SUPERALLOYS 2000

SUPERALLOYS 2000

Proceedings of the Ninth International Symposium on Superalloys sponsored by the TMS Seven Springs International Symposium Committee, in cooperation with the TMS High Temperature Alloys Committee and ASM International, held September 17-21, 2000, Seven Springs, PA.

Edited by

T.M. Pollock

R.D. Kissinger

R.R. Bowman

K.A. Green

M. McLean

S.L. Olson

J.J. Schirra

A Publication of

TMS

A Publication of The Minerals, Metals & Materials Society
184 Thorn Hill Road
Warrendale, Pennsylvania 15086-7528
(724) 776-9000

Visit the TMS web site at
http://www.tms.org

Printed in the United States of America
Library of Congress Catalog Number 00-133373
ISBN Number 0-87339-477-1

If you are interested in purchasing a copy of this book, or if you would like to receive the latest TMS publications catalog, please telephone 1-800-759-4867 (U.S. only) or 724-776-9000, EXT. 270.

Dedication

The Superalloys 2000 Symposium and these proceedings are dedicated to Mr. Wilford "Red" H. Couts, Jr. in honor of his pioneering contributions to the superalloy and gas turbine industries.

Preface

The purpose of the International Symposium on Superalloys is to provide a forum for researchers, producers and users to exchange recent technical information on the high temperature, high performance materials used in the gas turbine and related industries. Held once every four years, it is the goal of the conference to highlight new initiatives and future growth opportunities for superalloys, recent advances in the understanding of the behavior of these materials and progress in integrating these materials into new systems.

The First Symposium, held thirty-two years ago, emphasized phase instabilities in superalloys. Since then, the scope of the symposium has greatly expanded to cover all aspects of superalloy research, development, production and applications. The Eighth Symposium, held in 1996, was notable for the increased use of computer modeling, progress in understanding the constraints of alternate materials and for the degree of international participation. The Ninth Symposium, held at the turn of the century, once again covers recent research and new developments on all aspects of superalloys, with an increased emphasis on industrial gas turbines.

Starting with the Second Symposium in 1972, each symposium and corresponding published proceedings have been dedicated to individuals as a means of honoring that individual for his or her contributions to the superalloys industries. This Ninth Symposium is dedicated to Mr. Wilford "Red" Couts, Jr. He is author or coauthor of more than thirty technical and scientific papers and articles focusing on high temperature materials and their microstructure/mechanical property relationships and metalworking processing. Red Couts also holds four U.S. patents in the area of processing of high temperature materials, including forging processes, selective heat treatments applied to wrought products and the production of dual alloy disks featuring high quality bond lines. He was also a significant contributor to *Superalloys II* and *Superalloys, Supercomposites, and Superceramics*, both of which are important source references for the high temperature materials community. During his career, Red Couts worked at Battelle Memorial Institute in Columbus from 1954-1956, at General Electric Aircraft Engines in Evendale from 1956-1961 and at Wyman Gordon in North Grafton as a Research Engineer, Manager, and Senior Scientist - Superalloys from 1961 until his retirements in January of 1990. His significant technical contributions are related to wrought materials. At General Electric, Red was the specialist for A-286 and V-57. At Wyman-Gordon, he focused on the thermomechanical processing of a wide range of superalloys, from René 41 to Waspaloy to Astroloy. Perhaps most important are his contributions to the understanding of the microstructural/mechanical property relationships in IN-718 and its thermomechanical processing. Red Couts holds a number of honors from technical societies, is a Fellow of ASM and was Chairman, Technical Divisions Board, ASM International from 1987-1989.

Finally, this conference would not be possible without the efforts of the current and past members of the International Symposium on Superalloys Committee. The program committee, listed below, was responsible for development of the technical program, including critical review of abstracts and manuscripts for originality, technical content and industrial pertinence. The entire organizing committee, listed on the following page, devoted considerable effort to organizing all other aspects of the symposium.

July, 2000
T.M. Pollock
R.D. Kissinger
R.R. Bowman
K.A. Green
M. McLean
S. L. Olson
J.J. Schirra

Best Paper Award
(co-winners)

The papers listed below were selected by the Awards Subcommittee of the International Symposium on Superalloys as the Best Papers on the Ninth Symposium. The selection was based on the following criteria: originality, technical content, pertinence to the superalloy and gas turbine industries and clarity and style.

Novel Casting Processing for Single Crystal Gas Turbine Components

M. Konter, E. Kats, and N. Hofmann

Predicting Grain Size Evolution of UDIMET Alloy 718 during the "Cogging" Process through Use of Numerical Analysis

B.F. Antolovich and M.D. Evans

Ninth International Symposium on Superalloys
Committee Members

General Chairman.. Bob Kissinger
Secretary... Tim Howson
Treasurer.. Gern Maurer
Program Chairman.. Tresa Pollock
Program Committee.. Randy Bowman
                                                                                            Ken Green
                                                                                            Bob Kissinger
                                                                                            Malcolm McLean
                                                                                            Sonja Olson
                                                                                            Jack Schirra

Publication Chairman... Ken Green
Arrangements Chairman.. Jim Blair
International Publicity... Allister James
U.S. Publicity.. Sonja Olson
Awards Committee Chairman.. Doug Deye
Awards Committee.. Bob Stusrud
                                                                                            Chuck Kortovich
                                                                                            Lou Lherbier

TABLE OF CONTENTS

Keynote Address

Ingot, Powder and Deformation Processing

Blade Alloys

Disk Alloys

Mechanical Behavior

Coatings, Welding and Repair

Alloy Development

Keynote Address

SUPERALLOYS – THE UTILITY GAS TURBINE PERSPECTIVE

Brij B. Seth
Manager, Materials Engineering
Siemens Westinghouse Power Corporation

Abstract

The growing market for large utility gas turbines (UGTs) in recent years has rekindled competition for high efficiency, cost-effective large units. This, in turn, has accelerated the introduction of advanced superalloys and coatings narrowing the temperature gap with aeroturbines. The significant differences between aeroturbines and UGTs in size, operating environment and duty requirements create superalloy challenges that are unique to UGTs.

The superalloy related development needs for UGT blades, vanes and discs are discussed with reference to alloy composition, manufacturing processes, corrosion/oxidation, coatings, inspection, rejuvenation and repair. Minimum life cycle cost, the key market driver, dictates the need for advances in each of these fields. The current status and the approaches for achieving these advances are reviewed.

Background

Growth of Utility Gas Turbine Industry

Electric power consumption per capita has grown (Figure 1)[1] steadily in the USA and is projected to grow even more in future. Industrialization, electrical appliances, heating/air conditioning and more recently computerization have each been contributing factors. Worldwide growth in per capita consumption and population have increased the demand for electric power. Considering that several population intense countries such as China and India are still at the low end of per capita power consumption and advanced technologies are being introduced at a phenomenal pace, it is expected that the need for additional power generation will accelerate rapidly.

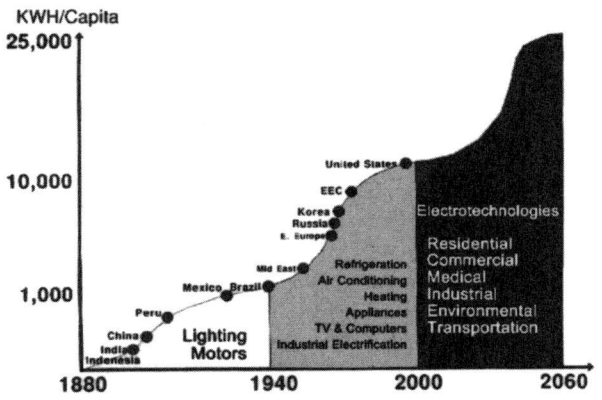

Figure 1: Growth of per capita power consumption in USA[1]

This power demand has been met through several different resources. Fossil fuel has dominated (Figure 2) and is expected to do so for the foreseeable future. Some growth is expected in hydro power worldwide, but the growth of nuclear power is expected to continue to be limited to a few countries. Solar and wind power, although politically popular, are not significant sources and their role is not expected to change.

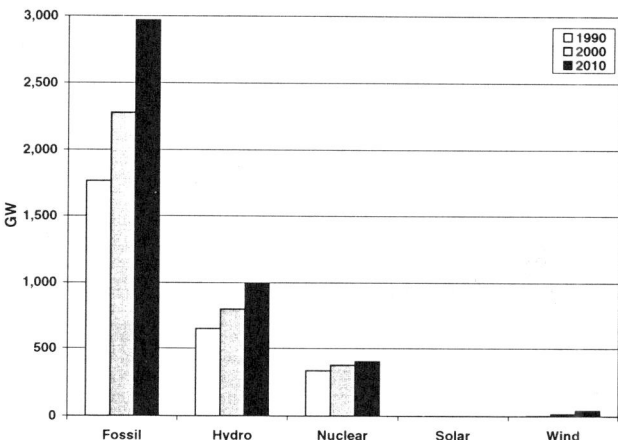

Figure 2: Sources of world power generation

Historically, steam turbines have dominated fossil power production with gas turbines filling only peak power needs because of their rapid start-up capability. Pollution concerns with coal, coupled with the abundant cost effective supply of gas and emergence of combined cycle technology, have, however, propelled gas turbines to the forefront as the major source of new base load capacity. The near 60% efficiency of combined cycle plants far exceeds that achieved by either steam turbine or gas turbine plants alone. Consequently, new gas turbine capacity is expected to grow rapidly (Figure 3) throughout the world with USA and Asia clearly dominating.

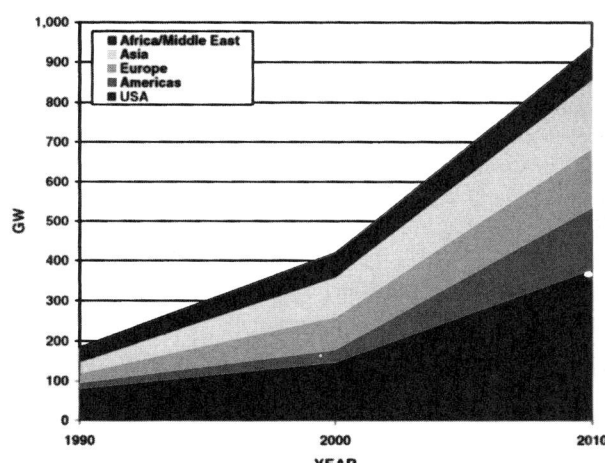

Figure 3: UGT growth by region

Superalloys 2000
Edited by T.M. Pollock, R.D. Kissinger, R.R. Bowman,
K.A. Green, M. McLean, S. Olson, and J.J. Schirra
TMS (The Minerals, Metals & Materials Society), 2000

The size of combined cycle plants is steadily increasing (Figure 4), approaching that of medium sized steam turbine plants. Also, gas turbine temperature (Rotor Inlet Temperature) and efficiency have almost doubled over the last forty years (Figure5) primarily as a result of better high temperature materials, more effective cooling designs and aerodynamically more efficient airfoils. This growth in size and efficiency are expected to continue.

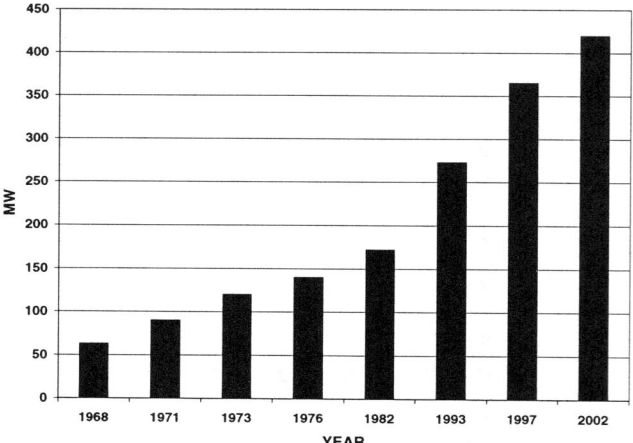

Figure 4: Evolution of UGT plant size

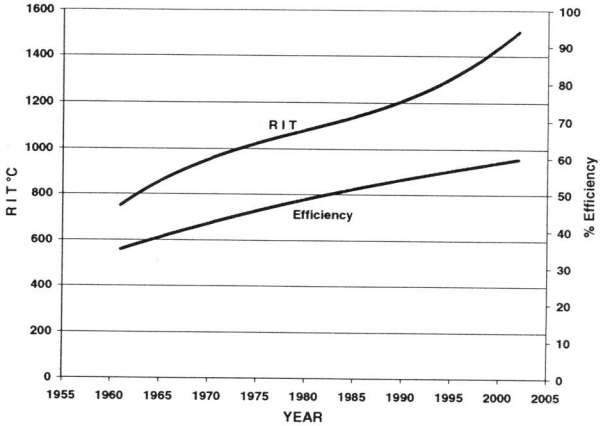

Figure 5: Historical increase in UGT efficiency and R.I.T.

The amount of superalloys used in UGT units has steadily increased (Figure 6) with the increase in unit size. This, combined with the increased demand for UGTs, has created a dramatic surge in the total quantity of superalloys consumed by UGTs. UGT sales[1a] and share of the total gas turbine market rose sharply in 1999 over the previous year. Sales increased from $8.5B to $13.0B and market share from 30% to 38%. By contrast, aeroengine market share[1a] actually declined (65% to 59%) despite a growth in the absolute value of sales ($18.0B to $20.0B). UGTs are expected to dominate the gas turbine market in the coming years. The focus of the superalloy industry, which historically has been only on aeroengines, must now shift towards UGTs. This shaft has already begun and it is significantly changing the outlook of superalloy manufacturers.

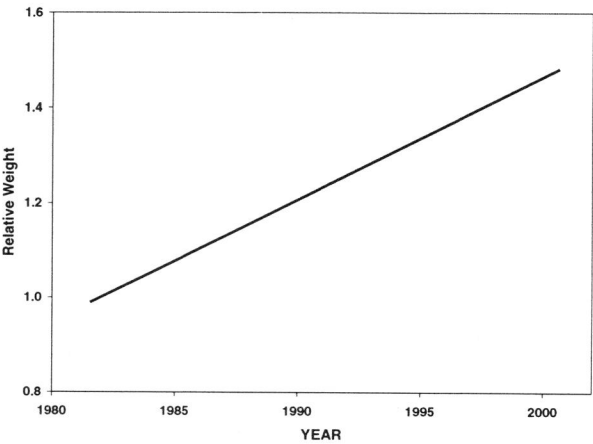

Figure 6: Increase in superalloy consumption per UGT

Customer Needs, Market Drivers and Materials Requirements for Utility Gas Turbines

Surplus new unit manufacturing capacity, despite greater demand for power production, has kept the market highly competitive. Competition is fierce due to deregulation and globalization. To succeed in this highly competitive market, engine manufacturers must have a clear understanding of customer needs, the associated critical market drivers and the resulting critical materials requirements. Needs and market drivers vary from customer to customer and country to country. The needs, market drivers and critical materials requirements common to most customers are summarized in Table I and discussed in this section.

Table I – Customer Needs, Market Drivers and Critical Materials Requirements for UGTs

Customer Need	Market Driver	Critical Materials Requirements
Regulations Compliance	Emission Levels	HT Materials to Increase Efficiency and Reduce Cooling Air
Low Plant Cost	Size	Manufacturability of Blades Vanes, Discs
Low Fuel Cost	Efficiency, Fuel Flexibility	HT Materials, Corrosion Control
Low Maintenance Cost	Reliability, Maintenance	Inspections, Life Extension
Short Project Time	Cycle Time	Material, Process and Component Behavior Modeling

Regulations Compliance - A critical regulation that customers must meet to obtain an operating permit concerns emission limits. While the percentage of emissions is controlled through the combustion process, improved engine efficiency reduces the amount of fuel needed per MW of power thereby reducing their total amount. Emissions can also be lowered if the materials allow the amount of cooling air to be reduced. The net result is a need for better higher temperature materials.

Low Plant Cost - Plant cost is affected by unit size. Generally, plant cost/MW decreases with increase in engine size, but increase in engine size means bigger components. The manufacture of large blades, vanes and discs from nickel based superalloys, particularly directionally solidified (DS, SC) blades and vanes, pose serious challenges of cost and quality.

Low Fuel Cost - With deregulation of the power industry, power producers will no longer be able to transfer fuel cost directly to consumers. The lowest fuel cost is achieved by the most efficient engines capable of operating with a wide variety of fuels. The implication of higher efficiency is the need for new materials of higher temperature capability. Fuel flexibility requires understanding the corrosion phenomenon and developing corrosion resistant coatings.

Low Maintenance Cost - Initial plant and fuel costs are very important, but to achieve the lowest life cycle cost, the equipment must be reliable and easy to maintain. In-service inspection, rejuvenation and repair are all critical elements of low maintenance cost.

Short Project Time - There is greater and greater emphasis on reducing the total project time for new plants. Return on investment begins only after the project is complete and commercialization has begun. While concurrent engineering has helped reduce project time, computer modeling of new alloys, their processing and component behavior will shorten the development cycle of new materials and lower costs by eliminating a large number of trial and error experiments.

The Use of Aerotechnology for Utility Gas Turbines

Since the aeroindustry has traditionally been the major user of superalloys, especially advanced superalloys, it has been the primary focus of the superalloy industry. The cold war drove new technologies for high performance military aeroengines and major advances were made in new superalloy compositions, processing technologies and solidification structures. The logical question that arises is, "Can the technology needs of UGTs be met through the advances already made in the aeroindustry?" The answer is not a simple "yes" or "no". The answer cannot be categorical because of the differing requirements of aeroengines and utility gas turbines. The key differences are summarized in Table II and discussed briefly below.

Table II – Key Differences Between the Requirements of Aeroturbines and Utility Gas Turbines

Parameter	Aeroengine	UGT
Weight	Very Important	Not Significant
Operating Time, Hours		
-- Steady State	25,000	> 100,000
-- Peak Temperature	< 1000	> 100,000
Cyclic Duty	Severe	Severe
Environment	Non Corrosive	Corrosive
Size	Small	Large

Weight - Weight is a critical parameter in aeroengines. The development and selection of aeromaterials is constrained by the need to consider alloy density, coating thickness and dimensional tolerances. Since weight is not of prime importance in UGTs, aeromaterials developed solely for weight considerations are of no direct relevance.

Operating Time - The highest temperature experienced by aeroturbines is during the brief take-off period and the total time at peak temperature over the life of the engine is less than 1,000

hours. During cruise, the aeroturbine actually operates at temperatures significantly lower than those during the steady state operation of UGTs. Cruise time over the life of the aeroengine is expected to total about 25,000 hours contrasting with the 100,000 + hours of steady state operation by UGTs. Thus, long term alloy stability, extrapolation of properties and interaction of creep and fatigue are much more critical for UGTs than for aeroengines.

Cyclic Duty - Both UGTs and aeroturbines must have adequate cyclic capability. While maximum temperature and strain are high, the hold period (take-off time) for aeroengines is very short (minutes). In contrast, UGTs operate at higher strain for days and weeks, especially for base load operation, making the interaction of creep and fatigue a much more severe damage mechanism.

Environment - Fuel quality in aeroengines is generally high and relatively free of corrosive elements such as vanadium, sulphur etc. Fuel quality in UGTs, however, varies considerably from plant to plant, region to region and country to country. These variations must be considered during design and operation in order to minimize corrosion and improve reliability.

Size - The most significant difference between aeroturbines and UGTs is size. Typical first stage blades for aeroturbines and UGTs are compared in Figure 7. All UGT dimensions are generally 2-3 times larger so that blades are 20-30 times heavier. Similarly, disc sizes (diameter and thickness) are also much bigger. This size/weight difference has an immense impact on component manufacturability and cost.

Figure 7: Comparison of typical Row 1 blades

Superalloy Challenges for Utility Gas Turbines

The UGT materials requirements discussed above pose major challenges to the superalloy industry. These include high temperature capability, processing SC materials for large blades and vanes, processing large disc materials, corrosion resistance, advanced thermal barrier coatings, advanced inspection techniques, life extension and computer modeling. The current status and implications of each of these challenges for the industry is discussed.

High Temperature Capability

The high-temperature oxidation resistance of Ni-Cr alloys was recognized in the early 1900's and the superalloy containing about 20% Cr has been a principal material for electrical heating elements ever since[2]. The discovery that these Ni-Cr alloys could be strengthened by the coherent precipitation of Ni_3 (Al, Ti) led to the development of the modern gamma prime (γ') strengthened superalloys. Increasing the volume fraction of γ' required the reduction of chromium. Although higher creep resistance could be achieved with reduced chromium levels, this resulted in both a loss of solid solution strengthening and oxidation resistance. The addition of molybdenum for solid solution strengthening and aluminum for oxidation resistance initially compensated for the reduced chromium, but it was quickly recognized that chromium levels around 15% were needed[2] to avoid the onset of hot corrosion. Molybdenum levels above approximately 3.5wt% were also found to be harmful to hot corrosion resistance, leading to the substitution of some molybdenum by other refractory metals such as tungsten, tantalum and niobium[3]. Grain boundary carbides play a major role in the control of creep and fracture behavior and can be modified by heat treatment and by the addition of minor constituents such as boron and zirconium. Hafnium is also a strong carbide former and is added to polycrystalline alloys to improve grain-boundary ductility[4].

Until recently, UGT hot section components where type II corrosion dominates, were made of superalloys containing high chromium (chromia formers). The lower chromium, higher aluminum containing alloys (alumina formers) were developed for aeroturbines, where gas stream temperatures are higher and oxidation concerns dominant. As firing temperatures of UGTs continued to increase, the emphasis shifted from modifying the composition of equiaxed alloys to controlling grain structure (Figure 8), as in the case of aeroturbines some twenty five years ago[5].

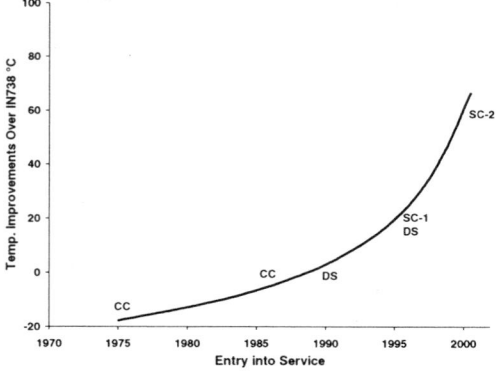

Figure 8: Increase in temperature capability of superalloys in UGTs

Directionally solidified (DS) and single crystal (SC) superalloys for blades and vanes are being introduced (Figure 9). Single crystals derive their beneficial properties from the absence of grain boundaries and, in directionally solidified nickel base superalloys, their preferred low elastic modulus in [100] orientation. The absence of grain boundaries eliminates preferred sites for crack nucleation under fatigue and thermomechanical fatigue conditions. Under creep conditions, the absence of grain boundaries not only removes the preferred sites for fracture initiation but also removes a potential mechanism of high temperature deformation, specifically grain boundary sliding.

Figure 9: Evolution of grain structures in UGT blades

The first generation single crystals were without any rhenium additions. Since then, second and third generation single crystals have been developed containing 3% and 6% Re respectively. Although single crystals find widespread applications in aeroturbines, the castability of large single crystal blades and vanes (discussed later) poses some unique challenges for UGTs.

Processing Single Crystal Materials for Blades and Vanes

The directionally solidified (DS & SC) structures deliver excellent properties but their processing is not without challenges. This is particularly true for the large blades and vanes used for UGTs. Because of these challenges efforts are ongoing to improve conventional SC casting processes and develop alternate processes.

This section discusses the conventional SC manufacturing process, its limitations, ways to overcome some of these limitations, and alternate processes for producing cost effective defect-free blades and vanes.

Conventional SC Manufacturing Process - The conventional Bridgman process for manufacturing single crystal blades and vanes has been used extensively for applications in aeroturbines. When used for large UGT parts, the result is very low yield due to distortion and cracking of the core, shell rupture, mold-metal reaction and numerous crystal defects. Figure 10 shows a large single crystal UGT blade with various defects. Continued improvement of the efficiency necessary for reducing emissions and lowering fuel costs depends on finding cost effective ways to manufacture these large blades and vanes in directionally solidified structures. Some of the approaches are discussed below:

Figure 10: Casting defects in a large single crystal UGT blade

Better Core and Shell Materials - Since a much larger volume of molten metal is required, there is a significant increase in the hydrostatic pressure on molds. Also, the longer solidification time increases mold-metal reactions and causes creep deformation of cores. To overcome these problems, it is necessary to develop shell systems and core materials that have greater high temperature strength and dimensional stability. The high shell strength has to be carefully balanced against the increased strain and corresponding propensity for recrystallized grains in castings.

Thermal Management - The crystal structure during directional solidification is controlled by growth rate and temperature gradient. Successful casting of single crystals therefore depends on proper thermal management during solidification. Historically this has been achieved through trial and error based on prior experience. The solidification modeling of the system allows optimization of the numerous parameters that control the growth rate and temperature gradient. Section size, mold clusters, mold wall thickness and conductivity, baffle design, withdrawal rates etc. all play an important role and the optimum combination of parameters for a given blade/vane geometry must be established.

Defect Tolerance - In producing real parts for engines there must be a trade-off between the need to accept only defect-free parts for optimum material performance and the need to allow some level of defects for realistic commercial production. Many different types of defects can be formed in a cast and heat treated single crystal part. Figure 11 illustrates typical defects that are found in such single crystal castings. Just as the preferred orientation and absence of grain boundaries enhance single crystal properties, the misorientation and presence of grain boundaries degrade them.

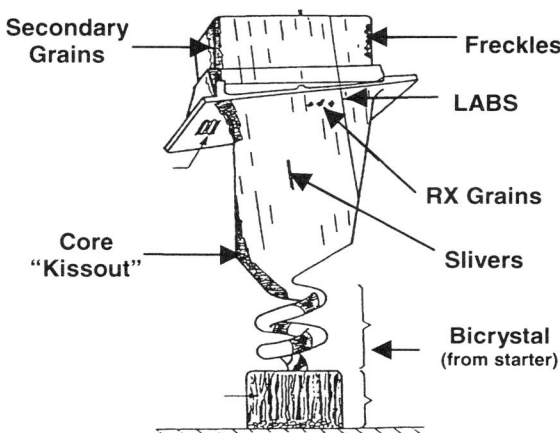

Figure 11: Typical defects in single crystal castings

In casting large single crystals, it is impossible to maintain perfect [001] crystal orientation. Mold geometry and the difficulty in controlling furnace baffling mean that [001] crystal growth cannot be maintained parallel to the desired axis at all locations within a part. Thus, parts must be accepted with orientations that lie close to but not exactly parallel to the preferred [001] axis.

High angle grain boundaries and multiple boundaries such as recrystallized regions and freckle chains, are obvious sources of weakness. These defects are particularly pernicious since the single crystal formulations do not contain the grain boundary strengtheners found in conventional alloys. In such cases the supposedly single crystal alloys will be even weaker than their conventionally cast counter-parts and the property deficit relative to expected single crystal properties can be catastrophic.

Considered less deleterious are the low angle grain boundaries (LABs) that form between essentially parallel single crystals. LABs are very difficult to avoid in large castings and for the sake of commercial viability some levels of LABs are always accepted. Their impact on properties is significant, although difficult to quantify precisely. Traditionally, the acceptance of LABs has been based on a philosophy similar to the effect of misorientation, the deleterious effects of LABs being related to the severity of the misorientation across the boundary.

The aeroindustry has established acceptance limits for various defects. Since minimum property requirements for UGTs differ from those for aeroturbines and critical property requirements vary from location to location within large UGT components, it seems prudent to reassess and indeed customize the prevailing aeroengine acceptance standards.

Alloy Optimization - The evolution of DS and SC alloys has been governed by their high temperature strength without much consideration for the manufacturability of large UGT blades/vanes. Alloys that are less susceptible to casting defects and less sensitive to grain defects need to be developed. The absence of any grain boundary strengtheners in SC alloys make them particularly sensitive to grain defects. Additions of small amounts of grain boundary strengtheners may provide a better alloy balance. Cannon Muskegon have developed[6,7] a second generation DS alloy called CM186. They have manufactured SC components from this alloy which contains C, B, Zr and Hf and found it to be less sensitive to heat treatment and low angle boundaries. However, under such conditions, the ability to fully solution heat treat has to be sacrificed. Effectively, a further trade-off between performance and yield has to be made.

Alternate SC Manufacturing Processes – Alternate manufacturing processes are being developed because advances in the conventional SC manufacturing process may not fully meet the quality and size demands of advanced UGTs. Two of these alternate processes, liquid metal cooling and transient liquid phase bonding, are reviewed.

Liquid Metal Cooling - Since many of the SC yield detractors in the conventional process result from low thermal gradient, slow growth rate and long solidification time, the use of liquid metal (Al, Sn) as a cooling medium holds great promise for casting large SC components. Figure 12 shows[8] a schematic comparison of the conventional Bridgman process and the liquid metal cooling (LMC) process. LMC has numerous advantages, the main one being increased thermal transfer due to conductive rather than radiative cooling. It has been shown [9] that heat transfer rates for LMC are three times faster than those obtained with radiation cooling in the Bridgman process. Also, since the liquid metal is in equal contact with every individual mold, view factor effects are minimized and greater part to part consistency is expected. The increased withdrawal rate due to the higher thermal gradient decreases the time for mold-metal reaction and increases the rate of production. Minimum mold spacing allows more pieces to be cast at one time, further improving the economics.

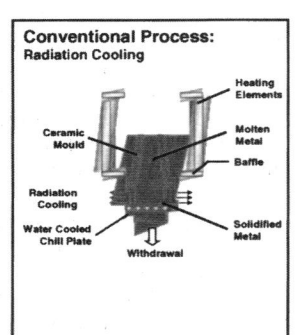 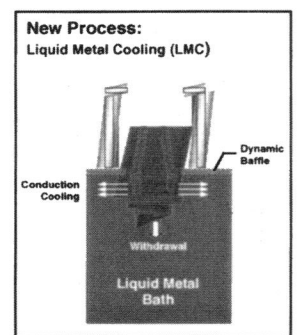

Figure 12: Conventional and liquid metal cooling processes[8]

Compared with parts cast by the Bridgman process, parts cast by LMC have[10] a much finer dendrite arm spacing (less than half). The benefits of finer arm spacing are reduced heat treatment time, a more uniform microstructure and improved mechanical properties due to less $\gamma - \gamma'$ eutectic.

Transient Liquid Phase Bonding - In this approach, small, easily castable segments of the blade/vane are produced as separate single crystal pieces and joined together to form the complete structure [11]. For this purpose, a very high quality joining process that can regenerate the optimum single crystal structure and properties across the bondline is required.

Transient Liquid Phase bonding is a high quality precision bonding technology that has been shown[12] to be capable of bonding Nickel-based superalloy single crystals for turbine applications. This process uses a bonding medium that is compositionally matched to the material to be bonded but containing a small amount of a melting point depressant element such as Boron. It is also critical that the melting point depressant exhibit

high solid state diffusivity in the alloy. By using only a very thin layer of bonding medium, a bond region can be formed by isothermal solidification which exhibits not only the same chemistry as the substrate, but also the same crystallography and fine scale microstructure. The principle of this process is shown in Figure 13 [13]. Figure 14 illustrates [12] how well matched bond zone and base material chemistries can be achieved when the Transient Liquid Phase bonding process is controlled.

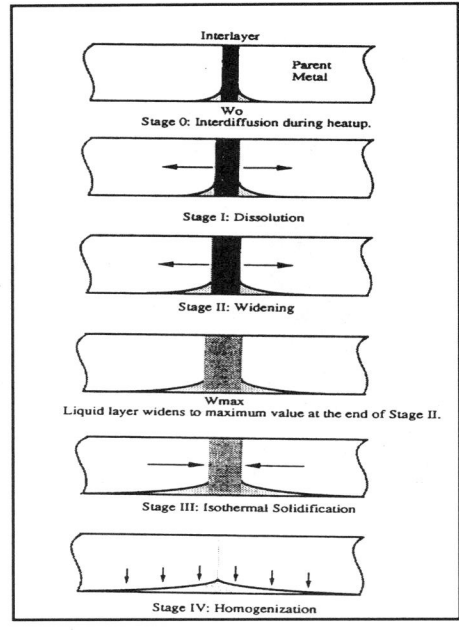

Figure 13: Principle of transient liquid phase bonding[13]

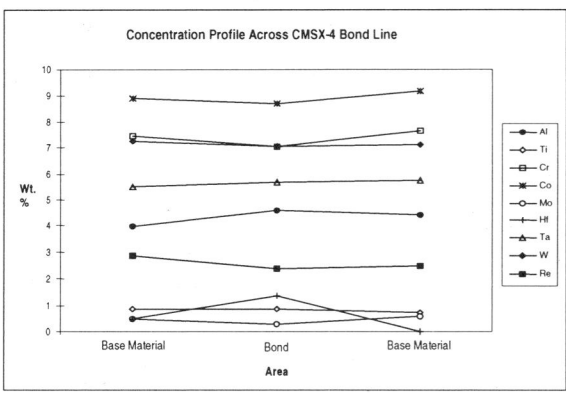

Figure 14: Microprobe analysis across the transient liquid phase bonded CMSX-4 single crystal bond region

For a bond region to demonstrate the same high level of mechanical properties as the original single crystal, not only must the bond region chemistry match that of the rest of the single crystal but the component must also be properly heat treated to produce the optimum $\gamma - \gamma'$ morphology. Specifically, for the second generation single crystal alloys the morphology of the γ' should consist of cuboidal particles of about 0.5 microns on edge and a secondary dispersion of finer spheroidal particles. Figure 15 demonstrates [12] that the appropriate post bonding solution heat treatment and subsequent precipitation aging treatment can create a bond line microstructure indistinguishable from the interior regions of the unbonded single crystals. Thus, proper bonding chemistry, bonding cycle and post bonding heat treatment not only generate the optimum microstructure in bonded materials but also produces mechanical properties close to those of the base material.[12]

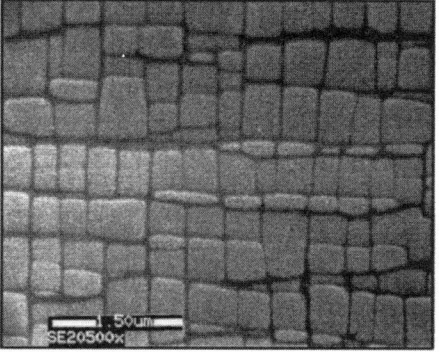 (a)

 (b)

Figure 15: Gamma-prime morphology of a) bond zone and b) base material in bonded CMSX-4 single crystal

While high levels of mechanical properties can be achieved in transient liquid phase bonded materials including single crystals, such properties are not always needed in real components. Most often, the full capabilities of advanced materials are needed at only a few critical locations in a blade or vane. At many other locations the imposed temperatures and stresses allow material properties to be well below those needed for critical locations. Thus, judicious placement of the bond in a low stress region will provide considerable margin in using transient liquid phase bonding.

Finally, process engineering practices for handling the large components used in UGTs are required. In contrast to aircraft gas turbines, the blade and vanes of UGTs can be over thirty inches in length and weigh several tens of pounds. Handling such parts for bonding is not a trivial task.

Processing Turbine Discs

Although superalloy discs are common in aeroengines, they are less frequently used in UGTs for a number of reasons[3]. First, rotating blades have a long shank between the platform and the root, keeping the dove-tailed root setion of the disc away from the hot gas path and enabling it to operate at significantly lower temperatures than those observed in aircraft engines. Second, although the utility turbine discs can have very large diameters, weight is not a consideration. Increased section size can therefore be used to reduce stress levels. Third, the difficulty of

manufacturing large superalloy ingots lowers yield and increases cost.

The higher firing temperatures and pressure ratios of the latest generation UGTs have led some engine manufacturers to use superalloy discs. Alloy 718 is the most widely used superalloy for aerospace disc applications, but high niobium levels create 'freckle' problems (regions of positive macrosegregation) in the large diameter ingots required for UGT discs. Alloy 706 has been developed to allow the production of ingots larger than those of alloy 718 without segregation problems. Ingots with diameters as large as 1,016mm (40 inches) have been successfully cast in alloy 706[14].

Corrosion Resistance

Although most gas turbines operate with clean gas fuel, variable oil quality and diverse ambient environments create potential for severe corrosion damage in hot parts. The corrosion concern is much more severe in UGTs than aeroturbines since the latter use high quality fuels. Also, the potential use of steam as a cooling medium in advanced UGTs requires thorough evaluation of its impact on oxidation and the effect of steam borne impurities. Details of steam cooling, corrosion resistant materials and oxidation/corrosion resistant coatings are discussed in this section.

Steam Cooling - The advanced UGT engines are expected to use closed-loop steam cooling for transitions and the early stages of vanes. Closed-loop steam cooling increases plant efficiency by eliminating the need for cooling air injection into the turbine flow path. Steam is a more effective cooling medium and the heat removed while cooling the components is recovered in the combined cycle steam turbine.

There are challenges to the successful use of closed-loop steam cooling. These include preventing corrosion and spalling of corrosion products over long periods of time, blockage of closed-loop cooling channels due to corrosion product outward growth, and loss of mechanical properties due to thickness loss. Also, common impurities in steam such as NaCl, Na_2SO_4, Na_3PO_4, SiO_2 and metal oxides may deposit inside the cooling holes leading to component failure from corrosion and / or blockage.

Steam passes through pipes, flanges, valves, blade ring and manifolds before being introduced into the vanes and transition cooling channels. Spallation of corrosion and oxidation products in these components carried by the steam may lead to cooling channel blockage in vanes and transitions. Therefore, in addition to vane and transition materials, the balance of plant materials need to be evaluated in high-temperature steam as well as in the presence of salt impurities.

Corrosion Resistant Materials - Until recently, high chromium alloys were commonly used to combat hot corrosion in UGTs. To meet the high temperature strength requirements of increased firing temperatures, the chromium contents have been gradually decreased and the inherent loss in corrosion resistance compensated by application of oxidation/corrosion resistant coatings. There are isolated cases of some high chromium single crystal alloys such as AF56 and PWA1483. More recently, Cannon Muskegon have developed[15] two new SC alloys CMSX11B and CMSX11C containing 12% and 14% chromium respectively. To the author's knowledge, these alloys have not yet been used.

Oxidation/Corrosion Resistant Coatings - In the early years of superalloy development, a majority of the coatings that provided oxidation and corrosion resistance were diffusion aluminide coatings. These were initially obtained by a pack cementation process[16] and more recently, by chemical vapor deposition processes[17]. The diffusion aluminide coatings possess excellent isothermal high temperature oxidation resistance but moderate to poor cyclic oxidation resistance, hot corrosion resistance and thermal fatigue resistance. To overcome the limitations of the diffusion aluminide coatings, overlay coatings were developed which were sprayed on to the part either by plasma spray processes or by physical vapor deposition. The overlay coatings provided a much larger aluminum reservoir for the oxidation resistance and also had superior mechanical properties [18,19]. The overlay coatings are denoted as MCrAlY, in which M represents a base metal of Ni, Co, Fe or some combination of these elements and Y represents a rare earth element, such as yttrium. In these alloys, Al provides the oxidation resistance, Cr increases the chemical activity of Al, Y improves adherence of the oxide scale to the bond coat and M provides compatibility with the substrate alloy. A typical microstructure of an MCrAlY coating on a superalloy substrate is shown in Figure 16. The outer surface consists of an oxide scale which, ideally, is slow growing and dense to prevent direct reaction of gases and deposits with the coating alloy. Since the protective scale on an MCrAlY coating is achieved by selective oxidation of Al in the alloy, the materials immediately beneath the scale will be depleted in Al content. A similar depletion zone is expected at the substrate/coating interface due to diffusion of Al into the substrate. Based on superior fatigue and oxidation data across several substrate-bond coat systems, these coatings are used in most hot section components in utility gas turbines.

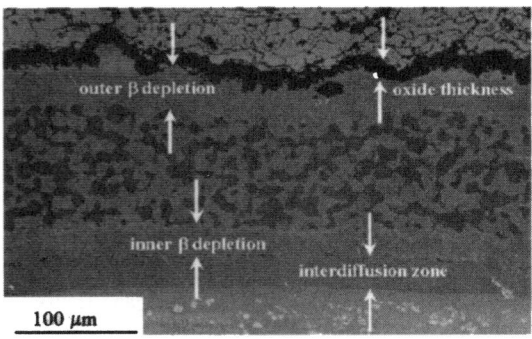

Figure 16: Typical MCrAly coating after thermal exposure

Advanced Thermal Barrier Coatings

As discussed earlier, significant increase in superalloy temperature capabilities has been achieved over the years through compositional modifications, process changes and structural controls. The third generation SC alloys are approaching the practical upper limit of temperature. Considering that manufacturing large UGT SC parts even with simpler alloys is a major challenge, the second and third generation alloys become virtually impossible to produce. It is therefore unlikely that additional advances in superalloy temperature capability can be realized by further modifications of superalloy compositions and structures. The next materials system for improved temperature capability is ceramic or ceramic matrix composites (CMCs). Despite many years of effort on ceramic systems, considerable research and development is still required before CMCs are commonly used in UGTs. The temperature capability gap between superalloys and CMCs has to be filled and the only viable means for doing so is the use of thermal barrier coatings (TBCs).

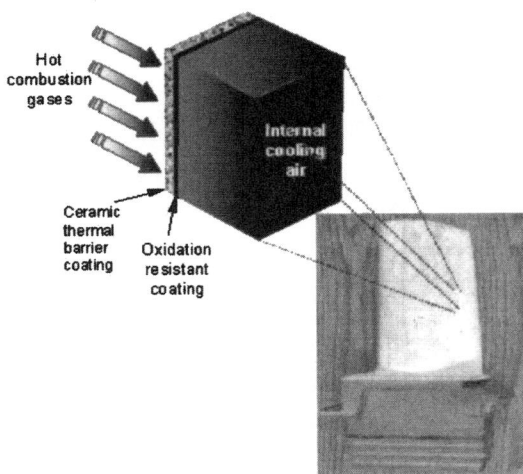

The thermal barrier coating system consists of a metallic bond coat and a ceramic thermal barrier coating (Figure 17). The bond coat provides oxidation and corrosion resistance by forming a slow growing adherent protective aluminum oxide scale, whereas the ceramic top layer reduces the substrate alloy temperature.

Figure 17: A thermal barrier coating system

Depending on the ceramic thickness and cooling effectiveness, substrate temperatures can be reduced by 100° - 300°F allowing a significant reduction in cooling air, and consequent increase in engine efficiency. The ceramic top coats are deposited on to the MCrAlY coatings, which function as a physical and chemical bond between the thermal barrier coating and the substrate (Figures 18-19). The current industrial standard for a ceramic thermal barrier coating is 8wt% yttria stabilized zirconia (8YSZ) deposited by air plasma spraying (APS) and electron beam physical vapor deposition (EB-PVD). Although, both coatings act as thermal insulators, their microstructures are fundamentally different from each other and so are their performances and application temperatures. APS coatings obtain their strain tolerance from gaps between the intersplat boundaries and vertical microcracks within the ceramic splats (Figure 18). A coating more tolerant to strains induced from thermal cycling can be obtained by EB-PVD. EB-PVD coatings have a columnar microstructure with intercolumnar gaps (Figure 19), which lend themselves to superior strain tolerance at temperatures higher than those of APS coatings. The higher performance of the EB-PVD coatings is, however, available only at an increased cost.

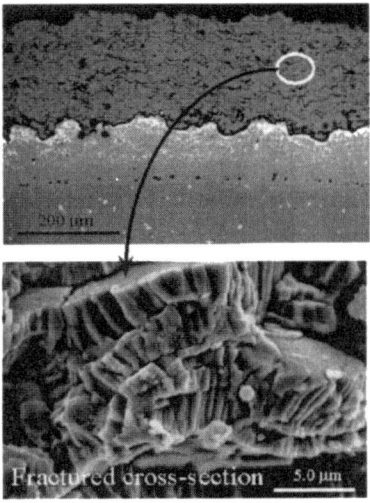

Figure 18: Air plasma sprayed thermal barrier coating

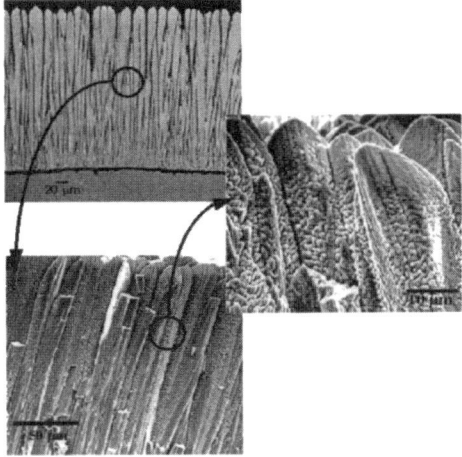

Figure 19: EB-PVD thermal barrier coating

With an increase in bond coat temperature, oxidation induced failure of the TBC system is expected to occur more rapidly. Depletion of aluminum in the bond coat is more rapid because of the increased growth kinetics of the aluminum oxide and interdiffusion with the substrate alloy. This severely degrades the ability of the MCrAlY coating to function as an oxidation resistant coating. The increased thickness of the thermally grown oxide (TGO) also results in the spallation of the TBC.

The thermal barrier coating, 8YSZ, is also vulnerable to the increased operational temperatures due to sintering of the intersplat boundaries and the microcracks within the splats. The sintering of the coating results in densification and loss in strain tolerance leading to spallation of the TBC. This failure occurs prior to spallation from bond coat oxidation - contrary to the previously observed failure modes. EB-PVD coatings are also expected to be susceptible to sintering resulting in a reduced time to spallation. In addition, at operating temperatures above 1200°C, the as-deposited non-transformable teteragonal phase destabilizes upon exposure (greater than 100h) [20-21], possibly leading to a loss of the mechanical integrity of the coating.

It is evident that future coating systems should overcome the limitations of both bond coat oxidation and poor sintering resistance of ceramic coatings. At expected higher temperatures, these new bond coats should form a slow growing adherent oxide scale. New TBC materials should possess a sintering resistance superior to that of 8YSZ and also maintain their phase stability at high temperatures. In addition, it is critical to understand that development of new bond coats for higher temperature capability is limited by superalloy design temperatures. Therefore, future benefits can be significant only if new TBC materials can simultaneously withstand higher surface temperatures and provide larger thermal gradients. These new TBC systems are required to reduce cooling air, increase reliability and extend component life.

Advanced Inspection Techniques

Deregulation is leading the power industry to push for the lowest operating costs, a key element of which is the lowest life cycle cost of parts. As more and more DS and SC components are used, component costs will rise significantly. To minimize costs, it is essential to rejuvenate and repair parts rather than replace them. To optimize rejuvenation and repair frequencies and minimize unit shutdowns it is necessary to develop inspection techniques to determine the amount of life consumed and on-line monitoring to detect the initiation and propagation of cracks.

Currently, the remaining life of components is assessed using analytical methods and destructive tests. These approaches have inherent limitations. Analytical methods must use worst case assumptions about stresses, material properties and operation so that results tend to be conservative. In the case of destructive tests, testing material from the most severely damaged location may not always be possible and there is usually insufficient time to test long term properties and insufficient material to test all the relevant properties.

Life Assessment Using NDE - It would be highly advantageous if components could be examined non destructively to determine the percent life consumed. Advanced NDE techniques are being explored for this purpose. Chou and Earthman[22] have demonstrated the feasibility of using laser light scanning to characterize low cycle fatigue damage in IN718. Goldfine and Clark[23] have reported an excellent correlation between the meandering winding magnetometer conductivity measurements and percent of total fatigue life of St. 304. Jeong K. Na, et al[24] have studied the accoustic properties of longitudinal velocity and non-linearity parameters and observed a linear correlation with log cycles in high cycle fatigue tests on St. 410 Cb material. The results from these NDE techniques are shown in Figure 20. These and any other promising approaches should be pursued further with the objective of developing practical tools for in-service applications.

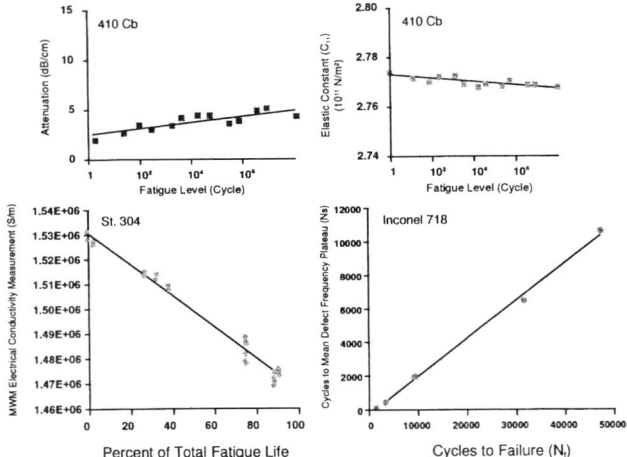

Figure 20: Fatigue damage detection by various NDE techniques[22-24]

On-line Monitoring - It would be even more advantageous if the above measurements could be taken while the turbine is running. Unit availability would significantly increase by eliminating the need for frequent shutdowns and openings. This may at first seem impossible since combustion turbines typically operate at 3000 or 3600 rpm with the blades moving at about 1200 miles per hour and completely encased in a pressure vessel. Further, the components of interest operate up to 1000°C with the gas temperatures up to 1540°C. Review of current sensor capabilities, however, reveals that the task is not impossible. Many of the sensors are based on electromagnetic or optical principles which have very fast measurement and response times. Relative to the speed of these sensors, the running turbine might as well be standing still. The real limiting factor becomes the ability to manage the data. Fortunately, the newest generations of personal computers have adequate capability.

Optical sensors could play an important role in real time materials measurements. The information content of the light scattering spectrum included in Raleigh, Brillouin and Raman scattering parameters is quite impressive. They are capable of measuring important properties like chemical composition, phase information, acoustic velocity, moduli, and thermal diffusivity. The measurements can be taken at light speed from a distant vantage point of milder conditions. Adding to the optimism, the most critical rotating components move past the sensor to create a self-scanned measurement map of the component, thus greatly reducing the number of sensors required to monitor many parts. Infrared imaging arrays could examine rapidly moving components in real time for thermal properties. Components with thermal barrier coating could also be examined to detect the formation and growth of defects.

Electromagnetic sensors could measure conductivity, permeability, or dielectric properties. The sensors must be used in close proximity to the component limiting measurements on rotating components. However, many useful measurements could be made on critical stationary components without these concerns.

Life Extension

It is important to distinguish between rejuvenation and repair although both are processes undertaken to extend the useful life of a component. Repairs can be thought of as 'external' processes that return the component to its original size and shape, or replace the protective coating. Rejuvenation can be defined as the regeneration of a microstructure leading to the restoration of mechanical properties equivalent to those of the original component prior to initial service.

Rejuvenation - The primary types of internal degradation observed in superalloys experiencing prolonged elevated temperatures are: gamma prime precipitate coarsening or overaging; changes in grain boundary carbides; cavity or void formation and the generation of Topologically Close Packed (TCP) phases. A conventional heat treatment involving complete solutioning, controlled cooling and subsequent aging should generally be sufficient to regenerate the original microstructure. It reverses the deleterious effects of gamma prime precipitate coarsening and changes in grain boundary carbides, and also sinters[25,26] cavities below a certain critical size.

The creep properties of relatively simple superalloys (such as Nimonic 80A) containing low volume fractions of gamma prime can be restored by merely annealing at temperatures below the γ' solvus temperature[27]. In this type of alloy where the degradation is caused by the formation of cavities, annealing at the creep temperatures without any stress is sufficient to sinter the cavities. The time required to sinter a cavity will obviously depend upon its size and the sintering temperature. Cavities formed during tertiary creep may be amenable to healing by heat treatment alone.

The creep properties of the more complex alloys containing higher volume fraction of γ' are controlled by microstructural changes such as overaging of γ' and carbide degeneration. Rejuvenation of this type of alloy can be achieved only by heat treatments which dissolve and reprecipitate the γ' in a distribution and size similar to those of the original microstructure[28].

Proper timing of a rejuvenation treatment is imperative if the maximum economic benefit is to be gained. It is generally agreed that to ensure the regeneration of creep properties the rejuvenation treatment should be applied at the end of the secondary or beginning of the tertiary state of creep. Creep strength can be recovered[29] after up to 1% creep deformation by heat treatment alone. Economical rejuvenation at later stages of creep may be possible if an HIP cycle is incorporated.

Conventional cast parts can be routinely rejuvenated using HIP and resolution heat treatment cycles to restore the microstructure and properties. The rejuvenation of directionally solidified (DS) and single crystal (SC) parts pose a greater challenge. Recrystallization is a major concern for DS and SC parts. Exposure to fatigue and creep loading in service may induce damage that can manifest itself as recrystallization during rejuvenation treatments at temperatures near to or above the γ' solvus. However, it is not only service exposure that can lead to recrystallization. Prior handling damage or machining induced strains at the manufacturing stage may also result in recrystallization during rejuvenation.

Recrystallization may be present on the surface or in the interior cooling channels. The conventional back reflection laue technique used in the aeroindustry could, in theory, detect surface recrystallization, but it is totally inadequate for detecting internal recrystallization. For this reason, a patented method[30] of through transmission diffraction (Figure 21) has been developed to detect small secondary crystals, as well as the orientation of the primary and secondary crystals relative to the component axis. A high flux, low energy, x-ray generator is used to provide the necessary photons for diffraction imaging of internal recrystallized grains. Figure 22 shows the diffraction pattern of a secondary internal crystal in a single crystal specimen, using this technique.

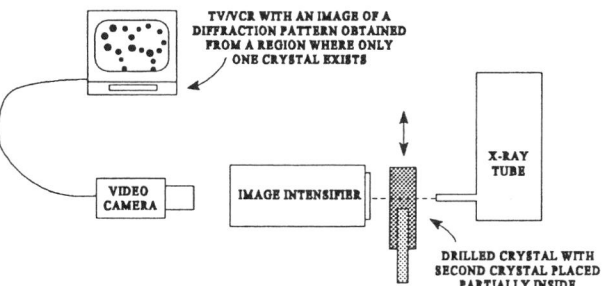

Figure 21: Schematic of SC through-transmission imaging

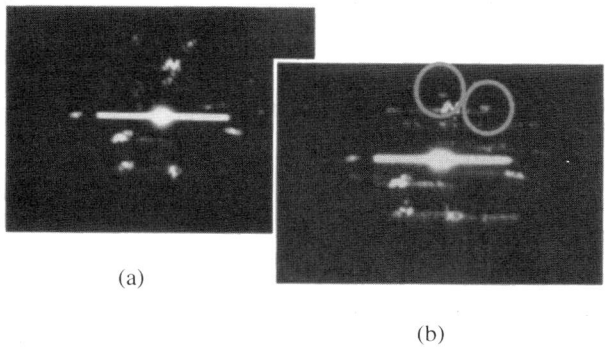

(a)

(b)

Figure 22: Through transmission SC image (a) no defect (b) internal misoriented grain

<u>Repair</u> - Repair of today's advanced superalloys poses significant challenges. As the industry developed new nickel based superalloys with higher temperature capabilities, the volume percent of γ' increased from 30-40% in the earlier forged and cast alloys, to 60-70% in some of today's DS and SC alloys. The high γ' content and the directional solidification of these latest alloys make repairs especially difficult. However, the high cost of these components dictates the need for developing cost effective repair procedures.

Earlier researchers have published[31] formulae (Figure 23) which correlate the blade material composition with its weldability. The high volume percent of γ' gives these alloys high strength at elevated temperatures, but very low ductility, which, combined with the residual stresses from welding, frequently results in cracking.

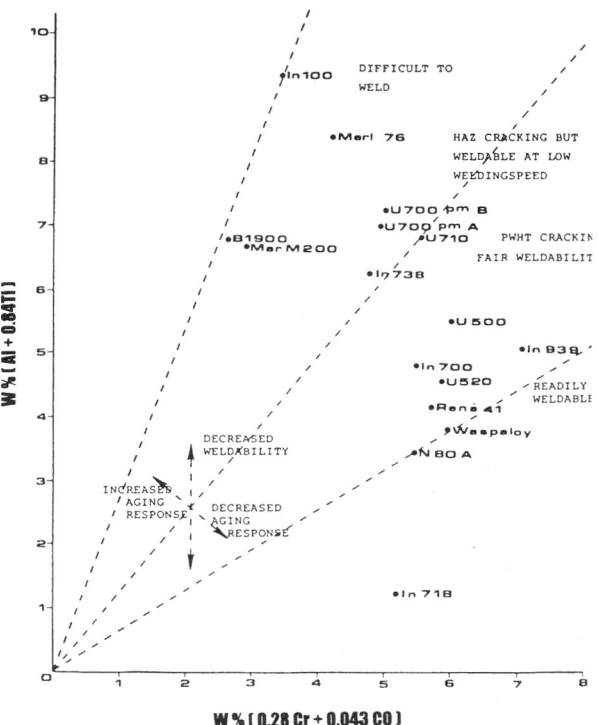

Figure 23: Weldability diagram for a number of nickel-base superalloys and related materials[31]

Currently weld repair is generally limited to the low stress tip regions of the components. Typically, welding is performed using manual Gas Tungsten Arc Welding (GTAW), and the weld filler material has been a low strength, solid solution strengthened nickel base alloy, such as Inconel 625. This high ductility filler material absorbs some of the welding stresses by yielding, and thereby reducing cracking. Still, some minor cracking is inevitable in the higher γ' alloys. The addition of braze filler to mend microfissuring is usually inadequate to meet engineered part life. Repairs that use higher strength or more oxidation resistant weld filler materials have been developed, but these repairs are also limited to the tip and platform areas.

To minimize cracking when welding high γ' alloys, the weld parameters must be carefully controlled. A highly concentrated heat source, lower heat input, automation and judicious application of preheat and controlled cooling are critical. Processes that provide a concentrated heat source include Laser Beam Welding – LBW (with powder or wire filler material), Electron Beam Welding – EBW, and Plasma Transferred Arc – PTAW (with powder or wire filler material).

The challenge is to improve the repair techniques for DS and SC alloys and to extend the repair envelope to higher stress regions elsewhere on the component. New and improved repair techniques must be developed. Some of these techniques are discussed briefly below.

<u>Material Conditioning</u> - Pre-weld heat treatments have been found[32] to improve the ductility and hence weldability of some alloys. These heat treatments usually overage the material, raising ductility and lowering strength.

High Preheat Welding - Crack-free welds were achieved by high preheat welding using the EBW process[33]. In high preheat welding, the component is heated to temperatures near the solution temperature during welding.

Blade Alloy and Filler Material Composition - Careful control of some of the elements that form low melting point compounds at the grain boundaries can improve weldability with little compromise in mechanical properties.

Transient Liquid Phase Bonding - The transient liquid phase bonding process discussed earlier[12] in connection with manufacturing large SC blades and vanes, yields excellent properties and can be used for repairs.

Liquid Phase Diffusion Sintering (LPDS) - LPDS uses a combination of a high melting point powder (typically a composition similar to the superalloy being joined) and a low melting point powder (typically a braze with boron as the melting point depressant). The powders are applied to the area to be repaired in the form of a paste, putty, or sintered preform. The repair is subsequently sintered in vacuum furnace below the melting point of the base metal substrate.

DS and SC alloys are among the most difficult materials to join. The challenge is to develop economical joining processes that achieve required strength. In the past, many of these alloys were considered unweldable, but today, some limited repairs are already possible. No one best process has yet been found, but success will be achieved by optimizing today's processes, selecting the best process for each specific application, and continuing the search for the new processes of tomorrow.

Material, Process and Component Behavior Modeling

The development of new alloys, their manufacturing process and evaluation of component behavior is an expensive and time consuming undertaking. It can take up to ten years and tens of millions of dollars to complete this cycle for a new alloy. To reduce cycle time, the experience-based trial and error approach of the past has to be replaced with a much shorter computer based modeling approach. An overall vision for modeling is shown in Figure 24. The conceptual materials model shown in Figure 25 is discussed below.

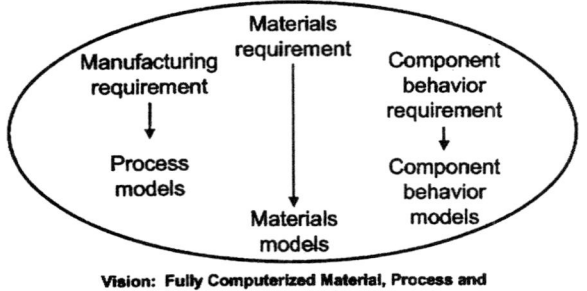

Figure 24: An overall vision for computer modeling

Successful modeling of advanced materials requires the ability to predict changes in the phase constitution as a function of composition, temperature and time, and translate this informa-

tion into properties targeted by the design engineer, such as creep strength, fatigue, oxidation and corrosion resistance.

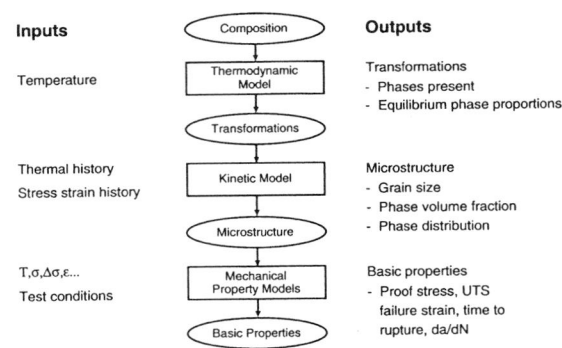

Figure 25: A conceptual material model

Several investigators[34-36] have reviewed the progress in modeling the phase stability of complex materials systems, especially multi-component alloys. They have focussed on predicting phase stability in thermodynamic equilibrium, a situation expected only upon long term exposure to high temperatures. Materials engineers, however, seek a heat treatment schedule which will optimize a specific mechanical property. To address such realistic aspects of materials processing, the models will require incorporation of kinetic aspects such as phase transformation and interdiffusion of components across several phases. Upon successful prediction of the phase compositions with temperature and time, it is necessary to correlate properties with composition, based on fundamental and empirical models.

In addition to predicting materials properties, it is important to evaluate susceptibility of advanced materials to the aggressive environments of the gas turbine, such as high temperature oxidation and corrosion. These may be modeled by using combined thermodynamic and kinetic models which include calculations for the activity and mass transport of the various components in the alloy, its oxides and surrounding environment. For reliable predictions, it is also imperative to know the correct physical and thermo-physical properties, such as elastic modulus, density, heat capacity, thermal conductivity etc. for each component. Predicting the behavior of the system (material and its environment) is definitely an important step for materials modeling and key to early identification of superalloy degradation during service.

Considering the progress already made on phase stability, composition-property correlations and environmental interactions, future efforts to combine these aspects in a unified model is no longer an unrealistic vision. This is especially true given the recent tremendous progress in efficient interfacing of software databases. A unified model will minimize the advanced materials development cycle and facilitate introducing new technology into the utility gas turbine industry.

Summary

1.   The growing demand for utility gas turbines and increased use of superalloys in them have created a dramatic surge in the market for superalloys.

2. The need to minimize the cost of electricity drives the need for increasing engine efficiency and reducing the life cycle cost of parts.

3. UGTs present unique challenges for transferring aeroturbine materials technologies and developing new UGT targeted materials. The key issues include longer life requirements, operation in hostile environments, cost effective manufacture of large components, inspection, rejuvenation and repair.

4. The cycle time for materials and process development for UGTs must be shortened by using knowledge-based systems and advanced computer modeling instead of trial and error.

5. The market for advanced superalloys with directional structures (DS and SC) will continue to flourish so long as the superalloy industry meets the power producers' need for high quality large components at low cost, short processing cycles, and rejuvenation and repair techniques applicable to the advanced superalloys used in UGTs.

Acknowledgements

The author gratefully acknowledges the discussions, input and constructive review of the manuscript by many of his colleagues. In particular, the contributions of Bill Trappen, Allister James. Dave Allen, Paul Zombo, Mike Taras, Frank Roan, Mike Burke, Gregg Wagner, Ramesh Subramanian and Steve Sabol are greatly appreciated. The author also wishes to thank Howmet Corporation for providing some of the data used in this paper. Finally, the author expresses his appreciation to Siemens Westinghouse Power Corporation for permission to publish this paper, and to Camille D'Amore for her patience in typing the manuscript.

References

1. K. E. Yeager, Presented at Westinghouse power generation symposium, Orlando, USA, February 14, 1994.

1a. Lee S. Langston, Gas turbine industry overview, 2000 IGTI Technology Report and Product Directory Land, Sea and Air, ASME, p. 7.

2. W. Betteridge and S.W.K. Shaw, Development of superalloys, Materials Science and Technology, Vol. 3 September 1987.

3. J. Stringer and R. Viswanathan, Gas turbine hot-section materials and coatings in electric utility applications, Advanced Materials and Coatings for Combustion Turbines, ed. V. P. Swaminathan and N.S. Cheruvu, 1993.

4. G. L. Erickson, Superalloy developments for aero and industrial gas turbines, Ibid.

5. G.L. Erickson, K. Harris and R.E. Schwer, Directionally solidified DS CM 247 LC - Optimized mechanical properties resulting from extensive gamma prime solutioning, Gas Turbine Conference and Exhibit, Houston, Texas - March 18-21, 1985.

6. G. M. McColvin, et al, Application of the second generation DS superalloy CM186LC to first stage turbine blading in EGT industrial turbines. Fourth International Charles Parsons Turbine Conference, Advance in turbine materials, design and manufacture, 4-6 November 1997. Newcastle upon Tyne, UK.

7. P. S. Burkholder et al, CM186LC Alloy single crystal turbine vanes, International Gas Turbine and Aeroengine Congress, Indianapolis, USA, June 7-10, 1999.

8. R. F. Singer, Advanced materials and processes for land-based gas turbines Materials for Advanced Power Engineering, Part II, eds. D. Coutsouadis et al, Kluwer Academic Publishers, London, 1994 p. 1707.

9. T. F. Fitzgerald and R. F. Singer, "Modelling of the directional solidification of turbine blades via liquid metal cooling", Proc. 2nd Pacific Rim Intl. Conf. on Modeling and Casting Solidification Processes, 1995 Ed. E. Niyama and H. Kodama.

10. J. Grossman et al - Investment casting of high performance turbine blades by liquid metal cooling - A step forward towards industrial scale manufacturing, Proc. Intl. Symp. on Liquid Metal Processing and Casting, Santa Fe, NM, February 6, 1999.

11. D. S. Duvall, W. A. Owczarski and D. F. Paulonis : "TLP bonding : A new method for joining heat resistant alloys" Welding Journal 53 1974 p. 203.

12. B. B. Seth, P. D. Freyer, M. A. Hebbar, T. Zagar, and M. A. Burke "Cost effective manufacturing of high-performance power generation combustion turbine component using the fabricated component method" in "Proceedings of the 4th Charles Parsons Turbine Conference" Eds. A. Strang, W. M. Banks, R. D. Conroy, M. J. Goulette. The Institute of Materials London, 1997.

13. I. Tuah-Poku, M. Dollar, T. B. Massalski, "A study of the transient liquid phase bonding process applied to Ag/Cu/Ag sandwich joint" Metallurgical Transactions, 19A 1988 p. 675.

14. A. D. Helms et al - Extending the size limits of cast wrought superalloy ingots, Superalloys, Ed. R. D. Kissenger et al, The Minerals, Metals and Materials Society, 1996.

15. G. L. Erickson, Superalloys resist hot corrosion and oxidation, Advanced Materials and Processes, March 1997, p. 27.

16. G. W. Goward and L. W. Cannon, Transactions of the ASME, 110, January 1988, p. 150.

17. Deposition technologies for films and coatings, Development and Applications Ed., R. F. Bunshah, Noyes publication, New Jersey 1982, p. 335.

18. A. Strang, E. Lang, Effect of coatings on the mechanical properties of superalloys, High Temperature Alloys for Gas Turbines. R. Brunetaud, D. Coustsordais, T. B. Gibbons, Y. Lindblorn, D. B. Meadowcroft, R. Stickler, Eds., D. Reidal Publishing Company, London, U. K., 1982, p. 469.

19. G. W. Goward, Protective coatings – Purpose, role and design, Proceedings of the Conference on Protective Coating Systems for High Temperature Gas Turbine Components," Mat. Sci and Tech., 2, 1972, p. 189.

20. R. A. Miller, R. G. Garlick and J. L. Smialek, Phase stability in plasma sprayed partially stabilized zirconia-yttria, Science and Technology of Zirconia, Vol. 3, Ed. A. H. Hueur and L. W. Hobbs, The American Ceramic Society, 1981, p. 241.

21. D. S. Suhr, T. E. Mitchell and R. J. Keller, Microstructure and durability of zirconia thermal barrier coatings, Advances in Ceramics, Vol. 3, Ed. A. H. Hueur and L. W. Hobbs, The American Ceramic Society, 1982, p. 503.

22. K. J. C. Chou and J. C. Earthman, Characterization of low cycle fatigue damage in inconel 718 by laser light scanning, J. Material Res. 12 no. 8, 1997, p. 2048.

23. N. Goldfine and D. Clark, Materials characterization using model based meandering winding eddy current testing, EPRI Topical Workshop, Electro-Magnetic NDE Applications in Power Industry, Charlotte, U. S. A., August 21-23. 1995.

24. K. Na. Jong et al, Linear and non linear ultrasonic properties of fatigue 410Cb stainless steel, Review of Progress and Quantitative Nondestructive Evaluation, Ed. D. O. Thompson and D. E. Chimenti, Plenum Press, New York, 1996 p. 1347.

25. P. W. Davies, J. P. Dennison and H. E. Evans, Recovery of a nickel-base high temperature alloy after creep at 750ºC, Journal of the Institute of Metals, 94, 1966 p. 270.

26. P. W. Davies, J. P. Dennison and H. E. Evans, The kinetics of the recovery of creep properties during annealing of Nimonic 80A after creep at 750ºC, Journal of the Institute of Metals, 95 1967, p. 231.

27. Cina and S. Myron, Means for extending the secondary creep life of nickel-base superalloys, Proceedings of the Second International Conference on Mechanical Behavior of Materials, 1976, p. 2025.

28. A. Balden, Rejuvenation procedures to recover creep properties of nickel-base superalloys by heat treatment and hot isostatic pressing techniques – A review Journal of Material Science, 26, 1991, p. 3409.

29. D. C. Steward et al – Rejuvenation of turbine blade material by thermal treatment, AFML-TR-79-4032, July 1979.

30. P. Zombo and F. Roan, System and method for using X-ray diffraction to detect subsurface crystallographic structure, U. S. Patent No. 6005913.

31. M. H. Haafkens and J. H. G. Matthey, A New Approach to the weldability of nickel-base as-cast and powder metallurgy superalloys, Welding Journal, Nov. 1982, p. 25.

32. D. S. Duvall and W. A. Owczarski, Heat treatments for improving the weldability and formability of Udimet 700, Welding Research Supplement, Sept, 1971, p. 401s.

33. B. Jahnke, High temperature electron beam welding of the nickel-base superalloy IN 738LC, Welding Journal Supplement, Nov. 1982, p. 343s.

34. Calculation of phase diagrams and thermochemistry of alloy phases, Ed. Y. A. Chang and J. F. Smith, Proceedings TMS-AIME, 1979.

35. Computer modeling of phase diagrams, Ed. L. H. Bennett, TMS, 1986.

36. Phase diagram calculations for Ni-based superalloys, N. Saunders, Superalloys 1996, Eds. R. D. Kissinger, et al, TMS 1996.

Ingot, Powder and Deformation Processing

CHARACTERIZATION OF FRECKLES IN A HIGH S[TRENGTH WROUGHT NICKEL SUPERALLOY]

[handwritten notes:] GW Gatorizable Waspaloy granda like GE PW mixed alloy. freckle has N₃Ti η. Root cause : proprietary

Paul D. Genereux and Ch[...]

Pratt & Whi[tney]
United Techno[logies]
400 Main St[reet]
East Hartford, Conne[cticut 06108]

Abstract

Nickel base superalloys, such as Gatorizable Waspaloy®, U-720®, and Rene 88® contain relatively high weight percents of hardening elements (e.g. Al & Ti), for a conventionally cast (VIM/VAR or VIM/ESR/VAR) and wrought (press and or rotary forged) product. Due to the higher volume fraction of gamma prime, this class of alloys has increased susceptibility for the formation of melt segregation defects, such as freckles. The propensity for freckle formation is further aggravated by the need for larger diameter ingots to support the trend toward higher thrust large diameter fan engines used on wide body twin engine aircraft. A review and assessment of production quality control techniques for freckle inspection are discussed. An in depth microstructural, chemical, and structural characterization is presented. The effects of freckle defects on key mechanical properties for the design of critical rotating engine components are shown. Finally, a brief discussion of a root cause investigation on the origin of freckle formation in this alloy class is given.

Gatorizable Waspaloy® is a trademark of United Technologies Corporation
U-720® is a trademark of Special Metals Company
Rene 88® is a trademark of the General Electric Corporation

Introduction

Vacuum arc remelting (VAR) is the most commonly used process for the remelting of superalloys. The VAR process involves the continuous melting of a consumable electrode, typically cast in a vacuum induction melting (VIM) furnace or the product of an electroslag remelt (ESR) process. Melting is achieved by the striking of a DC arc under vacuum between the electrode and the melt pool. The melt pool solidifies in a water-cooled copper crucible. VAR furnace design has remained largely unchanged since its inception in the late 1950s. The significant advances came from furnace controls and in the area of automation of the melting and process control through the selection of optimum melting parameters. The primary benefits to the vacuum arc remelting process are: elimination of pipe, the removal of dissolved gases, the reduction of oxide inclusions, minimization of detrimental trace elements due to their high vapor pressure, and chemical uniformity.

Due to the physics of the process, consumable electrode melting also is associated with unique melt related defects in the form of solute segregation such as tree ring patterns, white spots and

fre[ckles]. Segregation resulting in tree rings is not considered to have a significant effect on mechanical properties; however, the latt[er two] are generally considered to be detrimental. White spots, as [the name] suggests, appear as light etching regions and are gen[erally] lower in alloying elements (e.g. Ti or Nb). The form[ation] of white spots is presently considered to be caused by one [of the following] three elements: 1.) fall-in of shelf or crown material into the melt pool, 2.) fall-in of unmelted dendrites from the consumable electrode, and 3.) localized decrease or arrest in the solidification rate. White spots are considered to be an unavoidable consequence of VAR processing.[2] The primary goal for the melt source in controlling white spots is to minimize the occurrence and severity.

Freckles are characteristic of a deep melt pool causing interdendritic channeling of low-density element (e.g., Al and Ti) rich liquid. Other factors that may lead to freckle formation are disruptions in electromagnetic fields and variation in heat transfer rates in the melt pool and mushy zone. Freckles are generally continuous and longitudinal in orientation and show up as dark etched spots in a macro-etched condition.[3] They are most commonly, but not exclusively, found at an approximate mid-radial location in the ingot. Freckles consist of high concentrations of hardener elements (i.e. Ti or Nb) similar in chemistry to gamma prime (γ′). Detection of freckles in both ingot and partially or fully converted billet form can be very difficult due to poor etch contrast, density similarities to the base metal, and the tendency not to crack even after significant ingot reduction.

Significant work has been conducted to gain a better understanding of freckle formation during the VAR process.[4,5] This work has focused on the elimination of freckles in the ingot through the use of tight process controls. The primary concerns with freckle containing material are the microstructural and chemical variations, which may result in the formation and concentration of phases significantly harder than the base metal. This high concentration of hardening elements has the potential to be detrimental to mechanical properties because of the presence of local hard and possibly brittle regions that may act as crack initiation sites. Limited literature is available which quantifies these chemical composition and hardness differences and especially the property debit associated with freckle defects. Specifically, very little literature exists regarding this type of melt

Superalloys 2000
Edited by T.M. Pollock, R.D. Kissinger, R.R. Bowman,
K.A. Green, M. McLean, S. Olson, and J.J. Schirra
TMS (The Minerals, Metals & Materials Society), 2000

defect in the class of wrought, high volume fraction γ′ superalloys, which defines Gatorizable Waspaloy.

Industry trends over the past several decades have pushed the envelope for achieving higher strength wrought alloys at larger ingot diameters. These trends have been largely fueled by the inherent cost and cleanliness benefits that modern wrought alloys have over powder metal product and by improvements in conventional melting and conversion techniques.

Wrought nickel based superalloys, specifically those containing relatively high percentages (> 35% by volume of γ′) of hardening elements (e.g., Al and Ti), tend to be more susceptible to freckle formation due to the relatively narrow melt window for this alloy classification (e.g. Gatorizable Waspaloy, U-720, Rene 88, etc.). The inherent difficulty of freckle detection and the use of this product for critical rotating applications warrant an in-depth microstructural investigation and quantification of the effects of freckles on mechanical properties.

Background

In the early days of Gatorizable Waspaloy development, alloy melt difficulties were occasionally encountered in maintaining proper melt parameters resulting in the occurrence of freckles. This problem was further aggravated by the need for larger diameter ingots due to increased part size resulting from the trends towards larger fan diameter engines, and by the relatively high hardener element content for this material, as shown in Table I. Engines requiring this material, including the largest PW4000 models, have been developed to power a new generation of more fuel efficient wide body commercial twin engine aircraft.

Table I Nominal Composition of Conventional Waspaloy & Gatorizable Waspaloy

Element	Conventional Waspaloy	Gatorizable Waspaloy
Ni	Bal.	Bal.
Cr	19.5	15.0
Co	13.5	13.5
C	0.05	0.03
Ti	3.0	4.6
Al	1.4	2.2
Mo	4.1	4.1
B	0.007	0.007
Zr	0.08	0.08

During the process development program, a point indication (Figure 1) was discovered during final macro-etch inspection of a Gatorizable Waspaloy low pressure turbine (LPT) shaft. Replica metallography showed a structure, of what appeared to be, a high density of γ′ and large blocky ribbon-like precipitates of an unknown origin. This was believed to be a freckle melt segregation defect. This freckle indication had a diameter of 1.5 – 2.0 mm and was traced back to the 18 cm diameter billet running longitudinally equivalent to, or greater than 188 cm of billet length. This indication was located at a radial position of approximately 6 cm from the center line. The shaft was subsequently sectioned and the detailed metallurgical evaluation commenced.

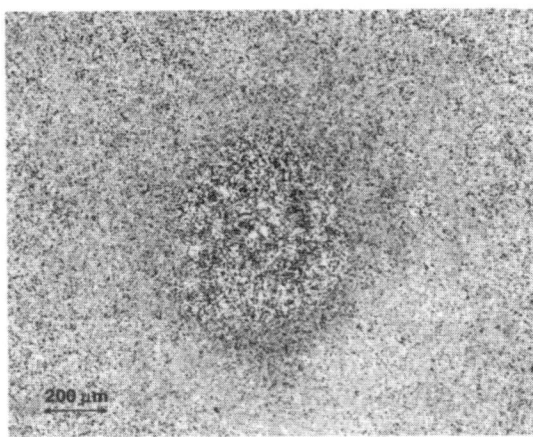

Figure 1: Photomicrograph of low density freckle found in Gatorizable Waspaloy LPT development shaft.

The freckle appeared darker than the matrix in a macroetched condition and microstructurally was found to contain a high volume fraction of large blocky ribbon-like and spheroidal precipitates that resembled overaged γ′. This was all contained in a matrix of what appeared to be fine γ′ in a gamma (γ) matrix.

During the production process macro-etched plates are the primary means of screening melt segregation defects at the billet stage. This is necessary because down stream etch and visual inspection can be compromised by material flow and the complexity of the square cut ultrasonic and finished machined shapes.

Conventional nondestructive inspection techniques may need some optimization for the fail-safe detection of freckles in billet material. Work was initiated to investigate alternate and potentially more reliable etching procedures to highlight structural differences on a macro scale. Standard production macro-etchants applied to billet slices and square cut ultrasonic parts, can at times, show poor contrast for detecting the segregation associated with freckles, making reliable detection heavily dependant on inspector interpretation and diligence.

Ultrasonic inspectability of alloys in this classification, with fine uniform grain size (typically ASTM 11), is generally very good due to the relatively low background noise. This fact, combined with the perceived hard and brittle nature of freckles due to local segregation and the significant reduction during conversion from ingot to billet, might be expected to result in cracking of the freckle defect, rendering it ultrasonically detectable. However, ultrasonic inspection of the part containing the freckle showed no evidence of any discontinuity despite conclusive evidence that the part contained a freckle defect.

A plan was formulated to conduct a thorough characterization of the defect. This plan involved full metallurgical evaluation including detailed metallography, macro and microhardness, macro-etch development trials and a controlled quantitative ultrasonic evaluation. Additionally, chemistry and structural information was obtained through the use of x-ray diffraction and electron microscopy employing energy and wavelength dispersive spectroscopy as well as selected area electron diffraction.

Analysis and Results

Based on the location and orientation of the defect in the billet and what appeared to be an increase in hardener elements, from a cursory chemical evaluation, the defect was most likely a freckle that formed due to melt segregation. The large blocky ribbon-like precipitates, contained within the freckle, were suspected to be eta (η). According to literature η phase (Ni_3Ti) exists in nickel based superalloys, particularly in alloys having high titanium/aluminum ratios, and it typically precipitates in a Widmanstatten appearing pattern. In contrast to face centered cubic (FCC) γ and γ', η is hexagonal and forms to near stoichiometric composition with little or no solubility for other elements. Eta (η) does not precipitate as rapidly as γ' but grows more rapidly into coarse large precipitates.[6] Plans were made to evaluate this phase for hardness, chemical, and structural character.

Most of the freckle evaluation conducted in this paper was done using freckles from the hot top region of a VAR ingot that experienced an identifiable melt disruption. These freckle defects showed a more well defined higher density of the suspected η phase and were more suited for the planned analyses. Although the suspected η phase within the freckle appeared denser than the indication found in the LPT shaft they appear visually identical in morphology.

Micro Metallography

Optical micro metallography was performed in both a transverse and longitudinal orientation to the axis of the VAR ingot. In the as polished condition, an increase in MC carbides in the location of the freckle was observed, as compared to the base metal matrix location. Figures 2 & 3 are photomicrographs of the freckle in the transverse and longitudinal orientation. Common γ' etchants clearly define all features within the freckle. The suspect η phase appears particularly Widmanstatten in the longitudinal etched condition.

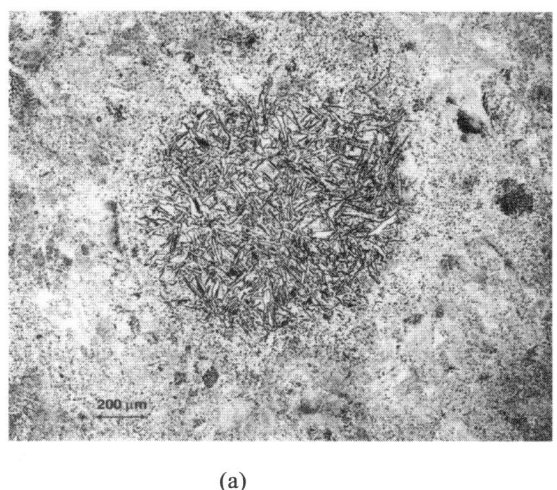

(a)

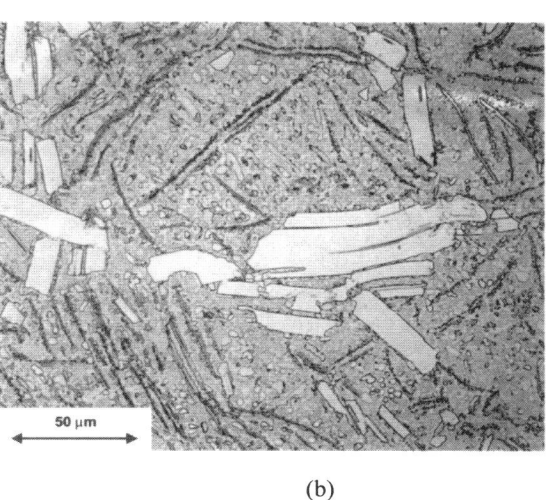

(b)

Figure 2: Photomicrographs of transverse section of freckle (a & b) from Gatorizable Waspaloy billet.

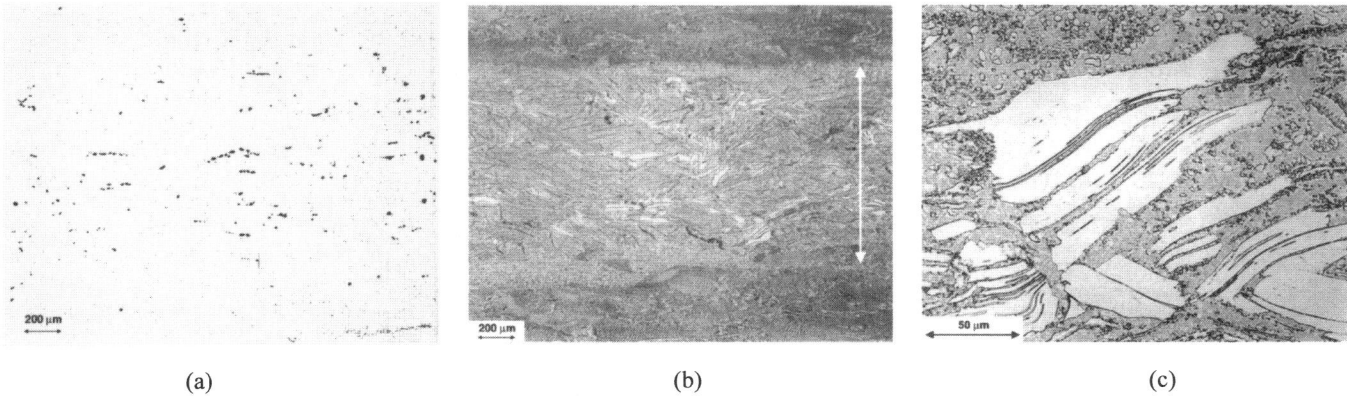

(a)   (b)   (c)

Figure 3: Photomicrographs of longitudinal section (a) as polished and (b & c) etched of freckle from Gatorizable Waspaloy billet. Arrow in (b) denotes freckle width.

21

Macro, Micro & Nano Hardness

Further characterization of the freckle was conducted through the use of hardness measurements. Macro, micro and nano hardness techniques were utilized at various regions within the freckle and outside the freckle in the base metal. Macro hardness results, which attempted to assess the overall bulk hardness of the freckle compared to the base metal, are show in Table II. The diamond indenter was positioned within the freckle with a sufficient load to insure the overall bulk hardness effect of the freckle was measured. The data indicates the bulk hardness within a freckle region does not differ significantly from the hardness outside the freckle in areas with normal microstructure.

Table II Macro hardness results taken within the freckle indication and outside the freckle to represent bulk hardness.

	VHN	HRC
Matrix	400	40.5
Freckle	390	39.8

When hardness measurements are performed on a micro scale as shown in Table III a dramatic difference is observed. Because the three dimensional morphology of the blocky ribbon-like phase was unknown (may be blocky or thin), and the difficulty of placing indentations in very small precipitates, nano hardness techniques were employed. Results from the nano hardness work show both phases within the freckle region, the blocky ribbon-like precipitates and the smaller spheroidal precipitates, exhibit similar high hardness values. Whereas, the matrix surrounding these precipitates, still within the freckle, the hardness is considerably lower than the precipitates and more inline with the typical bulk hardness for this material. The hardness readings of the precipitates within the freckle were approximately twice that of the matrix locations both within the freckle and in the base metal. The discrepancy in results, between the micro and nano hardness values, may be due to the size of the micro indenter relative to the precipitate size. This may have caused erroneous results due to the influence from the surrounding matrix. Thus, the nano hardness may be a better representation of the true hardness differences between the precipitate and the matrix. Table IV summarizes the hardness at several locations using nano techniques. Figure 4 shows a typical nano hardness indenter mark within the blocky ribbon-like precipitate.

Table III Micro hardness results within the blocky ribbon-like precipitates and in the surrounding base metal.

	VHN
Base Metal Matrix	460
Freckle Matrix	440
Blocky Precipitates in Freckle	550

Table IV Nano hardness results taken within the blocky ribbon-like precipitates, spheroidal precipitates, and matrix all within the freckle; in addition to the matrix away from the freckle.

	VHN
Base Metal Matrix	520
Freckle Matrix	550
Blocky Precipitates in Freckle	1120
Spheroidal Precipitates in Freckle	1030

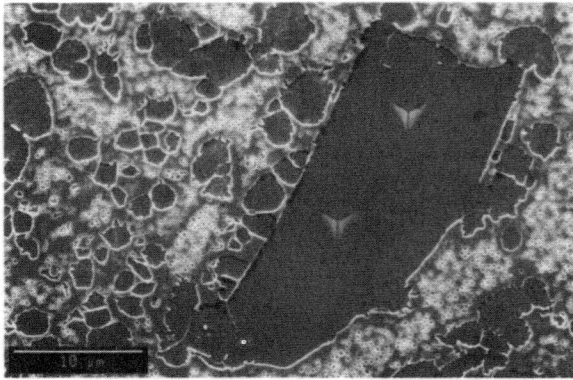

Figure 4: FESEM image of nano hardness impressions located in blocky ribbon-like precipitate within the freckle.

Macro-Etch Development

The most reliable technique for the detection of freckles in a routine production environment is macro-etch acid inspection. Using the standard macro-etch, a freckle indication typically appears as a dark gray spot in a lighter gray matrix. As previously stated, this single point indication was found during final etch inspection of the fully machined development part. This was the last of three inspections this part would receive having successfully cleared both the billet and post forge square cut sonic machined stages. This illustrates the criticality of reliable and definitive up-stream in-process macro-etch inspection methods.

Trials were conducted in an attempt to optimize the production macro-etch practice utilizing the same freckle with repeated repolished surfaces. Etching was performed on a surface that was grit blasted using 60 grit media and the test coupon was completely immersed in the acid solution. These trials consisted of evaluating several macro-etch candidates ranging from the standard production macro plate mill etch (diluted sulfuric, nitric, and hydrofluoric acids) to the finished machine part etch (anhydrous ferric chloride, nitric, and hydrochloric acids). Additionally, diluted HCl and HCl + H_2O_2 etches were evaluated. Figures 5a – 5d show a coupon from the billet slice that has been prepared and etched using the four macro-etch candidates.

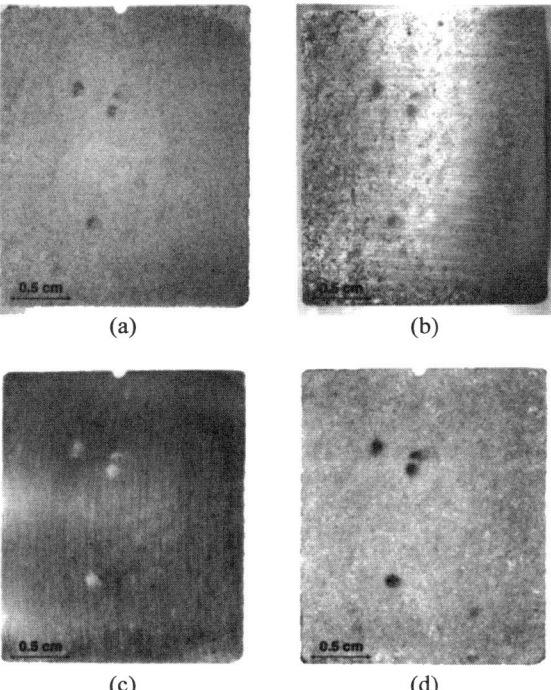

(a)   (b)

(c)   (d)

Figure 5: Coupon from billet containing freckles in macro etched condition: (a) Production mill etch, (b) finished machine part etch, (c) $HCl + H_2O$, and (d) $HCl + H_2O_2$.

The typical mill billet inspection etch turned the base metal uniformly gray while the indications appear darker. The freckle contrast can be poor and may go undetected, as was the case in the LPT development shaft. When the finished machined part etch was applied to a coupon, as is required for production, the matrix became spotty and non-uniform. The three freckle indications are difficult to see and can be confused with details within the matrix. The results of a diluted 2:1 HCl and water trial show the background has become very dark and the three freckle indications appear lighter. This high contrast makes detection relatively easy; however, good lighting is necessary to insure reliable detection. The $HCl + H_2O_2$ etch was similar in appearance to the production billet inspection etch but with better contrast. This etch effectively attacks the matrix while darkening the freckle. Generally, it was found that the surface roughness and amount of surface residual work influences the effectiveness of the macro-etchant.

Quantitative Ultrasonic Evaluation

The high hardness and possible brittle nature of this phase is evident in the micro and nano hardness work described above. Considerable material deformation, through press and rotary forging during the billet conversion process is imparted; it would be reasonable to suspect cracking or debonding in the interface between the parent matrix and the large blocky ribbon-like phase. However, during all inspections on this part, no significant difference in ultrasonic response was observed between the area associated with the freckle indication and an area without a freckle indication; indicating this type of melt defect is virtually

undetectable with the use of existing ultrasonic inspection techniques.

To better quantify inspection sensitivity, longitudinal and shear wave ultrasonic inspection techniques were evaluated. Blocks from the same production billet slice, at locations that contained freckles and at a location with normal microstructure free of freckles, were prepared for ultrasonic inspection. The blocks (measuring 5 cm x 5 cm x 10 cm thick) were polished on the sides to insure good surface finish and were machined parallel for optimum inspectability. The freckle indication was 2 mm in diameter and ran the entire length of the block. Because the grain size in this material is very fine (ASTM 11), no significant attenuation occurs and inspection to a #1 flat bottom hole (FBH) is possible. Longitudinal and shear wave ultrasonic inspection techniques were utilized. Comparing scans from the block containing freckles to block containing the normal microstructure, no differences in ultrasonic response was observed. This was consistent in scans taken from both the transverse and longitudinal orientations. Subsequent metallographic evaluation showed no cracking or debonding of the freckle. These results show that it is not possible to find indications of this type (intact freckles) through the use of current ultrasonic techniques.

Scanning Electron Microscopy and Dispersive Spectroscopy

Both field emission scanning electron microscopy (FESEM) and electron microprobe analysis (EMP) were employed to obtain high magnification high resolution photomicrographs. Additionally, energy dispersive spectroscopy (EDS) and wavelength dispersive spectroscopy (WDS) techniques were used in an attempt to get qualitative and semi-quantitative chemical data from the phases associated with the freckle indication. High resolution photomicrographs using the FESEM are shown in Figure 6a. Qualitative x-ray maps (Figures 6b – 6g) show that the blocky ribbon-like and spheroidal precipitates are enriched in nickel and titanium while depleted in chromium, cobalt, and molybdenum as compared to the base metal matrix. The level of aluminum is relatively constant throughout.

Using the EMP, a WDS chemical trace across the freckle highlights the concentration variations within this region. This semi-quantitative chemical analysis (Figure 7) illustrate the difference in the nickel, cobalt, chromium, molybdenum, titanium, and aluminum amounts within the freckle when compared to the base metal matrix.

X-Ray Diffraction

Phase and crystal structure identification was initially attempted using x-ray diffraction (XRD) on a polished slice containing the defect. The entire surface was masked off except for a 3 mm diameter circle exposing the freckle. Scan dwell times of 55 seconds over a range of 20° - 100° at increments of 0.02° were used for greater sensitivity. The interplanar spacings obtained were typical of a nickel base γ-γ′ alloy, with the exception of two small peaks located at 2θ values corresponding to d-spacing values of 0.21414 nm and 0.19560 nm. These values closely correspond to two η phase peaks, (201) and (202). Due to the relative low intensity of these two peaks and lack of associated peaks, confident resolution and subsequent phase identification was not possible. This is surprising due to the apparent significant volume of the suspect phase as compared with the matrix. This may suggest that the morphology of this phase is thin and ribbon-like resulting in a relatively small contribution of the total x-ray data collected.

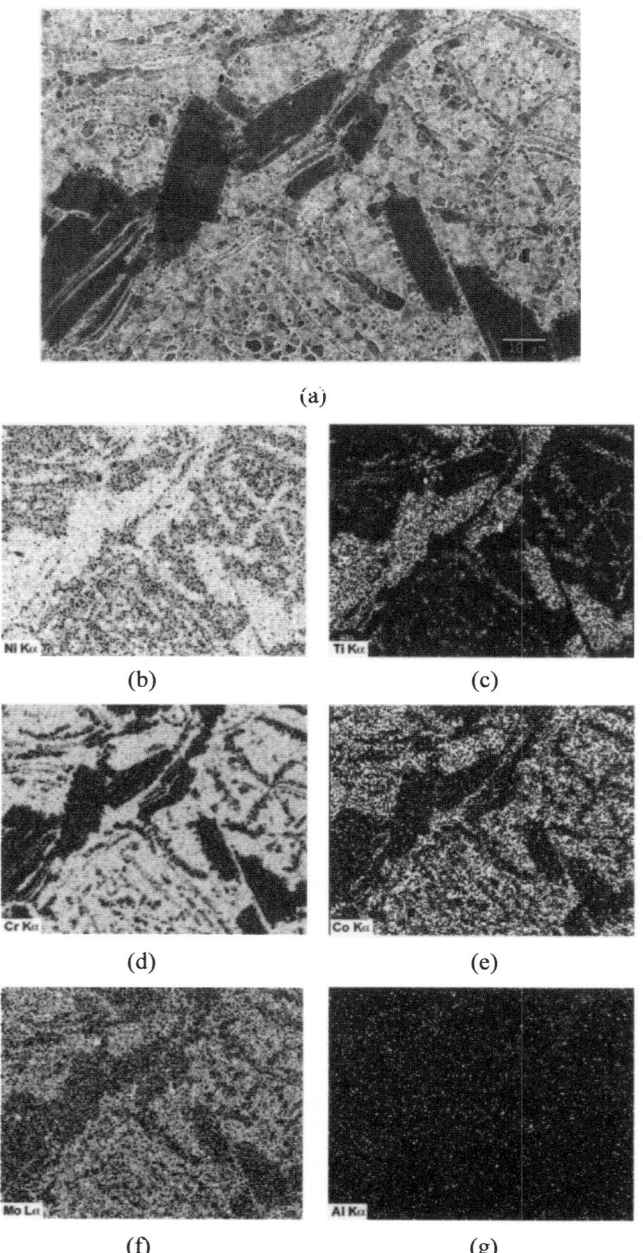

(a)

(b)

(c)

(d)

(e)

(f)

(g)

Figure 6: FESEM image (a) of freckle indication with associated x-ray maps for (b) nickel, (c) titanium, (d) chromium, (e) cobalt, (f) molybdenum, and (g) aluminum.

Transmission Electron Microscopy

Due to the inconclusive results from the XRD analysis, transmission electron microscopy (TEM), along with selected area electron diffraction (SAED), and micro-chemical analysis were utilized so accurate crystal structure and chemistry could be determined. TEM thin foil analysis is advantageous for thin

ribbon-like appearing phase identification. The foils are sufficiently thin and beam size sufficiently small to effectively isolate the phase from any matrix contributions.

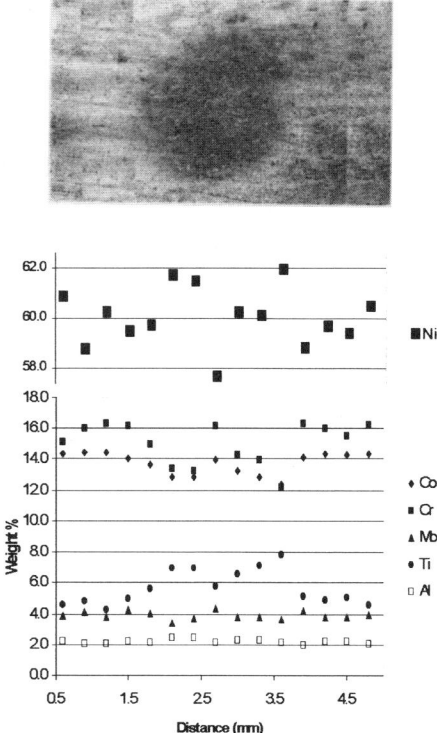

Figure 7: WDS compositional traces with corresponding image of area analyzed. Note the spikes in Ni and Ti and depressions in Co, Cr, and Mo at locations corresponding to the freckle indication.

Thin films were sliced from polished coupons, electro polished and ion milled to obtain an electron transparent surface for the analysis. TEM bright field images and electron diffraction patterns taken in the blocky ribbon-like precipitate are shown in Figures 8a – 8d. Electron diffraction pattern analysis indicate the blocky ribbon-like precipitate to have an ordered hexagonal crystal structure. This is significant in that it rules out the possibility of these precipitates being γ, γ', or carbide, all of which are FCC. Because of their smaller size, spheroidal shape and chemical composition the smaller round precipitates that showed slightly reduced hardness are believed to be overaged γ'. Further analysis is needed to confirm this.

Chemical analysis performed in the TEM using EDS (Figure 9) showed a small amount of solubility for cobalt, aluminum and chromium. This is in contrast to what was stated in literature, that η forms near the Ni_3Ti stoichiometric composition and should consist only of nickel and titanium. This may be alloy composition dependent and may not hold true in every alloy system.

24

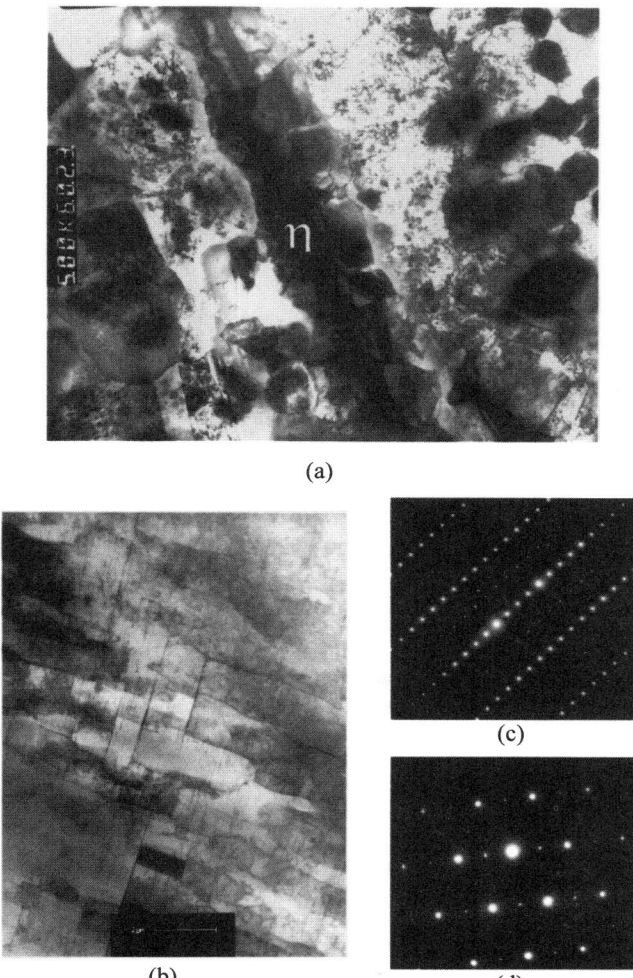

(a)

(b)   (c)

                       (d)

Figure 8: TEM bright field images (a & b) with associated electron diffraction patterns for the (b) 010, and (c) 10$\bar{1}$ zone axes. The η phase is designated on image (a).

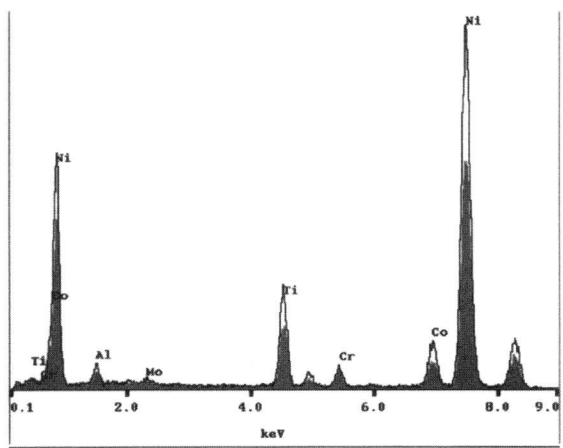

Figure 9: TEM EDS spectra of η phase within freckle.

Mechanical Properties

Mechanical test specimens were extracted with freckles positioned in an appropriate location normal to the applied stress. Cylindrical tensile and fatigue specimens were selected in an attempt to contain the full freckle diameter within the gage diameter. Both smooth and notched uniaxial fatigue specimens were extracted from the freckled area. Additional fatigue specimens were extracted from adjacent locations to insure the absence of freckles and achieve baseline property comparisons. Both tensile and fatigue testing were conducted in an effort to quantify the effects of freckles on critical design properties.

Tensile specimens were tested at room temperature. Results of the tensile testing show yield strengths equal to baseline and typical for this alloy. Ductility, however, was on average 20% of baseline and in some cases approached zero. This poor ductility resulted in the ultimate tensile strength being roughly equal to the yield strength. Figure 10 summarizes the tensile results as compared to baseline. Metallography of the tensile fracture in the area of the freckle is shown in Figure 11a. There is evidence of fracture at the blocky η precipitates as well as secondary cracking between this phase and the matrix as shown in Figure 11b.

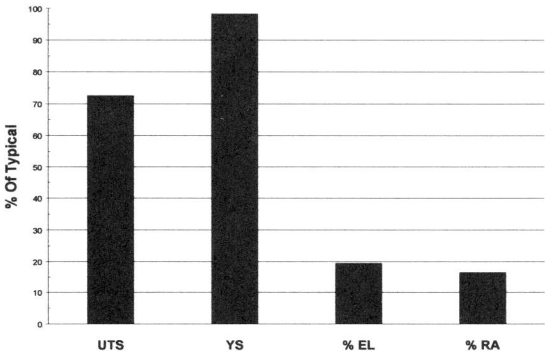

Figure 10: Tensile properties of specimens containing freckle as compared with specimens having normal Gatorizable Waspaloy microstructure.

Both smooth and notched low cycle fatigue specimens were tested to approximate engine relevant conditions. A typical fatigue fracture surface is shown in Figure 12, where the freckle location is clearly visible as a bisected cylinder through the gage section illustrating the brittle nature of this defect. Results of the fatigue testing indicate a smooth life debit of 4X, based on B0.1 lives, with failure origins at freckle sites. Figure 13 shows a plot comparing smooth fatigue life of specimens containing freckles vs. specimens free of freckles, both extracted from the same billet slice. It was not possible to assess the effect of freckle segregation on notch fatigue life on a B0.1 basis. Post test analysis show that the accuracy of locating the defect at the notch location corrupted the data. This likely contributed to the high scatter observed resulting in an artificially large debit. On a mean life basis an approximate debit of 2X, compared to baseline, is shown.

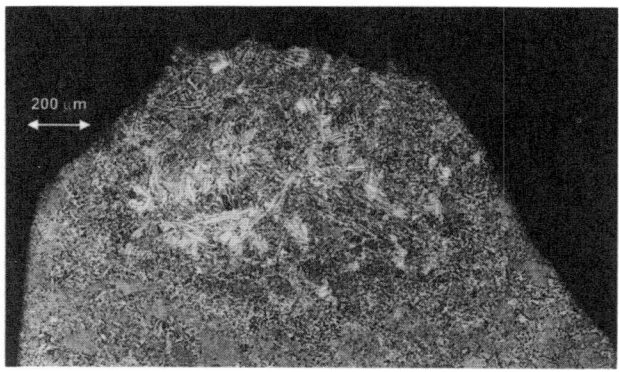

(a)

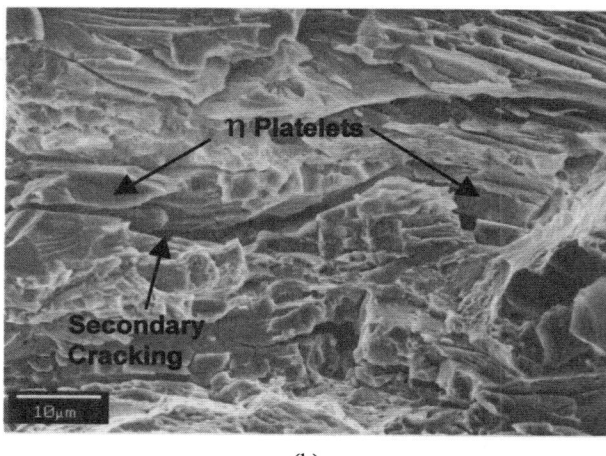

(b)

Figure 11: Photomicrograph in gage of failed tensile specimen (a) note freckle at fracture location, and SEM image (b) showing fracture occurred on η platelets and the presence of secondary cracking associated with η phase.

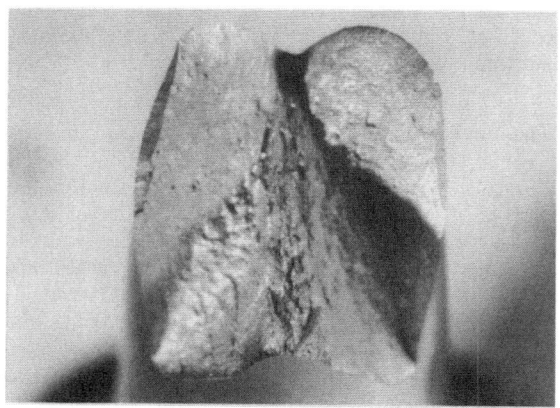

Figure 12: Typical fatigue fracture surface of specimen containing freckle defect.

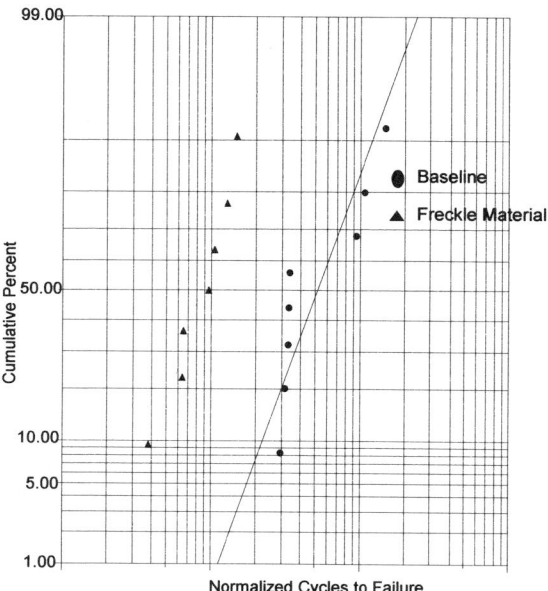

Figure 13: Lognormal plot of smooth fatigue life from specimens containing freckles compared to normal Gatorizable Waspaloy microstructure.

Conclusions and Recommendations

A thorough characterization of freckle melt defects found in Gatorizable Waspaloy product has been completed. A number of techniques were used in this characterization which include metallography, macro and nano hardness, macro-etch trails, ultrasonic inspection, x-ray diffraction and electron microscopy. Additionally, mechanical test specimens were utilized to assess the effects of freckle melt defects on key mechanical properties associated with aero engine LPT shaft designs.

Optical metallographic evaluation shows that the melt defect does appear to be a freckle. This indication contains what seems to be an increased concentration of carbides and γ' formers. The most distinguishing features are the large plate or ribbon-like precipitates that appear somewhat Widmanstatten in morphology.

Hardness measurements taken on a bulk or macro scale in the freckle indication and in the base metal matrix show no difference in hardness. On a nano scale, hardness values in the blocky ribbon-like precipitates and the large spheroidal overaged γ' appearing phase within the freckle were similar. These values were approximately double that of the γ matrix both inside and outside the freckle.

Evaluation of production inspection methods included both macro-etch and ultrasonic. An assessment of several alternative macro-etchants showed similar results, in terms of freckle delineation, when compared to the typical production billet inspection etch, used by the supplier. Additionally, some unique contrast enhancements were observed using alternate etchants, such as the HCl + H_2O_2 macro-etch. This etch produced a similar effect to the production billet etch but with more contrast making detection easier. However, due to the nature of this etch method, it was not consistently reproducible and not practical in a production environment. With the proper process control in place, the current production billet macro-etch should provide adequate and reliable detection of freckles in Gatorizable Waspaloy material.

Freckles were undetectable with the use of longitudinal and shear wave ultrasonic inspection techniques, despite low background noise. The conclusion that can be drawn from this observation is that the size and reduced density due to the increase in hardener content (Ti) is not sufficiently different to render the freckle ultrasonically detectable. Likely a crack or debond associated with the freckle would be necessary for detection. It is reasonable to consider that at the approximate mid radial location where freckles form, tensile strains are not sufficiently high during press or rotary forge to create cracks or debonds associated with the freckle. However, during the flange forging of a LPT shaft, where the use of deformation modeling predicts significant tensile strains, the freckle was not ultrasonically detectable and metallographic evaluation showed no cracks. This may be the result of improved high temperature ductility of the micro constituents within the freckle at the conversion or forge temperatures.

Chemical analysis of the blocky ribbon-like precipitates and spheroidal precipitates within the freckle, using x-ray mapping and EMP showed a higher concentration of both nickel and titanium precipitates within the freckle with a depletion of chromium, molybdenum and cobalt as compared to the base metal. Aluminum levels within the precipitates and in the base metal remained largely unchanged.

Phase and crystallographic identification using XRD was inconclusive for accurately identifying the large ribbon-like blocky precipitates within the freckle. As expected normal γ and γ' phases were identified. TEM analysis, including SAED, AEM, and EDS, was used to obtain structural and chemical information. The blocky ribbon appearing phase was conclusively identified as η (hexagonal Ni_3Ti) with evidence of ordering and with some limited solubility for elements other than nickel and titanium.

Mechanical property debits caused by the freckle defect were shown to be dramatic in both tensile and fatigue. Room temperature tensile ductility was reduced by 80%, as compared to baseline, and in one case approached zero. The ultimate tensile strength, at room temperature, was roughly equal to the yield strength, which did not differ significantly from baseline yield strength properties. Smooth fatigue properties of specimens containing freckles were debited by 4X, as compared to typical base metal results. On a mean life basis, notch fatigue specimens showed debits of 2X.

Because of the dramatic reduction in mechanical properties that were seen in this investigation, freckle defects of this type clearly must be eliminated in the VAR melted product. The only effective means for in process removal of freckles through inspection is with the use of billet macro plate etch inspection. The elimination of segregation defects such as freckles has principally been accomplished through tighter controls on the VAR melt process.

As a result of this event (a freckle found in a development part) a root cause investigation was initiated to review historical melt records and process parameters. Some of the VAR melt practices that were investigated include, but are not limited to: optimized melt rate (lower), electrode stress relief, improved ingot cooling with water and helium, use of electrode hot top practice, and attention to VIM electrode orientation.

In the author's opinion, implementing or optimizing some of the above listed melt process changes combined with increased inspector diligence and awareness training has effectively eliminated the risk of occurrence of freckle defects in Gatorizable Waspaloy. Some benefit may be derived from additional work to optimize macro-etch techniques for the reliable detection of freckles.

This paper has presented results from a detailed metallurgical characterization of freckle melt segregation defects found in a high volume fraction γ' large diameter wrought nickel based superalloy product. Little if any published data exists on the characterization of structural and chemical properties or in quantifying mechanical properties of freckle defects in this alloy classification. Results from this work illustrate the severity that freckle defects have on mechanical properties. Also highlighted is the difficulty of freckle detection using standard production inspection practices, which further emphasizes the need for the elimination of freckles through improved melting practices.

Acknowledgements

The authors would like to thank Laurence Jackman, Ph. D. and Christopher O'Brien from Allvac for insightful discussions on melting and conversion. Also, Joseph Bracale, Frederick Galli, Alexander Nytch, and David Snow, Ph. D. from Pratt & Whitney; with a special acknowledgement to William Lee and Gregory Levan for their many contributions.

References

1. L. W. Lherbier, "Melting and Refining," Superalloys II, ed. C.T. Sims, N.S. Stoloff and W.C Hagel, (New York: John Wiley & Sons, 1987), 403-405.

2. K. O. Yu, J. A. Domingue, G. E. Maurer and H. D. Flanders, "Macrosegregation in ESR and VAR Processes," Journal of Metals, 38 (1) (1986), 46-48.

3. R. J. Siddall, "Comparison of the Attributes of VIM + ESR and VIM + VAR Alloy 718," Superalloys 718, 625 and Various Derivatives, ed. E. A. Loria (Warrendale, PA: The Minerals, Metals & Materials Society, 1991) 34-35.

4. J. A. Van Den Avyle, J. A. Brooks, and A. C. Powell, "Reducing Defects in Remelting Processes for High Performance Alloys," JOM, 50 (3) (1998), 22-25.

5. J. M. Moyer, L. A. Jackman, C. B. Adasczik, R. M. Davis, and R. Forbes-Jones, "Advances in Triple Melting Superalloys 718, 706, and 720," Superalloys 718, 625, 706 and Various Derivatives, ed. E. A. Loria (Warrendale, PA: The Minerals, Metals & Materials Society, 1994) 43-45.

6. C. H. Lund and H. J. Wagner "Identification of Microconstituents in Superalloys" (Memorandum 160, Battelle Memorial Institute, 1962).

SIMULATION OF INTRINSIC INCLUSION MOTION AND DISSOLUTION DURING THE VACUUM ARC REMELTING OF NICKEL BASED SUPERALLOYS

W. Zhang, P.D. Lee and M. McLean
Department of Materials, Imperial College
London SW7 2BP, U~~K~~

R.J. Siddall
Special Metals Limited, Wiggin Wo~~rks, Holmer Road~~
Hereford HR4 9SL, U~~K~~

Abstract

Vacuum Arc Remelting (VAR) is one of the state-of-the-art secondary re-melting processes that are now routinely used to improve the homogeneity, purity and defect concentration in modern turbine disc alloys. The last is of particular importance since, as the yield strength of turbine disc alloys has progressively increased to accommodate increasingly stringent design requirements, the fracture of these materials has become sensitive to the presence of ever smaller defects, particularly inclusions. During the manufacture of these alloys extrinsic defects can be introduced in the primary alloy production stage from a variety of sources; these include ceramic particles originating from crucibles, undissolved tungsten (or tungsten carbide) and steel shot. In addition intrinsic inclusions can be produced during the VAR process itself such as those originating from the crown and shelf, often termed 'white spot'. VAR has proven to be effective in controlling inclusion content but optimization of the process, particularly for a new alloy composition, is a difficult and expensive exercise. Therefore, there is considerable interest in the use of numerical models to relate the process control variables to the final microstructure.

A transient macroscopic model of the VAR process has been developed and applied to simulate inclusion motion in the melt pool and the dissolution of these thermodynamically unstable phases in a liquid metal environment. The model has been applied to determine the trajectories, thermal history and

dissolution ~~of intrinsic particles for typical melt conditions of~~ INCONEL ~~718.~~

Introduction

The VAR p~~rocess has been investigated previously by both~~ experimental ~~and mathematical techniques. Early mathematical~~ approaches s~~olved for heat transfer and approximated the fluid~~ flow by usin~~g an enhanced thermal conductivity in the molten~~ region [1]. ~~The next stage of modeling complexity~~ incorporated ~~the fluid flow using a quasi-steady state~~ assumption [~~2,3]. More recently Jardy et al. [4] presented a~~ model that s~~imulated the transient nature of the flow and~~ macrosegregation. In the last few years some authors have used macromodels to investigate how the formation of defects such as freckles correlated to operational parameters [5,6]. The macromodels have also been coupled to micromodels to predict how these operational parameters alter the formation of grains and other microstructural features [7]. Another application of these macroscopic process models is the prediction of inclusion behavior. A preliminary study illustrating how particle tracking could be added to a process model was presented by the current authors [8].

This paper presents an extension of a transient model of the heat, mass and momentum transfer during VAR to include intrinsic particle dissolution. The origins of these particles are described first, followed by the theory of the particle tracking and dissolution model. The model is then applied to predict the thermal history and trajectories of a series of particle sizes,

---

[*] INCONEL is a trademark of Special Metals Corporation

Superalloys 2000
Edited by T.M. Pollock, R.D. Kissinger, R.R. Bowman,
K.A. Green, M. McLean, S. Olson, and J.J. Schirra
TMS (The Minerals, Metals & Materials Society), 2000

types and entry locations. The dissolution rate of these particles, and maximum size which can be safely dissolved, are explored.

Inclusion Origins

One of the practical benefits of vacuum arc remelting is its known ability to reduce the number of inclusions introduced in the primary alloy production stage. These extrinsic inclusions can come from a variety of sources including: ceramic particles from agglomerates of oxide and nitride particles; and undissolved tungsten (or tungsten carbide from machining tools). Between primary melting and VAR, steel shot used for cleaning may also become lodged in the outer surface of the electrode or trapped in the VAR crucible, turning into a potential source of inclusions. Finally, during the VAR process itself, intrinsic inclusions can be produced such as those originating from the crown and shelf, often termed 'white spots' because of their lighter etching appearance in billets. In this paper the inclusions will be subdivided by the location from which they originate: the crucible; the electrode; or the outer surface of the electrode.

From the Crucible

Waves or perturbations on the surface of the melt pool and metal vapor transport from the arc plasma can cause an area of rapidly solidified metal to form on the water-cooled crucible above the melt pool in the VAR process. This region is termed the 'shelf' and 'crown'. Oxide and nitride contaminants washed to the outside of the melt can also be entrained in the crown. Perturbations in the system may cause a fragment, especially of the crown, to detach from crucible wall and fall into the melt pool. The fragment may not completely remelt in the pool but instead it may be entrapped in the mushy zone, where it could form an inclusion (white spot), with or without ceramics particles, in the final ingot.

Electrode

Tungsten or tungsten carbide from cutting tools may exist as extrinsic inclusions in the consumable electrode. Although these are heavy particles that might not be efficiently removed or remelted, there was insufficient data on the dissolution rates to simulate these particles properly, and they have been left as a future topic for study. In this paper only intrinsic particles originating from the electrode will be studied, Nb lean electrode fall-in.

Vacuum induction melting (VIM) is used to obtain the correct composition and this liquid metal is then poured into a mold to produce the electrode for VAR. As the electrode solidifies, shrinkage occurs resulting in a pipe cavity towards the top of the electrode. In the shrinkage, dendrites of primary phase are often exposed as the liquid metal is drawn back to feed volumetric shrinkage closer to the mold wall. Some of this forest of large, solute lean dendrites may fall into the VAR melt pool. As the electrode is remelted in the VAR process, the interdendritic material will melt first due to its lower melting point, further exposing the dendrite spines. These dendrites

may melt faster near their trunks due to greater heat flow from conduction and the presence of increased solute causing them to fall in before being completely molten. If the dendrites do not completely melt in the pool before being entrapped in the mushy zone, they may cause solute depleted regions.

For some operating conditions a ridge or torus of metal will form at the outer edge of the electrode, thought to be dendrite fragments exposed by preferential melting [9]. This torus is solute lean. If it is undercut, falls into the melt, and survives until trapped in the mushy zone, it could form white spot similar to the 'pipe' dendrite clusters and crown fall-in.

Electrode Outer Surface Fall-in

As already mentioned above, the consumable electrode is produced by casting VIM melted metal into a mold. The metal is poured into the cold mold from a considerable height. When the molten stream of metal hits the bottom of the 2 m mold, it can splash up and freeze high on the wall. As the mold is further filled and the melt level passes above these frozen splashes, enveloping them, but not always completely remelting them. These splash particles may be detached from the electrode surface when heated by a cathode spot climbing up the electrode. If a sufficiently large mass of splash falls into the melt pool, it may not be completely remelted before entrapment in the mushy zone, becoming a region of different composition and microstructure in the final ingot.

Model Description

A transient finite volume model with a moving mesh was used to simulate the VAR ingot formation, solving the momentum, heat transfer and electric potential equations. This model has been presented elsewhere [3,7], hence only the development of particle trajectory and melting/dissolution theory is presented below.

Particle Motion

There are two well established methods of simulating the behavior of particles in molten metal. The first approach assumes a high density of fine particles, allowing the particles to be treated as a transported scalar - a continuum solute whose concentration is related to the particle density [10]. The second approach considers the behavior of a single independent particle in the fluid. Individual particle tracking provides an ideal way to investigate particle trajectories in the fluid and is particularly appropriate to follow the circulation of intrinsic inclusions; hence this approach was used in the present study.

The motion of an individual particle in a liquid is governed by many factors: the drag force; buoyancy; and capillary forces associated with solid-liquid and solid-solid interfacial interactions. Although the interfacial forces are important for particle agglomeration and adhesion, they are assumed to be negligible in this study. The particles were also assumed to be small compared to changes in the flow field allowing the fluid flow outside the particle boundary layer to be treated as uniform and particle rotation affects to be neglected.

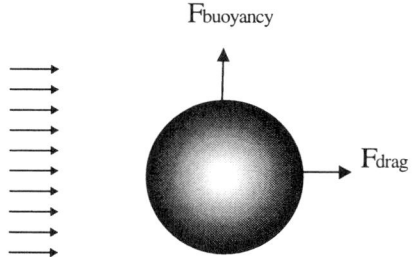

Figure 1.   Balance of forces on a particle in a fluid.

With these assumptions the general equation describing particle motion in the fluid can be derived by equating the acceleration of the particle to the buoyancy force, drag force and changes in particle mass, as shown schematically in Figure 1, giving:

$$M_p\left(\frac{d\vec{U}_p}{dt}\right) = V_p(\rho_p - \rho_l)\vec{g}$$
$$-\tfrac{1}{2}C_D A_p \rho_l(\vec{U}_p - \vec{U}_l)|\vec{U}_p - \vec{U}_l| + \frac{dM_p}{dt}\vec{U}_p \qquad 1$$

where M_p is the mass of particle, $\vec{U}_p$ is the velocity of particle, V_p is the volume of particle, ρ_p and ρ_l are the densities of the particle and liquid respectively, C_D is the dimensionless drag coefficient, $\vec{U}_l$ is the fluid velocity, D_p is the particle diameter and A_p is the cross-sectional area of the particle projected on a plane perpendicular to the direction of relative velocity, $\vec{U}_p - \vec{U}_l$. The velocity and properties of the fluid are calculated by interpolation from the macromodel grid. By solving equation 1 for the particle velocity in axial direction, u_p, and radial direction, v_p, the location of particle x_p and y_p can be calculated for each time step.

The drag coefficient, C_D, is calculated as a function of the Particle Reynolds Number, Re_p, as follows [11]:

$$C_D = \begin{cases} 24/Re_p, & 10^{-3} < Re_p < 2 \\ 18.5/Re_p, & 2 \le Re_p \le 500 \\ 0.44, & 500 < Re_p \le 2\times10^5 \\ 0.09, & 2\times10^5 < Re_p \end{cases} \qquad 2$$

where Re_p is defined as, $Re_p = \dfrac{\rho_l|\vec{U}_p - \vec{U}_l|D_p}{\mu}$.

Equation 2 assumes that the particles are spherical and hence is directly applicable only for some intrinsic inclusions. However, many intrinsic inclusions such as crown fall-in and electrode splash are plate-like or disk-like in shape. For non-spherical shapes Heywood [12] has suggested a multiplicative

correction factor for the drag force termed the volume shape factor, κ, given by:

$$\kappa = V / d_A^3, \qquad 3$$

where V is the volume of particle, d_A is the projected area diameter used to calculate the Re_p and C_D. He found this correlation was suitable for Reynolds numbers in the range of 1 to 1000. For lower Reynolds numbers creeping flow was assumed whilst for higher Reynolds numbers a sphericity correction was applied [13]. For most simulations in this paper Re_p was in the range where equation 3 was applicable.

Particle Melting/Dissolution

The rate at which a particle melts and/or dissolves in the molten nickel based superalloy is dependent upon both the heat and mass transfer between the particle and the surrounding fluid. Some inclusions with high melting points may only be dissolved by solutally driven dissolution; other particles may be dominated by thermally driven remelting. Both of these processes were simulated and are jointly termed 'dissolution' in this paper since they are solved simultaneously. The following assumptions were made in order to simulate the transient dissolution of inclusions during VAR:

1. The heat and mass transport is one dimensional. For spherical particles it is 1D in spherical coordinates; for plate-like particles it is 1D in Cartesian coordinates since the thickness is much smaller than any other dimension.

2. The solute and thermal distributions are uniform around the particle and boundary-layer.

3. The solute concentration at the interface of the film layer and bulk melt is constant, and the temperature of that interface equals that of the surrounding bulk liquid.

4. If a phase change occurs at the liquid and solid interface, thermodynamic equilibrium is established instantly.

With these assumptions, the thermal and solute profiles inside the inclusion are governed by the following partial differential equations. For thermal transport:

$$\frac{\partial(\rho h)}{\partial t} = \frac{1}{r^2}\frac{\partial}{\partial r}(\frac{k}{C_p}r^2\frac{\partial h}{\partial r}) + S_h, \qquad 4$$

Where S_h is a source term accounting for the phase change, ρ is the density, h is the enthalpy, t is the time, k is the thermal conductivity, C_p is the heat capacity and r is the spatial coordinate. For Cartesian coordinates ∂r becomes ∂x and the r^2 and $1/r^2$ terms are set equal to 1.

For solute transport:

$$\frac{\partial(\rho C_i)}{\partial t} = \frac{1}{r^2}\frac{\partial}{\partial r}(D_{i,s}r^2\frac{\partial C_i}{\partial r}) + S_{C_i}, \qquad 5$$

where S_{Ci} is a source term accounting for any reactions, C_i is the concentration of solute i and $D_{i,s}$ is the diffusion coefficient of solute i in the solid particle.

In order to solve these equations the boundary conditions must be determined. These boundary conditions are a strong function of the local flow characteristics, as presented separately below for the thermal and solute models.

<u>Thermal Boundary Conditions and Source Terms</u>   The heat transfer between the particle and bulk fluid can be approximated by a heat flux, q_b, dependent upon a solid-liquid heat transfer coefficient, h_{sl}:

$$q_b = A_{sp}h_{sl}\left(T_{bl} - T_{l\infty}\right),\qquad 6$$

where A_{sp} is the surface area of a particle; T_{bl} is the liquid temperature at the solid/liquid interface which is approximately equal to solid particle surface temperature T_{bs}; and $T_{l\infty}$ is bulk liquid temperature. The solid-liquid heat transfer coefficient is calculated from the Nusselt number, $Nu=h_{sl}D_p/k_l$. Nu is calculated assuming a sphere in forced convection [14]:

$$Nu \;=\; 2.0 \;+\; 0.6Re^{\frac{1}{2}}Pr^{\frac{1}{3}},\qquad 7$$

where Pr is the Prandtl number, defined as $Pr = \dfrac{C_p\mu}{k}$.

Equation 4 is solved in the domain of the solid particle using the boundary condition from equation 6 and a source term to account for the phase change given by:

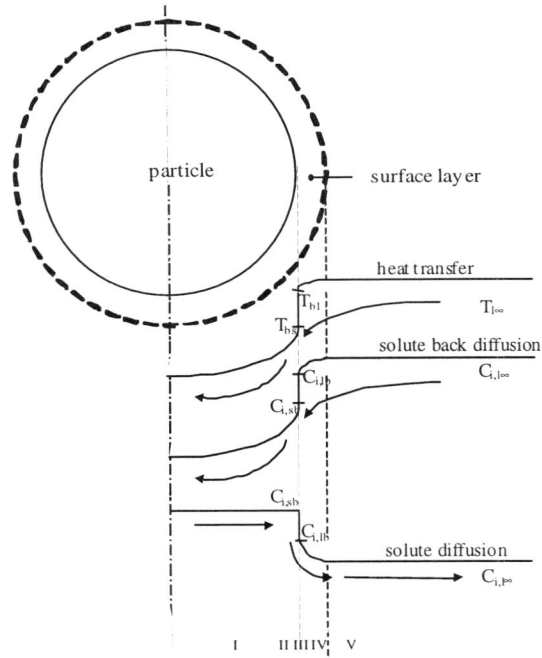

Figure 2.   Schematic of particle dissolution mechanisms.

$$S_h = \rho_p L_p \frac{df_s}{dt},\qquad 8$$

where L_p is the latent heat of the particle.

<u>Solute Boundary Conditions and Source Terms</u>   The dissolution of a solid particle in a liquid is governed by the rate at which solute molecules on the solid surface leave the face, reducing the mass of particle. The rate at which this process occurs can be controlled by any of the five steps [15,16] shown schematically in Figure 2 and listed below:

I.   Solid state diffusion of the solute molecules between particle core and surface.

II. Diffusion across a solid reaction product.

III. Transfer of the solute molecules across the solid-liquid interface to the liquid boundary layer.

IV. Solute diffusion within the liquid boundary layer.

V.   Solute transport from the liquid boundary layer through the bulk liquid by a combination of diffusion and convection.

Processes I and II are simulated over the domain of the particle by equation 5 whilst III and V are assumed to be negligible (i.e. there is no kinetic atomic detachment barrier and the bulk liquid is completely mixed). The thickness of solute boundary layer, IV, depends on both the particle properties and fluid properties and can be characterized by the Sherwood number, $Sh_p=h_{i,sl}d_p/D_{i,l}$, which relates an average mass transport coefficient between solute i in the solid and liquid, $h_{i,sl}$, to the molecular diffusivity of solute i in liquid phase, $D_{i,l}$ [17]. Using the mass transport coefficient the boundary condition of solute flux, $f_{i,b}$, for equation 5 can be determined:

$$f_{i,b} = h_{i,sl}A_{sp}\left(C_{i,l\infty} - C_{i,lb}\right),\qquad 9$$

where $C_{i,lb}$ is solute concentration of liquid at solid/liquid interface; and $C_{i,l\infty}$ is solute concentration of in bulk liquid.

For creeping flow the Sherwood number can be derived from a mass balance equation coupled to the Stokes velocity field [18], obtaining:

$$Sh_p \;=\; 1 + \left(1 + Pe_p\right)^{\frac{1}{3}},\qquad 10$$

where Pe_p is the Peclet number for a particle, defined as $Pe_p = \dfrac{\left|\vec{U}_p - \vec{U}_l\right|d_p}{D_{i,l}}$. For high Reynolds numbers the empirical correlation given by Clift [17] was used:

$$Sh_p = \begin{cases} 1+\left(1+\dfrac{1}{Pe_p}\right)^{\frac{1}{3}} Re_p{}^{0.41} Sc^{\frac{1}{3}}, \\ \qquad \text{for } 1 < Re_p \leq 100; \\[1.5em] 1+\left(1+\dfrac{1}{Pe_p}\right)^{\frac{1}{3}} 0.752 Re_p{}^{0.472} Sc^{\frac{1}{3}}, \\ \qquad \text{for } 100 < Re_p \leq 2\times10^3; \\[1.5em] 1+\left(1+\dfrac{1}{Pe_p}\right)^{\frac{1}{3}} \left(0.44 Re_p{}^{\frac{1}{2}} + 0.034 Re_p{}^{0.71}\right) Sc^{\frac{1}{3}}, \\ \qquad \text{for } 2\times10^3 < Re_p \leq 10^5. \end{cases} \qquad 11$$

Where Sc is the Schmidt number, defined as $Sc = \dfrac{\mu}{D_{i,l}}$.

Solution Technique and Simulation Details

In order to simulate the transient dissolution of inclusions in VAR, the thermal and solute profiles inside an inclusion were calculated using an explicit finite volume discretisation of equations 4 and 5 using the boundary conditions and source terms outlined above. A sufficiently small time step was used to ensure many interpolations of the particle motion (equation 1) per macromodel control volume.

The nominal VAR operation parameters used to simulate the macroscopic heat transfer and fluid flow during the INCONEL 718 ingot formation are listed in Table I. The initially quickly changing temperature profile and fluid flow becomes well established after the ingot height reached 0.8 m. Therefore, the thermal and flow field at this stage of the VAR process was used to investigate inclusion behavior.

Table I.   Nominal values used to simulate the VAR processing of INCONEL 718.

Parameter	Value	Units
Ingot Diameter	0.510	m
Final Ingot Height	2.00	m
Electrode Diameter	0.435	m
Current	6.3	kA
Volts	23	V
Melt Rate	6.47×10^{-2}	kg/s

In this study particle fall-in from three locations was studied: from the crucible, within the electrode; or on the outer surface of the electrode. Three types of intrinsic particles were considered: spherical; plate shaped; and disk shaped. The thermal physical properties for INCONEL 718 ingot and particle material are listed in Table II. For this study, the particles were assumed to have the same composition as the base material. In reality, the crown material may have an increased level of volatile solutes whilst electrode fall-in may be lean in solute (e.g. Nb). Both of these changes in composition will alter the liquidus and solidus temperatures; however, due to the paucity of experimentally measured compositions, these effects were not included in the model.

Table II.   Thermophysical property values for INCONEL 718 (both the melt and intrinsic particles).

Property	INCONEL 718	Units
ρ_l	7050	kg/m^3
ρ_p at 1000 K	7713	kg/m^3
ρ_p at solidus	7500	kg/m^3
ρ_p at liquidus	7050	kg/m^3
$T_{solidus}$	1533	K
$T_{liquidus}$	1609	K
L_p	2.72×10^5	J/kg
C_p	620	J/kg.K
k	25	W/m.K
μ	5.0×10^{-3}	kg/m.s
$D_{Nb,l}$	7.0×10^{-6}	kg/m.s
$D_{Nb,s}$	7.0×10^{-9}	kg/m.s

Results and Discussion

The results are grouped by the source of the particles, beginning with crown fall-in from the crucible. For each source and type of fall-in a sensitivity study was performed to determine how the temperature, size, exact location and height at which the particles drop affect the motion, melt rate and final state (fully melted or entrapped in the mushy zone forming an inclusion). The particle was assumed to become entrapped in the mushy zone once it reached a region with a fraction solid of greater than 0.01. For each set of conditions the maximum size that will fully melt was determined allowing the potential of such inclusions forming to be quantified. The range of conditions was based upon observations in industrial practice with a worst case factor added.

Crucible Fall-in

Fall-in of the crown formed on the crucible is thought to be the greatest source of 'white spot' in VAR ingots. The trajectories of crown fall-in treated as spheres with an initial diameter of 2 mm was simulated using three different initial temperatures, 1233, 1433 and 1533 K. Their predicted trajectories are plotted in Figure 3. All three particles are denser than the melt since they are below the liquidus temperature, hence they are accelerated downwards by gravitational forces. Near the edge of the melt pool a buoyancy driven flow from the top of the melt pool down along the edge of the mushy zone carries the particles down faster, with only the hottest and hence lightest particle escaping entrapment in the solid. The temperature history at the center of each of these three particles is shown in Figure 4. The particle with the highest initial temperature melts completely within 0.23 s, as shown by the particle temperature profile, 0.02, 0.12 and 0.22 s after entering the pool, plotted in Figure 5. Note that for the latter two cases each profile has a constant temperature region - the temperature of the surrounding liquid at that time/location since it has melted completely in this region. The particle with an initial temperature of 1433 K begins to melt but becomes entrapped in the mushy zone after 0.28 s; before it has completely

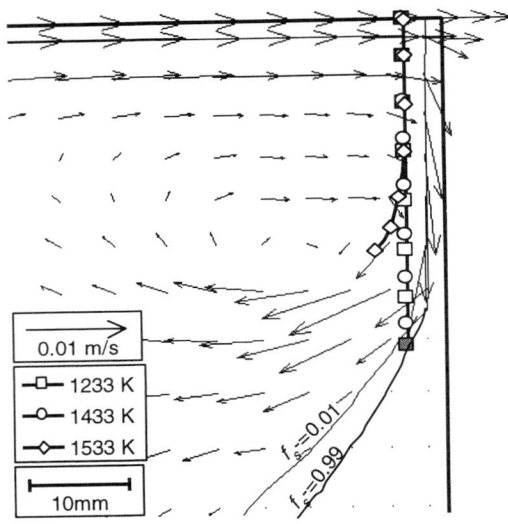

Figure 3. Simulated crown fall-in trajectories for 2 mm diameter spherical particles at three different initial temperatures, 1233, 1433 and 1533 K. (Markers are spaced by 0.04 s.)

melted. The diameter has reduced by 18.5%, a 46% volume reduction. The coldest particle reaches the mushy zone intact - the outer surface has become mushy but is not fully melted.

For all of these simulations and those that follow, it was found that the particle Reynolds number started at approximately 500 and decreased steadily once the particle began to melt. These values for Re_p are well above the creeping flow regime, hence transport across the surface boundary layer dominates the heat and solute transfer between particle and melt.

From an industrial perspective, the important questions are: what is the maximum size of particle that will completely remelt? And, if a particle does not remelt, where will it end up in the ingot? In order to answer these questions, a series of runs was performed on spherically shaped particles altering the size for three initial temperatures, three fall-in heights and three initial radial locations. The maximum size of spherical crown fall-in particles that dissolve completely for the three initial temperatures and three drop heights is shown in Figure 6. Increasing the initial temperature from 1233 K to 1533 K means the particles will melt faster, and are also less dense so that they do not sink as quickly. Therefore, as seen in Figure 6, the maximum size which will be melted increases for a given drop height.

The effect of increasing the drop height for crown fall-in particles from 0 to 30 mm on the maximum diameter that dissolves completely is also shown in Figure 6. The range of heights tested correlates to the industrial observation of melting 20 mm up the electrode plus an extra 10 mm for the arc gap. If the crown particles fall in from just above the melt pool surface, particles whose diameter is less than 3 mm will be safely dissolved if the initial temperature is above 1433 K. The safely dissolved size drops to 2 mm for the 200 K lower

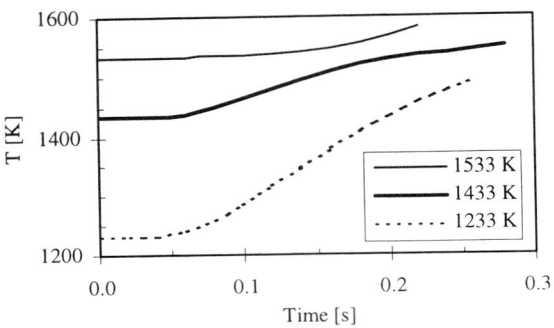

Figure 4. Thermal history at the center of each of the three crown fall-in particles whose trajectory is plotted in Figure 3.

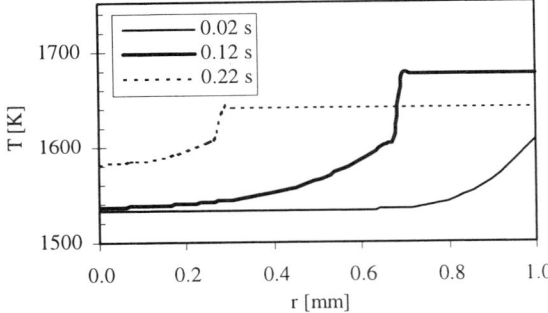

Figure 5. Thermal profile at three different times for the crown fall-in particle with an initial temperature of 1533K whose trajectory is plotted in Figure 3.

initial temperature. When the crown particles drop from 10 mm height, they enter the pool at 0.44 m/s, decreasing the safe diameter to 2.5 mm for an initial temperature of 1533 K; and reducing it further for colder particles or increased heights.

Figure 7 shows the effect of altering the fall-in location from: near the crucible wall ((r=250 mm); half way between crucible and electrode (r=236 mm); and near the electrode (r=219 mm). The trend is clear, the nearer the crucible wall, the faster a particle will be trapped in the mushy zone, and hence the smaller the initial safe diameter. The molten metal the particle is exposed to is also colder, further reducing the melt rate.

In summary, the simulations of spherical crown fall-in particles demonstrate that even small (2-3 mm diameter) crown fall in particles may not completely remelt.

Actual crown fall-in particles are more likely to be plate-like rather than spherical, perhaps spreading quite far circumferentially around the mould wall whilst remaining quite thin and close to the surface. Using a plate correction factor for the drag force and one dimensional melting through thickness in Cartesian coordinates, the motion and dissolution of plate-like fall-in was simulated. The maximum thickness for

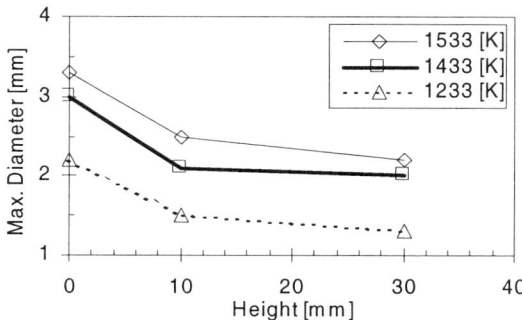

Figure 6. Maximum spherical crown fall-in size that dissolves completely for three initial temperatures and drop heights.

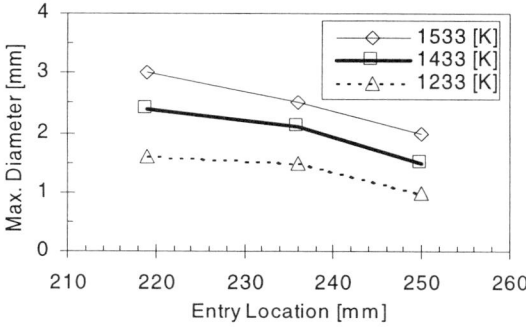

Figure 7. Maximum spherical crown fall-in size that dissolves completely for three different initial locations and temperatures.

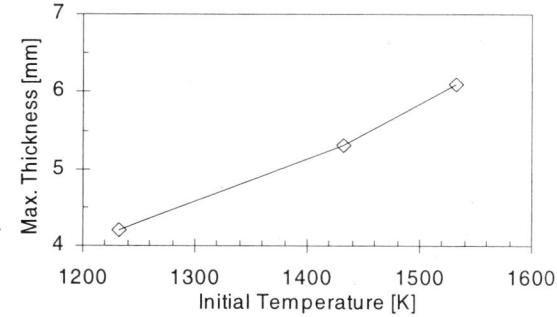

Figure 8. Maximum 40x20 mm plate-like crown fall-in thickness that dissolves completely for three different initial temperatures.

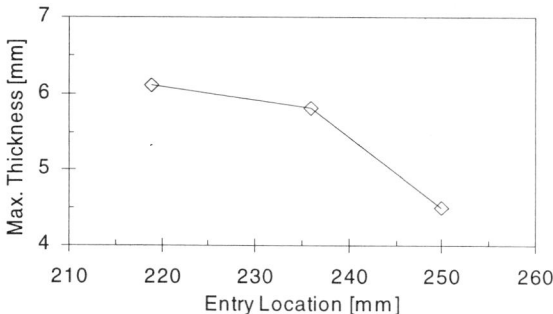

Figure 9. Maximum 40x20 mm plate-like crown fall-in thickness that dissolves completely for three different initial locations.

complete dissolution of plate-like crown particles sliding into the pool with no initial velocity for three different initial temperatures is given in Figure 8. These results are for a very large plate of dimensions 40 mm long, 20 mm high and of the thickness plotted. For these large plates, the assumption of the particle being small compared to the flow field may be less appropriate, hence the results are approximate. The plate-like structures do not sink as quickly and melt faster than equivalent spheres because the surface area to volume is greater. Even reasonably thick crowns of 4-6 mm, depending on initial temperature, will safely melt.

The maximum safe thickness for different entry locations into the melt pool is shown in Figure 9. Again, as for spherical particles, they are entrapped faster the closer they are to the crucible wall on entering the melt pool. A possible reason for crown fall-in to enter closer to the electrode is if they tip in hinged at the bottom. In which case they may have a velocity component radially inwards, which will increase their residence time in the pool, and hence increase the maximum safe size. The trajectories of 6 mm thick particles as a function of entry location are shown in Figure 10.

Electrode Fall-in

In the introduction the different types of particles that might fall-in from the electrode were outlined. A simulation of the trajectories and melting of spherical particles of INCONEL 718 from locations under the electrode allows a comparison with the crucible results. The maximum particle fall-in diameter that dissolves completely for combinations of initial temperatures and drop height are shown in Figure 11. The safe size is much larger than for particles at the outer edge of the pool, due to both higher molten metal temperatures and to greater pool depth. If the particle enters the pool with no initial velocity, superalloy particles or dendrite clusters will melt if the initial diameter is less than 6 mm and the initial temperature is close to the solidus. As the height from which the particles fall from increases the maximum safe size reduces, as expected. If the particles drop from further out the electrode, their chance of survival is greater, as illustrated in Figure 12.

In summary, particles falling in from the pipe in the center of the electrode are more likely to be completely melted than those entering from the crucible wall.

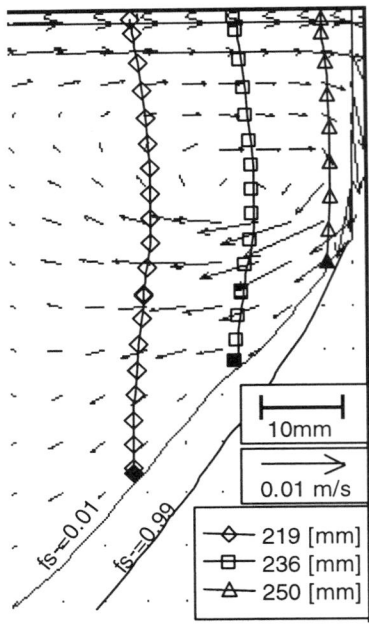

Figure 10. Trajectories of 40x20x5.5 mm plate-like crown fall-in for three different initial radial locations. (Markers are spaced by 0.1 s.)

Electrode Outer Surface Fall-in

The final location for particle fall-in studied was the outer surface of the electrode. In the introduction, the possibility of VIM splash melting off the outside and falling in was discussed. Such pieces of splash were simulated as disk-like shapes, 20 mm in diameter with a thickness of δ. The maximum thickness of such a disk-like splash fall-in that will dissolve completely is plotted in Figure 13 as a function of drop height. Even at the unlikely initial height of 100 mm a 3.5 mm thick disk will melt completely if its initial temperature is near the solidus. Actual pieces of loosely attached splash are unlikely to be very thick due to the way in which they are formed, therefore the simulations would suggest that their potential to be a source of 'white spot' inclusions is low.

For necessary simplicity ideal particle shapes such as spheres, plates and discs have been used for calculations. Also, larger sizes of potential features have been used in the simulations than is likely to be encountered in an industrial furnace to overestimate the potential of entrapment. Both of these assumptions, together with other factors such as increased drag and melting rates for real particles like dendrite fragments, mean the model will under predict the maximum safely dissolving particle size. Much larger fragments will be completely melted than these predictions indicate in an industrial VAR furnace. Therefore, the indicated trends have much relevance in allowing the relative hazards associated with each potential event to be assessed.

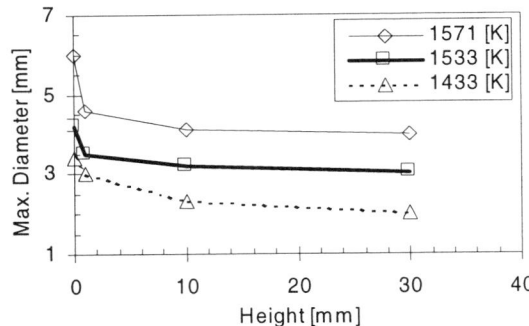

Figure 11. Maximum spherical electrode fall-in size that dissolves completely for three initial temperatures and drop heights.

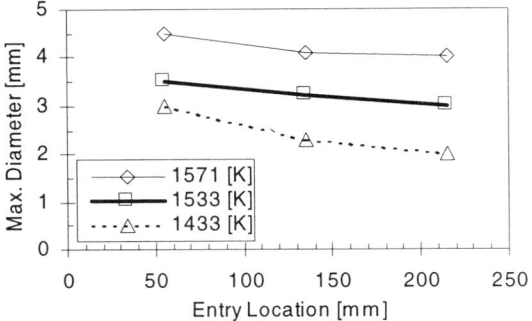

Figure 12. Maximum spherical electrode fall-in size that dissolves completely for three different initial locations and temperatures.

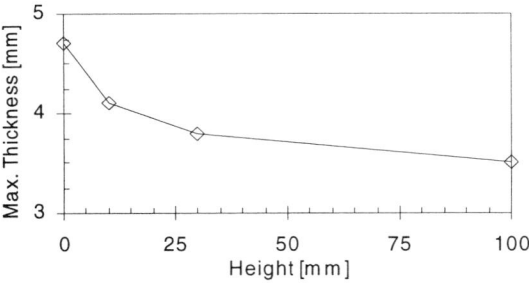

Figure 13. Maximum thickness of a 20 mm diameter disk-like splash fall-in that dissolves completely as a function of drop height.

Summary and Conclusions

A model to simulate the motion, melting and dissolution of particles in the melt pool of the vacuum arc remelting process for the production of INCONEL 718 ingots was developed. The model was used to predict the potential of particles to survive without completely remelting for a range of conditions: entry location; drop height; and particle shape.

Of all the particles studied, crown fall-in from the crucible was found to be the most likely source of 'white spot' inclusions in VAR ingots of INCONEL 718. The maximum safe diameter of spherical fall-in is only 2 to 3 mm if it falls directly in from the crucible wall. If the particles are plate-like instead of spherical, the maximum thickness of crown plate that will completely dissolve is 4 to 6 mm.

Spherical particles falling in from the pipe shrinkage at the center of the electrode were shown to melt completely before entrapment in the mushy zone if the diameter was less than 6 mm. Hence, they form an unlikely source of inclusions in the final ingot.

The detachment of splash from the outer surface of the electrode was also shown to be an unlikely source of inclusions; disk shaped particles up to a thickness of 3 to 6 mm completely remelt before possible entrapment in the mushy zone.

Acknowledgements

The authors would like to thank: Special Metals Wiggin Ltd., Rolls-Royce, DERA, and the EPSRC (GR/L57845 and GR/L86821) for financial support together with the provision of both materials and information. The authors also gratefully acknowledge National Physical Laboratory and the IRC Birmingham for their provision of material property data.

References

1. A.S. Ballantyne and A. Mitchell, "Modelling of Ingot Thermal Fields in Consumable Electrode Remelting Processes", Iron and Steelmaking 2 (1977), 222-239.

2. A. Jardy and D. Ablitzer, "On Convective and Turbulent Heat Transfer in VAR Ingot Pools", Mod. of Casting, Welding and Adv. Sol. Proc. V, Ed. M. Rappaz et al., (TMS 1990) 699-706.

3. P.D. Lee, R.M. Lothian, L.J. Hobbs, and M. McLean, "Coupled Macro–Micro Modelling of the Secondary Melting of Turbine Disc Superalloys", Superalloys 1996, Ed. R.D. Kissinger et al., (TMS 1996) 435-442.

4. A. Jardy, S. Hans and D. Ablitzer, "On the Numerical Prediction of Coupled Transfer and Solidification Process During Vacuum Consumable Arc Remelting of Titanium Alloys", Mod. of Casting, Welding and Adv. Sol. Proc. VII, Ed. M. Cross and J. Campbell, (TMS 1995) 205-212.

5. L.A. Bertram, J.A. Brooks and D.G. Evans, "Transient Melt Rate Effects on Solidification During VAR of 20 Inch Alloy 718", Proc. Int. Sym. Liq. Met. Proc. & Cast., Santa Fe, Feb 21-24, Ed. A. Mitchell, (AVS 1999) 156-167.

6. P. Auburtin and A. Mitchell, "Elements of Determination of a Freckling Criterion", Proc. Int. Sym. Liq. Met. Proc. & Cast., Santa Fe, Feb 21-24, Ed. Mitchell, (AVS 1997) 18.

7. X. Xu, W. Zhang, P.D. Lee and M. McLean, "The Influence of Processing Condition Fluctuations on Defect Formation During VAR of Nickel Based Superalloys", to be published in Proc. Mod. of Casting, Welding and Adv. Sol. Proc. IX, Ed. Sahm et al., (TMS 2000).

8. W. Zhang, P.D. Lee and M. McLean, "Inclusion Behaviour During Vacuum Arc Remelting of Nickel Based Superalloys", to be published in Proc. of EUROMAT 6, Intermetallics and Superalloys, Ed. Morris et al., 27-30 Sept., Munich (1999).

9. J.K Tien and T. Caulfield, Superalloys, Supercomposites and Superceramics, Acad. Press, San Diego, (1989) 80.

10. M.R. Aboutalebi. M. Hasan and R.I.L. Guthrie, "Coupled Turbulent Flow, Heat and Solute Transport in Continuous Casting Processs", Met. Trans. 26B (1995) 731-743.

11. J. Szekely, Fluid Flow Phenomena in Metal Processing, Acad. Press, (1979) 256-257.

12. H. Heywood, Symp. Interact. Fluids Part., Inst. Chem. Eng., London (1962) 1-8.

13. R. Clift, J.R.Grace & M.E. Weber, Bubbles, Drops and Particles, Acad. Press, New York, (1978) 161-162.

14. W.E. Ranz and W.R. Marshall, Jr., Chemical Engineering Progress 48, (1952) 141-146&173-180.

15. Y. Sahai and G.R. St. Pierre, Advances in Transport Process in Metallurgical Systems, Elsevier Sci., (1992) 15.

16. T. K., Sherwood, C. R. Wilke, Mass Transfer, McGraw Hill, N.Y., (1975) 151.

17. R. Clift et al., ibid, 117-120.

18. N. P. Cheremisinoff, Handbook of Heat and Mass Transfer Vol. 2, Mass Transfer and Reactor Design, Gulf Pub. Co., Houston, (1986) 59-109.

Best Paper Award

The following paper was selected by the Awards Subcommittee of the International Symposium on Superalloys as a co-winner of the Best Paper Award for the Ninth Symposium. The selection was based on the following criteria: originality, technical content, pertinence to the superalloy and gas turbine industries and clarity and style.

Predicting Grain Size Evolution of Udimet Alloy 718 during the "Cogging" Process through Use of Numerical Analysis

B.F. Antolovich and M.D. Evans

Predicting Grain Size Evolution of UDIMET® alloy 718 During the Cogging Process Through the Use of Numerical Analysis

Bruce Antolovich* Mike Evans

Special Metals Corporation,4317 Middle Setlemen... ...tford, NY*
Electralloy, 175 Main St., Oil Cit...

Abstract

A semi-automated finite element analysis program, in conjunction with user written subroutines, has been demonstrated to successfully predict thermo mechanical histories and grain size evolution. "Double cone" compression specimens were found to be highly efficient for generating data for the recrystallization behavior of UDIMET®[1] alloy 718. Recrystallization behavior was modelled as having two distinct regimes, dynamic and static; both of which were modelled using typical Mehl-Johnson-Avrami forms. This type of modelling capability is expected to improve the efficiency of the ingot-billet conversion process as well as making possible the development of unique products such as dual property billet.

Introduction

Historically, higher and higher levels of turbine engine performance have been achieved by increasing their operating temperature. Close control of grain size has been instrumental in allowing these increases in temperature; small grains near the hub are required for crack initiation resistance while large grains are preferred near the rim for creep resistance. Furthermore, ultrasonic inspectability is greatly improved through grain size reduction. Disk-to-disk variations in grain size must be kept to a minimum in order to fully exploit the possible material property and inspectability gains achieved through grain size control.

A typical manufacturing sequence for a turbine disk starts with the primary melting and consumable electrode remelting of an ingot followed by conversion of the ingot to a billet. The billet is then closed die forged into a disk blank which is

[1] UDIMET is a registered trademark of Special Metals Corporation

followed by final machining. Each of these steps is typically though not always, carried out by a separate manufacturer.

The conversion process of ingot to billet is called cogging and is accomplished by hot working the ingot, usually with open die forging to induce recrystallzation. In the last decade, the billet has become a controlled grain size product unto itself to enable improved inspectability and reduce operations and costs for the forgers of disks. This process will involve many reheats and forging passes. Its development and improvement can be very costly and time consuming. Reduction of the time and cost of this process can be achieved through numeric simulation of the cogging process and microstructural evolution prior to industrial trials and certification.

There are many well established grain size evolution models for nickel base superalloys including:

1. Mehl-Johnson-Avrami type models
2. dislocation based models
3. simple lookup tables

Most share the common elements of predicting grain size based upon prior grain size, temperature, strain, strain rate and hold time. A considerable amount of work has been conducted to integrate these models into finite element codes. This work has been quite successful for cases such as disk blanking where you can use an axisymmetric (2D) finite element analysis and only need to model a few deformation strokes. The case of cogging is quite a bit more complicated as you cannot take advantage of axisymmetry and the process typically involves thousands of deformation strokes and several reheats, each of which requires a separate analysis whose initial conditions are derived from the results of the previous analysis.

Superalloys 2000
Edited by T.M. Pollock, R.D. Kissinger, R.R. Bowman,
K.A. Green, M. McLean, S. Olson, and J.J. Schirra
TMS (The Minerals, Metals & Materials Society), 2000

If run manually, the analyst would perform a finite element analysis for a single deformation stroke. When the analysis of this deformation stroke was finished, the analyst would then invoke the preprocessor to read in the end results of the previous deformation stroke to be used as the initial condition for the next analysis. When any given deformation stroke can take between several minutes and several hours to complete, it is obvious that this process takes considerable amounts of time. If run in a manual mode by a single analyst, the computer would be incapable of computing 24 hours per day but would be restricted to those hours that the analyst is available; thereby introducing an artificial "slow-down" of over 50%. Furthermore, a mistake early in the process can lead to erroneous results at the end of an extremely long modelling effort (the author is aware of several modelling runs aborted after six weeks of effort). The use of a "template" in which the cogging parameters including number of reheats, number of passes per reheat, number of deformation strokes per pass are specified and then used to automate this process has been instrumental in carrying out these evaluations. Considerable effort has gone into making this template robust, easy to use yet sufficiently flexible to handle a wide variety of pass scheduling requirements.

This paper will explore modelling of two cogging processes. The first case is for cogging on a hydraulic radial forge machine in which grain size is directly predicted. The second is a process modification to improve homogeneity of final grain size in billed produced my more traditional open die press forging by changing cogging parameters to improve strain homogeneity throughout the billet. These two examples will show that direct prediction of grain size evolution is possible but that less sophisticated efforts can also yield significant product improvements.

For the case of the radial forge analysis, grain size will be predicted for a single reheat cogging sequence. For the open die forging, the effects of changing certain cogging parameters will be examined. The modifications were designed to "homogenize" the strain and strain rate distributions within the billet in an effort to reduce the variation in grain size.

Material

The material chosen for this study was UDIMET® alloy 718. It's nominal composition is shown in Table I.

Table I: UDIMET Alloy 718 Composition wt%

C	Cr	Fe	Mo	Nb + Ta
0.020	17.35	17.00	2.80	5.30

Ti	Al	B	Ni
0.85	0.40	0.0020	Bal

Recrystallization Models

The literature contains a great number of articles concerning recrystallization and cogging of nickel base superalloys. [1–6] These models have generally taken one of three forms as stated previously. Regardless of the model, there are two well accepted regimes of recrystallization along with a slightly controversial third type. In general, during load application, an original unrecrystallized grain may recrystallize dynamically. If 100% dynamic recrystallization is not achieved the remaining unrecrystallized portions of the original grain may undergo further recrystallization without additional strain input. Some authors call this "meta-dynamic" recrystallization since the principal driving force for recrystallization is the removal of dislocations introduced in the previous deformation. The third regime is static recrystallization and grain growth in which the principal driving force is the reduction of grain boundary energy. The factors affecting each of these types of recrystallization for any given material are :

1. Static
   - Hold time
   - Residual dislocation density
   - Temperature
   - Initial grain size
2. Dynamic
   - Strain
   - Strain rate
   - Temperature
3. Meta-Dynamic
   - Strain
   - Strain rate
   - Temperature
   - Initial grain size
   - Hold time

Regardless of the recrystallization model chosen, dynamically recrystallized and meta-dynamically recrystallized grain size may be reduced by increasing the total strain or strain rate. Increasing the temperature or hold time tends to increase the meta-dynamic or statically recrystallized grain size.

This author has chosen to model the recrystallization phenomenon by breaking it down into dynamic and static components without addressing the "meta-dynamic" possibilities. The modelling is Mehl-Johnson-Avrami based [7,8] with critical strains and strain rates to achieve static and dynamic recrystallization respectively.

Dynamic Recrystallization

Dynamic recrystallization will only occur if sufficient strain rates and strains are achieved. (*i.e.* if $\varepsilon > \varepsilon_{crit}$ and $\dot{\varepsilon} > \dot{\varepsilon}_{DRXCrit}$. If these conditions are achieved then the recrystallized fraction and grain size will be given by:

$$X_{dyn} = 1 - \exp\left[-\ln k\left(\frac{\varepsilon}{\varepsilon_{0.5}}\right)^n\right] \qquad (1)$$

$$d_{dyn} = C_1 Z^m \qquad (2)$$

Where ε is the applied strain, k, n, C_1 and m are material constants, $\varepsilon_{0.5}$ is the strain required to achieve 50% recrystallization and Z is the traditional Zener-Hollomon parameter given by:

$$Z = \dot{\varepsilon}\exp\left(\frac{Q}{RT}\right) \qquad (3)$$

Static Recrystallization

Static recrystallization will only ocurr if there has been sufficient accumulation of plastic strain. (*i.e.* $\varepsilon_p > \varepsilon_{SRXCrit}$) If this is achieved, then the fraction recrystallized and recrystallized grain size will be given by:

$$X_{sta} = 1 - \exp\left[-\ln k\left(\frac{t}{t_{0.5}}\right)^n\right] \qquad (4)$$

$$\text{with } t_{0.5} = t_{0.5}(d_o, \dot{\varepsilon}, Z)$$

$$d_{sta} = C_2 \varepsilon^{n_1} d_o{}^{n_2} Z^{n_3} \qquad (5)$$

Where t is the incremental time, $t_{0.5}$ is the time required to achieve 50% recrystallization, k, C_2, n are material constants and d_o is the initial grain size.

Experimental Procedures

Numeric Simulation of Cogging:

All thermomechanical simulation of the cogging process was done using the commercially available finite element package, DEFORM3®. This is a large 3D deformation code specialized for the forging environment. A "template" was developed by Scientific Forming Technologies Corporation[2] in order to make more tractable the problem of running many simulations sequentially. Essentially, this template sets up a batch job to run thousands of linked simulations with the output from one simulation serving as the input for the next simulation. In this template, the following parameters are specified:

[2] Scientific Forming Technologies Corporation
5038 Reed Road
Columbus, Ohio 43220-2514
(614) 451-8313
www.deform.com

1. Material
2. Heat exchange environment
3. Billet geometry
4. Die geometry
5. Die movement parameters
6. Reheat furnace temperature
7. Number of reheats
8. Number of passes per reheat
9. Billet advance per "bite" (travel increment across dies)
10. Draft per "bite"
11. Billet rotation per pass

After obtaining the complete thermomechanical history (strain, strain rate and temperature as a function of time) of a billet undergoing conversion, the grain size was predicted using Mehl-Johnson-Avrami type models. A user written subroutine was integrated into the DEFORM3 postprocessor. This subroutine is of a modular nature and thus will permit easy incorporation of different recrystallization models as they are developed and effectiveness proven.

Although there is a dependence of yield stress upon grain size, flow behavior was modelled to be only a function of temperature and strain rate and taken from material of intermediate grain size. Although this will cause errors in the prediction of adiabatic heating, the degree of error is relatively small and did not justify increasing either the complexity of the yield constitutive equation or the increase in computational time required. In other words, all predictions of grain size refinement were done on a "postprocessing basis."

Recrystallization Data Generation:

Generating recrystallization data for the three recrystallization modes requires samples with different initial grain size, temperature, hold time, strain and strain rate. In order to reduce the time and expense of testing, "double cone" compression specimens were chosen due to their ability to generate a wide variety of strains and strain rates within a single specimen. A typical "double cone" geometry and strain variation is shown in Figure 1. Similar variations are found for the strain rate as shown in Figure 2.

Typical microstructures for different specimen locations for a specimen tested at 1074°C with a post test hold time of 60 seconds are in Figures 3 and 4.

Analysis and Results

As mentioned in the introduction, two different types of cogging were modelled; traditional open die cogging and radial forge machine cogging in order to show that:

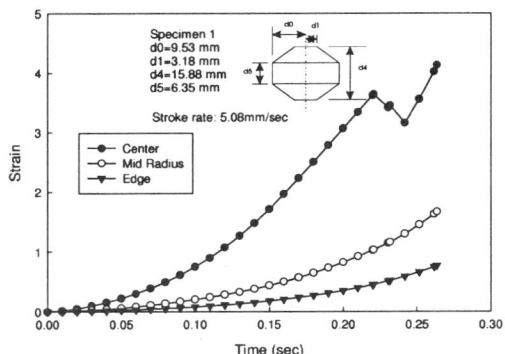

Figure 1: Strain evolution in double cone specimen.

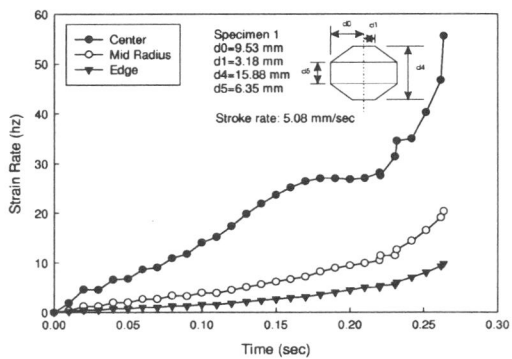

Figure 2: Strain rate evolution in double cone specimen.

1. Analysis of strain and strain rate fields often gives sufficient information to improve a process without the need for grain size refinement modelling.

2. The relatively simple breakdown of recrystallization phenomena into dynamic and static regimes give sufficient information to make good grain size predictions.

For the case of traditional open die forging, modifications were made to a set of existing forging sequences in order to decrease the variation in grain size as a function of position within the billet and to reduce the grain size. Changes were made to the cogging parameters for all reheats of the conversion process including the total reduction per reheats. The two cases modelled were cogging of a 430mm round corner square (RCS) billet to a 380mm octagon billet and a 460mm RCS billet to a 406mm octagon billet. These are practices 1 and 2 respectively. The total reduction in cross

Figure 3: Recrystallization near specimen edge with low deformation.

Figure 4: Recrystallization near specimen center with high deformation.

sectional areas are only very slightly different; ≈ 28% for practice 1 and ≈ 27% for practice 2. Most pertinent details of the cogging sequences are shown in Tables II and III.

Table II: Original reheat sequence

Pass	Forge to Size (mm)	Bite Advance (mm)	Rotation (deg)
1	381.0	190.5	90
2	381.0	190.5	90
3	381.0	190.5	90
4	381.0	190.5	90
5	419.1	190.5	45
6	419.1	190.5	90
7	381.0	190.5	90
8	381.0	190.5	90
9	381.0	190.5	45
10	381.0	190.5	90
11	381.0	190.5	-45
12	381.0	190.5	90

Table III: Modified reheat sequence

Pass	Forge to size (mm)	Bite Advance (mm)	Rotation (deg)
1	444.5	147.3	45
2	444.5	147.3	90
3	406.4	134.6	45
4	406.4	134.6	90
5	406.4	203.2	90
6	406.4	203.2	45
7	406.4	203.2	90
8	406.4	203.2	90
9	406.4	228.6	45
10	406.4	228.6	90
11	406.4	254.0	-45
12	406.4	254.0	90

For the case of the radial forge machine cogging, pertinent details of the cogging sequences are shown in Table IV.

Table IV: SMX-420 sequence

Pass	Shape	Forge to size (mm)	Bite Advance (mm)	Rotation per Bite
0	Octagon	355.0	0.0	0.0
1	Round	328.2	60.0	30.0
2	Round	295.5	60.0	30.0
3	Round	266.0	60.0	30.0
4	Round	266.0	25.0	15.0

Thermomechanical histories of open die cogging sequences

The evolution of strain, strain rate and temperature for various points in the billet for Practice 1 and 2 was predicted by finite element analysis. The final state of strain is shown graphically in Figures 5 and 6. *(Color versions of figures 5 & 6 appear on page 839.)*

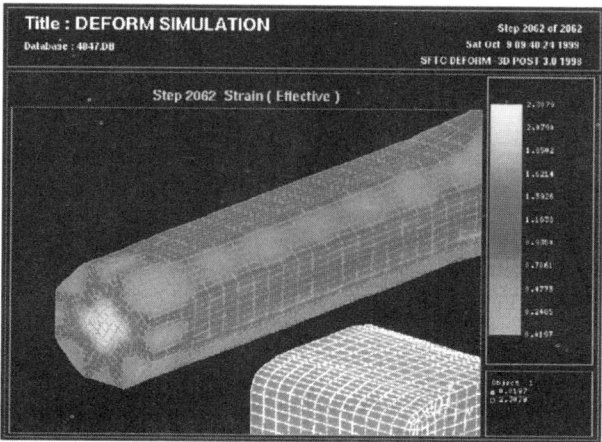

Figure 5: Practice 1: Final state of strain.

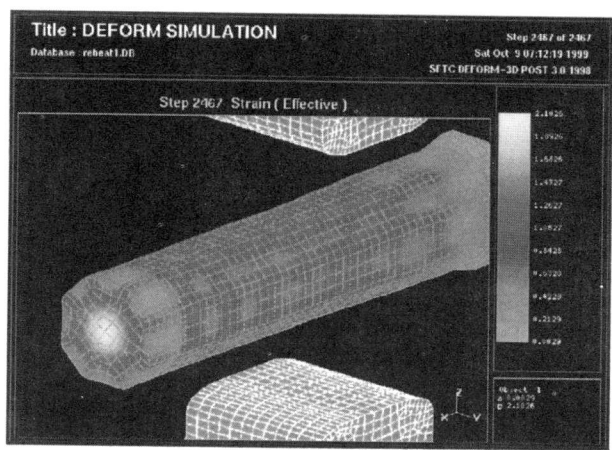

Figure 6: Practice 2: Final state of strain.

Three different line sections of the billet were examined for homogeneity of final cumulative strain. These sections are:

1. Along the centerline in a longitudinal direction in the middle (lengthwise) of the billet, away from end effects

2. 12.7 mm beneath the surface along a longitudinal direction line in the middle of the billet

3. In a radial direction from the billet centerline to the surface

The results are shown in Figures 7, 8, 9, 10, 11 and 12. It is quite clear that changing the pass schedules has substantially changed the thermomechanical history experienced throughout the billet. Practice 1 produced billets with significant variations in the edge grain size as one moved longitudinally along the billet whereas Practice 2 produced much more uniform edge grain sizes. This is clearly a result of changing the near edge strain distribution. The first sequence produced cumulative strains that varied between 0.55 and 0.80 whereas the modified sequence varied between 0.70 and 0.84. This is particularly noteworthy in light of the fact that Practice 1 had a greater overall reduction in cross sectional area of the billet. Not only was the variation reduced, the average cumulative strain experienced near the outer surface was increased which resulted in finer grain sizes. The *variation* in strain experienced along the centerline was increased somewhat for Practice 2 but was still quite low on an overall basis. Finally, Practice 2 clearly biased the deformation towards the surface of the billet whereas Practice 1 biased the deformation towards the center of the billet. This is clearly shown in Figures 11 and 12.

43

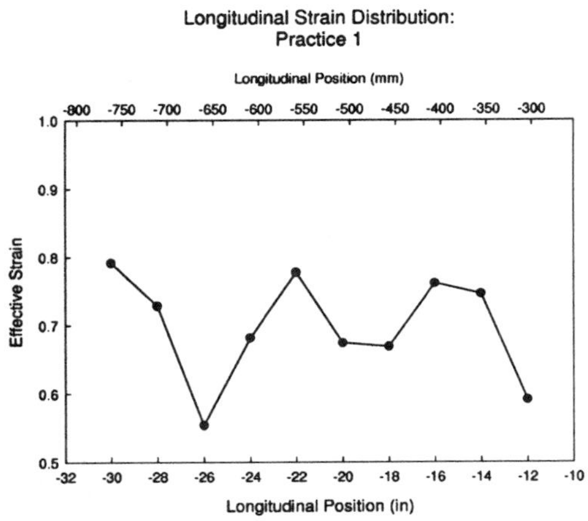

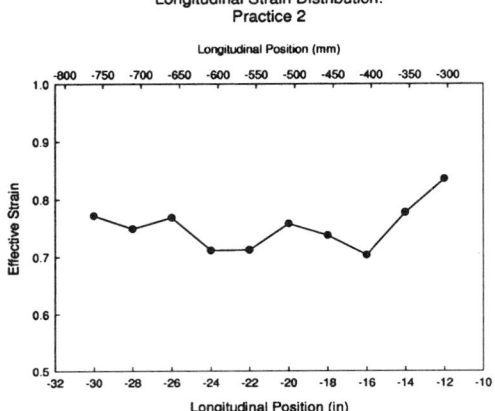

Figure 7: Practice 1: Longitudinal strain variations measured 25 mm from the surface.

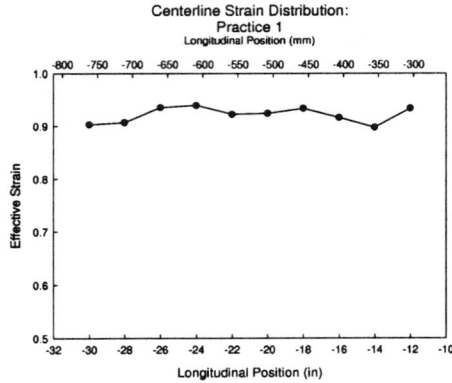

Figure 8: Practice 2: Longitudinal strain variations measured 25 mm from the surface.

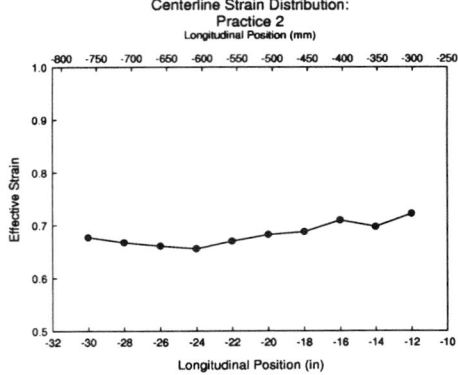

Figure 9: Practice 1: Centerline strain variations.

Figure 10: Practice 2: Centerline strain variations.

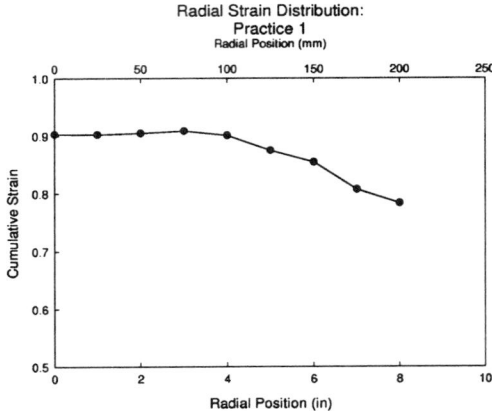

Figure 11: Practice 1: Radial strain variations.

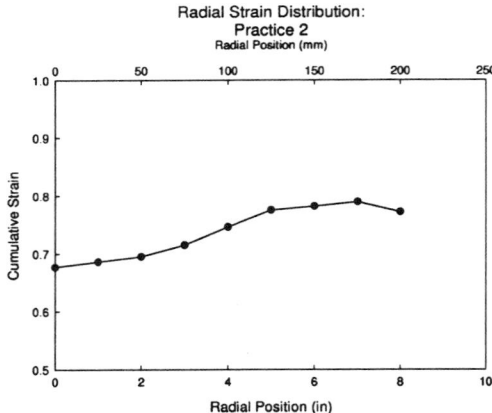

Figure 12: Practice 2: Radial strain variations.

The results of changing the cogging sequence upon final grain size are shown in Figure 13. The modified cogging sequence has clearly reduced the size of both the primary and as large as edge grain size while not significantly affecting the center grain size.

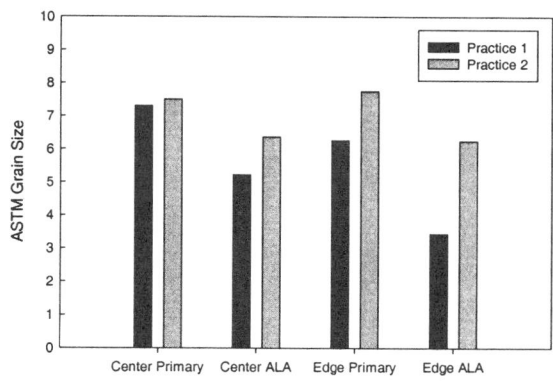

Figure 13: Grain size comparison for the two cogging sequences.

For the case of the cogging with a radial forge machine, predictions of statically recrystallized grain size, dynamically recrystallized grain size and percent fraction dynamic recrystallization are shown in Figures 14, 15 and 16 respectively. It must be noted that these are plots of billet prior to final grinding and polishing in which approximately 10–20mm of material in the radial direction is removed. Therefore, when comparing the predicted and measured grain sizes at the billet edge as shown in Table V, one must be careful to examine the predicted grain size on the finite element plots at approximately 10–20mm beneath the surface. *(Figures 14-16 appear in color on pages 839-840.)*

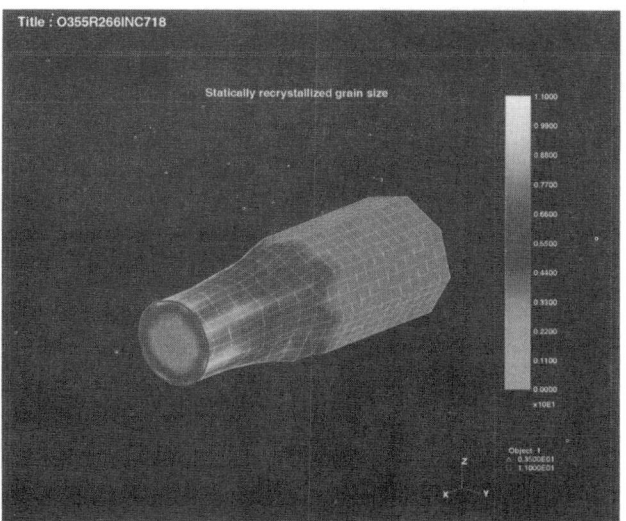

Figure 14: Statically recrystallized grain size. Predicted grain size ranges from ASTM 5.5 at center to 6.0 near edge.

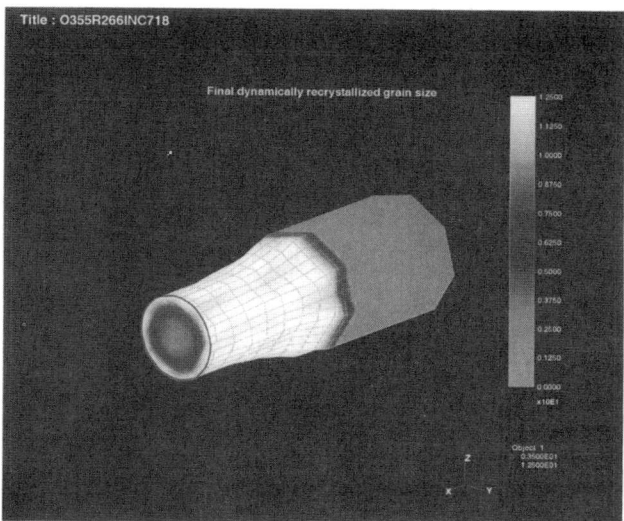

Figure 15: Dynamically recrystallized grain size. Predicted grain size ranges from ASTM 3.5 at center to 7.0 near edge where grain size measurements are made.

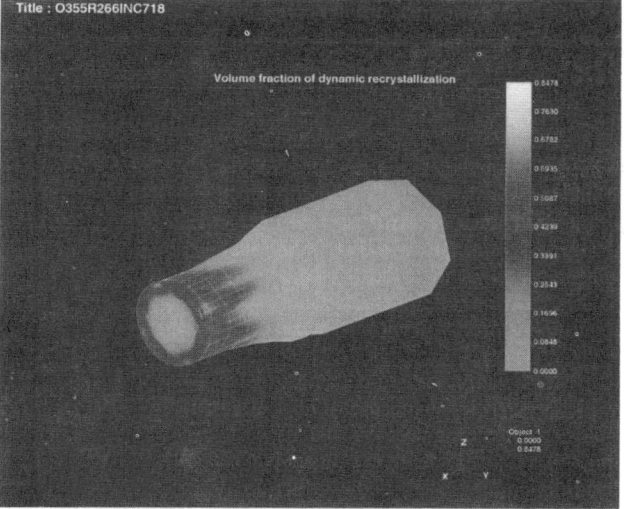

Figure 16: Dynamically recrystallized grain size fraction Rx

Grain size predictions for Rotary forge "cogging"

Using the previously discussed models, dynamically recrystallized grain size, statically recrystallized grain size and fraction dynamic recrystallization were predicted. Comparison between measured values and predicted values is shown in Table V.

Table V: Comparison of predicted and measured grain sizes

	Center (ASTM)	Mid-Radius (ASTM)	Near Edge (ASTM)
Prediction (SRX)	5.5	5.5	6.0
Prediction (DRX)	3.5	5.0	7.0 $V_f \approx 30\%$
Measured	6.0	7.0	3 15% 5.5

Examination of the finite element plots and tabulated date for static recrystallization, dynamic recrystallization and volume fraction of dynamic recrystallization shows that:

1. Near the center, there is insufficient strain to achieve significant volume fractions of dynamic recrystallization

2. Near the center, there is sufficient strain to achieve static recrystallization

3. Near the surface($\approx$ 10-15 mm subsurface, there is sufficient strain to achieve approximately 30-40% dynamically recrystallized grains of ASTM 7.

Applications

As is well known, there is a great desire on the part of engine manufacturers to produce so called "dual property" disks with small grains on the hub for LCF resistance and large grains on the circumference for good creep resistance. Efforts in the past to accomplish this have taken the form of selective induction heating on the disk circumference as well as welding different materials together to form a single disk. Both of these have obvious drawbacks including:

1. Extra processing on the part of the billet supplier to produce uniform fine grain which is subsequently removed

2. Difficulty in achieving uniformly large grain size in the geometrically complex blade attachment points

3. Crack initiation at the heat affected zone

Given the fact that the disk forger has areas in the disk which receive little deformation and little potential for grain size refinement, one elegant solution is to supply billet with an appropriate heterogeneous grain size. Through careful attention to reheat temperatures, bite advance, bite draft and die speed, it is theoretically possible to produce such a billet. The amount of industrial trials required in the past to perfect this process were prohibitive. In theory and practice, these billets may now be produced.

Conclusions

It has been shown conclusively that the use of finite element modelling can adequately predict billet thermo-mechanical histories and associated microstructural evolution during the cogging process. The development and use of a cogging "template" has been instrumental in allowing the analysis of "real world" cogging problems by allowing extremely complicated simulations with many "bites" per pass, multiple passes per reheat and multiple reheats. When used in conjunction with post processing based recrystallization models, sufficiently accurate grain size evolution predictions can be made to reduce the amount of required full scale testing to validate new cogging sequences. This process is now sufficiently robust, with sufficient ease of use, to be used on a regular basis as a critical tool in pass scheduling development. The trials conducted have also numerically confirmed and quntified the benefits of precise control of cogging parameters such as "draft", "bite", rotational orientation of workpiece, *etc.*, for the manufacture of billet for today's turbine engine disks.

References

[1] G. Shen, S.L. Semiatin, and R. Shivpuri. Modeling microstructural development during the forging of Waspaloy. *Metallurgical Transactions A*, **26A**,p1795–1803, 1995.

[2] C.A. Dandre, S.M. Roberts, R.W. Evans, and R.C. Reed. Microstructural evolution of Inconel 718 during ingot breakdown process modelling and validation. *Materials Science and Technology*, **16**,1,p14–25, 2000.

[3] F.J. Humphreys and M. Hatherly. *Recrystallization and Related Annealing Phenomena*. Oxford; New York; Yushimi: Pegramon Press, 1995, 1 edition, 1995.

[4] D. Zhao, S. Guillard, and A.T. Male. High Temperature Deformation Behavior of Cast Alloy 718. In *Superalloys 718, 625, 706 and Various Derivatives*, pages 193–204, 1997.

[5] Laurance A. Jackman, M.S. Ramesh, and Robin Forbes Jones. Development of a Finite Element Model For Radial Forging of Superalloys. In *Superalloys 1992*, pages 103–112, 1992.

[6] A.K. Chakrabarti, M.R. Emptage, and K.P. Kinnear. Grain Refinement in IN-706 Disc Forgings Using Statistical Experimental Design and Analysis. In *Superalloys 1992*, pages 517–526, 1992.

[7] Melvin Avrami. Kinetics of Phase Change. General Theory. *Journal of Chemical Physics*, **7**,p1103–1112, 1939.

[8] Melvin Avrami.   Kinetics of Phase Change.
Transformation-Time   Relations   for   Random
Distribution of Nuclei. *Journal of Chemical Physics*,
8,p212–224, 1940.

Control of Grain Size Via Forging Strain Rate Limits for R'88DT

Eric Huron
General Electric Engine Services

Shesh Srivatsa

General Electric Aircraft Engines

Ed Raymond
General Electric Aircraft Engines

Abstract

R'88DT is a powder metallurgy nickel-based superalloy used for rotating compressor and turbine disks. It was developed using a supersolvus heat treatment to achieve good creep and fatigue crack growth resistance. It was observed during development that the supersolvus heat treatment could result in large grains occuring in bands in the microstructure. Forging window studies were carried out using a specimen designed to produce a range of strain rates and strains for a given nominal specimen upset. These specimens were deformed over a range of nominal strain rates and the regions of large grains were correlated with strain rate. The presence of large grains was found to be related to a critical strain rate range depending on temperature. The large grains were termed Critical Grain Growth (CGG) because of this relationship. Deformation finite element analysis with commercial forging codes was used to establish a model to predict when CGG would occur and to design safe forging designs and process routes. This model has allowed the successful production of R'88DT forgings.

Superalloys 2000
Edited by T.M. Pollock, R.D. Kissinger, R.R. Bowman,
K.A. Green, M. McLean, S. Olson, and J.J. Schirra
TMS (The Minerals, Metals & Materials Society), 2000

Introduction:

Rene'88DT is an advanced high strength Ni-base powder metallurgy alloy used for rotating disks in gas turbines (References 1 and 2). It was developed with a nominal composition of 13Co-16Cr-4Mo-4W-2.1Al-3.7Ti-0.7Nb-0.03C (Reference 2).

This alloy is processed to a relatively coarse grain size (typically ASTM 6-10, or nominally 10 to 40 microns) to achieve good creep strength and damage tolerance. The processing route involves several major processes. First, argon atomized powder is produced and screened to desired mesh size (-270 mesh for R'88DT). The powder is extruded into billet, typically 20-30 cm in diameter. Cylindrical sections ("mults") are cut from the billet at target weights. The mults are isothermally forged in specially designed, controlled-environment forging presses. To achieve a consistent grain size, the forging condition must be carefully controlled. This paper describes the definition of the forging window using experiments designed to produce a controlled strain rate distribution in test coupons. The test coupons were forged, heat treated, and evaluated for grain size. Finite Element Modeling analysis was used to map grain size response vs. local strain rate to develop a model for the proper strain rate conditions. This model allows reliable forging of R'88DT hardware. Some of the other factors affecting the model are discussed in light of requirements for technology in forging of these alloys.

Procedure:

Test coupons were machined from extruded P/M R'88DT billet according to Figure 1. The development and composition of R'88DT are described in detail elsewhere (Reference 2). The coupons were tested on a servohydraulic testing machine equipped with a resistance furnace to control temperature and programmed to provide a nominal constant strain rate. Coupons were tested over a matrix of strain rates and temperatures from 954C to 1066C at strain rates from 0.0032/sec to 0.32/sec. Strain rates were programmed on a nominal value based on the overall specimen height to simplify the testing. Later, computer modeling was used to correlate the deformation conditions to actual local strain rate contours within the samples. After testing the coupons were heat treated with a supersolvus exposure of 1149C for 1 hour. The coupons were sectioned on radial planes and etched in Kalling's reagent to identify regions of grain size variation.

Some portions of the double cone matrix were replicated using standard right circular cylinder compression tests. The flow stress data from these specimens were used to calculate values of "m", the strain rate sensitivity factor, as a function of temperature and strain rate

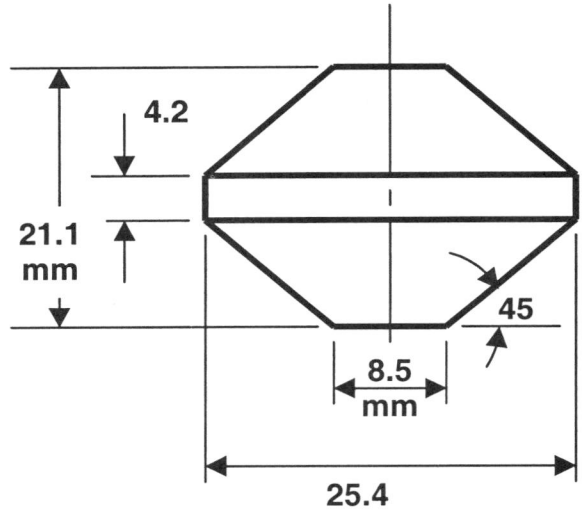

Figure 1: Double Cone Specimen Geometry. This specimen produces gradients of strain rate which simulate strain rate gradients caused by die radii and flow constriction in isothermal forging.

Results:

Most regions of the samples displayed the expected typical supersolvus microstructure of ASTM 6-10 grain size. However, regions of very large grains, ASTM 00, exceeding 500 microns in diameter, were observed. Figure 2 shows a typical uniform microstructure, and Figure 3 shows a region of the enlarged grains. The location of the enlarged grains was observed to change within the sample depending on the overall strain rate and temperature. Because it showed a definite relationship with a critical strain rate, this type of grain growth was termed Critical Grain Growth (CGG). For a given temperature, the location of CGG moved outward as nominal strain rate increased. Figure 4 shows the patterns of large grain regions observed in the samples. Generally the CGG occurred in bands within the samples. The grain size in the regions without CGG was generally ASTM 6-10. Within the CGG grains up to ASTM 00 were present. Maintaining the uniform ASTM 6-10 grain size is desirable to provide uniform mechanical properties. Although repetitive specimens did not always show CGG in exactly the same location, the same general patterns and locations of CGG were observed to occur. The emergent patterns observed over the repetitive sampling are shown schematically in Figure 5.

The main issue was then to transfer the results of the experiment into limits for production forging of R'88DT. This was accomplished by Finite Element Modeling (FEM) of the local strain rates in the coupons. Modeling was done using the commercial software package DEFORM. Flow stress data of R'88DT generated from 9" fine grain billet material was used. The models considered non-isothermal deformation with effects of adiabatic heating included. A shear friction factor of 0.2 corresponding to well lubricated specimens was used, and this value was independently verified by measurements in ring tests. The modeling results were verified by comparing measured and predicted load-displacement as well as the deformed specimen shapes.

The strain rates and the adiabatic heating experienced by all material points (finite elements) were tracked through the process and the maximum strain rate for each point was noted. The locations of CGG within the samples were matched visually to the modeled output to establish predictive maximum strain rate and adiabatic heating regions where CGG occurred. Using this approach, a strain rate window to predict CGG was established. If the maximum strain at a location occurs between

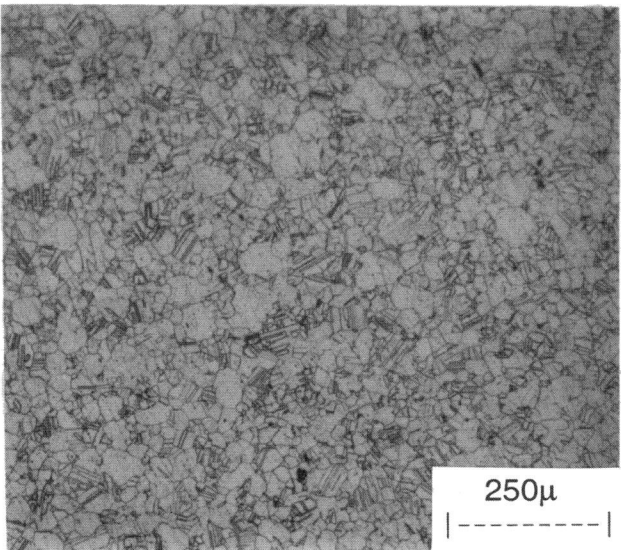

Figure 2: Typical Microstructure of R'88DT Away from CGG

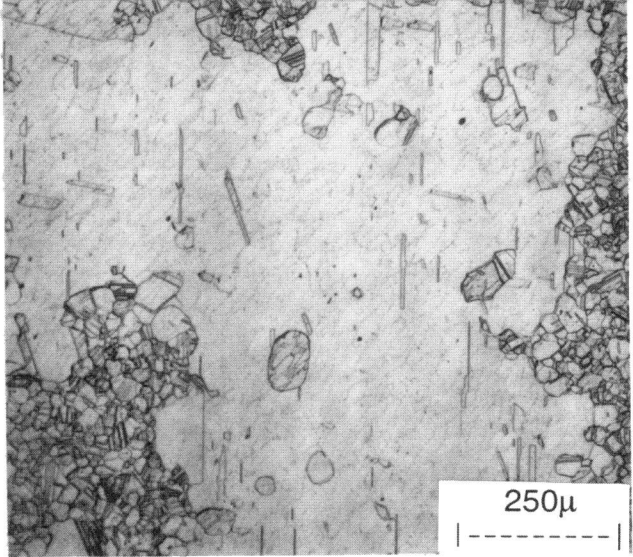

Figure 3: Typical Microstructure of R'88DT In Large Grain Band

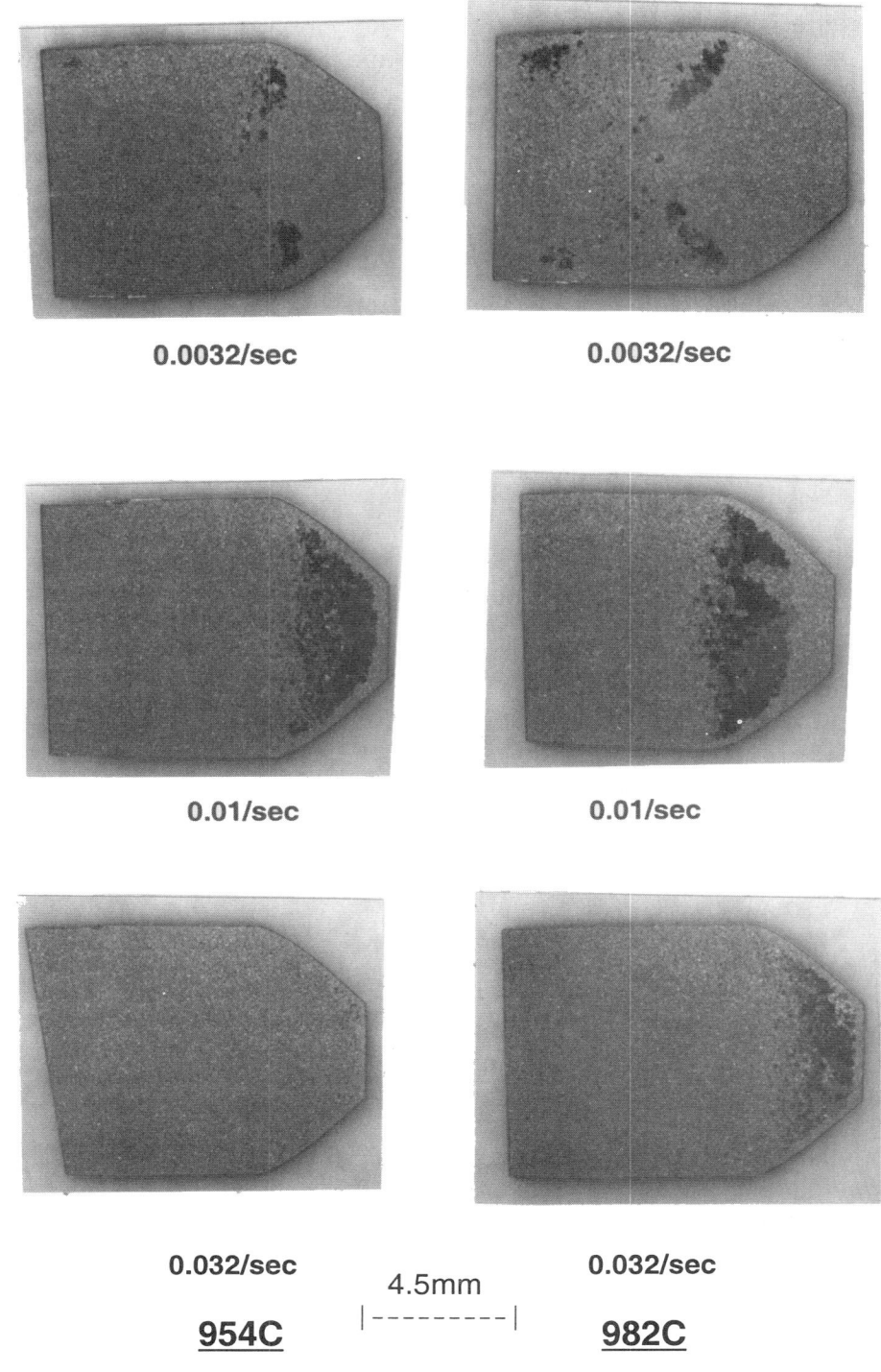

0.0032/sec   0.0032/sec

0.01/sec   0.01/sec

0.032/sec   0.032/sec

4.5mm

954C   **982C**

**Figure 4 : Photographs of Critical Grain Growth (CGG)
bands within deformed and supersolvus heat treated
R'88DT specimens**

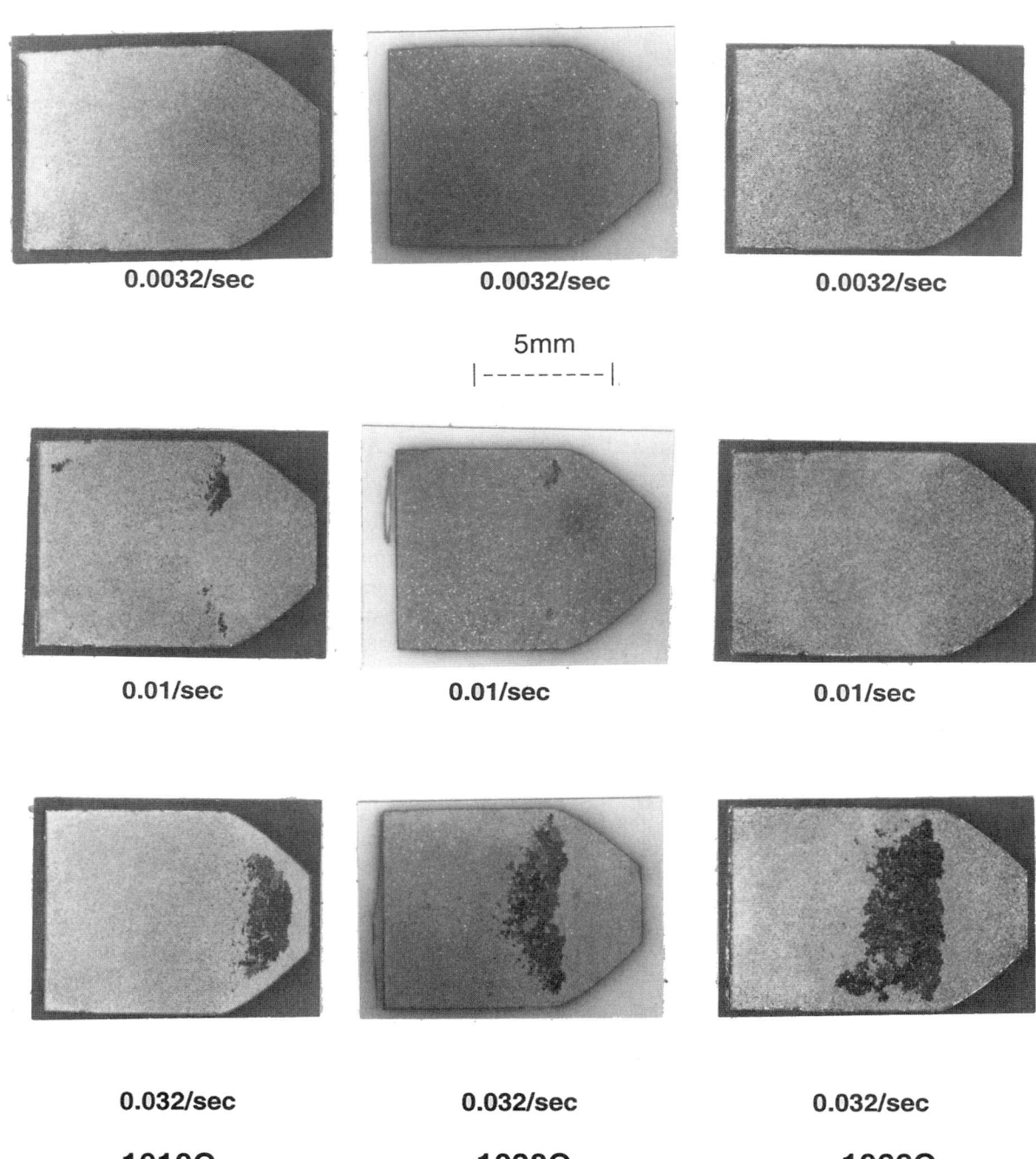

0.0032/sec   0.0032/sec   0.0032/sec

5mm

0.01/sec   0.01/sec   0.01/sec

0.032/sec   0.032/sec   0.032/sec

1010C   **1038C**   **1066C**

Figure 4 (continued): Photographs of Critical Grain Growth (CGG) bands within deformed and supersolvus heat treated R'88DT specimens

the limits of the CGG window, CGG is likely to occur. The strain rate window was found to be a strong function of temperature. At 954C, the maximum strain rate to avoid CGG was only 0.00244/sec, while at 1066C, a strain rate up to 0.0256/sec could be used. In addition, a strain rate above which CGG could be avoided was noted. For 954C, this strain rate was 0.01/sec, while at 1066C, it was much higher, 0.07/sec. The overall strain rate limits where CGG was observed are summarized in Table 1.

Table 1: Regions of Critical Grain Growth

Temperature (degrees C)	Strain Rates for Critical Grain Growth (/sec)
954	0.00244-0.01
982	0.00288-0.0128
1010	0.008-0.04
1038	0.010-0.05
1066	0.0256-0.07

The modeling results are summarized in Figure 6. The CGG regions are indicated by black color. The darker the black color, the closer is the max strain rate to the center of the CGG window. The predicted CGG locations agree well with the observed CGG (Figures 4 and 5) in all the specimens. This shows that the max strain rate is one of the variables controlling the occurrence of CGG. Further experimental tests and modeling are needed to establish the effects of gradients of strain, strain rate and temperature.

The strain rate regions where CGG was observed in the double cone specimens generally corresponded to regions of low "m" (strain rate sensitivity). Although the full analysis is beyond the scope of this paper, generally deformation under conditions of low "m" correspond to relatively low superplasticity. Instead of deforming superplastically (in theory, with little buildup of deformation debris and stored energy), the material deformed in this region is not highly superplastic and thus deformation debris and storage of energy could in fact occur.

Discussion:

The application of the testing to production forging has been fairly straightforward. The strain rate limits achieved by modeling analysis are used to guide forging design. Die design and press speeds are controlled so that no regions of the forging during isothermal forging are predicted to lie within the strain rate window. Exact details of the design are left to the forging supplier, but in general slower speeds or more generous die radii can be used to address any forging regions predicted to lie within the window. The model data have been provided to the commercial isothermal forging suppliers and have resulted in successful production of several thousand R'88DT isothermal forgings.

The model has been practically applied with great success even though the theoretical microstructural causes of CGG have not been fully determined. In extensive study at GE, the exact microstructural event leading to CGG has not been determined. If classical superplasticity is the cause, or rather, forging in regions of limited superplasticity, some microstructural evidence of stored energy should be expected. Superplasticity is commonly discussed in terms of the relative position along the classical stress vs. strain rate sensitivity curve, which is divided into three regions (References 3, 4). Region I, at low strain rates, is dominated by diffusion-accomodated grain boundary sliding with little stored energy. Region III, at high strain rates, is characterized by dislocation climb in addition to grain boundary sliding and extensive stored energy. Region II is an intermediate region with several possible contributing mechanisms. The present work for R'88DT appears to be associated with the upper end of Region II and the lower end of Region III. For IN100, as deformation conditions moved into this reason an increased dislocation density was noted and "m" decreased (Reference 5). This would correlate to increased tendency for CGG if it could be shown that the dislocation structure helps provide the driving force for CGG to occur. However, extensive Transmission Electron Microscope studies attempting to correlate measured dislocation densities in as-forged samples, with observed CGG in samples forged

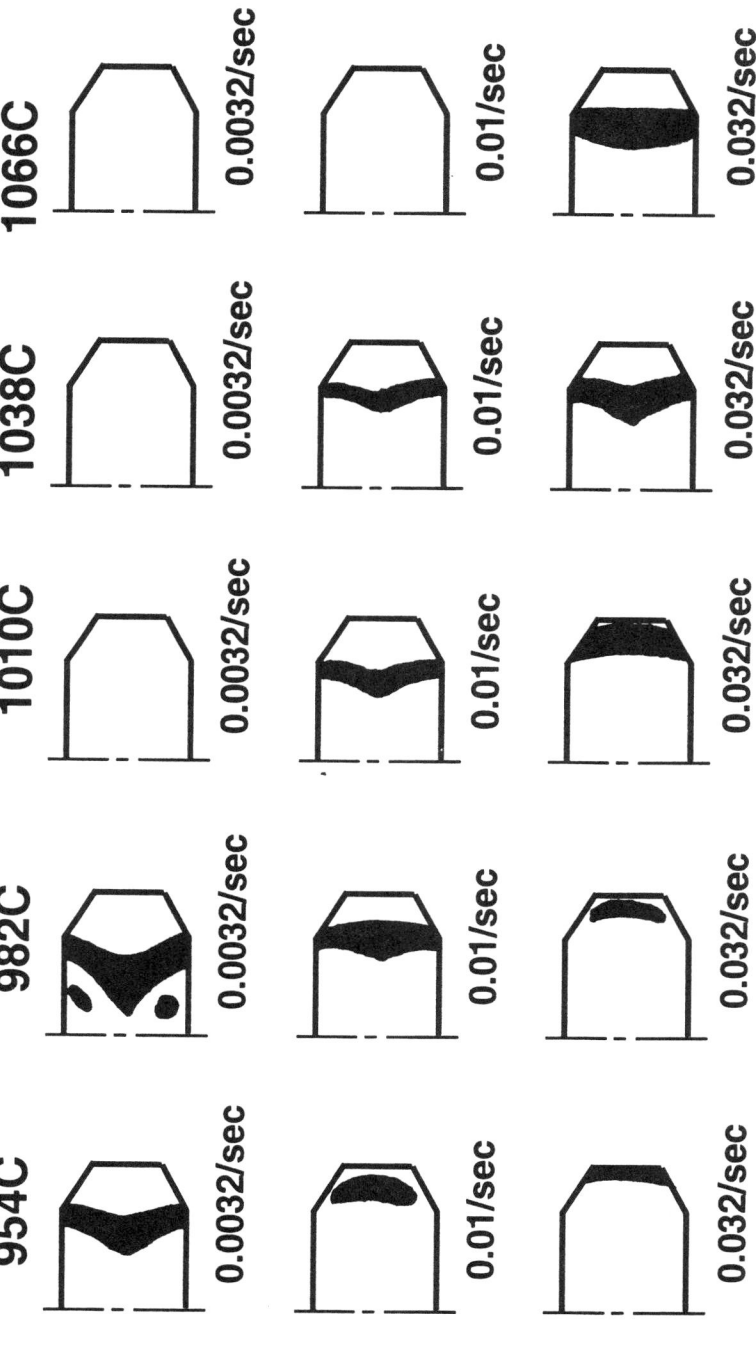

Figure 5: Schematic representation of Critical Grain Growth (CGG) patterns within deformed and supersolvus heat treated R'88DT specimens.

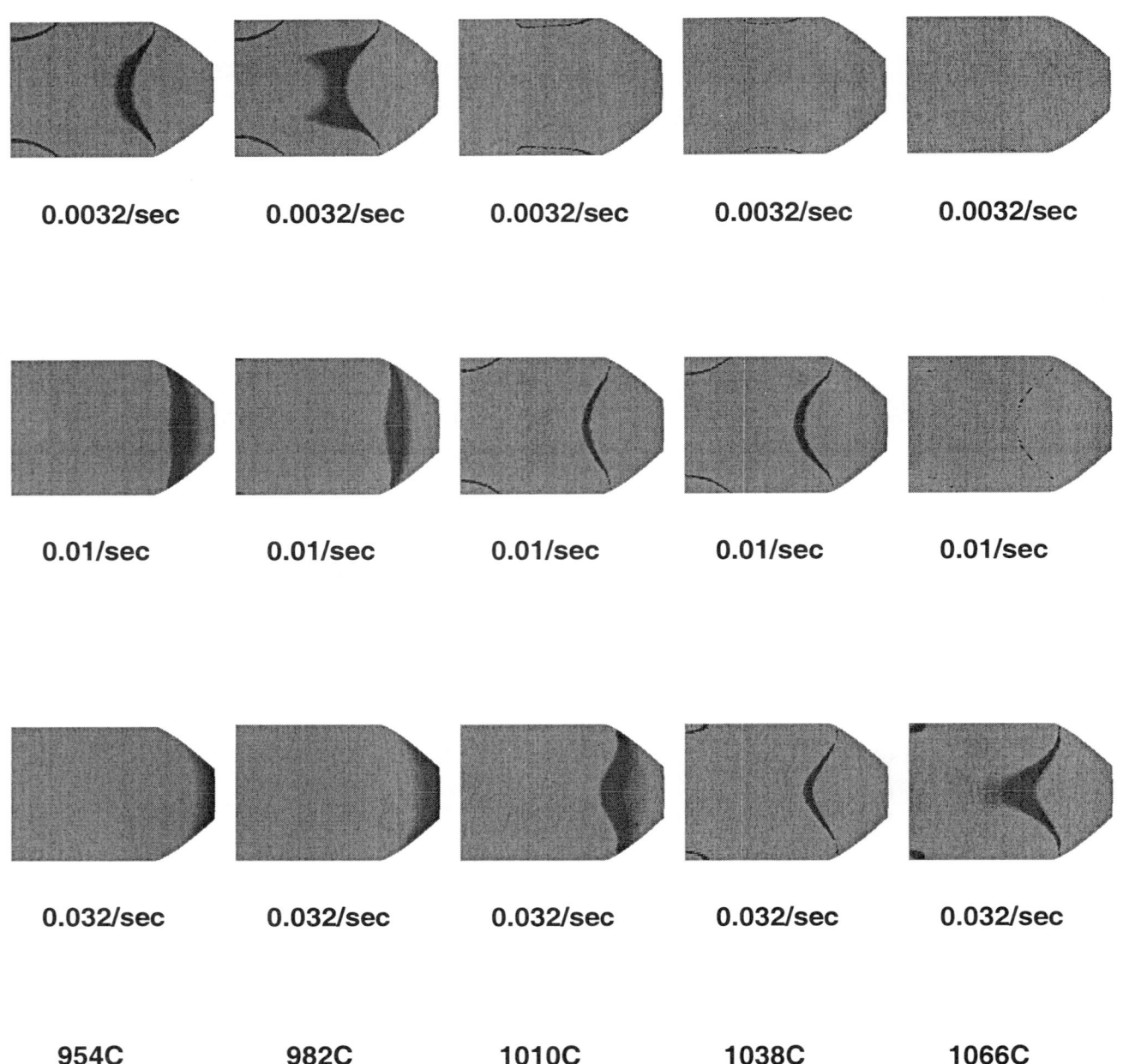

0.0032/sec   0.0032/sec   0.0032/sec   0.0032/sec   0.0032/sec

0.01/sec   0.01/sec   0.01/sec   0.01/sec   0.01/sec

0.032/sec   0.032/sec   0.032/sec   0.032/sec   0.032/sec

954C   982C   1010C   1038C   1066C

Figure 6 : Critical Grain Growth (CGG) Patterns predicted by modeling of the deformation process. These patterns agree well with the observed patterns shown in Figures 4 and 5.

at those conditions and then supersolvus heat treated, have failed to show a consistent correlation.

The results of CGG in R'88DT agree well with results reported for N18, another P/M Ni-base superalloy (Reference 6). This study also found that high temperature deformation, at a critical strain rate, produced large grains after annealing. The study also showed that the critical strain rate was a function of temperature. For N18, the critical strain rate was 0.0035/sec at 1100C, 0.008/sec at 1120C, and 0.01/sec at 1140C. Although the temperatures were higher, the trend of decreasing critical strain rate with decreasing temperature is consistent with the R'88DT results. The γ' content of N18 is higher than R'88DT (~55% vs. ~40%) and the γ' solvus is correspondingly higher (~1195C vs. ~1100C). The critical strain rate / temperature relationship may depend on the relative γ' amounts.

All of the results reported in this work are for a given billet microstructure and overall chemistry, and with controlled isothermal deformation. Most P/M billet is extruded to a fine grain superplastic structure. Billet microstructure influences the superplasticity and hence the energy storage and resultant grain size response. The results are beyond the scope of the present study, but billet microstructure must be maintained through consistent process controls. Recently new studies have explored alternatives to the most commonly used fine grained billet structure. Hot Isostatic Pressing (HIP) of P/M Udimet 720 was found to allow grain sizes up to ASTM 3, which could be hot worked to ASTM 10 (Reference 7). This work illustrates that minor phases as well as γ' control the grain size. Another study reports that hot die forging under non-isothermal conditions and at high strain rates promoted a different microstructural response (Reference 8). In this study forging was carried out to be deliberately above the critical strain rate, so that on subsequent annealing critical grain growth was intentional. Essentially a uniform high nucleation density promoted uniform grain size instead of a bi-modal distribution. The results did show that high strain rates, above the CGG window, produced uniform grain sizes within the samples. This suggests that forging throughput can be increased by using these higher strain rates. Unfortunately it is not possible to avoid local regions of sticking or low strain rate for areas of the forging in contact with the dies, so it is not practical to forge entirely above the CGG strain rate window.

Although much remains to be understood about the cause of CGG at the microstructural level, the present results and other studies suggest several practical approaches to forging design and practice. For uniform ASTM 6-10 microstructures, controlled isothermal forging, at strain rates below the critical strain rate, can be used to produce a uniform microstructure after supersolvus final heat treatment. Current commercial deformation models can easily be used to design forgings that have local strain rates within safe limits and within die loading and press capacity. Forging geometries and overall forging ram speed are modified iteratively until the models predict that all local regions are within the strain rate limits established by the coupon testing. This may require compromise on geometry radii or slower ram speeds, or both. This places a practical limit on throughput in isothermal forging, but by optimizing the models, good forging practice balancing speed, die design, and grain size control is achieved. Other grain sizes may be achievable by alternate processes, in particular the approach of forging entirely above the critical strain rate, but to date limitations on die capability have limited large-scale commercial use.

Acknowledgements:

All sample deformation testing for this study was done by Dr. Noshir Bhathena of Wyman-Gordon Forging Company, Houston, Texas. Metallographic specimen preparation and heat treatment were provided by Ron Tolbert and Bill Brausch of GE Aircraft Engines. Valuable technical consultation was provided by Drs. Micheal Henry and Charles P. Blankenship, Jr. of GE Corporate Research and Development.

References:

1. D. D. Krueger, R. D. Kissinger, R. G. Menzies, and C. S. Wukusick, U. S. Patent No. 4,957,567, General Electric Company.

2. D. D. Krueger, R. D. Kissinger, and R. G. Menzies, "Development and Introduction of a Damage Tolerant High Temperature Nickel Base Disk Alloy, Rene'88DT", in Superalloys 1992, ed. by S. D. Antolovich et. al., TMS-AIME, Warrendale, PA, 1992, 277-286

3. M. F. Ashby and R. A. Verrall, "Diffusion-Accomodated Flow and Superplasticity", Acta Met. 21, February 1973, pp. 149-173.

4. J. W. Edington, K. N. Melton, and C. P. Culter, "Superplasticity", Progress in Materials Science, 1976, 21, 1976, pp. 61-158.

5. R. G. Menzies, G. J. Davies, and J. W. Edington, "Microstructural Changes During Superplastic Deformation of Powder-Consolidated Nickel-Base Superalloy IN-100", Metal Science, 16, October 1982, pp. 483-494.

6. M. Soucail, M. Harty, and H. Octor, "The Effect of High Temperature Deformation on Grain Growth in a P/M Nickel-Base Superalloy", in Superalloys 1996, ed. by R. D. Kissinger et. al., TMS-AIME, Warrendale, PA, 1996, pp. 663-666.

7. G. Maurer, W. Castledine, F. A. Schweizer, and S. Mancuso, "Development of HIP Consolidated P/M Superalloys for Conventional Forging to Gas Turbine Engine Components, in Superalloys 1996, ed. by R. D. Kissinger et. al., TMS-AIME, Warrendale, PA, 1996, pp. 645-652.

8. C. P. Blankenship, Jr., M. F. Henry, J. M. Hyzak, R. B. Rohling, and E. L. Hall, "Hot Die Forging of P/M Ni-Base Superalloys", in Superalloys 1996, ed. by R. D. Kissinger et. al., TMS-AIME, Warrendale, PA, 1996, pp. 653-662.

SUB-SOLVUS RECRYSTALLIZATION MECHANISMS IN UDIMET® ALLOY 720LI

B. Lindsley, X. Pierron

Special Metals Corporation
Middle Settlement Road
New Hartford, NY, 13413

Abstract

Sub-solvus recrystallization mechanisms were studied in a nickel base superalloy and the initial microstructure was found to influence the operating mechanism. Processing maps were developed to describe recrystallization for 3 different initial microstructures. In the cast-wrought condition, recrystallization takes place by nucleation and growth along grain boundaries between γ' particles in a similar fashion to single phase alloys. For large grain, small γ' structures, growth of recrystallized grains was found to occur primarily by boundary looping of the γ', leaving the γ' misoriented with the new matrix. Electron backscatter diffraction was used to measure the orientation of the γ matrix and the γ'. After a heat treatment that resulted in static recrystallization, three γ' orientations were found in the recrystallized grain, including γ' with a twin relationship to the matrix. A mechanism is proposed to describe the formation of the cast-wrought structure consisting of fine grains and γ' misoriented with both the γ matrix and neighboring γ'.

Introduction

Recrystallization in multiphase microstructures is an important mechanism for conversion of nickel base superalloy ingot to wrought product. Typical cast-wrought nickel base superalloys contain a γ' volume fraction as high as 50% to increase high temperature mechanical properties. The presence of the coherent ordered γ' precipitates in the microstructure at deformation temperatures results in complex recrystallization micro-mechanisms which need to be understood if one wants to optimize the thermo-mechanical processes necessary to manufacture forged components. Second phase particles act as barriers to grain boundary motion. This effect, known as Zener pinning, plays a major role in retarding recrystallization, and may also affect grain nucleation. The volume fraction second phase (F_v) and particle size (r) are critical variables for grain boundary pinning, as they determine interparticle spacing. It is generally accepted that, if $F_v/r > 0.2$ μm^{-1}, recrystallization will be inhibited[1].

UDIMET®[*] alloy 720LI (720) is a challenging alloy as the γ' solvus temperature is quite high for a cast and wrought superalloy. Above the solvus temperature, grain growth is rapid and the workability of the alloy decreases dramatically; therefore the material is worked sub-solvus. The volume fraction of γ' at working temperatures is approximately 20%, so according to the above rule, recrystallization will be retarded when the particle radius is less than 1μm. Particle radii $< 0.5\mu m$ are typically

---

[*] UDIMET is a registered trademark of Special Metals Corporation.

Superalloys 2000
Edited by T.M. Pollock, R.D. Kissinger, R.R. Bowman,
K.A. Green, M. McLean, S. Olson, and J.J. Schirra
TMS (The Minerals, Metals & Materials Society), 2000

found in air cooled and furnace cooled specimens after supersolvus solutioning. Since grain boundary motion is suppressed by particle pinning, additional mechanisms must exist to enable boundary motion through a sub-solvus material.

The interaction of the grain boundary and the second phase during recrystallization in superalloys has been extensively studied(2-4). Several processes have been reported for Ni based superalloys:

1. Dissolution of the γ' precipitates after contact with the boundary and coherent re-precipitation of the γ' either at or behind the boundary;

2. Grain boundary pinning by the coherent γ' which have coarsened at the grain boundary, and subsequent nucleation of new grains to form a necklace structure about the original grain boundary;

3. Grain boundary cutting through a γ' precipitate and reorienting the precipitate to the orientation of the recrystallized grain (rarely seen).

The most common mechanism for Ni based superalloys was found to be dissolution and re-precipitation of the γ'.

Little work has been done on dynamic or static subsolvus recrystallization with a γ' diameter ≥ 0.5μm. The effect of γ' size and interparticle spacing has not been addressed within a given alloy system, nor has the associated mechanistic changes in recrystallization. The interparticle spacing also affects the competition between recovery and recrystallization. Recovery can be quite rapid since the working temperature is approximately $0.9T_m$, and the γ' pin grain boundaries and inhibit recrystallization. The current work attempts to expand the current understanding of subsolvus recrystallization in nickel base superalloys by examining the different recrystallization mechanisms within a single alloy system containing varying microstructures deformed under a range of conditions.

Experimental Approach

Three initial microstructures of UDIMET alloy 720LI were compression tested and the recrystallization response was evaluated using light optical microscopy, scanning electron microscopy and electron backscatter diffraction. The composition of the alloy used in this study is given in Table I. The alloy was heat treated to three different starting microstructures to evaluate the effect of γ' size and distribution on subsolvus recrystallization. The first microstructure was typical cast-wrought material rolled to 28.6mm (1 1/8") bar, containing large γ' and small grains (material A). For the other two microstructures, the as-rolled material was solutioned and furnace cooled to form a large grain, fine γ' microstructure (material B). This material was then aged below the solvus to coarsen the γ' to an intermediate size (material C). The γ' sizes given in the paper were evaluated using image analysis after a heat

treatment at 1110°C. Cylindrical compression samples were made from each microstructure and the samples were deformed 30% at several temperatures (1066°C (1950°F), 1110°C (2030°F) and 1132°C (2070°F)) and three strain rates (1, 0.1 and 0.01 sec^{-1}). The compression test resulted in a barrel shaped specimen with non-uniform deformation from the bottom to the center of the sample. A wedge shaped dead zone was found in each end of all samples where die lock occurred. Samples were then air cooled or oil quenched to room temperature and evaluated both after deformation and after subsequent heat treatments to evaluate dynamic and static mechanisms. The heat treatment temperature used for all samples was 1132°C. Samples were etched with either modified Kallings (100ml methanol, 100ml HCl, 50g CuCl$_2$) or Chromate (460ml H$_3$PO$_4$, 25ml H$_2$SO$_4$, 50g CrO$_3$). The resulting microstructures were analyzed using light optical and scanning electron microscopy.

Table I. Alloy composition in weight percent.

C	Al	Co	Cr	Mo
0.025	2.44	14.6	16.2	3.02
Ti	**W**	**Zr**	**B**	**Ni**
4.95	1.31	0.037	0.016	Bal.

The orientations of the γ' precipitates and the surrounding matrix were investigated using the electron backscattered diffraction technique (EBSD) in the SEM. This technique allows the orientation of the observed crystals to be described by indexing the diffraction patterns generated by the backscattered electrons on a detector. By mapping the orientations as a function of position on the sample, an orientation map or OM can be generated and analyzed. The data was collected by TSL Inc. on specimens which were first polished down to 0.05 micron using a silica solution. A backscattered electron image (BEI) of the scanned area was then taken in order to associate each phase with its orientation. The orientation data were analyzed using TSL software and various representations of the orientation data were constructed. The misorientation between two adjacent areas can be visualized by associating different colors to the two misoriented regions. In addition, regions of same orientation were highlighted in the orientation maps. Those orientation images were then superimposed with the backscattered electron images of the same area.

Results

Microstructures

Material A (as-rolied material) contained large γ' (2.8μm) and small grains (15μm). EBSD of this structure showed that the γ' are misoriented with the surrounding grains, Figure 1. An SEM image and an EBSD image of the same area are shown in Figure 1, and the two are overlayed in

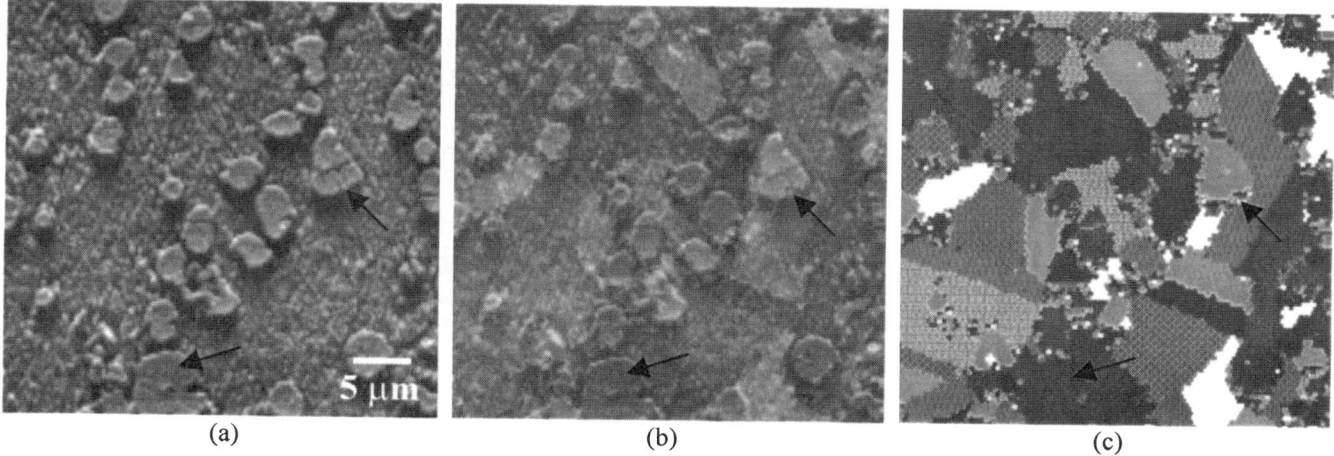

(a)   (b)   (c)

Figure 1. SEM image and EBSD orientation map of material A. (a) and (c) are SEM and orientation maps of the same region respectively. (b) is a superposition of image (a) and (c). The arrows indicate some γ' precipitates misoriented with the surrounding grains. *(A color version of this figure appears on page 841.)*

Fig. 1b. Each color represents a misorientation of greater than 15° with its neighbors, and the different colors of the γ' and surrounding γ grains reveal their different orientations. The solution and furnace cooled samples (materials B and C) had a large grain size after the supersolvus solution treatment (3 mm). Material B had an average γ' size of 0.7μm and material C had an average γ' size of 0.95μm. EBSD of the as-heat treated structure in material C showed that the γ' was coherent with the matrix, as expected in a non-deformed, as heat treated microstructure.

Processing Maps

Processing maps of strain rate vs. temperature were generated for the three microstructures, Figure 2. The process maps were created to evaluate the recrystallization response of the three microstructures to the testing variables that control recrystallization, namely strain rate and temperature (constant strain used for all tests). The result of the thermomechanical treatments to the material can generally be described by three responses: no recrystallization; a necklace of new grains at the original grain boundaries; or as some percentage of dynamic recrystallization. [Values for percent recrystallization are average values across the entire sample. The effect of changing strain and strain rates along the radius of the sample was therefore not addressed.] For material A, the processing map is separated into these three responses, although necklacing is the beginning of dynamic recrystallization for this microstructure. No dynamic recrystallization (DRX) was found at 1066°C or at low strain rates at 1110°C. Some recrystallization was found in the form of a necklace as the temperature and strain rate was increased, Figure 3a. The percent recrystallization is also given in Figure 2a. 100% recrystallization was found

at 1132°C and strain rates ≥ 0.1 sec^{-1}, Figure 3b. It should be noted that the dead zone of the compression sample was not included in the percent recrystallization values. The recrystallization of cast-wrought 720 was found to occur by new grains forming at the prior grain boundaries followed by grain growth until the deformed grains are consumed. This is a classic recrystallization mechanism in single phase alloys. Since the grain size of material A is on the order of the γ' spacing, little to no γ' exists within each grain, so on this scale the material behaves similarly to a single phase alloy.

The processing map for material B is given in Figure 2b. It was found that strain rate was a more important variable for dynamic recrystallization than temperature for this microstructure (as compared to material A). Again, the three responses were found in the material. No recrystallization was found at low temperature and low strain rate. At higher temperatures and strain rates ≤ 0.1 sec^{-1}, a necklace of fine grains 5 – 10 μm wide formed at the original grain boundaries. At a strain rate of 1 sec^{-1}, dynamic recrystallization was found for all three temperatures. The recrystallization was primarily within the grains and no longer associated with the boundaries. Preferential recrystallization was also found at nitrides within the matrix.

The processing map for material C is given in Figure 2c. Again, no recrystallization was found at low temperatures and strain rates. A necklace of recrystallized grains formed at the original grain boundaries at intermediate strain rates and temperatures. The band of recrystallized grains at the old boundary expanded in width to approximately 25 to 50μm at 1132°C and 0.01 sec^{-1}. Relative to material B, the region of the map containing dynamic recrystallization shifted toward higher temperatures. No recrystallization

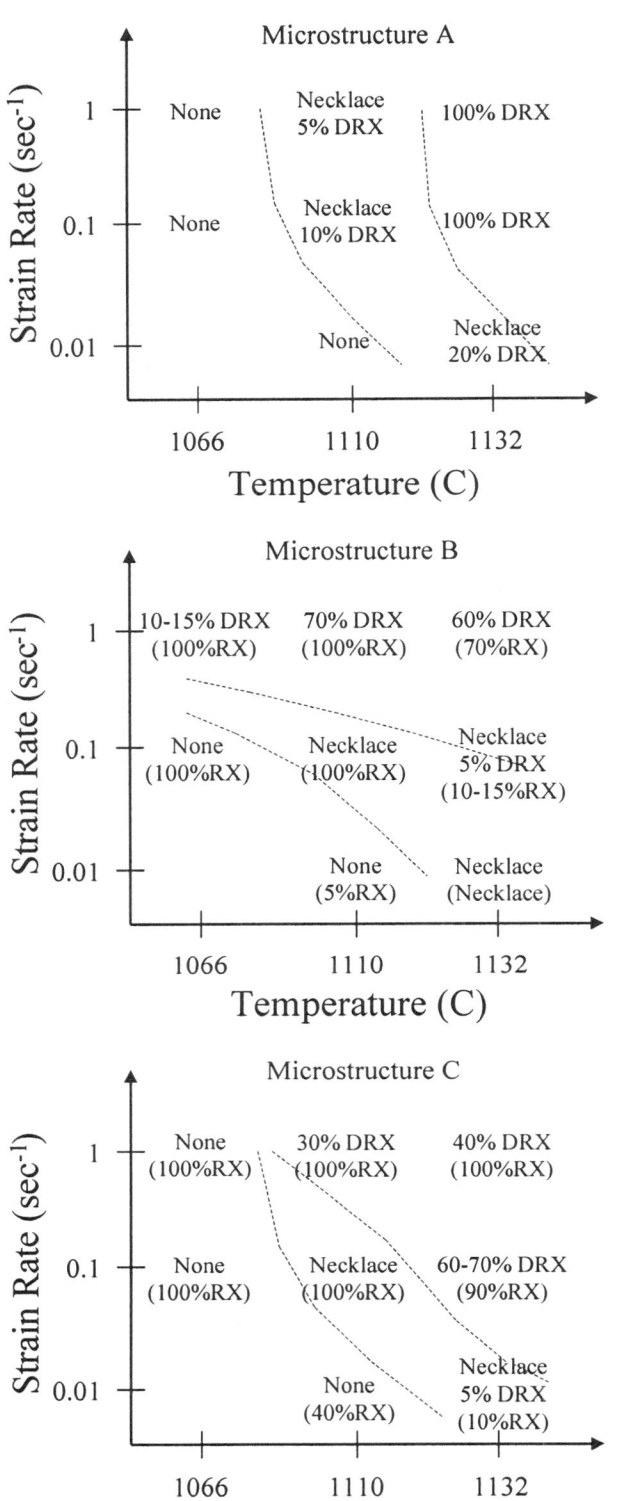

Figure 2. Processing maps for materials A, B, and C. The amount of dynamic recrystallization (DRX) is given for each deformation condition. Total recrystallization after an 1132°C heat treatment for 5 hours is given in parenthesis for B and C.

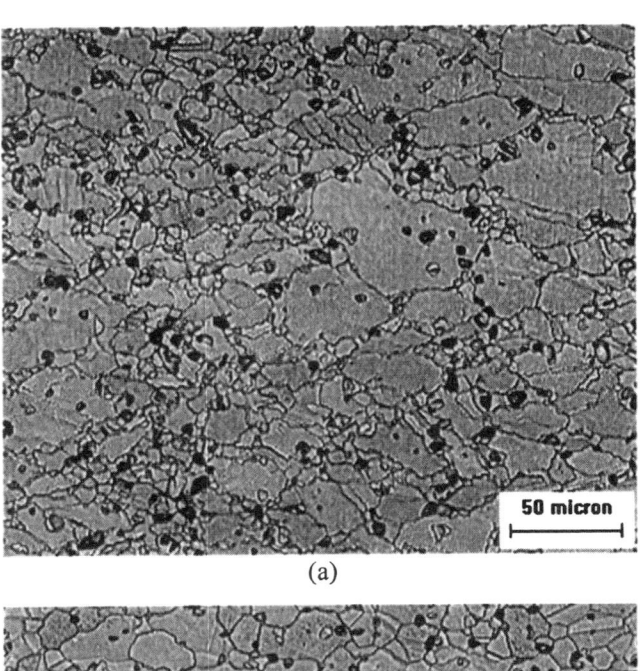

(a)

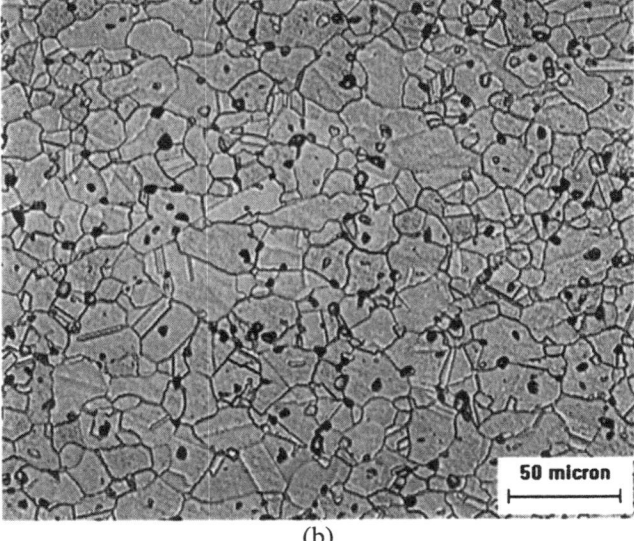

(b)

Figure 3. Recrystallized microstructures in material A: (a) necklace structure (b) fully recrystallized. Etched in modified Kallings.

was found at 1066°C and a strain rate of 1 sec⁻¹, as compared to 10-15% DRX for material B. Increased recrystallization at 1132°C and strain rates ≤ 0.1 sec⁻¹ was also found. Examples of dynamic recrystallization are given in Figure 4, where Figure 4a shows a finer grain size along the original grain boundary and larger grains towards the center of the grains, and Figure 4b shows the recrystallization front, with the newly recrystallized grain on the right. No change in the γ' size was found due to the dynamic recrystallization, although some evidence of the γ' inhibiting boundary motion was seen.

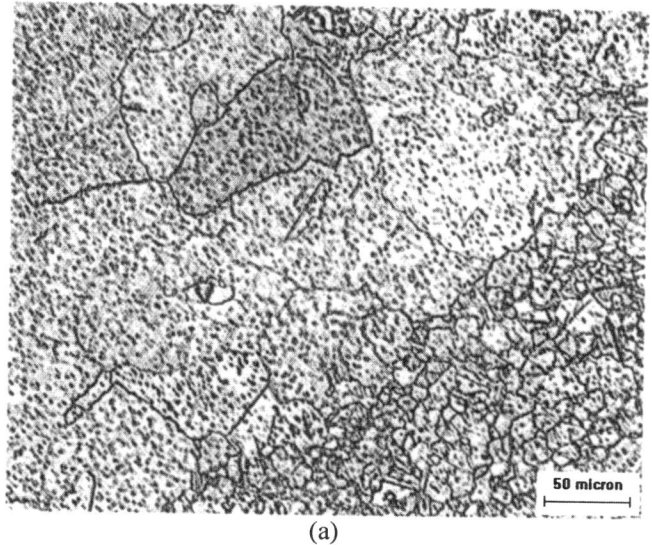

(a)

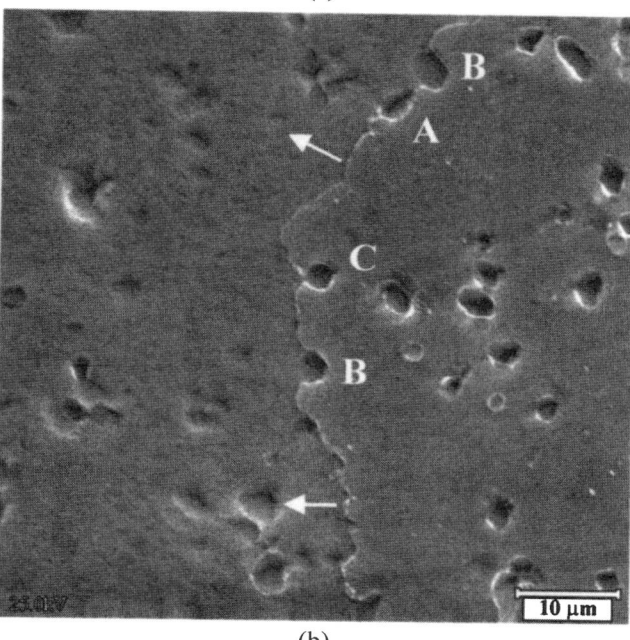

(b)

Figure 4. Dynamically recrystallized grains in material C. (a) Fine grains near the original grain boundary with coarser grains near the center of the original grain. (b) SEM image of a recrystallized grain boundary. Arrows indicate grain growth direction prior to quenching. Letters mark various stages of boundary looping around γ' precipitates. Etched in modified Kallings.

Effect of Heat Treatment

Samples from material B & C were heat treated at 1132°C for 5 hours and it was found that select samples statically recrystallized. The amount of recrystallization after the heat treatment is given in parentheses on the process maps (Figures 2b & c) for each sample. For both materials,

samples that were tested at temperatures ≤ 1110°C and strain rates ≥ 0.1 sec^{-1} were found to be 100% recrystallized after the heat treatment. Those samples tested at lower strain rates and higher temperatures tended to have less static recrystallization. Material B showed very little static recrystallization under these test parameters, whereas increased static recrystallization was observed for material C. The recrystallized grain size changed from the center to the edge of the prior grains. Near the prior grain boundary, the grain size ranged from 20 to 50 µm while in the center of the prior grains, grain sizes up to 500µm were found. Similar results were found during dynamic recrystallization for material C.

Figure 5 shows a boundary between a recrystallized grain (lower right) and an original grain (upper left). Pinning of the grain boundary by γ' particles is evident by the looping

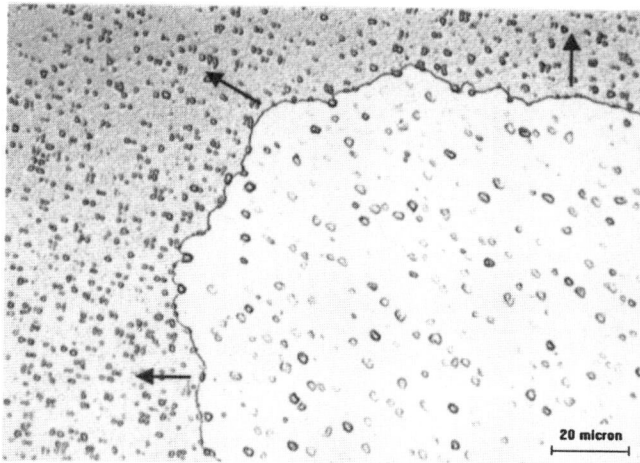

Figure 5. Statically recrystallized grain in material B after heat treatment. Arrows indicate grain growth direction prior to quenching. Etched in modified Kallings.

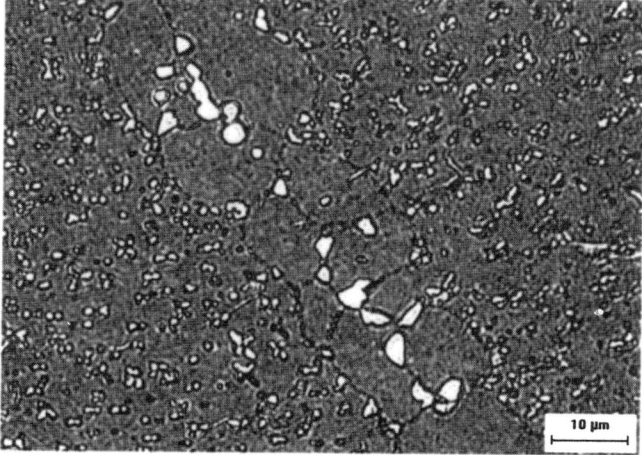

Figure 6. Necklace grain structure after heat treatment in material B. Larger γ' (white) lie primarily on the original grain boundary. Electrolytically etched in Chromate.

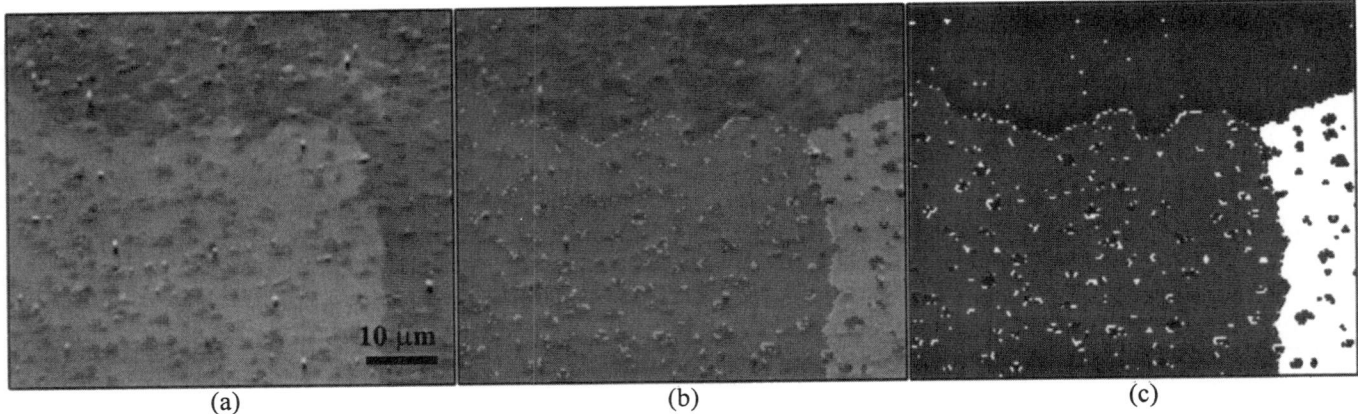

Figure 7. SEM image and EBSD orientation map of dynamically recrystallized material C. (a) and (c) are SEM and orientation maps of the same region respectively. (b) is a superposition of image (a) and (c). The red region is the original, unrecrystallized grain. *(A color version of this figure appears on page 841.)*

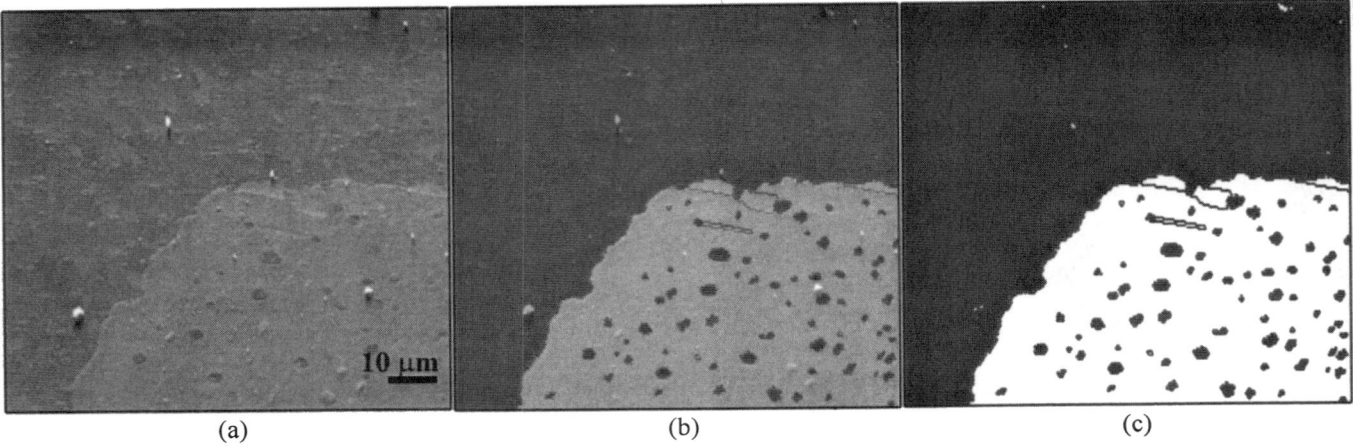

Figure 8. SEM image and EBSD orientation map of statically recrystallized material C. (a) and (c) are SEM and orientation maps of the same region respectively. (b) is a superposition of image (a) and (c). The red region is the original, unrecrystallized grain. *(A color version of this figure appears on page 841.)*

of the boundary about the particles. Significant coarsening of the γ' and an associated decrease in the number of γ' particles in recrystallized grains was observed relative to the γ' in both the unrecrystallized regions and in dynamically recrystallized grains. The γ' size increased from less than 1μm to 2-3μm. A necklace structure that formed after heat treatment of a low strain rate / high temperature sample is shown in Figure 6. Large γ' reside along the original grain boundary and the new necklace grains are primarily devoid of γ'.

Electron Backscatter Diffraction

EBSD was used to analyze the orientation relationship between the various phases and grains in order to provide insight on the dynamic and static recrystallization mechanisms that were observed in materials B & C.

An orientation map of a dynamically recrystallized region in material C is presented in Figure 7. Recrystallization

occurred after compression testing at 1132°C, with a strain rate of 0.1Hz. The region in Figure 7 contains two recrystallized grains that were growing toward the top of the image when the sample was quenched. One can note that the γ' size and distribution in the recrystallized grains and in the unrecrystallized grain at the top of the Figure are similar. The orientation map (OM) of the unrecrystallized grain shows that all the γ' precipitates have the same orientation as the matrix, as no change in color can be seen in the OM. In this Figure the misorientation treshold for color change was set at a 10° misorientation angle. In the recrystallized grains, the γ' is evidently misoriented with the new matrix. By highlighting all the regions with the same orientation as the unrecrystallized grain in red, it was found that all of the misoriented γ' in the OM had the same orientation as the original top grain (Fig 7c). A comparison of the microstructure and the EBSD results revealed that less than 5% of the γ' did not show up on the misorientation map. These γ' either have the same orientation as the matrix in the new grains or were too small to be detected during the EBSD scan.

In the same material, material C, the orientation relationship between the precipitates and the matrix was analyzed in the recrystallized region after deformation at 1110°C with a lower strain rate of 0.01Hz. After deformation, no recrystallized regions could be observed in this sample. The deformed material was then heat treated for 5 hours at 1132°C and recrystallized statically. The resulting microstructure can be seen in Figure 8. An unrecrystallized grain can be seen in the upper left part of the figure, and the orientation image shows that the γ' phase in this grain is oriented cube on cube with the surrounding matrix. By comparing the BEI image and the orientation image, it can be seen in the recrystallized grain that the majority of the γ' precipitates are misoriented with the surrounding matrix, as only a few precipitates are visible on the BEI image but not on the OM. By highlighting the phases with the same orientation, it was found that in the misoriented γ' in the recrystallized grain are of two kinds. The first kind (red in the OM) have the same orientation as their parent grain, the unrecrystallized top grain. This is similar to the precipitates found in the dynamically recrystallized grain. The second kind (blue in the OM) are not oriented with the original grain, but rather have a special orientation relationship with the recrystallized grain. A region containing the blue precipitates and the recrystallized matrix was selected and pole figures were generated with only these orientations, Figure 9. It follows from close examination of the pole figures that the blue γ' precipitates have a twin relationship with the matrix, and are rotated 60° around a [111] matrix

direction. The blue γ' and the matrix share one common (111) plane and three common [110] directions. Annealing twins were also present in the recrystallized grain and are visible through misorientation boundaries in the white region of Figure 8c. The four annealing twins share a common (111) twin plane with the matrix that is different to the (111) twin plane between the matrix than the blue γ'.

Discussion

It was found that the γ matrix and the γ' precipitates retained no residual orientation relationship in the as-rolled (cast-wrought) condition. Further, the γ' were misoriented with neighboring γ'. [It should be noted that the γ' referred to here only include those that precipitated at the working temperatures. γ' that precipitate during the final strengthening heat treatment will be a fine, cube on cube oriented precipitate.] However, when 720 is processed from ingot, the material is first homogenized, and after cooling, a large grain, fine γ' structure is produced, similar to material B, where all of the γ' have the same orientation within a large grain. It has been well established that such precipitates effectively inhibit grain boundary motion. Previous recrystallization mechanisms that have been reported in subsolvus superalloys include solutioning of the fine γ' ahead of the boundary and coherent re-precipitation in or behind the boundary, the boundary moving through the γ' precipitate, or the pinning of the original boundary by γ' and the formation of a necklace structure(2-4). All of these mechanisms may eventually lead to a random grain orientation, however, the γ' tend to maintain a common orientation within each grain. The questions of what recrystallization mechanism(s) occurs in the subsolvus condition and how the γ' eventually change from a cube on cube orientation with the matrix to being misoriented with both the matrix and neighboring γ' have not yet been answered.

The first step in answering these questions was to develop processing maps to understand recrystallization conditions in 720 with different microstructures. In the as-rolled condition, material A, the grain size is on the order of the γ' spacing. The γ' tend to reside at the grain boundary triple points and inhibit further grain growth, but few γ' reside within the grains or at the grain boundaries between triple points. Recrystallization via necklacing at the boundaries was not significantly inhibited by the γ'. This was found to be especially true at higher temperatures where the γ' spacing increased due to solutioning of the γ'. Further grain growth after necklacing leads to full recrystallization, which was also accelerated at high temperatures. Hence, once the material has been worked enough so that the grain size is on the order of the γ' size, recrystallization occurs in a similar manner to a single phase material.

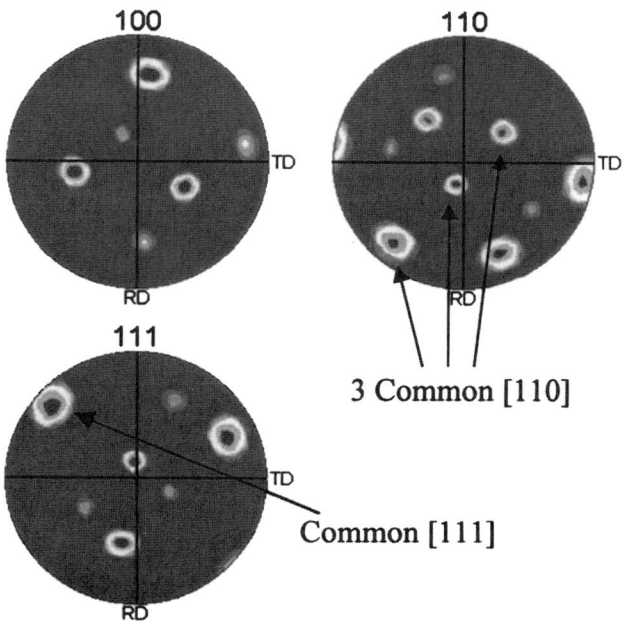

Figure 9. Pole figures for the recrystallized grain and the blue γ' in Figure 8c. (*This figure appears in color on page 842.*)

Material B has a fine γ' size, so for all test temperatures below solvus, the γ' spacing remains small. For this reason, temperature is a less significant recrystallization variable in this material, and strain rate appears to be the dominant recrystallization variable. It is interesting to note that dynamic recrystallization was found at a temperature of 1066°C and a strain rate of 1 sec⁻¹ for this condition, while none was found in the other materials. This is likely related to how the material accommodates the deformation, but is outside the scope of this paper and will not be addressed further.

The processing map for material C appears to be a hybrid between materials A and B. A large shift in the dymanic recrystallization regime was noted between materials B and C, with increased recrystallization occurring at hotter temperatures and lower strain rates for material C. The coarser γ' in material C equate to a larger interparticle spacing, which apparently encourages dynamic recrystallization at higher temperatures. However, the size of this shift was a bit surprising given that the starting microstructures are not vastly different for the two materials.

The heat treatment given to materials B and C resulted in 100% recrystallization for those samples tested at lower temperatures (≤ 1110°C) and higher strain rates (≥ 0.1 sec⁻¹). These test conditions lead to an elevated amount of stored work in the material, which upon heat treatment, resulted in full static recrystallization. Samples tested at higher temperatures and especially those tested at lower strain rates exhibited lower amounts of static recrystallization. This finding was attributed to recovery in the material. Materials that have high stacking fault energies, such as nickel, undergo recovery when deformed at high temperatures[5]. It has also been found that in low stacking fault materials, such as stainless steel, if recrystallization is inhibited by a second phase, then at sufficiently high temperatures, recovery can occur[1]. When the materials are deformed at lower strain rates, more time is available for dynamic recovery and cell wall formation. Reduced levels of static recrystallization in material B as compared to C under these conditions may be a result of increased boundary pinning by the smaller interparticle spacing.

For recrystallization to occur in materials B and C, the grain boundary must be able to move through an area containing γ'. As described previously, the γ' precipitates inhibit boundary motion. The mechanism by which dynamic recrystallization occurred in these materials was grain boundary looping of the γ'. The stages of looping can be seen in Figure 4b. The boundary initially contacts a γ' (A) and is halted at the γ'/recrystallized grain interface. As the rest of the boundary continues to move, the boundary begins to bow around the particle (B). Finally, the boundary loops around the particle (C) and continues on leaving γ' with their original orientation in its wake. The orientation map in Figure 7 clearly shows the γ' retain their original orientation after recrystallization. Both the microstructural and EBSD analysis support the boundary looping mechanism. It should be noted that grain boundary looping is not a new mechanism and often occurs for incoherent particles[1]. However, looping of coherent particles has rarely been reported. Looping of the coherent NbC phase in γ-Fe was found by Jones and Ralph[6] but has not been widely reported for nickel base superalloys.

One possible reason that looping is rarely seen for coherent, cube on cube oriented precipitates is that the coherent particles are more effective at inhibiting grain boundary motion than incoherent particles. It has estimated[2] that coherent precipitates are up to 4 times more effective in resisting grain boundary motion than incoherent particles, due to the additional interfacial energy required to transform the particle/matrix interface from coherent to incoherent. Interparticle spacing is also an important variable for this mechanism. The stress of the bowing boundary is inversely related to the radius of curvature of the boundary. If a very fine, uniform distribution of particles exists, the radius of curvature decreases between particles on the boundary and the resistance to boundary movement increases. As the particle spacing increases, the energy required for boundary movement decreases. It should be noted that once the first recrystallization front has passed through a region containing γ' and the particles become misoriented with the matrix, further recrystallization should be easier in this region. Any future boundary would encounter incoherent particles and would no longer need to supply the additional energy to transform the coherent interface to an incoherent interface.

In the case of static recrystallization, some looping of γ' is still observed. The red γ' in Figure 8 indicate some of the γ' retain their original orientation. In addition, boundary looping can be seen in Figure 5. However, additional processes occur during the heat treatment. Coarsening of the γ' occurs during the heat treatment, and is at least partially a result of accelerated growth on the grain boundary. The boundary is a fast diffusion path for Al and Ti hence larger γ' in the boundary will coarsen rapidly at the expense of smaller γ' that come in contact with the boundary. Some matrix diffusion ahead of the boundary can also occur. The coarsening of γ' in the boundary can be seen in Figure 5. In fact, in cases of low driving force, grain boundary movement may not occur without this coarsening mechanism. Coarsening of the γ' at the boundary reduces the number of precipitates and barriers to boundary motion. The γ' continue to coarsen at high temperature after the boundary has swept past (Oswald ripening) and the coarsening kinetics may be accelerated due to the higher interfacial energy of the misoriented γ' with the matrix. Regardless of whether the accelerated coarsening is due to the boundary interaction, the misoriented γ', or both, the coarsening is more rapid in the recrystallized region. The initial 48 hour heat treatment at 1132°C used to coarsen the structure in material B to material C resulted in a change in γ' size from 0.7μm to 0.95μm. After the 5 hours at 1132°C for the static recrystallization, γ' 2 to 3 μm in diameter were observed in the recrystallized regions.

Finding γ' with a twin relationship to the γ in the heat treated samples was surprising. To the authors' knowledge, γ' with a twin relationship to the γ have not been previously found in nickel based superalloys. The mechanism for their formation is unknown. One potential mechanism for their formation may be similar to the formation of annealing twins. As the boundary moves and solutions some of the smaller γ', new γ' may precipitate in the boundary, which has been documented by other investigators(2-4). It may be energetically favorable for the γ' to grow with a twin relationship to the new grain in order to minimize the total free energy of the resultant boundaries between the γ' and both the recrystallized grain and the original grain. In other words, if by forming the twin, the γ' lowers the interfacial energy between it and the old grain enough to offset the additional energy associated with the twinned interface, then this mechanism may be possible. For this mechanism to be viable, a special (low energy) orientation relationship should exist between the 'twinned' γ' and the old matrix which would minimize the interfacial energy. However, no special relationship was found between these γ' and the original matrix.

Another possible mechanism for the 'twinned' γ' could involve the dissolution of the misoriented γ' left behind from the looping process and the re-precipitation of γ' with some orientation to the matrix. The γ' with the twinned relationship represent a lower interfacial energy precipitate, and while not being the lowest energy precipitate (cube on cube orientation), may be metastable. Nucleation of such a particle may require a defect in the matrix, such as a pre-existing annealing twin. Both of the above mechanisms also allow for the formation of γ' oriented cube on cube with the matrix. This is an important point since γ' oriented with the matrix were also found.

The original question of how a structure with γ' that are misoriented to both the surrounding matrix and neighboring γ' can now be addressed. It was found that, for dynamic recrystallization, one passage of a grain boundary through the material lead to γ' misoriented with the matrix, but oriented with the surrounding γ'. Additional dynamic recrystallization will not change the orientation of the γ'. However, after heat treatment and static recrystallization, 3 γ' orientations were found. If another recrystallization front was to pass through this region via the looping mechanism, three populations of misoriented γ' would exist, and with additional holds at temperature, two more orientations of γ' (cube on cube and 'twinned' with the new matrix) could exist. It can be seen that repeated deformation and heat treatments could lead to condition where the γ' are now misoriented with neighboring γ'. In addition, as the material undergoes many repetitions of deformation and heat treatment, the γ' will coarsen to a point at which recrystallization can occur between the γ'. The process described above represents a possible mechanism to go from a large grain, cube on cube coherent γ' structure to a fine grain, misoriented γ' structure.

Conclusions

Analysis of the recrystallization mechanisms has shown how the starting microstructure influences the operating mechanism for a given set of processing conditions. In the cast-wrought condition, recrystallization occurs by nucleation and growth along grain boundaries between γ' particles in a similar fashion to single phase alloys. For a large grain, fine γ' structure, growth of recrystallized grains was found to occur primarily by boundary looping of the γ', leaving the γ' misoriented with the new matrix. After a heat treatment that resulted in static recrystallization, three γ' orientations were found in the recrystallized grain, including γ' with a twin relationship to the matrix. A mechanism is proposed to describe the transformation of a large grain, fine γ' structure containing γ' with a cube on cube orientation to the cast-wrought structure consisting of fine grains and γ' misoriented with both the γ matrix and neighboring γ'.

Acknowledgements

The authors would like to thank B. Antolovich for his insight during technical discussions and K. Sellitti who conducted SEM analysis and documentation.

References

1. F. J. Humphreys, M. Hatherly, *Recrystallization and Related Annealing Phenomena* (Pergamon, Elsevier Science Inc., Tarrytown, NY, 1996).

2. R. D. Doherty, Metal Science, **16** (1982) 1-13.

3. V. Randle, B. Ralph, Acta Metallurgica, **34** (1986) 891-898.

4. A. Porter, B. Ralph, Journal of Materials Science, **16** (1981) 707-713.

5. R. E. Reed-Hill, *Physical Metallurgy Principles* (PWS-Kent, Wadsworth Inc., Boston, MA, 1973).

6. A. R. Jones, B. Ralph, Acta Metallurgica, **23** (1975) 355-363.

THE MECHANICAL PROPERTY RESPONSE OF TURBINE DISKS PRODUCED USING ADVANCED PM PROCESSING TECHNIQUES

Anthony Bani
Special Metals Cor
Princeton, KY U

Kenneth A. Gr
Rolls-Royce Corp
Indianapolis, IN

ABSTRACT

Over the thirty years for which powder metal processing was applied in the production of superalloy turbine disk components, many advancements in the understanding of powder processing and applications have evolved. In this regard, a major shift to finer powder sizes has driven manufacturers to investigate advanced powder processing routes to minimize alloy costs. In addition, major advancements in understanding component temperature and strength capability have permitted improvements in designs that extend the ultimate performance of the turbine under extreme operating conditions. In this investigation, the powder UDIMET® Alloy 720 was produced using an advanced gas atomization nozzle in combination with process improvements consisting of hot isostatic pressing and low ratio extrusion to develop fine grain microstructures for subsequent isothermal forging to a low pressure turbine disk. The disks were evaluated both non-destructively to assess the ultrasonic inspection capability and destructively to demonstrate mechanical property capability.

Introduction

Powder metal (PM) superalloys have been utilized in gas turbine engines for over thirty years. Initial processing for powder metal superalloys utilized a hot compaction plus extrusion conversion sequence. High extrusion ratios were employed to develop a fine grain billet microstructure[1]. The fine grain billet microstructures provided starting material suitable for slow strain rate, isothermal forging techniques. The subsequent near-net shape forging was then heat treated to produce high strength, fine grain disks used initially in military turbines. As powder processing became better understood, powder metal superalloy applications expanded to include the most advanced commercial and military engines. These highly stressed components continue to utilize many of the powder billet processing

UDIMET is a registered trademark of Special Metals Corporation.

techniques established during the initial phases of powder development.

Concurrently with these improvements in powder processing, modified superalloy compositions were being developed to meet the demands of advanced turbine engine designs. One alloy, UDIMET Alloy 720, has been developed for both powder metal processing and conventional ingot metallurgy processing routes. Thus, the alloy provides an excellent baseline material to assess the benefits of advanced powder metal processing techniques compared to a baseline ingot metallurgy processing technique.

In the 1960's, the powder compaction and extrusion operations were established on powder material produced using conventional atomization processes. These processes utilized argon gas atomization techniques and provided powder distributions that ranged in size from –180 µm to +30 µm in diameter. The powder was then screened to restrict the coarse powder fraction and provide a means to control residual defect size in the material for life prediction methodologies. Typically, the powder was screened to provide –150 µm powder for consolidation and extrusion. High extrusion ratios provided refinement of the coarse powder structure and were thought to provide a dispersion of possible powder contaminants including organic and in-organic inclusions.

Low cycle fatigue tests have shown that the fatigue life of components can be significantly increased with the use of finer powder size. However, until recently, the costs for producing fine powder material has restricted applications to only the most critical hardware having the most demanding operating conditions relative to temperature and cyclic stress conditions.

An advanced atomization process incorporating a high yield nozzle (HYN) atomization process is presently being evaluated at Special Metals Corporation. The HYN has demonstrated the capability to increase the yields of fine superalloy powder in excess of 50%. The increase in fine powder yields results in a direct reduction of powder

Superalloys 2000
Edited by T.M. Pollock, R.D. Kissinger, R.R. Bowman,
K.A. Green, M. McLean, S. Olson, and J.J. Schirra
TMS (The Minerals, Metals & Materials Society), 2000

manufacturing costs. With the fine starting powder material, additional cost reductions associated with the consolidation and extrusion operations can also be incorporated.

Powder Characterization

The powder material was produced in a production gas atomization unit using the high yield nozzle and screened to –270 mesh (53 μm). Powder samples were obtained for characterization studies including chemistry, heavy liquid separation testing[2] and water elutriation testing. The baseline chemistry (Table I) of the powder was consistent with production UDIMET Alloy 720 material produced using conventional ingot metallurgy or powder metallurgy techniques.

Table I The powder was produced to the standard UDIMET Alloy 720 composition. (in weight percent unless otherwise noted.)

C	Cr	Co	Mo	Ti	Al
0.027	16.0	14.5	3.01	5.14	2.63
W	**Zr**	**B**	**Ni**	**O$_2$**	**N$_2$**
1.29	0.04	0.02	Bal.	190 PPM	9 PPM

Heavy liquid separation (HLS) testing was performed to assess the frequency of inclusions present in a 0.4 kg (0.5 lb.) powder sample. The particles are separated using a thallium formate solution followed by scanning electron microscopy to count and classify the inclusions. As indicated in Table II, the inclusions present were typical of those present during standard powder atomization.

Table II HLS testing revealed refractory type oxide particles.

Type	f	Area (mil^2) Min.	Area (mil^2) Max.	EDS (%) C	O	Na	Mg	Al	Si	Ca
Al	12	1.12	7.33	>0	>0	=0	=0	>0	>0	=0
Mg	2	3.36	7.48	>0	>0	=0	>=0			
Si	6	1.35	11.75	>0	>0	>0	<5	<5	>10	>0
Other	12	1.05	7.33							

Water elutriation testing was performed on the powder material and provided very erratic results. The water elutriation test was established for coarse powder fractions or –150 mesh material. With the present material of –270 mesh product, powder samples for elutriation testing required reconstructing the original powder distribution prior to performing elutriation testing. The subsequent powder exhibited a high frequency of powder contamination from the additional powder handling and the results were considered inconclusive.

Billet Manufacture

The powder was consolidated using a hot isostatic press consolidation process in combination with a 3:1 extrusion process to a 152 mm (6") diameter billet. Testing of the billet material included large bar tensile (LBT) testing, metallographic evaluations, electron beam button analysis and ultrasonic inspection.

The LBT evaluations were performed on test samples obtained from the lead and tail end of each extrusion. As indicated in Figure 1, the LBT test results indicated that the powder material contained refractory materials typical of vacuum melted and atomized product. The area of the material at the fracture initiation site was similar to those evident during HLS testing of the powder material. The morphology of the material however was elongated as a result of the extrusion operation.

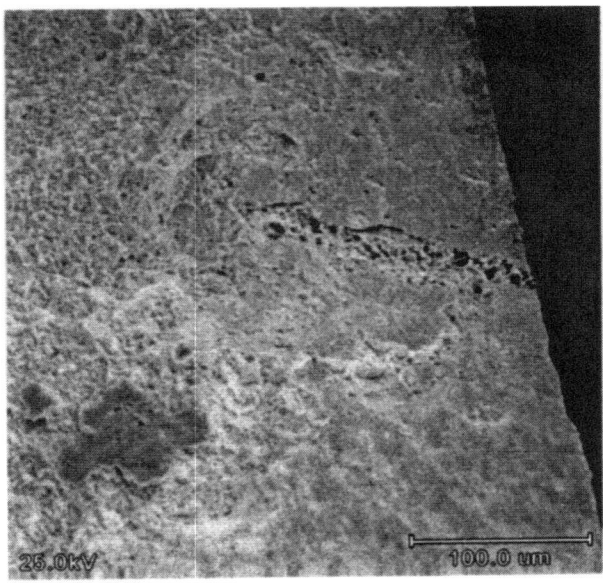

Billet	Location	Length (mil^2)	Width (mil^2)	Area (mil^2)
5306	Lead	110	45	5.6
	Tail	170	28	5.9
5307	Lead	105	20	2.9
	Tail	175	25	5
5311	Lead	155	40	6.3
	Tail	130	30	4.5

Figure 1 Powder processing tests revealed typical refractory materials associated with the primary melt operations that were elongated in the extrusion direction.

Test samples were obtained from each extrusion for electron beam button testing. The samples were approximately 1 kg (2 LB) each. The test was performed using a drip melting process followed by a gradual reduction in power to provide for directional solidification of the button prior to total reduction of the power. This process permits floatation of the low-density oxide materials to the top center of the button for subsequent analysis.

Each button sample was evaluated for the raft size and composition of the materials that were evident at the button surface. As indicated in Figure 2, the composition of the raft is similar to materials observed during previous tests performed on the powder material.

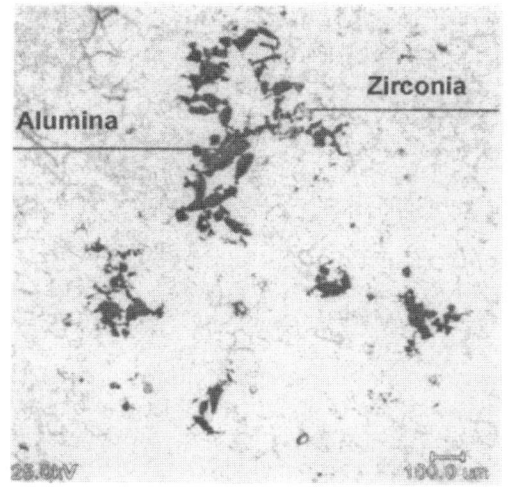

ID	Area (cm^2)	Cleanliness (cm^2/kg)
5306	0.21	0.07
5307	0.05	0.02
5311	0.04	0.01

Figure 2. Alumina and zirconia were evident on the button surface.

Metallographic evaluations and ultrasonic response were typical of fine grain powder metal billet product using high extrusion ratios as indicated in Figure 3. The grain size was assessed in the longitudinal direction at the lead and tail of each extrusion. The extruded product revealed ASTM 14 grain size at the center, mid-radius and edge locations after extrusion. The grain size of the low extrusion ratio product was similar to that obtained previously with an extrusion ratio of 6:1. Ultrasonic inspection of the billet material was performed to 10% of a No. 1 flat bottom hole (FBH) standard with no indications evident.

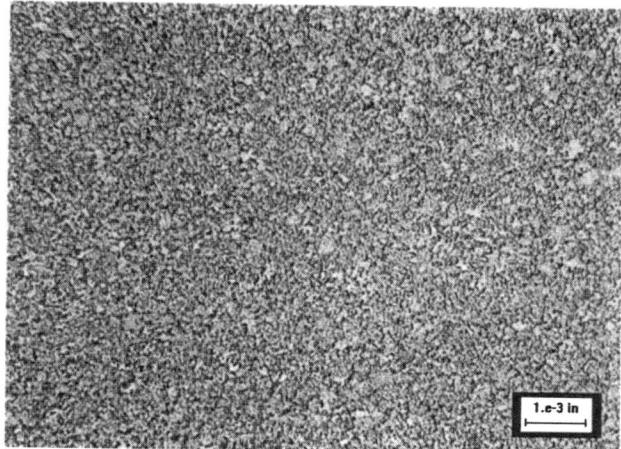

PM UDIMET Alloy 720 Extruded at 6:1 Reduction
152 mm (6") Diameter Billet

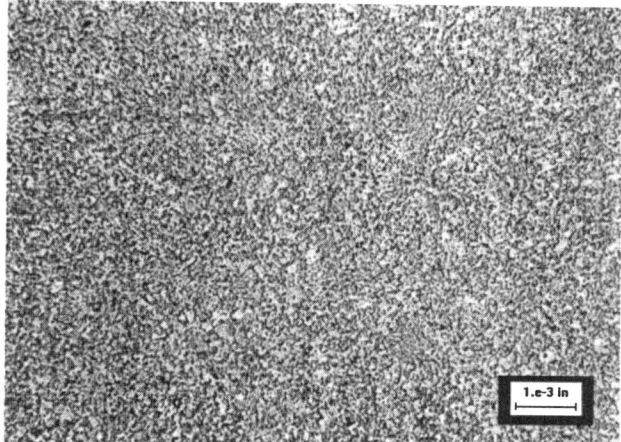

PM UDIMET Alloy 720 Extruded at 3:1 Reduction
152 mm (6") Diameter Billet

Figure 3. Metallographic evaluations indicated that the 3:1 reduction ratio provided a similar grain size to conventional product extruded at a ratio of 6:1.

The HYN powder billet material was subsequently isothermally forged by Ladish Company, Inc. into low-pressure turbine disks,[3] as shown in Figure 4. The thermo-mechanical processing schedule was tailored to produce a damage tolerant microstructure, which offered a balance of tensile yield strength and fatigue crack growth resistant properties. To accomplish this, a sub-solvus solution heat treatment in the range of 1121°C (2050°F) to 1149°C (2100°F) was used to produce a controlled coarsening of the grain structure to a uniform size of ASTM 11. A standard two-step age cycle of 760°C (1400°F)/8 hrs. + 649°C (1200°F)/24 hours produced the desired aging response to the gamma prime and carbide precipitates.

Figure 4. The powder billet was successfully isothermally forged into low-pressure turbine disks.

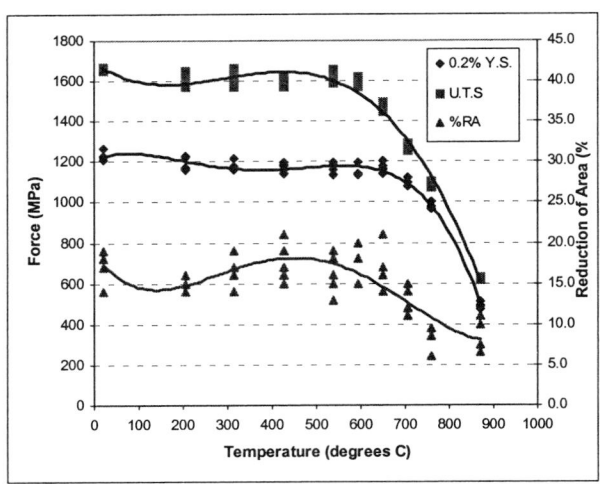

Figure 5. The PM UDIMET Alloy 720 disk exhibited excellent high strength tensile capability.

Results and Discussion

The disks were evaluated nondestructively to assess ultrasonic resolution. All disks passed the inspection based on rejection criteria of a 10% of No. 1 flat bottom hole (0.38 mm FBH) size. Mechanical property evaluations on the disks revealed the high yield nozzle atomized powder process in combination with the low extrusion ratio provided excellent high temperature tensile strength and ductility, as indicated by Figure 5.

Low cycle fatigue (LCF) testing was performed at elevated temperatures for several R-ratios. The results at 538°C (1000°F) for R = 0.0 are plotted in Figure 6 with results of PM UDIMET Alloy 720 produced by conventional atomization, screened to –150 mesh and heat treated to a grain size of predominantly ASTM 14[4]. The data of both materials are comparable, although a slight difference is noted at the high strain rates, where the ASTM 14 material is slightly better than the ASTM 11 material. At lower strain rates, the ASTM 11 material demonstrates longer life, possibly due to improved material cleanliness.

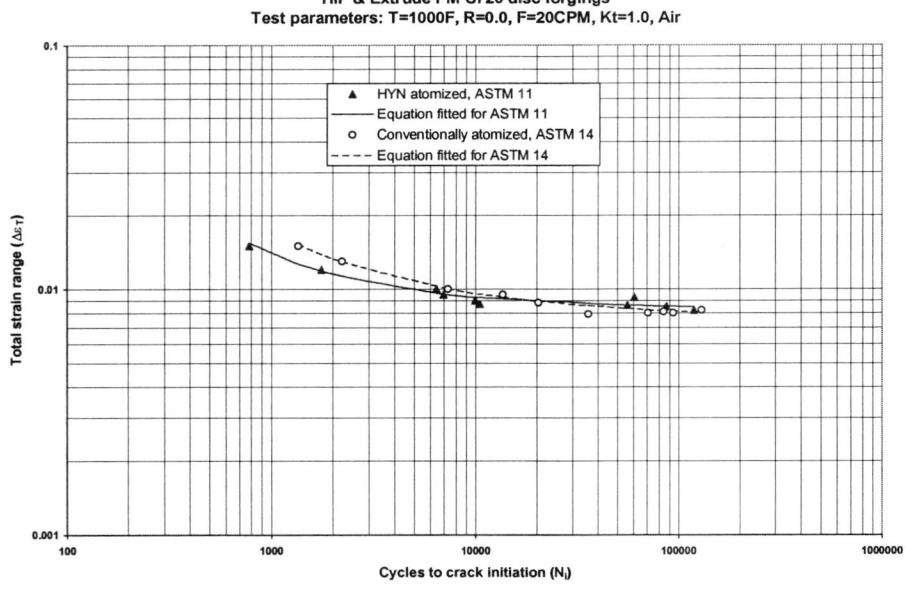

Figure 6. The difference in fatigue life at the higher strains can be attributed to the intentionally coarser grain size obtained by heat treatment.

An analysis of the fracture initiation sites performed on the LCF specimens show significant differences between the conventional atomized powder and the HYN atomized material. Table III compares the fracture initiation types and locations for both materials. As expected, the –270 mesh HYN material displays fewer non-metallic inclusions at the initiation sites. The predominant initiation feature for the –150 mesh conventionally atomized material, microporosity, is not evident in the HYN material as a result of the finer powder size. Only a few percent of the HYN specimens failed at non-metallic inclusions. The remaining HYN fatigue specimens failed at surface features, which is what one would expect for a very clean material. Another indication of material cleanliness is the fracture initiation location, also indicated in Table III. Only 2% percent of the initiation sites in the HYN material occurred internally, as compared with 16% internal initiation sites in the conventionally atomized material. The reduction in internal initiation sites is associated with fewer internal anomalies and is reflective as cleaner material. This information confirms the improved LCF capability of the HYN material.

Table III The initiation sites in the HYN material appears significantly cleaner when evaluating fracture initiation sites of fatigue samples.

			Conv. Atom. -150	HYN Atom. -270
Initiation Site Type	Type I	Discrete	8%	2%
	Type II	Agglomerated	15%	3%
	Type III	Microporosity	36%	0%
	Type IV	Surface, G.B.	41%	95%
Location	All Types	Internal	16%	2%
	All Types	Surface	84%	98%

Fatigue crack growth rate testing was conducted at 1200°F, R=0.05 and 10 Hz. Again the results are compared with conventionally atomized ASTM 14 material in Figure 7. The improvement in performance of coarser grain material is evident in the order of magnitude improvement in crack growth rate of the ASTM 11 material versus the ASTM 14 material. This trend is considered favorable for newer engine designs, which rely on damage tolerance as a major criteria for satisfying life requirements of critical rotating components.

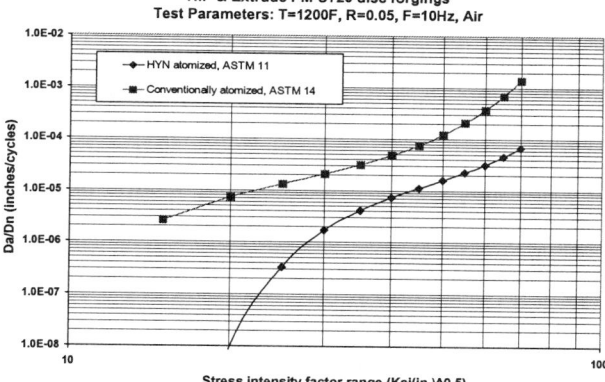

Figure 7 The uniform ASTM 11 grain size in the HYN material provided significantly improved fatigue crack growth response.

The advanced HYN powder process route provides a significant reduction in processing costs compared to traditional powder practices. In addition, the need for specialized extrusion equipment and high set-up costs is not required with the reduced extrusion ratios. When this approach for lower cost billet is coupled with near-net shape isothermal forging, significant cost reductions can be realized for forged components. Based on this information, Rolls-Royce Allison is aggressively pursuing these technologies for implementation in a wide range of engine applications.

Conclusions

1) Standard test practices for water elutriation testing resulted in extraneous inclusions when applied to the –270 mesh product.

2) Ceramic materials present in the powder and billet material produced using the high yield nozzle atomization process were similar in composition and lower in frequency to those present in conventional atomized product.

3) HIP plus low extrusion ratio (3:1) billet product was microstructurally and ultrasonically similar to conventional produced using a high extrusion ratio (6:1).

4) Isothermally forged PM UDIMET Alloy 720 material processed to a grain size of ASTM 11 exhibits attractive levels of elevated temperature tensile, low cycle fatigue and fatigue crack growth rate properties, capable of meeting low pressure turbine disk design requirements.

Acknowledgements

This work was performed under U.S. Navy Prime Contract No. N00140-92-C-BC49 at the National Center for Excellence in Metalworking Technology operated by Concurrent Technologies Corporation, under the direction of Dr. Gary Miller. The NAVAIR technical monitor was Dr. George Richardson. Technical assistance was provided by Karen Sellitti at Special Metals Corporation, Dr. Chi-An Yin and Mr. Dennis Overshiner at Rolls-Royce Corporation and Mr. Joe Lemsky at Ladish Co., Inc.

[1] J. B. Moore and R. L. Athey, *U.S. Patent 3,519,503,* (1970).

[2] J.C. Murray, P. G. Roth, J. E. Morra and J. M. Hyzak, *APMI / MPIF Conference,* 1994.

[3] G. Miller, *Workshop on High Temperature Structural Alloys,* West Virginia University, May 19-21, 1999.

[4] K. A. Green, J. A. Lemsky, and R. M. Gasior, "Development of Isothermally Forged P/M Udimet 720 for Turbine Disk Applications", Superalloys 1996 (Warrendale, PA: TMS, 1996), 697-703.

SEGREGATION AND SOLID EVOLUTION DURING
THE SOLIDIFICATION OF NIOBIUM-CONTAINING SUPERALLOYS

Wanhong Yang[*], Wei Chen[*], Keh-Minn Chang[*]
Sarwan Mannan[♀], John deBarbadillo[♀]

[*]West Virginia University, Morgantown, WV 26506-6106
[♀]Special Metals Corporation, Huntington, WV 25705

Abstract

A macrosegregation defect, commonly termed freckle, has limited the maximum diameter of remelted ingots of some Nb-strengthened nickel base superalloys. Two aspects of freckle formation were studied in three commercially important superalloys 718, 706 and 625: 1) the segregation characteristics which provide the driving force for freckle formation; 2) the solid development in the mushy zone which provides resistance to the flow inducing the defect. A Rayleigh number form freckle criterion that considers the effects of composition is derived. Freckle tendency of the alloys was evaluated using the experimental results.

Introduction

The production of remelted superalloy ingots has been driven by the increasing demand from the land based gas turbine industry. However, the size of remelted ingots in production has been limited because of some technical difficulties. One of them is the risk of forming freckles. Freckle is a channel-type macrosegregation, consisting of equiaxed grains and eutectic constituents[1]. Because of its macrosegregation nature, thermomechanical processing after remelting cannot eliminate the problem in billet products. The freckle defect is considered very harmful to the mechanical property of final products[2,3].

Freckle has been found in directionally solidified and single crystal airfoils, steel ingots and remelted superalloy ingots[1-5]. Similar channel type macrosegregation was also observed in Al and Ti alloys[6]. The formation of the channel-type defects is believed to be related to the density gradient in the solidifying mushy zone. The density gradient is caused by the segregation of solute elements into the interdendritic liquid. This density gradient is unstable under the action of gravity. As a result of the convective hydrodynamic instability, plume flow can occur in the mushy zone. This may cause misoriented grains or eutectic constituents to form in the freckle trail[7-10]. Several studies using transparent analog system NH_4Cl-H_2O[9], binary tin alloys[8,11] and computer simulation[12] have been successful in demonstrating freckle formation in upward unidirectional solidification and in static ingot casting. The formation of freckle during remelting of Nb-containing superalloys has also been widely studied and discussed[13-16]. A positive density gradient is thought to occur in the mushy zone with respect to the direction of gravity[13]. A proposed explanation of the freckle presence in this case is that the positive density gradient produces flow of interdendritic liquid that is moving downward along the sloped pool profile[14,15]. Some experimental evidence also supports the notion that the flow in the direction of secondary dendrites is easier to activate when the solidification front is tilted[16,17].

There are two basic approaches to control freckle formation in remelting. One is through the optimization of remelting practice. Improved electrode quality, intensified cooling, shallower pool and more stable melting reduce the possibility of freckles. However, in real melting the measures to intensify cooling is limited because the ingot size becomes so large that any forced cooling on ingot surface can not effectively penetrate to ingot interior. Melting system stability is limited because of the nature of remelting although significant improvements in melting control have been made.

Another approach is through alloy composition modification or alloy selection. The reason is that the density gradient inside the mushy zone is caused by solute segregation. If the segregation is so controlled that the density gradient during solidification is minimized, the risk of freckling can be reduced.

Alloy composition not only influences the segregation behavior, but also the solid evolution during solidification. Solid in mushy zone provides resistance to any interdendritic flow inside it. As more solid is formed, the possibility of interdendritic flow is reduced.

Superalloys 2000
Edited by T.M. Pollock, R.D. Kissinger, R.R. Bowman,
K.A. Green, M. McLean, S. Olson, and J.J. Schirra
TMS (The Minerals, Metals & Materials Society), 2000

The segregation and the solid formation behaviors of niobium-containing superalloys were studied with the purpose to quantitatively evaluate compositional effects. Three commercial alloys, 718, 625 and 706 were analyzed. A relative Rayleigh number accounting for the compositional effect was derived from Flemings' freckle criterion. Experimental results obtained from the three industrial alloys are incorporated into the relative Rayleigh number to quantitatively evaluate the freckle tendency of the alloys. The effects of alloying addition on density gradient and solid evolution are discussed.

Experimental

Materials

Bulk chemical composition of the materials used for the study was measured by wet chemical method and is listed in Table I. The solidification temperatures of the alloys are also listed in the Table. Niobium is added to all three alloys. Among them, alloy 718 has the highest Nb content and is also the most difficult to remelt. Freckle is a possibility in large diameter remelted ingots of the alloy.

Microsegregation analysis

A square mesh random sampling technique[18] was employed to measure the solute distribution in Ni-based alloys by EDS (Electron Dispersive Spectrometry). This method is essentially the same as the systematic point count method used in quantitative metallography for the measurement of phase volume fraction[19]. A polished sample was examined under SEM to get a backscattered image. A square mesh was overlaid on the image field. The grid size of the mesh was on the order of secondary dendrite arm spacing. Composition was taken at the cross points of the mesh. The change between the sampling points was achieved by digital beam control. Multiple fields were measured to accumulate enough points for high accuracy. Analysis indicated that a total of over 100 points was needed for each sample[20].

The analysis was conducted on a JEOL JSM-6400 Scanning Electron Microscope (SEM) equipped with a fully automated PGT EDS. The operating voltage was 20 kV and the current of the condensing lens was 0.3 nA. Live sampling time for each point was 30 seconds. Multi-element standards were used for the analysis since the bulk composition of the samples was already known.

Cast structures analyzed were either samples solidified after DTA (Differential Thermal Analysis) tests or cut from industrial ESR (Electroslag Remelting) or VAR (Vacuum Arc Remelting) ingots. DTA tests were done on TA-1600 using samples of about 170mg and the cooling rate of 20°C/min. The alloy 718 VAR ingot had a diameter of 510mm. The diameter of the alloy 706 ESR ingot was 915mm. All the samples were polished but not etched.

Measured compositions were sorted to get a profile of composition vs. solid volume fraction in the dendrite structure. Details of the data presentation can be found in Ref[18,20]

Solid fraction measurement

Solid fraction during solidification was obtained by quenching. A special kit shown in Figure 1 was designed and installed inside a DTA chamber for the test. The specimen was positioned on top of the flat end of the pure alumina rod. The specimen held its shape by the surface tension of the liquid when it was molten. When a designated quenching temperature was reached, the specimen was pushed off the top of the ceramic rod by the pure alumina tube, as shown in Figure 1. The specimen fell and was immediately quenched into the oil bath. Two differentially-connected thermocouples were positioned at the side of the alumina rod at a height just below the specimen. The thermocouples monitored the temperature and the phase transition in the sample. Figure 2 shows a curve recorded by the differential thermocouples, which clearly illustrates the melting points and the quench temperature. Actual tests show that the setup had very high reproducibility. The standard deviation in recorded melting point for an alloy is less than 1.5°C.

The quenching scheme was: heat to 1100°C at 100°C/min, change to 20°C/min and heat to a temperature 30°C over the alloy melting temperature (load the specimen when the temperature reaches 1150°C in the meantime), soak for 3 minutes, cool down at 5°C/min. The specimen was pushed off and quenched as soon as the cooling temperature had reached a set point. The specimen weighed about 100mg. Flowing argon was used to protect the specimen.

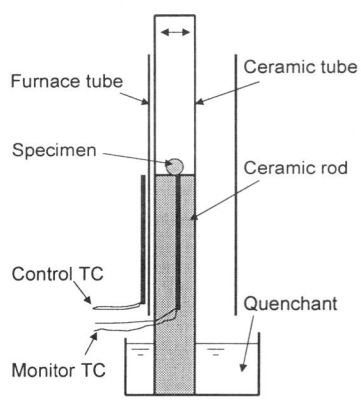

Figure 1 Test kit for the quenching of solidifying liquid

Table I Composition and solidification temperature of the alloys used in the study

Alloy	Raw Material	Composition, wt%									Transition temperature[¥], °C		
		Ni	Cr	Fe	Nb	Mo	Ti	Al	C	Si	T_L	T_S	T_E
718	VAR	53.52	18.19	17.77	5.41	3.01	0.98	0.56	0.039	0.03	1337	~1260	1164
706	ESR	41.39	16.30	37.03	2.98	0.12	1.70	0.19	0.012	0.05	1388	1336	1172
625	Billet	60.11	22.09	4.40	3.53	9.10	0.29	0.26	0.024	0.03	1352	1290	1155

[¥] Data from DTA tests with a scan rate of 20°C/min. T_L and T_E are from the cooling curves. T_S is from the heating curves.

Two methods were used to determine the solid fraction in the quenched samples. One was the systematic manual point count method[19]. The other was image processing. SEM backscattered electron images taken from an as-polished specimen were processed and analyzed by the software ScionImage. Both counting methods produced almost the same results. Therefore no mention is made for the method of analysis in presenting the data.

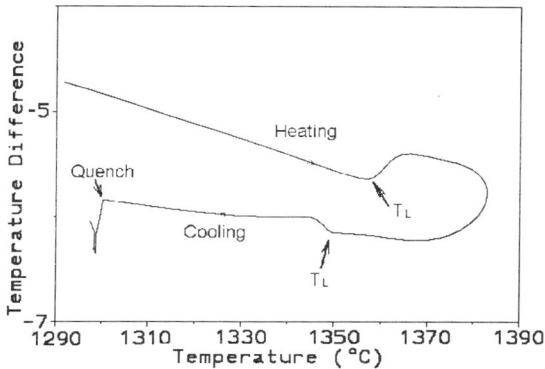

Figure 2 Typical heating and cooling curve recorded for the quenching test

Results

Microsegregation of alloys

A solidified dendrite microstructure can be clearly identified in the backscattered image of SEM, Figure 3. Because of the segregation of niobium and molybdenum into the interdendritic region, the dendrite cores are dark and the interdendritic areas are brighter. The brightest spots in the interdendritic region are Laves phase and carbide.

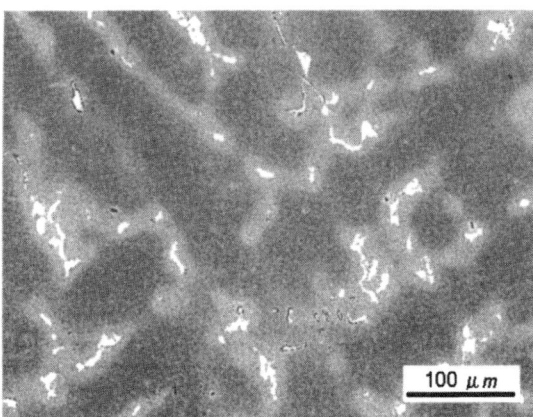

Figure 3 Dendrite structure of alloy 718 solidified at 20°C/min of DTA test

The segregation profile The measured Nb composition of every point is plotted against its sequence of sampling in Figure 4. This graph represents the random nature of the sampling method. Therefore, there is a wide scattering in the data. The Nb content in some sampling points is low, corresponding to points located at or

close to dendrite cores. Some points have extremely high Nb content, where the sampling points are located either on Laves phases or carbides.

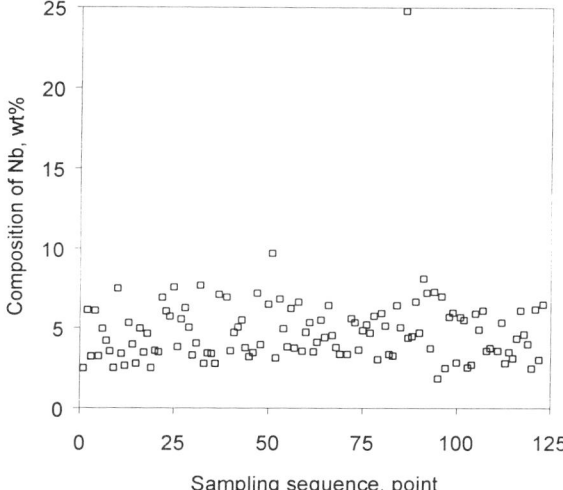

Figure 4 The composition of Nb vs. the sequence of sampling in VAR 718 alloy

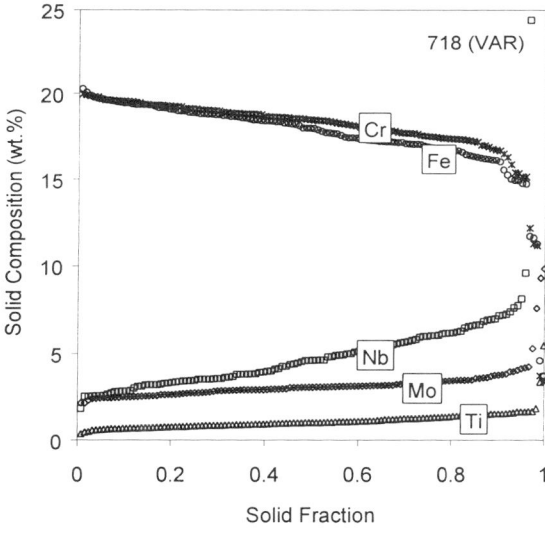

Figure 5 The solid segregation profile in VAR remelted 718 alloy

The apparently random compositions of all the points were sorted to obtain a segregation profile in the solid. Figure 5 is the sorted profile of Nb, and other solute elements for VAR alloy 718. It is proposed that this method is able to provide the complete segregation profiles across the whole solidification range.

The composition of the interdendritic liquid was calculated from the measured segregation profiles in solid. Assuming that the Scheil solidification condition is satisfied, i.e., no back diffusion in solid and complete solute mixing in the liquid, the composition of the interdendritic liquid at a solid fraction f_s for any solute element can be computed by the mass conservation principle[20]:

$$C_L = \frac{1}{1-f_s}(C_0 - \int_0^{f_s} C_s df_s) \qquad (1)$$

Where C_0 is the alloy composition, C_s is the solid composition at a solid fraction f_s, f_L is the liquid fraction and $f_s + f_L = 1$. Resultant liquid segregation profiles are illustrated in Figure 6. These liquid composition profiles will be used to evaluate the density gradient across the mushy zone.

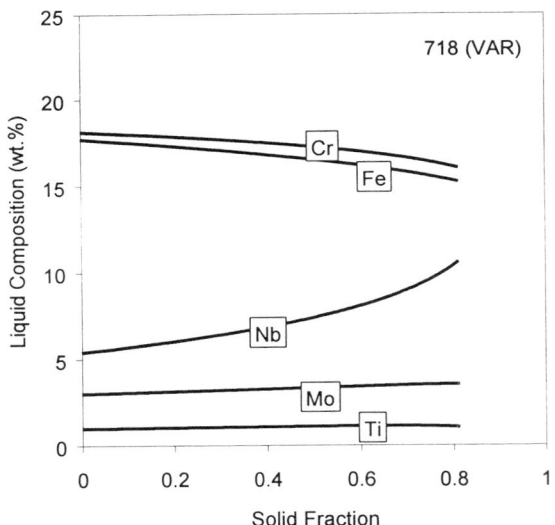

Figure 6 Computed composition profile of the interdendritic liquid during the solidification of alloy 718 in VAR remelting

The segregation characteristics The partition coefficient k for each element at a given solid fraction f_s, can be calculated as the ratio of the solid composition to the liquid composition, $k = C_s/C_L$. The results are shown in Figure 7 for alloy 718. The partition coefficients of Nb, Mo and Ti are smaller than 1. They are positive segregation elements, i.e., their contents in the liquid become higher as solidification proceeds (Figure 6). Nb has the smallest k among the three elements. It is more severely segregated in alloys. Cr and Fe are slightly negatively segregated ($k>1$) and therefore concentrated in dendrite cores, as shown in Figure 5. It is consistently observed that the segregation of Ti in the ESR or VAR structure is smaller than in the DTA solidified structure. This may be because the cooling rates in industrial alloys are much lower, 2°C/min in VAR/ESR ingots compared to 20°C/min in DTA test.

The partition coefficients are dependent on the base composition of the alloys. Figure 8 shows k values of Nb, Ti and Mo for alloy 625 and 706. Those of the VAR 718 alloy are also displayed in the figure for comparison purpose. The k of Nb and Ti decreases in the sequence from 625, 718 to 706. This follows the trend that alloy 625 has the least iron and alloy 706 has the most iron. This suggests the dependence of the coefficients with the iron contents in the base alloys.

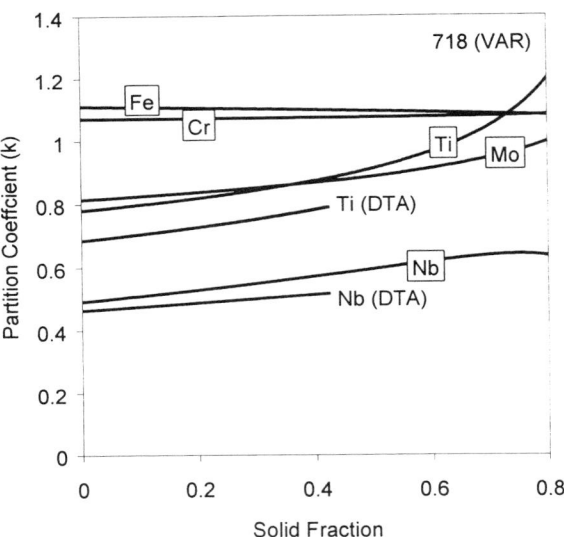

Figure 7 Partition coefficients of solute elements in 718 VAR ingot and DTA sample

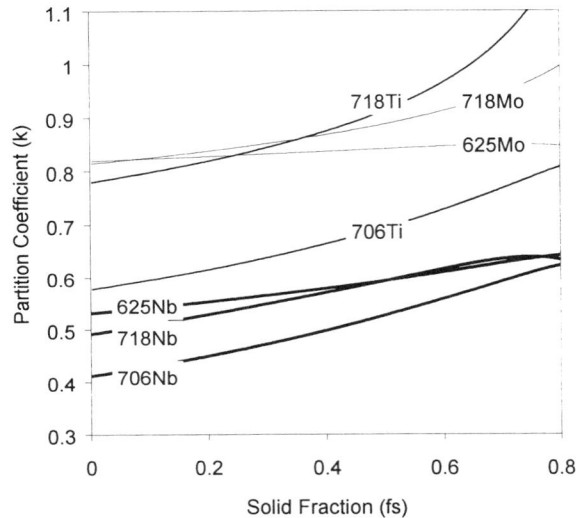

Figure 8 Comparison of elemental partition coefficients in the three alloys. 718 is measured from VAR structure, 706 is from ESR structure, 625 is from DTA sample

Precision of the analysis The above random sampling method also provides an accurate measurement of the segregation profiles. Figure 9 shows the accuracy analysis by a Monte-Carlo simulation using similar parameters as in the SEM/EDS random sampling experiment[20]. The standard deviations involved in the measured solid and the calculated liquid composition profiles change with the solid fraction. At smaller solid fraction, the deviation from actual composition is very small. The inaccuracy increases rapidly as the solid fraction approaches unity. This implies that the random sampling method can give accurate segregation results, unless the solid fraction is too high. The recommended limit of the solid fraction is $fs < 0.6 \sim 0.8$.

The random sampling has several advantages over some other methods. First it has very high accuracy. Second, it provides continuous segregation profiles covering most of the solidification range. Most other methods can only measure segregation at specific volume fractions, e.g., initial solid composition or eutectic composition. Therefore, the density gradient in the mushy zone can be accurately accessed by this method. Another merit of the method is that it uses conventionally solidified structure, no special specimen preparation is required.

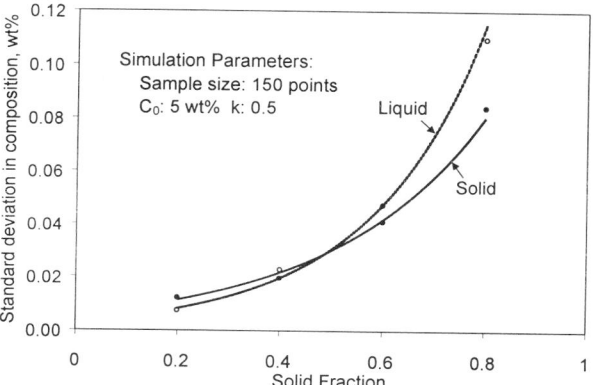

Figure 9 Standard deviation of the measured solid and calculated liquid compositions at different solid fractions in the SEM/EDS random sampling segregation profile analysis.

In addition to the accurate segregation profiles, the development of the segregation with temperature has to be determined to obtain the density gradient in mushy zone. This is because density changes across mushy zone come from both solutal and thermal effects, and that the thermal effect sometimes even forms a significant part of the overall density change in nickel-base alloys[15]. The temperature-segregation relation was investigated by the quenching study of solidifying liquid.

Solid development during solidification

Quenched microstructure of alloy 718 can be found in Figure 10. The rapid quenching chilled the interdendritic liquid and produced the fine microstructure located between the coarse dendrites formed during the slow cooling from the melting temperature. The interdendritic area looks brighter in SEM backscattered image because of the greater segregation of Nb and Mo.

The composition of the interdendritic areas and the dendrite cores were measured by SEM/EDS. The result can be found in Figure 11 and Table II, respectively. The measured composition of the interdendritic area is consistent with the calculated profiles by the random sampling of conventional cast structure. The dendrite core composition indicates there was some back diffusion of Ti during prolonged holding at high temperature. But the diffusion did not produce statistically significant composition change, especially for other solute elements. It seems that the cooling rate of 5°C/min was fast enough to suppress strong back diffusion in the solid.

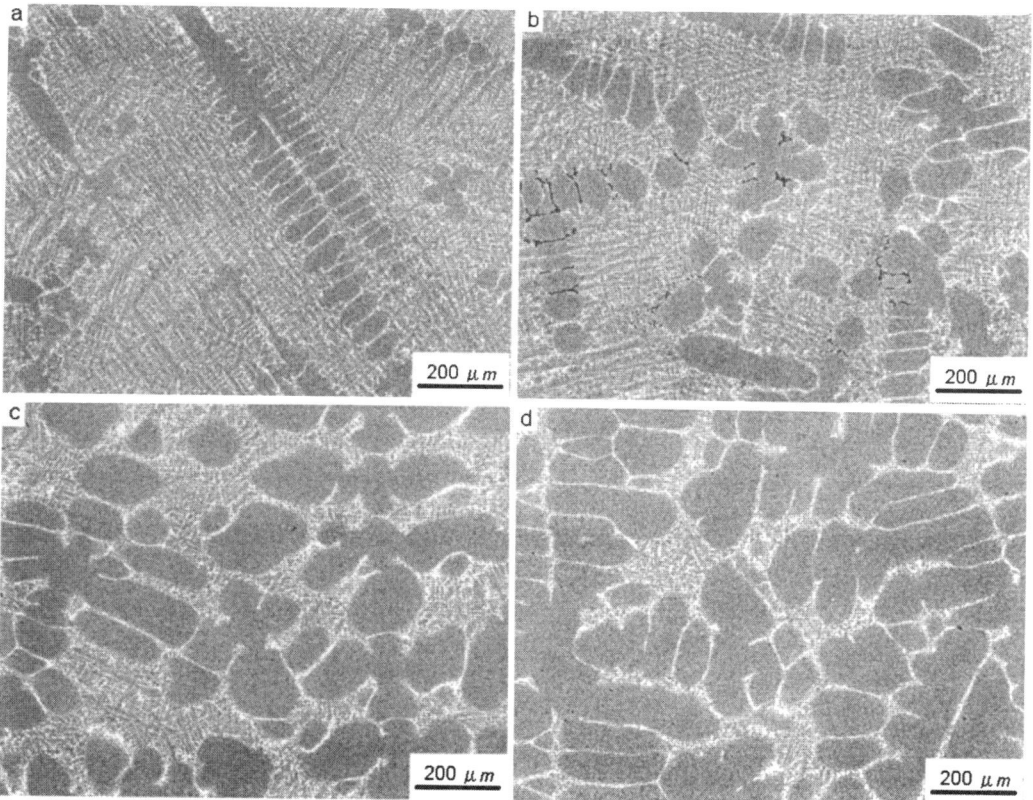

Figure 10 Quenched microstructure of alloy 718
a) ΔT=-2.6°C; b) ΔT=-7.6°C; c) ΔT=-27.6°C; d) ΔT=-47.6°C;

Table II Dendrite core composition of alloy 718 quenched at different cooling temperatures

ΔT	Ni	Cr	Fe	Mo	Nb	Ti	Al
-2.6	55.35	18.76	20.04	2.85	2.03	0.47	0.50
-7.6	55.45	19.13	19.75	2.72	1.85	0.52	0.57
-17.6	55.56	18.94	20.30	2.23	1.84	0.54	0.59
-27.6	55.14	19.11	19.94	2.70	2.05	0.52	0.54
-47.6	55.24	19.05	19.46	2.76	2.40	0.59	0.50
-97.6	55.29	19.32	19.57	2.18	2.35	0.66	0.64

$\Delta T = T - T_L$, T_L - alloy melting point, T – quench temperature

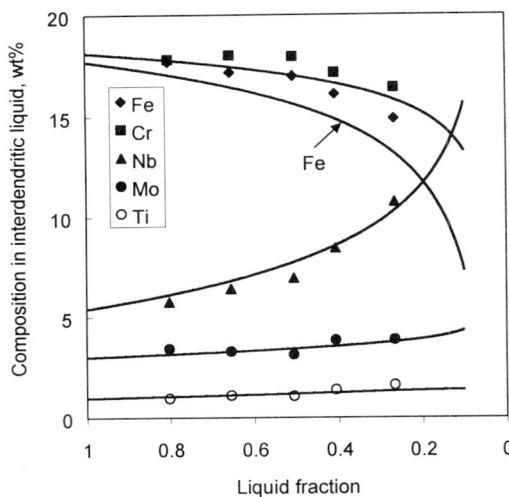

Figure 11 Comparison of the compositions of the interdendritic area in the quenched microstructure (dots) and those calculated from random sampling of DTA sample (lines).

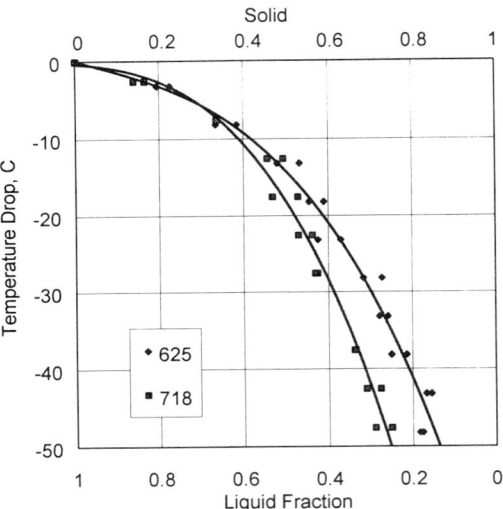

Figure 12 The transformation of liquid to solid during solidification

Solid fraction Figure 12 is the plot of the solid fraction vs. the temperature cooled from the alloy melting point, which is expressed as the temperature drop $\Delta T = T - T_L$. Here T_L is the averaged cooling melting point of each alloy. T is the quench temperature. The transformation of liquid to solid is relatively fast during the solidification. At a temperature drop of 20°C, more than 50% of the liquid has already solidified for both alloys. At the same temperature drop from the alloy liquidus, the solid content in alloy 625 is consistently higher than in alloy 718.

Prediction of the solid fraction Though experimental measurement of the solid fraction during solidification by quenching or other means is effective, it is very tedious. Theoretical calculation methods were explored to predict the quantity from alloy composition or from the segregation profile measured. These include Thermo-Calc and empirical alloy melting point equation.

In the Thermo-Calc analysis, Thermo-Calc M version and the all-element nickel-iron database were used[21]. Both equilibrium solidification and Scheil solidification modes were assumed in the calculation.

The melting temperature equation used is

$$T_L = T_0 + \sum_i m_i C_i \qquad (2)$$

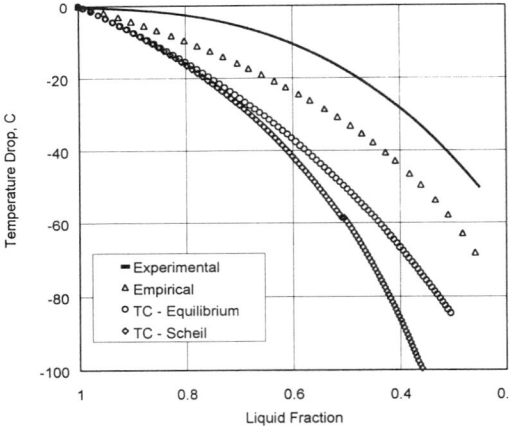

Figure 13 The solid formation in the mushy zone of alloy 718 predicted by Thermo-Calc and the empirical melting point equation.

where T_L is the melting point of an alloy, in °C. C is the composition of an element in nickel base alloy, in wt%. T_0 and m are constants (Table III). This equation is a linear regression of the melting points of a series of alloys based on Ni-20Cr[22]. The composition range covers Nb 0~14, Ti 0~5.5, Mo 0~9, Al 0~1.5, Co 0~18, Fe 0~36, C 0~0.36, Si 0~1.95. Eq.(2) offers a way to correlate the solid fraction, interdendritic liquid composition and temperature.

Element	Co	Cr	Fe	Nb	Mo	Ti	Al	C	Si	T_0
M	0.86	0.75	-0.16	-12.3	-5.24	-18	-14.63	-27.56	-29.45	1455

Figure 13 shows the calculated and the experimentally measured liquid fractions. The difference is large for Thermo-Calc results. Neither equilibrium nor Scheil solidification of Thermo-Calc gives results close to experiment. The deviation also enlarges as temperature drops, when the interdendritic liquid becomes more and more enriched with Nb and other solute elements. Therefore, the difference in Thermo-Calc calculation can not be narrowed by compensating the undercooling effect, back diffusion in the solid or other kinetic factors.

The interdendritic liquid composition in the random sampling method was input in the empirical melting point equation (2) to get the melting temperature. Obtained temperature - liquid fraction results are sketched in Figure 13. They are closer to experimental ones. The difference becomes stabilized after the initial transition at the low solid fractions. A temperature difference of about 12°C is observed between the calculated and experimental results. This temperature difference is of the same order as that of the undercooling. Therefore it is expected that the empirical melting point equation can be used to predict the solid evolution speed in the mushy zone if the undercooling correction is made.

The freckle tendency

Relative Rayleigh number

Many criteria were proposed in the literature for freckle prediction and prevention. Through careful examination of these criteria using experimental results, Flemings' macrosegregation criterion was found best for defining freckle phenomenon[23]. The explanation of the criterion is that if the flow of interdendritic liquid along crystal growth direction is faster than the crystal growth speed, freckle defect will be developed as a result of the structural instability[24]. Figure 14 is a schematic showing the crystal growth and the interdendritic flow at the tilted solidification front of remelting, where **V** is the velocity of the interdendritic flow, **R** is the crystal growth speed. According to Flemings' theory, freckle is expected to form in the tilted solidifying mushy zone if the derivative of V in the R direction is greater than R.

The velocity of the interdendritic flow, **V**, can be obtained by Darcy's law, which describes the liquid flow through porous media. However, for the remelting process, two factors make the computation more complex. One is the tilt angle of the solidification front, β. The other is the the angle α, which is needed to maintain a negative density gradient to drive the flow along the solidification front. Therefore the interdendritic flow velocity will have the following form if both angles are considered:

$$V = \frac{\Delta\rho g \Pi}{\nu f_L} f(\alpha,\beta) \qquad (3)$$

Where $\Delta\rho$ is the density difference, $\Delta\rho = (\rho - \rho_0)/\rho_0$, ρ_0 – density at alloy melting temperature, g – gravity, v- viscosity of

interdendritic liquid, Π – permeability of mushy zone, f_L – local liquid fraction. The tilt effect of the solidification front and the deviation of the flow from the front are included in the function $f(\alpha,\beta)$.

The comparison of the derivative of V in R direction results in the following criterion:

$$Ra = \frac{\Delta\rho g \Pi}{\nu f_L} \cdot \frac{1}{R} \cdot F(\alpha,\beta) \qquad (4)$$

This is a criterion written in the form of a Rayleigh number. But its physical meaning is that same as Flemings'. In Eq.(4), $F(\alpha,\beta)$ considers both the $f(\alpha,\beta)$ in Eq.(3) and taking the derivative of V in the R direction.

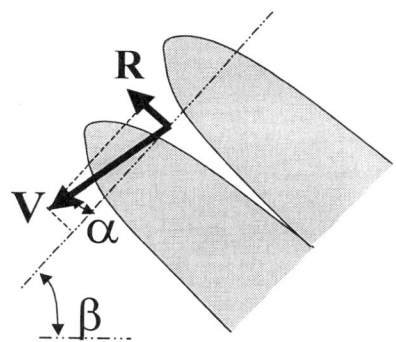

Figure 14 The interdendritic liquid flow pattern at the tilted solidification front of remelting. β is the slope angle of the solidification front, α is the deviation of interdendritic flow from the slope plane, V is the velocity of the interdendritic liquid flow, R is the crystal growth velocity.

Exact calculation of equation (4) requires, first of all, the definition of the slope effect, the function $F(\alpha,\beta)$, which is not understood at the present time. Accurate information of the dendrite spacing and physical parameters of interdendritic liquid is also required to calculate the permeability and other parameters. A reasonable approach to use the equation for freckle study is through some simplification.

For semi-quantitative analysis considering the composition effects, dendrite parameters, pool slope and physical parameters can be neglected, provided that the alloys studied have similar constituents and are processed under similar remelting schemes. If the permeability factor in equation (4) is split into two multiplying terms, $\Pi = \Pi(f_L) \cdot \Pi(G,R)$, with the first part considering the influence of solid fraction and the second part the processing (dendrite) effect, equation (4) can be simplified to:

$$Ra_r = \frac{\Delta\rho \cdot \Pi(f_L)}{f_L} \qquad (5)$$

This criterion considers only the composition effect, it can be used to compare the freckle tendency of similar alloys. Therefore we will call it relative Rayleigh number in the sections followed. In equation (5), $\Pi(f_L)$ is the overall permeability in mushy zone. For the flow along secondary dendrite arms, $\Pi(f_L)$ usually takes the form $\Pi(f_L) \propto f_L^{3.34}$ for a given liquid fraction[25]. Detailed discussion about its calculation can be found in Ref[23].

Freckle tendency of the alloys

With the segregation profile and the solid formation information, it is possible to quantitatively evaluate the freckle tendency of an alloy. Density of the interdendritic liquid was calculated by the molar addition method[26,27] using the software Metals. In the calculation, correction was made to include the mixing effect proposed by Sung et al[28]. Compared to the experimental measurements of the liquid density of several superalloys[28], the error in our calculation was less than 4%.

Obtained density changes of the three alloys studied are illustrated in Figure 15. The density difference is calculated relative to the alloy's density at its liquidus temperature. As the liquid fraction decreases, there are both changes in the composition of interdendritic liquid and in the temperature. Therefore, Figure 15 shows the overall density difference from both solutal effect and thermal effect. The interdendritic liquid density in alloys 718 and 625 increases as liquid fraction is reduced. Alloy 706 shows the opposite behavior. The absolute density gradient increases in the order 706, 625 and 718.

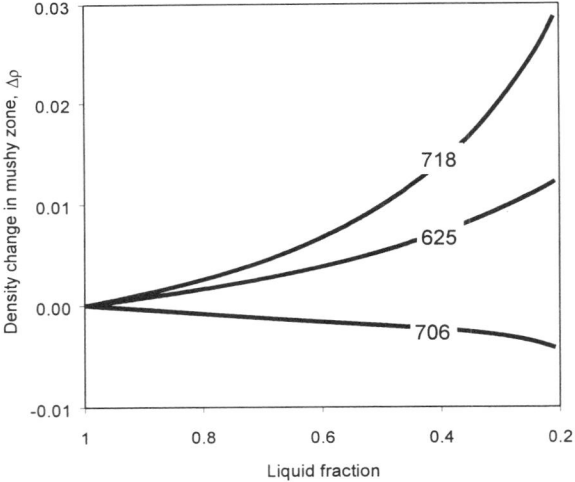

Figure 15 Density difference of interdendritic liquid in mushy zone of industrial alloys

Figure 16 shows the calculated relative Rayleigh number for the alloys. It is revealed that the maximum Rayleigh number value is not reached at the bottom of the mushy zone, where $f_L \rightarrow 0$. Instead, it is reached mid-way in the mushy zone. The reason is that there are two competing factors controlling overall Rayleigh number. One is the density, which increases monotonically as liquid fraction decreases, giving greater driving force for liquid flow.

However, the decrease in liquid fraction eventually overweighs this density difference effect, giving greater resistance to the flow. Therefore, a peak effect is observed. For 706 the maximum is reached at about 40% solid fraction. In 718 this is at about 60% solid fraction. Freckle flow most likely initiates in a place where the maximum Ra value is reached. Therefore, freckle flow starts deeper in the mushy zone in 718 alloy.

Comparing the maximum value of Ra_r for the three alloys, 718 has the highest Rayleigh number and alloy 706 has the lowest. This corresponds to industrial observations that alloy 706 is the easiest to remelt in the three alloys. Much larger ingots can be made from the alloys. Alloy 718 is the most difficult to process. Usually the ingot diameter is limited to about 510mm for high Nb rotor grade chemistry.

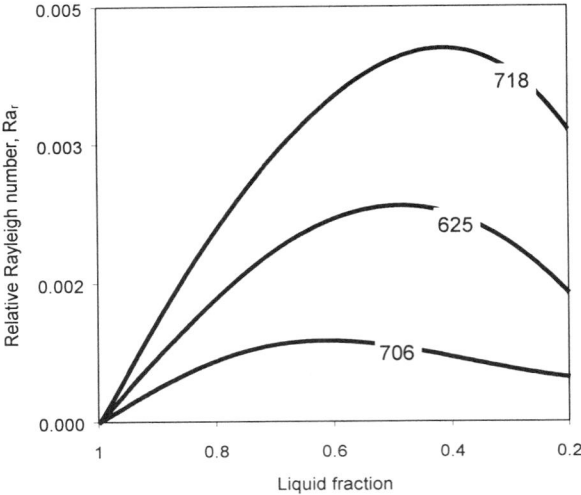

Figure 16 Estimated relative Rayleigh number for the industrial alloys

Alloying effect on freckle tendency

The above evaluation of the freckle tendency reveals why alloy 718 is prone to freckles. The alloy happens to have a combination of both high density gradient and a slow solid growth development as a result of the high content of Nb and its strong segregation nature. Generally, the trend in the Rayleigh numbers of the three alloys follows that of the density gradient, i.e., higher density gradient implies higher possibility of freckling. Therefore, the primary choice of alloying is to reduce the density gradient resulting from the elemental segregation. Among the major alloy strengtheners, Nb strongly segregates into the interdendritic liquid and makes the liquid there heavier. Ti, on the other hand, segregates into the same place. Therefore, one approach to reduce the segregation related density gradient is to balance the alloying of both elements. 706 is a good example of this balanced alloying. The alloy has a reduced content of Niobium and a higher Ti/Nb ratio. Even though the alloy contains a large amount of Fe, which makes the Nb segregate more severely, the tendency to freckle is reduced.

Conclusions

For the evaluation of the freckle tendency of an alloy, quantitative determination of the segregation behavior and the solid growth during the solidification are pertinent for the calculation of the density distribution and permeability inside the mushy zone. The study explored methods to obtain the necessary data by experimental measurement and theoretical calculation. Based on the precise measurement of the segregation behavior in cast structures and the solid formation process, the freckle tendency was examined for three commercial alloys. The trend in the calculated relative Rayleigh number follows the melting observations that alloy 718 is the most freckle prone and alloy 706 the least.

1. The segregation profiles in Nb-containing superalloys were obtained by SEM/EDS analysis using the random sampling method. This method has the capability to provide high precision continuous composition profiles from as-cast materials from industrial ingots or laboratory solidified samples.

2. The progressive development of solid was studied using quenching method. It was found that the speed of the solid evolution during solidification is very fast for the alloys studied. At the same degree of cooling, more solid is formed in alloy 625 than in alloy 718.

3. More accurate prediction of the solid formation during solidification can be accomplished by the empirical melting point equation. Further development in the thermodynamic analysis and the database is needed to improve its accuracy of calculation.

4. Density gradient and relative Rayleigh number calculation reveals that alloys 718 and 625 have a positive density gradient in the mushy zone, alloy 706 has a slightly negative density gradient. The freckle tendency predicted by the relative Rayleigh number decreases in the sequence 718 , 625 and 706. Among the three alloys, alloy 718 has the highest freckle potential.

Acknowledgment

This research was carried out under the project "Composition effects on macroscopic solidification segregation of superalloys", sponsored by Special Metals Corporation. Dr. Shailesh Patel, Director of R&D, is thanked for his continuous support of the project. The authors would like to acknowledge Mrs. Diane Schwegler-Berry and Mr. Joel Harrison, National Institute of Occupational Safety and Health, Morgantown, WV, for their help with SEM/EDS analysis. They are also grateful to Dr. Peter Quested, National Physics Laboratory, UK, for providing the software "Metals" for the calculation of liquid metal density. The public domain software ScionIamge was used for the quantitative image analysis. The authors would like to extend their appreciation to the software developers at the National Institute of Health (www.nih.gov). Prof. Pollock and the program committee of the 9th International Symposium on Superalloys are sincerely thanked for their reading and comments on the paper, which helped to avoid many technical and spelling mistakes.

References

1. K. O. Yu, J. A. Domingue, G. E. Maurer, et al. "Macrosegregation in ESR and VAR processes," JOM, (1993), 49-51.

2. A. D. Helms, C. B. Adasczik, L. A. Jackman, "Extending the size limits of cast/wrought superalloy ingots," Superalloys 1996, eds. R. D. Kissinger, et al., TMS, 1996, 427-433.

3. S. V. Thamboo, "Melt related defects in alloy 706 and their effects on mechanical properties," Superalloys 718, 625, 706 and Various Derivatives, ed. E. A. Loria, TMS, 1994, 137-152.

4. A. F. Giamei and B. H. Kear, "On the nature of freckles in nickel base superalloys," Metall. Trans., 1(1970), 2185-2192.

5. J. J. Moore, and N. A. Shah, "Mechanisms of the formation of A- and V- segregation in cast iron," Int. Met. Rev., 28 (12) (1983), 338-356.

6. J. A. Brooks, J. S. Krafcik, J. A. Schneider et al., "Fe segregation in Ti-10V-2Fe-3Al 30 inch VAR ingot β-fleck formation," Proc. 1999 Inter. Symp. On Liquid Metal Processing and Casting, eds. A. Mitchell et al, Vacuum Metallurgy Div., American Vacuum Soc., Santa Fe, New Mexico, Feb 21-24, 1999. 130-144.

7. S. M. Copley, A. F. Giamei, S. M. Johnson et al., "The origin of freckles in unidirectional solidification of nickel base crystals," Metall. Trans., 1(1970), 2193-2204.

8. J. R. Sarazin and A. Hellawell, "Channel formation in Pb-Sn, Pb-Sb, and Pb-Sn-Sb alloy ingots and comparison with the system NH$_4$Cl-H$_2$0," Metall. Trans. A, 19A(1988), 1861-1871.

9. C. F. Chen, "Experimental study of convection in a mushy layer during directional solidification," J. Fluid Mech., 293(1995), 81-98.

10. R. J. McDonald and J. D. Hunt, "Convective fluid motion within the interdendritic liquid of a casting," Metall. Trans., 1(1970), 1787-1788.

11. S. N. Tewari, R. Shah, "Macrosegregation during dendritic arrayed growth of hypoeutectic Pb-Sn alloys: influence of primary arm spacing and mushy zone length," Metall. Trans A, 27A(1996), 1353-1362.

12. S. D. Felicelli, D. R. Poirier and J. C. Heinrich, "Modeling freckle formation in three dimensions during solidification of multicomponent alloys," Metall. Mater. Trans. B, 29B(1998), 847-855.

13. P. Auburtin, S. L. Cockcroft, A. Mitchell et al., "Center segregation, freckles and development directions for niobium containing superalloys," Superalloys 718, 625, 706 and Various Derivatives, ed. E.A. Loria, TMS 1997, 47-54.

14. J. Van Den Avyle, J. Brooks, and A. Powell, "Reducing defects in remelting processes for high-performance alloys," JOM, 50 (3) (1998), 22-25.

15. W. Chen, W. H. Yang, K.-M.Chang et al. Proc. 1999 Inter. Symp. On Liquid Metal Processing and Casting, eds. A. Mitchell et al, Vacuum Metallurgy Div., American Vacuum Soc., Santa Fe, New Mexico, Feb 21-24, 1999, 122-129.

16. P. Auburtin, S. L. Cockcroft, and A. Mitchell, "Freckle formation in large superalloy single crystal airfoil castings," Proceedings: Materials for Advanced Power Engineering, v5, Part III, eds. J. Lecomte-Becker et al., 1998, 1459-1468.

17. N. Mori and K. Ogi, "Study on formation of channel-type segregation," Metall. Trans. A, 22A(1991), 1663-1672.

18. M. N. Gungor, "A statistically significant experimental technique for investigating microsegregation in cast alloys," Metall. Trans. A, 20A(1989), 2529-2533

19. Annual Book of ASTM Standards, "Standard Test Method for Determining Volume Fraction by Systematic Manual Point Count," vol. 03.01, Designation E 562-89, 1995, pp. 583-588.

20. W. H. Yang, W. Chen, K.-M. Chang, S. Mannan, J. deBarbadillo "Monte Carlo sampling for the microsegregation measurement in cast structures", accepted for publication in Mater. Metall. Trans. A.

21. Bo Sundman, User's Manual of Thermo-Calc, Ver.L, 1997.

22. Yang, L., Phase control in Nb-hardened superalloys, MS Thesis, West Virginia University, 1997.

23. W. H. Yang, W. Chen, K.-M. Chang, S. Mannan, J. deBarbadillo Freckle Criteria for the Directional Solidification of Alloys, submitted to Mater. Metall. Trans. A. for publication

24. R. Mehrabian, M. Keane, and M. C. Flemings, "Interdendritic fluid flow and macrosegregation: influence of gravity," Metall. Trans., 1(1970), 1209-1220.

25. D. R. Poirier, "Permeability for flow of interdendritic liquid in columnar-dendritic alloys," Metall. Trans. B, 18B(1987), 245-255.

26. A. F. Crawley, "Densities of liquid metals and alloys," Int. Metall. Rev., 19(1974), 32-48.

27. K.C. Mills, P. N. Quested, "Measurements of the physical properties of liquid metals," Liquid Metal Processing and Casting Conf. Proc., AVS, eds. A. Mitchell et al., Sept. 1994, 226-240.

28. P. K. Sung, D.R.Poirier, E. McBride, "Estimating densities of liquid transition-metals and Ni-base superalloys," Mater. Sci. and Engg., A231(1997), 189-197.

MICROSTRUCTURAL EVOLUTION OF NICKEL-BASE SUPERALLOY FORGINGS DURING INGOT-TO-BILLET CONVERSION: PROCESS MODELLING AND VALIDATION

C.A. Dandre, C.A. Walsh, R.W. Evans*, R.C. Reed and S.M. Roberts

Rolls-Royce University Technology Centre, Department of Materials Science and Metallurgy, University of Cambridge, Pembroke Street, Cambridge, CB2 3QZ, UK.

*IRC for Computer Aided Materials Engineering, Department of Materials Engineering, University of Wales Swansea, Singleton Park, Swansea, SA2 8PP, UK.

Abstract

A computer-based process modelling capability has been developed to simulate the evolution of microstructure during ingot-to-billet conversion of nickel-base superalloys. A unique feature of the model is the incorporation of rules describing the dynamic and static recrystallisation phenomena that govern microstructural development. Computational investigations consider the influence of process control parameters such as die velocity and displacement to obtain an optimal microstructure. Furthermore, a comparison is made of the effective strain profiles obtained by squeezing flat versus rounded sections of the workpiece.

Introduction

This paper is concerned with the development and application of a computer-based process model for the simulation of microstructural evolution [1-4], which occurs during the thermomechanical working of superalloys, e.g. Inconel®718[1] and Waspaloy, for the purposes of ingot-to-billet conversion. It would appear that only very little attention [5,6,7] has been paid to the analysis of the cogging process, despite its significance to the materials suppliers and its relevance to the manufacturers of gas turbines. Our initial studies were on Inconel®718 but have latterly been extended to encompass Waspaloy. Generically the strain and strain-rate distributions due to cog forging are considered similar for these alloys, although the relationship between the strain path and the microstructural evolution may differ slightly.

During the cogging process, the optimum amount of deformation must be imposed at the correct temperature within the right time-scale. A description of all the relevant phenomena has been attempted. The complex relationship between the processing parameters, the flow field and the microstructure is time dependent and strongly temperature dependent. The development of microstructure during ingot breakdown is governed by the various recrystallisation phenomena occurring. Under dynamic loading conditions, dynamic recrystallisation may occur. When the dies are removed, a significant period of dwell can be expected before the same volume of material is worked again. During this period static recovery, static (meta-dynamic) recrystallisation and grain growth may occur; it has been found that it is these phenomena which control the development of microstructure.

The details of the model are given with particular emphasis placed on the description of microstructural evolution. A comprehensive testing programme using a purpose built compression testing machine has been carried out to acquire the materials data required for the model. The analysis of the flow curves is described. The microstructural model has been implemented within the three-dimensional visco-plastic Forge3™[2] finite element software. In order to validate the methodology developed, interrupted cogging trials have been performed. The evolution of microstructure is explained in terms of the cogging model and a more optimal process procedure suggested.

Nickel-base superalloys for disc applications

This paper considers the nickel-base superalloys Inconel®718 and Waspaloy. Typical compositions of the two alloys considered are given in Table 1 [8]. These materials are used extensively for the production of high integrity gas turbine discs due to their excellent high temperature mechanical properties. These properties may be further enhanced by careful control and consideration of the processing route employed for disc manufacture. Ingot production involves a primary melting process e.g. vacuum induction melting (VIM) followed by a remelt procedure e.g. vacuum arc remelting (VAR) and/or electro slag refining (ESR). Following a homogenisation treatment to minimise the effects of macrosegregation, the ingot is then converted into a wrought product in billet form in order to breakdown the coarse, non-uniform as-cast grain structure. The resulting refined wrought structure is then suitable for supply to the forgemasters for near net shape disc forging. The final operations of ageing heat treatments and machining are not considered to affect the grain evolution.

The control of microstructure during the cogging process and during subsequent disc forging is essentially a compromise; higher temperatures enable larger amounts of strain to be imparted in a given step hence minimizing cost, at the same time however, higher temperatures have a detrimental effect on grain growth.

Table 1. Nominal alloy compositions (wt %) [8]

Alloy	Ni	Cr	Co	Mo	Fe	Al	Ti	Nb
Waspaloy	58	19.5	13.5	4.3	-	1.3	3.0	-
Inconel®718	52.5	19.0	-	3.0	18.5	0.5	0.9	5.1

---

[1] Inconel a trademark of Special Metals Corporation

[2] Forge3 a trademark of Transvalor

Superalloys 2000
Edited by T.M. Pollock, R.D. Kissinger, R.R. Bowman,
K.A. Green, M. McLean, S. Olson, and J.J. Schirra
TMS (The Minerals, Metals & Materials Society), 2000

The principal difference between the alloys is the substitution of iron and niobium (Inconel®718) for cobalt (Waspaloy). For Inconel®718, this leads to the formation of phases other than the ordered L1$_2$ gamma prime (γ') phase [9]. In particular the Ni$_3$Nb delta (δ) phase is often formed, although at lower temperatures the meta-stable gamma double prime (γ'') phase arises. As is well known, the 718 alloy is the workhorse of the nickel-base superalloys and accounts for 40-50% of all superalloys produced [9,10]. The high temperature capacity of the cobalt rich Waspaloy is dominated by the volume fraction of gamma prime (γ') formed, typically around 25%. In considering the forging sequences for these alloys it is necessary to consider the influence of the two major solvus temperatures. For Waspaloy, the γ' solvus has been reported to be 1030°C [11]. For Inconel®718, the δ solvus is 1000°C ±20°C, dependent upon the niobium content (4.7-5.5 wt%) [12].

The Cogging Process

The cogging process is an open-die thermomechanical forging operation that converts the as-cast ingot into a wrought product in billet form. The purpose of ingot-to-billet conversion is to break down the coarse as-cast, as-homogenised ingot microstructure to achieve a much refined and consistent microstructure. The open-die cogging process is a traditional approach to ingot breakdown, deriving from the introduction of the steam press in the 1840's.

The process involves passing the workpiece repeatedly through an open-die press to reduce the ingot diameter. The exact manner in which the workpiece is cycled through the press is varies from producer to producer, however the operation typically involves rotating the workpiece (e.g. through 45° or 90°) at the end of each pass in order to maintain a symmetrical octagonal or square cross-section. By working the sides down the length of the ingot via a series of narrow tool width ('bites'), strain is imposed within the deforming material by reducing the diameter at discrete time steps. The amount and rate of applied strain, together with the temperature, are key variables which influence the flow stress of the material.

However, due to the relatively large section sizes of the ingot (e.g. 380-520 mm diameters), full strain penetration may not be achievable. Instead, the cogging process comprises a number of stages, or heats, of limited deformation; after this, the workpiece is returned to the furnace for re-heat. As many as seven heats might be required to cog a 3 tonne ingot whilst maintaining uniformity of temperature and ease of forging. Figures 1(a-d) show the starting ingot size, two consecutive intermediate stages of the process, and the final as-cogged (rounded) billet section for a Waspaloy cogging.

The furnace temperatures are set in accordance with the stage of processing. Initial ingot forging is generally performed at a temperatures well above the carbide, δ and γ' solvii (Inconel®718 1120°C, Waspaloy 1170°C). Conversely final forging is performed close to the relevant solvii. For Inconel®718, final forging is performed at, or just above, the δ solvus at around 1010°C. As a consequence, δ-phase particles form which restrict subsequent grain growth by impingement of grain boundaries and which also act as nucleation sites for recrystallisation events. For Waspaloy, the final forging is performed around the γ' solvus, at approximately 1040°C.

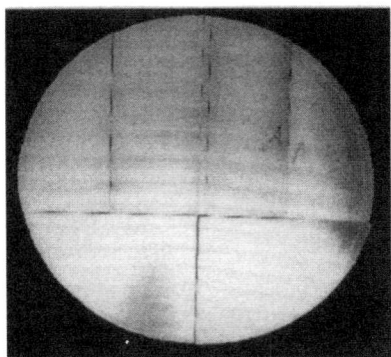

Figure 1(a). As-cast, as-homogenised section of Waspaloy.

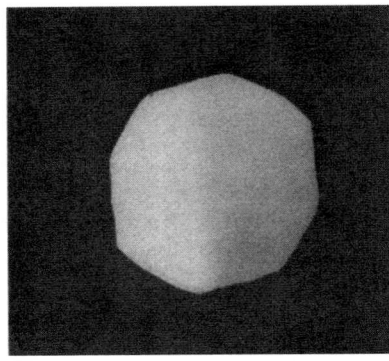

Figure 1(b). Intermediate stage 3 of Waspaloy.

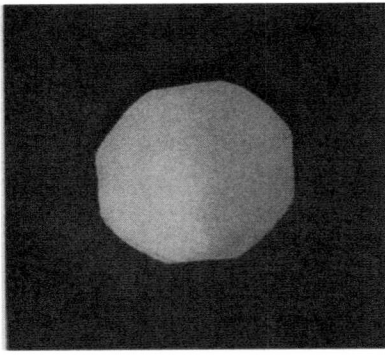

Figure 1(c). Intermediate stage 4 section of Waspaloy.

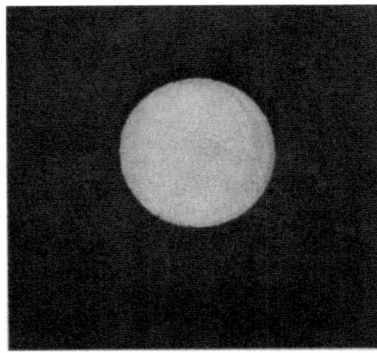

Figure 1(d). Final as-cogged section of Waspaloy.

The Process Model

As an example of the application of the model we consider a three-dimensional simulation of a Special Metals Wiggin Ltd. production schedule, which involves the reduction of a 2 tonne, 380mm diameter as-cast as-homogenised ingot of Inconel®718. Both radiative heat loss during transfer from furnace-to-press and heat conduction due to die contact have been modelled. Thermal predictions associated with radiative heat losses have been validated through pyrometric measurements of the ingot skin temperature, which cools from 1120 to 1060°C during furnace-to-press transfer. A shear friction factor of 0.38 is used to represent the interaction between die and ingot surfaces. The die speed has been set at a constant 40mm/s.

The constitutive model follows the methodology of Evans [1,2] and uses the Sellars-Teggart [13] inverse sinh relationship given in Equation 1:

$$\sigma = \frac{1}{\alpha} \sinh^{-1} \left[\left\{ \frac{\dot{\varepsilon}}{A} \exp\left(\frac{Q_{def}}{RT}\right) \right\}^{\frac{1}{n'}} \right] \tag{1}$$

This relationship may be used to define the yield stress (σ_1), maximum stress (σ_2) and steady-state stress (σ_3) in terms of both temperature and strain rate. These points are illustrated on the experimental stress-strain curves in Figure 2.

A further equation relates the strain (ε_2) at the maximum stress to the strain rate and temperature and is given in Equation 2:

$$\varepsilon_2 = B\dot{\varepsilon}^m \exp\left(\frac{Q_\varepsilon}{RT}\right) \tag{2}$$

Between σ_1 and σ_2 the work hardening portion of the flow curve is described by a second order polynomial. The softening portion of the curve is described by a linear fit and steady state conditions are assumed to ensue at 0.4 strain.

Unique sets of constants (α, A, n', Q_{def}, B, m and Q_ε) are required for each of the points corresponding to σ_1, σ_2, σ_3 and ε_2 [14]. A comprehensive compression testing programme was devised which considered a range of temperatures (950-1150°C) and strain-rates (0.001-10s⁻¹) to derive these constants. Specimens were machined into cylinders having 6mm diameter and 8mm height. The geometry of the specimens was selected to minimise the barrelling effects associated with friction. The specimen/platen surfaces were glass lubricated during testing. All tests were conducted under isothermal conditions and taken to a true strain of approximately 0.5 which resembles sufficiently closely the amount of deformation applied during the cogging operation. Compressed specimens were immediately water quenched upon removal from the testing furnace. The specimens were obtained from wrought product and were determined to have starting grain sizes of ASTM 5/6. Stress-strain curves for Inconel®718 at 1080°C are shown in Figure 2.

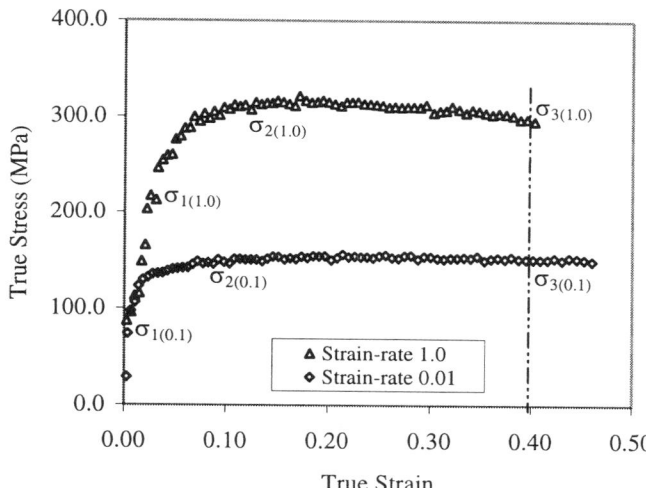

Figure 2. Typical stress-strain curves for Inconel®718.

The tool developed has been used to give an insight into the way in which microstructure evolves through careful study of the first and second passes.

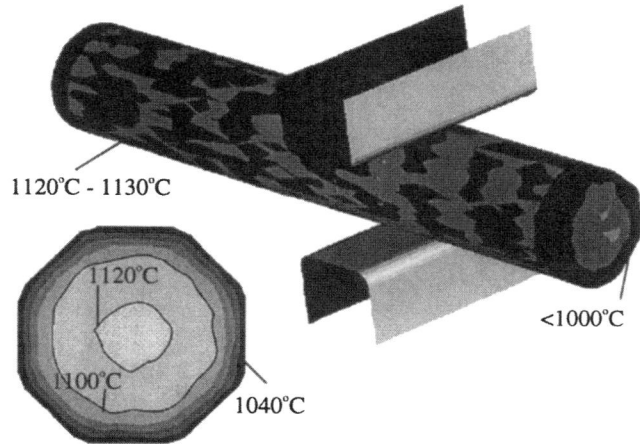

Figure 3. Numerical prediction of cogged billet temperatures.

Figure 3 shows both a three dimensional representation of the workpiece temperatures and a section at the mid-length. Adiabatic heating can be seen to have raised the centre section temperature above the initial 1120°C, whilst radiative and conductive losses have cooled the ingot surface. Towards the end of the ingot, the surface cooling is observed to be even more dramatic due to the extra surface area there. The value for the end section of between 1020°C at the centre, and 990°C at the corners, is in excellent agreement with the measured temperature of 1008°C. This has a significant effect on the variation of microstructure both through the cross section of the billet and along the billet length.

The Microstructural Model

A microstructural evolution model, which has been described elsewhere [1,2,4,14] and previously assessed [15], has been incorporated into the FORGE3™ finite element software package. Microstructural development is described using a number of purpose-built routines, which have been incorporated into the code. The dominant mechanisms for microstructural development during ingot breakdown are taken to be static recrystallisation and grain growth, which occur during periods of dwell between successive bites [16,17]. Previous observations assume that dynamic recrystallisation has a negligible influence on the evolving structure [18,19], which has been demonstrated to occur under specific conditions of temperature and strain-rate [20].

From an investigation of the flow curves and empirical measurements of microstructure three regimes have been identified; dynamic recrystallisation (DRX), dynamic recovery (DRV), and static (meta-dynamic) recrystallisation (SRX) with subsequent grain growth [1,2,14].

The softening behaviour exhibited by the material under applied load is indicative of DRX and/or deformational (adiabatic) heating. Upon closer examination of the Inconel®718 flow curves, there is considerable evidence to suggest that softening by DRX occurs only above both (i) a critical strain-rate of $0.1s^{-1}$ and (ii) a critical level of strain accumulation ($\varepsilon_{DRX} = \varepsilon_2$). At the onset of DRX, strain-free grains nucleate and appear with a distinct size governed solely by the instantaneous temperature and strain-rate. The nucleation rate is similarly governed by the instantaneous temperature and strain-rate.

At and below strain-rates of $0.1s^{-1}$ a plateau appears in the flow curve, with no further hardening nor any softening, which is indicative of dynamic recovery (DRV). Further straining results in the local generation of dislocations which are effectively annihilated to give a reduction in the local dislocation length [21]. The phenomena of DRV is not simulated in the model since static recovery, recrystallisation and grain growth are considered to overrule the effects of DRV.

If, under applied load, there has been insufficient work to drive dynamic deformation mechanisms, then the level of plastic strain accumulation may still be sufficient to drive SRX. Here, it is considered that SRX encompasses the processes of (a) static recovery which is followed by (b) subsequent recrystallization of strain-free grains (~1μm), which (c) grow at a rate dependent on the temperature. If the strain reaches a critical level ($\varepsilon_{SRX} = 0.8\varepsilon_2$), then SRX is assumed to occur upon removal of the applied load. If the level of strain accumulation is insufficient, then the strain energy is stored for future events. The model considers a static time variable which tracks dwell time between stages of SRX and which is used to determine the extent of static grain growth. The Avrami relationship is employed to describe the volume fraction of statically recrystallized material.

The recrystallisation logic applied is summarized in Figure 4. It is the belief of the investigators that, for a given temperature, there is a change in deformation mechanisms with decreasing strain-rates. At 'extreme' strain-rates ($10-20s^{-1}$), flow softening can be expected and is attributable to adiabatic heating. Dynamic recrystallisation may also play a significant role, whereby DRX grain growth may occur in the absence of δ-phase.

At 'high' strain-rates ($1-10s^{-1}$), flow softening may still occur, however this is almost exclusively due to dynamic recrystallisation (DRX).

At 'intermediate' strain-rates ($0.01-0.1s^{-1}$), DRX is less apparent and DRV becomes the dominant deformation mechanism, which is reflected by the plateau in the flow curve beyond peak stress.

At 'low' strain-rates ($<0.01s^{-1}$), a portion of DRV is considered to take place followed by softening which is attributable to grain boundary rotation and/or sliding. The frictional effects associated with grain boundary sliding are considered to break up material which experiences the greatest frictional forces. These regions accumulate sufficient strain to cause DRX. This is believed to result in the appearance of a necklace structure. This is illustrated in the microstructure observed in the Waspaloy ingot material after cogging (Figure 6). However, between strain-rates of 0.1 and $1s^{-1}$ there is a distinct change in the deformation mechanism which is supported by the transition in flow curves from a plateau to a softening mechanism. Generally, as strain-rate is increased from 0.1 to $1s^{-1}$, the deformation mechanism shifts from DRV to DRX. Consequently, a critical strain-rate of $0.1s^{-1}$ is incorporated into the microstructural model to simulate the restrictions of the DRX mechanism.

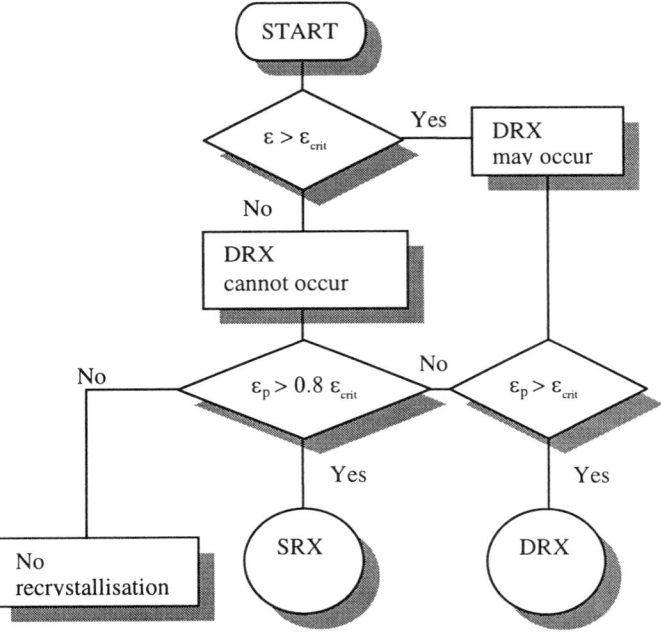

Figure 4. Recrystallisation logic.

Figure 2 provides good evidence that, for Inconel®718, flow softening occurs only above a critical strain-rate of $0.1s^{-1}$. Below this level, DRV events may occur. However, upon removal of the load, SRX will dominate microstructural evolution during dwell periods between blows. It is imperative that this dwell period is minimised in order to achieve a finer recrystallised microstructure.

Figure 5 shows a three-dimensional image of the computational process model for the Inconel®718 material: the predictions indicate the level of static grain growth within the workpiece at the end of the fourth pass. For a mid-length section, the model predicts ASTM 3/4 at the centre and mid-radius locations. The microstructure is much finer at the edge, ASTM 5/6. Such refinement is expected due to the lower temperature at the edges of the ingot compared to the adiabatically heated centre (Figure 3). A finer structure is predicted for the end of the workpiece since this material has just recrystallised. After some time, however, a grain size slightly finer than that of the bulk might be expected to due to the lower temperatures at the extremities of the workpiece. As discussed earlier, the lower temperatures offer less driving force for grain growth and, for temperatures below the δ solvus, grain growth becomes impinged. Table 2 illustrates the measured and predicted variation in microstructure [4] for the Inconel®718 material and clearly shows these phenomena.

Table 2. Microstructural results: observations versus predictions (Inconel®718).

Location	Mean RXD Grain Size (ASTM)			Volume Recrystallised (%)		
	Observed	Predicted		Observed	Predicted	
		Mid-length	End		Mid-length	End
Centre	3/4	3/4	6/7	86	100	84
Mid-radius	3/4	3/4	5/6	87	100	89
Edge	6	5/6	6-8	94	100	97

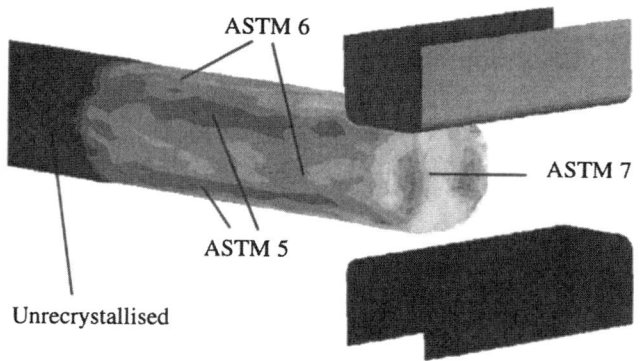

ASTM 6
ASTM 5
ASTM 7
Unrecrystallised

Figure 5. Prediction of SRX grain size as a function of position along the ingot length (Inconel®718)

Analysis of as-cogged billet section

Two forms of analysis were performed on cogged material. For the final stage of the as-cogged Inconel® 718 both chemical and microstructural analysis was performed. The Inconel® 718 microstructural results are recorded in Table 2 above. For the Waspaloy material only microstructural analysis has been performed so far, however this covers a number of cogging stages.

The chemical analysis in Inconel® 718, as can be seen in Table 3 below, indicated no significant macrosegregation across the radius of the ingot.

In Waspaloy, samples from the centre, mid-radius and periphery of the billet were characterized for stages 3, 4, 5, 6 and 7. Stages 3, 4 and 7 are shown in Figure 1, and a macroetched example of stage 5 is shown in Figure 8. Three images from different areas of each sample were analysed semi-automatically using a macro-program in conjunction with the Kontron Elektronik Imaging System KS300 v2.00 image analysis software package. The linear intercept technique was used as the basis for this analysis. A macro-program produced a series of lines in two orthogonal directions on each micrograph. The positions of the intercepts of these lines with the grain boundaries were then marked off manually using a mouse to position the cursor appropriately. The computer calculated and stored the intercept lengths in a database. Twin boundaries were not included in this analysis. The sum of the intercept lengths in each orthogonal direction was kept approximately equal. The total number of intercepts measured for each sample was typically between 500 and 1000. A mean ASTM grain size for each sample was obtained from the mean intercept length, l (mm) using the ASTM Standards [22] formula

$$G = 3.2877 - 6.643856 \log10 (l) \qquad (3)$$

The distribution of the intercept lengths for each sample as a function of fractional intercept length is shown as raw data. For each intercept measured this was calculated as

$$LFR = \frac{100 \times intercept\ length}{sum\ of\ all\ intercept\ lengths\ measured} \qquad (4)$$

This length fraction was then converted to the more familiar ASTM distribution of grain size as a fraction of intercept length.

The micrographs in Figure 6 are taken from the centre, mid-radius and edge locations of the as-cogged, stage 7 material. Microstructural analysis considered the distribution of intercept lengths for the samples associated with these regions.

Table 3. Results of chemical analysis for as-cogged section of Inconel®718 (Weight % except ppm - parts per million).

Location	C	Si	Mn	P	S	Al	Co	Cr	Cu	Fe	Mo	Nb	Ni	Ta	Ti	B
Centre	0.030	0.15	0.20	0.008	<0.001	0.48	0.18	19.3	0.07	17.0	2.97	5.17	53.2	<0.01	0.93	38 ppm
Mid-radius	0.025	0.14	0.20	0.008	<0.001	0.48	0.13	19.4	0.08	17.0	2.96	5.18	53.1	<0.01	0.93	37 ppm
Edge	0.028	0.14	0.20	0.008	<0.001	0.48	0.14	19.4	0.08	17.0	2.97	5.19	53.0	<0.01	0.94	39 ppm

CENTRE **MID-RADIUS** **PERIPHERY**

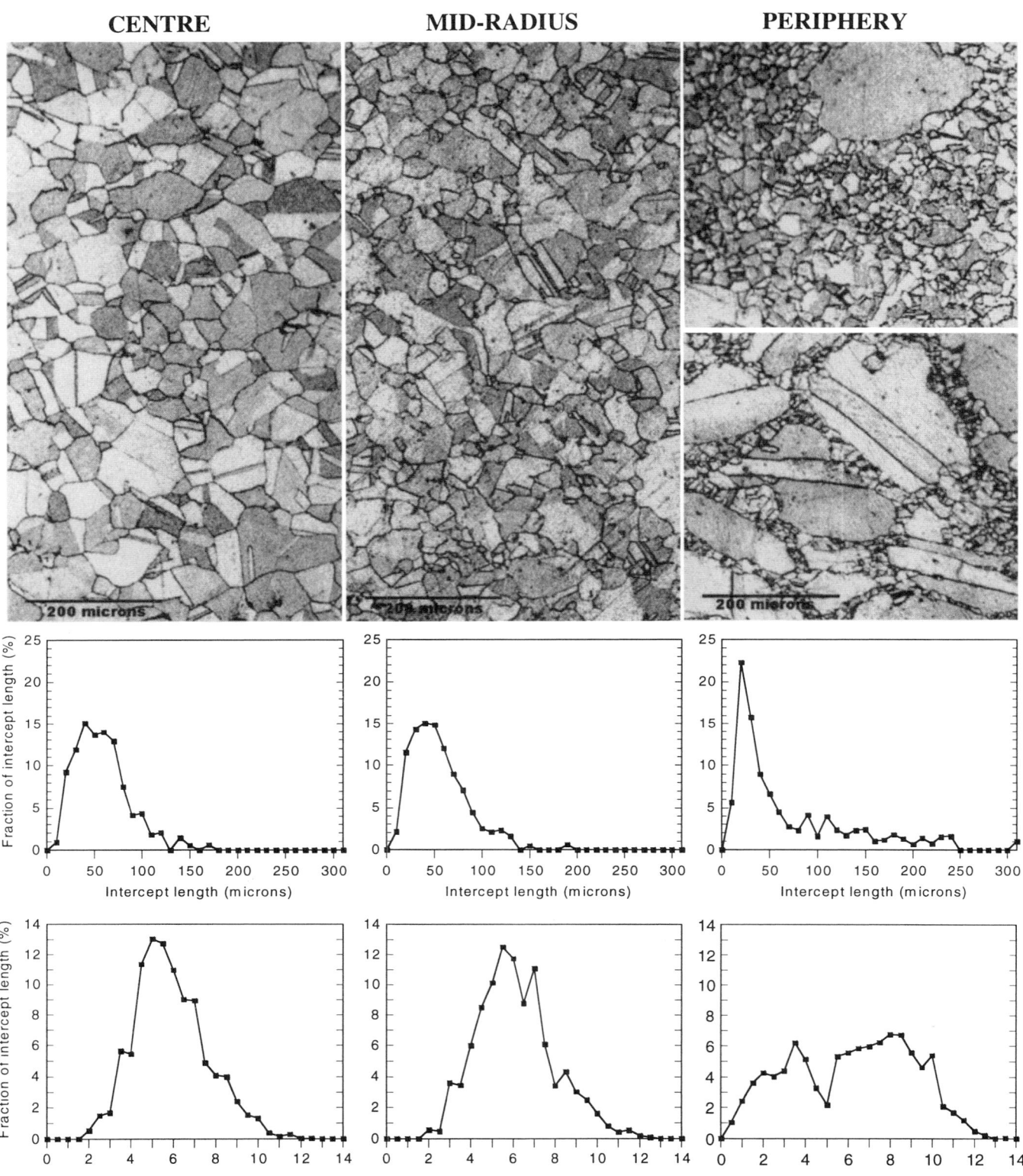

Figure 6. Micrographs and grain size distributions, measured using the linear intercept technique, from the centre, mid-radius and periphery of the final stage Waspaloy cogged billet. At the periphery two different microstructures were observed; micrographs of each are shown.

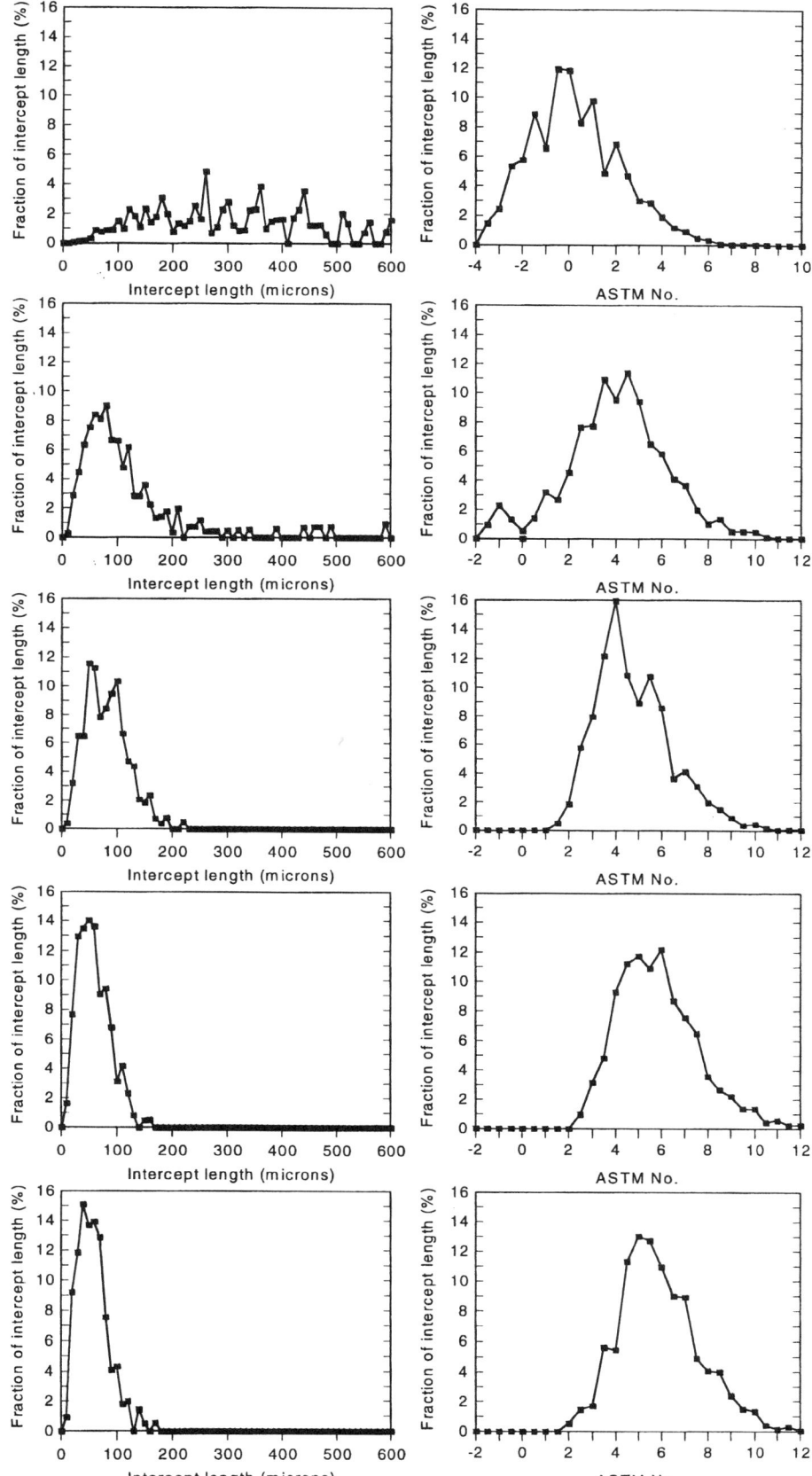

Figure 7.
Analysis of microstructure evolution at the centre of a Waspaloy ingot, during ingot-to-billet conversion. Grain size distributions were measured using the linear intercept technique, and taken for stages 3 to 7 (top to bottom respectively).

91

It can be seen from Figure 6 that a fairly uniform microstructure is achieved in both the centre and mid-radius locations of the billet material analysed. The slightly coarser nature of the centre material is due to the higher temperature associated with adiabatic heating, as demonstrated in the Inconel® 718 model (Figures 2 and 5). The periphery shows a mixed microstructure of two forms. The upper micrograph shows a true mixed mode, partially recrystallised structure, and is a result of insufficient strain accumulation to fully recrystallise this zone. The structure in the lower micrograph exhibits the classical necklace appearance as described in the previous section.

Of some concern would be the small percentage of grains with an ASTM of 1 or 2. Such grains are often associated with 'dead zones' in which the rate and amount of applied strain are insufficient to induce DRX. Figure 7 illustrates the steady improvement in microstructure brought about by the cogging procedure. The results represent the gradual refinement of the central ingot structure from ASTM 0 (stage 3) to ASTM 6 (stage 7).

<u>Dead zone evaluation</u>

One of the major objectives of process modelling is to enable the factory floor metallurgist to optimise actual forging schedules to refine the evolving microstructure. The underlying philosophy of this research considers that static grain growth is the dominant phenomenon governing microstructural development during ingot breakdown. A fundamental requirement for microstructural refinement, therefore, is the restriction of static grain growth. One may interrupt grain growth by ensuring DRX occurs during applied deformation.

The effect of variations in the process control parameters on the microstructure achieved is now examined, using the model as a predictive tool. Two parameters have been investigated, bite size and die speed. Process schedules combining slow die velocities with large bite sizes are compared to schedules combining fast die speeds with small bite sizes. We have found that the current combination of die speed and bite is near optimal. Small bites lack sufficient strain penetration at the billet centre thereby suppressing recrystallisation. Larger bites provide the necessary strain penetration to achieve full recrystallisation, however this is countered by excessive static grain growth times, and hence coarse microstructures, promoted by the slower die velocity. This is compounded by both the adiabatic heating in the centre which adds a further driver to the grain coarsening and excessive contact and radiative cooling to the periphery, resulting in a partially recrystallised grain structure. A small degree of process optimisation may be achieved through increases to both bite size and die velocity. This results in a more uniform microstructure with small levels of grain refinement achievable. In reality, such conditions may not be attainable since the hydraulic press is operated at near loading capacity.

The influence of squeezing the rounded sections of the octagonal section workpiece as opposed to the traditional approach of squeezing the flats has also been investigated. Figure 8(a) shows a macroetched section of Waspaloy billet, highlighting the areas of unrecrystallised material that are associated with the 'dead zones' created by forging on the flats. These dead zones are akin

Figure 8(a). Macroetched billet section showing coarse microstructure in dead zones.

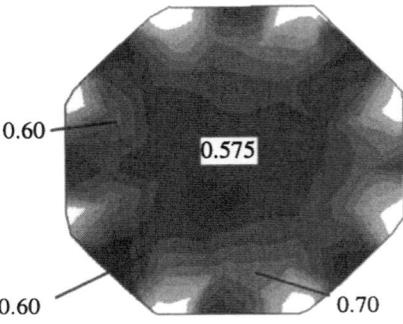

Figure 8(b). Effective strain profile associated with squeezing flats.

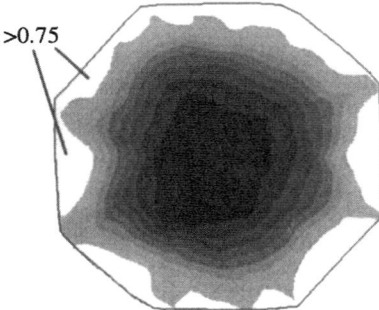

Figure 8(c). Effective strain profile associated with squeezing rounds.

to those seen in lubricant free compression tests. Surface friction effects at the die/workpiece interface are sufficient to prevent a significant volume of the material from deforming. The restrained material resides immediately under the die surface and exhibits a V-shape appearance. Figure 8(b) shows a typical effective plastic strain contour map taken from a two-stage process schedule simulation for Inconel®718. The first stage comprises four passes along one half of the ingot length to produce an octagonal cross-section. The workpiece is then rotated through 45° to ensure that the next four passes deliver blows to the flats of the workpiece.

An alternative simulation was performed to investigate the effect of squeezing octagonal section corners. This process schedule was identical to the previous model in all respects other than the rotation at the end of pass four. Here the workpiece was rotated through 22.5° thereby ensuring the delivery of subsequent blows to the rounded sections of the workpiece. The effective plastic strain profile for this approach is shown after the eighth pass (Figure 8(c)). It is clear that there is a significant rise in the level of strain accumulation around the periphery of the workpiece. This increase in strain is induced due to the wider section of material that must be worked in order to achieve the desired reduction. It is the authors' contention that such a methodology would eliminate the coarse 'dead zone' regions as seen in the macroetched section (Figure 8(a)). A possible disadvantage to this approach is the lower level of strain penetration attained at the centre of the ingot. However, given the microstructural analysis carried out in Figure 7, it is likely that there would still be sufficient strain penetration in the course of the whole cogging process to ensure full recrystallisation. In fact this slight reduction in the level of strain experienced at the centre of the ingot might prove beneficial, with its associated reduction in adiabatic heating and thus a lowering of the driving force for static grain growth.

Summary and Conclusions

A computer-based process model has been presented to simulate microstructural evolution of nickel-base superalloys during ingot-to-billet conversion. The theoretical approach has been validated by comparison to sectioned material taken from various stages in the cogging process. Analysis of macro-etched sections of the Waspaloy, indicate significant grain refinement through the cogging process, typically ASTM 0 to ASTM 6 at the center of the ingot. The microstructural analysis of the final stage Waspaloy material suggests that the bulk of the section is fairly uniform with only a slight variation in grain size from centre to mid-radius, the latter being slightly finer. This slight variation is attributed to the coarsening effect of the adiabatic heating at the centre of the ingot. Investigation of the periphery indicates two forms of partially recrystallised structure.

The microstructural model has been demonstrated to be effective in predicting the microstructures observed. A coarser recrystallised grain structure is predicted at the centre of the ingot, where a degree of adiabatic heating is modelled. Surface cooling effects due to radiation and die/billet interaction are also modelled successfully. From macroetched sections taken from an interrupted cogging schedule the effect of biting on the flats of a previous octagonal section are apparent. Such biting leads to a dead zone forming immediately below the die/ingot interface and extending someway into the cross section. Regions of unrecrystallised grains may be observed due to this lack of strain penetration. The process model accurately reflects these 'dead

zones' and was then used to derive a new schedule where the level of strain penetration was more uniform around the periphery of the ingot.

Acknowledgements

The authors are grateful to the sponsors of this work, Rolls-Royce and notably Special Metals Wiggin Ltd. for the supply of material and processing details. In particular useful discussions with Specials employees, past and present; Richard Siddall, Dave 'Rocky' Rochard, Dave Lambert, Andy McGilvray, Gern Mauer, Bruce Antolovich and Lesh Patel are acknowledged. Thanks also to George Durrant at Rolls-Royce for his continued enthusiasm for this work and to Jeff Brooks and Mike Henderson at DERA for their advice. Chris Dandre was, and Carrie Walsh is, employed on an EPSRC/DERA funded joint grant scheme. (GR/L 66397, 'Microstructural Evolution in Ni-base Superalloys Forgings: Process Modelling and Validation').

References

1) R.W. Evans, "Microstructural Modelling of Near Net Shape Forgings", (Paper presented at the EUROMAT'89 Conference, Aachen, FRG, 22-24 November 1989).
2) R.W. Evans, "Modelling of the Hot-Working of High Performance Alloys", Key Engineering Materials, 1993, 77-78, 227-240.
3) R.A. Jaramillo et al. "Evaluation of an Inconel® Alloy 718 Microstructural Evolution Model", Superalloys 718, 625, 706 and Various Derivatives, (ed. E.A. Loria, Warrendale, TMS, 1997), 257-266.
4) C.A. Dandre et al., "A Model Describing Microstructural Evolution for Nickel-Base Superalloy Forgings During the Cogging Process", Le Journal de Physique IV, v9, (1999), pp33-42.
5) D. Zhao et al., "Three dimensional simulation of Alloy 718 Ingot breakdown by Cogging", Superalloys 718, 625, 706 and Various Derivatives, (ed. E.A. Loria, Warrendale, TMS, 1997), 163-172.
6) C. Boyko, H. Henein, and F. Robert Dax, "Modelling of Open Die and Radial Forging Processes for Alloy 718", Superalloys 718, 625 and Various Derivatives, (ed. E.A. Loria, Warrendale, TMS, 1991), 107-124.
7) J.P. Domblesky et al, "Prediction of Grain Size During Multiple Pass Radial Forging of Alloy 718", Superalloys 718, 625, 706 and Various Derivatives, (ed. E.A. Loria, Warrendale, TMS, 1994), 263-272.
8) C.T.Sims, N.S. Stoloff, and W.C. Hagel, Superalloys II, (New York, NY:Wiley, 1987)
9) M. Durand-Charre, The Microstructure of Superalloys, (Gordon and Breach Science Publishers, 1997).
10) J.M. Zhang, Z.Y.Gou, J.Y. Zhuang, Z.Y. Zhong, 'Mathematical Modelling of the Hot Deformation Behaviour of Superalloy IN718', Met & Mat Trans A, (1999), 2701-2712
11) A.A. Guimaraes and J.J. Jonas, 'Recrystallisation and ageing effects associated with the high temperature deformation of Waspaloy and Inconel 718', Met. Trans. A (1981), 1655-1666.
12) R. Siddall private communications with authors, Special Metals Wiggin Ltd., Hereford, UK, February 2000.

13) C.M. Sellars and W.J. McG. Tegart, "Hot Workability", International Metallurgical Reviews., (1972), 17, 1-24.

14) C.A. Dandre et al., "Microstructural evolution of Inconel 718 during ingot breakdown: process modelling and validation", Mat Sci & Tech. v.16, (2000), 14-25

15) C.A. Dandre et al., "Exploring the Microstructural Evolution of Inconel 718 During Ingot Breakdown: Optimising the Cogging Process", (proceedings of COMPASS'99, University of Wales, Swansea, 1999).

16) M.C. Mataya and D.K. Matlock, "Effects of Multiple Reductions on Grain Refinement During Hot Working of Alloy 718", Superalloy 718 - Metallurgy and Application, (ed. E.A. Loria, Warrendale, TMS, 1989), 155-178.

17) M.C. Mataya, E.R. Nilsson and G. Krauss, "Comparison of Single and Multipass Compression Tests Used to Simulate Microstructural Evolution During Hot Working of Alloys 718 and 304L", Superalloys 718, 625, 706 and Various Derivatives, (ed. E.A. Loria, Warrendale, TMS, 1994), 331-343.

18) D. Zhao, S. Guillard and A.T. Male, "High Temperature Deformation Behaviour of Cast Alloy 718", Superalloys 718, 625, 706 and Various Derivatives, (ed. E.A. Loria, Warrendale, TMS, 1997), 193-204.

19) M.J. Weiss et al., "The Hot Deformation Behaviour of an As-Cast Alloy 718 Ingot", Superalloy 718 - Metallurgy and Application, (ed. E.A. Loria, Warrendale, TMS, 1989), 135-154.

20) N. Srinivasan and Y.V.R.K. Prasad, "Microstructural Control in Hot Working of IN-718 Superalloy Using Processing Map", Met. & Mat. Trans. A, 25A, (1994) 2275-2284.

21) W. Blum, and H.J. McQueen, "Dynamics of Recovery and Recrystallisation", Materials Science Forum, vols217-222, (1996), 31-42.

22) Annual book of ASTM Standards, v.03.01, E112, p227, 1996

Nomenclature

σ	- flow stress
ε_2	- strain at maximum stress
$\dot{\varepsilon}$	- strain-rate
DRX	- dynamic recrystallisation
Q_{def}	- activation energy for deformation
Q_ε	- activation energy for ε_2
R	- gas constant
T	- temperature
RXD	- recrystallised
RXN	- recrystallisation
SRX	- static recrystallisation
α, A, n', B, m	- material constants

REMOVAL OF CERAMIC DEFECTS FROM A SUPERALLOY POWDER USING TRIBOELECTRIC PROCESSING

J. M. Stencel, T.-X. Li, J. K. Neathery and L.W. Lherbier*

Center for Applied Energy Research, University of Kentucky, Lexington, KY 40511
*Dynamet Incorporated, 195 Museum Road, Washington, PA, 15301

Abstract

The high-pressure turbine is one of the most important components in aviation propulsion systems. The disks in a turbine work under the most severe conditions through which corrosive combustion gases are passed at high temperatures. Non-metal impurities in these disks, potentially introduced during processing of superalloy powders from which the disks are made, compromise their quality and generate discontinuities or defects, thereby becoming the location at which cracks or other mechanical failures can be initiated. Typically, sieving of the as-atomized superalloys through 200 mesh (75 μm) or 325 mesh (45 μm) screens is used as a way to decrease the concentration of the larger-sized ceramic particles. This paper describes fundamental studies using an alternative method - called triboelectric separation - to remove ceramic defects from superalloy powders.

A laboratory-scale apparatus, designed for fundamental investigations in the purification and processing of physical mixtures having particle diameters between 0-200 μm, was employed as the test platform. Through it were processed a series of alumina-seeded Udimet[R] - U720 - superalloy powders, the products from which were analyzed by sieving and scanning electron microscopy. The removal of the seed varied from 23% at a 5 wt. % seed concentration to 49% at a 0.01 wt. % seed concentration. Simultaneously, the recovery of purified superalloy powders was constant and near 92%. Although more investigations are needed, the data suggest that triboelectric separation processing may be useful for removing defects inherent to superalloys.

Introduction

The high-pressure turbine is one of the most important components in aviation propulsion systems. The disks in a turbine work under the most severe conditions through which corrosive combustion flue gases at the highest temperatures are passed. As a consequence, the life of these disks can determine the life of the whole engine. For decades, efforts have been made to improve the lifetime, or delay the failure of, turbine disks.

One advanced technique for high-performance engines, which is now common practice by companies such as Allison/Rolls Royce, General Electric, and Pratt & Whitney, is the use of powder metallurgy turbine disks. Superalloy powders are manufactured by melting and atomizing to create nearly spherical particles having diameters between 1-300 μm. These superalloy particles are consolidated into billets by hot isostatic pressing and/or extrusion. Subsequent forging of these billets into geometric shapes gives superior quality disks having isotropic and homogeneous characteristics[1-3].

However, it has been determined that non-metal impurities can be introduced during melting and processing of the superalloy powders. These so-called "inherent impurities" or "defects" are usually in the form of ceramic particulates. The incorporation of impurities generates discontinuities or defects in the turbine disk, which are believed to be the location at which cracks or other mechanical failures are initiated.

It has been suggested that a significant increase in the lifetime of turbine disks would be possible, even under conditions more severe than currently used, if the ceramic impurities could be eliminated from the superalloy powders. Unfortunately, they are almost impossible to eliminate because alumina crucibles are used to melt the alloy, which is then flowed through zirconia nozzles during powder formation. This process can lead to small pieces of ceramic nozzle being worn or chipped off and, as a consequence, being incorporated into the powdered alloy. In relative terms, the amount of such impurity is almost negligible – at a level near one part per

Superalloys 2000
Edited by T.M. Pollock, R.D. Kissinger, R.R. Bowman,
K.A. Green, M. McLean, S. Olson, and J.J. Schirra
TMS (The Minerals, Metals & Materials Society), 2000

million by volume or about 50 ceramic particles per pound of superalloy. Importantly, even at this extremely low level, a deleterious effect is expected on the quality and lifetime of the formed components.

A way to decrease the concentration of the larger-sized ceramic particles is to sieve the as-atomized superalloy powder using 200 mesh (75 μm) or even 325 mesh (45 μm) screens. However, sieving eliminates not only the larger ceramic particles but also the larger superalloy particles, thereby decreasing the superalloy yield per melt. This decrease causes significant wastage and increases costs.

An alternate approach to improving the process yield while increasing superalloy purity is to apply a physical separation technique that selectively removes ceramic impurities. This paper describes one such approach that is called triboelectric separation. It has been under development at the Center for Applied Energy Research, University of Kentucky. It uses gas transport of particles and an electric field to selectively separate physically distinct particles after they have been charged in a bipolar manner. Although energy materials have been the focus of its application, the difference between ceramic defects and superalloy particles suggested that they may be able to be separated under the conditions employed within the triboelectric separation apparatus. The following sections detail the experimental procedure and the data relating to the removal of alumina impurities which had been added to the Udimet superalloy powders.

Experimental

The Udimet powder, labeled as U720, was supplied as a -140 mesh (<106 μm) cut. Commercial sintered alumina having particle size less than 3 mm was purchased from Aldrich. It was crushed using a mortar and pestle, and then sieved to -50+75 mesh (-300 μm + 212 μm). The superalloy and the sieved alumina were mixed, by weight percentage, to five different concentrations, including 5%, 1%, 0.5%, 0.1% and 0.01% alumina-in-superalloy. Although substantially greater than the expected concentration of ceramic defects in the alloy, the alumina was difficult to accurately seed into the superalloy when the concentration was less than 0.01%. Properties of the superalloy and the alumina are presented in Table I.

Table I. Properties of the superalloy and alumina powders.

Table I. Properties of the Udimet 720 and alumina.

	U720	Al_2O_3
Size (mesh)	-140	-50+75
Density (g/cc)	8.3	3.97

The alumina powder is significantly larger than the Udimet superalloy because the method chosen to delineate seed removal was mechanical sieving. In other words, subsequent to passing the mixtures through the triboelectric separator, the rejected impurities and purified superalloy were sieved, the largest alumina particles of which remained on top of the screen. Then, the weight of these impurities could be easily measured.

The mixtures were used as the feed material in a laboratory-scale triboelectric separation system. A diagram of it is presented in Figure 1. It consists of a powder feed and transport section, a parallel plate electric field zone and an exhaust/filtering outlet.

Up to 50 g of seeded superalloy is placed into a vibratory feeder located within a sealed and pressurized container. The powder is dropped from the feeder into a 10 mm diameter tube leading to a N_2 gas eductor. The eductor entrains the powder in N_2, flowing it through a 6 mm diameter transport line leading to the top of a parallel plate separation chamber. The gas-entrained powder attains a bipolar charge by particle-particle interactions during this gas transport to the chamber.

The gas entrained particles enter at the top, middle of the separation chamber. A flow of gas, void of particles, surrounds these injected particles. As they enter the electric field, established by applying a voltage between Cu electrodes placed 10 cm apart, negative particles are attracted towards and attach to the positive plate whereas positive particles are attracted toward and attach to the negative plate. At the bottom of the chamber is a filter that retains particles which have either fallen off or were not attached to the electrodes. Gas flowing through the filter is exhausted by an induced draft fan.

Subsequent to an experiment, the positive electrode was coated with a visible film of white-shaded powder whereas the negative electrode was coated with a metallic-shaded film. After turning off the high voltage supply, the deposits were removed by

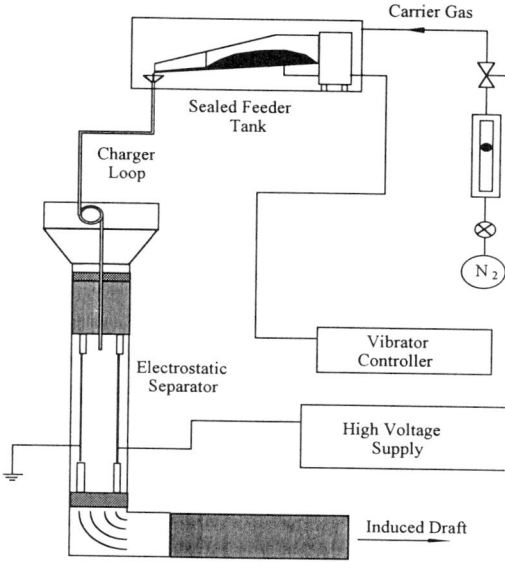

Figure 1. Laboratory triboelectrostatic separator used in pulverized coal beneficiation studies

gentle scraping and brushing into a small collection

vessel. The vessels with their contents were weighed and the mass compared to the mass of seeded superalloy that was in the feeder prior to the experiment. This comparison provided mass balance information, i.e. how much of the feed could be accounted for in the products.

The negative and positive electrode products were then sieved through a 140 mesh (106 μm) screen. Because of size differences between the alumina impurities and the superalloy particles, the alumina particles were retained on the top of the sieve, while the superalloy particles passed through the sieve. The mass on top of the screen from the positive electrode product was then weighed and compared to the feed mass. This ratio provides the % alumina reduction, i.e. the % Al_2O_3 removed from the feed during triboelectric processing. The ratio of the mass from the negative electrode product relative to the mass of the feed provides the alloy recovery rate.

The dc voltage applied to the electrodes was between 15kV-to-40kV; this implies the electric field strength was 150-400 kV/m. The current to the electrodes was small, typically less than 2 mA, because a relatively small mass of sample was used during each experiment. It is believed that, during application of the triboelectric method to continuous-feed

processing for superalloy cleaning, the current to the electrodes would also be small because the particles would not be collected by attachment to the electrodes. Rather, the electric field zone of the separator would be used to divert either purified superalloy or ceramic defects toward their respective collection outlets without the particles colliding with or attaching to the electrode.

Scanning electron microscopy (SEM) with energy dispersive spectroscopy (EDS) was used to determine the size distribution, morphology and particle/sample compositions. Products from both the negative and positive electrodes were examined.

Results and Discussion

The effect of voltage within the electric field cell on removing alumina from the superalloy for the 5% alumina-seeded sample is presented in Figure 2. The extent to which alumina was removed gradually increased from near 5%-to-10% as the voltage was increased from 15-to-30 kV. Then, it jumped rapidly to above 23% and was independent of voltage above 35 kV. The electric field is necessary for separating charged components from a physical mixture. Its optimal value, i.e. relative to impurity rejection efficiency, is dependent on size of the impurities to be rejected. Because the alumina seed was uniform in size, a voltage of 35 kV was used during the data acquisition for all other samples.

Table II summarizes the triboelectric separation results for the five seeded mixtures. The values in the table were the average of three or more tests. It is clear that, as the concentration of the alumina decreased, its removal rate increased from about 23% in the 5% alumina mixture to 50% in the 0.01% alumina mixture. Over this range of impurity concentrations, there was no change in the superalloy recovery rate. In all tests, about 7% of the superalloy mass reported to the electrode on which the alumina product reported, i.e. about 93% of the feed was recovered as purified superalloy.

The lowest alumina concentration used in the tests, 0.01% (wt), was equivalent to 100 ppm (wt) or 810 Al_2O_3 particle/lb of sample. At this level, the concentration of the alumina impurity is about 50 times greater than the anticipated concentration of ceramic defects inherent to some superalloys.

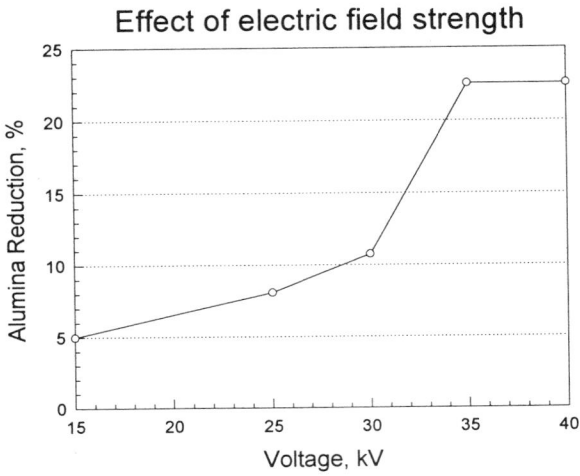

Figure 2. Alumina reduction from superalloy as the voltage on the electrodes is changed.

Table II: Summary of triboelectric separation results for alumina seeded U720 superalloy.

Feed Al$_2$O$_3$ Conc. (Wt. %)	# Al$_2$O$_3$ Particles per in^3	# Al$_2$O$_3$ Particles per lb	Al$_2$O$_3$ Removal Rate (%)	Alloy Recovery Rate (%)
5	72,500	405,000	23	93
1	14,500	81,000	31	92
0.5	7,250	40,500	36	92
0.1	1,450	8,100	43	92
0.01	145	810	49	93

However, plotting the data from Table II into Figure 3 shows that the separation efficiency was greater for samples with lower impurity concentrations. Coupled with a standard deviation analyses of the data (also presented in Figure 3) it is projected that the alumina removal rate would be nearly 70% for samples in which the impurity level would be as low as 0.0001% (1ppm or 8 Al$_2$O$_3$ particles/lb of superalloy). This concentration level is close to that anticipated for inherent ceramic defects introduced into some superalloys during atomization.

Prior to obtaining the data in Table II and Figure 3, it was anticipated that alumina seed removal from the superalloy either would become more difficult or remain constant as the seed concentration decreased. This rational was based on the fact that, in separation procession, it is generally more difficult to remove

impurities from gases, liquids and solids as their concentration decreases to near trace levels.

Additionally, in our gas transport, triboelectric experimentation it had been observed that impurities in other types of powdered materials are removed at relatively constant efficiency independent of their concentration over a range of about 3%-to-15%. However, because triboelectric separation depends on establishing bipolar charge on particles with distinct surface physical properties during particle-particle collisions, and because, during gas transport to the electric field cell, the probability increases that alumina particles will collide with superalloy particles as the concentration of the alumina is decreased, it is reasonable to anticipate that the alumina particles are more efficiently and highly charged at the lower seed concentrations. Hence, it is

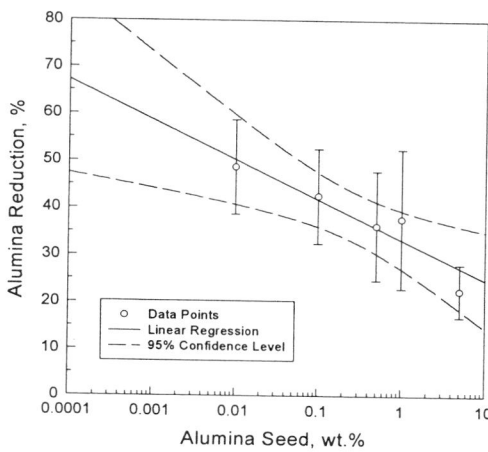

Figure 3. Alumina removal as a functional of seed concentration.

reasonable to anticipate that inherent defects, which are typically at levels near 0.0001%, would be more efficiently removed from the superalloy than is the seeded impurities.

SEM data were obtained on the products deposited on the negative and positive electrodes. These data detail morphology differences between the superalloy and the alumina impurities. The large impurity particles were absent from the negative electrode product and present in the positive electrode product. On the positive electrode were a combination of superalloy and alumina particles. These positive electrode products represent the 7% of the feed which did not report to the purified product (Table I). Also, it was evident that impurity particles were removed from the superalloy even though they were significantly larger than the superalloy particles.

Conclusions

It is possible to remove alumina particles seeded into a superalloy by use of gas transport, triboelectric separation. The removal of the impurity was dependent on its concentration; the lower the concentration, the greater was the removal. This trend suggests that the extent to which inherent defects can be removed from superalloys may be as great as 70% at defect concentration near 0.0001%, i.e.~8 particles per pound for the size of the impurities studied. Also, the results suggest that even larger sized defects are efficiently removed by

triboelectric processing. More studies are necessary to understand the potential of triboelectric separation as an alternative to sieving for controlling inherent defect concentrations in superalloys.

Acknowledgments

Assistance from and discussion with Bruce Ewing from Allison/Rolls Royce are gratefully acknowledged.

References

1. International Journal of Powder Metallurgy, American Powder Metallurgy Institute.
2. Advances in Powder Metallurgy, Compiled by T.G. Gasbarre and W.F. Jandeska, Metal Powder Industries Federation, Princeton, NJ. 1989.
3. ASM Metals Handbook, Vol. 7, Powder Metallurgy, MPIF, 1989.

PRODUCTION EVALUATION OF 718-ER® ALLOY

Wei-Di Cao and R.L. Kennedy

Allvac, An Allegheny Technologies Company
2020 Ashcraft Ave.
Monroe, NC 28110
USA

Abstract

Allvac 718-ER® alloy, for "extended rupture life," is a slightly modified version of alloy 718 with increased phosphorous (P) and boron (B) content. A full-scale production trial was conducted to evaluate 718-ER in comparison with standard alloy 718. The processing characteristics of both alloys in melting/remelting, homogenization and hot working were evaluated. Microstructure and mechanical properties were also compared in detail with those of standard 718. Results showed that 718-ER could be manufactured in nearly the identical manner as standard 718. Consistent with earlier subscale results, tensile properties changed very little, but nearly a 100% improvement in stress rupture and creep properties was achieved in a full scale production heat of 718-ER.

Introduction

Previous studies performed at Allvac[1-4] have shown that increasing the P content of alloy 718 significantly improved the stress rupture/creep properties, although P has generally been regarded as a detrimental element in superalloys. In addition, it was found that there was a strong synergistic effect between P and B, i.e. the stress rupture/creep properties could be further improved by controlled additions of both P and B. In comparison with standard alloy 718, the increase in stress rupture life may reach 100-200% at the optimum combination of 0.022%P and 0.011%B. Other mechanical properties of

_____

718-ER® is a registered trademark of ATI Properties, Inc.

718-ER, such as strength, ductility and low cycle fatigue, are basically identical to those of standard 718.

Subsequent studies[5-8], performed in China, have demonstrated similar effects. Tensile properties were not noticeably affected, but significant improvement of up to 300% in stress rupture life and creep rate was observed. The optimum P level reported in these studies was quite close to that found in references 1-4.

Detailed studies on 718-ER have been conducted on small pilot plant heats, but to fully determine the commercial potential, a production scale evaluation was critical. Therefore, a project was initiated to make a full-size heat of 718-ER and to evaluate its processing characteristics, macro and microstructures, and various mechanical properties. To make a better comparison, one standard 718 heat was made by identical processing, side by side, and evaluated together with a 718-ER heat.

Experimental Procedures

Manufacture of Test Alloys

A 9080-kg vacuum induction melted, 508mm Rd vacuum arc remelted 718-ER heat, JG61 was made using standard 718 procedures. Three heats of standard alloy 718 were selected for comparison. The majority of this study was conducted on standard heat JG60 that was made immediately prior to JG61. In some cases, two other standard 718 heats (EU03 and EW02) were used, mainly due to material availability for that specific test.

Superalloys 2000
Edited by T.M. Pollock, R.D. Kissinger, R.R. Bowman,
K.A. Green, M. McLean, S. Olson, and J.J. Schirra
TMS (The Minerals, Metals & Materials Society), 2000

All ingots were homogenized at identical conditions and forged in the same manner to 203mm diameter billet. The billets were peeled, ground and immersion-sonic inspected to a number two flat bottom hole. Macro plates were cut at different ingot locations for inspection. Additional test material was taken from hot rolled 16mm Rd bar and coupons (pancakes) upset-forged from 51mm cubes.

Segregation Behavior and Hot Workability

Studies[9,10] have shown that P and B may increase the tendency of microsegregation of Nb and other elements in alloy 718, and that increased segregation could adversely influence hot workability. Therefore, a detailed study was conducted to characterize the segregation behavior and hot workability both in as-cast and wrought conditions.

Microsegregation in the as-cast condition was judged by volume fraction of Laves eutectic determined by quantitative metallography. In the homogenized condition, the Nb content in areas corresponding to the center of interdendritic and dendrite arms was measured by Energy Disperse Spectroscopy (EDS), and the ratio was used as an indicator of microsegregation severity.

The hot workability of as-cast and as-homogenized ingot materials was investigated by the rapid strain rate hot tensile (RSRHT) test and reduction in area and elongation used as an indicator of hot workability. Differential thermal analysis (DTA) was also performed on both alloys to determine incipient melting temperatures which could be used as an indirect measure of hot workability and provide important clues for explaining the results obtained by RSRHT.

Evaluation in the wrought condition was done on the 203mm Rd forged billets. The distribution of Nb across the lightly banded structure revealed by etching was determined by EDS, and the microsegregation characterized again by the Nb ratio between "dark" and "light" bands that correspond respectively to dendrite arms and interdendritic areas in the as-cast structure. Due to detectability limits of EDS, a progressive solution treating method of evaluating delta solvus temperature was also used to determine the degree of microsegregation. This procedure involved incrementally solution treating micro samples from pancakes. The grain growth behavior was examined and grain size was plotted as a function of solution temperature.

Mechanical Property Tests

The mechanical properties, including room and elevated temperature tensile, stress rupture and creep properties, were tested on various product forms, including forged billet, pancakes and rolled bar. Since all previous pilot plant heats had been evaluated on the same size rolled bars, this test may provide the best comparison of the effect of P-B modification in pilot plant and production heats. All samples were subject to standard heat treatment (954°C x 1 hr., AC + 718°C x 8 hrs., FC at 56°C/hr. to 621°C and held 8 hrs., AC).

Microstructure and Fractographic Study

Samples were studied by optical and scanning electron microscopy to define the effect of P and B on the microstructure. A fractographic study was also conducted on broken stress rupture samples to check for changes in failure mode with increased P and B levels.

Experimental Results

Chemistry and Segregation

The chemistries of all materials tested in this program are listed in Table 1. The P level of JG61 (718-ER) was at the optimum level for 718-ER, but B was lower and C higher than optimum, as defined from prior work with pilot plant heats[1-4].

As-cast microstructures are shown in Figure 1. Quantitative metallographic study on ingot samples indicated that as-cast 718-ER contained about 20% more Laves phase eutectic, suggesting that P and B promote Nb segregation. This is consistent with prior studies[9]. However, essentially all Laves phase particles were eliminated after a standard homogenization treatment of both alloys. Results of EDS for Nb segregation are shown in Table II. These data suggest that residual Nb interdendritic segregation in the ingot, following homogenization, is higher in 718-ER than in the standard alloy. There is, however, significant scatter in the data.

In the wrought condition, EDS results still show slight evidence of segregation, but there does not appear to be a significant difference between the two alloys. This conclusion is supported by results of the progressive solution treatment test of as-forged samples. Grain size, plotted in Figure 2 as a function of solution temperature, shows almost identical grain growth behavior between alloy 718 and 718-ER.

Hot Workability

Percent elongation and reduction in area of data from RSRHT testing are listed in Table III. The data for as-homogenized ingots is plotted in Figure 3. From these results, it is apparent that the reduction of area for these two alloys is very comparable except at the very highest test temperature. Fortunately, 1204°C is well above the standard forging temperature for 718. Percent elongation results were higher for 718-ER at all test temperatures except 1204°C. These data are consistent with the actual processing results in that no difficulties were encountered during the forging of heat JG61 and no sonic defects were found on U/T inspection.

RSRHT tests of billets resulted in a similar trend as for ingots (see Table III and Figure 4). Surprisingly the fall off in hot ductility occurred at an even lower temperature in 718-ER billet. In general, ductility for billet was higher than for ingot, especially at the lower test temperatures and for elongation. As for the case with ingot, the lower hot ductility displayed for 718-ER at the higher temperatures should not be a problem in normal production since standard forging temperatures are in the range of 1121°C or less.

Table I. Chemistries of 718-ER® and 718 Heats

Element Wt.%	C	Mo	Cr	Fe	Mn	Si	Nb	Ti	Al	S*	P*	B*
JG61	0.023	2.92	17.9	17.5	0.06	0.08	5.39	0.99	0.57	4	220	77
JG60	0.022	2.90	18.0	17.4	0.08	0.09	5,36	0.96	0.57	4	60	40
EU03	0.026	2.93	18.1	17.1	0.06	0.10	5.44	0.97	0.49	4	80	37
EW02	0.023	2.90	18.1	17.2	0.08	0.11	5.37	0.94	0.49	5	50	41

* Contents of S, P and B are in ppm, others weight percent.

Table II. EDS Results of Microsegregation Analysis in Homogenized Ingot and Forged Billet

Alloy (Heat No.)	Nb Content in Homogenized Ingot			Nb Content in 203mm Rd Forged Billet		
	Interdendritic Area	Dendritic Arm	Nb Ratio	"Light" Bands	"Dark" Bands	Nb Ratio
718-ER (JG61)	5.77 +0.53/-0.32	4.18 +0.27/-0.19	1.380	4.77 +0.52/-0.59	4.51 +0.09/-0.11	1.058
718 (JG60)	–	–	–	4.98 +0.37/-0.86	4.62 +0.71/-0.71	1.078
718 (EU03)	5.22 +0.39/-0.46	4.39 +0.33/-0.41	1.189	–	–	–

Average of 12 independent measurements, weight percent.

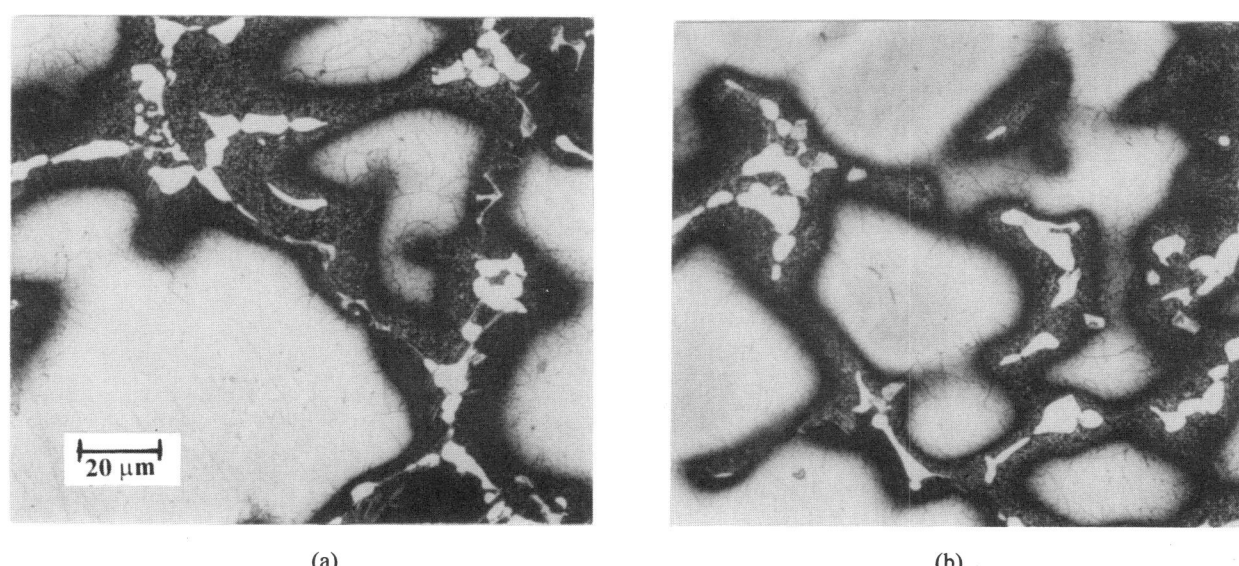

(a)   (b)

Figure 1. Cast Ingot Microstructure. (a) 718, EU03, (b) 718-ER®, JG61.

Table III. Results of Rapid Strain Rate Hot Tensile Tests of Ingot and Billet

	Hot Ductility (%)	Alloys	Test Temperature (°C)					
			927	1038	1093	1149	1177	1204
As-cast Ingot	EL	718 (EU03)	–	86.3	–	81.3	–	–
		718-ER (JG61)	–	70.7	–	6.5	–	–
	RA	718 (EU03)	–	95.9	–	80.3	–	–
		718-ER (JG61)	–	97.6	–	8.1	–	–
Homo-genized Ingot	EL	718 (EU03)	14.5	23.7	18.0	44.3	40.5	21.4
		718-ER (JG61)	28.8	37.8	67.9	67.5	54.8	1.4
	RA	718 (EU03)	44.6	72.1	88.2	93.7	84.1	82.0
		718-ER (JG61)	34.7	68.7	94.4	86.2	82.4	1.2
203mm Rd Billet	EL	718 (EU03)	36.9	50.7	47.3	52.0	53.7	12.3
		718-ER (JG61)	75.1	93.0	112.0	20.0	6.6	–
	RA	718 (EU03)	98.2	99.3	98.7	98.9	93.5	21
		718-ER (JG61)	98.0	98.9	98.9	20.0	0.8	–

Notes: 1. EL and RA are the elongation and reduction in area, respectively.
        2. Each data is the average value of two independent measurements.

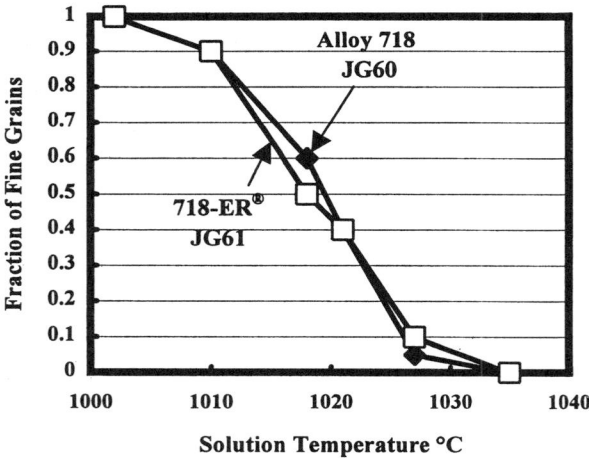

Fig. 2. Fraction of Fine Grains as Function of Solution Temperature in Progressive Solution Treatment.

Differential thermal analysis (DTA) showed that P and B in the range used had a minor effect on incipient melting (Laves eutectic) temperature of as-cast alloys. Only about a 6°C drop in temperature was seen in 718-ER (1143-1198°C to 1137-1188°C). There was almost no difference in incipient temperature of the wrought alloys (~1214°C). This seems somewhat inconsistent with the RSRHT test results.

Mechanical Properties

The mechanical properties of the various products tested are listed in Tables IV and V. Tensile properties within the three products tested were not significantly different for the two alloys. While ductilities were slightly lower at room temperature for 718-ER, they may be slightly higher at 649°C. The slightly higher strength for 718-ER billet most likely reflects a slight difference in structure.

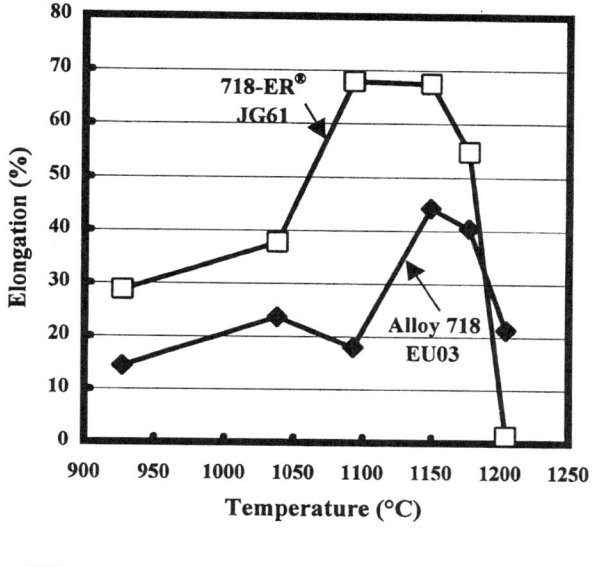

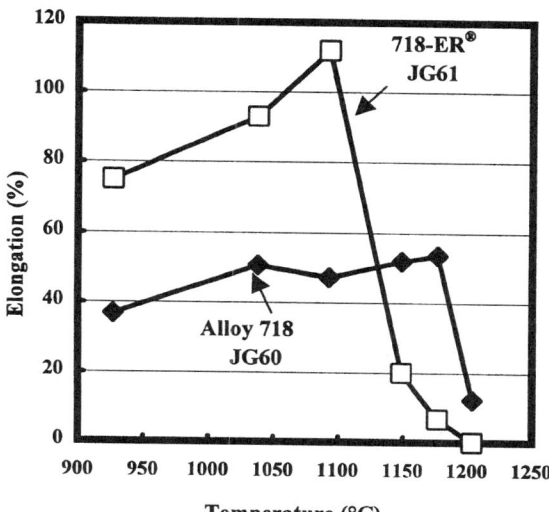

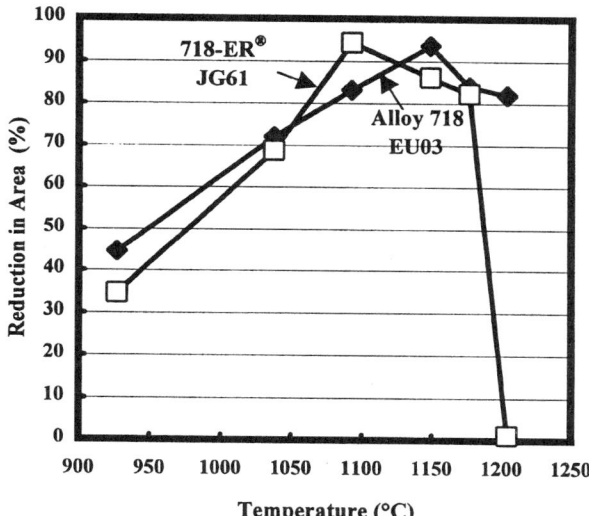

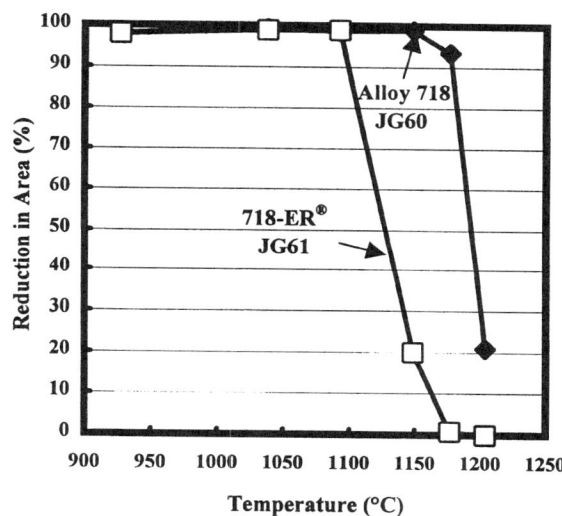

Fig. 3. Hot Ductility of Homogenized Ingots as Function of Test Temperature.

Fig. 4. Hot Ductility of 203mm Billets as a Function of Test Temperature.

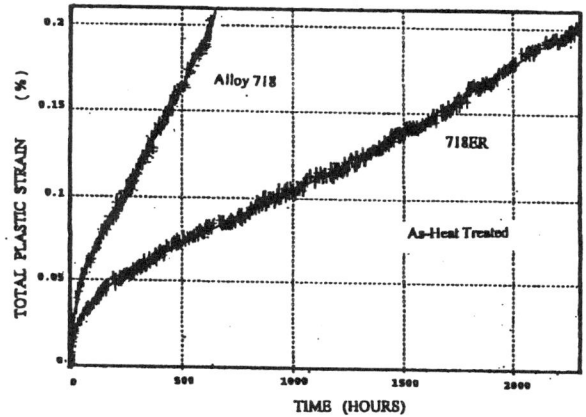

Fig. 5. Creep Curves of 718 (JG60) and 718-ER® (JG61) Pancake Samples, 621°C/621 MPa.

Table IV. Tensile Properties of Billets, Pancakes and Rolled Bars

Product Form	Alloy	Room Temperature Tensile				649°C Tensile			
		UTS MPa	YS MPa	EL %	RA %	UTS MPa	YS MPa	EL %	RA %
203mm Rd[(1)] Billet	718 (JG60)	1367	1134	20.8	40.2	1112	934	24.9	36.4
	718-ER (JG61)	1406	1237	17.7	38.0	1142	1038	23.8	43.0
Pancake[(2)]	718 (JG60)	1427	1108	18.8	28.0	1133	939	20.8	28.3
	718-ER (JG61)	1428	1096	16.5	19.7	1140	940	21.6	34.5
16mm Rd Bar	718 (EW02)	1459	1202	20.2	40.6	1174	1011	17.6	27.0
	718-ER (JG61)	1465	1216	19.6	38.5	1192	1004	23.9	57.9

All data points average of two tests:
(1) Mid-radius longitudinal orientation
(2) Transverse orientation

Table V. Stress Rupture and Creep Properties of Billet, Pancakes and Rolled Bar

Product Form	Alloy	Stress Rupture Test				Creep Test			
		649°C/759 MPa		649°C/690 MPa		621°C/621 MPa		649°C/621 MPa	
		Life (hrs)	EL (%)	Life (hrs)	EL (%)	Life to 0.2% (hrs)	Creep Rate (x 10^{-10})	Life to 0.2% (hrs)	Creep Rate (x 10^{-10})
203mm Rd[(1)] Billet	718 (JG60)	91	12.6	274	13.8	>2000	1.21	448	10.5
	718-ER (JG61)	166	11.2	528	17.2	>2000	0.67	824	5.93
Pancake[(2)]	718 (JG60)	54.8	22.5	158	22.5	866	5.51	57.5	73.6
	718-ER (JG61)	116.2	19.0	246	19.7	1889	2.50	169	28.4
16mm Rd Bar	718 (EW02)	96	19.0	233	18.5	–	–	–	–
	718-ER (JG61)	153	21.0	518	19.5	–	–	–	–

All data points average of two tests:
(1) Mid-radius longitudinal orientation
(2) Transverse orientation

As expected from previous work, there were substantial increases in stress rupture and creep performance for 718-ER compared to standard 718. Improvements ranged from 60% to 160%. All of these results agree very well with 718-ER alloys of similar P and B levels tested in previous studies[1-4]. Differences in rupture and creep properties for the different products tested (billet, pancake & bar) reflect differences in thermomechanical processing and structure (see microstructure and fractography section).

Figure 5 shows creep curves of standard 718 and 718-ER tested at 621°C/621 MPa. 718-ER was still in secondary creep well beyond 2000 hours, but 718 started tertiary creep at about 700 hours. The secondary creep rate of 718-ER was approximately 50% that of standard 718.

Microstructure and Fractography

The general features of billet microstructures of the two alloys were very similar, as shown by optical metallography and SEM studies, with two exceptions (Figure 6). The 718-ER billet had a slightly finer average grain size (JG61, ASTM 6 vs. JG60, ASTM 5), which is consistent with differences in tensile strength, and it appeared that the billet of JG61 had a more pronounced dark and light banding structure. The grain size in the light bands in JG61 was slightly finer and also contained more NbC particles. This indicates that JG61 had more residual microsegregation although EDS and the progressive solution treatment tests were unable to reveal it.

The grain size of pancakes was significantly refined in comparison to the billet grain size (ASTM 10-11 vs. 5-6). This finer grain size explains the lower stress rupture and creep life in pancake samples in comparison with billet samples.

The fracture surfaces of broken stress rupture samples of billet were analyzed, and their SEM photos are shown in Figures 7 (a) and (b). The fracture mode in 718-ER was completely dimple fracture, while some intergranular fracture facets can be seen in alloy 718. The intergranular failure in the standard alloy is clearly seen in the cross section of a broken rupture sample (Figure 7 (c)). This result is in agreement with previous work[4] and clearly illustrates that P and B additions increase grain boundary cohesion, and in turn the stress rupture resistance of 718-type alloys.

Discussion

These results clearly demonstrate the beneficial effect of P and B for stress rupture and creep properties of alloy 718 made on a full production scale. The improvements in stress rupture and creep properties seen in the 718-ER production heat were comparable with those reported on small-scale pilot plant alloys with similar P and B levels. On average, about 100% improvement in stress rupture life and creep resistance was achieved in production alloys of this study. As previously reported[4], pilot plant heats with similar P, B and C levels (0.022%P, 0.008%B and 0.025%C) showed similar increases in stress rupture life. Based on this work, it is reasonable to believe that further improvements in rupture and creep properties could be achieved in production scale heats by changing chemistry to the optimum levels for P, B and C (0.022%P, 0.011%B, <0.01%C).

Microstructural results suggest that P and B additions increase the amount of microsegregation, although the effect could not be measured by EDS or grain growth studies. This would suggest that an improved homogenization practice may be required for 718-ER.

Standard 718 processing conditions were used for all materials in this study and no difficulties were encountered with any of the resulting products. However, RSRHT testing clearly showed that P and B reduced the hot ductility of 718 at peak temperatures but may improve it at lower temperatures. The reduction in hot ductility was at temperatures (>1149°C) above the normal processing temperatures for 718. Nonetheless, this clearly points out the necessity of avoiding processing 718-ER at excessive temperatures. This effect most likely results from grain boundary segregation rather than liquation since the critical temperatures were much lower than the incipient melting temperature determined by DTA. Thus P and B segregation to grain boundaries may increase the grain boundary strength at lower temperature, but decrease it at higher temperature.

Conclusions

1. The beneficial effects of P and B additions on stress rupture and creep properties of 718, previously reported for pilot plant size heats, is fully scaleable to production size heats.

2. Stress rupture and creep properties observed in a production heat of 718-ER were on the order of 100% greater than for standard 718. Results from the production scale heat were in good agreement with data from small scale pilot plant heats of a comparable chemistry.

3. A full scale production heat of 718-ER was successfully processed to billet and bar using standard 718 thermomechanical processing conditions. Results show P and B have a detrimental effect on the hot ductility of 718, but only at very high temperatures, above those normally employed for hot working the alloy.

4. A higher degree of microsegregation was observed in 718-ER suggesting the need for improvement in homogenization practice.

References

1. W.D. Cao and R.L. Kennedy, Superalloys 718, 625, 706 and Various Derivatives, ed. E.A. Loria, TMS, 1994, p. 463.

2. W.D. Cao and R.L. Kennedy, Superalloys 1996, eds. R.D. Kissinger et al., TMS, 1996, p. 589.

3. R.L. Kennedy, W.D. Cao and W.M. Thomas, Advanced Materials & Processes, Vol. 150, No. 3, 1996, p. 33.

4. W.D. Cao and R.L. Kennedy, Superalloys 718, 625, 706 and Various Derivatives, ed. E.A. Loria, TMS, 1997, p. 511.

5. X.S. Xie et al., Superalloys 1996, eds. R.D. Kissinger et al., TMS, 1996, p. 599.

6. W.C. Yu et al., Superalloys 718, 625, 706 and Various Derivatives, ed. E.A. Loria,, TMS, 1997, p. 543.

7. W.R. Sun et al., Mater. Sci. Engr., Vol. 247 A, 1998, p. 173.

8. X.B. Liu et al., Mater. Sci. Engr., Vol. 270 A, 1999, p. 190.

9. Y.X. Zhu et al., Superalloys 1988, eds. S. Reichman et al., TMS, 1988, p. 703.

10. W.D. Cao and R.L. Kennedy. "Effect of Minor Elements on Microsegregation of Alloy 718," paper presented at 123rd TMS Annual Meeting, San Francisco, February, 1993.

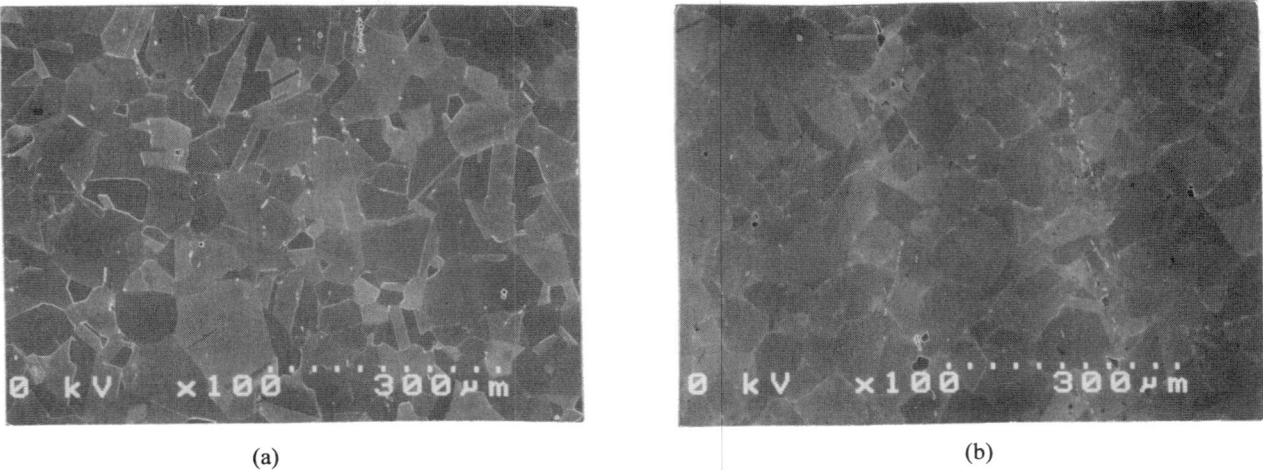

(a) (b)

Fig. 6. Microstructure of 203mm Rd Billets. (a) Alloy 718, JG60 and (b) 718-ER®, JG61.

(a) 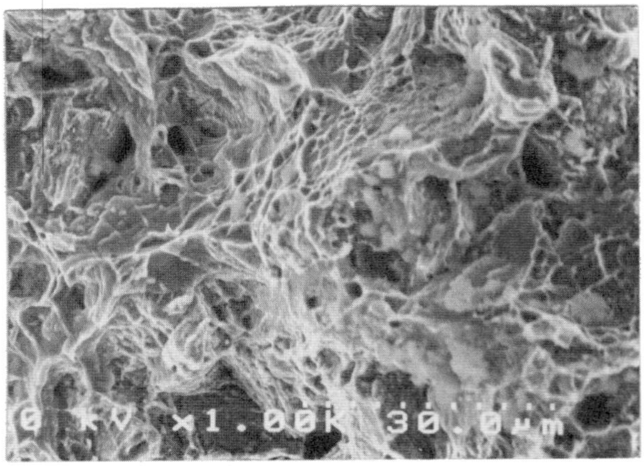 (b)

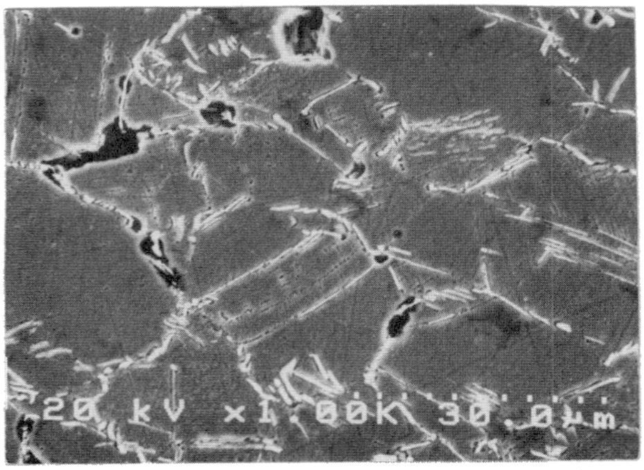

(c) 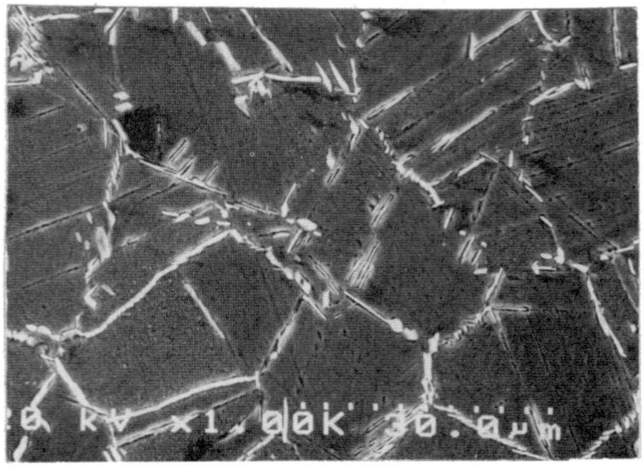 (d)

Fig. 7. SEM Photos of Fracture Surface and Microstructure of broken Stress Rupture Samples. (a), (c) Alloy 718, JG60, (b), (d) 718-ER®, JG61.

QUENCH CRACKING CHARACTERIZATION OF SUPERALLOYS USING FRACTURE MECHANICS APPROACH

Jian Mao[*], Keh-Minn Chang[*] and David Furrer[**]
* West Virginia University, Morgantown, WV 26506
**Ladish Co., Inc., Cudahy, WI 53110

Abstract

In the expectation of developing a new quench cracking criterion, an approach based on fracture mechanics has been investigated in order to improve computer modeling of quenching process. A fully automatic computer controlled data acquisition and processing system was set up to simulate the quenching process. Based on the study of the quench cracking resistance of several gamma prime strengthened superalloys, including U720Li, Rene'95, Rene88DT and HW model alloys, the mechanism of quench cracking was studied. Effects of the grain size, solution temperature, the composition of alloy as well as specimen size on quench cracking resistance have been investigated. Results show that quench cracking is featured with intergranular failure. The quench cracking toughness and the failing temperature is related to the quench fracture mode. There is a transition in fracture modes with the decrease of temperature. Grain size, composition and solution temperature were found to be the three major factors influencing quench cracking resistance. Fine grain structure can sustain more temperature drop and requires higher thermal stress to initiate the cracking. Intermediate grain structure, which was heated at either supersolvus or subsolvus temperatures, failed at higher temperature and hence developed lower quench toughness. γ' content in alloys only influences the cracking resistance of subsolvus quenched specimen, but does not show significant influence on that of supersolvus quenched one.

Introduction

PM superalloys and their application on turbine disk have matured since the 1970's. With the development of computer simulation technology, modeling of thermal and stress fields has further enabled improved process control in powder consolidation, forging, heat treatment and other processes [1-3]. Precise prediction of microstructure and mechanical properties has shown bright perspective [4-7]. However, the risk of quench cracking and distortion due to aggressive thermal stress still remains a concern in the superalloy community. Unexpected quench cracking on the practical disks usually results in a huge waste of money.

In the past few years, efforts in computer simulation of heat treatment have been focusing both on acquiring high strength in quenched parts and on avoiding quench cracking and severe distortion [8-11]. So far, the calculation of cooling rates and the prediction of mechanical properties in the quenching part are expanding. Comparing to practice in real situation, however, the simulation sometimes fails to produce reliable predictions on quench cracking. The problem is that the present computer simulation uses the cracking criterion that is based on strength consideration, i.e., if the thermal stress in a quenching part exceeds the strength limit, cracking is predicted to occur. No consideration about the surface defects on heat treated parts is included. In fact, several kinds of defects are unavoidably existent on the surface. These include, machining scratches, oxidation scales, inclusions, micro-cracks and other discontinuities. Therefore, modeling based on strength criterion cannot treat the scatter due to the variations in surface defects [12]. Moreover, an alloy with the highest strength actually shows the lowest cracking resistance [13].

Taking into account these surface defects and the weakness of the strength criterion, a new approach based on fracture mechanics has been investigated in the expectation of developing a new quench cracking criterion [13-15]. Based on the study of the quench cracking resistance of several gamma prime strengthened superalloys, including U720Li, Rene'95, Rene88DT and HW model alloys, the mechanism of quench cracking has been studied. Effects of grain size, solid solution temperature, composition of alloys as well as specimen size on quench cracking resistance have also been examined.

Superalloys 2000
Edited by T.M. Pollock, R.D. Kissinger, R.R. Bowman,
K.A. Green, M. McLean, S. Olson, and J.J. Schirra
TMS (The Minerals, Metals & Materials Society), 2000

Table I Composition of the superalloys, wt%

Alloy	C	W	Mo	Nb	Zr	Co	Al	Cr	Ti	B	Ni	γ'% at 650°C	Tγ's
PMU720Li	0.025	1.3	3.02		0.035	14.75	2.46	16.35	4.99	0.017	Bal	46.4	1150
CWU720Li	0.013	1.18	2.85		0.03	14.45	2.48	16.14	5.15	0.014	Bal	47.3	1161
Rene88DT	0.049	3.88	4.00	0.70	0.043	12.99	1.99	15.67	3.72	0.016	Bal	38.6	1100
Rene'95*	0.06	3.5	3.5	3.5	0.05	8.0	3.5	13.0	2.5	0.015	Bal	54.5	1150
HW1**			4.0		0.05	10.0	1.5	15.0	3.0	0.01	Bal	26.1	990
HW3**			4.0		0.05	10.0	2.0	15.0	4.0	0.01	Bal	36.7	1080

*Data is from Ref.12
**Nominal composition

Materials and Experimental Procedure

The materials studied were U720Li, Rene'95, Rene88DT and HW model alloys. Bulk composition of these alloys is listed in Table 1. PM U720Li, Rene88DT and Rene'95 are commercial powder metallurgy superalloys made by standard production procedure, HIP + extrusion + isothermal forging. Cast and wrought U720Li alloys were also investigated. HW alloys are cast & wrought model alloys designed for the study of γ' content effect. All these alloys are strengthened by the precipitation of γ'. The amount of γ' in alloys varies from 25% to 55%. γ' content and the γ' solvus temperature for those alloys were calculated by using Thermo-Calc software, vision M, NiFe database 3.0 with 13 elements. This calculated temperature was used to establish subsolvus and supersolvus treatments.

The tests were carried out on a servo-control MTS machine. A fully automated computer control and data acquisition system was set up for the machine. The machine is capable of heating and cooling the specimen accurately at designated rates because of the infrared radiant heating equipped furnace. The specimen has a dog-bone shape with a rectangular cross section of 6.35mm(0.25 inch) by 1.27mm(0.05 inch), and a gauge length is 44.45mm(1.75 inch). A single edge notch was made by EDM. The notch depth is 0.635mm(0.025 inches). Precracking was carried out at room temperature to avoid the influence of the starting notch geometry. Most of the precrack lengths were controlled to meet: $0.45 < a/w < 0.55$ (w: width of specimen, a: crack length) in this study.

During the quenching test, each specimen was held on the MTS machine and heated up to a designated solution temperature. After holding at this temperature for one minute, the quench process was activated with the setting of the grips at displacement control mode. Thermal load was then applied instantly due to the contraction of the specimen. When the on-cooling thermal stress (load) accumulated to a certain point, quench cracking happened abruptly. Based on the precrack length, the thermal stress was converted to toughness K according to the Tada's empirical equation, which is stated below, [16] and then the on-cooling K-T curve can be plotted. On the K-T curve, the temperature at which maximum K was reached was defined as a failing temperature (T_f). The value of K at this point was defined as quench cracking toughness K_Q.

$$K = \frac{P}{B\sqrt{W}} \frac{\sqrt{2\tan\frac{\pi a}{2w}}}{\cos\frac{\pi a}{2w}} \left[0.752 + 2.02\left(\frac{a}{w}\right) + 0.37\left(1 - \sin\frac{\pi a}{2w}\right)^3\right]$$

Where:
a — measured precrack length
W — width of specimen
B — thickness of specimen
P — maximum load while cracking

Metallographic samples were taken from broken quench specimens in the area adjacent to the fracture surface parallel to the stress axis. Chemical etching was used to study grain structures. The chemical etchant used is: $50\%HCL + 50\%H_2O + 0.5\%H_2O_2$. Electrolytic-etch was used to reveal the γ' morphology. This etchant is Chromium Acid Solution ($170ml H_3PO_4 + 16g CrO_3 + 10ml H_2SO_4$). Fracture morphology of broken specimens and precipitate distribution were observed using scanning electron microcopy (SEM).

Result and discussion

Simulation of the quench cracking process

Quench cracking is a type of abrupt failure. Figure 1 shows the scheme of a quench process and the plot of the cooling curve. Simulation of the process indicates that with the temperature decrease during cooling, the thermal load increases rapidly to a maximum value, then the catastrophic failure happens and the specimen is broken into two pieces.

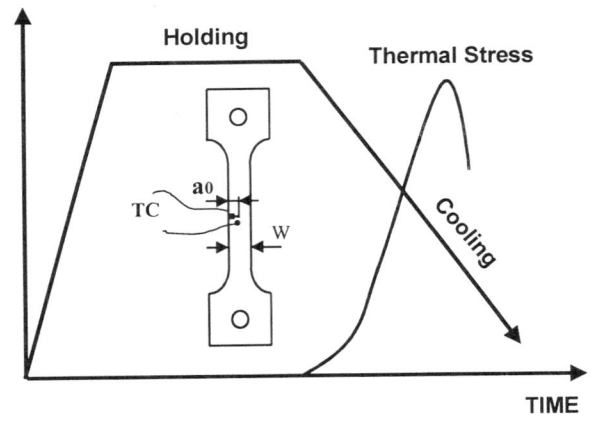

Figure 1: Scheme of quench process and plotting of on-cooling curve

Examination of the fracture surface indicates that quench cracking is characterized with a typical brittle intergranular fracture mode, as shown in Figure 2, which is consistent with the quench cracking failure mode in actual disks. This intergranular fracture feature differs from the precracking one, as shown on the right hand side of the Fig.2, which is typically transgranular. It is also different from that observed in the tensile or stress rupture tests at elevated temperatures, which is usually an intergranular dimple mode for most superalloys. This demonstrates that the quench simulating system based on the new fracture mechanics approach can well represent the real quench cracking process.

Quench cracking toughness and cracking mode

The measurement of the cracking toughness of alloys demonstrates that there is a relationship between the quench cracking toughness and the failing temperature. Figure 3 shows the results obtained from PMU720Li and CWU720Li alloys. Generally speaking, if the cracking occurs at a higher temperature, above 800°C, lower cracking toughness is obtained. In contrast, if material fails at a lower temperature, it usually results in higher cracking toughness. This relationship is associated with the quench fracture mode.

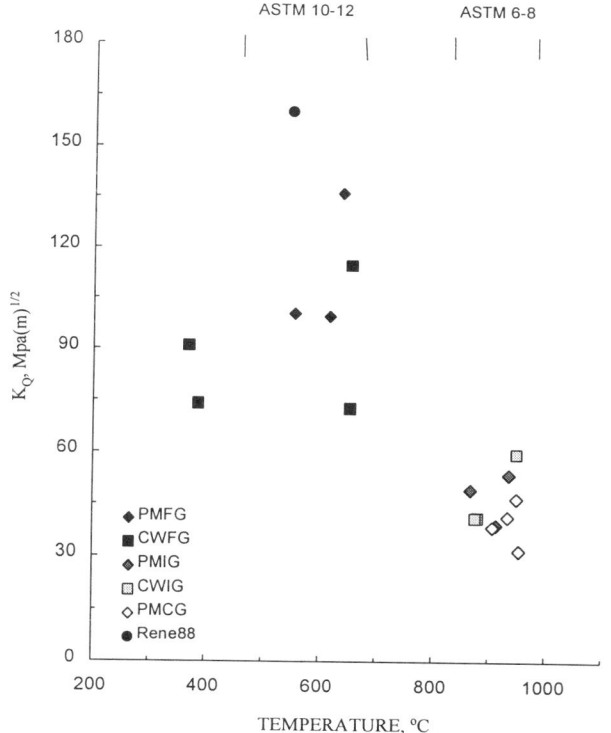

Figure 3: Relationship of failing temperature and quench cracking toughness

As mentioned above, intergranular fracture feature is a predominant cracking mode observed in quench cracking. This type of failure usually happens at a higher temperature ($T_f > 800°C$) and is featured with lower quench toughness ($K_Q = 40-60MPa\sqrt{m}$). In addition to that, two more types of cracking features were observed during the SEM examination. One is that the quench cracking initiates intergranularly and then transfers to a transgranular mode, as shown in Figure 4. In this case, the failing temperature drops down to about 500°C-700°C, cracking toughness is raised up to more than 70MPa√m. The third mode is completely transgranular, as shown in Figure 5, which looks like what we see in the tensile test. The failing temperature, in this case, is lower than 500°C, where the quench cracking toughness is observed to be even higher.

Fractography examinations indicate that there is a transition in fracture modes with the decrease of temperature during the quenching test. This temperature at which fracture mode switches bears great importance for the quench process control and is worth further investigation.

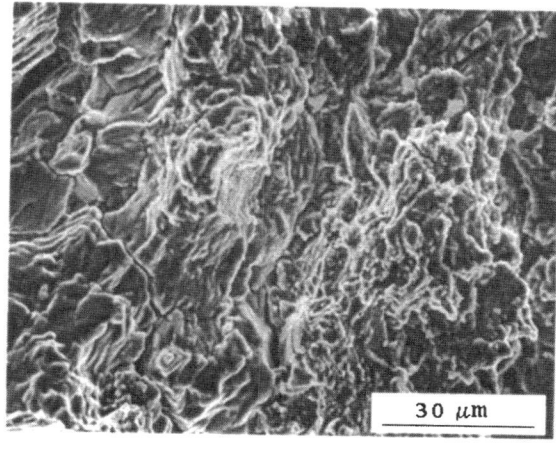

Figure 5. Completely transgranular failure in HW1 alloy, Solution temperature 1000°C, Cooling rate 600°C/min

Characterization of quench behaviors

As a characteristic parameter, quench cracking resistance varies with the microstructure, composition of alloys and solution temperature as well as cooling rate. Their influence is discussed respectively below.

Grain size effect The influence of grain size is accomplished through the deformation behavior and fracture mode at high temperatures. Figure 6 shows grain structures of PM U720Li and cast & wrought U720Li superalloys, where FG represents the fine grain structure with a grain size ASTM11-12. IG, CG represents a similar grain structure with a grain size ASTM 7-8, both obtained by supersolvus treatment before our quench tests. The difference is that CG was designated for a supersolvus solution in our quench test and IG was for a subsolvus solution. Grain size effects are demonstrated by comparing different structures from the same alloy, as shown in Figure 3. Specimens with fine grain structure tend to fail at a lower temperature and were characterized mainly with the II or III type fracture feature. As a result, higher quench cracking toughness is obtained. That indicates that fine grain structure can sustain more temperature drop and requires higher thermal stress to initiate the cracking. Specimens with coarse grain structures, no matter what temperature they were heated to, either supersolvus or subsolvus, all failed at higher temperatures and had the lower cracking toughness. The effect of the grain size on the quench crack toughness is related to the stress relaxation

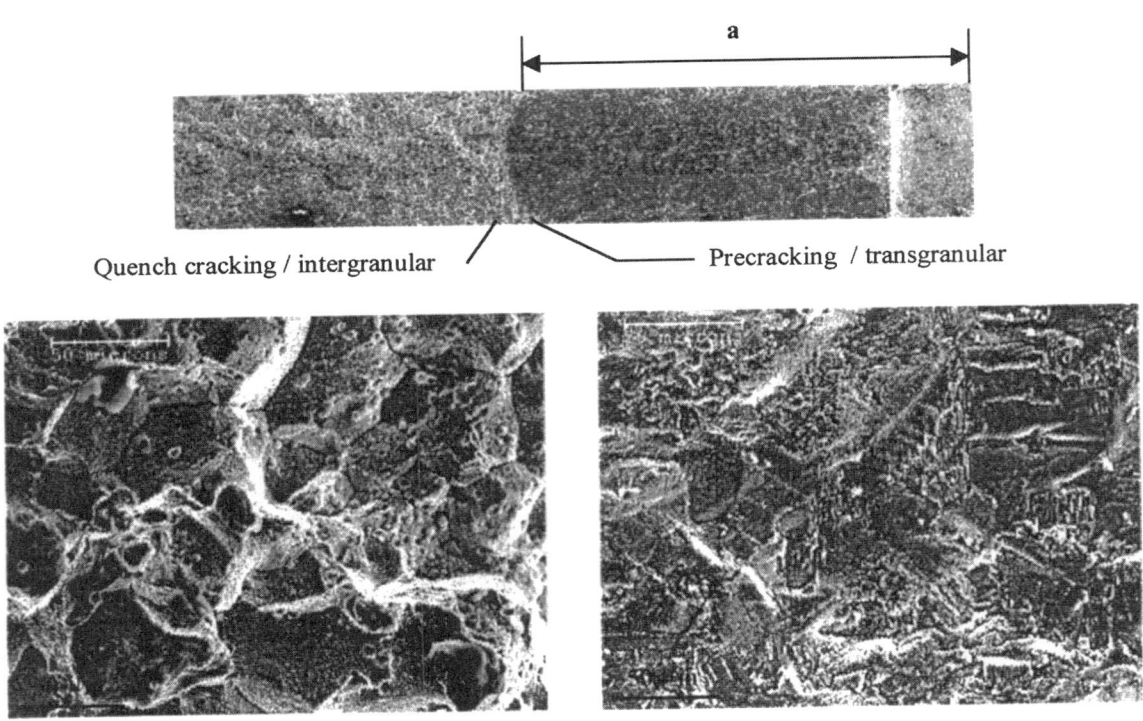

Figure 2: Typical fracture morphology observed in quench cracked specimen with coarse grain structure

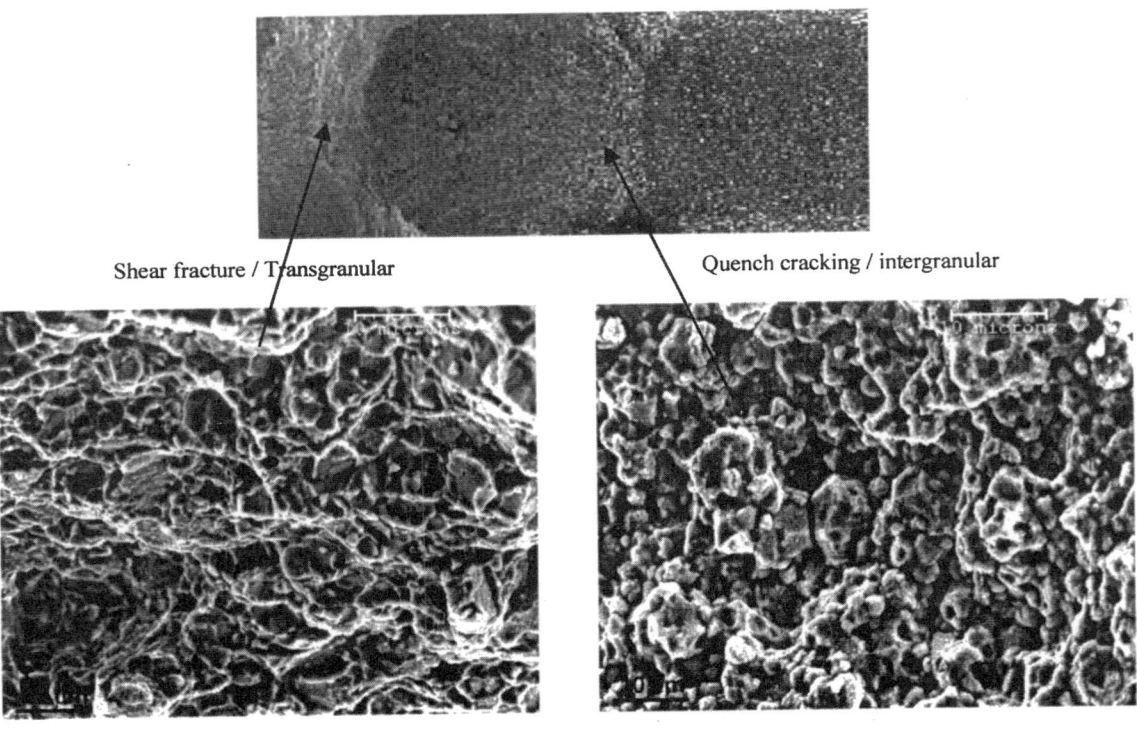

Figure 4: Transition of fracture mode from intergranular to transgranular in fine grain structure

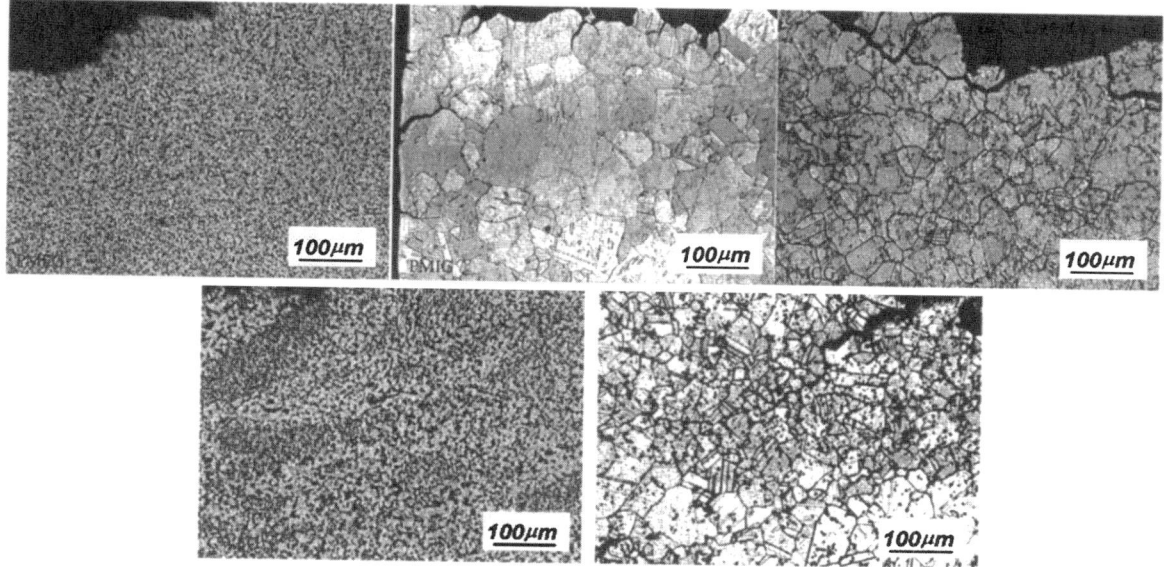

Figure 6 Grain structure of PM and CW Udimet 720Li superalloy after quenching test

capability of the structure in response to the thermal stress.

The topographic factor of crack propagation also contributes to the difference. In addition, a fractographic examination reveals that there are many secondary cracks found in the specimen with the fine grain structure. The development of abundant secondary cracks also contributes to a higher fracture toughness. It is believed that the value of the cracking toughness for different microstructures may vary with different alloys, but the grain size effect might always exist.

Solution temperature effect Solution temperature is another key factor to be considered in the study of the quench cracking behavior. Supersolvus solution provides alloys with preferred fatigue crack growth resistance [17-20]. Therefore, supersolvus heat treatment is favored for second or third generation PM superalloys. However, the quench cracking resistance under the supersolvus condition is inferior. Figure 7 shows a typical on-cooling K-T curve for supersolvus temperature quenching. All the specimens for the studying of solution temperatures and composition have almost the same grain size, ASTM 7-8. Supersolvus quenching carries lower quench cracking toughness than subsolvus treatment and usually fails abruptly at higher temperatures, above 800°C. The fracture feature is of an intergranular mode. Moreover, the quench cracking resistance seems to be somewhat alloy independent. All the specimens demonstrate the same cracking behavior regardless of alloy composition: cracking toughness is between 40-60 MPa√m, failing temperature above 800°C and each featured intergranular failure. It is believed that the normalization of microstructure and the weakening of the grain boundary at elevated temperatures plays the main role in composition independence for supersolvus quenching.

The subsolvus quench shows not only higher cracking toughness but also dependence on the composition of alloys.

Figure 8 shows some typical K-T curves for subsolvus quenching. The value of the quench cracking toughness varies from 30Mpa√m to 120Mpa√m with different alloys having almost the same grain structure and grain size ASTM 7-8. Solution temperature effect is believed to be associated with the balance of cracking resistance between the grain boundary and the intragranular area.

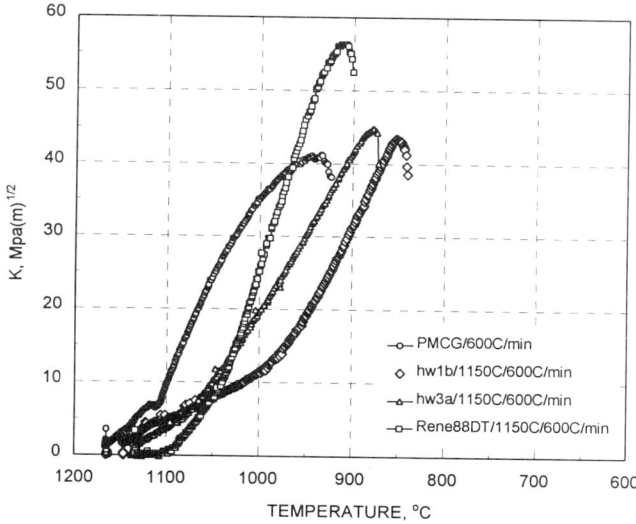

Figure 7: Typical on-cooling K-T curve for supersolvus temperature quenching

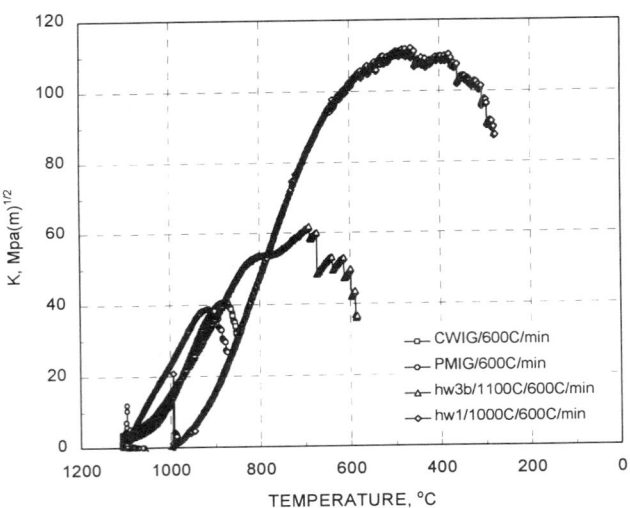

Figure 8 Typical on-cooling curve for subsolvus temperature quenching

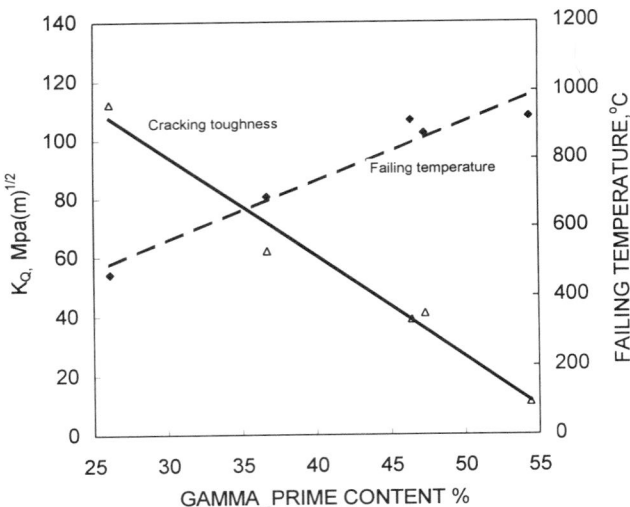

Figure 10: Effect of γ' content on quench cracking toughness and failing temperature

Composition effect Several superalloys are chosen in order to study the composition effect on quenching behavior. Composition differences cause a different volume fraction of γ' phase in the alloys. The volume fraction of gamma prime in the alloys was simulated by using Thermo-Calc software. The volume fraction of gamma prime is taken as the amount at 650°C according to the calculation, listed in Table I. Figure 9 displays the gamma prime solubility with the temperature in various superalloys. The HW1 is the alloy with lowest gamma prime content. The volume fraction of gamma prime is about 27%. Rene'95 has the highest gamma prime content, at about 56%. As we mentioned above, the alloy with different gamma prime content does not show a difference on cracking toughness during a supersolvus treatment. But during a subsolvus treatment, the effect of the gamma prime content and solubility of alloys on cracking behaviors are obvious. Figure 10 shows the relationship of γ' content and the quench cracking resistance as well as failing temperature. Data of Rene'95 alloy was taken from ref.13. If an alloy is highly alloyed

to obtain a higher γ' fraction, it will more likely exhibit a higher failing temperature and lower quench cracking toughness. A less alloyed alloy, having low γ' fraction, tends to fail at lower temperature and shows higher quench cracking toughness.

Specimen size effect Besides the influence of the above factors on quench cracking behaviors, the size of the specimen is another consideration. A narrow specimen with a cross section of 3.175mm(0.125") × 1.27mm(0.05"), which is half of the width of normal specimen, is used to study the specimen size effects. Alloys investigated are the same as used in the grain size study. As we can see in Figure 11, decreasing the width of the specimen usually lowers the cracking toughness, especially for fine grain structure. One of the reasons for this is that narrow specimen mean a shorter effective length for crack propagation and a limited accumulation of thermal stress. Before K can reach higher values, failure of the specimens already occurs. Another reason is that narrow specimen results in more stress relaxation. So it's hard to reach higher stress. The investigation of the specimen size effect is still underway.

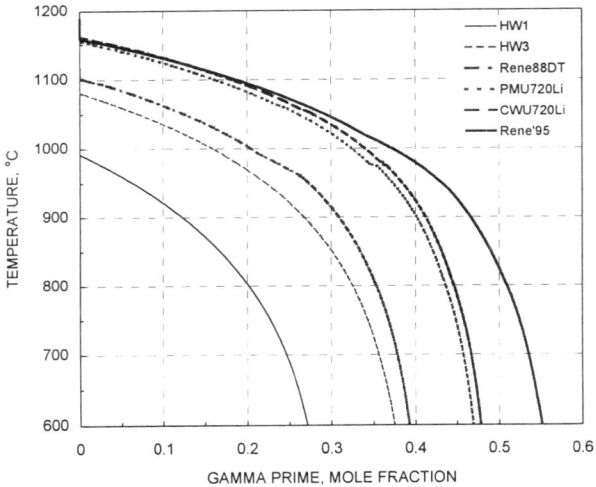

Figure 9 Gamma prime solubility with temperature in superalloys

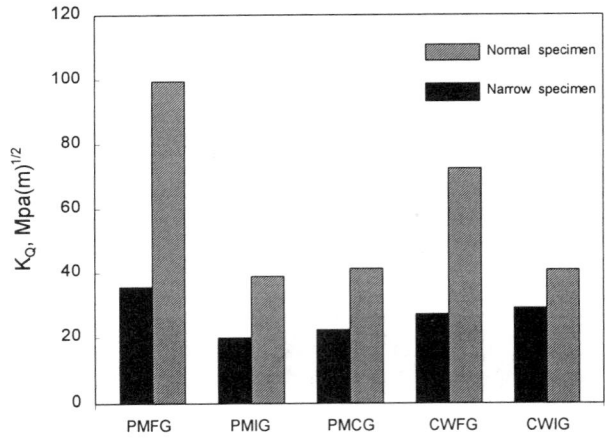

Figure. 11 Effect of the specimen size on cracking toughness
Discussion

114

Microstructure and deformation

Quench cracking behavior at elevated temperatures is believed to be associated with the deformation behavior of materials. Softening and hardening are considered as two typical processes during high temperature deformation. Figure 12 summarizes schematically the development of the on-cooling thermal load on specimens for various test conditions. At the very beginning of quenching, the rate of thermal stress increase is slow. The curvatures of thermal load vs temperature (L-T) curves are negative, which means that softening dominates the deformation process. With further decrease in temperature, the curvatures of L-T curves change from negative to positive, the hardening process gradually outweighs the softening process until cracking.

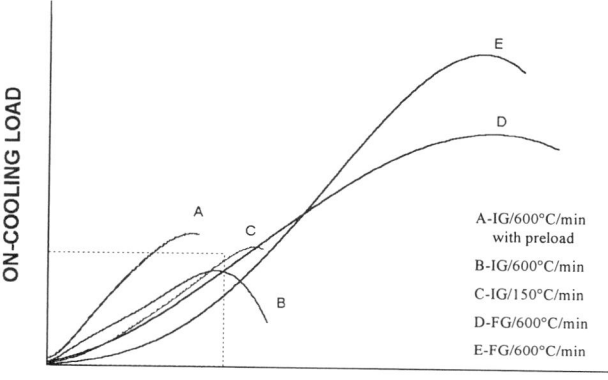

Figure 12: Scheme of the development of on-cooling thermal load

The curvature of L-T curves differs from fine grain structure to coarse grain structure. For all specimens, the curvatures on the L-T chart increase at beginning as the cooling rates decrease. As we know, fine grain is more flexible and compliant under stressful conditions. Therefore, softening in the fine grain structure is more severe than that in the coarse grain structure under the same cooling rate for the same material (see inside dashed line area in Figure 12). Furthermore, because the softening process consumes a lot of energy at the beginning of quenching, and the hardening process is postponed, so the specimen with fine grain structure can survive through the fracture mode transition temperature. The transition of fracture modes allows these specimens to resist more thermal stress before reaching a critical point. That is one of the reasons why fine grain structure shows higher cracking toughness and lower failing temperature for the same material.

The topographical factor of the crack path in the fine grain structure is another beneficial factor that brings in higher cracking toughness [13]. Once a crack initiates, the size and direction of grain features influence its propagation along the grain boundary. Therefore the fine grain structure can absorb much more cracking energy, taking more time to reach the critical fracture limitation before a catastrophic failure. In fact, when the temperature drops sufficiently, intergranular cracking is actually depressed by the precipitation of strengthening phases and the transition of the fracture mode takes place. Secondary cracks also appear to have some beneficial effects on energy release before final failure.

Lower cracking toughness of the coarse grain structure is due to its lack of compliance. Because of this, accumulation of thermal stress is faster than in the fine grain structure. As a result, hardening process dominates most of the cooling process. Before samples cool down to the transition temperature of failure modes, rapidly increasing thermal stress has already reached a critical value. On the other hand, because coarse grain provides less grain intersection obstruction to crack propagation and less grain boundaries to absorb cracking energy, once a crack initiates, unrestricted propagation along the grain boundary will continue to the end of the fracture path.

A lower cooling rate causes more softening at the beginning of the quenching process. Therefore, lower cooling rates result in higher quench cracking toughness in both fine and coarse grain structure samples.

If there are more grain boundaries, a lower cooling rate, a higher degree of softening, and more energy is released at the beginning of quenching, then there is more potential for the alloy to survive through the fracture mode transition temperature. A higher quench cracking toughness would be expected for this material. Grain size effect and brittle / ductile fracture transition was also found in the investigation of Rene'95 alloy [13]. In that study, a transition temperature of 700°C-850°C was suggested. The delayed transfer of production components from the heat treat furnace to the quench bath can help to avoid quench cracking. This may be due to the postponement of the hardening process, which allows the components to survive the transition temperature of fracture modes.

Balance of strengthening in grain boundary and matrix

As indicated before, supersolvus treatments usually feature intergranular failure, which results in lower cracking toughness. The reason is related to the balance of cracking resistance between the grain boundary and the matrix at high temperatures. During supersolvus treatment, all the γ' forming elements dissolve into the matrix. The material is highly solution strengthened. In this case, the grain boundaries become relatively weak, so it becomes the preferred path for crack propagating [12]. Moreover, since no γ' precipitates along the grain boundaries, it facilitates the crack propagation along the grain boundary. In addition, grain boundary micro-cracks due to oxidation at high temperatures may be another considerable factor [12].

Subsolvus treatment and some strengthening phases, like γ', precipitate from the matrix along grain boundaries, which weaken the solid strengthening of the intragranular area, but strengthen the grain boundaries. If grain boundaries are strengthened enough, it is possible for the intragranular area to become potential crack propagation paths when intergranular crack propagation is obstructed. This is proven by the observation of the fractography of the specimen after a subsolvus treatment. Most of the broken specimens for the subsolvus solution, regardless of the composition of alloys, are found undergoing the transition in fracture modes. Furthermore, for subsolvus treatment, once the fracture mode is transferred to the transgranular mode, the amount of primary or secondary γ', the size of γ', the mismatch of lattice and carbide content etc. will affect the fracture strength of the alloy. The lower γ' content alloy has less γ' forming elements which results in lower solution strengthening. Therefore, higher cracking resistance is obtained, as seen in Fig.10. That is why composition effects are obvious for subsolvus treatment. Topographic effect of primary γ' on the grain boundary is another important reason to slow down and block the cracking growth along the grain boundary.

Conclusions

A new fracture mechanics approach provides a way to quantitatively evaluate the quench cracking resistance of superalloys. Grain size, composition, and solution temperature are found to be the three major factors influencing quench cracking resistance and quench behavior.

The fracture mode affects the quench cracking resistance of alloys. Complete intergranular failure is usually concomitant with a lower cracking toughness. If an alloy can survive the intergranular cracking mode, or be able to cool down below the transition temperature of cracking modes, a higher cracking resistance is expected.

The alloy with higher γ' content shows lower quench cracking resistance than alloy with a lower γ' volume fraction under the same test condition and similar grain structure.

Solution temperatures influence the quench cracking resistance of γ' strengthened superalloys. Supersolvus solutions more likely cause intergranular cracking. Subsolvus solutions increase the possibility of fracture mode transition and show relatively higher quench cracking toughness. The influence of solution temperatures is associated with the balance of the fracture resistance between the grain boundary and the intragranular area. The topographic factor of primary γ' on the grain boundary and the grain boundary oxidation resistance may also affect quench cracking resistance.

Reference:

[1] T. Furman and R. Shankar, "Keys to Modeling the Forging Process," Advanced Materials & Processes, 154 (3) (1998) 45-46

[2] C.P.Blankenship, Jr., M.F.Henry, J.M.Hyzak, et al., "Hot-Die Forging of P/M Ni-Base Superalloys," Superalloy 1996, ed. R.D.Kissinger et al. The Minerals, Metals and Materials Society (1996) 653-662

[3] D.J.Bryant, Dr.G.McIntosh, "The Manufacture and Evaluation of A Large Turbine Disk in Cast and Wrought Alloy 720Li," Superalloy 1996, ed. R.D.Kissinger et al. The Minerals, Metals and Materials Society (1996) 713-722

[4] N.Gayraud, F.Moret, X.Baillin and P.E.Mosser, "Precipitation Kinetics in N18 P/M Superalloy: Experimental Study and Numerical Modeling," Journal de Physique III, 3 (1993) 271-276

[5] J. J. Schirra and S. H.Goetschius, "Development of An Analytical Model Predicting Microstructure and Properties Resulting from the Thermal Processing of A Wrought Powder Nickel-Base Superalloy Component," Superalloys 1992, ed. S.D.Antolovich et al. The Minerals, Metals and Materials Society (1996) 437-446

[6] M.P.Jackson, R.C.Reed, "Heat Treatment of Udimet 720Li: the effect of microstructure on properties," Materials Science and Engineering, 259 (A) (1999), 85-97

[7] David Furrer and H.Fecht, "Ni-Base Superalloys for Turbine Discs," JOM, 1999, no 1: 33-36

[8] Ronald A. Wallis, "Modeling Quenching: the state of the art," Advanced Materials & Processes, 1995, no.9: 42kk-42NN

[9] R.A. Wallis and I.W.Craighead, "Predicting Residual Stresses in Gas Turbine Components," JOM, 1995, no. 10: 69-71

[10] R. I. Ramakrishman and T. E. Howson, "Modeling the heat Treatment of Superalloys," JOM,44 (6) (1992), 29-32

[11] D.Persampieri, "Process Modeling for Improved heat Treating," Advanced Materials & Processes, 1991, no:3: 19-23

[12] D. L. Klastrom, "Heat Treatment Cracking of Superalloys," Advanced Materials & Processes, 1996, no.4: 40EE-40HH

[13] K.M.Chang and B.Wu, "A New Approach to Evaluate the Quench Cracking Resistance of P/M Superalloy," 1st International Non-Ferrous Processing and Technology, ed T.Bains and D.S.Mackenzie, ASM International (1997) 477-481

[14] Valerie L. Keefer, "Fracture Mechanics Approach in Studying the Quench Cracking Behavior of the High Strength Superalloy Udimet 720," (MS thesis, West Virginia University, 1998)

[15] J.Mao, V.L.Keefer, K.M.Chang and D.Furrer, "An Investigation on Quench Cracking Behavior of Superalloy Udimet 720 Using Fracture Mechanics Approach," Journal of Materials and Engineering Performance, (accepted, 1999)

[16] H. Tada, P. Paris, and G. Irvin, The Stress Analysis of Cracks Handbook, (Del Research Corp. 1973), 2.10-2.12

[17] Keh-Minn Chang, "Critical Issues of PM Turbine Disks," Acta Metallurgica Sinica, 9 (6) (1996), 467-471

[18] R.D.Kissinger, "Cooling Path Dependent Behavior of a Supersolvus Heat Treated Nickel Base Superalloy," Superalloy 1996 ed. R.D.Kissinger et al. The Minerals, Metals and Materials Society (1996) 689-695

[19] S.T.Wlodek, M.Keely and D.A.Alden, "The Structure of Rene88DT," Superalloy 1996, ed. R.D.Kissinger et al. The Minerals, Metals and Materials Society (1996) 129-136

[20] H.Hattori, M.Takekawa, D.Furrer and R.J.Noel, "Evaluation of P/M U720 for Gas Turbine Engine Disk Application," Superalloy 1996, ed. R.D.Kissinger et al. The Minerals, Metals and Materials Society (1996) 705-711

DEVELOPMENT AND CHARACTERIZATION OF A DAMAGE TOLERANT MICROSTRUCTURE FOR A NICKEL BASE TURBINE DISC ALLOY

R.M. Kearsey *, A.K. Koul *, J.C. Beddoes **, and C. Cooper ***

* Structures, Materials, and Propulsion Laboratory, Institute for Aerospace Research, National Research Council of Canada, Ottawa, ON, Canada, K1A 0R6
** Department of Mechanical and Aerospace Engineering, Carleton University, Ottawa, ON, Canada, K1S 5B6
*** Chief of Materials, Pratt & Whitney Canada, Longueuil, Quebec, Canada, J4G 1A1

Abstract

A modified heat treatment has been developed for the nickel-base superalloy, PWA 1113, to create a damage tolerant microstructure (DTM) using mechanistic microstructural design concepts. The DTM was designed with the aim of imparting improved fatigue crack growth resistance without forfeiture of other vital properties such as tensile strength, stress rupture life, and low cycle fatigue lifetimes. This was achieved by optimizing the material's grain size, grain boundary morphology, and the intragranular precipitate size and distribution. Mechanical testing demonstrated that when compared to the conventional microstructure (CM), the short crack growth rate for the DTM was slower by a factor of 3 at room temperature, and 2.2 times slower at 482 °C. Creep test results showed that at 690 MPa (100 ksi) and 705 °C, the creep-rupture life was extended by a factor of almost 4 for the new DTM. Tensile test results indicated minimal strength losses for the DTM with respective YS and UTS values of 80% and 90% of the CM baseline values at both test temperatures.

Introduction

Emerging safety standards for rotating engine components are now starting to demand properties in engineering materials, that may not meet the required specifications[1, 2]. In particular, the damage tolerance specifications imposed upon turbine disc materials have prompted researchers to investigate ways to improve crack growth properties of presently used alloys without compromising the traditional safe life limits. As a result, a microstructural design philosophy has been successfully developed that is aimed at improving the damage tolerance of conventional disc materials.[3]

Damage tolerance, by definition, is the ability of the disc material to exhibit greater resistance to the growth of inherent or service induced flaws under creep and/or fatigue loading conditions, while still maintaining adequate low cycle fatigue properties, tensile strength, as well as stress rupture strength.[4] Consequently, improving the damage tolerance of conventional Ni-base disc alloys has become an area of substantial interest amongst disc material manufacturers and engine designers alike.

Superalloys 2000
Edited by T.M. Pollock, R.D. Kissinger, R.R. Bowman,
K.A. Green, M. McLean, S. Olson, and J.J. Schirra
TMS (The Minerals, Metals & Materials Society), 2000

Design Lifing Methodologies

All rotating turbine components, such as discs and spacers, have been designed on the basis of the safe-life design philosophy since the late sixties. This safe-life design approach states that a component must be designed for a finite service life during which no significant fatigue damage can occur.[5] Specifically, disc life limits are set by the statistical number of low cycle fatigue (LCF) cycles required to form a detectable 800 μm surface crack, in 1 out of 1000 components tested.[6] This lifing methodology does not directly incorporate crack propagation behavior and therefore is unsuitable for damage tolerance applications.

Unlike traditional safe-life design methodologies, the damage tolerant design (DTD) philosophy includes the assumption that a flaw or a rogue defect is present during the initial design of the disc.[7] The initial size of this flaw, a_i, is assumed to be the minimum detectable crack size of the non-destructive inspection (NDI) technique used during engine maintenance. In contrast to the safe-life design methodology, that is dependant on pure statistical analysis of LCF tests, the DTD philosophy predicts disc lifetimes based on fatigue and/or creep crack growth using fracture mechanics principles. The fatigue crack growth rate (FCGR) of the material can be determined by performing a sufficient number of fatigue tests that reflect the in-service variables of stress amplitude, stress concentration, loading frequency and temperature. Once the FCGR of the material is known and a critical crack size, a_c, has been established, a safe inspection interval (SII) for the component can be determined. This SII is derived from the time it takes for a_i to propagate to a predetermined dysfunction crack size, a_m, based on the FCGR of the disc material. After each SII, the disc is removed from service for inspection. If no cracks are found the disc is returned to service, otherwise, it is retired. Not only does this methodology prevent catastrophic failure by assuming the presence of inherent flaws from initial service, but it also prevents unnecessary discarding of discs that are retired prior to their actual full-life capability. A schematic of this life cycle management process is provided in Figure 1.

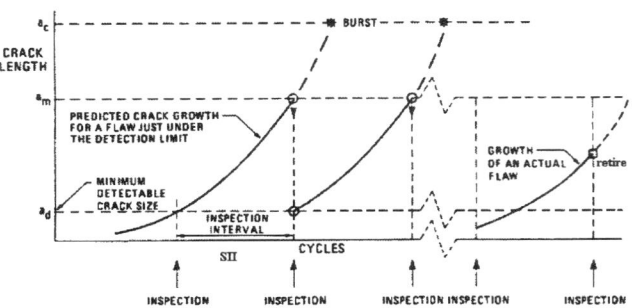

Figure 1: Safety Inspection Interval (SII) of DTD philosophy. [8]

It is apparent that the critical aspect of the DTD philosophy is the length of the SII. If the SII is too short, the maintenance downtime would make the process impractical. Therefore, materials that have consistently superior crack growth rate properties are essential for both the successful application of this life cycle management methodology and for meeting damage tolerance safety criteria. Hence, researchers must not restrict themselves solely to the area of damage tolerant microstructural design. They must also pay attention to the methods used to characterize the crack growth rate properties, which ultimately influence the SII predictions.

As the sensitivity of field NDI techniques improve, the study of short crack growth becomes important for the application of the DTD philosophy. This is largely due to the fact that, once nucleated, cracks in engineering components spend a large fraction of their lives as physically small cracks on the order of 5 to 1000 μm in depth.[9] In fact, the initiation and short crack growth periods generally account for a large portion of the total fatigue life of highly stressed components. Until recently, most fatigue research concentrated largely on the characterization of relatively long crack growth behavior of disc materials. Although this information is essential in itself, it only supplies the designer with a partial description of a material's overall fatigue behavior. Therefore, a damage tolerant microstructure (DTM) of a disc material must also retard short crack growth rates relative to its conventional microstructure (CM).

This paper describes the thermal processing route followed to create a DTM for PWA 1113, as well as the development of a crack monitoring technique that was developed for characterizing its short crack behavior. The influence of the new DTM on creep and tensile properties are also examined and compared with the mechanical properties of the CM at ambient as well as elevated temperatures.

Experimental Materials and Methods

Material

The conventional material used in this study was taken from an approximately 2-ft. diameter turbine disc that was provided by Pratt and Whitney Canada (P&WC). The nominal composition (wt%) for PWA 1113 is:

Cr	Co	Mo	Ti	Al	Zr	B	Fe	Ni
16.5	14.0	4.08	4.50	2.02	0.07	0.007	0.06	Bal.

The disc was produced by a forging method known as the Gatorizing process. This is a procedure that involves alloy powder consolidation through extrusion of a canned powder compact followed by isothermal forging. The CM contains a very fine microduplex γ-γ' grain size (ASTM 10) with a relatively straight grain boundary morphology and a, poorly distributed, coarse γ' precipitate structure, as seen in Figure 2.

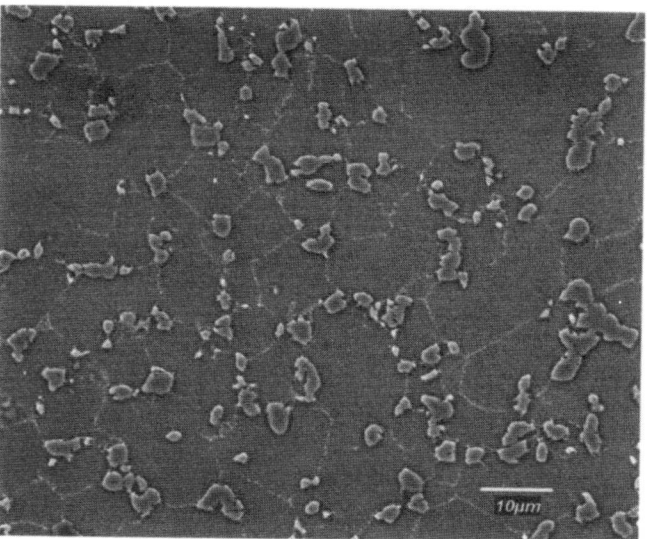

Figure 2: The CM of PWA 1113 showing large primary γ '.

DTM Development

The DTM design philosophy was based on successful work by Koul *et al.* who developed DTM's for Inconel 718 and MERL 76 with marked improvements in fatigue and creep resistance [3,7]. Based on their past success, a similar methodology was adopted that involved:

1. Grain size optimization by determining the solution treatment conditions required to dissolve all second phase γ' precipitates while preventing excessive grain growth.
2. Determination of a post solutioning heat treatment sequence that defines the controlled-cooling rate range required to promote the formation of a serrated grain boundary structure through controlled γ' precipitation along the grain boundaries.
3. Determination of an appropriate aging sequence that produces the ideal amount, size, and distribution of γ' precipitates required to strengthen the γ matrix.

Fatigue Testing

Short crack growth fatigue tests were performed using single edge notched (SEN) specimens in accordance with the ASTM-E647 specifications. Testing parameters included a loading frequency of 5 Hz, a zero stress ratio with a maximum load of 32 kN, and a triangular waveform loading pattern. As seen in Figure 3, the SEN notch radius was 1.2 mm (0.049 in.) to simulate the K_T factor of 2.68 that is commonly found in the critical regions of the disc. Three specimens from each microstructural condition were tested at room temperature and 482 °C.

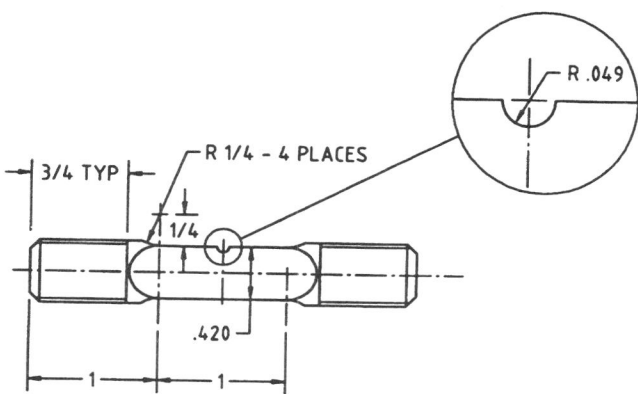

Figure 3: SEN specimen used for short crack fatigue testing.

Extensive experimental work was also performed to implement an alternating current potential difference (ACPD) technique required to continuously monitor the short crack growth rate properties without test interruption. Preparation involved meticulous placement of voltage probes and current leads to the specimen surface as seen in Figure 4. Ten calibration tests were performed over a wide crack size range (40 μm to 1500 μm) to determine the crack size-potential drop relationship. Figure 5 shows one such calibration test where a thumbnail crack of 748 μm in length corresponded to a 52 mV change in potential voltage. A 5-amp alternating current at a frequency of 130 kHz was used to give extremely high sensitivity of short crack detection to a depth of 1.5 mm below the specimen surface.

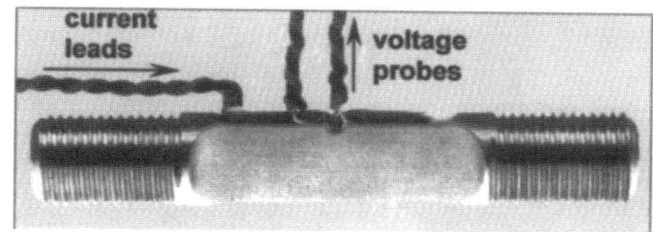

Figure 4: Specimen ACPD lead configuration.

Tensile Testing

Tensile tests were performed in accordance with ASTM-E8 at room temperature and 650 °C. Strain data was gathered using a high temperature quartz-rod extensometer while monitoring simultaneous load readings from the MTS load cell. Multiple tests were conducted for the conventional and damage tolerant microstructures at each test temperature.

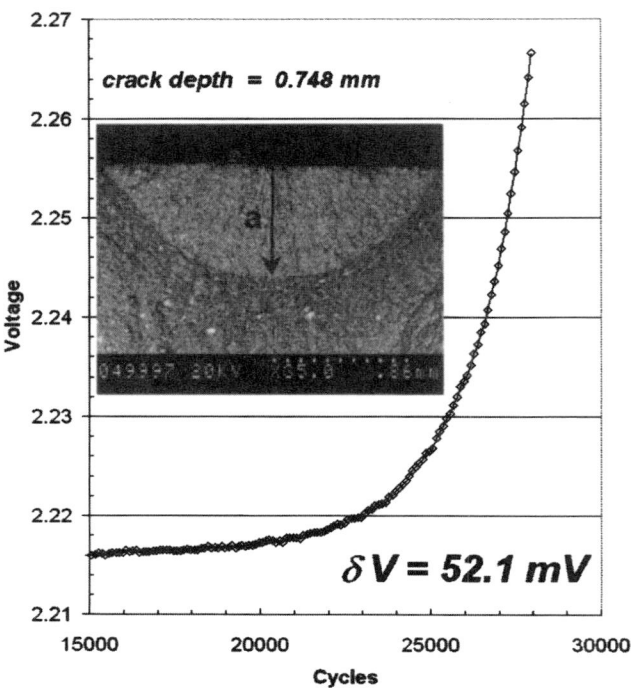

Figure 5: Sample crack calibration curve showing crack depth/potential drop relationship.

Creep Testing

Creep-rupture tests were performed in accordance with ASTM-E139 and were conducted at 705 °C under a constant load of 690 MPa (100 ksi). Both microstructural conditions were tested for comparison.

Results and Discussion

Grain Size Optimization

To determine the solutionizing temperature required to dissolve all γ′ precipitates without the occurrence of excessive grain growth, a detailed grain coarsening study was performed. Figure 6 shows the grain coarsening behavior of the material over a wide range of solutioning temperatures and hold times. It is apparent from this figure that a major shift in grain size occurs over a very short temperature interval (1130 °C to 1135 °C) for all solutioning hold times. It can therefore be deduced that, during this short interval, the γ′ precipitates are fully dissolved, effectively removing all grain boundary pinning agents and permitting grain growth to occur. Hence, the temperature chosen for solutioning the γ′ precipitate phase was at 1135 °C just past the inflection point on the curve to prevent excessive grain growth. The chosen solutioning temperature resulted in a final DTM grain size of ~70 μm (ASTM 4).

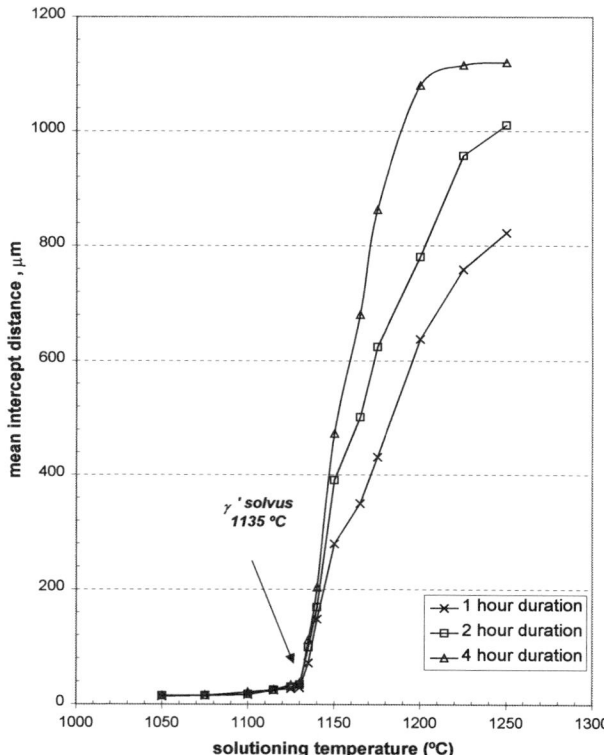

Figure 6: Grain coarsening behavior of PWA 1113.

Grain Boundary Serrations

Grain boundary serrations can be generated by controlled furnace cooling from the solution temperature. However, it has been determined by Koul *et al.* that only high volume fraction alloys with a sufficiently large lattice mismatch are capable of forming serrations during furnace cooling. [10] Furthermore, the proposed strain energy differential model by Koul and Gessinger, states that there is a minimum particle size required for serration formation based on the following equation [10],

$$ r_c = \frac{3\gamma_{st}}{4\mu\delta^2} \quad \dots\dots\dots\dots\dots\dots (1) $$

where γ_{st} is the interfacial energy due to the structural distortion caused by the misfit dislocations, δ is the particle-matrix misfit parameter, and μ is the shear modulus of the matrix. It should be noted that the model accounts for the varying effects of cooling rates on serration morphology if one considers the correlation between precipitate size and cooling time.

Once it has been established that the alloy is capable of forming grain boundary serrations the next logical step in DTM design is to determine the size of serrations required to optimize crack growth resistance. Wu and Koul have

120

developed a constitutive model that predicts the optimum grain boundary serration size that effectively inhibits grain boundary sliding.[11],[12] The model accurately describes the transient creep behavior of complex Ni-base superalloys with respect to the serration amplitude (h) and the serration wavelength (λ). The optimum serration geometry to prevent grain boundary sliding can be approximated by the simplified steady-state creep rate equation,

$$\dot{\varepsilon} = \phi A_0 \frac{D\mu b}{kT}\left(\frac{b}{d}\right)^q \left(\frac{l+r}{b}\right)^{q-1}\frac{(\sigma_a - \sigma_{ig})(\sigma_a - \sigma_{ic})}{\mu^2} \quad ....(2)$$

where A_0 is a non-dimensional constant, l is the grain boundary precipitate spacing, r is the radius of the grain boundary particle, d is the grain size, q is the grain size index, μ is the elastic shear modulus, σ_a is the applied tensile stress, σ_{ig} is the tensile stress opposing dislocation glide, σ_{ic} is the tensile stress opposing dislocation climb, and ϕ is the *grain boundary serration factor* defined by,

$$\phi = \frac{2}{\sqrt{1+\left(\frac{\pi h}{\lambda}\right)^2}} - 1 \quad .... (3)$$

There is an optimum range of serration amplitude and wavelength values required to reduce creep rate. From these equations, it is then possible to predict the γ' precipitate size required to form the appropriate serration h and λ, necessary for suppressing grain boundary sliding phenomena. Equations (1) through (3) define the basic microstructural design criteria. No attempt was made however, to optimize the grain boundary carbide precipitate size and distribution in the present study.

Numerous heat treatment sequences were used to determine the appropriate cooling rate range required to heterogeneously nucleate γ' precipitates along the grain boundaries. Extremely fast cooling rates resulted in relatively small grain boundary precipitates (<1 μm) that were ineffective in serration formation. In the other extreme, excessively slow cooling rates generated large precipitate growth that produced ineffective, oversized grain boundary serrations. The final cooling rate range was between 1°C/min and 3°C/min directly from the solvus temperature of 1135°C to a furnace removal temperature of 1060 °C. This sequence produced the optimum precipitate size (~1-3 μm) that generated grain boundary serrations throughout the microstructure's intergranular region, as seen in Figure 7. Similar to findings by Koul *et al.*, the majority of grain boundary γ' precipitates responsible for grain boundary serrations are elliptical in shape and are always located within the concave side of the serration [13].

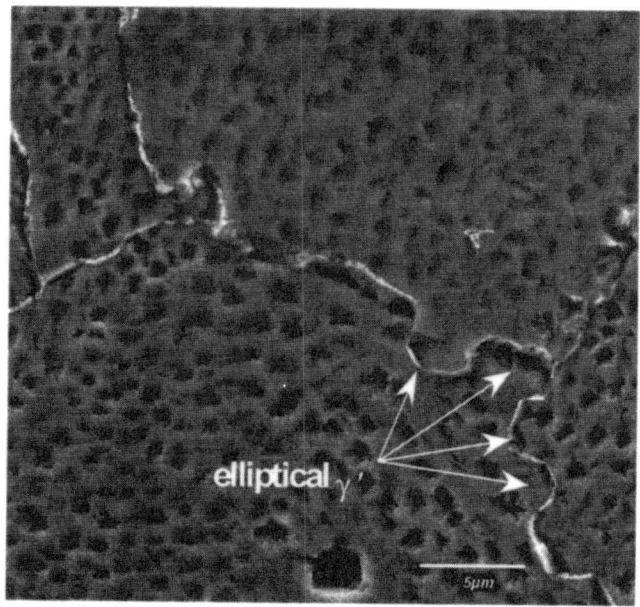

Figure 7: Grain boundary serrations in new DTM.

Morphology and Distribution of γ'

The size, volume fraction, and distribution of the γ' precipitates are the three main variables that influence the degree of precipitation hardening in Ni-base superalloys. It is quite obvious from Figure 8, that the primary γ' size and distribution for the CM is quite large and poorly distributed throughout the matrix. Primary γ' for the CM condition are up to 10 μm in diameter and are usually located along the grain boundaries, likely contributing little to the overall matrix strength. Precipitate optimization was accomplished by partially solutionizing the material at 1060 °C for 1 hour which partially dissolved the intragranular γ' without any loss of grain boundary serration morphology. This was then followed by a single-step aging sequence that entailed heat treating at 760 °C for 16 hours which produced the ideal amount, size, and distribution of γ' precipitates required to strengthen the γ matrix. Figure 9 shows the new precipitate morphology and distribution of the DTM. The new features include a smaller primary γ' precipitate size (~1μm) that is more homogeneously distributed throughout the intragranular region, as well as a much finer secondary precipitate size.

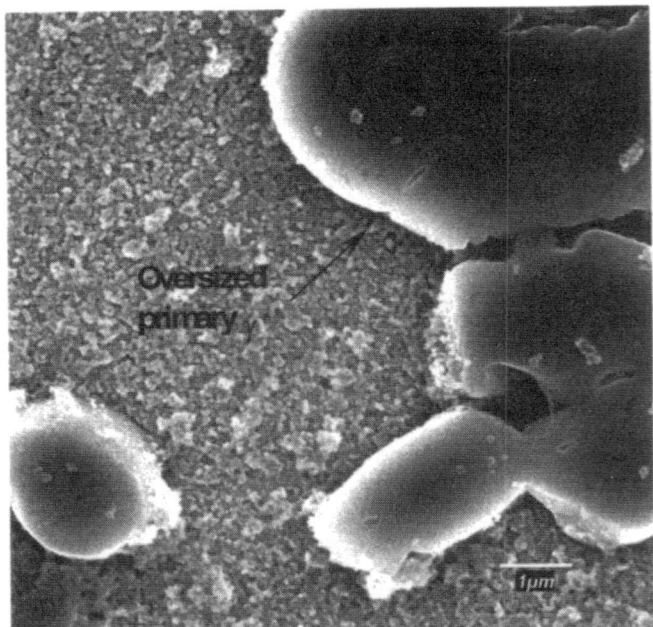

Figure 8: Morphology and distribution of γ′ for CM condition.

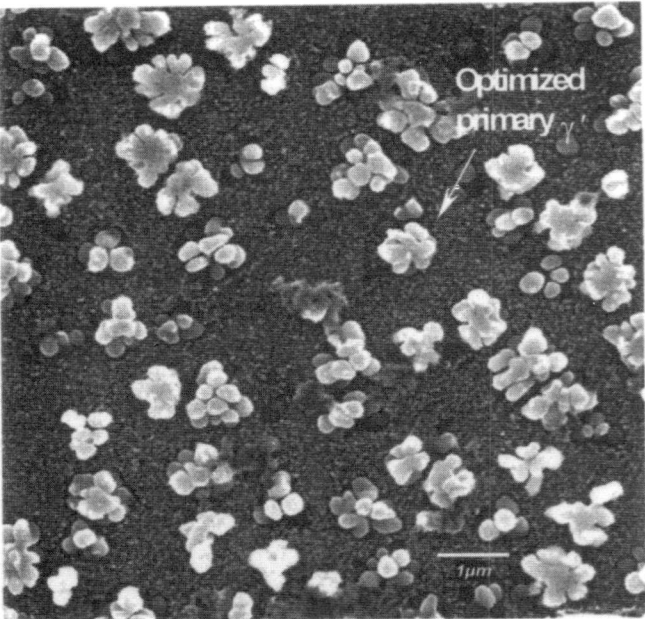

Figure 9: Morphology and distribution of γ′ for DTM condition.

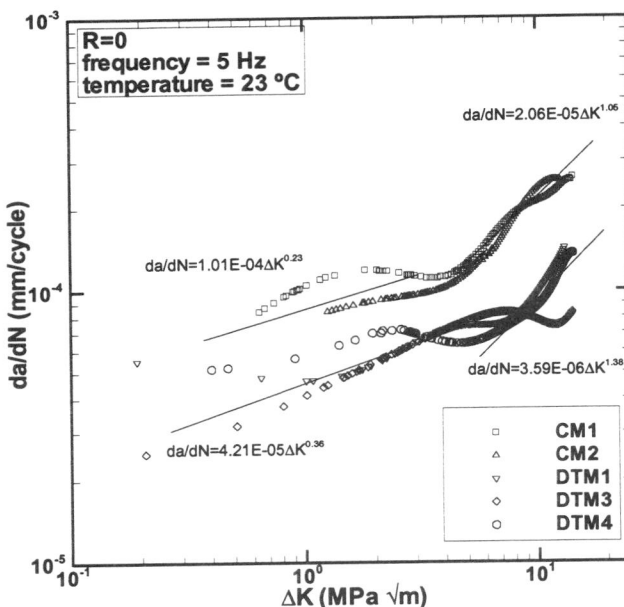

Figure 10: SCGR properties of CM and DTM at 23°C.

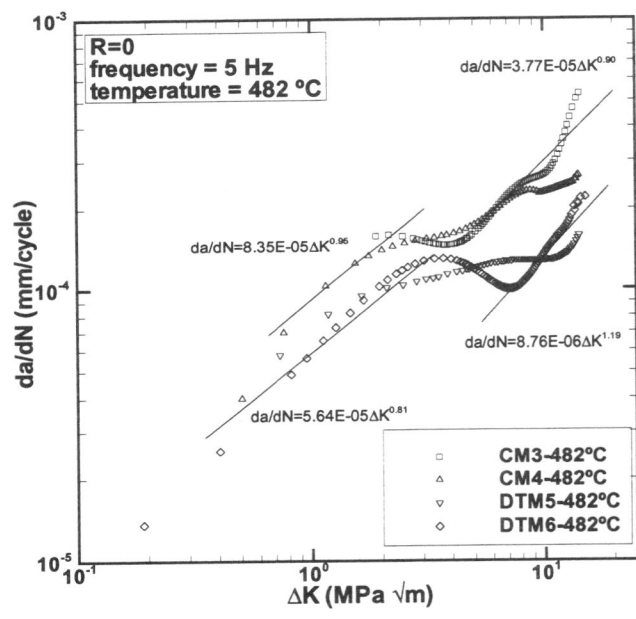

Figure 11: SCGR properties of CM and DTM at 482 °C.

Mechanical Properties

Short Crack Fatigue Properties

Fatigue test results revealed that the DTM possessed better short crack fatigue resistance than the CM at both room temperature and at 482 °C. For example, at a stress intensity (ΔK) of 10 MPa√m, the crack growth rate for the DTM is 2.7 times slower at room temperature and 2.2 times slower at 482 °C compared to the CM, as illustrated in Figures 10 and 11 respectively.

It is apparent from both the room and elevated temperature fatigue curves presented, that short crack growth behavior does not conform to traditional long crack growth characteristics. Generally, there are two main defining properties of short cracks that distinguish them from typical long crack growth features.

1. Short cracks have generally been observed to exhibit faster and more irregular propagation rates compared to long crack data.[14]

122

2. Short crack propagation occurs at stress intensity factors (ΔK) well below the traditional threshold value (ΔK_{th}) for long crack growth.[15]

These two distinguishing properties of short cracks have been reported by numerous researchers, all of whom have reported similar anomalous short crack behavior.[16] This is primarily due to the fact that microstructurally short crack growth, often deals with the propagation of cracks that do not abide by Linear Elastic Fracture Mechanics (LEFM) principles.

LEFM relies heavily on continuum mechanics principles for crack propagation. However, when dealing with short crack growth, the crack in the beginning is smaller than the actual grain size and microstructural features tend to influence crack propagation. This ultimately results in the breakdown of the material continuum and propagation exhibits a more discrete behavior. Due to the interference of short crack propagation by such features as grain boundaries and intragranular precipitates, the resulting growth rate is often irregular in nature[17]. When such an obstacle is encountered by the crack tip plastic zone, the growth rate tends to decrease followed by rapid acceleration upon traversing the barrier. This fluctuating crack growth behavior is visible in Figures 10 and 11, which show periods of deceleration and acceleration at relatively low ΔK levels. These inflection points on the curves occur at ΔK levels that are indicative of crack depths equivalent in length to the average microstructural grain size. This suggests that the initial deceleration is most likely due to dislocation pile-up at the impeding grain boundary, followed by rapid acceleration once sufficient stress intensity has been attained for slip band extension into the neighboring grain. As the crack length increases, microstructural features become less effective, crack tip deformation increases, and propagation behavior tends toward continuum mechanics principles (i.e long crack growth).[18]

The improvements in short crack growth rate properties can be explained on the basis of differences in the fracture modes observed in Figures 12 and 13. In Figure 12, it is obvious that the fracture mode of the CM is transgranular, with evidence of striations or faceted crack growth across the grains. There is also evidence of the ineffectiveness of the large overaged γ' precipitates in retarding crack growth. The DTM fracture surface also exhibits a transgranular fracture, however, due to the finer, more dispersed γ' precipitate morphology, crack growth is decreased within the intragranular region. This is most likely caused by the combination of crack deflection around the precipitates which prevents slip-band extension within the grain itself as well as the larger DTM grain size.

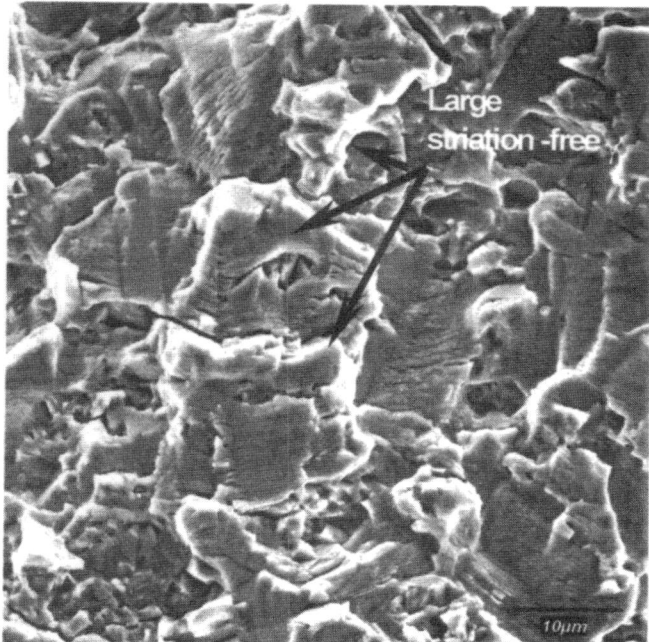

Figure 12: Fractograph of CM from 482 °C fatigue test.

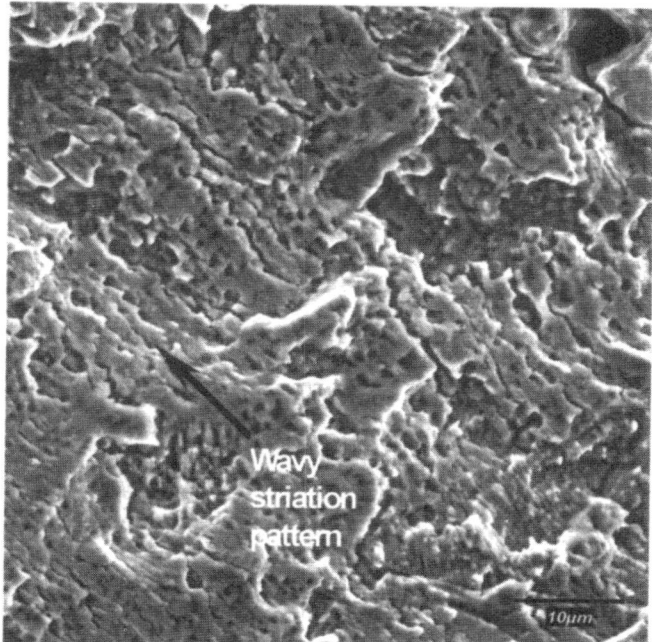

Figure 13: Fractograph of DTM from 482 °C fatigue test.

Creep Rupture Properties

In addition to the improved short crack growth rate properties of the DTM, the creep-rupture tests results demonstrate a marked improvement over the creep properties of the CM material. It was predicted that the DTM would have better creep resistance due to it's larger grain size and serrated grain boundary morphology. In fact, the test results shown in Figure 14 indicate that at an initial stress level of 690 MPa (100 ksi) and at a temperature of 705 °C, the creep-rupture life was almost 3.4 times longer for the DTM.

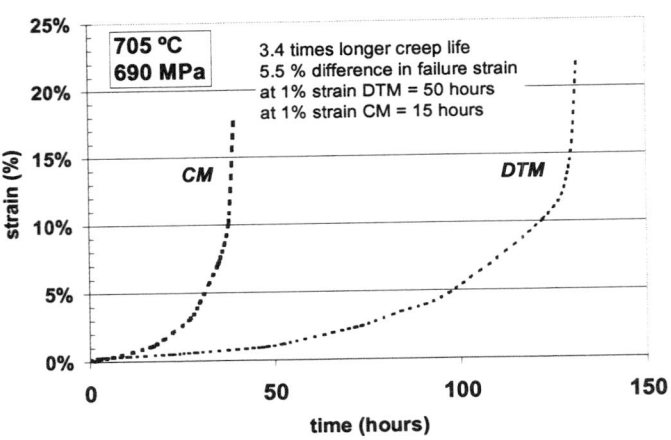

Figure 14: Creep-Rupture test results for CM and DTM.

Furthermore, it is apparent from the creep fractograph in Figure 15 that the fracture mode of the DTM is intergranular, as indicated by the irregular topography and grain boundary relief. Compared to the fracture surface of the CM, Figure 16, the DTM has a more highly featured surface profile. Although the fracture mode of the CM is also intergranular, it does not possess the same degree of rough topography due to its smaller grain size. This small grain size (~13 μm), results in a relatively smooth fracture surface almost indistinguishable from the transgranular overload section of the specimen as shown in Figure 16. The intergranular fracture in both materials suggests that the test temperature of 705 °C was indeed above the equicohesive limit. In other words, at this temperature the grain interiors retained superior creep resistance compared to the grain boundaries.

The data from Figure 14 also reflects two other important results,

1. There is a delayed onset of tertiary creep (~50 hrs) for the DTM as compared to the CM (~5 hrs)
2. The tertiary creep time interval is much longer for the DTM.

As for the superior creep resistance of the DTM, it may be attributed to a combination of a coarser grain size, the serrated grain boundary morphology, and the smaller γ′ precipitate size and more uniform γ′ distribution. The larger grain size (70-100 μm) of the DTM means that there is a lower grain boundary fraction present that could serve as creep fracture paths in the material. The serrated morphology of the DTM may also slow down surface creep crack propagation by presenting a more tortuous crack path,

as compared to the relatively planar boundaries of the CM. These microstructural features are known to suppress grain boundary sliding. The finer (1 μm) precipitate size, as well as their decoration along the grain boundaries, may also retard crack propagation. The higher fraction of grain boundary pinning, due to the more dispersed nature of these precipitates, results in higher intergranular strength. In contrast, the CM precipitate size is extremely large (10 μm) and poorly distributed playing no role in grain boundary strengthening.

Figure 15: Creep surface of DTM after 131.5 hours at 705 °C and 690 MPa.

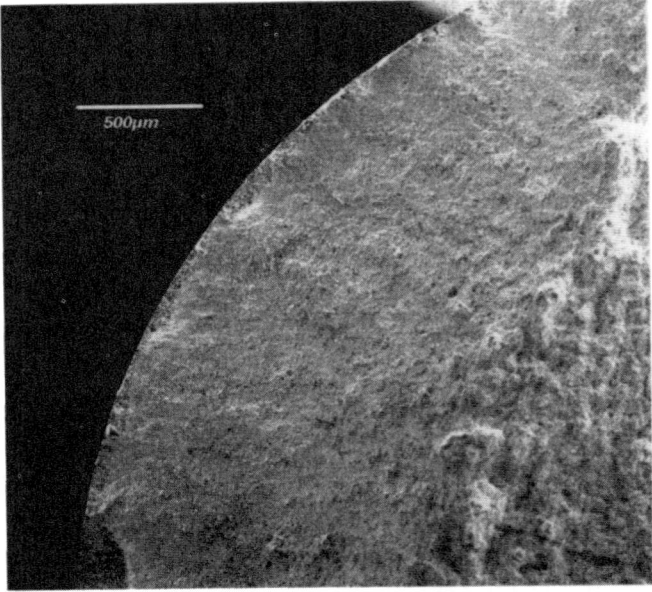

Figure 16: Creep surface of CM after 39.4 hours at 705 °C and 690-MPa.

124

Tensile Properties

Tensile tests were conducted at room temperature and at 650 °C. Figure 17 and Figure 18 show the tensile test results for room temperature and 650 °C respectively. Tensile test results indicated minimal strength losses for the DTM in terms of YS and UTS values with 80% and 90% of the CM baseline values for both room temperature and 650 °C. From these results it can be seen that the UTS and YS values of the new damage tolerant microstructure are comparable to that of the conventional microstructure. Yield and ultimate strengths are slightly lower in the new DTM as was to be expected due to the larger grain size present.

Room temperature and high temperature tests also showed an increase in specimen elongation for the new DTM. Average elongation values for room temperature tests were 12% and 18% for the CM and DTM respectively. Likewise, at 650 °C, the average elongation values were 12% for the CM and 15% for the DTM. This trend is consistent with the improvement in creep ductility of the DTM as shown in Figure 14. The increased ductility of the new DTM may also be accountable for the improved high temperature fatigue resistance observed. Bain *et al.* have developed an improved DTM for the Ni-base superalloy Udimet 720 by simply increasing the overall grain size. [19] Their findings have shown that at the expense of marginal losses in tensile strength, the increased ductility of the coarser grain sized material produces significant improvements in high temperature fatigue resistance as well as stress rupture life.

Conclusions

A successful DTM for the Ni-base superalloy, PWA 1113, has been designed. The DTM was produced by implementing a modified heat treatment sequence that optimized the grain size, grain boundary morphology, and γ′ precipitate size and distribution. The new heat treatment sequence imparted a controlled serrated grain boundary morphology, coarser grain size, and a finer, more uniformly distributed γ′ size. Short crack propagation rates were decreased by a factor of 3 at room temperature and a factor of 2.2 at 482 °C. Grain size had a profound effect on the shape of ΔK versus crack growth rate correlation. At 705 °C and 690 MPa, the creep rupture life of the DTM was extended by a factor of almost 4. Marginal losses in proof strength and ultimate strength were also observed during room temperature and high temperature tensile tests. Improved damage tolerance properties can be directly related to the improved grain boundary strength of the DTM that was invoked by controlled-cooling heat treatments. The finer γ′ precipitate size and distribution also contributed to slower intragranular crack propagation rates by impeding dislocation motion through the grain interiors. These findings for the new DTM are in good agreement to earlier damage tolerance studies for comparable Ni-base superalloys [3,7,13].

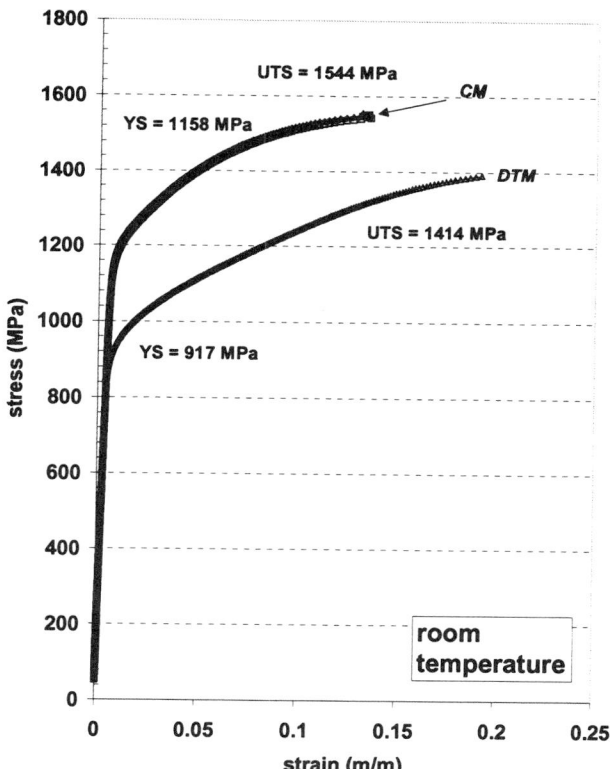

Figure 17: Room temperature tensile test results.

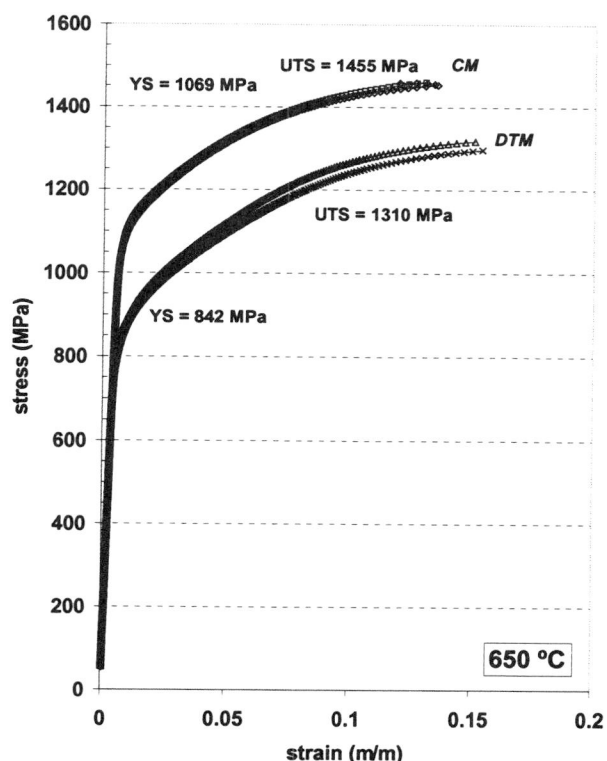

Figure 18: Elevated temperature tensile test results.

125

It was also proven that successful characterization of short crack properties is possible utilizing the ACPD crack monitoring technique. This is a significant result for future reference if damage tolerant lifing philosophies are to be implemented in turbine disc design.

Acknowledgements

This work was conducted at the Institute for Aerospace Research under IAR-NRC project #46-6JL-04. Financial assistance was also provided by Pratt and Whitney Canada and Carleton University.

References

1. MIL-STD-1783 (USAF): Military Standard, "Engine Structural Integrity Program (ENSIP)", by the Department of Defence, USA, Nov. 1984.

2. MIL-STD-1843 (USAF): Military Standard, "Reliability Centered Maintenance for Aircraft, Engine and Equipment", by the Department of Defence, USA, Feb. 1985.

3. Koul, A.K., Au, P., Bellinger, N., Thamburaj, R., Wallace, W., and Immarigeon, J-P., "Development of a Damage Tolerant Microstructure for Inconel 718 Turbine Disc Material", Superalloys 1988, (The Metallurgical Society, 1988), pp. 3-12.

4. Krueger, D.D., Kissinger, R.D., and Menzies, R.G., "Development and Introduction of a Damage Tolerant High Temperature Nickel-Base Disk Alloy, Rene 88DT", Superalloys 1992, (The Metallurgical Society, 1992), pp. 277-286.

5. Wanhill, R.J.H., "Short Cracks in Aerospace Structures", The Behavior of Short Fatigue Cracks, (Ed. by K.J. Miller and E.R. de los Rios, Mechanical Engineering Publications, London, 1986), pp. 27-36.

6. Evans, W.J., Smith, M.E.F., and Williams, C.H.H., "Disc Fatigue Life Predictions for Gas Turbines", AGARD-CP-368, 1984, paper 11.

7. Koul, A.K., Immarigeon, J.P., Wallace, W., "Microstructural Control in Ni-based Superalloys", Advances in High Temperature Structural Materials and Protective Coatings, (Ed. by A.K. Koul et al., National Research Council of Canada, Ottawa, Canada), Chapter 6.

8. Mom, A.J.A., Raizenne, M.D., "AGARD Cooperative Test Programme on Titanium Alloy Engine Disc Material", AGARD-CP-393, 1985, p.9.

9. Swain, M.H., "Monitoring Small-Crack Growth by the Replication Method", Small Crack Test Methods, Ed. by J.Larsen and J. Allison, ASTM STP 1149, 1992, pp.34-56.

10. Koul, A.K., and Thamburaj, R., "Serrated Grain Boundary Formation Potential of Ni-based Superalloys and its Implications", Metallurgical Transactions A, Vol. 16A, 1985, pp.17-26.

11. Wu, X.J., and Koul, A.K., "Grain Boundary Sliding at Serrated Grain Boundaries", Advanced Performance Materials, (Kluwer Academic Publishers, 1997), Vol.4, 1997, pp.409-420.

12. Wu, X.J., and Koul, A.K., "Grain Boundary Sliding in the Presence of Grain Boundary Precipitates During Transient Creep", Metallurgical Transactions A, Vol. 26A, 1995, pp.905-914.

13. Koul, A.K., and Gessinger, G.H On the Mechanism of Serrated Grain Boundary Formation in Nickel-Based Superalloys", Acta Metallurgica., Vol. 31, No. 7, July 1983, pp. 1061-1069.

14. Reed, P.A.S., and King, J.E., "Comparison of Long and Short Crack Growth in Polycrystalline and Single Crystal Forms of Udimet 720", Short Fatigue Cracks, (ESIS 13, Ed. by K.J. Miller and E.R. de los Rios, London, 1992), pp. 153-168.

15. Newman, J.C. Jr., "Fracture Mechanics Parameters for Small Fatigue Cracks", Small Crack Test Methods, (Ed. by J.Larsen and J. Allison, ASTM STP 1149, 1992), pp. 6-33.

16. Bolingbroke, R.K., and King, J.E., "A Comparison of Long and Short Fatigue Crack Growth in High Strength Aluminum Alloy", The Behavior of Short Fatigue Cracks, (Ed. by K.J. Miller and E.R. de los Rios, Mechanical Engineering Publications, EGF.1, 1986), pp.101-114.

17. Radhakrishnan, V.M., and Mutoh, Y., "On Fatigue Crack Growth in Stage I", The Behavior of Short Fatigue Cracks, (Ed. by K.J. Miller and E.R. de los Rios, Mechanical Engineering Publications, EGF.1, 1986), pp.87-99.

18. Blom, A., et al., "Short Fatigue Crack Growth Behavior in Al 2024 and Al 7475", The Behavior of Short Fatigue Cracks, (Ed. by K.J. Miller and E.R. de los Rios, Mechanical Engineering Publications, EGF.1, 1986), pp.37-66.

19. Bain, K.R. et al., "Development of Damage Tolerant Microstructures in Udimet 720", Superalloys 1988, (The Metallurgical Society, 1988), pp. 13-22.

The Microstructure Prediction of Alloy720LI for Turbine Disk Applications

T.Matsui*, H.Takizawa*, H.Kikuchi** and S.Wakita*

Mitsubishi Materials Co. Ltd

*1-297 Kitabukuro-cho, Omiya, Saitama, 330-8508, Japan

**1230 Kamihideya, Okegawa, Saitama, 363-8510, Japan

Abstract

In order to reduce the lead time of developmental stage and increase the reliability of the forging process of turbine disk, it is effective to predict the transformation of microstructure during the forging stage. In this paper, the microstructure prediction procedure for cast/wrought form alloy720LI is introduced. To determine the parameters quantitatively, two kinds of tests were carried out. One was isothermal heating test, the other was isothermal compression test. The relational expressions between microstructure and various parameters e.g. time, temperature and strain rate were formulated for both static grain growth model and dynamic recrystallization model. Static grain growth model and dynamic recrystallization model represents preheating process prior to forging and forging process itself respectively. The models were coupled to finite element analyzing system. Furthermore pancake forging and generic shaped disk forging were carried out to verify the effectiveness of proposed prediction system. It was confirmed that the proposed prediction system has good accuracy to apply to the actual disk forging process.

Introduction

High performance alloys are used for turbine disks and the requirement for its properties and reliabilities are increasing for advanced aircraft engines. Alloy720LI is one of the most effective alloy for turbine disk applications because of its high creep strength and low cycle fatigue (LCF) property at elevated temperature. In fact, several engine adopt alloy720LI as turbine disk material.

Alloy720LI was developed as for both cast/wrought alloy and powder metallurgy (P/M) alloy. Many investigations were performed in both regions[1-4]. Each has advantages and disadvantages. Generally cast/wrought form is more desirable compared to P/M form from the viewpoint of cost. Advanced aircraft engine requires not only improvement of properties but also reduction of cost. Therefore cast/wrought form is worthy of being investigated further.

As for property, cast/wrought process can give excellent creep and LCF property, much the same to P/M process. But it is difficult to give such high properties in actual parts. Alloy720LI include higher alloying elements than conventional alloys e.g. alloy718 and has greater temperature and strain rate dependence on various behaviors. Especially deformation resistance and microstructure behavior are important to produce the disk. The former restricts disk size and shape, the latter is related to final mechanical properties. Various parameters e.g. temperature, strain rate, total strain etc. have to be controlled carefully to obtain desirable mechanical properties through out any portion of disk. In case of cast/wrought process, the control of such

Superalloys 2000
Edited by T.M. Pollock, R.D. Kissinger, R.R. Bowman,
K.A. Green, M. McLean, S. Olson, and J.J. Schirra
TMS (The Minerals, Metals & Materials Society), 2000

TABLE I	Chemical composition of the billet											(mass%)
	Ni	Cr	Co	Mo	W	Al	Ti	Fe	Mn	Si	Zr	B	C
alloy720LI	Bal.	15.6	14.2	3.2	1.2	2.6	5.0	0.3	0.01	0.01	0.05	0.013	0.015

parameters is relatively difficult.

One of the effective ways to solve the problem is computer simulation. Generally computer simulation has been used to predict the load and deformed shape and to determine initial work size, preformed shape and so on. If the prediction of microstructure is possible, it enables to design the forging process that can obtain excellent and uniform properties through out any portion of disk and the reliability of the disk will be increased. Actually several microstructure modeling were proposed and reported in the past. But most of them were carried out on conventional alloys e.g. alloy718, Waspaloy[5,6]. In case of alloy720LI, large temperature and strain rate dependence on microstructure and dispersion of microstructure at billet condition make the prediction difficult. According to calculation of phase equilibria, the mole fraction of stable gamma phase changes about 10% under the range of temperature between gamma prime solvus and below that temperature by 40C, while only about 4% change is estimated in Waspaloy[7]. Such phase transition leads to drastic change of microstructure i.e. grain size, fraction of dynamic recrystallization and makes it difficult to recognize relationship between microstructure and various parameters. That is one of the reasons why accurate prediction in alloy720LI is difficult compared to other conventional alloys.

Furthermore, alloy720LI includes large amount of Al, Ti. Al and Ti are the major elements of gamma prime former and essential to obtain high strength but both elements tend to segregate in billet. Segregation of Al, Ti makes difficult to obtain complete uniform gamma prime distribution at billet stage. Actually two kinds of region are often observed. One has relatively high density of primary gamma prime particles i.e. gamma prime banding and the other has relatively low density of those. In case of disk production forging is carried out below the temperature of gamma prime solvus, the gamma prime distribution in billet condition effects strongly on the final microstructure. Therefore it is necessary to understand the behavior of each region for accurate prediction.

The objective of this investigation is to establish the microstructure prediction procedure of cast/wrought alloy720LI disk using computer simulation.

Materials and Experimental Procedure

Materials

Original material in this study was the billet produced from triple melted ingot (VIM+ESR+VAR). Chemical composition and typical microstructure of the billet are shown in TABLE I and Figure 1 respectively. The microstructure was controlled carefully to obtain equiaxed grain with average grain size of

ASTM 10-11 but gamma prime banding was observed on whole billet.

Isothermal Heating Test

Isothermal heating test was conducted to understand the static grain growth behavior. The specimens were put into furnace in which was controlled at specific temperature and were kept for set time. The range of temperature is 1100-1180C. Microstructure observation was performed by optical microscopy and grain size was estimated in accordance with ASTM E112.

Isothermal Compression Test

Isothermal strain rate controlled compression tests were carried out to understand the essential relationship between temperature, strain rate, strain and dynamically recrystallized grain size, fraction of dynamic recrystalization. The specimens were cylindrical shape with 8mm in diameter and 12mm in height. These specimens were heated up to test temperature by induction heating and compressed isothermally. Each range of temperature, strain rate and total strain is 950-1180C, 0.01-0.2s^{-1} and 5-70% respectively. The specimens were cooled immediately or after keeping test temperature for definite time to achieve quench and dwelled condition. Microstructure observation was carried out by optical microscopy for specific area with 1/2 in height and 2/3 in radius. The area was selected by finite element analysis as desirable observation point in which uniform conditions e.g. strain rate, temperature were achieved through compression. The results of microstructure observation were associated to test temperature, revised strain rate and total strain.

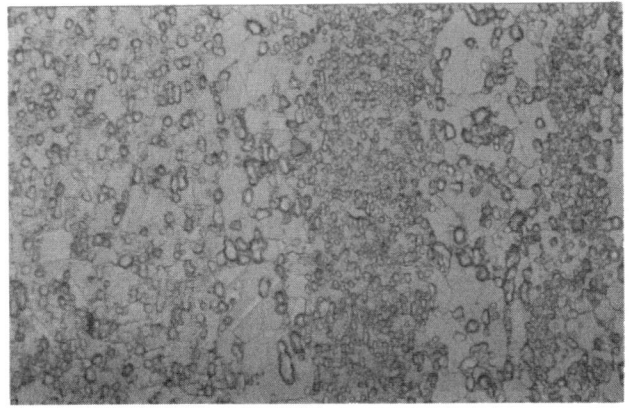

20 μm

Figure 1 : Typical microstructure of the billet.

Several kinds of heat treatment were done prior to machining of specimens to control grain size with care of remaining gamma prime banding. It is necessary to understand the effect of initial grain size on dynamic recrystalization behavior. Moreover the microstructure in as-billet condition is not always best to investigate dynamic recrystallization behavior. Stored strain to which was introduced through billet forging process will induce rapid grain growth during heat-up prior to compression and correct initial grain size will become indistinct. Excellent fine grain size may make it difficult to distinguish recrystallized grain from unrecrystallized grain and bring difficulty in correct measurement of recrystallized grain size and fraction of dynamic recrystalization.

<u>Modeling</u>

Static grain growth model which corresponded to preheating process prior to forging and dynamic recrystalization model which represented forging process itself were developed through quantitative analysis of the results which were obtained from isothermal heating tests and isothermal compression tests. Moreover constructed microstructure prediction models were coupled to the finite element analyzing system, DEFORM*. Estimation of average grain size and fraction of dynamic recrystallization in addition to general output e.g. temperature, strain and load on the whole area of disk was enabled by utilizing user subroutine developed under this study.

<u>Pancake Forging</u>

Hot die pancake forging were conducted with different three conditions to verify the effectiveness of proposed prediction system. Adopted forging conditions varied in temperature from 1125C to 1160C. Forged final shape had 200mm in diameter and 25mm in height.

<u>Generic Shaped Disk Forging</u>

For more accurate verification in practical shape and size, generic shaped disk forging were carried out with hot die condition. The component had 350mm diameter and 80mm height in boss region, 40mm height in rim region. Two kinds of preheating temperature condition, 1100C and 1120C were selected by proposed prediction system to obtain desirable grain size for disk applications.

Results and Discussion

<u>Static Grain Growth Model</u>

Static grain growth model represents mainly preheating process

---

*DEFORM is the trade name of Scientific Forming Technology Corporation.

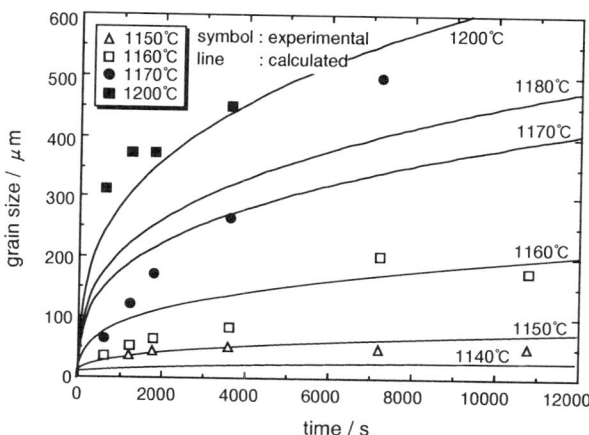

Figure 2: Temperature and time dependence of static grain growth.

prior to forging. The relationship between grain size and holding time is shown in Figure 2. The higher the temperature is, the larger the grain growth rate is. The tendency becomes obvious above gamma prime solvus temperature. Furthermore Figure 2 reveals that the relationship can be represented by following equation and the exponent n takes 3 in this case.

$$d_{i+1}^{n_{gro}} - d_i^{n_{gro}} = A_{gro}\, t\, \exp(-Q_{gro}/RT) \qquad (1)$$

where d_{i+1} is the grain size when time is t_{i+1}, d_i is the grain size when time is t_i, t is time, Q_{gro} is activation energy for grain growth, T is absolute temperature, R is gas constant and A_{gro}, n_{gro} are material constants. Generally the exponent n_{gro} varies from 2 to 4 with rate-controlling process of grain growth. In case of single phase structure, grain boundary migration controls the growth rate and the value n_{gro} takes 2. On dispersion structure condition in which dispersion particle plays as pinning of grain boundary, volume diffusion of atom which constructs dispersion particle controls not only growth of particle but also that of grain and n_{gro} takes 3. $n_{gro}=4$ is found in dual phase structure. According to differential thermal analysis, gamma prime solvus temperature varies from 1079C to 1164C in this material. Therefore primary gamma prime particle has great effect to grain growth at whole temperature range investigated. In addition to existence of gamma prime particle, carbide, nitride and boride are dispersed and remain in the structure even when the temperature is over gamma prime solvus. Consequently the estimated exponent 3 is supposed to be valid. However the equation (1) mentioned above represents essentially steady state grain growth with no internal strain. If the stored strain caused by billet forging process were extremely large or the distance among particles were extremely short compared to grain diameter, n_{gro} might not be the same. Actually activation energy Q_{gro} is effected by internal strain and initial dispersion of particles. As a result relatively large value is estimated while theoretical value is equivalent to activation energy for volume

129

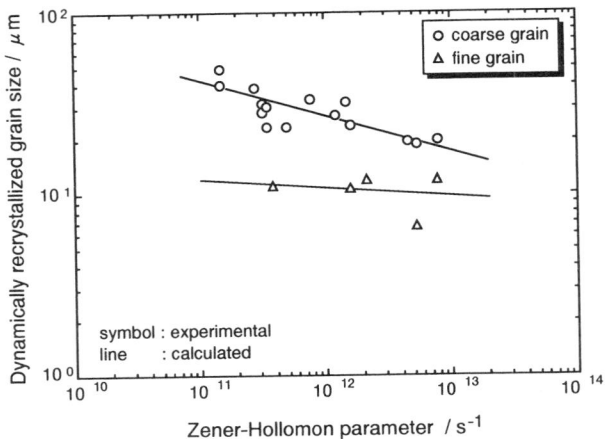

Figure 3 : The relationship between dynamically recrystallized grain size and Zener-Hollomon parameter.

diffusion of Ni or alloying element in Ni.

Dynamic Recrystallization Model

Dynamic recrystallization model corresponds to forging process itself. In case of cast/wrought forms, it is very important to control the occurrence of dynamic recrystallization. Dynamic recrystallization enables to bring desirable microstructure that consists of uniform, equiaxied and fine grain.

One of the most important factors to predict accurately is dynamically recrystallized grain size because dynamically recrystallized grain size has a great influence on average grain size. Average grain size is determined as mixture of initial grain size and dynamically recrystallized grain size. Figure 3 shows the relationship between dynamically recrystallized grain size and Zener-Hollomon parameter. Two kinds of relationship can be seen in Figure 3. One is observed in the area with relatively low density of primary gamma prime particles. The other is observed in the area with relatively high density of primary gamma prime particles i.e. gamma prime banding. The grain size tend to be coarse in the former condition and be fine in latter condition. Each relationship is represented with the following equation.

$$d_{dyn} = A_{dyn} Z^{n_{dyn}} \qquad (2)$$

where d_{dyn} is dynamically recrystallized grain size, Z is Zener-Hollomon parameter, A_{dyn} and n_{dyn} are material constants. Zener-Hollomon parameter means temperature-compensated strain rate and follows as

$$Z = \dot{\varepsilon} \exp(Q_{def}/RT) \qquad (3)$$

where $\dot{\varepsilon}$ is strain rate, Q_{def} is activation energy for deformation, T is absolute temperature, R is gas constant. Activation energy

for deformation Q_{def} is estimated by following equation that represents deformation behavior at high temperature.

$$\dot{\varepsilon} = \sigma^{n_{def}} \exp(-Q_{def}/RT) \qquad (4)$$

where σ is peak stress or steady stress, n_{def} is material constant. In this study, peak stress is adopted to estimate n_{def} and Q_{def} in equation (4). The exponent n_{def} takes constant value of 4.6 at temperature above gamma prime solvus and is larger value below the temperature. Corresponding activation energy Q_{def} is 360kJmol^{-1} or larger. The value of estimated activation energy is apparent activation energy and includes the effect of non-thermally activated process. Generally, larger value is obtained compared to volume diffusion of Ni or alloying element in Ni and similar tendency is observed in this study.

It is interesting that two kind of different relationships are shown in Figure 3 even when various parameters except primary gamma prime distribution are equivalent to each other. Dynamic recrystallized grain size can be described as a function of Zener-Hollomon parameter in each region. But different constants A_{dyn} and n_{dyn} in equation (2) are obtained. The absolute value of n_{dyn} in the area with relatively high density of primary gamma prime particles is smaller than that of low density area. And the fitting curve for high density gamma prime region is positioned below the curve for low density area. This fact means that primary gamma prime particles act effectively as nucleation site and obstacle for grain boundary migration. Consequently Zener-Hollomon parameter dependence becomes relatively small in gamma prime banding.

Precise prediction is required for fraction of dynamic recrystallization too. Equation (5) is the most common form which represents the behavior of fraction change.

$$X_{dyn} = 1 - \exp(-kt^n) \qquad (5)$$

where X_{dyn} is fraction of dynamic recrystallization, t is time, k and n are constants. But another Avrami type equation (6) was maintained in this study.

$$X_{dyn} = 1 - \exp\{-\ln2 \, (\varepsilon/\varepsilon_{0.5})^{n_{dynx}}\} \qquad (6)$$

where ε is strain, $\varepsilon_{0.5}$ is strain for 50% dynamic recrystallization, n_{dynx} is constant. $\varepsilon_{0.5}$ is function of Zener-Hollomon parameter and initial grain size. This equation shows that fraction of dynamic recrystallization is affected by temperature, strain rate, strain and initial grain size. Figure 4 shows the relationship between fraction of dynamic recrystallization and strain when initial grain size is 73 μm, test temperature is 1130C and 1150C. Obtained fitting curve is able to trace the actual tendency quantitatively. The higher the temperature is and the smaller the strain rate is, the higher the fraction of dynamic recrystallization is. The exponent n_{dynx} which reflects the mechanism of nucleation and growth takes 1.7

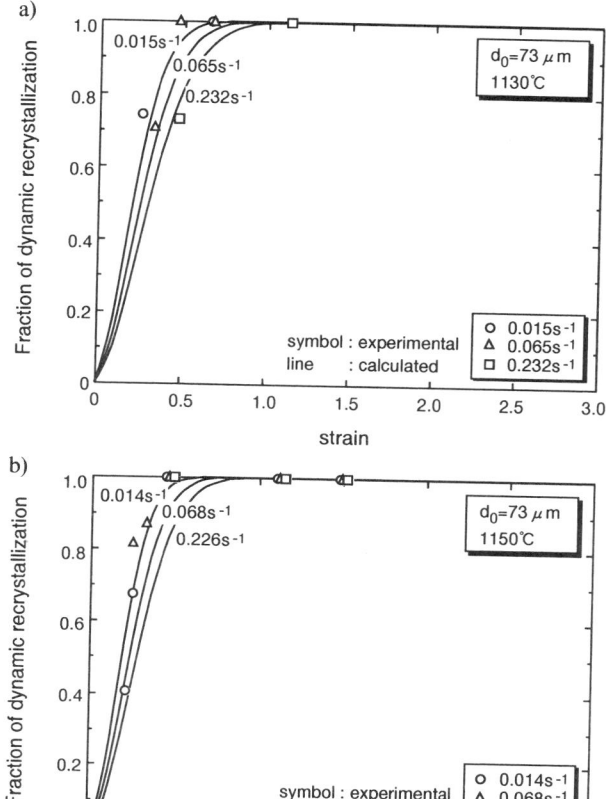

a)

b)

Figure 4 :The relationship between fraction of dynamic recrystallization and stain a)initial grain size; 73 μ m, test temperature; 1130C b)initial grain size; 73 μ m, test temperature; 1150C.

in 1130C and 1.9 in 1150C. However, the value of n_{dynx} changes as strain strictly. The change corresponds to various phenomena. Dynamic recrystallization occurs not only in unrecrystallized region but also in recrystallized region at the latter stage whereas it occurs only in unrecrystallized region at the early stage. Recrystallized grain is able to grow relatively free at early stage but the growth is restricted by next grains at later stage. Hence the estimated n_{dynx} is average value for prediction and dose not always represents the strict behavior at particular stage.

There are some amounts of possibility that grain growth occur in actual parts due to the deformation heat and stored strain. Therefore it is important to suppress the grain growth during and after forging in order to obtain the fine grain structure in whole disk. Moreover the degree of grain growth has to be predicted quantitatively when temperature exceeds the critical temperature. Hence accurate prediction of grain growth phenomena just after forging is required in particular. Figure 5 shows the grain growth behavior   of dynamically recrystallized grain   after isothermal

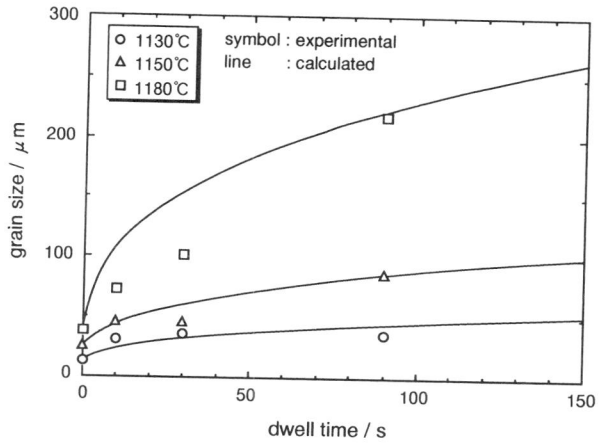

Figure 5 :Grain growth behavior after forging.

compression. Very rapid grain growth tendency is observed. The relationship between grain size and dwell time is described with following equation (7) similar to equation (1).

$$d_{dwe}{}^{n_{dgr}}-d_{com}{}^{n_{dgr}}=A_{dgr}\, t\, exp(-Q_{dgr}/RT) \qquad (7)$$

where d_{dwe} is the grain size after dwell, d_{com} is the grain size just after compression, Q_{dgr} is activation energy for grain growth in dwell condition, T is absolute temperature, R is gas constant and A_{dgr}, n_{dgr} are material constants. The exponent n_{dgr} takes 3 which is identical value in equation (1).

Pancake Forging

Figure 6 shows the predicted average grain size and actual observed one in case of pancake forging. The work shape is simple in this case, but the deformation is not uniform. For example, the strain distribution varied from low strain area near the die contact surface to high strain area at the center of the work. Therefore, distribution and history of temperature, strain rate and total strain seem to be complicated too. But in any forging temperature, the difference between predicted average grain size and observed one is relatively small and predicted values have good agreement with observed ones in spite of the large temperature dependence of microstructure change on alloy720LI. It does not need to say that precise prediction is required when the forging temperature is relatively low i.e. sub-solves temperature. However, it is also important to predict the degree of grain growth quantitatively when temperature exceeds the gamma prime solvus because there are some amount of possibility that grain growth may occur partially in actual large and complicated portion due to the deformation heat even if preheating temperature is sub-solvus one.

Generic Shaped Disk Forging

131

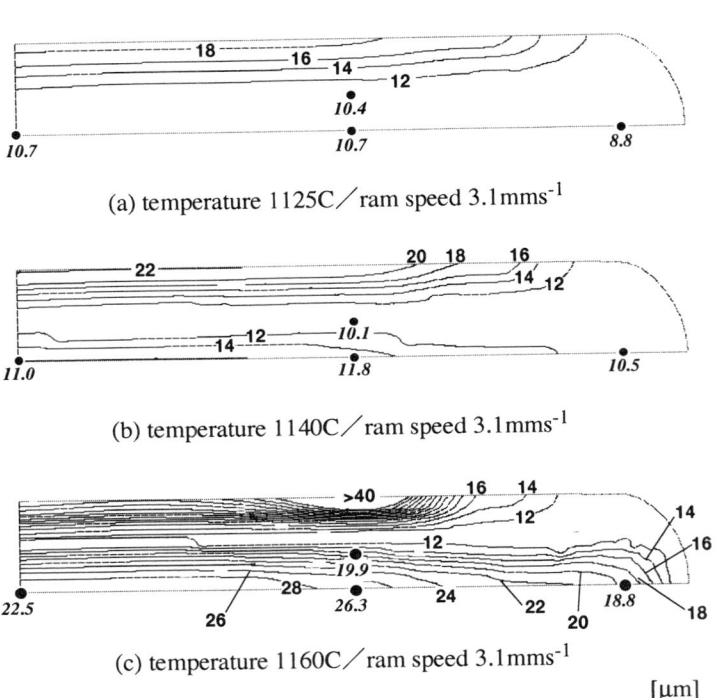

(a) temperature 1125C／ram speed 3.1mms^{-1}

(b) temperature 1140C／ram speed 3.1mms^{-1}

(c) temperature 1160C／ram speed 3.1mms^{-1}

[μm]

standard : **calculation**	
italic	: *experiment*

Figure 6 : Comparison between predicted and observed average grain size distribution in case of pancake forging (a)1125C with 3.1mms^{-1} ram speed (b)1140C with 3.1mms^{-1} ram speed (b)1160C with 3.1mms^{-1} ram speed.

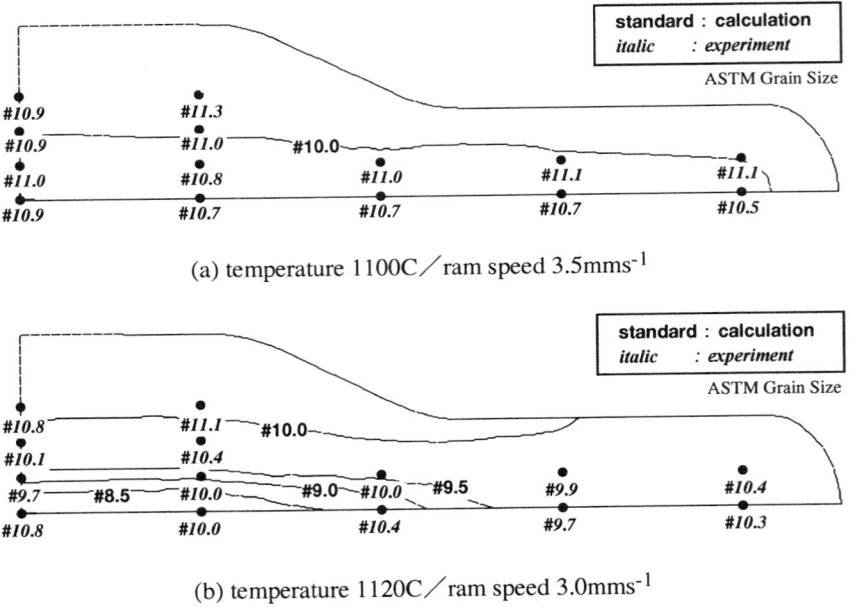

(a) temperature 1100C／ram speed 3.5mms^{-1}

(b) temperature 1120C／ram speed 3.0mms^{-1}

Figure 7 : Comparison between predicted and observed average grain size distribution (a)1100C with 3.5mms^{-1} ram speed (b)1120C with 3.0mms^{-1} ram speed.

The deformation is more complicated in this actual disk shape than in pancake one. Two forging conditions were chosen due to proposed simulation system. One gives the best microstructure distribution for disk, around ASTM No.10 at any portion of the work. The other gives some amount of distribution, from ASTM No.8.5 to 10. Figure 7 shows the comparison of prediction and observation in two conditions mentioned above. In each condition, good agreement between predicted values and observed ones was obtained in spite of complicated deformation.

Conclusions

The microstructure prediction procedure of cast/wrought form alloy720LI for turbine disk production has been developed. The essential relationship between microstructure and various parameters, e.g. temperature, strain rate, strain was revealed through isothermal heating tests and isothermal compression tests. Obtained relationship was formulated as static grain growth model and dynamic recrystallization model. The models were coupled to finite element analyzing system. In order to verify the effectiveness of proposed prediction system, pancake forging and generic shaped disk forging were conducted. It was confirmed that accuracy of prediction was enough to apply to practical forging process.

Acknowledgment

A part of achievement in this investigation was obtained under commission of The Society of Japanese Aerospace Companies, Inc. (SJAC). The authors would like to thank Ishikawajima-Harima Heavy Industries Co., Ltd. for collaborating on the work.

References

1. P.W.Keefe, S.O.Mancuso and G.E.Maurer, "Effects of Heat Treatment and Chemistry on the Long-Term Phase Stability of a High Strength Nickel-Based Superalloy," Superalloys 1992, (1992), 487-496.

2. K.A.Green, J.A.Lemsky and R.M.Gasior, "Development of Isothermally Forged P/M Udimet720 for Turbine Disk Applications," Superalloys 1996, (1996), 697-703.

3. H.Hattori et al., "Evaluation of P/M U720 for Gas Turbine Engine Disk Application," Superalloys 1996, (1996), 705-711.

4. D.J.Bryant, G.McIntosh, "The Manufacture and Evaluation of a Large Turbine Disc in Cast and Wrought Alloy 720Li," Superalloys 1996, (1996), 713-722.

5. G.Shen, J.Rollins and D.Furrer, "Microstructure Modeling of Forged Waspaloy Discs," Superalloys 1996, (1996), 613-620.

6. A.J.Brand, K.Karhausen and R.Kopp, "Microstructual Simulation of Nickel Base Alloy Inconel718 in production of turbine discs," Mat. Sci. and Tech., 12(1996), 963-968.

Unpublished Mitsubishi Materials data.

Characteristics and Properties of As-HIP P/M Alloy 720

J. H. Moll and J. J. Conway
Crucible Materials Corporation

Abstract

Alloy 720 is currently being used as a conventional cast and wrought material for turbine engine disks. Recently, the alloy was evaluated as an extrude plus isothermally forged (E+I) P/M material with excellent results. In the present work, P/M 720 was evaluated in the hot-isostatically-pressed plus heat treated (as-HIP) condition since this type of processing represents the potential for lower cost through elimination of the extrusion and forging steps to produce near-net shapes. Tensile, stress rupture and LCF properties were evaluated. The results compare favorably with those of (E+I) P/M material and indicate that as-HIP P/M Alloy 720 may be suited for engine application.

Superalloys 2000
Edited by T.M. Pollock, R.D. Kissinger, R.R. Bowman,
K.A. Green, M. McLean, S. Olson, and J.J. Schirra
TMS (The Minerals, Metals & Materials Society), 2000

Introduction

Alloy 720 is a high strength nickel based superalloy which was originally developed as a wrought turbine blade alloy for industrial turbines (1). The alloy is currently being used in the cast and wrought form as a turbine disk alloy. More recently, the alloy has been evaluated as a powder metallurgy (P/M) material processed using extrusion plus isothermal forging (E+I) (2,3). Reportedly, the alloy has excellent fatigue crack growth resistance and is being strongly considered for application in turbine disks for small to medium gas turbine engines. As-HIP refers to hot-isostatically-pressed plus heat treated material processed without further thermo-mechanical treatment. A major benefit of as-HIP processing is lower cost through the elimination of extrusion and isothermal forging and the reduction of input material through near-net shape capability. This approach has been successfully applied to Rene 95 and Low Carbon Astroloy with over 100,000 turbine engine components flying world wide (5). The objective of this work was to evaluate the mechanical properties of as-HIP Alloy 720 to determine if this alloy could be used in as-HIP condition in engine applications.

Billet Manufacture

Material used in this program was from 6.5-inch (165-mm) diameter billet produced using argon atomized Alloy 720 powder consolidated by hot-isostatic-pressing followed by heat treatment. The process is shown schematically in Figure 1.

In powder manufacture, vacuum induction melted metal is delivered to the atomizer as a thin stream which is impinged upon by high pressure argon gas. The metal stream breaks up via the transfer of kinetic energy to surface energy. Spherical droplets are formed due to surface tension and solidify in the range of 10^4 to 10^6 C/sec. For the current study, the powder was produced as several 5000 lb heats which were subsequently combined to make a master powder blend. Once atomized, the powder was screened to -270 mesh (-53 μm). All powder processing was performed in equipment and systems made from stainless steel. Specially designed valves and powder transfer bins were also used to insure powder cleanliness.

The powder was subsequently loaded into mild steel containers which were then outgassed and sealed. The containers were HIPed at 2065F (1129C)/15 ksi (103 MPa)/4 hr. The composition of the resulting P/M 6.5-inch (165 mm) diameter billet is given in Table I.

Table I Composition of P/M Alloy720

Composition - Wt%			
Carbon	0.010	Tungsten	1.28
Chromium	16.57	Boron	0.012
Cobalt	14.71	Zirconium	0.038
Molybdenum	3.00	Aluminum	2.49
Titanium	5.02	Nickel	Balance

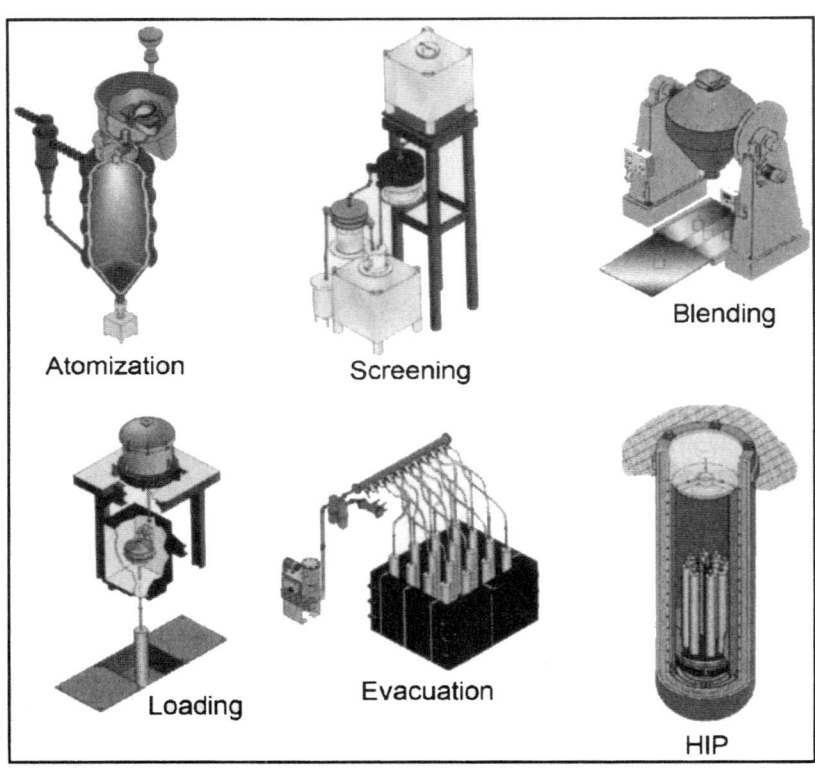

Figure 1. Schematic showing the steps involved in the as-HIP P/M process

136

Billet Evaluation

Microstructure

Initially, a solution treatment study was conducted in the temperature range of 2000F (1193C) to 2150F (1177C). Microstructural examination showed the γ' solvus temperature for the material to be between 2075F (1135C) and 2100F (1149C). This is congruent with that reported in the literature (5). As shown in Table II, heat treatment below the solvus temperature resulted in a fine uniform structure with a grain size of ASTM 10 and finer (Figure 2a). Heat treatment above the solvus resulted in a grain size of ASTM 8.5-9 which was stable up to at least 2150F (1177C) (Figure 2b). Based on these studies, 2040F (1116C) was selected as a sub-solvus solution treatment to maintain a fine uniform grain structure for high yield strength. A super-solvus treatment of 2150F (1177C) was selected to increase the grain size and maximum solutioning of second phase particles for optimal stress rupture properties.

Table II   Effect of Solution Treatment Temperature on Grain Size of As-HIP Alloy 720

Solution Treatment Temperature	Average Grain Size	
	ASTM No.	μm
2000F (1093C)	11.0	8.0
2025F (1107C)	10.5	9.4
2050F (1121C)	10.5	9.4
2075F (1135C)	10.0	11.0
2100F (1144C)	9.0	16.0
2125F (1163C)	8.5	18.9
2150F (1177C)	8.5	18.9

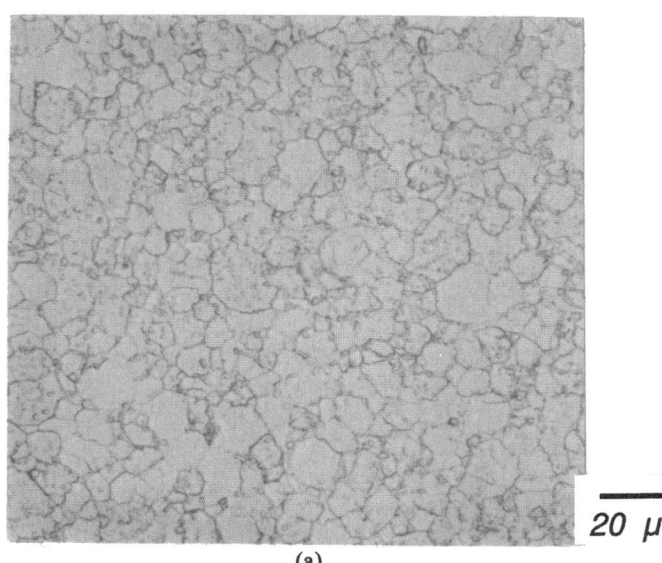

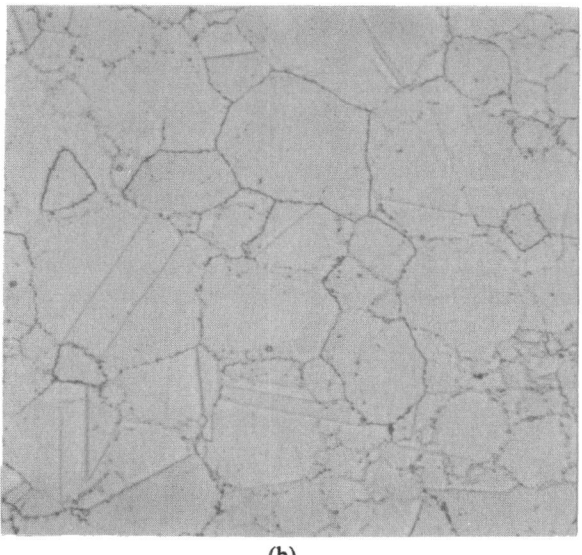

20 μm

(a)   (b)

Figure 2. Microstructure of as-HIP P/M Alloy 720 solution treated at (a) 2050F (1121C) and (b) 2150F (1177C)

Mechanical Properties

To determine the effect of grain size on the mechanical properties of as-HIP Alloy 720, tensile, creep-rupture and low cycle fatigue (LCF) properties were evaluated for both sub-solvus and super-solvus solution treated material. Following solution treatment, the materials were aged at 1400F (760C)/8 hr + 1200F (649C)/24 hr.

Tensile tests were conducted over the temperature range from room-temperature to 1600F (871C). Figure 3 shows the effect solution treatment temperature on the tensile properties of as-HIP 720 at room temperature and 1000F (538C), respectively. As expected, sub-solvus solution treatment results in higher strength and ductility than super-solvus solution treatment due to differences in grain size. The tensile properties from room temperature to 1600F (871C) for sub-solvus and super-solvus treatments are shown in Figures 4 and 5, respectively. As can be noted, as-HIP Alloy 720 retains excellent strength and ductility up to at least 1200F (1121C). As shown in Figure 6, the tensile properties also compare favorably

with E+I P/M material.

Stress-rupture properties were determined in the 1200F (1121C) to 1450F (788C) temperature range   The data for as-HIP along with those for E+I P/M Alloy 720 are summarized in Figure 7. As can be noted, sub-solvus treated as-HIP Alloy 720 is quite comparable to E+I P/M material over the entire temperature range. Super-solvus treatment resulted in a significant increase in rupture strength primarily due to the coarser grain size which results from this treatment.

Smooth bar LCF tests were conducted on as-HIP Alloy 720 at 1000F (538C). The tests were conducted at an R = 0 and a frequency of 20 cpm in strain control. The results are given in Table III and summarized in Figure 8. As shown in Figure 9, the strain for a 30,000 cycle mean initiation life of 0.9 % compares favorably with that reported for E+I P/M Alloy 720 and is higher than that of conventional cast and wrought Alloy 720 (2).

137

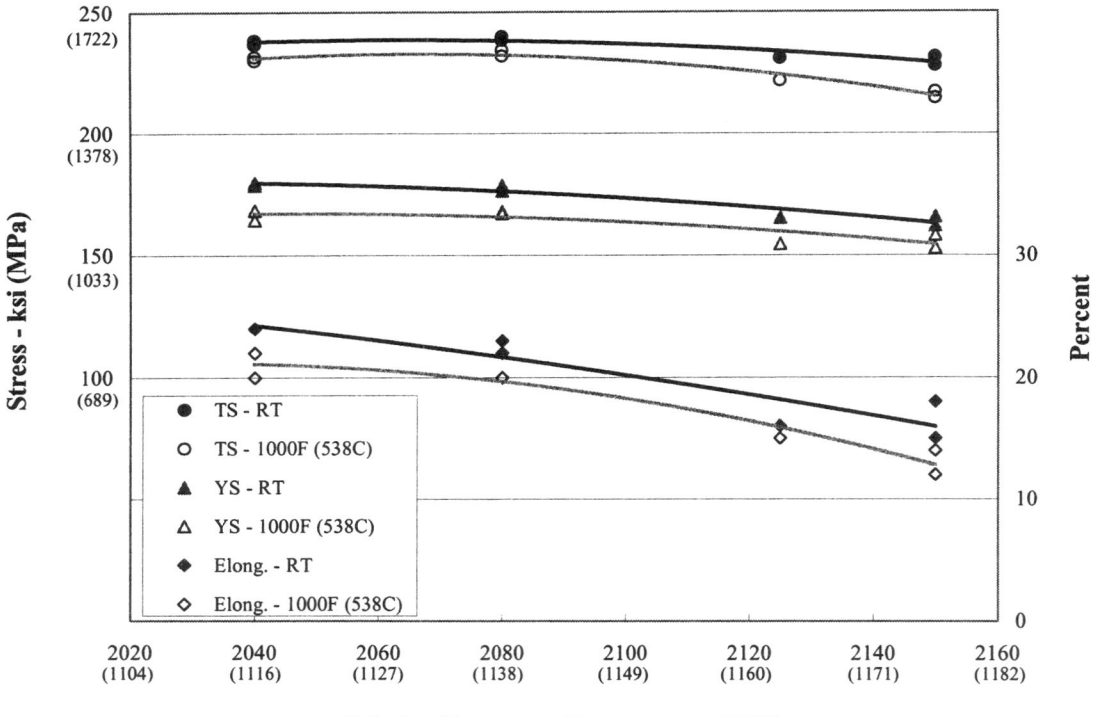

Figure 3. Effect of solution treatment temperature on room temperature and 1000F (538C) tensile properties of as-HIP Alloy 720. Material was solution treated at the indicated temperature and then aged at 1400 F (760C) / 8 hr + 1200 F (649C)/ 24 hr.

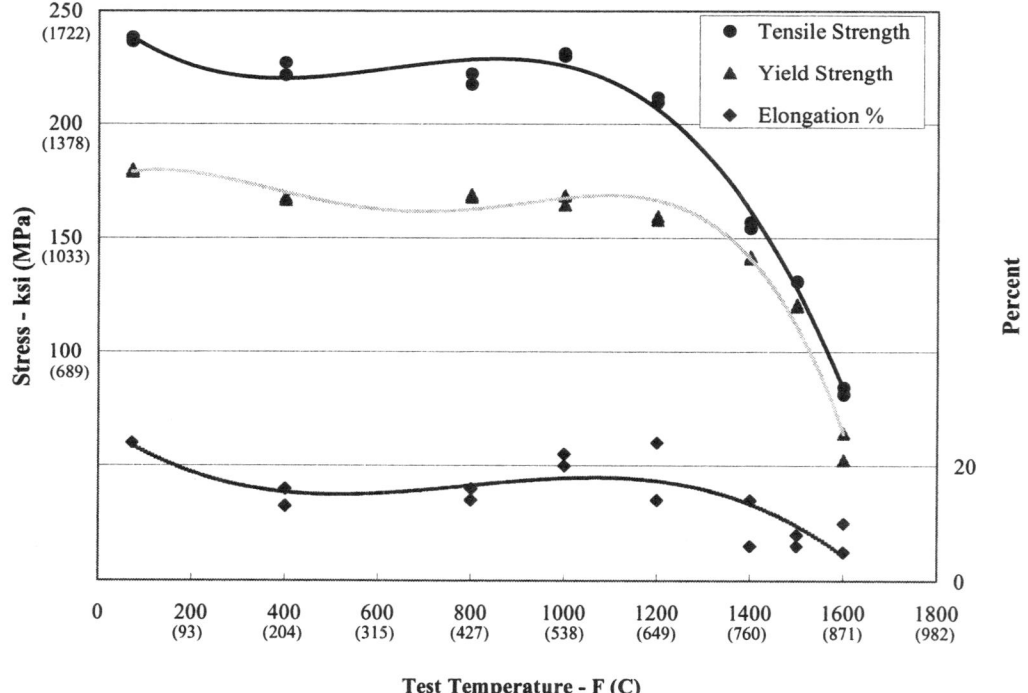

Figure 4. Tensile properties of as-HIP Alloy 720 sub-solvus solution treated at 2040F (1116C) and aged at 1400 F (760C)/ 8 hr + 1200 F (649C)/ 24 hr.

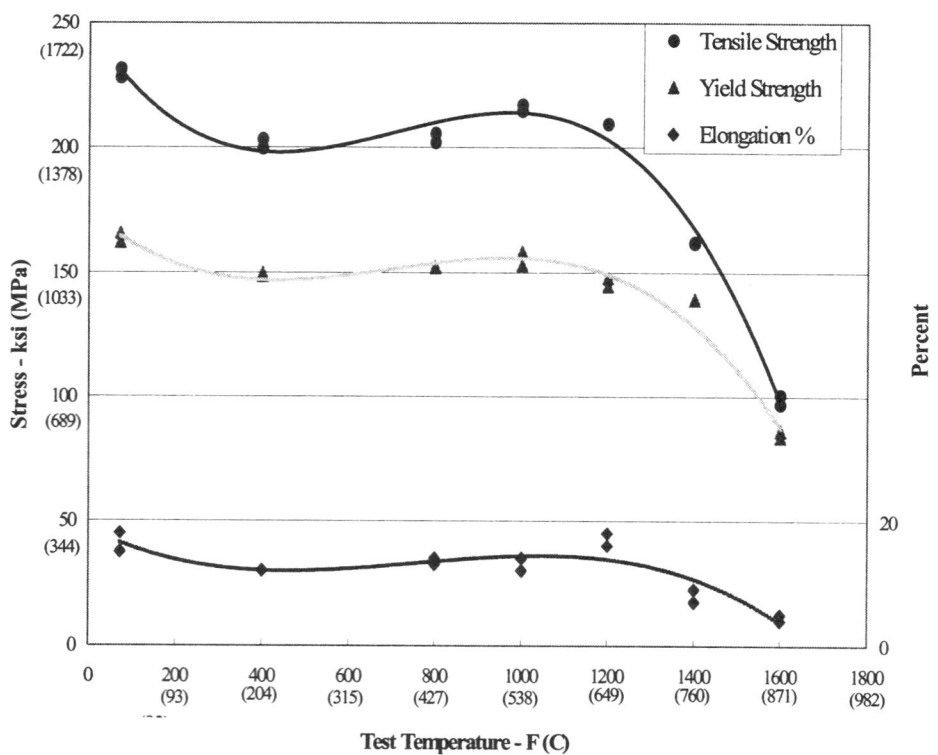

Figure 5. Tensile properties of as-HIP Alloy 720 super-solvus solution treated at 2150F (1177C) and aged at 1400 F (760C)/ 8 hr + 1200 F (649C)/ 24 hr.

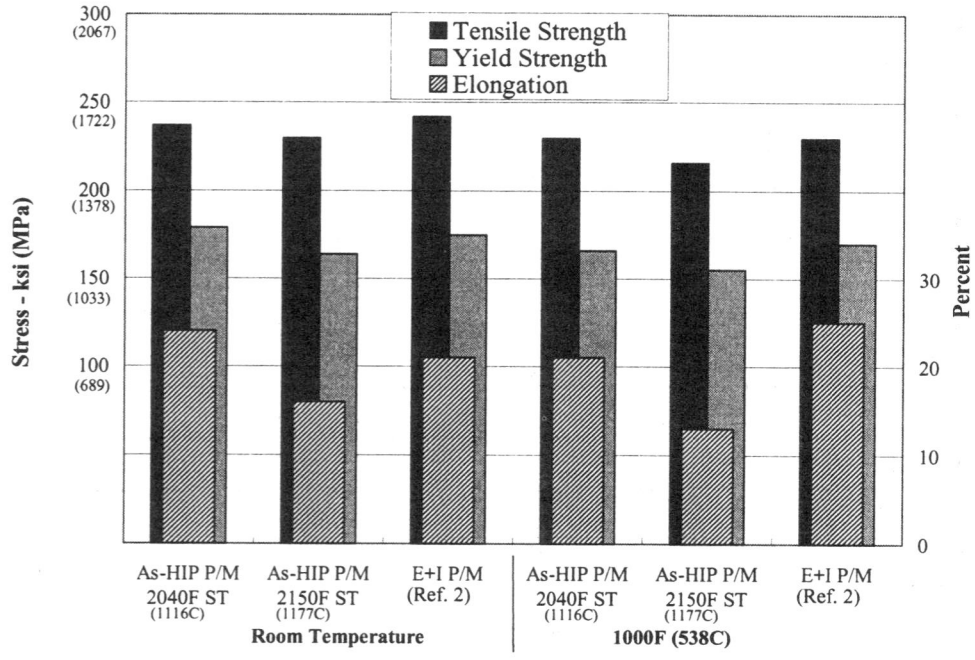

Figure 6. Comparison of the tensile properties of as-HIP Alloy 720 with E+I P/M material in the solution treated and aged condition.

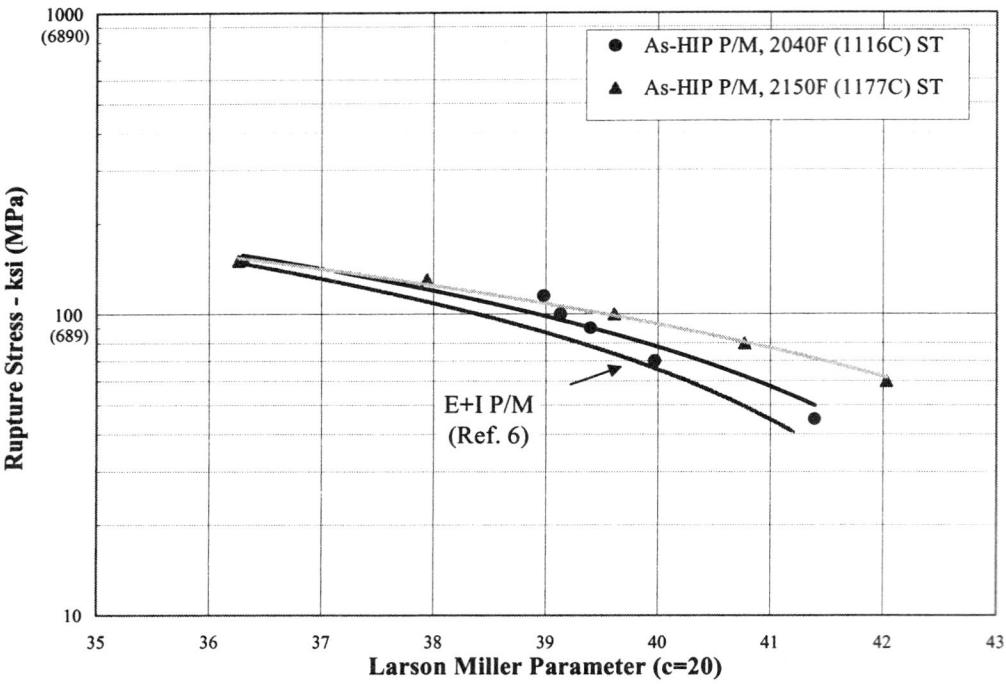

Figure 7. Stress rupture properties of as-HIP P/M and E+I P/M Alloy 720 in the solution treated and aged condition.

Table III Results of 1000F (538C) LCF Tests of As-HIP Alloy 720

Solution Temperature	Strain Range (%)	Cycles to Crack Initiation	Fracture Initiation Source		
			Size (milli-inches^2)	Type	Composition
2040F (1116C)	0.86	80,703	3.58	Ceramic	Al, O
2040F (1116C)	0.86	80,881	2.32	Ceramic	Al, O
2040F (1116C)	0.90	7,635	0.78	Ceramic	Al, O
2040F (1116C)	0.90	45,022	3.54	Ceramic	Al, Mg, Zr, O
2040F (1116C)	1.00	7,843	0.60	Pore	-
2040F (1116C)	1.00	8,130	0.48	Pore	-
2150F (1177C)	0.86	11,450	-	Grain	-
2150F (1177C)	0.92	11,118	-	Grain	-
2150F (1177C)	1.00	6,485	-	Grain	-

* Material tested at $K_t = 0$, R = 0, 20 cpm under strain control

140

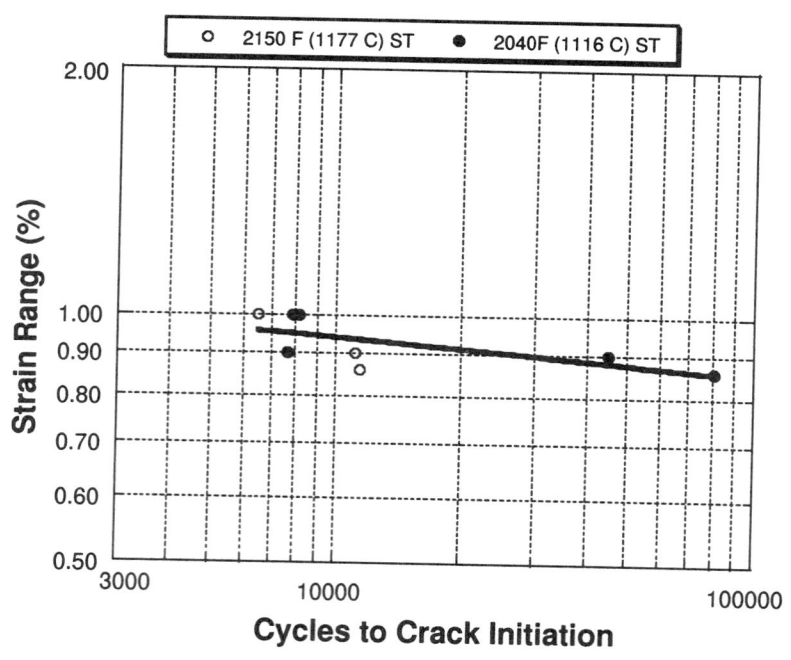

Figure 8. Low cycle fatigue properties of as-HIP P/M Alloy720 at 1000F (538C). Material was solution treated at the indicated temperature and aged at 1400 F (760C)/ 8 hr + 1200 F (649C) / 24 hr.

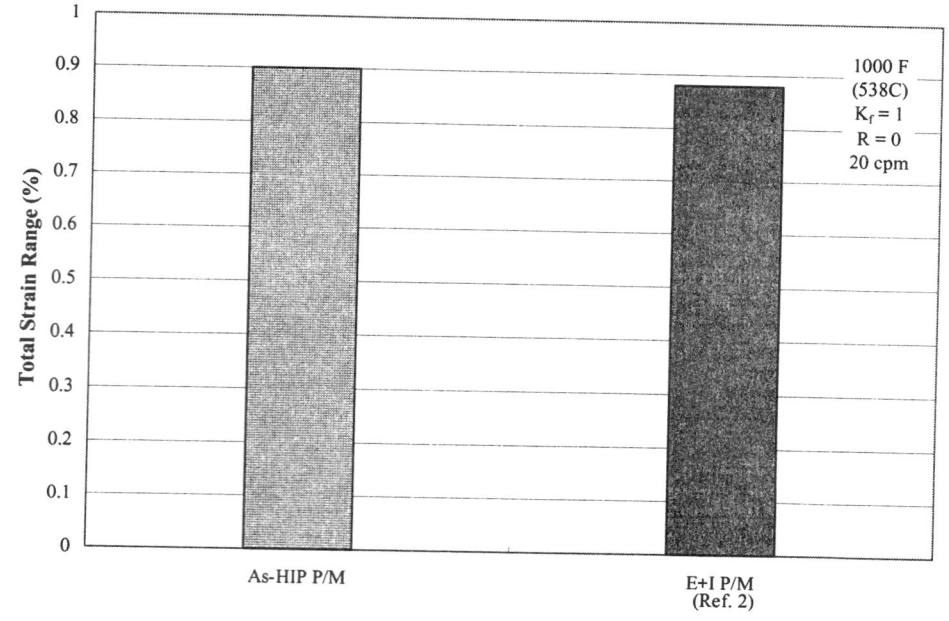

Figure 9. Comparison of as-HIP and E+I P/M Alloy 720 of total strain for 30,000 cycles mean initiation life.

Fracture surfaces of LCF specimens were examined to determine the features of fracture initiation sites. The results are included in Table III. The LCF bars from sub-solvus treated material failed at ceramic inclusions or pores while super-solvus treated material failed at grains. The inclusion size, type and composition observed as fracture initiation sites are similar to those observed in other E+I P/M superalloys produced from -270 mesh powder.

Summary and Conclusions

As-HIP Alloy 720 was evaluated for tensile, stress rupture and low cycle fatigue properties at room- and elevated-temperatures. The results show that the material exhibits excellent properties over a wide temperature range. The properties compare favorably with conventional ingot metallurgy and E+I P/M material.

The results of the work conducted to date indicate that as-HIP P/M Alloy 720 may be suited for engine applications. Additional testing is planned to further evaluate the low cycle fatigue behavior and fatigue crack growth resistance of this material to more fully evaluate its potential.

References

1. H. Pushnik, G. Zeiler, J. Fladisher, W. Eber and K. Keienburg, "Udimet 720 Turbine Blades Production and Properties", *Evolution of Advanced Materials,* Associazione Italiana di Metallurgia, Milano, 1989.

2. K. A. Green, J. A. Lemsky and R. M. Gasior, Development of Isothermally Forged P/M Udimet 720 for Turbine Disk Applications", *Superalloys 1996*, R. D. Kissinger, et al (eds), The Minerals, Metals & Materials Society, 1996.

3. H. Hattory, M. Takekawa, D. Furrer and R.J. Noel, Evaluation of P/M U720 for Gas Turbine Application", *Superalloys 1996*, R. D. Kissinger, et al (eds), The Minerals, Metals & Materials Society, 1996.

4. J. H. Moll, J. J. Conway and B. J. McTiernan, "As-HIP P/M Superalloys: A Technical and Commercial Success" *Advances in Powder Metallurgy and Particulate Materials*, MPIF, 1999.

5. J. Hyzak, R. Singh, J. Morra and T. Howson, "The Microstructural Response of As-HIP P/M U-720 to Thermomechanical Processing", *Superalloys 1992*, S. Antolovich et al, (eds), The Minerals, Metals & Materials Society, 1992.

6. "Powder Metal Udimet 720 Turbine Engine Components" ASAATCOM Report No. TR-96-F001, April 1996.

ENHANCED POWDER METALLURGY (P/M) PROCESSING OF UDIMET®* ALLOY 720 TURBINE DISKS – MODELING STUDIES

James J. Fisher, Albert Casagranda, Suhas P. Vaze, Frederick D. Arnold,
Shih-Yung Lin, Gopal Salvady, and Gary A. Miller

Concurrent Technologies Corporation
100 CTC Drive
Johnstown, PA 15904

Abstract

Enhanced P/M processing encompasses powder preparation, hot isostatic pressing (HIP) consolidation of powder, extrusion, isothermal forging, ultrasonic inspection, and machining technologies. This combination of technologies is being evaluated for producing P/M UDIMET alloy 720 turbine disks with improved quality at reduced cost.

The three modeling studies reported here were undertaken to support process optimization efforts to meet the project quality and cost objectives. Defect migration modeling supports the selective ultrasonic concept that can, once validated, reduce the "buy-to-fly" weight of P/M billets and also the cost of ultrasonic inspection. HIP modeling can be used to identify the most cost-effective path to achieve the desired as-HIPed density. Simulation of the machining processes used to convert a near-net shape isothermal forging to a finished disk can help to define the optimum machining sequence to minimize machining cost and disk rejects. This report describes the results of efforts to develop and apply these modeling and simulation techniques.

Introduction

The concept of enhanced P/M processing of superalloys for aircraft engine components is being undertaken as a U. S. Navy MANTECH project monitored by the NAVAIR Systems command. The project involves an extensive experimental effort by an integrated project team. The team members include: Rolls-Royce Allison, the engine manufacturer, Special Metals Corporation, the P/M billet producer, Ladish Co., Inc., the isothermal forging manufacturer, and Concurrent Technologies Corporation, the project manager and operator of the National Center for Excellence in Metalworking Technology. This project extends the work begun in the Army Comanche helicopter program [1].

project team by providing a means of rapidly evaluating processing alternatives to shorten the search for the optimum routes. Such efforts presuppose that suitable physical process models can be identified and validated prior to playing the "what if" optimization games. The three areas chosen for study using modeling and simulation techniques are material flow and defect migration, HIP processing, and machining.

Modeling Studies

Material Flow and Defect Migration

The use of fine powder (−270 mesh) enables ultrasonic inspection of the full volume of extruded billets. If no defects are created during isothermal forging, the forged disk need only be inspected in highly stressed areas. This concept, termed selective ultrasonic, can reduce the cost of ultrasonic inspection. It cannot be used for cast/wrought UDIMET alloy 720 material because the variation in grain size and composition submerges the defect signals within the background noise.

In order for the selective ultrasonic concept to be accepted by aircraft engine manufacturers, it is necessary to show that defects of concern can be detected reliably at the billet stage and tracked through isothermal forging. This is the reason for developing the defect migration model. This model requires thermophysical [1] and flow property data (upset tests on 0.5-inch diameter cylinders) for UDIMET alloy 720 in the hot working temperature range (950 – 1200 °C) at suitable strain rates (0.0001/sec to 10/sec).

Defect migration was followed using the point-tracking feature of DEFORM-2D™ going from the HIP canister to the 3:1 extrusion, and finally to the isothermal forging of the Stage 3 disk. This sequence is illustrated in Figure 1 for two widely separated planes of seeds.

---

* UDIMET is a registered trademark of Special Metals Corporation.

DEFORM is a registered trademark of Scientific Forming Technologies Corportation.

The modeling and simulation efforts reported here were undertaken to support the experimental work of the integrated

Superalloys 2000
Edited by T.M. Pollock, R.D. Kissinger, R.R. Bowman,
K.A. Green, M. McLean, S. Olson, and J.J. Schirra
TMS (The Minerals, Metals & Materials Society), 2000

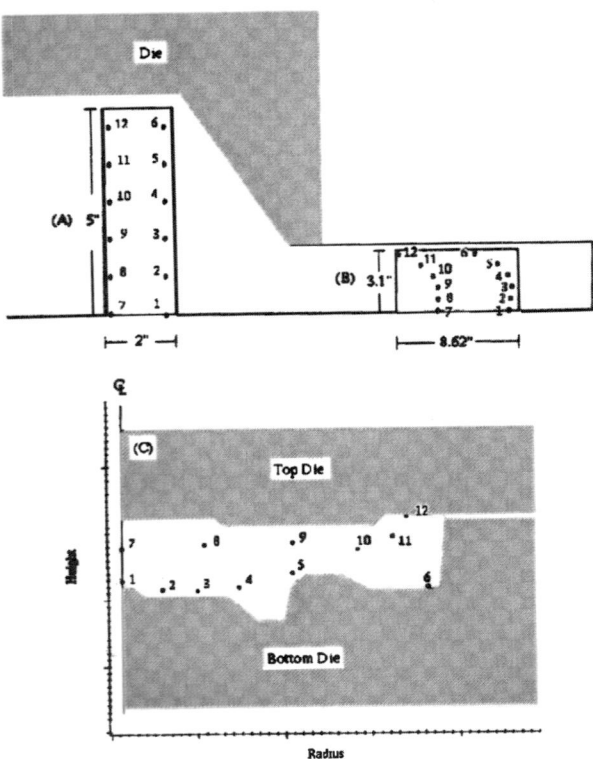

Figure 1: Defect migration simulation – A) HIP billet, B) 3:1 extrusion, C) Stage 3 disk

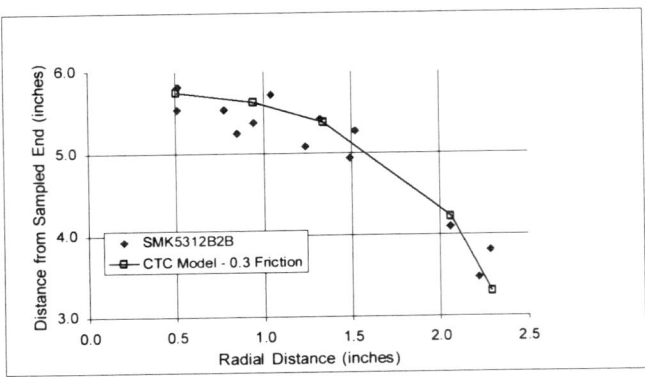

Figure 2: Comparison of predicted (finite element model) versus observed (SMK5312B2B) seed locations following HIPing and 3:1 extrusion

These planes were chosen to have the maximum separation while ensuring that the seed particles remain in the disk after isothermal forging. Points 1, 6, and 12 may have actually reached the surface.

In order to verify the numerical modeling techniques used to describe the material flow during extrusion and isothermal forging, a seeding experiment was devised. Alumina and silica particles (150 to 1000 μm in diameter) were deposited in a planar array of several layers in a HIP canister that was then extruded and sectioned. Special Metals Corporation used an ultrasonic immersion technique to find the seed particles after extrusion. A comparison of the defect migration predictions with the ultrasonic measurements, Figure 2, shows that the model appears to be a valid representation of these processes.

The curvature of the initially straight lines of seed particles (edge view) after extrusion, Figure 1B, matches results for cold extruded billets [2]. The validity of these results for extrusion, and ultimately for isothermal forging, depends on the validity of the assumptions. The defects of concern in UDIMET alloy 720 are likely to be refractory oxides. In modeling the movement of these particles it has been assumed that they (1) do not impede the flow and (2) do not change or interact with the superalloy powder particles. Based on extrusion results obtained to date, these assumptions seem reasonable. The final verification of the defect migration model will involve ultrasonic inspection, metallography, and comparison with model predictions for the disk shown in Figure 1C.

The existence of a valid defect migration model will enable a more efficient usage of billet material. If an inclusion (detected in the extruded billet) could end up in a highly stressed region of the disk, the location of the mult (within the billet) can be changed to prevent this from happening. This type of prescreening approach can avoid investing effort in a mult that could eventually be a discarded disk. The model will also be of use for detecting defect migration of particles producing ultrasonic signals below the threshold level of acceptance.

HIP Modeling

Although hot isostatic pressing (HIP) is an intermediate processing step within the enhanced P/M processing framework, considerable cost savings can be achieved through an optimization of the HIP process. The production HIP cycles used in industry are designed to attain the highest possible final density. However, since other processing steps (i.e., extrusion and forging) will be applied to the material during enhanced P/M processing, a density less than ~100% after HIP may be optimal. The goal of this effort is to develop an accurate HIP model for UDIMET alloy 720, so that the final density and microstructure can be accurately predicted. Using this model will allow an engineer to tailor the HIP cycle specifically for an application, thus reducing trial and error efforts and secondary operations.

The HIP model used in this program is based on a viscoplastic constitutive equation previously suggested for describing the densification of metal powders during HIP [3,4]. The powder aggregate is modeled as a compressible continuum with the following macroscopic potential ($\hat{\phi}$) describing its deformation

$$\hat{\phi} = S^2 + b(\rho)p^2 - c(\rho)s^2 = 0 \qquad (1)$$

where S is the magnitude of the deviatoric stress tensor, p is the pressure, and s is a measure of the deviatoric stress in the powder particle. In addition, $b(\rho)$ and $c(\rho)$ are functions of the relative density, ρ, and the form of these functions are determined by experiment.

To fully specify powder properties in the HIP model, one needs to determine the flow behavior of the powder particles, the functional forms of $b(\rho)$ and $c(\rho)$, the specific heat as a function of temperature for the fully dense powder, and the thermal conductivity of the powder as a function of both temperature and density [5]. Previously, a large set of HIP tests were required to calibrate the density dependence of the model, i.e., establish the functional form of $b(\rho)$ and $c(\rho)$. In the current project, these functions are determined using a smaller initial set of experiments and the forms of $b(\rho)$ and $c(\rho)$ based on other data from powders that have already been calibrated. The constants used in these functions are optimized for UDIMET alloy 720. This preliminary model will be used to prescribe additional HIP experiments to refine and validate the model.

The flow behavior of the powder is determined through a series of high temperature, constant strain-rate compression tests on samples over a range of density. Partially dense samples are obtained from interrupted HIP cycles and are used to calibrate the density dependence of the model. The partial HIP experiments also provide a means of validating the model after calibration. The thermophysical properties of the fully dense powder compact are normally determined using differential scanning calorimetry (specific heat) and the laser flash technique (thermal diffusivity). The density dependence of the thermal diffusivity must also be measured and is included in the model.

The test specimens for establishing the flow behavior of the powder are usually machined from as-HIPed material. However, in this case, enhanced P/M processed (i.e., HIPed, extruded, and forged) material was available and was used to save both costs and time during the initial model calibration. The experimental result of a series of compression tests on these fully dense UDIMET alloy 720 samples is shown in Figure 3.

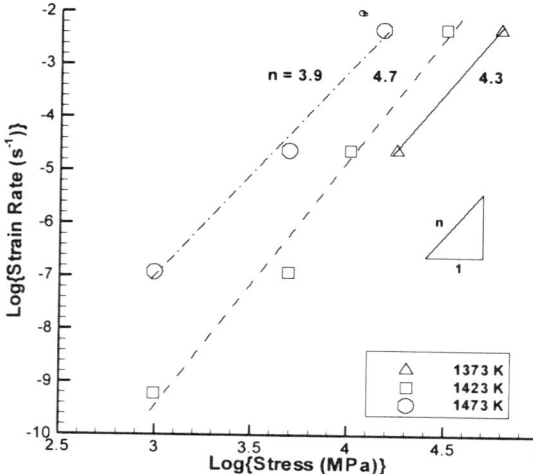

Figure 3: Creep stress exponent dependence on temperature for UDIMET alloy 720

A power law creep model, equation (2), was used to analyze the test results

$$\dot{\varepsilon} = A\sigma^n \exp\left(\frac{-Q}{RT}\right) \qquad (2)$$

where $\dot{\varepsilon}$ is the strain rate, A is a material constant, σ is the stress, n is the creep stress exponent, Q is the activation energy, R is the gas constant, and T is absolute temperature. Using the steady state results (true stress vs. true strain at constant strain rate) at different temperatures, the creep stress exponent can be determined. A least squares linear curve fit of the creep data for temperatures of 1100 °C, 1150 °C, and 1200 °C results in $n \sim 4$. In a similar fashion, the creep activation energy can be determined using the same set of compression tests as shown in Figure 4.

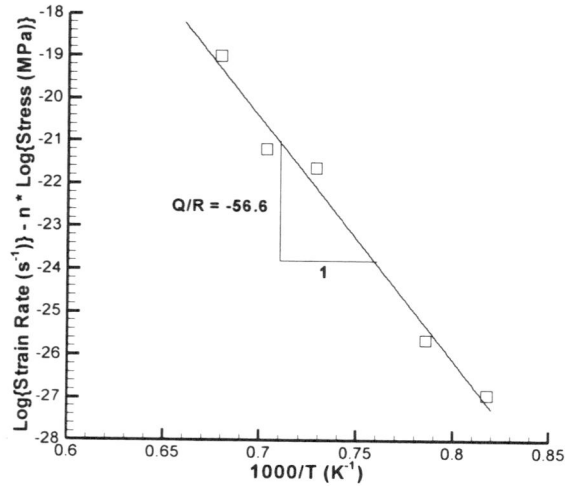

Figure 4: Creep activation energy for UDIMET alloy 720

In this case, the slope represents $-Q/R$, therefore $Q = 470$ kJ/mole. Finally, the material constant A is evaluated by plotting $\dot{\varepsilon}\exp(Q/RT)$ against $A\sigma^n$ on a log-log scale and determining the slope, which results in $A = 2.0E+08$. Additional testing will be required to more accurately establish these values for UDIMET alloy 720 over the full density range of interest. However, comparing the experimental data with the predictions of an ABAQUS™ simulation for these parameters shows good agreement for $\dot{\varepsilon}$ from 0.0001/sec to 0.1/sec, Figure 5.

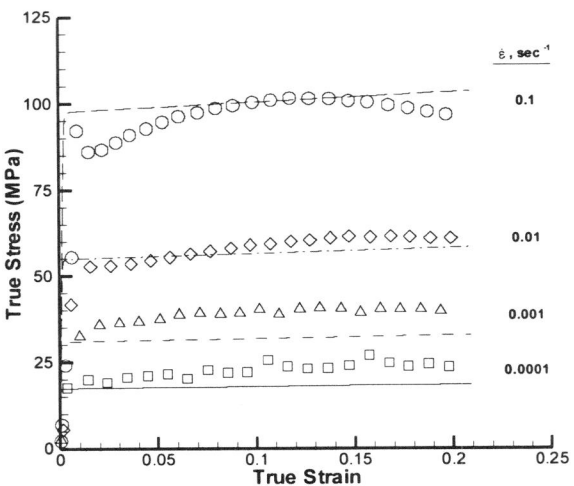

Figure 5: Comparison of experimental data (T = 1423 K) and ABAQUS simulation

Normally, the thermophysical property data are obtained from untested specimens machined from the set of HIP tests used to calibrate the model. Since the number of initial HIP experiments has been minimized in this case, another source for this information needed to be found. During the previous MANTECH program [1] on UDIMET alloy 720, measurements were made for specific heat and thermal conductivity as a function of temperature for fully dense material (HIP and extruded). The data were extrapolated to 1422 K for use in the HIP model (Figures 6 and 7).

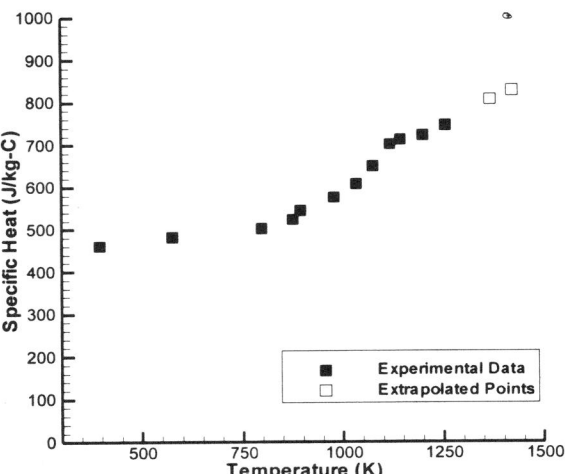

Figure 6: Specific heat as a function of temperature for UDIMET alloy 720

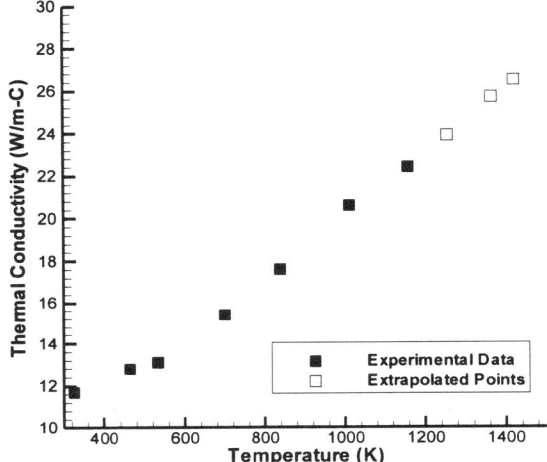

Figure 7: Thermal conductivity as a function of temperature for UDIMET alloy 720

The density dependence of the thermal conductivity for UDIMET alloy 720 is a required input of the HIP model. An estimate of this dependence was determined using thermal conductivity data for another Ni-based superalloy powder. Figure 8 shows the measured data for a range of temperatures and densities as well as the fitted equations used to interpolate the data. The exponential form of the curve fit worked well over the range of the data, since only the pre-exponential term varied with temperature. The following general form of the curve fit equation was used to scale the thermal conductivity (k) of fully dense UDIMET alloy 720 to reflect the influence of porosity

$$k = C(T) * \exp(4.0 * \rho)$$

where $C(T)$ is the temperature dependent term and ρ is the relative density. These estimated properties are suitable for the preliminary HIP model, but may be refined, if necessary, for the final model.

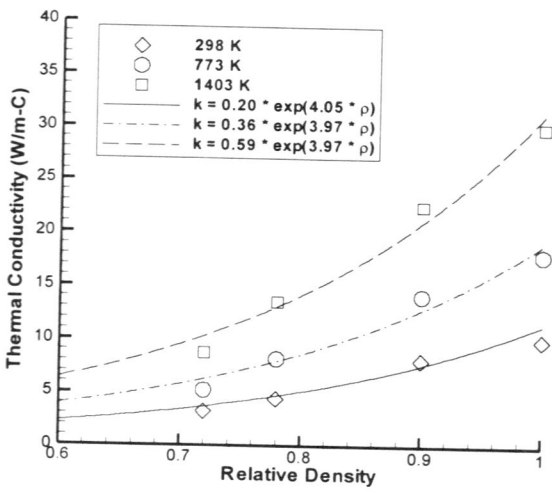

Figure 8: Thermal conductivity as a function of temperature and density for a Ni-based superalloy

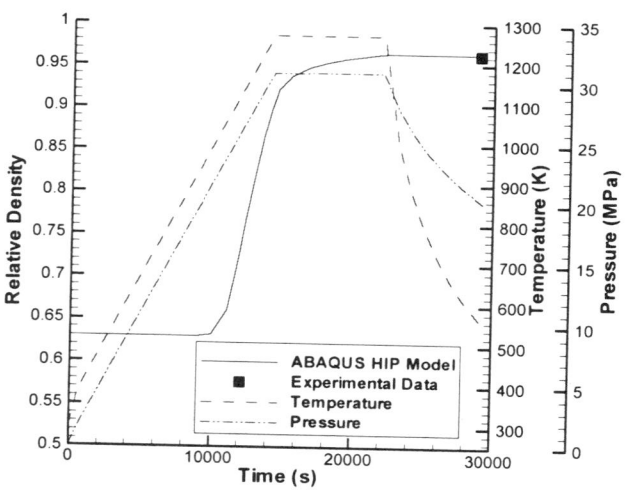

Figure 9: Relative density as a function of time for superimposed HIP cycle conditions

The densification rate during HIP is a strong function of both temperature and pressure. Therefore, the HIP conditions used to calibrate the model must be carefully specified to allow a range of final densities to be produced within a reasonable HIPing time. Once this HIP cycle has been determined, a series of interrupted HIP experiments are performed to produce specimens for further testing as well as to provide data for the model calibration.

Since a goal of this effort is to develop an accurate HIP model with as few experiments as possible, the initial attempts to determine a suitable partial HIP cycle were used to calibrate the preliminary HIP model. Figure 9 shows the model predictions and the experimental data point for the specified HIP cycle also shown. The preliminary model uses the following forms for the functions $b(\rho)$ and $c(\rho)$

$$b(\rho) = b_1 (1 - \rho)^{b_2}$$

$$c(\rho) = \rho^{c_1}$$

with the constants from a previously calibrated Ni-based superalloy as an initial estimate. The model used to predict the densification curve shown in Figure 9 was adjusted for UDIMET alloy 720 by scaling the parameter b_1 in order to agree with the data. This preliminary model was used to determine the partial HIP cycle parameters and additional HIP experiments required for model verification. These experiments are currently being performed.

Modeling of Residual Stresses and Distortions during Heat Treatment and Machining

This effort was undertaken because of the engine manufacturer's prior experiences with distortion occurring during machining of cast/wrought UDIMET alloy 720. Thus, distortion is caused by the redistribution of residual stresses resulting from differential cooling rate throughout the part during quenching. Subsequent aging treatments could reduce, but do not necessarily eliminate, these residual stresses.

A general non-linear finite element code (ABAQUS) was used to simulate the cooling behavior, predict residual stresses, and then model machining through removal of groups of elements. The first step in this method is to make a suitable mesh for the disk geometry. The second step is to simulate the quenching operation. This involves specifying a temperature-dependent, convection coefficient distribution on the disk surfaces and conducting a coupled thermo-mechanical analysis. Once the thermal simulation is complete, the residual stress distribution is available. After this, distortions are estimated during the simulated machining step. The removal of groups of elements changes the stresses in order to maintain static equilibrium. As it is currently configured, stress relaxation and the build up of residual stress due to the action of machining tools are not included in the model. Also, the disk is not fixtured or constrained during the simulated machining (removal of elements).

Figure 10 shows the forged geometry (Forged Shape) of the Stage 3 disk enclosing three intermediate machined shapes (M/C Shape 1, 2, and 3). Three operations are used to transform the forged shape to the final net shape, M/C Shape 3 (region 17). M/C Shape 1 results from removing regions 1 through 8; M/C Shape 2 from removing regions 9 through 15; and M/C Shape 3 from the removal of region 16.

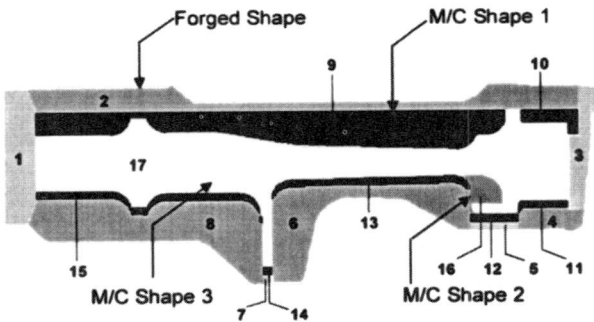

Figure 10: Shapes for disk : Forged and machined shapes

	1	2	3	4	5	6
Specification (s)	0.212	0.180	0.125	0.826	0.328	0.173
FEA Prediction (p)	0.214	0.179	0.149	0.804	0.319	0.170
Deviation (d)	x	x	+0.019	-0.017	-0.004	x

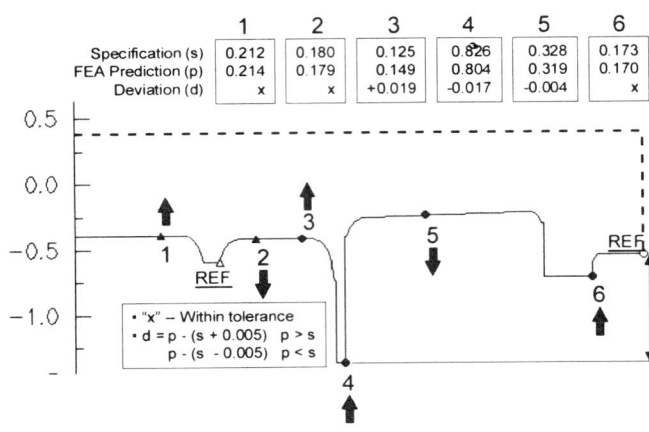

- "x" – Within tolerance
- d = p - (s + 0.005) p > s
     p - (s - 0.005) p < s

Figure 12: Specification and predicted locations for the bottom surface of M/C Shape 1. Specified tolerance limits (+/- 0.005 inch)

In an effort to validate the machining model and to compare the measured residual stress levels in P/M and cast/wrought UDIMET alloy 720, Moiré fringe strain measurements were made. These measurements ranging from −70 to −120 ksi showed that the circumferential and radial stresses were compressive and similar in magnitude to the average stresses for a disk. There appeared to be no significant difference in average residual stresses between P/M and cast/wrought UDIMET alloy 720. A comparison of the model predictions and the measured stresses are given in Figure 11.

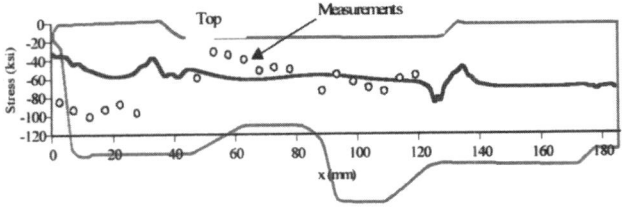

Figure 11: Comparison of Moiré fringe circumferential residual stress measurements and FEM predictions for the top surface for a cast wrought stage 3 disk

The best agreement is in the thinnest section near mid-radius. At the rim, the model predicts significantly lower stresses than were measured. The finite element predictions typically underestimated the Moiré fringe measurements by 15 to 50 percent. Experiments are currently underway to identify the source(s) of this disparity. Mesh size effects, the steepness of the stress gradients near the surface, the degree to which the thermal analysis represents the actual cooling conditions, and the surface preparations techniques for the Moiré fringe measurements, all could be contributors.

The engine manufacturer has set dimensional requirements for specific locations on the intermediate machined shapes in order to ensure that the final shape will be within tolerances. As an illustration, six locations were selected on the lower surface of the M/C Shape 1. The predicted vertical positions are compared to the specifications in Figure 12. The results show that predictions for locations 1, 2, and 6 are within the 0.005-inch tolerance while those for 3, 4, and 5 are out of tolerance. The results presented in Figure 12 are intended to illustrate the use of the machining model and are considered to be preliminary until they can be verified with actual machining data.

Conclusions

Based on the modeling and simulation approaches used to study defect migration, HIPing, and machining of P/M UDIMET alloy 720 turbine disk material, the conclusions are:

1. A model of oxide particle migration during HIPing and extrusion based on DEFORM-2D software provides accurate predictions of particle movement.
2. A model for describing the as-HIPed density as a function of HIP processing parameters has been developed for defining cost-effective processing.
3. A model of disk machining processes has been developed that has potential for identifying operation sequences to minimize distortion.

Acknowledgements

The authors acknowledge the support of the National Center for Excellence in Metalworking Technology (NCEMT), operated by Concurrent Technologies Corporation (CTC), under Contract N00140-92-C-BC49 to the U. S. Navy as part of the U. S. Navy Manufacturing Technology Program. The authors are grateful for the information supplied by, and the support of, the integrated project team – Mr. Kenneth A. Green of Rolls-Royce Allison, Mr. Joseph Lemsky of Ladish Co Inc., and Mr. Anthony Banik of Special Metals, Inc.

References

1. Kenneth A. Green, "Powder Metal Udimet 720 Turbine Engine Components," Light Helicopter Turbine Engine Company, (MANTECH Report No. TR-96-F-001, April 1996).

2. D.J. Blickwede, Metals Program, (Vol. 97, 1970), pp. 76-80.

3. R. Smelser, J. Zarzour, J. Xu, and J.R.L. Trasorras, "On the Modeling of Near Net Shape Hot Isostatic Pressing," (Paper presented at the International Mechanical Engineering Congress and Exposition, Chicago, Illinois, 6-11 November, 1994).

4. J.R.L. Trasorras, M.E. Canga, and W. Eisen, "Modeling the Hot Isostatic Pressing of Titanium Parts to Near-Net Shape," (Paper presented at PM²TEC, 1994) <u>Advances in Powder Metallurgy & Particulate Materials</u>, American Powder Metallurgy Institute, Princeton, NJ, 7 (1994), pp. 51-70.

5. J. Zarzour, J.R.L. Trasorras, J. Xu, and J. J. Conway, "Experimental Calibration of a Constitutive Model for a Hot Isostatic Pressing (HIP) of Metallic Powders" (Paper presented at PM²TEC, 1995) <u>Advances in Powder Metallurgy & Particulate Materials</u>, American Powder Metallurgy Institute, Princeton, NJ, 5 (1995), pp. 89-112.

Characterization and Thermomechanical Processing of Sprayformed Allvac® 720 Alloy

Henry E. Lippard and Robin Forbes Jones
Allvac
An Allegheny Technologies Company

Sprayforming is a promising route for production of highly segregation prone alloys such as the high γ′ fraction nickel alloys that are finding increased usage in the hot stages of aircraft turbines. Conventional ingot metallurgy (IM) is experiencing increasing difficulties with segregation and segregation-related defects such as freckles and white spots with the highly alloyed materials and probably will not be able to produce the next generation of hot stage alloys. Powder metallurgy (PM) is an expensive production route and the oxide cleanliness is inferior to IM. Sprayforming offers the segregation free microstructure of PM with the oxide cleanliness of IM through the clean metal sprayforming (CMSF) technique at a production cost intermediate between IM and PM.

Sprayformed Allvac® 720 Alloy was produced by nitrogen gas atomization and collected as 325 mm diameter billets. The nitrogen atomization eliminated the thermal induced porosity typical of powder metallurgy production using argon atomization and refined the initial structure relative to conventional cast material. The as-sprayed grain size is an equiaxed ASTM 6-7 and the carbonitride particle distribution is 1-2 μm. No segregation or segregation-related defects such as freckles or white spots were observed.

The response of sprayformed material to thermomechanical processing (TMP) is demonstrated by an experimental forging matrix measuring the effects of heat treatment, strain, and strain rate. The sprayformed material exhibits a strong recrystallization response and the only unrecrystallized grains are observed in low strain rate deformation conditions. The deformation temperature and strain rate are the primary variables affecting the flow stress and the prior heat treatment has a minor effect at lower temperatures in the hot working range.

Allvac is a registered trademark of ATI Properties, Inc.

Superalloys 2000
Edited by T.M. Pollock, R.D. Kissinger, R.R. Bowman,
K.A. Green, M. McLean, S. Olson, and J.J. Schirra
TMS (The Minerals, Metals & Materials Society), 2000

Introduction

The sprayforming production method is a competing manufacturing process to powder metallurgy (PM) and conventional ingot metallurgy (IM) for nickel-base superalloys. The sprayforming process in the simplest terms is atomization of a molten metal stream and collection of the resulting droplets in a "preform", which is analogous to an ingot. The atomization process is similar to powder metallurgy techniques but the droplets are not allowed to completely solidify before collection as a preform. The droplets are typically collected in a 60%-80% solid fraction range on a rotating starter plate that is withdrawn as the preform lengthens to maintain a constant deposition plane distance from the nozzle. The system complexity is quite high and the major variables are atomizing gas, atomizer design, gas:metal ratio, preform rotation speed, nozzle raster pattern, nozzle to preform distance, and molten metal pour rate.

Experimental Procedure

The preforms were sprayed in the United Kingdom by Osprey Metals Ltd* on pilot plant scale equipment with a 1200 kg melting capacity. The preforms were atomized with nitrogen gas to ensure any residual porosity or gas pockets would fully dissolve in the alloy and not reappear as thermal induced porosity at high temperatures. The nozzle experienced some clogging during the spray run that lowered the metal pour rate; thus the gas flow rate was reduced to maintain a constant gas:metal ratio. The as-sprayed preform sizes were 325 mm Ø x 1385 mm and 325 mm Ø x 880 mm length.

The preforms were sampled in steady-state regions near each end to provide the specimens for this study. The bulk of the preforms will be forged to billet based on these subscale processing results. The Gleeble compression specimens were heat treated to produce γ' dispersions A or B, then machined to 10 mm Ø x 15 mm length cylinders. The specimens were heated in an argon atmosphere at 10°C/s to the test temperature and then held for 5 minutes to equilibrate. The deformation was conducted in four steps of 0.2 true strain each with a one minute dwell time after each strain step. The specimens were water quenched after the final one minute hold to preserve the dynamic microstructure. The samples were sectioned, polished, and etched to reveal the centerline longitudinal plane for optical and SEM analysis.

Results and Discussion

The chemistry analyses listed in Table I demonstrate that no macrosegregation occurred during the sprayforming process. A microchemical analysis by energy dispersive spectrometry (EDS) compared the composition of the grain centers and boundaries to detect microsegregation, Table II. The composition differences between the grain centers and boundaries were not statistically significant above a 1σ range for the EDS measurement error nor did they exceed a customary definition of 5% difference for microsegregation. None of the major alloying elements are highly segregation prone and thus, Allvac® 720 Alloy is an ideal sprayforming candidate.

The high nitrogen level is consistent with absorption of the atomizing gas in the atomizing zone and as a hot preform. The center region remains higher in temperature due to the insulating effects of the outer preform layers and thus has higher nitrogen solubility. The top of the preform is also better insulated by the bottom of the preform relative to the initial starting conditions of

---

* Osprey Metals Ltd
  Red Jacket Works
  United Kingdom

the cold starter plate. The higher oxygen content in the preform bottom region is due to the initial air atmosphere in the spray chamber that becomes increasingly nitrogen rich as the spray time increases.

Table I Chemistry as a function of Preform Location

(wt%)	Top Surface	Top Center	Bottom Surface	Bottom Center
Ni	57.37	57.38	57.37	57.36
Cr	15.92	15.93	15.92	15.93
Co	14.50	14.49	14.52	14.54
Ti	5.00	4.98	5.00	4.99
Mo	3.03	3.03	3.03	3.03
Al	2.60	2.60	2.61	2.59
W	1.23	1.23	1.22	1.23
Fe	0.18	0.18	0.15	0.15
V	0.06	0.06	0.06	0.06
Zr	0.04	0.04	0.04	0.04
Ta	0.02	0.03	0.03	0.03
Nb	0.01	0.01	0.01	0.01
Si	0.01	0.01	0.01	0.01
Mn	0.01	0.01	0.01	0.01
Cu	0.01	0.01	0.01	0.01
B	0.015	0.014	0.015	0.015
C	0.016	0.016	0.014	0.016
N	0.0305	0.0320	0.0185	0.0257
O	0.0028	0.0026	0.0050	0.0047
P	0.003	0.004	0.003	0.003
S	<.0003	<.0003	<.0003	<.0003

Table II EDS Measurement of Microsegregation

	Avg. Grain Composition (wt%) Center	Avg. Grain Composition (wt%) Boundary	Avg. Grain Composition (wt%) GC-GB	EDS +/- 1σ range
Ni	57.05	57.68	-0.63	1.58
Cr	16.64	16.18	0.45	0.47
Co	15.00	14.62	0.38	0.96
Ti	4.74	4.95	-0.21	0.29
Mo	2.86	2.75	0.11	0.33
Al	2.23	2.31	-0.08	0.23
W	1.50	1.53	-0.03	0.34

The microstructure of the as-sprayed preform was similarly characterized as a function of location. Table III quantifies the grain size and carbonitride dispersions. Figure 1 shows the equiaxed grain structure inherent to the sprayforming process and illustrates the effects of the higher core temperature through the coarser grain size and decreased porosity at the preform center. The carbonitride dispersion is also coarser in the center and the particle aspect ratio is reduced.

Table III As-sprayed Microstructure of Alloy 720 Preform

	Surface	Center
Top		
Grain Size (ASTM)	7	6
Bottom		
Grain Size (ASTM)	7	6
Carbide Size (µm)		
Width, std. dev.	0.6 , 0.2	1.0 , 0.4
Length, std. dev.	1.1 , 0.4	1.3 , 0.5
Aspect Ratio, std. dev.	1.9 , 0.7	1.3 , 0.3

The γ' dispersions, A and B, are shown in Figures 2 and 3, respectively. The A γ' dispersion consists of fine, intragranular rounded cuboids ranging from 0.8 μm – 1.2 μm diameter and the grain boundaries are partially outlined with particles that range up to 10 μm in length and 3 μm width. The B dispersion has a coarser, intragranular butterfly type particle and the grain boundaries are outlined with coarser particles that reach 20 μm length x 6 μm width. The butterfly particles are ~4 μm in width and their shape is indicative of solid state dendritic arm growth. The coarser grain boundary γ' also increased the width of the precipitate free zone adjacent to the grain boundaries. Both dispersions were created by subsolvus slow cooling cycles with the same initial (2000°F) and final (1900°F) temperatures; however, the B specimens were cooled at a slower rate to produce the coarser microstructure.

The recrystallization response measured by grain size is listed in Table IV. The experimental variables covered the full range of practical thermomechanical processing parameters. The deformation temperature spans from near γ' (sub)solvus to the temperature required to produce fine grain forging billet material. The strain rate spans from the typical draw forging rates to the slowest upset forging rates. The heat treatment is designed to coarsen the γ' dispersion, which enhances the hot workability and recrystallization response.

The forging temperature was the only variable that significantly affected the recrystallization response. The preforging heat treatment and deformation strain rate did not consistently influence the recrystallized grain size. The decreasing deformation temperature increased the precipitated γ' volume fraction that pinned the recrystallized grains at smaller sizes. Figure 4 captures the microstructural evolution in a transient condition after forging the B γ' dispersion at strain rate 1.0 and 1066°C. The distinguishable grain size is ASTM 11 but many regions are in a non-equilibrium recrystallization and grain growth phase. These regions have many small, recrystallized grains less than 1 μm in diameter that consume one another until reaching an equilibrium grain size determined by the forging (holding) temperature. Figure 5 illustrates the microstructure evolution at higher temperatures (1121°C, strain rate 1.0, γ' dispersion A) where the kinetics are more rapid and the precipitated γ' volume fraction lower, which produces a larger equilibrium grain size. The striking feature is linear, submicron γ' precipitate structures extending across all or most of the grain boundary width. These structures may originate from the combination of metastable stacking faults and favorable γ' precipitation kinetics that preserve the stacking faults. No other specimens in the study exhibited this phenomenon.

All experimental conditions produced fully recrystallized structures except the A γ' dispersion specimens compressed at 0.03 strain rate in the 1066°C – 1093°C temperature range. Those three conditions had elongated, unrecrystallized grains present but at less than 0.05 area fraction. The coarser B γ' dispersion appears to enhance the fraction of recrystallization for a given amount of strain but the effect is relatively modest.

A secondary priority after the recrystallization response is minimization of the flow stress to lower the press tonnage required for forging. The peak flow stresses, summarized in Table V, show the strong influence of strain rate and a secondary effect from the γ' dispersion. Decreasing the strain rate from 1.0 to 0.03 produced an average 97 MPa decrease in the peak flow stress. The γ' dispersion effect is small (<10%) at near solvus temperatures and at the slower 0.03 strain rate. However, in the

1.0 strain rate and the 1066°C – 1079°C processing range, the B γ' dispersion flow stress is significantly higher than the A γ' dispersion. The complete flow stress relationship with true strain is shown in Figure 6 with examples to illustrate the effect of γ' dispersion, temperature, and strain rate. A feature common to each specimen is the discontinuity in the flow stress after the first 0.2 strain step. The second deformation step always starts at 40-50 MPa lower flow stress than the first step had been completed. This indicates some metadynamic recrystallization occurs during the one minute hold after the first deformation step. The initial flow stresses of the third and fourth deformation steps normally follow the trend established by the previous step or show a maximum discontinuity of 20 MPa.

Table IV Recrystallization of Gleeble specimens forged to 4x0.2 true strain with a one minute hold after each compression step

Temp.	Strain Rate	Heat Treatment	Grain Size	Avg. Grain Size @ Temp.
1066°C	0.03	A	12 ala 9	11
	0.03	B	10.5	
	1.0	A	10	
	1.0	B	11	
1079°C	0.03	A	11 ala 9	10.5
	0.03	B	9.5	
	1.0	A	10.5	
	1.0	B	10.5	
1093°C	0.03	A	10 ala 8	10
	0.03	B	9.5	
	1.0	A	10	
	1.0	B	9.5	
1107°C	0.03	A	8	8.5
	0.03	B	8	
	1.0	A	8.5	
	1.0	B	9	
1121°C	0.03	A	6.5	7
	0.03	B	5.5	
	1.0	A	7.5	
	1.0	B	8	

Table V Peak Flow Stress (MPa) from compression specimens

γ' Dispersion	A		B	
Strain Rate	0.03	1.0	0.03	1.0
1066°C	236	296	245	344
1079°C	194	297	212	321
1093°C	192	293	195	290
1107°C	172	282	159	270
1121°C	143	240	148	230

Conclusions

Sprayform production of Allvac® 720 Alloy yielded a high quality preform free from macro- and microsegregation. The as-sprayed structure is intermediate between conventional ingot metallurgy and powder metallurgy with an equiaxed ASTM 6-7 grain size and carbonitrides ~1 μm in diameter. These results demonstrate Allvac® 720 Alloy is an ideal candidate alloy for sprayform processing.

The fully recrystallized structures obtained in response to 0.8 true strain in the typical hot working temperature range indicate the thermomechanical processing advantages sprayformed material will possess relative to conventional ingot metallurgy material due to the finer initial structure.

Surface   **Center**

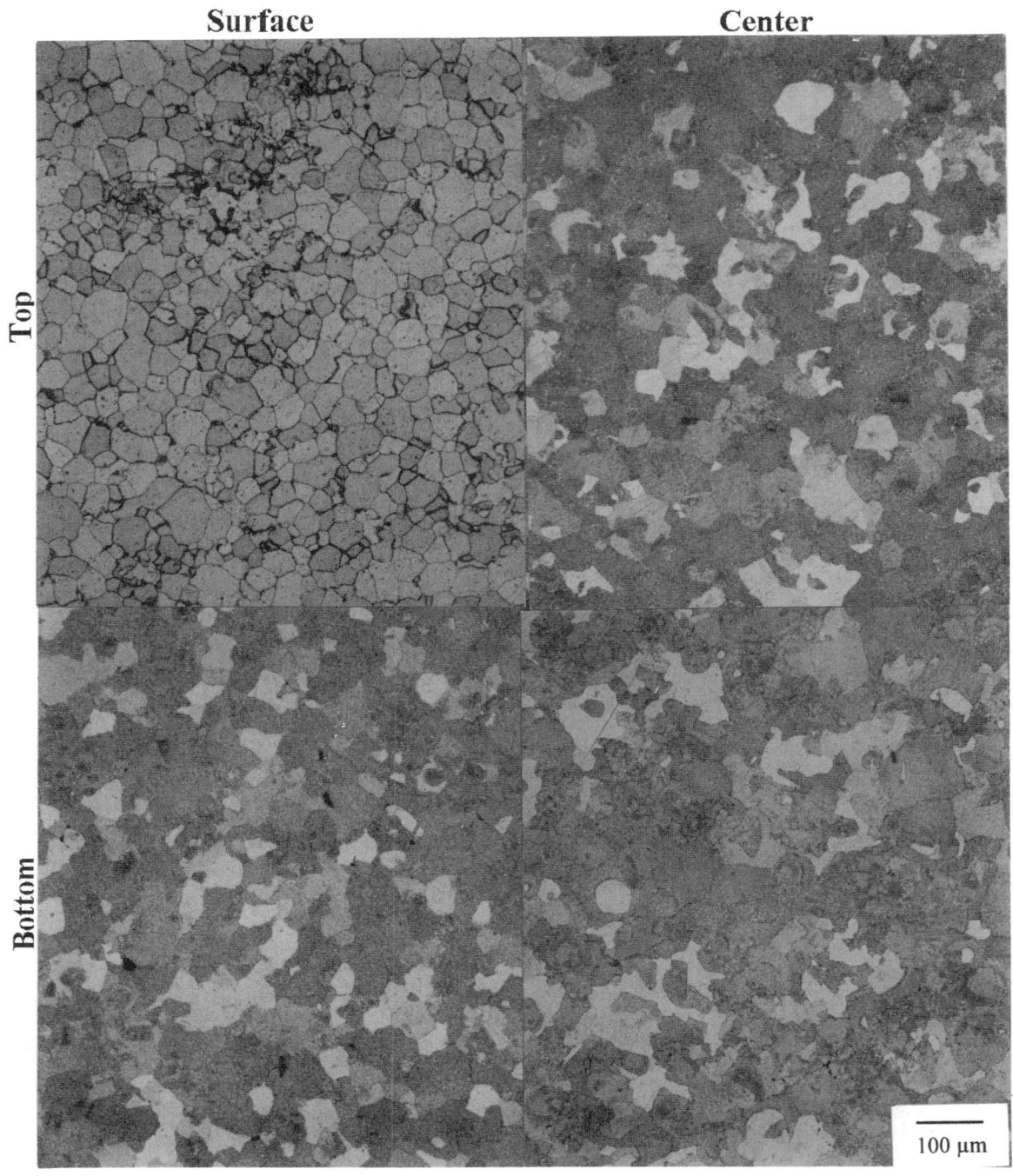

Figure 1: As-sprayed grain structure as a function of location.

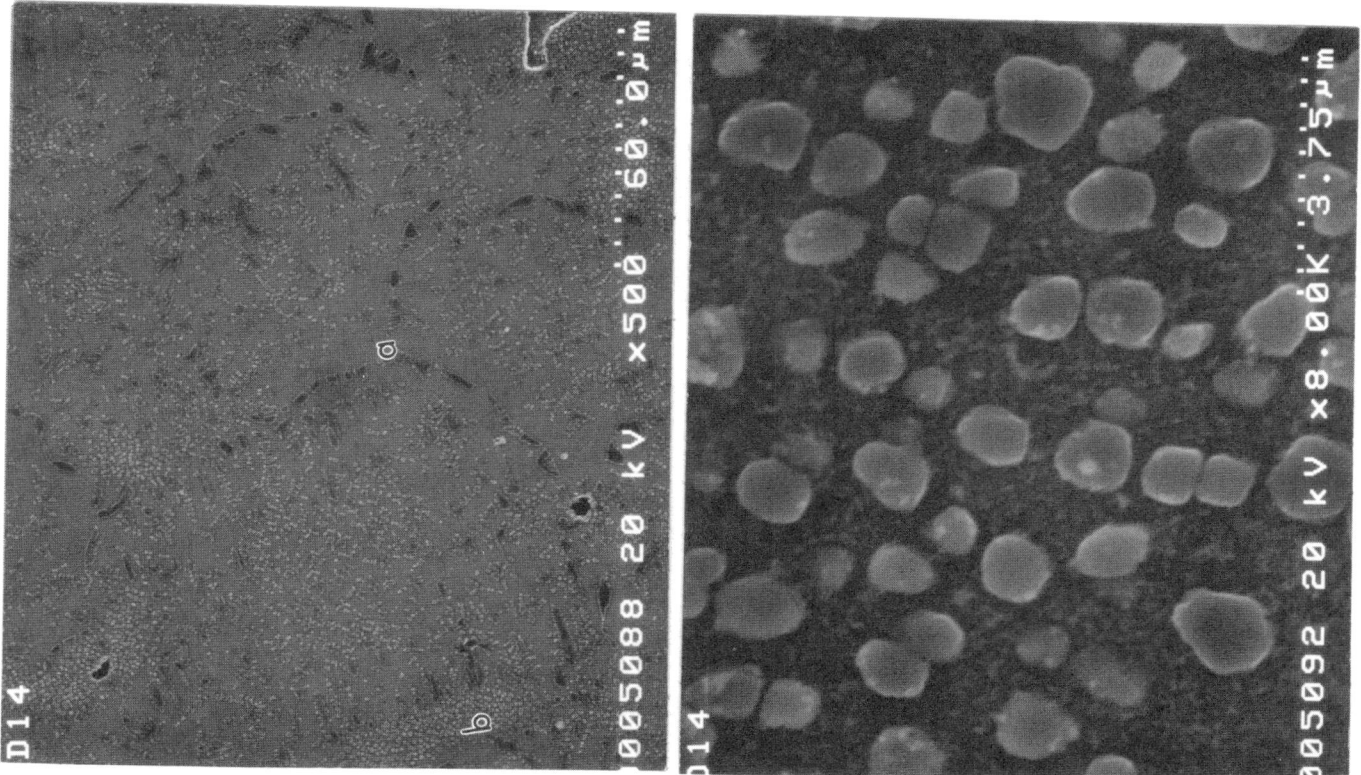

Figure 2: γ´ dispersion produced by heat treatment A.

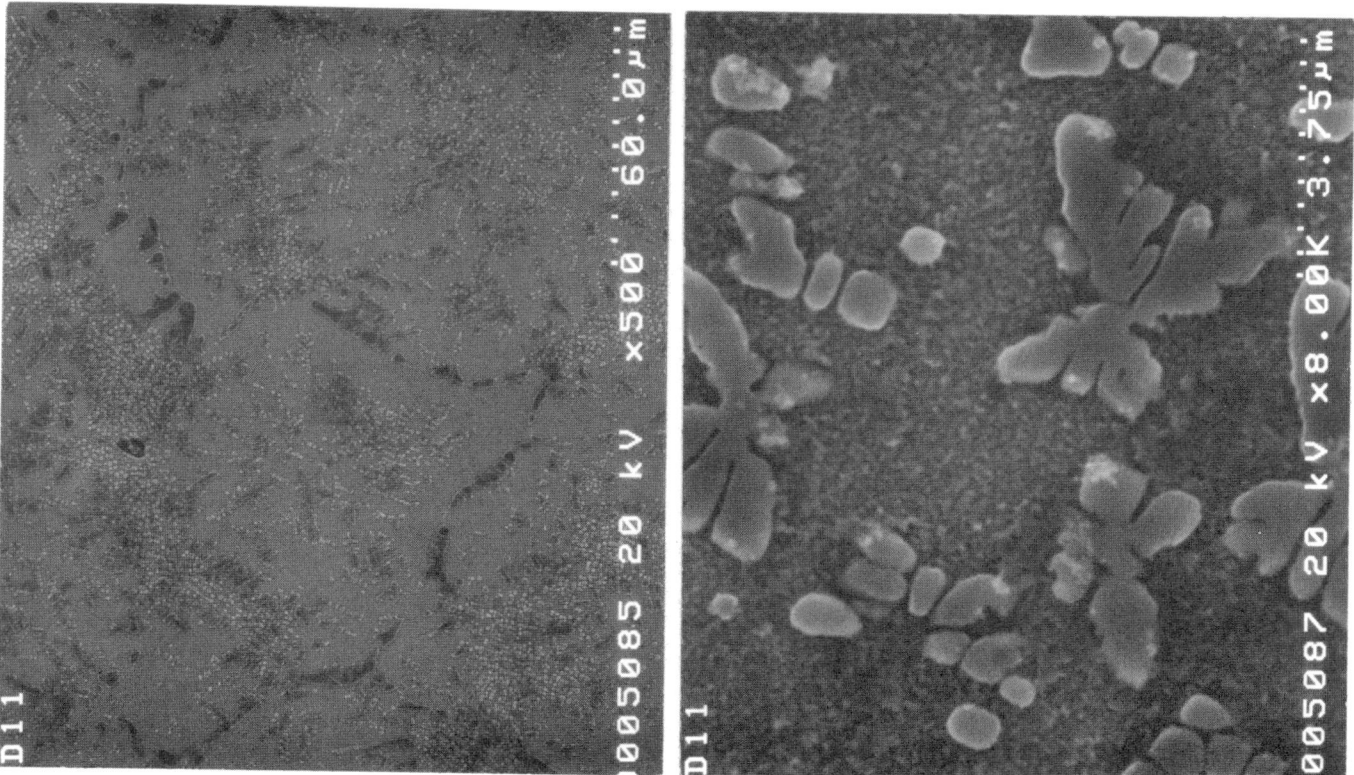

Figure 3: γ´ dispersion produced by heat treatment B.

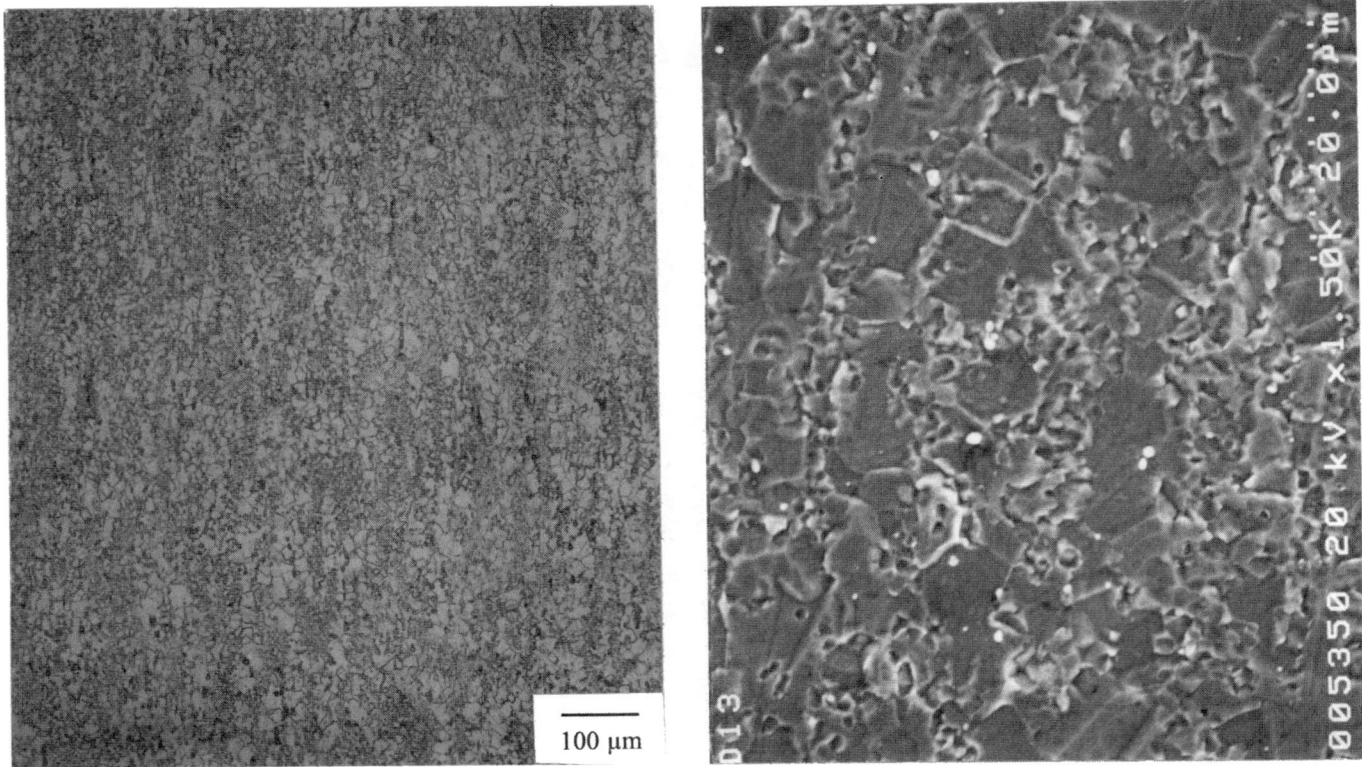

Figure 4: Recrystallization response of Gleeble specimen forged at 1066°C / strain rate 1.0 / γ′ dispersion B.

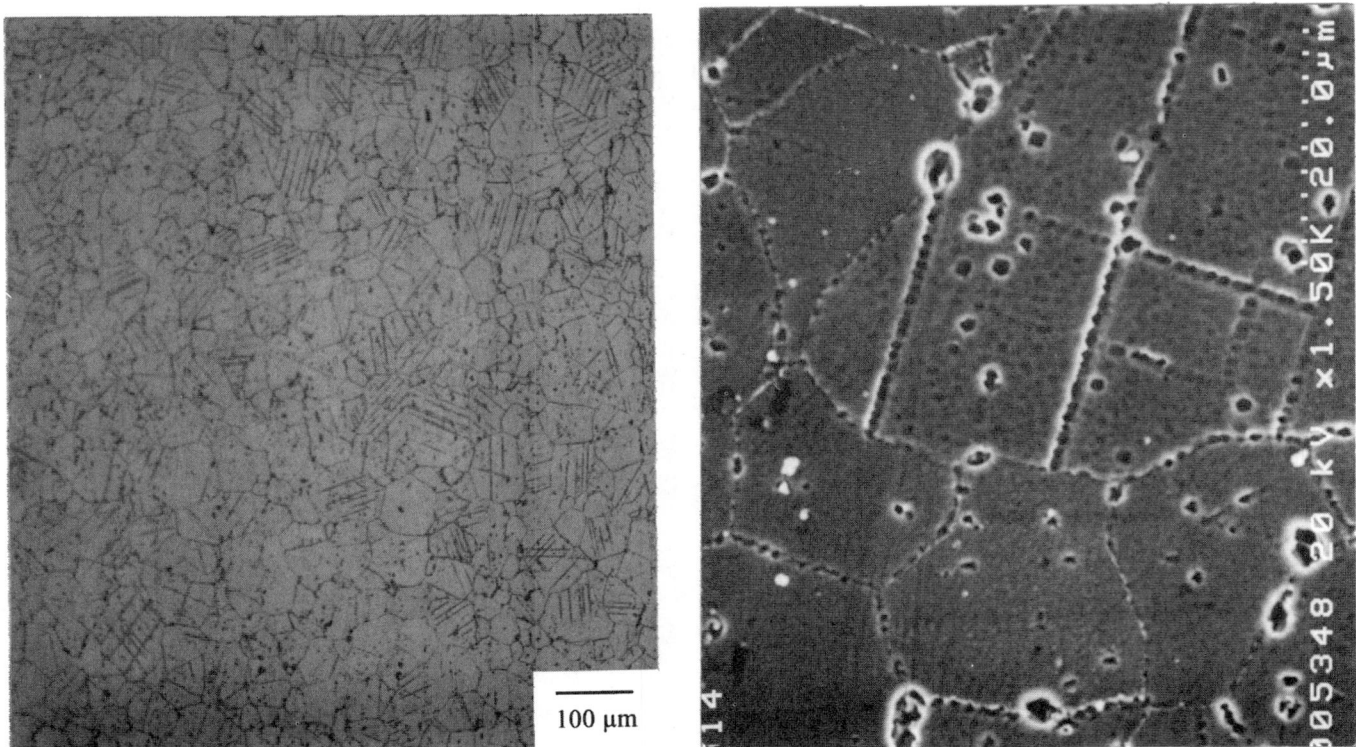

Figure 5: Recrystallization response of Gleeble specimen forged at 1121°C / strain rate 1.0 / γ′ dispersion A.

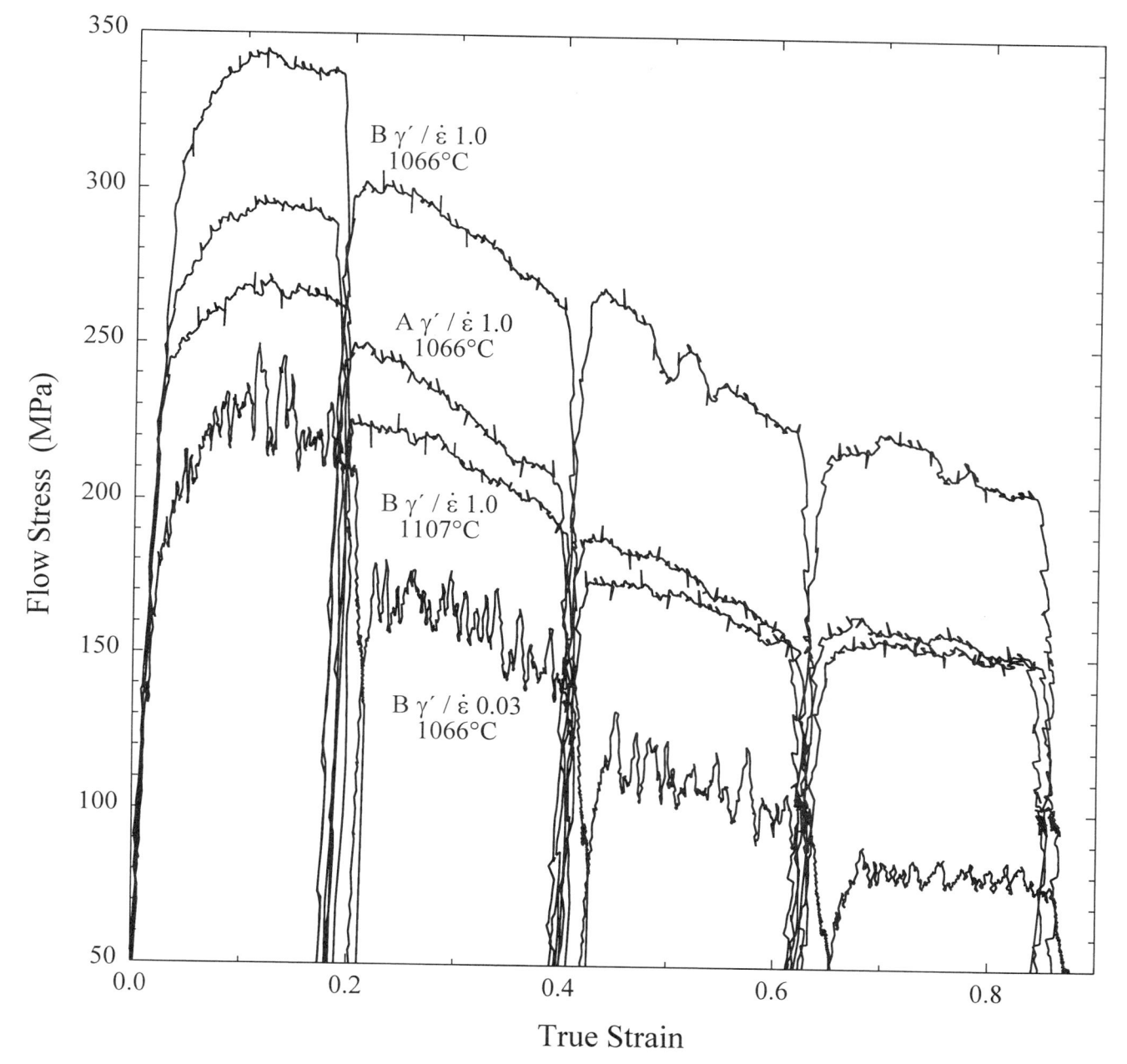

Figure 6: Flow stress as function of true strain from Gleeble compression tests.

Acknowledgements

This work is part of a US Air Force development program with Allvac and General Electric as industry partners. The work is partially funded by US Air Force Dual Use Science and Technology (DUS&T) Agreement Number F33615-99-2-5208 and administered by Mr. Rollie Dutton from Wright-Patterson Air Force Base.

Solidification and Casting Processing

PROPERTIES OF RS5 AN[...]
CAST USING THERMALLY CC[...]

M. L. Gambone[1], S. B. Shend[...]
M. N. Gungor[5], J. J. Vale[...]

[1]Rolls-Royce, Indianapolis, IN; [2]PCC Structurals, I[...]
[4]West Virginia University, Dept of Mechanical an[...]
[5]Concurrent Technologies Co[...]

[handwritten: Mary Lee Gambone / MAI / Rupprecht / anderson]

Abstract

This paper describes the evaluation of three superalloys—IN718, IN939, and RS5—cast using a new technology called thermally controlled solidification (TCS). The TCS casting technology enables the casting of large, complex, thin-walled components which have heretofore been impossible to cast with conventional investment casting methods. The characteristics of the candidate alloys in the TCS cast condition were analyzed for castability, weld reparability, microstructure, thermal stability, and mechanical properties. The RS5 alloy, recently patented, was also examined to optimize post-casting heat treatment. The goal of this evaluation was to select the best alloy with which to conduct a component demonstration as part of a U.S. Navy manufacturing technology program. While all the alloys demonstrated adequate castability and weld reparability, the alloy identified as the best candidate was RS5. This decision was based on the superior tensile properties of TCS cast RS5.

Introduction

Many aircraft propulsion systems include large, complex, fabricated components that are expensive to manufacture, inspect, and maintain. Conventional investment casting technology is not sufficient to produce castings to supplant these fabricated structures. Critical to economically producing such a one-piece casting is a newly developed technology, proprietary to PCC Structurals, Inc., called thermally controlled solidification (TCS). The TCS process enables the casting of large (diameters as great as 1 m), thin walled (~1.5 mm) superalloy components. The goal of the U.S. Navy manufacturing technology program that funded this work was to demonstrate TCS technology by replacing a fabricated component with a one-piece casting and, in doing so, create a significant cost savings. The demonstration component selected was the diffuser case for the AE 1107C engine manufactured by Rolls-Royce for the V-22 Osprey tiltrotor aircraft.

[...]e TCS process, which is critical in the success of the [pro]gram, i[...]volves preheating the invested mold inside a furnace, [po]uring t[...] molten alloy into the mold cavity, and withdrawing the heater. T[...] [...]plate is[...] [...] TCS furnace in such a manner that the lower half of the total mold height is at a temperature close to, but not below, the solidus temperature of the alloy to be poured in the mold cavity. In addition, the upper half section of the mold is heated in such a way that its temperature is close to the liquidus temperature of the alloy. The molten alloy is then introduced into the mold cavity through the inlet at the top of the mold, at the end of the preheat cycle.

Once the alloy is poured into the mold cavity, the heater assembly used to preheat the mold is withdrawn, while the mold remains stationary. The ratio of the gradient, G, on the mold to the rate of movement of the solidification front in the mold cavity, R, is closely monitored during the withdrawal process. G/R values of greater than 100 F-min/square inch are realized when the heater withdrawal rates range from about 7.5 in./hr to 30 in./hr. The achievement of adequate value of G/R is critical to the quality of the castings.

During solidification, the molten alloy decreases in volume. This reduction in volume of the alloy is compensated by the still molten alloy in the gates on the mold as well as by the molten alloy near the top end of the mold. Appropriate withdrawal speed of the heater ensures that this and the interdendritic shrinkage is fed, resulting in clean, shrinkage-free equiaxed castings. With the TCS process, geometrically complex structural castings are produced at reduced cost, and the costs associated with fabrication are eliminated. Both significantly reduce the total component cost.

Also critical to the program success is the demonstration of a superalloy that is castable via the TCS process and that has the required mechanical properties. The material from which the

Superalloys 2000
Edited by T.M. Pollock, R.D. Kissinger, R.R. Bowman,
K.A. Green, M. McLean, S. Olson, and J.J. Schirra
TMS (The Minerals, Metals & Materials Society), 2000

diffuser is currently fabricated is Ti-6242S. Superalloys are approximately twice as dense as titanium alloys; therefore, a direct material substitution of a cast superalloy for titanium would impose an unacceptable weight penalty in this application. Thus, the goal of this work was to select and characterize a superalloy with sufficient specific strength to allow a weight neutral substitution of a one-piece casting for the titanium alloy fabrication. Of the three nickel-based superalloys evaluated in this study—IN718, IN939, and RS5—IN718 was selected because it is commonly used in the conventionally cast condition and is well characterized. In addition, IN718 is readily weld repairable. The alloy IN939 is known to have improved strength at elevated temperatures, but it is not as amenable to weld repair as IN718. RS5 is a newly developed alloy, patented by Rolls-Royce plc, that has also demonstrated superior elevated temperature properties to IN718 in the conventionally cast condition and was designed to be weld repairable. None of these alloys had been thoroughly characterized in the TCS cast condition.

The aim of the research discussed in this paper was to determine the optimum alloy with which to continue the component development: the alloy most likely to enable achievement of the goal of weight-neutrality for the engine application. In this paper, the relative TCS castability of three candidate superalloys—IN718, IN939, and RS5—is evaluated. The mechanical properties and microstructure of the TCS cast alloys are analyzed as well. The response of the RS5 alloy to cooling rate and various heat treatment schedules has also been addressed.

Castability Evaluation

Experimental Procedure

Four different casting shapes, shown in Figure 1, were cast in all three alloys—IN718, IN939, and RS5—to examine their relative castability. The typical chemical composition of the alloys used is given in Table I. Molds produced from the wax patterns of the shapes were invested and cast using the TCS process. Each of the castings was non-destructively tested using x-ray and fluorescent penetrant inspection (FPI) techniques for surface and internal shrinkage porosity and shell-related and other casting defects such as hot tearing and cold shuts. These are standard techniques used to evaluate the soundness of investment castings.

After x-ray and FPI inspection and volumetric non-fill evaluation, the non-concentric ring castings of each alloy in the as-cast condition were cut-up at four different locations for microstructure evaluation. All the weld cavities on the weld test ring castings were ground, cleaned using standard techniques, and welded using the respective alloy filler wire. Welded test rings were heat treated, and the welded blanks were excised from the ring for microstructural evaluation and mechanical testing. The wagon wheel castings in all three alloys were welded in 10 different locations, heat treated, and sectioned for microstructure evaluation. There was no microstructure evaluation performed on the hot tear casting configuration.

Results of Castability Study

X-ray and FPI Evaluation FPI results for the non-concentric ring showed a few minor shell-related defects in all three alloys. This is not unusual in investment castings. There were no x-ray defects in the IN718 ring other than a minor amount (~ 0.5%) of non-fill. The IN939 ring revealed several hot tears and the greatest amount of non-fill (~ 0.2%) of all three alloys. The non-fill observed in all three alloys was in the thinnest section of the inner ring of the non-concentric ring casting. In FPI the RS5 ring demonstrated three minor hot tears and the least amount of non-fill (~ 0.2%) of the three alloys tested. These hot tears were not in the thin-wall section of the ring, but in the ribs connecting the inner and the outer ring. X-ray inspection of the RS5 ring did not show any additional defects. Based on these results, RS5 was the best alloy in filling thin wall molds, and IN718 was rated most hot tear resistant. Alloy IN939 performed the worst of the three alloys in both categories.

Table I Typical Chemical Composition in Weight Percent of the Alloys IN718, IN939, and RS5

Element	IN718	IN939	RS5	Element	IN718	IN939	RS5
C	0.05	0.15	0.08	B	0.003	0.009	0.006
Cr	18.5	22.4	16.0	Zr	0.006	0.010	0.005
Co	0.1	19.0	10.0	Fe	18.0	--	--
Mo	3.0	--	4.8	Mn	0.04	0.02	0.05
W	--	2.0	2.0	Si	0.07	0.01	0.05
Nb	5.0	1.0	4.8	S	0.003	0.003	0.003
Ti	0.8	3.7	2.7	P	0.007	0.005	.0.005
Ta	--	1.4	1.5	Ni	Bal.	Bal.	Bal.
Al	0.5	1.9	1.0				

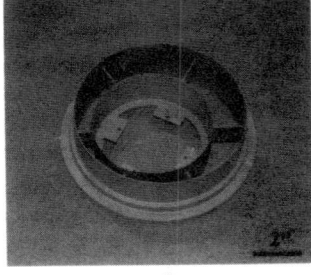

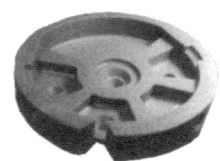

a   b   c   d

Figure 1: Castability test castings: (a) hot tear, (b) non-concentric ring, (c) weld ring, and (d) wagon wheel.

No x-ray or FPI defects were observed in any of the weld test rings inspected in the welded and fully heat treated condition. This indicated that all three alloys are weldable. IN939 was the most difficult alloy to weld, while IN718 was the easiest to weld, as anticipated. Similar to the weld test rings, no x-ray or FPI defects were found in the cavities in the wagon wheel in the fully welded and heat treated condition in any of the three alloys.

The x-ray and FPI results for the hot tear castings were in contrast to the results found in the non-concentric ring castings. Observation of the hot tear configuration showed that RS5 performed worst of the three alloys. The RS5 casting had 10.16 cm of hot tear length compared to 5.33 cm for IN939 and 9.84 cm for IN718. The RS5 casting also demonstrated the maximum amount of shrinkage in FPI (129.3 cm) but the least in x-ray (26.87 cm) compared with FPI and x-ray shrinkage of 49.78 cm and 42 cm, respectively, for IN718, and 60 cm and 31.34 cm, respectively for IN939.

<u>Microstructural Evaluation</u> Figure 2 shows a photomicrograph of the RS5 alloy microstructure in the as-cast condition. The needle-like structure in the RS5 alloy is believed to be the Nb-rich η-phase. Laves phases typically observed in IN718 in the as-cast condition were eliminated by heat treatment; no segregation was found in the IN939 alloy castings after full heat treatment. Typical grain size in the non-concentric ring castings ranged between 0.75 mm to 1.25 mm in all three alloys. In Table II the grain size of both thin and thick regions in the base metal of the wagon wheel casting is shown for all three alloys.

Results of the shrinkage evaluation of the non-concentric ring casting showed that the RS5 alloy exhibited the best filling characteristics compared to the other two alloys. The maximum percent shrinkage was 1.5% for RS5, 2% for IN939, and 5.1% for IN718; the maximum size of the shrinkage was 1.78 mm for RS5, 3.35 mm for IN939, and 4.14 mm for IN718. No internal shrinkage defects were found in the longitudinal sections of the welded and heat treated blanks sectioned from the weld test ring castings. No defects of any kind were observed in the base metal or the heat affected zone (HAZ) of the welded blanks in any of the alloys. A grain size of about 0.38 mm was measured in the weld and between 3.8 mm and 5.0 mm was measured in the base metal of all the alloy castings. These results indicate all these alloys develop a similar grain size in the TCS process for section thickness of about 1.4 cm and that grain size in the range of 5.0 mm in the base metal does not pose a problem in welding.

The 10 different locations on the wagon wheel that were welded simulated thick- and thin-wall welding and welding in fillets. No shrinkage defects were observed in any of the 10 welds in any of the alloy castings. It may be noted (Table II) that, in general, the grain size of the weld metal was significantly finer than that of

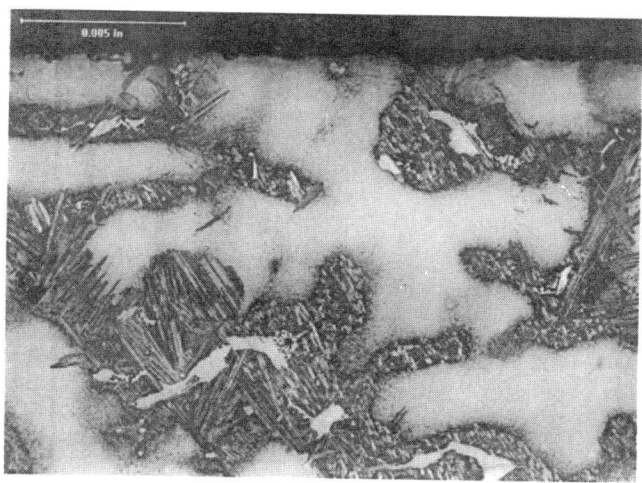

Figure 2: Microstructure of RS5 in the as-cast condition.

the base metal and that the grain size of the weld metal on thinner base metal was finer than that of welds on the thicker base metal. This effect is caused by the faster solidification rates that accompany welding. A more detailed discussion of these results can be found elsewhere [1,2].

RS5 Characterization

The RS5 alloy was patented by Rolls-Royce plc in 1994 for use in high temperature cast structures. Due to its relative immaturity in comparison to IN718 and IN939, more extensive characterization of RS5 alloy microstructure and property response to processing was conducted in this study. Several heat treatment variations were also evaluated to optimize the mechanical performance and stability of the alloy.

Continuous Cooling Transformation Results

Differential thermal analysis (DTA) and differential scanning calorimetry (DSC) techniques coupled with scanning electron microscopy energy dispersive spectroscopy (SEM-EDS) were utilized to study the phase transformations and develop a partial continuous cooling transformation diagram for the RS5 alloy. The liquidus of this alloy was 1330°C and 1312°C by cooling at rates of 1°C/min and 40°C/min, respectively. During solidification, the MC-type carbide forms at 50°C and 60°C below the liquidus cooling at 1 °C/min and 40°C/min, respectively. At approximately 1190°C, a eutectic reaction starts and finishes at 1140°C for the slower cooling rate and 1090°C for the faster cooling rate. This transformation is similar to that for a δ-eutectic phase, which has

Table II Grain Size Evaluation of Wagon Wheel Weld Metal and Base Metal in Alloys IN718, IN939, and RS5

Alloy/sample location	Avg. grain size, mm	Alloy/sample location	Avg. grain size, mm	Alloy/Sample location	Avg. grain size, mm
IN718 thin BM*	1.0	IN939 thin BM*	1.5	RS5 thin BM*	1.0
IN718 thin WM*	0.38	IN939 thin WM*	0.25	RS5 thin WM*	0.25
IN718 thick BM*	0.76	IN939 thick BM*	3.0	RS5 thick BM*	2.5
IN718 thick WM*	0.64	IN939 thick WM*	0.50	RS5 thick WM*	0.38

*BM and WM stand for base metal and weld metal, respectively.

been reported to occur in alloy René 220C [3]. X-ray diffraction of slower cooled samples revealed the presence of Laves phases; it is believed to have been formed in the same temperature range as the δ-like eutectic between 1190°C and 1140°C. Accurate-measurements of the solidus temperature on cooling were not possible; however, a solidus temperature of approximately 1179°C was estimated from the on-heating curves at 20°C/min. Further cooling of the as-cast microstructure indicated the precipitation of the plate-like η-phase between 1080°C and 980°C at a cooling rate of 5°C/min. The precipitation of the γ' was not observed in the DSC or DTA curves. Figure 3 shows the various solidification phases.

DSC analysis of hot isostatic pressed (HIPed), solution treated (1160°C), and aged RS5 specimens was used to develop the partial continuous cooling transformation diagram in a cooling rate range of 1 to 20°C/min shown in Figure 4. Two solid-state transformations were observed in this cooling rate range, one that begins to precipitate between 1050°C and 1010°C as the cooling rate increases and another phase transformation that occurs at approximately 980°C. The higher temperature precipitate is η-phase; its transformation temperature range appears to coincide with the range of solubility for the δ-phase in the René 220C alloy [3]. The lower temperature phase also appears to be η-phase, but its nucleation may have occurred in the matrix rather than at grain boundaries or "ghosts" of the interdendritic segrega-tion regions. Work is still in progress to elucidate this phenome-non. The γ and γ' phase precipitation fields were observed using DSC to extend from just below 930°C to about 580°C at a cooling rate of 5°C/min; at 20°C/min this transformation starts at 765°C and finishes at about 620°C.

RS5 Heat Treatment Optimization

Details of the standard heat treatment (HT#1) applied to as-cast RS5 are given in Table III. The heat treatment temperatures were determined on the basis of phase diagram modeling (Figure 5) for the nominal composition of the alloy. The HIP and solution heat treatment temperatures were chosen to ensure that full solutioning of all phases in the as-cast microstructure was achieved, while being sufficiently below the solidus temperature to avoid incipi-ent melting. A separate HIP and solution heat treatment have

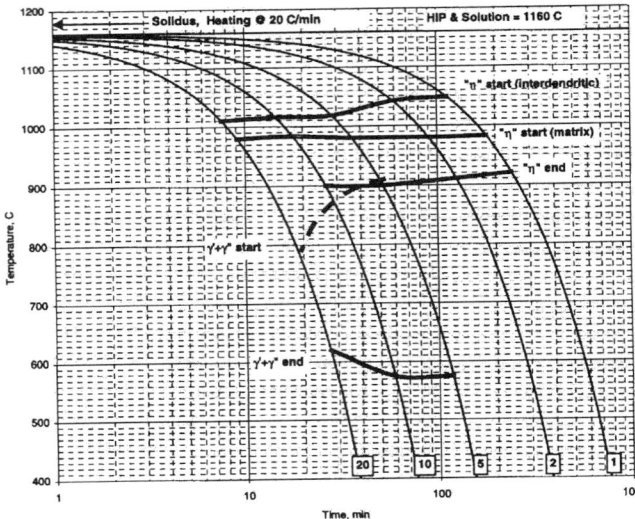

Figure 4: Continuous cooling transformation diagram for superalloy RS5.

been specified due to the limited availability of HIP facilities capable of achieving the required cooling rate of greater than 50°C per minute. This cooling rate is required in order to avoid η-phase precipitation during cooling from the solution temperature.

The HIP and solution heat treatment temperatures have been kept constant in this study, and the alternative heat treatments have concentrated on assessing the effect of applying an initial homog-enization treatment or alternative aging temperatures. The aim of HT#2, which included an 1100°C homogenization heat treatment, was to determine whether the additional lower temperature heat treatment was required to minimize compositional variation and prevent incipient melting in TCS cast RS5. The phase diagram model shows that 1100°C is still sufficient to fully solution the alloy. The aim of HT#3 and HT#4 was to assess alternative precipitation heat treatments. A 750°C aging temperature was chosen to produce a higher volume fraction of γ' and a lower

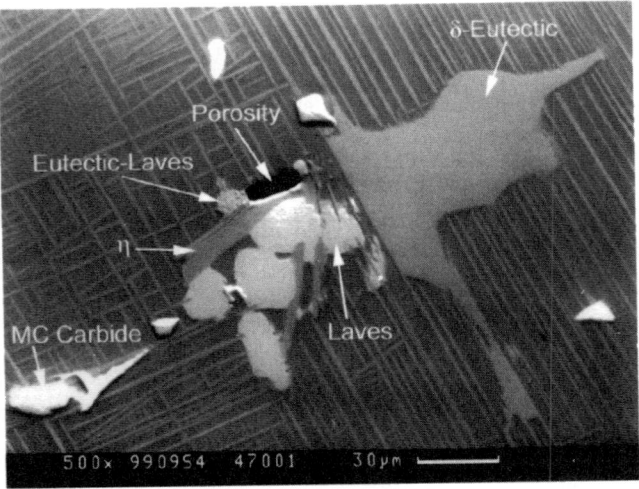

Figure 3: Solidification microstucture of an RS5 specimen cooled from liquid at 5°C/min.

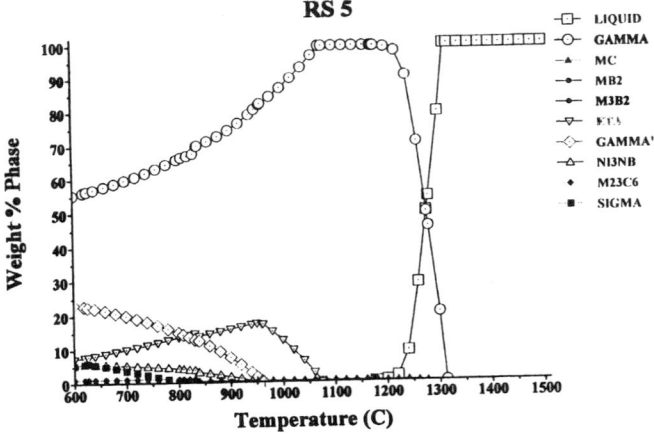

Figure 5: Phase diagram model for RS5.

164

Table III Heat Treatment Variations of the RS5 Alloy

Treatment	Temperature (°C)	Time (hr)	Pressure (MPa)	Cooling rate (°C/min)	Atmosphere
HT#1 – Standard Heat Treatment					
HIP	1160	4	103		Argon
Solution	1160	4	N/A	GFQ (>50)	Vac./inert gas
Age	800	16	N/A	Air cool	Air
HT#2 – Homogenization Heat Treatment					
Homogenization	1100	4	N/A	GFQ (>50)	Vac./inert gas
HIP	1160	4	103		Argon
Solution	1160	4	N/A	GFQ (>50)	Vac./inert gas
Age	800	16	N/A	Air cool	Air
HT#3 – Alternative Age Heat Treatment					
HIP	1160	4	103		Argon
Solution	1160	4	N/A	GFQ (>50)	Vac./inert gas
Age	750	16	N/A	Air cool	Air
HT#4 – Two-Step Age Heat Treatment					
HIP	1160	4	103		Argon
Solution	1160	4	N/A	GFQ (>50)	Vac./inert gas
Age	750	8	N/A	Air cool	Air
Age	650	8	N/A	Air cool	Air

volume fraction of η-phase than in the 800°C age used in HT#1 and HT#2. The purpose of a two-stage aging treatment in HT#4 was to determine whether, as for IN718, additional γ' of a finer particle size would precipitate at the lower temperature.

Microstructural Analysis

Examination of RS5 in the standard heat treated condition (HT#1) revealed that a ghost dendritic structure persisted after heat treatment of the TCS cast alloy (Figure 6). This indicates that full homogenization was not achieved. Secondary phases, such as carbide particles and irregular shaped plates that are assumed to be η-phase (Ni3Ti) or δ-phase (Ni3Nb), are concentrated within the interdendritic regions (Figure 7). The carbide particles are predominantly associated with primary grain boundaries, which form at the interface between dendrites in the cast structure. Hence, in the heat treated condition, grain size is strongly influenced by the dendrite spacing of the casting. Due to the nature of the dendrites, elongated grains that are as much as 1000 μm long and approximately 200 μm in diameter are formed. Precipitation of a phase, whose morphology is either needles or platelets (Figure 8), was observed on the primary grain boundaries. Transmission electron microscopic (TEM) examination of other

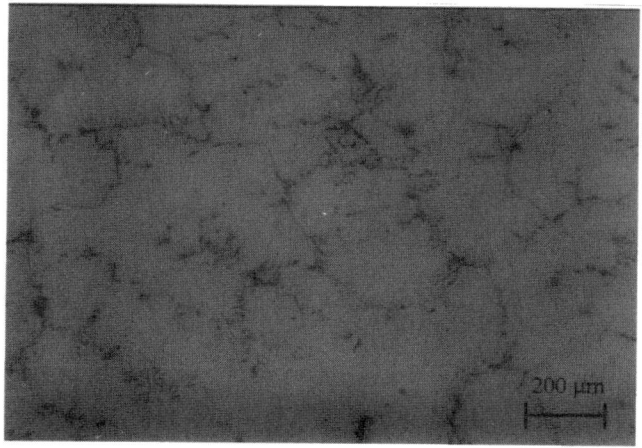

Figure 6: Optical micrograph of RS5 in the standard heat treatment condition (HT#1). Primary grain boundaries are evident at interface between dendrites in the original as-cast structure.

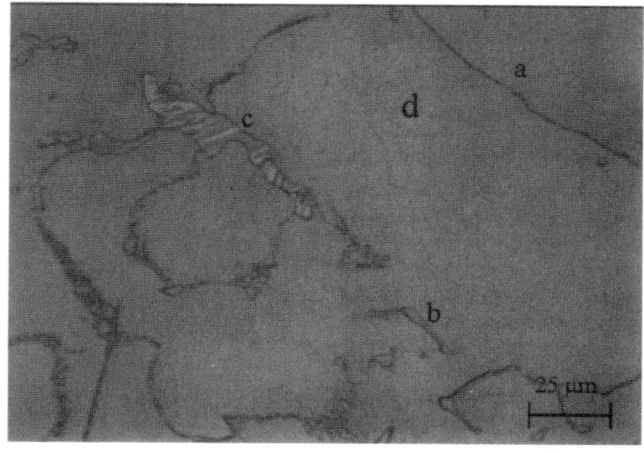

Figure 7: Optical micrograph of RS5 in the standard heat treatment condition (HT#1). Features present in the interdendritic region are primary grain boundaries (a), carbide particles (b), η or δ phase (c), and a subgrain structure (d).

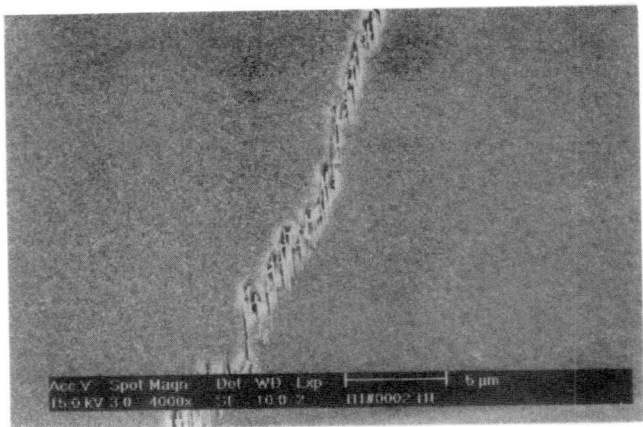

Figure 8: Secondary electron micrograph showing η-phase precipitates on a primary grain boundary in RS5, HT#1.

forms of RS5 has shown that this phase is η-phase. It is considered unlikely that the η-phase will have a detrimental effect on the strength of the grain boundaries as it is discreet and oriented approximately perpendicular to the grain boundary. Within the interdendritic regions a subgrain structure was also apparent (Figure 7). Gamma prime particles are present on these subgrain boundaries, which are taken to indicate that an orientation mismatch exists between these local areas in the structure. The elongated morphology of the γ' precipitates, which appear to form a basket weave structure, is also shown in Figure 9.

Compositional analysis of the dendrite cores and interdendritic regions using SEM-EDS showed that chromium, cobalt and tungsten concentrate in the former region and niobium, titanium, molybdenum and tantalum concentrate in the latter region. Analyses were also carried out on the carbide particles present in the structure and showed that the particles contained approximately 50 atomic percent carbon together with molybdenum or niobium and traces of tantalum, titanium and silicon. This analysis indicates that the carbides are of the MC type.

The homogenization heat treatment (HT#2) produced microstructures that were very similar to standard heat treatment (HT#1) with no obvious differences in the size, morphology, and distribution of phases being readily discernible. Compositional analysis of the dendrite cores and interdendritic regions showed that the compositional variation was similar to HT#1. These results indicate that the homogenization heat treatment was of limited benefit in minimizing compositional variation.

The RS5 samples given the alternative precipitation heat treatment (HT#3 and HT#4) both showed similar grain size to that of the standard heat treatment. They primarily varied from HT#1 and HT#2 in the morphology of the γ' and η-phase precipitates. The alternative precipitation heat treatment has produced γ' precipitates that are larger and spherical or cuboidal (Figure 10) rather than the elongated morphology formed in HT#1 and HT#2 (Figure 9). Also, the η-phase precipitates on the primary grain boundaries are longer and have a greater tendency to be aligned parallel with the grain boundaries. This orientation appears less favorable and suggests the possibility of a plane of weakness as compared with the essentially perpendicular orientation of the η-phase precipitates in HT#1 and HT#2. It was not possible at the resolution available on the SEM to determine whether the two-stage aging treatment (HT#4) had produced a bimodal γ' size distribution. Further studies using TEM will be carried out to determine if a finer γ' precipitate is present.

Mechanical Property Comparisons

The mechanical properties of the four different heat treatment variations of the TCS cast RS5 alloy were also evaluated and compared to determine the optimum heat treatment. Tensile tests were conducted over a range of temperatures from 20°C to 815°C, and creep tests were conducted to measure the time to 0.2% creep strain and rupture life between 650°C and 815°C. Strain-controlled low cycle fatigue (LCF) tests were also performed at an R-ratio of 0 at room temperature. The tensile results showed that the alternative aging treatments increased the yield and tensile strength of the alloy. At room temperature the yield strength of HT#3 and HT#4 material averaged about 35 MPa greater than that of HT#1 or HT#2, and the ultimate tensile strength (UTS) was also greater by about 40 MPa. The higher yield strength persisted at 426°C and 600°C but was negligible at higher temperatures. The same trend is evident with respect to UTS. The 0.2% creep strength of RS5 subjected to the standard heat treatment (HT#1) is greater than that of RS5 given the alternative aging treatments, and rupture strength shows the same trend. The effect is most noticeable at lower temperatures

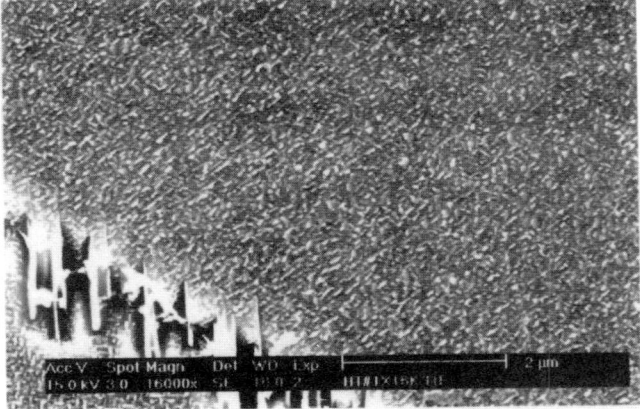

Figure 9: Secondary electron micrograph showing elongated nature of γ' precipitates in RS5, HT#1.

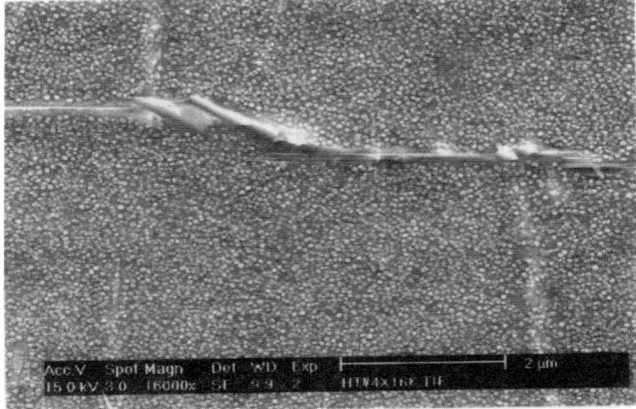

Figure 10: Secondary electron micrograph of sample HT#4. The structure contains spherical/cuboidal γ' precipitates and elongated η-phase precipitates along the primary grain boundaries.

(~650°C) and higher stresses. The four different heat treat conditions demonstrated no discernible differences in LCF behavior at room temperature.

Based on the results of the microstructural examination and mechanical property evaluation, the decision was made to characterize the RS5 alloy further only in the standard heat treatment condition (HT#1). The data shown in the remainder of the paper reflect the behavior of this heat treatment condition only.

Thermal Stability

Previous mechanical property evaluations of various forms of RS5 have revealed that during creep testing at the highest use temperature of the alloy a phase instability occurs in the microstructure [4]. Time, temperature, and stress are required to initiate this phase instability. To determine the extent to which TCS cast RS5 is prone to this phenomenon a series of creep tests were performed on the alloy for approximately 500 hr at 600°C/750 MPa, 700°C/400 MPa, and 800°C/150 MPa. On completion of the creep tests, the samples were longitudinally sectioned, mounted, polished, and etched for microstructural evaluation. The same creep tests and microstructural analysis were performed on TCS cast IN939 for comparison with the RS5 alloy.

RS5

Microstructural evaluation of RS5 following creep testing showed that at both 700°C and 800°C the ghost dendritic structure in alloy progressively decreased, and that coarsening of the γ' precipitates occurred, indicating that elemental diffusion, which is enhanced by the presence of stress, had occurred. The most dramatic change in the microstructure occurred on testing at 800°C. At this temperature an acicular phase was observed to precipitate in the microstructure (Figure 11). The density of the acicular phase was greatest in the interdendritic regions, with only a limited number of precipitates observed in the dendrites. Previous studies on a different form of RS5 have identified the precipitates as δ-phase. This phase was observed to precipitate on and grow from the η precipitates that are present on the grain boundaries in the alloy. The higher density of delta phase in the interdendritic region reflects the higher niobium content in this region, while the absence of precipitates in the dendrite cores indicates the lower niobium content and the absence of a composition gradient in this region of the structure.

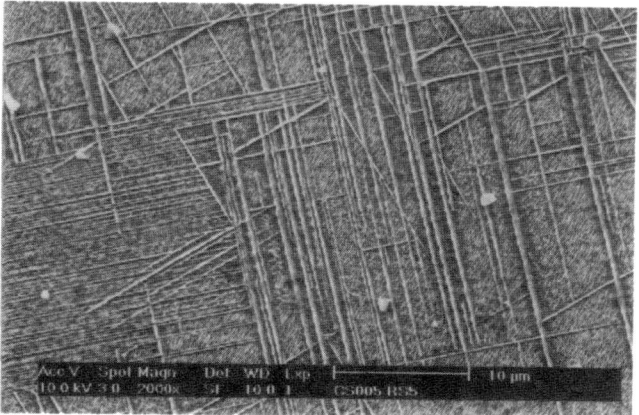

Figure 11: Acicular δ-phase precipitates in interdendritic region of RS5, HT#1 following creep testing at 800°C/150MPa for about 500 hr.

A temperature of 800°C is at the upper temperature limit at which RS5 was designed to operate. Although fundamentally the phase instability and acicular nature of the delta phase are undesirable, the predominant strengthening phase in the alloy, γ', has not been denuded by the phase change. Information in the literature for René220C [3] indicates that the δ-plate structure can be controlled to influence fatigue and effectively impede fatigue crack propagation. Additionally, preliminary studies to determine the residual strength of creep tested spray-formed RS5 [4] containing the δ phase have shown that the phase has minimal influence on the strength and ductility as compared with the base tensile properties for the alloy. This information indicates that, although undesirable, the formation of delta phase does not have a significant influence on the mechanical properties of the alloy.

IN 939

Microstructural evaluation of IN939 following creep testing showed that the ghost dendritic structure persisted in the microstructure on creep testing and that coarsening of the γ' precipitates did not occur. No phase instability was observed in the alloy in the temperature range studied. These observations with the mechanical property results, described in detail later, show that IN939, while having a lower strength, offers a higher temperature capability than RS5. Phase instabilities in IN939 are, therefore, likely to arise at higher temperatures than assessed in the current study.

Mechanical Behavior of TCS Cast Alloys

The purpose of the mechanical testing conducted in this study was, first, to compare the properties of the TCS cast alloys: IN718, IN939, and RS5; and, second, on a limited basis, to compare the TCS cast alloy properties with those for the same alloy conventionally investment cast.* The mechanical behavior of the TCS cast alloys was evaluated with respect to tensile, creep, LCF, and room temperature fatigue crack growth (FCG) behavior. In addition, the elevated temperature FCG performance of the RS5 alloy was characterized. Table IV shows the test matrix with all test conditions. The tensile, creep, and LCF specimens were cast-to-size hung-on-bars (HOBs) that were machined prior to testing. The FCG testing was performed on compact tension specimens, nominally 6 mm in thickness, that were sectioned from a TCS cast plate.

Tensile Properties

Figure 12 illustrates the yield and ultimate tensile strength of the three TCS cast alloys. The data shown are the average of three tests for each alloy at each temperature tested. The RS5 alloy demonstrated higher tensile strength than IN939 or IN718 across the entire temperature range investigated. The yield strength of RS5 was also superior to the other two alloys. However, while IN718 showed the lowest UTS, its yield strength was significantly greater (~150 MPa) than IN939. The yield and ultimate tensile strength of conventionally cast IN718 are included in Figure 12. The yield strength of the TCS cast IN718 is about 100 MPa greater than that for the conventionally cast alloy. This improvement is likely related to the finer grain size that results from TCS casting. The tensile ductility of the TCS alloys and conventionally cast IN718 was also measured. From room temperature to 600°C the reduction in area for RS5 and IN939 is between 8 and 10%, approximately half that for the TCS cast IN718, which is in turn about half that for the conventionally cast IN718 alloy.

---

* The data for the conventionally cast material was derived from Roll-Royce databases and was not produced as part of this program.

Table IV Mechanical Property Test Matrix for TCS Casting Process Assessment

Alloy	Condition	Test	Temperature (°C)	Tests/temp	Total tests
IN718	TCS + HIP + heat treat	Tensile	21, 427, 593, 704, 815	3	15
IN939	TCS + HIP + heat treat			3	15
RS5	TCS + HIP + heat treat #1			3	15
IN718	TCS + HIP + heat treat	Creep	650, 704, 760, 815	3	12
IN939	TCS + HIP + heat treat			3	12
RS5	TCS + HIP + heat treat #1			3	12
IN718	TCS + HIP + heat treat	LCF, R = 0, Kt = 1.0	21, 427, 650	5	15
IN939	TCS + HIP + heat treat	Freq. = 20 cpm		5	15
RS5	TCS + HIP + heat treat #1	Strain control		5	15
IN718	TCS + HIP + heat treat	FCG, R = 0.05	21	2	6
IN939	TCS + HIP + heat treat	Freq. =15 cpm	21	2	6
RS5	TCS + HIP + heat treat #1		21, 427, 650	2	6

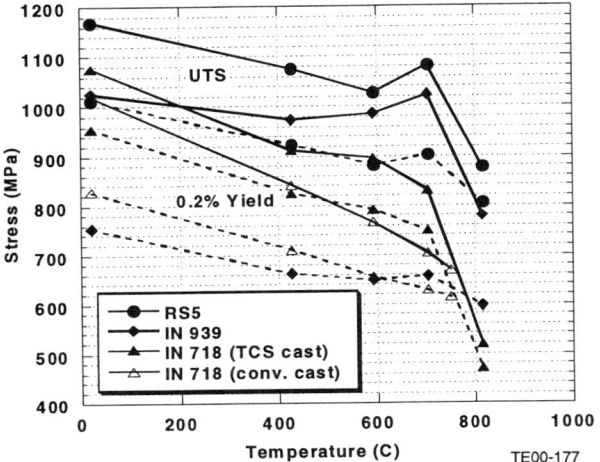

Figure 12: Yield and tensile strength of TCS cast alloys

Creep Properties

It is evident in Figure 13 that TCS cast RS5 clearly demonstrated superior creep resistance to 0.2% creep strain between 250 MPa and 800 MPa. At 700 MPa for a 1000-hr creep life to 0.2% strain, the TCS cast RS5 was approximately 75°C better than IN939 and 180°C better than conventionally cast RS5. At 500 MPa for a 1000-hr creep life to 0.2% strain, RS5 shows a 45°C improvement over IN939 and a 60°C improvement over TCS cast IN718. At 250 MPa IN939 and conventionally cast RS5 have equivalent 0.2% creep life to TCS cast RS5. In the case of stress-rupture (Figure 14), TCS cast RS5 is also superior to the other TCS cast alloys from 500 MPa to 800 MPa and to conventionally cast RS5 at stresses of 350 MPa and higher. At lower stresses and higher temperatures IN939 demonstrates superior stress-rupture properties to RS5. As would be expected, both conventionally and TCS cast IN718 show significantly reduced stress-rupture life across the entire temperature range of interest.

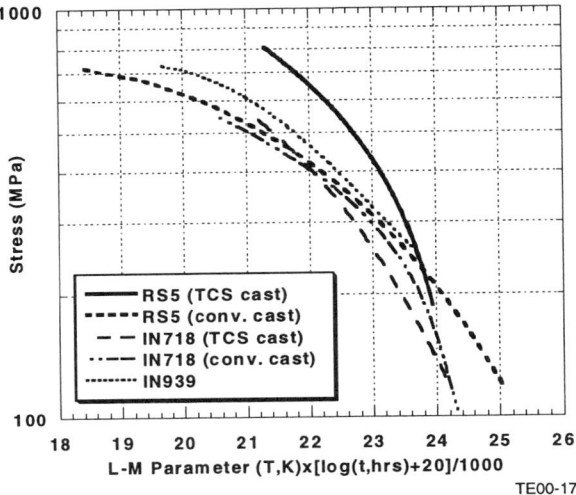

Figure 13: The 0.2% creep strain behavior of TCS cast alloys compared to conventionally cast IN718 and RS5.

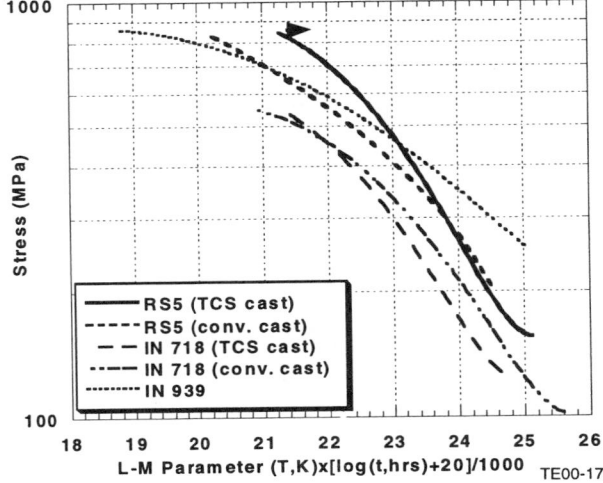

Figure 14: Stress-rupture behavior of TCS cast alloys compared to conventionally cast IN718 and RS5.

168

LCF Performance

LCF tests were conducted in strain control on unnotched (K_T=1.0) specimens at an R-ratio of zero. The test frequency was 20 cpm, and the wave form was triangular. Five tests were conducted for each alloy at each temperature. There was little difference in the room temperature LCF performance of the TCS cast alloys; Figure 15 illustrates the trend in room temperature LCF for RS5, IN939, and IN718. At the elevated temperatures tested, however, some differences in performance did emerge. At 427°C (Figure 15), it is evident that at strain ranges greater than 0.6% TCS cast IN718 demonstrated the longest life. At lower strain ranges and lives greater than 10,000 cycles TCS cast RS5 and conventionally cast IN718 were superior. At 650°C, the highest temperature tested (Figure 16), RS5 showed superior LCF life over all strain ranges tested, and IN939 demonstrated better performance than IN718, as expected.

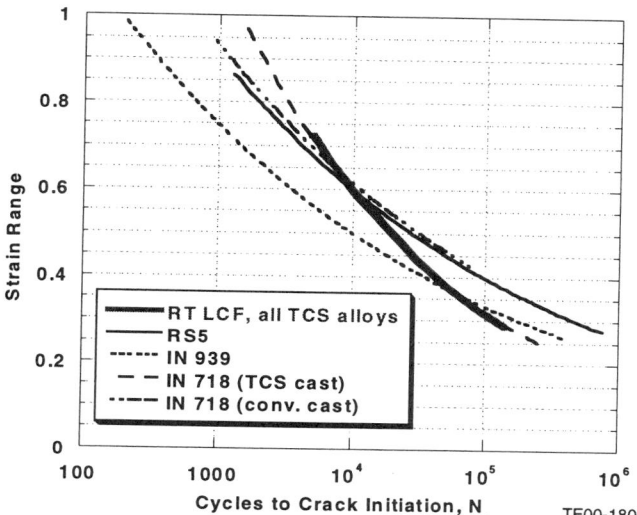

Figure 15: Strain-controlled LCF behavior of TCS cast alloys at room temperature and 427°C.

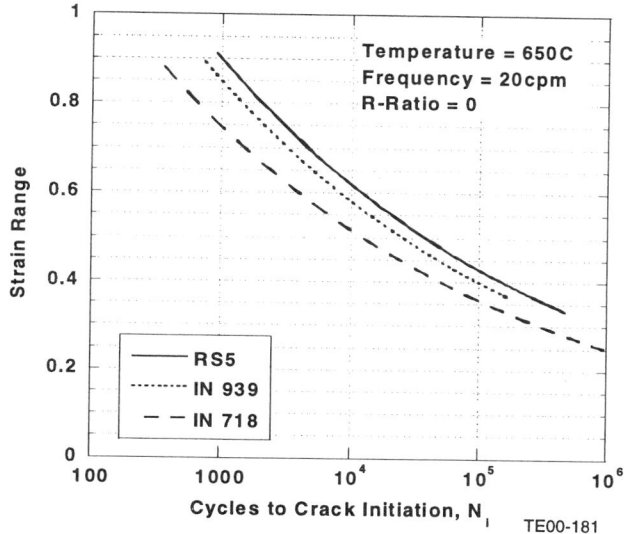

Figure 16: Strain-controlled LCF performance of TCS cast alloys at 650°C.

Fatigue Crack Growth Behavior

All the FCG tests, at both room and elevated temperature, were conducted using a trapezoidal wave form: 1 sec hold at minimum stress, 1 sec increase to maximum stress, 1 sec hold at maximum stress, and 1 sec unload to minimum stress. The R-ratio was constant at 0.05, and crack length was monitored both visually and using electropotential drop. All tests were conducted in accordance with ASTM specification E647-93 [5]. The alloys' steady-state FCG behavior at room temperature was similar; this behavior is as expected for cast superalloys with similar grain size. RS5 was the only alloy tested at elevated temperature, and the alloy demonstrated an increase in FCG rate at constant stress intensity with increasing temperature (Figure 17). At a stress intensity of 20 MPa(m)$^{1/2}$ there was an order-of-magnitude increase in crack growth between room temperature and 650°C.

Summary

The goal of the analyses presented here was to determine the optimum alloy with which to proceed with component development. The selection was made by evaluating the relative difficulty in casting and repairing the alloys via the TCS process, measuring thermal stability, and developing a preliminary mechanical property database as described previously. Castability and weld reparability are considered vital to production of these castings at minimal cost, the main thrust of the program. Thermal stability and adequate creep and LCF performance are required to maximize component life in the use environment. High tensile strength in the superalloy selected is necessary to produce a specific strength comparable to Ti-6242S.

The RS5 alloy was selected as the optimum alloy with which to continue the component demonstration. All three alloys evaluated demonstrated adequate castability and were weld repairable. The RS5 alloy demonstrated the best capability to fill thin walls, which is essential for the diffuser case being investigated in this

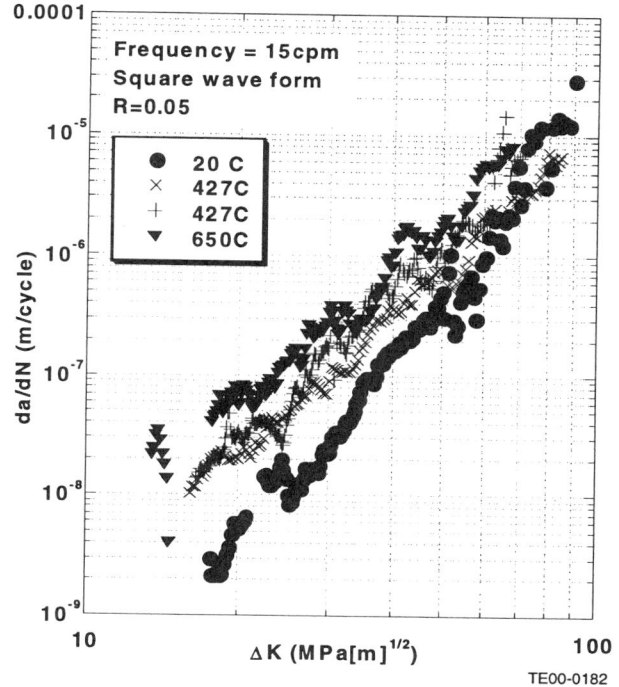

Figure 17: FCG behavior of TCS cast RS5 at room temperature, 427°C, and 650°C.

program and many other structural components. While RS5 showed a propensity toward hot tear in certain casting configurations, the non-concentric ring casting was more representative of the component under consideration, and RS5 demonstrated the minimal hot tear in that casting. The thermal stability of the RS5 alloy was also adequate in the temperature range of interest (400°C to 700°C), although the IN939 alloy was stable to significantly higher temperatures and would probably have been selected if the use temperature range was higher for the component. The most significant determiners among the mechanical properties evaluated were yield and tensile strength. The TCS cast RS5 alloy is clearly superior in strength to both IN718 and IN939 (Figure 12); it also has the highest specific strength of the three alloys and that most similar to Ti-6242S. RS5 was chosen because the goal of weight neutrality with the current component design is paramount to implementation of this technology in this program. Other structural applications with less stringent weight and lower temperature requirements would likely be less costly TCS cast in IN718. The RS5 alloy clearly fills a niche for parts with higher performance requirements than IN718 can fulfill and in a use temperature range below that for which IN939 is optimized. Because RS5 is TCS castable and weldable, the cost benefits derived through application of the process will be maintained with this new alloy.

Acknowledgments

The authors would like to acknowledge the U.S. Navy and the National Center for Excellence in Metalworking Technology, operated by Concurrent Technologies Corporation. This work was funded under contract No. N00140-92-C-BC49 to the U.S. Navy as part of their Manufacturing Technology Program.

References

1. S. B. Shendye, "Report on Task 3: Assessment of Thin-wall Castability of Candidate Alloys" (Report to Concurrent Technologies Corporation for Contract No. N00140-92-C-BC49, PCC Structurals, Inc., 3 August 1999).

2. S. B. Shendye et al., "The Castability and Mechanical Properties of Nickel Superalloys Cast using Thermally Controlled Solidification" (Paper presented at the TMS 2000 129th Annual Meeting and Exhibition, Nashville, TN, 12-16 March 2000).

3. S. T. Wlodek and R. D. Field, "The Structure of René 220C," Superalloys 1992, ed. S. D. Antolovich et al. (Warrendale, PA: The Minerals, Metals & Materials Society, 1992), 477-486.

4. P. Andrews, unpublished research, Rolls-Royce Derby, UK, 22 August 1997.

5. "Standard Test Method for Measurement of Fatigue Crack Growth Rates," Annual Book of ASTM Standards, vol. 3.01 (Philadelphia, PA: ASTM, 1993), 679-706.

ADVANCED SUPERALLOYS AND TAILORED MICROSTRUCTURES FOR INTEGRALLY CAST TURBINE WHEELS

R. C. Helmink*, R. A. Testin*, A. R. Price**, R. Pachman**, G. L. Erickson[†], K. Harris[†], J. A. Nesbitt[††], and J. F. Radavich[‡]

* Rolls-Royce Corporation, Indianapolis, Indiana; ** Howmet Research Corporation, Whitehall, Michigan;
[†]Cannon-Muskegon Corporation, Muskegon, Michigan; [††]NASA Glenn, Cleveland, Ohio;
[‡]Micro-Met Laboratories, West Lafayette, Indiana

Introduction

The life cycle costs of the Rolls-Royce Corporation Model 250 small turbine engines have not kept pace with those of international competitors. A significant contributor to this deficiency was the dated technology used in turbine section materials and manufacturing processes. The Advanced Materials for Small Turbine Engines (AMSTE) team, consisting of Rolls-Royce Corporation, Howmet Corporation, Cannon-Muskegon Corporation, Purdue Research Foundation, and the National Aeronautics and Space Administration (NASA) Glenn Research Center, was formed to significantly advance the state of the art in the manufacture of integrally bladed turbine wheels. The team was awarded a program by NASA for the Aerospace Industry Technology Program (AITP) designed to help maintain competitiveness in the production of small turboshaft engines and demonstrate an advanced material system for other gas turbine aerospace/surface applications.

The goals for this program were to develop the technology base needed to protect the small turbine engine industry by doubling or tripling the life of current gas generator turbine wheels and significantly reducing their life cycle cost. Specifically, this program addressed the development of advanced superalloy materials and manufacturing processes to make possible an affordable turbine wheel with significantly increased low cycle fatigue (LCF) life and improved airfoil stress rupture life.

To meet these goals, a technical effort was formulated to develop an integrally bladed turbine wheel casting technology that would provide equivalent or superior capabilities at a substantially lower cost when compared to conventional mechanically attached bladed

designs. The technical plan included a moderate risk "baseline" approach, featuring a state-of-the-art Grainex® (GX) casting process using Mar-M247 superalloy and an improved wheel design, and two higher risk approaches involving development/ evaluation of advanced superalloys and tailored cast microstructures. The two higher risk efforts are the subject of this paper.

Advanced Alloy Evaluations

The advanced alloy studies investigated the relative potential of derivative directionally solidified (DS) and single crystal compositions when utilized for investment cast integrally bladed turbine wheel designs. The alloys studied were compositional variants of the second-generation 3% rhenium (Re) containing DS alloy CM 186 LC®[1] and the third-generation 6% Re containing single crystal alloy CMSX®-10[2]. The nominal compositions of the alloys investigated within this program are provided in Table I. Previous attempts to make equiaxed castings using the CM 186 LC alloy with no modification in composition or heat treatment were unsuccessful because the creep rupture properties were inadequate[3]. The compositional modifications made in support of the present program focused on increasing grain boundary strength and ductility while maintaining microstructural stability.

The master alloys were used to produce Rolls-Royce Corporation Model 250 first-stage turbine wheels using Howmet's GX casting process to refine the grain structure in the web and hub of the wheels. The GX process utilizes oscillatory mold rotation to control convective fluid flow, the rate of heat transfer out of the casting and the resultant grain structure.[4-6] All GX wheel casting was performed at the Howmet LaPorte Casting Facility. The

Superalloys 2000
Edited by T.M. Pollock, R.D. Kissinger, R.R. Bowman,
K.A. Green, M. McLean, S. Olson, and J.J. Schirra
TMS (The Minerals, Metals & Materials Society), 2000

Table I. Nominal compositions of first, second, and third generation superalloys (weight %).

Alloy	Cr	Co	Mo	W	Re	Nb	Ta	Al	Ti	Hf	C	B	Zr	Ni
Mar-M246	9	10	2.5	10			1.5	5.5	1.5		0.15	0.015	0.05	
Mar-M247	8.4	10	0.65	10			3.1	5.5	1	1.4	0.16	0.015	0.05	
CM 186 LC	6	9	0.5	8	3		3	5.7	0.7	1.4	0.07	0.015	0.005	Bal
CM 186 Mod	5.9	9.4	0.4	8.5	3		3.3	5.7	0.75	1.5	0.09	0.019	0.01	Bal
CM 681	5.4	9.3	0.5	8.5	3		6.2	5.7	0.15	1.6	0.11	0.018	0.015	Bal
CM 681A	5	9.3	0.5	9	3		6.9	5.7		1.6	0.11	0.018	0.025	Bal
CMSX-10	2	3	0.4	5	6	0.05	8	5.7	0.2	0.03				Bal
CM 4670	4	3.4	0.5	5	5.3	0.05	8	5.7	0.13	1.2	0.09	0.017	0.015	Bal
CM 4670C	2.7	3.2	0.4	5	6	0.05	8	5.7	0.08	1.2	0.05	0.02	0.025	Bal
CM 4670D	1.7	3	0.35	5	5.7	0.05	8	5.7		1.7	0.1	0.018	0.015	Bal
CM 4670E	1.7	3	0.4	5	5	0.05	8.3	5.8		1.7	0.12	0.018	0.015	Bal

wheels were evaluated for compatibility with the GX casting and postcasting processing, microstructural stability during thermal processing and expected service conditions, and critical mechanical and environmental properties.

Iteration 1—Wheel Casting

The first casting iteration evaluated the CM 186 MOD, CM 681, CM 4670 and CM 4670C alloys. Macrostructure examination of integral and cross-sectioned wheels revealed uniform equiaxed grains in the rim, bore, and stub areas (Figure 1). The grain structure and porosity levels in these wheels were similar to production Mar-M247 cast wheels produced using the GX casting method. Since the porosity in the GX cast wheels was not surface connected, a full densification would be expected during subsequent hot isostatic pressing (HIP) processing.

Fluorescent penetrant inspection (FPI) revealed severe airfoil cracking in the CM 4670 and CM 4670C wheels. The cracks were classified as hot tear cracks and were intergranular. No airfoil cracking was observed in the CM 186 MOD and CM 681 wheels.

Microstructural examination of the as-cast wheels showed comparable grain sizes in the hub and rim areas for all four compositions. However, the airfoil grain size was somewhat finer in the CM 681 and CM 4670C wheels. A higher volume fraction of gamma/gamma prime eutectic phase was observed in the CMSX-10 variant wheels. This is credited to higher gamma prime forming element and refractory element content in the alloys.

HIP and Heat Treat Studies

One GX cast wheel from each of the four modified CM 186 LC and CMSX-10 compositions was processed at 1185°C/172 MPa/4 hr. This cycle is the standard production cycle used for GX cast Mar-M247 components. Metallographic examinations of the

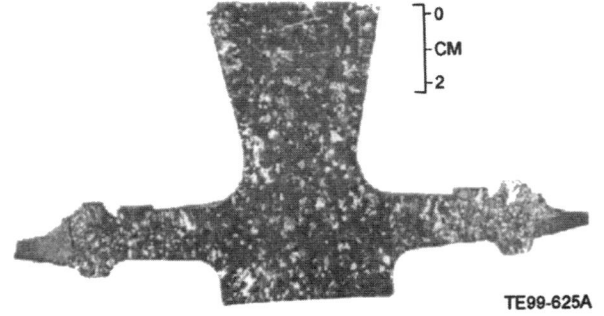

Figure 1. Macrostructure of Rolls-Royce Corporation Model 250 CM 681 GX wheel.

advanced alloy wheels revealed the presence of some residual porosity. It was suspected that adding rhenium increased the high temperature strength of the alloy and prevented full densification; therefore, an additional four wheels were HIP processed at a higher temperature and pressure (1232°C/200 MPa/4 hr). Metallographic examinations of this material revealed complete pore closure. The higher HIP temperature and pressure did not result in noticeable grain growth or in indications of incipient melting.

Optical microscopy and scanning electron microscopy (SEM) were performed on various sections of the wheels (airfoil, outer web, and inner web) in the as-cast and as-HIPed conditions and with post HIP resolutioning thermal treatments of 1232, 1246, 1260, and 1274°C for 2 to 5 hours. As-cast structures exhibited relatively uniform blocky gamma prime and no evidence of topologically close-packed (TCP) phases (Figure 2a). However, significant gamma prime coarsening and carbide degeneration occurred as a result of the high HIP temperature utilized to close the residual casting porosity and the subsequent high solution heat treat temperatures employed. In addition, the CMSX-10 variants exhibited some microstructural instabilities in the form of TCP needle phases from the postcasting thermal processing (Figure 2b). In an effort to restore the desired gamma prime morphology, high temperature solutioning treatments of 1271°C/3 hr for the CM 186 variants and 1288°C/5 hr for the CMSX-10 variants were selected followed by gas fan cooling (GFC) and double aging at 1093°C/2 hr/GFC + 871°C/20 hr/GFC.

Microstructural Characterization

SEM and energy dispersive spectroscopy (EDS) were employed to document the changes in carbide morphology occurring during the high temperature thermal treatments. The as-cast material showed many script MC carbides (Figure 3a). There are also a number of discrete carbides present near the primary gamma prime islands. Both the script and discrete carbides showed very high Ta content. The discrete carbides also showed some Hf.

The high temperature thermal treatments progressively break down the script TaC phase to release carbon to form HfC precipitates where Hf segregation exists, generally near primary gamma prime areas (Figure 3b). The carbon tied up as HfC cannot form the secondary Cr carbides at the grain boundaries and Cr concentrations can then lead to sigma formation. Since Hf is added to Ni base alloys to improve ductility at 760°C[7], loss of beneficial Hf in the gamma prime is detrimental. The Re appears not to take a significant part in primary carbide formation and is contained primarily in the gamma phase.[8]

Specimen Testing

The mechanical testing of HIP processed and fully heat treated GX cast wheels included room temperature and 538°C tensile tests, stress rupture tests, LCF tests at 538°C, and fatigue crack growth (FCG) testing at 538°C. These tests all were performed using material removed from the disk portion of the wheel. In addition, airfoil miniflat stress rupture tests were conducted.

The 0.2% yield strengths and ultimate tensile strengths of all four advanced alloys were equal or somewhat higher than the baseline GX Mar-M247 material.

Stress rupture results for the hub portion of the wheels are shown in Figure 4. The CM 186 variants were superior to the Mar-M247 baseline alloy at low stresses, but inferior at high stresses. The CMSX-10 variants were equivalent to Mar-M247 at low stresses and significantly worse at high stresses.

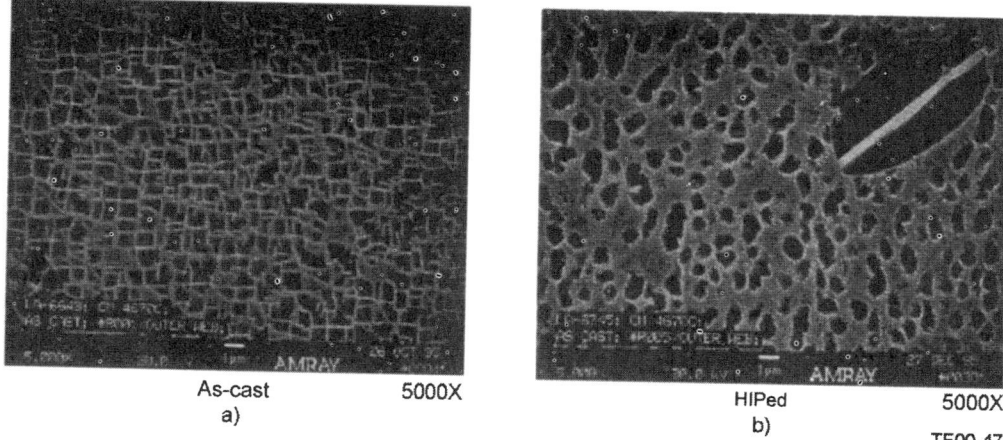

As-cast 5000X
a)

HIPed 5000X
b)

TE00-47

Figure 2. As-cast and as-HIPed microstructures from outer web area of CM 4670C GX wheel.

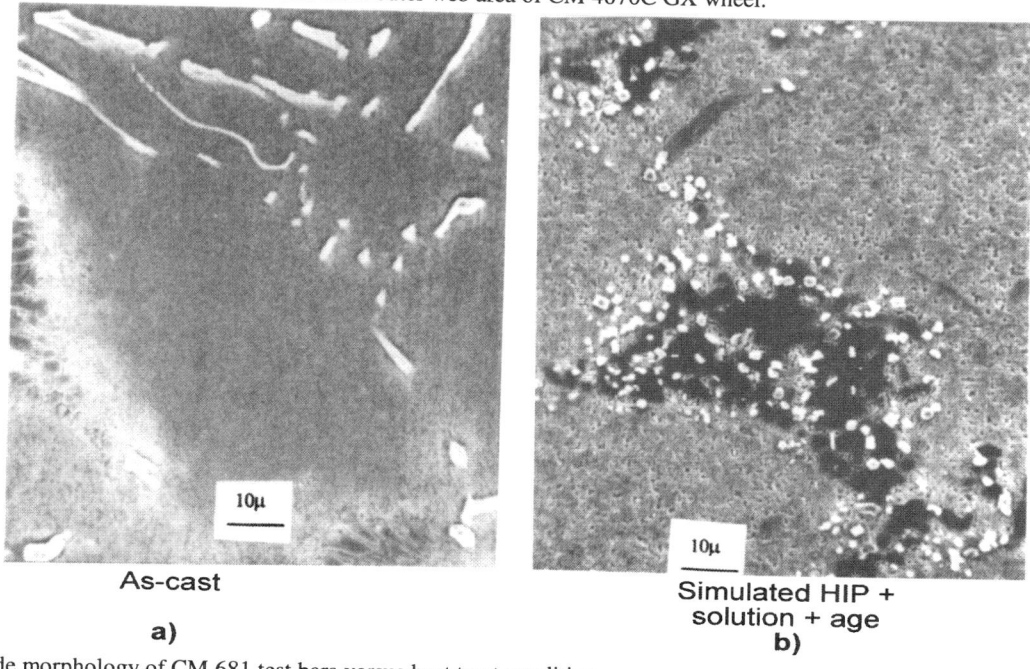

As-cast

a)

Simulated HIP +
solution + age

b)

Figure 3. Carbide morphology of CM 681 test bars versus heat treat condition.

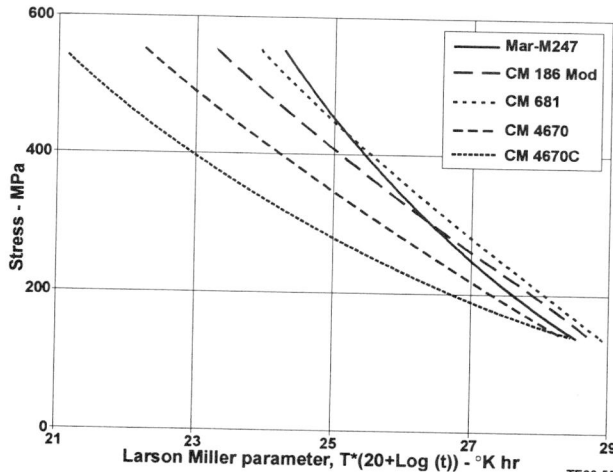

Figure 4. Hub stress rupture results from first-iteration equiaxed alloy variants.

The airfoil miniflat stress rupture lives of all four advanced alloys were roughly equivalent to each other and to the baseline Mar-M247 at the lowest stress level tested. At higher stresses, more divergence in rupture life was observed among the advanced alloys and all of the alloys were inferior to the baseline material.

The LCF capability of the CM 186 variants showed a crossover with Mar-M247 with the baseline alloy having an advantage at lower strain ranges. Since the critical portion of the wheel operates at high strain ranges, these curve shapes are favorable for the advanced alloys. The CMSX-10 variants showed reduced lives at all strain ranges.

The FCG test results are summarized in Figure 5. Surprisingly, all of the advanced alloys showed some advantage in FCG resistance over the critical stress intensity range of 20 to 40 MPa $\sqrt{m}$.

Burner rig sulfidation testing was performed by NASA Glenn. The test conditions were 899°C and Mach 0.3 air flow using JP5 fuel with 2 ppm by weight of synthetic sea salt. Each cycle consisted of

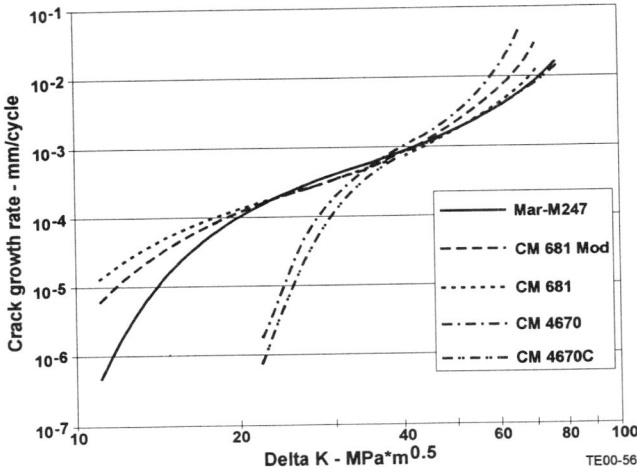

Figure 5. FCG summary from first-iteration of equiaxed alloy variants.

1 hr at temperature in a rotating carousel followed by a cold air blast. The weight change data are shown in Figure 6. All four advanced alloys performed considerably better than the baseline Mar-M247 alloy. Although the advanced alloys show improved corrosion resistance, a coating would be required because Model 250 engines often operate in corrosive environments.

Iteration 1 Conclusions

At the conclusion of the first iteration of advanced alloy evaluations, it was postulated the mechanical properties of the advanced alloys were being adversely affected by the thermal processing routes selected for the materials. Specifically, it was suggested the relatively slow cooling rates achieved during HIP processing caused excessive gamma prime coarsening. Subsequent heat treatment after HIP processing using a higher solution temperature and gas fan cooling did not restore the desired gamma prime sizes and morphology.

In addition, all of the high temperature processing was expected to have an adverse effect on carbide composition and morphology.

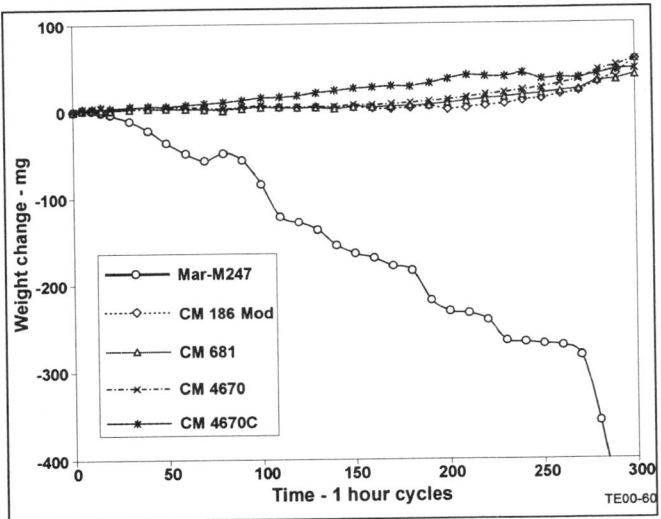

Figure 6. Burner rig hot corrosion test results from the Mar-M247 and modified CM 186 2c and CMSX-10 alloys.

Fractographic analysis on selected airfoil miniflat stress rupture bars revealed much smoother grain boundary fracture surfaces for the advanced alloys than the Mar-M247 baseline material. This suggests the grain boundary carbide morphology was less well developed during heat treatment for the advanced alloys than the Mar-M247 baseline material.

Also, it was clear the phasal stability of the modified CMSX-10 alloys was limited and they could not withstand the high temperature post casting thermal processing.

Advanced Alloy Test Bar Evaluations

To separate the effects of composition and thermal processing on mechanical properties, test bar castings were produced from each of the seven advanced alloy compositions and evaluated with various thermal processing conditions prior to initiating the second iteration of advanced alloy wheel casting.

The results of this testing can be summarized as:

- The phasal stability of the CMSX-10 variants was inadequate to withstand the high temperature postcasting thermal processing required for integrally cast GX wheels.
- The modified CMSX-10 alloys would be dropped from the program.
- CM 186 MOD alloy was noticeably weaker than the other two advanced alloys and also would be dropped from the program.
- The HIP conditions required to achieve pore closure and HIP cooling rate were key variables.

Iteration 2—Wheel Casting

Rolls-Royce Corporation Model 250 first-stage wheels were produced using CM 681 and CM 681A alloys and the Howmet GX casting process. The surface grain structures were uniform for wheels from both alloys. Grain sizes in the hub and rim appeared to be comparable to current production GX Mar-M247 cast wheels for both advanced alloy compositions.

HIP and Heat Treat Optimization

A series of wheels were HIPed at 200 MPa for 4 hr at temperatures ranging from 1185 to 1218°C. The initial metallographic examination of HIPed wheels for pore closure used specimens taken from the central hub region. The central hub is the thickest part of the casting and the last area to solidify; therefore, it was believed to be the area most prone to microshrinkage and the last area that HIP would close. Specimens removed from the central hub area of these wheels showed no evidence of residual microporosity. Subsequently, it was also decided to examine specimens from the web and rim areas for residual porosity, because small microshrinkage was occasionally observed on the fracture surfaces of the failed stress rupture bars. Surprisingly, several small pores with incomplete closure were located in the center of the rim area. Presumably, the greater susceptibility to microporosity in the rim area is related to the forced fluid flow during solidification associated with GX processing. The maximum pore size observed was 3 millimeters (mm) and was generally less than 1 mm. These results were reviewed with the materials applications engineers for the Model 250 engine and failure analysts at Rolls-Royce Corporation familiar with this engine. It was determined this small amount of residual porosity would be inconsequential to engine performance.

It was concluded from the HIP assessment studies that minimizing the HIP temperature was beneficial to mechanical properties with the advanced alloys. Accordingly, one GX wheel each of alloy CM 681 and CM 681A were HIPed at 1204°C/200 MPa/4 hr and a second CM 681 alloy wheel was HIPed at 1185°C/200 MPa/4 hr. One group of specimens from each wheel received the standard age of 1093°C/2 hr/GFC + 871°C/20 hr/GFC. A second group received a modified age of 1038°C/2 hr/GFC + 871°C/20 hr/GFC. A third group received a 1204°C/2 hr/GFC partial resolution followed by the modified double age.

The stress rupture lives at 843°C/552 MPa were 200 to 300% of baseline Mar-M247 lives for both advanced alloys and all three thermal processing conditions. The results from stress rupture tests conducted at 1038°C/138 MPa are presented in Figure 7. The lower temperature processing appeared to provide a significant improvement in the rupture life. The CM 681 alloy exhibited a somewhat higher rupture life than the CM 681A alloy. The LCF testing results are also shown in Figure 7. Most of the advanced alloy and thermal processing combinations provided improved LCF lives compared to the baseline Mar-M247 material examined. It also appears the resolutioning after HIP offers a benefit to fatigue life.

Overall, the 1185°C HIP followed by the modified age appeared to offer the best balance of properties, and this thermal processing was selected for the balance of the CM 681 and CM 681A wheels.

Specimen Testing

The test plan was identical to that employed for the first-iteration wheel evaluations. The testing included room temperature and 538°C tensile tests, stress rupture tests, LCF tests at 538°C, and crack growth testing at 538°C. The tests were all performed using material removed from the disk portion of the wheel. In addition, airfoil miniflat stress rupture tests were conducted.

The 0.2% yield strength and ultimate tensile strength of the CM 681 alloy was somewhat lower than the values achieved for this alloy in the first iteration and closer to the strength levels of Mar-M247. This represents the desired result, since a higher strength could disrupt the required burst sequence between the first-stage and second-stage turbine wheels and thereby force a redesign. No significant difference was observed in strength or ductility between CM 681 and CM 681A.

Stress rupture results for the hub portion of the wheels are shown in Figure 8. Both advanced alloys performed significantly better than the baseline Mar-M247 alloy at all stress levels. Compared to

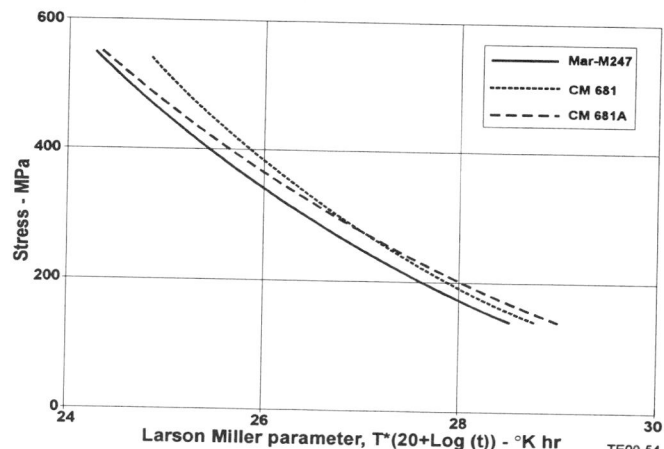

Figure 8. Hub stress rupture results from second-iteration equiaxed alloy variants.

the results of the CM 186 derivative alloys in the first iteration, it is evident the second-iteration thermal processing provides better performance in the high stress portion of the curve while maintaining an advantage over Mar-M247 in the low stress region. CM 681 performed slightly better than CM 681A at low stresses and CM 681A was superior at high stresses.

The airfoil miniflat stress rupture test results are provided in Figure 9. The advanced alloys are clearly superior to the baseline Mar-M247 alloy throughout the stress range investigated. This is in stark contrast to the first-iteration results in which the advanced alloys were dramatically inferior to the baseline material at high stresses. The CM 681A alloy exhibited a small advantage over the CM 681 alloy at higher stresses and a more distinct advantage in the low stress region.

The LCF test results are shown in Figure 10. The two CM 186 derivative alloys performed similarly. Both alloys were superior to Mar-M247 in the low life, high strain range portion of the curve and inferior to the baseline in the high life, low strain range region. Since the critical portion of the wheel operates at high strain ranges, these curve shapes are favorable for the advanced alloys. This is the same trend observed in the first-iteration results for the CM 186 derivative alloys, indicating the alternative thermal processing had only a minor effect on LCF properties.

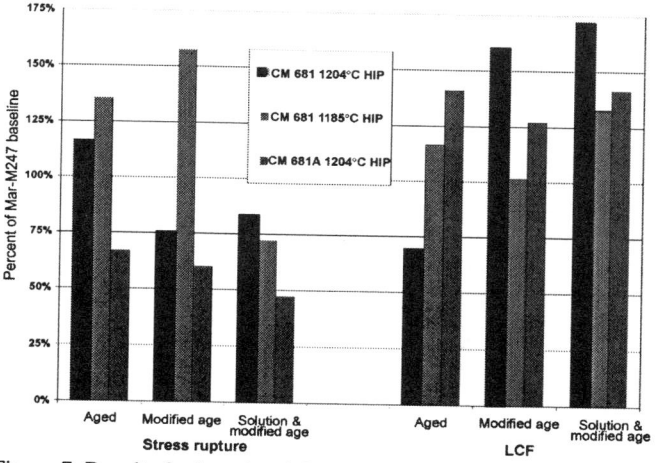

Figure 7. Results for iteration 2 843°C/552 MPa stress rupture and 538°C/R=0/Kt=1/Δε=0.95% LCF HIP and heat treat optimization study.

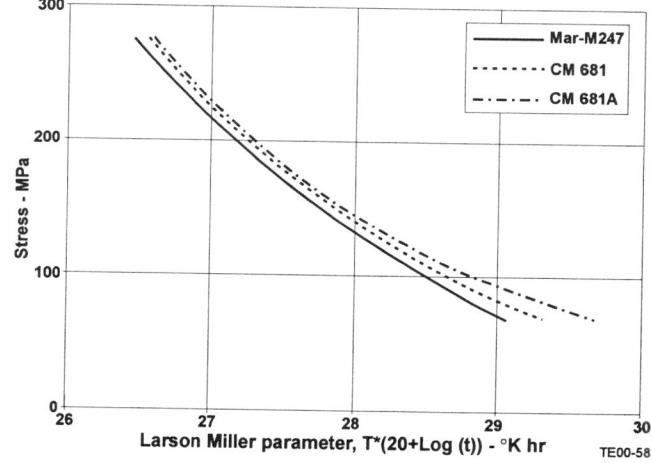

Figure 9. Airfoil miniflat stress rupture results from second-iteration equiaxed alloy variants.

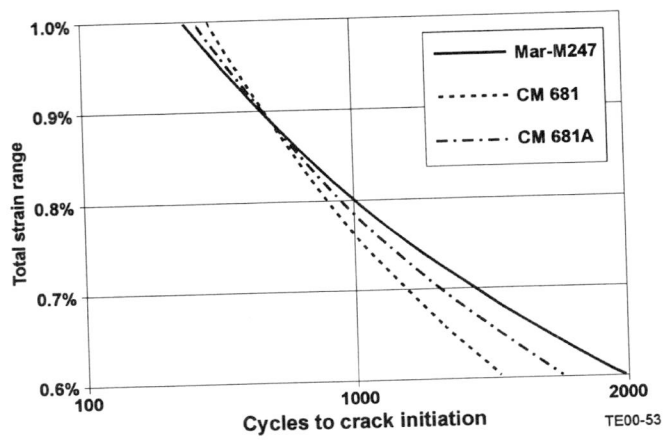

Figure 10. Hub LCF results from second-iteration equiaxed alloy variants.

The FCG test results are provided in Figure 11. The CM 681A alloy was similar to the baseline Mar-M247 material. The CM 681 alloy appears to offer a significant advantage in crack growth resistance compared to the baseline. Crack growth tests tend to be variable and the extent of testing conducted on this program was limited. Nevertheless, the CM 681 results were encouraging and would provide a major benefit to wheel life if this advantage is realized in engine testing.

In summary, both CM 681 and CM 681A exhibit significant advantages over the baseline Mar-M247 material. CM 681 was selected for the manufacturing scale-up because of its potential for greatly increased crack growth resistance.

Process Modeling and Validation of Tailored Cast Structures

The objective of the tailored cast microstructure effort was to produce DS airfoils in an integral turbine wheel that would then transition into a fine grain GX web and hub. Previous attempts by Howmet and others to produce integral wheels with DS airfoils were successful on occasion; however, process repeatability was lacking due to inadequate understanding and control of the key variables affecting the modified GX process.[9-11] Other technical difficulties included poor grain control in the airfoils, poor control of the coarse-to-fine grain structure transition, significant mold structural problems, and significant labor in put for each cast

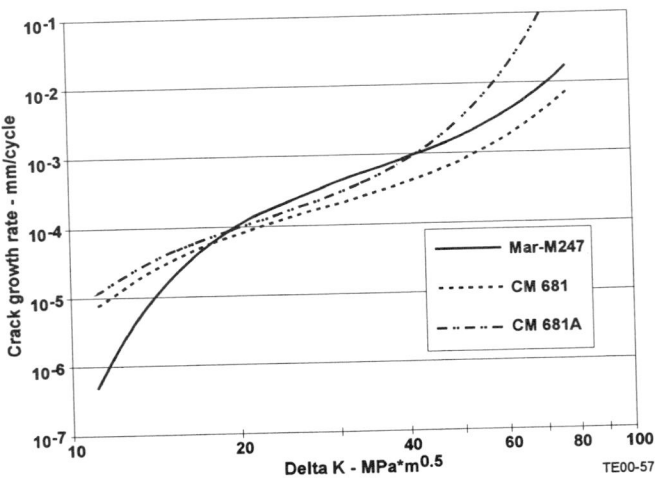

Figure 11. FCG summary from second-iteration equiaxed alloy variants.

wheel. In the present investigation Howmet relied heavily on their expertise with process modeling and manufacturing simulation tools to minimize trial and error experimentation.

The initial simulations and casting trials focused on improving the directionality of the airfoil grains. The goal of the simulations was to minimize axial and circumferential thermal gradients and isotherm velocities through the combined use of a radial heat sink, tailored pattern features, mold composition and thickness, wrapping procedures, and foundry parameters.

This iterative approach of model predictions followed by casting trials resulted in progressive improvement of the oriented grain structure with the later trials showing excellent orientation and directional solidification. The resultant process was designated DS/GX®. The initial casting trials also exhibited some mold splitting, but this problem was essentially eliminated with additional casting modifications. One defect that remained a concern was hot tears in the airfoil area. Figure 12 shows an example of the hot tears that occur in both the radial and tangential directions.

To address the airfoil hot tearing, stress modeling was used in conjunction with the fluid-thermal modeling to optimize the starter geometry. A number of thermal stress models were evaluated to simulate the relative hoop and tensile stress levels of the baseline and alternative geometries. Following the stress evaluation, fluid-thermal models were solved to assess the effects the modifications would have on the directional solidification of the airfoils. Tool modifications were made based upon the design showing the best improvement in both stress level and temperature gradient orientation. Figure 13 shows the three component temperature gradients for the revised configuration. The gradients were better aligned orthogonal to the airfoil stacking axis for the revised configuration than the baseline and resulted in better airfoil grain orientation as well as a significant reduction in overall stress levels.

Additional wheels were cast and overall the exceptional grain orientation quality was retained (Figure 14); however, airfoil hot tearing, while significantly reduced, was not eliminated. Since

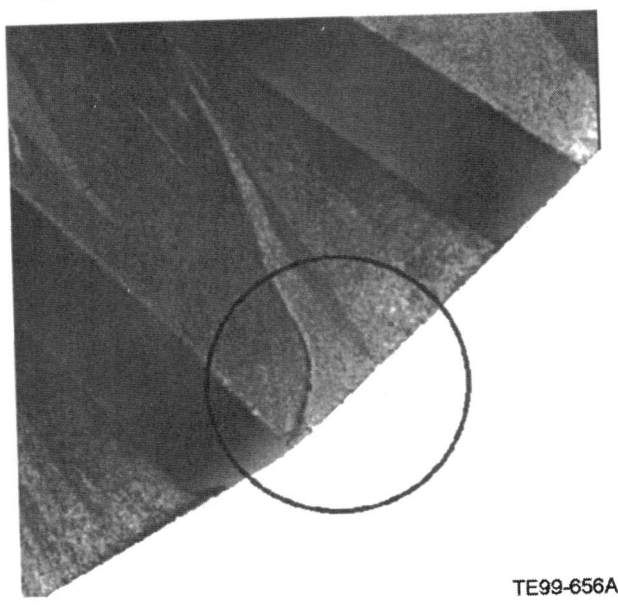

TE99-656A

Figure 12. Typical hot cracking in the airfoils of the DS/GX cast wheel.

176

Temperature Gradient - Allison DSGX Wheel

Vertical

Radial

Tangential

200 F/in

0

-200

TE00-95

Figure 13. Temperature gradients for the revised configuration for the modified DS/GX cast wheel.

TE99-661

Figure 14. DS/GX cast Mar-M247 wheel from the final casting trial showing directionally oriented grain in airfoils.

the airfoil hot tearing issue remained, a compromise process that eliminated the radial heat sink and utilized special mold configurations and casting processes to produce large, partially aligned grains in the airfoil was developed. This modified GX casting process has been termed Pseudo DS/GX casting. This process was demonstrated on a wheel shown in Figure 15.

DS/GX and Pseudo DS/GX Specimen Testing

Tensile and stress rupture testing was conducted on specimens taken from the columnar-to-equiaxed transition zone in several wheels. The tensile specimens all failed in the columnar grain region and exhibited base material strength and ductility. The transition area stress rupture specimens all failed in the equiaxed

TE99-654

Figure 15. Pseudo DS/GX cast Mar-M247 wheel from casting trial 3 showing enlarged grain in airfoils.

region with equivalent life versus the base material. Metallographic examination of transition areas revealed no unusual phases or microstructural features.

Testing also included tensile, stress rupture, and LCF tests using material removed from the disk portion of the wheel as well as airfoil miniflat stress rupture tests. The higher temperature foundry parameters associated with DS/GX and Pseudo DS/GX processing produced little or no reduction in yield and tensile strength and may have benefited ductility to a small extent.

The Pseudo DS/GX processing provided significant benefits to airfoil stress rupture life for both alloys and DS/GX processing adds another significant improvement in life as shown in Figure 16. The DS/GX and Pseudo DS/GX processing had little or no effect on hub stress rupture or LCF with Mar-M247, but caused some degradation in rupture life at high stresses and a substantially lower LCF life in the high strain range portion of the curve with CM 681. The LCF results are shown in Figure 17.

Apparently the higher processing temperatures caused enough microstructural damage in the CM681 alloy to partially offset the benefits obtained from the modified HIP, solution, and aging thermal treatments developed during this program.

Manufacturing Scale-Up

Wheel Casting

The AMSTE team selected CM 681 as the advanced alloy and Pseudo DS/GX for the advanced casting process, but concluded there were not yet sufficient data to make a final selection regarding the best combination of alloy and casting process.

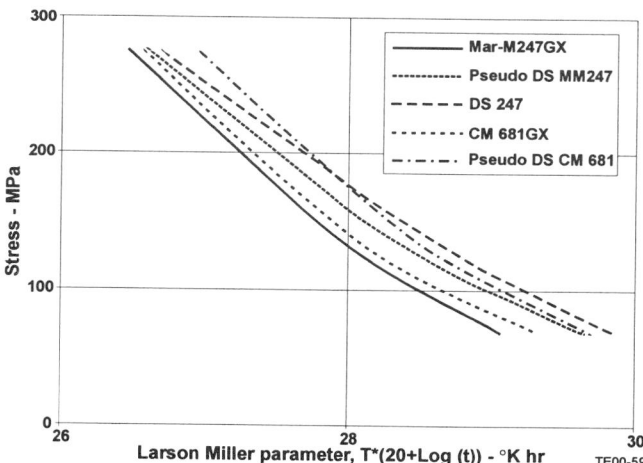

——	Mar-M247GX
‑ ‑ ‑ ‑	Pseudo DS MM247
– – –	DS 247
······	CM 681GX
–·–·–	Pseudo DS CM 681

Stress - MPa

Larson Miller parameter, T*(20+Log (t)) - °K hr   TE00-59

Figure 16. Airfoil miniflat stress rupture results from DS/GX and Pseudo DS/GX wheels.

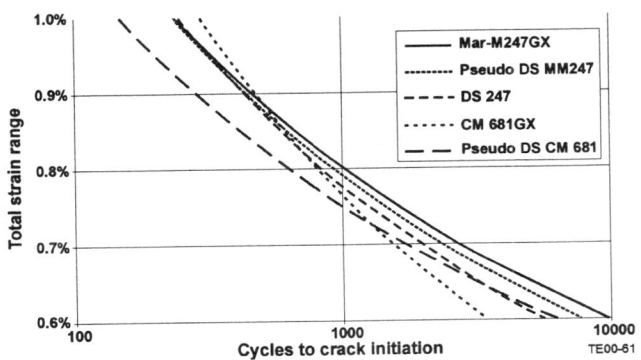

Figure 17. Hub LCF results from DS/GX and Pseudo DS/GX wheels.

Accordingly, both Mar-M247 and CM 681 alloys and GX and Pseudo DS/GX processing were included in the final scale-up plan.

Nine wheels were cast from Mar-M247 using the Pseudo DS/GX method. There were 16 wheels cast from CM 681, 8 each with the GX and Pseudo DS/GX casting processes. These castings were inspected per the Rolls-Royce Corporation production specification and therefore received a more detailed inspection than any other castings produced in this program to date.

Manufacturing/Affordability Study

The trade study was intended to assess the potential life cycle cost impacts of alternative designs, materials, and structures in the first-stage turbine wheel of the Model 250 Series IV engine. The inputs to this trade study were wheel casting costs, wheel life, and wheel airfoil temperature capability. Six variants were compared to the production Mar-M246 wheel. The wheel casting cost estimates were prepared by Howmet. The wheel life estimates were based on go-and-blow engine testing and mechanical testing (LCF and FCG). The airfoil temperature capability estimates were based on stress rupture testing of airfoil miniflats.

The first-stage turbine wheels developed under this program offer the two advantages of increased turbine wheel life and increased airfoil temperature capabilities compared to the Mar-M246 production configuration wheel. The increased turbine wheel life translates directly into increased time between overhaul for the engine hot section, with a subsequent reduction in the direct operating cost (DOC) of the engine. The increased airfoil temperature capabilities result in increased shaft horsepower output of the engine, which in turn leads to a potential increase in customer value. The potential for both the decreased DOC and increased net customer value for each alloy/structure combination, relative to the current production baseline first-stage turbine wheel, was assessed separately. Details of the trade study are shown in Table II.

The impact on engine DOC is a tradeoff between the increased turbine wheel life and cost. The Mar-M247 GX thick rim wheel shows a substantial improvement in engine DOC compared to the current production baseline configuration (-2.2%). The CM 681 GX provides an incremental improvement over the Mar-M247 GX wheel (-2.5% versus -2.2%). The Pseudo DS/GX and DS/GX processing showed an adverse effect on engine DOC, since these techniques increase wheel manufacturing cost and do not provide additional wheel life.

As with the DOC analysis, there is a tradeoff between the increased temperature capability of the turbine wheel and the increased cost of advanced wheel castings. For the alloy/structure combinations studied, however, the increase in customer value due to the increased temperature capability far outweighs the increase in casting cost. Thus, the wheels with the greatest increase in temperature capability also have the greatest potential for increased net customer value.

In summary, this trade study suggests that both the advanced alloy and the advanced processing, either alone or in combination, offer significant cost benefits over the current production practice.

Frequency/High Cycle Fatigue Blade Testing

Blade vibratory modes and blade high cycle fatigue (HCF) strength were determined from finish machined Mar-M247 Pseudo DS/GX and CM 681 Pseudo DS/GX wheels. The test is required to determine which blade vibratory modes lie within the engine operating range, to verify the consistency of fabrication, and to compare the fatigue strength to past Model 250 first-stage wheel experience. Blade frequencies were determined through bench frequency testing. Blade fatigue strength was determined at ambient temperature by securing an individual blade to a fixed base and exciting the airfoil with an acoustical siren that varies in excitation energy and frequency. The crystallographic orientation texture associated with the Pseudo DS/GX processing produces a 10 to 15% reduction in the effective elastic modulus of the airfoil in first bend mode, with a resultant 5 to 10% reduction in blade frequency in first bend. However, no detrimental engine order coincidences that would interfere with normal engine operation were observed. The Pseudo DS/GX processing reduced HCF blade strength 30 to 40%. Additional analysis would be required to assess this lower capability.

Table II. AMSTE manufacturing/affordability study.

Alloy	Wheel structure	Design	Casting acquisition cost	Change in turbine wheel list price	Life	Increased airfoil temperature	Increased output shaft horsepower	Change in DOC	Net increase in customer value
Mar-M246	Equiaxed	Baseline	Baseline	Baseline	1X	Baseline	Baseline	Baseline	Baseline
Mar-M247	GX	Thick rim	1.6X	26%	2X	+0°C	0	-2.2%	0
	Pseudo DS/GX	Thick rim	2.7X	71%	2X	+14°C	4.3%	-0.8%	8%
	DS/GX	Thick rim	8.0X	300%	2X	+24°C	7.3%	+5.9%	12%
CM 681	GX	Thick rim	2.7X	71%	3X	+4°C	1.3%	-2.5%	2%
	Pseudo DS/GX	Thick rim	4.0X	129%	3X	+21°C	6.2%	-1.4%	11%
	DS/GX	Thick rim	9.3X	357%	3X	+31°C	9.2%	+3.1%	15%

Spin Pit Burst Testing of Wheels

Burst speed of finish machined Mar-M247 Pseudo DS/GX, CM 681 Pseudo DS/GX, and CM 681 GX wheels were determined in spin pit testing at ambient temperature. This test is conducted in a spin pit capable of rotational speeds up to 85,000 rpm. Burst testing validates design calculations of both wheel burst speed and failure origin. All three wheels burst in a radial mode, which is considered normal, at speeds commensurate with the ultimate tensile strength of the materials as determined by specimen testing.

Cyclic Engine Testing

Historically, the most accurate means of evaluating crack initiation and growth in the Model 250 engine turbine wheels has been through cyclic engine testing known as go-and-blow testing. Only in engine operation are the actual rim-to-web thermal gradients achieved during various conditions such as start, acceleration, deceleration, and shutdown. The extensive experience of Rolls-Royce Corporation and the data base with this accelerated test allow for efficient screening of new designs, materials, and processes.

This test is more severe on the turbine wheel than an actual engine operating cycle (start-run-stop), producing LCF crack growth rates of approximately twice those experienced in actual service. The most important parameter in determining the life of the Model 250 turbine wheels is the length of the wheel rim cracks known as "B" cracks. A "B" crack is a crack that is initiated on the outer diameter of the wheel rim (the platform between the blades) and extends down the machined surface of the rim leading or trailing edges. Engine operational instructions define a maximum allowable "B" crack length for continued operation in service.

A CM 681 GX first-stage turbine wheel was fabricated and assembled into a Model 250-C47B engine. The test engine completed 2000 cycles of go-and-blow testing in 1000 cycle intervals. At the end of each 1000 cycle interval, the wheel was removed from the engine and inspected for cracks by FPI. The lengths of the cracks were measured and recorded during each inspection.

The results of the advanced alloy wheel go-and-blow engine test are compared to go-and-blow tests for a current production first-stage turbine wheel and the baseline Mar-M247 GX configuration in Figure 18. It is evident the baseline Mar-M247 GX

configuration offers a significant improvement (>2X) over the current production first-stage wheel. The advanced alloy wheel results were also promising. Additional engine testing of this wheel would be required to confirm the improved crack growth resistance of this alloy.

Acknowledgements

The authors thank Robert Miner, formerly at NASA Glenn, and Brian Griffin and Bruce Allmandinger at Howmet LaPorte Casting Corporation for their technical contributions to this program. This program was supported in part by the NASA Glenn Research Center under Cooperative Agreement NCC3-387.

References

1) K. Harris, G. L. Erickson, S. L. Sikkenga, W. D. Brentnall, J. M. Aurrecoechea and K. G. Kubarych, "Development of the Rhenium Containing Superalloys CMSX-4® and CM 186 LC® for Single Crystal Blade and Directionally Solidified Vane Applications in Advanced Turbine Engines," Superalloys, 1992 (Warrendale, Pennsylvania, TMS, 1992, pp 297-306).

2) G. L. Erickson, "The Development and Application of CMSX-10®," Superalloys, 1996 (Warrendale, Pennsylvania, TMS, 1996, pp 35-44).

3) Private communication from John Eridon, Cannon-Muskegon to Rolls-Royce Corporation, 1993.

4) C. T. Sims and W. C. Hagel, The Superalloys, (John Wiley & Sons, Inc., 1972, pp. 500-503).

5) J. L. Mallardi, "From Teeth to Jet Engines," (presented to Gas Turbine Society of Japan June 5, 1992) Howmet Corporation, 1992, Greenwich, CT.

6) Howmet Technical Bulletin No. 3000, "Grainex® Cast Mar-M 247 Alloy," 1989, Greenwich, CT.

7) C. T. Sims and W. C. Hagel, The Superalloys, (John Wiley & Sons, Inc., 1972, pp. 252-256).

8) R. W. Broomfield, et al., "Development and Turbine Engine Performance of Three Advanced Rhenium Containing Superalloys for Single Crystal and Directionally Solidified Blades and Vanes," (Paper presented at the 1997 ASME [IGTI] TURBO EXPO '97 Conference, Orlando, Florida, June 1997).

9) K. Harris, G. L. Erickson, and R. E. Schwer, "Development of a High Creep Strength, High Ductility, Cast Superalloy for Integral Turbine Wheels" (Paper presented at the 1982 AIME Conference, Dallas, Texas, February 1982).

10) B. A. Ewing, "A Solid to Solid HIP-Bond Processing Concept for the Manufacture of Dual-Property Turbine Wheels for Small Gas Turbines," Superalloys, 1980 (Warrendale, Pennsylvania, TMS, 1980, pp 169-178).

11) J.H. Moll, J.H. Schwertz, and V.K. Chandhok, "P/M Dual-Property Wheels for Small Engines," Conference Proceedings-Progress in Powder Metallurgy 1981, (Metal Powder Industries Federation, Vol. 37, 1982, pp. 303-319).

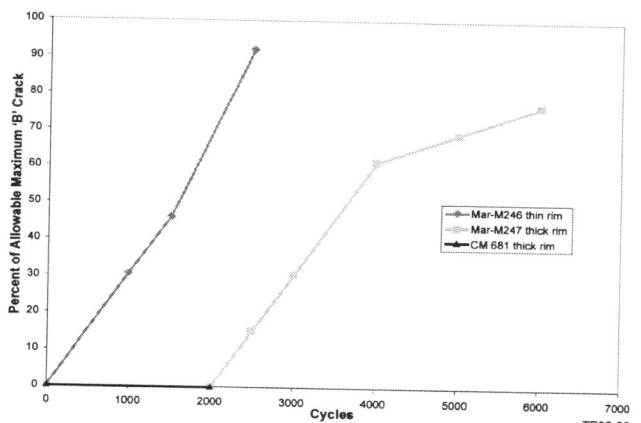

Figure 18. Maximum "B" crack length versus go-and-blow engine cycles.

IMPROVED QUALITY AND ECONOMICS OF INVESTMENT CASTINGS BY LIQUID METAL COOLING – THE SELECTION OF COOLING MEDIA

A. Lohmüller[1], W. Eßer[2], J. Großmann[3], M. Hördler[1], J. Preuhs[3] and R. F. Singer[1],

[1] Lehrstuhl Werkstoffkunde und Technologie der Metalle (WTM), 91058 Erlangen, Germany
[2] Siemens KWU, 45473 Mülheim a. d. Ruhr, Germany
[3] DONCASTERS Precision Castings - Bochum GmbH (DPC), 44793 Bochum, Germany

Abstract

A new directional solidification technique promising a more economic production of large blades is the Liquid-Metal-Cooling (LMC) process. The advantages of this technique are summarized in the paper. Direct temperature measurements as well as microstructural comparisons point towards an increase of the thermal gradient by a factor of two for large industrial turbine blades. This is at the present state of technical development, i.e. further improvements are expected eventually.

One of the most important factors in optimizing LMC is the choice of the cooling medium. In this paper tin and aluminium are compared. The advantages and disadvantages of these two cooling media are discussed in detail. Special attention is directed to the heat transfer potential and the possible dissolution of the cast component in the cooling bath in case of inadvertant contact. In addition information on the effect of tin as an intentional alloying element is given. Although tin looks rather favourable with respect to the points discussed, the long term performance on an industrial scale as compared to aluminium remains to be seen.

Introduction

Higher demands on turbine blades in modern industrial gas turbines (IGT's) have led to increasing use of directional solidification. The trend towards larger components and more complex alloy compositions reveales the limitations of the conventional Bridgman-technique (High-Rate-Solidification technique, HRS). Due to the limited temperature gradient only low withdrawal rates can be applied. An economic production based on radiation cooling becomes therefore quite difficult [1]. A promising alternative is the Liquid-Metal-Cooling (LMC) process [2,3]. LMC is expected to provide important improvements and may play a significant role in the industry in the near future [4].

The main advantages of the LMC technique have been well demonstrated for small scale, laboratory test pieces [5]. The higher thermal gradient and the increased solidification rate results in a finer microstructure as compared to the conventional Bridgman-technique. This reduces significantly the solution heat treatment soak period.

Superalloys 2000
Edited by T.M. Pollock, R.D. Kissinger, R.R. Bowman,
K.A. Green, M. McLean, S. Olson, and J.J. Schirra
TMS (The Minerals, Metals & Materials Society), 2000

Figure 1: Industrial scale LMC-furnace at DONCASTERS Precision Castings - Bochum GmbH, Germany. This furnace is in operation since mid of 1999.

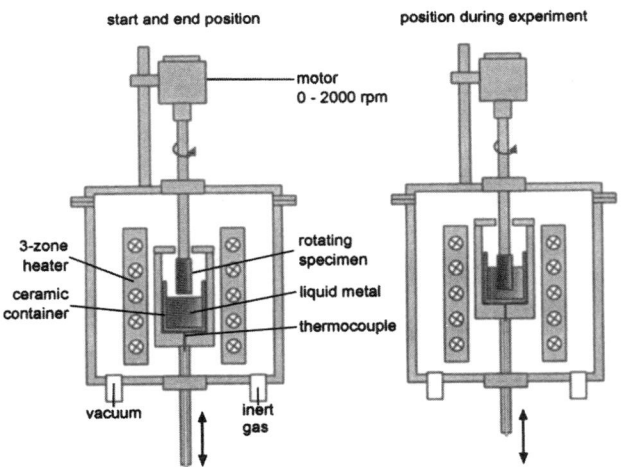

Figure 2: Schematic diagram of the apparatus used for the immersion and dissolution tests. During heating to temperature T the specimen and the liquid metal are not in contact (left). During the time t the rotating specimen is immersed into the liquid metal bath (right).

In 1994, a LMC pilot plant was installed at the WTM-institute in Erlangen [3]. By casting large test pieces and full scale, cored IGT single crystal turbine blades the feasibility of LMC was studied on a prototype scale. Some results from this program will be presented in the following together with some more general considerations that led recently to the construction of the worlds largest LMC production scale furnace (Figure 1) at DONCASTERS Precision Casting in Bochum. The main emphasis of the paper will be on the choice of a suitable cooling medium in LMC.

Experimental Procedure

LMC-casting trials have been carried out in the pilot plant at the WTM-Institute. The cooling bath with dimensions of 700 mm x 700 mm x 700 mm contains 1.7 t of liquid tin. Single crystal slabs (18 mm x 100 mm x 200 mm, two slabs per mold) and full scaled, cored IGT single crystal turbine blades were cast. The nickel base superalloys SC16 (16 wt.% Cr, 3,5 % Al, 3,5 % Ta, 3,5 % Ti, 3 % Mo and Ni as balance) and PWA 1483 (12,2 % Cr, 9 % Co, 4,1 % Ti, 3,6 % Al, 1,9 % Mo, 3,8 % W, 5 % Ta, 0,07 % C and Ni as balance) were used. The test pieces were solution heat treated and aged before mechanical testing.

In order to investigate the behaviour of the superalloys in direct contact with the cooling medium a laboratory test was developed (see Figure 2). A cylindrical, polished superalloy specimen (10 mm diameter) is immersed into a liquid metal bath (tin or aluminium) at constant temperature for an appropriate time. The specimen can be rotated at different speeds. In order to obtain reacting surfaces with well defined uniform flow conditions in the bath, the cylinder is coated at the bottom and near the fluid surface. In this way these surfaces are protected against dissolution. The bulk concentration c_t at the time t is calculated from the weight loss of the specimen.

Results

Advantages of the Liquid Metal Cooling-technique

Higher thermal gradients. It has always been assumed that the LMC-process has the advantage of producing a higher longitudinal thermal gradient during solidification. In the following, we want to explore the evidence for this when large IGT-geometries are considered.

The potential of the LMC process on a larger scale has been studied in an earlier investigation using an analytical model [6,7]. It was possible to predict the optimal withdrawal rate and the temperature gradient. Predominantly in large cross sections LMC was expected to double the gradient as compared to the HRS-technique. The influence of several parameters like temperature of the heater and the cooling medium, ceramic shell conductivity or mold thickness were examined. Depending on the parameter set, thermal gradients for LMC in the range between 2 K/mm and 8 K/mm were calculated.

The results of the earlier theoretical investigation will be now compared with the experimental results. At first we will derive the thermal gradient by evaluating the microstructure achieved during solidification for casting conditions that are typical for each process. In a first stage turbine blade (length ≈ 290 mm) LMC was found to reduce the primary dendrite arm spacing λ_1 by half [8]. At the same time LMC allows withdrawal rates twice as high as the HRS technique [8]. The use of the empirical formula [9]

$$\lambda_1 = 750\, G^{-\frac{1}{2}}\, v^{-\frac{1}{4}} \tag{1}$$

allows to estimate the temperature gradient present. For the HRS-technique ($\lambda_1 \approx 600\ \mu m$, v = 3.5 mm/min) a gradient of $G_{HRS} = 0.8$ K/mm can be calculated, while for LMC ($\lambda_1 \approx 300\ \mu m$, v = 8.0 mm/min) the gradient is $G_{LMC} = 2.2$ K/mm. It must be

noted, however, that the experimental data in [9] shows a large scatter. Therefore, the gradient is better represented by a range, e.g. from 1.5 K/mm to 3.8 K/mm for LMC and 0.7 K/mm to 1.1 K/mm for HRS.

To confirm the estimates based on the microstructure, temperature measurements with thermocouples have been performed during the LMC-process. Figure 3 shows a cooling curve for a first stage turbine blade in the middle of the airfoil. The cooling rate at the liquidus front was determined to be 40 K/min. For a withdrawal rate of 8 mm/min this is equivalent to a thermal gradient of 4.9 K/mm which is somewhat higher but close to the values based on equation (1).

Other LMC advantages. The LMC process has further advantages besides its capability for higher gradients. Higher withdrawal rates and reduced dendrite arm spacings have already been mentioned. In addition LMC-cast components show finer and more homogeneously distributed carbide precipitates and γ/γ'-eutectic. The γ'-precipitates are distributed more uniformly and cubic morphology is already achieved in the as cast condition. Due to the smaller dendrite arm spacings, concentration gradients are much higher and segregation is much more rapidly eliminated. The high temperature gradients can be achieved even for large blade clusters, since heat transfer is uniform even under these conditions. [8]

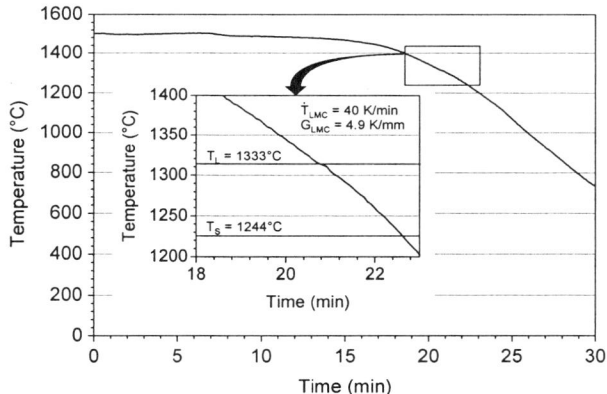

Figure 3: Cooling curve during withdrawal for an IGT blade (length ≈ 290 mm) solidified at a withdrawal rate of 8 mm/min using the LMC pilot plant at WTM-Institute. The thermocouple was situated in the middle of the airfoil. The thermal gradient is determined to be 4.9 K/mm. This value is in good agreement with values derived from evaluation of the microstructure as well as earlier analytical predictions. It is at least twice as high as in the corresponding HRS process.

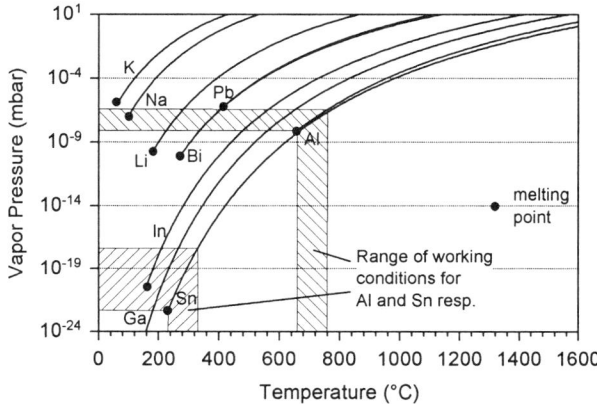

Figure 4: Vapor pressure of low melting point metals [18]. A low vapor pressure is a prerequisite for use as cooling medium in the LMC process. Both tin and aluminium are particularly suitable.

Choice of the Cooling Medium

<u>Physical properties and heat transfer.</u> One of the challenges in the optimization of the LMC technology on an industrial scale is to find a suitable cooling medium. Several requirements have to be fullfilled e.g. low melting temperature, low vapor pressure, high thermal conductivity, low viscosity, no toxicity, and economic efficiency.

Figure 4 presents the vapor pressure of low melting point metals as a function of temperature. The only liquid metals with suitable vapor pressure for a vacuum process at high temperatures are aluminium and tin. Gallium and indium are not considered due to their high price. Although the curves for aluminium and tin are nearly congruent, tin appears to be more favourable because the lower melting point allows a lower process temperature to be used.

One of the evident advantages of aluminium is its very high thermal conductivity ($\lambda = 104$ W/mK at a temperature of 660°C) promising a good heat transfer from the casting, see Table I. However there are other important factors in heat transfer that should not be overlooked. A simple pseudo-one-dimensional heat transfer model can give an idea of the magnitude of the heat flux possible. The situation is depicted in the insert drawing in Figure 5. For every temperature T_{SC} within the solidified cast component a local heat flux q perpendicular to the withdrawal direction can be calculated. The heat is transferred through a gap between the

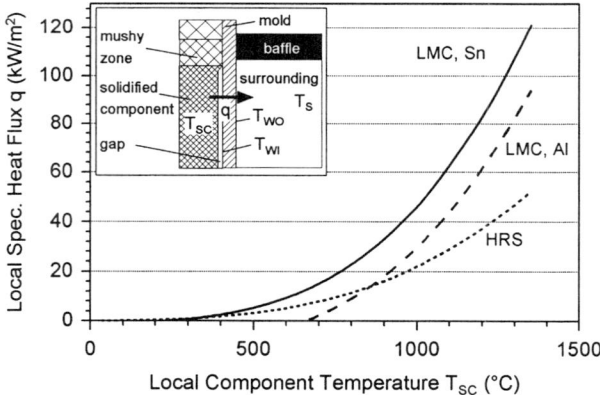

Figure 5: Comparison of the calculated local specific heat flux q for LMC with tin or aluminium as a cooling medium with the conventional HRS-technique. The heat flux for LMC with tin is significantly higher than for LMC with Aluminium. In general LMC leads to better cooling conditions than the HRS-technique.

Table I: Properties of liquid aluminium and liquid tin at their melting point [19].

Property	Unit	Aluminium	Tin
Melting point	°C	660	232
Density	kg/m³	2382	6980
Specific heat	kJ/m³ K	2597	1857
Thermal conductivity	W/mK	104	33.5
Thermal diffusivity	m²/s	40 x 10⁻⁶	18 x 10⁻⁶
Latent heat	kJ/m³	921834	418800
Kinematic Viscosity	m²/s	6,3 x 10⁻⁷	2,58 x 10⁻⁷

cast component and the inner mold surface at temperature T_{WI} by radiation. It was assumed that the heat transfer takes place between two infinite parallel plates [10]. For heat transfer through the shell wall (with thermal conductivity $\lambda_{md} = 4$ W/mK and thickness $d_{md} = 10$ mm) by conduction and heat transport from the outer mold surface at temperature T_{WO} into the surrounding liquid metal bath at temperature T_S by convection the overall heat balance takes the following form:

$$q = \frac{\sigma T_{SC}^4 - \sigma T_{WI}^4}{\frac{1}{\varepsilon_1} + \frac{1}{\varepsilon_2} - 1} = \frac{\lambda_{md}}{d_{md}}\left(T_{WI} - T_{WO}\right) = \alpha\left(T_{WO} - T_S\right) \quad (2)$$

with $\varepsilon_1 = \varepsilon_2 = 0.5$. The heat transfer coefficient α can be calculated, e.g. for flow over a flat plate, using the following formula [11]:

$$Nu = \frac{\alpha L}{\lambda_f} = 0{,}38\left(Re\ Pr\right)^{0,65} = 0{,}38\left(\frac{U_f L}{\nu_f} Pr\right)^{0,65} \quad (3)$$

where Nu is the Nusselt-number, Re is the Reynolds-number and Pr is the Prantl-number. Figure 5 shows the local heat flux q according to eq. (2) for an alloy with a liquidus temperature of 1350°C for both cooling media, aluminium ($T_S = 660$°C) and tin ($T_S = 232$°C). A flow velocity of $U_f = 0.1$ m/s and a characteristic length of $L = 0.1$ m for the cast component was assumed.

The most important parameter determining the heat flux in LMC is the temperature of the surrounding medium. Therefore, the heat flux with tin is significantly higher than with aluminium.

In addition, a curve for heat transport by radiation from the outer mold surface is given in Figure 5:

$$q = \frac{\sigma T_{SC}^4 - \sigma T_{WI}^4}{\frac{1}{\varepsilon_1} + \frac{1}{\varepsilon_2} - 1} = \frac{\lambda_{md}}{d_{md}}\left(T_{WI} - T_{WO}\right) = \frac{\sigma T_{WO}^4 - \sigma T_S^4}{\frac{1}{\varepsilon_1} + \frac{1}{\varepsilon_2} - 1} \quad (4)$$

with $\varepsilon_1 = \varepsilon_2 = 0.5$ and $T_S = 25°C$. It is obvious that the heat flux for both LMC-techniques is much higher than for the HRS-process.

There are other properties in Table I that have to be taken into account. For example the density of aluminium is only one third of tin thus faciliting contruction problems. On the other hand the energy ΔE needed for melting the cooling medium:

$$\Delta E = \int_{RT}^{T_S} C_P dT + \Delta H \qquad (5)$$

is more than 250 % higher for aluminium ($\Delta E = 2.47 \times 10^6$ kJ/m^3) than for tin ($\Delta E = 0.68 \times 10^6$ kJ/m^3).

Mechanical properties and effect of tin. A particularly important issue is the possible contamination of the superalloy with the cooling medium, e.g. when revert material is used. For aluminium, which is an alloying element in nickel base superalloys, this is not an issue. In the case of tin a review of the literature shows that tin is generally believed to be harmful. The removal of tin from the alloy using long melting times during production of the master melt is not possible [13, 14]. In order to clarify the effect of tin on alloy properties, test slabs with tin concentrations up to 4150 ppm were cast and heat treated. Creep rupture strength was tested for SC16 and PWA 1483.

A Larson-Miller-plot for SC16 with and without tin additions (Figure 6) illustrates that there is no important influence of tin on the creep strength. Creep rupture tests for PWA 1483 showed that this holds even for very high tin concentrations up to 4150 ppm that are extremely unlikely to occur in commercial practice. In Figure 7 SEM-micrographs of the microstructure of SC16 with and without tin doping are shown. There is no apparent effect of

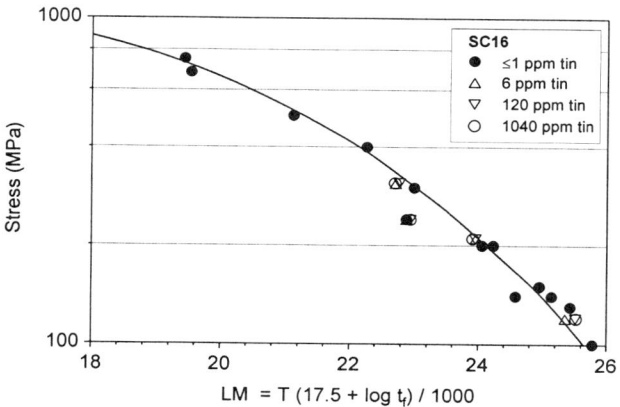

Figure 6: Creep strength of single crystal superalloy SC16. The Larsen-Miller parameter LM was calculated using temperature T in Kelvin and time t_f in hours. There is no important influence of added tin.

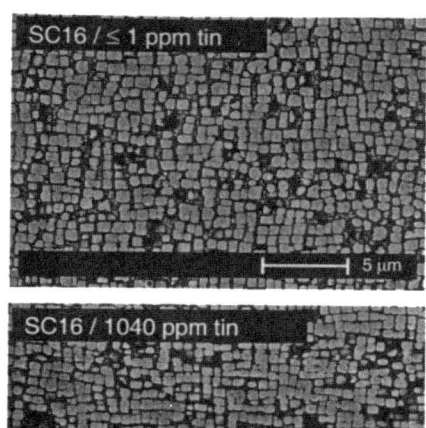
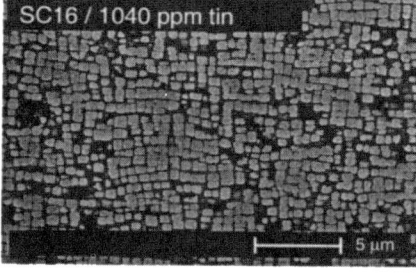

Figure 7: SEM-micrographs of the microstructure of SC16 with and without tin doping. There is no apparent effect of tin on shape, size or volume fraction of the γ'-precipitates.

tin on shape, size or volume fraction of the γ'-precipitates.

In addition low cycle fatigue has been tested for PWA 1483 with the same tin concentrations. In Figure 8 the results for 850°C and 950°C are presented. Again no influence on the mechanical properties was determined. The evaluation of other properties like corrosion resistance is presently under way.

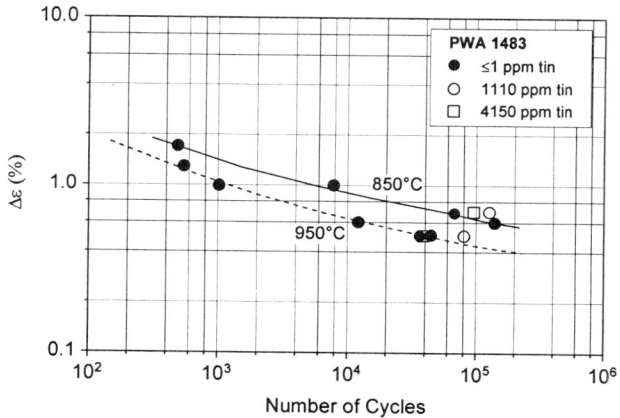

Figure 8: Low cycle fatigue strength of the single crystal superalloy PWA 1483. No significant influence of tin is found.

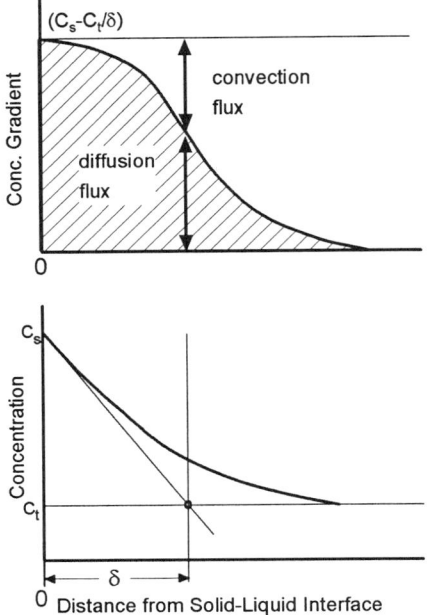

Figure 9: Schematic representation of the concentration profile in the case of the dissolution of a solid substance in a liquid. At the interface diffusive flux is rate limiting [15, 16].

Alloy dissolution. Further difficulties may appear if the solidified superalloy comes inadvertently into contact with the cooling medium. Reactions with the liquid metal of the cooling bath could lead to partial dissolution of the cast component. A related issue is the reaction between the liquid metal of the cooling bath and the bath container material and other bath components.

Several transport mechanisms might be limiting for the reaction rate in alloy dissolution, e.g. diffusion in the solid, mass transfer between solid and liquid, or the transport within the fluid. For the sake of the argument, let us assume in the following transport in the liquid phase limits the rate of the dissolution. The transport in the fluid occurs through a combination of diffusion and convection (see Figure 9) [15, 16]. The total flux is constant and does not change with the distance from the solid-liquid interface. Near the interface transport is by diffusion while further in the bulk convection becomes dominant. A boundary layer thickness δ can be defined as shown in Figure 9. Considering Fick's law and the time dependence of the diffusion flux, the rate of dissolution can be calculated by using the differential equation

$$\frac{dc}{dt} = D \frac{A}{V} \frac{c_s - c}{\delta} \qquad (6)$$

where D is the diffusion coefficient, V is the liquid metal volume, A is the area of the solid-liquid interface, c is the bulk concentration of the liquid metal bath and c_s is the saturation

concentration. For a transport controlled, one-dimensional reaction the solution of (6) can be given as:

$$c_t = c_s \left[1 - \exp\left(-\frac{D}{\delta} \frac{A}{V} t \right) \right] \qquad (7)$$

The term D/δ can be substituted by a solution rate constant k. The thickness of the boundary layer δ and therefore the constant k depends on the flow conditions.

Using a laboratory test with a cylindrical specimen rotating in a liquid metal bath (Figure 2) it was possible to investigate the kinetics of the dissolution of superalloys in liquid metals at various temperatures. Some results for the system SC16-Sn at 1000°C are presented in Figure 10. It is obvious that for a rotational speed of 80 rpm (Reynolds number Re = n d^2/ν = 620) corresponding to a laminar fluid flow there is only little difference to the static case, while a rotational speed of 300 rpm (Re = 2325) corresponding to a turbulent flow leads to a very fast dissolution. Higher rotating speeds lead to higher degrees of convection and therefore smaller boundary layers. The evident influence of the flow conditions in the bath is an indication for transport within the fluid to be limiting, rather than diffusion in the solid.

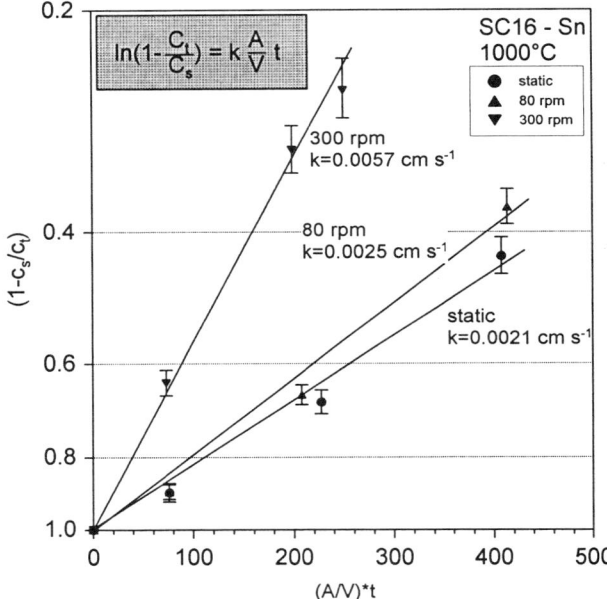

Figure 10: Plot of $\ln(1-c/c_s)$ as a function of (A/V) t for dissolution of superalloy SC16 in liquid tin, see equation (7) in the text. The slope of the curve is equivalent to the solution rate constant k. Stirring obviously speeds up the dissolution of the superalloy, thus confirming that transport in the bath is rate controlling.

Experiments at different temperatures and equal flow conditions allow to determine the activation energy ΔH using the formula:

$$k = k_0 \, e^{-\frac{\Delta H}{RT}} \qquad (8)$$

Figure 11 shows an Arrhenius-plot for the dissolution of SC16 in liquid tin. The activation energy is determined to be 56 kJ/mol. This rather low value is another indication that transport in the liquid is rate controlling. It is similar to other values found in the literature for dissolution of metals in liquids, where diffusion in the liquid controls the reaction rate. Processes like dissolution of silver in mercury (ΔH = 59.4 kJ/mol) [16] or of chromium in iron (ΔH = 44.8 kJ/mol) [17] are examples.

With respect to the dissolution of a superalloy, tin and aluminium will be compared below at their respective process temperatures. Figure 12 shows that an aluminium melt at 700°C (40 K above aluminium melting point) leads to considerable damage after only 15 minutes. For tin at 300°C (68 K above its melting point) no reaction is observed at all even after 300 minutes direct contact.

Conclusions

The Liquid Metal Cooling (LMC) process has several advantages in comparison with the conventional Bridgman-technique (high rate solidification, HRS). The rapid and uniform heat transfer allows higher withdrawal rates and larger clusters. At the same time a finer and more homogeneous microstructure is obtained.

An important factor for commercializing the process is the choice of the cooling medium. The discussion of the physical properties focusses on the two cooling media aluminium and tin. A heat balance shows the improved potential with both cooling media in comparison with the HRS-process. The maximum heat flux for tin was calculated to be significantly higher than for aluminium.

Measurements of the creep strength and low cycle fatigue strength for SC16 and PWA 1483 showed that intentional doping of the alloys with up to 4150 ppm tin has no significant influence on the mechanical properties.

Direct contact between the cast component and the cooling medium was examined. Diffusion in the fluid was found to be the limiting mechanism for a dissolution of the superalloy in the liquid metal. Tin and aluminium were compared at typical process temperatures. No reaction is observed for tin (300°C) after 300 min while aluminium (700°C) causes significant dissolution of the superalloy after 15 min.

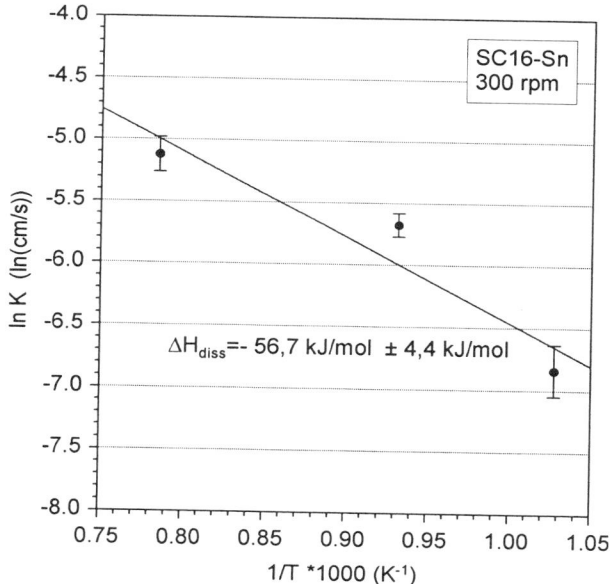

Figure 11: Arrhenius-plot for the solution rate constant k for SC16 in liquid tin. Specimens were rotated with 300 rpm. The activation energy was determined to be 56 kJ/mol which is typical for diffusion in the liquid metal to be rate limiting.

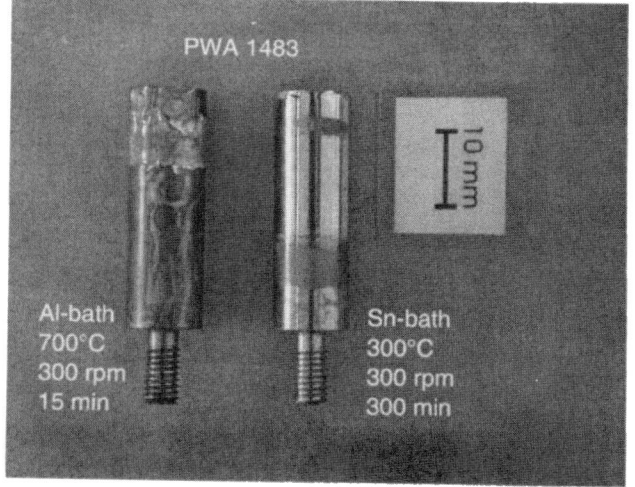

Figure 12: Comparison of dissolution of PWA 1483 in liquid aluminium or tin at typical process temperatures. No superalloy dissolution is apparent with tin, whereas significant dissolution occurs with aluminium.

References

1. M. Konter, N. Hoffmann, C. Tönnes, M. Newnham: "Influence of a Casting Process with High Cooling Rate on Structure and Properties of SX and DS components for Industrial Gas Turbines", (Paper presented at the 3rd ALD Symposium, Frankfurt, Nov. 1995)

2. J.S. Erickson, C.P. Sullivan, F.L. Versnyder, "Modern processing methods and investment casting of the superalloy family", in: High Temperature Materials in Gas Turbines, (ed. P.R. Sahm, M.O. Speidel), (New York, American Elsevier Publishing Company, 1974), pp. 315-340

3. R.F. Singer, "Advanced Materials and Processes for land-based Gas Turbines", in: Materials for Advanced Power Engineering, Part II (ed. D. Coutsousadis et al.), (Dordrecht, Kluwer Academic Press, 1994), pp.1707-1729

4. S. Olson, "A Market Forecast for Land-Based Gas Turbines & Superalloy Applications", (Paper presented at Superalloys 99, Gorham's Int. Business Conf., San Antonio, Texas, Nov. 1999)

5. A.F. Giamei, J.G. Tschinkel, "Liquid Metal Cooling: A New Solidification Technique", Met. Trans. A, Vol 7A (1976), pp. 1427-1434

6. T.J. Fitzgerald, R.F. Singer, "An Analytical Model for Optimal Directional Solidification using Liquid Metal Cooling", Met. Trans. A, Vol 28A (1997), pp. 1377-1383

7. T.J. Fitzgerald, R.F. Singer, "Modeling of the Directional Solidification of Turbine Blades via Liquid Metal Cooling", in: Proc. 2nd Pacific Rim Int. Conf. On Mod. of Cast. and Solid. Proc., E. Niyama & H. Kodama, eds., (Hitachi Ltd., Hitachi, 1995), pp. 201-216

8. J. Großmann, J. Preuhs, W. Eßer, R. F. Singer, "Investment Casting of High Performance Turbine Blades by Liquid Metal Cooling – A Step Forward Towards Industrial Scale Manufacturing", (Paper presented at the Int. Symp. on Liquid Metal Proc. and Cast., Sante Fe, New Mexico, Feb. 1999), pp. 31-40

9. D. Goldschmidt, "Einkristalline Gasturbinenschaufeln aus Nickelbasis-Legierungen, Teil I: Herstellung und Mikrogefüge", Mat.-Wiss. u. Werkstofftech., (1994), pp.311-320

10. M. N. Özişik, Basic Heat Transfer; (London, Mc-Graw-Hill Inc., 1981), p. 372

11. Kutateladze, S.S.; Borishanskii, V.M.; Novikov I.I.; Fedynskii, O.S.; Liquid Metal Heat Transfer; (New York Consultants Bureau, Ltd., 1959), p.73

12. P. Krug, "Einfluß einer Flüssigmetallkühlung auf die Mikrostruktur gerichtet erstarrter Superlegierungen", (Ph.D. Thesis, University of Erlangen, 1998)

13. J. Großmann, "Casting Single Crystal Superalloys for Land-based Gas Turbines", (Paper presented at Superalloys 99, Gorham's Int. Business Conf., San Antonio, Texas, Nov. 1999)

14. R.T. Holt, W. Wallace, "Impurities and trace elements in nickel-base superalloys", Int. Met. Rev., (1976), 21[1], pp.1-24

15. B. Ilschner, "Diffusion und Viskosität in Metallschmelzen", Z. Metallkde, (1966) H.3, 57, pp. 194-200

16. F.W. Hinzner, D.A. Stevenson, "Kinetics of Solution in Liquid Metals. Solution Rate of Zinc, Silver, and Tin in Mercury", J. Phys. Chem, (1963), 67, pp. 2424 ff.

17. W. Helber, "Über die Auflösung von Chrommetall und tiefgekohltem Ferro-Chrom in Eisenschmelzen", (Ph.D. Thesis, University of Aachen, Germany, 1969)

18. O. Kubashewski, C.B. Alcock, Metallurgical Thermochemistry, (Frankfurt a.M., Pergamon Press, 1979), pp. 358 ff.

19. R.I.L. Guthrie, Engineering in Process Metallurgy, (Oxford, Clarendon Press, 1989), p. 438

Best Paper Award

The following paper was selected by the Awards Subcommittee of the International Symposium on Superalloys as a co-winner of the Best Paper Award for the Ninth Symposium. The selection was based on the following criteria: originality, technical content, pertinence to the superalloy and gas turbine industries and clarity and style.

Novel Casting for Single Crystal Gas Turbine Components

M. Konter, E. Kats, and N. Hofmann

A NOVEL CASTING PROCESS FOR SINGLE CRYSTAL GAS TURBUNE COMPONENTS

M. Konter, E. Kats*, N. Hofmann*

ABB ALSTOM Power, CH 5401 Baden, Switzerland;
*ABB ALSTOM Power Technology, CH 5405 Baden-Dättwil, Switzerland

Abstract

The novel directional solidification process was developed at ABB for the manufacturing of large single crystal components. In addition to the radiation cooling, typical for the conventional Bridgman technology, an inert gas is injected directly below the furnace baffle. Experiments show that for the Gas Cooling Casting (GCC) process the heat transfer coefficients are significantly higher compared to those in the Bridgman process and are similar to the best LMC practice.

Introduction

The drive to higher efficiency land-based gas turbines has resulted in wide use of single crystal technology. Single crystal and directionally solidified blades are traditionally produced by a casting technique known as the Bridgman process.

In this process, the superalloy is molten in a vacuum chamber and poured into a heated ceramic mould. The mould is mounted on a copper chill, and the whole assembly is withdrawn from the melting/heating chamber into a cooling chamber so as to produce a vertical temperature gradient in the solidified (mushy) zone of the casting. While the first 40–60 mm of the casting are efficiently cooled by conduction of heat through the water-cooled bottom copper chill, the main part of the large IGT parts is cooled by a radiation heat transfer from the mould sides towards water-cooled vacuum chamber walls. The temperature gradient induces directional solidification, and either a single crystal seed or a complex spiral "pig tail" ensures that the blade is produced as a single crystal.

The economics of the single crystal casting process are directly related to the rate of grain defect production. Grain defects such as freckles, strain and spurious grains, and low angle boundaries limit the production yield. The tendency to grain defect formation increases as the size and mass of the component increases. This is due to the difficulty of maintaining a reasonably high temperature gradient at the liquid-solid metal interface by the heat radiation transfer in vacuum. This is illustrated by the fig.1, which schematically shows the defect–free region in thermal gradient – solidification rate coordinates. Improvement of casting yield for large SX blades has only a limited potential within an existing process design. Demand for a more efficient single crystal casting process led to rediscovery of the Liquid Metal Process (LMC) as well as development of the new technique, the Gas Cooling Casting (GCC).

Heat flux analysis

The conventional industrial single crystal casting process (so - called Bridgman technique) is described in the patent [1]. The mould is being preheated to the temperature above the metal liquidus and after the pouring of the molten metal is removed at a controlled withdrawal rate from the heating furnace to the cooling chamber. The heating furnace during withdrawing process is maintained at full power. The heating and the cooling chambers are separated by the baffle in order to maintain the thermal gradient as high as possible. The ceramic mould is placed on the chill plate, which is the part of withdrawing mechanism. In the beginning of the solidification process the heat is removed by conduction along the ingot to the chill plate. However, for large castings, conduction is mostly limited to the starter or a pig – tail, and the main part of the component solidifies by the process, where the predominant heat loss is a combination of:

- conduction through the portion of mushy and solidified metal placed above the baffle (in a case of a liquidus line being above the baffle) with the heat exchange coefficient α_{cm},

- heat transfer from the metal through the vacuum gap between the solid metal and the ceramic shell to the mould inner surface with the heat exchange coefficient α_{gap} (the metal – shell radiation);

- conduction through the porous ceramic shell mould to the mould outer surface with the heat exchange coefficient α_{cmd}, (more exactly, the heat transfer process here is a combination of the conduction with a

Superalloys 2000
Edited by T.M. Pollock, R.D. Kissinger, R.R. Bowman,
K.A. Green, M. McLean, S. Olson, and J.J. Schirra
TMS (The Minerals, Metals & Materials Society), 2000

radiation at pores, depending on the pore size and morphology);
- radiation heat transfer from the mould surface to the colder surroundings (water-cooled chamber walls) with the heat exchange coefficient α_ρ.

With certain approximation, the heat flux in the multistage cooling process can be described using a combined heat transfer coefficient α in the Newton heat transfer law

$$Q = \alpha\,(T - T_0), \qquad \{1\}$$

For an estimation of the potential of various casting processes, the α is defined as:

$$1/\alpha = 1/\alpha_{cm} + 1/\alpha_{gap} + 1/\alpha_{cmd} + 1/\alpha_r, \qquad \{2\}$$

though this approach is a stationary 1-D heat flux approximation, assuming that heat fluxes through the gap, mould and from the mold surface are the same (better would be consideration of steps in the heat flux rather than α).

For a large blade:

$$\alpha_{cm} = \lambda_m / \delta_m = 25 / 0.025 = 1000 \ \text{W/m}^2\text{K}, \qquad \{3\}$$

$$\alpha_{cmd} = \lambda_{md} / \delta_{md} = 1.5 / 0.012 = 125 \ \text{W/m}^2\text{K}, \qquad \{4\}$$

where λ_m and λ_{md} are the thermal conductivity of metal and mould, δ_m and δ_{md} are the thickness of the portion of metal described above (taken as 25 mm) and the mould thickness (12 mm) respectively (see also measurements reported below);

$$\alpha_{gap} = 4\varepsilon_{eff}\,\sigma\,T^3 = 4(1/(1/\varepsilon_1 + 1/\varepsilon_2 - 1))\,\sigma\,T^3$$
$$\alpha_{gap} = 230 \ \text{W/m}^2\text{K}, \qquad \{5\}$$

and
$$\alpha_r = \sigma(\varepsilon_3 T_1{}^4 - \varepsilon_4 T_0{}^4)/(T_1 - T_0) = 120 \ \text{W/m}^2\text{K}, \qquad \{6\}$$

where σ is the Steffan - Boltzmann constant, ε_1, T; ε_2, ε_3, T_1 and ε_4, T_0 are the emissivity and the temperature of the metal surface; mould inner and outer surface and the absorption and the temperature of the surroundings

respectively (($\varepsilon_1 = 0.4$, $\varepsilon_2 = 0.5, \varepsilon_3 = \varepsilon_4 = 0.45$, $T = 1580K$, $T_1 = 1500K$, $T_0 = 400K$).

Then,

$$1/\alpha = 1/1000 + 1/230 + 1/125 + 1/120 \qquad \{7\}$$

$$\alpha\ (\text{Bridgman process}) = 46 \ \text{W/m}^2\text{K}. \qquad \{8\}$$

Equation {7} shows that heat transfer process in Bridgman technique is limited by the heat exchange between the mold and environment but also by the mold and the vacuum gap.

The relatively low heat exchange in the Bridgman process limits the withdrawal speed w and the solidification front velocity v. Low v results in the coarse dendrite structure, followed by various grain defects, such as freckles and internal grains [2]. If withdrawal speed is increased in order to increase v, the solidification front moves into the cooling zone and the ratio between transverse and longitudinal temperature gradient on the liquidus front Gt/Gl exceeds the critical value range of 0.2-0.3, typical to formation of the spurious grains [3]. Often, the only remaining effective way to increase the yield is reduction in number of parts per mold, which increase the ε_3 in the equation {6}. This, however, significantly reduces process capacity and increases costs per part.

In order to increase the process effectiveness, several techniques have been developed. Patent [4] discloses a casting process in which the mould is immersed in a bath with liquid metal coolant (LMC process). The LMC process is described elsewhere [5-7] and is schematically shown on fig.2. The modification of this process with molten aluminum as a coolant is currently widely used in Russia. Though, when liquid aluminum is used, the distance between the solidification front and the heat extraction level is slightly increased, the last part of the equation {2} nearly disappears, being replaced by very high heat transfer coefficient of convection in the liquid metal. Floating baffle reduces the distance between the solidification front and the heat extraction level (in our example – LMC Sn). Now for a large blade:

$$1/\alpha = 1/\alpha_{cm} + 1/\alpha_{gap} + 1/\alpha_{cmd} + 1/\alpha_{lmc}. \qquad \{9\}$$

$$\alpha_{cm,baffle} = \lambda_m / \delta_m = 25 / 0.03 = 830 \ \text{W/m}^2\text{K}. \qquad \{10\}$$

$$\alpha_{cm,\,float.baf.} = \lambda_m/\delta_m = 25/0.022 = 1136 \ \text{W/m}^2\text{K} \qquad \{11\}$$

$$\alpha_{lmcAl} = Nu \, \lambda_{Al} \,/\, L = 11500 \ \text{W/m}^2\text{K}. \qquad \{12\}$$

$$\alpha_{lmcSn} = Nu \, \lambda_{Sn} \,/\, L = 6500 \ \text{W/m}^2\text{K}. \qquad \{13\}$$

where Nu is the Nusselt criteria, calculated using a combination of the Prandtl criteria and the Grassgoff criteria, and L is the length of the dipped mold. The thermal conductivity of the mold only slightly depends from the temperature in the temperature range $500° - 950°C$, and the α_{cmd} for the mentioned processes varies within +/- 5%, with mold thermal conductivity slightly lower at higher temperature (Al process).

$$1/\alpha_{lmcAl} = 1/830 + 1/230 + 1/120 + 1/11500 \qquad \{14\}$$

$$1/\alpha_{lmcSn} = 1/1136 + 1/230 + 1/125 + 1/6500 \qquad \{15\}$$

$$\alpha \ (\text{LMC Al}) = 72 \ \text{W/m}^2\text{K}. \qquad \{16\}$$

$$\alpha \ (\text{LMC Sn, float.baffle}) = 75 \ \text{W/m}^2\text{K}. \qquad \{17\}$$

Back to the heat flux in equation {1}, we can compare the potential effectiveness of various casting techniques:

$$Q \text{ Bridgman} = 46 \ (1700 - 400) = 60 \ \text{kW} \,/\, \text{m}^2 \qquad \{18\}$$

$$Q \text{ LMC Sn} = 75 \ (1700 - 550) = 86 \ \text{kW} \,/\, \text{m}^2 \quad \{19\}$$

$$Q \text{ LMC Al} = 72 \ (1700 - 950) = 54 \ \text{kW} \,/\, \text{m}^2 \quad \{20\}$$

where 1700K is the metal temperature on solidification front and 350K, 550K and 950K is the temperature of the "heat sink" in Bridgman process (cooled furnace walls), and in LMC process with liquid Sn and Al correspondingly.

One should anticipate, that despite an elimination of a bottle neck in the overall heat transfer process – the heat resistance of the vacuum chamber, the effectiveness of LMC process is limited by heat resistance of the mould and, to smaller extent, by heat resistance of the vacuum gap. On the other hand, in the real LMC process, especially one with the floating baffle, the real heat flux is quite close to the calculated potential. In a contrast, in Bridgman process a cooled mold faces significant heat input from the heat chamber through the baffle opening and from the central feeder/mould holder and adjoining parts in the cluster. This explains why, without having advantage in the cooling potential, the Al – LMC process in practice often shows better gradients compared to the Bridgman technique.

The novel directional solidification technique, developed by ABB ALSTOM Power, is targeted to significantly improve all three meaningful heat transfer processes: heat transfer through the gap, through the mold and heat remove from the mold surface. In the Gas Cooling Casting process (GCC) [8], in addition to the radiation cooling, the mould surface sees an impingement cooling by the inert gas, injected at high velocity directly below the furnace baffle (fig.3). In GCC process heat is effectively removed from the mold outer surface by high velocity gas flow and, additionally, by radiation heat transfer (mold surface is colder compared to the Bridgman process). The average heat transfer coefficient of the impingement gas cooling by argon – helium mix was conservatively taken as $510 \ \text{W/m}^2\text{K}$ (for an optimized process a higher value can be achieved):

$$\alpha_{GCC} = \alpha_{gas} + \alpha_r = 510 + 90 = 600 \ \text{W/m}^2\text{K}. \qquad \{21\}$$

Additionally, the high-pressure gas jet in the process of impingement cooling fills the porous mold and the vacuum gap with gas, predominantly helium, which increases their heat conductivity. The following values have been measured:

$$\alpha_{cmd} = \lambda_{md} \,/\, \delta_{md} = 2.0 \,/\, 0.012 = 166 \ \text{W/m}^2\text{K}, \qquad \{22\}$$

$$\alpha_{gap} = 300 \ \text{W/m}^2\text{K} \qquad \{23\}$$

Then,

$$1/\alpha_{GCC} = 1/830 + 1/300 + 1/166 + 1/600 \qquad \{24\}$$

$$\alpha_{GCC} = 81 \ \text{W/m}^2\text{K}. \qquad \{25\}$$

$$Q \text{ GCC} = 81 \ (1700 - 450) = 101 \ \text{kW} \,/\, \text{m}^2 \qquad \{26\}$$

From the heat flux point of view, GCC process has the highest potential among the considered techniques, followed by LMC with Sn bath and a floating baffle, while the LMC process with Al as a cooling media has no advantage compared to the Bridgman technique. However, as was already pointed out, in the casting practice, especially when multi-component mold is used, GCC as well as Bridgman process can not realize the full potential due to the additional radiation heat input. Industrial effectiveness of the various techniques is assessed in the Table 1 based on the casting process modeling results. The dendrite arm spacing (DAS), measured after industrial casting trials on large cored blades, confirms this ranking. Note, that an advanced furnace design, such as internal baffle and central heat sink [9], help to bring industrial effectiveness of the Bridgman (as well as GCC) process closer to the theoretical values.

Table 1. Assessment of the relative cooling effectiveness for casting techniques

Process	Bridgman	GCC	LMC Sn	LMC Al
Physical potential of the cooling effectiveness	1	1.7	1.45	1
Estimate of cooling effectiveness in an industrial process	0.6	1.5	1.5	1
DAS, μm for large cored blades; low airfoil.	430	320	(330) [10]	360

Experimental procedure

The GCC feasibility study and benchmarking of LMC and Bridgman casting processes have been carried out at ABB Corporate Research in an industrial-size furnace UVNK-8P, fully automated and modified for each of the mentioned techniques (fig.4). The furnace was equipped with induction crucible, two graphite resistance heaters, baffle and pull-down rod with a hanger for mould. Additionally, the bath with molten Al (LMC) or the nozzle system with gas supply (GCC) was placed below the baffle. The distance between Al surface and the baffle was varied from 20 to 40 mm. The distance between the baffle level and a projection of nozzle centerline on the mould surface was varied from 0 to 100 mm. The clusters of 2 shrouded solid blades, 240 mm in length and with airfoil wall thickness from 3 to 8 mm were cast with each technique (fig. 5). Alumina – based molds have been produced by the standard stucco-and-slurry technique with some of LMC molds additionally wire supported. The single crystal structure was provided by single crystal seeding technique and a 12 - 15 mm. long grain selector. The alloys used in experiments were CMSX-4 [11] with liquidus temperature 1410°C and MK-4 [12] with liquidus temperature 1407°C. Heaters temperature was carried at 1550°C for the upper and 1600°C for the lower heater. Al bath temperature was carried in the range of 680 – 700°C and the gas temperature was 100 – 170°C on the nozzle exit. As a gas media an Ar, He and their mixture in various proportions have been tested. For the process control additional thermocouples have been placed on mold, in the LMC bath, on the gas nozzles, in the heater chamber, on the inner wall of the vacuum chamber, on gas exhaust in front of and behind the intercooler.

After grain structure etching and visual inspection the blades have been cut in 2 locations in the airfoil and 2 locations in the root. Some of blades have been heat treated prior to etching. The heat treatment cycle for both alloys consisted of multi-step solutioning with the maximum cycle temperature of 1310°C [13], precipitation heat treatment at 1140°C and aging at 870°C. The primary dendrite arm spacing, porosity, γ'-phase size, concentration profiles of elements have been investigated using standard procedures of etching, optical and scanning (SEM) microscopy, EDX and VDDX microprobe analysis.

The fluid dynamic and heat transfer optimization of the GCC process has been carried on in a large vacuum chamber, where a 1:1 model of the induction heating chamber of industrial casting furnace was installed. The influence of various GCC process parameters on heat transfer values and distribution was studied on plaster model of the casting mold with 7 of GT24 low pressure 1st row blades. Heat transfer has been measured using thermocouples and the IR-scanner AGA-780. The thermocouples have been installed into the plaster model on various depths, so at the beginning the time at temperature, necessary for stabilization of the temperature profile has been defined. The plaster model has been heated up to 800°C for 30 min. Then the furnace with attached baffle and the nozzle ring beneath was lift up the certain height to section, where the measurements of heat transfer should be carried out. Than, the steady gas flow through the nozzles was activated. The recording of thermo - images by IR-scan and temperature by thermocouples was conducted prior and during the gas jet action, which allowed to use (with certain limitations) the heat conduction equation for a semi-infinite body with type III boundary conditions to describe the convective heat flux. The heat transfer coefficient was determined as follows:

$$\alpha = U \lambda / (a \tau)^{1/2}, \ W/m^2K, \qquad \{27\}$$

where λ and a are heat and temperature conductivity of plaster, τ - the time of jet action onto the model surface and U is the non-dimensional parameter, determined from the equation:

$$(T_t - T_0) / (T_s - T_0) = 1 - e^{U2} (1 - 2/\sqrt{\pi} \int_0^U e^{-t2} \, dt), \quad \{28\}$$

where T_t, T_0 and T_s are temperature of the mold surface at the moment t, initial mold temperature and stagnation temperature of the cooling gas.

The ceramic mold heat diffusivity was measured on a shell created by ACCESS e.V. (Aachen, Germany) by a laser flash method at the Institute of Ceramic Components (RWTH, Aachen). The measurement was performed under vacuum, air (1 bar), argon (150 mbar), helium (150 mbar) and a helium/argon mixture (20% He/ 80% Ar, 150mbar). The thermal diffusivity was measured in steps of 100°C in a temperature range between 20°C and 1600°C. At each temperature the diffusivity was measured 4 times and then averaged. The measurement error was approximately 3%. Measurement of thermal diffusivity described as:

$$\alpha = \lambda/(\rho c_p) \qquad \{29\}$$

[Heat Conductivity / (Density * Heat Capacity)]

The mold ceramic was Al_2O_3 with SiO_2 slurry. Samples from the rough and smooth surface of the shell, $\phi = 12.7$ mm, with a thickness of 3 mm, porosity 30%. Samples were covered with graphite to avoid transmission of the laser.

Mold conductivity in the gas cooling process

The thermal diffusivity of the ceramic shell mould was measured in a temperature range from 20°C to 1600°C. The purpose of this measurement was to study the impact of helium and argon on the thermal conductivity of the shell. The thermal diffusivity of the shell is conductivity to the amount of heat, which will be transported through the shell in the Bridgman process. The increase in the thermal diffusivity increases the thermal gradient and cooling rate.

The thermal conductivity is influenced by microstructure discontinuities such as cracks or pores, which will interfere with the direct flow of heat. In the case of cracks the volumetric heat capacity is maintained such that the heat storage capacity is unaffected.

Pores, in contrast, will lower the volumetric heat capacity in direct proportion to the volume fraction of the pores. Additionally the pore size and volume indicates that the specific heat is not affected by the combined effects of pore content and accompanying changes in the crystallographic phase content [14]. Although the volumetric heat capacity decreases with decreasing density, the *heat capacity per unit mass* of shell material is independent of the pore content.

On the other hand the thermal conductivity is strongly affected by the microstructure discontinuities like crack and pore size and distribution. There has been systematic experimental work on the methodological problems of determining the thermal conductivity of refractories [15]. Particularly the presence or absence of gas in the pores and cracks significantly influence the thermal conductivity. Most measurements on thermal conductivity are performed under atmospheric conditions and do not represent the thermal conductivity of porous ceramics under vacuum conditions like in a Bridgman process.

Figure 6 shows the thermal conductivity of the smooth surface of the shell. The thermal conductivity of the rough surface is approximately 1% higher than the smooth surface (not shown in the plot). The thermal conductivity of the shell in air at 1 bar is app. 50 % higher than the vacuum conductivity. For argon and the 80% argon + 20% helium mixture at 150 mbar the thermal conductivity is 30 % higher than in the vacuum. The use of pure helium at 150 mbar gives a limited improvement of the thermal conductivity compared to the gas mixture, from plus 30% over the vacuum in gas mix to plus 43 % compared to vacuum for the pure helium in the temperature range 800°- 1200°C.

The benefit of the GCC process is a significant improvement in the shell thermal conductivity (shell temperatures around 1000°C for GCC) of 30%. Consequently the heat flow, the cooling rate and the temperature gradient are increased by 30%.

Figure 7 shows the ratio of thermal conductivities between shell exposed to a given gas and vacuum thermal conductivity (conventional process). The figure illustrates that air has a decreasing influence on thermal conductivity with increasing temperature. Argon and helium in the temperature range between 800°C and 1200°C show the highest impact on the thermal conductivity of the shell. This is exactly the temperature range of the shell situated under the baffle in the casting process.

Influence of gas flow parameters

The following parameters have been optimized using fluid dynamic and heat transfer experiments:

- gas velocity in the jet;
- total jet pressure;
- nozzle diameter;
- level of initial turbulence;
- gas pressure in the vacuum chamber;
- gas volume;
- gas composition;
- distance from the nozzle to the mold;
- position of the jet centerline on the mold;
- jet inclination angle.

Jet inclination

Fig. 8 shows results on inclination angle experiments. The jet normal to the mold surface results in a relatively low cooling effectiveness due to vortex formation from reflected gas and poor gas flow along the profile. Effectiveness of the inclined jet is much higher, with a maximum in a wide range of inclination angles from 30° to 60°, showing heat transfer coefficient two times higher compared to one for a gas jet normal to the surface.

Experiments have demonstrated, that at inclination angle from 30° to 60° the gas flow follows the concave and convex profile of the cored large IGT blade mold. Almost no flow separation was detected even in the leading edge area. Distribution of heat transfer coefficients along the mold profile, measured with IR-scanner, was very homogeneous with α max / α min < 3. Taking into account significant homogenization of the temperature fields across the mold thickness (due to the low thermal conductivity of ceramic) the cooling inhomogenity along the blade profile is insignificant on the metal surface. Analysis of blade dendrite morphology after casting confirms this point.

Distance to the mold

Fig. 8 shows also an influence of distance from the nozzle to the mold on the cooling effectiveness. At selected flow parameters (velocity, pressure at the nozzle and in the chamber), cooling is still very effective even at distance up to 150 mm., with optimum values up to 100 mm.

Interestingly, the measured curve for heat transfer coefficient α vs. distance from the nozzle to the mold shows a significant difference compared to the curve which would be theoretically expected for vacuum from the law $\alpha \approx 1/x$ (where x is the distance to the mold). This is due to additional gas volumes, involved by turbulence in the gas jet from the chamber atmosphere.

Material structure

Macrostructure of the blades cast in various techniques shows a consistent difference in dendrite arm spacing, in porosity level and an insignificant difference in volume fraction of residual eutectic. Figure 9 illustrates the typical macrostructure after Bridgman, LMC Al and GCC process for the airfoil cross-section close to the platform. Figure 10 shows measurement results for dendrite arm spacing. The lower airfoil area is the most representative for assessment of the relative temperature gradient in vertical direction using the primary DAS parameter. This zone is characterized by stabilized solidification front velocity, quite close to the withdrawal speed. Comparison of GCC, LMC and Bridgman trials, with taking into account difference in withdrawal speed, confirms an advantage of GCC process.

Secondary and tertiary dendrite arms, strongly extended in one preferential direction on the figure 10 reflect a pronounced transverse gradient in Bridgman and, especially, LMC casting trials. In both trials withdrawal speed was kept as high as possible (5 mm/min and 8 mm/min respectively) in order to prevent freckle – induced grain defects. This, however, results in shifting solidification front beneath the baffle, strong inclination of the front line and, in a turn, fluctuation induced grain defects. This point is illustrated on the defect map on the figure 1. GCC process, in contrast, shows almost an ideal dendrite shape, typical for high longitudinal to transverse temperature gradient ratio and wide process window.

GCC process results (as a consequence of higher cooling rate and high longitudinal to transverse temperature gradient ratio) in a lower porosity. Comparison of porosity area fraction for three casting techniques is given in the Table 2.

Table 2. Area fraction of porosity.

Process	Bridgman	LMC-Al	GCC
Upper airfoil, %	0.32	0.18	0.06
Lower airfoil, %	0.24	0.19	0.08

More important for component life is, however, the maximum pore size. Pore size distribution is illustrated by figure 11. The largest detected pore had 70 mm, 60 mm and 45 mm maximum size for Bridgman, LMC and GCC process correspondingly. In general, GCC process results in much smaller amount of large pores, which provides a good potential for an extended fatigue life.

The area fraction of residual eutectic in as-cast condition does not show significant difference between various casting techniques (fig. 12). Finer morphology of eutectic islands after GCC process is beneficial for solutioning heat treatment.

There was no significant difference between casting techniques in dendritic segregation of elements in as-cast condition (fig. 13). A slight beneficial trend in segregation of Re may, however, result in better homogenization during solutioning heat treatment and, therefore, higher structure- and phase- stability during the long time high temperature operation.

Both LMC-Al and GCC process are characterized by significantly finer γ' –phase morphology in as-cast condition compared to the conventional casting technique (fig. 14). The γ' –size is determined by the cooling rate in the temperature range below γ' – solvus. In this range the liquid Al bath provides effective cooling, while radiation heat transfer from the mold is poor. With respect to GCC process, the low-temperature part of the casting is not directly exposed to the gas jets, however, to our opinion, the gas – filled mold and gap, as well as convection cooling through the chamber, result in the cooling rate comparable to LMC. Finer γ' – structure is potentially beneficial first of all for vane applications, if vanes are used in as-cast or partially solutioned condition in order to avoid recrystallization. Too high cooling rate at lower temperatures, however, results in residual stresses in casting and promotes recrystallization during solution heat treatment.

Casting process application

Industrial demand for the directional solidification process with an increased cooling rate is obvious. Wide application of the LMC process, however, requires high initial investments and, for the tin process, solution of certain mold technology problems. Gas cooling casting process does not request a replacement or a significant modification of the existing Bridgman equipment. The manifold with nozzles and attached gas supply can be easily installed in any DS furnace and investments in an additional gas and control equipment are marginal. The modified furnace always can function in the Bridgman process mode with only the manifold removed.

With regard to casting technology, experiments show that a flexible control over gas parameters during the withdrawing process allows optimization of casting structure in the most critical areas, such as shrouds and platforms. The typically higher process withdrawal speed also improves the plant economics.

Summary

The novel gas cooling casting process has been developed for production of single crystal and directionally solidified gas turbine components. The process has been validated on semi-industrial furnace and directly compared to the conventional and LMC single crystal casting technique. Theoretical analysis of heat flux during the various SX casting processes show a significant advantage of the gas cooling over existing techniques. Improvement in a heat flux during the solidification comes from the direct gas jet impingement cooling as well as from improvement in the mold heat conductivity. The sensitivity analysis of the heat transfer in the GCC process from various process parameters and gas system design had been performed on the test rig and allowed to design the gas system for industrial furnace. Casting trials show doubled temperature gradient compared to conventional Bridgman technique. Micro – and macrostructure of the GCC – cast blades show certain potential for easier post – processing and longer fatigue life.

References

1. F.L. Versnyder: US Patent No. 3 260 505, 1966.
2. T.M. Pollock, W.H. Murphy: The Breakdown of Single-Crystal Solidification in High Refractory Nickel-Base Alloys, Met. Trans., v. 27A, Apr. 1996
3. E. Kats, S. Korjakin, A. Amiljanchik, E. Spiridonov: Thermal process during formation of a SX structure in superalloys. Proc. Of the 1st International Heat-resistant Material Conference, 23 – 26 Sept. 1991, Fontana WI, 1991.
4. J.G. Tschinkel, A.F. Giamei and B.H. Kear: US Patent No. 3 763 926, 1973.
5. R.F. Singer: Directional and Single Crystal Solidification Using LMC. 3rd Symposium on Advanced Technologies and Processes for Metals and Alloys, Hanau, ALD Vacuum Technologies, 1995.
6. M. Konter, N. Hofmann, C. Tönnes, M. Newnham: Influence of a Casting Process with High Cooling Rate on Structure and Properties of SX and DX Componets for Industrial Gas Turbines. 3rd Symposium on Advanced Technologies and Processes for Metals and Alloys, Hanau, ALD Vacuum Technologies, 1995.
7. F. Hugo, U. Betz, J. Ren, S.-C. Huang, J. Bondarenko, V. Gerasimov: Casting of Directionally Solidified and Single Crystal Components Using Liquid Metal Cooling (LMC). International Symposium on Liquid Metal Processing and Casting, Santa Fe, VMD – AVS, 1999.
8. E. Kats, M. Konter, V. Lubenets, J. Rösler: US Patent No 5 921 310, 1999.
9. F. Hugo, R. Schumann, W.T. Carter Jr.: Advanced Vacuum Metallurgical Processes and Equipment for Gas Turbine Materials and Components. Gorham's Int. Business Conference, Tampa, Gorham Advanced Materials Inc., 1998.
10. J. Grossmann, J. Preuhs, W. Esser, R. Singer: Investment Casting of High performance Turbine Blades by Liquid Metal Cooling – A Step Foreward Towerds Industrial Scale Manufacturing. Proc. Of the 1999 int. Symposium on Liquid metal processing and Casting, Santa Fe, NM, Feb. 1999
11. K. Harris, L. Erickson: US Patent No 4 643 782, 1987.
12. M. Konter, C. Tönnes, M. Newnham: US Patent No 5 759 301, 1998.
13. M. Konter: US Patent No 5 882 446, 1999.
14. M.V. Swain, L.F. Johnson, R. Syed, D.P.H. Hasselman: Thermal diffusivity, heat capacity and thermal conductivity of porous partially stabilized zirconia, Journal of Material Science Letters 5, 1986, p. 799-802
15. V.V. Pustovalow: Thermal Conductivity of Refractory Materials, Translated from Russian, NASA Accession Number N71-15034 1971, Printed and Published by INSDOC, Delphi-1

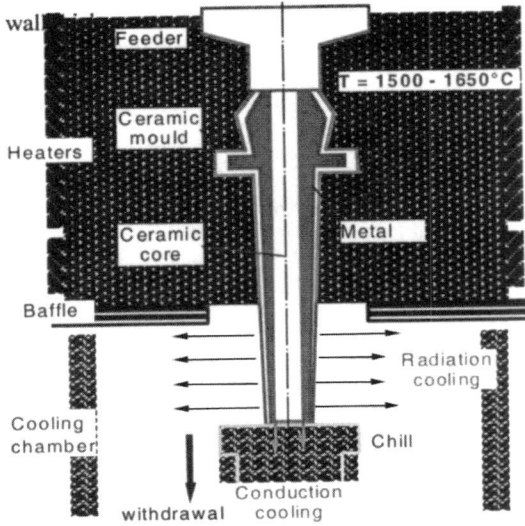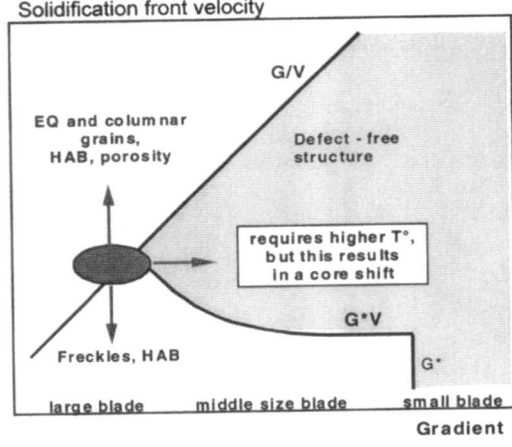

• Large IGT SX blades: low yield, high cost

Fig. 1: Bridgman process with grain defect map.

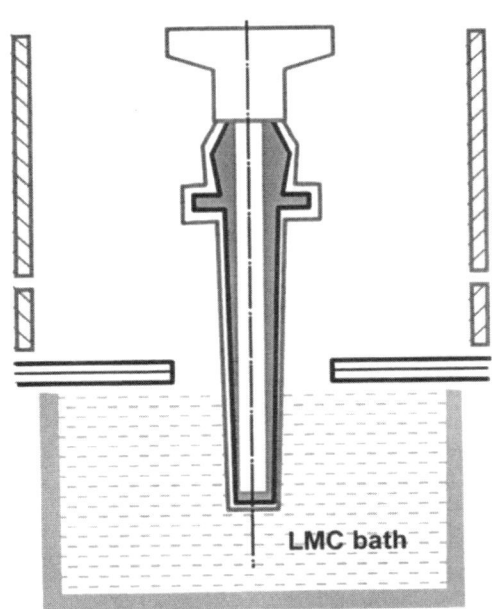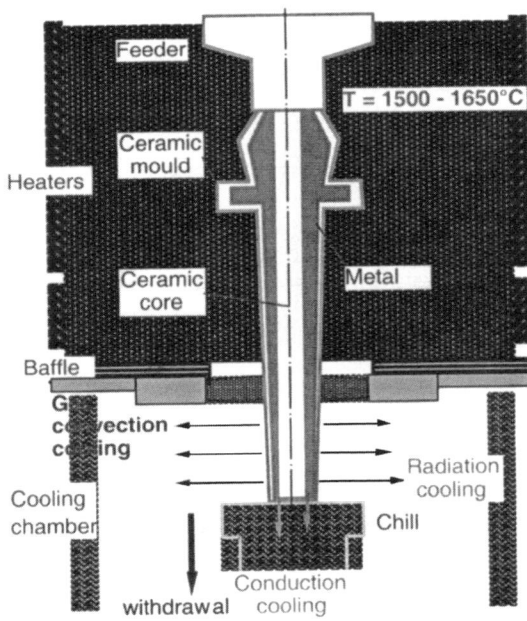

Fig. 2: Liquid metal casting process

Fig.3: Gas cooling casting process

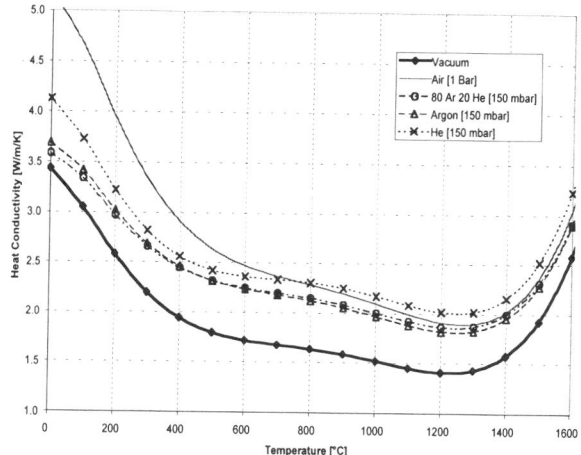

Fig. 4: Casting furnace at ABB ALSTOM Power used for trials with various cooling media.

Fig. 6: Thermal conductivity of shell material exposed to various gases.

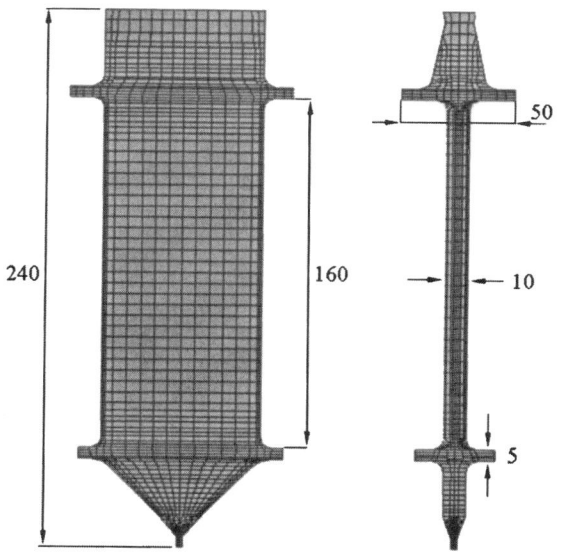

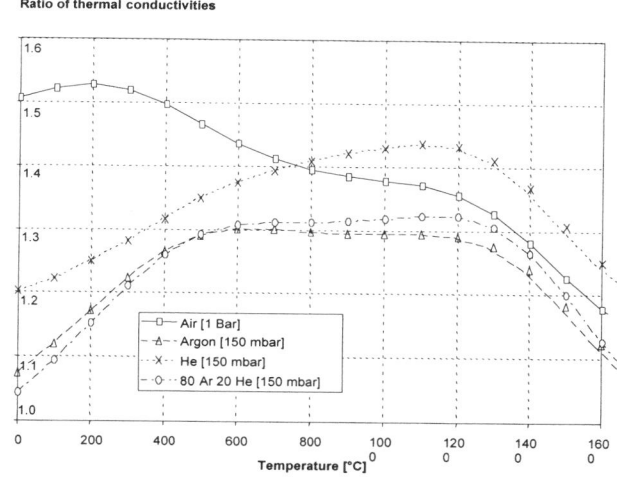

Fig. 5: Solid shrouded blade used for the initial casting experiments.

Fig. 7: Ratio of shell conductivity for mold exposed to a given gas to the conductivity of mold in vacuum.

197

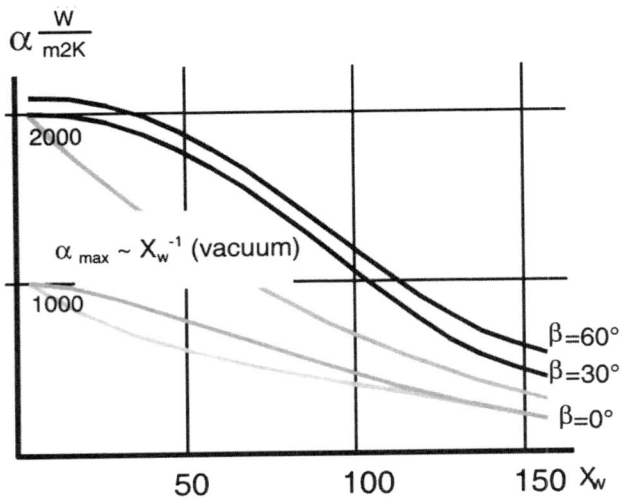

$\alpha \frac{W}{m2K}$

2000

$\alpha_{max} \sim X_w^{-1}$ (vacuum)

1000

$\beta = 60°$
$\beta = 30°$
$\beta = 0°$

50   100   150 X_w

Fig. 8. Influence of a distance between the nozzle and the mold (X_w, mm) and of the jet inclination (β) on the heat transfer coefficient α (Ar/He mix).

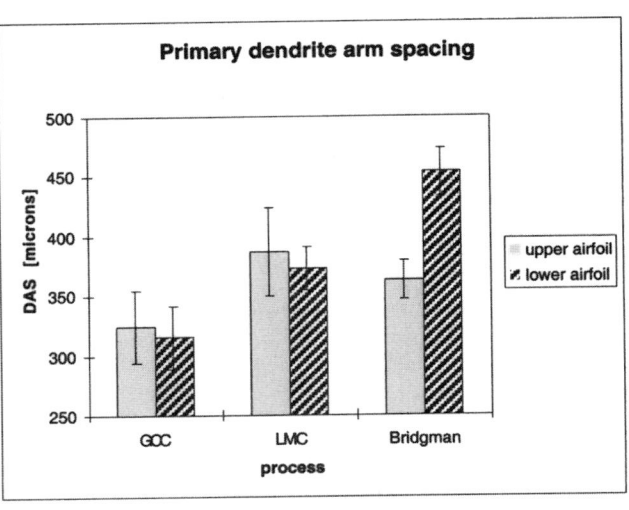

Fig. 9. Primary dendrite arm spacing after different casting processess.

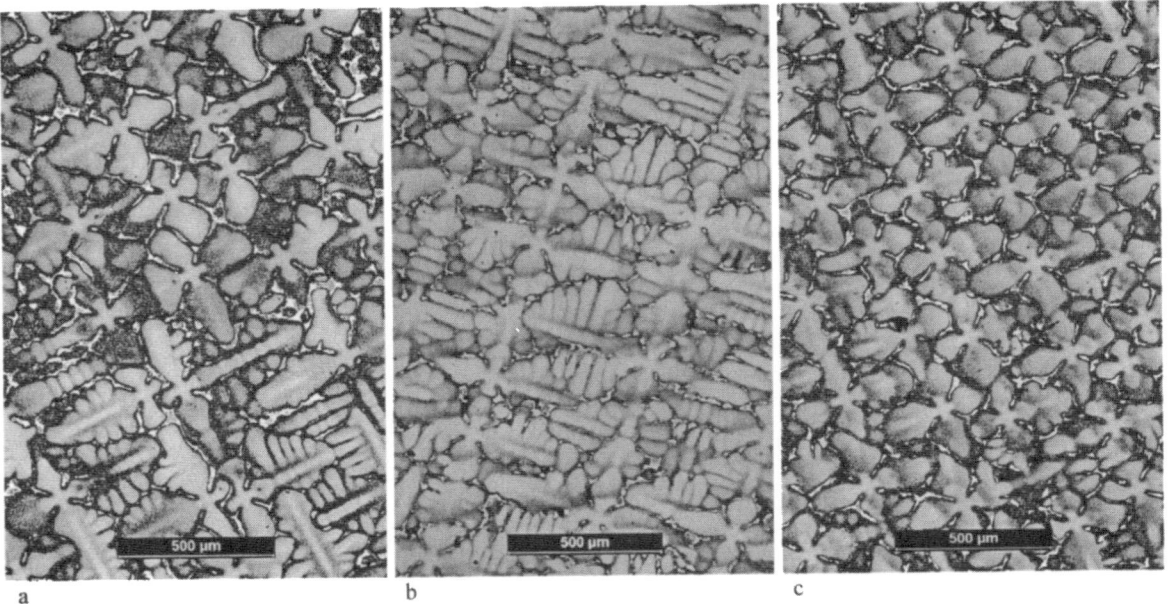

a   b   c

Fig. 10 . Macrostructure of SX alloy (MK4) in as cast condition after Bridgman (a), LMC – Al (b) and GCC (c) casting, lower airfoil.

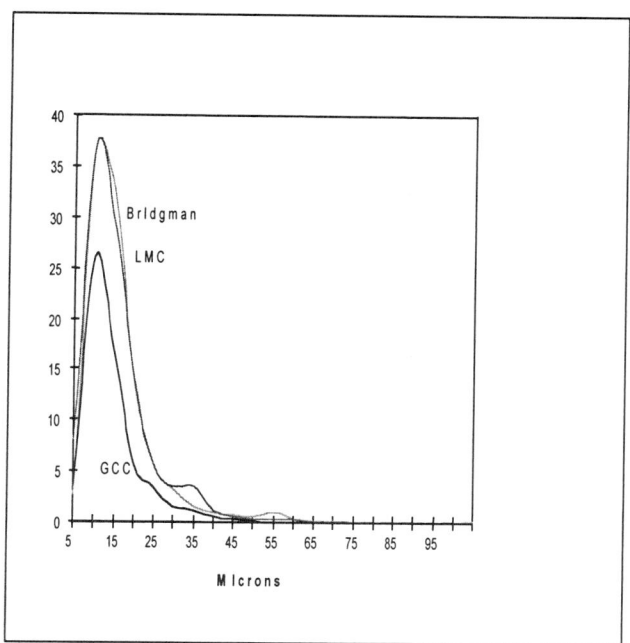

Fig. 11: Pore size distribution at lower airfoil cross-section.

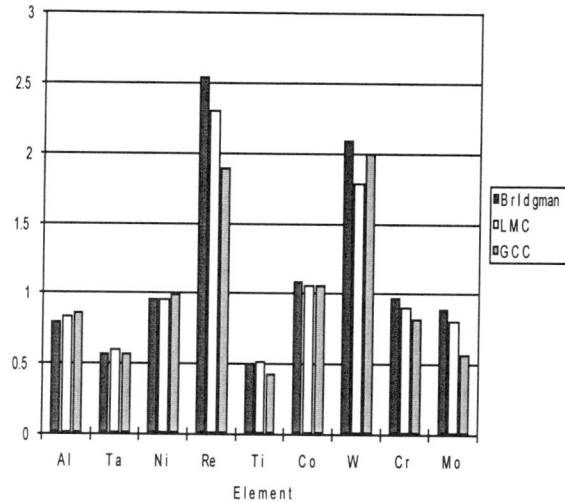

Fig. 13: Dendritic segregation of elements in as-cast condition vs. casting technique.

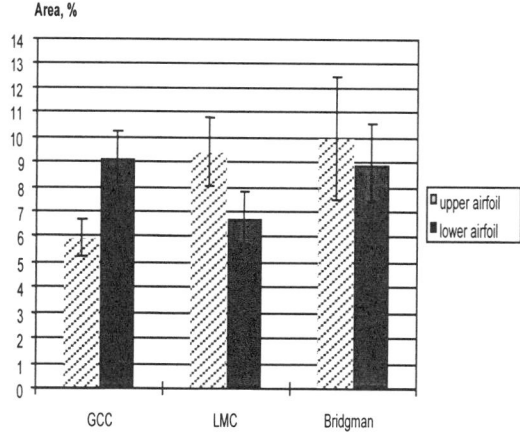

Fig. 12: Eutectic area fraction for different casting techniques.

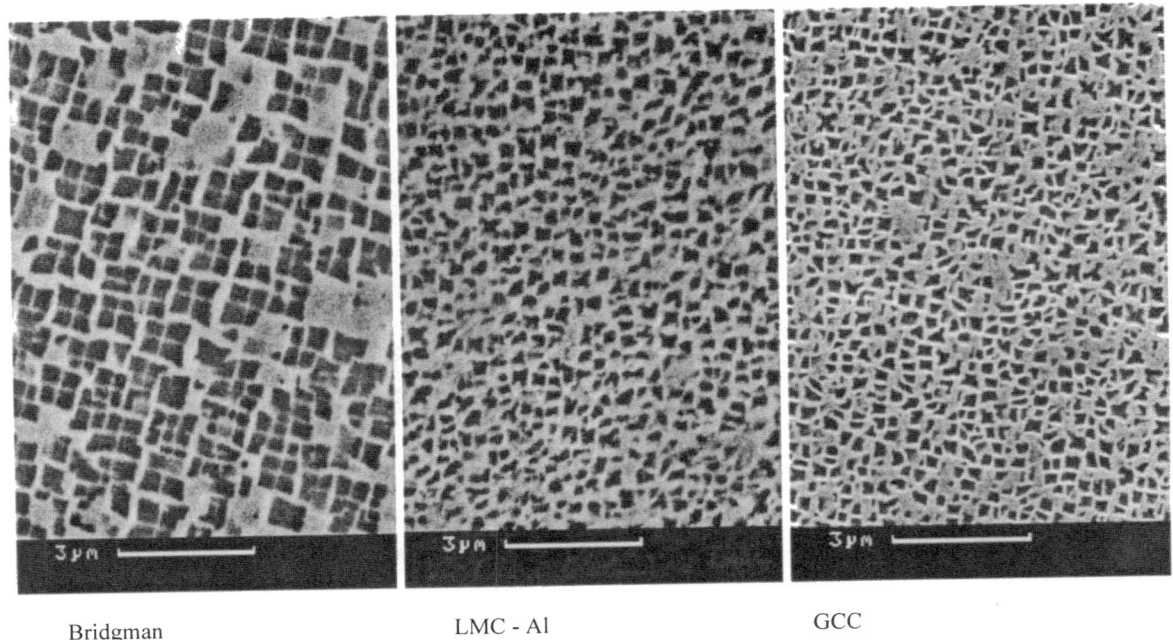

Bridgman   LMC - Al   GCC

Fig.. 14: Microstructure of as-cast SX alloy (MK-4) after various casting processes.

CARBON ADDITIONS AND GRAIN DEFECT FORMATION IN HIGH REFRACTORY NICKEL-BASE SINGLE CRYSTAL SUPERALLOYS

S. Tin, T.M. Pollock and W.T. King*
University of Michigan, Dept. of Materials Science and Engineering
Ann Arbor, MI 48109
*General Electric Power Systems

Abstract

Using two half fractional factorial experimental designs, the effect of alloying additions on the solidification characteristics of single crystal Ni-base superalloys has been studied in twelve distinct "2nd generation" type alloys. The experimental alloys, with compositional variations of C, W, Re, Ta, Al and Hf, were solidified as cylindrical bars in a large cluster mold. Under constant processing conditions, nominally similar experimental alloys containing additions of 0.125 wt.% C exhibited a decreased tendency to develop grain defects, such as freckle chains. The carbon additions resulted in the formation of three Ta-rich MC carbide morphologies which precipitate near the liquidus temperature of the alloy. Intentional carbon additions also affected the segregation behavior of the constituent elements. Comparison of distribution coefficients measured using a Scheil-type analyses revealed that less segregation of W and Ta occurred in the experimental single crystal alloys containing carbon. Preliminary observations suggest that carbon/carbide interactions may potentially change the segregation behavior of high density refractory alloying elements and affect the properties of the mushy zone in a manner which decreases the driving force for thermosolutal convection.

Introduction

As nickel-base single crystals are being implemented to increase efficiency in large, land-based, industrial gas turbine engines, grain defect formation during solidification has become an increasingly important problem. Typically, grain defects, such as freckles and misoriented grains, are caused by the onset of thermosolutal convective instabilities during dendritic solidification in these multicomponent alloys[1-6]. By achieving high thermal gradients at the growth front during solidification in the Bridgman process, these instabilities can be effectively suppressed[1-3,5,7-11]. However, this becomes a challenge when solidifying physically large single crystals required for land-based gas turbines. Thus, alloys with a low potential for developing these solute-induced density inversions are highly desirable.

Recent experiments with "3rd generation" type alloys have shown that increasing tantalum and/or decreasing rhenium and tungsten levels will have a beneficial effect in reducing the number of grain defects that develop under relatively low gradient solidification conditions[2-4]. Furthermore, it has also been shown that intentional carbon additions will also inhibit grain defect formation in certain high refractory alloys[12]. Improving the solidification characteristics of single crystal Ni-base superalloys through the addition of carbon represents a new alloying alternative with a great deal of potential. However, the effectiveness of these alloying approaches over a broader range of compositions has not yet been explored.

In this study, alloying approaches for the reduction of grain defects have been studied over a wider range of composition which covers "2nd generation" alloys such as René N5. Of particular interest was the influence of carbon on the segregation behavior of the constituent elements and the relationship between carbide precipitation and the tendency to form freckle defects. Results from detailed analyses of segregation and differential thermal analyses (DTA) of the alloys are reported. Based on these observations, a brief discussion regarding the potential interactions between carbon additions and freckling mechanisms follows.

Experimental Materials and Procedure

A total of twelve distinct alloys, listed in Table I, have been investigated in two half fractional factorial studies designed to reveal the effects of Al, Hf, Ta, Re, W and C on single crystal solidification. Each experiment was repeated to verify reproducibility. In an attempt to keep the levels of interstitial impurities constant, alloys were first sectioned from a solute-lean master heat then doped to the compositions listed in Table I. In the first matrix of experiments, referred to as Carbide Effects (CE), carbon, tantalum and hafnium levels for alloys varied form 0 to 0.125 wt.%, 4.0 to 8.0 wt.% and 0 to 0.3 wt.%, respectively. In the second matrix, or Main Effects (ME) experiments, levels of carbon, aluminum, tungsten and rhenium in the alloys

Superalloys 2000
Edited by T.M. Pollock, R.D. Kissinger, R.R. Bowman,
K.A. Green, M. McLean, S. Olson, and J.J. Schirra
TMS (The Minerals, Metals & Materials Society), 2000

Table I: Compositions of Experimental Single Crystal Alloys (wt.%)

	C	Al	Ta	Hf	W	Re	Mo	Cr	Co	Ni
ME-1	0.000	6.8	6.1	0.14	2.4	4.6	1.5	7.0	7.5	Bal.
ME-2	0.000	6.7	6.1	0.14	2.3	4.6	1.5	7.0	7.5	Bal.
ME-3	0.000	6.7	6.0	0.13	7.3	1.4	1.5	7.0	7.5	Bal.
ME-4	0.000	6.7	6.1	0.14	7.2	1.4	1.5	7.0	7.5	Bal.
ME-5	0.000	4.4	6.0	0.12	7.2	4.4	1.5	7.0	7.5	Bal.
ME-6	0.000	4.4	6.0	0.13	7.1	4.5	1.5	7.0	7.5	Bal.
ME-7	0.000	4.8	6.4	0.14	2.5	1.5	1.5	7.0	7.5	Bal.
ME-8	0.000	4.9	6.5	0.14	2.5	1.5	1.5	7.0	7.5	Bal.
ME-9	0.125	6.5	5.8	0.14	7.0	4.5	1.5	7.0	7.5	Bal.
ME-10	0.119	6.4	5.8	0.14	7.0	4.4	1.5	7.0	7.5	Bal.
ME-11	0.127	4.6	6.2	0.14	7.4	1.4	1.5	7.0	7.5	Bal.
ME-12	0.129	4.5	6.1	0.14	7.4	1.4	1.5	7.0	7.5	Bal.
ME-13	0.132	4.6	6.2	0.14	2.4	4.7	1.5	7.0	7.5	Bal.
ME-14	0.135	4.7	6.2	0.15	2.4	4.8	1.5	7.0	7.5	Bal.
ME-15	0.140	7.1	6.2	0.15	2.4	1.5	1.5	7.0	7.5	Bal.
ME-16	0.130	7.0	6.2	0.14	2.4	1.4	1.5	7.0	7.5	Bal.
CE-1	0.000	5.6	8.4	0.00	4.5	2.8	1.5	7.0	7.5	Bal.
CE-2	0.000	5.5	8.1	0.00	4.3	2.8	1.5	7.0	7.5	Bal.
CE-3	0.000	5.8	4.0	0.28	4.6	2.9	1.5	7.0	7.5	Bal.
CE-4	0.000	5.8	3.9	0.31	4.7	3.1	1.5	7.0	7.5	Bal.
CE-5	0.139	5.8	3.9	0.00	4.8	3.0	1.5	7.0	7.5	Bal.
CE-6	0.140	5.8	4.0	0.00	4.8	3.0	1.5	7.0	7.5	Bal.
CE-7	0.130	5.5	8.2	0.27	4.5	2.8	1.5	7.0	7.5	Bal.
CE-8	0.119	5.4	8.4	0.26	4.5	2.9	1.5	7.0	7.5	Bal.

ranged from 0 to 0.1 wt.%, 4.5 to 7.0 wt.%, 2.5 to 7.0 wt.% and 1.5 to 5.0 wt.%, respectively.

For each experimental alloy, fourteen cylindrical bars approximately 15cm in length with diameters of 12.7, 15.9 and 19.0mm were unidirectionally solidified in a large investment cluster mold by PCC Airfoils. Processing parameters used in these experiments were designed such that solidification occurred under relatively low thermal gradient conditions. The variations in bar diameters resulted in characteristic thermal gradients which allowed for the simulation of three different solidification conditions under a constant withdrawal rate of 20cm/hr.

Following solidification, as-cast specimens were macroetched in a hot nitric bath to reveal the presence of freckles and misoriented grains. Selected bars were then sectioned normal to the solidification direction to determine the characteristic primary dendrite arm spacings associated with the varying bar diameters.

For each of the carbon containing alloys, small rectangular samples (~5g) were sectioned from the bars and prepared for electrolytic extraction of the carbides. Using a small sheet of platinum as the cathode and a current density of 0.04 A/cm^2, the γ - γ' matrix was dissolved in a solution of 9:1 Methanol:HCl with 1 wt.% tartaric acid. The remaining carbides were then collected and analyzed in a Rigaku X-ray Diffractometer (XRD). A Phillips XL30 FEG scanning electron microscope (SEM) was used to characterize the carbide morphologies and perform qualitative energy dispersive spectroscopy (EDS) on the carbides.

A two step process was used to characterize segregation of individual alloying elements and to subsequently assess distribution coefficients, k, via a Scheil analysis. This technique is similar to the ranking technique reported by Gungor, Huang and Nastac[13-15]. As-cast samples were first sectioned normal to the <001> growth direction. Next, a line scan that crossed the dendrite core into the interdendritic region was used to determine whether individual elements segregate preferentially to the dendrite core or the interdendritic region. A second series of analyses were then conducted to measure composition at points in a square grid composed of 225 points. The grid was placed over a representative section of the dendritic microstructure, covering ~1.2 mm^2. The composition data acquired for the individual elements were then ranked in order according to their characteristic segregation behavior. For elements determined to have distribution coefficients greater than one, as determined previously by the linescan, the composition data was ranked in descending order and plotted as apparent fraction solid, yielding a Scheil plot. Composition data for elements segregating to the interdendritic regions were ranked in ascending order and plotted in the same manner. Distribution coefficients were then extracted by fitting the plots with a Scheil equation [1] and/or a modified Scheil analysis [2] which accounts for back diffusion,

$$C_s = kC_o(1-f_s)^{k-1} \qquad [1]$$

$$C_s = kC_o(1-(1-2ak)f_s)^{k-1/1-2ak} \qquad [2]$$

where C_s is the local composition of the solid, C_o is the nominal alloy composition, α is the Fourier number, and f_s is the fraction solid. This analysis assumes that the data collected was representative of the entire sample and that the distribution coefficients were constant. Steps were taken to determine the validity of these assumptions. First, the mean composition from the EDS point measurements were compared to the actual compositions of the different samples measured using X-ray Fluorescence (XRF). Experimental Scheil plots generated from the point measurements were also compared to theoretical Scheil plots with constant k values to detect any deviations between the two curves. All quantitative EDS measurements in these analyses were carried out on a Philips XL30 FEG SEM.

Finally, a SETARAM SETSYS 18 Differential Thermal Analysis unit was used to investigate the effects of alloying composition on alloy solidus, liquidus and carbide precipitation temperatures. Prior to testing of the experimental alloys, the DTA unit was calibrated with high purity Ni (99.99+ %) and Ag (99.9999%) at scanning rates of 3, 5, 10 and 20 °C/min using a platinum reference. All tests were conducted in a purged ultra-high purity argon atmosphere (flow rate ~70cc/min) using high purity alumina crucibles. Once calibrated, cylindrical samples (4mm diameter and ~3-4mm height) with masses ranging from 200-250mg were prepared from the various alloys. Due to the effects of undercooling, only the DTA traces collected during heating were evaluated. The heating cycle of the samples consisted of:

1) RT to 1000°C at 20°C/min
2) 10 min isothermal soak at 1000°C
3) 1000°C to 1550°C at 5°C/min

For this particular application, a scanning rate of 5°C/min was determined to yield the best combination of temperature accuracy and peak resolution. Additionally, this heating/cooling rate matched the calculated solidification rate of the 12.7mm diameter bars. Following the DTA experiments, the samples were metallographically prepared and examined to ensure that no major contamination or interaction with the alumina crucibles occurred during the analyses.

Results

Figure 1 shows as-solidified bars of alloy ME1 with diameters of 12.7, 15.9 and 19.0mm. The bars exhibit both freckles and larger, misoriented grains. Both types of defects have previously been shown to form due to the onset of thermosolutal convective instabilities during

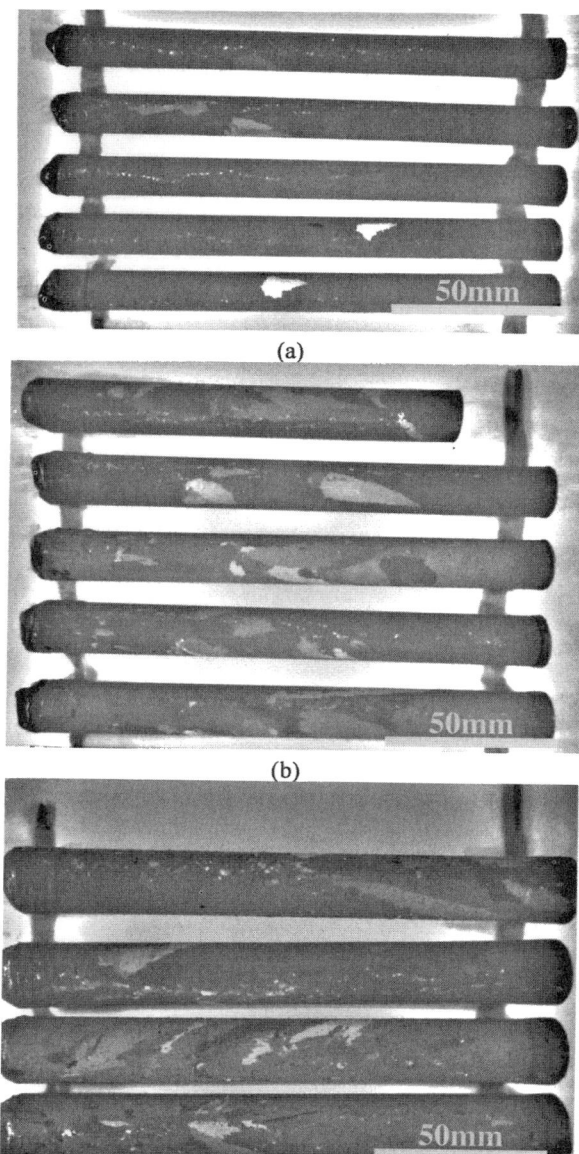

(a)

(b)

(c)

Figure 1: Photos of the macroetched castings exhibiting freckling and the development of misoriented grains in (a) 12.7mm, (b) 15.9mm and (c) 19.0mm diameter bars.

solidification[4]. From this, it is apparent that the number of grain defects increased as the cross-sectional area of the casting increased. In addition, micrographs of the as-cast solidified microstructure, Figure 2, revealed that the primary dendrite arm spacing measurements (PDAS) increased with the diameter of the bars. On average for all alloys, the PDAS of the 12.7, 15.9 and 19.0mm diameter bars were measured to be 340, 400 and 460μm, respectively. Figure 3 shows the average number of freckle-type grain defects as a function of alloy composition for the fractional factorial studies. The beneficial effects of both tantalum and carbon are evident

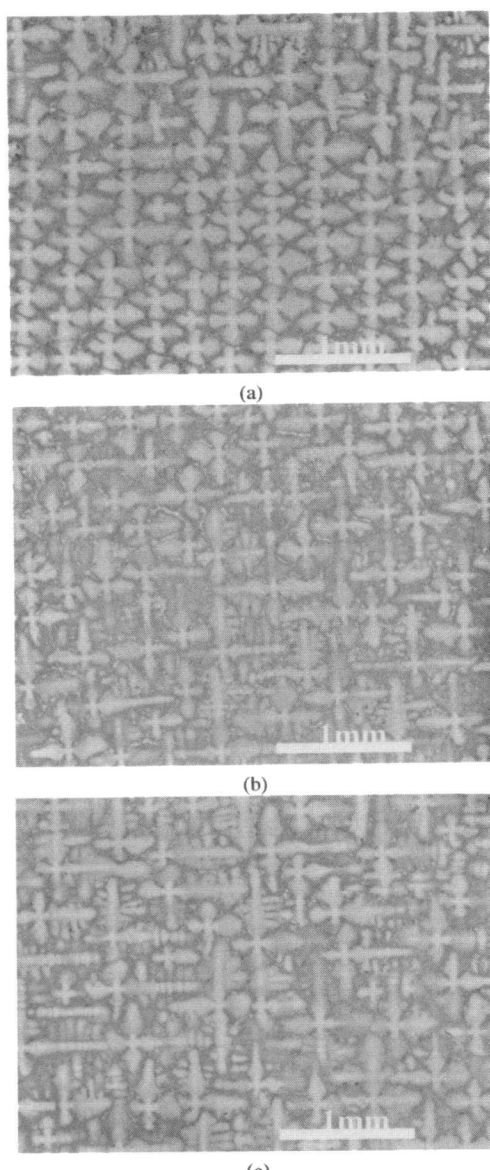

Figure 1: Typical dendritic microstructure of as-cast samples sectioned normal to the solidification direction with diameters of (a) 12.7 mm, (b) 15.9mm and (c) 19.0mm.

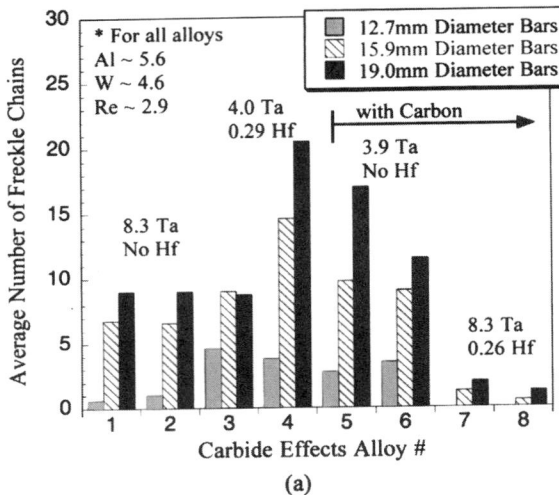

(a)

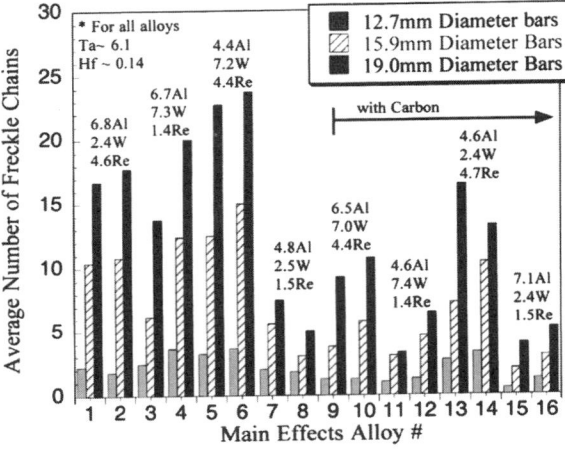

(b)

Figure 2: Results from the half fractional factorial solidification experiments involving (a) the CE and (b) the ME matrix of alloys.

from the solidification experiments involving alloys from the CE matrix, Figure 3a. All of these alloys were René N5-based with nominally identical levels of Cr, Co, Al, W, Re and Mo. For example, comparison of alloys CE1 and 2 to CE3 and 4 clearly shows that freckling was reduced when the level of Ta in the alloy increases from 4.0 to 8.3 wt.%. Interestingly, the beneficial effect from the addition of 0.125 wt.% C to the alloys varied dramatically between alloys CE5 and 6 compared to CE7 and 8. In alloys CE5 and 6, which contain ~3.9 wt.% Ta, the number of freckle defects observed on the surface of the castings decreased only slightly with the addition of carbon. However, the carbon addition was very effective in minimizing freckle defects in the alloys with 8.3 wt.% Ta. Comparison of alloys CE1 and 2 to CE7 and 8 shows that freckling was essentially eliminated in the 12.7mm diameter bars and was greatly reduced in the 15.9 and 19.0mm bars by the addition of carbon.

Results from the ME matrix of experiments are shown in Figure 3b. General trends indicate that increasing levels of W and Re promote the development of freckle defects while intentional C additions of 0.125 wt.% suppress them. Alloys used for these experiments were also René N5-based with nominally identical levels of Cr, Co, Mo, Hf and

204

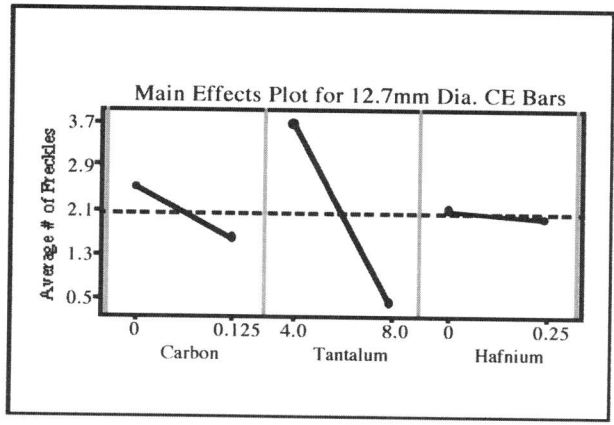

(a)

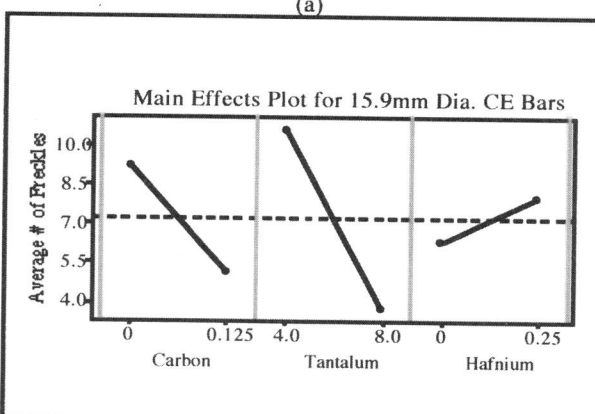

(b)

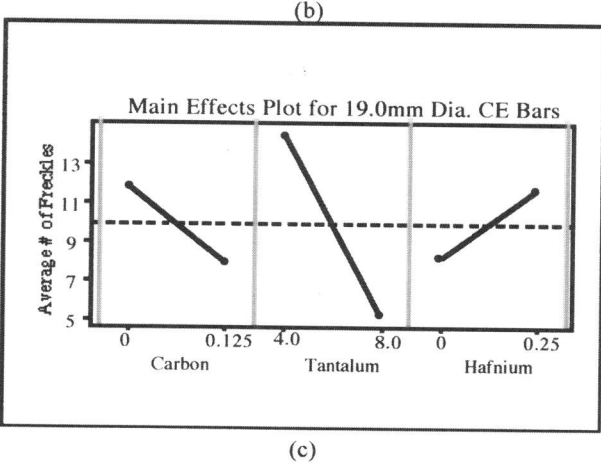

(c)

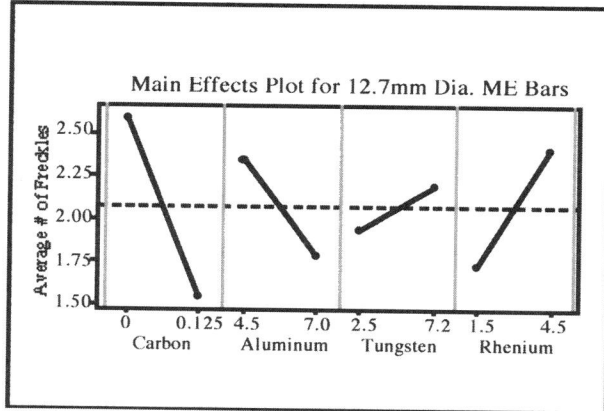

(d)

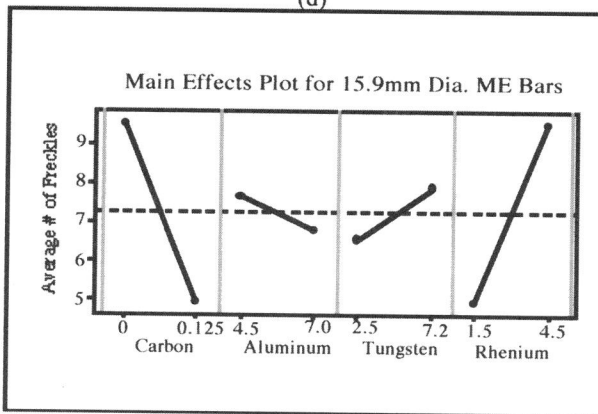

(e)

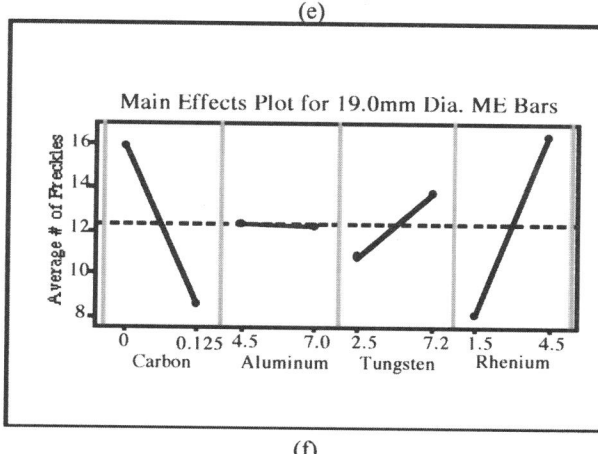

(f)

Figure 4: Main Effects plots generated using the results from (a) CE 12.7mm, (b) CE 15.9mm, (c) CE 19.0mm, (d) ME 12.7mm, (e) ME 15.9mm and (f) ME 19.0mm diameter bars.

Ta. Comparison of alloys ME1 through 8 reveals that the onset of convective instabilities was most prevalent in alloys with elevated levels of W and Re. In comparative instances, alloys with the intentional carbon additions, ME9 through 16, exhibited fewer freckle defects on the surface of the casting. However, the effectiveness of the carbon addition in suppressing grain defect formation again varied from alloy to alloy. In this instance, the beneficial effect of the carbon addition is shown to be much more significant in the alloys which contain elevated levels of W, ME9 through 12. The degree of freckling on the surfaces of the castings

with ~2.4 wt.% W, ME13 through 16, was observed to be only slightly lower than comparable alloys containing no carbon additions.

To statistically assess the influence of individual elements within the two sets of designed experiments, a commercially available statistical software package, MINITAB, was used to generate Main Effects plots, Figure 4.

In factorial experiments, such as these where the variables are systematically changed, these plots isolate the response of each variable. For the CE alloys, Figures 4a-c clearly indicate that increasing Ta levels in the alloys from 4.0 to 8.0 wt.% played a major role in inhibiting the onset of freckle formation. Due to a large two factor interaction between C and Hf in the CE matrix, their effects cannot be evaluated separately. However, results from the ME matrix, Figures 4d-f, reveal that the effect of C was indeed beneficial with respect to stabilizing against the onset of convective instabilities. In addition, freckle defects are shown to increase with elevated Re and W levels. Finally, changes in Al content become statistically insignificant as thermal gradients decrease in the larger diameter bars.

In all instances, carbon additions to the experimental alloys resulted it the formation of a small volume fraction of carbides located preferentially in the interdendritic regions, surrounded by a γ - γ' matrix. Extraction of carbides present in "2nd generation" alloys revealed morphologies that are identical to those observed in "3rd generation" René N6-based alloys. As seen in Figure 5, the three morphologies were characterized as blocky, script and nodular. Most of the experimental alloys contained a mixture of both script and nodular carbides. No correlation between carbide morphology and freckling was evident. XRD and EDS analyses identified these carbides to be Ta-rich MC carbides, with lattice parameters varying from 4.39 to 4.45 angstroms depending on the presence of minor amounts of Hf, Ni, Cr, Co, and Mo.

Due to a potential change in the solidification path of the alloy as a result of the precipitation of Ta-rich MC carbides, the segregation characteristics of the constituent elements were studied in the as-cast crystals. Quantitative EDS linescans through the dendrite core and interdendritic regions revealed that Ta and Al segregate preferentially to the interdendritic region, while W, Re, Co, Cr and Mo segregate to the dendrite core.

Comparison of the Scheil analyses of tantalum for a pair of nominally similar alloys with and without carbon is shown in Figure 6. Initially, the two curves are nearly identical, suggesting that solidification is progressing in the same manner for the two alloys. However, separation of the curves occurs at approximately 0.5 apparent fraction solid, corresponding to the onset of Ta-rich MC carbide precipitation in the mushy zone. The 1.5-2 wt.% (~0.6 at.%) deviation in the curves is consistent with the TaC reaction, consuming the 0.125 wt.% (0.6 at.%) carbon in solution. The difference in the overall slope of the two curves is captured by the fitted distribution coefficient, k. Using the modified Scheil equation to account for limited back diffusion (α = 0.01), the fitted distribution coefficients for CE1 and CE8 are 0.71 and 0.80, respectively. Although comparison of the

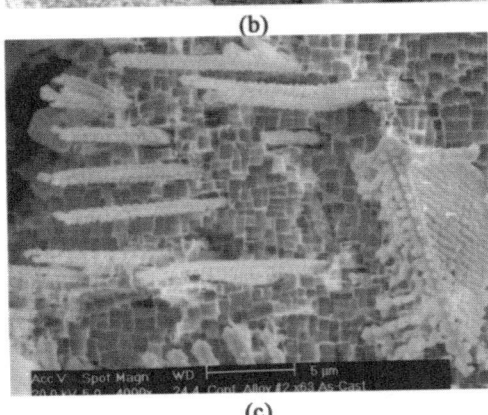

Figure 5: Photomicrographs of Ta-rich MC carbide morphologies seen in both René N5 and N6 - based alloys: (a) blocky, (b) script and (c) nodular.

segregation coefficients in Table II show that less segregation of the constituent elements occurred in alloys which contain carbon, the degree to which the elements were affected varied. Generally, the largest changes in segregation due to carbon were observed in Ta and W. Since carbides contain a significant amount of Ta, the change in Ta segregation was not unexpected. However, W segregation is altered by the presence of carbon in spite of the fact that no W could be detected in the carbides. As mentioned earlier, the ME alloys in which the carbon additions were determined to be most effective in suppressing the onset of convective instabilities also had elevated levels of W.

DTA analyses of the as-cast experimental single crystal alloys revealed information on the effects of alloy composition on solidus, liquidus and carbide dissolution temperatures. Comparison of alloys CE4 and CE5 in Figure 7a, which are essentially nominally similar alloys except for the addition of 0.125 wt.% C to CE5, reveal that carbon lowers the liquidus temperature of the alloy by ~7°C. Furthermore, it results in the dissolution of the Ta-rich MC carbides at approximately 24°C below the liquidus temperature of the alloy. As seen in Figure 7b, increasing the level of Ta dramatically lowers the liquidus temperature of the alloys by approximately 15°C. Although the liquidus temperatures of the alloys are significantly different, carbide dissolution in alloys CE5 through 8 occur at the same temperature, 1385°C. Collective results of the DTA analyses, Table III, show that the carbide dissolution temperatures (1377-1393°C) are much less sensitive to shifts in composition than the solidus (1331-1391°C) and liquidus (1381-1428°C) temperatures.

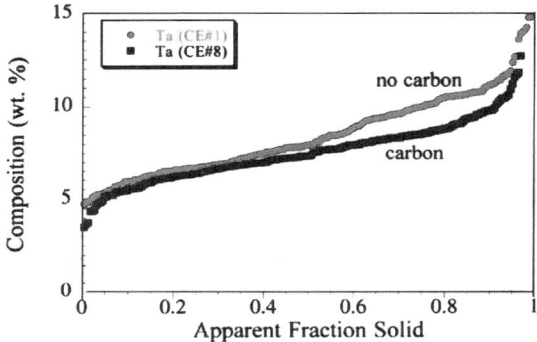

Figure 6: Comparison of tantalum distribution in alloys CE1 and CE8.

Discussion

The results of experiments reported here on "2nd generation" type alloys as well as results of previous experiments on "3rd generation" type alloys clearly show that intentional carbon additions to high refractory single crystal alloys assist in minimizing grain defect formation. However, the effects of carbon on the mechanisms which result in freckling are not yet well understood. One potential mechanism by which carbon improves solidification characteristics is by a change in segregation behavior of other constituent elements. Changes in segregation lead to slight compositional changes in the interdendritic liquid which may possibly lower the tendency for the onset of convective instabilities. Keeping in mind the variable factors in the ME alloys, comparison of the results from the solidification experiments show that alloys ME9&10 and ME11&12 have significantly improved solidification characteristics when compared to ME5&6 and ME3&4,

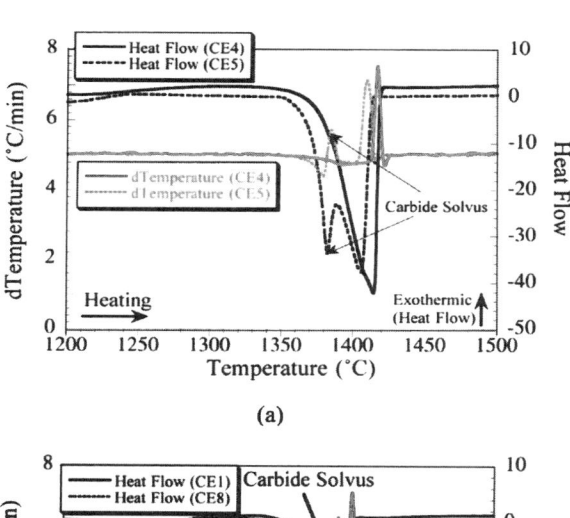

(a)

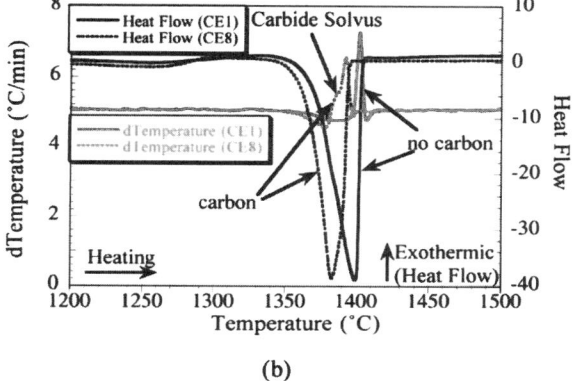

(b)

Figure 7: DTA traces from single crystal alloys (a) CE4 and CE5 (b) CE1 and CE8.

respectively. Since the effect of Al has been previously shown to be negligible, especially as thermal gradients decrease, these improvements can be attributed to the carbon addition. Relative to the effects of W, Re and Ta additions, previous studies involving "3rd generation" René N6-based alloys have also shown that minor variations in Al levels were statistically insignificant with respect to freckle formation[2,12]. Comparison of ME13&14 and ME15&16 with ME1&2 and ME7&8 show that the beneficial effects of the carbon addition are much less significant in alloys with low levels of W. Recalling the segregation analyses, the constituent elements most affected by the carbon additions were Ta and W. The alloys which benefit to the greatest degree from the C addition have the strongest segregation of Ta combined with the weakest segregation of W. Typically, elevated levels of W would promote freckling due to its tendency to segregate to the dendrite core and create a density inversion between the interdendritic liquid and the bulk melt[2,4]. However, in this case, the addition of carbon results in less segregation of W to the dendrite core thereby lowering the alloys' tendency to develop convective instabilities. In alloys ME1&2, ME7&8 and ME13 through 16 the presence of only ~2.4 wt.% W in the alloys would not seem likely to contribute significantly to the formation of

	CE#4	CE#5	CE#1	CE#8	ME#6	ME#10	ME#4	ME#11	ME#2	ME#14	ME#7	ME#15
Al	0.89	0.88	0.85	0.86	0.89	0.89	0.92	0.92	0.92	0.95	0.94	0.95
Cr	1.12	1.13	1.15	1.15	1.07	1.15	1.15	1.06	1.13	1.06	1.07	1.13
Co	1.10	1.08	1.10	1.10	1.05	1.09	1.09	1.04	1.07	1.04	1.05	1.09
Ni	0.97	0.98	0.97	0.97	0.96	0.94	0.96	0.98	0.97	0.97	0.98	0.98
Ta	0.67	0.76	0.72	0.77	0.79	0.87	0.76	0.89	0.78	0.89	0.78	0.84
W	1.39	1.25	1.39	1.35	1.38	1.31	1.36	1.28	1.54	1.35	1.58	1.40
Re	1.39	1.39	1.42	1.34	1.35	1.30	1.38	1.28	1.37	1.30	1.25	1.23
Mo	1.25	1.28	1.30	1.30	1.20	1.27	1.25	1.20	1.23	1.18	1.13	1.24

convective instabilities. Thus, a decreased degree of W segregation in these alloys apparently does not provide substantial benefits in improving the solidification characteristics. On the other hand, in alloys ME3 through 7 and ME9 through 12 (>7.0 wt.% W) where W would likely be considered to contribute heavily to freckle formation, changes in segregation could potentially improve solidification characteristics substantially. Similar changes in segregation behavior of W due to carbon have also been recently reported in experimental "3rd generation" single crystal superalloys[12].

Interestingly, the consumption of tantalum by the TaC carbide reaction during solidification is beneficial with respect to the solidification characteristics of the alloy. During solidification, tantalum segregates preferentially to the interdendritic regions and offsets the density imbalance which develops between the solute in the mushy zone and the bulk liquid. Previous studies have shown that decreasing the levels of tantalum in the mushy zone leads to an increased propensity for the onset of thermosolutal convection[2,4]. However, since the density of TaC (14.5 gm/cm^3) is almost twice that of the bulk liquid, the presence of these high density carbides in the mushy zone during solidification may also be beneficial in stabilizing convective fluid flow.

In the CE matrix of experiments, levels of Ta, C and Hf were varied in the alloys while all of the other constituents were held constant. The most obvious result from Figures 3a and 4a-c is that increasing the level of Ta in the alloy suppresses the onset of thermosolutal convection. In the results, the effect of Hf on the solidification characteristics was negligible. Although, Hf is a very potent MC carbide former, it was concluded that the presence of only ~0.3 wt.% Hf (~0.1 at.%) was insufficient to generate drastic changes, since it would consume only ~0.02 wt.% C (0.1 at.%) in the alloy. With ~4.5 wt.% W present in the CE alloys, the change in the segregation behavior of W due to carbon had little effect in reducing the number of freckles in alloys CE5 and 6. However, a substantial decrease in the number of freckles was observed between alloys CE1&2 and CE7&8. Clearly, in this instance, the carbon addition to alloys CE7 and 8 is affecting a mechanism of freckle

formation other than the segregation behavior of the constituent elements. Even though larger differences in W segregation were measured between alloys CE4 and CE5, no significant beneficial effect due to carbon was observed between alloys CE3&4 and CE5&6. Factoring in the DTA data, differences between the relative carbide dissolution temperatures of CE5&6 and CE7&8 are readily observed. Considering the sequence of events which begin with the initiation of the convective instabilities and ultimately result in freckle formation, it seems highly probable that carbide formation close to the liquidus temperature could potentially affect freckling mechanisms. Carbide dissolution in alloys CE7&8 and CE5&6 occur at ~8°C and ~24°C below their respective alloy liquidus temperatures. Referring back to the carbon containing alloys in the ME experiments, even though alloys ME9 and 10 had the highest potential for developing freckles based on composition, more freckle defects were observed on the castings from alloys ME13 and 14. In this case, the relative carbide dissolution temperatures in ME9&10 and ME13&14 were ~8°C and 22°C, respectively.

Along with the possibility of influencing the segregation behavior of the constituent elements, the resulting carbide precipitation may serve to alter the viscosity and permeability of the interdendritic fluid. Investigations of freckling in transparent and binary alloy systems have shown that the onset of thermosolutal convection occurs in the upper regions of the mushy zone where the fraction solid in no greater than ~0.5[5,6,16-18]. At higher fraction solid, the permeability of the dendritic network is too low for solute induced fluid flow to develop. Thus, changes to the upper regions of the mushy zone, such as the precipitation of carbides just below the liquidus, should theoretically lower the permeability and decrease the alloys' tendency to develop freckles.

To summarize, this study suggests that some of the inherent difficulties in casting single crystal blades can be lessened through slight modifications in alloy composition. Although freckles and other grain defects can potentially be reduced by optimizing the levels of Ta, Re, and W, this is unlikely to simultaneously benefit phase stability, corrosion, creep, and fatigue properties. Therefore, alloying approaches, which involve changes in "minor" alloying elements such as carbon, may be more useful for reducing grain defects without significantly compromising mechanical or physical properties. These results reveal that carbon additions are beneficial in lowering the alloys' tendency to develop freckle defects over a wide range of experimental "2nd generation" single crystal alloys. Although the addition of carbon resulted in an overall reduction of grain defects, certain alloy compositions benefited more than others. Carbon additions were determined to influence the segregation behavior of the constituent alloys and affect some aspect of the mechanisms which lead to solute-induced fluid flow. DTA analyses revealed a correlation between the relative carbide dissolution temperature and the number of observed grain defects. As MC carbide formation occurred closer to the liquidus temperature, the tendency for grain defect formation decreased. Thus, alloys designed with high levels of tantalum and carbide precipitation occurring near the liquidus provide the highest degree of resistance to the breakdown of single crystal solidification.

Table II: Liquidus, Solidus and Carbide Dissolution Temperature Measurements for Experimental Single Crystal Alloys With and Without Carbon.

Alloy	T(solidus) °C	T(carbide) °C	T(liquidus) °C
ME1	1335	-	1397
ME2	1333	-	1396
ME3	1331	-	1393
ME4	1331	-	1393
ME5	1393	-	1426
ME6	1396	-	1428
ME7	1390	-	1416
ME8	1391	-	1417
ME9	1355	1383	1390
ME10	1354	1378	1389
ME11	1376	1391	1412
ME12	1378	1391	1412
ME13	1378	1393	1416
ME14	1379	1393	1415
ME15	1355	1377	1381
ME16	1358	1378	1384
CE1	1369	-	1402
CE2	1370	-	1403
CE3	1381	-	1416
CE4	1379	-	1416
CE5	1371	1385	1409
CE6	1370	1385	1410
CE7	1366	1386	1393
CE8	1364	1385	1393

Conclusions

1) Under relatively low thermal gradients, elevated levels of Ta and intentional additions of 0.125 wt.% C improve single solidification characteristics over a wide range of compositions encompassing both "2nd and 3rd generation" alloys. Increasing the levels of W and Re in these alloys promotes the formation of freckle-type defects.

2) Intentional carbon additions result in the formation of Ta-rich MC carbides during solidification and alter the segregation behavior of the constituent elements.

3) Fewer freckle-type defects develop in carbon containing alloys where carbide precipitation occurs just below the liquidus temperature.

References

1. A.F. Giamei and B.H. Kear, "On the Nature of Freckles in Nickel-Base Superalloys," Metall.Trans., 1 (1970), p.2185-2192

2. T.M. Pollock, W.H. Murphy, E.H. Goldman, D.L. Uram, and J.S. Tu. "Grain Defect Formation During Directional Solidification of Nickel Base Single Crystals," Superalloys 1992. TMS. Seven Springs, PA: p.125-134

3. T.M. Pollock, "The Growth and Elevated Temperature Stability of High Refractory Nickel-Base Single Crystals," Mater. Sci. Eng., B32 (1995), p.255-266

4. T.M. Pollock and W.H. Murphy, "The Breakdown of Single-Crystal Solidification in High Refractory Nickel-Base Alloys," Metall. and Mat. Trans., 27A (1996), p.1081-1094

5. A.K. Sample and A. Hellawell, "The Mechanisms of Formation and Prevention of Channel Segregation During Alloy Solidification," Metall. Trans., 15A (1984), p.2163-2173

6. S.M. Copley, A.F. Giamei, S.M. Johnson, and M.F. Hornbecker, "The Origin of Freckles in Unidirectionally Solidified Castings," Metall.Trans., Vol. 1 (1970), p.2193-2204

7. P. Auburtin and A. Mitchell. "Elements of Determination of a Freckling Criterion," Proc. Symp. on Liquid Metals Processing 1996. AVS. Santa Fe, NM: p.18-34

8. S.D. Felicelli, J.C. Heinrich, and D.R. Poirier, "Simulation of Freckles During Vertical Solidification of Binary Alloys," Metall.Trans., 22B (1991), p.847-859

9. M.C. Schneider, J.P. Gu, C. Beckermann, W.J. Boettinger, and U.R. Kattner, "Modeling of Micro- and Macrosegregation and Freckle Formation in Single-Crystal Nickel-Base Superalloy Directional Solidification," Metall. and Mat. Trans., 28A (1997), p.1517-1531

10. N. Streat and F. Weinberg, "Pipe Formation in Pb-Sn Alloys," <u>Metall. Trans.</u>, 3 (1972), p.3181-3184

11. F.L. VerSnyder and M.E. Shank, "The Development of Columnar Grain and Single Crystal High Temperature Materials Through Directional Solidification," <u>Mater. Sci. Eng.</u>, 6 (1970), p.213-247

12. S. Tin, T.M. Pollock, and W.H. Murphy. "The Role of Carbon in Directionally Solidified Superalloys," Presentation at the 1999 *Fall TMS* Meeting Cincinati, OH. (Submitted to Metall. and Mat. Trans.)

13. M.N. Gungor, "A Statistically Significant Experimental Technique for Investigating Microsegregation in Cast Alloys," <u>Metall. Trans.</u>, 20A (1989), p.2529-2533

14. S.C. Huang, L. Peluso, and D. Backman. "Measurements of Solute Segregation Coefficients in Quaternary Ni-Based Superalloys,"*Solidification 1999* TMS. San Diego, CA: p.163-172

15. L. Nastac, L.S. Chou, and Y. Pang. "Assessment if Solidification-Kinetics Parameters for Titanium-Base Alloys,"*Liquid Metals Processing* Santa Fe, NM. p.207-223

16. J.R. Sarazin and A. Hellawell, "Channel Formation in Pb-Sn, Pb-Sb, and Pb-Sn-Sb Alloy Ingots and Comparisonn with the System NH_4Cl-H_2O," <u>Metall. Trans.</u>, 19A (1988), p.1861-1871

17. M.C. Schneider and C. Beckermann, "A Numerical Study of the Combined Effects of Microsegregation, Mushy Zone Permeability and Flow, Caused by Volume Contraction and Thermalsolutal Convection, on Macrosegregation and Eutectic Formation in Binary Alloy Solidification," <u>Inter. Jour. of Heat and Mass Trans.</u>, 38 (1995), p.3455-3473

18. D.R. Poirier, "Permeability for Flow of Interdendritic Liquid in Columnar-Dendritic Alloys," <u>Metall. Trans.</u>, 18B (1987), p.245-255

NEW ASPECTS OF FRECKLE FORMATION DURING SINGLE CRYSTAL SOLIDIFICATION OF CMSX-4

R. Schadt, I. Wagner, J. Preuhs*, P.R. Sahm
Gießerei-Institut, RWTH Aachen, Intzestr. 5, 52056 Aachen, Germany
*DONCASTERS Precision Castings – Bochum GmbH, Bessemer Str. 80, 44793 Bochum, Germany

Abstract

Freckle formation is a long time investigated but in its high complexity still not completely understood phenomenon. Some new aspects of this phenomenon especially concerning freckle formation inside single crystal castings and the simulation of freckling are alighted. A simple and near process criterion for the prediction of freckle threatened areas in single crystal components has been developed. The investigations reveal that there are still a lot of open questions concerning where freckle formation exactly takes place. Nevertheless the results presented offer new starting points concerning solutions and may inspire to intensify the research activities on this interesting phenomenon.

Introduction

Freckles are presently one of the main defects encountered in advanced directional solidification (DS) and single crystal (SC) casting technology of superalloys. They appear as long chains of equiaxed grains preferentially at the component's surface. A reasonable explanation of freckle formation is thermosolutal convection driven by a density inversion in the mushy zone. This is caused by alloy specific segregations, which is indicated by a noticeable shift of the freckles grain composition towards the alloy's eutectic composition. The convection evolution is influenced by the component geometry as well as by alloy composition and primary process parameters as temperature gradient G and solidification velocity v [1-3]. Today the most complete criterion available seems to be the Rayleigh criterion. It combines the influences of alloy composition and process parameters but does not take into account component geometry.

Sarazin and Hellawell [4] suggested to characterize fluid flow associated with freckle formation by using the Rayleigh number as follows:

$$\text{Freckling when:} \quad Ra = \frac{g\dfrac{d\rho}{dz}}{\dfrac{\eta D_t}{h^4}} > Ra^*$$

where the parameter h is a characteristic linear dimension of the system linked to the dendritic array in the mushy zone [3,4]:

$$h^4 = \lambda_1^4 \qquad \text{or} \qquad h^4 = K\lambda_1^2 .$$

Auburtin et al. [5] investigated the influence of the growth front angle on the freckle formation and modified the Rayleigh criterion. However this highly sophisticated criterion requires enormous efforts in measuring alloy specific thermophysical and thermochemical data.

In order to predict freckle threatened areas in complicated DS- or SC-components it is necessary to develop a criterion, which takes into account process parameters as well as geometrical effects but does not depend on the complete knowledge of alloys' chemistry.

Such an accurate criterion enables the production of specific freckle tainted specimens for an investigation of their lifetime relevant effects on DS- and SC-components.

Superalloys 2000
Edited by T.M. Pollock, R.D. Kissinger, R.R. Bowman,
K.A. Green, M. McLean, S. Olson, and J.J. Schirra
TMS (The Minerals, Metals & Materials Society), 2000

Experimental

The SC casting experiments were carried out in an industrial sized vacuum Bridgman furnace where solidification conditions are the same as in those furnaces used for production of DS/SC parts. The chosen solidification parameters were similar to those normally used in the production process. Withdrawal velocities from v = 0.5 mm/min up to 3.0 mm/min and heater temperatures T = 1500°C were applied. A round copper chill plate of 150 mm diameter was used for the cyclic clusters. Ceramic molds (Al_2O_3/SiO_2 based) were manufactured by standard investment process. Up to 20 thermocouples could exactly be positioned in the different specimens. The alloy CMSX-4 [6] was chosen for the investigations because due to its segregation behavior it is a typical freckle prone alloy. Furthermore there is great interest in widening the application range of CMSX-4 to large IGT-blades. The composition of the alloy is given in **Table I**.

Table I: Composition of the investigated alloy CMSX-4 (wt%) [6].

Cr	Co	Mo	W	Ta	Re	Al	Ti	Hf	Ni
6.5	9	0.6	6	6.5	3	5.6	1.0	0.1	bal

Several cylindrical geometries with constant or varying diameter were examined, **Figure1**. Freckle chains were characterized by Energy-Dispersive-X-Ray (EDX) and by Electron-Back-Scattering-Diffraction (EBSD). Composition and orientation distribution of the freckle chains in comparison to their counterparts in the single crystal matrix were examined in detail.

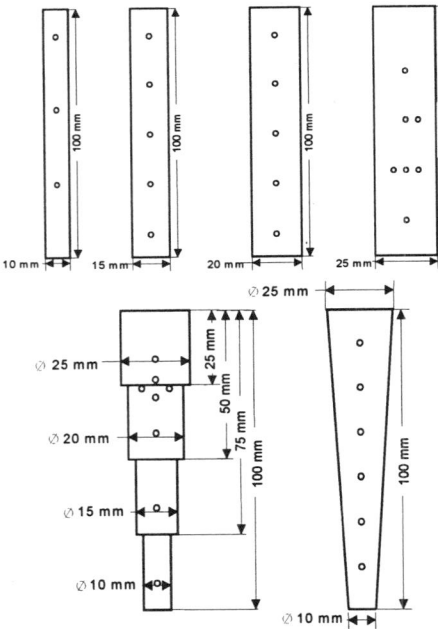

Figure 1: Outline of the investigated simple specimen geometries. The samples were solidified in clusters which consist of either four specimens with constant or four specimens with varying diameter. The circles indicate the position of the thermocouples.

All FEM-simulations were performed with the in-house program CASTS [7-10]. Simulations of grain growth were done with a cellular automaton (CA) algorithm [11].

Results and discussion

G*v-criterion

Based on temperature measurements with thermocouples exactly positioned in a ceramic shell mold a new, more simple and near application criterion for freckle formation has been developed. Late experiments prove, that the freckle chains do not always grow along the direction of the primary dendritic solidification front representing the crystallographic orientation, **Figure 2**. The comparison with FEM-simulations shows that the curved freckle chain which is visible on the surface of the turbine blade follows nearly exactly the curvature of the solidus line. According to the fact that freckle formation takes place during the last period of solidification, the measured velocity of the solidus isotherm and the temperature gradient just in front of this isotherm were used to generate a $G*v$-criterion suitable for practical use. It takes into account the varying local solidification conditions during the Bridgman process. The results of these experimental investigations are shown in **Figure 3**. It is revealed that freckling only occurs below the critical threshold value $G*v = 0.14$ K/s. In a transitional area, which cannot be determined exactly, most likely freckling mainly depends on geometrical influences.

Based on the temperature field calculations the experimental $G*v$-criterion could be transferred to the simulation. It was implemented as a post processing tool. Freckle threatened and certainly freckle free areas can accurately be predicted, **Figure 4**. This criterion cannot predict where freckling actually will take place. A new approach in the calculation of freckle formation using coupled nonequilibrium FEM-methods and cellular automaton calculations is part of the ongoing investigation. First results of a 3D-CA coupled with equilibrium FEM-calculations are presented later.

Expanding specimens

The microstructure analysis of the stepwise and continuous expanding specimens reveals no tendency for freckle formation inside the specimens. Even if a freckle was found at the surface of a specimen in front of a step the freckle chain could not penetrate the castings inside more than a few millimeters, **Figure 5** and **Figure 7**.

Experiments, in which specimens with stepwise and continuous expansion of cross-section were cast together in one cluster under the same process conditions show the same tendency. The specimen with continuously increasing cross-section shows one freckle chain starting at a diameter of 12.5 mm and growing along the surface of the whole specimen. The stepwise expanding specimen contains surface freckle chains at each diameter, but these chains always stop at a foot of a new step.

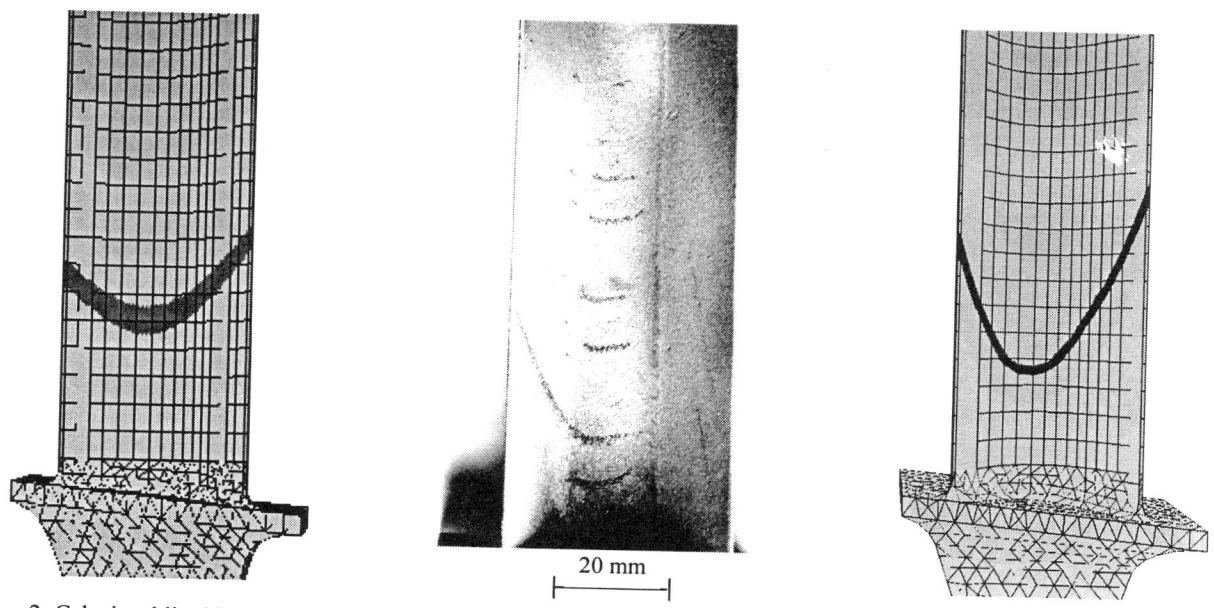

Figure 2: Calculated liquidus isotherm (left picture), turbine blade with extended freckling and calculated solidus isotherm (right picture) of an aero-engine turbine blade. The curvature of the freckle chains on the turbine blade seems to follow the solidus more than the liquidus line. The intermittent freckling structure occurs due to the non steady state conditions with locally varying growth velocities. The vertical temperature gradient G_v was slightly decreased, the lateral gradient G_l increased. Due to the torsion and the varying wall thickness not all freckle chains rises upward at the outer edges of the blade. Additionally should be remarked, that the liquidus and solidus isotherms have not the same propagation velocity. Simulations were performed with the FEM-program CASTS.

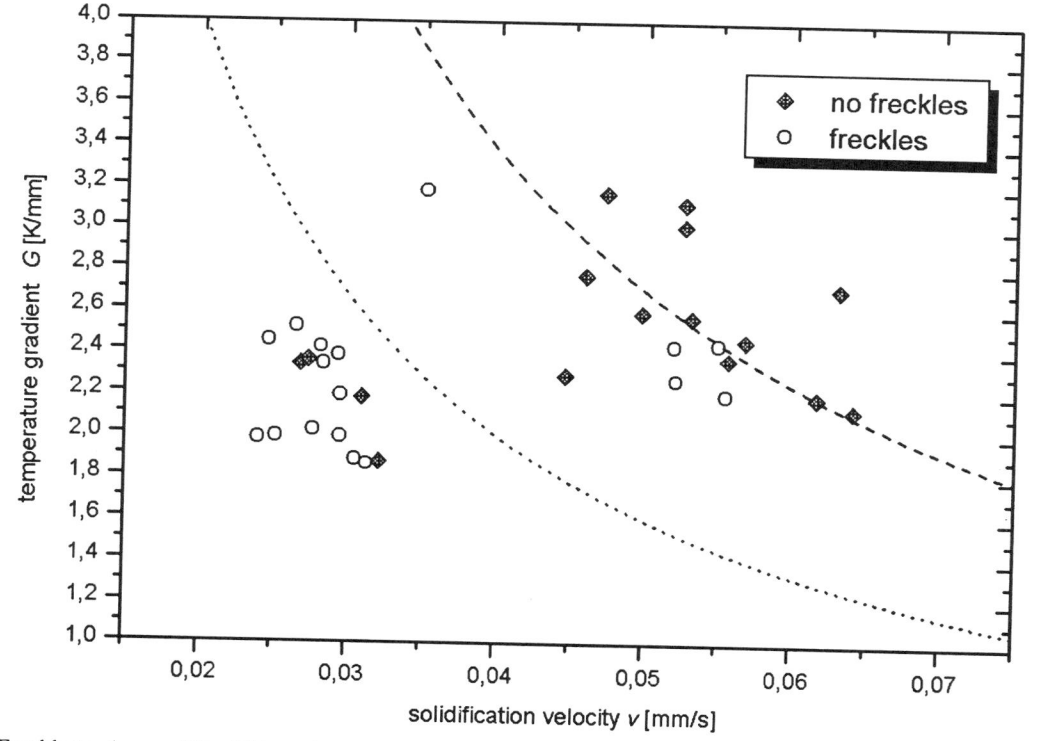

Figure 3: Freckle tendency of the Ni based superalloy CMSX-4. Above the critical threshold value $G*v = 0.14$ K/s (dashed line) no freckling occurs. In the transitional area (between the dashed and the dotted line) freckling seems to depend mainly on geometrical influences.

After that a new chain starts a few millimeters above the step at a different position according to the cluster's centerline. This indicates that the newly arising freckle chain is not correlated to the one that stopped at the step's foot. It does only occur because the conditions at the surface are prone for freckling, **Figure 6**.

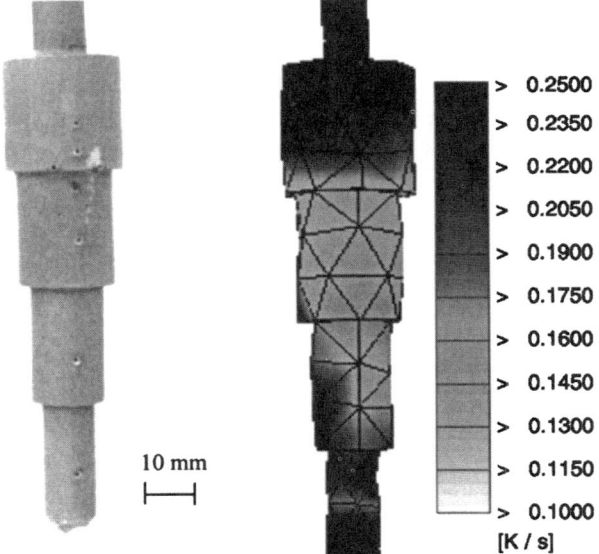

10 mm

> 0.2500
> 0.2350
> 0.2200
> 0.2050
> 0.1900
> 0.1750
> 0.1600
> 0.1450
> 0.1300
> 0.1150
> 0.1000
[K / s]

Figure 4: Comparison of a freckle tainted specimen with the simulated situation. The freckle chain appears in the region of the lowest calculated $G*v$-value. The lack of symmetry in the simulation is due to the inhomogeneous radiation conditions of the cluster configuration.

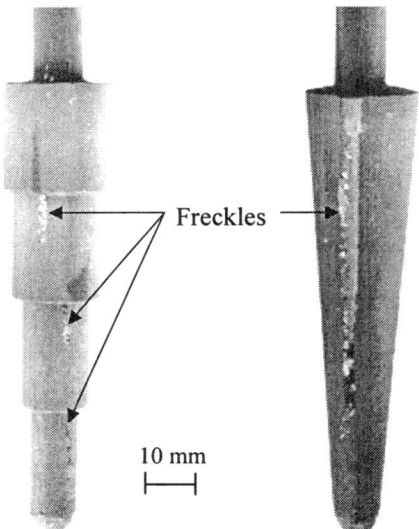

Freckles

10 mm

Figure 6: Specimens with stepwise and continuous increasing diameter. Whereas the step geometry shows different freckle chains at the foot of each step, the continuous expanding specimen contains one single freckle chain, which grew from a diameter of 12.5 mm to the top.

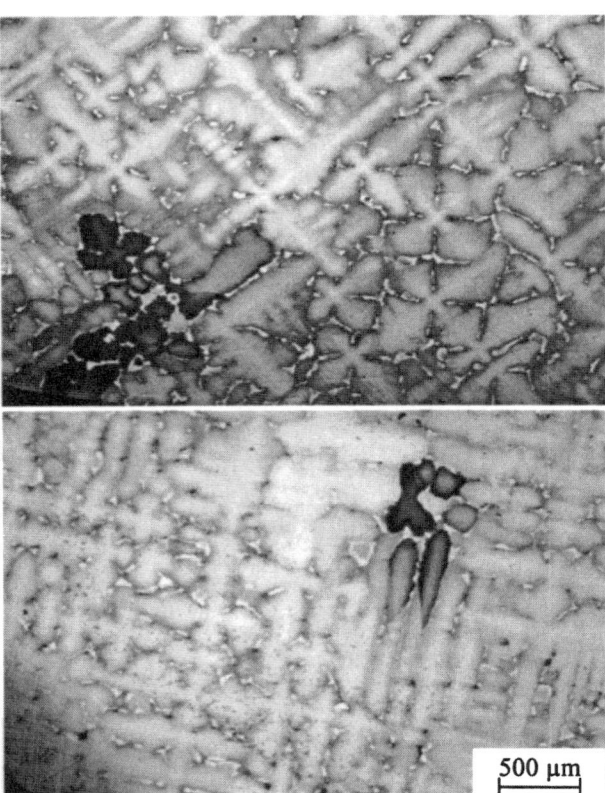

500 µm

Figure 5: Microstructure of a stepwise expanding specimen. A grain of a freckle chain is visible at the surface of the specimen in a transversal micrograph just below the step. 400 µm after the expansion only a small grain of the freckle chain is still visible in the single crystal matrix.

Additional investigations were performed on the question, if segregation channels, which are predicted by some authors [12-14] can be detected above the edge of these step specimens even if there was no freckling detected. 2D-simulations predict the formation of convection channels (plumes) in expanding domains [13]. Therefore freckle free specimens were investigated by EDX-analysis, **Figure 8**. Because the channels were expected to be in the dimension of dendrite axe spacing, a scan distance of 100 µm was chosen. The measured line scans show no significant concentration decrease or increase of any element in the area above the edge of the step, **Figure 9** and **Figure 10**. A few microns after the cross-sectional transition the distribution seems to be disturbed in the expanded area, **Figure 9**. But the distribution flattens with the distance from the edge, **Figure 10**. The high variations in the concentration of neighboring spots are due to the interdendritic segregations which are opposite for Ti and Ta on the one and Re and W on the other hand.

Figure 7: Freckle chain at the surface of a CMSX-4 stepwise expanding specimen. The freckle chain ends at the edge of the step probably due to a disturbance of the convection flow by the expansion. At the surface of the expanded part a slightly disoriented grain develops.

A characterization of the freckle chain in Figure 7 with EBSD-analysis gives an orientation distribution which indicates that there are no significant correlations between the single grains in the freckle chain, **Figure 11**. The polar plot shows the position of the [100]-orientations of the SC-matrix and the other measured grains. These orientations seem to be distributed quite coincidentally, but the high disorder is not only due to the different orientations in the chain but also to some fuzziness which is typical for the EBSD-analysis. The best focus is given in the center of the investigated specimen and the more the beam gets out of the center the more fuzzy the reflexes are. The matrix reveals a nearly perfect [100]-orientation, which is indicated by the dots in the polar plot.

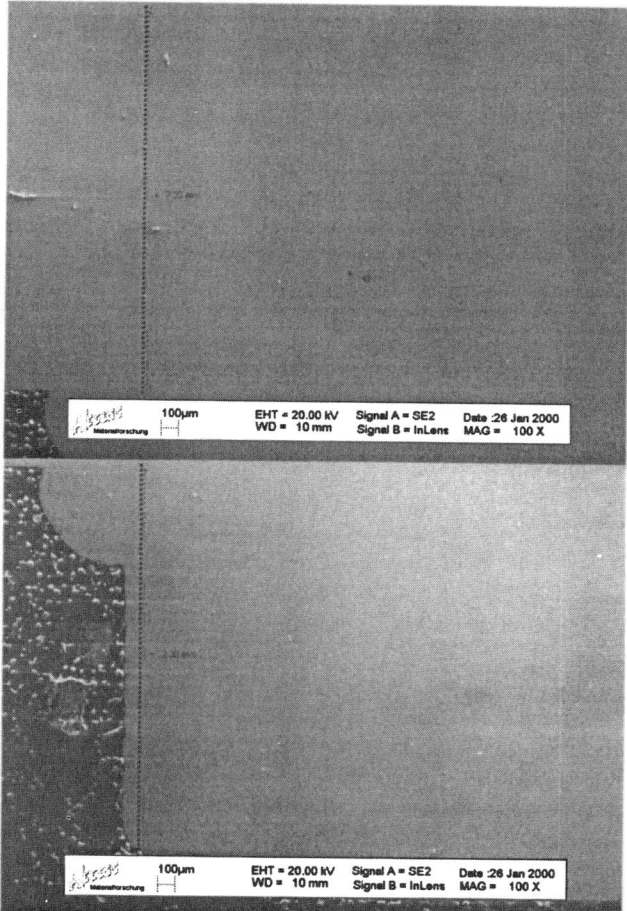

Figure 8. Expanding domain specimen investigated by EDX. The dashed line shows the path of the line scan. The distance between the measure spots was 100 μm, so that there were 46 measure points for each line scan. Two more line scans were performed at a distance of 7.5 mm and 12.5 mm from the edge.

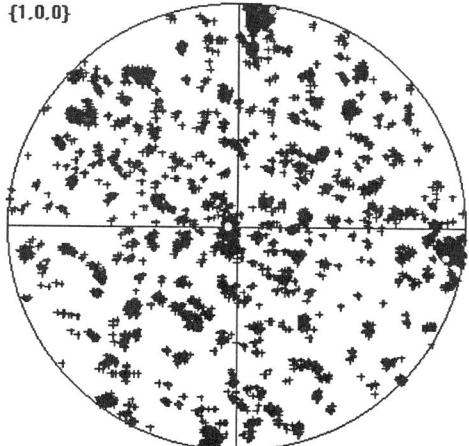

Figure 11: Polar plot of the measured [100]-directions of the SC-matrix (dots) and some freckle chain grains.

215

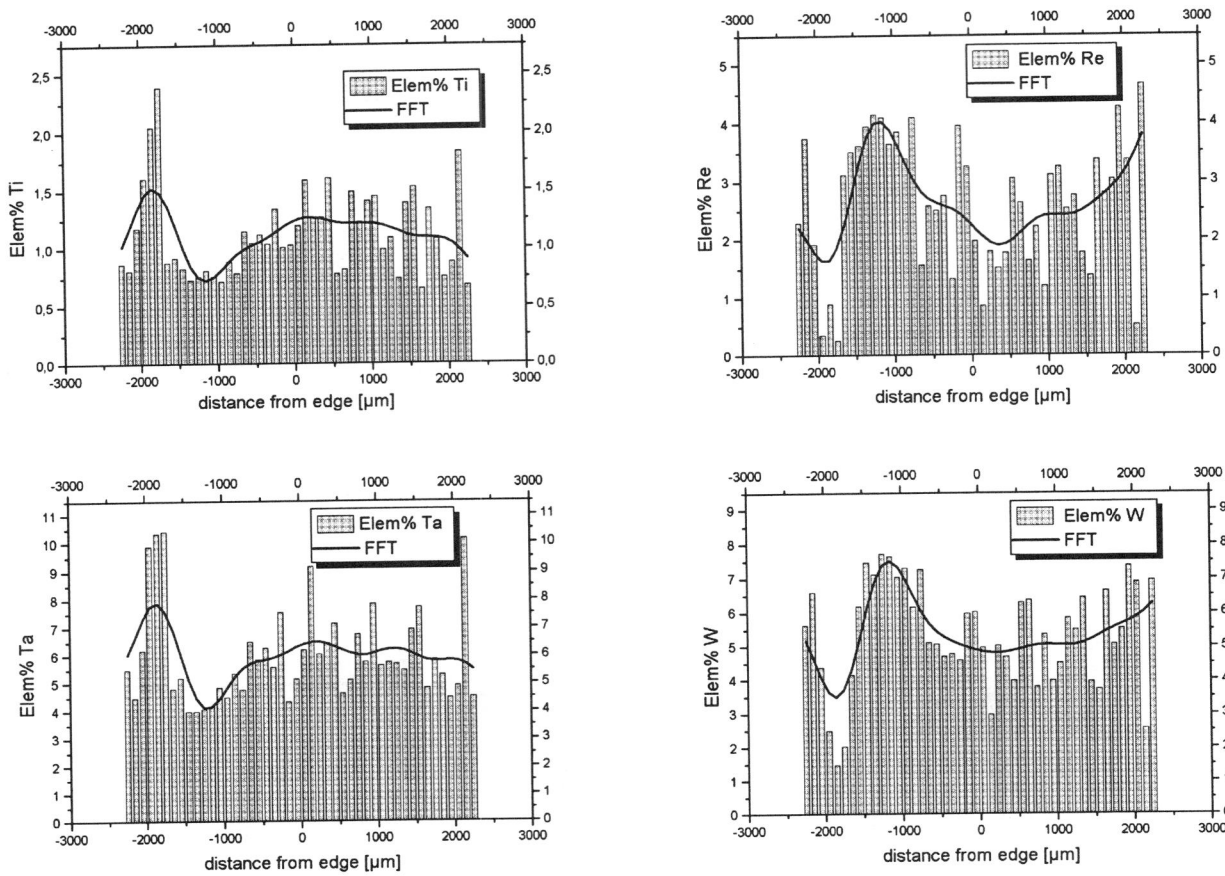

Figure 9: The EDX-line scan just above the edge of the expanding domain shows no significant concentration variation. It seems that the element distribution in the outer part of the domain (negative distance) is disturbed by the expansion. The columns display the concentration measured at each spot, whereas the lines represent an integral Fourier smoothing function term.

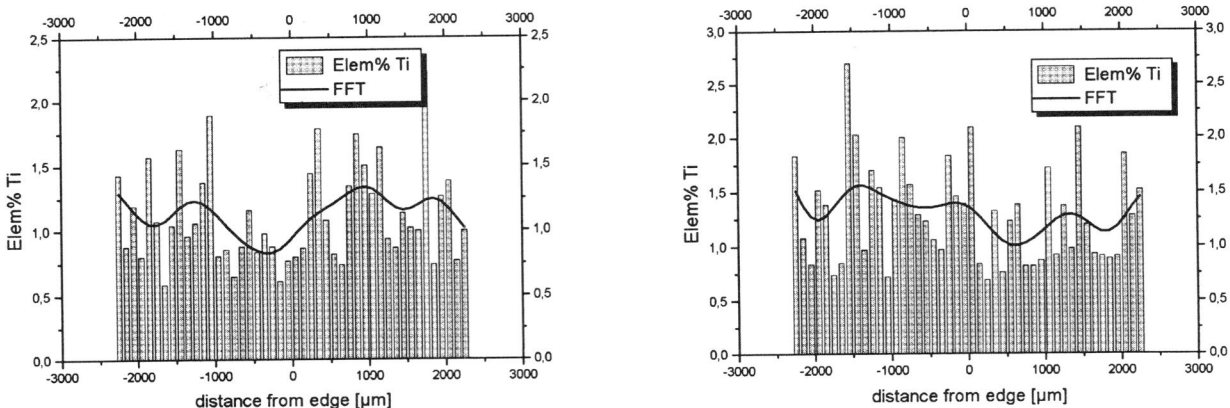

Figure 10: Titanium distribution 7.5 mm (left) and 12.5 mm (right) above the edge. The distribution seems to flatten with the distance above the edge. The high variations between neighboring spots are due to the interdendritic segregations.

Grain structure simulation

The simulation of the grain structure formation during DS- or SC-solidification was performed with a cellular automaton rotated grid algorithm. In this model the grain orientation is identified with the orientation of the grid. A detailed description of the rotated grid algorithm is given in [15]. 2D-calculations of grain selection development were performed. 3D-calculations are only done for the single crystal solidification. The grain growth of a single crystal in a simplified geometry of a turbine blade inner shroud was calculated, **Figure 12**. The results of the grain growth calculation with a grain orientation slightly rotated against the shrouds coordinate system (Euler angles $\vartheta = 0°$, $\psi = 0°$, $\varphi = 8°$) are shown in **Figure 13**. A transversal and two longitudinal (x- and y-plane) sections through the geometry at the same time step are displayed. The transversal section shows the slight rotation of the grain envelope which coincides with the crystallographic rotation. The longitudinal sections through the x- and y-plane look the same due to the crystal's symmetry. The same situation was calculated with the 3D-CA coupled with FEM temperature field simulation, **Figure 14**. Again a transversal and two longitudinal sections through this geometry at the same time step are shown. These pictures clarify the influence of the calculated temperature field on the grain envelope development. The anisotropic growth into the edges of the shroud is due to the faster cooling in the center of the cluster. The curvature of the surface in the longitudinal sections is caused by the macroscopic curvature of the isotherm in the temperature field simulation [11].

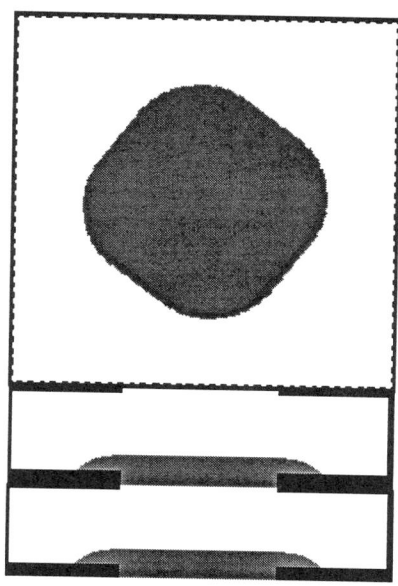

Figure 13: Transversal (upper picture) and longitudinal sections through the simplified geometry of an inner shroud. The gray areas represent the grain envelope of the single crystal whereas the white areas are still liquid. The snapshots were taken at the same calculation time step.

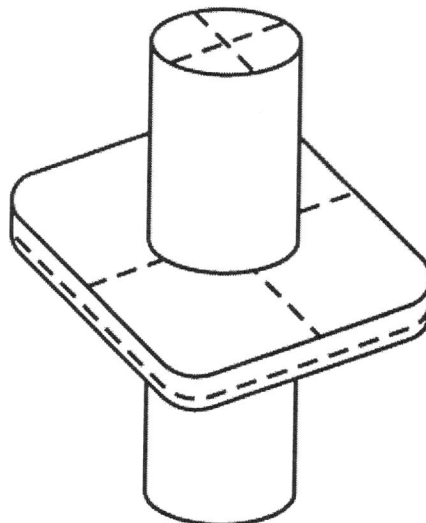

Figure 12: Principle sketch of the investigated simplified geometry of a turbine blade inner shroud. The lines indicate the sectional planes.

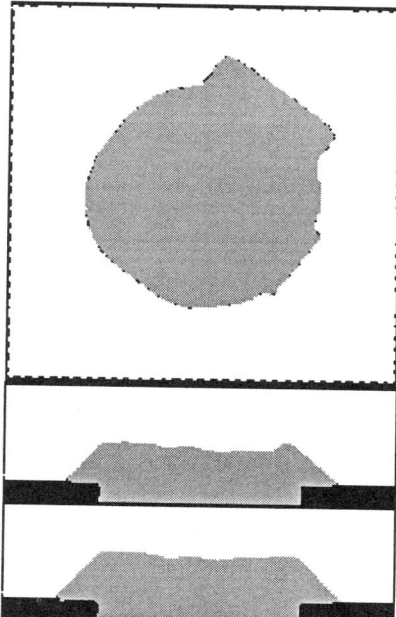

Figure 14: The same situation as in Figure 13 calculated with a 3D-CA coupled with FEM temperature field simulation. The anisotropic growth into the edges of the shroud is due to the faster cooling in the center of the cluster. The curvature of the surface in the longitudinal sections is caused by the macroscopic curvature of the isotherm in the temperature field simulation. The snapshots were taken at the same calculation time step.

To use this model for the prediction of freckling it will be necessary to include a fluid flow calculation and a nucleation model. This model has to consider undercooling as well as crystallographic and convection effects.

Conclusions

The main conclusions of the present work are as follows:

- A simple $G*v$-criterion for practical use has been developed which takes into account varying local solidification conditions near solidus temperature during Bridgman process.
- FEM temperature field simulation combined with the experimental $G*v$-criterion allows the prediction of component areas which are extremely freckle threatened and those areas, which are certainly freckle free.
- At solidification conditions typical of industrial production sized Bridgman furnaces freckles does not form inside the casting even at abrupt cross-section expansions.
- By now it is not possible to predict where freckle formation actually will take place. An improvement in freckle prediction by using a combination of FEM-simulation including fluid flow models with cellular automaton calculations is part of further investigation.

Acknowledgements

The authors would like to thank A. Schievenbusch, ACCESS e.V., and G. Nolze, BAM, for performing the EDX- respectively EBSD- Analysis. Partially the project subject matter of the present report has been promoted by funds of the German Federal Ministry for Education, Science, Research and Technology under the promotion reference 0327058C. Responsibility for the content of this publication lies with the authors.

References

1. A.F. Giamei, B.H. Kear, "On the Nature of Freckles in Nickel Base Superalloys", Met. Trans., 1 (1970) 2185-2192

2. S.M. Copley et al., "The Origin of Freckles in Unidirectionally Solidified Castings", Met. Trans., 1 (1970), 2193-2203

3. T.M. Pollock, W.H. Murphy, "The Breakdown of Single-Crystal Solidification in High Refractory Nickel-Base Alloys", Met. and Mat. Trans. A, 27A (1996) 1081-1094

4. J.R. Sarazin, A. Hellawell, "Channel flow in partly solidified alloy systems", Advances in Phase Transition, 10 (1987), 101-115

5. P. Auburtin, S.L. Cockroft, A. Mitchell, Freckle Formation in Large Superalloy Single Crystal Airfoil Castings", Materials for Advanced Power Engineering, eds. J. Lecomte- Beckers, F. Schubert and P.J. Ennis, (FZ Jülich, 1998), 1459-1468

6. G.L. Erickson, "The development and application of CMSX®-10", Superalloys, eds. R.D. Kissinger, D.J. Deye, D.L. Anton, A.D. Cetel, M.V. Nathal, T.M. Pollock, D.A. Woodford, (TMS, 1996), 35-44

7. P.R. Sahm, W. Richter and F. Hediger, "Das rechnerische Simulieren und Modellieren von Erstarrungsvorgängen bei Formguß", Gießerei-Forschung, 35 (2) (1983), 35-42

8. U. Reske et al., "Numerische Simulation der gerichteten Erstarrung nach dem Bridgman-Verfahren", Gießerei-Forschung, 43 (3) (1991), 101-106

9. N. Hofmann et al., "Numerische Simulation der gerichteten Erstarrung nach dem Bridgman-Verfahren II", Gießerei-Forschung, 44 (3) (1992), 113-120

10. F. Hediger, N. Hofmann, "Process Simulation for Directionally Solidified Turbine Blades of Complex Shape", Modeling of Casting, Welding and Advanced Solidification Processes, eds. M. Rappaz, M.R. Özgü, K.W. Mahin, (TMS, 1991), 611-619

11. J.L.L. Rezende, "Numerical Modelling of Microstructure Formation during Directional Solidification utilizing the Bridgman Process" (Dissertation, RWTH-Aachen, 1999)

12. M.C. Schneider et al., "Modeling of Micro- and Macro-segregation and Freckle Formation in Single-Crystal Nickel-Base Superalloy Directional Solidification", Met. and Mat. Trans A, 28A (1997),1517-1531,

13. C. Beckermann, private communication

14. S.D. Felicelli, D.R. Poirier and J.C. Heinrich, "Modeling Freckle Formation in Three Dimensions during Solidification of Multicomponent Alloys", Met. and Mat. Trans. B, 29B (1998), 847-855

15. N. Warnken et al., "Simulating Grain Structure Formation with a Rotated-Grid Cellular Automaton Algorithm", accepted for publication in, Modeling of Casting, Welding and Advanced Solidification Processes IX, eds. P.R. Sahm, P.N. Hansen, J.G. Conley, (2000)

Competitive Grain Growth and Texture Evolution during Directional Solidification of Superalloys

M.G.Ardakani, N.D'Souza, A.Wagner, B.A.Shollock and M.McLean

Department of Materials
Imperial College of Science, Technology and Medicine
Prince Consort Road, London, SW7 2BP, UK

Abstract

The development of crystallographic texture during directional solidification has been analysed quantitatively in columnar castings of the Ni-base superalloys, CMSX4 and CM186LC, produced with a range of cooling rates. The area density of grains having axial orientations within a specified scatter from <001> and the inverse pole figures (IPF's) were used to quantify the evolution of the axial texture. Increasing the solidification rate led to more effective competitive grain growth and a sharper <001> texture for a given alloy. However, there are significant differences in the solidification characteristics of alloys with minor compositional differences; sharper <001> textures were obtained in CM186LC than in CMSX4. The effects of solidification rate and composition on dendrite morphology have been investigated through quenched directional solidification experiments; the implications of these observations on the grain selection process and texture development are discussed.

1 Introduction

Casting of superalloy turbine blades in the directionally solidified and single crystal (SX) forms is now a well-established industrial process. However, recent developments in both alloy chemistry and in the dimension and complexity of castings have led to a concern about quality control; there is a trend to a more diffuse crystallographic texture during directional solidification, a wider range of single crystal orientations and more frequent occurrence of solidification defects such as spurious grains and freckles in Second and Third Generation alloys than in the earlier materials. The more important factors are:

• Recent developments in single crystal superalloy compositions have largely been directed to increasing the temperature capability of turbine blades and this has been achieved by progressively increasing the concentration of refractory solid solution strengthening additions, such as tungsten and rhenium. The improved high-temperature mechanical properties have been accompanied by an increase in the extent of microsegregation and significant changes in the solidification characteristics of the alloys.

• The dimensions of single crystal castings have progressively increased as the technology has been extended from small turbine blades for aero-engines to large castings for industrial gas turbines and to clusters of aero-engine parts. The larger mass of metal in these parts lead to quite different temperature distributions during casting. This leads to lower values of temperature gradient and solidification rate and to the possibility of there being a curved solid-liquid interface.

This paper addresses the role of the solidification rate and alloy chemistry on the evolution of texture; the role of interface curvature has been considered in earlier publications [1, 2]. The study has been carried out on two commercial nickel-base superalloys; CM186LC is a variant of the earlier single crystal alloy CMSX4 and is specifically designed for use in the directionally solidified form. It is also being considered for production of single crystal blades to ensure that significant creep strength is retained, even when spurious grains are formed. In the first part of this paper, the evolution of axial texture in CM186LC under different local solidification conditions is considered; the second part is concerned with the grain selection process as a function of the alloy composition for a fixed set of the solidification parameters.

2 Experimental

2.1 Directional Solidification Experiments

Directional solidification (DS) of pre-cast ingots of 12mm diameter encompassing a range of cooling rates was carried out using a Bridgman crystal growth apparatus with RF heating using graphite susceptors in an atmosphere of flowing argon. An important aspect of this investigation was the quantitative analysis of the competitive grain growth mechanism at curved liquidus isotherms which result from a radial component of the thermal gradient, in addition to the vertical component directed downward through the solid ingot to a water-cooled copper chill plate. The DS rig is shown schematically in Figure 1 and details of the solidification conditions are listed in Table 1. Directional solidification was achieved by translation of a liquid metal zone of ~ 30mm length along the length of the initial charge. Adjustment of the radiation baffles at the base of the hot zone, which also corresponds to the position of the solid–liquid interface, increased the radial component of heat flux from the solidified ingot.

Superalloys 2000
Edited by T.M. Pollock, R.D. Kissinger, R.R. Bowman,
K.A. Green, M. McLean, S. Olson, and J.J. Schirra
TMS (The Minerals, Metals & Materials Society), 2000

In industrial casting practice the entire charge is melted separately in a ladle in a vacuum or inert atmosphere and then poured into a shaped mould that incorporates a water-cooled chill. In the apparatus used for the present directional and single crystal solidification experiments, melting and solidification occur in a single chamber (Figure 1). The graphite susceptor, which lies within the central portion of the coils, directly couples with a copper induction coil to provide the heating source. However, because of the very high thermal gradients existing where the ingot contacts the chill, no melting occurred within 15-20mm of the chill. Consequently, the orientations of the pre-existing grains in the unmelted ingot were carried forward into the casting.

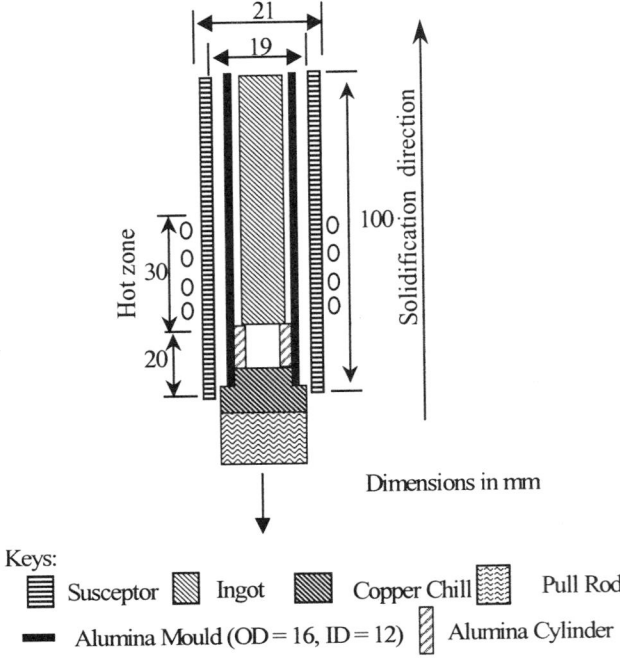

Figure 1. Schematic diagram showing configuration of solidification rigs used for directional solidification (Experiments 1-4, Table I).

In order to generate a truly random initial texture, similar to that occurring in industrial practice, the pre-cast ingot was supported on a hollow alumina cylinder of 12 mm diameter and 5 mm length to separate it from the chill. On melting the liquid metal flowed through the hollow gap and solidified on contact with the chill, thus nucleating a random axial crystallographic texture. The entire casting unit (mould with the charge and susceptor) was enclosed by a dense firebrick having a thermal conductivity of $0.1 Wm^{-1}K^{-1}$. The initial charges in experiments 1 to 4 (Table I) had square cross sections of $7 \times 7 mm^2$. The initial length of the charge in experiment 1 was 200mm, while it was 130mm for experiments 2, 3 and 4.

The thermal gradient, G_L at the solid-liquid interface and the solidification rate, V, are the most important parameters in determining the solidification structure. However, in practice, it is difficult to control these parameters independently in a DS apparatus. At steady state there is a balance between the heat supplied from the furnace, the latent heat generated on solidification and the heat lost to the chill. Considering only axial heat flow, this can be expressed:

$$k_S \, G_S = k_L \, G_L + \Delta H_F \, \rho \, V + \Delta E \, V \qquad (1)$$

where k is the thermal conductivity, the suffixes S, L represent solid and liquid respectively, G_L is the thermal gradient at the solidification front, G_S is the thermal gradient in the solid, ΔH_F is the latent heat per unit mass, ΔE is the external energy provided to unit volume of the melt, ρ is the density and V is the solidification rate. G_L and V are clearly inter-related quantities which are not readily calculated, particularly when radial contributions to heat loss must be considered.

In the present DS set-up there are two important contributers to changing solidification conditions:

a. In addition to the heat conducted through the solid to the water cooled copper chill plate, the unmelted solid lying above the molten zone can radiate heat reducing the superheat in the melt. However, as solidification progresses and the length of the unmelted ingot decreases, this contribution to the melt heat loss is eliminated. This is results in an increasing melt temperature (1580±20°C) in the final stage of DS, relative to that in the initial stages (1470±20°C). This results in higher values of the temperature gradient in the liquid G_L in the case of the shorter length ingots ($15-18 Kmm^{-1}$) than in the longer ingots ($11 Kmm^{-1}$).

b. As solidification progresses, the distance between the water-cooled copper chill and the liquidus front increases, thereby decreasing the axial temperature gradient in the solid G_S. In order to satisfy the energy conservation requirement, V and G_L must adjust to satisfy Equation 1.

The magnitude of the temperature gradient at the dendrite tips was estimated from measurements of the primary dendrite arm spacing (PDAS) λ_P and the solidification rate V by using the $G^{-0.25}V^{0.5}$ relationship that Quested and McLean [3] showed applied to a range of superalloys; this information is included in Table I. Solidification during single crystal production was interrupted by sudden quenching in a liquid metal bath in order to examine the development of the dendrite morphology in the mushy zone. Differential thermal analysis (DTA) was carried out on both alloys using a Stanton Redcroft STA-1780 Series simultaneous thermal analyser under an atmosphere of flowing argon.

Table I: Directional solidification experiments.

Exp. No.	Alloy	Nature	$V \times 10^5$ (ms^{-1})	Sol. Length (mm)	Primary dendrite spacing (μm)	G (°Kmm^{-1})
1	CM186LC	DS	2.8	55	236±7	5
2	CM186LC	DS	6.4	55	160±5	7
3	CMSX4	DS	2.8	35	197±27	7
4	CM186LC	DS	2.8	35	201±7	7
5	CMSX4	DS	1.9	40	185±5	10
6	CM186LC	DS	1.9	40	182±7	10

Table II: Alloy Composition (wt. %).

Alloy	Cr	Co	Mo	W	Ta	Re	Al	Ti	Hf	Ni	C	T_L°C	T_S°C
CMSX4	6.5	9	0.6	6	6.5	3	5.6	1.0	0.1	Bal	0	1387	1347
CM186LC	6.0	9	0.5	8	3	3	5.7	0.7	1.4	Bal	0.07	1380	1320

2.2 Determination of Crystallographic Orientations using Electron Back Scattered Diffraction (EBSD) Patterns

Specimens were prepared for metallographic and grain orientation characterisation by mechanical and electropolishing to remove any surface deformed layers produced during preliminary cutting, grinding and polishing. These samples were examined by means of the Electron Back Scattered Diffraction (EBSD) technique in a JEOL 840 scanning electron microscope, equipped with SINTEF hardware and CHANNEL+ software for automatic grain indexing and mapping. The maximum angular resolution was within ±1.5° and the spatial resolution was 1 μm parallel to the specimen tilt axis. All of the grain orientations were specified with respect to a co-ordinate system based on the ingot geometry; the Z axis was parallel to the macroscopic growth direction; the X, Y axes were parallel to arbitrary, orthogonal transverse directions that were identified on the grain maps that were constructed to show the spatial location of grains on a given section. The compositions of the Ni-based superalloys, CMSX4 and CM186LC, used in this study are listed in Table II.

3 Results

3.1 Evolution of the Axial Texture

3.1.1 Role of the Local Solidification Parameters

Figure 2 shows grain maps and inverse pole figures (IPF's) of surface normals of the grains on transverse sections taken at various distances along the length of a small commercial DS turbine blade of CM186LC. There is clearly both a progressive increase in grain size and the development of a sharp <001> texture. Examination of longitudinal sections shows the establishment of a columnar grain morphology. The IPF for a solidification length of 85mm shows that all of the grains have axial orientations within 5° of <001>. Even after 2mm of solidification the grain selection process has produced a diffuse texture with 90% of the axial orientations being clustered within 20° of <001>. The thermal history (solidification conditions) of this blade is not known. However, a progressive coarsening of the primary dendrite arm spacing along the length of the blade is indicative of a diminishing cooling rate. Primary dendrite arm spacing (λ_P) corresponding to solidification lengths of 2mm and 50mm were 90μm and 383μm respectively. These measurements lack statistical significance for the thin cross section at a height of 85mm because of an insufficient number of dendrites.

The purpose of the laboratory experiments described below is to characterise the grain morphology and texture development in different solidification conditions where the solidification parameters can be estimated. Figures 3a and d show grain maps from transverse sections after 55mm of solidification in Experiments 1 and 2. There is a significant difference in the cross-sectional areas of the columnar grains and this is accompanied by a sharper <001> texture being established in Experiment 2 than in Experiment 1 (Figures 3b and e). Figures 3c and f indicate the parts of the cross-section that have surface normals within 10° of <001> (95 and 38% for Experiments 1 and 2 respectively). The primary dendrite spacings, λ_p have been measured in both cases and are included in Table 1; the temperature gradients estimated from a knowledge of λ_p and V are also included in Table 1.

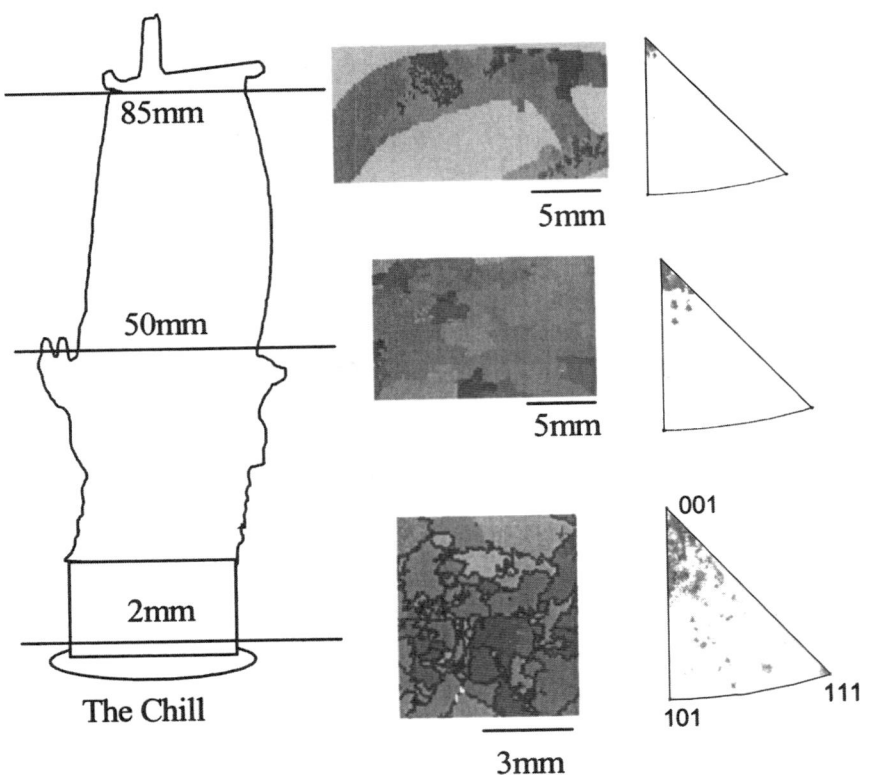

Figure 2. The transverse grain maps and inverse pole figures at different distances along a small directionally solidified turbine blade of CM186LC.

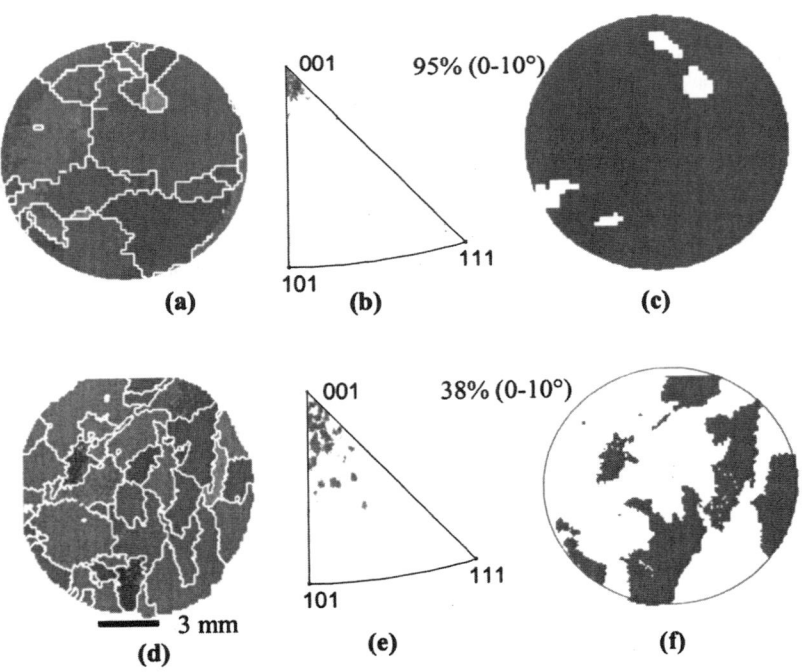

Figure 3. (a and d) Grain maps of transverse sections of CM186LC after directionally solidifying a length of 55 mm correspond to, Experiments 1 and 2 (Table I) respectively. The corresponding inverse pole figures (b) and (e) showing the axial orientation of the grains. (c) and (f) distribution of grains in (a) and (d) with axial texture within 10° from <001>.

3.1.2 Role of the Alloy Chemistry

The differences in composition of CM186LC and CMSX4 are quite minor (Table 2) and the physical constants that control heat flow and melting are also very similar. Consequently, it is expected that for the same ingot size and process conditions the principal solidification parameters will be identical. This expectation is supported by the observation of the same heating period before commencement of solidification for the two alloys; 18 minutes for Experiments 3 and 4. Consequently, any differences in grain or texture development is likely to be a consequence of the alloy constitution, rather than the solidification parameters, G_L and V.

The authors have previously shown that under identical solidification conditions involving high temperature gradients, CM186LC develops a sharp <001> texture whereas CMSX4 develops a diffuse texture (Figure 4 and Experiments 5 and 6 in Table 1). It is also noteworthy that for both alloys, the transverse grain diameters were 4 to 10mm, compared with 0.5 to 2 mm in Experiments 3 and 4; this indicates that grain selection was much more effective in Experiments 5 and 6. These earlier experiments were carried out without ensuring a totally random starting texture and consequently are not entirely comparable with Figure 4. Experiments 3 and 4, using the experimental procedures described above, were designed to investigate the grain selection process in solidification conditions that did not lead to a rapid establishment of a sharp <001> texture in CM186LC.

Figure 5 shows grain maps, IPFs and the area of the cross-section with axial orientations lying within 7° of <001> for CMSX4 and CM186LC after a solidification length of 35mm in Experiments 3 and 4. Unlike Figures 3b and e, the IPF's do not show a clear difference in the axial textures of the two alloys. A more detailed comparison of the distribution of axial orientations for the two alloys was undertaken. The grain maps in Figures 5c, f are sub-sets of Figures 5a, d respectively, which depict regions where the crystallites have axial orientations within a 7° scatter from <001>. The area densities of grains are plotted as cumulative frequency as a function of angular deviation from <001> in Figure 6a. The following deductions can be made:

1. A histogram plot of the area densities of grains as a function of axial misorientation shows a maximum for deviations of 10 to 15° from <001> (Figure 6b).

2. The plot of cumulative distribution of grain area as a function of misorientation shows that, even for the apparently diffuse textures obtained in Experiments 3 and 4, the proportion of grains falling within 10° of <001> is significantly greater for CM186LC than for CMSX4. For CM186LC 10% and 90% of the cross-sectional area is within 5° and 20° of <001> respectively. For CMSX4, only 5% of the area has misorientations of less than 5° and 30% of the area deviates from <001> by more than 20°.

There is a clear implication from these observations that grain selection during directional solidification under identical experimental conditions, is more pronounced in CM186LC than in CMSX4.

The DTA results (Figure 7) show that the CMSX4 and CM186LC have similar thermal characteristics.

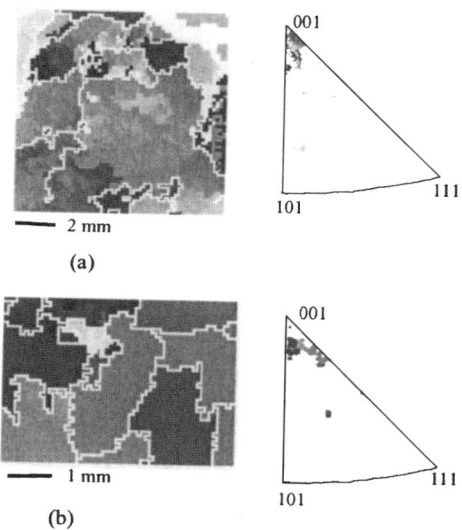

Figure 4. Grain maps and IPF's showing and axial orientation of grains on a transverse section after a solidification length of 40mm. (a) CM186LC (Experiment 6), (b) CMSX4 (Experiment 5).

CM186LC has a marginally lower liquidus temperature (by 7°C), a significantly lower solidus temperature (by 27°C) and a wider temperature range (by 20°C) than CMSX4. The most significant difference in the shape of the DTA curves is the kink at 1350°C that is thought to be associated with the formation of carbides. Other thermal characteristics, such specific heat and thermal conductivity, are almost identical for the two alloys. Similar forms and weight of charge were used in the directional solidification experiments to ensure that the melting sequences were identical. For both materials melting was initiated after a period of 18 minutes (Experiments 3 and 4, Table I). Thus, any difference in the evolution of the axial texture can be directly attributed to the alloy solidification characteristics rather than to the process history.

Figure 8(c) shows a longitudinal section of CMSX4 in which single crystal growth in the <001> direction has been interrupted by quenching in a bath of liquid metal. The ingot has been carefully aligned using the Laue Back reflection technique and then sectioned to ensure that the plane observed contains both the <001> growth direction and one of the <010> transverse orientations. The field shown in Figure 8(c) shows the primary dendrite propagating ahead of the solid/liquid interface and secondary dendrites developing further back in the semi-solid region.

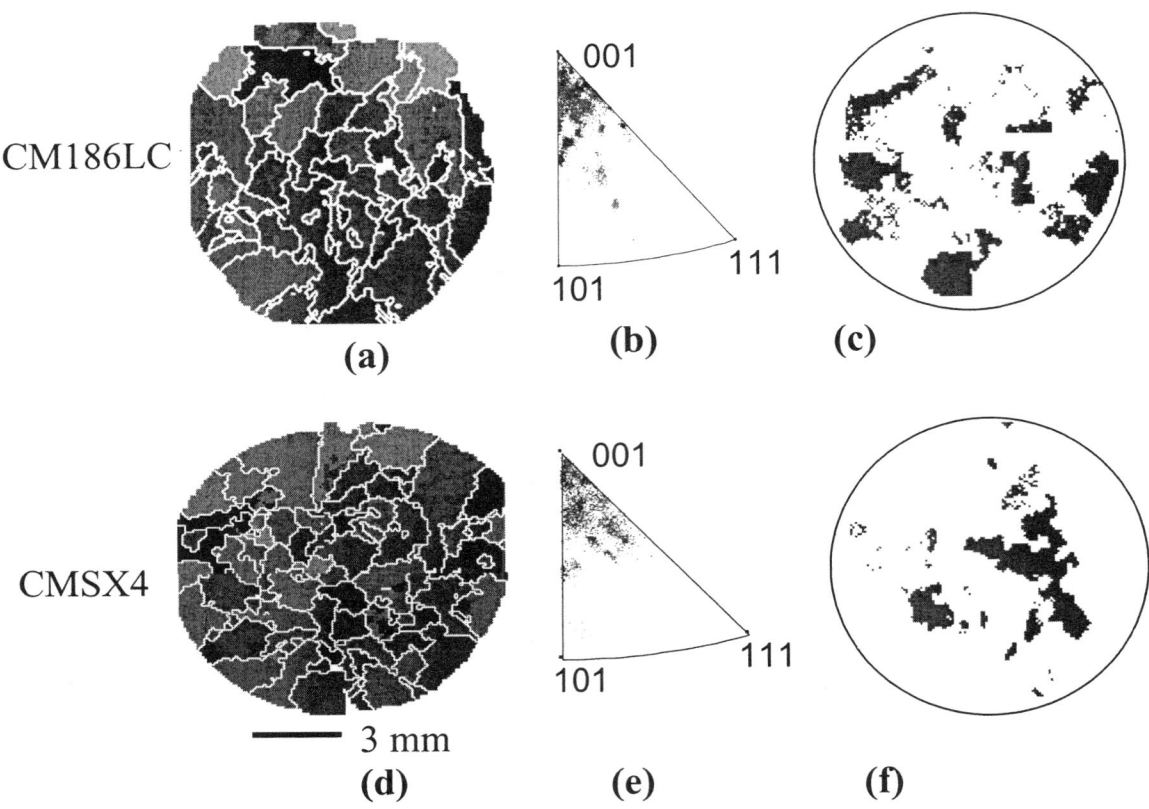

Figure 5. (a-c) grain map, IPF and distribution of grains showing spatial orientation of grains with respective axial orientations within 7° from <001>, for CM186LC, after solidification length of 35mm, experiment 4 (Table I). (d-f) similar to (a-d) for CMSX4, experiment 3, (Table I)

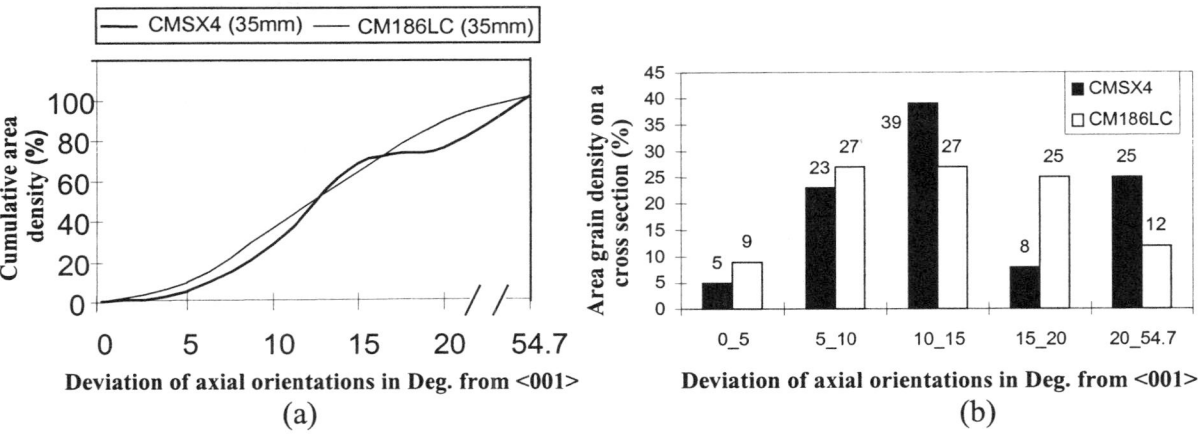

Figure 6. (a) Variation of the cumulative area density of grains with increasing angular deviation of the axial orientations from <001>, (b) Histogram plot showing the area density of grains in intervals of the angular deviation.

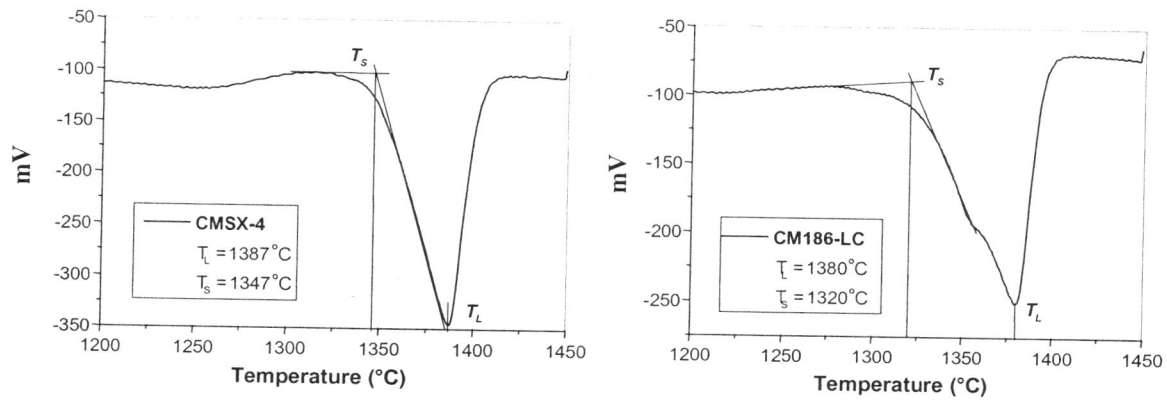

Figure 7. DTA curves for CM186LC and CMSX4 showing liquidus, solidus and carbide formation temperature.

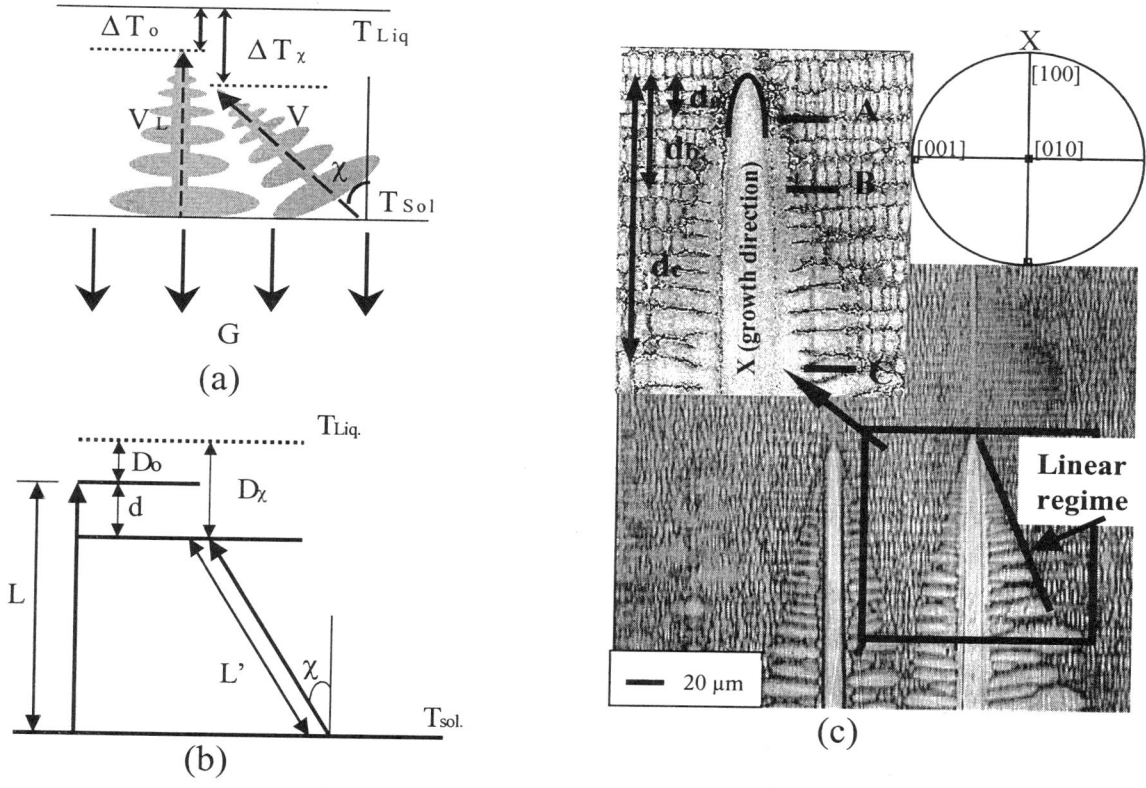

Figure 8. (a) Schematic diagram showing the tip undercoolings for favourably and unfavourably oriented dendrites with respect to the temperature gradient (G), (b) Estimation of the angle χ from geometrical and morphological constraints, (c) Optical and back scattered electron micrograph and (001) pole figure (inserted) showing the primary <001> dendrite and the two regimes in the secondary arm morphology.

4 Discussion

4.1 Nature of the Axial Texture

It has been reported extensively that dendritic growth in cubic systems is parallel to <001>, although significant deviations have also been reported in some cubic metals (e.g. Cr) [6]. Nevertheless, the classical experimental observations of Walton *et al* .[7], involving the directional growth of Al and Pb alloys, show a clustering of the axial orientations close to <001>. They concluded that the evolution of a preferred crystallographic orientation, leading to an axial grain texture, was a consequence of the dendritic growth mechanism, and in particular was associated with the dendrite morphology, particularly that close to the dendrite tip. Specifically, it has been suggested that fast growing near <001> grains block the development of slower growing mis-oriented grains by the growth of <010> secondary dendrite from which <001>

225

tertiary dendrites nucleate and grow in the <001> macroscopic growth direction.

4.2 Evolution of the Axial Texture
4.2.1 Role of the Local Solidification Parameters

A theoretical treatment of the growth of an array of dendrites or cells in a positive thermal gradient has been given by Burden *et al* [8]. Neglecting interfacial attachment kinetics they express the net undercooling at the dendrite tips as:

$$\Delta T = \Delta T_{Sol} + \Delta T_{Curv} \qquad (2)$$

where the first and second terms represent the solutal and curvature undercoolings respectively. The primary compositional contribution to the solutal undercooling is related to the diffusivity of the solute species in the melt ahead of the dendrite tip. Both alloys under consideration have similar concentrations of the highly segregating refractory solutes, rhenium and tungsten; they differ primarily in the levels of carbon and hafnium. It is likely that the solutal undercoolings will be dominated by the former, suggesting that there should be little difference between the alloys in this respect. The curvature undercooling, which has its origins in the Gibbs-Thomson parameter, depends on both the magnitude and anisotropy of the solid-liquid interfacial energy. Surface energies can be extremely sensitive to minor variations in alloy composition and these changes are not easy to predict. It is likely that this will be the cause of the significant differences in melting characteristics of the two, very similar alloys.

At steady-state, an unfavourably oriented dendrite with <001> deviating from the direction of the local thermal gradient (G) by an angle χ can advance with the same vertical component of velocity (V_L) as the liquidus isotherm, if it satisfies the condition:

$$V_L = V \cos\chi \qquad (3)$$

where V is the velocity of the primary dendrite parallel to its misoriented growth direction. Equation 3 can only be valid when $V > V_L$. In the presence of a positive thermal gradient the unfavourably oriented primary dendrite lags the more favourably oriented one, and therefore grows with greater undercooling. It is the solute and thermal field of the leading primary that retards the growth of its lagging neighbour [7]. In all isothermal and non-isothermal dendrite growth theories, the tip supercooling (ΔT) and solidification rate (V) can be expressed as [9]:

$$V = A (\Delta T)^n \qquad (4)$$

where A is a lumped material parameter that varies inversely with the solid-liquid interfacial energy and the exponent, *n* lies in the range: $2 \leq n \leq 3$.

For a given alloy, neglecting the temperature dependence of *A*, the difference in the undercooling between a primary dendrite tip aligned parallel to the local thermal gradient and another inclined at an angle χ, as a function of the solidification rate, can be written as

$$\Delta T_\chi - \Delta T_0 = \sqrt[n]{\frac{V_L}{A}} \left(\frac{1}{(\cos\chi)^{1/n}} - 1 \right) \qquad (5)$$

where ΔT_χ and ΔT_0 are the dendrite tip undercooling for the unfavourable and favourably oriented dendrite tips. These terms are shown schematically in Figures 8a in relation to the observed dendrite morphology. If G_L is known this can be translated into a distance between the tips of well-oriented and mis-oriented dendrites:

$$D_\chi - D_0 = \frac{1}{G_L} \cdot \sqrt[n]{\frac{V_L}{A}} \left(\frac{1}{(\cos\chi)^{1/n}} - 1 \right) \qquad (6)$$

It follows that for a given alloy (A constant), decreasing the growth rate (V) leads to a smaller difference in the undercoolings at the tips of well- and mis-oriented dendrites. This in turn will make overgrowth of the misoriented dendrite by a secondary from the well-oriented dendrite less likely and texture development more difficult.

Increasing the temperature gradient G_L influences both the dimensions of the dendrites through the $G^{-0.25} V^{0.5}$ dependence of λ_P and the morphology of the dendrite through the G/V dependence of the dendrite/cellular/plane-front solidification transitions. The present results suggest that, providing that solidification is well within the dendritic growth regime, the primary dendrite spacing provides a good indicator of the effectiveness of grain selection; the finer dendrites lead to sharper solidification textures for a given alloy. However, this does not explain the differences between alloys.

In the solidification experiments of Huang *et al* [11] involving the transparent organic analogue succinonitrile, the formation of the initial secondary arms at the primary dendrite tips and their spacing (prior to coarsening in the mushy zone when the diffusion fields overlap) is dependent on the dendrite tip undercooling. Therefore, instead of measuring the dendrite tip undercooling, the morphological aspects of the dendrite can be considered. This also assumes significance, since in the classical competitive grain growth process proposed by Walton *et al* [7], the growth of the tip of the unfavourably oriented dendrite, which lags the tip of the favourably oriented one, is retarded by the thermal and solutal environment of the latter. However, it is the secondary arms at the tip of the leading dendrite that eventually impair the further advancement of the lagging one beyond a certain length. Therefore, the morphology of these secondaries is crucial to the efficacy of the process of overgrowth. Therefore, from the perspective of competitive growth, the critical parameters would be:

a)   The distance along the dendrite stem measured from the unbranched tip that the secondary arms form. Huang *et al* [9] observed an inverse relationship.

b) The initial arm spacing (S_{Tip}), before overlapping of diffusion fields leads to coarsening in the mushy zone.

c) The dendrite tip radius (R), which is a measure of the supercooling.

Since the dendrite blocking mechanism depends on effects in the vicinity of the dendrite tips, it is of interest to use the observed dendrite morphologies in Figure 8c to estimate the threshold value of χ_{th} at which mis-oriented dendrites are overgrown. Li and Beckermann [10] have shown from a detailed measurement in succinonitrile that the side-branch evolution is divided into two regimes: an initial linear regime and a subsequent non-linear coarsening regime. Our preliminary measurements support that interpretation; Figure 8c shows that the projection of the newly formed secondary arms (line A) to line C does in fact lie on a straight line. However, beyond line B, some non-linear coarsening is evident. In the following analysis we restrict consideration to this linear regime.

Overgrowth of a misoriented dendrite will occur if the distance between its and the well aligned dendrite tips (D_χ-D_0) exceed the distance d along the dendrite at which significant secondary dendrites develop. Consequently a threshold value of χ can be identified below which no overgrowth will occur. Referring to the schematic diagram in Figure 8b:

$$\cos \chi_{th} = \frac{L-d}{L'} \cong 1 - \frac{d}{L} \qquad (7)$$

where L and L$'$ are the lengths of the two dendrites in the mushy zone (between T_{Liq} and T_{Sol}) and d is measured from the dendrite tip. For $d \ll L$, $L \approx L'$ and equals the mushy zone length There is some arbitrariness in determining d, but taking distances along the dendrite to positions A. B and C shown in Figure 6c leads to estimates of χ_{th}:

$$d_a = 35\mu m, \chi_A = 5°$$
$$d_b = 55\mu m, \chi_B = 6°$$
$$d_c = 110\mu m, \chi_C = 9°$$

There is now a solid body of evidence pointing to the solute and thermal fields of the secondaries of the leading dendrite retarding the growth of a lagging neighbour. The observed morphology in Figure 8c of the quenched dendrite in CMSX4, solidified with conditions (V=5x10^{-5} ms^{-1}, G = 4Kmm^{-1}, λ = 225μm, L $\approx$ 10mm) is consistent with a final axial texture having a scatter of up to 10° from <001>, which is reasonably consistent with casting experience.

The histogram shown in Figure 6b indicates that there is an increasing population of grains with increasing misorientation until χ is about 10^0. This is quite consistent with the above analysis. If all grains with χ<10^0 can grow

without being overgrown and if they originate from random distribution of grains then there are more possible orientations deviating from than with a precise <001> direction. For this regime of no selection the probability increases as (1-cosχ) which is approximately linear for χ<10^0. Grains with χ >10^0 are predicted to be overgrown leading to a reduction in frequency.

A detailed comparison of dendrite morphologies in a range of directionally solidified and single crystal alloys will be presented in a future publication.

4.2.2 Role of the Alloy Chemistry

The results presented above show that CM186LC and CMSX4 have a significant difference in axial texture resulting from identical processing conditions. The similarity in physical constants will lead to the same local solidification conditions (G_L, V). It follows that the observed difference in the axial textures can only be attributed to differences in thermodynamics and kinetics of dendrite growth, that is represented by the parameter, A. In the simplified model of Burden and Hunt. [8], the lumped material parameter, A varies inversely as $\frac{\gamma_{SL}}{D}$, where D is the solute diffusivity in the melt at the solidification front. For a multi-component system it is reasonable to take the diffusion coefficient of the slowest diffusing species, which in this case would be Ta, W and Re. The role of surface energy is more difficult to assess since both the absolute value and the anisotropy are very sensitive to minor changes in chemistry. CM186LC and CMSX4 have similar concentrations of refractory solutes, which segregate to the dendrites, suggesting that the solutal terms will be similar.

There has been a common view based on experimental evidence that the newer generation of SX superalloys (CMSX4, CMSX10) are less tractable to orientation control than some new alloys that have been specifically developed for use in the DS form (CM186LC). The major difference in the alloy composition, is the presence of Hf and C, both being added as grain boundary strengthening agents (Table II). However, C forms carbides and the liquid at the solidification front in CM186LC will be enriched in C only if the carbides are formed below the liquidus front. This is indeed the case from the DTA plot in Figure 4. The carbides form between 1350 and 1360°C, which for an average thermal gradient of 10Kmm^{-1}, is 2-3mm behind the dendrite tip. Therefore, the melt into which the dendrite tips are advancing is enriched in C. Experimental determination of γ_{SL} has usually been undertaken on transparent systems [11, 12] and such measurements at the high liquidus temperatures and in opaque systems are formidable and fraught with inaccuracy.

227

Conclusions

1. The effectiveness of grain selection during directional solidification of a superalloy is sensitive to the processing conditions. High temperature gradients and solidification rates lead to a sharper <001> texture and larger transverse grain diameters than low cooling rates. There is an inverse correlation between primary dendrite size and grain selection efficiency.

2. CM186LC develops a sharper <001> axial texture than does CMSX4 when processed under identical conditions. Even in conditions when both alloys produce diffuse textures, detailed analysis of area densities of grains with specific orientations shows that grain selection is more effective in CM186LC.

3. Analysis of the evolving morphology of the dendrite structure at a quenched solid-liquid interface indicates that misorientation of up to 10^0 will not be overgrown by secondary dendrites. This is consistent with the observation of a peak in orientations at about 10^0 misorientation.

4. DTA analysis shows that the carbides form 2-3mm behind the dendrite tips in CM186LC showing that the carbides have no significant role in the grain selection mechanism. There is a possible role of C being rejected into the melt at the dendrite tips modifying γ_{SL}.

Acknowledgements

The authors thank ALSTOM ENERGY (formerly European Gas Turbines) for provision of a studentship (ND'S) and EPSRC for support (Grant No.GR/L05433).

List of References

1. N. D'Souza, M. G. Ardakani, B. A. Shollock, M. McLean, *International Synposium on Liquid Metal Processing and Casting*, eds. A. Mitchell, L. Ridgway, M. Baldwin, American Vacuum Society, Santa Fe, 1999, p.1.

2. N. D'Souza, M. G. Ardakani, B. A. Shollock, M. McLean, *Met. Trans. A*, in press.

3. P.N. Quested and M. McLean: *Mat. Sci. Eng.* 1984, vol. 65, p. 171.

4. D. A. Porter, K. E. Easterling, *Phase Transformations in Metals and Alloys*, 2nd edition, 1992, Chapman and Hall, p. 204.

5. F. Giamei, J. G. Tschinkel, *Met. Trans,* Vol 7A, 1976, p. 1427.

6. D. N. Lee, K-H Kim, Y-G Lee, C-H Choi, *Materials Chemistry and Physics*, 1997, vol. 47, p. 154.

7. D. Walton and B. Chalmers: *Trans. Met. Soc. of AIME.* 1959, vol. 215, p. 447.

8. M. H. Burden and J.D. Hunt: *J. Crystal Growth.* 1974, vol. 22, p. 109.

9. M. E. Glicksman, R. J. Schaefer, J. D. Ayers, *Met Trans. A*, vol 7A, p. 1747.

10. Q. Li, C. Beckermann, *Acta Mater.*, 1999, vol. 47, p. 2345.

11. S-C. Huang and M.E. Glicksman: *Acta Met.* 1981, vol. 29, p. 717.

12. D. R. H., Jones, G. A. Chadwick, *Phil Mag.*, 1970, vol. , p. 291.

RECRYSTALLIZATION IN SINGLE CRYSTALS OF NICKEL BASE SUPERALLOYS

R. Bürgel *, P. D. Portella **, J. Preuhs ***

* Osnabrück University of Applied Sciences,
Albrechtstrasse 30, D–49076 Osnabrück, Germany
** Federal Institute for Materials Research and Testing (BAM),
Unter den Eichen 87, D–12205 Berlin, Germany
*** DONCASTERS Precision Castings–Bochum GmbH (DPC),
Bessemerstrasse 80, D–44793 Bochum, Germany
e-mail addresses: r.buergel@fh-osnabrueck.de; pedro.portella@bam.de; jpreuhs@doncasters.com

Abstract

The recrystallization behavior of some monocrystalline superalloys was investigated both in the temperature regime of solutioning as well as at lower temperatures in the γ+γ' field. At very high temperatures new grains already form after a small plastic deformation of about 1 %. They only nucleate at the surface in single-phase γ areas if the degree of cold work is not too high. Grain boundary motion is stopped by interdendritic γ/γ' areas but is barely retarded by carbides and residual eutectic islands. Even very extended recovery annealing procedures are not able to prevent recrystallization during subsequent full solutioning. For the process of cellular recrystallization in the γ+γ' field higher degrees of cold work are necessary. In this case a γ' free zone must be absent to initiate new grains. Along the moving grain boundary the γ' phase is taken into solution and reprecipitated immediately behind it. A thin recrystallized surface "jacket" leads to a higher crack density under LCF conditions but for the chosen parameters in this work the number of cycles to crack initiation is not affected.

Introduction

Recrystallization poses one of the major difficulties in post casting processing of directionally solidified, especially single crystal blades of nickel base superalloys. There are several possible sources for the necessary plastic deformation during manufacturing and processing of the new parts as well as during service and reconditioning: contraction stresses during cooling of the solid in the shell mold, removing the ceramic mold and core material mechanically, stamping identification marks, grinding the airfoil and the fir tree root to net shape, impact damage, removing of coating residues mechanically, etc. The deformation is either concentrated at the component surface or can extend into the bulk material. **Figure 1** shows an example of unacceptable recrystallization in the root-airfoil transition area after solution heat treatment of a new part.

The formation of new grains can take place during solutioning heat treatment even after relatively small degrees of work. Higher amounts of plastic deformation are needed to initiate recrystallization at lower temperatures, e.g. during age hardening or service exposure.

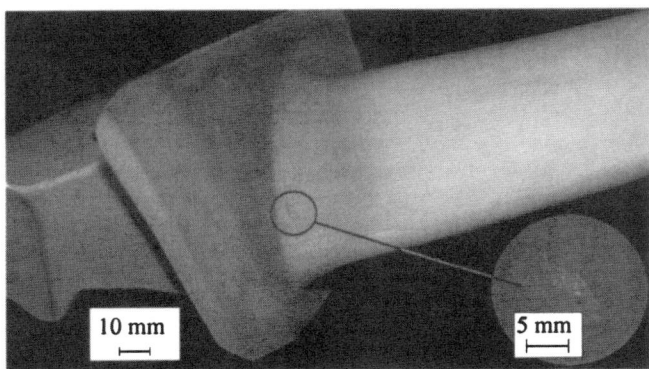

Figure 1: Recrystallization phenomena on a turbine airfoil after solution heat treatment.

Superalloys 2000
Edited by T.M. Pollock, R.D. Kissinger, R.R. Bowman,
K.A. Green, M. McLean, S. Olson, and J.J. Schirra
TMS (The Minerals, Metals & Materials Society), 2000

Recrystallization may dramatically reduce the fatigue life and also the stress rupture strength of directionally solidified components. Therefore, specifications typically limit recrystallized grains to a diameter of about 1 mm in the airfoil, with additional restrictions concerning maximum number of new grains and the distance between recrystallized areas.

Little information is available in the open literature on recrystallization of monocrystalline and columnar-grained components and possible methods of preventing it. Mostly surface deformation was stimulated, e.g. by hardness indentations or shot peening, and consequently recrystallization was initiated at the surface only [1-4]. The main purpose of the present investigation was to find out the mechanisms responsible for the formation and growth of new grains and whether recrystallization could be reduced or even suppressed. The work comprises two distinct parts: the first deals with recrystallization phenomena during solution heat treatment, i.e. in the γ region, and the second with recrystallization during annealing or service exposure in the γ+γ' field.

The alloys investigated were developed as blading material both for industrial gas turbines (CMSX–11B, PWA 1483) as well as for aircraft propulsion (SRR 99, CMSX–6).

Experimental

Alloys investigated
Table I shows the nominal chemical compositions of the alloys investigated. Recrystallization studies in the γ region were focused on CMSX–11B in comparison to PWA 1483 and SRR 99. For the behavior in the γ+γ' field CMSX–6 was chosen.

Table I Nominal chemical compositions in w%
of the alloys investigated (balance: Ni)

	Cr	Co	Mo	W	Ta	Nb	Al	Ti	C	Hf
CMSX–11B	12.5	7	.5	5	5	.1	3.6	4.2	–	.04
PWA 1483	12.2	9	1.9	3.8	5	–	3.6	4.1	.07	–
SRR 99	8.5	5	–	9.5	2.8	–	5.5	2.2	.02	–
CMSX–6	10	5	3	–	2	–	4.8	4.7	–	.08

Cylindrical bars were cast in Bridgman furnaces with their longitudinal axes in the [001] direction with not more than 10° deviation.

The heat treatments of the alloys are given in **Table II**.

Heat treatments
The heat treatments for investigation of the solutioning and recrystallization behavior in the γ field were performed under air atmosphere in a laboratory furnace which was accurately controlled by two calibrated thermocouples of type B and S. For comparison purposes some tests were run in a vacuum production furnace. In any case when a certain annealing treatment is mentioned in the text below, the steps of the solution heat treatment cycle according to the values given in Table II were run additionally up to the respective annealing temperature. The experiments in the γ+γ' field were performed under an argon/hydrogen atmosphere.

Table II Heat treatments of the alloys investigated

CMSX–11B	solutioning, SHT*	1204°C/2h + 1227°C/2h + 1249°C/3h + 1260°C/6h; heating with 1 K/min
	age hdn.	1120°C/5h +870°C/24h +760°C/30h
PWA 1483	solutioning*	1260°C/1h
	age hdn.	1090°C/4h
SRR 99	solutioning*	1270°C/0,5h + 1280°C/1h + 1290°C/2h + 1300°C/0,5h + 1305°C/0,5h; heating with 1K/min
	age hdn.	1080°C/4h + 870°C/16h
CMSX–6	solutioning*	1227°C/2h + 1238°C/2h + 1271°C/2h + 1277°C/3h + 1280°C/2h; heating with 1K/min
	age hdn.	1080°C/4h + 870°C/16h

* SHT: Solution heat treatment. Additional steps at lower temperatures may precede the solutioning to equilibrate the furnace.

Samples for γ' solutioning behavior
To quantify the amount of γ' precipitates as a function of temperature for CMSX–11B as-cast specimens were heated to temperature, held for 2 h, and then quenched in water to freeze the microstructure which had developed at the annealing temperature. This procedure was repeated in temperature intervals of 10 K up to the highest step of the solutioning heat treatment (see Table II).

The remaining volume fraction of the γ' phase, excluding eutectic γ', was determined through the area fraction in SEM micrographs of longitudinal sections.

Samples for recrystallization in the γ field
Samples of 12 mm diameter and 18 mm height were sectioned deformation-free and exactly parallel-sided by wire-guided electro discharge machining (EDM). Deformation of test pieces in the as-cast condition was applied by axial compression at room temperature, the degrees of cold work being given as relative height reduction.

Some samples of CMSX–11B were coated after cold work by a standard CVD aluminizing process used for airfoil coating in the aero turbine industry. The maximum temperature of the coating cycle was 1080 °C and a layer of about 60 µm thickness was produced.

For metallographic analysis longitudinal sections were prepared, except where otherwise stated. The specimens were cut along the long axis of the elliptical cross section which had formed and was slightly visible after cold work. All micrographs of longitudinal sections shown below are taken with the [001] direction vertical.

Samples for recrystallization in the γ+γ' field
The recrystallization kinetics in the γ+γ' field were determined using polished [001] oriented cross sections of CMSX–6. Indentations were produced with a prismatic punch of tungsten carbide [1], typically of 8 mm length and 0.1 mm depth. After annealing at 1080 °C (i.e. the first age hardening temperature), the recrystallized area was measured in several sections perpendicular to the indentation edge.

The influence of surface recrystallization on the mechanical behavior was investigated using solution-annealed cylindrical specimens which were machined to the nominal dimensions (gage diameter: 9 mm) before age hardening. The machining parameters were determined in order to generate a rather uniformly recrystallized surface layer of about 0.1 mm depth over the gage length after the final age hardening heat treatment.

Metallographic techniques

Polished samples were etched with one of two variants of the so-called Mo-reagent. The first consists of 100 ml H_2O, 100 ml HNO_3, 100 ml HCl, 6 g Mo-acid and is applied for about 5 s. The second one is a mixture of 60 ml of a concentrate (300 ml HCl, 300 ml H_2O, and 5 g Mo-acid) and 50 ml H_2O, 30 ml HNO_3, and 5 drops of a tensidic agent. This reagent should be prepared shortly before etching and be applied for about 20 s.

Electron microscopy

The conclusive identification of recrystallized grains as well as the characterization of the orientation distribution of grains in recrystallized areas was performed in a SEM by electron back-scattered diffraction (EBSD). Since in most cases the areas of interest lie near the specimen surface, the usual electrolytic polishing could not be applied. Instead, careful mechanical polishing alternating with slight chemical etching produces the required surface quality.

Transmission electron microscopy was employed for characterizing with very high spatial resolution the differences in dislocation structure, orientation, and chemical composition between different regions in the specimens. The near surface position of these regions required electrolytic deposition of a 3 mm thick nickel layer on the specimen surface, careful mechanical preparation of the disc and final twin jet thinning using a solution of 50 ml perchloric acid in 950 ml ethanol and 100 ml buthanol at -40 °C [5, 6].

Mechanical testing

LCF tests were carried out at 980 °C in air under total strain control with $R_\varepsilon = -1$ and ramps with $\dot{\varepsilon} = 10^{-3}$ s^{-1}. The tests were conducted either without hold time or with hold times of 300 s at ε_{min}. The testing systems have a capacity of 100 kN and are equipped with low bending moment grips and three-zone resistance furnaces. Creep tests were conducted in the same apparatus in air under constant load.

Results

γ' Solutioning behavior of CMSX–11B

Recrystallization in the γ field depends on the solutioning behavior of the γ' phase both as secondary precipitates as well as in the γ/γ' eutectic. This was quantified for CMSX–11B in the as-cast condition (**Figure 2**) which contains about 6 vol% eutectic. Solutioning is completed within the dendrite cores at about 1205 °C after 2 h dwell time whereas the interdendritic regions become single phase above about 1250 °C. Residual γ/γ' eutectic only dissolves at the highest solutioning step of 1260 °C after 6 h. After SHT the amount of retained eutectic is about 1 % to 2 %.

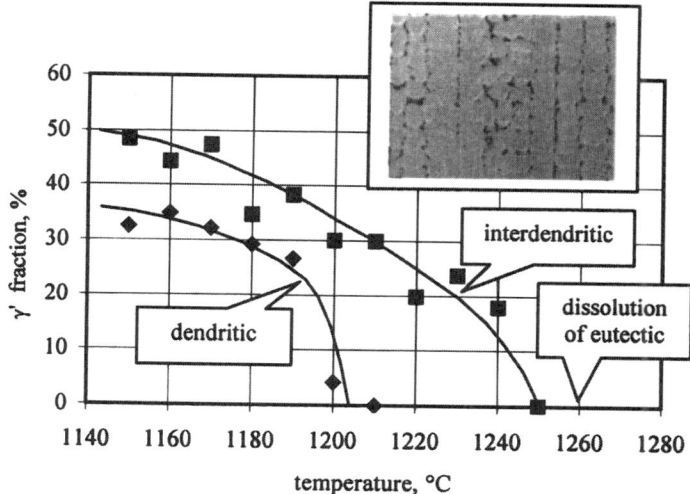

Figure 2: Solutioning behavior of the γ' phase in CMSX–11B (after [7]). The micrograph represents the as-cast start condition with about 6% eutectic.

Recrystallization behavior of CMSX–11B in the γ field

Critical strain   After about 1 % compressive strain new grains form in CMSX–11B during solution heat treatment (SHT). After 0.9 % height reduction some grains are located only in the specimen corners and some way along the circumference, **Figure 3**. After about 2 % strain the specimens fully recrystallize during SHT. Remarkably, *all* grains break the surface after these degrees of cold work and there are no new grains which have developed in the interior or which are entirely surrounded by other grains. To initiate nucleation in the bulk much higher deformation is needed in this alloy, in excess of about 10 %.

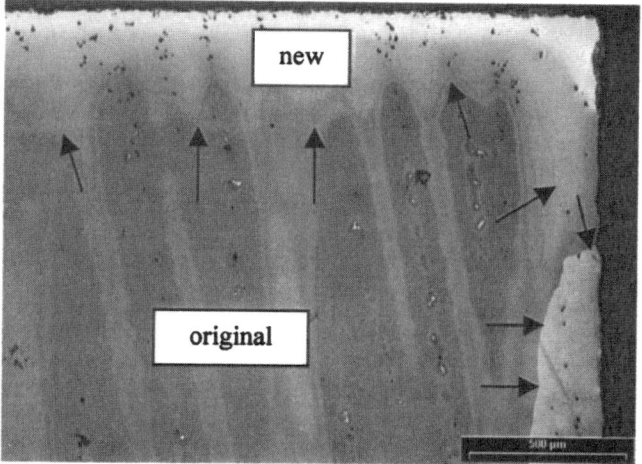

Figure 3: New grains along the surface after 0.9 % height reduction and solution heat treatment (SHT) in CMSX–11B.

Recrystallization temperature   Samples cold worked by 2 % or 3 %, which was chosen as the standard degree of deformation, are completely recrystallized after 1 h at 1249 °C (the step before last in the solutioning cycle, see Table II). After only 15 min at this temperature new grains develop along the surface. According to the common definition round 1250 °C represents the recrystallization temperature for CMSX–11B after 2 % to 3 % strain. All grains have access to the surface which means that they all have nucleated at the surface.

Recovery annealing   With 2 % to 3 % cold worked samples a test program was run with the aim to find a recovery procedure that sufficiently reduces the driving force for recrystallization.

Long term annealing at 1220 °C for up to 100 h, for example, does not prevent complete recrystallization after subsequent full SHT. The same holds true after a 1227 °C/10 h dwell period. After 10 h at 1230 °C the corresponding sample is for the most part recrystallized directly after this annealing (**Figure 4**).

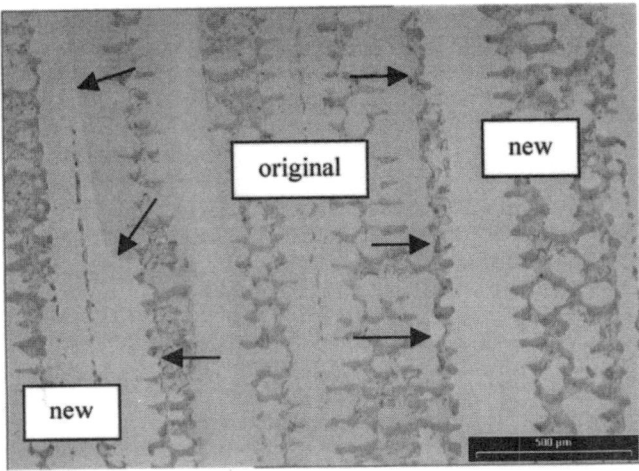

Figure 4: Grain boundaries (arrows) have overcome interdendritic γ/γ' obstacles in CMSX–11B (2.94 %; 1230 °C/10 h). The new grains grow from the surface. (Figure taken from [7]).

Extremely slow heating with 1 K/h from 1220 °C to 1250 °C again is not appropriate to avoid complete recrystallization during subsequent SHT. As stated above, in any case there was no indication that new grains have formed in the bulk, in other words they all break the surface.

At all recovery heat treatments some new surface grains develop after rather short exposure. This led to the supposition that these surface grains might be responsible for further growth into the specimen volume. Consequently, one sample was compressed by 2 %, annealed at 1220 °C for 68 h, and then reduced in diameter from 12 mm to 10 mm by spark erosion in order to remove surface recrystallization (its depth was checked before). It was subsequently annealed at 1249 °C for 15 min to see if new surface grains form again. Indeed, this is the case and complete recrystallization still occurs during full SHT.

To be sure that the observed effects are not influenced by axial compression straining some samples were prepared with 2 % tensile strain. A fully solution heat treated sample again com-

pletely recrystallizes with all grains breaking the surface. These findings coincide with those from compression testing.

Annealing in air causes depletion of oxide forming elements in the surface near region which in turn leads to γ' dissolution. The question arose whether this phenomenon is responsible for the observed surface grain formation. Therefore, some tests were run in a vacuum production furnace. A sample held for 10 h at 1220 °C, then slowly heated with 1 K/h to 1249 °C, followed by the last two steps of SHT (1249 °C/3 h + 1260 °C/6 h) exhibits only a very thin oxide layer but it fully recrystallizes as the air-tested sample does. Also from other comparison tests there is no indication that the results are significantly influenced by the annealing atmosphere.

As long as interdendritic γ' precipitates are present at a high volume fraction the grain boundaries of new grains nucleated at the surface are pinned along the first "row" of interdendritic γ/γ' spaces even after long annealing times, **Figure 5** (see also Figure 11 below). Only at higher temperatures when the fraction of interdendritic γ' is smaller grain boundary motion continues and the deformed microstructure will be consumed and recrystallized, Figure 4.

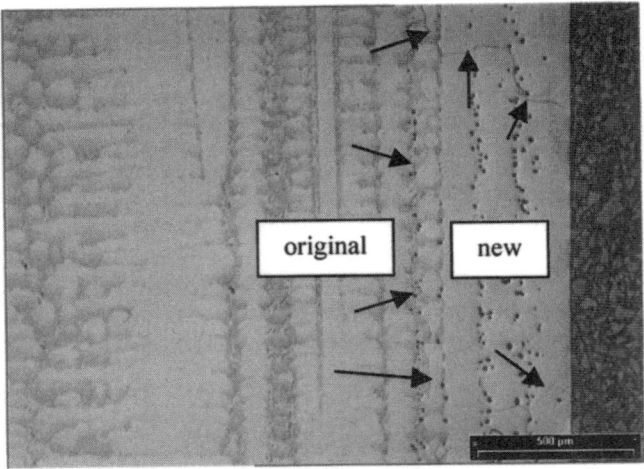

Figure 5: Grain boundaries (arrows) are pinned at the first "row" of interdendritic γ/γ' spaces in CMSX–11B (2.99 %; 1220 °C/97 h). Due to oxidation the near surface region is depleted in γ' and Kirkendall voids also occur. (Figure taken from [7]).

Cyclic recovery annealing   Following some indications in the literature [8, 9] cyclic recovery heat treatments were performed on 2 % strained samples. After ten cycles 1000 °C/30 min ↔ 1100 °C/30 min and final 1204 °C/1 h again new surface grains develop and the specimens fully recrystallize during SHT as before with static annealing. The most extended test series consisted of in total more than 100 cycles, divided into the following steps:
50 cycles 950 °C/5 min ↔ 1050 °C/5 min
+ 27 cycles 1050 °C/5 min ↔ 1150 °C/5 min
+ 29 cycles 1150 °C/5 min ↔ 1200 °C/5 min.

Finally, the SHT was added which again caused full recrystallization.

All together, no recovery heat treatment without any further manipulation could be found that was able to prevent recrystallization during subsequent SHT.

Coated specimens   To move the boundary of deformed material away from the surface and to avoid γ' dissolution due to oxidation different coating procedures were tried after 2 % cold work. An aluminide layer revealed the most meaningful results. After 1249 °C/1 h annealing only at some locations a few new grains develop underneath the coating, the rest of the specimen still exhibits the original single crystal, **Figure 6**. When a specimen is exposed to the full solutioning cycle the coating layer no longer remains continuous, **Figure 7**. This causes recrystallization only on a small scale and about ¾ of the original grain still appears.

Figure 6: Some new grains underneath an aluminide layer after 2 % cold work, CVD aluminizing, and 1249 °C/1 h annealing in CMSX–11B (arrows indicate the grain boundaries).

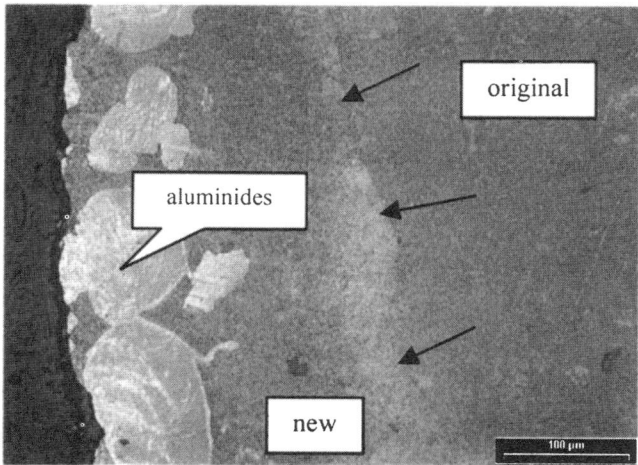

Figure 7: Discontinuous aluminide layer and grain boundary (arrows) after full solutioning cycle in CMSX–11B (2 % cold worked and CVD aluminized).

Another 2 % strained and aluminized specimen was annealed for 1249 °C/3 h and then a jacket was removed by EDM about 1.5 mm deep and 2 mm at the faces. Finally the test piece was fully solution treated. The aim of this experiment was to get rid of new grains which certainly have formed during 1249 °C annealing and to see whether recovery at this temperature eliminates the driving force for recrystallization in the remaining specimen core. After SHT the result was mixed: along the surface some small grains still nucleate but the majority of the volume remains unrecrystallized indicating that indeed bulk recovery dominates in case when surface recrystallization is inhibited.

Influence of carbides   To investigate the influence of dispersed carbides on the recrystallization kinetics an experimental derivative of CMSX–11B with 0.08 w% C was cast. It was shown that this alloy can be solution treated with the same parameters, i.e. the carbon content does not reduce the melting temperature below 1260 °C when all prior annealing steps are run. The carbides are concentrated in interdendritic regions and after SHT they appear rather coarse and blocky, sometimes bone-shaped, as primary carbides typically do. They were analyzed as Ti/Ta containing MC-type particles, as expected from the alloy composition.

A specimen of this alloy strained 1.88 % and fully solution treated exhibits the same features as the carbide free alloy. The whole volume is recrystallized and all grains have access to the surface. Particular pinning of grain boundaries by the carbides could not be detected.

Recrystallization behavior of PWA 1483 in the γ field

The fully solution treated condition of PWA 1483 looks very similar to that of the carbon version of CMSX–11B, i.e. relatively coarse primary carbides are visible of the MC type with Ti and Ta being the main metallic components.

A sample compressed by 2.11 % completely recrystallizes during SHT, **Figure 8**. All grains initiate from the surface, too. Heating a 2.12 % cold worked sample to 1249 °C with the same steps as for CMSX–11B and holding for 1 h at this temperature also produces full recrystallization. Thus it can be concluded that the behavior is very similar to that of CMSX–11B.

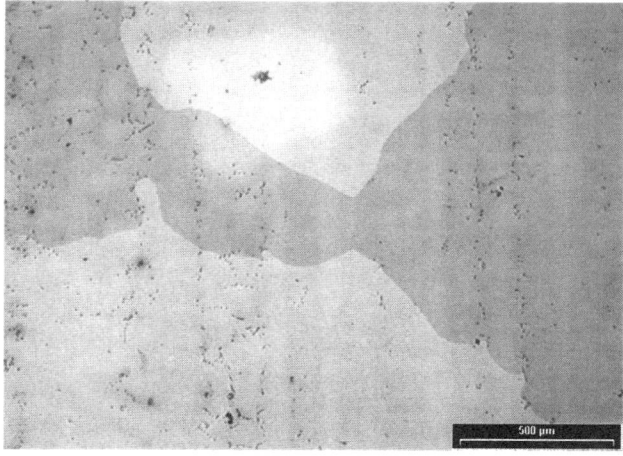

Figure 8: Full recrystallization in PWA 1483 after 2.11 % strain and SHT. The dark particles are MC type carbides.

Recrystallization behavior of SRR 99 in the γ field

SRR 99 shows pronounced glide bands indicating single slip and, with increasing strain, dual slip on {111}<110> slip systems. After 1.83 % strain and full solution heat treatment (see Table II) a band of new grains forms which runs diagonally through the specimen, **Figure 9**. The rest of the volume remains unrecrystallized. All grains have access to the surface at this degree of deformation. With increasing cold work the width of the recrystallization band increases. A 4.01 % compressed sample, for example, exhibits two crosswise bands with new grains as a consequence of dual slip.

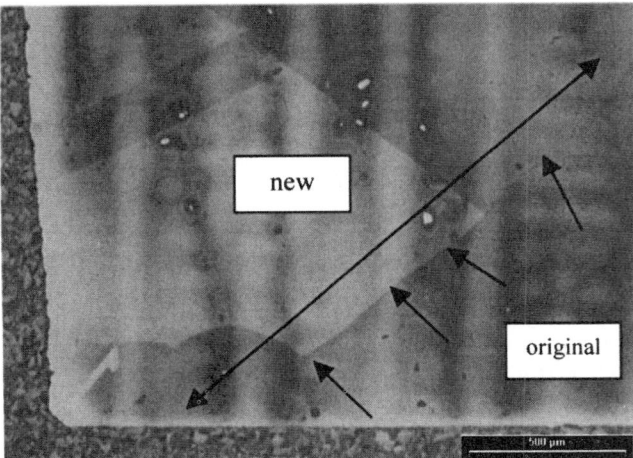

Figure 9: Recrystallization band starting at one corner and running diagonally through the specimen in SRR 99 (1.83 % and SHT).

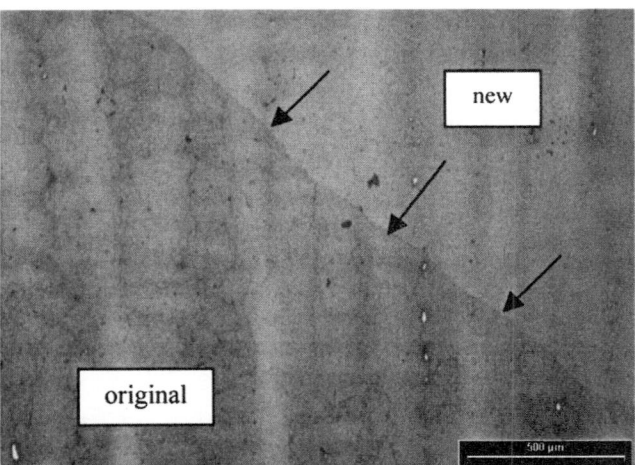

Figure 10: Boundary between a new grain and the original one in SRR 99 after 4.01 % and SHT. Within the original grain a feathery substructure can be detected.

In SRR 99 residues of the original crystal can be clearly distinguished from recrystallized areas by a feathery substructure, **Figure 10**. At higher resolution in a scanning electron microscope this feature emerges as somewhat coarser γ' particles which probably reprecipitate on low-angle grain boundaries du-

ring cooling from solutioning. Obviously this substructure remains stable also during SHT.

With SRR 99 recovery annealing was tried at 1250 °C for 10 h. At this temperature the dendrite cores are single phase γ but a high amount of γ' is still present in the interdendritic regions as well as eutectic γ/γ'. Where the deformation band runs out of the surface *and* a dendrite core breaks the surface new grains nucleate after 1250 °C/10 h, **Figure 11** (checked after 1.8 % and 3.24 % strain). Always when the SHT is added after 1250 °C annealing, recrystallization occurs in the way described above for the non-pretreated condition.

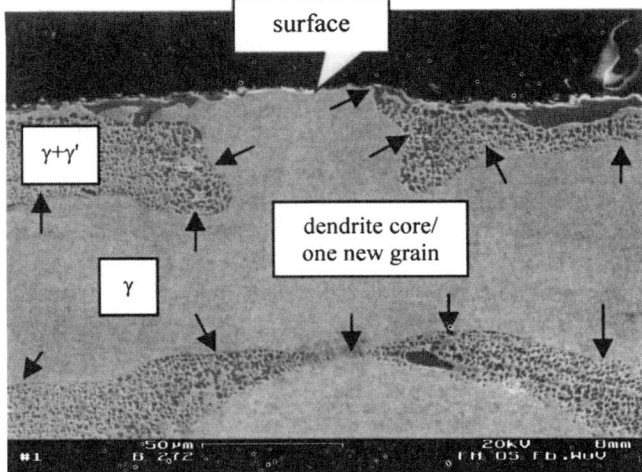

Figure 11: SEM micrograph of grain nucleation at the surface in a single phase dendritic area. The grain boundary (arrows) winds along the interdendritic γ/γ' spaces (SRR 99; 3.24 %; 1250 °C/10 h under vacuum; cross section).

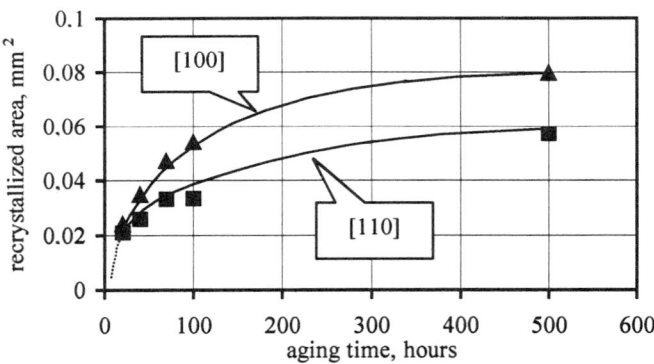

Figure 12: Recrystallized area in CMSX–6 as function of aging time at 1080 °C for two orientations of the prismatic indenter (evaluated section perpendicular to the indentation edge).

Recrystallization behavior of CMSX–6 in the γ+γ' field

Kinetic evaluation The time dependence of the recrystallized area in indented CMSX–6 specimens after annealing at 1080 °C is shown in **Figure 12**. The anisotropy of this process is clearly reflected by the effect of the edge orientation relative to the crystal directions [100] and [110], respectively. Similar results were obtained when determining the machining parameters for LCF specimens.

This experimental approach allows the comparison of different materials with respect to their proneness to recrystallization. The ranking we obtained in this way reflects the practical experience with this phenomenon. It is possible to reduce all the data by using a Johnson-Mehl-Avrami equation [10]. The incubation period depends on the applied indentation load, being shorter for higher loads as in Figure 12.

<u>Nucleation</u>   In order to nucleate new grains in a γ/γ' microstructure without partially dissolved γ' the following procedure was chosen. In a previous investigation [11] it was shown that tensile creep deformation produces a gradual change in the morphology of the interdendritic shrinkage pores. Creep fracture is induced by cracks originated at the surface of these pores. Cylindrical specimens of CMSX–6 with the complete heat treatment were creep deformed at 760 °C and 980 °C, respectively, to a strain of about 15 %, generating plenty of cracks around shrinkage pores without contact to the atmosphere. One may assume that the remaining oxygen partial pressure in the pores is very low. After interrupting the creep deformation by unloading, the control mode of the machine was changed and the samples were ruptured at 980 °C with a high strain rate. After annealing at 980 °C cellular recrystallization could be observed in many specimens at the crack tips, **Figure 13**, following the high plastic deformation in these regions.

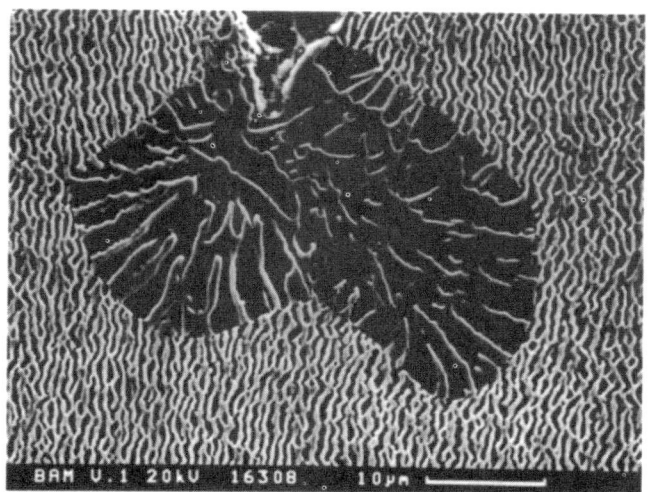

Figure 13: Recrystallized area at a crack tip in a CMSX–6 creep specimen (980 °C) after tensile rupture and annealing at 980 °C for 96 h.

<u>Microstructural characterization</u>   It could be shown that the γ/γ' microstructure in the new grains does not depend on the topological characteristics of the parent grain. In specimens creep tested at 760 °C, where γ' cuboidal precipitates are embedded in the γ phase, as well as in those creep tested at 980 °C, where γ platelets are involved by γ' phase, we observed the same γ/γ' microstructure in the recrystallized region. This microstructure has a much coarser distribution of γ/γ' than the parent grain, **Figure 14**. Due to the characteristic cells, this

process is known as cellular recrystallization. The analysis of the orientation distribution in SEM using the EBSD technique as well as the comparison of TEM diffraction patterns from the parent grain and adjacent recrystallized grains clearly reveal that the cells are grains with high angle boundaries and that there is no unique relationship between the parent and the new grains.

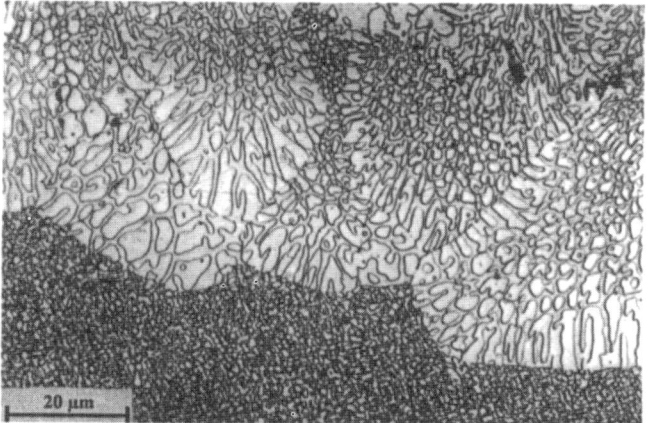

Figure 14: Change in the γ/γ' microstructure of a CMSX–6 specimen due to cellular recrystallization.

The γ/γ' microstructure of a recrystallized volume was registered in eight parallel sections by a serial polishing technique. The distance between neighboring sections lies between 0.5 μm and 1.5 μm. Comparison of the micrographs in different areas shows that the cells consist of a γ' continuum in which γ columns are embedded.

The chemical compositions of both phases in the parent grain and in a recrystallized area were determined by EDX in a TEM specimen, **Table III**.

Table III Chemical compositions of the γ and γ' phases in the parent grain and in recrystallized areas (RX) of CMSX–6

phase	grain	mass fraction in %					
		Ni	Cr	Co	Mo	Ti	Al
γ	parent	61.3	22.4	9.0	5.8	.9	.8
	RX	62.8	21.7	8.4	4.9	1.2	1.0
γ'	parent	77.3	8.3	4.7	2.9	5.6	1.4
	RX	81.7	3.3	3.9	2.1	7.2	2.0

<u>Influence of recrystallized areas on LCF behavior</u>   **Figure 15** depicts the influence of a 0.1 mm deep recrystallized surface layer on the LCF behavior at 980 °C and a total strain range of 0.7 %. The number of cycles to crack initiation, $N_{A(5\%)}$, does not differ significantly between virgin and recrystallized samples in both cycle modes (without and with hold times).

In comparison to specimens with no recrystallization, the density of cracks on the gage length of the recrystallized specimens is higher, especially in the <110> directions (which can be identified by looking at a polished specimen face, i.e. a (001) plane). The metallographic analysis of longitudinal sections and fractography confirm this higher crack nucleation rate of recrystallized specimens in comparison with single crystal ones.

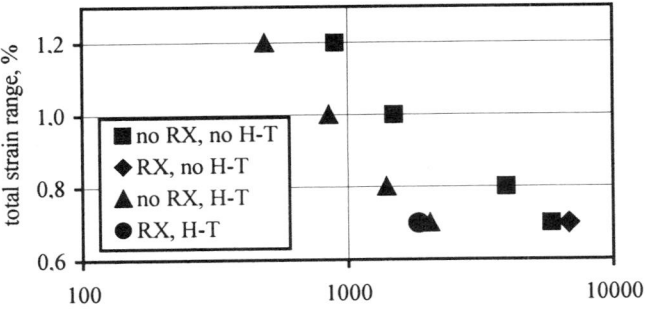

Figure 15: LCF test results of CMSX–6 at 980 °C.
Each symbol represents the average lifetime of two to five specimens. (RX: with cellular surface recrystallization;
H-T: hold time of 300 s at ε_{min}).

Discussion

Recrystallization in the γ field

The susceptibility of an alloy to recrystallization during solutioning treatment, i.e. when the single-phase γ field is approached or reached, is usually tested by hardness indentations or similar non-uniform surface deformation. In the present investigation uniform strain was applied to improve quantification of the relationship between deformation and recrystallization behavior, to improve the comparison between different alloys, and to determine the appearance of recrystallization when the bulk is deformed.

Cold work at room temperature is regarded as causing a dislocation structure representative of what develops by short-term high temperature deformation in the γ+γ' phase field, e.g. due to contraction differences between the blade and the shell mold and core. It was estimated that 2 % to 3 % are realistic strain values which can be imposed on the components when the mold and core do not comply sufficiently or crack.

After about 1 % plastic strain the first new grains appear at the surface and less than 2 % causes full recrystallization across the samples. This statement holds true for all alloys investigated for recrystallization in the γ phase field. *Two* conditions need to be satisfied simultaneously to initiate recrystallization after such low degrees of cold work: *i)* The γ' phase must be dissolved in the area of grain nucleation, and *ii)* a free surface must be available. Dissolution of γ' alone is not a sufficient prerequisite because then nucleation from internal dendrite cores would also occur as soon as these become single phase. This is definitely not the case.

In the bulk a grain nucleus has to overcome the energy barrier of a grain boundary around the whole grain boundary area whilst along a surface no new interface energy has to be generated. This increases the probability of heterogeneous nucleation

at the surface compared to the bulk formation obviously so much that only a small amount of cold work is needed for the former whereas much higher mechanical energy has to be introduced for volume grain nucleation (e.g. more than about 10 % strain in CMSX–11B). For blade assessment these findings mean that if recrystallization has taken place surface grains must in any case be present. If there are none there will be no recrystallization at all. Of course, for hollow blades and vanes the inspection of internal surfaces remains a problem.

With SRR 99 a deviation was observed compared to CMSX–11B and PWA 1483 in that slip during cold work is concentrated in prominent glide bands. Consequently, new grains only appear within these bands whereas for the other alloys investigated full recrystallization covers more or less the whole specimen volume. This feature might explain why in industrial practice components cast from SRR 99 may not generally be free of recrystallization but new grains often are small and within the specification.

Experiments with coated samples confirm the hypothesis of energetically favored grain nucleation at surfaces. Despite all difficulties with coating degradation at very high temperatures a clear difference of aluminized samples was observed compared to uncoated ones. If recrystallization from the surface is inhibited by a coating recovery can take place effectively in the bulk when the γ' precipitates are dissolved completely. However, this test series only served to improve understanding the nucleation mechanisms and the competition between recovery and recrystallization. In industrial practice it is not feasible to coat the blades to avoid recrystallization during SHT and to strip them afterwards.

All efforts to identify a recovery heat treatment that sufficiently reduces the driving force for primary recrystallization failed. Even unrealistic long annealing times and very low heating rates are not able to annihilate the dislocation structure noticeably below γ' solvus. TEM examinations of long-term annealed samples confirmed that the dislocation structure is extremely resistant to recovery as long as the dislocations are knitted into the γ/γ' interface. These dislocations supply the mechanical energy that is released by primary recrystallization as soon as the γ' phase disappears. Removal of surface grains after nucleation does not prevent complete recrystallization during subsequent SHT which means that indeed the strain energy is stored in the bulk and that the mechanism of primary recrystallization is operating.

For the production of copper crystals with an extremely low dislocation density cyclic recovery annealing can be applied [8]. This also forms the background of a patent for monocrystalline superalloys [9]. The idea is that dislocation annihilation is enhanced by repeated creation and absorption of vacancies to adjust the equilibrium concentration for the respective temperature. In the present investigation, however, very extended cyclic heat treatments could not avoid subsequent primary recrystallization either.

For most superalloys complete dissolution of γ/γ' eutectic requires significantly higher temperatures and/or longer times compared to dendritic and interdendritic γ' precipitates (see Figure 2). This temperature "window" is sometimes considered to be appropriate to solution treat the alloy completely, except eutectic, on the one hand and to inhibit recrystallization on the other hand because residual eutectic is believed to hinder grain boundary motion. No alloy examined in the present work confirms such a mechanism operating successfully. For example, CMSX–11B exhibits a recrystallization temperature of about 1250 °C (full recrystallization within 1 h after 2 % to 3 % plastic strain). At this temperature a fairly high amount of eutectic (more than 5 %) is still present.

Clearly, grain boundary motion is stopped or at least retarded by interdendritic γ/γ' spaces. A surface grain nucleated where a dendrite core breaks the surface will not grow further when the grain boundary encounters the next γ/γ' area (see Figure 11). Such a small grain might be acceptable in practice. At high enough temperatures and long times, however, these obstacles can be overcome (Figure 4). To utilize this effect would mean to dispense with complete γ' dissolution at the expense of strength.

Virtually no difference was detected between carbide free and carbide containing alloys. From the present investigations carbides do not seem to be an appropriate measure to inhibit grain boundary motion in the way these particles are present in the γ' solvus region.

Summarizing, the recrystallization behavior in the γ phase field does not differ much among the alloys investigated under the conditions applied. The critical degree of cold work for recrystallization initiation and the mechanisms of heterogeneous grain nucleation at the surface are similar. The extent of primary recrystallization is slightly different, being less for SRR 99 compared to the other alloys examined. This coincides with the experience that, for example, after stamping indentation marks and SHT the recrystallized grains are smaller with SRR 99 compared to other alloys. How much deformation is introduced into the blades during manufacturing is highly geometry-dependant and influenced by core and mold yielding and crushability as well as by the cooling conditions.

Recrystallization in the $\gamma+\gamma'$ field

Cellular recrystallization in superalloys with a high volume fraction of γ' precipitates occurs by solutioning this phase at the moving grain boundary [12]. It is not uncommon to detect this phenomenon in single crystal blades. In comparison to recrystallization in the γ field discussed above, a much higher degree of cold work is necessary is this case.

The onset of cellular recrystallization, examined for CMSX–6, does not require a γ' free zone in contrast to assumptions in the literature [13]. However, it was not possible to identify the mechanism of high angle grain boundary formation in the γ/γ' microstructure and whether a free surface also forms a prerequisite for nucleation as observed in the γ field.

Cellular recrystallization is determined by the movement of a grain boundary into the deformed parent material. Due to the crystal disorder in this boundary, it possesses a locally higher solubility and diffusivity than the bulk material [14]. The supersaturation leads to decomposition of the single phase immedi-

ately behind the moving boundary. This occurs by the formation of γ columns perpendicular to the grain boundary in a γ' continuum. A new grain nucleates and expands as a section of a sphere, the γ columns being homogeneously distributed in radial directions. A section through a recrystallized area shows cells with different shapes of the γ rods depending on their angle to the section plane. Since the γ/γ' microstructure in the recrystallized area is much coarser than in the parent phase (Figure 14), the movement of the grain boundary leads to a reduction of the γ/γ' interface energy per unit volume.

The reduction in the free energy associated with cellular recrystallization, ΔG_{RX}, may be decomposed into three terms:

$$\Delta G_{RX} = \Delta G_{mechanical} + \Delta G_{interface} + \Delta G_{chemical}$$

The mechanical term (sign –) reflects the annihilation of dislocations. The second component is related to the microstructural coarsening mentioned above (sign –) and the creation of high-angle grain boundary (sign +). The third term is assumed to be negligible since the chemical composition of both phases in the recrystallized area is nearly the same as in the parent material.

The crack formation under LCF loading was shown to be facilitated by the presence of a cellularly recrystallized surface layer. However, the number of cycles to crack initiation, $N_{A(5\%)}$, seems to be independent of the specific surface condition of the specimen used in this work, Figure 15. In the present case this is due to the small depth of the recrystallized layer relative to the specimen cross section. Therefore, the common load reduction criterion of 5 % to evaluate N_A can only be achieved if the cracks grow into the parent material [15]. In the case of thin-walled components the influence can be expected to be more pronounced.

Conclusions

The present investigations lead to the following conclusions:

(1)   Grain formation in the solutioning temperature regime requires only a small critical degree of cold work of about 1 % strain. In a rather wide range of plastic deformation, which in practice most certainly will not be exceeded, grain nucleation merely starts from the surface for energetic reasons in single-phase γ areas. If the deformed surface is subsequently shifted into the interior by an appropriate coating recrystallization can be suppressed. However, this treatment can hardly be utilized in practice.
(2)   Grain boundary motion is substantially hindered by interdendritic γ/γ' spaces but not noticeably by carbides or residual eutectic islands. Therefore, if a blade has been deformed to a critical degree it may only be heat treated considerably below the complete solvus of γ' to avoid larger grains.
(3)   No recovery heat treatment could be identified that sufficiently reduces the driving force for recrystallization due to very stable dislocation anchoring around the γ' precipitates.
(4)   Alloys in which deformation is concentrated within narrow glide bands, such as SRR 99 in the present work, exhibit localized recrystallization with possibly acceptable grain size and area fraction.
(5)   At lower temperatures in the $\gamma+\gamma'$ field cellular recrystallization occurs after much higher amount of cold work. Nuclea-

tion in this case is not bound to a γ' depleted zone. The subsequent γ' microstructure is much coarser than the initial one.

(6) A thin surface layer with cellular recrystallization was found to cause a higher crack density under LCF conditions but the number of cycles to crack initiation according to a load reduction criterion is not influenced under the chosen geometry parameters. For thin-walled components a reduction in LCF strength can be expected.

Acknowledgments

The following co-workers and students at Osnabrück University of Applied Sciences have contributed to the present work over the past two years: J. Düsterhöft, B. Fischer, H.-G. Kleinheider, K. Mey, T. Michaelis, U. Mussing, and S. Wilkens. The experiments at BAM were performed by A. Riepe and F. Menelao. T. Lübcke of Lufthansa Technik/Hamburg applied the CVD coatings. H. Rooch and W. Gesatzke of BAM, Dr. J. Penkalla of Research Center Jülich, and F. Pyczak of Erlangen University carried out TEM examinations. Prof. H. Mughrabi of Erlangen University gave some valuable advice. Dr. J. Blackford, formerly with Erlangen University, checked the manuscript. The authors gratefully acknowledge all these contributions. The work at Osnabrück University of Applied Sciences and at DPC was promoted by funds by German Ministry of Education, Science, Research, and Technology through Grant 03N2011A and B. The work at BAM was partially supported by Deutsche Forschungsgemeinschaft in the framework of the Priority Program "Microstructure and Mechanical Properties of Metallic High-Temperature Materials". The responsibility for the content of this publication lies with the authors.

References

1. S.D. Bond, J.W. Martin, "Surface Recrystallization in a Single Crystal Nickel-Based Superalloy", J. Mater. Sci., 19 (1984), 3867–3872.

2. T. Khan, P. Caron, Y.G. Nakagawa, "Mechanical Behavior and Processing of DS and Single Crystal Superalloys", J. Metals, July 1986, 16–19.

3. U. Paul, P.R. Sahm, D. Goldschmidt, "Inhomogeneities in Single-Crystal Components", Mater. Sci. Engg., A173 (1993), 49–54.

4. R.W. Salkeld, T.T. Field, E.A. Ault, "Preparation of Single Crystal Superalloys for Post-Casting Heat Treatment", US-Patent 5,413,648 (1995).

5. W. Österle, W. Gesatzke, W. Byrne, "TEM Study of the Effect of Machining on the Microstructure of a Soft Magnetic Ni76Fe17Cu5CrMo2 Alloy", Z. Metallkde., 82 (1991), 902–906.

6. W. Österle, P.X. Li, W. Niewelt, "Microstructural Changes Induced by Grinding of Ni-base Superalloy IN 738 LC and their Relationship to Machining Parameters", Z. Metallkde., 85 (1994), 20–27.

7. J. Düsterhöft, "Vermeidung von Rekristallisation gerichtet erstarrter Bauteile", Diploma Thesis, Osnabrück University of Applied Sciences, 1998.

8. S. Kitajima, M. Ohta, H. Tonda, "Production of Highly Perfect Copper Crystals with Thermal Cyclic Annealing", J. Crystal Growth, 24/25 (1974), 521–526.

9. W.J. Gostic, "Cyclic Recovery Heat Treatment", US-Patent 5,551,999 (1984).

10. F. Menelao, "Kinetik und Gefügebildung bei der zellularen Rekristallisation in der einkristallinen Nickelbasis-Superlegierung CMSX–6", Diploma Thesis, Berlin University of Applied Sciences and Economy, 1998.

11. P.D. Portella, C. Herzog, "Gefügeänderungen und Schädigungsentwicklung der einkristallinen Superlegierung SRR 99 unter Kriechbeanspruchung bei 980 °C", in: Vortragsveranstaltung Rastermikroskopie in der Materialprüfung, ed. V. Thien (DVM, Berlin, 1992), 51–60.

12. A. Porter, B. Ralph, "The Recrystallization of Nickel-Base Superalloys", J. Mater. Sci., 16 (1981), 707–713.

13. B. Ralph, C.Y. Barlow, B.A. Cooke, A.J. Porter, "The Recrystallization of High Performance Alloys", in: Proc. 1st Risø Int. Symp. on Recrystallization, ed. N. Hansen et al. (Roskilde: Risø Nat. Lab., 1980), 229–242.

14. K. Smidoda, W. Gottschalk, H. Gleiter, "Diffusion in Migrating Interfaces", Acta Met., 26 (1978), 1833–1836.

15. P.D. Portella, W. Österle, "Influence of Cellular Recrystallization on the Fatigue Behaviour of Single Crystal Ni-Based Superalloys", in: Microstructure and Mechanical Properties of Metallic High-Temperature Materials, ed. H. Mughrabi et al. (Weinheim: Wiley-VCH, 1999), 441–453.

STRUCTURE OF THE NI-BASE SUPERALLOY IN 713C AFTER CONTINUOUS CASTING

F. Zupanič, T. Bončina, A. Križman

University of Maribor, Faculty of Mechanical Engineering, Slovenia
Smetanova 17
SI-2000 Maribor, Slovenia

Abstract

In this work a Ni-base superalloy IN 713C was continuously cast in the form of rods with a circular cross section under a protective argon atmosphere using a vertical continuous caster. The structure of the rods cast at different casting parameters was investigated using different microstructural characterisation techniques. It was found that superalloy IN 713C could be continuously cast over a wide range of casting parameters. The continuously cast rods possessed a clean surface and excellent internal integrity without non-metallic inclusions and porosity. The microstructure consisted, regardless of the casting parameters used, of γ-grains, interdendritic γ/MC eutectic and coherent γ' precipitates distributed uniformly within the γ-matrix. The size, distribution, morphology and orientation of microstructural constituents depended strongly on local solidification conditions, which were greatly influenced by the nozzle/mould set up and the alternating drawing mode of continuous casting. The results of the work strongly suggest that IN 713C could also be continuously cast successfully using large-scale industrial equipment.

Superalloys 2000
Edited by T.M. Pollock, R.D. Kissinger, R.R. Bowman,
K.A. Green, M. McLean, S. Olson, and J.J. Schirra
TMS (The Minerals, Metals & Materials Society), 2000

Introduction

Continuous casting offers several advantages over the conventional casting of semi-finished metal products including lower production costs and higher productivity rates, as well as better surface quality and the excellent internal integrity of the products [1]. It has been widely used in the production of almost all alloys, but not yet in the manufacture of vacuum melted cast Ni-base superalloys [2]. It has the potential to replace conventional casting of remelting stick, which has high cost and labour intensive stages in the production of superalloys, although a large capital investment is unavoidable at the beginning. A further drawback for the introduction of continuous casting in the manufacture of superalloys arises from a lack of knowledge about the behaviour of particular superalloys when continuous cast. In order to answer the latter question, we initiated an investigation with the following objectives:

1) Evaluation of the castability of selected Ni-base superalloys in continuous casting using laboratory scale equipment

2) Finding a correlation between the adjustable casting parameters and the as-cast microstructure

3) Optimisation of the continuous casting parameters to obtain the best combination of production rate, process reliability and product quality

4) Determining the influence of the as-cast microstructure on the subsequent heat treatment stages, final microstructure and mechanical properties of the continuously cast products

5) Obtaining mostly unidirectional crystallisation of the Ni-base superalloys in a continuous caster.

So far, we have concentrated mainly on the first three objectives and have continuously cast with success three Ni-base superalloys. The aim of the present work was to carry out a characterisation of macro- and microstructure to evaluate the influence of casting parameters on the formation and development of microstructure during continuous casting. This information is critical in further optimisation of the casting parameters to obtain the desired goals.

The nickel-base superalloy IN 713C has been used for decades. Its application is in the form of precision cast parts for hot-end turbocharger wheels. It is interesting to note that the required properties are already attained normally in the as-cast condition, and therefore no heat treatment is necessary. The influence of casting parameters on the as cast microstructure of IN 713C was thoroughly investigated by Bhambri et al [3]. One of the most important reasons to experimentally continuously cast IN 713C lies in the fact that the annual production rate of this alloy is relatively high. In addition, the alloy possesses considerable high-temperature ductility and strength [4], which are usually an important factor in reliable continuous casting.

Experimental procedure

Continuous casting experiments were carried out using a pilot scale set-up consisting of a vacuum induction melting furnace (Leybold Heraeus IS 1.5) and a vertical continuous caster (Technica Guss; now SMS Meer, Demag Technica). Figure 1 a shows schematically the most important parts of the equipment. The alloy was melted in an alumina crucible. It was separated from the water-cooled Cu-Be mould by a ZrO_2 inlet nozzle. It is important to notice that the inner diameter of the ZrO_2 inlet nozzle (9.7 mm) was slightly smaller than the inner diameter of the Cu-Be mould (10 mm). It will be shown later that this feature greatly influenced the appearance of the rod and the formation of macroscopic in-homogeneities.

Table I   Chemical composition of the investigated alloy IN 713C*

Fe	Cr	Al	Ti	Nb	Mo	C	Ni
1.42	13.15	6.04	0.78	2.11	4.19	0.15	Rest

* all compositions in this work are given in weight percents unless otherwise stated

Table II   Conditions of continuous casting experiments

No. of a casting condition	#1	#2	#3	#4
Length of draw [mm]	5	7	8	8
Relative velocity of the drawing stroke [%]	30	35	35	35
Resting time [s]	0.15	0.15	0.15	0.10
Length of the reverse stroke [mm]	0.30	0.30	0.30	0.30
Relative velocity of the reverse stroke [%]	10	10	10	10
Casting speed [mm/s]	8.33	10.17	11.17	12.33
Melt temperature [°C]	1450	1420	1400	1420
Water flow rate [l/min]	20	20	30	30
Inlet water temperature [°C]	30	38	38	37
Outlet water temperature [°C]	30	43	43	43

In a typical trial around 14 kg of IN 713C, the composition of which is given in Table 1, was melted in the induction furnace in a vacuum of approximately 10^{-2} mbar. After heating up to 1420 °C the melt was continuously cast under a protective argon atmosphere at 1.030 bar through the Cu-Be mould, in which solidification occurred. The solid rod was pulled out of the mould using the predetermined "alternating drawing mode" (Fig. 1 b).

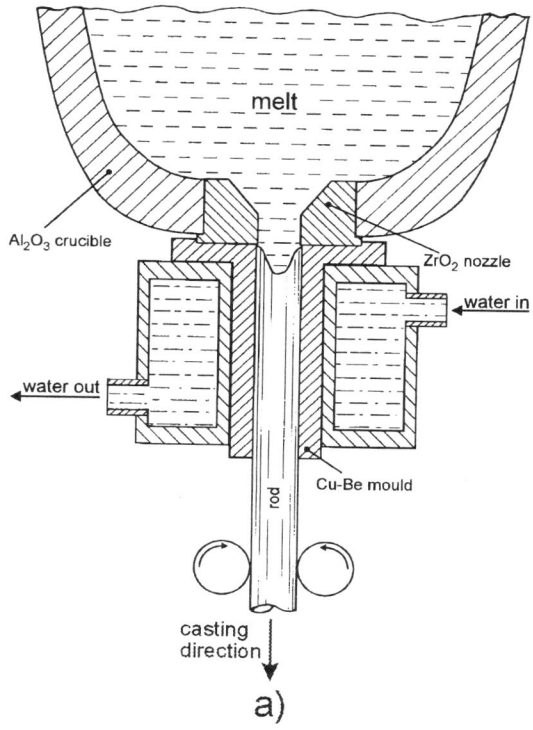

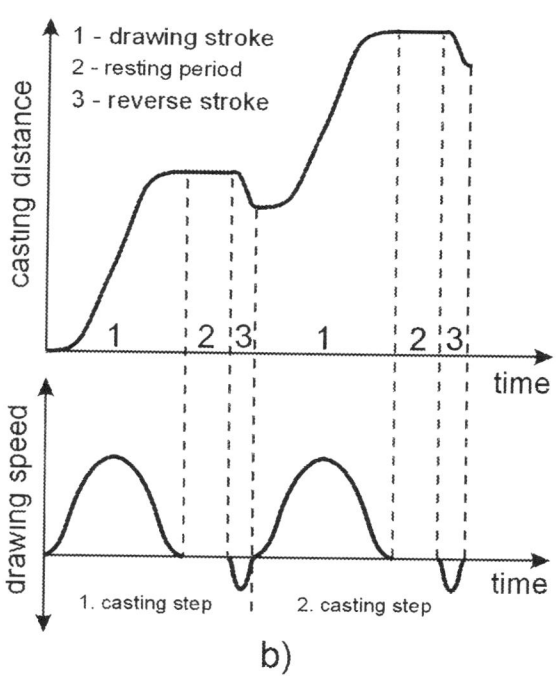

Figure 1: Schematic presentation of the continuous casting set-up (a); and the characteristics of the alternating drawing mode (b)

It consisted of three sequential stages: the drawing stroke, the resting period and the reverse stroke, which formed a casting step. It can be seen that during the drawing and the reverse stroke, the velocity increased from rest to the maximum velocity and decreased back to rest. During the reverse stroke the rod was moved in the direction opposite to the casting direction. This drawing mode is usually used when an alloy is continuously cast for the first time because it allows the most reliable continuous casting. Besides, the reverse stroke eliminates possible sticking between the bar and the Cu-Be mould.

During the trial, the length of draw was changed from 5 to 8 mm, the resting time between 0.10 and 0.15 s, but the length of the reverse stroke was always 0.3 mm. Different combinations of these parameters gave an average casting speed ranging from 8.33 to 12.33 mm/s. Details are given in Table II.

The continuously cast rods were prepared using standard metallographic procedures and investigated by different microstructural characterisation techniques, such as light (LM), scanning (SEM) and transmission (TEM) electron microscopy, and energy dispersive X-ray spectroscopy (EDS).

Results and discussion

Surface appearance of continuously cast rods

Continuous casting of the superalloy IN 713C took place without any breakouts of the rod or other deviations. Figure 2 shows some continuously cast rods. Their surface is bright, with a characteristic metallic shine. It is obviously free of thick oxide layers. Therefore subsequent grinding of the rods or even shot blasting, which are typical steps in the present production of remelting sticks, is unnecessary. Periodic markings on the surface are the consequence of the "alternating drawing mode". This is confirmed by the fact that the distance between them coincides nicely with the length of the casting step; which is defined as the difference between the length of draw and the length of reverse stroke. The diameter of the continuously cast rod at these markings is around 9.7 mm and thus smaller than in the other parts of the rod. It is suggested that there must be some relationship between it and the diameter of the inlet nozzle, which is also 9.7 mm.

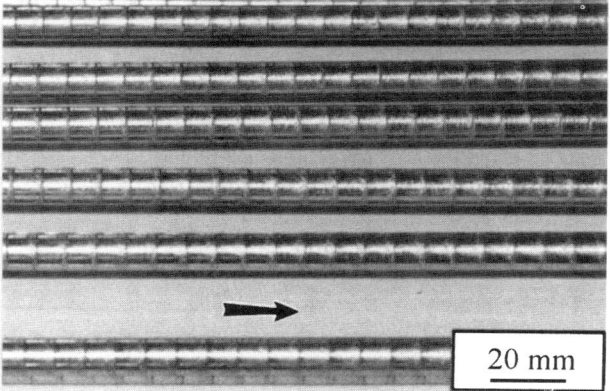

Figure 2: Continuously cast rods of IN 713C. Periodic markings are a result of the "alternating drawing mode"

Microstructural constituents in the continuous cast IN 713C

Metallographic analysis revealed that irrespective of the casting conditions the microstructure of the alloy consists of γ-grains with dendritic growth morphology, the MC/γ eutectic and coherent γ' precipitates distribution in the γ matrix are more or less uniform. On the other hand, it was observed that the size, distribution, morphology and orientation of these microstructural constituents depend on casting conditions. The γ/γ' eutectic, which was observed by Bhambri et al [3] in directionally solidified IN 713C, is not present in continuously cast rods. As a result of the alternating drawing mode a periodic pattern has formed with a wavelength almost equivalent to the length of a casting step (Fig. 3).

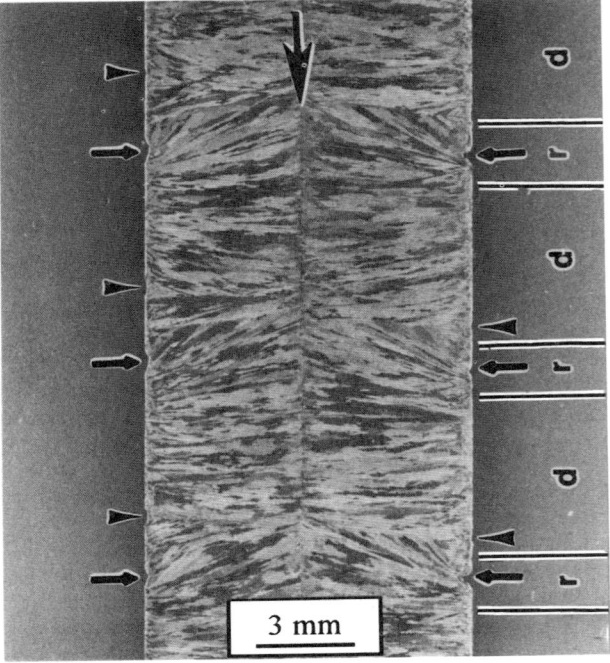

Figure 3: Macrostructure of a continuously cast IN 713C superalloy (r – the resting period and the reverse stroke, d – the drawing stroke). Casting conditions: the length of draw: 7 mm, the resting time: 0.15 s, the length of the reverse stroke: 0.3 mm, the average casting speed: 10.17 mm/s. The largest arrow indicates the casting direction

Macrostructure

Fig. 3 also shows the size and orientation of the γ grains. The γ grains are usually columnar and oriented in the radial direction. They must have formed on or very close to the mould wall and then grown toward the centre of the rod. In addition to columnar γ grains, a shallow layer of small equiaxed grains is present on the edge of the rod, and some equiaxed grains have also formed at the centre of the rod at the highest casting speed of 12.33 mm/s.

Within a casting step the orientation of γ grains changed from almost ideally perpendicular to the mould wall, towards orientations which were more closely aligned with the casting direction. The latter grains in fact grew in the opposite direction to the casting direction. It is likely that γ grains being perpendicular to the mould wall formed and grew during the drawing stroke, whereas the γ grains inclining to the cast direction grew during the resting period and the reverse stroke.

The periodic markings observed on the surface of the rod could be seen on the longitudinal cross sections in the form of large intrusions. They were positioned next to γ grains that were inclined to the casting direction; this clearly indicates that they must have formed during the resting period and the reverse stroke.

Besides these large intrusions we also observed smaller and less periodic intrusions (indicated by arrows ►). They were usually positioned close to the transition from the resting/reverse stroke and the drawing stroke. Their probable origin will be discussed later.

The drawing stroke

Despite the fact that the drawing velocity was changing all the time throughout the drawing stroke, the microstructure formed during the drawing stroke exhibited the same features over the whole drawing length. At this stage of continuous casting the melt was able to come into direct contact with the water-cooled Cu-Be mould. The heat was mostly extracted through the mould wall. This facilitated crystal grains having an [100] axis perpendicular to the mould wall to outgrow other less favourably oriented grains. Despite Bhambri et al. [3] claiming that the solidification of IN 713C should start with the primary crystallisation of MC carbide, the solidified microstructure shows characteristics, which are typical for primary crystallisation of the γ phase. It seems that carbides are present mainly within the MC/γ eutectic constituent in the interdendritic region.

Figure 4 compares microstructures formed at different distances from the edge of the rod during the drawing strokes. It is evident that microstructure is becoming coarser when moving toward the centre of the rod; the primary and secondary dendrite arm spacings are increasing, as well as the size and the amount of MC carbides. Close to the edge of the rod the MC carbides appear as separate particles almost parallel relative to the dendrite trunks, whereas in the middle of the rod they are present as a constituent of large MC/γ eutectic pools in the interdendritic region.

Bhambri et al. [3] determined the empirical relationship between the secondary dendrite arm spacing λ_2 and the local solidification time t_f:

$$\lambda_2 = At^n \qquad (1)$$

with $A = 6.79 \times 10^{-6}$ and $n = 0.43$. Using equation (1) and taking into account that the solidification range of IN 713C amounts to approximately 60 °C, we calculated local cooling rates on the basis of measured secondary arm spacings λ_2

(Table III). It can be seen that the local solidification rate $\dot{T}$ at a distance of 0.5 mm from the edge was 150-250 °C/s and ~20 °C/s at the centre of the rod. It is interesting to note that only minor differences were observed between different casting conditions.

Table III: Secondary dendrite arm spacing λ_2 and estimated cooling rates $\dot{T}$

		#2			#4		
		0.5 mm	2.5 mm	5 mm	0.5 mm	2.5 mm	5 mm
perpendicular	λ_2 [μm]	4.5	6.6	10.6	3.7	6.1	10.1
grains	$\dot{T}$ [°C/s]	156	64	21	246	77	24
inclined	λ_2 [μm]	5.2	6.2	10.7	5.8	7.7	11.3
grains	$\dot{T}$ [°C/s]	112	74	21	87	45	18

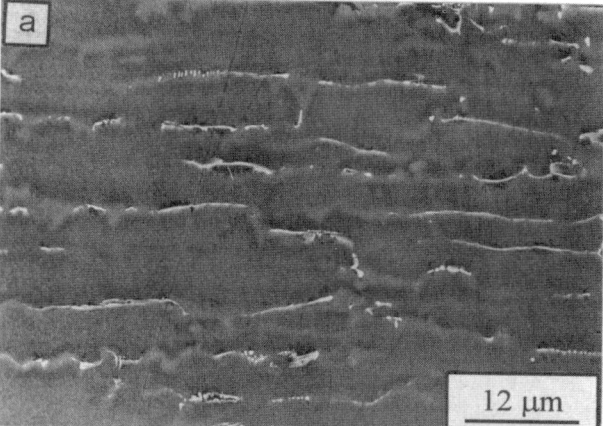

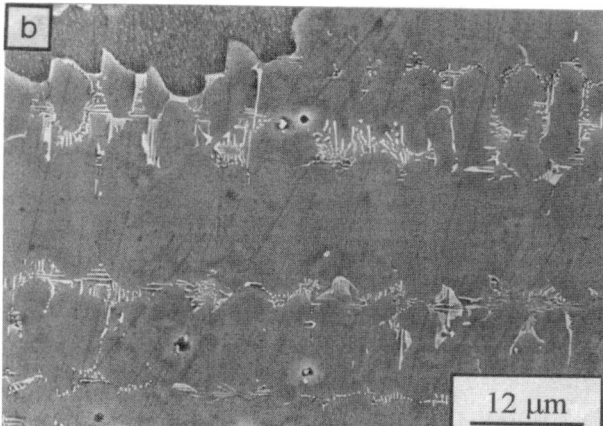

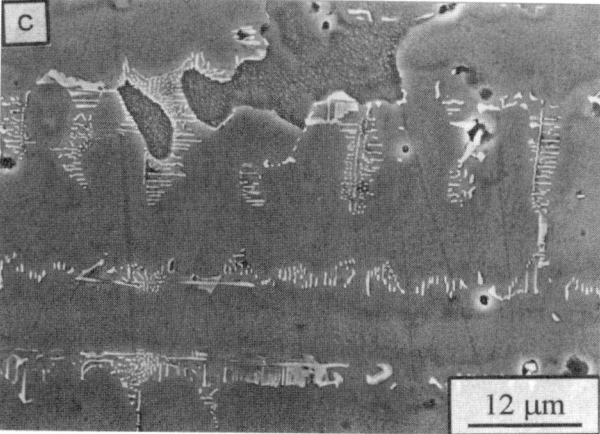

Figure 4: Microstructure of a grain growing perpendicular to the casting direction at different distances from the edge of the rod: a) 0.5 mm, b) 2.5 mm and c) 5 mm. (SEM, casting conditions #1). Dark γ-matrix, bright MC carbides

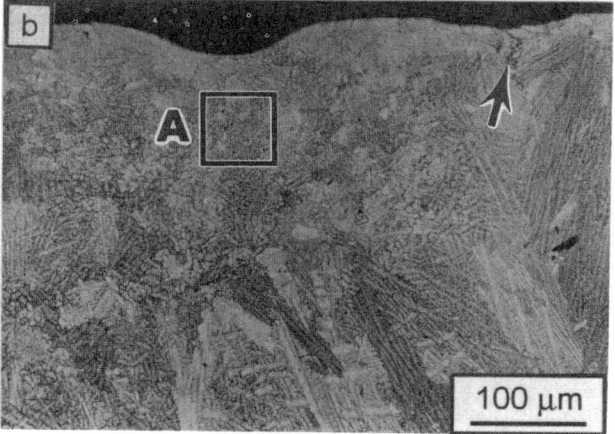

Figure 5: Light-optical micrographs of the region around a large intrusion, which have formed during the resting period and the reverse stroke a) casting conditions #1, b) casting conditions #4

The resting time and the reverse stroke

Less uniform microstructure has formed during the resting period and the reverse drawing stroke. Figure 5 shows this part of the rod at the smallest (conditions #1, 10.33 mm/s) and at the highest (conditions #4, 12.33 mm/s) casting speed. Just below the large intrusion in Fig. 5 a, we can observe that reorientation of γ grains (dendrite trunks) started to occur. This suggests that (1) in the course of the resting period and the reverse stroke the melt was not able to come into direct contact with the Cu-Be mould and (2) that the heat extraction direction had changed from being perpendicular to the mould wall towards the direction almost aligned with the rod axis. This could only happen if γ grains had started to grow into the inlet nozzle, which prevented dissipation of heat in the radial direction. Thus the relative amount of heat conducted in the axial direction increased considerably. In Fig. 5 a dendrite marked by the arrow is almost ideally aligned with the casting direction. It is approximately 150 μm from the rod edge. This clearly indicates that it must have grown along the inner wall of the inlet nozzle.

Primary and secondary dendrite arm spacing in the inclined grains are slightly larger than those of the perpendicular grains (Fig. 6 and Table III). It could therefore be inferred that cooling rates obtained during the resting time and the reverse stroke were smaller than those during the drawing stroke.

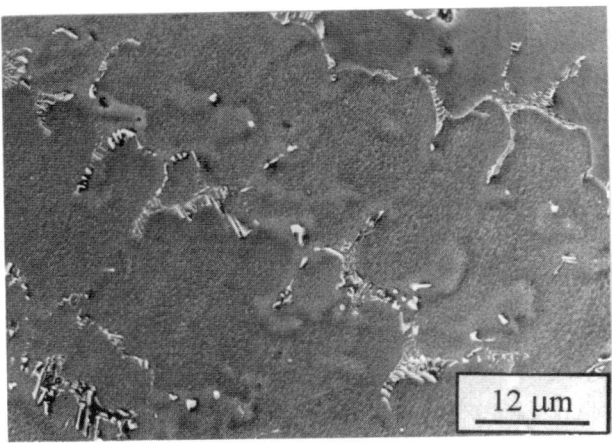

Figure 6: Microstructure of a grain growing inclined to the casting direction at a distance of 2.5 mm from the edge of the rod (SEM, casting conditions #1)

As a consequence of the growth of γ grains into the inlet nozzle and the fact that the inner diameter of the inlet nozzle was smaller than the diameter of the rod, a step has formed on the surface of the rod in the course of the resting time and the reverse stroke. At the initial stage of the following drawing stroke metallostatic pressure tried to force melt into the gap between the mould and the already solid shell. From Fig. 5 a it can be seen that the large marking is a result of incomplete filling of the mould; this means that the melt had solidified before completely filling the gap. With these processes, uniformly spaced large intrusions have formed. The microstructure formed in the gap consists of equiaxed γ grains with a much larger amount of MC/γ eutectic than in the other parts of the rod (Fig. 7 a). The latter observation indicates the melt which penetrated into the gap was already enriched by positively segregating elements; for instance Nb, C, Ti and Mo.

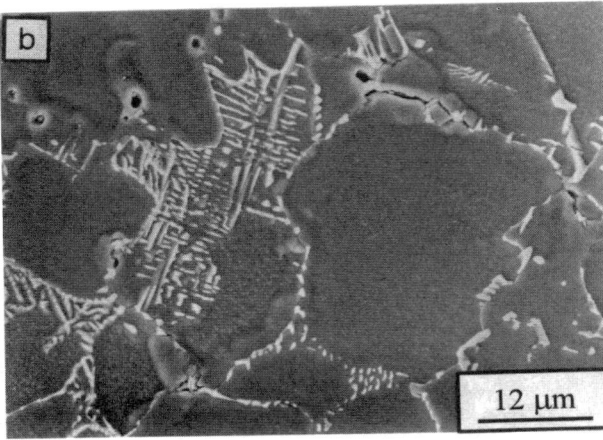

Figure 7: Scanning electron micrographs of a) the region marked by A in Fig. 5a, and b) the region marked by A in Fig. 5b

At casting speeds larger than 11.17 mm/s we were not able to observe dendrites growing parallel to the rod axis (along the inner wall of the inlet nozzle). On the other hand, the region of equiaxed grains next to large intrusions was considerably thicker than at casting conditions #1 and #2. This can be explained if we suppose that the temperature gradient in the liquid before the solid liquid interface increases with the growing casting speed. This would mean that at the beginning of the drawing stroke much hotter melt penetrates into the gap between the mould wall and the solid shell, causing partial remelting of the already solid rod in the vicinity of a large intrusion. Much coarser γ grains and γ/MC eutectic indicate smaller cooling rates compared to the casting conditions #1 and #2 (compare Fig. 7 a and b).

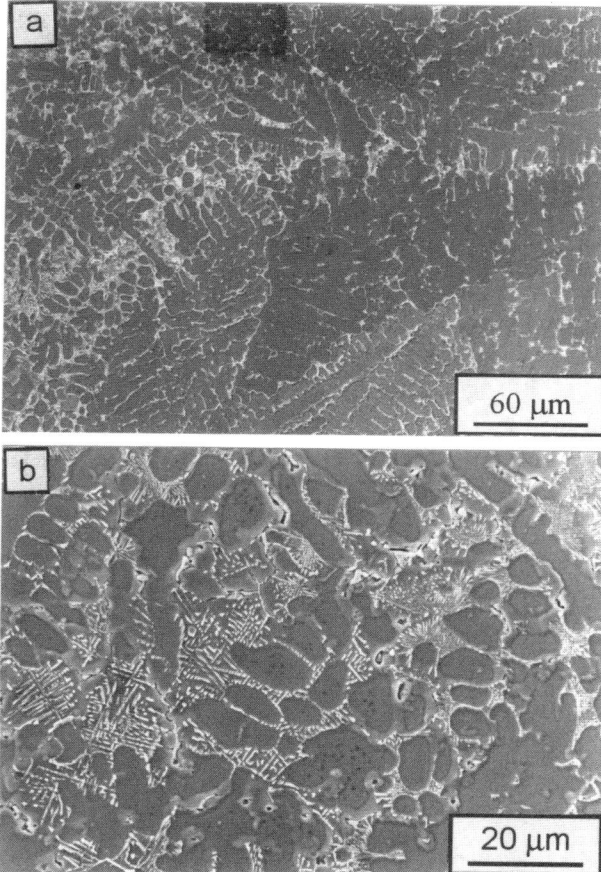

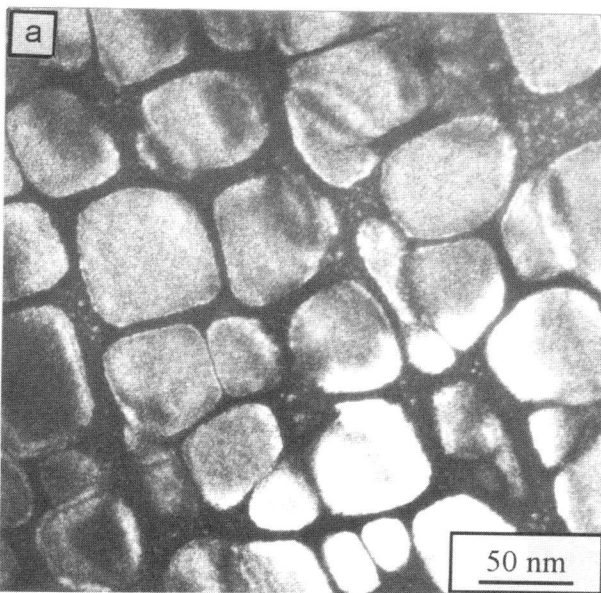

acceleration at the beginning of the drawing stroke. This and other possibilities are to be investigated in our future work.

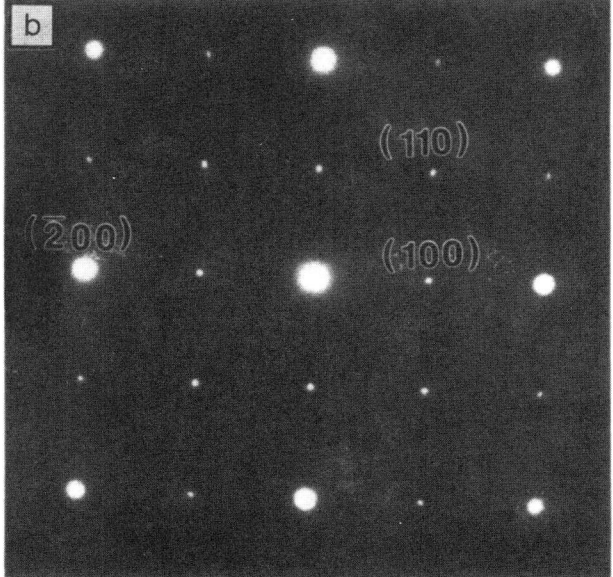

Figure 8: Scanning electron micrographs of a region under a shallow intrusion in Fig. 5a

Hot tearing

In Fig. 5 b we can observe a smaller intrusion, which is of the same type as those indicated by the smaller arrows in Fig. 3. Such intrusions are usually positioned at the beginning of the drawing stroke. With the small intrusion, a larger amount of MC/γ eutectic is present than elsewhere. This eutectic-rich band protrudes some 0.5 mm from the surface of the bar. It is believed that it forms because of hot tearing – formation of cracks by separating grains along their grain boundaries in the mushy zone. The occurrence of hot tearing can be explained as follows. Stresses arising from the acceleration of the rod during the initial stage of a drawing stroke resulted in a stretching of the thin solid shell and the separation of γ grains along the grain boundaries in the crack sensitive mushy zone. The melt enriched with the positively segregating elements filled the crack, and a eutectic band following the initial crack path has formed. Under the casting conditions applied, the appearance of hot tearing did not cause breakouts, but it was observed that the length of the cracks was increasing with casting speed, indicating that there exists a maximum casting speed, above which the reliability and safety of the continuous casting of IN 713C may be seriously endangered. The possibility of increasing production rate without decreasing safety and reliability would probably be a decrease of the

Figure 9: Transmission electron micrograph (a); and its selected area diffraction pattern (b). Small spots: a γ' precipitate, large spots: γ matrix.

Precipitation of γ'

The precipitation of γ' precipitates took place during cooling at temperatures under the γ'-solvus. So far, only a preliminary investigation has been carried out. The results indicate that duplex-sized γ' precipitates are present in the γ matrix (Fig. 9a). The average size of the larger population is around 60 nm, and the average size of the smaller population is around 5-10 nm. It is believed that the larger population has formed at higher temperatures and the smaller ones at lower temperatures. SADP (Selected Area Diffraction Pattern) of a large precipitate showed that it is coherent with the γ matrix (Fig 9 b), as usual in Ni-base superalloys [5]. Further work regarding this topic is in progress and will be reported later.

Conclusions

The results of the investigation show that a Ni-base superalloy IN 713C can be continuously cast over a wide range of casting parameters. The behaviour of IN 713C during the continuous casting of a small circular cross section in a vertical continuous caster strongly suggests that it would also be convenient for continuous casting using large-scale industrial equipment.

The continuously cast rods possessed a clean surface and excellent internal integrity. The microstructure of continuously cast rods consisted, regardless of the casting parameters used, of γ–grains, interdendritic γ/MC eutectic and coherent γ' precipitates distributed uniformly within the γ-matrix. The size, distribution, morphology and orientation of microstructural constituents depended strongly on local solidification conditions, which were greatly influenced by the nozzle/mould set up and the alternating drawing mode of continuous casting. The macrostructure changes periodically, and the wavelength of periodicity corresponds to the length of the casting step. According to the results of alloy characterisation some changes regarding the nozzle/mould set up and the alternating drawing mode will be necessary to attain all the goals of the investigation.

Acknowledgements

Persons at the National Centre for HREM at Delft University of Technology are gratefully acknowledged for investigations by transmission electron microscopy. We also thank Derek Hendley from Ross & Catherall, Sheffield, UK for providing the alloy and useful advice regarding the melting procedure.

References

1. E. Herrmann, D. Hoffman (eds), Handbook on Continuous Casting, (Aluminium-Verlag Düsseldorf 1980), V-VI

2. "INVESTMENT CASTING - IMPROVED QUALITY FOR SUPERALLOY BAR STOCK - Ross & Catherall combines vacuum induction melting with continuous casting", Foundry Trade Journal, 173 (1999), 3555, B, 52 / 4

3. A.K. Bhambri, T.Z. Kattamis, J.E. Morral: "Cast Microstructure of Inconel 713C and its Dependence on Solidification Variables", Metallurgical Transactions B, 6B, (1975), 523-537

4. Alloy Digest, Data on world wide metals and alloys, Inconel 713C, February, 1959

5. J.R. Davies (ed.), ASM Specialty Handbook, Heat Resistant Materials, ASM International, (1997), Metallurgy, Processing and Properties of Superalloys, 221-254

THE THERMAL ANALYSIS OF THE MUSHY ZONE AND GRAIN STRUCTURE CHANGES DURING DIRECTIONAL SOLIDIFICATION OF SUPERALLOYS

S. U. An, V. Larionov*, V. Monastyrski**, E. Monastyrskaia*, I. Grafas*,
J. M. Oh***, O. D. Lim, S. H. Kim, J. H. Lee****, D. Y. Seo*****

Division of Materials, Korea Institute of Science and Technology, Seoul 136-791, Korea
*Aerocast International, Taejon 306-230, Korea
**Russian Academy of Sciences, "IVTAN" Association, Moscow, 127412, Russia
and Taeseong Precision Industrial Co., Daegu 704-320, Korea
***Korea Electric Power Research Institute, Taejon 305-380, Korea
****R&D Center, Han Kook Tire Co., Taejon 305-343, Korea
***** Department of Aerospace and Mechanical Engineering
Carleton University, 1125 Colonel By Drive, Ottawa, KIS 5B6, Canada

Abstract

The influence of the casting parameters (such as heater temperature, withdrawal rate and thermal conductivity of the mould) on the mushy zone of directionally solidified (DS) casting by the Bridgeman's method have been studied by computer simulation. The grain widening phenomena in DS castings on two superalloys IN738LC and B1914 has been investigated. Based on the geometrical theory of moving grain interface, an empirical dependence of the columnar grains widening on the distance from the nucleation site has been obtained.

Introduction

The advantages of using columnar structured buckets in newly designed high temperature engines manifests itself by increased efficiency, life cycle and decreased overall running cost. The elimination of transverse grain boundaries increases the creep strength and ductility of the material at operating temperatures and its resistance to relative thermal-cycle fatigue. The growth structure conditions provide high density of casting material by decreasing micro porosity and segregation, and changing the shape and distribution of carbides. All these factors influence the structural stability at high operating temperatures.

To define the growth conditions of casting structure during DS process the following parameters are used: temperature gradient, cooling rate and solidification rate. The dependence of their values on technical casting parameters has been extensively studied experimentally and theoretically. It is known, that the increase of heater temperature results in an increase of temperature gradient, and the increase of withdrawal rate results in an increase of melt cooling rate. The change of growth conditions is esteemed as a consequence of change of heat flows transferred through the mushy zone. During DS process of superalloys, the extension of the mushy zone along a direction of the solidification can be some centimeters. In this case a change of technical casting parameters results in a change of heat flows through the mushy zone and also in a change of its configuration. The growth conditions in mushy zone are essentially different and can not be described by the values of the temperature gradient and cooling rate in any single point. The first part of the present paper describes the results of the theoretical study of growth conditions on the mushy zone boundaries - on the surfaces of solidus and liquidus.

Among the structural factors, which define the mechanical properties of a material, the grain structure shows an important feature. The columnar grains nucleated on the chill-plate change their transversal dimensions with increased distance from the chill-plate. The phenomenon of grain structure spreading is observed in DS bucket for industrial gas turbine engines. In previous papers the authors [1-3] have suggested a theoretical model of grain spreading process in conditions of unidirectional heat transfer. The model is based on the concept of mobility of grain boundaries during growth. It has shown that the increase of grain cross sections obeys the law of normal kinetic. The second part of the paper describes data of columnar grain structure investigation of real buckets and blades. To analyze the obtained experimental data, the geometrical theory of interface motion was used.

The mushy zone of casting

As is known, during the directional solidification (DS) of superalloys, macro- and microstructure forms in the mushy zone where dendrites of the solid phase and the melt coexist. The mushy zone boundaries are the isothermal surfaces $T_s = const$ and $T_L = const$.

Superalloys 2000
Edited by T.M. Pollock, R.D. Kissinger, R.R. Bowman,
K.A. Green, M. McLean, S. Olson, and J.J. Schirra
TMS (The Minerals, Metals & Materials Society), 2000

The mushy zone extent along the direction of the solidification can be from several millimeters to several centimeters depending on a temperature interval of alloy crystallization ($\Delta T = T_L - T_S$) and thermal gradient. The thermal gradient varies from point to point of the isothermal surfaces $T_L = const$ and $T_S = const$ and the values of thermal gradient can essentially be different.

The scheme of heat flows for the mushy zone is shown in Figure 1.

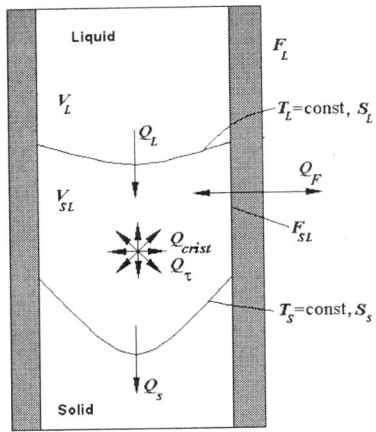

Figure 1: The scheme of thermal flows in the mushy zone of DS casting

The heat balance equation for a mushy zone can be written as:

$$Q_S = Q_L + Q_{crist} + Q_\tau + Q_F , \qquad (1)$$

where Q_S is the quantity of heat output from the mushy zone through the surface $T_S = const$; Q_L is the quantity of heat input to the mushy zone through the surface $T_L = const$; Q_{crist} is the crystallization heat; Q_τ is the quantity of heat output from the mushy zone as a result of metal cooling; Q_F is the quantity of heat input to (or output from) the mushy zone through the mould surface.

$$Q_S = \lambda_m(T_S) \int_{S_s} |gradT| dF = \lambda_S G_S S_S \qquad (2)$$

$$Q_L = \lambda_m(T_L) \int_{S_L} |gradT| dF = \lambda_L G_L S_L \qquad (3)$$

where $\lambda_m(T)$ is the thermal conductivity of the metal depending on temperature; G_L, G_S are the average thermal gradients at the surfaces of T_L and $T_S = const$; S_L, S_S are the areas of the surfaces T_L and $T_S = const$; λ_L, λ_S are the thermal conductivity at the temperature T_L and T_S.

When the mushy zone moves with a steady velocity w and its configuration remains constant while moving

$$Q_{crist} = \rho_L L w S_c \qquad (4)$$

where ρ_L is the melt density at the temperature T_L; L is the latent heat; S_S is the transversal section area of the casting, and

$$Q_\tau = w \int_{V_{SL}} \rho_m(T) C_m(T) |gradT| dV \qquad (5)$$

where $\rho_m(T)$ is the density of alloy; $C_m(T)$ is the specific heat of alloy; V_{SL} is the volume of the mushy zone.

The influence of last term of the equation (1) depends on the mushy zone position in the heater.

$$Q_F = \int_{F_{SL}} q_f dF \qquad (6)$$

where q_f is the heat flow through the lateral surface of the casting; F_{SL} is the lateral surface of the casting in the mushy zone (see Figure.1). Thus, the heat balance equation for the mushy zone (1) becomes

$$\lambda_S G_S S_S = \lambda_L G_L S_L + w \left[\rho_L L S_c + \int_{V_{SL}} \rho_m(T) C_m(T) |gradT| dV \right] + \int_{F_{SL}} q_f dF \qquad (7)$$

The equation for a heat flow through the isothermal surface of $T_L = const$ can be written similarly equation (7):

$$\lambda_L G_L S_L = \int_{V_L} \rho_m(T) C_m(T) \frac{\partial T}{\partial \tau} dV + \int_{F_L} q_f dF , \qquad (8)$$

where F_L is the surface of casting above the isothermal surface $T_L = const$.

The last terms of the equations (7) and (8) determine the quantity of heat transferred through the casting surface as a result of a heat exchange with a furnace. They depend on the mushy zone position in the heater and the heater temperature. Thus, the last term of equation (7) can change its sign depending on the mushy zone position into the hot or cold zone of the furnace. The second term of the equation (7) contains the solidification rate w, which is proportional to the mould withdrawal rate. The first term of equation (8) contains the cooling rate of the melt $\partial T/\partial \tau$, which also is proportional to the withdrawal rate.

Average thermal gradients G_L and G_S are proportional to the quantity of heat transferred through the corresponding isothermal surface in unit time. Therefore it is necessary to increase heat flows through the mushy zone to increase the thermal gradient. On the other hand, G_L and G_S are inversely proportional to the areas of the isothermal surfaces. The increase of heat flows will entail a displacement of the mushy zone in a heater, a flexure of its configuration and, hence, an increase of the isothermal surface solidus (liquidus) area S_L, S_S. In this case a change of the casting conditions will entail a reduction of the temperature gradients.

Thus, the equations (7) and (8) demonstrate a complex dependence of the thermal gradients G_L and G_S from major casting process parameters – the mould withdrawal rate and the temperature of the heater.

Mathematical model

To define forming conditions of the casting structure, computing experiments have been conducted. The software CASTDS^{2D} was used to simulate the thermal conditions of the Bridgman-type DS process. The effect of the heater temperature, the withdrawal rate and the mould thermal conductivity on the thermal gradient, solidification rate and the melt cooling rate in the mushy zone have been studied.

Mathematical basis of the CASTDS^{2D} model is a numerical solution of the unsteady heat conduction equation for the cylindrical mould.

In the CASTDS^{2D} software the furnace consists of a cylindrical heater, a cylindrical side cooler and a chill plate. Thermal properties of mould material and metal depend on the temperature.

The input data for the simulation are the temperature profiles along the furnace as a function of time, initial mould position and withdrawal rate, the pouring time point and the temperature of the metal poured into the mould.

Computing experiments consisted of the mould heating process simulation from a room temperature up to the pouring time point, the imitation of pouring process, calculation of the temperature distribution in the mould and solidifying metal during their moving out of heater.

The following factors have been taken into consideration:
- the radiation heat transfer from the surfaces of the heater to the surface of the ceramic mould and the melt surface;
- the radiation heat transfer between the ceramic mould surface and upper and lower surfaces of the baffle between hot and cold zones;
- the radiation heat transfer from the surface of the ceramic mould to the surfaces of the vacuum chamber and the side cooler;
- the convection and radiation heat transfer from the mould and metal to the chill plate. The chill plate temperature is established as a result of simulation of this heat process.

It was supposed that all elements of the furnace hot zone were made of graphite and their emissivity factor is equal to 1. The emissivity factors of ceramics and a melt in the mould are much less than 1. However the surfaces of the melt and the cylindrical mould are convex, so that it can be excluded from the account multiple reflections in the radiation heat transfer simulation.

The computing experiments were carried out for the cylindrical form casting with a diameter of 22mm and a length of 450 mm. In Figures 2a and 2b the typical curves of thermal gradients G_L and G_S, solidification rates w_L and w_S of the solidus and the liquidus isotherms are shown depending on their position in casting. In Figure 2d the position of isotherms T_L and T_S in casting depending on its position in a heater is shown. All results relate to an axis of casting.

It is possible to assume that the conditions in the mushy zone in initial time point mainly depend on metal pouring temperature and temperature of the heater. In a mushy zone the conditions are established in compliance with a given regime of the directional solidification during the initial stage of solidification. The mushy zone displaces in the heater to the position appropriated to the given withdrawal rate and the heater temperature. At this stage of solidification the velocity of the isotherms T_L and T_S movement can differ strongly from the withdrawal rate.

As shown in Figure 2, the initial stage is completed when it has reached stable conditions of solidification at constant values of the thermal gradient and the solidification rate. The time required to reach a steady state, and the duration of steady state solidification depends on the casting parameters and geometry of the casting. The final stage of a solidification of casting takes place, when an influence of the top of casting becomes significant. At high heater temperature the additional heat of metal through its free surface results to an increase of thermal gradient and displacement of the mushy zone in the direction of the cooler. At low heater temperature, insufficiently intensive heat of metal results in a displacement of the mushy zone in a direction of the heater and decrease of thermal gradient.

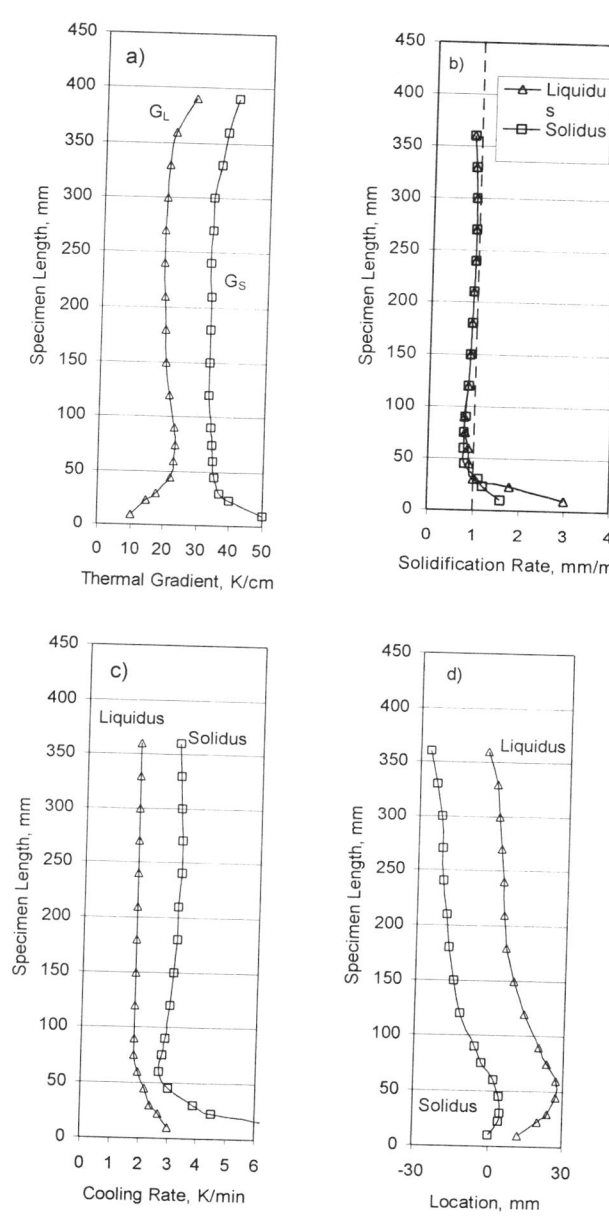

Figure 2: The typical calculated curves of the thermal gradient (a), solidification rate (b), cooling rate (c) and the mushy zone position (d) for simulated directional solidification process.

The results are related to a steady state regime of the solidification. The values G_L and w_L at the isotherm $T_L=const$ and its position x_L were determined at the time point, when casting axis temperature of section x=240 mm $T=T_L$. The values G_S, w_S and x_S, were determined at the same time point in the section, in which the casting axis temperature was $T=T_S$.

Influence of the withdrawal rate.

As shown in equations (7) and (8), there are additional sources of heat at a high rate of the directional solidification, which should supply higher temperature gradient in comparison with that of quasistationary conditions. However, when the withdrawal rate is increased, the temperature gradient grows at the solidus only (Figure 3).

At the low withdrawal rate, about several millimeters in an hour, the contribution of the non-stationary terms of the right part of the equation (7) is insignificant. The relationship between G_S and G_L is determined as the relationship of surfaces S_S and S_L.

results in an increase of the amount of heat left from the mushy zone through the solidus surface and an increase of the melt-cooling rate at the solidus U_S.

A change of the relationship between G_S and G_L due to a change of the mushy zone configuration, also results in a change of the relationship between G_S and G_L. The G_S growth with an increase of the withdrawal rate is decelerated with a displacement of the mushy zone in the side cooler area. Consequently, a sign of the last term is changed in equations (7) and (8).

On the one hand, the mushy zone cooling through the mould side walls reduces a part of the thermal energy took off through the surface S_S, and on the other hand, it results in an increase of the surface S_S area.

The influence of the withdrawal rate on conditions in the mushy zone will vary depending on the furnace design, thermal properties of the mould and casting geometry.

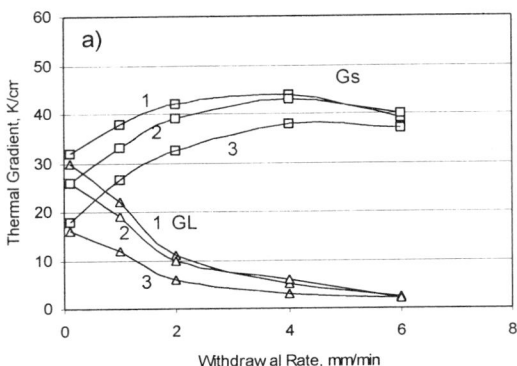

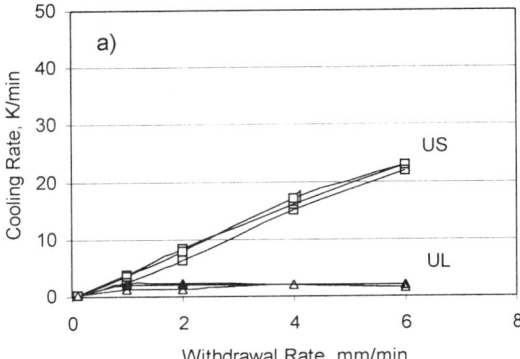

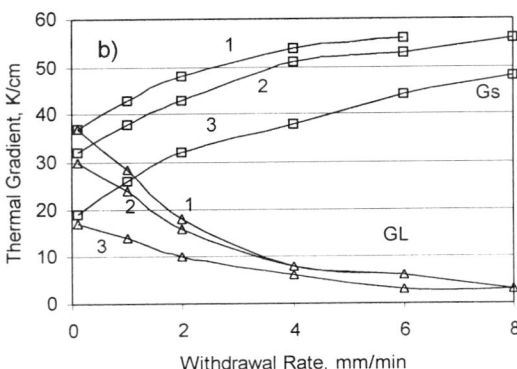

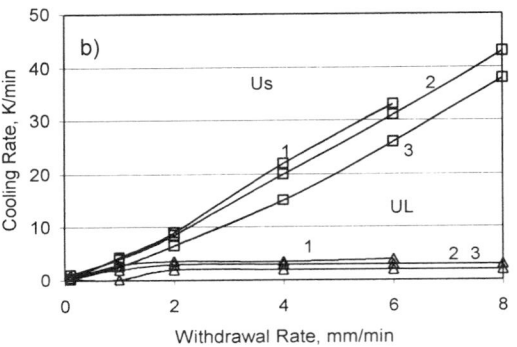

Figure 3: Calculated temperature gradient at the isothermal surfaces of solidus G_S and liquidus G_L depending on the withdrawal rate and heater temperature. (a) - low thermal conductivity of the mould λ_f=2.2$W/(mK)$; (b) - high thermal conductivity of the mould λ_f=20.2$W/(mK)$. 1 - the heater temperature T_h=1700°C; 2 - the heater temperature T_h=1600°C; 3 - the heater temperature T_h=1500°C.

Figure 4: Calculated melt cooling rate at the isothermal surfaces of solidus U_S and liquidus U_L depending on the withdrawal rate and heater temperature. (a) - low thermal conductivity of the mould λ_f=2.2$W/(mK)$; (b) - high thermal conductivity of the mould λ_f=20.2$W/(mK)$. 1 - the heater temperature T_h=1700°C; 2 - the heater temperature T_h=1600°C; 3 - the heater temperature T_h=1500°C

Influence of the thermal conductivity of the mould.

With increased withdrawal rate, the liberation of the latent heat results in heating of metal in the mushy zone. It reduces G_L and consequently the cooling rate of the melt U_L. Simultaneously, it

The results, shown in Figures 3-6, enable are to estimate a change of the thermal conditions in the mushy zone in a practical significant range of the mould thermal resistance. The calculations

250

are presented for the mould thickness of 14 mm and the thermal conductivity of ceramics $\lambda_f=2.2W/(mK)$ and $\lambda_f=20.2W/(mK)$.

The tenfold increase of the thermal conductivity results in some increase of the temperature gradient and the cooling rate in the mushy zone. The greatest growth (up to 30 %) of this value is observed at the high heater temperature (Figure 6) and at the high withdrawal rate (Figures 3 and 4).

The influence of thermal conductivity of the mould on the mushy zone position in the furnace is quite significant. The increase of the thermal conductivity has not changed the sensitivity of the mushy zone position to the temperature of a heater, but it has made the considerably less sensitive to the withdrawal rate.

According to the Figure 5a, for a mould with the thermal conductivity $\lambda_f=2.2W/m/K$, the withdrawal rate 5-6 mm/min fits to a very low mushy zone position practically at the exit of the side cooler. It is possible to assume that the further increase of the withdrawal rate will reduce to the violation of the unidirectional heat transfer, i.e. to the equiaxed solidification.

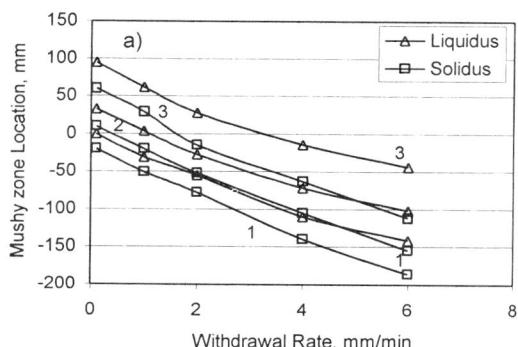

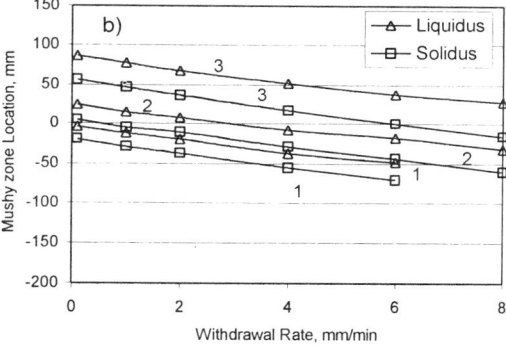

Figure 5: Calculated position of the isothermal liquidus and solidus surfaces depending on the withdrawal rate and heater temperature. (a) low thermal conductivity of the mould $\lambda_f=2.2W/(mK)$; (b) high thermal conductivity of the mould $\lambda_f=20.2W/(mK)$. 1- the heater temperature $T_h=1700°C$; 2 - the heater temperature $T_h=1600°C$; 3 - the heater temperature $T_h=1500°C$.

In case of solidification in a high thermal conductivity mould, an increase of the withdrawal rate has a smaller effect on the mushy zone position. The application of higher heat conductive ceramics allows to increase the withdrawal rate at the invariable position of the mushy zone in the furnace.

Materials and Experimental

Casting of buckets was undertaken with a commercial scale vacuum induction furnace. The Bridgmen's method was used as a base of casting technology. The furnace has the following technical parameters: weight of liquid metal is about 20kg, operating height of the solidification unit is 400mm, three- zones heater with individual regulation, maximal operating temperature of heater is 1650°C and diameter of chill-plate is 210mm.

The withdrawal mechanism allows to regulate velocity of mould moving from 0.01 mm/min up to 20 mm/min. The special transitions provide the possibility of using 20 and more thermocouples to monitor the temperature during casting.

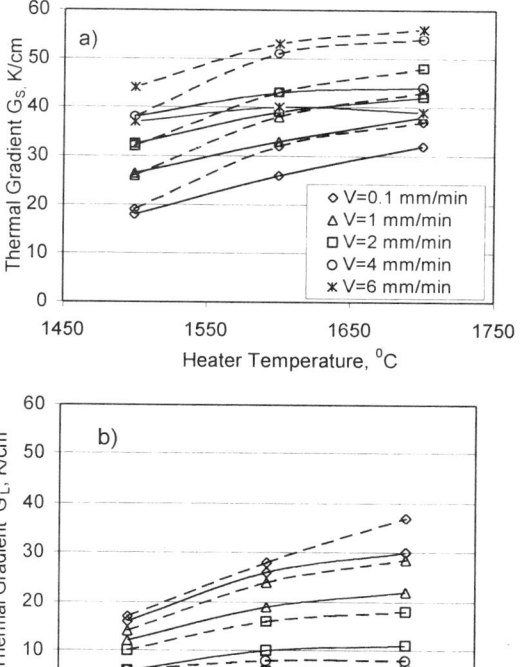

Figure 6: Calculated temperature gradient at the isothermal surfaces of solidus G_S (a) and liquidus G_L (b) depending on the heater temperature and the mould thermal conductivity. Solid lines mean low thermal conductivity of the mould $\lambda_f=2.2W/(mK)$; Dashed lines mean high thermal conductivity of the mould $\lambda_f=20.2W/(mK)$.

To investigate alloy"s composition effect, IN738LC (WILLAN Co., UK) and B1914 were used. The chemical compositions of these alloys are presented in Table 1.

Growth of the structure of blades and buckets was undertaken at the furnace temperatures of 1450-1500°C and under the withdrawal rate of 2.5-6.0 mm/min.

Transverse sections of the top, middle and transition to platform for DS-airfoil were examined by metallographic techniques.

TABLE I. Composition of the Alloys (mass %)

Element	B 1914	IN 738LC
Cr	10.10	15.78
Co	9.80	8.42
Mo	3.10	1.70
Ti	5.27	3.37
Al	5.58	3.32
C	0.001	0.10
Ta	-	1.70
W	-	2.55
B	0.11	0.008
Fe	0.11	10 ppm
Nb	-	0.70
Ni	Bal.	Bal.

The OMNIMET® 1" BUEHLER Image Analyze System was used to examine the grain sizes. The Standards Test Method for Determining Average Grain Size (E 1382-91 and E1181) was used. Each grain was measured in four different positions through the center of grain.

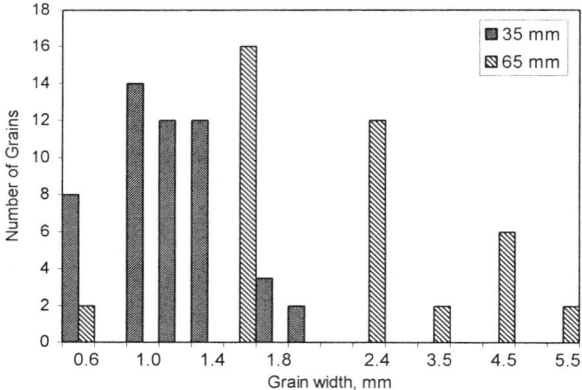

Figure 8: The distribution of measured grain width in two transversal sections of directionally solidified blades of B1914 alloys.

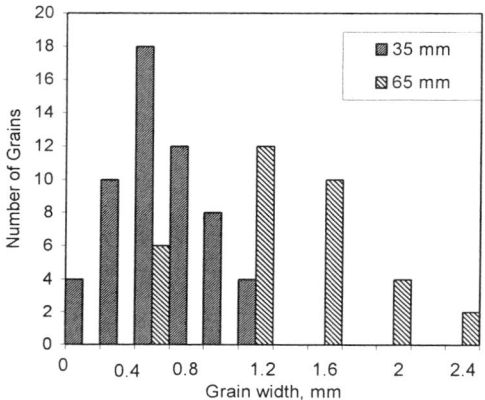

Figure 7: The distribution of the measured grain width in the two transversal sections of directionally solidified blades of IN738LC.

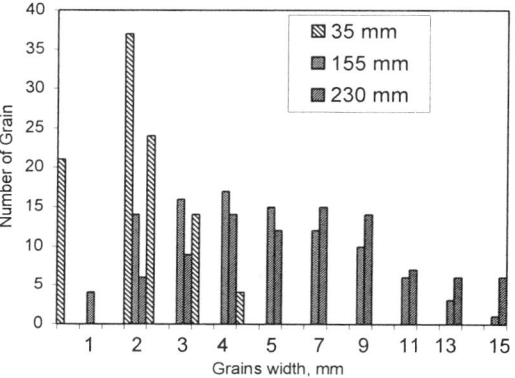

Figure 9: Grain size distribution versus distance from the chill-plate for the DS bucket of IN738LC alloy

Grain Size Measurement

The measurement data of grains from castings with columnar structure are presented on Figures 7 - 10 in the histogram fashion. The analysis of results shows that the grain spreading takes place in all cases of structure growth. The value of spreading is not identical for the studied alloys. The grains in castings of B1914 alloy spread more intensively.

In small size castings (total length is about 80 mm, thickness of airfoil is about 2-4mm, weight is about 0.1-03kg) the columnar structure is characterized with small sized grains cross section at the entrance into the airfoil. The dimensions of majority grains do not exceed 1.2mm with the maximum distribution curve at 0.6mm for IN738LC and 2.0mm with maximum of 0.8-1.0 mm for B1914 alloys.

The estimation of grains sizes in the sections, placed at the distance 30mm showed displacement of maximum of distribution curves to the right till 1,3mm for IN738LC and 1,5 – 1,7mm for B1914.

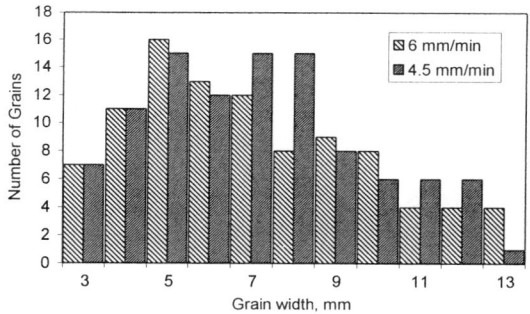

Figure 10 shows the measurement data of grains size in the cross section of the root of buckets solidified at the two different withdrawal rates.

The same tendency was noticed at the estimation of the grain dimensions in buckets of IN738LC. The grain structure loses the uniformity with the distance from chill-plate. The structure of the cross section at the distance of 35mm from chill-plate is rather uniform. The maximum of the distribution curve moves also to the right, (to the side of bigger grain dimensions - 4.0 mm at the distance 155mm and 5.5 mm at 235 mm respectively). It is necessary to mark the appearance of the very large grains with a cross section of 10-16 mm though in small quantities - no more than 4% of the total number of grains (Figure 9).

<u>Discussion</u>

As shown in Figures 7 – 10, the transverse size of columnar grains increases with the increase of a distance from the examined section to the chill-plate. To describe the growth of the columnar grain width quantitatively, the following physical model can be considered. All grains have nucleated on the chill plate and grow by the mushy zone movement. The average transverse grain size increases with the distance from the chill-plate. The increase of the transversal size of grains reached section X has taken place at the expense of their neighbors during competitive growth.

According to [1] the grain width grows under the normal growth law, independent of the growth morphology and can be described as following:

$$R^m = R_o^m + M\tau \qquad (9)$$

where τ is the growing time, R is the average transverse grain size at the time τ, R_0 is the average transverse grain size at time $\tau=0$; M is mobility of the grain boundary; m is the grain growth exponent.

It is obvious to assume that the transverse grain growth is possible only into the mushy zone. Then the time of the grain growth will be determined with the mushy zone moving velocity and its position in casting. Assuming that the mushy zone moving velocity is equal to the withdrawal rate, then the equation (9) can be rewritten as

$$R^m = R_o^m + M\frac{X}{V} \qquad (10)$$

where X is the mushy zone position at time τ, V is the withdrawal rate.

As the initial size of grain nucleated on the chill-plate $R \gg R_0$, the equation (10) can be transformed to

$$R = \left(M\frac{X}{V}\right)^{1/m} \qquad (11)$$

Thus, the experimentally observable increase of the transverse size of columnar grains in the directional solidification can be described by

$$R = A\left(\frac{X}{V}\right)^n \qquad (12)$$

where A and n are coefficient and exponent defined from experimental data.

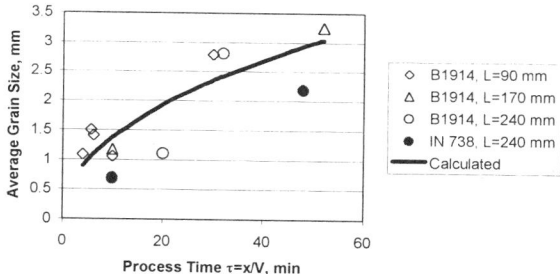

Figure 11: Comparison of calculated and measured columnar grain size of DS blades and buckets. L is the length of casting part.

Figure 11 shows the measured average transverse size of columnar grains in different cross-sections of blades and buckets with different withdrawal rate. The received data of B1914 alloy are approximated by the equation,

$$R = 0.46\left(\frac{X}{V}\right)^{\frac{1}{2}}, \qquad (13)$$

which is in good agreement with experimental data.

In equation (13) exponent $n=1/2$ and hence in the equations (9) - (11) exponent $m=2$. This value does not contradict the conventional notion when $2 \le m \le 3$ [4]. In addition, Wagner reported that $m=3$ for the diffusion controlled grain growth and $m=2$ for the interfacial-reaction controlled growth. However, it is not easy to debate the mechanism of interaction of grains at the competitive growth on the castings with a variable cross-section area - real turbine blades. The model of the competitive growth of columnar grains, suggested in [5] and [6], explains transversal growth of the grains at the expense of neighbors by the leading growth of the dendrites with preferable crystallographic orientation. Based on this model, it is possible to assume the process of transversal growth of columnar grains is controlled more dominantly by an interaction of the dendritic skeleton of grains than the diffusion processes in the mushy zone. Therefore, the exponent $m=2$ obtained from equation (13) seems more realistic. Therefore, we suggest in this work that the interfacial reaction process rather than the diffusion process control the grain widening in DS casting.

<u>Conclusions</u>

Modeling of different conditions of Bridgman's solidification allows one to investigate the influence of the technical casting parameters such as withdrawal rate, heater temperature and thermal conductivity of mould on the value of thermal gradient and cooling rate at the liquidus and solidus surfaces and on the mushy zone position.

The columnar grain structure of blades and buckets of B1914 and IN 738LC alloys solidified by Bridgman's process was investigated. It has been shown that the average transversal size of the columnar grains increases along the casting. The analysis of the obtained experimental data, based on the geometrical theory of interface motion, has allowed one to receive an approximate dependence of the columnar grains transversal size on the distance up to their nucleation site The result of approximation is in agreement with the experimental data.

References

1.   E. A. Holm, N. Zacharopoulos and D. J. Srolovitz, "Nonuniform and Directional Grain Growth Caused by Grain Boundary Mobility Variations", Acta Mater., 46 (3) (1998), 953-964.

2.   J. E. Taylor, J. W. Cahn and C. A. Handwerker, "Geometric Models of Crystal Growth", Acta Metall. Mater., 40 (7) (1992), 1443-1474.

3.   B. Szupar and R. W. Smith, "Monte Carlo Simulation of Solidification Processes; Porosity", Canadian Metallurgical Quarterly, 35 (1996) 299.

4.   V. Tikare and J. D. Cawley, "Numerical simulation of grain growth in liquid phase sintered materials", Acta Mater., 46 (4) 1998, 1343-1356.

5.   M. Rapaz and Ch.-A. Gandin, "Probabilistic modelling of microstructure formation in solidification processes", Acta Mater., 41 (2) (1993), 345-360.

6.   V. Monastyrsky and E. Kachanov. "The model of the competitive growth of columnar grain of multi-component alloy". Solidification and modeling. Transaction of the IV All Russia conference on problem on alloys solidification and computer modelling, Igevsk, October (1990), 10-24 (in Russian).

Freckle Formation in Superalloys

P. Auburtin, S. L. Cockcroft, A. Mitchell and T. Wang

Advanced Materials and Process Engineering Laboratory
University of British Columbia
Vancouver V6T 1Z4 BC
Canada

Abstract

The analysis of the formation mechanisms for freckling and for random grain formation in single crystal castings has resulted in a general understanding of the principles involved, but has not been tested extensively against results in actual superalloy castings. In this work, we present the results of experimental work on the production of freckling in superalloy castings, made directionally in a Bridgeman furnace capable of producing DS castings in which the directional axis can be rotated with respect to gravity. The results indicate that we may account quantitatively for freckle formation through an analysis based on fluid-flow as described by a modified Rayleigh criterion. The theoretical critical condition for freckle initiation is found to be a Rayleigh Number of unity; the experimental determination of this critical value is found to be in the region of 0.7 - 0.9 depending on the alloy examined. We ascribe the difference in these values to the secondary features of the dendrite morphology.

The relation to random grain nucleation is found to lie in the role of isotherm curvature in the solidifying region. This aspect is examined by the use of ProCAST computations in both castings and remelt ingots.

Introduction

The problem of freckle generation in castings is a very old one and has been studied by a number of workers (1 - 3). The general mechanism for the formation of the defect has been defined as bouyancy-driven flow in the liquid/solid region of the casting, and the analysis of this mechanism for low temperature systems, including aqueous analogues, has confirmed the fundamental parameters as a balance between the bouyancy driving-force and the resistance offered by the dendrite mesh in the mushy zone. The balance was quantified by Sarazin and Hellawell (4) in terms of the system's Rayleigh Number (Ra); a proposal which was verified by experimental work on a low melting alloy system. Further experimental work by Auburtin et al (5) on a range of segregation-sensitive superalloys confirmed the validity of this approach. In this latter work, several interesting features of the analysis were highlighted. First, it is interesting to note that the necessary bouyancy force is developed from extremely small density gradients in the segregating liquid, of the order of 0.001g/mm, and is thus quite difficult to determine experimentally or compute from existing thermophysical databases. Second, the Rayleigh Number as defined by the above workers contains a high order of dendrite spacing and is thus very sensitive to the structure developed during solidification. Although the superalloys all have very similar solidification characteristics in respect of fraction-solid as a function of temperature (6), small differences in dendrite morphology may have a large influence on the numerical value of Ra, and therefore on the development of freckling in a casting. Thirdly, the common industrial solution to freckling in directional castings, that of increasing the solidification temperature gradient, is shown to be reasonable in that its primary influence is on the dendrite spacing, but it is also seen to have clear limitations in the case of large thermal sections as would, for example, exist in large castings. Finally, the role of casting geometry and imperfect isotherm control during solidification is not at present part of the analysis and should be included for a full description of the phenomenon.

There is a putative link between freckling and the development of random grains in single crystal castings, based on two hypotheses. The first (7) relies on the existence of quite shallow temperature gradients in the vicinity of the liquidus surface which permits the development of substantial constitutional supercooling. At the same time, the liquid flow in the freckle channel is held to detach secondary or tertiary dendrite arms by a solution mechanism. The dendrite fragments are carried into the liquidus region and are able to nucleate new grain growth. The hypothesis is supported by the presence of both grain fragments and random grains in experimental castings which contain freckles. The second (8) links the two phenomena by the presence of isotherm curvature in the solidification region of the casting. The curvature, as shown below, provides an additional factor to the bouyancy driving force for freckle formation. At the same time, the curvature implies an increased undercooling in the surface region of the casting which heightens the probability of grain nucleation. This mechanism is supported by experimental work, reported below, on freckle formation, and also by the presence of random grains in castings which do not contain freckles, where the undercooling provoked by the curvature is sufficient for grain nucleation, but insufficient for freckle formation in the alloy composition concerned. The work reported in this paper is a contribution to these developments.

Experimental

The experimental basis for the work is a series of experimental castings made using a Bridgeman furnace of an unusual design. The furnace has been described previously (9), and is designed to cast cylindrical samples under carefully controlled conditions of temperature and withdrawal rate. The unusual design feature is that the furnace may be titled to an angle of 40° to the horizontal without changing the direction of the isotherms in relation to the axis of the casting. The alloys chosen for this study cover both casting and forging compositions as shown in Table 1.

Superalloys 2000
Edited by T.M. Pollock, R.D. Kissinger, R.R. Bowman,
K.A. Green, M. McLean, S. Olson, and J.J. Schirra
TMS (The Minerals, Metals & Materials Society), 2000

The alloys were solidified directionally as 2cm dia cylinders approximately 150mm long with a range of temperature gradients and rates, and the castings examined for freckle formation (10). The results are summarised in Table 2, demonstrating that the freckling phenomenon can be activated by increasing the angular driving force. For each of the alloys, the freckle composition was analysed (Table 3) in order to compute the liquid density; the average liquid density during solidification was also computed from a combination of previously-established segregation coefficients (9) and the software package "METALS" (11).

Discussion

Using the modified Rayliegh Number as developed by Auburtin et al (8)

$$Ra = \frac{g d\rho / dz}{\eta D_T}\left[\lambda_1\left(\frac{K}{K_y}\right)\right]^4$$

where

Ra = Rayleigh Number;
g = Gravitational Constant (ms^{-2});
$d\rho/dz$ = Density Inversion Term in Vertical Direction (kgm^{-4});
η = Dynamic Viscosity (kgm^{-1}s^{-1});
D_T = Thermal Diffusivity (m^{2}s^{-1});
λ_1 = Primary Dendrite Arm Spacing (m);
K = Permeability in Vertical Direction (m^2). (8)
K_y = Permeability Parallel to Primary Dendrite Arm (m^2). (8)

we may derive a threshold value for the freckle formation as shown in Table 2. It is seen from this analysis that whilst the threshold value can be used to predict areas in a casting which would be sensitive to freckle formation, there remains an alloy dependence which is not covered by parameters contained in the Rayleigh Number as formulated above. Since the relation between fraction solid and relative temperature between liquidus and solidus is approximately constant for all superalloys (Figure 1) we cannot assign this sensitivity to a difference in the volume fraction of dendrite present at the freckle site. However, dendrite shape is not constant throughout the superalloy composition range, and is it possible that the variation observed originates in a "shape factor" depending on the detailed shape of the secondary and tertiary dendrite arms.

The composition variations on a base of IN718 illustrate two points. First, that the flow system causing freckling is the same basic process whether the liquid rises or falls through the dendrite network. A threshold value can be derived for the type of freckling described (12) in the same way as for the classic mechanism. Second, the role of the silicon content is well demonstrated in that since it is a highly segregating light element, its concentration determines the direction of the bouyancy force and hence also the morphology of the resulting freckle. In principle, carbon can be shown to behave in the same manner, as has been observed in the case of the high-speed steels (8), but with the complication of concentration changes prompted by the precipitation of primary carbides.

Since the freckling mechanism is the same in both ingot and casting manufacture, the criteria developed above should be able to account for the morphology of freckles found in remelted ingots. In this case the freckles are found predominantly at the

mid-radius position, which is due to the balance described in the Rayleigh Number between the angular driving force and the resistance to flow offered by the dendrite structure. This balance results in a maximum freckle probability at the mid-radius position, as illustrated in Figures 2 - 4, indicated by the variation of Ra across the radius of the ingot. In this case, uncertainty in the computation of the precise temperature gradients at the liquidus surface leads to estimates of Ra which are higher than the expected values of approximately unity. The form of the radial variation of Ra, however, clearly indicates a maximum freckle probability at the mid-radius position.

Two further factors appear to be important in freckle formation, and are not accounted for by the above estimates. First, it is intuitively obvious that the freckle channel must draw liquid from a reservoir in the solidifying network, replenished by inflow from the bulk. One may speculate that this reservoir will have a minimum volume which is of the same order as that of the freckle channel itself. If this is the case, then freckles should not form if the "root" region lies in a section of the casting which is smaller than the required volume. Freckles should not develop, therefore, in casting areas where this minimum dimension is not available in all directions from the freckle "root". It is postulated that this factor is the reason why freckles are not seen on interior faces of the casting such as air-cooling passages in turbine airfoils. However, when the casting is large enough, the air channel faces will present more than the critical dimension and so we have the troublesome prospect of freckling on interior surfaces which cannot be inspected. The dimension, by reference to the freckles observed in industrial castings appears to be in the range of several mm. A second factor which is likely to play a role in DS and forging alloys is that of second-phase precipitation. When nitrides and/or carbides are precipitated in the interdendritic liquid, the composition changes in a manner which is predictable from phase relationships, and which can be included in the Rayleigh Number computation. However, since the precipitates are of the same order of size as the interdendritic spacing, they will have an influence on the development of interdendritic fluid flow. This effect has been demonstrated in the case of the formation of microporosity in castings on IN100 (18), and will play a role in the freckle formation mechanism through the structural component of Ra, by increasing the interdendritic flow resistance.

Although the above description of the freckle formation mechanism in terms of Ra appears to explain the features observed in practice, we must examine the extent to which the actual isotherm curvatures experienced in castings could give rise to the Ra values used. The role of furnace temperature and withdrawal conditions has been studied using "ProCAST" software (13) for the case of the present small cylinder and also the cases of larger castings and castings containing ceramic inserts. Some of the results obtained are presented in Figure 5 and clearly show that with relatively small changes in withdrawal conditions it is possible to produce substantial isotherm curvature. In an industrial system such changes could readily take place through small errors in estimating the relative positions of the actual liquidus front and the furnace profile. In the limiting case of very large section castings it is germane to question whether or not such curvature can be avoided at economically-viable withdrawal rates due to the thermal cross-section of the casting. It is also interesting to note that the addition of forced cooling to the system has the potential to increase the isotherm curvature if the correct withdrawal conditions are not precisely observed, as noted by Giamei (14) in his studies of liquid metal cooling.

The link with random grain formation through either of the two mechanisms noted above is through isotherm curvature. Estimations of the flow velocity in a freckle channel (8) lead to values at which it is feasible to envisage solid particles persisting against the dissolution process until they reach a position at the liquidus where they would melt. If the undercooling at that position is sufficient, such particles could, therefore nucleate fresh grains. The undercooling required, following the theories of Hunt et al (1515) is small, of the order of 2 - 3°C. However, it is still necessary to explain the formation of random grains in castings which do not exhibit freckling. In the case of random grain nucleation produced when isotherm curvature has produced sufficient undercooling to cause nucleation on features such as mold imperfections or inclusions, the same curvature should, in principle have given rise to freckling. It appears from observation of industrial castings results that the curvature required for the former condition is less than that required for the latter in the common SX alloys, leading to the conclusion that the undercooling necessary for heterogenous nucleation is quite small. This conclusion is difficult to reconcile with the lattice mismatch theory of Hunt, since the likely surface imperfections or inclusions all have lattice mismatches which lead to undercooling requirement estimates of at least 10°C (16).

Conclusions

We conclude that the freckle initiation condition can be represented by an appropriate form of the Rayleigh criterion which takes into account the isotherm curvature in the system. This curvature also appears to be a significant factor in the formation of random grains and may possibly be amenable to a similar treatment. As with many similar formulations in the field of solidification, the accuracy of this analysis is dependent on the accuracy with which the appropriate physical parameters of the alloy can be defined.

Acknowledgements

The authors would like to acknowledge the valuable support ad assistance received during the course of this work from Rolls-Royce plc and the Consarc Corporation. They are also grateful for the provision of alloy materials from Cannon-Muskegon Inc and the Special Metals Coroporation.

References

1. S.M. Copley, A.F. Giamei, S.M. Johnson, M.F. Hornbecker: *Met. Trans.*, 1970, Vol. 1, pp.2193-2204.

2 M.G. Worster: *Ann. Rev. Fluid Mech.*, 1997, 29, pp.91-122.

3. J.C. Heinrich, S. Felicelli, D.R. Poirier: *Num. Heat Transfer, Part B*, 1993, Vol. 23, pp.461-481.

4. J.R. Sarazin, A. Hellawell: *Met. Trans. A*, 1988, Vol. 19A, pp.1861-1871.

5. P. Auburtin, A. Mitchell: *Liquid Metal Processing and Casting AVS Conference* (Santa Fe, NM, February 1997), pp.18-34.

6. P. Auburtin: Ph.D. Thesis, University of British Columbia, Vancouver, BC, Canada, August 1998.

7. T.M. Pollock, W.H. Murphy: *Met. Trans. A*, 1996, Vol. 27A, pp.1081-1094.

8. P. Auburtin, T. Wang, S.L. Cockcroft, A. Mitchell: *Met. Trans. B*, 2000, to be published.

9. P. Auburtin: M.A.Sc Thesis, University of British Columbia, Vancouver, BC, Canada, August 1995.

10. T. Wang: M.A.Sc Thesis, University of British Columbia, Vancouver, BC, Canada, May 1999.

11. "METALS" software, developed by National Physical Laboratories, Teddington, Middlesex, UK.

12. W Chen, W. H. Yang, K-M Chang S K Mannan and J.J. deBarbadillo; *Proc. Int. Conference on Liquid Metals*, Santa Fe 1999, eds. A Mitchell et al, publ. AVS 1999, pp.122 – 130.

13. ProCAST software, User's Manual and Materials Database, developed by UES Inc., Dayton, OH, USA.

14. A.F. Giamei, B.H. Kear: *Met. Trans.*, 1970, Vol. 1, pp.2185-2192.

15. J. D. Hunt, *Materials Science and Engineering*, Vol. 65 (1984), pp.75-83.

16. J. Fernihough, *The Columnar to Equiaxed Transition in Nickel Based Superalloys AM1 and MAR-M200 Hf*, Ph. D Thesis, UBC, 1995.

17. A. S. Ballantyne, A. Mitchell and J-F Wadier *Proc. 6th Intl Vacuum Metallurgy Conference*, 1979, eds G K Bhat and R Schlatter, publ. AVS, pp.555-571.

18. J. Lecomte-Beckers and M. lamberigts: "High temperature Alloys for Gas Turbines 1986", eds. W. Betz et al, Part I, 1986, pp.745-756.

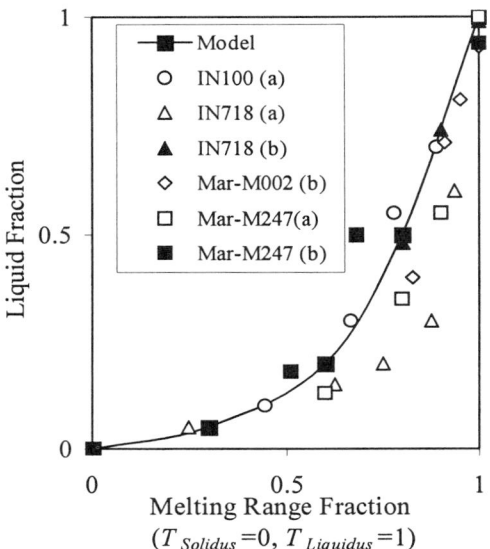

Figure 1 Typical liquid fraction profile along the mushy zone of nickel base superalloys.(experimental measurements : (a) from [9] and (b) from [17].)

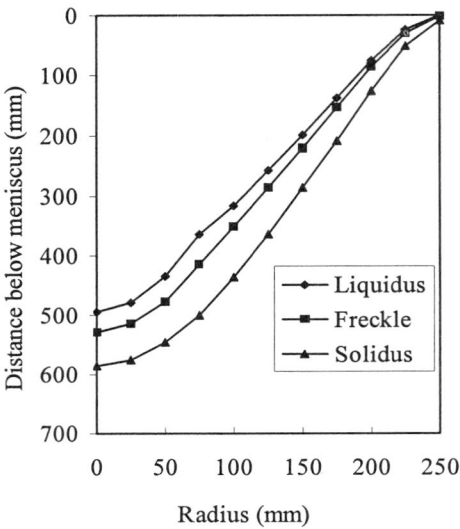

Figure 2 Pool depth vs. Radius in VAR IN718 shown by the isotherms for the liquidus, freckle and solidus temperatures. (260kg/hr melt rate) (after [17]).

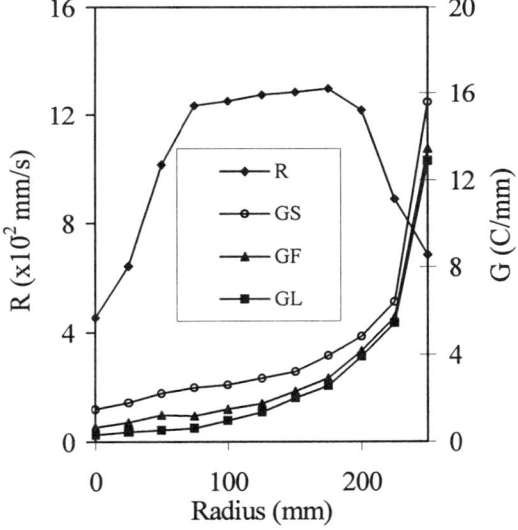

Figure 3 Thermal gradients at the liquidus, freckle and solidus temperatures and solidification rate in VAR IN718 (melt rate : 260kg/hr) (after [17]).

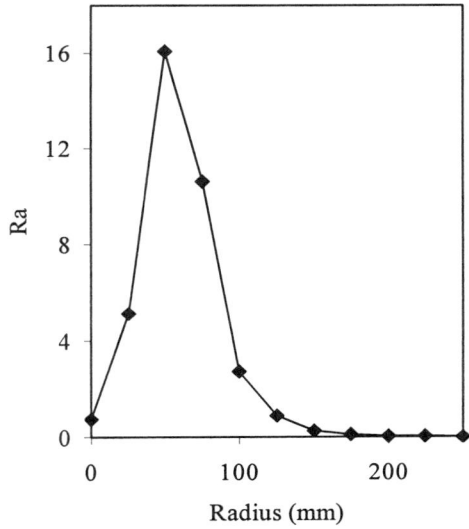

Figure 4 Rayleigh criteria profiles along the radius of VAR IN718 (melt rate: 260kg/hr.)

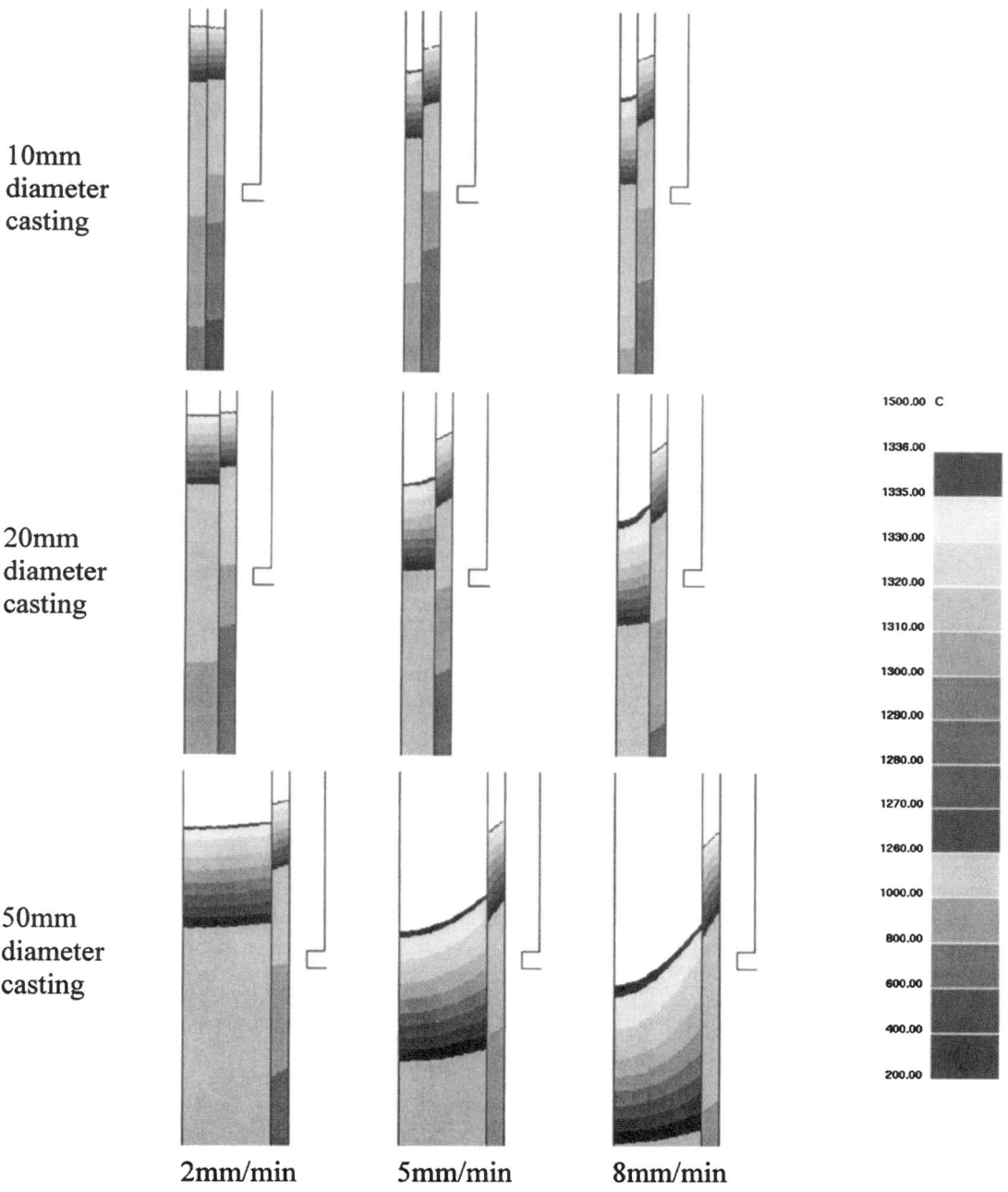

10mm diameter casting

20mm diameter casting

50mm diameter casting

2mm/min   5mm/min   8mm/min

1500.00 C

1336.00
1335.00
1330.00
1320.00
1310.00
1300.00
1290.00
1280.00
1270.00
1260.00
1000.00
800.00
600.00
400.00
200.00

Figure 5 Numerical simulation of the influence of casting cross-sections and withdrawal rates on the growth front angle found in DS/SX castings. (Control temperature 1435°C , after 130mm withdrawal) (ProCAST simulation)

259

Table 1 Compositions (in wt%) and melting range of chosen alloys

Alloy	Nominal Composition (wt%)	T_{Sol}-T_{Liq} (°C)
CMSX–11B	3.6Al, 7Co, 12.5Cr, 0.5Mo, 0.1Nb, 4.2Ti, 5W, 5Ta, 0.04Hf, Bal. Ni[45]	1275 – 1336*
René 88 (2)	2.1Al, 0.03C, 13Co, 16Cr, 4Mo, 0.7Nb, 3.7Ti, 4W, 0.015B, Bal. Ni[43]	1250 - 1355[43]
Nim 80A	1.4Al, 0.03B, 0.06C , 19.5Cr, 0.3Mn, 76.0Ni, 0.3Si, 2.4Ti, 0.06Zr[44]	1313 –1379*
IN718	0.46Al, 0.031C, 0.2Co, 18.12Cr, 2.96Mo, 5.27Nb, 53.46Ni, 1Ti, 0.08Si, 18.24Fe*	1253 - 1347*
IN718 – LSi	0.5Al, 0.008C, 0.001Co, 18Cr, 3Mo, 5.0Nb, 54.03Ni, 1Ti, 0.007Si, Bal. Fe	1280 – 1354*
IN718–HiSi		1175 – 1206*

*samples analyzed by the Special Metal Corporation.
** alloys provided by: Canon Muskegon Corp. (CMSX-11B); General Electric Corp. (René 88); Special Metal Corp. (IN718); Inco Alloys International (Nim 80A).

Table 3 Chemical analysis (measured by microprobe) of the matrix and freckles from some of the experimental samples. (wt%)

	IN718		IN718-0.4Si		IN718HiSi		CMSX-11B	
	Matrix	Freckle	Matrix	Freckle	Matrix	Freckle	Matrix	Freckle
Al	0.37	0.40	0.41	0.27	0.38	0.33	2.93	3.32
Si	0.23	0.27	0.50	0.97	0.62	1.35	-	-
Ti	1.08	1.31	0.95	1.61	0.90	1.39	3.93	7.08
Cr	18.48	17.98	18.67	15.90	18.58	16.08	11.34	5.59
Fe	18.63	17.41	18.82	15.06	19.45	16.33	-	-
Ni	53.94	53.0	53.42	53.06	53.41	51.88	59.02	61.75
Nb	4.66	6.09	4.21	9.07	4.05	8.86	0.13	0.11
Mo	3.07	3.54	3.04	4.05	2.86	3.79	0.48	0.26
Co	-	-	-	-	-	-	5.51	4.49
Hf	-	-	-	-	-	-	0.16	0.22
Ta	-	-	-	-	-	-	6.23	9.4
W	-	-	-	-	-	-	9.90	7.77

Table 2 Threshold values for the freckle formation in some superalloys

Alloys	Critical Rayleigh Number
CMSX-11B	0.88
René88	0.90
Nim80A	0.85
IN718-Si	0.65
Waspaloy	0.95
Mar-M247	0.86

Table 2 Summary of the directional solidification experiments carried out on the tiltable Bridgman furnace.

Alloy	Sample	Furnace T (°C)	Angle	$G_{Liquidus}$ (°C/cm)	$G_{Freckle}$ (°C/cm)	R (mm/min)	Freckling?[*]
CMSX-11B	#001	1435	0	23	27	1	no
	#002	1435	0	23	27	1	no
	#007	1500	0	30	33	1	no
	#008	1500	0	7	19	6	no
	#006	1435	20	23	27	1	yes
	#005	1435	20	7	18	6	no
	#012	1500	20	30	33	1	yes
	#010	1500	20	7	19	6	no
	#004	1435	35	23	27	1	yes
	#003	1435	35	7	18	6	no
	#011	1500	35	30	33	1	yes
	#009	1500	35	7	19	6	no
René 88	#101	1400	35	9	18	6	no
	#102	1465	35	26	32	1	yes
	#103	1400	20	9	18	6	no
	#104	1465	20	26	32	1	yes
	#105	1400	0	11	19	1	yes
Nim 80A	#201	1435	35	11	17	6	no
	#202	1500	35	26	30	1	yes
	#203	1435	20	11	17	6	no
	#205	1500	20	26	30	1	yes
	#204	1435	0	15	21	1	yes
IN718	#301	1400	35	10	16	6	no
	#302	1465	35	26	31	1	yes(underside)
	#303	1400	20	10	16	6	no
	#304	1465	20	26	31	1	yes(underside)
	#305	1400	0	14	20	1	yes
IN718LSi	#401	1435	35	15	22	6	no
	#402	1500	35	29	33	1	yes(underside)
	#403	1435	20	15	22	6	no
	#404	1500	20	29	33	1	no
	#405	1435	0	20	25	1	no
IN718HiSi	#501	1400	35	9	17	6	no
	#502	1465	35	35	38	1	yes
	#503	1400	20	9	17	6	no
	#504	1465	20	35	38	1	yes
	#505	1400	0	31	35	1	yes
IN718-0.1Si	#601	1465	35			1	yes(underside)
IN718-0.2Si	#701	1465	35			1	yes(underside)
IN718-0.3Si	#801	1465	35			1	yes(underside)
IN718-0.4Si	#901	1465	35			1	yes(underside)

*: "yes" means freckles occur on the upside surface of an inclined sample or the surface of a vertical sample;

"yes(underside)" means freckles occur on the underside surface of an inclined sample.

MODELLING OF THE MICROSEGREGATION IN CMSX-4 SUPERALLOY AND ITS HOMOGENISATION DURING HEAT TREATMENT

M.S.A. Karunaratne, D.C. Cox, P. Carter and R.C. Reed

University of Cambridge / Rolls-Royce University Technology Centre,
Department of Materials Science and Metallurgy, Pembroke Street, Cambridge CB2 3QZ, UK

Abstract

Microsegregation in the single crystal superalloy CMSX-4 has been studied using electron probe microanalysis, both in the as-cast condition and after solution heat treatment. In order to establish the solidification path, a statistical treatment of the data is proposed, which is based upon the local value of the quantity $C_{Ta} - C_{Re}$. Using differential scanning calorimetry and by appealing to a database of thermodynamic parameters, it is shown that the solidification path cannot be explained without acknowledging that backdiffusion occurs. Analysis of the microsegregation remaining after progressive heat treatment reveals that the dendritic and interdendritic regions homogenise at different rates, owing to the presence of the eutectic mixture; there is evidence of up-hill diffusion of solutes in the eutectic region during homogenisation. Simple expressions based upon a sinusoidal variation of composition are inadequate for the estimation of homogenisation times. A coupled thermodynamic/kinetic theory is able to explain most of the effects which occur, *i.e.* incipient melting, enrichment of residual γ' by Ta, Mo and subsequent γ' dissolution. Although more work needs to be carried out to better establish the thermodynamic and kinetic parameters required by the model, it is at this stage already useful for the assessment of new superalloy compositions and for the design of optimal heat treatments.

Introduction

The turbine blading for gas turbine engines [1] is now commonly cast from nickel-base superalloys in single crystal form, often with intricate channels included so that cooler air can be forced to flow within and along the blades during operation [2–4]. Since the capability for single crystal processing first became available, single crystal compositions have evolved considerably [5]. The addition of heavy elements such as Ta, W, Mo and particularly Re by the alloy designer has conferred considerable solid solution strengthening, with concomitant increases in the high temperature creep resistance. Such developments have, over the past twenty years or so, contributed significantly to a steady increase in the turbine entry temperatures of modern aeroengines, which has averaged about 5°C per year [6].

Less well appreciated however, are some disadvantages which have arisen as the single crystal compositions have become more heavily alloyed. For example, the heat treatment schedules required to homogenise the single crystal blades before they enter service have become longer, with tight control of the temperature/time sequence needed to avoid incipient melting. With this in mind, it is possibly surprising that the optimisation of the heat treatment of single crystal superalloys still involves a significant degree of empiricism. Possibly this is because reliable interdiffusion data pertinent to the superalloys have not been available, particularly for the heavy elements such as Re. Fortunately, this situation has recently been remedied [5] and for this reason it was decided to make a first attempt at developing a numerical model for the heat treatment process. This paper is concerned with CMSX-4, which is a second generation single crystal alloy which is finding widespread application in turbine applications; nonetheless it should be noted that the techniques reported here are quite generic and can therefore be applied to other grades of superalloy. In particular, the numerical model is highly suited to the assessment of experimental alloys, during the alloy design process.

Background

One should consider briefly the reasons why it is necessary to homogenise turbine aerofoils fabricated from single crystal superalloys, prior to their entering operation in the aeroengine. Single crystal superalloys have compositions with as many as twelve different elements present in significant proportions. On thermodynamical grounds alone one could then reasonably expect the freezing range to be rather wide; in practice however, there is a considerable kinetic contribution which arises as a consequence of the presence of the heavy elements, which do not diffuse rapidly in the solid which forms. For most single crystal compositions the very last stages of solidification occur via one or possibly more eutectic reactions, the nature of which differs from alloy to alloy [7]. These eutectic reactions give rise to a significant fraction of a coarse eutectic mixture, which usually contains a large volume fraction of the γ' phase.

Superalloys 2000
Edited by T.M. Pollock, R.D. Kissinger, R.R. Bowman,
K.A. Green, M. McLean, S. Olson, and J.J. Schirra
TMS (The Minerals, Metals & Materials Society), 2000

In practice therefore a homogenisation heat treatment is required, to prevent the occurrence of incipient melting of the eutectic regions during high temperature service. During this treatment, mass transport by bulk diffusion acts to reduce the scale and extent of the microsegregation and eutectic regions which are inherited from the casting process. Traditionally, the degree of microsegregation has been assumed to vary in a sinusoidal fashion, according to expressions such as [8,9]

$$C\{r,t\} = C_o + C_n \cos \left\{ \frac{2\pi r}{L} \right\} \exp \left\{ -\frac{t}{\tau} \right\} \qquad (1)$$

where C_o and C_n are constants and the time constant for homogenisation, denoted τ is given by

$$\tau = \frac{L^2}{4\pi^2 \tilde{D}} \qquad (2)$$

where $\tilde{D}$ is an interdiffusion coefficient. Diffusion is assumed to occur in one-dimension, over a characteristic distance denoted L; for the directional solidification of single crystal alloys it is probably appropriate to equate this to a distance somewhere between the primary and secondary dendrite arm spacing.

It has been suggested, particularly for binary alloys [e.g. 8] and for steels [10], that Equations 1 & 2 represent a useful first approximation for the way in which the degree of microsegregation changes during heat treatment. However, other studies [e.g. 11] have shown that the agreement with theory is poor. For single crystal superalloys, it would appear that the applicability of this simple approach has not yet been tested in an adequate sense. It should be appreciated that the approximations introduced might invalidate its use for this class of alloy. For example, (i) the assumption that the initial concentration profiles can be represented as sinusoidal waves might be unrealistic, particularly owing to the presence of the second phases or eutectic regions, (ii) the presence of large cross-terms in the interdiffusivity matrix might mean that the interdiffusion of the various elements are in fact coupled, with one of the heavy elements controlling the rate at which homogenisation occurs, and (iii) in practice, diffusion must occur in two or possibly three dimensions. It is then quite possible that a more complicated theory is to be preferred.

Experimental Details

Material

The CMSX-4 single crystal alloy used for the present work was supplied by Rolls-Royce plc in the as-cast condition, in the form of 10 mm diameter rods of approximate length 120 mm. These had been directionally solidified according to the methods described in [12], and therefore each rod had a < 001 > crystallographic direction within a few degrees of its long axis. The chemical composition of one of these bars was determined using Leco CS244/TC436AR furnaces and inductively coupled plasma optical emission spectroscopy (ICP-OES). The results are given in Table I.

Table I The chemical composition of the CMSX-4 material used for the present study.

	Al	Co	Cr	Hf	Mo	Ni	Re	Ta	Ti	W
wt%	5.7	9.5	6.4	0.09	0.7	Bal.	2.92	6.5	1.08	6.4

The rods were sectioned into 3 mm long slices. Each of these was subjected to an isothermal heat treatment at a temperature between 1250°C and 1315°C, in a vacuum furnace.

Electron Probe Microanalysis (EPMA)

After heat treatment, the specimens were polished to a 1 μm finish using traditional metallographic techniques. Concentration maps of Al, Co, Cr, Ni, Re, Ta, Ti and W were then obtained from each of the samples using a Cameca SX50 electron probe microanalyser (EPMA), in the following way. Measurements were made over a 1000 μm by 1000 μm square mesh at a spacing of 50 μm. The wavelength dispersive spectrometers (WDS) which were employed were equipped with crystals of TAP (used for Al, Ta and Re), PET (used for Ti, Mo and W) and LIF (used for Cr, Co and Ni) [13]. X-ray counts were recorded simultaneously for Kα (Al, Co, Cr, Ti), Kβ (Ni), Lα (Mo), Mα (Ta, W) and Mβ radiations. These were converted to concentration values using standard correction procedures [13,14] and signals from pure-element standards, except for Al for which an Al_2O_3 standard was used. The acceleration voltage, beam current and take-off angle for the analyses were 20 kV, 100 nA and 40° respectively and the counting time for each point was approximately 72 s.

Because of the nature of the analysis carried out, it became appropriate to characterise the microsegregation in the as-received, as-cast CMSX-4 with greater spatial resolution. For this purpose a 1000 μm by 1000 μm square mesh was used and data were collected at a resolution of 4 μm.

Differential Scanning Calorimetry

A Stanton-Redcroft model DSC1500 was employed for the measurements. To avoid sample contamination, sapphire crucibles were placed inside platinum ones. Cylindrical samples of 3.5 mm diameter and 2 mm thickness were found to yield an acceptable signal/noise ratio. The measurements were conducted at a cooling rate of 10°C/min under an atmosphere of scrubbed argon. The measurement range was 1500°C to 500°C. The reference pan had a similar arrangement of a sapphire crucible within a platinum one; runs were carried out with the reference pan empty, and also with a specimen of α-alumina of known mass and heat capacity. The specific heat capacity of the CMSX-4 sample was then determined using the ratio method, as described in [15]. Conversion of the data for specific heat capacity vs temperature to fraction solid vs temperature was carried out using a geometrical method [16].

Optical and Scanning Electron Microscopies

For the purposes of optical microscopy, the polished samples were etched in a mixture of 100 ml nitric acid, 500 ml hydrochloric acid, 25 g ferric chloride, 25 g cupric chloride and 400 ml water. It was found that this gave satisfactory contrast from the eutectic regions. Optical micrographs were taken with a Zeiss Axiotech reflecting light microscope using polarisation contrast.

In order to estimate the volume fraction of the eutectic mixture, use was made of a JEOL 6340 field-emission gun scanning electron microscope (FEGSEM); to minimise errors associated with the subsequent stereological analysis, backscattered electron signals were used. Digital image micrographs were taken with a magnification of 10,000 using 512×512 pixels, thus achieving a resolution of $\sim 0.02 \, \mu$m per pixel. Subsequently, the micrographs were subjected to image analysis using a commercial image analysis software package (Kontron Elektronik KS300 version 2.00).

Determination of Interdiffusion Coefficients and Atomic Mobilities

To make predictions of the rate of homogenisation of as-cast CMSX-4, it became necessary to make measurements of the interdiffusion coefficients in a number of model binary and ternary diffusion couples. Special emphasis was placed on the Ni-Re, Ni-W, Ni-Ta and Ni-Al-Ti systems. Diffusion couples were fabricated from high purity constituents which had been specially prepared using vacuum induction melting; bonding was achieved using a ThermecMaster-Z thermo-mechanical simulator [5]. These were heat-treated in a vacuum furnace for various times at temperatures between 1000°C and 1300°C. Electron probe microanalysis was then used to determine the extent of interdiffusion, and interdiffusion coefficients determined using a modified form of the Boltzmann-Matano analysis, following [5]. Values of the interdiffusion coefficients for Ni-Mo were estimated from [17,18] and for Ni-Co from [19,20]. For modelling purposes, atomic mobilities were evaluated using the method reported in [21], with care being taken to ensure consistency with a databank of thermodynamic parameters for the superalloys [22]. The mobilities given by Engstrom & Agren [23] for the ternary Ni-Cr-Al system were accepted. The expressions are tabulated in the Appendix. Figure 1 illustrates the way in which the interdiffusion coefficients in nickel vary as a function of temperature, for a number of elements pertinent to the superalloys.

Experimental Results and Analysis

On the Morphology and Scale of Dendritic Microstructure

In Figure 2, the morphology and scale of the dendritic structure on (a) transverse and (b) longitudinal sections is illustrated. Consistent with observations reported elsewhere [24–27], for this class of material and for the casting conditions used in a typical commercial investment casting foundry [12] the primary dendrite arm spacing L_p lies in the

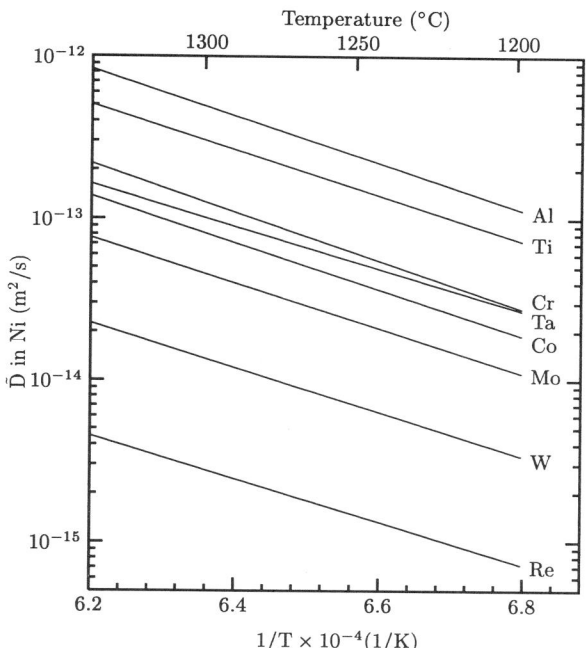

Figure 1: Variation of the interdiffusion coefficient in pure nickel of various elements pertinent to the superalloys, as a function of inverse temperature.

range 100 μm to 500 μm. The secondary arm spacing L_s is in the range 30 μm to 50 μm.

On the Microsegregation in CMSX-4 in the As-Cast Condition

Maps of local concentration, as determined on a transverse section using electron probe microanalysis, are given in Figure 3. It has been found that the elements Co, Cr, Re, W partition to the dendrite cores with Al, Ni, Ta, Ti segregating to the interdendritic regions. It is notable that the heavy elements Re, Ta and W segregate particularly strongly. For various reasons it is of interest to quantify the equilibrium partitioning ratios from the EPMA data. This can be done by assuming that no diffusion occurs in the solid during solidification and subsequent cooling, that the extent of undercooling is small, and that there is no macrosegregation [28]. The partitioning ratio of each component is then given by the ratio of the composition at the dendrite core to the mean composition determined. However, in our experience, the estimation of the chemical composition at the dendrite core requires some care because of the possible presence of statistical noise in the EPMA data, particularly when a large amount has been collected. It is helpful to have a procedure which helps with this situation.

For this reason we have adapted the statistical technique of Gungor [29], which was originally applied to binary alloys, to estimate the solidification path and thus the composition of the first solid to form. In a multicomponent alloy such as CMSX-4 there are several independent compositional variables and therefore one needs to make a decision concerning

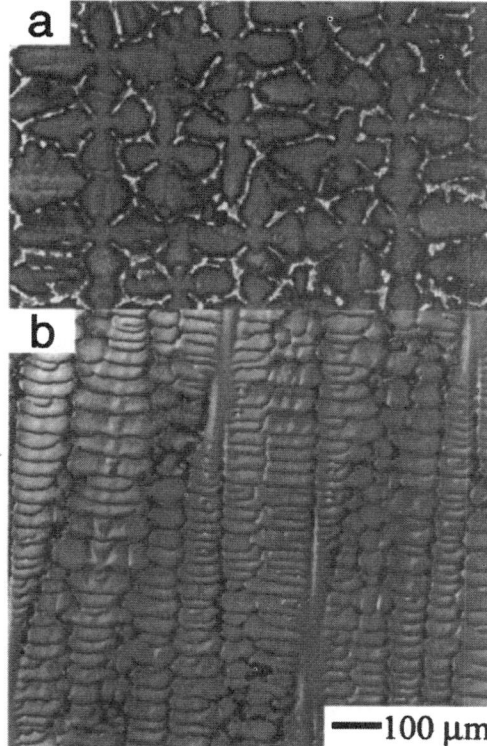

Figure 2: Optical micrograph of the as-cast structure of CMSX-4: (a) transverse section and (b) longitudinal section.

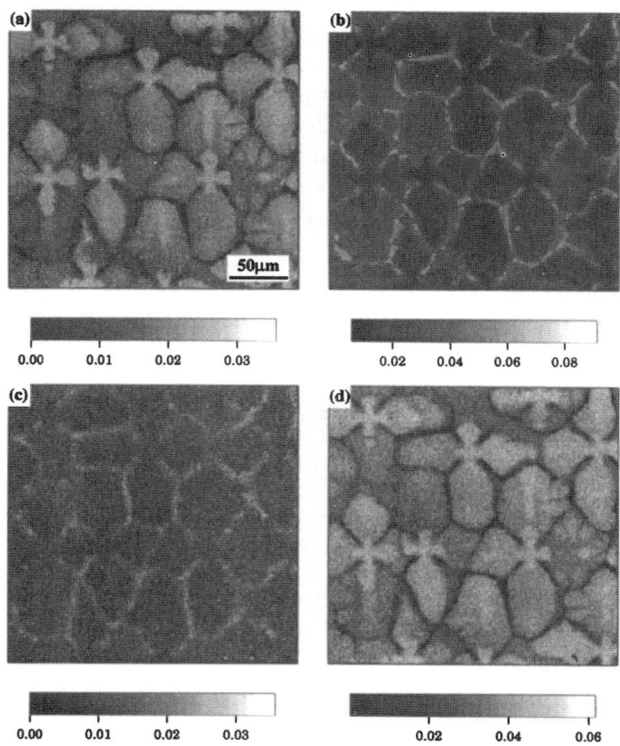

Figure 3: EPMA composition maps illustrating the interdendritic microsegregation in CMSX-4 in the as-cast condition, (a) Re, (b) Ta, (c) Ti and (d) W. Compositions expressed as weight fractions.

how to proceed. In the present case it is clear that Re and Ta segregate in different directions, so that the magnitude of the quantity $(C_{Ta} - C_{Re})$ can be used as a basis for estimating the solid fraction f_s to which it corresponds. A criterion based upon the Ta and Re concentrations is further supported by predictions made using a database of thermodynamic parameters [22], which indicate that the composition of these elements changes monotonically with increasing fraction of solid, even past the eutectic temperature, see the following section. The procedure employed [30] involves (i) the sorting of the datapoints into ascending order on the basis of the local value of $(C_{Ta} - C_{Re})$, (ii) the assignment of an integer i lying between unity and N on this basis, where N is the number of datapoints collected, and the quantity f_s placed equal to $(i-1)/(N-1)$ which is assumed to equate to the fraction solidified at this point and (iii) the computation of mean concentration values within intervals of f_s corresponding to 0.02, and the corresponding errors which are quoted as one standard deviation in the mean.

The concentrations of Re, Ta, Ti and W as a function of solid fraction f_s obtained in this way are given in Figure 4. The composition profiles are well behaved in the limit of zero f_s and this allows the composition of the first solid to form, and hence the partitioning ratios, to be estimated. The values determined in this way are listed in Table II. There is

reasonably good agreement between the values found here, and values reported elsewhere [31,32]. However, there is some discrepancy for Cr – for this element a partitioning coefficient k greater than unity was found here whereas other reported values are less than unity. The experiments reported by Ma & Grafe [32] employed liquid metal cooling and the directional solidification/quench (DSQ) method, so that the extent of any backdiffusion is likely to be much smaller than observed here; nevertheless any backdiffusion occurring in the present case would serve to increase our values of k, see Figure 4 and thus this effect cannot be used to explain this discrepancy.

Table II Equilibrium partitioning ratios determined for CMSX-4. The experimental values are determined at 1wt% solid, the calculated values at the liquidus temperatures.

Al	Co	Cr	Mo	Re	Ta	Ti	W	Method
0.86	1.08	1.05	0.86	1.66	0.67	0.86	1.31	Experimental[†]
0.91	1.11	0.92	0.68	1.62	0.56	0.69	1.18	Calculated[‡]

[†]Calculated at 1wt% solid
[‡]Calculated at liquidus temperature

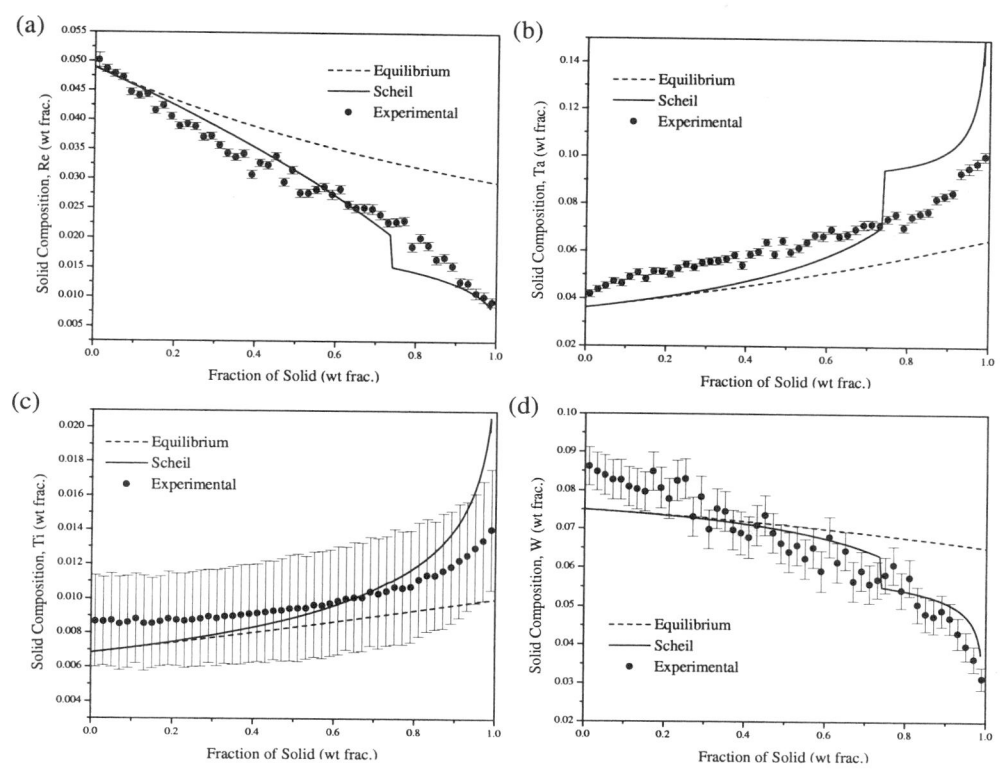

Figure 4: Solidification paths estimated by the statistical analysis of the EPMA data: (a) Re, (b) Ta, (c) Ti and (d) W. Also shown are the curves calculated using the equilibrium and Gulliver-Scheil assumptions.

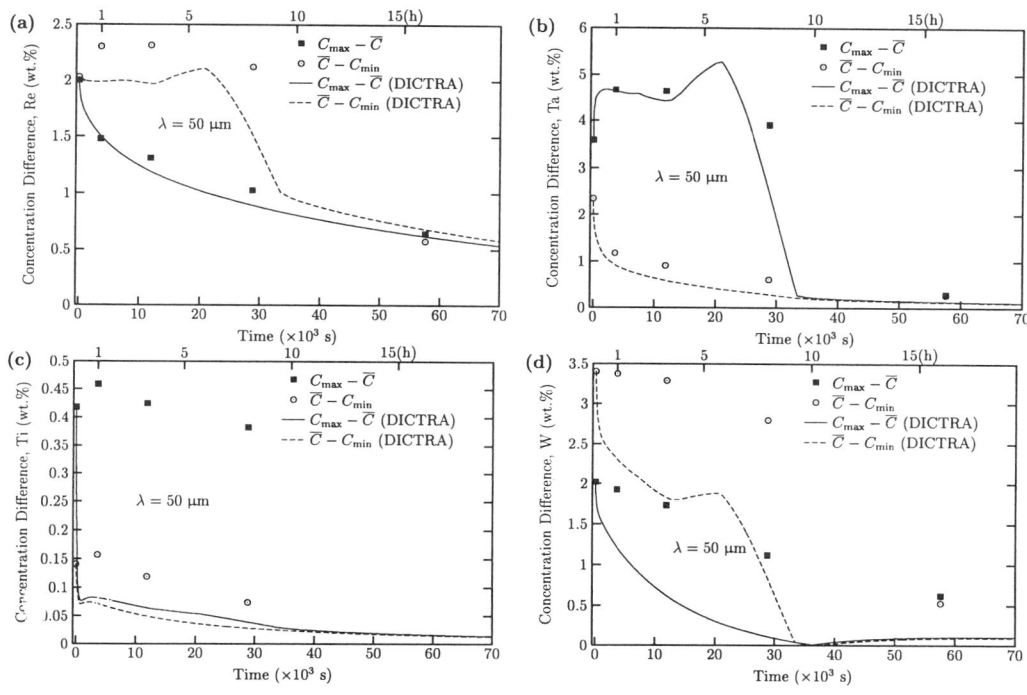

Figure 5: Variation of the quantities $(C_{max} - \overline{C})$ and $(\overline{C} - C_{min})$, during heat treatment at 1315°C: (a) Re, (b) Ta, (c) Ti and (d) W. Values predicted by the numerical model are also shown.

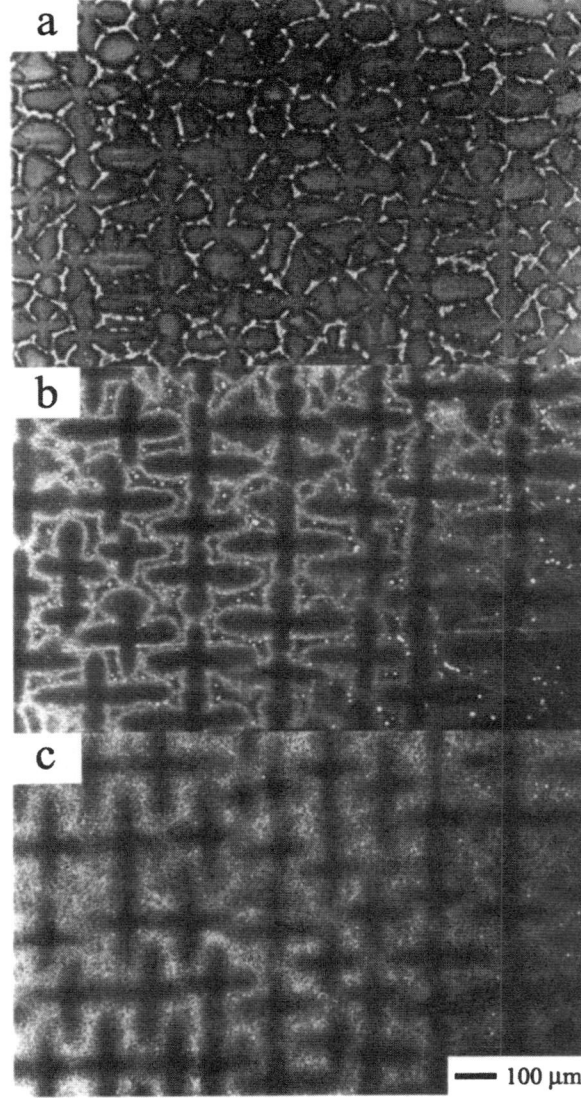

— 100 μm

Figure 6: Optical micrographs illustrating the microstructure of as-cast CMSX-4, and its evolution during heat treatment at 1315°C: (a) as-cast, (b) after 9 hours and (c) after 12 hours.

On the Homogenisation of Microsegregation in CMSX-4

We have found that the microsegregation which is inherited from the casting process, see Figure 3, and its subsequent homogenisation cannot be represented adequately by Equation 1. This can be seen most readily by examining the quantities $(C_{max} - \overline{C})$ and $(\overline{C} - C_{min})$, where C_{max}, C_{min} and $\overline{C}$ correspond to the maximum, minimum and average compositions seen on the solidification profiles; on Figure 5 these are the values corresponding to time zero. For CMSX-4, one can see that the quantities $(C_{max} - \overline{C})$ and $(\overline{C} - C_{min})$ differ significantly in magnitude. A further point concerns the rate at which the dendritic and interdendritic regions homogenise. Our data demonstrate beyond any reasonable doubt that the dendrite cores homogenise more rapidly than

the interdendritic regions. For some of the elements in the interdendritic region and particularly Ta, the composition rises at first, before falling again as homogenisation proceeds still further – thus there is evidence of the occurrence of up-hill diffusion. This behaviour requires further rationalisation.

Observations made on the eutectic mixture in the as-cast condition have indicated that it consists of a mixture of the γ and γ' phases, see Figure 6. Figure 7 shows how the fraction of the eutectic mixture evolves during heat treatment at 1315°C.

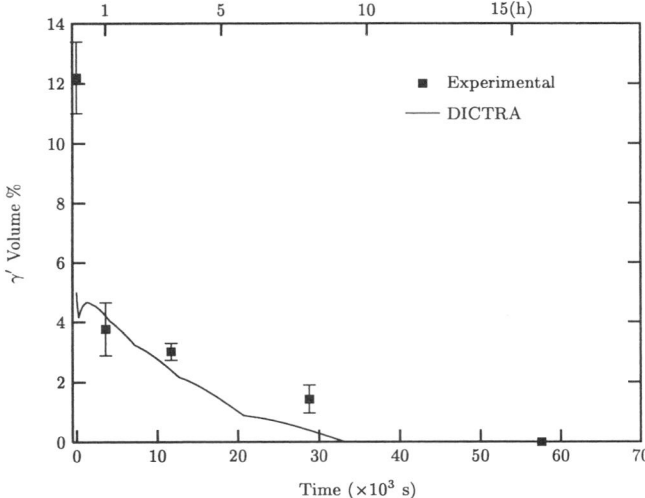

Figure 7: Variation of the fraction of coarse γ' in the eutectic mixture with time and temperature during heat treatment, as determined using image analysis. Values predicted by the numerical model are also shown.

Simulation of Microstructure Evolution

Modelling of the Evolution of Microsegregation

The experimentally determined solidification paths given in Figure 4 have been compared with the predictions made using the Thermocalc software [33] and a database of thermodynamic parameters [22]. For the purposes of the present analyses, one assumes that only the liquid, γ and γ' phases play a part in the reaction. On Figure 4 the equilibrium lines are shown, as are the lines predicted by the Gulliver-Scheil approach, determined using the temperature stepping scheme suggested by Sundman & Ansara [34]. Once the eutectic temperature is passed, the composition quoted corresponds to the weighted average of those corresponding to the γ and γ' phases, it being assumed that the EPMA method is incapable of resolving the compositions of the γ and γ' phases in the eutectic mixture.

From Figure 4 it can be seen that the models reproduce the observed sign of partitioning reasonably well. From the

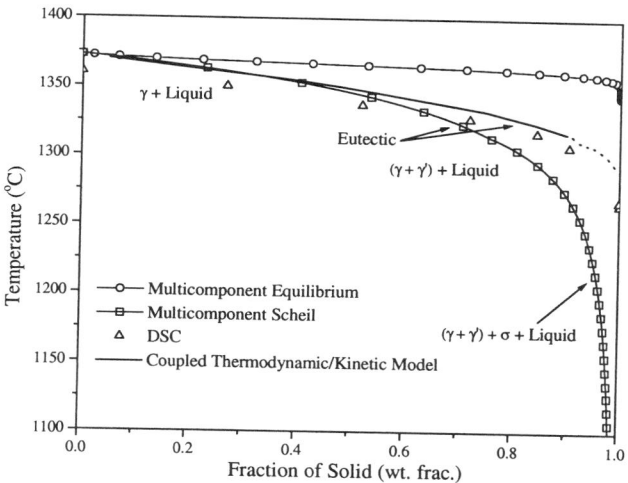

Figure 8: Variation of the solid fraction with temperature, as determined from the DSC measurements and from the numerical models.

composition of the first solid to form, one can see that the initial partitioning coefficients are also predicted reasonably, with the exception of W. However, the predictions at temperatures lower than the eutectic point are in rather poor agreement with experiment, with the sharp change in composition predicted by the database as the consequence of the eutectic reaction liquid→γ + γ' not being observed. This is probably a consequence of the limitations of the experimental techniques and the analysis method which has been employed.

From an examination of the data in Figure 4, it is not easy to confirm whether or not there is significant occurrence of back-diffusion during solidification. A better way of testing this to examine the manner in which the fraction of solid evolves with decreasing temperature, see Figure 8, as determined using differential scanning calorimetry. The solidification range predicted by the Gulliver-Scheil model is too wide and this proves conclusively that back-diffusion is indeed occurring.

Modelling of Homogenisation during Heat Treatment

The solution heat treatment of CMSX-4 has been simulated using the DICTRA software [e.g. 35] and the database of kinetic parameters given in the Appendix. All simulations have been made using the model for dispersed systems [36]. The labyrinth factor was taken to be unity. Thus long-range diffusion in the continuous matrix phase γ is accounted for. The formation and dissolution of dispersed phases γ' and liquid is allowed, but diffusion within them is not permitted. Hence, it is assumed that significant gradients in chemical potential exist only over distances much greater than the inter-particle spacing, with mass transport occurring in one dimension and planar co-ordinates. The phases in any given volume element are taken to be in local equilibrium

and hence their constitution at any given time during the simulation is estimated by solving the phase diagram problem [37]. The initial composition profiles representing the as-cast microsegregation were obtained from the statistical analysis described earlier, see Figure 4. Since this analysis does not yield a length scale, a planar region of $50\,\mu$m length has been assumed, which approximates to the secondary dendrite arm spacing, see Figure 2. The timestep Δt for the calculations was 0.5 s.

The results of the simulations are presented in Figure 5, for heat treatment at 1315°C which is approximately equal to the equilibrium solvus temperature of the alloy. Although the agreement is not perfect, one can see that the coupled thermodynamic/kinetic model accounts for the different homogenisation rates of dendritic and interdendritic regions, but also for the enrichment of the interdendritic regions with respect to Ta which occurs within the first few hours of heat treatment. This effect occurs because the interdendritic region forms at a temperature different from that at which heat treatment is performed; thus bulk mass transport by diffusion can occur giving rise to enrichment of interdendritic γ' by elements such as Ta. Quite clearly, the sinusoidal form implied by Equation 1 is valid only after these effects have occurred, about 10 hours. Unfortunately, the agreement between experiment and theory for Ti and W is not good, and this is probably because of tertiary interaction parameters which are not presently accounted for. On the other hand we have observed that the dissolution of the coarse γ' found in the eutectic mixture correlates well with the fraction of γ' predicted by the model, see Figure 7. The predictions of the rate of homogenisation at temperatures below the solvus are also in reasonable agreement with experiment, see Figure 9. It should be noted that the model is also capable of predicting the onset of incipient melting, i.e. the formation of interdendritic liquid and its subsequent resolidification as homogenisation progresses.

Summary and Conclusions

The following conclusions can be drawn from this work:

1. A statistical treatment of electron probe microanalysis data has been proposed, which appeals to the local value of the quantity ($C_{Ta} - C_{Re}$). The method allows the solidification path, the composition of the first solid to form and the partition coefficients to be estimated.

2. In CMSX-4 superalloy, the elements Co, Cr, Re, W segregate to the dendrite cores with Al, Ni, Ta, Ti partitioning to the interdendritic regions. It is notable that the heavy elements Re, Ta and W segregate particularly strongly. Some inconsistencies between theory and experiment have been noted.

3. By appealing to a database of thermodynamic parameters for the superalloys and the results from differential scanning calorimetry, it has been shown that the solidification path cannot be explained without acknowledging that some backdiffusion occurs during solidification.

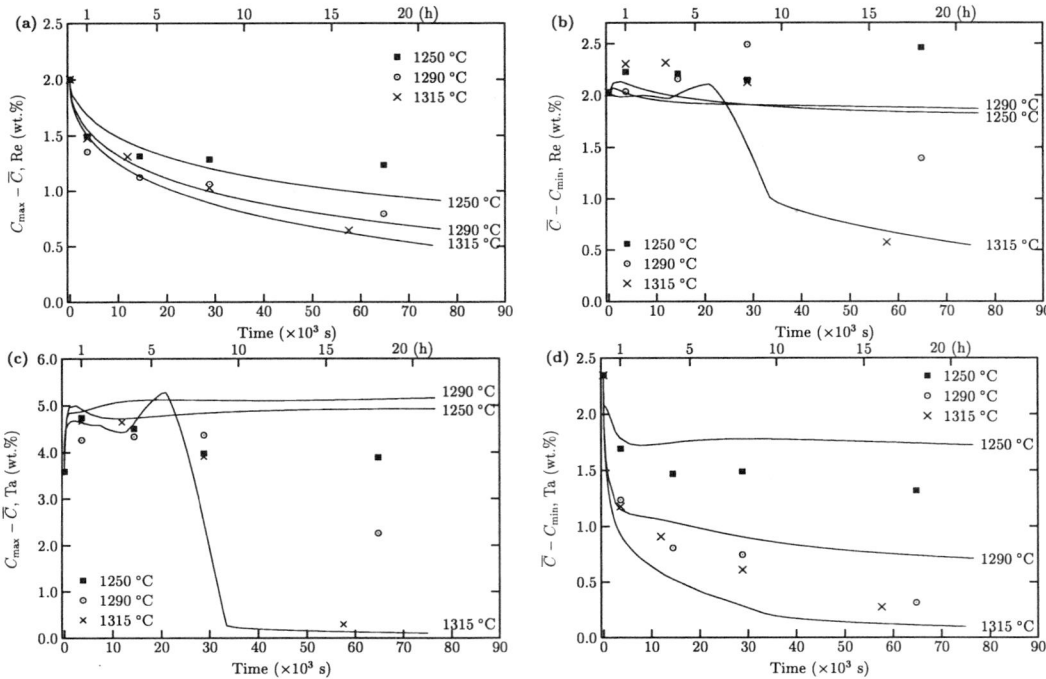

Figure 9: Response of the interdendritic segregation to heat treatment, (a) variation of $(C_{\max} - \overline{C})$ for Re, (b) variation of $(\overline{C} - C_{\min})$ for Re, (c) variation of $(C_{\max} - \overline{C})$ for Ta, (b) variation of $(\overline{C} - C_{\min})$ for Ta. Values predicted by the numerical model are also shown.

4. During solution heat treatment, the dendritic and interdendritic regions homogenise at different rates. Enrichment of some solutes, *e.g.* Ta, Co and Cr occurs in the eutectic region. Simple expressions which are based upon a sinusoidal variation of composition are unable to account for these effects. More complicated theory is required.

6. Kinetic parameters pertinent to the superalloys have been coupled with the thermodynamic parameters, so that the diffusion of solutes during homogenisation can be handled. It has been shown that the simulations are in reasonable agreement with the observations.

7. The model has reached the point where it can make a useful contribution to the process of alloy design, *e.g.* for the assessment of new superalloy compositions and for the design of appropriate heat treatments. However, more work needs to be carried out to better establish some of the thermodynamic and kinetic parameters required.

Acknowledgements

The authors thank the Cambridge Commonwealth Trust, the Engineering & Physical Sciences Research Council (EPSRC), Rolls-Royce plc and the Defence Evaluation & Research Agency (DERA) for sponsoring this work. The assistance of Dr. S.J.B. Reed with the EPMA analysis is gratefully acknowledged. Helpful discussions with Colin Small (Rolls-Royce), and Mike Henderson & Mike Winstone (DERA) are much appreciated. It is a pleasure to acknowledge the help and advice of Lindsay Chapman and Peter Quested of the National Physical Laboratory.

References

1. The Jet Engine, (The Technical Publications Department, Rolls-Royce plc, Derby, UK, Fourth Edition, 1992).

2. F.R.N. Nabarro and H.L. de Villiers, The Physics of Creep, (Taylor and Francis, 1995).

3. M. McLean, "Nickel-Base Superalloys: Current Status and Potential", Philosophical Transactions of the Royal Society London A, 351, (1995), 419–433.

4. R.D. Kissinger, D.J. Deye, D.L. Anton, A.D. Cetel, M.V. Nathal, T.M. Pollock and D.A. Woodford, eds., Superalloys 1996, (The Minerals, Metals and Materials Society, Warrendale, Pennsylvania, USA, 1996).

5. M.S.A. Karunaratne, P. Carter and R.C. Reed, "Interdiffusion in the Face-Centred-Cubic Phase of the Ni-Re, Ni-Ta and Ni-W Systems between 900 and 1300°C", Materials Science and Engineering, A281 (2000), 229–233.

6. J.K. Tien and V.C. Nardone, "The US Superalloys Industry – Status and Outlook", Journal of Metals, 36 (1984), 52–57.

7. M. Durand-Charre, The Microstructure of Superalloys, (Gordon & Breach Science Publishers, Amsterdam, The Netherlands, 1997).

8. G.R. Purdy and J.S. Kirkaldy, "Homogenisation by Diffusion", Metallurgical Transactions, 2 (1971), 371–378.

9. M. McLean, Directionally Solidified Materials for High Temperature Service, (The Metals Society, London, 1983).

10. R.G. Ward, "Effect of Annealing on the Dendritic Segregation of Manganese in Steel", Journal of the Iron and Steel Institute, 203 (1965), 930–932.

11. F. Weinberg and R.K. Buhr, "Homogenisation of a Low-Alloy Steel", Journal of the Iron and Steel Institute, 207 (1969), 1114–1121.

12. P. Carter, D.C. Cox, Ch.-A. Gandin and R.C. Reed, "Process Modelling of the Grain Selection during Solidification of Superalloy Single Crystal Castings", Materials Science and Engineering, A280 (2000), 233–246.

13. S.J.B. Reed, Electron Microprobe Analysis, (Cambridge University Press, Cambridge, Second Edition, 1993).

14. J.L. Pouchou and F. Pichoir, Proceedings of the 11th International Congress on X-ray Optics and Micro-analysis, J.D. Brown and R.H. Packwood, eds., (London, 1996).

15. M.P. Jackson, M.J. Starink and R.C. Reed, "Determination of the Precipitation Kinetics of Ni_3Al in the Ni-Al System using Differential Scanning Calorimetry", Materials Science and Engineering, A264 (1999), 26–38.

16. L. Chapman and P. Quested, National Physical Laboratory, Teddington, Middlesex, UK, Private Communication.

17. A. Davin, V. Leroy, D. Coutsouradis and L. Habraken, "Comparison of the Diffusion of Some Substitution Elements in Nickel and Cobalt", Cobalt, 19 (1963), 51–56.

18. C.P. Heijwegan and G.D. Rieck, "Diffusion in the Mo-Ni, Mo-Fe and Mo-Co Systems", Acta metallurgica, 22 (1974), 1269–1281.

19. T. Ustad and H. Sorum, "Interdiffusion in the Fe-Ni, Ni-Co and Fe-Co Systems", Physica Status Solidi, 20A (1973), 285–294.

20. A.B. Vladimirov, V.N. Kaygorodov, S.M. Klotsman and I.Sh. Trkhtenberg, "Bulk Diffusion of Cobalt and Tungsten in Nickel", Physics of Metals and Metallography, 46 (1978), 94–101.

21. B. Jonsson, "Assessment of the Mobilities of Cr, Fe and Ni in Binary Cr-Fe and Cr-Ni Alloys", Scandinavian Journal of Metallurgy, 24 (1995), 21–27.

22. N. Saunders, "Phase Diagram Calculations for High Temperature Structural Materials", Philosophical Transactions of the Royal Society of London A, 351 (1995), 543–561.

23. A. Engstrom and J. Agren, "Assessment of Diffusional Mobilities in Face-centered Cubic Ni-Cr-Al Alloys", Zeitschrift für Metallkunde, 87 (1996), 92–97.

24. J. Glownia and A. Janas, "Microsegregation in Dendritic Single Crystals of Nickel-Rich Alloys", Materials Science and Technology 3 (1987), 149–154.

25. V.A. Wills and D.G. McCartney, "A Comparative Study of Solidification Features in Nickel-base Superalloys: Microstructural Evolution and Microsegregation", Materials Science and Engineering, A145 (1991), 223–232.

26. S.N. Tewari, M. Vijayakumar, J.E. Lee and P.A. Curreri, "Solutal Partition Coefficients in Nickel-base Superalloy PWA-1480", Materials Science and Engineering, A141 (1991), 97–102.

27. D. Ma and P.R. Sahm, "Untersuchung des Erstarrungsvorgangs der gerichtet erstarrten Superlegierung CMSX-6", Zeitschrift für Metallkunde, 87 (1996), 640–644.

28. P.K. Sung and D.R. Poirier, "Liquid-Solid Partition Ratios in Nickel-base Alloys", Metallurgical and Materials Transactions, 30A (1999), 2173–2181.

29. M.N. Gungor, "A Statistically Significant Experimental Technique for Investigating Microsegregation in Cast Alloys", Metallurgical and Materials Transactions, 20A (1989), 2529–2533.

30. D.C. Cox, "Characterisation of the Microstructural Evolution in Single Crystal Nickel-Base Superalloys", (Ph.D Thesis, The University of Cambridge, 2000).

31. N. D'Souza, B.A. Shollock and M. McLean, "Quantitative Characterisation of Micro-segregation and Competitive Grain Growth in CMSX-4", Solidification Processing 1997, J. Beech and H. Jones, eds., (Department of Engineering Materials, University of Sheffield, UK, 1997), 316–320.

32. D. Ma and U. Grafe, "Microsegregation in Directionally Solidified Dendritic-Cellular Structure of Superalloy CMSX-4", Materials Science and Engineering, A270 (1999), 339–342.

33. THERMO-CALC, see for example, B. Sundman, Report D53, Summary of the Modules of Thermo-calc, (The Royal Institute of Technology, Stockholm, Sweden, 1984).

34. B. Sundman and I. Ansara, The SGTE Casebook, K. Hack, ed., (The Institute of Metals, London, UK, 1996), 94–98.

35. J.O. Andersson, L. Hoglund, B. Jonsson and J. Agren, in Fundamentals and Applications of Ternary Diffusion, G.R. Purdy, ed., (Pergamon Press, New York 1990), 153–163.

36. A. Engstrom, L. Hoglund and J. Agren, "Computer Simulation of Diffusion in Multiphase Systems", Metallurgical and Materials Transactions, 25A (1994), 1127–1134.

37. N. Matan, H.M.A. Winand, P. Carter, P.D. Bogdanoff and R.C. Reed, "A Coupled Thermodynamic/Kinetic Model for Diffusional Processes in Superalloys", Acta Materialia, 46 (1998), 4587–4600.

271

Appendix

In order to perform the calculations, it is necessary to estimate the diffusional mobilities of the components in CMSX-4. For simple disordered subsitutional phases such as the FCC phase, the mobility M_i of each component i can be represented by the following equation [36]

$$M_i = \frac{1}{RT} \exp\left\{\frac{-\Delta G_i^*}{RT}\right\} \qquad (A1)$$

where R is the universal gas constant and T the temperature. In an n component system the activation energy ΔG_i^* is expanded according to

$$\Delta G_i^* = \sum_{j=1,n}\left[x_j \Delta G_i^{*j} + 1/2 \sum_{j,k=1,n(j\neq k)} x_j x_k {}^0\Delta G_i^{*j,k}\right] \qquad (A2)$$

where x_i is the mole fraction of component i. The constants which were used in the calculations are as follows and are in units of J/mol.

Mobility for Al

$\Delta G_{Al}^{*Al} = -142000 + RT \ln(1.71 \times 10^{-4})$
$\Delta G_{Al}^{*Cr} = -235000 - 82T$
$\Delta G_{Al}^{*Ni} = -284000 + RT \ln(7.5 \times 10^{-4})$
${}^0\Delta G_{Al}^{*Al,Ni} = -41300 - 91.2T$
${}^0\Delta G_{Al}^{*Al,Cr} = 335000$
${}^0\Delta G_{Al}^{*Cr,Ni} = -53200$

Mobility for Co

$\Delta G_{Co}^{*Co} = -276731.84 - 80.06T$
$\Delta G_{Co}^{*Ni} = -277380.37 - 74.47T$
$\Delta G_{Co}^{*Al} = -174472.80 + RT \ln(4.64 \times 10^{-6})$
${}^0\Delta G_{Co}^{*Co,Ni} = 133337.99E - 66.98T$

Mobility for Cr

$\Delta G_{Cr}^{*Cr} = -235000 - 82.00T$
$\Delta G_{Cr}^{*Al} = -261700 + RT \ln(0.64)$
$\Delta G_{Cr}^{*Ni} = -287000 - 64.40T$
${}^0\Delta G_{Cr}^{*Cr,Ni} = -68000$
${}^0\Delta G_{Cr}^{*Al,Cr} = -487000$
${}^0\Delta G_{Cr}^{*Al,Ni} = -118000$

Mobility for Mo

$\Delta G_{Mo}^{*Mo} = -286273.92 - 56.46T$
$\Delta G_{Mo}^{*Ni} = -269133.21 - 84.35T$
$\Delta G_{Mo}^{*Al} = -54810.40 + RT \ln(1.04 \times 10^{-13})$
${}^0\Delta G_{Mo}^{*Mo,Ni} = -263315.40 + 172.55T$

Mobility for Ni

$\Delta G_{Ni}^{*Ni} = -287000 - 69.8T$
$\Delta G_{Ni}^{*Al} = -145900 + RT \ln(4.40 \times 10^{-4})$
$\Delta G_{Ni}^{*Co} = -272358.40 - 82.22T$
$\Delta G_{Ni}^{*Cr} = -235000 - 82.00T$
$\Delta G_{Ni}^{*Mo} = -286273.92 - 56.46T$
$\Delta G_{Ni}^{*Re} = -502659 - 76.96T$
$\Delta G_{Ni}^{*Ta} = -235390 - 86.73T$
$\Delta G_{Ni}^{*Ti} = -138456.95 - 132.72T$
$\Delta G_{Ni}^{*W} = -367004 - 64.50T$
${}^0\Delta G_{Ni}^{*Al,Ni} = -113000 + 65.50T$
${}^0\Delta G_{Ni}^{*Co,Ni} = 931556.43 - 627.82T$
${}^0\Delta G_{Ni}^{*Mo,Ni} = -288170.36 + 91.26T$
${}^0\Delta G_{Ni}^{*Ni,Re} = 1088526.75 - 702.62T$
${}^0\Delta G_{Ni}^{*Ni,Ta} = 1211176.04 - 1012.01T$
${}^0\Delta G_{Ni}^{*Ni,Ti} = -100716.30 + 100.45T$
${}^0\Delta G_{Ni}^{*Ni,W} = -366978.97 + 231.48T$

Mobility for Re

$\Delta G_{Re}^{*Re} = -502659 - 76.96T$
$\Delta G_{Re}^{*Ni} = -245037 - 127.54T$
${}^0\Delta G_{Re}^{*Ni,Re} = -168547.93 + 215.39T$

Mobility for Ta

$\Delta G_{Ta}^{*Ta} = -235390.00 - 86.73T$
$\Delta G_{Ta}^{*Ni} = -226313 - 109.20T$
${}^0\Delta G_{Ta}^{*Ni,Ta} = -629850.45 + 271.32T$

Mobility of Ti

$\Delta G_{Ti}^{*Ti} = -138456.95 - 132.72T$
$\Delta G_{Ti}^{*Ni} = -269267.07 - 68.81T$
${}^0\Delta G_{Ti}^{*Ni,Ti} = -260924.10 + 293.89T$

Mobility of W

$\Delta G_{W}^{*W} = -367004 - 64.50T$
$\Delta G_{W}^{*Ni} = -312426 - 64.39T$
${}^0\Delta G_{W}^{*Ni,W} = 1089264.53 - 923.05T$

ENHANCEMENT OF THE HIGH-TEMPERATURE TENSILE CREEP STRENGTH OF MONOCRYSTALLINE NICKEL-BASE SUPERALLOYS BY PRE-RAFTING IN COMPRESSION

U. Tetzlaff and H. Mughrabi

Institut für Werkstoffwissenschaften, Lehrstuhl I,

Friedrich-Alexander-Universität Erlangen-Nürnberg,

Martensstr. 5, D-91058 Erlangen, F.R. Germany

Abstract

In recent years, the development of γ'-hardened monocrystalline nickel-base superalloys focused on different aspects of alloy composition and microstructural refinement. In particular, the formation of so-called γ/γ'-raft structures, lying typically, in the case of a negative γ/γ'-lattice misfit δ, perpendicular to the stress axis during high-temperature tensile creep had attracted much attention. Unfortunately, this microstructural transformation leads in most cases to deteriorated creep properties. Earlier attempts to prevent rafting by suitable prior annealing treatments had been in so far unsuccessful, that while the applied annealing treatment retarded raft formation, it inevitably lead to microstructural coarsening accompanied by creep acceleration. The goal of the present extensive study is to show that suitable pre-rafting in compression, leading to a γ/γ'-raft structure parallel to the stress axis in the case of a negative lattice misfit, enhances not only the isothermal high-temperature fatigue strength, as shown earlier, but also the high-temperature creep properties. The observed macroscopic creep behaviour will be discussed with reference of microstructural observations.

Introduction

Monocrystalline nickel-base superalloys, hardened by a high volume fraction (up to ca. 70 %) of the ordered γ'-phase ($L1_2$-structure), which is coherently embedded in the fcc γ-matrix, are widely used as turbine blade material in aircrafts. In these high-temperature materials, so-called rafting (directional coarsening) has emerged as an important phenomenon both in laboratory tests [1,2,3] and in hot sections of turbine blades subjected to service conditions [4]. Furthermore, rafting has been observed after thermomechanical fatigue (TMF) [5]. A typical feature of the raft formation is, among other parameters, the strong dependence on the sign and magnitude of the applied stress and the γ/γ'-lattice mismatch $\delta = 2(a_{\gamma'}-a_{\gamma})/(a_{\gamma'}+a_{\gamma})$, where $a_{\gamma'}$ and a_{γ} are the lattice parameters of the phases γ' and γ, respectively. In the commonly used commercial nickel-base superalloys, the lattice mismatch is negative, and an external stress in tension (compression) produces a raft-like structure perpendicular (parallel) to the [001]-stress axis. The earliest model explaining this observation is due to Pineau [6]. According to the current understanding, which has been recently reviewed by Nabarro [7], the driving force for rafting, which is promoted by the combined action of dislocation activity and diffusion, stems from the fact that the superposition of the external stress and the coherency stresses introduces a gradient in elastic strain energy between the two types of γ-channels, which lie perpendicular and parallel to [001], respectively.

Since rafting was also detected in turbine blades subjected to service conditions [4,8], it was of considerable interest to study the effect of this modified microstructure on the fatigue and creep strengths. Therefore, Ott et al. [9,10] studied systematically the effect of pre-rafting on the isothermal fatigue behaviour of the alloys CMSX-6 and CMSX-4. For this purpose, they introduced raft structures parallel (by pre-rafting in compression) and perpendicular (by pre-rafting in tension) into the specimens by a small creep deformation (≤0.4 %) prior the fatigue tests. A typical result of this work is shown in Fig. 1 (stress amplitude $\Delta\sigma/2$ versus number of cycles N).

Superalloys 2000
Edited by T.M. Pollock, R.D. Kissinger, R.R. Bowman,
K.A. Green, M. McLean, S. Olson, and J.J. Schirra
TMS (The Minerals, Metals & Materials Society), 2000

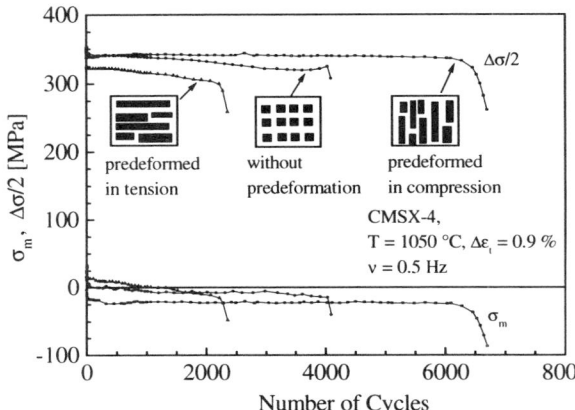

Fig. 1: Cyclic deformation curves of CMSX-4 crystals with the original cuboidal γ'-precipitates and after pre-rafting by tensile or compressive creep at 1050°C [9,10]. The initial microstructures are indicated schematically.

After an initial weak cyclic softening, the cyclic deformation curves show almost steady-state deformation until failure. The most important result is that while the specimens with the raft structure parallel to the stress axis exhibited the longest fatigue life, the specimens with the raft structure perpendicular to the stress axis showed the shortest fatigue life. The fatigue lives of the specimens with cuboidal γ'-particles were intermediate between these to extremes. It should be noted that the prior creep deformation induced mean stresses σ_m in the sense of the pre-deformation. In separate tests, it could be shown that while these mean stresses affected the fatigue behaviour, the basic result remained unaffected, when these mean stresses were compensated.

Over the last three decades, the effects of rafting on the creep behaviour have been discussed controversially in a number of papers [2,7,11,12,13,14]. It has to be mentioned that, in these investigations, all raft structures formed during tensile creep (for δ<0) and were oriented perpendicular to the stress axis.

In modern monocrystalline nickel-base superalloys with low negative misfit, such as SRR 99, CMSX-6, CMSX-4, rafting perpendicular to the stress axis causes under most experimental conditions a deterioration of the creep strength [2,15,16]. On the other hand, several authors suggest that a raft structure perpendicular to the stress axis can, under the condition of high temperatures and very low stresses, lead to an increase of the creep strength [2,7,13,17,18], as long as cutting of the rafts, which seems to be facilitated compared to cuboidal γ/γ'-microstructures, is avoided. In these cases, it is assumed, that under these conditions a raft structure becomes an effective barrier against dislocation by-passing by glide-climb motion along the γ/γ'-interfaces. Investigations of one-step heat treated SRR 99 specimens showed this behaviour in the case of very low stresses and a comparably high temperature [18,19]. Nevertheless, to the authors' knowledge, detailed experimental investigations of modern two-step annealed nickel-base superalloys, comparing well-aligned γ'-cuboids with pre-rafted material perpendicular to the stress axis, are still missing. A reason for this might be that rafting is unavoidable during service and therefore accepted as a fact of life.

For alloys with high negative misfit, improvements of the creep strength due to the formation of raft structures perpendicular to the stress axis have been found experimentally [20,21]. Pearson et al. [21] presented experimental evidence showing that a pre-rafted structure perpendicular to the stress axis, consisting of finely spaced γ/γ'-lamellas, enhances the creep strength remarkably. They concluded that for high temperatures and low stresses the operative creep mechanism involves dislocation motion primarily in the γ-phase with mobile dislocations circumventing or by-passing the γ'-phase by a combined glide and climb process, as firstly discussed by Carry and Strudel [22]. In an ideal situation, with γ'-rafts extending from one side of the crystal to the other in an almost perfectly regular manner, gliding/climbing around it by matrix dislocations is impossible. This conclusion is in accordance with opinions of other authors [1,13,21,23,24]. Another example for enhanced creep properties due to tensile pre-rafting can be found in [25]. In contrast, for lower temperatures and higher stresses, the same material used in [25], which had been pre-rafted in tension, exhibited a reduced creep strength after pre-rafting in tension, due to easier dislocation cutting of the γ'-rafts [26].

Summarising the observations on the creep strength of specimens with rafts perpendicular to the stress axis in material with high and low negative misfit, it can be proposed that rafts perpendicular to the stress axis may possibly have a higher potential for improving the creep strength in material with a high negative misfit, because the mobility and therefore the possibility for cutting of the rafts is reduced due to a smaller γ-channel width and more stabilised γ/γ'-interfaces consisting of finely spaced dislocation networks. Irrespective of these considerations, the formation of raft structures perpendicular to the stress axis leads in most cases to reduced creep properties and in all cases reported so far [9,10] to a reduced fatigue life. Therefore, it is concluded that pre-rafting in tension provides no promising potential for microstructural development.

The aim of the present work is to point out that suitable pre-rafting in compression not only improves the fatigue life, as reported earlier [9,10,27,28], but can also enhance significantly the creep strength of commercial monocrystalline superalloys, as suggested previously [10] and reported recently [28,29].

Theoretical background

In order to understand how different initial microstructures affect the creep behaviour, some simple theoretical considerations are helpful. The following deformation mechanisms are discussed:

- Circumvention of the γ'-particles by climb
- Circumvention of the γ'-particles by the Orowan mechanism
- Cutting of the γ'-particles

Firstly, we should consider the combined glide-climb motion along the γ/γ'-interface [30] in the spirit of the classical work by Carry and Strudel [22], which is dominant in the regime of low stresses and high temperatures, provided the threshold stresses for other mechanisms like the cutting or the circumvention of the γ'-particles by Orowan bowing are not exceeded.

According to calculations of both von Mises stresses [30,31,32] and the more realistic pertinent resolved shear stresses [30] as well as experimental observations [2,3,14,16], the largest local (shear) stresses develop in the case of an applied tensile load, initially in the γ-channels which lie perpendicular to the stress axis. Thus, tensile plastic yielding starts first in the horizontal γ-channels, whereas the dislocations are stopped after glide at the γ/γ'-interfaces (Fig. 2).

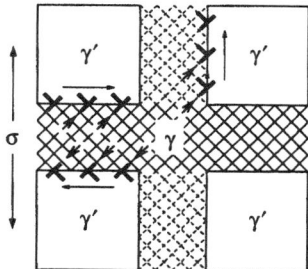

Fig. 2: Schematical drawing of dislocation accumulation and combined glide-climb motion of interfacial dislocations at γ/γ'-interfaces. The inclined lines represent the traces of slip planes, dashed lines indicate limited dislocation activity. The arrows parallel to the γ/γ'-interfaces indicate the direction of the glide-climb movement of dislocations. After [30].

The arrested interfacial dislocations whose Burgers vectors lie oblique to the γ/γ'-interface can move towards the edges of the γ'-particles by a combined glide-climb process, as originally suggested by Carry and Strudel [22]. In a next step, those dislocations which have reached the edge of the γ'-particle can glide across the vertical channel and build a dislocation arrangement at the vertical γ/γ'-interface, provided the local stresses are large enough. Those dislocations which are arrested at the vertical γ/γ'-interface can then move upwards in a similar manner by a combined climb-glide process toward the edge of the γ'-particle, and then glide into the next horizontal γ-channel, etc..

The authors propose, as in [22,30], that the climb process at the γ/γ'-interfaces becomes the strain-rate controlling mechanism.

It is of considerable interest to understand to what extent the γ/γ'-raft structures influence the creep properties. We approach this problem by calculating both the horizontal and the vertical climb components of the Peach-Koehler force acting on the arrested interface dislocations in the horizontal (001) and the vertical (100)/(010) planes, respectively [29]. (The stress axis is assumed to lie in the [001] direction.) The Peach-Koehler force is given by

$$\vec{F} = \left(\overline{\overline{\sigma}} \cdot \vec{b} \right) \times \vec{s} , \qquad (1)$$

where $\overline{\overline{\sigma}}$ is the stress tensor; $\vec{b}$ the Burgers vector and $\vec{s}$ the line direction.

It should be noted that in the Peach-Koehler formula, the stress tensor is the result of the superposition of the externally applied stress and the internal (coherency) stresses. During creep, this stress tensor can be modified by dislocation arrangements like networks which relax or increase the internal stresses. Further-

more, in these calculations it was assumed that only those dislocation segments which are arrested at the γ/γ'-interface after glide on an a/2 <110>{111} glide system contribute to the creep process. Based on geometric considerations, it is sufficient to examine only one plane of the {111} plane family, for example the (111) plane. On this plane, six Burgers vectors are possible: $\pm a/2[01\overline{1}]$, $\pm a/2[10\overline{1}]$ and $\pm a/2[1\overline{1}0]$. Considerations of the Schmid factor show that the slip systems with a Burgers vector $\pm a/2[1\overline{1}0]$ are not activated. For geometrical reasons, the Burgers vectors $\pm a/2[01\overline{1}]$ and $\pm a/2[10\overline{1}]$ are equivalent. Therefore, the calculations are only performed for dislocations with the $\pm a/2[01\overline{1}]$ Burgers vector on the (111)-plane. The geometry of the glide plane intersecting the γ/γ'-interfaces determines the line direction of the arrested dislocation segments [33]. Thus, dislocations with a $\pm a/2[01\overline{1}]$ Burgers vector at the (001)-γ/γ'-interface result in a $[1\overline{1}0]$ line direction (60°-dislocation) at the (010)-interface in a $[10\overline{1}]$ line direction (60°-dislocation) and for the (100)-interface in a $[01\overline{1}]$ line direction (pure screw). The screw dislocations do not contribute to the climb process.

Finally, after inserting the parameters discussed above in the Peach-Koehler formula, it follows that the climb force component in the horizontal channels acting in the appropriate directions in the horizontal (001) plane is higher than the climb force component in the vertical channels in the appropriate directions in the (100) or (010) planes [29]. Thus, the circumvention of γ' plates/particles by dislocation glide-climb motion along the corresponding γ/γ' interfaces is always easier for glide-climb motion along the horizontal (001) than along the vertical (010) or (100) interfaces. Therefore, it is proposed that the raft structure parallel to the stress axis should increase the creep strength compared to the initial cuboidal γ/γ'-microstructure or to a raft structure perpendicular to the stress axis, because the raft structure parallel to the stress axis provides the largest areas of vertical interfaces, where the climb process is impeded most strongly. Moreover, the raft structure parallel to the stress axis has only small areas of γ/γ'-interfaces perpendicular to the stress axis, where climb is more rapid.

Experimental

In the present study, rods of the two-step annealed γ'-hardened nickel-base superalloy SRR 99 with orientations, lying within 5° of [001], were used. The crystal orientations were determined by the Laue back reflection technique. The composition of the alloy is listed in Table 1:

Table 1: Composition of SRR 99.

Element	C	Si	Mn	Cr	Mo	Co
Wt.%	0.01	<0.1	<0.1	8.6	<0.03	5.02
Element	Ti	Al	Fe	B	Cu	S
Wt.%	2.22	5.61	0.03	<0.0013	<0.03	0.0006
Element	Zr	W	V	P	Ta	Ni
Wt.%	<0.005	9.66	<0.03	<0.005	2.83	bal.

After machining and electropolishing of the surface, the specimen had a gauge length of 12 mm and a diameter of 9 mm. The initial γ/γ'-microstructure consisted of fairly regularly aligned cuboidal γ' particles with a γ' edge length of 0.44 µm and a volume fraction of 72 %. The constrained misfit was determined by high resolution X-ray diffraction as $\delta \approx 1.4\times10^{-3}$.

The creep tests were performed on a servohydraulic test machine (MTS 880). For the investigations, a raft structure parallel to the stress axis was produced by pre-deformation in compression at a stress σ of -120 MPa and a temperature T of 1050° C. The required plastic creep strains were small and less than 0.4 %, to avoid prior damaging. For the verifying creep experiments, different applied stresses (150 to 300 MPa) and constant temperatures of 1000° C and 1050 °C were chosen. The deformed samples were investigated by scanning electron microscopy (SEM) as well as by transmission electron microscopy (TEM).

Results and Discussion

Creep at 1000°C

In this section, tensile creep curves (tensile plastic creep strain ε_{pl} versus creep time t) are compared for specimens with the initial cuboidal γ'-structure and the raft structure parallel to the stress axis for creep tests at different applied stresses (150-300 MPa) and a temperature of 1000 °C. These tests were limited to plastic strains of 3%, since larger strains are not tolerable in a component. The discussion of the macroscopic behaviour will be based on microstructural features.

Fig. 3. shows tensile creep curves ($\sigma = 150$ MPa, T = 1000°C) of two specimens in the as-aged condition and after pre-rafting in compression.

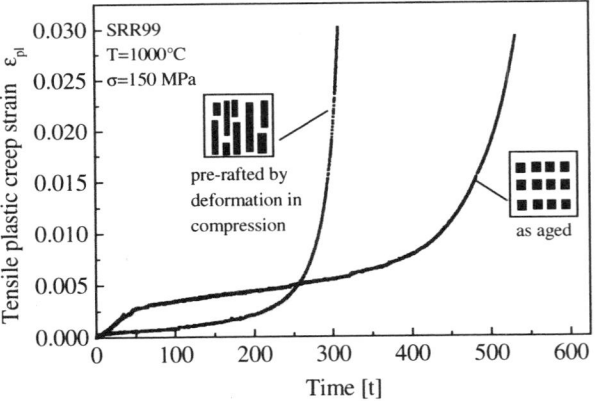

Fig. 3: Tensile creep curves ($\sigma = 150$ MPa, T = 1000 °C) up to 3% plastic strain of two monocrystalline SRR 99 specimens in the as-aged condition and after pre-rafting in compression, respectively.

It is obvious that up to a certain value of plastic creep strain the initially pre-rafted specimen shows a considerably enhanced creep strength compared to the specimens with the initial cuboidal γ/γ'-microstructure. This experimental result agrees with the conclusion of the previous theoretical considerations of the Peach-Koehler calculations, according to which rafts lying parallel to the stress axis should reduce the creep rate, if the climb at the γ/γ'-interfaces becomes the rate-determining process. In TEM-investigations of the two specimens after 3% plastic creep strain, this result could be confirmed in as much as that no cutting of the γ'-phase by dislocations could be found. After a creep time of approximately 260 h, the creep curves of the specimens with the initial cuboidal γ' and with the raft structure cross. This macroscopic behaviour can be explained through microstructural observations (see Fig. 4).

In the specimen with the initial cuboidal γ'-particles, a transformation to the typical (unfavourable) raft structure perpendicular to the stress axis occurs, as discussed before. On the other hand, the microstructure of the specimens with the initial well-aligned raft structure parallel to the stress axis develops into a rather irregular γ/γ'-microstructure, associated with the observed enhanced creep rate. It is evident that during tensile creep a raft structure parallel to the stress axis is energetically not as favourable as a γ' raft structure perpendicular to the stress axis. Therefore, the initial raft structure parallel to the stress axis tends to transform into a raft structure perpendicular to the stress axis. This transformation is, however, not completed within a plastic creep strain of only 3 %.

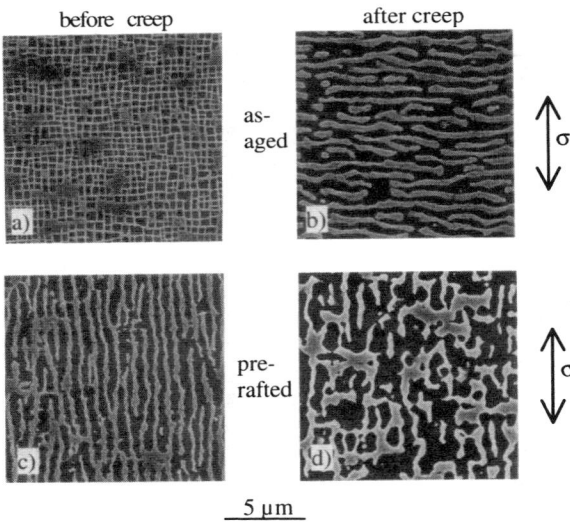

Fig. 4: SEM-micrographs of the microstructures in the initial state and after tensile deformation ($\sigma = 150$ MPa, T = 1000°C) a) as-aged b) as-aged, after 3% plastic creep strain in tension c) pre-rafted in compression d) pre-rafted in compression, after 3% plastic creep strain in tension.

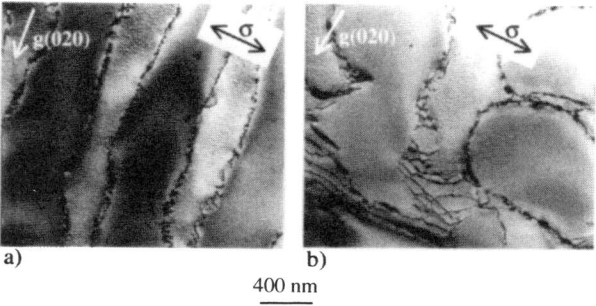

a)   b)

$\overline{\quad}$
400 nm

Fig. 5: TEM-micrographs of (010)-sections after tensile creep deformation at σ = 150 MPa/T = 1000°C. a) as-aged b) pre-rafted in compression.

At an increased stress of 200 MPa, pre-rafting in compression improves the tensile creep properties even more strongly, as shown in Fig. 6.

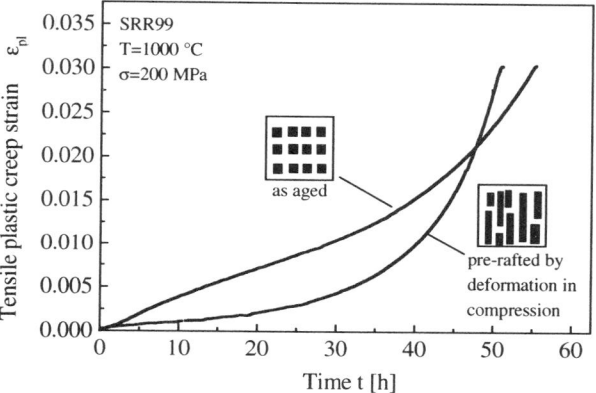

Fig. 6: Tensile creep curves (σ = 200 MPa, T = 1000 °C) up to 3% plastic strain of two monocrystalline SRR 99 specimens in the as-aged condition and after pre-rafting in compression, respectively.

As shown in Fig. 6, the creep strength of the specimen consisting of a raft structure parallel to the stress axis is improved appreciably up to a tensile plastic strain of about 2%, compared to the specimen with the initial cuboidal γ/γ'-microstructure. This improvement can be interpreted in the same manner as the creep experiments at 150 MPa, namely that climb as the rate-controlling process is remarkably impeded for a raft structure, which consists mainly of γ/γ'-interfaces parallel to the stress axis. At larger strains, however, this behaviour is reversed. Microstructural investigations (Fig. 7), show as for to the creep experiments performed at 150 MPa, that the enhanced creep rate for the specimen with the raft structure parallel to the stress is closely connected to the formation of an irregular γ/γ'-microstructure, providing deformation paths for facilitated glide and climb [13,24]. Due to the shorter creep times at 200 MPa, compared to 150 MPa, the relative time available for coarsening from a raft structure parallel to the stress axis toward a more stable raft configuration perpen-

dicular to the stress axis is reduced. Consequently the creep range of enhanced creep properties for the raft structure parallel to the stress is extended. This behaviour indicates that the processes of directional coarsening discussed here are more time-dependent than stress-dependent [12]. By contrast, the specimen with the initial cuboidal γ/γ'-microstructure transforms immediately into a well-aligned raft structure perpendicular to the stress axis. TEM-investigations of both specimens after 3% plastic creep strain show that cutting of the γ'-phase as a competing deformation mechanism can be excluded.

before creep   after creep

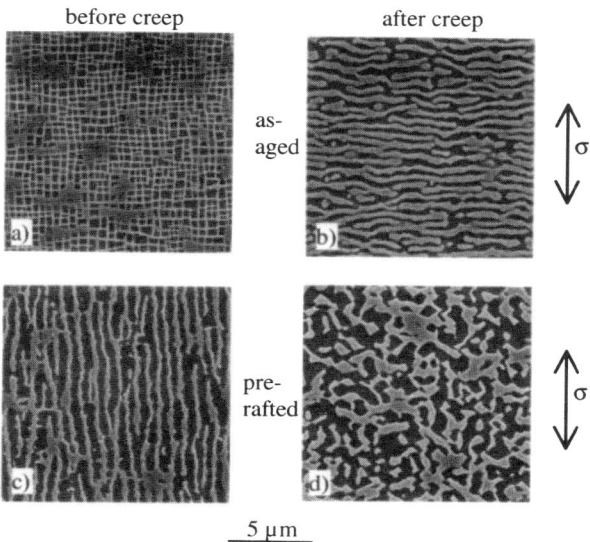

$\overline{\quad}$
5 μm

Fig. 7: SEM-micrographs of the microstructures in the initial state and after tensile deformation (σ = 200 MPa, T = 1000°C) a) as-aged b) as-aged, after 3% plastic creep strain in tension c) pre-rafted in compression d) pre-rafted in compression, after 3% plastic creep strain in tension.

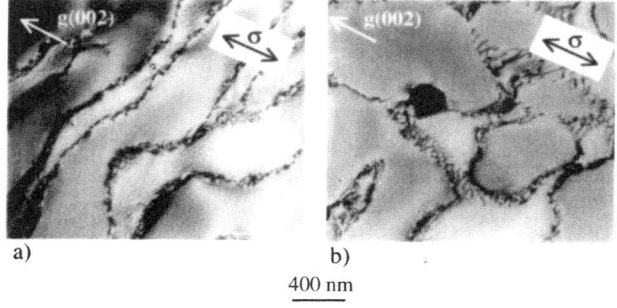

a)   b)

$\overline{\quad}$
400 nm

Fig. 8: TEM-micrographs of (010)-sections after tensile creep deformation at σ = 200 MPa/T = 1000°C. a) as-aged b) pre-rafted in compression.

At an even higher applied stress of 300 MPa, the creep behaviour of the specimen with a rafted structure parallel to the stress axis is significantly changed, compared to the creep test at 150 and 200 MPa (Fig. 9).

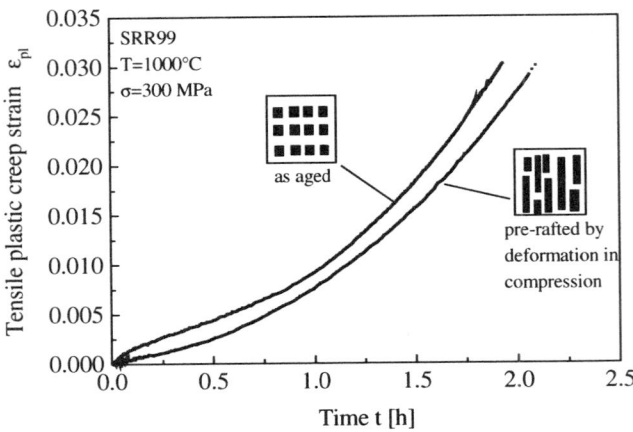

Fig. 9: Tensile creep curves (σ = 300 MPa, T = 1000°C) up to 3% plastic strain of two monocrystalline SRR 99 specimens in the as-aged condition and after pre-rafting in compression, respectively.

As shown in Fig. 9, the rafted structure parallel to the stress axis exhibits improved creep properties over the whole creep time.

Taking into account the microstructural investigations, it is obvious that the initial cuboidal γ/γ'-microstructure transforms immediately towards a raft structure perpendicular to the stress axis. More important, the initial pre-rafted structure parallel to the stress axis shows coarsening effects, but remains elongated parallel to the stress axis. Obviously, in this particular case, the time available for diffusion-controlled transformation of the initial raft structure parallel to the stress axis into an irregular γ/γ'-microstructure is too short; thus no crossing of the two creep curves can be observed. In fact, after 3% plastic strain, TEM-investigations revealed cutting of the γ'-particles in both types of initial γ/γ'-microstructures. Combining the microstructural observations and the macroscopic creep behaviour, it can be assumed that, in the beginning of the creep experiments, dislocation climb motion along the γ/γ'-interfaces controls the deformation rate. Furthermore, after successive strain hardening, when the sum of the external stress and the deformation-induced forward stress in the hard γ'-particles exceeds the local flow stress of the γ-particles, the ordered γ'-particles will be cut by dislocations. Thus, in a later stage of creep, cutting of the γ'-particles and the circumvention of the γ'-particles are competitive mechanisms controlling the creep rate.

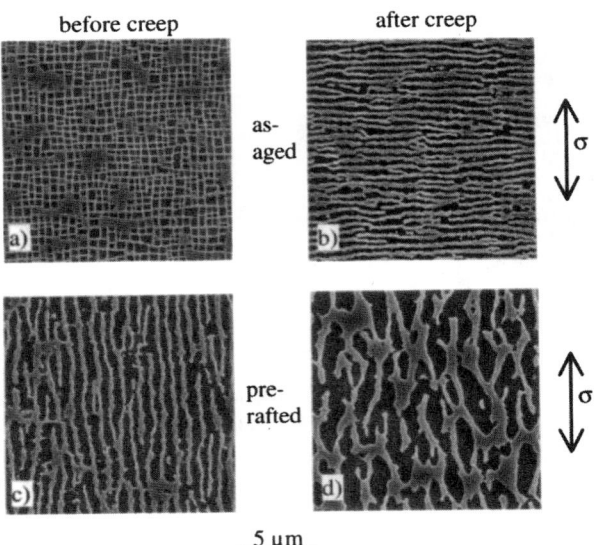

Fig. 10: SEM-micrographs of the microstructures in the initial state and after tensile deformation (σ = 300 MPa, T= 1000 °C) a) as-aged b) as-aged after 3% plastic creep strain in tension c) pre-rafted in compression d) pre-rafted in compression, after 3% plastic creep strain in tension.

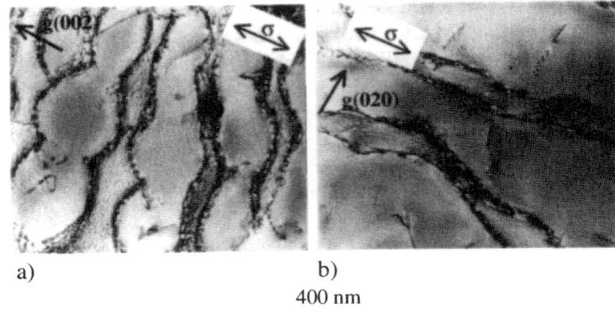

Fig. 11: TEM-microcraphs of (010)-sections after tensile creep deformation at σ = 300 MPa/T = 1000°C. a) as-aged b) pre-rafted in compression.

Creep at 1050°C

At an applied stress of 150 MPa, pre-rafting in compression leads to an improvement of the tensile creep properties, compared to the initial cuboidal γ/γ'-microstructure (see Fig. 12).

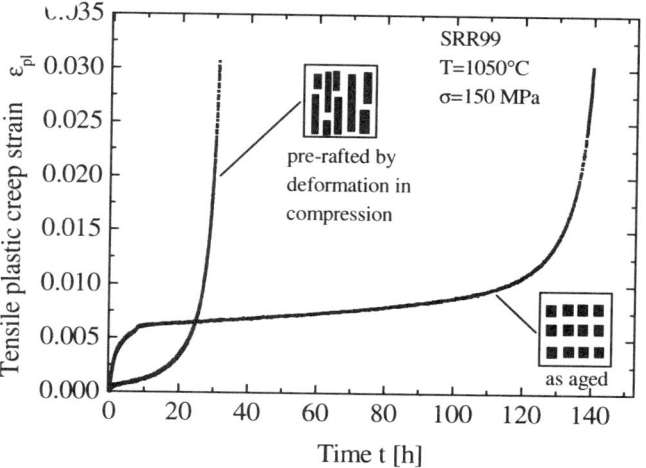

Fig. 12: Tensile creep curves (σ = 150 MPa, T = 1050 °C) up to 3% plastic strain of two monocrystalline SRR 99 specimens in the as-aged condition and after pre-rafting in compression, respectively.

This improvement by pre-rafting in compression can be attributed to the reduced climb force acting at the vertical γ/γ'-interfaces as discussed for creep at 1000°C.

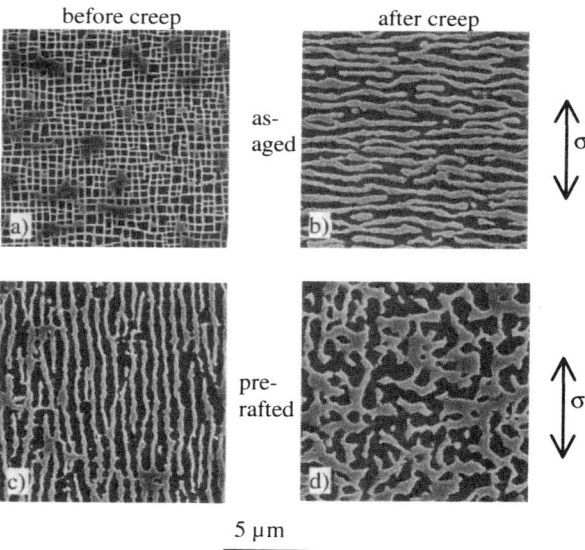

Fig. 13: SEM-micrographs of the microstructures in the initial state and after tensile deformation (σ = 150 MPa, T = 1050°C) a) as-aged b) as-aged after 3% plastic creep strain in tension c) pre-rafted in compression d) pre-rafted in compression, after 3% plastic creep strain in tension.

After a relatively short creep time, this behaviour is reversed. Comparing these creep experiments (σ = 150 MPa, T = 1050°C) with the creep experiments performed at the same applied stress of 150 MPa but at a lower temperature of 1000°C, the pronounced rapid increase of the creep rate at 1050°C is undoubtedly related to the facilitated diffusion-controlled transformation of the initial raft structure parallel to the stress axis to an irregular γ/γ'-microstructure at this higher temperature. This microstructural transformation is evident from Fig. 13 after 3% plastic creep strain.

Furthermore, TEM-investigations reveal that under these conditions (σ = 150 MPa, T = 1050°C), both types of initial γ/γ'-microstructure show, in contrast to the test at same applied stress at 1000°C, cutting of the γ'-phase by dislocations (Fig. 13). Thus, it can be concluded that cutting of the γ'-phase depends not only on the residual stresses but also on the temperature [15].

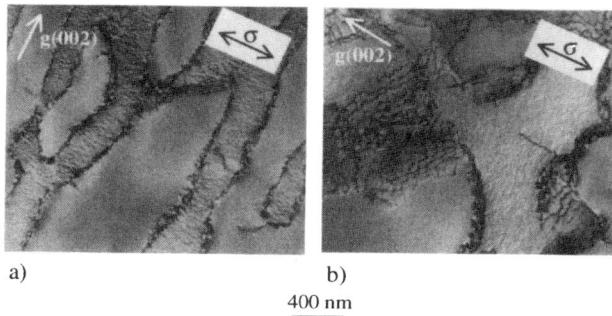

Fig. 14: TEM-micrographs of (010)-sections after tensile creep deformation at σ = 150 MPa/T = 1050°C. a) as-aged b) pre-rafted in compression.

At a higher applied stress of 200 MPa, a basically similar creep behaviour as discussed for a stress of 150 MPa is found (Fig. 15), the main difference being that the time at which the two creep curves cross is now shorter than at the lower stress (Fig. 12).

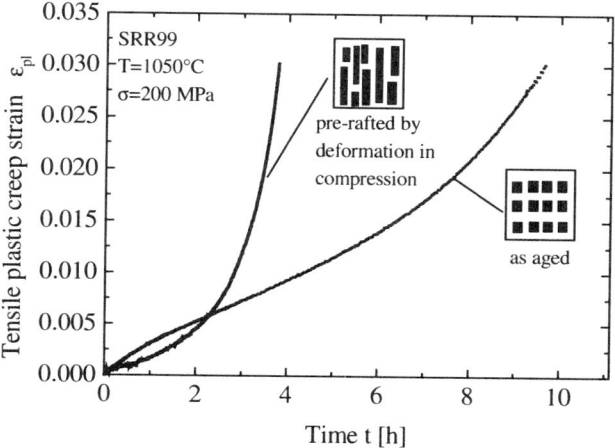

Fig. 15: Tensile creep curves (σ = 200 MPa, T = 1050°C) up to 3% plastic strain of two monocrystalline SRR 99 specimens in the as-aged condition and after pre-rafting in compression, respectively.

Subsequently, as in the previous examples, the behaviour of the initially pre-rafted specimen shows an enhanced creep rate, when the formation of an irregular γ/γ'-microstructure sets in (Fig. 16).

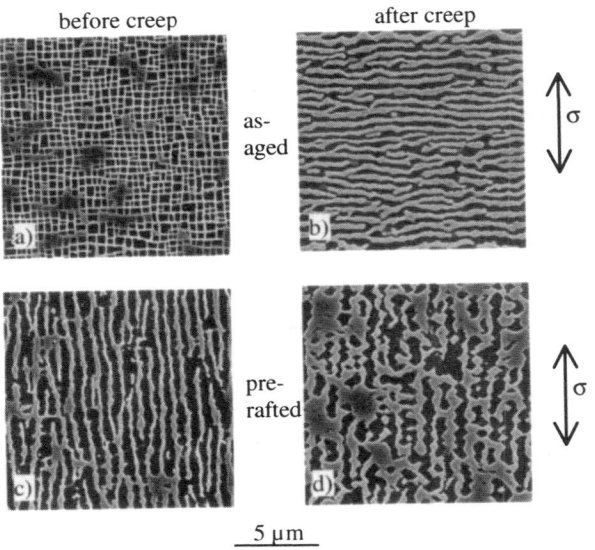

Fig. 16: SEM-micrographs of the microstructures in the initial state and after tensile deformation (σ = 200 MPa, T = 1050°C) a) as-aged b) as-aged after 3% plastic creep strain in tension c) pre-rafted in compression d) pre-rafted in compression, after 3% plastic tensile creep strain.

Here, it should be noted that the increased applied stress of 200 MPa results in enhanced cutting of the γ'-phase by dislocations for both initial γ/γ'-microstructures. As shown in Fig. 17, cutting of the γ'-phase is less pronounced for the initial cuboidal γ/γ'-microstructure which coarsened directly to an almost perfect raft structure perpendicular to the stress axis than for the initial raft structure parallel to the stress axis which transformed into an irregular raft structure. This experimental finding is in accordance with opinions of some authors [15,34], who pointed out that enhanced inhomogenity of the local deformation in irregularly shaped γ/γ'-microstructures leads to increased cutting of the γ'-particles by dislocations.

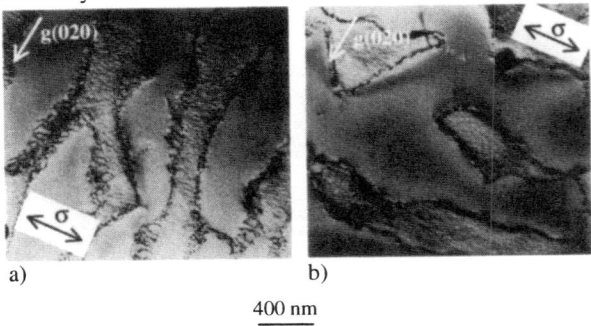

Fig. 17: TEM-micrographs of (010)-sections after tensile creep deformation at σ = 200 MPa, T = 1050°C a) as-aged b) pre-rafted in compression.

Influence of pre-rafting in compression on the minimum creep rate

In Fig. 18, a Norton-Plot of the minimum plastic creep rate $\dot{\varepsilon}_{pl.min}$ versus σ shows for all creep experiments presented previously a remarkable reduction of the minimum creep rate for the initial raft structure parallel to the stress axis, compared to the initial cuboidal γ/γ'-microstructure. Only in one particular case at a temperature of 1050°C and an applied stress of σ = 150 MPa, the minimum creep rate was not reduced by pre-rafting in compression.

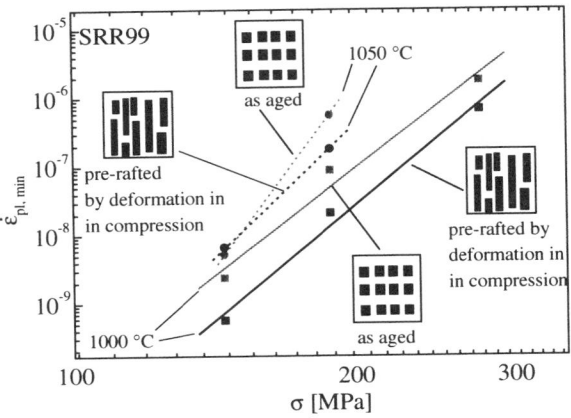

Fig. 18: Comparison of the minimum plastic creep rate of the initial cuboidal γ/γ'-microstructure and the raft structure parallel to the stress axis at different temperatures (1000°C, 1050°C) and different applied tensile stresses (150 - 300 MPa).

Conclusions

In the present work on the monocrystalline superalloy SRR 99, is was shown exemplarily that pre-rafting in compression is beneficial for the high-temperature creep strength. It is proposed that this conclusion will apply also to other monocrystalline nickel-base superalloys. The main result were as follows:

- Pre-rafting in compression is suitable to not only improve the isothermal fatigue strength, as shown earlier, but also the tensile creep strength.

- The observed enhancement of the creep strength is in good agreement with predictions based on the Peach-Koehler calculations of the rate-controlling dislocation climb force on vertical and horizontal γ/γ'-interfaces.

- At a given temperature, the range of enhanced creep strength due to pre-rafting depends primarily on the applied stress, which determines the time available for microstructural transformations of the rafts introduced prior to the tensile creep test by a small compressive creep strain.

- It is conceivable that the extent of the creep strain within which the creep strength can be enhanced could be extended by a suitable measure to stabilise the rafts lying parallel to the stress axis against transformation into rafts lying perpendicular to the stress axis.

References

1. T.P. Gabb, S.L. Draper, D.R. Hull, R.A. MacKay and M.V. Nathal, "The Role of Interfacial Dislocation Networks in High Temperature Creep of Superalloys", Materials Science and Engineering, A 118 (1989), 59-69.

2. H. Mughrabi, W. Schneider, V. Sass and C. Lang, "The Effect of Raft Formation on the High-Temperature Creep Deformation Behaviour of the Monocrystalline Nickel-Base Superalloy CMSX-4", in Strength of Materials, Proc. ICSMA 10, eds. H. Oikawa et al., (Sendai: The Japan Institute of Metals, 1992), 705-708.

3. T.M. Pollock and A.S Argon, "Directional Coarsening in Nickel-Base Single Crystals with High Volume Fractions of Coherent Precipitates", Acta Metallurgica et Materialia, 42 (1994), 1859-1874.

4. S. Draper, D. Hull and R. Dreshfield, "Observation of Directional Gamma Prime Coarsening during Engine Operation", Metallurgical Transactions A, 20A (1989), 683-688.

5. S. Kraft, I. Altenberger and H. Mughrabi, "Directional γ-γ' Coarsening in a Monocrystalline Nickel-Based Superalloy During Low-Cycle Thermomechanical Fatigue", Scripta Metallurgica et Materialia, 32 (1995), 411-416, and Scripta Metallurgica et Materialia, 32 (1995), 1903 (Erratum).

6. A. Pineau, "Influence of Uniaxial Stress on the Morphology of Coherent Precipitates During Coarsening-Elastic Energy Considerations", Acta Metallurgica, 24 (1976), 559-564.

7. F.R.N. Nabarro, "Rafting in Superalloys", Metallurgical and Materials Transactions, 27A (1996), 513-530.

8. H. Biermann, B. von Grossmann, T. Schneider, H. Feng and H. Mughrabi, "Investigation of the γ/γ' Morphology and Internal Stresses in a Monocrystalline Turbine Blade after Service: Determination of The Local Thermal and Mechanical Loads", Superalloys 1996, Proc. of 8th Int. Symp. on Superalloys, eds. R.D. Kissinger et al., (Warrendale, PA: The Minerals, Metals & Materials Society, 1996), 201-210.

9. M. Ott, U. Tetzlaff and H. Mughrabi, "Influence of Directional Coarsening on the Isothermal High-Temperature Fatigue Behaviour of the Monocrystalline Nickel-Base Superalloys CMSX-6 and CMSX-4", in Microstructural and Mechanical Properties of Metallic High-Temperature Materials / DFG research report, eds. H. Mughrabi et al., (Weinheim: Wiley-VCH, 1999), 425-440.

10. H. Mughrabi, M. Ott and U. Tetzlaff, "New Microstructural Concepts to Optimize the High-Temperature Strength of γ'-Hardened Monocrystalline Nickel-Base Superalloys", Materials Science and Engineering, A234-236 (1997), 434-437.

11. P. Caron, P.J. Henderson, T. Khan and M. McLean, "On the Effect of Heat Treatments on The Creep Behaviour of A Single Crystal Superalloy", Scripta Metallurgica, 20 (1986), 875-880.

12. R.A MacKay and L.J. Ebert, "The Development of γ-γ' Lamellar Structures in a Nickel–Base Superalloy during Elevated Temperature Mechanical Testing", Metallurgical Transactions, 16A (1985), 1969-1982.

13. M.V. Nathal and L.J. Ebert, "Elevated Temperature Creep–Rupture Behaviour of the Single Crystal Nickel–Base Superalloy NASAIR 100", Metallurgical Transactions, 16A (1985), 427-439.

14. A. Fredholm and J.-L. Strudel, "High Temperature Creep Mechanisms in Single Crystals of Some High Temperature Performance Nickel Base Superalloys", in High Temperature Alloys: Their Exploitable Potentials, Proc. of Petten Int. Conf, eds. J.B. Marriot et al., (London: Elsevier Applied Science, 1987), 9-18.

15. J. Svoboda and P. Lukáš, "Model of Creep in <001>-oriented Superalloy Single Crystals", Acta Materialia, 46 (1998), 3421-3431.

16. W. Schneider, J. Hammer and H. Mughrabi, "Creep Deformation and Rupture Behaviour of The Monocrystalline Superalloy CMSX-4—A Comparison With The Alloy SRR 99, in Superalloys 1996, Proc. of 7th Int. Symp. on Superalloys, eds. S.D. Antolovich et al., (Warrendale, PA: The Minerals, Metals & Materials Society, 1996), 589-598.

17. W. Schneider, "Hochtemperaturkriechverhalten und Mikrostruktur der einkristallinen Nickelbasis-Superlegierung CMSX-4 bei Temperaturen von 800°C bis 1100°C" (Doctorate Thesis, Universität Erlangen-Nürnberg, 1993), 155-156.

18. J. Hammer, "Kriech-und Zeitstandsverhalten der einkristallinen Nickelbasis-Superlegierung SRR99 unter besonderer Berücksichtigung mikrostruktureller Vorgänge und der Materialfehler" (Doctorate Thesis, Universität Erlangen-Nürnberg, 1990), 149-157.

19. H. Mughrabi, "γ/γ' Rafting and Its Effect on the Creep and Fatigue Behaviour of Monocrystalline Superalloys", in The Johannes Weertman Symposium, eds. R.J. Arsenault et al. (Warrendale, PA: The Minerals Metals and Materials Society, 1996), 267-278.

20. D.D. Pearson, F.D. Lemkey and B. H. Kear, "Stress Coarsening of γ' and its Influence on Creep Properties of a Single Crystal Superalloy, in Superalloys 1980, Proc. 4th Int. Symp. Superalloys, eds. J. K. Tien et al., (Metals Park, OH: ASM, 1980), 513-520.

21. D.D. Pearson, B.H. Kear and F.D. Lemkey, "Factors Controlling the Creep Behaviour of a Nickel–Base Superalloy," Creep and Fracture of Engineering Materials and Structures, eds. B. Wishire and D.R.J. Owen, (Swansea: Pineridge Press, 1981), 213-233.

22. C. Carry and J.-L. Strudel, "Apparent and Effective Creep Parameters in Single Crystals of a Nickel Base Superalloy - I. Incubation Period", Acta Metallurgica, 25 (1977), 767-777.

23. A. Fredholm and J. L. Strudel, "On the Creep Resistance of some Nickel Base Single Crystals", in Superalloys 1984, Proc. of 5th Int. Symp. on Superalloys, eds. M. Gell et al., (Warrendale, PA: Metallurgical Society of AIME, 1984), 211-220.

24. P. Caron and T. Khan, "Improvement of Creep Strength in a Nickel-Base Single-Crystal Superalloy by Heat Treatment", Materials Science and Engineering, 61 (1983), 173-184.

25. M.V. Nathal and R.A. MacKay, "The Stability of Lamellar γ - γ' Structures", Materials Science and Engineering, 85 (1987), 127-138.

26. M.V. Nathal, R.A. MacKay and R. V. Miner, "Influence of Precipitate Morphology on Intermediate Temperature Creep Properties of a Nickel-Base Superalloy Single Crystal", Metallurgical Transactions A, 20A (1989), 133-141.

27. M. Ott and H. Mughrabi, "Dependence of the Isothermal Fatigue Behaviour of a Monocrystalline Nickel-Base Superalloy on the γ/γ'-Morphology", in Proc. of FATIGUE '96, eds. G. Lütjering and H. Nowack, (Oxford: Elsevier Science Ltd, 1996), 789-794.

28. H. Mughrabi and U. Tetzlaff, "Microstructure and High-Temperature Strenghth of Nickel-Base Superalloys", in Advanced Engineering Materials, in print.

29. U. Tetzlaff, M. Nicolas und H. Mughrabi, "Can the High Temperature Tensile Strength of Nickel-Base Superalloys be Improved by pre-rafting?", in Proc. of EUROMAT '99, (Weinheim: Wiley-VCH, 2000), in print.

30. H. Mughrabi, H. Feng and H. Biermann, "On the Micromechanics of the Deformation of Monocrystalline Nickel-Base Superalloys", in Proc. of the IUTAM Symposium on Micromechanics of Plasticity and Damage of Multiphase Materials, eds. A. Pineau and A. Zaoui, (Dordrecht, Netherlands: Kluwer Academic Publishers, 1996), 115-122.

31. L. Müller, U. Glatzel and M. Feller-Kniepmeier, "Modelling Thermal Misfit Stresses in Nickel-Base Superalloys Containing High Volume Fraction of γ' Phase", Acta Metallurgica et Materialia, 40 (1992), 1321-1327.

32. T.M. Pollock and A.S. Argon, "Creep Resistance of CMSX-3 Nickel-Base Superalloy Single Crystals", Acta Metallurgica et Materialia, 40 (1991), 1-30.

33. T. Link und M. Feller-Kniepmeier, "Elektronenmikroskopische Untersuchung von γ/γ'-Phasengrenzen in der einkristallinen Nickelbasislegierung SRR 99 nach Hochtemperaturkriechen", Zeitschrift für Metallkunde, 79 (1988), 381-387.

34. M.V. Nathal, "Effect of Initial Gamma Prime Size on the Elevated Temperature Creep Properties of Single Crystal Nickel Base Superalloys", Metallurgical Transactions A, 18A (1987), 1961-1970.

Blade Alloys

Alloying Effects on Surface Stability and Creep Strength of Nickel Based Single Crystal Superalloys containing 12mass%Cr

Y. Murata [1], R. Hashizume [2], A. Yoshinari [3], N. Aoki [1], M. Morinaga [1], and Y. Fukui [3]

1) Department of Materials Science and Engineering, Graduate School of Engineering, Nagoya University, Furo-cho, Chikusa, Nagoya 464-8603, JAPAN,
TEL & FAX   +81 52 789 3232
E-mail : murata@numse.nagoya-u.ac.jp
2) Materials Research Section, Technical Research Center, The Kansai Electric Power Company Inc., 11-20 Nakoji 3-chome, Amagasaki 661-0974, JAPAN
3) Hitachi Research Laboratory, Hitachi Co. Ltd., 3-1-1 Ohmika, Hitachi, 317-8511, JAPAN

Abstract

A series of experiments is carried out with six nickel based single crystal superalloys for use in industrial gas turbines. Each alloy contains 12mass%Cr, but no Re. The alloy compositions are chosen with the aid of the d-electrons concept, so that any undesirable phases do not appear in the alloy. The creep rupture life is measured at a stress of 196MPa and a temperature of 1193K, which is close to the service temperature, 1150K, of industrial gas turbines in power plants. The measured life of two alloys in this condition is found to be comparable to that of the 2nd generation nickel based single crystal superalloy containing 3%Re alloy. Also, it is shown that the amount of 12mass%Cr is not necessary to get good hot-corrosion resistance, as long as both the compositional ratios of Ti/Al and Ta/(W+Mo) are high in the alloy. In addition, a dense TiO_2-rich layer is found on the surface of the alloy with good hot-corrosion resistance. In contrast to the hot-corrosion resistance, the oxidation resistance is lowered if the Ti/Al compositional ratio is high in the alloy. The possible mechanisms for hot corrosion and oxidation are discussed in this study.

Introduction

The importance of the combined cycle system in power plants has been increasing, because it has a great potential for the increase in the efficiency of power generation and also the attendant decrease in the emission of carbon dioxide into the air [1]. Nickel-based superalloys used for this system should be superior in hot corrosion resistance and oxidation resistance to the conventional alloys used for jet engines. This is because the examination interval for the maintenance of the system is much longer than that of jet engines. Another reason is partially due to the location of the power plant since it is often placed near the seaside, in particular, in Japan. There is a strong attack to the system when exposed to an NaCl corrosive atmosphere.

Furthermore, the nickel based superalloys for jet engines have been developed so as to increase the creep strength at the temperatures of more than 1300K [2,3]. However, the metal temperature of gas turbines is usually below 1170K. In addition, it has been found that the alloys with superior creep strength at high temperatures are not necessarily strong at low temperatures, as shown later. Thus, there are distinct differences in the targets for alloy design between jet engines and gas turbines.

The purpose of this study is to get fundamental information necessary for the design of single crystal superalloys for use in industrial gas turbines. The alloying effects are investigated experimentally on the surface stability and the creep strength of the alloys.

Experimental Procedure

Alloy chemistry

Six nickel-based single crystal superalloys containing 12mass%Cr, NKH71 ~ NKH76, are made with the aid of the

Superalloys 2000
Edited by T.M. Pollock, R.D. Kissinger, R.R. Bowman,
K.A. Green, M. McLean, S. Olson, and J.J. Schirra
TMS (The Minerals, Metals & Materials Society), 2000

Table 1 Chemical compositions and alloying parameters of experimental alloys (mol%/mass%)

Alloy	Ti	Cr	Co	Nb	Mo	Hf	Ta	W	Al	Ni	$\overline{Md}$	$\overline{Bo}$	Ti/Al	Mo+W	Ta/(Mo+W)
NKH71	1.51	13.99	0.00	0.00	0.30	0.00	1.91	2.52	11.02	bal.	0.985	0.688	0.137	2.82	0.68
	1.21	12.19	0.00	0.00	0.49	0.00	5.78	7.75	4.98						
NKH72	1.57	14.09	3.87	0.00	0.31	0.00	1.91	2.50	10.89	bal.	0.987	0.696	0.144	2.81	0.68
	1.26	12.27	3.81	0.00	0.50	0.00	5.78	7.70	4.92						
NKH73	2.99	14.05	0.00	0.06	0.36	0.01	2.17	1.98	9.01	bal.	0.985	0.694	0.332	2.34	0.93
	2.39	12.20	0.00	0.09	0.57	0.04	6.57	6.09	4.06						
NKH74	4.74	14.13	3.97	0.06	0.36	0.01	2.20	1.96	7.30	bal.	0.995	0.712	0.649	2.31	0.95
	3.77	12.20	3.88	0.09	0.57	0.03	6.61	5.97	3.27						
NKH75	2.98	14.07	3.93	0.06	0.30	0.01	1.99	1.86	8.76	bal.	0.980	0.697	0.340	2.16	0.92
	2.40	12.28	3.89	0.09	0.48	0.03	6.05	5.74	3.97						
NKH76	4.92	14.11	0.00	0.06	0.30	0.01	1.78	1.77	7.48	bal.	0.989	0.698	0.658	2.06	0.86
	3.97	12.36	0.00	0.09	0.48	0.03	5.43	5.47	3.40						
3%Re*	1.27	7.60	9.96	0.00	0.38	0.03	2.19	2.12	12.62	bal.	0.986	0.667	0.101	3.46**	0.88
	1.00	6.50	9.65	0.00	0.60	0.10	6.50	6.40	5.60						

*) 3%Re alloy contains 2.95mass%Re and is used as a reference alloy.   **) Mo+W+Re

Table 2 Heat treatment conditions employed in this study

Alloy	Solution	1st Step aging	2nd Step aging
NKH71	1573K/28.8ks	1373K/14.4ks, A.C.	1144K/72ks, A.C.
NKH72			
NKH73			
NKH74	1553K/14.4ks + 1558K/14.4ks		
NKH75	1573K/28.8ks	1393K/14.4ks, cooling to 1144K at a rate of 0.03K/s	
NKH76	1553K/14.4ks + 1558K/14.4ks		

d-electrons concept [4].   The advantage of the use of this concept is to predict precisely the alloying limits of refractory elements in view of the phase stability even in high Cr superalloys [4].

It is generally accepted that the temperature capability decreases with increasing Cr content in the alloy [5].   But here to get the high corrosion resistance, 12%Cr superalloys are prepared by modifying the alloy composition of TUT92 [4], which is one of the highest performance 2nd generation nickel based superalloys developed by us previously.   The Re content of TUT92 is 0.25mol% (0.75mass%), but there is no Re in the present 12%Cr superalloys.

Table 1 shows the chemical compositions of the six 12%Cr single crystal (SC) superalloys and one of the 2nd generation SC alloys containing 3%Re which is used as a reference alloy.   In the table, the upper row and the lower row represent the alloy composition in mol% and in mass%, respectively.   The compositional ratios of Ti/Al and Ta/(Mo+W) are varied in mol% units among these alloys.

$\overline{Md}$ and $\overline{Bo}$ values are also shown in the table.   They are obtained from the compositional average of the d-electron parameters, Md and Bo [4,6].   Both $\overline{Md}$ and $\overline{Bo}$ are known to be related to the phase stability of nickel based superalloys.   It is also known that high strength conventional nickel based superalloys including the 1st and the 2nd generation single crystal superalloys have special values, $\overline{Md}$: 0.975~0.995 and $\overline{Bo}$: 0.660~0.715 [4,7].   The $\overline{Md}$ and $\overline{Bo}$ values of the six alloys are set in these ranges by adjusting the compositions of the refractory elements such as Mo, W and Ta in the alloys.

Alloy preparation and heat treatments

First, the master ingots of these alloys are made in a vacuum induction furnace by controlling the contents of gas elements to be as low as possible.   Then, using those ingots, single crystals measuring 16mm in diameter and 150mm in length are grown by a directionally solidified method.   The conditions of solution heat treatment and precipitation heat treatment of the γ ' phase are shown in Table 2.   Every specimen is first heat treated following the conditions shown in Table 2 and then supplied for a series of experiments.

Microstructural observation

The microstructure of the SC alloys are observed with the SEM before and after the creep rupture tests.   The cross section of the specimens after hot corrosion tests and oxidation tests are also observed using a SEM equipped with EDX analyzer.   For these observations, the specimen surface is first polished mechanically with emery papers and then with buff together with water containing alumina powders.   Subsequently, the surface is etched chemically in an HCl-HNO₃ solution.

Creep rupture test

The heat-treated SC alloys are machined mechanically into the specimens for the creep rupture test.   The gauge length

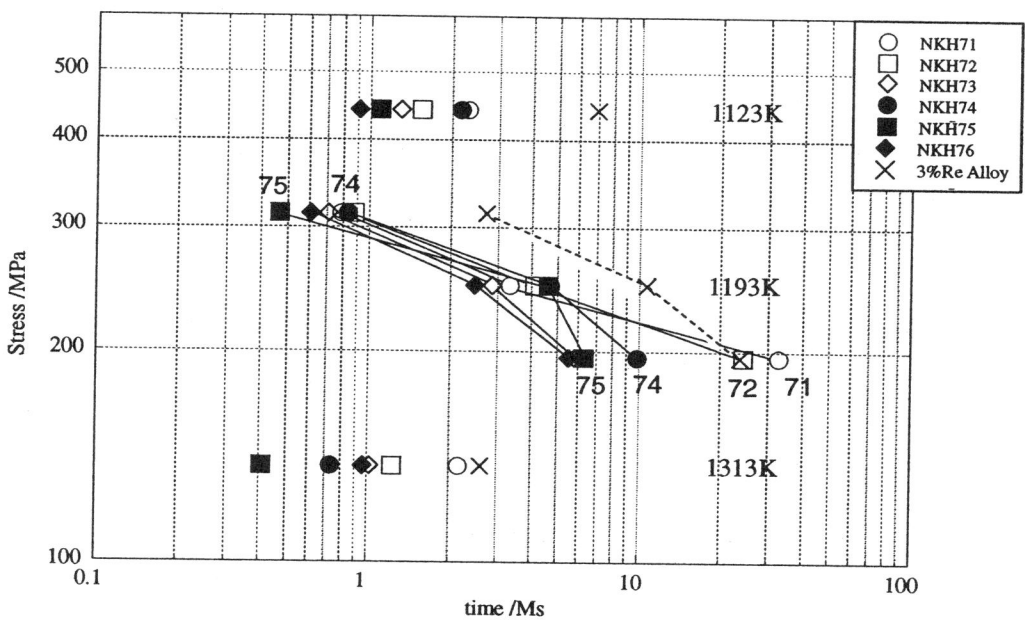

Fig.1 Results of creep rupture test.

of the specimen is 6mm in diameter and 30mm in length. The creep rupture tests are carried out under temperature/stress conditions of 1313K/137MPa, 1193K/314MPa, 248MPa, 196MPa, and 1123K/441MPa.

Oxidation and hot corrosion tests

The plate specimens measuring 10 x 25 x 1 mm are cut from the SC alloys by an electro-spark machine. Two kinds of oxidation tests are employed in this study. One is the cyclic oxidation test, in which the specimen is held in air at 1373K for 72ks followed by air cooling in each cycle. This cycle is repeated for 12 cycles, and the total exposure time at 1373K is 864ks. The other is the continuous oxidation test at 1313K for 2.16Ms.

In order to examine the hot corrosion resistance of the SC alloys, a burner-rig test is carried out at 1173K for 126ks with a specimen of 10mm in diameter and 30 mm in length. A fuel gas with atomized brine of 80ppm NaCl is used for this test. In addition, to prepare the specimens for the X-ray diffraction measurement, specimen surfaces of NKH71 and NKH76 are coated first with a solid solution of Na_2SO_4-25mol%NaCl salt, and then they are placed in a furnace at 1173K for 72ks.

Experimental Results

Creep strength

The results of the creep rupture test are shown in Fig.1. In the figure, the data on the reference alloy containing Re is represented by a cross mark. As mentioned earlier, all the experimental alloys have similar $\overline{Md}$ and $\overline{Bo}$ values. In the range of these values, the measured creep rupture lives are dependent on the alloy compositions as follows;
(1) Mo+W content is an important factor for increasing the creep rupture life in all the test conditions. For example as shown in Table 1, NKH71 and NKH72 alloys, both containing the highest amount of Mo+W among the experimental alloys, show longer rupture lives than other experimental alloys.

(2) When comparing the result of NKH71 with that of NKH72, Co is supposed to be an effective element in improving the creep rupture life in the high stress levels at 1193K, but it does not work anymore in the low stress, 196MPa, at the same temperature.

NKH74 and NKH75 exhibit longer rupture lives under 248MPa at 1193K than NKH71 and NKH72. However, this is not the case under 196MPa at the same temperature. In particular, the creep rupture lives of NKH74 and NKH75 under 196MPa are much lower than the values extrapolated from the creep rupture data at the higher stress levels at 1193K. These results indicate clearly that the Larson-Miller plot is inapplicable to the present experimental alloys.

The metal temperature of the turbine blade is controlled to be below about 1200K by the thermal barrier coating (TBC) and by air cooling. Also, the stress level used for the plant design is not as high as 200MPa. Since the Larson-Miller plot does not apply here, an accelerating creep test by increasing temperature or stress will give us an erroneous

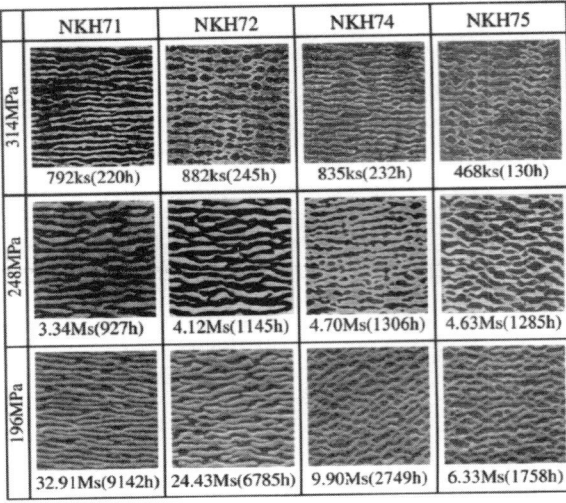

	NKH71	NKH72	NKH74	NKH75
314MPa	792ks(220h)	882ks(245h)	835ks(232h)	468ks(130h)
248MPa	3.34Ms(927h)	4.12Ms(1145h)	4.70Ms(1306h)	4.63Ms(1285h)
196MPa	32.91Ms(9142h)	24.43Ms(6785h)	9.90Ms(2749h)	6.33Ms(1758h)

Fig.2   Microstructures of the experimental alloys crept at 1193K under three stress levels.

Table 3   Aspect ratios of the γ' phase in the experimental superalloys crept at 1193K under three stress levels.

Alloy	Aspect ratio		
	314MPa	248MPa	196MPa
NKH71	49.2	34.5	21.9
NKH72	9.2	31.3	13.4
NKH74	50.0	14.1	5.9
NKH75	10.9	8.4	7.1

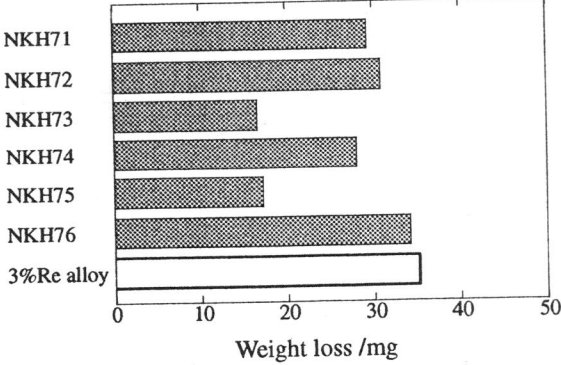

Fig.3   Weight loss of specimens after cyclic oxidation.

answer as to the temperature capability of the alloy.

Rafted structure

The typical microstructures of NKH71, NKH72, NKH74 and NKH75 alloys crept at 1193K are shown in Fig.2.   The rafted structure is observed in all the experimental alloys. The measured aspect ratios of the γ' phase obtained from the microstructures are listed in Table 3.

NKH71 and NKH72 alloys exhibit a regular rafted structure in all the stress conditions, whereas NKH74 and NKH75 alloys show somewhat irregular ones, except for NKH74 tested at 314MPa.   These results imply that the coherency between the γ phase and the γ' phase is lost in both NKH74 and NKH75 alloys, in particular, in the case of low stress conditions (longer rupture life).   This is related probably to the phenomena that the stress-time to rupture curve at 1193K becomes steep suddenly when the stress level is changed from 248MPa to 196MPa in these two alloys as shown in Fig.1.

Oxidation resistance

The results of the cyclic oxidation test are shown in Fig.3. The weight loss measured after 12 cycles is in the range of 17 to 35mg.   The weight of the specimen decreases due to the peeling off or spalling of surface scales.   All the experimental alloys are comparable or even superior in oxidation resistance when comparing to the reference alloy containing 3% Re.   Among the experimental alloys, NKH73 and NKH75, show an indication of good oxidation resistance, in which the Ti/Al compositional ratio is about 0.33.

The microstructures are observed in the cross section of the oxidation surface.   As shown in Fig.4, a layered structure is formed on the oxidation surface.   The outermost layer consists of mainly Al oxide phase, and the middle layer contains Al nitride phase, and the inner layer contains Ti nitride phase.   Some of these phases are identified as Al_2O_3, AlN and TiN by the EDX semi-quantitative chemical analysis.   The reason why this layered structure is formed will be discussed later.

The results of the continuous oxidation test of the SC alloys tested at 1313K for 2.16Ms are shown in Fig.5, and are very different from the results of the cyclic oxidation test shown in Fig.3.   The weight loss tends to increase approximately with the alloy number.   However, no difference is seen in the oxidation products between the cyclic and the continuous oxidation tests.   By comparing the results between NKH71 and NKH72, it is seen that Co is the element to deteriorate the oxidation resistance.   Also, there is a trend that the resistance decreases with increasing Ti/Al compositional ratio, except for the 3%Re alloy.

Hot corrosion resistance

The results of the burner-rig test at 1173K for 126ks are shown in Fig.6.   The weight loss decreases approximately with the alloy number.   This implies that the hot corrosion resistance is improved by increasing the Ti/Al compositional ratio.   This trend is opposite to the observation in the continuous oxidation test, as explained earlier.   Furthermore, by comparing the results between NKH71 and NKH72, it is found that the Co addition improves the hot corrosion resistance considerably.   This is also the reverse result of the continuous oxidation test.

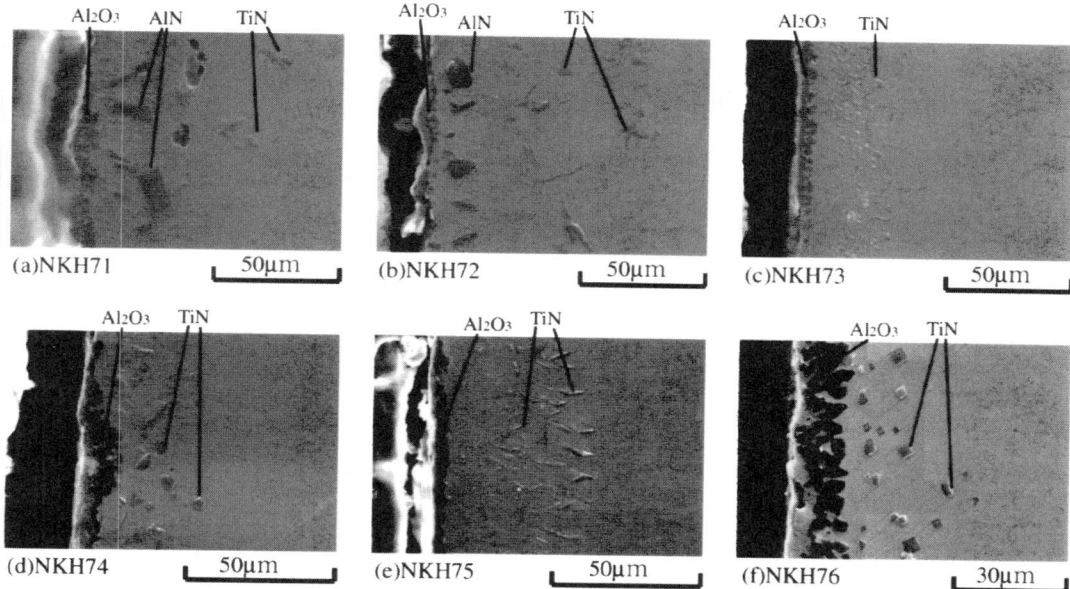

Fig.4   SEM microstructures in the cross section of specimens after the cyclic oxidation test.

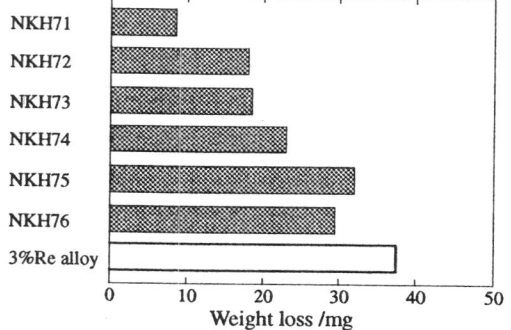

Fig.5 Weight loss of specimens after continuous oxidation.

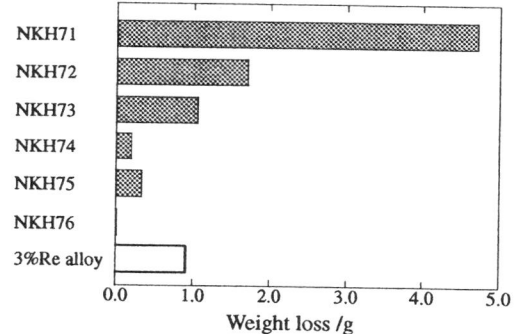

Fig.6 Weight loss of the specimens after the burner-rig test.

Among the experimental single crystal superalloys, the hot corrosion resistance is best for NKH76 and worst for NKH71. The main compositional differences between NKH71 and NKH76 are in Al, Ti and W contents.   In other words, high Ti/Al compositional ratio and low W content are conducive to better hot corrosion resistance in the alloys. Interestingly, in spite of a relatively high Cr content, 12mass%, in all the experimental alloys, the hot corrosion resistance is quite different among them.   This clearly indicates that the addition of 12mass%Cr is not necessarily needed to get good hot corrosion resistance.

After the burner-rig test, the SEM micrographs and the corresponding characteristic X-ray images are taken from a cross section of NKH71 and NKH76 alloys, and the results are shown in Fig.7 and Fig.8, respectively.   Schematic illustration is also drawn in each figure to show the layered structure of corrosion products.   Needless to say, the thickness of the corrosion layer is directly related to the weight loss shown in Fig.6, and is about 130 μm in NKH71

and about 30 μm in NKH76.   By comparing Fig.7 with Fig.8, it is apparent that the appearance of the layered structure formed in the corrosion surface resembles, for example, the sulfide layer existing beneath the oxide layer in both cases.   This is consistent with the previous result [8].   However, there is still a large difference between NKH71 and NKH76 in the state of the TiO_2 layer.   Namely, TiO_2 formed on the outermost surface is dense in NKH76 (Fig.8) but porous in NKH71 (Fig.7).   Furthermore, Cr sulfide is observed clearly in the corrosion products of NKH71, resulting in the existence of Cr depleted regions in this alloy.

The results of X-ray diffraction are shown in Fig.9 for (a) NKH71 and (b) NKH76, both measured after the hot corrosion test.   As described above, the main difference in the layered structure between NKH71 and NKH76 is whether a TiO_2–rich layer exists or not in the surface layer.   In fact, as shown in Fig.9 (b), TiO_2 is observed on the outer surface of good corrosion resistant NKH76.   However, as shown in

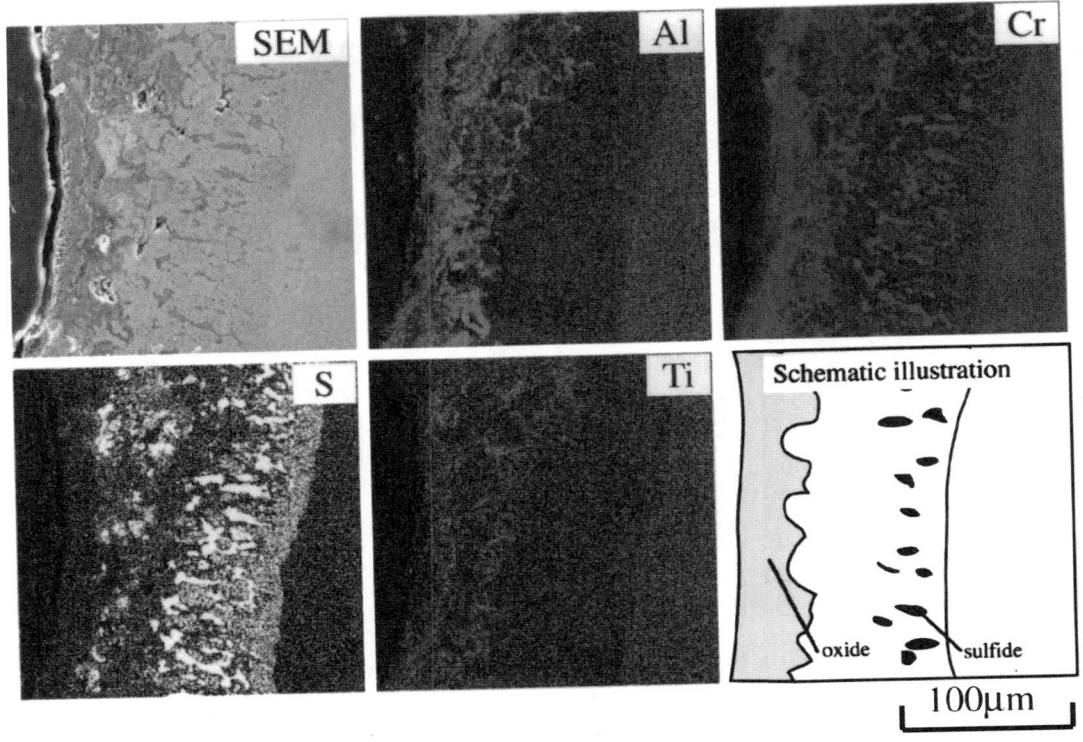

Fig.7   SEM image and the corresponding characteristic X-ray images taken from a cross section of NKH71 after
the burner-rig test.

Fig.9 (b), it is not detectable clearly on the surface of poor corrosion resistant NKH71.

Discussion

Creep strength

It is well known that the γ ' phase in nickel based superalloys changes its shape from cuboidal to rafted structure during the high temperature creep [9]. The mechanism of this phenomenon has been studied theoretically since the 1970's [10-12], and it is understood that both the lattice misfit and the elastic constant difference between the γ phase and the γ ' phase play an important role in the formation of the rafted structure [13]. As long as the coherency between the two phases is maintained, the elastic constant difference between them will operate to form such a rafted structure [11, 13].

NKH71 appears to keep the coherency between the γ phase and the γ ' phase up to about 33Ms at 1193K. This is because, once the coherency is lost in the alloy, the γ ' phase tends to coagulate and show an irregular microstructure, in order to lower the interfacial (chemical) energy by reducing the interfacial area in unit volume. In the case of NKH71, this never happens even after 33Ms, as shown in Fig.2.

On the other hand, the rafted structures in NKH74 and NKH75 are more irregular when compared to those in NKH71 and NKH72, despite the same creep conditions (1193K/196MPa) employed in these alloys. Therefore, it is supposed that there is a larger lattice misfit in NKH74 and NKH75 than in NKH71 and NKH72. As a result, the coherency between the γ and γ ' phases in NKH74 or NKH75 is lost considerably and misfit dislocations are introduced into the interface. The existence of such dislocations at the interface lowers the elastic energy between the two phases, so that the interfacial energy effect becomes more dominant than the elastic energy effect. As a result, the shape of the γ ' phase becomes round and irregular. In fact, the lattice mismatch of the NKH74 alloy is estimated to be 0.56% by using a relationship between the lattice constants and the chemical composition [14,15]. This lattice mismatch, 0.56%, for NKH74 is much larger than 0.38% for NKH71 and 0.35% for NKH72.

The relationship between the rafted structure and the creep strength depends on the creep temperature and the stress level as well. It is said that when dislocations do not cut the γ ' phase, the rafted structure gives a long way for dislocation climbing, resulting in the improvement of the creep rupture strength, but otherwise the creep strength tends to decrease by the rafting of the γ ' phase in the alloy [13].

Deviation from Larson-Miller plot

290

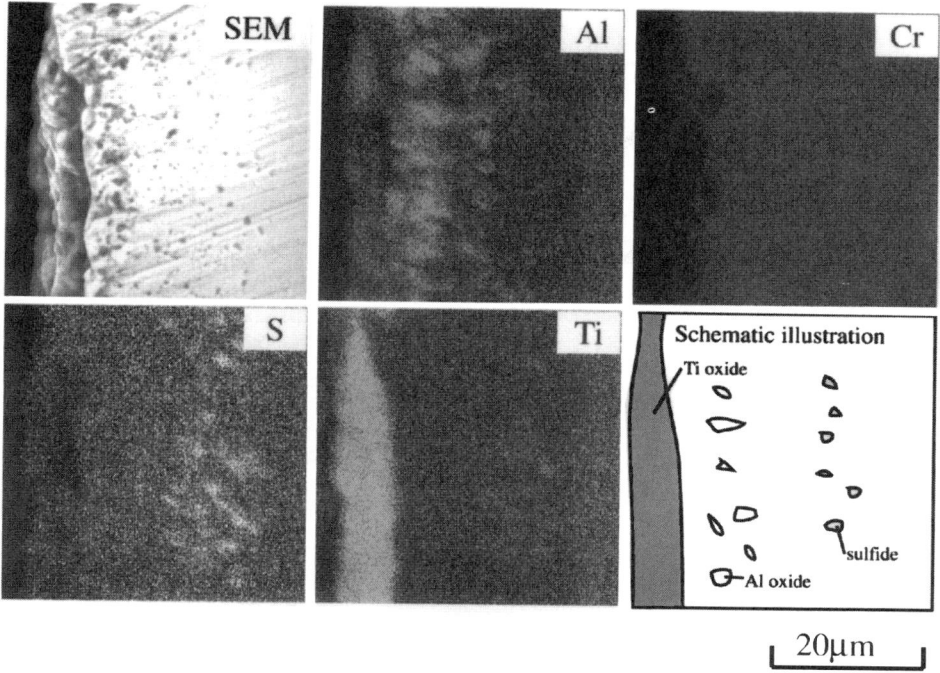

$20\mu m$

Fig. 8 SEM image and the corresponding characteristic X-ray images taken from a cross section of NKH76 after the burner-rig test.

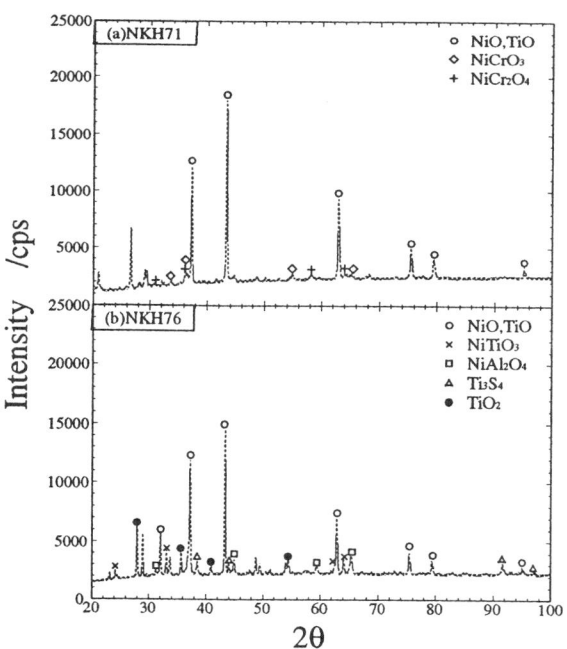

Fig.9 X-ray diffraction patterns of (a) NKH71 and (b) NKH76 after the hot corrosion test using the coating method.

It is stressed here that creep rupture data obtained at 1313K/137MPa are not usable for predicting the creep rupture life at 1193K/196MPa, by using a conventional Larson-Miller parameter. For example, Fig.10 is a Larson-Miller plot (C=30) which is drawn by using the creep data shown in Fig.1. In general, the Larson-Miller plot shows a smooth curve, when the microstructure in the alloy does not change during creep. However, as shown in Fig.10, the plot appears to give two different lines as indicated by A and B. Line A is obtained at high stress levels, whereas line B is obtained at low stress levels. It is likely that these two lines, A and B, are concerned with microstructural change during creep. In particular, the discrepancy between the A part and the B part in Fig.10 is larger in NKH74 and NKH75 than in NKH71 and NKH72. This is consistent with the results that the microstructures both of NKH71 and NKH72 are more stable during creep than those of NKH74 and NKH75, as shown in Fig.2.

Here, the data obtained at 1193K is located in the transition region from A to B. As mentioned before, the temperature, 1193K, corresponds to the upper limit temperature for the turbine blades to be serviced for a long term in industrial gas turbines. Also, stress operating on the turbine blades during the current service of gas turbines is not as high as 200MPa. Since the Larson-Miller plot is no longer valid in this temperature and stress range, creep experiments around 1193K/196MPa are important for testing the temperature capability of nickel based superalloys used in industrial gas turbines.

291

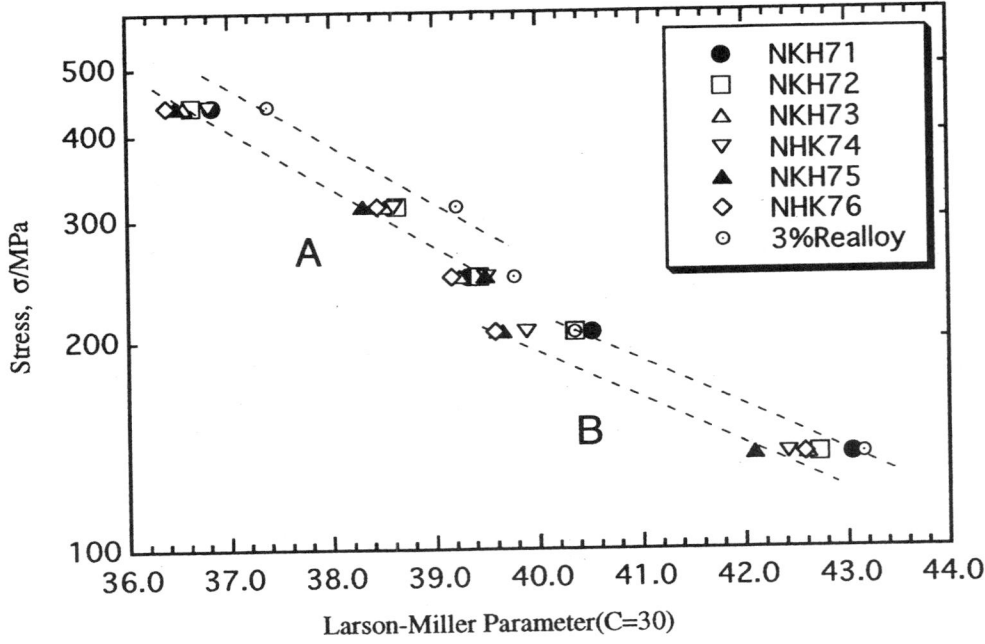

Fig. 10 Larson-Miller plot of the creep rupture data

Oxidation resistance and hot corrosion resistance

Both oxidation and hot corrosion resistance are important for materials used for blades and vanes in industrial gas turbines. In general, Al is known to be an effective element in improving high temperature oxidation resistance. Also, Cr is believed to be an effective element in improving hot corrosion resistance. However, in spite of the same Cr content, the hot corrosion resistance is considerably different among the present six experimental SC alloys. For example, the weight change is about two orders of magnitude larger in NKH71 than in NKH76.

In Fig.11 the results are plotted of both the burner-rig test and the continuous oxidation test. From this figure, it is clearly determined that the alloying effect is very different between the hot corrosion resistance and the oxidation resistance. As mentioned earlier, the increase in Ti/Al compositional ratio deteriorates the oxidation resistance but improves the hot corrosion resistance.

Possible mechanism for oxidation

In order to improve the oxidation resistance, it is necessary to form a dense oxide layer on the specimen surface. Both Ti and Al are strong oxide forming elements, so the oxidation occurs in a competitive way. It is known that the competitive reaction impedes the formation of a dense oxide layer [16]. Therefore, a purely dense oxide layer can not be formed on the specimen surface as long as the alloy contains comparable amounts of Al and Ti. Also, it is

considered that nitrides play an important role in the oxidation resistance, because nitrides are formed under the oxide layer in the oxidized specimen (see Fig.4). As explained below, this nitridation probably plays a role in the oxidation resistance in a similar way that the sulfidation does in the hot corrosion resistance.

The standard free energies of formation for TiN, TiO$_2$, AlN and Al$_2$O$_3$ are given as follows [17]:

$$2Ti + N_2 \rightarrow 2TiN \qquad \Delta G_f^0 = -425 kJ/g \cdot mol,$$
$$Ti + O_2 \rightarrow TiO_2 \qquad \Delta G_f^0 = -710 kJ/g \cdot mol,$$
$$2Al + N_2 \rightarrow 2AlN \qquad \Delta G_f^0 = -375 kJ/g \cdot mol,$$
$$2Al + 3/2 O_2 \rightarrow Al_2O_3 \qquad \Delta G_f^0 = -840 kJ/g \cdot mol.$$

At the beginning of oxidation in air, both oxides and nitrides are most likely to be formed. However, since oxides are more stable than nitrides, nitrides will be oxidized if the oxygen activity is higher than a certain level. For example, AlN is also one of the most stable nitrides, but still it is more unstable than Al$_2$O$_3$, judging from the standard free energy of formation. So, even if AlN is formed on the specimen surface, it turns to Al$_2$O$_3$ easily.

If a dense Al$_2$O$_3$ layer is formed predominantly on the specimen surface, the oxidation rate of the alloy will decrease greatly. This is plausible in those alloys (e.g., NKH71) which contain a high Al content. Following the above discussion, it will be necessary to hold free nitrogen

292

released by the oxidation of AlN. A small amount of Ti may be enough to capture such nitrogen, because TiN is relatively stable and it may be hard to change TiN into TiO_2. In fact, as shown in Fig.12, fine TiN precipitates are observed under the Al_2O_3 layer of NKH71 after the continuous oxidation test. Similarly, NKH73 and NKH75, both exhibiting a good indication of the cyclic oxidation resistance, have a dense Al_2O_3 layer and fine TiN precipitates under the layer, as shown in Fig.4

On the other hand, in the relatively Ti-rich alloys, TiN will form together with TiO_2 at the beginning of oxidation. But this TiN may not be oxidized to form TiO_2 so readily, judging from the small standard free energy difference between TiN and TiO_2. Therefore, the condensed TiO_2 layer may not be formed in the presence of TiN and Al_2O_3 in such alloys, resulting in poor oxidation resistance. Needless to say, if the Ti content is comparable to the Al content, the alloy can no longer form either dense Al_2O_3 nor TiO_2 as explained above. Thus, the Ti/Al compositional ratio influences the oxidation resistance greatly.

The oxidation mechanism mentioned here will be approximately correct, but there is a difference in the results between the cyclic and the continuous oxidation tests (Fig.3 and Fig.5). This is probably because the factor of adherence of the oxide layer to the specimen surface is more important in the cyclic oxidation than in the continuous oxidation. In other words, the thermodynamic consideration mentioned here is not applicable simply to the cyclic oxidation because of its dynamical phenomenon.

Possible mechanism for hot corrosion

As previously explained above, there is a large difference in the weight loss obtained from the burner-rig test among the alloys, despite their similar Cr content in them. It is well known that the addition of W and Mo deteriorates the hot corrosion resistance of nickel based superalloys by the acidic fluxing reaction [18]. NKH71 and NKH72 have a higher level of W and Mo content than the other experimental alloys. So, this is one of the reasons why these two alloys show poor hot corrosion resistance. By comparing the result of the hot corrosion between NKH71 and NKH72, it is seen that the Co addition improves the resistance, but the mechanism is still not clear at the moment. Also, as is evident from Table 1 and Fig.6, the compositional ratio, Ta/(W+Mo), needs to be set at about unity to increase the resistance, which is consistent with our previous result [4].

In the cross section of the specimen surface observed after the burner-rig test, a sulfide layer is present beneath the oxide layer. In general, it is known that sulfidation occurs prior to oxidation, and then sulfides are oxidized. Such sulfidation and subsequent oxidation reactions take place alternately and repeatedly, and the two layers penetrate into the base metal deeply. Therefore, the hot corrosion resistance will be improved if the sulfidation reaction is suppressed in some way. In NKH76, which shows the best hot corrosion resistance among the present experimental

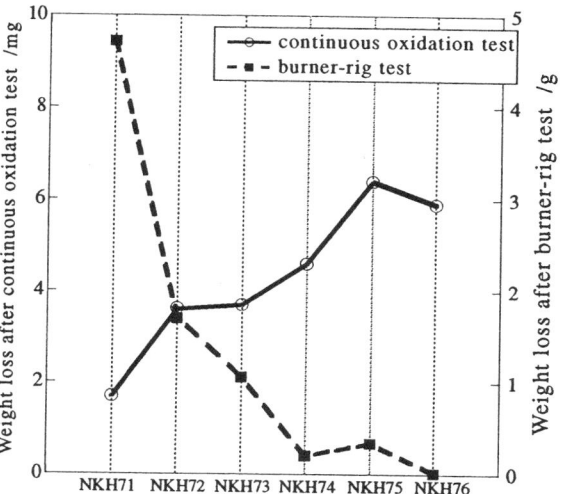

Fig.11 Comparison in the weight loss of the specimens between the continuous oxidation test and burner-rig test.

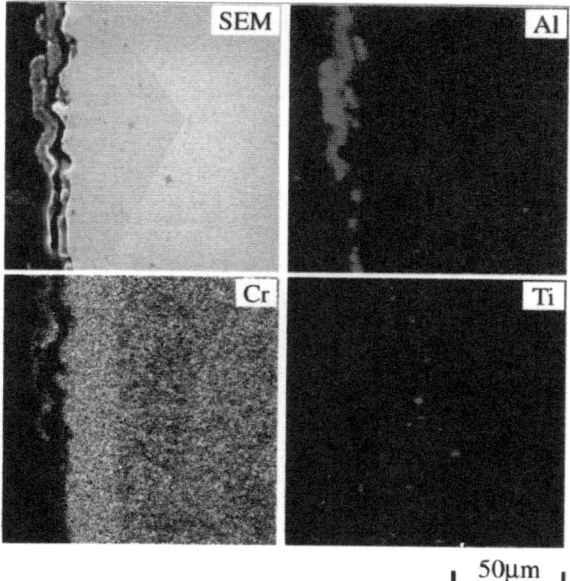

Fig.12 SEM image and the corresponding characteristic X-ray images taken from the cross section of NKH71 after the continuous oxidation test.

alloys, the dense TiO_2-rich layer is observed in the outermost layer of the corrosion products, as shown in Fig.8. From this result, it is deduced that the dense TiO_2-rich layer forms so quickly on the specimen surface when the alloy contains a certain level of Ti content because of the higher stability of TiO_2 than Cr_2O_3. In such a case, Cr is never oxidized in the presence of TiO_2 because of its higher dissociation pressure than TiO_2, but instead Cr can hold sulfur and form a sulfide under the oxide layer. As a result, sulfidation does not proceed into the base metal any longer, resulting in the maintenance of good hot corrosion resistance in NKH76.

Conclusion

In order to develop high performance nickel based superalloys for industrial gas turbines, 12mass% (14mol%)Cr and no Re alloys are made with the aid of the d-electrons concept, and the alloying effects are examined on the creep rupture strength, oxidation resistance and hot corrosion resistance. Among the experimental alloys, NKH71 exhibits the longest creep rupture life, 32.9Ms (9142h), at 1193K under a stress of 196MPa. NKH71 is superior in creep strength to the 2nd generation SC superalloy containing 3%Re. Also, it is shown that the Larson Miller plot is not suitable for presumption of longer rupture life in nickel based superalloys. In addition, the optimization of Ti/Al compositional ratio is found to be most important in improving both hot corrosion resistance and oxidation resistance at high temperatures. The amount of 12mass%Cr is not needed to get good hot corrosion resistance, as long as both the compositional ratios of Ti/Al and Ta/(W+Mo) are high in the alloy.

Acknowledgement

This research was supported in part by the Grant-in-Aid for Scientific Research (B) from the Ministry of Education, Science and Culture of Japan.

References

[1] R.Kehlhofer, "Power Engineering-Status and Trends", Materials for Advanced Power Engineering 1998, Eds. J. Lecomte-Beckers, et al., Forchungszentrum, Jülich, 1998, pp.3-17.

[2] W. S. Walston et al., "Rene N6: Third Generation Single Crystal Superalloy", Superalloys 1996, Eds. R.D. Kissinger et al., TMS, Warrendale, 1996, pp.27-34.

[3] G.L.Erickson, "The Development and Application of CMSX-10", ditto, pp.35-44.

[4] K. Matsugi et al. "Realistic Advancement for Nickel-Based Single Crystal Superalloys by the d-Electrons Concept", Superalloys 1992, Eds. S.D. Antolovich et al, TMS, Warrendale, pp. 307-316.

[5] R.F. Singer, "Advanced Materials and Processes for Land-Based Gas Turbines", Materials for Advanced Power Engineering 1994, Eds. D. Coutsouradis et al, Kulwer Academic Publishers, Dordrecht, 1994, pp. 1707-1730.

[6] M.Morinaga et al., "New PHACOMP and its Application to Alloy Design", Superalloys 1984, Eds. M.Gell et al, The Metall. Soc. Of AIME, (1984), pp.523-532.

[7] Y. Murata, M.Morinaga and R. Hashizume, "Trace of the Evolution of Ni-Based Superalloys by the d-Electrons Concept", Materials for Advanced Power Engineering 1998, Eds. J. Lecomte-Beckers, et al., Forchungszentrum, Jülich, 1998, pp. 1381-1390.

[8] G.Y. Lai, High Temperature Corrosion of Engineering Alloys, (Metals Park OH, ASM International, 1990), pp.117-144.

[9] J.K. Tien and S. M. Copley, "The Effect of Uniaxial Stress on the Periodic Morphology of Coherent Gamma Prime Precipitatate in Nickel-Base Superalloy Crystals", Metall. Trans., 2 (1971), pp.217-219.

[10] A. Pineau, "Influence of Uniaxial Stress on the Morphology of Coherent Precipitates suring Coarsening-Elastic Energy Considerations", Acta Metall., 24 (1976), pp.559-564.

[11] T. Miyazaki, K. Nakamura and H. Mori, "Experimental and Theoretical Investigations on Morphological Changes of γ ' Precipitates in Ni-Al Single Crystals During Uniaxial Stress-annealing", J. Mater. Sci. , 14(1979), pp.1827-1837.

[12] S. Socrate and D. M. Parks, "Numerical Determination of the Elastic Driving Force for Directional Coarsening in Ni-Superalloys", Acta Metall., 41 (1993), pp.2185-2209.

[13] F. R. N. Nabbaro, "Rafting in Superalloys", Metall. Mater. Trans. A, 27A (1996), pp.513-530.

[14] H. Harada and M. Yamazaki, "Alloy Design for γ' Precipitation Hardening Nickel-base Superalloys Containing Ti, Ta, and W", Tesu-to-Hagane, 65 (1979), pp.1059-1068.

[15] Y. Murata, S. Miyazaki and M. Morinaga, "Evaluation of the Partitioning Ratios of Alloying Elements in Nickel-Based Superalloys by the d-Electrons Parameters", Materials for Advanced Power Engineering 1994, Eds. D. Coutsouradis et al, Kulwer Academic Publishers, Dordrecht, 1994, pp. 909-918.

[16] Y. Murata et al., "Early Stage of High Temperature Oxidation of TiAl and FeAl at Low Oxygen Partial Pressures", J.Japan Inst. of Metals, 61 (1997), pp.702-709.

[17] S. Nagasaki ed.,"Metals Databook, 2nd ed.", Japan Inst. Met., 1983, pp.90-91.

[18] F.S. Pettit and G.H. Meier, "Oxidation and Hot Corrosion of Superalloys", Superalloys 1984, Eds. M.Gell et al, The Metall. Soc. Of AIME, (1984), pp.651-687.

EVALUATION OF PWA1483 FOR LARGE SINGLE CRYSTAL IGT BLADE APPLICATIONS

Dilip M. Shah and Alan Cetel
M/S114-43, Pratt & Whitney
400 Main Street, E. Hartford, CT 06033

Abstract

As a part of the Department of Energy "Land Based Turbine Casting Initiative" program, a high chromium containing, corrosion resistant single crystal alloy, PWA1483, was evaluated for large single crystal component application in industrial gas turbines (IGT). The objective of the program was to assess the impact of casting size on casting yield and structural performance of a large anisotropic body. For this purpose, large single crystal blades cast at two different growth rates, were evaluated for crystallographic quality, microstructure, and high angle boundary (HAB) type defects. Presence of HAB defects, is one of the principal factors, contributing to casting rejection. Testing consisted of 649°C tensile, and 760°C, 871°C, and 982°C creep, as a function of the deviation of the tensile direction from <001>. PWA1483 bicrystals were cast, and tolerance to HAB was evaluated in creep at 760°C and 982°C. To assess the potential benefit of minor element additions on HAB tolerance, single crystals and bicrystals of hafnium and boron modified PWA1483 were also evaluated.

The results show a linear decrease in tensile strength, and creep rupture life, and a corresponding increase in tensile elongation, with increasing deviation from <100>. As expected, for the single crystal alloy PWA1483, with no minor elements, a precipitous drop in creep capability occurred, when grain boundary misorientation exceeded 10°. However, in the modified PWA1483, with deliberate additions of hafnium and boron, surprisingly, there was no significant drop in creep-strength even with grain boundary misorientation as high as 25°. The potential role of other factors, in this unexpectedly high tolerance to HAB of the modified alloy, is discussed further. Overall, the results provide a good basis for determining tradeoffs between, tensile strength, temperature capability, and grain boundary tolerance, vis-a-vis potential increase in casting yields with improvement in casting technology.

Introduction

The objective of this effort was to support a broad program to assess and address the potential technical and economic barriers in transitioning single crystal technology to large industrial gas turbine (IGT) components. Among a variety of associated issues, single crystal casting acceptance standards and post-casting processing, were considered two major factors affecting casting

yield and design performance of large components. Discussion in this paper will be primarily focused on the factors influencing casting acceptance standards.

Casting Acceptance Standards

Mechanical and physical properties of components cast in equiaxed polycrystalline nickel base superalloys can be assumed to be isotropic for all practical purposes, when the best conventional casting practice is used. Directionally solidified (DS) components with a columnar grain structure can be treated as transversely isotropic, with no variation of properties in the plane normal to the growth direction. Here, the superior creep resistance and low modulus, parallel to the columnar grain, are well exploited when the component is DS along the stacking line of interest. However, a single crystal component is truly an anisotropic body, with significant variation in properties along different directions conforming to the crystal symmetry. To take optimum advantage of single crystal properties, control of crystallographic orientation with respect to the part geometry is required.

The additional orientation control is the price one has to pay to exploit the benefits stemming from the elimination of all grain boundaries. Consequently, it must also be assured that the objective of the removal of all grain boundaries is adequately achieved. Any inadvertent presence of HAB must be detected and quantified. Minor low angle boundaries (LAB), are acceptable when they result in no significant loss in performance.

To achieve optimum performance with single crystal nickel base superalloys, the grain orientation with respect to the part geometry, and the tolerance of the material to HAB type defects, must be specified and controlled within limits.

Crystallographic Characterization

While crystallographic control specifications are well in place for components currently used in aircraft or aeroderivative engines, a re-examination of the practice was deemed necessary, for heavy frame turbines, due to the 2-3X size increase of single crystal castings. The inevitable increase in total number of defects and variation in properties with increasing component volume, imparts additional cost and performance penalties. Furthermore, these issues must be addressed in the context of an altered operating scenario in which long term steady state durability is

Superalloys 2000
Edited by T.M. Pollock, R.D. Kissinger, R.R. Bowman,
K.A. Green, M. McLean, S. Olson, and J.J. Schirra
TMS (The Minerals, Metals & Materials Society), 2000

more critical than the number of start-up and shutdown cycles. Specific design of the components enters the picture as well. There is limited engine experience with large single crystal components. An iterative approach with emphasis on crystallographic characterization and property evaluation is necessary to achieve the most cost-effective transition of single crystal technology.

Anisotropy of Physical and Mechanical Properties

For nickel base superalloys, with cubic structure, physical properties such as thermal expansion and thermal conductivity are isotropic and do not vary with crystallographic direction. Elastic modulus on the other hand, being a fourth rank tensor can be anisotropic, and it does vary significantly with crystallographic orientation in nickel base alloys [1,2]. However, it is largely insensitive to specific alloy composition. In contrast, though, crystallographic slip dependent plastic properties, such as creep and tensile strength are not only anisotropic, but the extent of anisotropy is a strong function of alloy composition

Tensile Properties: There is ample evidence in the literature [1,3, 4] to suggest that tensile properties show significant anisotropic behavior. This variation is rooted in the competition between octahedral and cube slip behavior and therefore is also a complex function of γ ' size and alloy composition. The role of primary orientation, and dendrite splaying if any, on tensile properties must be recognized. From the turbine blade design standpoint, low temperature (~649°C) tensile properties and associated low cycle fatigue (LCF) life are most critical, for the durability of the root attachment.

Creep Resistance: Long term creep resistance is a life limiting factor for airfoils employed in heavy frame turbines for power generation, which undergo fewer start-up and shutdown cycles than aircraft engines. High temperature creep resistance in single crystal superalloys depends on a complex diffusion controlled slip-climb mechanism, which is highly sensitive to alloy composition. Not only is creep strength, for a given crystallographic orientation, composition sensitive, but the orientation dependence of creep strength can also vary significantly with alloy chemistry. Both major [1,5] and minor elements [6] can alter the nature of creep anisotropy.

The nature of creep anisotropy has important implications for large single crystal castings. In the current design system, variation in properties along the principal loading directions, must be accounted for. In addition, potential increases in spatial variability in primary orientation (splaying) must be considered as well. This is tied into the state of the art of casting technology, and economic viability of process modifications such as seeding. To maximize the future efficient application of large single crystal components, it will be necessary to understand the three dimensional aspects of loading. Understanding the nature of anisotropy will be a first step in that direction.

Effect of the High Angle Boundaries on Mechanical Properties

One of the principal motivations for pursuing directional solidification was the elimination of grain boundaries, normal to the principal stress direction. Successful application of columnar

grain Mar M200 (PWA 1422), required addition of hafnium (Hf) to maintain transverse strength. The maximum benefit of single crystal application occurred when all grain boundary strengthening elements were reduced to sufficiently low levels that segregation was decreased. A concomitant increase in incipient melting point allowed greater latitude in alloy design and solution heat treatment. It is important to recognize that alloys developed specifically for single crystal application are not tolerant to the presence of grain boundaries, especially under creep conditions. Yet from an engineering standpoint, given the complexity of investment casting, the inadvertent presence of low angle boundaries must be tolerated to maintain an economically viable casting yield.

The critical question is what level of grain boundary misorientation or low angle boundary (LAB), can be tolerated without any significant loss in mechanical performance. The misorientation in its simplest form is quantified as the minimum angle of rotation necessary to bring the crystallographic frame of reference in one grain to map on that of the second grain. Owing to cubic symmetry, 24 axes of rotation can be identified, but only one yields a minimum angle. If necessary, the mismatch angle can be separated into its twist and tilt components, but that requires knowledge of the orientation of the grain boundary with respect to the crystallographic frame of reference in one of the grains. However, the added characterization is of marginal value. Typically it is known that grain boundaries with misorientations beyond 10° result in a precipitous drop in creep-strength. It is interesting to note, that this is approximately the misorientation limit, beyond which grain boundaries can no longer be treated in terms of a dislocation network, and consequently resulting in an order of magnitude increase in diffusion compared to that expected transgranularly [7]. This correlation is another strong indication supporting the view that environmental interaction is a principal factor contributing to the lack of grain boundary ductility.

One of the obvious options to improve the grain boundary tolerance of single crystals, is to add a limited amount of grain boundary strengthening elements, and assess the trade-off between increases in casting yield and corresponding loss in high temperature performance.

For this study, PWA1483, a high chromium containing, single crystal alloy, specifically developed for IGT application was selected for evaluation. Large single crystal IGT blades were cast at two different growth rates, and were evaluated for orientation variability and low angle boundary type defects. The results showed little influence of growth rate on primary orientation. However, a reduction in the population of grain boundary defects was observed in crystals grown at the high growth rate.

The property characterization consisted of tensile (649°C), and creep (760, 871, and 982°C) properties as a function of orientation. To assess the tolerance of PWA1483 to low angle boundaries, bicrystals of known orientation were cast and evaluated in creep. To assess the potential benefit of minor element additions on high angle boundary tolerance, single crystal and bicrystal castings of hafnium and boron modified PWA1483 were also evaluated.

The results show a linear decrease in tensile strength, and creep rupture life, accompanied by an increase in tensile elongation, with increasing deviation from <100>. As expected, for the single crystal alloy PWA1483, containing no minor elements, a precipitous drop in creep strength occurred, when grain boundary misorientation exceeded 10°. However, in the modified PWA1483, with deliberate additions of hafnium and boron, no significant drop in creep-strength was observed, even with grain boundary misorientations as high as 25°. This was one of the most interesting findings of this evaluation. Broadly the results provide a good basis for determining tradeoffs between, tensile strength, temperature capability, grain boundary tolerance, and potential reduction in casting rejects, with improvement in casting technology.

Experimental Procedure

For this study, PWA1483 with 12% chromium, and a modification with hafnium and boron additions, were selected for evaluation. PWA1483 was specifically developed for marine and industrial gas turbine applications, requiring superior hot corrosion resistance. The advanced single crystal alloy PWA1484 was used for benchmarking casting yield in large components only. Nominal compositions for the three alloys are listed in Table I. The material procured for this study was cast by Howmet Corporation. Chemical analysis of all castings confirmed that the concentration of all elements but boron, was within 2% of the nominal. Boron was analyzed to be consistently higher than the nominal target of 0.008 weight %, and varied between 0.0098 and 0.018 weight %.

To benchmark casting behavior in large components, Siemens' large V84.4 blade tooling was used. Six large single crystal blades of PWA1483 and PWA1484 each, were cast in 2 molds at two different growth rates, differing by a factor of three. Maximum dimensions of these blades are approximately 21 x 14 x 4 cm.

Single crystal test material was variously cast as 15.9 mm and 25.4 mm diameter bars and 25.4 mm thick slabs. Molds of PWA1483 were obtained in all three configurations, whereas Modified PWA1483 was cast as slabs and large diameter bars.

For the evaluation of off-axis properties and to study the effect of HAB on mechanical properties, single crystal bars, and bicrystal slabs were cast using nominally <100>-oriented seeds. The bicrystal slab casting configuration is shown schematically in Fig. 1. The desired LAB or HAB misorientation was achieved by

rotating one of the seeds, around its axis, relative to the other. Bicrystals were cast, both in PWA1483 and Modified PWA1483 alloys.

PWA1483 single crystal castings received a solution heat treatment (SHT) of 1260°C/0.5 hr at the casting supplier, as this enhances subsequent grain etching, and facilitates visual inspection of grain boundary type defects. All material received in the as SHT condition was subsequently heat treated at 1079°C/4 hr to simulate a coating cycle. The Modified PWA1483 alloy was solution heat treated at 1204°C, to avoid incipient melting, and subsequently given the simulated coating cycle heat treatment. At 1204°C, only partial γ′ solutioning was achieved.

Standard optical and electron metallographic techniques were used to characterize the material. Grain boundary misorientation was quantified using the X-ray Laue determination of the crystallographic orientation of both grains. Based on this information, the misorientation angle was calculated.

All tensile tests were carried out at a strain rate of 0.5%/min, as per ASTM standard. The tensile testing was carried out at 649°C, reflecting a measure of root attachment strength. All creep tests were run per ASTM standard E139.

Creep tests were carried out at 760°C, 871°C, and 982°C. Based on prior experience, conditions for testing were selected to produce rupture lives of at least 100 hours. In many cases, the stress was iteratively increased or decreased, so that the stress for 100 hr rupture life could be defined.

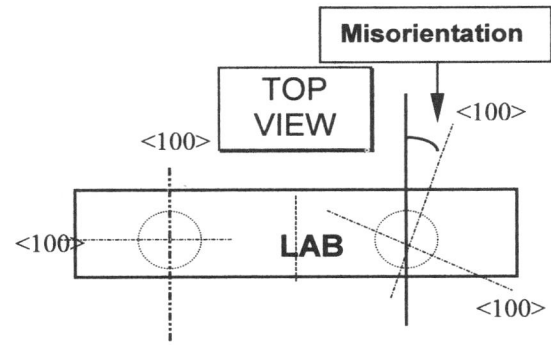

Figure 1: Schematic showing configuration of seeded bicrystal slabs.

Table I Nominal Composition of Alloys Used in P&W Study.

Alloy	Composition					Wt. %						
	Ni	Co	Cr	Mo	W	Re	Ta	Ti	Al	Hf	B	C
PWA1484	Bal.	10	5	1.9	5.9	3.0	8.7	-	5.65	0.10	-	-
PWA1483	Bal.	9.0	12.2	1.9	3.8	-	5	4.1	3.6	-	-	.07
Modified 1483	Bal.	9.0	12.2	1.9	3.8	-	5	4.1	3.6	0.5	0.008	.07

Results

Cast Microstructure Characterization of Large IGT Blades

Twelve large single crystal IGT (V84.4) blades were cast in 4 molds, at 2 different growth rates in PWA1483 and PWA1484. The objective of this effort was to assess the quality of single crystal castings, as affected by the casting size, growth rate, and alloy chemistry. Obviously with the small population of castings evaluated, no statistically meaningful conclusions can be drawn, but only major trends in casting quality can be discerned.

Grain defects: In both alloys, the faster growth rate appeared to be beneficial, in that fewer grain boundary type defects were revealed upon macroetching. At the slower growth rate, multiple defects were observed. Moreover, while all 3 blades of PWA1483, were virtually defect free at the faster growth rate, blades cast in PWA1484 were not. No defects were found in PWA1483 blade airfoils, whereas for PWA1484, one airfoil was cracked, and in two cases pits were observed at the tip end. One PWA1484 blade from each mold, showed orientation deviation of as much as 15° from the primary <100>.

Variation in primary orientation or Splaying: Four blades, representing two alloys, and two growth rates in each case, were mapped with Laue X-ray diffraction to determine primary orientation variation, sometimes referred to as splaying. In each alloy/growth rate group, a blade with the largest deviation in primary orientation was selected, since such orientations have the least crystallographic symmetry. The results indicated virtually no splaying for both alloys at either growth rate. A variation of 2-3° in primary and secondary orientation is considered natural, as it is comparable to the misorientation between neighboring dendrites. This can be verified using electron channeling pattern (ECP) mapping of dendrite orientation.

γ' size variation: Blades DS at the fast growth rate, were solution heat treated and sectioned in the root area, tip and mid-span of the airfoil. Optical metallography showed no evidence of under solutioning. To determine the finest γ' size present in the dendritic core, surface replicas were prepared and examined at high magnification, in a transmission electron microscope. As expected, the average γ' size was observed to be larger in the massive root area compared to that in the tip area. While the size varied from ~0.29 μm. to ~0.13 μm for PWA1484, less variation (~0.3 μm. to ~0.24 μm) was observed for PWA1483.

Dendritic Sub-structure: In order to discern any correlation between dendritic sub-structure, and rate of solidification, casting size, alloy chemistry, and casting defects; dendritic sub-structure of the available casting configurations was studied.

Surprisingly, the results showed no significant difference in the dendrite density in the large IGT blades cast at growth rates differing by a factor of 3X. The dendritic density was estimated to be around 400 /cm² in the root attachment and decreased to around 300 /cm² in the airfoil. In comparison, the density of dendrites was observed to be higher (550/cm²) in 25.4 mm thick bars and slabs, and almost a factor of two higher (861/cm²) in the case of 15.9 mm diameter bars. While the difference in the density of dendrites between 15.9 mm and 25.4 mm diameter bars

was small, there appeared to be an increased volume of γ/γ' eutectic phase in the latter. Another notable difference was the presence of secondary dendrite arms in Modified PWA1483 castings.

Evaluation of Off-axis Properties

Off-axis test specimens for tensile, and creep testing were machined from fully heat treated PWA1483, with the orientation of the specimen axis (α) varying from 10° to 29° from <001>. X-ray Laue data were further analyzed to define the actual crystallographic orientation of the specimen axis, within a standard stereographic triangle. However, the additional information proved academic, as there was insufficient data to discern any trend.

Tensile: All tensile testing was carried out at 649°C. Average results of four tests originating from a bar with a specific primary orientation (α) are plotted against the primary orientation angle α in Fig. 2. It is apparent that while 0.2% yield strength and ultimate tensile strength (UTS) decreases with increasing α, the % elongation increases. It is known that elastic modulus increases with increasing α. The results are not surprising, given that in the vicinity of <001>, single crystal nickel base superalloys deform by octahedral {111}<110> slip, and that strength is controlled by the critical resolved shear strength (CRSS) for the operative slip system. (See Reference 4 for further discussion.) For comparison, the average of duplicate test results for the modified alloy at 6.5°, are also plotted in Fig. 2.

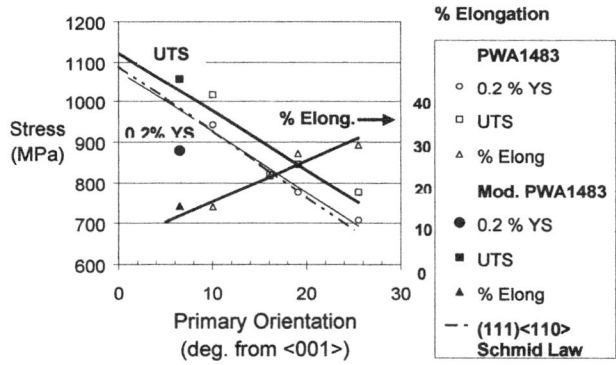

Figure 2: 649°C tensile properties of PWA1483, versus primary orientation, measured in α° from <001>. Results for Mod. PWA1483 are compared at α = 6.5°.

Creep: Off-axis creep behavior of PWA1483 was evaluated at two different stress conditions at 760°C, 871°C, and 982°C. A total of 8 bars with different orientations were used. To facilitate a direct comparison of data, at different temperatures and stress combinations, groups of 6 specimens machined from the same bar, with presumably identical orientation, were tested at six different temperature and stress conditions. Thus at a given creep condition (temperature and stress), creep behavior was evaluated for a maximum of 8 different orientations.

Creep test results at 760°C/793 and 655 MPa, 871°C/483 and 345 MPa, and 982°C/248 and 172 MPa are presented in Table II. Note that at all test conditions, there is a general tendency of rupture life decreasing with increasing deviation from <001> or angle α. Early failure of two specimens at 760°C, with α >20°, is attributed to plastic deformation prior to loading, suggesting that in these cases, the stress level may have exceeded the yield strength. Yield strength data presented in Fig. 2, corroborate this conclusion. At α=25°, the average 649°C yield strength was determined to be 113 MPa, and it is expected to be similar at 760°C. This suggests that relaxing the restriction on α beyond 20°, cannot be made without compromising some strength capability. It is also observed that at 760°C, there is a decrease in total creep ductility with increasing deviation from <001>.

Comparison of PWA1483 and Mod. PWA1483 in Creep

While Mod. PWA1483 was not evaluated extensively in the off-axis orientation, the creep properties were compared with PWA1483, at α = 6 ± 2°. The creep rupture data at 871°C and 982°C for both alloys are plotted in Fig. 3. It is clear that,

because of under-solutioning to prevent incipient melting with the minor element additions, there is a 5-7 MPa loss in creep strength for the modified alloy at 871°C and 982°C.

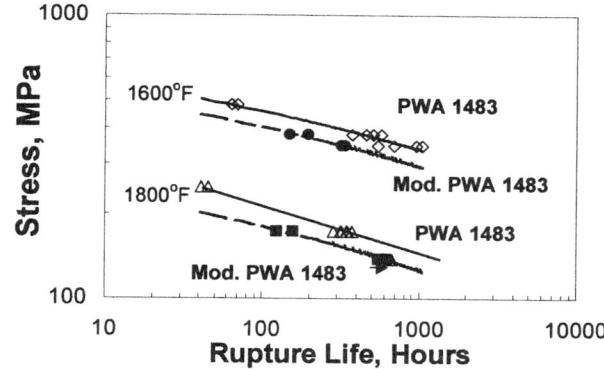

Figure 3: Creep-rupture comparison of PWA1483 and Mod. PWA1483 at 871°C and 982°C.

Table II Creep Test Results of Fully Heat Treated PWA1483 in Off-axis Orientations at 760°, 871°, and 982°C

Primary Orient.	Temp.	Stress	Time to 1 %	Time to 2%	Rupture Life	Elong.		Stress	Time to 1 %	Time to 2%	Rupture Life	Elong.	
α°	(C)	(MPa)	(hrs)	(hrs)	(hrs)	(%)		(MPa)	(hrs)	(hrs)	(hrs)	(%)	
10.0	760	793	1.8	8.5	59.6	17.0		655	69.0	183.0	647.2	13.4	
10.2	760	793	1.4	7.6	46.9	12.3		655	11.7	173.0	554.3	21.9	
14.7	760	793	7.9	18.8	62.8	16.4							
16.2	760	793	8.7	21.0	61.2	12.6		655	62.0	157.0	220.5	2.8	(c)
19.1	760	793	10.7	23.0	45.6	6.0		655	79.0	160.0	384.2	9.3	
19.2	760	793	20.0	33.0	57.0	7.0							
24.5	760	793	0.1	0.2	1.1	23.5	(a)	655	78.0	153.0	223.3	3.9	
26.0	760	793	0.2	0.4	1.9	21.9	(b)	655	79.0	161.0	307.8	6.4	
10.0	871	483	9.3	151.7	69.5	18.6		345	298.0	527.0	956.1	13.4	
10.2	871	483	8.2	20.0	63.9	12.7		345	377.0	591.0	1045.7	19.4	
14.7	871	483	10.2	22.3	65.2	20.5		345	264.0	475.0	824.3	15.9	
16.2	871	483	8.6	18.7	55.7	20.0		345	258.0	497.0	877.9	19.8	
19.1								345	275.0	488.0	810.5	14.8	
19.2	871	483	10.7	22.3	62.4	16.0		345	392.0	586.0	859.4	19.7	
24.5								345	295.0	477.0	661.2	7.5	
26.0	871	483	3.1	8.1	23.4	7.6		345	301.0	466.0	656.1	7.9	
10.0	982	248	13.1	151.7	41.3	18.2		172	137.0	269.0	375.9	26.6	
10.2	982	248	18.6	27.0	45.2	15.3		172	137.0	247.0	321.4	19.3	
14.7	982	248	11.0	20.6	37.5	18.4		172	188.0	228.0	266.6	14.4	
16.2	982	248	13.0	21.0	36.0	20.0		172	181.0	231.0	267.1	13.4	
19.1	982	248	24.0	31.0	43.5	15.3		172	182.0	204.0	223.2	9.5	
19.2	982	248	17.6	27.0	39.7	21.3		172	168.0	206.0	244.1	17.6	
24.5	982	248	16.1	24.0	33.2	11.7		172	132.0	150.0	165.6	8.4	
26.0	982	248	15.1	23.0	32.3	13.7		172	153.0	176.0	198.9	13.4	

(a) 10.26 % plastic creep on loading.   (b) 8.26 % plastic creep on loading.   (c) discontinued, no failure

Evaluation of Low and High Angle Boundaries (LAB & HAB)

A total of twelve (12) bicrystal slabs were cast, six (6) each from PWA1483 and Modified PWA1483. Casting yield for the bicrystals was very poor because of slab geometry, and the targeted grain boundary misorientation angles were not achieved with any precision. For the purpose of this evaluation, the average boundary misorientation measured for each bicrystal casting was assigned to all specimens from the same casting. Grain boundary misorientation of individual specimens was not characterized. A typical macrograph, presented in Fig. 4, confirms that both grain boundary misorientation, and primary orientation, were achieved as intended. Note that from each bicrystal slab, baseline specimens, with no grain boundaries, were always procured. They were machined from one of the two crystals, avoiding the grain boundary.

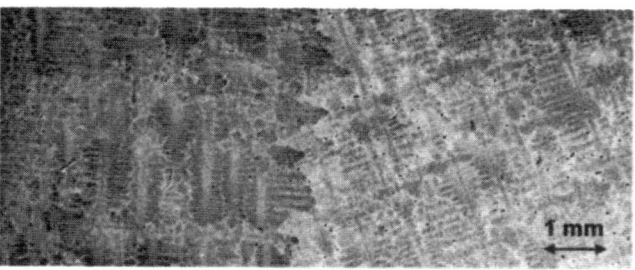

Figure 4: A typical macrograph showing orientation relationship between two grains of a bicrystal specimen.

Creep Strength: The bicrystal specimens were tested in creep at 760°C, and 982°C, at various stress levels so that stress for 100 hour rupture could be defined at each temperature for a given alloy. This required lowering the stress for specimens with high angle boundaries. The results for PWA1483 are presented in Table III, and Modified PWA1483 in Table IV. As noted earlier, the Modified PWA1483 alloy was cast with a late addition of 0.5 % Hf and 0.008 % B to PWA1483.

Comparison of data for PWA1483 and Modified PWA1483, clearly show that Modified PWA1483 is much more tolerant to grain boundary type defects. Specifically consider the short rupture life (38 and 97 hrs) of PWA1483 at 760°C/414 MPa for a 21.2° boundary. At the identical condition, a specimen of Modified PWA1483 had not ruptured after 4078 hrs. This is one of the most interesting findings of this investigation and will be discussed further.

Comparisons of fracture surfaces of creep specimens with and without high angle boundaries, showed a more or less featureless appearance in specimens with no grain boundaries, in contrast to an accentuated dendritic de-cohesion in specimens with high angle grain boundaries. Nonetheless, careful examinations of baseline specimens also revealed inter-dendritic failure. Since the artificially created boundaries in the present cases were largely tilt boundaries, the dendritic pattern was only uni-directional.

Discussion of Results

Quality of Large Single Crystal IGT Blades

Effect of Growth Rate: Our observations suggest that it is feasible to DS defect free large single crystal IGT blades at higher growth rates. At appropriate fast growth, not only are high angle grain boundary type defects suppressed, but crystallographic splaying is also virtually absent. However, assessment of how this translates to a large-scale production environment, is beyond the scope of this effort. There is minor influence of alloy chemistry, but no meaningful conclusions can be drawn from this limited study. Based on this study, it appears that PWA1483 is less prone to defects than PWA1484, but the difference is not considered significant. Such small differences can be optimized simply with minor changes in casting process parameters.

Based on our observations, it appears that either the casting technique has achieved considerable maturity, or the massive volume of the part is naturally helpful. Excessive splaying is generally expected in asymmetric parts with undesirable thermal gradients in the plane normal to the growth direction. However, proper clustering of the parts in a mold may mitigate this factor; and a larger mass of material requiring a larger thermal flux, can help flatten out the temperature gradients within the mold.

The Role of Cast Microstructure: To gain some understanding of the interrelationship between cast microstructure, casting process and configuration, and alloy chemistry, dendritic microstructure and sub-structure misorientation were analyzed. Results showed no major differences in dendrite density in blades cast at baseline and at 3X higher relative growth rate. In fact the 15.9 mm diameter bars normally used for laboratory evaluation of mechanical properties exhibited distinctly higher dendrite density. In a sense, 25.4 mm diameter bars more closely simulated the dendritic density of large IGT blades. Another notable difference was the presence of secondary dendrite arms in Modified PWA1483, where melting characteristics were inevitably altered with the addition of boron and hafnium. The presence of secondary arms lead to slightly increased dendrite arm spacing. This was also accompanied by a slight increase in sub-structure misorientation. Based on past experience, this may imply an increased susceptibility to grain boundary type defects, because of less tolerance to transverse thermal gradient. However, a limited comparison of single crystal and bicrystal slabs cast in both alloys did not reveal any strong differences in frequency of grain boundary type defects between the two alloys.

Variation in Off-axis Properties

Tensile Properties: The 649°C, tensile data are plotted against the primary orientation angle α in Fig. 2. As expected, while yield strength and UTS decrease with increasing α, the % elongation increases. At low α angle (<15°) the Schmid factor in the vicinity of <001> orientation, for the primary {111}<110> slip system can be approximated as $(1+0.0149\alpha°)/\sqrt{6}$.

Table III 760°C and 982°C Creep Test Results for PWA1483 Bicrystals

LAB MISORIENT (deg.)	Temp. (C)	Stress (MPa)	Time to 1 % (hrs)	Time to 2 % (hrs)	Rupture Life) (hrs)	Elong. (%) (%)	Temp. (C)	Stress (MPa)	Time to 1 % (hrs)	Time to 2 % (hrs)	Rupture Life) (hrs)	Elong. (%)
No bound.	760	793	5.0	21.0	106	13.7	982	207	70	103	158.3	13.9
No bound.	760	793	4.5	19.2	51.9	7.5	982	221	46	63.0	79.8	12.5
No bound.							982	221	22	50.0	97.1	16.1
4.2	760	690	269	338	499.7	10.3	982	207	96	134	168.2	14.8
4.2	760	758	13.8	38.0	134.3	10.3	982	207	38.0	75.0	131.8	16.2
4.2	760	793	4.0	16.6	88.6	11.6	982	221	25	54.0	104.9	17.2
4.2	760	758	6.8	26.0	110.2	9.2	982	221	21.0	49.0	98.9	14.3
7.6	760	690	17.2	53.0	67.3	2.9	982	207	14.1	29.0	62.9	13.8
7.6	760	690	18.3	50.0	249.2	21.0	982	207	12.3	32.0	41.3	3.1
7.6	760	621	61.0	175	300.8	3.6	982	172	62	---	117.9	2.9
16.9	760	414	---	---	977.4	0.21	982	207		---	12.7	1.60
16.9	760	552	---	---	6.5	0.60	982	138		---	243.1	1.80
16.9	760	483	---	---	25.4	0.28	982	124		---	227.6	0.80
16.9	760	483	---	---	73.3	0.40	982	138		---	240.7	1.0
21.3	760	414	---	---	38.3	0.19	982	207		---	3.6	0.37
21.3	760	414	---	---	97.4	0.70	982	138		---	58.4	0.45
21.3	760	345	---	---	+2034.2	0.00	982	124		---	153.3	0.90
21.3	760	345	---	---	868.1	0.5	982	124		---	420.7	0.80

+ Discontinued, no failure

Table IV 760°C and 982°C Creep Test Results for Modified PWA1483 Bicrystals

LAB MISORIENT (deg.)	Temp. (C)	Stress (MPa)	Time to 1% (hrs)	Time to 2% (hrs)	Rupture LIfe (hrs)	Elong. (%)	Temp. (C)	Stress (MPa)	Time to 1% (hrs)	Time to 2% (hrs)	Rupture LIfe (hrs)	Elong. (%)
No bound.	760	690	18.7	49.0	160.5	13.9	982	207	14.6	24.0	43.6	15.7
No bound.	760	690	30.0	66.0	187.8	13.8						
No bound.	760	690	35	77	176.2	12.6						
6.2	760	690	18.5	54.0	171.1	11.7	982	207	7.5	14.0	35.2	20.9
6.2	760	758	2.1	8.0	36.3	11.3	982	207	7.4	14.8	33.9	16.0
6.2	760	758	1.0	4.7	35.9	15.8	982	152	85.0	165	287	17.6
6.2	760	690	16.8	47	143.2	13.0	982	172	40.0	62	101.3	16.8
13.7	760	690	26.0	59.0	117.6	5.5	982	207	10.1	18	33.1	9.2
13.7	760	724	6.9	151.7	54.8	5.2	982	152	67.0	131	220.8	7.3
13.7	760	690	16.7	48.0	134.1	7.5	982	172	26.9	52.0	82.2	6.9
13.7	760	724	5.1	18.1	55.9	6.5	982	172	34.0	63.0	112	8.5
21.2	760	414	---	---	+4078.1	0.0						
21.2	760	552	388	814	1442.6	4.9	982	207	13.8	24.0	30.4	4.2
21.2	760	690	34.0	69.0	90.7	3.2	982	172	44.0	67.0	69.6	3.2

+ Discontinued, no failure

Thus yield strength can be linearly related to critical resolved shear strength (τ_c) by the relation

$$\sigma_y = \tau_c \sqrt{6} \, (1 - 0.0149\alpha°)$$

In Fig. 2, the 0.2% yield strength data reasonably fits this description, if it is assumed that τ_c = 445 MPa. Also plotted in Fig. 2, for comparison, are tensile data for Modified PWA1483 at 6.5°. It is evident that while there is little change in UTS, 0.2% yield strength is significantly reduced for the modified alloy.

Creep: As noted earlier, in all cases, except at 760°C/793 MPa, creep rupture life and time to 2% decrease with increasing deviation of α from <001>. The rate of decrease in rupture life was observed to be much higher than that for time to 2% creep. A Larson-Miller plot of the data, presented in Fig.5, depicts this trend in rupture life. In the figure, rupture life data at α=10° and α=25° are compared. Based on this plot a decrease of 207 in L-M parameter is determined, for increasing α from 10° to 25°. This translates into a 35% drop in rupture life at constant stress, or a 8°C temperature debit for the same creep strength, or a 30-40 MPa drop in creep strength, at the same temperature. Obviously, the decrease in creep capability will be less significant if time to 2% data were compared in Fig. 5. It is interesting to note that the loss in creep strength with a 15° deviation in primary orientation is comparable to the loss in creep-strength for the modified alloy with respect to PWA1483.

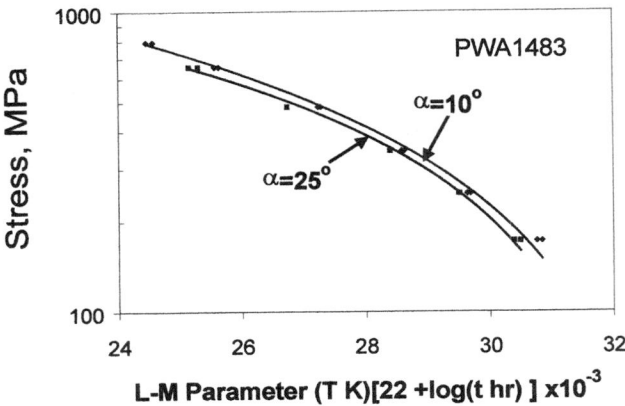

Figure 5: Larson-Miller comparison of creep rupture life of PWA1483 with primary orientation deviations(α) of 10° and 25°.

Grain Boundary Tolerance

From log-log plots of stress vs. rupture life, stress for rupture in 100 hrs was determined for PWA1483 and Modified PWA1483, respectively, at 760°C and 982°C, for various grain boundary misorientations. The results are depicted in Fig. 6, as a plot of stress for 100-hr rupture, versus grain boundary misorientation for both alloys at both temperatures. Note that while the stress drops precipitously for grain boundary misorientations greater than 10° for PWA1483, there is virtually no change in the case of Modified PWA1483. In fact, as reported in Table IV, testing of a specimen cast in Modified PWA1483 was discontinued at 760°C/414 MPa, after 4078 hours. The results clearly suggest that additions of minor elements, boron and hafnium, in the Mod. PWA1483, has significantly increased its tolerance to grain boundary type defects. It is apparent from Fig. 6, that there is almost a 50% debit in stress capability for PWA1483 (760°C) with HAB greater than 20°. However, for the Modified PWA1483 alloy, HABs do not significantly debit creep strength.

Stress debit is not the only measure of the impact of grain boundary type defects. Total elongation at failure is another measure of margin of safety in engineering applications. Creep rupture elongation measured after failure for both PWA1483 and Modified PWA1483, are plotted against grain boundary misorientation in Fig. 7. The elongation data at 760°C and 982°C for each alloy follow the same trend with grain boundary misorientation, and hence only one trend line is shown for each alloy. Note that even in the presence of a high angle boundary defect, the modified alloy exhibits ductility greater than 2%, compared to virtually none (<0.2%) in the base alloy. It is evident that addition of minor elements not only improves the stress capability, but also enhances ductility in the presence of HABs.

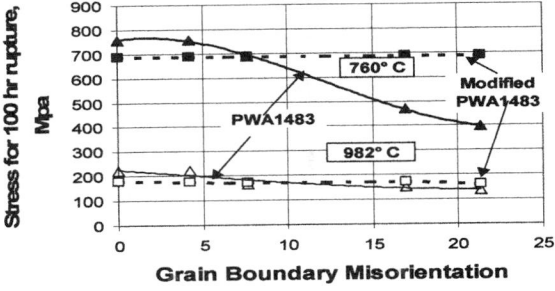

Figure 6: Stress for 100 hr rupture versus grain boundary misorientation at 760°C and 982°C for PWA1483 and Modified PWA1483.

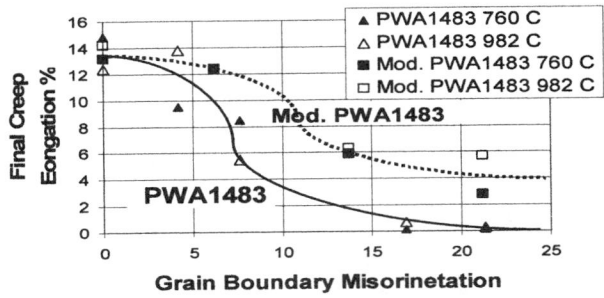

Figure 7: Creep rupture elongation at 760°C and 982°C, versus grain boundary misorientation for PWA1483 and Modified PWA1483. Trend in ductility behavior is almost identical at both temperatures.

While our results, on improving grain boundary tolerance with additions of B and Hf, is not totally surprising, the robust tolerance to HAB as large as 20°, is not in agreement with past experience. Ross and O'Hara [8], in their study of the alloy René N4, concluded that addition of minor elements at low level, enhance the grain boundary tolerance to 12°. Past testing of columnar grain alloys such as PWA1422, in transverse orientation, has always shown a significant creep strength debit in spite of even greater additions of minor elements. We shall return to this aspect later in the discussion.

As alluded to earlier, for alloys without minor elements, the tolerance to grain boundaries with 6-10° misorientation, coincides with a steep increase in grain boundary diffusion. This implies that without the minor elements, grain boundary failure may be further aggravated by environmental interactions. In the presence of minor elements, this aspect is largely mitigated. Thus, once the minor elements are added, the loss of creep strength may not originate entirely from the grain boundary per se, but may be dependent on transgranular creep behavior. Transgranular creep deformation is obviously controlled by the orientation dependence of creep behavior, as well as deformation compatibility in the vicinity of the grain boundary.

Role of Primary Orientation: When a bicrystal is tested in creep, with the grain boundary oriented normal to the stress axis, the grain boundary misorientation alone is not sufficient to define the creep behavior, especially for high angle boundaries. For example, a 20° bicrystal can be produced with a pure twist boundary with both grains oriented along either a <100> or <110> direction. Alternatively, a 20° tilt boundary can be produced with the orientation of each grain approximately 10° away from <100>, or one grain close to <100>, and the other off by 20° from <100>. Since the latter case was the way bicrystals were produced for this study, it is expected that creep properties will, as a minimum, degrade as the primary orientation of one of the crystal shifts away from <100>.

To assess this effect, 760°C/690 MPa rupture life is plotted in Fig. 8 for both bicrystals and off-axis specimens. Note that for PWA1483 bicrystals, the grain boundary misorientation obviously dominates. However, in the case of Mod. PWA1483, the inverse relation observed with grain boundary misorientation, is similar to the trend observed between rupture life and primary orientation for PWA1483. At 982°C/30 MPa, the misorientation relationship for the Mod. PWA1483 follows the same trend as primary orientation for PWA1483, although the PWA1483 rupture lives are consistently higher, as shown Fig. 9. This difference in creep strength is consistent with the creep-rupture comparison of both alloys (α=6°) presented in Fig. 3.

The data for both alloys (LAB < 6° and HAB 6-20°) is also compared on plots of stress vs. rupture life at 760 and 982°C (Figs. 10 and 11). At both test temperatures, PWA1483 creep strength is significantly debited by HAB, while the modified alloy shows much less of an effect. Even though the modified alloy has lower absolute creep-strength capability than PWA1483, the modified alloy shows a large advantage when comparisons are made with material containing HAB. As shown in Figs. 10 and 11, in material containing HAB, the modified alloy shows more than a 50% stress advantage at 760°C, and an ~10% stress advantage at 982°C, over PWA1483.

Role of <110> orientation: In a <100> oriented single crystal component, at least two secondary directions in the transverse plane will always coincide with <110>. This is known to be the weakest direction in creep, although single crystal components are used in practice in spite of that. The question is, whether a 20° boundary with <110> orientation would have performed as well. Unfortunately, no creep data were generated in <110> oriented PWA1483. However, if past results were any guide, rupture life would be expected to be very low. If 20° bicrystals were grown with average grain orientations near <110>, chances are that creep results would have been poor. This may represent the worst condition, similar to columnar grain material tested in a transverse direction. However, the key question is whether transverse properties are always poor because of the presence of grain boundaries or because of crystal orientation.

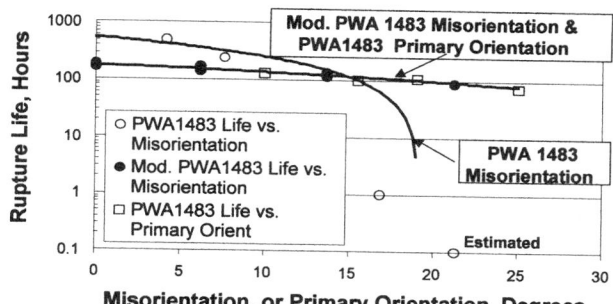

Figure 8: 760°C/690 MPa creep-rupture life of PWA1483 and Mod.PWA1483 vs. grain boundary misorientation, and primary orientation.

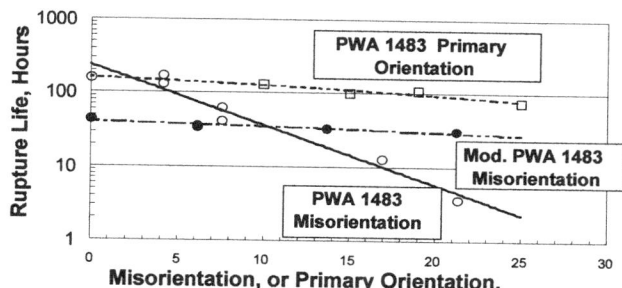

Figure 9: 982°C/30 MPa creep-rupture life of PWA1483 and Mod.PWA1483 vs. grain boundary misorientation and primary misorientation.

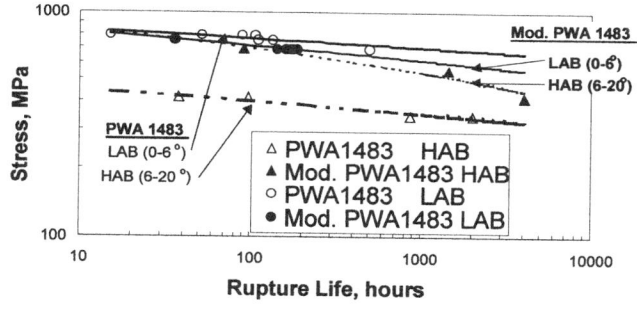

Figure 10: Effect of HAB on 760°C creep-rupture life of PWA1483 vs. Mod. PWA1483.

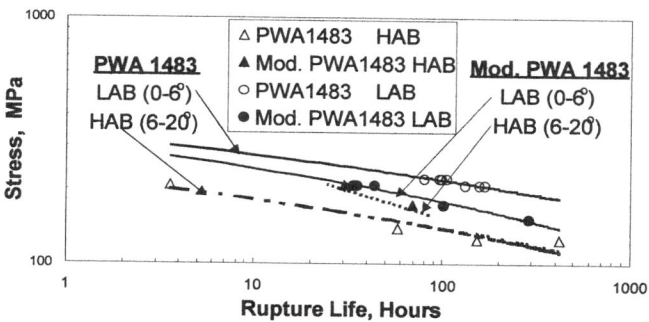

Figure 11: Effect of HAB on 982°C creep-rupture life of PWA1483 vs. Mod. PWA1483.

To answer this question, we revisited results for Mar M200+Hf, for which data in both columnar grain, and single crystal [4], form were available.

Mar M200 Experience: 760°C creep-rupture life for Mar M200+Hf, for columnar grain (CG) material in longitudinal and transverse directions, and single crystal material in <100> and <110> orientations, is presented Fig. 12. Linear parallel extrapolation is used to compare data for different orientations generated at different stresses. Clearly the wide scatter in data for the transverse orientation spans the range from the lowest rupture life for <100>, down to lower than <110> orientated single crystals. If what we have observed for Mod.PWA1483 were true for Mar M200+Hf, the large scatter in rupture life in the transverse orientation must be a consequence of the uncontrolled nature of primary orientation of grains in a transverse specimen. This suggests that there may be nothing unique about high Cr containing Mod.PWA1483. The high tolerance to grain boundary misorientation, may be a result of testing <100>-oriented bicrystals. Evaluation of HAB bicrystals with <110> primary orientation is needed to define the worst case scenario.

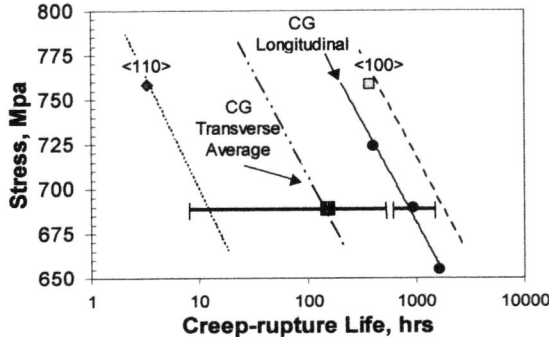

Fig. 12: Transverse 760°C creep-rupture life for columnar grain Mar M200+Hf spans range between longitudinal and <110> single crystal data at 690 MPa.

Summary

1. High growth rate seems to be helpful in suppressing grain boundary type defects as well as dendrite splaying in large single crystal castings. At the current development of casting technology, no significant difference in behavior of castable single crystal alloys could be discerned.
2. As expected, both low temperature tensile and high temperature creep strength of PWA1483 decrease with increasing deviation from the primary <001> direction.
3. Tolerance to high angle grain boundaries is significantly improved with minor element additions in creep at all temperatures. However, a creep strength debit is associated with this benefit at high temperatures.
4. In spite of a creep strength improvement with minor element additions, in the presence of high angle grain boundaries, creep ductility is limited compared to defect-free single crystal.
5. Comparison with data for Mar M200+Hf, suggests that the observed exceptional tolerance to high angle grain

boundaries with minor element additions, may be a result of the <100> orientation of the bicrystals evaluated.
6. Evaluation of <110> oriented high angle boundaries is necessary to assess the true benefit of addition of minor elements to single crystal alloys for improving tolerance to occasional grain boundary type defects.
7. With improvement in casting technology and better understanding of material behavior, and design tradeoffs, application of single crystal technology for large IGT seems economically viable.

Acknowledgement

Thanks are due to Mr. Allan Price of Howmet Research Corporation for supply of single crystal castings. Many thanks are due to Mr. Glenn Cotnoir of P &W for technical help.

References

1. R. P. Dalal, C. R. Thomas, and L. E. Dardi, "The Effect of Crystallographic Orientation on the Physical and Mechanical Properties of an Investment Cast Single Crystal Nickel-Base Superalloy" (SUPERALLOYS 1984, Edited by M. Gell, et al., TMS-AIME, Warrandale, PA., 1984) 185.

2. D. N. Duhl in Superalloys, Supercomposites, and Superceramics, (Edited by J. K. Tien and T. Caulfield, Academic Press, Boston) 149.

3. R. A. MacKay, R. L. Dreshfield, and R. D. Maier, "Anisotropy of Nickel-Base Superalloy Single Crystals" (SUPERALLOYS 1980, Edited by J. K. Tien et al., ASM, Metals Park, Ohio, 1980, p.385.

4. D. M. Shah and D. N. Duhl, "The Effect of Orientation, Temperature and Gamma Prime Size on the Yield Strength of a Single Crystal Nickel Base Superalloy" (SUPERALLOYS 1984, Edited by M. Gell et al., TMS-AIME, Warrandale, PA., 1984) 105.

5. D. M. Shah and A. Cetel, "Creep Anisotropy in Nickel Base γ, γ', γ/γ' Superalloy Single Crystals" (SUPERALLOYS 1996, Edited by R. D. Kissinger et al., TMS, Warrandale, PA) 273.

6. D. M. Shah and D. N. Duhl, "Effect of Minor Elements on the Deformation Behavior of Nickel-Base Superalloys" (SUPERALLOYS 1988, Edited by D. N. Duhl et al., TMS-AIME, Warrandale, PA., 1988) 693.

7. Shewman, Paul G., Diffusion in Solids, McGraw-Hill, New York, 1963, p.173.

8. E. W. Ross and K. S. O'Hara, "René N4: A First Generation Single Crystal Turbine Airfoil Alloy With Improved Oxidation Resistance, Low Angle Boundary Strength and Superior Long Time Rupture Strength" (SUPERALLOYS 1996, Edited by R. D. Kissinger et al., TMS, Warrandale, PA) 19.

EFFECT OF RU ADDITION ON CAST NICKEL BASE SUPERALLOY WITH LOW CONTENT OF CR AND HIGH CONTENT OF W

Yunrong Zheng, Xiaoping Wang, Jianxin Dong, and Yafang Han,
Institute of Aeronautical Materials, Beijing 100095, CHINA

H. Murakami and H. Harada
National Research Institute for Metals, 1-2-1, Sengen, Tsukuba, JAPAN

Abstract

The effect of Ru on two kinds of cast nickel-base superalloys in as-cast, heat-treated and oxidation tested specimens has been characterised by optical metallography and scanning electron microscopy. The evaluation of microstructure confirms the distribution of Ru in various phases, the beneficial role of Ru in stabilising microstructure, delaying formation of M_6C and increasing incipient melting temperature of the alloys. The role of Ru on oxidation resistance is different in various kinds of alloy; in Hf-bearing alloys, it has a beneficial effect, but in a 1.5% Cr alloy, the effect of Ru is not obvious.

Introduction

An important aim of developing new cast Ni-base superalloys is to increase the operating temperature of these alloys. One of the most effective ways to increase the stress rupture strength at temperatures above 1050°C is to decrease the content of Cr and increase the content of W in cast Ni-base superalloys. For example, in the third generation of single crystal superalloys, the content of Cr has been decreased to the level of 2-4wt%, and the total amount of refractory elements of W, Mo, Ta and Re is as high as 20wt% [1]. Among these refractory elements, Re is a unique element which can increase both the high temperature stress rupture strength and the oxidation resistance of the alloys. Re was first used in a conventional casting superalloy NASAII [2], and subsequently used in SC superalloys with Re content of 3-6wt%[1]. In BIAM, in order to develop a low cost Re-free casting Ni base superalloy 601, which has the stress rupture life of 50h under 1100 °C/118MPa, the content of W was increased to 16wt%, and the content of Cr was decreased to 1.5wt%[3]. However, the decrease of Cr content led to a decrease of oxidation resistance of this alloy, which should be compensated by addition of other oxidation resistant elements.

It has been found that Ru was a potential alloying element to increase the oxidation resistance, and can replace part of the Re due to its relatively lower density and cost compared with Re [4]. Since Ru belongs to the Pt family, it has high chemical stability and hence high oxidation resistance. Furthermore, Ru can strengthen both γ' and γ phases effectively due to its large atomic radius. Based on the above-mentioned considerations, this study was performed.

Experimental procedure

Two kinds of alloys were selected in order to study the role of Ru; their chemical compositions were listed in Table 1. Alloy 1 is a commercially cast superalloy K19H containing Hf, alloy 3 is a trial cast superalloy with low Cr, high W content selected on the basis of the stress rupture properties under the condition of 1100°C/118Mpa. Alloy 2 and alloy 4 are both Ru-bearing modified alloys from alloy 1 and alloy 3 respectively.

Superalloys 2000
Edited by T.M. Pollock, R.D. Kissinger, R.R. Bowman,
K.A. Green, M. McLean, S. Olson, and J.J. Schirra
TMS (The Minerals, Metals & Materials Society), 2000

Table 1 Chemical compositions of tested alloys, wt.%

Alloy	Co	Cr	Mo	W	Nb	Ru	Al	Ti	Hf	C	B	Ni
1	11.8	6.0	1.9	10.3	2.3	-	5.8	1.4	1.3	0.12	0.02	bal.
2	11.8	6.0	1.9	10.3	2.3	3.0	5.8	1.4	1.3	0.12	0.02	bal.
3	9.6	1.3	1.3	16.2	1.0	-	6.0	1.1	-	0.10	0.024	bal.
4	9.6	1.3	1.3	16.2	1.0	3.0	6.0	1.1	-	0.10	0.024	bal.

The distribution of Ru, dendrite segregation and phase precipitation were analysed using optical microscopy (OM) and SEM/EDX for all the alloys. The effect of Ru on the oxidation resistance was investigated by static oxidation test at the temperature of 1100°C, the composition and morphology of the surface oxide scale after oxidation testing were analysed by using the LV method of SEM. The microstructures after long-time exposure at 1100 °C for 100 hours were observed for all alloys. In order to determine the influence of Ru on the solid solutioning of γ' and the incipient melting point of alloys, the directionally solidified alloys 3 and 4 were heat-treated in the temperature range of 1260-1300°C, and thus a suitable solid solutioning treatment was determined.

Results and discussion

Composition selection of trial alloys

The chemical composition of alloy 3 in Table 1 was designed on the basis of the compositions of K19H alloy (6Cr-10W), by decreasing Cr content to 1.5wt% and increasing the total content of W+Mo+Nb+Ta up to 20wt%. The chemical compositions of these designed conventional cast alloys and their stress rupture lives are given in Table 2. It is shown from Table 2 that the strength level of alloy A is lower than that of alloys B and C. The precipitation of primary μ and α (W, Mo) phases

was observed in alloy D due to excessive alloying. The alloy C contains higher Ta which causes an increment of primary eutectic (γ+γ') phases (Fig.1) and leads to difficulty of solid solutioning treatment. Moreover, Ta is a high cost element; therefore, alloy B is selected as a trial alloy and named as alloy 3 in Table 1.

Distribution of Ru

The distributions of Ru in the precipitated phases, dendrite and interdendrite regions in alloys 2 and 4 are shown in Table 3.

The compositions of γ' and γ in alloy 4 were determined when γ' and γ were coarser after 1100°C /100h long-term exposure, and the compositions of M_6C carbides were determined from specimens after higher temperature treatment of 1300°C.

It can be seen from Table 3 that Ru hardly enters MC carbides, and the solubility of Ru in the M_6C carbides is higher than that in the MC carbides. However, Ru has a rather higher solubility in the γ' and Ni_5Hf phases. The Ru concentration in the eutectic γ' phase is higher than in the secondary γ' phase, and higher in the γ matrix compared with the secondary γ' phase. The partitioning ratio of (γ/γ') is near 1:0.7. Ru is uniformly distributed between dendritic and interdendrtic regions.

Table 2 Chemical compositions of trial alloys and stress rupture lives

Alloy	Cr	Co	W	Mo	Al	Ti	Nb	Ta	C	B	Ni	Stress rupture life (h) 118MPa	
												1080°C	1100°C
A	3.0	9.9	14.0	1.9	5.6	1.2	2.5	-	0.14	0.02	bal	61.4, 80.1	
B	1.3	9.6	16.2	2.1	6.0	1.1	1.0	-	0.10	0.024	bal		51.3, 52.4
C	1.5	9.8	16.2	2.0	4.9	1.1	1.1	4.4	0.10	0.034	bal		65.0, 52.1
D	1.4	9.6	16.0	2.0	5.0	1.1	1.0	6.7	0.10	0.02	bal		40.0, 41.4

Table 3 Distribution of Ru in as-cast alloys 2 and 4, wt%

Alloy	Eutectic γ'	γ'	γ	Ni_5Hf	MC	M_6C	dendrite	interdendrite
2	2.94	-	-	2.13	0.23	-	2.96	3.06
4	2.51	2.14*	2.97*	-	0.15	0.87**	2.68	2.78

*   After 1100°C/100h heat exposure
** After 1300°C/4h heat treatment

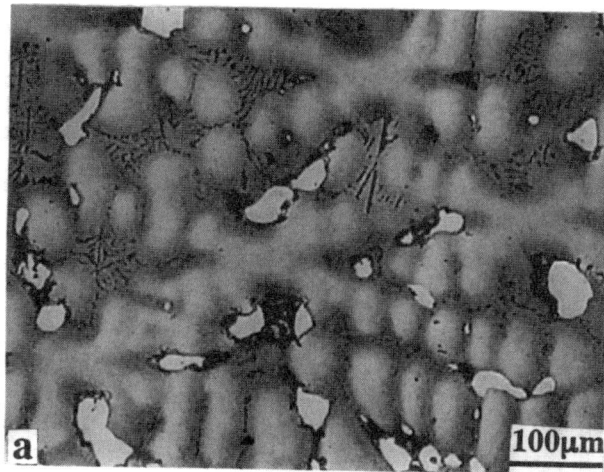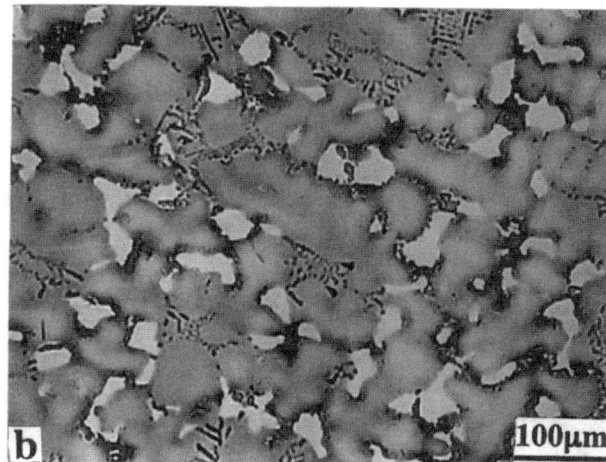

Fig.1 Comparison of eutectic (γ+ γ') in alloys; (a) alloy B, (b) alloy C.

Effect of Ru on microstructure

Addition of 3wt.% Ru in alloys 2 and 4 seems to have no effect on the as-cast microstructures, while it delays the precipitation of M_6C carbides. After heat treatment at 1100°C/50h for alloy 3, large amount of coarse-granular or a few short-plate M_6C carbide particles precipitated along grain boundaries and interdendritic regions (Fig.2). For alloy 4 exposed at 1100°C/100h, only at some grain boundary segments were M_6C particles detected (Fig.3a), and a strong raft tendency of the γ' phase was observed at the dendritic areas (Fig.3b).

Partial solid solution of γ' phases in dendrites takes place during heat treatment of 1260°C/4h of alloys 3 and 4. The secondary γ' phase was completely dissolved except for eutectic γ' phases in alloy 4 at treatment of 1280°C/4h. For alloy 3, a few secondary γ' phases existing in the vicinity of the eutectic γ' were observed, and some neck-like M_6C carbides enveloped by γ' film existed (Fig.4).

With increasing temperature to 1300°C for 4h, all kinds of γ' phases were dissolved, but the incipient melting phenomenon apparently occurred in two of the alloys. Compared to alloy 3, the degree of incipient melting of alloy 4 is lower and it takes place only in some eutectic γ'. For alloy 3, large parts of the original eutectic (γ+γ') region and some columnar grain boundary were melted. The morphology of the incipient melting zone in both alloys is shown in Fig.5. It can be seen from Fig.5 that some block shape or coarse plate M_6C carbides often appeared in the incipient melting zone and its adjacent zone, especially in alloy 3. Element W is the main composition of M_6C carbide and a typical content of

metal element in M_6C carbide of alloy 4 is 72.3W-6.7Mo-11.2Ni-3.4Co-1.7Al-0.9Ru-0.7Cr-0.2Ti, This proves that Ru is not a M_6C carbide former.

Oxidation resistance of alloys

Fig.6 shows the experimental results of oxidation tests for the four alloys, Fig.6b is a magnification of Fig.6a. The role of Ru is different in various kinds of alloys. For Hf-bearing alloys, Ru decreases the rate of weight gain slightly, but can drop the rate of weight loss during oxidation testing at 1100°C. The value of the weight gain decreases from $6.87g/m^2$ to $1.55g/m^2$ in alloy 1 compared with alloy 2 after oxidation test at 1100°C for 65h due to the decreased flaking of oxide scale.

Observation from SEM has shown that the oxide scale on the surface is porous and contains a crystallographic slope in the Ru-free alloy, as shown in Fig.7a, whereas, the scale of the Ru-bearing alloy is denser and there are nodules existing on some local regions (Fig.7b). On cross sections of the oxidation layer, the continuous scale was observed in alloy 2 (Fig.8a).

For low Cr alloys, the weight gain was increased during the first 3 hours; even alloy 4 has a larger weight gain. However, spalling of the scale in alloy 4 was later than that of alloy 3. Once spalling of the scale appeared five hours later, the weight loss rate was at the same level in alloys 3 and 4, about $8.1g/m^2$·h. The continuous scale did not appeared in alloys 3 and 4, this is apparent from Fig.8b, which shows the serious inter-oxidation in the low Cr alloy. The compositions of the oxidation layer at alloy 4 after the oxidation test of 1100°C/50h is listed in Table 4.

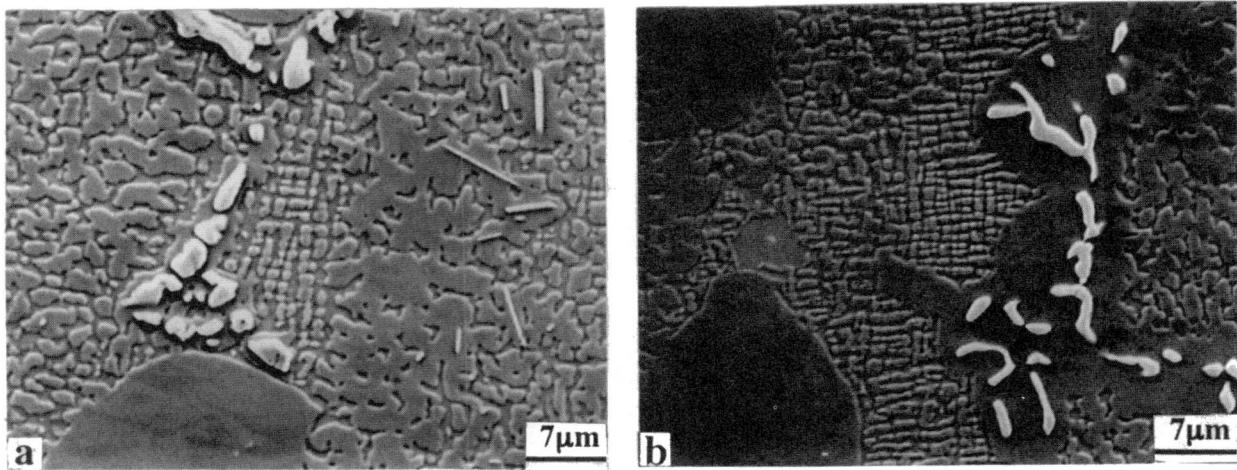

Fig.2 M_6C carbide precipitated from alloy 3 after 1100°C for 50h;
(a) granular and short-plate M_6C carbide at interdendrite, (b) M_6C along grain boundaries.

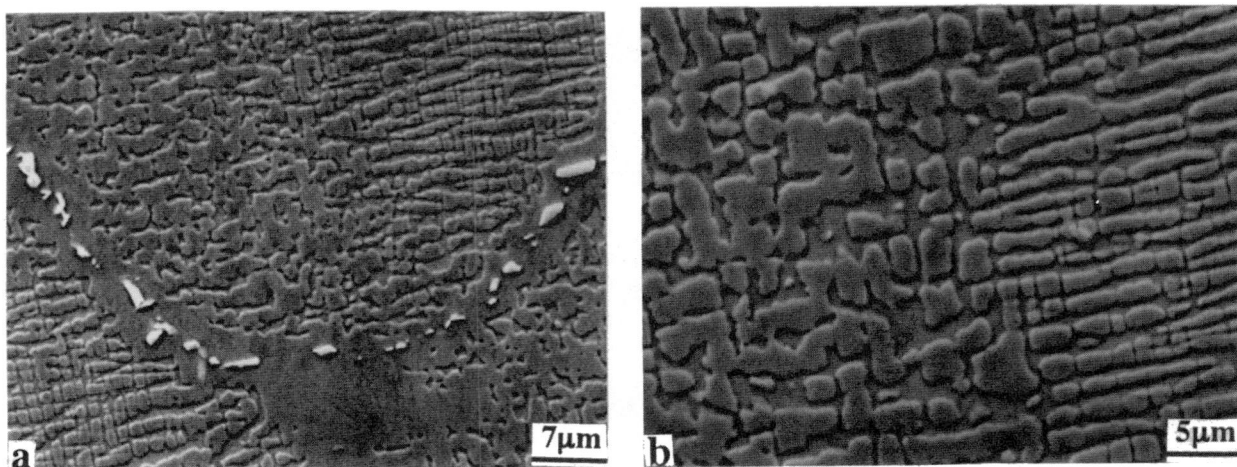

Fig.3 Microstructure of alloy 4 after 1100°C for 100h; (a) M_6C along grain boundary, (b) rafting γ' at dendrite.

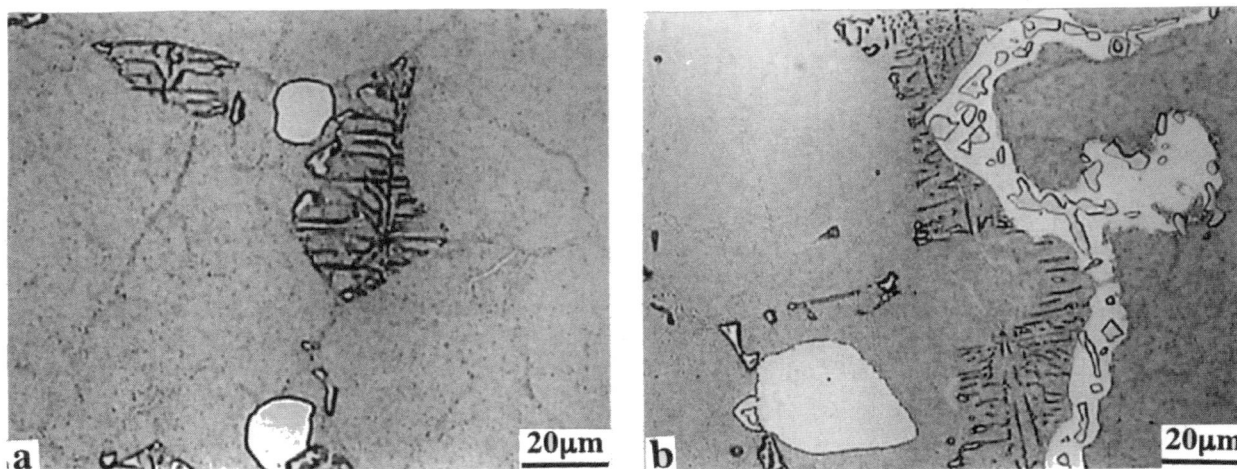

Fig.4 Microstructures after heat treatment at 1280°C/4h; (a) alloy 4 (b) alloy 3.

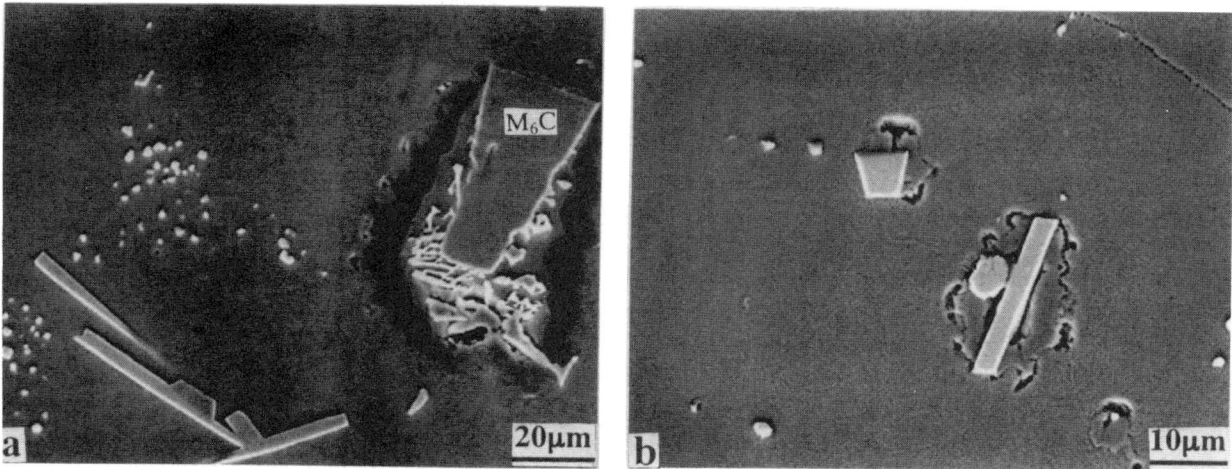

Fig.5 The morphology of incipient melting zones after 1300°C/4h solid solution treatment; (a)alloy 3, (b)alloy 4

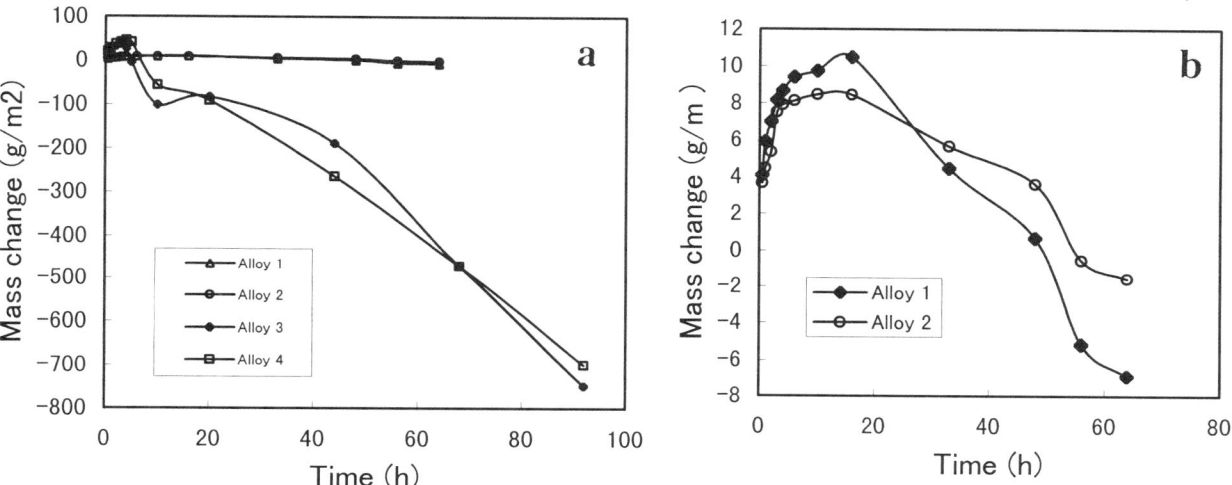

Fig.6 Weight change of alloys during oxidation test at 1100 °C

Fig.7 Morphology of scale after oxidation test of 1100°C/65h; (a) alloy 1 (b) alloy 2.

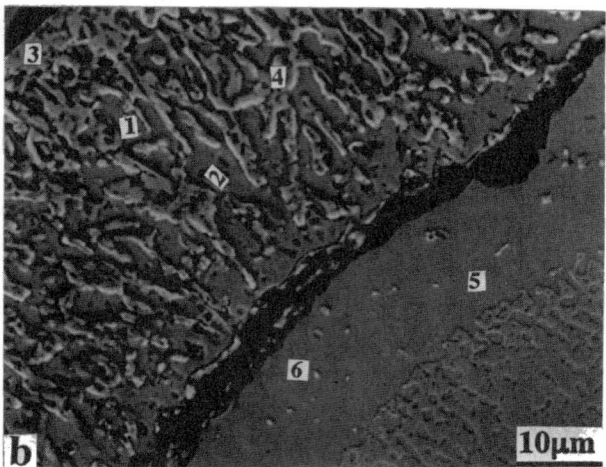

Fig.8 Morphology of oxide layer on alloys after oxidation test at 1100 °C/50h, (a) alloy 2 (b) alloy 4.

Table 4 Compositions of oxidation layer on alloy 4 after test of 1100°C/50h.

Point	O	Al	Nb	Mo	Ru	Ti	Cr	Co	Ni	W
1	2.26	0.21	0.36	1.35	4.26	0.22	0.35	2.69	84.67	3.63
2	2.22	0.17	0.22	0.83	3.57	0.29	0.30	1.51	87.28	3.62
3	17.98	0.88	0.42	0.68	0.34	0.11	0.76	4.01	15.74	59.07
4	15.40	1.01	0.36	1.04	0.42	0.36	0.61	3.57	15.87	61.35
5	2.09	2.33	0.20	1.77	1.86	0.28	1.09	9.45	58.19	22.63
6	2.26	2.69	0.60	1.87	1.87	0.52	1.16	10.23	58.78	20.04

The location of the data points is marked in Fig.8b. Indeed, large amounts of W-rich oxide particles exist in the Ni matrix, which separates the Ni matrix into small blocks and prevents the formation of a continuous scale.

Structural stability and oxidation resistance are the key points when we develop the low Cr high W casting superalloys applied at temperatures over 1100°C besides the strength considerations. The excessive alloying by W usually cause the precipitation of α (W, Mo) and μ phases [5] and it can be avoided by modifying the composition of alloys. Although Cr-free IC-6 alloy was applied [6], there is still worry about excessive reduction of the oxidation resistance due to the lower Cr content. Thus, how to affect the oxidation resistance and further improve high temperature strength is the key research topic motivating the substitution of Ru for Ni.

The experimental results indicate that the substitution 3wt%Ru for Ni did not cause the precipitation of any new phases. Ru is a very weak positive segregation element and restricts the formation of M_6C carbide, which is important for solid solution heat treatment and long time

service of high W alloys. For high W free-Ru alloys, large blocky M_6C carbides (up to 10μm) were formed after a short time when the temperature is over 1280°C. The content of W in M_6C carbide is up to 70% reducing the W content within matrix and γ' phases. Meanwhile, the presence of both coarsening of M_6C carbides and the formation of M_6C thin films at the grain boundaries and the interdendrite regions also reduce the high temperature stress rupture properties of high W alloys. Thus, it is beneficial for the addition of Ru to restrict and delay the formation of M_6C carbide.

The role of Ru on the improvement of oxidation resistance is not obvious. It slightly delays spalling of the scale, and does not significantly decrease the weight-loss and weight-gain rate. Especially for the alloy containing 1.5Cr, once the protective scale has not formed, the addition of Ru did not improve its oxidation resistance, while the addition of Y can improve the oxidation resistance of Cr-free alloy [7]. It is considered that the contribution to oxidation resistance of Y is superior to that of Ru.

Conclusions

1. Ru is a weak positive segregation element, which hardly enters into carbide, but has higher solubility in γ' and Ni_5Hf phases. The partitioning ratio of Ru in γ/γ' is about 1:0.7.
2. The addition of Ru in low Cr high W alloys can improve the structural stability and delay the M_6C carbide formation. Ru slightly increases the incipient melting temperature of alloys.
3. Ru can delay the oxide scale spalling, but does not reduce the oxidation rate for low Cr high W alloys. With respect to improvement of oxidation resistance, Y is superior to Ru.

Reference

1. W.S. Walston, K.S. O'Hara, E.W. Ross, T.M. Pollock and W.H. Murphy, "RENE N6: Third generation Single Crystal Superalloy", Proc. of the 8th Inter. Symp. on Superalloys, Ed. by R.D. Kissinger et al., TMS, (1996), 27-34.

2. H.E. Collins, "Effect of Thermal Exposure on the Microstructure and Mechanical Properties of Nickel-Base Superalloys", Metall. Trans., 6(1), 1974, 189-204.

3. Zheng Yunrong and Zhang Detang, Colour Metallographic Investigation of Superalloys and Steels (Beijing, National Defence Press, 1999), 2.

4. Xuan Nguyen-Dinh, "Phase Stable Single Crystal Materials", United Stated Patent Number 4, 935, 072, Jun. 19, 1990.

5. Zheng Yunrong, "Primary μ Phase in Cast Nickel Base Superalloys", Acta Metallurgical Sinica, 35(12) (1999), 1242-1245.

6. Y.F. Han, Z.P. Xing and M.C. Chaturvedi, 'Development and Engineering Application of a DS Cast Ni_3Al Alloy IC6", Structural Intermetallics 1997, Ed. by M.V. Nathal et al., TMS(1997), 713-719.

7. C. Xiao and Y.F. Han, "Effect of Yttrium on the Microstructure and Oxidation Resistance of Ni_3Al-Base Alloy IC6", Journal of Materials Engineering, 1998, no.6: 23-28.

PREDICTION AND MEASUREMENT OF MICROSEGREGATION AND MICROSTRUCTURAL EVOLUTION IN DIRECTIONALLY SOLIDIFIED SUPERALLOYS

B. Böttger, U. Grafe, D. Ma

ACCESS e. V., Intzestraße 5, D-52056 Aachen, Germany

A. Schnell

ABB Management AG, Research Center CH-5405 Baden-Dättwil, Switzerland

Abstract

Microsegregation and microstructural evolution in directionally solidified superalloys is calculated using a unit cell approach. The results are compared to experimental micrographs and composition profiles. CMSX-4 and Inconel 706 are used as examples for a single-crystal and a conventional superalloy respectively. Two micromodels are applied to simulate solidification of the two alloys. For CMSX-4 as a typical SX alloy the SX-Model is formulated based on the Scheil approximation. For conventional superalloys as IN706 however carbon as interstitial element leads to considerable solid-state back diffusion. For that reason the more sophisticated Shape-Constrained Phasefield Model (SCPF-Model) is presented, which applies for diffusion in all phases and includes precipitation of secondary phases. In both models thermodynamic databases assessed using the CALPHAD method are coupled using the Thermo-Calc software package.

Introduction

Solidification of multicomponent alloys such as Ni-base superalloys under practical conditions is always associated with solute redistribution of the alloying elements. Therefore formation of chemical inhomogenities and the consequent precipitation of interdendritic phases can be observed. Such inhomogeneous microstructures result in poor mechanical properties, even after heat treatment. For that reason it is required to predict segregation and secondary phase precipitation for different process parameters as well as for different alloy compositions.

Extensive work has been done to predict microsegregetion in dendritic microstructures. Frequently one-dimensional plate morphologies are used to describe the growing of secondary dendrite arms[1-4]. The half dendritic secondary arm spacing $\lambda_2/2$ is used as length of the unit cell. Some models include coarsening of secondary arms[3,4].

In processes like electroslag remelting (ESR) or vacuum arc remelting (VAR) dendritic growth of Ni-base superalloys normally is near to the cellular-dendritic transition region due to the high thermal gradients. No tertiary arms are observed and the dendrites form a more or less regular array. In such cases one-dimensional cylindrical models have been formulated, which use the primary dendritic distance λ_1 as the dimension for the unit cell[5,6]. Recently Ma[7] proposed a pseudo-2D model for cellular dendritic solidification of Ni-base superalloys. This approach uses a fourfold shape function for description of the isothermal cross-section perpendicular to the growth direction of the dendrite.

Superalloys 2000
Edited by T.M. Pollock, R.D. Kissinger, R.R. Bowman,
K.A. Green, M. McLean, S. Olson, and J.J. Schirra
TMS (The Minerals, Metals & Materials Society), 2000

Superalloys are complex multicomponent systems. Generally the correct thermodynamic description is much more important than sophisticated kinetic modelling to get results, which are comparable to experimental findings[8]. Multicomponent systems are often described by linearized or interpolated phase diagrams[9,10]. This is a crude approximation, especially because tielines are not specified. Thermodynamic databases instead provide Gibbs energy phase descriptions of the system, which are based on a large number of experiments and are assessed by the CALPHAD[11] approach. By total Gibbs energy minimization phase equilibria can be obtained using software packages such as Thermo–Calc[12]. In the case of superalloys such databases have been developed including up to 13 elements and a large number of phases of practical interest[13] as well.

The unit cell approach

The two models used here are based on a unit cell approach for directional dendritic solidification which makes use of a four-fold shape function for the approximation of the isothermal cross section of a dendrite, like that shown in Fig. 1, and was proposed by Ma[7]:

$$L(\beta) = L_0(t)(1 + A\cos 4\beta) \qquad (1)$$

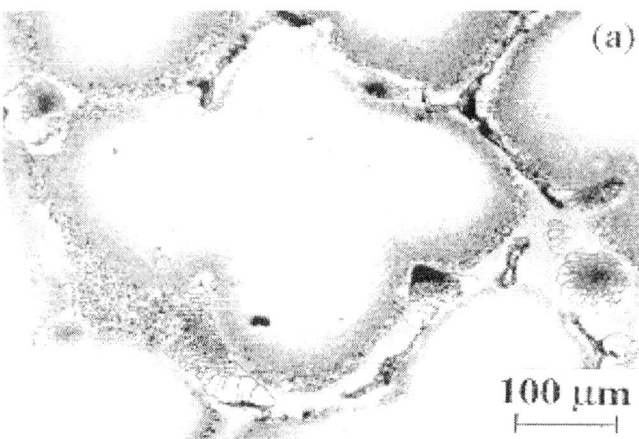

Fig. 1: Cross–section through a CMSX–4 dendritic array obtained by directional solidification.

The amplitude factor A determines the amount of anisotropy of the shape function. For $A=0$ a circular shape is observed corresponding to cellular growth, the maximum value is $A=1$. The unit cell is depicted in Fig. 2.

The function is chosen according to the description of the surface free energy of fcc crystals[14]. Under the conditions of high thermal gradients and low cooling rate no tertiary arms are observed and the shape function is a good geometric approach.

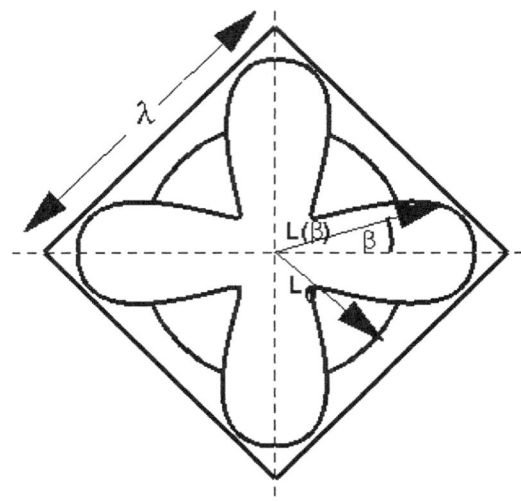

Fig. 2: Dendritic unit cell used for the SX– and the SCPF–Model.

The side length of the unit cell is correlated to the dendritic primary spacing λ_1 and can be estimated from the thermal gradient G and the solidification velocity v using the theory of Hunt[15]:

$$\lambda_1 = K G^{-1/2} v^{-1/4} \qquad (2)$$

According to Goldschmidt[16] the factor K can be considered as a unique constant for superalloys. Alternatively K can be estimated from experiments.

SX–Model: Scheil approximation for SX–alloys

Within the SX–Model back diffusion is neglected as no fast diffusing interstitial elements are present in single crystal alloys and diffusion in the liquid is considered to be complete (Scheil approximation). So the solid fraction f_s is only a function of temperature T which itself depends on solidification time t:

$$f_s = f(T) \qquad (3)$$

$$T = T_0 - G v t \qquad (4)$$

In Eq. 4 G denotes the thermal gradient and v the solidification velocity along this gradient.

Within the unit cell model f_s corresponds to the quotient of the dendrite area and the total area of the unit cell:

$$f_s = \frac{F}{\lambda_1^2} = \frac{1}{\lambda_1^2} \int 0.5 L^2(\beta) d\beta \qquad (5)$$

At later solidification times growth of the dendrite is hindered by neighboring dendrites. In this case the position of the solid-liquid interface may exceed the maximum length given by the primary dendrite arm spacing. Then the size of the shape function must be corrected properly[7] to get the correct value for f_s.

The amplitude factor A can be estimated using an empirical relation proposed by Ma[7]. The amplitude factor depends on the solidification velocity v and on the critical value v_e, at which the solidification front shows fully developed cells:

$$A = 1 - \left[\frac{v_c}{v} \right]^{\frac{1}{2}} \qquad (6)$$

In case of v being equal to v_e A equals to zero corresponding to a circular dendritic cross section. At $v=v_c$ the planar solidification front shows first instabilities, at a higher value v_t the transition from cellular to dendritic growth is observed. v_e is taken to lie halfway between v_c and v_t (Fig. 3).

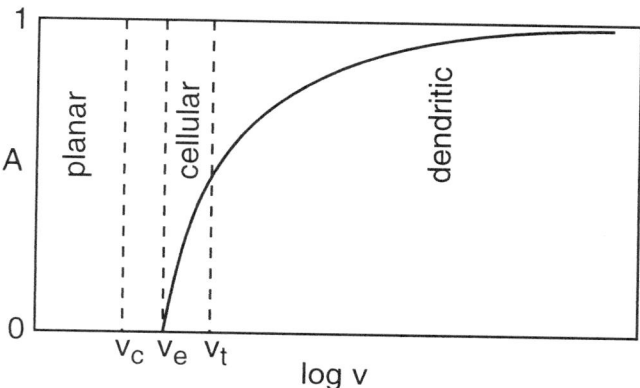

Fig. 3:   Relation between the amplitude factor A and the solidification velocity v

To model the evolution of the shape function with time, the temporal evolution of the fraction solid f_s is needed. For SX–superalloys back diffusion can be neglected and the multicomponent Scheil model is applied. It is implemented using a FORTRAN code, that calls subroutines of the TQ–interface provided by the thermodynamic software Thermo–Calc[12]. The thermodynamic description for the multicomponent CMSX–4 was developed by Thermotech Ltd. and Rolls–Royce plc. was used[13]. Further details about the model can be taken from elsewhere[7,17].

The Shape–Constrained Phasefield Model

For conventional superalloys the Scheil model is not suitable for prediction of microsegregation and secondary phase precipitation because of fast back diffusion of the interstitial elements. Therefore the more refined numerical Shape–Constrained Phasefield (SCPF–) Model is used.

Implementation of the shape function on a numerical finite differences grid

Although this model like the SX–Model is based on the unit cell approach (Fig. 2), the growing shape function is treated in a

numerical manner. The shape function is projected onto a finite difference grid (Fig. 4). Like in phasefield methods a diffuse interface is applied to avoid the necessity of front tracking and to allow the application of a single diffusion algorithm in the whole multiphase region. A phasefield parameter ϕ_α is defined, which denotes the phase fraction of all phases α. The diffuse interface region is created by application of a hyperbolic tangent function perpendicular to the interface (Fig. 4). The use of isolating boundary conditions at the border of the unit cell ensures proper treatment of the interactions with neighboring dendrites for a regular dendritic array. For reasons of symmetry only a quarter of the unit cell needs to be calculated.

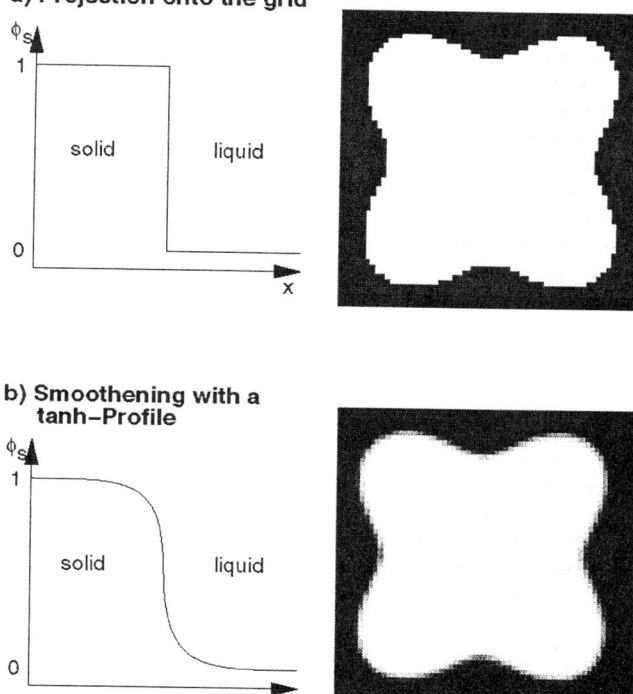

Fig. 4:   Implementation of the unit cell in the shape–constrained phasefield model on a finite differences grid with diffuse interface.

Multicomponent Diffusion

The diffusion algorithm used in the SCPF–Model is based on the Multi–Phasefield diffusion concept[18] and has been generalized for multicomponent systems. The total flux of component k dc^k/dt in the multiphase region is considered as the sum of the fluxes in the individual phases weighted by their phasefield parameter ϕ_α, which corresponds to the local volume fraction of phase α (volume average approach):

$$dc^k/dt = \nabla \sum_{\alpha=1}^{N} \phi_\alpha D_\alpha^k \nabla c_\alpha^k \qquad (7)$$

The diffusion coefficients D_α^k in all phases α are regarded as constant, but could also easily be implemented as temperature-dependent. The composition of phase α is obtained from the partition coefficients $K_{\beta\alpha}$ between phase α and all other phases β and the phasefield parameter ϕ_β:

$$c_\alpha^k = \frac{c^k}{\sum\limits_{\beta=1}^{N} \phi_\beta K_{\beta\alpha}^k} \quad (8)$$

Thermodynamic coupling

For IN 706 a commercial Ni database from Thermotech Ltd. with 11 elements was used and from there the partition coefficients and the undercooling of the interfaces are calculated.

As in the SX-Model thermodynamic coupling is done using the TQ-interface of the software Thermo-Calc. Partition coefficients must be evaluated for all interface cells to obtain the phase compositions needed for the diffusion algorithm (Eq. 8).

The growth kinetics of the shape function is determined using the average solutal undercooling $\Delta T_{solutal}$ of the interface, calculated from the average composition by Thermo-Calc subroutines. Additionally the curvature contribution $\Delta T_{curvature}$ is calculated to obtain the kinetic undercooling of the interface:

$$\Delta T = \Delta T_{solutal} + \Delta T_{curvature} \quad (9)$$

An exponential kinetic equation is used which is built up as a sum two terms for forward and backward motion of the solidification front:

$$v = v_0 \left[\exp\left(\frac{\Delta T}{\Delta T_0}\right) - \exp\left(\frac{-\Delta T}{\Delta T_0}\right) \right] \quad (10)$$

Here v_0 and ΔT_0 are numeric parameters which describe the mobility of the interface.

Separately the undercooling of the region of the secondary arm is calculated to obtain the growth velocity in the [001] direction of the dendrite:

$$v_{(001)} = v_0 \, \kappa \left[\exp\left(\frac{\Delta T_{(001)}}{\Delta T_0}\right) - \exp\left(\frac{-\Delta T_{(001)}}{\Delta T_0}\right) \right] \quad (11)$$

κ is the kinetic anisotropy factor which raises the mobility of the [001] direction. It is assumed to be 1.05 in the calculations below. With the two velocities v and $v_{[001]}$ L_0 and $L(0°)$ in Eq. 1 can be obtained and the amplitude factor A can be calculated for each time step:

$$A(t) = \frac{L(0°,t)}{L_0(t)} - 1 \quad (12)$$

By this way $A(t)$ is included as additional variable to allow the system to change from cellular to dendritic growth depending on the growth conditions, the diffusion coefficients and surface energy.

Precipitation of interdendritic phases

Precipitation of secondary phases is an important factor in solidification processes, as they can be detrimental for the mechanical properties or can raise the homogenization time necessary for their removal. In this model secondary phases are included as circular shape functions, i.e. the amplitude factor A is restricted to zero. Nevertheless after complete solidification the precipitates are not necessarily circular because they can be partially overgrown by the primary dendrites. Thus the exact shape of the precipitates cannot be predicted by the model, but the volume fraction and the spatial distribution in the unit cell. Interactions between solid phases are not included.

Nucleation is assumed to take place at the solid-liquid interface of the primary dendrite. For nucleation a certain undercooling ΔT_{nuc} is necessary. The local undercooling is calculated using Thermo-Calc and the thermodynamic database.

Experimental Details

For validation of the models two samples of CMSX-4 and INCONEL 706 were directionally solidified in a Bridgman furnace. The nominal compositions of the two alloys are shown in Tab. 1 and 2. The CMSX-4 sample was solidified at a velocity of 1.0 mm/min with a thermal gradient of 2.85 K/mm, the IN706 sample at $v=2$ mm/min and $G=15$ K/mm.

Tab. 1: Nominal and measured compositions of CMSX-4

	Ni	Co	Cr	Ta	W	Al	Re	Ti	Mo	Hf
nom.	bal.	9.0	6.5	6.5	6.0	5.6	3.0	1.0	0.3	0.1
EDX	bal.	10.02	6.32	6.04	5.83	5.0	2.78	1.0	0.36	0.51

Tab. 2: Nominal and measured compositions of IN706

	Ni	Fe	Cr	Nb	Ti	Al	Mg	Si	C	B
nom.	bal.	37.4	16.0	2.9	1.75	0.2	0.1	0.1	0,02	0.004
EDX	bal.	38.0	16.9	3.0	1.65	–	–	–	0.01	–

Composition measurements of IN706 were made using energy dispersive X-ray analysis (EDX, Type Oxford Link-Isis). For CMSX-4 microprobe measurements (WDX, Type Cameca SX 50) were necessary to distinguish between W, Re and Ta. For each alloy one linescan along the [001]-axis (secondary arm direction) and one in the [011]-direction was chosen. The exact positions are indicated in Fig. 5 and 6.

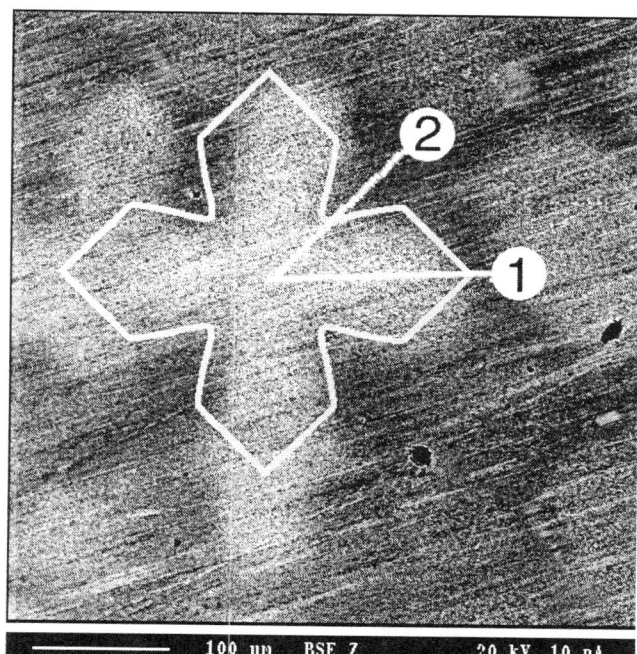

Fig. 5: Scan paths on the CMSX-4 sample. Scan 1 goes from the center along the [001]-axis, scan 2 crosses the region between the secondary arms.

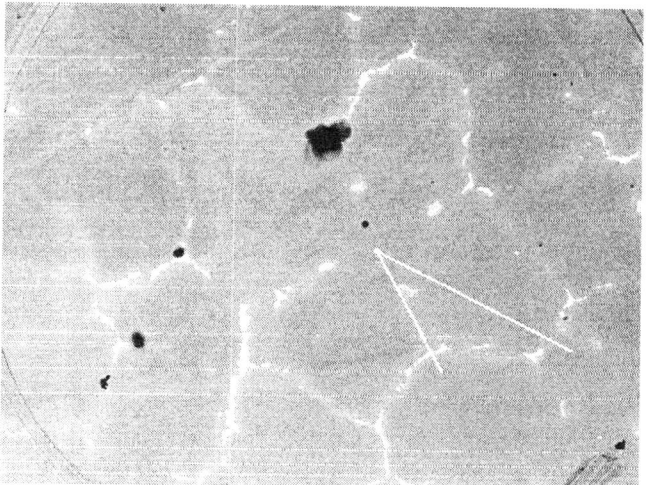

Fig. 6: Scan paths on the IN706 sample. Scan 1 goes from the center along the [001]-axis, scan 2 crosses the region between the secondary arms.

Experimental and Numerical Results

SX-Model

For the SX-Model the critical velocity v_c for the cellular-dendritic transition (Fig. 3) must be obtained from additional experimental data with very low solidification velocities. This was done for CMSX-4, where beginning front instabilities were found at v_c=0.009 mm/min. Fully developed cells were formed at v_c=0.0193 mm/min. From these values the amplitude factor A can be estimated to 0.87.

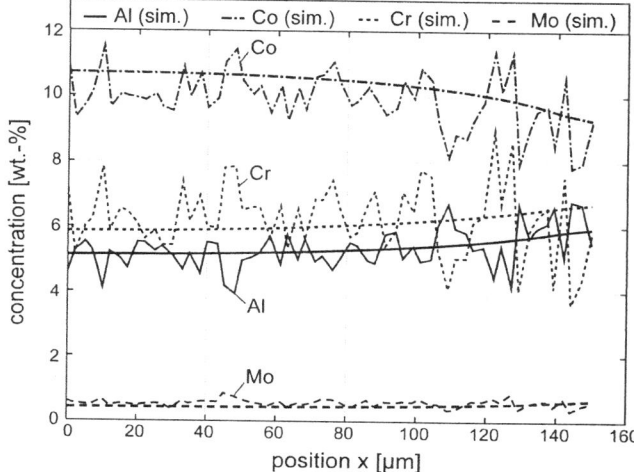

Fig. 7: Comparison between composition profiles of Al, Co, Cr and Mo obtained from scan 1 (Fig. 5) and those calculated using the SX-Model.

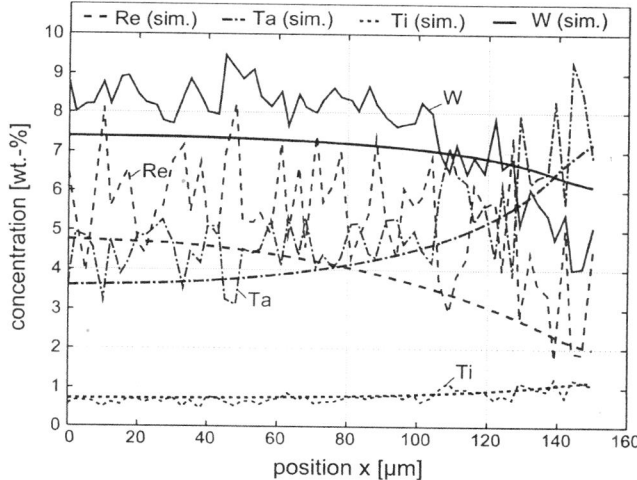

Fig. 8: Comparison between composition profiles of Ti, W, Re and Ta obtained from scan 1 (Fig. 5) and those calculated using the SX-Model.

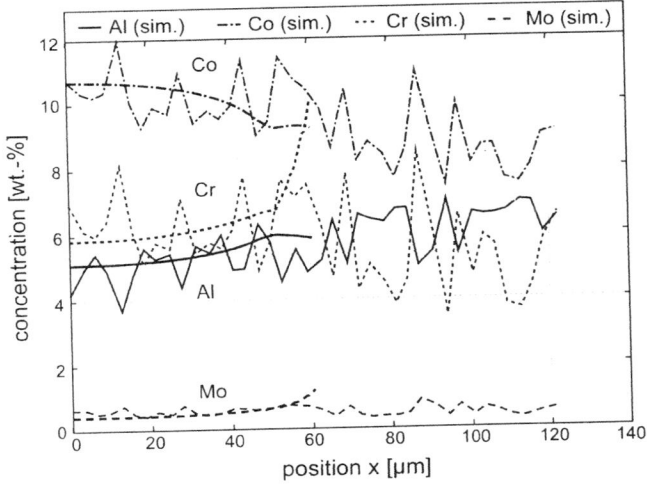

Fig. 9: Comparison between composition profiles of Al, Co, Cr and Mo obtained from scan 2 (Fig. 5) and those calculated using the SX–Model.

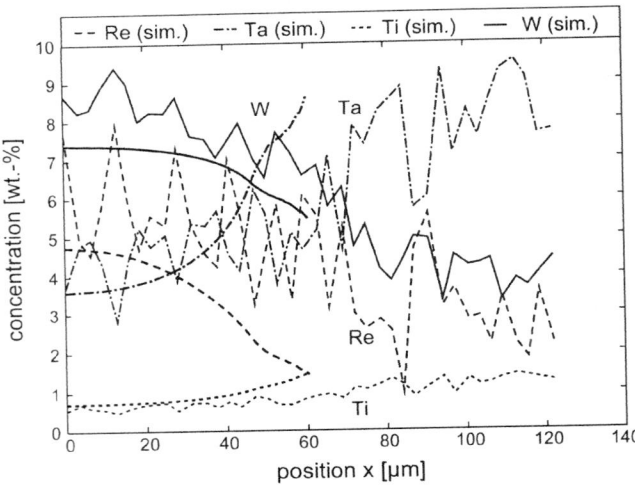

Fig. 10: Comparison between composition profiles of Ti, W, Re and Ta obtained from scan 2 (Fig. 5) and those calculated using the SX–Model.

For simulation a thermodynamic database developed by Thermotech Ltd. and Rolls–Royce plc. has been used. This database has been made available by ABB company, which holds exclusive rights on this data. All elements which are present in CMSX–4 were used in the simulation with exception of Hf .

The comparison between simulation and experimental results is shown in Fig. 7–10. The large oscillations of the microprobe curves are due to secondary γ' particles of about the same size as that of the electron beam used for measurements. The calculated curves match fairly well to the experimental composition profiles for Al, Co, Cr and Mo in Fig. 7 and for Ti and Ta in

Fig. 8 (Scan 1), while the values are too low for W and Re. The same tendency is observed for the [011] direction (scan 2).

Shape–constrained phasefield model

For calculation of IN706 solidification the Ni–Data from Thermo–Tech covering the elements Ni–Al–Co–Cr–Fe–Mo–Nb–Ti–Zr–B–C was used. To reduce computation time the system was reduced to Ni–Cr–Fe–Nb–Ti–C. The diffusion coefficients of all elements were set to $1.0 \cdot 10^{-9}$ m^2/s in the liquid and to $1.0 \cdot 10^{-12}$ m^2/s in all solid phases with the exception of carbon in γ' and MC carbides. For these phases the diffusivity was chosen to be $1.0 \cdot 10^{-10}$ m^2/s.

Nucleation was allowed for MC carbides, Laves and η phase using a critical undercooling of 5 K for carbides and 2 K for the other phases. Under the applied conditions only carbides and Laves phase occurred in simulation while a micrograph of the interdendritic region shows also some η needles at the border of the Laves phase (Fig. 11).

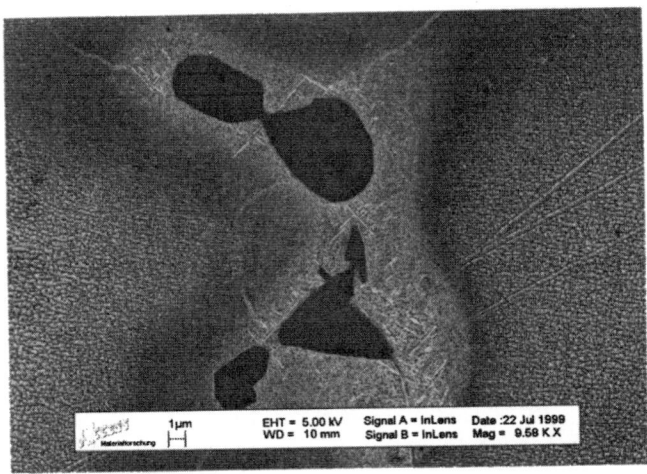

Fig. 11: Micrograph of the interdendritic region of a IN706 sample showing Laves phase and η needles.

Fig. 12 shows the calculated microstructure and composition distribution for C, Cr, Fe, Nb and Ti for a simulation which corresponds to the experimental conditions of the IN706 solidification experiments ($dT/dt=-0.5$ K/s, $\lambda_1=200$ μm). Due to their different chemical compositions carbides and Laves phase which formed at the end of solidification in the remaining interdendritic melt can easily be distinguished from the primary dendrite. In this simulation MC–carbides show the tendency to form diffuse patterns as they grow at the same velocity as the primary dendrite. This could be a hint for eutectic growth conditions.

In Fig. 13–16 the EDX composition profiles of the IN706 sample (Fig. 6) are compared to the calculation in Fig. 12. Although there is a qualitative agreement remarkable differences can be

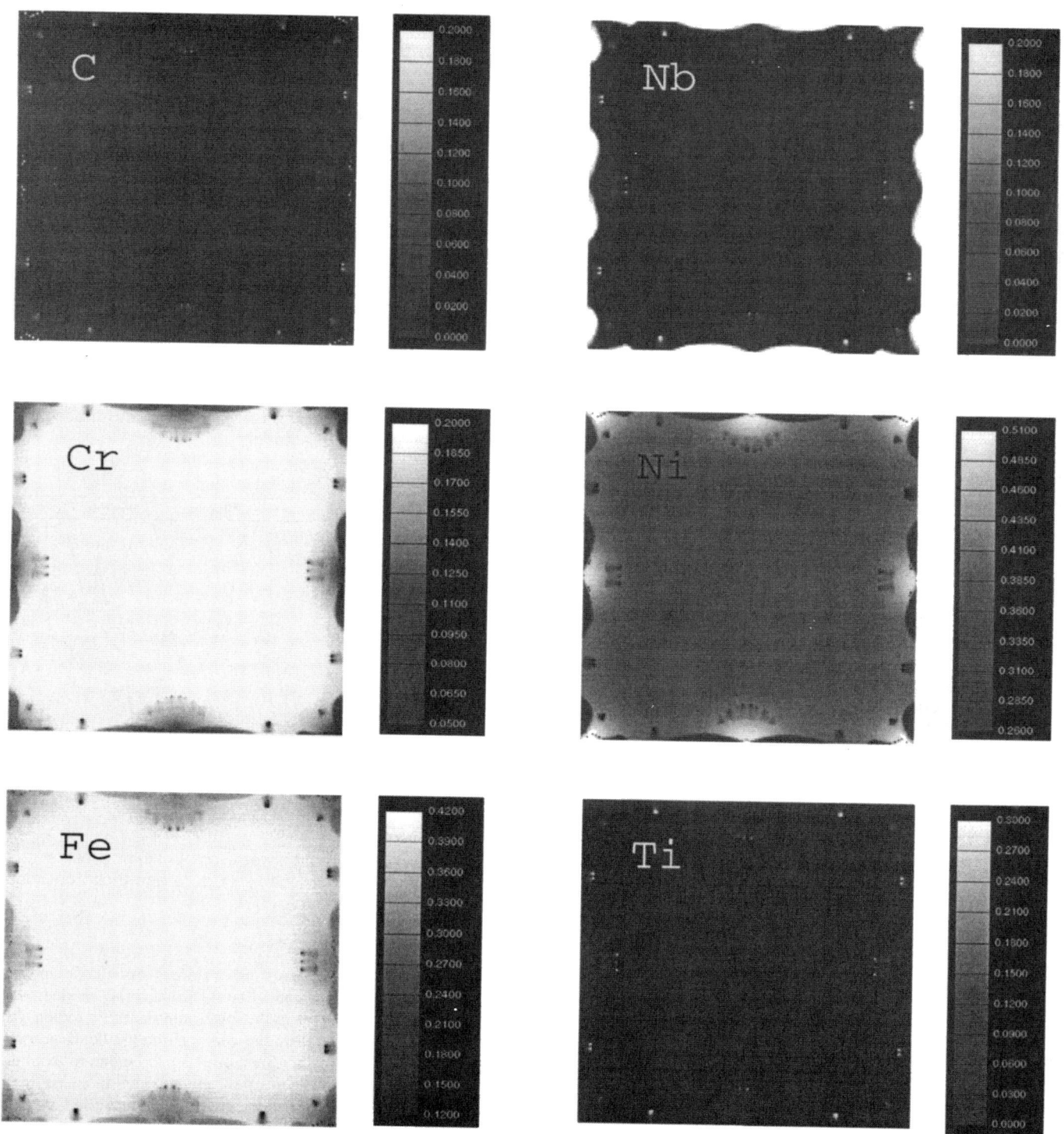

Fig. 12: Composition distribution of C, Cr, Fe, Nb, Ni and Ti in weight fractions calculated using from the SCPF–Model

seen between measured and calculated profiles. This is primarily caused by the position of precipitates, which heavily influence the composition due to their strong segregation. While the experimental Scan 1 does not cross any precipitate (Fig. 6) carbides and Laves phase are formed in this line in simulation. On the contrary Scan 2 shows two precipitation regions.

With EDX no carbides were found at all on the IN706 sample. The assumption for simulation, that carbides form at the γ–liquid interface with a nucleation undercooling of 5 K is very arbitrary, but affects strongly the simulation results.

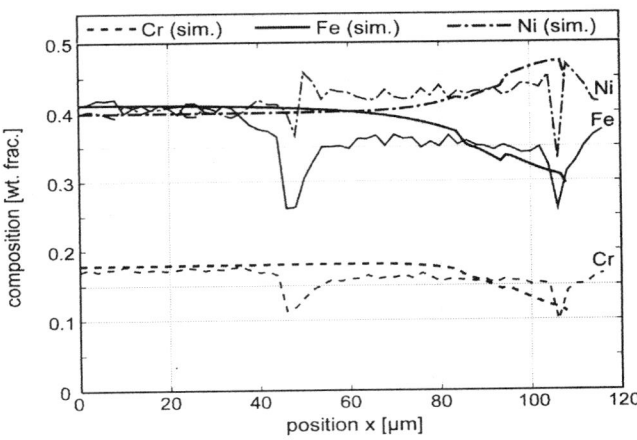

Fig. 15: Comparison between composition profiles of Ni, Fe and Cr obtained from scan 2 (Fig. 6) and those calculated using the SCPF-Model.

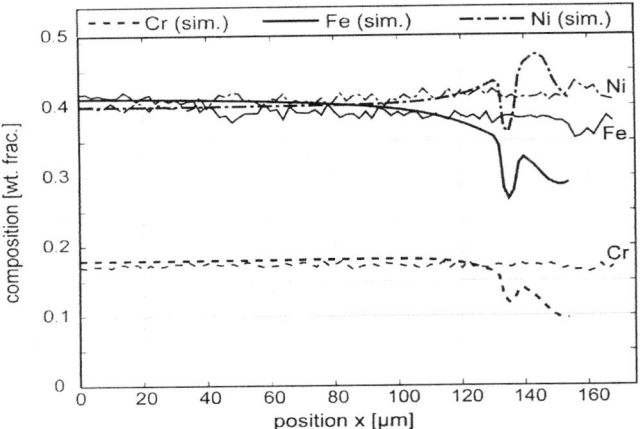

Fig. 13: Comparison between composition profiles of Ni, Fe and Cr obtained from scan 1 (Fig. 6) and those calculated using the SCPF-Model.

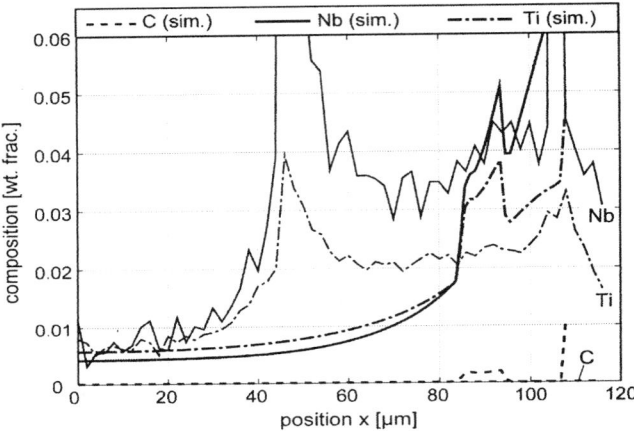

Fig. 16: Comparison between composition profiles of Nb and Ti obtained from scan 2 (Fig. 6) and those calculated using the SCPF-Model.

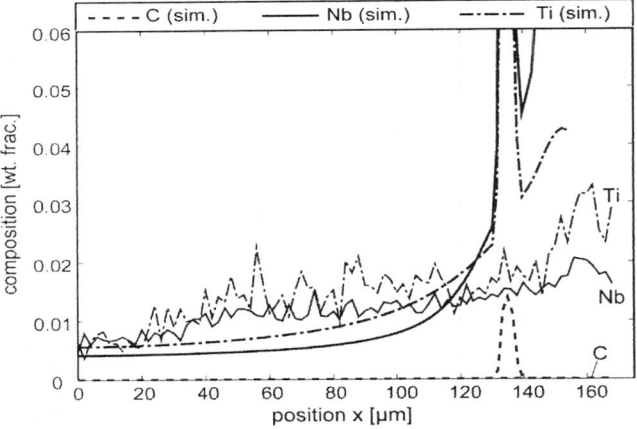

Fig. 14: Comparison between composition profiles of Nb, and Ti obtained from scan 1 (Fig. 6) and those calculated using the SCPF-Model.

In the experimental Scan 1 all elements except Cr show an unsteady behavior at about 40 µm which could be due to the formation of precipitates in the lower interdendritic region. Also at the edge of the dendrite the composition gradients generally seem to be overestimated. Such effects can be due to axial diffusion in the interdendritic melt and are not included in the two–dimensional SCPF–model. Another explanation could be a certain homogenization during the slow cooling process after complete solidification, which has not been taken into account in simulation.

Discussion

The agreement between the results of the SX−Model and the WDX profiles obtained from the CMSX−4 sample is quite good taking into account the oscillations in the WDX signal due to secondary γ' particles. This is ensured by using thermodynamic data which are based on a huge number of experiments and by incorporating a reasonable shape function for the dendrite into the model. The model is restricted to systems where multicomponent Scheil conditions hold, and formation of interdendritic precipitates is not included.

For systems like IN706, where precipitates of very different composition compared to the primary dendrite can heavily influence the microsegregation profiles and where interstitial elements exhibit considerable back diffusion the more refined Shape−Constrained Phasefield (SCPF−)Model is more adequate.

Due to the complexity of the system no perfect match of simulation and experiment should be expected. Too many factors like diffusion coefficients and nucleation undercoolings are unknown. Nevertheless a principially good agreement with the experimental data can be seen in Fig. 13−16 keeping in mind the strong effect of the precipitation position which are different in simulation and experiment and the fact that no carbides were found at all in the micrographs. They could have formed in the deeper interdendritic melt and reduced the carbon content in the measured cross section through axial diffusion. Other reasons for the deviations are unknown parameters, which have a strong influence on the results of the simulation. These are physical variables like diffusion coefficients, which are only estimated due to the lack of experimental data, the critical nucleation undercooling of the different phases or the interfacial energies, which determine the curvature undercooling. These parameters in future could partially be obtained by fitting the calculation results to the experiment.

Outlook

For validation of the SCPF−Model further experiments at different conditions and with different alloys will be done. Of special importance with IN706 is the role of carbides which will be studied with carbon−enriched samples. On this basis more reasonable nucleation conditions in simulation should lead to a better agreement between simulation and experiments.

Compared to the Phasefied Model[18,19] the use of the shape function concept in the SCPF−Model allows a dramatic reduction of the calculation effort. This is mainly due to the online coupling of thermodynamic data to obtain the driving force for the growing dendrite, which is considerably reduced by use of shape functions. But as the shape function approach is compatible to the Phasefield algorithm, the latter could be used just for secondary phases while the primary dendrite is further treated as shape function. With this combination homogenization and secondary precipitation after complete solidification could be included which would greatly enhance the capability of the model for microstructure prediction and alloy design.

Acknowledgment

The authors wish to thank DFG (Deutsche Forschungsgesellschaft) for financial support and ABB for supplying the database for CMSX−4 simulations.

References

1. H. D. Browdy and M. D. Flemings, "Solute Redistribution in Dendritic Solidification", Trans. AIME, 236 (1966) 615−624.

2. D. J. Allen, J. D. Hunt, "Diffusion in the Semi−Solid Region During Dendritic Growth", Metall. Trans. A, 10A (1979) 1389−1397.

3. A. Roósz, E. Halder and H. E. Exner, "Numerical Calculation of Microsegregation in Coarsened Dendritic Microstructures", Mater. Sci. Technol., 2 (1986) 1149−1155.

4. T. Kraft, M. Rettenmayr and H. E. Exner, "An Extended Numerical Procedure for Predicting Microstructure and Microsegregation of Multicomponent Alloys", Modelling Simul. Mater. Sci. Eng. 4 (1996) 161−177.

5. T. Himemiya and T. Umeda, "Solute Redistribution Model of Dendritic Solidification Considering Diffusion in Both the Liquid and Solid Phases", ISIJ International, 38 (1998), No. 7, 730−738.

6. D. Ma and P. Sahm, "Microsegregation in Cellular Microstructure", Metall. Trans. A, 23A (1992) 3377−3381.

7. D. Ma, "Modellierung des Dendritenwachstums und der Mikroseigerung bei gerichteter Erstarrung, Teil I: Entwicklung eines analytischen Mikromodells", Gießereiforschung, 1 (1998) 29−34.

8. T. Kraft and H. E. Exner, "Numerische Simulation der Erstarrung, Teil2: Mikroseigerung in ternären und höherkomponentigen Legierungen", Z. Metallkd. 87 (1996) 652−660.

9. A. J. W. Ogilvy and D. H. Kirkwood, "A Model for the Numerical Computation of Microsegregation in Alloys", Appl. Sci. Res. 44 (1987) 43−49.

10. A. Roósz and H. E. Exner, "Ternary Restricted−Equilibrium Phase Diagrams − I. A First Report: General Principles and Definitions; II. Practical Application: Aluminium−Rich Corner of the Al−Cu−Mg System", Acta metall. mater. 38 (1990) 2003−2008, 2009−2016.

11. N. Saunders and A. Miodownik: CALPHAD, Calculation of Phase Diagrams, A Comprehensive Guide; Elsevier S. Ltd. 1998, ISBN 0−08−042−129−6.

12. B. Sundman, B. Jansson and J. O. Anderson, "The Thermo-Calc Databank System", CALPHAD, 9 (1985) 153−190.

13. N. Saunders, "Phase Diagram Calculations for Ni−Based Superalloys", Superalloys 1996, (Warrendale, PA: Met. Soc. AIME., 1996), 101−110.

14. D. A. Kessler, J. Koplik and H. Levine, "Steady–State Dendritic Crystal Growth", Phys. Rev. A, 31 (1986) 3352–3357.

15. J. D. Hunt, Solidification and Casting of Metals, Vol. 192. The Metal Society, London.

16. D. Goldschmidt, "'Einkristalline Gasturbinenschaufeln aus Nickelbasis–Legierungen, Teil I", Mat.–wiss. u. Werkstofftech., 25 (1994) 311–320.

17. U. Grafe et al., "Calculation of Microsegregation for the Directionally Solidified Superalloy CMSX 4 Using a Pseudo 2–Dimensional Model", Modelling of Casting, Welding and Advanced Solidification Processes VIII, The Minerals, Metals & Materials Society, 1998, 227–234.

18. J. Tiaden et al., "The multiphase–field model with an integrated concept for modelling solute diffusion", Physica D, 115 (1997) 73–86.

19. I. Steinbach et al., "A Phase Field Concept for Multiphase Systems", Physica D, 94 (1996) 135–147.

DEVELOPMENT OF A THIRD GENERATION DS SUPERALLOY

T.Kobayashi, M.Sato, Y.Koizumi, H.Harada, T.Yamagata,
A.Tamura*, and J.Fujioka*

High Temperature Materials 21 Project, National Research Institute for Metals
1-2-1 Sengen, Tsukuba Science City, Ibaraki 305-0047, Japan
*Kawasaki Heavy Industries, Ltd.
2-4-1 Hamamatsu-cho, Minato-ku, Tokyo 105-6116, Japan

Abstract

A third generation DS superalloy with creep strength as high as the second generation SC superalloys was developed. The alloy TMD-103 was designed by simply adding grain boundary strengthening elements, carbon and boron, to a third generation SC superalloy TMS-75 containing 5wt% Re. The alloy was cast to columnar grained directionally solidified structure with solidification rate at 200mm/h. Creep test was performed after solution and aging heat treatments. Typical creep-rupture lives at 900°C-392MPa and 1040°C-137MPa being 519h and 884h, respectively, which were equivalent with CMSX-4. Burner rig tests showed good hot corrosion and oxidation resistance. A DS hollow blade of a 2000KW class industrial gas turbine was successfully cast with TMD-103.

Introduction

The production cost of the advanced single crystal (SC) superalloys are becoming higher due to the complicated casting and heat treatment conditions and the resulting lower yields. It is practically very important to develop superalloys with higher cost-performance. The production cost of directionally solidified (DS) columnar grained superalloys is considerably lower than that of SC superalloys. However the creep strengths of the present DS superalloys are not competitive with the advanced SC superalloys. The so-called second generation DS superalloys with 3wt% Re, for example CM186LC [1], which is the strongest DS superalloy existing, are not stronger than the first generation SC superalloys.

In the present paper, a new class of DS superalloy with a high cost-performance, namely, the third generation DS superalloy is reported. The microstructures, creep and fatigue properties, and hot corrosion and oxidation properties are evaluated comparing with the present DS and SC superalloys.

Alloy Design

The third generation SC superalloy TMS-75, that was developed by some of the authors [2], was used as the base alloy. This alloy has excellent processability and phase stability as well as high creep strength and hot corrosion resistance. In this study, we simply added 0.07wt% carbon (C) and 0.015 wt% boron (B) to the TMS-75 as the grain boundary strengthening elements. The chemical compositions of the alloys are given in Table I with some typical DS and SC superalloys as references.

The carbon forms carbides in superalloys. It is estimated by our alloy design program (NRIM-ADP) that MC and M_6C carbides are formed in TMD-103. It is also predicted that the MC carbide contains Ta as the main part of 'M'. In the M_6C, W and Mo are predicted to occupy 'M'. The behavior of Re can not be predicted due to the lack of experimental data.

The carbide formations result in slight decreases of the solid solution strengthening elements in the γ and γ' phases. Also the γ' fraction in the TMD-103 is predicted to become slightly smaller than TMS-75. However, the effects of these microstructural changes on the creep strength were estimated to be small.

Superalloys 2000
Edited by T.M. Pollock, R.D. Kissinger, R.R. Bowman,
K.A. Green, M. McLean, S. Olson, and J.J. Schirra
TMS (The Minerals, Metals & Materials Society), 2000

Table I Chemical compositions (wt%) of TMD-103 and the base alloy TMS-75, with those of typical DS and SC superalloys.

	Ni	Co	Cr	Mo	W	Al	Ti	Ta	Hf	Re	C	B	Zr
TMS-75 (3rd SC)	Bal.	12.0	3.0	2.0	6.0	6.0	-	6.0	0.1	5.0	-	-	-
TMD-103 (3rd DS)	Bal.	**12.0**	**3.0**	**2.0**	**6.0**	**6.0**	-	**6.0**	**0.1**	**5.0**	**0.07**	**0.015**	-
CMSX-4 (2nd SC)	Bal.	9.0	6.5	0.6	6.0	5.6	1.0	6.5	0.1	3.0	-	-	-
CM186LC (2nd DS)	Bal.	9.0	6.0	0.5	8.0	5.7	0.7	3.0	1.4	3.0	0.07	0.015	0.005
CM247LC (1st DS)	Bal.	9.2	8.1	0.5	9.5	5.6	0.7	3.2	1.4	-	0.07	0.015	0.015
IN792Hf (1st DS)	Bal.	8.9	12.3	1.8	4.4	3.4	3.9	4.0	1.0	-	0.12	0.014	0.050

Casting and Heat Treatment Characteristics

A TMS-75 master heat bar was melted and, after the C and B additions, cast to DS bars of 10mm diameter with solidification rate at 200mm/h. The microstructure was examined after single-step solution treatment at temperatures between 1225 and 1300°C for 2h to find the optimum solution temperature. The DS bars were solution heat treated under thus selected standard condition; the solution treatment was performed at 1275°C for 5h after heating at 1225°C for 1h as a pre-homogenization to avoid any chance of incipient melting. A two-step aging treatment was performed, first at 1150°C for 4h and second at 870°C for 20h, both followed by air-cooling.

Figure 1 shows the microstructures after solution heat treatment at 1275°C and 1300°C. It is shown that almost all the γ' precipitates are dissolved at 1275°C and re-precipitated as fine γ', except for the eutectic γ' and the grain boundary γ'. At 1300°C, a perfect solution including the eutectic and grain boundary γ' is achieved. However, a small amount of incipient

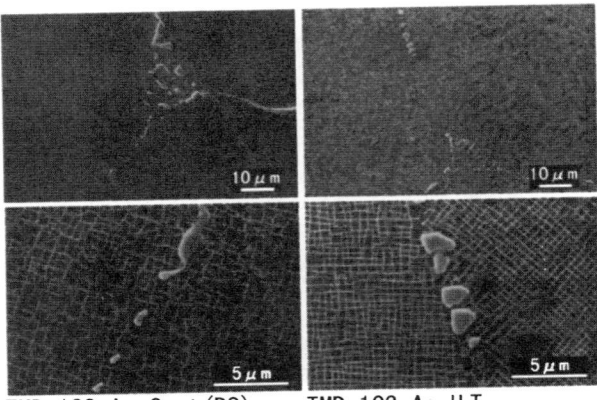

TMD-103 As Cast (DS)   TMD-103 As H T

Figure 2: SEM micrographs of as cast and as standard heat treated TMD-103 DS samples.

melting started to occur involving carbides, resulting in cavities formed during the following solidification. From these results, 1275°C was selected as the standard solution heat treatment temperature.

Figure 2 shows the microstructures in as cast and as standard heat treated (including aging) samples. The γ' precipitates became finer and their alignment became better after the standard heat treatment. The γ' size in as cast sample is rather small and of fairly regular shape. This must be due to the high Re content; Re reduces the γ' size because of its low diffusibity and high γ/γ' partitioning coefficient, e.g., 10:1 [4], both effectively prevent the γ' growth. Carbide(s) with bright contrast are observed mostly at grain boundaries and also in the interdendritic regions. The bright contrast suggests that the carbide(s) contain elements having higher atomic numbers, such as Ta, W, Mo and possibly Re.

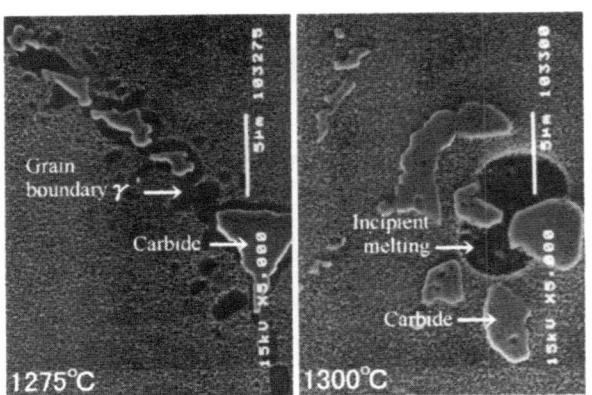

Figure 1: SEM micrographs of TMD-103 DS after solution heat treatment at 1275°C and 1300°C.

The castability was evaluated with a DS hollow blade for a 2000KW class industrial gas turbine as shown in Figure 3. The

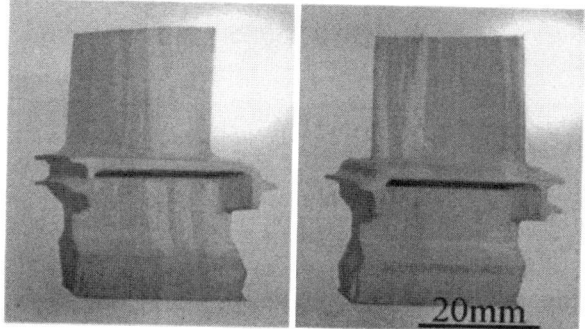

Figure 3: A DS hollow blade for an industrial gas turbine, cast with TMD-103.

solidification rate was again 200mm/h. The normal inspection was carried out to examine the castings. It was found that sound castings were obtained without any defects. There are no difficulties so far in production of such DS blades.

Mechanical Properties

Creep tests were performed in short-term and long-term to estimate the strength level and phase stability with

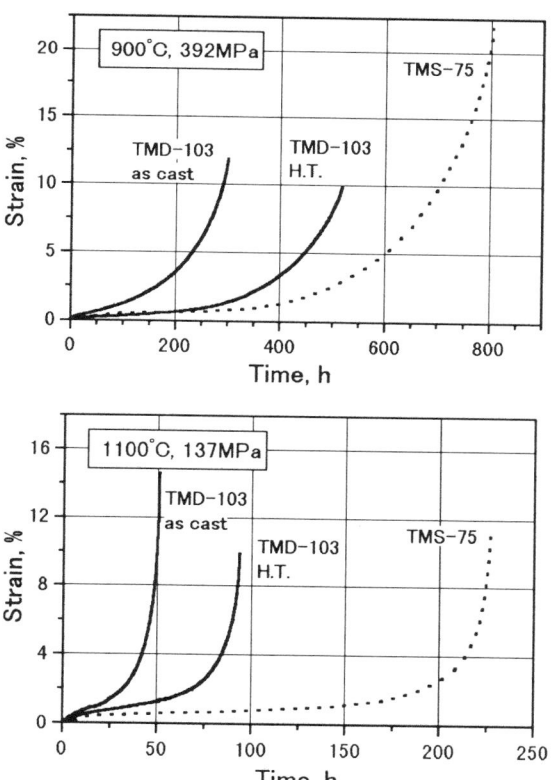

Figure 4: Creep curves of as cast and heat treated (HT) TMD-103 DS longitudinal samples, with those of TMS-75 SC.

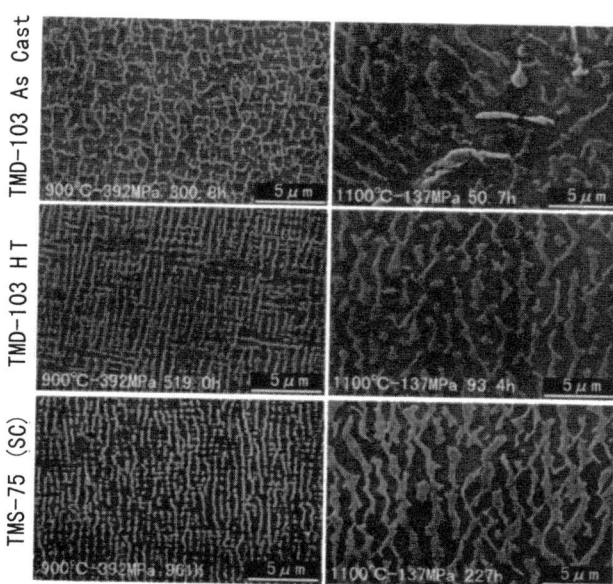

Figure 5: Microstructures in creep ruptured samples.

temperatures ranging 900-1100°C. A newly developed optical monitoring system was used for measuring the high temperature creep strain. Low cycle fatigue tests were also conducted at 950°C in air with R ratio of -1 and strain rate of 0.1%/sec.

Creep Test (Short-Term)

In Figure 4 the creep curves of as cast and heat treated TMD-103 DS longitudinal samples are compared with those of TMS-75 SC. The creep-rupture lives of heat treated DS samples are about 40% to 60% of those in TMS-75.

Figure 5 represents the microstructures in the creep-ruptured samples. The creep conditions and rupture lives are given in the figure. The microstructures in TMD-103 as cast samples

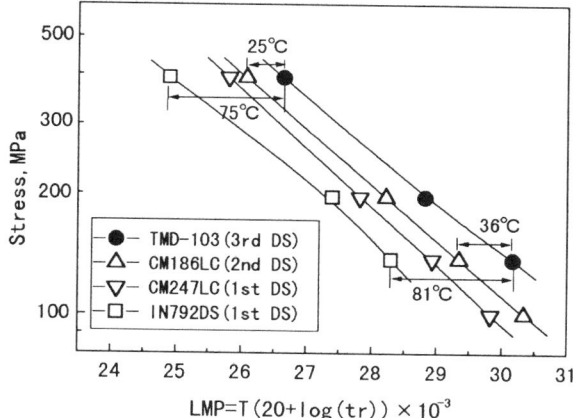

Figure 6: Comparison of creep-rupture strengths among first, second, and third generation DS superalloys.

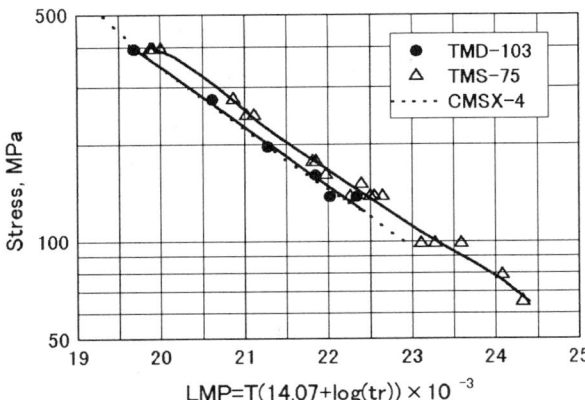

Figure 7: Comparison of creep-rupture strengths including long-term data with re-fitted Larson Miller Parameter (C=14.07).

are coarser than the others in the both creep conditions, which must cause the shorter creep-rupture lives. The heat treated DS sample has very similar microstructures to those in TMS-75 SC but has shorter creep-rupture lives. This should be attributed to the remaining undissolved eutectic γ' phase and the presence of grain boundaries in the TMD-103 DS samples. In Figure 6 a comparison of creep-rupture strengths is made among first, second, and third generation DS superalloys. It is demonstrated that the third generation DS superalloy TMD-103 has 25 to 36°C higher temperature capability than the second generation DS superalloy CM186LC. When comparison is made with IN792Hf DS the capability of our alloy is higher by 75 to 81°C.

Creep Test (Long-Term)

In Figure 7 the creep-rupture results of TMD-103 and TMS-75 including some long-term data over 1000h are compared with the CMSX-4 data with re-fitted Larson Miller Parameter at

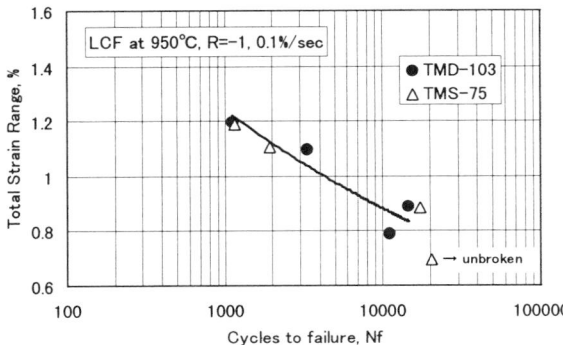

Figure 8: Low cycle fatigue failure data of TMD-103 DS and TMS-75 SC at 950°C.

C=14.07 [5]. The longest data of TMD-103 and TMS-75 are about 6300h and 7800h, respectively. The creep-rupture strength of TMD-103 was found to be as strong as the second generation SC superalloy CMSX-4 for thousands hours.

Low Cycle Fatigue Test

Figure 8 shows the low cycle fatigue failure results of TMD-103 and TMS-75. Although data points were not enough and scattering, the low cycle fatigue property of TMD-103 was almost as good as SC superalloy TMD-75.

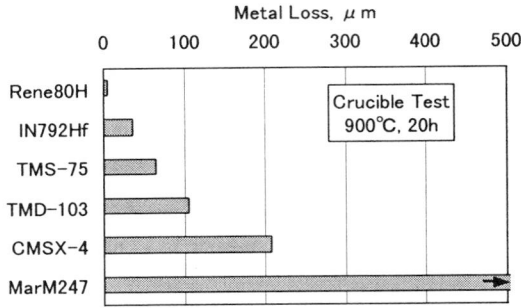

Figure 9: Crucible test results at 900°C, 20h with 75%Na$_2$SO$_4$ + 25%NaCl mixed salt.

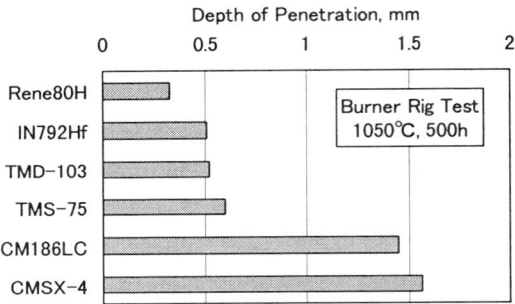

Figure 10: Burner rig hot corrosion test results at 1050°C, 500h with kerosene fuel and corrosive ingredients (S, Cl, Na, etc.).

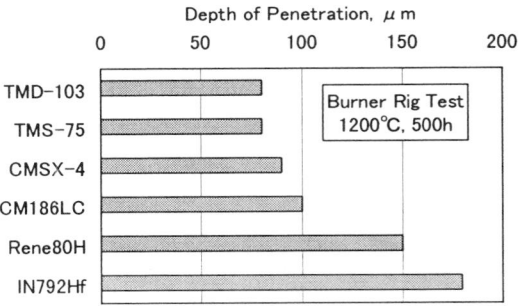

Figure 11: Burner rig oxidation test results at 1200°C, 500h with kerosene fuel only.

Environmental Properties

In high temperature corrosion tests, varied results are often given at the different condition. One crucible test and two burner rig tests were conducted to evaluate the environmental properties of TMD-103 relatively.

Crucible test were performed at 900°C for 20h with 75%Na$_2$SO$_4$ + 25%NaCl mixed salt in air atmosphere.

Burner rig tests were performed with a hot corrosive condition and an oxidative condition. In the hot corrosion test, to simulate a severe hot corrosive environment in a heavy duty industrial gas turbine, some sulfuric oil and artificial seawater were added into kerosene burner at gas temperature of 1050°C and exposure of 500h. The oxidation test was run at 1200°C for 500h without the addition of corrosive ingredients.

Alloys IN792Hf, CMSX-4, etc., were also tested under the same condition as references.

Crucible Test

In Figure 9 the crucible test results are compared with some reference alloys. TMD-103 showed a good corrosion resistance, forming a protective scale like a Cr$_2$O$_3$-former type in spite of the low Cr content.

Burner Rig Test (Hot Corrosion)

In Figure 10 the burner rig hot corrosion test results are represented with several DS and SC superalloys. There is a general tendency that the alloy with higher Cr content has higher hot corrosion resistance. However, TMD-103 which has only 3wt% Cr showed the hot corrosion resistance comparable with IN792Hf containing 12.5wt% Cr. This is due to the 5wt% Re content. As already reported [6], Re is effective in improving hot corrosion resistance as well as creep strength.

Burner Rig Test (Oxidation)

Figure 11 shows the burner rig oxidation test results. At higher temperature, alloys with lower Cr and higher Al content tend to provide the good resistibility. On the other hand, most of corrosion-resistible alloys with higher Cr content, as Rene80H and IN792Hf, show poor resistibility at such conditions. These behaviors are generally explained with the balance of oxidation and sulfidation, and with the scale type of Al$_2$O$_3$-former or Cr$_2$O$_3$-former.

In this condition TMD-103 showed also a good performance. These results represent that TMD-103 has a good resistibility in both the hot corrosive and oxidative environment.

Conclusions

A third generation DS superalloy TMD-103 was designed as a high cost-performance superalloy and evaluated experimentally. The following results have been obtained;

(1) TMD-103 has about 36°C higher temperature capability than the second generation DS superalloy CM186LC at stress level of 137MPa.

(2) Long-term creep-rupture strength of the alloy is equivalent with second generation SC superalloy CMSX-4.

(3) Low cycle fatigue property of the alloy is almost the same as that of TMS-75 SC at 950°C in air.

(4) Hot corrosion resistance of the alloy is found to be equivalent with IN792Hf (12.5wt%Cr) in spite of the low (3wt%) Cr content.

(5) Oxidation resistance of the alloy is as good as or better than CMSX-4.

(6) Heat treatment of the alloy is simple, and a DS hollow blade for an industrial gas turbine was successfully cast with the alloy.

TMD-103 was thus thought to be promising as a high cost-performance turbine blade material for industrial gas turbines, marine gas turbines, and aero-engines.

Further work on modification of TMD-103 is being conducted.

Acknowledgments

This work is a part of the KHI-NRIM collaborative research in the High Temperature Materials 21 Project.

The authors wish to thank Dr.N.Akikawa, Mr.T.Takenaka, and Mr.H.Ishihara of KHI (Akashi Works) for their conducting the burner rig tests.

The authors also acknowledge the ROSS & CATHERALL LTD. for supplying the melting stock.

References

1. K.Harris, et al., "Development of the Rhenium Containing Superalloys CMSX-4 & CM186LC for Single Crystal Blade and Directionally Solidified Vane Applications in Advanced Turbine Engines", Superalloys 1992 (Warrendale, PA: TMS, 1992), 297-306.

2. T.Kobayashi, et al., "Design of High Rhenium Containing Single Crystal Superalloys with Balanced Intermediate and High Temperature Creep Strengths", Advances in Turbine Materials, Design and Manufacturing (Newcastle upon Tyne, U.K., 4-6 November 1997), 766-773.

3.   H.Harada, et al., "Alloy Design for Nickel-base Superalloys", High Temperature Alloys for Gas Turbines 1982 (Liege, Belgium, 4-6 October 1982), 721-735.

4.   H.Harada, et al., "Phase Calculation and Its Use in Alloy Design Program for Nickel-base Superalloys", Superalloys 1988 (Warrendale, PA: TMS, 1988), 733-742.

5.   C.K.Bullough, et al., "The Characterization of The Single Crystal Superalloy CMSX-4 for Industrial Gas Turbine Blading Applications", Materials for Advanced Power Engineering 1998 (Liege, Belgium, 5-7 October 1998), 861-878.

6.   G.L.Erickson, "The Development and Application of CMSX-10", Superalloys 1996 (Warrendale, PA: TMS, 1996), 35-44.

The Development and Long-Time Structural Stability of a Low Segregation Hf Free Superalloy - DZ125L

Yaoxiao Zhu[1], John F. Radavich[2], Zhi Zheng[1], Xiuzhen Ning[1], Langhong Lou[1], Xishan Xie[3], Changxu Shi[4]

(1) Institute of Metal Research, Chinese Academy of Sciences, Shenyang 110015
(2) Micro-Met Laboratories, Inc., West Lafayette, IN 47906, USA
(3) University of Science & Technology Beijing, Beijing 100083, China
(4) National Natural Science Foundation of China, Beijing 100083, China

Abstract

Solidification studies of cast superalloys carried out at the Institute of Metals Research showed that the trace elements P, Si, B, and Zr had a profound effect on the resultant as-cast structure. These trace elements lowered the liquidus and incipient melting temperatures as well as enhanced solidification segregation.

The control of the P, Si, B, and Zr levels, called low segregation technology, can raise the incipient melting temperature as to provide a higher service temperature of 20-25°C and reduce the segregation to enhance long-time stability at 900°C.

A new Hf free DS alloy has been developed by the low segregation technology called DZ125L. It shows rupture strengths two to three times longer than the corresponding conventional alloys either at high or medium temperatures in both longitudinal or transverse directions. These properties are attributed to less shrinkage porosity due to a smaller liquidus-solidus gap and lower segregation.

Introduction

At the present time and in the future, superalloys will still be the "workhorse" of materials for jet engines and industrial gas turbine engines. Current studies are underway to raise the service temperature of existing alloys by melt practice, design, or new coatings. However, the melting point limits the temperature of application of superalloys and the addition of large amounts of refractory elements for high temperature strength lowers the melting point and increases segregation which affects the stability of the alloy during service.

Solidification studies of superalloys have been carried out since the sixties at the Institute of Metal Research. It was found that some trace elements, notably P, Si, B, and Zr, affected the segregation of these elements resulting in lower incipient melting points. A number of new superalloys have been developed based on the principle of trace element control (1-3). A most recent application of low segregation technology has been in the field of directionally solidified alloys.

DS superalloy blades were developed in the early sixties to take advantage of no grain boundaries perpendicular to the direction of applied stress, which resulted in higher creep and rupture strengths. However, because of the high refractory element content, the DS blades showed erratic third stage creep, low ductility, and exhibited transverse cracks along the grain boundaries. This detrimental behavior was solved by the addition of 1-2% Hf (4,5).

Hf is a rare and expensive element and reacts strongly with oxygen to form HfO either during the melt or remelt cycles. A number of studies have been carried out on the effects of Hf on the microstructure and mechanical properties of a number of superalloys (6,7). The objective of this paper is to describe the results of applying the low segregation technology to the development of a Hf free DS superalloy.

329

Superalloys 2000
Edited by T.M. Pollock, R.D. Kissinger, R.R. Bowman,
K.A. Green, M. McLean, S. Olson, and J.J. Schirra
TMS (The Minerals, Metals & Materials Society), 2000

Alloy Development

To study the effects of trace elements P and Si on the solidification and mechanical properties of a Ni base alloy with high refractory content, a series of alloys were made with varying amounts of P and Si and stress-rupture tested at 760°C and 804 MPa. The results of the tests are given in Table 1.

Based on these results, a low segregation alloy called DZ125L was developed similar in composition to René 125 except the Hf and Zr were removed and the B, P, and Si contents were reduced to very low levels as shown in Table 2.

The nominal composition of René 125 is shown in Table 2. Although René 125 was put in a DS form in order to compare it to a Hf free composition, DZ125L, René 125 is a Ni base superalloy developed by GE and is used in equiaxed turbine blades. The development and application of René 125 has been described by P. Aldred (8).

The René 125 and DZ125L compositions were directionally solidified in a conventional DS furnace. Samples of DS René 125 and DZ125L alloys were stress-ruptured tested in the longitudinal and transverse directions. The results are given in Table 3.

The results show a much longer life in stress-rupture for the DZ 125L alloy. To understand and explain the reasons for the longer stress rupture life of the DZ 125L alloy compared to that of the DS René 125 alloy, a solidification study was carried out on both alloys as well as a long time structural stability program.

Solidification Behavior of DZ125L and DS René 125

In the solidification study, a 10 mm cube of each alloy was melted in a graphite boat and heated to 1420°C in a SiC tube furnace and held for 5 minutes to ensure a total molten condition. It was then cooled down slowly to a preset temperature, held for 10 minutes and water quenched. The percent of liquid and $(\gamma+\gamma')$ eutectic structure of the solidified material was determined under an optical microscope. Table 4 shows the results of this study.

The data in Table 4 shows that DS René 125 begins solidification below 1350°C and $(\gamma+\gamma')$ precipitates below 1230°C. The alloy is finally solidified at 1100°C. For the DZ125L alloy, the liquidus is 1370°C, the $(\gamma+\gamma')$ eutectic precipitates from 1250°C, and the solidus is at 1230°C from which γ' precipitates. The difference in the solidus temperatures of the two alloys can range up to 130°C.

The $(\gamma+\gamma')$ eutectic formation range for the DZ125L is about 30°C while in the DS René 125 alloy the eutectic keeps forming until 1100°C. The amount of eutectic measured is 14.5% for the DS René 125 alloy to only 5% for the DZ125L alloy.

An electron probe study of the final liquid to freeze in the DS René 125 alloy, Table 5 shows the greatest enrichment takes place in the P content followed by Zr, B, and Si. These elements can form low melting point eutectics such as Ni-P and Ni-Zr. Samples that were quenched from 1350°C and 1230°C show the degree of MC formation at 1350°C is more advanced in the DZ125L alloy, Figure 1a, while a large amount of segregation is still present in the DS René 125 alloy at 1230°C, Figure 1b.

The nature of the segregation is quite different in the DS René 125 at 1290°C as compared to the DZ125L alloy as shown in Figure 2. The segregation area shown for the DS René 125 contains small discrete borides and Hf enrichment while the segregation areas in the DZ125L tend to show high Ti liquid giving rise to an MC structure.

Structure Analysis

As-Cast Condition

The as-cast microstructure of the DZ125L alloy consists of two MC phases. The main MC phase is a script form with lessor amounts of a discrete carbide. The MC carbide in the DS Rene 125 is a discrete shaped carbide.

Table 1. Effect of P on the stress rupture life (hr).

P(wt%)	0.0005	0.0022	0.0034	0.0052
760°C/804MPa	336	218	187	126

Effect of Si on the Stress Rupture Life (hr).

Si(wt%)	0.03	0.1	0.2
980°C/235MPa	117	111	59

Table 2. Composition of the Experimental Alloys (wt%)

Alloy	C	Cr	W	Al	Ti	Mo	Co	Ta	Hf	B	Zr	P	Si	Ni
A*	0.11	9	7	4.8	2.5	2	10	3.8	1.8	0.015	0.1	0.005	0.1	Bal
B*	0.11	9	7	4.8	2.5	2	10	3.8	--	0.008	0.05	0.0005	0.05	Bal

Table 3. Stress Rupture Life of DS René 125 and DZ125L Alloys (hr).

Alloy	Rupture Strength (hours)		
	980°C/235MPa(L)**	760°C/804MPa(L)**	760°C/725MPa(T)***
A*	25-50	72-104	28-39
B*	98-135	218-354	62-97

A*= conventional DS Rene'125   B*= low segregation DS Rene'125 without Hf.
L**= longitudinal   T***= transversal

Table 4. Solidification of Two Alloys.

Temperature °C	DS Rene' 125		DZ125L	
	L(%)	(γ+γ')(%)	L(%)	(γ+γ')(%)
1370	100		100	
1350	100		70	
1330	60		35	
1310	40		25	
1290	20		10	
1270	19		6	
1250	18		3	2
1230	17	1	Trace	5
1215	10	7		
1200	8	10		
1180	6	11		
1160	5	12		
1140	3	13		
1120	2	14		
1100	Trace	14.5		

Table 5. Enrichment in the Final Residual Liquid in DS Rene' 125 Alloy.

Element	P	B	Zr	Hf	Si
Alloy(wt%)	0.0029	0.019	0.10	1.15	0.21
In final residual liq.(wt%)	0.8/1.3	0.6/1.2	3.8/5.7	15/21	1.4/3.3
Enrichment rate	275/48	31/63	38/57	13/18	6/15

Table 6. Content of M in MC

Element		Ti	Ta	Hf	W	Mo	Cr	Ni	Co
DZ125L	wt%	19.9	56.7	-	15.9	2.6	1.0	3.4	0.6
	at%	44.6	33.7	-	9.3	3.0	2.1	6.3	1.0
DS	wt%	17.0	62.5	2.1	11.7	1.7	0.7	4.1	0.5
	at%	40.1	39.2	1.3	7.2	2.0	1.4	7.0	1.0

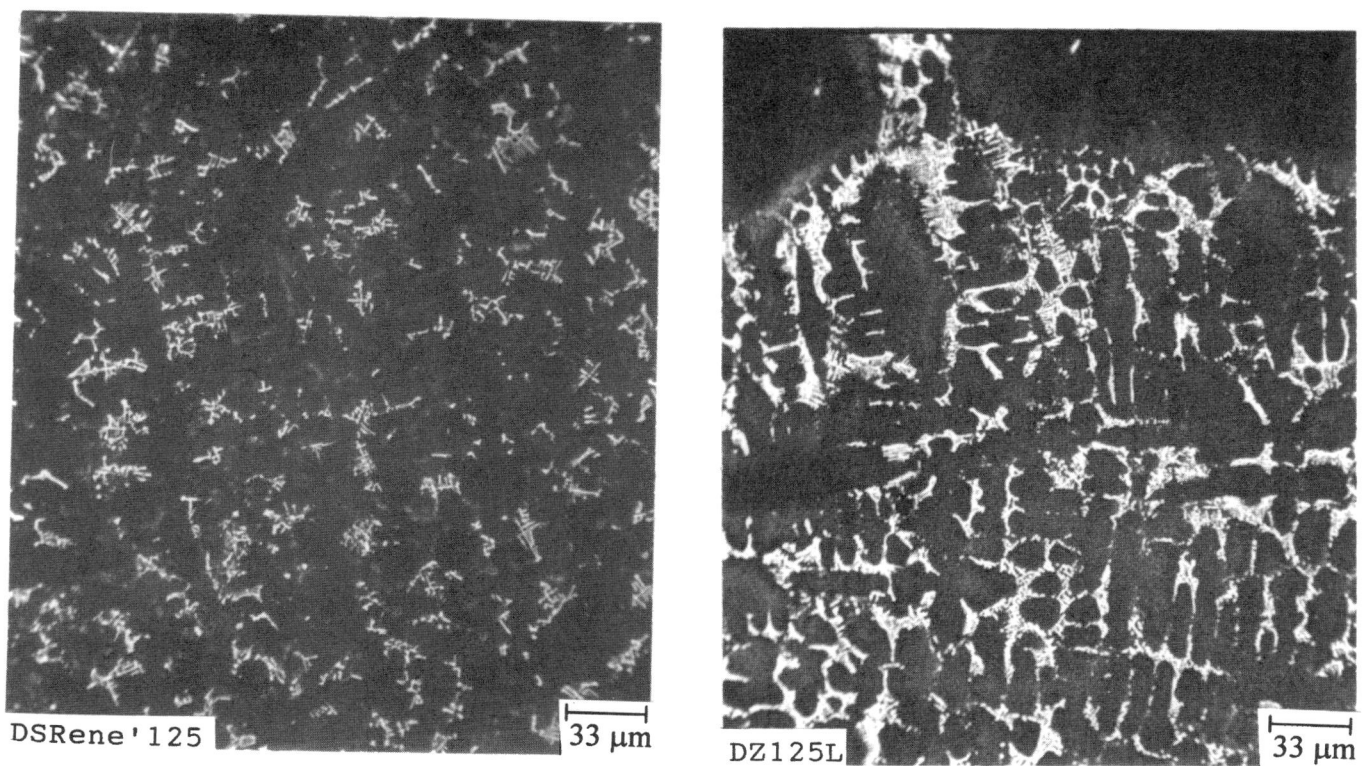

Figure 1a: MC precipitation at1350°C

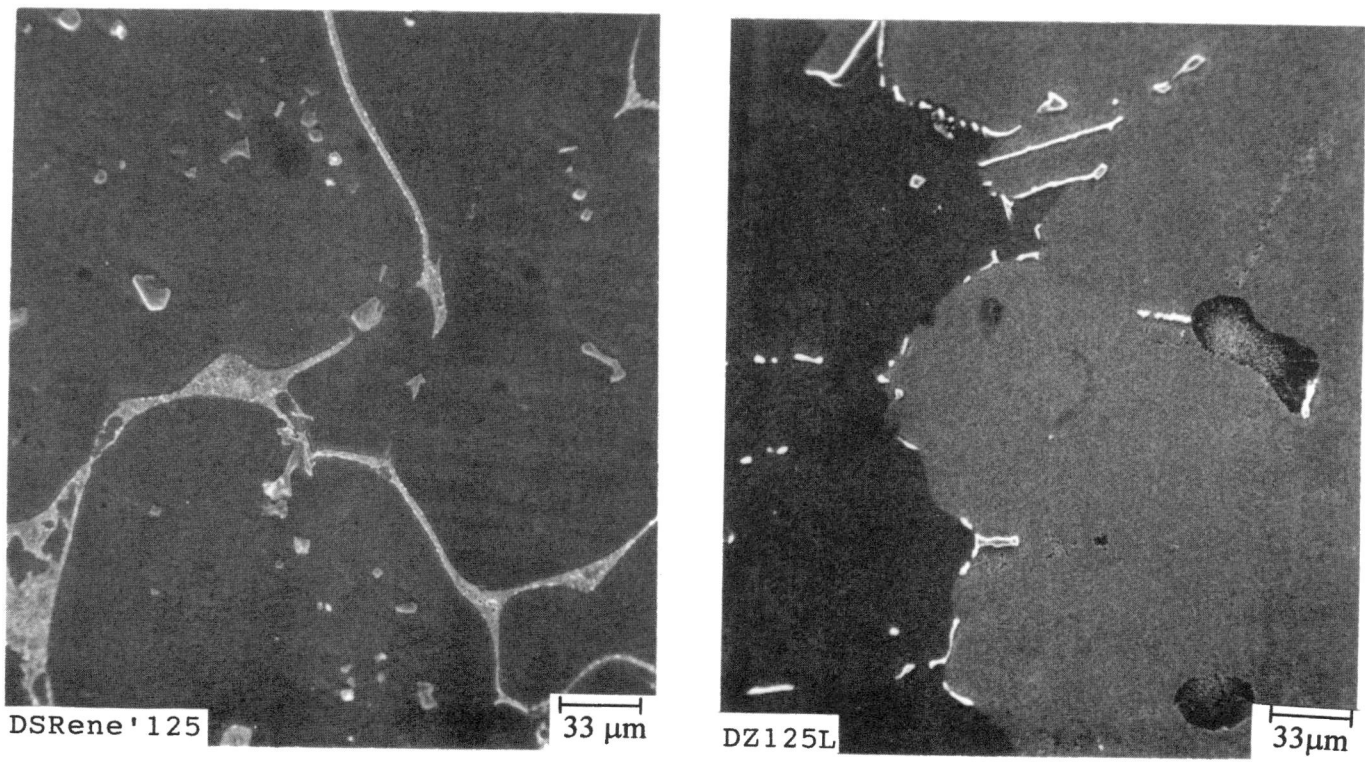

Figure 1b: Segregation structure at 1230°C

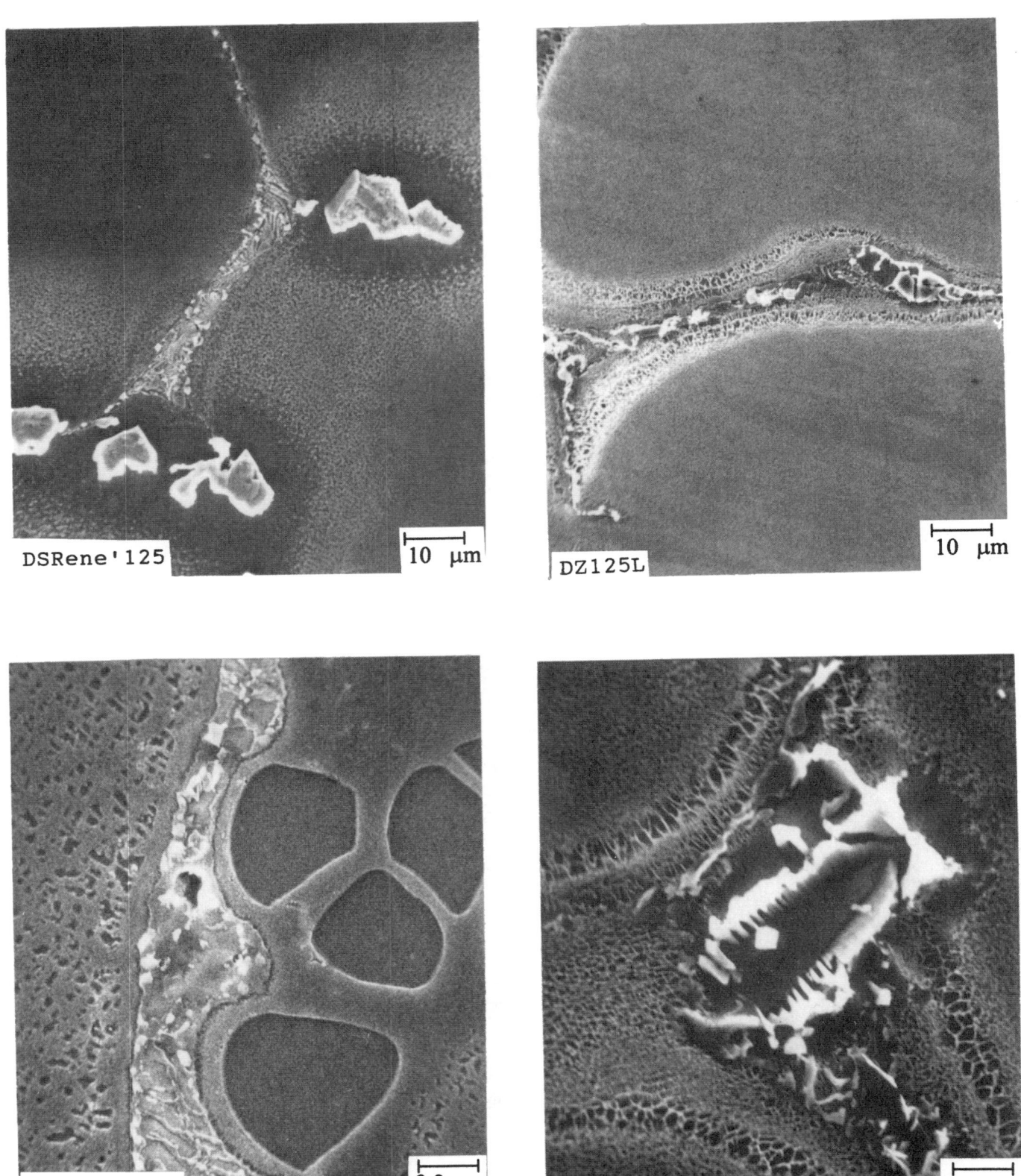

Figure 2: Segregation areas at 1290°C and 1140°C

334

Both alloys contain eutectic or primary γ' islands but the amount of these island on M23s is much greater in the DS René 125 alloy, Figure 3. There appears to be one γ' size in the DS René 125 alloy while two sizes of γ' appear in the DZ 125L in a dendrite pattern.

The amount of γ' forming elements are the same in both alloys but the presence of a large amount of primary γ' islands in the DS René 125 alloy would indicate higher segregation and less possible γ' for strengthening in the DS René 125 alloy. The periphery of the primary γ' islands are known areas where borides and Hf rich carbides form upon high temperature heat treatments.

The chemical analyses of the carbides in both alloys are given in Table 6. High Ti and Ta are present in the carbides in both alloys with minor amounts of Cr, Ni, and Co. However, the carbides in DS René 125 also contain Hf and lessor amounts of W+Mo than those in the Hf free DZ125L. Additional micro probe studies show the discrete carbides in the DZ125L alloy are richer in Ti than the script carbides which would indicate a greater tendency to decompose during heat treatments and/or long-time exposures.

Heat Treatment

Both alloys were given the same heat treatment before testing and long-time stability tests at 900°C. The heat treatment consisted of 1220°C/2 hours-A.C. + 1080°C/4 hours A.C. + 900°C/16 hours A.C.

The resultant structures are shown in Figure 4a. The DS René 125 alloy shows some small precipitation near primary γ' islands due to the 1220°C temperature, $M_{23}C_6$ at the grain boundaries and a uniform γ' phase. Near the primary γ' islands there are areas of finer γ' present. The main γ' precipitation is due to the γ' solutioning at 1220°C and re-precipitation of γ' at 1080°C.

The γ' has not totally solutioned at 1220°C in the DZ125L alloy. The γ' precipitated at the 1080°C temperature is more carboidal in shape than the unsolutioned γ'.

The grain boundaries also show some $M_{23}C_6$ precipitates.

900°C Stability

After 1000 hours exposure at 900°C, the DS René 125 shows more fine precipitates near regions of primary γ'. Some carbides show evidence of decomposition and additional $M_{23}C_6$ is seen at the grain boundaries, Figure 4b. An occasional needle or plate is present in random areas.

In the case of the DZ 125L, the discrete carbides show decomposition resulting in a rosette like formation of $M_{23}C_6$. More $M_{23}C_6$ is present at the grain boundaries and no plate structure is seen.

After 2000 hours exposure at 900°C, there is more discrete $M_{23}C_6$ precipitation in the grain boundaries of each alloy. The γ' phases do not appear to have changed due to the additional 1000 hour exposure as shown in Figure 5a. The most notable change appears to be as large additional formation of the plate phase in the DS Rene 125 and only a few small areas showing plates in DZ 125 L as shown in Figure 5b.

A micro-probe analysis of the plate phase gives in wt.% 39.1W, 15.3Mo, 13.4Cr, 16.9Ni, 11.7Co, 1.3Al, 1Ti and 1.4Ta. The composition suggests the Mu phase which by extraction and x-ray analysis confirms the phase to be Mu.

Phase Identification

The inert carbide and other phases present after 2000 hours at 900°C were extracted in a 10% HCl-methanol solution and analyzed by x-ray diffraction.

The x-ray chart for the DZ125L shows only $M_{23}C_6$ and an MC whose lattice parameter is 4.36Å. The presence of only one $M_{23}C_6$ phase indicates that the grain boundary carbide is $M_{23}C_6$ and the rosette structure due to the discrete breakdown is also $M_{23}C_6$.

The x-ray pattern of the DS René 125 is more complex. Two MC phases are present with lattice parameters of 4.37Å and 4.52Å, a $M_{23}C_6$ phase, and a Mu phase.

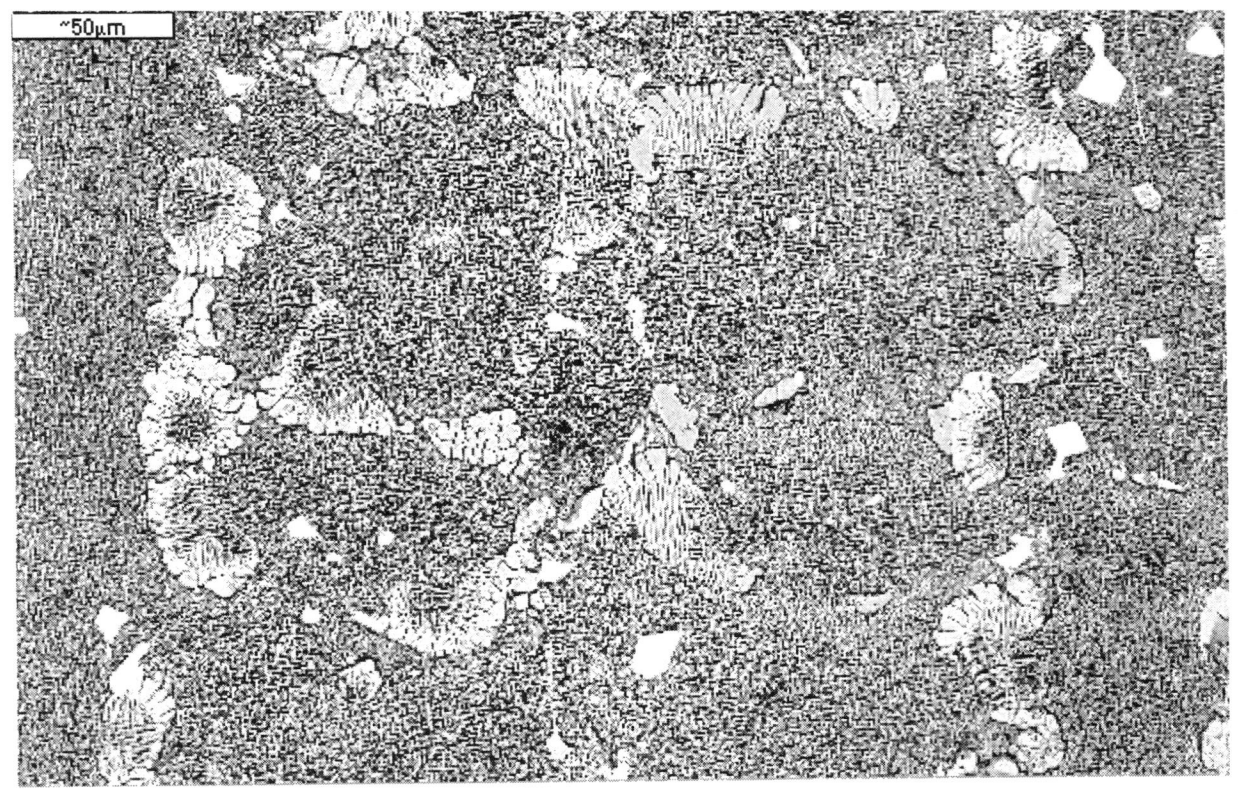

DS René 125   **(γ+γ′) · Eutectics   14%**

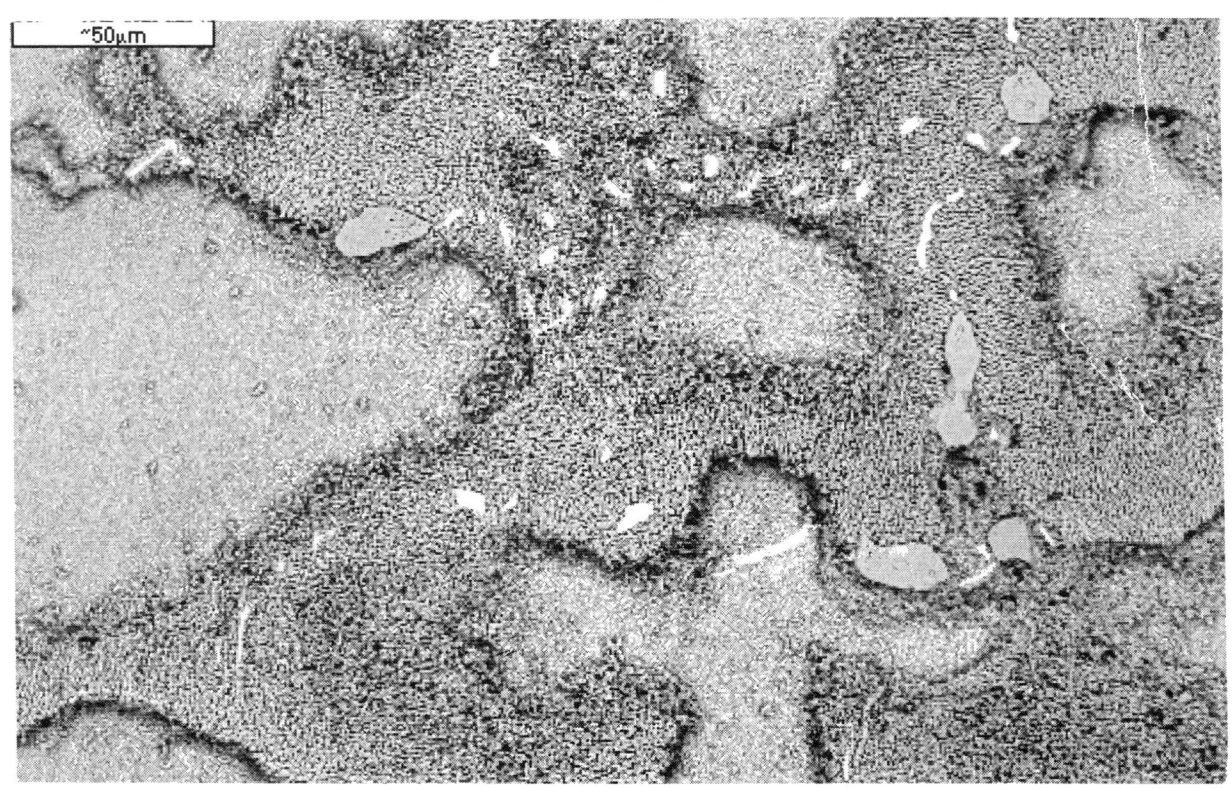

DZ 125L   **(γ+γ′) Eutectics   5%**

Figure 3: A-cast (γ+γ′) eutectic structure

336

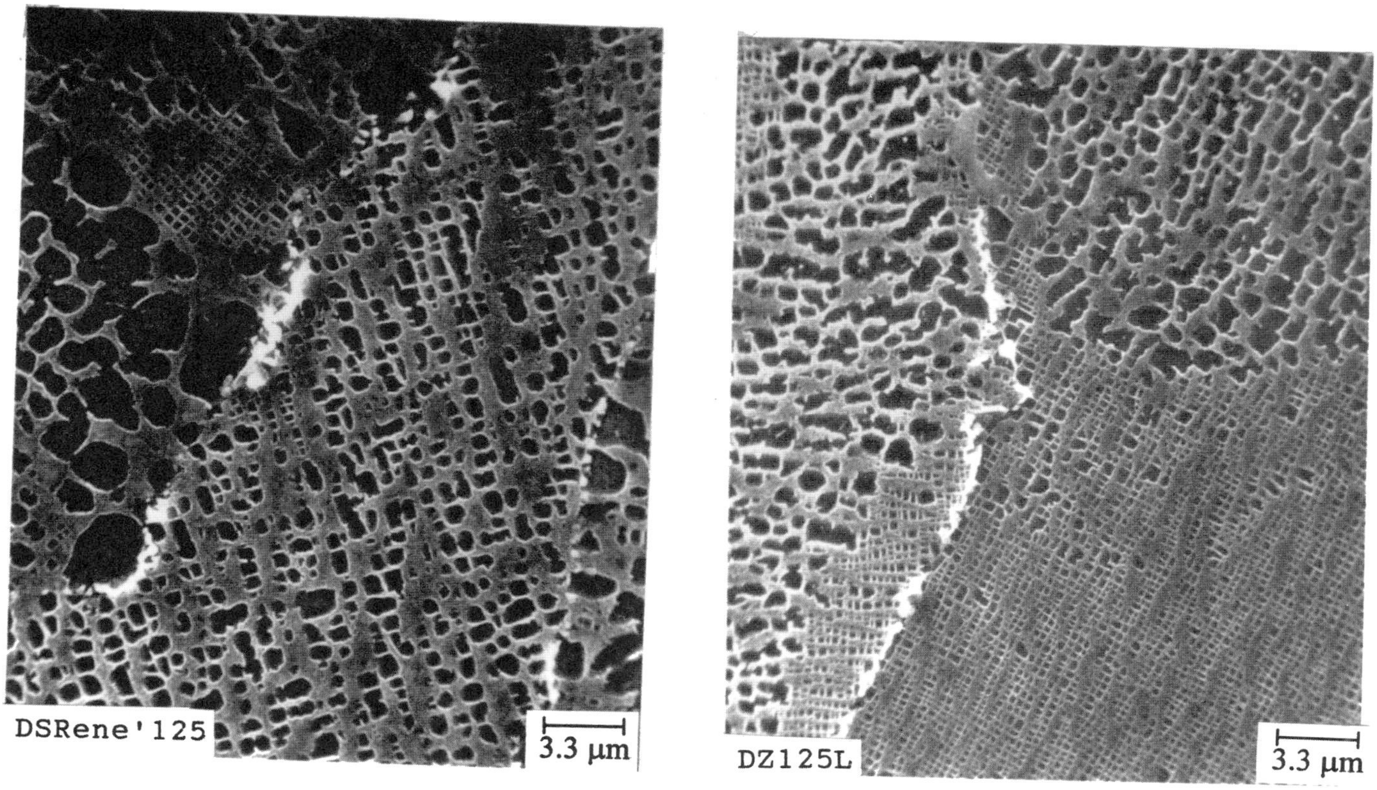

Figure 4a: As heat treated structures

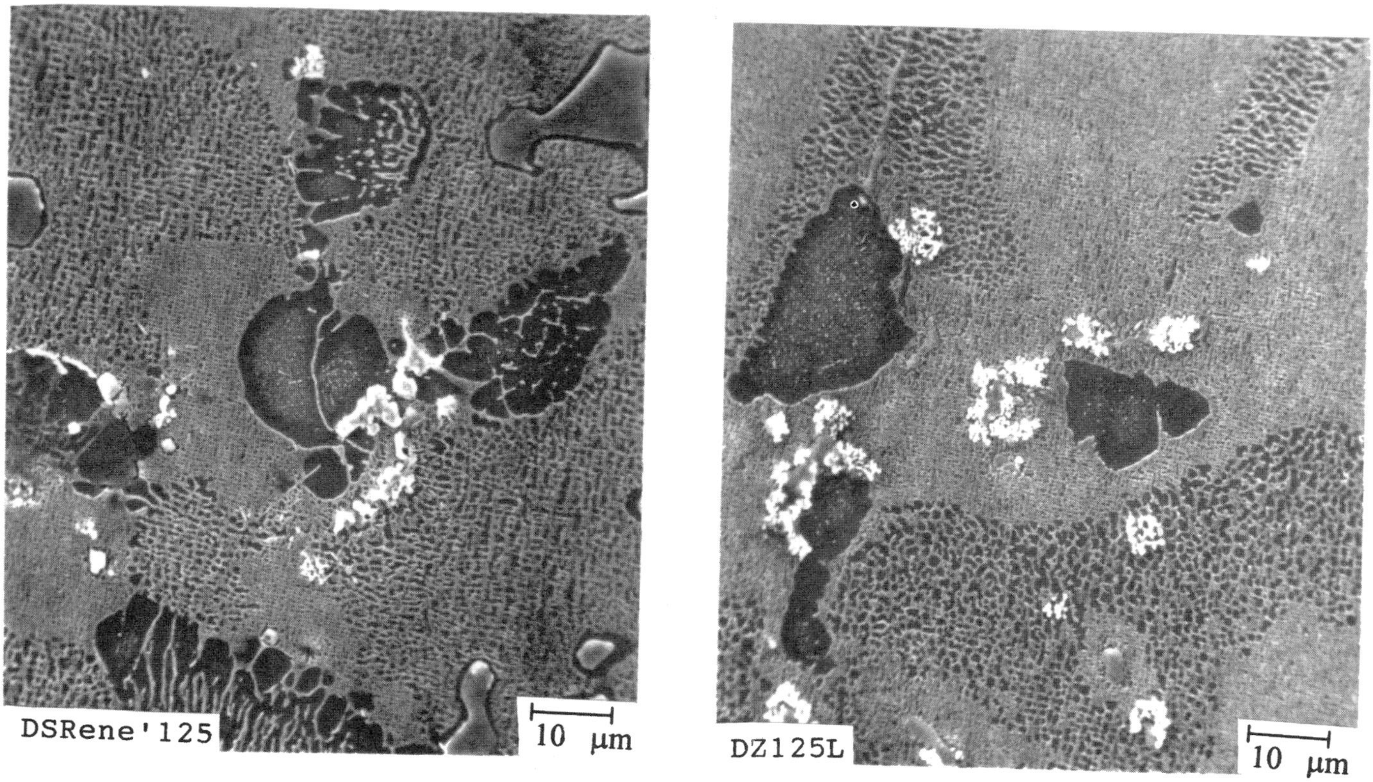

Figure 4b: Microstructures after 1000 hrs at 900°C

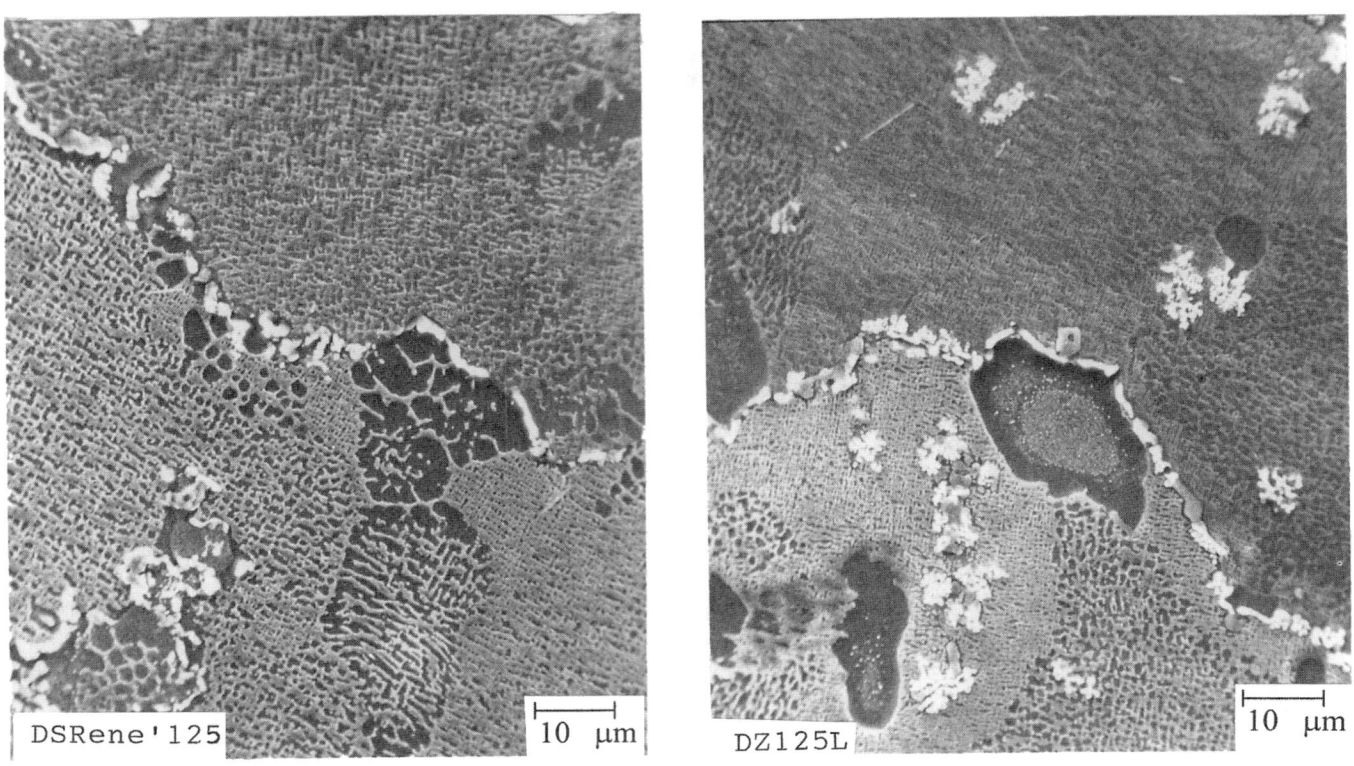

Figure 5a:Microstructures after 2000 hrs at 900°C

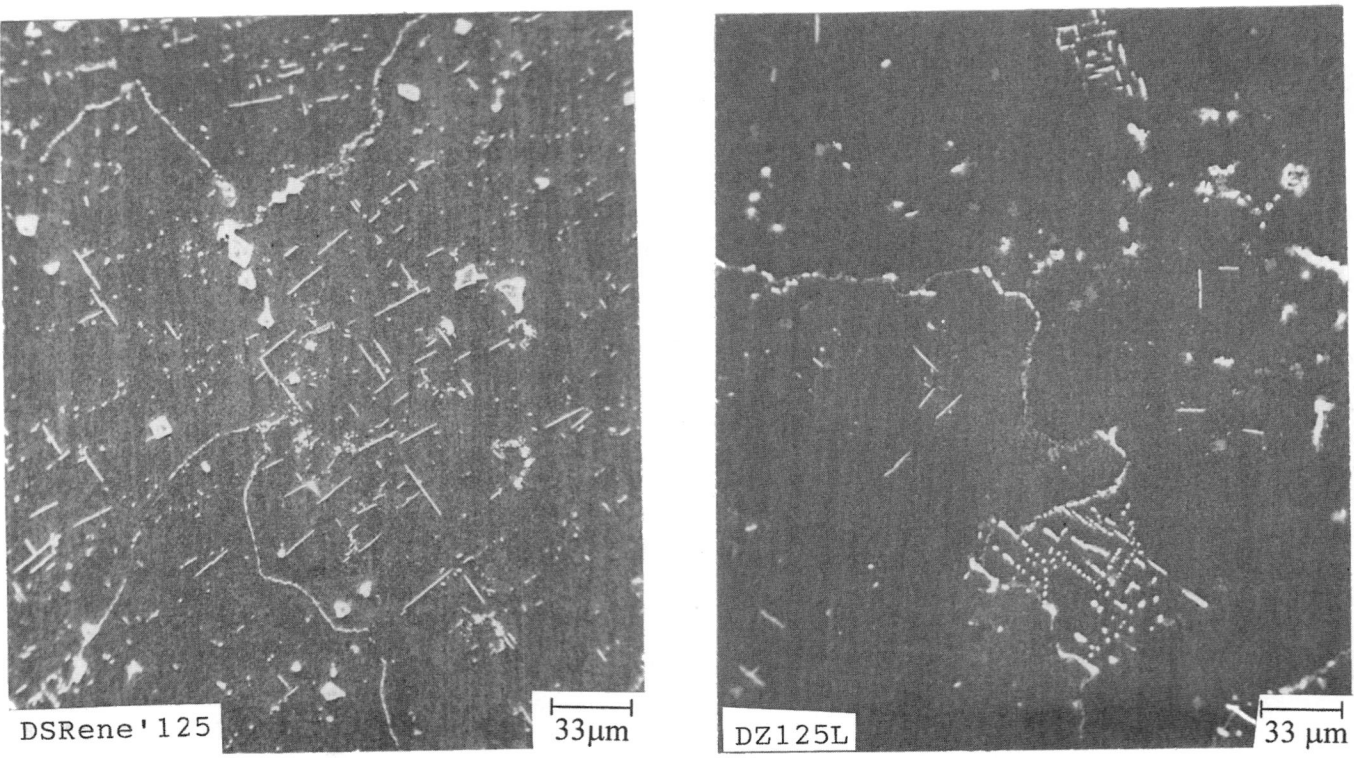

Figure 5b: Mu phase precipitation after 2000 hrs at 900°C

Conclusions

A low segregation alloy DZ125L similar in base composition to DS René 125 was developed in China. It contains low P, Si, B and no Hf or Zr. Solidification studies show DZ125L to have a much narrower solidification gap compared to DS René 125 and has much fewer primary γ' islands (5% vs. 14%). As the Al+Ti in primary γ' islands does not contribute to strengthening, it is felt the higher stress-rupture properties of DZ125L compared to DS Reneé 125 is attributed to more Al+Ti available because of the fewer primary γ' islands.

As cast DS René 125 contains segregation areas on the periphery of many γ' islands. These areas are rich in Mo and Hf and they form borides and Hf rich carbides during high temperature processing or heat treatments. In contrast, the areas of segregation found along some primary γ' islands in the DZ125L alloy are richer in Ti and revert to TiC during processing.

High segregation areas in alloys with high γ' and refractory elements form plate structures when the solubility of the matrix is exceeded by more alloying or long-time exposures. DS René 125 develops appreciable Mu plates after 2000 hours at 900°C while only a small amount of isolated Mu phase is found in DZ125L alloy in the same exposure time. Because the solidification gap in DZ125L is small, the degree of porosity is minimized and columnar growth during solidification is improved. Porosity enhances the tendency for longitudinal cracking and lowers the transverse properties at intermediate temperatures.

References

1. Y. X. Zhu et al., "Superalloys with Low Segregation," D. N. Duhl et al., eds., Proceedings of Superalloys 1988, TMS (1988) 703.

2. Y. X. Zhu et al., "A New Way to Improve Superalloys," S. D. Antolovich et al., eds., Proceedings of Superalloys 1992, TMS (1992) 145.

3. Y. X. Zhu et al., "Superalloys with Low Segregation Controlled by Trace Elements," An award paper presented at the Conference on Innovations in Real Materials, Washington, D.C., July 21, 1998.

4. A. W. Cochardt, "High-Temperature Alloys," U.S. Patent 3,005,705, October 24, 1961.

5. Martin Metal's brochure (Ductility Improvement Discovery, MM004) See also: C.H. Lund, J. Hockin and J.J. Woulds, "High Temperature Castable Alloys and Casstings," U.S. Patent 3,677,747, July 18, 1972.

6. J.M. Dahl, W.F. Danesi and R.G. Dunn, "The Partitioning of Refractory Metal Elements in Hf Modified Cast Nickel Base Superalloys," Met. Trans. 4, p. 1087, April 1973.

7. J.E. Doherty, B.H. Kear and A.F. Giamei, "On the Origin of the Ductility Enhancement in Hf Doped Mar-M-200," Journal of Metals, 23, (11), p. 59, 1971.

8. P. Aldred, "René 125 Development and Application," Paper presented at the National Aerospace engineering Meeting, Culver City, Los Angeles, Nov. 17-20, 1975.

The Growth of Small Cracks in the Single Crystal Superalloy CMSX-4 at 750 and 1000°C

F. Schubert, T. Rieck and P.J. Ennis
Research Centre Juelich
Institute for Materials and Processes in Energy Systems IWV-2
52425 Juelich
Germany

Abstract

Creep-fatigue crack growth experiments have been carried out on alloy CMSX-4 single edge notched specimens at 750°C (maximum root-temperature) and at 1000°C (maximum aerofoil temperature). At 750°C and below the cracks, controlled by the K-concept, followed a zigzag line due to alternate gliding on the {111} and {100} planes. An influence of the test atmosphere was observed, the tests in vacuum exhibiting higher crack growth rates than those measured in air during the early stages of testing. At high temperatures, such as 1000°C, the crack propagation was along the {100} planes.

Introduction

During the last decade, there has been a dramatic development in heavy-duty, stationary industrial gas turbines for electrical power generation. The application of cooling systems, the introduction of single crystal technology, both originally developed for aircraft engines, and the use of thermal barrier coatings have enabled gas temperatures as high as 1400°C to be reached. Figure 1 indicates the development line for turbine blading materials. One of the currently favoured blading materials is the second-generation single crystal alloy CMSX-4, with which long term material temperatures up to 950°C can be achieved. This alloy exhibits excellent creep resistance at these high temperatures due to hardening by γ' precipitates (more than 70% by volume) with optimized distribution and morphology and due to solid solution hardening by 3 wt.% Re addition. Single crystal alloys, however, exhibit a strongly anisotropic deformation behaviour.

Only limited information exists in the literature regarding the behaviour of technical cracks in single crystal Ni-base superalloys (mainly observed with CT specimens). Up to test temperatures of about 800 to 850°C, the crack propagation in polycrystalline Ni-superalloys is correlated with the stress intensity factor K_I. Numerical and finite element (FE) analyses have shown differences between the crack growth behaviour in isotropic and anisotropic Ni-superalloys /1, 2/. Therefore, the use of similar K_I values for the description of crack growth phenomena in anisotropic single crystalline Ni-superalloys is totally uncertain, because the role of gliding systems for the crack propagation is unknown. However, numerical estimations /3/ have demonstrated that similar K values are obtained for cracks perpendicular to the stress direction and cracks with deviations of up to 30° from the stress direction.

The typical temperature dependence of yield and rupture strength of superalloys and the different deformation mechanisms (dislocations and particles interactions) control the temperature dependence of the growth of macroscopic flaws. Only limited results for fatigue and creep-fatigue crack growth of small cracks in single crystal alloys, especially for CMSX-4, are available in the literature, e.g. /4-9/. Unfortunately, there are some results in the open literature that are contradictory and need clarification.

Estimation of allowable operational time and allowable numbers of operational cycles requires information concerning the initiation and growth of small flaws in the area of root (about 750°C) and the blade (about 1000°C). Therefore, a study of the growth of small cracks at 750 and 1000°C was undertaken for the alloy CMSX-4. The possible influence of test atmosphere was also studied, by comparing the results of crack growth tests in air with those from tests carried out in vacuum.

Experimental details

Material

Single crystalline superalloys, first developed for aero gas turbines, exhibit a significant improvement in creep and fatigue resistance, compared with conventionally cast superalloys, e.g. IN 738 LC, so that a potential increase of 80C in the materials operating temperature may be obtained. From the application in aero gas turbines, three classes of single crystalline superalloys may be denoted: the first generation with reduced grain boundary hardening alloy elements; the second generation with additions of about 3 wt.% Re; and the third generation alloys which have somewhat higher Re contents (about 5 to 6 wt%).

Alloy CMSX-4 is a typical second-generation single crystal material hardened by about 70 vol% γ' with solid solution strengthening of the γ channels in-between the cuboidal γ' particles due to the addition of 3 wt.% Re. The nominal chemical composition of CMSX-4 in wt% is as follows: Ni bal, 5.6 Al, 1.0 Ti, 6.5 Ta, 6.5 Cr, 0.6 Mo, 6.0 W, 9.0 Co, 3.0 Re, 0.1 Hf, less 0.002 C. The heat treatment consists of: 6 h solid solution treatment at 1305°C with controlled cooling rate, followed by a two stage ageing treatment of 6 h at 1140°C and 20 h at 871°C, both with controlled cooling rates. The material was as delivered as single crystalline plates, directionally solidified in the <001> direction and fully heat treated.

Small cracks in turbine blades were simulated by single edge notched specimens (SEN) with artificial edge or corner cracks. Figure 2 shows the details of the test pieces used for crack growth experiments.

341

Superalloys 2000
Edited by T.M. Pollock, R.D. Kissinger, R.R. Bowman,
K.A. Green, M. McLean, S. Olson, and J.J. Schirra
TMS (The Minerals, Metals & Materials Society), 2000

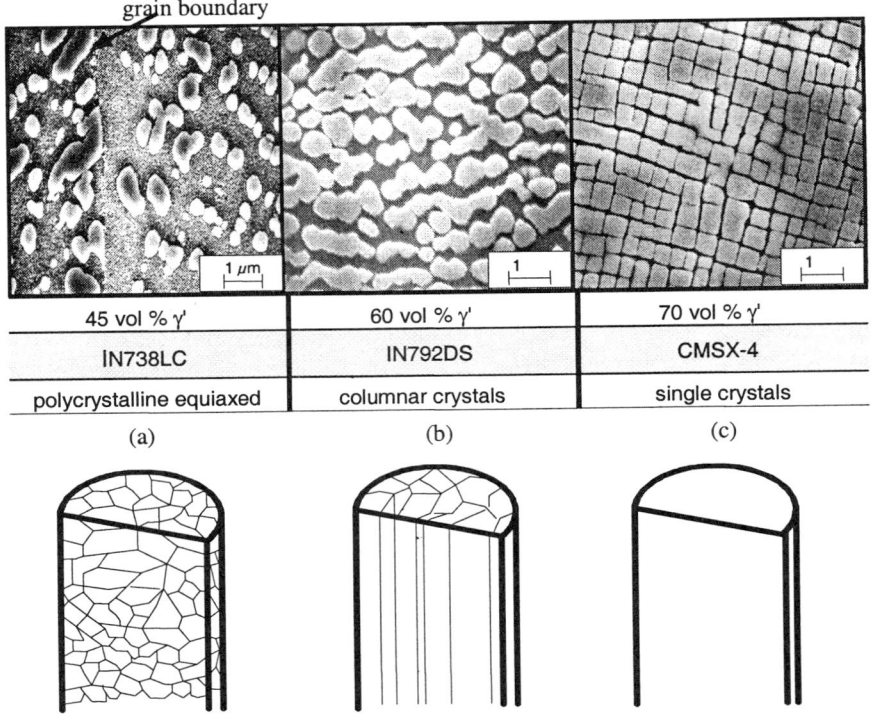

grain boundary

45 vol % γ'	60 vol % γ'	70 vol % γ'
IN738LC	IN792DS	CMSX-4
polycrystalline equiaxed	columnar crystals	single crystals
(a)	(b)	(c)

Figure 1
Schematic diagram of alloy development for superalloys: (a) polycrystalline, equiaxed conventionally cast IN 738 LC,
(b) columnar grained directionally solidified IN 792 and (c) single crystalline alloy CMSX-4.

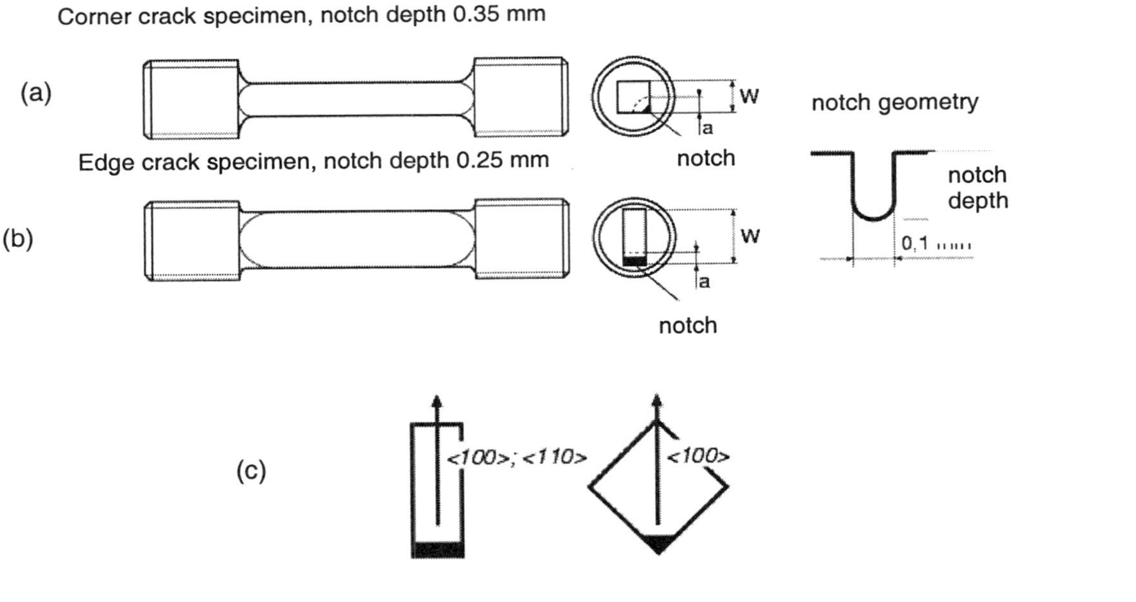

Corner crack specimen, notch depth 0.35 mm

(a)

W

a

notch

notch geometry

Edge crack specimen, notch depth 0.25 mm

(b)

W

a

notch

notch
depth

0,1

(c)

<100>; <110>

<100>

Figure 2
Geometries of SEN specimens; (a) corner crack, cross-section 8 x 8 mm, (b) edge crack, cross-section 4.5 x 12 mm,
(c) secondary crack orientations

Fatigue test

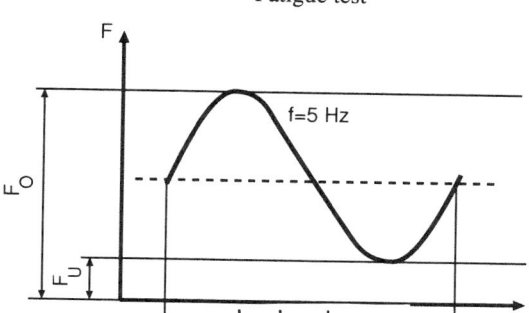

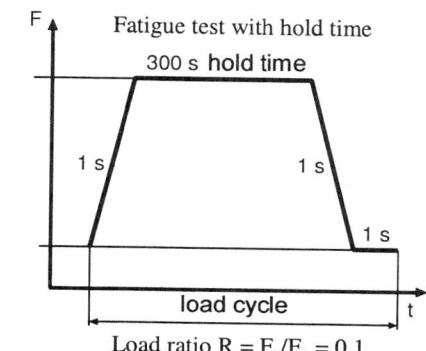

Load ratio R = F_u/F_o = 0.1

Figure 3
Schematic presentation of the loading cycles for fatigue tests with and without hold times

Some fracture mechanics equations

The K_I function for SEN specimen under stress is normally given by

$$K_I = \sigma \sqrt{\pi a \cdot c} \qquad (1)$$

where K_I is the stress intensity, σ the applied stress, a the crack length and c a correction factor that depends on the crack length a and the specimen geometry. For small cracks, $c = 1.12$ and for larger cracks, c is given by

$$1.99 - 0.41\left(^a/_w\right) + 18.7\left(^a/_w\right)^2 - 38.5\left(^a/_w\right)^3 + 53.8\left(^a/_w\right)^4 \qquad (2)$$

where w is the specimen width, for $0 \le \left(^a/_w\right) \le 0.6$.

SEN specimens with corner cracks exhibit a much more complicated stress distribution ahead of the crack tip than edge crack specimens and represents the realistic crack geometry within a component due its three-dimensionality. Mathematical results are given in /10, 11/, in which a square edged crack surface area is estimated. Thereby one differentiates between the stress intensity factors along the surface of the specimen and in the direction of 45°. Then the stress intensity factor across the whole crack surface may be estimated by

$$K_{I\,mean} = \frac{K_{I\,45°} + K_{I\,surface}}{2} \qquad (3)$$

$$= \left(0.97 - 0.09\left(\frac{a}{w}\right)^2\right) \cdot K_{I\,surface} \qquad (4)$$

The approximations help in understanding the crack propagation of a corner edged crack. Further estimations as well as the experimental observations indicate that the K_I-concept may be used for both test temperatures.

Test conditions

The fatigue and creep fatigue experiments were carried out at 750 and 1000°C in air and in vacuum (ca 2.5 x 10⁻⁵ bar) using a servohydraulic test machine. The fatigue loading was sine-wave with a frequency of 5 Hz and the creep fatigue load cycle was trapezoid with a hold time of 300 s at maximum load and 1 s at minimum load, as shown in Figure 3. To obtain a sharp crack the notched specimens were subjected to a 10 Hz sine-wave load 50% above the ΔK value and a load ratio of 0.1 at the test temperature before the actual test. The induced cracks were 0.4 – 0.5 mm long in the edge crack specimens and 0.3 – 0.4 mm in the corner crack specimens. The test were stopped at a crack length to specimen width ratio of 0.5.

Results and discussion

Figure 4 shows the results of crack growth experiments carried out at 750 and 1000°C in air and in vacuum. At 750 and 1000°C, the fatigue crack growth behaviour of the specimens resulted in the expected functional behaviour ("Paris-Erdogan") of da/dN versus ΔK_I (the cyclic stress intensity factor). At 750°C, the threshold values were higher and the slope of the Paris equation not so steep compared to the values at 1000°C. The influence of crystal orientation seemed to be more marked at 750 than at 1000°C. Specimens with a <100> crack orientation came to a rapid fracture by a spontaneous change to the {111} sliding planes. The <110> crack orientation did not show this behaviour.

As a consequence, one may assume that the position of the crack front to the {111} sliding planes is responsible for the occurrence of the observed rapid fracture. At 1000°C and for low ΔK values and low crack growth rates, the crack growth behaviour may be understood as typical crack behaviour of small cracks. This behaviour could be explained by the crack closure because of plastic deformation at the crack tip, oxidation and depletion of the crack surface areas and the start of the γ' rafting process. These influences blunt the crack tip and the stress singularity decreases, so that the crack could be stopped or slowed down.

In CMSX-4, oxidation at the crack tip became more important at the higher test temperatures. The comparison of the crack growth experiments in air and vacuum showed that the oxidation process influences significantly the crack initiation point or the initial stage of crack growth (see Figure 4).

Fractographic examinations using scanning electron microscopy (SEM) indicate for CMSX-4 (high volume fraction of γ') a slightly different behaviour compared to equiaxed Ni alloys with γ' volume fractions below 50%.

The crack surface of CMSX-4 at 750°C in air followed at low ΔK-values the γ channels or the γ/γ' interface region. At high ΔK-values, a change in the crack surface growth to the {111} plane was observed and cutting of γ' precipitates occurred.

Metallographic examinations to reveal the mechanisms of crack growth (see Figure 5) showed that:

- at 750°C and low and medium ΔK values, the propagation followed a zigzag line by changing the orientation along the {111} and {100} planes, mainly following γ/γ' interface. At higher ΔK values sliding along the {111} planes by cutting of the γ' precipitate occurred.
- at 1000°C, crack propagation along the {100} planes was mainly observed, which means propagation by cross sliding on {100} planes. The cutting of γ' pre-

cipitates became difficult and the crack path followed the γ/γ' interface areas along {100} planes.

Figure 6 compares the fatigue and the creep-fatigue behaviour at 1000°C. The edge crack specimen showed the same threshold values for both types of test, but the creep crack curves did not exhibit any changes in the crack growth rate. Therefore one may expect that creep-fatigue is more influenced by the deformation at the crack tip than by oxidation. If K is used as the stress intensity factor controlling the creep crack behaviour, the fatigue and the creep-fatigue results lie in the same range. At high K values and crack growth rates, the differences between fatigue and creep-fatigue became more significant. Because of these observations, one may conclude that creep dominates the crack growth process at low ΔK or K values, and fatigue at high ΔK or K values.

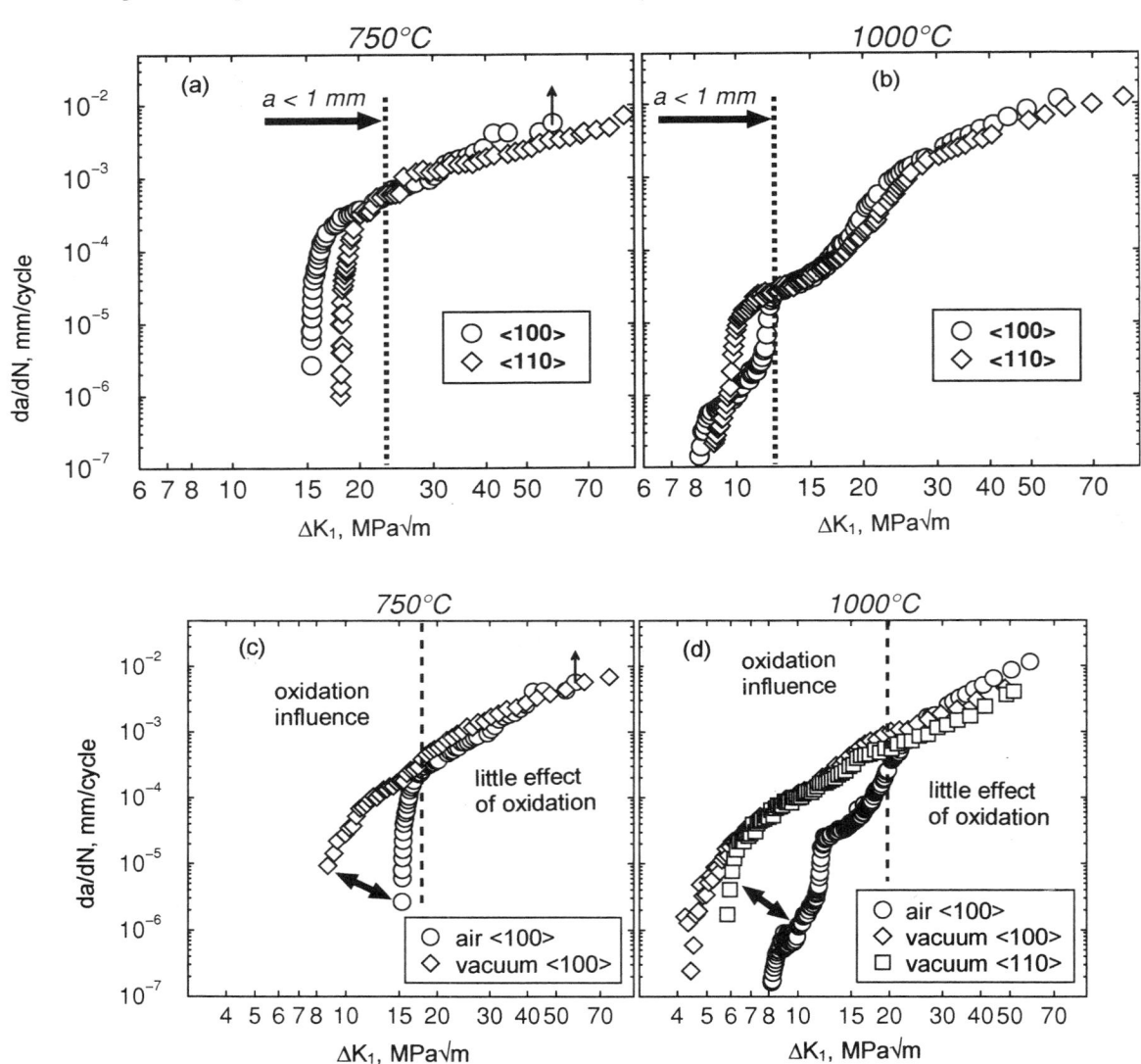

Figure 4
Fatigue crack growth of edge crack specimens; (a) and (b) <100> and <110> orientations tested in air, at 750 and 1000°C; (c) comparison of <100> orientations tested in air and vacuum at 750°C; (d) <100> and <110> orientations tested in vacuum at 1000°C compared with <100> orientation tested in air

(a) 750°C

(b) 1000°C

Figure 5
Crack surfaces of corner edge specimens of CMSX-4 at (a) 750 and (b) 1000°C in air

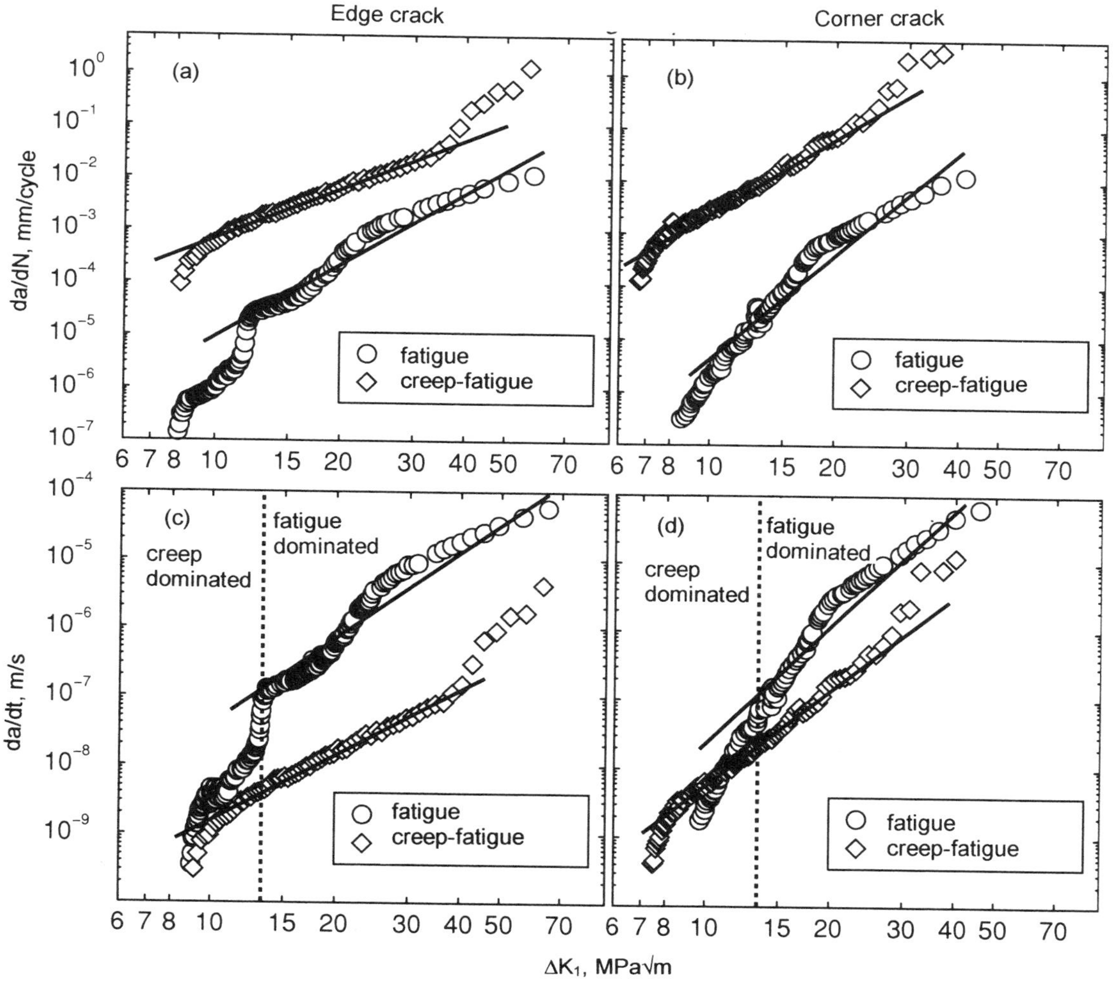

Edge crack

Corner crack

(a)

da/dN, mm/cycle

10^0
10^{-1}
10^{-2}
10^{-3}
10^{-4}
10^{-5}
10^{-6}
10^{-7}

○ fatigue
◇ creep-fatigue

(b)

○ fatigue
◇ creep-fatigue

(c)

da/dt, m/s

10^{-4}
10^{-5}
10^{-6}
10^{-7}
10^{-8}
10^{-9}

creep
dominated

fatigue
dominated

○ fatigue
◇ creep-fatigue

(d)

creep
dominated

fatigue
dominated

○ fatigue
◇ creep-fatigue

6 7 8 10   15   20   30 40 50 70

ΔK_1, MPa√m

Figure 6
Fatigue and creep/fatigue crack growth of CMSX-4 specimens tested in air; (a) edge crack, ΔK fatigue parameter,
(b) corner crack, ΔK fatigue parameter, (c) edge crack, K creep crack parameter, (d) corner crack, K creep crack parameter

345

Conclusions

At 750°C, the crack growth threshold values are higher and the slope of the Paris equation not so steep compared to the values at 1000°C.

The oxidation behaviour at the crack tip becomes more important at the higher test temperatures. The oxidation process influences significantly the crack initiation point and the initial stages of crack growth.

At 750°C and low/medium ΔK values, the crack follows a zigzag line by alternating between the {111} and {100} planes, mainly following γ/γ' interface. At higher ΔK values, sliding along the {111} planes by cutting the γ' precipitate occurs.

At 1000°C, crack propagation along the {100} planes is observed, which means propagation by cross sliding at {100} planes. The cutting of γ' precipitates becomes difficult and the crack path follows the γ/γ' interface areas along {100} planes.

At high K values and high crack growth rates, the differences between fatigue and creep-fatigue become more significant. It may be concluded that creep dominates the crack growth process at low ΔK or K values, and fatigue at high ΔK or K values.

Acknowledgement

The support of this work by Deutsche Forschungs-gemeinschaft in the Priority Programme "Microstructure and Mechanical Properties of Metallic High Temperature Materials" is gratefully acknowledged. This was a preliminary investigation for the co-ordinated research work with the RWTH Aachen, SFB 561: "Thermomechanical behaviour of highly cooled, porous, multilayer structures for combined cycle plants."

References

/1/   K.S. Chan, T.A. Cruse, Engineering Fracture Mechanics, 23, 863-874 (1986)

/2/   T. Rieck, F. Schubert, Bruchmechanische Bewertungskonzepte im Leichtbau, 173-182, DVM-Bericht 231, (1999)

/3/   K.S. Chan, T.A. Cruse, Journal of Engineering Fracture Mechanics, 23, 863 (1982)

/4/   B.F. Antolovich, A. Saxena, S.D. Antolovich, Superalloys 1992, 727-736, The Minerals, Metals & Materials Society, Warrendale (1992)

/5/   A. Defrense, L. Rèmy, Materials Science and Engineering A, 45 (A129), 55-64 (1990)

/6/   A Sengupta, S.K. Putatunda, Journal of Materials Engineering and Performance, 2, 57-68 (1993)

/7/   A. Sengupta, S.K. Putatunda, M. Balogh, Journal Engineering and Performance, 3, 540-550 (1994)

/8/   O.R. Murphy, J.W. Martin, Proc. of the 7th European Conference on Fracture, Budapest (1988), Engineering Materials Advisory Services Ltd, Halesowen UK, 1988, 1138-1140

/9/   V. Lupinc, G. Onofrio, G. Vimercati, Superalloys 1992, 717-726, The Minerals, Metals & Materials Society, Warrendale (1992)

/10/   T. Rieck, F. Schubert, "Wachstum kleiner Risse bei hohen Temperaturen und Zug-Schwellbeanspruchungen in den einkristallinen Superlegierungen CMSX-4 und SC 16", Diss. RWTH Aachen, Nov. 1999

/11/   J. Newman, I. Raju, Stress intensity factor equations for cracks in three-dimensional finite radius subjected to tension and bending loads, NASA Langley Research Center, Hampton, TM 85793, Washington DC, 1984

The Influence of Load Ratio, Temperature, Orientation and Hold Time on Fatigue Crack Growth of CMSX-4

Steffen Müller*, Joachim Rösler, Christoph Sommer* and Walter Hartnagel*

Technical University Braunschweig, Institut für Werkstoffe, Germany

*ABB ALSTOM POWER Technology Ltd., Heidelberg, Germany

ABSTRACT

CMSX-4 is a widely used single crystal material for critical components like gas turbine 1st stage blades and vanes. For lifetime prediction of such components knowledge on defect behavior is of particular interest. For that reason, results of a systematic study, characterizing the influence of load ratio (R=0.1/0.7), temperature (T=550°C/950°C), orientation (<001> and <011> parallel to load axis) and hold times on fatigue crack growth rates will be presented in this paper. It was found, that increasing R and decreasing temperature leads to a reduction of the threshold value for fatigue crack propagation, whereas the fatigue crack growth rate in the Paris region seems to be weekly temperature- and R-dependent. Changing the load axis from <001> to <011> did not influence the fatigue crack growth rate. Independent of temperature, R and loading direction a smooth mode I fracture was found. Room temperature precracking lead to fatigue crack growth along {111} planes. Introducing hold times during fatigue crack growth at 950°C retards the crack propagation rates only at higher stress intensity levels which is attributed to creep deformation during hold times.

Additionally to fatigue crack propagation, isothermal low cycle fatigue (LCF) as well as out-of-phase thermomechanical fatigue (OP-TMF) tests have been performed on test bars with idealized casting defects acting as crack starters. OP-TMF tests compare well with LCF results with compressive dwell time at maximum temperature of TMF tests. LCF tests without dwell time and at lower TMF temperature lead to longer lifetimes. By in-situ observation of crack propagation starting from these defects da/dN data were obtained and compared to those obtained from standard CT specimen. Both for 550 and 950°C, a good agreement between crack growth rates in LCF tests without dwell times and CT results was found.

Superalloys 2000
Edited by T.M. Pollock, R.D. Kissinger, R.R. Bowman,
K.A. Green, M. McLean, S. Olson, and J.J. Schirra
TMS (The Minerals, Metals & Materials Society), 2000

INTRODUCTION

Single crystal (SC) superalloys are widely used for blade and vane applications in aeroengines and, more recently, in land based gas turbines. They offer improved creep and fatigue resistance compared to conventionally and directionally solidified materials due to the lack of grain boundaries, elimination of low melting point elements and anisotropy of elastic modulus. Within the family of SC superalloys, CMSX-4 is a widely used second generation, rhenium containing alloy [1].

For lifetime prediction of SC components such as 1st stage blades and vanes, knowledge on defect behavior, especially under fatigue loading, is of particular interest. Therefore, this article deals with measurement of CMSX-4 fatigue crack growth (FCG) resistance. As systematic studies, characterizing the influence of load ratio, temperature, SC orientation and hold time on FCG, are not available in the open literature for this material, da/dN measurements on standard CT specimen were conducted in <001> and <011> loading direction at two temperatures (550°C, 950°C) and R-ratios (R=0.1, 0.7). By introducing a hold time at the peak stress of the fatigue cycle (300 sec after every 1000 cycles), the influence of creep and/or oxidation effects was determined. Furthermore, low cycle fatigue (LCF) tests at 550°C and 950°C as well as out-of-phase thermomechanical fatigue (TMF) tests (temperature range 550-950°C) have been performed on test bars with idealized casting defects acting as crack starters. These defects, realized with ceramic inclusions with a defined geometry and orientation to the load axis, should give information about the influence of defects in SC material on LCF and TMF lifetime. Besides the characterization of fatigue life of CMSX-4 with and without defects, also fatigue crack propagation has been measured in these tests by means of continuous video monitoring.

EXPERIMENTS

The material used for all tests was taken from a single melt (chemical composition see Table I). From that melt near-net-shape rods or plates were grown by grain selector technique at Doncasters Precision Castings-Bochum GmbH (DPC), Germany. From these ingots, specimens were machined out for the <001> direction parallel to the solidification direction and for <011> direction tilted by 45° to solidification. In both cases <010> was the crack propagation direction. FCG tests were performed at two different laboratories: ABB ALSTOM POWER Ltd. Baden, Switzerland (named AA) and Technical University Braunschweig, Institut für Werkstoffe, Germany (named TUBS). Two different specimen geometries were used: ¾" CT standard specimen at TUBS (thickness ¾") and 1" CT standard specimen at AA (thickness ½") according to ASTM E-647. The crack propagation was monitored continuously by means of potential drop technique in both laboratories. The measured potential signal was converted to crack lengths with Johnson's formula (see ASTM-E647). Whereas the tests at

TUBS were performed with a fixed frequency of 5 Hz, the test frequencies at AA ranged from 5 (at high ΔK) to 50 Hz (at low ΔK) within one test. The specimens were heated with a chamber furnace as well as with HF conduction heating.

Table I: Nominal chemical composition of CMSX-4 (in wt.-%, Ni bal.).

Al	Ti	Ta	Cr	Re
5.6	1.0	6.5	6.5	3.0
Mo	**W**	**Co**	**Hf**	
0.6	6.0	9.0	0.1	

TMF and LCF tests were carried on closed-loop servohydraulic testing machines which were equipped with induction heating. Out of the near-net-shape rods cast by DPC cylindrical specimens (diameter 10 mm, gauge length 15 mm) were machined. Loading direction was <001>. The specimens contained rectangular ceramic pins (cross section 1 x 1 mm) in the center of the gauge length, c.p. Fig. 1. The pins have been introduced during the near-net-shape casting process. They penetrated 1 to 4 mm from the surface into the specimens.

LCF tests were performed according to ASTM E 606 under total strain control with a constant strain rate of 6 %/min. Temperature was controlled by a ribbon-like Pt-PtRh thermocouple within the gauge length (see Fig. 1). Testing temperatures were 550°C and 950°C with R = -1. Part of the LCF tests at 950°C included compressive dwell periods of two minutes. The temperature range in the out-of-phase TMF tests was 550°C to 950°C with 2 min compressive dwell times at maximum temperature. The TMF tests have been performed with temperature rates of 3 K/sec and constant mechanical strain rates. Mechanical strain ranges were 0.6 to 0.8%. The number of cycles to crack initiation in the LCF tests was determined at a decrease of the tensile stress amplitude by 5 %. For the evaluation of the TMF tests a 5 % decrease of the maximum tensile stress was used.

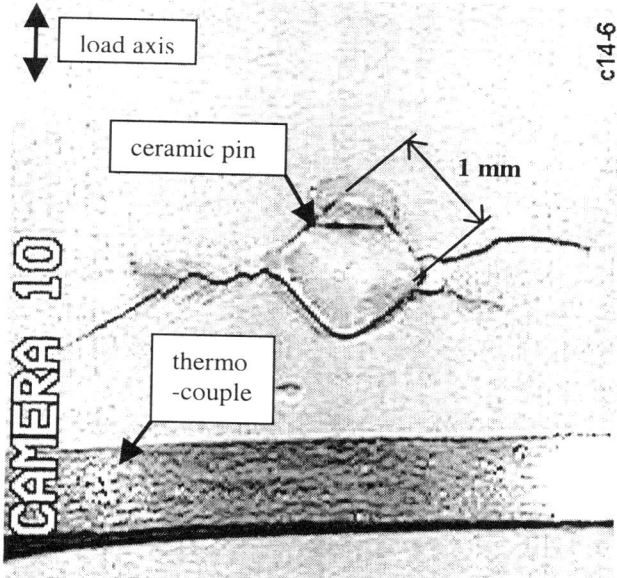

Fig. 1: Micrograph of the ceramic pin within a LCF specimen at 950°C. The picture was taken during testing with the video monitoring system.

The ceramic pins, c.p. Fig. 1, which acted as crack starters were monitored by video observation. After fixed numbers of cycles pictures were stored at maximum strain in the LCF tests and at minimum temperature in the TMF tests. Crack length vs. cycle number data were smoothened by a second order polynom and evaluated crack growth rates compared to fracture mechanics data. According to [2] stress intensities were calculated by

$$\Delta K_I = Y \cdot \Delta\sigma \cdot \sqrt{\pi \cdot a} \qquad (1)$$

with

$$Y = \sqrt{\frac{d+a}{a}} . \qquad (2)$$

$\Delta\sigma$ means tensile stress of the LCF and TMF test, a denotes crack length (measured from pin edge to crack tip) and d denotes half pin width.

RESULTS

Fatigue Crack Growth at 950°C

The fatigue crack growth rates of CMSX-4, loaded in <001> direction, for R ratios of 0.1 and 0.7 are shown in Fig. 2 and 3, respectively. The curves show the typical form with a threshold region and the Paris regime of stable fatigue crack growth. The data for R=0.1 compare well with previous results for CMSX-4 [3] and René-N4 [4]. In agreement with literature findings [5,6], increasing the R-value increases the fatigue

crack growth rate for a given ΔK. According to ASTM-E647 the threshold values ΔK_{th} for fatigue crack propagation were determined. The results are summarized in Table II. An increase of R from 0.1 to 0.7 decreases the threshold value for fatigue crack growth by 40%, whereas the FCG rate is accelerated by a factor of about 5 in the Paris regime.

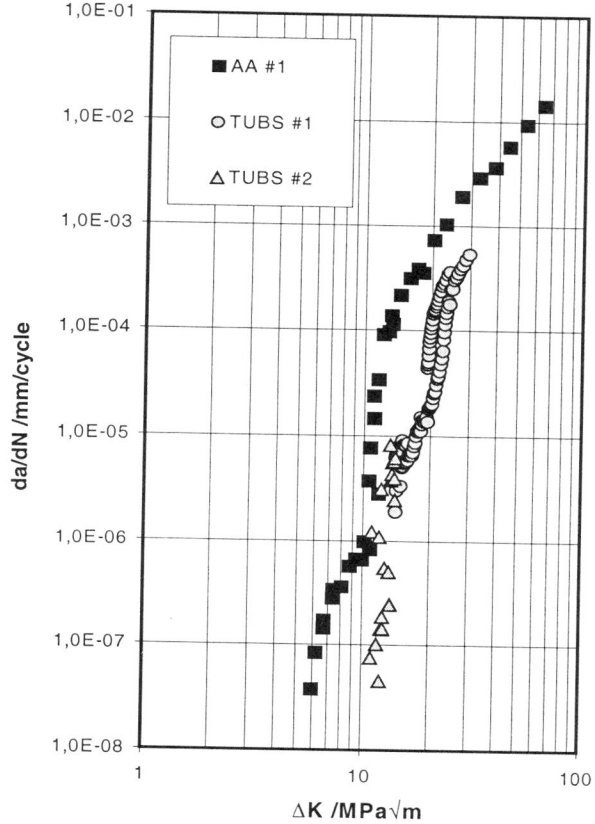

Fig. 2: Fatigue crack growth rates of CMSX-4 at 950°C, <001> load direction and R=0.1 (AA and TUBS refers to test laboratory).

Table II: Threshold values for fatigue crack growth in CMSX-4 at 950°C. For details see text.

T	R=0.1	R=0.7
950	11.0 MPa√m/ 5.5 MPa√m	6.8 MPa√m

Whereas the results for R=0.7 are in a narrow scatter band in the lower stress intensity region (Fig. 3), a strong difference between single measurements for R=0.1 are visible (Fig. 2). One test resulted in a tail at very small ΔK values which predicts a threshold value below 6 MPa√m (compared to 11.0 MPa√m). An explanation for that may be the different test frequencies for the different curves: The tail part of the test named AA was run at frequencies between 40 and 50 Hz, whereas the curves named TUBS were obtained at a much lower test frequency (5 Hz). At lower test frequencies much more time is spent for oxidation at the crack tip. This oxidation results in crack tip blunting, decreases the stress intensity ahead of the crack tip and leads to a shift of ΔK_{th} to higher ΔK-values [7].

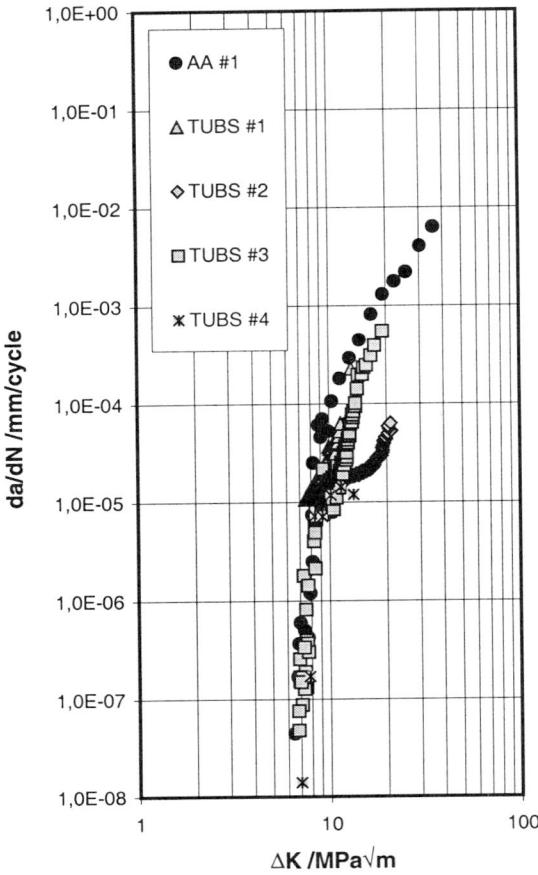

Fig. 3: Fatigue crack growth rates for CMSX-4 at 950°C, <001> load direction and R=0.7 (AA and TUBS refers to test laboratory).

Fractographic investigations of the samples tested at 950°C revealed a macroscopic and microscopic smooth fracture surface (mode I) independent of cyclic stress intensity factor (Fig. 4). This behavior is well-documented in the literature and attributed to the plastic deformation along {100}-planes [8,9]. It is assumed that the homogenous deformation is performed within the narrow γ matrix channels between γ' particles. Additionally, an oxide scale is visible which is probably Al_2O_3 [10]. The thickness of this scale varies only with exhibition time to high temperatures. No influence of stress intensity level or load ratio was found.

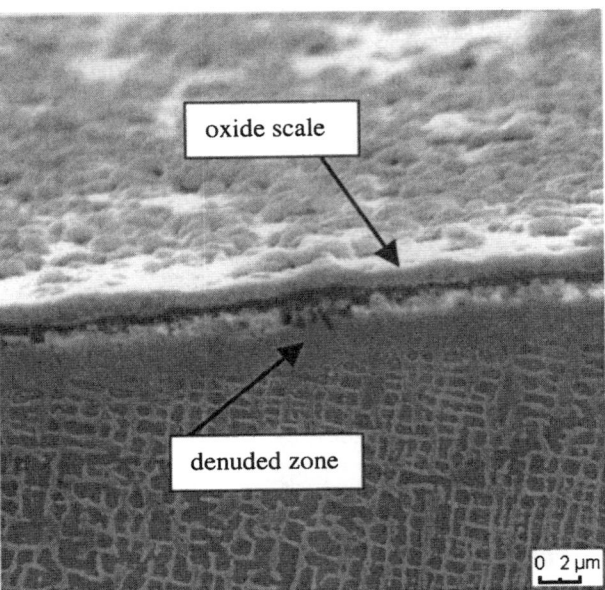

Fig. 4: Fracture path of fatigue crack at medium ΔK at 950°C (R=0.1). A smooth fracture surface with an oxide scale and the γ' denuded zone is visible.

In contrast to high temperature fatigue cracking, fatigue precracking at room temperature leads to crystallographic crack propagation along {111} planes (Fig. 5). These planes are glide planes for CMSX-4 at lower temperatures with the highest shear stresses when the crystal is loaded in <001> direction [3]. This type of fracture was independent of R-value.

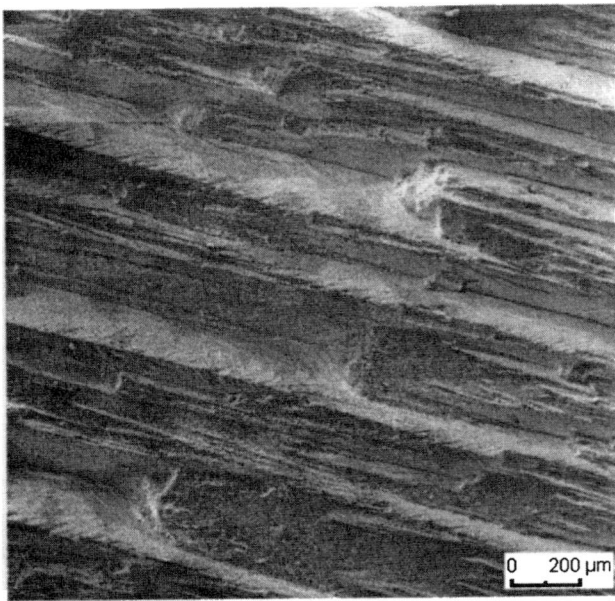

Fig. 5: Fracture surface of CMSX-4 after fatigue crack growth at room temperature (during precracking). A crystallographic fracture along {111} planes is visible.

Fatigue Crack Growth at 550°C

The fatigue crack growth rates for CMSX-4 at 550°C for both load ratios tested (R=0.1 and 0.7) are shown in Fig. 6. Compared with the results for 950°C (Fig. 2 and 3) the following differences were found:

- Surprisingly, testing at 550°C results in smaller threshold values for fatigue crack growth (see Table III) than at 950°C (Table II). This contrasts literature findings where with decreasing temperatures increasing threshold values for fatigue crack growth were found [3,10]. An explanation may be seen in the less intensive oxidation at 550°C. Whereas the growing oxide scale is filling the crack tip and thus, reducing the effective stress intensity factor at 950°C, this mechanism is not possible to act at lower temperatures [11].

- Whereas the Paris region seems to be independent of R, strong differences occur in the threshold region: For both load ratios tested a very strong dependence of the fatigue crack growth rate on the ΔK-value was found. This means, that small changes in the cyclic stress intensity factor leads either to fatigue crack propagation above 10^{-6} mm/cycle or no crack propagation was observed even after 10^6 cycles. Thus, fatigue crack growth rates below 10^{-6} mm/cycle were hardly measurable. This "sharp" threshold at 550°C contrasts the finding of a "smooth" threshold behavior at 950°C (see Fig. 2 and 3).

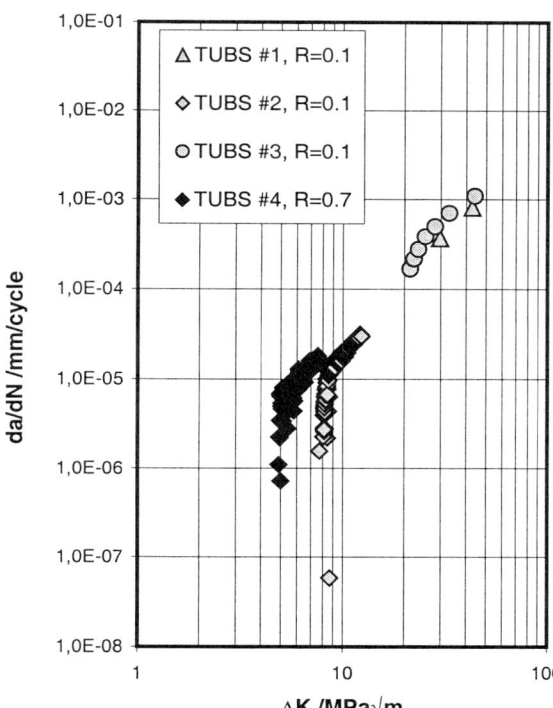

Fig. 6: Fatigue crack growth rates for CMSX-4 at 550°C, <001> load direction, R=0.1 and R=0.7 (TUBS refers to test laboratory).

Table III: Threshold values for fatigue crack growth in CMSX-4 at 550°C.

T	R=0.1	R=0.7
550	8.0 MPa√m	5.0 MPa√m

The fractographic observations made at 550°C were similar to that at 950°C: A macro- and microscopically smooth fracture (mode I on {001} planes) with an oxide scale was found at low and intermediate cyclic stress intensities.

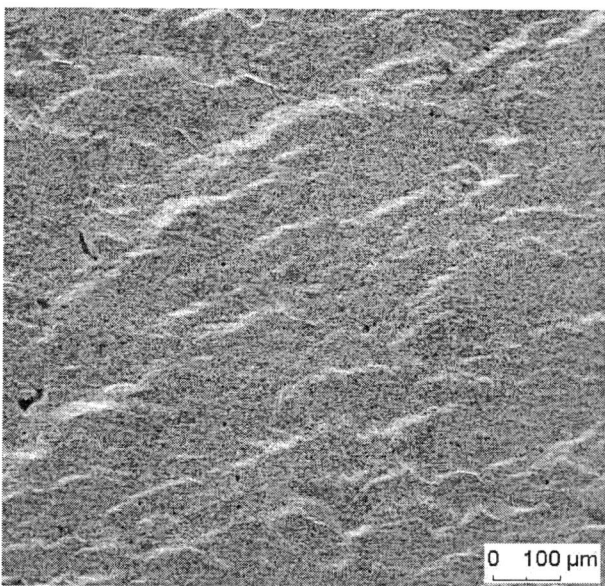

Fig. 7: Fracture surface near the specimen sideface after fatigue crack propagation at higher ΔK and 550°C (R=0.1). A smooth fracture surface with an oxide scale is visible.

At higher cyclic stress intensity factors a deviation from mode I is visible. Especially at the specimen sideface crack propagation along {111} planes was found which could be attributed to the more easily shearing of γ' particles at high stresses [10]. Also a step-wise crack propagation along (macroscopic) {100} and {111} planes was found (Fig. 7).

Influence of Hold Times

In order to study the effect of hold time on fatigue crack propagation, experiments with a hold time at the maximum stress intensity level were performed. Because oxidation and creep effects are expected contributions to fatigue crack growth with hold times, these test were run at 950°C (R=0.7) only. After a package of 1000 cycles (at 5 Hz) a hold time of 300 sec was introduced. The results are given in Fig. 8.

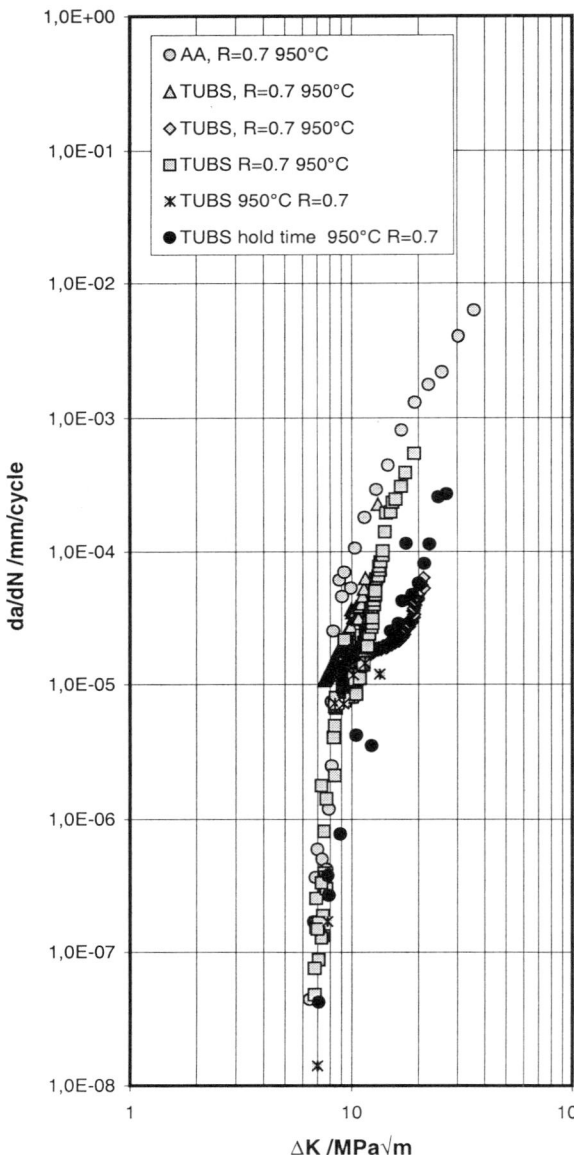

Fig. 8: Influence of hold time on fatigue crack growth rate in CMSX-4 at 950°C (<001> load direction, R=0.7). The hold time of 300 sec was introduced at the peak stress after every 1000 cycles at 5Hz.

As expected from literature findings [12], the introduction of a hold time leads to a retardation of the fatigue crack growth rate by one order of magnitude. But this observation is only true for higher stress intensities. Near the threshold value (ΔK<9MPa√m) no differences between tests with and without hold times were found. In principle, hold time effects may be explained by two contributions: crack tip blunting due to creep deformation and crack closure caused by oxidation. As the oxidation time pre crack growth increment increases with decreasing crack growth rate, the latter mechanism is expected to dominate at low ΔK values. In contrast, crack tip blunting by creep deformation requires sufficiently high stress intensities. Thus, it appears that the observed hold time effect is a consequence of creep deformation rather than oxidation.

The conclusion is also supported by metallographic cross sections of the fracture path after fatigue crack growth with hold times. As shown in Fig. 9 and 10, a wavy structure is apparent where each wave can be a attributed to a hold time. During this hold time, the material creeps and secondary cracks are introduced at higher stress intensities.

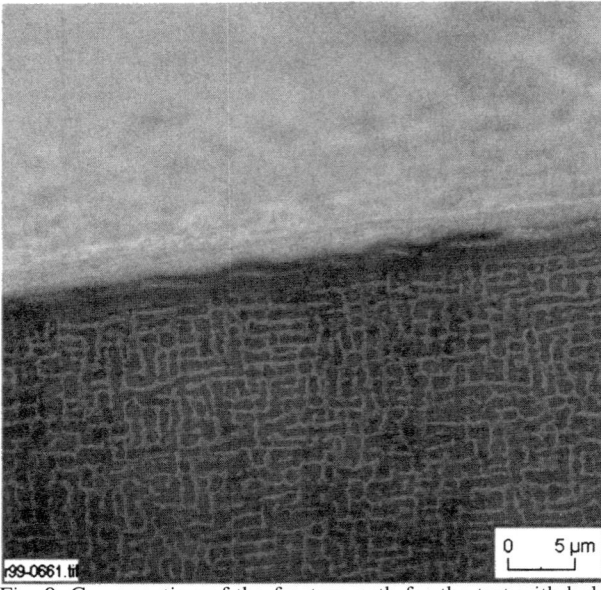

Fig. 9: Cross section of the fracture path for the test with hold times at low ΔK. A wavy structure is apparent.

The above observation is schematically illustrated in Fig. 11. As shown in Fig. 11a, pure fatigue crack propagation is accompanied by a very sharp crack tip. Fig. 11b shows the situation after 300 sec creep hold time: due to creep the crack tip is blunted. This creep time is followed by 1000 pure cycles, leading again to crack tip sharpening and growth (Fig. 11c). The crack length increment is according to the fatigue crack growth rate at the given ΔK level over 1000 cycles. In consequence, a wavy fracture path results where each "step" is associated with hold time creep and the crack advance per fatigue package is given by the distance between "steps".

352

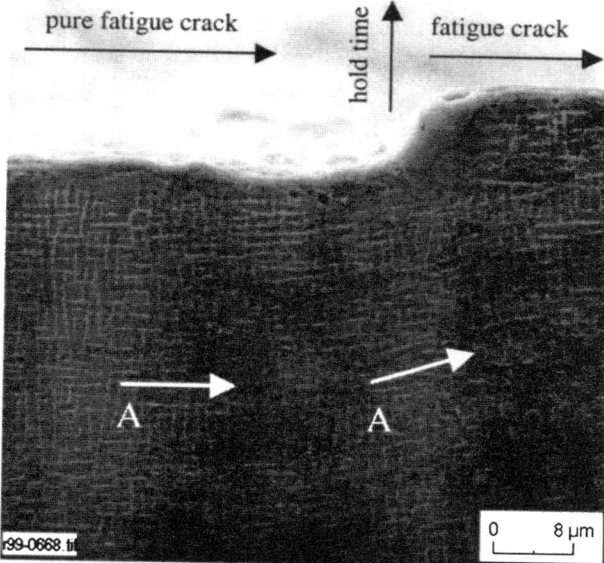

Fig. 10: Cross section of the fracture path for the test with hold times at high ΔK. The wavy structure is clearly visible. The end of each wave is accompanied by creep deformation, which can be seen from deformation of γ' (see arrows A).

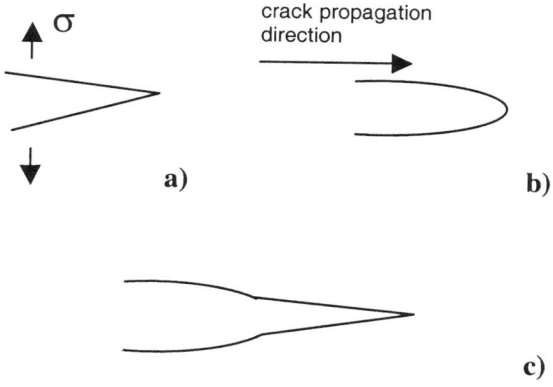

Fig. 11: Schematic illustrating the formation of wavy fracture path during fatigue crack growth with hold times. For details see text.

Influence of Specimen Orientation

No differences between <001> and <011> as loading direction were found at 550°C (R=0.7) as well as at 950°C (R=0.1 and 0.7). All <011> fatigue crack growth curves were in the scatter band of <001> direction with the same curve shape. The fracture surface was again macroscopically smooth and oriented perpendicular to the load axis. The same independence of FCG rates on crystal orientation with respect to load axis was reported for MAR-M200 [13,14].

Low Cycle Fatigue and Thermomechanical Fatigue Tests

In Fig. 12 lifetimes of the TMF tests are compared to those of the LCF tests performed at minimum and maximum temperature. Out-of-phase TMF lives coincided very well with those of LCF dwell time tests at maximum temperature. LCF loading at 950°C without dwell resulted in lifetimes being longer by a factor of approximately 8 while isothermal fatigue at the minimum temperature of the TMF test revealed a further increase of life by a factor of 3 to 4.

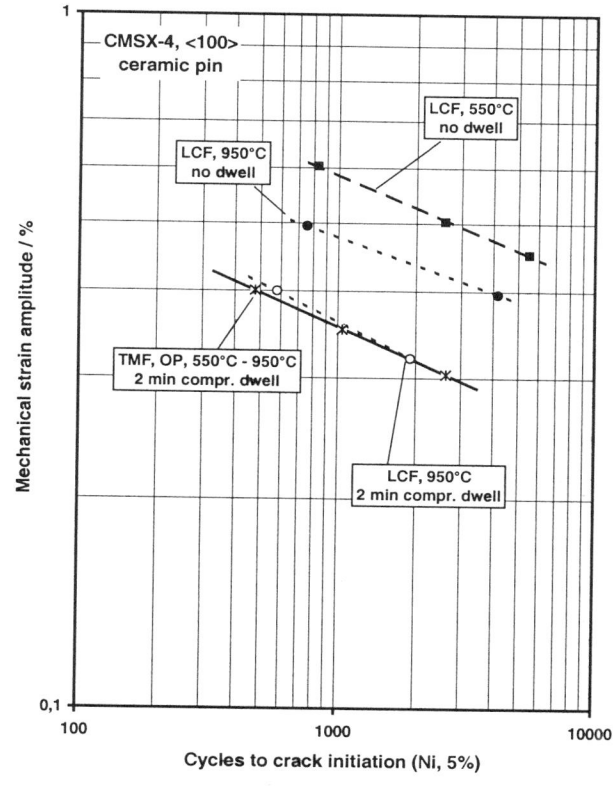

Fig.12: Lifetimes of TMF tests compared with LCF tests performed at minimum and maximum temperature, specimen with ceramic pin as crack starter.

353

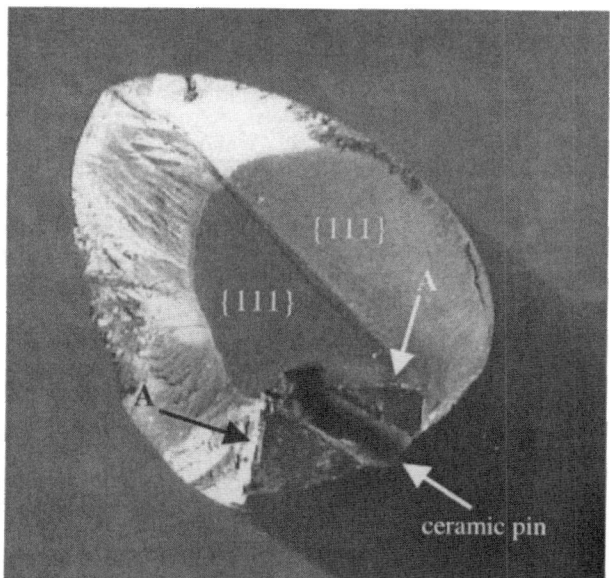

Fig. 13: Fracture surface of a LCF specimen tested at 550°C ($\Delta\varepsilon$=1%, no dwell time, N_f=2600). For details see text.

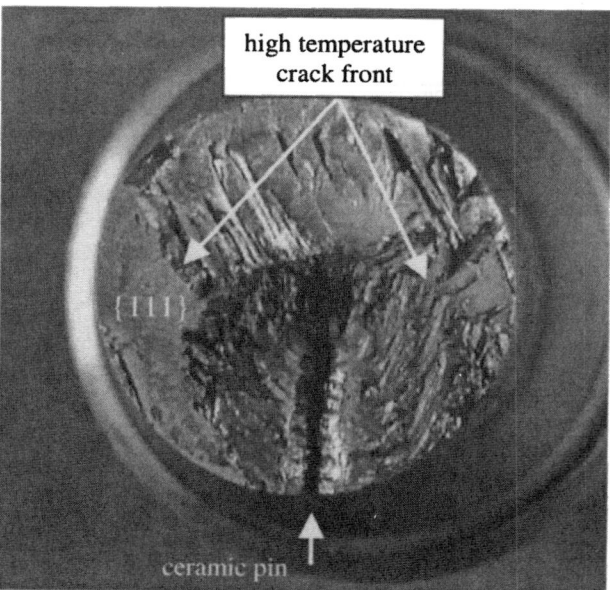

high temperature crack front

ceramic pin

Fig. 14: Fracture surface of a LCF specimen tested at 950°C ($\Delta\varepsilon$=0.8%, 2 min compression dwell time, N_f=720). For details see text.

The fracture surface of a broken LCF sample after testing at 550°C ($\Delta\varepsilon$=1%, no dwell time, N_f=2600) is shown in Fig. 13. Starting from the edges of the ceramic pin the crack propagated macroscopically in Mode I for about 2 mm. This part of fracture, designated with arrows A in Fig. 13, has a smooth surface and is comparable to the fracture type of CT specimen at 550°C. But, as clearly visible in Fig. 13, the crack

deviated to grow along two {111} planes. It could be concluded that fracture along {111} is the preferred one because this type, only evident on one side of the ceramic pin, led to complete fracture of the sample. The crystallographic fracture type was never found in CT specimen to such an extend. Only at very high ΔK values some cracking along {111} planes at the surface of the CT specimens was evident which is in agreement with other authors [15]. Probably LCF crack growth rates along {111} planes correspond to very high ΔK values at FCG in CT specimen.

In Fig. 14 the fracture surface of a LCF specimen tested at 950°C ($\Delta\varepsilon$=0.8%, 2 min compression dwell time, N_f=720) is shown. Compared with the tests at 550°C the following features were observed:

- The crack propagates primary at Mode I, but this fracture is accompanied by a microscopically rough fracture surface.
- As at 550°C some crystallographic fracture was observed at 950°C. This type is located at the surface of the sample which may be explained by the plane stress state.

Surface Fatigue Crack Propagation During LCF and TMF Testing

The surface crack length observed at the ceramic pins in dependence on cycle number is plotted in Fig. 15. Comparable crack growth was observed for TMF dwell and LCF dwell at 950°C for comparable strain ranges. Crack growth was reduced for 950°C LCF loading without dwell and lowest crack growth was observed for 550°C LCF tests without dwell. But it must be mentioned, that fast crack growth along {111} planes as shown in Fig. 14 was not included in these crack propagation measurements.

Crack growth rates vs. range of stress intensity (calculated according to Eqn. 1 and 2) are presented in Fig. 16 for the 550°C LCF tests and in Fig. 17 for the 950°C LCF and the TMF tests. Included in both plots are the scatterbands determined by the related CT specimen tests of Fig. 6 and Fig. 2 with the load ratio R = 0.1. For the 550°C tests a good correlation between the standard fracture mechanics fatigue crack propagation test results and the video monitoring of surface crack growth in the LCF tests was observed at high stress intensity ranges. The higher apparent threshold stress intensity ΔK_{th} observed for the LCF tests can be readily interpreted as a consequence of the different R-values. At 950°C LCF crack propagation measured in tests without dwell corresponded well to fracture mechanics data. LCF tests with dwell periods and TMF tests (also with dwell periods) revealed higher crack growth rates. This was attributed to the dwell periods in each cycle which were not included in the fatigue crack propagation tests on standard CT specimens.

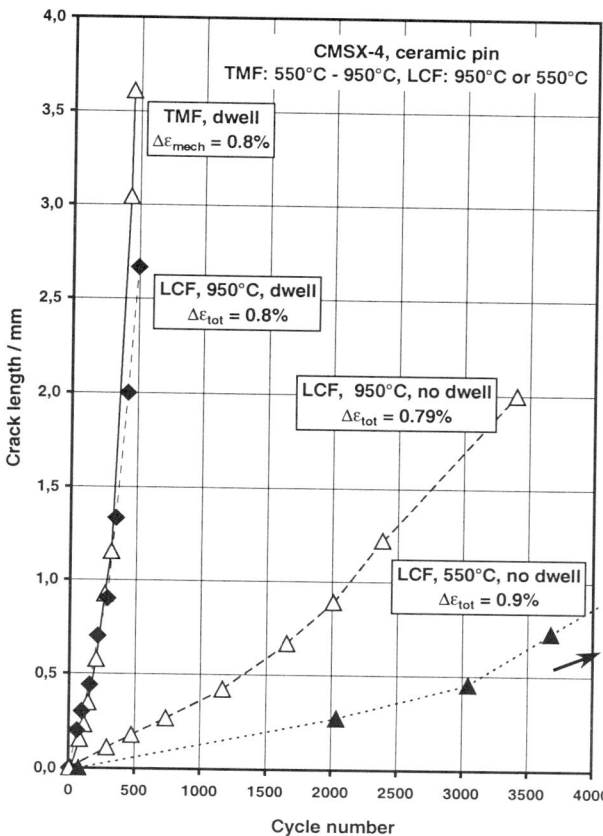

Fig. 15: Crack lengths at the ceramic pins as a function of cycle numbers for TMF tests and LCF tests at minimum and maximum temperature.

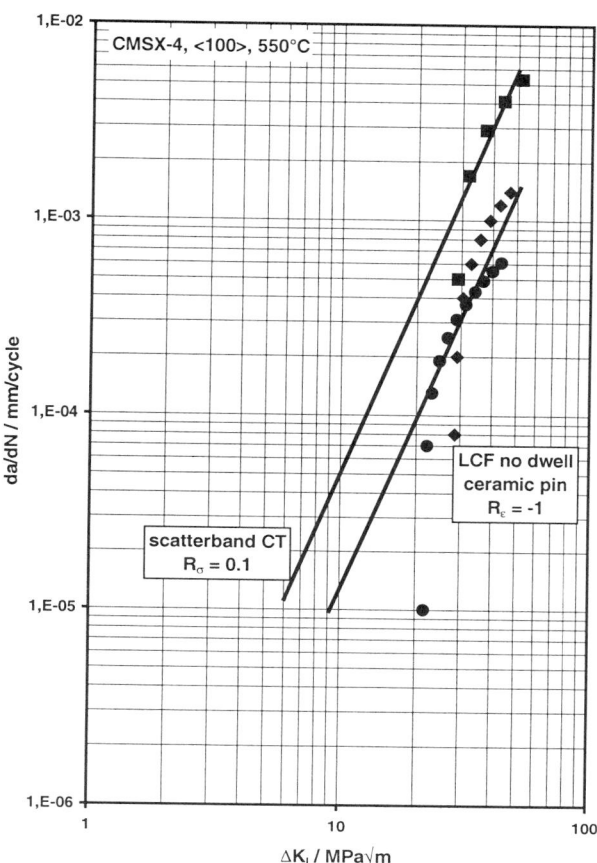

Fig. 16: Crack growth rates vs. range of stress intensity factor for LCF tests at 550°C, LCF specimen with ceramic pin.

CONCLUSIONS

- The fatigue crack growth rate of CMSX-4 in <001> load direction/<010> crack propagation direction showed the following characteristics:
  - Increasing R increases the threshold value ΔK_{th} and the FCG rate in the Paris regime of da/dN at both temperatures.
  - Increasing the temperature from 550°C to 950°C increases the threshold for FCG. Whereas a "sharp" threshold region was found at 550°C, a "smooth" transition from threshold to the Paris regime was found at 950°C.
  - Within the Paris regime, the scatter band of 550°C falls within the band of 950°C data.
- Changing the load axis from <001> to <011> did not influence the FCG rate.
- For all temperatures, R values and orientations a mode I fracture along {001} planes was observed. Room temperature precracking lead to stage I FCG along {111} planes.

- Introducing hold times after 1000-cycle packages during FCG at 950°C/R=0.1, the FCG rate is lowered at higher ΔK due to creep deformation whereas at lower ΔK no influence was found. In contrast, introducing a compressive hold time during LCF/TMF testing reduces the lifetime. Whereas the tensile hold time during FCG leads to creep deformation, the compressive hold time during LCF/TMF testing does not lead to creep, but shifts the mean stress of the cycle to higher tensile stresses.
- In contrast, a compressive dwell time during LCF testing (specimen with ceramic pin) at 950°C leads to a reduction in lifetime by factor 8.
- Out-of-Phase TMF test results compare well with LCF (dwell time at maximum temperature of TMF) data. In contrast, LCF tests at minimum temperature and LCF tests at maximum temperature without dwell time lead to longer lifetimes.
- By comparing the da/dN data from CT specimen with crack propagation starting from defects under LCF/TMF loading the following observations were made:

355

- At 550°C a good correlation between LCF tests without dwell times and CT results was found.
- At 950°C the LCF data without dwell time fit to data from CT tests, whereas LCF tests with dwell times as well as TMF tests showed higher FCG rates than CT tests. This is somehow surprising since LCF/TMF tests were performed with negative R-values. Probably, short cracks may grow faster than one would expect from long crack growth measurements from CT specimen. This difference seems to manifest itself also in different fracture modes in LCF/TMF and FCG testing.

ACKNOWLEDGEMENTS

The work was supported by BMBF grant No. 0327058A and is gratefully acknowledged.

REFERENCES

1. Harries, G.L. Erickson, S.L. Sikkenga, W.D. Brentnal, J.M. Aurrecoechea and K.G. Kubarych, in: Superalloys 1992, S.D. Antolovich et al. (Eds.), 297-306, TMS 1992.
2. O.L. Bowie, J. Math. and Phys., 35, (1956) 60-71.
3. A. Sengupta and S.K. Putatunda, Scripta Met., 31, (1994) 1163-1168.
4. B.A. Lerch and S.D. Antolovich, Met. Trans. A, 21A (1990), 2169-2177.
5. J.S. Crompton, J.W. Martin, Met. Trans. A, 15A (1984) 1711.
6. A. Sengupta, S.K. Putatunda, M. Balogh, J. Mat. Eng. and Perf., 3 (1994) 540-550.
7. E. Affeld, J. Timm, A. Bennet, in: Fatigue under Thermal and Mechanical Loading, J. Bressers et al. (Eds.) 159-169 (Netherlands) Kluwer Academic Publishers 1996.
8. A. Diboine, J. Peltier, R. Pelloux, In: Proc. of MECAMAT Int. Seminar on High Temperature Fracture Mechanisms and Mechanics, Part –IV, P. Bensussan et al. (Eds.), 71-94, Dourdan, France (1987).
9. V. Lupinc, G. Onofrio, G. Vimercati, in: Proc. of the Conf. on Superalloys 1992, S.D. Antolovich et al. (Eds.), 717-726, TMS 1992.
10. T. Rieck, F. Schubert, Berichte des Forschungszentrums Jülich, Nr. 3706, Jülich 1999.
11. M. Henderson, J. Martin, In: Superalloys 1992, S.D. Antolovich et al. (Eds.), 707-716. TMS (1992).
12. G. Onofrio, S. Ai, V. Lupinc, G. Vimercati, Conf.: Fracture Behaviour and Design of Materials and Structures - ECF8, Torino, Italy, 1231-1236.
13. K.S. Chan, J.E. Hack, G.R. Leverant, Met. Trans. A, 18A, (1987) 581-591.
14. K.S. Chan, J.E. Hack, G.R. Leverant, Met. Trans. A, 18A, (1987) 593-602.
15. J. Telesman, L.J. Ghosn, Eng. Frac. Mech., 34, (1996) 1183-1196.

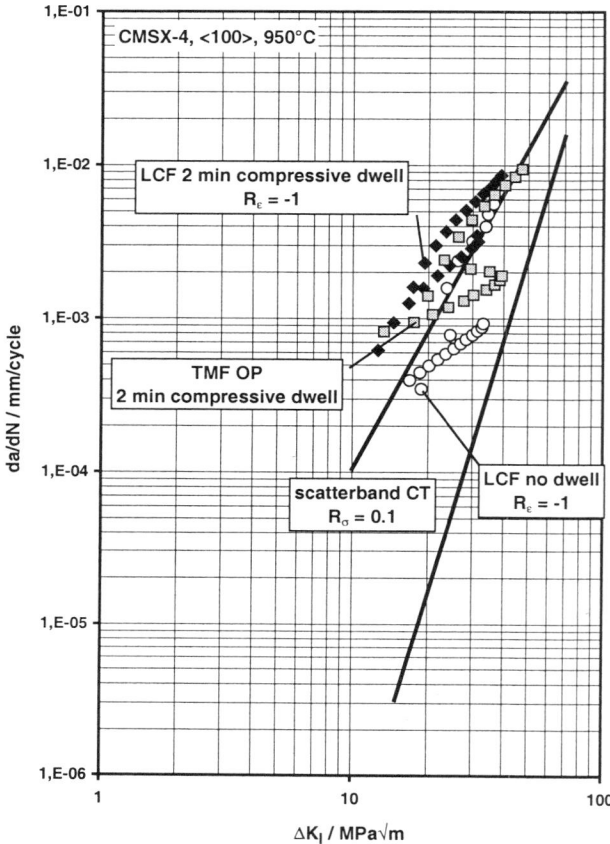

Fig. 17: Crack growth rates vs. range of stress intensity factor for TMF tests and LCF tests at 950°C.

MODELLING THE ANISOTROPIC AND BIAXIAL CREEP BEHAVIOUR OF NI-BASE SINGLE CRYSTAL SUPERALLOYS CMSX-4 AND SRR99 AT 1223 K.

D.W. MacLachlan, L.W. Wright, S.S.K. Gunturi, and D.M. Knowles

Rolls Royce University Technology Centre
Department of Materials Science & Metallurgy
University of Cambridge, UK.

Abstract

The use of single crystal turbine blade components has made many traditional models for creep unsuitable for application in certain loading situations. Such models generally assume deformation is isotropic. Since the introduction of single crystal components a generation of crystal plasticity or slip system models has emerged. Amongst the challenges of such models is to explain the orientation dependence of uniaxial creep behaviour and to attempt to make correlation between uniaxial and multiaxial stress states so that confidence can be gained as to the performance of a model when it is applied to complex loading conditions. In this paper an attempt is made to address these issues. A slip system based finite element creep model has been fitted to uniaxial data at 1223 K in a range of crystallographic orientations. Based on experimental evidence the operative slip systems incorporated are the {111}<101> and {111}<112> families of slip system.

Analysis of the data and a consideration of the dislocation mechanisms likely to occur suggest that the significant component of the hardening matrix is latent hardening by the {111}<101> systems on the {111}<112> systems, although hardening between the {111}<101> systems may also occur. The model can describe creep deformation as a function of orientation to a reasonable degree of accuracy. It is also shown to have reasonable predictive capability when used to analyse the results of thin cylinder biaxial creep tests on CMSX-4 and SRR99. The reason for this success is explained in terms of activated slip systems and the magnitude of the cumulative shear strain rate on different types of slip system as a function of orientation and stress state.

Superalloys 2000
Edited by T.M. Pollock, R.D. Kissinger, R.R. Bowman,
K.A. Green, M. McLean, S. Olson, and J.J. Schirra
TMS (The Minerals, Metals & Materials Society), 2000

Introduction

Single crystal Ni-base superalloys are commonly employed for use as high temperature creep and oxidation resistant blade alloys in the early stages of modern gas turbine aero-engines. CMSX-4 was developed for such an application by the Cannon-Muskegon Corporation [1]. The excellent high temperature creep properties of this superalloy are derived directly from its microstructure, consisting of 70% volume fraction of coherent, cuboidal γ'-precipitates in a matrix of solid solution strengthened γ-phase, which act as barriers to dislocation motion at high temperature. The large atomic radii elements Mo, W and Re in CMSX-4 preferentially partition to the γ-matrix [2]; causing the relative expansion of the γ-lattice parameter compared with that of the γ'-precipitates, hence CMSX-4 is a negative misfit alloy.

The use of high strength, two phase, single crystal superalloys for blade materials has led to complex relationships between load, temperature and deformation. This is exacerbated by the complex geometry and anisotropy of blade components as well as the multiaxial stress states they experience. Accurate modelling of deformation occurring under such conditions has been highlighted as a requirement for optimising the efficiency in design of components, with respect to both preventing failure and avoiding over-design.

Materials and Methods

The standard creep specimens used in this study were machined from cast bars of CMSX-4 supplied by Rolls-Royce plc. with compositions as given in Table I. The bars were seeded so as to give a number of different axis orientations. The orientations listed in table II and illustrated in figure 1 were determined using a backscattered Laue method [3].

Table I Nominal compositions (wt. %) of CMSX-4.

Ni	Co	Cr	W	Mo	Re	Al	Ti	Ta	Hf
Bal	9.7	6.5	6.4	0.6	2.9	5.6	1.0	6.5	0.1

The bars had undergone a standard three-step solution heat treatment and two-step ageing treatment. The solution and homogenisation heat treatments are designed to minimise residual dendritic microsegregation and maximise phase stability [4]. Figure 2 shows a bright field transmission electron microscopy (TEM) micrograph, typical of the fully heat-treated material. Standard creep specimens were machined from this material with a gauge length of 28.0 mm and diameter of 5.6 mm.

The tests were performed in air on a 20 kN constant load creep machine. The full loads were applied gradually within a minute using a self-levelling arrangement. The operating temperature was controlled by platinum/13% rhodium-platinum thermocouples to ±2 K. The creep strain was measured by averaging two high temperature extensometers with linear displacement voltage transducers (LDVTs). The signals generated were logged on a PC.

Table II Creep specimen orientations.

Test Specimen	Nominal Orientation	θ† Degrees	ρ† Degrees	Test Stress [MPa]
A	[Ī11]	53.2	45.0	250
B	[Ī11]	53.5	42.9	300
C	[Ī11]	52.2	41.4	325
D	[Ī11]	51.5	42.5	355
E	[Ī11]	50.8	43.9	400
F	[Ī11]	53.3	42.5	450
GΨ	[001]	-	-	480
H	[011]	44.9	1.0	350
I	[Ī 19 23]	41.2	3.0	250
J	[Ī 4 10]	23.0	13.6	250
K	[Ī 4 10]	23.0	13.6	400
L	[Ī12]	34.0	40.0	300
M	[2̄67]	42.0	19.3	350
N	[2̄67]	42.0	19.3	400
O	[Ī 25 32]	38.1	2.3	300

† The terms θ and ρ are defined in figure 1.
Ψ This specimen was interrupted before rupture and its orientation was not measured. Additional creep data from the Rolls-Royce database, for nominal < 001 > orientated specimens, were used for model development.

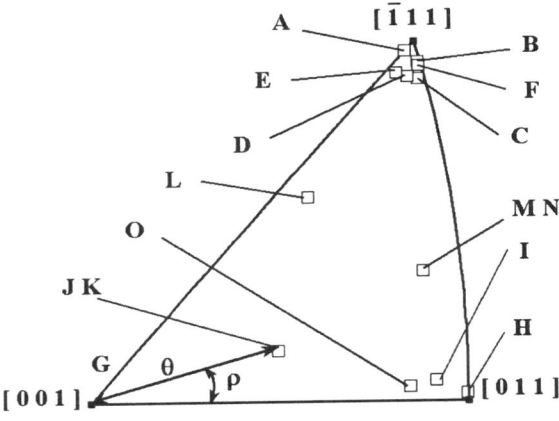

Figure 1: Standard [001] stereographic projection showing the axis orientations of the specimens tested.

Foils taken at a distance greater than 3 mm from the fracture surface were prepared for TEM examination. The ground TEM foils were electropolished in 10% perchloric acid in ethanol in a conventional jet polisher using 30 V at 260 K. TEM examination was conducted on a JEOL 2000 FX with an operating voltage of 200 kV.

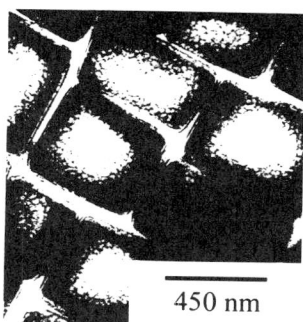

Figure 2: Bright field TEM micrograph of fully heat-treated CMSX-4, foil plane (100). The faces of the γ′-cubes are aligned with the {001} planes. The γ-channels lie in < 001 > directions. The γ-channels width is about one-tenth of the γ′-cube width.

The lattice rotations that occurred along the specimen gauge length during creep deformation were calculated using electron back scatter diffraction patterns (EBSP) [5,6]. Small bars of 3 x 4 mm^2 cross-section were machined along the axial direction of the specimen gauge length. These bars were then mechanically ground and polished using colloidal silica suspension. A JEOL 5800LV scanning electron microscope was then used to obtain Kikuchi patterns at 0.5 mm intervals along the length of the bars. TSL Inc. OIM software, was used to index the crystal orientation by averaging values obtained from scanning a 225 μm^2 area at each position.

Mechanisms

Mechanisms that occur in γ/γ′ superalloys have been reported in numerous works, [7,8]. This section details plastic deformation mechanisms which have been observed at 1223K for the alloy CMSX-4 in the orientations tested (Table II).

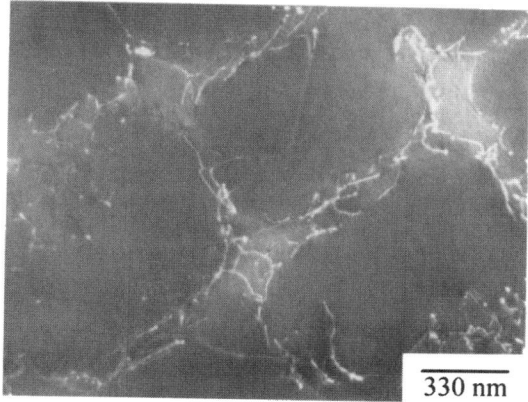

Figure 3: Weak beam TEM micrograph of a specimen orientated close to [001] after 9 hours of creep testing at 480 MPa initial stress, the specimen had accumulated 1.5 % strain.

Towards the [001] pole, networks of $a/2<110>$ dislocations were observed in the γ-phase. Figure 3, taken from a test interrupted after 9 hours orientated close to the [001] pole, shows a typical $a/2<110>$ network of creep dislocations principally confined to the γ-phase. Interfacial dislocation formation around the γ′-precipitates has also occurred.

For specimens tested near to the [111] pole deformation twinning was also observed. Figure 4(a) shows the resultant microstructure after twinning in a specimen with an initial nominal orientation of [$\bar{2}$67]. This mechanism is typical of the deformation occurring in specimens orientated towards [111] (although not observed near [001] orientations). Deformation twinning in superalloys at high temperature has been reported previously. For example in Alloy 718 after low frequency fatigue at 823 K [9] and in alloys CMSX-2 and SRR99 during elevated temperature creep [10] and [11] respectively.

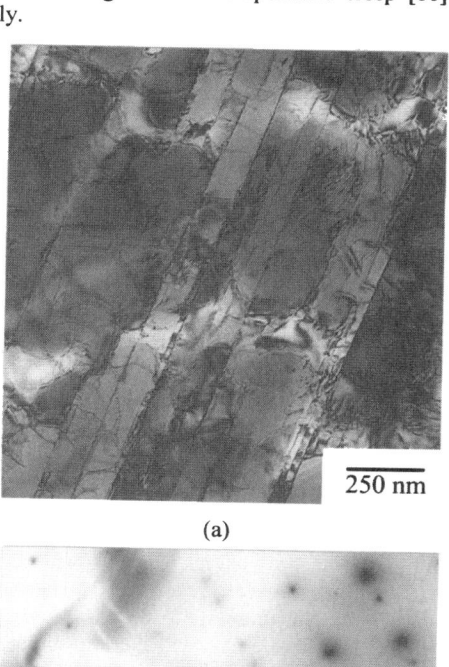

(a)

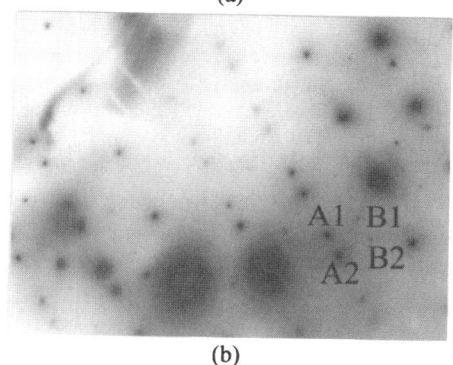

(b)

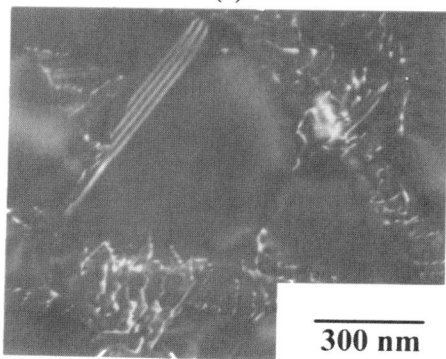

(c)

Figure 4: (a) TEM micrograph of CMSX-4 specimen M with nominal orientation [$\bar{2}$67] tested at 350 MPa. (b) Electron diffraction pattern for (a), twin diffraction spots from the γ-phase, (A1 and A2) and from the γ′-phase, (B1 and B2) are shown. (c) Stacking faults present in specimen N, ([$\bar{2}$67], 400 MPa).

Common with slip deformation, the shearing of a crystal resulting from deformation by twinning enables it to undergo a plastic deformation. The fixed magnitude of the twinning shears shown in figure 4(a) results in the reorientation of parts of the crystal. The orientation relationship of the original crystal and the twinned parts can be obtained from the electron diffraction pattern figure 4(b). The pairs of diffraction spots in the pattern indicate that twinning has occurred. Additionally, the larger pairs of diffraction spots, (e.g. A1 and A2) show that twinning has occurred in the γ-phase, while the fainter pairs of diffraction spots, (e.g. B1 and B2) show that twinning has also occurred in the γ'-phase.

The deformation twins thus extend through the γ/γ' microstructure twinning both the γ and γ' phases. For twining in a FCC, face centred cubic, phase a deformation twin occurs by sliding adjacent (111) planes by a/6[112] resulting in a shear s of magnitude $1/\sqrt{2}$. In a L1$_2$ structure however twinning requires the passage of a dislocation with twice this magnitude, giving $s=\sqrt{2}$ [12]. Such a shear also produces a twin in the FCC lattice, but is not generally observed for energetic reasons. As the twins observed in CMSX-4 are continuous through both phases it would appear that they are the result of the passage of a/3<112> dislocations through the whole structure. The amount of longitudinal plastic strain resulting from deformation twinning is: [13]

$$\varepsilon_p = V_f \left(\sqrt{\left(1 + \gamma^2 \cos^2 \phi + 2\gamma \cos \phi \cos \lambda\right)} - 1 \right) \quad (1)$$

where: ε_p is the plastic strain associated with twinning; V_f is the volume fraction of twins in the crystal; γ is the shear strain; ϕ is the angle between the tensile axis and slip plane normal and λ is the angle between the tensile axis and the slip direction. From Table II, for specimen M, the relation between the nominal tensile axis [$\bar{2}$67] and the most highly resolved system for this orientation (111)[$\bar{2}$11], gives $\phi = 47.7$ ° and $\lambda = 42.6$ °, with γ as $\sqrt{2}$, thus from equation 1, every one percent volume fraction of twins in the specimen contributes 0.8 % to the longitudinal strain.

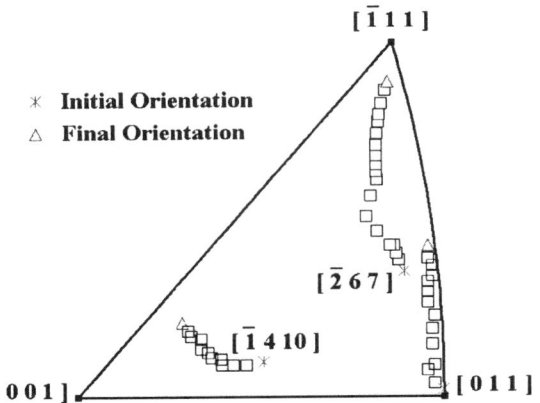

Figure 5: EBSP measurements for specimens H, J, and M plotted within a standard [001] pole figure.

Figure 5 shows the lattice rotations resulting from creep testing of three different initial orientations. For specimen J, with an initial orientation of [$\bar{1}$ 4 10] tested at a stress of 250 MPa, no twin deformation was observed, the rotation shown is consistent with slip towards the [101] pole as a result of the a/2<101> dislocations observed in the γ-phase. For specimen H, with an initial orientation of [011] tested with an initial stress of 350 MPa the rotation is along the [011]-[111] boundary towards the [$\bar{1}$11] pole. For specimen M, with an initial orientation of [$\bar{2}$67] the lattice rotation generally proceeds towards the [211] pole. The failure strain for this specimen was 35 %. The volume fraction of twins in this specimen is approximately 15-20 % which using equation 1 leads to a longitudinal strain of approximately 12–15 %. Twining is clearly a significant deformation mechanism in this orientation but is not responsible for the total axial strain observed.

<u>Discussion</u>

A significant problem in the use of slip system models to predict anisotropic creep at relatively high temperatures (above 1173 K) has been their inability to explain the low creep resistance of the [111] orientation when compared to the [001] orientation. The [111] orientation is approximately a factor of two stronger than the [001] orientation, though the relative resistance increases as the applied load is reduced [14]. If deformation is assumed to occur in both orientations by the operation of {111}<101> slip, resistance in the [001] orientation should be orders of magnitude stronger, based on the relative Schmid factors for the two orientations. For brevity {111}<101> and {111}<112> slip will from here on be referred to as type I and type II slip respectively. It is widely known that deformation at 1223 K in the [001] orientation occurs by type I slip [8]. The test results suggest that a different mechanism is operating in the [111] orientation. Previous slip system approaches have addressed this issue by asserting that cube slip is responsible for the apparent lower resistance in the [111] orientation [15, 16]. There is, however very little evidence for such cube slip in single crystal superalloys [17] and its use has been called into question [18]. Many slip system models using cube slips invoke a macroscopic cube slip mechanism as being similar to type I slip restricted to the matrix channels that lie in cubic orientations. However, as there is no direct microstructural evidence of cube slip, whilst type II slip is clearly active, it is important to investigate the effects of incorporating the latter into a macroscopic model.

Experimental evidence given in the previous section shows that type II slip systems may be operative in orientations towards the [111] pole. Figure 5 shows the results of EBSD measurements of the change in crystallographic orientation with strain of an initially orientated [$\bar{2}$67] specimen located close to the [011]–[111] boundary. The rotations are similar to those that would occur if the primary type II slip system was active. The exact rotations that would occur in this situation are shown in figure 6 taken from MacKay and Maier [19]. TEM analysis of a specimen deformed in this orientation has shown that a major mechanism of deformation is by twinning, figure 4, which is derived from type II slip and is consistent with incorporating such a mechanism into a slip system model. Similar rotations to those in figure 5 could occur if the two cubic slip systems (001)<110> and (010)<101> were active, but in light of extensive twinning associated with creep in this orientation it is more likely due to type II slip. From the previous section it is clear that the total global strain experienced by the specimen did not come from deformation twinning and additional strain may be due to type I or type II slip. In either case the reduced creep resistance (relative to an analysis based on type I Schmid factors) as one approaches the [011]-[111] boundary and the [111] pole may be due to the operation of an additional deformation mechanism.

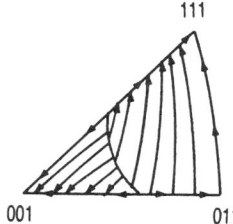

Figure 6: Ideal crystallographic rotations that would occur if deformation resulted from slip on the most highly stressed type II system, taken from [19].

If type II systems are considered to be operative towards the [111] orientation a similar problem is encountered as for modelling with the type I systems mentioned above. This can be seen with reference to figure 7 (taken from [19]) which shows the highest type I and II Schmid factors as a function of orientation in the standard stereographic triangle. The Schmid factor for type II slip is much higher in the [001] pole than it is in the [111] pole so if type II slip is active in the latter why should it give way to type I slip in the former?

Creep deformation on type I slip systems is confined to the continuous matrix phase for a significant portion of a creep test and does not usually occur within the precipitates before the onset of tertiary creep. The reason for this is primarily the high energy APB that would be created if an a/2[110] dislocation entered the precipitate. It is well known that during creep type I dislocations percolate throughout the matrix, which has a very low initial dislocation density [8]. As this process proceeds, interfacial dislocations are laid down at the phase boundary relieving coherency stresses. This situation is in contrast to that which occurs when type II dislocations are active. Type II dislocations pass through both matrix and precipitate, in order for type II dislocations to fully restore the L1$_2$ lattice the burgers vector must be of magnitude a<112>. This large vector may dissociate in various ways as it shears both matrix and precipitate [7, 20]. The important point however is that there is much evidence which suggests that such a dislocation can pass through and completely restore the structure whilst leaving no deformation or dislocation debris in its wake. Such a dislocation reaction has been reported at lower temperatures [21] where the passage of an a[112] dislocation was composed of a/3[112] + 2a/6[112] + a/3[112] acting in the form of a ribbon. In certain situations [20] loops or intrinsic stacking fault are seen left behind after such dislocation motion, but these are frequently forms of superlattice stacking fault and are confined to the precipitates (as in figure 4c).

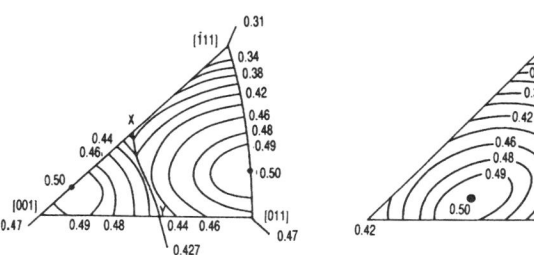

Figure 7: Schmid factor contours for the most highly stressed type II and type I slip system respectively, taken from [19].

In order for significant strain to be associated with specific slip system activity the dislocations need to be able to travel through large portions of the lattice. Type I dislocations reside chiefly within the interconnected γ-channels and although slip is hindered by the γ′-precipitates, climb and cross slip processes provide mechanisms for continued progress through the structure. For deformation via type II slip the dislocations must be able to pass through both the γ and γ′ phases. This may occur by dislocation pairs expanding or contracting in the two phases, but it necessarily requires that they can proceed through the precipitate interfaces and γ channels relatively unhindered. If this is not the case their activity will be confined to individual precipitates leading to reduced strain levels.

Although the passage of type II dislocations through the lattice will lead to little disruption of the structure which affects type I activity, the same cannot be said of the converse case. If type I slip systems are highly activated their progress though the material will leave a wake of dislocation networks at the precipitate interfaces and a further build up of dislocation networks in the γ-channels. If conditions are such that type II dislocations are activated they will interact with these networks on entering or leaving the precipitates leading to the rapid formation of jogs and tangles which will render them sessile, terminating type II activity. It can be concluded that type II activity will have very little influence on type I slip systems whose deformation is chiefly confined to the γ channels. However if conditions are such that the type I slip systems are stressed significantly the subsequent dislocation network which ensues in the γ channels will be sufficient to inhibit significant type II deformation leading to rapid hardening.

Modelling

Uniaxial Creep

Following the reasoning of the previous section uniaxial creep behaviour at 1223 K can be rationalised as follows. In the [001] orientation deformation proceeds rapidly on type I slip systems due to the high Schmid factors they experience, this causes the formation of a network of dislocations throughout the matrix, located primarily at the interface. Such a network prevents precipitate shear by type II dislocations and hence restricts activity of type II systems. As the orientation changes from [001] towards [111] the activity on type I slip systems decreases preventing hardening and thus allowing long term activation of the type II systems.

The model used in this analysis is based on the established theory of finite deformation single crystal plasticity [22]. Deformation is analysed using the FE package ABAQUS, the implementation of the FE slip system model has been described previously [14]. Earlier studies [23, 24] have incorporated slip system models into FE codes primarily to analyse the effect of type 1 and cube slip. The current work aims to use a similar macroscopic technique to investigate the effect of type I and type II shear on macroscopic behaviour. In the analysis the full specimen is simulated, which enables accurate application of loads and boundary conditions and also the analysis of biaxial stress states. Half of the solid specimen model is shown in figure 8 from which the internal and external mesh can be seen. During creep in unstable crystallographic orientations elastic rotation causes bending of the specimen resulting in non-uniform stresses and an elliptical cross section. Such features can be incorporated into the creep analysis using a solid model [23]. As a qualitative comparison figure 9 shows the undeformed creep specimen and mesh as well as the deformed shape following actual and simulated creep in an unstable orientation where rotation occurs.

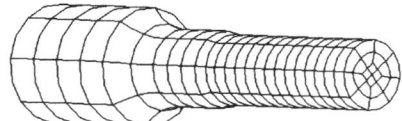

Figure 8: One-half of the solid specimen used in the FE model showing internal and external mesh.

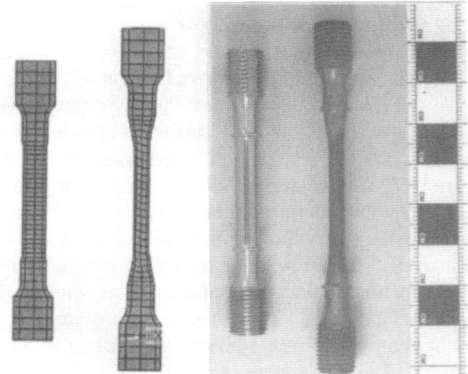

Figure 9: Comparison of creep test specimen and FE model specimen before and after uniaxial creep deformation in an unstable orientation.

The equations used to describe the shear strain rates on the different slip systems are given by [14, 25]:

$$\dot{\gamma}_\alpha = E_{101} \times \left(\frac{\tau_\alpha}{\tau_{C101} - \tau_\alpha} \right)^{u101} \quad \ldots \quad \alpha=1 \text{ to } 12 \text{ (Type I)} \quad (2)$$

$$\dot{\gamma}_\alpha = E_{112} \times \left(\frac{\tau_\alpha}{\tau_{C112} - \tau_\alpha} \right)^{u112} \quad \ldots \quad \alpha=13 \text{ to } 24 \text{ (Type II)} \quad (3)$$

α is the slip system number, E and u are material constants describing shear strain rate and τ_c is the critical resolved shear stress determined from tensile tests. Whilst it is straightforward to conclude that shear strain should accumulate on specific slip

systems it is not evident how damage should evolve as a function of stress or strain. It is clear that damage does not occur on, and should not be related to specific slip systems as the accumulation of plastic shear strain does not cause damage (it may cause hardening) but the perfect restoration of the crystal lattice (large strain may result in intense shear bands but such localised processes are not considered here). Uniaxial rupture life cannot be correlated with stress based parameters such as the stress tensor or von Mises equivalent stress as they predict a rupture life independent of orientation. It has been shown [14] that rupture life also does not correlate directly with Schmid factors on active slip systems. Furthermore, as will be shown later, biaxial rupture life does not correlate with von Mises effective stress. As such stress based parameters are inappropriate damage accumulation is taken as a function of the strain rate tensor derived from equations 2 and 3 giving the additional equations:

$$\dot{\omega}_{101} = C_{101} \times \left[\frac{\dot{\varepsilon}_{101}}{E_{101}} \right]^{v101/u101} \quad (4)$$

$$\dot{\omega}_{112} = C_{112} \times \left[\frac{\dot{\varepsilon}_{112}}{E_{112}} \right]^{v112/u112} \quad (5)$$

$$\dot{\omega} = \dot{\omega}_{101} + \dot{\omega}_{112} \quad (6)$$

$$\dot{\varepsilon}_{101} = \sum_\alpha \dot{\gamma}_\alpha \mathbf{b}_\alpha \mathbf{n}_\alpha \quad \ldots \quad \alpha=1 \text{ to } 12 \quad (7)$$

$$\dot{\varepsilon}_{112} = \sum_\alpha \dot{\gamma}_\alpha \mathbf{b}_\alpha \mathbf{n}_\alpha \quad \ldots \quad \alpha=13 \text{ to } 24 \quad (8)$$

C and v are further material parameters and $\mathbf{b}$ and $\mathbf{n}$ are the slip system direction and slip plane normal respectively. The subsequent effect of global damage on stress and strain rates is factored into the model in the conventional damage mechanics method using an effective stress, equation 9 from which the resolved shear stresses in equations 2 and 3 are calculated.

$$\sigma = \frac{\sigma_0}{1 - \omega} \quad (9)$$

Using strain rate rather than stress to determine a global damage state allows the combined effects of load and crystallographic orientation to be factored into the rate of damage accumulation. The effect of resolved shear stresses on the various slip systems is therefore incorporated into the equations as a global rather than a local effect. Two tensor measures of damage are used which describe how slip on different types of slip system effect the material. Thus the effects of different origins of shear strain on the material is incorporated into the model at the macroscopic level. If the model is to be used consistently to describe strain evolution in a range of orientations, including the [001] and [111] poles, it is necessary to utilise equations describing the hardening mechanisms discussed earlier.

In classical single crystal plasticity, interaction (or hardening) between different slip systems is introduced through the hardening matrix. One set of equations which has been used [26] is:

$$\dot{\gamma}_\alpha = \dot{a}_\alpha \left(\frac{\tau_\alpha}{g_\alpha} \right)^{\frac{1}{m}} \quad (10)$$

362

$$\dot{g}_\alpha = \sum_\alpha h_{\alpha\beta} \dot{\gamma}_\beta \qquad (11)$$

$$h_{\alpha\beta} = qh + (1-q)h\delta_{\alpha\beta} \qquad (12)$$

$$h = h_o Sech^2\left(\frac{h_o\gamma}{\tau_s - \tau_o}\right) \qquad (13)$$

$$\dot{g}_\alpha = \sum_\beta h_{\alpha\beta} f(\dot{\gamma}_\beta) \qquad (15)$$

$$f(\dot{\gamma}_\beta) = \left\{ 1 + tanh\left[a_0 \ log\left(\frac{\dot{\gamma}_\beta}{\dot{\gamma}_{ref}}\right)\right]\right\} \qquad (16)$$

a_0 and $\dot{\gamma}_{ref}$ are material parameters.

In these equations $\dot{a}$ and m are material constants and g characterises the current strain hardened state of the crystal. The rate of material hardening is specified by the evolution equations for $\dot{g}_\alpha$. The constant h_o represents an initial hardening rate and τ_s denotes a saturation strength, γ in equation 13 is the sum of accumulated shear strain on all slip systems. The use of the Kronecker delta in equation 12 makes the hardening matrix symmetric such that the degree of self hardening for all systems is equal and the rate of latent hardening by all slip systems on each other is equal, the difference between self and latent hardening is set by the parameter q.

When different types of slip systems are active the equations and especially the hardening matrix take a slightly different form. Firstly, there are two different intrinsic types of slip system, twelve of each, between which hardening will be different, hence the hardening matrix is not symmetric. The use of the hardening matrix to describe interactions between similar and dissimilar slip systems has been investigated elsewhere [24, 27]. In the current analysis the effect of hardening is restricted to hardening by <101> systems on <112> systems. Equation 12 is therefore replaced with the hardening matrix:

$$h_{\alpha\beta} = \begin{array}{c|cccccc} & 1 & . & 12 & 13 & . & 24 \\ \hline 1 & 0 & . & 0 & 0 & . & 0 \\ . & . & & . & . & & . \\ 12 & 0 & . & 0 & 0 & . & 0 \\ 13 & H_{13,1} & . & H_{13,12} & 0 & . & 0 \\ . & . & & . & . & & . \\ 24 & H_{24,1} & . & H_{24,12} & 0 & . & 0 \end{array} \qquad (14)$$

Where $H_{ij}=H_{kl}$ for $13 \le i,k \le 24$ and $j,l \le 12$. Equation 13 is retained in our analysis as it provides a sensible function by which hardening should saturate with accumulated strain, in this case the accumulated shear strain, γ, is restricted to that from the <101> systems.

The attainment of stable interfacial dislocation networks surrounding precipitates is a very rapid process [8, 14, 25] and is highly sensitive to the rate of shear on octahedral slip systems; this is evident by the appearance of these networks under the TEM in creep specimens which have been interrupted very early into the creep life. It has been shown that the interfacial dislocation network was almost fully formed in a creep specimen interrupted after 20 minutes and 0.5 % creep strain during testing at 950 °C and 480 MPa in the [001] orientation [14]. The rapid formation of this hardening structure is thought to be responsible for the lack of significant primary creep in CMSX-4 at 1223 K. The speed of formation of interfacial networks is sensitive to temperature and at lower temperatures significant primary creep can be seen as hardening occurs at a lower rate. The linear dependence of hardening rate on shear strain rate in equation 11 is not sufficient to describe the high sensitivity of hardening and is replaced by the equations:

The procedure used to calibrate the model described thus far has been to fit equations 2-5 describing strain and damage associated with shear on the type I and II slip systems to data in the [001] and [111] orientations respectively. This is first done for the type I systems and a correction is then made for the small amount of strain occurring on the type I systems in the [111] orientation before fitting the type II systems. The hardening equations are then introduced and the degree of type I on type II hardening is set so that it is sufficient to negate the effect of type II slip in the [001] orientation and also to optimise the prediction of further test data in unstable orientations. The overall amount of hardening is controlled through the initial hardening rate and the rate of saturation of hardening with accumulated shear strain.

Figures 10 to 12 show the results of creep simulations using the model presented here. Figures 10 and 11 show available data in the [001] and [111] orientations respectively whilst figure 12 contains data from unstable orientations inside the stereographic triangle. It can be seen that reasonable agreement is achieved between test and simulation and also that the desired accuracy of the fits in the [001] and [111] poles is maintained.

The model described is a macroscopic approximation to the likely microscopic deformation mechanisms. The effect of the two phase microstructure and of different precipitate orientations and morphologies on flow of the matrix material is being considered in a separate project, it has not yet been possible to combine a slip system analysis with microstructural features. However activating two different types of slip system which operate primarily in the two different phases allows the effect of the microstructure to be considered in a continuous fashion. One area where the model may be refined further is in the contribution of type I slip to creep in the [111] orientation. It was discussed earlier that the twin deformation observed in orientations near [111] could not be responsible for all of the macroscopic strain. The proportion of the remaining strain derived from type I and type II slip systems is not known. Superlattice stacking faults were observed in these orientations but they also contain a reasonably high incidence of type I dislocations in the matrix. On a strict Schmid factor basis it is difficult to activate any significant type I slip in the [111] orientation due to the very low Schmid factor relative to that in the [001] orientation. However there are two possible mechanisms by which type I slip may be activated in the [111] orientation. The first is due to the different orientation of the precipitates where there is less constraint on flow of the matrix material which may provide a geometrical weakening for type I slip [28]. This argument has been used previously to advocate the use of a macroscopic cube slip mechanism [15,16]. The second is the lack of hardening between the type I slip systems. In the [001] orientation eight type I slip systems are highly activated whilst in the [111] orientation, six type I slip systems are weakly activated. The effect of hardening in the [001] orientation due to several highly activated slip systems may result in strengthening of this orientation relative to [111]. Work is ongoing to address these issues.

The extent of transient creep is greatly reduced in single crystal superalloys at the higher temperature of 950 °C and as such it has been omitted from the analysis thus far. However, such transient creep is pronounced at lower temperatures and has been incorporated into the crystallographic model at low temperature through the use of a softening parameter for type II slip [29].

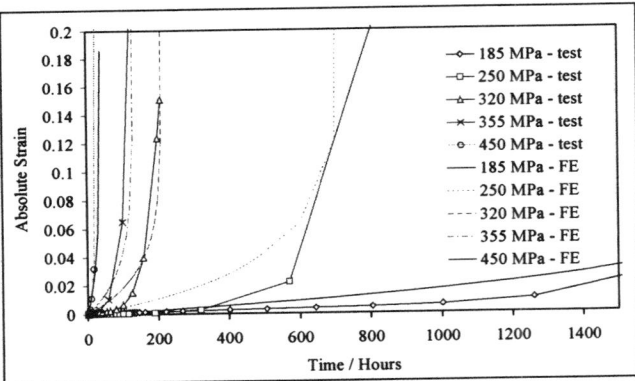

Figure 10: <100> creep data and model simulations.

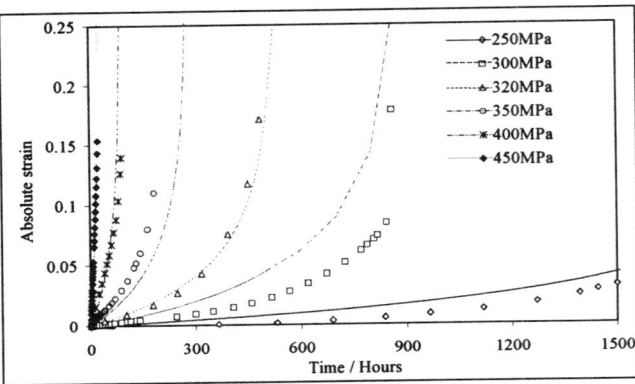

Figure 11: <111> creep data and model simulations.

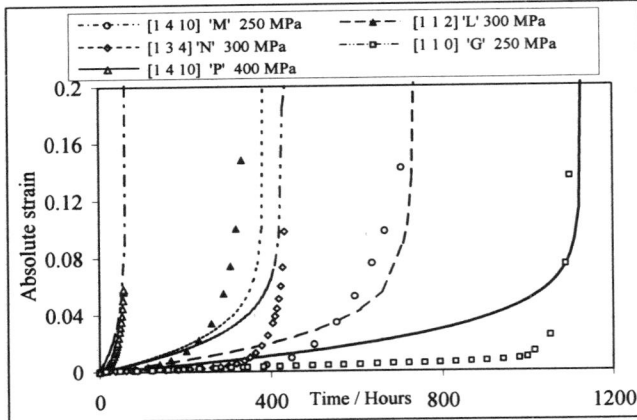

Figure 12: Creep test data and model simulations in various unstable orientations.

Biaxial Creep

Thin cylinder tension / torsion creep tests have been performed at the University of Swansea on servo-hydraulic machinery. Tests were performed at 1223 K on a specimen with a 10 mm gauge length and internal and external radii of 3 and 4 mm respecively. Tensile and torsional strains were measured from the machine remote from the specimen gauge length. Such biaxial creep testing can be characterised by two parameters, the equivalent stress and the stress angle [30], given by:

$$\sigma_{eq} = \left[\sigma^2 + \left(\sqrt{3}\tau\right)^2\right]^{\frac{1}{2}} \qquad (17)$$

$$\theta = Tan^{-1}\left(\frac{\sigma}{\sqrt{3}\tau}\right) \qquad (18)$$

In these equations σ_{eq} is the equivalent stress, θ is the stress angle, σ in normal stress and τ is shear stress. A number of tension / torsion tests were performed on single crystal superalloys SRR99 and CMSX-4. Tests on the former were performed at an equivalent stress of 270 MPa and at stress angles of 90, 53, 30 and 0 degrees and on the latter at an equivalent stress of 360 MPa and stress angles of 90, 30 and 0 degrees. For a given equivalent stress the results show that the material is least resistant in tension, the resistance increases up to a certain combination of tension and torsion and then decreases back to pure torsion which has a resistance slightly higher than pure tension. Comparable results have been found previously for the biaxial behaviour of a similar single crystal superalloy [30].

When considering the behaviour of single crystals subjected to biaxial tension / torsion creep there are two primary factors which need to be taken into account. The first is the change in cumulative shear strain rate on the relevant slip systems as the stress angle is changed progressively from tension to torsion, the second is the change in shear strain rate, for a given stress angle, as one rotates around the surface of the thin cylinder. With respect to the latter the shear strain rate remains the same for pure tension but changes around the circumference if a torsional stress is present. The circumferential angle around the surface of the test cylinder, starting at a point in a cubic crystallographic direction is denoted ϕ. Figure 13 shows the variation with θ of the combined shear strain rate derived from all slip systems, for each of the type I and type II slip systems, and for different values of the parameter ϕ. These profiles are obviously dependent on the stress exponents u_{101} and u_{112}. For the values these exponents take of 3.2 and 3 respectively, there is little effect of ϕ. The cumulative shear strain rate on both types of slip system, decreases with decreasing θ, for any value of ϕ. This would suggests that creep resistance in torsion should be stronger than that in tension, however as mentioned previously the resistance is highest for some combination of tension and torsion (stress angle approximately 30 degrees).

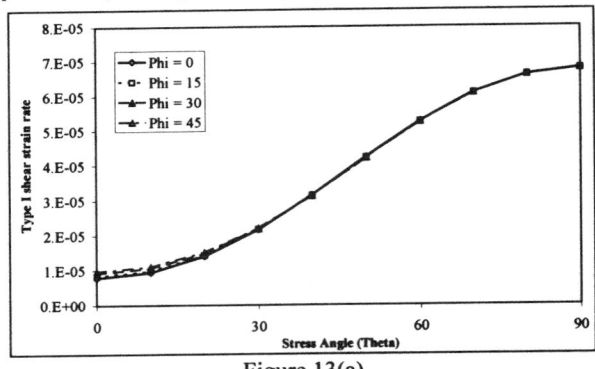

Figure 13(a)

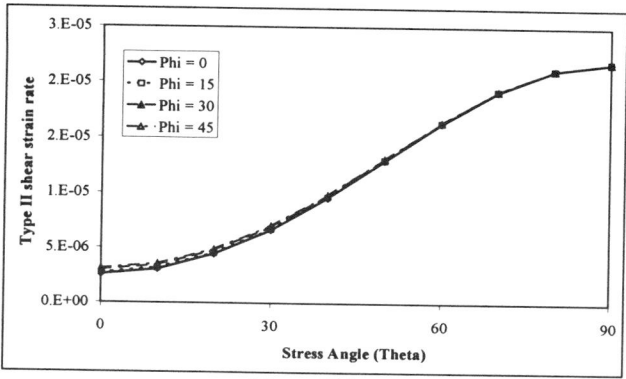

Figure 13(b)

Figure 13: Plot of the cumulative shear strain rate, on the type I, figure (a), and type II, figure (b), systems as a function of the stress angle for different values of the parameter ϕ.

The situation in progressing from tension to torsion is similar to that which occurs in uniaxial creep when the tensile orientation changes from [001] to [111] – the type I Schmid factors decrease but the strength does not increase accordingly. In terms of modelling the behaviour the same phenomenology can be applied to the two situations. In tension, deformation along [001] occurs on type I slip systems and the rate of such deformation is sufficient to harden the type II slip systems. As the stress angle is reduced, and the torsional stress component increased, the creep rate initially decreases due to the decreasing type I shear strain rate. With further reduction in stress angle the shear strain rate on type I slip systems becomes too low to permit hardening of the type II slip systems, which become active. As the stress angle is reduced to zero degrees the type II slip systems experience less hardening and the creep resistance decreases.

The hardening functions and parameters obtained from fitting the model to unaixial data at 1223 K have been found sufficient to give reasonable predictions of creep behaviour under combined tension / torsion testing. The comparison between model simulations and available test data for SRR99 and CMSX-4 are shown in figure 14.

Conclusions

1. A finite element slip system based damage mechanics model has been implemented to analyse creep of single crystal superalloys SRR99 and CMSX-4. Mechanical testing and microstructural analysis has shown that $<101>\{111\}$ (type I) slip systems are dominant in orientations close to the [001] pole, but that $<112>\{111\}$ (type II) slip systems are dominant in orientations close to the [111] pole. These deformation mechanisms have successfully been included in the model.
2. A significant component of the model is a hardening matrix, through which latent hardening by the type I systems on the type II systems is introduced. This hardening explains the inactivity of the latter in the [001] orientation. There is strong microstructural evidence for such hardening, which also allows the relatively low anisotropy between the [001] and [111] orientations at 950 °C to be explained.
3. Due to the orientation and stress state dependence of rupture life it is clear that damage evolution is dependent on strain rather than stress. This has been incorporated into the model by

including two tensor measures of damage which evolve as a function of the strain rate tensor derived from slip on the different types of slip system. Thus the effect of crystallographic slips at the material level can be incorporated into the macroscopic constitutive model.

4. Tension torsion testing on SRR99 and CMSX-4 has shown these materials follow similar trends to biaxial testing performed on another single crystal superalloy [30]. The materials are weakest in pure tension, the resistance increases for combined tension torsion testing up to a maximum at a stress angle of approximately 30 °, and then decreases again to pure torsion. Life in torsion is approximately a factor of two stronger than in tension. These biaxial stress effects have been predicted from the uniaxial model. Type I slip prevails in tension, but in torsion type I activity is insufficient to harden the type II systems which dominate deformation.

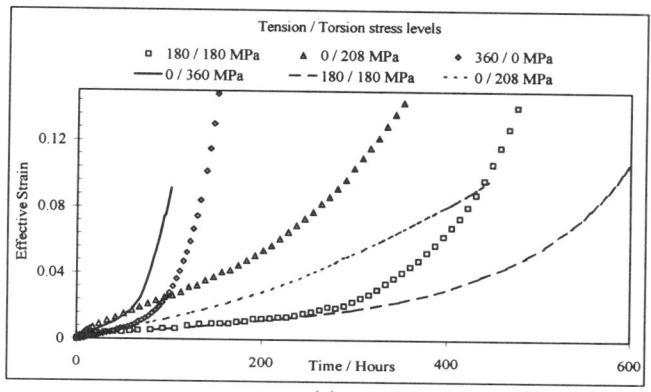

(a)

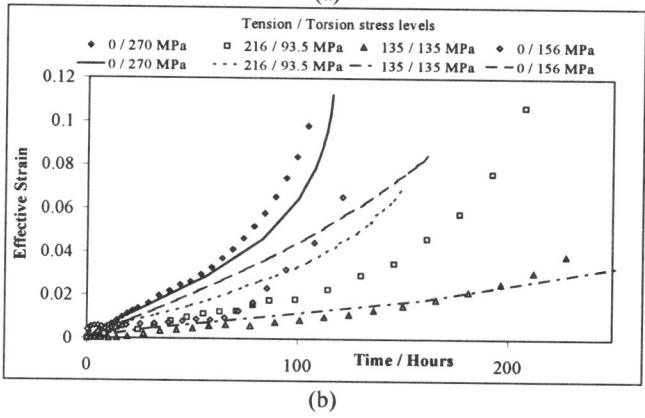

(b)

Figure 14: Comparison of biaxial test behaviour and model simulations. The effective strain is plotted against time for different stress angles and constant effective stresses, symbols are test data and lines are model results. (a) shows CMSX-4 data at an effective stress of 360 MPa, (b) shows SRR99 test data at an effective stress of 270 MPa.

Acknowledgements

The authors are grateful to Professors A. Windle and C.J. Humphreys for provision of the research facilities. They would also like to thank Rolls-Royce plc and DERA for funding for this work.

References

1. K. Harris, G. L. Erickson, R. E. Schwer, "Development of an Ultra High Strength Single Crystal Superalloy CMSX-4 for Small Gas Turbines," TMS-AIME Fall Meeting, (Philadelphia, Pennsylvania, 1983).

2. H. Murakami, P. J. Warren, H. Harada, "Atom-Probe Microanalyses of some Ni-Base Single Crystal Superalloys," 3rd International Charles Parsons Turbine Conference – Materials Engineering in Turbines & Compressors, (Newcastle UK, 1995), 343-350.

3. M. J. Goulette, P. D. Spilling, R. P. Arthey, "Cost Effective Single Crystals," Proceedings of the Fifth International Symposium on Superalloys, ed. M. Gell, C. S. Kortovich, R. H. Bricknell, W. B. Kent, J. F. Radavich, (Warrendale, PA: The Metallurgical Society AIME, 1984), 167-176.

4. R. W. Broomfield et al., "Development and Turbine Engine Performance of Three Advanced Rhenium Containing Superalloys for Single Crystal and Directionally Solidified Blades and Vanes," Journal of Engineering for Gas Turbines and Power, 120 (1998), 595-608.

5. S. S. K. Gunturi, D. W. MacLachlan, D. M. Knowles, "Anisotropic Creep in CMSX-4 in Orientations Distant from <001>," Mat. Sci. Eng. A, Accepted for publication.

6. R. A. Schwarzer, "Automated Crystal Lattice Orientation Mapping Using A Computer-controlled SEM," Micron, 28 (3) (1997), 249-265.

7. D. P. Pope, S. S. Ezz, "Mechanical Properties of Ni_3Al and Nickel-Base Alloys with High Volume Fraction of γ'," Int. Metals Rev., 29 (3) (1984), 136-167.

8. T. M. Pollock, A. S. Argon, "Creep Resistance of CMSX-3 Nickel Base Superalloy Single Crystals," Acta metall. mater., 40 (1) (1992), 1-30.

9. M. Clavel, A. Pineau, "Frequency and Wave-Form Effects on the Fatigue Crack Growth Behavior of Alloy 718 at 298K and 823K," Metallurgical Transactions A, 9A (1978), 471.

10. R. Bonnet, A. Ati, "Evidence for Microtwinning as a Deformation Mechanism in CMSX-2 Superalloy," Scripta Metallurgica et Materialia, 25 (1991), 1553-1556.

11. M. G. Ardakani, M. McLean, B. A. Shollock, "Twin Formation during Creep in Single Crystals of Nickel-Based Superalloys," Acta mater., 47 (9) (1999) 2593-2602.

12. J. W. Christian, T. B. Massalski, "Deformation Twinning", Progress in Materials Science, 39, (1995), 1-157.

13. C. N. Reid, Deformation Geometry for Materials Scientists, (Pergamon Press Ltd., 1973), 179-202.

14. D. W. MacLachlan et al, "Constitutive Modelling of Anisotropic Creep Deformation in Single Crystal Blade Alloys SRR99 and CMSX-4", Int. J. Plasticity, accepted for publication.

15. R. N. Ghosh, R. V. Curtis, M. McLean, "Creep Deformation of Single Crystal Superalloys - Modelling the Crystallographic Anisotropy," Acta metall. mater., 38 (10) (1990), 1977-1992.

16. P. E. McHugh, R. Mohrmann, "Modelling of Creep in a Ni base superalloy using a single crystal plasticity model," Computational Materials Science, 9 (1997), 134.

17. M. Kolbe, A. Dlouhy, G. Eggeler, "Dislocation reactions at γ/γ' interfaces during shear creep deformation in the macroscopic crystallographic shear system (001)[110] of CMSX-6 superalloy single crystals at 1025°C," Materials Science & Engineering A: Structural Materials: Properties, Microstructure and Processing, A246 (1998) 133-142.

18. M. Kolbe, K. Neuking, G. Eggeler, "Dislocation reactions and microstructural instability during 1025°C shear creep testing of superalloy single crystals," Mat. Sci. Eng. A, A234 (1997) 877-879.

19. R. A. MacKay, R. D. Maier, "The Influence of Orientation on the Stress Rupture Properties of Nickel-Base Superalloy Single Crystals", Metallurgical Transactions A, 9A (1982), 1747-1754.

20. B. H. Kear, J. M. Oblak, "Deformation modes in γ' precipitation hardened Nickel-base alloys," Journal de Physique, C7, 12 (1974), 35.

21. C. Rae, N. Matan, R. C. Reed, "Stacking Fault Shear in the Primary Creep of CMSX-4 at 750° C", Mat. Sci. Eng. A, Accepted for publication.

22. D. Peirce, R. J. Asaro, A. Needleman, "An Analysis of Non-uniform and Localised Deformation in Ductile Single Crystals," Acta Metall., 30 (1982), 1087-1119.

23. D. Nouailhas, G. Cailletaud, "Tension-Torsion Behaviour of Single-Crystal Superalloys: Experiment and Finite Element Analysis," Int. J. Plas.. 11 (4) (1995), 451-470.

24. L. Meric, P. Poubanne, G. Cailletaud, "Single Crystal Modelling for Structural Calculations: Part 1 – Model Presentation," J. Eng. Mat. Tech., 113 (1991), 162-170.

25. D. W. MacLachlan and D. M. Knowles, "Creep Behaviour Modelling of Single Crystal Superalloy CMSX-4", Met Trans, in press.

26. D. Peirce, R. J. Asaro, A. Needleman, "Material Rate Dependence and Localised Deformation in Crystalline Solids: Overview No. 32," Acta Metall., 31 (12) (1983) 1951-1976.

27. H. Brehm, U. Glatzel, "Material Model Describing the orientation Dependent Creep Behavior of Single Crystals Based on Dislocation Densities of Slip Systems," Int. J. Plas.,15 (1998) 285-298.

28. L. W. Wright et. al., "High Temperature Creep of CMSX-4 – Effect of Precipitate Orientation and Morphology", Parsons 2000, 5th International Charles Parsons Turbine Conference, (Cambridge, UK: IOM, 2000).

29. D.W. MacLachlan, G.S.K. Gunturi, D.M. Knowles, "Modelling the Uniaxial Creep Anistropy of Nickel Base Single Crystal Superalloy CMSX-4 at 1023 K," submitted to Scripta Met.

30. N. Ohno, T. Takeuchi, "Anisotropy in Multiaxial Creep of Nickel-Based Single-Crystal Superalloy CMSX-2: Experiments and Indentification of Active Slip Systems," JSME International Journal, Series A, 37 (2) (1994) 129-137.

CBED-MEASUREMENT OF RESIDUAL INTERNAL STRAINS IN THE NEIGHBOURHOOD OF TCP-PHASES IN NI-BASE SUPERALLOYS

Florian Pyczak, Horst Biermann, Haël Mughrabi, Andreas Volek and Robert F. Singer

Institut für Werkstoffwissenschaften, Universität Erlangen-Nürnberg,
Martensstr. 5, D-91058 Erlangen, Fed. Rep. Germany

Abstract

The influence of TCP-phases on the internal strains in the surrounding γ- and γ'-phase in two different Ni-base superalloys was investigated. One of the superalloys contained rhenium as an alloying element, while the other one did not. In the first alloy, rhenium is strongly enriched in the TCP-phases, and the phase possesses a σ- or P-type lattice structure. In the second alloy, the TCP-phase contains high amounts of the alloying elements tungsten, molybdenum and especially chromium and is of σ-type lattice structure. The method of convergent beam electron diffraction was used to determine the shear strains in the γ- and γ'-phase in an area of about one square micrometer around the TCP-phases.

Introduction and Objectives

In recent years, increasing amounts of different refractory elements were introduced in Ni-base superalloys to improve the mechanical properties. Rhenium proved to be an especially effective solid solutioning strengthener and significantly enhanced the mechanical strength of so-called second and third generation superalloys [1]. While the solid solution strengthening effects are desirable, refractory elements have also a tendency to precipitate as so-called topologically closed-packed phases (TCP-phases). It is commonly known that the presence of TCP-phases can degrade the mechanical properties of a Ni-base superalloy [2,3]. For this reason, different measures to avoid the precipitation of TCP-phases, such as careful balance of the alloy composition or homogenizing heat treatments up to 50 hours [1], are used to produce Ni-base su-peralloys of the second and third generation. Nevertheless, after long-term service at high temperatures, commercial alloys tend to precipitate small amounts of TCP-phases [4]. It seems that in many modern Ni-base superalloys minor contents of TCP-phases have to be accepted.

The σ-phase is the best known TCP-phase found in Ni-base superalloys until today. The lattice structure is similar to the well known σ-phase which occurs in steels, but the composition differs as tungsten, molybdenum and other alloying elements are also enriched in the σ-phase in the case of Ni-base superalloys in addition to chromium [5]. In the Ni-base superalloys of the second generation which contain rhenium, σ-phase is found and reported to contain larger contents of rhenium, but the amount of chromium is remarkably lowered. The composition is identical to another kind of TCP-phase, the P-phase which was first obtained near rhenium-containing coatings in Ni-base superalloys [6]. As mentioned above, the compositions of P- and σ-phase in Re-containing alloys are identical, but, in contrast to the tetragonal lattice cell of the σ-phase, the lattice cell of P-phase is orthogonal. The space groups and lattice parameters for the two phases are shown in Table I, and due to their structure and lattice parameters, parts of P-phase type lattice cell can be found together with parts of σ-type lattice cell in the same TCP-phase. The so-called μ-phase is the last type of TCP-phase which is frequently observed in Ni-base superalloys. The lattice structure data for μ-phase are also given in Table I. Additionally, many other types of TCP-phases, like δ-, β- or H-phase, are reported to occur in ternary systems with compositions similar

Superalloys 2000
Edited by T.M. Pollock, R.D. Kissinger, R.R. Bowman,
K.A. Green, M. McLean, S. Olson, and J.J. Schirra
TMS (The Minerals, Metals & Materials Society), 2000

Table I: Crystal structure information on σ-, μ- and P-phase [5].

TCP-phase	Space Group	Lattice Parameter (nm)
σ-phase	$P4_2/mnm$	a=0.930, c=0.486
P-phase	Pnma	a=1.720, b=0.486, c=0.920
μ-phase	$R\overline{3}m$	a=0.900, α=30.8°

to the Ni-base superalloy γ-phase [7], but none of these has been reported in the common Ni-base superalloys.

The purpose of the present work is the investigation of the effects of TCP-phases on the strains in the surrounding γ-matrix and in the coherently embedded γ'-particles. Interactions between γ–matrix and γ'-particles through the internal coherency stresses due to the γ/γ'-lattice mismatch have been thoroughly investigated during recent years and remain an important topic of present investigations in the field of Ni-base superalloys [8,9,10]. In this regard, the method of convergent beam electron diffraction (CBED) as a tool to measure lattice parameters and lattice misfits with high precision and an extremely high lateral resolution, is widely applied in Ni-base superalloys [11,12,13,14,15,16]. Kaufman et al. extended the usage of CBED to the investigation of strains induced by α-Mo precipitates in surrounding γ'-particles in a ternary Ni-Al-Mo alloy [17]. In the present work, CBED was used to characterize lattice distortions in both the γ- and the γ'-phase in the vicinity of TCP-phases. These lattice distortions induced by the TCP-phases are localized in areas of about 1 square micrometer. Thus, CBED is the ideal method to investigate these strain fields.

Experimental

Specimens

Two experimental directionally solidified Ni-base superalloys were investigated. Both are developed for the use as turbine blade material in stationary gas turbines, and their composition is based on that of the commercial alloy IN792. One alloy, later referred to as alloy A, contains up to 3wt.% of the element rhenium, and the second, subsequently referred to as alloy B, contains no rhenium. To enhance the properties of these experimental alloys compared to IN792, the contents of the elements Mo, W, Ti and Ta were also increased by 1 to 2 wt.%.

The specimens investigated in this work were aged for 500 h at 1000°C in air. Under these conditions, TCP-phases precipitate in both alloys and can be clearly recognized in the scanning electron microscope pictures (SEM), as shown in Figure 1. The shape of the precipitates is needle- or plate-like with an orientation of 45° relative to the γ'-cubes in a {001}-section. This results from certain lattice orientation relationships which are found between the different types of TCP-phases and the surrounding γ- and γ'-phase, as reported by several authors [5,18]. TEM-specimens of both alloys were produced from sections perpendicular to the solidification direction (lying close to [001]) by mechanical grinding and subsequent electrochemical polishing in an electrolyte of perchloric acid and acetic acid. By the use of chemical polishing, it is possible to avoid the introduction of strains during specimen preparation which is sometimes the case, if specimens are produced by ion milling.

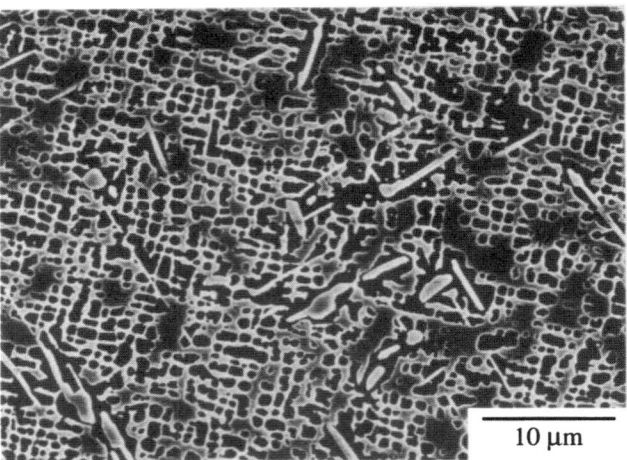

Figure 1: TCP-phases in alloy B in a scanning electron microscope image of a {001}-section.

Experimental set-up

All TEM-measurements were performed on a Philips CM200 TEM at an acceleration voltage of 120 kV or 200 kV, respectively. For CBED-measurements, a Gatan cooling stage was used, and the specimens were cooled to the temperature of liquid nitrogen to enhance the contrast in the CBED-patterns. All CBED-measurements were performed at an acceleration voltage of 120 kV. Spot sizes between 7 and 9, which result in nominal beam focuses between 27 and 15 nm, were utilized. In γ'-particles, in which relatively homogenous strain states were encountered, broader spots resulting in a higher intensity, were used, while the small focus sizes were necessary for the measurements in narrow γ–matrix channels or in the direct vicinity of TCP-phases. An incident beam parallel to the [001]-direction was chosen, because only <001>-orientations or orientations very near to <001> allow measurements in the very narrow γ-matrix channels. Furthermore, tetragonal lattice distortions are clearly distinguishable from shear distortions in <001> zone axis CBED-patterns.

All other TEM-investigations, including energy dispersive spectrometry (EDS) and the recording of selected area diffraction patterns (SAD-patterns) for the different phases, were performed using an acceleration voltage of 200 kV. The contamination of the specimens, while performing the EDS-measurements, was reduced by use of the Gatan-cooling stage. The EDS-measurements were done in the scanning mode of the TEM, working with a nominal spot diameter of about 4 nm.

Evaluation of CBED-patterns

The experimental patterns were compared with simulated patterns, calculated by using the kinematic routines of the EMS-simulation package by Stadelmann [19]. A set of patterns for pure shear distortions in a range of 0.15° were calculated for comparison with the experimental patterns. From the analysis of these simulated patterns we found that the length ratio between the two orthogonal axes parallel to the <110>-directions in the <001> zone axis CBED-pattern, which will be later referred to as R-ratio (see Figure 2), is very sensitive to changes in the angle γ of the lattice cell,

368

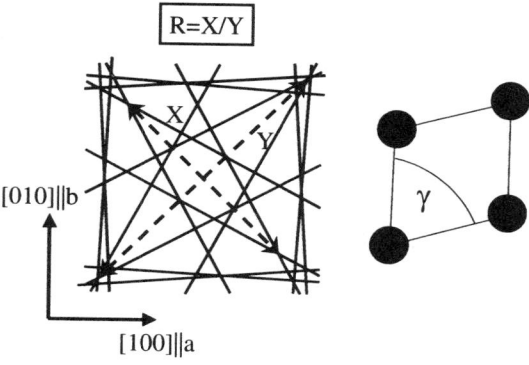

Figure 2: Simulated CBED-pattern of a shear-distorted lattice cell and a sketch of the lattice cell. The axes used to calculate the R-ratio are marked.

but is in no way recognizably influenced by pure tetragonal distortions of the lattice cell. This was also checked by kinematic simulations of CBED-patterns. In the following, a coordinate system with the lattice parameter c parallel to the [001]-direction is used. Hence, γ is the angle in the plane of the (001)-oriented specimens between the lattice axes a and b as drawn in Figure 3. It is obvious that, if the TCP-phase induces a compressive strain along the [1$\bar{1}$0]-direction, a shear component, which decreases the angle γ in the way shown in Figure 3 always exists. These shear components of the induced strain have been investigated in this work and later are simply termed as shear strains.

The CBED-patterns can be oriented due to the magnetic rotation of the TEM. Hence, directions in reciprocal space and the appropriate direction in image space can be identified as shown in Figure 4. The R-ratios of the experimental CBED-patterns were measured by using an image analysis program. By comparing the measured

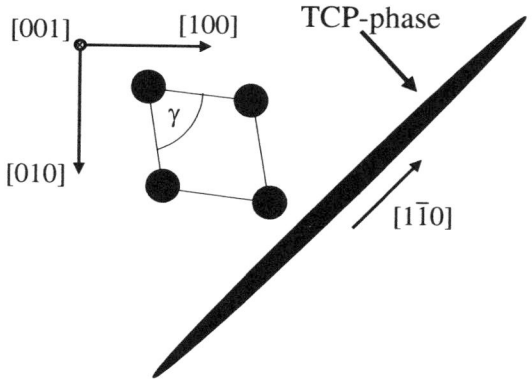

Figure 3: Schematic diagram of the used coordinate system. The lattice cells in both alloys are always under compressive strain along that <110> direction which is parallel to the long axis of the TCP-phase, as shown in the figure.

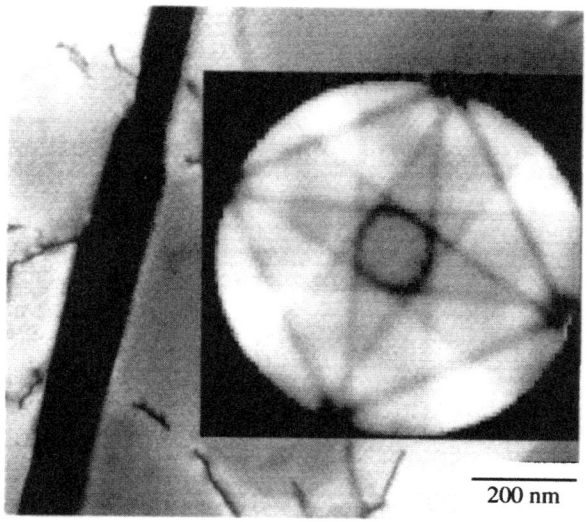

Figure 4: TEM-image of a TCP-phase together with an appropriately oriented <001> zone axis CBED-pattern.

R-ratios with a calibration curve derived from the simulated patterns, it was possible to determine the angle γ of the lattice cell [20] with an accurracy of better than 0.01°. Due to the 45°-inclination of the TCP-phases with respect to the edges of the γ'-cubes, their presence should induce a recognizable shear distortion in the <001> zone axis CBED-patterns, in their direct vicinity. As mentioned above, these shear distortions are distinguishable from tetragonal distortions, which could also be caused by the lattice misfit between γ- and γ'-phases [8,10,14]. This is shown in Figure 5, by examples of CBED-patterns of a tetragonally distorted and a tertagonally and, additionally, shear-distorted lattice cell.

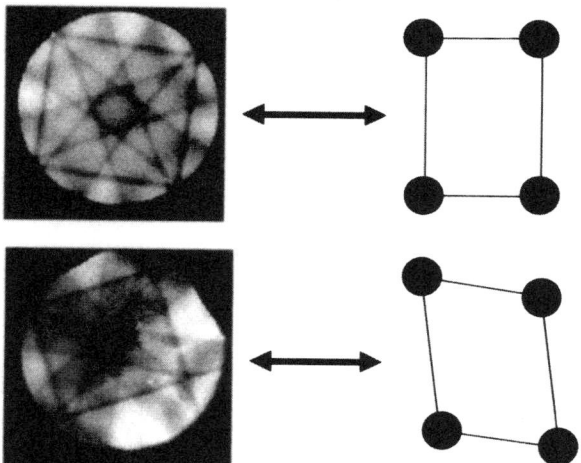

Figure 5: Schematics of a tetragonally distorted lattice cell together with the appropriate {001}-CBED-pattern and the same for a tetragonally and, additionally, shear-distorted lattice cell

369

Results

Microstructural investigations and SAD-measurements

The TCP-phases have been investigated by scanning electron microscope (SEM) and conventional TEM-imaging. As mentioned above, the TCP-phases in both alloys appear in SEM-images as needle-shaped particles, which are always oriented at an angle of 45° relative to the edges of the γ'-cubes. For alloy A, the concentration of TCP-phases in dendritic cores seems to be higher than in interdendritic regions (see Figure 6), in correspondence with the well-known tendency of rhenium to segregate in the dendritic cores [21]. In both alloys, the widths of the TCP-phases vary between 0.2 and 0.5 μm. Their lengths range from 10 to 20 μm in the case of alloy A and up to 100 μm in the case of alloy B.

No further microstructural features of the TCP-phases such as for example, faulting were detectable in the TEM brightfield-pictures in addition to the results obtained already in the SEM.

SAD-patterns of the TCP-phases in both alloys were taken to identify the type of investigated TCP-phase. In both cases, the TCP-phases were oriented in such a way that diffraction patterns of high symmetry were obtainable. The orientations in both cases were between the <112>- and <113>-orientations of the γ/γ'-phase, which was determined with the help of Kikuchi patterns, as shown in Figure 7 and Figure 8. From the diffraction patterns, the TCP-phase in alloy B (see Figure 7) was identified as σ-phase. The calculated lattice parameters of 0.884 nm for the a and b directions is in fair agreement with the lattice data known from literature [5,7,22,23].

For the TCP-phase in alloy A, the identification is not so unambiguous. Appropriate indexing of the diffraction patterns was possible for a P-type lattice cell as well as for a σ-phase type lattice cell (see Figure 8). Kikuchi patterns of the TCP-phase supplied no further information, since due to the thickness of the phase no clearly visible Kikuchi bands were observable.

EDS-measurements

The TCP-phases in both alloys consist mainly of elements which are enriched either in the γ-matrix, such as chromium or evenly distributed between the γ/γ'-phases, such as tungsten. Elements

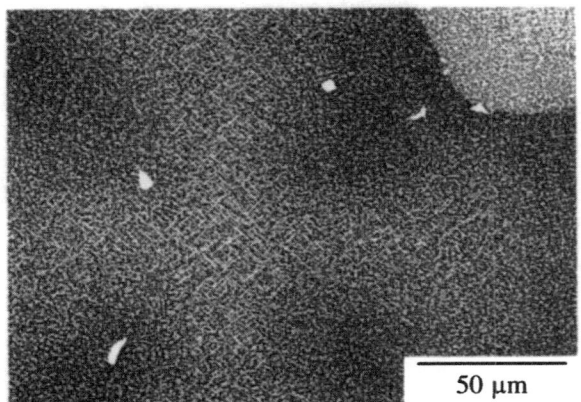

Figure 6: The SEM-picture shows clearly the higher concentration of TCP-phases in the dendritic cores in a (001) section of alloy A.

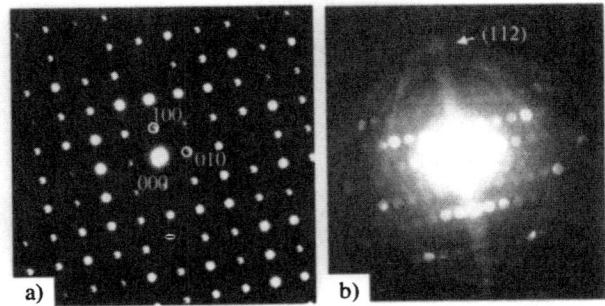

Figure 7: SAD-pattern of the TCP-phase (a) and the Kikuchi pattern of the same orientation for the neighbouring γ/γ'-phases (b) for alloy B.

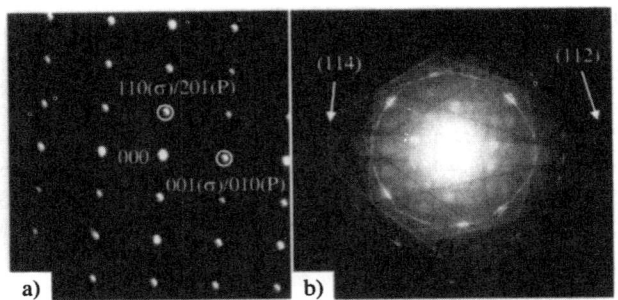

Figure 8: SAD-pattern of the TCP-phase with an indexing for σ- and P-type lattice cell marked (a). Kikuchi pattern for the neighbouring γ/γ'-phases in alloy A (b).

known as γ' builders, like titanium or tantalum, are not encountered in significant amounts in the TCP-phases. The exact compositions for both TCP-phases are shown in Table II. The refractory elements tungsten and rhenium in alloy A and tungsten in alloy B especially tend to be enriched in the TCP-phases, while the level of nickel is reduced relative to the surrounding matrix. In the TCP-phases of alloy A, high amounts of rhenium and tungsten are concentrated, while the level of chromium corresponds to that found in the matrix phase. The level of molybdenum is also enhanced compared to the γ-matrix composition.

In the TCP-phases in alloy B, mainly chromium and a remarkable amount of tungsten and molybdenum were found. However, the tungsten content in the TCP-phases of alloy B is not as high as in alloy A. No γ'-builders like tantalum or titanium were found in the TCP-phases.

Table II: Composition of the TCP-phases in the two different alloys in wt.% (elements which only occur in contents below 1% are omitted).

	Cr	Co	Ni	W	Mo	Re
Alloy A	18.8	5.0	10.0	29.1	8.4	26.6
Alloy B	32.4	11.8	23.0	20.3	12.6	0

CBED-measurements

In the present work, shear-distortions are only found near TCP-phases. They are clearly identified in the CBED patterns and are easy to distinguish from pure tetragonal distortions. The latter are recognizeable without difficulty in CBED-patterns of <001>-zone axes by examining the length inequality between the two principal axes parallel to <001>-directions [10].

In both alloys, the lattice spacings oriented parallel to the TCP-phases are compressed. This can be verified by aligning the CBED-patterns according to the magnetic rotation of the TEM. Then, the directions in the diffraction patterns correspond directly to directions in the images (see Figure 4).

It is very difficult to deduce direct connections between the induced shear strain and the distance from the TCP-phase. Obviously, in both alloys, shear distortions far above the level encountered elsewhere are measured in the direct vicinity of TCP-phases. These shear distortions vanish at larger distances from the TCP-phase which normally lie in the range of 400 to 800 nm in alloy A and 800 to 1000 nm in alloy B. However, the exact dependence of the shear distortions on the distance from the TCP-phase seems to be strongly modified by the presence of microstructural features such as dislocations or the shape of γ′-particles and matrix channels. Taking this into account, it is necessary to observe the microstructure near all individual sites of measurement, in order to understand the effects on the measured shear strains.

In alloy A, the strong shear distortions in the γ′-particles in the direct vicinity of the TCP-phases result in angles γ of about 89.95° which are encountered at distances between 70 to 300 nm away from the TCP-phases. At distances above 600 nm, the angles γ differ between 89.99° and 90.00° for all series of measurements. This corresponds to negligible shear distortions, since the shear strains measured in areas of the alloy far away from TCP-phases are in the same order of magnitude. In the γ-matrix, the tendency is similar to the situation in γ′-particles. Up to 400 nm away from the TCP-phase, shear distortions are clearly recognizable. At a distance of 600 nm from the TCP-phase, the shear distortions in the γ-matrix also vanish totally.

Two examples of series of measurements for alloy A are discussed in the following. In the first case (see Figure 9 and Figure 10), measurements in a γ′-particle directly connected to the TCP-phase and in a neighbouring matrix channel were performed. The slightly rafted γ′-particle is oriented at 45° relative to the TCP-phase. This is the situation normally found in the alloy due to the orientation relationships between TCP-phases and γ′-particles, as mentioned above. The relationship between shear strain and distance from the TCP-phase is very similar in the γ′-particle and the matrix, but the range of induced shear strains in the matrix seems to be greater. In both cases, the strong shear distortions remain, up to distances of ca. 300 nm. At larger distances, the shear distortions decrease rather abruptly and become similar to those encountered in other regions of the specimen, where no TCP-phases are present.

The series of measurements in an irregular shaped γ′-particle shown in Figure 11 reveals another picture. Though the maximum shear distortion in the immediate vicinity of the TCP-phase is sim-

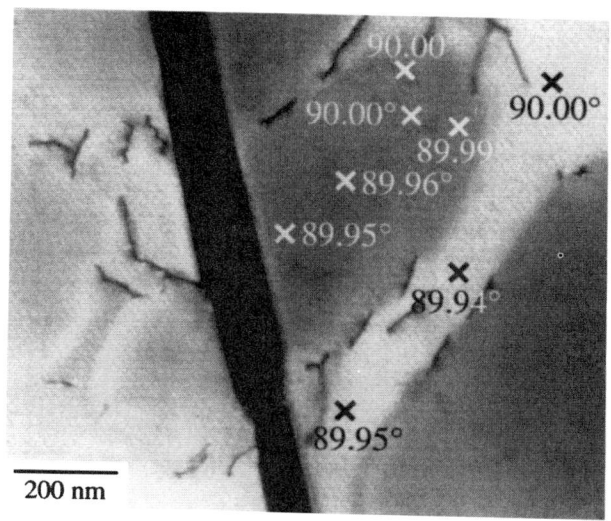

Figure 9: A series of CBED-measurements in alloy A: The measurement positions and the measured angles γ are marked.

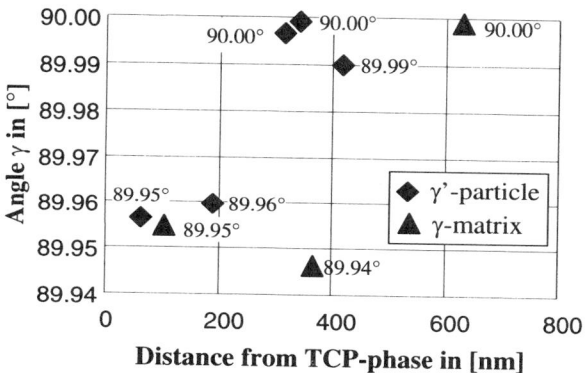

Figure 10: Angle γ plotted versus distance from TCP-phase for the series of measurements in γ- and γ′-phase in Alloy A shown in Figure 9. Triangles mark measurements in γ-matrix and squares measurements in γ′-particles.

ilar to the case presented above, the decrease of shear strain with increasing distance from the TCP-phase occurs smoothly and not as abruptly, as in the previous example, until an undistorted strain state is found at about 800 nm distance from the TCP-phase (see Figure 12).

In alloy B, the sphere of influence of the TCP-phases on the shear strains extends to much larger ranges. Measurements in γ′-particles directly connected with the TCP-phase result in angles γ between 89.84° and 89.92°. It is remarkable that, even in γ′-particles which are not directly connected to the TCP-phase but separated by one γ-matrix channel, shear distortions of the lattice cell of the same level as in direct vicinity to the TCP-phase in alloy A are encountered.

As in the case of alloy A, we shall now discuss several measurements in alloy B in more detail. As a first example, we take a series of measurements in the γ′-phase in a precipitate directly connected to the TCP-phase and a precipitate separated by one γ-matrix chan-

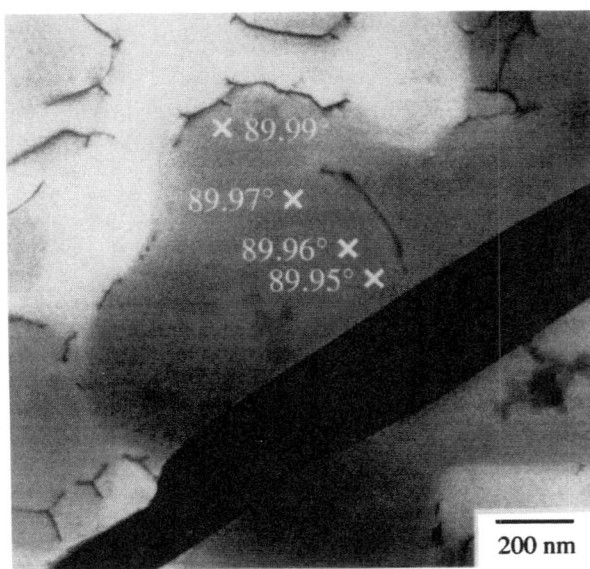

Figure 11: Series of CBED-measurements in a γ'-particle in alloy A. The beam positions and the measured angles γ are marked in the image.

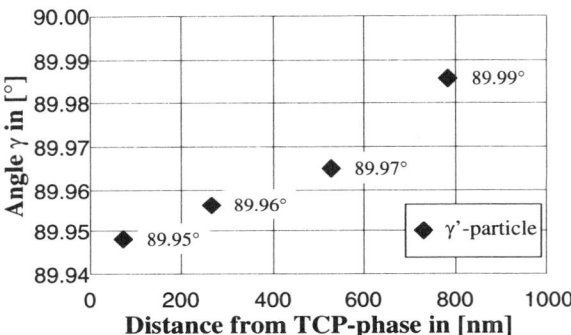

Figure 12: Angle γ plotted versus distance from TCP-phase for a series of measurements in the γ'-phase in alloy A shown in Figure 11. Compared with the results shown in Figure 10, the shear distortions have a greater lateral range and a smoother decay over the distance.

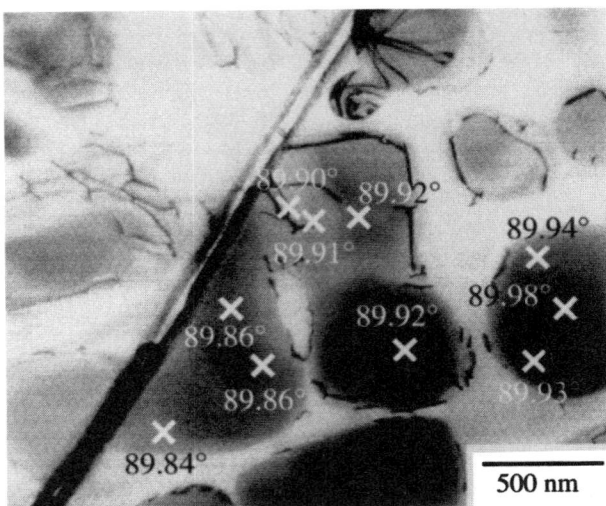

Figure 13: CBED-measurements in alloy B in a γ'-particle directly connected with the TCP-phase and a γ'-particle separated from the TCP-phase by one γ-matrix channel. Even at the measurement positions in the separated γ'-particle a shear distortion is recognizable.

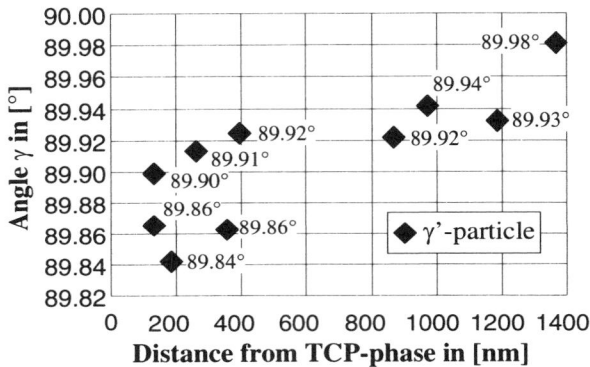

Figure 14: Angle γ plotted versus distance from TCP-phase for the series of measurements in Alloy B shown in Figure 13.

nel (Figure 13). The microstructure near the TCP-phase is very inhomogeneous in this measurement, because of some dislocations in the γ'-particle which is directly connected to the TCP-phase, and a γ-matrix channel penetrating partially into that γ'-particle. Regarding the measured shear distortions plotted in Figure 14, it is quite obvious, that stronger shear distortions occur in the areas of the γ'-particle, where no dislocations are present, compared to the level of shear distortions measured in areas, where dislocations are present. Even in the γ'-particle, which is separated by a matrix channel from the TCP-phase, shear strains are measurable at a position more than one micrometer away from the TCP-phase. It is unlikely that these shear distortions are caused by lattice mismatch between γ- and γ'-phase, because the position is in the centre of the γ'-particle, where strain states due to γ/γ'-lattice mismatch should be of hydrostatic nature.

In a second example, measurements in the γ- and the γ'-phases near two closely neighbouring TCP-phases are shown (Figure 15). The respective changes of angle γ are shown in Figure 16. In the matrix phase, shear distortions near the TCP-phase are more intense than in alloy A, reaching an angle γ of 89.91° at a distance of about 200 nm from the TCP-phase. Also at a distance of 600 nm, the shear strains are much more intense than in alloy A. The picture in the γ'-phase is surprising at first sight, as slight shear distortions, leading to an angle γ of 89.97°, are encountered in a measurement position which is only 600 nm away from the TCP-phase. At this distance, shear distortions of a much higher level were measured in the series presented above (see Figure 13). But, when the microstructure of these two measurement series are compared, it is obvious that in the latter case, there is always a γ-matrix area between the measurement position and the TCP-phase, which could be responsible for the decrease of the shear strains.

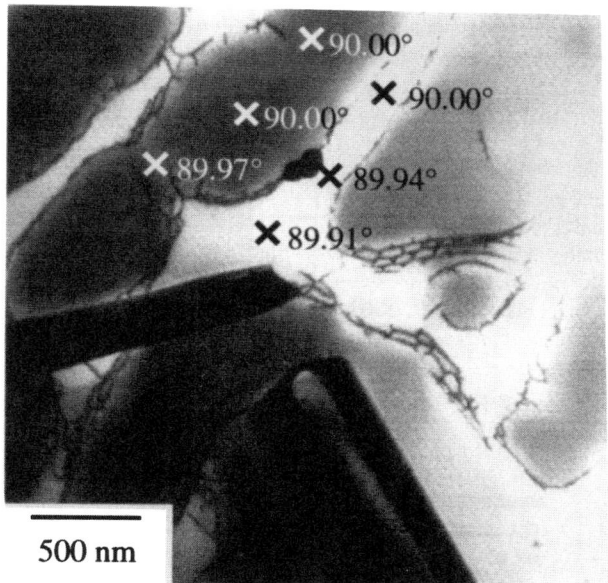

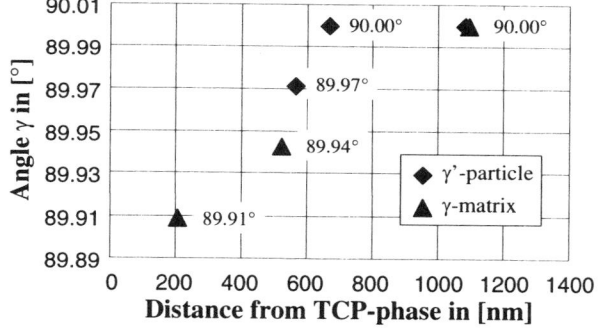

Figure 15: Two series of CBED-measurements in alloy B, one in a γ'-particle and the other in a γ-matrix channel. For all three positions in the γ'-particle the direct junction to the TCP-phase is separated by a γ-matrix channel.

Figure 16: Angle γ plotted versus distance from TCP-phase for the measurement in alloy B shown in Figure 15. Triangles mark measurements in γ-matrix and diamonds measurements in γ'-phase.

Discussion

Type of the TCP-phase

The presented identification of the TCP-phase occurring in alloy B as a σ-phase is considered to be relatively reliable, although no useful diffraction patterns of other high symmetry orientations were obtainable from the (001)-oriented specimens used in this work, to cross-check the orientation determination. The lattice parameters obtained are in good agreement with literature data [5,7,22,23], especially if one takes into account, that the lattice parameters normally are influenced by the composition of the nonstoichiometric TCP-phases. The occurrence of σ-phase is not unexpected as σ-phase is commonly found in Ni-base superalloys which contain no rhenium and have an alloy composition which promotes the precipitation of TCP-phases. The chemical composi-

tion of the phase also supports the identification as σ-type as discussed in the next section.

The identification of the TCP-phase in alloy A is ambiguous, because an identification as σ-phase as well as P-phase is reasonable, based on the diffraction patterns. It can be excluded that the investigated TCP-phase possesses a μ-type lattice structure, because the experimentally recorded diffraction pattern and a diffraction pattern of μ-type lattice structure in any possible orientation show no correspondence.

It is presently unclear how strongly the uncertainties regarding a full identification of the investigated TCP-phases, impair the appreciation of the results, concerning the induced strains due to the presence of TCP-phases. Even for the same type of lattice structure, the lattice parameters for TCP-phases, which can be found in the literature [5,7,22,23,24], vary in a remarkable range. Also very little is known about the properties of the interface between the γ/γ'-phase and the different TCP-phases. All these questions have to be answered first, before it can be judged, whether for example the lattice structure or the chemical compositions and their influence on the lattice parameters are the most important effects which affect the strains in γ/γ'-phase.

Composition of the TCP-phases

The compositions measured by EDS correspond well with data known from the literature. Dariola et al. also report [5] that the chromium content is greatly enhanced in σ-phases which contain no rhenium. This is also true for the σ-phase in alloy B. The chromium contents of TCP-phases in rhenium-containing alloys, for σ-phase as well as P-phase, is reported to be lower and similar to the chromium-content in γ-phase [5]. This was also found for the TCP-phases in alloy A.

The refractory elements rhenium and tungsten are strongly enriched in the TCP-phases in alloy A, and their combined amount reaches more than 50 % which is three times more than in the γ-phase. Compared with the amount of tungsten in the TCP-phases in alloy B, which exceeds the tungsten content in the γ-phase at most by a factor of two, the prominence of the refractory elements in the TCP-phases of alloy A is obvious. Rhenium and tungsten levels of the same magnitude are also reported in the work of Dariola et al. [5].

Unfortunately, the measured TCP-phase compositions give no hint on the type of TCP-phase, which is encountered in alloy A, because the compositions of rhenium-containing σ-phase are reported to be identical with that of P-phase and sometimes both phases are found together in the same alloy. The phase composition in alloy B corresponds well with the data known from literature for σ-phase in rhenium-free alloys and thus confirms the results obtained by phase-identification via SAD-diffraction.

Shear strains induced in γ- and γ'-phase

It is obvious that, due to the presence of TCP-phases, shear strains in an area of about one square micrometer are induced in the surrounding γ-matrix and γ'-particles. This can be seen in Figure 17, where all measured shear distortions for both alloys are plotted versus the distance from the TCP-phase. It is obvious, that the shear distortions obtained in alloy B are for the most cases higher than in alloy A. Some deviations can be explained by the influence of microstructural features near the measurement positions (see Figure 15 and Figure 16) and are discussed in detail below. The difference between shear strains in the matrix phase and in the γ'-particles is not significant in both alloys. Especially at measurement positions in direct vicinity of the TCP-phase, the induced shear strains in matrix phase and γ'-phase are of the same magnitude.

The magnitudes of the measured shear strains appear reasonable, if one considers the stresses associated with strains of similar magnitude. Assuming a shear modulus of the γ'-phase of 120 GPa, then shear stresses of 90 MPa in alloy A and 350 MPa in alloy B are found for the maximum shear strains measured in both specimens. These values are below the flow stress of both alloys at the temperature of measurement.

The different series of measurements presented above suggest that the correlation between the induced shear strains and the distance between TCP-phase and measurement position is not unique. The shear distortions are obviously partially influenced by locally present microstructural features, such as dislocations or γ'-particles of different shapes. The measured shear strains in the γ'-particle in Figure 15 are almost negligible and lower than in the γ-phase, suggesting that the specific distortion energy is lower than in the γ'-phase. The presence of dislocations also seems to decrease the intensity of induced shear strains, but no regular dislocation arrangements, comparable to the kind developing at γ/γ'-interfaces [26], are observed to compensate the lattice mismatch. The data available at present is insufficient to conclude whether dislocations really reduce the induced shear strains in a systematic manner.

It is commonly known that TCP-phases are not the only source of internal strains in Ni-base superalloys. The strains due to lattice mismatch between γ- and γ'-phase have been investigated thoroughly in the past [8,9,10,11,12,13,14,15,16,17]. In order to measure solely the effects of the TCP-phases, one has to find a method to separate strains induced by γ/γ'-mismatch from the strains in-

duced by the TCP-phase. The observation, that intense shear distortions are encountered near TCP-phases, is helpful in this regard, because it is reasonable and confirmed by measurements, that the γ/γ'-mismatch induces mainly tetragonal distortions. Only near γ/γ'-interfaces, at corners of γ'-particles and in narrow matrix channels some shear distortions are present [10]. However, in the two alloys investigated, these shear distortions are far below the levels observed in the vicinity of TCP-phases. Hence, one can assume that shear distortions above a certain level are caused mainly by the presence of TCP-phases. Nonetheless, we suppose that TCP-phases also induce tetragonal distortions in the surrounding γ- and γ'-phase which can, however, not be distinguished from the effects due to the γ/γ'-mismatch.

On the one hand, CBED is the only method known to the authors that is able to measure such localized strain states [25]. On the other hand, the value of the information obtained is uncertain with respect to any conclusions regarding the properties of a bulk material at service temperature. The strain and stress states in a thin TEM-specimen are not equivalent to the situation in the bulk material. Hence, it is difficult to infer reliably from information obtained from thin specimens the strain states in bulk material. Another difficulty of CBED-measurements arises from the fact that the specimen must be cooled down to the temperature of liquid nitrogen in order to obtain CBED-patterns of useful quality. As the service temperature for Ni-base superalloy applications lies normally between 600 and 1100°C, results obtained at liquid nitrogen temperature provide no direct information on the influence of certain internal strains and stresses on the properties of the material under service conditions [9]. Thus, even marked differences between the measured intensities of the induced shear strains near the TCP-phases in the two different alloys do not, at present, permit any valid conclusions as to how this would affect the mechanical properties under service conditions. At the present state of this study, it is premature to decide whether the data obtainable by CBED-measurements could provide a potentially useful basis for simulations in order to model the situation at service temperature.

Significance of TCP-phases for damage initiation

The presence of a third phase with elastic moduli and thermal expansion coefficients which deviate strongly from that of γ-matrix and γ'-particles may have strong effects on the mechanical properties and on the damage behaviour of the superalloy. The elastic hardness of the TCP-phases is assumed to be high compared with the surrounding two-phase compound. Hence, externally applied stresses cause elevated internal stresses in the TCP-phases. The fracture of the brittle TCP-phase under these circumstances could degrade the fatigue properties of the superalloy, because the broken TCP-phases could act as initiation points for cracks. The differences in thermal expansion coefficients can also affect the properties of a Ni-base superalloy in service. Temperature changes from ambient temperature to elevated service temperatures and back to ambient temperature or temperature changes during service often occur in a Ni-base superalloy turbine blade, e.g. at the start-up of a jet engine turbine. During these temperature changes, the differences in thermal expansion coefficients cause thermally induced internal strains in the TCP-phases and in the surrounding superalloy. These internal strains are either compensated by elastic deformation of the TCP-phase and the superalloy, leading to the

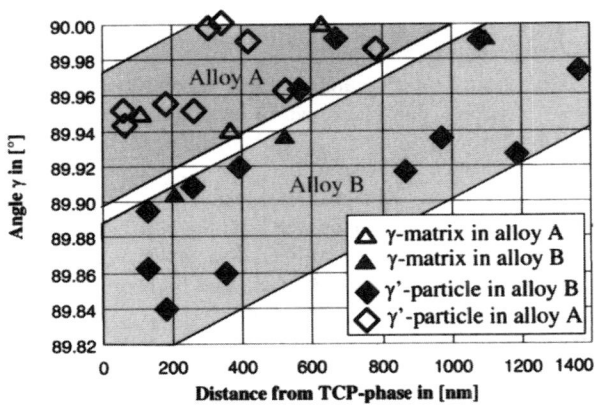

Figure 17: All measured shear strains in both alloys plotted versus distance from the TCP-phase.

corresponding internal stresses, or by plastic deformation of the superalloy at the interfaces between TCP-phase and γ/γ'-compound. In this investigation, no evidence of plastic deformation of the γ/γ'-compound, i.e. a higher dislocation density near the TCP-phases compared with other regions, has been found. If thermally induced internal stresses occur due to the presence of TCP-phases, they are superimposed on the externally applied stresses and increase the superalloy's probability of failure due to the fracture of the TCP-phases. As these internal stresses are not homogeneously distributed over the whole superalloy, but concentrated near the TCP-phases, these stress concentrations will probably cause fracture of the TCP-phase or decohesion between the TCP-phase and the surrounding γ/γ'-compound.

As outlined above, it is important to be concerned about the effects of TCP-phases on the mechanical properties of a Ni-base superalloy not only at service temperature but also during temperature changes. Although the data presented allow no conclusions, whether one of the two TCP-phases affects the mechanical properties of the superalloy at service temperature more than the other by inducing internal strains in the surrounding γ/γ'-compound, it can be determined from the results, that the effects of the two TCP-phases on the mechanical properties must be different with regard to temperature changes. The effects of the TCP-phases in alloy A and B are different at the temperature of the measurement, where the strains in alloy A are significantly smaller. Even if at one temperature, and by chance this may even be the service temperature, the induced strains caused by the different TCP-phases are equal, the situation in both alloys during heating up to service temperature or cooling down after a thermal cycle must be different. Hence, the TCP-phase in the Re-containing alloy A obviously must have different effects on the mechanical properties of the Ni-base superalloy compared with the classical σ-phase in alloy B, as the strains induced by both TCP-phases are of significantly different magnitude. This result is especially relevant, as the importance of rhenium as an alloying element has strongly increased in recent years in modern Ni-base superalloy development.

Conclusions

In the present work, the distortions in the vicinity of TCP-phases of two experimental superalloys were investigated by CBED. The most important findings are the following:

1) The lattice structures of the two investigated types of TCP-phases in both alloys have been partially identified by the evaluation of SAD-patterns. Both TCP-phases are of σ- or P-type. The identification of the TCP-phase in the rhenium-free alloy B as σ-type is quite unambiguous and in good correlation with the measured chemical composition and the shape of the TCP-precipitates. The TCP-phase in alloy A could either be of σ- or P-type.

2) The compositions of the TCP-phases in both alloys were measured by the use of EDS in the TEM. In alloy A, mainly rhenium and tungsten and in Alloy B chromium and a minor content of tungsten are the main elements encountered in the TCP-phases. These compositions are very similar to data published in literature for both kinds of phases.

3) The presence of TCP-phases in a Ni-base superalloy induces recognizable shear strains in the surrounding γ- and γ'-phase. These shear distortions are distinguishable from internal strains caused by other microstructural features like γ/γ'-mismatch, because their magnitude exceeds the level of shear strains encountered in areas of the specimen, where no TCP-phases are present. The shear strains were confined to areas of about one square micrometer around the TCP-phase.

4) In the present work, CBED was applied to measure these shear strains, because CBED is the only method which permits measurements of lattice parameters with the necessary high lateral resolution. The shear strains can be separated from pure tetragonal distortions by investigating CBED-patterns of <001>-zone axes, and evaluating the length ratio of the axes parallel to the <110>-directions.

5) The measurements presented for two different Ni-base superalloys showed significant differences in the magnitude of the induced shear strains. Re-containing TCP-phases cause shear distortions of a much lower level in their direct vicinity than TCP-phases which consist mainly of the elements tungsten and chromium and are free of rhenium. It was impossible to deduce exactly, how the induced shear distortions are correlated with the distance from the TCP-phase, because of the interference of other microstructural features which obviously modified the local strain states.

6) Since the measurements were performed at about -180° C and because of the modified strain states due to the relaxation in the thin TEM-specimens, it is difficult to draw conclusions regarding the effects in bulk materials at service temperature.

7) Further work is necessary in order to substantiate whether data of the type presented can provide a reliable basis for simulations to calculate the internal strain and stress states caused by TCP-phases under service conditions.

Acknowledgments

This work was performed within the framework of a larger collaboration project which financially supported by the Bundesministerium für Bildung und Forschung (BMBF). This support is acknowledged gratefully. The authors also wish to thank their cooperation partners, especially DONCASTERS Precision Casting (DPC) in Bochum for the casting of the two investigated alloys.

Literature

1.   G. L. Erickson, "The development and application of CMSX-10", Superalloys 1996, ed. R.D. Kissinger, D.J. Deye, D.L. Anton, A.D. Cetel, M.V. Nathal, T.M. Pollock and D.A. Woodford (Warrendale, PA: The Minerals, Metals &Materials Society, 1996), 35-44.

2.   G. Chen, C. Yao and Z. Zhong, "The effect of σ phase on the mechanical properties in Ni-Cr-Co base wrought superalloys", Superalloys 1980, ed. J.K. Tien, S.T. Wlodek, H. Morrow, M. Gell and G.E. Maurer, (Warrendale, PA: American Socierty for Metals, 1980), 355-364

3.   M. Simonetti and P. Caron, "Role and behaviour of μ phase during deformation of a nickel-based single crystal superalloy", Materials Science and Engineering A, 254 (1998), 1-12.

4.   W. Schneider and H. Mughrabi, "Investigation of the creep and rupture behaviour of the single-crystal nickel-base superalloy CMSX-4 between 800° C and 1100° C", Proceedings 5th International Conference on Creep and Fracture of Engineering Materials and Structures, ed. B. Wilshire and R.W. Evans, (London: The Institute of Materials, 1993), 209-220.

5.   R. Dariola, D.F. Lahrman and R.D. Fields, "Formation of topologically closed packed phases in nickel base single crystal superalloys", Superalloys 1988, ed. S. Reichmann, D.N. Duhl, G. Maurer, S. Antolovich and C. Lund (Warrendale, PA: The Metallurgical Society, 1988), 255-264.

6.   W.S. Walston, J.C. Schaeffer and W.H. Murphy, "A new type of microstructural instability in superalloys - SRZ", Superalloys 1996, ed. R.D. Kissinger, D.J. Deye, D.L. Anton, A.D. Cetel, M.V. Nathal, T.M. Pollock and D.A. Woodford (Warrendale, PA: The Minerals, Metals &Materials Society, 1996), 9-18.

7.   E. Gozlan et al., "Topologically close-packed precipitations and phase diagrams of Ni-Mo-Cr and Ni-Mo-Fe and of Ni-Mo-Fe with constant additions of chromium", Materials Science and Engineering A, 141 (1991), 85-95.

8.   H.A. Kuhn, H. Biermann, T. Ungar and H. Mughrabi, "An X-ray study of creep-deformation induced changes of the lattice mismatch in the γ'-hardened monocrystalline nickel-base superalloy SRR 99", Acta Metallurgica et Materialia, 39 (1991), 2783-2794.

9.   H. Biermann, M. Strehler and H. Mughrabi, "High-temperature measurements of lattice parameters and internal stresses of a creep-deformed monocrystalline nickel-base superalloy", Metallurgical and Materials Transactions, 27A (1996), 1003-1014.

10.   B. von Grossmann, H. Biermann and H. Mughrabi, "Measurement of service-induced internal elastic strains in a single-crystal nickel-based turbine blade with convergent beam electron diffraction", Philosophical Magazine A, in press.

11.   A.J. Porter et al., "Coherency strain fields: magnitude and symmetry", Journal of Microscopy, 129 (1983), 327-336.

12.   D.F. Lahrman, R.D. Field, R. Darolia and H.L.Fraser, "Investigation of techniques for measuring lattice mismatch in a rhenium containing nickel base superalloy", Acta Metallurgica, 36 (1988), 1309-1320.

13.   D. Mukherji and R.P. Wahi, "On the measurement of lattice mismatch between γ and γ'-phases in nickel-base superalloys by CBED technique", Scripta Materialia, 35 (1996), 117-122.

14.   R.R. Keller, H.J. Maier, H. Renner and H. Mughrabi, "Local lattice parameter measurements in a creep-deformed nickel-base superalloy by convergent beam electron diffraction", Scripta Metallurgica et Materialia, 27 (1992), 1167-1172 and Scripta Metallurgica and Materialia, 28 (1993), 661 (Errata).

15.   R.C. Ecob et al., "The application of convergent-beam electron diffraction to the detection of small symmetry changes accompanying phase transformations", Philosophical Magazine A, 44 (1981), 1117-1133.

16.   R. Völkl, U. Glatzel and M. Feller-Kniepmeier, "Measurement of the lattice misfit in the single crystal nickel based superalloys CMSX-4, SRR99 and SC16 by convergent beam electron diffraction", Acta Materialia, 46 (1998), 4395-4404.

17.   M.J. Kaufman, D.D. Pearson and H.L. Fraser, "The use of convergent-beam electron diffraction to determine local lattice distortions in nickel-base superalloys", Philosophical Magazine A, 54 (1986), 79-92.

18.   M. Pessah-Simonetti, P. Donnadieu and P. Caron, "T.C.P. phase particles and prediction of the orientation relationships", Scripta Metallurgica et Materialia, 30 (1994), 1553-1558.

19.   P.A. Stadelmann, "EMS - a software package for electron diffraction analysis and HREM image simulation in material science", Ultramicroscopy, 21 (1987), 131-146.

20.   F. Pyczak and H. Mughrabi, "CBED-measurement of residual internal strains in the neighbourhood of TCP-phases in Ni-base superalloys", Proceedings of EUROMAT conference 1999, in press.

21.   U. Brückner, A. Epishin and T. Link, "Local X-ray diffraction analysis of the structure of dendrites in single-crystal nickelbase superalloys", Acta Materialia, 45 (1997), 5223-5231.

22.   J.R. Mihalisin et al., "Sigma - its occurrence, effect and control in nickel-base superalloys", Transactions of the Metallurgical Society of AIME, 242 (H) (1968), 2399-2414.

23.   A. Proult et al., "Identification of the fault vectors of planar defects in the sigma phase in a nickel-based superalloy", Philosophical Magazine A, 72 (1995), 403-414.

24.   J. Burslik, "The existence of P phase and Ni_2Cr superstructure in Ni-Al-Cr-Mo system", Scripta Materialia, 39 (1998), 1107-1112.

25.   H.J. Maier et al., "On the unique evaluation of local lattice parameters by convergent-beam electron diffraction", Philosophical Magazine A, 74 (1996), 23-43.

26.   T.P. Gabb, S.L. Draper, D.R. Hull, R.A. MacKay and M. V. Nathal, "The role of interfacial dislocation networks in high temperature creep of superalloys", Materials Science and Engineering, 118 (1989), 59-69.

The Influence of Dislocation Substructure on Creep Rate during Accelerating Creep Stage of Single Crystal Nickel-based Superalloy, CMSX-4

Nobuhiro Miura, Yoshihiro Kondo and Narihito Ohi*

The National Defense Academy, Yokosuka 239-8686, Japan
*Ishikawajima-Harima Heavy Industries Co., Ltd. Tokyo 188-8555, Japan

Abstract

The evolution of the dislocation substructure during the accelerating creep stage in the single crystal nickel-based superalloy, CMSX-4, was investigated through the microstructural observation of interrupted creep specimens at 1273K-160MPa. The dislocation substructure evolved, not within γ', but at the γ/γ' interface and in the γ channel. The dislocation density at the γ/γ' interface increased with increasing creep deformation. The correlation between the dislocation density and the creep rate during the accelerating creep stage led to an equation relating the creep rate to the fifth power of the dislocation density. This result was quite different from the supposition proposed by Dyson et al. that the creep rate during the accelerating creep stage would be directly proportional to the dislocation density. They supposed that the dislocations at the γ/γ' interface acted as mobile ones. The thickness of the γ channel increased with creep, and the creep rate during the accelerating creep stage was proportional to the fifth power of the thickness of the γ channel. The increase in the thickness of the γ channel meant the increase in the radius of dislocation curvature in the γ channel, then the loss of creep resistance during the accelerating creep stage would be interpreted by an increase in the radius of dislocation curvature. This same correlation between the creep rate during the accelerating creep stage and the thickness of the γ channel at 160MPa was confirmed in the wide stress range of 100 to 400MPa. Using the creep rate-γ channel thickness curves, the creep rate at the certain thickness of the γ channel were estimated, and the stress-creep rate curve at the certain thickness of the γ channel could be drawn. Comparing the measured stress-creep rate curve with the estimated stress-creep rate curve (at the certain thickness of the γ channel), the lager creep rate was obtained at a lower stress level, and this was also interpreted by an increase in the thickness of the γ channel. Consequently, the present experimental evidence, showing the effect of the thickness of the γ channel on the creep rate was confirmed.

Superalloys 2000
Edited by T.M. Pollock, R.D. Kissinger, R.R. Bowman,
K.A. Green, M. McLean, S. Olson, and J.J. Schirra
TMS (The Minerals, Metals & Materials Society), 2000

Introduction

Kondo et al. compared the creep rate of prior-creep tested specimens of the single crystal nickel-based superalloy, CMSX-4, with that of the as-heat treated one, using a stress-enhanced creep test[1)-4)]. Stress-enhanced creep tests were conducted at 1273K-250MPa, using the specimen interrupted prior-creep tests at 1273K-160MPa in the wide creep range from the transient to the accelerating creep stage. The microstructural observations were done for creep interrupted specimens and the rafting process was examined. And it was indicated that the rafting of the γ′ phase acted as a creep weakener and that the loss of creep resistance was caused by the widening of the γ channel. The extending of the γ channel led to an increase in the radius of dislocation curvature in the γ channel. This supposition was substantially accepted to interpret the accelerating creep. However, it is important to directly investigate the correlation between the thickness of the γ channel and the creep rate during the accelerating creep stage.

This study was undertaken to show that the origin of the onset of the accelerating creep in the single crystal nickel-based superalloy was caused not by mechanical damage like voids and cracks, but by microstructural changes. Dyson et al. also indicated that the onset of the accelerating creep in the nickel-based superalloy was caused by a microstructural change, that is, an increase in the dislocation density at the γ/γ′ interface[5)]. They determined the dislocation at the γ/γ′ interface acted as the mobile dislocation, so an increase in the dislocation density meant an increase in the creep rate according to the Orowan equation of $\dot{\varepsilon} = \rho bv$, where $\dot{\varepsilon}$ is the creep rate, ρ is the dislocation density, b is the burgers vector and v is the velocity of the dislocation. This conception was based on the dislocation glide model. In contrast to the model proposed by Dyson et al., the supposition proposed by Kondo et al. was based on the dislocation climb model.

Experimental procedure

A single crystal alloy of CMSX-4 (having the chemical composition in weight percent ; 6.4Cr, 9.3Co, 5.5Al, 0.9Ti, 6.3Mo, 6.2Ta, 6.2W, 2.8Re, 0.1Hf, balance Ni) was prepared in the form of bars 13mm in diameter by a directional precision casting. The exact orientations were determined by the Laue back-reflection technique ; longitudinal axes of single crystals selected for this study were within 5deg of the [001] orientation. After employing an eight step solution treatment (1550Kx7.2ks → 1561Kx7.2ks → 1569Kx10.8ks → 1577x10.8ks → 1586Kx7.2ks → 1589Kx7.2ks → 1591Kx7.2ks → 1594Kx7.2ks → GFC) and a three step aging heat treatment (1413Kx21.6ks → 1353Kx21.6ks → 1144Kx72ks → AC), the specimens for the creep tests with a gauge diameter of 6mm and a gauge length of 30mm were prepared.

The creep tests were carried out at 1273K-160MPa (creep rupture life is 3.60×10^6s) under constant load. Creep strain was measured automatically through linear variable differential transformers (LVDT's) attached to extensometers. The creep tests were interrupted at the certain times ranging from the transient creep stage to the accelerating creep stage. All the creep interrupted specimens were followed by cooling under load.

Microstructural examinations by a scanning electron microscopy (SEM ; Hitachi S-4200) and by a transmission one (TEM ; Hitachi H-8000) were carried out on the specimens sectioned parallel to (100), (101) and (111) planes. Specimens for the SEM observation were prepared metallographically and electroetched with a supersaturated phosphoric acid-chromic acid solution. TEM foils were electropolished with a 5% solution of perchloric acid in alcohol. The thickness of the γ channel was measured by use of the image processor-analyzer.

Results

Microstructure of the as-heat treated specimen

The microstructure of the as-heat treated CMSX-4 is shown in Fig. 1. The cuboidal γ′ phase is regularly arrayed in the γ matrix. The average edge length of the cuboidal γ′ phase is

Fig. 1. Scanning electron micrograph of a single crystal nickel-based superalloy, CMSX-4.

about 0.5μm. The mean thickness of the γ channel is approximately 0.1μm. The volume fraction of the γ' phase is about 80 percent. There were no eutectic γ' phase.

Creep interrupted test

The creep rate-time curve of CMSX-4 at 1273K-160MPa is shown in Fig. 2. As described in the previous reports[1)-3)], the creep curve at 1273K-160MPa consists of the transient creep stage and the accelerating creep stage. The conception that creep in single crystal nickel-based superalloys consisted of the extended steady state and the accelerating creep stage and the ratio of the transient creep stage to rupture life was small, was perfectly disappeared. In this case, the ratio of the transient creep stage to rupture life occupied more than 10 percent. The open circles corresponded to time where the creep tests were interrupted to prepare the specimens for the microstructural examinations. Creep interrupted tests were done over a wide range from the latter half of the transient creep stage to the end of the accelerating creep stage.

Microstructural changes during creep

The scanning electron micrographs of the specimens tested for 1.08×10^5, 1.08×10^6 and 3.24×10^6s are shown in Fig. 3. The stress axis was the normal direction of these photos. By subjecting the creep testing time of 1.08×10^5s (Fig. 3a), at the latter half of the transient creep stage, the γ' phase is remained cuboidal, whereas some of the cuboidal γ' phase contact each other. After connecting to each other, the lamellar γ/γ' structures, that is, rafted γ' structures, form

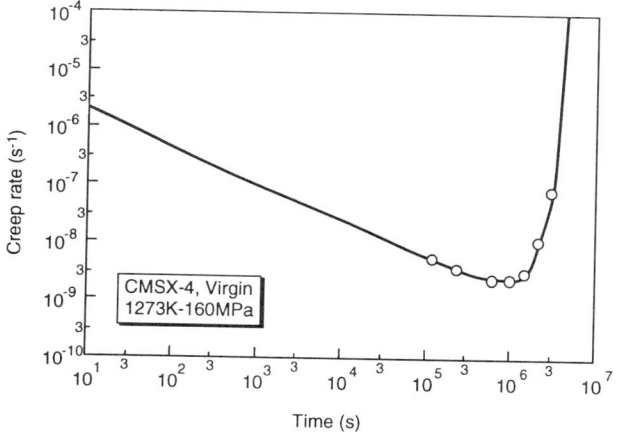

Fig. 2. Creep rate-time curve of CMSX-4 at 1273K-160MPa. The open circles correspond to time where the creep tests were interrupted for the micro-structural examinations.

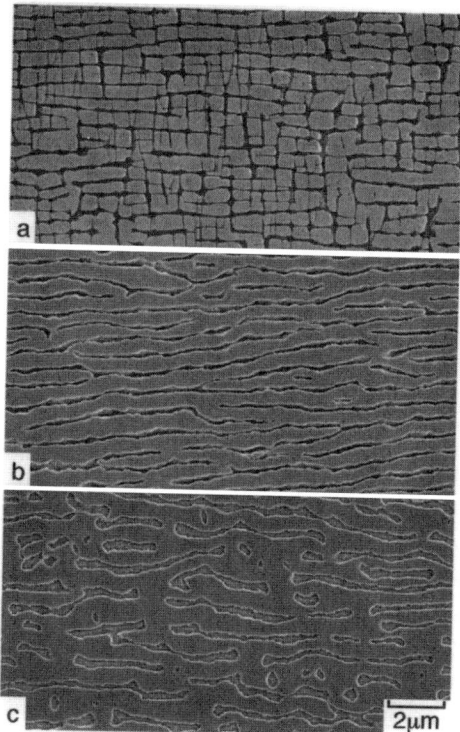

Fig. 3. Scanning electron micrographs of the specimens tested at 1273K-160MPa for (a) 1.08×10^5, (b) 1.08×10^6 and (c) 3.24×10^6s. Stress axis is vertical in these photos.

perpendicular to the stress axis as shown in Fig. 3b (the specimen creep tested for 1.08×10^6s). Further coarsening of the γ' phase is detected in the specimen tested for 3.24×10^6s (Fig. 3c), at the end of the accelerating creep stage. Regularity of the rafted γ' diminishes.

The transmission electron micrograph of the specimen tested for 1.08×10^5s, at the end of the transient creep stage, is shown in Fig. 4, where the electron beam direction, B,

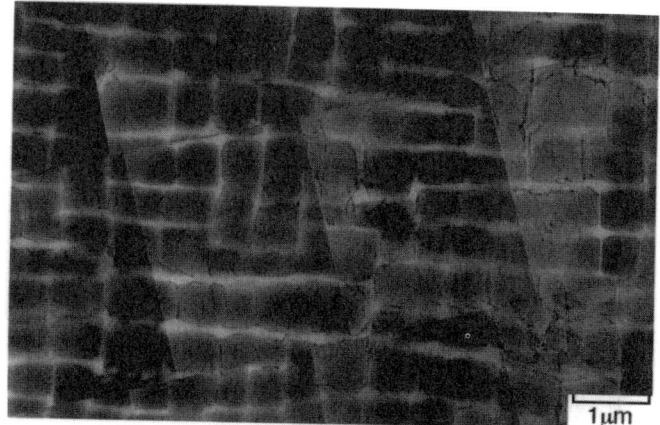

Fig. 4. Transmission electron micrograph of the specimen tested at 1273K-160MPa for 1.08×10^5s, where B=[100]. Stress axis is vertical in this photo.

379

was close to [100]. Few dislocations are observed in the γ matrix and the γ′ phase, while a small number of dislocations are present at the γ/γ′ interface.

For the specimen tested for 1.08×10^6s, at the minimum creep rate, Fig. 5, where B=[100], few dislocations are observed in the γ matrix and the γ′ phase, similar to the specimen tested for 1.08×10^5s. However, a number of dislocations are observed at the γ/γ′ interface.

For the specimen tested for 3.24×10^6s, at the end of the accelerating creep stage, Fig. 6, where B=[100], few dislocations are observed in the γ matrix and the γ′ phase, similar to the specimens creep tested for 1.08×10^5 and 1.08×10^6s. A large number of dislocations are observed at the γ/γ′ interface.

Thus few dislocations existed within the γ′ phase and there was no evidence that the dislocations cut through the γ′ phase. The dislocation density at the γ/γ′ interface was found to increase with increasing creep deformation. However, TEM observation where B=[100] did not indicate accurate information about the dislocation substructure, such as the dislocation density, at the γ/γ′ interface. Consequently, it was necessary to carry out TEM observations where B=[101].

The transmission electron micrograph of the specimen tested for 1.08×10^5s, at the latter half of the transient creep stage, is shown in Fig. 7, where B=[101]. A few dislocations are observed at the γ/γ′ interface.

In the specimen tested for 1.08×10^6s, at the minimum creep rate, Fig. 8, where B=[101], most of the dislocations are aligned. The dislocation density at the γ/γ′ interface is higher than that of the specimen tested for 1.08×10^5s (Fig. 7).

In the specimen tested for 3.24×10^6s, at the end of the accelerating creep stage, Fig. 9, where B=[101], the dislocations are tangled with each other. The dislocation density at the γ/γ′ interface is considerably higher than that

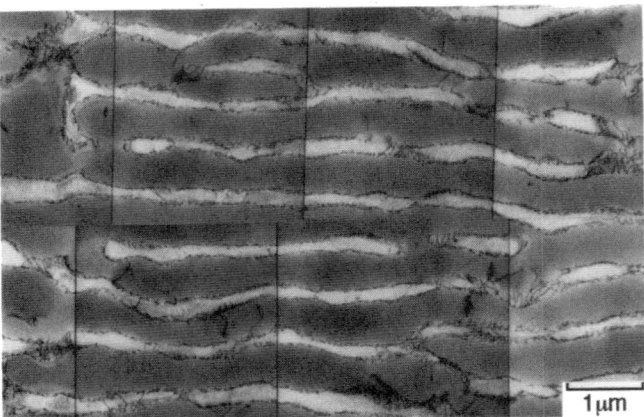

Fig. 5. Transmission electron micrograph of the specimen tested at 1273K-160MPa for 1.08×10^6s, where B=[100]. Stress axis is vertical in this photo.

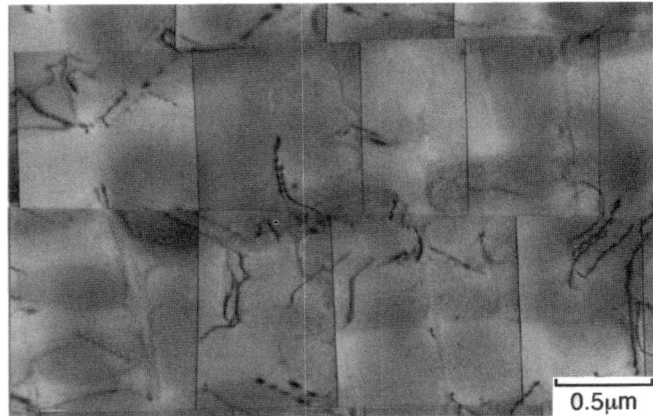

Fig. 7. Transmission electron micrograph of the specimen tested at 1273K-160MPa for 1.08×10^5s, where B=[101].

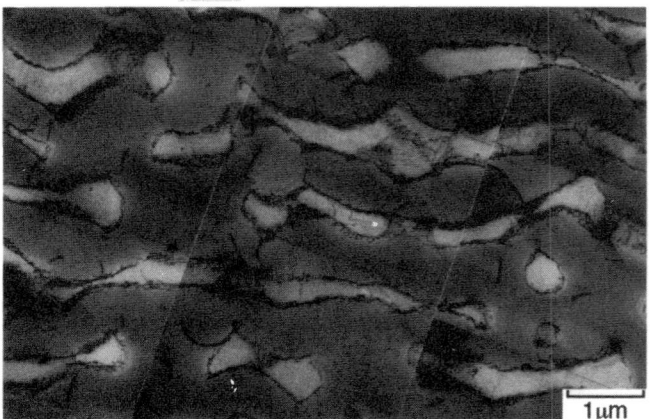

Fig. 6. Transmission electron micrograph of the specimen tested at 1273K-160MPa for 3.24×10^6s, where B=[100].

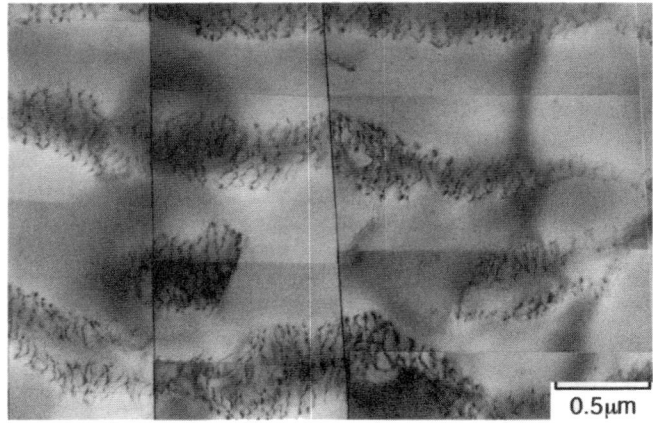

Fig. 8. Transmission electron micrograph of the specimen tested at 1273K-160MPa for 1.08×10^6s, where B=[101].

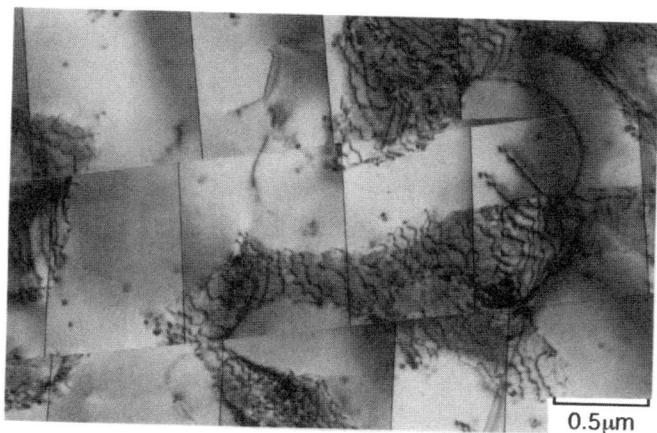

Fig. 9. Transmission electron micrograph of the specimen tested at 1273K-160MPa for 3.24x10⁶s, where B=[101].

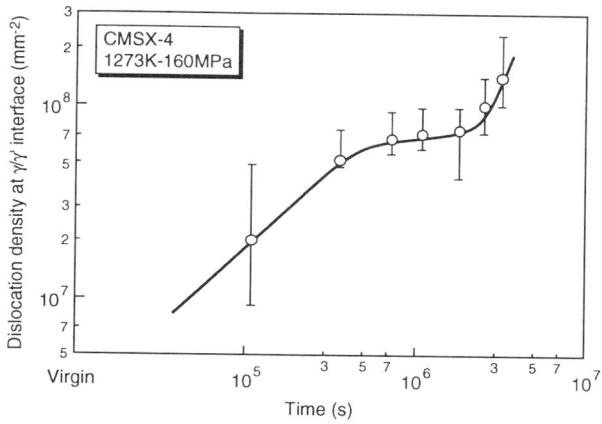

Fig. 10. Change in the dislocation density at the γ/γ' interface with the creep time.

of the other specimens, and therefore, increased remarkably with increasing creep deformation.

Discussion

Dyson et al. suggested that the rate-controlling process of the creep deformation of the nickel-based superalloys was the dislocation glide, since an increase in the creep rate during the accelerating creep stage was interpreted by an increase in the mobile dislocation density as mentioned before[5]. In the present study, attention is paid to the dislocation density at the γ/γ' interface which changed remarkably during creep deformation. One more important point is that the dislocation density at the γ/γ' interface was measured quantitatively on the specimens selected parallel to (101) planes, according to the Hirsch's manner[6], thereby more detailed observations could be done in order to evaluate the suggestion proposed by Dyson et al.

The change in the dislocation density at the γ/γ' interface with the creep testing time is shown in Fig. 10. The dislocation density increased after subjecting creep tests up to 3x10⁵s, and leveled off at 7x10⁷mm⁻² at 10⁶s. However, with further increasing time, the dislocation density increased markedly, corresponded to an increase in the creep rate during the accelerating creep stage, Fig. 2.

The creep rate during just the accelerating creep stage was plotted as a function of the dislocation density at the γ/γ' interface, as indicated in Fig. 11. The correlation between the creep rate during the accelerating creep stage and the dislocation density at the γ/γ' interface showed linear behavior on a log-log scale, the slope being five. In case that the creep rate was controlled by the dislocation glide suggested by Dyson et al., the linear slope would be one.

Consequently, it is difficult to state that the increase in the creep rate during the accelerating creep stage results from an increase in the dislocation density at the γ/γ' interface. In general, an increase in the creep rate during the accelerating creep stage of the polycrystalline heat-resistant alloys has been thought to be mainly attributable to the stress increase by the reduction of the cross-sectional area due to mechanical damage such as initiation and propagation of cracks[7,8]. However, no cracks and voids were

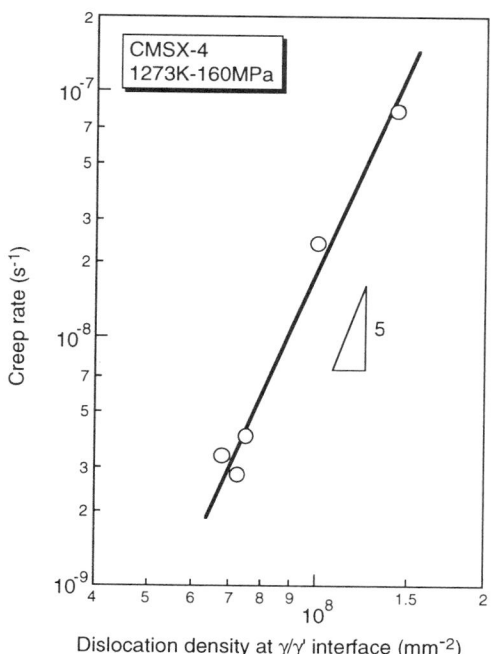

Fig. 11. Relation between the creep rate during the accelerating creep stage and the dislocation density at the γ/γ' interface.

381

observed even in the scanning electron micrograph of the specimen tested for 3.24x10⁶s, at the end of the accelerating creep stage (Fig. 3c). Therefore, it was difficult to claim that the creep rate during accelerating creep stage of this single crystal nickel-based superalloy was attributable to the stress increase due to mechanical damage. On the other hand, it was reported that an increase in the creep rate during the accelerating creep stage of the polycrystalline heat-resistant alloys was caused by the loss of creep resistance through the microstructural changes[9)-18)].

Kondo et al. revealed that the minimum creep rate of the single crystal nickel-based superalloy directly correlated with the radius of dislocation curvature which was the reflection of the thickness of the γ channel, independent of the shape and the size of γ' phase[1)]. Then, the change in the thickness of the γ channel with creep deformation was examined. From the SEM observation of the specimens tested for 1.08x10⁵, 1.08x10⁶ and 3.24x10⁶s as shown in Fig. 3, the thickness of the γ channel seemed to increase with an increase in the creep testing time.

The thickness of the γ channel of the interrupted specimens was plotted as a function of the creep testing time as shown in Fig. 12. The thickness of the γ channel increased immediately after loading and increased markedly after 10⁶s, when γ' phase rafting occurred. This increase in the γ channel corresponded to the increase in the creep rate during the accelerating creep stage, Fig. 2.

The creep rate during the accelerating creep stage was plotted as a function of the thickness of the γ channel in Fig. 13. The correlation between the creep rate during the accelerating creep stage and the thickness of the γ channel was linear on a log-log scale, the slope being five. Consequently, an increase in the creep rate during the

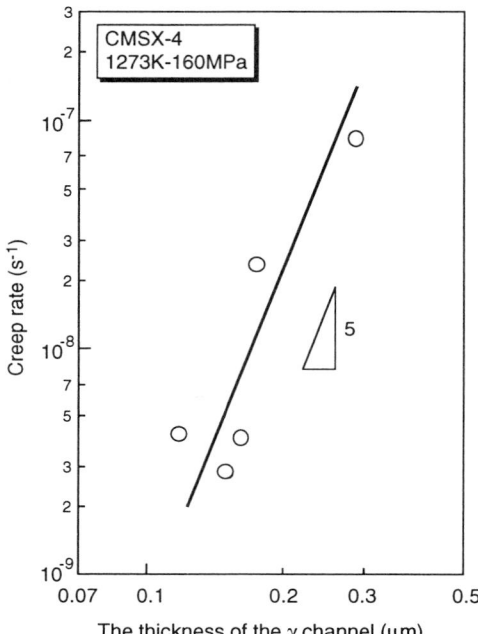

Fig. 13. Relation between the creep rate during the accelerating creep stage and the thickness of the γ channel of the interrupted specimens.

accelerating creep stage was caused by the loss of creep resistance through an increase in the thickness of the γ channel.

High magnification transmission electron micrographs of the specimens creep tested for 1.08x10⁵, 1.08x10⁶ and 3.24x10⁶s was shown in Fig. 14, where B=[100]. The bent dislocation with the small radius of curvature was observed within the γ channel of the specimen creep tested for 1.08x10⁵s (Fig. 14a), at the latter half of the transient creep stage. The radius of dislocation curvature of the specimen creep tested for 1.08x10⁶s (Fig. 14b), at the minimum creep rate, was larger than that of the specimen creep tested for 1.08x10⁵s and smaller than that of the specimen creep tested for 3.24x10⁶s (Fig. 14c), at the end of the accelerating creep stage. The radius of dislocation curvature of these specimens cannot be quantitatively compared with that of the other specimens because the electron beam direction was not perpendicular to (111) planes which were the slip plane of FCC metals. But the radius of dislocation curvature seemed to increase with increasing the creep testing time. Here, the radius of dislocation curvature, R_d, is inversely proportional to the shear stress, $\tau = A/R_d$ (A ; constant), therefore the smaller radius of dislocation curvature resists a larger shear stress[19)-21)]. Conversely, an increase in the radius of dislocation curvature facilitates dislocation motion. The measurements of the radius of dislocation curvature

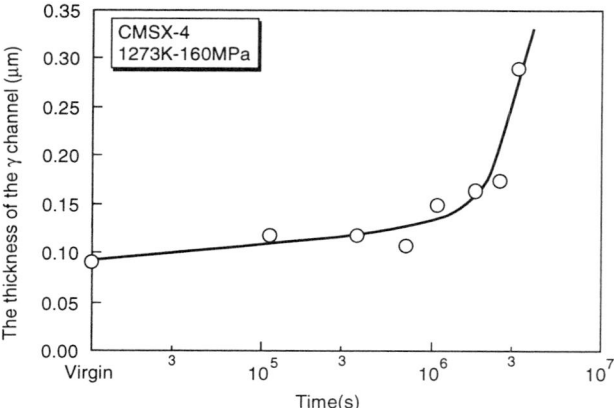

Fig. 12. Change in the thickness of the γ channel of the interrupted specimens with the creep time.

382

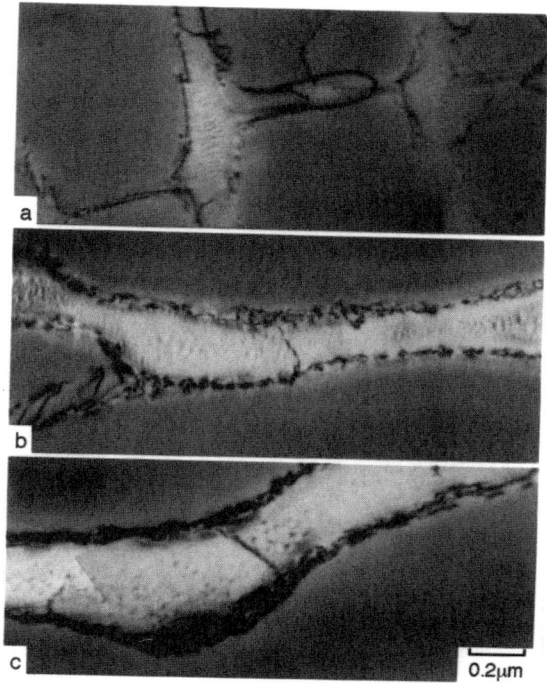

Fig. 14. High magnification transmission electron micrographs of the specimens tested at 1273K-160MPa for (a) 1.08×10^5, (b) 1.08×10^6 and (c) 3.24×10^6s. Stress axis is vertical in these photos.

were made on the specimens sectioned parallel to (111) planes which are the slip planes of FCC metals.

The relationship between the radius of dislocation curvature and the thickness of the γ channel was shown in Fig. 15, where the thickness of the γ channel was converted to the spacing between the two neighboring γ' phases on (111) planes. The radius of dislocation curvature was directly

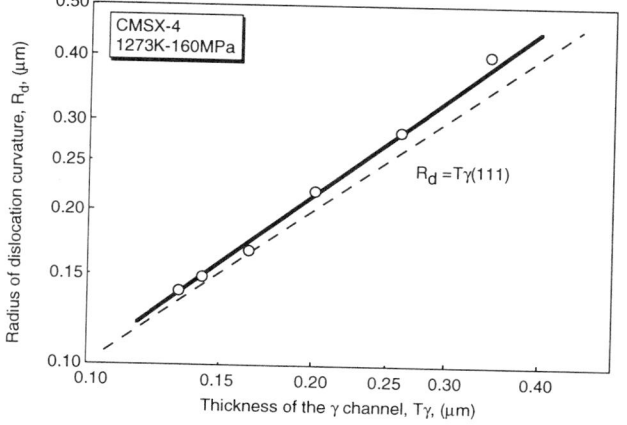

Fig. 15. Relation between the radius of dislocation curvature and the thickness of the γ channel of the interrupted specimens.

proportional to the thickness of the γ channel. From these results, it was concluded that an increase in the creep rate during the accelerating creep stage was caused by the loss of creep resistance through an increase in the radius of dislocation curvature.

Kondo et al. examined the stress exponent of the minimum creep rate, n-value, of the single crystal nickel-based superalloy, CMSX-4, crept at 1273K in the stress range of 100 to 400MPa, and suggested that a decrease in n-value from 7 at the high stress level to 3 at the low stress level resulted from a decrease in creep resistance due to the formation of the rafted structure of γ' phases[4),22),23)]. In this wide stress range, creep rate of the single crystal nickel-based superalloy may be expected to be dependent upon the thickness of the γ channel. Therefore, the thickness of the γ channel of CMSX-4 was measured on the creep interrupted specimens used in previous works[1)-4),22)-27)].

The minimum creep rate and the creep rate during the accelerating creep stage were used as the creep rate. The data at 1273K for the range of 100 to 400MPa was plotted as a function of the thickness of the γ channel in Fig. 16.

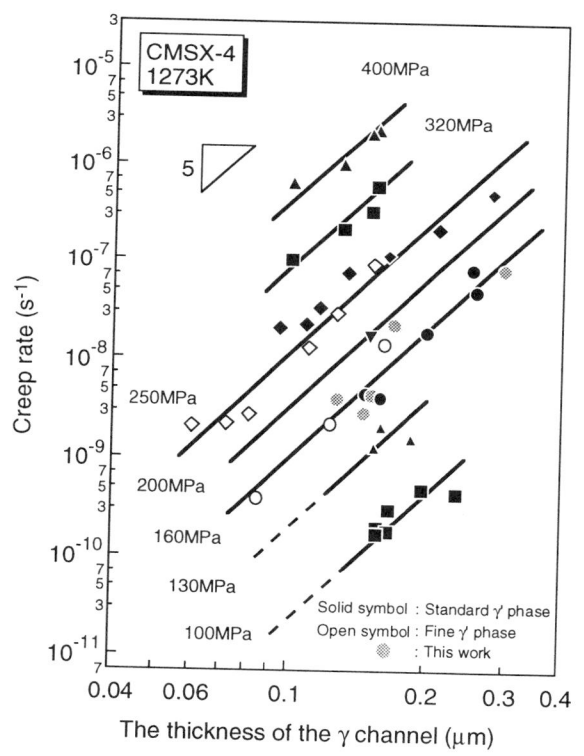

Fig. 16. Relation between creep resistance, that is, the minimum creep rate and the creep rate during the accelerating creep stage, and the thickness of the γ channel of CMSX-4 crept at 1273K in the stress range from 100 to 400MPa.

The correlation between the creep rate and the thickness of the γ channel was linear on a log-log scale; the curves at the stresses of 400, 320, 250, 160 and 100MPa have the same slope of five. The curves with the slope of five were drawn at 130 and 200MPa.

From seven straight lines in Fig. 16, stress-creep rate curve under the constant thickness of the γ channel could be drawn, as shown in Fig. 17. The solid line was the estimated one at the thickness of the γ channel of 0.1μm, which was the initial thickness of the γ channel. As shown in Fig. 17, the solid line was bent at the stress of 250MPa. Under the stresses less than 250MPa where the cuboidal γ' phase turned to the rafted structure, the solid line had the slope with the value of five. For the stresses larger than 250MPa, the cuboidal γ' phase remained at the minimum creep rate. The estimated stress-creep rate curve was compared with the measured one, the dotted line. The estimated creep rate was smaller than the measured one under the stresses less than 250MPa. In this case, the rafted γ' structures acted as a creep weakener, whereas the measured creep rate was approximately equal to the calculated one at the higher stresses than 250MPa where the γ' phase remained cuboidal. From these results, it was concluded that an increase in the creep rate at the lower stresses caused by an increase in the thickness of the γ channel due to the γ' phase rafting.

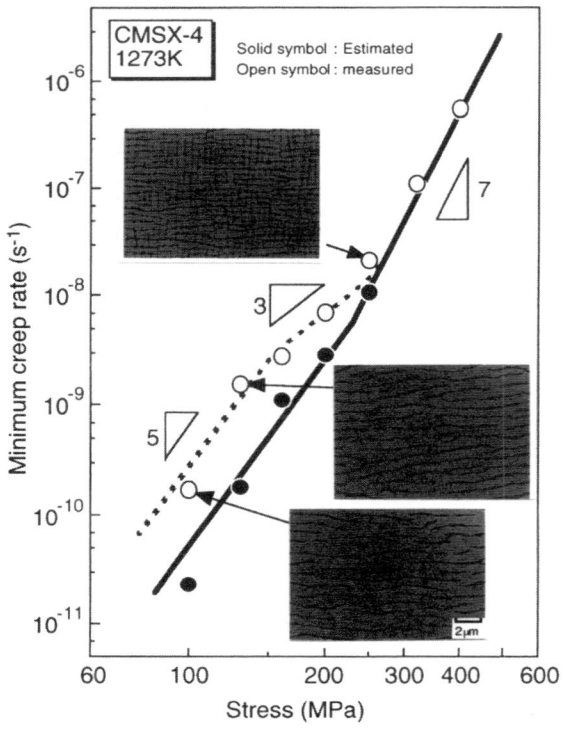

Fig. 17. Estimated and measured stress-minimum creep rate curves of CMSX-4 at 1273K.

Conclusions

Creep resistance, especially the creep rate during the accelerating creep stage crept at 1273K-160MPa, of the single crystal nickel-based superalloy, CMSX-4, was investigated in connection with the dislocation substructure. The results can be summarized in the following.

1) The γ/γ' lamellar structures were formed perpendicular to the stress axis at the minimum creep rate, but regularity of the rafted γ' diminished at the accelerating creep stage.

2) The dislocation substructure was not formed in the γ phase, and the dislocation density at the γ/γ' interface increased with the creep deformation.

3) The correlation between the creep rate during the accelerating creep stage and the dislocation density at the γ/γ' interface was linear on a log-log scale, the slope being five. The slope was different from the supposition proposed by Dyson et al., that an increase in the creep rate during the accelerating creep stage resulted from an increase in the mobile dislocation density.

4) The thickness of the γ channel increased with creep deformation and the correlation between the thickness of the γ channel and the creep rate during the accelerating creep stage was linear.

5) The radius of dislocation curvature was directly proportional to the thickness of the γ channel.

6) Consequently, an increase in the creep rate during the accelerating creep stage was attributable to the loss of creep resistance due to an increase in the radius of dislocation curvature caused by an increase in the thickness of the γ channel.

7) The correlation between creep resistance, $\dot{\varepsilon}$, and the thickness of the γ channel, t_γ, of the single crystal nickel-based superalloy, CMSX-4, crept at 1273K in the stress range from 100 to 400MPa showed $\dot{\varepsilon}$ t_γ^5 behavior.

8) Comparing the practical stress-creep rate curve with the estimated stress-creep rate curve at the certain thickness of the γ channel, the estimated creep rate was smaller than the measured one under the stresses less than 250MPa where the cuboidal γ' phase turned to the rafted structure.

9) From these results, it was concluded that an increase in the creep rate at the lower stresses caused by an increase in the thickness of the γ channel due to the γ' phase rafting.

Reference

1) Y. Kondo et al., "Effect of Morphology of γ' Phase on Creep Resistance of A Single Crystal Nickel-Based Superalloy, CMSX-4," Proc. of the 8th Inter. Conf. Superalloys '96,

(1996), 297-304.

2) N. Kitazaki et al., "Effect of Aging and Stress Aging on Creep Resistance of Single Crystal Ni-base Superalloy, CMSX-4," 123rd Committee on Heat Resisting Metals and Alloys Rep., 34 (1993), 173-181.

3) Y. Kondo et al., "Effect of Aging and Stress Aging on Creep Resistance of Single Crystal Ni-base Superalloy CMSX-4," Tetsu-to-Hagané, 80 (1994), 568-573.

4) Y. Kondo and T. Matsuo, "Creep of Single Crystal Superalloys," 123rd Committee on Heat Resisting Metals and Alloys Rep., 38 (1997), 269-286.

5) B. F. Dyson and M. McLean, "Particle-coarsening, σ_0 and Tertiary Creep" Acta Metall., 31 (1983), 17-27.

6) P. B. Hirsch et al., Electron Microscopy of Thin Crystals (London: Butterworths, 1965), 422.

7) D. A. Woodford, "Creep Damage and the Remaining Life Concept," J. Eng. Mater. Technol., 101 (1979), 311-316.

8) N. Shin-ya and S. R. Keown, "Correlation between Rupture Ductility and Cavitation in Cr-Mo-V Steels," Mater. Sci., 13 (1979), 89-93.

9) K. R. Williams and B Wilshire, "Effect of Microstructural Instability on the Creep and Fracture Behavior of Ferritic Steels," Mater. Sci. Eng., 28 (1977), 289-296.

10) K. R. Williams and B Wilshire, "Creep behavior of 1/2Cr 1/2Mo 1/4V Steel at Engineering Stresses," Mater. Sci. Eng., 38 (1979), 199-210.

11) C. N. Bolton, B. F. Dyson and K. R. Williams, "Metallographic Methods of Determining Residual Creep Life," Mater. Sci. Eng., 46 (1980), 231-239.

12) L. P. Stoter, "Thermal Aging Effects in AISI Type 316 Stainless Steel," J. Mater. Sci., 16 (1981) 1039-1051.

13) R. A. Stevens and P. E. J. Flewitt, "The Effects of γ' Precipitate Coarsening During Isothermal Aging and Creep of the Nickel-base Superalloy IN-738," Mater. Sci. Eng., 37 (1979), p.237-247.

14) J. M. Leitnaker and J. Bentley, "Precipitate Phases in Type 321 Stainless Steel after Aging 17 Years at ∼ 600℃," Metall. Trans. A, 8 (1977), 1605-1613.

15) J. H. Hoke and F. Eberle, "Experimental Superheater for Steam at 2000Psi and 1250 F - Report after 14,281 Hours of Operation," Trans. ASME, 79 (1957), 307-317.

16) M. Tanaka et al., "Study on Microstructure and Mechanical Properties of SUS 304 Used for a Long Time at Elevated Temperature," 123rd Committee on Heat Resisting Metals and Alloys Rep., 24 (1983), 373-384.

17) Y. Kondo et al., "Effect of Recovery Treatment on the Mechanical Properties of SUS 304 Serviced for Prolong Time at Elevated Temperature," 123rd Committee on Heat Resisting Metals and Alloys Rep., 26 (1985), 133-139.

18) Y. Yamaguchi et al., "Microstructures and Mechanical Properties of SUS304 Serviced for Prolonged Time at 520 ∼ 610℃," Tetsu-to-Hagané, 24 (1984), B-61

19) A. Kelly and R. B. Nicholson, Strengthening Methods in Crystals (Amsterdam: Elsevier, 1971),9.

20) F. R. N. Nabarro, Theory of Crystal Dislocations (London: Oxford, 1967), 53.

21) J. Weertman and J. R. Weertman, Elementary Dislocation Theory (New York, NY: Macmillan, 1964)

22) N. Kitazaki et al., "Creep Resistance of Single Crystal Ni-base Superalloy, CMSX-4, at 1273K," 123rd Committee on Heat Resisting Metals and Alloys Rep., 35 (1994), 353-360.

23) K. Ishibashi et al., "Long-term Creep Rupture Properties of Single Crystal Ni-base Superalloy, CMSX-4," 123rd Committee on Heat Resisting Metals and Alloys Rep., 37 (1996), 1-10.

24) N. Kitazaki et al., "Applied Stress Dependence of Transient Creep in Single Crystal Ni-based Superalloy," CAMP ISIJ, 7 (1994), 739.

25) N. Kitazaki et al., "Creep Resistance of Single Crystal Ni-based Superalloy CMSX-4 at 1273K," CAMP ISIJ, 6 (1993), 1725.

26) Y. Hoshizashi et al., "Change in the Radius of Dislocation Curvature with Creep Deformation in Single Crystal Ni-base Superalloy, CMSX-4," 123rd Committee on Heat Resisting Metals and Alloys Rep., 36 (1995), 123-129.

27) Y. Hoshizashi et al., "Dislocation in γ Channel in Single Crystal Ni-base Superalloy, CMSX-4, with Creep Deformation," 123rd Committee on Heat Resisting Metals and Alloys Rep., 36 (1995), 331-341.

Oxidation Improvements of Low Sulfur Processed Superalloys

Tammy M. Simpson and Allen R. Price

Howmet Research Corporation
1500 South Warner Road
Whitehall, Michigan, 49461-1895

Abstract

At the relatively high operating temperatures at which many superalloys are utilized, oxidation is of great concern. Nickel based superalloys tend to form alumina scales that protect the material from further oxidation. The oxidation resistance is reduced, however, when sulfur segregates from the bulk alloy and degrades the alumina layer. Reducing the sulfur level in the alloy material can prevent this from occurring. Howmet Dover Alloy has developed methods for lowering the sulfur content in alloys to less than one ppm using a liquid refining method during the master heat formulation.

Both standard production and low sulfur modified single crystal superalloys were subjected to oxidation testing at 1093°C (2000°F) and 1177°C (2150°F) in order to determine to what extent the sulfur affected the oxidation resistance. Alloys were tested both in the coated and the uncoated condition. The three coatings that were evaluated were LDC-2E, MDC-150, and MDC-150L. The LDC-2E coating is the product of a pack-cementation process. The MDC-150 coating is a platinum aluminide coating developed at Howmet, and MDC-150L coating is a low sulfur platinum aluminide coating that was also developed at Howmet. CMSX-4®, and CMSX-10®, Rene' N5®, and PWA 1484® alloys were evaluated.

In general, the low sulfur alloys were more oxidation resistant than their production counterparts in both the uncoated and the coated states for all of the alloys evaluated. Reducing the amount of sulfur present in the nickel based alloys did increase the oxidation resistance especially at 1177°C (2150°F). All three coatings were effective at limiting alloy oxidation at 1093°C (2000°F) up to 2000 hours.

Superalloys 2000
Edited by T.M. Pollock, R.D. Kissinger, R.R. Bowman,
K.A. Green, M. McLean, S. Olson, and J.J. Schirra
TMS (The Minerals, Metals & Materials Society), 2000

Introduction

At the relatively high operating temperatures at which many superalloys are utilized, oxidation is of great concern. Nickel based superalloys tend to form alumina scales that protect the material from further oxidation[1]. The oxidation resistance is reduced, however, when sulfur segregates from the bulk alloy and degrades the adhesion of the alumina layer[2]. Lowering the sulfur level in the alloy material can prevent this from occurring[3]. Howmet Dover Alloy has developed methods for lowering the sulfur content in alloys to less than one ppm using a liquid refining method during the master heat formulation[4]. Typical sulfur levels in single crystal superalloys are between two to ten ppm.

Technical Approach

The alloys, coatings, and testing temperatures used in this experiment are summarized in Table I. Both low sulfur and standard production heats of PWA 1484 and Rene' N5 were produced at Howmet Dover Alloy (HDA). Multiple master heats of the two alloys were produced in sizes ranging from 180 to 2270 kg (400 to 5,000 pounds) having bulk sulfur levels of < 0.5ppm. In addition, blends of CMSX-4 and CMSX-10 were purchased with different sulfur levels. Each alloy was tested at both temperatures with each coating as well as in the uncoated state. Alloys with the "LS" designation are low sulfur alloys. Major chemistry differences of the alloys are highlighted in Table II. None of the alloys contained the reactive element yttrium.

Table I: Parameters for the Oxidation Experiment

Alloy*	Coating	Parameters
Rene' N5	LDC-2E	A) 1093°C
Rene' N5 LS	MDC-150	100 hours
PWA 1484	MDC-150L	at temperature
PWA 1484 LS	Uncoated	
CMSX-4 Blend		B) 1177°C
CMSX-4 Virgin		50 min heat-up &
CMSX-10 Virgin		10 min cool-down

* LS stands for low sulfur

The LDC-2E coating is the platinum aluminide product of a pack-cementation process. The MDC-150 coating is a platinum aluminide coating developed at Howmet, and the MDC-150L coating is a low sulfur platinum aluminide coating that was also developed at Howmet. These coatings are deposited via chemical vapor deposition. The LDC-2E and MDC 150 are high activity coatings, whereas the MDC-150L is a low activity coating[5]. The aluminum activity affects the direction of diffusional growth[5].

Table II: Main Elemental Differences Among the Alloys

Element	Rene' N5	PWA 1484	CMSX-4	CMSX-10
Al	6.2	5.7	5.6	5.7
Co	7.5	10.0	10.0	3.3
Cr	7.0	5.0	6.5	2.3
Mo	1.5	1.9	0.6	0.4
Re	3.0	3.0	3.0	6.2
Ta	6.5	8.7	6.0	8.4
Ti	0.04 max	-	1.0	0.22
W	5.0	5.9	6.0	5.5

Coupons were prepared from panels cast in each of the alloys. Sulfur levels were measured below 1ppm using a technique developed at Howmet Research Corporation that compares well with glow discharge mass spectrometry (GDMS). Four coupons from each alloy were tested. The coupons were grit blasted prior to coating in order to clean the surface. The coupons were either left in the uncoated state or coated with one of the three coating types.

After coating, the coupons were subjected to a heat treatment to stabilize the coating. Cyclic oxidation testing at 1177°C (2150°F) consisted of 50 minutes of hot time at the designated temperature and 10 minutes of cooling for a total of 2000 hours of testing. At 1093°C (2000°F), the only cooling period occurred when the coupons were weighed. Past oxidation research efforts have shown that the cyclic testing at 1177°C (2150°F) is representative of what is observed in normal operating conditions, only at an accelerated rate. The 1093°C (2000°F) testing condition was chosen because other research has been conducted at that particular temperature.

Periodically the coupons were cooled to room temperature and weighed. Samples were oxidized for a total of 2000 hours. The average weight change was monitored throughout the oxidation test. The weight change was normalized to the surface area of the coating. Depletion zone measurements were made on the uncoated samples. The oxidation resistance of the low sulfur alloy version was compared to the production version.

Results and Discussion

Measurements were made of the depletion zone depth for each of the alloys that completed the uncoated oxidation testing. Table III is a summary of the average depletion zone thickness and final 2000 hour weight change for the 1093°C (2000°F) cycle. This table is not intended to relate alloy performance with depletion zone depth but to show the ability of the lower sulfur containing alloys to form a protective scale that is less susceptible to spallation. As the rate of protective scale spallation is reduced, both the depth of alloy depletion and alloy weight loss are decreased.

Table III: Results of 1093°C (2000°F) Oxidation Trials After 2000 hours

Alloy	Sulfur Level (ppm)	Avg. Depletion Zone Depth	Avg. Final Wt. Change (mg/cm^2)
Rene' N5	3.5	0.0011"	-10.661
Rene' N5 LS	0.8	0.0005"	0.616
PWA 1484	2.5	0.0023"	-8.856
PWA 1484 LS	0.3	0.0006"	-0.001
CMSX-4 Blend	4.1	0.0040"	-16.753
CMSX-4 Virgin	2.6	0.0017"	-4.553
CMSX-10 Virgin	1.8	0.0021"	-24.329

The depletion zone data for the 1177°C (2150°F) uncoated oxidation coupons are listed in Table IV. The results were similar to those seen in the 1093°C (2000°F) oxidation cycle, only more pronounced at the higher temperature.

Table IV: Results of 1177°C (2150°F) Oxidation Trials After 2000 hours

Alloy	Sulfur Level (ppm)	Avg. Depletion Zone Depth	Avg. Final Wt. Change (mg/cm^2)
Rene' N5	3.5	0.0074"	-15.396
Rene' N5 LS	0.8	0.0034"	-5.611
PWA 1484	2.5	0.0100"	-328.878
PWA 1484 LS	0.3	0.0038"	-7.332
CMSX-4 Blend	4.1	0.0110"	-390.581
CMSX-4 Virgin	2.6	0.0074"	-105.827
CMSX-10 Virgin*	1.8	0.0048"	-567.276

* Coupons were removed after 1650 hours due to severe oxidation.

Figures 1 and 2 are optical micrographs of uncoated PWA 1484 after oxidation for 2000 hours at 1177°C (2150°F). Production PWA 1484 is shown in Figure 1, and low sulfur PWA 1484 is shown in Figure 2. The reduced sulfur level helps to maintain the integrity of the protective oxide layer for a greater amount of time.

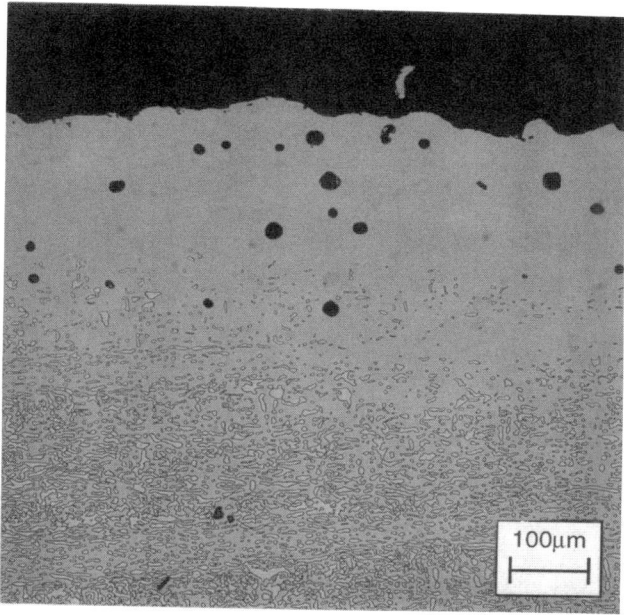

Figure 1: Optical photograph of production PWA 1484 after oxidation for 2000 hours at 1177°C (2150°F).

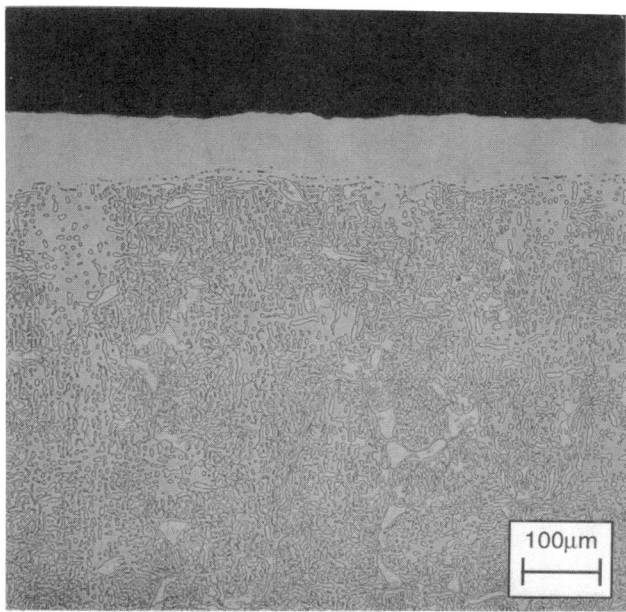

Figure 2: Optical photograph of low sulfur PWA 1484 after oxidation for 2000 hours at 1177°C (2150°F).

The results for the coated alloys tested at 1177°C (2150°F) are shown in Figures 3 to 6. Each graph compares the relative oxidation resistance as a measure of weight change for both the low sulfur and the production version of each alloy for all three coating types. Since the various coatings are not the same thickness, weight change results were normalized to the thickness of the coating after post-coat heat treatment to be able to more accurately compare coating effectiveness. Failure was designated as a weight change of zero as compared to the original weight of the coated coupon. Please note that the legends are ordered according to the amount of weight loss at 2000 hours. In addition, "LS" stands for low sulfur processed material and "Prod" stands for production alloy.

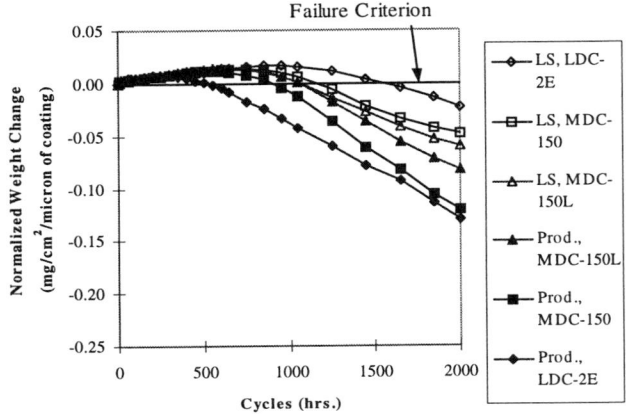

Figure 3: Cyclic oxidation results at 1177°C for production and low sulfur versions of coated Rene' N5.

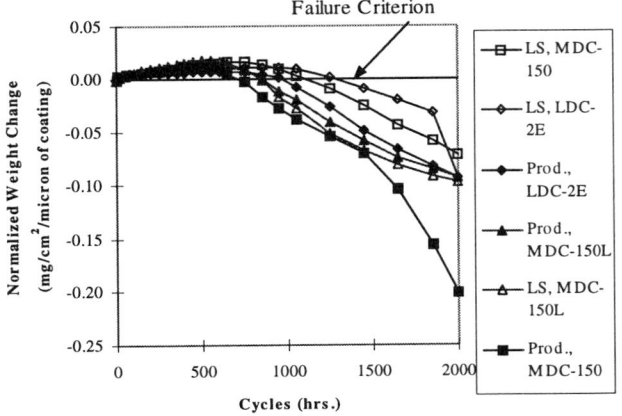

Figure 4: Cyclic oxidation results at 1177°C for production and low sulfur versions of coated PWA 1484.

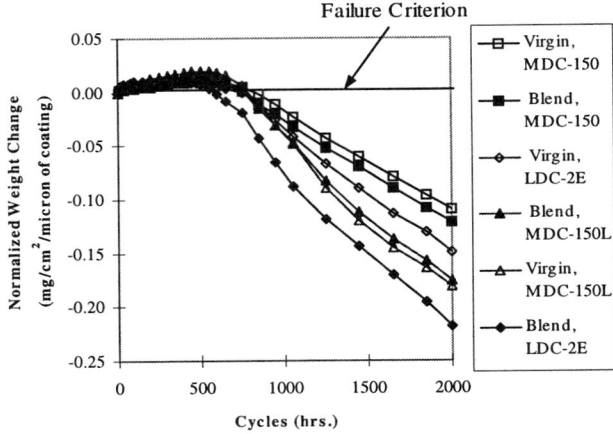

Figure 5: Cyclic oxidation results at 1177°C for virgin and blend versions of coated CMSX-4.

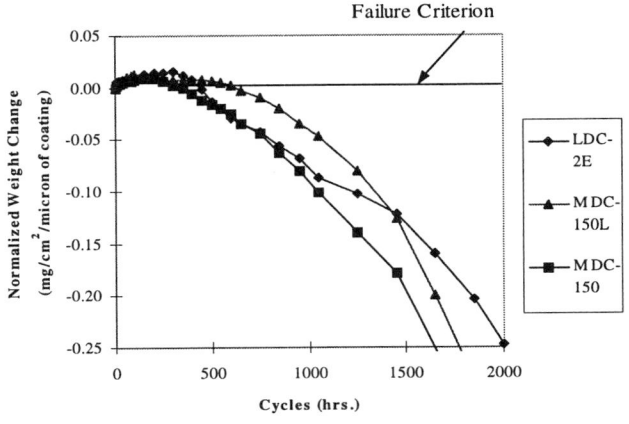

Figure 6: Cyclic oxidation results at 1177°C for coated CMSX-10.

Oxidation resistance results at the 1177°C (2150°F) testing temperature are detailed in Table V. Failure was designated as the number of cycles that corresponded to a weight change of zero as compared to the original weight of the coupon prior to oxidation testing. The number of cycles to failure was then divided by the original coating thickness to determine the number of hours of oxidation resistance per micron of coating. In general, the low sulfur alloys were more resistant to oxidation for a longer time period than the production alloys. That trend is more apparent with the LDC-2E and MDC-150 coated samples. However, no differences could be determined between the production and low sulfur versions of the MDC-150L coated samples.

Table V: Cycles to Failure at 1177°C for Each Alloy

Alloy	Sulfur Level (ppm)	Cycles to Failure/Micron of Coating (Hours/Micron of Coating)		
		LDC-2E	MDC-150	MDC-150L
Rene N5	3.5	6	10	13
N5 LS	0.8	18	14	14
PWA 1484	2.5	12	9	12
1484 LS	0.3	15	14	12
CMSX-4 Blend	4.1	7	7	6
CMSX-4 Virgin	2.6	8	8	6
CMSX-10	1.8	5	3	5

Most of the alloy systems exhibited excellent oxidation resistance at 1093°C (2000°F) even up to 2000 hours. The only cooling period occurred when the coupons were to be weighed. Periodically (every 100 hours), the samples were cooled to room temperature and weighed. The 1093°C (2000°F) oxidation results for Rene' N5 are detailed in Figure 7. As can be seen, all coated coupons experienced weight gain up to the 2000 hour time period.

Similar results were found with PWA 1484 and CMSX-4 (Figures 8 and 9). The coated CMSX-10 coupons (shown in Figure 10) exhibited a very different behavior. The MDC-150 coated coupons had significant weight loss after about 750 hours. The LDC-2E coating had begun to experience a minor amount of weight loss, while the MDC-150L coating continued to gain weight.

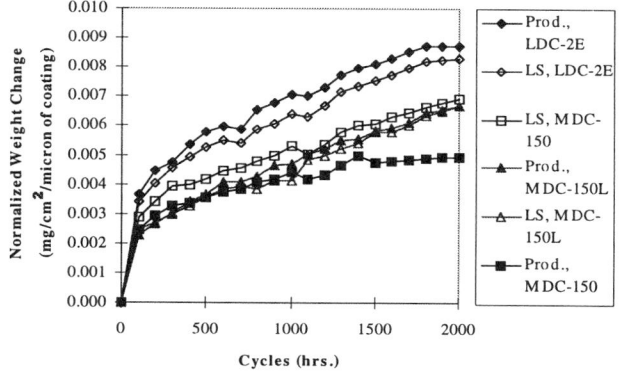

Figure 7: Cyclic oxidation results at 1093°C for production and low sulfur versions of Rene' N5.

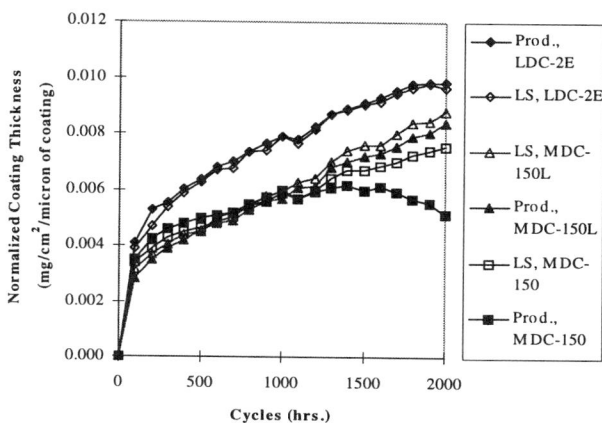

Figure 8: Cyclic oxidation results at 1093°C for production and low sulfur versions of PWA 1484.

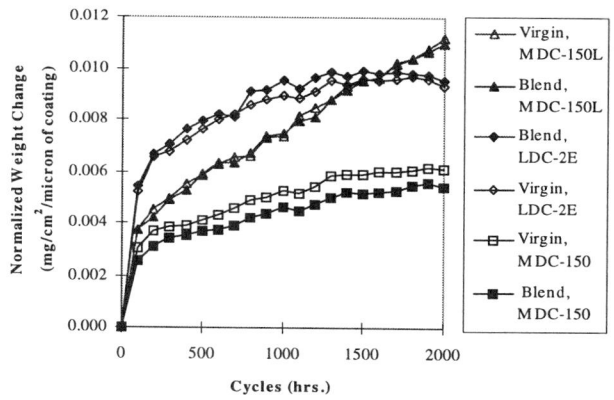

Figure 9: Cyclic oxidation results at 1093°C for virgin and blend versions of CMSX-4.

391

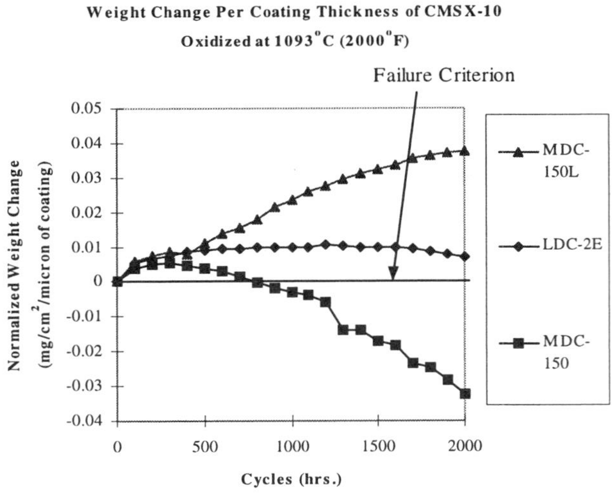

Figure 10: Cyclic oxidation results at 1093°C for CMSX-10.

In general, it is evident that the Howmet low sulfur processing and lower sulfur test coupons provided improved oxidation performance for the alloys tested. At 1177°C (2150°F) the low sulfur alloys experienced a smaller weight loss than their production alloy counterparts that have a higher sulfur content. However, at 1093°C (2000°F), the driving force for oxidation was not as great, and therefore protective layer spallation was not a problem up to 2000 hours. Thus, most of the alloys at that testing temperature experienced only weight gain. The sulfur content may not affect the oxidation resistance of the alloys at 1093°C (2000°F) until exposed for a longer time period.

Conclusions

Alloys with a sulfur level of less than one ppm were produced using a liquid refining method developed at Howmet Dover Alloy. Reducing the amount of sulfur present in the nickel based superalloys did increase the oxidation resistance especially at 1177°C (2150°F). The increase in oxidation resistance with reduced sulfur content was more apparent with the LDC-2E and MDC-150 coatings rather than the MDC-150L coating. All three of the coatings used in the evaluation appeared to be especially effective at limiting alloy corrosion at 1093°C (2000°F) up to 2000 hours.

Acknowledgments

The authors would like to acknowledge Dr. Bruce Warnes of Howmet Thermatech Coating for informative technical discussions. In addition, Rebecca Boczkaja is acknowledged for running all of the oxidation tests and evaluating the oxidation coupons.

References

1.   Chester Sims, Norman Stoloff, and William Hagel, Superalloys II – High Temperature Materials for Aerospace and Industrial Power (New York, NY: John Wiley and Sons, 1987), 301.
2.   C. Sarioglu et al. "The Control of Sulfur Content in Nickel-Base, Single Crystal Superalloys and Its Effects on Cyclic Oxidation Resistance," Superalloys 1996, 71-80.
3.   James Smialek, "Toward Optimum Scale and TBC Adhesion on Single Crystal Superalloys," Electrochemical Society Proceedings, 98-99.
4.   Jeffrey Irvine et al., "Ultra Low Sulfur Superalloy Castings and Method of Making," U.S. Patent # 5,922,148 July 13, 1999.
5.   B. Warnes, "Improved Pt Aluminide Coatings Using CVD and Novel Platinum Electroplating" (Paper presented at International Gas Turbine and Aeroengine Congress and Exhibition, Sweden, 1998 ASME 98-GT-391).

Disk Alloys

OPTIMISATION OF THE MECHANICAL PROPERTIES
OF A NEW PM SUPERALLOY FOR DISK APPLICATIONS

D. Locq, M. Marty and P. Caron

Office National d'Etudes et de Recherches Aérospatiales (ONERA)
BP 72 – 92322 Châtillon Cedex – France

Abstract

The roles of the subsolvus solutioning temperature and of the ageing temperature on the microstructure and on tensile and creep properties were evaluated on a new PM superalloy, designated NR3. A single-stage ageing treatment was applied instead of the usual double-stage ageing treatment. The principal objective was to improve the creep resistance of this superalloy while keeping a high tensile strength by using a simplified heat treatment.

The solution treatment essentially affected the respective proportions of primary and secondary γ' precipitates and the grain size. On the other hand, the major influence of the increase of the ageing temperature was to coarsen the tertiary γ' particles. These microstructural effects were successfully combined to obtained a significant increase of the creep strength associated with a slight rise of the tensile properties.

Introduction

Nickel base superalloy disks processed by the powder metallurgy (PM) route exhibit a high level of mechanical properties due to their homogeneous fine-grained microstructure and due to high levels of alloying elements which can be added. A number of alloys for disk applications were thus developed with the aim to get optimum properties, mainly by optimisation of their chemistry.

Another way to improve the mechanical properties of these materials is to control their microstructure by varying thermal and/or thermomechanical parameters. The main microstructural characteristics are the grain size, the volume fraction and the distribution of the strengthening γ' phase precipitates and the intergranular carbide/boride precipitation. In a number of studies,

these parameters were modified essentially by choosing between subsolvus or supersolvus solution treatments leading respectively to fine-grained or coarse-grained microstructures, and/or by varying the cooling rate after this high temperature treatment, thus controlling the size of the secondary γ' precipitates. On the other hand, the effects of the ageing treatments usually applied after the solution treatment were only scarcely studied.

The alloy selected for this work was NR3 (1), recently developed at ONERA in order to satisfy the requirements for high temperature disk applications. This alloy was selected amongst a series of experimental alloys derived from the reference N18 alloy (2, 3). The experimental chemistries were defined in order to avoid the precipitation of intergranular and intragranular topologically close-packed (TCP) phase particles that occurs in the N18 alloy during exposures at temperatures above 650°C (4), while showing comparable or higher mechanical properties (5). The NR3 alloy had a coarse-grained microstructure obtained through a supersolvus solution treatment which improves the creep strength at high temperatures. In these conditions the NR3 alloy therefore exhibited a satisfactory balance between tensile and creep properties, while showing no propensity for TCP phase precipitation at temperatures above 650°C.

The purpose of the present work was to identify an optimised γ' particle distribution obtained through an appropriate ageing treatment procedure in order to further improve the creep strength, while keeping a fine-grained microstructure with the aim to maintain a high level of tensile properties.

Alloy Development

The NR3 alloy was the result of the study of a series of nine experimental superalloys designed, processed and evaluated at

Superalloys 2000
Edited by T.M. Pollock, R.D. Kissinger, R.R. Bowman,
K.A. Green, M. McLean, S. Olson, and J.J. Schirra
TMS (The Minerals, Metals & Materials Society), 2000

ONERA (5). Its common features with the N18 alloy were :
- a γ' atomic fraction maintained at 0.50-0.55
- a Co content of 15 wt. %
- a Cr content fixed at 25 at. % in the γ matrix to maintain a good environmental resistance at high temperatures especially during crack propagation
- minor elements contents (C, B, Zr and Hf) close to those of N18 alloy.

The Mo content was lowered in order to decrease the $\overline{Md}$ parameter and in this way to avoid the precipitation of TCP phases. The stability criterion $\overline{Md}$ was calculated using the New Phacomp method (6) and a critical value for TCP phase precipitation was determined to be 0.915 by previous experiments. To counterbalance the loss of solid solution strengthening, the Ti/Al ratio was increased to 0.9 (vs 0.6 for N18 alloy (at. %)). The chemistry of the NR3 alloy was balanced by using a program based on the method developed by Watanabe for calculation of the fraction of γ' phase and the composition of the γ and the γ' phases from the alloy chemistry (7). The density of this alloy was estimated using the formula developed by Hull (8). After supersolvus solution treatment, the NR3 alloy exhibited tensile and creep properties very close to or better than those of N18 alloy. At last, even after a long-time exposure of 10 000 hours at 750°C, no intragranular TCP phases were observed in NR3 alloy.

Experimental Procedures

An as-forged pancake of NR3 alloy was processed by Tecphy and SNECMA using the N18 industrial route : i) vacuum induction melting, ii) argon atomisation, iii) screening (ϕ < 75 μm (-200 mesh)), iv) hot extrusion and v) isothermal forging. The chemical composition is reported in Table I. The density of NR3 alloy was measured to be 8.05 g.cm^{-3}.

Table I NR3 Alloy Composition (wt. %)

Ni	Co	Cr	Mo	Al	Ti	Hf	B	C	Zr
bal.	14.65	11.8	3.3	3.65	5.5	0.33	0.013	0.024	0.052

Solution heat treatments were conducted in argon below the γ' solvus temperature. The subsequent cooling was performed at a controlled rate of about 100 K.min^{-1} that is representative of the industrial conditions.

Primary γ' area fraction and grain size were determined by image analysis of pictures acquired by optical microscopy. Microstructural observations of γ' precipitates and dislocation structures were performed by transmission electron microscopy (TEM) using a JEOL 200 CX microscope operating at 200 kV. Thin foils were prepared by electrochemical polishing in a twin-jet polisher using a solution of 45 % acetic acid, 45 % butylcellosolve and 10 % perchloric acid, cooled to 263 K and using a potential of 25 V.

Vickers microhardness tests were performed in air at room temperature under a load of 1 kg. The results presented are the average of 5 to 10 measurements. Tensile tests were conducted in air at 400 and 700°C on 4 mm diameter cylindrical specimens with an initial strain rate of $1.7 \times 10^{-3} s^{-1}$. Tensile creep tests were conducted in air at 700°C on 3 mm diameter cylindrical specimens.

Experimental Results

Heat Treatment and Microstructure Selection

Introduction. A standard heat treatment ("SHT") based on that applied to N18 alloy has been defined for NR3 alloy. This treatment consists of a subsolvus solution treatment (30 K below the γ' solvus temperature) followed by two ageing treatments. The γ' solvus temperature was measured to be close to 1205°C (by isothermal treatments and optical metallography). So, the NR3 SHT is :

$$1175°C/4h + 700°C/24h + 800°C/4h.$$

The necessity for two ageing treatments is not clearly justified and understood and some authors even asked themselves about the influence of each ageing treatment (4, 9). On this basis and with the goal of a better understanding of the ageing effect on the microstructure, single-stage ageing treatments were tested after the solution heat treatment.

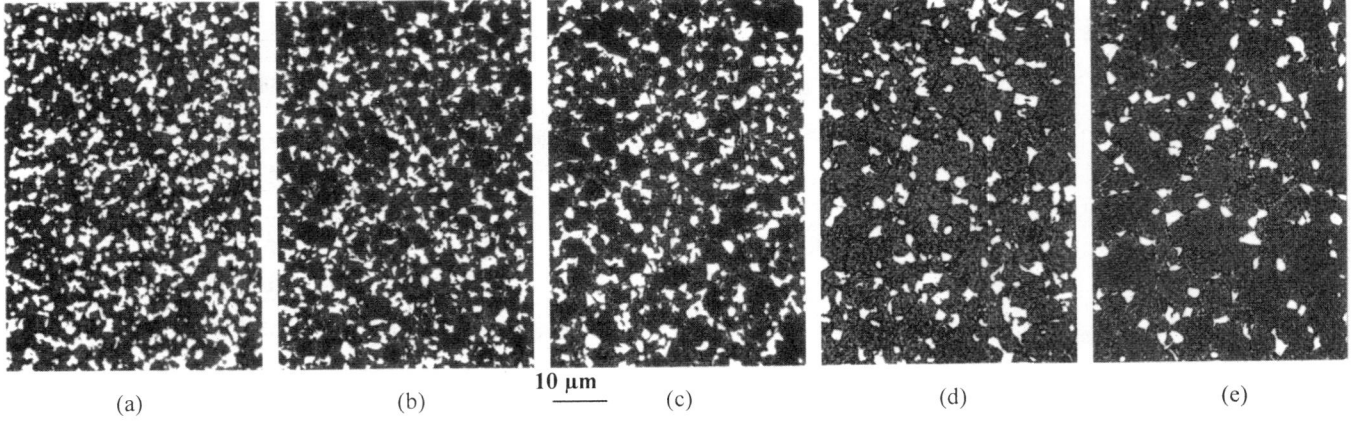

| (a) | (b) | 10 μm | (c) | (d) | (e) |

Figure 1: Variation of the area fraction of the primary γ' with the solutioning temperature (optical micrographs) : (a) 1145°C/4h, (b) 1165°C/4h, (c) 1175°C/4h, (d) 1185°C/4h and (e) 1195°C/4h.

<u>Effect of the Subsolvus Solution Treatment</u>. Six solutioning temperatures were chosen between the forging temperature and the γ′ solvus temperature (i.e. between 1120 and 1205°C). The duration of these treatments was 4 hours.

The micrographs of Figure 1 show the variation in the primary γ′ fraction with the solutioning temperature. The area fraction of primary γ′ and the average grain size *vs* the solutioning temperature are plotted in Figure 2. It can be seen that even 4 K below the γ′ solvus temperature, 5 % of primary γ′ remained. This gradual dissolution of the primary γ′ led to a relatively slow growth of the grain size especially in the range 1145-1185°C. Above 1185°C, the average grain size increased more quickly because there were no more enough primary γ′ precipitates to pin the grain boundaries which begin to move.

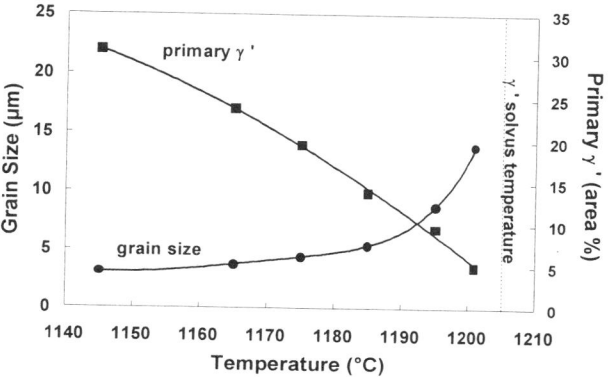

Figure 2: Variation of the grain size and the area fraction of primary γ′ with the solutioning temperature.

Another microstructural parameter affected by the solutioning temperature is the average size of the secondary γ′ precipitates. Actually, the average dimension of the edge of these precipitates was about 130 nm after 4 hours at 1145°C and moved to about 190 nm after 4 hours at 1195°C (see Figure 3). The distribution and the size of the tertiary γ′ precipitates were not significantly affected by this heat treatment.

Microhardness tests were performed at room temperature on some specimens. This is a simple test commonly used for mechanical characterisation of microstructure evolution (4, 9). The average values are presented in Table II.

Table II Effect of Solutioning Temperature on Microhardness

Solutioning temperature	Microhardness HV 1 kg
As-forged	430
1145°C/4h	422
1175°C/4h	411
1195°C/4h	407

A slight decrease of the microhardness was observed with the increase of the solutioning temperature. The increase of the fraction of secondary and tertiary γ′ precipitates which contribute to the strengthening of the alloy was very likely balanced by the increase of the grain size and the secondary γ′ precipitate size.

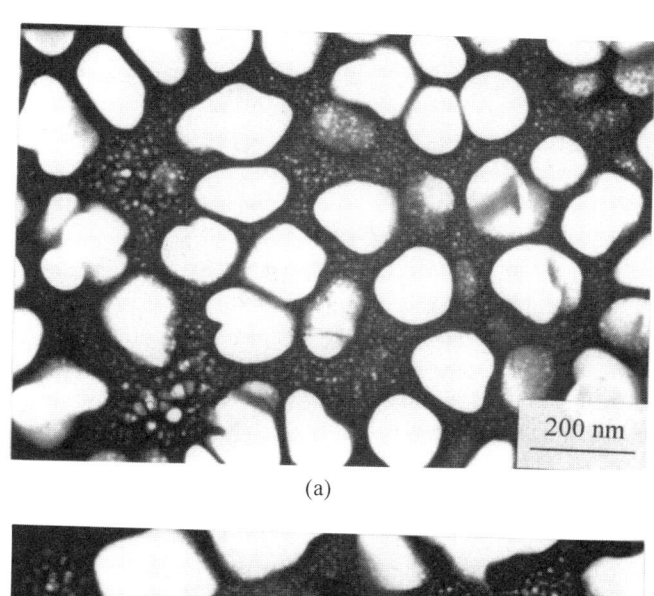

(a)

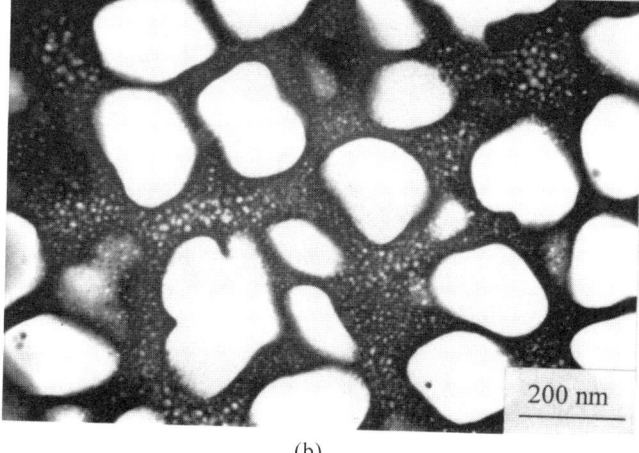

(b)

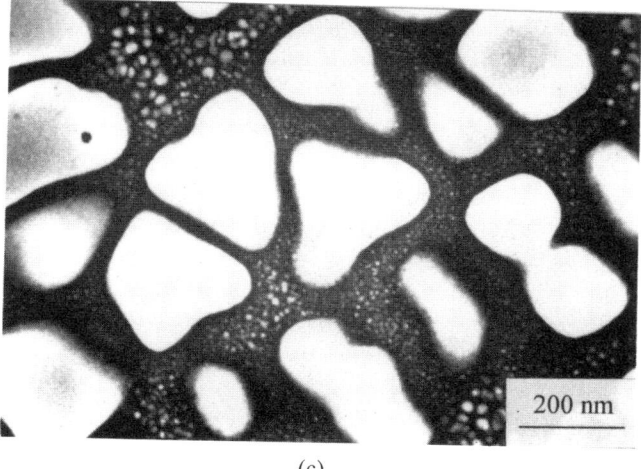

(c)

Figure 3: Effect of the solutioning temperature on the γ′ precipitation (+ ageing treatment : 750°C/4h), (TEM dark field micrographs using γ′ diffraction spots) : (a) 1145°C/4h, (b) 1175°C/4h and (c) 1195°C/4h.

397

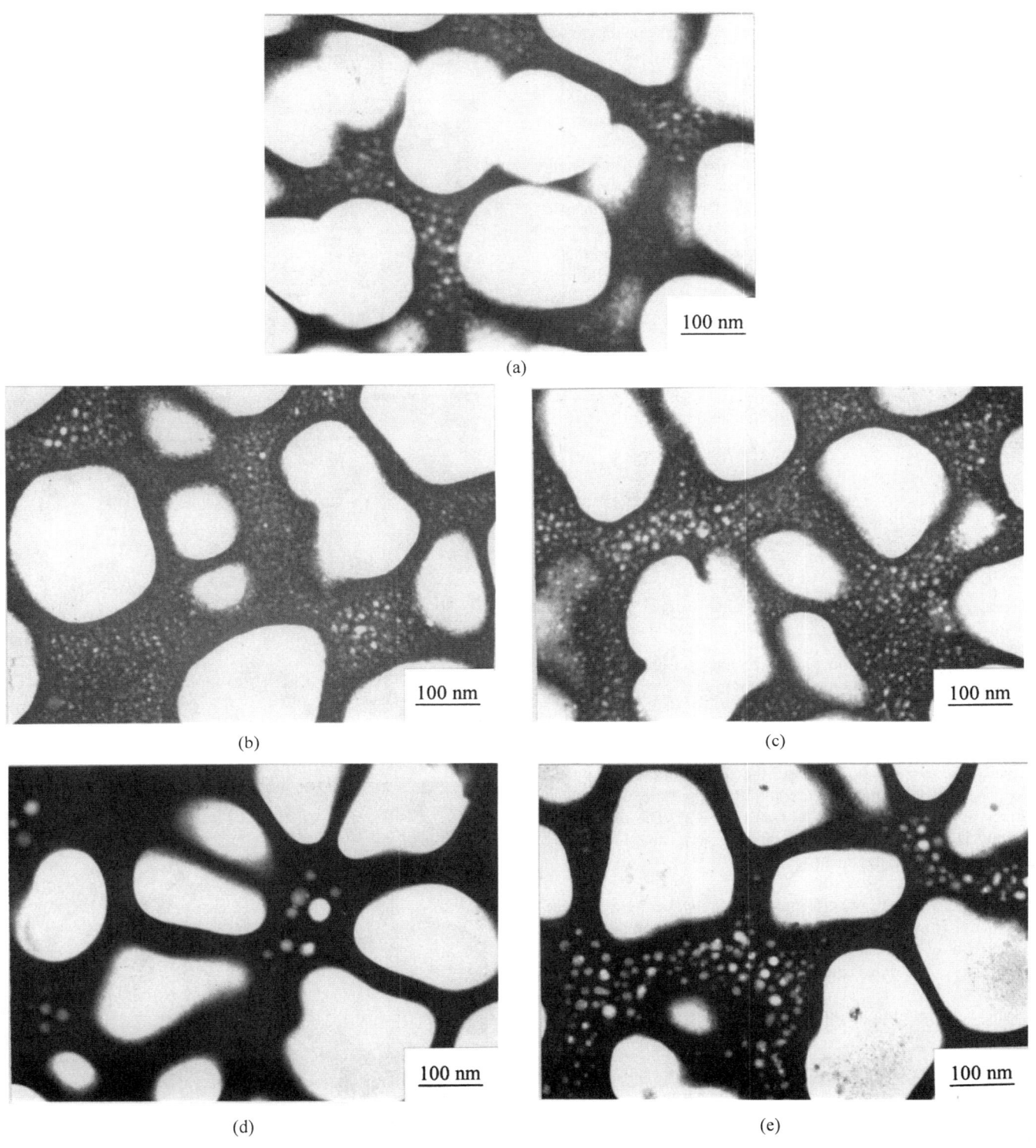

Figure 4: Effect of the ageing sequence on the γ 'precipitation (solution treatment 1175°C/4h (100 K.min⁻¹ cooling)), (TEM dark field micrographs using γ ' diffraction spots) : (a) as-solutioned, (b) 650°C/4h, (c) 750°C/4h, (d) 850°C/4h and (e) 700°C/24h + 800°C/4h.

Effect of the Ageing Treatment. The use of a single-stage ageing treatment instead of the double-stage ageing treatment usually applied for the major disk superalloys presented two advantages : i) a direct relationship between the heat treatment and the effect on microstructure and mechanical properties should be drawn up more easily and ii) it could be technically and economically interesting for the manufacturer. The duration of the ageing treatment was fixed to four hours which could also be an industrial benefit. The effect of the ageing treatment was studied in the range 650-900°C. Three solutioning temperatures were chosen for this study, 1145, 1175 and 1195°C. This choice will be discussed latter.

Clearly, the ageing treatment had no effect on the dissolution of primary γ' precipitates and so on the grain size. Three samples were treated at 1175°C during 4 hours and aged at 650, 750 or 850°C for microstructural observations. Two additional microstructures were also observed : as-solutioned and the SHT microstructures. The TEM micrographs of Figure 4 showed that none of the these ageing treatments induced the coarsening of the secondary γ' precipitates. It could be also noted that tertiary γ' precipitates were already present in the γ channels after the cooling following the solution treatment (Figure 4a). The ageing treatment at 650°C led to a higher particle density of the tertiary γ' population (Figure 4b). These spherical particles generally have a diameter lower than 10 nm with some bigger ones (15-20 nm) located in the largest γ channels or at intersections of channels. After the 750°C ageing treatment the tertiary γ' was on average slightly coarser (Figure 4c). Moreover, after the ageing treatment at 850°C the tertiary γ' precipitates were dramatically less numerous than after ageing at lower temperatures (Figure 4d). The finest γ channels were free from these particles and the largest ones contained some coarser precipitates (15 to 30 nm). The second effect resulting from the ageing at 850°C was a spare intergranular precipitation (Figure 5). These particles ($M_{23}C_6$ type carbides) were about 0.1 μm thick. The microstructure of the NR3 alloy was also studied after the SHT. The particle density and the size of the tertiary γ' precipitates were an intermediary state between those observed after an ageing at 750°C and 850°C (Figure 4e). Very sparse and fine intergranular carbides were also observed.

Microhardness tests were performed on samples solution heat treated at 1145, 1175 or 1195°C for 4 hours. For the three solutioning temperatures, the maximum of microhardness (peak ageing) was obtained for an ageing temperature close to 750°C (Figure 6). It was also observed that the variation of the microhardness with the ageing temperature is higher for the solution treatment at 1175°C than for the two other solutioning temperatures. Finally, one can see that a single-stage ageing treatment after a solution treatment at 1175°C can lead to a microhardness equal or superior to the value measured on a SHT sample (1175°C/4h+700°C/24h+800°C/4h).

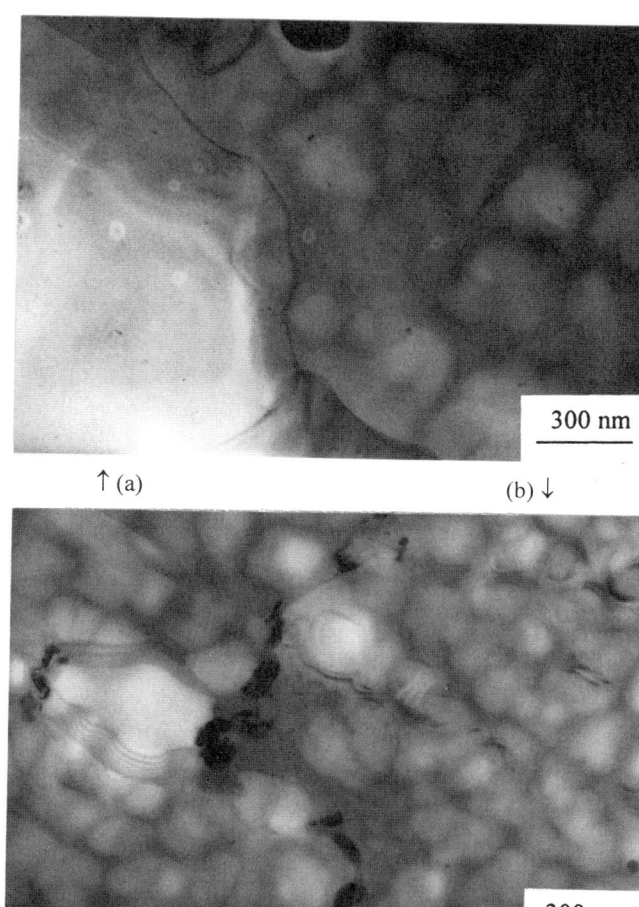

↑ (a)   (b) ↓

Figure 5: Variation of the grain boundary microstructure with the ageing temperature (solution treatment 1175°C/4h (100 K.min⁻¹ cooling)) (TEM bright field) : (a) 750°C/4h and (b) 850°C/4h.

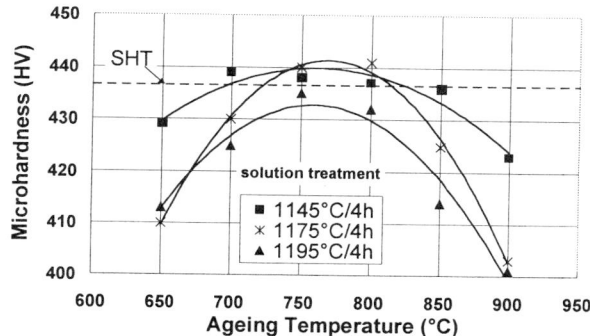

Figure 6: Variation of the microhardness at room temperature with the solutioning temperature and the ageing temperature.

Selection of the Heat Treatments for Mechanical Assessment.
Six heat treatments were selected for tensile and creep tests on the basis of the microstructural observations and of the microhardness results :

- 3 solutioning temperatures with the same ageing temperature
- 4 ageing treatments with the same solutioning temperature.

These different heat treatments are presented in Table III.

Table III Heat Treatments Selected for Mechanical Assessment

Solution treatment	Ageing treatment			
	650°C4h	750°C/4h	850°C/4h	700°C/24h + 800°C/4h
1145°C/4h		X		
1175°C/4h	X	X	X	X
1195°C/4h		X		

The common solutioning temperature selected for the study of the influence of the ageing temperature was 1175°C because it was the solutioning temperature of the SHT and also the temperature which presented the best potential of hardening with the ageing treatment (see Figure 6). The 1145°C solution treatment was selected because the fine grain size and the fine secondary γ′ precipitate size are open to improve the tensile properties. On the contrary, the 1195°C solution treatment was chosen because the coarse grain size and the large fraction of secondary and tertiary γ′ (low fraction of primary γ′) should be propitious to creep strength improvement.

The common ageing temperature selected for the study of the effect of the solutioning temperature was 750°C because the best microhardness levels were obtained around this temperature (see Figure 6). The ageing temperatures of 650°C and 850°C were chosen to estimate the influence of a very fine tertiary γ′ population (650°C-underageing) and of a sparse and coarse tertiary γ′ population (850°C-overageing) on the mechanical properties. The double-stage ageing treatment was also applied after the 1175°C solution treatment for the mechanical characterisation of the SHT microstructure.

Mechanical Properties

Tensile Properties. In Tables IV and V are compared the tensile properties of the NR3 alloy at 400°C and 700°C with the selected microstructures.

At 400°C, there was a significant effect of the solution treatment on the yield stress (Y.S.) and the ultimate tensile stress (U.T.S.). Finer were the grain size and the secondary γ′ precipitate size and higher were the tensile properties. On the other hand, for a given solutioning temperature, the maximum of tensile properties was reached with the 750°C ageing treatment. These two effects were qualitatively predicted by the hardness tests at room temperature.

Table IV Influence of the Heat Treatment on Tensile Properties at 400°C of NR3 Alloy (average of 2 tests)

Heat treatment	0.2% Y.S. (MPa)	U.T.S. (MPa)	Elongation (%)
1145°C/4h+750°C/4h	1043	1535	24.0
1175°C/4h+650°C/4h	1000	1484	23.8
1175°C/4h+750°C/4h	1036	1515	23.3
1175°C/4h+850°C/4h	970	1454	22.6
1195°C/4h+750°C/4h	987	1472	23.5

Table V Influence of the Heat Treatment on Tensile Properties at 700°C of NR3 Alloy (average of 2 tests)

Heat treatment	0.2% Y.S. (MPa)	U.T.S. (MPa)	Elongation (%)
1145°C/4h+750°C/4h	1043	1255	23.2
1175°C/4h+650°C/4h	1002	1250	18.1
1175°C/4h+750°C/4h	1034	1252	19.9
1175°C/4h+850°C/4h	981	1198	22.6
1195°C/4h+750°C/4h	1004	1273	20.2
SHT	1022	1237	21.2

These two influences were also observed for the yield stress at 700°C but only the effect of the ageing temperature was verified for the UTS. Coarser grain size and secondary γ′ precipitate size seem to be auspicious for this property at 700°C. At last, the samples treated with the SHT present tensile properties intermediate between those of the 1175°C/4h+750°C/4h and the 1175°C/4h+850°C/4h treatments.

Globally, the most efficient heat treatment to improve the tensile properties was a low solutioning temperature associated with the ageing temperature corresponding to the peak ageing condition. All these results also showed that the presence of tertiary γ′ precipitates in the γ channel is an essential condition for high tensile properties.

Creep Rupture Properties. Creep rupture tests were conducted at 700°C with an initial load of 700 MPa in order to assess the effects of the different microstructures. All the creep curves are plotted in Figures 7 and 8 and some creep data are presented in Table VI.

Table VI Influence of the Heat Treatment on the Typical Creep Data of NR3 Alloy (700°C / 700 MPa) (average of 2 tests)

Heat treatment	Time to 0.2% creep (hours)	Rupture time (hours)	Minimum creep rate 10^{-9} s^{-1}
1145°C/4h+750°C/4h	203	460	1.4
1175°C/4h+650°C/4h	246	491	1.6
1175°C/4h+750°C/4h	329	854	0.8
1175°C/4h+850°C/4h	26	171	21.7
1195°C/4h+750°C/4h	156	510	2.7
SHT	79	317	7.8

The influence of the solutioning temperature on the creep properties is shown in Figure 7. It appeared that the best behaviour was obtained with the 1175°C solution treatment. The two other solutioning temperatures led to the same range of rupture time but with different creep behaviours. The minimum creep rate associated with the 1195°C solutioning temperature is 2 times those obtained with the 1145°C solutioning temperature but the increase of the creep rate during the tertiary stage is slower. The higher value of the minimum creep rate should be associated with the coarser secondary γ ' precipitate size (and accordingly with the larger γ channels) while the slower increase of the creep rate in the tertiary creep should be attributed to the larger grain size obtained with the high solutioning temperature.

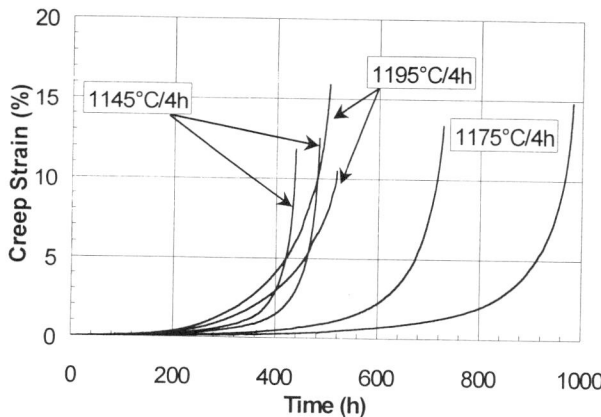

Figure 7: Influence of the solutioning temperature on the creep behaviour of NR3 alloy (700°C/700 MPa) (with an ageing treatment at 750°C/4h).

The effect of the ageing treatment on the creep behaviour is illustrated in Figure 8. As seen for the tensile properties, the best creep behaviour was obtained with the ageing treatment at 750°C. In the same way, the overageing treatment (850°C/4h) is drastically deleterious for the creep behaviour (rupture time 5 times lower and minimum creep rate about 25 times higher in comparison with the best results). On the other hand, the underageing treatment (650°C/4h) led to a creep behaviour significantly worse than for the 750°C ageing. However, the 650°C ageing treatment induce better creep properties than those obtained with the double-stage ageing treatment (SHT).

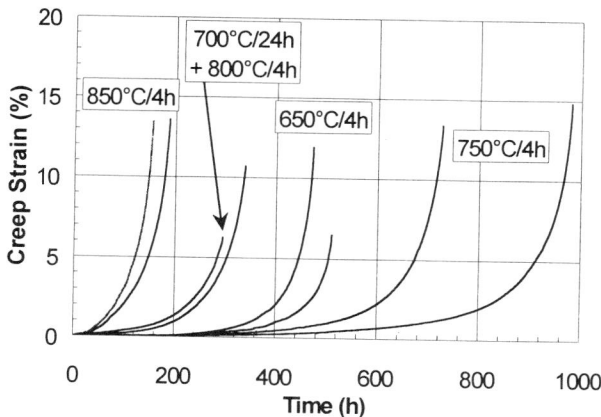

Figure 8: Influence of the ageing temperature on the creep behaviour of NR3 alloy (700°C/700 MPa) (with a solution treatment at 1175°C/4h).

Discussion

Heat Treatment – Microstructure Relationship

The increase of the subsolvus solutioning temperature led to the following microstructural modifications :
- decrease of the fraction of the primary γ ' precipitates which implies the increase of the secondary and tertiary γ ' fraction and the increase of the average grain size
- increase of the secondary γ ' precipitate size associated to a lower precipitate density.

Some dilatometric experiments were conducted in order to understand this second phenomenon. The starting precipitation temperature (Tsp) was measured during the cooling following the thermal cycle "room temperature-isothermal subsolvus solution treatment (4 hours)-room temperature". The heating and cooling were conducted at a rate of 10 K.min^{-1}. The dilatometric experiments were performed on two experimental superalloys of the same series (NR3 and NR6 alloys (5)). The results presented in Figure 9 showed that the Tsp decreased with the solutioning temperature. A gap of about - 30 K was observed between the Tsp and the solutioning temperature in the temperature range studied.

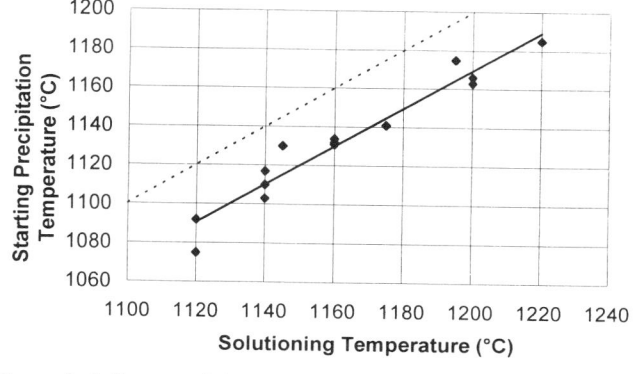

Figure 9: Influence of the solutioning temperature on the starting precipitation temperature of the NR3 and NR6 superalloys.

The microstructural changes associated to the cooling rate increase (which led to the Tsp decrease) in several superalloys were an increase of the secondary γ ′ precipitate density and a decrease of their mean size (10, 11). In the present case, the Tsp decrease was not due to a higher cooling rate but to a lower solutioning temperature.

The increase of the ageing temperature led to the increase of the mean size of the tertiary γ ′ precipitates and to the decrease of their particle density. The solvus temperature of the tertiary γ ′ is close to 850°C for an ageing duration of 4 hours. This value is in good agreement with those determined by Wlodek for the N18 alloy and the René 88 DT alloy (4, 12). On the other hand, a sparse intergranular precipitation of carbides was observed after the double-stage ageing and the 850°C treatments.

Microstructure – Mechanical Property Relationship

Globally, the highest tensile properties of the NR3 alloy at 400 and 700°C were obtained with a microstructure presenting the following features :
- a fine grain size
- a fine secondary γ ′ size
- and a slightly coarsened tertiary γ ′ precipitates.
The highest UTS at 700°C was reached with coarse grain size (12 µm instead of 4 µm) and secondary γ ′ precipitate size but the previously described microstructure led however to an UTS value not so lower. The lower tensile properties were always associated to the almost complete dissolution of the tertiary γ ′ precipitates. So, it appears that a secondary γ ′ precipitate size as fine as possible is required for high tensile properties. On the other hand, as for the hardness, improved tensile properties could be reached by optimising the tertiary γ ′ size by appropriate ageing treatment.

Our experimental results are in agreement with those published by Jackson and Reed [9] on the Udimet 720Li alloy. They demonstrated indeed that by using a single-stage ageing treatment, it was possible to obtain an optimum γ ′ precipitate size leading to comparable to or even better tensile and hardness properties than those produced through a standard two-stage ageing procedure. Applying standard precipitate hardening theories (13, 14) Jackson and Reed showed that this optimal γ ′ size corresponds to the transition between γ ′ cutting mechanisms involving pairs of weakly coupled a/2<110> dislocations and pairs of strongly coupled dislocations (9). A similar analysis could be therefore applied to the results obtained on NR3, but further TEM observations would have to be performed in the future to completely validate it.

As the creep behaviour of NR3 is concerned, the effects of the size and distribution of the tertiary γ ′ precipitates were even more spectacular than for the tensile properties. In particular, the almost complete dissolution of the tertiary γ ′ particles resulting from the ageing treatment for 4 hours at 850°C led to poor creep characteristics. TEM observations were carried out on thin foils prepared from specimens aged at 650, 750 and 850°C, then creep strained at 700°C. The three samples exhibited similar dislocation substructures : stacking faults extended through the γ and γ ′ phases and isolated a/2<110> dislocations homogeneously distributed between the secondary γ ′ precipitates (see Figure 10a).

The extended stacking faults are produced by {111}<112>-type glide involving shearing of the γ ′ precipitates by a/3<112> super-lattice partials as observed for instance during creep at 650°C of the PM René 95 alloy (15). The sinuous shape of the matrix dislocations is typical of a motion combining slip and climb deformation mechanisms around the secondary γ ′ precipitates. Such a deformation process was also identified during creep of René 95 (15) (Figure 10b).

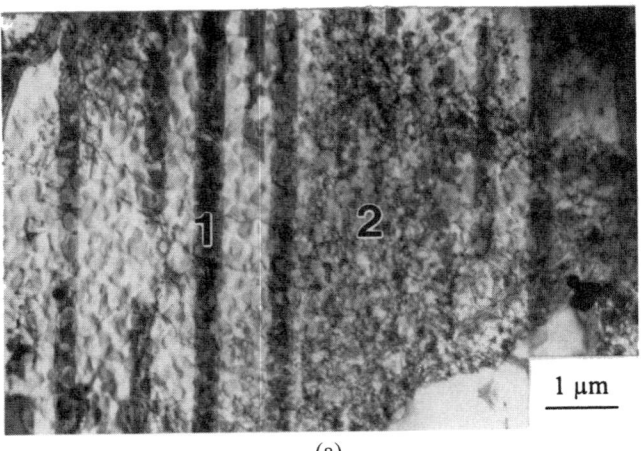

1 µm

(a)

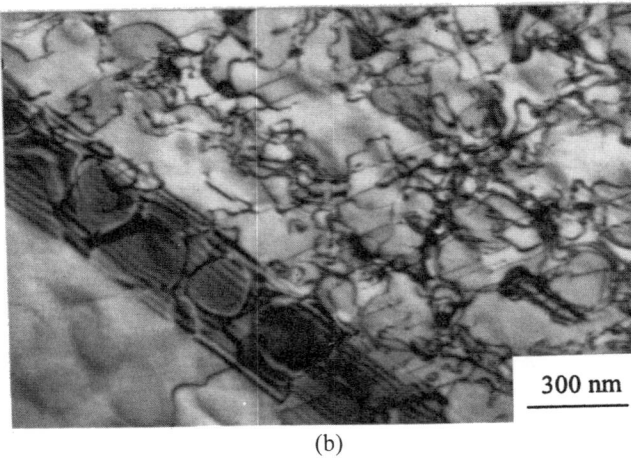

300 nm

(b)

Figure 10: Dislocation substructures of a specimen aged at 650°C/4h and then creep tested at T=700°C and σ = 700 MPa (0.32 % strain) (TEM bright field) : (a) zone 1 : extended stacking faults and zone 2 : homogeneous distribution of matrix dislocations and (b) sinuous shape of a/2<110> dislocations between the secondary γ ′ precipitates.

Due to their large extents, the stacking faults appeared as prominent features of the deformation structure in the crept NR3 samples. The corresponding <112>-type dislocation density is however rather low compared to the a/2<110> matrix dislocation density and the dominant creep mechanism in the tested specimens is thought to be the motion of a/2<110> dislocations between the secondary γ ′ precipitates. As the motion of these dislocations involves the by-passing of the tertiary γ ′ particles, it is obvious that their size and density will have significant effects on the creep behaviour, that was effectively experimentally observed. As no paired dislocations were observed in the creep strained samples, it is possible to rule out the cutting mechanism of the γ ′ precipitates by strongly coupled a/2<110> dislocations as observed during tensile deformation of Udimet 720 Li (9). On the other hand, γ ′ shearing by loosely coupled a/2<110> pairs is likely for the fine tertiary precipitates, but not easy to evidence within the rather narrow channels separating the secondary γ ′ precipitates. As the stress needed for this cutting mechanism increases with the size of the precipitates (16), that may explain the decrease of the creep rate observed when the temperature of the single-stage ageing treatment increases from 650°C to 750°C. The creep strength decrease observed when applying the ageing treatment at 850°C could be interpreted by a transition to an Orowan-type mechanism combining slip and climb processes with expansion and looping of a/2<110> dislocations around the coarser and less numerous remaining tertiary γ ′ precipitates. More accurate dislocation structure analyses will have however to be carried out to identify precisely the tertiary γ ′ particle/dislocation interactions and to confirm these hypotheses.

Conclusion

A significant improvement of the creep strength of the new PM superalloy NR3 was obtained while keeping high tensile properties. This was reached through the application of a simplified heat treatment sequence. A single-stage ageing treatment was indeed applied after the subsolvus solution treatment instead of the usual double-stage ageing treatment. The solution treatment as well as the ageing treatment were conducted in order to optimise the size and the distribution of the secondary and tertiary γ ′ particles. These results have shown the important influence of, firstly, the presence of tertiary γ ′ precipitates in the γ channels and secondly, of their size and particle density which were modulated by the ageing treatment

Acknowledgements

This work was supported by the French Ministry of Defence. The authors would like to acknowledge Mrs S. Baudu, F. Passilly and C. Ramusat for their assistance in this study.

References

1. M. Marty et al., French Patent 2 737 733, "Superalliages à Base de Nickel Stables à Hautes Températures," 1997.

2. J-Y. Guedou, J-C. Lautridou, and Y. Honnorat, "N18, PM Superalloy for Disks : Development and Applications," Superalloys 1992, ed. S.D. Antolovich et al. (Warrendale, PA : TMS-AIME, 1992), 267-276.

3. A. Walder et al, "N18, a New High Strength, Damage Tolerant PM Superalloy for Turbine Discs Application," (Paper presented at the 16th congress of the International Council of the Aeronautical Sciences (ICAS), Jerusalem, Israel, 28 August - 2 September 1988).

4. S.T. Wlodek, M. Kelly, and D. Alden "The Structure of N18," Superalloys 1992, ed. S.D. Antolovich et al. (Warrendale, PA : TMS-AIME, 1992), 467-476.

5. D. Locq, A. Walder, M. Marty, and P. Caron "Development of New PM Superalloys for High Temperature Applications," (Paper presented at the EUROMAT 99 Congress, Munich, Germany, 27 - 30 September 1999).

6. M. Morinaga et al., "New Phacomp and its Applications to Alloy Design," Superalloys 1984, ed. M. Gell et al. (Warrendale, PA : TMS-AIME, 1984), 523-532.

7. R. Watanabe, and T. Kuno, "Alloy Design of Nickel-base Precipitation Hardened Superalloys," Transactions ISIJ, 16 (1976), 437-446.

8. F.C. Hull, "Estimating Alloy Densities," Metal Progress, n°11 (1969), 139-140.

9. M.P. Jackson, and R.C. Reed, "Heat Treatment of Udimet 720Li: The Effect of Microstructure on Properties," Mat. Sci. and Eng., A259 (1999), 85-97.

10 T. Grosdidier, A. Hazotte, and A. Simon, "Precipitation and Dissolution Processes in γ / γ ′ Single Crystal Nickel-base Superalloys," Mat. Sci. and Eng., A256 (1998), 183-196.

11 H. Loyer Danflou, "Serrated Grain Boundaries and Resulting Properties at 750°C of a P/M Nickel Base Superalloy" (Ph.D. thesis, Paris XI-Orsay, 1993)

12 S.T. Wlodek, M. Kelly, and D. Alden "The Structure of RENE' 88 DT," Superalloys 1996, ed. R.D. Kissinger et al. (Warrendale, PA : TMS, 1996), 129-136.

13 L.M. Brown, and R.K. Ham, Strengthening Methods in Crystals , ed. A. Kelly and R.B. Nicholson (London, : Applied Science Ltd., 1971).

14 W. Hüther, and B. Reppich, Z. Metalk., 69 (1978), 628.

15 P.R. Bhowal, E.F. Wright, and E.L. Raymond, "Effects of Cooling Rate and γ ′ Morphology on Creep and Stress-Rupture Properties of a Powder Metallurgy Superalloy," Met. Trans. A, 21A (1990), 1709-1717.

16 B. Reppich, "Some New Aspects Concerning Particle Hardening Mechanisms in γ ′ Precipitating Ni-Base Alloys - I. Theoretical Concept," Acta Met., 30 (1982), 87-94.

γ′ FORMATION IN A NICKEL-BASE DISK SUPERALLOY

Timothy P. Gabb*, Daniel G. Backman**, Daniel Y. Wei**, David P. Mourer**, David Furrer***, Anita Garg*, and David L. Ellis****

*NASA Glenn Research Center
**General Electric Aircraft Engines
***Ladish Co., Inc.
****Case Western Reserve University

Abstract

A streamlined, physics-based kinetic model was formulated, calibrated, and validated on a nickel-base superalloy. It was designed to be simplified and streamlined, to speed up calculations and facilitate linkage to finite element modeling thermal codes. The model describes and predicts the formation of γ′ precipitates during the heat treatment process. Standardized experimental methods were employed to calibrate and initially validate the microstructure model. The model was then validated on an oil-quenched generic disk. Predictions of primary cooling γ′ size and area fraction agreed well with experimental measurements over a large range of values and cooling rates.

Introduction

Many mechanical properties of nickel-base superalloys are strongly influenced by the morphology of the strengthening γ′ precipitates (1,2). Both the sizes and area fractions of different populations of precipitates are influential and therefore need to be quantified and modeled. These microstructural characteristics are strongly affected by the heat treatments a component undergoes (3-5). Previous work has often modeled precipitate size through heat treatments using cubic coarsening equations (5). For physically accurate treatments, it is necessary to describe the kinetics of nucleation and subsequent growth and coarsening of γ′ precipitates as these processes occur within a given heat treatment. This requires tracking multiple populations of precipitates through heat treatments containing multiple steps.

The NASA-industry Integrated Design and Process Analysis Technologies (IDPAT) program had the objective of developing computational tools that could be applied together to improve the product and process development cycle for aircraft engine components. It was recognized that this required a "fast-acting" γ′ precipitation model for superalloy components, which could rapidly perform the necessary computations while integrated with material processing and mechanical property models. This would enable prediction of γ′ precipitate microstructure and resulting mechanical properties from the processing. This paper describes the work performed in development of this fast-acting γ′ precipitation model.

The objective of this work was to model γ′ precipitation for the case of a powder metallurgy disk superalloy. A physics-based kinetic model was formulated, calibrated, and validated on an advanced disk superalloy. The model describes and predicts the nucleation, growth, and coarsening of γ′ precipitates during the heat treatment process. However, it was designed to be simple and streamlined, in order to speed up calculations and facilitate linkage to finite element thermal and structural codes. Standardized experimental methods were employed to calibrate and initially validate the microstructure model. The model was then further validated using an oil quenched disk. γ′ precipitate morphologies were extensively quantified as functions of thermal histories and compared to model predictions. Model calibration and validation efforts described here were focussed on the primary cooling γ′ precipitates, which form during the first (primary) burst of nucleation during the solution quench.

Model Formulation

During the quench of a nickel-base superalloy from solution temperatures, supersaturation of γ′ phase solutes builds up in the γ phase. When the supersaturation builds up to a sufficient degree, a burst of nucleation of new γ′ precipitates occurs. This greatly reduces the supersaturation of γ′ solutes. Both the new and prior precipitates continually grow by interface-diffusion controlled growth. The precipitates also undergo coarsening, where smaller precipitates dissolve to provide solute for larger precipitates to grow. A precipitation model of the γ′ phase in nickel-base superalloys had been previously developed by General Electric Aircraft Engines (6) using classical nucleation and diffusion controlled growth. That model encompasses the full consideration of thermodynamic excess free energy, supersaturation buildup, nucleation, growth and coarsening. In contrast, the present work was focused on a precipitation model that is fast-acting such that it can be integrated with finite element model thermal analyses. However, this precipitation model must still encompass the entire precipitate physics including supersaturation buildup, nucleation, growth and coarsening, to ensure the streamlined model is capable of predicting the γ′ structure under the full spectrum of heat treatment conditions.

The fast-acting γ′ precipitation model has two basic requirements: i) to generate high-fidelity predictions and provide explanations for subtle but important relationships between γ′ size distribution and the heat treat schedule, and ii) to provide a computationally efficient algorithm that can be easily mated to FEM thermal codes without burdening the overall computational effort.

Superalloys 2000
Edited by T.M. Pollock, R.D. Kissinger, R.R. Bowman,
K.A. Green, M. McLean, S. Olson, and J.J. Schirra
TMS (The Minerals, Metals & Materials Society), 2000

The fast-acting precipitation model includes the following elements to meet these requirements:

a. A pseudo-binary solvus formulation that describes the solvus curve for the γ' formers.

b. A kinetic model that predicts the minimum size of nuclei as a function of supersaturation. This model employs a simple kinetic criterion to determine the onset of nucleation in lieu of a sophisticated thermodynamic model to compute the free energy and strain energy for nucleation.

c. Classical nucleation expressions that predict nucleation rate as a function of supersaturation and time.

d. A precipitate growth relationship based on an asymptotic solution that accounts for diffusive transport driven by the interfacial concentration gradient. These gradients form because of the concentration difference between the supersaturated matrix and the precipitate.

e. Particle coarsening relationships that predict changes in the average particle size of a size distribution based on the functional dependency of particle solubility upon particle radius.

These elements were integrated into a computer code that numerically integrates the effect of these relationships as a function of time and tracks mean size as shown in Figure 1 in contrast to a conventional approach. For each time step, the calculation starts by invoking mass conservation to calculate the supersaturation. If sufficient supersaturation is present, nucleation of new γ' precipitates occurs, and these precipitates are added to the size distribution. Precipitates are allowed to grow, driven by interface diffusion based on the level of supersaturation. Coarsening is then allowed to modify the existing particle size distribution through mean size and γ' volume conservation. This mechanism allows some precipitates to dissolve, removing them from the distribution.

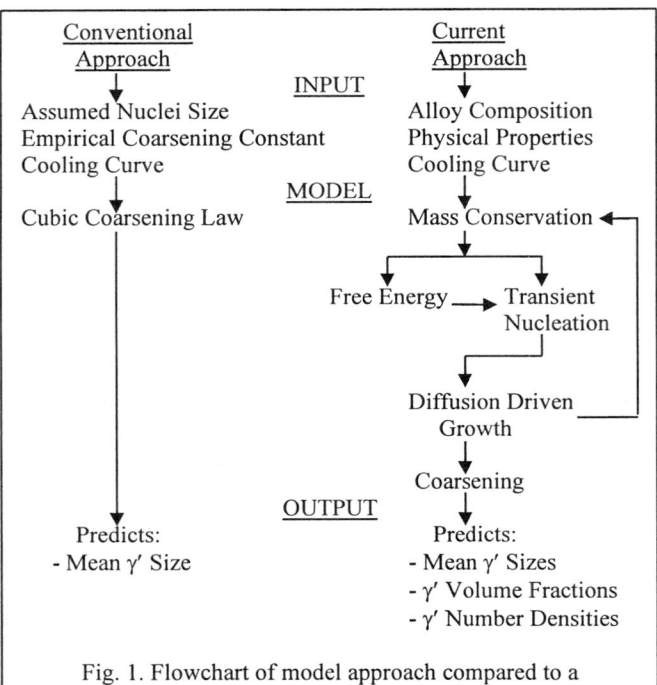

Fig. 1. Flowchart of model approach compared to a conventional approach.

Each of the model's elements is elucidated in the following:

a. Equilibrium γ' solvus

The solvus line in a binary system of limited terminal solubility can be expressed by the following equation (7)

$$C_\gamma = C_0 \exp\left(-\frac{Q}{R}\left(\frac{1}{T} - \frac{1}{T_{solvus}}\right)\right) \qquad (1)$$

where: C_γ = γ solubility
C_0 = initial composition of γ' former (e.g., Al + Ti + Ta)
Q = excess free energy per mole of γ' former in γ solution
R = gas constant
T_{solvus} = solvus temperature

The lever rule can be stated as:

$$C_\gamma = \frac{V_{\gamma'} C_{\gamma'} - C_0}{V_{\gamma'} - 1} \qquad (2)$$

where $V_{\gamma'}$ is the γ' volume fraction. Equation (1) can be restated as:

$$V_{\gamma'} = \frac{C_0\left(1 - \exp\left(-\frac{Q}{R}\left(\frac{1}{T} - \frac{1}{T_{solvus}}\right)\right)\right)}{C_{\gamma'} - C_0 \exp\left(-\frac{Q}{R}\left(\frac{1}{T} - \frac{1}{T_{solvus}}\right)\right)} \qquad (3)$$

The γ' composition was considered independent of temperature. Therefore, a series of measurements of equilibrium γ' volume fraction at temperatures up to the γ' solvus was used to correlate the constant Q and approximate the solvus line of γ' formation $C_{\gamma'}$ as a function of temperature.

b.-c. Nucleation

The classical solid state homogenous nucleation theory (8-12) defines the free energy ΔG* required for a spherical nucleus as:

$$\Delta G^* = \frac{16\pi\sigma^3}{3(\Delta G_v + \Delta G_s)^2} \qquad (4)$$

where: ΔG_v = volume free energy
ΔG_s = induced strain energy in the precipitates
σ = precipitate/matrix interfacial energy

For the case where ΔG_V, ΔG_S, and σ are independent of radius, the critical radius r* of nuclei can be expressed as:

$$r^* = \frac{2\sigma}{(\Delta G_v + \Delta G_s)} \qquad (5)$$

In the fast acting model, this time-intensive thermodynamic calculation was avoided. The critical size r* of surviving γ'

406

nuclei was defined to satisfy the following stability criterion employing an adjustable nucleation scaling factor S:

$$r^* = \frac{2\sigma V_m}{RT} \frac{S}{\ln(C_s / C_e)} \qquad (6)$$

where C_S is the saturated solute concentration in the γ matrix, C_e is the equilibrium solute concentration in the γ matrix and $V_m = a^3/4$ is the average atomic volume in the γ' precipitate using a, the lattice parameter. The interfacial energy σ and nucleation scaling factor S in equation 6 are determined through iterative estimation of these constants, followed by comparison to measurements from calibration heat treatments. For each time step in the precipitation model algorithm, r* is first calculated using equation 6. r* is then used in equation 5 to obtain $(\Delta G_v + \Delta G_s)$. ΔG^* is then obtained from equation 4.

Formally, the isothermal nucleation rate J^* is given by nucleation theory as (11):

$$J^* = NZ\beta^* \exp(-\Delta G^* / kT) \exp(-\tau / t) \qquad (7)$$

in which N is the density of active nucleation sites per unit volume, Z is the Zeldovich non-equilibrium factor which corrects the equilibrium concentration of critical nuclei for the loss of nuclei to growth and thus to the concentration present at steady state, β^* is the rate of atomic attachment to the critical nuclei, ΔG^* is the activation energy for the formation of a critical nucleus as defined above, k and T have their usual meaning, and τ is the incubation time to form embryos before nucleation. However, to keep the entire calculation manageable in the fast-acting model, the latter incubation effect was instead related to r*. For an f.c.c. or $L1_2$ lattice, the frequency factor β^* was defined as:

$$\beta^* = (C_S D / \lambda^4) \, 4 \, \pi r^{*2} \qquad (8)$$

where $\lambda = a/\sqrt{2}$ is the atomic spacing, D the diffusion coefficient of the rate limiting species in the matrix, and r* was determined before. The Zeldovich factor was calculated using:

$$Z = [\Delta G^* / (3\pi k T)]^{1/2} \qquad (9)$$

where ΔG^* was determined as previously described from equation 4. A volume-related time dependent nucleation rate was derived, given that as nucleation proceeds, a portion of the volume will have been transformed into γ' precipitates. Within this volume and its neighborhood further nucleation is not possible. Therefore, it was assumed that the stationary nucleation rate can proceed only in the remaining (non-depleted) volume, using the expression:

$$V(t) = V_0 - \sum_{i=1}^{k} 4/3 \pi n_i (r_i + \Delta r_i)^3 \qquad (10)$$

where V_0 is the total volume considered, k is the number of nucleation bursts, and V(t) the remaining non-depleted volume.

The number of nuclei ΔN_v at time interval Δt and available volume V(t) is then:

$$\Delta N_v = J^* \, \Delta t \, V(t) \qquad (11)$$

d. Growth

Once a precipitate nucleates, its rate of growth in a supersaturated matrix is determined by [13-17]:

$$\frac{dr}{dt} = \frac{D \frac{\partial C_\gamma}{\partial r} |_{\gamma'/\gamma}}{(C_{\gamma'} - C_{\gamma|\gamma'/\gamma})} \qquad (12)$$

where the varying concentration field in the matrix is determined by solving the diffusion equation with above moving boundary:

$$D \nabla^2 C = \partial C / \partial t \qquad (13)$$

Now total solute mass conservation requires:

$$C_s(t) = \frac{C_0 V - C_{\gamma'} V_{\gamma'} V_r}{(V - V_{\gamma'} V_r)} \qquad (14)$$

at any instant, where $C_s(t)$ is the solute composition in the matrix, V is the total volume considered in the simulation, $C_{\gamma'}$ is the solute composition in the precipitates, $V_{\gamma'}$ is the total volume of precipitates at any instant, V_r is molar volume ratio between matrix and precipitate, and C_0 is the total solute composition in the alloy.

The solute concentration C_γ at the γ/γ' interface for a precipitate with radius r can be determined using the equation:

$$C_{\gamma|\gamma/\gamma'} = C_e \exp(\frac{2\sigma V_m}{RT r}) \qquad (15)$$

where R is the gas constant and V_m is precipitate molar volume per atomic site.

The interface kinetics are likely to be rate limiting during the very early stage of growth, since the diffusion distance tends to be negligible. However, diffusion is likely to become dominant, as the precipitates grow larger. For the growth of spherical precipitates, an invariant size approximation solution can be stated as:

$$\lambda = \frac{-\Omega}{2\pi^{1/2}} + (\frac{\Omega^2}{4\pi} - \Omega)^{1/2} \qquad (16)$$

where λ is a non-dimensional parameter related to growth rate and Ω is a saturation index factor. The saturation index factor Ω

describes the level of supersaturation of γ' formers in the matrix, driving both nucleation and growth:

$$\Omega = \frac{2(C_\gamma - C_0)}{(C_{\gamma'} - C_0)} \qquad (17)$$

The resulting growth rate relationship is then defined as:

$$\Delta r = 2\lambda (D\Delta t)^{1/2} \qquad (18)$$

where D is estimated as described before, Ω and λ are determined through equations 16 and 17.

e. Coarsening

Particle coarsening is driven by the higher solubility of small particles, which tend to dissolve as solute diffuses to larger particles. The driving potential is provided by the reduction of total interfacial energy while conserving the total volume fraction of precipitates. From coarsening theory (18-23), the general form of precipitate coarsening can be written as:

$$\frac{dr}{dt} = \frac{2DC_e\sigma V_m}{RT r(C_{\gamma'} - C_e)}\left(\frac{1}{\bar{r}} - \frac{1}{r}\right)(1+br) \qquad (19)$$

where C_e is the equilibrium solute content, and b is correction factor for the case where volume fraction of precipitates is high (19,20). The correction factor b tends to broaden the precipitate size distribution, which is not tracked by the current mean size approximation. So this correction factor was neglected during the coarsening calculation. All other terms in this equation had previously been estimated.

Summarizing, these elements and associated equations were integrated into a computer code that numerically integrates the effect of these relationships as a function of time and tracks mean size as shown in Figure 1. For each time step, the calculation starts by invoking mass conservation (eq. 14) to calculate the supersaturation. Nucleated particles (eq. 6) are then added to the size distribution if sufficient driving force (eq. 4) is present. Precipitates are allowed to grow based on the level of supersaturation (eq. 17) using the growth relationship (eq. 18). The coarsening law (eq. 19) is finally applied to modify the existing particle size distribution through mean size and γ' volume conservation. This did allow small precipitates to dissolve, removing them from the distribution.

Experimental Procedure

The powder metallurgy nickel-base disk superalloy CH98 (24) was selected for evaluation. This alloy has a nominal composition in weight percent of 4Al-0.03B-0.03C-18Co-12Cr-4Mo-3.8Ta-4Ti-0.03Zr-bal. Ni. Alloy powder was atomized, then hot compacted, extruded, and isothermally forged. A variety of heat treatment experiments were first performed on the alloy to help formulate, calibrate, and initially validate the γ' model. Small cube specimens were heated to a variety of temperatures and soaked for a sufficient period of time to determine the equilibrium area fraction of γ' as a function of temperature, in order to calibrate the solvus formulation. Other small specimens were supersolvus solution heat treated, cooled to temperatures below the solvus for short dwells of up to 20 minutes, and then rapid quenched in water to determine the kinetics of γ' formation near the solvus, for the kinetic model. These rapid quenches were accomplished via a drop quench furnace, where a specimen was directly dropped from the hot zone of the furnace into a chamber of water for a transfer time of less than 1 second. Standardized Jominy bar end quench tests and sampling techniques were adapted to establish a uniform and reliable test for the kinetics of γ' formation in rapid quenching conditions. One end of thermocoupled bars 25.4 mm in diameter and 63 mm in length was quenched with water. Sections of the bars were extracted using precision numerical controlled electro-discharge machining and prepared from three locations selected to span the range of cooling rates expected in oil quenched disks. Other small specimens with attached thermocouples were cooled in a programmed furnace to assess slower cooling rates than that achieved in the Jominy bars.

The model was further validated on a generic disk via measuring gamma prime sizes at specific locations where cooling rates were measured by thermocouples. Disks were subsolvus solution heat treated at 1149°C/2h, and then quenched in oil. Thermocouples were embedded within one disk to gather accurate temperature-time data at selected disk locations during the oil quenching process. Another disk was then heat treated in the same process with no thermocouples attached. Microstructural specimens were excised from bore and rim locations of the second disk.

Specimens evaluated in the SEM were first metallographically polished and then electrolytically etched to leach away the γ matrix to expose γ' particles. Specimens evaluated in the TEM were electrochemically thinned. At least two TEM foils were examined from each selected specimen. Multiple weak beam dark field images were consistently taken from near the [001] zone axis, selecting grains which required tilting of less than 30°. Precipitate size measurements included major and minor axis widths, equivalent circle (feret) diameter, and several shape parameters. Size-frequency histograms were then generated for primary cooling and tertiary γ' precipitates. Over 100 precipitates were measured to estimate each distribution. Precipitates were then grouped into 3 classes for area fraction measurements by point counting: remnant γ' particles > 1.0 μm in diameter that were not solutioned during subsolvus solution heat treatments, cooling γ' precipitates between 0.05 and 1.0 μm, and tertiary γ' precipitates less than 0.05 μm in diameter.

Results and Discussion

Model Calibration

After formulation, the model had to be initially calibrated with heat treatment experiments. Equilibrium γ matrix solubility was determined as part of the kinetic model calibration. Specifically, the solubility curve for the γ/γ' pseudo binary system was correlated against values of γ' volume fraction measured at several temperatures up to the solvus temperature. This solubility relationship was used to model γ' solubility for temperatures that ranged from the solvus temperature down to

870 °C. The empirically derived solubility curves were then used to compute the solute supersaturation levels in the γ matrix. These experiments indicated a γ' solvus of approximately a 1200°C for CH98.

The objective of the rapid quench coupon tests was to determine the time to onset and resulting fraction of γ' precipitation within the temperature range between the alloy solvus temperature and 80 °C below the solvus. All samples water quenched in these tests had a "background" quantity of γ' precipitates less than 0.015 μm in size. These were present even in specimens directly water quenched in less than 1 sec. This indicated that nucleation could not be fully suppressed here in this alloy. Samples quenched before nucleation took place in the near-solvus dwell periods exhibited no coarse γ', whereas samples quenched following nucleation near the solvus had coarse primary γ'. Coarse γ' precipitates were observed after 1200 s dwells at 1160°C or lower temperatures, and after 180 s dwells at 1150 °C or lower. The morphology of γ' precipitates formed near these temperatures had a roughly cuboidal morphology, with some dendritic growth of the cube corners. Precipitates aged at lower temperature exhibited a more cuboidal morphology. These observations are consistent with γ' morphological transitions reported in the literature (13). The results of these experiments were then mapped on to a γ' time-temperature-transformation diagram shown in Figure 2 for CH98. These observations are essential to calibrate the nucleation portion of the kinetic model.

The fast acting model contains four additional adjustable kinetic parameters that also required calibration:
i)   the nucleation scaling factor S which is assumed to be temperature independent,
ii)   the value of the γ/γ' interfacial energy σ,
iii)   the diffusion activation energy Q of γ' formers,
iv)   the diffusivity D for the γ' formers.

Model calibration and validation efforts were focused on the primary (larger) cooling γ' sizes. The model's cooling rate response was calibrated using cooling curve and microstructural measurements from a programmed furnace cooling experiment run at a cooling rate of 1.1°C/s. A series of size estimates at intermediate times during the quench were also generated by both the fast-acting and full thermodynamic model (6) after initial calibrations. This was performed to model the thermal path dependence of the γ' microstructure. A non-linear regression analysis was then conducted to obtain the values of the four kinetic parameters in the fast-acting model that gave the best data fit and most reasonable predictions of intermediate sizes.

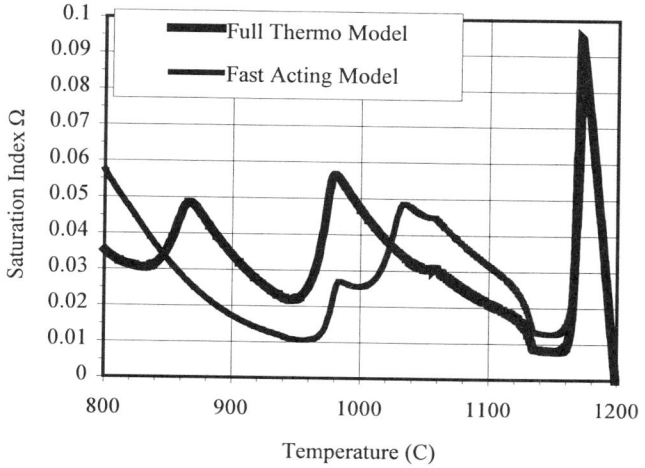

Fig. 3. Saturation levels predicted using the models.

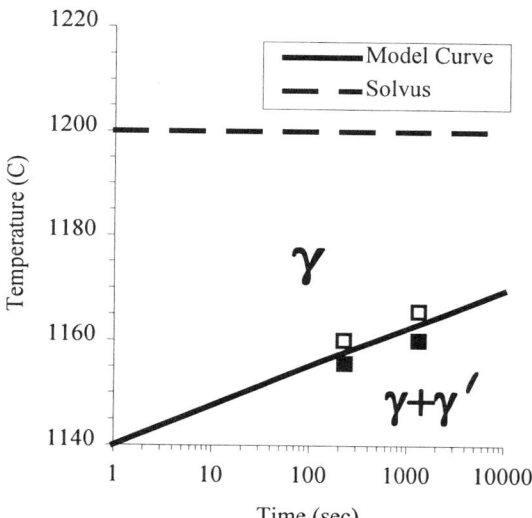

Fig. 2. Time-temperature-transformation data and model cu near the γ' solvus temperature, open symbols indicate no co; γ'was present, solid symbols indicate the presence of coarse

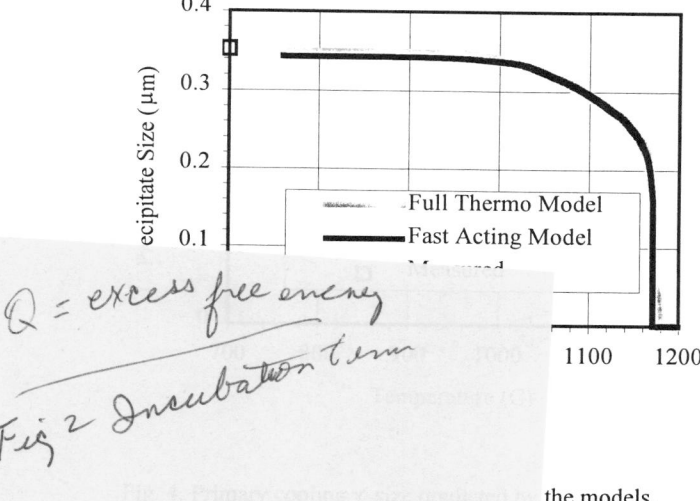

Q = excess free energy

Fig 2 Incubation term

the models.

Figures 3 and 4 show the calibration results, comparing fast acting model and full scale thermodynamic predictions for CH98. Figure 3 shows good agreement of predicted supersaturation levels, as indicated by the saturation index Ω, during quench for the full scale and fast acting models. As can be clearly observed, the fast-acting model is capable of tracking the balance between nucleation and growth in accordance with the predictions of the full scale model for the primary burst of cooling γ' nucleation. There is room for further improvements in predictions of subsequent bursts of precipitate nucleation. However, predictive capability for the subsequent precipitation bursts was not a prime goal for the fast acting model. Figure 4 compares the primary cooling γ' sizes generated by the two models. Excellent agreement was achieved between the experimental measurements, the fast acting model predictions, and full scale model calculations. The full scale model has previously been demonstrated to provide good agreement with experimental observations on other alloys.

Model Validation

The model was initially validated using controlled cools of other small specimens and standardized Jominy bar end-quench tests. Cooling curves from the various validation experiments are compared in Fig. 5. The Jominy end quench tests were used to initially validate the model for fast cooling conditions in CH98. A slow programmed furnace cooling heat treatment was used to validate the model for slow cooling paths. The γ' distributions produced in the Jominy bar at distances of 1.9cm and 5.7cm after water quenching from supersolvus solutioning are compared to the slow furnace cooling case in Fig. 6. Histograms illustrating the primary cooling γ' size distributions for these cases are also shown. The size distributions were considered normally distributed and a normal curve fit is shown. The mean size of the primary cooling γ' slightly increased with decreasing cooling rate, while the area fraction remained comparable. Excellent agreement was obtained between predicted and experimentally measured mean sizes of the primary cooling γ' particles for these cases.
Due to the promising results with the supersolvus solution heat treatment quench tests, subsolvus Jominy tests were also performed. The precipitate distributions developed in the Jominy bar after water quenching from subsolvus solutioning conditions are compared in Fig. 7. The subsolvus cases produced much smaller primary cooling γ' sizes, as predicted by the model. Model predictions of precipitate sizes again agreed well with observed results.

A CH98 disk of generic shape shown in Fig. 8 was then subsolvus solution treated and oil-quenched for further model validation. Thermocouple temperature versus time data collected during the disk oil quenching from the bore, web, and rim locations indicated the bore location clearly cooled slower than the web and rim, which had similar cooling responses. The temperature-time responses were employed for model predictions at several locations. Typical SEM images of bore and rim edge microstructures from an as-quenched disk are compared in Fig. 9. Predicted and measured sizes of primary cooling γ' at the bore hole, bore, and rim edge locations agreed well, as shown in the comparisons of Fig. 10.
Another disk was oil quenched and then aged 760°C/16h. Typical microstructures from the bore, mid rim, and rim edge of this disk are compared in Fig. 9. Predicted and measured sizes

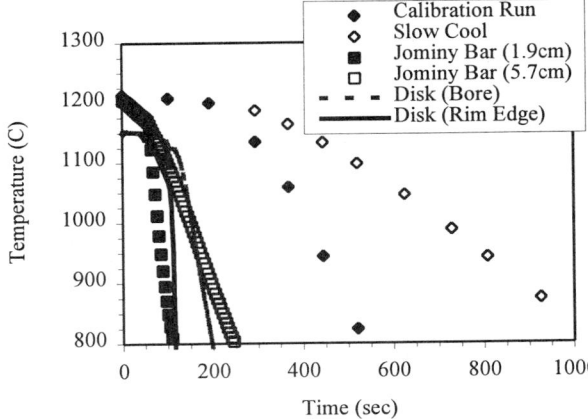

Fig 5: Comparison of representative cooling curves from heat treatments.

of primary cooling γ' at bore, web, and several rim locations are compared in Fig. 10 - 11. The model predicted the trend of the measurements, but predicted consistently lower sizes for these cases. This may be due to the lack of any calibration experiments for aging effects.

Comparisons of measured and predicted primary cooling γ' precipitate sizes and area fractions for various validation experiments are shown in Fig. 10 and 11. The model gave good predictions of primary cooling γ' size for the heat treat coupons, jominy bar, and generic disks, over a very broad range of precipitate sizes. The model's predictions of area fractions of primary cooling γ' were also reasonable, but showed room for further improvements.

Differences between predicted and measured γ' sizes and quantities can be attributed to model inaccuracies as well as measurement difficulties. As can be seen in the γ' images of Fig. 4-6, several groups γ' precipitates having different sizes and morphologies can be present in the same microstructure, as predicted by the model. Also, tips and corners of the somewhat rectangular precipitates can be sectioned in a metallographic section or TEM foil. This effect is most notable for slow cooled primary cooling γ', where the cube corners show enhanced growth (19). The size vs. frequency distribution of all these precipitates was combined together for measurements. However, the size distributions of each of these groups sometimes could not be separated or deconvoluted in the combined measurements. Objective experimental analyses were capable of grouping precipitates into 3 classes for size and area fraction measurements: remnant γ' particles that were not dissolved during subsolvus solution heat treatments which were greater than 1.0 μm in diameter, cooling γ' precipitates between 0.05 and 1.0 μm, and tertiary γ' precipitates less than 0.05 μm in diameter. Attempts to subdivide and measure relative quantities and area fractions of each of several groups of cooling γ'

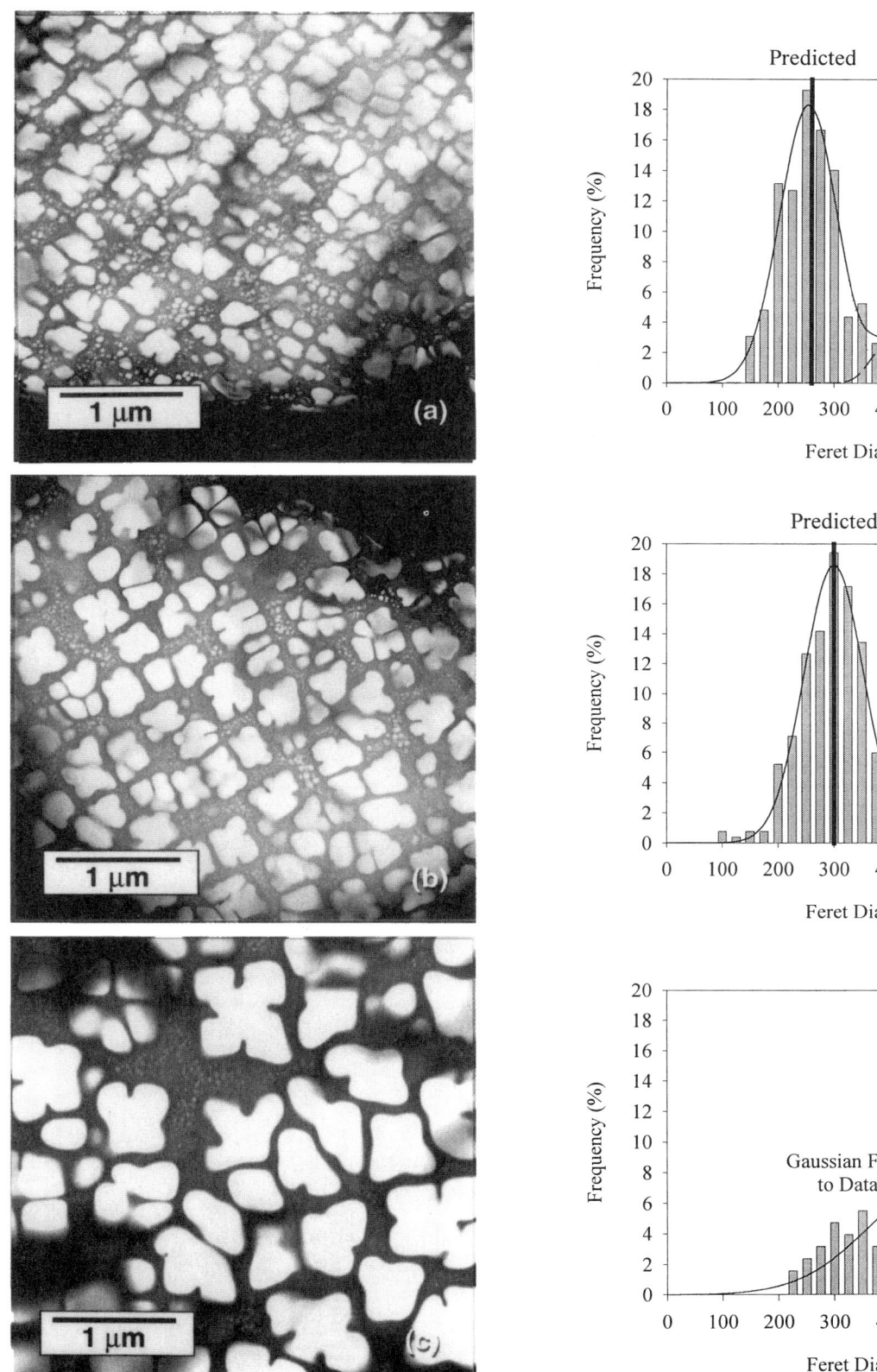

Fig. 6: Comparison of γ′ microstructures and size histograms after supersolvus heat treatments: (a) Jominy bar 1.9cm and (b) Jominy bar 5.7cm from the quenched end, (c) slow furnace cool.

411

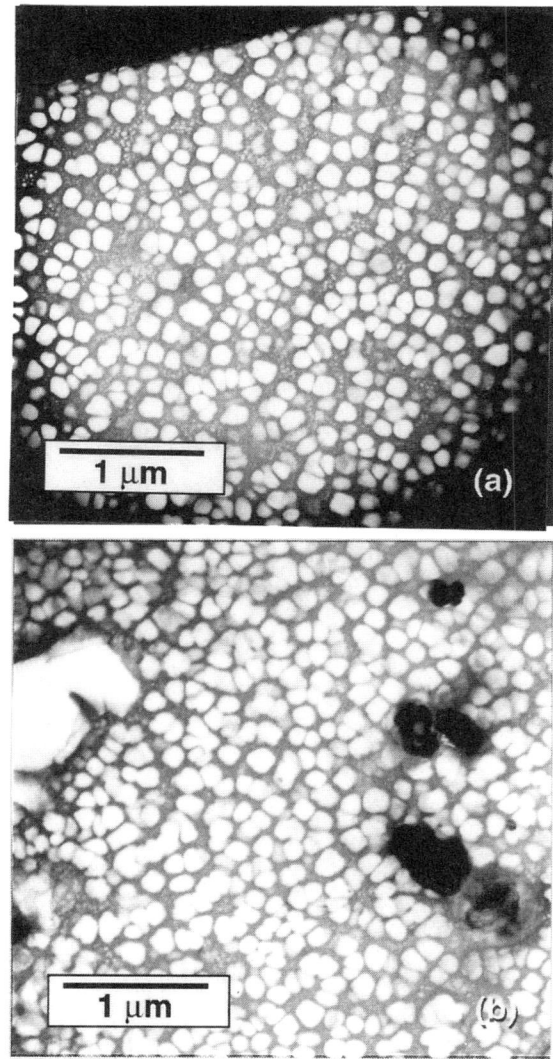

Fig. 7. Comparison of the γ′ microstructures in subsolvus solution heat treated Jominy bars: (a) 1.9cm and (b) 5.7cm from the water quenched end.

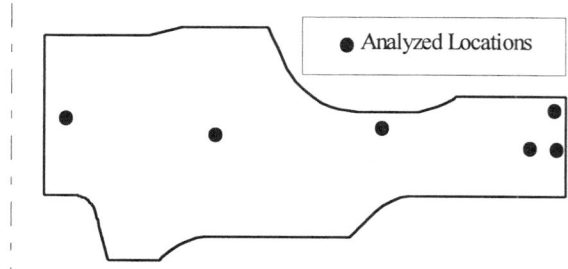

Fig. 8. Generic disk shape used for further model validation.

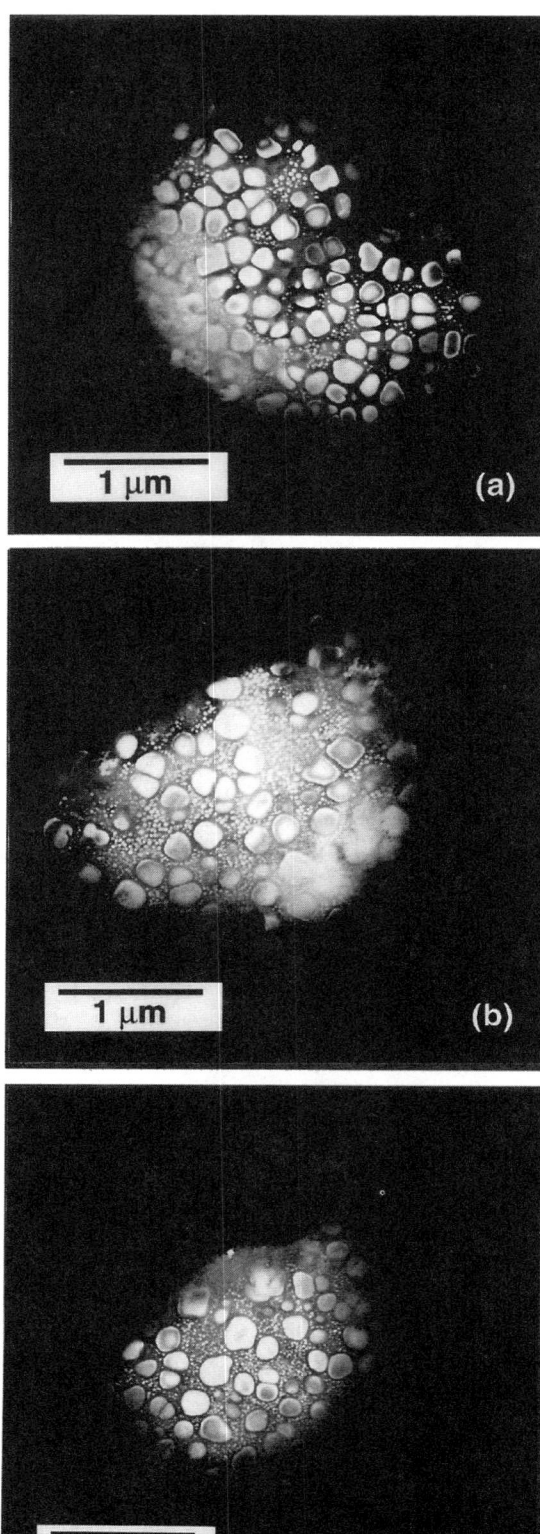

Fig. 9. γ′ microstructures produced in oil quench and aged generic disk: a) bore, b) mid rim, c) rim edge

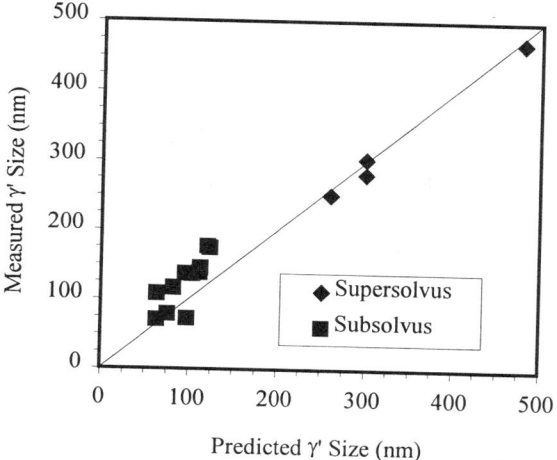

Fig. 10. Comparison of predicted and measured primary cooling γ' sizes.

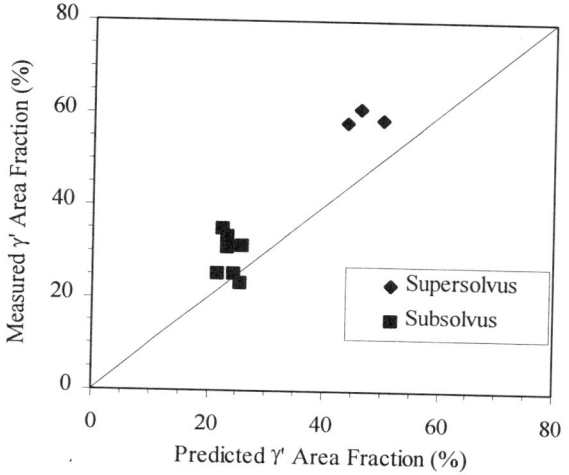

Fig. 11. Comparison of predicted and measured primary cooling γ' area fractions.

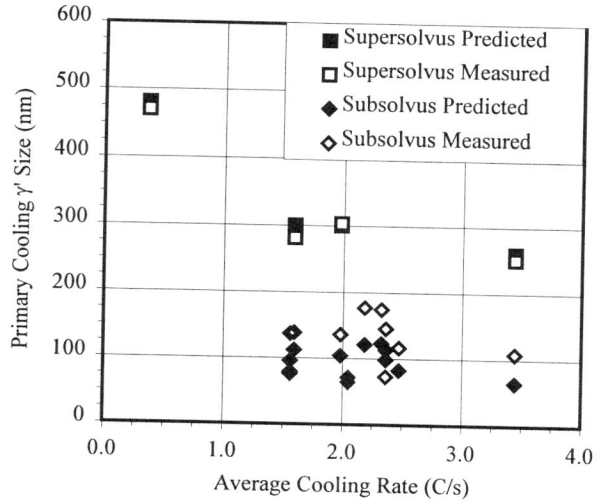

Fig. 12. Comparison of predicted and measured primary cooling γ' as a function of average cooling rate from solution temperature to 870°C.

precipitates between 0.05 and 1.0 μm in size would require more subjective selections among precipitates. Yet, the model was capable of predicting the precipitation of several groups of cooling γ', through later bursts of nucleation and subsequent growth during the quench. Two groups of cooling γ' were in fact predicted for the case of the disk rim edge. There were possible indications of this in the associated micrographs and histograms, but this could not be fully proven due to the above mentioned limitations. However, the model would benefit from further refinements in predictions of the formation of these later groups of secondary cooling γ' precipitates.

The ability of this model to track both size and quantity of multiple distributions of γ' particles can be very important in fully relating microstructures to mechanical properties in disk superalloys. Inspection of Fig. 9-12 indicates the mean size of the primary cooling γ' precipitates did not decrease linearly with cooling rate as often expected. However, their area fraction decreased in the web and rim, allowing a larger area fraction of tertiary γ' precipitates <0.02μm in size. Mechanical testing underway indicates these subtle changes produce a significant increase in tensile and creep properties of the rim and web, over the bore. Both primary cooling and tertiary γ' need to be quantified for mechanical property modeling in such cases

Further model refinements are contemplated to refine area fraction prediction capabilities and enhance secondary γ' prediction capabilities. Further calibration experiments are planned to improve the thermal path modeling and include more effects of multi-step stress relief and aging heat treatments used on many disk alloys. Integration trials have already successfully linked this model to typical processing codes for processing sensitivity studies. This allows prediction of microstructures and mechanical properties resulting from processing inputs. This enhanced prediction capability allows reducing the forging design cycle time and expands modeling capabilities to improve the product and process development cycle for aircraft engine components.

Summary and Conclusions

1. A streamlined physics-based model describing the evolution of γ' precipitates during the heat treatment process was formulated.
2. The resultant calibrated model gave reasonable predictions of primary cooling γ' size over a very broad range of conditions, agreeing well with measurements from experiments run on coupons, Jominy bars, and a generic disk.
3. Improvements in predictions of primary γ' area fraction and secondary sizes and area fractions are planned in future work.
4. The streamlined γ' model was successfully linked with a commercial heat treat code enabling prediction of γ' sizes and area fractions across a component in a computationally efficient manner.

Acknowledgements
This work was supported by the NASA IDPAT program, managed by Douglas Rohn and John Gayda. The authors also wish to acknowledge the work of Howard Merrick, Honeywell Engines and Technologies (Allied-Signal Engine Co.) and Timothy Howson, Wyman-Gordon Forgings Co.

References

1. R. F. Decker, "Strengthening Mechanisms in Nickel-Base Superalloys", Steel Strengthening Mechanisms Symposium May 5-6, 1969, Zurich, Switzerland, Climax Molybdenum Co., 1969.

2. N. S. Stoloff, "Fundamentals of Strengthening", in Superalloys II, ed. C. T. Sims, N. S. Stoloff, W. C. Hagel, John Wiley & Sons, New York, NY, (1987), 61-96.

3. R. A. Wallis, P. R. Bhowal, "Property Optimization in Superalloys Through the Use of Heat Treat Process Modeling", in Superalloys 1988, Proceedings of the International Symposium at Seven Springs Mountain Resort, PA, TMS-AIME, ed. D. N. Duhl, G. Maurer, S. Antolovich, C. Lund, S. Reichman, 1988, 525-534.

4. J. R. Groh, "Effect of Cooling Rate From Solution Heat Treatment on Waspaloy Microstructure and Properties", in Superalloys 1992, Proceedings of the International Symposium at Seven Springs Mountain Resort, PA, TMS-AIME, ed. R. D. Kissinger, D. J. Deye, D. L. Anton, A. D. Cetel, M. V. Nathal, T. M. Pollock, D. A. Woodford (1992), 621-626.

5. Schirra, J., in Superalloys 1992, Proceedings of the International Symposium at Seven Springs Mountain Resort, PA, TMS-AIME, ed. R. D. Kissinger, D. J. Deye, D. L. Anton, A. D. Cetel, M. V. Nathal, T. M. Pollock, D. A. Woodford, 1992.

6. D. G. Backman, D. Y. Wei, unpublished research, 1998.

7. P. Gordon, " Principles of Phase Diagrams in Materials Systems", in Materials Science and Engineering Series, Published by McGraw-Hill, Inc. (1968), 142-145.

8. P. Haasen and R. Wagner, "High-Resolution Microscopy and Early-Stage Precipitation Kinetics", Met. Trans. A, 23 (1992), 1901-1914.

9. S. Q. Xiao and P. Haasen, "HREM Investigation of Homogenous Decomposition in a Ni -12 at. % Al Alloy", Acta Met., 39(4) (1991), 651-659.

10. H. Wendt, and P. Haasen, "Nucleation and Growth of γ'-Precipitates in Ni-14 at. % Al", Acta Met., 31(10) (1983), 1649-1659.

11. H. I. Aaronson and F. K. LeGoues, " An Assessment of Studies on Homogeneous Diffusional Nucleation Kinetics in Binary Metallic Alloys", Met. Trans. A, 23 (1992), 1915-1945.

12. W. F. Lange, III. M. Enomoto, and H. I. Aaronson, "Precipitate Nucleation Kinetics at Grain Boundaries", Int. Mat. Rev., 34(3) (1989), 125-150.

13. R. A. Ricks, A. J. Porter and R. C. Ecob, "The Growth of γ' Precipitates in Nickel-Base Superalloy", Acta Met., 31 (1983), 43-53.

14. H. I. Aaronson, "Atomic Mechanisms of Diffusional Nucleation and Growth and Comparisons with Their Counterparts in Shear Transformations", Met. Trans. A, 24 (1993), 241-276.

15. H. I. Aaronson, C. Laird and K. R. Kinsman, in Phase Transformation, A. S. M. Metals Park, Ohio, and Chapman & Hall, London, p. 313.

16. R. D. Doherty, in Physical Metallurgy, edited by R. W. Cahn and P. Haasen, 3rd edition, North Holland, Amsterdam, 1983, 934.

17. M. F. Henry, Y. S. Yoo, D. Y. Yoon, and J. Choi, "The Dendrite Growth of γ' Precipitates and Grain Boundary Serration in a Model Nickel-Base Superalloy", Met. Trans. A, 24 (1993), 1733-1743.

18. G. W. Greenwood, "Particle Coarsening", Acta Met., 4 (1956), 243.

19. C. S. Jayanth and P. Nash, "Review Factors Affecting Particle-Coarsening Kinetics and Size Distribution", J. Mater. Sci. Engng., 24 (1989), 3041-3052.

20. R. A. MacKay and M. V. Nathal, " γ' Coarsening in High Volume Fraction Nickel Base Alloys", Acta Met., 38(6) (1990), 993-1005.

21. E. H. Van Der Molen, J. M. Oblak, and O. H. Kriege, "Control of γ' Particle Size and Volume Fraction in the High Temperature Superalloy Udimet 700", Met. Trans. A, 2 (1971), 1627-1633.

22. G. I. Rosen, S. F. Direnfeld, M. Bamberger and B. Prinz, "Computer Aided X- Ray Analysis for Determining Growth Kinetics of γ' Phase in Nickel-based Wrought Superalloys", High Temp. Mater. and Proc., 12(4) (1993), 183-191.

23. P. W. Voorees, "Ostward Ripening of Two-Phase Mixture", Annu. Rev. Mater. Sci., 1992, 197-215.

24. K.-M. Chang, U.S. Patent 5,662,749

25. M. L. Macia, T. H. Sanders, Jr., "A Study on the Dendritic Growth of Gamma Prime in Astroloy", Proc. of the 2nd Int. Conf. On Heat-Resistant Materisls, Gatlingberg, TN, Sept. 11-14, 1995, 163-170.

MICROSTRUCTURE AND MECHANICAL PROPERTY DEVELOPMENT IN SUPERALLOY U720LI

David Ulrich Furrer and Hans-Jorg Fecht*

Ladish Co., Inc., PO Box 8902, Cudahy, WI 53110, USA
*Universität Ulm, Albert-Einstein-Allee 47, D-89081 Ulm, Germany

Abstract

Superalloy Udimet 720LI™ has been studied and implemented into many production applications. [1,2] Further work has been conducted to assess the nature of the microstructural evolution and mechanical property development in this alloy. This effort is focussed on thermal processing cycles to tailor gamma-prime distribution and grain boundary morphologies, and subsequent mechanical properties. The results of this work have been compared to previous studies and assessed with regard to basic physical metallurgy understanding of the observed microstructure/property relationships. A new thermal processing route has been established which will allow for an optimum balance in mechanical properties for this material for potential future applications.

Experimental Procedure

A three-phased experimental program was undertaken to study U720LI material. A laboratory phase was conducted on heat treated microspecimens. A property screening phase, which incorporated mechanical property test specimens to evaluate heat treatment cooling rates, was performed. A full scale forging demonstration and validation phase was also conducted to more completely assess the microstructure and mechanical property interactions in U720LI.

Laboratory Heat Treat Studies

To study the microstructural evolution in superalloy U720LI, laboratory experiments where performed to characterize the microstructure and phases present after various thermal treatments. Initial studies were performed on billet samples from both powder metallurgy (P/M) and cast and wrought (C&W) (i.e. ingot metallurgy) materials. Laboratory thermal treatment studies focused on both subsolvus and supersolvus solution heat treatment cycles and aging treatments.

The specific chemistry for the U720LI materials investigated in this program is listed in Table I. The gamma-prime solvus for the P/M and cast and wrought U720LI billet material was determined to be 1155°C, and 1160°C respectively.

Table I. Chemistry of the U720LI material studied (w%).

	Cr	Co	Ti	Al	Mo	W	Zr	C	B	Ni
P/M	16.26	14.73	5.05	2.50	3.01	1.27	0.036	0.023	0.018	Rem.
C&W	16.06	14.52	5.04	2.54	3.08	1.20	0.047	0.013	0.018	Rem.

Billet slices of each material were sectioned by wire electric discharge machining (EDM) to produce 12.7 x 12.7 x 25.4 mm heat treatment blocks. Holes 1.6 mm in diameter were drilled in the center of the 12.7 x 12.7 mm face to a depth of 12.7 mm for insertion of a friction fit sheathed thermocouple.

Solution heat treatment studies were performed on the heat treatment blocks with thermocouples attached. Controlled cooling at different cooling rates was achieved by imposing different cooling media, such as air, oil and water, and by wrapping the blocks with various layers of insulating media to retard cooling to simulate large turbine disc-type components. An extensive matrix of heat treatment conditions were performed on the program samples, which included supersolvus and subsolvus solution cycles.

After solution heat treating, the samples were sectioned at the middle of the block, normal to the thermocouple hole for metallographic analysis and further aging studies. Aging studies were performed on select samples to determine the changes in phase types, quantities and morphologies.

Superalloys 2000
Edited by T.M. Pollock, R.D. Kissinger, R.R. Bowman,
K.A. Green, M. McLean, S. Olson, and J.J. Schirra
TMS (The Minerals, Metals & Materials Society), 2000

Property Screening Phase

In addition to the heat treated metallographic samples, laboratory test bar samples were processed, mechanically tested and characterized. These samples are similar to the smaller block samples outlined above, but are much longer in size to accommodate machining mechanical test specimens from them after laboratory heat treatment. These cooling-rate test bars also had thermocouples inserted into one end to allow measurement of the exact cooling rate profile the bars experienced during the imposed heat treatment cycle. These samples were used to evaluate the effects of solution heat treat cycle and cooling rate on final elevated temperature tensile and creep properties.

Both cast and wrought and powder metallurgy U720LI were tested in this phase. Supersolvus solution treatments were performed to allow assessment of intermediate (P/M) and coarse (C&W) structures. Table II lists the processing conditions given to each series of test bars.

Table II. Solution heat treatment cycles used for the mechanical property screening test samples. All samples were given a 760°C/8hrs./AC + 649°C/24hrs./AC age treatment.

Series ID	Solution Heat	Solution Cooling
C15 / P15	1168°C	Continuous Cooled
C26 / P26	1168°C	Continuous Cooled
C37 / P37	1168°C + 1143°C	Continuous Cooled
C48 / P48	1168°C	Step-Cooled

Full-Scale Forging Evaluation

The forging phase of this program was designed to develop and expand upon the microstructure / mechanical property relationships for P/M U720LI material. C&W U720LI

processed by a supersolvus solution cycle did not show a capability of developing a good balance of tensile and creep properties and was not pursued after the property screening phase.

For this effort, 229mm diameter by 132mm tall P/M U720LI billet material was isothermally forged into a 427mm diameter x 38mm thick pancake material. The chemistry of this material is listed in Table III.

Table III. Chemistry of the P/M U720LI billet material used for the full-scale forging efforts (wt.-%).

Cr	Co	Ti	Mo	Al	W	Fe	Zr
16.35	14.78	5.01	3.00	2.41	1.35	0.09	0.031

B	C	Mn	Si	Cu	P	S	N2
0.017	0.021	<0.02	<0.06	0.02	<0.010	0.0003	0.0017

Four different heat treatments were given to the full-scale forged material. These heat treatments were aimed at assessing the development of optimum grain boundary and gamma-prime morphologies. All solution heat treat cycles are aimed at developing a uniform ASTM 7 grain size. The solution heat treatment cycles are graphically shown in Figure 1. Each solution cycle was numbered to correspond to the forged sample serial number. Solution cycle 1001 was aimed at developing large grain boundary serrations along with primary gamma-prime. Solution cycle 1002 was aimed at developing little grain boundary serrations and very fine secondary gamma-prime. Solution cycle 1003 was aimed at developing serrated grain boundaries with no primary gamma-prime. Solution cycle 1004 was aimed at developing serrated grain boundaries with fine gamma-prime.

Three different aging practices were performed on each solution heat treated material to further assess the effect of aging treatments on mechanical property response. Table IV lists the aging cycles used.

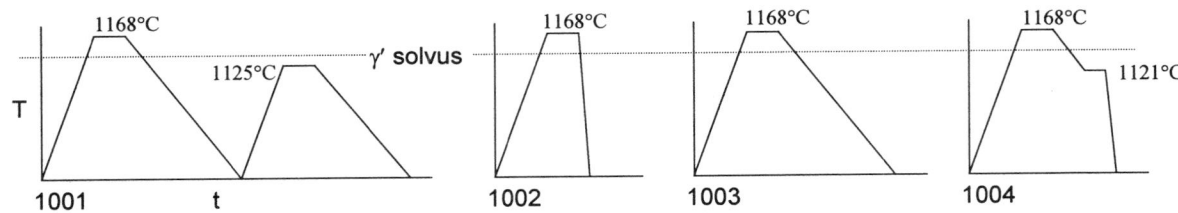

Figure 1. Schematic time temperature plots that show the solution heat treatment cycles relative to the gamma-prime solvus, which were utilized for the full-scale forging phase.

Table IV. Aging cycles given to the full-scale forged
samples.

Age	Cycle 1	Cycle 2
A	843°C/8hrs./FAQ	760°C/16hrs./AC
B	760°C/16hrs./AC	- -
C	649°C/24hrs./AC	760°C/16hrs./AC

Results

Laboratory Heat Treat Studies

Optical and electron microscopy was performed on the
laboratory heat treat samples. Grain size, gamma-prime
size and morphology, and grain boundary morphology
features were fully characterized as functions of solution
heat treatment temperature and cooling rate. The grain
growth behavior of the powder metallurgy and ingot
metallurgy materials are distinctly different. Figure 2
shows grain growth plots for both materials as a function
of solution temperature.

Light and electron optic metallography performed on the
laboratory heat treatment specimens revealed a strong
cooling rate / gamma-prime size dependence. Figure 3
shows example photomicrographs, which show this
correlation. Secondary gamma-prime size measurements
were performed on carefully prepared scanning electron
microscope specimens. Supplemental measurements
were conducted on select samples using transmission
electron microscopy. The TEM measurements confirmed
the measurements performed by the SEM evaluation
technique.

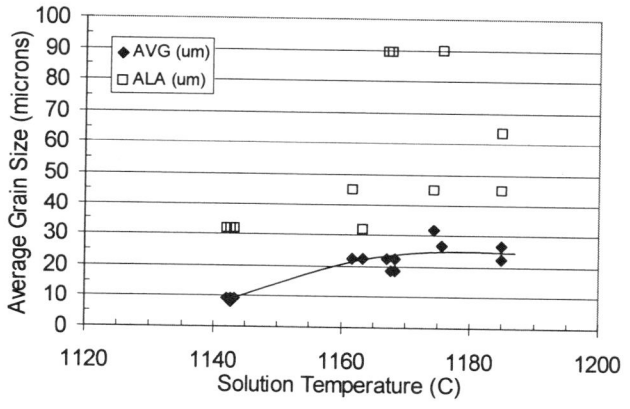
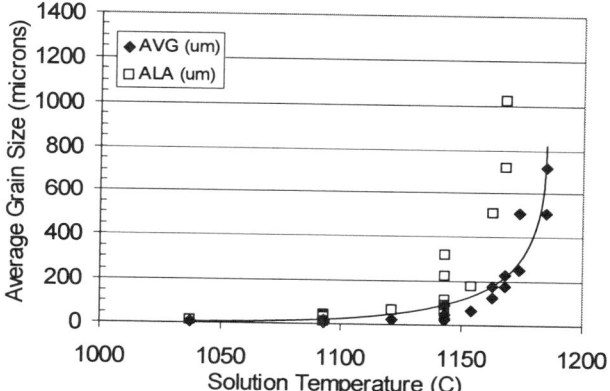

Figure 2. Plot of grain size as a function of solution heat treatment temperature for A) P/M and B) C&W
U720LI material.

The grain boundary morphology of supersolvus solution
heat treated samples was seen to change as a function of
cooling rate. Figure 4 shows photomicrographs of P/M
U720LI grain boundaries after two different cooling rates.
All of the grain boundaries of supersolvus solution heat
treated material exhibited serrations. The faster the cooling
rate, the smaller the amplitude and period of the serrations.
Figure 5 shows a plot of serration morphology as a function
of cooling rate.

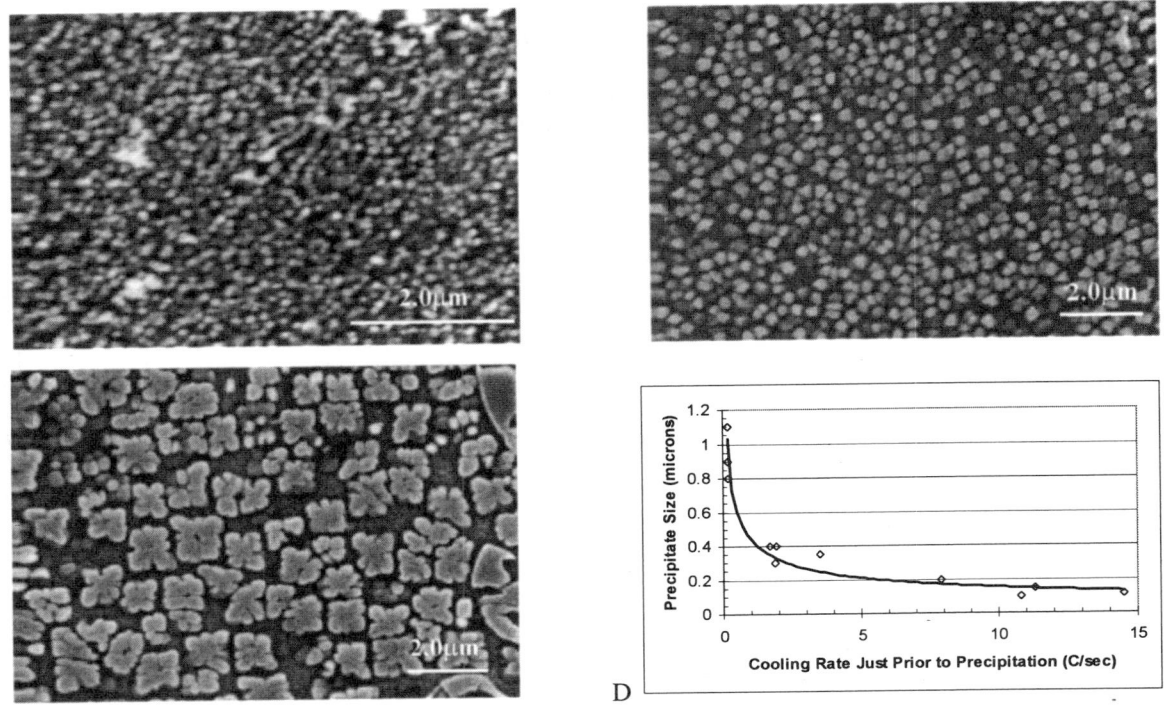

Figure 3. Photomicrographs of supersolvus solution heat treated P/M U720LI samples cooled at A) 12.7°C/sec, B) 1.3°C/sec, C) 0.12°C/sec, and D) a plot expressing gamma-prime size as a function of cooling rate.

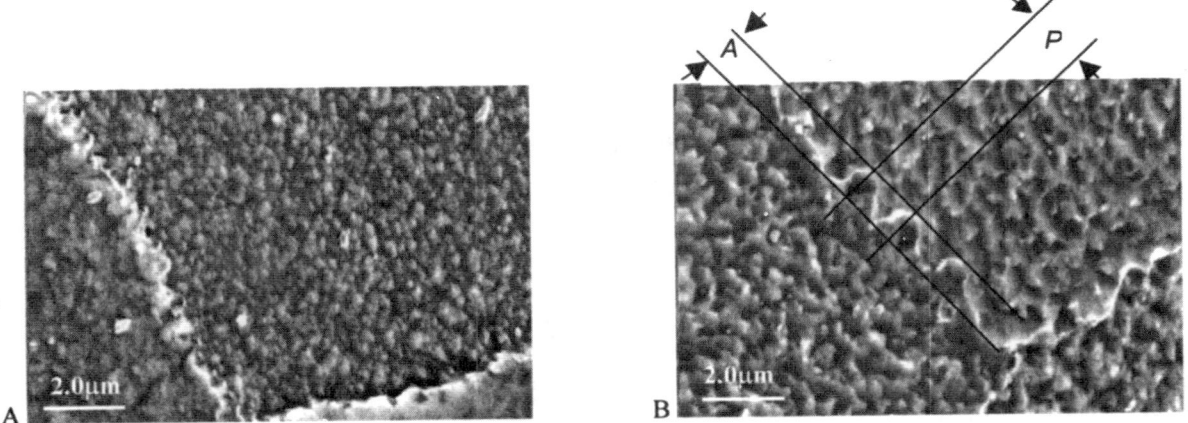

Figure 4. Serrated grain boundary morphologies of P/M U720LI material supersolvus solution treated and cooled at A) 12.7°C/sec and B) 1.3°C/sec. *A* and *P* represent the measurements taken to assess serration amplitude and period respectively.

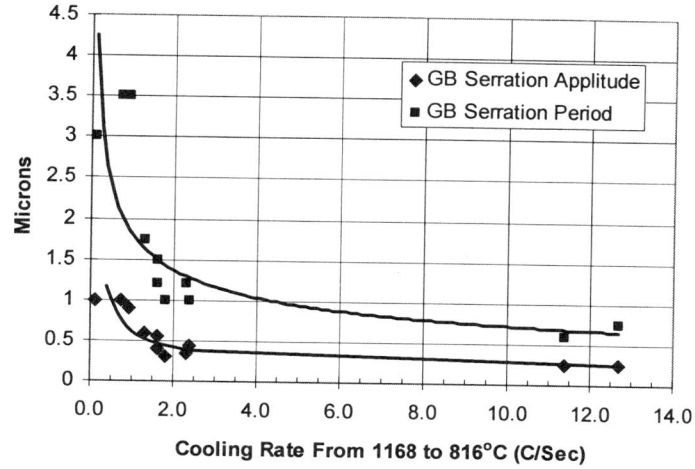

Figure 5. Plot of grain boundary serration amplitude for supersolvus solution treated P/M U720LI as a function of cooling rate.

Electrolytic extractions were performed on all laboratory heat treatment samples to assess the formation of carbide and boride phases as a result of the various thermal treatments. Samples were evaluated in the solution heat treated as well as aged condition. The as-solution heat treated samples allowed an estimation of the $M_{23}C_6$ and M_3B_2 phase solvus temperatures as being between 1030°C-1040°C and 1060°C respectively.

The cast and wrought U720LI material contained large primary carbides [1], whereas the powder metallurgy material only exhibited relatively fine and uniform carbide precipitates. The X-ray diffraction patterns from the cast and wrought material also exhibited two primary carbide phases with distinctly different lattice parameters. It is believed that one is an MC carbide and the other an M(C,N) carbo-nitride phase.

Relative fractions of carbide and boride phases were estimated by comparison of the heights of the strongest peak from each phase diffraction pattern. For all as-

solution heat treated samples, MC carbides were the most predominant. The amount of $M_{23}C_6$ and M_3B_2 increased with decreased cooling rate, and was seen to exceed the amount of primary carbide at the very slowest solution cooling rates examined. Aging also increased the amount of $M_{23}C_6$ and M_3B_2, but MC was still the most predominant carbide/boride phase after each aging treatment except for the 843°C age, where $M_{23}C_6$ phase content exceeded that of the primary carbide.

Property Screening Phase

The mechanical property results for the test bar samples are shown in Table V. There is a strong mechanical property dependence on solution cooling rate, as seen for both materials. The tensile properties of the P/M material are clearly superior to those of the cast and wrought samples, but conversely the cast and wrought material exhibited the best creep resistance for the same thermal treatments.

Table V. Results from the mechanical property screening tests. The tensile tests were conducted at 538°C and the creep-rupture condition used was 704°C/690MPa.

ID	Cooling Rate (°C/sec)	Grain Size (ASTM #)	YS (MPa)	UTS (MPa)	%E	%RA	Hrs. to Rupture
C15	0.68	0.5	755	1138	14.7	16	509
C26	5.25	1.5	948	1217	9.7	13.8	912
C37	0.67/0.9	1.5	810	1258	16.6	19.9	635
C48	0.74	1.0	807	1258	15.8	18.5	554
P15	0.68	7.5	914	1431	20.6	25.8	250
P26	4.46	7.5	1100	1569	13.2	16	391
P37	0.5/0.73	7	938	1458	22	24.5	252
P48	0.74	8	965	1472	18.5	19.2	256

The grain size that resulted after the heat treatment of each specimen was also measured and is listed in Table V. These average grain sizes are consistent with those observed in the laboratory heat treatment microspecimens. It is easy to understand that the very large difference in the mechanical properties between the P/M and C&W samples stems from the large difference in grain size. Because cast and wrought material can not be effectively processed to intermediate grain sizes (ASTM 6-8), and the results from property screening tests were so poor, this material pedigree was not carried into the full-scale forging phase of this program.

Full-Scale Forging Evaluation

The microstructure from the full-scale forged P/M U720LI materials is shown in Figures 6 and 7. The intentional variations in gamma-prime size and grain boundary morphology between each sample can be clearly seen.

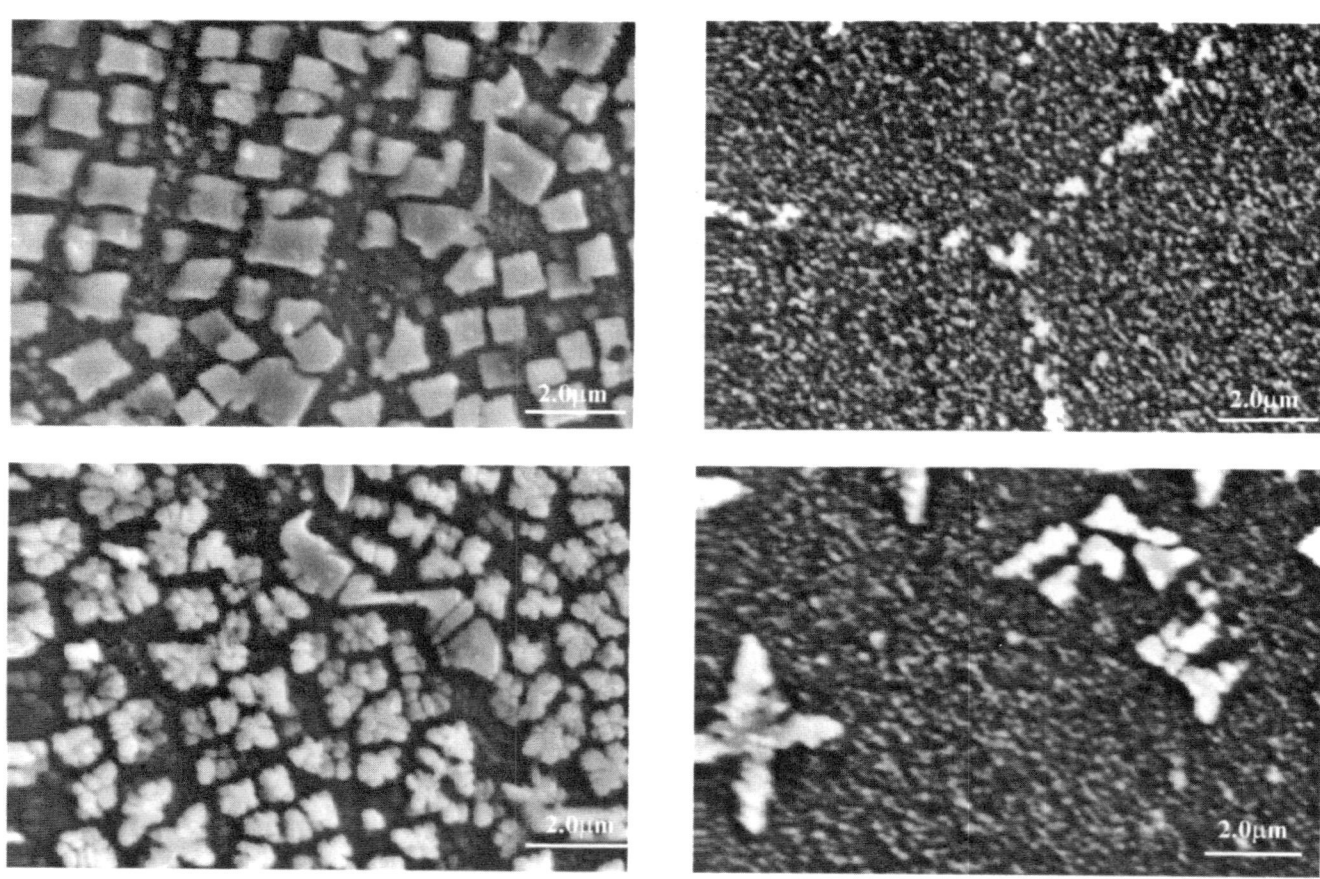

Figure 6. Photomicrographs of the secondary gamma-prime particles within S/N-1001 (A), S/N-1002 (B), S/N-1003 (C), and S/N-1004 (D).

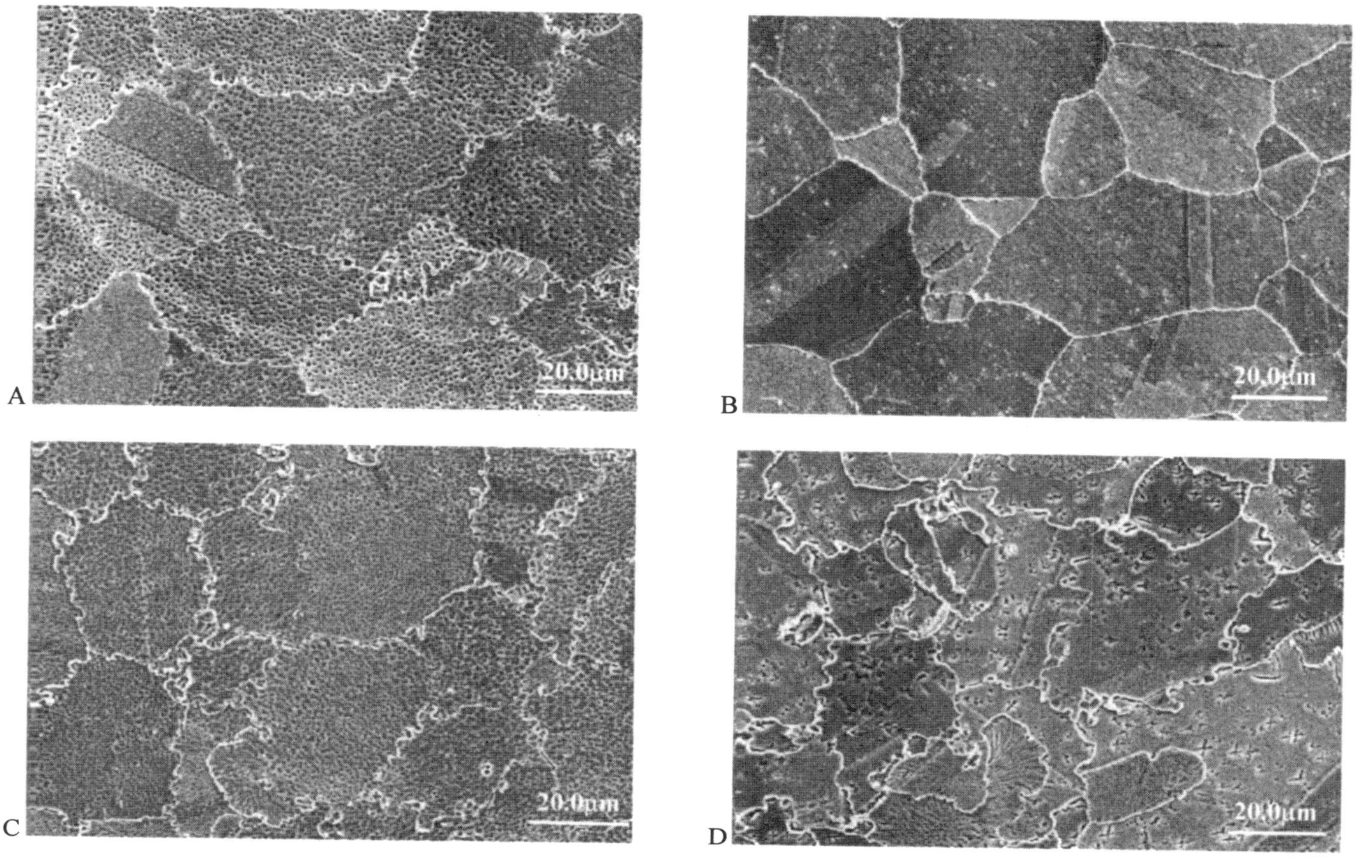

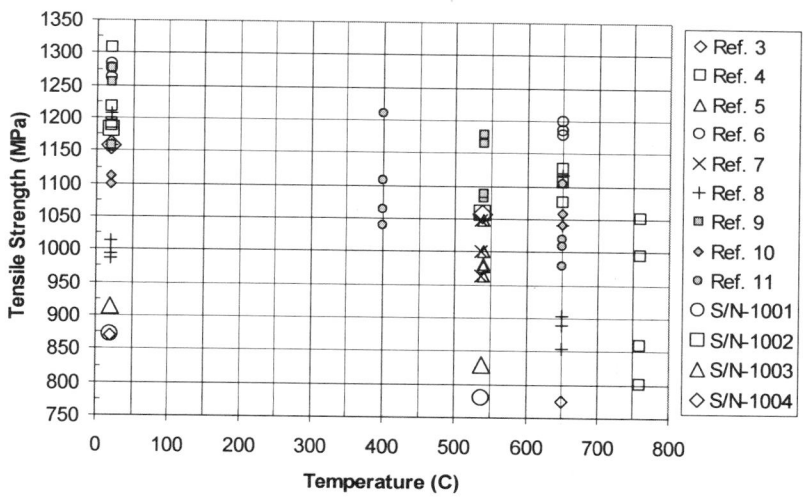

Figure 7. Photomicrographs of the grain boundary morphologies for S/N-1001 (A), S/N-1002 (B), S/N-1003 (C), and S/N-1004 (D).

Figure 8. Average tensile properties from the full-scale forged and heat treated U720LI samples, along with literature data on U720 and U720LI material.

The average tensile properties from each of the full-scale samples are shown in Figure 8. The average creep-rupture properties from each sample are presented in Figure 9. This Larsen-Miller plot shows that the material optimized for maximum grain boundary serration and finest secondary gamma-prime size results in the best creep properties. This material is seen to have greater creep-rupture resistance compared to other U720LI and U720 disc materials and is even approaching the properties of coarse grain U720 blade material.

To assess the fracture resistance behavior of the full-scale samples, room temperature charpy impact tests were conducted. Figure 10 shows the results from this dynamic crack propagation screening test.

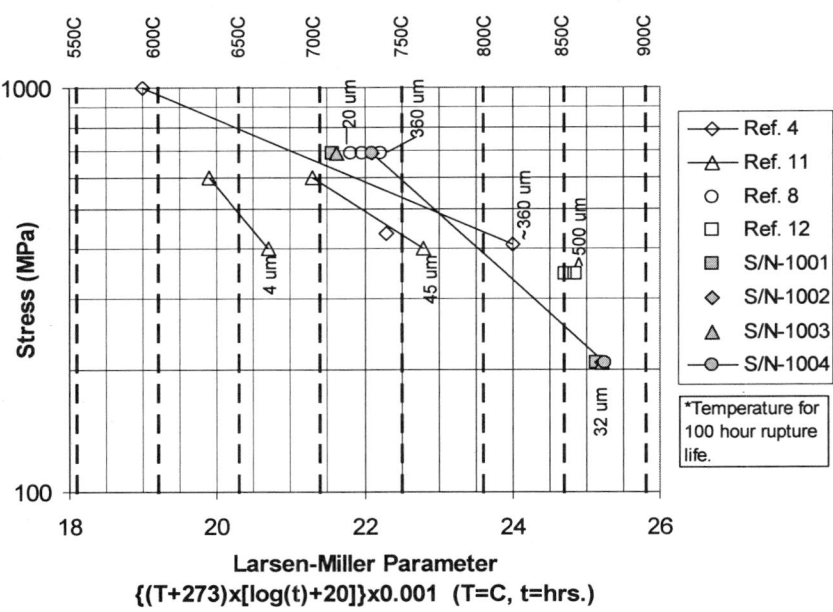

Figure 9. Larsen-Miller creep-rupture plot of the full-scale forged and heat treated U720LI samples, along with literature data for U720 and U720LI material.

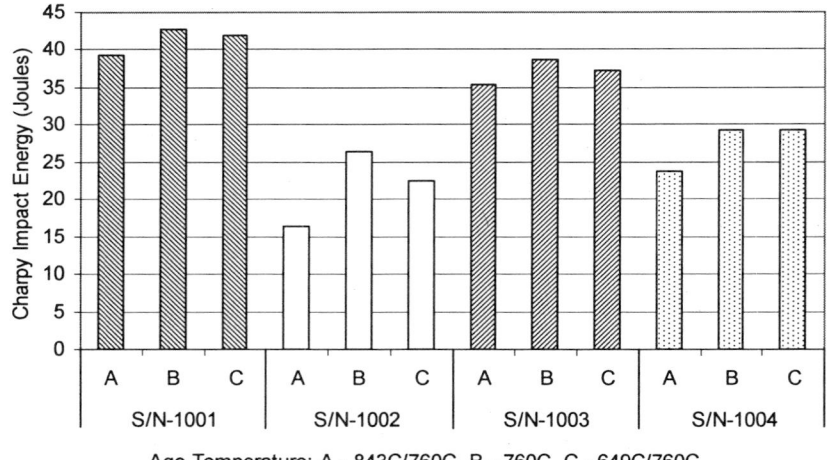

Figure 10. Graph of the average charpy impact energies for each full-scale forged sample.

422

Discussion

Equations in the form of the arrhenius rate equation can been developed to describe the grain growth behavior of C&W and P/M U720LI. The equations developed in this program are as follows:

$$d^{1.5} - d_o^{1.5} = 5.95 \times 10^{45} t \exp(-1273900/RT) \quad [C\&W] \quad (1)$$

$$d^{5.4} - d_o^{5.4} = 2.37 \times 10^{91} t \exp(-2432900/RT) \quad [P/M] \quad (2)$$

These equations are shown in Figures 2, along with the experimental data. It can be seen that the developed equations fit the observed gain growth behavior well.

The gamma-prime size was seen to be closely related to cooling rate. It has been previously reported that very little difference in precipitation undercooling was observed for this material. [2] It is believed that the secondary gamma-prime grows after initial nucleation and therefore the gamma-prime size must be a diffusion controlled process. Since diffusion distance is proportional to the square root of the diffusivity and time, an expression of this form has been developed which shows extremely good correlation with the observed data. In this expression, cooling rate contains a factor of time, while the remaining factors are related to a function of the material diffusivity. This equation is shown in Figure 3 along with the measured experimental data.

$$d = 0.4425 \times (C)^{-0.4506} \quad (3)$$

The formation of serrated grain boundaries in U720LI is very interesting. It was seen from detailed metallographic analysis that the serrated grain boundaries are caused by the formation and growth of grain boundary gamma-prime. The relationships between the grain boundary serration and gamma-prime sizes versus cooling rate are also very similar, lending further evidence that growth of grain boundary gamma-prime is the cause of the grain boundary serrations.

The mechanical properties of U720LI can be linked to gamma-prime size through an analysis of dislocation interactions and predictions of critical resolved shear stress. This approach has been previously performed on subsolvus processed C&W U720LI. [13] Using this approach with supersolvus processed U720LI results in a similar relationship between gamma-prime size and theoretical critical resolved shear stress. Theoretical estimations of tensile strength have been calculated based on knowledge of the critical resolved shear stress. A proportionality factor (the Taylor factor) relates the critical resolved shear stress to observed tensile strength values.

From this approach, it has been seen that theoretical and experimental strength values correlate well.

The results from the creep and charpy impact tests show that the gamma-prime size have a large influence on these mechanical properties as well. Additionally, grain boundary serrations appear to also play a role in these properties. While S/N-1002 and S/N-1004 had nearly identical tensile properties, S/N-1004 (sample with large grain boundary serrations) exhibited improved impact resistance.

The age process used during the heat treatment of U720LI has been shown to be critical in the development of optimum mechanical properties. The single step aging treatment exhibited the best properties for all solution heat treatments and cooling conditions. This is believed to be due to a combination of optimum growth of the fine gamma-prime particles and optimum distributions of borides and carbides. It was seen that the high temperature aging cycle resulted in extensive $M_{23}C_6$ formation, and gamma-prime particles growth. These changes are seen to be very deleterious to the final component mechanical properties.

Conclusions

- This program resulted in a better understanding of the U720LI phases, phase relationships and formation kinetics, and microstructure mechanical property relationships.

- A model has been developed for the prediction of gamma grain growth, gamma-prime precipitate size and grain boundary serration morphology.

- A correlation between microstructure and mechanical properties has been established.

- A new thermal processing cycle has been developed to produce an intermediate grain size U720LI material with reduced residual stresses, and optimized mechanical properties.

- The new thermal processing cycle includes a supersolvus solution cycle, a stepped cooling rate, and a single aging treatment.

References

1. D. Furrer and H. Fecht, "Superalloys for Turbine Disc Applications", <u>JOM</u>, January, 1999, pp. 14-18.

2. D.U. Furrer and H.-J. Fecht, "γ' Formation in Superalloy U720LI", <u>Scripta Mat.</u>, Vol.40, No.11, pp. 1215-1220, 1999.

3. L.S. Buslavsky, et. al., "Improvement of the Granulated Nickel Alloy EP741NP for the Purpose of an Increase in the Mechanical and Operating Characteristics", <u>Technology Of Light Alloys,</u> Vol. 2, VILS, Russia, 1997, pp. 24-26.

4. F.E. Sczerzenie and G.E. Maurer, "Development of Udimet 720 for High Strength Disk Applications", **Superalloys 1984**, eds., Gell, et. al., TMS, 1984, pp. 573-582.

5. J.F. Radavich and J. Hyzak, "Effect of Processing and Thermal Treatment on Alloy 720", **Proceedings of the Tenth International Conference on Vacuum Metallurgy**, Vol. 1, Specialty Melting, Beijing, China, June 11-15, 1990.

6. S.Bashir and M. Thomas, "Effect of Interstitial Content on High Temperature Fatigue Crack Propagation and Low Cycle Fatigue of Alloy 720", **Superalloys 1992**, eds., S.D. Antolovich, et. al., TMS, 1992, pp. 747- 755.

7. J.M Hyzak, et. al., ., "The Microstructural Response of As-HIP P/M U-720", **Superalloys 1992**, eds., S.D. Antolovich, et. al., TMS, 1992, pp. 93-102.

8. K.R. Bain, et. al., "Development of Damage Tolerant Microstructures in Udimet 720", **Superalloys 1988**, eds., S. Reichman, et. al., TMS, 1988, pp. 13-22.

9. K.A. Green, J.A. Lemsky and R.M. Gasior, "Development of Isothermally Forged P/M Udimet 720 for Turbine Disk Applications", **Superalloys 1996**, eds., R.D. Kissinger, et. al., TMS, 1996, pp. 697-703.

10. D.J. Bryant and G. McIntosh, "The Manufacture and Evaluation of a large Turbine Disc in Cast and Wrought Alloy 720LI", **Superalloys 1996**, eds., R.D. Kissinger, et. al., TMS, 1996, pp. 713-722.

11. H. Hattori, M. Takekawa, D. Furrer and R. Noel, "Evaluation of P/M U720 for Gas Turbine Engines Disk Application", **Superalloys 1996**, eds., R.D. Kissinger, et. al., TMS, 1996, pp. 705-711.

12. I. Tsuji and H. Itoh, "Long Term Stability of Ni-Base Superalloy U720 Bar and Blade for Gas Turbine Blades", <u>Transactions of the Iron and Steel Institute of Japan</u>, Vol. 22:4, 1982, p. B-112.

13. M.P. Jackson, R.C. Reed, <u>Material Science and Engineering</u>, A259, 85, (1999).

SUB-SOLIDUS HIP PROCESS FOR P/M SUPERALLOY CONVENTIONAL BILLET CONVERSION

X. Pierron, A. Banik, G. E. Maurer
J. Lemsky*, D. U. Furrer*
S. Jain**

Special Metals Corporation, 4317 Middle Settlement road, New Hartford, NY
Ladish Co., Inc., 5481 South Packard Ave, Cudahy, WI 53110*
Rolls-Royce Corporation, 2001 South Tibbs, Indianapolis, IN 46206**

Abstract

Recent studies have identified thermo-mechanical conditions for powder UDIMET®alloy 720[1] which resulted in significantly higher ductility of HIP compacts[1]. In this process, the compact is produced by hot isostatic pressing (HIP) at temperatures just below the solidus of the alloy (slightly sub-solidus HIP or SS-HIP). The workability of the resulting as-HIPed material is high enough to allow billet conversion using conventional forging. This new process has the potential of reducing the cost of powdered metal (P/M) billet manufacturing. The results of the present study provide some insight on the mechanisms leading to the enhanced forgeability of the SS-HIP processed UDIMET alloy 720. In addition, a full-scale trial has shown that this new processing route has potential application for future production of high performance turbine disks, with equivalent microstructural and mechanical properties as parts manufactured with the conventional compact plus extrude plus isothermal forging route.

Introduction

UDIMET alloy 720 is a high strength nickel base superalloy currently produced in cast and wrought and P/M form for turbine disk applications. UDIMET alloy 720 was developed originally as a wrought turbine blade material for power plant turbines[2]. A new generation of the alloy was developed in order to increase its high temperature stability and fatigue resistance to make it suitable for turbine disk applications. The modified cast and wrought alloy, UDIMET alloy 720LI, has since been recognized as having outstanding strength and fatigue resistance compared to other advanced turbine disk materials[3].

UDIMET alloy 720LI is one of the most highly alloyed cast and wrought disk materials, making the production of billet larger than 6.5" diameter very challenging. On the other hand, the powder metallurgy route allows the production of very homogeneous material, without the segregation caused by melting and re-melting processes. The grain size of P/M material is in general finer and more uniform than cast and wrought material, resulting in better defect detection capability by ultrasonic inspection. In addition, the P/M version of the alloy in the compact plus extrude plus isothermal forge form offers greater microstructural flexibility and control during post forging heat treatments[4].

[1] UDIMET is a registered trademark of Special Metals Corporation

Superalloys 2000
Edited by T.M. Pollock, R.D. Kissinger, R.R. Bowman,
K.A. Green, M. McLean, S. Olson, and J.J. Schirra
TMS (The Minerals, Metals & Materials Society), 2000

Powder metal superalloy billets are conventionally processed by hot compaction of powder in sealed cans followed by an extrusion process. During the hot compaction process, phases precipitate at the particle surface, and the resulting continuous second phase network or prior particle boundaries (PPB), was found to be responsible for the observed low ductility of the compact[3, 4].

The effects of PPB's in P/M superalloys have been extensively reviewed by Thamburaj et al.[5]. Their review showed that the PPB's particles form as a result of carbon segregation during consolidation and particle contamination at the PPB's. The PPB particles were identified as carbides, nitrides or oxy-carbo-nitrides, and the PPB particle volume fraction was found to depend greatly on the atomization and handling method. Alloy designers have attempted to reduce the amount of PPB precipitation by chemistry and HIP cycle modification in alloys such as Astroloy, but the benefit of putting the PPB particles in solution was offset by an increase in grain growth and reduced workability during final forging[6].

The detrimental effects of the PPB particles in the microstructure can be greatly reduced if the material goes through a heavy warm working process which provide high strains within the component to ensure the dispersion of the prior particle boundary phase network. Specialized extrusion processes were developed to induce the required large strains without extensive cracking. The extrusion process also develops a fine grain microstructure suitable for the slow strain rate, isothermal forging of high performance jet engine turbine disks[3].

One disadvantage of the conventional compaction plus extrusion route to manufacture P/M superalloys billet is its cost. The process calls for specialized equipment to provide the high tonnage necessary for extrusion, and the setup and maintenance of the specialized tooling requires significant costs relative to initial procurement, on-going maintenance and operations. Contrary to the compaction plus extrude consolidation route, the consolidation of powders using a hot isostatic pressing (HIP) route can be performed using a single piece process flow. In addition, HIP consolidation is readily available throughout the world with over 1,000 units operating[7].

Initial attempts to convert as-HIPed P/M superalloys on conventional press conversion equipment were unsuccessful due to poor workability of the as-HIPed material compared to ingot material, making conventional press forging with the available press speeds, reduction and thermal control not cost effective. Special Metals Corporation has recently identified thermo-mechanical

conditions for powder UDIMET alloy 720 which resulted in significantly higher ductility of the HIP compact[1, 8]. The compact was produced by hot isostatic pressing at temperatures just below the solidus of the alloy (slightly sub-solidus HIP or SS-HIP). The SS-HIP temperature range is define as $T_{solidus} - T_{SSHIP} < 28°C$. The as SS-HIP material possessed enough workability to be processed using conventional press conversion equipment. A 3000kg, 50cm diameter, SS-HIP consolidated UDIMET alloy 720 ingot was successfully press forged in Special Metals Corporation forging facility in Dunkirk, NY, to produce a 16.5cm billet.

In order to understand the micro-mechanisms leading to the enhanced workability of the SS-HIP material, a study of the effect of HIP temperature on microstructure and mechanical properties was undertaken on sub-scale containers filled with -140 mesh gas atomized powder. In addition, material from the SS-HIP consolidated and press forged billet was sent to Ladish Co. Inc.. Forged components were produced by Ladish Co. Inc. to compare the properties of the SS-HIP UDIMET 720 material with conventional P/M and cast and wrought material.

Sub-scale study

Experimental Procedures

Lab-scale experiments consisted of a detailed study of the HIP temperature effects on microstructure and mechanical properties of the as-HIPed and HIP plus rolled UDIMET alloy 720. The composition of the alloy can be found in Table I.

Table I: UDIMET alloy 720 P/M composition

	Ni	Co	Cr	Al	Ti
Wt.%	57.59	14.4	16.15	2.55	4.9
	C	B	Zr	Mo	W
Wt.%	0.022	0.018	0.03	3	1.34

Sub-scale 304 stainless steel 4.6 cm diameter containers were filled with -140 mesh gas atomized powder. HIP was performed under 103.4 MPa pressure for 3 hours. The HIP temperature was varied from 1120°C (sub-solvus) to 1270°C (slightly below the solidus temperature of the alloy) with a controlled cooling rate of 0.8°C /min.

The microstructure was examined using optical, scanning and transmission electron microscopy techniques (SEM and TEM respectively). In order to observe the PPB's by optical microscopy, the material was first heat treated for 5 min. at 1170°C (just above the γ' solvus), removing most of the γ' precipitates which mask the fine PPB particles.

Extraction and analysis of the secondary phase particles were performed by Micro-Met Laboratories, Inc[2]. The extraction was done using a 10% HCl in methanol electrolytic solution. This solution dissolves the γ and γ' phases, leaving in the extracted residue the carbides, nitrides, borides, oxides and oxy-carbo-nitrides. Extraction of secondary phase particles from the HIP compact was performed on material which was solution heat treated for metallographic observation of the PPB networks. The particles in the extracted residue were observed using SEM techniques and phase identification was performed by X-ray diffraction.

Workability was evaluated by performing elevated temperature tensile tests (1093°C) at a strain rate of 2 sec^{-1} and measuring the reduction of area at failure. In addition to testing the as-HIPed material, HIP materials were rolled down and tested along the rolling direction to study the effect of microstructural refinement possible by press forging and its affect on workability. Roll down operations were performed at 1093°C , for a final height reduction of 30%.

Microstructural analysis

PPB particles could clearly be observed in the microstructure of the as-HIPed materials, Figure 1. The particles are believed to precipitate as a result of surface contamination during the atomization, and carbon segregation to the particle free surface in the first stage of the consolidation process[6]. The PPB particles formed a network of secondary phases in the consolidated material throughout the microstructure, following the location of the powder particle surfaces. In this study, the particle density on the PPB's was found to decrease with increasing HIP temperatures, Figure 1. This is in accord with the second phase extraction investigation, which found that the amount of residue collected decreased for increasing HIP temperatures.

The particle size in the collected extraction residue from the HIP trial materials was approximately 0.2 μm. The few particles extracted from the SS-HIP material were on average larger than for the other HIP conditions. The larger particles in the SS-HIP material were up to 0.5 μm diameter. Those larger particles are believed to be TiC which have precipitated inside the grains but not at the PPB's. X-ray diffraction performed on the residue

[2]Micro-Met. 209 North Street, West Lafayette, IN 47906

(a) 1120°C

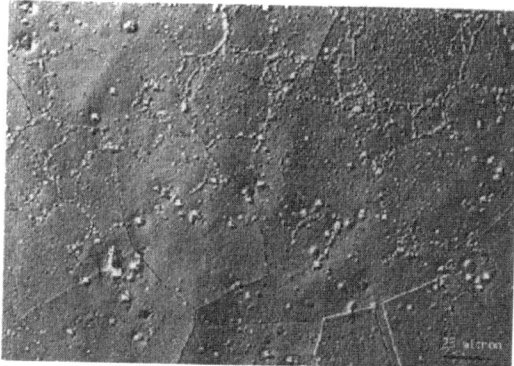

(b) 1265°C

Figure 1: View of the PPB particle network and grain size after a 3 hour HIP cycle at (a) 1120°C and (b) 1265°C .

extracted from material HIPed at 1120°C (below solvus), 1230°C and 1265°C (SS-HIP condition) gave rise to diffraction patterns fitting MC carbides, and zirconium oxide ZrO_2 or possibly ZrCaTiO. The lattice parameter of the MC carbides was 42.8 nm. EDAX on the residue observed by SEM showed that the particles were enriched in Ti, indicating that the MC carbides should be TiC.

The TEM study of the microstructure of the as-HIPed materials showed discrete particles aligned along what could have been the free surface of a prior powder particle, Figure 2. No continuous phase or film was observed at the PPB's. Grain boundaries were not always associated with the particles. However, some particles were observed to pin grain boundaries and seem to impede grain growth. EDAX analysis performed in the TEM showed that some of the particles were enriched in Ti or in Zr, in accord with the extraction results. In the case of SS-HIP material, very few particles could be found in the TEM thin foils.

These results indicate that, in the HIP temperature range investigated, the majority of the particles forming on the PPB's were Ti rich MC carbides and Zr rich oxides.

Similar studies[9] performed on Rene' 95, a P/M nickel based superalloy containing 3.5 wt.% Nb and 0.05 wt.% Zr, have found Nb and Ti rich MC carbides and ZrO_2 oxides at the PPB's. The precipitation of Zr oxides rather than similar common superalloy oxides such as Al and Mg oxides could be due to a higher stability of the Zr oxides or to a faster segregation of Zr to the free surface of the powder particle in these alloys. As the HIP temperature increased, some second phase particles went into solution. It seems unlikely that the oxides, due to their inherent stability, could go in solution at the HIP temperatures investigated, but the MC carbides seem to be substantially dissolved.

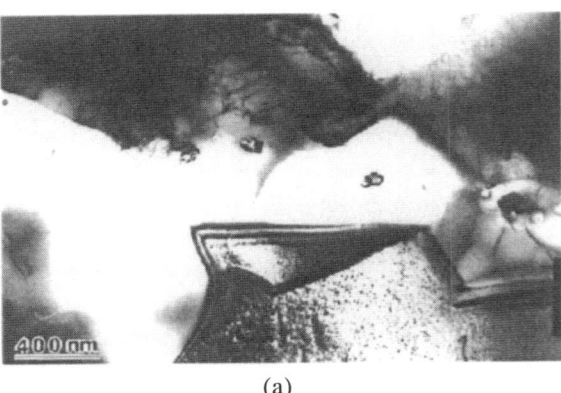

(a)

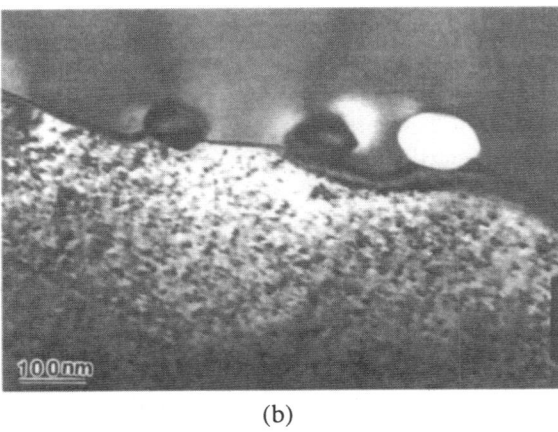

(b)

Figure 2: TEM bright field micrographs showing (a) the discret particles along the PPB's, and (b) particles pinning a grain boundary.

Figure 3 shows that the secondary γ' precipitates in material HIPed below the solvus temperature was much finer than the secondary γ' observed in material HIPed at temperatures above solvus. This is due to the very different precipitation conditions of the γ' phase in those two cases. For super solvus HIP, the γ' phase precipitates as the material cools slowly in the HIP unit at the end of the HIP cycle. This results in large γ' precipitates, occasionally forming a fanlike structure[10]. In the case

of sub-solvus HIP, very fine, well dispersed primary γ' precipitates in the supersaturated matrix quenched from the liquid during the atomization process, Figure 3.

The effect of HIP temperature on the grain size of P/M UDIMET alloy 720 is presented in Figure 4. The results suggest that grain growth occurs in stages, corresponding to temperatures where grain growth barriers, such as γ', nitride, carbides or oxy-carbo-nitrides, go in solution. The driving force for grain growth in the case of HIP should be primarily grain boundary surface minimization, as after consolidation, the isostatic pressure would not induce sufficient plastic strain to provide enhanced driving force for grain growth.

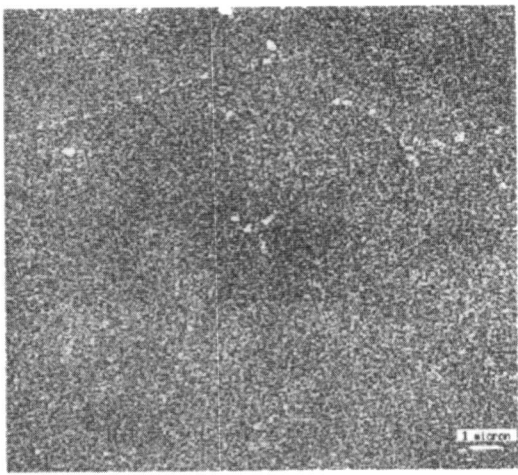

(a) 1120°C

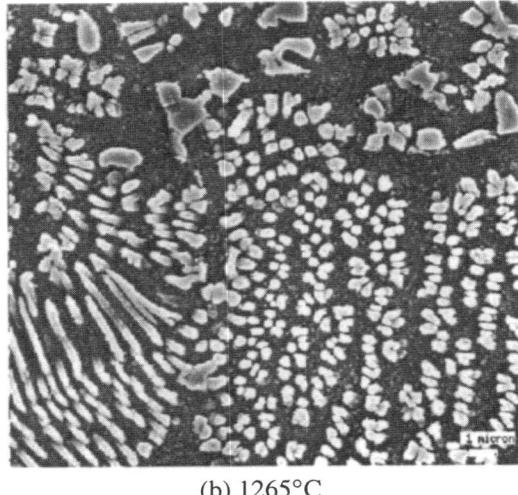

(b) 1265°C

Figure 3: γ' precipitates after HIP at (a) 1120°C and (b) 1265°C

Below the γ' solvus, grain growth was restricted within the prior particle by the γ' precipitates and the PPB particles. The resulting grain size was in general smaller than the prior powder particles. Above γ' solvus, grains started to coarsen as a result of γ' solutioning but the PPB network was still present, preventing extensive grain growth. As the HIP temperature was increased, some of the PPB particles dissolved, and the grains were able to grow to larger sizes before being pinned again on clusters of larger or more densely packed PPB particles. It was only in the SS-HIP temperature range, where extensive dissolution of the PPB's network was observed, that grains coarsened clearly beyond several prior particles to reach ASTM 3-4, Figure 1b. The drastic increase in grain size in the SS-HIP temperature range indicates that the rate of dissolution of the MC carbides increased readily at those temperatures.

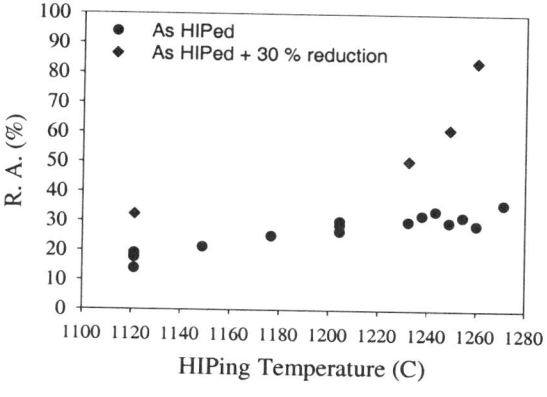

(a)

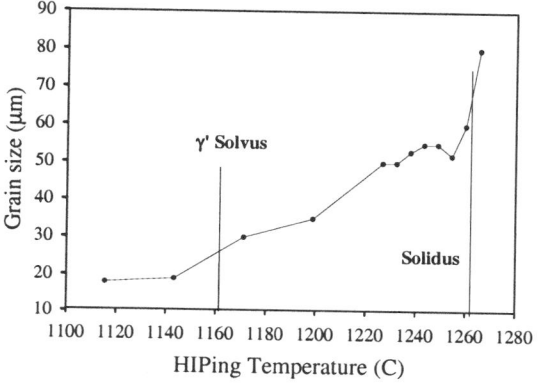

Figure 4: Effect of HIP temperature on grain size of as-HIPed microstructure.

Aging treatments of the SS-HIP material were performed for 3 hours at temperatures below the SS-HIP temperature in order to study the eventual re-precipitation of the carbides at the PPB's. No new PPB particles re-precipitated in the microstructure after aging. However aging at temperatures above approximately 1200°C at atmospheric pressure created thermal induced porosity (TIP), i.e. the entrapped argon in the material during gas atomization and the consolidation expands and formed visible pores.

Mechanical testing

Figure 5 shows the evolution of the workability of the as-HIPed material as a function of HIP temperature. The workability at 1093°C increased steadily with the HIP temperature and reached 30% R.A. for HIP temperatures above 1230°C . For HIP temperatures above the solidus, the workability drops as incipient melted areas start to form at grain boundary triple points in the microstructure.

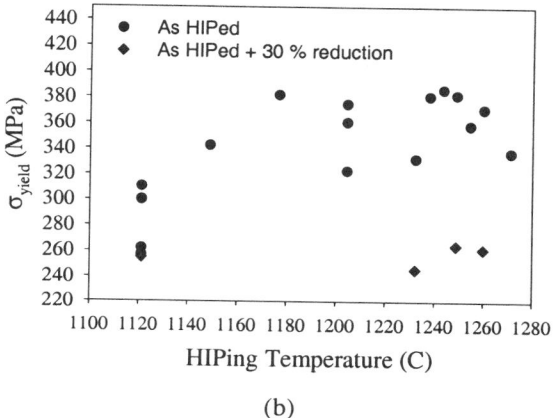

(b)

Figure 5: Effect of HIP temperature on (a) workability and (b) flow stress. Testing was done at 1093°C with a strain rate of 2 sec^{-1}.

The fractured specimens were sectioned such that the fracture initiation sites under the fracture surface could be observed, Figure 6. Fracture initiated at voids found around PPB's particles in SS-HIP and non SS-HIP material. In the non SS-HIP material, the PPB particle network and the grain boundaries were in general close to each other as grain growth during HIP was contained by the PPB particles. As a result, occasionally a void initiated at a PPB particle could initiate a crack into a neighboring grain boundary. For SS-HIP material, the voids were generally isolated, coalescing with neighboring voids where the PPB's particle inter-spacing was low. Some voids were also observed at grain boundary triple points.

After rolling, the grain size (ASTM 11-12) was independent of HIP temperature and uniform. The workability increase that could be obtained by press forging was found to be highly dependent on the HIP temperature,

429

Figure 5. For HIP temperatures just above the γ' solvus, the workability after rolling increased to just above 30%. However, as the HIP temperatures get closer to the SS-HIP condition, the workability of the press forged material increased steeply and exceeded 85% RA for the SS-HIP material. The 1093°C flow stress at high strain rate (2 sec^{-1}) of the as-HIPed and as-HIPed plus rolled material can be seen Figure 5. For HIP temperatures between the γ' solvus and the solidus, the flow stress of the as-HIPed and as-HIPed plus rolled material did not vary significantly. Rolling of the as-HIPed material reduced the flow stress in the rolling direction from approximately 350 MPa to 242 MPa.

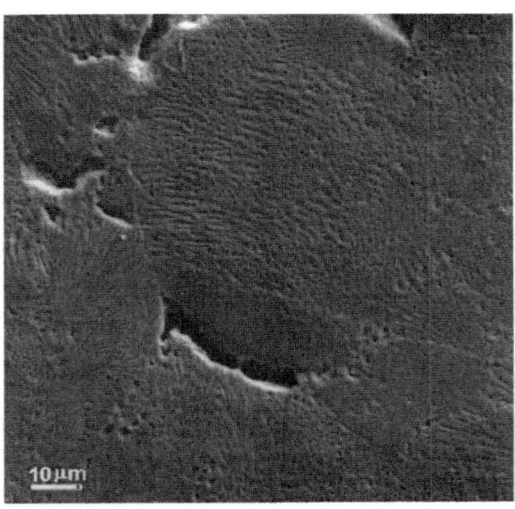

Figure 6: View of the fracture initiation sites at the PPB particle under the fracture surface for material HIPed at 1260°C and tested in tension at 1093°C with a strain rate of 2 sec^{-1}.

Workability is highly controlled by grain size, and usually increases as grain size decreases. In the case of P/M superalloys, the workability is also influenced by the PPB particles, at which voids initiate during high temperature and high strain rate deformation. In the case of P/M UDIMET alloy 720, as HIP temperatures increase, the grain size increases and the PPB particle volume fraction decreases. Figure 5 indicates that for increasing HIP temperature up to the solidus, the workability stayed unchanged at 30%, showing that the detrimental effect of grain growth on the workability was balanced by the beneficial dissolution of some of the PPB particles. After refining the microstructure by rolling, all the HIP materials had the same grain size, i.e. ASTM 11-12, and the workability would only be controlled by the PPB particles. This is indeed what can be observed Figure 5, where the workability increases with the HIP temperature, i.e. as the PPB particle volume fraction

decreases.

For SS-HIP UDIMET alloy 720, despite a grain size of ASTM 3-4 in the as-HIPed condition, the material had sufficient workability to be thermo-mechanically processed using conventional forging techniques used to break down ingot structure. After some grain refinement, the workability of the SS-HIP material plus press forge was comparable to cast and wrought material. Those results indicated that a UDIMET alloy 720 SS-HIP powder can could be forged to produce a highly homogeneous and workable billet using conventional open die press forging equipment.

Large Scale Trials

A 3000 kg, 50 cm diameter SS-HIP consolidated UDIMET alloy 720 ingot was press forged in Special Metals Corporation forging facility in Dunkirk, NY. The billet was forged at 1080-1110°C to 17.5 cm using a direct draw conversion sequence. The sonic inspectability of the 16.5 cm diameter billet was tested and exhibited an ultrasonic noise level of 15-18% on a 0.4 mm diameter reference hole (#1FBH), a much better sonic response than what can be accomplished with a cast and wrought billet. The microstructure was very uniform, ASTM 10-11 grain size, across the billet cross section.

Small lot quantities were provided to Ladish in diameters of 16.5 and 23 cm for forging and characterization studies. Producing two billet diameters with a relatively small lot size represents a distinct advantage of the SS-HIP process relative to standard powder metal conversion by extrusion for which minimum lot sizes are quite large.

A desirable attribute of powder metal superalloys is the ability to tailor the microstructure over a large range of grain size. In this fashion, a single alloy can meet widely varied engine disk design requirements by altering the specific TMP cycle and subsequently the grain size. In contrast, cast and wrought materials have less grain size flexibility and often require working within a difficult or impractical process window to obtain desired grain size requirements. The grain coarsening response of the SS-HIP UDIMET alloy 720 material was investigated.

Ladish Co. Inc. performed controlled sub-solvus and super-solvus (1107°C - 1177°C range) thermo-mechanical processing trials in which a variety of grain size targets were successfully achieved in forged components. The results of this study showed that the SS-HIP material revealed good grain size flexibility and control. Table II shows grain size results from this study.

Table II: Grain size control evaluation of the SS-HIP processed billet

Target grain size ASTM, average	Resulted Grain Size ASTM (average, ALA)
	11, 5 (input billet)
4-6	4.5, 2
6-8	7, 5
8-10	9.5, 8
10-12	11, 9

Component Demonstrations

Large, complex, closed-die shapes were produced with the SS-HIP material and examined for forgeability, microstructure and mechanical properties. Two shapes that were produced are shown in Figure 7. These components were produced by isothermal forging to a uniform fine grain size requirement. The turbine disk shown in Figure 7a was further machined to a rectilinear shape and ultrasonic tested. The test showed substantially improved ultrasonic transparency when compared to the same part produced from cast and wrought UDIMET alloy 720LI material.

The forging shown in Figure 7b is a multi-part disk in which seven individual turbine disks are excised from one forging. This contoured near-net shape forging measuring approximately 70 cm. in diameter was produced in an efficient single-step isothermal forge process. The excellent chemical and microstructural homogeneity associated with the powdered metal product enabled an isotropic product on this non-axisymmetric forging[11].

The results previously shown in Table II reflect individual forgings each produced to a single microstructure, *i.e.* uniform grain size throughout the volume of the forging. The same grain coarsening behavior of the SS-HIP plus press forged product can also be locally applied and controlled to produce a component with dual microstructure and property characteristics. Figure 8 shows a cross-section of a disk that was processed to a dual microstructure. It is the combination of the forge and the post forge thermal practices that control the grain coarsening response.

Another example of a disk processed to provide a dual microstructure is shown in Figure 9. The component was solution heat treated such that the rim maintained a nominal temperature of γ' solvus + 28°C , and the bore maintained a nominal temperature of γ' solvus - 28°C . The disk was then stabilized at 760°C for 8 hours followed by an air cool, and aged at 649°C for 24 hours followed by an air cool. The forging was cross-sectioned and specimens were removed from the

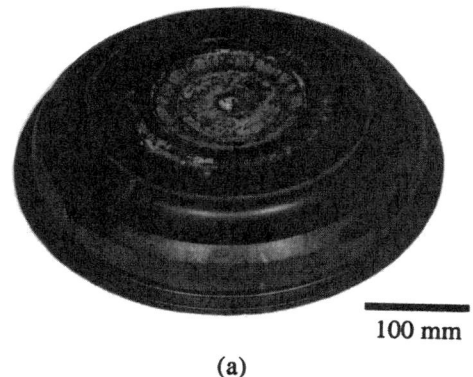

100 mm

(a)

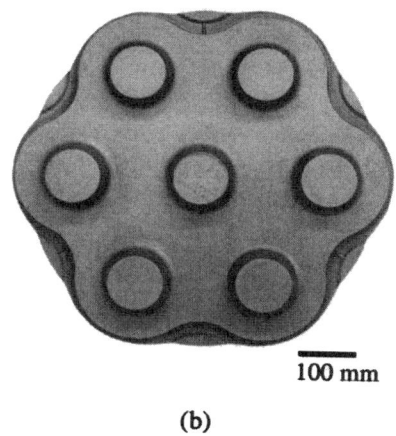

100 mm

(b)

Figure 7: Closed die forging shapes were produced with SS-HIP processed material, (a) and (b).

coarse grain rim region and the fine grain bore region. The tensile properties for this P/M dual microstructure forging are shown on Table III, with a comparison to cast and wrought forgings also made into this forged shape.

A machineability trial was performed on SS-HIP plus press forged and cast and wrought billet utilizing a well established finish machine production plan at Rolls-Royce Allison. Two AE3007 second stage turbine disks from SSHIP plus press forged material were successfully machined to blueprint requirements as shown in Figure 10. The machineability of powdered metal U720 was shown to be better than cast and wrought due to a more homogenous chemistry and microstructure.

Table III: Average tensile properties from a turbine wheel forging utilizing SS-HIP and cast and wrought UDIMET alloy 720 billet processed to a dual fine grain (FG) and coarse grain (CG) microstructure.

Billet	Grain size	Temp.	UTS (MPa)	σ_{Yield} (MPa)	Elong. (%)	R.A. (%)
SS-HIP	FG Bore	Room	1631	1190	22	23.5
SS-HIP	CG Rim	Room	1537	1082	21	25
SS-HIP	FG Bore	538°C	1564	1131	20.5	23.5
SS-HIP	CG Rim	538°C	1376	989	18.5	23
C & W	FG	Room	1634	1203	21	22
C & W	FG	538°C	1558	1128	19.5	24

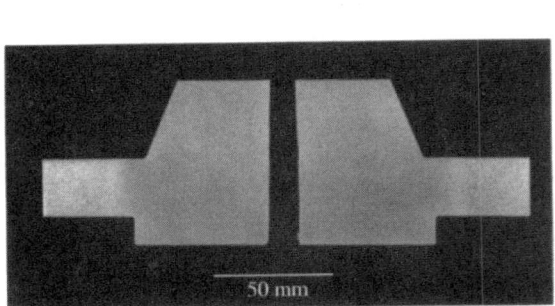
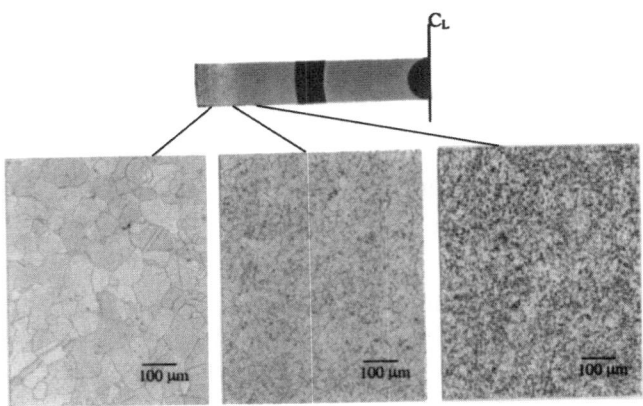

Figure 8: A closed die forging shape produced with SS-HIP processed material and thermomechanically processed to a dual microstructure.

Figure 9: A turbine wheel produced with SS-HIP processed material and thermomechanically processed to a dual microstructure.

Figure 10: Finish machined turbine wheel, P/M SS-HIP + press forged input stock

Conclusions

1. In UDIMET alloy 720, a large fraction of the PPB particles formed during powder atomization and HIP can be eliminated by HIP in the temperature range $T_{solidus} - T_{SSHIP} < 28°C$, well within HIP unit capabilities.

2. SS-HIP powder metal UDIMET alloy 720 can be converted to billet using conventional press conversion practices.

3. SS-HIP material has superior ultrasonic inspectability compared to cast and wrought material.

4. Conventional press conversion practices employed with SS-HIP processed materials allow for increased production flexibility and reduced billet cost.

5. Grain size response of the SS-HIP processed material can be controlled using a combination of forging and heat treatment processes.

6. Turbine disks were successfully produced by Ladish Co., Inc. for Rolls-Royce Allison from the SS-HIP processed billet.

Acknowledgements

The authors would like to thank Prof. John Radavich and Sam Mancuso for their helpful discussions on the analysis of the second phases found in our study. We would also like to thank Paul W. Keefe and Bruce A. Lindsley for their insights on the SS-HIP billet process.

References

1. G. E. Maurer, W. Castledine, F. A. Schweitzer, and S. Mancuso. "Development of HIP Consolidated P/M Superalloys for Conventional Forging to Gas Turbine Engine Components". In Superalloys 1996 (Kissinger et al., eds.), (Warrendale, PA), The Minerals, Metals and Materials Society, (1996), 645–652.

2. F. E Sczerzenie and G. E. Maurer, "Development of UDIMET 720 for High Strength Disk Applications", In Superalloys 1984 (M. Gell et al., eds.), (Warrendale, PA), The Minerals, Metals and Materials Society, (1984), 573–582.

3. K. R. Bain et al., "Development of Damage Tolerant Microstructures in UDIMET 720",In Superalloys 1984 (M. Gell et al., eds.) (Warrendale, PA), The Minerals, Metals and Materials Society, (1984), 13–22.

4. R. Chang, D. D. Krueger, and R. A. Sprague, "Superalloy Powder Processing, Properties and Turbine Disk Applications". In Superalloys 1984 (M. Gell et al., eds.), (Warrendale, PA), The Minerals, Metals and Materials Society, (1984), 245–273

5. R. Thamburaj, A. K. Kroul, W. Wallace, and M. C. deMalherbe. "Prior Particle Boundary Precipitation in P/M Superalloys". In 1984 International Powder Metallurgy Conference, Metal Powder Metallurgy, (1986), 635–674

6. G. A. J. Hack, J. W. Eggar and C. H. Symonds, "A comparaison of APK-1 Consolidated at HIP Temperature Above and Below the Boride Solvus". In Powder Metallurgy Superalloys Conference, Zurich, (1980), 20.1–20.50

7. A. Lazzaro, Bodycote IMT, personnel communication.

8. US Patent No 5,451,244. "High Strain Rate Deformation of Nickel-Base Superalloy Compact". W. Castledine. Special Metals Corporation.

9. Y. Wanhong et al., "Effect of Heat Treatment on Prior Particle Boundary Precipitation in Powder Metallurgy Nickel Base Superalloy". In Advanced Performance Materials, Vol. 2, (1995), 269–279.

10. D. U. Furrer, H. J. Fecht, "γ' Formation in Superalloy U720LI", Scripta Materialia, vol. 40, (1999), 1215–1220.

11. K. A. Green, J. A. Lemsky, R. M. Gasior, "Development of Isothermally Forged P/M Udimet 720 for Turbine Disk Applications". In Superalloys 1996 (Kissinger et al., eds.), (Warrendale, PA), The Minerals, Metals and Materials Society, (1996), 697–703.

EFFECT OF OXIDATION ON HIGH TEMPERATURE FATIGUE CRACK INITIATION AND SHORT CRACK GROWTH IN INCONEL 718

T. Connolley, M. J. Starink and P. A. S. Reed

Materials Research Group, School of Engineering Sciences
University of Southampton
Highfield, Southampton, SO17 1BJ, United Kingdom.

Abstract

Fatigue tests at 600°C in air were performed on U-notch specimens of wrought IN718, using 1-1-1-1 and 1-20-1-1 trapezoidal cycles. SEM examination and acetate replication was used to study crack initiation and growth. In support of the fatigue tests, thermal exposure experiments were performed on unstressed specimens to study surface and sub-surface primary carbide oxidation. Compared to a 1 second dwell, a dwell time of 20 seconds at maximum load had a beneficial effect on fatigue lifetime. Polishing the U-notch to remove broaching marks also had a beneficial effect on fatigue life. Multi-site crack initiation along the root of the U-notch was observed in fatigue specimens. Many of the cracks initiated at bulge-like features, associated with the oxidation of sub-surface primary carbides. Crack initiation was first observed as early as 12% of the total fatigue life, with further crack initiation occurring as tests progressed. Short cracks in the U-notches grew at a constant rate, except when interactions between parallel cracks resulted in crack arrest or coalescence. Significant crack coalescence occurred towards the end of the fatigue life, producing a dominant defect which propagated rapidly to failure. Surface eruptions and localised surface deformation were observed in fatigue and unstressed thermal exposure specimens, demonstrating a considerable volume expansion when primary $(Nb,Ti)C$ carbides oxidised. It is proposed that the misfit strains due to primary carbide oxidation were superimposed on the plastic strain field in the U-notch due to external loading. This would create local strains high enough to cause rupture of the matrix, hence initiating a fatigue crack.

Introduction

Inconel 718 (IN718) is a precipitation-strengthened nickel-based superalloy with many applications, such as components for gas turbines, liquid-fuelled rockets, high temperature plant and cryogenic systems. IN718 has good oxidation and creep resistance at service temperatures up to 650°C, combined with high strength and fatigue resistance. These properties mean IN718 is widely used for gas turbine discs, which experience long periods at high temperatures in an oxidising environment. Varying engine speeds generate cyclic loads, producing conditions in which low cycle fatigue (LCF) can occur. The high temperature fatigue properties of IN718 have been extensively studied [1], paying particular attention to the effect of mechanical variables [2, 3], microstructure [4, 5, 6] temperature and environment [4, 7, 8, 9] on the propagation of long fatigue cracks. Less is known about fatigue crack initiation and subsequent propagation when cracks are microstructurally and/or mechanically short. Use of long crack propagation data can lead to non-conservative estimates of component lifetimes [10, 11], so there is a continuing requirement for short crack initiation and growth data for use in life prediction models. There is also a need for further understanding of high temperature crack initiation and growth in notch stress fields, such as the fir tree root fixings in the rims of gas turbine discs.

It is generally accepted that high temperature crack initiation can be caused by oxidation of primary carbides at the surface of components, either by the creation of a micro-notch or the formation of oxide intrusions along grain boundaries.

Superalloys 2000
Edited by T.M. Pollock, R.D. Kissinger, R.R. Bowman,
K.A. Green, M. McLean, S. Olson, and J.J. Schirra
TMS (The Minerals, Metals & Materials Society), 2000

For superalloys in general, primary carbides are of the form "MC", where "M" is Ti, Nb, Ta, Hf, Th or Zr. For IN718, the main primary carbide formers are Nb and Ti. Enhanced crack initiation and growth due to preferential primary carbide oxidation has been observed in cast superalloys such as MAR-M 509 [12, 13] and IN100 [14, 15]. In cast alloys, primary carbides often have a script-like morphology, which react to form oxide intrusions from which fatigue cracks can grow. In wrought IN718, primary carbides usually have a globular morphology and the mechanism of fatigue crack initiation by carbide oxidation is likely to be different. In a study of notch-rupture sensitivity of IN718 [16], volume expansion of oxidising primary carbides was identified as a mechanism of environmentally assisted crack propagation, due to the stress intensifying effect of expanding particles at or near crack tips. In the same study, surface eruptions of oxidised phases were seen on polished specimens exposed for 1 hour at 650°C [16].

Primary carbide oxidation has been proposed as a mechanism of environmentally-assisted creep crack growth, due to the release of niobium onto grain boundaries, forming an embrittling Nb_2O_5 niobium oxide phase [17]. Recently, eruption-like features caused by oxidation of primary carbide particles were frequently observed at fatigue crack initiation sites in U-notch specimens machined from an extruded IN718 bar [18]. Further work provided evidence that volume expansion of oxidised primary carbides caused fatigue cracks to initiate U-notch specimens of IN718 at 600°C [19].

Published information on high temperature crack growth in IN718 for blunt notch geometries is limited. Bache et al. [20] studied the growth of pre-initiated semi-elliptical cracks of initial surface length $2c \approx 1$ mm growing in a U-notch with a stress concentration factor 2.23. Very rapid, accelerating crack propagation rates were observed at 600°C in air, producing critical crack lengths within 2000 cycles. Assuming production turbine discs do not contain millimetre-size defects prior to entering service, it is desirable to obtain information on the natural initiation of cracks and their growth rates in the sub-millimetre size range from 10-1000 μm. This information, combined with data available for longer cracks, will enable more accurate prediction of turbine disc fatigue lives.

In this paper, results of observations of high temperature crack initiation and growth in a U-notch geometry are presented and discussed for IN718 specimens taken from a turbine disc forging. The results also provide further information about the role of primary carbide oxidation in fatigue crack initiation.

Material

Material used for this study was taken from an Inconel 718 turbine disc forging, the composition of which is given in Table I. The forging was solution treated at 955°C for 1 hour followed by an air cool. It was then aged at 720°C for 8 hours, cooled at 50°C/h to 620°C, then aged at 650°C for a further 8 hours, finishing with an air cool.

The disc material had a non-equiaxed grain structure, with larger elongated grains (diameter approximately 30-70 μm) surrounded by extensive regions of small, more equiaxed grains (diameter approximately 5-20 μm). Grain boundaries were extensively decorated with δ phase particles of various morphologies, ranging

from globular to almost continuous films. An SEM micrograph of the typical microstructure is shown in Figure 1. Primary (Nb,Ti)C carbides exhibited a globular morphology and were distributed in clusters throughout the microstructure. Some primary (Ti,Nb)N nitrides were also present. The mean area fraction of primary carbides and nitrides determined by image analysis on polished microsections was 0.48%. Mean Vickers hardness (30kg) measured on specimens extracted from the disc was 444 ± 11 H$_v$.

Table I: Chemical Composition of IN718 Material Studied.

Element	wt%
B	0.003
C	0.031
Mg	0.002
Al	0.46
Si	0.08
P	0.01
S	0.0004
Ti	1.02
Cr	18.0
Mn	0.06
Fe	18.6
Co	0.43
Ni	52.9
Cu	0.10
Mo	3.02
Nb + Ta	5.22

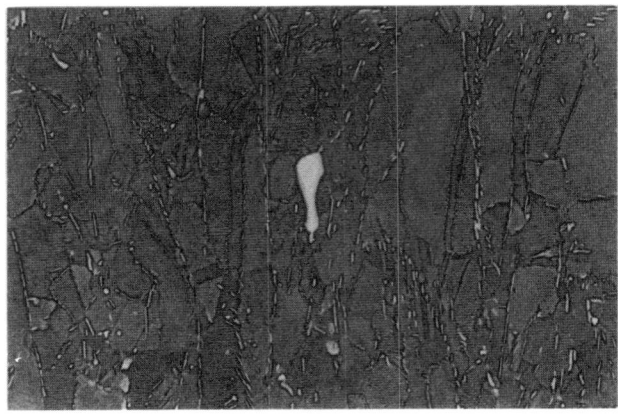

■ 10μm

Figure 1: Backscattered electron image of typical microstructure. The bright feature at the centre of the image is an (Nb,Ti)C primary carbide.

Experimental Procedures

Fatigue tests were performed on U-notch specimens extracted from the disc forging with their long axes tangential to the circumference and the notch axis parallel to the disc axis, as shown in Figure 2. The geometry of the U-notch specimen, shown in Figure 3, was designed to produce a stress concentration factor of 2.23. Notches were machined using a broaching process similar to that used to produce blade root fixings in the rims of production discs. The broaching process left long, straight marks running parallel to the notch axis.

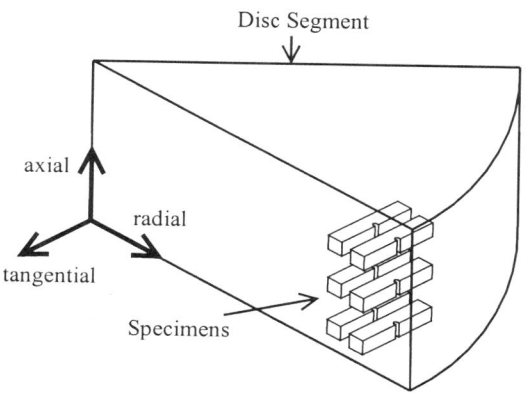

Figure 2: Orientation of U-notch specimens relative to disc forging.

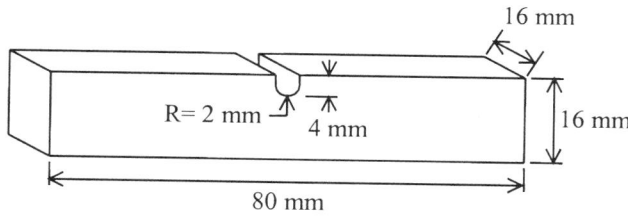

Figure 3 : U-notch specimen geometry.

Fatigue testing was conducted inside an ESH high temperature chamber mounted on an Instron 8501 servo-hydraulic testing frame. All tests were performed at 600 +/- 2°C in an air environment. Specimens were heated using four high intensity quartz lamps and the temperature was monitored and controlled via an R-type thermocouple attached to the front face of each specimen. Specimens were loaded in three point bend so that the maximum net section stress σ_{max} below the notch was 750 MPa, with an R-ratio of 0.1. Tests were conducted under load control using 1-1-1-1 or 1-20-1-1 trapezoidal waveforms (see Figure 4).

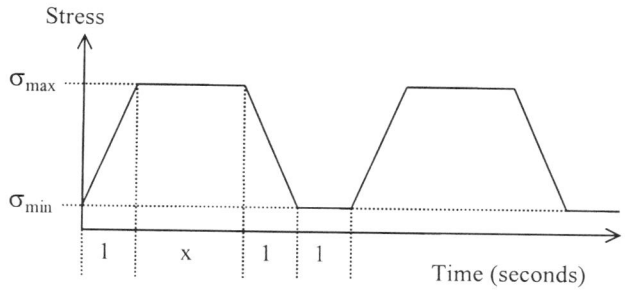

Figure 4: Trapezoidal waveform used for high temperature fatigue testing. x = dwell time at maximum load: either 1 or 20 seconds.

Three fatigue tests are described in detail in this paper. Two of the tests were performed on specimens in the as-broached condition, using trapezoidal 1-1-1-1 and 1-20-1-1 waveforms to study the effect of dwell time at maximum load on fatigue behaviour. The tests were interrupted periodically to enable scanning electron microscope (SEM) examination of the U-notch surfaces. A third

1-1-1-1 test was performed on a specimen with a polished U-notch, taking cellulose acetate replicas every 2000 cycles so that crack initiation and growth behaviour could be studied in more detail. Prior replication trials demonstrated that it was difficult to reliably distinguish between cracks and broaching marks in the U-notch. Therefore, the broaching marks were removed by polishing the notch to a 1μm diamond finish. To minimise enlargement of the notch, which can occur during manual polishing, the specimen was clamped in a specially-designed jig and polished using soft 4mm x 8mm cylindrical dental felts mounted in a pillar drill.

SEM examination of the specimens during interrupted tests and after fracture was used to identify fatigue crack initiation sites and to investigate carbide oxidation. Energy-dispersive X-ray spectroscopy (EDX) was used to determine the chemical composition of features found at crack initiation sites.

In addition to the fatigue tests, thermal exposure experiments in air at 550°C were carried out on flat, unstressed samples of the IN718 disc material. Prior to exposure, the surface of each thermal exposure sample was polished to a 1μm diamond finish. Sample exposure times ranged from 1 to 256 hours. After exposure, specimens were examined using an SEM to look for signs of carbide oxidation on the surface. Cross-sections from selected specimens were nickel plated and prepared for metallographic observation to look for further evidence of surface and sub-surface carbide oxidation. The flat polished surface was used to provide a reference against which any deformation associated with sub-surface carbide oxidation could be checked.

Results

U-notch Specimen Lifetimes

Total fatigue lifetimes for the U-notch specimens are presented in Table II.

Table II : Specimen Lifetimes

Notch Surface Condition	Waveform	Lifetime (cycles)
As-broached	1-1-1-1	25311
As-broached	1-20-1-1	50756
Polished	1-1-1-1	51279

The introduction of a 20 second dwell at maximum load resulted in almost double the fatigue life of the baseline test with 1 second dwell. (At the time of writing, a specimen with a 30 second dwell at maximum load had also exceeded double the life of the 1 second dwell test). This trend of increasing lifetime with increasing dwell is similar to that observed by the authors in their work on U-notch specimens extracted from extruded IN718 bar [18, 19]. Polishing the notch to remove the broaching marks resulted in a significant increase in fatigue life for a 1-1-1-1 cycle at 600°C. This is similar to previously published observations for as-broached and polished U-notch specimens tested using a 10 Hz sawtooth waveform in air at 600°C [18]. SEM observations of the interrupted as-broached tests and examination of replicas taken from the polished notch showed that crack initiation occurred within the first 12-20% of the overall fatigue life. After initiation, a significant proportion of the fatigue life was taken up by growth of numerous short cracks, with further crack initiation occurring as the tests progressed. Eventually, coalescence of cracks produced a dominant defect which propagated rapidly, assisted by further crack coalescence, until specimen failure occurred. Crack initiation and growth behaviour is described in more detail in the following sections.

Short Crack Initiation

Multi-site crack initiation along the root of the U-notch was seen in all fatigue tests. Three types of crack initiation site were observed: (i) sites with split, bulge-like features on the notch surface; (ii) sites where oxidised primary carbides had erupted out of the notch surface and (iii) sites with no distinct surface feature. An example of a crack initiation site in the notch root is shown in Figure 5.

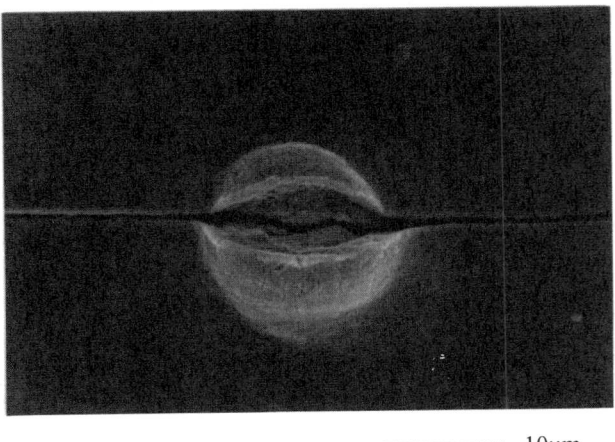

_____ 10μm

Figure 5 : SEM image in plan view of crack initiated at a bulge-like feature.

_____ 10μm

Figure 6 : Post-fracture SEM image of main fracture surface, showing bulge and associated sub-surface particle at a crack initiation site.

During post-fracture examination, it was found that the bulge-like initiation sites had sub-surface particles associated with them, for example Figure 6. EDX analysis of the particles below the bulges confirmed they were oxidised primary carbides.

The replica record from the 1-1-1-1 polished notch test confirmed the early onset of crack initiation at bulge–like features. Cracks were first observed after 6000 cycles, which is within the first 12% of the overall fatigue life of 51279 cycles. Figure 7 shows the variation in the number of cracks observed as the test progressed. In the first half of the test, the number of surface cracks in the U-notch increased linearly with the number of cycles. Between 10000 and 30000 cycles the rate of nucleation of cracks from bulge-like features was calculated by linear regression to be 0.003 cracks/cycle with a correlation coefficient 0.98. After 30000 cycles the number of cracks observed approached a plateau. Beyond 44000 cycles, substantial crack coalescence led to an overall decrease in the number of cracks observed.

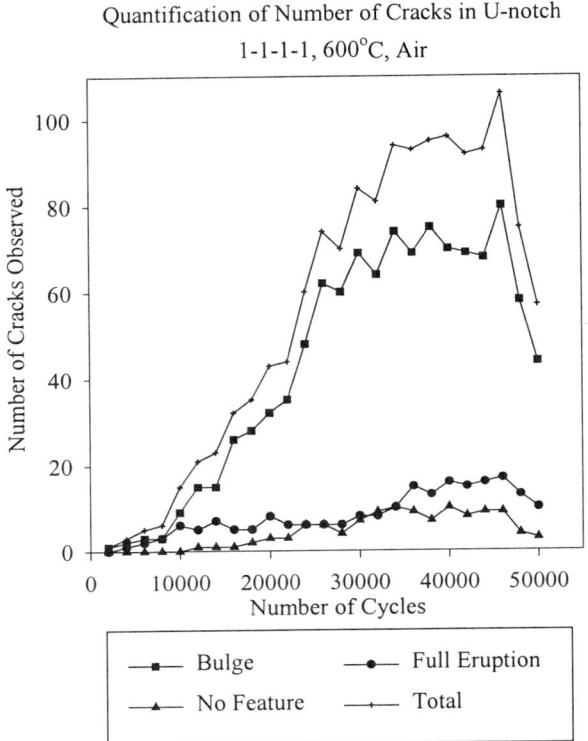

Figure 7 : Variation in number of surface cracks observed on replicas. 1-1-1-1, 600°C, polished U-notch.

Not all of the cracks which initiated in notch roots coalesced to form the dominant final through-thickness defect. Results of post-fracture SEM quantification of cracks which formed the through-thickness defect are listed in Table III. For the 1-1-1-1 test with a polished U-notch, 38 short cracks were identified which contributed to the final defect, whereas the replica record showed that over 100 cracks in total had initiated in the notch root.

438

Table III : Quantification of cracks which coalesced to form the final through-thickness defect in U-notch specimens.

Waveform & Notch Surface Condition	Initiation Site Type			Total no. of cracks
	Bulge	Full eruption	No feature	
1-1-1-1 As-broached	12	0	14	26
1-20-1-1 As-broached	16	3	9	28
1-1-1-1 Polished	20	7	11	38

The data in Table III provide further evidence of the importance of oxidised carbides as crack initiation sites, particularly the sub-surface particles associated with bulges in the notch surface.

Short Crack Growth

The convention used for describing the length of cracks observed is shown in Figure 8. Surface cracks were defined as having a total length $2c$, divided into two segments c_1 and c_2. Segment c_1 was the distance from the first crack tip to the middle of the initiation site and c_2 the distance from the second crack tip to the middle of the initiation site, so that $2c = c_1 + c_2$.

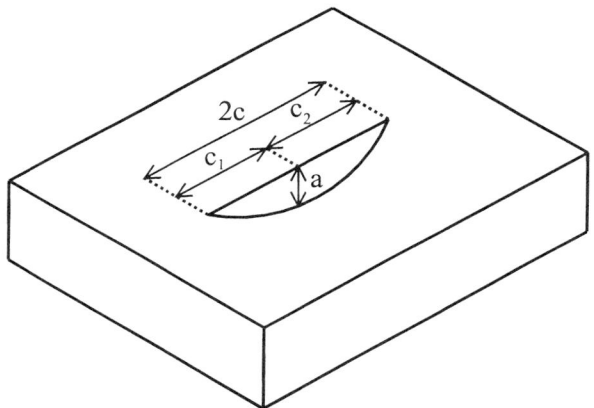

Figure 8: Convention used for describing crack lengths.

Crack length vs. cycle data for four typical surface cracks are presented in Figure 9. These four cracks were chosen because they initiated early in the fatigue life and exhibited behaviour typical of other cracks observed.

Crack I was first observed after 6000 cycles, in a large, shallow bulge-like feature. Its initial growth rate was slow, even arresting at one point, but then the growth rate accelerated until it reached an approximately constant value. Later in the test Crack I became asymmetric about its initiation point, with segment c_1 growing more slowly than c_2. This retardation of segment c_2 probably occurred due to shielding by other cracks growing parallel to Crack I.

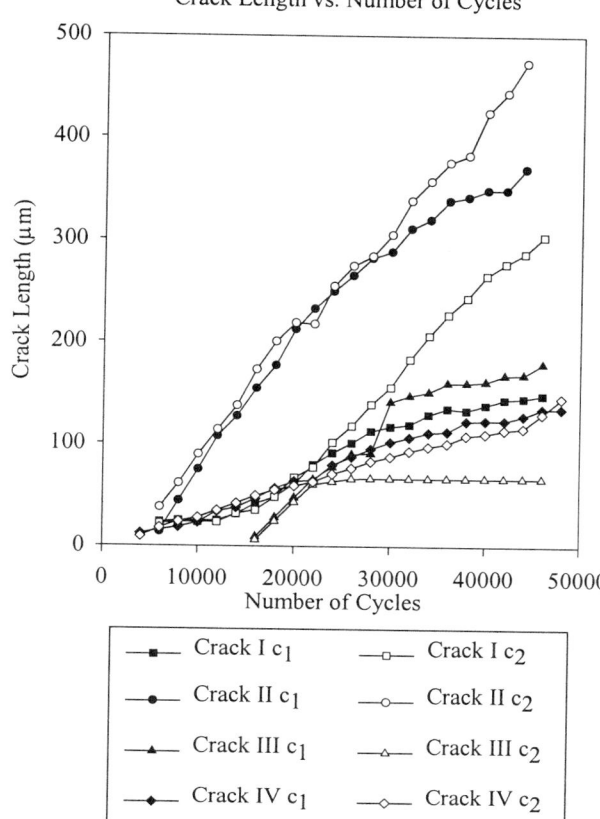

Crack Length vs. Number of Cycles

—■— Crack I c_1		—□— Crack I c_2	
—●— Crack II c_1		—○— Crack II c_2	
—▲— Crack III c_1		—△— Crack III c_2	
—◆— Crack IV c_1		—◇— Crack IV c_2	

Figure 9: Crack length vs. number of fatigue cycles for four typical surface cracks. Waveform 1-1-1-1, 600°C, air environment.

Crack II initiated at a small surface eruption after 6000 cycles and was the fastest growing of all the cracks observed, averaging $d(2c)/dN = 2.1 \times 10^{-8}$ m/cycle. Crack II reached a length of $2c = 840$ μm at 44000 cycles, after which coalescence with other cracks occurred and a dominant through-thickness crack formed. Like Crack I, Crack II became asymmetric about its origin later in the test, due to shielding of segment c_1 by other parallel cracks and possible interaction of their crack tip stress fields.

Crack III was first observed after 16000 cycles and initiated at a bulge-like feature. The development of this bulge can be seen in the sequence of images in Figure 10. Segment c_1 of crack III displayed a jump in length after 28000 cycles due to coalescence with a smaller crack directly in its path. The growth of crack III was suppressed at longer crack lengths, probably due to interaction with the crack tip stress fields of other cracks, with which crack III eventually coalesced. Segment c_2 of Crack III completely arrested until it coalesced with another crack after 48000 cycles.

Crack IV was first observed at a bulge-like feature after 4000 cycles and grew until coalescence with other cracks after 48000 cycles.

439

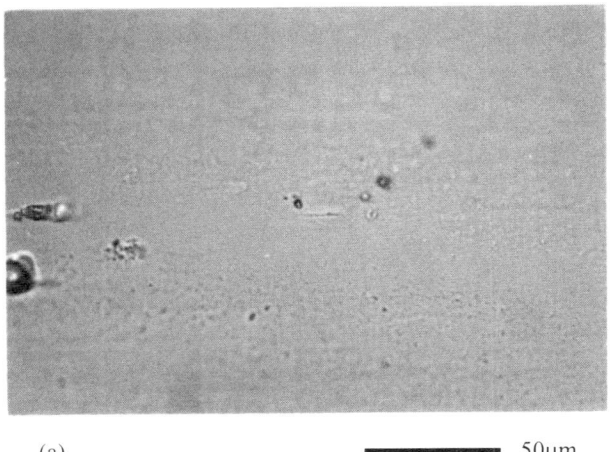

(a)   ▬▬▬▬▬ 50μm

(b)   ▬▬▬▬▬ 50μm

(c)   ▬▬▬▬▬ 50μm

Figure 10 : Sequence of images from acetate replicas showing development of bulge at a crack initiation site. (a) 16000 cycles; (b) 18000 cycles; (c) 20000 cycles.

The growth rates of total surface crack lengths, d(2c)/dN are plotted in Figure 11 for cracks I to IV. The cracks appear to have approximately constant growth rates across a wide range of crack lengths (ignoring crack arrest events). This is in contrast to the usual trend of increasing growth rate with increasing crack length observed for long crack tests [1-9]. The surface cracks in the U-notches were all remarkably straight, with no deflections except those occurring during crack coalescence. It therefore appeared that surface crack propagation was transgranular, though SEM fractography indicated that propagation in the "a" direction into the bulk of the specimen became mostly intergranular beyond a distance of 30-50 μm below the notch surface, shown in Figure 12.

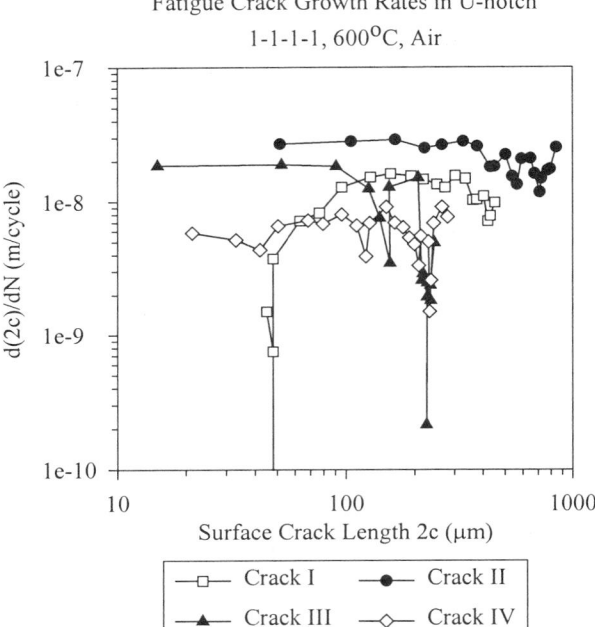

Figure 11: Growth rates of typical surface cracks in a U-notch. Waveform 1-1-1-1, 600°C, air environment.

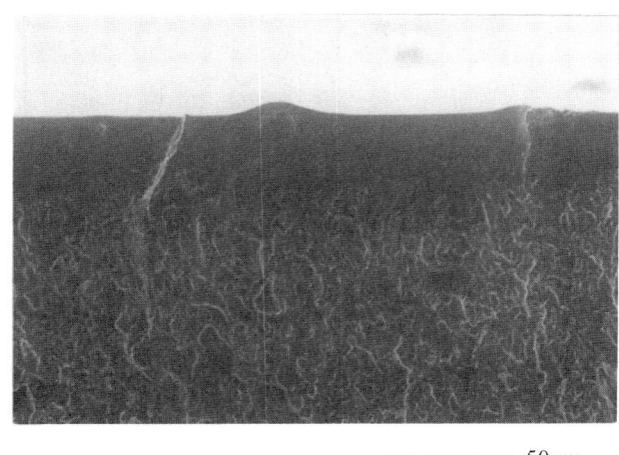

▬▬▬▬▬ 50μm

Figure 12: SEM image of fracture surface showing transition from very flat, transgranular fracture to intergranular fracture. Note the bulge at the crack initiation site.

440

Carbide Oxidation in Thermal Exposure Specimens

Eruptions due to primary carbide oxidation were observed by SEM on the surfaces of all thermal exposure specimens, confirming the observations made by Sjoberg et al. [16]. When the specimens were tilted to view the eruptions from the side, many of the eruptions were seen to have a remarkable tower-like morphology (Figure 13). Other eruptions were more irregular in shape (Figure 14). Optical and SEM microscopy on cross-sections through the thermal exposure specimens showed that some sub-surface oxidation of primary carbides had occurred, with associated bulges in the surface above (Figure 15). These bulges were similar to those observed at crack initiation sites in the U-notch fatigue specimens (Figure 6). Matrix deformation was also observed in the vicinity of some of the oxidised carbides which had erupted onto the surface. EDX confirmed that the surface eruptions and sub-surface oxidised particles were oxidised primary carbides.

Further optical microscopy and SEM performed on cross-sections through the thermal exposure specimens showed that in most cases, only carbide particles intersecting the surface had oxidised, even for the longest exposure time of 256 hours. Occasionally, sub-surface oxidised carbides like those shown in Figure 15 were observed, while other primary carbides at the same depth below the surface (2-3μm) were unoxidised. This suggests that a preferential diffusion path, such as a grain boundary, was required to produce sub-surface carbide oxidation.

——— 5μm

Figure 14: SEM image of irregular morphology of oxidised primary carbide erupting from the polished surface of a thermal exposure specimen. 16 hours exposure at 550°C.

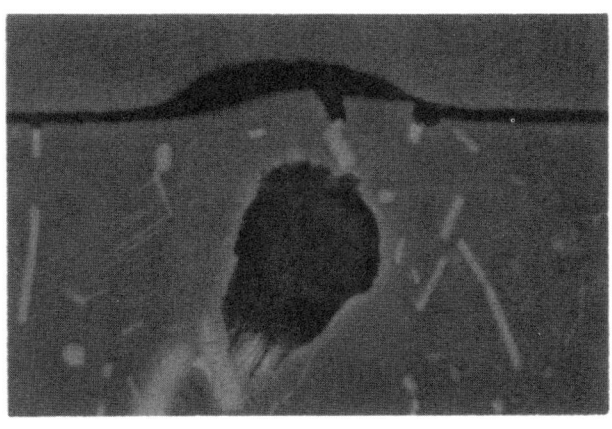

——— 5μm

Figure 15: SEM backscatter image of sub-surface oxidised primary carbide in a thermal exposure specimen, showing associated local deformation of surface. 128 hours exposure at 550°C.

——— 5μm

Figure 13: SEM image of tower-like morphology of oxidised primary carbide erupting from the polished surface of a thermal exposure specimen. 16 hours exposure at 550°C.

Discussion

Short Crack Initiation

The results of the interrupted tests on U-notch specimens provide some insights into the role of primary carbides in fatigue crack initiation in IN718. Crack initiation commenced early in the fatigue life, with the majority of cracks observed on replicas initiating at bulge-like features. At first, crack initiation from bulge-like features occurred at a constant rate, before reaching a plateau which can be attributed to: (i) exhaustion of potential crack initiation sites; (ii) reduction in the local strain at initiation sites due to shielding by neighbouring cracks and (iii) crack coalescence. Post-fracture examination confirmed that a significant number of short cracks forming the main fracture initiated at bulge-like features with oxidised primary carbides below them.

The surface bulges associated with oxidised sub-surface primary carbides in the fatigue specimens and thermal exposure specimens are evidence that plastic deformation of the surrounding matrix occurred during particle oxidation. In order to produce the observed bulges, the mismatch strains generated by carbide oxidation must be sufficiently high to cause matrix deformation. Pure NbC is known to oxidise rapidly between 490°C and 600°C to form Nb_2O_5 [21]. Estimating the volume expansion for the transformation of NbC to Nb_2O_5 in the temperature range 500-600°C is problematic because Nb_2O_5 is polymorphic, and has been synthesised and observed in several different crystalline and amorphous forms [22]. It is not known what form is adopted by the oxide eruptions observed in IN718, though a crystalline oxide form is considered to be more likely. Using available crystallographic data [22, 23], a maximum volume expansion factor of 2.28 was estimated for the transformation of NbC to Nb_2O_5. This estimate appears reasonable given the amount of volume expansion observed experimentally. Analytical solutions for the misfit stresses around a spherical particle in an infinite matrix [24, 25] were used to estimate the plastic zone size around an NbC particle which had transformed completely to Nb_2O_5. Using estimated mechanical property data for IN718, NbC and Nb_2O_5 at 600°C, it was calculated that volume expansion by a factor of 2.28 would produce a maximum plastic zone radius of 4.2 times the particle radius. Although this simple estimate is for an infinite matrix and does not consider the situation of a particle close to the surface, the result demonstrates the potential for matrix deformation during primary carbide oxidation, as observed experimentally. Another phenomenon to consider is the evolution of CO or CO_2 gas during carbide oxidation, but it is not known at this stage whether gas entrapment was a significant factor in bulge formation.

In the U-notch fatigue specimens, the misfit strains due to carbide oxidation were superimposed on the strain field due to loading. This may have created local strains high enough to cause rupture of the matrix in the vicinity of an oxidising particle, hence initiating a fatigue crack. The high incidence of bulges at crack initiation sites suggests that expansion of oxidising primary carbides was a significant mechanism of fatigue crack initiation in the notch root. The stress concentration effect of the notch root probably promoted stress-assisted oxygen diffusion, causing more rapid oxidation of the primary carbides than that observed for the unstressed thermal exposure specimens. Once a crack had initiated, the crack itself would provide another fast oxygen diffusion route to a sub-surface particle.

The total number of cracks initiated by oxidation of primary carbides and their initiation rate will be controlled by several factors. The number of primary carbides which are potential crack initiation sites will depend upon the number density of primary carbides in the material, their distribution relative to the notch surface and their distribution relative to preferential oxygen diffusion pathways such as grain boundaries. The rate of crack nucleation will depend upon the rate of primary carbide oxidation, which in turn is affected by the depth of individual carbides below the surface and the rate of oxygen diffusion to them. In addition, the plastic strain in the notch root due to external loading is not uniform, so that the local matrix strain around an oxidising particle will vary according to its position relative to the notch axis.

Short Crack Propagation

Preliminary results for high temperature short crack growth in U-notches indicate that the majority of the specimen life was taken up by growth and interaction of numerous short fatigue cracks, until coalescence of several cracks generated a dominant defect which propagated rapidly to failure. Following the discussion about the factors affecting crack initiation in the previous section, it is likely that the fatigue lifetimes of specimens tested under the same conditions will vary, depending upon: (i) the number of short cracks which initiate; (ii) the rate of crack initiation; (iii) the relative positions of the short cracks relative to the notch axis and to each other; (iv) the rate of crack growth and (v) the degree to which cracks interact with one another via arrest or coalescence events. Hence, a life prediction model for U-notch geometries incorporating short crack data may have to take into account microstructural factors such as primary carbide distribution, because of its effect on the number and relative positions of cracks initiated from carbides.

The remarkable straightness of surface cracks in the U-notches and their apparent transgranular path near the surface was probably a consequence of the high plastic strain field caused by the stress concentration effect of the notch. At the maximum stress of 750 MPa, the theoretical elastic stress in the notch root would be 1672.5 MPa, which is well above the yield stress and ultimate tensile strength of IN718 at 600°C. In reality, plastic yielding of material around the notch occurred during testing, so that the surface cracks grew in a situation where they were mechanically short relative to the size of the notch plastic zone. It is thought that the high level of notch plasticity resulted in transgranular Stage II type crack growth at the surface, with a fairly constant fatigue crack growth rate. This implies that the local crack tip driving force was constant for short cracks growing in isolation, until it was reduced for longer crack lengths by crack shielding and interactions between the crack-tip stress fields of neighbouring cracks.

Effect of Hold Time and Surface Condition

Tests performed on U-notch specimens at 600°C in air show a beneficial effect of dwell times at maximum load on overall fatigue life. A 1-20-1-1 cycle increased the fatigue life by a factor of 2 compared to a 1-1-1-1 cycle. This is the same trend as that observed previously for U-notch specimens taken from an extruded bar of IN718 [18]. Crack initiation occurred within the first 20% of the overall fatigue life, so that a significant proportion of life was taken up by short crack growth and coalescence. A speculative explanation for the increased lifetime is that the dwell at maximum load retarded the growth of short cracks in the notch root, rather than affecting time to crack initiation. However, this cannot be confirmed until short crack growth rate measurements are made on U-notch specimens with a dwell at maximum load. The retardation mechanism may be stress relaxation at the crack tips in the high plastic strain field of the notch, or possibly oxide-induced crack closure.

A beneficial effect on fatigue life of polishing out broaching marks was observed, consistent with previously published results on uninterrupted tests [18]. Surface roughness is known to influence fatigue crack initiation through the creation of local stress concentrators, but the surface replica results show that even for a polished U-notch, crack initiation readily occurs.

442

Alternatively, the broaching marks parallel to the U-notch axis could have influenced crack propagation. Residual stresses effects introduced by the polishing process are considered to be unlikely, since they will be negligible compared to the plastic stress-strain field generated by the notch.

Morphology of Erupted Carbides

Volume expansion of oxidised primary carbides has already been discussed in relation to crack initiation. Volume expansion can also explain the formation of the various eruptions seen on the polished thermal exposure specimens, shown schematically in Figure 16.

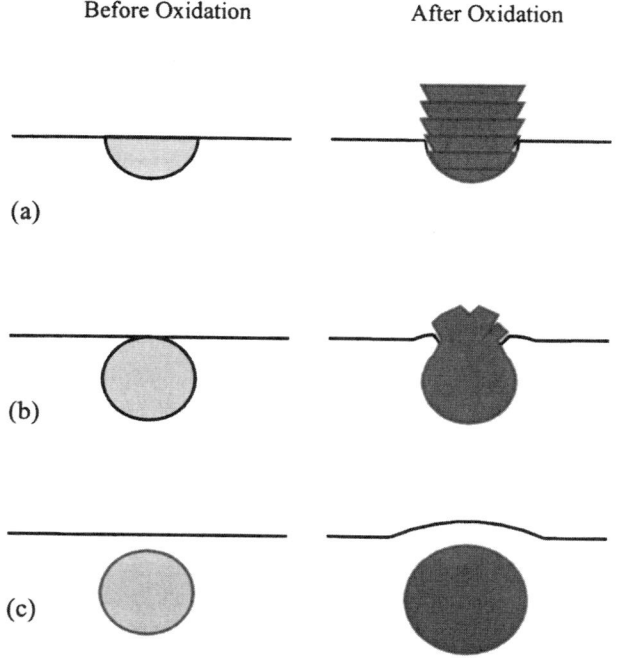

Figure 16: Schematic diagram of different primary carbide oxidation scenarios. (a) Exposed carbide on polished surface; (b) Carbide just intersecting surface; (c) Sub-surface carbide causing matrix deformation.

The layered, tower-like eruptions observed on polished surfaces are possibly the result of a lamellar oxidation process. The flat surface of a carbide exposed on the polished surface is easily oxidised. Volume expansion due to oxidation raises the oxidised material proud of the surface. The oxide/carbide interface created then provides a preferential oxygen diffusion path for further oxidation. Another layer of oxide forms and expands, forming another oxide/carbide interface. This process continues, building the observed layered eruption (Figure 16a). Alternatively, the layered appearance could be crystallographic in nature, because the primary carbides have a cubic crystal structure, while the various forms of Nb_2O_5 are reported to be orthorhombic or monoclinic [22]. For carbides only just exposed at the surface, constraint by the matrix prevents free expansion of the oxide, resulting in a more irregular surface eruption and some matrix deformation (Figure 16b). This situation is also thought to apply for surface eruptions observed on rougher, machined surfaces like a broached U-notch. The third scenario is where a sub-surface carbide oxidises, and mismatch stresses due to the volume change between carbide and oxide are sufficiently high to cause matrix deformation (Figure 16c).

Conclusions

Fatigue crack initiation and growth in U-notch specimens taken from an IN718 turbine disc forging was studied at 600°C in air, using 1-1-1-1 and 1-20-1-1 cycles. SEM examination and cellulose acetate replication enabled detailed study of crack initiation and subsequent crack growth. Thermal exposure experiments on unstressed specimens provided further information about the role of primary carbide oxidation in crack initiation.

SEM observations on interrupted tests in air at 600°C and an acetate replica study showed that crack initiation in U-notch specimens occurred early within the overall fatigue life, so that a significant proportion of life was taken up by short crack growth and coalescence. Short crack growth rates in a U-notch specimen tested using a 1-1-1-1 loading cycle at 600°C in air tended to be constant, except when crack-crack interactions resulted in arrest or coalescence. This has implications for life predictions of U-notch specimens, since the overall life will vary according to the distribution of crack initiation sites such as primary carbides, the rate at which cracks nucleate and the rate at which they grow and coalesce to form a dominant defect.

Compared to a 1 second dwell, a dwell time of 20 seconds at maximum load had a beneficial effect on the fatigue lifetimes of U-notch specimens. Polishing the U-notch to remove broaching marks also had a beneficial effect on fatigue life under the test conditions used.

Oxidation of (Nb,Ti)C primary carbides in IN718 readily occurred at temperatures of 550-600°C. The volume expansion due to oxidation was sufficient to cause substantial plastic deformation of the surrounding matrix. Surface eruptions of oxidised material were observed in thermal exposure and fatigue specimens.

For crack initiation in U-notch specimens of IN718, it is proposed that the misfit strains due to primary carbide oxidation were superimposed on the plastic strain field in the U-notch due to external loading. This would create local strains high enough to cause rupture of the matrix, hence initiating a fatigue crack. Examination of crack initiation sites in U-notch fatigue specimens confirmed that localised matrix deformation due to sub-surface carbide oxidation was a significant mechanism of fatigue crack initiation in the notch root.

Acknowledgements

This work was funded by the United Kingdom Engineering and Physical Sciences Research Council and ABB-ALSTOM Power Generation UK Limited. The authors wish to thank Dr. S. J. Moss and M. Hughes of ABB-ALSTOM Power Generation for valuable technical discussions and provision of the IN718 material. We also wish to acknowledge the use of the EPSRC's Chemical Database Service at Daresbury.

443

References

[1] L. A. James "Fatigue Crack Propagation in Alloy 718: A Review" Proc. Conf. Superalloy 718 - Metallurgy & Applications, Ed. E. A. Loria, TMS, Warrendale PA, USA (1989), 499-515

[2] M. Clavel & A. Pineau "Frequency and Wave-Form Effects on the Fatigue Crack Growth Behavior of Alloy 718 at 298K and 823K" Metall. Trans. A 9 (1978), 471-480

[3] H. Ghonem, T. Nicholas & A. Pineau "Elevated Temperature Fatigue Crack Growth in Alloy 718-Part I: Effects of Mechanical Variables" Fatigue Fract. Eng. Mater. Struct. 16 (1993), 565-576

[4] J. P. Pedron & A. Pineau "The Effect of Microstructure and Environment on the Crack Growth Behaviour of Inconel 718 Alloy at 650°C Under Fatigue, Creep and Combined Loading" Mater. Sci. & Eng. 56 (1982), 143-156

[5] L. A. James "The Effect of Grain Size upon the Fatigue-Crack Propagation Behavior of Alloy 718 under Hold-Time Cycling at Elevated Temperature" Eng. Fract. Mechanics 25 (1986), 305-314

[6] S. P. Lynch et al. "Fatigue Crack Growth in Nickel-Based Superalloys at 500-700°C. II: Direct-Aged Alloy 718" Fatigue Fract. Eng. Mater. Struct. 17 (1994), 313-325

[7] S. Floreen & R. H. Kane "An Investigation of the Creep-Fatigue-Environment Interaction in a Ni-Base Superalloy" Fatigue Eng. Mater. Struct. 2 (1980), 401-412

[8] H. H. Smith & D. J. Michel "Effect of Environment on Fatigue Crack Propagation Behavior of Alloy 718 at Elevated Temperatures" Metall. Trans. A 17 (1986), 370-374

[9] H. Ghonem, T. Nicholas & A. Pineau "Elevated Temperature Fatigue Crack Growth in Alloy 718 - Part II: Effects of Environmental and Material Variables" Fatigue Fract. Eng. Mater. Struct. 16 (1993), 577-590

[10] K. J. Miller "The Short Crack Problem" Fat. Eng. Mater. Struct. 5 (1982), 223-232

[11] J. Lankford J. & S. J. Hudak (Jr.) "Relevance of the Small Crack Problem to Lifetime Prediction in Gas Turbines" Int. J. Fatigue 9 (1987), 87-93

[12] J. Reuchet & L. Remy "High Temperature Low Cycle Fatigue of MAR-M 509 Superalloy I: The Influence of Temperature on the Low Cycle Fatigue Behaviour from 20 to 1100°C" Mater. Sci. & Eng. 58 (1983), 19-32

[13] J.Reuchet & L. Remy "High Temperature Low Cycle Fatigue of MAR-M 509 Superalloy II: The Influence of Oxidation at High Temperatures" Mater. Sci. & Eng. 58 (1983), 33-42

[14] M. Reger & L. Remy "High temperature Low Cycle Fatigue of IN 100 Superalloy I: Influence of Temperature on the Low Cycle Fatigue Behaviour" Mater. Sci. & Eng. A 101 (1988), 47-54

[15] M. Reger & L. Remy "High Temperature Low Cycle Fatigue of IN 100 Superalloy II: Influence of Frequency and Environment at High Temperatures" Mater. Sci. & Eng. A 101 (1988), 55-63

[16] G. Sjoberg, N-G. Ingesten & R. G. Carlson "Grain Boundary δ Phase Morphologies, Carbides and Notch Rupture Sensitivity of Cast Alloy 718" Proc. 2nd Int. Symp. Superalloy 718, 625 & Various Derivatives Ed. E. A. Loria, TMS, Warrendale PA, USA (1991), 603-620

[17] M. Gao, D. J. Dwyer & R. P. Wei "Niobium Enrichment and Environmental Enhancement of Creep Crack Growth in Nickel-Base Superalloys" Scr. Metall. & Mater. 32 (1995), 1169-1174

[18] P.A.S. Reed et al. "Creep-Fatigue Initiation and Early Crack Growth in Inconel 718" Proc. 8th Int. Conf. Progress in Mechanical Behaviour of Materials (ICM8), Eds. F. Ellyin & J. W. Provan, Fleming Printing Ltd. Victoria, BC, Canada, vol. 1 (1999), 418-423

[19] T. Connolley, P. A. S. Reed & M. J. Starink "A Study of the Role of (Nb,Ti)C Carbides in Fatigue Crack Initiation in Inconel 718 at 20°C and 600°C" (To be published in Proc. 5th International Charles Parsons Turbine Conference, Cambridge UK, 3rd – 7th July 2000)

[20] M. R. Bache, W. J. Evans & M. C. Hardy "The Effects of Environment and Loading Waveform on Fatigue Crack Growth in Inconel 718" Supplement to Int. J. Fatigue 21 (1999), S69-S77

[21] S. Shimada & M. Inagaki "A Kinetic Study on Oxidation of Niobium Carbide" Solid State Ionics 63-65 (1993), 312-317

[22] H. Schäfer, R. Gruehn & F. Schulte "The Modifications of Niobium Pentoxide" Angew. Chem. Internat. Edit. 5 (1966), 40-52

[23] D. A. Fletcher, R. F. McMeeking & D. Parkin "The United Kingdom Chemical Database Service" J. Chem. Inf. Comput. Sci. 36 (1996), 746-749

[24] J. K. Lee et al. "Plastic Relaxation of the Transformation Strain Energy of a Misfitting Spherical Precipitate: Ideal Plastic Behavior" Metall. Trans. A 11 (1980), 1837-1847

[25] M.J. Starink, P. Van Mourik & B. M. Korevaar "Misfit Accommodation in a Quenched and Aged Al-Cu Alloy with Silicon Particles" Metall. Trans. A 24 (1993), 1723-1731

THE EFFECTS OF PROCESSING ON STABILITY OF ALLOY 718

G. Shen*, J. Radavich**, X. Xie***, and B. Lindsley****

*Ladish Co., Inc. 5481 S. Packard Ave., Cudahy, WI 53110-8902, USA
** Micro-Met Laboratories, Inc., 209 North Street, West Lafayette, IN 47906, USA
*** University of Science & Technology Beijing, Beijing 100083, China
****Special Metals Corporation, New Hartford, NY 13413, USA

Abstract

A program was undertaken to (1) investigate the degradation of mechanical properties of Alloy 718 processed in four different forging/heat treatment process combinations under the conditions of post forging thermal exposures and the combination of post forging cold work and thermal exposures and (2) to identify the initial microstructural changes which are associated with the mechanical property degradation in a temperature range of 593°C to 677°C.

It was found that a significant impact degradation occurred after exposures at 593°C and 650°C but not as much as seen for exposures at 677°C for 1000 to 2000 hours. The Charpy impact energies after exposure were higher for samples which were given the standard Solution and Age (Super) while the hot tensile and notch rupture properties were better for samples given the Direct Age.

Embrittlement at temperatures of 650°C and higher is the result of large amounts of αCr phase and delta at the grain boundaries and overaging of the γ" phase. The αCr precipitation is accelerated when the exposure temperature is above 650°C. Residual strain from cold work was found to accelerate the αCr formation and promote Charpy property degradation. The αCr phase is found to precipitate at the γ-delta phase interface and is located mainly at the grain boundaries. The formation of αCr is attributed to the depletion of Ni and an enrichment of Cr as a result of Cr rejection during delta phase formation.

Introduction

Previous structural studies have shown that when Alloy 718 was exposed for long times at temperatures of 593°C to 760°C, αCr formed in addition to delta phase [1,2]. Significant amounts of αCr have been found in retired engine disks. The degree of αCr formation has been found to be associated with the delta phase that formed during processing, heat treatment or during engine service. A significant drop in impact energy occurs when a large amount of αCr forms.

Recent studies [2,3] on Alloy 718 have shown that large amounts of residual strain produced by cold work will form αCr in short exposure times. Some highly strained Alloy 718 bolts, which have failed prematurely, have shown αCr to be present as a result of the initial heat treatment [3]. It has long been known that residual strain accelerates phase reactions. Varying amounts of residual strain in Alloy 718 disks produced by different forging processes may affect the rate of αCr formation and hence the structural stability. Understanding the formation of αCr and the associated property degradation is of great interest for aero and land based turbine applications.

Superalloys 2000
Edited by T.M. Pollock, R.D. Kissinger, R.R. Bowman,
K.A. Green, M. McLean, S. Olson, and J.J. Schirra
TMS (The Minerals, Metals & Materials Society), 2000

A research program was undertaken to study the stability of Alloy 718 under thermal exposure up to 2000 hours. A variety of forging and heat treatment conditions were used in processing the 718 disks used in this program. The objectives of the program were:

(1) To investigate the degradation of a variety of properties of the 718 disks processed in four different forging/heat treatment process combinations under the conditions of post forging thermal exposures and the combination of post forging cold work and thermal exposure;
(2) To investigate the microstructure deterioration associated with the mechanical property degradation.

Experimental Procedures

Processes and Thermal Exposure Experiments

Triple melt Alloy 718 was forged into a simple turbine disk configuration (Figure 1) using four different processes as shown in Table 1. The 718 had a composition (in weight %) of 0.021 C, 0.08 Mn, 0.07 Si, 17.89 Cr, 53.88 Ni, 0.37 Co, 2.99 Mo, 0.94 Ti, 0.49 Al, 0.0029 B, 0.0003 S, 0.01 P, 0.05 Cu, 0.01 Ta, 5.36 Nb and the balance Fe. Two heat treatments were applied to these disks. One was direct age (DA) from forging operation. The other was post forging solution and age. The disks were then cut into sections for thermal exposures up to 2000 hours. The actual thermal exposure experiments are shown in Table 2. To detect the possible early structure changes, a 1000-hour exposure was used for all of the three thermal exposure temperatures and a 500-hour exposure was used for the exposure at 677°C. To address the issue of the cold work/thermal exposure interaction, a set of samples was cold worked 9-16% and exposed at 650°C for 1000

hours. Two disks were processed for each of the four conditions listed in Table 1.

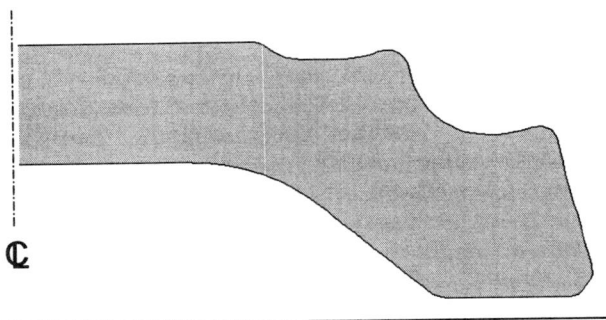

Figure 1. Half cross section of the disk (diameter=280 mm) used in this research.

Mechanical Tests

Hardness and Charpy tests were run on all conditions listed in Table 2. Hot tensile and notched rupture tests (notch factor Kt=3.6) were run on all (except the cold work/thermal exposure) conditions listed in Table 2. Compact tension tests were run on selected samples.

Structural Evaluations

Modified electrolytic techniques were developed and used to reveal the following: 1) αCr phase in relief, 2) preferential etching of the αCr, or 3) both the αCr and delta phases as separate particles. Examples of such selective actions on αCr are shown in Figure 2. SEM studies were carried out on samples in all conditions listed in Table 2. X-ray diffraction of extracted residues was used to confirm the presence of the αCr phase.

Table 1. 718 Processed in Four Different Conditions

Disk ID	Classification	Processes	Grain Sizes
1,2	DA-718	Fine grain forging with Direct Age heat treatment	ASTM 11-12
3,4	Super-718	Fine grain forging with solution + age heat treatment	ASTM 11-12
5,6	Super-II 718	Ladish proprietary process (Special fine grain processing + DA)	ASTM 10-11
7,8	Standard-718	Medium grain forging with solution + age heat treatment	ASTM 7-9

Table 2. Thermal Exposure Experiments

	0 hour	500 hours	1000 hours	2000 hours
593°C	Disk 2,4,6,8		Disk1,3,5,7	Disk1,3,5,7
650°C			Disk1,3,5,7 Disk1,3,5,7 (cold work)	Disk1,3,5,7
677°C		Disk2,4,6,8	Disk2,4,6,8	Disk2,4,6,8

However, the SEM metallographic technique was the primary tool to detect the αCr in the early stages of formation.

Results

Mechanical Properties

Hardness: There was no significant difference in hardness among the four groups of samples after exposures at 593°C, 650°C, and 677°C for up to 2000 hours. The mean hardness of the four groups before the exposure was 45HRc. After exposure at 593°C for 2000 hours the mean hardness of the four groups increased to 48HRc (age hardened). The mean hardness was 42HRc after exposure at 650°C for 2000 hours. The largest decrease in mean hardness appeared in samples exposed at 677°C for 2000 hours. The change was from 45HRc to 40HRc.

Impact: The decrease in Charpy impact energy is dramatic. For an exposures up to 2000 hours at 593°C, the Charpy impact energy dropped from 17.6J (DA-718) and 21.6J (Super-II 718) to 13.5J (DA-718) and 14.9J (Super-II 718), Figure 3(a), which represented a 23% and 31% reduction from the values obtained under unexposed condition for DA-718 and Super-II 718 respectively. After a 2000 hours at 677°C, the lowest Charpy impact energies were 13.5J, 12.2J, 10.8J, and 8.1J for Super-718, Standard-718, Super-II 718, and DA-718 respectively. The Charpy value for DA-718 was the lowest both before and after the thermal exposure. The addition of cold work prior to exposure reduced the impact energy further. With a 9-16% amount of cold work and a 1000-hour thermal exposure at 650°C, the Charpy impact energies of the samples are worse than samples without cold work exposed for 2000 hours at the same temperatures (Figure 3 (b)).

Hot Tensile: The hot tensile strength did not have any noticeable reduction after 2000 hour exposures at 593°C and 650°C. However, a 19% decrease in hot tensile strength was found after exposure at 677°C for 2000 hours for all of the four group samples (Figure 4). The DA-718 and Super-II 718 had a better hot tensile strength both before and after the thermal exposures than the Super-718 and the Standard-718.

Notched Rupture: A decrease of more than 80% in notched rupture time was found after the exposure at 677°C for 2000 hours (Figure 5). The time to rupture for DA-718, Super-II 718, Standard-718, and Super-718 were 34, 22, 15, and 8 hours respectively. DA-718 and Super-II 718 had longer notched rupture time than the Super-718 and the Standard-718 both before and after the thermal exposures.

Structural Evaluations

593°C: The etching out of the αCr technique was used to detect and follow the αCr formation with increasing exposure times and temperatures. αCr was most closely associated with the delta plates in the early stages of formation and then it was found also in the grain boundaries. After 1Kh of exposure, the αCr does not form in the standard 718 but does form in the DA-718 and the Super-II 718. More αCr is seen in the later two samples after 2Kh at 593°C while no αCr is found in the former sample.

650°C: After 1Kh of exposure, more αCr was found in the DA-718 and Super-II 718 samples. αCr was found in the Standard-718 and the Super-718 after 2Kh but not as much as was found in the DA-718 and the Super-II 718. The γ" phase also grew to a larger size comparing to that observed from samples exposed at 593°C.

The cold worked samples all showed greater αCr formation compared to their counterparts that were not cold worked. Figure 6 shows the effect of cold work on αCr formation in the case of the Super 718 samples. The cold work also causes the growth of γ"/γ' phases.

677°C: At this temperature αCr was detected after 500 hours in all samples, but greater amounts were found in the DA-718 and Super-II 718 samples. The αCr continued forming during the 1Kh and 2Kh exposures. While the αCr was forming, the γ" phase was coarsening. The size of the γ" after 500 hours at 677°C appears to be the same size as the γ" found in 2Kh at 650°C. Figure 7 shows the increased αCr amounts associated with the delta phase after 1Kh exposures at 593°C, 650°C, and 677°C. The coarsening of the γ" is also clearly seen with increased temperatures of exposures.

Conclusions

Alloy 718 processed in four different ways showed property degradation after exposures at 593°C, 650°C, and 677°C for up to 2000 hours. Among the three exposure temperatures, the 677°C gave the worst property degradation. The Charpy impact energies after exposure were higher for the Super-718 and Standard-718 than the DA-718 and Super-II 718, while the DA-718 and Super-II 718 had better hot tensile and notched rupture properties before and after exposure. Cold work was found to accelerate the Charpy property degradation as the cold worked samples showed much lower Charpy impact energies than those samples without cold work. These mechanical property changes appear to occur in shorter times than indicated by the published data of Radavich and Korth [1].

There is an interaction of the αCr and the delta phase as the αCr is found to precipitate at the γ-delta phase interface at temperatures as low as 593°C. The initial αCr formation has not been previously shown in its early stages due to the fact that in metallographic preparation the top surface of the delta phase with its αCr layer is polished off and only the αCr on the sides and ends of the delta phase is retained.

The formation of αCr is thought to be due to the effects of residual strain and microsegregation of Cr as the delta phase forms. Similar αCr formation behavior in Ni base alloys having a large fraction of γ' was reported by Lemaire, Fornwalt, and Kear [4].

The αCr formation is accelerated when the temperatures of exposure go above 650°C as large amounts of Cr diffuse to the grain boundaries to form large discrete αCr while the γ'' coarsens in the matrix. Continued exposure at 650°C tends to form continuous film of αCr similar to $M_{23}C_6$ carbides in Ni base alloys. Embrittlement of Alloy 718 is caused by large amount of αCr and the over aging of γ''/γ' phases.

References

[1] J. F. Radavich and G. E. Korth, "Effects of Very Long Time Aging in Alloy 718", TMS Annual Meeting, San Diego, CA 1992.

[2] B. Lindsley, X. Pierron, G. Maurer, and J. F. Radavich, "α-Cr Formation in Alloy 718 During Long Term Exposure : The Effects of Chemistry and Deformation", TMS Annual Meeting, San Diego, CA 1999.

[3] J. F. Radavich, Unpublished Research.

[4] L. Lemaire, D. E. Fornwalt, and B. H. Kear, "Co-Precipitation of γ' (Ni_3Al) and αCr in a Nickel-Base Alloy", MICROSTRUCTURES, Vol. 2, No. 1, December/January 1971.

LONG TERM THERMAL STABILITY OF INCONEL ALLOYS 718, 706, 909, AND WASPALOY AT 593°C AND 704°C

Sarwan Mannan, Shailesh Patel, and John deBarbadillo
Special Metals Corporation, 3200 Riverside Drive, Huntington, WV 25705

Abstract

This paper presents a study of the isothermal stability of INCONEL* alloys 718, 706, 909, and WASPALOY**. Standard annealed and aged materials were exposed at 593°C up to 10,000h and at 704°C up to 5,000h. The exposed materials were tested for room temperature tensile, room temperature impact, and high temperature tensile. The strength of alloy 909 degraded on 593°C exposure but 706 and 718 retained their strength. At 704°C exposure, WASPALOY had the best stability in terms of high temperature strength whilst 909 was the poorest and retained only 50% of its strength. On 704°C exposure, alloy 706 and WASPALOY retained their room temperature impact strength better than alloy 718. The tested materials were subjected to optical microscopy to develop an understanding of the reasons for the above results.

* INCONEL is a trademark of the Special Metals Family of Companies
** WASPALOY is a trademark of the United Technology Corporation

Superalloys 2000
Edited by T.M. Pollock, R.D. Kissinger, R.R. Bowman,
K.A. Green, M. McLean, S. Olson, and J.J. Schirra
TMS (The Minerals, Metals & Materials Society), 2000

Introduction

Materials used in gas turbines must have good high temperature microstructural stability. To utilize their maximum potential, precipitation strengthened superalloys are used in the turbine at intermediate temperatures where phase transformations and growth kinetics are rather sluggish. Since the evaluation of intermediate temperature stability requires long term exposure and extensive testing, this data is rare. Although thermal stability of alloy 718 is extensively characterized [1-4], the literature lacks comparative stability data on the Ni-base superalloys.

Dispersion strengthened materials are prone to the coarsening of their precipitates on intermediate exposure, which degrades their strength. The coarsening rate of γ' precipitates in Ni-base alloys was found to relate to coherency strain [5]. Increasing the Ti/Al ratio was found to increase the coherency strain and the coarsening rate in a Ni-based alloy [6]. In a multicomponent system like alloy 718, almost all the alloying elements were analytically detected in γ'/γ'' precipitates [7]. Since the effect of various elements on coherency strain at different atomic levels is not known, it is difficult to accurately predict thermal stability γ'/γ'' in a multicomponent system.

Experimental Procedure

Commercially produced 101.5 mm hot finished round bar of alloy 718 was used as a starting material. For alloys 909, 706 and WASPALOY, commercially produced forging stocks were hot rolled to 12.5 mm flats in the laboratory. Hot worked materials were subjected to the standard annealing and aging procedures as shown in Table I. The AC and FC in Table I stands for air cool and furnace cool. 706-3 and 706-2 denote 3-step and 2-step aged alloy 706 respectively. The final grain size of alloy 909, 718, 706, and WASPALOY were ASTM # 9.5, 7.5, 5, and 8 respectively. The chemical compositions of these alloys are given in Table II.

Table I. Annealing and aging conditions for the tested alloys

Alloy	Annealing	Aging
718	982°C/1h/AC	718°C/8h, FC 55°C/h to 621°C, hold 621°C/8h, AC
706-2	982°C/1h/AC	Same as alloy 718
706-3	982°C/1h/AC	843°C/3h, AC + same age as alloy 718
909	982°C/1h/AC	Same age as alloy 718
WASP	1020°C/4h, Oil Quench	850°C/4h, AC + 760°C/16h, AC

Table II. Chemical Compositions of the tested alloys in weight percent.

Alloy	Ni	Fe	Cr	Co	Mo	Cb	Al	Ti	C	Si
718	53.4	18.0	18.3	0.13	2.97	5.45	0.59	0.92	0.037	0.10
706	41.5	37.3	16.2	0.03	0.05	3.00	0.24	1.52	0.021	0.05
909	38.3	41.4	0.1	12.9	0.03	5.10	0.05	1.56	0.005	0.42
WASP	57.8	0.40	19.6	13.8	4.25	0.00	1.44	2.92	0.018	0.04

Rough-machined, annealed plus aged specimens were exposed at 593°C up to 10,000h and at 704°C up to 5,000h. This was followed by room temperature tensile (RTT) and high temperature tensile (HTT) testing. Specimens exposed at 593°C were HTT tested at 649°C and the specimens exposed at 704°C were HTT tested at 704°C. The specimens exposed at 704°C were also subjected to room temperature impact testing.

The tested specimens were examined by optical microscopy to characterize the microstructure. Alloys 718, 909, and WASPALOY were immersion-etched using Kalling's reagent (100ml methanol, 5gm cupric chloride, and 100ml hydrochloric acid). Alloy 706 was swab-etched using Seven-acids etchant (30ml hydrofloric acid, 60ml acetic acid, 60ml phosphoric acid, 60ml nitric acid, 300ml hydrochloric acid, 30ml sulphuric acid, 30gm anhydrous ferric chloride, and 300ml water).

Results and Discussions

593°C Exposure:

Figure 1 shows room temperature and 649°C yield strength of alloys 718, 706-3, 706-2, and 909 following 593°C exposure. Room temperature yield strength of alloy 718 increases up to 8,000h exposure but falls back to almost its as-produced strength on further exposure, Table III and Figure 1. Although the yield strength after 10,000h exposure is higher than the as produced yield strength, it is lower than 8000h exposed yield strength by 55MPa. This is in contrast to a reported study where room temperature yield strength continued to be higher for 50,000h exposure at 593°C [4].

Alloy 909 begins to lose its strength after 2000h at 593°C. On 10,000h exposure at 593°C, the loss in strength of alloy 909 is more than 200 MPa, Table I. Although the as produced 649°C yield strength of 706-2 is 70 MPa lower than 706-3, the 649°C yield strength of 593°C/10,000h exposed 706-2 is 100 MPa higher than the exposed 706-3, Table III. This is related to the additional 843°C/3h heat treatment that alloy 706-3 undergoes following annealing, Table I. Room temperature and high temperature elongation of any of the alloys is not significantly affected by 593°C exposure, Table III.

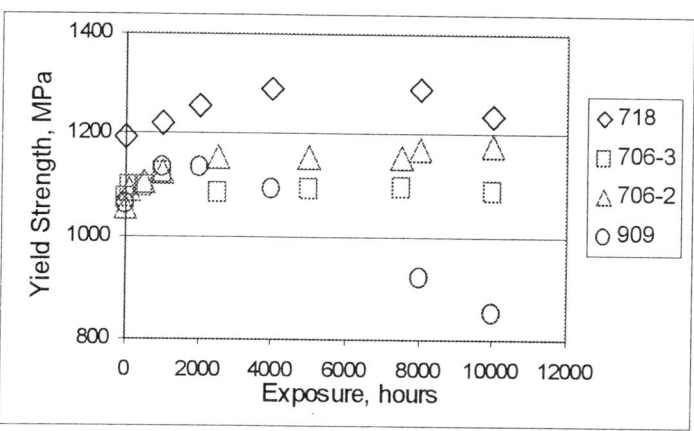

Figure 1a: Room temperature yield strength of materials exposed at 593°C.

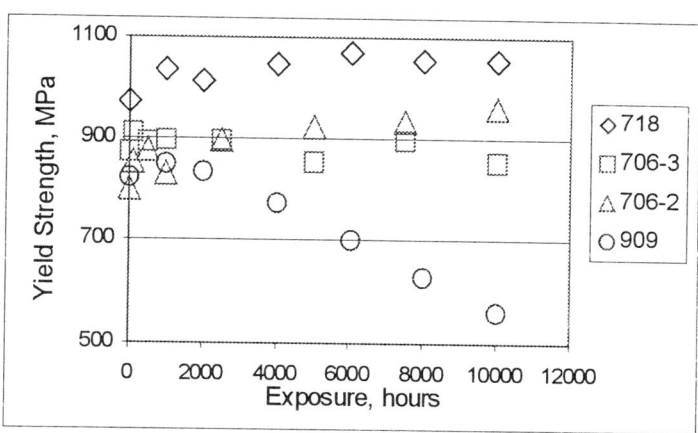

Figure 1b: 649°C yield strength of materials exposed at 593°C.

Table III: Room temperature and 649°C tensile properties following 593°C exposure. Yield strength, tensile strength, percentage elongation, and percentage reduction of area are denoted by YS, UTS, %El, and %RA respectively. The data are the average of two tests. The average values are rounded-off to the whole number.

Alloy	Exposure Condition	Room temperature tensile				649°C tensile			
		YS, (MPa)	UTS (MPa)	%El	%RA	YS, (MPa)	UTS (MPa)	%El	%RA
718	As-produced	1193	1455	16	36	972	1179	31	60
	1000h	1220	1489	14	35	1034	1200	23	62
	2000h	1255	1482	15	35	1014	1200	24	61
	4000h	1289	1496	15	32	1048	1220	22	58
	6000h					1069	1241	21	59
	8,000h	1289	1503	15	34	1055	1241	23	61
	10,000h	1234	1503	14	34	1055	1234	23	60
706 3-Step	As-produced	1077	1319	18	32	874	1025	25	54
	100h	1100	1345	18	36	909	1046	23	54
	500h	1097	1346	17	35	890	1023	25	55
	1000h	1124	1345	18	33	895	1026	23	55
	2500	1085	1346	17	33	896	1026	24	54
	5,000h	1095	1356	16	26	854	985	27	56
	7,500h	1098	1354	17	33	894	1008	24	51
	10,000h	1091	1349	18	32	854	997	25	51
706 2-Step	As-produced	1056	1289	25	53	799	981	31	54
	100h	1091	1313	25	53	852	1030	29	60
	500h	1106	1310	24	55	877	1040	29	61
	1000h	1123	1305	24	53	831	1022	29	61
	2,500h	1153	1337	23	51	894	1046	26	58
	5,000h	1154	1342	22	51	920	1054	27	58
	7,500h	1172	1356	22	51	932	1066	25	57
	10,000h	1178	1357	23	50	960	1064	24	57
909	As-produced	1062	1351	12	33	821	972	26	54
	1,000h	1138	1414	11	34	848	972	19	45
	2,000h	1138	1386	11	35	834	986	19	66
	4,000h	1096	1365	11	31	772	931	23	62
	6,000h					703	890	23	61
	8,000h	924	1262	11	28	628	821	27	69
	10,000h	855	1214	10	22	559	766	29	71

704°C Exposure: Figure 2 shows room temperature and 704°C yield strength of 718, 706-3, 706-2, 909, and WASPALOY exposed at 704°C for up to 5,000h. All the alloys lose their strength after 704°C exposure. Alloy 909 has the poorest thermal stability of all these alloys. After 500h exposure, the loss of strength for alloy 909 was more than 300 MPa, Figure 2. WASPALOY had the best stability in terms of retention of strength. Loss of yield strength at 704°C for WASPALOY and 718 following 704°C/5,000h exposure was 100 MPa and 350 MPa respectively, Table IV. Yield strengths at 704°C of the 704°C/5,000h exposed 706-3 and 706-2 were comparable. Elongation of all the alloys at 704°C increases with the initial 1000h exposure and then saturates on further exposure, Figure 3a. Room temperature elongation is essentially unaffected by exposure at 704°C, Figure 3b.

The decrease in strength of alloys 718, 706, 909, and WASPALOY after 704°C exposure is presumably related to the coarsening of γ'/γ'' precipitates. Coarsening of γ'/γ'' precipitates is a diffusion controlled process. The coarsening rate depends on matrix/particle lattice parameter mismatch, which is responsible for γ/γ' and γ/γ'' interface energy [8]. The higher the lattice mismatch, the higher the growth rate [8]. In the present study, the better stability of WASPALOY compared to alloy 718 is probably related to its lower lattice mismatch. The lattice mismatch of WASPALOY and alloy 718 are +0.30% and +0.80% respectively [9]. Increasing the Ti/Al ratio was found to increase the lattice mismatch and the coarsening rate in a Ni-base superalloy [6]. The decrease in coarsening rate on increase of Cr from 10 wt% to 37 wt% in a Ni-Cr-Ti-Al alloy was attributed to a high

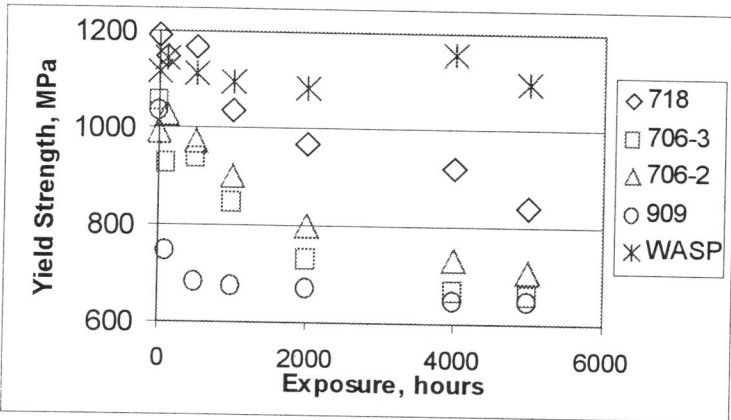

Figure 2a. Room temperature yield strength of the materials exposed at 704°C.

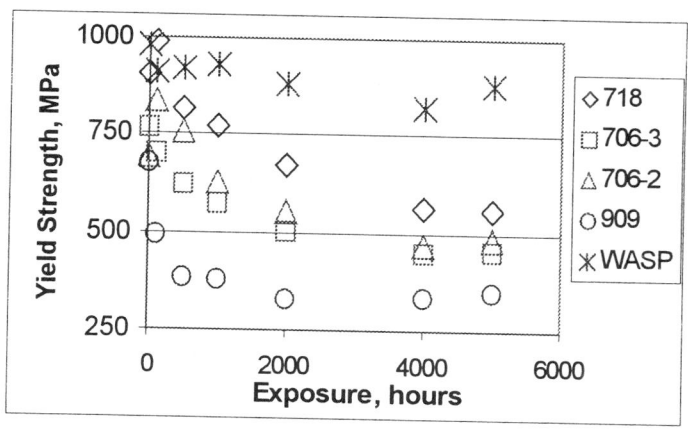

Figure 2b. 704°C yield strength of the materials exposed at 704°C

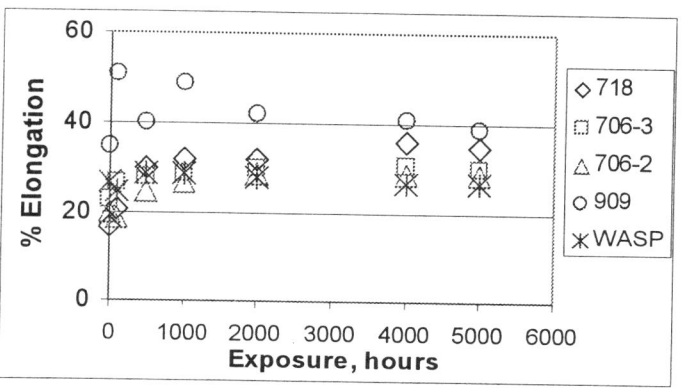

Figure 3a: 704°C elongation of materials exposed at 704°C.

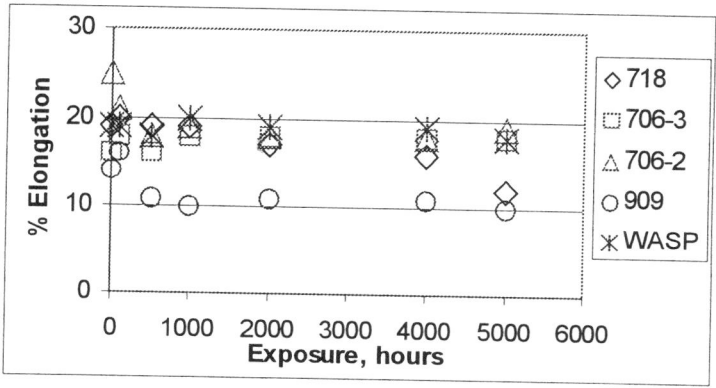

Figure 3b. Room temperature elongation of the material exposed at 704°C.

partitioning coefficient and reduction of coherency strain [5]. A very high partitioning coefficient for an element for γ' reduces the equilibrium concentration in the matrix which decreases the driving force for growth [5]. The addition of 5.5 wt% Mo was found to decrease the coherency strain and retard the coarsening rate of γ' in a Ni-Cr-Ti-Al alloy [10]. The lattice mismatch of alloys 706 and 909 are not known. Thermal stability data of alloy 706 and 909 presented in this investigation will be useful in future to see if the lattice mismatch trend can be extended to these alloys.

Impact Strength Following 704°C Exposure:
Table V shows room temperature impact strength of alloys 718, 706, 909, and WASPALOY following 704°C

exposure. The impact values of as produced and 704°C/5000h exposed alloy 909 are 15 and 10 joules respectively. The impact strength of 706-3 and WASPALOY are essentially unaffected by the exposure. Alloys 718 and 706-2 lose their impact strength on exposure. The impact strength of as produced 706-2 is 111 joules. The impact value of the 704°C/5000h exposed alloy 706-2 is 39 joules which is quite respectable for a precipitation strengthened material. However, the as produced and 704°C/5000h exposed impact strength of alloy 718 are 50 and 9 joules respectively. The low impact value of exposed alloy 718 of 9 joules could be a concern in certain applications. The impact strength versus exposure time plot is shown in Figure, 4.

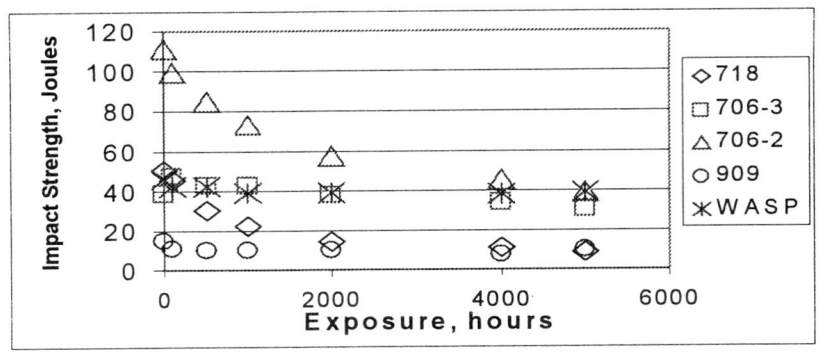

Figure 4. Room temperature impact strength of the materials exposed at 704°C.

Table IV: Room temperature tensile and 704°C tensile properties following 704°C exposure. Yield strength, tensile strength, percentage elongation, and percentage reduction of area are denoted by YS, UTS, %El, and %RA respectively. The data are the average of two tests. The average values are rounded-off to the whole number.

Alloy	Exposure Condition	Room temperature tensile				704°C tensile			
		YS, (MPa)	UTS (MPa)	%El	%RA	YS, (MPa)	UTS (MPa)	%El	%RA
718	As-produced	1192	1352	19	49	904	998	17	29
	100h	1150	1358	20	49	991	1082	21	45
	500h	1167	1377	19	38	816	987	30	68
	1000h	1037	1325	19	44	771	944	32	72
	2000h	969	1289	17	36	671	876	32	66
	4000h	921	1249	16	29	561	811	36	80
	5000h	840	1223	13	17	556	817	35	76
706 3-Step	As-produced	1055	1283	16	30	767	885	23	50
	100h	927	1233	18	39	703	816	27	56
	500h	941	1236	16	38	623	763	28	60
	1000h	847	1177	18	34	572	732	29	58
	2000h	733	1111	18	30	501	685	30	61
	4000h	667	1063	18	29	449	642	31	63
	5000h	662	1056	18	30	457	643	30	60
706 2-Step	As-produced	992	1231	25	56	690	845	20	23
	100h	1029	1264	21	52	834	881	19	47
	500h	974	1236	18	50	759	847	25	62
	1000h	902	1199	19	46	627	767	27	65
	2000h	801	1140	18	35	552	714	29	62
	4000h	752	1097	18	32	468	670	29	59
	5000h	709	1078	19	30	485	679	29	60
909	As-produced	1035	1341	14	39	676	776	35	72
	100h	747	1102	16	29	491	625	51	92
	500h	685	996	11	19	383	556	40	94
	1000h	677	965	10	18	377	530	49	95
	2000h	672	963	11	18	330	510	42	89
	4000h	650	949	11	18	332	476	41	95
	5000h	649	942	10	15	348	495	39	93
WASP	As-produced	1118	1461	19	31	979	1105	27	53
	100h	1145	1489	19	37	910	1129	25	39
	500h	1112	1478	18	29	922	1112	29	55
	1000h	1095	1461	20	38	929	1102	29	57
	2000h	1086	1462	19	37	879	1086	28	50
	4000h	1156	1486	19	49	823	1077	27	56
	5000h	1098	1462	18	36	883	1088	27	52

Table V Room temperature impact strength (joules) of the materials exposed at 704°C

Exposure Time	Materials				
	718	706-3	706-2	909	WASPALOY
0	50	39	111	15	46
100h	45	48	99	11	42
500h	30	43	85	10	42
1000h	22	43	73	10	39
2000h	14	38	57	10	39
4000h	11	34	45	8	38
5000h	9	31	39	10	39

Optical Metallography: As produced and 704°C/5,000h exposed specimens of alloy 718, 706, 909, and WASPALOY were examined with optical microscopy to understand the impact properties. Since microstructural development for these alloys is well documented in the literature [1, 2, 4, 12, 13, 14], morphological similarity was used as the only identification for the observed precipitates. The microstructures of as produced and exposed WASPALOY are comparable, Figure 5a and 5b. This explains its good retention of impact strength on exposure. The grain boundaries of as produced 706-2 are completely devoid of precipitates whereas the grain boundaries of exposed material contains acicular precipitates, presumably Eta phase [12, 13], Figure 5c and 5d. These precipitates would have been responsible for lowering the impact strength from 111 joules to 39 joules,

Table V. Further, intragranular areas of exposed 706-2 have a mottled appearance (Figure 5d) probably due to the coarse γ'/γ'' precipitates. As produced 706-3 contains intergranular Eta phase [12, 13], Figure 5e. On exposure, it grows (Figure 5f) resulting in marginal degradation in the impact strength, Table V. Figure 5g and 5h show the microstructure of as produced and 704°C/5,000h exposed alloy 718. Extensive growth of Delta phase colonies [1, 2, 4], relatively higher carbon (0.037wt% as compared to 0.021wt% in alloy 706, Table II), and formation of α-Cr [15] may have been responsible for lower impact strength of exposed alloy 718. Low impact strength of as produced and exposed alloy 909 is due to the presence of globular Laves phase and acicular Epsilon phase [14], Figure 5i and 5j.

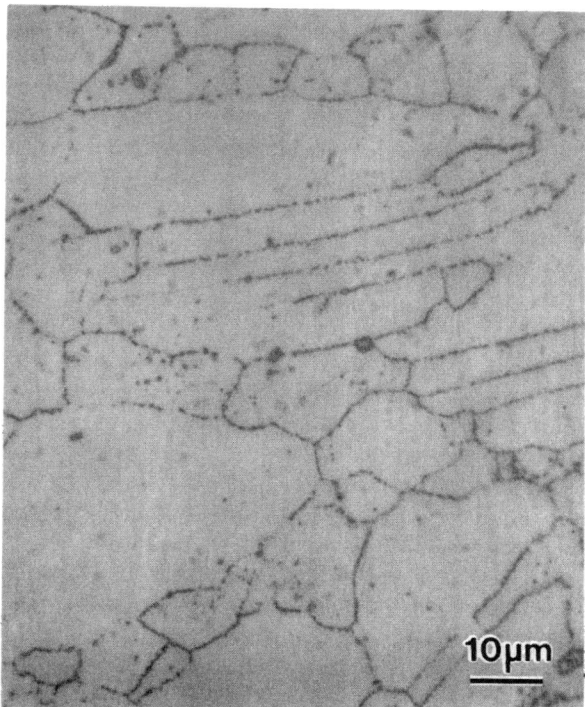

Figure 5a: Optical photomicrograph of as produced WASPALOY.

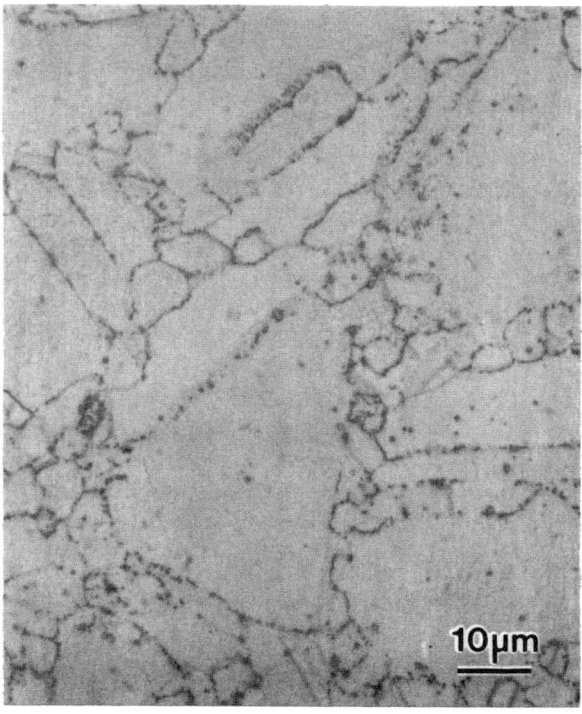

Figure 5b: Optical photomicrograph of WASPALOY exposed at 704°C for 5,000h.

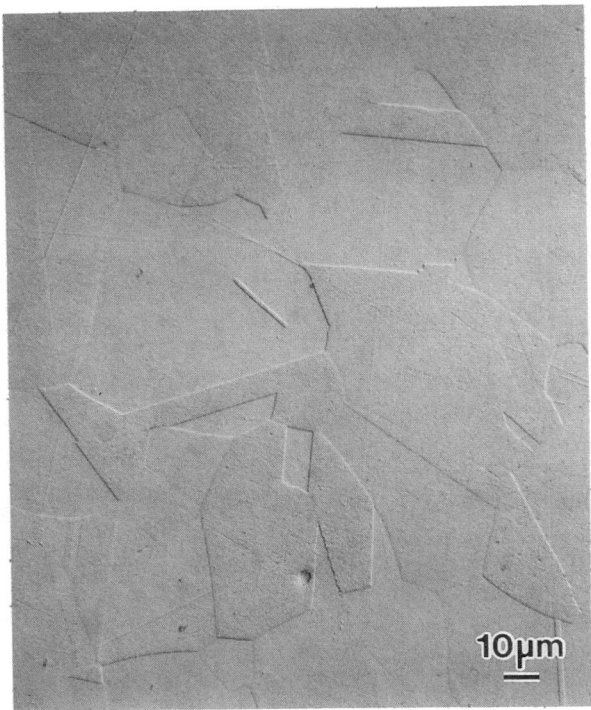

Figure 5c: Optical photomicrograph of as produced 706-2.

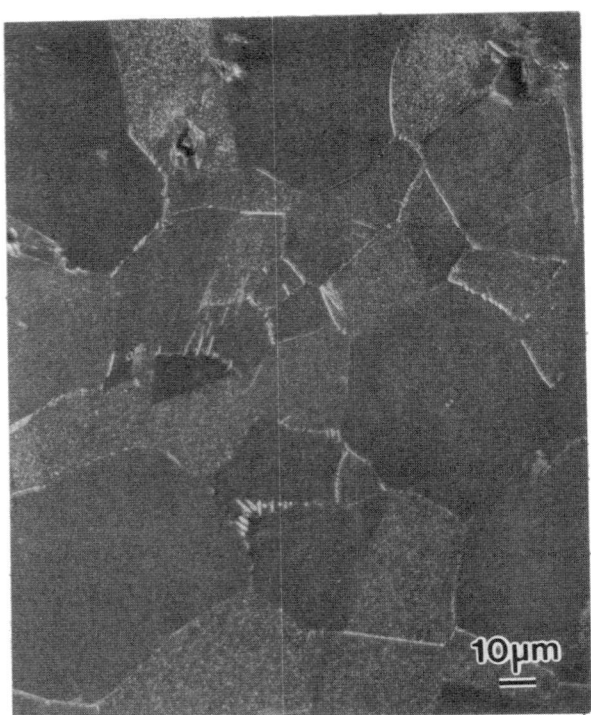

Figure 5d: Optical photomicrograph of alloy 706-2 exposed at 704°C for 5,000h.

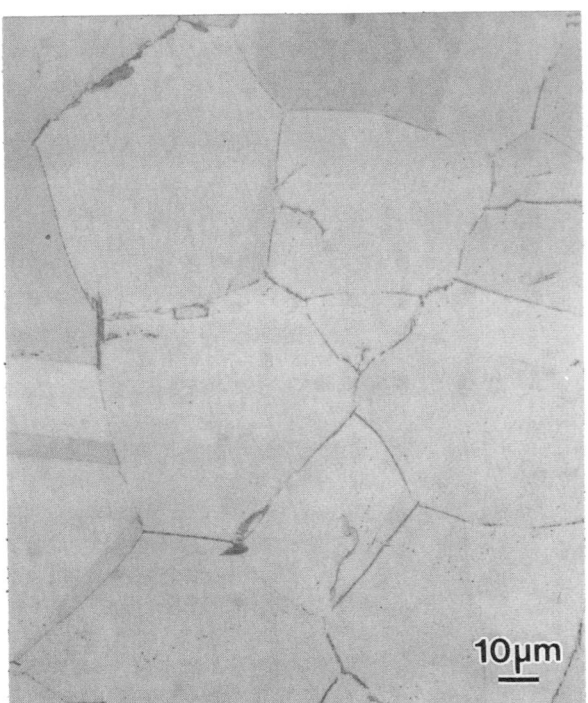

Figure 5e: Optical photomicrograph of as produced alloy 706-3.

Figure 5f: Optical photomicrograph of alloy 706-3 exposed at 704°C for 5,000h.

456

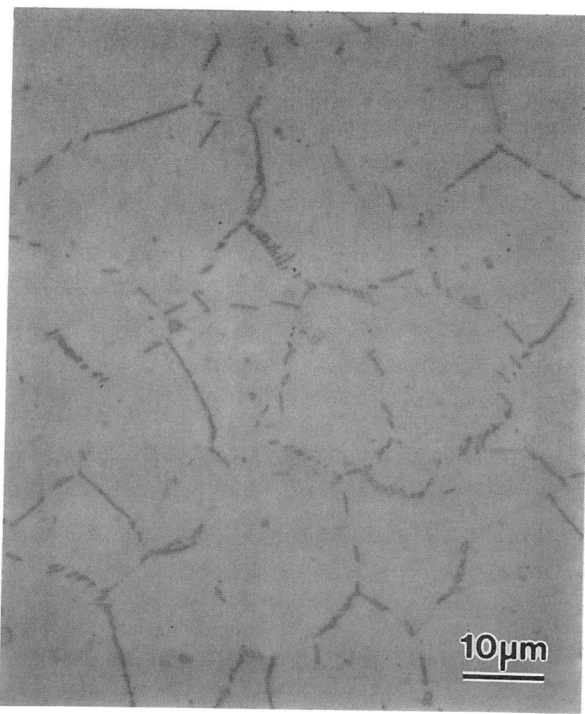

Figure 5g: Optical photomicrograph of as produced alloy 718.

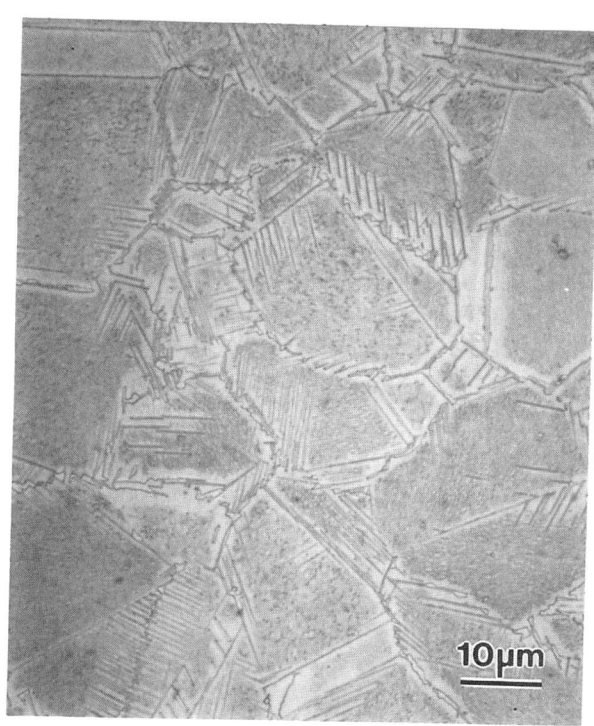

Figure 5h: Optical photomicrograph of alloy 718 exposed at 704°C for 5,000h.

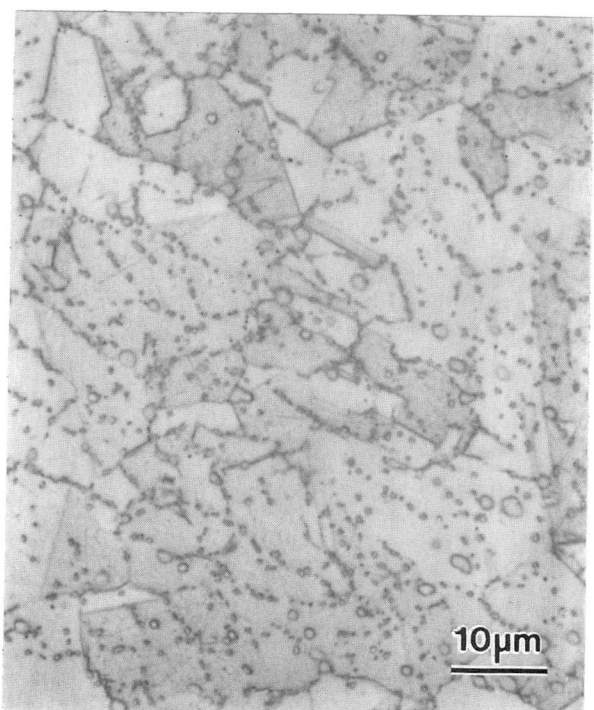

Figure 5i: Optical photomicrograph of as produced alloy 909 showing globular Laves phase.

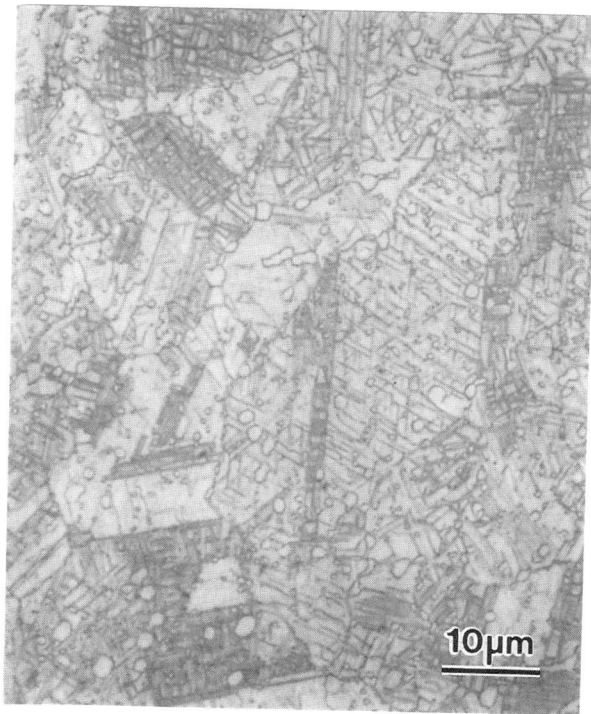

Figure 5j: Optical photomicrograph of alloy 909 exposed at 704°C for 5,000h showing acicular Epsilon phase.

457

Conclusions

1. Alloy 718 and 706 retain their strength upon 10,000h exposure at 593°C. Alloy 909 begins to lose its strength after 2,000h exposure. Room temperature and 649°C yield strength of 593°C/10,000h exposed alloy 909 is approximately 200 MPa lower than the yield strength of as produced material. Room temperature and 649°C elongations of these materials are not degraded with exposure.

2. Alloy 718, 706, 909, and WASPALOY lose their strength upon 704°C exposure. WASPALOY is the most stable, alloy 909 is the least stable, and alloy 718 and 706 are comparable for their retention of strength on exposure. Room temperature and 704°C elongations of these alloys were not adversely affected after exposure.

3. Room temperature impact strength of WASPALOY and 706-3 (3-step aged) was essentially unaffected upon 704°C/5,000h exposure. Alloy 706-2 (2-step aged) did lose its impact strength but the retained impact strength following 5,000h exposure was respectable (39 joules). Room temperature impact strengths of as produced and 704°C/5,000h exposed alloy 718 were 50 joules and 9 joules respectively. The observed impact strength values were correlated with the microstructure.

References

1. J. W. Brooks and P. J. Bridges, " Metallurgical Stability of INCONEL Alloy 718", Superalloys 1988, Edited by S. Reichman, D. N.Duhl, G. Maurer, S. Antolovich, and C. Lund, TMS, 1988, 33-42

2. J. F. Radavich, "Long Term Stability of a Wrought Alloy 718 Disc", Superalloy 718-Metallurgy and Applications, Edited by E. A. Loria, TMS, 1989, 257-268.

3. S. D. Antolovich, "The effect of Metallurgical Instabilities on the Behavior of IN 718", ibid, 647-653.

4. G. E. Korth and C. L. Trybus, "Tensile Properties and Microstructure of Alloy 718", Superalloys 718, 625 and Various Derivatives, Edited by E. A. Loria, TMS, 1991, 437-446.

5. R. F. Decker, "Strengthening Mechanisms in Nickel-Base Superalloys", Paper presented at Steel Strengthening Mechanisms Symposium, Zurich, Switzerland, May 5-6, 1969, 1-24.

6. E. A. Fell, "The Effect of Thermal Treatment on the Constitution of 80-20 Nickel-Chromium Alloys Hardened With Titanium and Aluminum, Metallurgia, Vol. 63, 1961, 157-166

7. M. G. Burke, M. K. Miller, "Precipitation in 718: A Combined AEM and APFIM Investigation", Superalloys 718, 625 and Various Derivatives, Edited by E. A. Loria, TMS, 1991, 337-350

8. T. Tkach et al, "The role of Alloying Elements on γ' Phase Kinetics in Ni-Base Alloys", High Temperature Materials and Processes, Vol. 15, No 3, (1996), 195-200.

9. M. Prager and C. S. Shira, "Welding of Precipitation – Hardening Nickel Alloys", (Welding Research Council Bulletin, # 128, February 1968)

10. G. N. Maniar, J. E. Bridge Jr., "Effect of Gamma Prime Mismatch, Volume Fraction, and Gamma Prime Morphology on Elevated Temperature Properties of Ni, 20Cr, 5.5Mo, Ti, Al Alloys", Metallurgical Transaction, 2 (1971), 95-102.

11. C. E. Jordan, R. K. Rasefske, and A. Castagna, "Thermal Stability of High Temperature Structural Alloys", Long Term Stability of High Temperature Materials, Edited by G. E. Fuch, K. A. Dannemann, T. C. Deragon, TMS, 1999.

12. K. A. Heck, "The Time-Temperature-Transformation Behavior of Alloy 706", Superalloys 718, 625, 706 and Various Derivatives, Edited by E. A. Loria, TMS, 1994, 437-446.

13. J. H. Moll, G. N. Maniar, and D. R. Muzyka, "The Microstructure of 706, a New Fe-Ni-Base Superalloy, "Metallurgical Transactions, 2 (1970) 2143-2151.

14. K. A. Heck, et al., "The Physical Metallurgy of a Silicon – Containing Low Expansion Superalloy", Superalloys 1988, Edited by S. Reichman, D. N. Duhl, G. Maurer, S. Antolovich, and C. Lund, TMS, 1988, 151-160

15. B.A. Lindsley et al., "α-Cr Formation in Alloy 718 During Long Term Exposure: The Effect of Chemistry and Deformation", Special Metals Corporation, Report # TR-99-008, May 25, 1999.

Effects of Microstructure and Loading Parameters on Fatigue Crack Propagation Rates in AF2-1DA-6

J. Mason[*], J. Lemsky[**], T. Stuhldreher[*] and D. Furrer[**]

[*]Aerospace and Mechanical Engineering
University of Notre Dame
365 Fitzpatrick Hall
Notre Dame, IN 46556

[**]Ladish Co., Inc.
5481 S. Packard Ave.
Cudahy, WI 53110

Abstract

Fatigue crack growth is examined in AF2-1DA-6 over a wide spectrum of growth rates from 10^{-8} to 10^{-6} m/cycle for two mean stress values (R = 0.7 and 0.05) and two frequencies (10 and 2 Hz) at 649 °C. The effects of one processing parameter, the cooling rate from the solution heat treatment, on Paris law growth rates is examined. In addition, the effects of changes in grain size on the same parameters are examined. No effect of cooling rate from solution heat treatment or of grain size on Paris law crack growth is seen for high R ratio tests at 10 Hz. However, low R ratio tests at lower frequency, 2 Hz, indicate that crack growth rates can vary with microstructure. Namely, crack growth rates are higher for a slow cooling rate of 34 °C/min for both the ASTM 10 ½ and 6 average grain sizes. The effect is more dramatic for the smaller grain size. Crack growth rates in that material, i.e. AF2-1DA-6 cooled at the slow rate with a grain size of ASTM 10 1/2, exhibited the highest growth rate of all when R=0.05 and the sinusoidal load frequency was 2 Hz. Some implications of these results on forging of such alloys are discussed.

Introduction

The forging of aerospace components such as turbine disks or airframe components has increased in sophistication over the last few years. It is now possible to a control desired material properties as well as final shape with a great degree of accuracy. Process modeling using the finite element method (FEM) has been the biggest contributor to these recent advancements.(1) FEM can be used to predict material behavior during forging processes as long as characteristics of materials, tools and processes are well known beforehand. Without good material characterization, the complex, coupled thermo-mechanical deformation of the material cannot be accurately predicted and all accuracy is lost. Prediction of the material deformation during the process is only one of the goals of forging modeling using FEM---prediction of the *properties* of the completed component is a goal, as well. With good metallurgical models of material response, grain size and strength of the forged component can also be predicted.(1) Therefore, it is important to develop sophisticated material models for use in FEM models of forging.

Typically, in forging, customers specify the desired mechanical properties for a rotating part such as a turbine disk to be manufactured. The forging engineer then determines whether the particular part can be manufactured and estimates the cost and time of that manufacture, if possible. Since the materials used in such aerospace applications are often made of expensive elements, material is not to be wasted and therefore the margin for error is minimal if it exists at all. The answers must be accurate. It is not possible to produce such an accurate response to the customer's request without an accurate data base of material properties, without experienced modeling engineers or without proven process models that can predict the time, cost and feasibility of producing the requested component with the grain size needed for the mechanical properties specified. Clearly, the data base of material properties should include information characterizing the effect of forging upon all important material properties, and the engineer should have a good understanding of what factors in the process affect each mechanical property most.

It is known, for example, that forging strain and cooling rates are critical in determining the final mechanical properties of a forging.(2) These parameters control the grain size for the final material and consequently can have an effect upon the yield strength and ductility of the finished product. However, while higher cooling rates may improve the strength of the material, they may also have detrimental effects on the fatigue resistance. Until recently, little information about the effects of cooling rate upon the fatigue properties of the material has been incorporated into the design and modeling of forgings. Such information is of significant value to the aerospace engine industry where damage tolerant design---design based on the growth of cracks or flaws assumed to exist in the component---is common. Further sophistication in the modeling of the forging of aerospace components can be achieved if the effects of grain size and cooling rate on the fatigue properties of a forging material can be measured. However, the measurement of fatigue properties of aerospace forging materials can be a costly undertaking. There are many alloys used in a plethora of components exposed to a wide range of operating environments. For each material the fatigue response over a range of microstructures and environments may need to be evaluated leading to large testing programs and lengthy testing times. On the other hand, gathered data can be invaluable in producing customer specified properties and may be used in multiple models reducing the average cost. It is, therefore, important that the effects of forging process parameters upon fatigue crack growth resistance be measured and utilized for future FEM models of forge processing.

In this study, the fatigue crack propagation rate is measured in a turbine disk material that has been exposed to a range of cooling

459

Superalloys 2000
Edited by T.M. Pollock, R.D. Kissinger, R.R. Bowman,
K.A. Green, M. McLean, S. Olson, and J.J. Schirra
TMS (The Minerals, Metals & Materials Society), 2000

rates. The desired effect is to imitate the forging process and to gather data that can be used in the modeling of forging of these materials. Microstructure is known to have an effect upon the fatigue crack growth rate in P/M U720 at high temperatures.(3) This investigation intends to determine if similar effects are seen in the P/M alloy AF2-1DA-6.

Experimental Procedure

Fatigue crack propagation rate tests were performed following ASTM 647 as described below. The purpose of the tests was to reproduce the characteristic crack growth curve, da/dn vs. ΔK, for each material. Following the collection of the data, a Paris law (4) relationship of the form,

$$\frac{da}{dn} = C\Delta K^m, \qquad (1)$$

was fit to the curve in the Paris Law regime giving the values of the material constants C and m for the material and microstructure tested. The materials were tested in several conditions as described below.

Materials

The material used for this investigation was P/M alloy AF2-1DA-6, that was isothermally forged, cut into sized blanks, and heat treated to specific thermal profiles prior to machining of compact tension specimens and program fatigue crack growth testing. The nominal chemistry for AF2-1DA-6 is shown below:

Table I: Nominal Composition for alloy AF2-1DA-6 (wt %)

C	Cr	Co	Mo	Ti	Al	B	W	Ni	Solvus
0.04	11.6	10.0	2.7	2.8	4.6	0.015	6.5	Bal	1195°C

Specimens were heat treated to two distinct grain size groups; a coarse grain ASTM 6 and a fine grain ASTM 10 ½. Within each grain size group, specimens were heat treated to establish distinct cooling profiles. Figure 1 is a schematic representation of the program heat treatments for both the fine and coarse grain specimen groups.

The grain size was established by an initial high temperature cycle: the coarse grain material was obtained by heating above the γ' solvus, while the fine grain size was obtained by heating near-solvus. After establishing the grain size, specimens of both coarse and fine grain groups were re-treated at a lower 1149°C temperature and cooled in a controlled manner for distinct γ' morphology generation. Then, ageing was carried out at 760°C.

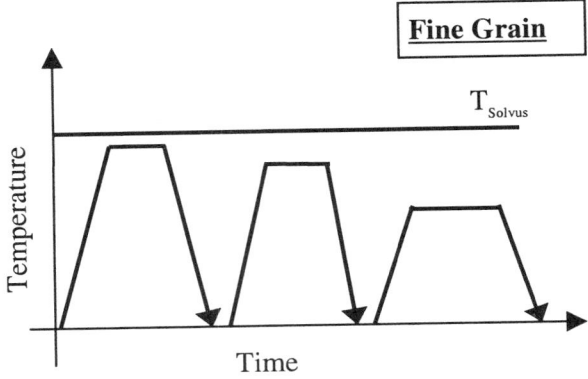

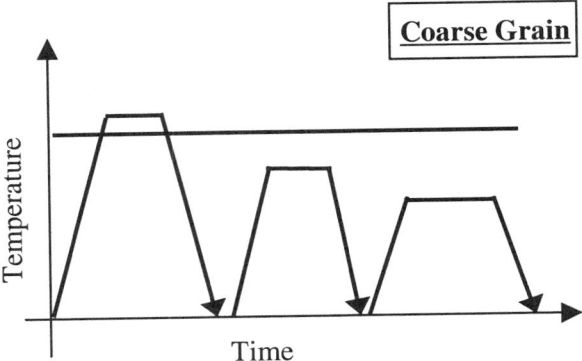

Figure 1. Schematics of heat treatments applied to AF2-1DA-6

The heat treatment cycles given advanced nickel-base superalloys, such as alloy AF2-1DA-6, has a marked influence on gamma prime size, distribution and morphology. Supersolvus processed material resulted in very fine, uniformly distributed primary γ' upon reheating to the second subsolvus solution temperature. Cooling from this temperature resulted in very fine secondary γ' precipitates between the primary γ' particles at fast cooling rates (Figures 2a). At slower cooling rates, (Figures 2b, 2c), significant primary γ' coarsening is observed with greatly reduced content of secondary γ'. The slowest cooled, coarse grain samples resulted in γ' particles which coarsened to a size which resulted in a morphology change from spherical to cuboidal due to the misfit strains in this alloy.

The fine grain material resulted in distinctly different γ' evolution and morphologies. This material consisted of 3-5μm primary γ' particles which were spaced at distances approximate to the grain size. The secondary γ' that formed in this material after the second solution cooling cycle is shown in Figures 3a and 3b. Fast cooling resulted in very fine spherical precipitates, while slow cooling resulted in a degeneration to a butterfly precipitate morphology.

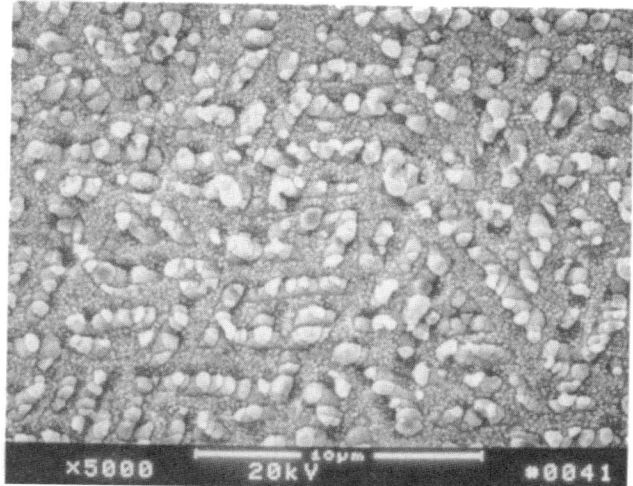
Figure 2a - 315 °C/min

Figure 3a - 315 °C/min

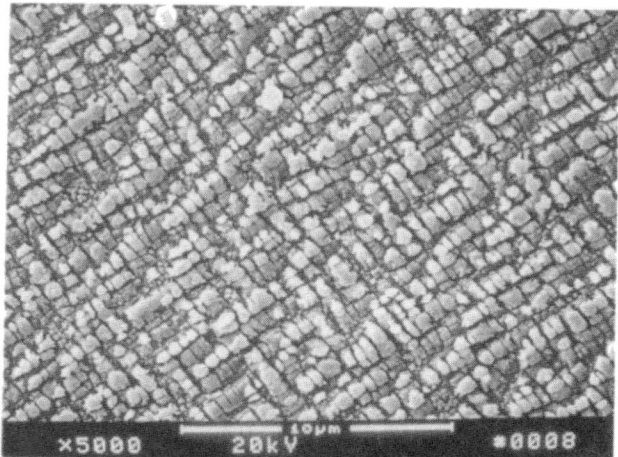
Figure 2b - 100 °C/min

Figure 3b - 100 °C/min

Figure 3: SEM microstructures of secondary ϒ' particles
after fine grain heat treat processing

Figure 2c - 34 °C/min

Figure 2: SEM microstructures of ϒ' particles after coarse
grain heat treat processing

461

Fatigue Crack Propagation Tests

Fatigue crack propagation tests were performed following the guidelines of *ASTM Standard E 647* (5) for crack growth rates from 10^{-8} to 10^{-6} m/cycle using a compact tension specimen. A personal computer is employed to control the stress intensity factor range at the crack tip directly. This is achieved through measurement of the load on the specimen and the crack length in the specimen in real time during the test. These two measurements are then used to calculate the stress intensity factor range using a known solution for the stress intensity factor and the geometry of the specimen. Measurements of the specimen size are taken and entered at the beginning of the test; typical specimen dimensions are given Figure 4. Crack length is measured using the electrical potential drop method.(5) In this method a constant current of 3 Amp is passed through the specimen, entering at point A and exiting at point B in Figure 4, through wires spot-welded to the specimen surface. As the crack length increases the path of the current from point A to point B increases and the resistance to the current increases accordingly. Because of this increase in current path length and resistance, the voltage across points C and D increases as well. Through a careful calibration performed at Notre Dame, the change in voltage can be correlated with the change in crack length. Thus, the voltage across points C and D is measured automatically and is converted into a crack length. Any thermocouple effects due to the difference between the voltage measurement wires and the specimen were removed by current switching. That is, the polarity of current supply was automatically, periodically, switched so that current alternately entered from point B then point A. The voltage for each polarity was measured and the two measurements were compared to each other. Any difference in magnitude was removed by manual adjustment of the amplifier and by computer averaging in a redundant system. Tests were carried out in laboratory room air (relative humidity, 20-50%) at 649°C ($\pm$ 3 °C) with a sinusoidal load frequency of 10 Hz or 2 Hz. The temperature of the specimen was monitored throughout the experiments by a thermocouple spot welded directly to the side of the specimen.

As per the standard, each test began with 1 mm of crack growth at a growth rate of about 10^{-9} m/cycle. Thus, the effects of crack initiation and of the notch tip geometry were minimized. By the end of 1 mm of growth the crack growth rate stabilized to a constant value. Then the crack growth rate was decreased using an automated load shedding procedure. The equation for load shedding developed by Saxena et. al. (6) was used in this procedure,

$$\Delta K = \Delta K_o e^{c(a-a_o)} \qquad (2)$$

where $\Delta K = K_{max} - K_{min}$ is the nominal stress intensity range, a is the instantaneous crack length, a_o is an initial crack length, ΔK_o is the initial stress intensity range and c is a constant. The intial crack length was taken to be the crack length acheived after the preliminary 1 mm of growth away from the notch tip. A value of -0.10 mm^{-1} was used for c. This value is slightly more negative than the value recommended in ASTM E-647, -0.08 mm^{-1}. However, it is noted in the ASTM standard that more negative values of c can be used depending upon the load ratio, test material and environment. Here, verification of the obtained curves was established by increasing ΔK from a value near

threshold with c=0.10 mm^{-1}. Both increasing and decreasing curves were coincident. Crack growth rates are presented in the following section in terms of the nominal stress intensity range which was computed for the compact tension specimen as per *ASTM E-399*.(7)

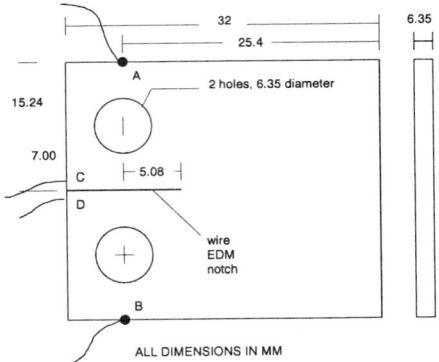

Figure 4. Compact tension, C(T), specimen used in these studies.

The upper range of crack growth rates was investigated. Typically, a stress intensity factor range 20% below the initial ΔK in the load shedding procedure, i.e. $0.8\Delta K_o$, was chosen as a starting point, then the stress intensity factor range was increased using equation (2) with a positive value for c. Some of the data for the previous load shedding procedure was reproduced for growth rates slightly below 10^{-9} m/cycle and then new data was generated for growth rates above 10^{-9} m/cycle. The overlapping data provided confirmation of the accuracy of the tests. Once confirmation of the lower growth rate data was obtained, the specimen was sacrificed as the higher growth rate data was obtained. The crack was allowed to continue growing until the specimen failed.

TABLE II. Fit parameters for fatigue crack propagation rate data of AF2-1DA-6 for two grain sizes at several different cooling rates from the solution heat treatment.

Grain Size (avg.) ASTM	Cooling Rate °C/min	Exponent m	Multiplicative Constant $C\ (\times 10^{-12})$	ΔK_{th} $Mpa\sqrt{m}$
10.5	315	2.86	2.70	5.60
10.5	180	2.71	3.88	5.38
10.5	100	2.37	9.51	5.49
10.5	55	2.47	6.86	5.18
10.5	34	2.35	10.12	4.62
6	315	2.85	2.06	5.49
6	180	2.92	1.79	5.49
6	55	2.52	5.06	4.13
6	34	2.61	4.93	3.90

Tests were performed in this study at two R ratios, R= 0.7 and 0.05 where

$$R = \frac{K_{min}}{K_{max}}. \qquad (3)$$

The high R ratio was initially chosen to minimize the effects of crack closure near the threshold stress intensity range. However, the second round of tests was performed at R=0.05 to better match the loading in typical applications for this material and to better match standard testing procedures in the superalloy industry.

Experimental Results and Discussion

Tests on the AF2-1DA-6 material at R=0.7 with sinusoidal loading at a frequency of 10 Hz revealed little variation in the Paris law crack growth with microstructure or processing parameters. As seen in Figure 5 and Table I, the fatigue crack propagation rates in all materials are very similar in the Paris law regime regardless of the grain size or the cooling rate from solution heat treatment. All of the curves fall within in a tight band of uncertainty. Fits of the Paris law for fatigue crack growth, equation (1), to the data resulted in the values for the emperical constants, C and m, given in Table I. Regardless of grain size, the exponent, m, falls in the range from 2.35 to 2.92 and the multiplicative constant, C, falls in the range from 1.79×10^{-12} to 10.12×10^{-12}. The average value of the exponent, determining the slope of the Paris law regime in the figures, is 2.63 with all values falling within 10% of the average.

Threshold measurements are tabulated in Table I. These values range from 3.90 $MPa\ m^{1/2}$ to 5.60 $Mpa\ m^{1/2}$. Measurement between duplicate tests did show some scatter. The exact reason for the scatter is not known at this time. Threshold measurements can be sensitive to numerous factors including microstructure, mean stress and environment, and the mechanisms leading to crack arrest---and the subsequent measurement of threshold under load shedding conditions such as these---which can be statistical in nature. That is, the precise cause of the threshold value may depend upon stochastic variables and therefore the measured value may vary in a stochastic fashion. To be sure, more investigation of the threshold mechanisms is needed through the use of further fatigue crack propagation tests and microscopy of the cracked material when growth rates are very low, i.e. near threshold. In the interest of saving time and due to limited resources for this project, such investigations were not performed here.

TABLE III. Fit parameters for fatigue crack propagation rate data of AF2-1DA-6 for two grain sizes at several different cooling rates from the solution heat treatment.

Grain Size (avg.) ASTM	Cooling Rate °C/min	Exponent M	Multiplicative Constant $C\ (\times 10^{-12})$
10.5	315	2.77	8.33
10.5	100	3.33	5.81
10.5	34	2.93	26.7
6	315	2.86	13.1
6	100	2.76	19.0
6	34	2.83	24.1

Keeping in mind that the measurements of ΔK_{th} should be considered preliminary due to the scatter between tests and the incomplete information about the mechanisms of crack arrest in these tests, one might make some cursory observations about the

results. First, by examining the results shown in Figure 5 it is possible to surmise that the materials cooled at 55 °C/min and 34 °C/min have lower threshold values than those cooled at higher rates. In fact, further testing may show that the highest threshold may be achieved by the highest cooling rate. Decreasing the cooling rate seems to decrease ΔK_{th} as seen in Table I.

Tests were also performed at lower frequency and lower R ratio to further investigate the effects of grain size and cooling rate on FCGR in AF2-1DA-6 material. The coarse grain material tested in these conditions (Figure 6) appears to show a slightly slower FCGR as compared to the fine grain samples.

Tests at lower frequency and lower R ratio revealed that a cooling rate of 34 °C/min can result in Paris Law crack growth rates significantly higher than that for materials with higher cooling rate under the same loading conditions. The results of the testing are presented in Figure 6 and Table II where data for tests on AF2-1DA-6 with two grain sizes are presented. It is easy to see that the crack growth rate in the material with a grain size of ASTM 10 1/2 cooled at 34 °C/min is about three times higher than that in all the materials cooled at 100 or 315 °C/min. The fatigue crack growth rate in this same material is also about 1.7 times higher than the crack growth rate in the material with a grain size of ASTM 6 cooled at 34 °C/min.

Fitting of the Paris law to the data resulted in the values given in Table II. The exponent for all materials range from 2.76 to 2.97, with an average value of 2.85, in good agreement with the tests at 10 Hz and R=0.7. The results are roughly of the same slope at both frequencies of loading.

Conclusions

Fatigue crack propagation rate tests were performed on AF2-1DA-6 for various cooling rates from the solution heat treatment temperature and for two different grain sizes. Growth rates over a wide spectrum were measured. As expected a region of growth governed by the Paris law was observed for a domain of stress intensity factor ranges, ΔK. In the high frequency tests, for lower values of ΔK, a threshold stress intensity factor range was observed, ΔK_{th}, below which long-crack growth rates were vanishingly small or zero. For higher values of ΔK, a maximum stress intensity factor range was observed above which immediate failure of the specimen could be expected.

At R=0.7 and a loading frequency of 10 Hz, fatigue crack growth rates in the Paris law regime for AF2-1DA-6 were insensitive to changes in the cooling rate from the solution heat treatment, but they were also insensitive to changes in grain size from ASTM 6 to 10.5. However, it was found that the threshold stress intensity factor for long-crack growth rate quite possibly *was* sensitive to cooling rates during processing and that higher cooling rates lead to the more desirable higher threshold stress intensity factor range, ΔK_{th}.

It is not surprising that Paris law governed growth was not sensitive to changes in the processing parameters for the higher frequency of loading reported here. Changes in cooling rate will lead to changes in precipitate microstructure as seen in Figures 2 and 3, and it has long been known that Paris law growth is usually insensitive to changes in microstructure except for special values of frequency and special environmental conditions.(8)

It appears, however, that for low frequencies on the order of 2 Hz and low R ratios special conditions exist where Paris law growth *is* a function of microstructure, especially for materials tested at 649 °C. In previous work, it has been observed that P/M U720 exhibited a change in crack growth rate for different processing parameters. Likewise, AF2-1DA-6 as tested here demonstrated a dependence of Paris law crack growth rate upon processing parameters. Namely, for the lowest cooling rate, 34 °C/min, the crack growth rates were significantly higher in both the material with a grain size of 6 and the material with a grain size of 10. These growth rates were higher than the same materials cooled at much higher rates. The worst case was the material with a grain size of 10 cooled at 34 °C/min which had the highest crack growth rate of all. Materials cooled at lower higher cooling rates had lower crack growth rates. These crack growth rates were very similar to the crack growth rates observed in tests at 10 Hz as can be seen in Tables II and III.

Of interest is the behavior of the materials cooled at moderate cooling rates, 100° C/min. For a grain size of 10, a moderate cooling rate resulted in crack growth similar to the lowest cooling rate, for the same grain size. However, for a grain size of 6, the moderate cooling rate resulted in crack growth similar to the highest cooling rate.

It is not precisely clear why these materials exhibit a dependence of Paris law crack growth rate upon microstructure at low frequency and low R ratio, but is certainly worth noting to the metallurgist and manufacturing engineer. The mechanism must account for the difference in behavior of materials cooled at 100 C/min for the two grain sizes tested. At present, no definitive mechanism can be identified, however it appears that the slowest cooling rates lead to the worst properties in terms of high cycle fatigue crack growth response for these testing conditions.

References

1. T. Furman and R. Shankar, Advanced Materials and Processes 154, (9) (1998) 45-46

2. D. Furrer, Advanced Materials and Processes, 155, (3) (1999) 33-36

3. P.C. Paris, M.P. Gomey and W.P. Anderson, The Trend in Engineering, 13, (1961) 9-16

4. H. Hattori, M. Takekawa, D. Furrer, and R.J. Noel, Superalloys 1996, Kissinger et al. Eds., TMS Pubs, (1996) 705-711

5. ASTM Annual Book of Standards, E 647-95, Vol. 3.01, (1995) 578-614

6. A. Saxena, S.J. Hudak, Jr., J.K. Donald and D.W. Schmidt, J. Testing and Evaluation, 6, (1978) 167

7. ASTM Annual Book of Standards, E 399-90, Vol. 3.01 (1995) 412-442

8. R.O. Ritchie, in the Encyclopedia of Materials Science and Engineering M.B. Bever, ed. Pergamon Press, Oxford and New York, 3 (1986) 1650-1661

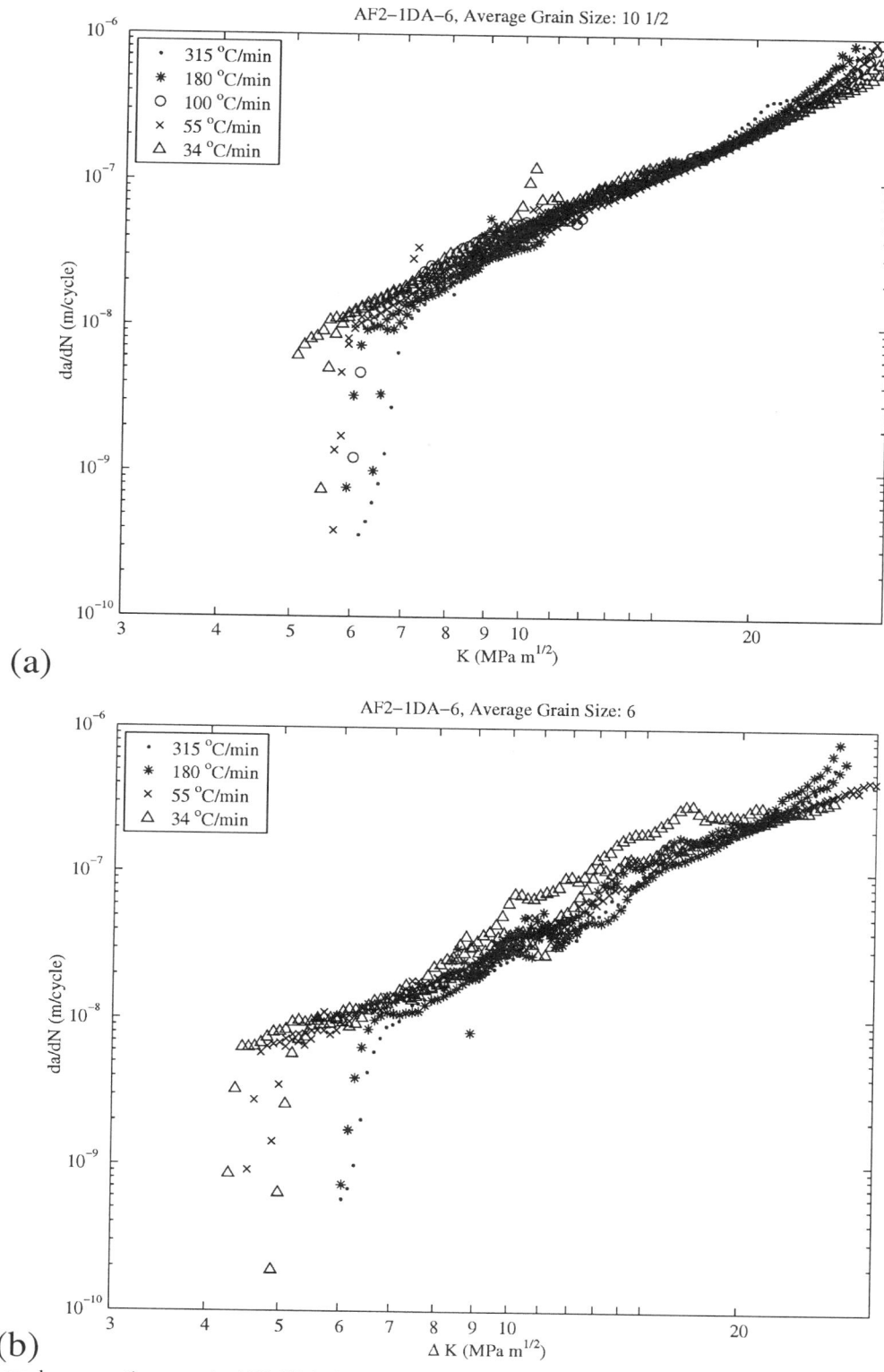

Figure 5. Fatigue crack propagation rates in AF2-1DA-6 tested at 1200 F with sinusoidal load at 10 Hz and R=0.7.

465

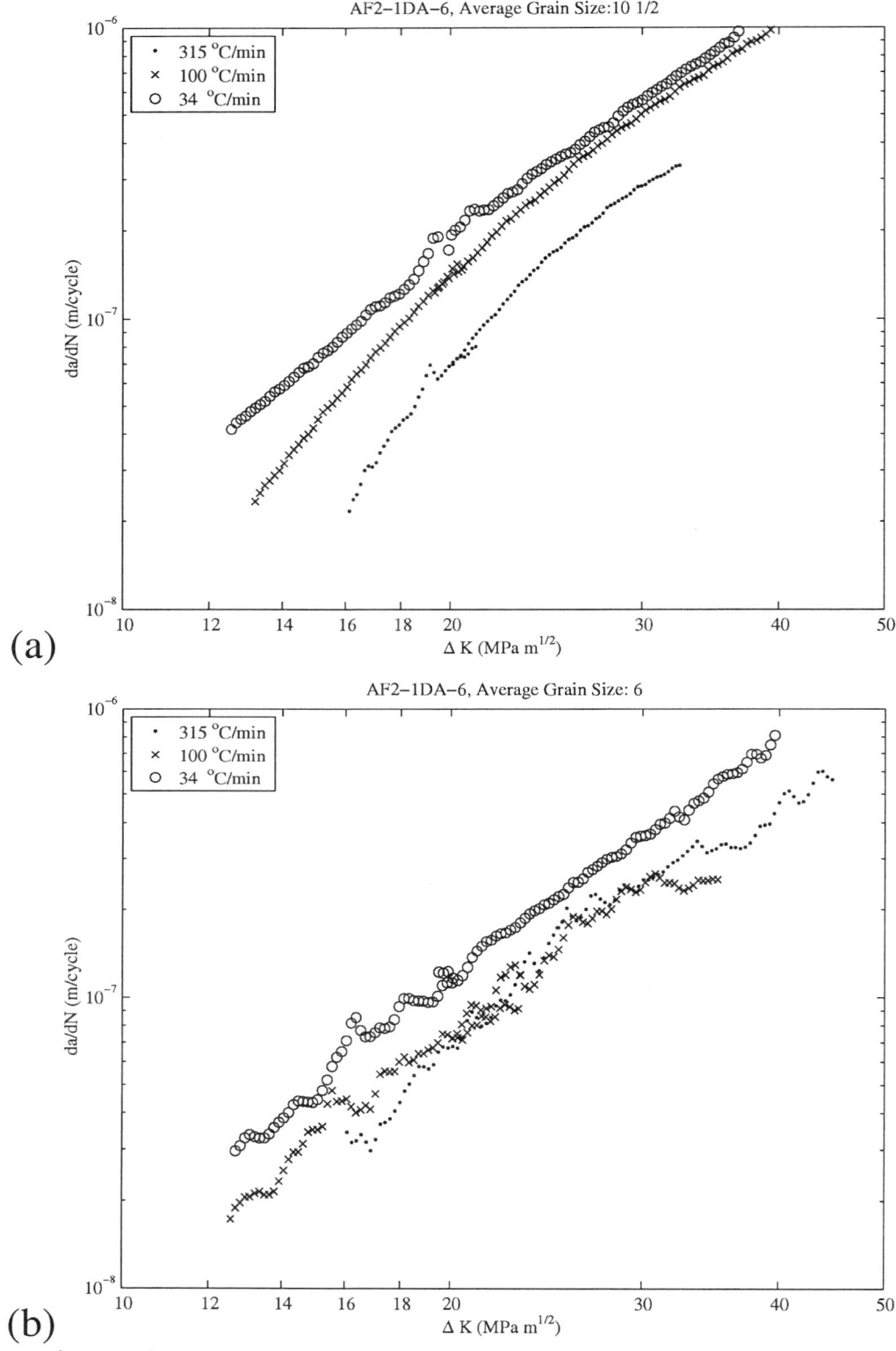

Figure 6. Fatigue crack propagation rates in AF2-1DA-6 tested at 1200 F with sinusoidal load at 2 Hz and R=0.05.

466

THE COMMON STRENGTHENING EFFECT OF PHOSPHORUS, SULFUR, AND SILICON IN LOWER CONTENTS AND A PROBLEM OF A NET SUPERALLOY

W. R. Sun, S. R. Guo, J. T. Guo, B. Y. Tong, Y. S. Yang, X. F. Sun, H. R. Guan, Z. Q. Hu

Department of Superalloys and Special Castings,

Institute of Metal Research, Chinese Academy of Sciences

72 Wenhua Road, Shenyang 110015, P. R. China

Abstract

The effects of phosphorus, sulfur, and silicon on the microstructure and the stress rupture properties of a Ni-Fe base superalloy were studied. Phosphorus had no effect on the grain size of the alloy, but sulfur reduced and silicon enlarged the grain size of the alloy. The 'net' alloy with the lowest amount of the minor elements had a problem of short rupture life. It was determined by analysis that the dislocations in the 'net' alloy were not firmly pinned by the segregation of the minor elements and could move easily, which increased the creep rate and reduced the rupture life of the alloy. In this case, a small addition of any minor element resulted in the same effect, significantly increasing the rupture life but drastically decreasing the rupture elongation. The concentration of the minor elements at the dislocations might reach the saturation point at a very low addition level. When the addition of the minor elements was above the level, their effect of inhibiting the dislocation movement did not increase again, and other influencing mechanisms by the elements took effect. Phosphorus had an effect to resist the intergranular ingression of the environmental oxygen. But the oxygen ingression along the grain boundaries was not the dominant factor leading to the failure of the 'net' alloy. Phosphorus at 0.016 wt. % strengthened the alloy by a perfect combination of the three influencing mechanisms, including the inhibition of the dislocation movement, the improvement of the intergranular precipitation, and the resistance to the intergranular ingression of the environmental oxygen. And hence the alloy had the longest life. Sulfur and silicon seemed having only one beneficial effect of pinning the dislocations, and hence they increased the rupture life of the alloy in the very small addition ranges. Sulfur had an abnormal effect to improve the rupture elongation of the alloy because it reduced the grain size noticeably.

Introduction

Minor elements play diverse and great roles in determining the microstructure and performance of Ni-base superalloys. Great efforts have been made throughout the history of superalloys to reduce the negative roles and allow full play to the positive roles of the minor elements. Several review papers [1-3] have summarized the major results up to 1984. Phosphorus, sulfur, and silicon are all classified as detrimental elements in these review papers. The effects of sulfur have been extensively studied and well documented. Generally speaking, sulfur reduces the ductility

Superalloys 2000
Edited by T.M. Pollock, R.D. Kissinger, R.R. Bowman,
K.A. Green, M. McLean, S. Olson, and J.J. Schirra
TMS (The Minerals, Metals & Materials Society), 2000

of superalloys by concentrating at the grain boundaries or forming the brittle Y-M_2CS phase [3]. Silicon has some solubility in Ni-base superalloys and is detrimental to malleability even in small quantities. It reduces the workability of the alloys by forming the M_6C phase or segregating at the grain boundaries. Si also forms the low melting point phases, such as Ni-Hf-Si, in superalloys [1]. Phosphorus has not been studied in detail and rare information can be retrieved from the literature up to 1984. The maximum solubility of phosphorus in nickel is very small (0.32%) and a eutectic exists at 875°C. The addition of phosphorus to Inconel 600 causes the susceptibility to intergranular attack in strongly oxidizing media.

In the late eighties, some minor elements were suggested to be controlled as low as possible to get premium quality superalloys with homogeneous microstructure and mechanical properties. Because the elements, including phosphorus, sulfur, and silicon, tended to be segregated in the interdendritic areas and drop the finishing solidification temperature of the alloys [4-7]. But the study shows that the segregation tendency is different not only for different elements in the same alloy but also for the same element in different alloys [8]. For example, phosphorus can be dissolved into Laves phase and its effect on the solidification of Inconel 718 alloy can be neglected when its content is lowered to some extent.

Recently, great attentions have been paid to the influencing mechanism of phosphorus on superalloys, because phosphorus was found significantly prolonging the rupture life of some alloys [8-10]. Phosphorus tends to be segregated at the grain boundaries [9, 11]. A strong interaction between phosphorus and boron has been found, and the interaction is an important, or even dominant factor that controlling the rupture life of the alloys [12, 13]. For example, the effect of phosphorus in alloy Waspaloy directly relies on the addition of boron [12]. A beneficial effect of phosphorus on the rupture life of the alloys comes from its improvement of the intergranular precipitation, by lowering the surface energy of the grain boundaries and interacting with carbon and boron [14, 15]. In Inconel 718 alloy, phosphorus was found to resist the environmental oxygen attack efficiently [15]. In China, Xie's group carried out a series of works to study and compare the effects of phosphorus in Ni-Cr-Fe, Ni-Cr-Fe-Mo and Ni-Cr-Fe-Mo-Nb-Ti-Al systems [16-18]. The co-segregation of phosphorus, chromium and molybdenum has been detected at the

grain boundaries [11], and a new point was raised that the abnormal effect of phosphorus might result from the effect of P-GB-Mo or P-GB-M (metal element) complex on the binding strength and ductility behavior of the grain boundaries. The assumption is supported by the results of the theoretical analysis and calculations [19]. The results and assumption demonstrate that phosphorus is not so poisonous in superalloys as in steels and supply with us the basis of employing phosphorus as a strengthening element but not a harmful one.

The effects of phosphorus, sulfur, and silicon on the microstructure and stress rupture properties of the alloy were investigated and compared in this study. The goal of this paper is to manifest the influencing mechanism of these elements.

Materials and Experimental

The materials were prepared in a vacuum induction furnace using high-purity raw materials. To minimize the compositional variations among the heats, the master alloy was prepared, and its composition (wt. %) was C0.011, Cr12.89, Ni43.54, W3.12, Mo1.59, Al1.71, Ti3.64, B0.003, Si0.059, S0.003, and Fe balance. And the master alloy was remelted to give ten 10kg ingots. The ingots were added into different contents of phosphorus, sulfur, and silicon, with the same content of carbon and boron. The remelting also made the alloy much more homogeneous. The contents of carbon, boron, phosphorus, sulfur, and silicon were analyzed and are listed in Table 1.

The ingots were forged into bars of 45-mm-square section at 1120°C, and rolled into round bars of 18mm in diameter. These materials were then given a standard heat treatment of 1120°C/2h, water cooled; 850°/4h, air cooled; and 750°C/24h, air cooled.

The microstructure was observed by using optical microscopy, transmission electron microscopy (TEM), and scanning electron microscopy (SEM). Phase identification was mainly carried out using selected area diffraction. And the trace phases were analyzed by x-ray diffraction (XRD) after anodic matrix dissolution. An electron microprobe was used to measure the composition of precipitates.

The stress rupture lives and elongation of the alloys were measured at 650°C under constant load of 637 MPa, and the fracture surfaces of the test pieces were examined by SEM.

Table 1. Contents of Carbon, Boron, Phosphorus, Sulfur, and Silicon of the Alloys (wt. %)

Alloy	P	S	Si	C	B
1	0.0005	0.002	0.070	0.038	0.0049
2	0.0009	0.002*	0.070*	0.039	0.0044
3	0.016	0.002*	0.070*	0.041	0.0045
4	0.051	0.002*	0.070*	0.036	0.0047
5	0.0005*	0.0085	0.070*	0.040	0.0047
6	0.0005*	0.017	0.070*	0.042	0.0045
7	0.0005*	0.046	0.070*	0.042	0.0052
8	0.0005*	0.002*	0.083	0.041	0.0050
9	0.0005*	0.002*	0.38	0.047	0.0042
10	0.0005*	0.002*	0.91	0.039	0.0045

*not added when remelting and kept at the lowest level as in the no.1 alloy

Results and Discussion

As shown in Fig.1, phosphorus scarcely affected the grain size of the alloys (see Figs. 1a and 1b). Comparatively, sulfur noticeably decreased the grain size (see Figs. 1a and 1c), and silicon increased the grain size when its content was as high as 0.91 wt. % (see Figs. 1a and 1d). The precipitate lines were observed sometimes lying in the rolling directions. Phosphorus had little or no effect on the appearance of the precipitate lines. However, sulfur increased, and silicon decreased the appearance of the precipitate lines.

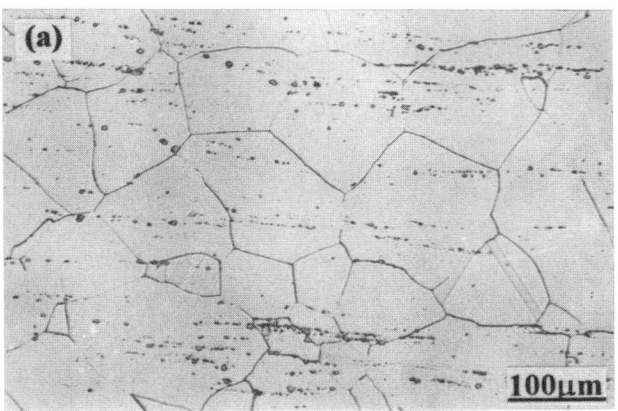

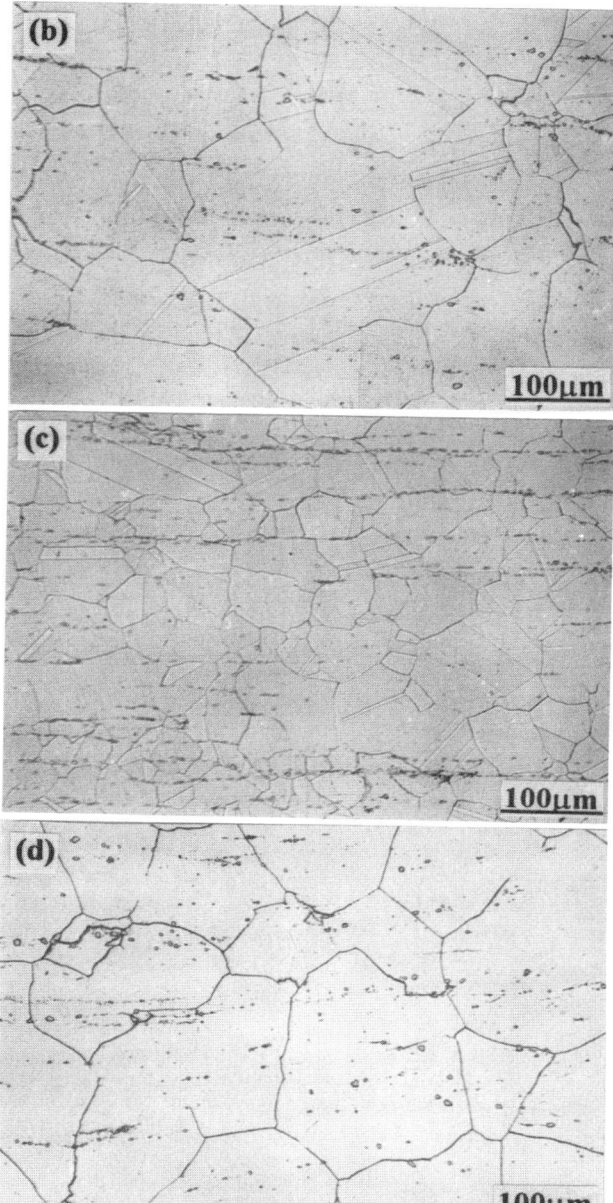

Fig.1 Effect of phosphorus, sulfur and silicon on the microstructure of the alloys. (a) no.1 alloy; (b) no.4 alloy; (c) no.7 alloy; (d) no.10 alloy

By using XRD, it was determined that the trace phases were made up of the MC carbide, and a small amount of M_3B_2 boride and Y-M_2CS sulfide. Phosphorus increased the precipitation of $M_{23}C_6$ carbide. The amount of $M_{23}C_6$ carbide was so low in the no.1 alloy that it was not detected by XRD. Sulfur naturally increased the precipitation of the Y-M_2CS phase. Silicon increased the precipitation of Laves phase and decreased the precipitation of MC carbide.

469

The MC carbide precipitated at 1380°C during the solidification of the alloy. The influence of phosphorus and sulfur on the MC carbide crystallization could be neglected because the amounts of the residual liquids were large and hence the concentrations of the two elements in the liquids were lower at the temperature the carbide precipitated [8]. Silicon also had insignificant effect on the carbide precipitation when its content was lower than 0.38 wt. %. However, the addition of silicon at 0.91 wt. % decreased the carbide precipitation, and the primary Laves phase precipitated at 1100°C in the solidification of this alloy [8]. The $Y-M_2CS$ phase precipitated at 1280°C in the solidification of the no.1 alloy, and the precipitation temperature increased with the increasing sulfur content [8].

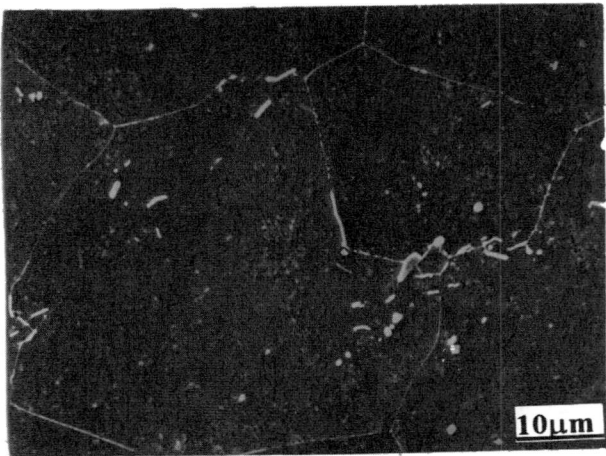

Fig.2 Distribution of $Y-M_2CS$ phase in the no.7 alloy

The MC carbides were stable at the temperature that the thermal mechanical processing and the heat treatment were performed. They were broken into pieces by forging and rolling and distributed within the grain interior and on the grain boundaries. The recrystallization was inhibited by the MC carbide particles. Phosphorus did not affect the grain size of the alloys because it did not affect the precipitation of the primary MC carbide. The $Y-M_2CS$ phase was also stable and the small rods of the broken pieces were also distributed within the grain interior or at the grain boundaries as shown in Fig.2. For this reason, sulfur noticeably reduced the grain size of the alloys. Comparatively, the primary Laves phase was not stable at the temperature of the thermal mechanical processing and the heat treatment, and it was dissolved while heating at 1120°C and failed to inhibit the recrystallization. As a result, silicon increased the grain size of the alloys. The Laves particles could only be observed on the grain boundaries, indicating that they were not the primary precipitates formed during the solidification.

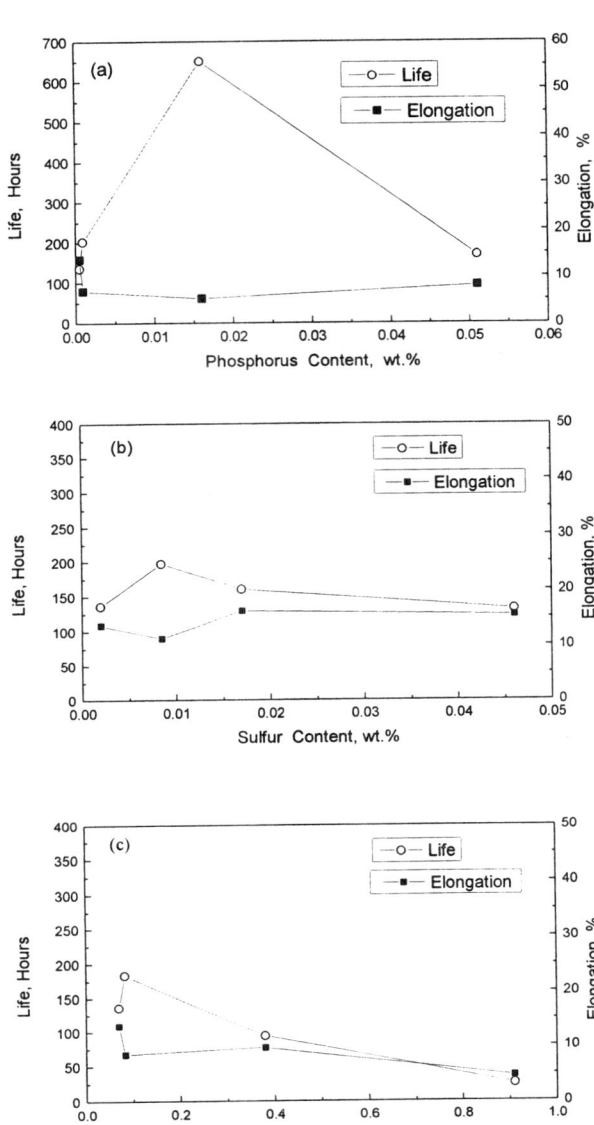

Fig. 3 Effect of phosphorus, sulfur, and silicon on the stress rupture properties of the alloys.

The effects of phosphorus, sulfur, and silicon on the stress rupture properties of the alloys are shown in Fig. 3. In some content ranges, phosphorus, sulfur and silicon all increased the rupture lives and simultaneously reduced the rupture elongation of the alloys. For the alloys added with phosphorus and sulfur, the longest lives are corresponded to the smallest elongation. For the alloys added with silicon, the situation is the same when the silicon content is lower than 0.35 wt. %. It should be noted that all of the life lines go up with a significantly steep slope in the lower addition ranges (phosphorus 0.0005-0.0009 wt. %; sulfur 0.002-0.0085; silicon 0.070-0.083 wt. %) of the three elements, and correspondingly the elongation lines go down drastically. The results indicate that the three elements in their lower addition ranges might have a common influencing mechanism on the stress rupture properties of the alloys, though their influencing mechanisms are in general different from each other in the higher addition ranges.

The nickel-base superalloys obtain their high temperature strength mainly from three types of strengthening, including precipitation strengthening, solution strengthening and grain boundary strengthening. And the minor elements influence the stress rupture properties of the alloys mainly by modifying these strengthening mechanisms. The amount of the γ' phase was about 20wt. % in this alloy, and phosphorus, sulfur and silicon slightly increased the particle size and precipitation amount of the γ' phase [14]. In the lower addition ranges, the effects of these elements on the γ' precipitation were insignificant and could be neglected. Therefore, it could be determined that the special significant effect of the three elements in the lower addition ranges was not originated from their influence on the precipitation strengthening mechanisms. Phosphorus, sulfur, and silicon are all surface-active elements, they tend to be segregated at the interfaces of various phases and the grain boundaries, and their concentrations in the γ matrix are generally very small even when they are highly added. That is to say that the influence of the three elements in the lower addition ranges on the solution strengthening of the alloys was also not the reason for their common acute effect. The segregation of phosphorus, sulfur and silicon at the grain boundaries of superalloys has been verified by many previous studies [9, 11], and it was most possible that the great effects of the three elements in the lower addition ranges resulted from their modification on the grain boundary strength. However, this influencing mechanism was also impossible. For example, sulfur is typically a harmful element reducing grain boundary strength [3]. Phosphorus has a bad reputation to damage the grain boundary strength of steels. But both of them presented the same dramatic effect on the stress rupture properties in the lower addition ranges. The minor elements can also modify the grain boundary strength by affecting the grain boundary precipitation.

In the previous study of this alloy, phosphorus was found to increase the intergranular precipitation of $M_{23}C_6$ carbide and M_3B_2 boride, and it has been taken as one of the main reasons for phosphorus to prolong the rupture life of the alloy [14]. However, the observation showed that the intergranular precipitation of the alloy was scarcely changed when the phosphorus content is in the range of 0.0005 - 0.0009 wt. %. The addition of silicon in the range of 0.070-0.083 also made no difference on the grain boundary precipitation. Because silicon increased the intergranular precipitation of the harmful Laves phase, it seemed impossible that the rapid increment of the rupture life was caused by the effect of silicon on the intergranular precipitation even if silicon did noticeably increase the grain boundary precipitation within such a small addition range. The effect of sulfur on the intergranular precipitation was more noticeable than that of phosphorus and silicon because it directly took part in the formation of $Y-M_2CS$ phase. However, compared with phosphorus and silicon in their lower content ranges, the life line goes up and the elongation line goes down more slowly when the sulfur content is in the range of 0.002 - 0.0085 wt. % (see Fig.3). Apparently, the modification of the three elements on the intergranular precipitation was also not the reason for their common dramatic influencing mechanism on the stress rupture properties in the lower addition ranges.

The testing condition or the service environment is an important factor for determining the long-time properties, such as the stress rupture properties of the alloys. Phosphorus has been found to resist the grain boundary ingression of the environmental oxygen attack [15]. The similar effect was also found in this alloy, phosphorus greatly reduced the formation of the oxides at the grain boundaries on the surfaces of the specimens. However, it was at least not the main reason for phosphorus to increase the rupture life in the range of 0.0005 - 0.0009 wt %. If the intergranular ingression of the environmental oxygen was the predominant reason causing the premature failure of the stress rupture specimens, the cracks would form and propagate quickly along the grain boundaries and the deformation of the no.1 alloy should be smallest. However, the elongation of the no.1 alloy was in fact the largest among the alloys added with phosphorus. The real process of the failure was that the opening of the surface grain boundaries caused the oxygen ingression and the oxidation, but not the contrary. Sulfur had no effect on the grain boundary

471

oxidation. Although silicon efficiently resisted the oxidation of the alloys, its effect in the range of 0.070 – 0.083 wt. % was not noticeable. In addition, the silicon addition also reduced the rupture elongation of the alloys as phosphorus did. Therefore, the improvement of the environmental oxygen attack resistance by the three elements in the lower content ranges can also be determined not to be the reason for their common significant effect.

With the exception of the grain boundaries and the interfaces between the phases, the loose structures such as dislocations are also the places where the minor elements tend to be segregated. It is well known that phosphorus is prone to form 'atmosphere' at the dislocations in steels and cause temper brittleness. Because other possible influencing mechanisms have been excluded as above, the concentration of the three elements at the dislocations might be the true reason leading to their common significant effect in the lower content ranges. Because the total amount of the minor elements in the 'net' no.1 alloy was so low, the concentration of the minor elements at the dislocations might not reach the saturation point. Without the pinning by the concentration of the minor elements, the dislocations moved easily, leading to the large deformation of the metals, the quick formation and propagation of the cracks at the grain boundaries, and the premature failure of the specimens. This explains the high rupture elongation of the no.1 alloy. In this case, a small increment of the minor elements could greatly increase their concentrations at the dislocations. The pinning effect was thus quickly enhanced and the movement of the dislocations was efficiently inhibited. As a result, the deformation became more difficult and the creep rate was correspondingly slowed down. This accords well with the results shown in Fig.3 that the rupture lives were greatly increased and the rupture elongation largely reduced by a small addition of any of the three elements.

In addition, phosphorus and silicon existed most likely as free atoms instead of forming the precipitates when their contents were lower, and subsequently a small addition increment could result in a greater concentration increase at the dislocations. Sulfur formed the Y-M_2CS phase even when its content is as low as 0.002 wt. % in the no.1 alloy, and therefore the increment of sulfur content led to a relatively small increase of the sulfur

concentration at the dislocations. As a result, the effect of sulfur was slighter than that of phosphorus and silicon in their lower content ranges (see Fig.3). This may be an extraneous evidence of the above assumption.

The concentration of any element at the dislocations resulted in the same pinning effect, this explains the common influence of the three elements in their lower content ranges.

As shown in the Fig.3a, after the very sharp increase of the rupture life in the range of 0.0005 – 0.0009 wt. %, the increasing rate of the life line becomes lower when the phosphorus content is in the range of 0.0009 – 0.016 wt. %. Correspondingly, the elongation line shows an inverse tendency. Actually, the concentration of phosphorus at the dislocations might quickly reach the saturation point with a small increment of the phosphorus addition. With the further increment of the phosphorus addition, its effect on retarding the dislocation movement ceased to increase. In this case, other influencing mechanisms began to take effect. Except the effect of improving the resistance to the intergranular ingression of the environmental oxygen, phosphorus increased the intergranular precipitation noticeably [14].

Different from phosphorus, the rupture lives of the alloys added with sulfur and silicon began to reduce after the rapid increase in the lower content ranges. Sulfur had no effect on the intergranular ingression of the environmental oxygen, and the Y-M_2CS phase was in general a harmful phase. Therefore, different from phosphorus, sulfur did not increase the rupture life of the alloys after its concentration at the dislocations reached saturation point. Silicon also reduced the rupture life of the alloys at a higher addition level because it increased the precipitation of harmful Laves phase.

Sulfur is typically a harmful element to reduce the plasticity of the alloys. However, it presented an abnormal effect to increase the elongation of this alloy as shown in Fig.3b. Sulfur mainly formed the Y-M_2CS phase. This greatly reduced its concentration at the grain boundaries and hence the plasticity of the alloy was not damaged largely by the segregation mechanism. But the formation of the Y-M_2CS phase largely decreased the grain size, and hence the rupture elongation was increased.

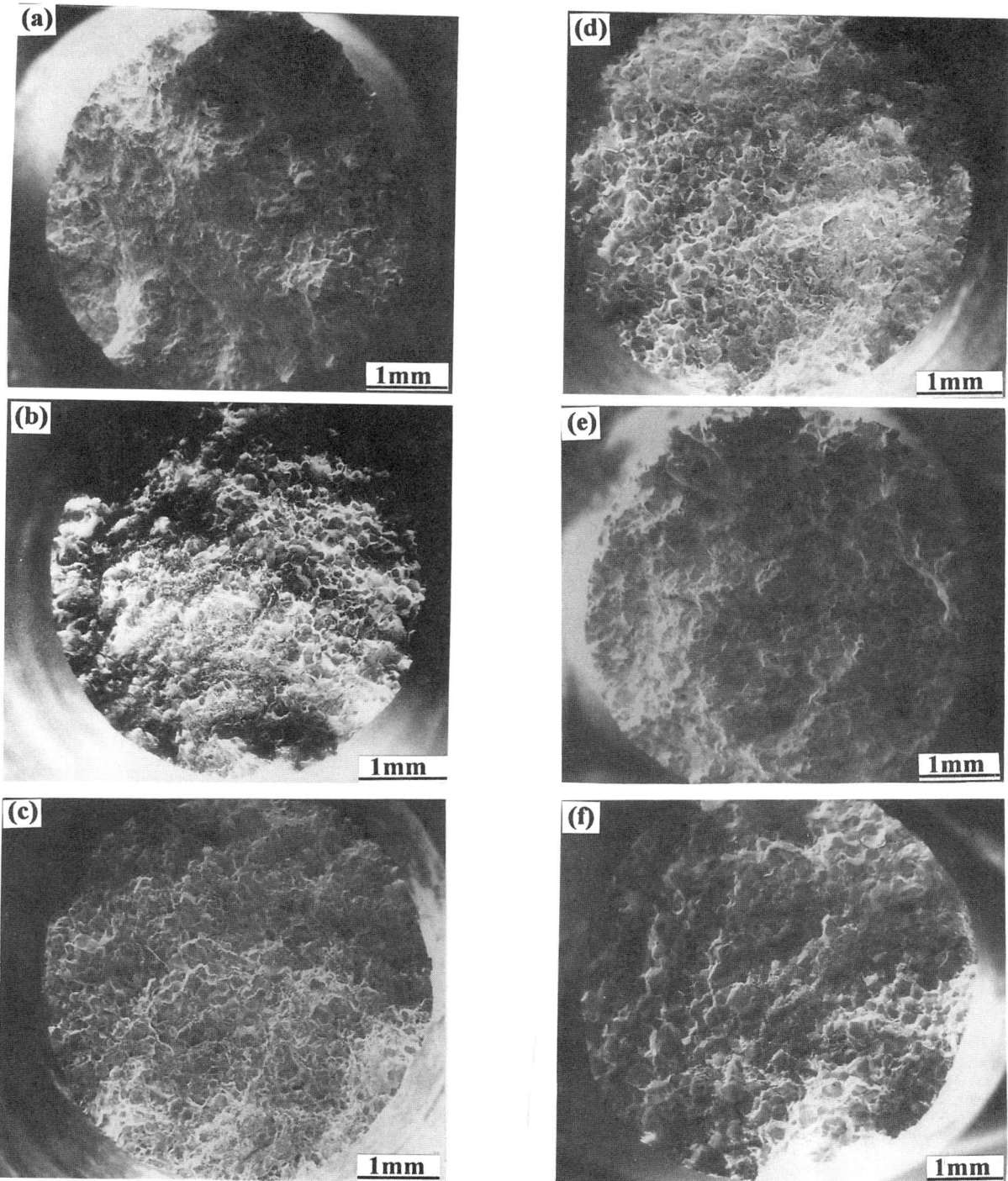

Fig.4 Effect of phosphorus, sulfur, and silicon on the fracture mode of the alloys

(a) no.1 alloy; (b)no.2 alloy; (c) no.3 alloy

(d) no4 alloy; (e) no.9 alloy; (f) no10 alloy

It is apparent that the 'net' superalloy had a problem of short life. When the alloy was very net or the total amount of the minor elements was very small, the dislocations could move easily. The cracks were quickly formed due to the rapid deformation of the grains, and they propagated rapidly because the grain boundaries were not sufficiently strengthened by the intergranular precipitation [14]. In this case, the increased addition of any minor element that tends to be segregated at the dislocations can efficiently prolong the rupture life but simultaneously reduce the rupture elongation of the alloy.

The fractographs of the alloys added with phosphorus are shown in Figs. 4a - 4d. The fracture surface can be divided into two areas. One is the 'undulated area' with smooth exposing grain boundary surfaces. And the other is the 'flat area' where the detail character is not clearly shown at the lower magnification of Fig.4. The fine microstructure of the 'flat area' is shown in Figs. 5a and 5b, there are many small pits with a plastic tearing character on the area. In addition, in the flat area some of the grain boundaries perpendicular to the fracture surface are widened due to the oxidation (see Fig.5b). It is no doubt that the grain boundaries parallel to the fracture surfaces were more prone to be oxidized under load. Then the small pits on the fracture surface might not be the plastic tearing but the places where the particles of the oxides were formed and fell off when the samples were broken. The flat area was liable to be taken as the area that was finally pulled apart, because there are many small pits with a character looked like the plastic tearing. But it was actually the area where the grain boundaries were initially opened, and then oxidized while soaking at 650°C under the load.

It is shown in Fig.4c that most area of the fracture surface of the no.3 alloy is ' undulated area' with exposing grain boundary surface, indicating that the grain boundaries of the alloy were well prevented from the environmental oxygen ingression. But in Fig.4d, the fraction of the 'flat area' of the no.4 alloy increased again. Considering that phosphorus increased the resistance to the environmental oxygen attack, the 'flat area' formed by the oxidation of the grain boundaries in the no.4 alloy should be less than that in the no.3 alloy. Actually, the grain boundaries in the no.4 alloy were easy to open due to the excessive precipitation caused by the high addition of phosphorus [14]. The opened grain boundaries facilitated the oxygen ingression and hence led to the increased fraction of the flat area in the no.4 alloy. This reflects that the improvement of the intergranular precipitation was the predominant influencing mechanism for phosphorus to prolong the rupture life of the alloy [14]. This also indicates that the segregation of phosphorus atom at the grain boundaries might not damage but increase the boundary strength [16-18]. Considering that the rupture life of the no.3 alloy was more than 600 hours, the improvement of the resistance to the intergranular oxygen ingression must be a necessary factor for phosphorus to prolong the rupture life of the alloys. Apparently, phosphorus at 0.016 wt. % maximally strengthened the alloy by a perfect combination of its three influencing mechanisms, including the inhibition of the dislocation movement, the improvement of the intergranular precipitation, and the resistance of the environmental oxygen attack. And hence the no.3 alloy had the longest rupture life among all of the alloys studied.

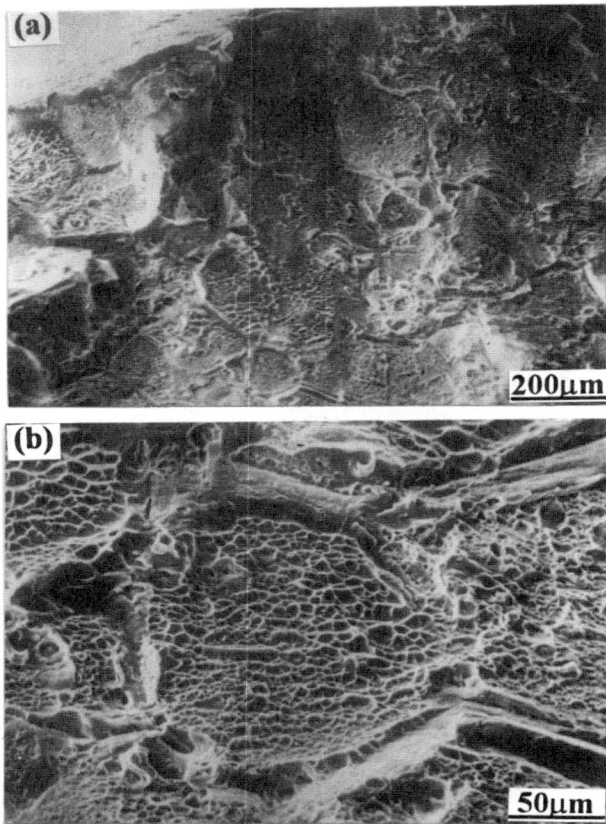

Fig.5: (a) Oxidized area on the fracture surface of the No.2 alloy

(b). An area from figure (a) at high magnification

474

Because sulfur had no effect to resist the environmental oxygen attack, the fractographs of the alloys added with sulfur were similar with that of the alloys with a lower phosphorus addition. Because silicon increased the Laves precipitation, the cracks were easy to form at and propagate along the grain boundaries. The opened grain boundaries facilitated the oxygen ingression and hence the 'flat area' of the no.9 alloy was not reduced (see Fig.4e), though silicon had a noticeable effect to resist the oxidation. However, the 'flat area' of the no.10 alloy was reduced (see Fig.4f), though the cracks could form at and propagate along its grain boundaries more easily. This is for two reasons, one is that the rupture life of the no.10 alloy was very short, the other was that the resistance of silicon at 0.91 wt. % to the oxidation is very high. As a result, the fracture surface of No.10 alloy was very smooth as shown in Fig.6.

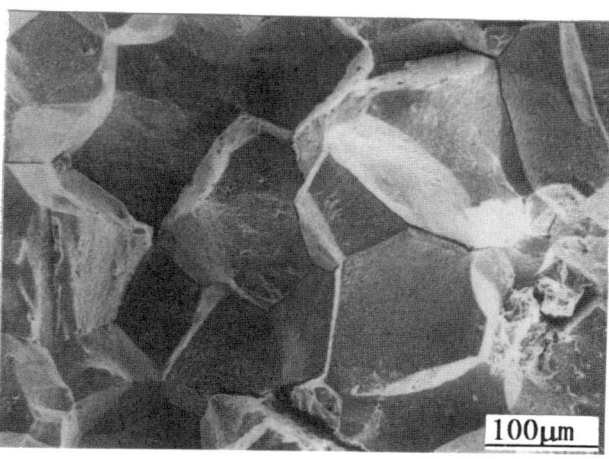

Fig.6 Smooth grain boundaries on the fracture surface of the no.10 alloy

Conclusions

1. The recrystallization of the alloys was inhibited by the MC carbide. Phosphorus scarcely influenced the MC carbide precipitation and hence had no effect on the grain size. Silicon decreased the MC carbide precipitation and hence enlarged the grain size of the alloys. Sulfur reduced the grain size because the $Y-M_2CS$ phase inhibited the recrystallization of the alloy.

2. The net superalloy with the lowest amount of the minor elements had a problem of short rupture life, because the dislocations in the alloy were easy to move without the pinning by the concentration of the minor elements. The addition of the minor elements in some lower content ranges significantly increased the rupture lives and drastically decreasing the rupture elongation of the alloys. The common significant effect of the minor elements was caused by their concentration at the dislocations which largely inhibited the dislocation movement.

3. Phosphorus influenced the rupture properties of the alloys mainly by three mechanisms, including the inhibition of the dislocation movement, the enhancement of the intergranular precipitation, and the resistance to the environmental attack. The alloy with 0.016 wt. % phosphorus addition had the longest life due to the perfect combination of the three strengthening effect of phosphorus at this content.

4. Sulfur increased the rupture elongation because it reduced the grain size of the alloy. Silicon increased the intergranular precipitation of Laves phase but resisted the oxidation.

Acknowledgement

This work was founded by National Natural Science Foundation of China under the contract of No. 59801013

References

1. R. T. Holt and W. Wallace, "Impurities and trace elements in nickel-base superalloys," International Metals Reviews, 21(1976), 1-24.

2. M. McLean and A. Strang, "Effect of trace elements on mechanical properties of superalloys," Metals Technology, 11(1984), 454-464.

3. G. W. Meetham, "Trace elements in superalloys-an overview," Metals Technology, 11 (1984), 414-418.

4. Y. Zhu et al., "A new Way to Improve the Superalloys," Superalloys 1992, Eds. S. D. Antolovich et al., (Warrendale, PA: The Minerals, Metals & Materials Society, 1992), 145-154.

5. H. Q. Zhu et al., "The effect of silicon on the microstructure and segregation of directionally solidified IN738 superalloy," Materials at High Temperatures, 12 (4) (1994), 285-291.

6. W. R. Sun et al., "Effect of sulfur on the solidification and segregation in Inconel 718 alloy," Materials Letters, 31 (1997), 195-200.

7. C. Chen, R. G. Thompson and D. W. Davis, " A Study of Effects of Phosphorus, Sulfur, Boron and Carbon on Laves and Carbide Formation in Alloy 718,"Superalloys 718, 625, 706 and Various Derivatives, Ed. E. A. Loria, (Warrendale, PA: The Minerals, Metals & Materials Society, 1991), 81-90.

8. Z. Q. Hu, W. R. Sun and S. R. Guo, "Effect of P, S and Si on the solidification, segregation, microstructure and mechanical properties in Fe Ni base superalloys," Acta Metallurgica Sinica, 9 (1996), 443-449.

9. W. D. Cao, and R. L. Kennedy, "The Effect of Phosphorus on the Mechanical Properties of Alloy 718," Superalloys 718, 625, 706 and Various Derivatives, Ed. E. A. Loria, (Warrendale, PA: The Minerals, Metals & Materials Society, 1994), 463.

10. Xishan Xie et al., "THE ROLE OF PHOSPHORUS AND SULFUR IN INCONEL 718," Superalloys 1996, Eds. R. D. Kissinger et al., (Warrendale, PA: The Minerals, Metals & Materials Society, 1996), 599-606.

11. J. X. Dong et al., "The segregation of sulfur and phosphorus in nickel-base alloy 718," Acta Metallurgica Sinica, 10 (6) (1997), 510-514.

12. W. D. Cao and R. L. Kennedy, "Phosphorus and Boron Interaction in Nickel – Base Superalloys," Superalloys 1996, Eds. R. D. Kissinger et al., (Warrendale, PA: The Minerals, Metals & Materials Society, 1996), 589-597.

13. Wei-Di Cao and R. L. Kennedy, "Effect and mechanism of phosphorus and boron on creep deformation of alloy 718," Superalloys 718, 625, 706 and Various Derivatives, Ed. E. A. Loria, (Warrendale, PA: The Minerals, Metals & Materials Society, 1997), 511-520.

14. W. R. Sun et al., "Effect of phosphorus on the microstructure and stress rupture properties in an Fe-Ni-Cr base superalloy," Metallurgical and Materials Transactions, 28A (1997), 649-654.

15. W. R. Sun et al., "Effect of phosphorus on the δ-Ni$_3$Nb phase precipitation and stress rupture properties in alloy 718," Materials Science and Engineering, A247 (1998), 173-179.

16. Xishan Xie et al., "Segregation behavior of phosphorus and its effect on microstructure and mechanical properties in alloy system Ni-Cr-Fe-Mo-Nb-Ti-Al," Superalloys 718, 625, 706 and Various Derivatives, Ed. E. A. Loria, (Warrendale, PA: The Minerals, Metals & Materials Society, 1997), 311-542.

17. Xishan Xie et al., "An abnormal effect of phosphorus on mechanical properties in nickel-base superalloys of Ni-Cr-Fe, Ni-Cr-Fe-Mo and Ni-Cr-Fe-Mo-Nb-Ti-Al system," The Third Pacific Rim International Conference on Advanced Materials and Progressing (PRICM 3), Eds. M. A. Imam et al., (Warrendale, PA: The Minerals, Metals & Materials Society, 1998), 215-222.

18. Xingbo Liu et al., "Investigation of the abnormal effects of phosphorus on mechanical properties of Inconel 718 superalloy," Materials Science and Engineering, A270 (1999), 190-196.

19. Xingbo Liu et al., "Molecular dynamics simulation on phosphorus behaviors at Ni grain boundary", (University of Science and Technology Beijing, to be published).

Simulation of microstructure of nickel base alloy 706 in production of power generation turbine discs.

J. Huez, J-F. Uginet

FORTECH
BP 173 – 09102 PAMIERS cedex - FRANCE

Abstract

In die forging the conditions of strain and of the subsequent heat cycle have a significant influence on the structure and properties of the final product, that is why the determination and modelling of grain size development during and after deformation is of primary interest.

In this paper we present the simulation procedure and the methodology used to develop a model which can predict microstructural changes in alloy 706 during hot forming.

High temperature compression test samples were used to develop a constitutive equation for the alloy 706 which was then applied via a finite element model to firstly simulate industrial forging process and finally optimize it.

Important parameters such as critical deformation to initiate recrystallization have been determined.

The model was found to accurately predict the recrystallized grain size and the percentage of recrystallization on compression test samples as well as on industrial parts.

Superalloys 2000
Edited by T.M. Pollock, R.D. Kissinger, R.R. Bowman,
K.A. Green, M. McLean, S. Olson, and J.J. Schirra
TMS (The Minerals, Metals & Materials Society), 2000

Introduction

High temperature nickel base alloy 706 is used for power generation applications requiring excellent mechanical properties at elevated temperatures.

These properties, principally high strength and good low cycle fatigue resistance, necessitate the process being controlled accurately to obtain the necessary microstructures and to ensure similar mechanical properties for every forging.

But the progress in high pressure turbine discs is measured not only in technical terms (mechanical properties, parts shape and size) but also in economical terms (input material, manufacturing cycles, press power).

In this context, numerical simulation becomes an essential tool to improve product quality and process efficiency.

The finite element method, used in Forge2® [1], gives the strain and thermal parameters of each instant of the operation. Through the analysis of these results it is possible to determine the evolution of the microstructure in relatively simple cases, but for more complex ones, including re-heating and cooling sequences, it would appear that we need to control the prediction of the microstructure. It is also the only way to optimize a microstructure and thus the process.

The aim of this work is to present the simulation results obtained with a model which incorporates microstructural phenomena taking place during hot forging of alloy 706.

Material

The chemical composition of alloy 706 used in this investigation is given in Table I.

The microstructure of alloy 706 is governed by the fcc lattice of the γ matrix associated with characteristic precipitates such as the aluminum rich fcc γ' phase. The presence of this precipitate is the essential strengthening mechanism in this alloy. γ' will transform to a stable hexagonal η phase during elevated temperature exposure. The solvus temperature of the η phase in the present material is ~ 954°C (1750°F).

TABLE I Chemistry of alloy 706 used. Weight Percentage.

Ni	Fe	Cr	Nb	Ti	Al	Mn	C	Si	S
40.55	38.74	15.83	2.84	1.71	.18	.14	.034	.03	.006

The main steps of the processing route of alloy 706, have been extensively describe elsewhere [2], we will summarize here the essential elements.

The alloy 706 is triple melted: the primary melting is a Vacuum Induction Melting (VIM), followed by an Electro Slag Remelting (ESR), and a final Vacuum Arc Remelting (VAR).

An upsetting and drawing of the ingot is carried out well above the recrystallization temperature before the close die forging. Generating a uniform level of strain throughout the workpiece is key to producing a uniform microstructure . Grain size of ASTM 0-1 is obtained after billetizing (see figure 1).

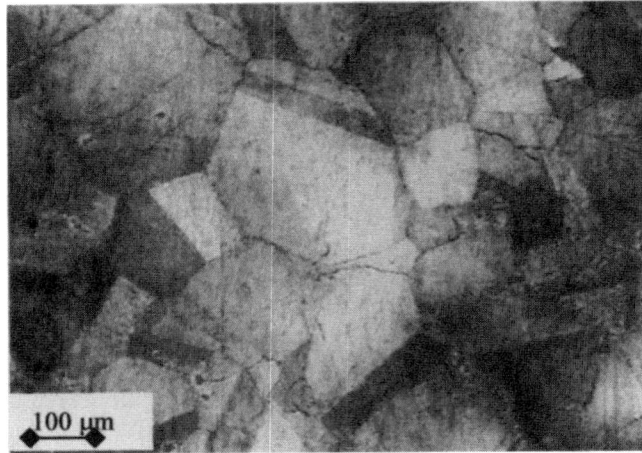

Figure 1 : billet microstructure

Modelling

By the use of the finite element method the prediction of mechanical, thermal and structural conditions is possible but material data has first to be well known.

Knowledge of both the thermomechanical behavior and thermophysical properties of the material as well as that of the kinetic tools is essential to give an accurate description of the deformation process.

The constitutive law

In the first phase, a rheological model of the material was defined through high temperature compression tests.

Typical flow curves were obtained and analyzed.

The constitutive law used is the Norton-Hoff model, expressed as follow :

$$\sigma = \sqrt{3}^{m+1} K(T) \dot{\varepsilon}^{m} (\varepsilon_0 + \varepsilon)^n$$

with σ the stress flow
$K(T) = K_0 \exp(Q/RT)$ the strength constant of the material,
ε the strain
ε_0 the strain hardening regularization term
$\dot{\varepsilon}$ the strain rate
n the strain hardening coefficient
m the strain rate hardening coefficient

The temperature is not only studied within the range of the forging window but also until the ambient temperature, to simulate correctly the heat treatment.

Interfacial and thermophysical properties of alloy 706

Ring tests are performed to reach a compromise between friction reduction and thermal insulation but, above all to adapt good, effective Coulomb friction coefficient for the calculation. The knowledge of the influence of the friction coefficient on the deformation process permits the increase or the decrease of the minimum strain in the workpiece which is a very important parameter leading to the microstructure and to partial recrystallization.

The thermal constants are also determined such as calorific capacity, thermal conductivity, emissivity, and global transfer coefficient : heat transfer coefficient between die and workpiece and exterior area interface. So losses by radiation and convection are taken into account as are conduction effects and adiabatic heating produced during deformation.

The kinetic tools

In a first approach it can be considered a constant velocity during the simulation, but in reality the press velocity decreases when a force limit and/or power limit is reached.

Thus for an accurate calculation, we are taking into account a driving velocity application simulating hydraulic press.

Microstructure evolutions

To take full advantage of controlled deformation processes it is necessary to understand the interactions of the forging parameters with the microstructure developed.

Thus the effects on microstructure of strain, strain rate and temperature during forging and after heat treatment have been studied in details by Plisson [3]. The aim of his study was to find a semi empirical equation describing the microstructural evolution, essentially the grain size and the recrystallized fraction evolutions, of alloy 706.

Recrystallization has been studied through compression tests : miniature cylinders of 80 mm in diameter and in length are tested in compression along their axial direction, in different conditions of strain rate and temperature, so that nominal strain varying between 0.1 and 1.2 is obtained (see figure 2). It appears from Plisson studies that in alloy 706 dynamic recrystallization does not take place in that strain domain and that the evolution of microstructure is essentially governed by static recrystallization and grain growth phenomena.

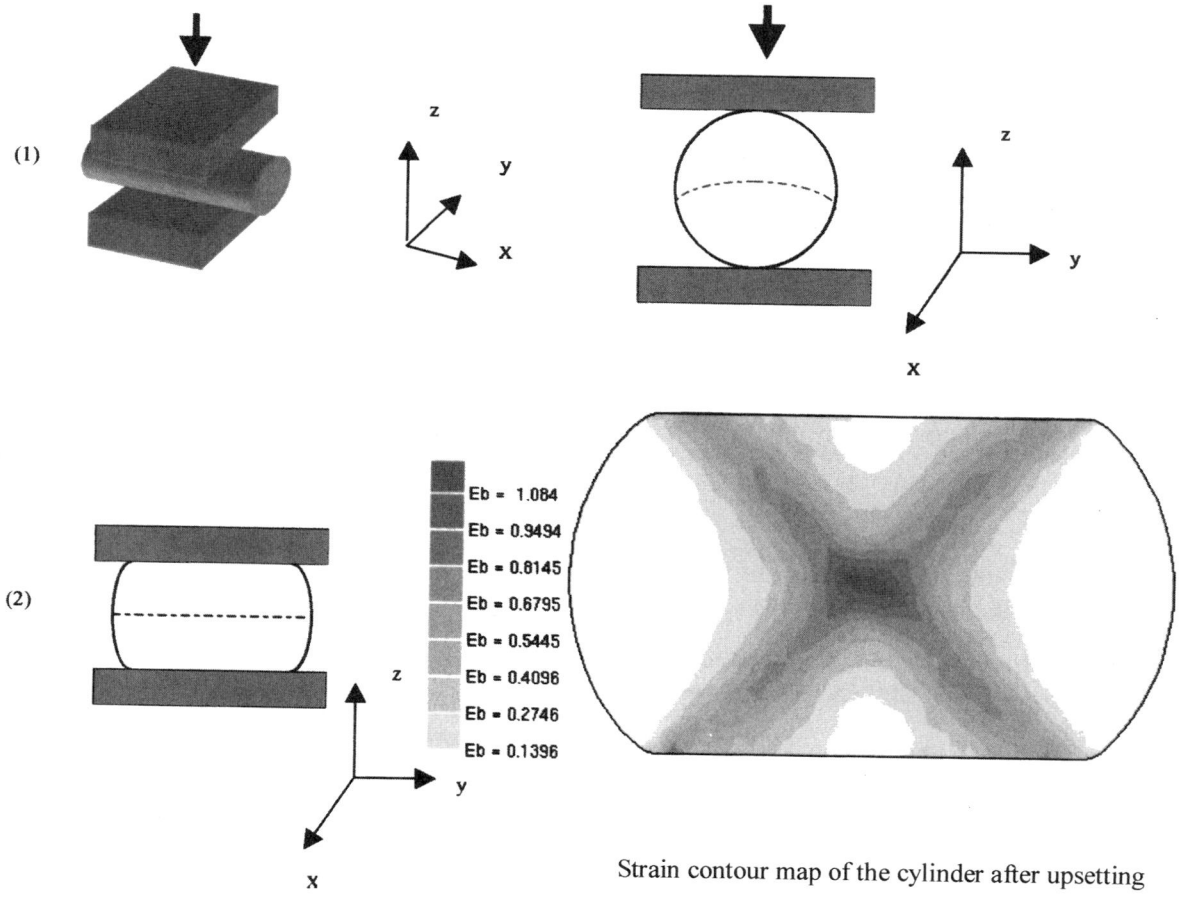

Figure 2 : Compression test on cylinder of 80 mm in diameter and in length. Strain rate, temperature and forging ratio varying.

Concerning the static recrystallization, an Avrami-Sellars [4,5,6,7] analysis allowed the establishment of the equation of the kinetic evolution which after discretization [8], is directly usable by numerical simulation. The expression of the recrystallized fraction and of the grain size after static recrystallization are :

$$X = 1 - \exp\left[\ln(0.5).\left(\frac{t}{t_{0.5}}\right)^k\right]$$

were X is the recrystallized fraction, t the time, $t_{0.5}$ the necessary time to have 50% of the structure recrystallized and k the Avrami exponent.

$$t_{0.5} = \alpha\varepsilon^a Z^b D_0^c . \exp\left(\frac{Q_{rex}}{RT_{rex}}\right) \qquad \text{and}$$

$$Z = \dot{\varepsilon} . \exp\left(\frac{Q_{def}}{RT_{def}}\right)$$

$$d_{rex} = \beta\varepsilon^n D_0^m Z^l$$

with Z the Zener-Hollomon parameter, R the universal gas constant, D_0 and d_{rex} respectively the initial and the final recrystallized grain size (in mm), Q_{def} the activation energy for hot deformation, Q_{rex} the activation energy for static recrystallization, T_{def} the temperature of deformation and T_{rex}, the temperature of subsequent annealing. α, β, a, b, c, n, m, and l are constant.

Concerning the grain growth mechanism, the experimental results allow finding the parameters of the expression

$$A^n = A_0^n + B . \exp\left(\frac{Q_{gg}}{RT}\right).t$$

with A the mean area of the grain at t, A_0 the mean area of the initial grain (in mm^2), Q_{gg} the activation energy for grain growth, B a constant.

The experimental study shows that the recrystallization is static and that the dynamic recrystallization is not significant in the industrial shaping process of alloy 706. Thus microstructure evolutions will be calculate separately from the thermomechanical calculation.

The simulation starts by a mechanical calculation which gives strain, stress, strain rate and temperature, these last two being averaged on the calculation time. The temperature distribution at the end of the mechanical calculation initializes the thermal calculation which will follow.

All these values are re-entered at the beginning of the thermal calculation. For a given node, the material is divided in several microstructural elements, to take into account the structure heterogeneity. If it is the first thermal calculation, the microstructural values such as grains size are initialized. If not, the program reads in a file the characteristics of each microstructural element (recrystallized fraction, mean grain size in the recrystallized and in the worked part of the disc).

For each step of the thermal calculation, recrystallized fraction at each point increases. If the material does not reach the critical strain needed to recrystallize, the grain growth mechanism is applied to the worked grains. As soon as all the material is recrystallized the grain growth law is brought into action.

Each mechanism is limited by a value experimentally determined:

- no microstructural evolution starts until a minimum temperature T_{min} is reached
- recrystallization starts only if the critical strain (ε_c) is reached . The value of ε_c depends on the deformation temperature.
- the recrystallization is considered to be finished, when 95% of the structure is recrystallized and when the grain growth can take place. Then the recrystallized fraction appears to be 100%.

The microstructure parameters are calculated assembling the information of the totality of the microstructure elements of each node. So the recrystallized fraction, the mean grain size, the standard deviation between grain size and the maximal grain size are estimated; these two last parameters permit the estimation of the heterogeneity of the microstructure.

To validate the calculation development, forging tests have been carried out on identical cylinders as those described before. Figure 3 presents one of these results.

The predicted grain size differs from experimentation by only half a unit on the ASTM grade, while predicted recrystallized volume fraction differs from experimentation by 5 %.

Finally, the variety of the microstructure (deformed grains and recrystallized grains) is also accurately predicted : the center of the material has a slightly coarser grain size than that near the surface. This is a result of the non uniform deformation but also of the relative difference in heating rates between center and surface location.

A good match exists between predicted and actual values, confirming the correspondence between the parameters and the analytical description of the evolution of the microstructure introduced in the Forge2® software.

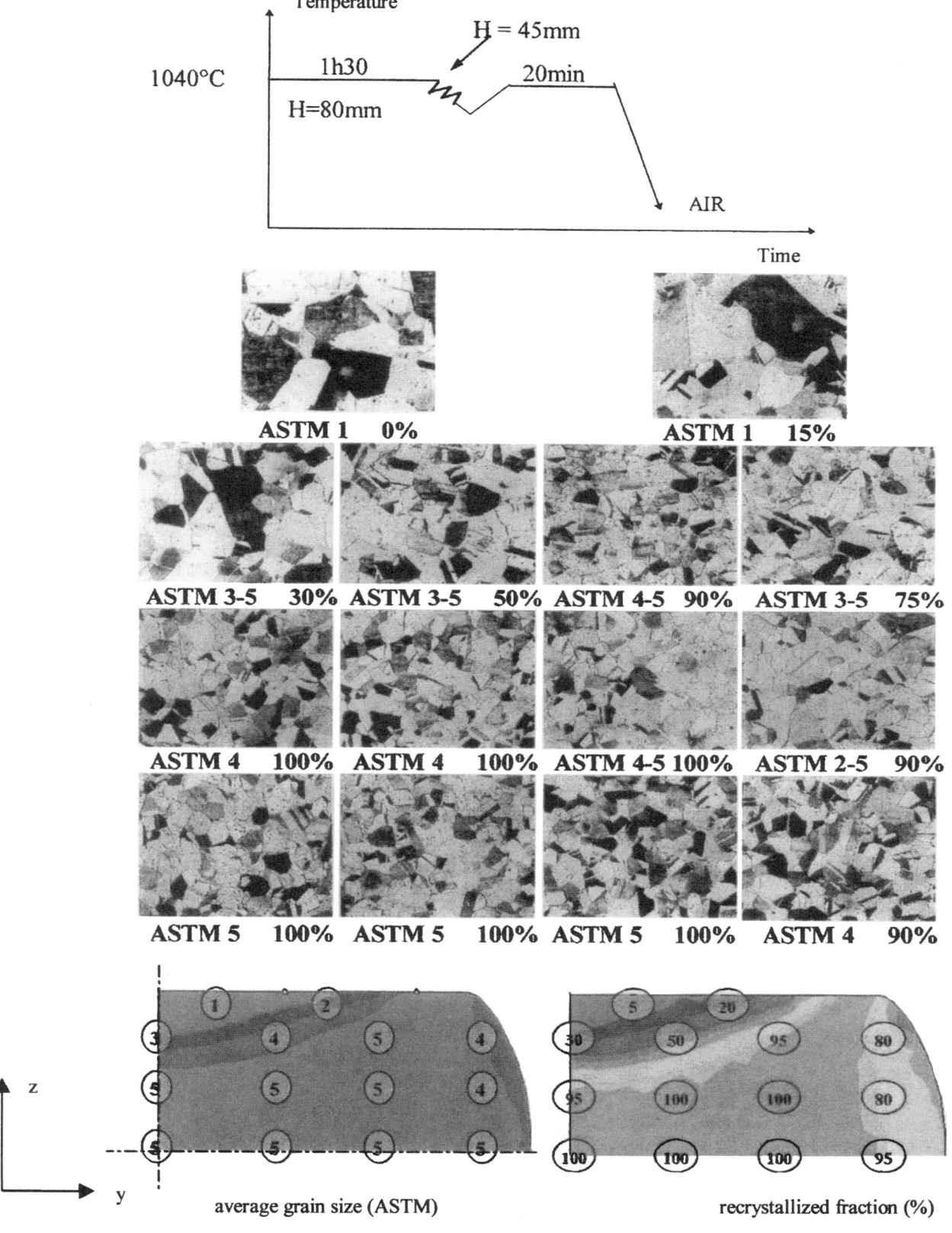

Figure 3 : Numerical results

Industrial application

Given these positive results, the model was used for the industrial production of gas turbine discs.

In the following, results of the improvement and of the optimization of a die forging process are presented. Figure 4 shows the scheme of the thermomechanical process studied and simulated. The dimensions of the billet are in the 0.7 to 0.9 m range concerning the diameter, with a length to diameter ratio of approximately two.

The 65000 tonnes press at Interforge adequately meets requirements necessary to optimize both temperature/reduction rate parameters. At this step of the process it is important to know the strain distribution in the preform, as achieving uniform recrystallization is also dependent upon generating sufficient strain throughout the part.

The first forging process consists of an upsetting followed by punching and then die forging. The simulation of this process shows a lack of strain just below the surface of the billet. Indeed, during the first heat, the work does not penetrate efficiently from surface to mid-length of the part. Thus, adequate strain is not achieved, which results in a partial recrystallization, with negative effect on the final properties of the product. To avoid the problem of die-lock area a punching operation was added. Figure 5 shows the strain, the recrystallized fraction and the average grain size

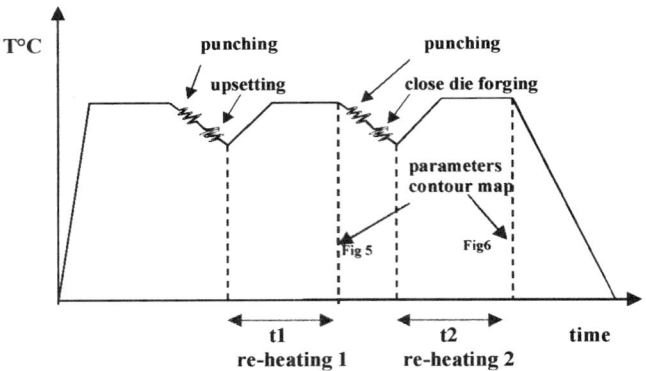

Figure 4 : Scheme of the industrial process studied and simulated

contour map inside the preform resulting from an upsetting plus a re-heating (a) and from a punching followed by an upsetting plus re-heating (b), to reach the same thickness.

The stored energy of deformation during the punching allows recrystallization to take place during thermo-mechanical processing, which follows : upsetting to thickness H followed by re-heating 1 at temperature T (°C) during a time t1.

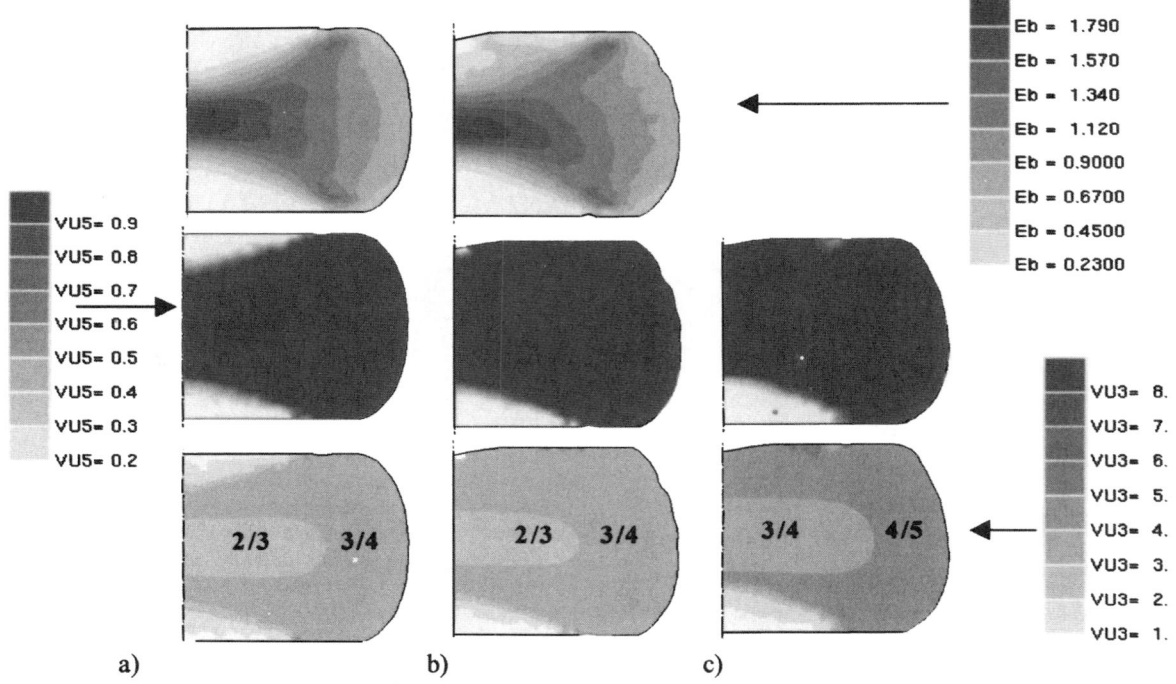

Figure 5 : First heat . Strain (top), recrystallized fraction in percent (center) and average grain size in ASTM grades (bottom) predicted after   (a) an upsetting to H followed by a re-heating during a time t1

         (b) a punching plus an upsetting to H followed by a re-heating during a time t1

         (c) a punching plus an upsetting to H followed by a re-heating during a time t1'

The same operation, punching, is repeated on the other face of the part, before the final close die forging and a re-heating (2) followed by air cooling.

To avoid inverting the mesh in the simulation, the second punching is made on the bottom face. Thus on the final product contour maps, the real top face is at the bottom and vice et versa.

The industrial production process, subdivided into forging a preform and finish close die forging, is simulated. Figure 6 shows the final microstructure obtained, which corresponds to that observed on the workpiece.

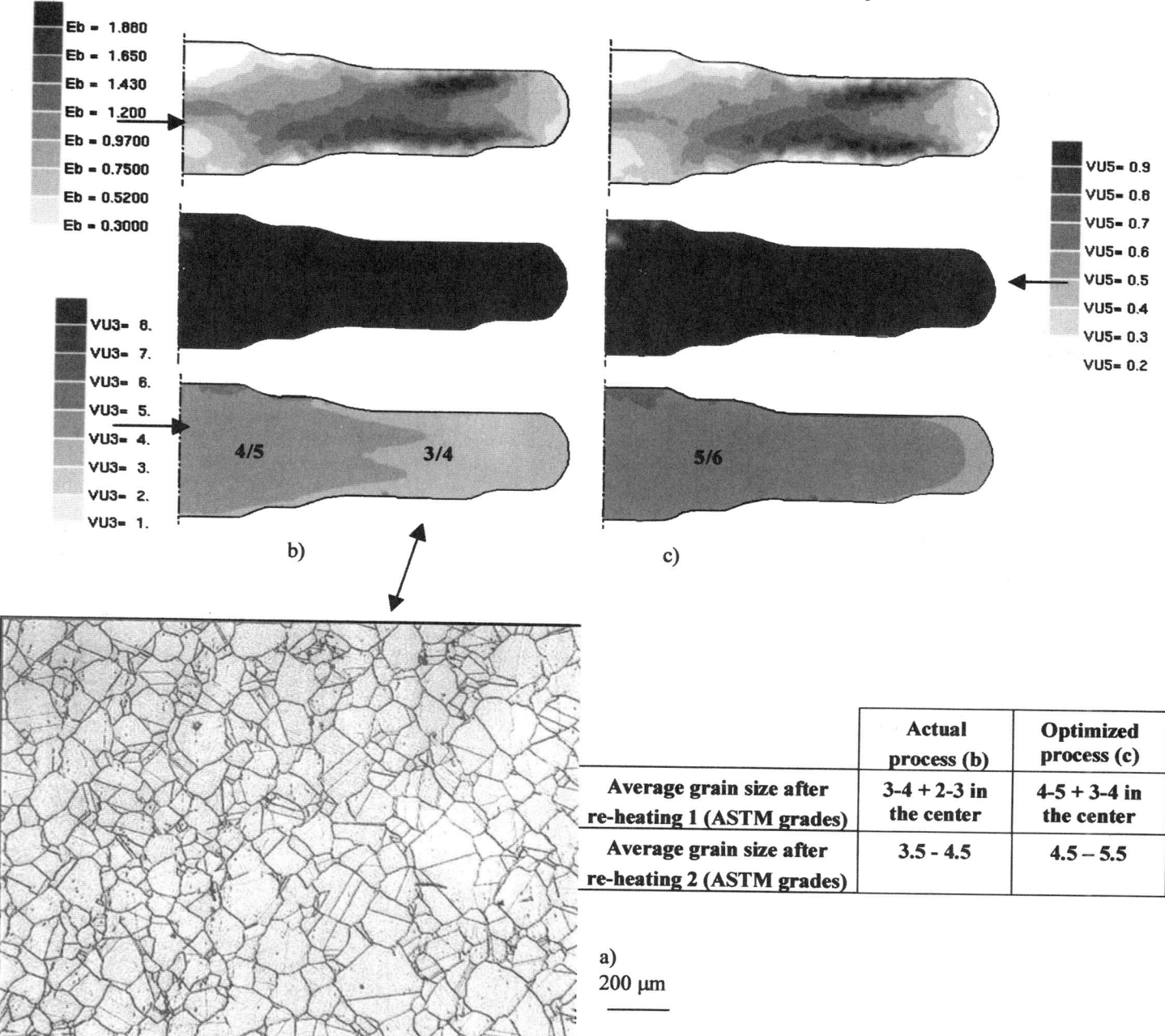

	Actual process (b)	Optimized process (c)
Average grain size after re-heating 1 (ASTM grades)	3-4 + 2-3 in the center	4-5 + 3-4 in the center
Average grain size after re-heating 2 (ASTM grades)	3.5 - 4.5	4.5 – 5.5

Figure 6 : Second heat. Strain (top), recrystallized fraction in percent (center) and average grain size in ASTM grades (bottom) predicted on the final product after a re-heating during the time t2 (b) and t2' (c). (a) microstructure observed on the final product.

483

The simulation gives, at each step of the process, similar grain size and morphology (deformed or recrystallized) to those obtained on the workpiece. Once again, these results demonstrate that the laws which govern the physical phenomena surrounding alloy 706 are well identified.

Thus, by varying the re-heating time between two forging steps and/or the temperature of the heat treatment, the optimization of the forging procedure can be obtained through computer techniques of modelling.

The present study also investigated an alternative production route, consisting of reducing the re-heating time between two forging steps. The same simulation was carried out with t1 and t2 reduced. Below we will speaking about an actual production process (time t1 and t2) and the optimized process (reduced time t1' and t2').

The different strain, grain size and recrystallized fraction contour map for both process are summarized in figures 5 and 6.

The microstructure obtained after the re-heating 1, presents an average grain size of 4-5 and of 3-4 ASTM grade respectively for the actual process and the optimized one. A coarser grain 1 to 2 ASTM grades larger is observed in the center of the part because the strain and the temperature levels and thus the adiabatic heating leads to a rapid recrystallization followed by the grain growth. The finer structure predicted by the simulation for the optimized process is due to shorter time for grain growth before the subsequent forging step is applied. The recrystallized volume fraction remains unchanged.

The microstructure predicted by simulation on the final product, still has the difference of 1 to 2 ASTM grades on the grain size, for the same fraction of recrystallized grains.

The difference between the two processes of final average grain size has no significant influence on the mechanical properties, a good recrystallization is also achieved in both cases. So the reduction of the re-heating time leads to a reduction of cost and increase of press availability.

Simulation allows to correctly adapt initial and finished shapes to ensure to have sufficient strain in all of the disc and associated to the close-die forging procedure with a precise time and temperature sequence, it leads to a controlled microstructure and to the required mechanical properties.

Conclusion

Using compression test samples, the microstructural evolutions and especially static recrystallization and grain growth of alloy 706 have been identified and introduced into 2D modelling software. The model has been validated and is now used for industrial process optimization. This study illustrates that now the finite element model is not only used to correctly predict the material flow occurring in the part but today numerical simulation has become an essential tool for process optimization and improvement of microstructure.

The knowledge of constitutive equations that involve microstructural parameters and allow to predict microstructural evolution during forging are of first importance.

Of course the identification of the laws which governs the microstructure evolution is not easy especially when there is not only static recrystallization but also dynamic one, but it is the next step in pushing back the blacksmiths' boundaries and to decrease the cost and manufacturing cycles.

[1] Forge2®, a modelling software for the shaping of axisymmetrical and /or two-dimensional parts, Transvalor, BP037, 06901 Sophia Antipolis Cedex - France.

[2] D. Rayne, J.F. Uginet, " Fabrication of large components in 706 Alloy for gas turbine application ", 11th International Forgemasters Meeting Terni /Spoleto, Italy, 1991.

[3] C.Plisson, " Mise en forme du superalliage base nickel 706 ", (Internal Report 1997).

[4] Avrami, " Kinetics of phase change I, general theory ", Journal of Chemical Physics, vol. 7, (1939).

[5] Avrami, " Kinetics of phase change II, transformation-time relations for random distribution of nuclei ", Journal of Chemical Physics, vol. 8, (1940).

[6] Avrami, " Kinetics of phase change III, granulation, phase change and microstructure ", Journal of Chemical Physics, vol. 7, (1939).

[7] Glovers,. Sellars, " Static recrystallisation after hot deformation in alpha-iron ", Metallurgical Transactions A, vol. 3, (1972).

[8] P.Audrerie, "Etude du comportement rhéologique au cours de traitements thermomécaniques d'alliages de titane", (Ph. D., Ecole Nationale Supérieure des Mines de Paris, 1993).

Mechanical Behavior

INFLUENCE OF LONG TERM EXPOSURE IN AIR ON MICROSTRUCTURE, SURFACE STABILITY AND MECHANICAL PROPERTIES OF UDIMET 720LI

Dietmar Helm and Olaf Roder

DaimlerChrysler Aerospace, MTU Motoren- und Turbinen-Union München GmbH
Dachauer Strasse 665
80995 Munich, Germany

Alloy Udimet 720LI was designed for high temperature application of aircraft-engine disks up to about 730°C. Operating at such high temperatures for long times raises the question of thermodynamical stability of the alloy (bulk stability) and surface integrity. Subject of the present paper was to investigate the influence of long time exposure up to 1000 hours at high temperatures up to 845°C in air on microstructure and phase stability as well as on surface integrity. It could be shown by appropriate microscopy (LM, SEM, TEM) that coarsening of the γ'-phase occurred with increasing time of exposure at temperatures of 760°C and 845°C. Additional formation of topological closed packed (TCP) phases (σ) could be observed. It was found that this lack of stability exhibited an impact on mechanical properties such as tensile strength, creep resistance, and low cycle fatigue strength, especially when a dwell time was superimposed. Furthermore, the reaction of the alloy in the near-surface area of the specimen exposed to air at high temperatures with superimposed stresses was investigated by light microscopy (LM) and SEM. A quantitative analysis by EDX investigation in a SEM of the thus formed surface oxide layer and the subsurface area with a depletion of alloying elements was performed.

Superalloys 2000
Edited by T.M. Pollock, R.D. Kissinger, R.R. Bowman,
K.A. Green, M. McLean, S. Olson, and J.J. Schirra
TMS (The Minerals, Metals & Materials Society), 2000

I. INTRODUCTION

UDIMET* 720LI (U720LI) is a high strength corrosion
*UDIMET is a trademark of Special Metals Corporation
resistant nickel-based superalloy, precipitation strengthened by γ', Ni₃(Al,Ti), and solution strengthened by Mo, W, Cr and Co, being particularly used in aero-engines for highly stressed rotating components, such as compressor and turbine disks. U720LI (low inclusion chemistry) is a derivative from UDIMET 720 (U720, conventional chemistry), an alloy originally designed for optimum oxidation and sulfidization resistance for turbine blades operating in the harsh environment of land-based turbine engines. The chemistry of U720LI is optimized for improved phase stability and reduced carbide and boride stringers for cast and wrought processed material. Table I shows the nominal compositions of U720 and U720LI.

Table I Nominal Composition of U720 and U720LI

Alloy	Cr	Co	Mo	W	Ti	Al	C	B	Zr
U720	18	14.7	3.0	1.25	5.0	2.5	0.035	0.033	0.03
U720LI	16	14.7	3.0	1.25	5.0	2.5	0.010	0.015	0.03

U720LI as well as U720 have received increased interest due to their good balance of mechanical properties and economic potential. It has been reported that cast and wrought U720LI is less costly than comparable competitive P/M alloys [1]. The original chemistry of U720 has a problem with phase stability, especially in the fine grained subsolvus heat treated condition. The high chromium content of the conventional U720 chemistry makes it prone to sigma phase formation. Figure 1 (adopted from [2]) shows the effect of composition on the precipitation of sigma phase in subsolvus heat treated material for U720 by comparing sigma phase formation "C"-curves of U720 with U720LI.

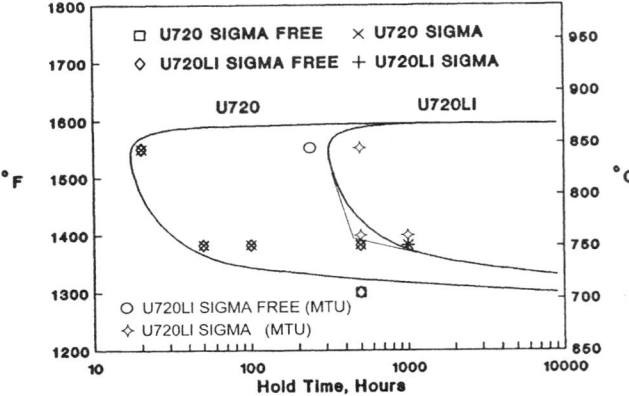

Figure 1: Effect of composition on the precipitation of σ-phase in subsolvus heat treated U720 [2]

Mechanical property and microstructure changes with in-service thermal exposures have shown that sigma phase formation in U720 is very deleterious [2]. Recent studies of the stability of U720LI have shown that this material is susceptible to sigma phase formation, though to a much lower degree as compared to U720 [3].

The objective of this study was to investigate the influence of long time exposure up to 1000 hours at high temperatures up to 845°C in air on microstructure and phase stability as well as on surface integrity for U720LI.

II. EXPERIMENTAL PROCEDURE

For the purpose of the present work specimen blanks were taken from a regular cast and wrought U720LI disk, which was heat treated as listed below:
- 4 h 1100°C / Oil Quench
- 24h 650°C/AC + 16h 760°C/AC

This condition was used to determine the baseline microstructure and mechanical properties to which all other conditions were compared. Table II shows the various conditions being produced by different exposure heat treatments of the baseline condition. All exposure heat treatments were performed in air in a laboratory furnace followed by furnace cooling.

Table II Performed Exposure Heat Treatments

Condition	Heat Treatment	Investigation
760-S	500 h 760°C	LM, SEM, TEM
760-L	1000 h 760°C	LM, SEM, TEM, MT
845-S	120 h 845°C	LM
845-M	240 h 845°C	LM, SEM
845-L	500 h 845°C	LM, SEM

Characterization of the microstructural changes as a function of thermal exposure heat treatment were performed by optical microscopy (LM), scanning electron microscopy including energy disperse X-ray measurements (SEM) and transmission electron microscopy. As the thermal exposure of 1000h at 760°C seems to be a reasonable upper bound limit for microstructural changes under a practical point of view (maximum accumulation of time at this temperature during the whole engine life), mechanical testing was performed on the baseline condition and on condition 760-L. Mechanical testing (MT) included ambient and hot tensile testing up to 750°C, creep testing under two test conditions, 625°C at 730MPa and 670°C at 650MPa, and load controlled fatigue testing at 700°C (k_t = 2.1) without dwell time and with a dwell time of 1 minute. All mechanical tests were performed in laboratory air. Most specimens were machined after thermal exposure to study the influence of bulk stability (designation for example 760-L). Additionally, some specimens were heat treated with finished surface to determine surface integrity effects superimposed to bulk effects on mechanical properties (designation for example 760-L Surface).

III. RESULTS AND DISCUSSION

The effect of heat treatment on the stability of U720LI is illustrated in Figures 2 to 4 by comparing optical, SEM and TEM micrographs showing the microstructural changes of conditions 760-S, 760-L, 845-M and 845-L.

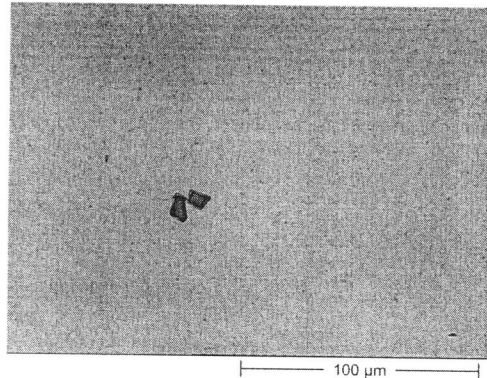

Figure 2a: Baseline Condition, no σ-phase

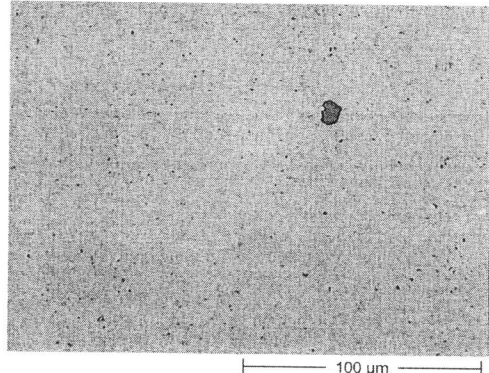

Figure 2b: Condition 760-S, small amount of σ-phase

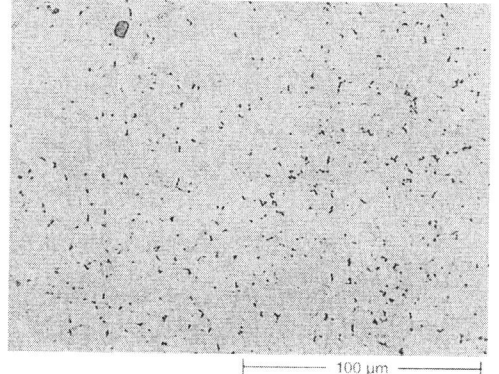

Figure 2c: Condition 760-L, σ-phase present

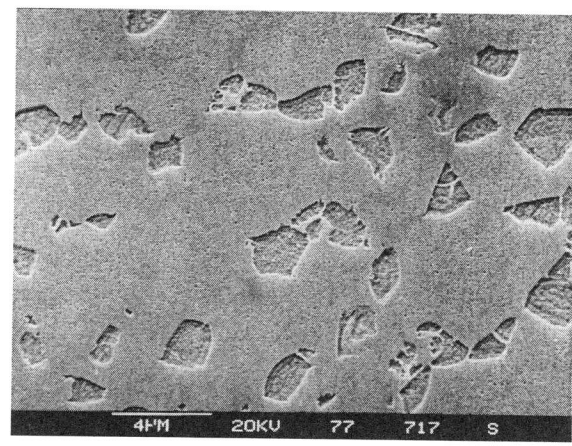

Figure 3a: Baseline Condition, SEM

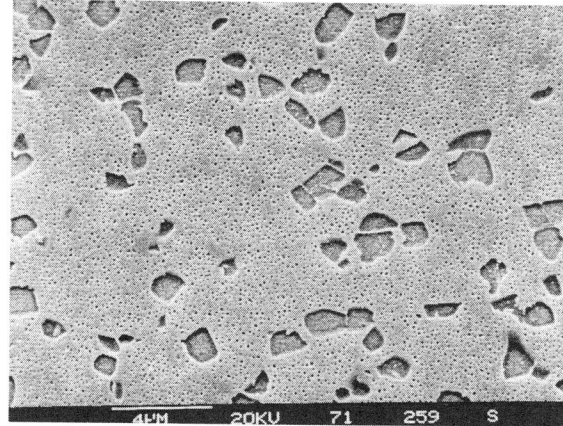

Figure 3b: Condition 845-M, SEM

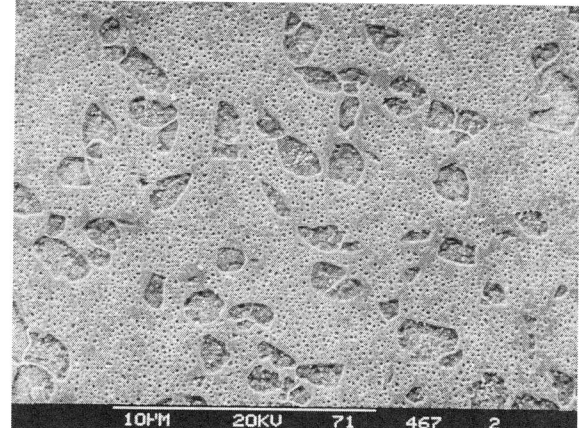

Figure 3c: Condition 845-L, SEM

489

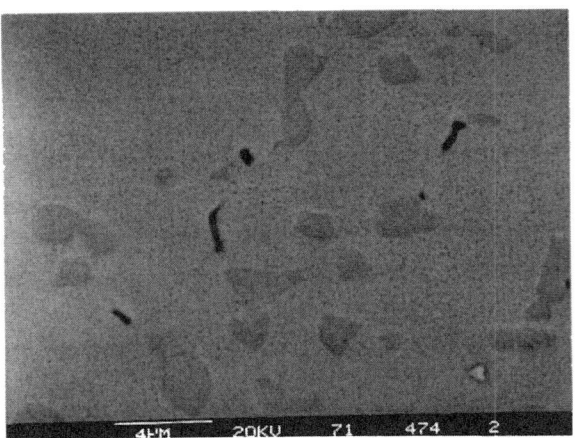

Figure 3d: Condition 845-L, σ-phase, SEM

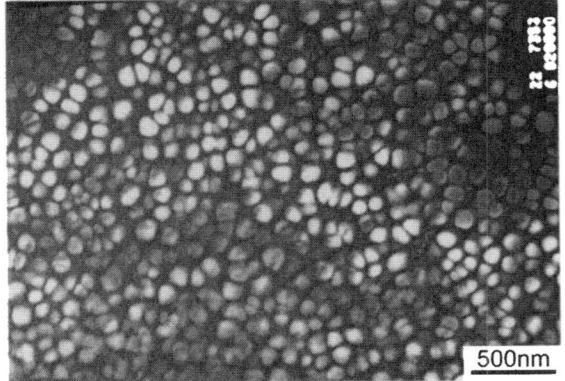

Figure 4a: Baseline condition, size of secondary γ', TEM

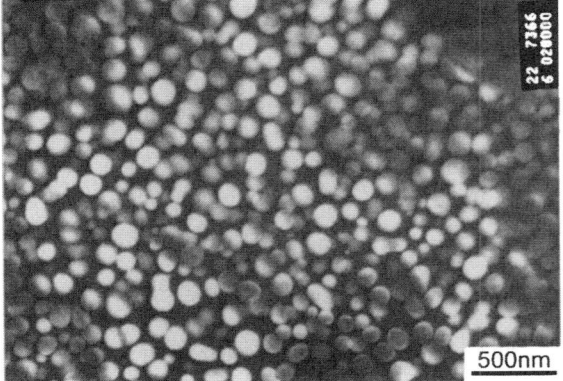

Figure 4b: Condition 760-L, size of secondary γ', TEM

The exposure of U720LI to high temperatures lead on one hand to a coarsening of the secondary γ', the size of these particles grows from about 70nm in the base-condition (Fig. 3a, Fig. 4a) to sizes of about 150nm in condition 760-L (Fig. 4b) and 845-L (Fig. 3c). Table III lists the size of the secondary γ' particles for all six conditions. On the other hand the high temperature exposure lead to the

formation of TCP-phases in conditions 760-S, 760-L and 845-L (Fig. 2b, c, Fig. 3d). The phases in conditions 760-S, 760-L and 845-L supposed to be σ-phase were analyzed using quantitative energy disperse X-ray (EDX) measurements. These measurements showed an increase in Cr, Mo and Co and a depletion of Ni for these phases compared to the analysis of the surrounding matrix thus indicating that these phases are σ-phase, the measured composition being about $(Cr_{0.5}, Mo_{0.1}),(Ni_{0.2}, Co_{0.2})$ and matching the suggested composition in [4]. The information on the presence of σ-phase is incorporated into figure 1 confirming the existing data from [2] mainly. Our data suggest that the slope of the C-curve for U720LI at lower temperatures is a little smaller and the upwards turn starts at lower temperatures (see updated Fig. 1).

Table III Size of secondary γ'

Condition	Secondary γ' size [nm]
Base	70 - 100
760-S	~125
760-L	~ 150
845-S	not investigated
845-M	~ 110
845-L	~ 150

Furthermore some investigations were performed to study the influence of long time high temperature exposure on the surface integrity of U720LI. Figure 5 shows as an example a SEM micrograph of condition 845-L being exposed to 845°C for 500h without machining the surface afterwards.

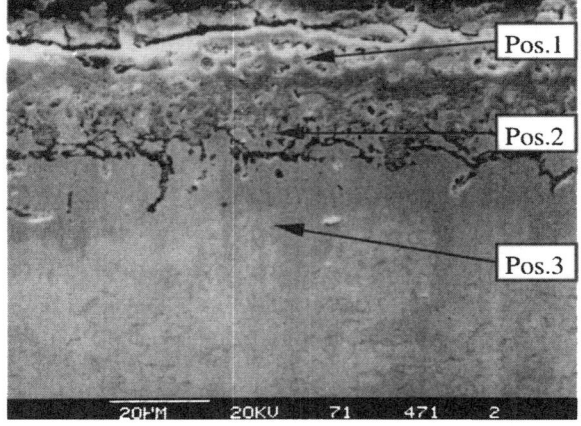

Figure 5: Condition 845-L, near surface microstructural changes

Typical microstructural changes near and at the surface after long time high temperature exposure can be described as a relatively dense zone on the surface (Fig. 5, Pos.1), followed by a porous zone (Fig. 5, Pos. 2) just below the surface and a zone below these two zones which shows a depletion of the γ' particles (Fig. 5, Pos.3). EDX-analysis of the three zones are listed in table IV.

Table IV   EDX-analysis in wt.-% of near surface area in condition 845-L with surface exposed to high temperature
Position (Pos) see Figure 5

Pos	O$_2$	Al	Ti	Cr	Co	Mo	Ni
1	41	3,8	11,2	28,9	3,4	0,9	10,3
2	11,5	4,0	3,1	13,0	15,2	4,3	55,2
3	8,8	1,6	4,9	13,7	14,1	3,3	53,1
Nom		2,5	5,0	16	14,7	3,0	57,5

Mechanical properties were determined for the baseline condition which did not show any presence of σ-phase at all and for condition 760-L with σ-phase in the microstructure. Figure 6 shows the tensile properties as a function of test temperature for the above mentioned conditions.

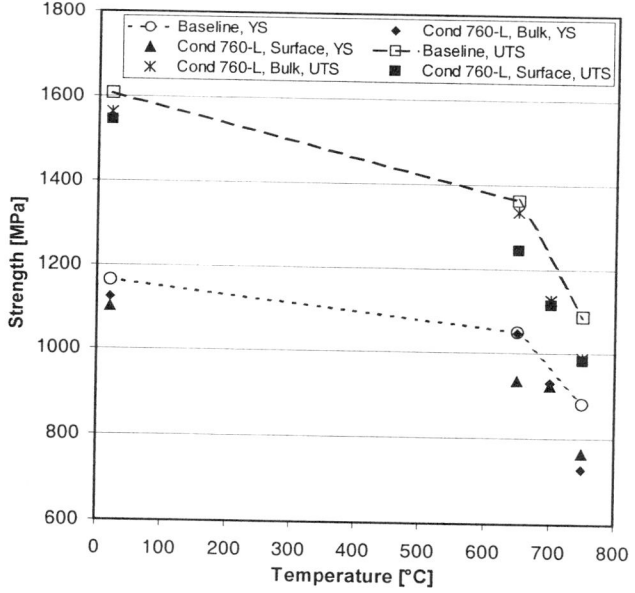

Figure 6a: 0.2% Yield Strength (YS) and Ultimate Tensile Strength (UTS) as a function of test temperature for baseline condition (NO σ-phase) and condition 760-L (σ-phase present)

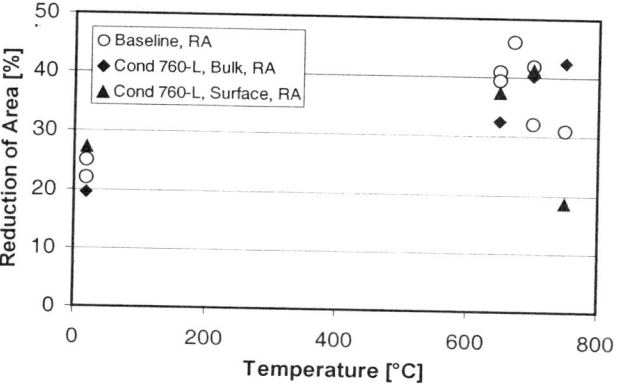

Figure 6b: Reduction of Area (RA) as a function of test temperature for baseline condition (NO σ-phase) and condition 760-L (σ-phase present)

The tensile data presented above (Figures 6a and 6b) illustrate several important features of property changes due to coarsening of secondary γ' and σ-phase precipitation.

The strength values of condition 760-L are lower over the whole temperature range as compared to the baseline condition. The drop in strength is higher for higher temperatures. No differences were observed between the two variations of condition 760-L (760-L and 760-L Surface).

The strength of γ' hardened nickel-based superalloys arises from a combination of hardening mechanisms. On one hand there are alloying additions in solid solution to provide strength, creep resistance and/or resistance to surface degradation and on the other hand nickel-base alloys contain elements that form small coherent particles of an intermetallic compound of the type Ni$_3$(Al,Ti) after an appropriate thermo-mechanical processing or heat treatment, contributing to strength, creep and fatigue resistance [5]. On this basis the decrease in strength of condition 760-L as compared to the baseline condition can be explained:

1.   The coarsening of the secondary γ' particles leads to a change of the interface of these particles and by that to a change of the dislocation – particle interaction. In the baseline condition the secondary γ' particles are coherent and moving dislocations will pass the particles by a cutting mechanism. With increasing time of high temperature exposure (condition 760-L) the secondary γ' particles grow and become semi-coherent or in-coherent, forcing moving dislocations to pass by an Orowan-looping mechanism. This would be a reasonable explanation why at a constant volume fraction of γ' an increasing particle size (secondary γ') and corresponding increasing particle spacing leads to a decrease in strength.

2. Additionally the depletion of the γ-matrix from Cr, Co, and Mo due to the precipitation of σ-phase will reduce solid solution strengthening by these elements and therefore lead to a further decrease in strength. The decrease in strength due to depletion of solid solution strengthening elements is smaller compared to the mechanism described under 1., due to the small volume fraction of σ-phase precipitated in condition 760-L. As Co is a stronger solid solution strengthening element at high temperatures, the depletion of Co might be responsible for the bigger decrease in strength at higher temperatures.

Values of reduction of area do not show any differences between the baseline condition and both variations of condition 760-L. Above temperatures of about 650° the data scatter more in comparison to lower temperatures and show a decreasing tendency after reaching a maximum at about 670°C (Fig. 6b). This behavior is probably due to an increasing contribution of grain boundary sliding to the fracture process at temperatures above 670°C in U720LI. Condition 760-L does not show a different behavior compared to the baseline condition, indicating that the volume fraction of σ-phase is too small to influence the fracture behavior in short term mechanical tests. Furthermore, similar tensile test results of both variations of condition 760-L (without and with superimposed surface effects) indicates, that the surface and near-surface changes of the microstructure (Fig. 5) due to long time high temperature exposure do not lead to an embrittlement of the material.

Table IV compares the results of the creep testing for condition 760-L with typical results of the alloy.

Table IV Results of creep testing condition 760-L

	$t_{0,2\%}$ [h]	$\varepsilon_{pl,100h}$ [%]
Test Conditions: 625°C, 730MPa		
Typical	230	0.1
760-L, Bulk	32	0.4
Test Conditions: 670°C, 650MPa		
Typical	50	0.3
760-L, Bulk	5 - 8	1.14 – 2.15

The creep results show a drastic drop in creep resistance for condition 760-L compared to the baseline condition, expressed by an increase in time to 0.2% plastic strain and/or plastic strain after 100h, respectively.

Creep in nickel-based superalloys is very dependent of a number of microstructural parameters. Thus, several of the

above described changes in the microstructure are to a certain extent responsible for the deterioration of the creep resistance of condition 760-L compared to the baseline condition. Besides the decrease of the creep resistance due to the overall decrease of high temperature strength as a result of the coarsening of the secondary γ' particles and the depletion of solid solution strengthening elements, it is well known [6] that the appearance of σ-phase noticeably impairs the creep strength. The loss in properties associated with the formation of phases such as σ usually is either the result of the depletion of alloying elements contributing to solid solution strengthening, or is due to cracking in the new phase or at the new phase-matrix interface.

Figure 7 presents the results of the load controlled fatigue tests of notched specimens ($k_t = 2.1$) on the baseline condition and condition 760-L, performed at 700°C without a dwell time and with a dwell time of 1 minute.

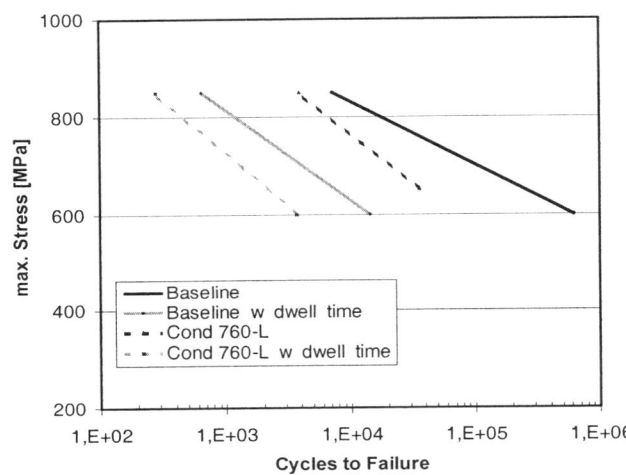

Figure 7: Results of fatigue testing, 700°C, R = 0.1, $k_t = 2.1$

For the baseline condition the tests with a dwell time of 1 minute per cycle show a decrease in life (cycles to failure) by a factor of about 10 for high maximum stresses and of about 15 for lower stresses as compared to the tests without a dwell time. Condition 760-L shows a further drop of life by a factor of 2-3 as compared to the baseline condition for both test conditions (without a dwell time as well as with a dwell time).

At 700°C fatigue life becomes dependent not only on the number of cycles but also on the time per cycle. The accumulated time especially at high loads contributes significantly to the total life. At high maximum stresses the total number of cycles to failure is relatively small and the difference in total life between samples tested without a dwell time and those tested with a dwell time is mainly due to the accumulated time at maximum stress and the resultant creep damage. At lower maximum stresses the number of cycles to failure becomes bigger and thus the time at maximum stress increases accumulating additional creep damage. This may explain the increasing difference in life time with increasing total life for tests performed with a dwell time and those performed without a dwell time for the baseline condition.

The difference in total life for condition 760-L when compared to the baseline condition can be explained by the difference of the creep resistances. As described above the creep resistance of condition 760-L is much lower than the creep resistance of the baseline condition and therefore exhibits a lower total fatigue life at 700°C with accumulated creep damage contributing significantly to the total life.

IV. CONCLUSIONS

Long time high temperature exposure of the nickel-based superalloy UDIMET 720LI in a subsolvus heat treated fine grained condition for aero-engine disk application lead to the changes in the microstructure listed below and as a result of this to significant changes of important mechanical properties:

- Long time exposure at temperatures up to 845°C and times up to 1000h increases the size of the secondary γ' particles and under certain combinations of time and temperature σ-phase is precipitated.
- σ-phase formation in the microstructure is observed in U720LI material exposed for 500h to 845°C, 500h to 760°C and 1000h to 760°C.
- From a practical point of view (maximum accumulation of time at this temperature during the whole engine life) thermal exposure at 760°C for 1000h seems to be a reasonable upper bound limit for estimating the maximum possible deterioration of mechanical properties.
- Tensile strength from room temperature to 750°C (YS as well as UTS) was reduced by 5 to 10% when σ-phase was present.
- Creep resistance was reduced significantly by a factor of about 5.
- Load controlled fatigue tests performed at 700°C without a dwell time as well as with a dwell time of 1 minute showed a decrease in total life (cycles to failure) by a factor of 2 to 3 as compared to their baseline counterparts.

These results show that long time high temperature exposure of U720LI leads to a significant deterioration of the important mechanical properties. Furthermore these results indicate that the practical maximum temperature for long time usage should be below 730°C.

ACKNOWLEDGEMENTS

This work was supported by the Bundesministerium für Bildung und Forschung (BMBF).
Prof. G. Lütjering of the Technical University of Hamburg-Harburg is kindly acknowledged for performing TEM work.

REFERENCES

1. G. Kappler et. al., "Conventionally Procecessed High Performance Disc Material for Advanced Aeroengine Design", 4th European Propulsion Forum 1993, Bath, UK, pp. 9.1-9.9

2. P. W. Keefe, S. O. Mancuso, G. E. Maurer, "Effects of Heat Treatment and Chemistry on the Long-Term Phase Stability of a High Strength Nickel-Based Superalloy", in Superalloys 1992, Eds. S.D. Antolovich et. al., The Minerals, Metals & Materials Society, 1992, pp. 487-496

3. Y. S. Na, et. al. "Quantification of Sigma Precipitation Kinetics in Udimet 720LI", Proc. of the 4th Intl. Charles Parsons Turbine Conference, Newcastle, UK, November 4-6, 1997, pp. 685-697

4. R. C. Reed, M. P. Jackson, and Y. S. Na, "Characterization and Modeling of the Precipitation of the Sigma Phase in UDIMET 720 and UDIMET 720LI", Metallurgical and Materials Transactions A, Volume 30A, March 1999, pp. 521-533

5. C. T. Sims, N.S. Stoloff, and W. C. Hagel, "Superalloys II", Chapter 3, John Wiley & Sons, New York 1997, pp.61-96

6. C. T. Sims, N.S. Stoloff, and W. C. Hagel, "Superalloys II", Chapter 9, John Wiley & Sons, New York 1997, pp.241-262

EFFECTS OF GRAIN AND PRECIPITATE SIZE VARIATION ON CREEP-FATIGUE BEHAVIOUR OF UDIMET 720LI IN BOTH AIR AND VACUUM

[1]N.J. Hide, [2]M.B. Henderson and [1]P.A.S. Reed

[1]Materials Research Group, School of Engineering Sciences, University of Southampton, Southampton, SO17 1BJ, U.K.
[2]Mechanical Sciences Sector, DERA Farnborough, Farnborough, GU14 0LS, U.K.

Abstract

The high temperature fatigue characteristics of U720Li have been investigated over the temperature range 650°C to 725°C under imposed dwell times (at maximum load) of 1 and 20 seconds in vacuum and air conditions. The effect of varying grain size and coherent precipitate size under these conditions has been assessed.

Testing in air resulted in oxidation dominated intergranular crack growth at all temperatures and dwell times with the slope (m-values) of the crack growth rate curves remaining constant. Increased crack growth rates are seen at the higher temperatures and at longer dwells, although no effect of dwell was observed at 650°C in the as-received fine grained variant. In vacuum crack growth rates were much lower than in air and a purely cyclic dependent regime was evident at 650°C. As temperature and dwell time at maximum load was increased, m-values increased and were accompanied by a change in crack growth mechanism from transgranular to intergranular cracking. This indicated that true, time-dependent, creep-fatigue processes were occurring.

The large grain variant of the U720Li showed little advantage in crack growth rates within the cyclic dependent and creep-fatigue regime, but did show a significant increase in resistance to crack growth in the time dependent (oxidation-fatigue) regime. The effect of the large precipitate variant was to give similar or worse crack growth resistance than the baseline U720Li at temperatures up to 725°C (1 second dwell) but improved crack growth resistance when oxidation processes predominated at 725°C in air with an imposed 20 second dwell.

Introduction

Engine operating temperatures are being pushed ever higher with target temperatures for nickel-base disc materials such as U720Li now approaching 725°C to improve aero-engine efficiency. Service conditions for these alloys are severe due to operating temperature, oxidation and the fluctuating loads imposed during engine start-up and shut-down cycles. These conditions pose problems in terms of high temperature fatigue, oxidation assisted fatigue and (when sub-surface defects occur which grow in an effective vacuum) creep-fatigue. Optimisation of turbine materials so that operating temperatures can be maximised thus requires a fundamental understanding of the mechanisms involved in high temperature fatigue and oxidation/creep-fatigue as well as an assessment of the optimised microstructure.

Nickel-base superalloys largely comprise an austenitic f.c.c. matrix (γ phase) and coherent nickel rich γ' precipitates, from which the material derives most of its strength and creep-rupture properties. As such the characteristics of the γ grains and γ' precipitates can significantly alter the fatigue and oxidation/creep-fatigue properties of the material. At elevated temperatures material properties such as E and σ_y decrease and creep deformation and environmental effects become important (1). Slip character also changes as cross-slip and climb processes become easier. Cycle dependent behaviour is observed at intermediate temperatures, where crack growth rates are independent of frequency, but may be increased due to decreases in E and σ_y increasing the ΔCTOD. Such crack growth is generally characterised by a transgranular crack propagation mechanism. At higher temperatures frequency effects start to become apparent, with higher crack growth rates observed at lower frequencies (time-dependent behaviour). The transition from transgranular to intergranular crack growth is widely acknowledged to indicate the onset of time dependent behaviour (2,3,4). This may be caused by environmental effects (5) such as grain boundary embrittlement or from true creep effects, which are usually observed under vacuum (2,6) and are facilitated by triple point cracking or grain boundary sliding and consequent cavity formation. The point at which intergranular crack growth occurs (under any environment) may be linked to the amount of stress experienced by the grain boundary and the integrity of the grain boundary in relation to the ease with which cracks can propagate transgranularly. The amount of stress present at a grain boundary can be associated with the slip character of the material, which in turn is linked to the coherent γ' characteristics (2,7,8). Fine γ' precipitates have been associated with slip of a highly planar and heterogeneous nature (1), and whilst this may be beneficial for increasing resistance to crack growth at room temperatures (through increased σ_y) the intersection of intense planar slip bands at

495

Superalloys 2000
Edited by T.M. Pollock, R.D. Kissinger, R.R. Bowman,
K.A. Green, M. McLean, S. Olson, and J.J. Schirra
TMS (The Minerals, Metals & Materials Society), 2000

grain boundaries may occur. At high temperatures, a high build-up of stress concentration through this mechanism can cause crack propagation through wedge cracking or grain boundary sliding. Increasing the coherent γ' size has been cited by a number of authors as decreasing high temperature crack growth rates (in air) through increasing the degree of slip homogeneity within nickel based superalloys and subsequently lowering the stress at grain boundaries giving a resultant longer rupture life (9,10). However, this effect must also be balanced against the decrease in hardness and flow stress that can occur if γ' are grown into the overaged condition and the concomitant effects on crack growth. In an oxidising environment, the nature of the slip damage can also be important with respect to the diffusion of oxidising species to grain boundaries along intense slip bands. The inference being that a microstructure producing a more intense slip band might well produce an easier path for diffusion of oxidising species (10,11).

Increased grain size has been linked to lower crack growth rates at elevated temperature. Merrick and Floreen (9) when testing P/M Astroloy at 650°C noted lower fatigue crack growth rates in air when the grain size was increased from ASTM 8 (22.5μm) to ASTM 5 (63.5μm). Creep strains in U720Li at 600°C showed a decrease when the grain size was increased from 22μm to 420μm which was linked by Torster et al (12) to a classical grain boundary sliding mechanism. It was however found that the L.C.F. strength **decreased** with increasing grain size (in-line with variations in yield strength). Grain boundary morphology has also been linked to variation in high temperature crack growth rates with serrated or "wavy" grain boundaries producing lower crack growth rates in air (9,10). This was attributed to an increase in the resistance to grain boundary sliding. It has also been reported that producing wavy grain boundaries increased stress/rupture properties at >500°C in a nickel base alloy (3). The morphology and distribution of secondary phases (e.g. $M_{23}C_6$ and primary γ') on grain boundaries may also be expected to affect intergranular crack growth rates. Garimalla et al (7) state that $M_{23}C_6$ are brittle and deleterious to intergranular crack growth resistance if they occur continuously along grain boundaries, but reduce the propensity for grain boundary sliding and void growth when intermittently decorated along grain boundaries with consequently improved creep/rupture properties. In this context the distribution of intragranular carbides were found to be relatively unimportant. Runkel and Pelloux (13) have also looked at the effects of $M_{23}C_6$ and primary/grain boundary γ' distributions on L.C.F. behaviour in Astroloy between 400°C and 700°C and noted that: Large blocky γ' gave best resistance to crack growth, a combination of g.b. γ' and $M_{23}C_6$ was beneficial, fine g.b. γ'; and a combination of intermittent $M_{23}C_6$ + some blocky MC provided the least resistance to crack growth.

Hence we need to assess the effects of grain size/morphology in addition to grain boundary character (including distribution of grain boundary primary γ') and coherent γ' characteristics to identify high temperature fatigue and creep-fatigue resistant microstructures. In this study a fundamental examination of the effects of microstructure on the high temperature fatigue and oxidation/creep fatigue of U720Li has been carried out. The aim is to optimise the alloy's microstructural characteristics for maximum resistance to crack growth under these conditions.

Three material variants were studied, namely a standard as-received form of powder metallurgy U720Li (base-line material) as well as a large grain (U720LG) and large precipitate variant (U720LP) of U720Li. The U720Li microstructural variants were produced by increasing the solution temperature (to increase grain size) and by slowing the cooling rate from the solution temperature (to increase the coherent precipitate size). The heat treatment routes used for each material are listed in Table I.

Alloy Variant	Solution Treatment	Cooling medium
U720Li	4hr @ 1105°C	oil quench
U720LG	4hr @ 1135°C	oil quench
U720LP	4hr @ 1105°C	insulated air cool

Table I. Heat treatment conditions of alloy variants. All alloys received a subsequent two stage ageing treatment of 24hr @ 650°C (air cool) followed by 16hr @760°C (air cool)

The materials were characterised by etching with Fry's reagent and subsequently image analysed to determine the grain size, incoherent (grain boundary) γ' size and volume fraction. Coherent γ' characterisation was performed using TEM carbon extraction replicas of each material so that size and size range distributions of the secondary and tertiary γ' could be found. Vickers hardness tests were also performed on each of the materials so that the effect of heat treatment could be assessed in terms of mechanical strength (flow stress variations).

High temperature fatigue testing was conducted in both air and vacuum between 650°C and 725°C at an R ratio=0.1 using a trapezoidal 1-X-1-1 waveform where X was the dwell time at maximum load of either 1 or 20 seconds. All vacuum testing was conducted on an Instron 8501 servohydraulic testing machine fitted with an ESH vacuum furnace attachment using single edged notch bend (SENB) specimens. Air environment testing was conducted at DERA Farnborough using compact tension (CT) specimens. Standard d-c potential drop techniques were used to monitor crack growth over the ΔK range 15-40MPa√m. Constant ΔK tests were also conducted at a ΔK of 20 (±2) MPa√m with cracks grown for approximately 2mm under test conditions. Fracture surface and etched fracture surface sections were observed under the SEM to determine the crack propagation mode.

Results and discussion

Microstructural characterisation

The composition of the U720Li was analysed by wet chemical analysis and is given in Table II

Cr (wt%)	Co(wt%)	Ti(wt%)	Mo(wt%)	Al(wt%)
15.92	14.57	5.18	2.98	2.44

W(wt%)	Fe(wt%)	Zr(wt%)	B(wt%)	C(wt%)
1.35	0.08	0.042	0.016	0.023

Table II Chemical composition (wt%) of U720Li, bal. Ni

The microstructural characteristics and Vickers hardness of the alloy variants used in this study are presented in Table III.

Alloy Variant	Grain Size (μm) ±S.D.	Primary γ′ (μm) ±S.D. [Vf (%)]	Secon-dary γ′(nm) ±S.D.	Tertiary γ′ (nm) ±S.D.	Vickers Hardness ±S.D.
U720Li	5.9 ±2.5	2.6±1.0 [18.2%]	102± 51	16±22	459±2.9
U720Li (LG)	17.4 ±8.3	2.9±1.2 [6.9%]	190± 110	17±17	445±2.7
U720Li (LP)	10.1 ±3.4	2.9±1.0 [23.2%]	254± 135	27±25	425±2.3

Table III. Microstructural characteristics of the alloy variants

From Table III it is apparent that the variation in solutionising temperature and cooling rates of U720Li have produced significant variations in grain size and coherent γ′ sizes. The large grain variant (higher solutionising temperature) has a grain size three times that of the as-received alloy whilst coherent γ′ sizes have increased somewhat. The volume fraction of grain boundary pinning primary γ′ has decreased from 18.2% to 6.9%. For all three variants the primary γ′ size has remained similar. The range in grain size (reflected in the standard deviation) is greater for the larger grain size. The slower cooling rate has produced coherent secondary precipitates that are over twice the size of the as-received alloy. However, some grain growth has occurred as well as a somewhat counter-intuitive increase in the volume fraction of primary γ′. All microstructural variants were produced from the same turbine disc material, and specimens were cut tangential to the disc rim, the variations achieved in grain size and primary γ′ distribution may therefore also reflect microstructural variations within the disc (possibly due to differing heat schedules experienced in different sections of the disc.) Hardness testing indicates that increasing γ′ size decreases flow stress, whilst increasing grain size has little effect (with similar implications for yield strength variations). This trend has been reported elsewhere (6) for more extreme variations in grain size and coherent γ′ size in the same alloy.

Fatigue testing of as-received U720Li in air and vacuum

The fatigue crack propagation (f.c.p.) rates obtained in air and vacuum for the as-received U720Li are shown in Figure 1. Testing temperatures of 650 and 725°C with 1 second and 20 second dwells have been investigated. Very similar f.c.p. rates and m-values have previously been reported with no effect of dwell in vacuum at 600°C and 650°C (6). In air crack growth rates at 650°C are considerably higher, but also show no apparent dwell effect. SEM examination of the fracture surfaces and etched fracture surface sections obtained from growth out and constant ΔK tests (Figure 2a-b) obtained under vacuum at these temperatures showed flat, featureless cracking that was highly transgranular in nature. The similarity of the f.c.p. rates at 600°C and 650°C and the smooth, transgranular crack growth mode indicate that crack growth under these conditions may be of a purely cyclic nature. Tests in air at 650°C revealed a mixed (transgranular and intergranular) crack growth mode both with and without dwell (Fig 2c)

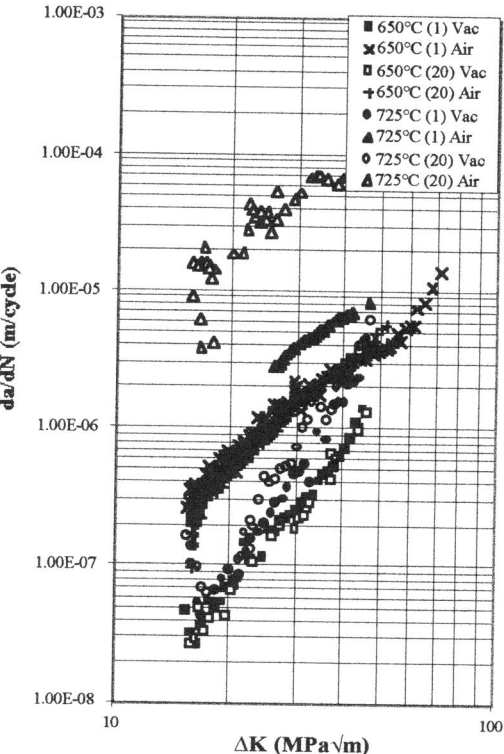

Figure 1 Comparison of behaviour in air and vacuum in U720Li (as-received)

Fig 2a Fracture surface section showing transgranular crack path obtained under 1-1-1-1 loading at 650°C in U720Li

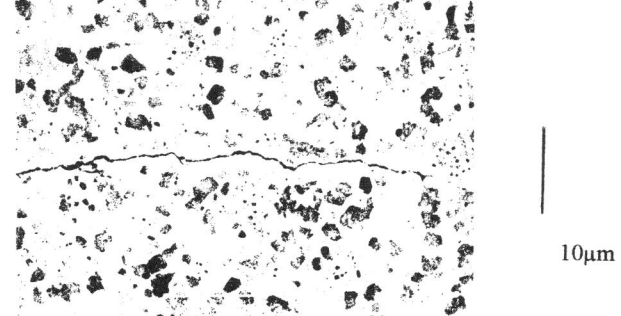

Fig 2b Sectioned constant ΔK test; U720Li, 650°C (1-1-1-1), ΔK = 20 MPa√m, crack tip region

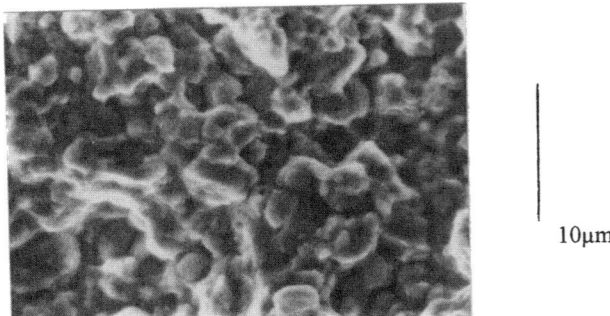

Fig 2c Fracture surface in air; U720Li, 650°C (1-1-1-1)

The f.c.p. rates and m-values for U720Li in vacuum are seen to increase as test temperature and dwell time at maximum load are increased. SEM observations of tests conducted at 725°C in vacuum indicated much rougher fracture surfaces (c.f. lower temperature tests) with the failure modes becoming increasingly intergranular in nature as dwell times were increased (Figure 3a).

Fig 3a Fracture surface section of U720Li 725°C (1-20-1-1) showing typical intergranular cracking

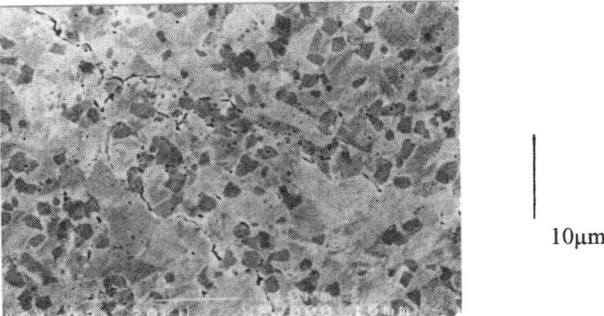

Fig 3b Sectioned constant ΔK test; U720Li, 725°C (1-20-1-1), ΔK = 20 MPa√m, crack tip region

Crack tunnelling was also evident for tests conducted under these conditions. The change in crack growth mode coupled with the increasing f.c.p. rates and m-values noted may indicate the presence of time dependent (creep-fatigue) behaviour. Analysis of constant ΔK tests using backscattered SEM showed that at 725°C voids were appearing ahead of the crack tip (Figure 3b) and were seen to be more prevalent as dwell time was increased. This further indicates the presence of static creep modes under these conditions.

The effect of environment can also be seen in Figure 1 where the vacuum data is compared with CT tests carried out in air at DERA Farnborough. In an air environment, crack growth rates are much higher under all testing conditions with increases noted as temperature and dwell times are increased. Of significance are the slopes of the crack growth curves (m-values), which remain relatively constant over all the testing conditions. Intergranular cracking modes were also noted to predominate under all testing temperatures and dwell times, indicating that the mechanism of crack growth remains constant throughout. This is in contrast to the vacuum data where m-values increased and the crack growth mode changed from transgranular to intergranular as test temperatures and dwell times were increased. In an air environment the results indicate that oxidation embrittlement along grain boundaries dominates the nature of crack growth under all testing conditions. This would explain the constant m-values and the consistently mixed or intergranular nature of the crack growth observed. However, in vacuum the effects of oxidation are effectively removed and at high temperatures with lower frequencies (increased dwell), the change in m-values and crack growth mode would indicate the onset of true creep processes. It is unlikely that any true creep processes can occur in an air environment as the oxidation assisted crack growth occurs at such a rate that there is insufficient time for creep processes to occur, in addition they would be insignificant with respect to the effect of oxidation.

Fatigue testing of the large grain and precipitate microstructural variants in air and vacuum

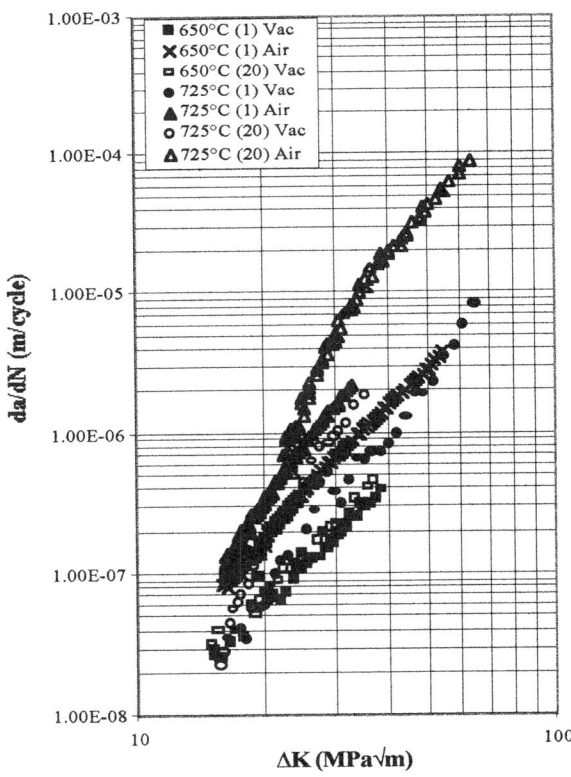

Figure 4: Comparison of f.c.p. rates in U720LG in air and vacuum

498

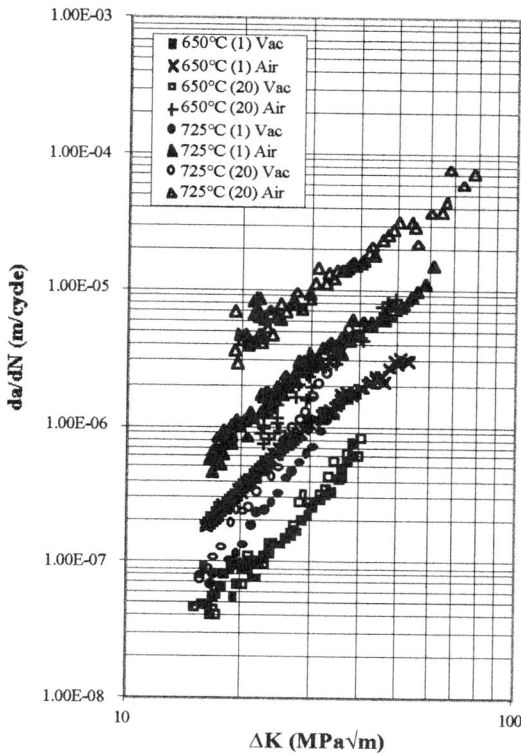

Figure 5: Comparison of f.c.p. rates in U720LP in air and vacuum

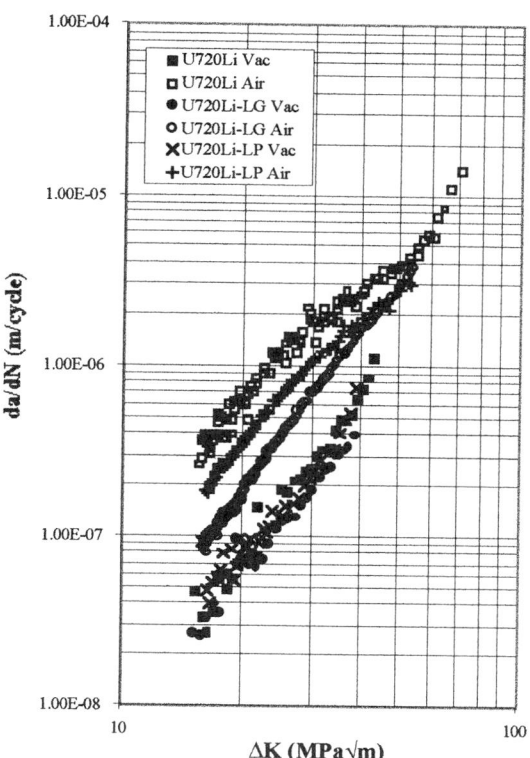

Figure 6: Comparison of f.c.p. rates 1-1-1-1 cycle at 650°C in U720Li, U720LG and U720LP in air and vacuum

Similar trends are seen in both the U720LG and LP microstructural variants in air and vacuum at 650°C and 725°C. Lower crack growth rates are seen in vacuum than in air, with no dwell effect observed at 650°C in vacuum. The U720Li variants with fine γ' precipitates (U720Li and U720LG) show no effect of dwell in air at 650°C, whereas the coarse γ' variant (U720LP) shows an approximate twofold increase in da/dN as the dwell increases from 1s to 20s. Increasing temperature leads to increased crack growth rates for all conditions in both air and vacuum and increased dwell at 725°C also leads to increased crack growth rates in both air and vacuum. Transgranular crack growth modes were observed at 650°C in vacuum for all microstructural variants, whilst at the higher temperatures and dwells increasing proportions of intergranular failure were observed in vacuum. In air mixed or intergranular crack growth modes were observed.

<u>Comparison of the effect of grain size and γ' precipitate size on high temperature fatigue in U720Li</u>

The microstructural analysis work indicates that increasing grain size appears to have less of an effect on flow stress than increasing γ' precipitate size. In Figures 6-9 a comparison of the effect of an increase in grain size (from 5.9μm to 17μm) on the high temperature fatigue behaviour of U720Li in air and vacuum can be seen. At 650°C there appears to be only a very slight advantage (if any at all) observed in terms of f.c.p. resistance of the large grain size variant over the base-line material in vacuum (with or without dwell), whereas in air the benefits of increased grain size at 650°C are far more evident and more marked for the longer dwell. The slight expected reduction in flow strength at 650°C for the larger grain size

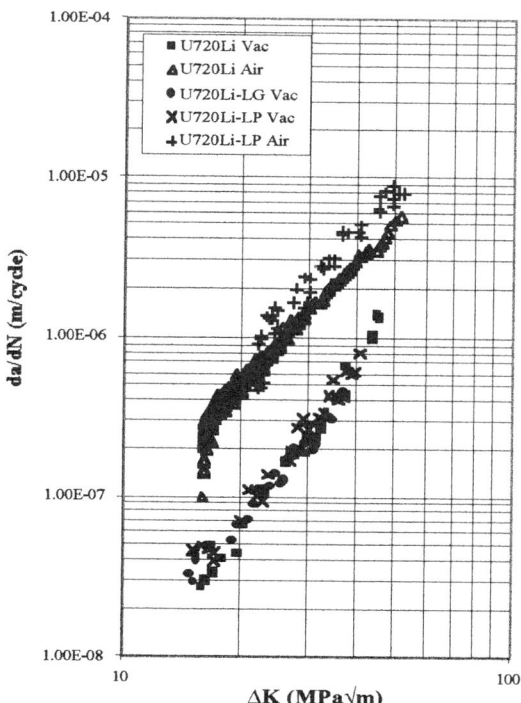

Figure 7: Comparison of f.c.p. rates 1-20-1-1 cycle at 650°C in U720Li, U720LG and U720LP in air and vacuum

does not seem to have reduced the crack propagation resistance. In the purely cyclic dependent regime (650°C in vacuum) there appears to be little effect of increased grain size, indicating that the reduced damage accumulation expected for larger grain sizes has not contributed to improving fatigue crack growth resistance. In air, where a mixed transgranular/intergranular crack growth mechanism predominates and oxidation embrittlement of the grain boundaries accelerates crack growth rates, the benefit of the larger grain size is much more pronounced. This can be linked to the reduced amount of grain boundary area for the larger grain size, limiting the extent of the oxidation embrittlement. In addition the larger grain sized material will have more tortuous and deflected crack growth, yielding an increased shielding effect, and hence a lowered ΔK will be experienced at the crack tip compared with the finer grain sized variants which will have a less tortuous crack path profile. The U720LP variant has a slightly coarser grain size (10.1µm c.f. 5.9µm) and this yields slightly improved f.c.p. resistance at 650°C in air with a one second dwell, but at a longer dwell (more oxidation embrittlement, Figure 7) the f.c.p. resistance is lowered. Previous studies (9,10) have reported an improvement in oxidation fatigue resistance with an increase in coherent γ' size, but this is not supported by our findings. At 725°C the effect of grain size is quite marked in air both with and without dwell (Figures 8 and 9), with the slowest crack growth again observed for the larger grain sized material. The vacuum data however shows only a very slight improvement in f.c.p. resistance even in the creep-fatigue regime (725°C, 20 second dwell). The large precipitate variant showed similar or slightly worse f.c.p. resistance than the baseline, as-received U720Li at 725°C in vacuum with and without dwell and in air without dwell. In air at 725°C with dwell, the U720LP shows improved f.c.p. resistance over the as-received U720Li (although the U720LG still shows the best fatigue resistance). The flow stress of the U720LP material is likely to be somewhat lower than that of the U720LG and U720Li variants, and this, together with the higher proportion of grain boundary γ' as well as the coarser coherent γ' should be taken into account in considering these results.

At 725°C in air with a 20 second dwell intergranular cracking modes appeared to predominate (indicating time-dependent oxidation behaviour) and it is this growth mode that is thought to favour the LP microstructure in terms of increasing resistance to crack growth. It should also be noted that the grain size and volume fraction of primary γ' had been increased somewhat over that of the base-line U720Li which may also contribute to increased creep-fatigue crack growth resistance as these features can help to restrict grain boundary sliding. It is therefore instructive to consider the nature of the grain boundary in more detail, particularly the nominal line fraction of primary grain boundary γ' along the grain boundaries.

Line-fraction estimation of grain boundary γ' along the grain boundaries

We have assumed a hexagonal-type array of equi-sized grains, with equivalent area and circularity to the average grain size determined by image analysis for each microstructural variant. The circularity is defined as:

$$circularity = \frac{4\pi.area}{perimeter^2} \qquad (1)$$

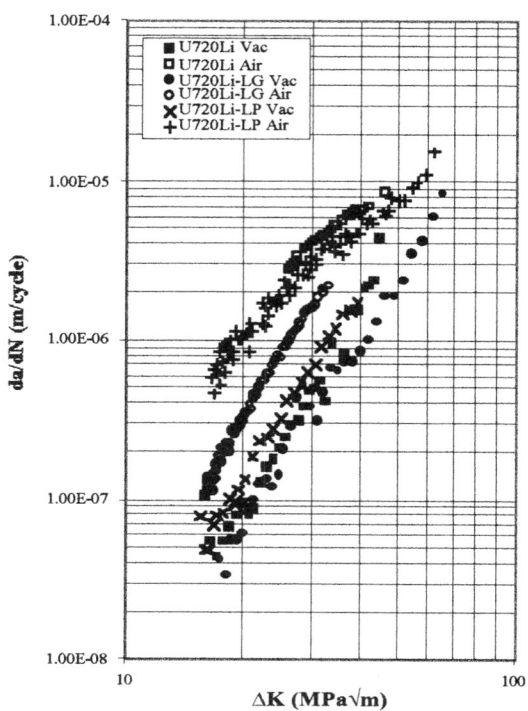

Figure 8: Comparison of f.c.p. rates 1-1-1-1 cycle at 725°C in U720Li, U720LG and U720LP in air and vacuum regime.

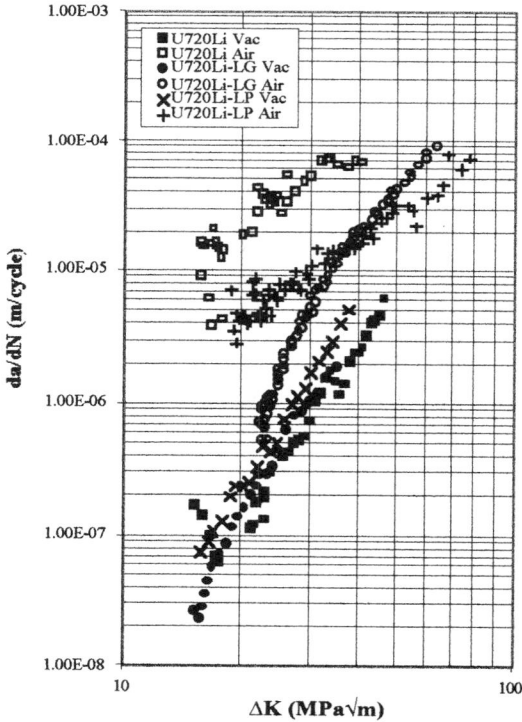

Figure 9: Comparison of f.c.p. rates 1-20-1-1 cycle at 725°C in U720Li, U720LG and U720LP in air and vacuum regime.

500

Hence yielding a circularity of 1.0 for a circle, the average circularity for the three grain sizes was found to be 0.66. The grain dimensions and the grain array considered in this estimation are shown in Figure 10

All internal angles
are assumed to be 120°C

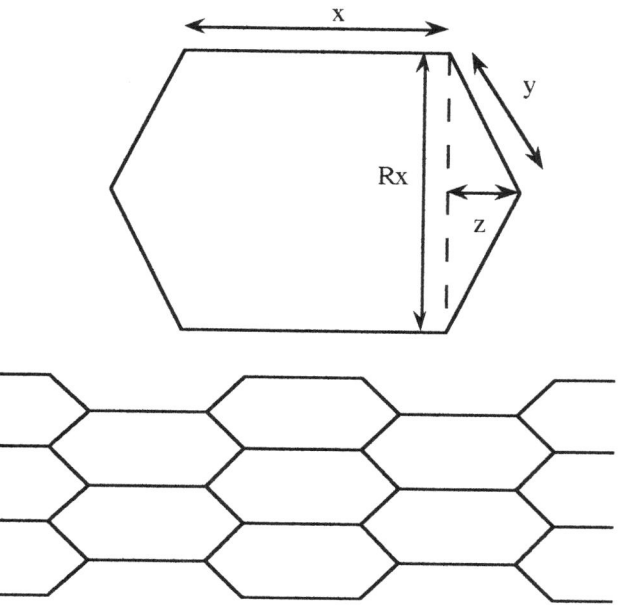

Figure 10 - Grain dimensions and grain array considered in line fraction estimation

An elongated hexagonal grain shape was chosen (as shown in Figure 10) R was chosen to be 0.4 to yield a circularity of 0.66. All internal angles have been taken as 120°, yielding the following relationships for the grain dimensions:

$$y = \frac{Rx}{2\cos 30°} \qquad (2)$$

$$z = \frac{Rx\sin 30°}{2\cos 30°} \qquad (3)$$

where x is chosen to yield an equivalent area to the grain size area for each microstructural variant. The total grain boundary length within the grain array is then:

$$g.b.length = 24y + 18x = \left(\frac{12R}{\cos 30°} + 18\right)x \qquad (4)$$

whilst the area of the grain array is:

$$array area = 12(Rx^2 + \frac{R^2 x^2 \sin 30°}{2\cos 30°}) + 3Rx^2 \qquad (5)$$

The length of grain boundary per m^2 is then simply eqn (4) divided by eqn (5). The number density of primary grain boundary γ′ per m^2 is estimated as the ratio of the V$_f$ γ′ (equivalent to the area fraction) and the estimated average γ′ area (making a simple circular assumption). Then, assuming the primary γ′ always sits on a grain boundary the line fraction of primary γ′ along the grain boundaries can be seen to be:

$$Line fraction = \frac{no.density \times \gamma' diameter}{g.b.length / m^2} \qquad (6)$$

Table IV lists the various estimates for the three microstructural variants

Alloy	Grain Area (m^2)	x (m)	array g.b. length (m)	array area (m^2)	No. dens- ity γ′	Line Fract- ion
U720 Li	2.734 x10^{-11}	7.82 x10^{-6}	1.841x 10^{-4}	4.015 x10^{-10}	3.428 x10^{10}	0.194
U720 LP	8.012 x10^{-11}	13.4 x10^{-6}	3.155x 10^{-4}	1.177 x10^{-9}	3.512 x10^{10}	0.380
U720 LG	2.378 x10^{-10}	23.1 x10^{-6}	5.438x 10^{-4}	3.494 x10^{-9}	1.045 x10^{10}	0.195

Table IV - line fraction estimates

It seems that the lower area fraction of grain boundary γ′ in the large grain sized variant compared with the as-received material can be directly related to the reduction in grain boundary area. A very similar line fraction of grain boundary γ′ is estimated for both the as-received U720Li and the U720LG. The U720LP had an increased area fraction of grain boundary γ′ compared with the as-received material, despite having a somewhat larger grain size, and this is reflected in the higher estimated line fraction of grain boundary γ′ (almost twice as high). It is clear that the U720LP material has significantly more primary grain boundary γ′ along the grain boundaries than the other two variants as well as coarser coherent γ′. The nature of the grain boundary is clearly different for the U720LP and this may go someway towards explaining it's improved performance under the most severe conditions (725°C in air 1-20-1-1 cycle). Under these conditions the crack growth behaviour is dominated by the nature of the grain boundary since intergranular crack growth modes predominate. The increased amount of primary g.b. γ′ is thought to reflect the fact that the original U720Li material, prior to the γ′ coarsening heat treatment, may have experienced a slightly different solution treatment due to it's position within the turbine disc.

Attempting to systematically vary the powder metallurgy microstructure is a complex business. Altering grain size requires longer solutionising times and temperatures to remove some of the grain boundary pinning primary γ′ which will then allow grain growth. Significant coarsening of grain size in this way is not easily controllable, and can lead to very anisotropic grain structures containing blown grains, when some grains grow extremely fast. Even when similar grain boundary character has been retained (as is the case for our U720Li and U720LG variants) with similar line fractions of primary γ′ along the grain boundaries, the amount of coherent γ' will have been altered. To illustrate this we have estimated the proportion of coherent γ' in each microstructure, based upon equilibrium predictions of the total amount of γ' and our experimental measurements of primary γ'. The equilibrium predictions have been made at Cambridge University using Thermo-calc, a commercially available thermodynamic prediction programme, using a proprietary Rolls Royce thermodynamic database.

Material	Total γ' predicted at 250°C	Primary γ' from micrographs	Remaining (predicted) coherent γ'
U720Li	48.3%	18.2%	30.1%
U720LP	48.3%	23.2%	25.1%
U720LG	48.3%	6.9%	41.4%

Table V - Predicted amounts of coherent γ' in each variant

It can be seen that the large grain sized variant (U720LG) is predicted to have the highest proportion of coherent γ' whilst the U720LP has the lowest. Thus changing grain size also affects the proportions of coherent and incoherent (primary) γ'. In the case of the U720LP, the γ' coarsening heat treatment has resulted in an unexpected variation in grain size and proportion of primary γ' (which may reflect solution heat treatment variations within the disc). Hence it is hard to unambiguously resolve the relative contributions of grain size, grain boundary character (e.g. line-fraction of grain boundary primary γ') coherent γ' size and proportion in the high temperature fatigue and oxidation/creep fatigue behaviour. In addition, the effect of these microstructural variables on flow stress may affect the crack tip strain range experienced at equivalent applied stresses.

Normalising for these effects will allow the effect of micromechanism changes due to microstructure variation to be more clearly established. Assuming the ratio of room temperature hardness values gives a reasonable estimate of the ratio of difference in yield strength at the test temperatures of interest, no change in E, and:

$$\Delta CTOD = \frac{\Delta K^2}{\sigma_y E} \qquad (7)$$

then taking the hardness of U720Li as the baseline, ~1.5% increase in "effective" ΔK experienced at the crack can be estimated for the U720LG case, and ~4% increase in "effective" ΔK experienced at the crack can be estimated for the U720LP case. Such changes will increase the normalised ΔK levels by negligible amounts in the log-log plots. Clearly the true high temperature flow stress behaviour should be ascertained for each microstructural condition to establish whether the microstructural changes have affected global mechanical properties.

This work indicates that a larger grain size material has the best high temperature fatigue, creep fatigue and oxidation fatigue resistance, but it should be noted that there are inherent problems with reproducibility of a consistent microstructure when attempting to produce a significant increase in grain size in this class of materials.

Summary and conclusions

The time-dependent crack growth regime in air is primarily due to oxidation assisted crack growth, since any underlying "true" creep-fatigue behaviour, such as that witnessed at 725°C in vacuum is swamped by rapid oxidation effects.

The effect of varying grain size (whilst maintaining similar γ' sizes) on f.c.p. rates in vacuum has been evaluated. A larger grain size offers little benefit in resisting creep-fatigue and cycle-dependent fatigue. This indicates that there is little reduced intrinsic driving force for fatigue crack growth for an increased grain size. Our results do not support the argument for less damage accumulation (14,15) due to increased slip reversibility to higher crack tip stress levels as crack tip plasticity is contained within a larger grain. The beneficial effect of larger grain size expected in the "true" creep-fatigue regime in vacuum, where the decreased number of grain boundaries might be expected to offer more resistance to intergranular grain boundary creep failure mechanisms, has not been shown. The larger grain sized material does however show improved oxidation fatigue resistance over all temperatures in air, where an intergranular oxidation embrittlement process occurs. Altering grain size was found to have little effect on the room temperature hardness of the material, whereas altering the coherent γ' size gave far greater effects. Normalising for the expected variation in yield strength at test temperature indicated a negligible effect of this mechanical property variation on high temperature fatigue resistance.

In terms of f.c.p. behaviour, larger coherent precipitates do not offer improved performance when static creep failure modes appear to dominate in vacuum or under pure cyclic fatigue conditions in vacuum. In air the U720LP material showed slightly improved f.c.g. resistance at 650°C with a 1-1-1-1 cycle, although this effect disappeared with the introduction of a dwell and was not evident for the 725°C 1-1-1-1 cycle. Under the most severe conditions (725°C 1-20-1-1 in air) the U720LP variant showed improved f.c.g. resistance c.f. the baseline material (although the large grain sized variant still offered the best performance). The slightly larger grain size and volume fraction of primary γ' may also be a factor within this regime where intergranular failure modes are important.

Further Work

Further work is planned to establish the tensile behaviour of all microstructural variants at the test temperatures of interest. The line fraction estimates of grain boundary γ' will be confirmed by experimental measurements using image analysis. Representative measures of crack deflection will be taken and the effect of crack path in producing extrinsic shielding effects assessed. This will allow the assessment of micromechanistic contributions aside from extrinsic shielding effects or reduced ΔCTOD due to yield stress variations. All the work described here relates to long crack propagation, and a further programme of work is planned to assess the high temperature fatigue performance of the microstructural variants in terms of crack initiation and short crack growth which may have more relevance to the turbine disc service application.

Acknowledgements

The work in this paper has been supported by the Mechanical Sciences Sector, DERA Farnborough under DTI-CARAD MOD Applied Research Package funding whose support is gratefully acknowledged. Rolls Royce plc are thanked for the supply of materials, use of equipment and technical discussion. The assistance of T.Powell and A.Tucker in carrying out the air testing at DERA Farnborough is gratefully acknowledged. D.Hunt is also thanked for doing the Thermo-calc predictions at the University of Cambridge.

References

1. J.E. King, "Fatigue crack propagation in nickel-base superalloys-effects of microstructure, load ratio, and temperature", Mat. Sci. Tech., 3, (1987), 750-764.

2. H.H. Smith and D.J. Michel, "Effect of environment on fatigue crack propagation behaviour of alloy 718 at elevated temperatures", Met. Trans. A, 17A, (1986), 370-374.

3. R.P. Skelton (ed.), Fatigue at high temperature, (Elsevier Applied Science, Oxford, 1983), 224.

4. L. Davis (ed.), ASM speciality handbook: Heat resistant materials, (ISBN 0871705966, ASM, 1997), 500.

5. E. Andrieu, R. Molins, H. Ghonem and A. Pineau, " Intergranular crack tip oxidation mechanism in a nickel-based superalloy", Mat. Sci. Eng., A154 (1992), 21-28.

6. N.J. Hide, M.B. Henderson, A. Tucker and P.A.S. Reed, "The effects of microstructure and environment on high temperature fatigue and creep-fatigue mechanisms in a nickel base superalloy" Progress in mechanical behaviour of materials ed. F. Ellyin and J.W. Provan (ICM8, Victoria, Canada, 1999), I, 429-434

7. L. Garimella, P.K. Liaw and D.L. Klarstrom, "Fatigue behaviour in nickel-based superalloys: A literature review", Journal of Metals, 49(7), (1997), 67-71.

8. T. Isomoto and N.S. Stoloff, "Effect of microstructure and temperature on high cycle fatigue of powder metallurgy Astroloy", Mat. Sci. Eng., A124, (1990), 171-181.

9. H.F. Merrick and S. Floreen, "The effects of microstructure on elevated temperature crack growth in nickel-based alloys", Met. Trans. A, 9A, (1978), 231-236.

10. H. Ghonem, T. Nicholas and A. Pineau, "Elevated temperature fatigue crack growth in alloy 718-part II: Effects of environmental and material variables", Fat. & Fract. Engng Mat. & Struct., 16 (6) (1993), 577-590.

11. E. Andrieu, G. Hochstetter, R. Molins and A. Pineau, "Oxidation mechanisms in relation to high temperature fatigue crack growth properties of alloy 718", Superalloys 718, 625, 706 and Various Derivatives (ed. E.A. Loria), (The Minerals, Metals and materials society, 1994), 619-631.

12. F. Torster, G. Baumeister, J. Albrecht, G. Lütjering, D. Helm and M.A. Daeubler, "Influence of grain size and heat treatment on the microstructure and mechanical properties of the nickel-base superalloy U720Li", Mat. Sci. Eng. A, A234-236, (1997), 189-192.

13. J.C. Runkel and R.M. Pelloux "Micromechanisms of low-cycle fatigue in nickel based superalloys at elevated temperatures" In Fatigue Mechanism (ed. J.T. Fong), ASTM STP 675, (1979) 501-527

14. A. Lasalmonie and J.L. Strudel, "Influence of grain size on the mechanical behaviour of some high strength materials" J. Mat. Sci., 21, (1986) 1837-1852

15. D.D. Krueger, S.D. Antolovich and R.H. Van Stone, "Influence of grain size and precipitate size on the fatigue crack growth behaviour of alloy 718 at 427°C" Met. Trans. A, 18A (1987) 1431-1448

EFFECT OF LOCAL CELLULAR TRANSFORMATION ON FATIGUE SMALL CRACK GROWTH IN CMSX-4 AND CMSX-2 AT HIGH TEMPERATURE.
- For Refurbishment Technology -

Masakazu Okazaki, Tomoyuki Hiura and Toshio Suzuki

Department of Metallurgy, School of Engineering
The University of Tokyo
Hongo 7-3-1, Bunkyou-ku, Tokyo 113-8656
JAPAN

Abstract

In order to get fundamental understanding to establish a refurbishment technology for advanced gas turbine components, the cellular formation associated with a γ/γ' microstructure coarsened in lamellar or equiaxed arrays in single crystal Ni-base superalloys, CMSX-2 and CMSX-4, have been studied, supposing the case in which they were previously subjected to a damage associated with local plastic deformation, followed by a re-heat treatment. During this study special attention was paid to understand the nucleation and the growth of the transformation from viewpoints of crystallo-plasticity and thermo dynamics. The experimental evidence indicated that the transformation originated and developed with high anisotropy, being influenced by the following factors: the strain field produced by local plastic deformation, the crystallographic orientation, the re-heat treatment temperature and time, and the microsegregation in the material. It was shown that the transformation was reproduced in material previously subjected to fatigue and thermo-mechanical fatigue damage, in which the shearing of γ' precipitates resulting from the activation of $\{111\}<1\bar{1}0>$ slip systems was significant. It was shown that diffusion was controlling the growth rate of the transformation, accompanied with an activation energy of 36 kcal/mol. Furthermore, the effect of the local cellular transformation on the high temperature small fatigue crack propagation was quantified experimentally.

Superalloys 2000
Edited by T.M. Pollock, R.D. Kissinger, R.R. Bowman,
K.A. Green, M. McLean, S. Olson, and J.J. Schirra
TMS (The Minerals, Metals & Materials Society), 2000

Introduction

Recently industrial gas turbine power plant have been the major baseload source of power for many countries. These applications require a highly reliable, long-term load machine which can operate using cheap, available fuels [1]. Nobody doubts that the directions of the development are not only to achieve high thermal efficiency but also cost reduction for production and maintenance. Thus, repair, recoating and refurbishment technology of the hot section components in service under extremely severe conditions will be inevitable in very near future. Nevertheless, no standards, or no sophisticated technologies have not been established yet.

As an example, let us suppose a simple refurbishment process in which single crystal alloys have been subjected to actual damage (e.g., fatigue, creep-fatigue and thermo-mechanical fatigue) during the servicing period and then a reheat treatment is given for damage recovery as is normal for pollycrystalline alloys. What re-heat treatment should be given? Is the same treatment condition as for the virgin material applicable? Is there a problem of local recrystallization in this case? There seems to be little understanding on the above factors. Some researchers have studied the phenomenon of local recrystallization and abnormal precipitate growth [2-5]. According to these investigations, abnormally growing cellular microstructures are found to occur. Because single crystal alloys are generally free from grain boundaries strengthening elements, it is not difficult to suppose harmful effects on the mechanical properties, once local or general recrystallization is reached. This may not be unusual. For example, a shot blasting

fatigue, creep-fatigue and thermo-mechanical fatigue damage? Note that these damages are more or less associated with plastic deformation.

This work has been conducted to clarify the factors leading to the abnormal microstructure, or cellular transformation, resulting from the local plastic deformation in single crystal Ni-base superalloys, CMSX-2 and CMSX-4. During the study special attention was paid to understand the nucleation and growth behavior from viewpoints of crystallography and thermo dynamics. The relationship between the transformation and the actual failure damage was also explored. Furthermore, the effect of the local cellular transformation on high temperature fatigue strength was investigated.

Materials

The materials tested in this work are single crystal Ni-base superalloy, CMSX-4 and CMSX-2. The chemical compositions and the conditions of heat treatments are given in Table I. At the stage of the heat treatments, the volume fraction and the size of cubic γ' precipitates are about 0.5 μm and 65 % in the former, and 0.5 μm and 62 % in the latter.

Results

Outline on Formation of Cellular Microstructure.

What can happen in the single crystal alloys from a metallurgical viewpoint, when they receive refurbishment and recoating processing after a service period? In order to

TABLE I: Chemical Compositions and The Conditions of Heat Treatments.

Chemical compositions (wt. %)

	Cr	Co	Mo	W	Ti	Ta	Re	Hf	Al	Ni
CMSX-4	6.4	9.7	0.6	6.4	1.0	6.5	2.9	0.1	5.7	bal.
CMSX-2	7.9	4.6	0.6	7.9	1.0	-	-	0.1	5.6	bal.

Heat treatments:
CMSX-2: 1250°C x 15 min. +1276°C x2 h.+ 1296°C x3 h. → Ar. F.C.
1080°C x 4 h. → Ar. F.C. ....870°C x 4 h.→ A.C.
CMSX-4: 1277°C x 2 h. +1288°C x 2 h.+ 1296°C x 3 h. + 1304°C x 3 h. + 1313°C x2 h.
+1316°C x 2 h. + 1318°C x2 h. + 1321°C x2 h. → Ar. F.C.
1140°C x 6 h. → Ar. F.C. ....870°C x 20 h. → A.C.

procedure on Ni-based superalloys before coating, which is often conducted to improve the adhesion between the film/substrate, is also a straining process [2]. More recently, Walston et al. [5] have studied microstructural instability in the second and third generation single crystal superalloys including higher levels of refractory elements, or the precipitation of Topologically Close Packed (TCP) phases. They have confirmed that local straining due to shot-blasting accelerated the transformation accompanied with TCP phase formation at the substrate/coating film interface, and that it could result in significant reduction of creep properties. What happens in these kinds of superalloys when exposed to

TABLE II: Conditions of Reheat Treatments.

	Reheat treatment	Aging treatment
Condition A	1080 °C x 20 hr.*	1080 °C x 4 hr. + 870 °C x 4 hr.*
Condition B	1220 °C x 1.5 hr.*	1080 °C x 4 hr. + 870 °C x 4 hr.*
Condition C	1250 °C x 1.5 hr.*	1080°C x 4 hr. + 870°C x 4 hr.*
Condition D	1280 °C x 1.5 hr.*	1080°C x 4 hr. + 870°C x 4 hr.*

* heating rate: 14 °C/min. in vac..
cooling rate: 30°C/min. in vac..

explore this situation, plastic strain was introduced at room temperature by Vickers indentation on CMSX-4, and then some re-heat treatments summarized in Table II were given in Ar atmosphere. The former and the latter are planned to simulate the process of damage during servicing, and that of recovery as usual in high temperature components, respectively. From another point of view the former can be reproduced by shot-blasting process before the coating [2]. The Vickers indentations with a pyramidal angle of 136 deg. were punched into the CMSX-4 plate specimen, whose surface was within 5 degree from the (100) crystallographic plane, under a load of 1000 kN. The indentations were punched at equal intervals of 300 μm in the 1.5x1.5 mm square region: 5x5 in the longitudinal and transverse directions, respectively. On the other hand, the heating rate during the re-heat treatment was 15℃/min. Hereinafter the condition of reheat treatment will be represented according to the notations in Table II.

The metallographs after the heat treatments are shown in Figures 1 and 2. At the macroscopic level the region associated with different morphology from the surroundings are clearly identified in Figure 1 in the both cases. Note that the geometrical bowl of the indentations was completely polished away. When the material was exposed to the condition D treatment (see Table II, Fig. 1 (b) and Figure 2), local recrystallization has been achieved, which can be confirmed from such a feature in Figure 2 that γ′ precipitates align individual orientation in each microstructural unit. This was also confirmed by the X-ray Laue pattern. It is important to note that the recrystallization has begun at the temperature

clearly lower than the normal solution heat treatment temperature for the bulk alloy (about 1315 ℃). These results indicate that simple reheat treatment for single crystal alloy subjected to previous plastic strain would be dangerous. In Figure 2 the white precipitates are also found at the grain boundaries, which were identified as being W-rich by using an electron probe micro analyzer (EPMA).

On the other hand, when the condition B treatment was applied, recrystallization is not seen at least at the macroscopic level, and the microstructurally abnormal regions were formed corresponding to the prior indentation traces (Fig. 1 (a)). This transformed region exhibits a cellular structure, which consist of the coarsened γ and γ′ precipitates (Figure 2). It is worth noting that the general profile of the transformed region reveals a shape of pollygonization, suggesting that this would be a recovery process accompanied with strain energy release. The shape of the cells is equi-axied at the center and lamellar at the interface of the transformed/normal regions, respectively (Figure 2). These geometries must result from the influence of stress fields due to the indentation [10]. Some researchers have been also confirmed the same type of transformations as Figure 2 in Ni-base superalloys [2-5]. Depending on their research background, they call them "reaction zone", "cellular colony", "cellular precipitation" and "cellular recrystallization". In the present work they will be called "cellular transformation", and will be the focus of this work.

It is worth noting that morphological changes occur drastically between Figs. 1(a) and (b). In order to study this change, differential thermal analysis during the heating process was carried out in the sample in which the 5x5 Vickers indentations were introduced; Figure 3. For the analysis a TGD 9600 (Shinku-Riko Co., Japan) was used, in which an Al₂O₃ block was employed as a reference. The heating rate was 40℃ /min up to 800℃, and 15℃ /min between 800-1320℃. In Figure 3 a clear exothermic reaction is found to occur at about 1250 ℃, which may corresponds to the γ′ solvus [6]. A comparison between Figure 3 with Figure 2 indicates that once γ′ precipitates are re-solutioned the local recrystallization developed dramatically. A gradual exothermic reaction which can be seen between 1050℃ and 1250 ℃ in Figure 3, must correspond to the microstructural change as shown in

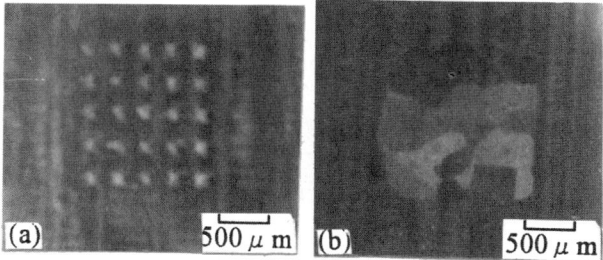

Figure 1: Macroscopic metallographs after the re-heat treatments; (a) at 1220℃, and (b) at 1280℃.

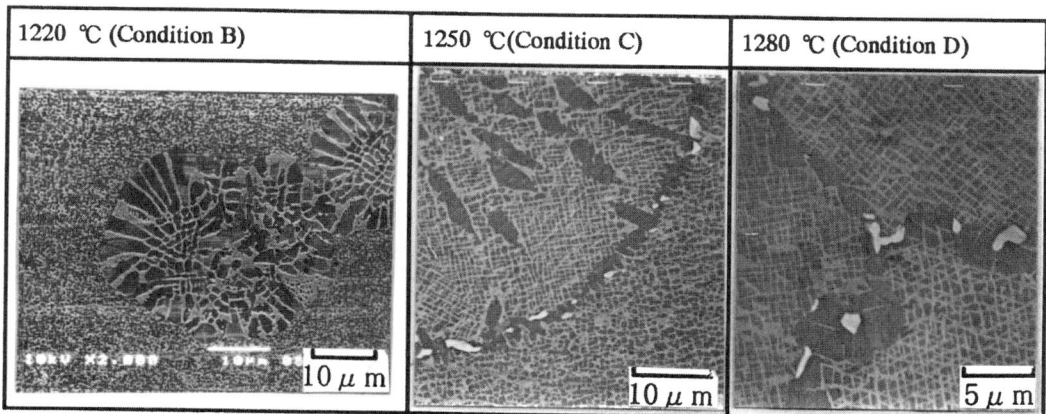

Figure 2: Change of microstructure with the reheatimg temperature in CMSX-4 (Reheating time in 1.5 hrs.).

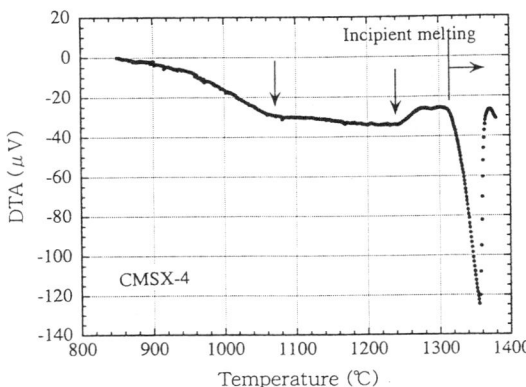

Figure 3: The results of differential thermal analysis (CMSX-4).

and (111) planes, and then the re-heat treatment B was performed (heated at 1220℃ for 1.5 hrs., then aged at 1080℃ for 4 hrs. and at 870℃ for 4 hrs.).

Micrograghs around the indentation before and after the re-heat treatments are summarized in Figure 4. It is found that after the indentation the characteristic slip pattern are activated depending on the crystallographic planes. What is the most exciting in Figure 4 is that the nucleation exhibits a significant crystallographic anisotropy: the predominant directions are <110> directions on the (001) plane indentation, [100] directions on the (011), and <112> directions on the (111) plane, respectively. It is worth noting that the predominant direction was strongly dependent on the crystallographic indentation plane. The mechanism on the anisotropy will be discussed later.

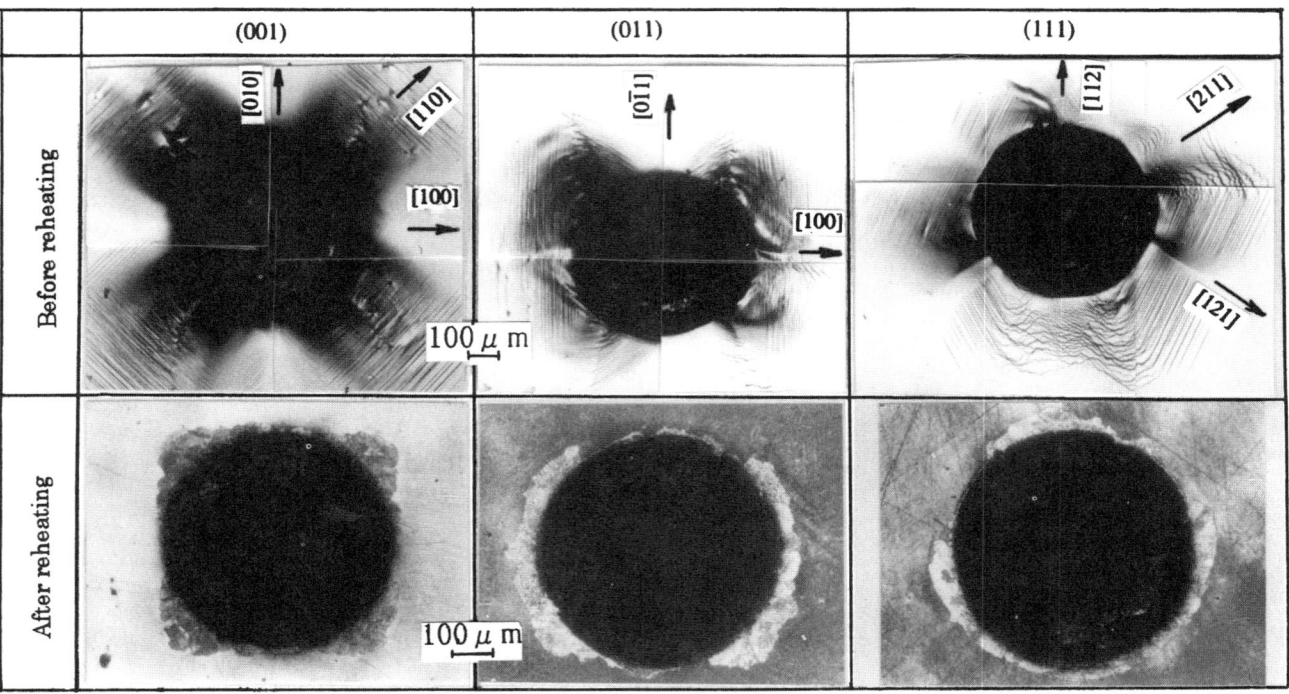

Figure 2.

Characteristics of Cellular Transformation

Nucleation and Anisotropy of Cellular Transformation.
Since stress and strain fields are singular at the pyramidal corner of Vickers indention, the cellular transformation should have predominantly nucleated there. This was confirmed experimentally in the previous section. In the other words, it is difficult to study on the anisotropy in the nucleation and development of the cellular transformation, by the procedure employing the Vickers indentation. In order to avoid this problem, a Brinell indenter with a sphere tip whose radius was about 400 µm was employed, and the plastic strain was introduced at room temperature on some crystallographic planes of CMSX-4; planes within 5 degree from (001), (011)

Relation between γ′ Rafting and Cellular Transformation
The rafting phenomenon [9] is famous as a morphological change of γ′ precipitates. In order to discuss the relation between γ′ rafting and the cellular transformation, the microstructure around the Brinell indentation after the re-heat treatment A (see Table II) was observed: Figure 5. The region accompanied with the rafting can be identified independently, neighboring to the cellularly transformed region. This implies that there is a difference between them.

The volume fraction of the γ′ phase in the cellularly transformed region is compared with that in the rafted and in the bulk areas (Figure 6). On the measurement the transformed region was classified into two parts: equi-axed

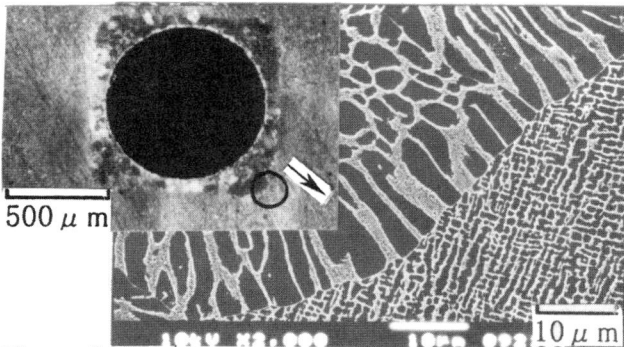

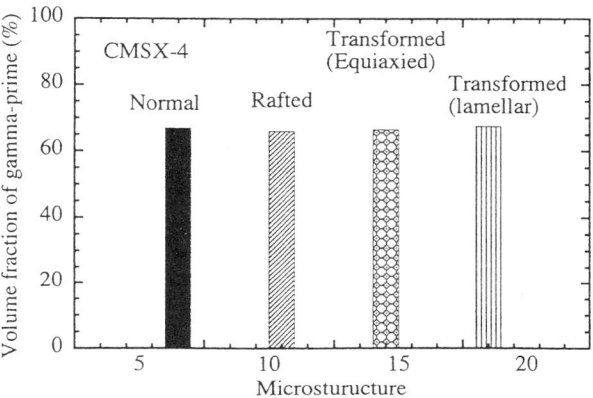

Figure 5: Microstructure around the transformed region (CMSX-4). Indented on (001), and then exposed to the re-heat treatment A in Table II.

Figure 6: Comparison of γ′ volume fraction in the cellular transformed region with that in the other regions (CMSX-4).

region at the center and lamellar region near the interface. No significant differences are found in Figure 6 at any areas, although the measurement at the lamellar region possibly involves large scatter due to their highly elongated, or anisotropic morphology. From these observations it is reasonable to consider that the cellular transformation is different from in some aspects but similar to the rafting phenomenon in other ones.

Effect of Microsegregation on the Cellular Transformation.
The indentation methods employed in the previous section is not always suitable to quantitatively clarify the criterion under which the cellular transformation is developed, because the plastic strain distribution has a high gradient around the indentation. Thus, a uniform distribution must be better. Bi-axial, or tri-axial plastic straining is preferable to uniaxial one, because the actual components generally experience the former type deformation. In order to realize this, cold working was applied to the cylindrical specimen of CMSX-4 as illustrated in Figure 7, by die forging at room temperature. The total strain in the radial direction by this method is about 8.3 % in compression. After die forging the material was exposed to the re-heat treatment C (see Table II).

Metallographs in the longitudinal and transverse sections of

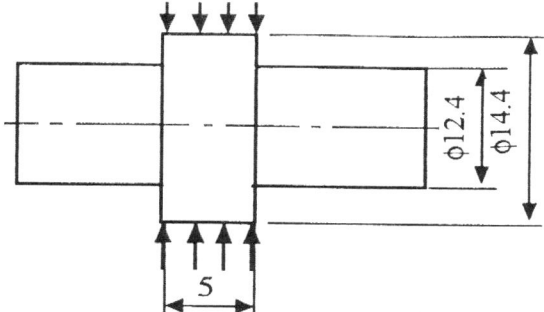

Figure 7: Geometry of specimen die-forged at room temperature.

the sample are given in Figures 8(a) and (b). It is found that while the surface region has been fully recrystallized (Figure 8(c)), the midsection has been partially recrystallized (Figure 8(d)). At the sample core any drastic morphology changes have not occurred (but the loss of γ/γ′ coherency) (Figs. 8(a) and (b)). The above inhomogeneous morphological changes may be resulted from the gradient of plastic strain introduced by die-forging. It is worth noting that the inhomogenity of local recrystallization at the mid-section of the sample (Figs. 8(a), (b) and (d)): the local recrystallization grows predominantly along the dendritic core under the same plastic strain.

The EPMA measurement on the distribution of the main chemical elements along the arrow mark in Figure 8(d) is shown in Figure 9. It is found that whereas the γ-side elements (e.g., W, Cr, Co) are relatively rich at the dendritic core, the γ′-side elements (e.g., Al, Ti, Ta) are rich at the interdeditric region, respectively. This result indicates that microsegregation, which depends on the distribution coefficient on phase diagram during the solidification process and on the post-solidified heat treatment(s), play an important role in the preceding inhomogeneous recrystallization.

Relation between Cellular Transformation and Actual Engineering Damage. Does the cellular transformation appear only when the plastic deformation is applied by the Vickers or Brinell indentations ? In order to explore this, the re-heat treatment C was exposed to the CMSX-2 specimen which had been previously subjected to the following three kinds of actual engineering damages: one was due to fatigue at 600℃ [7], the second was due to so-called creep-fatigue loading with hold time on the tension side at 950 ℃ [8], and the third was due to the thermo-mechanical fatigue between the temperature of 400-900 ℃ under the out-of-phase condition. All the damage was introduced into the CMSX-2 specimens in air under uni-axial loading condition, where the specimen axis laid within 5 degree from <001> direction. The experimental details are given elsewhere [7,8].

The metallograph near the crack face of the CMSX-2 specimen is given in Figure 10 (a), that had been previously subjected to the fatigue crack propagation test at 600℃ and then re-heated under the condition C treatment (i.e., reheated at 1250℃ for 10 hrs, and then aged at 1080℃ for 4 hrs,

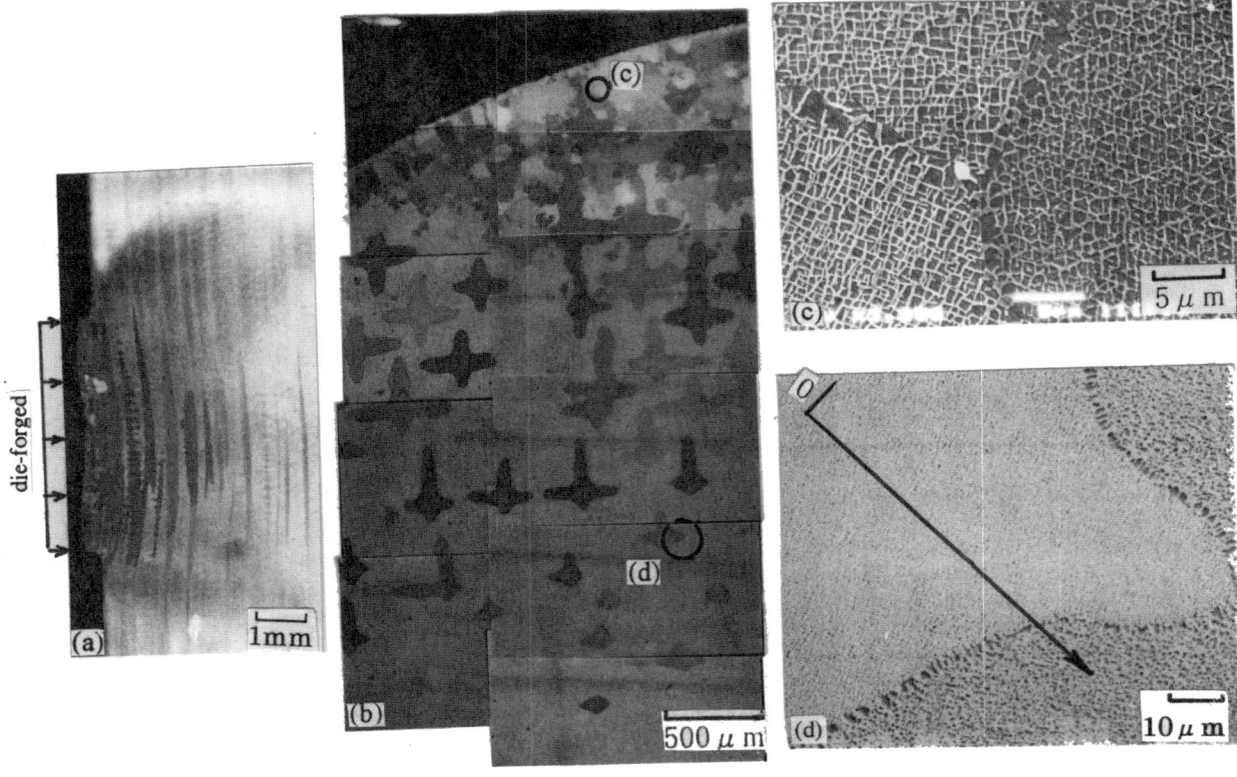

Figure 8: Metallographs of the sample die forged at room temperature, and then reheated by the treatment C (CMSX-4).
(a) longitudinal section, (b) transverse section, (c) microstructure near the surface, (d) microstructure inside the sample

followed by aged at 870℃ for 4 hrs.) During the prior fatigue process the fatigue crack had initiated at the center portion perpendicular to the loading axis and then propagated along the crystallographic {111} slip planes [7]. It is found from Figure 10 (a) that the cellularly transformed region is found to be developed along the fatigue fracture surface, and does not develop apart from there, after the reheat treatment. The transformation seems to grow along the dendritic core, which may be due to the effect of microsegregation (see the previous section). The same phenomenon can be seen along the crack plane in the specimen which had been previously exposed to the thermo-mechanical fatigue damage, as shown in Figure 10(c). It is worth noting that the cracks in both cases were propagated mainly by the activation of {111}<1$\bar{1}$0> slip system, which can be identified by the shearing phenomenon of the γ' precipitates [7].

On the other hand, no cellular transformation was found to occur in Figure 10(b), in which the CMSX-2 was exposed to the creep-fatigue damage. Instead, the γ' rafting is observed near the fracture surface, which had been already confirmed before the reheat treatment [8]. In this case the {100}<1$\bar{1}$0> system may be activated more predominantly than the {111}<1$\bar{1}$0> [12].

These experimental results indicate that the cellular

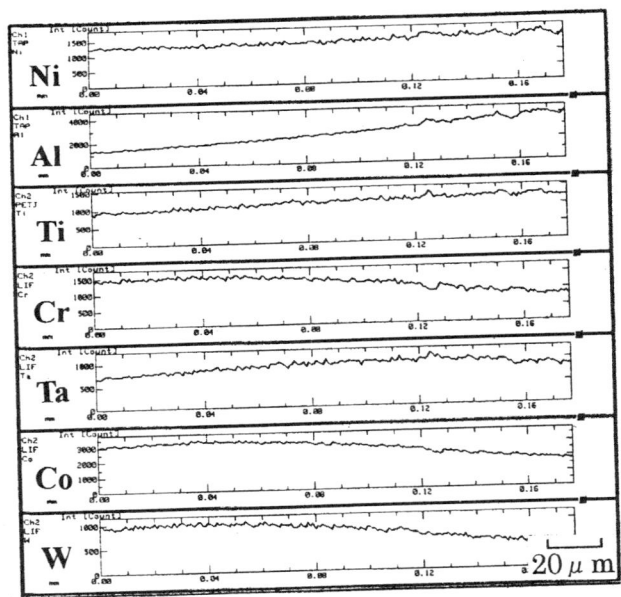

Figure 9: Microsegregation of the main elements in the dedritic structure (CMSX-4).

510

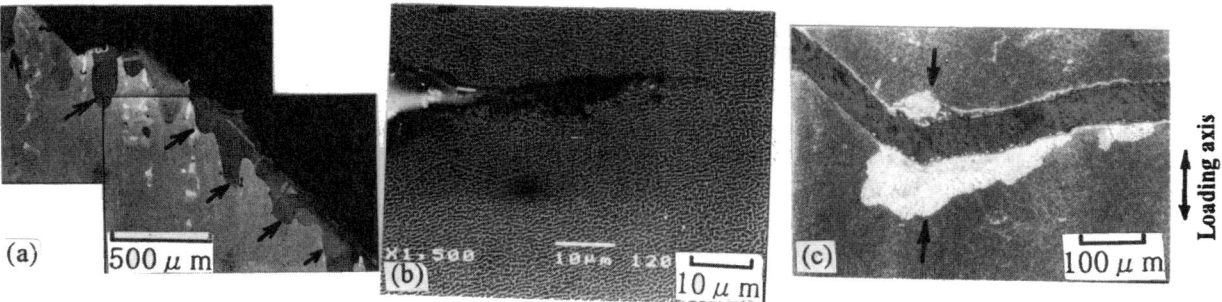

Figure 10: Metallograph of the CMSX-2 primarily subjected to; (a) fatigue damage at 600℃, (b) creep-fatigue damage with hold time on tension side at 950 ℃, and (c) thermo-mechanical fatigue damage between the temperature of 400-900 ℃ under the out-of-phase condition, followed by the reheat treatment, C.

transformation can be realized in actual components. Furthermore, the above features strongly suggest that the cellular transformation must originate only when the inelastic strain inducing the shearing of γ' precipitates is introduced. The experiments obtained in earlier sections support this interpretation. It is also important to note that this appeal could be a potential tool for failure analysis.

Effects of Cellular Transformation on Small Fatigue Crack Propagation at High Temperature.

It is not difficult to suppose that once the cellularly transformed region is formed in single crystal Ni-base superalloys, it is harmful for the high temperature strength, because single crystal alloys are generally free of grain boundary strengthened elements. The effect of local cellular transformation on fatigue strength is investigated in this section. For the experiments, solid cylindrical smooth specimens with locally cellular transformed areas were prepared, according to the following procedures: after the cold working was applied to the CMSX-4 specimen with same geometry and by the same procedure as Figure 7 at room temperature, the re-heat treatment C was performed, followed by the machining into the solid cylindrical smooth specimen of which gauge section diameter and length were 5 mm and 13 mm, respectively. At the center of the specimen gauge section a transformed region was formed over about 5 mm in length (See Figure 6). The fatigue tests were carried out at 950 ℃ in air at a frequency of 10 Hz under a load ratio of -1.

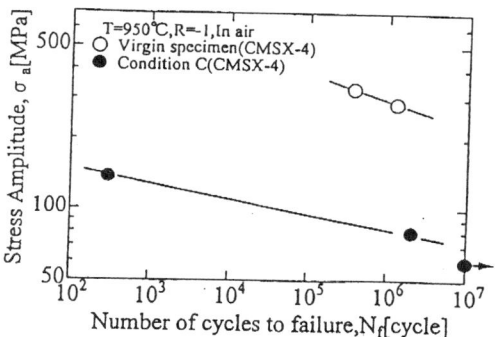

Figure 11: Comparison of the fatigue lives between the locally transformed and the bulk materials at 950 ℃ (CMSX-4).

Figure 11 is a comparison of the fatigue lives between the locally transformed and the bulk materials, which indicates the former exhibits the fatigue strength remarkably lower than the latter. This is a direct evidence indicating the seriousness of the problem.

As already shown in Figure 2, the cellularly transformed region reveals some different morphologies depending on the re-heat treatments. The effect of the morphology on the small fatigue crack propagation is a next project. For this purpose a 5x30x1.5 mm plate specimen with a locally transformed region with different morphologies was prepared: after Vickers indentation was punched into the CMSX-2 plate, the heat treatments; B and D, were carried out, followed by the polishing until the geometrical bowl of the indentation completely disappeared. Any mechanical notches were not introduced in the specimen. The transformed region whose size was about 0.4x0.4 mm in area and 50 μm in depth, respectively, existed at the center portion of the plate specimen. The specimen axis lies within 5 degree from the <001> direction. The small fatigue crack propagation tests were carried out at 950 ℃ in air at a frequency of 10 Hz under a load ratio of 0.05. The naturally initiated surface small fatigue crack growth was monitored, by periodically replicating the specimen surface. The small crack length measured was about from 20 μm to 200 μm.

Figure 12 shows a small fatigue crack in the smooth specimen with a locally cellular transformed region, which indicates that predominant crack initiation is from the transformed/normal interface. The propagation rates of these naturally initiated small cracks are compared with those in the bulk material in Figure 13, as a function of stress intensity factor range, ΔK, where ΔK was evaluated from the Raju-Newman equation by assuming a semi-circular geometry for the surface cracks [13]. The data points in the Figure exhibit the rates of multiple small cracks near the transformed region, which were obtained by using multiple specimens. From the Figure it is found that the small cracks near the transformed region exhibit the rates significantly higher than those in the bulk material, and they generally grew at lower ΔK level. It is important to note that this trend can be seen not only in the cracks running at the locally recrystallized area (which was exposed to the re-heat treatment D) but also in the cracks running at the cellularly transformed area. These experiments show that it is a serious concern from industrial point of

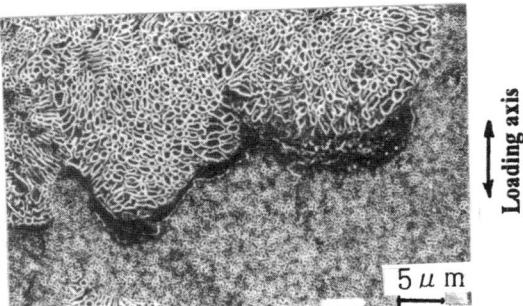

Figure 12: A fatigue crack nucleated in the CMSX-4 specimen with a locally cellular transformation region.

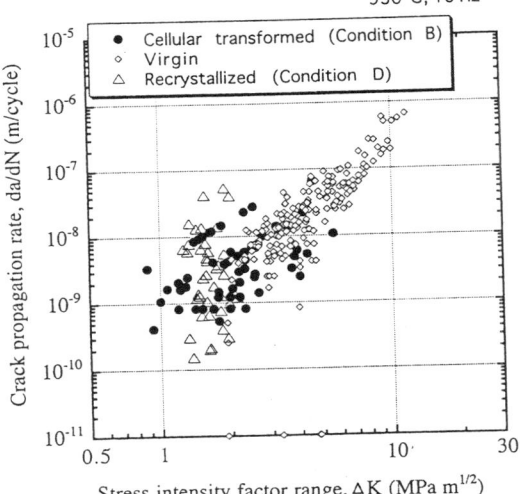

Figure 13: Small fatigue crack propagation in the CMSX-2 specimen with a locally cellular transformed region.

view to get understanding on the transformation and to develop the prevention method.

<u>Discussion</u>

Nucleation of Cellular Transformation.

It is natural to suppose from the results of the previous section that the main driving force for the cellular transformation should be strain energy stored in the material [2-4]. On the other hand, the nucleation of the cellular transformation was found to reveal significant anisotropy, in which the shearing of γ precipitates resulting from the activation of $\{111\}<1\bar{1}0>$ system may play an intrinsic role. Thus, let's consider the slip system activated in the Brinell indentation process.

The stress field around the Brinell indentation can be modeled and analyzed by applying the famous analysis on a sphere cavity expanding in infinite body. The stress components at the surface of the cavity, or Brinell indentation, are expressed by

$$\sigma_{rr} = -p$$
$$\sigma_{\theta\theta} = \sigma_{\phi\phi} = p/2 \qquad (1)$$

where p is a internal pressure. The r, θ and ϕ are in spherical coordinates, which are taken with reference to the specimen into which the Brinell indentations is punched (see Figure 14). These stress components are converted to σ_{ij} (i,j =1,2,3) on the basis of crystallographic cubic axis (see Figure 14), through the tensor transformation:

$$\sigma_{ij} = l_{si} l_{tj} \sigma_{st} \qquad s, t = r, \theta, \phi \text{ and } i,j = 1,2,3 \qquad (2)$$

where l_{si} and l_{tj} are the directional cosine between the two sets of coordinate axes. Based on Eqs. (1) and (2) the shear stresses, τ, on the twelves series of $\{111\}<1\bar{1}0>$ slip systems can be determined. It is natural to consider that the slip system accompanied with the maximum τ value, τ_{max}, of all the slip systems should be primarily activated at a given point.

The distribution of the τ_{max} normalized by the indentation pressure, p, is illustrated in Figure 15, together with the primary slip system activated, where the calculation was performed for $\theta = \pi/3$ rad. (See Figure 14). Figure 15 is illustrated so that the value of τ_{max} corresponds to the distance from the origin. Note the specific τ_{max} distribution dependent on the crystallographic indentation planes: 4-fold symmetric axes on (001) indentation, 2-fold ones on (011), and 3-fold ones on (111), respectively. There is an interesting relationship between these symmetries and the profile of the cellular transformation given in Figure 4.

Comparing Figure 15 with Figure 4, at first, note a common feature that the nucleation sites always involve the site where the most dominant slip system is changed from one to another in all the cases: e.g., [112] and [211] directions in the (111) indentation, [110] and [$\bar{1}\bar{1}$0] directions in the (001) indentation, and [100] and [0$\bar{1}$1] directions in the (011) indentation, respectively. However, there are some exceptions: e.g., [211] direction in the (111) indentation, [100] direction in the (001) indentation, and [0$\bar{1}$1] direction in the (011) indentation, respectively. Accordingly the above

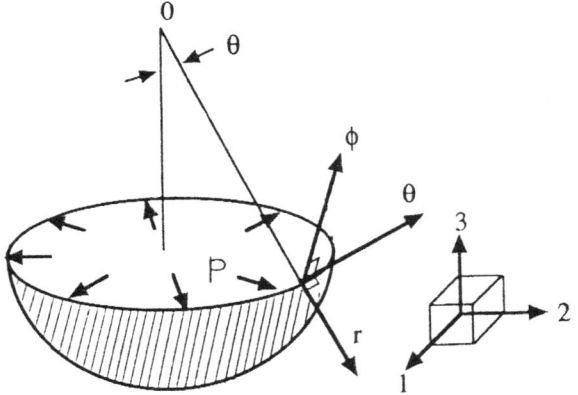

Figure 14: Illustration on spherical coordinates around the Brinell indentation, and the crystal coordinates.

condition is not sufficient. Relating to the nucleation site, the second common feature found in Figure 15 is that the changing ratio in τ_{max} along with orientation (which may correspond to the gradient of stress and strain) is very high at the nucleation site, which seems to be another requirement

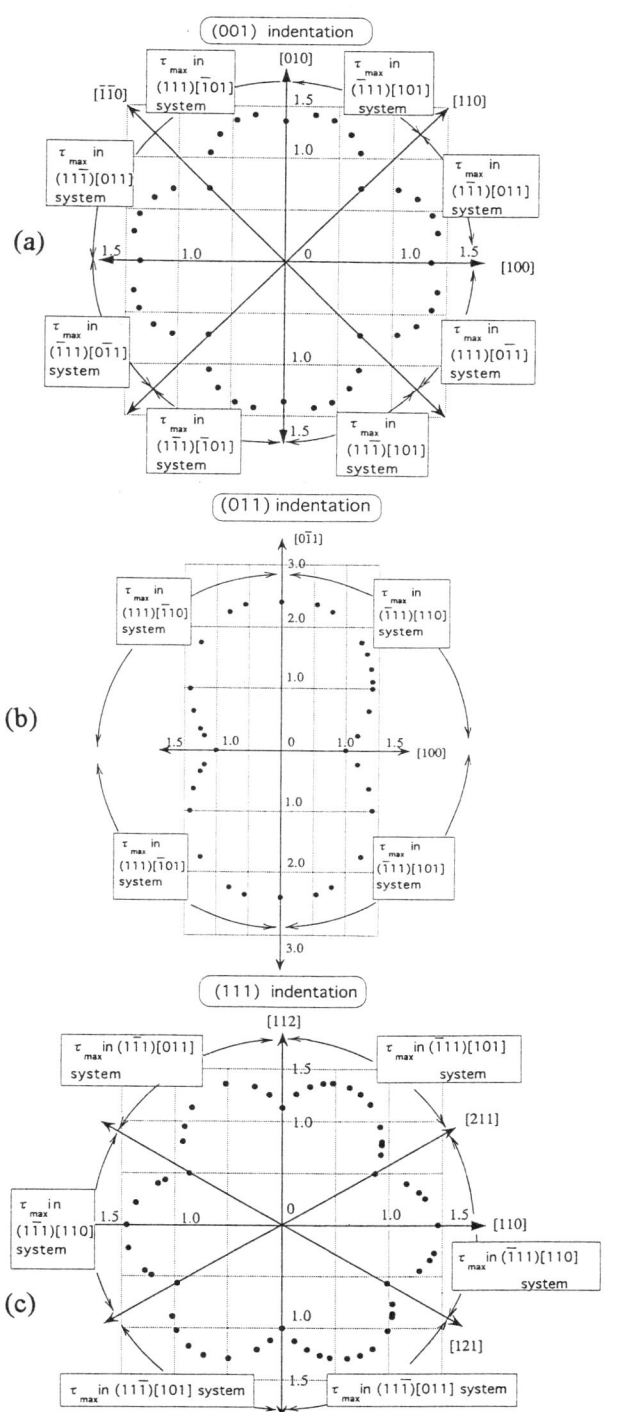

(a)

(b)

(c)

Figure 15: The distribution of the τ_{max} around the Brinell indentation, and the primary slip systems activated.

condition for the nucleation site. Remember that micro shear, or kinked bands where the gradient in stress and strain is very high, is a preferential site for recrystallization in general materials [14]. The reason why the $[0\bar{1}1]$ direction in the (011) plane indentation was not the nucleation site (Figure 4) is reasonably explained by the above second condition, because the changing ratio in τ_{max} is very low there. One exception which does not satisfy the second aspect is found in the [211] direction in the (111) indentation. According to the calculation in Figure 15, only slip direction but no slip plane is supposed to be changed between this direction. This would reduce the singularity of stress field there, resulting in non-nucleation there.

Growth of Cellular Transformation.

Regarding as general recrystallization, Johnson-Mehl type equation is famous. The recrystallized ratio, r(t), at a given time, t, is generally expressed by

$$r(t) = 1 - \exp(-At^m) \qquad (3)$$

where A and m are constants, respectively [14]. Employing new constants C and n, Eq. (3) can be converted into

$$R(t) / R_f = 1 - \exp(-Ct^n) \qquad (4)$$

where R(t) and R_f are the size of cellular transformation zone at re-heat treatment time, t, and that finally reached after infinite heat treatment time. The latter would be a function of transformed/bulk interface energy. It is interesting to study the growth of cellular transformation, according to Eq.(4).

The change of R(t) with the reheating time, t, is shown in Figure 16, when the Brinell indention was punched on (001) plane. Because the transformation was highly anisotropic as shown in the previous section (see Figure 4), R(t) was represented and measured by the size between the Brinell circle and the transformed/bulk interface to <110> direction.

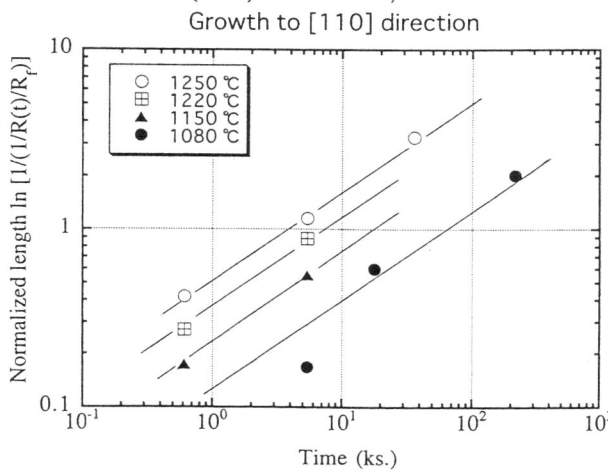

Figure 16: The change of the cellular transformation size with the heat treatment time.

513

(001) indentation,
Growth to [110] direction

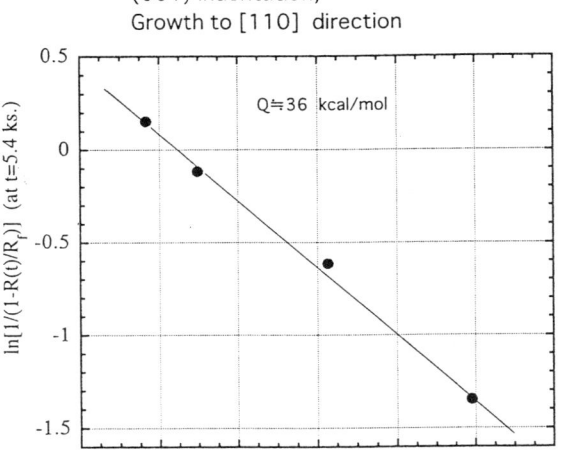

Figure 17: Temperature dependence in the cellular transformation size.

The value of R_f was tentatively approximated by 135 μm. It is found from Figure 16 that the growth can be successfully expressed by Eq. (4). Note that the value of n can be well approximated by 0.5, which indicates that the diffusion was controlling the growth rate.

It is also interesting to investigate the temperature dependence in the transformation, by employing Arrhenius type activation process in the constant, C. The Arrhenius plot at t=5.4 ks. is given in Figure 17, which shows an activation energy of about 36 kcal/mol. The activation energies have been measured 26-28 kcal/mol for Ni self diffusion, and 64 kcal/mol for diffusion of Al atoms in Nickel by other researchers [11]. Interestingly the present value is intermediate between them. This indicates that the cellular transformation must be a diffusion process of Al and Ni atoms.

Conclusions.

The conclusions obtained are summarized as follows:
(1) When CMSX-2 and CMSX-4 were subjected to damages associated with local plastic deformation, followed by the re-heat treatments, the cellular transformation associated with γ/γ' microstructure coarsened in lamellar or equiaxed array was originated. This was not a special phenomenon, but it was actually reproduced in the material subjected to fatigue and thermo-mechanical fatigue damages.
(2) The transformation was originated and developed with high anisotropy, being influenced by the following many factors: i.e., the strain field by the local plastic deformation, the crystallographic orientation, the re-heat treatment temperature and time, and the microsegregation in the material. The γ'-solvus in the reheating process drastically accelerated the local recrystallization.
(3) The shearing of γ' precipitates in the straining process may play an intrinsic role in the cellular transformation: when the $\{111\}<1\bar{1}0>$ systems were activated, the formation was

significant. When the $\{100\}<1\bar{1}0>$ slip systems were activated such as in creep failure at higher temperature, on the other hand, it was negligible.
(4) The growth of the transformation was successfully represented by the Johnson-Mehl type equation. The growth of the transformation must be controlled by diffusion process of Al and Ni, where activation energy was determined by about 36 kcal/mol.
(5) The small fatigue crack propagation rate was remarkably enhanced at the transformed region.

Acknowledgement:

Financial supports by the Ministry of Education, Japan, as Grant-in-Aid for Scientific Research (No: 10550097) and by Zairyo-Kagaku Foundation are greatly acknowledged.

References

1. For example, J. Stringer and R. Viswanathan, "Gas Turbine Hot Section Materials nd Coatings in Electric Utility Applications", (Paper presented at *ASM 1993 Congress Materils Week, Pittsburgh, 1993*) , 1-8.
2. S.D.Bond and J.W.Martin, Surface Recrystallization in A Single Crystal Ni-based Superalloy, J. Mater. Sci., 19, (1984), 3867-3872.
3. A Porter and B. Ralph, The Recrystallization of Ni-base Superalloys, J. Mater. Sci., 16, (1981), 707-713.
4. P. Portella, Influence of Cellular recrystallization on the fatigue Behavior of Single Crystal Ni-based Superalloys, " MicroStructure and Mechanical Properties of Metallic High Temperature mnaterials, Eds. by H. Mughrabi, DFG, (1999), 441-453.
5. W.S. Walston, J.C., Schaeffer and W.H. Murphy, Super-alloys '96, A New type Microstructural Instability in Super-alloys -SRZ, (1996), 9-18.
6. U. Paul and P.R. Sahm, Inhomogenities in Single Crystal Components, Mater. Sci. & Engg., A173, (1993), 49-54.
7. M. Okazaki, H. Yamada and S. Nohmi, Temperature Dependence of the Intrinsic Small Fatigue Crack Growth Behavior in Ni-Base Superalloys Based on Measurement of Crack Closure, Metall. & Mat. Trans., 26, (1996), 1021-1033.
8. M. Okazaki, Creep-Fatigue Small Crack Propagation of a Single Crystal Ni-Base Superalloy: Microstructural and Environmental Effect , Int. J. Fatigue, 20 (1999), 79-86.
9. F,R,N. Nabarro, Rafting in Superaloys, Metall. Trans., 27-A, (1996), 513-530.
10 A. Pineau, Influence of Uniaxial Stress on the Morphology of Coherent Precipitates During Coarsening, Acta Met., 24, (1976), 559-567.
11. K. Smidoda, W. Gottschalk and H. Gleiter, Diffusion in Migrating Interfaces, Acta Met., 26, (1978), 1833-1836.
12. For example, V. Paidar, D.P. Pope and V. Vitek, A Theory of the Anomalous Yield Behavior in L1$_2$ Ordered Alloys, Acta. Met., 32, (1984), 435-448.
13. I.S.Raju and J.C. Newman: Eng. Frac. Mech., 11, (1979), 817-835.
14. For example, F.J. Humphrey and M. Hathery, Recrystallization and Related Annealing Phenomena, (Pergamon Press, 1995).

Multiaxial Creep Deformation of Single Crystal Superalloys: Modelling and Validation.

H.C.Basoalto, R.N.Ghosh[1], M.G.Ardakani, B.A.Shollock and M.McLean

Department of Materials
Imperial College of Science, Technology and Medicine
Prince Consort Road, LONDON, SW7 2BP, UK

Abstract

The creep behaviour of the single crystal nickel-base superalloy CMSX-4 has been studied under multiaxial loading, using double Bridgman-notch specimens with <001> and <111> nominal axial orientations, at 850°C and net section stresses of 600 to 850 MPa. The creep lives measured were an order of magnitude greater than those obtained with the same load in uniaxial loading. The disposition of the triaxial stress state with respect to crystallographic orientation varies around the perimeter of the notch. The creep-deformed material has been examined by electron back-scatter diffraction (EBSD) to characterise local changes in crystal orientation resulting from creep strain. It is shown that there is a spatial distribution of crystal rotation across the specimen diameter that is consistent with the activation of <111>{1$\bar{1}$0} and <001>{110} slip for <001> and <111> specimens respectively. A model of anisotropic creep, developed at Imperial College over a number of years that assumes viscous slip on such a restricted set of slip systems, has been implemented in a commercial finite element code to simulate the notch-creep behaviour. There is good agreement between the computer simulation and the experimental results.

1 Introduction

The development of single crystal technology for the production of blades for gas turbines has been one of the most important developments in aero-engine technology in the past twenty-five years or so. Single crystal turbine blades are now state-of-the-art in modern aero-engines and are being considered for application in industrial turbines for electricity generation. There is now a good understanding of the creep behaviour of these materials under uniaxial loading. In conventional design procedures elastic stress analysis is used and the creep performance is represented by simple measures, such as rupture life or the time to achieve a specific strain. The evolution of engineering design makes additional demands on the formulation of elevated temperature deformation and fracture behaviour. Here, it is necessary to represent the full shape of the creep curve mathematically and to account for the effects of multiaxial loading through appropriate constitutive laws that can be integrated into advanced numerical design procedures. Empirical representation of an extensive database accounting for all appropriate combinations of stress (including multiaxiality), temperature and orientation is not practicable. Rather, there is a need to develop approaches to interpolation and extrapolation, both to long times and to more complex loading, from relatively sparse databases dominated by steady, uniaxial loading. This requires a detailed appreciation of the life-limiting deformation and fracture processes.

The design of gas turbines for aerospace and industrial applications must satisfy the increasingly stringent and different demands for performance and efficiency, which have resulted in increasingly complex blade designs and service cycles. Consequently, the blades will experience complex multiaxial states of stress, which will vary during service. Models of anisotropic creep that have been developed to represent uniaxial creep data cannot be automatically extended to account for multiaxial and/or variable loading. It is important to characterise experimentally the effects of multiaxial state of stress both on the creep performance and on the operative deformation mechanisms. This will establish whether or not the constitutive relations for anisotropic creep that have been successful in accounting for deformation in uniaxial loading can be extended to multiaxiality in a straight forward manner. The model can then be incorporated into a finite element routine to simulate the behaviour of a multiaxial test piece or a real component.

This paper presents the results of multiaxial creep tests using Bridgman notch specimens. It is particularly concerned with the characterisation of the changes in crystallography resulting from creep deformation due to the multiaxial stresses and of the implications of these measurements for the active slip systems controlling creep. These measurements are used to validate the extension of the Imperial College crystallographic-slip model of anisotropic creep to account for multiaxial loading.

2 The Creep Model

Continuum damage mechanics (CDM) has proved to be a useful means of describing the high temperature deformation of complex engineering alloys. The Imperial College model has its origins in the work of

---

[1] Permanent address: Deputy Director
National Metallurgical Laboratory, Jamshedpur, Bihar, India

Superalloys 2000
Edited by T.M. Pollock, R.D. Kissinger, R.R. Bowman,
K.A. Green, M. McLean, S. Olson, and J.J. Schirra
TMS (The Minerals, Metals & Materials Society), 2000

Ion *et al.* [1], who developed an isotropic continuum damage mechanics model that accounted for the dominant tertiary creep behaviour of a range of engineering alloys by a strain softening mechanism associated with the accumulation of mobile dislocations. In order to describe the anisotropic creep behaviour of single crystal superalloys, deformation was assumed to take place by viscous glide along specific glide systems [2]. The ability of the model to represent an extensive uniaxial database and to predict the creep response under complex uniaxial loading at high temperatures has been demonstrated on a number of single crystal superalloys [3-5]. A principal objective of the present work is to further extend the current model to a multiaxial formulation. The model has been presented in some detail previously; only the main elements are described below.

For nickel-base superalloys it is assumed that two families of slip systems are operative that give rise to the anisotropic creep behaviour of single crystals. These are octahedral and cube systems denoted by the sets $G_1 = \{111\} <110>$ and $G_2 = \{001\} <110>$, respectively. The activity of each k-slip system is determined by the magnitude of the resolved shear stress and the shear strains, $\gamma^{(k)}$, are determined by solving the evolution equations with appropriate boundary conditions. These equations take the general form,

$$\dot{\gamma}^{(k)} = \dot{\gamma}_0^{(k)}(1 - S^{(k)})(1 + \omega^{(k)})$$
$$\dot{S}^{(k)} = \dot{\gamma}_0^{(k)} H^{(k)}(1 - S^{(k)}/S_{ss}^{(k)})$$
$$\dot{\omega}^{(k)} = \beta^{(k)}\dot{\gamma}^{(k)} \tag{1}$$

where $S^{(k)}$ and $\omega^{(k)}$ are hardening and strain softening state variables, respectively. The hardening describes the partitioning of stress between the hard γ' particles and the matrix, leading to a description of primary creep (Figure 1). The damage state variable $\omega^{(k)}$ is related to the strain softening effect of an increasing mobile dislocation density, which gives rise to a tertiary response on the strain-time curve.

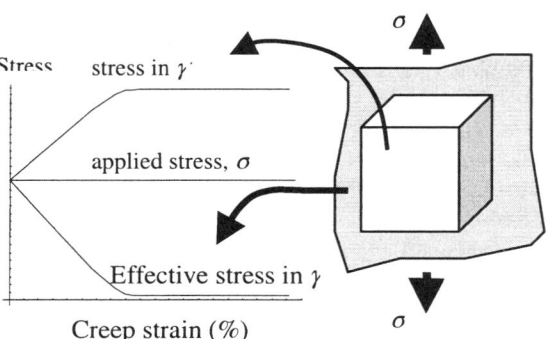

Figure 1

Once the individual $\gamma^{(k)}$'s on all active slip systems have been calculated it is possible determine the macroscopic strain along any direction (e.g. in the direction the tensile load or in any transverse direction). In order to do this we need to determine the deformation rate tensor, $\dot{F}_{ij}$, for an infinitesimal material element. For the k-slip system let the slip plane normal and direction be represented by $n^{(k)}$ and $m^{(k)}$. It can be readily shown that the deformation rate tensor has the following form [6],

$$\dot{F}_{ij} = \sum_{k \in G} \dot{\gamma}^{(k)} m_i^{(k)} n_j^{(k)} \tag{2}$$

and using $G = G_1 \cup G_2$ then,

$$\dot{F}_{ij} = \sum_{k \in G_1} \dot{\gamma}^{(k)} m_i^{(k)} n_j^{(k)} + \sum_{k \in G_2} \dot{\gamma}^{(k)} m_i^{(k)} n_j^{(k)} \tag{3}$$

The last result is a simply a statement of superposition, that is, the total deformation rate tensor is the sum of contributions of all active octahedral and cube slip systems. Hence, we can write Equation (3) in a more compact manner,

$$\dot{F}_{ij} = \dot{F}_{ij}^{G_1} + \dot{F}_{ij}^{G_2} + \cdots \tag{4}$$

The importance of $\dot{F}_{ij}$ is that it describes the deformation of a material element. Thus, once $\dot{F}_{ij}$ is known (determined from the accumulation of shear strains on the active slip systems) then the macroscopic creep response can be obtained. This is achieved by calculating the equivalent strain using a Mises- type relation. In addition to accounting for strain in arbitrary directions, the model predicts local changes in crystal orientation that result from these anisotropic strains.

Isotropic creep models accounting for multiaxial loading in finite element (FE) codes are generally based on an assumption that deformation takes place by shear and, consequently the von Mises formulation for individual elements can be used. However, in the isotropic case there is no restriction on the choice of slip systems. An isotropic material is assumed to deform on any plane on which the resolved shear stress has the maximum value. This is always equal to half of the effective stress. In terms of crystallographic deformation this means that the Schmid factor is always 0.5. On the other hand, for anisotropic materials the Schmid factor can have any value between 0 and 0.5, depending on the choice of slip system and orientation of the stress axis. The main advantage of this approach is that it allows the use of the uniaxial creep law even for multiaxial loading provided the stress and strains are replaced by an effective stress ($\bar{\sigma}$) and effective strain ($\bar{\varepsilon}$). For example, in power law creep the creep rate is described by the relation $\dot{\varepsilon} = A\sigma^n$, which become $\dot{\bar{\varepsilon}} = A\bar{\sigma}^n$ for multiaxial loading. The expressions for effective stress and effective strain rate ($\dot{\bar{\varepsilon}}$) for isotropic material are as follows,

$$\bar{\sigma}^2 = \frac{1}{2}\left[(\sigma_{11} - \sigma_{22})^2 + (\sigma_{22} - \sigma_{33})^2 + (\sigma_{33} - \sigma_{11})^2\right]$$
$$+ \frac{1}{3}\left[\sigma_{23}^2 + \sigma_{31}^2 + \sigma_{21}^2\right] \tag{5}$$

$$\bar{\varepsilon}^2 = \frac{9}{2} \left[(\varepsilon_{11} - \varepsilon_{22})^2 + (\varepsilon_{22} - \varepsilon_{33})^2 + (\varepsilon_{33} - \varepsilon_{11})^2 \right]$$
$$+ \frac{1}{3} \left[\varepsilon_{23}^2 + \varepsilon_{31}^2 + \varepsilon_{21}^2 \right] \qquad (6)$$

For anisotropic material, both of these relations will be functions of the co-ordinate system that defines the state of stress. In the case of a cubic crystal, because of its symmetry the above expressions remain valid if the states of stress and strain are represented with respect to its normal crystallographic axes. Therefore, as long as such a practice is followed, there is no problem in interfacing the slip-based formulation through a user-defined subroutine for the calculation of incremental creep strain at each time-step defined by the commercial finite element code. In the present study the ABAQUS finite element package has been used, where the model was interfaced with the software through a modified creep user-subroutine. This approach is simpler and less computing intensive than incorporating the creep law into ABAQUS through a user-material routine (U-MAT) [7]. Calculations have been carried out for circumferentially notched specimens using the full CDM crystallographic slip model. Previous work has mainly been restricted to Norton-type steady state creep relations [8, 9].

An experimental programme, which has been carried out to begin to evaluate the FE calculations, is the focus of the present work. For this creep tests were carried out using circumferential Bridgman notch specimens in tension, and using electron back-scatter diffraction (EBSD) for characterisation of the deformation under multiaxial stresses [10].

3 Experimental

The experimental programme has been carried out on the commercial single crystal superalloy CMSX4; the chemical composition is given in Table 1. Single crystal castings were supplied by PCC Aerofoil Corporation, using state-of-the-art commercial casting technology. The single crystals were supplied in the fully heat-treated condition. The standard commercial heat treatment (solutioning for 1h/1280°C, 2h/1290°C, 6h/1300°C and ageing for 4h/1140°C, 16h/870°C) produced cuboidal γ' particles with size of about 0.5 μm occupying a volume fraction of about 80%. The initial orientation of each casting was determined by the X-ray Laue back reflection technique; the axial orientation of each cylindrical casting used was within 4° of <001> or <111>.

Double notch Bridgman creep specimens, Figure 2, were machined from the single crystal castings. They were tested at 850°C with net-section stresses in the range 600-850MPa. A shadowgraph technique was used to measure local strain [11]. Fracture surfaces of all test-pieces were examined by optical and scanning electron microscopy. Longitudinal and transverse section of the uniaxial and multiaxial creep tested samples were electropolished using 10% perchloric and 45% acetic acid in butan-1-ol at 25 V at -5 °C. The EBSD (electron back-scattered diffraction) technique was employed in conjunction with a JEOL 840 scanning electron microscope (SEM) equipped with SINTEF hardware to map the spatial distribution of crystal orientation over these sections. CHANNEL software was used to analyse the three Euler angles φ_1, ϕ, and φ_2, measured by indexing Kikuchi bands. These Euler angles are used by the software for displaying orientation data in the form of inverse pole figure (IPF), pole figure (PF) or orientation map (OM). For details of the procedure see Ref. [10, 12].

Table 1. CMSX4 alloy composition in wt%.

Cr	Co	W	Ta	Al	Ti	Re	Mo	Ni
6	9	6	7	5.6	1	3	0.6	Bal

4. Results

4.1 Creep Rupture life

Figure 3 compares the times to rupture for uniaxial and the multiaxial creep tests of <001> and <111> single crystal of CMSX4 at 850°C. This clearly shows the effect of the multiaxial stress state is to extend the creep life by about an order of magnitude.

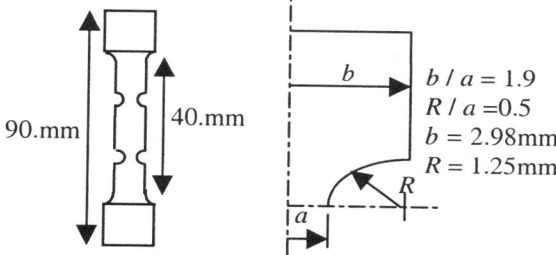

Figure 2 Geometry of double Bridgman notch

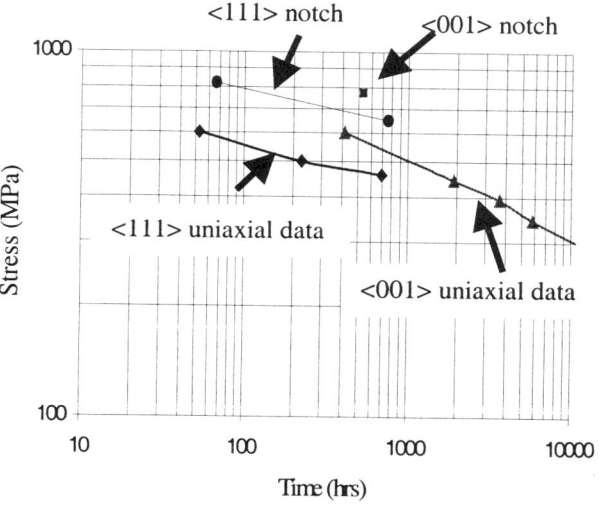

Figure 3 Lifetime data for CMSX4 <111> and <001> orientations under uniaxial and multiaxial loading conditions at 850°C.

517

4.2 Fracture behaviour

The fracture surfaces for the <001> and <111> orientated notch specimens were studied. Figure 4(a) shows the fracture surface for a <001> crystal tested at 775MPa/850°C, clearly showing crystallographic facets with four-fold symmetry consistent with deformation having occurred on the {111} planes that are symmetrically inclined to the load axis. In Figure 4(b) the fracture surface for the <111> orientated crystal shows crystallographic facets with three-fold , cube-on-edge symmetry that is consistent with deformation being constrained by the (001) faces of the γ′ particles.

The fracture surfaces of the multiaxial creep specimens for both <001> and <111> orientated crystals have two quite distinct regions:

- The central zones are jagged and macroscopically perpendicular to the tensile axis,

- There is a smoother outer ring showing less obvious crystallographic character; the width of this zone decreases with increasing load. Figure 5 shows the fracture surfaces for <111> tests.

4.3 Lattice Rotations

The present authors have used EBSD extensively to measure lattice rotations resulting from creep deformation of uniaxially loaded specimens [11]. Large rotations are found in specimens that deviate significantly from the high symmetry <001> and <111> orientations. In the case of near <001> and <111> there is some spread in orientation as a result of creep strain, but the mean orientation remains unchanged.

Figure 6(a) show the orientation map (OM), Inverse pole figure (IPF) and (111) pole figure obtained from transverse section of as received [001] single crystal. Although the OM exhibits a contrast change across the section, the maximum misorientation measured is ± 0.5°.

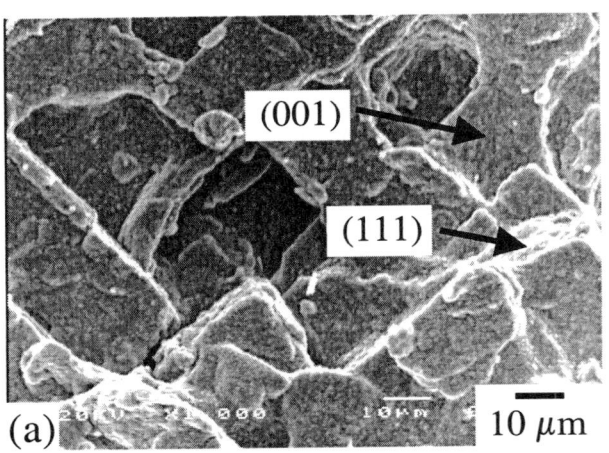
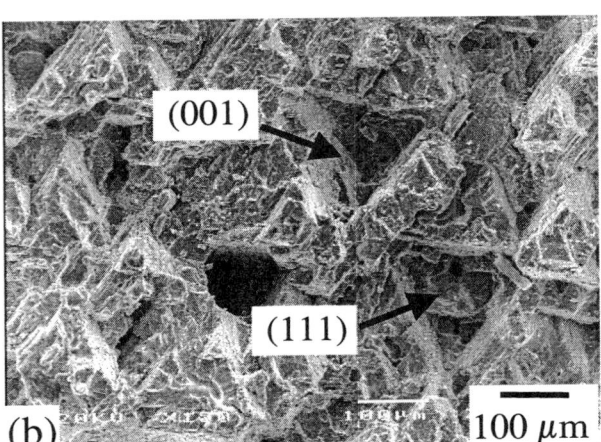

Figure 4 SEM of the fracture surfaces for (a) CMSX4<001> at 775MPa/850°C, and (b) CMSX4 <111> at 650MPa/850°C.

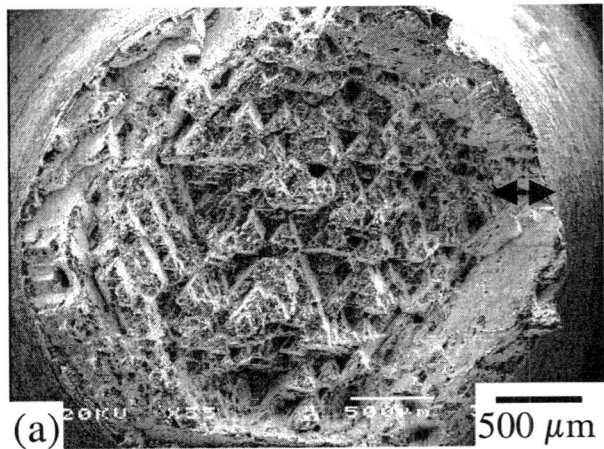
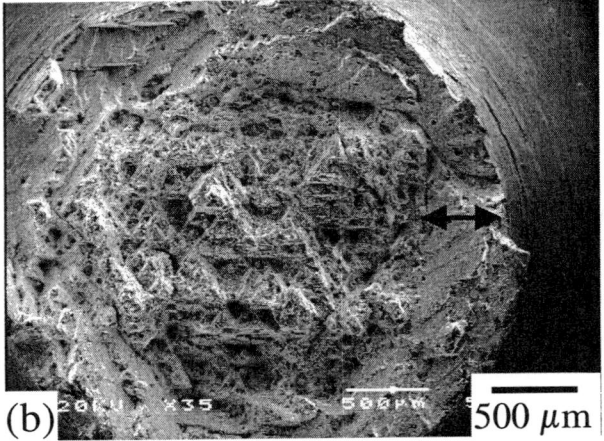

Figure 5 Fracture surfaces of CMSX4 <111> crystals tested at 850°C

(a) 820MPa, and (b) 650MPa.

This is clearly shown by the spread of orientation in figure 6(b), (IPF) and (111) pole figure. Figure 7 shows a transverse section of a uniaxially loaded <111> specimen tested to fracture; no systematic trend in crystal rotation is observed. The OM shows a uniform contrast and the maximum misorientation is about ±2°. Thus, in uniaxial tests it is found that the crystal rotation for the stable <001> and <111> single crystals is minimal. As will be shown, this is found to be in marked contrast when looking at the case of multiaxial deformation. Figure 8 shows the range of orientations observed on a longitudinal section of a Bridgman-notch creep specimen tested to fracture at 820MPa/850°C. It shows that the deformation is highly localised and that there are regions where very significant lattice rotations have occurred from Y to Y1 (see Figure 8(d)). In the zone of most intense deformation there is clear evidence of mechanical twinning having occurred.

Transverse sections through the minimum diameter of the notch were prepared by polishing one fracture surface until a flat surface was obtained. This was then electro-polished as described above. Figure 9 shows the results of EBSD analysis of the transverse section of a <001> specimen tested at 755MPa/850°C. In Figure 9(a) orientation map, inverse pole figure (IPF) and (111) pole figure for the entire section are shown. The orientation map clearly shows the four-fold symmetry associated with the <001> axis. The map was generated using a grid of 150×150 with a step size of 20μm, and a beam spot of approximately 1μm. There has been a spread of orientation of ±15° from the initial near-<001> orientation of the initial specimen. Further analysis of the data shows that the perimeter of the section at diameters pointing in <110> directions shows a rotation from <001> towards <111> on the IPF (Figure 9(b)). However, at the perimeter of diameters pointing in the transverse cube directions (<100> and <010>) the rotation is from <001> to <101>, see Figure 9(c). The Pole Figures show that the (111) poles rotate in different directions, depending on position on the section perimeter, forming cruciforms.

Similar data for a <111> notch specimen creep tested to fracture at 820MPa/850°C are shown in Figure 10. Here the three-fold symmetry of the <111> axis is clearly apparent in the orientation map. The overall deviation from <111> is ±20°. However, again the crystal rotations observed vary systematically around the circumference of the section. In transverse radii pointing towards <211> the rotation on the IPF is from <111> to <001> (Figure 10(b)) and for radii pointing to <011> the rotation is from <111> towards <101> (Figure 10(c)). At the centre of the section there is a small spread of orientations, but they are all within ±5° of <111> (see Figure 10(d)).

Furthermore, it can be seen from Figures 9 and 10 that the <001> crystal has been divided into approximately 100 subgrains, whereas for the <111> crystal less than 10 subgrains have developed. The reasons for this marked difference are currently under investigation.

4.4 Finite Element simulation

The crystallographic slip model, with parameters established by analysis of a uniaxial creep database, has been used to simulate the deformation of Bridgman notch specimens. The FE analysis was carried using the commercial package ABAQUS, and the model was interfaced using a modified creep user-subroutine [7]. An example of the output is shown in Figure 11, which shows a transverse section through the minimum notch diameter for a <001> specimen tested at 400MPa/850°C. The four-fold symmetry is readily apparent. In the example shown there is the accumulation of significant strains at the perimeter of the section (1.5 to 2.5%) but very low strains (0.5%) at the centre.

5 Discussion

The creep rupture data show that there is significant notch-strengthening that is similar to the level observed in polycrystalline superalloys. For isotropic materials the notch is considered to play two roles:

- It locally amplifies the axial component of stress σ_{ZZ} at the notch root giving rise to localised deformation and redistributing load to the specimen centre, and
- The radial component of stress σ_{RR} reduces the resolved shear stress and inhibits deformation.

The observations on the fracture surfaces are quite consistent with deformation being concentrated at the section perimeter. The smooth outer ring is typical of a creep failure. As the load-bearing section is reduced, the stress on the interior of the specimen increases to a level where fast fracture occurs; this is characterised by the jagged crystallographic facets that are observed.
As indicated in Section 2, the normal assumption of shear occurring in the direction of maximum shear stress, which is central to the von Mises formulation, is not valid for single crystals where the influence of specific slip systems is thought to be important. Clearly, the spatial distribution of lattice rotations observed by EBSD and illustrated in Figures 9 and 10 confirm this view.

Assuming elastic isotropy, which should be valid for a cubic system, the initial elastic stress field around the Bridgman notch in polar form is given by the expressions [13]:

$$\sigma_{zz} = F\left[1 + \log\left(\frac{a^2 + 2aR - r^2}{2aR}\right)\right] \quad (7)$$

$$\sigma_{rr} = F\log\left(\frac{a^2 + 2aR - r^2}{2aR}\right) \quad (8)$$

$$\sigma_{rr} = \sigma_{\theta\theta} \quad (9)$$

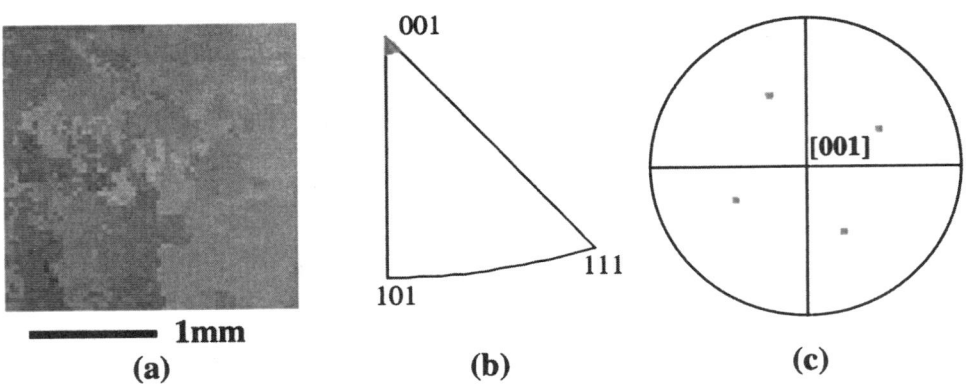

Figure 6 Transverse section of as received [001] single crystal (a) Orientation map, (b) (IPF) and (c) (111) pole figure.

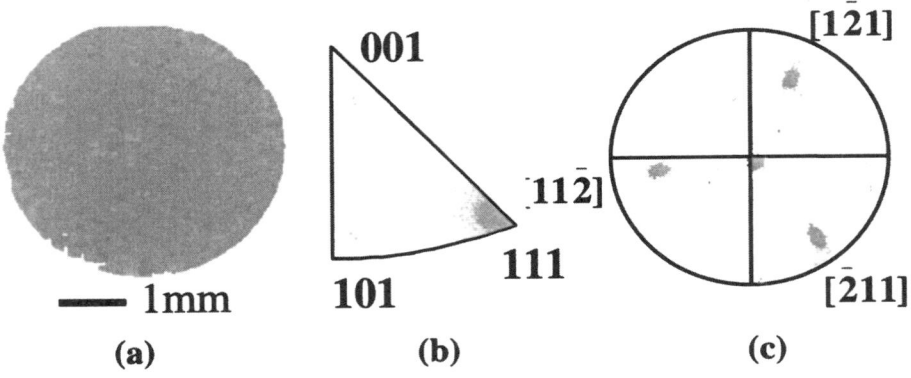

Figure 7 Transverse section, near fracture surface, of uniaxially loaded [111] single crystal creep tested at 850°C/460MPa. (a) orientation map, (b) inverse pole figure (IPF), (c) (111) pole figure.

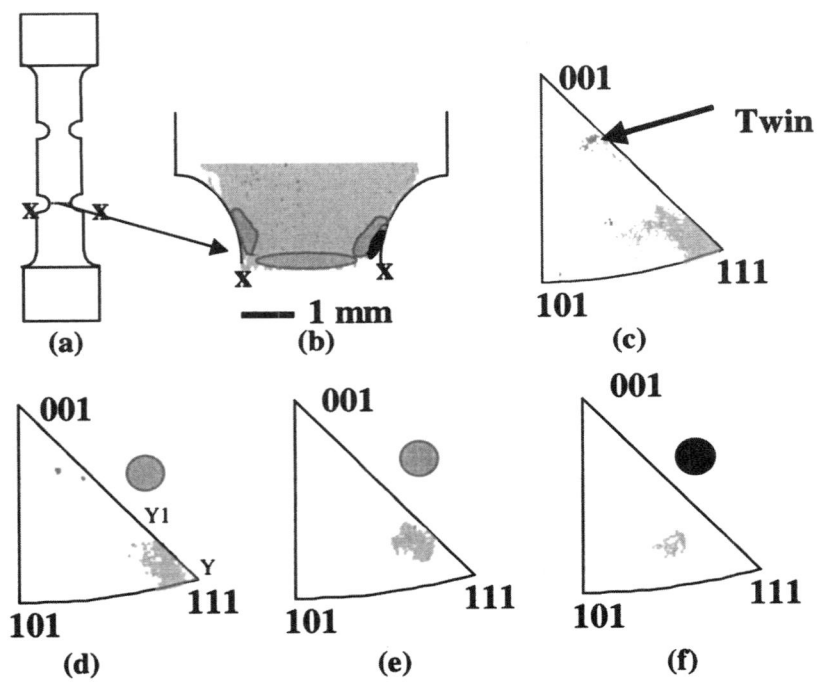

Figure 8 Longitudinal section of [111] single crystal creep tested at 850°C/820 MPa showing, (a) the position of fractured area (b) orientation map, (c) overall orientations of the map, (d, e and f) orientation of subsets shown in (b).

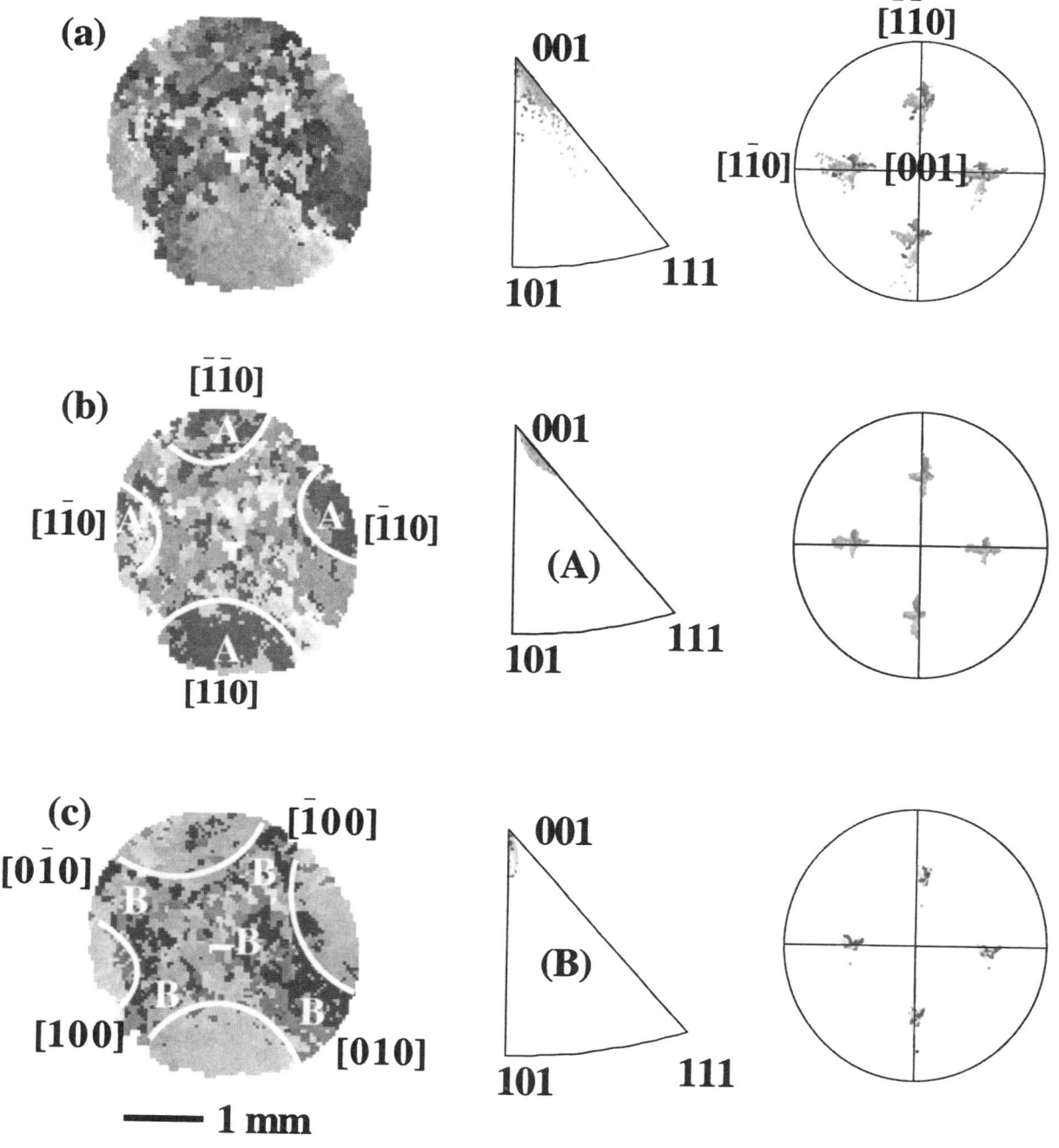

Figure 9 Transverse section, near fracture surface, of notched [001] single crystal tested at 850°C/775 MPa. (a) Overall orientation map, inverse pole figure (IPF) and (111) pole figure. (b) Orientation map showing four-fold symmetry of deformation at (A), IPF and (111) pole figure show the local orientation of subsets marked (A). (c) Orientation map, IPF and (111) pole figure showing the local orientation of subsets marked (B).

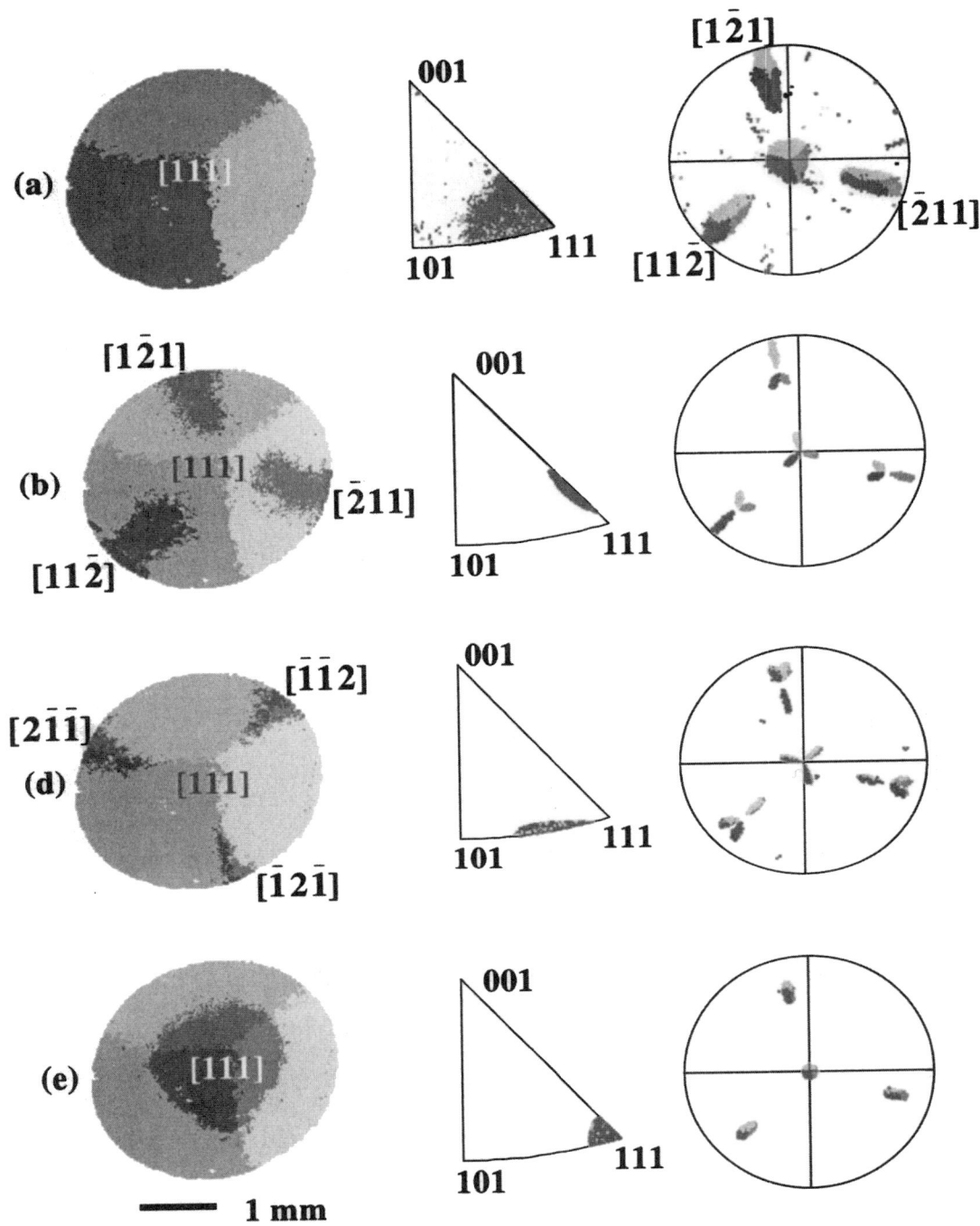

Figure 10 Transverse section, near fracture surface, of notched [111] single crystal tested at 850°C/820 MPa. (a) Overall orientation map, inverse pole figure (IPF) and (111) pole figure. (b, c) Orientation map showing three-fold symmetry of deformation, IPF and (111) pole figure showing the local orientation of (dark contrast). (e) Orientation map, IPF and (111) pole figure showing the local orientation of subsets at the centre.

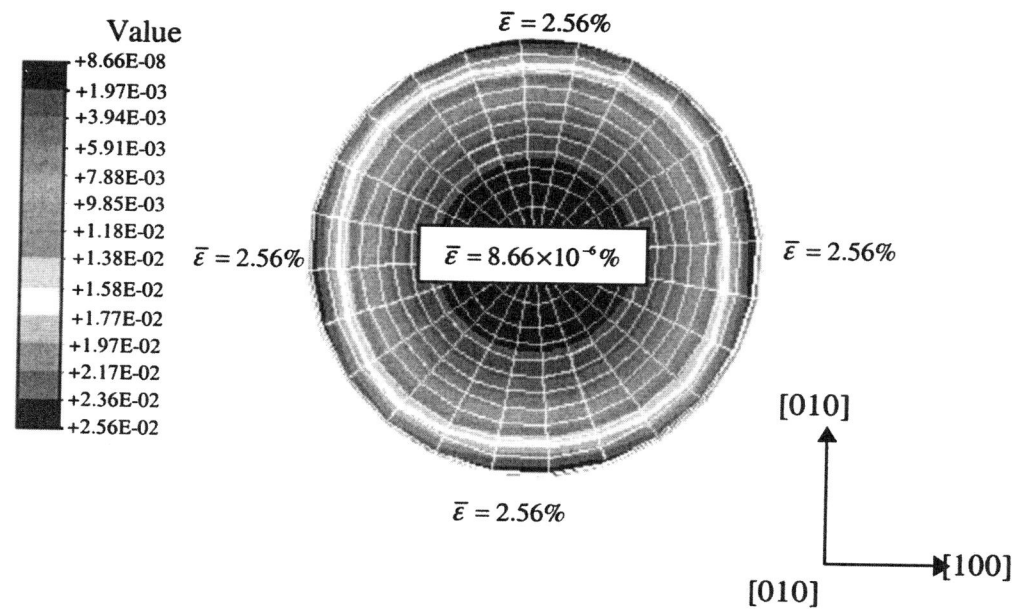

The color scale bar reads:

Value
- +8.66E-08
- +1.97E-03
- +3.94E-03
- +5.91E-03
- +7.88E-03
- +9.85E-03
- +1.18E-02
- +1.38E-02
- +1.58E-02
- +1.77E-02
- +1.97E-02
- +2.17E-02
- +2.36E-02
- +2.56E-02

$\bar{\varepsilon} = 2.56\%$

$\bar{\varepsilon} = 2.56\%$ $\bar{\varepsilon} = 8.66\times10^{-6}\%$ $\bar{\varepsilon} = 2.56\%$

$\bar{\varepsilon} = 2.56\%$

[010]
[100]
[010]

Figure 11 Finite element calculation of the equivalent strain, $\bar{\varepsilon}$, using anisotropic slip model for a <001> orientated crystal at 400MPa/850°C.

where F is the applied load, a the outside radius of the cross section of neck, R radius of curvature at the neck, r position along the cross section.

For exact <001> and <111> orientations the axial σ_{ZZ} component leads to equal and symmetrically distributed resolved shear stresses on the active slip systems so that deformation retains the original orientation. Any deviation from the exact orientation amplifies the shear stress on one or two systems and reduces it on others. This is exactly the situation for uniaxially stressed specimens, except that there is a stress concentration in the case of the notch. The σ_{RR} component which is at right angles to the axial stress has the same general effect as a misorientation in a uniaxial specimen in that it amplifies some resolved shear stresses and reduces others. An important difference is that σ_{RR} amplifies different slip systems in different positions of the specimens.

For <001> specimens the axial load leads to a maximum resolved shear stress for octahedral slip. There has been debate on whether this occurs with <110> or <211> Burgers vector; however, there appears to be a consensus that the latter predominately occurs at low temperatures and high stresses where γ' cutting is possible [14]. The effect of the radial component of stress on the possible slip systems is shown as maximum Schmidt factors in Figure 12. The two octahedral systems have similar values each varying by about 20% with a periodicity of 45^0. The radial

component in the direction <010> will enhance the shear stress on $(1\bar{1}1)(011)$ and lead to a rotation of the specimen axis from <001> towards <011>. On the other hand, the radial component in the direction $\langle\bar{1}10\rangle$ will enhance the shear stresses on both $(1\bar{1}1)(011)$ and $(1\bar{1}1)(\bar{1}01)$ and this will lead to a rotation of <001> towards $\langle\bar{1}11\rangle$. This is exactly the behaviour that is observed.

For <111> specimens, the effect of σ_{RR} is quite different at opposite ends of a diameter of a transverse section. Here the maximum shear stress due to the axial load is on (100)<011> type slip systems. The σ_{RR} at the $\langle11\bar{2}\rangle$ radius enhances the $(001)(110)$ slip vector and consequently leads to a rotation of <111> towards <110>. However, at the opposite end of the diameter, towards $\langle\bar{1}\bar{1}2\rangle$, two cube slip systems are enhanced-(100)<011> and (010)<101>. The combined effect is a rotation of <111> towards <001>. Again, this is quite consistent with the EBSD observations.

This analysis, based on the initial elastic stress field, can only be indicative. In practice, there will be a redistribution of stress as a result of creep deformation at the notch root. It requires a full time-dependent plastic analysis to account for this effect. Nevertheless, the present measurements of micro-crystallographic changes and the approximate analysis provide strong support for the basic assumptions of the crystallographic slip model, which incorporates both octahedral, and cube slip. This gives confidence in extending the uniaxial model to account for deformation in multiaxial stresses and to implement in FE simulations of specimen or component behaviour.

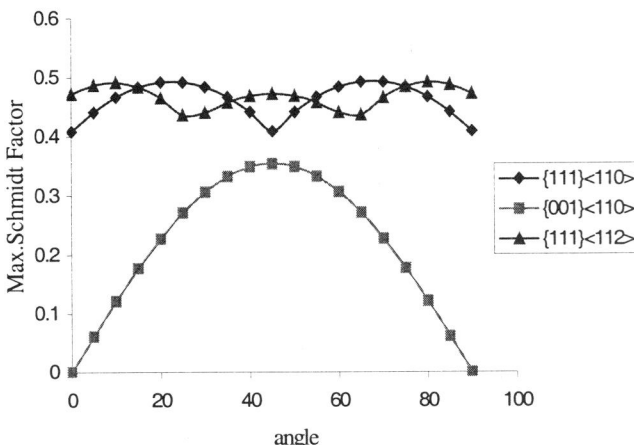

Figure 12 Maximum Schmidt factor due to radial stress component σ_{RR} calculated for [001] direction..

6. Conclusions

1. The multiaxial stress state produced by a circumferential Bridgman notch increases the creep rupture lives of CMSX4 by an order of magnitude relative to the net-section tensile uniaxial stresses in the <001> and <111> directions.

2. Creep deformation of <001> and <111> specimens leads to changes in orientation of up to 20^0 that vary across the specimen section.

3. Crystallographic analysis shows that the observed orientation changes are consistent with the radial component of the multiaxial stress state modifying the resolved shear stress for crystallographic slip such that the $(111)\langle\bar{1}10\rangle$ and $(001)\langle110\rangle$ systems operate on <001> and <111> loaded specimens respectively.

4. The results are consistent with the Imperial College model of anisotropic creep which has been implemented in a commercial FE code to simulate the creep behaviour of notch-creep specimens. There agreement between experimental observation and FE simulation is promising.

7. Acknowledgements

The work was made possible by support from the Engineering and Physical Science Research Council (Grant Numbers GR/J02667, GR/K19358; Visiting Fellowship GR/L67042) and BRITE EURAM III Project BE 96-3911. The experimental assistance of Ms Anne Ellis of Princeton University during a student internship at Imperial College is gratefully acknowledged.

8. List of References

1. J.C. Ion et al., Report DMA A115, The National Physical Laboratory, Teddington 1986

2. R.N. Ghosh and M. McLean, Acta Metal. Mater., 1990, Vol. 40 , p. 1977

3. R.N. Ghosh, R.V. Curtis, and M. McLean, Acta metall. mater., 1990, Vol. 38, p.1997.

4. L-M.Pan, B.A.Shollock and M.McLean, Proc. R. Soc. Lond. A, 1997, **453**, 1689-1715.

5. H.Basoalto, M. Ardakani, R.N. Ghosh, B.A Shollock and M. McLean,2000, Key Eng. Materials, Vol. 171-174, p.545.

6. D.Peirce, R.J.Asaro, and A.Needleman, Acta metall., 1982, Vol. 30, p. 1087.

7. ABAQUS user's manual, version 5.4.

8. N. Ohno, T. Mizuno, H.Kawaji, and I.Okada, Acta metall. mat., 1992, Vol. 40, p. 567.

9. L.Meric,P.Poubanne, and G.Cailletaud, Trans. ASME, 1991, vol. 113, p.162.

10. D. J. Dingley and V. Randle, J. Mater. Sci. 1992, Vol. 27,p. 4545.

11. M.B.Henderson, L.M.Pan, B.A.Shollock and M.Mclean, Proc. Euromat, 1995, Symp. D, 25/9-28/9, Pdua/Venice,1-10

12. F. J. Humphreys. J. Microscopy, 1999, Vol. 195, p.170.

13. B.P.W. Bridgman, *Large Plastic Flow and Fracture*, Metallurgical Series, McGraw-Hill Book Company Inc., 1952.

14. B.A. Shollock, M. Ardakani, R.N.Ghosh, H.Basoalto and M.Mclean, 1999, Proc. of Plasticity '99, Neat Press, p.193

INVESTIGATIONS ON THE ORIGIN AND EFFECT OF ANOMALOUS RAFTING

Horst Biermann, Ulrich Tetzlaff, Haël Mughrabi, Berthold von Grossmann[1], Stefan Mechsner[2] and Tamas Ungár[3]

Institut für Werkstoffwissenschaften, Lehrstuhl I, Universität Erlangen-Nürnberg,
Martensstr. 5, D-91058 Erlangen, Fed. Rep. Germany
[1]now at: AUDI AG, D-85045 Ingolstadt, Fed. Rep. Germany
[2]now at: Siemens AG, D-97076 Würzburg, Fed. Rep. Germany
[3]Institute for General Physics, Eötvös University, P. O. Box 32, H-1518 Budapest, Hungary

Abstract

Anomalous rafting is a morphological transformation of the phases γ and γ' occurring in monocrystalline nickel-base superalloy turbine blades which is observed under the surface of the turbine blades after service. The origin of anomalous rafting is discussed, based on microstructural investigations of turbine blades before and after service and of laboratory samples after shot peening and ageing. Microscopical investigations are completed with microhardness and X-ray diffraction measurements. Residual macrostresses and long-range internal microstresses are discussed as driving forces of anomalous rafting, induced by the plastic deformation of subsurface layers of the material, e.g. during peening. Finally, the effect of anomalous rafting is discussed on the basis of fatigue and creep experiments.

Introduction and Objectives

Rafting of monocrystalline nickel-base superalloys has been investigated experimentally and modeled in detail in recent years [1,2]. The change of the originally cuboidal shape of the γ' precipitates towards a γ/γ' plate structure occurs at temperatures above about 1150 K during deformation or in samples predeformed at a lower temperature during subsequent (stress-free) ageing treatments [3-5]. In the regime of plastic deformation above plastic strains of ε_{pl} = 0.001 to 0.002, the orientation of the plate structure depends on the lattice mismatch δ (δ is the relative difference of the lattice parameters of the two phases γ' and γ) and on the direction of the externally applied stress σ. Detailed experimental investigations have shown

that the modification of the internal stress state during deformation is responsible for the type of rafting observed, i.e. whether the plate structure has an orientation parallel or perpendicular to the stress axis [3-6]. Thus, it was shown that the type of rafting is governed strongly by the sign of the plastic deformation [3-6].

In the hot regions of the airfoils of turbine blades, a γ/γ' raft structure develops which has an orientation perpendicular to the stress axis [7-9], because the lattice misfit of technically applied turbine blade superalloys is negative at the high service temperatures, and the external stress axis acting on a blade during service is tensile due to the centrifugal stresses. In regions near the surface, however, a different microstructure, called anomalous raft structure, with an orientation of the plates parallel to the surface has been found repeatedly [7,8,10,11] which was in some cases attributed to the influence of a surface coating [10,11].

In the present work, we consider different types of anomalous rafting, occurring in
- real turbine blades during service,
- initially unused turbine blades during ageing and
- shot-peened laboratory samples during ageing.

The origin of anomalous rafting occurring in different cases is discussed on the basis of microhardness and microfocus X-ray diffraction investigations. Furthermore, its effect on the high-temperature strength properties of nickel-base superalloys is investigated in fatigue and creep experiments which are part of a more extended study of the influence of different γ/γ' raft structures on mechanical high-temperature properties [12-15].

Superalloys 2000
Edited by T.M. Pollock, R.D. Kissinger, R.R. Bowman,
K.A. Green, M. McLean, S. Olson, and J.J. Schirra
TMS (The Minerals, Metals & Materials Society), 2000

Experimental

Material

Turbine blades Three different NiAl-coated turbine blades made of the nickel-base superalloy CMSX-6 with orientations near [001] were available ([001] is defined in the following to be that <100> direction which is nearly parallel to the turbine blade or specimen axis, respectively). Before the (developmental) coating treatment, the blades were peened. The following investigations were performed on these standard two-stage heat treated blades: i) one turbine blade was studied in the as-manufactured state. In addition, sections of this turbine blade were aged under vacuum at temperatures of 1273 K and 1323 K, respectively, for times between 200 h and 1000 h. ii) The other two turbine blades were exposed to accelerated mission tests in a ground-based test engine for different durations (for details see references [8,9]; it has to be noted that the total test times of service in the engine tests were different. The microstructure of the turbine blade with the higher service time was described in detail earlier [8,9]). The engine tests were performed under continuous maximum temperature conditions with a higher temperature than under normal operation of the engine.

Laboratory samples The deformation experiments (i.e. shot peening, fatigue and creep tests) were carried out on monocrystalline samples of the nickel-base superalloys SRR 99. The rods were supplied in the standard state after a two-stage heat treatment. More details are given in references [14,15].

Shot peening

Shot peening was applied on monocrystalline samples of the nickel-base superalloy SRR 99 with an orientation near [001], cf. reference [16]. Electrolytically polished rods of 10 mm diameter and 150 mm length were peened with steel shot of specification S170, an Almen intensity of 0.42 mmA and a peening coverage of 200 %. After peening, a thin surface layer of about 5 µm was removed electrolytically in order to avoid iron contamination. Subsequently, sections of the rods were aged in vacuum at temperatures between 1173 K and 1353 K for times between 4 h and 300 h. The microstructure of these samples was investigated by scanning electron microscopy (SEM).

After preliminary ageing experiments performed on the shot-peened rods, samples for mechanical fatigue and creep tests were shot-peened under the same conditions as the rods. Some samples were aged in vacuum for 24 h at a temperature of 1323 K and some for 200 h at 1223 K, respectively. These conditions were found to give an "anomalous" raft structure, as will be shown later. The second heat treatment was performed to obtain anomalous rafts without strong coarsening of the γ' particles in the bulk of the samples (the ageing conditions are mentioned in the corresponding sections appropriately). Most peened and aged samples showed a recrystallized surface layer, as will be discussed later.

Microcharacterization

Scanning electron microscopy Sections of the turbine blades and of the laboratory specimens were cut parallel to {100} lattice planes and polished mechanically up to a final diamond powder grade of 1 µm. Subsequently, the sections were etched with a solution of H_2O, HCl, MoO_3 and HNO_3 which etches preferentially the γ' phase and finally sputtered with gold. Scanning electron microscopy was performed at an acceleration voltage of 20 kV in secondary electron contrast.

Microhardness testing The microhardness after Vickers (HV0.05) was determined as a function of the distance from the surface on different shot-peened samples of the nickel-base superalloy SRR 99. One sample was investigated before ageing, two samples after ageing under conditions which produced an "anomalous" raft structure (24 h at 1323 K and 100 h at 1273 K), and one after ageing, but without rafting (24 h at 1273 K).

X-ray diffraction The shift of the centre of gravity of X-ray peak profiles indicates the effect of residual stresses arising from shot peening which will be called macrostresses in the following. Such peening stresses are of biaxial compressive character directly under the surface of peened samples. They decrease with increasing distance to the surface. In addition to the peening (macro-)stresses, long-range internal microstresses acting between the two phases γ and γ' were investigated. These microstresses which were repeatedly found (e.g. after creep deformation [17] or in turbine blades after service [8,9]) cause counteracting shifts of the individual subprofiles of the phases γ and γ'. Thus, the peak profiles are asymmetric and consist of two subprofiles. The asymmetry is of opposite character for peak profiles measured parallel and perpendicular to the deformation axis.

In order to study X-ray peak profiles of the samples with adequate (high) lateral resolution, a new microfocus synchrotron radiation X-ray diffraction technique was applied (for details see [18,19]). A monochromatic, focussed X-ray beam with a line focus of $2 \times 50 \ \mu m^2$ (wavelength of the radiation $\lambda = 0.06886$ nm) was used to study the local variation of the lattice parameter distributions under the surface of shot-peened samples before and after an ageing treatment (24 h at 1323 K). The samples were mounted with the line focus parallel to the outer shot-peened surface of the samples. In addition, complementary peak profiles were measured with a "conventional" double-crystal X-ray diffractometer (wavelength $\lambda = 0.15406$ nm), cf. [17] which has a focal spot on the sample of about $0.5 \times 0.5 \ mm^2$.

Deformation tests

Fatigue Symmetrical total strain controlled fatigue tests were carried out at a temperature of 1223 K at two different total strain ranges of $\Delta \varepsilon_t = 0.01$ and 0.008. Reference samples were tested in the standard state after electropolishing at both strain ranges (samples A: $\Delta \varepsilon_t = 0.01$ and B: $\Delta \varepsilon_t = 0.008$). Additional samples were tested after different pretreatments: At the larger strain range, sample C was tested in the shot peened state after removal of a thin layer by mechanical and electrolytical polishing (approximately 10 µm) to remove the roughness; sample D was peened and aged for 24 h at 1323 K. A thin surface layer of about 30 µm was removed, but the sample still showed recrystallized grains on the surface. Samples E and F were aged for 200 h at 1223 K. Subsequently, a layer of about 80 µm was removed by electropolishing. These samples had "anomalous" rafts on the surface without recrystallized grains after polishing. Sample E was tested at the larger and F at the smaller strain amplitude, respectively.

<u>Creep</u> Tensile creep tests were performed on shot-peened samples with a diameter of 3 mm under a constant stress of 200 MPa at a temperature of 1273 K up to plastic strains of about 3 %. One undeformed reference sample and two shot peened samples after ageing at different conditions were investigated.

Anomalous rafting in turbine blades

Initial state

Already in the unused (peened and coated) state of a turbine blade of the superalloy CMSX-6, a small amount of anomalous directional coarsening with γ/γ' plates oriented parallel to the surface was observed beneath the interdiffusion zone between the NiAl coating and the bulk of the superalloy, see Fig. 1.

Anomalous rafting during service

During service of the turbine blades in a test engine, a change of the γ/γ' microstructure occurred in the hot regions of the airfoil at the leading and trailing edges, and a "normal" γ/γ' raft structure developed with an orientation perpendicular to the [001] blade axis. Under the surface of the blade exposed to service for the shorter time, a completely different microstructure with "anomalous"

rafts, i.e. γ/γ' plates with an orientation parallel to the surface, was observed, as shown in Fig. 2. Such a microstructure, however, was not found in the second turbine blade with the double service time.

In addition to the anomalous rafts, a γ' enriched zone and topologically closed-packed phases appeared under the interdiffusion zone between aluminide coating and superalloy material. A more detailed description of the phases occurring under an aluminide coating was given earlier, e.g. by Moretto and Bressers [10].

Rafting during stress-free ageing

The γ/γ' microstructure of a section of the unused turbine blade aged for 200 h at 1273 K is given in Fig. 3. The micrograph shows a γ/γ' plate structure in a region of up to 30 μm thickness. The plates have an orientation parallel to that {100} plane which has the smallest deviation from the outer surface of the turbine blade. Similar γ/γ' morphologies were also obtained after longer ageing times up to 1000 h or after ageing at a higher temperature (1323 K). However, it has to be mentioned that the plate structure shown in the figure was only observed under the convex surface of the aged turbine blade. Under the concave surface, separate γ' particles of cuboidal shape or coarsened due to the ageing treatment persisted up to the γ' enriched zone.

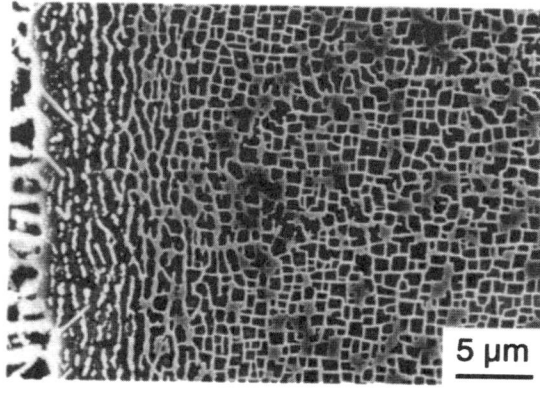

Figure 1: γ/γ' microstructure under the coating of an unused turbine blade, nickel-base superalloy CMSX-6. The interdiffusion zone under the aluminide coating is seen on the left.

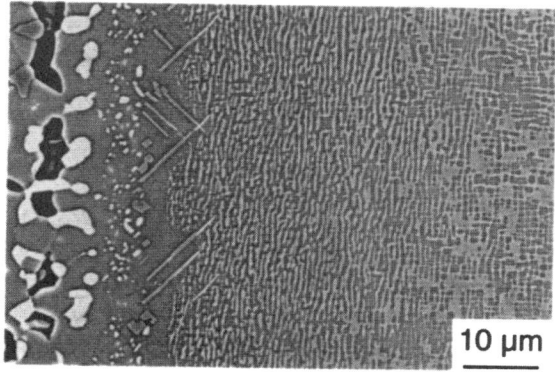

Figure 3: Microstructure of a specimen of the unused turbine blade (nickel-base superalloy CMSX-6) after ageing for 200 h at 1273 K.

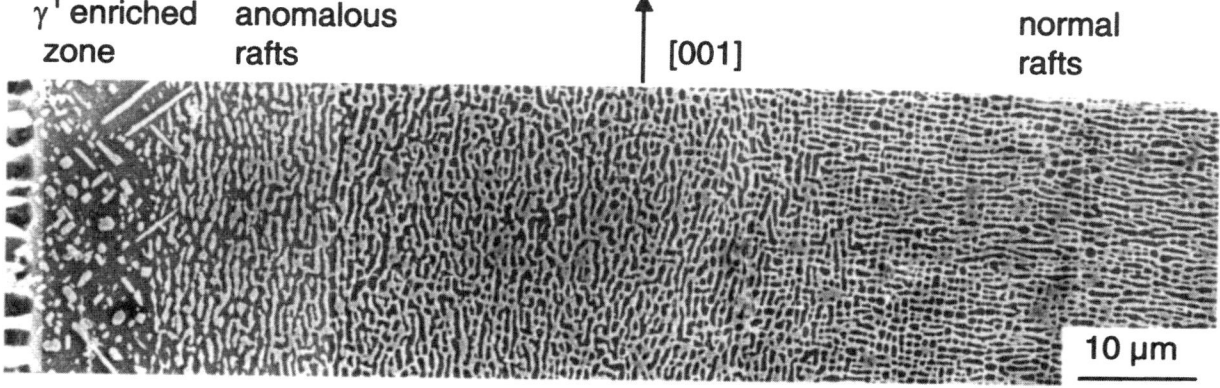

Figure 2: γ/γ' microstructure under the surface of a turbine blade after an accelerated mission test, nickel-base superalloy CMSX-6. On the left side, the interdiffusion zone under the NiAl coating is visible. Subsequently, a γ'-enriched zone (with precipitates of Cr-rich and topologically closed-packed phases) and a zone of "anomalous rafts" follow.

Shot-peened samples

To obtain information on the origin of the microstructural changes occurring under the surfaces of nickel-base superalloy turbine blades during high-temperature applications, samples were investigated after shot peening and after shot peening and an additional ageing treatment. The results of microstructural investigations, microhardness tests and X-ray diffraction experiments are presented in the following sections.

Rafting

After the ageing treatments, some of the samples showed a marked raft structure under the surface with an orientation of the γ/γ' plates parallel to the surface. Fig. 4 shows the microstructure of a sample aged for 24 h at 1323 K. Directly at the surface, a layer of cellular recrystallization was obtained. Under this recrystallized layer, a region with γ' particles of unregular shape and a γ/γ' plate structure with an orientation parallel to the surface follow. In the bulk of the samples, cuboidal γ' particles were still present. Due to the ageing treatment, these γ' particles were coarsened. In those regions with an orientation of the surface of the rods parallel to {110}, the γ/γ' plates had orientations parallel to both corresponding {100} lattice planes. A summary of the cases of anomalous rafting occurring during ageing for different times at different temperatures is given in Table I. It is clear that the process of rafting occurred after shorter times at higher temperatures (e.g. after 4 h at the highest temperature of 1353 K) than at lower temperatures. At the lowest temperature of 1173 K, only a small amount of rafts was found after 300 h. This temperature dependence is clearly due to the thermal activation of diffusion which is necessary for rafting.

In order to decrease the width of the recrystallized region or to avoid recrystallization at all, different heating rates were applied: Some of the samples were put into the preheated furnace. Other samples were put into the cold furnace which was then heated at a rate of 50 K per hour to the ageing temperature with a holding time of 2 h at 1073 K. These heating procedures, however, didn't yield a difference in the recrystallization behaviour.

Finally, the results obtained on an additional sample shall be mentioned: After shot peening, a surface layer of 50 μm was removed before the ageing treatment (24 h at 1323 K). This polishing should remove the highly damaged outer material which is under the action of the largest compressive macrostresses after peening. As a result of this procedure, a rafted microstructure comparable to that shown in Fig. 4 was found after ageing, but no recrystallization was obtained.

Table I: Microstructure of shot-peened samples after ageing. The symbols "×" indicate the cases of a marked γ/γ' raft structure with an orientation parallel to the surface, "≈" the cases of a small microstructural change and "—" the cases without rafting, respectively.

	4 h	10 h	24 h	50 h	100 h	200 h	300 h
1173 K					—		≈
1223 K			—			×	
1273 K	—	—	—	≈	×		
1323 K		×	×				
1353 K	×						

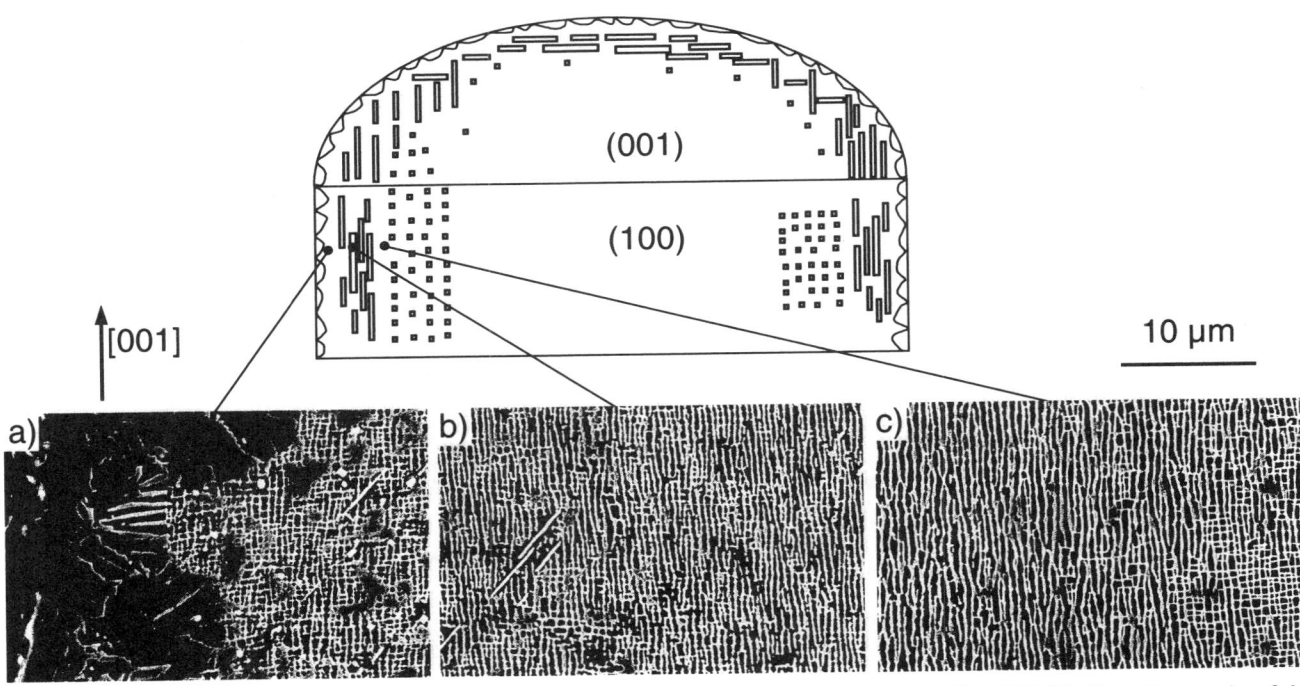

Figure 4: Microstructure of a shot-peened sample after ageing for 24 h at 1323 K, nickel-base superalloy SRR 99. The micrographs of the (100) section show a) cellular recrystallization directly under the surface, b) γ/γ' rafts with an orientation parallel to the surface and c) the transition from the rafts to cuboidal γ' precipitates at a distance of about 200 μm from the surface.

Microhardness

The courses of the microhardness under the surface of two shot-peened samples before and after ageing, respectively, are shown in Fig. 5. In the interior of the samples, an average hardness of 430 HV0.05 was obtained. An increase of the hardness under the surface was found prior to ageing, starting at a distance from the surface of about 200 μm up to an average value of 600 HV0.05 at a distance of 30 μm from the surface. This increase can easily be explained to be a result of the peening treatment.

The results of microhardness tests obtained on a sample aged for 24 h at 1323 K are also shown in Fig. 5. An approximately constant value comparable to that of the undeformed interior of the material was found. The increase of hardness induced by shot peening vanished completely during the ageing procedure. A similar behaviour was also obtained after ageing at 1273 K, irrespective of the duration of the ageing procedure (24 h or 100 h). This is noteworthy, because the microstructure of the sample aged for 24 h at 1273 K exhibited no rafts. Thus, it can be concluded that the ageing treatment and the respective annealing is responsible for the decrease of the peening-induced hardness, and not the process of rafting.

X-ray peak profiles

Two shot-peened samples were investigated by microfocus and "conventional" high-resolution X-ray diffraction, respectively, one before and one after an ageing treatment (24 h at 1323 K). The samples were cut parallel to a {100} lattice plane (which was parallel to the rod axis) which is called (100) in the following, and perpendicular to the [001] rod axis. The normal of the peened surface was parallel to [010]. In the following sections, the results are presented, divided into peak position (macrostresses), peak broadening (deformation) and internal strains (microstresses).

Peak position The centres of gravity of the peak profiles are a measure of the peening-induced macrostresses. The results of the measurements of (004) and (400) peak profiles obtained on a peened sample (without ageing treatment) are shown in Fig. 6. These two sets of investigations were measured with the focused synchrotron

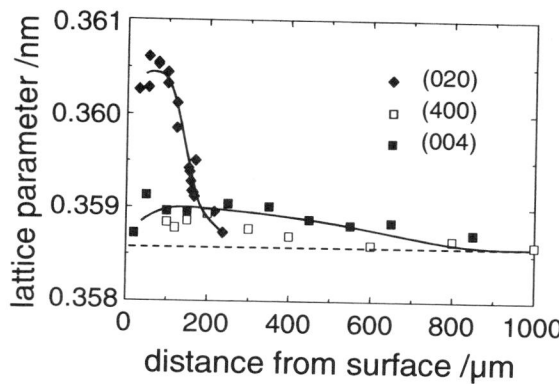

Figure 6: Local variation of the centres of gravity of the peak profiles (004) and (400) measured with synchrotron radiation X-ray diffraction and (020) measured with a home-lab double-crystal diffractometer. The dashed line indicates the stress-free lattice parameter.

radiation X-ray diffraction technique. The two Bragg reflections (004) and (400) are equivalent to another with respect to the stress state: Both reflections indicate the distribution of lattice parameters which have an orientation perpendicular to the normal of the shot-peened surface (which has an orientation of (010)). Therefore, it is not surprising that the results of the two Bragg reflections (004) and (400) shown in Fig. 6 are in good agreement. In contrast, the peak profiles of the type (020) (measured on the "conventional" double-crystal diffractometer) were measured on a tangential cut of the sample, thus yielding the lattice parameter distribution normal to the surface of the peened sample. In the shot-peened and aged sample, however, significant deviations of the centres of gravity from the stress-free state of undeformed material (dashed in Fig. 6) were not observed.

Peak broadening Fig. 7 shows the variation of the integral breadths of the X-ray peak profiles obtained on the sample without ageing, indicating the strong increase which is due to the plastic deformation induced by the peening treatment. As in the case of the microhardness tests, the increase of the integral breadths was found to occur in a region extending up to 200 μm from the surface. The peak profiles were nearly symmetrical without a significant separa-

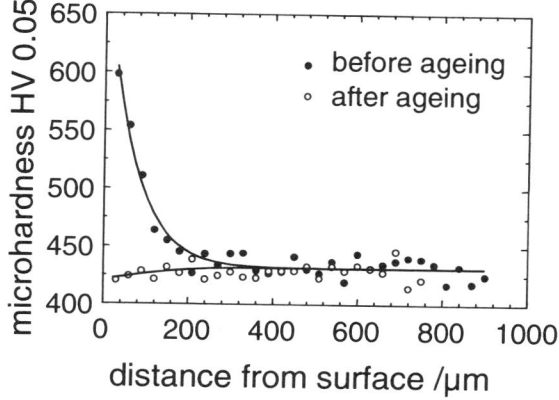

Figure 5: Microhardness after Vickers (HV0.05) for peened samples before and after ageing (24 h at 1323 K) vs. distance from the surface.

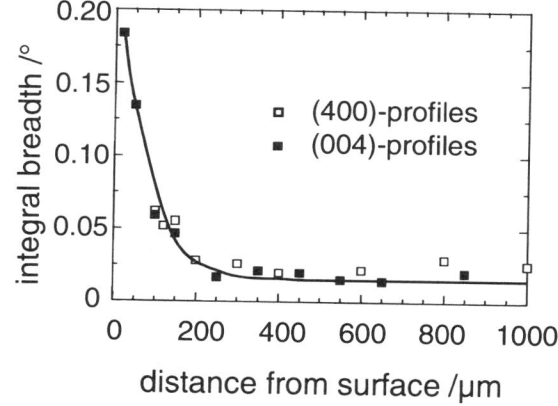

Figure 7: Integral breadth of X-ray peak profiles of a shot-peened sample before ageing vs. distance from the surface, nickel-base superalloy CMSX-6.

tion into the subprofiles of γ and γ'. The peak profiles of the aged sample, however, showed no significant broadening, but a strong separation into the subprofiles of γ and γ', as described in the next section. Therefore, the peak broadening of the aged sample is not shown here.

Internal strains The peak profiles of the shot-peened sample before ageing could not be separated into the subprofiles of γ and γ', as mentioned above. Therefore, only results obtained on the peened and aged sample (24 h at 1323 K, the microstructure is shown in Fig. 4) are presented in the following.

The main results obtained on the aged sample by microfocus X-ray diffraction are shown in Fig. 8. In the interior, a constant lattice parameter was determined for both phases γ and γ'. The peak profiles were nearly symmetrical, in agreement with peak profiles of undeformed material of the superalloy CMSX-6. Towards the surface, a separation of the (400) and (004) profiles was obtained, with a larger peak corresponding to γ' and a smaller peak corresponding to γ. The lattice parameter values of γ increase and those of γ' decrease with decreasing distance from the surface, respectively. The relevant changes of the lattice parameters in the directions [001] and [100] occur in that region of the sample in which a rafted microstructure was obtained (see Fig. 4).

Mechanical behaviour

In most cases, in which the phenomenon of anomalous rafting was described [7,8,10,11], the relevance of this microstructural transformation for the mechanical behaviour remained unclear. Therefore, in the present work, the effect of this kind of rafting was investigated in more detail. In particular, it was of interest whether a rafted surface layer has effects on the mechanical high-temperature properties under fatigue and creep loading.

Fatigue lives Cyclic deformation curves of the samples tested at the larger total strain range of $\Delta\varepsilon_t = 0.01$ are shown in Fig. 9. As reference, sample A was tested in the initially untreated (fully heat-treated) state. The shot-peened sample C which was not aged at all exhibits a clearly shorter fatigue life compared to the reference sample. However, the peened and aged sample E with an

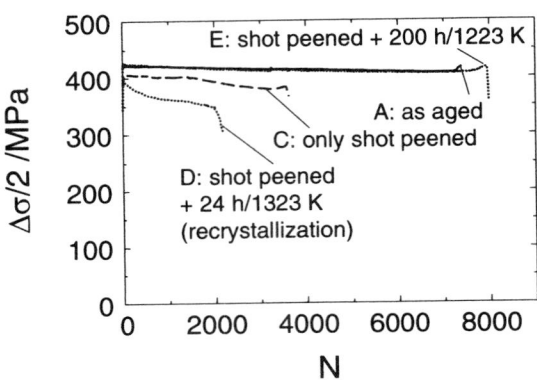

Figure 9: Cyclic deformation curves of the stress amplitude $\Delta\sigma/2$ vs. number of cycles N for the samples tested at a temperature of 1223 K at the larger total strain range of $\Delta\varepsilon_t = 0.01$. After [15].

"anomalous" raft structure (ageing for 200 h at 1223 K) showed a cyclic deformation curve and a fatigue life comparable to the reference sample. A comparable result was also obtained in the fatigue tests performed at the smaller strain range. It has to be noted that the recrystallized surface layers of these samples were totally removed by electropolishing before the fatigue tests.

It is also clear from figure 9 that sample D which still had a recrystallized surface layer which was not removed totally by polishing showed the shortest fatigue life. A network of cracks had formed on the surface of this sample along the grain boundaries. Thus, the short fatigue life is well understood as a consequence of the surface grain structure.

Creep strength Creep curves of three different samples are plotted in Fig. 10 as the plastic creep rate $\dot\varepsilon_{pl}$ versus the plastic strain ε_{pl}. The creep behaviour of one of the peened and aged samples (24 h at 1323 K) is clearly worse than the creep behaviour of the reference material. The peened sample aged for 200 h at 1223 K shows a lower minimum creep rate, but then a faster creep acceleration after the minimum than the reference sample. The times to a strain of 3 % were shorter in both aged samples compared to the reference sample.

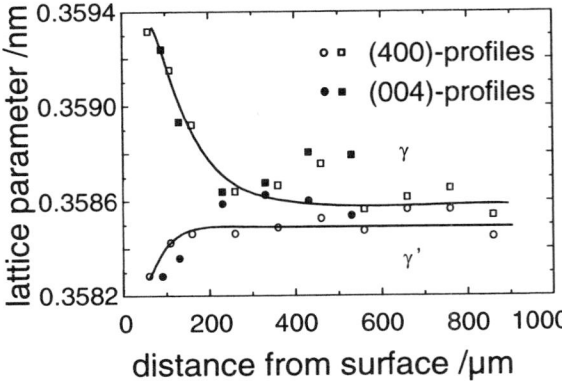

Figure 8: Variation of the lattice parameter values of γ and γ' of a peened and aged sample in the directions [001] and [100] (ageing conditions: 24 h at 1323 K).

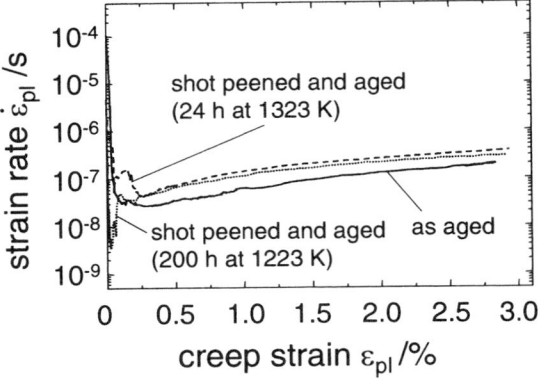

Figure 10: Creep curves of samples of the nickel-base superalloy SRR 99 in the as aged condition and after shot peening and ageing, respectively, for deformation at $\sigma = 200$ MPa and T = 1273 K, cf.[15].

Discussion

Origin of anomalous rafting

The experimental results obtained on the different samples concerning the origin of anomalous rafting can be summarized as follows:

1. Anomalous rafting was repeatedly obtained in turbine blades after service. However, the occurrence in different turbine blades seems to be unsystematical.
2. Anomalous rafting was observed already in the unused state of an aluminide-coated turbine blade.
3. After ageing of sections of an unused turbine blade, anomalous rafting was found to be very pronounced. However, the rafted zone was obtained only on one side of the blade.
4. In shot-peened specimens, a phenomenon similar to anomalous rafting in turbine blades was observed after ageing under appropriate conditions (high temperature above 1173 K).
5. The region in which anomalous rafting was observed in the shot-peened samples was correlated to the region in which severe plastic deformation was evidenced by the measurement of X-ray peak broadening and an increase of microhardness.
6. In a surface layer of shot-peened samples, a complex stress state was determined, indicating biaxial compressive stresses acting parallel to the surface.

These observations can be explained in two ways: i) A first hypothesis is that the modification of the internal microstress state between γ and γ' by plastic deformation is responsible for the microstructural transformation occurring during subsequent ageing treatments at sufficiently high temperatures. ii) The second hypothesis is that the deformation of a surface layer induces a complex state of residual macrostresses which act as driving force for anomalous rafting during a subsequent ageing treatment. In the following, these two possibilities are discussed.

Deformation-induced long-range internal microstresses There are several hints indicating that the internal microstresses induced by plastic deformation are the driving force for anomalous rafting. These long-range internal microstresses have been repeatedly shown to cause rafting, at least in laboratory samples after uniaxial deformation. The most significant examples are the experiments of Véron et al. [3,4] and Fährmann et al. [5]. These two research groups observed that rafting occurred in a predeformed sample during a subsequent stress-free ageing treatment in the manner which one would expect for the sense of predeformation. Even in a case, in which a plastically predeformed sample was subsequently aged under the action of a stress with the opposite sign compared to the predeformation, rafting was determined by the preceding plastic deformation process [5]. Also in the case of thermomechanical fatigue experiments, the sense of the plastic strain (and therefore the change of the internal microstress state) and not the external stress has been shown to determine the occurring type of rafting [6].

In the case of the present work, this possible origin for rafting can explain all discussed cases: Turbine blades are surface treated by grit blasting or shot peening before the coating process. This procedure will probably cause plastic deformation of a surface layer.

It has to be noted that this treatment is not always systematically applied, since it is performed in some cases only when necessary, and sometimes the surfaces are treated manually. Subsequently, at the high temperature occurring during the coating procedure, a first rafting process can occur, see the micrograph of an unused turbine blade in Fig. 1. Finally, after service at sufficiently high temperatures applied for some time, rafting will inevitably occur under the surfaces with an "anomalous" orientation of the γ/γ' plates parallel the surfaces, whereas the "normal" rafting process will occur in the interior of the material perpendicular to the external [001] stress axis.

With this explanation, it can be understood why the anomalous rafting was found only in one of the operated turbine blades: That turbine blade had probably a pretreatment which was more severe and which therefore induced (more) plastic deformation, leading to anomalous rafting during service. This would explain the observations on the aged sections of the unused turbine blade: These sections had only one side which showed "anomalous" rafts. The reason is probably that only this side was treated severely enough to induce anomalous rafting, and the other side was treated less severely.

The lattice parameter changes between γ and γ' observed after rafting (shown in Fig. 8) are finally a consequence of cooling to room temperature after ageing and can be explained as follows: During the ageing treatment (i.e. during rafting) the two phases γ and γ' were at high temperature and, because of the dynamic transformation of the morphology, tended to reduce their elastic strain energy. It can therefore be assumed that the two phases were approximately free of stresses after completion of the rafting procedure. During cooling, the plates of the two phases were contracted thermally. Due to the difference of the coefficients of thermal expansion of the two phases γ and γ', the plates would separately experience different thermal contraction strains. However, there is a strong constraint in the material, and thermally-induced internal microstresses arise. Similar thermally-induced microstresses and -strains have experimentally been observed in rafted creep-deformed samples [20] and calculated by finite element modelling [21].

There exists at present no micromechanical plasticity model which explains how the anomalous rafting is induced as a consequence of plastic deformation (and the change of the internal microstresses) by e.g. shot peening. In order to formulate such a model, transmission electron microscopic investigations of the respective zone of potential anomalous rafting before ageing will be necessary.

Shot-peening induced macrostresses As a second possibility, the peening-induced macrostresses have to be discussed. These internal stresses are known to be of biaxial compressive character acting parallel to a peened surface [22]. The investigation of the peened, but unaged sample by X-ray diffraction (Fig. 6) indicates the presence of such a complex stress state under the surface of the sample. The large increase of the lattice parameter in the [010] direction (i.e. parallel to the surface normal) is the corresponding Poisson strain of the peening-induced macrostresses. The measured increase of the lattice parameters in the directions [100] and [001] parallel to the surface plane cannot be fully explained in the present work. A possible reason for this increase is the relaxation of the large peening macrostresses during cutting of the peened specimen. As an example, on the (010) surface, the peening compressive macrostresses act in the directions [100] and [001]. During preparation, the sample was cut parallel to the (100) plane. Then, the stress acting in [100] direction is relieved, and only the compressive stress acting in the [001] direction remains. As a consequence, this stress will increase the lattice parameter in the [100] direction by the Poisson strain. A similar effect occurs in the case of a (001) cut.

Now, it shall be discussed how the peening-induced macrostresses could cause the observed cases of anomalous rafting. Under the assumption that during surface treatments like peening or grinding compressive stresses arise, we can compare the effect of these biaxial compressive stresses with the distortion induced by an external tensile stress acting in the normal direction on the surface of peened material. Such a tensile stress, however, is known to induce in alloys of negative lattice misfit (like SRR 99 and CMSX-6 which have a negative lattice misfit at the high service temperatures [8,20]) rafts perpendicular to this (tensile) stress state [1]. Thus, the anomalous rafts could also be explained by biaxial compressive stresses acting under a mechanically treated surface.

However, there is an argument against this hypothesis. Khadhraoui et al. investigated experimentally and by modelling the relaxation of shot-peening residual stresses [22]. After ageing for up to 100 h at temperatures of 873 K and 923 K, a strong recovery of the peening stresses up to 50 % of the initial value was reported. This result makes the peening-induced macrostresses as the driving force for anomalous rafting improbable, because the temperatures necessary for rafting are much higher than those applied by Khadhraoui et al. [22]. Therefore, the peening-induced macrostresses will probably vanish early during the ageing treatment.

In addition, the microstructural results obtained on the sample with a removed surface layer of 50 μm indicate that the peening-induced macrostresses are of minor relevance. The effect of "anomalous rafting" observed in this sample was similar to that observed in the other samples aged at comparable conditions. However, it can be assumed that the peening-induced macrostresses were substantially modified and decreased by the removal of a layer of 50 μm. Therefore, the macrostresses seem to be of minor importance for "anomalous" rafting.

The fact that the mean lattice parameters of the peened and aged sample with "anomalous" rafts indicate a macroscopically stress-free state does not give an answer to the question whether the peening-induced micro- or macrostresses were the origin for anomalous rafting. The investigated sample corresponds to the already rafted microstructure which is not expected to contain macrostresses.

As a summary of the two possible origins of anomalous rafting, we conclude that it seems to be more probable from the presented results that the effect of anomalous rafting is induced by a mechanical surface treatment, at least in those cases discussed in this work. The specific question whether the microstresses acting between γ and γ' or the peening-induced macrostresses were the main driving forces for the microstructural transformation cannot be answered

finally. Nevertheless, there are strong hints that the microstresses are the responsible microstructural reason for anomalous rafting.

Anomalous rafting during engine operation

The results of the present work on anomalous rafting in turbine blades seem different from earlier observations. Draper et al. [7] observed anomalous rafting under the surface of uncoated, monocrystalline turbine blades after a "performance test" (P) and after a "high cycle fatigue test" (cycle A) or after combinations of other tests with one of these two test conditions during ground-based engine tests. Both test conditions (P) and (A) consist of stepwise varying power settings between idle and cruise power levels. After other test cycles, e.g. constant operation at maximum temperature ("stress rupture, cycle B"), "low cycle fatigue, cycle C" or a "mission mix, cycle D", or combinations of these, no anomalous rafting was found, although, in some cases, normal rafting occurred. The authors discuss that anomalous rafting is caused by a "complex stress state at the surface of the blade" during the respective tests which stems from blade resonant conditions at an intermediate power level.

In the present case, however, it is clear that the observed rafting is caused by other processes: i) We have investigated coated blades which were peened before the coating procedure. ii) In the unused blade before as well as after the (stress-free) ageing treatment, the blades had not been subjected to engine operation at all. In the blade operated in an accelerated mission test, a test condition similar to the "stress rupture test, cycle B" of Draper et al. was applied. However, in contrast to the results of Draper et al., we observed anomalous rafting after operation under this condition.

These reasons indicate that, at least in the present case, the origin of anomalous rafting is not the operation condition in the test engine. In addition, the present observation of anomalous rafting on only one side of the aged blade shows that the reason seems to be in some way unsystematical.

Recrystallization

The ageing treatments applied to the shot-peened samples yield interesting results on the recrystallization behaviour of nickel-base superalloy single crystals. The appearance of recrystallization is very similar to the cellular recrystallization described earlier [23,24] (see also the recent work of Bürgel et al. [25] on the origin of recrystallization in superalloys). Recrystallization could not be avoided by the application of a drastically reduced heating rate (50 K per hour) and a holding time during heating (2 h at 1073 K). Therefore, the recovery occurring during the slow heating and the holding time at 1073 K were not sufficient to reduce the driving force for recrystallization. Furthermore, it has to be noted that the removal of the surface roughness by electropolishing did also not influence recrystallization.

However, recrystallization could be largely avoided in one sample, from which a surface layer of 50 μm was removed by polishing. This is worth mentioning, because X-ray peak broadening (Fig. 7) and microhardness investigations (Fig. 5) showed the zone of plastic deformation to have a depth of about 200 μm. It can be concluded from this observation that there is a critical deformation occurring in the outer 50 μm which is necessary for recrystallization in the nickel-base superalloy SRR 99. At lower deformation levels, no recrystallization was obtained.

Effect of anomalous rafting on the mechanical behaviour

Fatigue Recently, Mughrabi et al. showed that samples with a rafted microstructure with an orientation of the γ/γ' plates parallel to the stress axis (induced by compressive predeformation) have increased life times under isothermal high-temperature fatigue loading than the initial microstructure with cuboidal γ' particles [2,12]. Such an effect was an original motivation for the present work. However, the thickness of the rafted layer induced by shot peening and ageing was only about 100 μm. The cyclic deformation curves are identical for the reference samples with cuboidal γ' precipitates and for the samples with a rafted surface layer (the deformation curves obtained with the higher applied strain range are shown in Fig. 9). It can therefore be concluded that this zone is not thick enough to influence the mechanical behaviour or the fatigue life of the total sample appreciably.

The fatigue lives of the peened samples C (only peened) and D (peened and aged, but with recrystallized grains) were significantly shorter than the fatigue life of the reference material. This is partially in contrast to results obtained earlier [24]. However, a reason for this difference cannot be given yet.

Creep The increase of the creep rate in the case of the aged samples is due to the deterioration of the creep resistance caused by coarsening of the microstructure in the bulk of the samples, similar to the results obtained by Mughrabi et al. on aged samples [13-15]. The enhancement reported by the authors for the case of samples which had a γ/γ' raft structure parallel to the tensile deformation axis in the whole specimen induced by compressive predeformation was not obtained in the present samples with a thin surface layer of similar rafts. The reason is that the area fraction of rafted material with respect to the total cross section is small, and a possible enhancing effect is overcompensated by the deterioration due to the coarsening effect in the bulk.

Microhardness The investigations of the peened and aged samples by microhardness testing yield information on the room temperature strength of the different microstructures. The values obtained on the differently aged samples, however, were equal within some scatter. Only a small decrease of the microhardness was found in the rafted region which was smaller than the scatter. Therefore, it cannot be considered significant.

Conclusions

The phenomenon of anomalous rafting was investigated in real turbine blades and in shot-peened laboratory samples by scanning electron microscopy, microhardness testing and high-resolution X-ray diffraction. From the results, conclusions were drawn on the origin of this microstructural transformation. In addition, the effect of anomalous rafting on the mechanical high-temperature properties was investigated in fatigue and creep tests. The following conclusions can be drawn:

1. Anomalous rafts were obtained in aluminide-coated turbine

blades of the nickel-base superalloy CMSX-6 before and after service. In addition, this kind of rafting was observed after ageing of unused sections of a turbine blade.

2. A similar effect was found in shot-peened samples of the superalloy SRR 99 after shot-peening and ageing.

3. Two different possibilities were discussed to explain anomalous rafting: deformation-induced internal microstresses acting between the two phases γ and γ' and compressive macrostresses, respectively. Experimental results indicate that the internal microstresses are the main driving force for rafting.

4. A layer of rafts with an orientation parallel to the surface (i.e. anomalous rafts induced by shot peening and subsequent ageing) is more or less indifferent for the fatigue strength and life time.

5. The creep behaviour is not affected by thin surface layer of anomalous rafts.

6. Recrystallization of shot-peened samples can only be avoided by the removal of the highly deformed outer surface layer, not by recovery during a heating procedure. However, it is not necessary to remove the whole plastically deformed material to avoid recrystallization in the case of the nickel-base superalloy SRR 99.

Acknowledgments

The authors thank Dr.-Ing. D. Goldschmidt (formerly MTU, Munich, now Siemens AG/KWU, Mülheim), and the company MTU for providing the investigated turbine blades. The authors are grateful to V. Schulze, Institut für Werkstoffkunde I, Universität Karlsruhe, for shot-peening, to A. Snigirev, A. Souvorov and M. Drakopoulos, ESRF, Grenoble, for their competent technical assistance at the microfocus X-ray experiments performed at the optics beam-line (BM 5/ID 10) and to K. Zinn, Institut WTM, Universität Erlangen-Nürnberg, for microhardness testing. Parts of this work were supported by Deutsche Forschungsgemeinschaft (Mu502/12), by Volkswagen-Stiftung (I/69 362), and by ESRF, Grenoble.

Literature

1. F.R.N. Nabarro, "Rafting in Superalloys", Metall. Mater. Trans., 27A (1996), 513-530.

2. H. Mughrabi, "γ/γ' Rafting and its Effect on the Creep and Fatigue Behaviour of Monocrystalline Superalloys", in The Johannes Weertman Symposium, eds. R.J. Arsenault, D. Cole, T. Gross, G. Kostorz, P.K. Liaw, S. Parameswaran and H. Sizek, (Warrendale: PA, The Minerals, Metals and Materials Society, 1996), 267-278.

3. M. Véron, Y. Bréchet, and F. Louchet, "Directional Coarsening of Nickel Based Superalloys: Driving Force and Kinetics", in Proc. 8th Int. Symp. on Superalloys (Superalloys 1996), eds. R.D. Kissinger, D.J. Deye, D.L. Anton, A.D. Cetel, M.V. Nathal, T.M. Pollock and D.A. Woodford, (Warrendale, PA: The Minerals, Metals and Materials Society, 1996), 181-190.

4. M. Véron and P. Bastie, "Strain Induced Directional Coarsening in Nickel Based Superalloy: Investigation on Kinetics using Small Angle Neutron Scattering (SANS) Technique", Acta mater., 45 (1997), 3277-3282.

5. M. Fährmann, E. Fährmann, O. Paris, P. Fratzl, and T.M. Pollock, "An Experimental Study of the Role of Plasticity in the Rafting Kinetics of a Single Crystal Ni-Base Superalloy", in Proc. 8th Int. Symp. on Superalloys (Superalloys 1996), eds. R.D. Kissinger, D.J. Deye, D.L. Anton, A.D. Cetel, M.V. Nathal, T.M. Pollock and D.A. Woodford, (Warrendale, PA: The Minerals, Metals and Materials Society, 1996), 191-200.

6. S. Kraft, I. Altenberger and H. Mughrabi, "Directional γ/γ' Coarsening in a Monocrystalline Nickel-Based Superalloy during Low-Cycle Thermomechanical Fatigue", Scripta metall. mater., 32 (1995), 411-416, and Scripta metall. mater., 32 (1995), 1903 (Erratum).

7. S. Draper, D. Hull and R. Dreshfield, "Observations of Directional Gamma Prime Coarsening during Engine Operation", Metall. Trans., 20A, (1989), 683-688.

8. B. von Grossmann, "Mikrostrukturelle Bestimmung der lokalen Belastungen in einkristallinen Turbinenschaufeln aus Nickelbasis-Superlegierungen" (Doctorate Thesis, Universität Erlangen-Nürnberg, 1998, Aachen, Germany: Shaker Verlag, 1999).

9. H. Biermann, B. von Grossmann, T. Schneider, H. Feng and H. Mughrabi, "Investigation of the γ/γ' Morphology and Internal Stresses in a Monocrystalline Turbine Blade after Service: Determination of the Local Thermal and Mechanical Loads", in Proc. 8th Int. Symp. on Superalloys (Superalloys 1996), eds. R.D. Kissinger, D.J. Deye, D.L. Anton, A.D. Cetel, M.V. Nathal, T.M. Pollock and D.A. Woodford, (Warrendale, PA: The Minerals, Metals and Materials Society, 1996), 201-210.

10. P. Moretto and J. Bressers, "Thermo-Mechanical Fatigue Degradation of a Nickel-Aluminide Coating on a Single-Crystal Nickel-Based Alloy", J. Mat. Sci., 31 (1996), 4817-4829.

11. J. Bressers, D.J. Arrell, K.M. Ostolaza and J.L. Vallés, "Effect of an Aluminide Coating on Precipitate Rafting in Superalloys", Mater. Sci. Engng A, 220 (1996), 147-154.

12. M. Ott, U. Tetzlaff and H. Mughrabi, "Influence of Directional Coarsening on the Isothermal High-Temperature Fatigue Behaviour of the Monocrystalline Nickel-Base Superalloys CMSX-6 and CMSX-4", in Microstructure and Mechanical Properties of Metallic High-Temperature Materials, eds. H. Mughrabi, G. Gottstein, H. Mecking, H. Riedel and J. Tobolski, (Weinheim: Wiley-VCH, 1999), 425-440.

13. H. Mughrabi and U. Tetzlaff, "Microstructure and High-Temperature Strength of Monocrystalline Nickel-Base Superalloys", Adv. Eng. Mater., (2000), in print.

14. U. Tetzlaff, M. Nicolas and H. Mughrabi, "Can the High Temperature Tensile Strength of Nickel-Base Superalloys be Improved by Pre-Rafting?", in Proc. EUROMAT '99, Vol. 10, eds. D.G. Morris et al., (Weinheim: Wiley-VCH, 2000), in print.

15. U. Tetzlaff, "Gezielte gerichtete Vergröberung (Floßbildung) des γ/γ'-Gefüges und Auswirkungen auf die mechanischen Hochtemperatureigenschaften einkristalliner Nickelbasis-Superlegierungen (Doctorate Thesis, Universität Erlangen-Nürnberg, 2000).

16. H. Biermann, U. Tetzlaff, B. von Grossmann, H. Mughrabi and V. Schulze, "Rafting in Monocrystalline Nickel-Base Superalloys induced by Shot Peening", Scripta mater., accepted for publication.

17. H.-A. Kuhn, H. Biermann, T. Ungár and H. Mughrabi, "An X-ray study of creep-deformation induced changes of the lattice mismatch in the γ'-hardened monocrystalline nickel-base superalloy SRR 99", Acta metall. mater., 39 (1991), 2783-2794.

18. H. Biermann, B. von Grossmann, S. Mechsner, H. Mughrabi, T. Ungár, A. Snigirev, I. Snigireva, A. Souvorov, M. Kocsis and C. Raven, "Microbeam Synchrotron Radiation Diffraction Study of a Monocrystalline Nickel-Base Turbine Blade after Service", Scripta mater., 37 (1997), 1309-1314.

19. H. Biermann, B. von Grossmann, T. Ungár, S. Mechsner, A. Souvorov, M. Drakopoulos, A. Snigirev and H. Mughrabi, "Determination of Local Strains in a Monocrystalline Turbine Blade by Microbeam X-ray Diffraction with Synchrotron Radiation", Acta mater., 48 (2000), 2221-2230.

20. H. Biermann, M. Strehler and H. Mughrabi, "High-Temperature Measurements of Lattice Parameters and Internal Stresses of a Creep-Deformed Monocrystalline Nickel-Base Superalloy", Metall. Trans. A, 27A (1996), 1003-1014.

21. H. Feng, H. Biermann and H. Mughrabi, "3D Finite Element Modelling of Lattice Misfit and Long-Range Internal Stresses in Creep-Deformed Nickel-Base Superalloy Single Crystals", Mater. Sci. Engng. A, 214 (1996), 1-16.

22. M. Khadhraoui, W. Cao, L. Castex, and J.Y. Guédou, "Experimental Investigations and Modelling of Relaxation Behaviour of Shot Peening Residual Stresses at High Temperature for Nickel Base Superalloys", Mat. Sci. Technol., 13 (1997), 360-367.

23. S.D. Bond and J.W. Martin, "Surface Recrystallization in a Single Crystal Nickel-Based Superalloy", J. Mater. Sci., 19 (1984), 3867-3872.

24. P. D. Portella and W. Österle, "Influence of Cellular Recrystallization on the Fatigue Behaviour of Single Crystal Ni-Based Superalloys", in Microstructure and Mechanical Properties of Metallic High-Temperature Materials, eds. H. Mughrabi, G. Gottstein, H. Mecking, H. Riedel and J. Tobolski, (Weinheim: Wiley-VCH, 1999), 441-453.

25. R. Bürgel, P.D. Portella and J. Preuhs, "Recrystallization in Single Crystals of Nickel Base Superalloys", in Proc. 9th Int. Symp. on Superalloys (Superalloys 2000), (Warrendale, PA: The Minerals, Metals and Materials Society, 2000), this volume.

Stress-Rupture Behavior of Waspaloy and IN-738LC at 660°C (1112°F) in Low Oxygen Gaseous Environments Containing Sulfur

David C. Seib

Dresser-Rand Company
P.O. Box 560
Olean, NY 14760

Abstract

Stress-rupture lives of precipitation strengthened Nickel base Superalloys are detrimentally effected when subjected to atmospheres consisting of a low oxygen partial pressure and a high sulfur partial pressure. Short time failures, originating at mechanical notches, in these atmospheres have been experienced under both service and laboratory conditions. This notch sensitive behavior has previously been documented.[1,2] During laboratory testing of various specimens, trends have emerged to suggest that grain size can influence notch behavior in corrosive gaseous environments.

Introduction

Fluidized Catalytic Cracker Units (FCCU) are used by refineries to convert crude oil into lighter hydrocarbons such as propane, kerosene, gasoline, and fuel oil. During this process a catalyst is used to strip away some of the carbon atoms from the crude. The spent catalyst is heated in a regenerator to burn-off the carbon, and then the rejuvenated catalyst is returned to the reactor to repeat the cycle again. Exhaust gasses from the regenerator are directed through a FCCU Expander, i.e. a power turbine in the exhaust flue, to reclaim waste energy.

Regenerator atmospheres at refineries are divided into two groups, complete combustion and partial combustion. Thermodynamically calculated equilibrium diagrams can be used to compare these two atmospheres (Fig. 1). It can be seen that while the complete combustion process contains sufficient oxygen to predict the formation of protective chromium oxide and nickel oxide scales, the partial combustion process contains less oxygen, and therefore would be expected to produce chromium oxide and nickel sulfide scales.

Expanders powered by partial combustion atmospheres have become increasingly susceptible to failure as the sulfur content of available crude oils is increasing and the demand for higher horsepower continues. Service failures of FCCU Expanders in this atmosphere have revealed very similar metallurgical features. In all but one case, fracture had originated from a blade or disc firtree serration and progressed by intergranular separation. When the fractures / cracks were viewed in cross-section, a two layered scale was present. The scale immediately adjacent to the base metal had a gray colored appearance and consisted primarily of chromium, oxygen, and sulfur. The scale along the crack centerline had a gold colored appearance and consisted primarily of nickel and sulfur.

These sulfidation fractures are notably different from low temperature type II (layer-type) hot corrosion. Fractures resulting from this mechanism: exhibit an intergranular topography, occur below the 635°C (1175°F) nickel sulfide eutectic temperature, and contain no salts in the scales. It has been evident since the first occurrence in the early 1980s, that this sulfidation fracture mode was due to the combined effects of stress, temperature, time, and a low oxygen atmosphere containing sulfur.

A paper published by Mr. Ken Natesan, of Argonne National Laboratory, entitled "Effect of Oxidation / Sulfidation on Creep Behavior of Alloy 800" revealed similar two layered scales in the test range of 650°C to 927°C.[3] For Alloy 800, the scales were identified as iron sulfide and either chromium oxide or chromium sulfide depending upon test atmosphere. Mr. Natesan concluded that the alloy suffered substantial reductions in creep life when the test environment pO_2 was less than the transition boundary pO_2 between chromium oxide / chromium sulfide formation.

Superalloys 2000
Edited by T.M. Pollock, R.D. Kissinger, R.R. Bowman,
K.A. Green, M. McLean, S. Olson, and J.J. Schirra
TMS (The Minerals, Metals & Materials Society), 2000

Temperature, Equipment, & Procedure

The test temperature of 600°C (1112°F) was selected to promote stress-rupture failure of Waspaloy in air, and stay below the 635°C (1175°F) nickel sulfide eutectic temperature. A direct comparison between stress-rupture results in air and the corrosive atmosphere was desired.

Laboratory testing was performed using mixed gases of hydrogen and hydrogen sulfide slowly bubbled through distilled water to produce the desired atmospheres. Atmosphere 1 was achieved by bubbling a 51.3 % H_2S – 48.7 % H_2 mole percent mixture through room temperature water. Atmosphere 2, 3, or 4 was produced by bubbling a 5.9 % H_2S – 94.1 % H_2 mole percent mixture through room temperature, 45°C, or 80°C water. At the test temperature of 600°C, these atmospheres were (Fig.1):

Atmosphere 1:	Log O_2 = - 26.3	Log S_2 = - 5.6
Atmosphere 2:	Log O_2 = - 26.9	Log S_2 = - 8.0
Atmosphere 3:	Log O_2 = - 25.9	Log S_2 = - 8.0
Atmosphere 4:	Log O_2 = - 24.0	Log S_2 = - 8.1

Equipment used for the tests consisted of a stress-rupture frame, pressurized cylinders of mixed gases, distilled water, and a stainless steel reaction chamber. Connection of the gas system was accomplished by a 304 stainless steel supply tube from the flow meters on the pressurized cylinders to the bottom of the pot of water. Another 304 stainless steel supply tube extended from the top of the water pot to the lower end plate of the reaction chamber. Variac heating tape was used to prevent the condensation of water vapor in the later supply tube. The reaction chamber was 304 stainless steel schedule 40 pipe approximately 48 cm long x 5.3 cm inside diameter (19 inches long x 2.1 inches ID) with flanges welded on both ends, and end plates bolted to the flanges. Thermo-well tubes of 304 stainless steel entered the reaction chamber just above and below the specimen shoulders, approximately 90 mm (3.6 inches) apart. Temperature control was maintained by a thermo-couple placed in the upper thermo-well. Specimen gauge temperatures of 600+/-2°C (1112+/-4°F) were recorded during equipment calibration.

The stress-rupture frame was of a standard Satec design, incorporating a 20:1 specimen load to pan weight ratio. Furnace around the reaction vessel was of a clam shell design operated by a single zone controller.

A sixteen hour nitrogen gas purge was used to vent oxygen from the system prior to each test. The furnace was turned on, and the corrosive gas mixture was flowed at 0.1 SCFM. After about four hours, the temperature of the upper thermo-well indicated the specimen had achieved 600°C (1112°F). Test temperature was stabilized for 10 minutes prior to applying load.

Alloys Tested

Waspaloy received the greatest amount of testing during this project. Other forging alloys tested were X-750 and U-720Li.

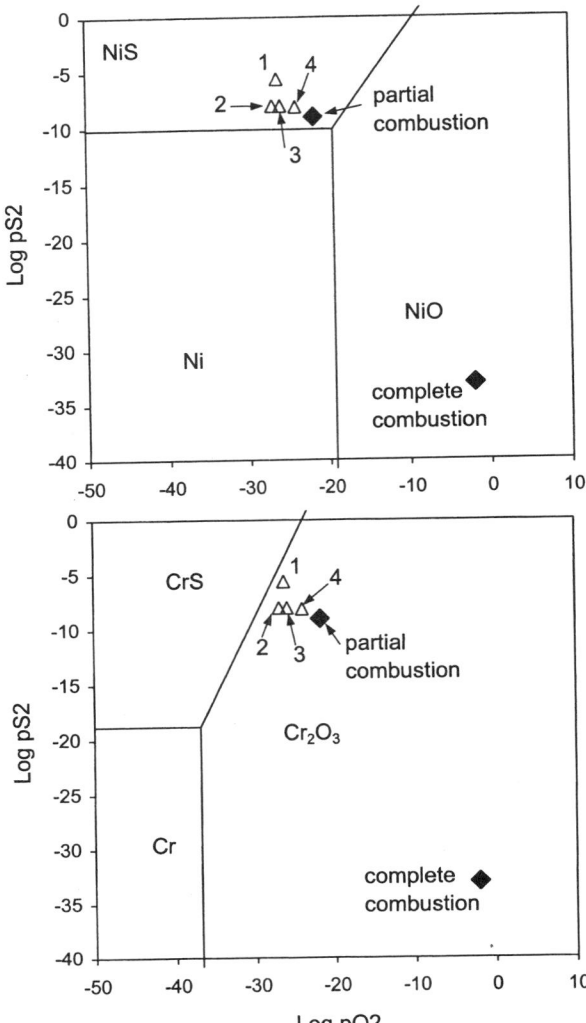

FIG. 1 Thermodynamically calculated equilibrium diagram for (a) the Ni-O-S systems and (b) the Cr-O-S systems at 600°C. Numbers in the pictures corresponds to the atmospheres.

Cast alloys tested were equiax IN-738LC, directionally solidified IN-738LC, and an experimental high chromium directionally solidified alloy (hereafter designated NiCr DS).

Derivatives of Waspaloy with modified chemistries were also tested. It seemed appropriate to modify the elements that effect grain boundary chemistry rather than bulk alloy chemistry, since the failures experienced in service progressed along grain boundaries. The majority of these Waspaloy modifications involved the reduction of carbon and increase of boron. Previous testing of U-720 in molten salt by the Westinghouse Research and Development Center had suggested that boron compounds along grain boundaries were more corrosion resistant than carbides. [4]

Waspaloy Bar, the first item listed in Table I, was used as a comparative standard during the testing. Approximately 258 kg

(569 pounds) of this heat was purchased per AMS 5704. Diameter of this bar was 15.8 mm (5/8 inch).

Waspaloy 11 through 41 were experimental ingots approximately 70 mm (2 ¾ inch) diameter rolled-down into 12 mm (1/2 inch) thick plates.

Waspaloy ER1 and ER2 were two heats provided by Allvac in the form of 15.8mm (5/8 inch) diameter bar.

All Waspaloy samples were solution treated, water quenched, aged at 845°C (1550°F) for 4 hours, and aged at 760°C (1400°F) for 16 hours.

U-720Li was 1104°C (2020°F) solution treated, water quenched, aged at 760°C (1400°F) for 8 hours, then aged at 649°C (1200°F) for 24 hours. The bar was 12mm (1/2 inch) diameter.

IN-738LC equiax and DS were cast as 16 mm (5/8 inch) diameter bars, 1185°C (2165°F) HIP processed, 1121°C (2050°F) solution treated, and 843°C (1550°F) aged for 24 hours. HIP of the DS alloy was not performed.

NiCr DS was directionally cast as 19 x 102 mm (0.75 x 4 inch) plate, 1204°C (2200°F) homogenized, 1079°C (1975°F) solution treated, and aged at 871°C (1600°F) for 20 hours. HIP was not performed.

TABLE I   Chemistries of Alloys, Weight Percent

Alloy	C	Cr	Co	Mo	Al	Ti	B
Waspaloy Bar	0.068	19.21	13.50	4.10	1.40	3.13	0.006
Waspaloy 5	0.034	19.20	13.25	4.34	1.34	3.07	0.005
Waspaloy 6	0.032	19.18	13.25	4.03	1.32	3.08	0.004
Waspaloy 11	0.018	18.9	14.0	4.1	1.36	3.10	0.005
Waspaloy 12	0.018	18.9	14.0	4.1	1.35	3.10	0.030
Waspaloy 22	0.008	19.0	14.0	4.1	1.36	3.15	0.065
Waspaloy 41	0.010	19.0	14.0	4.1	1.36	3.15	0.011
Waspaloy ER1	0.039	19.72	13.43	4.32	1.30	3.00	0.015
Waspaloy ER2	0.034	19.71	13.35	4.27	1.32	2.96	0.013
X-750	0.04	16.7	0.14	0.07	0.8	2.42	
U-720 Li	0.011	16.3	14.1	2.9	2.43	5.0	0.014
IN-738LC	0.10	16.06	8.36	1.85	3.48	3.35	0.010
IN-738LC DS	0.09	16.20	8.7	1.8	3.4	3.5	0.009
NiCr DS		15	6		3.6	4.5	

Mechanical Properties

Solution heat treatment temperatures, grain sizes, room temperature tensile, and 732°C – 552 MPa (1350°F - 80 ksi)

combination bar stress-rupture testing (Kt = 3.9 notch) results for the specimens were as shown in Table II. A "F" or "C" after the alloy name indicates that a sub-solvus fine grain (F) or super-solvus coarse grain (C) version of the chemistry was tested.

TABLE II   Room Temperature Tensile and 732°C–552MPa (1350°F - 80 ksi) Stress-Rupture Properties in Air

Alloy	Solution Temp.	Grain Size	RT UTS (MPa)	RT Yield (MPa)	RT Elong.	SR Life	SR Elong.
Wasp. Bar	1038°C	5	1324	972	24 %	58 hr.	15 %
Wasp. 5	1025°C	6	1372	931	27 %	47 hr.	22 %
Wasp. 6-F	1004°C	9	1407	993	26 %	58 hr.	38 %
Wasp. 6-C	1052°C	4	1282	834	30 %	26 hr.	11 %
Wasp. 11	1024°C	4 & 9	1365	1007	30 %	75 hr.	29 %
Wasp. 12	1032°C	9	1441	1041	28 %	53 hr.	19 %
Wasp. 22	1032°C	9	1413	1000	28 %	25 hr.	9 %
Wasp. 41	1004°C	4 & 9	1365	979	29 %	36 hr.	23 %
Wasp. ER1	1018°C	9	1411	1024	28 %		
Wasp. ER2-F	1004°C	9	1379	965	26 %	49 hr.	35 %
Wasp. ER2-C	1052°C	3	1269	800	29 %		
X-750	Unknown	2	1117	662	16 %		
U-720 Li	1104°C	10	1703	1317	21 %		
IN-738LC	1121°C		1013	827	10 %		
IN-738LC DS	1121°C		1110	855	8 %		
NiCr DS	1079°C		1007	876	4 %		

Specimen Geometry & Sulfidation Test Results

Notched (N) gauge and Combination (C) gauge specimens of 3.8 mm (0.151 inch) diameter, Kt = 2.4, were used throughout this testing. Test atmospheres were air, hydrogen bubbled through room temperature water, and the mixed gas atmospheres 1, 2, 3, and 4 (See "Temperature, Equipment, & Procedure" heading). The atmosphere for each test is listed in the "Atm." column. Specimens that did not break display a "+" in the Life column. Results were as shown in Table III.

TABLE III Stress-Rupture Properties at 600°C in Selected Atmospheres

Test # (gauge)	Alloy	Atm.	Stress		Life	Fracture
1 (C)	Waspaloy Bar	Air	896 MPa	(130 ksi)	348 hr.	Smooth
2 (C)	Waspaloy Bar	Air	896 MPa	(130 ksi)	217 hr.	Smooth
3 (N)	Waspaloy Bar	1	517 MPa	(75 ksi)	169 hr.	Notch
4 (N)	Waspaloy Bar	1	689 MPa	(100 ksi)	55 hr.	Notch
5 (N)	Waspaloy Bar	1	689 MPa	(100 ksi)	35.5 hr.	Notch
6 (N)	Waspaloy Bar	1	689 MPa	(100 ksi)	43.6 hr.	Notch
7 (N)	Waspaloy Bar	1	689 MPa	(100 ksi)	43.6 hr.	Notch
8 (C)	Waspaloy Bar	1	689 MPa	(100 ksi)	33.9 hr	Notch
9 (C)	Waspaloy Bar	1	689 MPa	(100 ksi)	35.8 hr.	Notch
10 (C)	Waspaloy 5	1	689 MPa	(100 ksi)	148.5 hr.	Notch
11 (C)	Waspaloy 5	1	827 MPa	(120 ksi)	14.5 hr.	Notch
12 (N)	Waspaloy 11	1	827 MPa	(120 ksi)	300+ hr.	
13 (N)	Waspaloy 12	1	689 MPa	(100 ksi)	115+ hr.	
14 (N)	Waspaloy 22	1	827 MPa	(120 ksi)	300+ hr.	
15 (N)	Waspaloy 41	1	827 MPa	(120 ksi)	285 hr.	Notch
16 (C)	Waspaloy ER1	1	827 MPa	(120 ksi)	300+ hr.	
17 (N)	X-750	1	517 MPa	(75 ksi)	76 hr.	Notch
18 (N)	X-750	1	517 MPa	(75 ksi)	88 hr.	Notch
19 (C)	IN-738LC	Air	896 MPa	(130 ksi)	400+ hr.	
20 (N)	IN-738LC	1	689 MPa	(100 ksi)	8 hr.	Notch
21 (N)	IN-738LC	1	689 MPa	(100 ksi)	8.1 hr.	Notch
22 (N)	IN-738LC DS	1	689 MPa	(100 ksi)	3 hr.	Notch
23 (N)	NiCr DS	Air	517 MPa	(75 ksi)	500+ hr.	
24 (N)	NiCr DS	1	517 MPa	(75 ksi)	0.1 hr.	Notch
25 (N)	NiCr DS	H_2	517 MPa	(75 ksi)	200+ hr.	
26 (C)	Waspaloy Bar	2	689 MPa	(100 ksi)	18.4 hr.	Notch
27 (C)	Waspaloy Bar	2	689 MPa	(100 ksi)	16.2 hr.	Notch
28 (C)	Waspaloy Bar	3	689 MPa	(100 ksi)	106.0 hr.	Notch
29 (C)	Waspaloy Bar	3	758 MPa	(110 ksi)	39.6 hr.	Notch
30 (C)	Waspaloy Bar	3	758 MPa	(110 ksi)	42.4 hr	Notch
31 (C)	Waspaloy Bar	4	724 MPa	(105 ksi)	311.2 hr.	Smooth
32 (C)	Waspaloy Bar	4	758 MPa	(110 ksi)	51.9 hr.	Notch
33 (C)	Waspaloy Bar	4	793 MPa	(115 ksi)	54.2 hr.	Notch
34 (C)	Waspaloy Bar	4	793 MPa	(115 ksi)	47.2 hr.	Notch
35 (C)	Waspaloy Bar	4	862 MPa	(125 ksi)	25.4 hr.	Notch
36 (C)	Waspaloy Bar	4	862 MPa	(125 ksi)	26.2 hr.	Notch
37 (C)	Waspaloy Bar	4	862 MPa	(125 ksi)	55.7 hr.	Notch
38 (C)	Waspaloy Bar	4	862 MPa	(125 ksi)	85.2 hr.	Notch
39 (C)	Waspaloy Bar	4	896 MPa	(130 ksi)	23.1 hr.	Notch
40 (C)	Waspaloy 6-F	4	827 MPa	(120 ksi)	52.3 hr.	Smooth
41 (C)	Waspaloy 6-F	4	862 MPa	(125 ksi)	114.5 hr.	Smooth
42 (C)	Waspaloy 6-F	4	931 MPa	(135 ksi)	47.5 hr.	Smooth
43 (C)	Waspaloy 6-C	4	862 MPa	(125 ksi)	2.4 hr.	Notch
44 (C)	Waspaloy 22	4	862 MPa	(125 ksi)	2.0 hr.	Notch
45 (C)	Waspaloy 41	4	862 MPa	(125 ksi)	16.4 hr.	Notch
46 (C)	Waspaloy ER2-F	4	862 MPa	(125 ksi)	27.4 hr.	Smooth
47 (C)	Waspaloy ER2-F	4	862 MPa	(125 ksi)	45.8 hr.	Smooth
48 (C)	Waspaloy ER2-C	4	862 MPa	(125 ksi)	15.2 hr.	Smooth

49 (C)	X-750	4	689 MPa	(100 ksi)	9.8 hr.	Smooth
50 (C)	U-720 Li	4	862 MPa	(125 ksi)	14.5 hr.	Notch
51 (C)	U-720 Li	4	931 MPa	(135 ksi)	23.5 hr.	Notch
52 (C)	IN-738LC	4	586 MPa	(85 ksi)	54.4 hr.	Notch
53 (C)	IN-738LC	4	689 MPa	(100 ksi)	4.9 hr.	Notch
54 (C)	IN-738LC	4	793 MPa	(115 ksi)	4.6 hr.	Smooth
55 (C)	IN-738LC	4	862 MPa	(125 ksi)	0.1 hr.	Smooth
56 (C)	NiCr DS	4	689 MPa	(100 ksi)	0.1 hr.	Notch

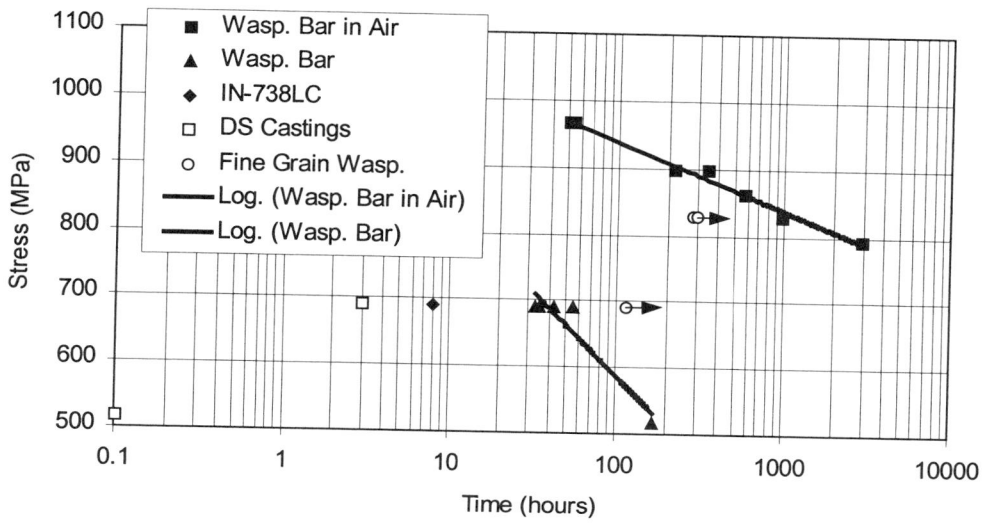

FIG.2. Semi-log plot of stress-rupture data at 600 °C (1112 °F) in atmosphere 1, i.e., log O_2 = -26.3 and log S_2 = -5.6, except where stated in air.

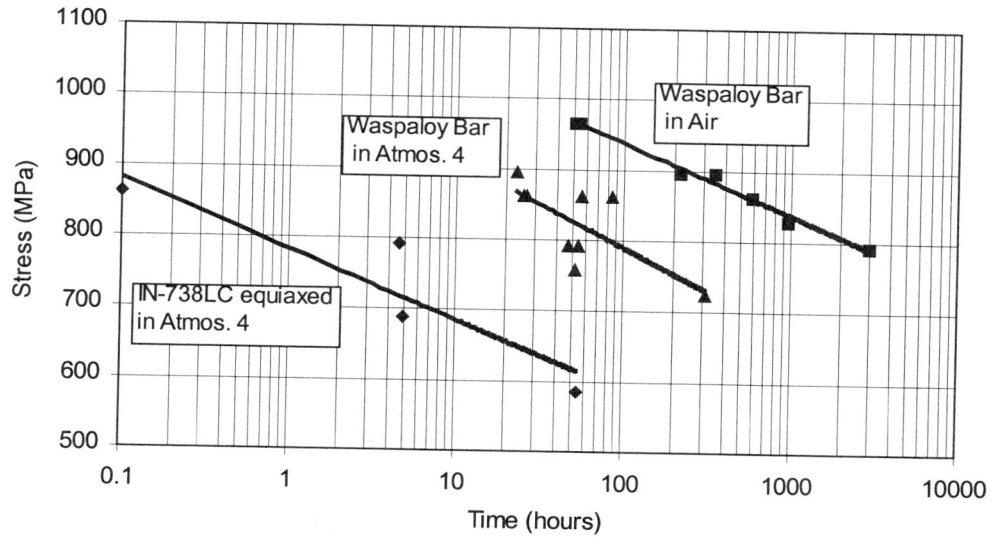

FIG.3. Semi-log plot of stress-rupture data at 600 °C (1112 °F) in air and atmosphere 4, i.e., log O_2 = -24.0 and log S_2= -8.1.

The semi-log plots produced from the rupture data revealed some interesting observations (Figs. 2 & 3). The most striking result was that the wrought alloys outperformed the cast alloys. The high performers in atmosphere 1 were fine grained wrought alloys. In atmosphere 4 these same fine grained alloys produced varied results. Rupture life data scatter was also larger in atmosphere 4.

Analysis of Waspaloy Surface Scales

Surface scales on the Waspaloy specimens were visually examined (Fig. 4). Atmosphere 1 produced gold colored bright crystalline surface scales, occasionally peppered by small gold colored nodules. Atmosphere 4 produced gold colored surface scales that were frequently covered with thick gray-green powdery nodules. The gold scale of atmosphere 1 had the color of pyrite, while the gold scale of atmosphere 4 displayed a tint of olive green. Both gold scales were crystalline and seemed epitaxial. The crystals formed in atmosphere 1 could easily be misinterpreted as intergranular fracture features (Fig. 5). The gray-green scale formed in atmosphere 4 was more porous and had the appearance of sponge coral composed of fine crystals (Fig. 6).

Energy Dispersive Spectroscopy (EDS) revealed the gold scales, gold nodules of atmosphere 1, and gray-green powdery nodules of atmosphere 4 were composed of 48-60% nickel, 27-41% sulfur, 6-12% cobalt, and 0-6% oxygen in atomic percent. As throughout this paper these percentages were determined by standardless quantitative analysis using an EDAX DX' system.

Many of the specimens were longitudinally sectioned, and a few were examined in the SEM. Longitudinal sectioning revealed the characteristic two layered scale (Fig. 7). Appearance of the inner scale immediately adjacent to the base metal was light gray, and appearance of the outer scale was gold. The gray-green nodules of atmosphere 4 not only had the same chemistry as the gold surface scales, but were revealed to be the peaks of the gold scales. Optical appearance of the gray-green nodules had apparently been effected by the fine crystal size. Further examination using backscatter SEM imaging revealed the outer scale consisted of two phases, a Ni_xS_y phase and a $(Ni,Co)_xS_y$ phase.

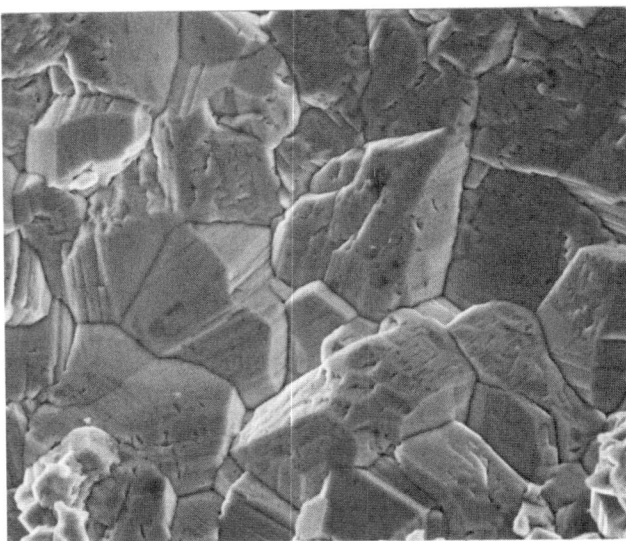

FIG. 5 SEM image of the gold colored surface scale produced in atmosphere 1. (Mag. 1,000X)

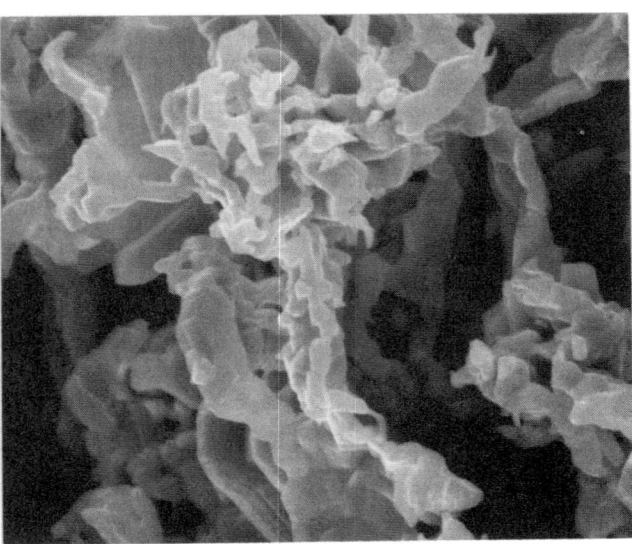

FIG. 6 SEM image of the gray-green porous scale produced in atmosphere 4. Sharp corners on the crystals confirmed that 635°C was not exceeded during the test. (Mag. 3,000X)

FIG. 4 Two Waspaloy Bar specimens after fracturing through the notch during testing. The top specimen was subjected to Atmosphere 1 for 43.6 hours. The bottom specimen was subjected to Atmosphere 4 for 26.2 hours. (Mag. 10X)

EDS of the inner scale revealed different chemistries for atmospheres 1 & 4. Atmosphere 1 inner scale consisted of approximately 22% chromium, 52% sulfur, 14 % oxygen, 5% titanium, 4% aluminum, 2% nickel, and 1% cobalt in atomic percent. Atmosphere 4 inner scale consisted of approximately 28% chromium, 18% sulfur, 33% oxygen, 5% titanium, 5% aluminum, 6% nickel, and 4% cobalt in atomic percent. The higher sulfur and lower oxygen content of atmosphere 1, did produce an inner scale with a higher sulfur : oxygen ratio.

Sulfur to oxygen ratios of the inner scale for Waspaloy were 3.7:1 in atmosphere 1, and 0.5:1 in atmosphere 4.

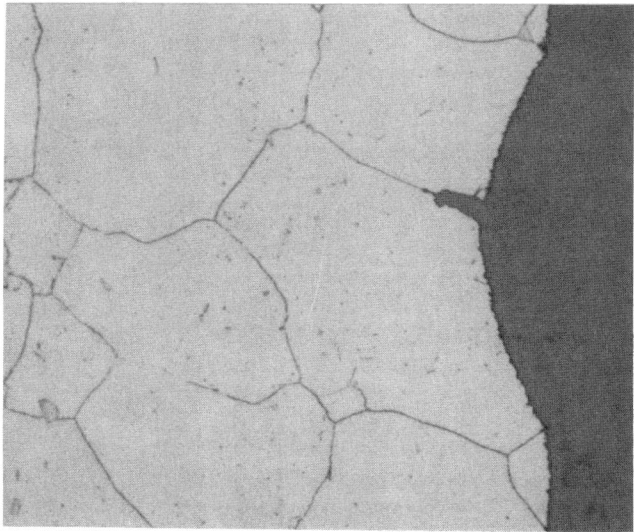

FIG. 8   Longitudinal section of a Waspaloy Bar specimen. The beginning of grain boundary attack is evident. (Mag. 500X)

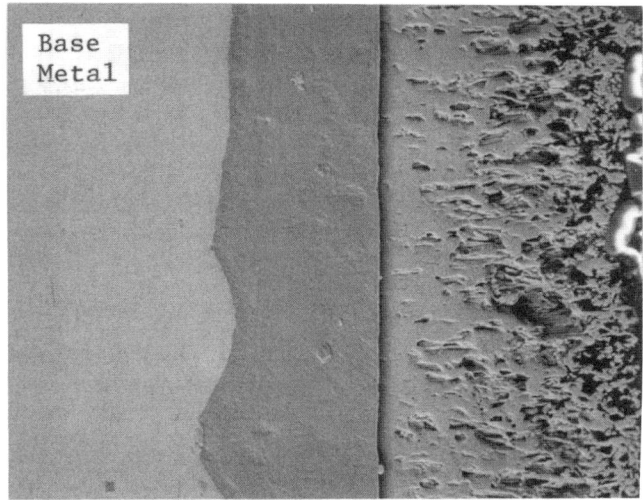

FIG. 7 Longitudinal section of a Waspaloy Bar specimen. SEM backscatter imaging reveals the outer scale as a continuous growth extending outward from the inner scale. Base metal carbides were evident in the inner scale. (Mag. 250X)

Carbides visible in the inner scale suggested that the interface between the inner scale and the outer scale was the original gauge diameter of the specimen prior to testing. The Waspaloy Bar specimen that had survived for 311.2 hours in atmosphere 4 was cross-sectioned. Diameter of the specimen prior to testing was 3.84 mm (0.151 inch). After testing, diameter of the inner scale / outer scale interface was 3.76 mm (0.148 inch), and the diameter of the base metal / inner scale interface was 3.63 mm (0.143 inch). Cross-section reduction as a result of sulfidation had resulted in a 10% increase in specimen stress during the duration of the test.

The inner scale / base metal interface was not perfectly straight. Blunt intrusions of inner scale were evident along the length of each specimen. A few of these intrusions exhibited the beginnings of a very localized attack into the grain boundaries (Fig. 8). No grain boundary attacks greater than 0.025 mm (0.001 inch) were detected, thereby suggesting that once preferential grain boundary attack began it continued very rapidly until failure.

Discussion of Waspaloy Specimens

Waspaloy Bar, the comparative standard, fractured through the notch in every sulfidation test except one. The one test that produced a smooth gauge failure was in atmosphere 4 at 724 MPa (105 ksi). Longitudinal sectioning of the Waspaloy Bar specimens confirmed the fracture mode in all cases was intergranular (Fig. 9).

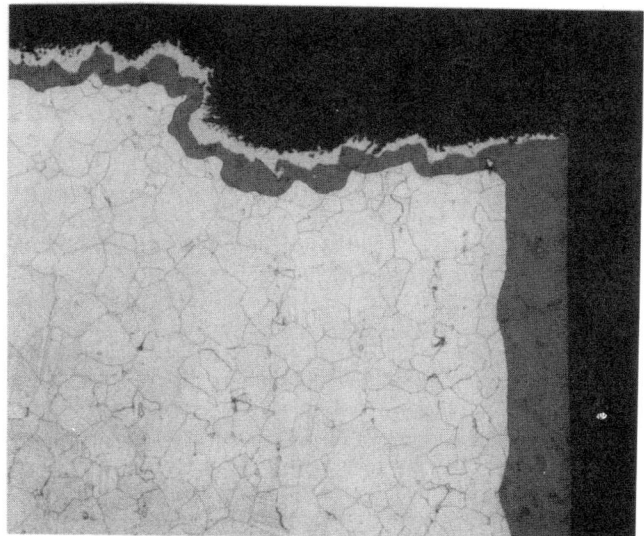

FIG. 9   Longitudinal section of a Waspaloy Bar specimen. Fracture was intergranular. Outer scale had flaked off gauge diameter. Grain size ASTM E112 No. 5. (Mag. 100X)

Waspaloy 5 and Waspaloy 6 had nearly identical chemistries within the normal acceptance limits for Waspaloy. The grain size ASTM E112 No. 4 version fractured through the notch in 2.4 hours in atmosphere 4. The grain size ASTM E112 No. 6 and 9 versions performed equal to or better than Waspaloy Bar in atmospheres 1 & 4.

Discussion of Modified Waspaloy Specimens

Chemically modified Waspaloy specimens were tested in atmosphere 1 and revealed encouraging results. These modified chemistries, and the standard chemistry of Waspaloy 11, had all been sub-solvus heat treated and were either a uniform ASTM E112 No 9 grain size or were of a necklace microstructure containing ASTM E112 No. 9 grains. Grain size in this atmosphere played a larger role in sulfidation resistance than alloy chemistry. It could be speculated that spreading out the desirable or undesirable grain boundary constituents, such as carbon or selenium, among more and more grain boundaries, resulted in grain boundaries that were less susceptible to sulfidation attack.

Fine grain size did not enhance stress-rupture life in atmosphere 4. The only benefit a fine grain size provided in this environment was a tendency toward smooth gauge rupture. The modified chemistry most detrimentally effected by atmosphere 4 was the carbon lean, high boron, Waspaloy 22. This fine grained alloy survived 300 hours in atmosphere1, but fractured after only 2 hours in atmosphere 4.

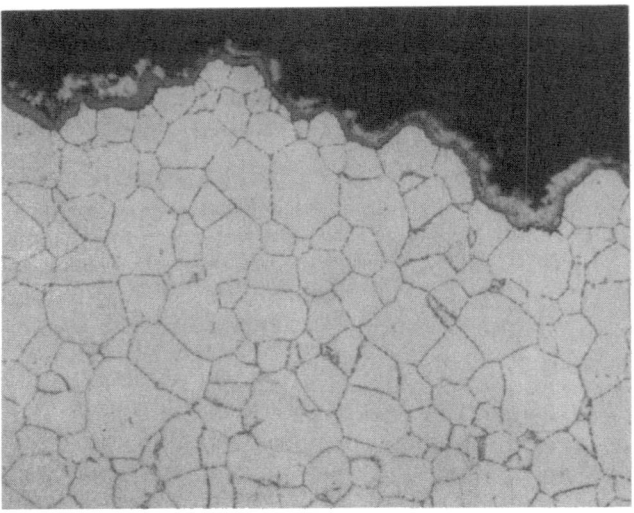

FIG. 10 Longitudinal section of a Waspaloy ER specimen. Fracture was intergranular. (Mag. 500X)

Waspaloy 41 and the two Allvac ER heats, ER1 and ER2, had similar chemistries, with the most significant difference being that Waspaloy 41 had approximately 1/3 the carbon content of the Waspaloy ER grades. This lean amount of carbon seemed to detrimentally effect stress-rupture life in atmosphere 4. The

Waspaloy ER grade was also interesting in that fracture always occurred through the smooth gauge rather than the notch, no mater what grain size was tested (Fig. 10).

Discussion of X-750 Specimens

X-750 has lower stress-rupture strength compared to Waspaloy in air, and was not expected to outperform the Waspaloy material in the sulfidation atmospheres. Stress-rupture lives of X-750 specimens in atmospheres 1 & 4 were shorter than that of the Waspaloy Bar specimens at equivalent stress. Fracture of X-750 in both test atmospheres was intergranular.

Discussion of U-720Li Specimens

Udimet 720Li is a high strength, fine grain, superalloy with lower carbon and higher boron content than commercially available Waspaloy. Westinghouse research on the corrosion behavior of Udimet 720 (higher carbon, chromium, and boron version of this alloy) in molten salt had revealed superior sulfidation resistance.[4] During this testing, however, stress-rupture life of U-720Li in atmosphere 4 seemed no better than Waspaloy Bar. Fracture of U-720Li was intergranular and originated at the notch (Fig. 11).

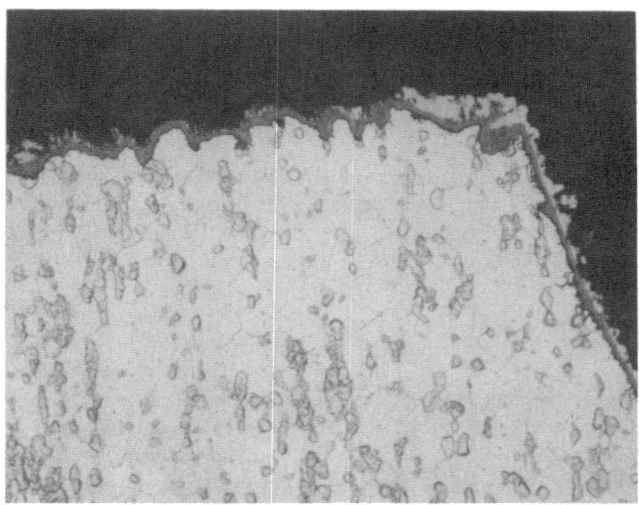

FIG. 11 Longitudinal section of a U-720Li specimen. Fracture was intergranular. (Mag. 500X)

Discussion of IN-738LC Equiax Cast Specimens

Conventionally cast IN-738LC performed exactly the same in atmosphere 1 & 4. Fracture of these specimens occurred along cast grain boundaries (Fig. 12). Fracture topography was dendritic. EDS of the outer scales revealed 42-57% nickel, 40-46% sulfur, 3-5% cobalt, 3-5% chromium, and 0-3% oxygen in atomic percent.

EDS of the inner scale formed in atmosphere 1 was approximately 15% chromium, 47% sulfur, 15% oxygen, 5% titanium, 12% aluminum, 2% nickel, 2% tungsten, and 2% tantalum. EDS of the inner scale formed in atmosphere 4 was approximately 17% chromium, 34% sulfur, 23% oxygen, 6% titanium, 10% aluminum, 5% nickel, 2% tungsten, and 2% tantalum.

Sulfur to oxygen ratios of the inner scale for IN-738LC were 3.1:1 in atmosphere 1, and 1.5:1 in atmosphere 4.

FIG. 13   Fracture surface of the NiCr DS specimen tested in atmosphere 4. Fracture initiation was flat, along what appears to be crystallographic planes, then shifted to dendritic. (Mag. 200X)

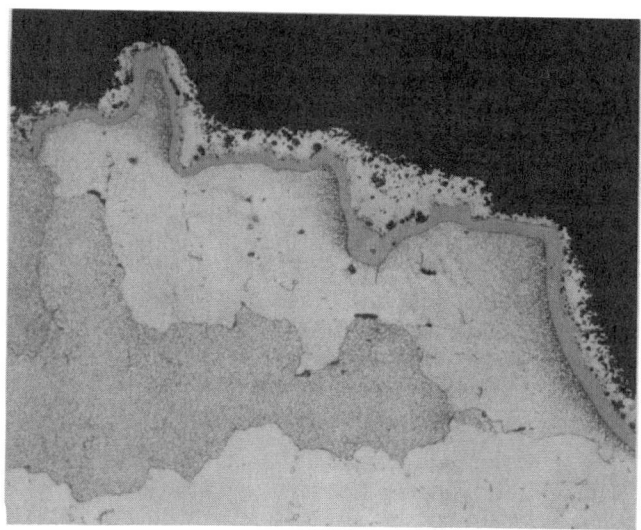

FIG. 12   Longitudinal section of an IN-738LC equiax specimen. Fracture occurred along the cast grain boundaries. (Mag. 100X)

Discussion of DS Cast Specimens

It was anticipated that significant stress-rupture life in the sulfidation atmospheres could be achieved by aligning the grain boundaries parallel to the direction of stress. Two high chromium directionally solidified alloys were selected. These alloys were IN-738LC and NiCr, a proprietary high chromium alloy. Neither alloy was HIP processed. Both alloys performed very poorly. Fracture of these specimens occurred through the notch in 2 minutes to 3 hours. The fractures initiated and progressed a short distance perpendicular to the direction of applied stress, then transitioned to dendrite arm boundaries (Figs. 13 & 14). Several origins were evident around the gauge diameter of the specimens.

To determine if these short failure times were due solely to the presence of atmospheric sulfur, a test using pure hydrogen bubbled through room temperature water was performed. This hydrogen test did not cause failure in 200 hours. It was concluded that DS alloys have a crystallographic plane extremely susceptible to sulfidation in atmospheres 1 & 4.

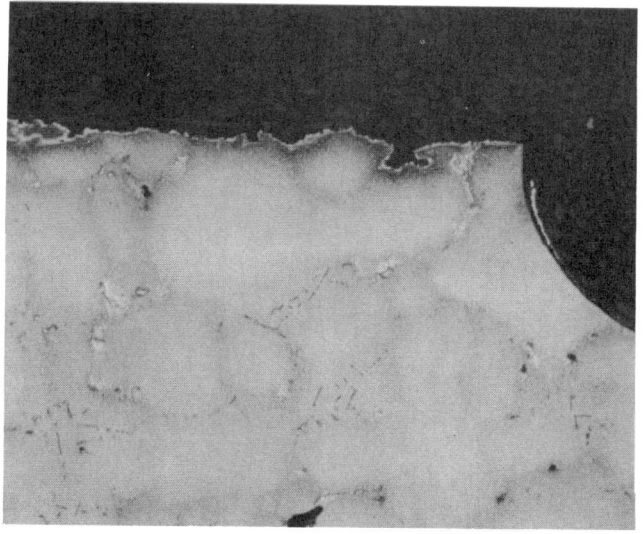

FIG. 14   Longitudinal section of the IN-738LC DS specimen tested in atmosphere 1. Fracture initiated perpendicular to the direction of applied stress, then progressed along dendrite arm boundaries. (Mag. 100X)

Conclusions

1. Nickel base precipitation strengthened superalloys can suffer significant stress-rupture life reductions at temperatures below the 635°C (1175°F) nickel sulfide eutectic. These life reductions are possible in low pO_2 environments containing significant pS_2.

2. Surface scales that formed on the test specimens in atmospheres 1 & 4 consisted predominantly of an outer scale of nickel sulfide, and a mixed inner scale of chromium sulfide / chromium oxide. The ratio of sulfur : oxygen in the inner scale of each alloy was effected by the ratio of sulfur : oxygen in the test atmosphere.

3. Stress-rupture life of the wrought alloys, Waspaloy and U-720Li, was notably longer than the cast alloys, IN-738LC and IN-738LC DS, in the sulfidation atmospheres.

4. Grain size refinement of Waspaloy significantly increased the 600°C (1112°F) stress-rupture life of specimens tested in atmosphere 1.

5. Current understanding of this sulfidation fracture mode is that it begins with the outward diffusion of nickel from the base metal to form a crystalline nickel sulfide outer surface scale. Sulfur and oxygen replace the depleted nickel to form an inner scale of chromium sulfide / chromium oxide. Shallow spikes then preferentially attack the grain or dendrite boundaries and cause rapid separation.

References

1. P. Dowson, D. Rishel, N. Bornstein, "Factors and Preventive Measures Relative to the High Temperature Corrosion of Blade / Disc Components in FCC Power Recovery Turbines, Proceedings of the Twenty-Fourth Turbomachinery Symposium, Texas A&M University, College Station, Texas, (1995), pp. 11-26.

2. F. Pettit, G. Meier, D. Seib, "Simulation Testing of Alloys and Coatings for Gas Expander Applications", Proceedings of the Third International Turbo-Expander Users' Council, (1995).

3. K. Natesan, "Effect of Oxidation / Sulfidation on Creep Behavior of Alloy 800", Conference Proceedings of the 2nd International Conference on Heat-Resistant Materials, (1995), pp. 353-360

4. G. Whitlow, C. Beck, R. Viswanathan, E. Crombie, "The Effects of a Liquid Sulfate / Chloride Environment on Superalloy Stress Rupture Properties at 1300F (704C)", Metallurgical Transactions A, Volume 15A, (Jan. 1984).

Thanks to Mr. Michiel Brongers and Mr. Arun Agrawal, CC Technologies, for performing the tests.

Special thanks to Dr. Gernant Maurer, Special Metals Company, & Dr. Fred Pettit, University of Pittsburgh, for technical assistance.

Within Dresser-Rand, the assistance of Mr. William Wilber III, Mr. David Vitale, Mr. Jiandong Shi, Mr. Steve Knight, and Mr. Dale Thibodeau is appreciated.

ISOTHERMAL AND THERMOMECHANICAL FATIGUE OF SUPERALLOY C263

Y.H. Zhang and D.M. Knowles

Rolls Royce University Technology Centre
Department of Materials Science and Metallurgy
University of Cambridge, England

Abstract

A study has been conducted to examine the behaviour of isothermal low cycle fatigue (LCF) and thermomechanical fatigue (TMF) of Ni-base superalloy C263 over a temperature regime from 25 to 950°C for several strain ranges. The macroscopic performance was correlated by detailed microstructural analysis using TEM and SEM. Initial cyclic hardening in isothermal LCF occurred at all temperatures. At room temperature, slight cyclic softening occurred following the hardening stage. This effect is more evident with increasing strain amplitude. TEM analysis revealed that this was related to dislocation shearing of γ' precipitates. Continuous cyclic hardening and considerable softening were observed respectively at 600 and 800°C. As at room temperature, planar slip was pronounced, but more slip systems operated resulting in a higher dislocation density as temperature increased. A gradual change of fracture mode from planar transgranular at room temperature to more wavy tansgranular failure at 800°C was observed. At 950°C, the load carrying capacity of C263 alloy was significantly reduced and the material was essentially cyclically stable. During LCF at this temperature, carbides precipitate in the matrix, promoted by plastic deformation, while the γ' phase was dissolved. IP and OP TMF tests suggested that considerable high temperature creep deformation occurred followed by heavy plastic deformation at low temperatures. Plastic recovery and recrystallization produced a microstructure with elongated grains and subgrains which is different from that in isothermal fatigue. In all TMF tests and isothermal LCF at 950°C, fracture was predominantly via an intergranular mechanism.

Superalloys 2000
Edited by T.M. Pollock, R.D. Kissinger, R.R. Bowman,
K.A. Green, M. McLean, S. Olson, and J.J. Schirra
TMS (The Minerals, Metals & Materials Society), 2000

Introduction

Nimonic C263 is a polycrystalline Ni-base superalloy that is solid solution strengthened by chromium, cobalt, molybdenum and $Ni_3(Ti,Al)$ γ' precipitates. It is widely used in stationary components such as combustion chamber, casing, liner, exhaust ducting and bearing housing in aeroengines. The choice of this material is based primarily on a combination of excellent fabricability, weldability and good resistance to oxidation [1]. When it is used in combustion chambers, it suffers from cyclic thermal and mechanical stresses during the start-up, steady-state and shut-down operations. Furthermore, standard laboratory isothermal testing can not include all the damage and failure processes that may occur under service conditions [2,3]. One example of such changes is the high temperature recovery of plastic deformation introduced at lower temperatures.

In recent years, there has been an increasing interest in combustor lifing. Combustor lifing differs markedly from blades and discs because the material can creep to such an extent at the high temperature end of the cycle that the residual stress at low temperatures causes severe deformation by plasticity. Therefore, the lifing for combustors represents challenges not seen for turbine aerofoil and disc materials.

In order to generate physically realistic constitutive equations for C263 combustor material, a comprehensive understanding of both macroscopic cyclic fatigue behaviour and microscopic deformation and fracture mechanisms under LCF and TMF is essential. Currently, the research in this field for this material is limited in the open literature.

Materials and Experiments

The C263 alloy was provided by Rolls-Royce plc as 21.0mm diameter bars in a standard heat treatment. This treatment involves solution at 1150°C for 2 h, quenching in water and ageing at 800°C for 8h followed by air cooling. Microstructural observation revealed that the material has a mean-linear-intercept grain size of about 104 μm, an average γ' precipitate size of about 22 nm, a small quantity of undissolved primary MC carbides and many annealing twins. Fine discontinuous $M_{23}C_6$ carbides are present along grain and twin boundaries and are coherent with the neighbouring grains. The chemical composition of the alloy is given in Table 1.

Fully-reversed (R=-1), total axial strain controlled isothermal LCF tests were carried out on 6.4mm dia. and 14.0 mm parallel gauge length specimens at temperatures in the range 25°C to 950°C. A symmetric 0.25 Hz trapezoidal cyclic waveform with 1 second ramps and holds was used. The axial strain ranges were from 0.5% to 2.0%. TMF tests were performed under total strain control on a computer controlled, servohydraulic test system with a load cell capacity of 50 kN. TMF specimens in the form of hollow cylindrical test-pieces having a gauge length of 12.5 mm, external diameter of 6.4 mm and a wall thickness of 1.0 mm were heated by an induction coil and cooled by forced air to achieve linear heating and cooling rates. In-phase (IP) and out-of-phase (OP) TMF tests were conducted with a cycle period of 130 seconds: 45 seconds for heating up and 85 seconds for cooling down. The temperature limits were T_{min}=300°C and T_{max}=950°C with a

mechanical strain range, $\Delta\varepsilon$, of 0.6% and strain ratio R=0 for IP and R=- ∞ for OP TMF tests. IP and OP are defined in Fig. 1. Strains and temperatures were monitored by contacting extensometry and optical pyrometry, respectively. The thermal expansion for the material was first recorded under zero load in order that it could be subtracted from the total strain during testing in order to impose a given mechanical strain.

Table 1. Nominal compositions (wt. %) of C263 alloy.

Ni	Co	Cr	Fe	Mo	Mn	Si	Ti	Al	C
Bal.	20	20	0.7	5.8	0.6	0.4	2.15	0.45	0.06

Both as-received material and some selected tested specimens were examined in a scanning electron microscope (SEM) and a transmission electron microscope (TEM) to identify the deformation and damage mechanisms. For isothermal LCF specimens, TEM foils were sliced along a plane vertical to the loading direction, while for TMF, the OP specimen was sectioned along the plane parallel to the loading direction. TEM foils were examined in a JEOL 2000 FX electron microscope with an operating voltage of 200 kV.

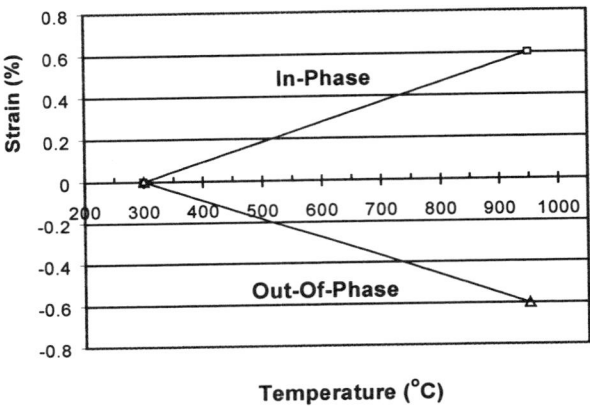

Figure 1: Schematic showing the definition of IP and OP TMF and the temperature and strain ranges applied.

Results and Discussion

Isothermal LCF

Table 2 summarises the isothermal LCF and TMF test results. Fig. 2 shows the influence of temperature on peak stress and fatigue lives when compared at a strain range of 0.75% (since tests were carried out at R=-1 and tensile and compressive stresses exhibited symmetry, only the maximum tensile stresses developed during isothermal LCF are plotted). All specimens initially exhibited cyclic hardening. At lower temperatures it took longer to reach a given saturated stress range. At room temperature, the specimen exhibited slight softening following the cyclic hardening stage. At 600°C, a consistent, moderate cyclic hardening was observed before crack initiation and propagation.

When the temperature was increased to 800°C, the initial maximum stresses were comparable with that at 600°C during hardening stage, but a significant softening then followed. At 950°C, a short period of limited hardening followed by very gradual cyclic softening can be seen with the general load carrying capacity significantly decreased. The hysteresis loops of strain against stress of the first 10 cycles were shown in Fig. 3 where it can be seen that significant creep deformation occurred near both strain extremes.

Generally, the shape and evolution of the stress/strain hysteresis LCF curves of C263 alloy are more dependent on temperature than strain range. The general features of the deformation such as cyclic hardening/softening and creep at each temperature were not affected by strain range, but the magnitudes of hardening or softening behaviour have been found to be dependent on the range, especially at room temperature as shown in Fig. 4. At this temperature, with increasing strain range, cyclic softening begins earlier and is more evident. When the strain range was increased to 2.0%, a marked cyclic hardening followed by softening can be seen. This is in agreement with the work on Nimonic 90 superalloy [4] where increasing strain range resulted in earlier cyclic softening and higher stress maximum at room temperature.

Table 2. The isothermal LCF and TMF testing conditions and results.

Specimen No	Temperature (°C)	Strain range (%)	Cycles to failure
LCF123	20	0.75	6250
LCF132	20	1.00	5090
LCF142	20	2.00	600
LCF221	600	0.75	2985
LCF232	600	1.00	1333
LCF241	600	1.50	360
LCF311	800	0.50	3680
LCF321	800	0.75	1059
LCF331	800	0.90	680
LCF411	950	0.50	800
LCF421	950	0.75	390
IP TMF	300-950	0.60	280
OP TMF	300-950	0.60	210

Thermomechanical Fatigue

Each TMF test started from the lower temperature and a shake-down or shake-up of stresses occurred. Stress-strain response was stabilised once the temperature reached and passed the maximum temperature in the first cycle. This implies that the initial strain R ratio will not play a significant role in the subsequent stress-strain behaviour in the TMF test provided little damage occurred in the first cycle. When approaching the maximum temperature, both tensile and compressive stresses dropped significantly to much less than the maximum tensile and compressive stresses experienced in the isothermal fatigue at 950°C. A typical temperature-stress hysteresis loop for each type of TMF test is shown in Fig. 5. At certain temperature ranges, especially during cooling from high temperature, a stress fluctuation was seen. This

stress discontinuity is considered to be the effect of dynamic strain ageing resulting from interaction between dislocations and diffusing alloying atoms [2]. This phenomenon was not observed during isothermal fatigue tests.

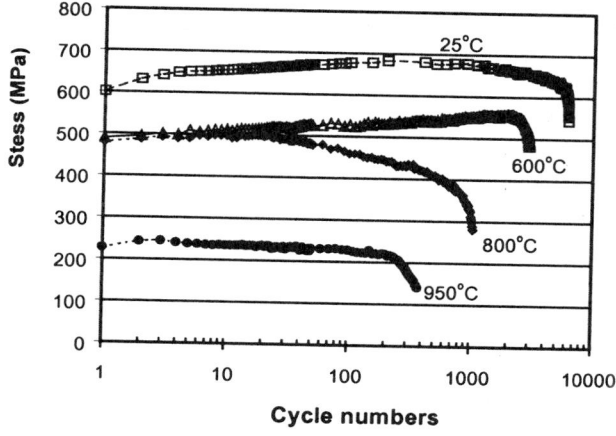

Figure 2: Development of the maximum tensile stresses with increasing cycle numbers at several temperatures, Δε=0.75%.

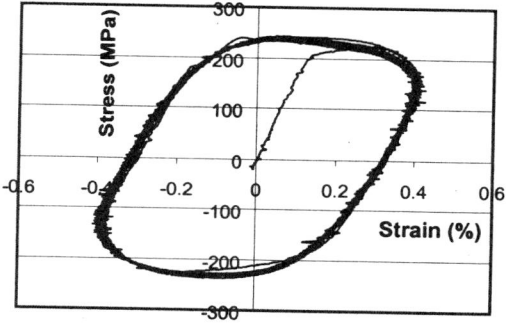

Figure 3: Cyclic hysteresis loops of strain against stress at 950°C, showing creep deformation under tensile and compressive loads.

Microstructural analysis

Isothermal LCF After fatigue failure, several specimens, covering different temperature and strain ranges, were prepared for microstructural analysis. Generally, dislocation density increases with increasing strain ranges. Stacking faults have also been observed in all specimens tested from 25°C to 800°C. In addition, both dislocation and grain boundary structure were strongly influenced by temperature.

Typical TEM microstructures obtained at different temperatures but similar strain ranges are compared in Fig. 6. At room temperature it was found that slip was predominantly planar and concentrated in local parallel slip bands which have the highest resolved shear stress. To investigate the reason for cyclic softening at room temperature, detailed TEM analysis was carried out on the γ′ morphology near the deformed region. Away from the plastic

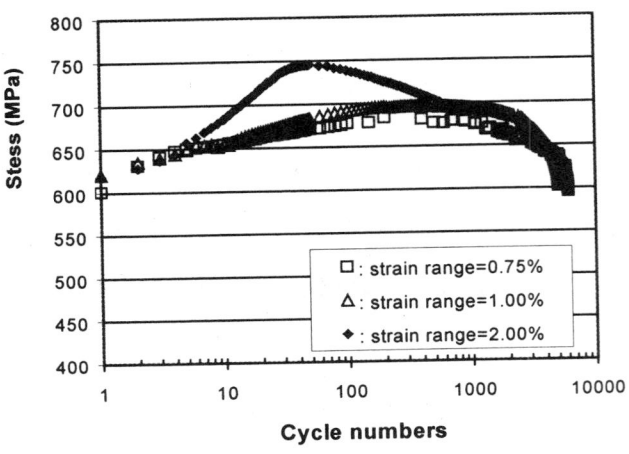

Figure 4: Comparison of maximum tensile stresses with cycle numbers at three strain ranges at room temperature. They exhibit cyclic hardening followed by cyclic softening.

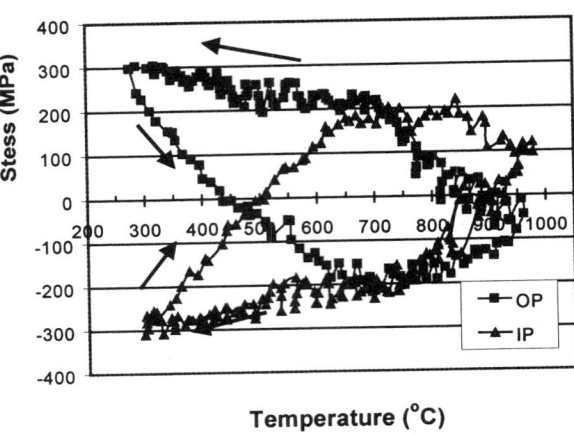

Figure 5: Typical stress-temperature hysteresis loops during IP and OP TMF (taken from the 20th cycle).

deformation area, the γ' precipitate morphology was unchanged and still exhibited a round shape; but within the slip bands, it was often found that γ' phase was sheared by dislocation movements, see Fig. 7. This might be linked with the cyclic softening at room temperature. It suggests that repetitive cutting of the fine precipitates by the motion of dislocations results in the reduction in the size of γ' and gradual cyclic softening. This phenomenon has also been observed in Nimonic alloys PE16 [5], Nimonic 80 [6] and Nimonic 90 [4] and was attributed to the shearing of γ' precipitates by dislocations during cyclic fatigue at room temperature. Cyclic stress response depends on the internal resistance to deformation of a material. This resistance comes from pinning of dislocation movements by either other dislocations (work hardening) and/or particles (precipitation hardening). When the dislocation structure reaches a saturated state, the reducing resistance to deformation is due to the decreased precipitation contribution owing to the reduced precipitation cross section. With increasing strain range, the dislocation density increases; severe dislocation interaction leading to dislocation tangles have been observed in a specimen tested at $\Delta\varepsilon=2.0\%$. With such strong cyclic softening occurring in this strain range, it was expected that clear interaction between γ' precipitates and dislocation deformation would be seen. However, the high deformation in this specimen made it impossible to see the diffraction contribution of γ' phase, and therefore not possible to view their interaction.

As observed at room temperature, deformation by planar slip of dislocations was still predominant at 600°C and 800°C and stacking faults were readily seen. However, as temperature increased, more slip systems operated leading to frequent slip band intersection and dislocation tangles. Plastic deformation became more uniform and stacking fault lengths shortened. Small amounts of $M_{23}C_6$ carbides precipitated on dislocations and were coherent with matrix. The grain boundary carbides maintained coherency with one of their neighbouring grains. Slip systems were identified as $\{111\}\frac{a}{2}<110>$, as at room temperature. It appears that the gradual cyclic hardening at 600°C was due to the

increased strength gained from more extensive dislocation interaction. Cyclic softening at 800°C is attributed to thermal activation allowing climb or cross slip process to overcome such interaction barriers.

The previous discussion on deformation mechanisms is supported by fractographic analysis. Scanning electron microscopy was used in failure analysis of fracture surfaces and cross sections of LCF tested specimens. It revealed that multi-crack initiation was the main feature in all specimens. At room temperature, cracks initiated by a transgranular localised slip process and fracture surfaces exhibited planar fracture characteristic. An example of such crack initiation and propagation sites is given in Fig. 8. When the temperature was increased to 800°C, a transgranular mode of crack initiation still persisted but a more wavy fracture surface was observed (Fig. 9), indicating a reduction in the planarity of slip. The fracture surface of the specimens tested at 600°C exhibited a mixture of the features shown at these two temperatures. It should be noted that creep cavitation at grain boundaries leading to intergranular fracture has been observed at 800°C for several stress levels [7]. The transgranular fracture mode and the lack of cavitation observed here indicate that in the present cycle waveform, fatigue played an important role in deciding the failure mechanism at 800°C.

Although C263 superalloy is designed for application at a temperature less than 800°C, it does experience temperatures as high as 950°C in service. At such temperature, it is expected that all the γ' phase is dissolved as the solvus temperature of γ' phase in C263 alloy is between 910-925°C [1]. When cooled down after LCF testing at 950°C, however, the specimen was found to contain dispersed γ' phase in the matrix. In order to determine if precipitates were present at 950°C, a piece of C263 material, which was held at 950°C for 1.5h and then quickly air cooled, has been examined under TEM. No γ' was observed after this treatment. The γ' precipitates in the isothermal LCF tested specimen at 950°C were therefore simply due to the re-precipitation of γ' phase during the slow cooling down process when the specimen failed.

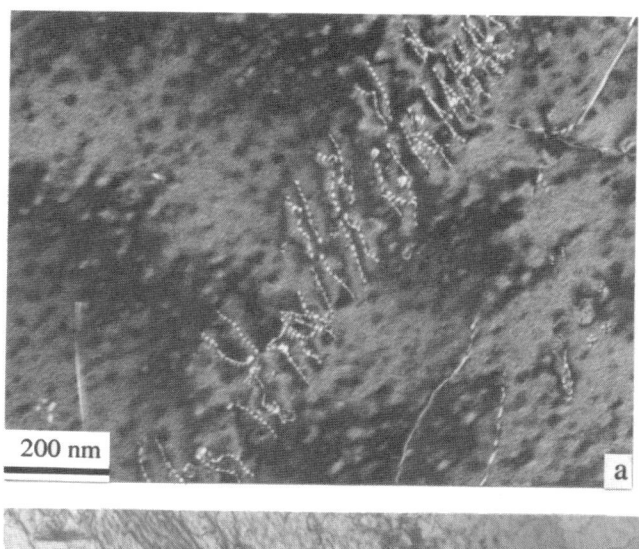

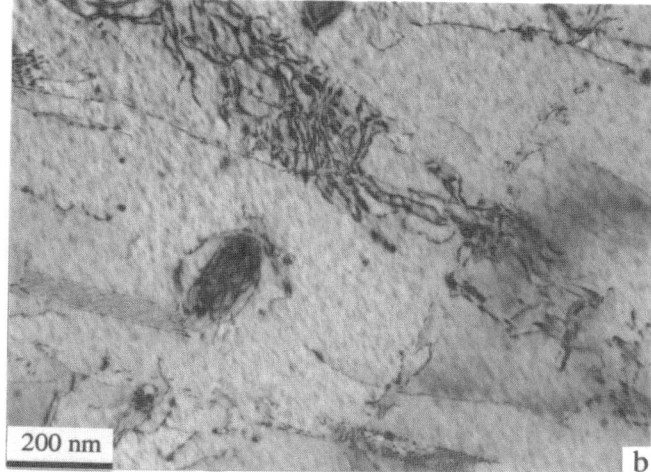

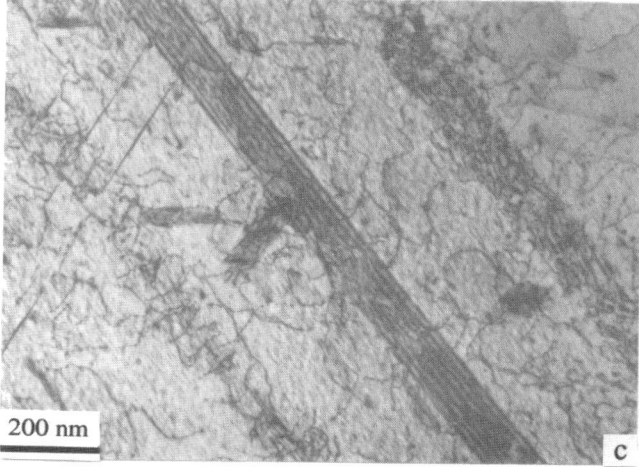

Figure 6: TEM micrographs illustrating the deformation microstructure at (a), room temperature, $\Delta\varepsilon=0.75\%$, (b), 600°C, $\Delta\varepsilon=0.75\%$, (c), 800°C, $\Delta\varepsilon=0.75\%$ and (d), 950°C, $\Delta\varepsilon=0.50\%$.

After a short period of hardening for the majority of fatigue life slight softening was observed. At this temperature, the grain boundary $M_{23}C_6$ carbides became coarse and more discontinuous. In some regions, they lost their original coherency with neighbouring grains. Uniform dislocation distribution was observed within grains. The dislocation Burgers vector was still $\frac{a}{2}<110>$, and the majority of the dislocations observed were in the {111} planes, however some were found to reside on {100} planes. Dynamic recovery led to subgrains and cells. Secondary $M_{23}C_6$ particles precipitated preferably on subgrain boundaries and slip bands. One example of $M_{23}C_6$ carbides precipitating on slip bands is shown in Fig. 10 where these carbides precipitated in a linear fashion. Trace analysis of several such sites were carried out and found that these lines with carbide precipitates lay in {111} planes. Although these carbides can act as barriers to dislocation movement, it is unlikely that they can play a significant role in increasing the strength of the material because of their small quantity. Cavity induced cracks have been found at grain boundaries but failure initiated mainly from grain boundaries

near specimen surfaces presumably because of the influence of oxidation, see Fig. 11, leading to intergranular fracture.

Thermomechanical Fatigue TEM observation of the OP TMF tested specimen in a longitudinal section revealed that the specimen has suffered from heavy plastic deformation. Elongated grain and sub-grains can be seen, Fig. 12. Heavy plastic deformation leads to recovery and recrystallization. Fig. 13 shows a grain boundary migrating towards a high density dislocation region, leaving behind original grain boundary carbides within a grain. Creep cavities were observed at grain boundaries indicative of their weakness during TMF, Fig. 14. Compared to the LCF specimen tested at 950°C, the dislocation density was higher and only a very small quantity of carbides was observed within grains. Fracture was primarily via an intergranular mode and was more prominent in IP TMF test.

Turbine components experience thermal and mechanical fatigue. Traditionally, design of these components against fatigue is based on the isothermal fatigue data at the highest temperature with the largest stress or strain. However, there is a strong likelihood that

549

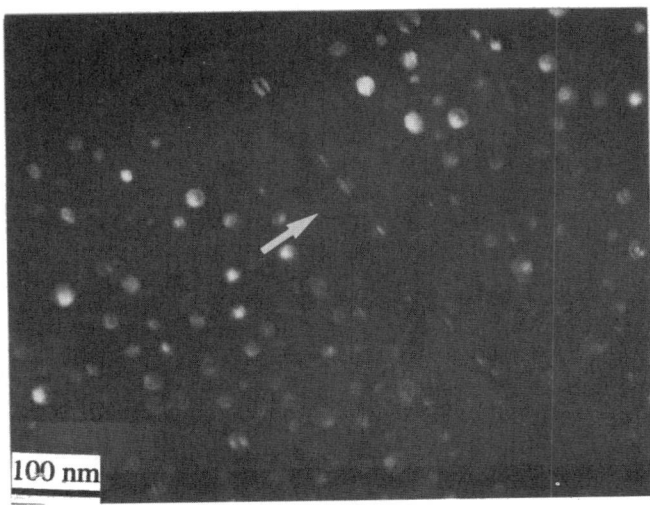

Figure 7: TEM micrograph from room temperature LCF tested specimen, showing sheared γ' precipitates in the slip band (arrowed), $\Delta\varepsilon=0.75\%$.

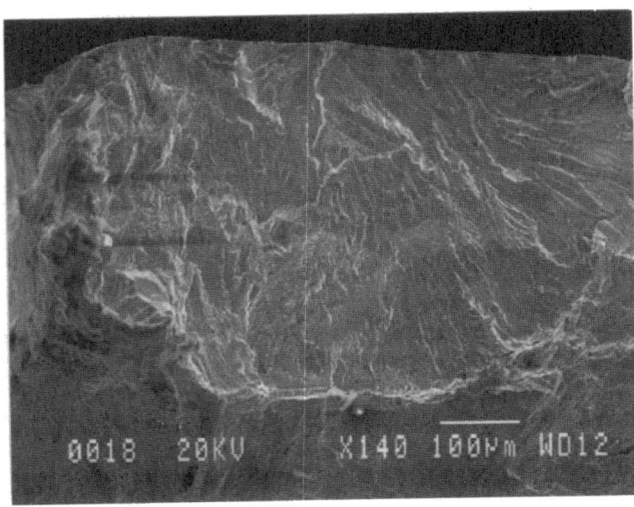

Figure 9: SEM fractograph showing quasi-cleavage fracture surface seen in isothermal LCF at 800°C, $\Delta\varepsilon=0.75\%$.

Figure 8: SEM fractograph showing trangranular cleavage crack initiation and propagation characteristics at room temperature, $\Delta\varepsilon=0.75\%$.

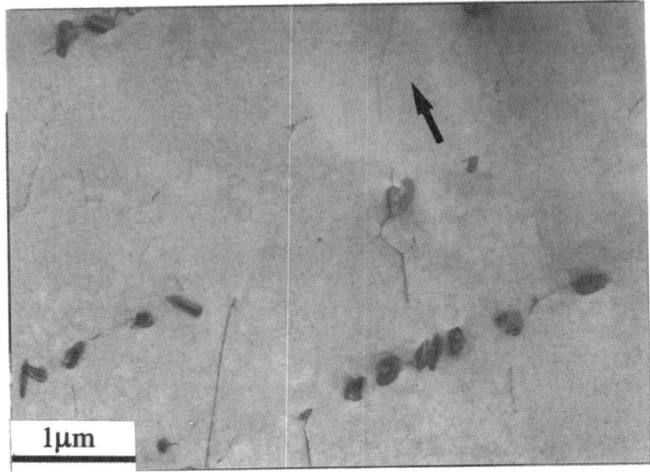

Figure 10: TEM micrograph showing $M_{23}C_6$ carbides precipitating along {111} slip bands in lines during isothermal LCF at 950°C ($\Delta\varepsilon=0.5\%$), $\mathbf{g}=111$ (arrow direction).

the deformation mechanisms are different under TMF which cast doubt over the feasibility of above methodology. The difference between isothermal fatigue and TMF both macroscopically and microscopically have been found in other combustion materials [2,3]. Present TMF tests revealed the following differences in micromechanisms in deformation from LCF at 950°C: 1. High density dislocations and recrystallization. In TMF testing, because of severe creep deformation at high temperature, large reversed plastic deformation occurred at lower temperature causing a high density of dislocations. When the specimen was heated to high temperature, those heavily deformed regions either recovered or recrystallized, leading to elongated grains or subgrains; 2. Carbide precipitation. In isothermal fatigued specimen at 950°C, many carbides precipitate within grains. They can act as barriers for dislocation movements leading to precipitation hardening although

their contribution might not be significant due to their small quantity. In the TMF specimen, little such carbide precipitation was observed. This is probably because of the shorter exposure time of the specimen to higher temperature during a TMF test.

Furthermore, solutionising of the microstructure in a TMF test might be the reason for the stress discontinuity seen in TMF tests. The phenomenon of dynamic strain ageing occurs in solid solutions where solute atoms are particularly free to diffuse through the parent lattice. It is preferable for these either interstitial or substitutional solutes to occupy sites in the vicinity of dislocations where they form Cottrell atmosphere around dislocations and hinder their movement. Since the microstructure of TMF tested specimens exhibited a solutionised structure with solute atoms dissolved in the γ matrix, it is not unexpected that dynamic strain ageing occurs during TMF. Indeed, in a similar

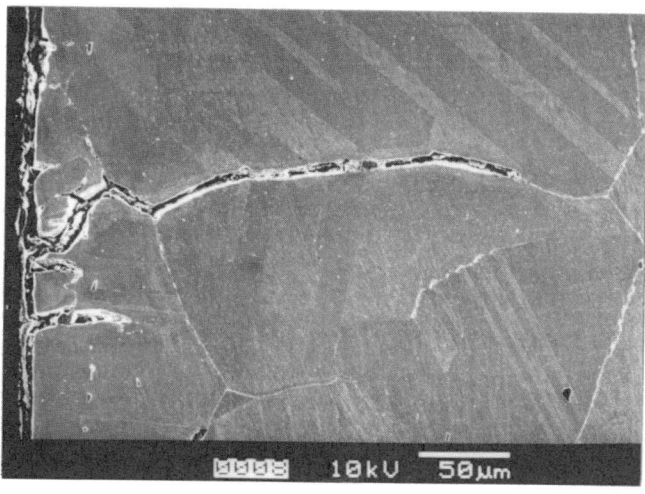

Figure 11: SEM micrograph taken from the cross section of a LCF tested at 950°C with Δε=0.5% showing intergranular cracking from specimen surface.

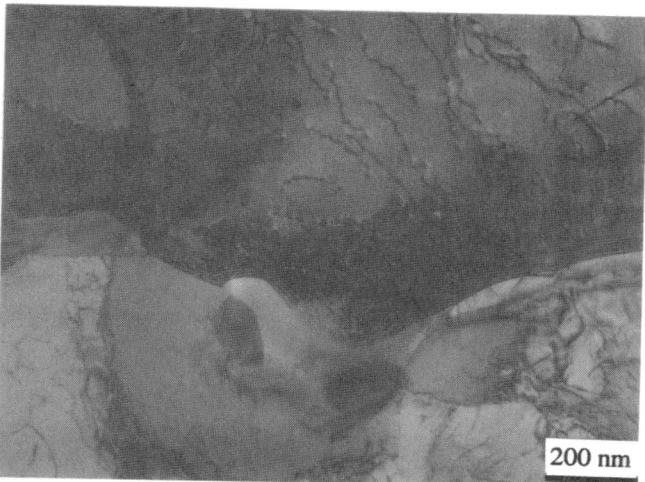

Figure 13: TEM micrograph showing shift of a grain boundary (upwards) during OP TMF. It moves towards a heavily deformed region, leaving original grain boundary carbides behind within a grain.

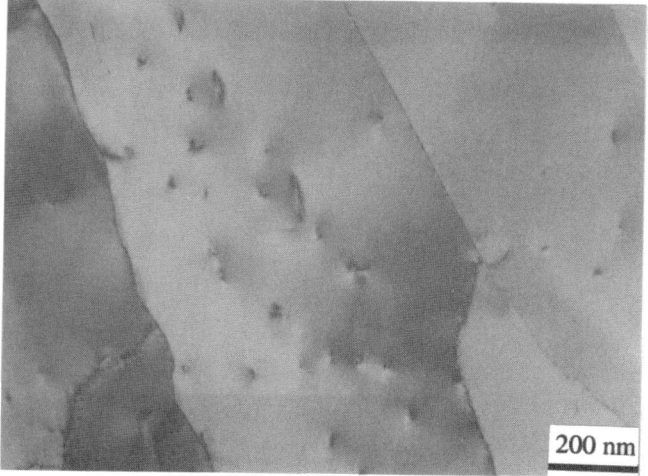

Figure 12: TEM micrograph showing elongated grain and subgrain structure developed during OP TMF.

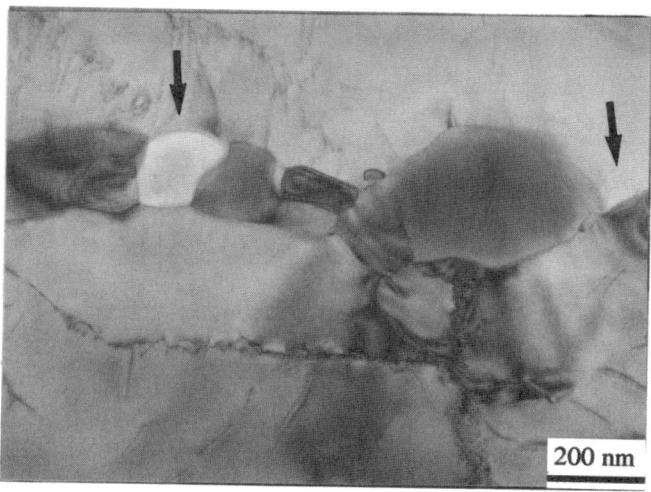

Figure 14: TEM micrograph showing cavity formation at a grain boundary during OP TMF.

superalloy, it has been found that the precondition for the appearance of dynamic strain ageing (discontinuous yielding) is that the maximum temperature must be high enough [8]. Although RF induction heating can induce a significant electrical noise leading to such stress fluctuation [9], it appeared that the stress discontinities are the true effect of dynamic strain ageing in the present study on the basis that the most severe stress discontinity occurred during cooling, the same observation as found in another material [8].

Conclusions

1, Initial cyclic hardening was observed at all tested temperatures in isothermal fatigue. It took less cycles to reach stress saturation with increasing temperature.

2, Cyclic hardening/softening depends on temperature. Hardening and softening was observed at 25°C, the levels being dependent on the strain range. At 600°C continual hardening occured. At 800°C and 950°C there was very limited hardening. Softening at 800°C was significant, but at 950°C the material was essentially cyclically stable with the stress bearing capacity significantly reduced.

3, Dislocation planar slip is predominant at temperatures up to 800°C. More slip systems operate with increasing temperature. Wavy slip occurred at 950°C and plastic recovery produced subgrains and cells.

4, TMF tests exhibit different deformation mechanisms from isothermal fatigue. High temperature creep deformation leads to heavy low temperature plasticity, which in turn results in significant recovery and recrystallization during the next cycle.

5, Fracture is mainly transgranular from room temperature up to 800°C in isothermal LCF. It became predominantly intergranular in TMF and isothermal LCF at 950°C.

Acknowledgements

The authors wish to thank EPSRC for financial support for Y.H.Z. and Rolls-Royce Plc for the provision of materials. Helpful discussions with Mr. S. Williams (Rolls-Royce) are appreciated. The provision of laboratory facilities in the Department of Materials Science and Metallurgy at the University of Cambridge by Professor A.H. Windle is acknowledged.

References

1. W. Betteridge and J. Heslop ED., The Nimonic Alloys (Edward Arnold Publishers Limited, London 1974), 18.

2. M.G. Castelli, and K.B.S. Rao, "Cyclic Deformation of Haynes 188 Superalloy Under Isothermal and Thermomechanical Loadings", Proceedings of the 8th International Symposium on Superalloys, ed. R.D. Kissinger, D.J. Deye, D.L. Anton, A.D. Cetel, M.V. Nathal, T.M. Pollock and D.A. Woodford, (Warrendale, PA: The Metallurgical Society AIME, 1996), 375-382.

3. M.G. Castelli, R.V. Miner and D.N. Robinson, "Thermomechanical Deformation Behavior of a Dynamic Strain Ageing Alloy, Hastelloy X", Thermomechanical Fatigue Behaviour of Materials, ed. H. Sehitoglu (ASTM STP 1186, Philadelphia, 1993) 106-125.

4. V.S. Sarma, M. Sundararaman and K.A. Padmanabham, Effect of Size on Room Temperature Low Cycle Fatigue Behaviour of a Nickle Base Superalloy, Materials Science and Technology, 14 (1998) 669-675.

5. M. Valsan, D. Sundararaman, S.K. Ray and S.L. Mannan, "Low Cycle Fatigue Deformation Behaviour of a Nimonic PE-16 Superalloy in a Double-Aged Microstructural Condition", Trans. Indian. Inst. Met., 49 (1996) 471-477.

6. B.A. Lerch and V. Gerold, "Room Temperature Deformation Mechanisms in Nimonic 80A", Metall. Trans., 18A (1987) 2135-2141.

7. Y.H. Zhang and D.M. Knowles, "Micromechanism of Creep Deformation of C263 Superalloy" (Interim Report 2/AT29, UTC, Department of Materials Science and Metallugy, Cambridge University, Jan. 2000).

8. B. Kleinpass, K.H. Lang, D. Lohe and E. Macherauch, "Thermal-Mechanical Fatigue Behabiour of NiCr22Co12Mo9", Fatigue Under Thermal and Mechanical Loading, ed. J. Bressers, L. Remy and M Steen and J.L. Valles (Kluwer Academic Publishers, The Netherlands, 1996) 327-337.

9. M.G. Castelli and J. Ellis, "Thermomechanical Fatigue Behaviour of Materials", ed. H. Sehitoglu (ASTM STP 1186, American Society for Testing and Materials, Philadelphia, 1993), 195-211.

STRUCTURE/PROPERTY INTERACTIONS IN A LONG RANGE ORDER STRENGTHENED SUPERALLOY

M. F. Rothman[1], D. L. Klarstrom[1], M. Dollar[2], and J. F. Radavich[3]

[1] Haynes International, Inc., P. O. Box 9013, Kokomo, IN 46904-9013
[2] Illinois Institute of Technology, 10 W. 32nd St., EL243, Chicago, IL 60616-3793
[3] Micro-Met Laboratories, 209 North St., West Lafayette, IN 47906

Abstract

An investigation of the long range ordering and secondary phase reactions in HAYNES® 242™ alloy, and their relationships to the mechanical properties of this age-hardenable material, has been conducted. The structure and property responses of the material to various intermediate temperature heat treatments interposed between a solution treatment and final aging treatment have been evaluated for a range of commercial heat chemistries utilizing elevated temperature tensile testing, as well as metallographic, TEM, SEM and extraction/x-ray diffraction techniques. Results of this work indicate that these interposed intermediate temperature heat treatments have major effects upon nature of the ordered domains formed in the material during final aging, the nature and morphology of secondary phases found, and the final mechanical properties exhibited by the alloy. Results also indicate that small variations in heat chemistry can exert a measureable influence upon both structure and properties.

Introduction

Utilization of long range order as a primary mechanism for providing enhanced elevated temperature strength in commercial wrought nickel-chromium-molybdenum alloys is a relatively recent development [1]. The development of long range ordered domains in these alloys after long exposure at intermediate temperatures, with an attendant increase in elevated temperature strength, has been a well known phenomenon for some time [2]. It is only in the last 10 years, however, that compositions in the nickel-chromium-molybdenum system have been identified which allow the strengthening associated with the formation of these long range ordered domains to be achieved with a commercially viable aging time.

HAYNES 242 alloy is one such material. Containing nominally 25% molybdenum and 8% chromium by weight, with nickel as a base, this alloy exhibits up to a doubling of its yield strength over the annealed condition with aging at 650°C (1200°F) for from 24 to 72 hours. This strengthening is accompanied by a modest reduction in ductility, as might be expected. The increase in strength is attributable to the observed rapid formation of very small long range ordered domains with an $Ni_2(Mo,Cr)$ stoichiometry, very similar to those which have previously been observed to form in nominally Ni-16%Cr-16%Mo type alloys following hundreds of hours of aging [2]. The domains exhibit a lenticular shape, similar to that of γ" particles in 718 alloy,

HAYNES is a registered trademark, and 242 is a trademark, of Haynes International, Inc.

with a typical initial long axis dimension of about 10-20 nm. Previous work [3] has shown that these ordered domains tend to coarsen slowly with long-term exposure at temperatures above about 540°C (1000°F), with the coarsening rate increasing as exposure temperature increases. Long range order (and related strengthening) appears to be lost above about 760°C (1400°F) [4], although domains have been observed to persist at or near these higher temperatures for a short exposure time [3].

Since its commercial introduction in the early 1990's, 242 alloy has been selected for a variety of gas turbine engine component applications, many of which are made from forged or rolled rings. From the onset of production-scale reforge billet and ring manufacturing, the alloy's properties have exhibited marked sensitivity to a seemingly small degree of normal compositional and thermo-mechanical processing variability. This has been particularly the case for elevated temperature tensile properties, for which extremes of high strength/ low ductility and high ductility/ low strength have been observed.

In the present work, efforts have been made to comprehensively and painstakingly characterize the nature of the long range ordering and secondary phase reactions in the alloy in an attempt to provide the fundamental understanding needed to remediate this problem. Our purpose was to examine the influence of minor composition variation upon those reactions, as well as the effect of short duration thermal exposure at temperatures at or above 760°C (1400°F), and tie the structural observations back directly to the observed tensile property effects.

Materials and Experimental Details

For these studies, reforge billets from 242 alloy commercial heats were selected with major element compositions (molybdenum and chromium) representing the high ("rich") and low ("lean") sides of the normal variation in production. Chemical analyses for study heats are given in Table 1. It should be noted that although the differences in molybdenum and chromium content among these heats might be considered minor in the context of other alloys, it was anticipated that this magnitude of variation in 242 alloy could have a significant effect upon the ordering reaction. Most of the work being presented in this paper was performed for Heat A and Heat D. Only selected illustrative structures from Heat B and Heat C are included.

Full cross section, 51-mm (2-inch) thick samples were taken from each billet. These were hot rolled in the laboratory in the billet transverse direction, in several sessions between 1120°C and 1205°C (2050°F and 2200°F), to yield 9.5-mm (0.375-inch) thick

Superalloys 2000
Edited by T.M. Pollock, R.D. Kissinger, R.R. Bowman,
K.A. Green, M. McLean, S. Olson, and J.J. Schirra
TMS (The Minerals, Metals & Materials Society), 2000

TABLE I Chemical Composition of Study Heats
(Weight %)

Heat	Ni	Mo	Cr	Fe	Mn	Al	W	Si	C	B
A	Bal.	24.84	8.07	1.05	0.32	0.28	0.12	0.07	0.003	0.003
B	Bal.	24.73	8.27	1.08	0.36	0.23	0.11	0.11	0.006	0.002
C	Bal.	25.10	8.66	1.32	0.38	0.20	0.15	0.10	0.005	0.002
D	Bal.	25.24	8.45	1.11	0.33	0.24	0.09	0.13	0.003	0.003

Note: Other analyzed elements found to be present in substantially similar trace amounts among the four heats.

plates. These plates were finish annealed at 1175°C (2150°F) for 30 minutes, and then water quenched. While the commercial annealing temperature range for 242 alloy is normally 1065 to 1095°C (1950 to 2000°F), this higher annealing temperature was selected in order to insure that a consistent, fully recrystallized grain structure was obtained for all heats, and that any and all secondary phase precipitates would be fully dissolved. In fact, grain sizes obtained for all of the laboratory rolled plates were in the range of ASTM 2½ - 3½, and as-annealed microstructures were observed to be essentially devoid of significant second phase precipitation upon the grain boundaries.

In order to assess the structure and property effects associated with thermal exposure at temperatures below 1175°C (2150°F), samples of the annealed plates were exposed for both 30 minute and 4 hour durations at intermediate temperatures between 760°C and 1040°C (1400°F and 1900°F) in 55°C (100°F) increments, and between 1040°C and 1150°C (1900°F and 2100°F) in 28°C (50°F) increments, followed by a water quench. A sample with no intermediate thermal exposure was included as a control.

Following the various intermediate heat treatments (IHT's), all samples were given a final aging treatment of 72 hours at 650°C (1200°F), and air cooled. The commercial age hardening heat treatment for 242 alloy is normally 24 to 48 hours at this temperature. Aging for 72 hours was selected in order to at least partially offset the effects of the coarser than normal grain size, and result in a tensile yield strength level somewhat on the same order as that for commercially produced material.

Most of the initial evaluation of the overall matrix of samples produced was centered upon the materials given a 4 hour IHT, with selective work done for the 30 minute exposed samples. Tensile testing was performed in duplicate at 650°C (1200°F), which is the Q/C test condition for commercial material. Samples were oriented in the direction of plate rolling. Corresponding specimens were subjected to detailed examination using light metallography, TEM and SEM. Selected materials were subjected to phase extraction and x-ray analysis.

Tensile Test Results

Four Hour IHT's

Results of 650°C (1200°F) tensile tests (average of duplicates) for samples given the 4 hour IHT's are plotted in Figures 1-4. *Strength and ductility values shown for the 650°C (1200°F) intermediate heat treatment temperature are actually the results*

obtained for control samples not given an intermediate heat treatment, but just given the 72 hour final aging treatment at 650°C (1200°F).

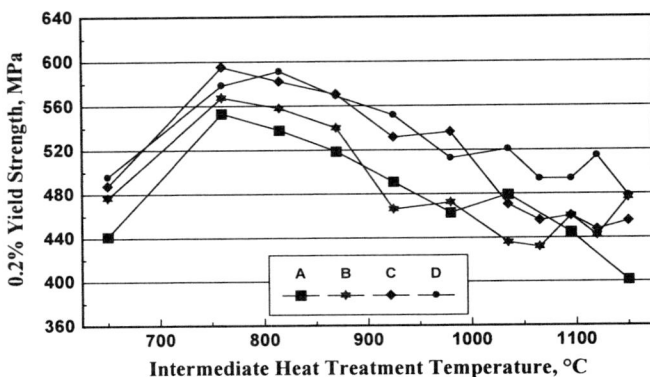

Figure 1: Yield Strength at 650°C (1200°F) for Heats A, B, C and D as a function of 4-hour IHT temperature.

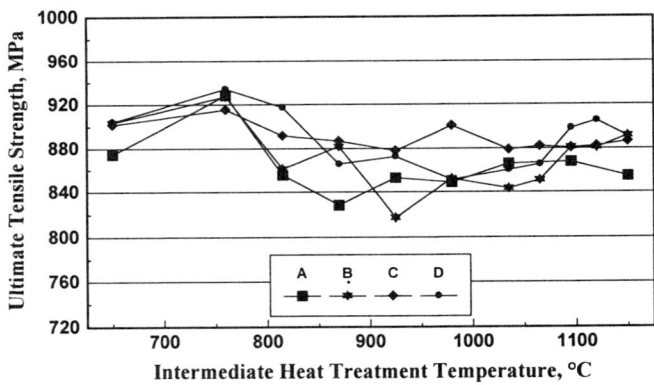

Figure 2: Ultimate Tensile Strength at 650°C (1200°F) for Heats A, B, C and D as a function of 4-hour IHT temperature.

Yield strength results for the control samples with no IHT ranged from an average of 440 MPa (64 Ksi) for Heat A ("lean") to 495 MPa (72 Ksi) for Heat D ("rich"), with Heats B and C in between. Ultimate Tensile Strength results exhibited a similar pattern. Tensile ductilities ranged from 48% elongation and 57% RA for Heat A, to 38% elongation and 47% RA for Heat C ("rich"), with Heats B and D in between. Consistency of the duplicate test

554

results was very good in all cases. As anticipated, this data for the no-IHT control tests indicates that minor composition variations do exert an influence upon the tensile properties of the alloy.

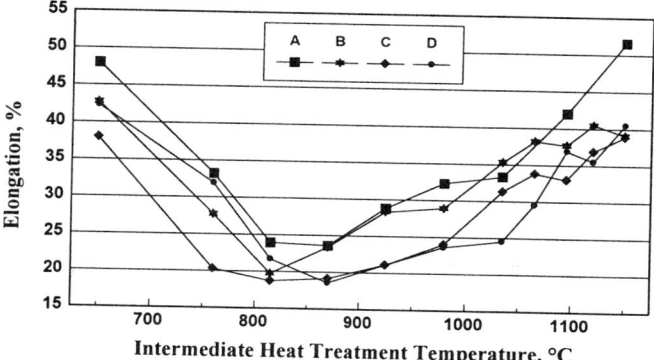

Figure 3: Tensile Elongation at 650°C (1200°F) for Heats A, B, C and D as a function of 4-hour IHT temperature.

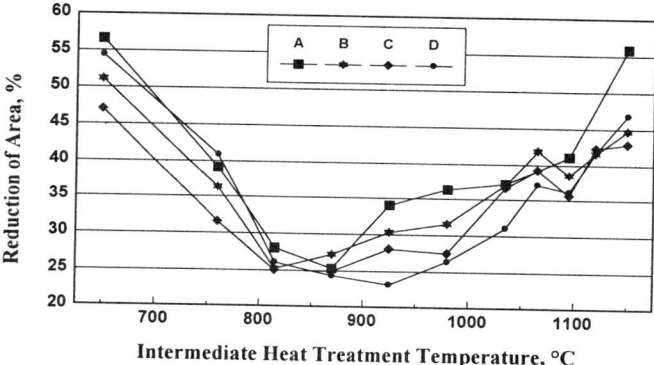

Figure 4: Tensile Reduction of Area at 650°C (1200°F) for Heats A, B, C and D as a function of 4-hour IHT temperature.

As mentioned earlier, it has been reported [4] that the disordering temperature for the Ni₂(Mo,Cr) long range ordered domains in 242 alloy is in close proximity to 760°C (1400°F). Kumar and Vasudevan [3] found that, for exposures up to about 10 hours at 750°C (1382°F), hardening associated with relatively large ordered domain formation was still observed, although longer exposures at this temperature resulted in some coarsening of domains ... although ... volume fraction, and a loss of hardening ... Accordingly, some ... at 760°C (1400°F) must ... as anticipated. On the other hand ... expected that an IHT at 815°C (1500°F) or above would ... exhibit such an effect.

Results presented in Figure 5, however, indicate an increase in 650°C (1200°F) yield strength of about 100 MPa (60 Ksi) was obtained with the imposition of IHT's in the range of 760-815°C (1400-1500°F) for four heats. Moreover, as IHT temperatures were increased to about 815°C (1500°F), increased yield strength continued to be observed, although a gradually decreasing extent, until IHT temperature reached about 900-1035°C (1800-1900°F).

Domains are not forming at 760°C contrary to paper

As may be seen in Figure 2, Ultimate Tensile Strength did not exhibit similar behavior. Except for a slight increase observed for four hour IHT at 760°C (1400°F), test results for samples given an IHT were generally the same *or lower* in value as those for the no-IHT control tests. This behavior may, in part, reflect the tensile ductility effects brought on by the imposition of the IHT's, which are illustrated in Figures 3 and 4.

As may be seen from the plots of both Elongation and RA, signifi-cant loss of tensile ductility resulted from the imposition of IHT's as compared to the no-IHT control samples. The magnitude of this ductility drop was generally worst for IHT at 815-870°C (1500-1600°F), and overall ductlity levels even on the order of the control samples were only observed for IHT's at or above about 1095°C (2000°F). Similar behavior was exhibited among the four study heats, although Heat A results could be judged to be somewhat better than those for the other heats.

Thirty Minute IHT's

Results of 650°C (1200°F) tensile tests for samples given the 30 minute IHT's are plotted in Figures 5-8. Again, strength and ductility values shown for the 650°C (1200°F) IHT temperature are actually the results obtained for control samples just given the 72 hour final aging treatment at 650°C (1200°F).

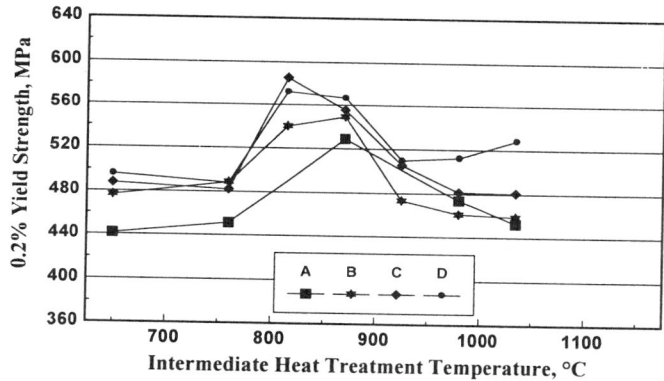

Figure 5: Yield Strength at 650°C (1200°F) for Heats A, B, C and D as a function of 30-minute IHT temperature.

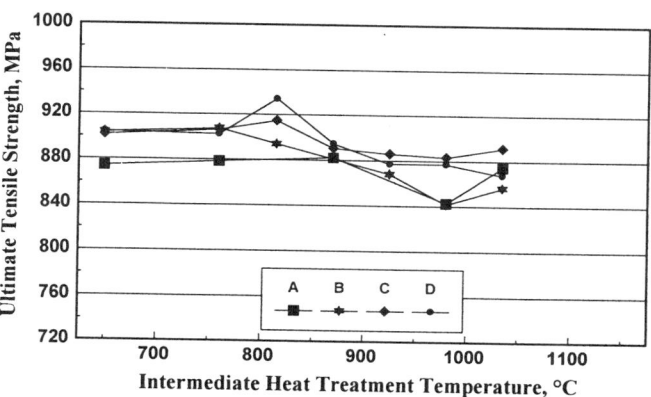

Figure 6: Ultimate Tensile Strength at 650°C (1200°F) for Heats A, B, C and D as a function of 30-minute IHT temperature.

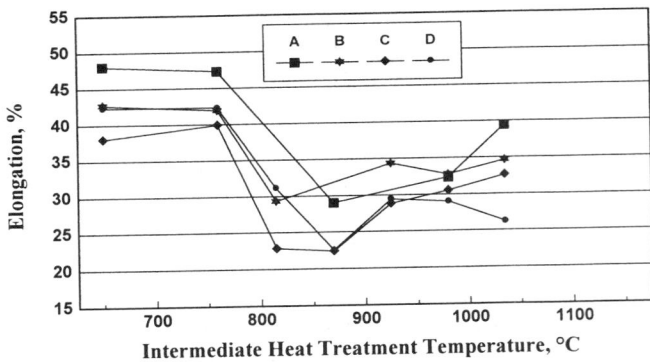

Figure 7: Tensile Elongation at 650°C (1200°F) for Heats A, B, C and D as a function of 30-minute IHT temperature.

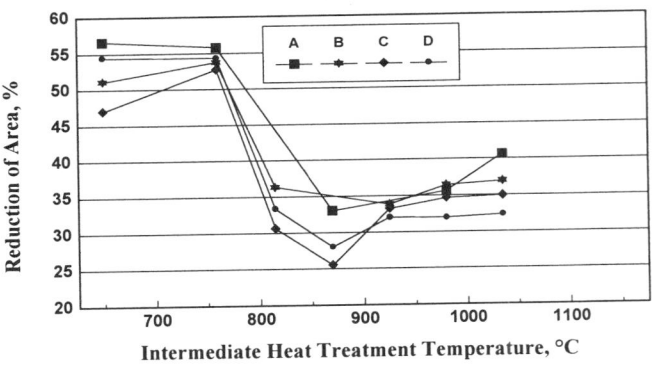

Figure 8: Tensile Reduction of Area at 650°C (1200°F) for Heats A, B, C and D as a function of 30-minute IHT temperature.

While the highest 30-minute IHT temperature tensile tested was only 1035°C (1900°F), the results, as given in Figures 5-8, were very similar to those obtained for the 4-hour IHT's, with a few exceptions. First, and most importantly, the 760°C (1400°F) 30-minute IHT, unlike the corresponding 4-hour IHT, produced no significant change in properties compared to the no-IHT control test results. Secondly, the ductility losses exhibited for the 30-minute IHT's in the 815-870°C (1500-1600°F) range appeared somewhat less severe than those for the 4-hour IHT's. Lastly, the ultimate tensile strength fall off in the 815-870°C (1500- 1600°F) range also appeared less pronounced for the shorter IHT.

Metallography, SEM, Extraction & X-Ray Analysis

Light Metallography

Light metallography was performed for all four of the heats in all of the finished heat treatment condition iterations utilized in this study. In all cases, structures consisted of fully recrystallized equiaxed grains, in the range of ASTM 2½ - 3½ in size, with isolated patches of what appeared to be groups of large, blocky primary carbides spread throughout. In some cases, evidence of decoration of prior grain boundaries with small particles was found. The condition of newly recrystallized grain boundaries was found to vary from essentially clean to heavily and continuously precipitated, apparently dependent upon both the heat treatment imposed and the heat of material in question. A visual

summary of the grain boundary conditions observed as a function of heat and heat treatment imposed is presented in Figure 9.

IHT	Heat A	Heat B	Heat C	Heat D
None				
760°C/4-hours				
815°C/4-hours				
870°C/4-hours				
925°C/4-hours				
980°C/4-hours				
1035°C/4-hours				
1065°C/4-hours				
1095°C/4-hours				
1120°C/4-hours				
1150°C/4-hours				
760°C/30-minutes				
815°C/30-minutes				
870°C/30-minutes				
925°C/30-minutes				
980°C/30-minutes				
1035°C/30-minutes				
1065°C/30-minutes				
1095°C/30-minutes				
1120°C/30-minutes				
1150°C/30-minutes				

Clean/Light Discrete   Lt.-Mod. Semi./Cont.
Moderate Discrete   Mod. Semi./Cont.
Light Semi./Cont.   Heavy Semi./Cont.

Figure 9: Grain boundary precipitate conditions found for Heats A, B, C, and D, as a function of IHT, using 500X microscopy.

Heavy Semi Continuous

[...]ed in Figure 9 [...] quantitative [...] the discrete, [...] ain boundary [...] ven at 500X. [...] al key points.

[...] in boundaries [...])-IHT control [...] 1400°F) also [...] boundaries. [...] 00°F) for four [...] s, resulted in [...] t the grain [...] the possible [...] heaviest for [...] ontinuous to

The final key point revealed by Figure 9 is that, for the higher temperature IHT samples, a clear difference in behavior was observed for the four study heats. Heats A and B, the "lean" materials, exhibited grain boundaries that were clean or had light, discrete precipitation in samples given IHT at 1035 or 1065°C (1900 or 1950°F) or higher. On the otherhand, Heats C and D, the "rich" materials, only exhibited such behavior for IHT's at 1095 or 1120°C (2000 or 2050°F) or higher. This indicates that the secondary precipitation reaction(s) in 242 alloy are indeed influenced by these compositional variations.

Scanning Electron Microscopy

Comprehensive SEM evaluation of no-IHT control samples and samples given 4-hour IHT's was performed. Samples were prepared by electropolishing in a 20% H_2SO_4 - methanol solution, and then electroetching in a 10% HCl - methanol solution. Some selective SEM work was done for 30-minute IHT samples from Heats A and D as well.

Evaluation of grain boundary structures by SEM yielded results which are in reasonably good qualitative agreement with the observations made for light microscopy. In the main, the no-IHT control samples had clean or light discrete second phase precipitated grain boundaries, although some continuous film-like structure was observed near the isolated large primary carbide clusters.

Samples given 4-hour IHT's at 760 to 870°C (1400 to 1600°F) tended to exhibit significant amounts of continuous secondary precipitation at the grain boundaries in all heats, apparently consisting of two separate phases . For 4-hour IHT between 925

and 980°C (1700 and 1800°F), somewhat more semi-continuous precipitate morphology was found, particularly for the "lean" heats A and B. The difference in this behavior is well illustrated by the structures shown for Heats A and D in Figure 10.

For 4-hour IHT's at 1035°C (1900°F) and above, patches of the continuous and semi-continuous precipitate structure were observed less and less frequently, apparently replaced with a significantly lower density of discrete particles of apparently one phase. For 4-hour IHT's at 1095°C (2000°F) and above, only discrete particles were observed.

Behavior consistent with the above was observed for the limited amount of work performed upon the 30-minute IHT samples from Heats A and D, with one exception. For the 30-minute IHT at 760°C (1400°F), no evidence of continuous or semi-continuous second phase precipitation at the grain boundaries was found (again, away from the isolated primary carbide clusters). In general, for all samples examined, IHT at about 815°C (1500°F) produced the heaviest precipitation at the grain boundaries, with that in Heats C & D somewhat heavier than that in Heats A & B.

Extraction and X-Ray Analysis

A limited amount of phase extraction work was performed in this study, using a 5% H_3PO_4 plus 10% HCl - methanol extraction

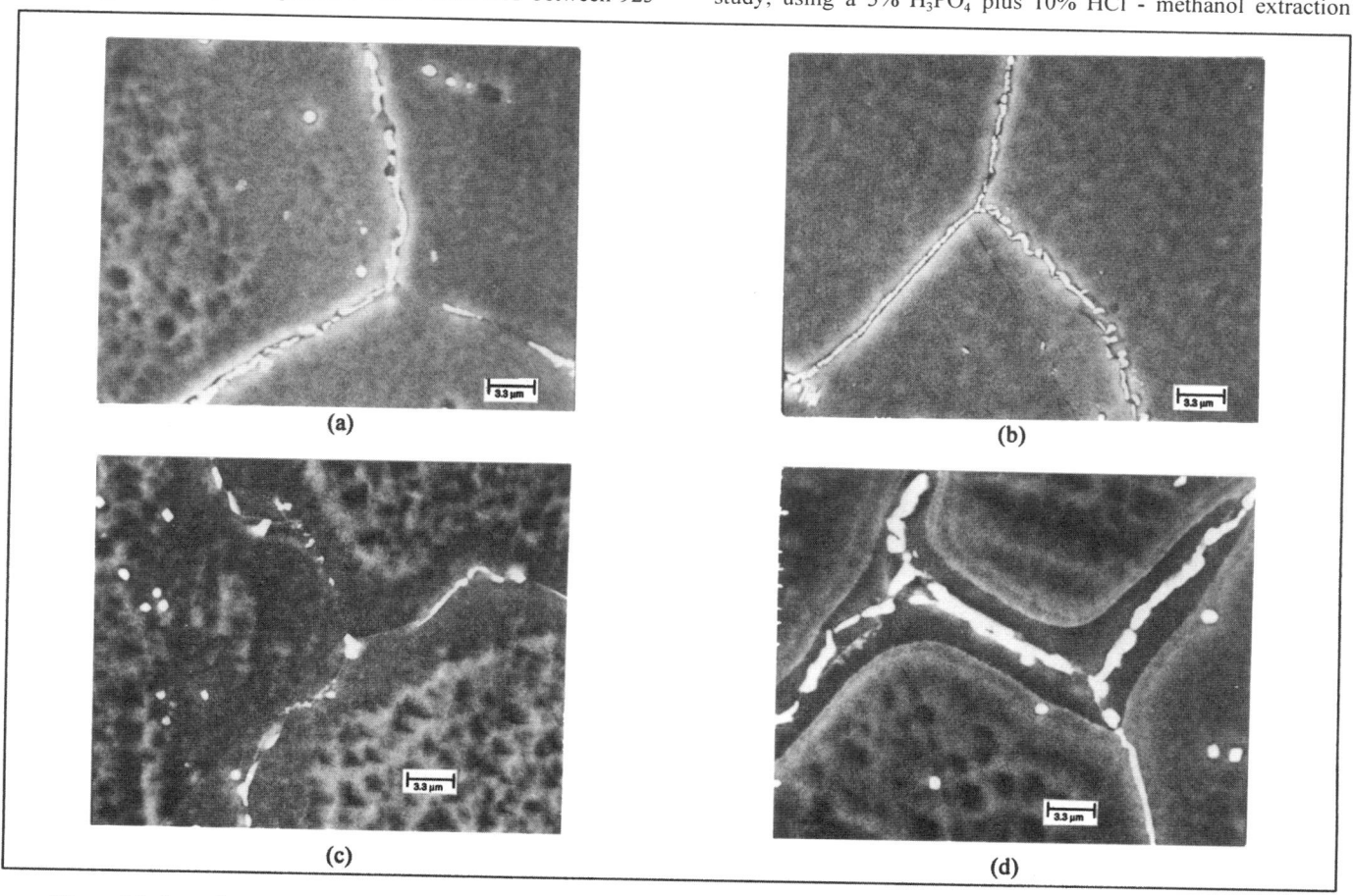

(a)

(b)

(c)

(d)

Figure 10: Scanning electron micrographs for samples of Heats A and D: (a) Heat A, 815°C (1500°F)/4-Hour IHT; (b) Heat D, 815°C (1500°F)/4-Hour IHT; (c) Heat A, 980°C (1800°F)/4-Hour IHT; (d) Heat D, 980°C (1800°F)/4-Hour IHT.

media. Film portions of the extracted residues were concentrated by ultrasonically removing the residue every five minutes and decanting the floating films into a separate sample. This residue was used for x-ray diffraction studies, while small portions were used for SEM studies.

Samples used in the extraction were rinsed in alcohol, and re-examined in the SEM. This type of sample provided a 3-D picture of the true morphology of the grain boundary films, and confirmed that the films were being extracted, not preferentially dissolved. An example is shown in Figure 11 for Heat C given an IHT of 870°C (1600°F) for four hours. EDS analysis of this extracted grain boundary film indicated it to be molybdenum-rich, with lesser amounts of chromium and nickel present.

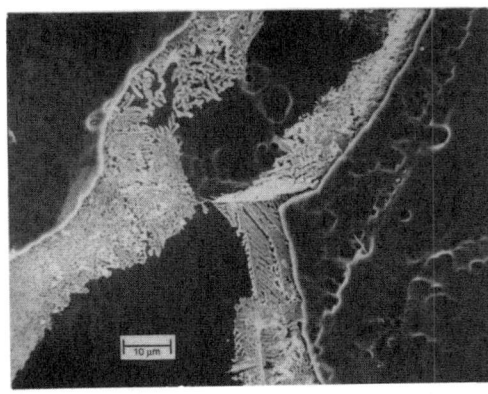

Figure 11: In-situ SEM view of the grain boundary films extracted from a sample of Heat C given a 4-hour IHT at 870°C (1600°F).

By way of contrast, a view of the extracted residue taken from the no-IHT control sample of Heat C, shown in Figure 12, appears to indicate that the residue was comprised mostly of large and small discrete particles, with little evidence of film structures.

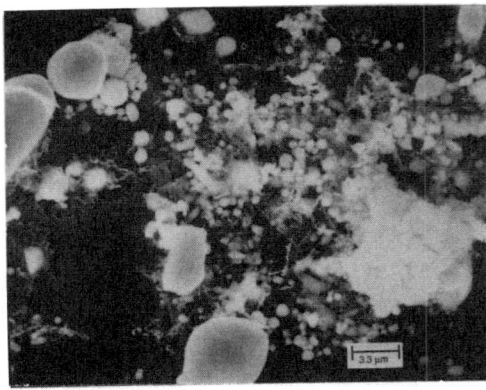

Figure 12: SEM view of the residue extracted from the no-IHT control sample of Heat C.

X-ray diffraction analysis performed upon the residues extracted in this study revealed two patterns, one for M_6C carbide, and the other best identified as a variant of mu phase. One strong diffraction line for the latter was missing; however, this may have been due to the particular orientation of the films in the residue.

Transmission Electron Microscopy

Transmission electron microscopy was performed upon thin foils prepared from the no-IHT control samples from all four heats, and most of the samples given 4-hour IHT's. Ellipsoidal long range ordered domains on the order of 10-30 nm in length were observed to be present in all samples. As anticipated from prior work [5], selected area diffraction patterns for these domains revealed the structure to be similar to an Ni_2Mo type phase. The size of the domains did not permit EDS analysis using a TEM nano- probe; however, in previously unpublished work by Dollar on 242 alloy exposed for 4000 hours at 650°C (1200°F) [6], domains up to 200 nm in length were observed, and EDS analysis using the TEM nanoprobe was performed. Results, given in Table II, confirmed the $Ni_2(Mo,Cr)$ stoichiometry.

Table II Composition of Ordered Domains in 242 Alloy

Element	$Ni_2(Mo,Cr)$ Domain		Matrix	
	Weight %	Atomic %	Weight %	Atomic %
Ni	56.80	64.81	65.85	71.78
Mo	34.58	24.15	24.31	16.22
Cr	7.86	10.12	8.63	10.63
Fe	0.77	0.92	1.20	1.38

A number of key observations were made from the results of the TEM evaluations. Firstly, despite the use of IHT's prior to final aging at 650°C (1200°F), the $Ni_2(Mo,Cr)$ ordered domains found in grain interiors in all samples were observed to be of a relatively uniform size within the sample. Secondly, the size of the domains found was significantly increased relative to the no-IHT control samples when 4-hour IHT's at 760 to 815°C (1400 to 1500°F) were used. This increased domain size was observed to a lesser and lesser degree as IHT temperature increased above 815°C (1500°F), with sizes roughly equivalent to those found for the no-IHT control samples generally observed with IHT at 980°C (1800°F) or above. These observations are illustrated by the TEM dark field images presented in Figures 13 to 15.

Figure 13: TEM dark field image of ordered domains in the no-IHT control sample from Heat D.

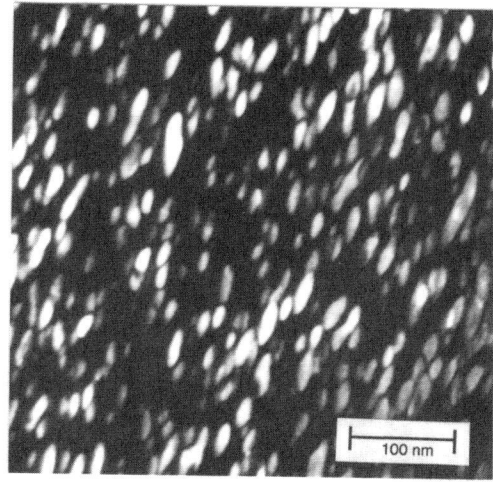

Figure 14: TEM dark field image of ordered domains in the sample from Heat D given a 4-hour IHT at 760°C (1400°F).

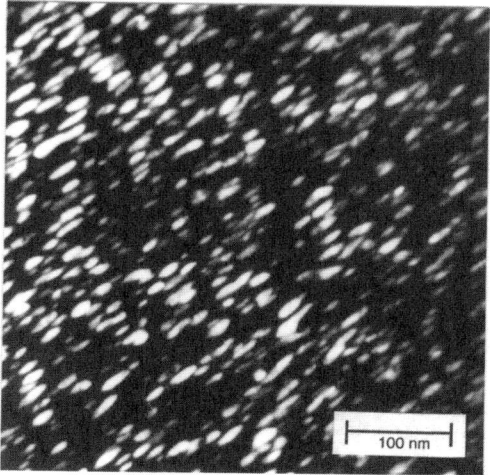

Figure 15: TEM dark field image of ordered domains in the sample from Heat D given a 4-hour IHT at 870°C (1600°F).

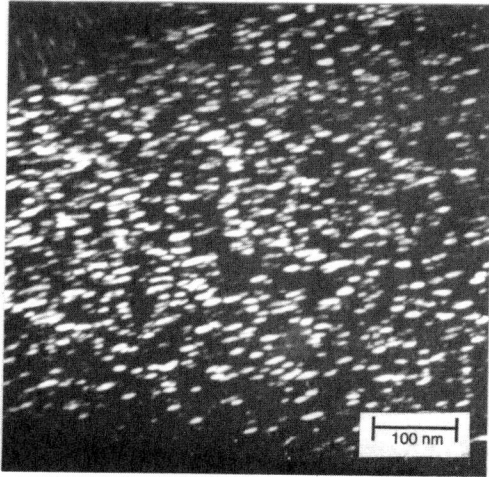

Figure 16: TEM dark field image of ordered domains in the no-IHT control sample from Heat A.

A third important observation made from the TEM evaluations was that the general size of the ordered domains appeared to vary as a function of the heat examined. The domain size observed in the "leaner" materials, Heats A and B, appeared to be larger than that observed in the "richer" materials, Heats C and D, for the same heat treatment condition. This is illustrated by comparing the image shown in Figure 13 for Heat D for the no-IHT control sample, to that in Figure 16 for same condition for Heat A. While the image in Figure 16 is at a lower magnification than that of Figure 13, the domains in the Figure 16 are clearly larger in size.

In addition to the study of the ordered domains, TEM was also used to examine the secondary phase precipitation at the grain boundaries. Confirming the work done with SEM, extraction and x-ray analysis, two grain boundary phases were identified by selected area diffraction techniques — cubic structured M_6C carbide and hexagonal structured mu phase. The large, isolated primary particles were also confirmed to be M_6C carbide. The dominant grain boundary phase found in samples given 4-hour IHT's at between 760 and 980°C (1400 and 1800°F) was found to be the hexagonal mu phase. Above 980°C (1800°F), mu phase was not really found to be in evidence.

Discussion

Structures and Tensile Ductility

It is not surprising to find that primary and secondary M_6C carbides were present in the materials studied. Even though the carbon contents of the study heats were only 0.002 to 0.006%, the extremely low solubility of carbon in nickel-molybdenum alloys is well known [7]. The minor loss of tensile ductility that occurs in many nickel-base alloys as a result of precipitation of carbides at grain boundaries during exposure at intermediate temperatures is also well known.

There is a hint of increased ductility loss with IHT at intermediate temperatures for the two highest carbon heats, B and C, associated perhaps with a bit heavier grain boundary precipitation; however, since it was found in the TEM work that the dominant phase precipitating at grain boundaries during IHT between 760 and 980°C (1400 and 1800°F) is mu phase, the additional relative contribution of the secondary carbide precipitation is hard to assess. Above 980°C (1800°F), it is likely that the M_6C secondaries are related to the somewhat lower ductility observed.

It is likely, in any case, that the continuous films of predominantly mu phase formed at the grain boundaries below 980°C (1800°F) are responsible for the majority of the tensile ductility losses shown in Figures 3, 4, 7 and 8. It is also instructive to note that continuous grain boundary films were not observed for the 30-minute IHT at 760°C (1400°F), but were found for the 4-hour IHT. There was no ductility loss observed for the former, and substantial loss for the latter.

All of the foregoing is not to say that some of the observed loss in ductility isn't simply associated with the higher strength levels achieved with IHT at these temperatures, which is due to a completely separate mechanism. General experience with 242 alloy would indicate that this might account for about 5% or so lower elongation or RA values in absolute terms.

Since the formation of continuous grain boundary films was observed in this study for IHT exposures of as little as 30 minutes, it implies that even shorter exposures may be significant. In fact, in unpublished work by Wu [8], it has been found that simply cooling through the intermediate temperature regime at a slow enough rate following annealing can yield results similar to those found here. In his work, the critical cooling time between about 1095 and 650°C (2000 and 1200°F) was on the order of about two to three minutes or longer.

Structures and Strength Properties

Considering the qualitative observations made with TEM, it would appear that the imposition of an IHT between the solution treatment and 650°C (1200°F) aging steps of the heat treatment for 242 alloy can result in a significant increase in the size of the ordered domains observed following the aging step, depending upon the IHT temperature. This increase in domain size apparently causes the observed increase in yield strength behavior illustrated in Figures 1 and 5. A mechanistic explanation for this behavior involves particle cutting by dislocations, and is described more fully by Kumar and Vasudevan [9].

Attempts were made to perform actual domain size measurements for several of the samples in this study in order to better quantify the relationship among IHT, domain size and yield strength. It must be recognized that measurements such as these are highly complicated by factors such as domain orientation to the plane of viewing, amongst others. Nevertheless, the results of these domain measurements were in generally good agreement with the observed yield strength behavior, as illustrated for Heat D data shown in Figure 17.

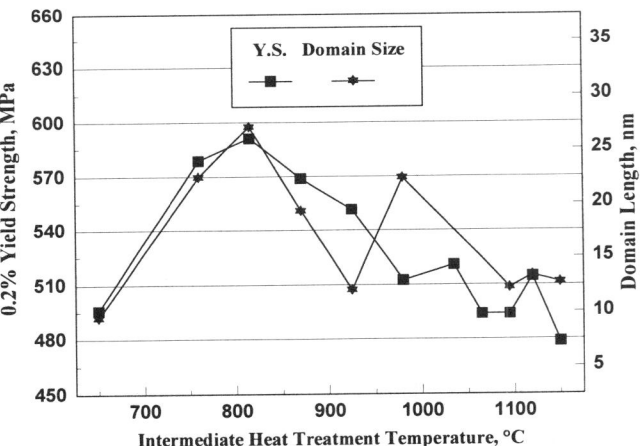

Figure 17: Yield Strength at 650°C (1200°F) & average measured $Ni_2(Mo,Cr)$ domain size for Heat D as a function of 4-hour IHT temperature.

Kumar and Vasudevan [4] have cited 766°C (1410°F) as the $Ni_2(Mo,Cr)$ disordering temperature in 242 alloy, and have observed the continued existence of $Ni_2(Mo,Cr)$ ordered domains in the alloy after exposure at 750°C (1382°F) [3]. It is therefore conceivable that ordered domains were actually formed during the 4-hour IHT's at 760°C (1400°F) used in this study. If this was

indeed the case, the observed absence of a bimodal distribution of domain sizes in the grain interiors indicates that subsequent aging at 650°C (1200°F) served only to grow those domains formed during the IHT at 760°C (1400°F) to a size greater than that achieved with aging at 650°C (1200°F) alone. Again, this size difference may be discerned for Heat D, as an example, by comparing Figures 13 and 14.

For IHT at 815°C (1500°F) and above, long range ordered $Ni_2(Mo,Cr)$ domains should not have been produced, and yet an effect of such IHT's upon both yield strength and domain size following final aging at 650°C (1200°F) is clearly observed. Moreover, the magnitude of the effect appears to gradually diminish as the IHT temperature is increased from 815°C (1500°F) to 980°C (1800°F).

One explanation for this behavior might be that the IHT exposure serves to establish a greater degree of short range order in the alloy relative to that produced by the solution treatment. The magnitude of the increase in short range order could be expected to greatest just above the $Ni_2(Mo,Cr)$ disordering temperature, and to diminish as IHT temperature increases. A greater degree of short range order thus present in the material prior to aging at 650°C (1200°F) might accelerate the kinetics of formation and/or growth of the $Ni_2(Mo,Cr)$ ordered domains during aging. This would produce the observed increased domain size and attendant increased yield strength.

It should be mentioned that direct measurement of *the degree* of short range order present in the samples was not attempted in the present study, and may in fact be exceedingly difficult to do. Further work in this area is planned, together with a more definitive examination of structures in the as-intermediate-heat-treated condition.

Once again, the fact that IHT effects upon strength were observed for as little as 30-minute exposures implies that even shorter exposures may be significant. In the previously referenced unpublished work by Wu [8], cooling rates between the solution annealing temperature and the 650°C (1200°F) aging temperature for 242 alloy were found to have an influence upon yield strength (in addition to that upon tensile ductility) of a similar magnitude to that observed in the present study for isothermal IHT.

Structure, Strength and Chemistry Issues

One of the key observations made in the TEM portion of this work was that the ordered domain size found in the "leaner" materials, Heats A and B, appeared to be larger than that found in the "richer" materials, Heats C and D, for the same heat treatment condition. In light of the fact that yield strengths for Heats C and D were observed to be higher than those for Heats A and B for the same heat treatment condition, this appears to be at odds with the conclusion that yield strength increases with increasing domain size.

One possible explanation for this may be that the strengthening effect of the ordered domains is not only a function of the size of the ordered domains, but also a function of their volume fraction in the alloy. Such a dual dependency would be consistent with the behavior of most particle-strengthened materials. If the volume fraction of Ni₂(Mo,Cr) ordered domains in 242 alloy increases with molybdenum content, for example, it might be resonably expected that such a larger volume fraction of domains, coupled with a smaller domain size, might net out to a higher level of strength for the "richer" heats C and D than that for the "leaner" heats A and B.

Traditional phase extraction techniques that would permit direct measurement of Ni₂(Mo,Cr) domain volume fraction in 242 alloy have yet to be developed, and estimates based upon TEM measurements are impractical for a variety of reasons. Thus, it is only possible at this point to demonstrate indirectly that Ni₂(Mo,Cr) domain volume fraction in 242 alloy increases with increasing molybdenum content. Such indirect evidence is presented in Figures 18 and 19, and in Table III.

The structure shown in Figure 18 is that for Heat C given a 4-hour IHT at 870°C (1600°F). Note that the grain boundary shown is heavily precipitated with mu and M_6C carbide phases, which have depleted the adjacent matrix in molybdenum, resulting a zone devoid of Ni₂(Mo,Cr) domains next to the boundary. With increasing distance away from the grain boundary and towards the the grain interior, molybdenum content is known to increase. The domains encountered nearest to the grain boundary are larger in size and apparently lower in volume fraction than those encountered further into the grain interior, where the molybdenum content is higher. This is consistent with both the domain size behavior observed in the bulk study alloys as a function of heat chemistry, and the postulated volume fraction effect of heat molybdenum content.

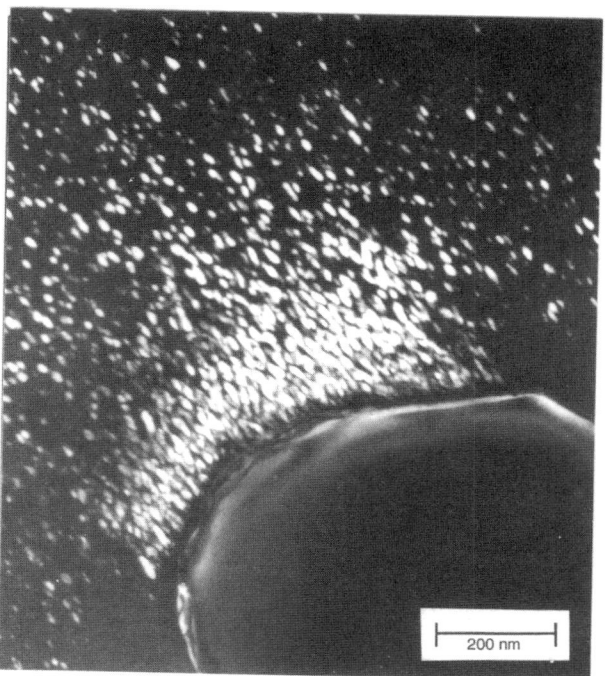

Figure 19: TEM dark field image of structure near a primary M_6C carbide in the sample from Heat A given a 4-hour IHT at 980°C (1800°F).

The structure shown in Figure 19 is that for Heat A given a 4-hour IHT at 980°C (1800°F). Note the increase in volume fraction of Ni₂(Mo,Cr) domains in proximity to the large M_6C carbide which has apparently undergone some slight dissolution, rejecting molybdenum into the surrounding matrix, as shown in Table III. Again, this supports the postulated effect of heat molybdenum content upon Ni₂(Mo,Cr) domain volume fraction.

Table III TEM EDS Analyses Relative to Figure 19

Element	Near M_6C Carbide		Away From Carbide	
	Weight %	Atomic %	Weight %	Atomic %
Ni	61.49	68.46	63.41	69.82
Mo	29.05	19.79	24.31	16.22
Cr	7.86	9.88	8.41	10.46
Fe	1.60	1.87	1.52	1.76

Summary and Conclusions

A comprehensive evaluation of the relationships that exist among the various aspects of metallurgical structure, heat treatment, composition and elevated temperature tensile properties of 242 alloy has been conducted. As a result of this work, a much clearer picture now exists of the factors which control the level of performance properties the material will exhibit in commercial components.

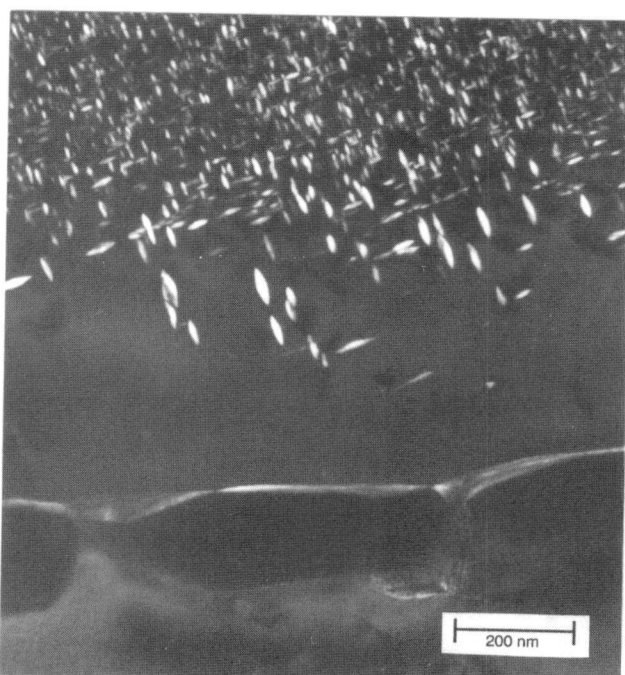

Figure 18: TEM dark field image of grain boundary structure in the sample from Heat C given a 4-hour IHT at 870°C (1600°F).

Major conclusions supported by this work are as follows:

1) 242 alloy achieves its strength from the formation of about 10-30 nm long ellipsoidal, ordered domains with a structure similar to that of Ni_2Mo phase, and a stoichiometry of Ni_2X, where X equals about 0.7 Mo atoms and 0.3 Cr atoms.

2) The 650°C (1200°F) yield strength of 242 alloy increases with increasing $Ni_2(Mo,Cr)$ domain size for a given heat chemistry.

3) $Ni_2(Mo,Cr)$ domain size in 242 alloy after aging at 650°C (1200°F) will be increased if the final aging is preceded by exposure for as little as 30 minutes at 815 to 980°C (1500 to 1800°F), with the size increase diminishing as exposure temperature increases in this range. It is believed that the effect of such exposure is to establish a greater degree of short range order in the material than is present following solution annealing, and thus accelerate the rate of formation and/or growth of the long range ordered domains during final aging.

4) A similar increase in final domain size will be observed if final aging is preceded by a 4-hour exposure at 760°C (1400°F), but not if a 30-minute exposure is used. This effect may be due to the actual formation of ordered domains at 760°C (1400°F) in four hours, which are simply grown in size during final aging. The 30-minute exposure may not be long enough to either enhance short range order or form long range ordered domains.

5) It was found that heats with higher molybdenum and chromium contents will exhibit higher yield strengths than those with lower such contents, yet have smaller $Ni_2(Mo,Cr)$ domain sizes. Indirect evidence indicates that this is a result of increased volume fraction of $Ni_2(Mo,Cr)$ domains with higher molybdenum content.

6) In addition to the $Ni_2(Mo,Cr)$ domains, phases to be found in 242 alloy include primary M_6C carbides, secondary M_6C carbides, and secondary mu phase. Primary and secondary carbides will be observed even at bulk carbon contents as low as 0.002%.

7) Secondary mu phase and, to a lesser extent, M_6C carbides will precipitate in 242 alloy, largely in a semi-continuous or continuous grain boundary film morphology, when the alloy is exposed for 30 minutes or longer in the temperature range from 815 to 980°C (1500 to 1800°F). Similar structures will be observed for 4-hour exposure at 760°C (1400°F), but not for a 30-minute exposure. *When present, these grain boundary films will have a strong deleterious effect upon 650°C (1200°F) tensile ductility of fully aged samples.*

8) Grain boundaries of the alloy aged directly at 650°C (1200°F) following an effective water quench from the solution treatment temperature will exhibit little or no precipitation which, when found, will normally only be in a discrete particle morphology.

9) Mu phase will be largely absent for exposure temperatures over 980°C (1800°F). Secondary M_6C carbides will precipitate as discrete particles at the grain boundaries for exposure temperatures above about 980°C (1800°F), and up to 1150°C (2100°F), with an apparent mild adverse effect upon tensile ductility of fully aged samples.

10) Within the context of this work, the extent of grain boundary precipitation that will occur in 242 alloy does not appear to be strongly dependent upon heat chemistry; however, grain boundary precipitate film morphology will apparently persist to somewhat higher exposure temperatures for heats with higher molybdenum and chromium contents.

Acknowledgements

The authors would like to acknowledge the considerable assistance of M. Farooqi of IIT, and M. Richeson and J. Cotner of Haynes International, in the performance of this work.

References

1. M. F. Rothman and H. M. Tawancy, "Low Thermal Expansion Superalloy", U. S. Patent No. 4,818,486, April 4, 1989.

2. H. M. Tawancy, "Long-Term Aging Characteristics of Some Commercial Nickel-Chromium-Molybdenum Alloys", Journal of Materials Science, 16 (1981), 2883-2889.

3. M. Kumar and V. K. Vasudevan, "Ordering Reactions in an Ni-25Mo-8Cr Alloy", Acta Mater., 44 (4) (1996), 1591-1600.

4. M. Kumar and V. K. Vasudevan, "Short-Range to Long-Range Ordering Reactions in a Ni-25Mo-8Cr Alloy", Mater. Res. Soc. Symp. Proc., 213 (1991), 187-192.

5. S. K. Srivastava and B. E. Lewis, "The $Ni_2(Mo,Cr)$ Ordering Transformation", Kinetics of Ordering Transformations in Metals, ed. H. Chen and V. K. Vasudevan, (Warrendale, PA: TMS, 1992), 141-149.

6. M. Dollar, private communication with the author, Illinois Institute of Technology, 1998.

7. V. L. Gernets, I. A. Tomilin, and T. V. Svistunova, "The Solubility of Carbon in Ni-Mo Alloys", IZV. Akad. Nauk. SSSR, Met (2) (1976), 200-204.

8. R. Wu, private communication with the author, Haynes International, Inc., 1999.

9. M. Kumar and V. K. Vasudevan, "Mechanical Properties and Strengthening of a Ni-25Mo-8Cr Alloy Containing $Ni_2(Mo,Cr)$ Precipitates", Acta Mater., 44 (12) (1996), 4865-4880.

MICROSTRUCTURAL CHANGES IN MA 760 DURING HIGH TEMPERATURE LOW CYCLE FATIGUE

A. Hynnä and V.-T. Kuokkala
Tampere University of Technology
Institute of Materials Science
P.O.Box 589, FIN-33101 Tampere, FINLAND

Abstract

The HTLCF properties and related microstructures of MA 760, a mechanically alloyed oxide dispersion strengthened Ni-base superalloy, were studied using specimens of two different grain orientations. The push-pull fatigue tests with total strain as a control parameter were conducted at two different strain rates, $\dot{\varepsilon} = 0.005$ s^{-1} and $\dot{\varepsilon} = 0.0005$ s^{-1}, at the temperatures of 650 °C, 950 °C, 1000 °C and 1050 °C. The fracture surfaces and the dislocation microstructures of the fatigued specimens were studied using optical microscopy, scanning electron microscopy (SEM) and analytical electron microscopy (AEM). In this paper, the changes in the microstructure of MA760 during strain cycling at high temperatures are reported and discussed.

Introduction

To extend the good corrosion resistance and high strength values of conventional superalloys to even higher temperatures, new processing techniques and new types of microstructure have been developed during the past two decades. Dispersion strengthened (ODS) superalloys form one of such new materials groups, increasing the useful service temperature of, e.g., gas turbines, by almost hundred degrees (Ref. 1). ODS superalloys are produced by mechanically alloying the elementary constituents and a master alloy with pre-processed oxide powder (Ref. 2), followed by conventional processing and heat treatments, such as extrusion, hot rolling, recrystallization- and γ'-precipitation heat treatments. The fine dispersion of stable, incoherent oxide particles formed during the mechanical alloying process is, to a large extent, responsible for the excellent high temperature mechanical properties of ODS- alloys by acting as strong obstacles to the mobile dislocations (Ref. 3). ODS Ni- base superalloys are commonly strengthened also by coherent γ'-precipitates which, however, start to lose their strengthening ability at around 1000 °C, above which the dispersion strengthening becomes the dominant strengthening mechanism (Ref. 4).

The fatigue endurance of crystalline materials depends primarily on their ability to resist plastic deformation which, in turn, depends on the type, strength, distribution and stability of the obstacles that the mobile dislocations meet on the slip planes. In high temperature fatigue, the changes in the obstacle structure may be caused by both temperature as well as by cyclic deformation. For example, formation of dispersion poor areas may weaken the material substantially by allowing local plastic deformation which, under suitable conditions, may lead to the initiation and growth of fatigue cracks and finally to the failure of parts or components. The research results of microstructural changes during strain cycling at high temperatures are therefore of great importance when assessing the performance and reliability of high temperature materials in actual service conditions.

Material and experimental techniques

The mechanically alloyed Ni-base ODS superalloy MA 760 was produced in the form of bars with two different rectangular cross-sections, SCS (200mm×600mm) and LCS (320mm×950mm). Compositions of the delivered bars are given in Table I. The final heat treatment of the SCS bar was 0.5 h solution treatment at 1100 °C, furnace cooling with 60 °C/h to 600 °C, followed by uncontrolled furnace cooling. The heat treatment procedure for the LCS bar was the same with the exception that the solutioning temperature was 1120 °C.

Table I Compositions of the delivered bars

Element	SCS, [wt-%]	LCS, [wt-%]
C	0.043	0.042
Si	0.04	0.05
Fe	1.02	1.04
Cr	19.66	19.79
Al	5.97	5.93
Mo	1.92	1.96
S	0.003	0.003
Zr	0.14	0.14
N	0.284	0.286
W	3.5	3.5
B	110 ppm	110 ppm
P	<0.005	<0.005
Ni	Bal.	Bal.
Y$_2$O$_3$	1.03	1.03

563

Superalloys 2000
Edited by T.M. Pollock, R.D. Kissinger, R.R. Bowman,
K.A. Green, M. McLean, S. Olson, and J.J. Schirra
TMS (The Minerals, Metals & Materials Society), 2000

The SCS material was tested in the longitudinal (L) direction and the LCS material in the long-transverse (LT) direction. The orientation of the fatigue specimens relative to the grain orientation and the extrusion direction are shown in Figure 1.

The total-strain controlled push-pull fatigue tests with a symmetrical triangular wave form (R = -1) were conducted at four different temperatures in air using a strain rate of $0.5\% \text{ s}^{-1}$ (fast, F) or $0.05\% \text{ s}^{-1}$ (slow, S). Details of the testing conditions for each specimen can be found in Table II.

For microstructural studies of the reference material and the fatigued samples, optical microscopy, scanning electron microscopy and transmission electron microscopy were applied. Most fatigued specimens were cut parallel to the stress axis but in a few cases they were sectioned perpendicular to the applied stress. The reference material of both the SCS- and LCS-bars was investigated from both longitudinally and perpendicularly cut surfaces with respect to the extrusion axis.

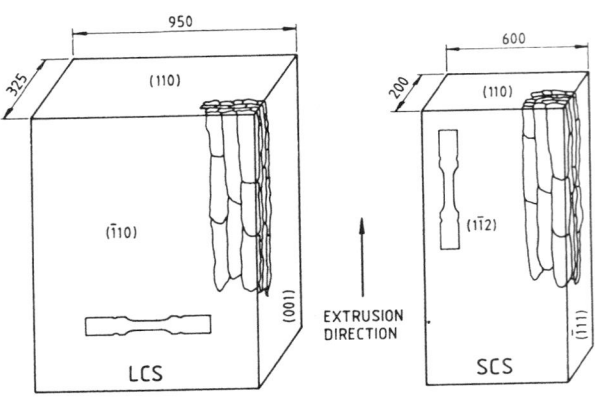

A summary of the fatigue test parameters and mechanical test results is shown in Table II. The stress amplitudes σ_{max}, σ_{min} and $\Delta\sigma/2$ were measured from the cycle recorded at half life $N_f/2$. As a failure criterion, a 20 % stress drop was used. The interrupted tests mentioned in Table II were performed to study the dislocation structures free from the possible changes caused by cooling down under decreased stress. In these tests the specimens were cycled for 10 loops, straining was stopped at maximum total tensile strain of 0.5 %, and the cooling down was conducted under the tensile stress thus obtained.

Figure 1. Cutting of the specimens from the extruded bars. Specimen gage length is 12 mm and diameter 7 mm.

Table II Numerical values recorded in the fatigue tests. Tests 28-31 were interrupted after 10 cycles.

Test No.	Bar ID	T [°C]	$\dot{\varepsilon}$ [%s^{-1}]	$\Delta\varepsilon_t$ [%]	$\Delta\varepsilon_{pl}$ [%]	$\Delta\varepsilon_e$ [%]	σ_{max} [MPa]	σ_{min} [MPa]	$\Delta\sigma/2$ [MPa]	N_f [cycl.]
1	SCS	650	0.5	1.0	0.1	0.9	850	-950	900	792
2	SCS	650	0.5	0.8	0.04	0.76	705	-780	742.5	3925
3	SCS	650	0.5	0.6	0.01	0.59	560	-575	567.5	16320
4	SCS	950	0.5	1.0	0.51	0.49	310	-340	325	570
5	SCS	950	0.5	0.8	0.35	0.45	275	-305	290	960
6	SCS	950	0.5	0.6	0.2	0.4	250	-270	260	4575
7	SCS	950	0.05	1.0	0.49	0.51	290	-300	295	415
8	SCS	950	0.05	0.8	0.34	0.46	270	-285	277.5	880
9	SCS	950	0.05	0.6	0.19	0.41	240	-265	252.5	3945
10	SCS	1050	0.5	0.8	0.55	0.25	160	-180	170	395
11	SCS	1050	0.5	0.8	0.55	0.25	160	-180	170	595
12	SCS	1050	0.5	0.6	0.35	0.25	170	-195	182.5	1025
13	SCS	1050	0.5	0.35	0.11	0.24	150	-160	155	6925
14	SCS	1050	0.5	0.25	0.01	0.24	160	-180	170	67300
15	SCS	1050	0.05	0.8	0.55	0.25	170	-190	180	305
16	SCS	1050	0.05	0.6	0.36	0.24	160	-180	170	620
17	SCS	1050	0.05	0.35	0.1	0.25	160	-170	165	3350
18	LCS	950	0.5	1.0	0.66	0.34	285	-300	292.5	600
19	LCS	950	0.5	0.6	0.3	0.3	250	-275	262.5	2105
20	LCS	950	0.5	0.45	0.15	0.3	240	-260	250	6770
21	LCS	1050	0.5	1.0	0.8	0.2	140	-160	150	105
22	LCS	1050	0.5	0.6	0.42	0.18	145	-155	150	585
23	LCS	1050	0.5	0.45	0.25	0.2	145	-150	147.5	1125
24	LCS	1050	0.05	1.0	0.79	0.21	140	-145	142.5	110
25	LCS	1050	0.05	0.6	0.4	0.2	135	-155	145	580
26	LCS	1050	0.05	0.45	0.26	0.19	145	-150	147.5	1280
27	SCS	1000	0.05	0.45	0.21	0.24	190	-195	192.5	5980
28	SCS	1000	0.05	0.45	N/A	N/A	N/A	N/A	N/A	10 (int.)
29	SCS	650	0.5	1.0	N/A	N/A	N/A	N/A	N/A	10 (int.)
30	SCS	950	0.5	1.0	N/A	N/A	N/A	N/A	N/A	10 (int.)
31	SCS	1050	0.5	1.0	N/A	N/A	N/A	N/A	N/A	10 (int.)

Results and discussion

Grain and boundary structures

The grain structure in both the SCS- and LCS-bars is very coarse. In general, the grains are several centimeters long in the L-direction and a few millimeters wide in both the ST- and LT-directions, as illustrated in Fig. 1. According to Martin and Tekin, (Ref. 5), most of the boundaries are of the low angle type.

The boundaries between adjacent grains are covered by agglomerated γ' giving rise to the contrast effect presented in Figure 2. Also discontinuous precipitation of chromium carbide along the boundaries is present. After cyclic straining at 650 °C, no visible changes in the general appearance of the boundaries could be detected. At 950 °C, the agglomerated γ' has, at least partly, disappeared. However, the most dramatic changes appear in the boundary structure of the samples fatigued at 1050 °C. This is due to the fact that 1050 °C is above the γ' solidus of MA 760, thus causing the dissolution of the agglomerated γ', leaving only a discontinuous carbide structure in the boundaries.

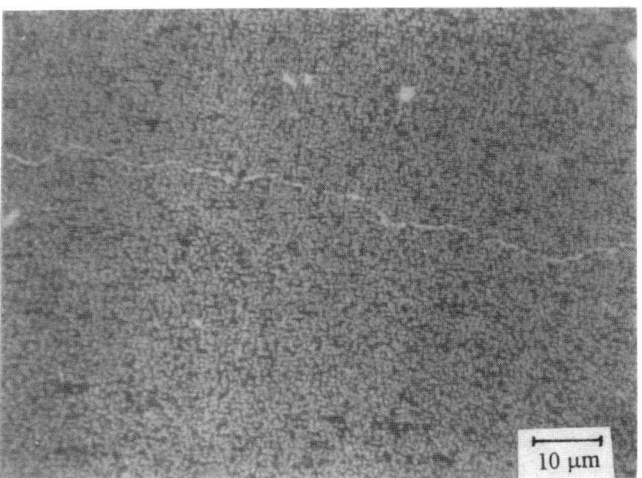

Figure 2. Optical micrograph of the longitudinal boundary in the SCS-bar reference sample.

The susceptibility of the boundary to the opening phenomenon was evident in the tests conducted at 1050 °C, as illustrated in Figure 3. Although this phenomenon was observed in the fracture surfaces in all of the conducted tests, the opening of the boundaries is not so evident when studying the longitudinal sections of the lower temperature samples, especially in the case of tests at 650 °C. The susceptibility of the boundaries to opening is caused by oxidation effects, as stated in Ref. 6. The precipitated carbide particles at the boundaries make this opening even more pronounced. A clear evidence of oxidation is the γ' depleted zone beside the crack, which was observed also near the air exposed surface of the samples. This effect is more pronounced at higher temperatures and was not observed at 650 °C at all. The formation of these γ' depleted zones is caused by the diffusion of Al to form Al_2O_3 at the crack, leaving an insufficient amount of Al to form γ' in the vicinity of the crack. This

phenomenon, as well as the accompanying formation of Kirkendall porosity, is illustrated in Figure 4. It should be pointed out that in the cracks observed inside the specimen which were not connected to the surface of the specimen, the formation of γ' depleted zones was not observed. The effects of transverse grain boundaries on the fatigue lives of LCS samples are discussed in an earlier paper (Ref. 15).

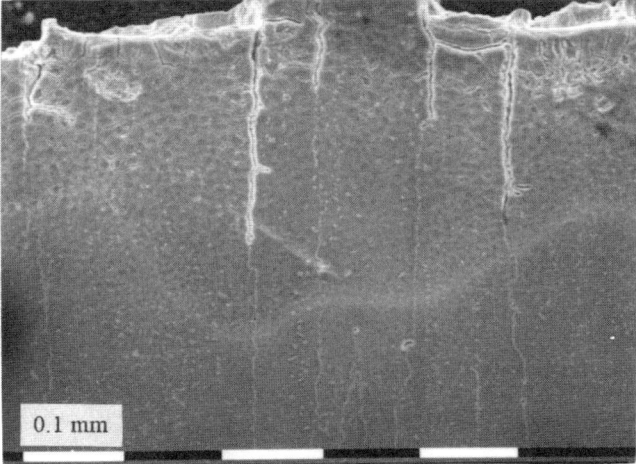

Figure 3. Opened longitudinal grain boundary starting from the fracture surface (sample 12).

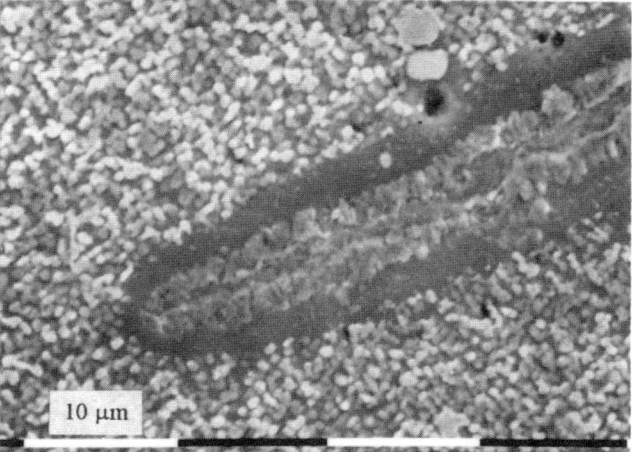

Figure 4. Tip of the opened grain boundary in sample 8.

γ'-structure

In the reference materials, the γ' appeared as cuboidal precipitates, the cube sides oriented roughly along the extrusion direction. However, a considerable variation from grain to grain existed, especially in the small grains. The average side length of the precipitates was 0.5 μm. No large difference in the size and orientation of γ' between the SCS- and LCS-bars was observed.

In the samples fatigued at 650 °C, no observable changes in γ' size or shape could be detected, when compared with the reference material. After the exposure at 950 °C, the precipitates had clearly rounded, but their average size had not changed markedly; the average diameter was still measured to be 0.5 μm. However, considerable size variation is present with the diameter of the largest observed precipitates approaching 1.0 μm. This observation suggests that at these temperatures the process of Ostwald ripening, where small precipitates are dissolving and the larger ones are thus growing at the expense of smaller ones, is operative. The same process is also observed at 1000 °C, although the average size of the precipitates was reduced to 0.4 μm and the variation in sizes was larger than at 950 °C. Finally, at 1050 °C, the precipitates seem to have more or less completely dissolved during the heating up period in each test. The small round precipitates found after testing are assumed to be due to the reprecipitation during cooling.

Defect structure

In general, MA 760 was found to be quite free of defects when compared with the previous versions of Ni-base MA alloys, e.g., MA 6000. However, several types of defects could be identified, such as small grains or groups of small grains, stringers of inclusions and individual inclusions. Generally, these defects were not critical when considering their ability to initiate propagating fatigue cracks.

In the same ways as the boundaries between large grains, the boundaries between large and small grains are covered by continuous agglomerated γ' film embedded with chromium carbide, the amount of which seems to be higher than in the boundaries between large grains. The orientation of γ' is very different in the small grains compared to the adjacent large grains, suggesting that these boundaries are of the high angle type. This observation seems to explain the increased susceptibility of the small grain boundaries to the boundary opening phenomenon as compared to the low angle boundaries. Also the shape of γ' inside small grains was sometimes, but not always, disturbed. This could be caused, e.g., by local variations in the element concentrations in the small grain area, since chemical composition has an effect on γ' morphology. In the case of disturbed γ' structures, possibly the heavy precipitation of carbides has, at least partly, been involved in the composition modification.

In addition to small grain areas, other types of defects were also observed. The most common of these defects were stringers of chromium carbide inclusions formed in the extrusion process. These inclusions were generally quite small, the average size being of the order of 3 μm. Also, a large variety of mainly individual Cr-, Al- and Ti-rich inclusions, containing small amounts of Cu, S and Si were occasionally observed. All the elemental analyses of these inclusions were carried out using an EDS analyzer. The shapes of these types of defects varied largely from a blocky appearance to long elongated rods. Also the variation of their sizes was large, starting at 1 μm and extending up to 0.5 mm.

Generally all the defects, except stringers of chromium carbides, were found to initiate internal cracks which did not extend to the surface of the sample. These cracks, however, were secondary- or non-propagating and therefore they do not have an effect on the

fatigue life. Two main types of damage processes were associated with these defects; pore formation at the boundaries of the defects and cracking of the inclusion and subsequent crack growth to the matrix.

Pore formation

The pore formation in the fatigued samples falls into two main categories; pore formation ahead (on the boundary) or near a longitudinal crack and pore formation near the air exposed surface, including the fracture surface, of the sample. Concerning the latter category, it should be mentioned that there does not exist a general acceptance of the reasons for this phenomenon. In this research it was found that the amount of pores increased as a function of increasing temperature and increasing fatigue time. At 650 °C, no pore formation was observed at all, and still at 950 °C the capability of pores to form was very limited. Thus, the formation of pores is clearly restricted to the higher temperatures. A general observation in the fatigued samples was the concentration of stringers of pores in certain regions along the longitudinal direction. The pores lay in a surface region of 0.1 mm maximum depth, which decreases with decreasing fatigue time. In the longest tests, a tendency of the pore stringers to coalesce and to form longitudinal cracks was clearly observed. The pore size did not markedly depend on temperature or fatigue time, which only affected the number of the pore rich regions and the amount of individual pores inside these regions. The average pore size was of the order of 3 μm, that is, approximately the same size as the carbide particles in stringers. Very often, but not always, Cr was found to be present in the pores. Also Al-content, as measured by an EDS-analyzer, decreased towards the surface resulting in the formation of γ'-depleted zones. The contents of the other alloying elements did not markedly vary as a function of depth. The appearance of the pore rich regions is presented in Figure 5. The above mentioned concerns only the L-samples. The picture changes dramatically when we look at the LT-samples. In this case the stringer formation is not observed at all and individual pores are lying completely in the γ' depleted zone of approximately 10 μm in depth. The average pore diameter is approximately 1.5 μm.

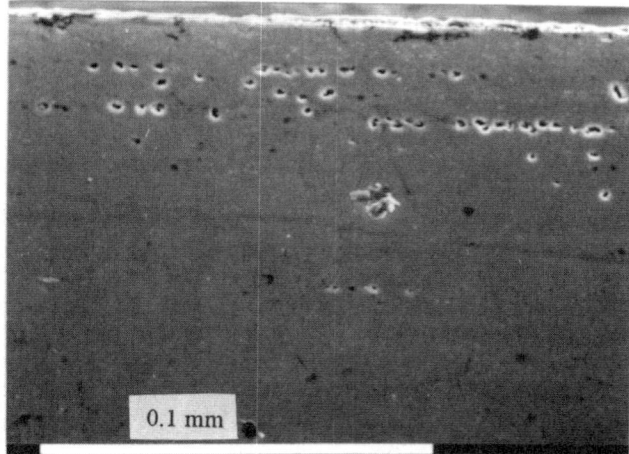

Figure 5. Pore formation in sample 14. Air exposed surface at the top.

It is quite reasonable to assume that in spite of Al diffusion to the surface, the stringers of Cr-carbides are also the reasons for the observed pore formation in the L-samples. The pores may be, at least partly, due to the specimen preparation, during which the carbide particles are pulled out of the matrix. Another possible source of the pull-out particles are the boundaries between the large grains or the boundaries around small grains, if suitably situated near the surface of the sample. The reasons for the pull-out phenomenon are partly unclear, but interfacial decohesion between the carbide particles and the matrix takes place during deformation (Ref. 7). Also, the effects related to the cyclic straining are not alone responsible for the pore formation, because in the annealing tests of ODS alloys the phenomenon has also been observed (Ref. 8). In order to check whether adequate temperature is the only requirement for pore formation in MA 760, a sample was heat treated for 144 hours at 1050 °C. After the treatment no pore stringers were observed and therefore it is concluded that at the temperatures and fatigue times used in this study, cyclic straining is required to achieve sufficient loss of adherence of the boundary between the carbides and the matrix. In Ref. 8 it is stated that grain boundaries and scale/metal interface both act as a sink for vacancies. It is proposed that in this case the carbide/matrix interface at least partly acts as a sink for vacancies and therefore the interface looses its adherence to the matrix.

The pore formation associated with the grain boundary opening was also generally observed. The formation of pore strings takes place in the matrix beside the crack no further than 0.1 mm from the crack surface. In fact, this subcategory resembles almost identically the case discussed above and therefore the formation mechanisms of the pore strings can be suspected to arise from the same basis. It should be noted that in this case pore string formation at most 0.05mm in front of the crack tip was also occasionally observed. These strings of pores were not at the boundary itself, rather they were in the same line than the pore strings observed beside the crack.

TEM studies of the reference material

Both longitudinal and transverse slices were prepared from the SCS- and LCS-material. No major differences could be detected either between the bars or between the slices. The microstructure was found to consist of approximately 50% cuboidal γ' with the average edge length of 500 nm. To achieve a comprehensive picture of the size distribution of the dispersoids, over 200 randomly distributed dispersoid particles were analyzed. Generally the shape of the particles was almost round and the dispersoid diameters ranged from 2 nm to 120 nm, the average being 35 nm. With the exception of some clusters of dispersoids, the distribution was relatively homogeneous. The dispersoid structure was nearly the same both in γ and γ'. Microtwins, both in γ and γ', with a width variation from 100 nm to 300 nm and a length variation from 700 nm to 2500 nm were frequently observed. The twin boundaries were often irregular due to the presence of dispersoid particles.

In the as-delivered condition, the alloy did not exhibit extensive dislocation structures. Generally the observed dislocations lay either at the γ/γ' boundaries or were pinned to dispersoid

particles. Figure 6 presents a TEM-image of the microstructure of the as-delivered material.

Dispersion-dislocation interaction, usually referred to as back-side pinning, is a frequent observation in many studies (e.g. Refs. 13 and 14). The phenomenon is illustrated in Figure 7, which shows a dislocation in a position where it has already surmounted a particle by climb (or pulled from the matrix to the interface) and adheres to the "departure side" of the dispersoid because of the attractive force between the dispersoid and the dislocation. The curvatures of the dislocation in the matrix and at the particle, showing the direction of the effective shear stress, have the same sign. An explanation for the attractive force is that a dislocation in the vicinity of an incoherent boundary surface can reduce a part of its strain field (and thus achieve a lower energy configuration) at high temperatures by diffusion. The dislocations also seem to be in intimate contact with the dispersoid and exhibit a sharp bend at the point of leaving the particle, which supports local climb theories.

Figure 6. Microstructure of the SCS reference material.

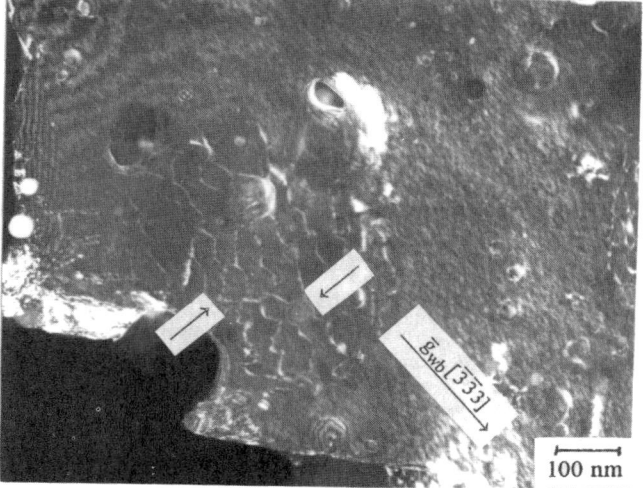

Figure 7. Weakbeam image from sample 26 illustrating the back-side pinning phenomenon (arrows).

Occasionally dispersion free areas were observed also in the as-delivered material. These defects originate from manufacturing processes and most probably they are milling defects. In the absence of pinning points for dislocations these defects have relatively low dislocation densities, which was also observed in the fatigued specimens.

TEM studies of the 650°C tests

In the tests conducted at 650 °C, the slip behavior of the alloy was rather homogeneous and, e.g., no signs of slip concentration to slip bands was observed. The most striking observation was that in γ' dislocations were travelling in pairs. This, in turn, indicates that the operative deformation mechanism is γ' cutting. Generally the dislocations formed loosely spaced tangles. The back-side pinning was not observed as often as in the samples fatigued at higher temperatures, rather it seems that at this temperature the Orowan looping dominates over back-side pinning. When one hundred dislocation/dispersoid interactions were analyzed, it was found that 80% of the interactions were of the looping type in accordance with the observations in Ref. 9 for MA 754. The dislocations were more numerous in γ', although there were dislocations also at the γ/γ' interface and in γ.

No dramatic effects in the size or in the shape of γ' could be detected in the samples fatigued at 650 °C when compared to the reference samples. This observation applies also to the dispersoid structure. Also in dislocation densities no marked differences could be detected. This seems quite natural because of the lack of cyclic hardening or softening, and therefore one expects a stable dislocation density to be generated during the initial cycles. In the interrupted test, the dislocation pairs cutting γ' were more readily observed, and generally the dislocations were not so curved as in test 3. This suggests that obviously some changes do happen in the dislocation microstructure during the crack growth (effective stress drops as a function of crack length) and cooling down (effective stress is zero) periods. Especially the amount of dislocation pairs cutting γ' was lower in the test 3 than in the interrupted test. From these observations it can be concluded that the possible rearranging mechanisms (during the absence of effective stress) may be thermally activated glide or the climb of dislocations to the γ/γ' interfaces and the dislocation/dispersoid interaction reactions. The TEM-images from the samples fatigued at 650 °C are presented in Figure 8.

TEM studies of the 950°C tests

The marked feature at 950 °C was the size and shape changes of γ'; the shape had become a little more irregularly rounded, the average diameter varying in the 200 nm to 500 nm range. Obviously the misfit parameter had changed, because clear evidence of the formation of interfacial dislocations between γ and γ' was detected. In the Y_2O_3 size distribution, no noticeable changes could be detected. However, the dispersoid particles were observed to be less homogeneously distributed, i.e., the amount of particle clusters, as well as the amount of areas of low particle concentrations, had increased markedly. The above mentioned concerns test 6, whereas in the interrupted test the changes were not visible. These phenomena became more

pronounced in test 27 conducted at 1000 °C, and this point will be discussed in detail in the next chapter.

When considering the dislocation interaction with the dispersion particles, roughly 80% of the dislocations were found to be attached to the particles. This observation is consistent with Ref. 9. Also γ' shearing by paired dislocations was evident. However, most of the dislocations appeared to have concentrated at the γ/γ' interfaces and in γ. As in the 650 °C case, in the interrupted test the γ' cutting was more evident, indicating the necessity for cooling down under stress in order to avoid the relaxation phenomena. A typical microstructure of the samples fatigued at 950 °C is presented in Figure 9.

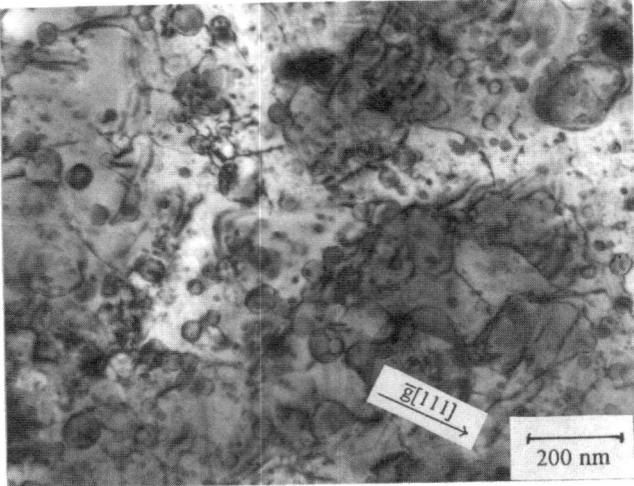

Figure 8. TEM-image from the sample 3.

Figure 9. TEM-image from the sample 6.

TEM studies of the 1000°C tests

In test 27 conducted at 1000 °C, the formation of dispersoid depleted zones in γ' (particle free zones, PFZ) was apparent, as presented in Figures 10 and 11. It was observed that the average size of the rounded γ' has increased to 750 nm. At the same time the amount of γ' particles had decreased to 10 vol-% and the wavy contrast of the matrix had became clearly visible. This observation may be regarded as an indication of small γ' nuclei below the resolution limit all around the matrix (Ref. 10). These phenomena are a clear indication of Ostwald ripening, in which process some of the precipitates are growing at the expense of others in order to achieve energetically the most favorable structure.

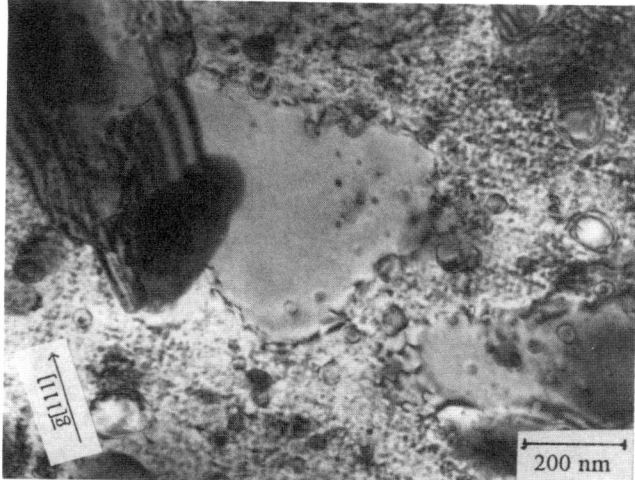

Figure 10. Particle free zone in sample 27

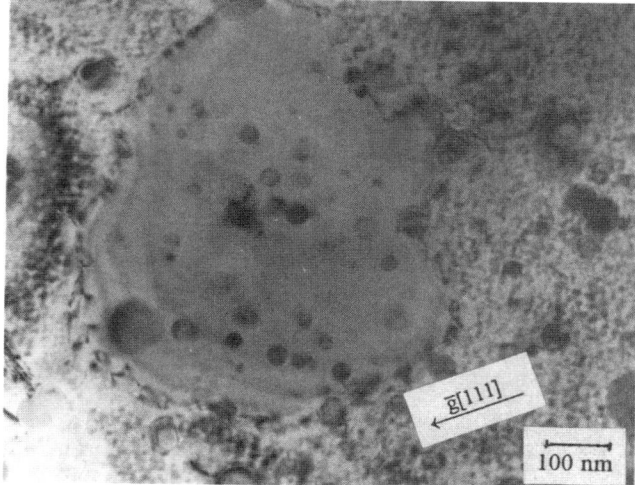

Figure 11. Particle free zone in sample 27.

A highly inhomogeneous distribution of the dispersoid particles is evident, as compared to the material in the as delivered condition. The same phenomenon can be observed inside the γ', as well as in the matrix. In fact, in the γ', where the dispersoids are more clearly visible, the particles seem to have arranged themselves along some loosely defined curves, as illustrated in Figure 11. In the following, it is supposed that the moving γ/γ' boundary exerts a dragging force on the dispersoid particles thus transporting them with the boundary and creating dispersoid free areas. Based on the latter, the explanation for the particles lying along curves (in fact, lying on surfaces), may be twofold. First, it

is possible that the moving γ/γ' boundary can escape the dispersoid particles thus leaving them inside the growing precipitate. The second explanation is based on the assumption that the particles are showing the maximum dimension of some earlier precipitate, which has reached its maximum size and has started to dissolve. Considering the second alternative we have to suppose that the dragging force of the growing boundary is higher, when compared to the case of dissolutioning. Possibly both mechanisms operate; support for the second alternative is given in Figure 11, where the dotted line indicates a curve containing particles and curved in the 'wrong' direction, if supposed to be formed during the growth of the present precipitate.

Models for the interaction of second-phase particles with moving boundaries, especially with high angle grain boundaries are well established, both theoretically, e.g. in Ref. 11, and also experimentally (e.g. Ref. 12). When the moving boundary is interacting with particles, several phenomena can exist: 1) the boundary may be pinned by the particles, 2) the particles may be dragged by the boundary, 3) the particles may coarsen until the boundary can escape from them and 4) the boundary may migrate through the field of particles, if the driving force is sufficiently high. When considering phase transformations, the driving pressures P for boundary motion according to Ref. 11 are very high, lying in the 10^7 - 2×10^{10} N/m^2 range. A boundary detachment will occur, if

$$P \geq \frac{3f_v \psi}{2r} \quad ,$$

where f_v = volume fraction of the particles, ψ = boundary energy and r = particle radius.

In the case when the driving pressure is insufficient to detach the boundary from the particles, the boundary tends to drift slowly, dragging the particles with it. If we consider a particle moving with a velocity v under a force $F_p = A_p P$ (A_p = the mean area of boundary per particle), we can define to particle mobility M by

$$v = MF_p$$

The mobility, depending on the temperature and the particle size, is determined by the relative rates of several processes illustrated in Figure 12. In this case it might be thought that the particle could advance as a rigid body through the solid by the diffusive motion of the surrounding matrix. Then its mobility would be determined by the fastest of the three alternative paths in the upper part of Figure 12: 1) diffusion through the matrix, 2) diffusion along the interface and 3) diffusion through the particle itself, which in this case can be ruled out. Because of the high stability of the dispersoid, the diffusional processes involving the movement of the atoms of the particle (lower part of Figure 12), can also be excluded. According to Ref. 11, the crystallinity of the matrix imposes a constraint on the mobility of the particle. Removing a single matrix atom (black square, Figure 12) does not permit the remaining matrix atoms to rearrange so that the particle moves forward. The crystal lattice is still well defined and all the remaining atoms occupy defined positions. To overcome this problem, it was assumed that a dislocation-like defect climbs in the interface, but with a mobility which is limited by diffusion in the particle. Another possibility is the advance by the diffusive motion of a particle atom into the vacancy left by the removal of (m). However, because of the high stability of the yttria particle, it seems probable that in this case the necessary diffusional processes to move the particle take place in the matrix or at the particle/matrix interface.

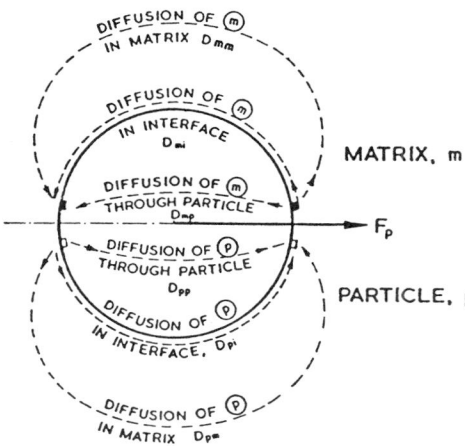

Figure 12. Diffusion paths (Ref. 11).

When analyzing more carefully the movement of the growing γ' boundary, it is clear that there is a net flux of γ'-forming elements through the boundary and into the precipitate. Clearly there exists a dragging force on the boundaries, but unfortunately it is very difficult to establish any numerical values of the particle- or the boundary mobility itself because there are too many unknown factors involved in the process. Most important, it is not clear, whether a specific γ' precipitate has existed or has nucleated and grown from the beginning of the test. There is strong evidence that the case is not so, but the Ostwald ripening detected in this study is a dynamic process.

At 1000 °C, the formation of well developed networks of interfacial γ/γ' dislocations was clearly observed as shown in Figure 13. The tendency for the formation of these dislocation networks is increased with decreasing coherency of the γ' and their function is to reduce the matrix/precipitate strain energy (Ref. 13). For this reason, these networks are not supposed to be damaging. According to Ref. 14, for the most efficient relief of misfit, the dislocations should be of edge character with Burgers vectors lying in the plane of the interface. This leads to a complicated three-dimensional model of misfit dislocations based exclusively on a/2⟨110⟩ type dislocations. The network extends over the 14 faces of a tetrakaïdecahedron: the eight {111} faces are covered with hexagonal grid patterns and the six {100} faces with square and octagonal dislocation arrangements. It may be thought that these dislocation networks help the diffusion around the dispersoid particles at the moving γ/γ' interface by pipe diffusion, and with the aid of this mechanism the dispersion

mobility is increased leading to the increased susceptibility for the PFZ formation.

TEM studies of the 1050 °C tests

In the tests at 1050 °C, the γ' had dissolved more or less completely during the heating up period. During the cooling down period, reprecipitation occurs. The precipitates are round with a diameter of approximately 50 nm. Also in this case, as well as in the case of sample 30, a general roughening of the background appears, as illustrated in Figure 14. This can be explained by assuming that most of the precipitates are smaller than the practical resolution limit of precipitates in TEM, approximately 5 nm (Ref. 10). This is supported by the observation that superlattice reflections of γ' are present in the diffraction pattern. The dispersoid structure resembles closely that detected in the tests conducted at 950 °C, although the structure appears to be more homogeneous. The dislocations are totally pinned to the dispersoids forming tangled structures.

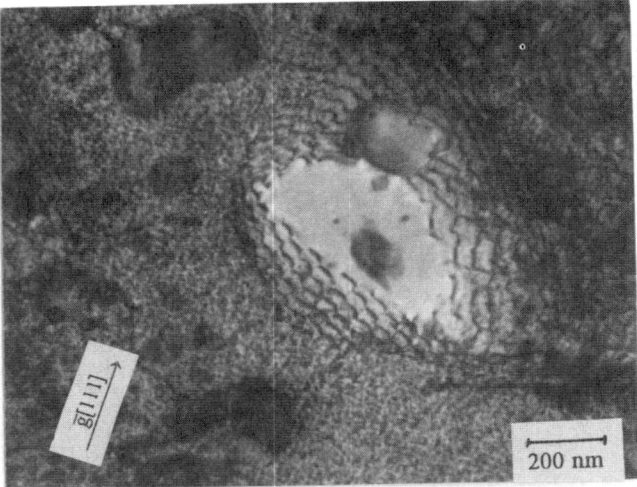

Figure 13. γ/γ' interfacial dislocations in sample 27.

Figure 14. TEM-image from sample 31.

570

Summary of Results

1. The γ' structure did not markedly change during the tests conducted at 650 °C, but at 950 °C the precipitates became clearly rounded compared with the original cuboidal shape. At 1000 °C, the process of Ostwald ripening was strongly operative and at 1050 °C, the γ' precipitates dissolved more or less completely, possibly already during the heating-up period.

2. At 650 °C and 950 °C, cutting is the dominant dislocation interaction mechanism with γ'. At 950 °C, the dislocations concentrated more into γ and γ/γ' interfaces than into γ', which seems to be the case at 650 °C.

3. Departure side pinning of dislocations to dispersoid particles was a commonly detected phenomenon. However, at 650 °C, Orowan bowing seems to dominate, the case being opposite at 950 °C and at higher temperatures.

4. The formation of dispersoid free γ' areas was detected at 1000 °C. This phenomenon is caused by Ostwald ripening, where some of the precipitates grow at the expense of the others by the diffusion of the γ' forming elements through the γ/γ' boundaries. The boundaries exert a dragging force on the dispersoid particles, thus moving them with the boundary and creating dispersoid free areas.

References

1. G. W. Meetham, "Superalloys in Gas Turbine Engines", Met. and Mat. Tech., 14(1982), 387-392.
2. J. S. Benjamin, "Mechanical Alloying", Sci. Am., 234(5)(1978), 40-48.
3. Y. Kaieda, "Trends in Development of Oxide-Dispersion-Strengthened Superalloys", Trans. Nat. Res. Inst. Metals, 28(3)(1986), 18-24.
4. M. J. Fleetwood, "Mechanical Alloying - The Development of Strong Alloys", Mat. Sci. Tech., 2(1986), 1176-1182.
5. J. W. Martin and A. Tekin, "COST 501-2, Project No. UK7, WP 1 - ODS", (Report No. 2; Jan. 89 - Jun. 90, Oxford University, Department of Materials, Oxford, UK, 1990).
6. A. Tekin and J. W. Martin, "Fatigue Crack Growth Behaviour of MA6000", Mat. Sci. Eng., (96)(1987), 41-49.
7. R. F. Decker and C. T. Sims, "The Metallurgy of Nickel-Base Alloys", in The Superalloys (eds. C. T. Sims and W. C. Hagel), (John Wiley & Sons, New York, USA, 1972), 33-77.
8. A. H. Rosenstein, J. K. Tien and W. D. Nix, "Void Formation in INCONEL MA-754 by High Temperature Oxidation" Metall. Trans. A, (17A)(1986), 151-162.
9. R. S. Herrick, J. R. Weertman, R. Petkovic-Luton and M. J. Luton, "Dislocation/particle Interactions in an Oxide Dispersion Strengthened Alloy", Scripta Metall., (22)(1988), 1879-1884.
10. M. von Heimendahl, Electron Microscopy of Materials, (Academic Press, New York, 1980), 198-199.
11. M. F. Ashby, "The Influence of Particles on Boundary Mobility", in Recrystallization and Grain Growth of Multi Phase and Particle Containing Materials, Proc. of the 1st Risø Int. Symp. on Metallurgy and Mat. Sci. (eds. N. Hansen, A. R. Jones, T. Leffers), (Risø Nat. Lab., Roskilde, Denmark, 1980), 325 - 336.
12. M. F. Ashby and R. M. A. Centamore, "The Dragging of Small Oxide Particles by Migrating Grain Boundaries in Copper", Acta Metall., (16)(1968), 1081-1092.
13. S. D. Antolovich, S. Liu and R. Baur, "Low Cycle Fatigue Behavior of René 80 at Elevated Temperature" Metall. Trans. A, (12A)(1981), 474-481.
14. A. Lasalmonie and J. L. Strudel, "Interfacial Dislocation Networks around γ' Precipitates in Nickel-Base Alloys", Phil. Mag., (32)(1975), 937-949.
15. A. Hynnä, V.-T. Kuokkala and P. Kettunen: "Crack Initiation Studies of MA760 During High Temperature Low Cycle Fatigue", Superalloys 1996, ed. R. D. Kissinger et al. (Warrendale PA: The Minerals, Metals & Materials Society, 1996), 369-374.

HIGH-TEMPERATURE LOW-CYCLE FATIGUE BEHAVIOR OF HAYNES® 230® SUPERALLOY

L. J. Chen, Y. H. He, and P. K. Liaw

Department of Materials Science and Engineering
The University of Tennessee Knoxville
Knoxville, TN 37996-2200

J. W. Blust and P. F. Browning

Solar Turbines Inc.
2200 Pacific Highway, P. O. Box 85376, MZ R-1
San Diego, CA 92186-5376

and

R. R. Seeley and D. L. Klarstrom

Haynes International, Inc.
1020 West Park Avenue, P. O. Box 9013
Kokomo, IN 46904-9013

Abstract

Total strain-controlled low-cycle fatigue tests and stress intensity factor range (ΔK)-controlled crack growth tests without and with hold times, as well as stress relaxation tests were conducted at high temperatures for a nickel-chromium-tungsten-molybdenum HAYNES 230 superalloy sheet material. The influences of test temperature and hold time on low-cycle fatigue and crack growth behavior were determined. The low-cycle fatigue life shows a strong dependence on test temperature and hold time. At all test conditions used in this investigation, the alloy exhibits strain hardening followed by softening. The crack growth rate per cycle was found to increase with longer hold time. Scanning electron microscope (SEM) observations show that crack initiation occurred in either transgranular or intergranular mode, while the crack propagated in either transgranular or mixed transgranular and intergranular mode. These cracking modes were dependent on test temperature and hold time.

Introduction

HAYNES 230 alloy is a solid-solution strengthened nickel-chromium-tungsten-molybdenum superalloy that possesses excellent high temperature strength, outstanding resistance to oxidizing and nitriding environments, very good long-term thermal stability, and excellent forming and welding characteristics. The primary solid solution strengthening element is tungsten. The alloy is used for making high-temperature components in the aerospace, power, chemical process, and industrial heating industries. These high-temperature components generally operate under very severe conditions. Frequently, they are subjected to cyclic loading at elevated temperatures, which can lead to high-temperature, low-cycle fatigue failure. Therefore, the problem of high-temperature, low-cycle fatigue is of great importance to the design of these high-

Superalloys 2000
Edited by T.M. Pollock, R.D. Kissinger, R.R. Bowman,
K.A. Green, M. McLean, S. Olson, and J.J. Schirra
TMS (The Minerals, Metals & Materials Society), 2000

temperature components.

It has become apparent that the fatigue life under high-temperature, low-cycle fatigue conditions depends not only on testing temperature but also on waveform. Generally, in studies concerning the total strain–controlled high-temperature, low-cycle fatigue, the fatigue life was observed to decrease with increasing test temperature[1-3]. A great amount of studies indicated that the incorporation of hold time may have a pronounced effect on the cyclic life of nickel base superalloys[4-8]. Moreover, many investigations concerning the fatigue crack initiation and propagation modes have shown that at elevated temperatures, damage was usually caused by fatigue and creep mechanisms, depending on the number of cycles and duration of hold time[9-13]. Both transgranular and intergranular cracking have been observed in nickel-based superalloys after testing.

Extensive research studies show that crack growth in various materials at high temperatures can be cycle-dependent, time-dependent, or a combination of both, depending on the type of the alloy and the test conditions. Such factors as test temperature, test frequency, hold time, loading ratio, initial microstructure, and environment can significantly affect crack growth. In the studies on wrought nickel-based WASPALOY superalloy[14], it was found that the general effect of increase in temperature was to give faster fatigue crack growth rates by a factor of about three from 550 to 700°C. Moreover, for 300 seconds hold cycling, the effects of temperature on the fatigue crack growth rate were very dramatic. With increasing temperature from 550 to 650°C, there was a severe increase in crack growth rates by up to a factor of 13. However, a reduction in da/dN values was observed with increase in test temperature from 650 to 700°C. Fatigue crack growth rates have also been found to increase generally with the decrease in test frequency or the addition of hold time[15-17]. In the tests on INCONEL® alloy 718, Diboine and Pineau[18] found that the introduction of a one minute hold time resulted in an increase in fatigue crack growth rates which were twice the growth rates measured under continuous cycling. Gayda, Gabb, and Miner [19] showed a systematic increase in da/dN with the test frequency ranging from 0.02 to 5 Hz in René 95 fatigue tested in air at 650°C.

In this investigation, high-temperature, low-cycle fatigue and crack growth tests using a range of cyclic periods ranging from one second to five hours, as well as stress relaxation experiments were conducted to determine the influence of testing temperature and hold time on the low-cycle fatigue behavior of HAYNES 230 alloy. Another goal of this research was to develop improved life prediction methods and data for the design of combustor liners manufactured from sheet metal products.

Test Material and Procedures

The HAYNES 230 alloy was supplied by Haynes International,

Inc. and received in the solid-solution annealed condition. The final annealing temperature was 1,232°C. The nominal chemical composition of the alloy is given in Table I[20]. The examination of the microstructure was carried out using optical microscopy. The micrographs of metallographic sections of the alloy are shown in Figure 1. It can be noted that the alloy is essentially composed of austenitic grains with some annealing twins. In addition, a small amount of white second phase particles can be also observed, as indicated by arrows in Figure 1(b). It has been

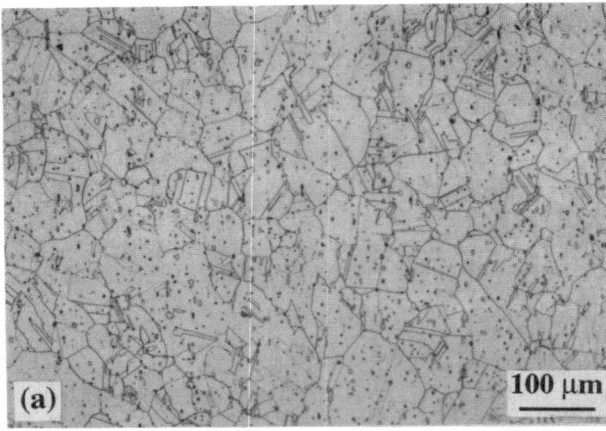

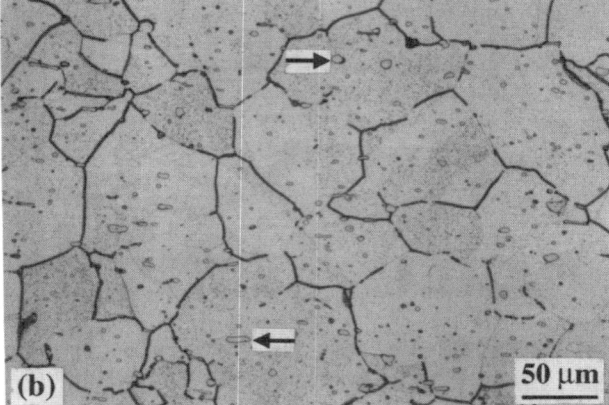

Figure 1: The initial Microstructure of HAYNES 230 alloy (as-received)

determined that these second phase particles are M_6C carbides in which M is primarily tungsten[21].

The low-cycle fatigue specimens used in this investigation were rectangular, with a thickness of 3.2 mm. The geometry of specimens is shown in Figure 2. All fully reversed pull-push low-cycle fatigue tests with and without hold times were performed in laboratory air employing a computer-controlled Material Test System (MTS) servohydraulic testing machine. A

Table I Nominal Chemical Composition of HAYNES 230 alloy (Weight Percent)

Element	Ni	Cr	W	Mo	Co	Fe	B
Content	57[a]	22	14	2	5*	3*	0.015*

[a] As balance   *Maximum

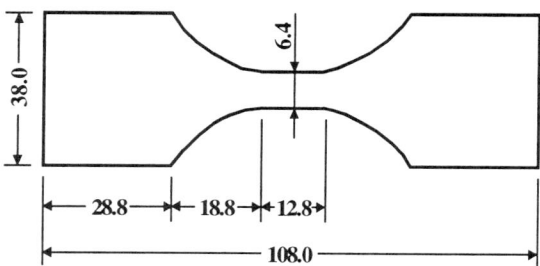

Figure 2: Geometry of low-cycle fatigue specimen (mm)

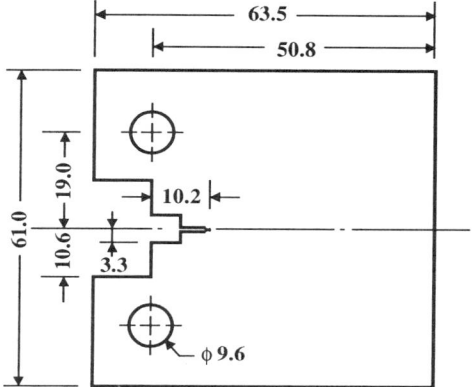

Figure 3: Geometry of compact tension specimen (mm)

high frequency induction generator was used to heat the specimen. The fluctuation of test temperature was maintained within a range of ±2°C. An axial total strain-range control mode was applied. A high-temperature extensometer was used to measure the axial strain. The extensometer was spring-loaded, and has two ceramic legs which were in direct contact with the gage-length area of the specimen. The extensometer was cooled by the forced air during high-temperature, low-cycle fatigue testing. The cyclic frequency of 1 Hz was used. For the high-temperature, low-cycle fatigue tests without hold time, the test temperatures of 760°C to 982°C were used, and the imposed axial total strain range was varied in the range of 0.25 to 1.0%. In the case of hold-time tests, the chosen test temperatures were 816°C and 927°C, respectively. In addition, a 120 seconds, 600 seconds, or infinite tensile hold time were introduced at the maximum tensile strain of each cycle. The imposed total strain range was 1.0%. All tests were run to failure.

All crack growth data were generated using the compact-tension specimen shown in Figure 3. The specimens have a thickness of 3.2 mm. Creep-fatigue crack growth specimens were prepared from the sheet materials in accordance with American Society for Testing and Materials (ASTM) standards E647-86[22]. Creep-fatigue crack propagation tests were conducted on an Instron servo-controlled, hydraulically actuated, and closed-loop test machine. An electric potential system, consisting of a constant current supply and a multiple gain amplifier, was used to continuously monitor the physical crack length on the specimens. Electric potential probes were attached to the front face of the specimen on either side of the specimen notch. Current leads were attached to the top and bottom of the specimen in order to provide a uniform current flux in the specimen gage section. The crack length was obtained by passing a constant current through the specimen, and measuring the potential drop between the two potential probes. The crack length is related to the change in the potential drop by a relationship known as Johnson's equation[23]:

$$\frac{U}{U_0} = \frac{\cosh^{-1}\left[\dfrac{\cosh\left(\pi y / W\right)}{\cos\left(\pi a / W\right)}\right]}{\cosh^{-1}\left[\dfrac{\cosh\left(\pi y / W\right)}{\cos\left(\pi a_0 / W\right)}\right]} \tag{1}$$

Where U_0 and a_0 are the initial values of the potential and crack length, U and a the instantaneous values of the potential and crack length, y one half the probe gage length, and W the specimen width. In addition, a high-temperature clip gage was placed across the crack mouth opening to measure the crack opening displacement. The specimens were precracked to approximately 1.27 mm. The final ΔK used during precracking was 20 MPa $\sqrt{m}$. The precracking was performed at a R-ratio of 0.1, using a triangular waveform at a frequency of 10 Hz. All crack growth tests were conducted in laboratory air at a temperature of 816°C under the constant ΔK control mode. A ΔK level of 27.5 MPa $\sqrt{m}$ and a R-ratio of 0.05 were employed. A triangular waveform and a frequency of 0.333 Hz were used in the crack growth test without hold time. Crack growth experiments with various hold times ranging from 3 to 18,000 seconds were conducted using a baseline frequency of 0.333 Hz with superimposed hold times at maximum load. Crack growth rates at different test conditions were determined through computer analyses of the crack length versus either number of cycles or time data using an apparent linear fitting program.

The fracture surface of some specimens subjected to low-cycle fatigue and creep-fatigue crack growth tests were examined using a Cambridge S-320 scanning electron microscope (SEM) in order to determine the crack initiation and propagation mode.

<u>Test Results and Analysis</u>

<u>Strain Fatigue Life Behavior</u>

<u>The Effect of Temperature</u>   For high-temperature, low-cycle fatigue without hold time, the curves of the fatigue life versus total strain range at different test temperatures are shown in Figure 4[24]. It can be seen that the low-cycle fatigue properties of HAYNES 230 alloy exhibit some dependence on the test temperature and imposed total strain range. At total strain ranges less than 0.7%, the low-cycle fatigue life of the alloy decreased significantly with increasing test temperature in the range of 760°C to 982°C. At a total strain range of 1.0%, it was found that the fatigue life of the alloy did not decrease with increasing test temperature, as observed in other superalloy systems. In the

temperature range of 760°C to 871°C, increasing the test temperature led to a significant reduction in the fatigue life of the alloy, while the fatigue life of the alloy appeared to increase with test temperature in the 871°C to 927°C range.

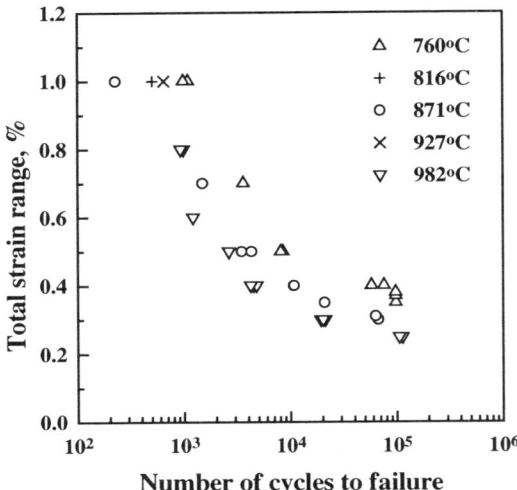

Figure 4: Fatigue life as a function of total strain range at different temperatures

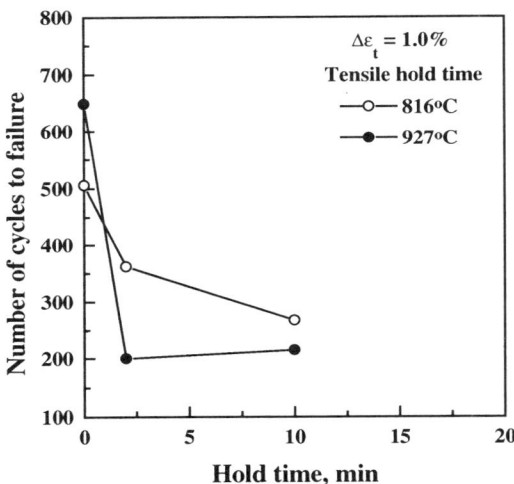

Figure 5: Fatigue life as a function of strain hold time

The Effect of Hold Time The low-cycle fatigue life of the alloy as a function of hold time is illustrated in Figure 5. It can be noted that the introduction of the tensile strain hold time resulted in a reduction of the fatigue life of the alloy. The amplitude of reduction in fatigue life appeared to be related to the test temperature and duration of tensile strain hold. Although the alloy showed a longer fatigue life at 927°C than at 816°C under the continuous cycling conditions, all low-cycle fatigue tests with hold times gave the shorter fatigue lives at 927°C than at 816°C.

Cyclic Stress Response Behavior

The cyclic stress response behavior of nickel-based superalloys under continuous cycling has been reported extensively in the literature. It has been demonstrated that their cyclic stress response is closely related to factors like the type and initial microstructure of alloy, test temperature, and strain rate. Some micromechanisms have been suggested to explain the cyclic hardening and softening phenomena occurring during cyclic deformation[25-27].

Vecchio, Fitzpatrick, and Klarstrom[21] found in their study on low-cycle fatigue properties of HAYNES 230® alloy at three different temperatures of 760, 871, and 982°C that the alloy usually exhibited cyclic hardening. It appears that the degree of cyclic hardening, which was defined as the midlife stress amplitude divided by the first-cycle stress amplitude, would decrease with increasing test temperature from 760 to 982°C.

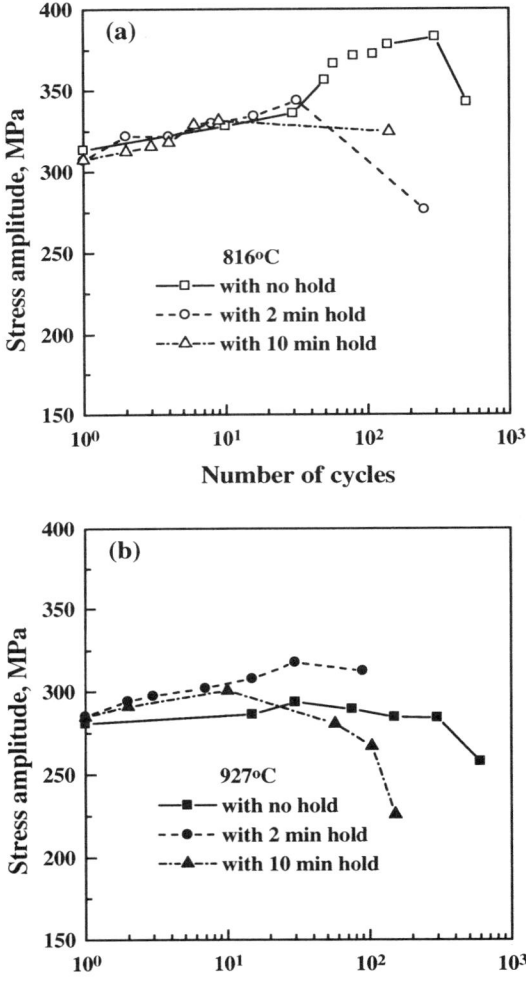

Figure 6: Cyclic stress response curves of HAYNES 230 alloy during low-cycle fatigue with and without hold time at (a) 816°C and (b) 927°C

576

Figures 6(a) and (b) illustrate the cyclic stress response curves for the strain hold times at 816°C and 927°C, respectively. It can be noted from the figures that the alloy showed an initial strain hardening to a peak stress followed by a period of softening. It is apparent that the number of cycles to reach the peak stress decreased due to the addition of a hold time. At 816°C, the cyclic stress response behavior of the alloy was similar in the early stage of cyclic deformation for low-cycle fatigue tests without and with hold time. However, it is notable that at 927°C, both tests with the tensile hold time showed a higher level of cyclic stress amplitude than continuous cycling in the initial stage of cyclic deformation. For the low-cycle fatigue test with a hold time of 2 minutes, the cyclic stress amplitude remained at a higher level throughout the test as compared to continuous cycling and the test with a 10 minute hold time.

Stress Relaxation Behavior

The stress relaxation curves of HAYNES 230 alloy at both temperatures of 816°C and 927°C during low-cycle fatigue with infinite hold time are shown in Figure 7. It can be observed that HAYNES 230 alloy showed similar stress relaxation behavior at both test temperatures used in this investigation. At the beginning of an infinite hold, a very rapid drop of the tensile stress occurred, while during the remaining period, comparatively less additional stress relaxation took place. The kinetics of stress relaxation during infinite hold was also similar at temperatures of 816°C and 927°C, with the majority of stress relaxation occurring within the initial 50 seconds.

In order to clearly show the dependence of the stress relaxation behavior on test temperature, the amount of stress relaxation

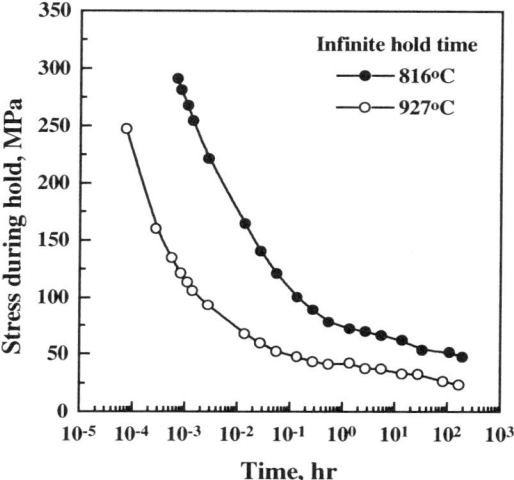

Figure 7: Stress relaxation curves of HAYNES 230® alloy

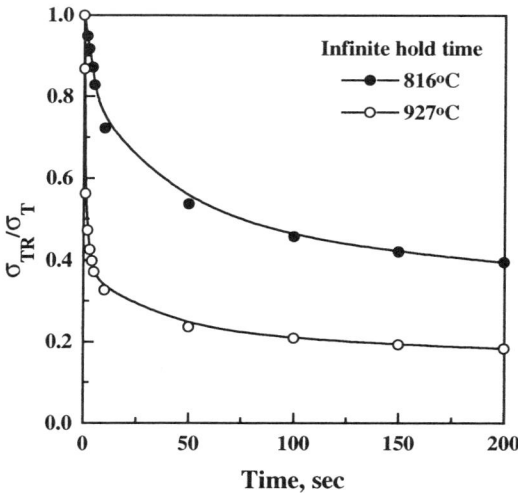

Figure 8: The amount of stress relaxation relative to time in the early stage of hold

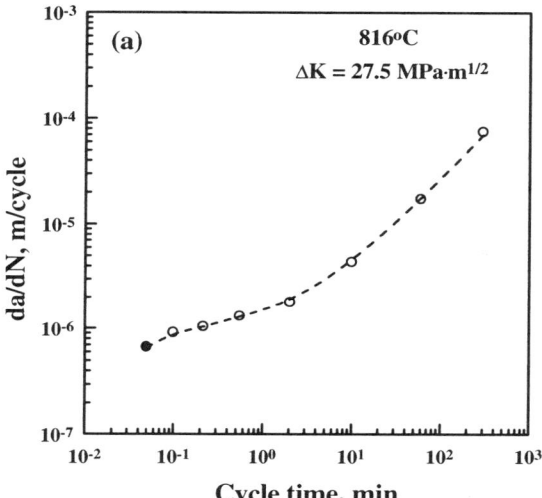

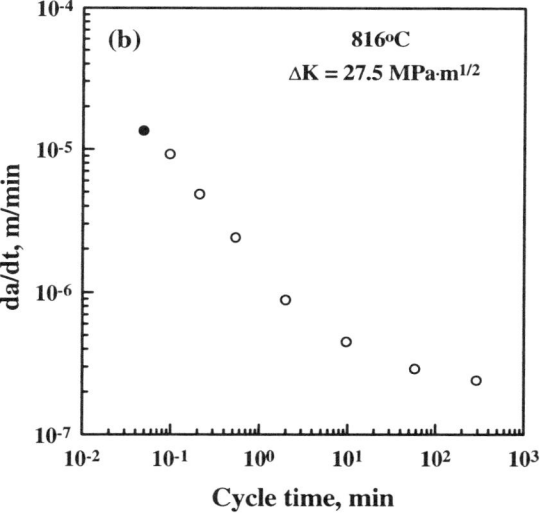

Figure 9: Crack growth rates in HAYNES 230 at 816°C. Solid symbol refers to cycling with no hold time.

occurring in the early stage of infinite hold tests relative to time is depicted in Figure 8. In the figure, σ_T represents the initial maximum tensile stress, and σ_{TR} is the relaxed stress at different times. It is obvious that increasing test temperature resulted in greater amounts of stress relaxation.

Influence of Hold Time on fatigue Crack Growth Rate

The crack growth rate data obtained from tests conducted at 816°C and different cycle times are presented in Figure 9. As seen in Figure 9(a), the introduction of a hold time at maximum loading led to a significant increase in the crack growth rate per cycle compared with that under continuous cycling. As expected, the crack growth rate per cycle increased with increasing hold time. It was also noted that at hold times less than 2 minutes, the increase in da/dN with hold time was lower. However, at the hold times above 10 minutes, da/dN showed a rapid and nearly linear increase.

When the crack growth per unit time (da/dt) is plotted versus cycle time, the crack growth rate is found to decrease with increasing hold time, as shown in Figure 9(b). The crack growth rate per unit time decreases rapidly at the shorter hold times and begins to reach a stable value at the longer hold times, indicating that time-dependent effects are beginning to dominate the rate of crack propagation at the longer hold time.

There existed an immediate relationship between da/dN and da/dt, which could be expressed as follows:

$$da/dN = \left(\frac{1}{60f} + t_h\right)(da/dt) \qquad (2)$$

where da/dN is measured in m/cycle, da/dt in m/min, f is the test frequency for continuous cycling and measured in Hz, and t_h is the hold time per cycle and measured in minutes. Using equation (2), the conversion between da/dN and da/dt can be easily done. Dash line in Figure 9(a) represents the corresponding da/dN values at different hold times calculated from equation (2) using actually obtained da/dt data given in Figure 9(b). It is obvious that the measured value of da/dN is considerably consistent with the calculated value.

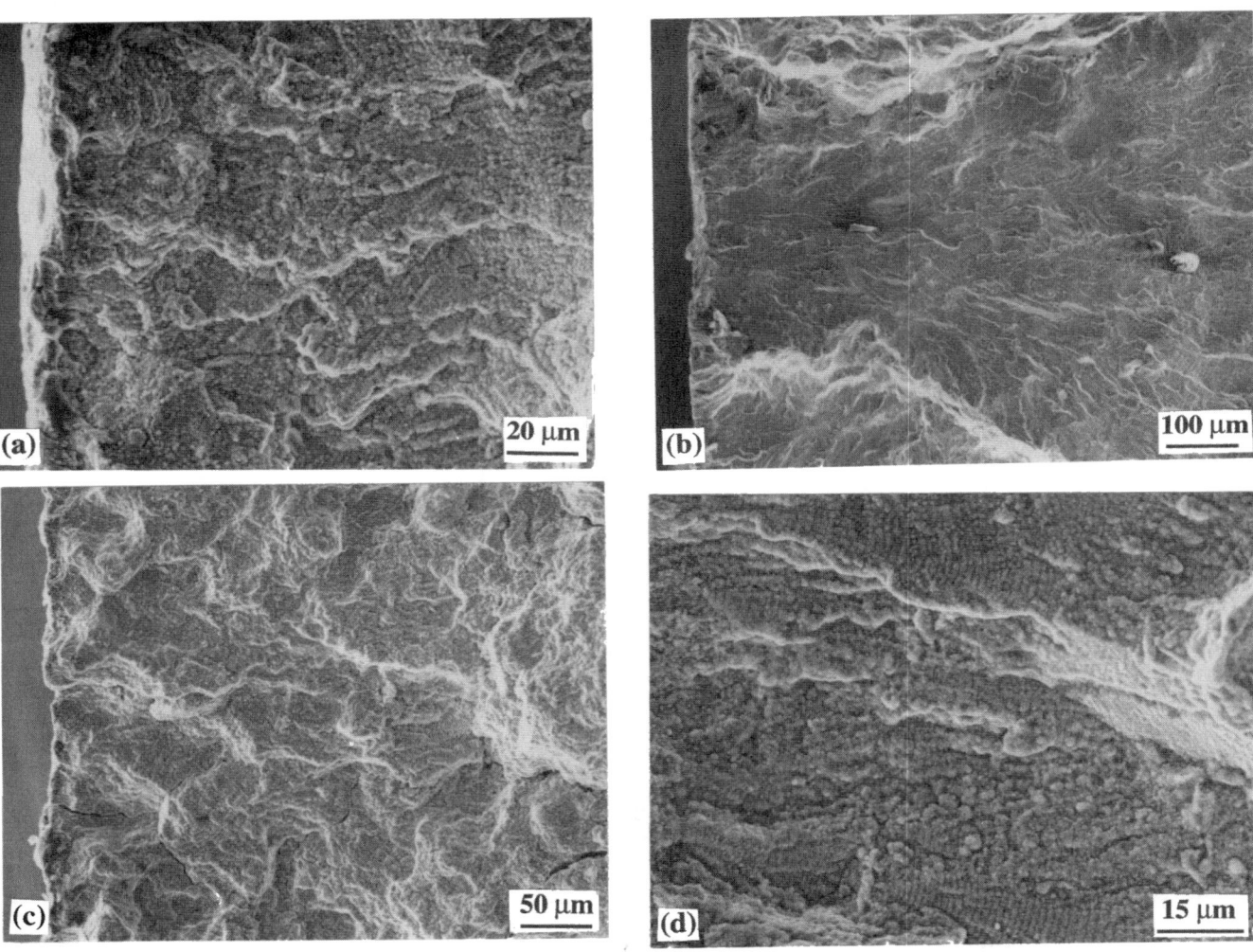

Figure 10   The typical SEM photographs of fracture surfaces for low-cycle fatigue tests
(a) 816°C, 10 minute hold time;   (b) 927°C, without hold time
(c) 927°C, 10 minute hold time;   (d) 816°C, 2 minute hold time

578

Examination of the fracture surfaces of the low-cycle fatigue and creep-fatigue crack growth tested specimens was performed in a scanning electron microscope. Figure 10 shows the typical SEM photographs of fracture surfaces obtained under different test conditions. There existed multiple crack initiation sites on all fracture surfaces, which originated entirely from the surface of specimens. For all of low-cycle fatigue tests with and without hold times performed at 816°C, it was observed that crack initiated in a transgranular mode, as exemplified in Figure 10(a) for the test with a 10 minute hold time. At 927°C, the crack initiation mode was also found to be transgranular in low-cycle fatigue tests without hold time and with a 2 minute hold time, as exemplified in Figure 10(b). However, for the test with a 10 minute hold time, it can be seen that crack initiated intergranularly, as shown in Figure 10 (c).

For all low-cycle fatigue tests carried out in this investigation at both temperatures of 816 and 927°C, crack propagation occurred

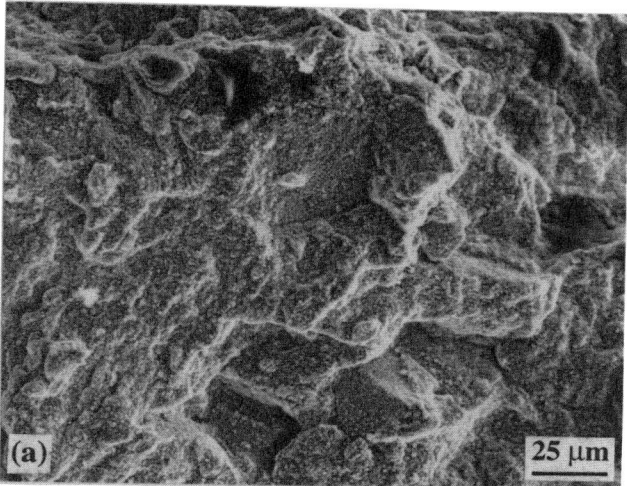

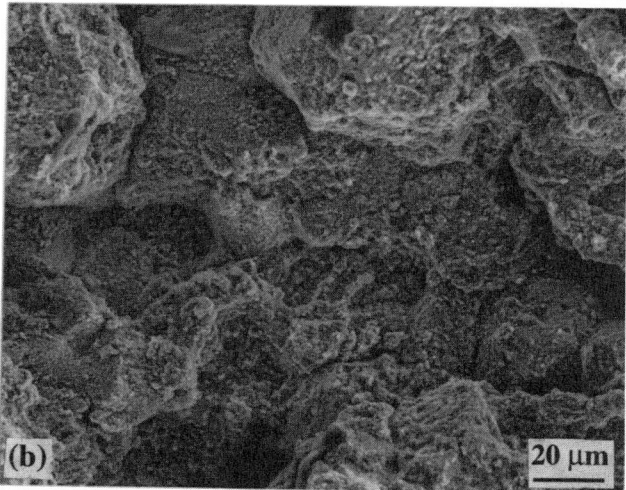

Figure 11 The typical SEM photographs of fracture surfaces for crack growth tests
(a) 816°C, 1 hour hold time;   (b) 816°C, 5 hour hold time

predominately in a transgranular manner, even if the hold time of 10 minutes was introduced. As shown in Figure 10(d), well-defined fatigue striations can be seen on the fracture surfaces, which are considered to be characteristic features of transgranular crack propagation.

The SEM observations on creep-fatigue crack growth specimens tested at 816°C showed the crack propagated transgranularly when the hold times were shorter than 10 minutes. However, if hold times of one and five hours were introduced at the maximum tensile stress, the crack propagation mode became the mixed transgranular and intergranular, as shown in Figure 11(a) and (b), respectively. It was also noted that a greater proportion of intergranular cracking was observed in the test with a hold time of five hours than for one hour.

<u>Discussion</u>

Such time-dependent damages as creep and oxidation can be used to interpret the changes in fatigue life with test temperature and tensile hold time. Oxidation damage is responsible for the reduction in fatigue life especially at higher test temperatures because it tends to accelerate crack initiation and to assist the propagation process of a fatigue crack. Creep taking place at higher temperatures and during the tensile hold time can also promote propagation of a fatigue crack and lead to a decrease in fatigue life. The significant decrease of the fatigue life in the 2 minute hold time test performed at 927°C can be at least considered to be related to the particular cyclic stress response behavior. The observed higher response of stresses developed during cyclic deformation can result in a large stress concentration at the crack tip, which will account for the accelerated crack growth, and hence, a significantly reduced number of cycles in the crack propagation stage.

The phenomenon, that hold time will increase crack growth rates, can be in part attributed to the stress relaxation and environmental degradation, especially oxidation. The stress relaxation that occurred during hold time led to an increase of the accumulated strain in the crack tip region, which will induce crack growth during the hold time. In addition, the crack tip stress relaxation from the previous hold time can increase the amount of crack tip sliding and accelerate the crack propagation during subsequent fatigue cycling. The additional increase in the crack growth rate per cycle due to the introduction of one and five hour hold time may be associated with the intergranular fracture mode.

The intergranular crack initiation behavior exhibited at 927°C and 10 minute hold time may be associated with oxidation damage. It was well known that grain boundaries are the most favorable sites for oxygen penetration and oxidation due to the lattice mismatch and presence of oxidizable elements, such as Cr, at grain boundaries. The oxidized grain boundary apparently weakened, and had lower strength than the matrix. Consequentially, the crack would be localized on the inherently weak grain boundary. Because oxidation is both time-dependent and thermally-activated processes, it is logical to conclude that intergranular crack initiation should occur at the combined condition of higher temperatures and longer hold times.

Conclusions

Based above analysis and discussion, the following conclusions can be made:

1. During low-cycle fatigue with tensile hold times, HAYNES 230 alloy usually exhibited cyclic hardening followed by softening.

2. The fatigue life of HAYNES 230 alloy generally decreased due to the introduction of the tensile hold time.

3. The stress relaxation behavior showed that at the beginning of strain hold, the stress dropped very quickly and then decreased very slowly with increasing time.

4. The creep-fatigue crack growth rates of HAYNES 230 alloy exhibited a strong dependence on the tensile hold time. With increasing tensile hold time, da/dN increased, while da/dt decreased.

5. Cracks usually initiated in either trangranular or intergranular mode at the surface of the specimens. For tests without hold time and with hold times less than 10 minutes, the crack propagation mode was transgranular. For tests with hold times above 1 hour, crack propagation occurred in a mixture of transgranular and intergranular modes.

Future Work

The work reported here shows the initial results of an ongoing research effort to characterize the effects of time-dependent processes on crack initiation and propagation in Haynes 230. Future work will focus on characterizing the effects of longer hold times, up to five hours, on LCF life. Additional work will be performed to characterize effects at other strain ranges and temperatures, and to study statistical variation in test results by performing multiple tests at selected test conditions. These results will be combined with microstructural and fractographic observations to produce life prediction methods that allow for incorporation of time-dependent effects in the prediction of hot section component lives in industrial gas turbines.

Acknowledgements

This work is supported by the Solar Turbines Inc., Haynes International, Inc., the University of Tennessee, and the U. S. Department of Energy's Advanced Turbine Systems Program. We also acknowledge the financial support of the National Science Foundation, the Division of Design, Manufacture, and Industrial Innovation, under Grant No. DMI-9724476, and the Combined Research-Curriculum Development Program, under EEC-9527527, to the University of Tennessee, Knoxville, with Dr. D. Durham and Ms. M. Poats as contract monitors, respectively. Low-cycle fatigue testing was performed at the University of Tennessee and Metcut, Inc. Creep-fatigue crack growth testing was carried out at Westmoreland Testing, Inc.

References

1. H. F. Merrick, "The Low Cycle Fatigue of Three Wrought Nickel-Base Alloys," Metallurgical Transactions, 5 (1974) 891-897.

2. D. Fournier and A. Pineau, "Low Cycle Fatigue Behavior of Inconel 718 at 298K and 823K," Metallurgical Transactions, 8A (1977) 1095-1105.

3. M. A. Burke and C. G. Beck, "The High Temperature Low Cycle Fatigue Behavior of the Nickel Base Alloy IN-617," Metallurgical Transactions, 15A (1984) 661-670.

4. M. Y. Nazmy, "High Temperature Low Cycle Fatigue of IN 738 and the Application of Strain Range Partitioning," Metallurgical Transactions, 14A (1983) 449-461.

5. D. C. Lord and L, F. Coffin, Jr., "Low Cycle Fatigue Hold Time Behavior of Cast René 80," Metallurgical Transactions, 4 (1973) 1647-1654.

6. M. D. Mathew, V. Singh, W. Chen, and R. P. Wahi, "Life Prediction for a Nickel Base Alloy Nimonic PE 16 Under Low Cycle Fatigue Loading at 923K," Acta Metallurgica Materialia, 39 (1991) 1507-1513.

7. H. Tsuji and T. Kondo, "Strain-Time Effects in Low-Cycle Fatigue of Nickel-Base Heat-Resistant Alloys at High Temperature," Journal of Nuclear Materials, 190 (1987) 259-265.

8. W. J. Ostergren, "A Damage Function and Associated Failure Equations for Predicting Hold Time and Frequency Effects in Elevated Temperature Low Cycle Fatigue," Journal of Testing and Evaluation, 4 (1976) 327-339.

9. S. D. Antolovich, S. Liu, and R. Baur, "Low Cycle Fatigue Behavior of René 80 at Elevated Temperature," Metallurgical Transactions, 12A (1981) 473-481.

10. B. A. Lerch and N. Jayaraman, "A Study of Fatigue Damage Mechanisms in Waspaloy from 25 to 800°C," Materials Science and Engineering, 66 (1984), 151-166.

11. K. B. S. Rao, H. Schiffers, H. Schuster, and H. Nickel, "Influence of Time and Temperature Dependent Processes on Strain Controlled Low Cycle Fatigue Behavior of Alloy 617," Metallurgical Transactions, 19A (1988) 359-371.

12. C. J. McMahon and L. F. Coffin, Jr., "Mechanisms of Damage and Fracture in High-Temperature Low-Cycle Fatigue of a Cast Nickel-Based Superalloy," Metallurgical Transactions, 1 (1970) 3443-3450.

13. S. Bashir, P. Taupin, and S. D. Antolovich, "Low Cycle Fatigue of As-HIP and HIP + Forged René 95," Metallurgical Transactions, 10A (1979) 1481-1490.

14. J. Byrne, R. Hall, and L Grabowski, "Elevated Temperature Fatigue Crack Growth Under Dwell Conditions in Waspaloy," International Journal of Fatigue, 19 (1997) 359-367.

15. J. Pédron and A. Pineau, "The effect of Microstructure and Environment on the Crack Growth Behavior of Inconel 718 Alloy at 650°C under Fatigue, Creep, and Combined Loading," Materials Science and Engineering, 56 (1982) 143-156.

16. B. A. Cowles, D. L. Sims, J. R. Warren, and R. V. Miner, Jr., "Cyclic Behavior of Turbine Disk Alloys at 650°C," Journal of Engineering Materials and Technology, 102 (1980) 356-363.

17. P. Shahinian and K. Sadananda, "Crack Growth Behavior Under Creep-Fatigue Conditions in Alloy 718," 1976 ASME-MPC Symposium on Creep-Fatigue Interaction, (New York, NY: American Society of Mechanical Engineers, 1976), 365-390.

18. A. Diboine and A. Pineau, "Creep Crack Initiation and Growth in Inconel 718 Alloy at 650°C," Fatigue and Fracture of Engineering Materials and Structures, 10 (1987) 141-151.

19. J. Gayda, T. P. Gabb, and R. V. Miner, "Fatigue Crack Propagation of Nickel-Base Superalloys at 650°C," Low Cycle Fatigue, ASTM STP 942, ed. H. D. Solomon, G. R. Halford, L. R. Kaisand, and B. N. Leis (Philadelphia, PA: American Society for testing and Materials, 1988), 293-309.

20. Haynes International Technical Bulletin H-3000F, HAYNES 230® Alloy.

21. K. S. Vecchio, M. D. Fitzpatrick, and D. L. Klarstrom, "Influence of Subsolvus Thermomechanical Processing on the Low-Cycle Fatigue Properties of Haynes 230 Alloy," Metallurgical and Materials Transactions, 26A (1995) 673-689.

22. ASTM Standard E647-86, 1986 Annual Book of ASTM Standards, ASTM, Philadelphia, PA, 1986, Vol. 3.01, pp. 714-736.

23. H. H. Johnson, "Calibrating the Electric Potential Method for Studying Slow Crack Growth," Materials Research and Standards, 5 (1965) 442-445.

24. R. R. Seeley and V. R. Ishwar, "Fatigue in Modern Nickel-Base Alloys for Gas Turbine Applications," Proceedings of Life Assessment of Hot Section Gas Turbine Components, October 5-7, 1999, Edinburgh, UK.

25. S. K. Hwang, H. N. Lee, and B. H. Yoon, "Mechanism of Cyclic Softening and Fracture of An Ni-Base γ'-Strengthened Alloy Under Low-Cycle Fatigue," Metallurgical Transactions, 20A (1989) 2793-2801.

26. R. E. Stoltz and A. G. Pineau, "Dislocation-Precipitate Interaction and Cyclic Stress-Strain Behavior of a γ' Strengthened Superalloy," Materials Science and Engineering, 34 (1978) 275-287.

27. J. B. Lerch and V. Gerold, "Room Temperature Deformation Mechanisms in Nimonic 80A," Acta Metallurgica, 33 (1985) 1709-1716.

High-Cycle Fatigue of ULTIMET® Alloy

L. Jiang,[1] C. R. Brooks,[1] P. K. Liaw,[1] and D. L. Klarstrom[2]

1. Department of Materials Science and Engineering
The University of Tennessee Knoxville
Knoxville, TN 37996-2200

2. Haynes International, Inc.
Kokomo, IN 46904-9013

Abstract

ULTIMET® alloy, a relatively new commercial Co-26Cr-9Ni superalloy, was studied to examine its fatigue behavior at low and high frequencies and in different environments. The microstructure of the alloy in the solution-annealed condition consisted of face-centered-cubic, single phase, relatively fine, and uniform grains with annealing twins. Stress-controlled fatigue tests were performed at room temperature with an R-ratio (the ratio of the applied minimum stress to the maximum stress levels) of 0.05. The high-frequency (1,000 Hz) high-cycle fatigue tests gave results comparable to the low-frequency (20 Hz) tests. Interestingly, there was a plateau region in the S-N (maximum applied stress level versus number of cycles to failure or run-out) curves of this alloy regardless of the testing frequencies and environments. At 20 Hz, the maximum stress level of the plateau was around 600 MPa, which is approximately equal to the yield strength of 586 MPa, and at 1,000 Hz the plateau region occurred at 540 MPa. The microstructural characterization of tested specimens revealed that the evolution of stacking faults played an important role in the fatigue process. The results of fractographic studies indicated that cracking was initiated either on the specimen surface or subsurface depending on test conditions, and the crack-initiation sites were cleavage-like in nature, typical of stage I crack initiation. The presence of oxygen and/or moisture in the test environments was found to have significant effects on the fatigue-crack initiation behavior and fatigue life of ULITMET alloy. The fatigue-fracture surfaces were observed to have a crystallographic appearance for all of the conditions investigated.

Introduction

Fatigue-crack initiation and propagation have been identified as one of the critical areas in determining the safety, durability, and reliability of structural components. To our knowledge, little work has been done on the fatigue behavior of ULTIMET alloy. In the present work, fatigue-crack initiation and propagation characteristics of ULTIMET alloy were investigated.

The availability of fatigue data in ULTIMET alloy will provide new commercial opportunities. In addition, the systematic study of the effects of applied maximum stress levels, frequencies, and environments on the fatigue lives of ULTIMET alloy offers pertinent information for engineering applications. It also facilitates the understanding of fatigue-crack nucleation and propagation behavior in light of microstructural characterizations by x-ray diffraction, optical microscopy (OM), scanning electron microscopy (SEM), transmission electron microscopy (TEM), etc.

This paper presents the first comprehensive description of the high-cycle fatigue behavior of ULTIMET alloy tested at room temperature under two frequencies, 20 Hz and 1,000 Hz, and various environments, including typical air with 20% to 24% relative humidity, 0 grade dry air with 0% relative humidity, and vacuum of 10^{-6} to 10^{-7} torr. Based on the S-N curves, the effects of environments and frequencies on fatigue life were examined.

Furthermore, subsurface crack-initiation sites were observed for ULTIMET alloy undergoing high-cycle fatigue. Internal fatigue-crack-initiation origins in most materials have invariably been associated with the presence of inclusions, but ULTIMET alloy is a "clean" material due to Argon-Oxygen Degassing (AOD) and Electric Slag Remelting (ESR) refining processes. Observations were made of the effect of environmental conditions and testing frequencies by microscopic techniques in order to understand the formation of internal fatigue-crack-initiation origins.

Experimental

Material

Wrought cobalt-based alloys including ULTIMET alloy provide excellent elevated-temperature strength, and resistance to various forms of wear and sulfidizing environments.[1-5] The nominal chemical composition of ULTIMET alloy, developed by Haynes International, Inc., is shown in Table I.[6] It possesses high tensile strength combined with excellent impact toughness and ductility.[6] The typical room-temperature yield strength of the alloy plated tested was 586 MPa, the tensile strength was 1,000 MPa, and the elongation was 39%. ULTIMET alloy is an appropriate material for fabricating structural components, such as agitators, blenders, spray nozzles, screw conveyors, and valve parts.

Table I: Nominal Chemical Composition of ULTIMET Alloy (Weight Percent).

Co	Cr	Ni	Mo	Fe	W	Mn	Si	N	C
54[a]	26	9	5	3	2	0.8	0.3	0.08	0.06

[a] as balance

The element cobalt exhibits two crystallographic forms. Below 400 °C, the stable structure is hexagonal-close-packed (HCP or ε); above this temperature, it is face-centered-cubic (FCC or γ), as shown in Figure 1.[7] This phase transformation makes cobalt a unique alloy base.[1] For most of the cobalt-based alloys, the cooling rates from the solution annealing temperature are high enough to avoid the FCC to HCP transformation, and the lack of sufficient thermal energy prevents the transformation from occurring at room temperature. However, the transformation can be triggered at room temperature by mechanical deformation. For Co-27Cr-5Mo alloys, the FCC to HCP transformation can be easily triggered by mechanical deformation.[5,8] Haynes 25 alloy, a

Superalloys 2000
Edited by T.M. Pollock, R.D. Kissinger, R.R. Bowman,
K.A. Green, M. McLean, S. Olson, and J.J. Schirra
TMS (The Minerals, Metals & Materials Society), 2000

Co-20Cr-15W-10Ni alloy, can result in the FCC to HCP transformation by 20% cold reduction.[5,8] For MP35N alloy, a Co-35Ni-20Cr-10Mo alloy, the stress (or strain) induced FCC to HCP transformation does not occur at room temperature, however, extensive twinning has been observed.[5,8] Therefore, it is likely that ULTIMET alloy may undergo the γ-ε transformation during fatigue testing.

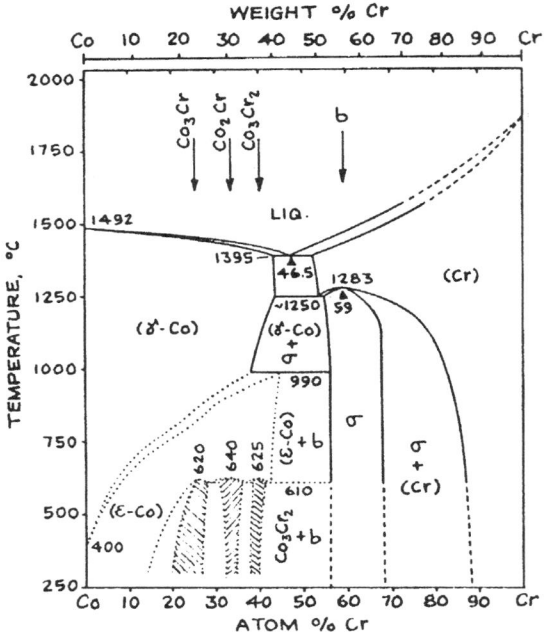

Figure 1: Cobalt-Chromium Binary Phase Diagram.[7]

The specimens of ULTIMET alloy, used for high-cycle fatigue testing, were produced as follows: (a) Plate feedstock was reduced in thickness from 30.48 cm to 1.27 cm in a 4-pass, cross-rolling sequence at 1,200°C, and (b) The material was solution-annealed at 1,120°C for about 20 to 30 minutes, then water-quenched to room temperature in order to retain the FCC structure.

Fatigue Testing

Uniaxial high-cycle fatigue tests were performed using smooth round-bar specimens. The test procedures were in accordance with the American Society for Testing and Materials (ASTM) E466 for "Conducting Constant Amplitude Axial Fatigue Tests of Metallic Materials."[9] The gage section of the test specimens was 5.08 mm in diameter and 19.05 mm in length. The final polishing operation produced fine marks parallel to the longitudinal axis of the test specimen, and the surface finish of a surface roughness smaller than 0.2 μm was in compliance with the ASTM standard.

For low-frequency (20 Hz) fatigue testing, a newly developed electrohydraulic material test system, MTS® 810, was used. This system was equipped with an advanced CENTORR/VI TESTORR™ furnace, such that tests could be conducted in different environments, including air, vacuum, argon, etc. The high-vacuum environment was obtained by first using a mechanical pump to obtain a rough vacuum, and then the furnace was evacuated with a diffusion pump to reach 10^{-6} to 10^{-7} torr vacuum within a few hours.

For high-frequency (1,000 Hz) fatigue testing, a state-of-the-art material test system, MTS® 810 1,000 Hz, was employed. This material test system had a Teststar II controller® and operation software. It had high-speed and accurate data acquisition capabilities. The "voice coil" servovalve enabled the material test system to conduct tests at 1,000 Hz. This high-frequency material test system provided the capabilities to conduct the fatigue test under accurate control and complete a fatigue test of life up to 10^9 cycles with in a reasonable time period (11 days).

High-cycle fatigue tests were performed under load control using a sinusoidal waveform at 20 Hz and 1,000 Hz. Loading was from tension to tension, with an R-ratio of 0.05, and in different environments, i.e., regular air, dry air, and vacuum. All tests were conducted at ambient temperature (25 °C). The specimens were cyclically loaded until failure or up to approximately 10^7 cycles as run-out for 20 Hz test, up to 10^9 cycles as run-out for 1,000 Hz test.

Microstructural Characterization

The as-received and tested specimens were subjected to detailed microstructural characterizations using X-ray diffraction, OM, SEM, and TEM, and SEM for fractography.

The metallographic specimens prepared for the OM and SEM observations were mechanically polished according to standard procedures. Prior to etching, the mechanically-polished specimens were electropolished. A solution of 730 ml ethanol, 78 ml perchloric, 90 ml distilled water, and 100 ml butylcellosolve was used for electropolishing. The mechanically-polished specimens were placed on a Buehler Electromet®4 Polisher/Etcher at a direct current (DC) potential of 20 volts with a stainless steel cathode for 5 to 10 seconds in order to remove the mechanically-deformed surface by electropolishing.

For etching, a solution composed of 95 ml of reagent grade hydrochloric acid (HCl) and 5 gram oxalic acid powder was used. The specimen was contacted with a stainless steel probe as the anode. A graphite rod was used as the cathode, and it was placed in the electrolyte near the specimen. Etching was conducted at a DC potential of 1 volt for 5 to 10 seconds.

The specimens for the TEM study were sliced 0.1 mm thick and 3 mm diameter, and then mechanically polished. They were electropolished by a Struers Tenupol-3 electrolytic polishing apparatus at a DC potential of 20 volts, using the solution mentioned above.

Temperature Measurement

In the present study, an advanced, high-speed, and high-sensitivity Raytheon Radiance HS® infrared (IR) imaging system was used to record the temperature changes during high-cycle fatigue. The IR thermography is the process of detecting the invisible infrared radiation and converting the energy detected into visible light. The resultant image depicts and quantifies the energy being radiated and reflected from the object viewed, and can be transformed into temperature maps. IR thermography is a convenient technique for developing digital temperature maps from the invisible radiant energy emitted from stationary or moving objects at any distance. There is no surface contact or any perturbation of the actual surface temperature of the objects investigated.

The IR camera employed had a 256 x 256 pixels focal-plane-array (FPA) InSb detector, and was sensitive to 3–5 μm wavelength thermal radiation. The camera operates in a snapshot mode at a frequency up to 144 Hz, and can be externally triggered. The temperature resolution of the camera is 0.015 °C at 23 °C. Before fatigue testing, a thin submicroscopic graphite coating was applied to the specimen in order to reduce IR reflections. The temperature measurement of the IR camera was calibrated by a thermocouple, and the relationship of IR radiant-energy intensity and temperature was established by a best-fit line of the measured results.

Results and Discussion

High-Cycle Fatigue Stress-Life Data

The stress vs. fatigue life in terms of number of cycles to failure or run-out (S-N) curves, including 20 Hz in regular air, dry air, and vacuum, and 1,000 Hz in regular air, are shown in Figure 2. The regular air environment contained the typical content of about 20% oxygen and 80% nitrogen (volume percent). In the laboratory, the relative humidity of the regular air was 20% to 26%. The test environment of dry air contained the same amount of oxygen and nitrogen but with 0% relative humidity. For the vacuum environment, the amount of oxygen and moisture was negligible under the vacuum of 10^{-6} to 10^{-7} torr.

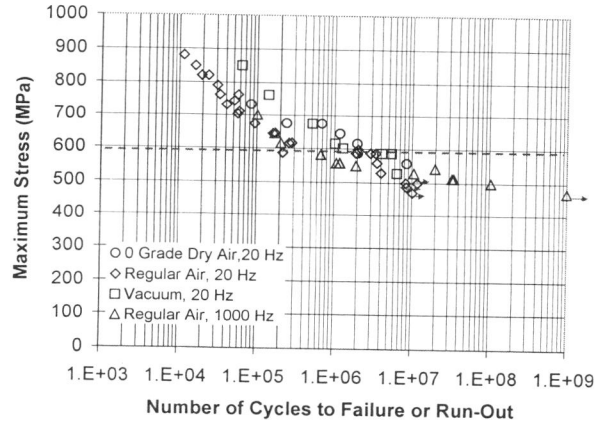

Figure 2: High-Cycle Fatigue Testing Results, S-N Data, of ULTIMET Alloy. Note That the Arrow Indicates a Run-Out without Failure, and the Dashed Line Represents the Applied Maximum Stress of the Plateau Region.

The general trend of the S-N data was that the fatigue life increased with decreasing maximum stress level, as is normally observed. As can be seen from Figure 2, the fatigue life in vacuum environment at 20 Hz was much longer than that in the regular air environment at 20 Hz. Generally, the fatigue life in the vacuum environment was 4 times greater than that in the regular air environment. It can also be noticed that the fatigue life in dry air at 20 Hz was longer than that in the regular air. This trend indicates that the fatigue life of the ULTIMET alloy was greatly affected by the oxygen and moisture in the environment.

It has been generally found that there is a great increase in temperature at high-frequency, 1,000 Hz.[10-13] In the present study, the temperature increase was measured by an advanced IR camera. Depended on the stress level, at low frequency, 20 Hz, it

was about 0 to 60 °C. At high frequency, 1,000 Hz, it was about 0 to 400 °C. As shown in Figure 2, within the experimental error, the S-N curves of 20 and 1,000 Hz in the regular air environment were in reasonably good agreement. Specifically, at the applied maximum stress level above 600 MPa, the test results at 20 and 1,000 Hz in the regular air environment grouped together. However, at the applied maximum stress level below 600 MPa, but above 530 MPa, the fatigue life at 20 Hz was greater than that at 1,000 Hz. At the applied maximum stress level below 530 MPa, the fatigue life at 1,000 Hz was longer than that at 20 Hz

Regardless of the testing frequencies and environments employed, there was a plateau region on the S-N curve. For the 20 Hz fatigue test results, the maximum stress level at the plateau region was near the yield strength of 586 MPa, and slightly higher than that at 1,000 Hz, 540 MPa. For the tests in dry air environment, there was another plateau region, which was at 670 MPa.

It is likely that the plateau region was due to the phase transformation induced by the plastic deformation. The shift of the plateau region with testing frequencies might be explained by temperature rises, which accelerated the phase transformation. The presence of oxygen and moisture in the testing environment was found to be harmful to the high-cycle fatigue properties of the ULTIMET alloy.

Microstructure Characterization

Figures 3 and 4 show the microstructure of ULTIMET alloy in the as-received, solution-annealed condition. The microstructure consisted essentially of a single FCC phase with a large number of annealing twins, and the grain size varied from 50 to 250 μm, with an average of approximately 80 μm as shown in Figure 3. The high-magnification SEM micrographs in Figure 4 show the Widmanstatten-type morphology, which is more clearly represented in the back-scattered-electron image, Figure 4(a), than the secondary-electron image, Figure 4(b). The single-phase, FCC, characteristic of ULTIMET alloy was also confirmed by the x-ray diffraction analysis,[14] which indicated that the amount of secondary phase, HCP, was very small or not present.

ULTIMET alloy has a low stacking fault energy (SFE). The stacking faults were abundant in the as-received material, as evidenced in Figure 5(a) and confirmed by TEM in Figure 5(b). In the high-magnification TEM images shown in Figure 5(b), the characteristic features of stacking faults, which were created by the movement of partial dislocations, are very clear. In fact, the Widmanstatten structure in Figure 4 and Figure 5(a) were composed of these stacking faults. The stacking faults were uniformly scattered in each grain.

It has been observed that twinning contributes significantly to the plastic deformation of low SFE metals at relatively large strains (about 10%).[15] Formation of the HCP phase or FCC twins of macroscopic size may be considered as a nucleation and growth process. The transformation of FCC to HCP can be achieved by the passage of a partial dislocation on every second {111} plane in an FCC crystal, i.e. transforming ABCABC to ABABAB, where A, B, and C are the stacking sequences of an FCC crystal. While the passage of a partial dislocation on every adjacent {111} plane results in the formation of an FCC twin, i.e. transforming ABCABC to ACBACB. Once the HCP phase or FCC twin embryo is formed by dislocation interactions, it may then thicken by overlapping of other nuclei formed on the {111} planes parallel to its habit planes. Dislocation density is an important factor for the formation of the HCP phase and FCC twins.

For the high-cycle fatigue tests of ULTIMET alloy, the strain was relatively small (less than 5%), and the important microstructural feature observed was the development of abundant stacking faults. The interaction of the stacking faults may form the FCC twins and/or HCP phase. As presented in Figure 6, the stacking faults formed a crisscrossed network over the entire deformed area in the gage section.

The dark-field image of Figure 7(a) exhibited the FCC twins oriented in the 10 o'clock direction which contributed to the "a" Bragg reflection of Figure 7(d). The bright FCC twins along the 2 o'clock direction of Figure 7(b) provided the contribution of "b" Bragg reflection of Figure 7(d). In the bright-field image, Figure 7(c), the dislocation structure was lying along directions and planes corresponding to the easy glide systems {111}<110>. Apparently, during fatigue, the interaction of the great amount of stacking faults formed FCC twins and had strong influence on the fatigue behavior of ULTIMET alloy.

After specimen failed, one broken half was sectioned in the longitudinal direction and mechanically- and electro-polished. As shown in Figure 8(a), below the fracture surface, there was a secondary crack. In the high-magnification SEM back-scattered-electron image 8(b), at the area around the tip of the secondary crack in Figure 8(a), there were a large number of microvoids which were aligned along the twin boundaries. It is suspected that the linkage of these microvoids may have created the macrocrack.

The stacking faults were found to play an important role in the fatigue process of ULTIMET alloy. As fatigue progressed, the stacking faults tended to form a network structure, and the density and thickness of stacking faults increased. The large number of stacking faults formed in the material could also serve as nuclei for the FCC-HCP transformation. Although the HCP phase was not found in the as-received and fatigue-fractured specimens examined, it is suspected that the phase transformation may have occurred in the region close the fracture surface. Microvoids formed along the interface of twins where deformation triggered slip systems on both sides of the interface. The linkage of the microvoids eventually led to the formation of the crack which caused the failure of the material.

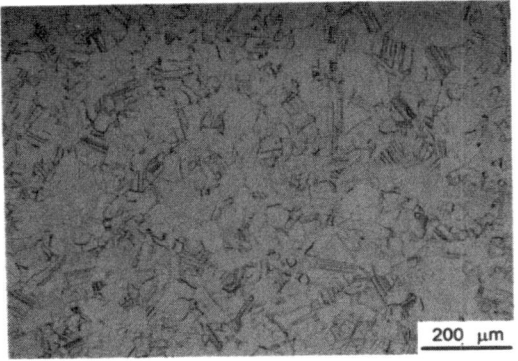

Figure 3: Optical Micrograph Showing the Characteristic Microstructure of ULTIMET Alloy.

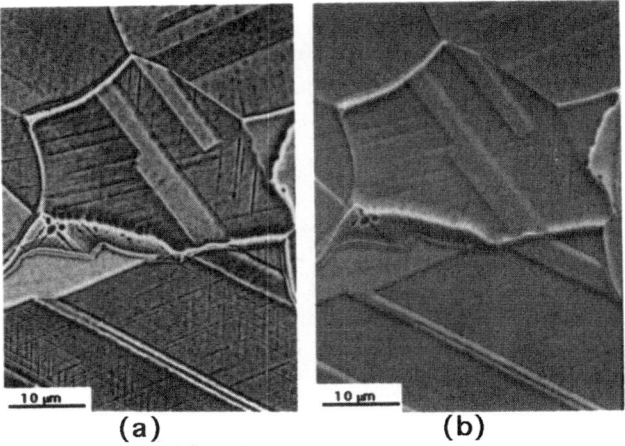

(a)   (b)

Figure 4: SEM Micrographs Showing the Widmanstaten-Pattern in the Grains: (a) Back-Scattered Electron Image and (b) Secondary-Electron Image.

(a)

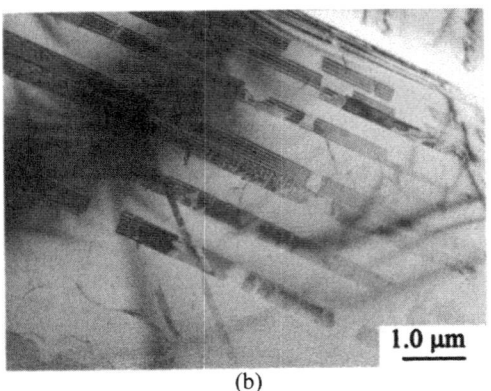

(b)

Figure 5: Abundant Stacking Faults of ULTIMET Alloy in the As-Received Condition: (a) SEM Back-Scattered-Electron Image and (b) High-Magnification TEM Image.

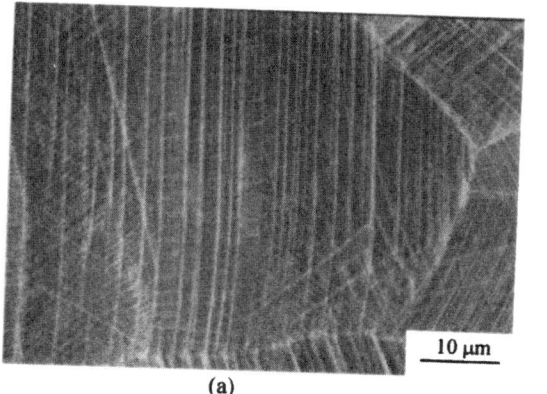 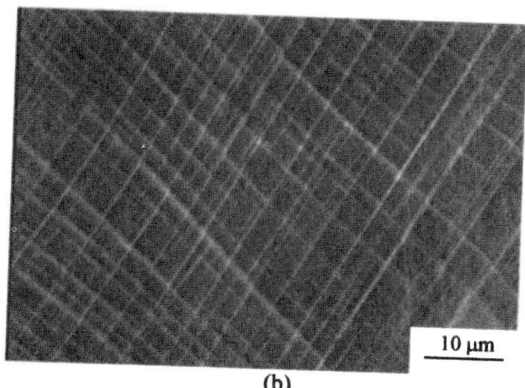

Figure 6: Abundant Stacking Faults of ULTIMET Alloy in the Cyclically-Deformed Condition: (a) Fatigue Tested at the Maximum Stress Level of 586 MPa, (b) Fatigue Tested at the Maximum Stress Level of 703 MPa.

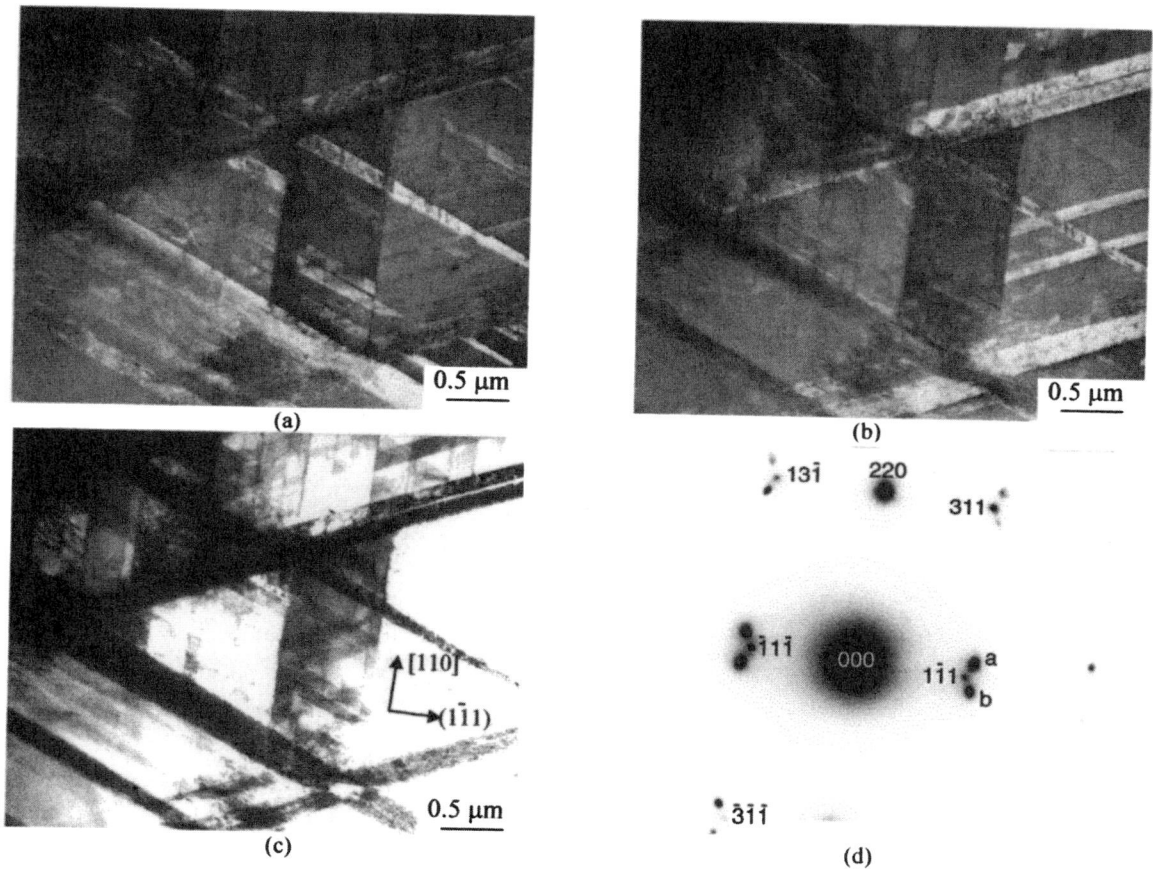

Figure 7: TEM Micrographs of ULTIMET Alloy in the Fatigue-Fractured Condition Showing FCC Twins: (a) Dark Field Image of One Set of FCC Twins Contributing to Bragg Reflection "a" in Figure 7(d), (b) Dark Field Image of Another Set of FCC Twins Contributing to Bragg Reflection "b" in Figure 7(d), (c) Bright Field, and (d) diffraction Pattern from Figure 7(c).

Fractography

The fracture surfaces of the failed specimens were examined by SEM. Figure 9 presents the fracture surface of a specimen fatigued at a relatively high applied maximum stress of 821 MPa, with an R-ratio of 0.05, in air and at room temperature. The topography of the fracture surface was uneven and possessed very crystallographic features as shown in Figure 9(a). The crack-initiation sites were identified by tracing back along the fracture-surface markings, which radiated from the point of origin. There were five fatigue-crack-initiation sites, as indicated by circles, and all of them were at the specimen surface. A representative high-magnification micrograph of the initiation sites presented in Figure 9(b) reveals a cleavage-like fracture surface typical of stage I crack initiation. The size of the crack-initiation sites was on the order of the grain size, ranging from 100 to 200 μm. At a lower applied maximum stress levels, the fracture surface showed the same appearance, but there were fewer crack-initiation sites.

After the crack-initiation stage, the cracks propagated through the gage section of the specimen. The characteristic feature of the crack-propagation region was the "stair step" appearance, as revealed by SEM examination in Figure 9(c). This trend indicated that the crack propagation was very crystallographic. In previous work, Shen et al.[16] pointed out that the striation spacing was small, and the propagation speed was slow.

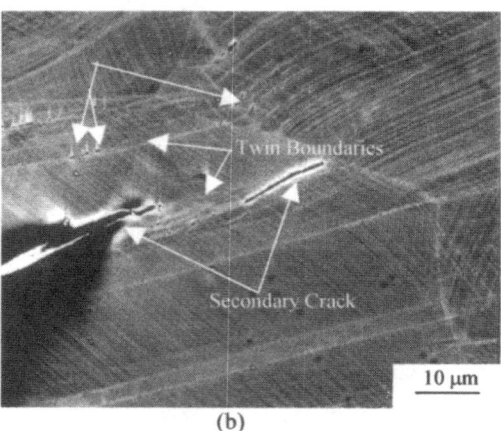

(a)   (b)

Figure 8: SEM Micrographs of ULTIMET Alloy in the Fatigue-Fractured Condition (a) Secondary-Electron Image Showing the Fracture Surface and Secondary Crack, (b) Back-Scattered-Electron Image Showing the Microvoids around the Tip of a Secondary Crack in Figure 8(a) and Their Linkage to Form a Crack.

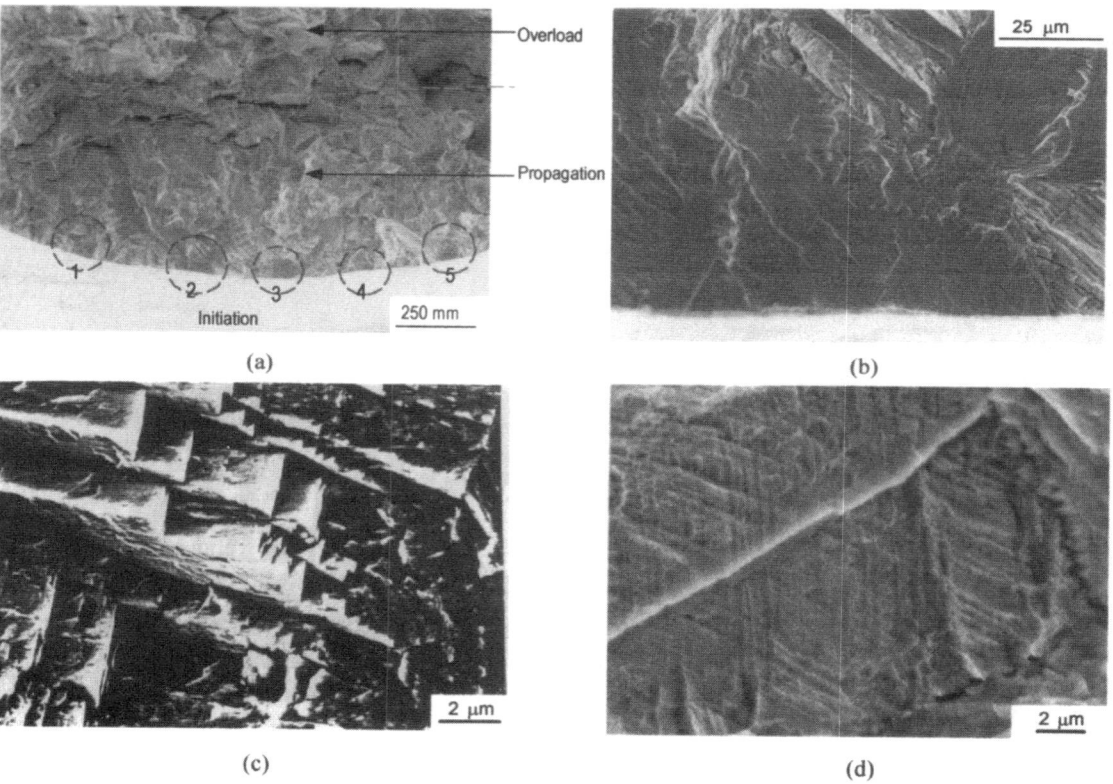

(a)   (b)

(c)   (d)

Figure 9: SEM Fractographs of a Fatigue-Fractured Specimen: (a) Overview of the Fractured Surface and the Crack-Initiation Sites Noted by the Circles with Numbers, (b) SEM Fractograph of the Initiation Site 2 from Figure 9(a), (c) 'Stair Step' Feature of the Crack-Propagation Region, and (d) Dimpled Fracture Surface of the Final Overload Region. (σ_{max} = 821 MPa, R = 0.05, Room Temperature, in Regular Air, Fatigue Life of 23,568 Cycles, 20 Hz)

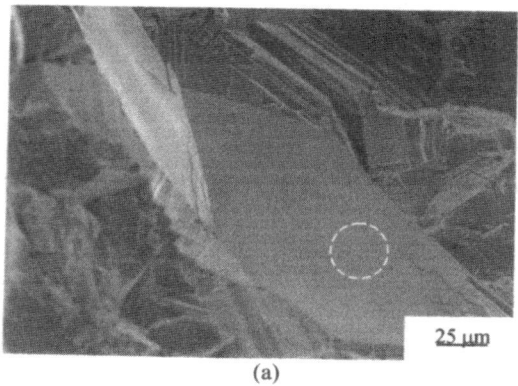
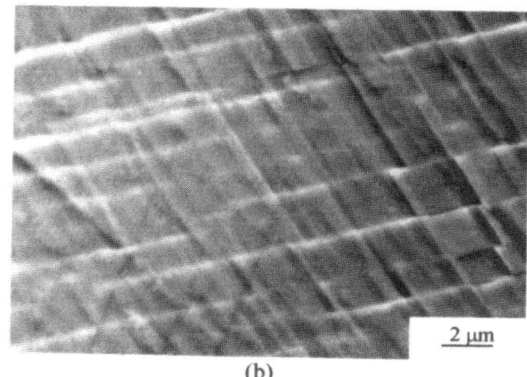

(a)   (b)

Figure 10: SEM Fractographs of a Fatigue-Fractured Specimen: (a) Crack-Initiation Site, (b) High-Magnification Micrograph from the Circled Region in Figure 10(a). (σ_{max} = 674 MPa, R = 0.05, Room Temperature, in Dry Air, Fatigue Life of 695,280 Cycles, 20 Hz)

The fractograph shown in Figure 9(d) illustrates the final overload region. The features were similar to that of a typical tensile overload. Interestingly, the regularity of the microstructure of the fracture surface in Figure 9(d) is consistent with the networked stacking-faults structure in Figures 6 and 7.

The SEM micrographs in Figure 10 show the crack-initiation site of a specimen fatigue tested at 20 Hz with an applied maximum stress of 674 MPa in the dry air. The fatigue crack initiated from an internal origin of the specimen, as shown in Figure 10(a), but it possessed the same feature as the crack-initiation site in Figure 9(b). The high-magnification view of the cleavage-like crack-initiation site, Figure 10(b), shows the presence of slip bands. The structure of the slip bands on the crack-initiation site is consistent with the networked stacking-faults structure shown in Figure 6. This feature also provided the evidence that the crack may initiate from the interface of twins which experienced large plastic deformation during fatigue, as shown in Figure 8(b).

Under all conditions, the high-cycle fatigue behavior of ULTIMET alloy exhibited a typical two-stage fatigue-crack-growth process: (a) stage I fatigue-crack initiation in which the cracks formed on those planes most closely aligned with the maximum shear-stress direction in the grains of the fatigue specimen, and (b) stage II fatigue-crack growth in which the maximum principal tensile stress controlled crack propagation in the region of the crack tip. It was noted that the stage I crack was confined within one grain and the fracture appearance was cleavage-like. Generally, for the crack initiation from the surface, as shown in Fig. 9(b), the persistent slip bands on the surface led to the stage I crack propagation, but for the crack initiation from the internal origin, as presented in Figure 10, the crack initiation resulted from the evolution of the stacking faults.

Crack-Initiation Location

Typically, the fatigue-crack initiation occurs on the specimen surface as a result of the irreversible process of extrusion and intrusion formation through slip deformation. However this was not generally true for the fatigue-crack initiation of ULTIMET alloy. The crack-initiation site was found to vary from surface to subsurface depending on test variables. Figure 11 summarizes the fatigue-crack-initiation sites observed under various testing conditions of frequency and environment. In the regular air

environment, for the lower-frequency (20 Hz) high-cycle fatigue at higher applied maximum stress levels greater than 600 MPa, the fatigue cracks initiated from the surface, as shown in Figure 9(b). However, at peak stresses lower than 600 MPa, corresponding to the plateau region, subsurface or near surface crack initiation was observed. This can be seen from Figure 12(a), for a peak stress of 586 MPa. The crack-initiation sites presented in Figures 12(b) and (c) indicate that the cleavage-like features were retained.

Similarly, in the regular air environment, for the high-frequency (1,000 Hz) high-cycle fatigue tests at the applied maximum stress levels greater than 600 MPa, the fatigue cracks initiated at the surface. For the applied maximum stress levels below 600 MPa the fatigue cracks initiated from the subsurface.

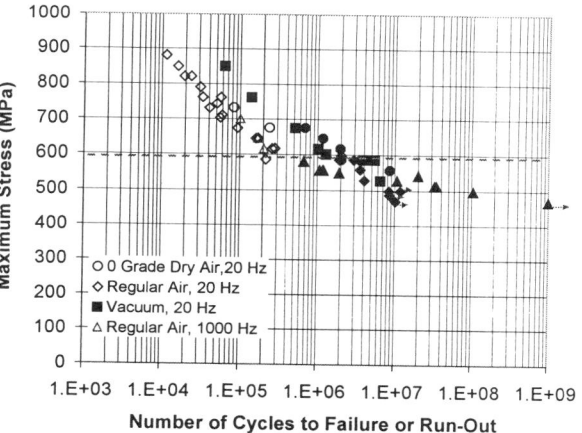

Figure 11: High-Cycle Fatigue Testing Results, S-N Data, of ULTIMET Alloy. Note That the Opened Symbol Represents the Fatigue Crack Initiated from the Surface, the Solid Symbol Stands for the Fatigue Crack Initiated from the Subsurface, the Arrow Indicates a Run-Out without Failure, and the Dashed Line Represents the Applied Maximum Stress of a Plateau Region.

For tests conducted in the dry-air environment, fatigue cracks initiated at the surface when the applied maximum stress levels were greater than 674 MPa. At stress levels below 674 MPa, the fatigue cracks originated from the subsurface, as exhibited in Figure 10. In the vacuum environment, all the cracks initiated at subsurface locations regardless of the maximum applied stress level at the test frequency of 20 Hz.

For both surface and subsurface crack initiation, the common features of the crack-initiation sites were: (a) flat facets with the size on the order of 100 to 200 μm at the origin, (b) the absence of precipitates or inclusions. All of the fracture surfaces examined retained a crystallographic appearance, and the initiation sites exhibited the cleavage-like feature.

Internal fatigue-crack-initiation origins have been attributed to the presence of inclusions in some materials.[17] In ULTIMET alloy, no inclusions were observed to cause crack initiation. Careful investigation of the material by SEM showed that it was very clean. It has been reported that in the titanium alloy system, e.g., Ti-6Al-4V,[17-19] and stainless steel system, such as a typical martensitic stainless steel,[18] the crack-initiation site changed from the surface to subsurface in specimens undergoing high-cycle fatigue subjected to either low peak stresses or long fatigue lives.

The fatigue-crack-initiation site was found to be dependent on the environmental conditions. All of the specimens tested in vacuum experienced subsurface crack initiation. For specimens tested in dry air, the fatigue cracks started to form internally at a higher applied maximum stress levels than that of the regular air.

For the specimens in the as-received condition, all surfaces were mechanically polished. Therefore, the surface was harder than the interior of the specimen. At high applied stresses, the surface was more susceptible to damage from oxygen and moisture, and surface crack initiation was favored. At low applied stresses, the harmful effects of oxygen and moisture were reduced because fewer slip bands were formed, and internal crack initiation was favored. At 1,000 Hz, this effect was not as pronounced. The change of the crack-initiation site from the surface to subsurface might also help to explain the occurrence of the plateau.

Another reason for the presence of the plateau region might be related to the FCC to HCP phase transformation induced by plastic deformation. All of the tests conducted in air at 20 Hz had the same plateau region at the applied maximum stress level of 600 MPa. The plateau region for the high-frequency fatigue tests in air shifted down to 540 MPa. This change was probably due to the fact that the specimens tested at 1,000 Hz experienced a higher temperature rise than those tested at 20 Hz. Further work is planned to gain a better understanding of the plateau region by conducting strain-controlled tests at room and elevated temperatures.

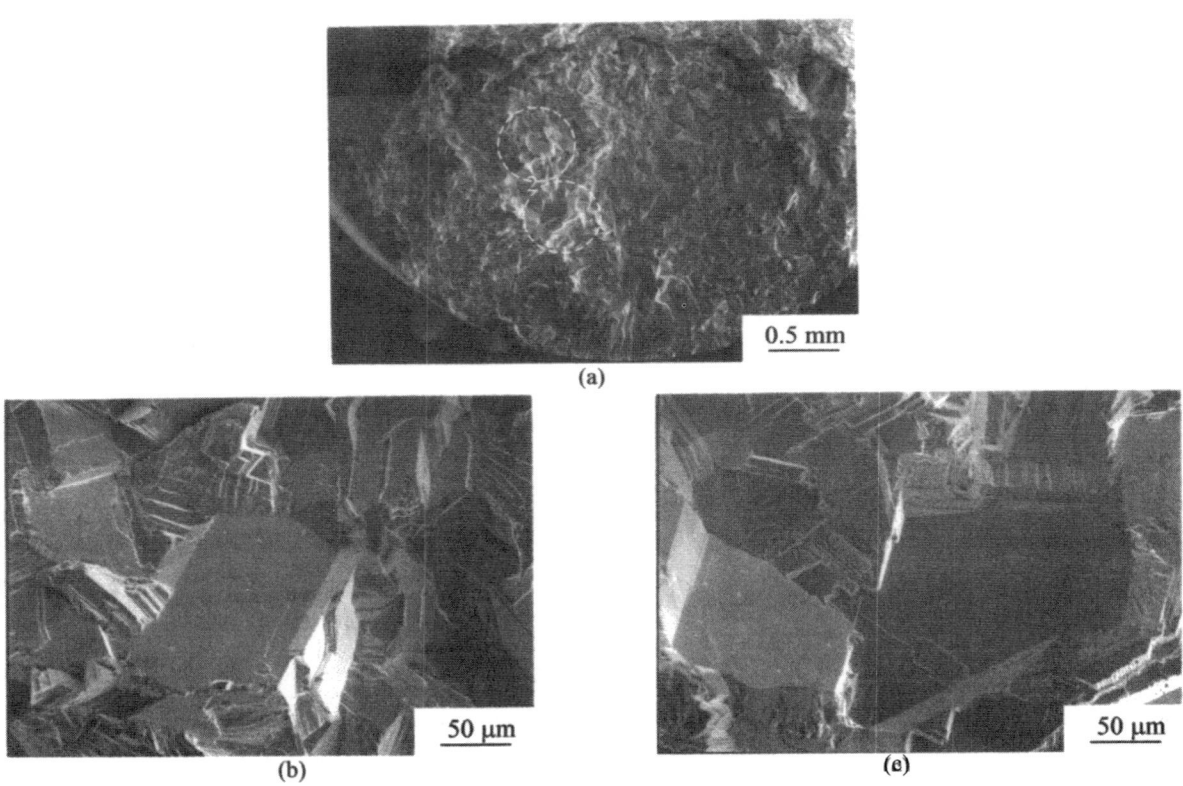

Figure 12: SEM Fractographs of a Fatigue-Fractured Specimen: (a) Overview of the Fractured Surface and the Crack-Initiation Sites Noted by the Circles with Numbers, (b) SEM Fractograph of the Initiation Site 1 from Figure 12(a), and (c) SEM Fractograph of the Initiation Site 2 from Figure 12(a). (σ_{max} = 586 MPa, R = 0.05, Room Temperature, in Regular Air, Fatigue Life of 2,941,100 Cycles, 20 Hz)

Conclusions

1) High-cycle fatigue data were developed for the ULTIMET alloy. The fatigue lives decreased as the applied maximum stress levels increased, as is normally observed. The high-frequency (1,000 Hz) high-cycle fatigue tests gave results comparable to the low-frequency (20 Hz) tests.

2) Moisture and oxygen were found to have significant detrimental effects on the fatigue lives. The absence of moisture and/or oxygen increased fatigue life.

3) Plateaus were found in all of the S-N curves. For tests at 20 Hz, the plateau occurred at an applied maximum stress level near the yield strength of the alloy, 596 MPa. For tests at 1,000 Hz, the plateau occurred at a maximum stress level of 540 MPa. It is likely that the plateau region was due to the FCC to HCP phase transformation induced by the plastic deformation and the change of the crack-initiation site from the surface to subsurface.

4) The microstructural characterization of the test specimens revealed that the evolution of stacking faults is a key fatigue-damage process. Stacking faults were formed in a network structure. The number of stacking faults increased as cycling progressed.

5) The fractographic studies showed that there was a change in the crack-initiation site from the surface to subsurface, depending on the applied maximum stress level and environment. The presence of oxygen and moisture tended to promote crack initiation at the surface. The change of the crack-initiation site from the surface to subsurface was associated with a long fatigue life.

6) Regardless of the crack origin, the fracture surface at the crack-initiation site was cleavage-like, and its size was on the order of the grain size, 100 μm to 200 μm. The crack-propagation fracture surface was crystallographic in nature.

Acknowledgements

This work is supported by the Haynes International, Inc. We also acknowledge the financial support of the National Science Foundation, the Division of Design, Manufacture, and Industrial Innovation, under Grant No. DMI-9724476, and the Combined Research-Curriculum Development Program, under EEC-9527527, to the University of Tennessee, Knoxville, with Dr. Delcie R. Durham and Ms. Mary Poats as program managers, respectively. We appreciate the financial support of the Center for Materials Processing and Office of Research Administration at the University of Tennessee, Knoxville. The great help of Dr. H. Wang, from Oak Ridge National Laboratory on the temperature measurements during high-cycle fatigue testing, is highly appreciated. Many thanks are due to Mr. Doug Fielden, Mr. Greg Jones, and Mr. Larry Smith at the University of Tennessee for their great help in setting up the electrohydraulic machines and excellent technical support.

References

1. D. L. Klarstrom, "Wrought Cobalt-Base Superalloys," Journal of Materials Engineering and Performance, No. 2, 1993, pp. 523-530.

2. P. Crook and D. L. Klarstrom, "Advances in Alloy Design for Aggressive Wear Service," ADVMAT' 91, San Diego, CA, June 1991.

3. P. Crook and A. V. Levy, "Cobalt-Base Alloys," Metal Handbook, Vol. 18, 1992, pp. 766-771.

4. A. V. Levy and P. Crook, "The Erosion Properties of Alloys for the Chemical Processing Industries," Wear, Vol. 151, 1991, pp. 337-350.

5. H. M. Tawancy, V. R. Ishwar, and B. E. Lewis, "On the fcc-hcp Ttransformation in a Cobalt-Base Superalloy (HAYNES Alloy No. 25)," Journal of Materials Science Letter, Vol. 5, 1986, pp. 337-341.

6. Haynes International, ULTIMET® Alloy, 1994, p. 3.

7. W. G. Moffatt, Binary Phase Diagrams Handbook, General Electric Company, 1976, p. 78.

8. H. Farhangi, R. W. Armstrong, and W. F. Regnault, "Transmission Electron Microscopy Detection of Cyclic-Deformation-Induced fcc to hcp Transformation in a Cobalt-Based Prosthetic Device Material," Materials Science and Engineering, Vol. 114, 1989, pp. 25-28.

9. ASTM Standard Practice (E466) for "Conducting Constant Amplitude Axial Fatigue Tests of Metallic Materials," 1995.

10. H. Wang, L. Jiang, P. K. Liaw, C. R. Brooks, and D. L. Klarstrom, "Infrared Temperature Mapping of ULTIMET® Alloy During High-Cycle Fatigue Tests," Metallurgical and Materials Transactions A Vol. 31A, 2000, pp.1307-1310.

11. P. K. Liaw, H. Wang, L. Jiang, B. Yang, J. Y. Huang, R. C. Kuo, and J. G. Huang, "Thermographic Detection of Fatigue Damage of Pressure Vessel Steels at 1,000 Hz and 20 Hz," Scripta Metallurgica et Materiala, Vol. 40, 2000, pp. 389-395.

12. L. Jiang, H. Wang, P. K. Liaw, C. R. Brooks, and D. L. Klarstrom, "Characterization of Temperature Evolution During High-Cycle Fatigue Tests, Part I, Experiments," submitted to Acta Metallurgica et Materialia.

13. L. Jiang, H. Wang, P. K. Liaw, C. R. Brooks, and D. L. Klarstrom, "Characterization of Temperature Evolution During High-Cycle Fatigue Tests, Part II, Modeling," submitted to Acta Metallurgica et Materialia.

14. L. Jiang, C. R. Brooks, P. K. Liaw, D. L. Klarstrom, and B. Muenchen "Phenomenological Aspects of the High-Cycle Fatigue of ULTIMET® Alloy " to be submitted to Materials Science and Engineering.

15. S. Asgari, E. El-Danaf, S. R. Kalidindi, and R. D. Doherty, "Strain Hardening Regimes and Microstructural Evolution during Large Strain Compression of Low Stacking Fault Energy Fcc Alloys That Form Deformation Twins," Metallurgical and Materials Transactions A, Vol. 28A, 1997, pp. 1781-1795.

16. E. Y. Shen, C. R. Brooks, P. K. Liaw, D. L. Klarstrom, and J. Y. Huang, "Fatigue Behavior of a Commercial Co-26Cr-9Ni (wt. %) Base Alloy (ULTIMET®)," High Cycle Fatigue of Structural Materials, Soboyejo, W. O. and Srivatsan, T. S., eds, TMS, 1997, pp. 313-326.

17. D. F. Neal and P. A. Blenkinsop, "Internal Fatigue Origins in α-β Titanium Alloy," Acta Metallurgica, Vol. 24, 1976, pp. 59-63.

18. A. Atrens, W. Hoffelner, T. W. Duerig, and J. E. Allison, "Subsurface Crack Initiation in High Cycle Fatigue in Ti6Al4V and in a Typical Martensitic Stainless Steel," Scripta Metallurgica, Vol. 17, 1983, pp. 601-606.

19. O. Umezawa and K. Ishikawa, "Phenomenological Aspects of Fatigue Life and Fatigue Crack Initiation in High Strength Alloys at Cryogentic Temperature," Materials Science and Engineering, Vol. A176, 1994, pp. 397-403.

THE EFFECT OF STRAIN RATE AND TEMPERATURE ON THE LCF BEHAVIOR
OF THE ODS NICKEL-BASE SUPERALLOY PM 1000

Martin Heilmaier[1], Michel Nganbe[1], Frank E.H. Müller[2], Ludwig Schultz[1]

[1]Institute of Solid State and Materials Research Dresden, D-01171 Dresden, Germany
[2]Plansee GmbH Lechbruck, D-86983 Lechbruck, Germany

Abstract

The high temperature LCF behavior of the oxide dispersion strengthened (ODS) nickel-base superalloy PM 1000 was studied using single-step tests with constant total strain amplitudes ranging from 0.2 to 0.7%. Specifically, the effect of varying strain rate and temperature on the cyclic lifetimes as well as on the fatigue crack initiation and propagation was studied. It was found that a decrease of the applied strain rate by a factor of 100 drastically reduces cyclic lifetime whereas an increase of temperature by 150 K decreases N_f only moderately. This behavior can be qualitatively understood on the basis of post-mortem microstructural investigations of fatigued specimens which reveal pronounced void formation and coalescence at transverse grain boundaries followed by cleavage at longitudinal grain boundaries leading to failure.

Introduction

Oxide dispersion strengthened (ODS) superalloys produced by mechanical alloying (MA) technique (1,2) have recently received renewed attention from the scientific community and industry. This interest originates mainly from the potential of ODS superalloys for structural applications, notably in gas turbine components, such as vanes or nozzles, where outstanding high temperature creep and fatigue resistance at temperatures of 1000 °C and higher are desirable.

In essence, PM 1000[*] is a powder-metallurgical nickel-chromium solid solution additionally strengthened by nominally 1 vol.% of incoherent and thermally stable nano-scale Y_2O_3 particles dispersed in the metallic matrix by solid state reaction processing, i.e. mechanical alloying. Powder consolidation is subsequently followed by an extensive thermomechanical processing including hot extrusion and hot rolling. A final annealing treatment above $0.9\ T_m$ promotes secondary recrystallization leading to coarse and elongated grains up to several mm in length. However, the transverse grain boundaries have been considered to be weak links during high temperature deformation as they are the most probable locations for void formation and growth during service. Consequently, this highlights the major importance of the grain aspect ratio (GAR) in this class of alloys (3). Additionally, fine equiaxed grains – so called "recystallization defects" – are present in the material which may – depending on their volume fraction – further weaken the materials

---

[*] Trademark of Plansee GmbH, Lechbruck, Germany

Superalloys 2000
Edited by T.M. Pollock, R.D. Kissinger, R.R. Bowman,
K.A. Green, M. McLean, S. Olson, and J.J. Schirra
TMS (The Minerals, Metals & Materials Society), 2000

deformation resistance(3-5). While the creep resistance of these ODS nickel-base superalloys is known to be outstanding – even superior to that of single crystal superalloys at temperatures around and beyond 1000°C – only a few reports deal with their low cycle fatigue (LCF) behavior (6,7).

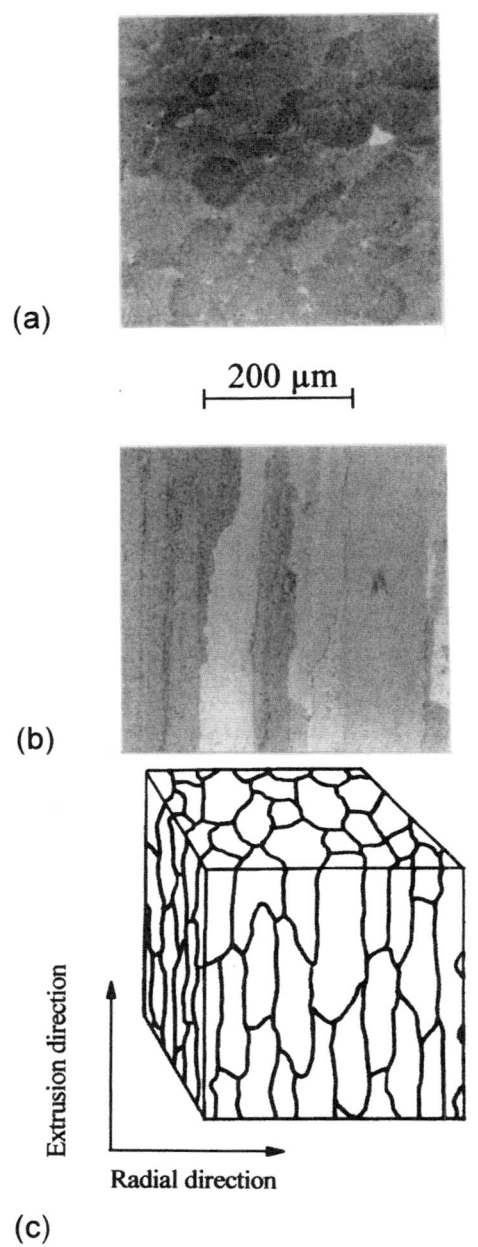

(a)

200 μm

(b)

(c)

Figure 1: Grain structure of the investigated bar material of PM 1000; (a) cross-section, (b) longitudinal plane, (c) 3-dimensional sketch.

In this paper we have extended our previous work on the influence of orientation and grain structure (8) by including the effect of strain rate and temperature on the LCF behavior of PM 1000. Specifically, we will focus on the effect of void formation and growth upon damage and cyclic lifetime.

Experimental

Material

PM 1000 having the nominal chemical composition (wt.%) of Ni-20Cr-0.5Al-0.3Ti-0.6Y$_2$O$_3$ was supplied in the fully recrystallized condition (final heat treatment: 1h @ 1315 °C) in form of a hot extruded bar material (diameter 50 mm). The grain parameters were inspected by optical microscopy (OM) and the average length of the coarse grains was found to be about 2.6 mm in longitudinal direction. In the transverse direction the diameter was about 0.25 mm, yielding a GAR of about 10, see Fig. 1. The orientation of individual grains was measured by scanning electron microscopy (SEM, Zeiss DSM 962) using electron back scatter diffraction (EBSD) as well as neutron diffraction. Both methods concurrently yield a strong <100> fiber texture parallel to the extrusion (longitudinal) direction (Fig. 2).

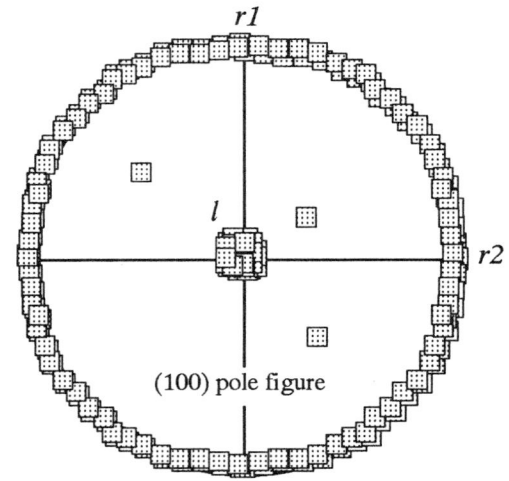

Figure 2: (100) pole figure of the PM 1000 bar material, obtained by SEM and EBSD.

Transmission electron microscopy (TEM, Jeol CM 200 X) was used for investigating the particle and dislocation microstructure before and after fatigue deformation. Quantitative stereology was applied to determine the relevant particle parameters, namely the mean diameter d, the mean planar center-to-center spacing L, and the volume fraction f. The thickness of the TEM foils was determined by convergent beam electron diffraction. From micrographs such as Fig. 3 one obtains values of particle diameter d = 14 nm, interparticle spacing L = 100 nm, and volume fraction f = 1.0 % nearly identical to that observed in Inconel MA 754 of comparable chemical composition (5). While the arrangement of most of the nano-scale dispersoids appears to be rather homogeneous or random in Fig. 3, relatively large inclusions in the micro-meter range were also detected. These inclusions were identified as titanium carbonitrides and cuboidal alumina. Though these large inclusions play little or no role in pinning dislocations, their role in pinning grain boundaries and in serving as nucleation sites for grain boundary cavities must be considered (5).

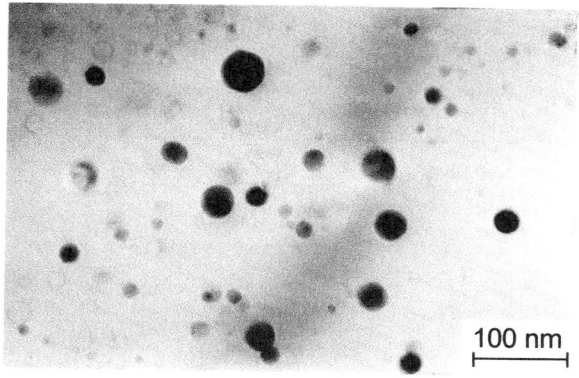

Figure 3: Bright field TEM micrograph of PM 1000 bar material, as-received condition.

Mechanical testing

For fatigue testing cylindrical specimens with threaded ends having a gage length of 12.5 mm and a gage diameter of 6 mm were produced by electro-erosion and subsequent machining such that the loading axes were parallel to the longitudinal direction. Prior to mechanical testing in air they were ground and polished to remove surface scratches. Cyclic symmetrical (R=-1) push-pull tests with triangular wave shape were carried out at the two different temperatures of 850° and 1000°C and at strain rates of 10^{-3} and 10^{-5} s^{-1}. All tests were performed under total strain control representing the loading condition of engineering components like vanes or nozzles in gas turbines undergoing a certain degree of structural constraint. Total strain amplitudes $\varepsilon_{a,t}$ were varied from 0.2 to 0.7%.

Results and discussion

Cyclic stress-strain response

Cyclic hardening/softening curves of PM 1000 at a strain rate of 10^{-3} s^{-1} with different total strain amplitudes $\varepsilon_{a,t}$ ranging from 0.2 to 0.6% are plotted in Fig. 4a. In Fig. 4b

the cyclic stress response at a strain rate of 10^{-5} s^{-1} is displayed for $\varepsilon_{a,t}$ values ranging from 0.3 to 0.7%.

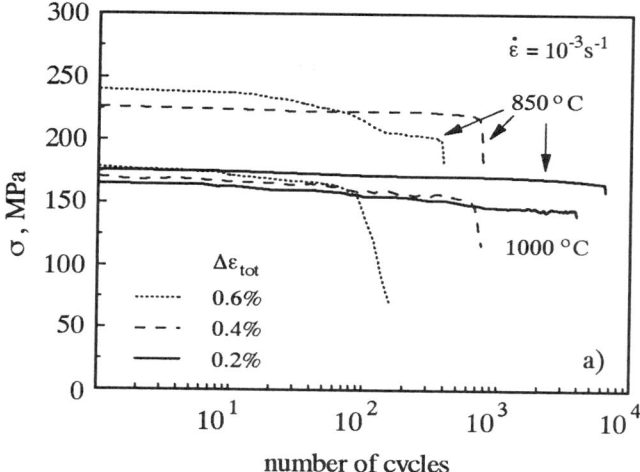

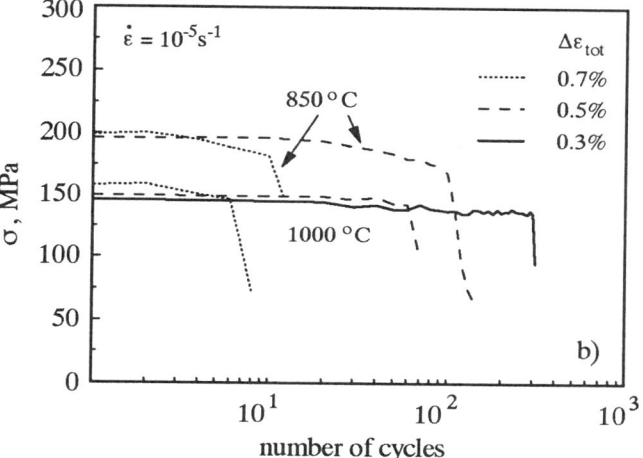

Figure 4: Cyclic hardening / softening curves for different applied total strain amplitudes and strain rates of 10^{-3} s^{-1} (a) and 10^{-5} s^{-1} (b), respectively.

At 850°C curves with strain amplitudes less than 0.6% show initial cyclic softening which is, however, only weakly pronounced, followed by a region of fairly constant stress amplitudes which may be defined as the saturation stress level $\sigma_{a,s}$ (9). In contrast, specimens cycled at the higher temperature of 1000°C generally show continuous cyclic softening from the start of the test like those at 850°C with strain amplitudes of 0.6% and higher. In this case "saturation" was defined at half life before the transition to a steeper decline of the curve indicates the onset of external necking in the specimen. Several points are obvious from Figs. 4a and b: first, at a given constant strain rate the saturation stress level $\sigma_{a,s}$ for cyclic deformation decreases with increasing temperature. Second, at a given temperature $\sigma_{a,s}$ decreases with decreasing strain amplitude. However, this effect is relatively weak. Third, the decrease of the applied strain rate leads to a dramatic reduction of the cyclic lifetimes N_f (note the different scale of the x-axis in Figs. 4a and b).

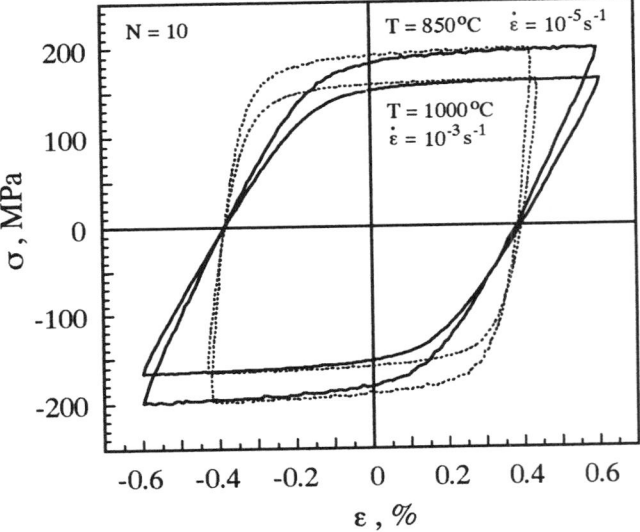

Figure 5: Selected stabilized σ-ε-hysteresis loops with a total strain amplitude of $\varepsilon_{a,t}$ = 0.6% (solid curves); dashed: σ-ε_{pl}-hysteresis loops; testing temperatures and strain rates as indicated.

Fig. 5 depicts representative hysteresis loops obtained from fatigue tests conducted at $\varepsilon_{a,t}$ = 0.6% in the form of stress (σ) vs. total strain (solid line) and plastic strain (dashed line), respectively. As discussed in detail in Ref. (10), severe damage, which affects cyclic stress-strain response, occurs early in fatigue life under these tests conditions. Therefore, the hysteresis loops plotted in Fig. 5 were recorded after ten cycles in order to show the true cyclic stress-strain response of the (undamaged) material. Plastic strain (ε_{pl}) was obtained by subtracting the elastic strain (ε_{el}) from the measured total strain. Time dependent deformation processes affect the stress-strain response at high temperatures, therefore the elastic modulus (E) is not reliable when determined from the slope of the unloading part of the hysteresis loop as usually done. Consequently, the dynamic elastic modulus was used, and E was determined to be 101 GPa and 92 GPa for temperatures of 850° and 1000°C, respectively. In general, the shape of the hysteresis loops appear fairly similar under the chosen conditions of $1000°C/10^{-3}s^{-1}$ and $850°C/10^{-5}s^{-1}$. Assuming the data collected at $850°C/10^{-3}s^{-1}$ to be a upper reference limit for the $\sigma_{a,s}$ (see the dotted curve in Fig. 4a), Fig. 5 clearly reveals that the effect of a temperature increase by 150°C is more significant on the saturation stress level than the reduction of the applied total strain rate by a factor of 100.

Cyclic lifetime

Fig. 6 shows the cyclic lifetime data of PM 1000 as a function of strain rate and temperature. First, we address the influence of temperature on the dependence of the number of cycles to failure N_f on the applied (total) strain amplitude. Obviously, an increase of temperature by 150°C leads to a lifetime reduction (e.g. compare the open and full circles in Fig. 6). The course of the data points may be approximated by straight lines with the slope decreasing slightly with increasing temperature. Second, decreasing the strain rate by a factor of 100 at the same temperature of 850°C yields a further substantial reduction of cyclic lifetime which is even more emphasized at the higher temperature of 1000°C. Moreover, a pronounced bend in the course of the data points is visible.

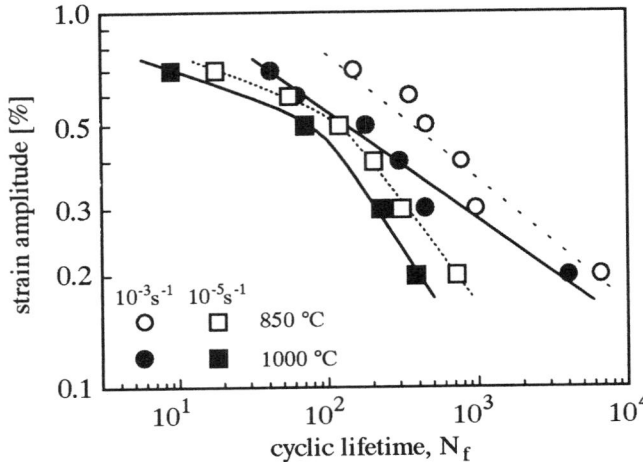

Figure 6: Total strain – cyclic lifetime curves of PM 1000 as a function of temperature and strain rate.

In the present case, the strain controlled LCF tests fatigue life should be characterized on the basis of the plastic strain amplitude according to Coffin and Manson (11). They noted a linear relationship between the plastic strain amplitude $\varepsilon_{a,pl}$ and the number of load reversals to failure $2N_f$ when plotted in a double-logarithmic form, i.e.

$$\varepsilon_{a,pl}\Big/_2 = \varepsilon_f \cdot \left(2N_f\right)^c \qquad (1)$$

where ε_f is the fatigue ductility coefficient and C is the fatigue ductility exponent, see Figs. 7a and b. Again, the data points (open and full circles in Figs. 7a and b) at the higher strain rate of $10^{-3}s^{-1}$ can be satisfactorily approximated by straight lines yielding comparable values for C and ε_f as defined in equation (1). These values are listed in Table I.

Table I: Fatigue ductility exponent C and fatigue ductility coefficient ε_f as evaluated from the Coffin-Manson plots in Figs. 7a and b

T /°C	850		1000	
Strain rate /s^{-1}	10^{-3}	10^{-5}	10^{-3}	10^{-5}
C	-0.75	-0.77	-0.49	-0.83
ε_f	0.368	0.195	0.048	0.052

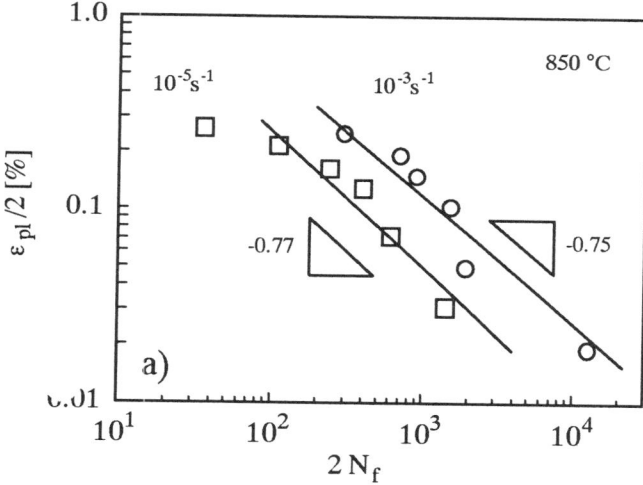

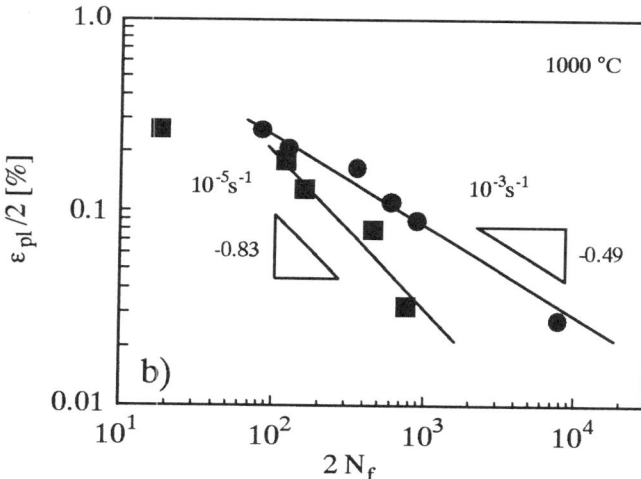

Figure 7: Plastic strain amplitude $\varepsilon_{a,pl}/2$ vs. number of load reversals to failure 2N$_f$ of PM 1000 as a function of strain rate at 850°C (a) and 1000°C (b).

In contrast, linear regression analysis of the open and full squares in Figs. 7a and b representing the data points obtained at the lower applied total strain rate of 10^{-5} s^{-1} would seriously overestimate the cyclic lifetime at service-relevant large cycle numbers. Hence, the data points for the highest plastic (and total) strain amplitude which caused the bend of the continuous curves in Fig. 6 were disregarded for the least squares fit. In doing so the quality of the fit was significantly improved (viz.

correlation coefficients R > 0.93). Also, the lifetime prediction at cycle numbers > 10^4 is significantly enhanced.

Further, the plot shown in Fig. 7 yields two interesting aspects: first, the fatigue ductility exponents C at 850°C are fairly similar while their counterparts at 1000°C deviate strongly from each other (Table I). However, except for the one determined at 1000°C and 10^{-3} s^{-1} (C = –0.49) these values considerably exceed the range of –0.5 to –0.7 commonly found for many engineering metals and alloys (12). Second, the fatigue ductility coefficients ε_f are found to be roughly equal to the fracture ductilities observed in monotonic tensile tests at the same temperatures and nominal strain rates (13).

Fractography

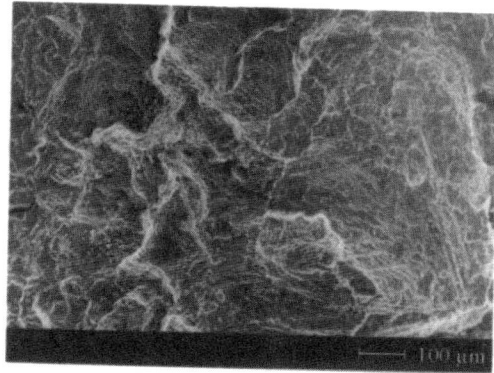

Figure 8: SEM micrograph of the fracture surface of a LCF specimen fatigued at 1000°C and 10^{-3} s^{-1}. The strain amplitude of 0.4% favors predominantly transgranular failure.

The observed dependence of the LCF lifetime can be elucidated on the basis of the following microstructural observations. First, we address the effect of strain rate and amplitude on the crack mode at constant temperature. At the lower temperature, predominantly transgranular crack propagation occurs at low applied strain amplitudes and high strain rates, see Fig. 8. Both, transverse and longitudinal grain boundaries were found to be nearly undamaged. By contrast, lowering the strain rate by a factor of 100 and increasing the strain amplitude favors superimposed creep effects. As fatigue specimens subjected to the lower strain rate of 10^{-5} s^{-1} remain longer in the tensile part of the hysteresis loop by a factor of 100, diffusional growth and coalescence of voids forming short cracks on transverse grain boundaries by the acting nominal stress are enhanced. In turn, cyclic lifetime is severely reduced by a decrease of the applied strain rate which is obvious from the experimental results shown in Figs. 6 and 7.

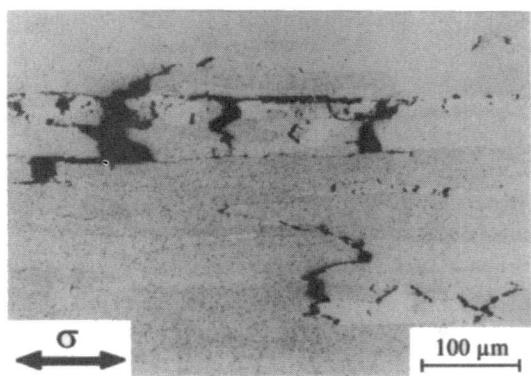

Figure 9: Optical micrograph of the longitudinal plane of a LCF specimen fatigued at 850°C to about 90% of cyclic lifetime. The strain rate of 10^{-5} s^{-1} and the strain amplitude of 0.7% favor void formation and coalescence at transverse grain boundaries.

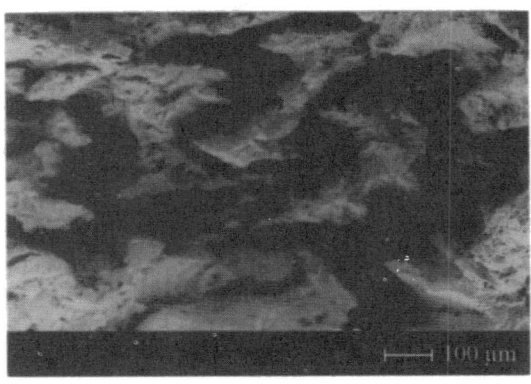

Figure 10: SEM micrograph of the fracture surface of a LCF specimen fatigued at 1000°C. As compared to Fig. 8, the lower strain rate of 10^{-5} s^{-1} and the strain amplitude of 0.7% favors intergranular failure.

At a later stage of fatigue, i.e. prior to final failure, there is pronounced internal necking which includes separation of longitudinal grain boundaries. Fig. 9 gives a representative view on this scenario obtained for a specimen fatigued to about 90% relative lifetime with a large strain amplitude of 0.7% at the lower strain rate of 10^{-5} s^{-1}. Obviously, the strong damage present in Fig. 9 may also explain the deviation of the data points at the largest strain amplitude of 0.7% in the Coffin-Manson lifetime prediction for the strain rate of 10^{-5} s^{-1} (see Figs. 7a and b). From the damage scenario displayed in Fig. 9 a mixed mode of trans- and intergranular failure should be expected. Fig. 10 supports this conclusion revealing mainly intergranular failure for a specimen fatigued to final failure under the same deformation conditions as that shown in Fig. 9.

Finally, it should be pointed out here that these observations on fatigue fracture coincide nicely with earlier reports (3-5,14,15) on creep of ODS nickel-base alloys like MA 754 and MA 6000. In particular, a mixed (inter- and transgranular) mode of failure was found by Singer and Arzt in MA 6000 for GAR values of around 10 as also determined for PM 1000 in the present investigation.

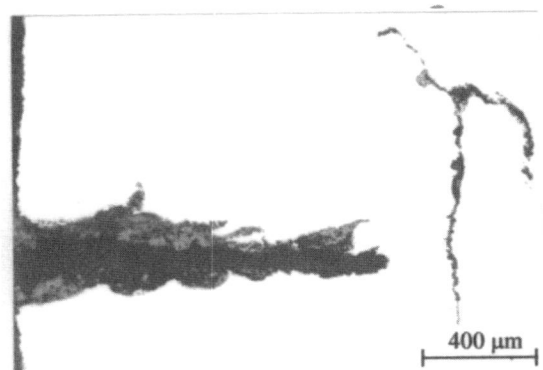

Figure 11: Optical micrograph (non-etched) of the longitudinal plane of a specimen fatigue deformed at 1000°C, strain rate 10^{-5} s^{-1}, strain amplitude 0.3%.

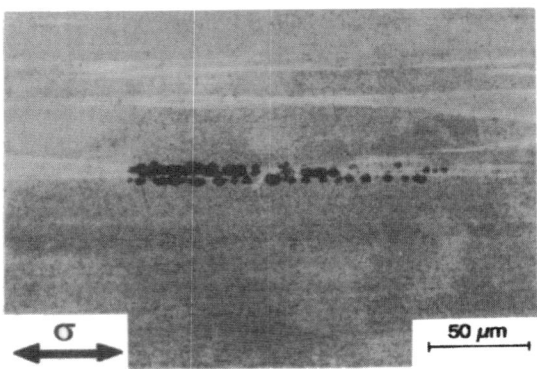

Figure 12: Optical micrograph of the longitudinal plane of a specimen fatigue deformed to about 25% relative lifetime. Deformation conditions: temperature 1000°C, strain rate 10^{-3} s^{-1}, strain amplitude 0.4%.

In contrast to the effect of a strain rate decrease the influence of a temperature change on cyclic lifetime is of minor importance. In essence, Figs. 6 and 7 demonstrate for both strain rates applied a rather marginal further lifetime reduction. We believe this tendency to be caused by the accelerated oxidation on the surface of the samples. A typical example for this is shown in Fig. 11. As expected the mean thickness of the oxide layer was found to be significantly increased at the higher deformation temperature thus giving rise to a higher probability for crack initiation from the specimen surface due to strain incompatibilities between substrate and oxide.

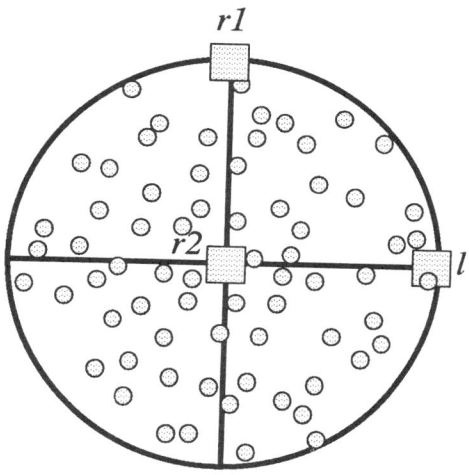

Figure 13: (100) pole figure of a fine-grained section of PM 1000 (SEM and EBSD). The squares indicate the orientations of the coarse elongated recrystallized grains whereas the circles represent the orientations of the fine grains. *r1* and *r2* denote two orthogonal radial directions in cross-section of the round bar.

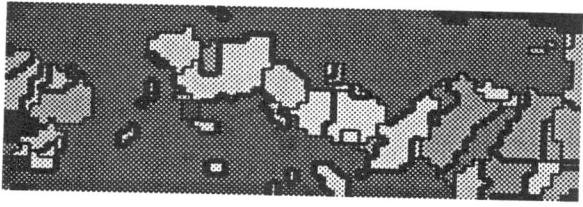

Figure 14: Grey-scale image of the grain structure corresponding to Fig. 13.

While Figs. 8 to 11 elucidate the damage state of PM 1000 subjected at final failure, Fig. 12 shows that internal damage already occurs relatively early in fatigue life. In particular, the sample prepared for optical microscopy in Fig. 12 was taken from a test interrupted after around 25% relative life, i.e. still in the region of stabilized cyclic deformation (Fig. 4a). Thorough metallographic preparation and inspection of several specimens interrupted in the saturation state prior to failure revealed (i) that the voids present in Fig. 12 were not artefacts from the grinding and polishing and (ii) that they were former positions of fine equiaxed grains with a diameter of 10 to 20 μm. These grains were found to be mainly situated in pockets along longitudinal grain boundaries of the coarse recrystallized grains. In the present heat these equiaxed grains occupy about 2% of the areal fraction of the longitudinal plane. Both, the mean diameter as well as the

areal fraction are in quantitative agreement with the results on Inconel MA 754 reported by Stephens and Nix (5).

Closer SEM examination of the grain orientation by EBSD proved these fine grains to be randomly distributed. Figures 13 and 14, respectively, show a stereographic projection of a (100) pole figure and the corresponding grey-scale image of the grain structure of a fine-grained region of the longitudinal plane in the non-deformed state. Besides the peaks of the coarse elongated grains (squares in Fig. 13) which are centered around the (100) poles a random distribution of the peaks resulting from the fine grains (circles) is observed. This result is in obvious contradiction to the observation of Stephens and Nix (5) who believed the fine grains to be recrystallized grains which were impeded in their growth by high local particle and inclusion concentrations. Though the presence of inclusions next to fine grains in PM 1000 was confirmed by Estrin et al. (16) which supports the assumption of grain growth impedement, we rather conclude from our observations that fine grains have undergone normal growth not preceded by recrystallization. Hence, we suggest to call them "recrystallization defects".

Considering a grain assembly with pockets of small equiaxed grains aligned between coarse grains elongated in longitudinal direction (Fig. 12) to be strained parallel during low cycle fatigue, it is conceivable that stresses arise between differently oriented grains due to plastic incompatibilities at the grain boundaries. These incompatibilities may originate from the experimentally verified anisotropy of the elastic constants in PM 1000 (10) which gives rise to different plastic strain amplitudes during total strain controlled LCF of differently textured PM 1000 crystals (8,9). Hence, the grain boundaries between the fine grains themselves as well as between fine and coarse grains can act as stress concentrator and become potential sites for void formation (Fig. 12). Consequently, the load-bearing cross-sectional area of the fatigue specimens is reduced and internal damage is initiated. It should be pointed out here that this effect occurs irrespective of whether the grain boundaries of the coarse elongated grains follow the extrusion direction quite closely or not. However, it is amplified in the case of fine grain pockets aligned along grain boundaries inclined at an angle with the extrusion direction.

Conclusions

The investigation of the high temperature LCF behavior of a powder-metallurgical ODS nickel-base superalloy and, in particular, the comparison with the creep behavior reported in the literature yields the following main conclusions:

1. In total-strain controlled LCF tests cyclic softening occurs which is more pronounced at higher temperatures.

2. A strain rate decrease severely reduces the cyclic lifetimes achievable.

3. The increase of testing temperature plays a minor role in decreasing cyclic lifetime.

4. Fatigue damage predominantly occurs due to mechanically assisted void formation and coalescence at transverse grain boundaries as well as in regions with fine-grain pockets.

5. The fine equiaxed grains were confirmed to be non-recrystallized. Rather, they underwent normal grain growth after the MA process. Consequently, they should be called "recrstallization defects".

Acknowledgments

This work has been carried out within the framework of the priority program "Microstructure and mechanical behavior of metallic materials at high temperatures" of the Deutsche Forschungsgemeinschaft (DFG). Funding from the DFG under contract no. He 1872/3 is gratefully acknowledged. One of the authors (M.H.) would like to thank the Alexander von Humboldt-Stiftung for financial support through the Feodor-Lynen program. Thanks are also due to Prof. Dr. H. J. Maier for stimulating discussions, Dr. B. de Boer for the help in texture analysis and H.-J. Klauss for experimental support during LCF testing.

References

1. J. S. Benjamin, "Dispersion strengthened superalloys by mechanical alloying", Metall. Trans., Vol. 1 (1970) 2943-2951.

2. J. S. Benjamin and T. E. Volin, "The mechanism of mechanical alloying", Metall. Trans., Vol. 5 (1974), 1929-1934.

3. E. Arzt, "Creep of dispersion strengthened materials: a critical assessment", Res Mechanica, Vol. 31 (1991), 399-453.

4. R. F. Singer and E. Arzt, "Structure, processing and properties of ODS superalloys", in Proc. Conf. "High Temp. Alloys for Gas Turbines and other Applications", edited by W. Betz et al., Reidel Publ. Comp., Dordrecht, 1986, 97-125.

5. J. J. Stephens and W. D. Nix, "The effect of grain morphology on longitudinal creep properties of INCONEL MA 754 at elevated temperatures", Metall. Trans. A, Vol. 16A (1985), 1307-1324.

6. M. Y. Nazmy, "High-temperature low cycle fatigue behavior and lifetime prediction of a nickel-base ODS alloy", in "Low Cycle Fatigue", ASTM STP 942, 1988, 385-397.

7. V. Banhardt, M. Nader and E. Arzt, "High-temperature low-cycle fatigue of an iron-base oxide-dispersion strengthened alloy: grain structure effects and lifetime correlations", Metall. Mater. Trans. A, Vol. 26A (1995), 1067-1077.

8. F. E. H. Müller, M. Heilmaier and L. Schultz, "The influence of texture and grain structure on the high temperature low cycle fatigue behaviour of the ODS nickel-based superalloy PM 1000", Mater. Sci. Eng. A, Vol. A234-236 (1997), 509-512.

9. M. Heilmaier, H.J. Maier, M. Nganbe, F.E.H. Müller, A. Jung and H.-J. Christ, "Cyclic stress-strain response of the ODS nickel-base superalloy PM 1000 under variable amplitude loading at high temperatures", Mater. Sci. Eng. A, Vol. A281/1-2 (2000), 37-44.

10. F. E. H. Müller M. Nganbe, H.-J. Klauss and M. Heilmaier, "Monotonic and cyclic high temperature deformation behavior of the ODS nickel-base superalloy PM 1000", J. Advanced Mater., Vol. 32 (2000), 9-20.

11. L. F. Coffin, "A study of the effects of cyclic thermal stresses on a ductile metal", Trans. ASME, Vol. 76 (1954), 931-950.

12. S. Suresh, "Fatigue of Materials" (New York, NY: Cambridge University Press, 1992), 136-140.

13. F. E. H. Müller, "Mikrostrukturelle Modellierung des monotonen und zyklischen Verformungsverhaltens der ODS Nickelbasissuperlegierung PM 1000 bei hohen Temperaturen" (Ph.D. thesis, Technical University of Dresden, 2000).

14. J. D. Whittenberger, "Creep and tensile properties of several oxide dispersion strengthened nickel base alloys", Metall. Trans. A, Vol. 8A (1977), 1155-1163.

15. T. E. Howson, J. E. Stulga and J. K. Tien, "Creep and stress rupture of oxide dispersion strengthened mechanically alloyed Inconel MA 754", Metall. Trans. A, Vol. 11A (1980), 1599-1607.

16. Y. Estrin, M. Heilmaier and G. L. Drew, "Creep properties of an oxide dispersion strengthened nickel-base alloy: the effect of grain orientation and grain aspect ratio", Mater. Sci. Eng. A, Vol. A272 (1999), 163-173.

EFFECT OF THERMOMECHAINCAL PROCESSING ON FATIGUE CRACK PROPAGATION IN INCONEL ALLOY 783

L. Z. Ma*, K. M. Chang*, S. K. Mannan**, S. J. Patel**
* West Virginia University, Morgantown, WV
** Special Metals Corporation. Huntington, WV

Abstract

Recently developed INCONEL® alloy 783 (nominal composition of Ni-34Co-26Fe-5.4Al-3Nb-3Cr) is precipitation strengthened by Ni_3Al-type γ' and NiAl-type β phases. Due to its low co-efficient of thermal expansion, high strength, and good oxidation resistance, alloy 783 is used for casings and bolting applications in gas turbines. During thermomechanical processing (TMP), the high Al content in alloy 783 results in the formation of β phase in an austenite matrix. The size and distribution of β phase governs the microstructure of the alloy. Therefore, optimization of TMP is critical for microstructure and mechanical properties. This study presents the effect of TMP on fatigue crack growth in alloy 783, especially in time-dependent condition.

Initially, the material was hot rolled in the mill at 1095°C. This was followed by rolling in the laboratory to 50% reduction at temperatures of 870°C, 1010°C and 1150°C. Materials rolled in the mill and laboratory were direct aged without solution anneal-ing. The aged materials were subjected to fatigue crack growth testing at room temperature, 300 °C, 450 °C and 600 °C and constant stress intensity mode. The results exhibit that time-dependent fatigue crack propagation resistance of alloy 783 can be dramatically improved by TMP. Comprehensive microstruc-ture characterization and fractographic analysis suggests that the enhancement of fatigue crack growth in time-dependent condition is associated with the change of fracture mode. The SAGBO (Stress Accelerated Grain Boundary Oxidation) mechanism and damage zone model were applied to explain these phenomena.

Introduction

Over the past twenty years, high strength low coefficient of thermal expansion (low CTE) superalloys, such as INCOLOY® alloy 903, 907, and 909, have been used for compressor cases, rings and shrouds to enable higher efficiency of gas turbine engines through tighter control of blade tip clearances over the range of turbine operating conditions [1-3]. The ferromagnetic characteristic below the Curie points of these alloys is responsible for thermal expansion coefficients lower than observed in paramagnetic alloys. These Ni-Fe-Co base alloys also have very low chromium content, as added Cr lowers the Curie temperature and thereby increase overall thermal expansion rate [4]. However, low content of Cr makes these alloys susceptible to stress accelerated grain boundary oxidation (SAGBO). INCONEL alloy 783 was developed to have low CTE and superior SAGBO and surface oxidation resistance than comparable alloys [1-3].

Relative to the INCOLOY Alloy 900-series of low CTE superalloys, alloy 783 with Al above 5% employs a three-phase microstructure (γ-γ'-β) to control mechanical properties, SAGBO and oxidation resistance. Fine γ' (Ni_3Al, Nb) precipitates form at low temperature aging treatments and provide adequate alloy strength [5]. The formation of the unique β phase (NiAl) is responsible for the improvement in SAGBO and oxidation resistance. The β phase is made possible by the high Al content in this alloy. Since β particles are heterogeneous and precipitate both in the grain boundaries and intragranularly in the dislocation networks, thermomechanical processing (TMP) can be utilized to control their precipitation kinetics and thereby control the properties and microstructure of this alloy. Early work has shown that TMP has a significant effect on microstructure and room temperature properties of alloy 783 [6]. The present study investigates how TMP affects fatigue crack propagation in this alloy.

® INCONEL and INCOLOY are trademarks of the Special Metals family of companies.

Material and experimental procedure

The INCONEL alloy 783 used in this study was supplied from regular production material by Special Metals Corporation, Huntington, WV. An ingot of 500mm diameter was vacuum induction melted (VIM) and vacuum arc remelted (VAR) using standard superalloy melting practices. The ingot chemical analyses are given in Table I

Superalloys 2000
Edited by T.M. Pollock, R.D. Kissinger, R.R. Bowman,
K.A. Green, M. McLean, S. Olson, and J.J. Schirra
TMS (The Minerals, Metals & Materials Society), 2000

Table I. Chemical Composition (wt%) of INCONEL Alloy 783

Element	Analyzed Results	Nominal Composition
Ni	28.21	28
Fe	24.88	Bal.
Co	34.39	34
Al	5.32	5.4
Nb	3.11	3.0
Ti	0.32	0.2 max
Cr	3.24	3.0

The precipitation temperature of two major phases in alloy 783, γ' (Ni₃Al) and β (AlNi), is respectively from 600°C to 800°C and from 700°C to 1140°C [1, 2, 6]. Therefore, TMP of this alloy is usually performed in the β precipitation temperature range to employ the β precipitates for structure control. Upon homogenization, the ingot was converted into billet in flat bar form at 1120°C. The received flat bar for this study was rolled to 19mm at 1095°C, which was designated as the primary rolling (HR0). This piece was used as a reference for the other three TMP procedures, which received a final rolling at specified rolling temperatures. Once the rolling temperature was reached, the plate was hot rolled at a 50% reduction ratio to 9.5mm thick. The first TMP rolling (HR1) was set at 870°C, which is above the γ' solvus, but still at the bottom of the β precipitation zone. The second rolling temperature (HR2) was chosen at 1010°C, which falls in the middle of the β precipitation zone. A last section (HR3) was rolled above the β solvus at 1150°C. Table II summarizes all TMP schemes used in the present work.

Specimens were machined from all the plates receiving the different TM schedules in the longitudinal direction, and received a direct aging heat treatment as follows: 843°C/4h, air cool + 720°C/8h, furnace cool at 55°C/hr to 620°C, hold at 620°C for 8 hours then air cool. The direct age, which eliminates the solution annealing, can markedly show the effect of TMP processing.

Table II. TMP Parameters

HR0	Primary roll at 1095°C
HR1	Primary roll + 50% reduction at 870°C
HR2	Primary roll + 50% reduction at 1010°C
HR3	Primary roll + 50% reduction at 1150°C

Fatigue crack propagation (FCP) tests were conducted at room temperature, 300°C, 450°C and 600°C respectively by using single edge-notched tension specimens. Crack length was monitored continuously by a D.C potential drop system. A MTS servohydraulic system equipped with the quartz light heaters heated the specimens to temperature to an error of around ±6°C. This test system can be completely monitored by a computer during the FCP test for either constant stress intensity factor mode or constant load mode. A sinusoidal fatigue cycle of 10 Hz was adopted to examine the crack propagation under pure fatigue. A slower fatigue cycle of 1/3 Hz was used to measure the effect of frequency on FCP. A third fatigue waveform consisting of a 1/3 Hz sinusoidal fatigue cycle and 100 seconds hold time at the maximum load of each cycle was also used, which usually results in time-dependent crack growth for several superalloys during elevated temperature fatigue [7-9].

FCP tests for every TMP specimen were conducted at constant stress intensity factor (ΔK) mode, Kmax = 28 MPa√m. The maximum-minimum load ratio (R) was kept at 0.05 in all waveforms. FCP tests were performed continuously from high temperature to low temperature, and from high frequency to low frequency. All specimens were precracked under fatigue condition of Kmax = 28 MPa√m, R= 0.05, f = 10 Hz and room temperature.

Each specimen was broken into halves when the test was finished. Many beach marks associated with the change of test temperature were observed on the fracture surface as seen in Figure 1. The measurement of beach marks was made for crack-length and ΔK value calibration. The linear regression was employed to calculate crack growth rate based on crack length vs. number of cycle's results.

Metallographic samples were prepared using conventional mechanical grinding and polishing procedures according to standard laboratory practices. Kalling's reagent consisting of 6 gm CuCl₂, 100 ml HCl, 100 H₂O, and 100 ml CH₃OH was used to reveal the microstructure. Scanning electron microscopy was used to examine the fracture surfaces of every sample.

Results

Microstructure

Figure 2 illustrates the microstructure of the alloy after TMP and direct aging. As expected, due to the high rolling temperature and no secondary deformation, the sample hot rolled at 1095°C (HR0) shows typically equiaxed recrystallized grains, with a grain size of ASTM 12-13, as seen in Figure 2(a). However, samples hot rolled at 870°C (HR1) and 1010°C (HR2) exhibit hot worked microstructure containing elongated β particles and stringers, as seen in Figure 2 (b) and (c). The sample hot rolled at 1150 °C (HR3) displays a dynamically recrystallized serrated grain boundary microstructure containing intergranular and transgranular β phase particles, as seen in Figure 2(d).

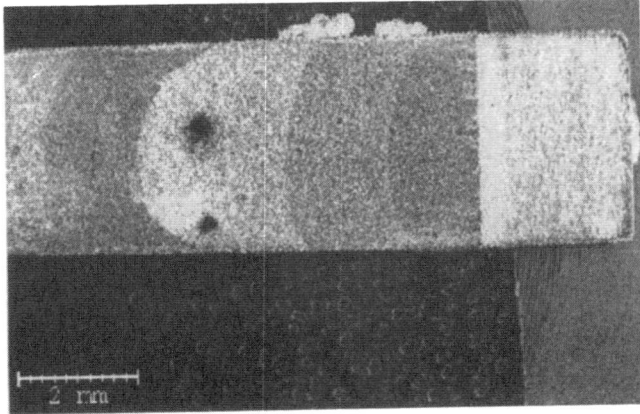

Figure 1: Specimen profile after FCP testing

Fatigue Crack Propagation

Constant ΔK crack propagation curves at 600°C for two hot rolled materials, HR1 and HR3, are presented in Figures 3 (a) and

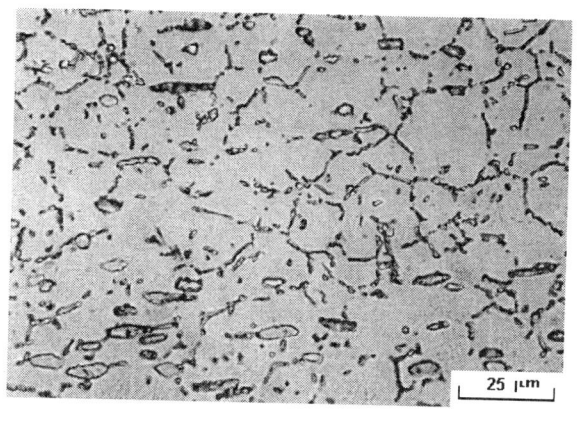

Figure 2 (a): Hot rolling at 1095°C (HR0)

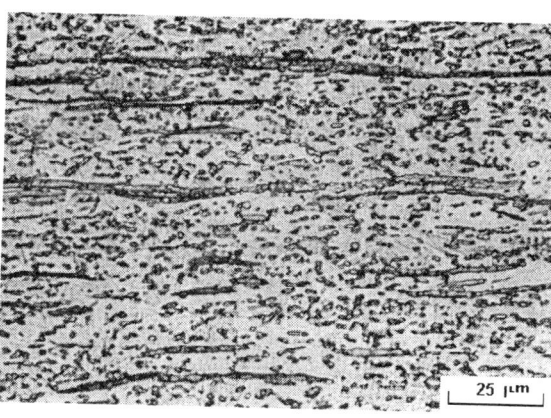

Figure 2 (b): Hot rolling at 870°C (HR1)

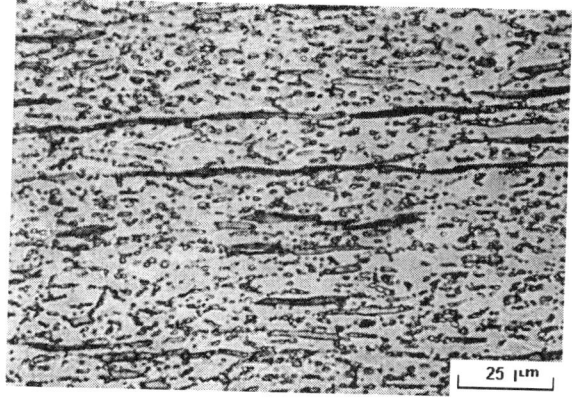

Figure 2 (c): Hot rolling at 1010°C (HR2)

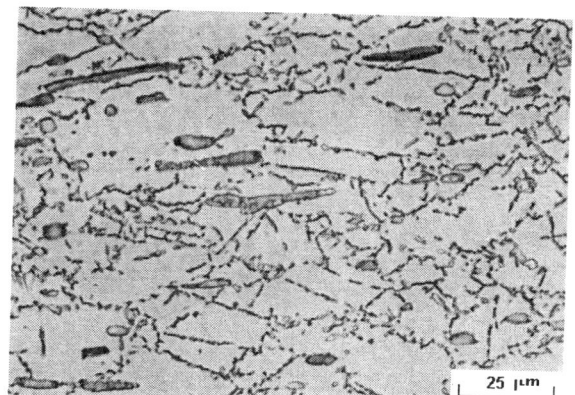

Figure 2 (d): Hot rolling at 1150°C (HR3)

Figure 2: Microstructure after TMP and direct aging

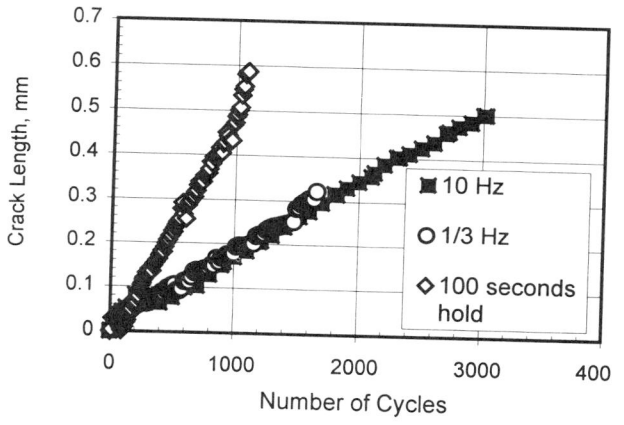

Figure 3 (a): Hot rolling at 870°C (HR1)

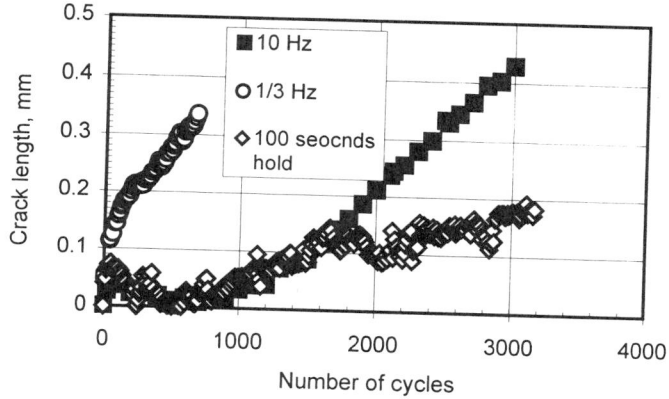

Figure 3 (b): Hot rolling at 1150°C (HR3)

Figure 3: Fatigue crack growth at ΔK=28 MPa$\sqrt{m}$ and 600°C

603

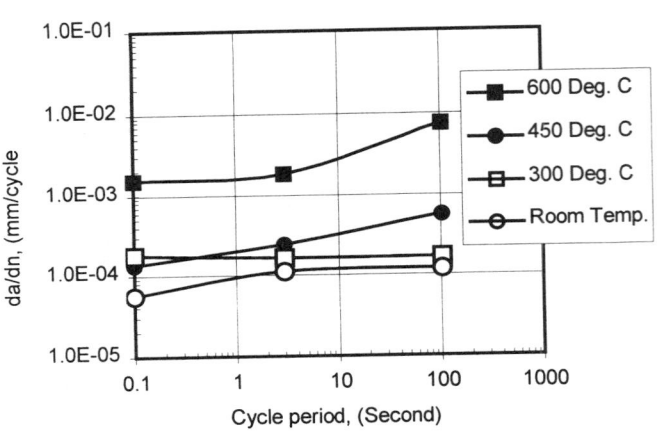

Figure 4 (a): Hot rolling at 1095°C(HR0)

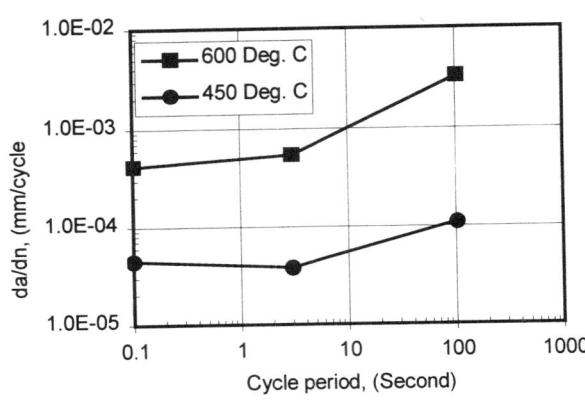

Figure 4 (b): Hot rolling at 870°C (HR1)

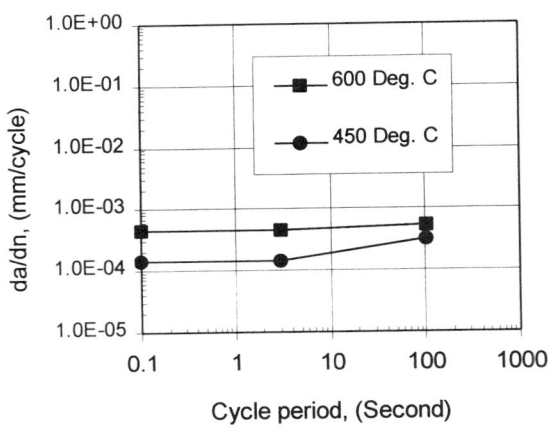

Figure 4 (c): Hot rolling at 1010°C (HR2)

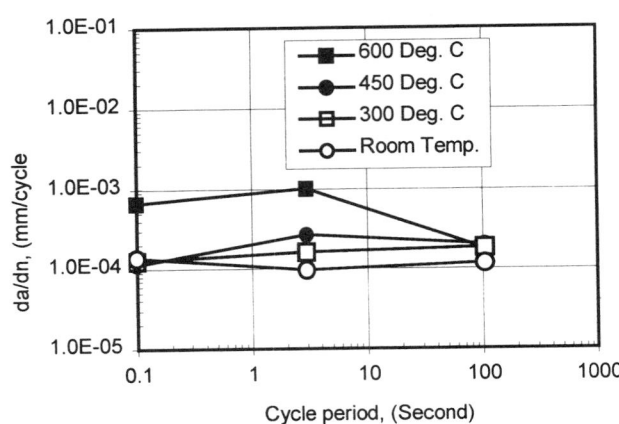

Figure 4 (d): Hot rolling at 11500°C (HR3)

Figure 4: Fatigue crack growth of TMP specimens at $\Delta K=28$ MPa$\sqrt{m}$

Table III Fatigue crack growth (mm/cycle) at $\Delta K=28$ MPa$\sqrt{m}$

Temperature (°C)	Frequency (Hz)	HR0	HR1	HR2	HR3
600	10	1.53×10^{-3}	4.17×10^{-4}	4.59×10^{-4}	6.72×10^{-4}
	1/3	1.84×10^{-3}	5.43×10^{-4}	4.50×10^{-4}	1.02×10^{-3}
	100 seconds Hold	7.51×10^{-3}	3.32×10^{-3}	5.45×10^{-4}	1.77×10^{-4}
450	10	1.36×10^{-4}	4.51×10^{-5}	1.43×10^{-4}	1.14×10^{-4}
	1/3	2.41×10^{-4}	3.80×10^{-5}	1.47×10^{-4}	2.67×10^{-4}
	100 seconds Hold	5.65×10^{-4}	1.11×10^{-4}	3.13×10^{-4}	2.01×10^{-4}
300	10	1.79×10^{-4}			
	1/3	1.65×10^{-4}			
	100 seconds Hold	1.67×10^{-4}			
Room Temp.	10	5.60×10^{-5}			
	1/3	1.12×10^{-4}			
	100 seconds Hold	1.21×10^{-4}			

(b). The slope of the linear portion of each curve represents the crack growth rate, which can be calculated by linear regression analysis. For the 870°C rolled material (HR1) as seen in Figure 3 (a), the crack growth rate for the 100 second hold test is 5 times higher than for the 10 Hz and 1/3 Hz tests, wherein the crack growth rate keeps constant. This is so called time-dependent fatigue crack growth and the phenomenon has been documented well in other alloys [10-12]. However, for the 1150°C rolled material (HR3), the crack growth rate for the 1/3 Hz test is dramatically higher than for the other two conditions as illustrated in Figure 3 (b). These observations will be discussed later.

Crack growth rates at different temperatures as a function of cycle period are shown in Figures 4 (a) to (d). Crack growth rates at 450°C and 600°C for material tested using the 100 second hold time rate is 5-10 times higher than for the 10 Hz and 1/3 Hz for the materials rolled at 1095°C (HR0) and 870°C (HR1). The difference in growth rate at those test temperatures is less pronounced with varying cycle times for the specimens rolled at 1010°C (HR2). However, for 1150°C rolled material (HR3), the crack growth rate at 600°C during the 100 second hold time test is significantly lower than in the case of the 10 Hz and 1/3 Hz tests. Further, the crack growth rate of 1150°C rolled specimen (HR3) during the 100 second hold test is also considerably lower than that for the specimens rolled at 1010°C, 1095°C, and 870°C. These findings demonstrate that an increase in the hot working temperature significantly decreases the time-dependent fatigue crack growth at 600°C for alloy 783. Crack growth rates at 300°C and room temperature for the specimens rolled at 1095°C and 1150°C were comparable for all three test conditions. This illustrates that at lower temperatures, crack growth is independent of rolling temperature, test frequency, and hold time. Fully understanding the relationship between microstructure and crack growth can provide the potential method to improve component life and enhance service temperature. Table III shows the data used in Figure 4.

Discussion

The effects of TMP on the microstructure and room tensile strength of alloy 783 as well as the TTT diagrams have been well documented [1-6]. In addition to the strengthening γ' precipitates developed during low temperature aging treatments, many precipitate phases may exist in the hot rolling temperature range. The equilibrium precipitation phase, β phase, has a solvus temperature around 1140°C. Since the β phase would not be coherent with the fcc matrix phase, they prefer to nucleate and grow along the defect sites in the alloy. Therefore, for HR1 and HR2, grain boundaries and dislocation networks are commonly decorated by the numerous β precipitates as seen in Figure 1 (b) and (c). As a result, the recrystallization will not easily take place during hot rolling because of retardation of β phases, and the structure consists of elongated, uncrystallized grains. However, when the hot rolling temperature is close to or above the β solvus, the recrystallization will occur during or after hot rolling. For HR3, dynamic recrystallization produced a very fine grain size with a serrated grain boundary structure. Some residual strains still remain due to the high deformation amount. In HR0, since the hot rolling reduction ratio is relatively low, there is a very fine equiaxed grain structure without deformation flow lines and with a serrated grain boundary structure.

Comparison of crack growth rates among four TMP microstructures are replotted in Figure 5, wherein transverse axis represents the temperature of TMP, with the origin point representing the HR0 processing. At a cycle-dependent stage, e.g. 10 and 1/3 Hz frequencies, four TMP specimens have a similar crack growth rate, which is simply determined by the pure mechanical driving force such as crack opening distance (COD) [13]. However, at hold time loading, the crack growth rate is drastically decreased with an increase of TMP temperature, i.e. the microstructure can have an extensive effects on FCP under the time-dependent condition.

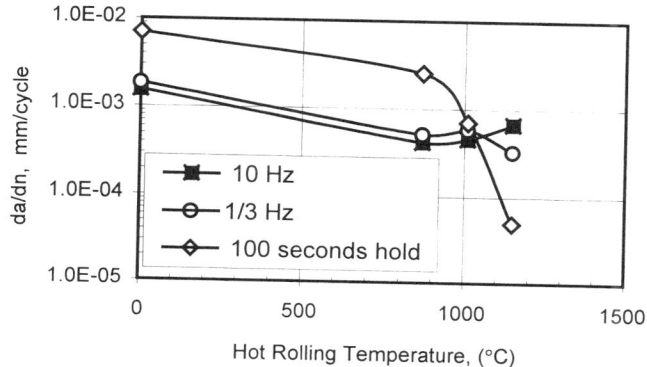

Figure 5: Fatigue growth rate against TMP Temperature at 600°C and ΔK=28 MPa√m

The fundamental mechanism causing the time-dependent crack propagation has been studied extensively. Conventionally three categories of theories, which are environmental effects, creep cracking effects, and strain rate effects respectively [13-17], have been proposed to explain the acceleration of crack growth under the time-dependent testing condition. Of course, complete separation of these three mechanisms may be impossible, while one of them generally becomes dominant in some alloys. Based on data generated in vacuum and other aggressive or inert gaseous environment for some superalloys, [11-13, 18-19] the environmental effect is believed to play the major role at elevated temperature crack growth. But the detailed mechanism of environmental effects is still not clear. In air, oxygen is the particular species responsible for the environmental effect. Presumably slow fatigue frequency allows oxygen atoms to be transported through the crack tip and to embrittle the grain boundaries ahead of the crack front. As a consequence, crack propagation is accelerated due to oxygen diffusion into grain boundaries ahead of the crack tip, and then time-dependent crack propagation occurs. It is important to note, however, that an environment, which can increase the crack growth rate under fatigue and creep loads, usually, does not produce significant general attack in unstressed materials. Thus, the stress assisted grain boundary oxidation (SAGBO) mechanism is considered to be responsible for time-dependent crack growth in this current study. It has been verified in alloy 718 that the SAGBO mechanism can cause the formation of a damage zone ahead of the crack tip during hold time of a sustained load. The damage zone, the size of which is a function of hold time, temperature, oxygen partial pressure, and stress intensity factor, can induce intergranular fracture and accelerate the crack growth rate [10-12, 20]. Therefore, based on the SAGBO mechanism and damage zone model, a phenomenological explanation for the observed crack

Therefore, based on the SAGBO mechanism and damage zone model, a phenomenological explanation for the observed crack growth behaviors in different TMP specimens of alloy 783 is proposed. When the fatigue frequency is high, e.g., 10 and 1/3 Hz, the crack growth rate is cycle-dependent and is determined by the cyclic stress intensity (ΔK), with only a limited effect of microstructure [13]. Lowering the frequency will enable oxygen diffusion to the grain boundaries producing a damage zone ahead of the crack tip. The damage zone causes intergranular fracture during crack growth and therefore accelerates the growth rate. This may explain the faster crack propagation rate observed with hold time for the 1095°C and 870°C hot worked specimens (HR0 and HR1) tested at 450°C and 600°C. Fractographs of these specimens show completely intergranular fracture as seen in Figure 6. It should be noted that the fracture surface of the HR1 specimen in Figure 6 (b) exhibits large amounts of precipitates along the grain boundary areas and fluctuated fracture paths. These precipitates are believed to be β phases, which may retard the crack propagation during intergranular fracture and decrease the effects of oxidation on the grain boundary and matrix [1-2, 21]. Also, the residual strain structure referred to as stringers, which are dislocation networks and segregation of β phase, can provide the preferential sites for oxygen diffusion during the build-up of the damage zone. The combination of the above effects is expected to enhance the time-dependent fatigue crack growth resistance. Thus, the crack growth rate of the HR1 specimen would be lower relative to the HR0 specimen during time-dependent fracture.

It is interesting to note that specimens rolled at 1010°C had many secondary parallel cracks as seen in Figure 7 (a). These secondary cracks were in the primary crack propagation direction with the interspacing about 100μm, which is almost equal to the distance between parallel β phase stringers as shown in Figure 1(c). These β phase stringers presumably provide preferential oxygen absorption sites resulting in embrittlement and secondary cracking along the stringers. In this type of cracking, stress concentration at the crack tip can be released to form multiple parallel cracks, which will not show a time-dependent behavior under cyclic-loading [22]. High magnification fractograph of the secondary crack areas show that the fracture mode is predominantly transgranular, and the fracture surface is covered by many particles (mostly β phase) and is relatively rough as shown in Figure 7(b).

Secondary cracks were also found in the specimen hot rolled at 1150°C (HR3) as shown in Figure 8. Interspacing of these cracks was approximately 5-10μm. The fracture surface demonstrates an obvious border between tests of hold time at 600°C and high frequency (10 Hz) at 450°C. There exist many secondary cracks in hold time specimen tested at 600°C. High magnification micrographs of these areas exhibits an intergranular fracture path and a ductile appearance with many dimples as seen in Figure 8 (b). However, in the high frequency specimen tested at 450°C, the fracture mode is typically of a microvoid nature. The ductile fracture feature in material hot rolled at 1150°C (HR3) in hold time condition could be ascribed to the serrated boundaries in this material induced by TMP as shown in Figure 2(d). Serrated grain boundaries are thought to restrict grain boundary sliding which increases the fracture path resulting in better crack growth behavior [23-24]. This could explain the better crack growth

resistance of the 1150°C hot worked material tested at high temperature under hold time mode.

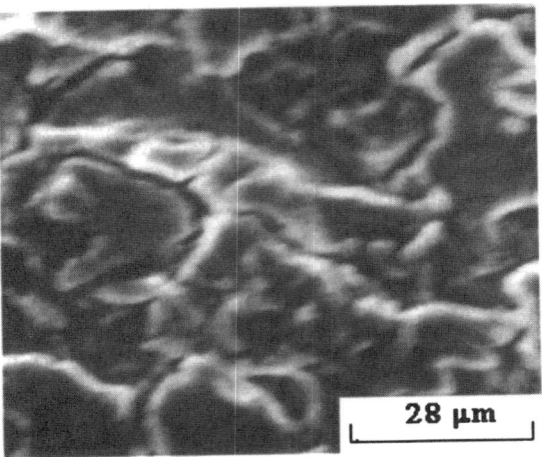

Figure 6 (a): Hot rolling at 1095°C (HR0)

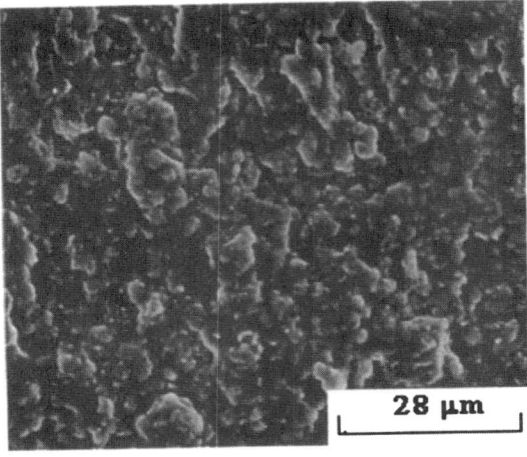

Figure 6 (b): Hot rolling at 870°C (HR1)

Figure 6: Fractograph of SEM at $\Delta K=28$ MPa√m, 600°C, and 100 seconds hold

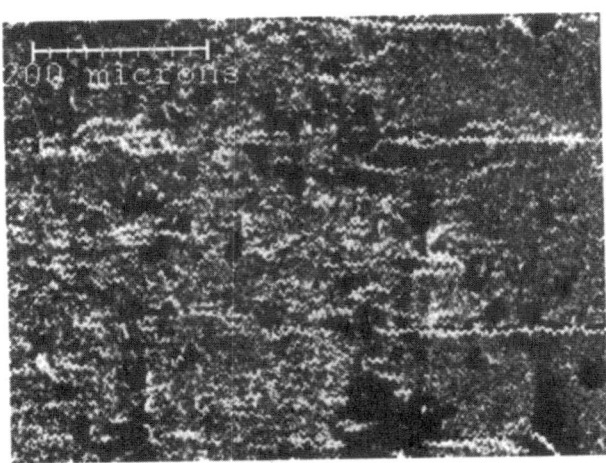

Figure 7 (a): Secondary crack in HR2

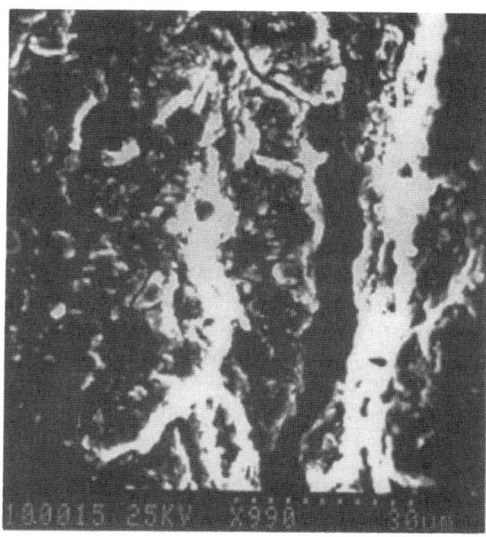

Figure 7 (b): High magnification of (a)

Figure 7: Fractograph of SEM for hot rolling at 1010°C
at ΔK=28 MPa√m, 600°C, and 100 seconds hold

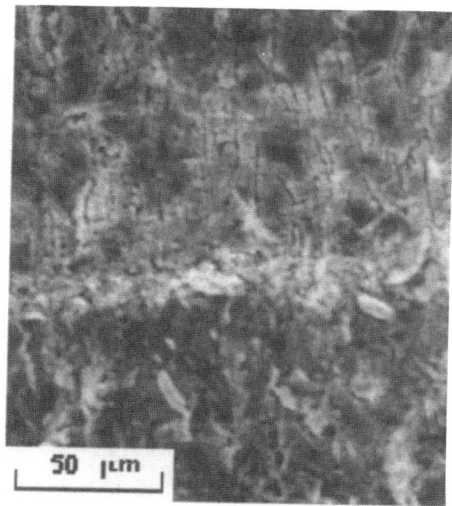

Figure 8 (a): Fracture surface in HR3

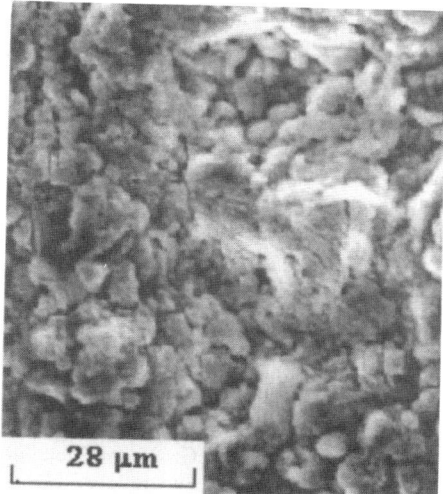

Figure 8 (b): High magnification of (a)

Figure 8: Fractograph of SEM for Hot rolling
at 1150°C under ΔK=28 MPa√m and 600°C

Conclusions

Four different microstructures of INCONEL Alloy 783 have been identified as the result of different TMP parameters and their fatigue crack propagation behaviors have been evaluated under time-dependent testing conditions. The data suggests the following:

1. Microstructures induced by the thermomechanical processing (TMP) have substantial effects on time-dependent fatigue crack propagation behavior in alloy 783, but little influence on cycle-dependent crack growth rate.

2. A deformed grain structure without recrystallization and with a large number of β phase particles distributed within the grain increases the resistance to crack propagation under time-dependent condition. The appearance of secondary cracks weakens the effects of the time-dependent behavior on the crack growth rate, together with a change of fracture mode.

3. The introduction of grain boundary serration produced by TMP can significantly improve the resistance to fatigue crack propagation under time-dependent condition.

4. SAGBO and an environmentally induced damage zone at the crack tip are believed to play a major role in time-dependent fatigue propagation in alloy 783 at elevated temperatures.

Acknowledgments

The authors would like to thank Specials Metal Corporation at Huntington, WV for supplying the INCONEL alloy 783 used in this study and contributing insight on this project. Many thanks are also due to F. J. Veltry and J. M. Davis at Specials Metal Corporation for their technical support in metallographic sample preparing. This study was sponsored by West Virginia EPSCOR program.

Reference

1. J. S. Smith and K. A. Heck, "Development of Low Thermal Expansion, Crack Growth Resistant Superalloy" Superalloys 1996, edited by R. D. Kissinger, D. J. Deye, D. L. Anton, A. D. Cetel, M. V. Nathal, T. M. Pollock, and D. A.Woodford, The Minerals, Metal & Materials Society, Seven Spring, PA, USA, 1996, 91-100.

2. K. A. Heck, J. S. Smith, R. Smith, "Inconel alloy 783: An oxidation Resistant, Low Expansion Superalloy for Gas Turbine Applications," International Gas Turbine and Aeroengine Congress & Exhibition, Birmingham, UK, June 10-13, 1996: 1-8.

3. K. A. Heck, D. F. Smith, M. A. Holderby and J. S. Smith, "Three-phase Controlled Expansion Superalloys with Oxidation Resistance" Superalloy 1992, edited by S. D. Antolovich, R. W. Stusurd, R. A. MacKay, D. L. Anton, T. Khan, R. D. Kissinger, D. L. Klarstrom, The Minerals, Metal & Materials Society, Seven Spring, PA, USA, 217-226.

4. M. M. Morra, " Stress Accelerated Grain Boundary Oxidation of INCOLOY Alloy 908 in High Temperature Oxygenous Atmosphere" (Ph.D. thesis, Massachusetts Institute of Technology, 1995), 16-17.

5. S. Mannan and J. deBarbadillo, "Long Term Thermal Stability of INCONEL Alloy 783," International Gas Turbine and Aeroengine Congress & Exhibition, Stockholm, Sweden, June 2-5, 1998, 1-10.

6. B. E. Higginbotham, K. M. Chang, S. Mannan, J. J. deBarbadillo, "Microstructure and Property Control of INCONEL 783 Through Thermomechanical Processing (TMP)," Proceedings of the 1st International Non-Ferrous Processing and Technology Conference, St. Louis, Missouri, 10-12 March 1997, 483-489.

7. P. Shahinian and K. Sadananda, "Effects of Stress Ratio and hold-Time on Fatigue Crack Growth in Alloy 718," The American Society of Mechanical Engineers (Proc.), June 25-29, 1979, 1-9.

8. K. Sadananda and P. Shahinian, "Effect of Microstructure on Creep-fatigue Crack Growth in Alloy 718, Res. Mechanica. Vol. 1, 1980, 109-128.

9. H. Ghonem and D. Zheng, "Frequency Interactions in High-Temperature Fatigue Crack Growth in Superalloys," Metall. Trans. A, Vol. 23 A, Nov. 1992, 3067-3072.

10. K. M. Chang, M. F. Henry and M. G. Benz, "Metallurgical Control of Fatigue Crack Propagation in Superalloys," JOM, Vol. 42, No.12, Dec. 1990, 29-35.

11. K. M. Chang, "Elevated Temperature Fatigue Crack Propagation After Sustained Loading," Effects of Load Thermal Histories on Mechanical Behavior of Materials (Proc.), TMS Spring Meeting, Denver, CO, 1987, ed., P. K. Liaw et al., AIME-TMS, 13-26.

12. K. M. Chang, "A Phenomenological Theory for Time-Dependent Fatigue Crack Propagation in High-Strength Superalloys," (G.E R & D Report, No: 91CRD066, March 1991, 1-19.).

13. D. L. Davidson and J. Lankford, "Fatigue Crack Growth in Metals and Alloys: Mechanisms and Micromechanics," International Mater. Rev. Vol. 37, No: 2, 1992, 45-75.

14. K. Sadananda and P. Shahinian "High Temperature Time Dependent Crack Growth, Micro and Macro mechanics of Crack Growth," Proceeding of Symposium, Metallurgical Society of AIME, Warrendale, Pa, 1982, 119-130.

15. D. A. Woodford and R .H. Bricknell, "Environmental Embrittle of High Temperature Alloys by Oxygen," Treatise on Materials Science and Technology, Vol.25, 1983, 157-196.

16. S. Floreen and R. H. Kane, "A Critical Model for the Creep Fracture of Nickel-Base Superalloys," Metall. Trans. A, Vol.7 A, Aug. 1976, 1157-1160.

17. J. Z. Xie, Z. M. Shen, and J. Y. Hou, "Fatigue Crack growth Behaviors in Alloy 718 at High Temperature," Superalloy 718 625 and Various Derivatives, edited by E. A. Loria, The Minerals, Metal & Materials Society, 1997, 687-594.

18. S. Floreen and R. H. Kane, "Effects of Environment on High-Temperature Fatigue Crack Growth in a Superalloy," Metall. Trans. A, Vol. 10 A, Nov. 1979, 1745-1751.

19. K. Sadananda and P. Shahinian, "Effect of Environment on Creep Crack Growth Behavior in Austenitic Stainless Steels under Creep and Fatigue Conditions," Metall. Trans. A, Vol. 11A, Feb. 1980, 267-276.

20. W. Carpenter, B.S.-J. Kang and K. M. Chang, "SAGBO Mechanism temperature Cracking Behavior of Ni-base Superalloys," Superalloy 718 625 and Various Derivatives, edited by E. A. Loria, The Minerals, Metal & Materials Society, 1997, 679-688.

21. J. S. Lyons, A. P. Reyonlds and J. D. Clawson, "Effect of Aluminide Particle Distribut-ion on the High Temperature Crack Growth Characteristics of A Co-Ni-Fe Superalloy," Scripta Mater., Vol.37, No:7, 1997, 1059-1064.

22. K. M. Chang, "Improving Crack Growth Resistance of IN 718 Alloy Through Thermomechanical Processing," (G.E. Report, No. 85CRN187, Oct. 1985, 1-12).

23. R.Raj and M.F.Ashby: On Grain boundary sliding and diffusion creep, Metall. Trans. A, Vol. 2, April 1971: pp. 1113-1127.

24. J. M. Larson and S. Floreen, "Metallurgical Factors Affecting the Crack Growth Resistance of a Superalloy," Metall. Trans. A, Vol. 8 A, Jan. 1977, 51-55.

The Ductility of HAYNES® 242™ Alloy as a Function of Temperature, Strain Rate and Environment

Stephen D. Antolovich
Professor of Mechanical and Materials Engineering
Washington State University
Pullman, WA 99164-2920

Dwaine L. Klarstrom
Manager of Product Research and Development
Haynes International
1020 West Park Ave.
POB 9013
Kokomo, Indiana 46904-9013

John F. Radavich
President
Micro-Met Laboratories
209 North Street
West Lafayette, IN 47906

Abstract

HAYNES 242 is an alloy that is used when good combinations of strength, toughness and corrosion resistance are required. Since it is intended to be used for long time applications, the effects of temperature, time of exposure, and strain rate are all important factors that must be considered. In this study, the effects of prior exposure, temperature and strain rate were investigated as to their effects on mechanical properties such as yield strength (YS), ultimate tensile strength (UTS), and ductility (%RA). It is shown that the ductility and UTS increase with increasing strain rate when tested in air, in contrast to what is usually expected. When tested in vacuum, the ductility decreased with increasing strain rate in accord with usual observations. These contrasting behaviours are interpreted in terms of an environmental effect which causes changes in the fracture mechanism as well as the properties. It is noted that while there is an environmental effect, the absolute values of ductility remain very high making this material suitable for applications where long-term exposure and impact loading are considerations.

I.   Introduction

HAYNES 242 alloy has an excellent combination of mechanical and physical properties and finds application at high temperatures in gas turbine seal rings where a good combination of ductility, strength, toughness, and low coefficient of thermal expansion is important [1]. Deformation has been determined in the past to occur through complex mechanisms including dislocation glide and twinning in a matrix containing very fine, ordered coherent $Ni_2(Mo,Cr)$ domains [2-6]. The object of this study was to determine the effects of temperature, strain rate, environment and prior exposure at high temperature on the strength and ductility of this important commercial alloy.

II.   Experimental Procedure

A.   Material

The material used in this investigation was HAYNES 242 Alloy whose nominal composition (wt %) is:

Ni [bal], Mo [25], Cr [8], Co [2.5 max], Mn [0.80 max], Si [0.80 max], Al [0.50 max], C [0.03 max], B [0.006 max], Cu [0.50 max] .

The material tested in this program was taken from heat 8422-3-7566 in the form of 12.7mm (0.5 inch) thick plate. Specimens were aged in air for either 24h or 1000h at 649C (1200F) after annealing in order to simulate potential service conditions.

Superalloys 2000
Edited by T.M. Pollock, R.D. Kissinger, R.R. Bowman,
K.A. Green, M. McLean, S. Olson, and J.J. Schirra
TMS (The Minerals, Metals & Materials Society), 2000

B. Tensile Testing

Specimens were fabricated in the form of standard ASTM specimens with a 6.35 mm diameter and a 35.6 mm gage length. Tests were carried out at 427C and 649C (800F and 1200F) using nominal strain rates of 0.5, 5.0, and 50%/min (in one case the strain rate was 500%/min). Testing was done in air at 427C and 649C (800F and 1200F) and in vacuum at 649C (1200F). The vacuum was about 0.1×10^{-5} torr. The specimens tested in air had an extensometer directly attached and the strain rate could be controlled directly. The specimens tested in vacuum did not have an extensometer attached and the strain was computed from the crosshead displacement. Since there is considerable deflection in the load train, the crosshead displacement does not represent the displacement in the specimen. The strain rate was controlled using a pacing technique. Since the load/displacement records were in terms of cross head displacement, they had to be corrected to reflect actual specimen strain. At low loads, the displacement was made up entirely of elastic deflection of the specimen and elastic deflection of the machine. However, Young's Modulus of the HAYNES 242 alloy was known and the deflection of the specimen could be subtracted from the total elastic deflection to yield machine deflection. Since the machine deflection will *always* track linearly with load, a curve for machine deflection v. load could be established. This then permitted the machine deflection to be subtracted from the total deflection at *any* load. Thus actual engineering stress/strain curves could be constructed. The stress/strain curves for all of the vacuum tested materials were determined in this way. A partial check of the validity of the technique is to note the measured longitudinal strain and to compare it with the strain computed from the stress/strain curves determined as described above. In all cases the agreement was quite good.

C. Optical Microscopy SEM Examinations

Specimens were examined via optical metallography and SEM to gain some insight into the microstructure, fracture features and the mechanisms of failure. Standard metallography and SEM procedures were used.

D. TEM Examination

Representative specimens were examined via TEM to determine if there were substantial changes in the deformation mode, which could account for the ductility changes that were observed. Foils were taken near the fracture surfaces (for the most part)

and were polished using standard techniques. A 200 kV TEM with analytical capability was used.

III. Results and Discussion

A. Tensile Test Results

The results of the tensile tests are tabulated in Tables I and II where yield strength, tensile strength, elongation, and per cent reduction in area are given.

Table I - Tensile Test Results for All Specimens Tested at 649C Air and Vacuum

Spec ID	Strain Rate (%/min)	0.2% YS (MPa)	UTS (MPa)	Elong (%)	RA (%)	Comments Vac (Torr)
Aged at 649C for 24 hrs - Alloy 8422-3-7566						
24-12	0.5	630.0	928.9	14.0	22.0	
24-3-V	0.5	626.5	1043.0	47.0	70.0	0.1×10^{-5}
24-4-V	0.5	619.5	1043.0	44.0	73.0	0.1×10^{-5}
24-4	5.0	574.0	994.0	20.0	22.0	
24-10	5.0	581.0	1001.0	19.0	26.0	
24-1-V *	5.0	612.5	1085.0	25*	23*	1×10^{-5}
24-2-V	5.0	605.5	1064.0	36.0	64.0	0.1×10^{-5}
24-5-V	5.0	602.0	1071.0	37.0	61.0	0.1×10^{-5}
24-6-V	50.0	595.0	1036.0	33.0	58.0	0.1×10^{-5}
24-13-V	50.0	595.0	1057.0	34.0	54.0	0.1×10^{-5}
24-2	50.0	651.0	1064.0	31.0	49.0	
24-1	50.0	591.5	1057.0	33.0	53.0	
24-9	50.0	574.0	1050.0	35.0	58.0	
24-3	500.0	640.5	1043.0	32.0	56.0	
Aged at 649C for 1000 hrs - Alloy 8422-3-7566						
1000-10	0.5	714.0	861.0	3.0	2.6	Fail in thrd
1000-11	0.5	721.0	903.0	3.7	3.5	Fail in thrd
1000-13-V	0.5	728.0	1099.0	43.0	69.0	0.1×10^{-5}
1000-4-V	0.5	714.0	1141.0	34.0	60.0	0.1×10^{-5}
1000-3	5.0	728.0	1015.0	9.0	16.0	
1000-4	5.0	721.0	994.0	9.5	17.0	
1000-12-V	5.0	728.0	1134.0	30.0	56.0	0.1×10^{-5}
1000-3-V	5.0	707.0	1148.0	29.0	54.0	0.1×10^{-5}
1000-2	50.0	704.9	1092.7	18.0	24.0	
1000-1	50.0	702.8	1117.9	22.0	30.0	
1000-1-V	50.0	686.0	1127.0	27.0	55.0	2.0×10^{-5}
1000-2-V	50.0	689.5	1120.0	26.0	50.0	0.1×10^{-5}

V = tested in vacuum. Vacuum level is indicated in torr
* = ductility data not valid - stopped before failure -pulled out of adapter

The results in Table I show that there is a large effect of strain rate, environment and aging time on ductility with less of an effect of these variables on the yield and ultimate strengths. Similar conclusions can be drawn about the air testing carried out at 427C as seen in Table II.

Table II - Tensile Test Results for 427C - Air Only

Spec ID	Strain Rate (%/min)	0.2% YS (MPa)	UTS (MPa)	Elong (%)	RA (%)
Aged at 649C for 24 hrs - Alloy 8422-3-7566					
24-11	0.5	650.3	1214.5	38.0	44
24-8	5.0	651.0	1190.0	39.0	48
24-7	5.0	651.0	1183.0	38.0	49
24-6	50.0	654.5	1155.0	41.0	46
24-5	50.0	658.0	1169.0	39.0	51
Aged at 649C for 1000 hrs - Alloy 8422-3-7566					
1000-9	0.5	728.0	1295.0	29.0	47
1000-7	5.0	770.0	1295.0	31.0	42
1000-8	5.0	777.0	1281.0	31.0	43
1000-6	50.0	753.2	1257.2	31.0	42
1000-5	50.0	758.9	1242.5	29.0	43

A typical stress/strain curve is shown in Fig. 1.

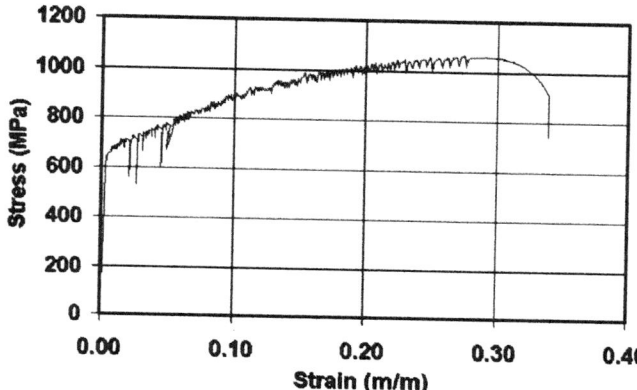

Fig.1. Stress/strain curve for specimen 24-2. Aged 24 hrs at 649C and tested at 649C and 50%/min in air.

The features seen in Fig. 1 were similar for air and vacuum. The serrations on the stress/strain curve are associated with either dislocation "bursts" or deformation by twinning. It has been shown in the literature that deformation occurs by both mechanisms. [2-5]. It is not likely that there is a strain aging effect here since the drops are too large and the range of conditions for which this behavior are observed are too extensive for such a mechanism to be operating. A stress strain curve for specimen 1000-V-4, tested and aged under different conditions is shown in Fig. 2. Specimens tested under these conditions also exhibited serrated stress strain curves. However, due to limitations of the equipment, testing was done in cross head control and machine deflection was subtracted out analytically. The test method and the data reduction procedure both operated to artificially reduce the appearance of the serrations.

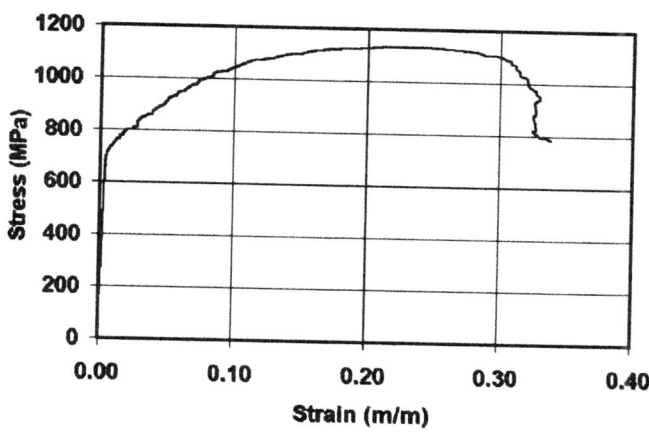

Fig. 2 . Stress strain curve for specimen 1000-V-4. Aged 1000 hrs at 649 C and tested at 649C at 0.5%/min

For completeness, a stress strain curve for specimen 1000-7, aged 1000 hrs at 649C and tested at 427C in air is shown in Fig. 3.

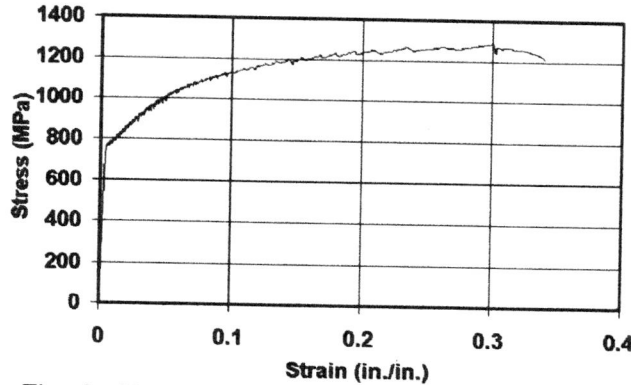

Fig. 3. Stress strain curve for specimen 1000-7. Aged 1000 hrs at 649C and tested at 427C at 5%/min in air.

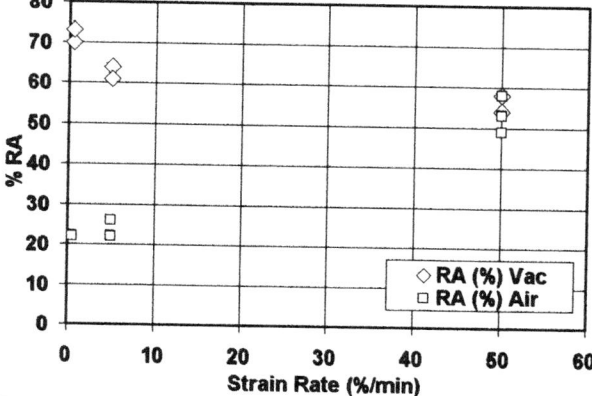

Fig. 4. % RA v strain rate for material aged for 24 hrs at 649C and tested in air & vacuum at 649C.

611

Figure 4 (24 hr age material) demonstrates that the ductility in vacuum is significantly higher than it is in air at the low strain rates. It is significant that:

1. ductility decreases with increasing strain rate for the vacuum tests , an expected result while,
2. ductility *increases* with strain rate in air, a rather uncommon and unexpected result.
3. the ductilities for both conditions converge at a strain rate of about 50%/min.

Similar behavior is observed in Fig. 5 for material that was aged for 1000 hrs at 649C and tested at 649C. The difference between these conditions is that the ductilities at a given strain rate are significantly lower and convergence appears to be shifted to higher strain rates (with a lower ductility). The lower ductility may be understood in terms of a larger embrittled zone. Reasons for the shift to convergence at higher strain rates is less intuitive.

seen that the ductility is relatively constant (somewhat lower for the longer term age).

Note that the data scatter is small in all cases making it possible to draw unambiguous conclusions.
Figures 7-9 shows the strength properties as a function of strain rate. Here we see that the yield strength for both air and vacuum are virtually identical while there is a modest increase of the UTS for the material tested in vacuum over that tested in air at the low and intermediate rates.

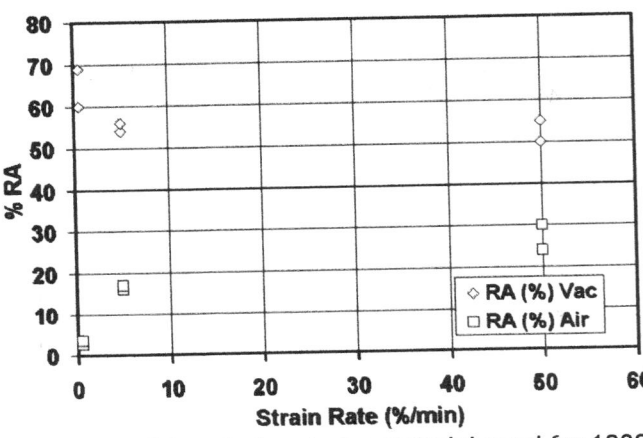

Fig. 5. % RA v strain rate for material aged for 1000 hrs at 649C and tested in air or vacuum at 649C

The ductility at 427C for ageing at 649C either 24 or 1000 hrs is shown in Fig. 6. In this figure it can be

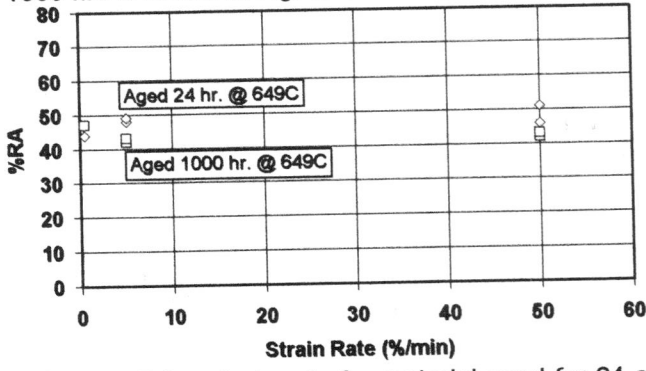

Fig. 6. % RA v strain rate for material aged for 24 or 1000 hrs at 649C and tested in air at 427C.

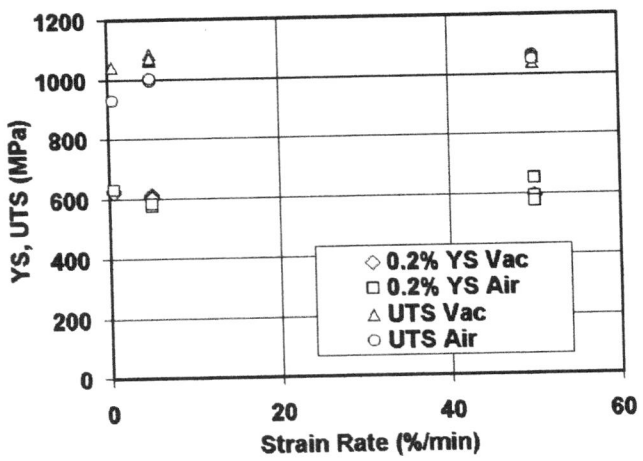

Fig. 7. Yield and tensile strength v strain rate for material aged for 24hrs at 649C and tested in air or vacuum at 649C.

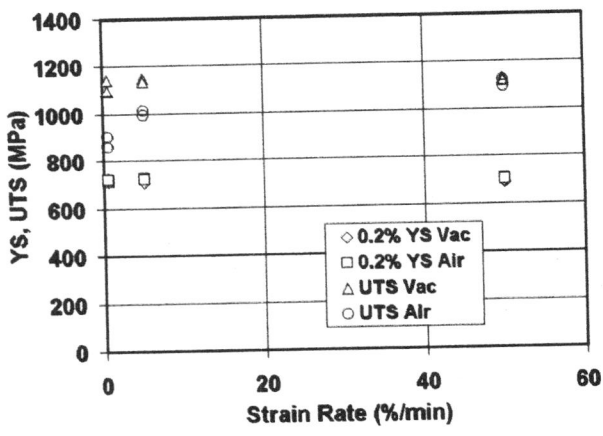

Fig. 8. Yield and tensile strength v strain rate for material aged for 1000 hrs at 649C and tested in air or vacuum at 649C.

All of these observations are fully consistent with an environmental effect. The increased ductility of the vacuum-tested material over the air-tested material at

649C can be understood on the basis of reduced ductility or toughness in air. In other words, in terms

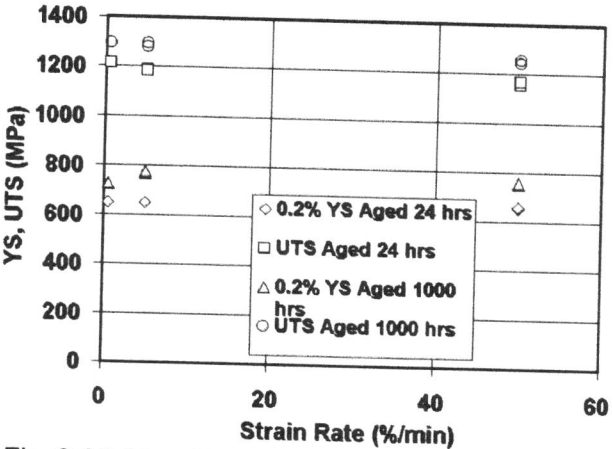

Fig. 9. Yield and tensile strength v strain rate for material aged for 24 or 1000 hrs at 649C and tested in air at 427C.

of an environmental effect. However, the ductility in air *increases* with strain rate (an unusual but not unique effect) because with increasing strain rate the time for a deleterious environmental effect (e.g. diffusion of oxygen) is reduced. Thus even though the intrinsic ductility would be expected to decrease (as seen in the vacuum tests) this is offset by the effect of the environment. The decrease in ductility with increasing strain rate observed in the vacuum tests is simply the usual trend observed in most metallic alloys. The strength data supports this argument as well. The yield strength is not much affected by environment since it is a measure of the first dislocation movement. Hence it is not surprising to see the vacuum and air yield strengths essentially the same. However, the UTS is affected by environment; during the time it takes to reach the UTS there is time for environmental interactions and hence a reduced toughness which is reflected in a lower UTS for the air tested material. It is well known that increased strain rate will increase the flow stress. Thus if there were an environmental effect we would expect to see similar yield strengths, higher tensile strengths for the vacuum tested material and increasing tensile strengths for each with convergence to a common value at some point. To repeat, this convergence is seen at a strain rate of about 50%/min. The environmental explanation fits very well with the specimens tested in this program.

The same trends are observed for the material that was aged for 1000 hrs. before testing except that the elongation and %RA did not converge at 50%/min. while the yield and ultimate strengths did converge. The ductility was reduced only modestly for the

vacuum tested material compared to the material which was aged for 24 hrs. This observation may be taken as an indication that extensive ageing doesn't significantly affect the *intrinsic* ductility that is measured in the vacuum tests. The ductility for the air tests showed a very large drop from the 24 hr. age, especially at the lowest strain rates. This would indicate that there is a strong environmental effect but that there may be significant structural changes when aged for 1000 hrs. which promote even more environmental damage in the course of the test. This would seem to follow from the fact that independent of how the testing was done, all starting material was aged under similar conditions in air prior to machining samples from the interior. These results indicate that there is a dynamic environmental effect associated with the deformation mode which may be influenced by structural changes during aging. Such a conclusion is strengthened by considering the yield and ultimate tensile strength results. As the ageing time is increased from 24 to 1000 hrs, there is a small but definite increase in the yield strength and a smaller increase in the tensile strength as seen in comparing the results in Figs. 7 and 8. These strength differences are consistent with modest changes in the underlying microstructure.

Testing at 427C did not reveal any rate effects, either on ductility or strength, as can be seen in Figs. 6 and 9. There were, however, modest ageing effects; the ductility of the material aged for 1000 hrs was modestly lower than that aged for 24 hrs and the strengths for the 1000 hr age were higher. These observations are fully consistent with the preceding comments on structural changes and the lack of an environmental effect at 427C. The absence of a rate effect can be understood in terms of an elementary diffusional model which follows the Arrhenius relationship and in which the activation energy is assumed to be unaffected by stress. These assumptions result in an equation of the form:

$$\frac{1}{\tau} = A \exp{-\left(\frac{Q}{RT}\right)} \qquad \ldots (1)$$

where τ = characteristic process time (sec)
Q = activation energy (cal/mol)
R = gas constant (1.987 cal/mol-K)
T = temperature (K)

If an activation energy of 50,000 cal/mol is assumed (reasonable for many processes in these kinds of alloys including oxygen diffusion) then the ratio of the characteristic times at 427C (700K) and 649C (922K) is calculated to be about 5740. This means that a process that requires 0.1 sec at 649C, say, will require just about 10 minutes at 427C. Given this large difference it is not surprising that the rate effect is masked at the lower test temperature. Another way of looking at this is that the material is practically impervious to environmental degradation at temperatures of 427C and lower.

B.   Optical Metallography

The microstructure after annealing is shown in Fig. 10 and after ageing in Fig. 11. It can be seen that the microstructure contains intermediate sized grains, twins, and stringers. The stringers and grain boundary precipitates have been identified as carbides and μ phase. The stringers and precipitates hold the grain size down and determine the grain size between them.

Fig. 10. Optical micrograph of sample annealed at 1066C for 30 min. and water quenched.

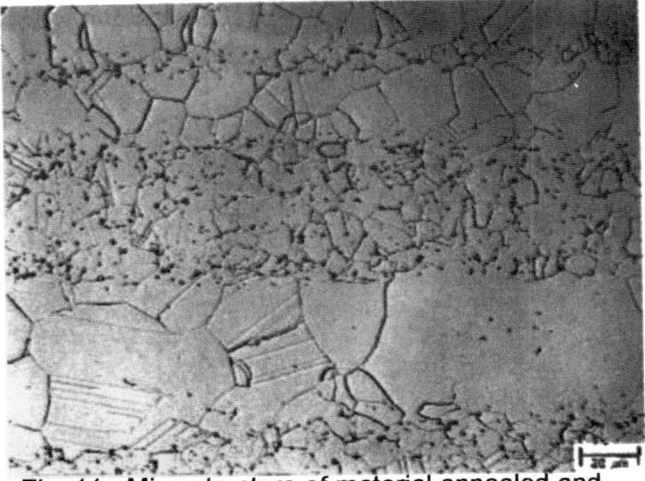

Fig. 11. Microstructure of material annealed and aged for 1000 hrs at 649C.
The effect of the precipitates is clearly seen in Fig. 11.

The grains are very large where the precipitate density is low and very small where the precipitate density is high. This is an excellent practical example of Zener pinning.

C. Scanning Electron Microscopy (SEM)

Specimens were examined via SEM to see if there were any significant differences in the fracture features as a function of pre-treatment or testing mode. In general, specimens tested in air at low rates and high temperatures tended to show secondary cracking (i.e. cracking on the side surfaces of the specimen) and significant areas in which the cracking mode was intergranular. A typical SEM micrograph of the surface of a specimen tested in air is shown in Fig. 12.

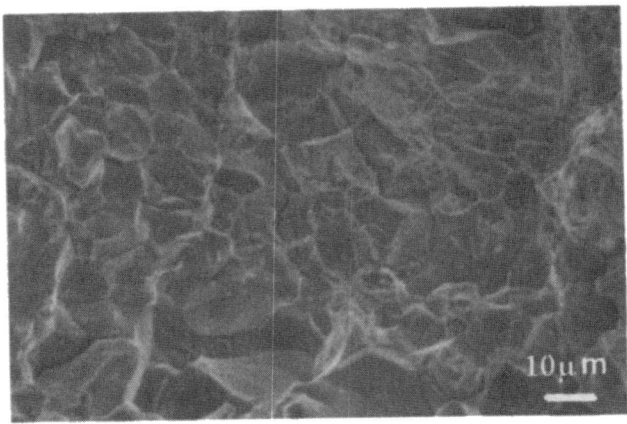

Fig. 12. Fracture surface of specimen tested in air at 649C and 0.5%/min. Note the large area of intergranular fracture along with some ductile dimpling.

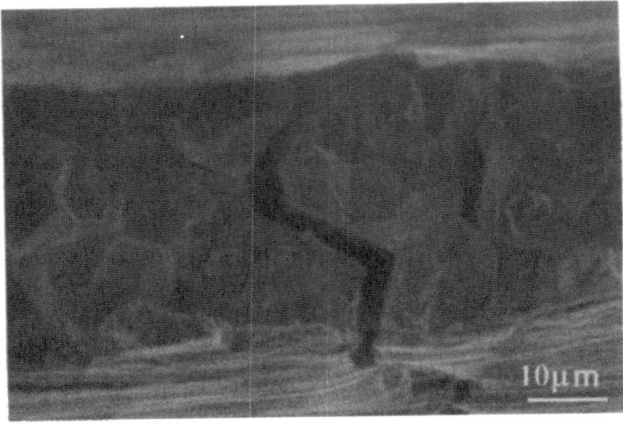

Fig. 13. Intergranular cracking on the side of same specimen as in Fig. 12.
The initiation of intergranular fracture on the side of the specimen is seen clearly in Fig. 13 for the same

614

specimen. Features similar to those seen in Figs. 12 and 13 were seen for the air tested specimens with the amount of intergranular cracking decreasing as the strain rate was increased. However, for specimens tested in vacuum, the situation was completely different. The fracture features of a vacuum tested specimen are seen in Figs. 14 and 15.

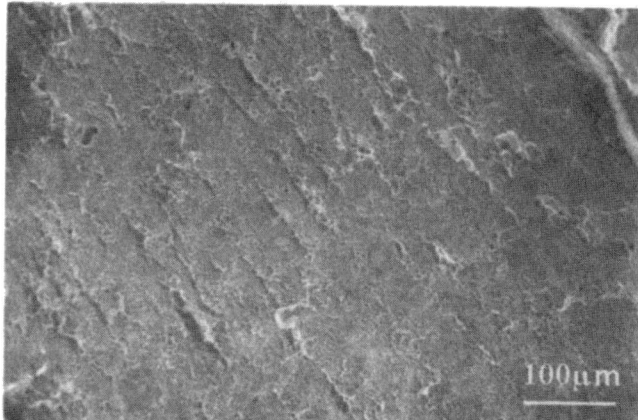

Fig.14. Low magnification fracture surface of specimen tested in vacuum at 649C and at 0.5%/min. Note the aligned stringers.

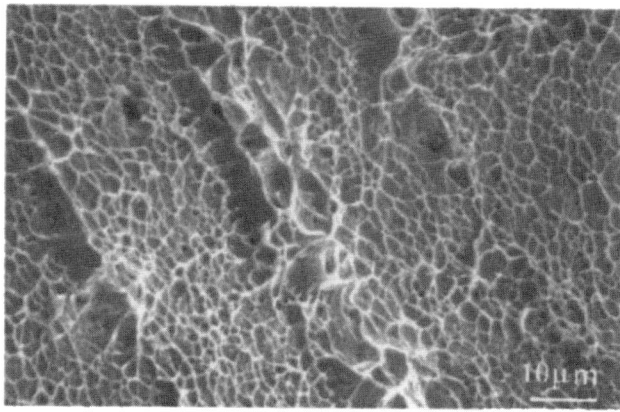

Fig. 15. Higher magnification of fracture surface shown in Fig. 14. Note the shallow ductile dimples and effect of stringers.

The fracture surfaces did not exhibit any intergranular cracking nor were there any significant cracks on the sides of the specimen. Furthermore, the great majority of the fracture surface consisted of "ductile dimples" albeit of a rather shallow nature. In short, under the same test conditions, the fracture features are essentially ductile in vacuum and brittle in air. Figure 14 shows macroscopic aligned "stringers" features which were commonly seen on fracture surfaces of vacuum tested and air tested materials. In these regions, we have demonstrated that there are large particles whose composition is close to the overall-composition. When these were analyzed in the TEM it was found that they were essentially large precipitate particles (with perhaps some carbides). This is evidence of a macroscopic processing effect which may have to do with solidification and subsequent mechanical working, much like banding in steels. These regions are involved in the fracture process. However, their role in crack initiation and in providing easy paths for rapid diffusion of a degrading species is not clear.

These observations are entirely consistent with the tensile test results where the macroscopic ductility was greater for the vacuum tests than for the air tests. The grain boundaries of some specimens which were given a prolonged thermal exposure were examined to see if the environmental effect could be explained by grain boundary precipitation effects. In no case that was examined was it possible to find any evidence of enhanced grain boundary precipitation of carbides or μ phase. We thus tentatively conclude that the environmental sensitivity is not a result of the formation of a new, deleterious phase.

D. Transmission Electron Microscopy (TEM)

Specimens typical of all of the conditions tested were examined by TEM in a effort to determine if there were significant differences in the deformation mode with strain rate, temperature, and aging time. The major result was that the deformation mode was essentially independent of environment (air or vacuum), temperature, strain rate or aging conditions over the range of conditions investigated in this study. Representative TEM micrographs are shown in Figs. 16-20. Figure 16, taken from specimen 24-5V (24 hr age and tested at 5%/min in vacuum at 649C) shows a deformation substructure that was generally characteristic of all conditions studied; planar deformation with a high density of slip bands and/or twins. Figure 17 is a dark field image taken from the same region as Fig. 16. This image was formed using a beam diffracted from the coherent precipitates and demonstrates unambiguously that the precipitates are being sheared (the local order is destroyed by shearing and the deformation band shows up dark while the precipitates outside of the band are bright). Figure 18 shows the deformation substructure of specimen 1000-9 (1000 hr age and tested at 0.5%/min in air at 427C). It has a structure similar to that seen in Fig. 16 although the ageing and test conditions were quite different. Figure 19 is a dark field image of the region showed in Fig. 18. Shearing of the precipitates is again observed. In addition, it is clear that ageing for 1000 hrs significantly increases the precipitate size.

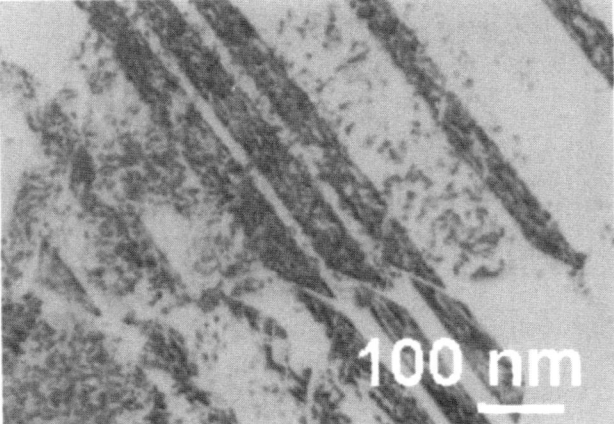

Fig. 16. TEM micrograph of specimen 24-5V. Aged 24 hrs at 649C and tested at 5%/min in vacuum at 649C. Note planar deformation bands.

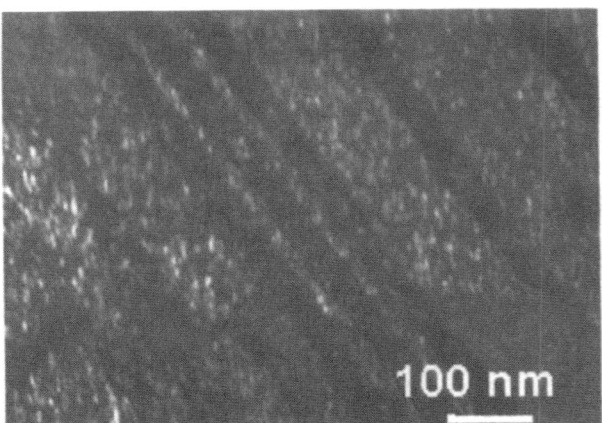

Fig. 17. Dark field TEM micrograph of same area as seen in Fig. 16. Note the fine precipitates and strong evidence of precipitate shearing.

Fig. 18. TEM micrograph of specimen 1000-9. Aged 1000 hrs at 649C and tested at 0.5%/min in air at 427C. Planar deformation bands are again present.

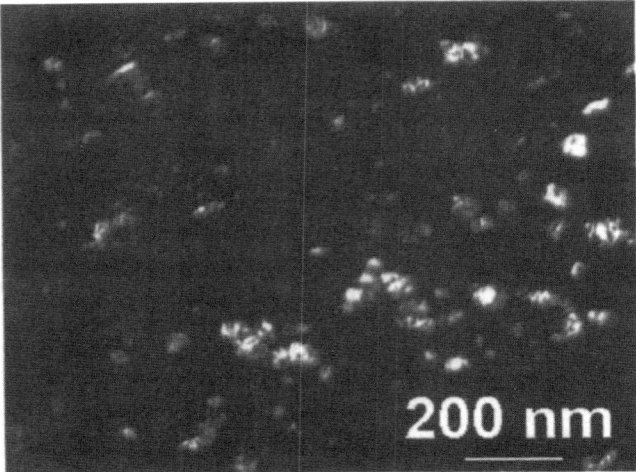

Fig. 19. Dark field TEM micrograph of same area as seen in Fig. 18. Note the significantly coarsened precipitates and precipitate shearing.

It is well known that as long as precipitate shearing remains operative, the strength increases with increasing precipitate size. Thus the increase in strength that was observed with increased ageing time is completely consistent with the microstructure. This increase in strength is also consistent with a reduction in ductility, especially in air. The higher stresses that can develop in the long-term aged material coupled with environmental ingress (which is a zone of weakness) gives rise to a reduction in the toughness or ductility. These ideas have been verified for many superalloys by one of the authors of this paper [7-15].

Figure 20 is a TEM micrograph taken from a low temperature/high rate test. Again planar deformation features, similar to those observed in all other specimens, are seen. Thus it appears as if *changes* in the deformation mode are not responsible for the drop in ductility. Deformation consisted of planar arrays of slip bands and twins (quasi twins to be precise). Both of these deformation modes would be expected to produce more sensitivity to environmental effects than wavy glide. The slip steps at the surface could break any protective film and could also provide energetically favorable sites for the dissociation of the damaging species.

Fig. 20. TEM micrograph of specimen 1000-6. Aged 1000 hrs at 649C and tested at 50%/min in air at 427C. Planar deformation bands are again present.

The picture that develops is that at low rates the environment, aided by the planar deformation mode, has time to penetrate into the specimen and to reduce the ductility by what is probably a very complex mechanism involving changes in the interatomic bonding forces. As the strain rate increases, the time for environmental interactions is decreased and the behavior of the material approaches that characteristic of vacuum. Prolonged ageing causes a moderate decrease in the ductility through a precipitate coarsening mechanism. However, the fact that the ductility is high on an absolute basis and increases for impact loading in air implies that this alloy has a significant degree of damage tolerance and is suitable for many engineering applications.

IV.   Summary and Conclusions

Based on the results of the experimental program the following conclusions can be drawn:

1.   The ductility of this alloy is substantial for all temperatures, strain rates and prior exposures that were investigated.

2.   The ductility in air increased with increasing strain rate while in vacuum it decreased. The ductilities for materials that were tested at 649C (1200F) in air and vacuum after exposure at 1000F for 24 hrs converged at a strain rate of about 50%/min.

3.   The ductility for specimens aged for 1000 hrs at 649C (1200F) followed the same trends as for the 24 hr aged materials except that

convergence was apparently shifted to higher strain rates.

4.   The ductility at 427C (800F) was essentially independent of strain rate. This was explained on the basis of a thermally activated damage mechanism whose rate was calculated (assuming Arrhenius behavior and an activation energy of 50 [k-cal/mol]) to be reduced by several orders of magnitude at the low temperature.

5.   Increasing the ageing time from 24 to 1000 hrs had the effect of increasing the precipitate size, increasing the strength and decreasing the ductility, all in accord with established physical metallurgy principles.

6.   The fracture surfaces of those specimens tested in air exhibited a significant degree of intergranular cracking, especially at the lower strain rates.

7.   The deformation substructure consisted of linear arrays of dislocations and/or twins.

8.   The strength was generally lower in air than in vacuum for equivalent exposures, temperatures and strain rates.

9.   The previous four conclusions are consistent with an environmental effect , similar to that observed in some other high temperature alloys.

10.   The fact that the ductility is quite substantial and increases with increasing strain rate implies that this alloy is suitable for applications in which impact loads may be encountered.

References

1.   S.K. Srivastava and G.Y. Lai: Paper presented at ASME Gas Turbine and Aeroengine Congress, Toronto (1989), Paper 89-GT-329.

2.   S. Dymek, M. Dollar, and D.L. Klarstrom: Scripta Met., 10, (1991), 865-869, 1991.

3.   M. Kumar and V.K. Vasudevan: Acta. Met., 44, (1996), 3575-3583.

4.   M. Kumar and V.K. Vasudevan: Acta Met., 44, (1996), 4865-4880.

5.   M. Kumar: Ph.D. Dissertation, University of Cincinnati, Cincinnati, OH, (1995)

6. S.K. Srivastava: <u>Superalloys 1992</u>, S.D. Antolovich, R.W. Stusrud, R.A. MacKay, D.L. Anton, T. Khan, R.D. Kissinger, D.L. Klarstrom eds., TMS, Warrendale, PA, (1992), 227-236.

7. Stephen D. Antolovich, P. Domas and J. L. Strudel: Met. Trans., vol. 10A, (1979),1859-1868.

8. Stephen D. Antolovich. S. Liu and R. Baur: <u>Superalloys 1980</u>, Tien, Wlodek, Gell and Maurer, eds. ASM, Metals Park, OH (1980), 605-615.

9. Stephen D. Antolovich, S. Liu and R. Baur: Met. Trans., vol. 12A, (1981),473-481.

10. Stephen D. Antolovich and N. Jayaraman: <u>Fatigue: Environment and Temperature Effects</u>, J. J. Burke and V. Weiss, eds., Plenum Press, NY, (1983),119-144.

11. Stephen D. Antolovich and J. E. Campbell: <u>Superalloys Source Book</u>, ASM, (1984),112-169.

12. B. A. Lerch, Stephen D. Antolovich and N. Jayaraman: Mat. Sci. and Engr., (1984),151-165.

13. Stephen D. Antolovich: <u>Superalloy 718-Metallurgy and Applications</u>, ed., E.A. Loria, TMS, Warrendale, Pennsylvania, (1989), 647-653.

14. Bruce F. Antolovich, Ashok Saxena, and Stephen D. Antolovich: <u>Superalloys 1992</u>, S.D. Antolovich, R.W. Stusrud, R.A. MacKay, D.L. Anton, T. Khan, R.D. Kissinger, D.L. Klarstrom eds., TMS, Warrendale, PA, (1992), 727-736.

15. M.P. Miller, D.L. McDowell, R.L.T. Oehmke, and S.D. Antolovich: <u>Thermomechanical Fatigue Behavior of Materials</u>, ASTM STP 1186, Ed. H. Sehitoglu, (1993), 35-49.

Coatings, Welding and Repair

PROCESSING EFFECTS ON THE FAILURE OF EBPVD TBCs ON MCrAlY AND PLATINUM ALUMINIDE BOND COATS

N. M. Yanar , M.J. Stiger, G. H. Meier and F. S. Pettit

Materials Science and Engineering Department, University of Pittsburgh,
Pittsburgh, PA, 15261

Abstract

Thermal barrier coatings (TBCs) composed of yttria stabilized zirconia (YSZ) deposited via electron beam physical vapor deposition (EBPVD) on MCrAlY and platinum diffusion aluminide bond coats on the superalloy Rene N5 have been compared in cyclic oxidation in air at 1100°C. The MCrAlY bond coats were fabricated by using several different plasma spray processes. The TBCs on the MCrAlY bond coats had shorter lives compared to the TBCs on the platinum aluminide bond coats. This difference in TBC performance is proposed to be caused by the presence of other oxides than alumina on the MCrAlY bond coats which caused the toughness of the TBC-TGO-bond coat region to be reduced. Processing variables which affect TBC performances are discussed.

Superalloys 2000
Edited by T.M. Pollock, R.D. Kissinger, R.R. Bowman,
K.A. Green, M. McLean, S. Olson, and J.J. Schirra
TMS (The Minerals, Metals & Materials Society), 2000

Introduction

It has been documented in numerous publications that thermal barrier coatings (TBCs) consisting of a $6-8$ wt.% Y_2O_3 stabilized ZrO_2 (YSZ) are an effective method to increase the useful lifetimes of high temperature components such as combustion liners and vanes and blades in gas turbines (1,2). Such coatings are usually applied in conjunction with a bond coat, as shown in Figure 1, to superalloy hardware. While it is clear that TBCs are effective coatings, the fabrication procedures that will result in optimum lives require better definition. The mechanics by which TBCs fail are not clear. The paper by Marcin and Bose provides a chronological description of the progression of TBC technology (1). Generation I TBCs consisted of air plasma spray (APS) MCrAlY bond coats and air plasma spray YSZ TBCs. Failure was caused by thermally induced stresses and occurred in the air plasma sprayed bond coats. Improved performances were obtained in Generation II TBCs by using low pressure plasma spray (LPPS) MCrAlY bond coats and air plasma spray YSZ TBCs. The failure of the Generation II TBCs occurred predominantly in the YSZ TBC. Generation III TBCs consisted of a LPPS MCrAlY bond coat, or a platinum modified aluminide bond coat, and a YSZ TBC deposited by using electron beam physical vapor deposition. In Generation III TBCs failure was no longer in the TBC but rather along the interface between bond coat and the thermally grown oxide (TGO) that forms during exposure at elevated temperatures. The progression of technology from Generation I to Generation III TBCs certainly represent the development of improved TBC systems, however numerous questions remain unanswered concerning the mechanisms of TBC failures. Generation II and III TBCs consist of YSZ TBCs deposited via different processes, Figure 2, and the TBCs are usually of different thicknesses. For example, the TBCs deposited via APS are usually on the order of 300 μm whereas the TBCs prepared by EBPVD are between 100 and 150 μm in thickness. The effect of this TBC thickness must be considered in comparing Generation II and Generation III TBC failures. Generation III TBCs may use either MCrAlY bond coats, deposited via LPPS or some other type of spray process that does not cause oxidation of the MCrAlY powder, or platinum modified diffusion aluminide bond coats. There are data in the literature to indicate that TBCs with platinum aluminide bond coats are better than those on MCrAlY bond coats (3). Nevertheless, there are TBCs being used in service that have MCrAlY bond coats. Moreover, results are also available that show in some cases TBC performance using MCrAlY bond coats can be exceptional (4). In order to understand the effects of different process variables on TBC performance it is necessary to investigate and describe in detail TBC failure mechanisms. This paper is concerned with the failure of EBPVD YSZ TBCs on MCrAlY and platinum aluminide bond coats after cyclic oxidation at 1100°C in air.

Characterization of As Processed TBCs

The TBC specimens, Figure 1, were prepared by depositing YSZ via EBPVD on either MCrAlY or platinum aluminide bond coats. The superalloy substrate was Rene N5 (Ni-7.5 Co-7.0 Cr-1.5 Mo-5.0 W-3.0 Re 6.5 Ta-6.2 Al-.15 Hf-0.05 C – 0.004 B-0.01 Y in wt.%). All of the TBCs were prepared in the same coater. The YSZ had a segmented structure with a cellular structure within segments, Figure 3. The cellular structure became more developed upon exposure to elevated temperatures. At times

defects were evident in the TBC, Figure 4, but such defects were not numerous and have not been found to contribute to failures.

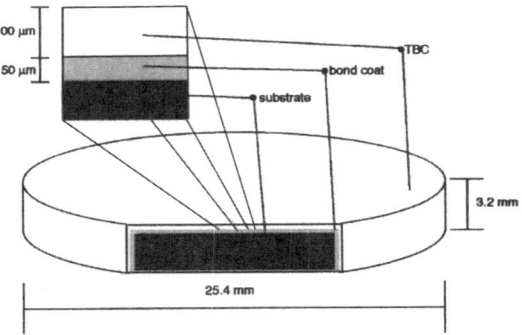

Figure 1. Schematic to show the relationship between TBC, bond coat and superalloy on specimens used in this program

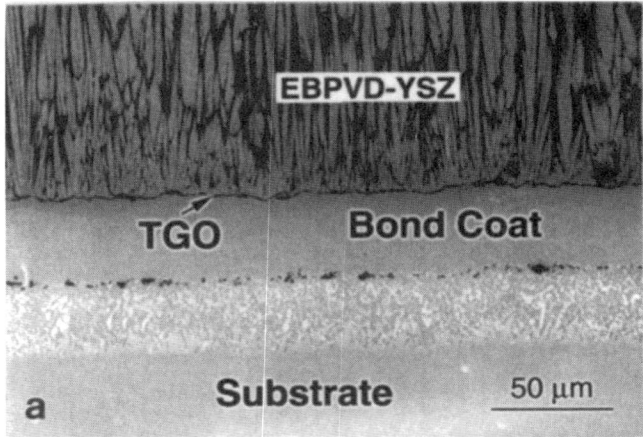

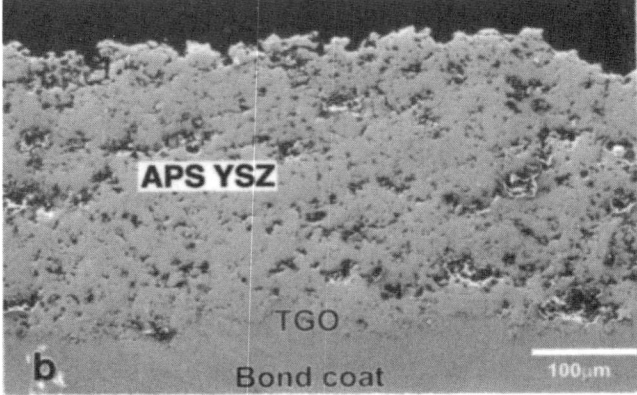

Figure 2. Photomicrographs showing typical electron beam physical vapor deposited (EBPVD) and air plasma sprayed (APS) yttria stabilized zirconia thermal barrier coatings. The EBPVD TBC consists of segmented,columnar grains whereas the APS TBC is composed of a particle like structure due to fusing of the YSZ powders.

Three types of MCrAlY bond coats were used. The compositions of these coatings, as well as the technique via which they were deposited, are presented in Table I. Three platinum aluminide bond coats were studied. All three bond coats were fabricated via the same general approach, namely a platinum coating was electrolytically deposited on the N5 substrate followed by aluminizing using a high temperature low aluminum activity process. Bond coat A was deposited by one manufacturer and bond coats B and C by another. The difference between bond coats B and C was that two different electrolytic techniques were used to deposit the platinum coatings. The surface of all bond coats were prepared by grit blasting using alumina powder prior to TBC deposition.

Photographs showing the as processed TBCs with MCrAlY bond coats are presented in Figures 5, 6 and 7. A thin TGO was evident as well as β (Ni, Co, Al) and γ (nickel solid solution) phases in the coatings beneath the TGOs. In all of these coatings nickel and chromium in addition to aluminum was detected in the TGOs on these TBCs. However, the TGOs were very thin and the nickel and chromium may have come from the bond coat. The TBCs on the platinum aluminide bond coats all had similar structures, Figure 8a. However, the aluminide coating thickness on A was less than on B and C, and the platinum content at the bond coat-TGO interface was greater in C than in B. All of these TBCs contained a thin TGO, Figure 8b, for which the thickness varied by a substantial amount. Such variations in the TGO thicknesses were also observed on the MCrAlY bond coats and may have been caused by grit blasting. The TGOs on the platinum aluminide bond coats were usually thicker than those on the MCrAlY bond coats and contained only aluminum and oxygen.

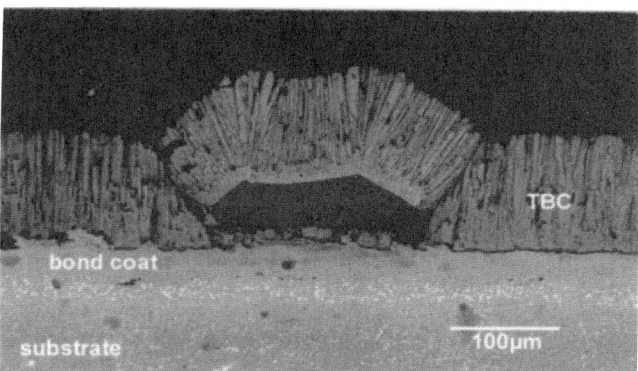

Figure 4. As processed YSZ TBC on a platinum aluminide bond coat showing a defect that was present in the TBC.

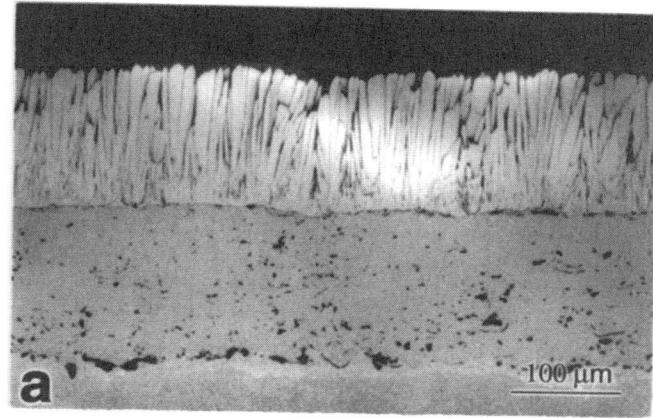

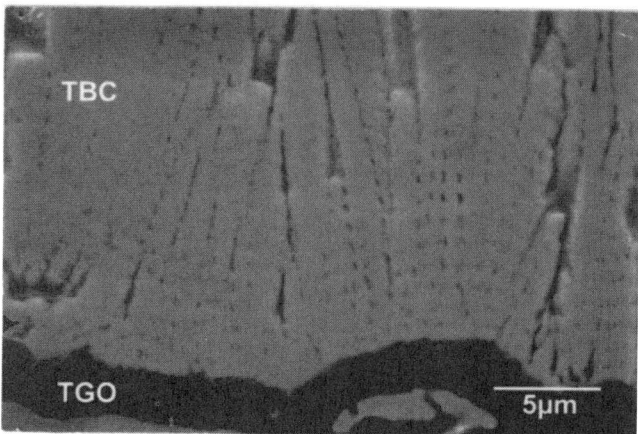

Figure 3. Scanning micrograph of a TBC showing cellular structure that developed in the TBC upon exposure for 60 hrs at 1200°C

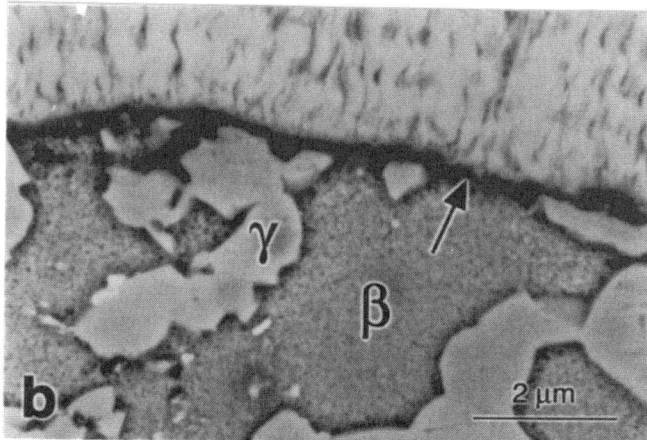

Figure 5. The as processed LN-11 bond coat with a TBC was uniform in thickness and contained some porosity and oxide inclusions, (a). A thin TGO was evident between the TBC and the bond coat, arrow (b). The bond coat contained β and γ phases.

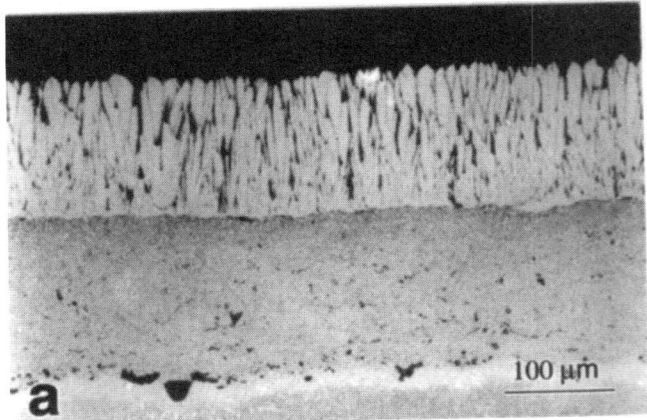

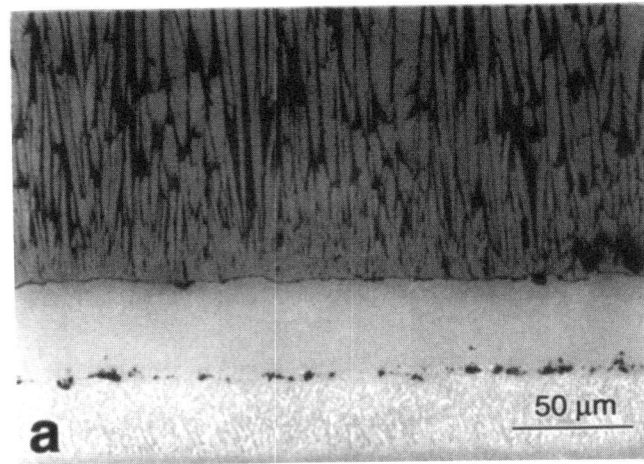

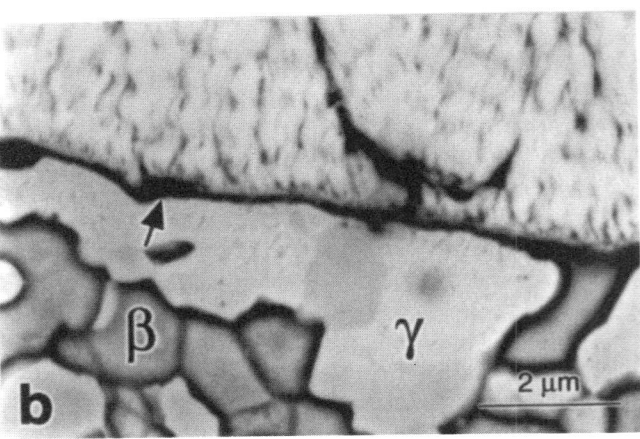

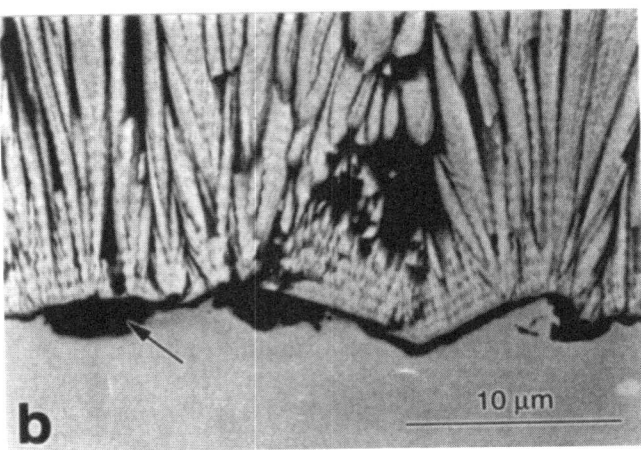

Figure 6. The TBC on the LN72 coating was uniform and some porosity and oxide inclusions were evident in the bond coat, (a). A TGO was evident between the TBC and bond coat in which some variations in thickness were evident, arrow, (b).

Figure 8. A typical as processed TBC system is shown in, (a). The TGOs on the platinum aluminide bond coats exhibited a variable thickness, arrow, (b), and were thicker than the TGOs on the as - processed MCrAlY bond coats.

Experimental Oxidation Conditions

All of the TBCs were exposed to cyclic oxidation conditions in a bottom loading furnace at 1100°C in air. This furnace had a platform upon which all of the specimens could be exposed simultaneously. The test cycle consisted of 10 minutes heating to temperature, 45 minutes at temperature followed by 10 minutes of forced air cooling. All of the specimens were examined at various time intervals by using optical metallography. The observations were directed at determining when and where cracking was evident in the TBCs. Upon termination of testing the fracture surfaces of failed specimens were characterized by using scanning electron microscopy. Sections through the TBCs and bond coats were prepared for examination via optical metallography and the SEM.

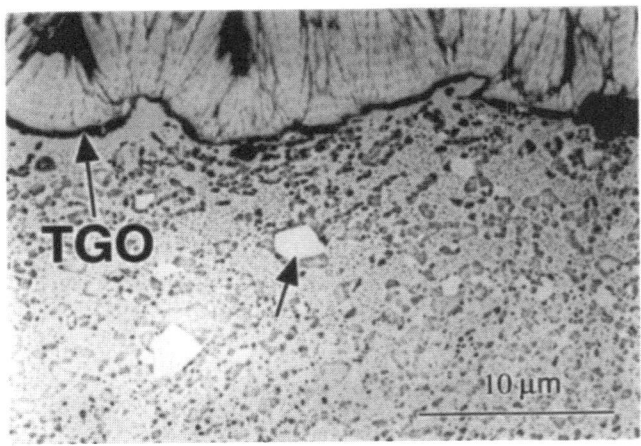

Figure 7. The TBC on the as processed BC52 bond coat contained an irregular TGO. Cr-Re particles were evident in this bond coat (arrow) along with both β and γ phases

624

Table I. Compositions of MCrAlY Coatings

Coating	Ni	Co	Al	Cr	Hf	Y	Si	O	C	Fe	Re	Ta
D-Gun (LN-72)	44.4	22.95	13.82	16.59	-	0.57	-	0.9	0.58	-	-	-
Shrouded Argon (LN-11)	48.0	21.78	12.58	16.45	-	0.43	-	0.16		-	-	-
Shrouded Argon (BC-52)	55.4	9.95	6.55	17.31	0.58	0.28	1.19	0.012	-	-	2.03	6.11

Table II. Exposure Cycles to Failure of TBCs

Bond Coat	Exposure Cycles
Shrouded Argon Plasma MCrAlY (LN-11)	76, 102
D-Gun MCrAlY (LN-72)	102, 139
Shrouded Argon Plasma MCrAlY (BC-52)	291
Platinum Aluminide (A)	1005
Platinum Aluminide (B)	1212
Platinum Aluminide (C)	1280

Failure Characteristics of the Cyclically Oxidized TBCs

The cycles to failure of the TBCs with different bond coats are presented in Table II. The TBCs on MCrAlY bond coats deposited via the LN-11 and LN-72 processes failed after essentially the same number of cycles. The BC 52 composition bond coat had a slightly longer life. The TBCs on platinum aluminide bond coats B and C had longer lives than the TBCs on bond coat A.

The fracture surfaces exposed upon failure of the TBCs with MCrAlY bond coats are shown in Figures 9a, b and c for LN-11, LN-72 and BC-52, respectively. The fractures have proceeded along the TGO-bond coat interfaces with numerous excursions into the TGOs and TBCs. The excursions into the TGOs occurred where the TGO extended into the bond coat due to oxidation of elements such as yttrium, hafnium and tantalum, Figure 10a, b and c. The TGOs that formed on the MCrAlY bond coats were not pure alumina but consisted of other oxides such as spinels ($NiCr_2O_4$, $NiAl_2O_4$), and oxides containing yttrium, hafnium and/or tantalum. Aluminum depletion of the MCrAlY bond coats was not the cause of TBC failure since the aluminum rich β phase was present in these bond coats at the time of TBC failure, Figure 11.

Specimens of the TBCs on the platinum aluminide bond coats were examined prior to failure after 758 cycles at 1100°C, and after failure. In Figure 12 a cross section of the TBC on platinum aluminide C is presented after 758 cycles. The TGO was pure alumina and some deformation of the aluminide bond coat was evident as indicated by undulations that developed at points where the TBC fractured, Figure 12b. The aluminide bond coat contained predominantly β phase but γ' was evident in the bond coat adjacent to the TGO, Figure 12c. Similar features were evident in the TBCs on platinum aluminide bond coats A and B, but the TBCs became detached during metallographic preparation. Failures occurred at the TGO-bond coat interfaces with numerous

excursions into the TGOs and TBCs, Figures 13a and b. At times a zone was evident in the outer part of the TGO that contained a mixture of alumina and YSZ, Figure 13b. Such zones have been observed in TGOs previously (5).

A typical surface of a platinum aluminide bond coat after spalling of the TBC is shown in Figure 14. It is evident that fracture has occurred in the TBC and the TGO, as well as along the TGO-bond coat interface but since failure occurred during testing, the exposed bond coat has been oxidized.

Mechanisms of TBC Failures

The results clearly show that Generation III TBCs do not fail by cracking solely along the TGO-bond coat interface. In the case of the MCrAlY bond coats the TGOs in the as processed condition may have contained spinels, and other oxides than alumina developed during oxidation. The absence of relatively pure alumina TGOs is believed to be responsible for the short lives of TBCs on the MCrAlY bond coats. As processed TBCs on MCrAlY bond coats were delaminated by using an indent test (7), but this test did not cause delamination of the TBC on as processed platinum aluminide bond coats. These results show that the interfacial toughness in the TBC-TGO-bond coat region is greater for the TBCs on the platinum aluminide bond coats compared to the MCrAlY bond coats. More work is required to determine where fracture of the TBC initiates, but it appears to involve in some way either the spinel phases and/or the oxides of tantalum and hafnium. Therefore fracture may initiate in the TGO. The lives of TBCs on MCrAlY bond coats should be increased if pure alumina TGOs are developed on these bond coats. Work is in progress to test this hypothesis.

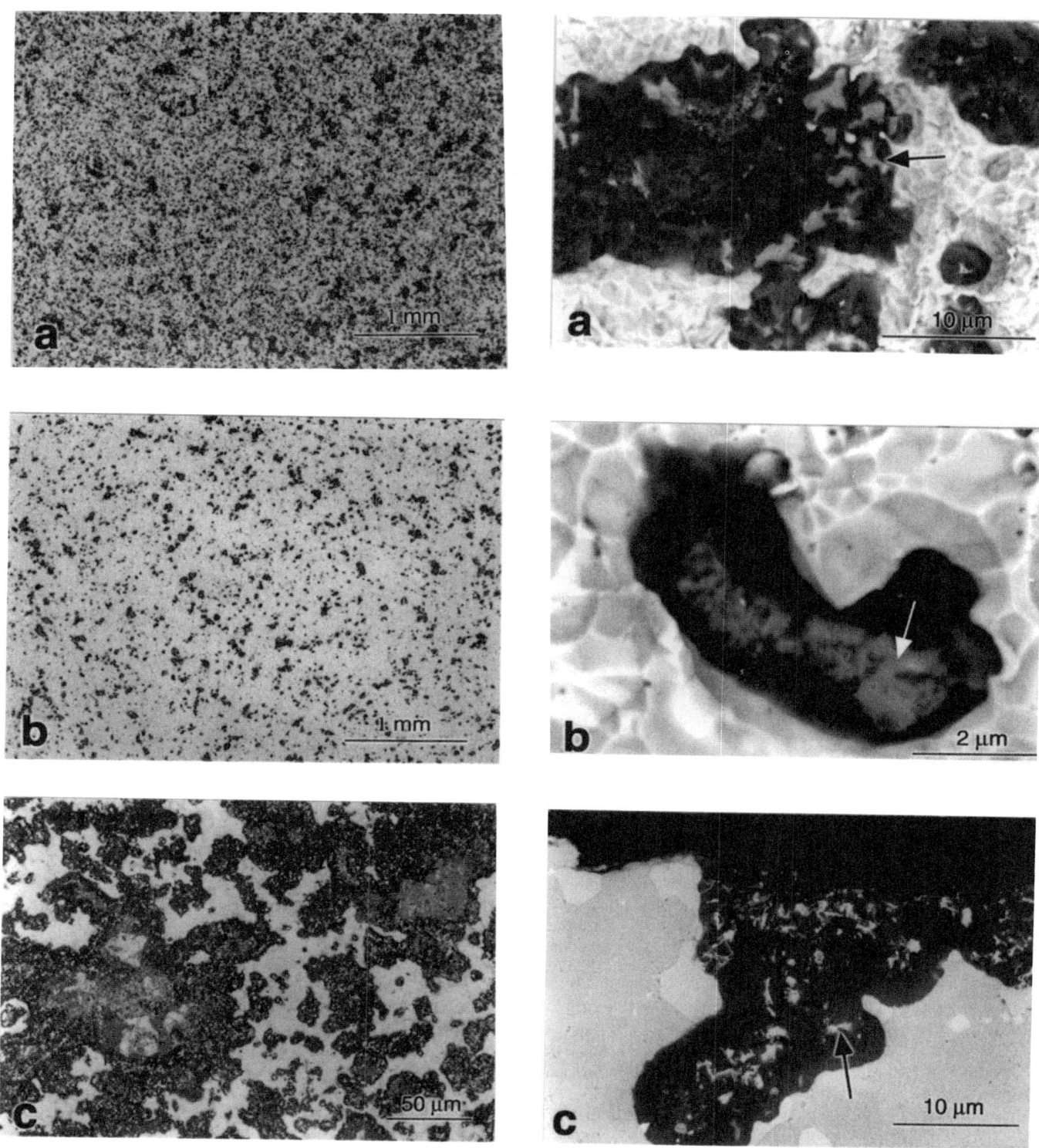

Figure 9. Surfaces exposed upon failure of the TBCs on LN-11, (a), LN-72, (b) and BC-52, (c). The failures have exposed bare bond coat (white areas) with numerous excursions into the TGOs and TBCs.

Figure 10. Some of the dark areas in Figure 9 are shown at higher magnifications for LN-11, (a), LN-72, (b), and BC-52, (c). The TGO on LN-11 contained alumina with oxides of yttrium and hafnium, arrow (a). The TGO on LN-72 bond coat contained ytrrium oxides, arrow (b), and the TGO in BC-52 was permeated with oxides of hafnium and tantalum, arrow (c).

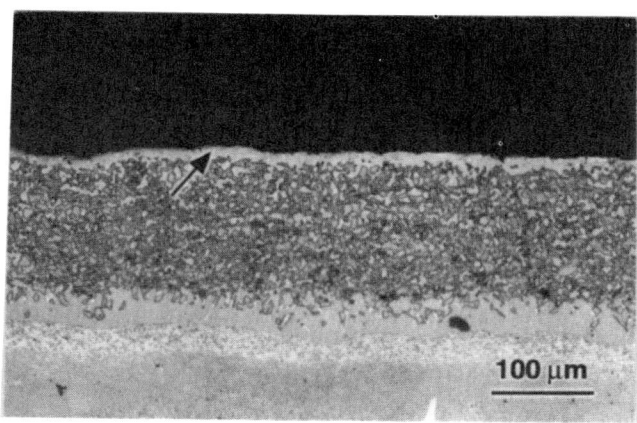

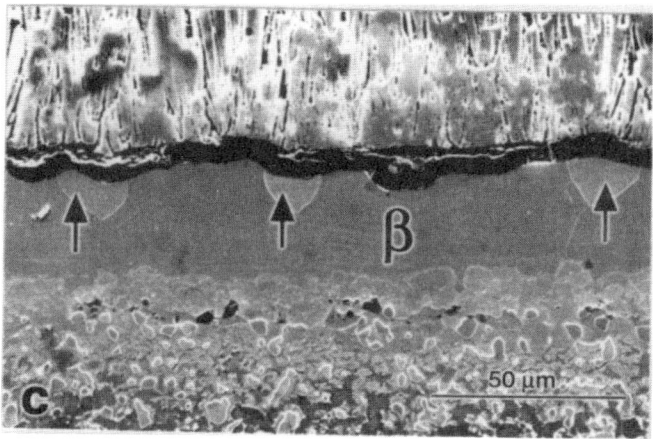

Figure 11. Photomicrograph showing the bond coat on LN-72 after failure of the TBC (139 hrs at 1100 °C). The β phase has been depleted from a very thin layer (arrow) of the coating. A similar zone was evident after failure of the TBC on LN-11 whereas the β phase was removed from 50% of the BC-52 bond coat.

Figure 12. Scanning micrographs are presented for aluminide C after 758 cycles at 1100°C. The TBC had not failed. Undulations are evident in the TGO, arrows, (a) and (b), where it is evident that some of the fragments of TBC have been broken off from the coating. The aluminide bond coat contains γ' phase, arrow (c), and β phase.

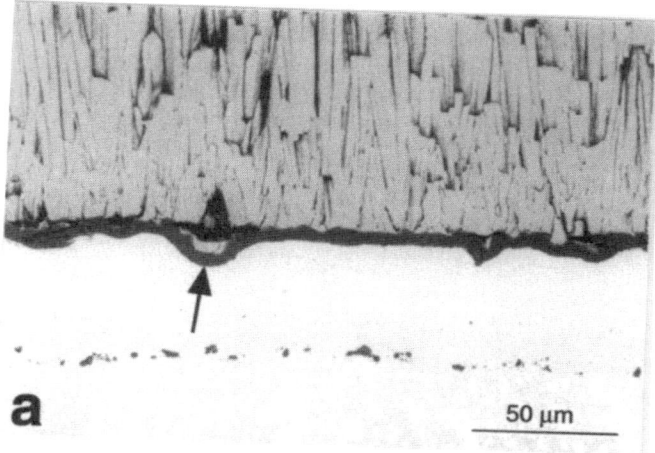

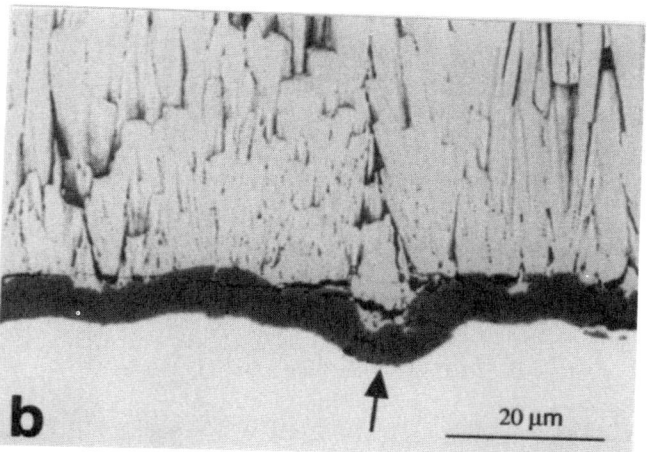

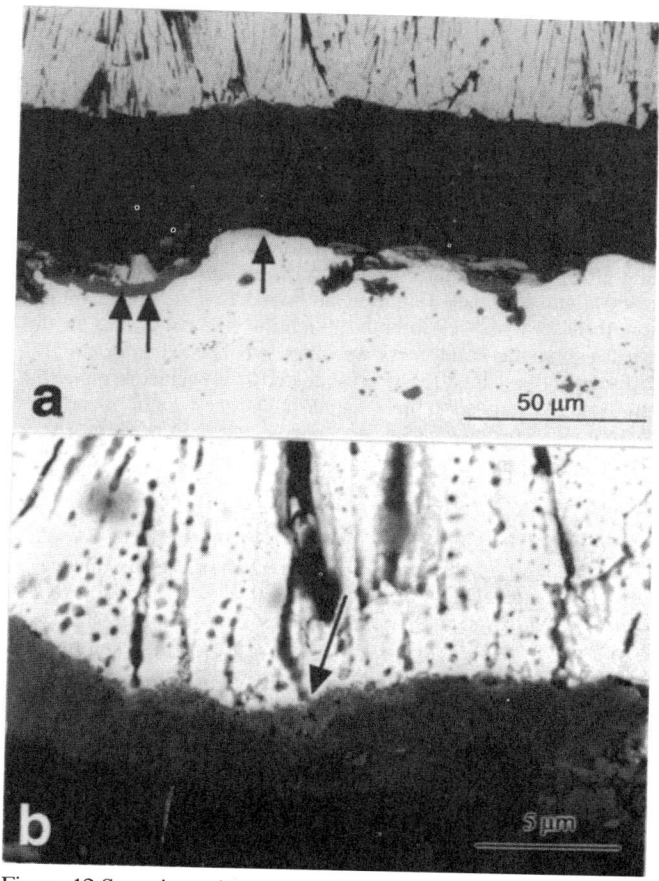

Figure 13. Scanning micrograph of platinum aluminide bond coat A after 758 cycles at 1100°C. The TBC did not fail during testing, but became detached upon metallographic preparation. It is evident that failure has involved fracture at the TGO-bond coat interface arrow (a), as well as fracture in the TGO and TBC, double arrows, (a). At times an outer zone was apparent in the TGO extending into the TBC, arrow, (b).

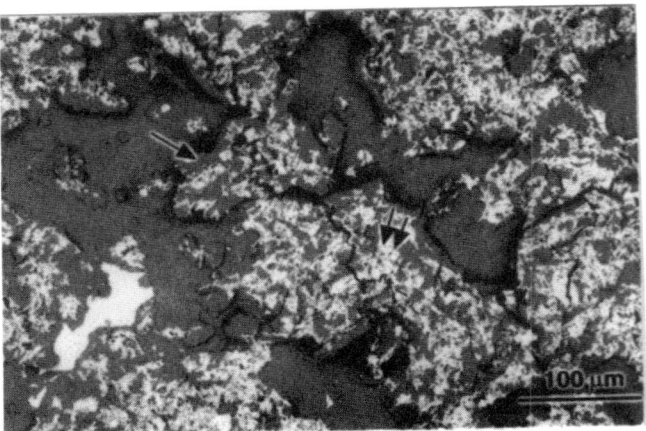

Figure 14. Scanning micrograph of typical platinum aluminide bond coat surface exposed after spalling of the TBC. The fracture occurred in the TGO(arrow) and the TBC (double arrows) as well as along the TGO-bond coat interface. The failure occurred during cyclic oxidation and the bare bond coat has been oxidized.

The failure of the TBCs on the platinum aluminide bond coats are believed to initiate in the TBC at ratchets in the TGO, Figure 12b. The cracks then propagate in the TBC and TGO as well as in the TGO-bond coat interface. Wrinkling of the bond coat may be a factor in initiating such failures, and more strong aluminide bond coats could extend coating lives. While aluminum depletion has been observed in these bond coats, it is not obvious that aluminum depletion, nor the presence of substrate elements, is playing a role in the failures of the TBCs on the platinum aluminide bond coats. Gell et al (8) have found that TBCs on platinum aluminide bond coats failed by crack initiation at boundaries in the aluminide bond coat at which preferential oxidation had occurred. In this investigation the bond coats were not grit blasted prior to TBC deposition, in contrast to the present study. It has been found that grit blasting inhibits the preferential oxidation of grain boundaries in aluminide bond coats (5).

Conclusions

The lives of TBCs on MCrAlY bond coats exposed to oxidizing conditions were substantially shorter than those on platinum aluminide bond coats. It is proposed that this may have been caused by the presence of other oxides than alumina on the as processed MCrAlY bond coats. It is not clear that failures of Generation III TBCs occur due to decreased interfacial toughness of the TGO-bond coat interface. Other factors such as ratcheting of the bond coat and the TGO may play significant roles in some failures. Processing conditions play a crucial role in TBC performances. Consequently, there probably are a number of different mechanisms by which Generation III TBCs fail. TBC performances should be increased by utilizing proper processing procedures.

Acknowledgements

This work was suppoted by AFOSR(contract number F49620-981-0221). The authors are grateful to GE Aircraft Engines(Ro Grylls and D. Wortman) and to Praxair Surface Technologies (A. Bolcavage and T. Taylor) for specimen preparation and numerous technical discussions.

References

1. S. Bose and J. DeMasi-Marcin, "Thermal Barrier Coating Experience in Gas Turbine Engines at Pratt & Whitney" Thermal Barrier Coating Workshop,, Nasa Lewis Research Center, 21000 Brookpared Rd, Cleveland, OH, 44135. NASA Conference Publication 3312, 63 – 77, (1995).

2. A. Maricochi et al, "PVD TBC Experience on GE Aircraft Engines" ibid, 79 – 89.

3. Z. Mutasim, C. Remlinger and W. Brentnall, "Characterization of Plasma Sprayed and Electron Beam Physical Vapor Deposited Thermal Barrier Coatings," The American Society of Mechanical Engineers, 97 – GT – 531, June (1997).

4. J. Schaeffer, " The Effect of Alumina Phase Transformations on Thermal Barrier Coating Durability", TBC 1997 Workshop, Nasa Lewis Research Center, 21000 Brookpared Rd, Cleveland, OH, 44135. 99-108, (1997)

5. M. J. Stiger, et al ,"Thermal Barrier Coatings for the 21st Century," Zeitschrift für Metallkunde, 90 (1999), 12, pp. 1069 – 1078.

6. A. Vasinonta and J. L. Beuth, "Measurement of Interfacial Toughness in Thermal Barrier Coating Systems by Indentation," Submitted to: Journal of the Mechanics and Physics of Solids, April 1999.

7. M. Gell et al, "Mechanism of Spalling in Platinum Aluminide/Electron Beam Physical Vapor Deposition Thermal Barrier Coatings," Met. and Mat. Trans. A, 30A, (1999), pp. 427 – 435.

COMPOSITIONAL EFFECTS ON ALUMINIDE OXIDATION PERFORMANCE: OBJECTIVES FOR IMPROVED BOND COATS

B. A. Pint, J. A. Haynes, K. L. More, I. G. Wright and C. Leyens*

Oak Ridge National Laboratory
Metals and Ceramics Division
Oak Ridge, TN 37831-6156
*on leave from DLR - German Aerospace Center
Institute of Materials Research
51170 Cologne, Germany

Abstract

In order to achieve long thermal barrier coating lifetimes, underlying metallic bond coats need to form adherent, slow-growing Al_2O_3 layers. A set of guidelines for developing aluminide bond coat compositions is proposed in order to maximize oxidation performance, i.e. forming a slow-growing adherent alumina scale. These criteria are based on results from cast, model alloy compositions and coatings made in a laboratory-scale chemical vapor deposition facility. Aluminide coatings are thought to have more long-range potential because of their lower coefficient of thermal expansion compared to MCrAlYs. The role of Pt in improving alumina scale adhesion and countering the detrimental role of indigenous sulfur is discussed. However, the improvements associated with Pt are minimal compared to reactive element doping. One strategy which has great promise for improvement is to incorporate Hf into the coating. From an oxidation standpoint, this would preclude the need for Pt in the coating and also reduce the scale growth rate. While excellent oxidation performance was observed for cast Hf-doped NiAl, its benefits can be compromised and even eliminated by co-doping with elements such as Cr, Ti, Ta and Re. Creating a pure Hf-doped NiAl is one promising approach for improving the oxidation performance of bond coats.

Superalloys 2000
Edited by T.M. Pollock, R.D. Kissinger, R.R. Bowman,
K.A. Green, M. McLean, S. Olson, and J.J. Schirra
TMS (The Minerals, Metals & Materials Society), 2000

Introduction

Formerly, the role of metallic coatings on Ni-base superalloys was simply to limit environmental attack of the underlying substrate. Now, a new paradigm has been established for metallic coatings that have been adapted as bond coats for thermal barrier coating (TBC) systems. It is no longer sufficient that the metallic coating simply minimize the corrosion rate. The metallic coating must form an adherent, slow-growing external Al_2O_3 layer beneath the overlying low thermal conductivity ceramic top coat. The ability of the coating to reform a protective thermally grown oxide or scale in the event of spallation is no longer the key. It is much more important that the scale be developed with a minimum of transient oxides and that it have near-perfect adhesion to limit spallation of the ceramic top coat, thereby achieving a long TBC lifetime.

While oxidation is not the only concern in complex TBC systems, it is, however, a primary factor in developing the next generation of bond coats with longer lifetimes or for engines with higher firing temperatures. Therefore, a set of compositional guidelines for coatings is proposed in order to assist in maximizing oxidation performance. These criteria are based on results from tests in which cast alloy compositions were used to quantify and understand possible improvements as a basis for further investigations using coatings made by chemical vapor deposition (CVD). Experimental work involved furnace cycle testing, isothermal kinetic measurements and in-depth characterization of the alumina scale, including transmission electron microscopy (TEM).

Experimental Procedure

Cast alloys (16mm diameter) were fabricated by vacuum induction melting and solidifying in a water-chilled copper mold. Chemical compositions were measured by inductively coupled plasma analysis and combustion analysis on as-cast material, Table I. All compositions are listed in atomic percent. After casting, alloys were annealed for 4h at 1300°C in quartz ampules. Oxidation coupons (1-1.5mm thick, typically 15mm diameter) were polished to 0.3μm alumina and cleaned in acetone and methanol prior to oxidation. Aluminide coatings with and without Pt were fabricated by a CVD process on Y-free, René N5 substrates (a General

Electric single crystal superalloy). This low-sulfur coating process is described in detail elsewhere.[1,2]

Isothermal exposures were performed in dry, flowing O_2 and mass gain was measured using a Cahn Instruments model 1000 microbalance. Cyclic testing was performed in two ways.[3] Long-term cycles (100-500h) were conducted in air with specimens in individual pre-annealed alumina crucibles. Crucibles and specimens were cooled to room temperature for >1h before weighing. Short-term (1-10h) cycles were performed in dry, flowing oxygen with specimens attached to alumina rods with Pt-Rh wires. Specimens were cooled for 10min between cycles and the cycle time is the time at temperature. Specimen mass changes were measured using a Mettler model AG245 balance. Hot corrosion testing was performed at 950°C and used 1h thermal cycles. Specimens were coated with $1.0mg/cm^2$ Na_2SO_4 after 1h and 100h. This testing is described in detail elsewhere.[4]

After oxidation, specimens were characterized by field emission gun, scanning electron microscopy (FEG-SEM) equipped with energy dispersive x-ray analysis (EDX). Specimens also were Cu-coated and sectioned for metallography. Scanning transmission electron microscopy (FEG-STEM) equipped with EDX was used to analyze cross-sectional specimens fabricated by focused ion beam milling (FIB).[5] During FIB specimen preparation, a W layer is deposited to protect the gas interface of the scale. Thermal expansion measurements were performed on specimens (3mm diameter, 1cm or 2.5cm long) of several alloys up to 1200°C on a Theta Industries dual push rod differential dilatometer.

Results

These results focus on nickel aluminide compositions. While MCrAlY-based coatings represent an important class of coatings for current applications, results from this lab suggest that scale adhesion on MCrAlYs is inherently inferior to that on aluminides at 1100°-1200°C [6,7] and therefore this coating approach does not represent the most fruitful option for developing high temperature bond coats. Figure 1 compares the temperature dependence of the coefficient of thermal expansion (CTE) of René N5 (Table I) and castings of generic bond coat compositions: NiAl+Hf and NiCoCrAlY. While the CTEs for René N5 and NiAl+Hf are reasonably similar, that for NiCoCrAlY was found to be significantly greater at temperatures above 600°C. These results

Table I. Chemical composition (in atomic%) of cast alloys determined by inductively coupled plasma analysis and combustion analysis.								
	Ni	Al	Hf	Pt	Cr	C	S(ppma)	Other
Ni-40Al	59.68	40.27	<0.01	--	<0.01	0.04	<4	
Ni-42.5Al	57.34	42.61	<0.01	--	<0.01	0.04	<4	
Ni-50Al	49.91	50.05	<0.01	--	<0.01	0.04	<4	
Ni-51Al	48.67	51.22	<0.01	--	<0.01	0.04	4	0.01Fe, 0.06Si
Ni-40Al+Hf	59.79	40.11	0.05	<0.01	<0.01	0.04	<4	
Ni-42.5Al+Hf	58.00	41.90	0.05	<0.01	<0.01	0.04	<4	
Ni-50Al+Hf	49.83	50.07	0.05	--	0.01	0.04	<4	
NiAl-2Pt	47.36	50.21	0.005	2.35	<0.01	0.04	4	0.01Fe, 0.01Cu
NiAl-5Pt	45.10	49.60	<0.01	5.20	0.05	0.04	9	
NiAl-2Pt+Hf	47.78	49.71	0.05	2.35	0.03	0.04	4	0.01Cu
NiAl-2Cr+Hf	48.28	49.63	0.05	<0.01	2.00	0.04	4	
NiAl-5Cr+Hf	47.76	47.20	0.05	<0.01	4.95	0.04	8	
NiAl-10Cr+Hf	45.13	45.07	0.05	<0.01	9.70	0.04	5	0.01Fe
NiAl-1Re+Hf	48.37	50.53	0.04	--	<0.01	0.04	1	1.00Re, 0.02Si
NiAl-1Ta+Hf	48.67	50.15	0.05	--	<0.01	0.04	4	0.92Ta, 0.14Fe, 0.01Co
NiAl-1Ti+Hf	48.58	50.33	0.05	--	<0.01	0.04	1	1.00Ti, 0.01Fe
René N5	64.85	13.88	0.05	--	7.79	0.25	7	2.1Ta,7.3Co,1.6W,1.0Re,.01Y
NiCoCrAlY	40.38	23.49	<0.01	--	17.43	0.04	8	18.48Co, 0.15Y, 0.01Ti

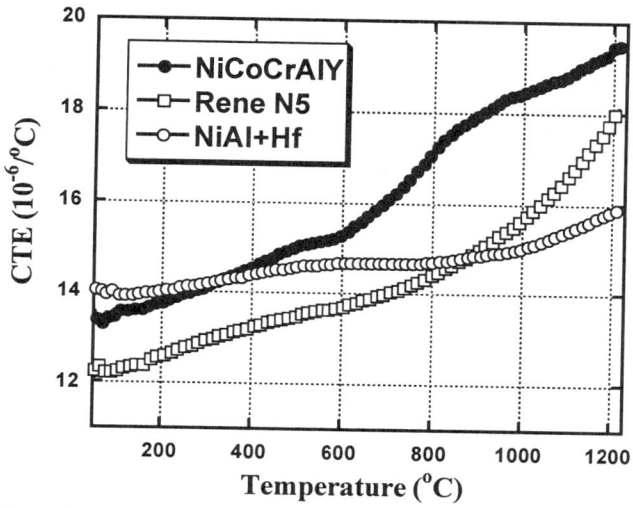

Figure 1. Coefficient of thermal expansion data for 3 alloys from 25°-1200°C. The aluminide specimen had the lowest expansion from 900°-1200°C.

are similar to earlier work.[8] Since the thermal stress in the alumina scale is related to the CTE difference, the additional thermal expansion compared to that for alumina ($\approx 9 \times 10^{-6}$°C^{-1}) represents a significant increase in thermal stress. Such a problem cannot be solved by reactive element (RE) doping or impurity control (e.g. S) and thus is an inherent limitation of the MCrAlY class of materials. Therefore, this work attempts have been made to understand and optimize the oxidation performance of aluminide coatings.

Current Compositions: The Pt Effect

Current aluminide bond coats often contain Pt, which is now commonly added in conjunction with a low activity CVD aluminization process which results in a single phase β-(Ni,Pt)Al coating.[9] The role of Pt in oxidation has not been fully explained but it is clear from recent results on both cast aluminides[10] and CVD aluminide coatings[2] that Pt improves alumina scale adhesion. This beneficial effect is shown at 1100°C for cast aluminides, Figure 2. With the addition of 2 or 5 at%Pt (10 or 20wt%), very little scale spallation was observed after 1000 cycles. Under the same conditions, undoped NiAl showed significant scale spallation. When Pt was added to CVD aluminide coatings, Pt mitigated the detrimental role of sulfur, present in the superalloy (Y-free René N5) substrate, on scale adhesion, Figure 3. Previous work on CVD NiAl coatings (without Pt) had shown that substrate sulfur content had a significant effect on alumina scale adhesion.[1] This result suggests that Pt somehow counters the detrimental role of S. One suggested mechanism for the effect of S on scale adhesion is that it increases the growth of interfacial voids.[11,12] The addition of Pt to CVD NiAl appeared to have reduced or eliminated this type of void growth, thereby improving contact between the metal and scale.[13]

The comparison of simple and Pt-modified CVD aluminide coatings on René N5 also demonstrated that Pt did not (1) reduce the alumina scale growth rate, (2) alter the diffusion of heavy elements into the coating during deposition or oxidation, or (3) alter the coating Al content before or after oxidation.[2,13].

Another area where the addition of Pt did not appear to play a strong role was hot corrosion resistance.[14] Figure 4 shows

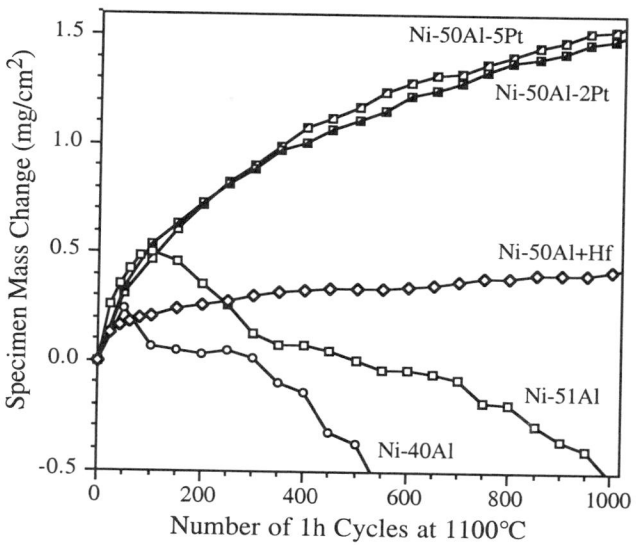

Figure 2. Specimen mass changes for various Ni-Al alloys during 1h cycles at 1100°C in O$_2$. The addition of Pt significantly improved scale adhesion compared to simple aluminides without Hf. The addition of Hf improved scale adhesion and reduced the scale growth rate.

results from testing at 950°C with a 1.0 mg/cm^2 Na$_2$SO$_4$ coating applied after 1h and 100h. Additions of Pt did not alter the behavior compared to NiAl+Hf. Only the addition of Cr improved performance. While Pt-modified aluminide coatings may be superior to simple aluminide coatings in hot corrosion due to a second order effect of improved alumina scale adhesion, there is no chemical effect of Pt in improving hot corrosion resistance. The Cr content of the coating is the critical component.

To provide additional information on the role of Pt on scale adhesion, TEM cross-sections were prepared of the scale formed on NiAl-5Pt after 2h at 1200°C in O$_2$. These conditions were selected to develop a 1-2μm thick oxide layer which is fully α phase and to allow comparison with previous work. In general the scale was 1-1.5μm thick and no transient Ni-rich oxide was observed. As surmised from the good scale adhesion, there were no interfacial voids observed in the thin area, Figure 5. The scale

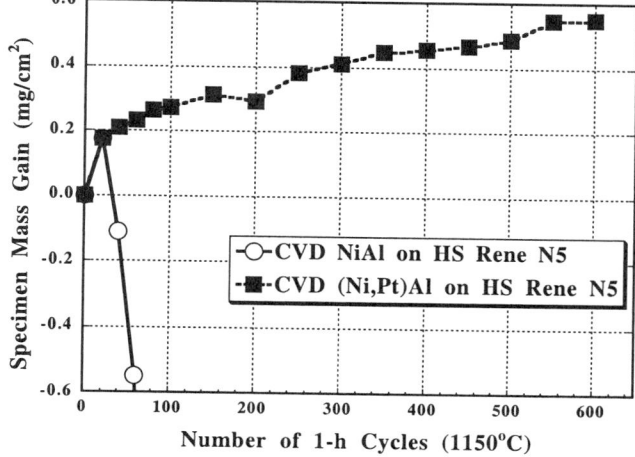

Figure 3. Specimen mass changes during 1h cycles at 1150°C for CVD aluminide coatings with and without Pt on a Y-free René N5 substrate that was not de-sulfurized (4 ppma S).

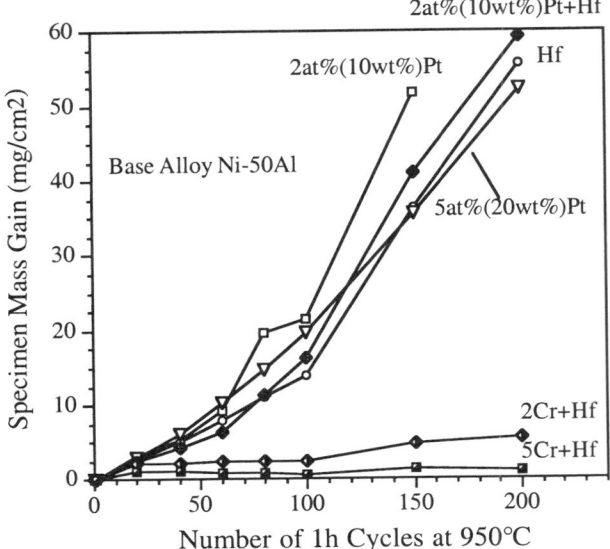

Figure 4. Hot corrosion specimen mass gains at 950°C for cast NiAl alloys with various modifications. All specimens were coated with 1.0 mg/cm² Na₂SO₄ after 1h and 100h. Chromium had a beneficial effect but Pt and Hf showed no benefit.

was generally two grains thick with the outer grains containing more voids than the inner, more columnar grains.

The most interesting feature of the scale was the roughness of the metal-scale interface, particularly the metal protrusions extending into the oxide. Protrusions have been observed previously for both cast Pt-Al alloys [15] and for CVD Pt aluminide coatings.[2] Each protrusion was centered on a single alumina grain and extended into it. This is seen clearly in a STEM annular dark field image, Figure 6. As transport occurs primarily along alumina grain boundaries at this temperature, one explanation for the formation of these protrusions is that new oxide is forming rapidly at the

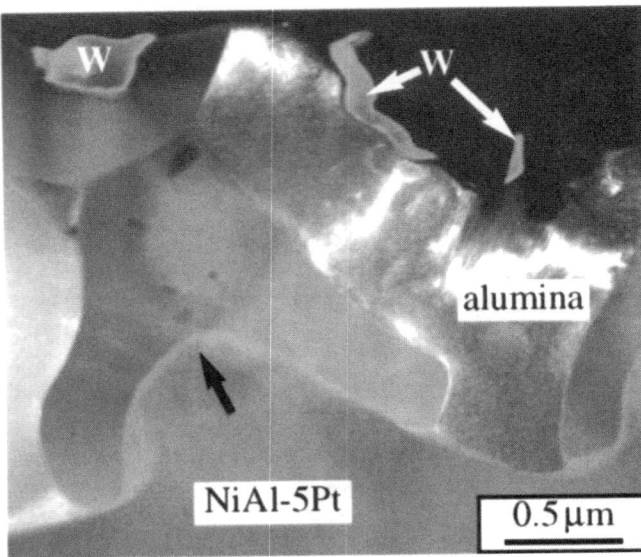

Figure 6. STEM annular dark field image of the alumina scale formed on NiAl-5Pt after 2h at 1200°C. Above the metal protrusion (arrow) is a single alumina grain. Small remnants of the FIB-process W coating remain at the gas interface of the scale.

juncture of oxide grain boundaries and the metal. If lateral oxygen diffusion does not keep up with this process, the interface will not remain flat. Inward growing ridges have been observed on undoped NiAl [16] and when the oxide loses contact with the metal due to interfacial void formation.[12] The ridges have been proposed to grow by a similar mechanism where lateral diffusion cannot keep up with the inward grain boundary diffusion of oxygen.[12] In this case where no voids form, the ridges appear to grow into the substrate leaving metal protruding into the center of the oxide grain. While the structure looks like the metal grew into the oxide, it may actually be metal remaining from a non-uniform oxidation front. The gas interface of the scale also was rough. This

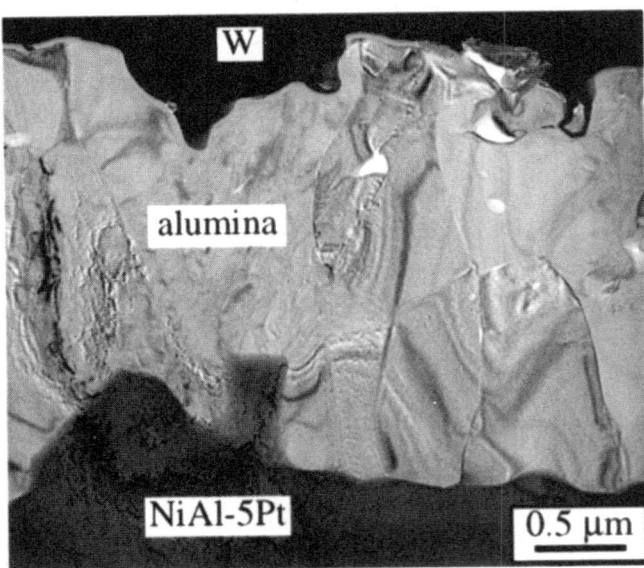

Figure 5. TEM bright field image of the alumina scale formed on NiAl-5Pt after 2h at 1200°C. In general, the scale is two grains thick and neither the gas interface or metal interface of the scale is smooth. The W coating is part of the FIB specimen preparation.

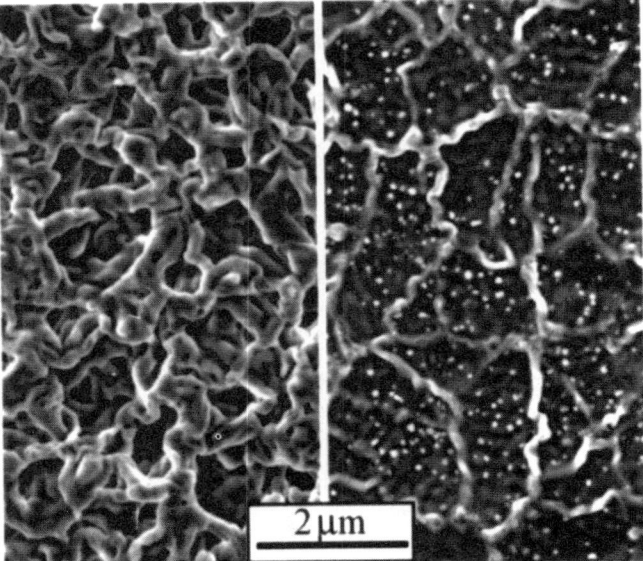

Figure 7. FEG-SEM plan-view images of the alumina scale formed after 2h at 1200°C on (a) NiAl-5Pt and (b) NiAl-2Pt+Hf. The "intrinsic" ridges in (a) are much wider and form due to outward Al transport. The "extrinsic" ridges in (b) remain from the θ−α phase transformation.

is typical of undoped alumina scales (those formed on substrates without a RE dopant) which form a network of closely spaced "intrinsic" ridges (Figure 7a) at this temperature due to the outward diffusion of Al.[17,18] Thus, the adherent alumina scales on (Ni,Pt)Al are attributed to the elimination of voids at the metal-scale interface and the metal protrusions into the scale acting as a mechanical interlock.

Ideal Alumina-Formation: The Hf Effect

While Pt is beneficial to alumina-scale adhesion, its overall influence on oxidation behavior is minimal compared to that of an optimized RE addition, particularly 0.05%Hf.[10] (Yttrium is not considered a viable alternative to Hf for aluminides because it readily forms NiY_x precipitates, which internally oxidize and disrupt scale formation.[19]) Hafnium not only improves alumina scale adhesion but also has been shown to reduce the scale growth rate by an order of magnitude compared to undoped β-NiAl [10,20]. This effect is evident in Figures 2 and 8 when comparing the cyclic oxidation behavior of Pt-doped NiAl to Hf-doped NiAl. Both alloys form adherent alumina scales at 1100°C but, because the scale on NiAl+Hf grows more slowly, it has a significant long-term advantage. There is some critical scale thickness where the thermal cooling strains will exceed the interfacial strength and scale spallation cannot be avoided. Empirical models have suggested critical scale thicknesses in the 10-15μm range (corresponding to ≈2-3mg/cm²), e.g. Ref.21. After 1000h at 1100°C, this level has not been reached. However, at 1150°C, this limit is attained for Pt-doped NiAl and spallation was observed to occur after less than 1000, 1h cycles, Figure 8. With the slower growing scale on Hf-doped NiAl, the critical scale thickness will not be reached for thousands of hours. No additional benefit was observed from combining Pt- and Hf-doping, Figure 8.

The scale microstructure formed on Hf-doped NiAl was investigated by TEM. On this substrate, the scale was thinner and had fewer internal voids than that formed on NiAl-5Pt (compare Figures 5 and 9). The addition of Hf is believed to act like Zr and inhibit the outward Al boundary transport,[17] resulting in growth

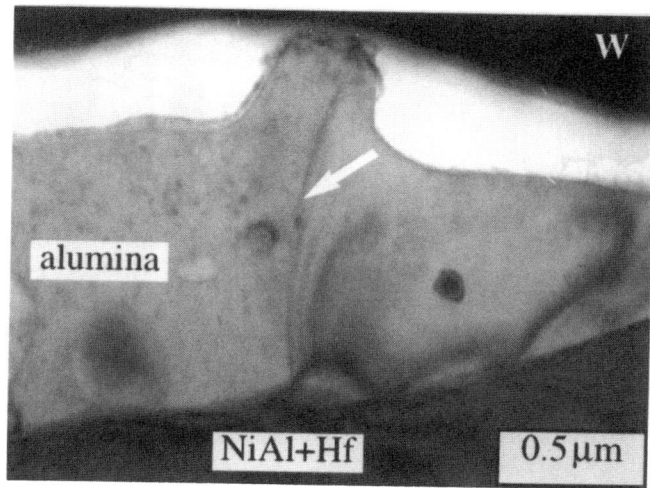

Figure 9. TEM bright field image of the alumina scale formed on NiAl+Hf after 2h at 1200°C. Grain boundaries (arrow) are associated with the ridges at the gas interface and the scale is only one grain thick. The W coating was part of the FIB preparation.

primarily by inward grain boundary diffusion of oxygen. This eliminates the intrinsic ridges and leaves only the "extrinsic" ridges[18,22] from the θ–α phase transformation, Figure 7b. At each of the extrinsic ridges, a grain boundary was observed as predicted.[22] Using STEM/EDX, Hf ions were detected as segregants to the scale grain boundaries and the metal-scale interface. Similar behavior has been observed for all RE additions.[7,10,17,23-30] The segregation of Hf ions is the most likely explanation for the low scale growth rate; however, its benefit over other dopants (e.g. Y and Zr), which also segregate at similar levels, has not been explained.

Thus, a promising approach to improved coating performance would be to incorporate Hf at the appropriate level in an aluminide coating, thereby achieving a significant improvement in alumina scale adhesion and making the addition of Pt unnecessary, at least from the viewpoint of oxidation behavior.

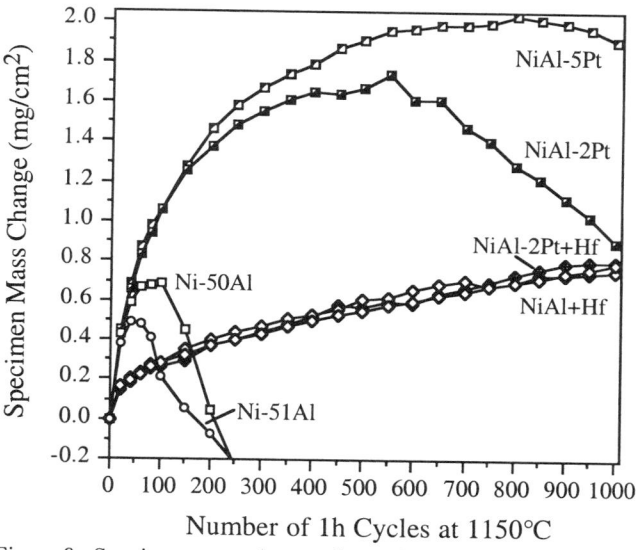

Figure 8. Specimen mass changes for various Ni-Al alloys during 1h cycles at 1150°C in O_2. The addition of Pt significantly improved scale adhesion compared to simple aluminides without Hf. The addition of Hf improved scale adhesion and reduced the scale growth rate.

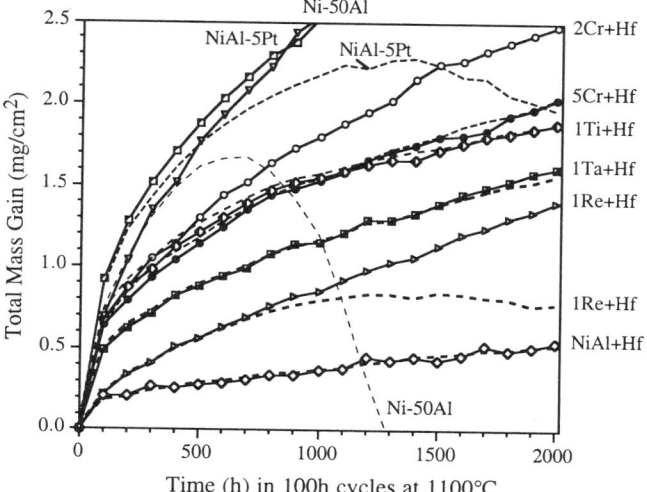

Figure 10. Total mass gain (specimen + spall) for various alloys during 100h cycles in alumina crucibles at 1100°C. Dashed lines show specimen mass changes. Higher total mass gains indicate higher scale growth rates or more scale spallation.

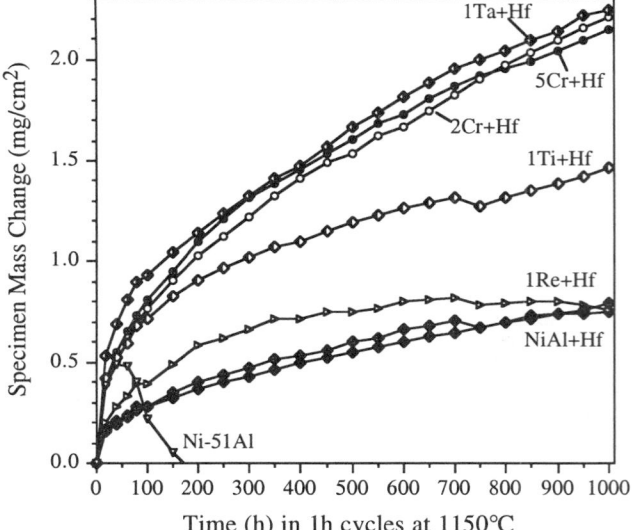

Figure 11. Specimen mass changes for various Ni-Al alloys during 1h cycles at 1150°C in O_2. The addition of Hf improved scale adhesion and reduced the scale growth rate. When secondary dopants were added, the scale growth rate increased.

<u>Real World Coatings: Co-Doping Effects</u>

The major problem with the Hf-doping approach is that current aluminide coatings contain numerous other elements besides Ni and Al due to interdiffusion with the superalloy substrate.[1] The benefits of Hf can be compromised if additional elements are also incorporated into the coating. Strong detrimental effects were found from the addition of 1%Re and 2-5%Cr to Hf-doped NiAl, Figures 10-12. Lesser negative effects were observed for additions of 1%Ti or Ta. The data shown in Figure 10 for 1100°C are from 100h thermal cycles which were used to simulate the longer duty cycles used for land-based gas turbines for power generation. The total mass gain reflects the metal wastage by summing the

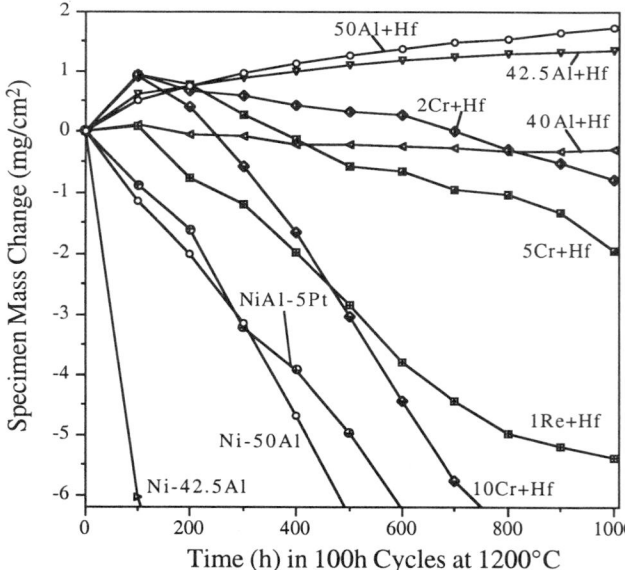

Figure 12. Specimen mass changes during 100h cycles at 1200°C in air. While Hf improves Ni-Al scale adhesion, additions of Re and Cr degrade it.

specimen mass gain and spalled scale inside the alumina test crucible. The difference between the total (solid line) and specimen (dashed line) mass gains reflects the amount of scale spallation. The lowest total mass gain was for NiAl+Hf. Higher mass gains were measured for NiAl-1Re+Hf due to some scale spallation and for NiAl with other co-dopants because of an accelerated scale growth rate (and little scale spallation). A similar observation is made for 1h cycles at 1150°C, Figure 11. Very little scale spallation was observed for Cr- or Ta-doped NiAl+Hf but the scale growth rate was 3-4X that for NiAl+Hf. At 1200°C, the additions of Cr and Re were found to cause significant scale spallation during 100h cycles, Figure 12. Thus, the improvements associated with Hf-doping can be compromised to various degrees by other dopants.

After 1000, 1h cycles at 1150°C, various co-doped NiAl specimens were sectioned, Figure 13. As expected from the mass change data, the scale on NiAl+Hf was thinnest. The addition of Ta, Ti and Cr resulted in a thicker but adherent scale with an indication of

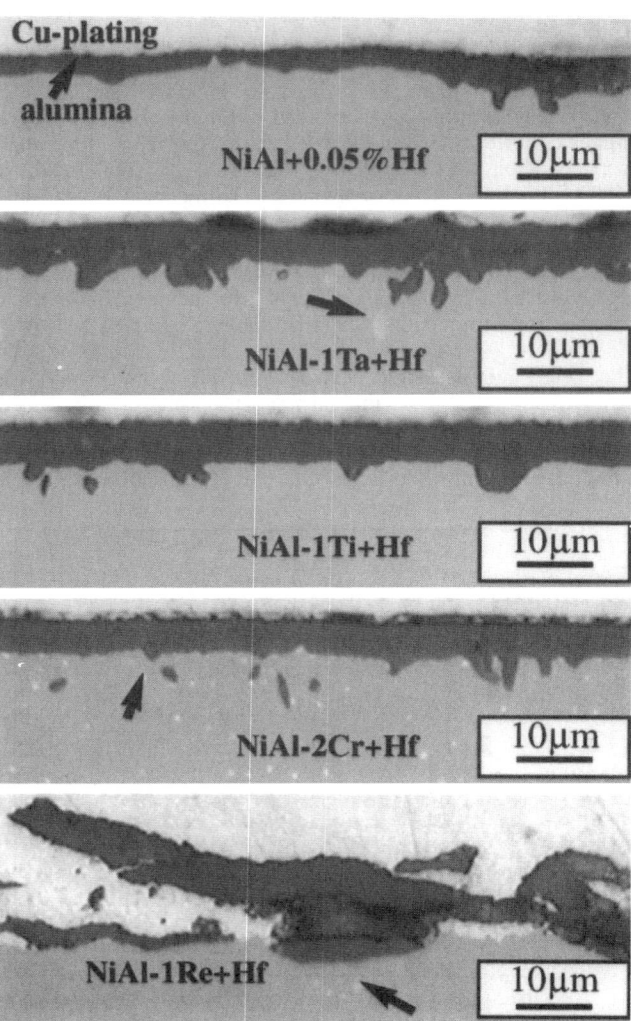

Figure 13. Copper-plated metallographic cross-sections of surface scale after 1000, 1h cycles at 1150°C on (a) NiAl+Hf; (b) NiAl-1Ta+Hf, (c) NiAl-1Ti+Hf (d) NiAl-2Cr+Hf and (e) NiAl-1Re+Hf. The scale on Hf-doped NiAl was thinner as expected from the mass change data. Precipitates (arrows) are noted with the addition of Ta, Re or Cr.

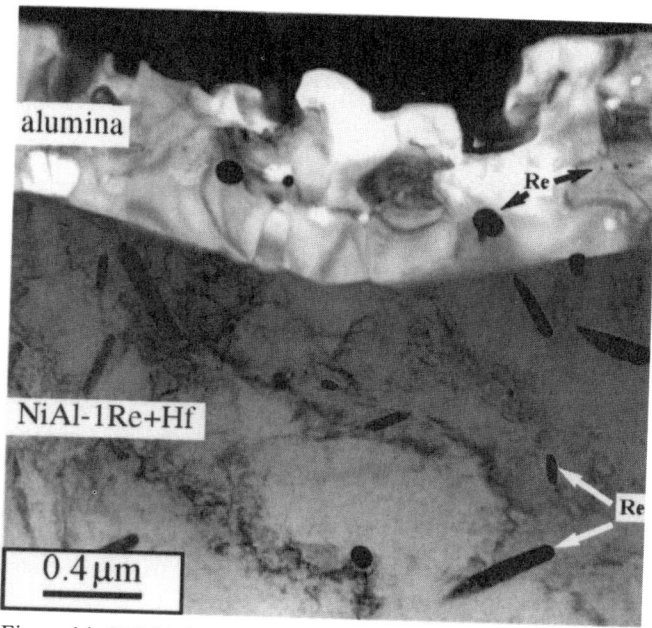

Figure 14. TEM bright field image of the alumina scale formed on Ni-50at%Al-1Re-0.05Hf after 2h at 1200°C in O₂. Re-rich precipitates are observed in the metal and in the oxide.

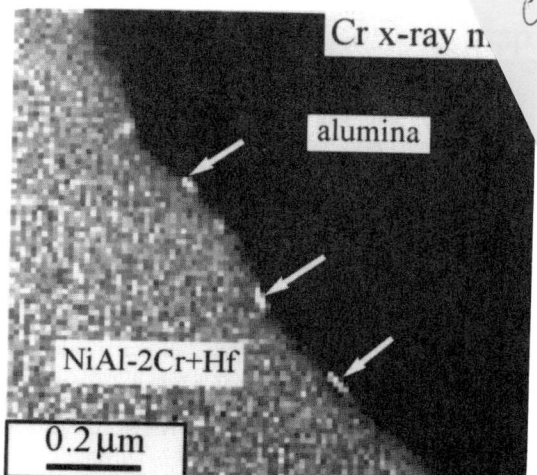

Figure 15. High resolution STEM Cr x-ray map of the scale formed on Ni-49Al-2Cr-0.05Hf after 2h at 1200°C in O₂. Arrows mark α-Cr precipitates at the metal-oxide interface.

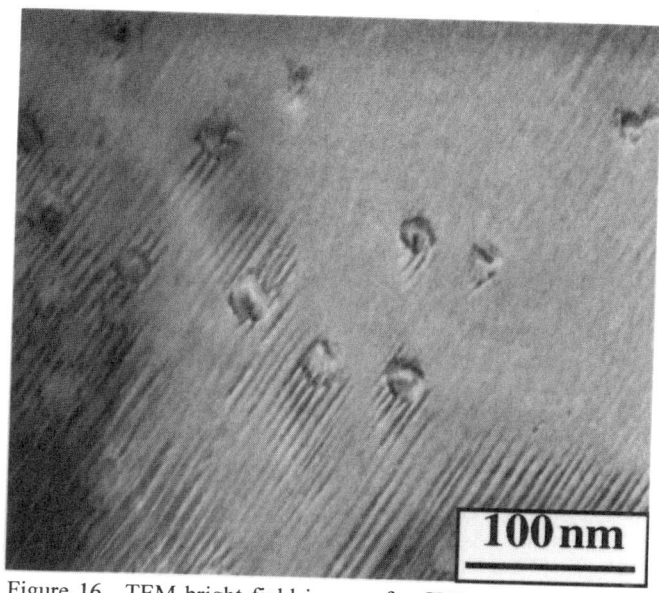

Figure 16. TEM bright field image of a CVD simple aluminide coating on a René N5 substrate after 100h at 1150°C. Cr-rich precipitates are observed throughout the coating.

more oxide penetrations into the substrate in each case, Figures 13b-d. Some penetrations contain additional Ta- or Ti-rich oxide particles. The scale on NiAl-1Re+Hf showed significant scale spallation which also was evident in the mass change data, Figure 11. Additions of Ta, Re and Cr all resulted in precipitate form in the substrate (arrows in Figures 13b, d & e).

The same alloys were sectioned for TEM an for 2h at 1200°C. Compared to the scale o Ti and Ta additions showed little effect on t under these conditions. Using STEM/EDX, were found as boundary segregants along wi Re-doped specimens were of more interest promoted scale spallation at 1200°C, Figure 12 formed α precipitates in the β-matrix which app scale adhesion. For NiAl-1Re+Hf, α-Re particles

[handwritten annotation:] Cr does help slow hot corrosion — why is Cr coming out as α particles? — processing can help?? ??

in the metal and in the oxide, Figure 14, suggesting a low solubility of Re in the β phase. After a 100h exposure at 1200°C, interfacial voids were observed by SEM but, after only 2h, there was no indication of interfacial void formation in the TEM section. In general, the microstructure was very different with coarse thickness, much like that on Pt-doped The scale formed on NiAl-2Cr+Hf had a very to that formed on NiAl+Hf and Hf ions were ary and metal-oxide interface only difference noted was the presence of small α-Cr particles at the metal-scale interface. These are shown in the STEM-EDX Cr x-ray map in Figure 15. observations of other Cr-doped alloys, appear to increase in size with exposure time and temperature (e.g. Figure 13e). These precipitates are not observed as-cast NiAl-2Cr+Hf. Their nucleation may be a result of Cr becoming enriched in the near-surface region as it is rejected from the oxidation front. Since the CTE of α is much lower than β, the precipitates may cause additional strain during cooling. Similar Cr-rich α precipitates were observed in both simple and Pt-modified aluminide coatings, after 100h at 1150°C, Figure 16. Thus a similar mechanism is likely operating in CVD aluminide coatings with these α precipitates having a negative effect on scale adhesion.

An additional issue in the development of CVD aluminides is that typically the coatings do not contain 50at%Al (31.5wt%) as did the cast aluminides. A coating will also lose Al during high temperature exposure by diffusion into the substrate. After only 100h at 1150°C, γ' formation was noted by TEM at CVD Pt-modified aluminide grain boundaries, indicating significant Al loss.[13] Ni₃Al has significantly worse scale adhesion than NiAl, even when RE-doped.[19,32] This also has been attributed to its higher CTE than NiAl.[22] In order to explore the effect of Al content within the β phase field, Hf-doped Ni-Al alloys were cast with Al contents down to 40at%Al, Table I. Doping with Hf significantly improved the performance of undoped, low Al content (40-42.5at%) NiAl, Figure 12. However, the mass loss observed for Ni-40Al+Hf suggests that it may be more prone to spallation than higher Al content NiAl. This aspect is still being explored with microstructural characterization and lower (35%) Al content alloys. Because castings, unlike coatings, do not lose Al by back diffusion, results from castings may not be entirely representative of coating behavior.

Discussion

These results have been compiled in order to suggest some composition guidelines that may lead to bond coats with improved performance. The guidelines relate to oxidation performance and may be altered by mechanical property or other considerations involved in maximizing TBC performance. They are intended as a starting point for an overall discussion of TBC system design and performance.

From an oxidation standpoint, an aluminide coating appears to have greater potential than MCrAlYs for use as a bond coat. Though not well recognized, the alumina scale adhesion on NiCrAlY-type alloys is decidedly inferior to that formed on RE-doped NiAl or FeCrAlY.[6,7] Cast NiCrAlY spalls readily at 1100°C while RE-doped NiAl only spalls after >1000 1h cycles at 1200°C. While Pt does improve scale adhesion in aluminides, its potential is limited compared to Hf additions. (One attractive feature of Pt is that it is easily incorporated into a CVD aluminide coating, unlike Hf.) Many of the beneficial effects often attributed to Pt, such as reducing heavy element diffusion into the coating or increasing the Al content of the coating, have not been confirmed experimentally.[2,13] One issue that requires further study is the interaction between Pt and S. Understanding this mechanism may provide useful insight into the role of S in high temperature oxidation.[33,34]

The incorporation of Hf into aluminide coatings offers significant improvements in performance and could preclude the use of Pt. The 0.05Hf level used in this study is sufficient to be uniformly included in the alloy while avoiding excessive internal oxidation associated with higher Hf contents. The best performance would likely be achieved with a coating that only contained Ni, Al and Hf. Incorporation of additional elements would inhibit the potential of Hf to improve scale adhesion and, more importantly, to reduce the scale growth rate. The ultimate benefit of proper Hf-doping is the order of magnitude drop in the scale growth rate. The reduction in rate allows longer time at temperature or higher temperature operation before a critical scale thickness is formed where spallation occurs.

The weak point of a pure NiAl+Hf coating would be hot corrosion. Chromium was the only element which appeared to improve the hot corrosion resistance of NiAl (Figure 4). However, Cr increased scale spallation and the isothermal scale growth rate compared to NiAl+Hf, Figures 10-12. These effects are in good agreement with previously reported work.[35-37] Based on the available data for cast aluminides, it appears that hot corrosion resistance and exceptional scale spallation resistance are mutually incompatible goals for coating performance.

A pure NiAl+Hf coating will be difficult to produce by traditional CVD coating processes. Since elements detrimental to scale growth and adhesion, such as Cr and Re, incorporate into both inward and outward growing aluminide coatings, the coating essentially must be formed by depositing Ni as well as Al, with a uniform, low level of Hf in the coating. This strategy would allow the formation of an initial scale doped solely by Hf and would require diffusion of substrate elements (e.g. Ta, Cr, Re) to the gas side of the coating before they could negatively impact oxidation performance. Of course, a stable diffusion barrier between the

substrate and coating would further improve this strategy by reducing outward and inward diffusion, but the development of such a barrier appears unlikely.

Attempting to incorporate Hf by surface modifications of the coating is not an appropriate strategy. Ion implantation[38] and RE surface oxide coatings[39] have been shown to be ineffective alternatives to a uniform RE distribution in the substrate.[40] A model for the role of RE dopants on alumina scales suggests that a uniform, constant flux of RE ions diffusing from the substrate into the oxide scale is necessary to maintain the RE benefit.[25]

A final point is that there appears to be no inherent problem with a thick (125μm, 5 mil) ceramic coating adhering to a metallic substrate. A 125μm-thick, physical vapor deposited, Y_2O_3-stabilized ZrO_2 coating on Zr-doped β-NiAl[41,42] had a 20% coating spallation lifetime of 3600, 1h cycles at 1150°C and has not reached 20% spallation after 800, 2h cycles at 1200°C. These observations indicate that an extended coating lifetime should result from the formation of an adherent alumina layer. Current bond coats simply do not form alumina scales with sufficient adherence. The next generation of bond coats will require more careful compositional control in order to achieve significant improvements in performance.

Summary

Based on experimental results from cast aluminides and laboratory-scale CVD aluminide coatings, it appears that single-phase aluminides have a greater potential for the development of improved bond coats than do MCrAlY-based compositions. A series of guidelines is proposed for aluminide coating compositions:
(1) The main features desired in a bond coat to maximize oxidation performance are:
- rapid formation of an α-Al_2O_3 scale with no transient oxides
- excellent scale adhesion
- low rate of scale growth
- minimal transport of additional elements into the oxide scale
(2) Pt incorporation provides improved scale adhesion and increases time to first spallation apparently by countering the detrimental role of indigenous sulfur.
(3) Reactive element (Hf) doping of aluminides results in a reduced rate of scale growth
(4) No synergistic benefits are anticipated from co-doping Pt and Hf.
(5) Transport of elements such as Cr, Re, Ti and Ta to the oxide scale can lead to an increase in oxidation rate and a decrease in resistance to scale spallation.
(6) Within the β-phase, Al contents of 50at%Al show better scale adhesion than lower (40at%) Al contents.

Acknowledgements

Much of the coatings work is part of the thesis work of Y. Zhang. The authors would like to thank K. M. Cooley, G. Garner, L. D. Chitwood, M. Howell, D. W. Coffey, W. D. Porter and K. S. Trent at ORNL for assistance with the experimental work and P. F. Tortorelli, D. F. Wilson and J. H. DeVan at ORNL for their comments on the manuscript. Work at ORNL was supported by the U.S. Department of Energy (DOE), Assistant Secretary for Energy Efficiency and Renewable Energy, Office of Industrial Technologies, as part of the Advanced Turbine Systems Program under contract DE-AC05-96OR22464 with Lockheed Martin Energy Research.

References

1. W. Y. Lee, Y. Zhang, I. G. Wright, B. A. Pint and P. K. Liaw, "Effects of Sulfur Impurity on the Scale Adhesion Behavior of a Desulfurized Ni-based Superalloy Aluminized by Chemical Vapor Deposition," Met. Trans. A, 29A (1998), 833.

2. Y. Zhang, W. Y. Lee, J. A. Haynes, I. G. Wright, B. A. Pint, K. M. Cooley and P. K. Liaw, "Synthesis and Cyclic Oxidation Behavior of a (Ni,Pt)Al Coating on a Desulfurized Ni-base Superalloy," Met. Trans. A, 30A (1999), 2679-87.

3. B. A. Pint, P. F. Tortorelli and I. G. Wright, "Effect of Cycle Frequency on High Temperature Oxidation Behavior of Alumina- and Chromia-Forming Alloys," in Cyclic Oxidation of High Temperature Materials, ed. M. Schütze and W. J. Quadakkers (London, UK: The Institute of Materials, 1999) 111-132.

4. C. Leyens, I. G. Wright and B. A. Pint, "Hot Corrosion of Ni-Base Alloys in Biomass Derived Fuel Simulated Atmosphere," in Elevated Temperature Coatings: Science and Technology III, ed. J. M. Hampikian and N. B. Dahotre (Warrendale, PA: The Minerals, Metals & Materials Society, 1999), 79-90.

5. K. L. More, D. W. Coffey, B. A. Pint, K. S. Trent and P. F. Tortorelli, "TEM Specimen Preparation of Oxidized Ni-Based Alloys Using the Focused Ion Beam (FIB) Technique," (Paper presented at Microscopy and Microanalysis 2000, Philadelphia, PA, August 2000).

6. B. A. Pint and I. G. Wright, "Cyclic Oxidation Behavior of cast NiCrAl Alloys," in High Temperature Corrosion and Materials Chemistry, Electrochem. Soc. Proc. v.98-9, ed. P. Y. Hou, M. J. McNallan, R. Oltra, E. J. Opila and D. A. Shores (Pennington, NJ: The Electrochemical Soc., 1998) 263-74.

7. B. A. Pint, K. L. More, I. G. Wright and P. F. Tortorelli, "Characterization of Thermally Cycled Alumina Scales," (Paper presented at Microscopy of Oxidation 4, Cambridge, UK, September 1999).

8. C. E. Lowell, R. G. Garlick and B. Henry, "Thermal Expansion in the Ni-Cr-Al and Co-Cr-Al Systems to 1200°C Determined by High-Temperature X-Ray Diffraction," Met. Trans., 7A (1976), 655-60.

9. B. M. Warnes and D.C. Punola, "Clean Diffusion Coatings By Chemical Vapor Deposition," Surf. Coat. Technol., 94-95 (1997), 1-6.

10. B. A. Pint, I. G. Wright, W. Y. Lee, Y. Zhang, K. Prüßner and K. B. Alexander, "Substrate and Bond Coat Compositions: Factors Affecting Alumina Scale Adhesion," Mater. Sci. Eng., 245 (1998), 201-11.

11. H. J. Grabke, D. Weimer and H. Viefhaus, "Segregation of Sulfur During Growth of Oxide Scales," Appl. Surf. Sci., 47 (1991), 243-50.

12. B. A. Pint, "On the Formation of Interfacial and Internal Voids in α-Al_2O_3 Scales," Oxid. Met., 48 (1997), 303-28.

13. A. Haynes, Y Zhang, W. Lee, B. Pint, I. Wright and K. M. Cooley, (1999) "Effects of Pt Additions and S Impurities on the Microstructure and Scale Adhesion Behavior of Single-Phase CVD Aluminide Bond Coatings," in Elevated Temperature Coatings: Science and Technology III, ed. J. M. Hampikian and N. B. Dahotre (Warrendale, PA: The Minerals, Metals & Materials Society, 1999), 79-90.

14. C. Leyens, B. A. Pint and I. G.Wright, "Effects of Composition on the Hot Corrosion Resistance of NiAl and (Ni,Pt)Al," NACE Paper 00-260, Houston, TX, presented at NACE Corrosion 2000, Orlando, FL, March 2000.

15. E. J. Felten and F. S. Pettit, "Development, Growth, and Adhesion of Al_2O_3 on Platinum-Aluminum Alloys," Oxid. Met., 10 (1976), 189-223.

16. H. M. Hindam and W. W. Smeltzer, "Growth and Microstructure of α-Al_2O_3 on β-NiAl," J. Electrochem. Soc., 127 (1980), 1630-5.

17. B. A. Pint, J. R. Martin and L. W. Hobbs, "^{18}O/SIMS Characterization of the Growth Mechanism of Doped and Undoped α-Al_2O_3," Oxid. Met., 39 (1993), 167-95.

18. B. A. Pint, M. Treska and L. W. Hobbs, "The Effect of Various Oxide Dispersions on the Phase Composition and Morphology of Al_2O_3 Scales Grown on β-NiAl," Oxid. Met., 47 (1997), 1-20.

19. J. D. Kuenzly and D. L. Douglass, "The Oxidation Mechanism of Ni_3Al Containing Yttrium," Oxid. Met., 8 (1974), 139-178.

20. B. A. Pint, "The Oxidation Behavior of Oxide-Dispersed β-NiAl: I. Short-Term Cyclic Data and Scale Morphology," Oxid. Met., 49 (1998), 531-60.

21. J. T. DeMasi-Marcin, K. D. Sheffler and S. Bose, "Mechanisms of Degradation and Failure in a Plasma-Deposited Thermal Barrier Coating," J. Eng. Gas Turb. & Power, 112 (1990), 522-7.

22. J. Doychak, "Oxidation Behavior of High Temperature Intermetallics," in Intermetallic Compounds, Vol.1: Principles, ed. J. H. Westbrook and R. L. Fleischer (New York, NY: John Wiley & Sons, 1994) 977-1016.

23. B. A. Pint, A. J. Garratt-Reed and L. W. Hobbs, "The Reactive Element Effect in Commercial ODS FeCrAl Alloys," Mater. High Temp., 13 (1995), 3-16.

24. E. Schumann, J. C. Yang, M. Rühle and M. J. Graham, "High Resolution SIMS and Analytical TEM Evaluation of Alumina Scales on β-NiAl Containing Zr or Y," Oxid. Met., 46 (1996), 37-49.

25. B. A. Pint, "Experimental Observations in Support of the Dynamic Segregation Theory to Explain the Reactive Element Effect," Oxid. Met., 45 (1996), 1-37.

26. C. Mennicke, E. Schumann, C. Ulrich and M. Rühle, "The Effect of Yttrium and Sulfur on the Oxidation of FeCrAl," Mater. Sci. Forum, 251-4 (1997), 389-96.

27. B. A. Pint and K. B. Alexander, "Grain Boundary Segregation of Cation Dopants in α-Al_2O_3 Scales," J. Electrochem. Soc., 145 (1998), 1819-29.

28. B. A. Pint, A. J. Garratt-Reed and L. W. Hobbs, "Possible Role of the Oxygen Potential Gradient in Enhancing the Diffusion of Foreign Ions on α-Al_2O_3 Grain Boundaries," J. Amer. Ceram. Soc., 81 (1998), 305-14.

29. E. C. Dickey, B. A. Pint, K. B. Alexander and I. G. Wright, "Oxidation Behavior of Platinum-Aluminum Alloys and the Effect of Zr-Doping," J. Mater. Res., 14 (1999), 4531.

30. K. Y. Kim, S. H. Kim, K. W. Kwon and I. H. Kim, "Effect of Yttrium on the Stability of Aluminide-Yttrium Composite Coatings in a Cyclic High-Temperature Hot-Corrosion Environment," Oxid. Met., 41 (1994), 179-201.

31. K. Fritscher, C. Leyens and M. Peters, "Development of a Low-Expansion Bond Coating for Ni-base Superalloys," Mater. Sci. Eng., A190 (1995), 253-8.

32. B. A. Pint, A. J. Garratt-Reed and L. W. Hobbs, "Analytical Electron Microscopy Study of the Breakdown of α-Al_2O_3 Scales Formed on Oxide Dispersion Strengthened Alloys," submitted to Met. Trans. A,.

33. J. G. Smeggil, A. W. Funkenbusch and N. S. Bornstein, "A Relationship Between Indigenous Impurity Elements and Protective Oxide Scale Adherence Characteristics," Met. Trans., 17A (1986), 923-32.

34. J. L. Smialek, D. T. Jayne, J. C. Schaeffer and W. H. Murphy, "Effects of Hydrogen Annealing, Sulfur Segregation and Diffusion on the Cyclic Oxidation Resistance of Superalloys: A Review," Thin Solid Films, 253 (1994), 285-92.

35. R. A. Rapp, "Chemistry and Electrochemistry of Hot Corrosion of Metals," Mater. Sci. Eng., 87 (1987), 319-27.

36. J. G. Smeggil, "The Effect of Chromium on the High

Temperature Oxidation Resistance of Ni:Al," <u>Surf. Coat. Tech.</u>, 46 (1991), 143-53.

37. D. L. Ellis, "Hot Corrosion of the B2 Nickel Aluminides," (Report CR 191082, NASA Lewis Research Center, Cleveland, OH, 1993).

38. G. Fisher, P. K. Datta, J. S. Burnell-Gray, W. Y. Chan and J. C. Soares, "The Effect of Active Element Additions on the Oxidation Performance of a Platinum Aluminide Coating at 1100°C," <u>Surf. Coat. Tech.</u>, 110 (1998), 24-30.

39. P. Y. Hou, Z. R. Shui, G. Y. Chuang and J. Stringer, "Effect of Reactive Element Oxide Coatings on the High Temperature Oxidation Behavior of a FeCrAl Alloy," <u>J. Electrochem. Soc.</u>, 139 (1992), 1119-26.

40. B. A. Pint, "Limitations on the Use of Surface Doping for Improving High-Temperature Oxidation Resistance," <u>Mater. Res. Soc. Bull.</u>, 19 (10) (1994), 26-30.

41. B. A. Pint, B. A. Nagaraj, and M. A. Rosenzweig, "Evaluation of TBC-Coated β-NiAl Substrates Without a Bond Coat," in <u>Elevated Temperature Coatings: Science and Technology II</u>, ed. N. B. Dahorte, J. M. Hampikian, and J. J. Stiglich (Warrendale, PA: The Minerals, Metals & Materials Society, 1996), 163-74.

42. B. A. Pint, I. G. Wright and W. J. Brindley, "Evaluation of TBC Systems on Novel Substrates," <u>J. Thermal Spray Technol.</u>, in press (2000).

MODELLING AND NEUTRON DIFFRACTION MEASUREMENT OF STRESSES IN SPRAYED TBCs

J.A. Thompson[1], J. Matejicek[2] & T.W. Clyne[1]

[1] Department of Materials Science & Metallurgy
University of Cambridge, Pembroke St,
Cambridge, CB2 3QZ, UK.

[2] Department of Materials Science & Engineering
State University of New York, Stony Brook,
NY 11794-2275, USA.

Abstract

Thick TBC deposits have been deposited onto mild steel substrates. A CoNiCrAlY bond coat was applied by vacuum plasma spraying (VPS), while a ZrO_2-8wt% top coat was deposited by air plasma spraying (APS). An *in-situ* curvature monitoring technique was used, in conjunction with thermal histories and deposit and substrate properties, to assist in the running of an existing numerical process model for the prediction of residual stresses in thermally sprayed coatings. Neutron diffraction experiments were performed independently on the same samples, in order to obtain the through thickness stress profiles. The agreement between the results obtained via these two techniques was excellent.

Introduction

With current superalloy technology rapidly approaching its theoretical limiting temperature, TBCs are now considered essential for raising the operating temperature of the next generation of aerospace engines(1,2). A coating 250 μm in thickness can raise the maximum engine temperature by about 100°C(3), an increase which represents significant savings in terms of cooling air and fuel reduction. Thermal spraying of such coatings results in a ceramic deposit of low thermal conductivity(4) and stiffness(5). Current procedures involve zirconia containing 6-8wt% Y_2O_3 being sprayed in air, after deposition of a vacuum plasma sprayed bond coat. The latter, commonly a CoNiCrAlY alloy, provides a dense active barrier to substrate oxidation, in addition to the rough surface necessary for improved bonding of the ceramic top coat(6).

Experimental Methods

Plasma Spraying

The powder compositions sprayed are given in Table I.

Vacuum plasma spraying was performed using a Plasma Technik

AG VPS system. The spray parameters are given in Table II. Air plasma spraying of the partially stabilised zirconia top coat could be carried out in the VPS chamber by operating the plasma gun without prior evacuation of the chamber. Mild steel was used for the substrate because it is very suitable for systematic neutron diffraction measurements, being virtually free of the complications which arise in many alloys during heat treatments as a result of changes in phase constitution.

The samples deposited are described in Table III

Table II: Plasma spray parameters for bond coat and top coat powder deposition.

Material	CoNiCrAlY	Zirconia - 8 wt % Y_2O_3
Spray type	VPS	APS
Spray distance (mm)	270	120
Arc current (A)	500	500
Voltage (V)	55	60-70
Ar flow rate (SLPM)	50	80
H_2 flow rate (SLPM)	10	18
Gun speed (mm s^{-1})	85	50
Chamber pressure	200 mbar	atmospheric
Nozzle diameter (mm)	8	8

Table III Specimen ID table

Sample ID	Substrate thickness (mm)	Bond Coat thickness (mm)	Top Coat thickness (mm)
B1	6.35	1.2	-
B2	6.35	0.8	2.07
B4	2.00	1.2	-
B5	2.00	0.8	1.43

Table I: Composition (by weight) of powders for plasma spraying.

Powder details	Content by weight %.						
Coating Type	Material	Ni	Co	Cr	Al	Y	ZrO₂	Y₂O₃

Coating Type	Material	Ni	Co	Cr	Al	Y	ZrO_2	Y_2O_3
bond coat	CoNiCrAlY	32	bal	21	8	0.5	-	-
top coat	PSZ	-	-	-	-	-	bal	8

Superalloys 2000
Edited by T.M. Pollock, R.D.,Kissinger, R.R. Bowman,
K.A. Green, M. McLean, S. Olson, and J.J. Schirra
TMS (The Minerals, Metals & Materials Society), 2000

Diffraction techniques

The emergence in recent years of neutron diffraction as a viable technique has opened up the possibility of obtaining accurate stress profiles through the thickness of thermally sprayed samples, due to the much greater penetration depth of neutrons, compared to that of traditional X-rays(7,8). In this sudy, neutron diffraction experiments were performed by J. Matejicek at the NIST facility, Gaithersburg, USA.

Table IV: Data used in the neutron diffraction analysis performed on TBC samples.

Material	hkl peak analysed	d spacing (nm)	E (GPa)	Poisson's ratio
Fe	110	0.2033	216	0.28
CoNiCrAlY	111 (γ')	0.2076	233	0.29
ZrO$_2$	112	0.2101	200	0.30

Curvature based methods

Evaluation of curvature data is a potentially powerful method for the determination of residual stress and has been used with success in the past(9). However, care must be taken when employing this technique, particularly with regard to the interpretation of data.

In this study, the curvature is recorded continuously, via a video camera mounted alongside the spray chamber (Figure 1). The defection of the free end of the sample is recorded and converted to a curvature, since the length of the deflecting strip is known. This provides a powerful means of validation for the numerical process model described below, since the curvature history recorded during spraying of the coating can be easily compared with that predicted by the process model. The thermal history during spraying is also recorded, via a thermocouple spot welded to the rear face of the sample.

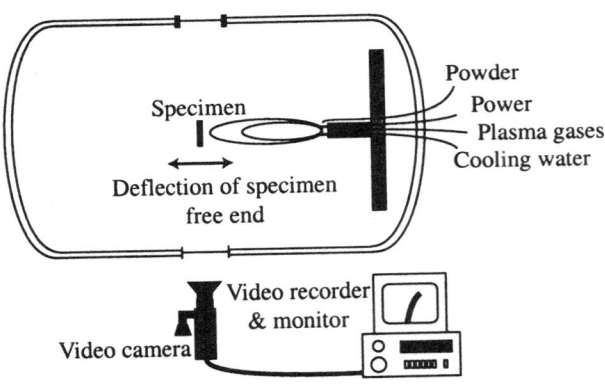

Figure 1: Arrangement of apparatus for *in situ* curvature monitoring.

Substrate Characterisation

The residual stress levels in the substrate material were evaluated prior to spraying by neutron diffraction.

Due to the wide range of properties exhibited by mild steel, dependent upon the precise composition and manufacturing process, the yield stress and creep properties of the substrate material were evaluated experimentally. The substrate yield

stress was measured as a function of temperature. Substrate creep data were obtained using a non-contacting laser scanning extensometry (Lasermike) technique(10). Creep tests were performed under conditions of constant load and the steady state creep rate fitted to the Norton creep law (Equation 1).

$$\dot{\varepsilon} = A\sigma^n \exp\left(\frac{-Q}{RT}\right)$$

Equation 1

where A is the pre-exponential constant, σ is the applied stress, n is the stress exponent for creep, Q is the activation energy for creep, R is the molar gas constant and T is the absolute temperature.

Numerical modelling

An existing numerical model(11-14) has been used in this study. The main points and assumptions behind this model can be summarised as follows:

i.) Heat flow is assumed to be one dimensional, through the thickness of the sample. Gaussian distributions of heat and mass are assumed and the gun movement during spraying can be modelled as required.

ii.) An equal biaxial stress state is assumed in the sample and through-thickness stresses are taken as negligible. Mechanical bonding at the substrate / coating interface is assumed to be perfect.

iii.) Each incoming molten splat generates unbalanced stress (the quenching stress) upon arriving at the surface.

iv.) The stress calculation is based on the determination of a relaxed width for each element in a stack through the thickness of the modelling point, according to temperature and mechanical properties. All elements are then constrained to have the same relaxed width (obtained using a force balance) and are then allowed to adopt a constant strain gradient through the thickness (obtained using a moment balance).

The thermal and curvature histories can be outputted by the model and compared with those obtained experimentally. This provides dual validation of the model. Experimental curvature histories can only be obtained for thin substrates, since deflections are difficult to measure for thicker substrates.

However, if the quenching stress can be measured from the compared curvature histories for coatings on thin substrates, the stress predictions for coatings deposited on thick substrates should be valid.

Results

Substrate Characterisation

The initial substrate residual stress levels are shown in Figures 2 & 3. While the substrate stress levels are low in the case of the thin substrate, they are high for the thicker samples. It was therefore necessary to consider these residual stresses in the numerical simulations of stress development during spraying and subsequent cooling.

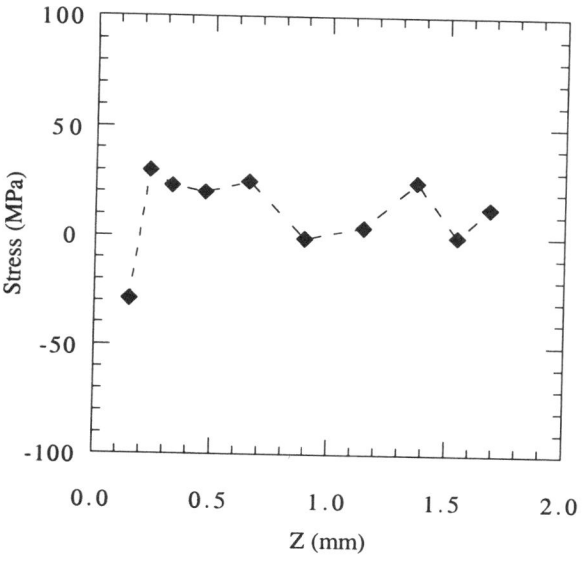

Figure 2: Residual stress state in thin steel substrate, measured by neutron diffraction.

The initial stress levels in the thick substrates were treated by altering the initial relaxed element width, such that, when constrained by the force and moment balances, the resulting stress state approximated to that measured by neutron diffraction. It is important that the changes in initial relaxed element width do not violate the conditions of the model. A stress distribution which is symmetrical about the neutral axis and exhibits equal amounts of tension and compression will satisfy the criteria of the model, such that the associated curvature is equal to zero (Figure 4).

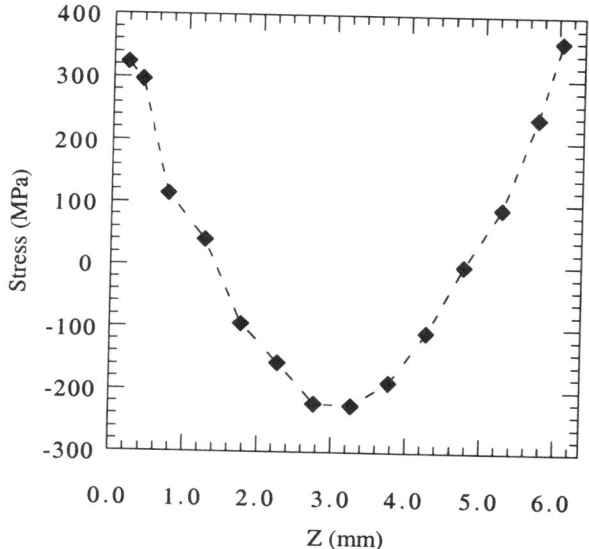

Figure 3: Residual stress state in thick steel substrate, measured by neutron diffraction.

The measured yield stress of the mild steel substrate is shown as a function of temperature in Figure 5. It is clear that, at the deposition temperature, the yield stress of the substrate material is significantly reduced. In addition, the substrate will undergo significant stress relaxation by creep processes at such temperatures. The creep data are summarised in Table V. This will result in non-elastic behaviour of the substrate during processing. It is important that this be accurately simulated in the residual stress calculations.

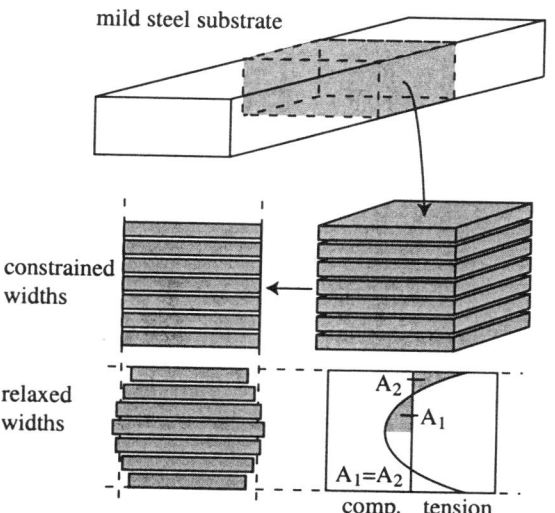

Figure 4: Schematic representation of the way in which the initial substrate residual stress state is treated by the numerical model.

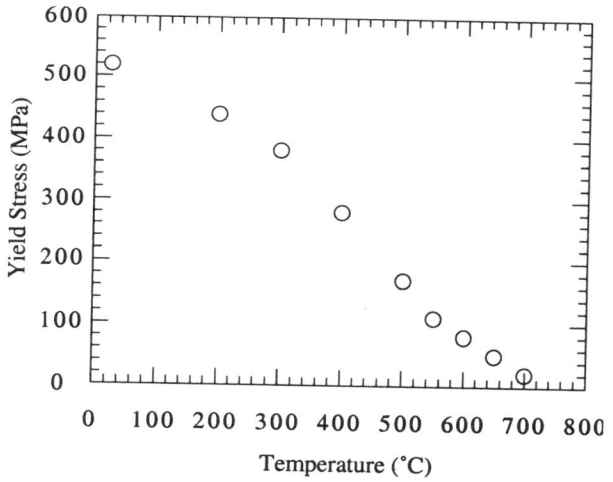

Figure 5: Variation of mild steel substrate yield stress with temperature.

641

Table V: Summary of mild steel substrate yield and creep data.

Yield Stress Data									
T (°C)	25	200	300	400	500	550	600	650	700
σ_y (MPa)	520	440	380	280	170	110	80	50	20
Creep Data									
n	3								
Q (kJ mol^{-1})	400								
A (s^{-1} Pa^{-n})	4.7×10^{-5}								

Table VI: Numerical modelling input data(15).

Property	CoNiCrAlY	ZrO$_2$ 8wt% Y$_2$O$_3$	mild steel
Thermal conductivity (W m^{-1} K^{-1})	12.0 (293K) 20.0 (1000K)	0.72 (283K) 2.164 (1352K)	49.6 (300K) 40.0 (673K) 23.6 (1073K)
Specific heat capacity (J kg^{-1} K^{-1})	350.0	480 (293K) 610 (873K)	483 (293K) 596 (873K)
Latent heat of fusion (kJ kg^{-1})	300	300	-
Melting point (K)	1673	3080	-
Droplet temperature (K)	1673	3080	-
Density (kg m^{-3})	7.00	5.28	7.85
Thermal expansivity $\times 10^6$ (K^{-1})	13.0 (298K) 18.5 (773K) 20.1 (873K) 21.5 (973K)	10.5 (283K) 11.5 (811K) 11.8 (1144K)	12.0 (298K) 13.3 (573K) 15.9 (573K)
Young's modulus (GPa)	125 (293K) 85 (1073K)	10.0 (293K) 5.0 (1293K)	208 (366K) 186 (590K) 117 (1033K)
Poisson ratio	0.30	0.30	0.30
Elastic Limit (MPa)	∞	∞	520 (293K) 170 (773K) 70 (873K) 20 (973K)
Quenching Stress (MPa)	130 (T_s = 550°C)	5.0	-
Creep parameters n Q (kJ mol^{-1}) A (s^{-1} Pa^{-n})	2.9 361 4.6×10^{-10}	- - -	3 400 4.7×10^{-5}

Numerical Modelling

The input data for the numerical model are summarised in Table VI.

The effect of creep on the predicted curvature history is significant. Figure 6 shows the experimental curvature history during deposition and subsequent cooling for sample B4. Figure 7 shows the predicted curvature histories for the same spray run, with and without creep included in the simulation.

The effect of creep is particularly clear during the first pass of the gun, where the increase in curvature is underpredicted when creep is neglected. The effect is most apparent here because the flexural stiffness of the beam is low (the sample is thin and predominantly made from the rapidly creeping substrate material, due to the low coating thickness at this time). During subsequent passes, the thickness of the substrate / coating sample increases, raising the flexural stiffness. In addition, the greater creep resistance of the CoNiCrAlY coating, in comparison with that of the substrate, lowers the specimen deflection during later passes of the gun.

Figures 8 & 9 show a comparison between predicted and measured temperature profiles during bond coat and top coat deposition (sample B5). It can be seen that the agreement is good. This is also the case in Figures 10 & 11, which compare the experimental curvature output with that predicted by the model. Both plots provide a means of validation for the model. Further compelling evidence is to be found in Figures 12-15, which show comparisons between residual stress levels obtained by neutron diffraction and those predicted by the model. Agreement is in general very good.

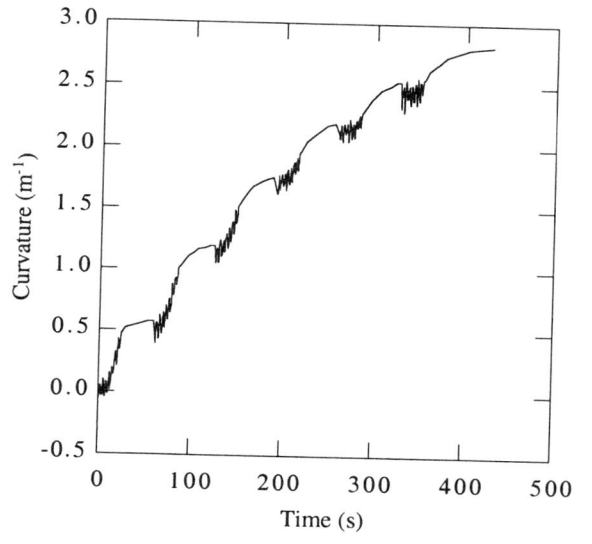

Figure 6: Experimental curvature history during spraying of CoNiCrAlY bond coat onto a thin mild steel substrate (sample B4).

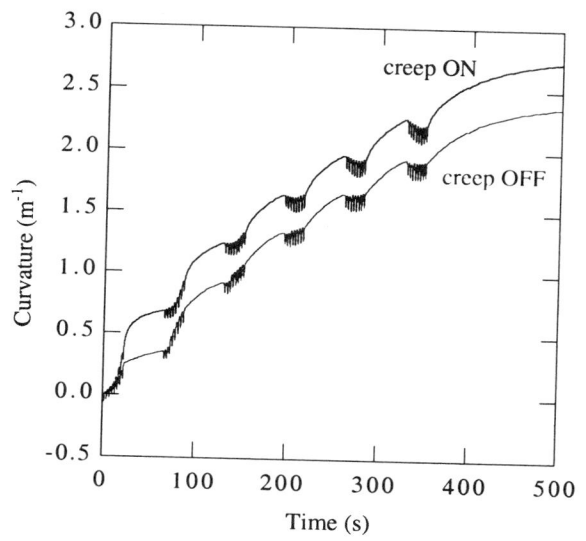

Figure 7: Predicted curvature histories for CoNiCrAlY deposition onto a thin mild steel substrate (sample B4), showing the effect of creep being incorporated in the calculation.

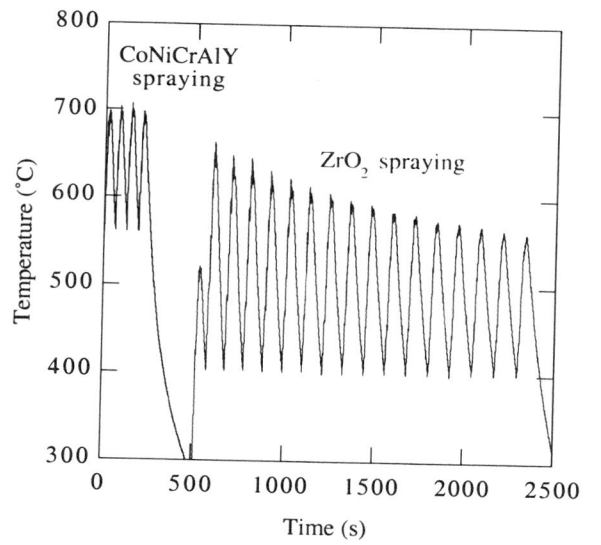

Figure 8:   Experimental temperature history for sample B5.

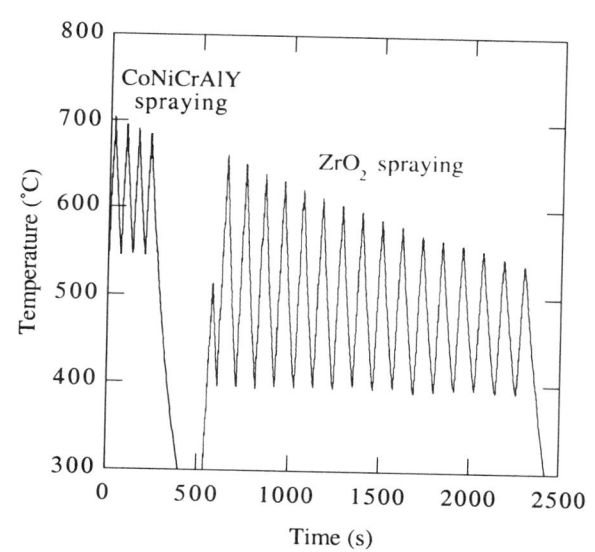

Figure 9:   Predicted temperature history for sample B5.

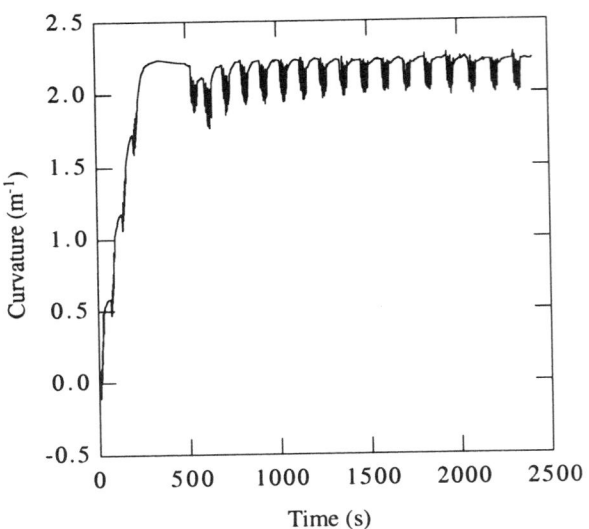

Figure 10: Experimental curvature history during spraying of sample B5.

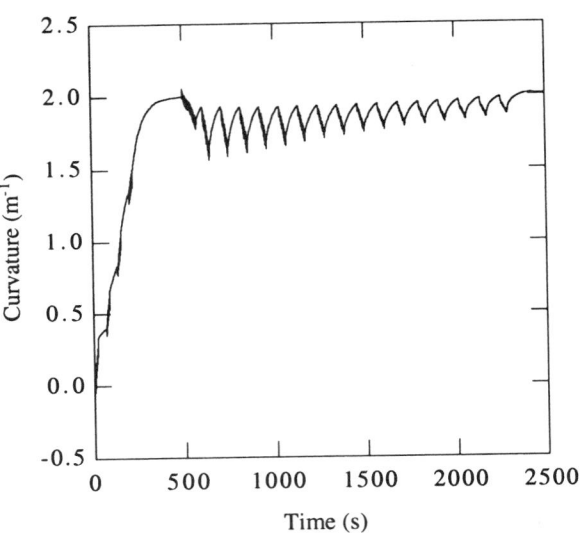

Figure 11: Predicted curvature history during spraying of sample B5.

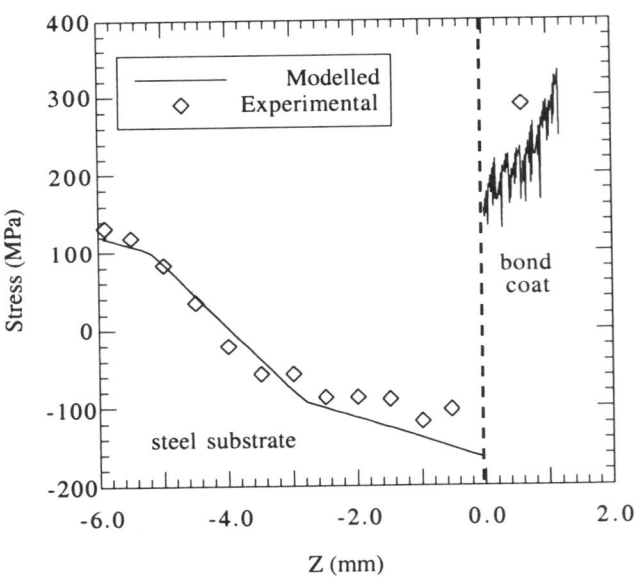

Figure 12: Residual stress comparison for sample B1 .

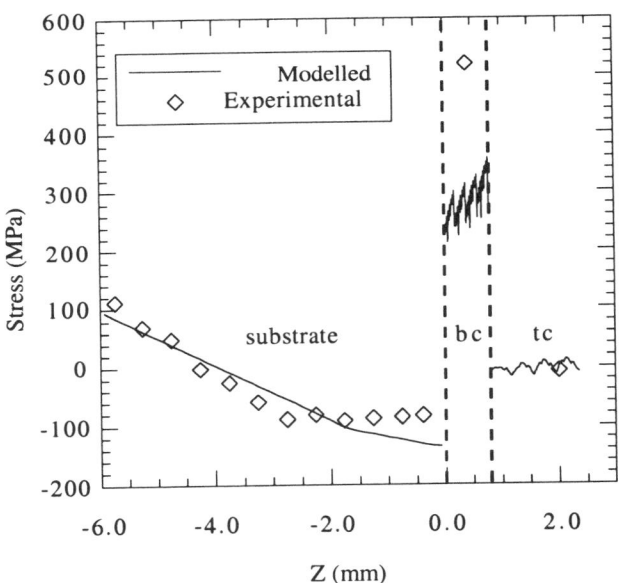

Figure 13: Residual stress comparison for sample B2.

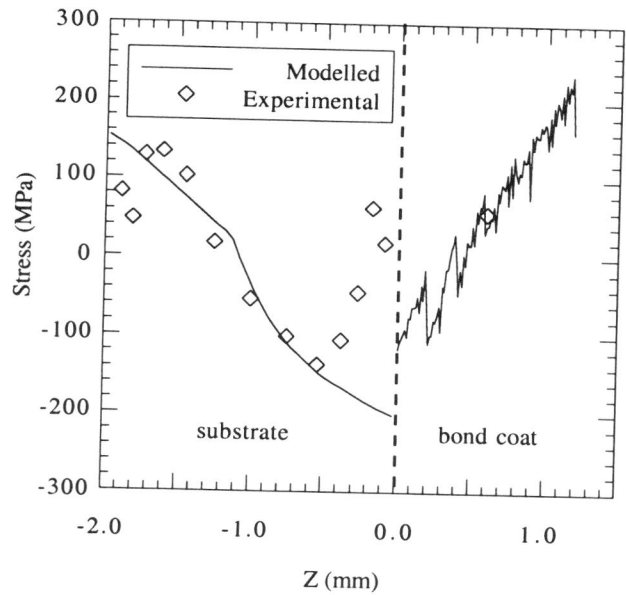

Figure 14: Residual stress comparison for sample B4.

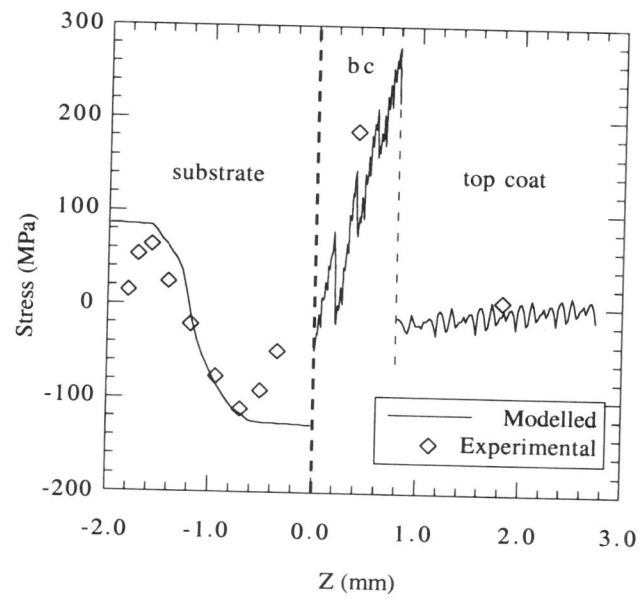

Figure 15: Residual stress comparison for sample B5.

Unfortunately, due to the requirements in terms of gauge volume needed to give acceptable count rates during neutron diffraction, through-thickness profiling was not possible in the bond coat (despite the artificially high thicknesses deposited) or in the zirconia.

While substrate stress levels predicted for the case of the thick steel substrates (B1/2) are in excellent agreement with experiment, there is a discrepancy in the case of the thin substrates (B4/5). In the regions close to the substrate free surface and the substrate / bond coat interface, the neutron diffraction data indicate a reduction in the magnitude of the measured stress. This aspect of the stress distribution is not predicted by the numerical model.

Microstructural investigation (Figure 16) demonstrates that there is no significant change in the grain structure in these regions. Therefore recrystallisation of the substrate can be ruled out as a possible explanation for this discrepancy. It seems more likely that the treatment of edge effects during the analysis of the neutron diffraction data may be responsible for the deviation from of the predicted stress distribution. However, it is not clear why this behaviour was seen for the thin substrates, but not in samples deposited on thicker substrates.

The stress distribution in the as-sprayed bond coat shows good agreement with the model prediction (samples B1/4). However, after the top coat was deposited, the model gives a lower average stress in the bond coat than that measured by neutron diffraction. This may be due to bond coat creep having a stronger effect than that simulated in the model. (Although the effects of CoNiCrAlY creep are included in the model, the creep rates assumed are steady state values. It may be the case that creep is occurring faster in the initial stages - i.e. that primary creep is significant.)

Also of interest is the predicted stress gradient through the bond coat. Coatings on both thin and thick substrates exhibit a pronounced stress gradient. However, the effect is stronger for thin substrates. This can be understood by considering the deposition process (Figure 17). (1) A thin layer impinges on substrate and contracts on cooling (2) to the substrate temperature. Force and moment balances are applied (3,4) to give the stress state and curvature. A second layer then impinges, freezes and cools to the substrate temperature (5,6). New relaxed widths are calculated (7), before the force and moment balances are reapplied (8,9). This process is repeated until the coating reaches the desired thickness. A positive stress gradient in the coating arises as a result of the incremental deposition process, demonstrating the importance of the simulation occurring progressively in order to accurately represent the true physical effects. Where the substrate is thicker, the change in initial element width with subsequent layer deposition will be smaller (due to the increase in sample flexural stiffness) and the coating stress gradient is reduced.

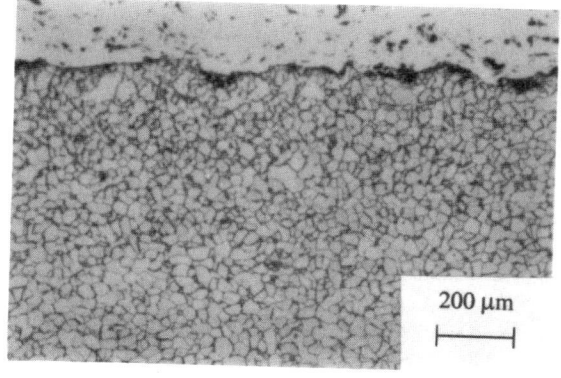

Figure 16: Optical micrograph of mild steel substrate / bond coat interface . Note that there is no change in substrate grain size in the region near the interface.

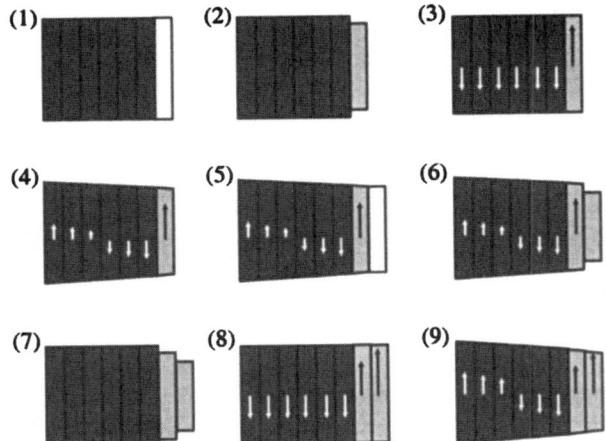

(1) (2) (3)
(4) (5) (6)
(7) (8) (9)

Figure 17: Schematic representation of stress development during spraying.

The top coat stresses are low in all cases. This is due to the low stiffness of the APS zirconia coating, which arises as a result of the dense network of microcracks which form during quenching of incoming zirconia splats to the substrate temperature (Figure 18). Further evidence of low top coat stresses can be found in the experimental curvature history during top coat deposition (Figure 10). The overall changes in curvature during top coat spraying are small, indicating low stresses.

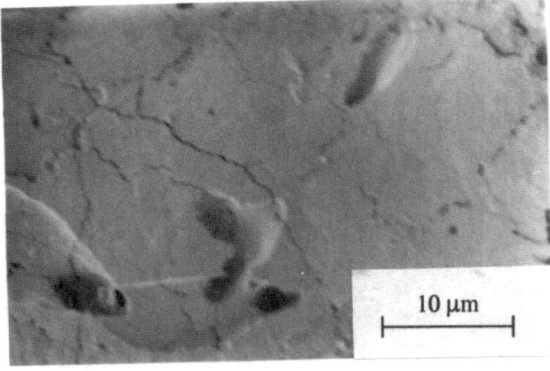

Figure 18: SEM micrograph of the top surface of an as-sprayed ZrO_2 8wt% Y_2O_3 coating, showing the presence of microcracks which contribute to the low coating stiffness.

Conclusions

These results show the power of the numerical process model to predict residual stresses with a high level of reliability, given the correct input data for the process. Samples were prepared on both thick and thin substrates and the model predictions were accurate in both cases, proving that the quenching stress values obtained for spraying onto thin substrates were applicable for the case of the thicker substrate.

Development of residual stresses in service is of considerable importance when considering the possibility of coating failure,

usually by spallation of the top coat. This study provides useful confirmation that the modelling assumptions are valid. The same calculation process (relaxed width method) can be used to simulate the changes in residual stress during service.

However, the availability of suitable material data is an important issue. Given these data, the numerical model can provide a useful insight into the relative importance of various phenomena (e.g. bond coat creep, top coat sintering), which might contribute to eventual coating spallation.

References

1. A. Bennett, "Properties of Thermal Barrier Coatings", Mater. Sci. Technol., 2 (1986), 257-261.

2. A. Bennett, F. Toriz and A. Thakker, "A Philosophy for Thermal Barrier Coating Design and its Corroboration by 10000h Service Experience on RB211 Nozzle Guide Vanes", Surface & Coatings Technology, 32 (1987), 359-375.

3. H. E. Eaton and R. C. Novak, "Sintering Studies of Plasma Sprayed Zirconia", Surf. Coat. Technol., 32 (1987), 227-236.

4. T. Taylor, "Thermal Propertries and Microstructure of Two Thermal Barrier Coatings", Surface & Coatings Technology, 54/55 (1992), 53-57.

5. P. Scardi, M. Leoni and L. Bertamini, "Residual Stresses in Plasma Sprayed Partially Stabilised Zirconia TBCs - Influence of the Deposition Temperature", Thin Sol. films, 278 (1996), 96-103.

6. G. W. Meetham, "Coating Requirements in Gas Turbine Engines", J. Vac. Sci. Technol, A3 (1985), 2509-2515.

7. O. Kesler et al., "Measurement of Residual Stress in Plasma-Sprayed Metallic, Ceramic and Composite Coatings", Mat. Sci. & Eng., (1998),

8. J. Matejicek et al., "Quenching, Thermal and Residual Stress in Plasma Sprayed Deposits: NiCrAlY and YSZ Coatings", Acta Mater, 47(2) (1999), 607-617.

9. S. Kuroda, T. Fukushima and S. Kitahara, "Generation Mechanisms of Residual-Stresses in Plasma-Sprayed Coatings", Vacuum, 41(224) (1990), 1297-1299.

10. Y. C. Tsui, J. A. Thompson and T. W. Clyne, "The Effect of Bond Coat Creep on Residual Stresses and Debonding in Plasma Sprayed Thermal Barrier Systems", in Thermal Spray: Meeting the Challenges of the 21st Century. Proceedings of the 15th Int. Thermal Spray Conf. (1998), Nice, France, C. Coddet (Ed.) Vol. 2, ASM International, 1565-1570.

11. S. C. Gill and T. W. Clyne, "Stress Distributions and Material Response in Thermal Spraying of Metallic and Ceramic Deposits", Met. Trans., 21B (1990), 377-385.

12. S. C. Gill and T. W. Clyne, "The Effect of Substrate Temperature and Thickness on Residual Stresses in Plasma Sprayed Deposits during Thermal Spraying", in

2nd European Conference on Advanced Materials and Processes, Euromat'91 (1992), Cambridge, UK, T. W. Clyne and P. J. Withers (Eds.), Vol. 1, Institute of Materials, 289-297.

13. S. Gill and T. W. Clyne, "Thermomechanical Modelling of the Development of Residual Stress during Thermal Spraying", in 2nd Plasma Technik Symposium (1991), Lucerne, Switzerland, H. Eschenauer, P. Huber, A. R. Nicoll and S. Sandmeier (Eds.), Vol. 3, Plasma Technik, 227-238.

14. Y. C. Tsui, S. C. Gill and T. W. Clyne, "Simulation of the Effect of Creep on the Stress Fields during Thermal Spraying onto Titanium Substrates", <u>Surf. Coat. Technol.</u>, 64 (1994), 61-68.

15. Y. C. Tsui, "Adhesion of Plasma Sprayed Coatings", (PhD Thesis, University of Cambridge, 1996), Pages.

INTERDIFFUSION BEHAVIOR IN NiCoCrAlYRe-COATED IN-738 AT 940° AND 1050°C

K. A. Ellison, J. A. Daleo and D. H. Boone

BWD Turbines Ltd.
1-601 Tradewind Dr., Ancaster, Ontario L9G 4V5

Abstract

Interdiffusion between a NiCoCrAlYRe coating and an IN-738 blade alloy substrate was investigated. Tests were conducted by exposing coated blade specimens in still air at 940°C and 1050°C for times of up to 9720 h. The interdiffusion zone microstructures were characterized by scanning electron microscopy and energy dispersive x-ray spectroscopy. It was found that at 1050°, the central NiCoCrAlYRe coating was comprised of γ, β and α phases and that a continuous α-Cr layer formed along the original substrate interface. At 940°C a more complex sequence of interdiffusion microstructures was observed, including the formation of a continuous γ' + α zone adjacent to the blade alloy substrate. The NiCoCrAlYRe coating was also comprised of four phases at the lower exposure temperature: γ, β, α and σ. It is shown that the complex interdiffusion behavior requires the use of diffusion path analysis for proper interpretation. The implications for development of service temperature estimation models based on coating/substrate interdiffusion in this system are discussed.

Introduction

Overlay coatings of the NiCoCrAlY type are used extensively for protection of hot section components in gas turbine engines.[1] Commercial NiCoCrAlY coatings are based on the γ-FCC and β-(Ni,Co)Al phases, but they may in addition contain phases such as γ', α-Cr and σ-(Cr_xCo_y).[2,3] Rhenium alloying is reported to improve the oxidation resistance and stability of NiCoCrAlY coatings.[2] The nickel-based superalloys used for rotating blades are primarily comprised of the γ and γ'-(Ni_3Al) phases. Interdiffusion between the coatings and substrates results in complex microstructural transformations which vary with the alloy compositions, temperature and time.[4]

Recently, the rates of interdiffusion between a Co-29Cr-6.5Al-0.5Y (wt %) coating and a nickel-base superalloy substrate were quantified and used to predict the service operating temperatures of a combustion turbine blade airfoil.[5] The metallurgical temperature estimates are used to calibrate finite element heat transfer models that support turbine hot section component life and repair calculations in an integrated approach to life management.[6,7,8] It would be quite useful if this type of temperature estimation model could be extended more generally to other commercial coating / substrate systems. However, a qualitative description and understanding of the diffusional interactions and microstructural transformations that occur at turbine operating temperatures is required before this can be accomplished. In this investigation, the interdiffusion behavior in NiCoCrAlYRe-coated IN-738 was evaluated at 940° and 1050°C.

Experimental Procedures

The chemical compositions of the IN-738LC blade and NiCoCrAlYRe coating alloys used in this investigation are given in Tables I and II. The commercial name for the NiCoCrAlYRe coating is SICOAT 2453.[*] The sample material was sectioned from commercially processed gas turbine blades. The as-received coating was approximately 200 μm thick. After transferred arc cleaning and deposition by low pressure plasma spray (LPPS), the coating was diffusion bonded to the alloy substrates by annealing in vacuum at 1120°C for 2-4 hours. The coated blade was then aged at 843°C for 24 hours.

Table I Nominal composition of IN-738LC (weight percent).

Ni	Co	Cr	Al	Ti	Mo	W	Ta	Nb	Zr	B	C
Bal.	8.5	16.0	3.4	3.4	1.75	2.6	1.75	0.85	0.12	0.012	0.13

Table II Chemical composition of NiCoCrAlYRe coating (weight percent).

Ni	Co	Cr	Al	Re	Y
Bal.	8.8	24.4	10.7	2.3	0.5

The coating samples were isothermally exposed in still air for up to 9,720 hours at 940° and 1050°C. The samples were removed from the furnaces and rapidly quenched in air to preserve the high temperature microstructures. The coating interdiffusion zones were examined by optical and scanning electron microscopy. The microstructural features were revealed by backscattered electron (BSE) imaging and after etching in a 1% chromic acid solution at 5VDC. The various phases in the coatings were identified on the basis of energy dispersive spectroscopy (EDS) results obtained from unetched specimens using a LINK Analytical QX2000 analyzer attached to a Phillips 515 scanning electron microscope operated at 20 keV.

---

[*] SICOAT 2453 is a designation of Siemens AG, Power Generation (KWU), Mülheim, Germany

Results

Interdiffusion at 1050°C

A coating sample exposed for 1300 h at 1050°C is shown in Figure 1. Compared with CoCrAlY / Ni-base and CoNiCrAlY / Ni-base samples studied previously[5,9], the coating was relatively intact and retained a substantial amount of beta phase within the central coating layer. Large Kirkendall-type pores were observed along the substrate interface.

A magnified view of the coating/substrate interdiffusion zone is shown in Figure 2. The central coating region was comprised of three principal phases: the γ matrix, β-(NiCo)Al and Cr,Re-rich α precipitates, Table III. Most of the rhenium partitioned to the α phase, which appeared white in BSE imaging mode. Interdiffusion between the NiCoCrAlYRe coating and IN-738 alloy resulted in the formation of a β-depletion zone above the original substrate interface leaving only the γ and α phases. Along the original interface, the α precipitates formed an essentially continuous layer. This zone was followed by a single-phase γ zone immediately adjacent to the γ + γ′ IN-738 alloy.

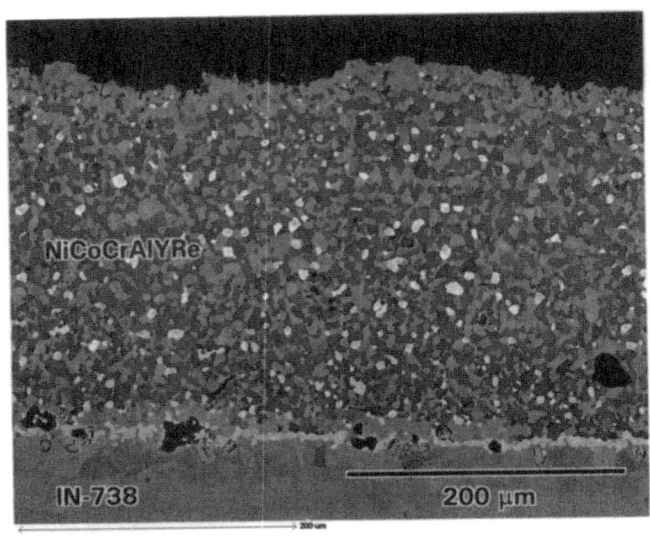

Figure 1: NiCoCrAlYRe coating after 1300 h at 1050°C (BSE image, unetched).

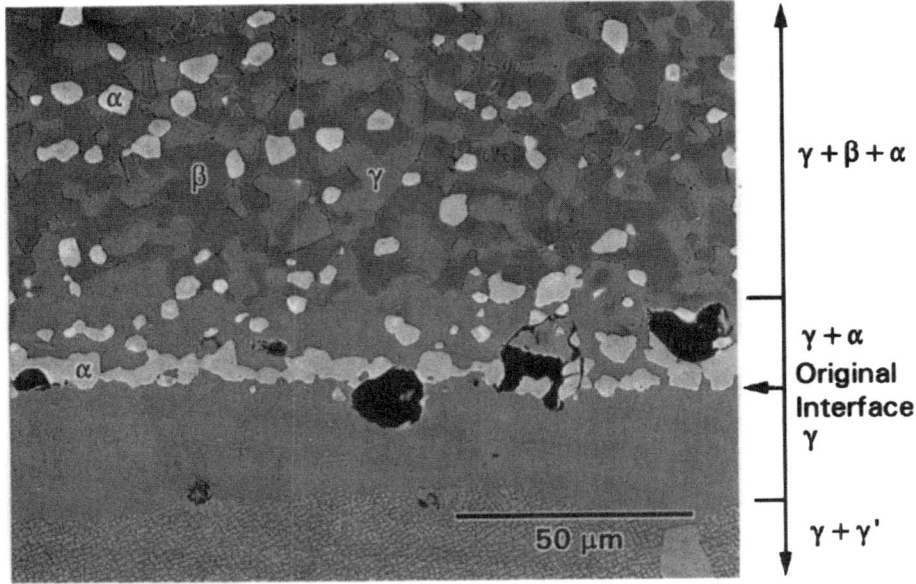

Figure 2: Interdiffusion zone microstructure after exposure for 1300 h at 1050°C (BSE image, 1% chromic).

Table III EDS results for individual phases identified in Figure 2 (atomic percent).

Ni	Co	Cr	Al	Ti	Ta	Mo	Re	Zone	Phase I.D.
45	5	8.5	40	1				γ + β + α	β
43	11	32	13					γ + β + α	γ
3.5	1.5	90	3.5				1	γ + β + α	α
3	1.5	90	4			0.5	0.5	γ + α	α
42.5	10.5	31.5	13.5	1		0.5		γ	γ

Interdiffusion at 940°C

The NiCoCrAlYRe coating exposed for 9,720 h at 940°C is shown in Figure 3. Once again, the coating was relatively undepleted after this length of exposure, compared to CoCrAlY / Ni-base and CoNiCrAlY / Ni-base samples examined in previous investigations.[5,9] A magnified image of the coating / substrate interdiffusion zone is shown in Figure 4. At this temperature, the central coating zone was comprised of four phases: the γ matrix, β-(NiCo)Al, Cr-rich α and σ precipitates, Table IV. The rhenium partitioned to the α and σ phases, which each contained about 1 at % of this element. There was very little contrast difference between the two in BSE imaging mode.

The diffusion zone between the NiCoCrAlYRe coating and IN-738 substrate was more complex at 940°C as compared to that observed at 1050°C. Moving from the central coating zone towards the alloy substrate, the EDS results showed that the α precipitates disappeared, leaving a three-phase, γ + β + σ zone. This was followed by a zone in which the β and σ precipitates also disappeared and were replaced by γ' and α (white in BSE mode). The three-phase γ + γ' + α zone extended down to the original coating/substrate interface. Below the original interface, a two-phase γ' + α structure was observed. In this zone the α phase precipitates were dark in BSE imaging mode and EDS analysis showed that that they contained almost no rhenium. The interface between the two-phase γ' + α zone and the γ + γ' substrate was non-planar.

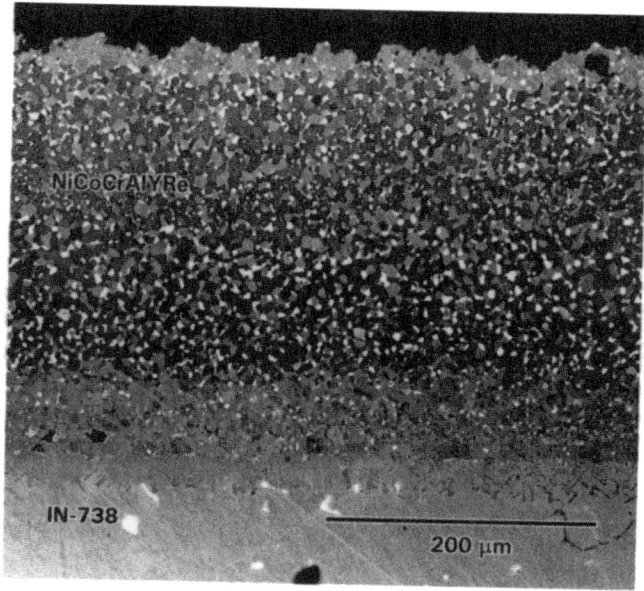

Figure 3: NiCoCrAlYRe coating after 9720 h at 940°C (BSE image, unetched).

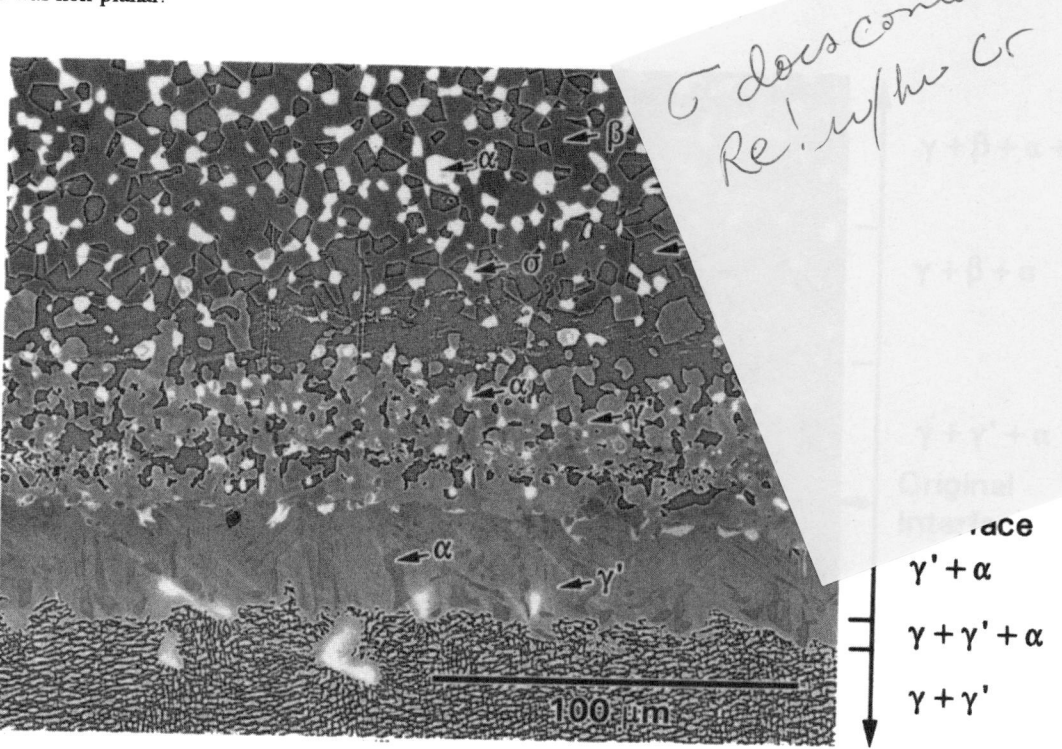

Figure 4: Interdiffusion zone microstructure after exposure for 9720 h at 940°C (BSE image, 1% chromic).

651

Table IV EDS results for individual phases identified in Figure 4 (atomic percent).

Ni	Co	Cr	Al	Ti	Ta	Mo	Re	Zone	Phase I.D.
50	5	8	37					$\gamma + \beta + \alpha + \sigma$	β
40	12.5	36	11					$\gamma + \beta + \alpha + \sigma$	γ
3.5	1.5	92	2				1	$\gamma + \beta + \alpha + \sigma$	α
12.5	12.5	70	3.5				1	$\gamma + \beta + \alpha + \sigma$	σ
58	6.5	10.5	23	2				$\gamma + \gamma' + \alpha$	γ'
4.5	2	89.5	2.5				1.5	$\gamma + \gamma' + \alpha$	α
62	5.5	5	22	5			0.5	$\gamma' + \alpha$	γ'
7.5	2	85	1.5	0.5		2.5		$\gamma' + \alpha$	α

Discussion

Interpretation of the Interdiffusion Zone Microstructures

The multi-phase diffusion zones at the coating / substrate interfaces can be represented in a compact way and related to the governing phase equilibria by plotting diffusion paths on the appropriate phase diagrams.[10] A schematic isothermal phase diagram for Ni-Cr-Al at 1050°C is shown in Figure 5. This diagram is hypothetical but is based on the closest available Ni-Cr-Al[11,12] section at 1025°C. A schematic diffusion path has been drawn to represent the NiCoCrAlYRe / IN-738 system after 1300 h at 1050°C. The multi-component NiCoCrAlYRe coating alloy was approximated as a Ni-Cr-Al ternary alloy by treating the chromium concentration of the α and σ phases as an equivalent (Cr_{eq} = Cr + Re). This is a reasonable approximation since the rhenium tends to partition strongly with the chromium in the α and σ phases.[2,3] Likewise, an equivalent aluminum concentration was used for the IN-738 alloy (Al_{eq} = Al + Ti + Ta + Nb) since these elements tend to partition to and stabilize the γ' phase in $\gamma + \gamma'$ alloys.[13] Note that, the presence of minor phase constituents (e.g. MC, $M_{23}C_6$, Ni_5Y and Y_2O_3) which are present in the commercial coating and substrate alloys cannot be represented within the Ni-Cr-Al ternary diagram.

The diffusion path of Figure 5 proceeds from the outer coating interface (point 1) to the IN-738 substrate (point 4). Oxidation at the outer coating surface of the NiCoCrAlYRe coating resulted in depletion of the α and β phases, so the initial part of the diffusion path is represented as residing closer to the γ, single phase field (point 1). The path then moves upward as it passes through the central coating region, representing the higher aluminum concentrations and higher volume fractions of β phase within the "bulk" coating layer (point 2). Moving in the direction of the coating / substrate interface, the path drops down into the two-phase $\gamma + \alpha$ phase field created by the depletion of aluminum from the coating into the alloy substrate. Enrichment of the α phase along the coating/substrate interface is represented by the loop towards the α single-phase field on the left of the diagram (point 3). The path then crosses back through the $\gamma + \alpha$ field. Since no distinct $\gamma + \alpha$ region was visible below the α particles, this portion of the diffusion path would lie on or close to a tie line through the $\gamma + \alpha$ field. The diffusion path then enters the γ

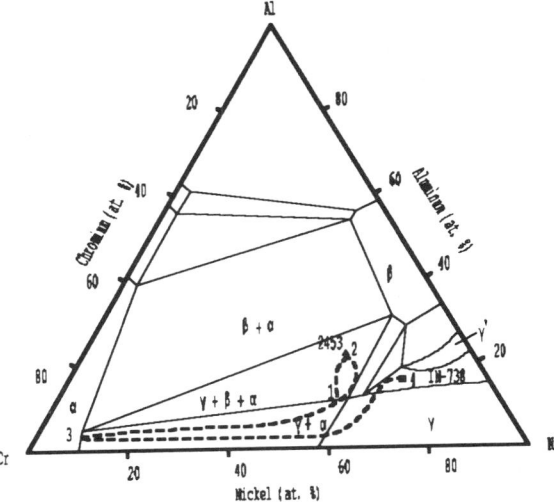

Figure 5: Schematic diffusion path (dashed line) for 1050°C interdiffusion between the NiCoCrAlYRe coating and IN-738 alloy substrate.

single-phase field before rising upwards into the $\gamma + \gamma'$ field and terminating at the equivalent IN-738 alloy composition (point 4).

The microstructural features produced by 940°C interdiffusion between the NiCoCrAlYRe coating and the IN-738 alloy substrate were more complex than at 1050°C and could not be represented on a single ternary phase diagram. According to the Gibbs phase rule, the four-phase $\gamma + \beta + \alpha + \sigma$ equilibrium observed in the central coating zone cannot be represented on anything less than a quaternary diagram. However, Sims[14] has noted that, "... no observation of a 'quaternary' phase between metals has been reported .... thus, regardless of the number of (alloying) elements involved, all phase relations applying to any nickel- or cobalt-base superalloy can be expressed in terms no more complicated than quaternary diagrams."

Inspection of the available ternary diagrams shows that the Ni-Co-Cr-Al quaternary system contains all of the major phase constituents (e.g. γ, γ', β, α and σ) observed in the present study. Until recently, no quaternary diagrams had been published in the Ni-Co-Cr-Al system.[9] Too few data are available to permit speculation on the appearance of the interior regions of the

Ni-Co-Cr-Al diagram at 940°C. However, the faces of the pyramidal Ni-Co-Cr-Al quaternary diagram at 940°C can be represented to a first approximation by using the closest available Ni-Cr-Al[11,12], Ni-Co-Cr[15], Co-Cr-Al[16,17] and Ni-Co-Al[18] ternary diagrams. Note that, the Ni-Cr-Al diagram at 850°C was used since there is an invariant transformation ($\gamma + \beta \leftrightarrow \gamma' + \alpha$) at approximately 990°C. Using these diagrams, the pyramidal Ni-Co-Cr-Al section has been unfolded into a planar view, as shown in Figure 6. The Ni-Co-Al ternary section is not shown in Figure 6 since it is not critical to the explanation.

Diffusion paths that go through the interior of the quaternary pyramid cannot be mapped onto the unfolded Ni-Co-Cr-Al diagram. However, to a first approximation, the sequence of major phase fields observed between the bulk NiCoCrAlYRe coating and the IN-738 substrate traversed the Ni-Co-Cr-Al diagram close to the ternary faces. This is shown by the schematic diffusion path superimposed on Figure 6. The path originates at a point within the Ni-Cr-Al diagram representing the equivalent IN-738 composition (point 1). Moving towards the coating, the path climbs to higher aluminum concentrations and passes through the $\gamma + \gamma' + \alpha$ and $\gamma' + \alpha$ phase fields before reaching the original coating/substrate interface. It then crosses back through the $\gamma + \gamma' + \alpha$ field, identified above the coating/substrate interface. Next, the path crosses the $\gamma + \alpha$, $\gamma + \alpha + \sigma$ and $\gamma + \sigma$ phase fields, before entering the $\gamma + \sigma + \beta$ field on the Co-Cr-Al side of the diagram (point 2). Note that, the $\gamma + \alpha$, $\gamma + \alpha + \sigma$ and $\gamma + \sigma$ zones were not distinct in the diffusion zone microstructure, but a transition region between the $\gamma + \gamma' + \alpha$ and $\gamma + \sigma + \beta$ zones was apparent, as noted above. From the $\gamma + \sigma + \beta$ field, the path leading to the bulk coating composition would need to turn towards the interior region of the diagram, containing an adjacent $\gamma + \sigma + \beta + \alpha$ four-phase field. As noted earlier, this final segment of the path cannot be shown on the unfolded diagram.

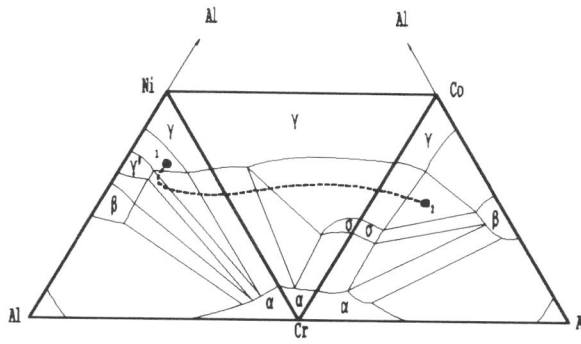

Figure 6: Schematic diffusion path (dashed line) for 940°C interdiffusion between the NiCoCrAlYRe coating and IN-738 alloy substrate.

Implications for Temperature Estimation

As described in Ref 5, the temperature estimation model for service-exposed CoCrAlY (GT-29) coatings on Ni-based (GTD-111) substrates was based on the rate of growth of the well-defined $\gamma + \beta$ zone below the original interface. This general feature of the interdiffusion zone was present over the temperature range of interest to the analysis of service-exposed turbine blades and growing at a rate sufficient to allow temperature estimates to be made with reasonable accuracy. A review of the Ni-Co-Al ternary system[18] confirms that the $\gamma + \beta$ phase field is, in fact, stable over a very wide temperature range (150°C to 1300°C), although the extent of the two-phase region varies considerably with temperature. The latter effect was accounted for in the temperature estimation model by incorporating a constitutional term within the total activation energy for $\gamma + \beta$ zone growth.

Over the range of temperatures used in this investigation, significant constitutional changes occurred within the NiCoCrAlYRe coating and throughout the coating/substrate interdiffusion zone. These microstructural transformations prevent the calibration of a temperature estimation model based on growth of a single zone over a comparable range of temperatures. The rates of intermediate zone growth (γ, $\gamma' + \alpha$) were also qualitatively slower in the NiCoCrAlYRe / Ni-base system than for $\gamma + \beta$ growth observed in the CoCrAlY / Ni-base and CoNiCrAlY / Ni-base systems examined previously.[5,9] The latter effect may be partly related to the formation of continuous α (at 1000°C and 1050°C) and $\gamma' + \alpha$ (at 940°C) zones between the NiCoCrAlYRe coating and IN-738 substrate. The transport of coating and substrate elements across these layers is expected to be slower than through intermediate layers comprised of a γ matrix phase, as observed in the CoCrAlY / Ni-base and CoNiCrAlY / Ni-base systems. The reduced rates of intermediate zone growth in the NiCoCrAlYRe / Ni-base system would lead to greater uncertainty in temperature estimates based on this phenomenon.

As an alternative to layer growth, it might be possible to make use of constitutional changes within the coating and coating/substrate interdiffusion zones to establish benchmark temperature estimates for service exposed turbine components. For example, a change from γ to $\gamma' + \alpha$ formation was observed between 940°C and 1000°C below the original NiCoCrAlYRe / IN-738 interface. The diffusion path analysis showed that this was consistent with the presence of an invariant $\gamma + \beta \leftrightarrow \gamma' + \alpha$ transformation in the Ni-Cr-Al system at approximately 990°C. Additional transformations may occur within the coating and/or interdiffusion zones at lower temperatures. Further investigations are planned to explore this possibility in greater detail.

Conclusions

The interdiffusion behavior in the NiCoCrAlYRe-coated IN-738 system is complex and requires the use of diffusion paths for proper interpretation. Constitutional differences were observed both within the coating and coating/substrate interdiffusion zone at the two temperatures used in this investigation. The observed constitutional changes will make it necessary to develop alternative approaches for metallurgical temperature estimation based on coating/substrate interdiffusion.

Acknowledgements

Financial support for this work was provided, in part, by the National Research Council of Canada, Industrial Research Assistance Program.

References

1 Proceedings of the Conference on "Protective Coating Systems for High-Temperature Gas Turbine Components", Materials Science and Technology, 2 (1986), 193-326.

2 W. Beele, et al., "Long-term Oxidation Tests on a Re-containing MCrAlY Coating", Surface and Coatings Technology, 94-95 (1997), 41-45.

3 N. Czech, F. Schmitz and W. Stamm, "Microstructural Analysis of the Role of Rhenium in Advanced MCrAlY Coatings", Surface and Coatings Technology, 76-77 (1995), 28-33.

4 P. Mazars, D. Manesse, and C. Lopvet, "Degradation of MCrAlY Coatings by Interdiffusion" (Paper No. 87-GT-58, presented at the ASME/IGTI Gas Turbine Conference and Exhibition, Anaheim, California, May 31 - June 4, 1987).

5 K. A. Ellison, J. A. Daleo and D. H. Boone, "Metallurgical Temperature Estimates Based on Interdiffusion Between CoCrAlY Overlay Coatings and a Directionally Solidified Nickel-Base Superalloy Substrate", Materials for Advanced Power Engineering 1998, ed. J. Lecomte-Beckers et al. (Jülich, Germany: Forschungszentrum Jülich GmbH Central Library, 1998), 1523 – 1534.

6 Daleo J.A., Boone D.H., "Metallurgical Evaluation Techniques In Gas Turbine Failure Analysis And Life Assessment", Failures 96, Risk, Economy and Safety, Failure Minimization and Analysis, ed R.K Penny (Rotterdam: Balkema, 1996, ISBN 90 5410 8231), 187-201.

7 Woodford D. A., Daleo J.A., "Stress Relaxation as a Basis For Blade Creep Life Assessment" (Paper presented at the IOM conference "Life Assessment of Hot Section Gas Turbine Components", James Watt Centre, Heriot Watt University, Edinburgh, UK, Oct. 5 - 7 1999).

8 Daleo J. A., Ellison K. A., Boone D. H., "Metallurgical Considerations for Life Assessment and the Safe Refurbishment and Re-qualification of Gas Turbine Blades" (Paper presented at the ASME/IGTI TURBO EXPO 2000, Munich, Germany May 8-11, 2000).

9 K. A. Ellison and J. A. Daleo, "Microstructural Evaluation of MCrAlY / Superalloy Interdiffusion Zones" (Paper presented at the IOM conference "Life Assessment of Hot Section Gas Turbine Components", James Watt Centre, Heriot Watt University, Edinburgh, UK, Oct. 5 - 7 1999), 31.

10 J. S. Kirkaldy and L. C. Brown, "Diffusion Behaviour in Ternary, Multiphase Systems", Canadian Metallurgical Quarterly, 2 (1963), 89 - 115.

11 A. Taylor and R. W. Floyd, "The Constitution of Nickel-Rich Alloys of the Nickel-Chromium-Aluminum System", J. Inst. Met., 81 (1952-1953), 451-464.

12 S. M. Merchant and M. R. Notis, "A Review: Constitution of the Al-Cr-Ni System", Materials Science and Engineering, 66 (1984), 47-60.

13 C. C. Jia, K. Ishida and T. Nishizawa, "Partition of Alloying Elements Between (A1), '(L1$_2$) and (B2) Phases in Ni-Al Base Systems", Metallurgical and Materials Transactions A, 25A (1994), 473 - 485.

14 C. T. Sims, "Prediction of Phase Composition", in Superalloys II, ed. C.T. Sims, N.S. Stoloff, and W.C. Hagel (New York: John Wiley & Sons, 1987), 217-240.

15 W. D. Manly and P. A. Beck (NACA Technical Note 2602, February 1952).

16 B. G. Livshits and N. N. Myuller, "Investigations of the Phase Equilibria in the Co-Cr-Al System" (in Russian), Nauch. Doklady Vyssh. Shkoly, 3 (1958), 201-206.

17 B. G. Livshits and N. N. Myuller, "Phase Equilibria in the Co-Cr-Al System" (in Russian), Sbornik Moscov. Inst. Stali, 39 (1960), 267-283.

18 J. Schramm, "Das Dreistoffsystem Nickel-Kobalt-Aluminium", Zeitschrift fur Metallkunde, 33 (1941), 403-412.

654

EFFECT OF COATING ON THE TMF LIVES OF SINGLE CRYSTAL AND COLUMNAR GRAINED CM186 BLADE ALLOY

S.D. Peteves, F. De Haan, J. Timm & J. Bressers
Institute for Advanced Materials, JRC, European Commission
Petten, The Netherlands
&
P. M. Hughes & S. J. Moss
ABB ALSTOM POWER UK Ltd, Leicester, UK
&
P. Johnson & M. Henderson
DERA, Farnborough, UK

Abstract

CM186, a Hf containing nickel-base superalloy is an attractive alloy, in terms of castability and thus manufacturing costs, for aero engine and land based gas turbine applications. Mechanical property and coating characterisation studies of this alloy are sparse, whilst thermo-mechanical fatigue (TMF) data do not exist in the literature. In service however, engine components experience severe cyclic thermal gradients and mechanical loads and as a consequence, TMF is a major lifing factor. This study investigates the TMF behaviour of single crystal and directionally solidified CM186 alloy and the effect of a silicon modified aluminide coating, Sermalloy 1515, on the TMF lives. Strain controlled TMF tests on coated and uncoated samples are carried out with an out-of-phase 180° difference between mechanical strain and temperature at 3 different strain ranges. In-situ monitoring of microcrack initiation and propagation and post-mortem fractographic analysis are implemented to understand the damage mechanisms and coating performance.

Introduction

High pressure gas turbine blades are frequently made from single crystal nickel-based alloys, microstructurally designed to retain strength up to high temperatures. The absence of an expensive solution heat treatment, along with high casting yields and achievable high temperature properties, makes columnar grained (CG) (directionally solidified) CM186 an attractive alternative alloy for aero engine and land based gas turbine applications.[1,2] The life of such a CG component is primarily determined by the interaction of mechanical and thermal loads and environmental contributions. Various protective measures are taken to guarantee efficient operation of the blades up to service temperatures. [3] Different coatings have been developed to protect the blades against environmental degradation. The failure modes, which must be taken into account for lifetime assessment of such coated systems include oxidation, corrosion, erosion and microstructural discontinuities caused by interdiffusion between the substrate and the coating. However, the major cause of failure is thermo-mechanical fatigue due to mechanically applied and thermally induced stresses, resulting from the thermal strains generated by the strong temperature gradients experienced during heating and cooling.

In this work, the results of TMF tests on single crystal (SC) and CG CM186 blade alloys in the uncoated and Sermalloy 1515 coated conditions are reported. Strain controlled thermo-mechanical fatigue cycles simulating the thermal and mechanical loading situation at critical locations of the blades were applied, and the damage evolution on the samples was in-situ monitored throughout the TMF cycle.

Superalloys 2000
Edited by T.M. Pollock, R.D. Kissinger, R.R. Bowman,
K.A. Green, M. McLean, S. Olson, and J.J. Schirra
TMS (The Minerals, Metals & Materials Society), 2000

Experimental

Specimens were machined from single crystal (SC) and columnar grained (CG) (directionally solidified) Ni-based superalloy CM186 bars with the orientation of the long axis of the samples in the [001] crystallographic direction. The nominal chemical composition in wt % of CM 186 is given in Table 1.

Table 1. Chemical composition of CM 186 (in wt. %)

Co	W	Al	Cr	Ta	Re	Hf	Ti	Ni
9.25	8.5	5.7	6	3.4	2.95	1.4	0.7	bal

Threaded end TMF specimens with a solid rectangular cross-section of 12 x 3 mm^2 and a parallel length of 9mm were machined. The rectangular cross section was adopted to enable observation of the flat surface by means of a computer vision system during testing, as discussed later. The parallel section was ground and polished in the longitudinal direction to a final surface finish of $R_z = 0.05$ μm. The edges of the gauge length section were rounded to promote good adherence of the coating.

Some of the SC and CG CM186 samples were coated with a modified aluminide coating, Sermalloy 1515, via an Al/Si slurry diffusion process, with an approximate thickness of 80μm.

The strain-controlled TMF tests were carried out in air on a computer controlled electro-mechanical closed loop testing machine of 100 kN capacity. Tests with a minimum to maximum strain ratio, $R_\varepsilon = -\infty$. at three different strain range levels of 1%, 0.8% and 0.7% and a constant strain rate during the cycle were performed. The temperature varied symmetrically and linearly with time between 350 °C and 950 °C with a cycle period of 200 s. A schematic picture of the applied cycle is shown in Fig. 1.

The mechanical strain was varied synchronously and linearly with time, but with a phase shift of 180° with respect to the temperature. This cycle imposes maximum strain/stress at minimum temperature and minimum strain/stress at maximum temperature. The mechanical strain rate was $|d\varepsilon_m/dt| = 10^{-2} \Delta\varepsilon$, [%s^{-1}] constant at any instant in time within the cycle. Strain was measured and controlled by means of a water cooled, high temperature axial extensometer with ceramic rods attached to the parallel length of the specimen. The original gauge length at 350 °C was 8 mm. The loads were measured by means of a strain gauged load cell located in the load train of the testing machine. Specimens were heated by means of electromagnetic induction, using a commercially available 6 kW high frequency induction heater and an induction coil with optimised geometry, so that the temperature gradient over the gauge length was within ±15°C at any given time during the cycle. The sample temperature was measured and controlled by means of thermocouples spot-welded on the specimen surface. An IR-pyrometer was also used for additional monitoring of the temperature during the test.

Control of the testing machine, the high frequency induction heating system and data acquisition was performed by means of a dedicated computer system using LabVIEW 5.1. Both the cyclic stress-strain response and the thermo-mechanical fatigue life were determined. Tests were conducted until failure, or stopped when the cyclic stress range had dropped beyond 50% of the stabilised value. TMF life was then defined as the cycle number at which the stress range Δσ had decreased to 50 % of the stabilised or maximum value.

During TMF testing, images of the front face of the specimen were taken at pre-selected cycle numbers by means of a computer controlled video imaging system, allowing contact-less, in-situ and fully automated monitoring of the information related to the evolution of damage. To enhance the detection of initiating and growing cracks (with a surface length ≥ 15μm), the CCD camera with a flash unit was synchronised with the TMF-cycle and images were taken when the imposed stress on the specimen exceeded 80% of the maximum tensile stress of the previous cycle. The images taken covered the specimen width and 11.6 mm of the parallel length of the specimen. All the images were digitised and stored for post-processing and analysis.

Fractographic analysis of the tested samples was performed to obtain an understanding of the fracture in terms of initiation sites, crack propagation modes and morphologies and the relationship between fracture surface features and the microstructure of the coating and substrate alloy. Polished cross-sections of selected samples were metallographically prepared and observed under SEM and analysed by means of EDS.

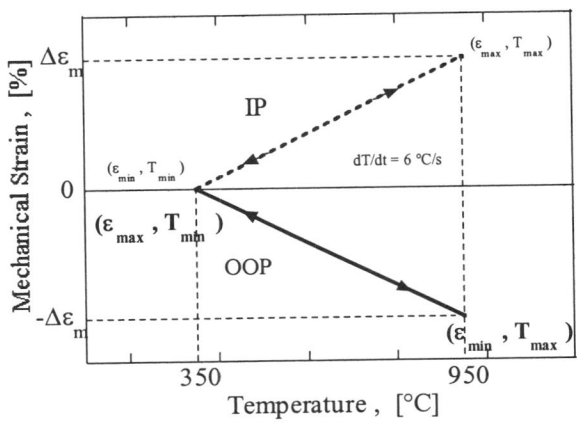

Figure 1: TMF cycle shapes in terms of mechanical strain ε_m versus temperature (IP: in phase, OOP:out-of-phase, applied in this study).

Isothermal strain controlled fatigue tests were also carried out on both solid and hollow specimens of CM186, in the CG and SC form. The solid specimens were cylindrical in section with a gauge diameter of 7mm. The hollow specimens were manufactured to the same external dimensions, but were cast with a hollow interior giving an as-cast nominal wall thickness in the gauge section of 1mm. Final machining included honing of the inner surface to provide a smooth finish. Testing was carried out on a servo-hydraulic machine at a strain rate of 0.1%s^{-1}. A constant temperature was maintained with a radiant heat furnace. Testing was carried out at several strain ratios and temperatures up to 950°C.

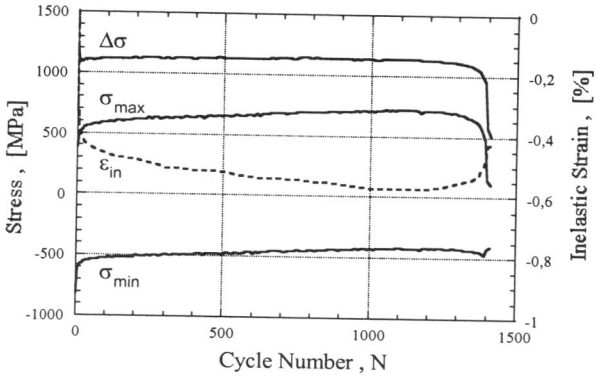

Figure 2: Evolution of the characteristic cycle stresses and accumulated inelastic strain with cycle number, N, measured on CG CM186 uncoated, at $\Delta\varepsilon_m = 1\%$.

Results & Discussion

Stress-strain-temperature response

The typical evolution of the maximum and minimum cycle stresses and the stress range with cycle number N during OOP TMF testing is shown in Fig. 2 for the uncoated CG material at a mechanical strain range of $\Delta\varepsilon_m = 1 \%$.

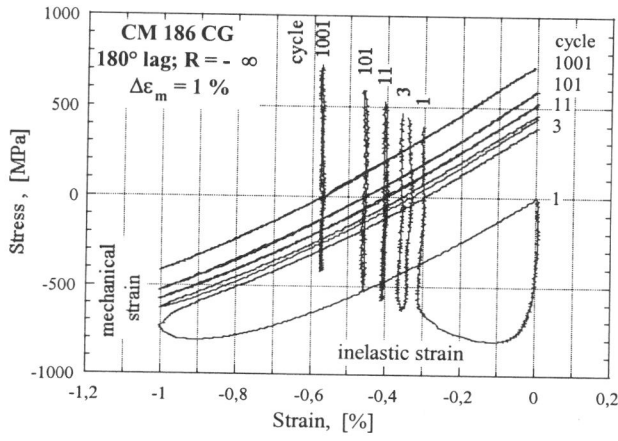

Figure 3: Stress-strain loops at three different stages during a test on CG CM186 uncoated, at $\Delta\varepsilon_m = 1\%$.

The material does not exhibit any strong cyclic evolution. This is supported by the results from strain controlled isothermal LCF tests at several temperatures. There is a rapid shift in stress during the early TMF cycles. The pace of this process decreases with continued cycling, with the stress cycle exhibiting a positive mean value. Subsequently, extensive cracking influences the compliance. This process is entirely consistent with the dominance of plasticity and more importantly creep at the compressive/high temperature end of the cycle. This is further emphasised by the evolution of inelastic strain (Fig. 2). Note that the shift of ε_{in} with cycling is depicted from the stress-strain loops shown in Fig. 3. After cycle 1, in which a significant amount of hysteresis is present, the build-up of ε_{in} with cycling slows down. This is also commensurate with the isothermal data.

From about $N / N_f \approx 0.1$ onwards, creep in the high temperature part of the cycle with stress in compression is the dominant deformation process responsible for a continuation of steady accumulation of inelastic strain. This results in a gradual shift of the compressive stress to smaller absolute values and a similar shift of the tensile stress to higher values. The shift continues until the amount of creep deformation in the compressive part of the cycle is compensated by an equivalent inelastic deformation in the tensile part of the cycle at lower temperature. However a decrease of tensile stress might already set in before this state of equilibrium is reached if the compliance of the sample increases due to crack formation and growth.

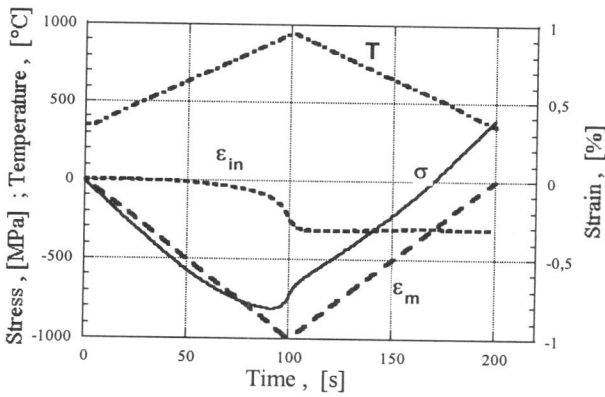

Figure 4: Stress (σ), temperature (T), mechanical strain (ε_m) and inelastic strain (ε_{in}) as a function of time for the first cycle; CG CM186 uncoated tested at $\Delta\varepsilon_m = 1\%$.

Most of the inelastic strain is primarily created during the heating phase of the first cycle in compression. Fig. 4 shows the stress, temperature, mechanical and inelastic strain as a function of time in the first TMF cycle, when the CG material in the uncoated condition is tested at a strain range of 1 %. In the first part of the cycle a purely elastic increase of the compressive stress can be observed. With increasing mechanical strain and temperature, plastic flow of the material sets in. Once the yield strength of the material is reached at about 800 °C, plastic deformation occurs leading to a decrease in compressive stress with further increase of mechanical strain and temperature. The peak of the compressive stress does not correspond with the minimum mechanical strain and maximum temperature. Stress in compression was minimum at about 820 °C while the mechanical strain reached its minimum at the maximum temperature of the cycle of 950 °C. In the second half of the cycle, with decreasing mechanical strain and temperature, the sample deforms almost exclusively elastically while the inelastic strain remains constant till the end of the cycle. The inelastic strain imposed on the specimen during the compression period should be balanced by a corresponding and equivalent deformation in tension. In this regime of low temperature, the yield strength of the material is not reached and consequently the tensile stress increases almost exclusively elastically. By the third and fifth cycles a different picture has emerged.

measured on specimens in the uncoated and coated condition at different mechanical strain ranges. The results show that the general deformation behaviour of the material appears to be essentially independent of whether the material is of SC or CG structure. The similar cyclic behaviour of the uncoated and coated CM186 suggests that the stress-strain response of the bulk is not influenced by the presence of the coating. Figures 5 and 6 show that, in compression all the stress data for all strain ranges coincide. At all strain ranges the yield strength of the material is reached in the high temperature compressive part of the first cycles. Thereafter, the compressive stress level is the same for all strain levels, as there is no plastic deformation in the low temperature tensile part of the cycle. Further inelastic deformation is due to creep and approximately the same for all tests. Fig. 7 summarises the evolution of the stress range with cycle number for all tests performed. For both material structures in the bare condition and the mechanical strain ranges considered, the stress range assumes an almost constant, may be slightly increasing value up to the onset of final fracture. The result for the coated SC, at $\Delta\varepsilon_m = 1$ %, is somewhat different, showing a progressive reduction in stress range from 20% of TMF life.

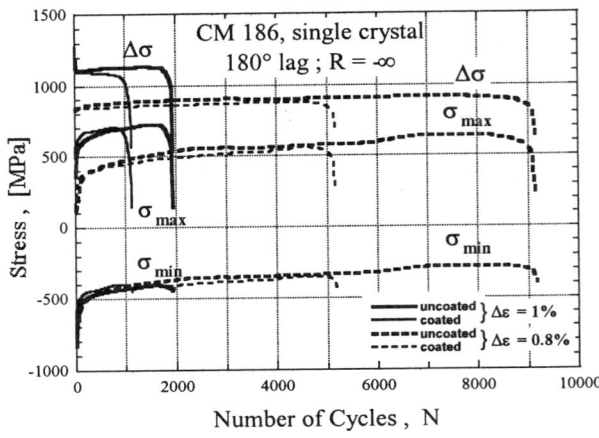

Figure 5: Evolution of stress range ($\Delta\sigma$), maximum stress and minimum stress with TMF cycles; uncoated and Sermalloy 1515 coated SC CM186.

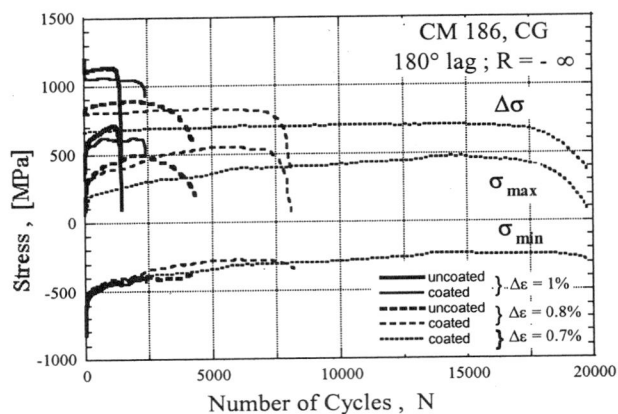

Figure 6: Evolution of stress range ($\Delta\sigma$), maximum stress and minimum stress with TMF cycles; uncoated and Sermalloy 1515 coated CG CM186.

The minimum of the compressive stress almost coincides with the minimum of mechanical strain. No pronounced yielding is observed and the total plastic deformation during the entire cycle is less than 0.06 % and 0.02 % in the third and fifth cycle, respectively. This behaviour indicates that the hysteresis loops are almost stabilised after a few cycles of the test. The pronounced inelastic deformation early in life results in a shakedown of the cyclic stresses to values close to the elastic regime, resulting in very narrow and almost closed stress-strain loops during the major part of life.

Figures 5 and 6 display the evolution of the maximum and minimum cycle stresses and the stress range with cycle number, N, for the SC and CG material respectively,

Microstructural Analysis

As-received state The main microstructural differences between the SC and CG materials are the presence of high angle boundaries and coarse carbides, having a Chinese script morphology, being present in the grain boundaries of the latter.

Figure 8 shows the microstructure of a coating in a state assumed "as-received" (cross-section well outside the gauge length of the shortest lived sample). The coating is approximately made up of 3 layers, each about 25μm thick, separated by 2 interlayers (ca. 5μm). The dark coloured matrix of the layers is made up from Al and Ni, with additional elements that have diffused from the substrate, while the bright precipitates and interlayers are rich in Si. At the coating/substrate interface a substrate-affected zone can be seen, in which the γ matrix is richer in Ta, W and Re than the substrate.

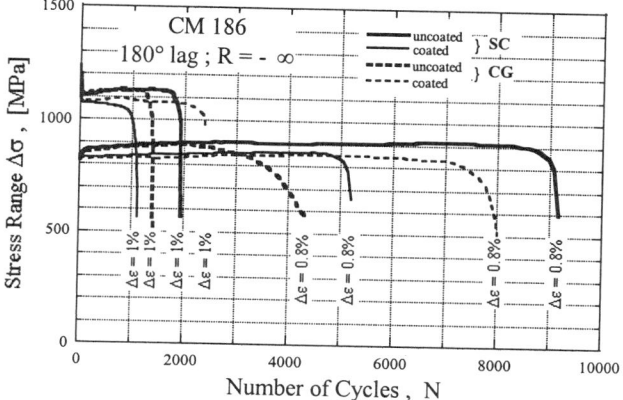

Figure 7: Evolution of the stress range with the number of TMF cycles; uncoated and Sermalloy 1515 coated SC and CG CM186.

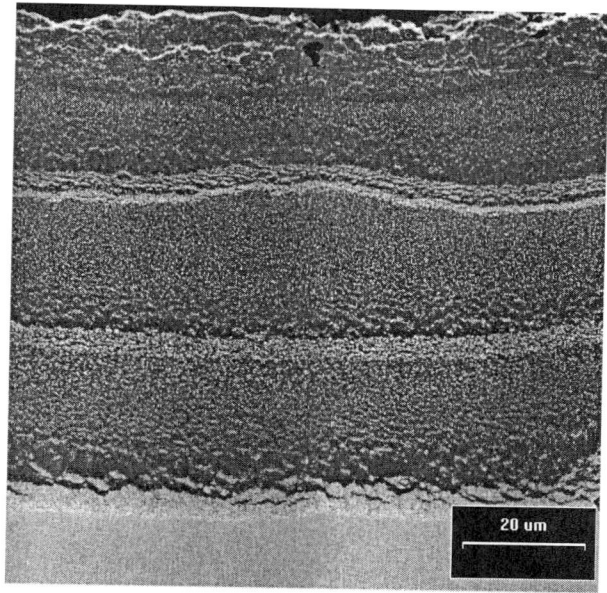

Figure 8: SEM image, (back scattered mode) of the Sermalloy 1515 coating in a nearly as-received state.

TMF Lives

In Fig. 9, the TMF lives of the SC and CG CM186 in the uncoated and Sermalloy 1515 coated condition respectively are plotted as a function of the mechanical strain range $\Delta\varepsilon_m$. As mentioned earlier, life to failure is defined as the cycle number at which the stress range $\Delta\sigma$ has decreased to 50 % of the stabilised or maximum value. In the uncoated form, the SC material appears to have a longer life when compared with the CG material. Fig. 9 also reveals the influence of the presence of the coating on the TMF lives. The presence of the coating on the SC material results in a reduction in life by a factor of about two when compared with the results from the SC material tested in the uncoated condition. The CM186 material with a CG structure shows the opposite behaviour. Sermalloy 1515 coating of the material leads to an increase in life of a factor of about two. However, the total fatigue life of the specimens can be considered to consist of two parts: A crack initiation phase and crack propagation to failure. An important aspect of this work has been the use of an in-situ video imaging system to monitor the first appearance and early growth of fatigue cracks at the specimen surface. Initiation is defined here as the cycle number N_i where a crack with a length (surface projection) of approximately 30 μm is first detected. This length is arbitrary and dictated by the resolution of the imaging system.

The computer vision records reveal that crack initiation takes place in relative early stages of the life below $N/N_f = 0.25$ and 0.1 when the bare material was tested at a mechanical strain range of $\Delta\varepsilon_m = 1$ % and 0.8 %, respectively. These results are also shown in Figure 9 as crack initiation lives, N_i. In terms of crack initiation, the CG material exhibits shorter lives than the SC in the uncoated form, whereas the coated CG material has a longer N_i life than the coated SC.. Furthermore, the presence of the coating leads to shorter TMF lives for the SC and longer for the CG material.

The N_i, N_f, and N_i/N_f TMF data are given in Table 2.

Table 2. Summary of TMF data

	N_i	N_f	N_i/N_f	N_i	N_f	N_i/N_f	N_i	N_f	N_i/N_f
SC	600	1941	0.308	1000	9165	0.109			
SC coated	200	1121	0.178	300	5196	0.058			
CG	400	1412	0.283	600	4500	0.133			
CG coated	300	2404	0.124	400	7985	0.05		19656	
$\Delta\varepsilon_m$ (%)	1			0.8			0.7		

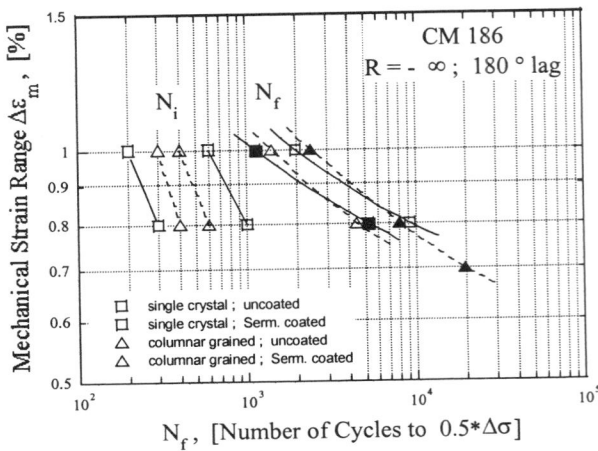

Figure 9: Cycles to failure (N_f) and crack initiation lives (N_I) of uncoated and Sermalloy 1515 coated SC and CG CM186 as a function of mechanical strain range.

Crack Initiation and Early Growth

The results show that, for all material and test conditions the smaller part of TMF life is spent in initiating microcracks while the larger fraction of life is spent in crack growth. Two different types of cracks are initiated and propagate during cycling as discussed also elsewhere. (4) One of these types is associated primarily with the uncoated specimens, and the other with the coated specimens.

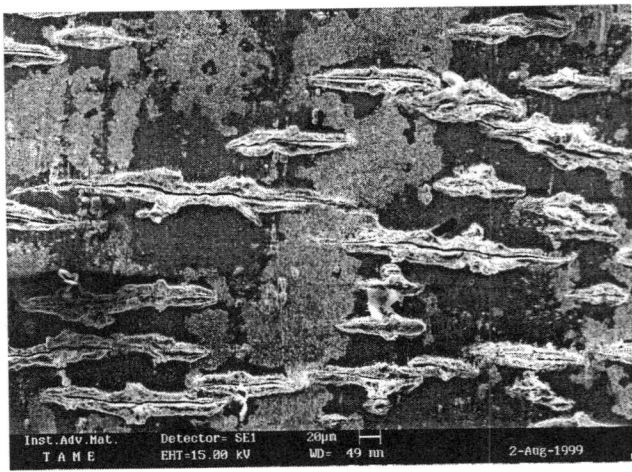

Figure. 10: Point surface initiation cracking of uncoated SC material TMF tested at a strain range of $\Delta\varepsilon_m = 1$ %.

Uncoated Specimens In the case of the uncoated material, multiple cracks initiate at coarse microstructural features; also found were a number of examples of cracks apparently initiated at casting pores. In all cases, initiation sites are associated with cracked oxide hillocks and thus, with preferential oxidation of microstructural features. At these preferred sites, oxide hillocks protruding from the surface are formed which finally crack as shown in Fig. 10.

Small cracks are initiated with fracture surfaces approximately perpendicular to the stress axis, this is apparently invariant with $\Delta\varepsilon_m$. The picture is similar for the SC and CG material structure, no obvious difference in initiation sites, initiation mode and density of initiated cracks could be observed. All cracks grow individually by inward extension along a semi-circular or semi-elliptical front and a crack path perpendicular to the stress axis in the radial direction. In a later stage of TMF life, coalescence of adjacent cracks starts to contribute to the extension of the cracks. Fig. 11 displays the fracture surface with some typical point-initiated, semi-elliptical surface cracks, when a CG sample was TMF tested at a mechanical strain range of $\Delta\varepsilon_m = 1$ %.

Fractographic examinations of isothermally fatigued uncoated LCF specimens tested at 950°C have revealed similar results. Multiple initiation takes place at surface features rich in nickel and oxygen. This contrasts with the observations of lower temperature isothermal LCF tests where fatigue crack initiation is limited to either a few or single sites located at the surface or at sub-surface positions. These are generally found to be either pores or surface breaking brittle phases rich in hafnium, tantalum and tungsten. It can therefore be concluded that the crack initiation sites found in the TMF specimens are characteristic of those found in high temperature LCF tests for this material.

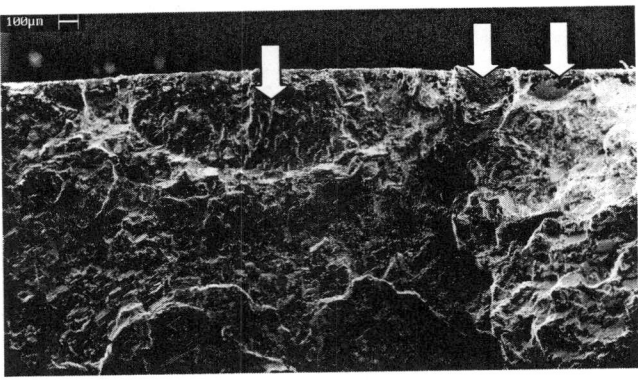

Figure 11: Fracture surface of uncoated CG material tested at a strain range of $\Delta\varepsilon_m = 1$ % with typical semi-elliptical surface cracks

Branching of cracks was found in the case of the CG material as shown in Figure 12, in which an example of a crack interacting with a precipitate is shown. On the other hand in the SC material, cracks grow straight and normal to the loading axis.

(a)

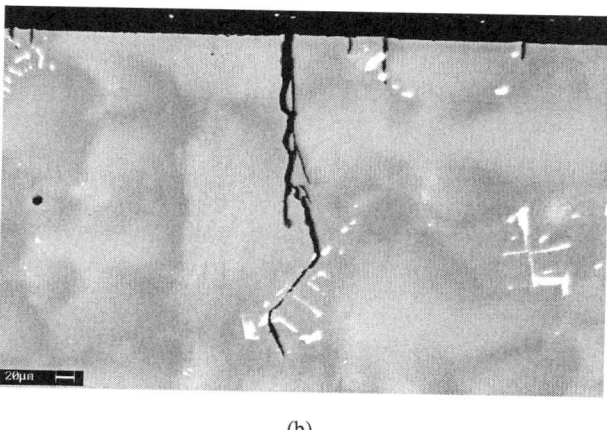

(b)

Figure 12: Observed crack paths in uncoated SC (a) and CG (b) material showing interaction with precipitates.

The fraction in life spent for initiation and propagation in the uncoated material is comparable for the SC and the CG structure but cracks appear to be initiated at a smaller number of cycles in case of the CG material when compared with the SC structure, see Fig. 9. The crack initiation earlier in life probably accounts for the lower TMF life of the CG material. There is no evidence that differences in the microstructure account for the difference in TMF lives. Further analysis of the computer vision system data with respect to evolution of crack length and crack density as a function of cycle number N may reveal any differences that exist between the initiation lives and crack propagation rates for the CG structure and the SC material, respectively.

Coated Specimens   For the coated material, brittle fracture of the coating in a line mode is observed at all strain range levels tested. Long shallow cracks are initiated over relatively large distances relative to the specimen width, Fig. 13. It is generally agreed that the applied cycle which imposes a compressive strain on the coating at high

temperature and tension at low temperature is the most damaging one.(5)   Due to the accumulation of inelastic deformation in the high temperature compressive part of the cycle an increasing tensile stress is built up in the low temperature part of the cycle.   Brittle cracking of the coating resulting in the formation of line cracks occurs once the limit of ductility of the coating below the ductile to brittle transition temperature, DBTT, is exceeded. (6-8)

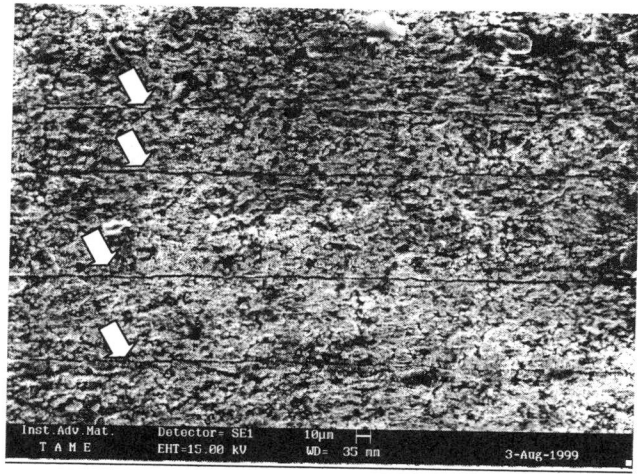

Figure 13:   Surface view of Sermalloy 1515 coated CG material TMF tested at a strain range of $\Delta\varepsilon_m = 0.8$ % showing a system of parallel line initiated straight cracks spanning the entire surface.

Cracks are always initiated in the coating or at the surface of the coating and propagate into the substrate by Mode I extension producing a crack front almost parallel to the specimen surface and a crack path perpendicular to the loading direction, see Fig. 13. (8,9)   Crack nucleation sites could not be discerned by means of fractographic investigation of the fracture surface.   Whether the crack starts at the outer coating surface or in the coating itself could not be identified.   The overall picture of crack growth mode and morphology as shown in Fig. 13 is similar when the SC and the CG materials are tested in the coated condition.

Observations of cross-sectioned coated samples show that most of the cracks stop at the coating/substrate interface, suggesting an abrupt change of mechanical properties at this site, Figure 14.   There appear to be more shallow cracks, i.e. not propagating as deeply in the substrate, in the CG material than in the SC, which could justify the enhanced TMF life of the CG coated material.   Other mechanisms that could reduce ΔK_{eff} and thereby reduce overall crack growths in the coated CG materials could not be discerned. The crack initiation lifetimes, Fig. 9, and the differences of maximum stress at half-life for the same applied mechanical strain are unlikely to produce significant differences in crack growth rates during the propagation phase.

SC CM186 coated   CG CM186 coated

Figure 14. Optical microscopy images of longitudinal cross sections of Sermalloy 1515 coated SC and CG CM186 alloys TMF tested at a strain range of $\Delta\varepsilon_m = 1$ %

For both material structures, crack initiation is observed to occur after a life fraction N_i/N_f of 0.15 and 0.05 for the strain range of $\Delta\varepsilon_m = 1$ % and 0.8 %, respectively. In terms of absolute cycle numbers crack initiation appears to take place earlier in life for the coated SC material relative to the CG material, see Fig. 9 and Table 2. As for the SC material the coating results in a reduction of the cycles to crack initiation N_i by a factor of three for both strain range levels investigated. The formation of line cracks earlier in life and the more detrimental effect of propagating line cracks relative to semi-circular or semi-elliptical cracks (in terms of ΔK) explains the life reducing effect of the coating.

In contrast, a life lengthening effect of the coating is observed in the case of CG material. The number of cycles to crack initiation is only slightly lower for the coated material relative to the uncoated. No explanation is found yet for the longer lives of the coated material. In a number of studies into the TMF behaviour of nickel-based superalloys the presence of the coating is reported as detrimental or beneficial to TMF life. (5, 7, 10-13) The lives to failure of the coated CG TMF specimens are shown, alongside the lives of isothermal LCF specimens tested under similar mechanical loads in Figure 15. This result, if not coincidental, is interesting in that it provides some support to the TMF results for the CG coated material in terms of absolute number of cycles to failure.

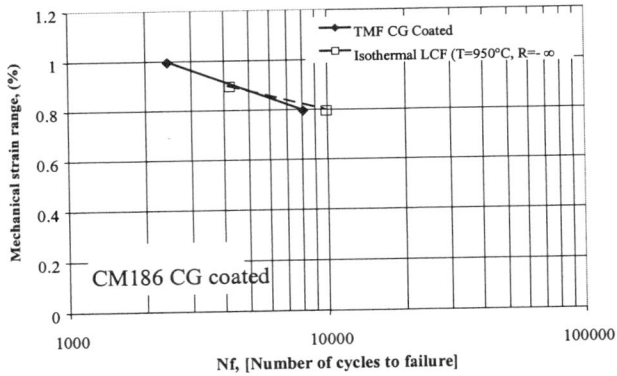

Figure 15: Comparison of TMF and isothermal LCF (T=950°C) results in terms of cycles to failure for coated CG CM186.

Conclusions

TMF tests on single crystal and columnar grained CM186 alloy in the uncoated and Sermalloy 1515 coated state were performed. The results suggest that the SC material has a longer TMF life than its CG counterpart. However, the effect of the presence of the coating is different for the two materials, namely enhancing the TMF life of the CG materials and degrading that of the SC. In-situ monitoring of the initiation and early growth of surface cracks have demonstrated that valuable information can be obtained on the basis of crack initiation and, thus, for lifing purposes. Initiation life represents a small fraction (0.1 to 0.25) of the total TMF life.

The type of crack initiation is similar in the SC and CG forms of CM186, even though the numbers of cycles to initiation are different. Brittle cracking of the coating manifests itself by continuous planar surface cracks, whereas in the uncoated material cracks develop from points in a semi-elliptical form.

ACKNOWLEDGEMENTS

Part of this work is carried out within the Commission research and development programme. PMH, SJM, PJ and MH would like to acknowledge support by the Department of Trade and Industry (UK) under the CARAD (civil aircraft research and demonstration) initative.

REFERENCES

1. K. Harris, G.L. Erickson, S.L. Sukkenga, W.D. Brentall, J.M. Aurrecoechea and K.G. Kubarych, "Development of the rhenium containing superalloys CMSX-4 & CM186LC for single crystal blade and directionally solidified vane applications in advanced turbine engines," TMS-AIME, 7th International Symposium on Superalloys, Seven Springs, Champion, PA, USA, September 20-24, 1992.

2. P.S. Burkholder, M.C. Thomas, R. Helmink, D. J. Frasier, K. Harris and J. B. Wahl, " CM 186 LC® Alloy Single crystal turbine blades," ASME, 99-GT-379

3. J.R. Nicholls, "Designing Oxidation-Resistant Coatings" J. of Metals, 52(2000), 28-35

4. J. Bressers, J. Timm, S. Willimams, A. Bennett, E. Affeldt, "Effects of Cycle Type and Coating on the TMF Lives of a Single Crystal Nickel-based Gas Turbine Blade Alloy", Thermo-mechanical Fatigue Behaviour of Materials, 2nd Volume, eds. M.J. Verrilli and M.G. Castelli (ASTM STP 1263, American Society for Testing and Materials, Philadelphia, 1995).

5. G.A. Swanson, I. Linask, D.M. Nissley, P.P. Norris, T.G. Meyer and K.P. Walker, 2nd Annual Status Report, NASA-CR-179594, 1987.

6. J.M. Martinez-Esnaola, M. Arana. J. Bressers, J. Timm, A. Martin-Meisozo, A. Bennett, E. Affeldt, "Crack Initiation in an Aluminide Coated Single Crystal during TMF", Thermo-mechanical Fatigue Behaviour of Materials, 2nd Volume, eds. M.J. Verrilli and M.G. Castelli (ASTM STP 1263, American Society for Testing and Materials, Philadelphia, 1995).

7. J. Bressers, J.M. Martinez-Esnaola, A.M. Martin-Meisozo, J. Timm, M. Arana, "Effects of Cycle Type and Coating on the TMF Lives of a Single Crystal Nickel-based Gas Turbine Blade Alloy", Thermo-mechanical Fatigue Behaviour of Materials, 2nd Volume, eds. M.J. Verrilli and M.G. Castelli (ASTM STP 1263, American Society for Testing and Materials, Philadelphia, 1995).

8. Zhang Detang, "Effect of ductile-brittle transition temperature of Al-Si coating on fatigue properties of Ni-base superalloys", Acta Metall. Sin. Series A 3, 435 – 438, (1990).

9. M.I. Wood, "Mechanical interaction between coating and superalloys under conditions of fatigue", Surface and Coatings Technology, 39/40, 1989, pp. 29-42.

10. K.R. Bain, "The Effect of Coatings on the Thermo-mechanical Life of a Single Crystal Turbine Blade Material", (Paper presented at the AIAA/SAE/ASME/ASEE 21st Joint Propulsion Conference, Monterey, Ca, July 8-10, 1985)

11. H. Bernard, L. Remy, "Thermal-Mechanical Fatigue Damage of an Aluminide Coated Nickel Base Superalloy", Advanced Materials an Processes, eds. H. Exner and V. Schumacher, (EUROMAT, 1989), 529-534.

12. J.C. Lautridou, J.-Y. Guedou and J. Delautre, "Comparison of Single Crystal Superalloys for Turbine Blades through TMF Tests", Fatigue under Thermal and Mechanical Loading : Mechanisms, Mechanics and Modelling, eds. J. Bressers, L. Rémy, M. Steen and J.L. Vallés (Kluwer Academic Publishers, 1996), 141-150.

13. J.-Y Guedou and Y Honnorat, "Thermo-mechanical fatigue of turboengine blade superalloys", Thermomechanical Fatigue Behaviour of Materials, ASTM STP 1186, ed. H Sehitoglu, San Diego, Oct. 1991, pp. 157-175

PROCESS MODELLING OF THE ELECTRON BEAM WELDING OF AEROENGINE COMPONENTS

R. C. Reed, H.J. Stone, D Dye and S.M. Roberts

Department of Materials Science and Metallurgy/Rolls-Royce University Technology Centre
University of Cambridge, Cambridge, CB2 3QZ, UK

S. G. McKenzie

Rolls-Royce plc.,
PO Box 31, Elton Road, Derby, DE24 8BJ, UK

Abstract

A numerical process model is described for the prediction of residual stresses and distortion arising from the electron beam welding of Waspaloy. Simulations have been conducted with this model for welds performed at a number of focal positions of the electron beam. Comparison is made between the model predictions and experimental measurements of the thermal cycles, the residual stresses and distortion. Good agreement is observed between the model and experiment for the thermal cycles and residual stresses. Distortion can also be predicted but some discrepancies have been noted.

Introduction

Electron beam welding (EBW) [1] is used widely by gas turbine manufacturers for the fabrication of compressor assemblies. A photograph of a typical component and a close-up of the weld is given in Figure 1. Successful processing relies on the careful generation and manipulation of a high power electron beam, which provides power densities at the workpiece surface in excess of $10^{10}\,\mathrm{W\,m^{-2}}$ [2]. At these power densities, a near cylindrical cavity of vapour or 'key-hole' is produced which may extend to considerable depths into the work-piece [3]. This characteristic enables autogenous welding to be accomplished on thick sectioned components in a single welding pass. Furthermore, as energy may be deposited almost uniformly through the depth of the key-hole, the post-welding distortions produced are typically smaller than those produced by more conventional fusion welding processes [4].

Nevertheless, considerable care is required when using electron beam welding for the manufacture of compressor assemblies, particularly in nickel-base alloys, for a number of reasons. First, the distortions that are generated by welding, whilst small, can be close to the limits that can be tolerated. Excessive distortion necessitates costly reworking or scrapping of the components. Second, as the structural integrity of these components must be ensured, reliable estimates of the residual stresses generated by welding need to be obtained for calculations of projected component life. Such estimates are often made via residual stress measurement techniques which are destructive. Finally, selection and optimisation of process parameters such that post-weld distortions are minimised, has traditionally involved a considerable amount of experimentation on a trial-and-error basis. Such strategies are highly reliant on the expertise of welding engineers and can often be inefficient, particularly when high strength alloys such as superalloys are employed.

The field of computational weld mechanics now offers the possibility of using numerical models to perform weld investigations virtually, *i.e.* on the computer. These models employ finite element codes to obtain a piece-wise polynomial solution to the differential equations governing heat transfer and the mechanical response of the component geometry to the applied thermal and mechanical loads. However, it should be noted that considerable experimentation is required during the development of these models for the purpose of calibration and validation. In the present paper, a numerical process model for the electron beam welding of Waspaloy is presented. The results obtained are compared with data obtained with a number of experimental techniques. The advantages and limitations of the approach are identified.

Numerical Modelling

The numerical model was constructed with the aim of providing a predictive capability for the residual stresses and distortions that arise as a consequence of the electron beam welding of Waspaloy. In such a model, the desire to incorporate as much of the process physics as possible must be balanced against the implementation and validation costs. As a consequence, a pragmatic decision was made to circumvent the complex physical processes that occur in the weld pool and to approximate the deposition of heat into the work-piece by an arbitrary volumetric heat generation rate.

Superalloys 2000
Edited by T.M. Pollock, R.D.,Kissinger, R.R. Bowman,
K.A. Green, M. McLean, S. Olson, and J.J. Schirra
TMS (The Minerals, Metals & Materials Society), 2000

Figure 1: Photograph of an aeroengine compressor assembly with a close-up of the electron beam weld.

The computations have been performed as sequential thermal and mechanical analyses so that the transient thermal cycles generated by the moving heat source are used as the input for an elastic-plastic continuum mechanics-based solution for the strains, displacements and stresses. The division of the computation into separate thermal and mechanical analyses is deemed to be acceptable as the heat generated by the adiabatic heating of the material from the temperature induced deformations is insignificant compared to the heat deposited by the heat source [5].

Mesh Generation

One of the principal difficulties encountered with the successful implementation of a weld model is that of scale [6,7]. The element size at the location of the heat source has to be sufficient to ensure that the amount of energy deposited into the plate in reality is matched by the model. Furthermore, the form of the heat source is important and the mesh should be sufficiently fine to represent this. The implication of such a criterion is that the size of elements required around the heat source is likely to be small compared to the length of the weld path. If the entire geometry were to be discretised with elements of this size, the resultant mesh would possess a very considerable number of elements. Analysis of such a model would be likely to prove intractable. To avoid this difficulty a form of adaptive remeshing has been devised in which the necessary refinement around the heat source is provided along a short section of the total length only with larger elements beyond [8]. This methodology relies on the division of the total analysis into a series of sub-analyses, each with their own associated mesh. In each sub-analysis, the heat source is traversed along the short distance with high local element refinement, thereby allowing the energy deposited by the heat source to be accurately captured. For the thermal analysis, the results of the last time step on a given mesh are then interpolated onto the mesh of the subsequent sub-analysis as the starting conditions for the new

sub-analysis and the heat source allowed to traverse along the next refined region. For the mechanical analysis, in addition to passing the results obtained at the end of the previous sub-analysis, the thermal results for that sub-analysis are also required. Three consecutive meshes are shown in Figure 2.

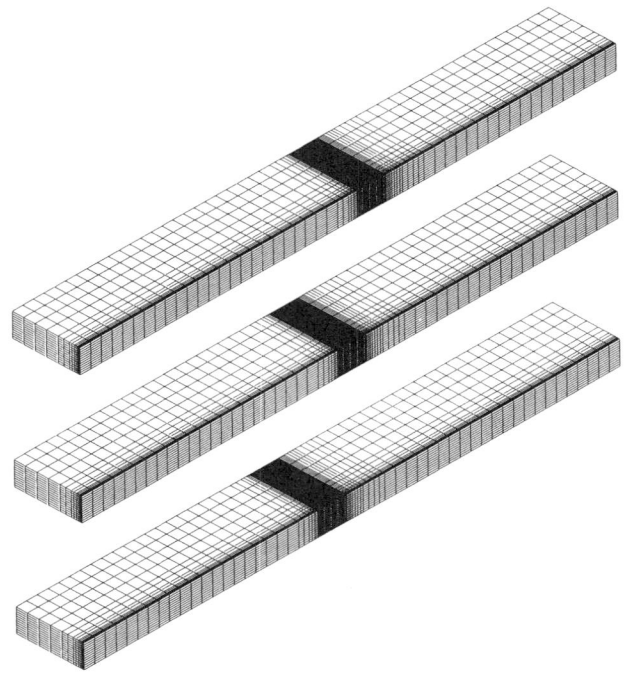

Figure 2: Three consecutive meshes used for the numerical analyses of welded test-plates.

To avoid incompatibility between the thermal and mechanical strain fields [9], linear elements have been used in the thermal analyses and quadratic elements with reduced integration in the mechanical analyses. The use of such elements in the mechanical analyses has the added advantage that it avoids the problems of volumetric and shear locking encountered with other element types [10].

Thermal Analysis

The deposition of heat into the work-piece geometry has been modelled by a linear combination of two distributed heat generation rates; that of a uniform circular surface flux and a uniform conical volumetric flux. Given a rectilinear plate such as that used in this investigation, with dimensions X, Y and Z in the Cartesian co-ordinate system (x, y, z) and with the sources moving at a constant velocity, v, in the yz-plane in the X-direction, the two source strengths, $\dot{Q}_{\text{surface}}$ and $\dot{Q}_{\text{volume}}$ are given by

$$\dot{Q}_{\text{surface}}\{\xi, y, 0\} = q_{\text{surface}} \tag{1}$$
$$\text{for } \xi^2 + y^2 \leq a_{\text{surface}}^2$$

and

$$\dot{Q}_{\text{volume}}\{\xi, y, z\} = q_{\text{volume}} \tag{2}$$

for $\xi^2 + y^2 \leq (mz + a_{\text{volume}})^2$ and $z \leq d$

where $\xi = x - vt$, in which t is the welding time, q_{surface} and q_{volume} are the surface and volume fluxes corresponding to the two sources, a_{surface} and a_{volume} are the characteristic radii of the two terms, m is the gradient of the conic decrease in width of the volume term and d is the penetration depth of the electron beam, which equals the thickness of the plate in the fully penetrating condition.

From the conservation of energy, the total power delivered by the two heat source terms must equal that provided by the electron beam. If a partitioning ratio, r, is defined as the proportion of the total power delivered to the surface term then

$$q_{\text{surface}} = \frac{r\eta VI}{\pi a_{\text{surface}}^2} \tag{3}$$

and

$$q_{\text{volume}} = \frac{(1-r)\eta VI}{\pi d \left(\frac{1}{3}m^2 d^2 + mda_{\text{volume}} + a_{\text{volume}}^2\right)} \tag{4}$$

where VI is the power supplied to the electron beam and η is the experimentally determined coupling efficiency.

The parameters a_{surface}, a_{volume}, m and r cannot be measured directly and instead must be optimised by comparison with the weld fusion boundary profiles and thermocouple data. To achieve this, a series of single step analyses were performed in the Eulerian reference frame. The solidus isotherm from the model was then compared with the fusion boundary profile and the parameters refined until good agreement between the two was achieved.

Once all of these modelling parameters were acquired, the full thermal transient analysis could be performed. This was conducted using the Broyden-Fletcher-Goldfarb-Shanno inverse update (BFGS) iteration method [11]. The calculation was continued until the model had achieved a uniform saturation temperature; after which, forced cooling of the model to ambient temperature (25°C) was performed.

Mechanical Analysis

On completion of the thermal analysis, the nodal temperatures were transferred to the mechanical model. The model was performed as an elasto-plastic analysis with linear isotropic hardening. As with the thermal analysis, the BFGS iteration method was used.

The mechanical constraints imposed on the model consisted of: (i) a symmetry condition across the plane of nodes along the weld path; (ii) a pin joint at one end of the plate; (iii) a simple support at the other. These constraint conditions are consistent with the 'unconstrained' bead on plate welds performed, allowing free deformation of the plates during welding.

Weld Manufacture and Characterisation

Plate test pieces of the dimensions $200 \times 50 \times 9$ mm were machined from Waspaloy. After machining, the plates were subjected to a vacuum heat treatment consisting of a 5 hour dwell at 760°C, followed by furnace cooling to ambient temperature. The purpose of this heat treatment was to alleviate the residual stresses generated by the prior manufacturing operations.

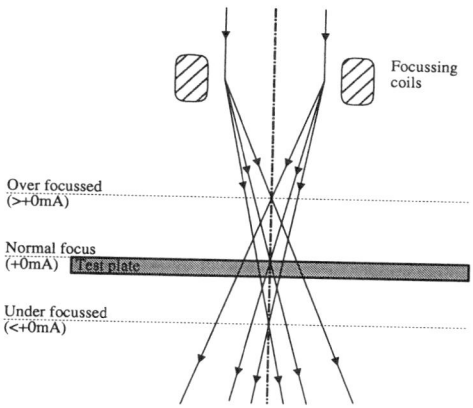

Figure 3: Schematic illustration of overfocussed, normal focussed and underfocussed electron beams

Autogenous bead-on-plate welding passes were conducted along the centre of each test-piece in the long direction of the plate using a Steigerwald K-100 electron beam welding machine with computer numeric control. The welds were conducted at ten different focal positions of the electron beam between the under-focussed and over-focussed conditions (Figure 3), with the remaining welding parameters being kept constant. The welding parameters used are given in Table I. In the following sections the samples have been referenced by the difference between the current in the focussing coils at which they were welded and the current required to achieve a normal focus.

Tabel I Welding parameters used for the fabrication of the test-pieces investigated in this study

Voltage (kV)	Beam current (mA)	Velocity mm s^{-1}	Defocus (mA)
150	23	8.4	+90
150	23	8.4	+70
150	23	8.4	+50
150	23	8.4	+40
150	23	8.4	+30
150	23	8.4	+20
150	23	8.4	+10
150	23	8.4	+0
150	23	8.4	-20
150	23	8.4	-40

Measurement of Thermal Transients

In-situ monitoring of the thermal transients generated during the welding process was performed using surface mounted thermocouples attached to commercial data-logging equipment via a shielded extension cable. Additional noise reduction was achieved with the use of filtering in the data-logging software.

Metallographic Sectioning

Metallographic sections were prepared of each of the welds in the plane perpendicular to the direction of welding. Microstructural relief was achieved using Kalling's reagent (5 g $CuCl_2$, 100 ml HCl, 100 ml ethanol).

Characterisation of Weld-Induced Distortion

The post-weld distortion of the welded plates were determined from measurements made of the surface topography of the welded samples. These measurements were performed using a Leitz PMM12106 three-dimensional needle probe co-ordinate measurement device. By this method it was possible to identify and characterise the two principal distortion modes of the test-pieces: (a) angular distortion or 'butterflying', which is manifest as a bending of the test-pieces along the weld path; and, (b) camber distortion, occurring as a curvature of the test-pieces along the length of the weld. These distortion modes are shown schematically in Figure 4.

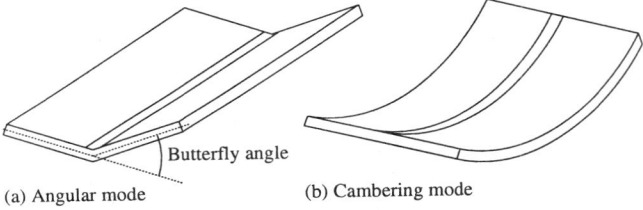

(a) Angular mode   (b) Cambering mode

Figure 4: Schematic diagram of two principal distortion modes (a) angular distortion and, (b) camber distortion.

Residual Stress Determination

Measurements of the residual stress state in the plate welded at +20 mA defocus were conducted using neutron diffraction, X-ray diffraction and incremental hole-drilling methods. Comparisons made of the applicability of each of these three techniques to the measurement of residual stresses around electron beam welds in Waspaloy suggest that these techniques should be considered as complementary [12].

Neutron Diffraction Neutron diffraction measurements were conducted at the ISIS facility, Rutherford Appleton Laboratory, UK. Measurements were performed in a step-wise fashion across the centre of the test-piece perpendicular to the welding direction as illustrated in Figure 5. For the full determination of the strain tensor, six independent measurements of strain are required [12]. However, to minimise the time required at each measurement position, it was assumed that the principal strain directions lay along the x, y and z directions of the plate (see Figure 5). This necessitated the

measurement of strain in these directions only. The experimental set-ups for the simultaneous measurements of the strains in the y & z and x & z directions (ε_y & ε_z and ε_x & ε_z) are shown in Figure 6. Spatial resolution was provided

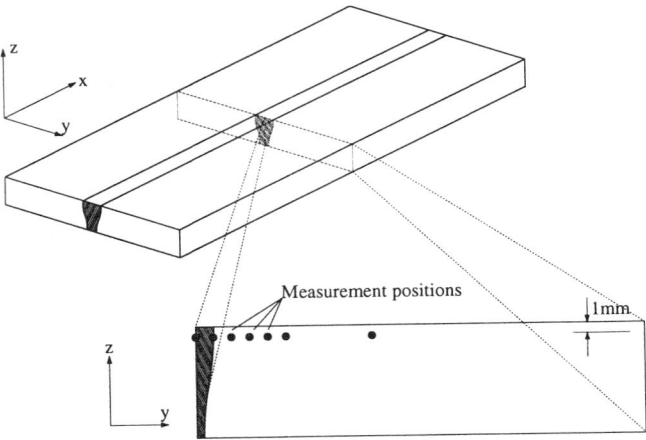

Neutron diffraction measurement positions at 0.8, 1.6, 2.4, 3.2, 4.0 and 8.0mm from the weld centre line

Figure 5: Schematic diagram of the test plate welded at +20 mA defocus illustrating the measurement locations used for the neutron diffraction measurements.

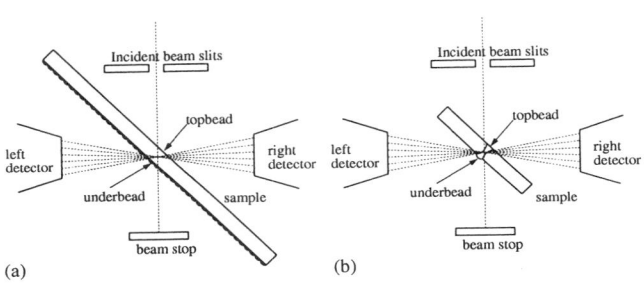

Figure 6: Experimental set-up for measurement of (a) ε_x & ε_z and (b) ε_y & ε_z.

by the use of slits 1×10 mm in the incident beam path for the measurements performed in the transverse and through-thickness directions and slits 2×2 mm for the measurements performed in the longitudinal and through thickness directions. These configurations gave illuminated gauge volumes of $1 \times 1.7 \times 10$ mm and $2 \times 1.7 \times 2$ mm respectively. The use of slits with different sizes for each of these measurement orientations was necessitated by the need to maximise the gauge volume in the sample and hence minimise the data collection time whilst ensuring that the spatial resolution remained sufficient to identify sharp variations in the stress field. The measurements were performed at a depth of 1 mm beneath the sample surface in order to ensure that the gauge volume was fully immersed in the sample in order to avoid edge effects. As the ISIS facility is a time-of-flight source, spectra containing all lattice reflections were obtained. The resultant spectra are composed principally of the overlaid diffraction patterns generated by the γ matrix and the γ' precipitates. As a result of the small lattice mismatch between these phases and the large intrinsic line widths it was not possible to resolve the reflections into their various phase constituents. However, it is believed that the treatment of the composite diffraction pattern as a single pattern does not generate erroneous results [13]. Consequently, the average strain in the sample was obtained by performing a Reitveld refinement on each full spectrum assuming only a single phase material. Conversion of the measured strains to stress could then be achieved using the bulk elastic modulus and Poisson's ratio. It is believed that this method provides an accurate measure of the macroscopic residual stress state, whilst overcoming the problems created by the presence of microstresses in the material [14].

<u>X-ray Diffraction</u> Measurements were also conducted by X-ray diffraction using a TEC 1610-2b dedicated residual stress measurement system. This system utilises a position sensitive proportional counter to capture the Bragg reflections from which the measurements are to be made. Calculation of the lattice strains is achieved using the $\sin^2\psi$ technique and the appropriate plane-specific diffraction elastic constants [15]. For the measurements performed in this study Cr-Kα radiation was employed, thus allowing access to the γ/γ' {220} lattice reflection at a fixed 2θ angle of approximately $128°$. At each measurement position 6 ψ-tilts were used. Spatial resolution was provided by a 2 mm diameter circular collimator. The measurement positions used are illustrated schematically in Figure 7.

<u>Hole-Drilling</u> Incremental hole drilling measurements were carried out according to ASTM standard E-837-94 [16] by Stresscraft plc, Shepshed, Leicestershire, UK. In order that the region of relaxed stresses around a hole does not interfere with the next measurement made, a separation of greater than 15 mm was used between adjacent measurements. As a consequence, it was only possible to perform four individual measurements on the welded plate within the region in which the stresses are believed to be invariant [12].

Results and Discussion

<u>Thermal Analysis</u> The metallographic sections obtained from the welded samples are given in Figure 8. Acquisition of appropriate heat source parameters, a_{surface}, a_{volume}, m and r to model these welds was achieved by the manual variation of the parameters and comparison with the fusion boundary profiles, presented in Figure 9. The heat source parameters obtained in this fashion are given in Table II. As these heat source parameters were selected to provide the optimum agreement with the experimentally determined fusion boundary profiles, it is not surprising that good agreement was achieved between the model and the experiments.

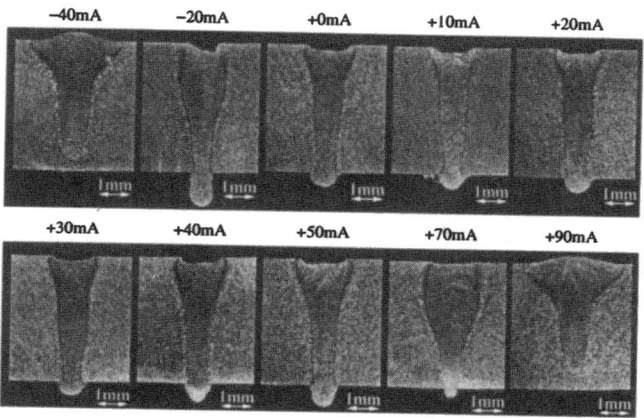

Figure 8: Metallographic sections from the bead-on-plate welds manufactured at different defocusses of the electron beam.

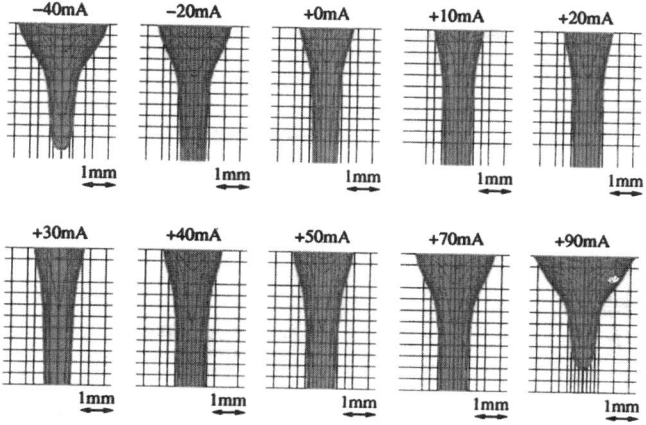

Figure 9: Predicted fusion boundaries obtained from the numerical model.

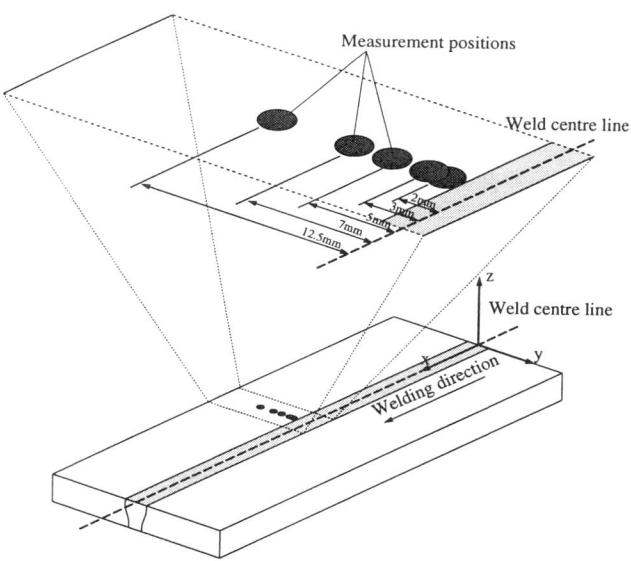

X-ray diffraction measurement positions at 2.0, 3.0, 5.0, 7.0 and 12.5 mm from the weld centre line

Figure 7: Schematic diagram of the test plate welded at +20 mA defocus illustrating the measurement locations used for the X-ray diffraction measurements.

Table II Parameters used in the thermal model

Defocus (mA)	η	a_{surface} (mm)	a_{volume} (mm)	m	r	d (mm)
+90	0.96	3.50	1.05	-0.025	0.35	7.7
+70	0.97	2.52	1.02	-0.0225	0.20	9
+50	0.88	1.76	1.0	-0.02	0.10	9
+40	0.88	1.79	1.0	-0.0175	0.10	9
+30	0.81	1.58	0.95	-0.015	0.075	9
+20	0.80	1.54	0.95	-0.013	0.075	9
+10	0.80	1.57	0.95	-0.013	0.075	9
+0	0.79	1.65	1.0	-0.015	0.10	9
-20	0.93	2.20	1.02	-0.02	0.20	9
-40	0.97	3.05	1.05	-0.025	0.30	8.7

A more stringent test of the thermal model is provided with comparison to the data acquired from the thermocouple measurements. The results obtained from the plates welded with normal focussed (+0 mA) and highly underfocussed (-40 mA) beams are presented in Figures 10 and 11. For the FEM results, two curves are plotted alongside each of the corresponding thermocouple results. These are representative of the upper and lower bounds of the predicted temperatures based upon the fact that the position of the thermocouple junctions could only be determined to within ±0.2 mm of these measured values. For all of the welding conditions investigated, excellent agreement was found between the FEM model and the thermocouple measurements. It can however be seen that the thermocouples appear to respond slower than the FEM results to the arrival of the heat source. This is likely to arise as a result of the finite separation of the thermocouple junctions. The measured temperature will therefore be some average of the temperature between these two points.

With these data it was also possible to extract an estimate of the thermal efficiency of the process, via the measurement of the uniform saturation temperature which each test-piece attained. The efficiencies obtained in this fashion are given in Table II.

For electron beam welding, it has been reported that the efficiency of energy coupling between the electron beam and the work-piece is typically 90-95% [1]. However, the efficiencies obtained from the thermocouple measurements, presented in Table II, are observed to decrease from 98% with the most defocused electron beams to 80% at the normal focus condition. This observation may be attributed to the fact that at all focal positions other than the most defocused, a fully penetrating weld cavity is produced. With full penetration, a proportion of the power supplied by the electron beam will pass through the work-piece and impinge on the fixture beyond. Additionally, as the normal focus condition is approached, so the incident power density will increase, and hence, a greater proportion of the power of the electron beam will be lost through the work-piece. Only in the cases of the non-penetrating welds generated with the most defocused electron beams is the true efficiency of energy coupling between the electron beam and the work-piece measured by this method.

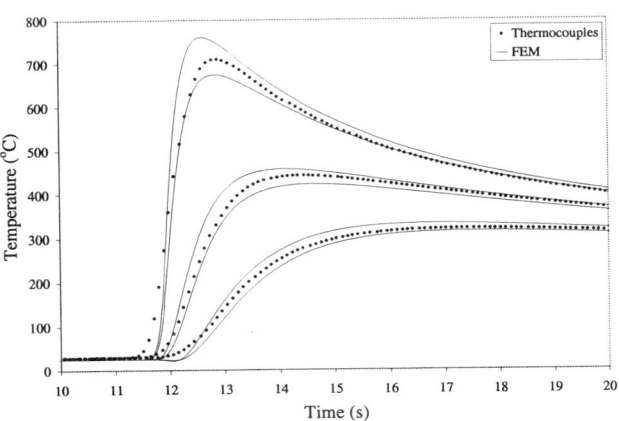

Figure 10: Thermal transients measured by surface mounted thermocouples and the associated upper and lower bound predictions from the associated numerical model for the test plate welded at +0 mA defocus.

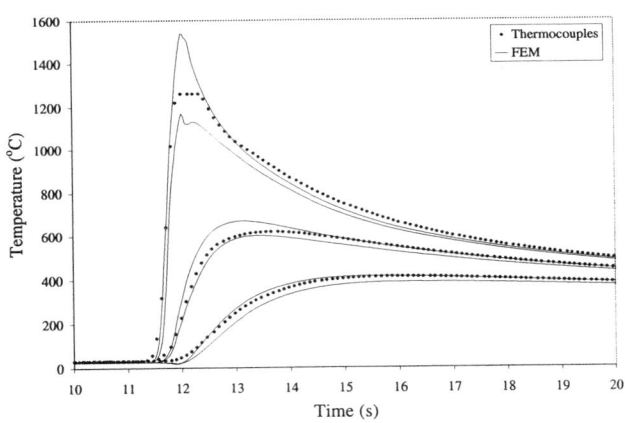

Figure 11: Thermal transients measured by surface mounted thermocouples and the associated upper and lower bound predictions from the associated numerical model for the test plate welded at -40 mA defocus.

Mechanical Analysis Contour plots of the longitudinal and transverse stress distributions predicted by the numerical model in the plane perpendicular to the welding direction at the centre of the plate are given in Figure 12. It can be seen that both stress components exhibit variations with depth into the sample. With the longitudinal stress component, the region of tensile stress is observed to be wider than with a lower peak stress at the surface of the plate than at depth. The transverse stress component exhibits greater variation still with compressive regions being seen at the surfaces of the plate in the weld and tensile stresses at depth.

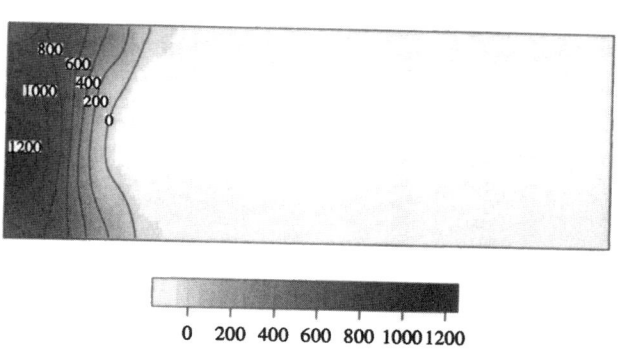

Longitudinal residual stress (MPa)

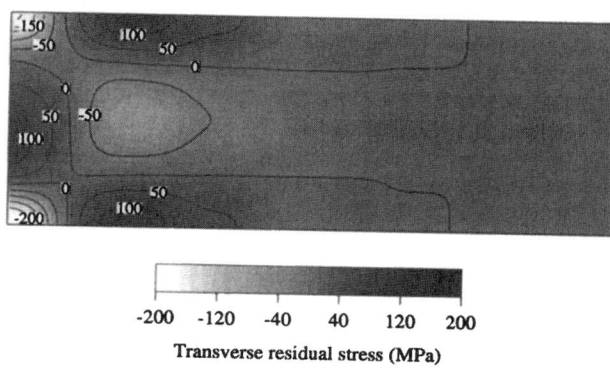

Transverse residual stress (MPa)

Figure 12: Contour plots of the longitudinal and transverse residual stresses across the centre of the model perpendicular to the weld.

The predicted and measured residual stresses across the plate perpendicular to the welding direction are shown in Figures 13 and 14. With the variations with depth identified in Figure 12 it is not appropriate to compare directly the residual stresses measured by the neutron diffraction technique at 1 mm into the sample with those obtained from the surface by X-ray diffraction and incremental hole-drilling. As a result, two lines have been plotted from the predictions made with the numerical model. The first line corresponds to the stresses predicted at the top surface of the plate, from which comparison with the stresses measured by the X-ray and hole-drilling methods may be made and the second line corresponds to those stresses predicted at a depth of 2.25 mm below the top surface. This has been se-

lected as providing the approximate extent of penetration of the gauge volume used in the neutron diffraction measurements. If the upper limit of the gauge volume is assumed to lie at the surface of the sample, the neutron diffraction results would be expected to be bounded by the two lines.

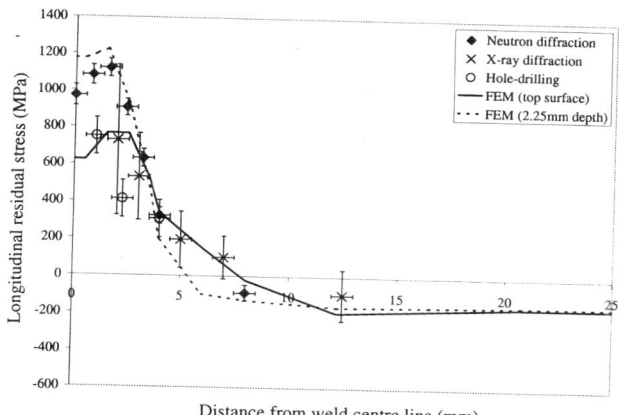

Figure 13: Plot of the measured and predicted longitudinal stress as a function of distance from the weld centre line.

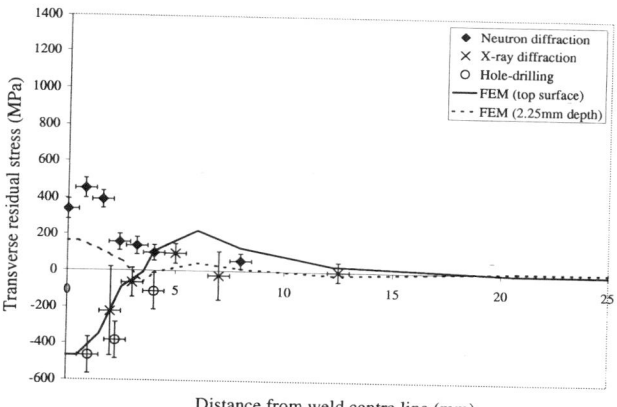

Figure 14: Plot of the measured and predicted transverse stress as a function of distance from the weld centre line.

In Figure 13, large tensile stresses are seen in the longitudinal direction in the vicinity of the weld. These stresses decrease in magnitude with increasing distance from the centre line, becoming compressive in the far field. As discussed earlier, smaller peak stresses are observed at the surface of the sample than at depth. Good agreement is seen between the measurements made with the X-ray diffraction and hole-drilling methods and also with the predictions made by the numerical model at the surface of the sample. Similarly, the measurements made by neutron diffraction are seen to lie between the two lines plotted from the numerical model as expected.

With the transverse residual stresses the numerical model predicts that little or no stress exists in the far-field away from the weld (Figure 14). As the weld is approached, small tensile stresses (~200 MPa) are predicted at the surface of the sample giving way to compressive stresses in and immediately adjacent to the weld. Reasonable agreement is seen between this model prediction and the results obtained by X-ray diffraction and incremental hole-drilling. A different pattern of behaviour is predicted at greater depths into the sample; the stresses are predicted to remain approximately zero up to the weld, after which tensile stresses develop. Again, this behaviour is replicated by the experimental measurements, however, considerably larger stresses are measured in the vicinity of the weld than are predicted by the numerical model. It is likely that this discrepancy arises as a result of the strong crystallographic texture present in the fusion zone. In this region, the lack of crystallites which fulfil Bragg diffraction criteria results in a reduced contribution to the measured reflection from this region. The effect will therefore be similar to the partial immersion of the gauge volume in the sample and will lead to an artificial and systematic shift of the results into tension [17]. If such an effect is indeed being encountered, the magnitude of both the longitudinal and transverse residual stresses may be in error in this region. In both cases it is possible that the actual stress may be smaller than shown.

Comparison between the predictions made for the other beam defocuses showed only small variations between the residual stress profiles at these depths. Indeed, the differences that were observed were smaller than the errors associated with the residual stress measurements given in Figures 13 and 14.

The measured and predicted angular distortions along the length of each of the plates in the investigation of beam defocus are shown in Figures 15 and 16 respectively. In all cases, the angular distortion is seen to remain approximately constant over the majority of the plate with slight variations at the edges arising as a result of the change in constraint on the weld in these regions. The magnitude of the angular distortion can be seen to increase as the defocussing of the electron beam is increased. This behaviour is displayed more clearly in Figure 17 in which the angular distortion obtained from the centre of each plate is plotted against the defocus at which it was welded. It can be seen that the minimum distortion condition is achieved with a slightly overfocussed electron beam. As the defocus is increased from this condition, a corresponding increase in the angular distortion is also observed, with the largest distortions being generated with the greatest defocuses of the electron beam. This behaviour arises as more material is melted above the neutral axis of the test pieces than below it. The subsequent transverse contraction of this material on solidification and cooling pulls the upper half of the weld together by a greater amount than the lower half and hence gives rise to a positive angular distortion, the magnitude of which is dictated by the asymmetry of the weld pool through the thickness of the test piece. Thus those welds which exhibit the greatest asymmetry also display the greatest angular distortion.

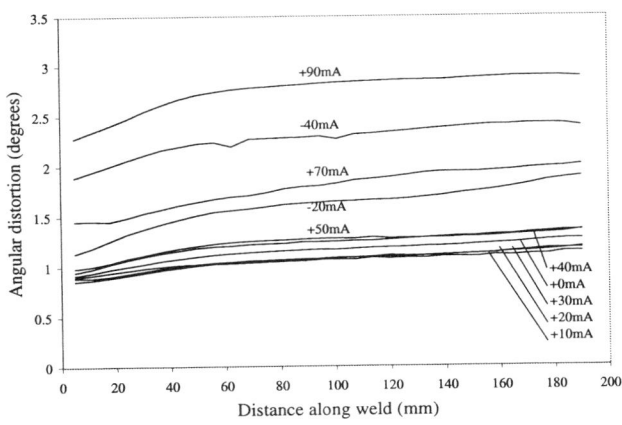

Figure 15: Plot of the angular (butterfly) distortions measured from the samples welded at different defocusses of the electron beam.

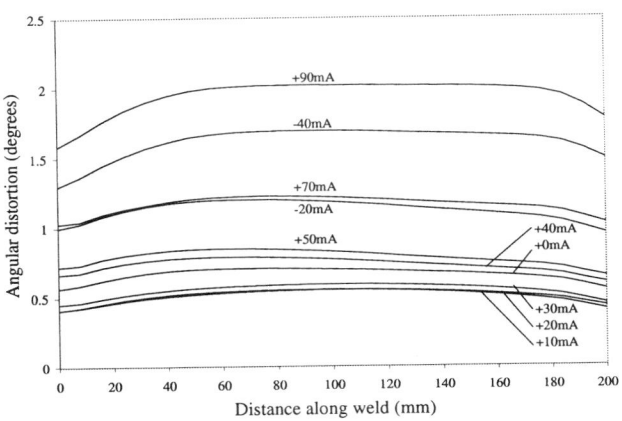

Figure 16: Plot of the angular (butterfly) distortions calculated with the numerical model for each of the samples welded at different defocusses of the electron beam.

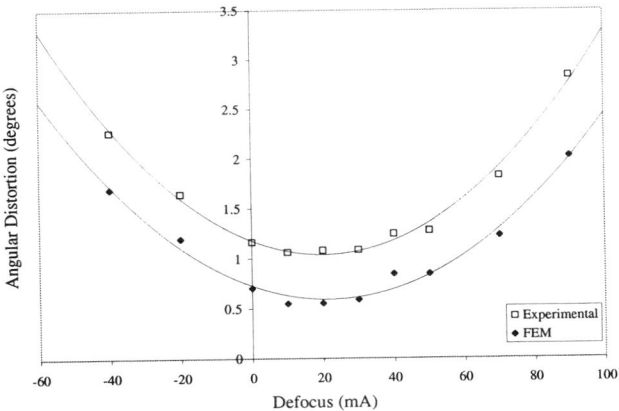

Figure 17: Plot of the measured and predicted angular (butterfly) distortions at the centre of the bead on plate test pieces as a function of the defocus of the electron beam. The lines are a guide to the eye only.

Beyond the range of defocusses investigated, it may be expected that the magnitude of the angular distortions will decrease again as the more diffuse electron beams produced will lead to a reduction in the penetration of the beam and a shallower weld pool.

It can be seen from Figure 17 that whilst the numerical model adequately captures the trends in this distortion mode, it underpredicts the magnitude of this distortion and as will be seen later, overpredicts the magnitude of the camber distortion.

Plots of the measured and predicted camber deflection of the plates along the length of the weld are presented in Figures 18 and 19 respectively. Whilst a systematic variation in this distortion mode is predicted with the numerical model, no such systematic variation is evident in the experimentally determined data. Indeed, it can be seen that adjacent plates in the defocus series exhibit camber distortions that are of the opposite sense, *i.e.* n-shaped rather than u-shaped. The corresponding camber distortions predicted by the numerical model show only u-shaped distortion patterns. These variations may be observed more readily in Figure 20 which shows the peak deflection along each weld as a function of the defocus of the electron beam.

If such distortion arises principally as a result of the differential transverse shrinkage of the weld pool through the thickness of the work piece then it would be anticipated that only a u-shape camber distortion would result, as all of the welds investigated display larger proportions of the weld pool above the neutral axis of the plate than below it. Indeed, at the largest defocuses of the electron beam where the most asymmetric welds are produced, u-shaped camber distortion patterns are observed. It therefore appears likely that the plates welded with more highly focussed electron beams are subject to an instability condition. The numerical model cannot predict such behaviour and poor agreement must necessarily be accepted for these welding conditions.

It was observed from the predicted residual stresses across the top surface of the plates that the variations in the distributions produced at the various defocuses of the electron beam were smaller than the estimated errors associated with the measurements. In contrast, the variations in the distortions produced with the different focal positions of the electron beam were sufficiently large to be measured reliably. It is therefore suggested that the measured post-welding distortion offers a more rigorous test of such a numerical model than the residual stresses.

Summary and Conclusions

The details of a numerical model suitable for the prediction of residual stresses and distortion arising from electron beam welding of aeroengine components has been described. From this work, the following conclusions have been drawn:

1. The magnitude and extent of the residual stresses predicted by the numerical model were shown to be in rea-

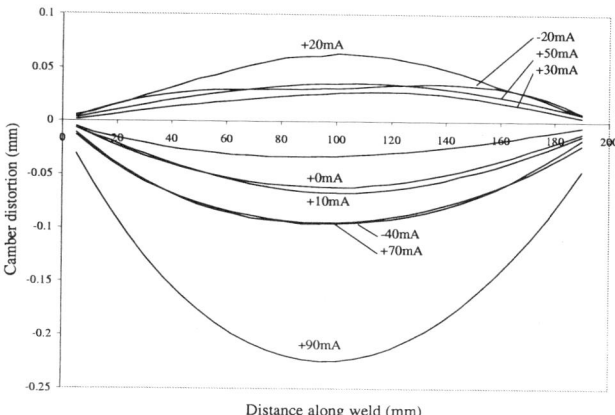

Figure 18: Plot of the camber distortions measured along the samples welded at different defocusses of the electron beam.

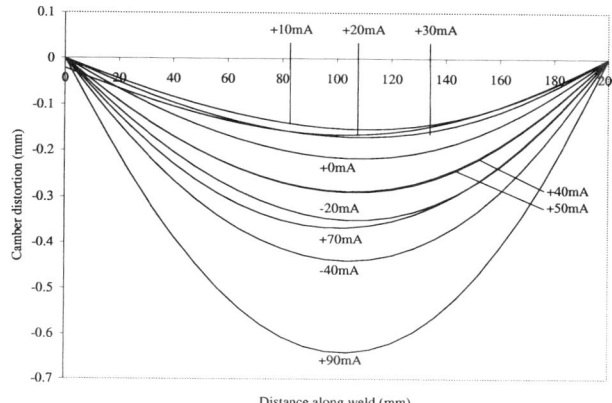

Figure 19: Plot of the camber distortions calculated with the numerical model along each of the samples welded at different defocusses of the electron beam.

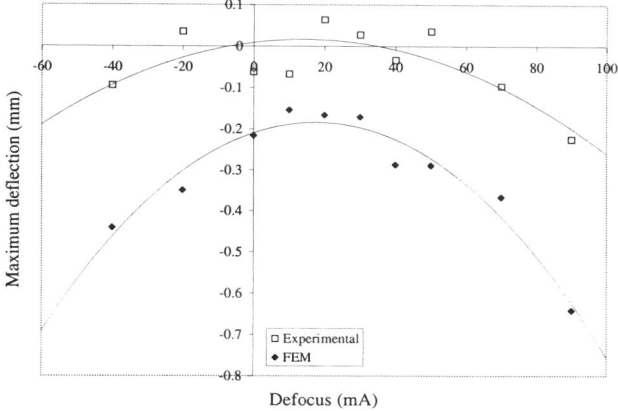

Figure 20: Plot of the peak deflection in the camber distortion mode measured and predicted from the samples welded at different defocusses of the electron beam. The lines are a guide to the eye only.

sonable agreement with those obtained experimentally. As expected, a peak residual stress of the order of the uniaxial yield strength of the material was observed in the longitudinal direction in the vicinity of the weld. The focal position of the electron beam is predicted to have only a small effect on the surface residual stress distribution.

2. The angular distortion generated by this welding process shows a systematic variation with the degree of defocus of the electron beam. A minimum distortion condition was achieved with a slight over-focussing of the beam. The numerical model was found to replicate this behaviour in form, with the predictions being approximately half of their experimentally determined counterparts.

3. The sign of the camber distortion was found to be either positive or negative, with little systematic variation being displayed. This was particularly true for welds made near the normal focus. This was caused by an instability condition arising from the near-uniform deposition of energy through the workpiece. The model as currently formulated is incapable of predicting this phenomenon.

4. The computer simulation of welding processes now provides a virtual capability for the prediction of residual stresses and distortion. However, it is believed that further efforts are required to improve the accuracy of the predictions. This will be possible as computer hardware and software mature.

Acknowledgements

This work was sponsored by the Engineering and Physical Sciences Research Council (EPSRC), Rolls-Royce plc and the Defence & Evaluation Research Agency (DERA). The support of Steve McKenzie & Paul Spilling (Rolls-Royce) and Mike Winstone & George Harrison (DERA) is appreciated. Helpful discussions over a number of years with Moyra McDill, Alan Oddy and John Goldak at Carleton University, Ottawa, Canada are acknowledged.

References

1. H. Schultz, Electron Beam Welding (Abington Publishing, Abington Hall, Cambridge, England, 1993).

2. Edited by J. F. Lancaster, The Physics of Welding (Pergamon Press, Headlington Hill Hall, Oxford, UK, 2nd edition, 1986).

3. H. Schwarz, "Mechanism of high-power-density electron beam penetration in metal", Journal of Applied Physics, 35(7) (1964), 2020–2029.

4. ASM Handbook Volume 6: Welding, Brazing and Soldering (ASM International, 10th edition, 1993).

5. J. H. Argyris, J. Szimmat and K. J. Willam, "Computational aspects of welding stress analysis", Computer Methods in Applied Mechanics and Engineering, 33 (1982), 635–666.

6. J. Goldak, M. McDill, A. Oddy, R. House, X. Chi and M. Bibby, "Computational heat transfer for weld mechanics", In S. A. David, editor, Advances in Welding Science and Technology - Proceedings of an International Conference on Trends in Welding Research, (ASM International, 1986), 15–20.

7. J. Goldak, V. Breiguine and N. Dai, "Computational weld mechanics: A progress report on the ten grand challenges", In H. B. Smart, J. A. Johnson, and S. A. David, editors, Trends in Welding Research - Proceedings of the 4th International Conference, (ASM International, 1995), 5–11.

8. S. M. Roberts, H. J. Stone, J. M. Robinson, P. J. Withers, R. C. Reed, D. R. Crooke, B. J. Glassey and D. J. Horwood, "Characterisation and modelling of the electron beam welding of Waspaloy", In H. Cerjak, editor, Mathematical Modelling of Weld Phenomena 4, (The Institute of Materials, 1998), 631–648.

9. A. S. Oddy, J. M. J. McDill and J. A. Goldak, "Consistent strain fields in 3D finite element analysis of welds", Journal of Pressure Vessel Technology, 112 (1990), 309–311.

10. O. C. Zienkiewicz and R. L. Taylor, The Finite Element Method: Solid and Fluid Mechanics, Dynamics and Non-linearity, volume 2, (McGraw-Hill Book Company, Maidenhead, Berkshire, England, 4th edition, 1994).

11. Edited by E. Hinton, NAFEMS Introduction to non-linear finite element Analysis, (NAFEMS, Birniehill, East Kilbride, Glasgow, UK, 1992)

12. H. J. Stone, P. J. Withers, T. M. Holden, S. M. Roberts and R. C. Reed, "Comparison of three different techniques for measuring the residual stresses in an electron beam-welded plate of Waspaloy", Metallurgical and Materials Transactions A, 30A (1999), 1797–1808.

13. H. J. Stone, T. M. Holden and R. C. Reed, "On the generation of microstrains during the plastic deformation of Waspaloy", Acta Materialia, 47(17) (1999), 4435–4448.

14. M. R. Daymond, M. A. M. Bourke, R. B. Von Dreele, B. Clausen and T. Lorentzen, "Use of Rietveld refinement for elastic macrostrain determination and for evaluation of plastic strain history from diffraction spectra", Journal of Applied Physics, 82(4) (1997), 1554–1562.

15. H. J. Stone, T. M. Holden and R. C. Reed, "Determination of the plane specific elastic constants of Waspaloy using neutron diffraction", Scripta Materialia, 40(3) (1999), 353–358.

16. ASTM E 837, "Standard Test Method for Determining Residual Stresses by the Hole-Drilling Strain-Gage Method", 1994.

17. Edited by J. Lu, Handbook of measurement of residual stresses (The Fairmont Press, 700 Indiana Trail, Lilburn, GA, US, 1996)

Novel Techniques for Investigating the High Temperature Degradation of Protective Coatings on Nickel Base Superalloys

A.A.Alibhai, D.P Garriga-Majo, B.A Shollock, D.S. McPhail

Department of Materials, Imperial College of Science, Technology and Medicine,
Prince Consort Road, London, SW7 2BP

Abstract

In order to facilitate improved fuel efficiency and power to weight performance in gas turbines, components are subjected to increasing operating temperatures. The demanding operating conditions have lead to the development of coating systems, which protect the gas turbine components from environmentally induced attack. The effectiveness of the coating relies on the mechanism by which a healing oxide scale is formed and replaced during thermal cycling. Therefore, in order to optimise coating performance, it is vital to develop a clear understanding of the oxidation mechanisms involved during scale formation. [1]

Novel approaches involving $^{18}O_2$ isotopic tracer experiments, secondary ion mass spectrometry (SIMS) and focused ion beam (FIB) in-situ milling, have been employed to investigate the high temperature oxidation mechanisms of various commercial protective coatings on industrial gas turbines and other nickel base superalloy systems. These techniques lend themselves well to the study of high temperature degradation, as they are able to provide essential information on the diffusion pathways of oxygen anions and metal cations during oxide formation [2-3].

In the first part of this study, the use of a two-stage tracer oxidation technique and Secondary Ion Mass Spectrometry have been employed to investigate the oxygen diffusion mechanisms occurring during scale formation on uncoated superalloys CMSX4 and CM186LC at temperatures between 826°C to 1100°C.

The cation depth profiles obtained give information regarding the chemical nature of the oxides formed and these results provide a baseline against which the oxidation performance of coatings may be assessed.

In the second part of this study, oxides formed on a platinum aluminide coating MDC150L which had been oxidised at 870°C, have been analysed by micromaching cross sections of the oxide layer using a Ga$^+$ focused ion beam (FIB-FEI Europe) instrument. These cross sections were analysed on the FIB, using secondary ion and electron imaging, and oxygen mapping.

Results from this study have provided essential information regarding the microstructural evolution of the oxide layer during high temperature exposure, revealing the oxide thickness and isotopic oxygen distribution. For uncoated superalloy CM186LC, outward cation diffusion was observed for all temperatures between 820°C and 1100°C. In the case of superalloy CMSX4, outward cation diffusion occurred at temperatures up to 996°C. Above 996°C a change in mechanism was observed and combination of outward cation and inward oxygen diffusion occurs. For the coating system MDC150L, outward cation diffusion is observed at 870°C, for extended oxidation periods up to 82 hours, and the coating formed an effective alumina scale.

These encouraging results obtained have highlighted the usefulness of the focused ion beam technique and secondary ion mass spectroscopy in elucidating the high temperature degradation of protective coatings and nickel base superalloys.

Superalloys 2000
Edited by T.M. Pollock, R.D. Kissinger, R.R. Bowman,
K.A. Green, M. McLean, S. Olson, and J.J. Schirra
TMS (The Minerals, Metals & Materials Society), 2000

Introduction

Thermal barrier coating (TBC) systems are typically applied to hot section components of gas turbine engines in order to reduce part temperatures and improve component durability. TBC systems involve a ceramic top coat, which is most commonly yttria–stabilised zirconia (YSZ) due to its poor thermal conductivity and good coefficient of thermal expansion. An underlying metallic bond coat is applied between the superalloy substrate and the ceramic top coat. This bond coat must be resistant to high temperature oxidation as the ceramic top coat does not act as a barrier to oxygen and hence does not protect the substrate alloy. The metallic bond coat is designed to form a protective and adherent α-Al_2O_3 scale. Spallation of this thermally grown alumina scale leads to the loss of the ceramic top coat and ultimately failure of the TBC system. Various studies [4–6] have shown that the oxidation mechanism of the metallic bond coat is a key aspect in determining the useful lifetime of TBC systems.

The use of isotopic tracer experiments and conventional Secondary Ion Mass Spectrometry (SIMS) has extensively been reported for the study of high temperature oxidation [7], especially for alumina forming systems [2,8-12]. These studies, combined with results obtained from other techniques such as X-Ray diffraction, Rutherford Backscattering Spectrometry and Transmission Electron Microscopy [13-14], have emphasised the possible complexity of the oxide scale growth mechanism, which may involve oxygen and/or cationic diffusion through different pathways and also phase transformations within the oxide scale.

Focused Ion Beam (FIB) technology has recently been applied to the investigation of the high temperature degradation of industrial turbine blades [15] and coatings [16]. FIB technology is routinely used within the semiconductor industry and has only recently been used in other application areas such as specimen preparation for biological electron microscopy [17], multilayer optical oxide/nitride films, hard carbide coating on tool steels [18], and Al-Li-Cu quasicrystal investigations [19]. A major advantage of the application of the FIB system to the study of high temperature oxidation is its capability to perform *in situ* micro-machining (ion milling) of the specimen and to prepare cleaned surfaces for high resolution SIMS analysis and imaging.

The recent developments in high Z-resolution microscopes such as the Atomic Force Microscope and Interferometric Optical Microscopes have made it possible to obtain valuable information regarding the oxide surface properties as well depth calibration of the SIMS craters.

In this paper, novel approaches towards investigating the high temperature oxidation of uncoated superalloys and coating systems are introduced. Investigation of uncoated substrate alloys is useful as it provides information regarding the durability of the bare superalloy in the event of coating failure. It also provides a foundation on which the oxidation performance of coatings may be assessed. In this study, the high temperature isothermal oxidation of uncoated single crystal CMSX4 (Cannon Muskegon) and directionally solidified alloy CM186LC (Cannon Muskegon) were investigated using the two-stage isotopic tracer technique and SIMS analysis. Oxide scales formed on a platinum modified aluminide coating MDC150L (Howmet-Thermatech) were analysed using a focused ion beam system (FIB200) and micromaching cross-sections of the oxide. Analysis of the oxide

layer was performed using secondary electron and secondary ion imaging, oxygen mapping and depth profiling.

Experimental Procedure

Materials

Uncoated superalloys CMSX4 and CM186LC were studied at oxidation temperatures between 826°C and 1100°C for a total of 22 hours. These alloys have virtually the same chemical composition, with small differences in elements added to aid the directionally solidified process. Table 1 gives the chemical composition of these superalloys.

A platinum modified aluminide coating MDC150L applied to the directionally solidified alloy CM186LC using a low aluminium activity- chemical vapour deposition process [20] was also investigated for various oxidation periods at 870°C. This coating has a single phase (Ni,Al)Pt additive layer. Beneath this is a well developed diffusion zone, which is almost one third of the total coating thickness, and is characteristic of an outward grown coating. Typically, the total coating thickness (post heat treatment) is 77.5μm and the coating composition is 19.5wt.%Al and 19.0wt.%Pt.

Specimen preparation

In the case of the uncoated superalloys samples were cut with specimen orientation normal to the <100> plane. For alloy CM186LC, specimen orientation was determined via X-Ray Diffraction texture analysis using the Phillips X'Pert Materials Research Diffractometer (MRD). In the case of the single crystal alloy CMSX4, back reflection Laue was performed using a 30kV voltage and a 30mm film to specimen distance. Samples were then sliced normal to the <100> plane using the Accutom 5 cutting machine (Struers). Samples were then polished to 1 μm and cleaned ultrasonically in acetone for 20 minutes.

The MDC150L/CM186LC system was received as a fatigue test specimen, which had previously been oxidised for 20 hours in air at 870°C. Small samples were cut for analysis as described above.

Oxidation Study

A two-stage oxidation treatment was performed on all the samples. In this tracer oxidation technique, the sample was first oxidised in one isotope of oxygen with an $^{18}O_2$ partial pressure of g_1, for a period of time t_1, thus forming the first oxide layer. The sample was then oxidised in the tracer isotope with an $^{18}O_2$ partial pressure of g_2 for a period of time t_2, thus forming the new oxide. Uncoated samples CMSX4 and CM186LC were first oxidised for 18 hours in $^{18}O_2$ (enriched to 50%, g_1=0.5), followed by 4 hours in unlabelled $^{16}O_2$ (g_2=0.002), at temperatures between 826°C and 1100°C. The coated samples MDC150L on alloy CM186LC were first oxidised in air (g_1=0.002) for 20 hours at 870°C (as received condition). The samples were then further oxidised in $^{18}O_2$ (enriched to 87%, g_2=0.87) for 4 hours and 62 hours at the same temperature. All oxidations took place at 200 millibar pressure and the sample tube was briefly evacuated between anneals. Table 2 summarises the oxidation treatments and the SIMS and FIB experiments performed on the uncoated superalloys CMSX4 and CM186LC and the coating MDC150L.

ALLOY	COMPOSITIONAL DATA (wt %)											
	Ni	Hf	C	Cr	Co	Mo	W	Re	Ta	Al	Ti	B	Zr
CM186LC	bal	1.4	0.07	6.0	9.0	0.5	8.4	3.0	3.4	5.7	0.7	0.015	0.015
CMSX4	bal	0.1	0.1	6.5	9.0	0.6	6.0	3.0	6.5	5.6	1.0	-	-

Table 1: Compositional Data (wt%) for alloys CMSX4 and CM186LC, which were investigated in the two-stage isotopic oxidation study.

Secondary Ion Mass Spectroscopy

SIMS depth profiling is an extremely useful technique for analysing the tracer oxygen diffusion mechanisms in metal systems, as it is able to distinguish all the isotopes in the periodic table. Although this is a destructive surface analysis technique, it combines very high depth resolution with exceptional detection limits (typically parts per million). This technique involves a mono-energetic focused beam of primary ions (typically 0.25keV to 15keV), bombarding the surface of a solid sample in a vacuum. The momentum transfer between the incoming ions and the constituent atoms leads to the emittance of secondary species from the surface of the material. This process is known as sputtering. The sputtered species may be neutral or charged particles. The charged particles are detected and analysed using conventional mass spectrometry analysis. As the ion beam continually rasters across the surface of the sample, it sputters material from the surface thus forming a square crater.

In this study SIMS depth profiling measurements were conducted on the oxidised superalloy samples using the ATOMIKA 6500 Ion Microprobe. A Xenon (Xe^+) primary ion beam was used to raster the surface of the sample at near normal incidence, with a primary beam energy of either 8keV or 15keV. Negative secondary ions of $^{16}O^-$ and $^{18}O^-$ were monitored as a function of time for all samples. During analysis of the uncoated samples, metal ions such as $^{27}Al^-$, $^{59}Co^-$, $^{58}Ni^-$ and $^{52}Cr^-$ were also monitored. Scan widths of between 250μm and 300μm were employed and secondary ions were gated to 25% of the total raster area in order to avoid any crater wall effects.

The Focussed Ion Beam System

The focussed ion beam system (FIB200-FEI UK Ltd.) works in essentially the same manner as conventional SIMS, but in this case the FIB200 employs a gallium Ga^+ liquid metal ion source, which operates at 30keV. The size of the beam can be pre-programmed depending on the beam current used and typically a beam diameter of just a few nanometers can be achieved using a current of approximately 1 picoampere. Typically, high currents (few nA) are used to mill different shapes of crater and smaller currents (few pA) permit high resolution SIMS analysis. In order to prevent surface charging a thin layer of platinum (typically 1μm) may be deposited in-situ.
The FIB200 system is equipped with different detectors:
- a secondary electron detector.
- a channeltron (CDEM) used to detect secondary ions and high energy electrons for imaging.

- a quadrupole mass spectrometer for SIMS analysis (mass spectrum, depth profiling, elemental mapping).

The Ga+ primary beam generates a low secondary ion signal compared to conventional ion sources such as O_2^+ and Xe^+, but it offers a very high spatial ability. The different motions of the stage X, Y, Z, rotation and tilt can be followed using an optical camera.

In this study oxides formed on the coated sample MDC150L were analysed by ion milling cross sections of the oxide and performing secondary electron imaging and isotopic oxygen mapping. The oxide surface was coated with platinum in order to avoid any surface charging. Depth profiling using the FIB200 was also performed on all the oxides grown on the coated sample.

Depth Calibration

Depth calibration of the profiles was performed using a Z-high resolution microscope (Zygo) and an Atomic Force Microscope (Quesant Resolver), by measuring the total depth of the craters and assuming a constant sputter rate throughout the oxide scale.

Determination of Tracer Enrichment

In order to be able to compensate for any undesirable variations in secondary ion intensities due to surface charging, the isotopic fractions ir_{18} and ir_{16} are plotted as a function of depth. Neglecting any contribution from $^{17}O^-$, which has a natural abundance of 0.037%, the isotopic fractions for oxygen 18 and oxygen 16 are given as: $ir_{18} = \dfrac{^{18}O^-}{^{18}O^- + ^{16}O^-}$ and $ir_{16} = 1 - ir_{18}$ respectively.

These isotopic fractions reflect the actual isotopic concentration variations in the oxide scale. In order to quantify the relative isotope tracer enrichment in the oxide scale, the parameter ν was estimated for the oxidation treatments performed and is given by:

$$\nu = \frac{ir_{18} - g_1}{g_2 - g_1}$$

where g_1 and g_2 are the $^{18}O_2$ partial pressure in the first and second oxidation atmospheres respectively. By plotting the tracer enrichment parameter ν as a function of depth, it is possible to distinguish the regions where the new oxide has grown. Regions of pure new oxide growth have a ν-value of unity, whereas regions of old oxide growth have values of zero. Regions that contain a mixture of new and old oxide have a ν-value between 0 and 1.

Material	Oxidation Treatment					SIMS Depth Profile			
	Temperature (°C/°F)	t_1 (hours)	g_1	t_2 (s)	g_2	SIMS Machine*	Conditions	Sputtering Time (s)	Zygo crater depth (µm)
CMSX4	826/1519	18	0.50	4	0.002	Atomika	$Xe^+/8keV$	1.434×10^4	4.39
CMSX4	906/1663	18	0.50	4	0.002	Atomika	$Xe^+/8keV$	0.936×10^4	3.95
CMSX4	996/1825	18	0.50	4	0.002	Atomika	$Xe^+/8keV$	1.638×10^4	8.15
CM186LC	826/1519	18	0.50	4	0.002	Atomika	$Xe^+/8keV$	1.392×10^4	4.32
CM186LC	906/1663	18	0.50	4	0.002	Atomika	$Xe^+/8keV$	1.638×10^4	7.29
CM186LC	996/1825	18	0.50	4	0.002	Atomika	$Xe^+/8keV$	2.232×10^4	12.91
CM186LC	1100/2012	18	0.50	4	0.002	Atomika	$Xe^+/8keV$ [†]	4.096×10^4	20.58
CM186LC+MDC150L	870/1598	20	0.002	-	-	Atomika	$Xe^+/15keV$	0.252×10^4	0.58
CM186LC+MDC150L	870/1598	20	0.002	4	0.87±0.02	Atomika	$Xe^+/15keV$ [†]	0.324×10^4	0.75
CM186LC+MDC150L	870/1598	20	0.002	62	0.87±0.02	FIB200	$Ga^+/30keV$	0.084×10^4	1.60
CM186LC+MDC150L	870/1598	20	0.002	62	0.87±0.02	FIB200	$Ga^+/30keV$	0.072×10^4	1.05

Table 2: Summary of oxidation treatments and SIMS/FIB analyses performed on alloys CMSX4, CM186LC and coating system MDC150L. [†] An electron gun with a current of 1.2 A was used to ensure charge compensation on these samples. *Atomika refers to the Atomika 6500 Ion Microprobe and FIB200 refers to the Focused Ion Beam 200 system (FEI UK Ltd.)

Depending on the oxidation temperature and the superalloy substrate, the diffusion of the tracer and the further growth of the oxide may take place at one of these three locations:

1. At the gas/oxide interface with a new oxide growth of thickness d_g.
2. Within the oxide formed during the first thermal treatment, the thickness of this new oxide layer is called d_o.
3. At the oxide/metal interface with a new growth of thickness d_m.

The total oxide thickness is typically determined by the depth over which the sum of the two negative oxygen secondary ion intensities falls to half its maximum value and is denoted as $d_{0.5}$. An example of this is given in figure 1 for alloy CMSX4 oxidised at 996°C and values of $d_{0.5}$ are given in table 3.

The cation diffusion mechanisms are presented as normalised secondary ion intensities. These are determined by dividing the negative metal secondary ion count rate at profiling time t over the maximum negative secondary ion count rate for that metal species. This method allows the relative variation of each metal species in the oxide to be displayed. Examples of cation depth profiles for CM186LC and CMSX4 oxidised at 996°C for total oxidation periods up to 22 hours are given in figures 2 and 4.

Results

Uncoated Superalloys

Values for the total oxide thickness $d_{0.5}$ and regions of new and partially new tracer oxide thickness at the gas/oxide interface (d_g), metal/oxide interface (d_m) and within the existing oxide (d_o) are given in Table 3, as well as the predominant oxidation mechanisms that occur in superalloys CMSX4 and CM186LC which have been oxidised for 18 hours in $^{18}O_2$, followed by 4 hours in $^{16}O_2$. The average oxide growth rate for these superalloys for each temperature has been determined by dividing the total oxide thickness $d_{0.5}$, by the total oxidation time, which is 1320 minutes (22 hours) in all cases. The oxygen depth profile for alloy CMSX4 oxidised at 996°C, for 18 hours in $^{18}O_2$, followed by 4 hours in $^{16}O_2$ is shown in figure 1.

The depth over which the sum of the oxygen 18 and oxygen 16 ions falls to half its maximum value, is one possible method of determining the total oxide scale.

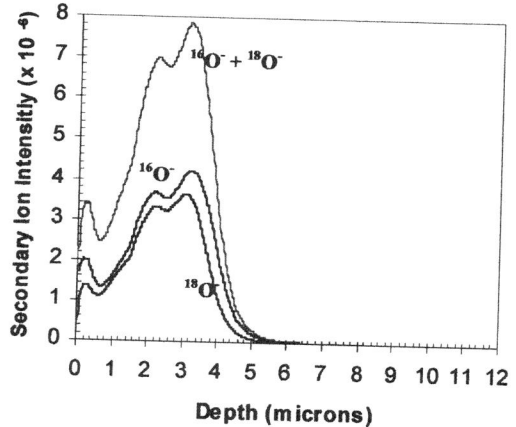

Figure 1: Oxygen depth profile for sample CMSX4 oxidised at 996°C. The total oxide thickness $d_{0.5}$ occurs at 4.2μm, which is the depth at which the sum of the $^{18}O^-$ and $^{16}O^-$ falls to half its maximum

Alloy	Oxidation Temperature (°C)	Total Oxide Thickness $d_{0.5}$ (μm)	Average Growth Rate (nm/min)	New Oxide Growth			Oxidation Mechanism
				d_g (μm)	d_o (μm)	d_m (μm)	
CMSX4	826	1.2	0.91	0.1	-	-	
CMSX4	906	3.2	2.42	0.15	1.0	-	Outward Cation Diffusion
CMSX4[†]	996	4.2	3.71	0.5	-	1.0	Combination of Outward Cation and Inward Oxygen Diffusion
CM186LC	826	1.4	1.06	0.2	-	-	
CM186LC	906	5.5	4.17	1.2	-	-	Outward Cation Diffusion
CM186LC[‡]	996	7.0	5.30	1.6	-	-	
CM186LC	1100	12.0	9.0	4.8	-	-	

Table 3: Summary of results for CMSX4 and CM186LC showing where the tracer oxide has formed and predominant oxidation mechanism. [†] Cation depth profiles and tracer enrichment profiles for CM186LC for these conditions are shown in figures 2 and 3 respectively. [‡] Cation depth profiles and tracer enrichment profiles for CMSX4 for these conditions are shown in figures 4 and 5 respectively.

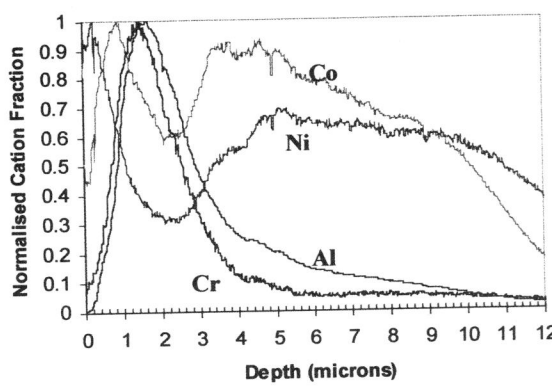

Figure 2: Normalised cation depth profile for CM186LC oxidised at 996°C for 18 hours in $^{18}O_2$ and 4 hours in $^{16}O_2$.

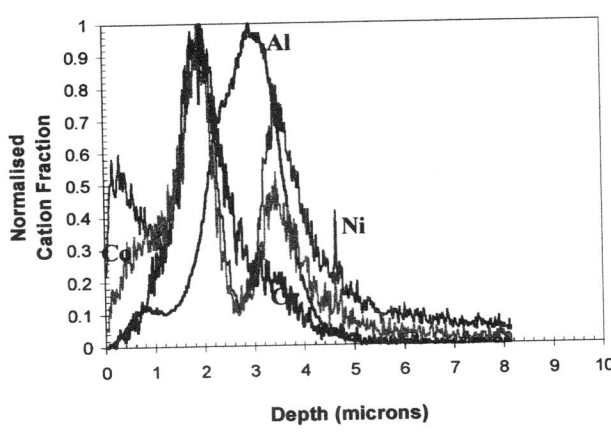

Figure 4: Normalised cation depth profile for CMSX4 oxidised at 996°C 18 hours in $^{18}O_2$ and 4 hours in $^{16}O_2$.

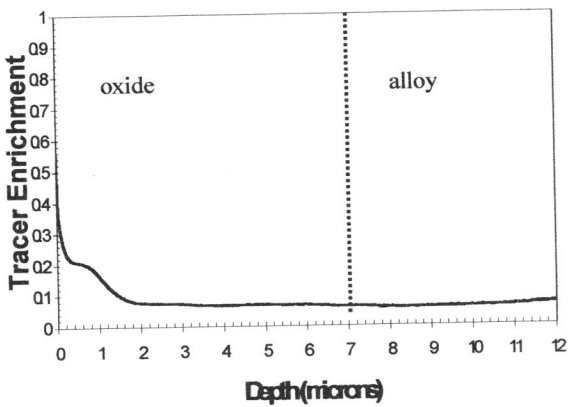

Figure 3: Tracer enrichment profile for CM186LC oxidised at 996°C as for figure 2.

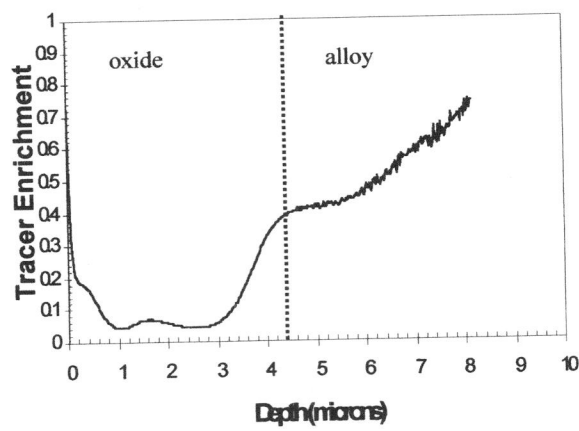

Figure 5: Tracer enrichment profile for CMSX4 oxidised at 996°C as for figure 4.

Coating MDC150L

Total oxide thickness $d_{0.5}$, average oxide growth rate and the predominant oxidation mechanism that occurs for the coating system MDC150L at 870°C for oxidation in air for 20 hours and various periods up to 62 hours in $^{18}O_2$ are given in Table 4.

Oxidation Time (hours)	Total Oxide thickness (μm)	Average Growth Rate (nm/min)	Predominant Oxidation Mechanism
20 hours air	0.5	0.41	Outward Cation Diffusion
20 hours air + 4 hours $^{18}O_2$	0.55	0.39	Outward Cation Diffusion
20 hours air + 4 hours $^{18}O_2$	0.95	0.19	Outward Cation Diffusion

Table 4: Summary of results obtained for coating MDC150L oxidised at 870°C.

Imaging results from the focused ion beam study for the coating MDC150L which had been oxidised in air for 20 hours and then 62 hours in $^{18}O_2$ at 870°C are given in figures 6 and 8 to 11. The oxide morphology is shown in figure 6, which is a secondary electron image of the crater wall formed after depth profiling. Figure 7 is a schematic diagram showing the direction the gallium beam during in-situ ion milling and imaging which was used to obtain figures 8 to 11. Figure 8 is a secondary ion image of the surface of the oxide formed on the coating, and the position of the wedge crater, which was formed using the Ga+ beam. The oxide surface above the crater was coated with platinum in order to avoid any surface charging effects. This also helps to clearly define the top of the oxide surface. Figure 9 is a secondary electron image and is an enlargement of the grain boundary area shown figure 8. Figures 10 and 11 are $^{16}O^-$ and $^{18}O^-$ maps respectively and correspond to the same region shown in figure 9.

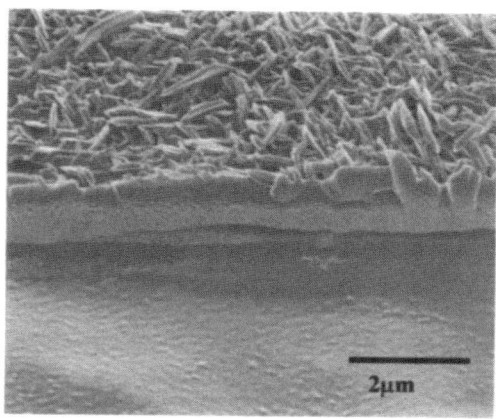

Figure 6: Secondary electron image of the crater wall formed after depth profiling revealing a θ-Al_2O_3 whisker morphology on the oxide surface.

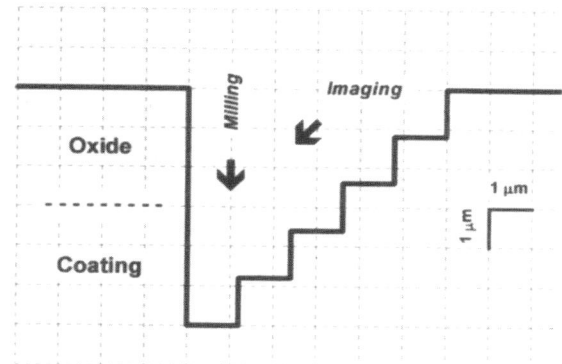

Figure 7: Schematic diagram showing the direction of the Ga^+ ion beam during in-situ ion milling and imaging.

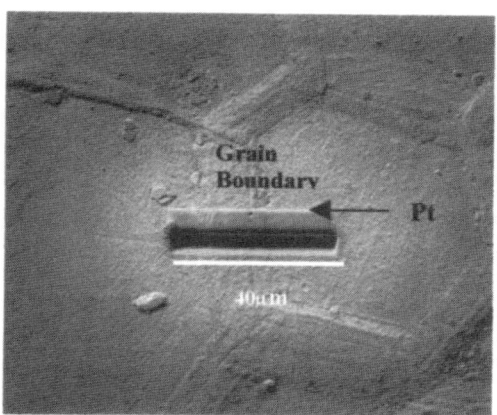

Figure 8: Secondary ion image of the ion-milled crater made on the oxide formed on coating MDC150L oxidised at 870°C for 20 hours in air plus 62 hours in $^{18}O_2$. Platinum was deposited above the crater in order to avoid any surface charging effects. The hexagonal grain structure of the coating can also be seen.

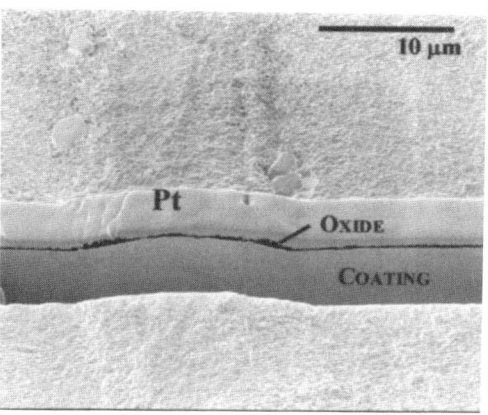

Figure 9: A general view of the milled cross-section. (Secondary electron image, tilt 45°).

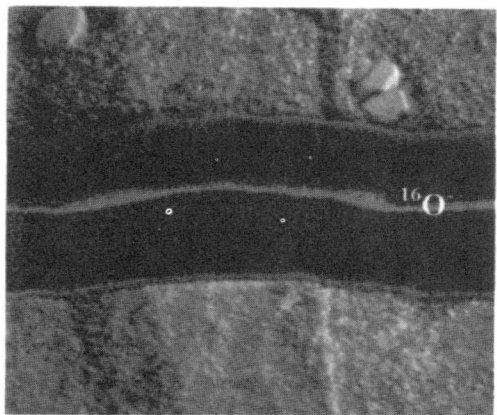

Figure 10: $^{16}O^-$ elemental map over the grain boundary region shown in figure 6.

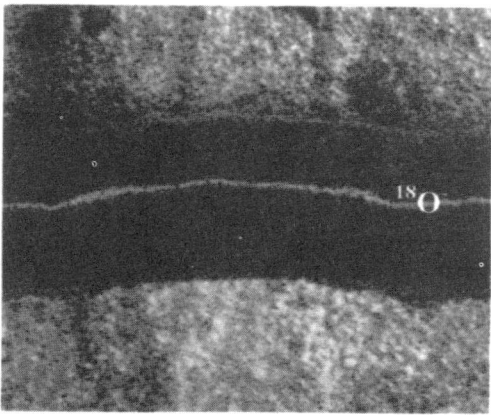

Figure 11: $^{18}O^-$ elemental map over the same region as shown in figure 8.

Discussion

Uncoated Superalloys

With respect to the tracer oxidation study made on the uncoated superalloys CMSX4 and CM186LC (given in figure 2 and 3), a number of key observations can be made. Results for alloys CM186LC show that outward cation diffusion is the predominant oxidation mechanism for all temperatures between 826°C and 1100°C. This is shown in the example given in figure 3. The tracer enrichment parameter v is high at the gas-oxide interface indicating possible surface exchange and new oxide growth. Also, with the exception of the oxide formed at 826°C, the total oxide thicknesses are significantly greater for the directional solidified alloy CM186LC compared to those grown on the single crystal alloy CMSX4. Samples for both these superalloys have been cut normal to the <100> orientation and hence this result may be attributed to fast oxygen diffusion down grain boundaries in the directional solidified alloy CM186LC. Such a result is comparable to hot oxidation tests performed by others [22-23] on these alloys.

With respect to the total oxide thickness values $d_{0.5}$ given in tables 3 and 4, it is important to note that this value often gives an underestimate of the true oxide thickness as can be seen for alloy CM186LC, which has been oxidised at 996°C. The $d_{0.5}$ value for this sample is found to be 7.0μm, although the cation diffusion profile at this point is still in the nickel and cobalt plateau. An alternative method for determining the total oxide thickness would be to evaluate the regions of new and old oxide growth at the gas-oxide, old oxide, and oxide-metal interface from the tracer enrichment profile. However, as it is not always possible to determine the oxide growth at the oxide-metal interface (as shown in figure 3), this method is limited. Hence, the total oxide thickness $d_{0.5}$ found from the sum of the oxygen profiles remains an adequate method for determining the total oxide thickness.

The cation diffusion profiles for this alloy also provide interesting information with regard to the nature of the oxides that form. Figure 2 is an example of a normalised cation diffusion profile obtained for this alloy at 996°C. This shows that the oxide at the gas-oxide interface is nickel rich and most likely to be NiO. A middle oxide layer is cobalt rich and the rest of the oxide near the oxide-alloy interface is rich in aluminium and chromium. The nickel plateau observed at depths greater than 5μm indicates a third extended oxide region, which is also rich in cobalt. Although the SIMS technique does not provide quantitative information on the oxide composition, such cation diffusion depth profiles are very useful as they provide information regarding the nature of the oxide formed.

The results obtained for CMSX4 show that a change in oxidation mechanism is observed at 996°C. At lower temperatures, this alloy displays outward cation diffusion. However, at 996°C the mechanism is altered and a combination of inward oxygen and outward cation diffusion through the first oxide layer is observed. Figure 5 shows the tracer enrichment profile for this alloy, which has been oxidised at 996°C. This shows an area of high tracer enrichment at the gas–oxide interface and represents a new growth in this region. It is also interesting to note that in figures 3 and 5 the tracer enrichment does not fall to zero, which may simply represent background enrichment. However, it is plausible

that this may correspond to the tracer having exchanged into the previous oxide grains or formed a new phase within the previous oxides grain boundaries. Such behaviour is as predicted by others [21], who have performed such two stage isotopic tracer experiments on alumina forming alloys. The tracer enrichment value then increases to a plateau with a v-value of 0.42, which indicates a region of new oxide growth within the existing oxide. At the oxide-metal interface the tracer enrichment value begins to rise as $^{16}O_2$ gas has been used as the tracer gas. The noise in the rising tail of the profile is characteristic of profiling into the substrate alloy.

The normalised cation depth profile in figure 4 for alloy CMSX4, which has been oxidised at 996°C shows that the region closest to the gas oxide surface is nickel rich, indicating nickel cation diffusion through the oxide to this interface. The central oxide is a mixture of nickel, cobalt and chromium oxide, beneath which a large region of an aluminium rich oxide exists. All the normalised cation fractions tail off towards zero and indicate profiling into the substrate. This is due to the fact that during the SIMS depth profiling, ionisation of the negative cation species takes place more readily in the oxide compared to the superalloy substrate and so low secondary ion counts are obtained in the substrate.

A key observation to be made from these results is that for both superalloys investigated in this study, neither alloy formed a protective alumina or chromia scale and therefore they are likely to degrade rapidly without a protective coating.

Coating MDC150L

Comparison of the oxide scales formed on the coating system MDC150L and uncoated alloy CM186LC reveals that the oxide layer is much reduced for the coating system, even for extended total oxidation periods up to 82 hours. At 870°C, the thickness of the oxide growth on the coating was found to be 0.95μm for an oxidation period of 82 hours, whereas at 826°C, the uncoated CM186LC formed an oxide of thickness 1.2μm for an oxidation period of 22 hours. This is because the coating system is able to form a protective Al_2O_3 scale, which acts as a barrier against the highly damaging environment. Table 4 shows that the average oxide growth for the alumina scale that forms on the coating drops from 0.50nm/min for a total oxidation period of 20 hours, to 0.19nm/min for a total oxidation period of 82 hours. This drop in oxide growth rate indicates the effectiveness of the alumina scale to as a barrier to further oxidation.

The high resolution imaging and isotopic oxygen mapping performed using the FIB200 provided direct characterisation of the oxidation growth mechanisms that occur at 870°C in 20 hours in air and 62 hours in oxygen 18. Figure 6 is a secondary electron image of the crater wall that has formed after depth profiling. This reveals the whisker like surface morphology which is characteristic of θ-Al_2O_3 and has been observed by others [24]. Figure 8 is a secondary ion image showing the ion milled crater on the oxidised surface and the platinum deposit. The oxygen 16 and oxygen 18 maps are shown in figures 10 and 11

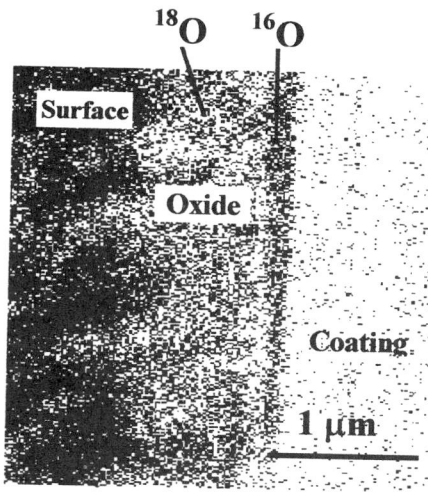

Figure 12: MDC150L Isotopic Ratio Map
Oxidation for: 20 hours in air + 62hours in $^{18}O_2$ at 870°C.

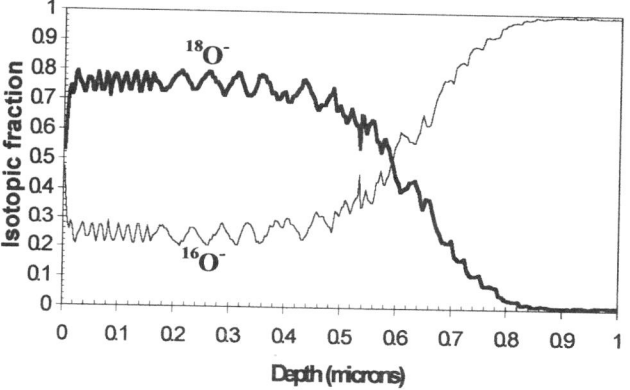

Figure 13: Isotopic Ratio Profile for coating MDC150L obtained via depth profiling for sample oxidised as above. The isotopic fractions are given by: $ir_{18} = \dfrac{^{18}O^-}{^{18}O^- + {^{16}O^-}}$ and $ir_{16} = 1 - ir_{18}$.

The isotopic ratio map that is shown in figure 12 above was determined directly from the oxygen 16 and oxygen 18 maps. The FIB oxygen mapping technique provides very localised information regarding the oxidation mechanism of a very specific area of interest and can be used to compare any isotopic oxygen difference in the oxide scale formed at the coating grain and grain boundary. The results obtained from the isotopic ratio maps are in good agreement with the isotopic fraction profile that was obtained by depth profiling, which is shown in figure 13. This shows that for oxidation treatments at 870°C, the oxygen 18 isotopic ratio is high at the gas–oxide interface and hence relates to outward cation diffusion. This is true for all the oxidation treatments irrespective of oxidation time.

Conclusion

These promising results obtained have emphasised the usefulness of Secondary Ion Mass Spectroscopy and the Focused Ion Beam technique in elucidating the high temperature oxidation of nickel base superalloys and protective coatings. Alloy CM186LC displayed outward cation diffusion for all temperatures between

826°C and 1100°C, whereas CMSX4 has shown a change in mechanism. At temperatures between 826°C and 906°C, this alloy displays outward cation diffusion, whereas at 996°C a combination of inward oxygen and outward cation diffusion is observed. Neither alloy formed a protective oxide scale. The coating system MDC150L displayed outward cation diffusion at 870°C for oxidation periods up to 82 hours and formed a effective alumina scale. The tracer enrichment profiles that were obtained for the uncoated superalloys provide useful information regarding the oxidation mechanisms that occur during oxide evolution. Their respective cation depth profiles provide detailed information regarding the compositional nature of the thermally grown scale. The high resolution and in-situ milling capabilities of the FIB system have been particularly useful in characterising the oxide formed on the coated system MDC150L. Oxygen isotopic ratio maps give direct information regarding the diffusion mechanism and are free from errors associated with depth calibration of depth profiles.

This encouraging application of secondary ion mass spectrometry and the focused ion beam technique has clearly established the usefulness of such analyses in determining the high temperature oxidation of superalloy materials and protective coatings.

Acknowledgements

The authors wish to acknowledge R.Chater for his help and advice with the SIMS analyses and J.Walker from FEI-Europe for his assistance with the FIB experiments. Financial support for this project was provided by the Engineering and Physical Science Research Council, UK.

References

[1]T.N. Rhys-Jones, "Protective Oxide Coatings on Superalloys and Coatings Used in Gas Turbine and Vane Applications." Mat. Sci. & Tech., 4, (1988), 421-430.

[2]J.Jedlinski, G.Borchardt and S. Mrowec, "Transport Properties of Alumina Scales on the β-NiAl Intermetallic" Solid State Ionics, 50, (1992), 67-74

[3]B.A. Pint, J.R. Martin and L.W Hobbs, "^{18}O SIMS Characterisation of the growth mechanism of doped and undoped α-Al₂O₃," Oxidation of Metals, 39, (1993), 167

[4]H.M Tawancy, N. Sridhar and N.M Abbas, "Failure Mechanism of a Thermal Barrier Coating on a Nickel-Base Superalloy" Journal of Mat.Sci, 33, (1998), 681-686.

[5]A.M Freborg et al., "Modelling Oxidation Induced Stresses In Thermal Barrier Coatings" Mat Sci & Engineering A, A245, (1998), 182-190.

[6]B.A. Pint et al., "Substrate and Bond Coat Compositions: Factors Affecting Alumina Scale Adhesion" Mat Sci & Eng A, VolA245, (1998) pp.201-211.

[7]H Caspari,. "Protective coatings extend gas turbine life and efficiency." Reprinted from Power Generation Technology (1995) for Sermatech International Inc.

[8]P Sahoo and G.W Goward.: "On the suitability and application of MCrAlY coatings under various operating conditions". Thermal Spray Industrial Applications – (paper presented at the 8th National Thermal Spray Conference, Houston, Texas, September 11-15, 1995), 539-544.

[9]M.J Graham.: "The study of oxide scales using SIMS", Microscopy of Oxidation. Institute of Materials, No. 500, 10-18.

[10]J Doychak.: "Oxidation behaviour of high-temperature intermetallics.," Intermetallic Compounds, (John Wiley & Sons, 1994), vol. 1, chap. 43.

[11] W.J Quadakkers. et al, "Effect of the θ alumina formation on the growth kinetics of alumina-forming superalloys". Oxidation of Metals. 46, (5/6), (1996), 465-480.

[12] G Borchardt., J Jedlinski. and W Wegener, "Oxygen-18 tracer diffusion as a tool to study high temperature oxidation of metals". (In Proceedings of the 8th. Secondary Ion Mass Spectrometry, Amsterdam, The Netherlands, September 15-20th 1991), 737-740.

[13] J Doychak. and J.L Smialek. "Transient oxidation of single crystal β-NiAl". Metal. Trans. 20A, (1989), 499.

[14] P.T Mosely et al,. "The microstructure of the scale formed during the high temperature oxidation of Fecralloy steel" Corrosion. Science., vol. 24, 1984, p. 547.

[15] D.Garriga-Majo et al., "Novel strategies for evaluating the degradation of protective coatings on superalloys" Journal of Inorganic Materials, 1, (1999), 325-336

[16] A.Alibhai, D.McPhail, B.A Shollock, "Use of isotopic tracers and SIMS analysis for evaluating the high temperature degradation of protective coatings on nickel based superalloys" (In Proceedings of the 12th. Secondary Ion Mass Spectrometry, Brussels 1999).

[17] K Robinson et al. "Applications of focused ion beam milling in biological electron microscopy". Microsc. and Analysis., (1997), 15-16

[18]N. Presser and M.R Hilton, "Applications of focused ion beam machining to the characterisation of carbide, nitride and oxide films," Thin Solid Films, 308, (1997), 369-374

[19]F Wang, P Garoche and L Dumoulin, "Focused ion beam imaging of grains in Al-Li-Cu quasicrystal," Journal. Of Physics-Condensed Matter, 10, (16), (1998), 3479-3488.

[20]B.M. Warnes and D. Punola, "Clean diffusion coatings by chemical vapour deposition," Surface and Coatings Technology, 94-95, (1997), 1-6.

[21]S.N.Basu and J.W. Halloran, "Tracer isotope distribution in growing oxide scales," Oxidation Of Metals, 27, (3/4), (1987), 142-155.

[22]P.S. Burkholder et al., "CM186LC alloy single crystal turbine vanes," (International Gas Turbine and Aeroengine congress and exhibition, Indianapolis, Indiana, 7-10 June 1999).

[23]P.S. Korinko, M.J. Barber and M. Thomas, "Coating Characterisation and Evaluation of Directionally Solidified CM186LC and Single Crystal CMSX4," (ASME Turbo EXPO 1996, NEC, Birmingham, UK).

[24]G.C. Rybicki and J.L Smialek, " Effect of the θ-α-Al_2O_3 transformation on the oxidation behaviour of β-NiAl +Zr," Oxidation Of Metals, 31,(3/4), (1989), 275-304.

SINTERING OF THE TOP COAT IN THERMAL SPRAY TBC SYSTEMS UNDER SERVICE CONDITIONS

J.A.Thompson, W.Ji, T. Klocker and T.W.Clyne

Department of Materials Science & Metallurgy
Cambridge University
Pembroke Street
Cambridge CB2 3QZ, UK

Background

Improved reliability of Thermal Barrier Coating (TBC) systems, under conditions in which the temperature at the free surface may be required to reach values as high as 1400°C, is a critical aim for the next generation of gas turbine engines. It is now becoming clear that a significant contributory factor in the tendency for top coat spallation to occur in TBC systems during service is related to sintering effects in the zirconia top coat. There is already evidence[1] that the stiffness of detached top coats can rise substantially during heat treatments, even after relatively short periods at temperatures as low as about 1100-1200°C. This increase in stiffness will lead to enhanced differential thermal contraction stresses during subsequent temperature changes and hence to an increase in the driving force for detachment of the coating[2]. It seems likely that this phenomenon is at least partially responsible for many of the observed cases of debonding during service.

Part I - Conventional Plasma-Sprayed Coatings

It has been proposed[1] that stiffening can occur in thermally sprayed coatings, at temperatures well below those needed for conventional consolidation of zirconia powder compacts, as a consequence of certain features of the as-deposited structure. This consists of overlapping splats, with microcrack networks within individual splats and relatively poor inter-splat bonding. Strengthening of inter-splat bonds and healing of microcracks can occur as a consequence of local diffusional processes, with diffusion distances appreciably shorter than those necessary to allow a powder compact to consolidate.

A factor which has hitherto received little attention is the effect of the stress state in the top coat on its sintering behaviour. In particular, it might be expected that the presence of in-plane tensile stresses in the top coat would inhibit stiffening by opening up microcracks and preventing the diffusional transport across them expected to be necessary for the crack healing process. There is thus interest both in experimental measurement of stiffening behaviour in the presence of defined stress levels within the top coat and in simulation of the stress state under service conditions. In this paper, an experimental investigation is made of the sintering characteristics with and without the top coat adhering to the substrate. A previously-developed numerical process model[3-5] is used to predict the effect of the imposition of a thermal gradient through the system.

Experimental Procedures

Plasma Spraying

TBC systems were produced using a Plasma Tech Vacuum Plasma Spray (VPS) System. Bond coats (~100 μm) were sprayed at low pressure, while ZrO_2-8wt%Y_2O_3 top coats (~400 μm) were air plasma sprayed (APS). Nimonic 80A substrates (thickness = 1.24 mm) were used. Powder compositions and spray parameters are given in Tables I and II.

Table II: Plasma spray parameters for bond coat and top coat powder deposition.

Material	CoNiCrAlY	Zirconia - 8 wt % Y_2O_3
Spray type	VPS	APS
Spray distance (mm)	270	120
Arc current (A)	500	500
Voltage (V)	55	60-70
Ar flow rate (SLPM[†])	50	80
H_2 flow rate (SLPM[†])	10	18
Gun speed (mm s⁻¹)	85	50
Chamber pressure	200 mbar	atmospheric
Nozzle diameter (mm)	8	8

[†]SLPM = Standard litres per minute

Specimen Heat Treatment

Free-standing top coat samples were heat treated in air at temperatures in the range 1000-1400°C. A Lenton UAF 15/5 furnace was used for this purpose. TBC samples with the substrate attached were sealed in Argon filled silica tubes prior to heat treatment, in order to reduce oxidation of the metallic components. Top coat removal was achieved by immersion in a bath of warm HCl.

Table I: Composition (by weight) of powders for plasma spraying.

System Component	Content by wt%								
	Material	Ni	Co	Cr	C	Ti	Al	Y	ZrO₂	Y₂O₃
Substrate	Nimonic 80A	bal	-	19.5	0.06	2.4	1.6	-	-	-
Bond coat	CoNiCrAlY	32	bal	21	-	-	8	0.5	-	-
Top coat	PSZ	-	-	-	-	-	-	-	bal	8

Superalloys 2000
Edited by T.M. Pollock, R.D.,Kissinger, R.R. Bowman,
K.A. Green, M. McLean, S. Olson, and J.J. Schirra
TMS (The Minerals, Metals & Materials Society), 2000

Cantilever Bend Testing

Coating stiffness was measured using a cantilever bend arrangement on detached top coats, using a scanning laser extensometer (Lasermike) to establish the beam deflection (Figure 1). The Young's modulus can then be deduced from the load / displacement plot using Equation 1, where I is the sample second moment of area and y is the measured deflection.

$$E = \frac{P}{yI}\left\{ \frac{Lx^2}{2} - \frac{x^3}{6} \right\} \qquad \text{Equation 1}$$

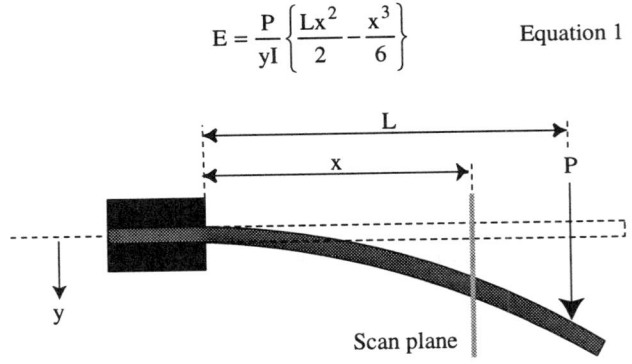

Figure 1: Experimental arrangement for the measurement of coating Young's modulus by cantilever bend technique.

Results

Free-standing coatings

The data in Figure 2 show that, while the stiffness of as-sprayed top coats is typically about 10 GPa, this approximately doubles after a few hours at temperatures as low as 1100°C. The low Young's modulus of the as-sprayed top coat is attributed to the presence of a large number of flaws in the sprayed microstructure, in particular interlamellar pores and through thickness microcracks (Figures 4 and 5).

While little further increase in stiffness occurs with holding at 1100°C, a further progressive rise, to values of the order of 60-70 GPa, occurs at 1300-1400°C. The initial sharp increase is attributed to improvements in inter-splat bonding, whereas the progressive rise is a consequence of microcrack healing.

Figure 3 illustrates the proposed sintering mechanism schematically. At low temperatures, only defects with a small separation can be healed (i.e interlamellar pores where material from adjacent splats is in very close contact). At temperatures in excess of 1200°C, however, diffusion starts to occur considerably more rapidly and the healing of larger defects (through thickness microcracks) becomes possible. The nature of the sprayed microstructure, with two relevant flaw types, thus results in the two stage stiffening observed in Figure 2. The stiffness changes measured will result in higher stress levels in the ceramic top coat after exposure to high temperature. This, in turn, raises the interfacial strain energy release rate, the driving force for top coat debonding and means that the probability of coating failure is increased(6).

Microstructures illustrating the proposed sequence of sintering are shown in Figures 6-9. Note that after 100 hours at 1100°C, microcrack healing is not observed (Figure 7), but there is evidence of grain growth across splat boundaries (Figure 6), giving Stage I stiffening. However, after a similar period of time at 1300°C, both mechanisms are observed (Figures 8 and 9).

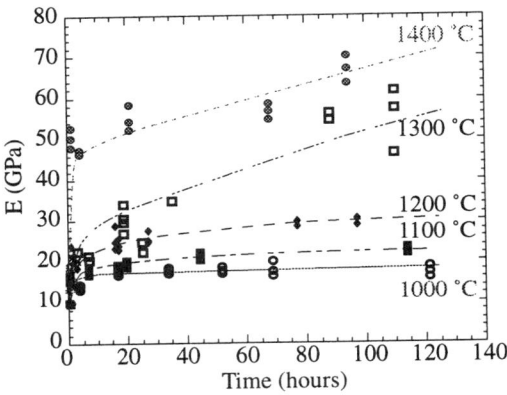

Figure 2: Measured stiffness changes of PSZ top coats after heat treatment (detached from the substrate) at various temperatures.

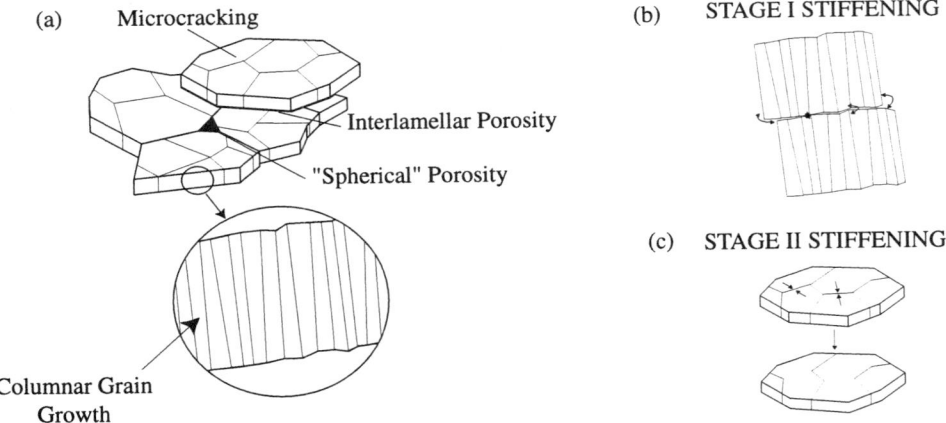

Figure 3: Schematic representation of the proposed stiffening mechanism occurring at high temperarure. The as-sprayed microstructure incorporates three types of pore, (a). Stage I stiffening (b) involves healing of interlamellar porosity, where the separation between material is very small, while at longer times and at higher temperature, Stage II stiffening (c) occurs by microcrack healing.

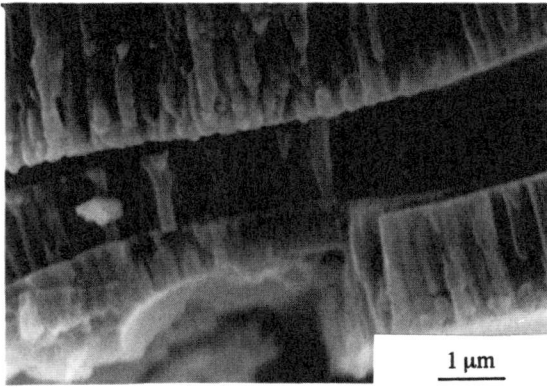

Figure 4: SEM micrograph of the fracture surface of APS-PSZ in the as-sprayed condition.

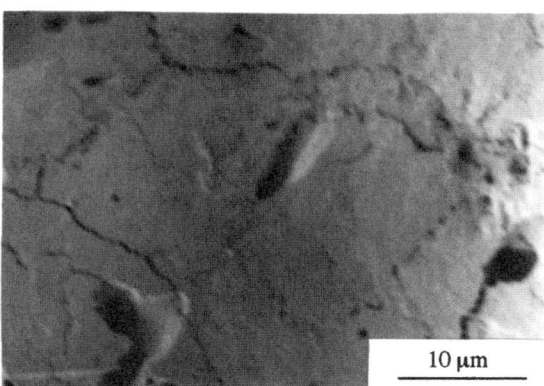

Figure 5: SEM micrograph showing the network of microcracks on the top surface of as-sprayed APS- PSZ.

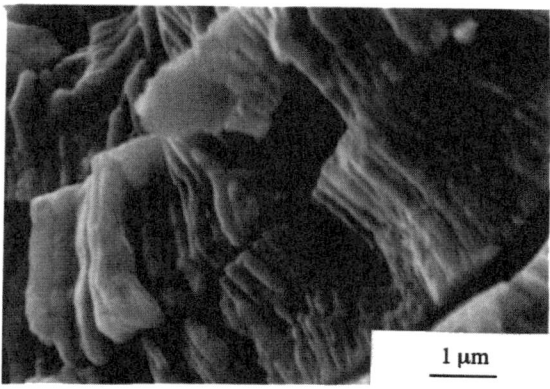

Figure 6: SEM micrograph showing the fracture surface of APS-PSZ after 100 hours at 1100°C. Note where grain growth has occurred across splat boundaries.

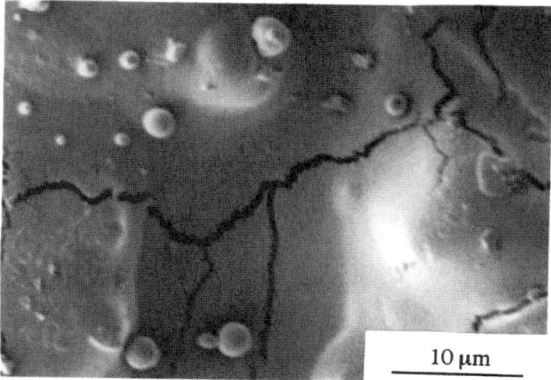

Figure 7: SEM micrograph of the free surface of APS-PSZ after 100 hours at 1100°C. There is no evidence of microcrack closure.

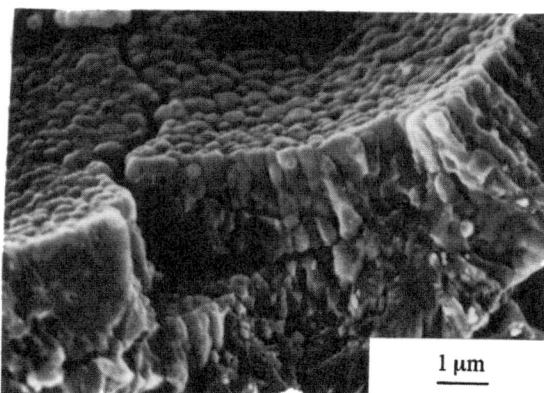

Figure 8: SEM micrograph of the fracture surface of APS-PSZ after 114 hours at 1300°C. Again note where grain growth has occurred across splat boundaries.

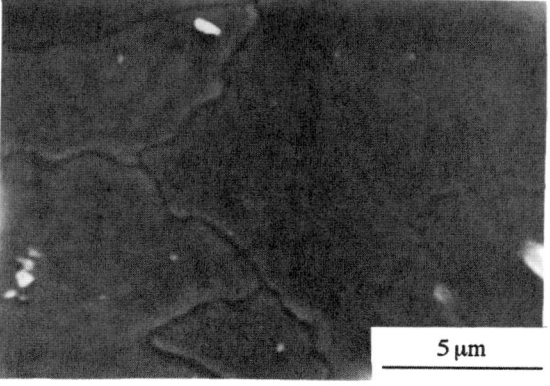

Figure 9: SEM micrograph of the free surface of APS-PSZ after 114 hours at 1300°C. Microcrack closure can be clearly seen.

Attached Coatings

The preceding results apparently indicate that the stiffness of the top coat can rise very sharply during relatively short periods of exposure to temperatures of the order of 1200-1300°C, which would be expected to cause substantial increases in the driving force for debonding as differential thermal contraction stresses rise during subsequent cooling. However, repetition of the cantilever bend test measurements, for specimens which remained attached to the substrates during the heat treatment, revealed that the rates of stiffening were thereby significantly reduced (Fig. 10).

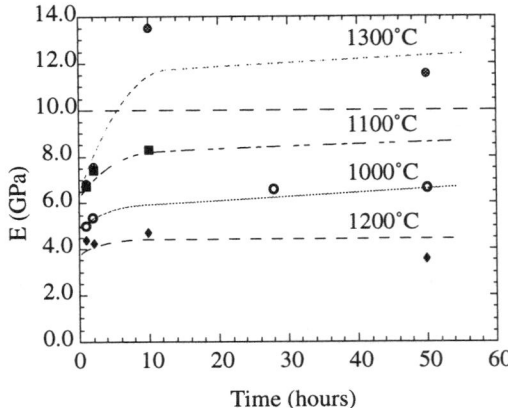

Figure 10: Variation in (room temperature) top coat Young's modulus after various heat treatments (coating attached to substrate during exposure to high temperature).

On comparing these data with those in Figure 2, it is clear that, not only has the presence of the substrate reduced the rate of top coat stiffening, but the heat treatment has actually generated a reduction in the Young's modulus after treatment at low temperature and for short times at the higher temperature (1300°C). It is concluded that, due to the higher coefficient of thermal expansion of the metallic substrate and bond coat, microcracks are held open at elevated temperatures. Thus it becomes clear that the sintering kinetics are a function, not only of the holding temperature and top coat microstructure, but also of the stress state at temperature. This is in principle an unsurprising result, since compressive stress is often applied to powder compacts in commercial sintering operations in order to increase the rate of densification(7), although the magnitude of the effect in this case is certainly striking.

There may be some growth of microcracks in tension, or sintering may occur at temperature such that the cracks becoming locked in in the open position, resulting is a more compliant structure upon cooling to room temperature. Evidence for this is seen in Figure 11, where, in addition to the usual microcracking pattern, a wider crack can also be seen. The presence of such defects will act to reduce the coating stiffness.

However, at higher temperatures (1300°C), the coating stiffness increases to a level above that found in the as-sprayed coating. Due to the increased rate of diffusional processes at such temperatures, sintering can apparently start to raise the coating stiffness. Evidence of microcrack healing is seen in Figure 12.

Numerical Modeling

A limitation of the experimental procedure used in this study is that the sample is isothermal during heat treatment. This should be contrasted with the in-service conditions, when the presence of cooling air channels within the blade, and high incident heat fluxes on the free surfaces, results in pronounced through-thickness thermal gradients within the top coat during most of the time spent at elevated temperatures. Such conditions are very difficult to recreate in the laboratory. However, it is important to examine the likely effect of such temperature gradients.

A previously-developed numerical model(3-5) was used to explore the expected stress distributions in service, with and without an imposed thermal gradient. The results of these simulations are shown in Figures 13 and 14. It can be seen that the presence of the substrate leads to a tensile stress of about 40 MPa for the isothermal case, but this is reduced to a range of about 20 MPa (free surface) to 30 MPa (at interface with bond coat) when the through–thickness thermal gradient is introduced.

That such stresses are expected to have a significant effect on sintering of these coatings can be deduced by recognising that they will generate strains of the order of a few millistrain, thus opening microcracks spaced a few tens of microns apart by increments of the order of 0.1 μm. Since observed typical crack opening flank separations at room temperature are around 0.1 μm, this change is expected to be significant in terms of inhibiting microcrack healing.

However, for the case in which a thermal gradient was imposed, the top coat stresses are reduced significantly in comparison with the isothermal case, particularly at the free surface. It is expected therefore, that the rate of sintering (and thus stiffening) will be fastest at the free surface of the top coat. This is not only due to the higher temperature there, but also to the reduced tensile stress levels arising as a result of the thermal gradient. In order to investigate experimentally the sintering kinetics of the top coat in this regime, using an isothermal arrangement, it would apparently be necessary to either artificially apply a stress of 20 MPa to a detached coating during the heat treatment or to use a substrate having a lower mismatch in thermal expansivity with the top coat.

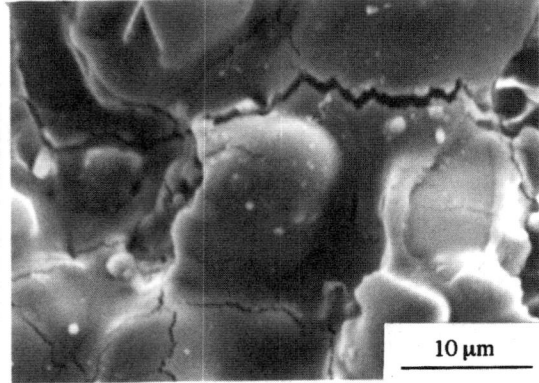

Figure 11: APS zirconia coating after 10 hours at 1100°C. Note how the crack opening displacement has apparently increased for at least one of the microcracks after heat treatment, as a consequence of the tensile stresses present at high temperatures.

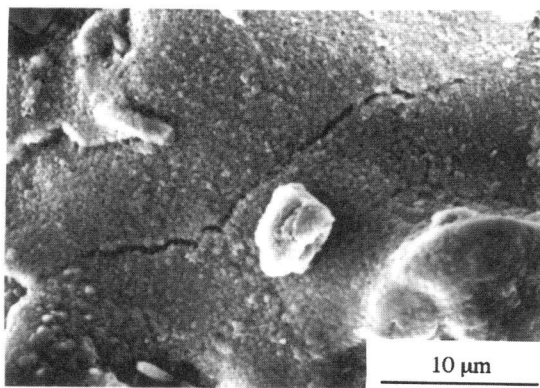

Figure 12: APS zirconia coating after 50 hours at 1300°C. Note the partial healing of microcracks which has occurred.

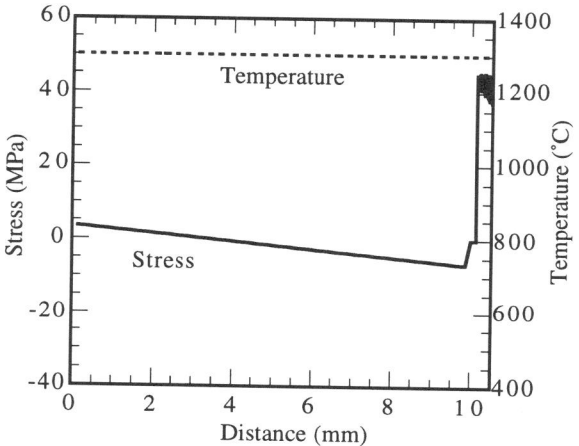

Figure 13: Predicted through-thickness stress (and temperature) distribution for a TBC system after 5 minutes spent at 1300°C (isothermal heating).

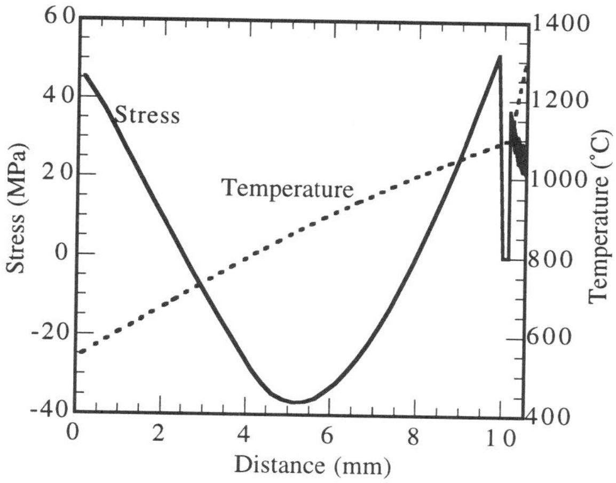

Figure 14: Predicted through-thickness stress and temperature distribution obtained with an increase in the inputted rear face heat transfer coefficient.

Conclusions

It has been shown that the conventional air plasma sprayed micrstructure exhibits desirable strain tolerance (low stiffness) in the as-sprayed state, largely as a result of the network of small defects present. However, it has also been shown that healing of these flaws at typical service temperatures can significantly reduce the strain tolerance of the coating, raising the likelihood of top coat spallation. The presence of the substrate during high temperature exposure has been shown to hinder the sintering process, but it is still considered likely than top coat sintering can play an important role in the in-service failure of thermal barrier coatings.

Part II - Co-spraying of dense and hollow ZrO₂ powder particles

Background

There is thus considerable interest in the production of coatings which are, not only relatively compliant, but also resistant to stiffening during prolonged exposure to elevated temperature. Of course, some factors are known to promote sintering. These include the presence of various impurities(8,9), particularly those which generate low melting point segregate material (e.g. silica). Use of high purity powders is likely to lead to at least some improvement in sintering resistance, but this approach may be unacceptable in terms of the associated economic penalties. The development of microstructures which are inherently less susceptible to sintering is a more attractive concept. This requires that the mechanisms by which stiffening occurs in conventional thermally sprayed coatings should be identified. This has been addressed in recent publications(1,10). It was concluded that the processes primarily responsible for stiffening are: (a) locking together of poorly-bonded overlapping splats and (b) healing of the network of microcracks which forms within each splat during deposition. The type of microstructural feature which is expected to confer a high compliance and resistance to stiffening is relatively large scale pores(11), without the regions of high interfacial curvature which promote rapid local sintering.

Experimental Procedures

Plasma spraying

TBC systems were produced by vacuum plasma spraying of a CoNiCrAlY bond coat (~100 μm) and a ZrO_2-8wt%Y_2O_3 top coat (~400 μm) onto Nimonic 80A substrates (thickness = 1.24 mm). Spray deposition of the top coat was carried out both using conventional, fully dense zirconia powder particles, of about 20 μm diameter, and also by co-spraying this powder with larger (60 μm), highly porous particles (Figure 15). Co-spraying was performed by VPS in order to increase the probability of collisions between molten ZrO_2 particles and unmelted hollow spheres. Under suitable conditions (Table III), this resulted in deposit microstructures exhibiting relatively large pores, associated with the presence in the spray cone of the large, porous particles.

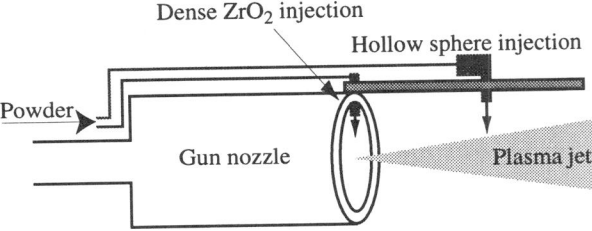

Figure 15: Schematic representation of the powder injection positions for co-spraying of dense and hollow zirconia powder particles.

Table III: Plasma spray parameters for co-spraying of dense and hollow zirconia particles.

Material	ZrO_2 - 8wt% Y_2O_3 (Dense particles)	ZrO_2 - 8wt% Y_2O_3 (Hollow spheres)
Powder feed rate (%max)	15	45
Average particle size (m)	20	60
Spray type	VPS	
Chamber pressure (mbar)	200	
Spray distance (mm)	330	300
Arc current (A)	600	
Voltage (V)	80	
Ar flow rate (SLPM[†])	60	
N2 flow rate (SLPM[†])	15	
Gun speed (mm s^{-1})	50	
Nozzle diameter (mm)	8	

[†] SLPM = Standard litres per minute

Results

As-sprayed microstructure

Co-spraying of dense and hollow powders produced coatings with very different microstructures to those exhibited by conventional plasma-sprayed coatings (Figure 16). It can be seen that the regular lamellar "splat" structure has been disrupted. This is believed to be due to the high proportion impacting unmelted or partially melted hollow particles. Although some intact hollow spheres are trapped in the coating (Figure 17), it was more common to find fragments of broken particles which remained in the coating. The result of this is to raise the overall porosity of the coating. The typical defect size is also increased. It is likely that this will have an effect upon the rate of stiffening due to sintering processes. It should, however, be mentioned that the suitability of such a microstructure with regard to other properties (thermal conductivity, erosion resistance) would require attention before its use could be seriously considered in practice.

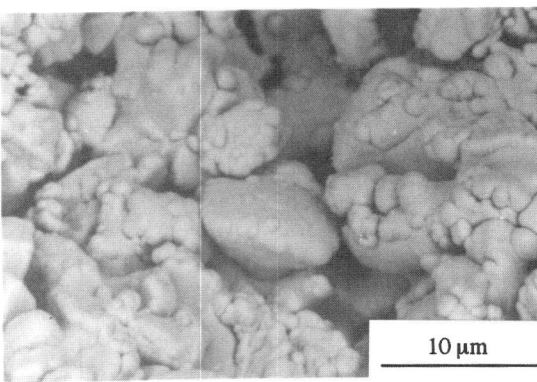

Figure 16: SEM micrograph showing typical morphology and distribution of porosity in the co-sprayed deposits.

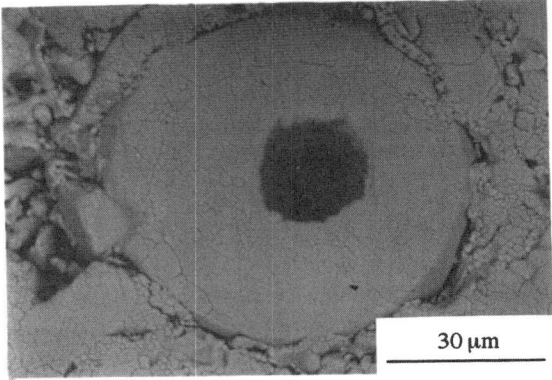

Figure 17: SEM micrograph of a polished top coat cross section for a co-sprayed deposit, illustrating the incorporation of a hollow zirconia particle in the coating.

Co-sprayed Coating Stiffness

An indication of the stiffening characteristics is given in Figure 18. It can be seen that (a) the stiffness of the as-sprayed co-spray material (E ~ 5-15 GPa) is appreciably lower than that of the conventionally-produced coatings (E ~ 10-30 GPa) and (b) the rate of stiffening during heat treatment is considerably lower for the co-sprayed material, at least for the temperatures studied.

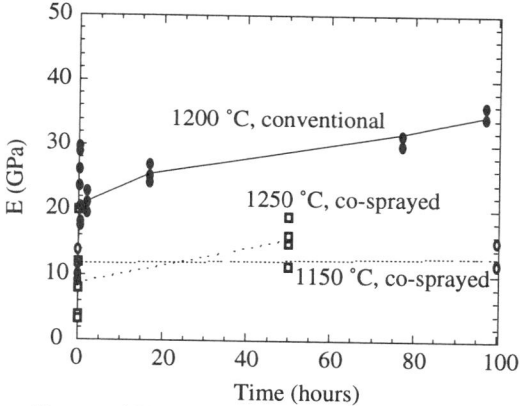

Fig.18 Measured Young's modulus values for ZrO_2-8wt%Y_2O_3 coatings, produced either by conventional spraying or by co-spraying of dense and hollow powder particles, after heat treatments at different temperatures, carried out while detached from the substrate.

These improvements in the high temperature sintering behaviour of plasma sprayed thermal barrier layers are due to the microstructural changes described. Typical pore sizes in the co-sprayed material are of the order of a few microns, while in conventional plasma-sprayed coatings, microcrack opening (in the as-sprayed state) are of the order 0.1-1 μm, while interlamellar distances may be much less than this. Clearly, the healing of such large defects requires the transport of large amounts of material over long distances. However, there is evidence of such processes occurring. Figure 19 shows sintering necks present in a sample heat treated at 1150°C for 100 hours. These necks will lock up the structure in the same way as microcrack closure and interlamellar pore healing in conventional plasma-sprayed coatings.

Figure 19: SEM micrograph showing sintering necks in a co-sprayed ZrO_2 - 8wt% Y_2O_3 coating.

However, the increases in stiffness exhibited by co-sprayed coatings are much lower than those shown by the conventionally sprayed material. A preliminary attempt has been made to model this effect. Using the diffusion coefficient data of Jiminez-Melendo et al(12), for cubic $ZrO2$ - 8wt% Y_2O_3, the √Dt diffusion distances can be calculated and compared with typical microstructural defect sizes (Figure 20). A grain size of 2 μm and

grain boundary thickness of 1 nm were assumed. These diffusion distances are comparable with the defect sizes observed in the sprayed microstructures, and it is clear that the increase in pore size observed in the as-sprayed material is likely to result in a significant increase in the maximum operating temperature of the material. However, the data should be treated with some caution, since the diffusion coefficients used apply to the cubic zirconia system, rather than the metastable tetragonal phase which predominates in the plasma-sprayed material.

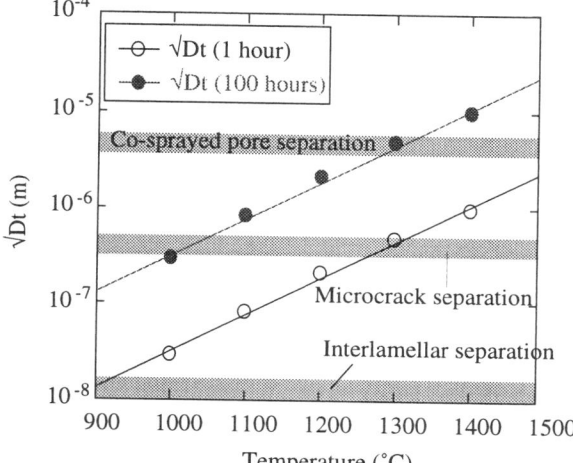

Figure 20: Calculated √Dt diffusion distances for zirconium ions in ZrO_2 (including contributions from both grain boundary and lattice diffusion), compared with typical defect sizes.

Conclusions

Co-spraying of dense and hollow zirconia powder particles, with a downstream injection position for the hollow spheres has resulted in the deposition of a novel top coat micrstructure. Some intact hollow particles are incorporated into the coating, but most break up on impact. The pore distribution is altered significantly in the co-sprayed material, with a higher density of larger flaws.

The stiffening behaviour of the new material shows an improvement over the conventional plasma sprayed material as a result of the new pore distribution. However, some sintering is observed and the growth of necks can be seen over quite large distances (>1 μm) after moderate high temperature exposure (100 hours, 1150°C).

The deposition of novel microstructures for improved resistance to high temperature stiffening by sintering mechanisms has shown some promising preliminary results. However, a greater understanding of the precise mechanisms controlling top coat sintering is required, particularly with regard to the role of coating impurities. In addition, the suitability of such microstructures for use in an engine environment needs addressing. For example, a microstructure exhibiting increased sintering resistance would be of limited use if it were accompanied by a reduction in erosion resistance or a dramatically increased thermal conductivity.

References

1. J. A. Thompson and T. W. Clyne, "The Stiffness of Plasma Sprayed Zirconia Top Coats in TBCs", in United Thermal Spray Conference (1999), Dusseldorf, E. Lugscheider and P. A. Kammer (Eds.), DVS, 835-840.

2. T. W. Clyne and S. C. Gill, "Residual Stresses in Thermally Sprayed Coatings and their Effect on Interfacial Adhesion - A Review of Recent Work", J. Thermal Spray Technol., 5 (1996), 1-18.

3. S. C. Gill and T. W. Clyne, "Stress Distributions and Material Response in Thermal Spraying of Metallic and Ceramic Deposits", Met. Trans., 21B (1990), 377-385.

4. Y. C. Tsui, S. C. Gill and T. W. Clyne, "Simulation of the Effect of Creep on the Stress Fields during Thermal Spraying onto Titanium Substrates", Surf. Coat. Technol., 64 (1994), 61-68.

5. S. C. Gill and T. W. Clyne, "Investigation of Residual Stress Generation during Thermal spraying by Continuous Curvature Measurement", Thin Solid Films, 250 (1994), 172-180.

6. J. A. Thompson, W. Ji and T. W. Clyne, "The Effect of Top Coat Sintering on Ceramic Spallation in Plasma-Sprayed Thermal Barrier Coatings", in 13th Int. Conf. on Surface Modific. Techn.(SMT XIII) (1999), Singapore, K. A. Khor (Ed.) NTU, in press.

7. W. D. Kingery, H. K. Bowen and D. R. Uhlmann, Introduction to Ceramics, Wiley and sons, 1976), Pages

8. M. Verkerk, A. Winnubst and A. Burggraaf, "Effect of Impurities on Sintering and Conductivity of Yttria-Stabilised Zirconia", J Mat Sci, 17 (1982), 3113-3122.

9. H. E. Eaton and R. C. Novak, "Sintering Studies of Plasma Sprayed Zirconia", Surf. Coat. Technol., 32 (1987), 227-236.

10. J. Ilavsky, G. G. Long and A. J. Allen, "Evolution of the Microstructure of Plasma Sprayed Deposits During Heating", in !5th International Thermal Spray Conference (1998), Nice, France, C. Coddet (Ed.) Vol. 2, ASM International, 1641-1644.

11. J. L. Shi et al., "Sintering Behavior of Fully Agglomerated Zirconia Compacts", J.Am. Ceram. Soc., 74(5) (1991), 994-997.

12.   M. Jimenez-Melendo et al., "Cation Lattice Diffusion in Yttria-Stabilized Zirconia deduced from Deformation Studies", Materials Science Forum, 239-241 (1997), 61-64.

OVERALUMINISING OF NiCoCrAlY COATINGS BY Arc PVD ON Ni-BASE SUPERALLOYS

Lucjan Swadzba, Adolf Maciejny, Boguslaw Mendala

Silesian Technical University
Krasinskiego 8b
40-019 Katowice, Poland

Abstract. Investigations of depositing high temperature coatings on Ni base superalloys by the Arc-PVD method, using exothermic reaction processes between Ni and Al with NiAl intermetallic formation are presented in the article. By diffusion heating at 1323K, a NiAl high temperature diffusion coating containing 21 % at. Al and 50 μm thick was obtained. In the next stage coatings with more complex chemical composition- NiCoCrAlY were formed. The NiCoCrAlY coatings were made with two targets. Good consistence between the chemical composition of the targets and a uniform distribution of elements in the coatings was shown. Then the surface was also covered with aluminium by the Arc-PVD method. In the vacuum chamber of the equipment a synthesis reaction between NiCoCrAlY and Al with the formation of NiAl intermetallics of high Co, Cr, Y content was initiated. The final heat treatment of coatings was conducted in air and vacuum at 1323K. Strong segregation of yttrium into the oxide scale in the specimens heated in air was shown. It was possible to obtain NiAl intermetallics phase coatings modified by Co, Cr and Y by the Arc-PVD method. The coatings were formed on cast Ni base superalloy ZS6K and single crystal CMSX-4. An example of the application of this method for the aircraft engine turbine blades was presented.

Superalloys 2000
Edited by T.M. Pollock, R.D. Kissinger, R.R. Bowman,
K.A. Green, M. McLean, S. Olson, and J.J. Schirra
TMS (The Minerals, Metals & Materials Society), 2000

Introduction

Diffusion coatings with an aluminide base are widely used to protect stationary and aircraft gas turbines against high temperature corrosion [1-5]. One of the limitations characterising those coatings is that during their formation, great amounts of elements, included in the coated alloy composition are introduced into their chemical composition. Even small amounts of elements such as molybdenum, tungsten or titanium can decrease the corrosion resistance of the coatings significantly at high temperature. There is a need, therefore, to obtain coatings free of these detrimental elements by forming coatings of which both the chemical composition and the protective properties would be independent of the substrate chemical composition. That was the reason for development of a protective system by forming the MeCrAlY coatings by the EB-PVD method, plasma spraying and the modified methods of PtAl and CrAl coating formation [5]. The development of modern technology and SIP (sputter ion plating) and Arc-evaporation equipment and their modification increased the possibility of producing the MeCrAlY coatings more easily [6-9]. The SIP equipment has certain attractive features such as electron beam evaporation. Furthermore, the equipment is simple. The estimated cost of SIP is also attractive compared with that of plasma sprayed coatings. This is partly because the total cost of SIP is low and also because some processing steps such as post deposition polishing have been eliminated [10]. The best results, however, have been obtained by combining methods of coating based on the MeCrAlY coating deposition followed by the diffusion aluminising [3,11,23]. The investigations of protective coatings and the deposition process by Arc and SIP (sputter ion platting) PVD methods for nickel-base alloys and γ-titanium aluminides were realised in the COST 501 program.[4,12]. The PVD method has been used to deposit PtAl$_2$ heat resistant coatings on TiAl alloys [13,14]. Intensive investigations on obtaining coatings with low sulphur content (clean coatings) are being conducted [15-18]. The purity of (Ni,Pt)Al is directly related to its oxidation resistance. Over aluminizing of MCrAlY coatings can be one of the ways to obtain coatings without detrimental impurities [18]. Arc-PVD methods allow obtaining clean overaluminising MeCrAlY coatings with very small or without active impurities such sulphur, which are present in standard aluminide coatings The Arc-PVD method gives a lot of possibilities for its industrial application [19-23].

Experimental

Materials

The nickel base superalloys with the following chemical composition (wt.%): ZS6K - 9% Cr, 5% Co, 4.5% Al, 5% W, 10,5% Mo, <0,1% C, balance Ni and CMSX-4 - 6.54Cr, 10.34Co, 1.03Ti, 4.69Al, 4.06Re, 5.92Ta, 5.9W, 0.5Mo, Ni-balance were used. Targets for Arc-PVD were produced by a melting method in a Leybold-Heraeus vacuum furnace IS III/5. Then they underwent homogenising in the protective atmosphere at 1373K and their chemical composition was determined. The assumed chemical composition of MCrAlY targets was following (wt.%): Co-32; Ni-38; Cr-21; Al-8; Y-1. The Al and Al-Si targets were also used. The coatings were made on samples and finally on rotor turbine blades of aircraft engine.

Coatings

Coatings were obtained by use of two targets in the form of discs with 100 mm diameter in PVT-550 equipment (Plasma und Vacuum Technik GmbH). The samples were made in the form of rod slices with 10 mm diameter. Before the process the samples were ground on the abrasive papers and then rinsed carefully with detergents at 343K. After that they were degreased in trichloroethylene with use of ultrasonic and dried carefully. During coating the samples rotated on the special plate, and in addition to it around their own axis. The shortest distance from the target was 150 mm. On the basis of earlier tests the basic coating parameters: current of 80A, ion cleaning at a bias from 600V to 50V for 5 minutes, were determined during the whole process. The aluminium deposition process was conducted in the following way: 5 min - 600V, 80 min. - 50V. The sample temperature was 473K during coating. Then the formed coatings were subjected to the diffusion heat treatment process, at 1323K, 4 hours. The obtained coatings underwent examinations of morphology, EDX, structure on a SEM microscope. The distribution of chemical elements was analysed by microprobe analysis.

Results

Coatings on ZS6K alloy

Depositing the heat resistant coatings on the nickel base superalloys by the Arc-PVD method was divided into two steps: NiCoCrAlY deposition and the aluminium deposition on the NiCoCrAlY coating. After the aluminium deposition process the temperature was increased until the moment of the exothermic reaction initiation between Al and Ni contained either in NiCoCrAlY or in the Ni-base superalloys. Those investigations were preceded by an analysis of phenomena during the aluminium deposition on the pure nickel in order to form intermetallic phases by the Arc-PVD method. The final structure formation of the coating takes place in the heat treatment process at 1323K.

Aluminide coatings. Aluminide coatings, which can be classified into HTLA (High Temperature Low Activity), were obtained. The structure and the EDX analysis results of the coating are presented in fig. 1 (a-c).
The outer zone matrix is a β-NiAl phase with the substochiometric Al content. In the precipitates (point of the analysis) greatly increased molybdenum and tungsten concentration and low aluminium and chromium content was found. Between the β-NiAl outer coating and the substrate there is a transient zone with the elongated precipitates with the increased content of the high-melting elements. This zone was formed during the heat treatment as a result of dominant diffusion of nickel in β-NiAl<50at.%Al.

Overaluminising NiCoCrAlY. The developed technological principles of obtaining and characteristic of the aluminide and NiCoCrAlY coatings were the basis of developing the principles of the technology of obtaining the complex MeCrAlY+Al coatings in the Arc-PVD process. The NiCoCrAlY coatings were formed by the Arc-PVD method and then this surface was also covered with aluminium by the same method. That system of coatings was subjected to diffusion heating at 1323K for 4h in vacuum. The structure and EDX results of the obtained coating is presented in fig.2.

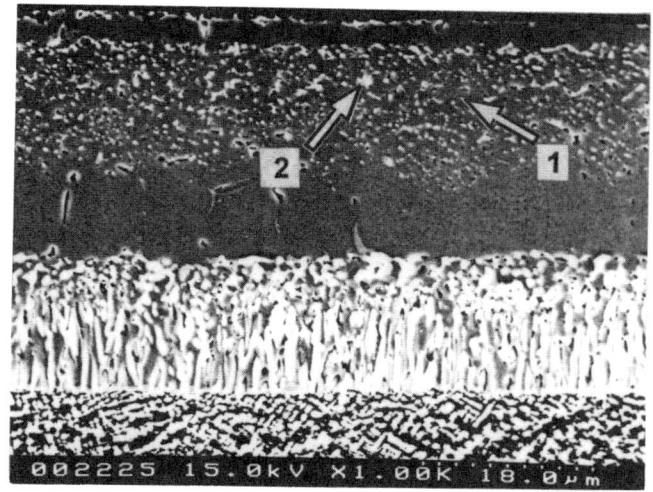

a.

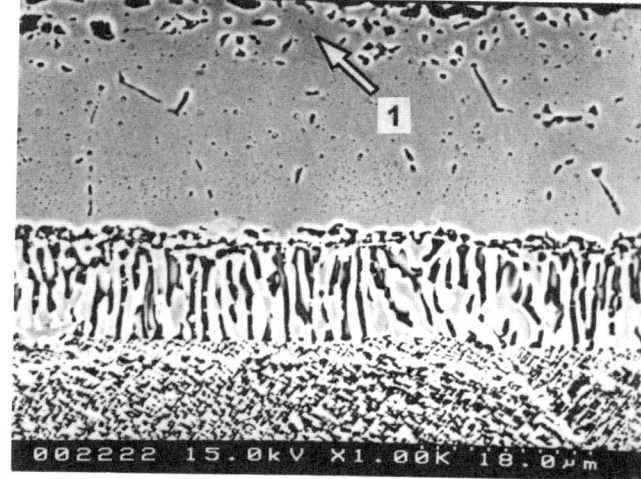

a.

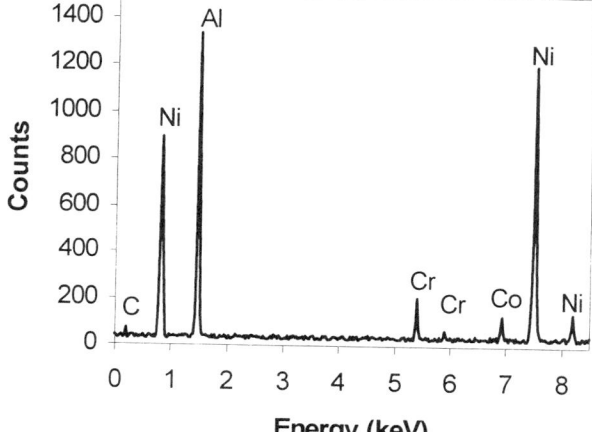

b.

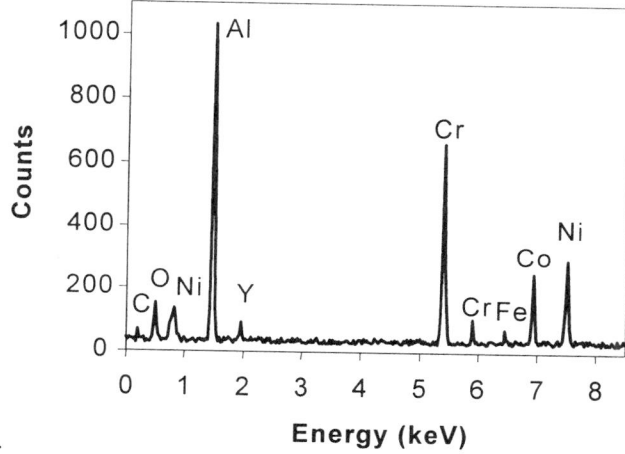

b.

Figure 2: The structure of the NiCrAlY+Al coating obtained by the Arc-PVD method on the ZS6K superalloy: the structure of the coating (a), the EDX analysis in point 1 (b).

The coating formed was about 35 µm thick. The analysis of the chemical composition of elements in the microareas showed that in spite of heat treatment of the specimens in the vacuum, on the surface there was a very thin oxidised zone in which oxygen, aluminium and yttrium are concentrated. In the 2-4 µm zone below the oxidised zone the amount of aluminium and yttrium decreased and the content of nickel, cobalt and chromium increased.

Coatings on CMSX-4

Investigations of NiCoCrAlY and Al coating deposition on the ZS6K cast superalloy experiments and formation of coatings on the CMSX-4 single crystal alloy were made. The coatings were made in the following series: a) NiCoCrAlY deposition on CMSX-4, heat treatment at 1323K, 4h, b) NiCoCrAlY deposition on CMSX-4, Al-Si deposition on NiCoCrAlY, exothermic reaction, heat treatment at 1323K, 4h. The coatings obtained in the particular batches differed in structure and chemical composition.

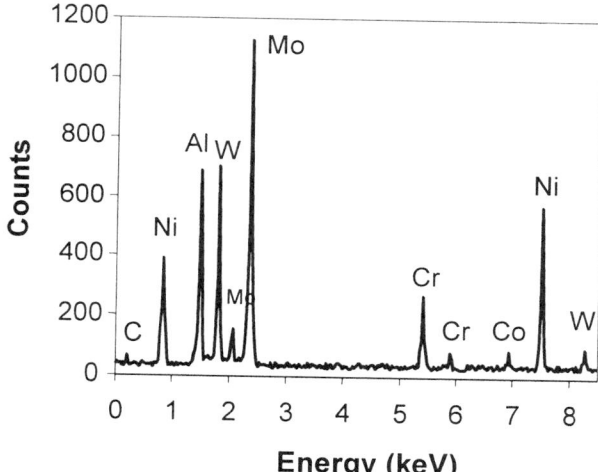

c.

Figure 1: The structure of the Al coating obtained by the Arc-PVD method on ZS6K alloy, heat treated at 1323K/4h (a) and the EDX analysis in point 1(b) and point 2(c).

NiCoCrAlY coatings. The structure of the NiCoCrAlY coating obtained on the CMSX-4 alloy is presented in Fig.3. The concentration of elements in the characteristic microareas of the coating is presented in table 1. The coating thickness is 36 μm. The deep etching revealed characteristic features of the coating structure. It has been shown that the coating consists of elongated discontinuous zones, which are parallel to the surface. Such a structure of the surface results from the deposition process. Specimens were rotated around their own axis and around the axis of the table on which they were placed. Therefore, rotation rate and deposition intensity have significant influence on the coating structure. Single zones have thickness of about 0.5 to 3 μm. Particular zones have a good bond between each other and they were free of any defects, Between the NiCoCrAlY layer and the substrate there is a thin continuous diffusion zone ensuring good adhesion of the coating to the substrate. This zone is formed in the first stage of the process during ion cleaning, when the deposition parameters have the maximal values and temperature of the specimen surface increases.

The EDX quantitative analysis presented in Table I shows that it is possible to deposit, by the Arc-PVD method, all elements which were present in the target. On the basis of the evaluation of the mean concentration of elements in the coating measured on the whole thickness of the coating and width 3mm, presented in table I. The following observations were made. The aluminium concentration in the coating is lower than the concentration of this element in the target, the yttrium concentration is also lower, the amount of cobalt increased, whereas amount of nickel is slightly lower. Good compatibility was obtained with chromium content. The average concentration of this element in the target was 21 wt.% whereas measured in the coating it was 21.82 wt.%, table I. Those results constitute important information for selection of the chemical composition of the target.

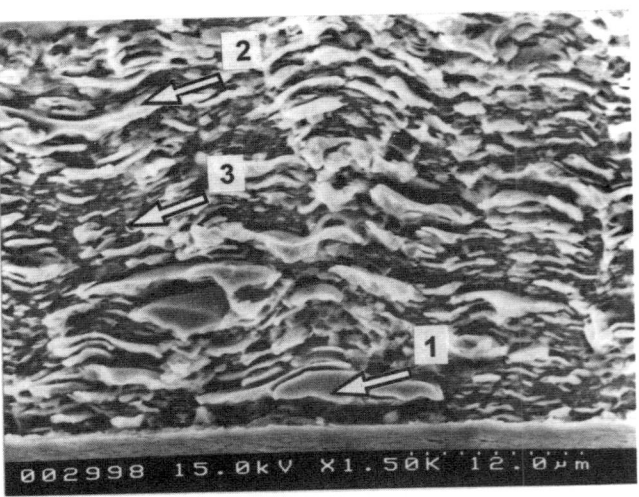

Figure 3: Structure of NiCoCrAlY coating obtained by Arc-PVD method on CMSX-4 alloy.

Aluminide coatings. Aluminide coatings on the CMSX-4 alloy were formed by the Arc-PVD method by deposition of Al-Si from the target with the eutectic composition (11 wt. % Si). After forming the Al-Si coating by changes in the deposition parameters the temperature was increased on the specimen surface above the melting temperature of the Al-Si coating to initiate the exothermic reaction between nickel and aluminium. A coating, being a mixture of phases of strong segregation of elements, with 30 μm thickness was obtained. The X-ray diffraction analysis showed that the $NiAl_3$ phase was the matrix. The characteristic feature of this coating is a clear interface with the substrate. The structure of the coating is presented in fig.4.

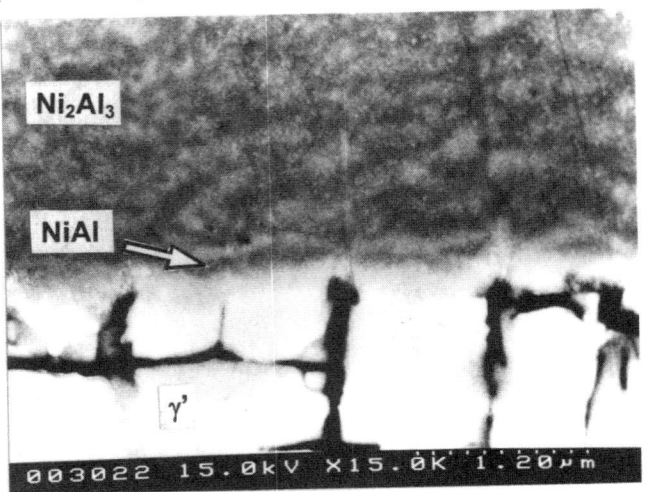

Figure 4: Structure of Al-Si coating on CMSX-4 alloy after the deposition process and exothermic reaction.

The structure of the Al-Si coating obtained by the Arc-PVD method on the CMSX-4 alloy after heat treatment is presented in Fig. 5(a) and (b). The EDX analysis results are shown in Fig. 5(c) and (d). Quantitative analysis results of elements in the microareas are presented in table II.

The structure, presented in fig.5a, has three clear zones formed as a result of diffusion processes in the outer zone and on the boundary with the base alloy. The outer zone is characteristic for high activity coatings of which the matrix of coating is NiAl>50at.% or Ni_2Al_3 phase with high concentration of fine precipitates. The morphology of precipitates in the outer zone with the marked points of the quantitative analysis is presented in fig.5b.

The diffusion zone is formed as a result of dominant outward nickel diffusion. Morphology and size of precipitates differs considerably from one observed in the outer zone. They are bigger and have more irregular shapes. The similar structure was observed in the coatings on cast alloy. However, below the diffusion zone there is a zone which was absent in the aluminium coatings both on traditional cast superalloys and coatings on the CSMX-4 alloy before the heat treatment, fig.1and fig.4.

Table I The concentration of elements in the NiCoCrAlY coating on the the CMSX-4 in the point marked in fig. 3.

Measurement Points	Al.		Y		Cr		Co		Ni	
	at. %	wt. %	at. %	wt. %	at. %	wt. %	at. %	wt. %	at. %	wt. %
1	10.44	5.19	0.47	0.77	16.97	16.26	33.11	35.97	38.29	41.43
2	17.06	8.86	0.48	0.82	15.39	15.41	34.84	39.54	30.41	34.38
3	2.61	1.26	-	-	29.72	27.61	37.02	38.98	30.65	32.15
Average	8.61	4.26	0.30	0.48	22.92	21.82	35.60	38.42	32.57	35.02

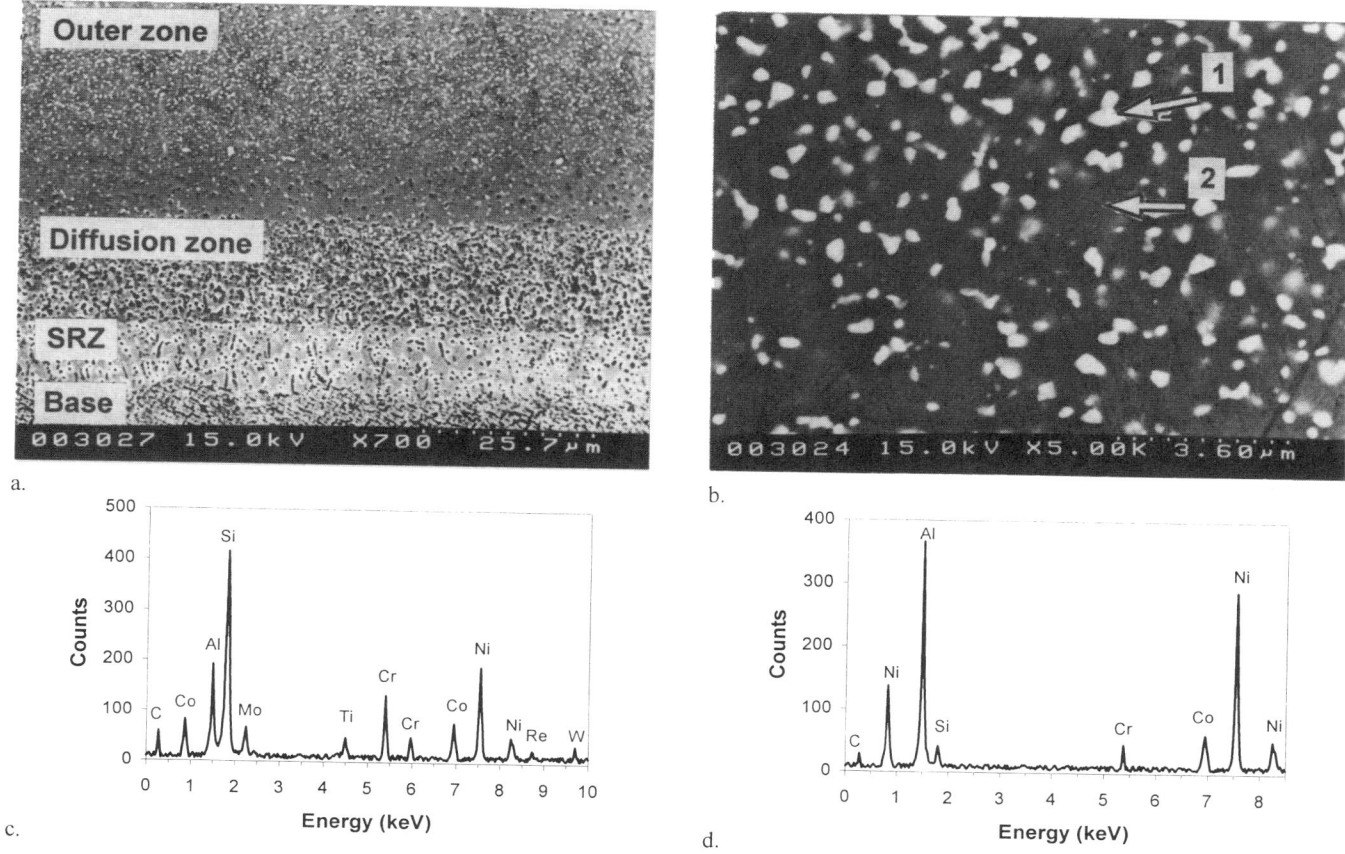

a.

b.

c.

d.

Figure 5: Structure and chemical composition of aluminide coatings obtained by Arc-PVD method and heat treatment at 1323K, 4h: the structure of the coating (a), a fragment of the structure with the marked points of the EDX and quantitative analysis (b), the EDX analysis results in point 1(c) and point 2(d).

Results of qualitative and quantitative analysis of elements in the coating are presented in fig. 5 (c, d). The analysis of elements in precipitates (point 1 fig. 5(b)) showed very high contents of Si, W, Mo and Re. The Re concentration is particularly high and equals to 15.81%. In the coating matrix besides precipitates the chemical composition is completely different. It is the NiAl phase composition without such elements as W, Mo, Re with the small amount of Cr and Si.

Overaluminising of NiCoCrAlY. The structure of coating obtained in the process of the NiCoCrAlY deposition in the first stage and the aluminium deposition in the second stage, initiation of the exothermic reaction and performing the heat treatment, is presented in fig. 6(a)-(d). As can be seen in the figure 6(a) the coating structure consists of three zones: outer zone bordering with the specimen surface, the diffusion zone bordering with the substrate and the zone belonging to the base alloy with the structure of which characteristic features will be presented in the discussion of results. In the outer zone shown in fig. 6(b) there are precipitates of irregular shapes and size at the surface. The structure of the boundary between the outer zone and diffusion zone is presented in fig. 6(c).

The results of analysis of elements in the microareas of overaluminising of the NiCoCrAlY coating are presented in fig.7 and table III. Precipitates in the outer zone show very high chromium content, being 86 wt.% in point 1 in fig. 6(a) and table III, and in other precipitates in this zone above 59 wt.%. In the coating matrix in point 2, fig. 6(b) and table III, the aluminium concentration is 15 wt.%, whereas there was a significant increase in the nickel content and decrease in chromium and cobalt content in comparison with their content in the initial NiCoCrAlY coating. In this zone chromium was found in the amount of 1.31 wt.%. Besides large precipitates having elevated Cr concentration there are also fine spherical precipitates which contain not only basic elements but silicon and yttrium as well, fig. 6(b) point 3, fig. 7(c) and table III point 3. Near the diffusion zone there are also present very fine (white) precipitates marked in fig. 6(c) as point 5. Their approximated chemical composition due to accuracy of the microanalysis at such small precipitates, is presented in fig. 7(d) and table III. The analysis of chemical composition on the boundary with the diffusion layer, point 9, and between precipitates, point 7 fig. 6 (c) and 7 (e, f) and table III revealed significant diversification neither in the qualitative nor in quantitative composition. The white precipitates in the diffusion zone had elevated concentration of tantalum titanium and silicon.

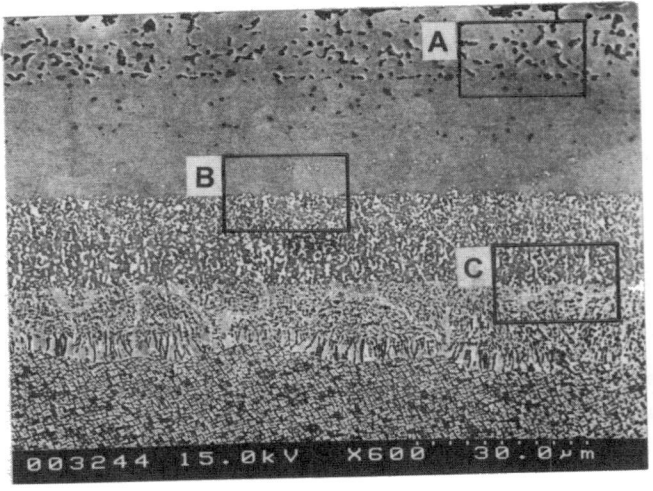

a.

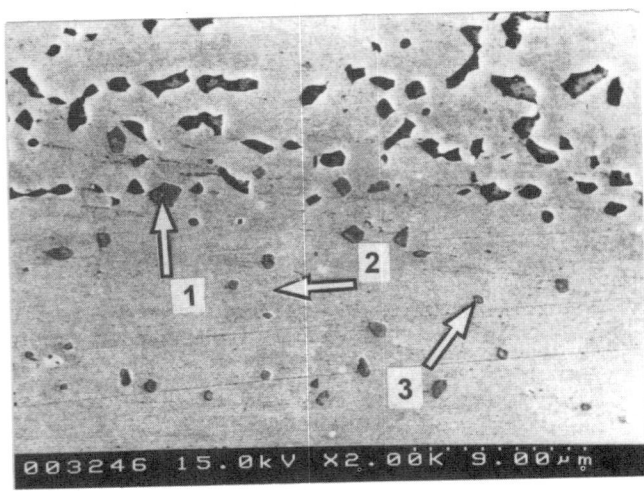

b.

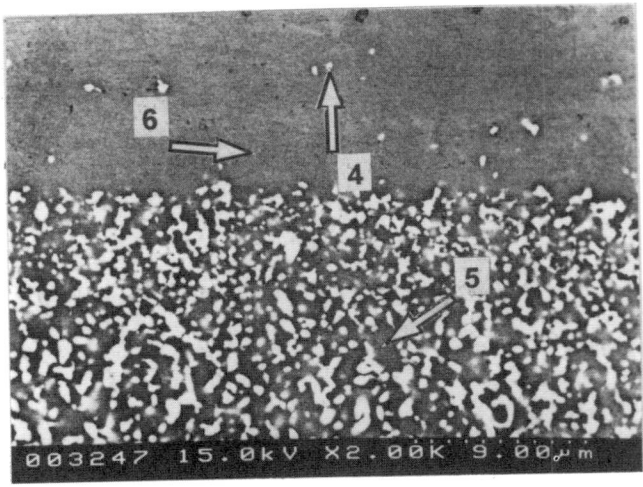

c.

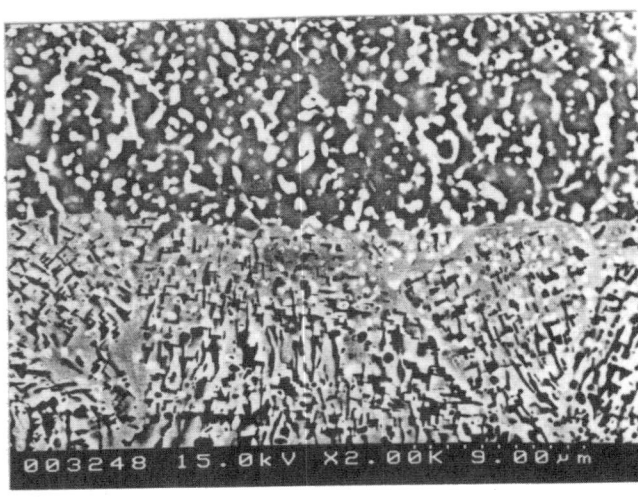

d.

Figure 6: Structure of overaluminising of NiCoCrAlY coatings after heat treatment at 1327K, 4h (a), fragment A (b) fragment B (c) fragment C (d).

Table II Concentration of elements in the aluminide coating shown in fig. 5(b).

point	Al		Si		W		Ti		Cr		Co		Ni		Re		Mo	
	at. %	wt. %	at. %	wt. %	at. %	wt. %	at. %	wt. %	at. %	wt. %	at. %	wt. %	at. %	wt. %	at. %	wt. %	at. %	wt. %
1	13.94	6.40	22.41	10.71	4.19	13.10	1.19	0.97	9.40	8.31	9.61	9.64	32.97	32.93	4.99	15.81	1.31	2.13
2	36.88	21.53	1.95	1.18	-	-	-	-	3.11	3.50	6.21	7.92	51.85	65.87	-	-	-	-

Table III Concentration of elements in the NiCoCrAlY+Al coating in the points marked in fig. 6(b) and (c).

point	Al		Si		Y		Ti		Cr		Co		Ni		Ta	
	at. %	wt. %	at. %	wt. %	at. %	wt. %	at. %	wt. %	at. %	wt. %	at. %	wt. %	at. %	wt. %	at. %	wt. %
1	3.13	1.61	5.86	3.14	-	-	-	-	51.18	50.65	21.72	24.37	18.11	20.24	-	-
2	27.04	14.92	2.28	1.31	-	-	-	-	8.45	8.98	13.74	16.56	48.49	58.23	-	-
3	27.35	15.02	2.37	1.35	1.20	2.16	-	-	8.46	8.95	13.31	15.96	47.32	56.55	27.22	59.80
4	10.61	3.48	15.28	5.21	-	-	11.94	6.94	4.57	2.89	7.30	5.22	23.09	16.46	-	-
5	26.27	14.57	3.81	2.20	-	-	1.03	1.02	6.86	7.33	10.68	12.93	51.36	61.96	-	-
6	27.72	15.35	2.32	1.34	-	-	0.73	0.71	6.17	6.58	10.90	13.18	52.16	62.84	-	-

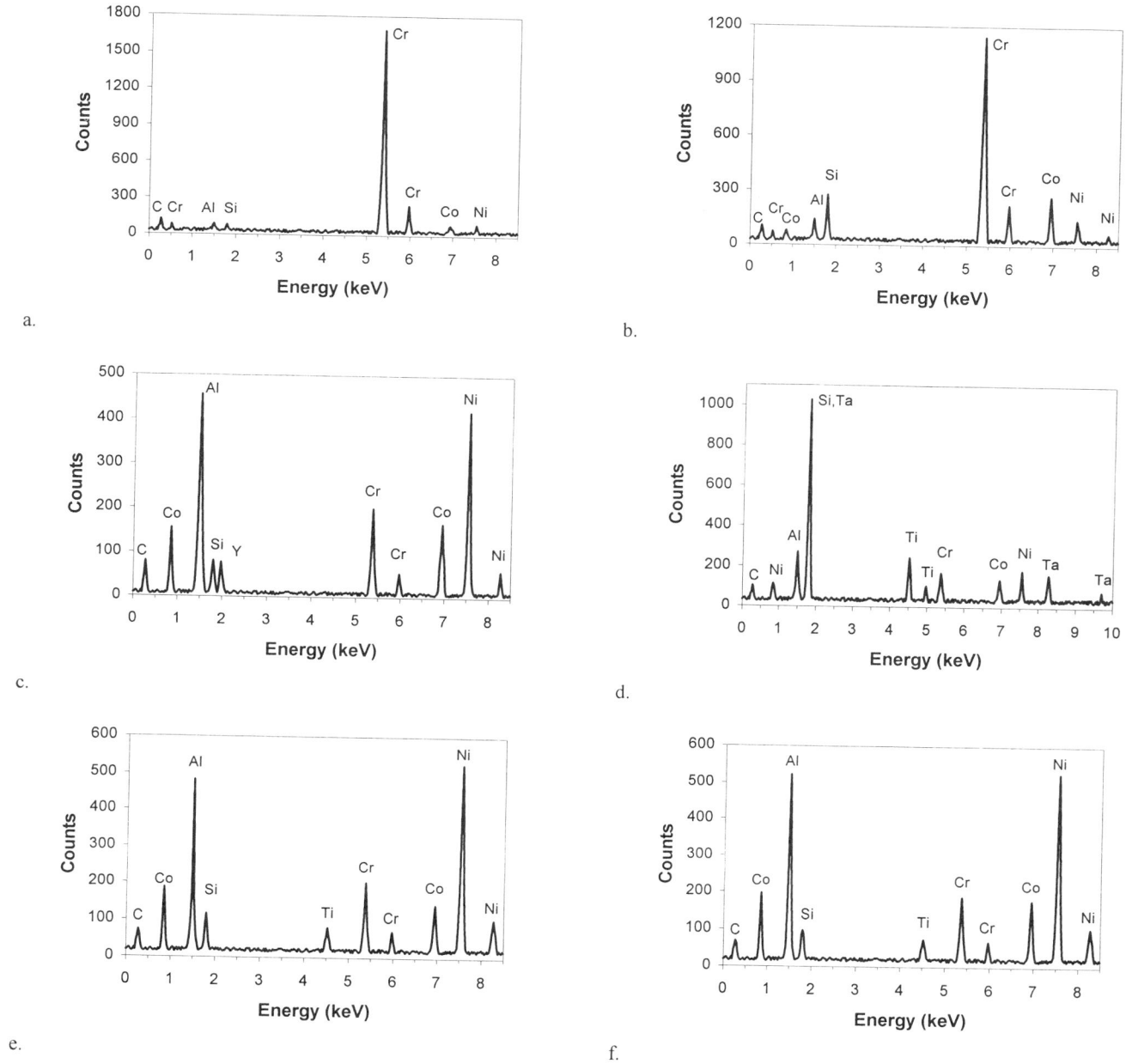

Figure7: Results of EDX analysis in the marked point shown in fig. 6(a), 6(b) and 6(c): point 1(a), 2(b), 3(c), 4(d), 5(e), 6(f).

Discussion

The process of overaluminising of MeCrAlY consists of the following mechanisms assuming as a criterion of classification, the physical phenomena influencing shaping of the structure: 1. the stage of sputtering and deposition of nickel cobalt, chromium, aluminium and yttrium from the target having these elements; 2. the stage of aluminium and silicon deposition from target with eutectic composition; 3. the stage of the exothermic reaction between aluminium and nickel contained in the alloy; 4. the stage of the diffusion growth during diffusion heating in the vacuum furnace.

The earlier investigations showed that it was possible to select such deposition parameters at which the content of Ni, Cr and Co in the coating diverges slightly from the chemical composition of targets. In the selection of deposition parameters J.A. Thornton's model [24], which formulates relationship of the coating structure with the ratio of the substrate temperature to the melting temperature of the deposited element (T_s/T_m) and argon pressure, can be used. Those parameters have an influence on the structure of the obtained coatings. Using those conclusions resulting from that model is needed and possible only in the initial period of deposition when the structure with fine thickened crystallite influencing adhesion is required.

In the next stage when high efficiency of the process is required to obtain appropriate coating thickness (to 25 μm) the deposition process parameters diverge greatly from the requirements resulting from Thornton's model. Shaping the MeCrAlY coating structure is conditioned by a system of the specimen rotation. Both the rate of the table rotation around main axis and the rotation of specimens around their own axis decides about configuration of the structure. Another factor influencing the MeCrAlY coating structure is a number of sources from which evaporation takes place. Two sources ensure high effectiveness and uniformity of coating.

The Al and AlSi deposition by the Arc-PVD method constituted a certain problem, though greater stability of the process was observed during the AlSi alloy deposition.

The Al or Al-Si coating immediately after the deposition was remelted on the surface of the alloy or MeCrAlY coating by an increase in the temperature. Magnitude of thermal effect after initiating the exothermic reaction depended on size of specimens and amount of the deposited aluminium (coating thickness). It has been shown that the NiAl coating structure can be shaped by changes in time of maintaining high temperature of the surface in which aluminium can react totally with the surface or the reaction can be interrupted by which the phase composition of coatings can be shaped. However after finishing the exothermic reaction in which heat came only from this reaction supplied from the external source, thin layer of the pure aluminium was always found on the surface.

An analysis of the structure of the coating obtained after the exothermic reaction showed from the surface presence of: phases- $NiAl_3$, Ni_2Al_3 and very thin NiAl and Ni_3Al phase zones on the boundary with the substrate. The further analysis of the Ni-base alloy aluminide coating deals with heat treatment in the vacuum at 1323K. Firstly, melting of aluminium takes place, after exceeding temperature of 933K, $NiAl_3$ melting after exceeding of 1127K and diffusion transformation leading to an increase in Ni_2Al_3 phase thickness. Further process of the increase in the diffusion layer in the system of Ni_2Al_3 should be considered. In a short time as a result of the Al diffusion to the substrate, the Ni_2Al_3 phase changes into NiAl>50 %at. Al phase in which there is also dominant Al diffusion and the layer growths as a result of Al diffusion to the substrate. When the NiAl<50 %at. Al phase occurs direction and kinetics of the growth change. In the NiAl <50 %at. Al phase nickel diffusion is dominant but the rate of this process is lower than in high aluminium phases [25,26]. From this moment formation process of the so called diffusion zones starts. The nickel diffusion to the substrate changes aluminium and nickel concentration in the outer zone and leads to changes in the phase composition. The temperature and time decided about final structure of the coating. An important problem to be explained is high diversification in the structure of the diffusion zone in the aluminide coatings obtained on the ZS6K and CMSX-4 alloys and the structure of the zone below the diffusion zone observed in the CMSX-4 alloy which was not seen in the coating on the ZS6K alloy.

Microphotograph of aluminide coating presented in fig. 1 shows high thickness of the diffusion zone in comparison with outer one and precipitates perpendicular to the surface and enriched in Cr, Mo , W and Ti. Such structure resulted from thermal treatment at 1323K for 3 h. The total thickness of the coatings is 41 μm. The ratio of outer zone thickness to the transient zone can be measure of growth by Al diffusion inward and Ni diffusion outward. The ratio of the outer zone thickness to the transient zone in the aluminide coating on the ZS6K alloy was 1.6.

The structure of the coating on the CMSX-4 alloy presented in fig.3 is different from one on the ZS6K alloy for several reasons, among others: it is a Al-Si coating, the thickness of the preliminary deposited Al-Si coating was higher than Al on the ZS6K alloy, time of heat treatment of the coating on the ZS6K alloy was 3h whereas on the CMSX-4 alloy time of heat treatment was 4 h. The factors above had an influence on difference in the ratio of the outer zone thickness to diffusion one. The ratio of thickness in the coatings on the CMSX-4 alloy was 2.75 what shows significant dominance of the aluminium diffusion to the substrate over the outward diffusion of nickel.

Diversification of the structure of the zone directly beneath the diffusion zone from the substrate side requires explanation. As it is seen in fig. 1, in the coating on the ZS6K alloy directly under the diffusion zone there is a reinforcing γ phase whereas in the coating on the CMSX-4 alloy the structure of this zone is different what fig. 5 shows. This is a structure similar to one on the aluminide coatings on the single crystal alloys of the third generation. This type of microstructural instability has been found in superalloys containing high rhenium content. This instability has been termed second reaction zone (SRZ) [27] because it typically occurs beneath the diffusion zone of coatings that are typically applied to superalloys. As it is seen in fig. 5 the SRZ zone beneath the aluminide coating is about 10 μm.

However, it was not the subject of the detailed analysis in this article, it seems that the SRZ structure is also present beneath overaluminising of NiCoCrAlY coating on the CMAX-4 alloy, fig. 6 (a,c).

Conclusion

The experimental investigations of which results are presented in this work show possibility of forming the diffusion coatings on the nickel base superalloys by the Arc-PVD method.

The concentration of elements in the MeCrAlY coatings differed insignificantly from their concentration in targets. There was a segregation of yttrium into the outer zone in the air thermally treated coatings.

The aluminide coatings, obtained in the Arc-PVD process and heated in vacuum at 1323K had the structure similar to the coatings obtained by the pack cementation process. NiAl with precipitates of phases with high concentration of Mo, W, Cr constituted the coating matrix.

In the overaluminising of NiCoCrAlY obtained by Arc-PVD and heated in vacuum the (Ni,Co,Cr,Y)Al phase contained very low amounts of detrimental elements such as Mo and W.

Forming the MeCrAlY coatings by the Arc-PVD method followed by diffusion aluminising or Al-Si process allowed to obtain coatings with the chemical composition without influence of the chemical composition of the base superalloys.

In the coatings obtained by the Arc-PVD method on the CMSX-4 alloy, presence of the SRZ zone (secondary reaction zone) was found and this zone was absent beneath the diffusion zone of coatings on a typical casting ZS6K superalloy.

References

1. G.W.Goward, <u>Materials Science and Technology</u>, March 1986, v. 2, 194-199.
2. R.Mevrel, C.Duret and R.Pichoir, <u>Materials Science and Technology</u>, March 1986, v. 2, 201-205.

3 E.Restal, M.Malik and L.Singheiser, "High Temperature Alloys For Gas Turbines and Other Application", Proc. of Conf. held in Liege, Belgium, 6-9 Oct. 1986, (Part.I, ed. by W.Betz, Dordrecht, 1986), 357-404.

4 L.Peichl, D.F.Betridge, "Materials for Adv. Power Engineering", Proc. of a Conference held in Liege, Belgium, 3-6 October 1994, (Part I, Ed. be D. Coutsouradis, 1994), 717-724.

5 R.Streiff, D.H.Boone, "Corrosion Resistance Modified Aluminide Coatings", Conf. on Coatings and Bimetalics for Energy Systems and Chemical Process Environments, Hilton Head, South Calif. 1984, 159-169.

6 Pat. USA 3978251, Aug 31, 1976.

7 Pat. USA 3754903, Aug 28, 1973.

8 J.R.Nicholls, P.Hancock and L.H.Al Yasiri, Mat. Sc. and Technology, Aug 1989, v. 5, 799-805.

9 J.R.Coad, J.E.Restal, Metals Technology, Dec 1982, v.9, 499-503.

10 J.E.Restal, M.I.Wood, Materials Sci. and Technology, March 1986, Vol. 2, 225-231.

11 P.Hancock, Materials Science and Technology, Mar 1986, 310-313.

12 D.F.Bettridge, R.Wing and S.R.J.Saunders, Materials for Advanced Power Engineering, Proceedings of the 6th Liege Conference vol. 5, (Part II, eds. J.Lecomte –Beckers, F.Schubert, P.J.Ennis, 1998), 961–976.

13 UK Patent Application No. GB 9302978.3, Jan 1993.

14 M.J. Deakin, J.R. Nichols, "Surface coatings on Titanium Alloys to Limit Oxygen Ingress", Materials Science Forum Vol. 251-254 (1997) pp. 777-784.

15 Y.Zhang et.al., "Synthesis and Cyclic Oxidation Behavior of a (Ni,Pt)Al Coating on a Desulfurized Ni-base Superalloy", Metallurgical and Materials Transaction A, V. 30A, October 1999-2679.

16 J.A.Haynes et. al., "Effects of Platinum Additions and sulfur Impurities on the Microstructure and Scale Adhesion Behavior of Single –Phase CVD Aluminide Bond Coatings", Elevated Temperature Coatings: Science and Technology III, (ed. by J.M. Hampikian and Dahotre, The Materials, Metals & Society, 1999), 185-196.

17 W.Y.Lee et al., "Effects of Sulfur Ompurity on the Scale Adhesion Behavior of a Desulfurized Ni-Based Superalloy Aluminizes by Chemiczal Vapuor Deposition", Metallurgical and Materials Transaction A, v. 29A, March 1998-833.

18 B.M.Warnes, D.C.Punola, "Clean diffusion coatings by chemical vapor deposition", Surface and Coatings Technology 94 – 95 (1997) 1-6.

19 J.Vetter et al, "MCrAlY Coatings Deposited by cathodic vacuum arc evaporation", Surf. and Coatings Technology, 68/69 (1994) 27-31.

20 O.Knotek et al, "Arc evaporation of multicomponent MCrAlY cathodes", Surface and Coatings Technology, 74-75 (1995) 118-122.

21 L.Swadźba et al, "MeCrAlY Coatings Obtained by Arc PVD and Pack Cementation Processes on Nickel Base Superalloys", in High Temperature Corrosion and Protection of Materials 4, Part 2, (R.Streiff et al., Editors, Materials Science Volumes, Trans Tech Publications,1996) , 793.

22 L.Swadźba, B.Formanek and A.Maciejny, "Modified heat-resistant protective coatings for nickel-base superalloys", International Journal of Materials and Product Technology, v.8, No 2,3,4, 1993.

23 L.Swadźba et al., Materials for Advanced power Engineering, Proceedings of the 6th Liege Conference vol. 5, Part III, (eds. J.Lecomte –Beckers, F.Schubert, P.J.Ennis, 1998), 1513-1522.

24 J.A. Thornton, Journal of Vacuum Science and Technology, A4(6), 1986, 3059-3065.

25 M.M.P.Janssen, Metallurgical Transactions, v. 4, June 1973, 1623.

26 M.M.P.Janssen, G.D.Rieck, Trans. TMS-AIME, 1967, vol. 239, 1372-85.

27 W.S. Walston et al., "A New Type of Microstructural Instability in Superalloys – SRZ.", Superalloys 1996, ed. R.D.Kissinger et al., (TMS 1996), 9-18.

THE INFLUENCE OF B, P AND C ON HEAT-AFFECTED ZONE MICRO-FISSURING IN INCONEL TYPE SUPERALLOY

S. Benhaddad, N.L. Richards, U. Prasad, H. Guo and M.C. Chaturvedi
Department of Mechanical and Industrial Engineering
The University of Manitoba
Winnipeg, MB
Canada R3T 5V6

Abstract

The influence of minor elements, viz., C, B and P, present either individually or in combination, on HAZ micro-fissuring susceptibility of Inconel 718 Superalloy was studied. The micro-fissuring susceptibility was evaluated by hot ductility testing in a Gleeble 1500 system and by the measurements of cracks in HAZs around electron beam welds. It was then correlated to the microstructure of the alloys and the segregation of minor elements to the grain boundaries. It was observed that B affected the HAZ micro-fissuring susceptibility very adversely, but the influence of C was only marginal. The addition of P to the B containing alloy made it more susceptible. However, the addition of C to both the B, and B and P containing alloy was beneficial in reducing the HAZ micro-fissuring susceptibility. The influence of minor elements could be reasonably explained by the segregation and/or formation of precipitates on grain boundaries.

Superalloys 2000
Edited by T.M. Pollock, R.D. Kissinger, R.R. Bowman,
K.A. Green, M. McLean, S. Olson, and J.J. Schirra
TMS (The Minerals, Metals & Materials Society), 2000

Introduction

An understanding of the effect of minor elements to the base composition of Inconel[1] 718 is essential to improving the mechanical properties of the alloy. The effect of elements such as C, B and P on the properties of Superalloys is complex, with their influence being dependent on the chemical composition of the alloy and heat treatment history. Carbon provides a source of carbides, whereas the role of Boron in Superalloys has been generally believed to be beneficial, especially for mechanical properties such as creep parameters. P is generally considered a tramp element, though concentrations of up to 0.015 wt. % in Inconel 718 have been reported, [1-3]. Allvac has also concluded that the effect of a combined addition of B and P, on creep strength of IN718, is much greater than the sum of individual effect of P and B. The role of the elements B, C and P on weldability has been less clear, though initially Pease [4] considered B to be harmful. C has been considered to be either detrimentally affecting the weldability [5], or innocuous [6]. The effect of P has been generally reported to be detrimental to weldability. Savage [7] for example indicated a tripling of the total crack length (TCL) values in Inconel 600 for an increase in P from 0.001 to 0.01 wt.%.

Previous work by the authors has shown significantly adverse influence of B on the micro-fissuring susceptibility of weld HAZ in IN718 [8-10], however to-date, no detailed studies have been published on the effect of P on weldability of IN718. In addition, the influence of the combined addition of C, B and P on the micro-fissuring behavior of IN718 has not been studied. This communication, therefore, presents the results of a study of the effect of controlled additions of C, B and P individually, and in combination, on the weldability of 718 type alloys.

Experimental Methods and Materials

16.0 mm diameter hot rolled bars, of Inconel 718 type alloys with varying B, C and P concentrations (Table 1) in a base of 53.16% Ni, 19.38% Fe, 17.8% Cr, 2.87% Mo, 5.06% Nb+Ta, 0.92% Ti, 0.63% Al, 0.10% Si,, 0.004 % S, 0.0017% Mg, were provided by Allvac. The as-received bars were swaged in two steps, with two intermediate annealing treatments of 1 hour each at 1200°C and 1050°C to produce 6 mm diameter rod suitable for testing in a Gleeble thermal simulator. The specimens were given a final solution heat treatment at 1050°C for 45 minutes followed by two cooling rates, viz. air cooling 11°C/sec and water quenching 284°C/sec.

Scanning electron microscopy with energy dispersive x-ray (EDX) analysis and analytical transmission electron microscopy were used for metallography and phase analysis. Secondary Ion Mass Spectroscopy (SIMS) was carried out to determine the grain boundary segregation of the minor elements.

The weldability of all the alloys was evaluated by hot ductility testing in Gleeble 1500 thermo-mechanical simulation system and by electron beam welding. For hot ductility testing 6.00 mm dia.x100mm long specimens were heated to a peak temperature of 1210°C, held there for 5 seconds and then cooled to the test

temperature and tensile tested to failure. The ductility of the specimens at the test temperature was measured by the reduction in area. The Ductility Recovery Temperature (DRT) and the Liquidation Cracking Temperature Range (LCTR) was used to evaluate the Heat Affected Zone (HAZ) liquation cracking tendency.

Table 1: Concentration of C, P and B in various IN718 Base Alloys

Alloy	C (wt. %)	B (wt. %)	P (wt. %)
988 Base	0.008	< 0.001	< 0.001
982 Base + C	0.031	< 0.001	< 0.001
989 Base + B	0.005	0.013	< 0.001
991 Base+ C+ B	0.031	0.012	< 0.001
T91 Base+B+ P	0.006	0.010	0.022
726 Base + C+ P	0.030	< 0.001	0.022
727 Base +B+C+P	0.033	0.011	0.022

Electron beam welding was carried out at 40 kV, 30 ma and 77 cm/min. Ten metallographic sections were polished and etched and the length of individual micro-fissures measured on an SEM. The results are reported as the average of the Total Crack Length per section (Av. T.C.L.)

Results

Microstructural Analysis

The microstructure of the base alloy (988) was essentially single phase FCC of about 90 μm grain size with a few particles dispersed within the grains and at grain boundaries, as shown in a typical SEM micrograph in Fig. 1(a).

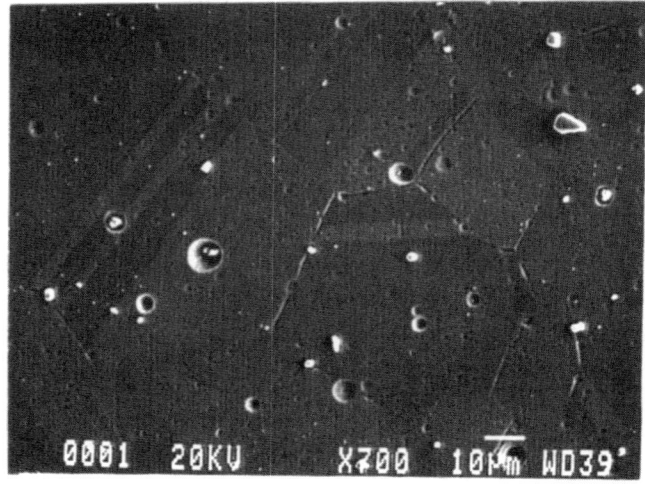

Fig 1 (a) S.E.M. Image

[1] Inconel is a trademark of INCO Alloy International, Huntington, WV.

These particles (78 at. % Nb/16 Ti) were observed to be rich in B and C as shown in an EDX spectrum in Fig. 1b. Due to the software limitation it was not possible to quantify the B and C concentrations. The analysis of selected area diffraction patterns (SADP), obtained by TEM examination of these particles extracted on carbon replicas revealed them to correspond to MC type particles with FCC crystal structure and a lattice parameter of about 0.45 nm. The microstructure of the higher carbon alloy (982), after air-cooling as well as water quenching, was essentially the same as that of 988 except that a few TiN precipitates were also observed.

The microstructure of high B and low C alloy (989) contained two types of precipitate. One was Nb-rich carbide containing B, as was observed in the base alloy (988) and another 38 Nb/31 Mo/23 Cr/5Fe (at. %) containing precipitates; in addition about 3 at. % P was observed in some of the precipitates. The latter particles were also extracted on a carbon replica and a representative TEM micrograph is shown in Fig. 2.

and c = 0.22 nm. The crystal structure and lattice parameter of these particles is the same as was observed by Vincent in the HAZ of Inconel 718 [11]. The alloy containing C and B, (991), also had the same general microstructure with Nb-rich carbide particles containing B, along with some Ti-rich carbides. The EDX spectra revealed the presence of both C and B in the precipitates, but the Nb/Mo/Cr/Fe tetragonal boride was not observed in this alloy.

The microstructure of the high B and high P alloy (T91) was, in general, similar to those of the other alloys, with Nb-rich B containing carbide precipitates being present. However, some P (up to 5 at.%) was also observed in a few Nb/Ni/Cr containing particles that were present in this alloy. In addition, 15-20% of the precipitates examined were of M_3B_2 type borides whose chemical composition was the same as that of the particles observed in alloy 989. Fig 3 (a) shows an S.E.M. image of a collection of Nb/Mo/Cr/Fe boride particles, along with an E.D.X. spectrum from one of the precipitates, Fig 3 (b) and X-ray maps for B. C, P, Ni and Mo in Fig 3 (c).A few Ti-rich carbides, with less than 5% of the other elements in them, were also observed.

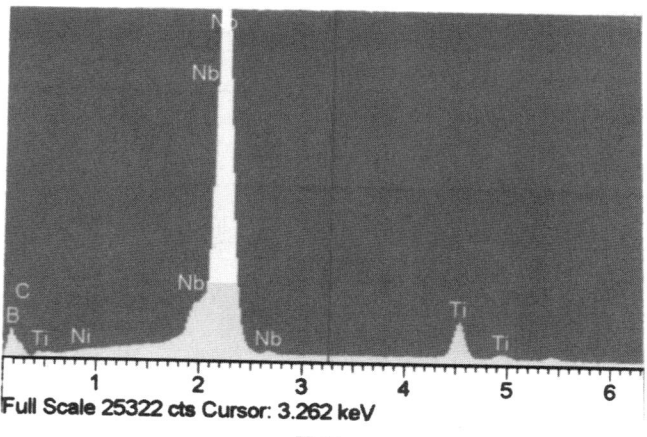

Fig 1 (b) E.D.X. Spectrum of Nb Rich MC Carbide

Fig 2 T.E.M. Micrograph of M_3B_2 Boride

The SADP analysis of these particles revealed them to be tetragonal M_3B_2 particles with a lattice parameters of a = 0.57 nm

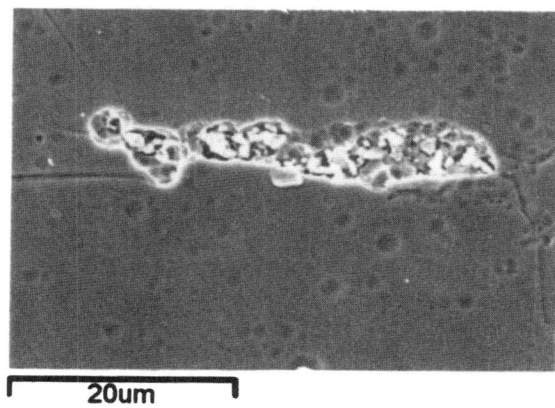

Fig 3 (a) S.E.M. Image of M_3B_2 Boride, T91

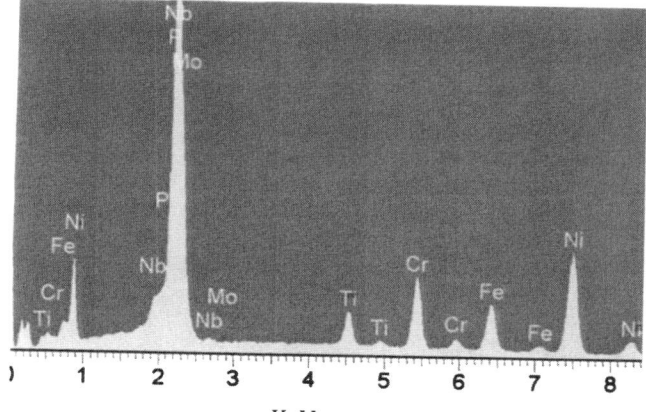

Fig 3(b) E.D.X. Spectrum of Boride, T91

705

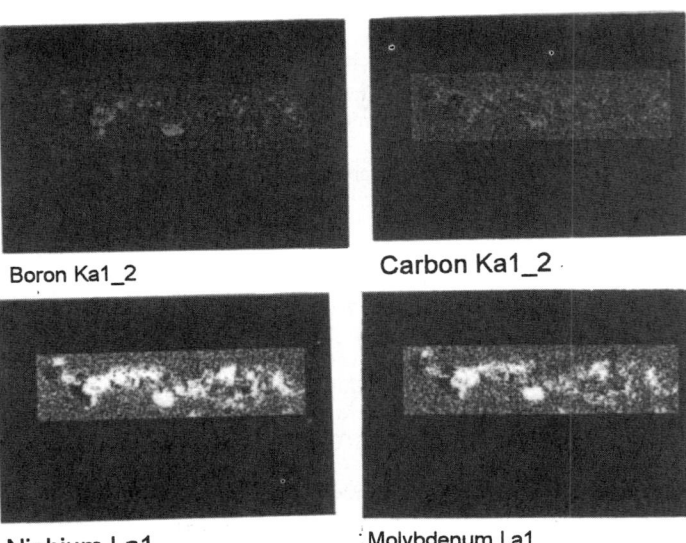

Boron Ka1_2   Carbon Ka1_2

Niobium La1   Molybdenum La1

Fig 3 (c) X-Ray Maps of B, C, Nb and Mo from Boride

The two remaining alloys, high C and high P (726) and high C, high B and high P alloy (727) also exhibited the same general microstructure. The high C and high P alloy (726) contained mainly Nb-rich B containing carbide particles of varying sizes. Minor amounts of Nb/Ni/Ti/Cr/Fe-rich precipitates were also present. This phase was also found to be FCC with a lattice parameter of 0.45 nm, i.e. it was a compositional variation of the MC phase. Fine carbides (<1.0 μm) were observed in the highly alloyed (B, C, P) material (727), which were of the Nb-rich type. No B was observed in any of the EDX spectra; however, some Ti-rich carbides were present.

Segregation Analysis by Secondary Ion Mass Spectroscopy

The segregation of B, C and P was examined by a Cameca Secondary Ion Mass Spectrometer (SIMS). The analysis of alloy 988 and 982 only showed the presence of carbon as precipitate, Fig 4(a), corroborating the observations by SEM and TEM, while a very small degree of segregation of B at the grain boundaries was also observed in the A.C. alloy.

SIMS analysis of alloy 989, containing high B concentration, revealed the presence of B in particles as well as on grain boundaries (Fig. 4 b). However, the degree of segregation of B at grain boundaries was smaller in the WQ specimens as compared to the air-cooled (AC) specimens, as was also observed by the authors in their previous studies [9-10]. The high C and high B alloy (991) exhibited similar images, especially in the AC specimen, though the TEM studies did not reveal the presence of boride particles. The B-ion images in SIMS are thus likely to be from Nb-rich carbide particles instead of boride particles. In the AC high B and high P alloy (T91), segregation of B was observed at the grain boundaries as well as in the precipitate particles (Fig. 4c). However, a significantly small degree of segregation was observed in the water quenched specimen, as would be expected if segregation of B was to occur by non-equilibrium segregation mechanism. It should be noted that in the high C and high P alloy when B concentration was less than 10 ppm, some B segregation was observed at the precipitates. This is corroborated by the SEM/EDX observations of B and C segregating to the precipitates.

The high C, high B and high P alloy (727) exhibited the segregation of B at the grain boundaries (Fig. 4d) in the AC specimens but only a few B containing particles were observed. In the water-quenched specimens the segregation of B to carbides was even more limited. This is consistent with the SEM/EDX observations, where only carbon was observed to be present in the particles and B was not observed. The SIMS analysis of P in high P containing alloy (T91) did not reveal any segregation of P either to the grain boundaries or to the precipitate particles, although some P was detected in a few particles by SEM/EDX analysis. It is, therefore, likely that P was generally present as monolayer in the grain boundaries but could not be detected by the SIMS used in this study. Horton et al [3] detected segregation of B, C and P to grain boundaries, but only a few M_3P precipitates of < 50 nm in size were observed.

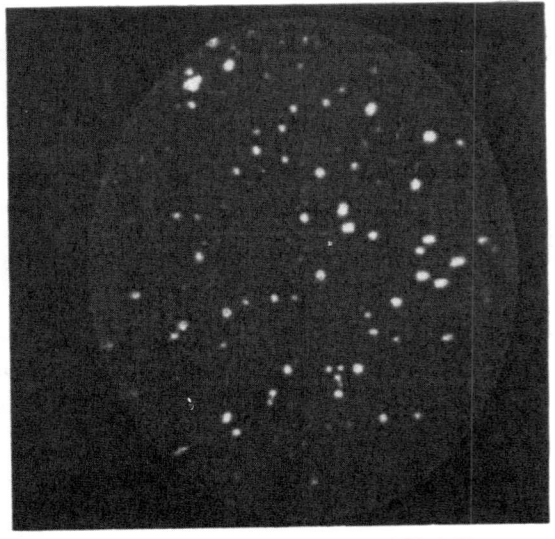

Fig 4(a) S.I.M.S. C Image, 988 A.C.

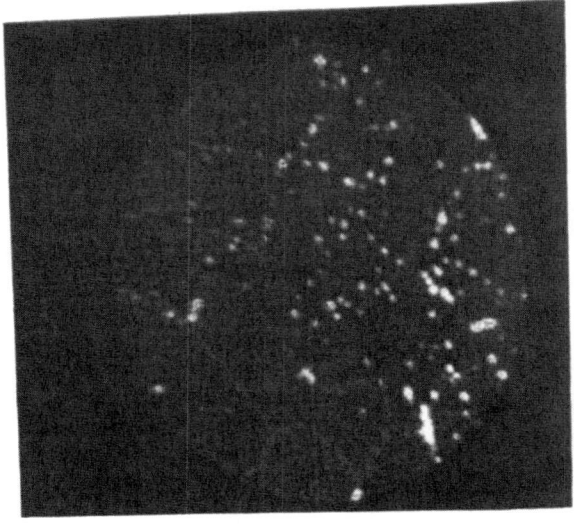

Fig 4 (b) SIMS B Image, 989 A.C.

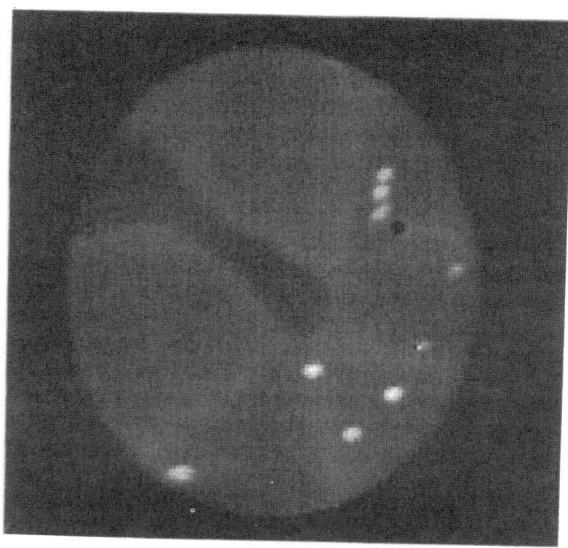

Fig 4 (c) S.I.M.S. B Image T91 A.C.

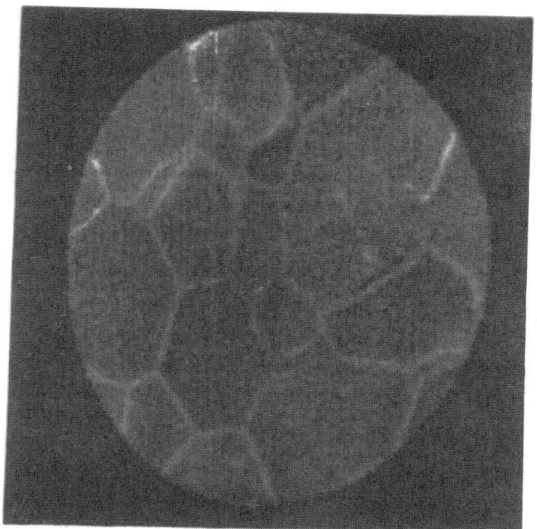

Fig 4 (d) S.I.M..S. B Image 727 A.C.

Hot Ductility Testing by Gleeble 1500

HAZ microfissures during welding usually form because of a lack of ductility in the material near its melting point. Therefore hot ductility measurements in a Gleeble thermo-mechanical simulator are conducted to study the weldability of a material. One of the commonly used parameter is called ductility recovery temperature (DRT). It is defined as the temperature at which a material recovers its ductility on cooling from the melting point or peak temperature[12]. The DRT values of various alloys were measured by hot ductility testing in a Gleeble 1500 system.. The 6 mm dia. X 100 mm specimens were heated at a rate of 150°C/s to 1210°C, held there for 5 seconds then cooled to various temperatures and tested to failure. The effect of C, B and P, individually and in combination with each other, on the weldability was analyzed by plotting the DRT values in four different graphs shown in Fig. 5 a-d. As seen in Fig. 5a the base

alloy had the highest DRT temperature of 1200°C i.e., its weldability was highest.

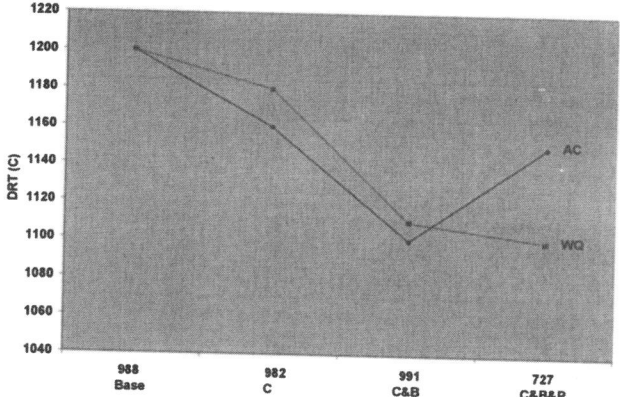

Fig 5 (a)   D.R.T. vs. B, C and P Additions

The addition of C reduced the DRT value by 30-40°C with the value of AC specimens being somewhat lower than that of the WQ specimens. The addition of B to the C containing alloy (991) reduced the value of DRT even further by 60-80°C. Again, the DRT value of the AC specimens was smaller than that of the WQ specimens by a small, but noticeable, amount. It was also observed that the addition of P to the alloy containing C and B (i.e. alloy 727) raised the DRT value of the AC specimens but that of the water-quenched remained almost unchanged. The influence of the addition of P to the C containing alloy (726) is

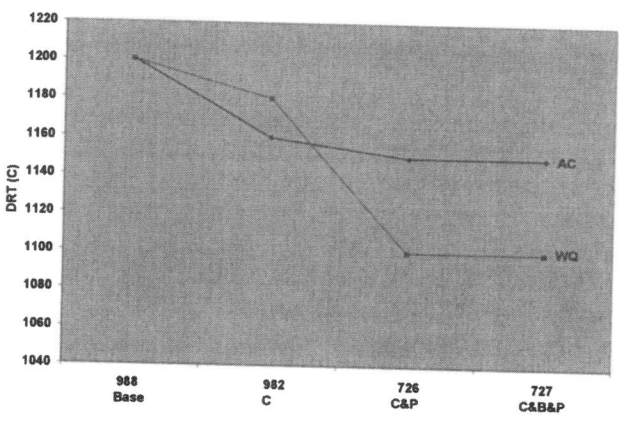

Fig 5 (b)   D.R.T. vs. B, C and P Additions

shown in Fig. 5b. It is seen the addition of P had no influence on the DRT value of the air-cooled specimens but reduced it in the WQ specimens from 1180°C to 1100°C. The addition of B to this alloy (727) had, curiously, no influence.

Fig. 5c shows the effect of B on the DRT value. It is seen that B has the greatest influence reducing the DRT value of the base alloy from 1200°C to about 1090°C in both the air-cooled as well as WQ condition. The addition of C to the B containing alloy improved the DRT values to a small extent. However, as shown in Fig. 5d, when P was added to the B containing alloy the DRT values of the alloy reduced even further (alloy T91). The ternary addition of C to this alloy (727) increased the DRT values in both the WQ and air-cooled specimens, with improvement being greater in the air-cooled specimen (Fig. 5d).

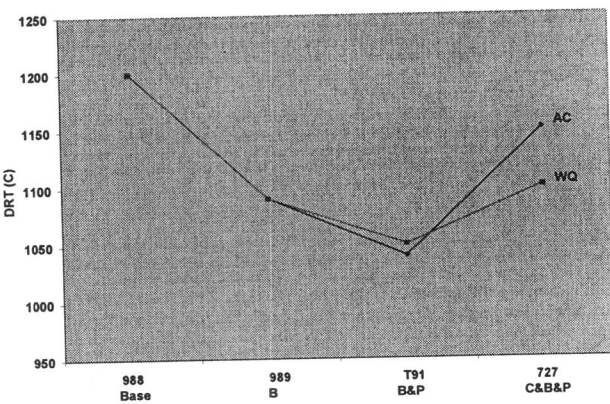

Fig 5 (d) D.R.T. vs. B, C and P Additions

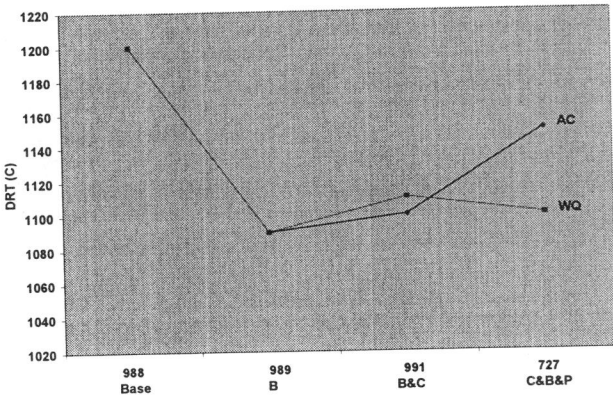

Fig 5 (c) D.R.T. vs. B, C and P Additions

(727) showing varying effect with the cooling rate, Fig. 6 (a). Fig. 6(b) show that in the sequence base, base+C, base+C+P, base +C+P+B, there is a continual increase in TCL with minor alloy addition. On the other hand, in the sequence, base, base+B, base+B+P, base+B+P+C, the detrimental effect of P is evident when C is low, Fig. 6, in both the AC and WQ conditions. With C present, Fig. 6, the detrimental effect of P is somewhat mitigated.

These results suggest that the adverse effect of B on the DRT values and weldability is most severe. The addition of P to B alloys reduces the weldability even further. However, the addition of C to the B, as well as to the B and P containing alloy is beneficial, although when only C is present the weldability of B and P free alloy is reduced to a small extent.

Evaluation of Micro-fissuring in EB Welds by Total Crack Length Measurements

The weldability of various alloys was also determined by measuring the lengths of micro-fissures in cross-sections of EB welded samples in a SEM. The average of the sum of crack length measured in 10 cross-sections was designated as the average total crack length or TCL. The variation in the values of TCL by the additions of B, C and P is shown in Fig. 6 (a-b).

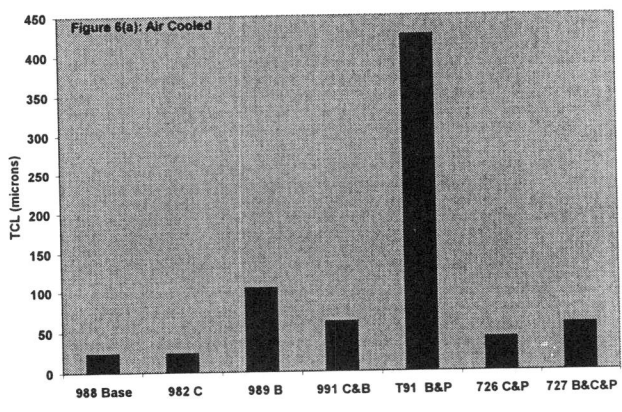

Fig 6 (a) T.C.L. vs. Alloy, for A.C. and W.Q.

In general, the values of TCL followed the same trend that was exhibited by the DRT values. The lowest TCL values were seen in the base alloy (988) with a TCL of 24 μm. The addition of C (alloy 982) did not significantly affect the TCL in either the AC or WQ condition. In contrast, the addition of high B (989), did increase the TCL to 105 μm and 72 μm in the AC and WQ condition, respectively.

Adding B (991) to the base plus C alloy (982), increased the TCL values in both the AC and WQ conditions with the addition of P

Discussion

To attempt to understand the effects of the additions of C, B, P, individually and in combinations, on weldability through their effect on the DRT and TCL parameters, it is necessary to briefly review current theories on the influence of secondary and tertiary addition on segregation of minor element to grain boundaries. Two main theories have been proposed, one by Guttmann [13] and another by Erhart and Grabke [14]. Guttmann [13] considered a ternary addition to a binary alloy, where the ternary

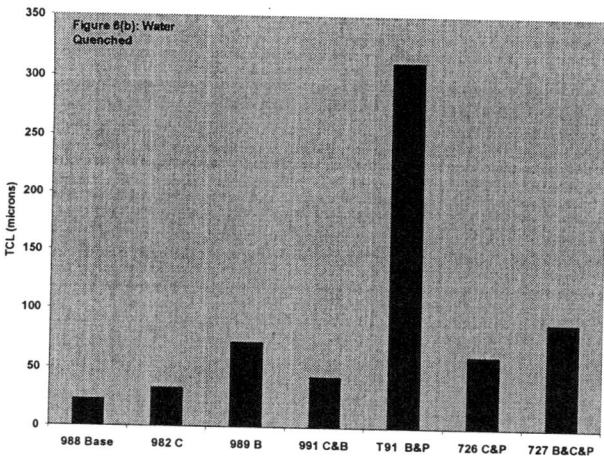

Fig 6(b) T.C.L. va. Alloy, for A.C. and W.Q.

element was able to influence the solubility of the tramp element such as S and thus increase the grain boundary segregation.
It should however be noted that Guttmann's theory is more likely to be applicable to higher alloying additions (> 1%), than used in the present study.

Site competition was however proposed by Erhart and Grabke [14], in a systematic investigation of P segregation in plain C and Cr alloyed steels. They showed that C for example could displace P from grain boundaries. That is, site competition rather than an intensification of the segregation of a minor element by a ternary addition is responsible for the increased concentration of minor elements to the grain boundary. The addition of Cr to a steel, resulted in the Cr reacting with C, which reduced the activity of C, allowing P segregation to occur on the grain boundary.

In analyzing the effect of minor elements on weldability another factor that needs to be considered is the potential effect of increased C additions on the solidification path. Thompson [5] has shown that increasing the C level can influence the solidification path via increased fractions of carbides, ensuring that the liquid film is retained to lower temperatures relative to a smaller amount of liquid. Similarly, Dupont et al [15], have shown that C additions of > 0.1 wt % can increase the initiation of the reaction, L -> γ + Laves, and a decrease in the temperature interval of the primary L -> γ reaction. Considerable improvements in solidification cracking behavior at higher C levels was observed during Varestraint testing, using the maximum crack length as the failure criterion.

Effect of the Chemical Composition

The discussion on the effect of minor elements on microstructure and HAZ micro-fissuring is divided in three groups: individual effect, interaction of two and of three elements.

Individual effect: The effect of C can be evaluated by comparing the behavior of the base alloy (988) and the higher C alloy (982). A carbon addition to the base alloy increased the volume fraction of carbides, though the overall increase in C level from 0.008 wt. % to 0.03 % had little effect on the TCL values. This is in agreement with Kelly's observation [6] that C additions up to 0.1 wt % had no effect on the TCL values in cast Inconel 718. Thompson (5) on the other hand found that the TCL values

increased by 24% in cast 718 on raising the C level from 0.02 wt % to 0.06 wt.%. In the present situation the increase in C level from 0.008 wt. % to 0.031 wt % has had little effect on the TCL values (Fig 5), though the DRT was reduced by up to 40°C. This is probably related to the amount of carbide particles present and the potential availability of liquated phase during the thermal cycle.

The effect of B on the weldability parameters from the present work is seen by comparing the base alloy (988) to the B alloy (989), (Fig. 5). The addition of B led to the precipitation of boride particles in the solution treated condition and to significant segregation of B to the grain boundaries, especially in the A.C. condition. This modification of the microstructure affected the ductility of the material. The high B alloy (989) showed increased micro-fissures after EB welding (Fig. 5) compared to alloy 982, especially in the A.C. specimen, and a low DRT in Gleeble testing (Fig.5 c & d). The segregation of B on grain boundaries lowered the solidification temperature of the boundaries and led to the formation of micro-fissures due to cooling stresses after welding [8-10].

From a purely weldability point of view therefore, B is considered detrimental to the weldability of Inconel 718 as demonstrated through its influence on the hot ductility DRT parameter and the TCL values.

Interaction of two elements: 1. Carbon-Boron: This interaction can be evaluated by considering the DRT and TCL values of alloys 988, 989 and 991.   Summarizing from the results section, the addition of B to the 0.03 wt. % alloy (982), resulted in a reduction in the value of DRT and a slight increase in the TCL values (i.e. reduction in weldability).   The addition of C to the B alloy, i.e. alloy 991, showed a slight improvement in DRT and a moderate improvement in TCL values (i.e. improvement in weldability).   The detrimental influence of the addition of B to the 0.03% C alloy on the weldability parameters is related to boron's well known effect in reducing weldability, as explained by Huang et al [8-10] and outlined in the previous section, i.e., B exacerbates the effect of C additions in the range of from as low as possible to up to about 0.1 wt %, via a combination of equilibrium and non-equilibrium segregation.

In the case of the C addition to the B alloy, since the C addition is not large enough to effect the solidification path (> about 0.1 wt % needed according to Dupont et al, [15]), it is possible that the C can compete with B for grain boundary sites and thus mitigate to some extent the deleterious effect of B. In addition, with more carbides present, some B is incorporated into the carbides and consequently reduces the effectiveness of B on the grain boundaries in influencing weldability of the alloy.   Borides were not observed when C was present in the alloys.

2. Carbon-Phosphorous: The interaction of C with P is given by alloys (982) and (726). The effect on the DRT value was variable depending on cooing rate, Fig 4(b), whereas the TCL value was adversely affected by the P addition, Fig 5. The microstructure of alloy 726 did not contain any new precipitates such as phosphide or Laves phase, neither was P detected on the grain boundaries using SIMS. It may be that P segregated to the boundaries as a monolayer, which was difficult to detect by SIMS.   From Paju's research, [16,17], C is evaluated as influencing the segregation of P in austenite by mutual displacement, with both elements

segregating to grain boundaries. Similar results were obtained by Erhart and Grabke [14], where it was proposed that the activity of C was such that C can displace P from grain boundaries, even in a high P material. Consequently in an alloy containing a sufficient level of C as in alloy 726, the effect of P is likely to be mitigated by the 0.03 wt.% C present.

3. Boron-Phosphorous: The effect of B and P on weldability is seen by comparing alloys (989) and (T91). The P addition to the high B alloy dramatically increased the TCL values and reduced the DRT values by 40-50°C, to the lowest values observed during the investigation.

In the air-cooled alloy containing B and P, (T91), the Nb/Mo/Cr boride formed, which was observed to be detrimental to micro-fissuring resistance of the alloy (Fig.4d & 5). As is to be expected, if non-equilibrium segregation was occurring, the SIMS image for B would be stronger in the AC than in the WQ condition, as was the case; however the SIMS analysis did not show any segregation of P. Again the P may be segregating by equilibrium segregation or the B may be influencing the grain boundary segregation. This has been also observed by Paju [16, 17] where B additions were found to reduce the segregation of P to grain boundaries. In the present set of results however the combination of P and B in a low C alloy was observed to be detrimental to weldability, showing the lowest DRT values and the highest TCL values for the entire experimental set.

Interaction of three elements

The effect of C, B and P, on micro-fissuring in HAZ, is observed in the B, C and P alloy (727). The most dramatic effect of the three element combination is seen in Fig 5 where the addition of 0.03 wt. % C in alloy 727 reduced the TCL values from the B+P alloy (T91) by a factor of about 80%, in both AC and WQ conditions. Similarly the DRT values improved by 50 to 110°C for the AC and WQ conditions (Fig. 4d).

The microstructure of alloy 727 showed only carbides, contrasting with alloy T91 where borides were observed. The location of the borides on the grain boundaries in alloy T91 considerably affected the weldability parameters. The C addition in alloy 727 however eliminated the borides and considerably improved weldability of the alloy.

Following Paju's observations [16, 17], one can assume that the 0.03 wt. % C addition was sufficient to reduce segregation of P on grain boundaries to a level where the precipitation of phosphides was prevented. It is likely, however, that some P segregation still exists in the alloy, but it's effect is severely mitigated by the C activity arising from the 0.03 wt. % addition. As mentioned previously, Horton et al have observed C, B, and P elemental segregation to the grain boundaries of modified alloy 718, using the Atom Probe, [3].

Effect of the Cooling Rate

Some general trends on the interrelation between the effect of the cooling rate and the alloy composition on the DRT and TCL values were noticed.

There was no effect of cooling rate in the base alloy (988). This is to be expected since the alloy did not contain a high concentration of any minor element, which may segregate or precipitate on grain boundaries. The high C alloy (982) had very similar weldability results to alloy 988, after AC or WQ considering inherent errors in the measurement process.

The B containing alloy (989) and the C and B alloy (991) had similar hot ductility results, though the slower cooling rate in air allowed B to segregate to grain boundaries, increasing the micro-fissuring as shown by the TCL values.

In the alloys containing P, (T91, 726 and 727), the general trend in the TCL values indicated an improvement in weldability in the AC rather than in the WQ condition. The DRT values also reflected this trend in two of the three alloys (726 and 727).

Conclusions

1. The addition of minor elements, either singly or in combination, can significantly influence the weldability of IN 718 alloy.

2. C additions of 0.03 wt. % to the base alloy (0.008 wt. % C) had little effect on the TCL, but adversely affected the DRT up to 40°C in the AC condition

3. B additions to the base alloy had a significantly detrimental effect on the TCL and DRT values.

4. C additions to the B containing alloy somewhat mitigated the detrimental effects of B. This can be explained by site competition as postulated by Erhart and Grabke.

5. B+P additions gave the worst combination of TCL and DRT values through the formation of borides.

6. The addition of C to a P containing alloy reduced the adverse effect of P, which could also be attributed to the site competition..

7. Similarly in a C+ P+B alloy, C mitigated the detrimental effects of B and P, with the elimination of borides, the B being incorporated into the Nb rich carbides.

8. Borides were not observed in the microstructure when C was present.

Acknowledgements

The authors would like to thank Natural Sciences and Engineering Research Council of Canada and a consortium of Manitoba aerospace industries for the financial support. The authors are extremely grateful to Dr. W.D. Cao and Allvac for providing the material, to Dr. G. McMahon of Materials Technology Laboratory of CANMET, Ottawa, for the SIMS analysis. Thanks are also due to Dr. Q. Xu of Industrial Technology Centre, Winnipeg, for his assistance with the TEM analysis.

References

[1] W.D. Cao and R. L. Kennedy, Superalloys 718, 625, 706 and various derivatives, Ed. E. A. Loria, Minerals, Metals and Materials Society, (1994), 463 – 477

[2] R. Kennedy, W.D. Cao and W. M. Thomas, Adv. Mater. and Processes, 149, n 3 (1996), 33-35

[3] J.A. Horton, C.G. McKamey and M.K. Miller, Superalloys 718, 625, 706 and various derivatives, Ed. E. A. Loria, Minerals, Metals and Materials Society, (1997), 401 – 408

[4] G.R. Pease, Weld. J., (1957), . 330-s –334-s.

[5] R. G. Thompson, D. E. Mayo and B. Radhakrishnan,, Metall. Trans. A, vol. 22 A, Feb, (1991), 557 – 567

[6] T. J. Kelly, Advances in Weld. Science and Tech., (1986), 623-627.

[7] W. F. Savage et al, Welding Res. Supplement, 08 (1977), 245s-253s .

[8] X. Huang, N. L. Richards and M. C. Chaturvedi, 3rd Int. S.A.M.P.E. metals and Metals Processing Conf., Eds. F. Froes, W. Wallace, R. Cull and E. Struchholt, Toronto, Canada, 1992, M231-242. SAMPE.

[9] X. Huang, M.C. Chaturvedi and N. L. Richards, Metall. Trans. 1993, 24A, 819-831

[10] X. Huang, M.C. Chaturvedi, N. L. Richards and J. Jackman, Acta Mater., 1997, 48, 3095-3107

[11] R. Vincent, Acta. Metall., vol. 33, n 7, (1985), 1205-1216.

[12] W.Lin, J.C.Lippold and W.A.Baeslack III, Welding J., (1993),72,135s-153s.

[13] M. Guttmann, Surface Sci, (1975), 53, 213-227.

[14] H. Erhart and H. J. Grabke, Metal Sci, (1981), 15, 401-408.

[15] J. N. Dupont, C. V. Robino and A.R. Marder, Weld. J, Oct. (1998), 417-s – 431-s.

[16] M. Paju and R. Moller, Scripta Metall., 18 (1984), 813-815.

[17] M. Paju and H. J. Grabke, Mater. Sci. and Tech., vol. 5, Feb. (1989), 148-154.

Improving Repair Quality of Turbine Nozzles Using SA650 Braze Alloy

Wayne A. Demo, Stephen Ferrigno, David Budinger, and Eric Huron
GE Engine Services, Cincinnati, OH

Abstract:

Due to competitive pressures, airlines are requesting that engine manufacturers reduce cost of ownership for their products. Developing repairs instead of replacement, and enhancing repairs already available, are key methods to do this. Airfoil components of high pressure and low pressure turbines are subject to wear and replacement due to thermal fatigue and oxidation. These are the highest cost items for maintenance and replacement and thus repairs for these components were among the first to be developed. Initially these parts were repaired by Gas Tungsten Arc (GTAW) welding, however, yields were low and part performance was degraded by the metallurgical degradation of the weld heat affected zone. The next improvement in repair was the development of the brazing process known as Activated Diffusion Healing (ADH). By cleaning and brazing the cracked areas, many parts were successfully repaired and returned to service. Now, GE Engine Services has developed a more advanced brazing alloy, SA650, which is a significant improvement in strength over older brazing alloys while offering superb resistance to high temperature oxidation. This provides higher tensile and creep strength in the repaired component, allowing improved performance and increased time on wing before cracking re-occurs and repair again becomes necessary. This is an example of the overall strategy of GE Engine Services to expand the number and quality of repairs offered to its customers.

Superalloys 2000
Edited by T.M. Pollock, R.D. Kissinger, R.R. Bowman,
K.A. Green, M. McLean, S. Olson, and J.J. Schirra
TMS (The Minerals, Metals & Materials Society), 2000

Introduction:

Airlines are operating in an increasingly competitive environment. In response, gas turbine engine manufacturers must offer solutions to reduce cost of ownership for their products. One of the key methods to do this is to provide improved repairs for key gas turbine components. Airfoil components of High Pressure and Low Pressure turbines (HPT and LPT) are subject to wear and replacement due to thermal fatigue and oxidation. In the very early stages of repair development, these parts were repaired by Gas Tungsten Arc (GTAW) welding, however, yields were low and part performance was degraded by the metallurgical degradation of the weld heat affected zone. Next, GE developed a brazing process known as Activated Diffusion Healing (ADH) (Reference 1). By cleaning and brazing the cracked areas many parts were successfully repaired and returned to service. Now, GE Engine Services has introduced the Partitioned Alloy Component Healing (PACH) process (Reference 2) braze process, and has used this process to apply a more advanced brazing alloy, SA650. SA650 is a significant improvement in strength and wide gap fill capability over older brazing alloys while offering superb resistance to high temperature oxidation. This provides higher tensile and creep strength in the repaired component allowing improved performance and increasing time on wing before cracking re-occurs and repair again becomes necessary. This improved brazing material is just one example of the overall strategy of GE Engine Services to expand the number and quality of repairs offered to its customers.

Effect of Engine Exposure on Turbine Nozzle Condition:

The operating environment within a gas turbine engine presents severe temperature and environmental conditions. Cobalt-base and Ni-base superalloys are used for these components; however, even components made from these alloys are subject to degradation. A common condition is cracking due to thermal fatigue. Figure 1 shows a typical example of this condition in a HPT nozzle. The cracks occur in part to relieve stresses generated by the thermal fatigue; they typically reach stable lengths and do not propagate to lengths sufficient to cause failure. However, after the engine reaches its next service increment these cracks need to be repaired to allow the part to be returned to service.

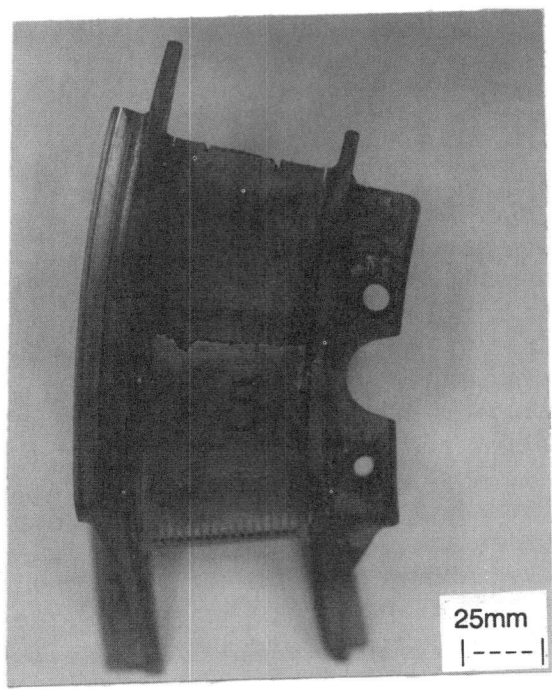

Figure 1: CFM56-3 Stage 1 HPT Nozzle showing typical thermal-mechanical fatigue cracks at airfoil trailing edge region

Another common condition is oxidation, sometimes with sulfidation. The major impact of oxidation is to coat the surface of the part with an inert barrier that prevents any braze repair alloy from bonding to the metal; this condition is most severe within the tight confines of small cracks. Under severe exposures or after long service lives and multiple service-repair-service-repair cycles, base metal erosion can occur. The geometries of the nozzle segments are carefully controlled to provide the proper aerodynamic conditions in order to achieve design performance of the gas turbine. Thus the impact of cracking on part geometry and the buildup of oxidation or impact of erosion on the nozzle surfaces must be corrected by the repair process.

The requirements of the repair process are clear: the part must be cleaned thoroughly before beginning the repair process; the oxidation within cracks must be removed and cracks must be healed with adequate strength to resist premature cracking in subsequent operation; the smooth surface and dimensions of the airfoils must be restored to provide desired performance.

Repair Process Description:

The part most commonly repaired with this process is the Stage 1 High Pressure Turbine Nozzle (HPT) of engines such as the CFM56 and CF6-6, -50, and –80-series engines. This part is subjected to high temperatures and frequent thermal cycling during engine acceleration and takeoff. Thermal-mechanical fatigue cracks occur in the airfoil regions of the part. The typical repair process for this part will be used to show how brazing is used to repair these parts.

The repair process (Figure 2) begins with incoming inspection. The part is characterized for location and size of defect. Parts with cracks over

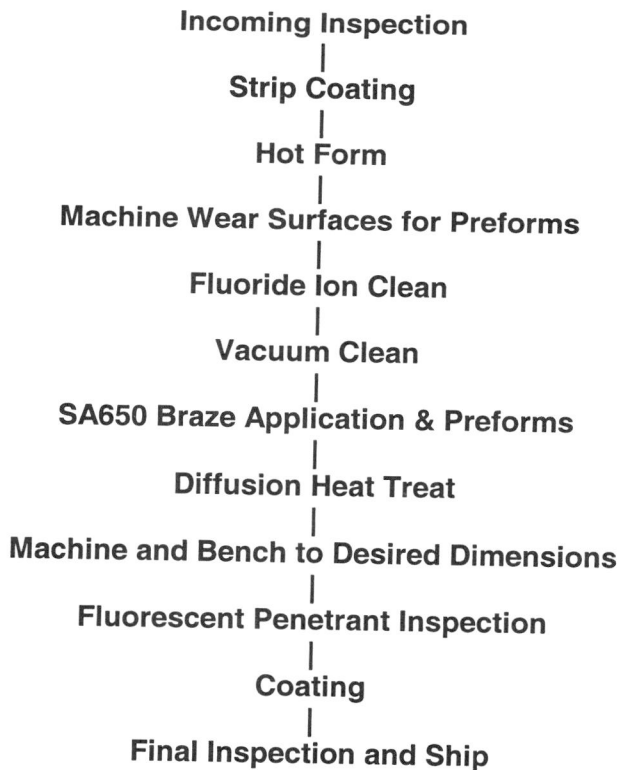

Incoming Inspection
|
Strip Coating
|
Hot Form
|
Machine Wear Surfaces for Preforms
|
Fluoride Ion Clean
|
Vacuum Clean
|
SA650 Braze Application & Preforms
|
Diffusion Heat Treat
|
Machine and Bench to Desired Dimensions
|
Fluorescent Penetrant Inspection
|
Coating
|
Final Inspection and Ship

Figure 2: Typical High Pressure Turbine nozzle process map

repairable limits are rejected; parts with repairable cracks are submitted into the production repair process. Some HPT nozzles have various coatings; these are removed by chemical stripping. The part may be hot-die formed to correct any trailing edge bowing or nozzle platform distortion. Wear surfaces on the nozzle platform are machined in preparation for re-buildup of key dimensional features.

One of the most important factors influencing specific braze alloy development and braze repair of cracks in general is the cleanliness of the cracks. GE developed the dynamic Fluoride Ion Cleaning (FIC) process to address this problem (References 3 and 4). The early work on FIC developed the basic chemical process. The process consisted of combining ammonium fluoride with chromium powder to produce chromous fluoride, CrF_2 and ammonia. Then at higher temperatures the CrF_2 combined with hydrogen to produce HF, which acted as the cleaning gas to reduce oxides of chromium, titanium, and aluminum, effectively cleaning the cracks. Subsequent work refined the process including the development of specialized furnace equipment and methods for safely handling the hydrogen and fluoride gases to allow direct introduction of HF gas. This produced the desired cleaning without the complications of the CrF_2 powder. This process is now fairly common in the superalloy repair industry, although process parameters still require optimization depending on the alloys being cleaned. After FIC, a vacuum clean then removes any residue and prepares the part for brazing.

After all initial cleaning and preparation for final dimensional restoration steps are repaired, the part is ready for brazing. The braze alloy is applied as a powder mixture, within an organic binder to make a paste. Figure 3 shows an example of a part after braze application. The paste is applied to the cracked areas. Pre-sintered preforms for wear dimensional buildup are applied as well. The entire part is brazed using a cycle that allows diffusion of the braze alloy into the part. After brazing, final dimensional features are re-machined into the part. Fluorescent Particle Inspection (FPI) is performed to ensure successful crack repair. The part is re-coated as appropriate, final inspected for dimensional quality, and readied for shipment.

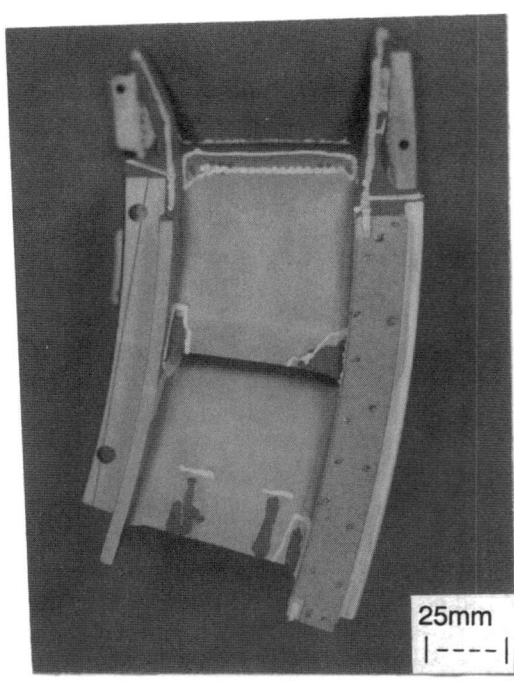

Figure 3: Appearance of part after braze application. Cracks are repaired with braze paste. Stop off paste is applied to prevent braze run in areas not needing restoration.

Characteristics of SA650 Braze Alloy:

GE has used an alloy called ADH (Activated Diffusion Healing) for many years (Reference 1). ADH alloys have various compositions depending on the base alloy being repaired, but work on the same principle of a two-component system. A slurry is applied to the part consisting of a high melting point superalloy powder, usually the same composition as the alloy being repaired, and the ADH alloy, which has lower melting point (achieved with B or Si) powder. The slurry is mixed together and suspended in typical industry-standard organic-based brazing binders. GE's ADH alloys have achieved their low melting point primarily using boron.

The boron level must be balanced between a minimum level for braze flow, acceptable crack filling, and reasonably low braze process temperature on one side, against excessive impact on mechanical properties on the other side. The initial ADH alloy development (Reference 5) evaluated a range of Boron content alloys. When AMS 4778 powder (nominal 3.2 w/o

boron) was mixed with R'77 powder, mechanical properties were too low for successful engine test. However, a GE-developed composition, D15, with 2.3 w/o boron, showed good properties when mixed with R'77. D15-Rene'77, D15-Rene'80, D15-X40, and D15 mixtures with other base alloys were highly successful and were validated on a number of components. The nominal composition of D15 is 15.3Cr-10.3Co-3.45Al-3.4Ta-2.3B, balance Ni (Reference 4).

This process was highly successful, however, over time as HPT and LPT nozzles were cycled through multiple service/repair/service cycles, the boron content of the repaired crack regions (which often occur in the same locations repeatedly) led to reduced strength because of boron buildup.

GE undertook the development of SA650 to build on the success of the demonstrated ADH technique with improved strength and to improve yields for multiple repair cycles. The overall goal of the SA650 braze is to produce a final brazed joint with improved high temperature properties due to reduced boron content. SA650 was also designed with application for single crystal and directionally solidified airfoils, and the low boron content results in minimal impact on single crystal and directionally solidified base alloys. The nominal composition of SA650 can be tailored for the specific application, in particular depending on whether the part to be repaired is a Co-based part or a Ni-base part. SA650 has a similar base composition to D15 but has additional refractory elements for strength, and lower boron level (Reference 2). The increased strengthening element content and the lower B content result in substantially improved high temperature properties, but unfortunately these composition modifications negatively impacted brazeability.

The need to achieve good brazeability with a composition higher in refractory elements and reduced boron required some additional complication in the powder mixture. The ADH alloys were simple two-component blends: usually D15 blended with alloy powder of the same composition as the part being repaired. When attempting to develop a lower overall boron content, however, achieving good flowability was not straightforward. Instead, SA650 was developed as a Partitioned Alloy Component Healing (PACH) alloy, having at least three components. The nominal SA650 composition range for Ni alloys is 15-25Co, 2-30Cr, 0.5-4.0Al, 2-11Ti, 1-3Mo, 1.5-4W, 0.1-0.3Hf, 0.5-6Nb, 1-4Ta, and up to 1 B (Reference 2). The compositions of the components and their relative amounts in the mixture are tailored depending on the composition of the alloy being brazed and other factors such as temperature cycle limitations due to coating or other process steps. In general the first component has a high melting composition and is rich in W and Mo. The second component is lower in melt point and typically contains Cr, Al, and Ni or Co. The third component contains a eutectic alloy with a much lower melting temperature. The elements Si or B are concentrated in the eutectic alloy. For some applications more than four components have been found to improve the balance of flowability vs. final strength. The separation of the elements into multiple powder components allowed the eutectic alloys to promote flowability of the higher melt point powders into cracks, followed by melting and diffusion of those alloys, in a process similar to liquid phase sintering. The end result was adequate flowability for crack filling with lower overall boron content. Si can also be added. The multi-component nature of SA650 has other advantages as well. The delay in melting of the high melt alloy allows relatively wide gaps to be brazed. This significantly enhances repair yields.

Typical microstructures of cracked regions repaired by SA650 are shown in Figure 4. Note the excellent fill of braze alloy along the crack and minimal porosity.

Extensive property testing has been done to evaluate SA650 mechanical properties. SA650 is considerably stronger than the earlier generation ADH composition (Figure 5). In fact, SA650 approaches the strength of the R'80 base alloy (Figure 6). For some alloys such as X40 the braze joint exceeds the strength of the parent metal. This is verified by test bars which break away from the braze joint.

Similarly, stress rupture shows the same trends as tensile data (Figure 7). SA650 has significantly improved rupture performance over the earlier ADH braze compositions, and comes very close to the stress rupture properties of the base substrate

R80 alloy. This allows restoration of cracks with properties much closer to the original new make part and enhances life after component repair.

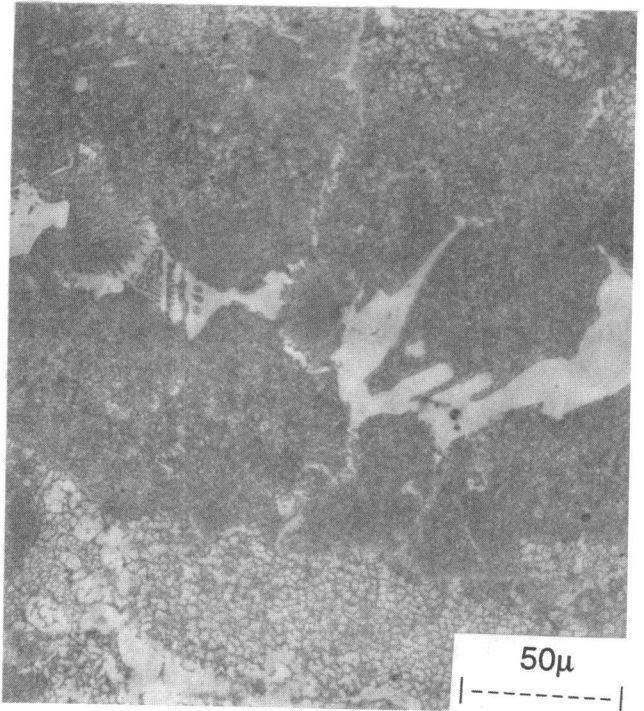

Figure 4: Typical microstructure of crack repaired using SA650 braze repair. Note excellent alloy fill and minimal porosity.

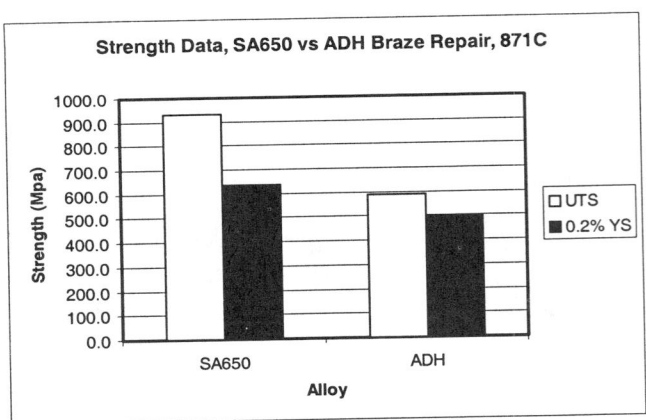

Figure 5: Comparison of SA650 Braze Strength with ADH Braze Strength. SA650 is considerably stronger.

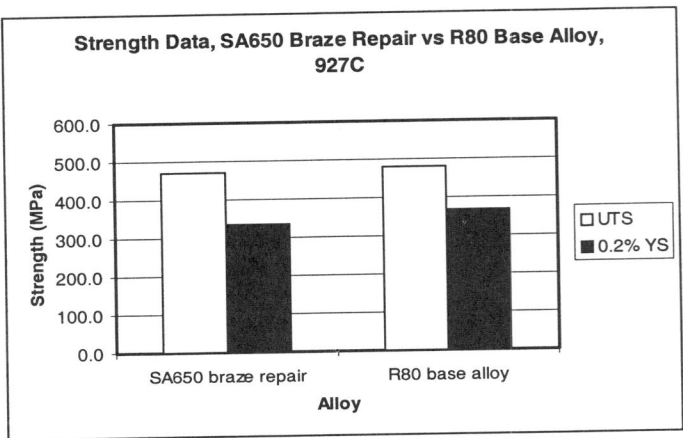

Figure 6: Strength comparison between SA650 braze repair alloy and baseline R80 properties. SA650 braze strength in the repaired regions of he part approaches that of the substrate alloy being repaired.

Results of fatigue testing are presented in Figure 8. Here the data is compared to R'80 base metal and properties approach common equiaxed airfoil alloys for the narrow gap test specimens which had braze gaps up to 0.13 mm, while properties were slightly lower for the wide gap specimens. The excellent fatigue performance is achieved despite the minor porosity typical of braze alloy shrinkage.

LCF failures typically occur at pores. This is common in braze repair alloys (Reference 6) due to the inherent nature of the process. Boron is a

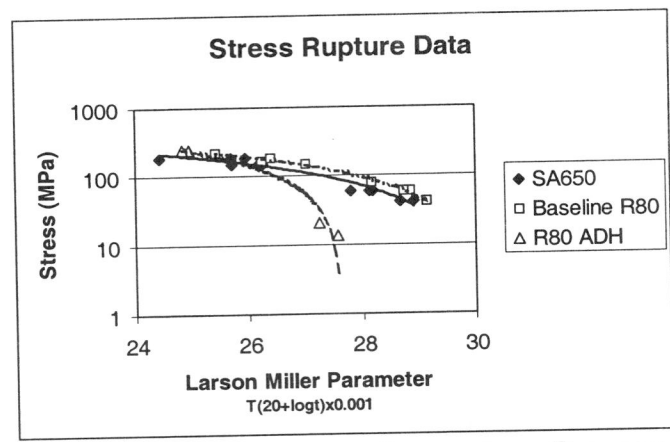

Figure 7: Stress Rupture Comparison Between SA650 braze alloy, ADH braze alloy, and R80 Base metal. The improvement offered by SA650 is clear.

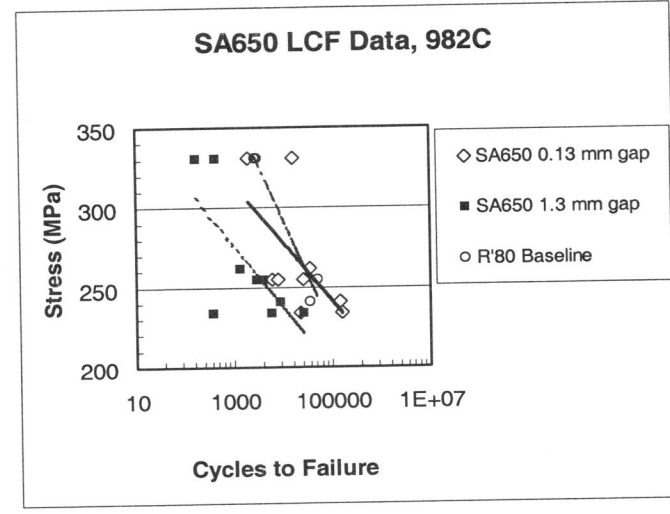

Figure 8: Low Cycle Fatigue properties of SA650 braze repair compared to baseline R80 properties

key melt point depressant used in braze alloys. The boron level of LPM[TM] IN738 (Reference 6) is similar to that of SA650, and SA650 displays similar failure modes. Boron has been found to influence porosity in Ni-base powder superalloys (Reference 7 and 8), thus it would be of interest to compare LCF properties with brazeability across boron levels for the same base braze composition. However, extensive testing, characterization, and field experience have shown that typical porosity levels associated with these alloys allow good performance.

Quality control evaluations on PACH-repaired hardware include metallurgical cutup to ensure proper braze flow and acceptable porosity for each batch. A typical repaired part is shown in Figure 9. General Electric often subjects parts to Furnace Cycle Testing (FCT) at aggressive cycle conditions to verify performance prior to release for engine application. Figure 10 shows a typical part after FCT test. The regions in the part that had prior cracks that were repaired did not fail prematurely in the test. Other regions of the part that had not been repaired showed distress and ultimate failure occurred away from brazed areas. This validates the ability of brazing, with proper alloy selection, to produce good component life after a repair. SA650 has proven itself in repair with over 100,000 airfoil components repaired successfully. The reduced metallurgical impact to the base alloy, made possible by the lower boron content, both improves the properties of the parts after the initial repair and allows multiple repairs to the components to be made, greatly reducing operating costs for airlines. Parts repaired by SA650 are now performing successfully on a daily basis.

20mm
|-----|

Figure 10: Results of FCT testing of an advanced braze alloy. This part had been engine run in service, repaired using techniques described in the text, and Furnace Cycle tested. Eventual part failure occurred at a crack away from any previous repaired regions, demonstrating the capability of braze repair.

25mm
|----|

Figure 9: Typical part at completion of repair process.

References:

1. Wayne A. Demo and Stephen J. Ferrigno, "Brazing Method Helps Repair Aircraft Gas-Turbine Nozzles", Advanced Materials and Processes, March 1992, pp. 43-45.

2. S. J. Ferrigno, M. Somerville, and W. R. Young, U. S. Patent 4,830,934, May 16, 1989.

3. D. L. Keller and W. R. Young, Cost Effective Repair Techniques for Turbine Airfoils – Volume 1, Air Force Report AFWAL-TR-81-4009, Volume 1, Contract F33615-78—5134, Wright-Patterson AFB, Ohio, 45433, June 1981).

4. D. L. Keller and D. L. Resor, U.S. Patent 4,098,450, Superalloy Article Cleaning and Repair Method, July 4, 1978.

5. D. L. Keller and W. R. Young, Cost Effective Repair Techniques for Turbine Airfoils –

Volume 2, Air Force Report AFWAL-TR-81-4009, Volume 2, Contract F33615-78—5134, Wright-Patterson AFB, Ohio, 45433, April 1982)

6. Keith A. Ellison, Joseph Liburdi and Jan T. Stover, "Low Cycle Fatigue Properties of LPM™ Wide-Gap Repairs in Inconel 738", in Superalloys 1996, ed. by R. D. Kissinger et. al., TMS-AIME, 1996, pp. 763-773.

7. E. S. Huron, R. L. Casey, M. F. Henry, and D. P. Mourer, "The Influence of Alloy Chemistry and Powder Production Methods on Porosity in a P/M Nickel-Base Superalloy, in Superalloys 1996, ed. by R. D. Kissinger et. al., TMS-AIME, 1996, pp. 763-773.

8. E. S. Huron, U.S. Patent No. 5,584,948, Method for Reducing Thermally-Induced Porosity in a Polycrystalline Nickel-Base Superalloy Article, December 17, 1996.

IMPROVING PROPERTIES OF SINGLE CRYSTAL TO POLYCRYSTALLINE CAST ALLOY WELDS THROUGH HEAT TREATMENT

A.E. Kolman

General Electric Power Systems, Schenectady, NY 12345

Abstract

Joining between single crystal and polycrystalline cast materials is desirable for advanced steam cooled land based gas turbine airfoil components. Electron beam weld joints between single crystal Rene N5 and polycrystalline GTD-222 have been characterized in terms of microstructure, crystallographic orientation, low cycle fatigue behavior, and chemistry across the weld centerline. The weld centerline was shown to be the vulnerable area of the joint due to carbide segregation and crystallographic mismatch. In addition, the weld fusion zone hardness was significantly higher than that of either base material. The effect of several proposed heat treatments on weld hardness was examined. All of the heat treatments significantly reduced weld hardness as compared to a standard GTD-222 solution treatment and age cycle. Scanning electron microscopy analysis revealed the hardness drop was due to coarsening of gamma prime in the weld fusion zone.

Background

Directionally solidified or single crystal cast nickel based alloys are key materials for high efficiency land based gas turbine components. While one piece castings are commonly utilized for rotating airfoils, fabricated assemblies are possible for stationary parts, offering the opportunity to use lower cost polycrystalline structures for portions of the assembly away from the hot gas path. Joining studies of single crystal to polycrystalline material have mainly explored diffusion bonding techniques[1,2]. Challenges for this technique include determining the optimal bonding HIP/heat treatment cycle for two dissimilar materials, and practical joining of non-planar components.

Studies of fusion welding of single crystal or polycrystalline gamma prime strengthened alloys have focused on causes of weld cracking[3,4], often from strain age cracking[5,6]. Little activity has been reported on the metallurgy of single crystal to polycrystalline fusion weld joints. Weld properties of single crystal to polycrystalline superalloy joints can be less than optimal due to abrupt changes in crystallographic orientation near the weld centerline, the propensity for carbide precipitation near the weld centerline, and the high hardness of the weld compared to the base materials, all contributing to early centerline failure in low cycle fatigue (LCF) testing. These contributing factors are examined in the following study for Rene N5 to GTD-222 electron beam welded joints.

Experimental Procedures

Compositions of GTD-222 and Rene N5 are shown in Table I. Cast slabs of GTD-222 and Rene N5 were welded together using an autogenous electron beam weld process. The slab thickness was reduced in the area of the weld to 0.76 cm. The welded slabs were subjected to the GTD-222 solution and age heat treatment cycle (1150°C for 4 hours, and 800°C for 8 hours), followed by a fluorescent penetrant (FPI) and x-ray inspection. Curved strips were cut from the slabs, with the weld in the center of each strip, yielding small disks between each pair. Round bar LCF specimens were machined from the curved strips, excluding any areas containing indications. The specimens were 11.7 cm long, with a 1.5 cm grip diameter and 0.76 cm gage diameter. The gage length was 1.5 cm, with the autogenous weld at the center of the gage section. FPI inspection was performed again after specimen machining to exclude test bars with indications from LCF testing. The disk remainders were mounted, polished and etched for microstructural evaluation.

Table I: GTD-222 and Rene N5 Compositions

	Ni	Cr	Co	Mo	W	Nb	Al	Ti	C	Ta
GTD-222	balance	22.5	19.1	-	2.0	0.8	1.2	2.3	0.08	0.95
Rene N5	balance	7.0	7.5	1.5	5.0	-	6.2	-	0.05	6.5

Superalloys 2000
Edited by T.M. Pollock, R.D. Kissinger, R.R. Bowman,
K.A. Green, M. McLean, S. Olson, and J.J. Schirra
TMS (The Minerals, Metals & Materials Society), 2000

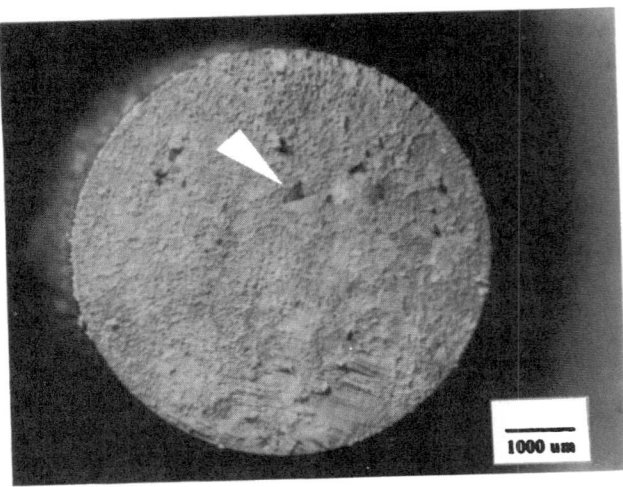

Figure 1: Fracture surface of 870°C LCF test specimen, Rene N5 side, with pyramid indentations.

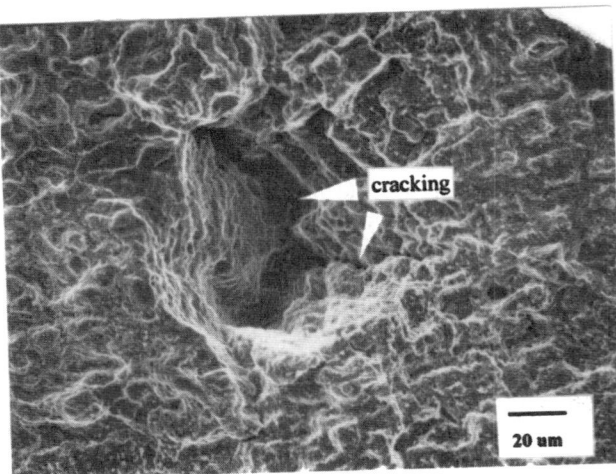

Figure 2: Pyramid indentation on Rene N5, with sidewall and apex cracking.

Figure 3: Fracture surface of 760°C LCF specimen at the GTD-222 fusion line with secondary intergranular cracking.

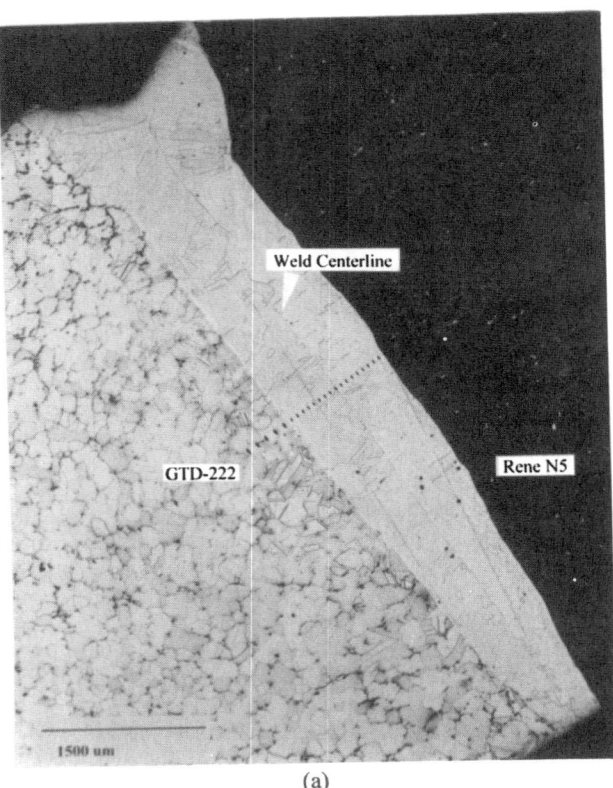

(a)

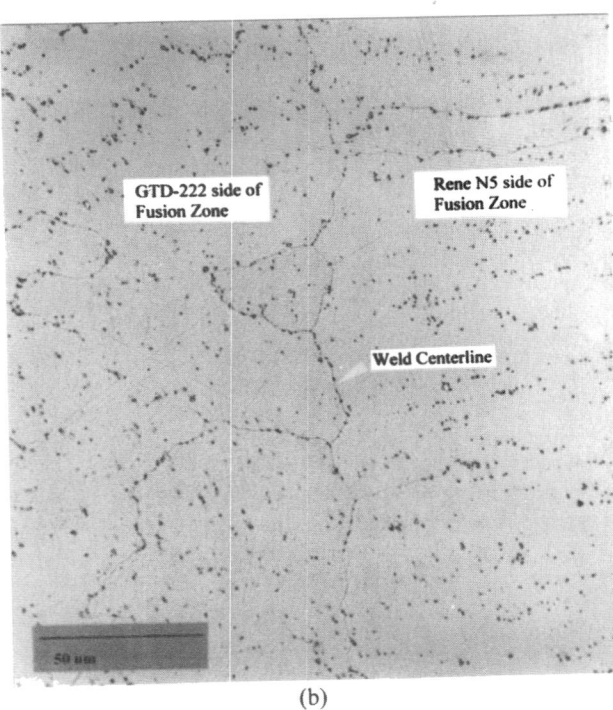

(b)

Figure 4: Cross-section of autogenous Rene N5 to GTD-222 electron beam weld at a) low magnification b) high magnification.

722

LCF specimens were tested with a two minute hold time in tension under strain control at 650°C and 760°C. A two minute hold time in compression was necessary for 870°C testing, because excessive yielding of the material occurred under tensile hold conditions at 870°C.

Results and Discussion

LCF Behavior

A specimen tested at 870°C, 0.40% strain achieved only 6 LCF cycles prior to failure, with flat, interdendritic fracture occurring at the weld centerline (figure 1). Small pyramid structures were observed on both fracture surfaces, but primarily on the GTD-222 side. Examination by electron microprobe revealed these structures were not titanium or tantalum-rich, as would be expected for MC carbides. Rather, the pyramids were identical in composition to the rest of the fusion zone. The pyramid indentation in figure 2 shows a convergence of dendrite arms at the apex, with sidewall cracking that led to pyramid separation.

A higher number of cycles to crack initiation were achieved at the lower temperature, lower strain test conditions. Bars tested at 760°C with 0.30% and 0.50% strain achieved 1237 and 2 cycles, respectively, while bars tested at 650°C with 0.55% and 0.72% strain failed after 4908 and 414 cycles. Specimens tested at the lower temperatures exhibited more intergranular fracture surfaces than the 870°C specimens, and typically had some or all of the fracture occur at the GTD-222 fusion line (figure 3).

Microstructure Characterization

Remainder disks from the welded slab were cross-sectioned, mounted and examined. The weld centerline was linear and highly decorated with tantalum- and titanium-rich carbide precipitates. MC-type carbides were also evident within the fusion zone grain boundaries (figure 4).

Element plot traces across the fusion zone were generated by electron microprobe (figure 5). To improve clarity of lower weight percent elements, Cr was omitted from the plot. The fusion zone composition is biased towards the GTD-222 composition (table II).

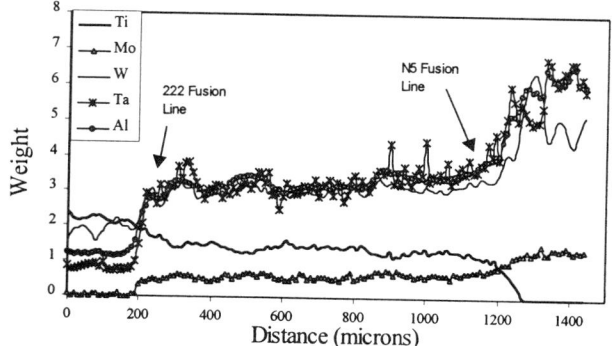

Figure 5: Microprobe traces across the Rene N5 to GTD-222 weld after GTD-222 solution treatment and age.

Table II: Fusion Zone Composition Comparison

	GTD-222	Rene N5	Calculated Mean Composition	Measured Fusion Zone Average Composition
Cr	22.5	7.0	14.75	16.3
Ti	2.3	0	1.15	1.5
Mo	-	1.5	0.75	0.75
W	2.0	5.0	3.5	3.0
Ta	1.0	6.5	3.75	3.3
Al	1.2	6.2	3.7	3.3

Crystallographic Orientation

Coupon disks were metallographically mounted and analyzed by Electron Backscatter Pattern (EBSP) analysis. This technique provides real time diffraction patterns on bulk specimens. The EBSP plot of the weld fusion zone (figure 6) shows crystallographic orientation differences in the single crystal Rene N5 to equiaxed GTD-222 weld joint. The weld centerline is an interface between low and high angle boundary regions, corresponding to single crystal and polycrystalline regions, respectively. The Rene N5 side of the fusion zone formed epitaxially from the base material. The GTD-222 side of the weld is comprised of random high angle grain boundaries. The GTD-222 base material has a mixture of random and twin boundaries. The sharply defined crystallographic boundary at the center of the fusion zone predisposes centerline failure.

Weld Hardness

Knoop hardness traverses across the fusion zone were measured on coupon disks. Hardness readings were significantly higher in the fusion zone compared to either base material after GTD-222 solution and age heat treatments (figure 7). Deleterious effects of excessive hardness in the fusion zone have been shown in laser welded Waspaloy[7] and laser welded Hastelloy X to Mar-M247[8]. In the autogenous Waspaloy welds, the heat affected zone (HAZ) remained hard even after solution and age heat treatments. During subsequent LCF testing, cracks initiated predominantly at the fusion zone / HAZ boundary. In the case of wrought Hastelloy X welded to cast Mar-M247, a direct correlation was observed between hardness of the fusion zone and fusion zone cracking. This tendency was reduced by intentional off-center welding into the Hastelloy X base material.

Effect of Heat Treatment

Additional coupons from the welded slab were subjected to several heat treatment sequences (table III). The sequences were selected to examine the effects of a Rene N5 age cycle, a Rene N5 stress relief cycle, and fabrication heat treatments. Fusion zone hardness of these test specimens was significantly lower than the hardness of welds which received only GTD-222 solution and age heat treatments (figure 8). Two of the heat treatment sequences lowered the hardness of the Rene N5 base material.

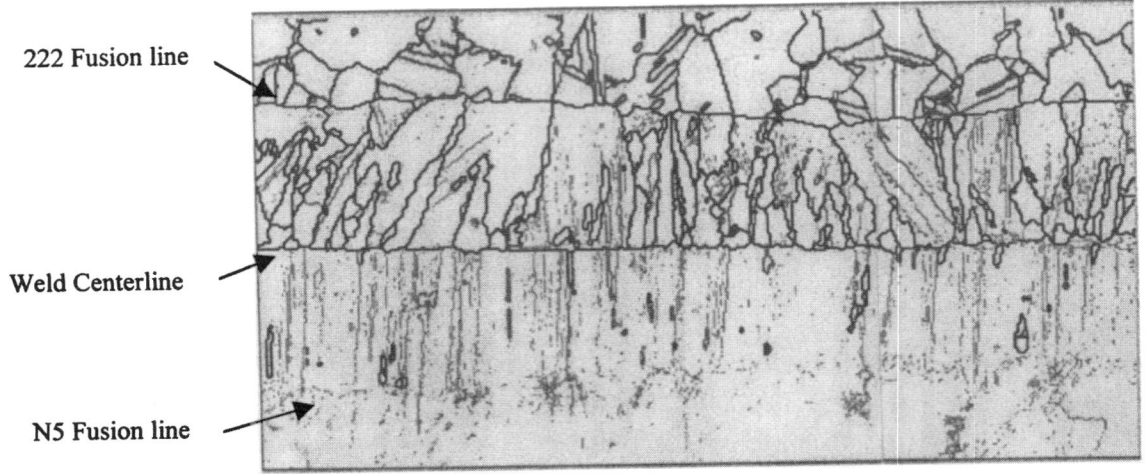

Figure 6: EBSP image of Rene N5/GTD-222 EB weld joint

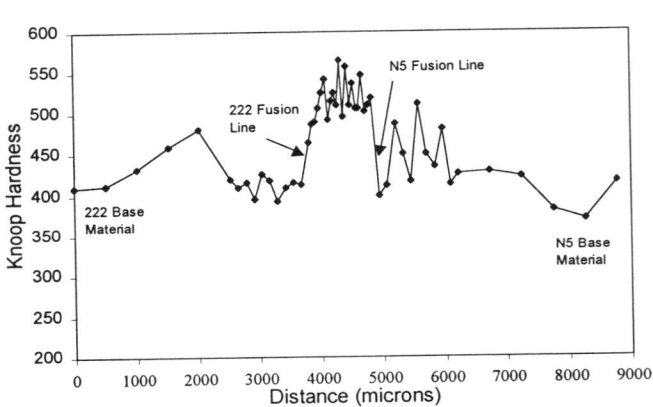

Figure 7: Knoop hardness measurements across the fusion zone of GTD-222 / Rene N5 weld after GTD-222 solution treatment and age.

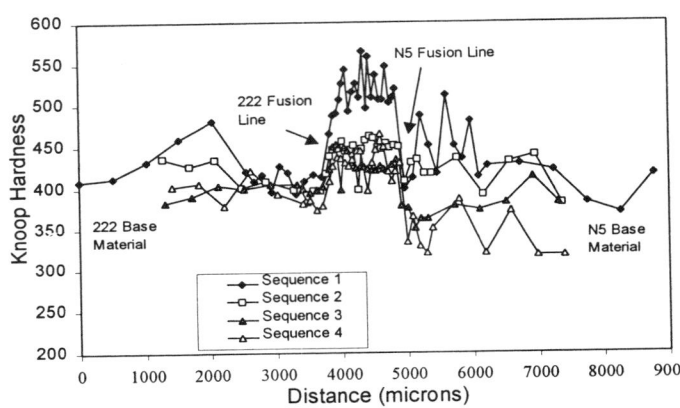

Figure 8: Knoop hardness measurements across the fusion zone of GTD-222/Rene N5 welds following heat treatment sequences 1, 2, 3, and 4.

Table III: Heat Treatment Sequences

Heat Treatment Sequence	1	2	3	4
Initial slab HT (GTD-222 soln & age): 1150°C / 4 hrs., 800°C / 8 hrs.	√	√	√	√
1150 / 2 hrs.		√	√	√
480°C / 45 min.		√	√	√
>1040°C / 9 hrs.		√	√	√
222 age: 800°C / 8 hrs.	√			
N5 stress relief: 1210°C / 1 hr				√
N5 age: 900°C / 4 hrs.			√	√

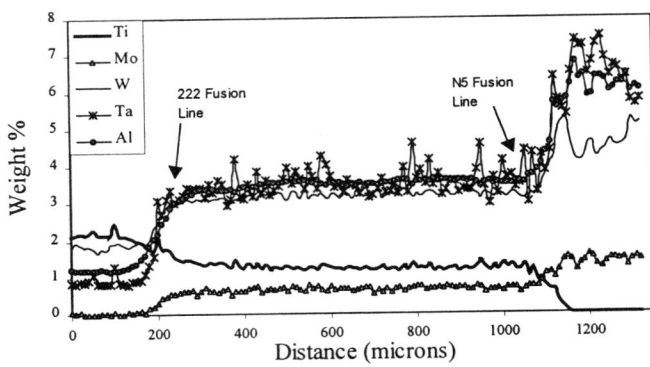

Figure 9: Microprobe traces across the Rene N5 to GTD-222 weld after sequence 2 heat treatment.

724

Chemistry differences were not detectable between samples that received different heat treatments. Microprobe traces across the fusion zone of extended heat treatment samples were virtually identical to those of samples treated with the GTD-222 solution and age cycles (figure 9).

Examination of the weld microstructure by scanning electron microscopy revealed the hardness drop was attributable to an increase in gamma prime size. The typical size of gamma prime precipitates in the extended heat treatment specimens was 0.33 µms, compared to 0.10 µms for GTD-222 solution and age treated specimens (figure 10). Agglomerated gamma prime particles of 1 µm size were observed in the extended heat treatment specimens.

In another study, overaging Rene 41 prior to welding has been shown to be beneficial in reducing strain age cracking during post weld heat treatment[9]. In the present study, overaging after welding improved weld ductility through reduced hardness. Because the fusion zone was enriched in aluminum from Rene N5 dilution, the gamma prime solvus temperature of the weld was greater than that of GTD-222. The standard GTD-222 solution heat treatment cycle was too low to return gamma prime particles to solution. Subsequent heat treatment sequences caused these particles to grow and overage.

The preferred extended heat treatment is sequence 2, because of the minimal effects on Rene N5 base material hardness.

Conclusions

Examination of single crystal to polycrystalline welds showed a high degree of carbide precipitation along a straight centerline boundary, as well as abrupt changes in crystallographic orientation. Fracture during 870°C LCF testing occurred predominantly at the weld centerline. Lower LCF test temperature specimens which had at least a portion of fracture at the fusion zone/base material interface achieved more cycles to crack initiation. The weld hardness after standard GTD-222 solution and age heat treatment was significantly higher than that of the base materials. Extended heat treatment cycles which effected gamma prime overaging resulted in reduced weld hardness. Critical material property requirements of the weld joint and base materials would need to be determined for the specific application prior to employing the overaging strategy.

Acknowledgements

The author would like to acknowledge Tymm Schumaker at GE Power Systems for SEM analysis; Lou Peluso, Charles Mukira, and Shyh-Chin Huang at the GE Corporate Research and Development Center for electron microprobe and EBSP analysis; and Jack Wood and John Murphy for helpful discussions.

(a)

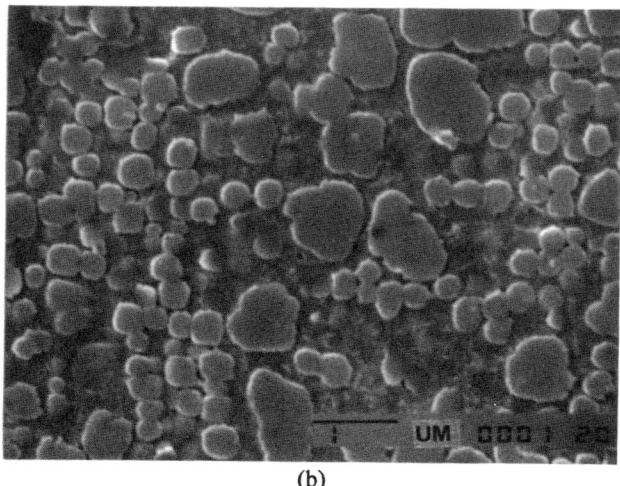

(b)

Figure 10: Gamma prime particles in weld after a) GTD-222 solution treatment and age b) sequence 2 heat treatment. (Scale bar in both photos is 1 µm).

References

1. Y. Bienvenu et al., "Diffusion Bonding of Nickel Base Superalloys to Manufacture Turbine Components with a Graded Microstructure", FGM 94 Proceedings of the 3rd Intl. Symp. on Structural and Functional Gradient Materials, Lausanne Switzerland, Oct 1994, 487-494.

2. R. Larker, J. Ockborn, B. Selling, "Diffusion Bonding of CMSX-4 to Udimet 720 Using PVD-Coated Interfaces and HIP", Journal of Engineering for Gas Turbines and Power, 121 (July 1999), 489-493.

3.   S. A. David et al., "Welding of Nickel Base Superalloy Single Crystals", Science and Technology of Welding and Joining, 2, (2) 1997, 79-88.

4.   S. A. David, S. S. Babu, J. M. Vitek, "Weldability and Microstructure Development in Nickel-base Superalloys", Mathematical Modeling of Weld Phenomena 4 (1998), 269-289.

5.   C. Sims, N. Stoloff, W. Hagel, ed., Superalloys II, (New York, NY, John Wiley & Sons 1987), 511-516.

6.   D. L. Olson et al., volume eds., ASM Handbook Volume 6 Welding, Brazing and Soldering (1993), 572-574.

7.   Z. Li, et al., "Metallurgical Investigation of Laser Welds in Wrought Waspaloy", Science and Technology of Welding and Joining, 3, (1) (1998) 1-7.

8.   Z. Li, S. L. Gobbi, K. H. Richter, "Autogenous Welding of Hastelloy X to Mar-M 247 by Laser", Journal of Materials Processing Technology, 70 (1997), 285-292.

9.   T. F. Berry, W. P. Hughes, "A Study of the Strain-Age Cracking Characteristics in Welded Rene 41- Phase II", Welding Journal, 48, (1969), 505S-513S.

Alloy Development

DEVELOPMENT OF A NEW SINGLE CRYSTAL SUPERALLOY FOR INDUSTRIAL GAS TURBINES

T.Hino, T.Kobayashi, Y.Koizumi, H.Harada and T.Yamagata

High Temperature Materials 21 Project, National Research Institute for Metals

1-2-1 Sengen, Tsukuba Science City, Ibaraki 305-0047, Japan

Abstract

In order to raise the turbine inlet gas temperature to improve thermal efficiency of gas turbines, turbine blade and vane materials are required to have high creep rupture strengths. In the present study, a new single crystal superalloy with a moderate Re addition (2.4wt%) has been developed. The alloy shows a higher creep rupture strength than the second and even the third generation single crystal superalloys.

The composition of the developed single crystal superalloys were designed with a computer aided alloy design program which was developed in National Research Institute for Metals (NRIM-ADP) and finally one alloy was selected experimentally. The developed single crystal superalloy had a stress rupture temperature advantage over 30°C in comparison with the second generation single crystal superalloys at the 137MPa/10^5hours condition. The developed alloy has a large negative lattice misfit. The large negative lattice misfit enhances the formation of continuous γ' platelets, the so-called raft structure, and a fine interfacial dislocation network during creep tests. These are considered to prevent the movement of dislocations and thus decrease the creep strain rate.

Introduction

Many combined cycle power plants have been in operation due to high thermal efficiency and good operatabilty in contrast with steam turbine power plants[1]. The combined cycle power plant is mainly composed of gas turbines, steam turbines, and heat recovery steam generators. The thermal efficiency of the combined cycle power plant can be improved by raising the inlet gas temperature of gas turbines. To increase the inlet gas temperature, the materials for turbine blades and vanes are required to have higher creep rupture strength.

Ni-base single crystal (SC) superalloys have higher creep strength in comparison with conventionally cast and directionally solidified superalloys and are now used in the new generation of gas turbine plant[2][3]. Creep rupture strength of SC superalloys are reported to be improved by adding Re; second generation SC superalloys contain about 3% Re[4][5][6] and third generation SC superalloys contain 5 to 6 % Re [6][7][8]. However, it is also reported that adding Re to SC superalloys tends to cause Re-rich Topologically Closed Packed (TCP) phase precipitation[9][10] which is known to reduce creep rupture strength.

Superalloys 2000
Edited by T.M. Pollock, R.D. Kissinger, R.R. Bowman,
K.A. Green, M. McLean, S. Olson, and J.J. Schirra
TMS (The Minerals, Metals & Materials Society), 2000

In the present study, it was intended to develop a new SC superalloy which has excellent phase stability and creep strength in comparison with second and even third generation SC superalloys with a moderate amount of Re addition.

Alloy Development

Alloy design was conducted with the NRIM-ADP. This program can estimate mechanical properties by equations derived empirically from mechanical properties of SC superalloys, volume fraction of γ' phase, lattice misfit between γ and γ' phases, and concentrations of solid solution elements in these phases[11]. Schematic flow of this program is shown in Figure 1.

The alloy was designed to have superior creep rupture strength by an effective use of negative lattice misfit ($a\gamma > a\gamma'$).

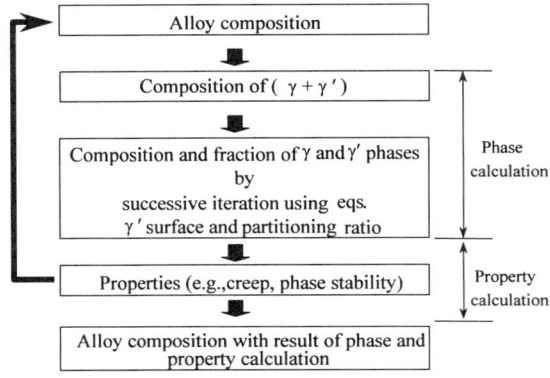

Figure 1: Schematic flow chart of alloy design program.

Other properties such as long-term phase stability, corrosion/oxidation resistance and castabilty were also considered. Creep rupture life at 1100°C/137MPa was designed to be longer than 200h (e.g., 100h for CMSX-4). Re content was designed to be less than 2.5% which prevents precipitation of TCP phases and make the alloy cheaper than present second and third generation SC superalloys. The alloy density was set below 8.9g/cm³ and the solution heat treatment window was designed to be larger than 50°C. The phase stability was estimated by a Solution Index (SI) value. SI values is defined as, SI = Σ (Ci/CLi), where Ci is an atomic fraction of i-th element in γ' phase and CLi is a solubility limit of i-th element in γ' phase of the Ni-Al-i-th element ternary system. If the SI value exceeds 1.25, precipitation of TCP phases is predicted.

Alloy A was designed to have a volume fraction of γ' phase of about 60% and maximum content of solid solution elements without exceeding the SI value 1.25. As Re tends to segregate in dendrite core areas and SI value happened to exceed the critical value in some areas, then, alloy B and C were made by diluting alloy A with Ni-8wt%Al. Chemical compositions of these alloys are shown in Table I.

Experimental Procedure

Single crystal bars of 10mm dia. were cast with the directionally solidified furnace in NRIM. After checking that the longitudinal axes of these single crystal bars were within 15deg from the [001] orientation, heat treatments were conducted with the sequences shown in Figure 2 and Table II.

Solution heat treatments were considered with temperatures ranging from 1280 to 1360°C, to dissolve γ' phase into γ phase without incipient melting.

Table I   Chemical composition of tested and reference alloys.

	Co	Cr	Mo	W	Al	Ti	Ta	Hf	Re	Ni	wt% Solution Index
Alloy A	8.0	5.0	2.0	9.0	5.2	0.5	6.2	0.1	2.5	Bal.	1.24
Alloy B (TMS-82+)	7.8	4.9	1.9	8.7	5.3	0.5	6.0	0.1	2.4	Bal.	1.20
Alloy C	7.5	4.7	1.9	8.5	5.4	0.5	5.8	0.1	2.4	Bal.	1.16
CMSX-4	9.0	6.5	0.6	6.0	5.6	1.0	6.5	0.1	3.0	Bal.	1.12
TMS-75	12.0	3.0	2.0	6.0	6.0	-	6.0	0.1	5.0	Bal.	1.10

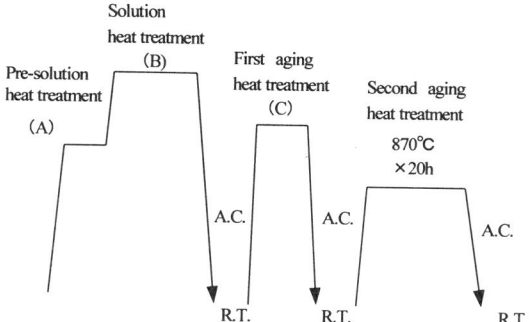

Figure 2 : Heat treatment cycle of tested alloys.

Table II   Heat treatment condition of each cycle.

	Pre-solution heat treatment (A)	Solution heat treatment (B)	First aging heat treatment (C)
Alloy A	-	1300°C×4h	1100°C×1h
Alloy B	1300°C×1h	1320°C×5h	1150°C×4h
Alloy C	1300°C×1h	1320°C×5h	1150°C×4h

After the heat-treatments, creep test specimens (4mm dia. with 22mm gage section length) were machined from these single crystal bars. Creep tests were performed at 900°C/392MPa, 1000°C/196MPa and 1100°C/137MPa. In addition to these conditions, alloy B (TMS-82+) was further tested at 1100°C/158MPa, 1100°C/98MPa and 1000°C/ 245MPa.

To investigate the high temperature creep mechanism, specimens for interrupted creep tests at 64hours under 1100°C/137MPa were prepared with alloy B (TMS-82+) and TMS-75. TMS-75 is a third generation SC superalloy also developed in NRIM[7]. Microstructural examination was conducted by transmission electron microscopy (TEM) and scanning electron microscopy (SEM). These specimens were cut to thin slices both normal and parallel to the longitudinal [001] directions. The thin foil specimens for TEM observation were prepared using an electro-polishing method with a reagent consisting of 50 mℓ HClO₄ and 250 mℓ C₂H₄O₂ at 5°C. Specimens for SEM observation were mounted in molds, polished and etched using an aqua regia reagent consisting of 10 mℓ HNO₃ and 30 mℓ HCl diluted by 40 mℓ C₃H₈O₃. High temperature phase stability of these alloys were evaluated with the microstructure in end section of creep specimens after tests.

The Results and Discussion

Selection of Alloy

Figure 3 shows the creep rupture strengths plotted against SI values. Except for a low temperature and high stress condition such as 900°C/392MPa, alloy B is the strongest in the three alloys. Microstructural examination showed many precipitates of a TCP phase in the creep specimen of alloy A, especially near the rupture portion, and this is the reason why alloy A is weaker than alloy B. The creep rupture strength of alloy C is also lower than that of alloy B since alloy C is not strengthened enough by solution hardening. Thus we selected alloy B as a final alloy. This alloy was designated as TMS-82+ and further examination was conducted.

The Heat-Treatment Capability of TMS-82+

Figure 4 shows the microstructures of TMS-82+ in the as-cast condition and after heating for 2 hour between 1280°C and 1360°C. Most of the γ' precipitates are dissolved into the γ phase at 1280°C. At 1360°C, incipient melting occurs. These show that TMS-82+ has a heat treatment window over 60°C wide providing a very good heat treatment processability.

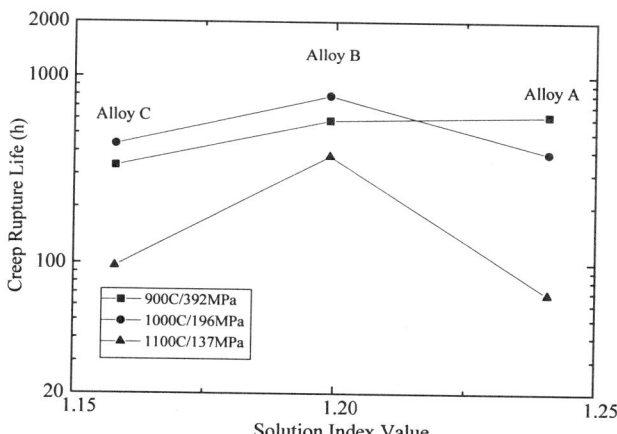

Figure 3: Creep rupture lives of alloy A,B and C plotted against the Solution Index value.

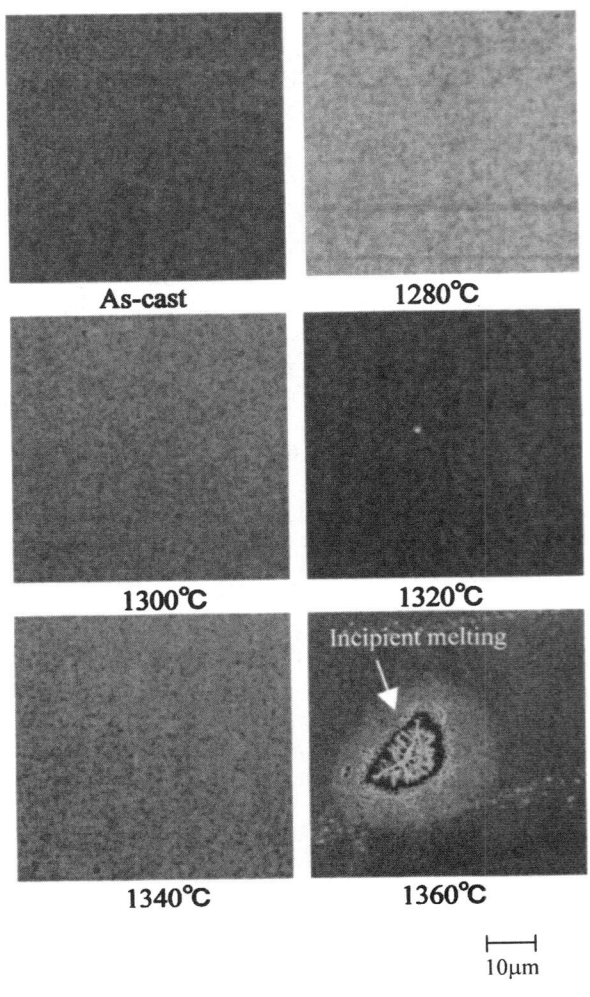

Figure 4: The result of solution treatment trials at different temperature for 2hours with TMS-82+.

Creep Property

Figure 5 shows the creep rupture curves of TMS-82+, CMSX-4[4], TMS-75[7], Rene'N5[5], Rene'N6[5] and a patent alloy (alloy 11) containing Ru[12]. The creep rupture strength of TMS-82+ is superior to those of the second generation SC superalloys such as CMSX-4 and Rene'N5 in all the stress and temperature range, the temperature capability at 137MPa/10⁵ hours of TMS-82+ is 50-60°C higher than Rene'N5. Moreover, in the higher temperature and lower stress range, TMS-82+ is stronger than the third generation SC superalloys such as TMS-75 and Rene'N6, and even the Ru containing US patented alloys[12].

Figure 6 shows creep curves of TMS-82+ and the third generation SC superalloy TMS-75. Cross section micrographs of creep interrupted specimens cut along the longitudinal direction in the middle of a specimen are shown in Figure 7. A so-called raft structure is observed both in TMS-82+ and TMS-75; γ' precipitates are connected with each other normal to the stress axis. Third generation SC superalloys tend to precipitate the TCP phase which is known to reduce the creep rupture strength[9][10]. However, there is no precipitates of TCP phase, the creep strain rate of TMS-75 is, nevertheless, larger than that of TMS-82+. This suggests that the large creep strain rate of TMS-75 compared with TMS-82+ in this condition is not attributed to precipitation of TCP phase.

As for the morphology of the raft structure, more continuous γ' platelets are observed in TMS-82+ compared with TMS-75. This structure improves the creep resistance effectively by providing effective barriers to dislocation climb around γ' platelets[13].

Table. III shows the lattice misfit of TMS-82+ and TMS-75 measured at 1100°C by X-ray diffraction techniques[14] .The lattice misfit of TMS-82+ is negative and the absolute value is larger than that of TMS-75. The large negative misfit enhances the γ' rafting[15] which improves creep strength of TMS-82+.

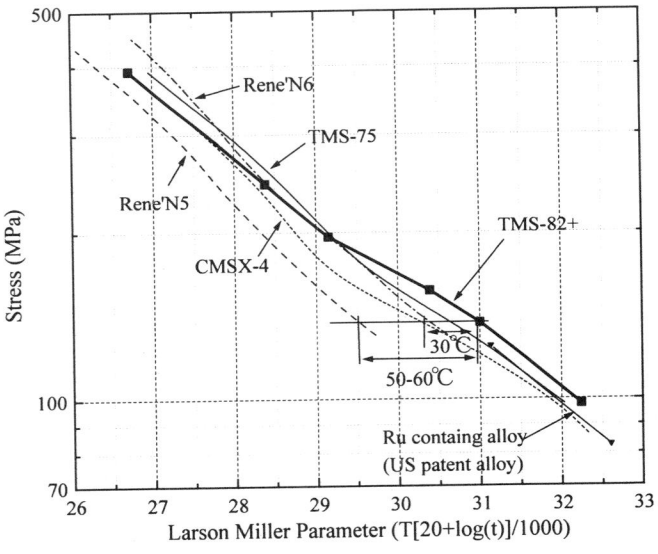

Figure 5 : Creep rupture strengths of TMS-82+, Rene'N5, CMSX-4, Rene'N6, TMS-75 and alloy11 (Ru containing alloy).

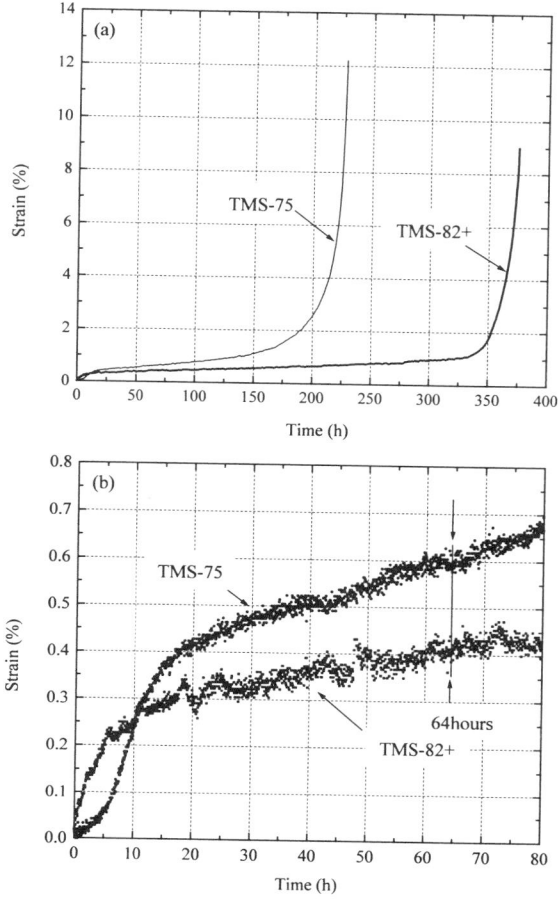

Figure 6: Creep curves of TMS-82+ and TMS-75 at 1100°C /137MPa condition.

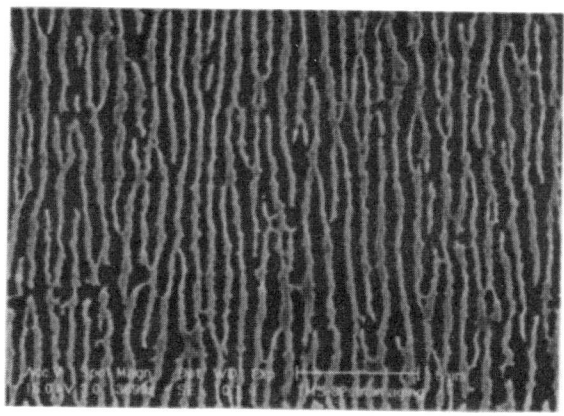

(a) TMS-82+

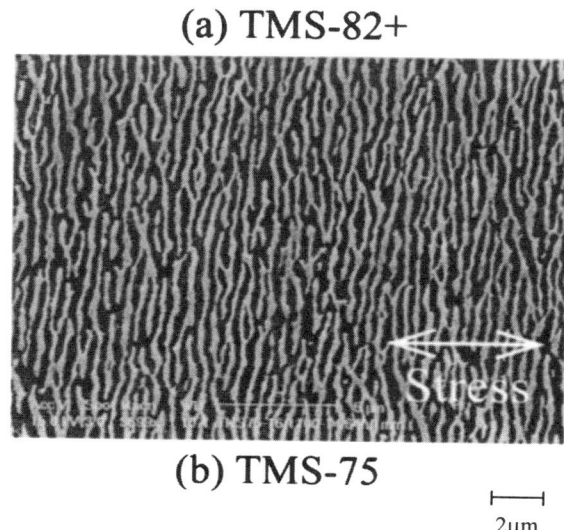

(b) TMS-75

⊢———⊣
2µm

Figure 7: Raft structures in (a)TMS-82+ and (b)TMS-75 ; creep interrupted at 64 hours under 1100°C/ 137MPa condition.

Table III   Lattice misfit, creep strain and creep strain rate of TMS-82+ and TMS-75 at 64 hours creep under   1100°C/137MPa.

	Lattice misfit (1100°C)	Creep strain	Creep strain rate
TMS-82+	-0.24	0.41%	2.4×10^{-3}%/h
TMS-75	-0.08	0.59%	3.4×10^{-3}%/h

Figure 8 shows TEM images of interfacial dislocation networks between γ and γ' phases in the creep interrupted specimen of TMS-82+ and TMS-75 after 64 hours at 1100°C/137MPa. The interfacial dislocation network is formed due to the misfit strain between γ and γ' phases and additionally due to the applied stress leading to creep. The size of the dislocation network in TMS-82+ is finer than TMS-75 as shown in Figure 8. It is considered that the finer dislocation network in TMS-82+ is formed mainly by the larger negative lattice misfit which improves the creep strength by preventing dislocation movement especially the cutting motion into γ' precipitate.

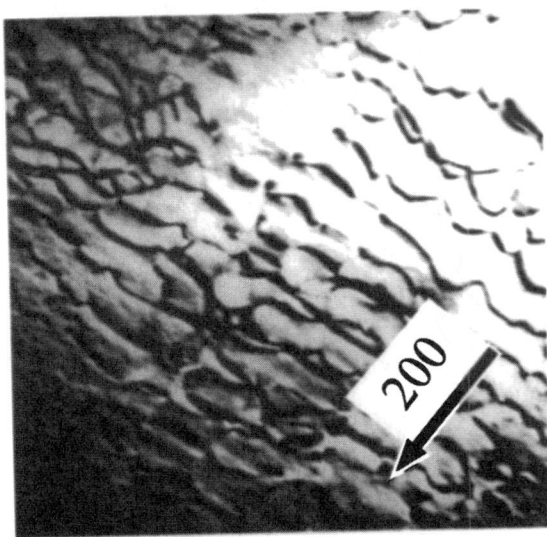

(a) TMS-82+

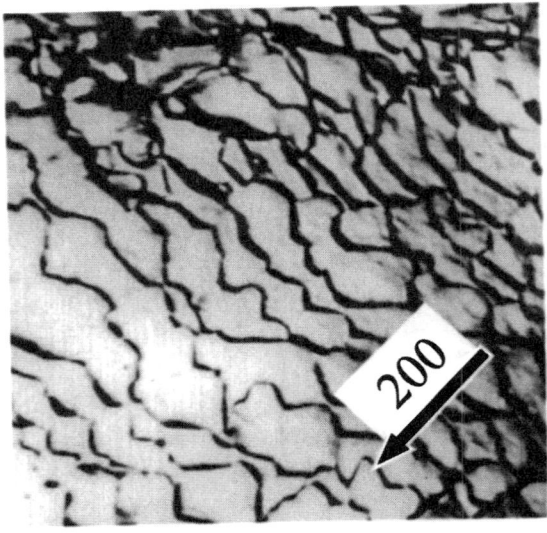

(b) TMS-75

|———————|
100nm

Figure 8: Dislocation network on γ/γ'interface in (a)TMS-82+ and (b)TMS-75; creep interrupted at 64 hours under 1100°C/137MPa condition, foil prepared perpendicular to stress axis, k_0=[001], g=(200).

This is another reason that TMS-82+ has high creep rupture strength compared with other second and third generation SC superalloys. Figure 9 shows creep strain rates plotted against dislocation network spacings along g=[200] direction. This indicates that creep strain rate decreases as the network spacing becomes fine. Once the good rafted structure is established as shown in TMS-82+, dislocation climb becomes very difficult. In this condition dislocation cutting into γ' platelet is forced to be the predominant creep mechanism and then the finer dislocation network can act as a very effective barrier to this.

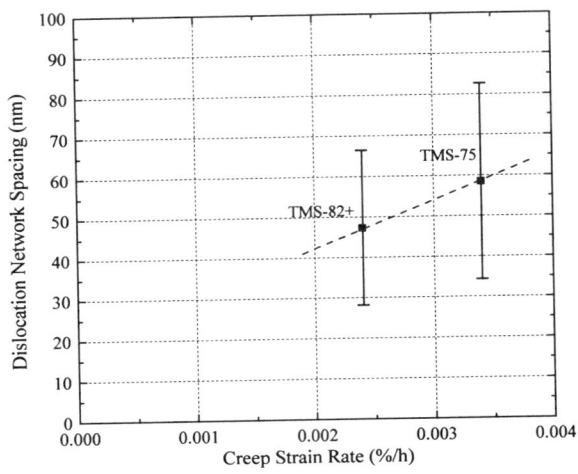

Figure 9: Dislocation network spacing of dislocation networks of TMS-82+ and TMS-75 at 1100°C /137MPa for 64hours plotted against their creep strain rate.

Phase Stability of TMS-82+

Figure 10 shows the end section of TMS-82+ creep rupture specimen tested at 1000°C/196MPa for 790hours. A very small amount of TCP precipitates are observed. In other conditions such as 1100°C/137MPa and 1100°C/98MPa, TCP particles were also precipitated, but the amounts are small. This reveals that TMS-82+ has a good phase stability.

⊢————⊣
10μm

Figure 10: Microstructure of TMS-82+ after 790h aged at 1000°C.

We developed a new SC superalloy which has excellent creep rupture properties compared with present second and third generation SC superalloys. Now, other properties, such as a longer term phase stability, hot corrosion resistance, high and low cycle fatigue properties are being examined. A castability test using a dummy mold of gas turbine blade is being conducted. After the laboratory tests, a turbine rotating test in an actual gas turbine is scheduled in mid 2000 year.

Conclusion

In this study, it was intended to develop a new SC superalloy which has excellent creep properties compared with present second generation SC superalloys. The following results were obtained.

1. We developed a new SC superalloy with NRIM-ADP and optimized the chemical composition through experimental evaluation. The creep strength of the developed alloy, TMS-82+, is higher than that of third generation SC superalloys at high temperature and low stress condition.

2. In microstructual observation, the length of raft structure perpendicular to the stress axis of TMS-82+ is longer than that of third generation SC superalloy, TMS-75. The dislocation network in TMS-82+ is finer than that of TMS-75. These are considered to be the reasons for the creep properties of TMS-82+ being superior to that of other SC superalloys especially in high temperature and low stress condition.

3. TMS-82+ has a large negative lattice misfit between γ and γ' phases which accelerated the formation of raft structure and very fine dislocation network, attributing the high creep strength.

Acknowledgements

We would like to express sincere thanks to Dr.M.Maldini, Mr.T.Yokokawa, Dr.H.Murakami, Dr.Y.Yamabe-Mitarai, Mr. S.Nakazawa, Dr.M.Osawa and Mr.M.Sato of National Research Institute for Metals for their advice. We would like to express sincere thanks to Dr.P.E.Waudby of Ross & Catherall ltd. for making the master ingot and analyzing the compositions of developed alloys.

References

1.T.Aizawa,"The out line of Yokohama ACC and Future Plan of ACC Power Generation ,"Proc.of 1995 Yokohama int. gas turbine congress, (Tokyo : Gas Turbine Society of Japan, 1995), No3: 341- 348.

2.H.Yokoyama,"Advanced Technology of ABB Gas Turbine", 28th gas turbine seminar text, (Tokyo: Gas Turbine Society of Japan, 2000), 77-86.

3.I.Myougan,"Single Axis Combined Cycle Plant of Siemens", 28th gas turbine seminar text, (Tokyo : Gas Turbine Society of Japan, 2000), 101-110.

4.A.D.Cetel and D.N.Duhl, "Second-generation Nickel-base Single Crystal Superalloy",Superalloys1988,ed.D.N.Duhl, et.al. (Warrendale, PA : The Minerals, Metals & Materials, Society, 1988), 235-244.

5. G.L.Erickson and K.Harris, "DS and SX Superalloys for Industrial Gas turbines", Material for Advanced Power Engineering 1994, ed. D.Coutsouradis et.al. (Kluwer Academic Publishers, 1994), Part I :1055- 1074.

6.W.S.Walston et.al.,"ReneN6 : Third Generation Single Crystal Superalloys", Superalloys1996, ed. R.D.Kissinger et.al. (Warrendale, PA : The Minerals, Metals & Materials, Society, 1996), 27-34.

7.G.L.Erickson, "The Development and Application of CMSX-10", Superalloys1996, ed. R.D.Kissinger et.al. (Warrendale, PA : The Minerals, Metals & Materials, Society,1996) , 35-44.

8.Y.Koizumi et.al., "Third Generation Single Crystal Superalloys with Excellent Processability and Phase Stability", Material for Advanced Power Engineering, ed. J.Lecomte-Beckers et.al (Kluwer Academic Publishers, 1998), Part II, 1089-1098.

9. R.Darolia et.al., "Formation of Topologically Closed Packed Phases in Nickel Base Single Crystal Superalloys" Superalloys1988, ed.D.N.Duhl et.al. (Warrendale, PA : The Minerals, Metals & Materials, Society 1988), 255-264.

10.T.Hino et.al. , "Design of High Re Containing Single Crystal Superalloys for Industrial Gas Turbines", Material for Advanced Power Engineering, ed. J.Lecomte-Beckers et.al (Kluwer Academic Publishers, 1998), Part.II: 1129- 1137.

11.H.Harada et.al., "Phase Calculation and Its Use in Alloy Design Program for Nickel-Base Superalloys", Superalloys1988, ed.D.N.Duhl et.al. (Warrendale, PA : The Minerals, Metals & Materials, Society 1988),733-742.

12.K.S.O'hara et.al., U.S.Patent 5,482,789 "Nickel Base Superalloy and Article".

13.MacKay,R.A. et.al., "Factors which Influence Directional Coarsening of γ' during Creep in Nickel-base Superalloy Single Crystals", Superalloys1984, ed. M.Gell,et.al., (Warrendale, PA : The Minerals, Metals & Materials, Society, 1984), 135-144.

14.T.Yokokawa and M.Osawa, private communication with author, National Research Institute for Metals, 25 Februrary 2000.

15.D.D.Pearson et.al., "Stress Coarsening of γ' and Its Influence on Creep Properties of a Single Crystal Superalloy ", Superalloys1980, ed.J.K.Tien et.al. (Warrendale, PA : The Minerals, Metals & Materials, Society, 1980) ,513-520.

HIGH γ' SOLVUS NEW GENERATION NICKEL-BASED SUPERALLOYS
FOR SINGLE CRYSTAL TURBINE BLADE APPLICATIONS

P. Caron

Office National d'Etudes et de Recherches Aérospatiales (ONERA)

BP72 - 92322 Châtillon - France

Abstract

A series of single crystal nickel-based superalloys containing additions of rhenium and ruthenium were designed with the aid of time-saving formulae allowing to predict some of their physical characteristics. This alloy design programme succeeded in identifying the new generation single crystal alloy MC-NG suited for gas turbine blade and vane applications. This alloy exhibited a density-corrected high temperature creep strength comparable to that of high-rhenium containing alloys, but with a lower density and no propensity to form deleterious TCP phases.

Introduction

The continuous demand of the gas turbine engine manufacturers for an increasing turbine inlet temperature have pushed the alloy designers to develop nickel-based superalloy chemistries with high rhenium contents for single crystal blade applications. This alloying element was indeed shown to have a strong beneficial effect on the high temperature mechanical properties, especially as the creep strength is concerned. The level of rhenium was typically increased up to 6 wt. % in the alloys CMSX-10 [1] and René N6 [2].

These so-called third generation alloys however suffer from a number of drawbacks mainly due to their high content of rhenium: a high density, up to 9.1 g.cm^{-3}, a propensity to form deleterious topologically close packed (TCP) phases during exposures at high temperatures and a propensity to form grain defects during single crystal casting.

We therefore decided to explore the possibility of developing new generation nickel-based superalloys for single crystal turbine blade applications with a temperature capability significantly increased as compared to that of MC2 [3] which is considered as the best single crystal superalloy without rhenium. The ultimate target was to attain a creep strength comparable to that of high rhenium containing third generation alloys, but with reduced density and no propensity to form TCP phases.

One of the main idea driving this work was to study alloys containing some amounts of ruthenium substituting partially for rhenium. The density of this refractory element is close to that of molybdenum, therefore approximately one-half of that of rhenium. Moreover, preliminary studies showed that alloys containing both rhenium and ruthenium were less prone to precipitation of TCP phase than alloys containing only rhenium.

Alloy Design

A series of nickel-based experimental alloys was defined by varying the content of different alloying elements, particularly the refractory elements Mo, W, Re, Ru and Ta, in order to fulfil several requirements : i) a high γ' solvus temperature; ii) a high amplitude of γ/γ' mismatch; iii) a density as low as possible; iv) a good phasial stability.

The chemistries of the experimental alloys were defined with the help of time-saving formulae allowing to estimate the values of the γ' solvus temperature, of the γ/γ' mismatch and of the density. The New PHACOMP method [4] was used in order to avoid the precipitation of TCP phases.

An additional objective was to get an acceptable environmental resistance (hot corrosion and high temperature oxidation) taking into account that the protective coatings usually applied on the blades may crack or spall during service. Additions of 0.1 wt.% of silicon and 0.1 wt.% of hafnium were therefore made in all the experimental alloys in order to improve the adherence of the

Superalloys 2000
Edited by T.M. Pollock, R.D. Kissinger, R.R. Bowman,
K.A. Green, M. McLean, S. Olson, and J.J. Schirra
TMS (The Minerals, Metals & Materials Society), 2000

Table I Aimed Chemical Compositions of Some Relevant Experimental Alloys Compared to that of Reference Alloys (wt.%)

Alloy	Ni	Co	Cr	Mo	W	Re	Ru	Al	Ti	Ta	Others
MC2	Bal.	5	8	2	8	-	-	5	1.5	6	-
MC533	Bal.	-	7	-	5	3	3	6	-	6	0.1 Si; 0.1 Hf
MC534	Bal.	-	4	4	5	3	4	5.8	-	6	0.1 Si; 0.1 Hf
MC544	Bal.	-	4	1	5	4	4	6	0.5	5	0.1 Si; 0.1 Hf
MC645	Bal.	-	5	-	6	4	5	6	0.5	5	0.1 Si; 0.1 Hf
MC653	Bal.	-	4	1	6	5	3	5.3	1	6.2	0.1 Si; 0.1 Hf
CMSX-10M	Bal.	1.75	2	0.4	5.4	6.5	-	5.78	0.24	8.2	0.08 Nb
René N6	Bal.	12.5	4.5	1.1	5.75	5.35	-	6	-	7.5	0.15 Hf; 0.05 C; 0.004 B
Alloy #11	Bal.	12.65	5	1.55	5.65	4.50	1.60	6	0.4	6.4	0.15 Hf; 0.05 C; 0.004 B

protective alumina scale formed at high temperature, that would be beneficial to the high temperature oxidation resistance as previously demonstrated for the AM3, AM1 and MC2 single crystal superalloys [5]. In order to preserve the hot corrosion resistance, only moderate levels of Mo and Ti were added to the experimental alloys.

More than twenty experimental alloys were thus defined, primarily deriving from MC2, all containing additions of Re and some of them containing both Re and Ru additions. Small-scale laboratory heats of these alloys were melted in a high vacuum induction furnace and single crystal rods were directionally cast by the withdrawal process using <001> oriented seeds. Two additional reference alloys reproducing the chemistries of the high-rhenium containing alloys CMSX-10M [6] and René N6 [2] and one alloy reproducing the chemistry of the Alloy #11 designed by General Electric [7] and containing both Re and Ru were also cast as single crystals for the sake of comparison.

As it will be shown later in this paper, the best balance of properties were obtained with alloys containing both Re and Ru additions. The chemistries of five of these most relevant experimental alloys are compared in Table I to that of the reference alloys. MC544 will be ultimately designated MC-NG.

γ solvus temperature

An important objective fixed for this work was to obtain alloys with a γ solvus temperature as high as possible in order to promote a very high creep strength at temperatures above 1100°C. The creep strength of the γ'-strengthened nickel based superalloys is obviously strongly dependent on the volume fraction of fine γ' phase precipitates. As typically shown in the case of the AM1 single crystal superalloy [8], the solutioning rate of the γ' phase is rather low when the temperature increases up to typically 1100°C, whereas it increases rapidly above this temperature. In the high temperature range, the volume fraction of residual γ' phase is therefore strongly sensitive to the temperature and it was supposed that increasing significantly the γ' solvus temperature will allow to keep a higher volume fraction of γ' phase at elevated temperatures.

The values of the γ' solvus temperatures of the various alloys were primarily estimated using a formula deduced from a multiple linear regression analysis of experimental data obtained previously at ONERA on a series of nickel-based superalloys [9]. This analysis was continuously refined as and when new data

were acquired along the development work of the new generation single crystal superalloys. This formula is currently :

$$T_{\gamma \, solvus} \, (°C) = 1299.315 - 2.415 \, wt.\% \, Co - 6.362 \, wt.\% \, Cr - 2.224 \, wt.\% \, Mo + 3.987 \, wt.\% \, W + 0.958 \, wt.\% \, Re + 2.424 \, wt.\% \, Ru - 2.603 \, wt.\% \, Al - 4.943 \, wt.\% \, Ti + 3.624 \, wt.\% \, Ta.$$

Figure 1 compares these estimated data to experimental data obtained by thermal analysis using a high temperature dilatometer. The specimens were heated up to the corresponding solutioning temperature determined by preliminary heat treatment studies. The full solutioning heat treatment was performed within the dilatometer furnace. The specimens were then cooled to the room temperature. The heating and cooling rates were 3.6×10^{-2} °C.s^{-1}. The onset of the γ' precipitation, evidenced at cooling by a significant contraction of the specimen, was taken as the γ' solvus temperature. For instance, the estimated value of 1265°C for MC2 fitted perfectly the corresponding measured value.

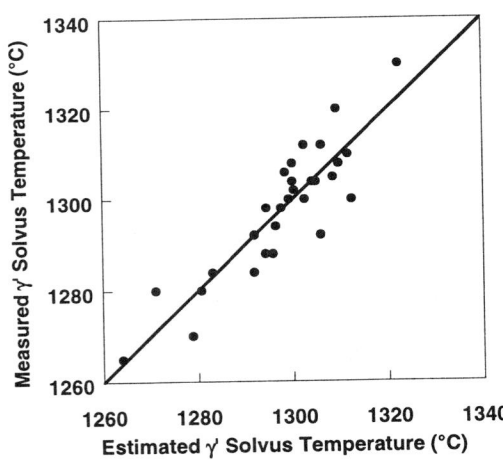

Figure 1: Comparison of estimated and measured values of the γ' solvus temperature of a series of experimental nickel-based superalloys.

More generally, there was a satisfying agreement between the estimated and experimental values, the largest difference being 14°C, which corresponds to a relative error of 1%. These results demonstrate the usefulness of such a formula for the prediction of the γ' solvus temperature knowing the chemistry of the alloy.

Table II Estimated and Measured Physical Properties

| Alloy | γ' solvus temperature (°C) | | Density (g.cm⁻³) | | δ at R.T. | New PHACOMP parameters | | Prone to TCP phase |
	Estimated	Measured	Estimated	Measured	Estimated	$\overline{Md_t}$	$\overline{Md_\gamma}$	
MC2	1265.1	1265	8.63	8.62	0.34%	0.987	0.918	-
MC533	1290.2	1292	8.638	8.64	0.21%	0.966	0.883	No
MC534	1304.9	1312	8.80	8.79	-0.66%	0.980	0.924	Yes
MC544	1305.3	1292	8.737	8.75	-0.24%	0.975	0.893	No
MC645	1308.0	1320	8.789	8.75	-0.31%	0.977	0.903	No
MC653	1311.0	1308	8.928	8.93	-0.15%	0.981	0.899	Yes
CMSX-10M	1322.7	1330	9.037	9.02	0.56%	0.965	0.843	Yes
René N6	1278.7	1270	8.87	8.87	0.05%	0.984	0.898	Yes
Alloy #11	1271.4	1280	8.791	8.78	-0.48%	0.995	0.937	Yes

The estimated and experimental values of the γ' solvus temperature for the alloys of Table I are compared in Table II. The experimental alloys exhibited γ' solvus temperatures much higher than that of MC2, that agreed with the initial target. For the experimental alloys of Table I, the increment varied between 27 and 55°C. It must be pointed out that the highest value was measured for the CMSX-10M alloy, presumably due to the low levels of Cr and Co which are γ' solvus temperature depressant elements and to the high content of Ta which increases this temperature. On the contrary, the Alloy #11 and René N6 exhibited quite low values of γ' solvus temperature owing to their high Co content.

To validate the hypothesis that a dramatic increase of the γ' solvus temperature would lead to a significant increment of the volume fraction of residual γ' phase at temperatures higher than 1100°C, some quantitative measurements were performed on the MC2, MC544 and MC653 alloys. In order to facilitate the γ' area fraction evaluation, measurements were made on pre-rafted specimens. Interrupted creep tests were thus performed at 1050°C on cylindrical specimens of the three alloys. Samples were machined from the specimen gauges, then heat treated in argon for one hour at 1150, 1200°C and 1250°C, with salted iced water quenching in order to freeze the high temperature microstructure. Samples for scanning electron microscopy (SEM) were prepared with sections parallel to the <001> tensile axis to obtain side views of the γ/γ' rafted microstructure. Using the back-scattering electron mode allowed to perfectly differentiate the two phases.

Moreover, supposing that the γ' rafts were infinitely long as compared to their thickness, it was assumed that the area fraction of γ' phase measured on such a section was close to its volume fraction. Such examples of microstructures observed after a hold at 1200°C are illustrated by the micrographs of Figure 2. Curves of Figure 3 show the variation with temperature of the volume fraction of γ' phase in MC2, MC544 and MC653 alloys. Between 1150°C and 1250°C, the γ' volume fraction is significantly higher in the new alloys, MC653 showing the highest content of γ'. There is effectively a good correlation between the residual content of γ' phase and the γ' solvus temperature.

Density

When designing the chemistry of the experimental alloys, a particular attention was devoted to the control of the density. An increase of the density will be indeed penalising for the rotating blades and for the disk at which the blades are attached. Particularly, when adding heavy refractory elements such as Re in the most recently developed alloys, a fraction of the potential creep strength improvement is lost due to the parallel increase of the density. The creep strength advantage offered by the third generation alloys as compared to the older ones is thus reduced when evaluated on a specific strength basis.

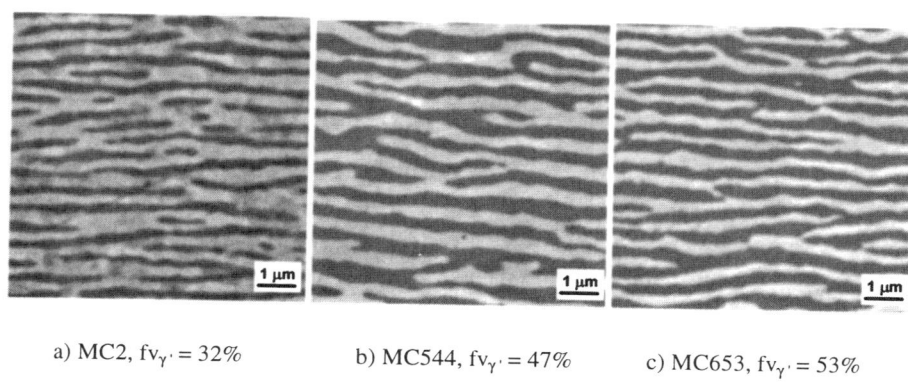

a) MC2, $fv_{\gamma'}$ = 32% b) MC544, $fv_{\gamma'}$ = 47% c) MC653, $fv_{\gamma'}$ = 53%

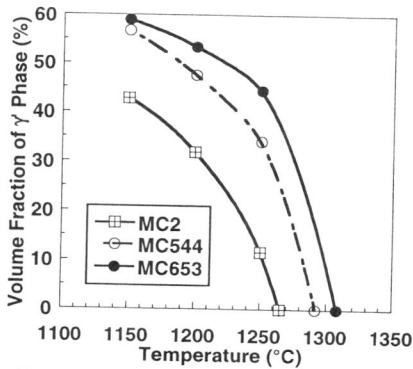

Figure 2: γ/γ' microstructures of pre-rafted single crystal alloys after a hold for 1 hour at 1200°C, followed by iced water quenching (SEM using the back-scattering electron mode : bright phase : γ, dark phase : γ').

Figure 3: Evolution with temperature of the γ' volume fraction. The temperatures for nil γ' phase are the measured γ' solvus temperatures.

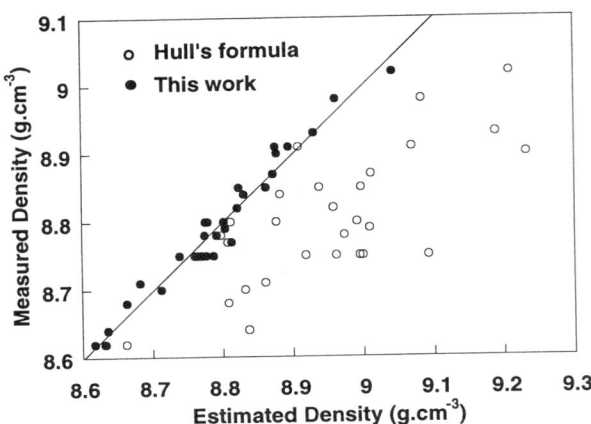

Figure 4: Comparison of estimated and measured values of the density of a series of experimental nickel-based superalloys.

The Hull's formula [10] was primarily applied to predict the density of the experimental alloys from their compositions. This formula contains one term resulting from the application of the rule of mixtures and several additional correction terms deduced from regression analyses performed on data collected on 235 alloys. However, this formula does not contain any correction terms for Re and Ru. Application of this formula to the whole series of experimental and reference alloys gave the results illustrated in Figure 4. When compared to experimental values calculated from volume and weight measurements, a significant deviation appeared between the estimated and the calculated values, which increased with the density. This therefore showed the necessity to include correction terms related to Re and Ru in the Hull's formula.

Instead of trying to modify the Hull's formula, a multiple linear regression analysis of experimental data obtained on the experimental and reference alloys was run. As for the γ' solvus temperature evaluation, this analysis was refined continuously along the design procedure of the new alloys and the resulting formula is currently :

d (g.cm^{-3}) = 8.29604 - 0.00435 wt.% Co - 0.0164 wt.% Cr + 0.01295 wt.% Mo + 0.06274 wt.% W + 0.0593 wt.% Re + 0.01811 wt.% Ru - 0.06595 wt.% Al - 0.0236 wt.% Ti + 0.05441 wt.% Ta.

The values of the density calculated using this formula for the series of experimental alloys and for the reference alloys are compared in Figure 4 to the corresponding measured values. There was a very good agreement between the predictions and the measurements, the largest relative error being 0.45%. Estimated and measured values for the alloys of Table I are compared in Table II.

γ/γ' mismatch

A survey of the literature show that the lattice mismatch $\delta = 2(a_{\gamma'} - a_\gamma)/(a_{\gamma'} + a_\gamma)$ is generally considered as a factor which could play a significant role on the elevated temperature creep strength of the single crystal superalloys [11-15]. There is however still a debate about the optimum value which must be aimed for improving the high temperature creep behaviour of these alloys [15]. Anyway a

common consent seems to be that a large negative value of the mismatch would be preferable because it will promote a fast evolution of the microstructure towards a stable and creep resistant γ/γ' rafted microstructure, and also dense dislocation networks forming at the γ/γ' interfaces and acting as efficient barriers against mobile dislocations, therefore reducing the creep rate [12-15]. When designing the chemistries of the experimental alloys, we have therefore tried to attain a negative value of the mismatch at high temperature with an amplitude as large as possible in order to optimise this property. We have also aimed to attain various values of mismatch in order to try to evidence some effects of this parameter on the creep behaviour of the experimental alloys.

According to the method used by Watanabe [17], the room temperature (RT) mismatch values of the experimental and reference alloys were calculated from γ and γ' lattice parameters estimated using the following formulas :

$$a_\gamma = a_{Ni} + \Sigma V_i C_i \qquad (1)$$

$$a_{\gamma'} = a_{Ni3Al} + \Sigma V'_i C'_i \qquad (2)$$

where a_{Ni} and a_{Ni3Al} are the lattice parameters of pure Ni and pure Ni$_3$Al, V_i and V'_i are the Vegard's coefficients for the element i in Ni and Ni$_3$Al respectively, and C_i and C'_i are the atomic fractions of element i in γ and γ' phases.

The lattice parameters for Ni and Ni$_3$Al were taken as 3.524 Å and 3.570 Å respectively [11]. The Vegard's coefficients for Co, Cr, Mo, W, Al, Ti, Ta and Nb in γ and γ' phases were get from Mishima et al. [18]. As no data existed for Re, we have deduced the Vegard's coefficient of Re in Ni from X-ray diffraction experiments performed on Ni-Re solid solutions containing 1at.% and 2at.%Re respectively. The resulting value of V_{Re} was 0.441. The Vegard's coefficient for Ru in Ni was taken as 0.3125 [19]. No data were found in the literature for the Vegard's coefficients for Re and Ru in Ni$_3$Al. To calculate these values, we have used the following relationship established by Mishima et al. [18] in the case where the alloying element i substitutes for Al in Ni$_3$Al :

$$V'_i = V_i - V_{Al} \qquad (3)$$

As atom probe analyses have shown that the refractory elements Re, Mo and W substitute preferentially for Al in Ni$_3$Al [20], we assumed that Ru exhibited the same behaviour, and applying the equation (3) gave $V'_{Re} = 0.262$ and $V'_{Ru} = 0.1335$.

The following formulae were finally obtained :

a_γ (Å) = 3.524 + 0.0196 C_{co} + 0.110 C_{Cr} + 0.478 C_{Mo} + 0.444 C_W + 0.441 C_{Re} + 0.3125 C_{Ru} + 0.179 C_{Al} + 0.422 C_{Ti} + 0.7 C_{Ta} + 0.7 C_{Nb}.

$a_{\gamma'}$ (Å) = 3.57 - 0.004 C'_{Cr} + 0.208 C'_{Mo} + 0.194 C'_W + 0.262 C'_{Re} + 0.1335 C'_{Ru} + 0.258 C'_{Ti} + 0.5 C'_{Ta} + 0.46 C'_{Nb}.

The compositions of the γ and γ' phases were estimated by using a method derived from that developed by Watanabe [17] which itself derived from the PHACOMP method [21]. The values of the calculated lattice mismatch δ for the relevant experimental alloys and for the reference alloys are compared in Table II.

Knowing that the thermal expansion coefficient of the γ' phase in the nickel-based superalloys is less than that of the γ phase [13, 22, 23], a positive mismatch at room temperature will become negative when the temperature will increase above a given value, and a negative mismatch at RT will remain negative with an increasing amplitude [13]. Our method thus allowed to propose a relative ranking of the alloys in respect of the lattice mismatch. For instance, it was foreseen that the alloy MC534 would exhibit a higher amplitude of δ at high temperature than MC544. It must be pointed out that all the experimental alloys of Table II would exhibit higher amplitude of δ than MC2, at the relevant temperatures.

New PHACOMP parameters

In order to predict and to control the occurrence of deleterious TCP phases during exposure at high temperatures of the experimental alloys, we have applied the New PHACOMP method devised on the basis of the molecular orbital calculations of electronic structures [4]. As done by Yukawa et al. [24, 25] and Zhang et al. [26], average values of the Md parameter, $\overline{Md}_t$ and $\overline{Md}_\gamma$, were respectively calculated from the total composition and from the estimated γ matrix composition of each experimental and reference alloy using the following formula :

$$\overline{Md} = \Sigma \, C_i \, (Md)_i$$

where C_i is the atomic fraction of the element i in the alloy or in the γ matrix, and $(Md)_i$ is the Md value for the element in Table III.

This method aims at predicting the occurrence of TCP phases such as σ, μ or Laves phases when the Md values become larger than critical values deduced from experience. The values of $\overline{Md}_t$ and $\overline{Md}_\gamma$ for the relevant experimental alloys and the reference alloys are reported in Table II.

Table III   List of Md Values [4].

Element	Md	Element	Md
Ti	2.271	Hf	3.02
Cr	1.142	Ta	2.224
Co	0.777	W	1.655
Ni	0.717	Re	1.267
Nb	2.117	Al	1.9
Mo	1.55	Si	1.9
Ru*	1.006*		

*Courtesy from Pr. M. Morinaga

Experimental Results

Microstructures and Heat Treatments

Some typical as-cast dendritic microstructures of experimental and reference alloys are illustrated in Figure 5. All these alloys contained higher fractions of interdendritic γ/γ' eutectic nodules than observed in first generation single crystal superalloys, and even in second generation single crystal superalloys containing Re contents close to 3wt.%. This is mainly due to the higher total level of γ'-former elements such as Al, Ti, Ta and eventually Nb. However, adequate solution heat treatments were identified for the experimental alloys to eliminate all or almost all of the γ/γ' eutectic nodules without incipient melting as illustrated in [ ]. For example, a two-stage solution heat-treatment including a hold for 3 hours at 1315°C, then a heating ramp at a rate of [ ] and a final treatment for 3 hours at 1340°C led to a complete reshaping of the γ/γ' eutectic and obviously of the γ' alloy γ' precipitates in the MC544 alloy. All the alloys of Table I have been fully homogenised using such heat treatment procedures. As Rene N6 and Alloy #11 contained voluntary additions of carbon some carbide particles remained within the interdendritic areas as shown in Figure 5c.

[handwritten note, partially overlapping text]: #4 New Phacomp Ru Md =1.006 Ru Superalloys 84

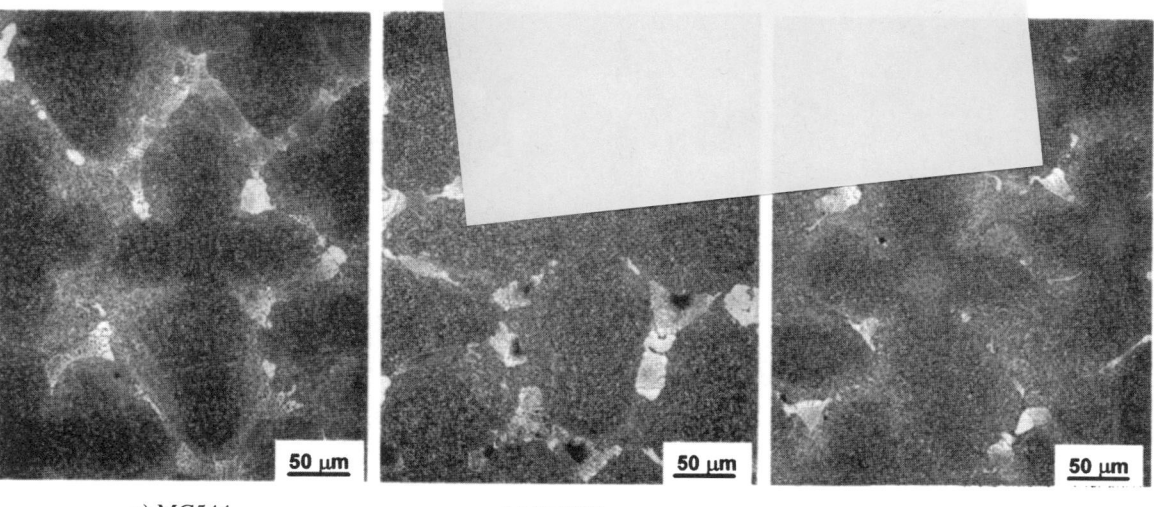

a) MC544   b) MC653   c) Alloy #11

Figure 5: As-cast dendritic structures of <001> single crystals.

741

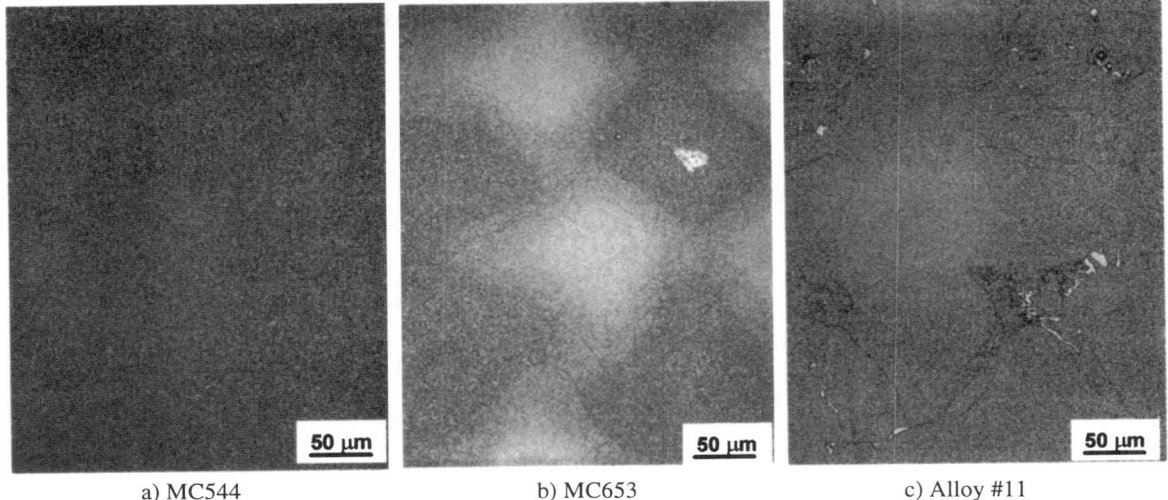

| a) MC544 | b) MC653 | c) Alloy #11 |

Figure 6: Solution heat-treated microstructures of <001> single crystals

After complete solutioning, all the experimental alloys were applied an ageing heat treatment for 4 hours at 1100°C followed by a treatment for 24 hours at 850°C in order to produce a homogeneous distribution of γ' precipitates. Depending on the alloy chemistry the mean size of the γ' particles varied within the range 260-400 nm. As shown in Figure 7, the shape of the precipitates depends on the alloy and it was possible to evidence a relationship between this precipitate shape and the value of the estimated room temperature mismatch. These observations agreed with the results of a number of studies showing that the shape of the γ' precipitates depend strongly on the value of δ at the ageing temperature [27, 28]. A transition from sphere to cube occurred when the amplitude of δ increases. Such an evolution was observed for our experimental alloys, that demonstrates the validity of our predictive method, if we extrapolate the RT δ values at 1100°C following the argument developed previously.

Creep properties

Constant load tensile creep tests were performed in air in the temperature range 950°C-1150°C on cylindrical specimens (20 mm gauge length, 3 mm diameter) machined from fully heat treated single crystals. Typical stress-rupture lives obtained on the most relevant experimental alloys and on the reference alloys are compared in Table IV. Typical creep curves are reproduced in Figures 8 to 10.

Table IV   Stress-Rupture Lives (in hours)

Alloy	Creep conditions		
	T = 950°C σ = 300 MPa	T = 1050°C σ = 150 MPa	T =1150°C σ = 100 MPa
MC2	198	485	6
MC533	298	401	52
MC534	266	372	60
MC544	458	486	151
MC645	404	499	185
MC653	456	726	194
CMSX-10M	690	637	137
René N6	531	434	65
Alloy #11	370	466	45

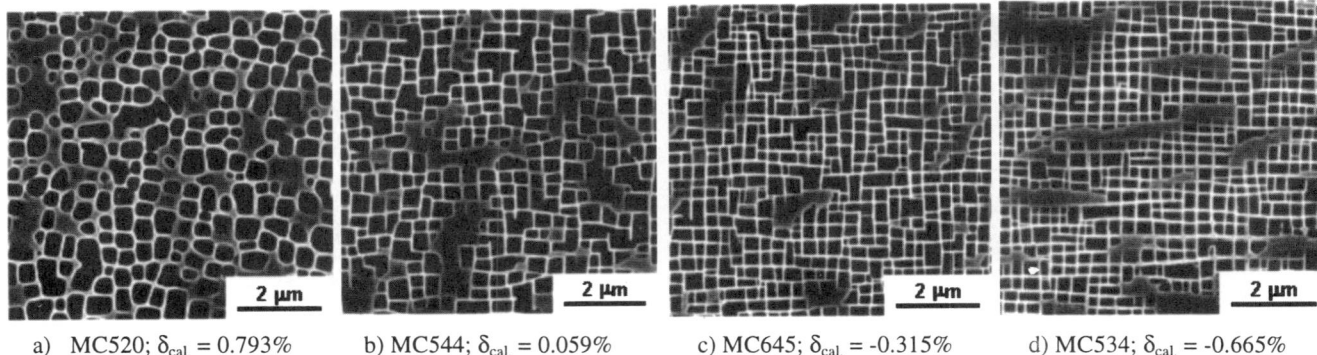

| a) MC520; $\delta_{cal.}$ = 0.793% | b) MC544; $\delta_{cal.}$ = 0.059% | c) MC645; $\delta_{cal.}$ = -0.315% | d) MC534; $\delta_{cal.}$ = -0.665% |

Figure 7: Relationship between the RT estimated lattice mismatch δ and the shape of the γ' precipitates.

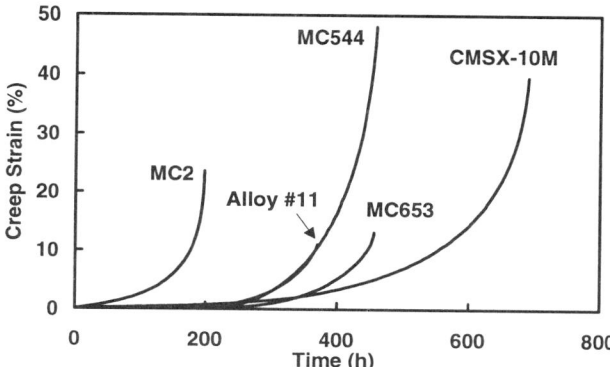

Figure 8: Typical creep curves at 950°C and 300 MPa.

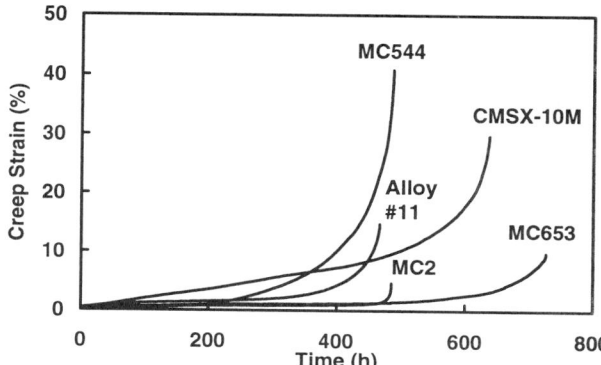

Figure 9: Typical creep curves at 1050°C and 150 MPa.

The studies aiming at understanding in details the effects of chemistry modifications on the creep behaviour of these new alloys are still in progress, but it is already possible to draw some hypotheses to explain the significant differences observed here between the various alloys.

Compared to MC2, all the Re-containing alloys exhibited significantly improved creep lives at 950°C. This advantage is thought to be due to the addition of Re that slows down all the diffusion-controlled processes which influence the creep behaviour of these alloys [29]. The coarsening rate of the γ' precipitates is thus strongly reduced, and the mobility of the matrix dislocations which move essentially by a diffusion-controlled climb process at the γ/γ' interfaces is also decreased. At this temperature, the two most resistant alloys are effectively CMSX-10M and René N6 which contain 6.5 wt.% Re and 5.35 wt.% Re respectively. Considering the creep curves, the most striking feature relative to the Re-containing alloys is a very low creep rate during a long period before entering the tertiary creep stage, whereas this stage started practically as soon as the load was applied in the case of MC2 which behaviour is representative of Re-free single crystal superalloys.

At 1050°C, only the MC653 and CMSX-10M alloys exhibited a significant creep life improvement as compared to MC2, while the other alloys showed at best a comparable stress-rupture life. These results therefore show that adding Re is not a panacea when searching at improving the creep strength of single crystal superalloys at every temperature.

The creep behaviour at 1050°C of this class of alloys is specific in the sense where the strengthening γ' precipitates generally experience during the primary creep stage a dramatic evolution towards a γ' rafted morphology, which remains stable during all the secondary creep stage, before to be destabilised during the tertiary creep. It is therefore obvious that the creep behaviour of these alloys will primarily depend on the more or less quick transition from cube to plate morphology of the γ' particles, and on the morphological stability of the resulting γ/γ' rafted microstructure, which in turn will depend on parameters such as the γ/γ' mismatch and the bulk diffusion rate.

As shown in Figure 9, the various shapes of the creep curves observed for instance for the MC2, MC544 and CMSX-10M alloys thus suggest significant variations of the γ' rafting behaviour between these three alloys. Preliminary investigations

of the microstructural variations during creep at 1050°C of the MC2 and MC544 alloys have shown that the cube-raft transition is delayed in MC544 compared to MC2, but that the resulting rafted microstructure is less stable in MC544 than in MC2 that correlates with the earlier initiation of the tertiary creep stage [30]. Further investigations are in progress to completely describe and understand the creep behaviour of such alloys at 1050°C, and with the aim to improve it.

At 1150°C, the creep strength advantage exhibited by the Re-containing over MC2 was spectacular (Figure 10). Typically, when MC544 is compared to MC2, the creep life is multiplied by a factor of 25. This dramatic effect is attributed mainly to the larger amount of γ' phase remaining in the high solvus MC544 and MC653 alloys at this temperature as compared to MC2 (see Figure 3). The Alloy #11 which exhibited a significantly lower creep strength than MC544, MC653 or CMSX-10M, is characterised by a lower γ' solvus temperature that underlines the importance of this parameter as the very high temperature creep strength is concerned. That obviously does not exclude other beneficial effects resulting from a decrease of the γ' coarsening kinetic or from solid solution strengthening effects associated to the increase of the refractory element contents. However the increase of the residual fraction of γ' phase is thought to be the main reason for the very high creep strength at elevated temperatures of the new alloys. That therefore fully justified the initial choice to develop alloys with a high γ' solvus temperature.

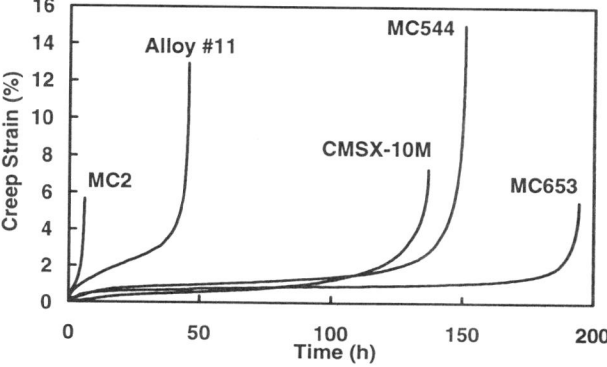

Figure 10: Typical creep curves at 1150°C and 100 MPa.

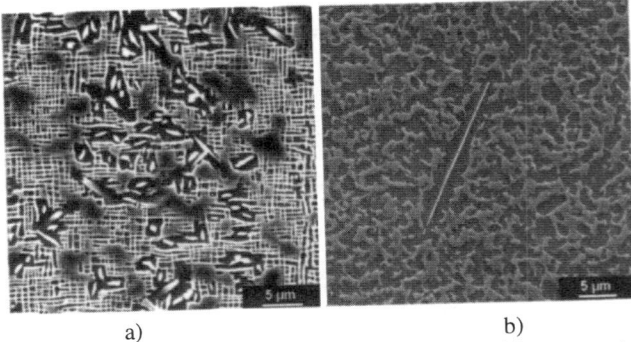

a)   b)

Figure 11: Re-rich TCP phase precipitates in dendrite cores: a) Alloy #11, 1050°C/200h; b) MC653, 1100°C/130 MPa/216h.

Microstructural Stability

Some ageing heat treatments for 200 hours at 950, 1050 and 1150°C were performed on homogenised samples of the whole series of experimental alloys and of the Re-containing reference alloys in order to check their phasial stability. Microstructural assessments were also performed on stress-ruptured specimens of all the alloys. Some of the experimental alloys were observed to be prone to the precipitation of needle-like Re-rich particles in the dendrite cores whereas the other ones did not contain TCP phases (Figure 11).

Attempt was made to determine the critical values of $\overline{Md_t}$ and $\overline{Md_\gamma}$ above which TCP phases precipitated. The diagram of Figure 12 summarises the results of the microstructural observations. It appeared in fact impossible to determine unique critical values for the $\overline{Md}$ parameters. Analysis of all the results showed that in any case the values of $\overline{Md_t}$ and $\overline{Md_\gamma}$ must not go beyond 0.980 and 0.903 respectively to avoid the precipitation of Re-rich TCP phase particles. These values were therefore considered as maximum allowable values when designing the most relevant of our experimental alloys, i.e. MC544, MC645 and MC653 .

However, it was observed that TCP phases may occur for values of $\overline{Md_t}$ and $\overline{Md_\gamma}$ as low as 0.965 and 0.843 respectively as in the case of CMSX-10M. A more accurate analysis of the results showed that the critical values of the Md parameters decreased as the Re content increased, but at the contrary increased as the Ru level increased in the alloys.

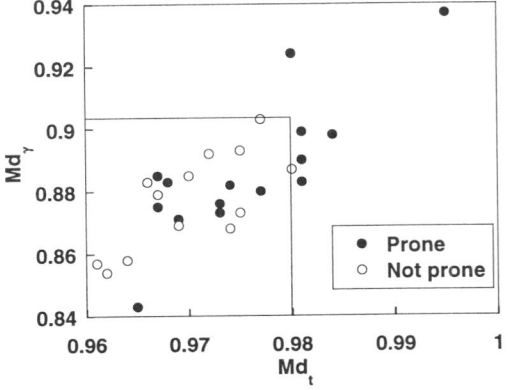

Figure 12: Proneness to TCP phase precipitation in single crystal superalloys as a function of Md_t and Md_γ parameters.

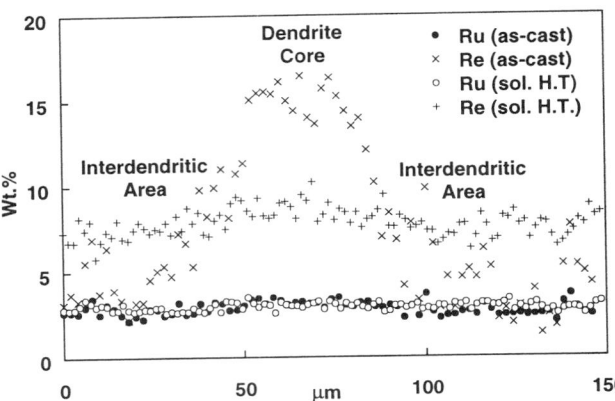

Figure 13: Variations of Ru and Re concentrations (wt.%) throughout dendrites of the MC393 alloy, before and after a solution heat treatment.

The sensitivity of the critical $\overline{Md}$ values to the Re concentration is due in fact to the residual segregation of this element in the dendrite cores, even after the solution treatment. Figure 13 illustrates the results of electron probe analyses performed through dendrites of the experimental alloy MC393 containing 8.2 wt.% Re, before and after a homogenisation heat treatment (1330°C/3h, ramp at 3°C.h^{-1} up to 1360°C, then 1360°C/18h/air cooling). These analyses showed that Re segregated strongly within the dendrite cores during solidification, and that, even after a 31h solution heat treatment on this experimental alloy, the Re concentration may still attain 9.5 wt.% in the dendrite cores. It is this local Re concentration which would have in fact to be considered when computing the values of the $\overline{Md}$ values for the prediction of TCP phase precipitation in the dendrite cores. Higher is the Re content in an alloy, higher will be its concentration within the dendrite cores, even after a long solution heat treatment, and lower will be the critical values for $\overline{Md_t}$ and $\overline{Md_\gamma}$.

Another interesting feature evidenced by these analyses is that Ru did not segregate in the alloy during solidification, that is an advantage compared to Re, because it did not induce local over-concentration. This feature could partly explain why the alloys containing both Re and Ru were observed to be less prone to TCP phase precipitation than alloys containing only Re additions. This beneficial effect could be also partly due to a modification of the partitioning ratio of various alloying elements such as molybdenum or rhenium between the γ and γ' phases as suggested by O'Hara et al. [6], that would render the γ matrix less prone to the formation of TCP phase particles.

Environmental Resistance

High temperature cyclic oxidation tests were performed in air at 1100°C using one hour-cycles on the experimental alloys. Their performance were comparable to that of AM3, AM1 or MC2 alloys containing both additions of Si and Hf. No significant effect of the major alloying elements was evidenced. On the other hand, cyclic hot corrosion tests performed at 850°C with addition of Na_2SO_4 on some of the experimental alloys and on reference alloys revealed various behaviours depending on the chemistry. Some of these results are illustrated in Figure 14.

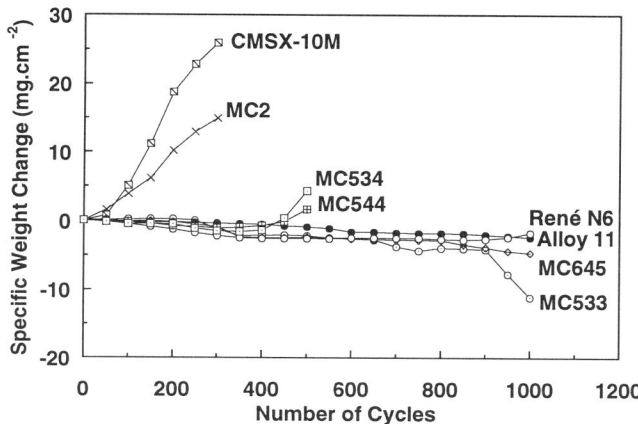

Figure 14: Cyclic corrosion behaviour at 850°C of single crystal superalloys.

The experimental alloys all exhibited a hot corrosion resistance superior to that of MC2 and CMSX-10M. The poor corrosion resistance of MC2 can be attributed to its rather high level of W and Mo, elements known to be deleterious for this property. It can be pointed out that the most corrosion resistant experimental alloys, MC645 and MC533, are free of Mo. The weak corrosion resistance of CMSX-10M is very likely due to its low level of chromium. It thus appeared that a minimum content of 4wt% of chromium must be kept in the alloys in order to ensure an acceptable level of corrosion resistance, even in the case of high rhenium containing alloys. On the other hand, no significant role of the ruthenium was evidenced.

Discussion and Conclusions

Whereas a number of studies have still to be performed to completely understand the effects of chemistry variations on the creep behaviour of the most recently developed single crystal superalloys, it has been shown that, using rather simple hypotheses and with the aid of computation methods aiming at predicting various physical parameters from the alloy chemistry, it was possible to identify promising new generation alloys which offer a unique combination of properties. It was thus demonstrated that the concept of high γ solvus alloy was particularly relevant when searching at improving the very high temperature creep strength of these materials. The new PHACOMP method aiming at predicting the occurrence of deleterious TCP phases must be carefully used because it does not take into account the dendritic segregations of low diffusivity elements such as Re and W. As the lattice parameter mismatch δ is concerned, its influence on the creep behaviour of the experimental alloys is still not clear, even if the most creep resistant of these alloys are predicted to show a high amplitude of δ at the testing temperatures. As discussed elsewhere [16], the role of the γ/γ' mismatch is complex, particularly at the temperatures where γ' rafting occurs and further work would have to be performed on this topic to define exactly the optimum value of δ for given creep conditions.

On a density-corrected basis, the best experimental alloys exhibited a creep strength equivalent to that of the third generation alloys CMSX-10 and René N6 as evidenced by the Larson-Miller plots of Figure 15.

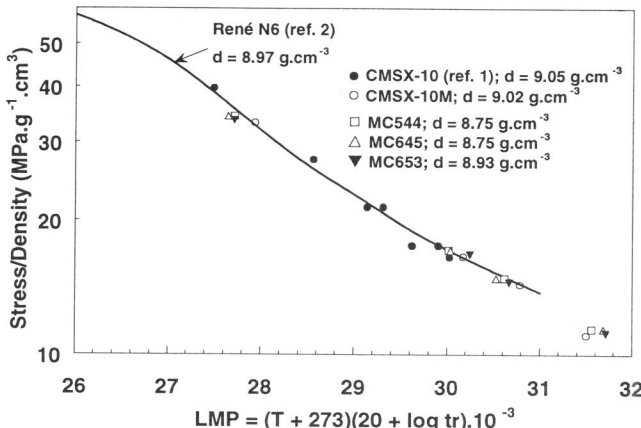

Figure 15: Comparative Larson-Miller stress-rupture data for various single crystal superalloys.

Amongst these various experimental alloys, MC544 exhibited the best balance of overall properties with a density of 8.75 g.cm^{-3} and a specific creep strength comparable with that CMSX-10 and René N6, but without any propensity to form rhenium-rich TCP phases. These results demonstrated that it exists a promising alternative to the third generation superalloys for increasing the temperature capability of the single crystal turbine blades without being penalised by an excessive density or by microstructural instability features. Further evaluation of the MC544 alloy under the name of MC-NG is in progress at an industrial scale in order to confirm its interest for various turbine blade and vane applications [31].

Acknowledgements

This work was funded by the French Ministry of Defence. The author is grateful for the experimental support given by J.-L. Raffestin, S. Navéos, C. Ramusat and F. Passilly.

References

1. G.L. Erickson., "The Development and Application of CMSX-10," Superalloys 1996, ed. R.D. Kissinger et al. (Warrendale, PA: The Minerals, Metals & Materials Society, 1996), 35-44.

2. W.S. Walston, K.S. O'Hara, E.W. Ross, T.M. Pollock, and W.H. Murphy, "René N6: Third Generation Single Crystal Superalloy," Superalloys 1996, ed. R.D. Kissinger et al. (Warrendale, PA: The Minerals, Metals & Materials Society, 1996), 27-34.

3. P. Caron, and T. Khan, "Development of a New Nickel Based Single Crystal Turbine Blade Alloy for Very High Temperatures," Advanced Materials and Processes, Vol. 1, ed. H.E. Exner and V. Schumacher (Oberursel, Germany: DGM Informationsgesellschaft mbH, 1990), 333-338.

4. M. Morinaga, N. Yukawa, H. Adachi, and H. Ezaki," New PHACOMP and its Application to Alloy Design", Superalloys 1984, ed. M. Gell et al. (Warrendale, PA: The Metallurgical Society of AIME, 1984), 523-532.

5. P. Caron, S. Navéos, and T. Khan, "Improvement of the Cyclic Oxidation Behaviour of Uncoated Nickel Based Single Crystal Superalloys," Materials for Advanced Power Engineering 1994 - Part I, ed. D. Coutsouradis et al. (Dordrecht, Holland: Kluwer Academic Publisher, 1994), 1185-1194.

6. G.L. Erickson, Cannon-Muskegon Corporation, US Patent # 5,540,790 (1996).

7. O'Hara K.S., Walston W.S., Ross E.W., Darolia R., General Electric Company, US Patent # 5,482,789 (1996).

8. T. Grosdidier, A. Hazotte, and A. Simon, "Precipitation and Dissolution Processes in γ/γ′ Single Crystal Nickel-Based Superalloys," Mat. Sc. and Engin., A256 (1998), 183-196.

9. M. Marty, private communication, ONERA, 1995.

10. F.D. Hull, "Estimating Alloy Densities," Metal Progress, November 1969, 139-140.

11. T.M. Pollock, and A.S. Argon, "Creep resistance of CMSX-3 Nickel Base Superalloy Single Crystal," Acta metall. Mater., 40 (1) (1992), 1-30.

12. R.A. MacKay, and L.J. Ebert, "The Development of γ-γ′ Lamellar Structures in a Nickel-base Superalloy during Elevated Temperature Mechanical Testing," Met. Trans. A, 16A (1985), 1969-1982.

13. M.V. Nathal, R.A. MacKay, and R.G. Garlick, "Temperature Dependence of γ-γ′ Lattice Mismatch in Nickel-base Superalloys," Mat. Sc. Engin., 75 (1985) 195-205.

14. M.V. Nathal, and R.A. MacKay, "The Stability of Lamellar γ-γ′ Structures," Mat. Sc. Engin., 85 (1987) 127-138.

15. R.A. MacKay, M.V. Nathal, and D.D. Pearson, "Influence of Molybdenum on the Creep Properties of Nickel-Base Superalloy Single Crystals," Met. Trans. A, 21A (1990) 381-388.

16. F.R.N. Nabarro, "Rafting in Superalloys," Met. and Mat. Trans. A, 27A (1996) 513-530.

17. R. Watanabe, and T. Kuno, "Alloy Design of Nickel-Base Precipitation Hardened Superalloys," Trans. ISIJ, 16 (1976), 437-446.

18. Y. Mishima, S. Ochiai, and T. Suzuki, "Lattice Parameters of Ni(γ), Ni₃Al (γ′) and Ni₃Ga (γ′) Solid Solutions with Additions of Transition and B-subgroup Elements," Acta metall., 33 (6) (1985), 1161-1169.

19. W. Pearson, A Handbook of Lattice Spacings and Structures of Metals and Alloys, Vol. 2 (Oxford, UK: Pergamon Press, 1967), 1135.

20. D. Blavette, P. Caron, and T. Khan, "An Atom-Probe Study of Some Fine-Scale Microstructural Features in Ni-Based Single Crystal Superalloys," Superalloys 1988, ed. D.N. Duhl et al. (Warrendale, PA: The Metallurgical Society, 1988), 305-314.

21. L.R. Woodyatt, C.T. Sims, and H.J. Beatty, Jr., "Prediction of Sigma-Type Phase Occurrence from Compositions in Austenitic Superalloys", Trans. Met. Soc. AIME, 236 (1966), 519-527.

22. S. Yoshikate, T. Yokokawa, K. Ohno, H. Harada, and M. Yamazaki, "Temperature dependence of γ/γ′ Parameter Misfit for Nickel-Base Superalloys," Materials for Advanced Power Engineering 1994, Part I, ed. D. Coutsouradis et al. (Dordrecht, Holland: Kluwer Academic Publisher, 1994), 875-882.

23. L. Müller, T. Link, and M. Feller-Kniepmeier, "Temperature Dependence of the Thermal Lattice Mismatch in a Single Crystal Nickel-Base Superalloy Measured by Neutron Diffraction," Scripta Met. et Mat., 26 (1992) 1297-1302.

24. N. Yukawa, M. Morinaga, H. Ezaki, and Y. Murata, "Alloy Design of Superalloys by the d-Electrons Concept," High Temperature Alloys for Gas Turbines and Other Applications 1986 - Part II, ed. W. Betz et al. (Dordrecht, Holland: D. Reidel Publishing Company, 1986), 935-944.

25. N. Yukawa, M. Morinaga, Y. Murata, H. Ezakin, and S. Inoue, "High Performance Single Crystal Superalloys Developed by the d-Electrons Concept," Superalloys 1988, ed. S. Reichman et al. (Warrendale, PA: The Metallurgical Society, Inc., 1988), 225-234).

26. J.S. Zhang, Z.Q. Hu, Y. Murata, M. Morinaga, and N. Yukawa, "Design and Development of Hot Corrosion-Resistant Nickel-Base Single-Crystal Superalloys by the d-Electrons Alloy Design Theory: Part I. Characterisation of the Phase Stability," Met. Trans. A, 24A (1993), 2443-2450.

27. R.A. Ricks, A.J. Porter, and R.C. Ecob, "The Growth of γ′ Precipitates in Nickel-Base Superalloys," Acta. metall., 31 (1983), 43-53.

28. M. Fährmann, P. Fratzl, O. Paris, E. Fährmann, and W.C. Johnson, "Influence of Coherency Stress on Microstructural Evolution in Model Ni-Al-Mo Alloys," Acta. metall. mater., 43 (3) (1995), 1007-1022.

29. A.F. Giamei, and D.L. Anton, "Rhenium Additions to a Ni-Base Superalloy: Effects on Microstructure," Met. Trans. A, 16A (1985), 1997-2005.

30. P. Caron, M. Benyoucef, A. Coujou, J. Crestou, and N. Clément, "Creep Behaviour at 1050°C of a Re-containing Single Crystal superalloy," Paper presented at the International Symposium on Materials Ageing and Life Management (ISOMALM 2000), Kalpakkam, India, 3-6 October 2000.

31. D. Argence, C. Vernault, Y. Desvallées, and D. Fournier, "MC-NG: a 4th Generation Single-Crystal Superalloy for Future Aeronautical Turbine Blades and Vanes", this Conference.

Distribution of Platinum Group Metals in Ni-Base Single-Crystal Superalloys

H. Murakami, T. Honma, Y. Koizumi and H. Harada

National Research Institute for Metals, 'High Temperature Materials 21' Project,

1-2-1, Sengen, Tsukuba, 305-0047, Japan

Abstract

The effects of platinum group metals (PGMs) addition on microscopic characteristics of Ni-Al-PGM ternary and Ni-base single crystal superalloys were investigated. Several Ni-19at%Al-Xat% PGM (PGM: Ru, Rh, Pd, Ir and Pt) ternary alloys were prepared by arc-melting in order to understand the distribution of PGMs in γ and γ' two phases. It is found from differential thermal analysis (DTA) that among PGMs, Ir and Ru increase the melting points whereas Pt and Pd decrease the melting points. These alloys were successfully solution treated at 1300°C for 4h. After ageing treatment, all Ni-19Al-1PGM alloys form γ and γ' two-phase structure, while 9at% addition of Ir, Ru, Rh and Pd leads to the precipitation of PGM-enriched phase which is different from both γ and γ' phases, except for a Ni-19Al-9Pt alloy which maintains γ and γ' two-phase structure. SEM-EDX analysis revealed that Ir and Ru have a preference to partition to the γ phase while Pt, Pd and Rh have a preference for the γ' phase. Microscopic distribution of Ir and Ru in multi-component Ni-base single crystal superalloys were also investigated. Atomprobe field ion microscopy (APFIM) revealed that partitioning behaviour of Ir in multi-component Ni-base single crystal superalloys depends on alloy composition. For Re-free alloy TMS-79, Ir atoms have a small preference to be located in the γ phase, while for Re-containing TMS-80, Ir tends to partition to the γ' precipitates. On the other hand, Ru is found to prefer the γ phase in Re-containing TMS-91. These results suggest that among alloying elements, 'competition of partitioning' takes place in multi-component alloys. The order of alloying elements having stronger preference to the γ phase can be summarized as Re > Cr > (Mo, Ru, Co) > W > Ir.

Introduction

Ni-base single crystal superalloys, which have remarkable mechanical properties at elevated temperatures, have been designed for use as turbine blades in aeroengines. In order to develop higher efficiency engines, considerable efforts are still devoted to enhance the temperature capabilities of superalloys. In addition to Ni and Al, superalloys usually contain several alloying elements, such as Cr, Co, Mo, W, Ta. Among these alloying additions, much attention has recently been paid to Re since the development of CMSX-4[1] because of its contribution to improve both the creep and corrosion properties. However, the solubility limit of Re in Ni is rather low due to the difference in crystallographic structure between Re (hcp) and Ni (fcc). Also, since Re has a low diffusibility, Re

Superalloys 2000
Edited by T.M. Pollock, R.D. Kissinger, R.R. Bowman,
K.A. Green, M. McLean, S. Olson, and J.J. Schirra
TMS (The Minerals, Metals & Materials Society), 2000

atoms tend to be segregated in matrices. The addition of Re thus assists the formation of a topologically close-packed (TCP) phase, resulting in deteriorating the creep properties. On the other hand, the addition of platinum group metals (PGMs) in Ni-base superalloys is of particular interest. Since PGMs usually have high melting points and high hot corrosion resistance, addition of PGM is expected to enhance high temperature mechanical properties and hot corrosion resistance. Moreover, except for Ru and Os which have hcp structure, PGMs and Ni have the same fcc crystal structure and they form a complete solid solution system. Therefore we expect that a fairly high amount of PGMs can be alloyed in contrast to Re. However, there are few reports investigating the effect of PGM addition in Ni-base alloys[2-6]. The aim of this study is to investigate the distribution of PGMs, such as Ru, Rh, Pd, Ir and Pt, in Ni-base γ and γ' two phase alloys. This paper deals with two kinds of alloy systems. As a fundamental investigation, Ni-Al-PGM ternary alloys having γ and γ' two phases were examined. The microscopic distribution of Ir and Ru in multi-component Ni-base single crystal superalloys were investigated as a next step, with discussing the change in partitioning behaviour of alloying elements as a function of alloy composition.

Table I Chemical composition of alloys (in at%)

Alloy	Ni	Al	Ir	Ru	Rh	Pd	Pt
1Ir	Bal.	18.1	0.95	-	-	-	-
3Ir	Bal.	18.5	2.81	-	-	-	-
5Ir	Bal.	18.0	4.72	-	-	-	-
9Ir	Bal.	18.1	8.60	-	-	-	-
1Ru	Bal.	18.2	-	0.90	-	-	-
5Ru	Bal.	18.3	-	4.88	-	-	-
9Ru	Bal.	18.4	-	8.81	-	-	-
1Rh	Bal.	18.3	-	-	0.94	-	-
5Rh	Bal.	18.7	-	-	4.73	-	-
9Rh	Bal.	19.1	-	-	8.11	-	-
1Pd	Bal.	18.2	-	-	-	0.92	-
5Pd	Bal.	18.5	-	-	-	4.69	-
1Pt	Bal.	18.2	-	-	-	-	0.91
5Pt	Bal.	18.5	-	-	-	-	4.50
9Pt	Bal.	18.4	-	-	-	-	8.22

Ni-Al-PGM ternary alloys

Experimental procedure

In order to obtain alloys having γ and γ' two phases, several Ni-19at%Al-Xat%PGM alloys were arc-melted under Ar in a vacuum chamber. Chemical composition of the as arc-melted alloys was determined by fluorescent X-ray spectroscopy, and the result is listed in Table I. Hereafter, the alloy name is denoted as 1Ir, 5Pt etc. The SEM-EDX analysis for the as arc-melted alloys confirmed that in all the alloys, Ni, Al and PGM are uniformly mixed without segregation. DTA was then conducted to obtain melting points of the alloys. DTA runs were all performed at a heating / cooling rate of 10K / min starting from room temperature to 1550°C. The alloys were then heat-treated at 1300°C for 4h followed by water quenching and then aged at 1100°C for 100h with water quenching to investigate microstructural evolution. The heat-treated specimens were examined by optical microscopy (OM) and SEM-EDX analysis.

Results and Discussions

The melting temperatures of the as-cast ternary alloys were measured using DTA. Figure 1 summarizes the relationship between PGM content and liquidus points. It is found that liquidus temperatures of Ir-containing alloys linearly increase with increasing Ir content. The increment rate is roughly estimated to 4.7K / 1at%Ir.

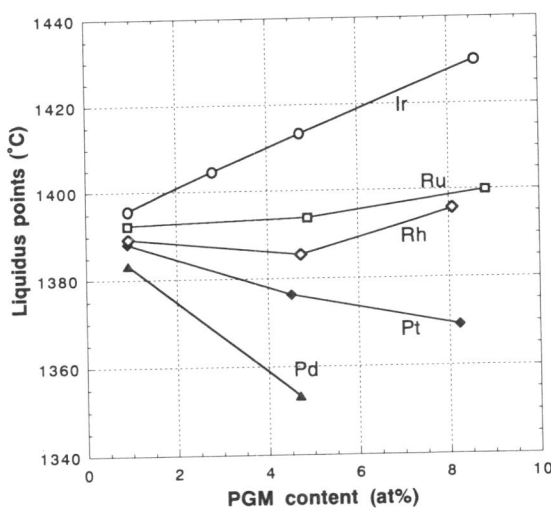

Figure 1: The relationship between liquidus points and PGM content.

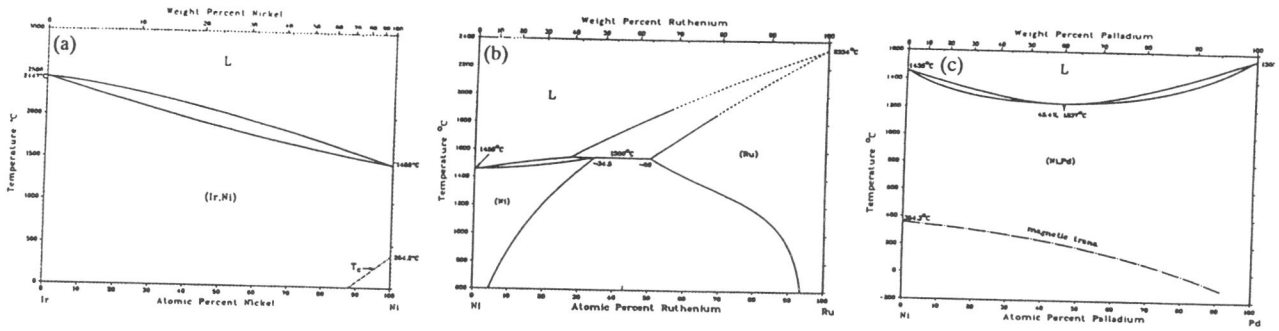

Figure 2: the (a) Ni-Ir, (b)Ni-Ru and (c)Ni-Pd binary phase diagrams[7].

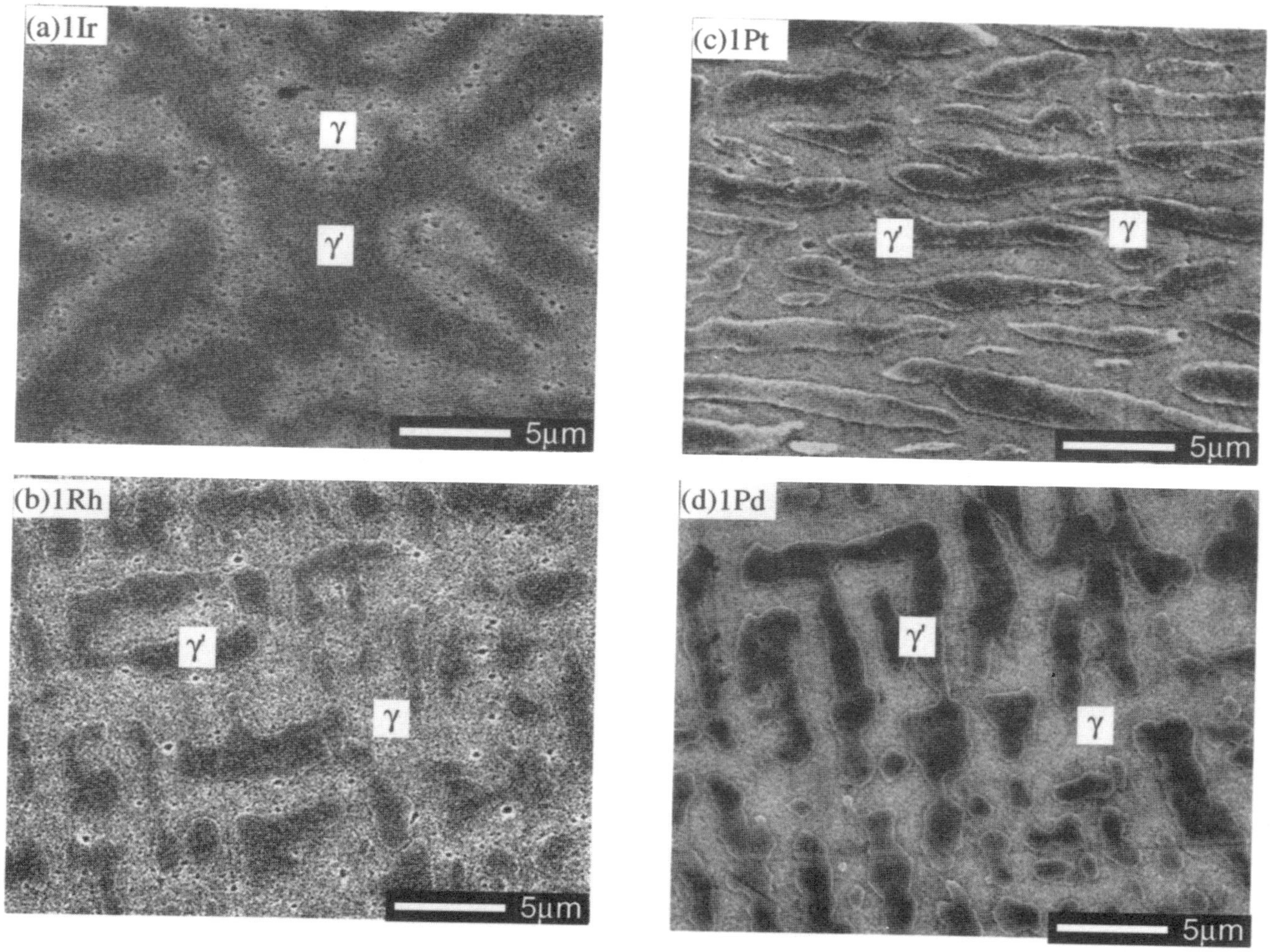

Figure 3: Typical SEM images of (a)1Ir, (b)1Rh, (c)1Pd and (e)1Pt alloys aged at 1100°C for 100h.

749

While Ru addition also increases melting points, Pt and Pd addition is found to decrease melting temperatures. These results can be explained from the Ni-PGM binary phase diagrams. Figure 2 shows the (a)Ni-Ir, (b)Ni-Ru and (c)Ni-Pd binary phase diagrams[7]. The phase diagram of the Ni-Rh system is similar to the Ni-Ir system and the Ni-Pt being similar to the Ni-Pd system, respectively. Ir, Rh, Pt and Pd all have identical crystal structure to Ni and they all have complete solid solubility with Ni at high temperatures. However, while Ni-Ir and Ni-Rh have a continuous series of (Ni, Ir) and (Ni, Rh) solid solution system and the melting points steadily increase with increasing Ir and Rh content, Ni-Pt and Ni-Pd systems exhibit a minimum in the liquidus. The decrease of melting points in Ni-Al-Pt and Ni-Al-Pd systems can be caused by this eutectic-like reaction of Ni-Pt and Ni-Pd. In designing Ni-base alloys, elemenets which increase melting temperature are favourable because at a given temperature, an alloy having a higher melting point generally has better mechanical properties. From this point of view, it is suggested that Ir, Ru and Rh are promising elements to increase the creep properties.

The alloys were heat treated at 1300°C for 4h followed by water quenching. OM and SEM analysis revealed that except for 5Rh, 9Rh, 9Ir and 9Ru, all the alloys showed almost uniform microstructure, indicating that the alloys can be fully solution-treated at 1300°C. Following the solution treatment, ageing treatment at 1100°C for 100 h was conducted. Figure 3 shows the typical SEM images of aged (a)1Ir, (b)1Rh, (c)1Pd and (d)1Pt alloys. All the alloys have γ and γ' two-phase structure, confirming that for all PGMs, 1at% of alloying addition is within the solubility limit in the γ and γ' two-phase region. On the other hand, 9at% of PGM addition except for Pt exceeds the solubility limit for the γ and γ' two-phase region and leads to precipitate other phase as shown in Figure 4. It should be noted that 9Pd alloy has not yet been investigated. EDX analysis indicated that the third phase shown in this figure can be identified as

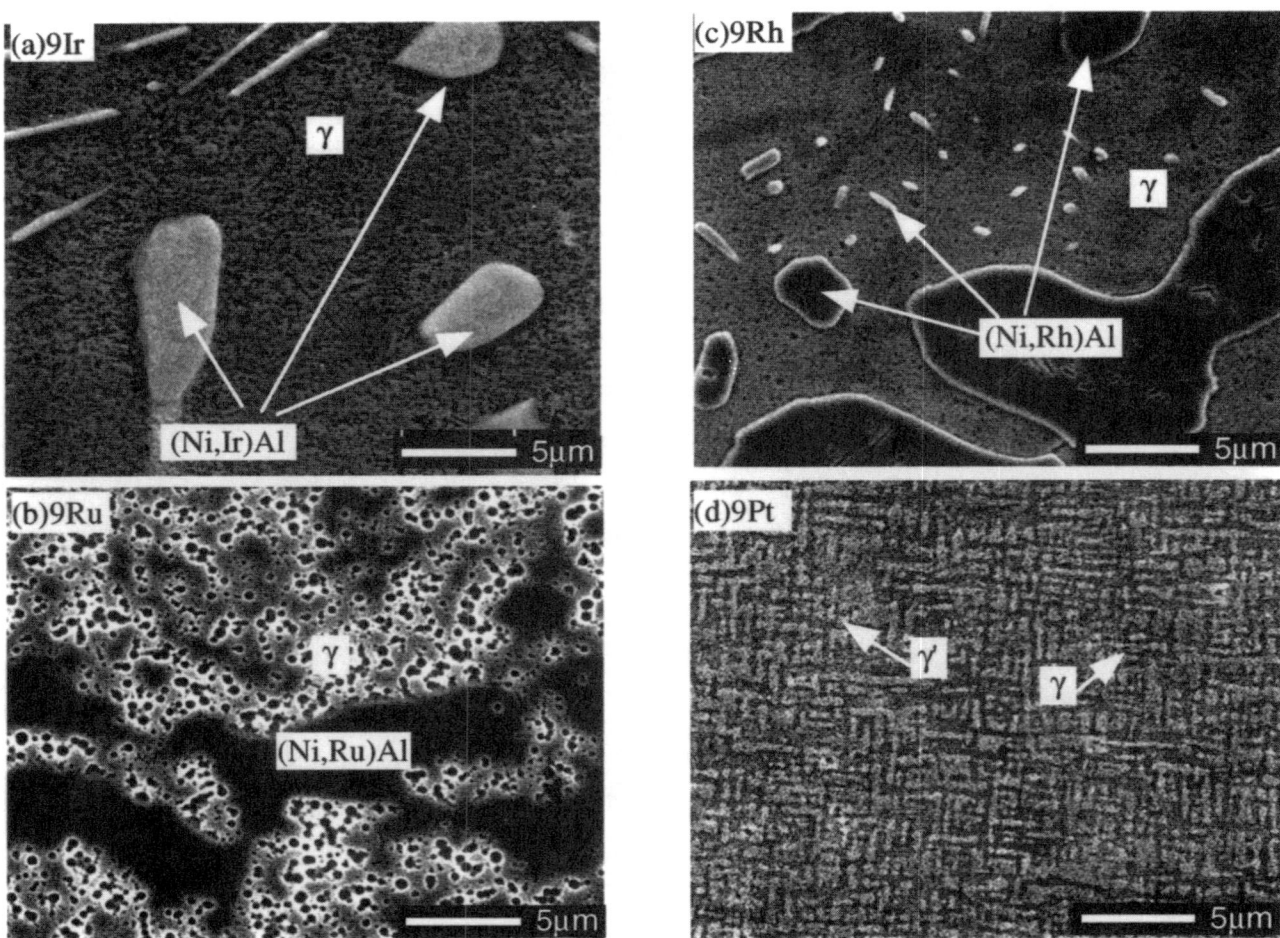

Figure 4: SEM images of (a)9Ir, (b)9Ru, (c)9Rh and (d)9Pt alloys after aged at 1100°C for 100h.

the β phase having B2 structure, or its higher order phase, with composition being approximately $(Ni,PGM)_{50}Al_{50}$.

According to Al-PGM phase diagrams[7], only Pt and Al form the Pt_3Al phase with $L1_2$ structure. It is thus likely that Ni_3Al may equilibrate with Pt_3Al to form $(Ni,Pt)_3Al$. In this case, unlimited amount of Pt could be substituted for Ni in the γ and γ' two-phase region. Other PGMs such as Ir, Ru, Rh do not form $L1_2$ structure, they only have B2 structure with Al in Al rich region. If this β phase seems to be chemically stable, increasing PGM content would change the microstructure from γ and γ' two-phase to β + γ or β + γ + γ' etc. To confirm this hypothesis, however, further investigation will be required.

In order to determine the partitioning behaviour of alloying elements, EDX analysis was conducted for alloys having γ and γ' two phases. Whether an alloying element 'i' partitions to the γ phase or γ' phase can be clearly shown by using a partitioning parameter k_i which is defined as

$$k_i = c_{i\gamma} / c_{i\gamma'}$$

where $c_{i\gamma}$ and $c_{i\gamma'}$ are the concentrations of an alloying element 'i' in the γ and γ' phases, respectively. Thus, for example, when $k_i > 1$, the element 'i' has a tendency to be located in the γ phase and vice versa. In Figure 5, partitioning parameters of PGMs are summarized as a function of k_i. Ru and Ir have a preference to partition into the γ phase while Pt, Rh and Pd have a preference to partition into the γ' phase. Elements to have larger atomic radii and to enrich in the γ phase are of particular interest because they preferentially increase the lattice parameter of γ phase and thus leads the lattice misfit $(a_{\gamma'} - a_{\gamma})/a_{\gamma}$ towards negative direction[8]. It is also expected that alloys having a preference for γ phase play a role as a solid solution strengthner of the γ phase. Since PGMs all have larger atomic radii than Ni, (When normalising the atomic radius of Ni as 1, the atomic radii of Rh, Ru, Ir, Pt and Pd are 1.079, 1.063, 1.089, 1.113 and 1.104, respectively[9].) it is predicted that Ru and Ir lead the lattice misfits to negative direction while Pt, Rh

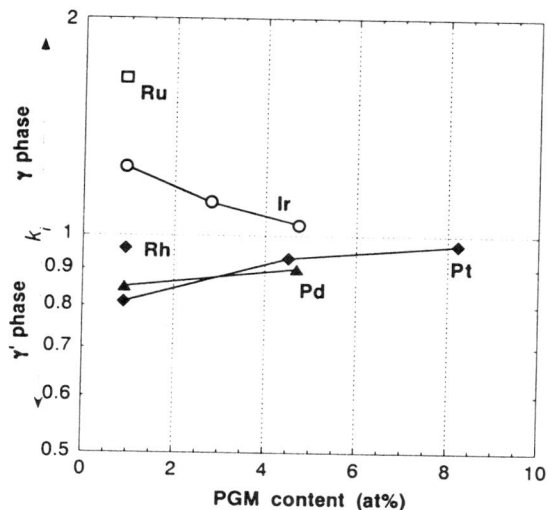

Figure 5: Partitioning behaviour of PGM in Ni-19Al-PGM ternary alloy systems.

and Pd leading to positive.

From these experimental results, it is expected that Ir and Ru, which increase the melting temperatures and partition to the γ phase in ternary alloy systems, will play a favourable role in enhancing creep properties of Ni-base superalloys. The effects of Ir or Ru addition to multi-component Ni-base alloys are therefore discussed in the following section.

Multi-component Ni-base single-crystal superalloys

Experimental procedure

The compositions of alloys investigated here are listed in Table II. Among these alloys, composition of TMS-63 and TMS-75 were determined by using 'alloy design program' (ADP)[10,11], which is the regression analysis based computer program developed by the authors, for designing Ni-base superalloys. The mechanical properties and microstructures of TMS-63[10,11] and

Table II Nominal composition of alloys (in at%)

Alloy	Ni	Al	Cr	Co	Mo	Ta	W	Re	Ru	Ir
TMS-63	Bal.	12.8	7.8	-	4.6	2.8	-	-	-	-
TMS-79	Bal.	12.6	7.7	-	4.6	2.7	-	-	-	1.8
TMS-75	Bal.	13.7	3.6	12.6	1.3	2.1	2.0	1.7	-	-
TMS-80	Bal.	13.6	3.5	12.4	1.3	2.0	2.0	1.7	-	1.0
TMS-91	Bal.	13.3	3.5	12.3	1.2	2.0	2.0	1.6	0.9	-

751

TMS-75[12] have already been reported in previous papers. The Re-free TMS-79[2,3] was obtained by the 1.8at% of Ir addition to the base material TMS-63, and Re-containing TMS-80 and TMS-91 were obtained by the 0.8at% of Ir addition, and the 0.9at% of Ru and 1.5at% of Ni addition to the base material TMS-75, respectively. The alloys were solution treated at 1300°C for 1h followed by 1320°C for 5h and gas fan cooled. After then, TMS-91 was aged at 1150°C for 4h followed by gas fan cooled, and the other two alloys were aged at 1100°C for 5h and water quenched to form coherent cuboidal γ' precipitates. The materials were then examined using SEM, TEM and atom-probe field ion microscopy (APFIM).

Results and Discussions

Figure 6 shows the typical SEM images of (a)TMS-75, (b)TMS-80[4] and (c)TMS-91, respectively. SEM observation revealed that the as-aged alloys all have a coherent γ and γ' two-phase structure with γ' precipitates being ~0.5μm in size. Although the size of the γ' precipitates is somewhat different among the alloys, the shape of the γ' precipitates is almost identical, suggesting that Ir or Ru addition does not drastically change the microstructure of the alloys. It should also be noted that there are no detrimental phases such as TCP phases observed in all the alloys.

Since the average size of γ' precipitates of the alloys is less than 1μm after heat treatment, it is very difficult to identify γ and γ' phase compositions using SEM-EDX analysis. TEM and APFIM analyses were therefore conducted to understand the microscopic distribution of alloying elements. Figure 7(a) shows the dark-field transmission electron micrograph of heat-treated TMS-91. In this figure, primary precipitates with 100~400nm in size, together with the small secondary γ' precipitates being ~10nm in size are clearly observed. It should be noted that the secondary γ' precipitates are not observed in the narrow γ channels, but they are observed in the central region of wide γ channels. Since the APFIM analysis is chemical identification of atoms successively field evaporated from the surface, if the analysing region is such like Figure 7(b), chemistry change between a primary γ' precipitate and a γ matrix in the narrow channel can be obtained. Also, if the APFIM specimen is prepared like Figure 7(c), chemistry change between secondary cooling γ' precipitates and a γ phase in the wide channel can be obtained. Microscopic partitioning behaviour of alloying elements can be determined in this manner. In order to visualise the partitioning behaviour

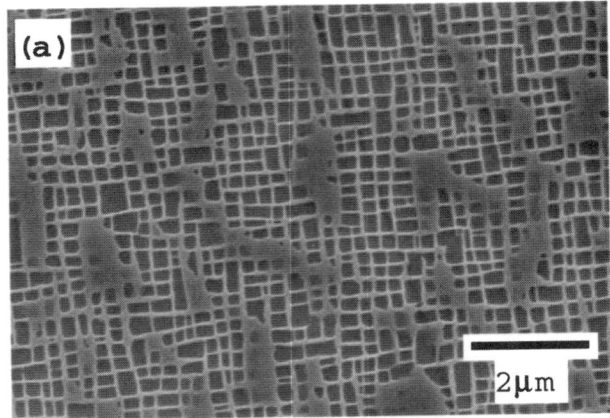

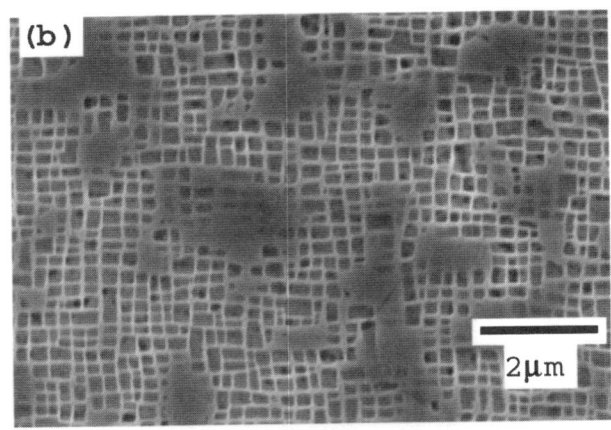

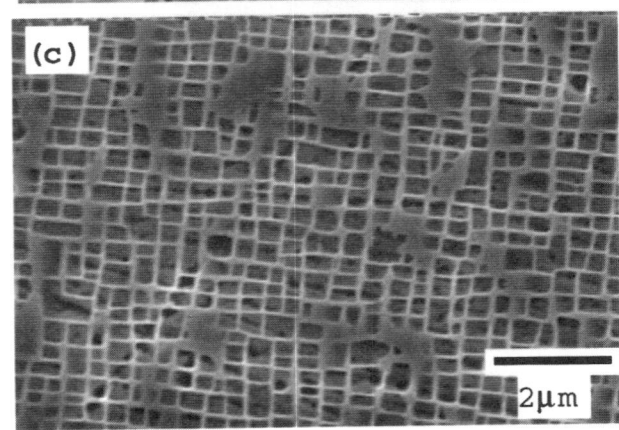

Figure 6: Typical SEM images of (a) TMS-75, (b) TMS-80[4] and (c)TMS-91.

of alloying elements, so-called 'ladder diagram' is used, where the cumulative number of each solute atom is plotted against the total number of detected atoms. Hence, the horizontal axis includes depth information and the gradient of each curve corresponds to the local concentration of each element, respectively. Figure 8 is the typical ladder diagram of TMS-91 showing the

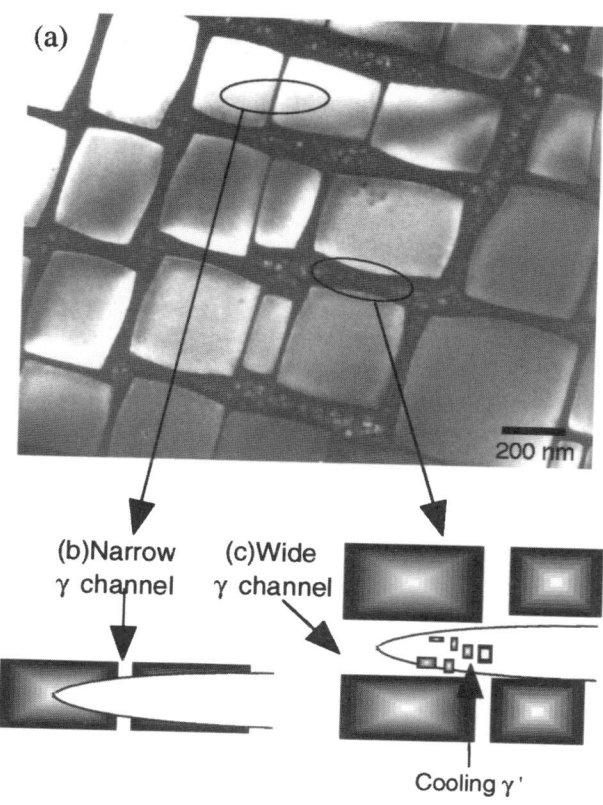

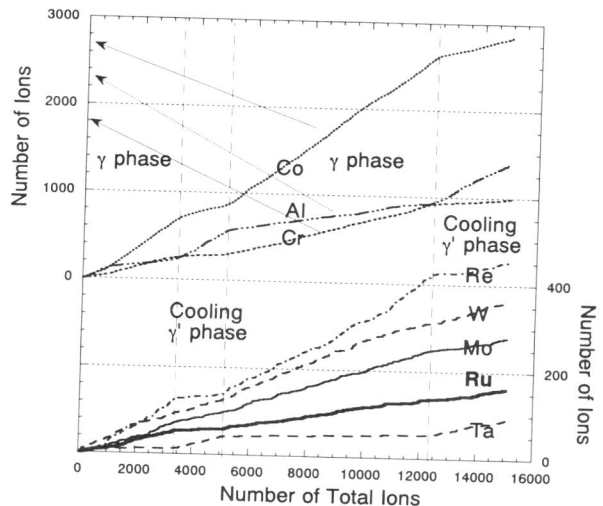

Figure 8: Ladder diagram of TMS-91 showing the chemistry change in a wide γ channel.

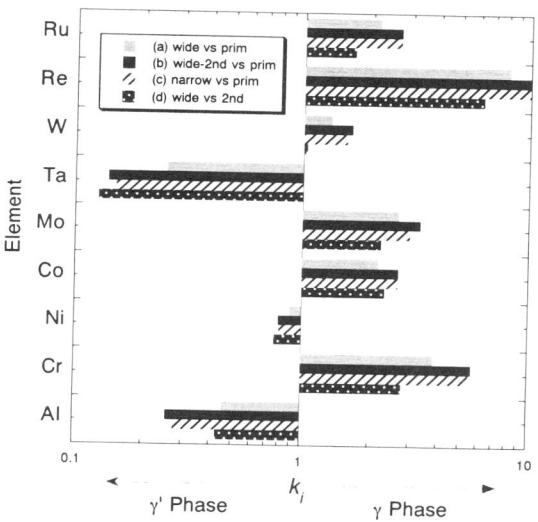

Figure 9: Partitioning behaviour of alloying elements in TMS-91 as a function of k_i.

Figure 7: (a)Dark field TEM image of TMS-91 and schematic diagrams of APFIM specimen for analysing (b)narrow γ channel and γ', and (c)wide γ channel and cooling γ'.

chemistry change in a wide γ channel. The specimen used in this analysis corresponds to Figure 7(c). This figure indicates that Ru, together with Cr, Co, Mo, W and Re, have a preference to partition into the γ phase whereas Ta having a preference for the secondary γ' precipitates. In Ru-containing TMS-91, partitioning behaviour of alloying elements between (a) primary cuboidal γ' phase and wide γ channel with secondary γ' precipitates, (b) primary γ' phase and wide γ channel without secondary γ' phase, (c) primary γ' phase and narrow γ channel, and (d) secondary cooling γ' precipitates and wide γ channel, are determined in this manner and they are summarised in Figure 9. This figure indicates that (1) Ru always partitions to the γ phase with $k_{Ru} \sim 1.6$ to 2.5. (2) partitioning of case (d) is less prominent than the other cases. (3) partitioning parameters of alloying elements for case (b) are almost identical to those for case (c), suggesting that alloying elements in the γ phase almost uniformly distribute either in a narrow channel or a wide channel, provided that secondary cooling γ' phase is omitted. From these results, the diffusion of alloying elements during cooling process can be explained as follows. Under equilibrium condition, γ' volume fraction increases as temperature decreases. Therefore, when the alloy is cooled, supersaturated solute atoms in the γ phase at γ and γ' interface tend to dissolve into the growing γ' precipitates. On the other hand, in the middle of wide γ channel, there remains the supersaturated γ region where nucleation of secondary γ' precipitates occurs. Once secondary γ' phase is precipitated, solute atoms in the wide γ channel would either dissolve into the primary γ' precipitates or secondary cooling γ' precipitates. In this way, concentration of alloying elements in the γ phase becomes uniform. It is thus suggested that case (a) should be used to describe the partitioning behaviour of alloying elements at the aged temperature. It should also be noted that even though chemical identification of extracted γ' phase is regarded as the most accurate way of determining γ' phase composition, it does not always describe the composition of γ' phase at a certain temperature if the extracted γ' includes both primary and secondary precipitates.

The partitioning behaviour of Ir in Ni-base single crystal superalloys has also been investigated. Figures 10 and 11 are the ladder diagrams of TMS-79[2] and 80[4] showing composition change at γ / γ' interfaces, respectively. As for partitioning, all the alloying elements except for Ir behaves similarly to TMS-91, that is, Cr, Co, Mo, W and Re have a preference to partition into the γ phase whereas Ta having a preference for the γ' phase. Note that while having a slight preference for the γ phase in the Re-free alloy TMS-79, Ir tends to partitione to the γ' phase in the Re-containing alloy TMS-80. Experimental results for Ni-Al-Ir ternary alloys confirmed that Ir inherently has a preference to partition to the γ phase rather than the γ' phase. However, since the partitioning tendency of Re to the γ phase is much stronger than that of Ir, Ir atoms are rejected from the γ phase due to the presence of Re. These results suggest that among alloying elements, 'competition of partitioning' takes place in Ni-base superalloys.

Partitioning behaviour of alloying elements in TMS-75, Ru-containing TMS-91 and Ir-containing TMS-80 are summarized by the partitioning parameter k_i, as shown in Figure 12. In TMS-91, Ru prefers to partition to the γ phase with $k_{Ru} \sim 2.0$. When compared with TMS-75, k_W decreases from 1.6 to 1.3 while k_{Re} keeps constant by the addition of Ru. O'Hara et al reported in their patent document that the addition of Ru to their alloys causes the partition of Al and Ti to γ phase, while Re, W and Cr in concentration in the γ' phase[6]. Although our study almost supports O'Hara's report, there are some discrepancies. For instance, partitioning behaviour of Re should be re-considered. In TMS-80, on the other hand, Ir addition does not affect the partitioning behaviour of

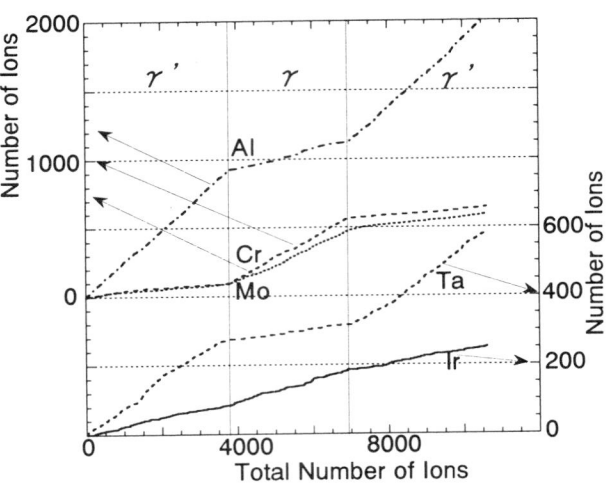

Figure 10: The ladder diagram of TMS-79 showing composition change at γ / γ' interfaces[2].

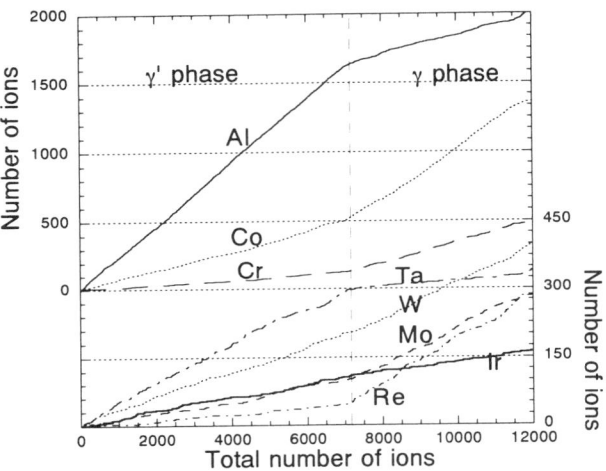

Figure 11: The ladder diagram of TMS-80 showing composition change at the γ / γ' interface[4].

other alloying elements. From these results, it is suggested that there is an order of alloying elements having stronger preference to partition to the γ phase, which can be determined from the partitioning parameters as Re > Cr > (Mo, Ru, Co) > W > Ir. Further investigation is now being conducted to investigate the distribution of other PGMs in multi-component Ni-base superalloys and their mechanical properties.

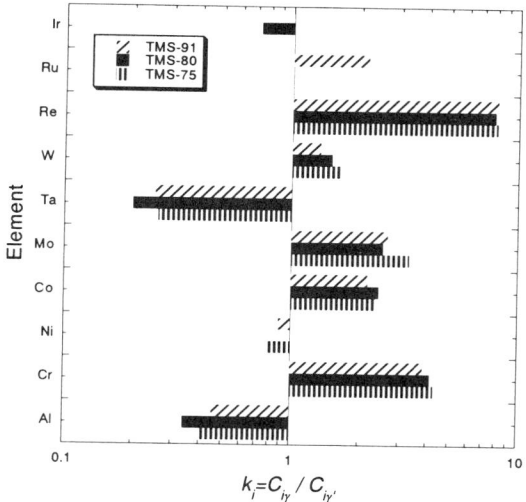

Figure 12: Partitioning behaviour of alloying elements in TMS-75, 80 and 91.

Conclusions

The distribution of PGMs in Ni-base γ and γ' two phase alloys was investigated and the following results have been obtained

1. For Ni-Al-PGM ternary alloys:

(1) addition of Ir and Ru increases, whereas addition of Pt and Pd decreases the melting temperatures.

(2) The third phase enriched in Al and PGM is precipitated in 5Rh, 9Rh, 9Ru and 9Ir alloys. This phase may be attributed to β phase with composition being approximately 50at% of (Ni+PGM) and 50% of Al.

(3) For alloys having γ and γ' two phase structure, Ru and Ir have preference to partition to the γ phase while Pt, Pd, and Rh prefer the γ' phase.

2. For Ru- and Ir-containing multi-component Ni-base single crystal superalloys:,

(1) Ru has a preference to partition into the γ phase while partitioning behaviour of Ir is composition dependent, in the Re-free alloy TMS-79, Ir behaves similarly to ternary alloys, whereas in Re-containing TMS-80, Ir has a preference to partition to the γ' phase.

(2) The order of alloying elements having stronger preference to the γ phase can be determined from the partitioning parameters as Re > Cr > (Mo, Ru, Co) > W > Ir.

Acknowledgement

This work is carried out in the scheme of 'High Temperature Materials 21' project. The authours are also grateful to Dr. S. Ito and Mr. H. Yamaguchi for analysing the composition of Ni-Al-X ternary alloys.

References

1. D.J. Frasier et al., "Process and Alloy Optimization for CMSX-4 Superalloy Single Crystal Airfoils," Proceedings of the conference "High Temperature Materials for Power Engineering 1990", Liège, Belgium (1990), 1281-1300.

2. H. Murakami et al., "Atom Probe Microanalysis of Ir-bearing Ni-based Superalloys," Mat. Sci. Eng., A250 (1998), 109-114.

3. H. Murakami et al., "The location of Atoms in Ir-Containing Ni-Base Single Crystal Superalloys," Proceedings of the conference "High Temperature Materials for Power Engineering 1998", Liège, Belgium (1998), 1139-1145.

4. H. Murakami et al., "The Distribution of Ir in Ni-base Single-Crystal Superalloys", Proceedings of the international symposium of iridium, Ed. By E. K. Ohriner et al., Nashville, Tennessee, USA, (2000), 121-128.

5. T. Kobayashi, et al., "Design of High Rhenium Containing Single Crystal Superalloys with Balanced Intermediate and High Temperature Creep Strengths", Proceedings of the Fourth International Charles Parsons Turbine Conference, Ed. by A. Strang et al, Newcastle upon Tyne, U.K. (1997), 766-773.

6. K.S. O'Hara et al., "Nickel Base Superalloy and Article", U. S. Patent 5,482,789, (1996).

7. T. B. Massalski, Binary Alloy Phase Diagrams, (ASM International, 1990).

8. H. Murakami, H. Harada and H.K.D.H. Bhadeshia, "The Location of Atoms in Re- and V- containing Multicomponent Nickel-base Single-crystal superalloys," Appl. Surf. Sci., 76/77(1994), 177-183.

9. M. Winter, WebElements, The periodic table on the World-Wide Web, http://www.webelements.com.

10. H. Harada et al., "Design of High Specific-strength Nickel-base Single Crystal Superalloys," Proceedings of the conference "High Temperature Materials for Power Engineering 1990", Liège, Belgium (1990), 1319-1328.

11. H. Harada et al., "Computer Analysis on

Microstructure and Property of Nickel-base Single Crystal Superalloys," Proceedings of the 5th International Conference on Creep and Fracture of Engineering Materials and Structures, Swansea, U.K. (1993), 255-264.

12. K. Koizumi et al., "Third Generation Single Crystal Superalloys with Excellent Processability and Phase Stability," Proceedings of the conference "High Temperature Materials for Power Engineering 1998", Liège, Belgium (1998), 1089-1097.

Development of A Low Angle Grain Boundary Resistant Single Crystal Superalloy YH61

H. Tamaki, A. Yoshinari, A. Okayama and S. Nakamura

Hitachi Research Lab., Hitachi, Ltd., 7-1-1 Ohmika, Hitachi, Ibaraki, 319-1292, Japan (E-Mail: hidekit@hrl.hitachi.co.jp)

K. Kageyama, K. Sato and T. Ohno

Metallurgical Research Lab., Hitachi Metals, Ltd., 2107 Yasugi, Yasugi, Shimane, 692-8601, Japan

Abstract

In this paper, a new single crystal (SC) superalloy YH61 is proposed to solve the problem of grain defects which are present in single crystal buckets or vanes of industrial gas turbines. First, the effects of solution heat treatment on various properties of the SC with and without grain defects were evaluated. Increasing the solution-heat-treated area was found to have a positive effect on the strength of a defect-free SC but harmful for the strength of a grain boundary and growth of recrystallization. The solution heat treatment condition for YH61 was determined after considering properties of both the defect-free SC and the SC with grain defects. Secondly, the effects of grain boundary misorientation angle (α) on the mechanical properties of grain boundaries were examined at several temperatures. No significant decrease in strength was found with increasing α at temperatures below about 800°C, although the fall-off of strength was observed with increasing α at temperatures above about this temperature. Even when a significant decrease in strength was not observed, unexpected deformation behavior was found with increasing α because the Schmid factor for the slip system was also increased with increasing α. In a strain-controlled LCF test, another type of deformation was observed for higher α values. This deformation caused significant fall-off of life as a function of α. From the results of long term creep-rupture tests, YH61 was found to show similar creep-rupture strength to a second generation single crystal superalloy under certain conditions although YH61 contains higher levels of grain boundary strengthening elements.

Introduction

Application of single crystal (SC) buckets and vanes in aero-engines has significantly improved the engine performance. In the field of industrial gas turbines (IGT), application of SC components is necessary for improving the efficiency of IGT due to increases in gas-firing temperature. Although this demand is common to all IGT manufacturers, directionally solidified (DS) buckets and vanes are still the mainstream technology for IGT especially in the case of heavy-duty machines. Although there have been considerable efforts for adopting SC components for IGTs [1][2], only two IGT manufacturers have announced that they have introduced SC buckets or vanes into their machinery [3][4].

The main reason why SC components have not been widely adopted for IGT is the casting problems of the SC for IGT. Grain defects such as low angle grain boundaries (LAB), high angle grain boundaries (HAB) and recrystallization tend to occur in IGT buckets or vanes, since their sizes are larger and their shapes are more complicated than those of the aero-engines' components. Because of these reasons, the casting yield of the SC for IGT is significantly lower and their cost is higher than the DS components.

If the resistance of SC superalloys to LAB or HAB is increased, higher yields and lower costs can be realized for SC buckets and vanes in IGT. In order to meet these demands, a new SC superalloy, YH61 (Table 1) [5] has been developed. Under certain conditions, creep-rupture strength of the defect-free YH61 is similar to that of a second-generation SC superalloy [6], although YH61 contains higher amounts of boron, carbon and hafnium and requires only a partial solution heat treatment (Figure 1).

Superalloys 2000
Edited by T.M. Pollock, R.D. Kissinger, R.R. Bowman,
K.A. Green, M. McLean, S. Olson, and J.J. Schirra
TMS (The Minerals, Metals & Materials Society), 2000

Table 1 Nominal Composition of YH61, mass%.

Table 1 Nominal Composition of YH61, mass%.

Cr	Co	W	Re	Mo	Ta	Nb	Al	Hf	C	B	Ni
7.2	1.0	8.8	1.4	0.9	8.8	0.8	5.0	0.25	0.07	0.02	Balance

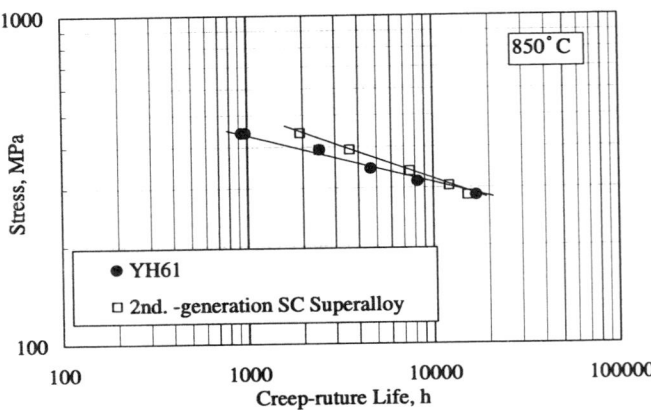

Figure 1: Stress-rupture curves for YH61 and a second-generation SC superalloy [6] at 850°C. Creep-rupture strength of YH61 is similar to that of the second-generation superalloy when creep-rupture life is longer than 10000h.

In order to utilize YH61 in IGT, the following targets should be achieved:

(1) A proper solution heat treatment condition which is compatible with both defect-free SC strength and LAB or HAB strength,

(2) The degree of α which can satisfy the design requirement for each part should be maximized.

It is generally accepted that higher creep-rupture strengths along the solidification direction of SC and DS can be achieved by increasing the volume percent of fine γ' [7]. Higher solution heat treatment temperatures are also known to increase the volume percent of fine γ'. For these reasons, solution heat treatment temperatures close to the incipient melting points have been adopted for conventional SC superalloys. Because such higher solution heat treatment temperatures can be achieved by removing grain boundary strengthening elements such as boron, hafnium and zirconium from the alloys, the conventional SC alloys contain a limited range of LAB. Ross and O'Hara [8] have reported a unique SC superalloy that has moderate LAB strength but higher grain boundary strength than this alloy required for IGT.

For alloys with grain boundary strengthening elements, Cetel and Duhl pointed out that the creep-ductility for DS in the transverse direction is decreased with increasing volume percent of fine γ' [9]. This increase was also found to decrease the creep-rupture life of DS in the transverse direction even in an alloy which contains grain boundary strengthening elements [5]. Especially, a higher volume percent causes a significant decrease in the creep-rupture life of DS

in the transverse direction. Thus, for achieving higher strength of LAB and HAB, both the addition of the grain boundary strengthening elements and optimizing solution heat treatment conditions seem to be necessary. This paper discusses optimization of the solution heat treatment condition for SC superalloys with grain defects.

Ross and O'Hara [8] and Burkholder et al. [10] studied the relationship between grain boundary misorientation angle (α) and LAB or HAB strength. Although they discussed the fall-off (or retention) of creep-rupture life as a function of α, the change of deformation mechanism with increasing α should also be considered. If one grain is rotated towards another in order to increase α, the Schmid factor for the rotated grain changes. When investigating the effect of α on the strength of LAB or HAB bearing SC, both the effect of α on grain boundary strength and on the rotated grain strength should be considered since the strength of the rotated grain is also affected by the changed Schmid factor. This paper discusses this effect of α on the strength of LAB or HAB bearing SC at various temperatures and stresses by considering the above mentioned issues.

Experimental Procedure

Effect of Solution Heat Treatment Conditions

Seven 150 kg - master alloys and one 1.5 ton - master alloy of YH61 were prepared for this study. SC bars (φ15 x 165 mm) and DS slabs (100 x 15 x 250 mm) were cast from the master alloys by a mold withdrawal method.

The conditions of the multistep solution heat treatment adopted for this alloy are listed in Table 2.

Table 2   Solution Heat Treatment Conditions

Condition No.	Condition for Final Step
A	1250°C/4h/AC
B	1260°C/4h/AC
C	1270°C/4h/AC
D1	1280°C/1h /AC
D	1280°C/4h /AC
D8	1280°C/8h /AC
D20	1280°C/20h /AC
E	1290°C/4h /AC

The aging condition for all specimens was 1080°C/4h/AC followed by 871°C/20h/AC.

Specimens for evaluating the perfect SC's mechanical properties were machined from SC bars. The stress axis of the specimens was parallel to the growth direction of SC bars and was within 10° from the <100> direction. Specimens machined from DS slabs were used for evaluating mechanical properties of the SC with grain defects such as HAB. The stress axis of the specimens was the DS transverse direction which was perpendicular to grain boundaries.

A schematic of the specimen used for evaluating the effect of solution heat treatment conditions on the growth of recrystallized grain is illustrated in Figure 2. In this study, an alloy SC610 [11] (Ni - 7.5Cr - 7.2W - 1.4Re - 0.8Mo - 8.8Ta - 1.7Nb - 5.0Al - 0.1Hf - 1.0Co, mass%) was adopted for the evaluation for the following reasons: (a) a carbon free alloy was selected to evaluate the obstacle effect of only the non-solutioned γ' on the growth of recrystallized grains, (b) additionally, the γ' solvus temperature of the selected alloy should be similar to YH61. Specimens were cut perpendicular to the growth direction from as-cast SC bars. Deviation of the <100> direction from the growth direction was within $10°$. The wire-cutting method was utilized in order to have low residual stresses. One side of the cut surface was polished by #1200 grade polishing paper followed by 0.06μm Al_2O_3 slurry polishing. A nucleus for recrystallization was made by loading the sample in a Brinell hardness tester. A load of 9800N was applied on the polished surface for 30 second, and then the specimens were solution-heat-treated. After the solution heat treatment, the maximum depth of the recrystallized grain was evaluated.

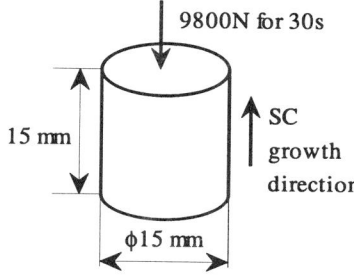

Figure 2: Schematic of the specimen for evaluating the effect of solution heat treatment on the growth of recrystallization.

The Effect of Grain Boundary misorientation angle

Bi-crystalline slabs described in Figure 3 were cast from the same master alloys mentioned above. The growth direction of both seeds was <001>, which corresponded to the longitudinal direction of the slab. The primary seed was set in order to align the <110> direction parallel to the transverse direction of the slab. The <110> direction of the second seed was rotated from transverse direction by α. Consequently, a grain boundary with a misorientation angle α was present along the center of the slab. Specimens were machined from the slabs parallel to the transverse direction. The grain boundary lay perpendicular to the stress axis in the middle of the specimen. The stress axis of the primary grain was parallel to the <110> direction, and the stress axis of the second grain deviated from the <110> direction by α in {100} plane. The maximum value of α was $20°$ in this study. Specimens were later solution-heat-treated according to the condition D listed in Table 2 followed by aging for which conditions were given above.

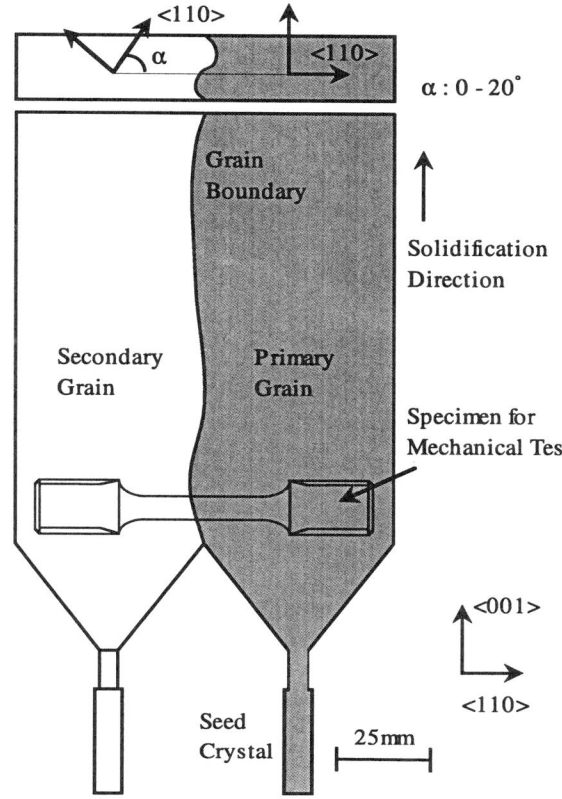

Figure 3: Schematic of the bi-crystalline slab for evaluating effect of α on strength of SC with LAB or HAB. Misorientation angle (α) was defined as rotation angle of the secondary grain against the primary grain on the {100} plain.

<u>EBW conditions</u>
Accelerating Voltage: 150kV
Beam Current: 5.0mA
Welding Velocity: 0.5m/ min
Focal Position: Under Focused
Temperature: 900°C

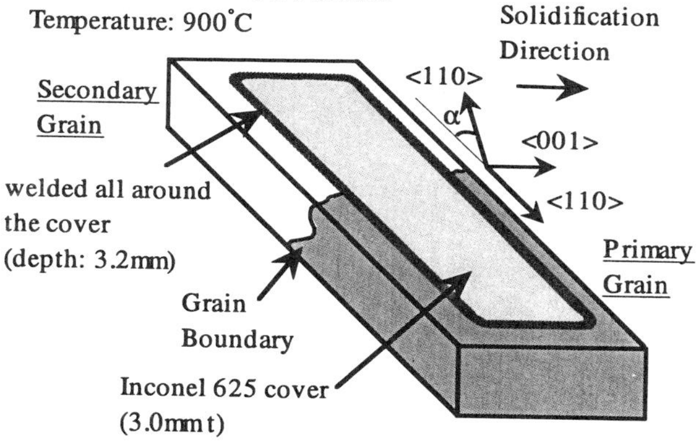

Solidification Direction

<u>Secondary Grain</u>

welded all around the cover (depth: 3.2mm)

Grain Boundary

Inconel 625 cover (3.0mm t)

<110>
α
<001>
<110>

<u>Primary Grain</u>

Figure 4: EBW direction was arranged perpendicular to the grain boundary whose misorientation angle was α.

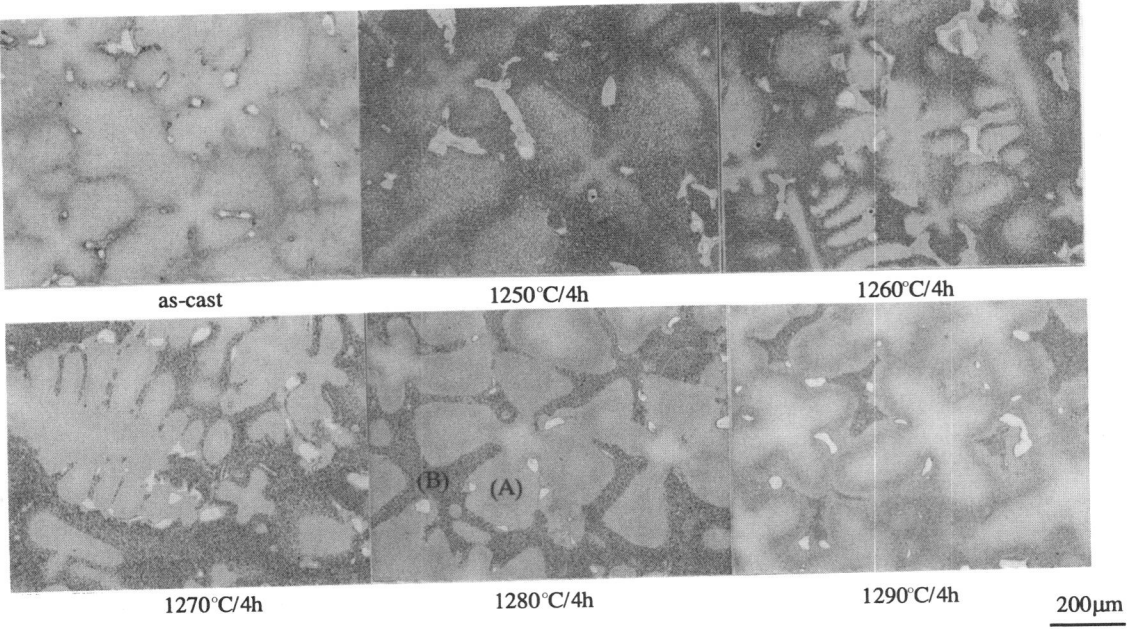

as-cast   1250°C/4h   1260°C/4h

1270°C/4h   1280°C/4h   1290°C/4h   200μm

Figure 5: Typical microstructure of YH61 after various solution heat treatment and that of as-cast. (A) indicates the region where γ' was completely solutioned and (B) is the region where γ' was not solutioned.

A specimen was machined from the above-mentioned slabs for evaluating electron beam (EB) weldability of SC with a grain boundary. Electron beam welding (EBW) was carried out after the solution heat treatment. A schematic of the sample and conditions of EBW are shown in Figure 4. Before (after the EBW) and after aging, existence of cracks was investigated by the liquid penetrant test method.

Results and Discussion

The Effect of Solution Heat Treatment Conditions

Although the design approach for the alloy chemistry of YH61 was detailed in another publication [5], highlights of its chemistry is summarized below:

(1) Higher amounts of refractory elements such as Ta, W and Re are added in order to realize a similar level of creep-rupture strength as the second-generation SC superalloys without full solution

(2) A higher content of boron compared to conventional DS alloys was used to keep an adequate level of boron at grain boundaries after solution heat treatment.

Typical microstructures of YH61 after several solution heat treatment conditions are shown in Figure 5. The effect of solution heat treatment temperature on the volume percent of solutioned γ' is shown in Figure 6. In this figure, data for a final solution heat treatment time of 4h are plotted. The term " volume percent of

solutioned γ' " is used to describe the volume percent of the region in which γ' is completely solutioned at the solution heat treatment and precipitated as fine particles during aging.

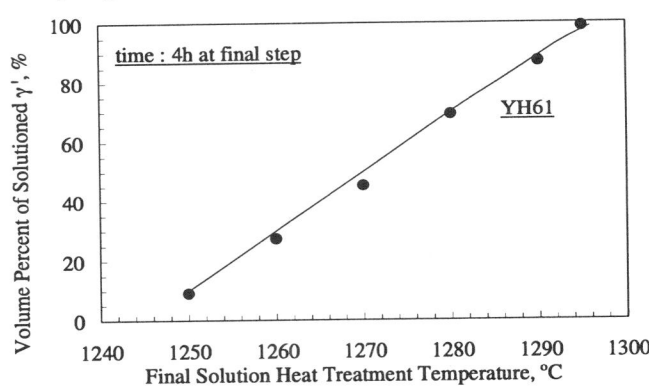

Figure 6: Volume percent of solutioned γ' increased linearly with increasing solution heat treatment temperature in this temperature range.

A Linear relationship between the volume percent of solutioned γ' and temperature was observed between 1250 and 1295°C. Although almost 100% solutioning with no incipient melting can be achieved in this alloy, residual eutectic γ-γ' colonies cannot be dissolved without incipient melting due to the higher content of boron in this alloy.

Figure 7 describes the relationship between the solution heat treatment time and the volume percent of solutioned γ' at 1280°C. At this temperature, the volume percent of solutioned γ' increased with time and was saturated at about 70%.

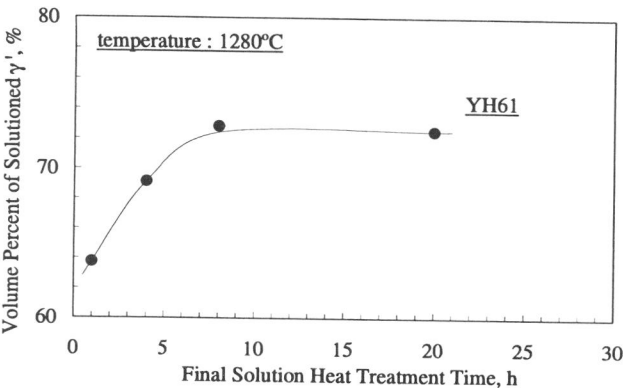

Figure 7: Volume percent of solutioned γ' was saturated to about 70% with increasing solution heat treatment time at 1280°C.

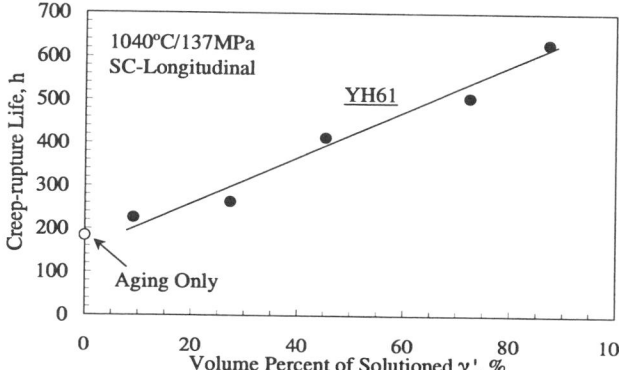

Figure 8: Creep-rupture life of SC longitudinal direction increased with increasing volume percent of solutioned γ'.

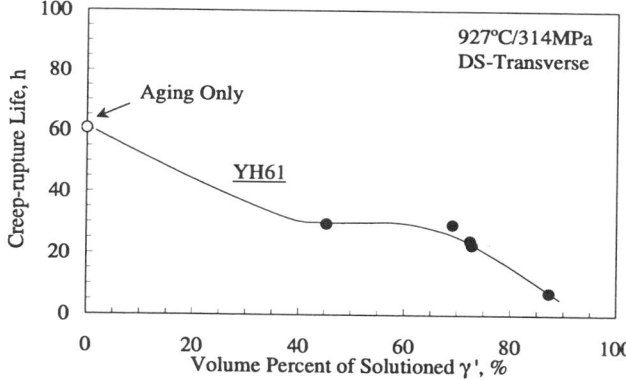

Figure 9: Creep-rupture life of DS transverse direction decreased with increasing volume percent of γ'. DS transverse direction should be considered as substitution for HAB.

The effects of solution heat treatment on the creep-rupture life of defect-free SC longitudinal direction and DS transverse direction are shown in Figures 8 and 9, respectively. The creep-rupture life in the DS transverse direction was evaluated for simulating the creep-rupture life of SC with HAB. The creep-rupture life of the defect-free SC longitudinal direction was improved by increasing the volume percent of solutioned γ'. Although this result has

already been established, it must be noted that the creep-rupture life in the DS transverse direction decreased with increasing the volume percent of solutioned γ'. It was pointed out in our previous study [5] that the grain boundary surrounded by the non-solutioned γ' (Figure 10-a) still contains higher levels of boron after solution heat treatment. On the other hand, the grain boundary exposed to the solutioned γ' (Figure 10-b) was found to contain lower amounts of boron after solution heat treatment. If the solution heat treatment temperature or the treatment time is increased, the increase in the volume percent of solutioned γ' also increases the percentage of the exposed grain boundary. Thus, these could be considered as reasons why creep-rupture life in the DS transverse direction decreased with increasing the volume percent of solutioned γ'.

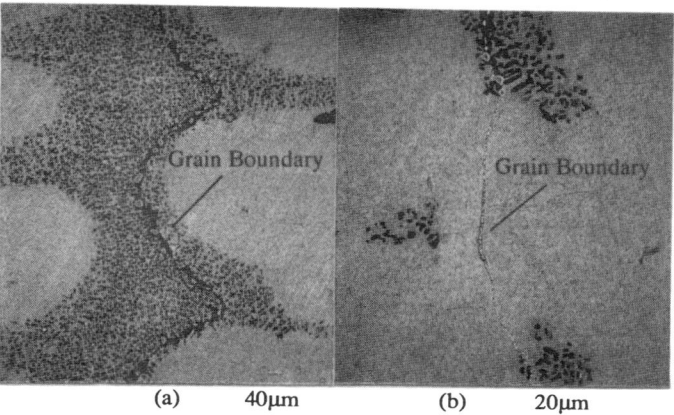

Figure 10: Two typical grain boundary structures after solution heat treatment (Final step: 1290°C/4h).

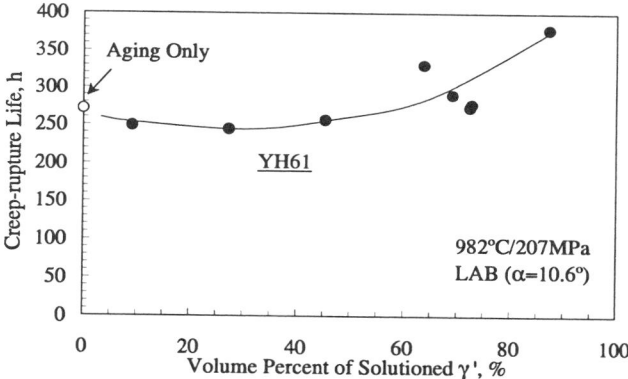

Figure 11: Decrease in creep-rupture life with increasing volume percent of solutioned γ' was not observed for LAB.

In the case of LAB (α = 10.6°), no significant degradation of SC transverse creep-rupture life was observed with increasing the volume percent (Figure 11). Additionally, a moderate LAB creep-rupture life was retained up to about 6° in SC610 which was full solutioned and did not contain boron (Figure 18). It can be deduced from these results that the presence of boron at grain boundaries does not play an important role in LAB while boron is essential for HAB strength. Therefore, it follows that the solution heat treatment condition would not influence the LAB strength significantly since diffusion of boron from the grain boundary during solution heat treatment is not a major issue for LAB strength. However, it must be noted that the addition of boron increases the misorientation angle at which there is a fall-off of strength, or rather, the addition of boron extends the range of LAB.

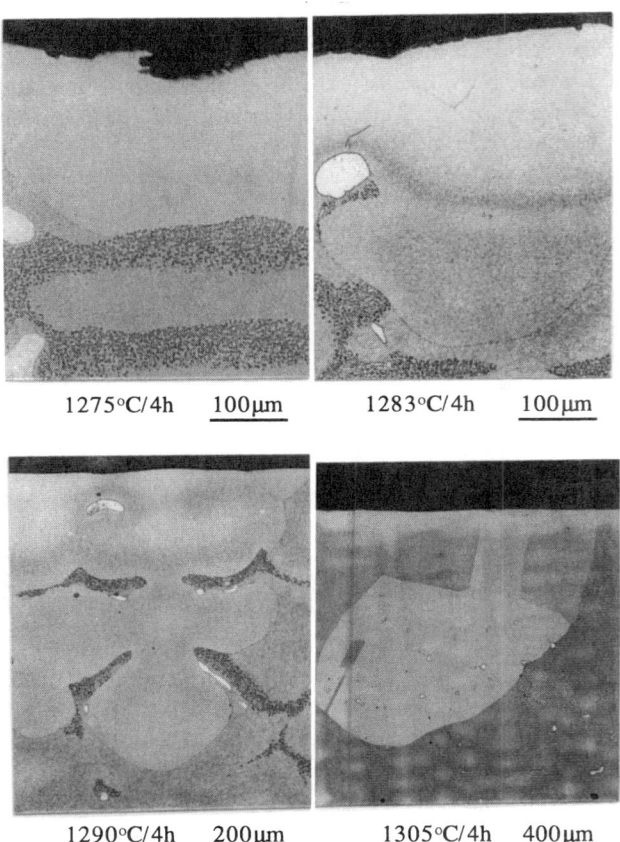

| 1275°C/4h | 100μm | | 1283°C/4h | 100μm |

| 1290°C/4h | 200μm | | 1305°C/4h | 400μm |

Figure 12: Depth of the recrystallization as a function of solution heat treatment temperature in SC610

Figure 12 shows typical micrographs of recrystallized grains in SC610. In order to apply this result to YH61, Figure 13 describes the maximum depth of recrystallization as a function of the volume percent of solutioned γ'. It is reasonable to suppose that the relationship between the volume percent and the depth can apply in principle to YH61 as well as other SC superalloys, because growth of recrystallization was observed to depend on the existence of a non-solutioned γ' barrier but to be independent of the solution heat treatment temperature as long as non-solutioned γ' existed. At temperatures below the solidus, recrystallization only occurred in γ'-denuded layer which was caused by the formation of a surface oxide layer. Although the recrystallized grain grew into the γ' solutioned area at temperatures above the solidus, non-solutioned γ' still hindered the growth. The recrystallized grain could not grow thermodynamically until full-solutioning was completed. These observations suggested that severe recrystallization cannot occur as long as non-solutioned γ' exists. Walston et al.[12] proposed another effective prevention method for similar crystal

growth by borides or carbides. However, a continuous network of the non-solutioned γ', which can be formed in a higher volume percent than borides or carbides, would be necessary for preventing the severe recrystallization observed in IGT castings since their large and complicated features introduces higher casting residual stress than aero-engine turbines. It follows that a partial solution heat treatment should be adopted for IGT castings while the alloy's potential for strength can not be fully exhibited in this condition.

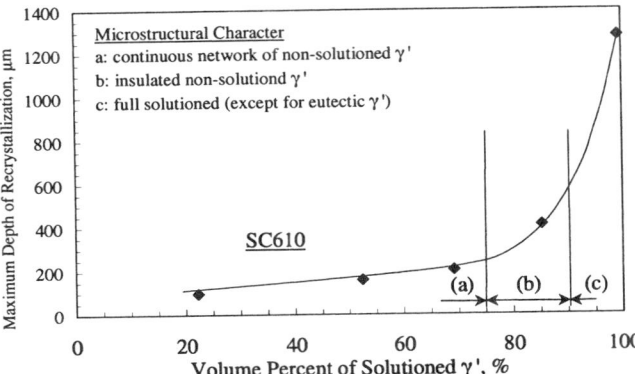

Figure 13: The effect of volume percent of solutioned γ' on growth of recrystallized grains.

These results have led to the conclusion that determining the solution heat treatment condition is one of the most important procedures for manufacturing reliable IGT buckets and vanes with low cost. Conditions to control the volume percent of solutioned γ' should be determined by taking account of the followings:

(1) Defect-free SC strength
(2) LAB strength
(3) HAB strength (in case of allowing HAB)
(4) Preventing severe recrystallization

It must be noted that the condition for a defect-free SC strength conflicts with that required for HAB strength and preventing severe recrystallization. A compromise condition should be selected in order to meet the design criterion for each part. Finally, it must be noticed that an extreme insufficiency of solutioning causes lower hot corrosion resistance and phase stability, which is not discussed in this paper.

The Effect of Grain Boundary misorientation angle

The relationship between grain boundary misorientation angle (α) and LAB or HAB strength is one of the major concerns for IGT manufacturers. From the viewpoint of alloy metallurgy, the effect of α on deformation behavior is also interesting.

Tensile property. Figure 14 shows the relationship between α and 0.2% yield strength at 700°C. No significant fall-off of the 0.2% yield strength was observed. Dependency of 0.2% yield strength on α at room temperature exhibited a tendency similar to that shown in this figure. Although the increase in α caused only a slight decrease in the 0.2% yield strength, fracture can be divided into three types depending on their α values (Figure 15). Specimens, whose α was less than about 4°, deformed uniformly and finally fractured at the center of gauge length (Type A). For α values between about 4 and 10°, significant deformation occurred in the rotated grain whereas apparent deformation was not observed in the primary grain (Type B). In this case, sheared fracture occurred in the rotated grain. In specimens whose α was more than about 10°, brittle fracture at the grain boundary was observed (Type C). Although unilateral deformation was also observed in the rotated grain, it did not lead to fracture.

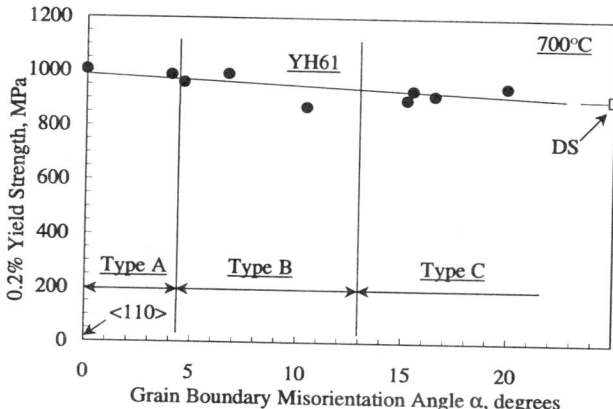

Figure 14: Relationship between grain boundary misorientation angle and 0.2% yield strength.

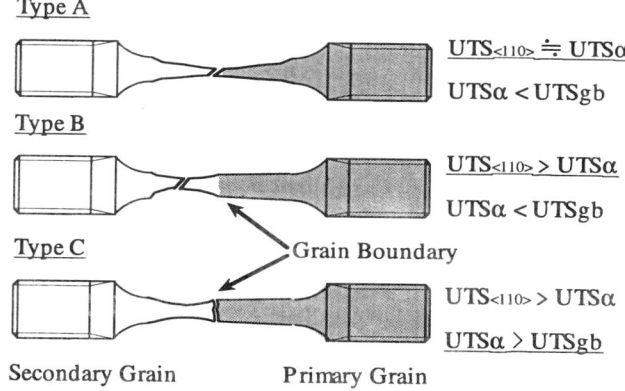

Figure 15: Schematic illustration of deformation and fracture types for YH61 with LAB or HAB.

These obesvations can be explained by the orientation dependence of 0.2% yield strength. Shah and Duhl [13] pointed out that orientations along the <001> to <110> boundary of the standard stereographic triangle deform by octahedral slip, but do not obey the Schmid's law since <001> and <110>, whose Schmid factors are equivalent, exhibit different yield strengths. Figure 16 shows the relationship between α and the 0.2% yield strength of defect-free SC YH61 as well as the data of Shah and Duhl. These data correspond to the 0.2% yield strength for the secondary grain itself. The difference of the 0.2% yield strength between <001> and <110> was also observed in YH61. Although it is not clear

whether the orientation dependence of the yield strength obeys the Schmid's law, it must be noted that the yield strength decreases with increasing α for α values less than about 20°. Thus, the yield strength of the secondary grain decreases with increasing α while that of the primary grain is not affected by α.

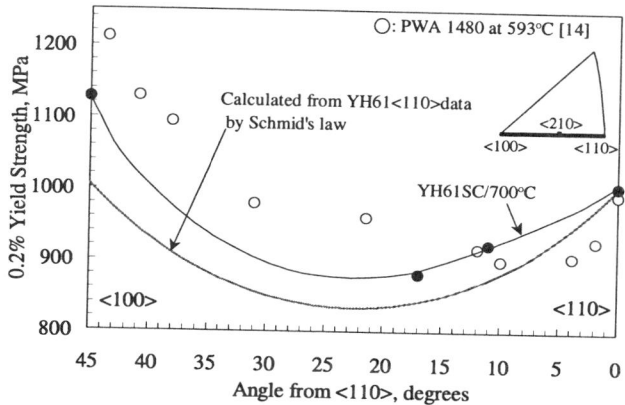

Figure 16: For angles less than about 20°, 0.2% yield strength decreases with increasing the angle. This range corresponds to the rotation angle of the secondary grain.

It seems reasonable to suppose from these results that the different behavior observed for the three deformation types can be explained by comparing ultimate tensile strength value of the primary grain (UTS$_{<110>}$), the rotated grain (UTSα) and the grain boundary (UTSgb). For type A, it can be considered that deformation behavior is the same as defect-free SC. In this range, it must be noted that shear occurred across the grain boundary. Therefore the existence of a grain boundary is thought to be insignificant for these α values, whereas a grain boundary can be considered as an obstacle for shear at greater α values as discussed later. For type B, deformation was localized in the rotated grain since UTSα becomes lower than UTS$_{<110>}$ with increasing α. Considering that the specimen was not fractured at the grain boundary, type B deformation can be included in a SC deformation. Classified with regard to the deformation behavior alone, "LAB" may correspond to α values for type A and B. For type C, although UTSα still decreases with increasing α, it can be supposed that UTSgb fell rapidly and therefore it became lower than UTSα. Thus, the specimen fractured at the grain boundary. In this range, even if the yield strength for higher α values retained more than about 90% of the yield strength for <110>, the deformation can no longer be regarded as a SC deformation. This α may correspond to "HAB". It follows from these results that the argument, whether life across LAB or HAB is comparable to that of a defect-free SC, is not valid unless their crystallographic orientations are clearly specified. It is considered that both the grain boundary strength and strength anisotropy of the rotated grain should be taken into account for evaluating the life of the SC with LAB or HAB.

763

Creep-rupture property. Figures 17 and 18 show the relationship between α and creep-rupture life for different test temperatures. Although a significant fall-off of the 0.2% yield strength for α was not observed, creep-rupture life exhibited apparent fall-off at higher α values. Furthermore, the magnitude of the fall-off increased with increasing temperature. For 760°C/490MPa, the creep-rupture life of the specimen whose α was 15.5° retained more than 70% of SC$_{<110>}$ life. On the other hand, the creep-rupture life of this specimen decreased to 30% of SC$_{<110>}$ life at 982°C/207MPa.

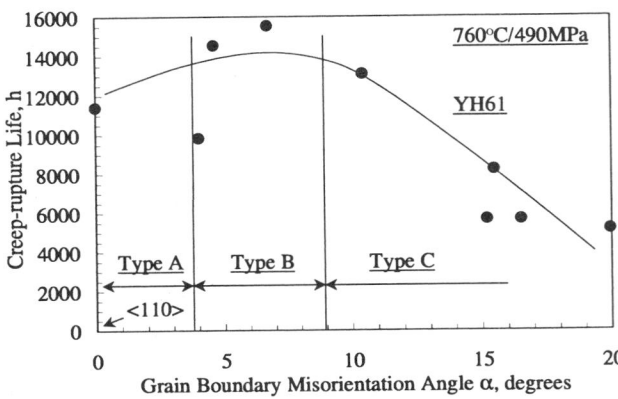

Figure 17: The effect of grain boundary misorientation angle on creep-rupture life for 760°C/490MPa.

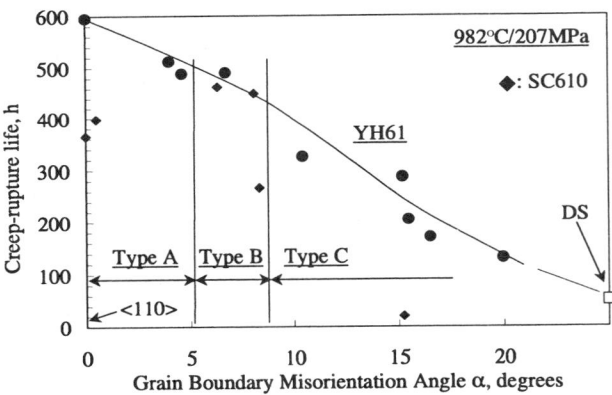

Figure 18: The effect of grain boundary misorientation angle on creep-rupture life for 982°C/207MPa.

It is found that the transition angle of which deformation type changes for A to B and B to C do not depend on temperature or stress. The transition angle for the change from type A to B is about 4° for each creep-rupture and for each tensile test condition. The same tendency was observed for the transition angle for the change from type B to C at about 10°. It follows from these results that classification with respect to the deformation type will help understanding the LAB or HAB tolerability of an alloy since the transition angle hardly depends on the test conditions whereas the magnitude of the fall-off strongly depends on the conditions.

Although the fall-off of creep-rupture life across LAB or HAB was observed for YH61, the absolute value was comparable to the transverse creep-rupture life of a second-generation DS superalloy [6] even for α of about 15°. This result leads to the conclusion that an optimized solution heat treatment condition and the presence of grain boundary strengthening elements in YH61 make it possible to realize a good balance between its LAB or HAB strength and defect-free SC strength.

Low cycle fatigue (LCF) property. A strain-controlled LCF test provided another interesting result for LAB or HAB strength. Deformation was localized in the rotated grain, which was the weaker grain, for the tensile test or the creep-rupture test, while every specimen fractured at the grain boundary during the LCF test. Although the life of the 10°-misoriented specimen was comparable to that of the SC$_{<110>}$ specimen in the LCF test, a rapid fall-off of the life was observed when α increased above about 12° (Figure 19).

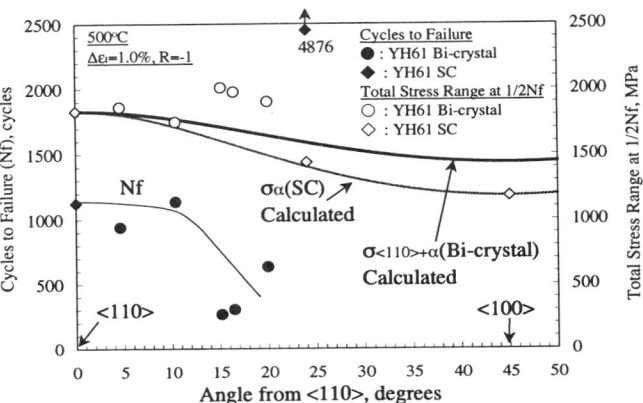

Figure 19: The effect of α on LCF life for bi-crystal(●) and single crystal(◆). For the bi-crystal, the angle from <110> corresponds to the grain boundary misorientation angle α. Observed total stress ranges at 1/2 Nf for each specimen are also plotted(○,◇). The lines σ$_{<110>+α}$ and σ$_α$ describe the theoretical total stress for the bi-crystal and single crystal, respectively. These values are calculated from measured elastic moduli.

In order to understand this particular deformation mechanism for the strain-controlled LCF test, experimental results for a total stress range at 1/2 Nf are compared with theoretical values (Figure 19). Although the total stress range is the sum of the elastic and plastic stress ranges, only the elastic stress is taken into account because no plastic stress was observed at 1/2 Nf. Strain for the primary grain (ε$_1$), the secondary grain (ε$_2$) and total strain of the bi-crystal (ε$_t$) have following relationship,

$$\varepsilon_t = 1/2 \cdot (\varepsilon_1 + \varepsilon_2), \tag{1}$$

and total stress for the bi-crystal (σ$_{<110>+α}$) can be written as

$$\sigma_{<110>+\alpha} = \varepsilon_1 E_1 = \varepsilon_2 E_2 = 2 \cdot E_1 \cdot E_2/(E_1 + E_2) \cdot \varepsilon_t, \tag{2}$$

where E_1 is the Young's modulus for the primary grain and E_2 for the secondary grain. E_1 is equivalent to $E_{<110>}$, and E_2 can be described as E_α while ε_t is equal to 0.01 in this study. Therefore,

$$\sigma_{<110>+\alpha}=2 \cdot E_{<110>} \cdot E_\alpha/(E_{<110>}+E_\alpha) \cdot 0.01, \qquad (3)$$

where $E_{<uvw>}$ is the Young's modulus for $<uvw>$ direction. The Young's modulus of a cube single crystal for any direction can be written as [14] [15]

$$E(\theta,\phi)=[s_{11}-2(s_{11}-s_{12}-1/2 \cdot s_{44})$$
$$\times \sin 2\theta(\cos 2\theta+1/4 \cdot \sin^2\theta \cdot \sin^2 2\phi)]^{-1}, \qquad (4)$$

where s_{11}, s_{12} and s_{44} are elastic compliance, θ the rotation angle from [001] axis and ϕ the angle between the projection on (100) plane and [100] axis [15]. Since the stress axis is only rotated on the (100) plane in this study, θ is equal to 0 and ϕ can be taken as $(45-\alpha)$ by using rotation angle from [110] axis. Then, equation (4) can be written in terms of α (degrees) as

$$E(\alpha)=[s_{11}+\sin^2(45-\alpha) \cdot \cos^2(45-\alpha) \cdot (2 \cdot s_{12}+s_{44}-2 \cdot s_{11})]^{-1}. \qquad (5)$$

Moreover, for these stress axes, the above elastic compliances can be described as

$$s_{11}=1/E_{<100>}, \quad s_{44}=1/G_{<100>}, \quad s_{12}=2/E_{<110>}-s_{11}-1/2 \cdot s_{44}, \qquad (6)$$

where $G_{<100>}$ is the shear modulus for $<100>$ direction. For YH61, experiments showed that $E_{<100>}=118$GPa, $E_{<110>}=183$Gpa and $G_{<100>}=102$GPa are valid at 500°C. Also, the relationship between the total stress range for the bi-crystal and α can be calculated from equations (3) and (5) (Figure 19). The total stress range for SC rotated by α from $<110>$ (σ_α) is also plotted in Figure 19. It is obtained from the equation;

$$\sigma_\alpha=E_\alpha \cdot \varepsilon_t . \qquad (7)$$

In Figure 19, a significant fall-off of the life was observed for the bi-crystal whose α was higher than about 12° as mentioned above. It is remarkable that the bi-crystal which suffered from the lower life showed a higher experimental total stress than the calculated value while the stress range should decrease with increasing α because of the decrease in the Young's modulus as calculated in equation (3). For α values below about 12° and for a whole range of SC, the experimental value followed the calculated value. In a SC rotated 24° from $<110>$, a lower total stress increased the life about four times that of $SC_{<110>}$ and the bi-crystal showed a comparable life to the $SC_{<110>}$ within the range where the experimental value followed the calculated value.

From these observations, the following conclusions are considered as applicable. When α is less than about 12°, deformation of the bi-crystal took place in accordance with equation (1) because no significant deviation of the experimental stress range from the calculated value was observed. For these values of α, the bi-crystal retains the life of $SC_{<110>}$ or shows a slightly higher life

because the total stress decreases with increasing α. Therefore, deformation of the bi-crystal can be considered as being the same as SC in this range. When α is higher than about 12°, higher stresses must have significantly decreased the life. It is thought that the existence of the grain boundary causes the difference between the experimental and the calculated stress ranges. The grain boundary, which lies perpendicular to the stress axis, can be regarded as an obstacle to shear deformation across the grain boundary. Since the stress increased to the determined strain range in the strain-controlled test, the resultant high stress must have been used to elongate the grain boundary. Since the grain boundary of high strength superalloys suffers from poor ductility [16], the additional stress, which was not observed for lower α, must have caused the significant shortening of life at higher α values. It should be noted that thermal stresses, which are more important to advanced IGT than centrifugal stress, is a strain-controlled stress. It is concluded from this discussion that close attention must be paid to LAB or HAB at the parts or the sections where thermal stress can be of a major concern. It is also thought that the strain-controlled test is most essential test method to evaluate LAB or HAB strength of a SC superalloy.

Electron beam weldability. In order to allow the existence of LAB or HAB in IGT castings, the effects of α on weldability as well as mechanical properties of SC superalloys should be investigated.

Grain boundary cracks caused by EB welding were not observed when α was lower than about 10°. For α above this value, both cracked and non-cracked grain boundaries were observed. Although the presence of grain boundary cracks was investigated after a post weld heat treatment in this study, almost all of the cracks had been observed after welding. Welding defects other than grain boundary cracking were not observed.

When a weldment cools to ambient temperature, tensile stress is generated near the weldment. Typically, the magnitude of this stress may be as high as the yield stress of the parent material [17]. Therefore, the grain boundary may elongate under this stress if the welding is performed across the LAB or HAB. Since the stress can be considered as a kind of thermal stress, it is possible that the grain boundary cracking can be prevented if LAB or HAB have enough ductility to elongate to strain values which are balanced by the thermally induced strain. The higher LAB resistance observed for YH61 is considered to be necessary for IGT castings from the viewpoint of weldability as well as mechanical properties.

Summary

Two important issues for the adaptation of SC castings to IGT were studied for YH61.

The effect of solution heat treatment condition
(1) Increasing the solutioned area decreased the HAB strength and increased the depth of recrystallization.
(2) A compromise solution heat treatment condition was selected

for the SC with some degree of grain defects since the best solution heat treatment condition for defect-free SC strength conflicts with that required for HAB strength and that for necessary to preventing severe recrystallization.

The effect of grain boundary misorientation angle (α)

(3) The effect of α on 0.2% yield strength, creep-rupture life and low cycle fatigue strength were studied. The magnitude of the fall-off as a function of α increased with increasing test temperatures for tensile and creep tests.

(4) In the bi-crystal test, an increase in α caused a decrease in yield strength of the rotated grain as well as the grain boundary strength. Based on crystal orientations, the strength of the rotated grain may be lower than that of resultant grain boundary. Therefore, both the grain boundary strength and strength anisotropy of the rotated grain should be taken into account when evaluating the life of SC with LAB or HAB.

(5) Different deformation behaviors were observed for load-controlled tests such as the creep test and strain-controlled tests such as the LCF test. The fall-off of life for higher α values was remarkable for the strain-controlled test due to its specific deformation mechanism.

Although the fall-off of life as a function of α was observed in YH61, the absolute life of higher α values was still comparable to that of conventional DS superalloy's transverse direction. An optimized solution heat treatment condition and grain boundary strengthening elements provide a higher level of compatibility for both defect-free SC strength and HAB strength. Since effects of α on alloy mechanical properties are influenced by temperature, stress level and strain-mode, a practical α-tolerance should be determined for each component or each section after considering their operation conditions.

Acknowledgments

The authors would like to thank Dr. B. Önay, Hitachi Research Lab., Hitachi, Ltd., for his reviewing the manuscript. The authors gratefully acknowledge helpful discussions with Dr. N. Matsuda, Hitachi Engineering Co., on the fatigue properties of YH61. The authors acknowledge the many metallurgical tests for YH61 which were performed at Hitachi Research Lab. by M. Kobayashi, N. Watanabe, T. Kashimura and K. Sato.

References

1. D. Goldschmidt, "Single-Crystal Blades", Materials for Advanced Power Engineering 1994, Part I, ed. D. Coutsouradis et al., (Netherlands: Kluwer Academic publishers, 1994), 661-674.

2. M. Cybulsky and P.E.C. Bryant, "Application of Aero-Engine Turbine Materials Technology to Industrial Gas Turbines", Advanced Materials and Coatings for Combustion Turbines, ed. V.P. Swaminathan and N.S. Cheruvu, (ASM, 1993), 23-27.

3. Thomas Barker, "Siemens' New Generation", Turbomachinery International, 1995, Jan/Feb: 20-22.

4. Hans-Jürgen Kiesow and Dilip Mukherjee, "The GT24/GT26 Family Gas Turbine: Design for Manufacturing", Advances in Turbine Materials, Design and Manufacturing, ed. A. Strang et al., (London, UK: The Institute of Materials, 1997), 159-172.

5. H. Tamaki et al., "Development of A New Ni-Based Single Crystal Superalloy for Large Sized Buckets", Materials for Advanced Power Engineering 1998, Part II, ed. J. Lecomte-Beckers et al., (Germany: Forschungszentrum Jülich GmbH, 1998), 1099-1110.

6. M. Sato, H. Tamaki et al., "High Temperature Strength of Large Size SC and DS Buckets for Industrial Gas Turbines", ASME Paper 95-GT-365, (ASME, 1995).

7. J.J. Jackson et al., "The Effect of Volume Percent of Fine γ' on Creep in DS Mar-M200 + Hf ", Metall. Trans. A, 8A (1977), 1615-1620.

8. Earl W. Ross and Kevin S. O'Hara, "René N4: A First Generation Single Crystal Turbine Airfoil Alloy with Improved Oxidation Resistance, Low Angle Boundary Strength and Superior Long Time Rupture Strength", Superalloys 1996, ed. R.D. Kissinger et al., (TMS, 1996), 19-25.

9. A.D. Cetel and D.N. Duhl, "Second Generation Columnar Grain Nickel-Based Superalloy", Superalloys 1992, ed. S.D. Antolovich et al., (TMS, 1992), 287-296.

10. Phil S. Burkholder et al, "CM186LC® Alloy Single Crystal Turbine Vanes", ASME Paper 99-GT-379, (ASME, 1999).

11. K. Sato, H. Tamaki et al., U.S. Patent 5,916,382, "High Corrosion Resistant High Strength Superalloy and Gas Turbine Utilizing The Alloy", 1999.

12. W.S. Walston et al., "A New Type of Microstructural Instability in Superalloys – SRZ", Superalloys 1996, ed. R.D. Kissinger et al., (TMS, 1996), 9-18.

13. D.M. Shah and D.N. Duhl, "The Effect of Orientation, Temperature and Gamma Prime Size on The Yield Strength of A Single Crystal Nickel Base Superalloy", Superalloys 1984, ed. Maurice Gell et al., (TMS, 1984), 105-114.

14. J.F. Nye, "Physical Properties of Crystals", (London, UK: Oxford University Press, 1960).

15. Tadashi Hasebe et al., "High Temperature Low Cycle Fatigue and Cyclic Constitutive Relation of MAR-M247 Directionally Solidified Superalloy", J. Eng. Mat. & Tech., 114 (1992), 162-167.

16. B.J. Piearcey and B.E. Terkelsen, "The Effect of Unidirectional Solidification on the Properties of Cast Nickel-Base Superalloys", Trans. TMS-AIME, 239 (1967), 1143-1150.

17. J.M. Robinson et al., "Residual Stress and Distortion Associated with Electron Beam Welding of Waspaloy", Joining and Repair of Gas Turbine Components, ed. Donald Tillack and Raymond Thompson, (ASM, 1997), 101-107.

TOPOLOGICALLY CLOSE PACKED PHASES IN AN EXPERIMENTAL RHENIUM-CONTAINING SINGLE CRYSTAL SUPERALLOY

C.M.F. Rae, M.S.A. Karunaratne, C.J. Small*, R.W. Broomfield*, C.N. Jones*
and R.C. Reed

University of Cambridge / Rolls-Royce University Technology Centre
Department of Materials Science and Metallurgy
Pembroke Street, CAMBRIDGE CB2 3QZ, UK

* Rolls Royce plc
PO Box 31, Derby, DB24 8BJ

Abstract

The kinetics, morphology and composition of the formation of TCP phases in an experimental alloy containing no tungsten is studied. At high temperature P phase forms after 20 h, whereas below 950°C the phases μ and R occur. At lower temperatures a polycrystalline form of σ phase is observed which is meta-stable but acts as a nucleation site for the other phases. The phase occurrence and compositions are compared with a thermodynamic model using a rhenium-containing database, and reasonable agreement is found for the P, R and σ phases. However the model underestimates the stability of the μ phase.

Introduction

Over the last 25 years additions of the elements tungsten, molybdenum and rhenium to single crystal superalloys have increased greatly and this situation is largely responsible for the improvements in the high temperature creep and rupture properties that have been achieved [1, 2]. However, the amounts of these elements which can usefully be introduced is limited by the formation of the group of phases known as TCP or Topologically Close-Packed phases [3]. The crystal structures consist of close-packed layers of relatively smaller atoms, such as nickel and chromium, with the larger atoms, such as rhenium, tungsten and tantalum, residing in 14, 15 or 16 co-ordinated sites between the layers. The number of these sites, centred on one of the so-called Kasper polyhedra [4], is characteristic of a given TCP phase, and, as a consequence of this, a wide range of compositions are displayed. In general, the close-packed layers have a high degree of coherency with the {111} planes of the γ and γ′ phases. Frequently, the TCP phases form in an extensive plate-like morphology parallel to the four {111} planes. This results in the characteristic needle-like appearance when sectioned in the {100} plane. Table 1 lists the crystallography of the phases observed in this work. TCP phases have been shown to have an adverse effect on the mechanical properties of alloys [5, 6], but, although they are brittle, there is conflicting evidence that they are directly involved in the failure and fracture of single crystal alloys [7,8,9]. A major threat posed to these alloys by the formation of TCP phases comes from the loss of the TCP-forming elements, and their potential strengthening effect, from the gamma phase.

The prediction of alloy stability is an important part of the alloy design process and recently the use of thermodynamic modelling of these multi-component alloys has proved to be a very powerful technique [10, 11,12,13,14]. However the accuracy of this method in predicting interactions in systems of up to 14 components depends on the availability of accurate experimental data on the equilibrium phases. Although TCP phases have been studied for many years, and their presence carefully monitored in a great number of alloys, data on the specific nature of the phases formed and their compositions in Re-containing single crystal superalloys are limited in extent [7, 8, 9, 15]. Recently Re has been added to the existing Ni-based superalloy database [16], and the work described here forms part of a larger study performed to validate the predictions of this database with respect to TCP formation and to provide further data for development of the model. The alloy described here, RR2071, is a second-generation single crystal superalloy containing ~3 wt % of rhenium comparable with the commercial superalloys such as CMSX-4. It is unusual in that it contains no tungsten and exhibits all four TCP phases typically found in single crystal alloys. This paper describes the TCP phases formed in the alloy, their composition and structure and demonstrates the role of sigma nucleation in the formation of the other TCP phases. Comparison with CMSX-4 and Alloy 800 [7,15] allows the effect of the replacement of tungsten by molybdenum to be explored.

Experimental Details

The alloy RR2071 used in this study is an experimental rhenium-containing single crystal alloy. Table 2 gives its composition together with that of CMSX-4 and Alloy 800 [15]. In RR2071 (in comparison with CMSX-4) the tungsten is replaced by a similar atomic percent of the lighter element molybdenum with the aim of reducing the density of the alloy. Other modifications are a slightly increased titanium percentage and the addition of a small amount of niobium.

Single crystal bars RR2071 were supplied by Rolls-Royce plc, (Derby, U.K.) in the fully heat-treated condition which involved: (i) solutioning at 1295°C for 6 hours, (ii) a primary age of 4 hours at 1130°C and finally (iii) a secondary age of 16 hours at 870°C. Samples were sealed under vacuum in silica tubes after flushing with argon and annealed for various times and temperatures. Some samples were annealed for shorter times in air and this did not have any effect on the formation of the TCP phases.

Specimens for TEM examination were polished using 20% perchloric acid in acetic acid at 10°C and 50V. A JEOL 2000FX

Superalloys 2000
Edited by T.M. Pollock, R.D.,Kissinger, R.R. Bowman,
K.A. Green, M. McLean, S. Olson, and J.J. Schirra
TMS (The Minerals, Metals & Materials Society), 2000

electron microscope was used to identify and analyse the phases using selected area diffraction and Energy-Dispersive Spectrometry. For the X-ray analysis the TCP phases were extracted from the γ/γ′ matrix using the techniques described in Reed et. al. [17].

Bulk specimens were prepared by mechanical abrasion finishing with a polish using colloidal silica for at least 20 minutes. The samples were examined in the JEOL 6340F, field emission gun scanning electron microscope using back-scattered electron contrast.

TCP Phase	System	Space Group	Space Group No.	Atoms per Unit Cell	Lattice Parameter	α
σ	Tetragonal	P4$_2$/mnm	136	30	a = b = 0.878nm, c = 0.454nm	90°
P	Orthorhombic	Pnma	62	56	a = 1.698nm, b = 0.475nm c = 0.907nm	90°
μ	Rhombohedral	R-3m	166	13 (hex. 39)	a = b = c = 0.904nm a = b = 0.4755nm, c = 2.583nm	30.5° 120°
R	Rhombohedral	R-3	148	53 (hex. 159)	a = b = 1.093nm, c = 1.934nm	120°

Table 1: Crystallography of the TCP Phases.

Wt %	Ni	Co	Cr	Mo	W	Re	Al	Ti	Ta	Nb	Hf
RR2071	62.0	9.50	6.60	4.50	0.00	2.80	5.60	1.30	7.30	0.30	0.10
CMSX-4	61.7	9.0	6.50	0.60	6.00	3.00	5.6	1.00	6.50	0.00	0.10
Alloy 800	63.6	7.5	8.0	1.5	4.0	3.0	5.8	1.5	5.0	0.0	0.0

Table 2: Alloy Compositions in Wt %.

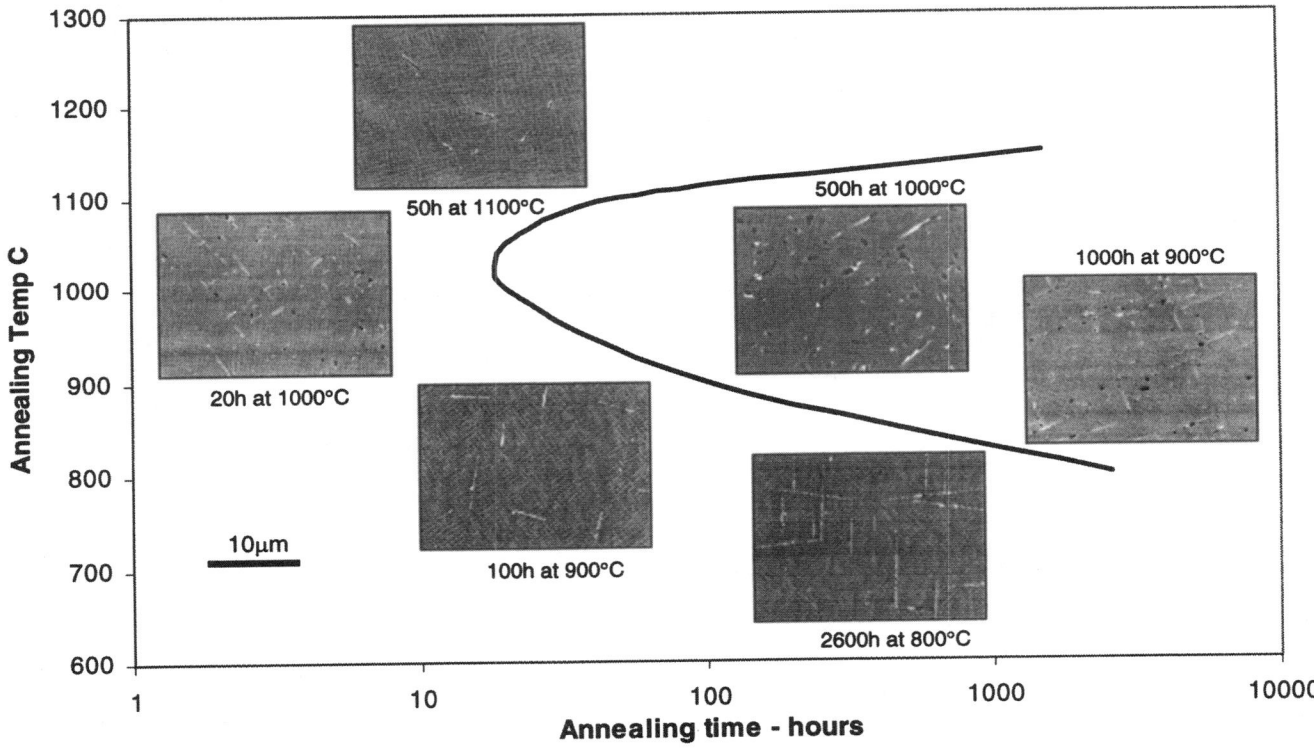

Figure 1: TTT curve for the formation of TCPs including backscattered electron SEM micrographs showing the morphology of the phases.

Results and Discussion

Transformation Kinetics

Figure 1 shows the experimentally determined TTT curve for the onset of the formation of TCPs in RR2071. TCPs are apparent after 20 hours annealing times at temperatures of 1000°C - 1050°C, but are not observed after prolonged annealing at 1150°C. The TCPs are more abundant in the dendrite cores corresponding to the retained segregation of rhenium to the core despite the full solution heat treatment [18], Figure 2.

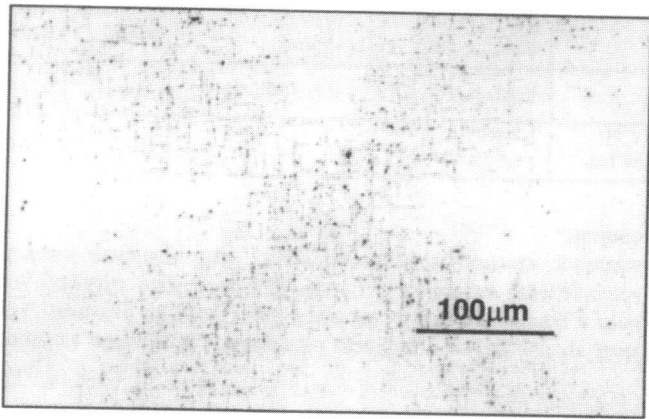

Figure 2: RR2071 annealed for 500h at 1000°C showing the formation of TCPs at the dendrite cores. (Dark-field optical micrograph, printed in negative.)

Precipitate Morphology

The morphology of the precipitates changes with both isothermal annealing temperature, and time. This is illustrated in Figure 1, which includes back-scattered electron images of the microstructure superimposed at the appropriate position. At low temperatures (e.g. 800°C) the precipitates are needle-like; closer examination shows the needles to consist of a series of discrete precipitates, accurately aligned over a distance of typically 10-20μm. These are also seen at 900°C, in the early stages of the precipitation, (Figure 3), together with two other distinct morphologies: fine planar precipitates interspersed with more cylindrical needles and separate larger blocky precipitates. With increasing annealing time the fine planar precipitates gradually disappear leaving individual precipitates aligned over distances of 20μm. The fine planar precipitates are not seen when the annealing temperature is raised to 1000°C and above. After 20h at 1000°C precipitates are visible as coarse single needles. Subsequent annealing shows that these needles break up and coarsen, as shown, for example, in Figure 1 after 500h at 1000°C. At temperatures above 1000°C and after long annealing times at 1000°C, the alignment of the precipitates disappears and they coarsen to become blocky. It is interesting to note that pores appear adjacent to the TCP phases as the precipitates grow; compare, for instance, the microstructure at 900°C after 100h and 1000h. Pore formation is most prolific at temperatures around the nose of the TTT curve, i.e. 1000-1050°C, but is apparent as low as 800°C.

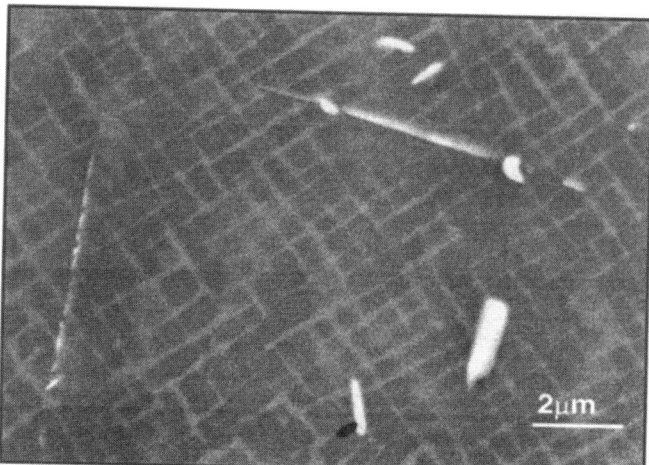

Figure 3: SEM micrograph (back-scattered electron image) of the precipitate morphology after annealing for 100h at 900°C

TEM analysis of specimens annealed at 800°C and 900°C identified the planar precipitates as thin (~30nm) plates of σ phase and the co-planar, prismatic needle phases as μ, Figure 4(a), 4(b), 4(c). The R phase was observed to form sporadically, in association with the σ phase, see Figure 4(d).

The structure of the σ phase in this alloy is unusual in being polycrystalline resulting in the mottled/feathery appearance of the precipitates (Figure 4(b)). Figure 5 shows the fine structure of the σ phase sectioned in the plane of the precipitate, i.e. with the foil normal being $[1\bar{1}1]_\gamma$. All the σ grains show the orientation relationship with respect to the γ/γ' matrix first reported in [15]. The σ tetragonal axis $[001]_\sigma$ is parallel to the $[1\bar{1}1]_\gamma$ axis and the $[110]_\sigma$ axis parallel the $[110]_\gamma$, $[011]_\gamma$ or $[10\bar{1}]_\gamma$ axes. This gives a total of three different orientations for the σ with respect to the γ/γ' matrix for each $\{111\}_\gamma$ plane. Each tiny grain of σ is mis-oriented from the next by 30° and the grain size is approximately 150nm. Three areas are indicated in Figure 5(a) where the orientations are clearly visible and rotated by 30°. The complex diffraction pattern from this form of sigma results from the superposition of the three σ [001] patterns and this is shown in Figure 5(b).

We have found that this form of σ nucleates very rapidly in the early stages of precipitation. We believe that this is due to a particularly low nucleation barrier for the formation of σ resulting from the low energy of the σ/γ interface on the $(1\bar{1}1)_\gamma$ plane in this orientation, thus enabling it to form as a meta-stable phase. The μ phase precipitates form as long, prismatic needles, usually in the plane of the σ plates, and consume the σ phase as they grow. Figure 6 shows polycrystalline σ and μ needles in RR2071 sectioned on the $(1\bar{1}1)$ matrix plane. The σ phase is on several $\{111\}_\gamma$ planes including that of the specimen. The long axis of the μ phase needle lies along the three-fold axis of symmetry, the $[001]_\mu$ axis, which is aligned within the $(1\bar{1}1)_\gamma$ plane of the γ/γ'matrix, close to the $[10\bar{1}]_\gamma$ direction, Figure 6(b). The deviation can be seen in the micrograph by comparing the edge of the μ phase needle with the sigma plate adjacent on an intersecting

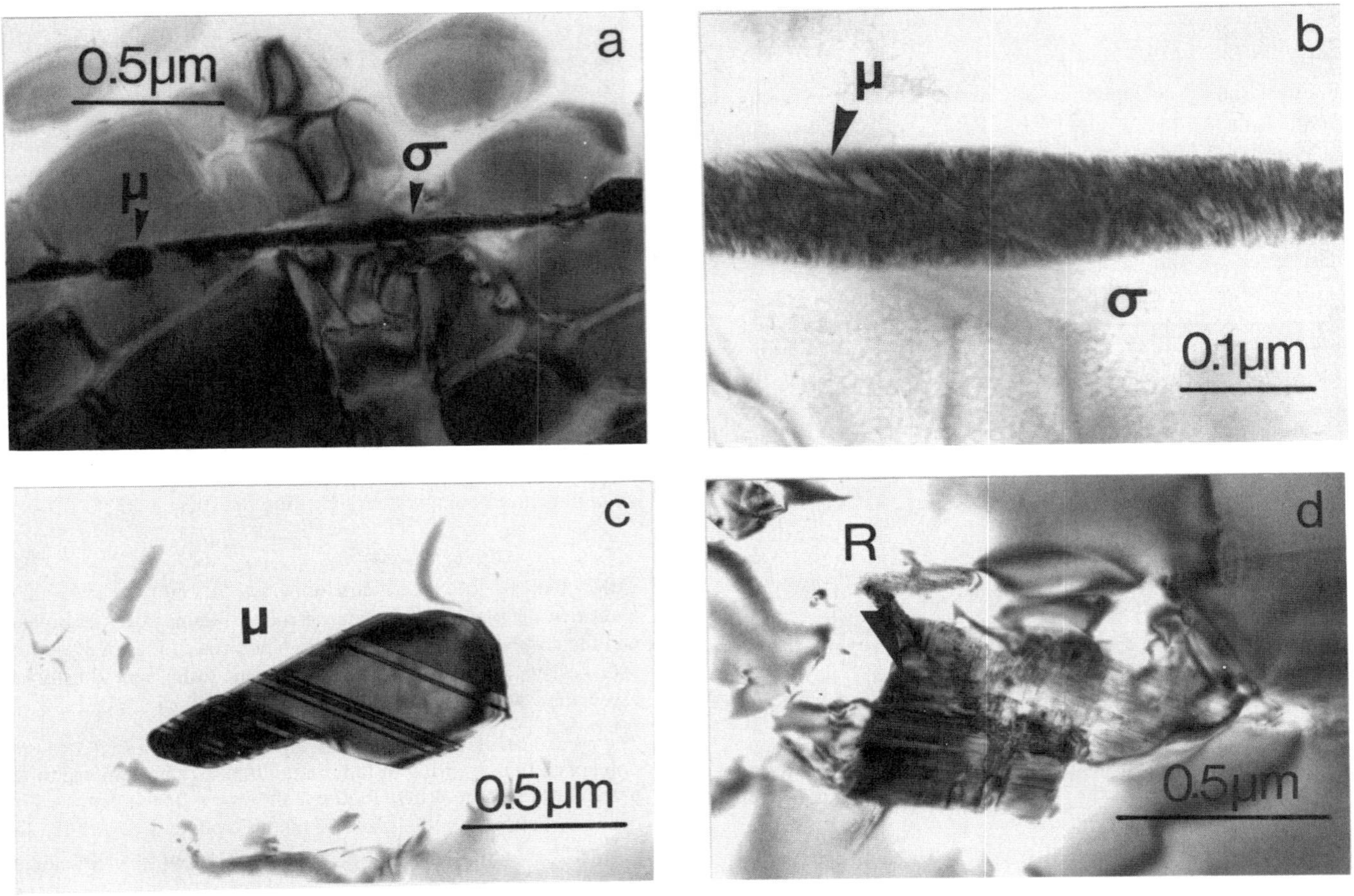

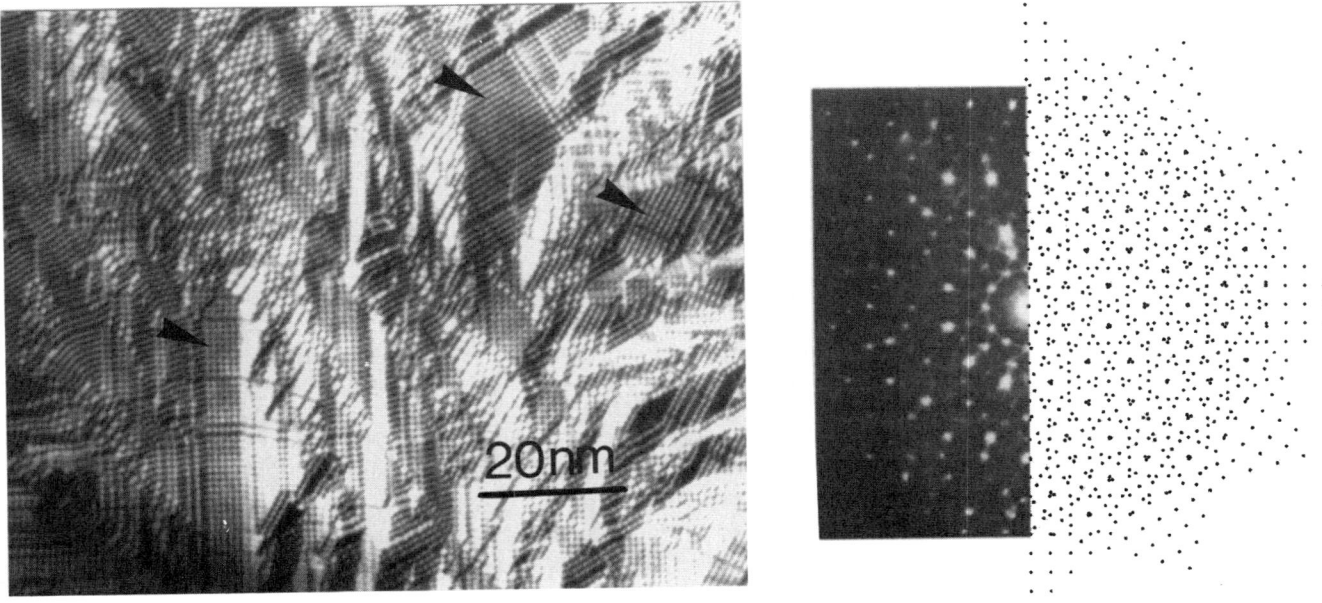

Figure 5: TEM micrograph of polycrystalline σ sectioned on the $(001)_\sigma/(111)_\gamma$ plane showing the structure of the small grains. The diffraction pattern results from the superposition of three [001] σ patterns rotated by 30°.

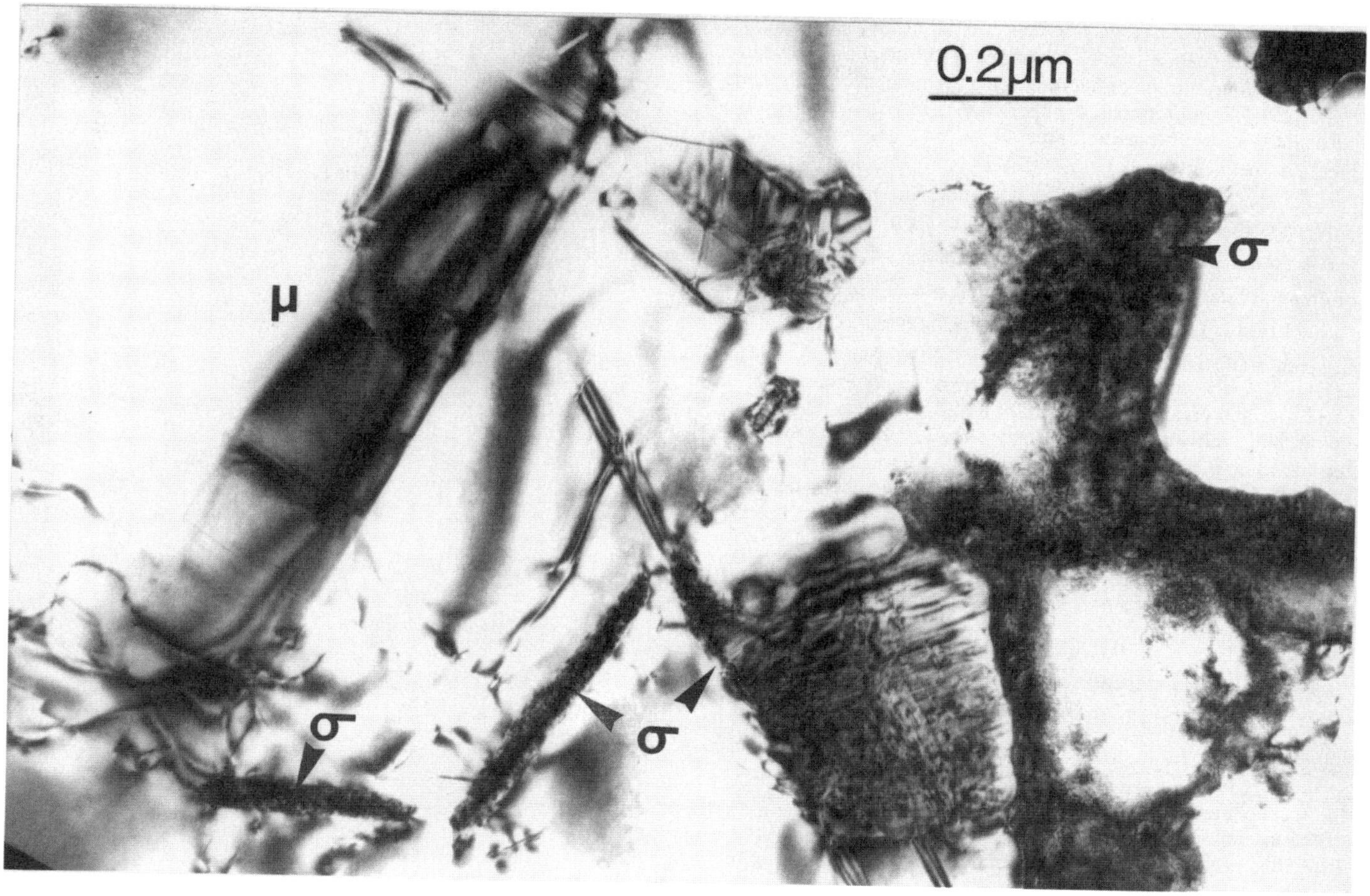

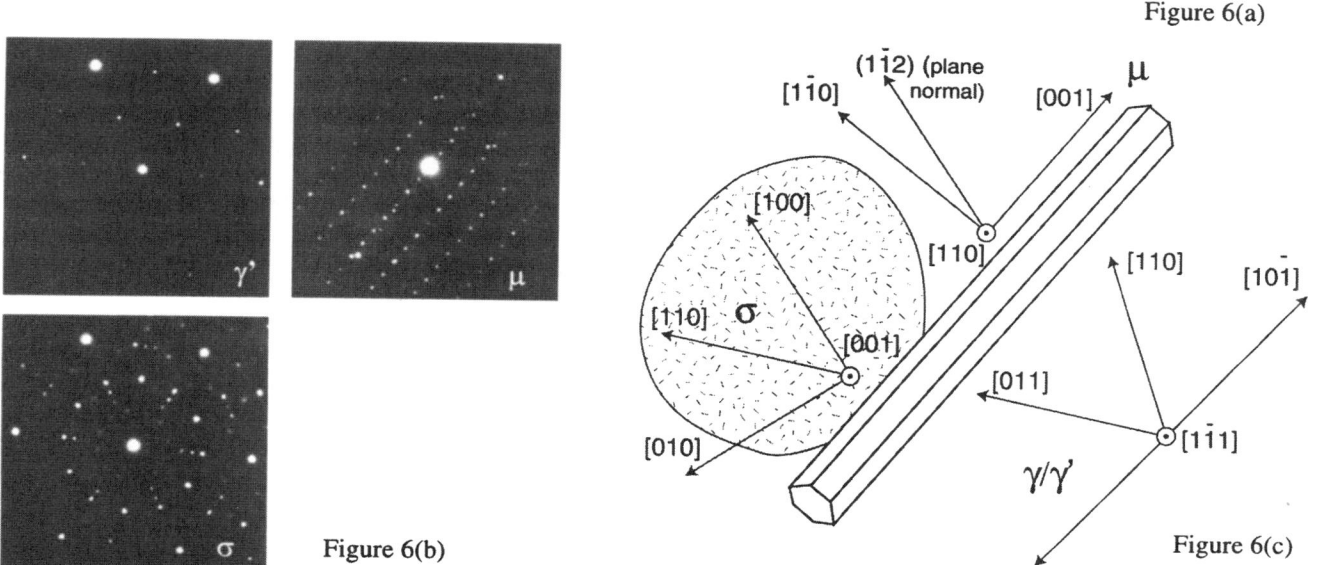

Figure 6(a)

Figure 6(b)

Figure 6(c)

Figure 6: (a): Micrograph of RR 2071 showing planar precipitates of polycrystalline σ on {111}γ planes and needle of μ phase aligned in the (1$\bar{1}$1) γ plane, 3° from [10$\bar{1}$].

(b): Diffraction patterns showing orientation relationships between γ, σ and μ phases.

(c) Schematic diagram showing orientation relationship between μ, σ & γ.

{111} plane. The observed orientation relationship between the μ phase and the matrix can be explained if the μ phase nucleated from the σ phase in the orientation relationship observed by Zhou et.al. [19] in an iron based superalloy. In this orientation a particularly close fitting (and potentially low energy) interface is produced between the μ and σ phases with the interface plane parallel to the $[001]_\sigma$ axis, i.e. perpendicular to the sheet of polycrystalline σ. The TCP phases are aligned with $[001]_\sigma$ and $[110]_\mu$ directions normal to the $(1\bar{1}1)$ plane of the γ/γ'. The orientation relationship reported by Zhou places the $(010)_\sigma$ plane parallel to the $(\bar{1}1\bar{2})_\mu$ plane. Given the orientation relationship between σ and γ, the $[110]_\gamma$ lies within 3° of the $[001]_\mu$ axis. This agrees with the orientation relationship reported Proctor [7] in CMSX-4, and presently, in the alloy RR2071. The small twinned grains indicated in Figure 4(b) may show an early stage of the formation of μ within the polycrystalline σ. The three orientation relationships are listed below and are illustrated schematically in Figure 6(c).

$(100)_\sigma$ // $(1\bar{1}2)_\mu$ and $[001]_\sigma$ // $[110]_\mu$.   equiv. to Zhou et. al.[19]

$(001)_\sigma$ // $(1\bar{1}1)_\gamma$ and $[110]_\sigma$ // $[011]_\gamma$ Darolia[15]

$[110]_\mu$ // $(1\bar{1}1)_\gamma$ and $[001]_\mu \sim^*$ // $[011]_\gamma$ equiv. to Procter [7]

          *Deviation measured as 4°.

The morphology and orientation of the μ phase indicate that it nucleates from within the meta-stable σ and grows at the expense of it. After 1000h at 900°C only traces of the planar σ are left, but the μ precipitates remain aligned with the original $\{111\}_\gamma$ planes of the matrix. Of the two examples of the R phase observed in the alloy, one was found adjacent to a sheet of σ, thus indicating that it is also possible that the R phase can nucleate from the σ phase.

At 950°C the precipitates were shown to be P phase with only a few examples of the μ phase. At 1000°C all the precipitates analysed were P phase. No trace of σ was seen at either temperature. The P phase grows as large rectangular precipitates, elongated in the $[001]_P$ direction, Figure 7. The same orientation relationship as noted by Darolia et. al. [15] was observed:

$$(010)_P \text{ // } (1\bar{1}1)_\gamma \text{ and } [10\bar{2}]_P \text{ // } [110]_\gamma$$

The precipitates are faulted and twinned. The fringes spaced at 1.69nm correspond to the largest of the three unit cell dimensions. The P phase precipitates are much larger than the μ phase seen at the lower temperatures and are typically 0.5μm in cross section.

The macroscopic form of the precipitates can be seen from SEM micrographs of the extracted residues used for X-ray analysis. The extraction process has left some of the γ phase supporting the TCPs. At 1000°C the cylindrical precipitates form a three-dimensional network of rods, Figure 8(a). These are beginning to thin at intervals as they break up into shorter, thicker lengths. At 850°C dense mats of interwoven precipitates are seen as the μ phase forms from the planar σ, Figure 8(b).

At higher temperatures the equilibrium phase, P, is able to form without the σ phase or before it has grown to any appreciable size; the result is a random nucleation of precipitates and their growth in the orientation relation described above. Figure 9 shows a

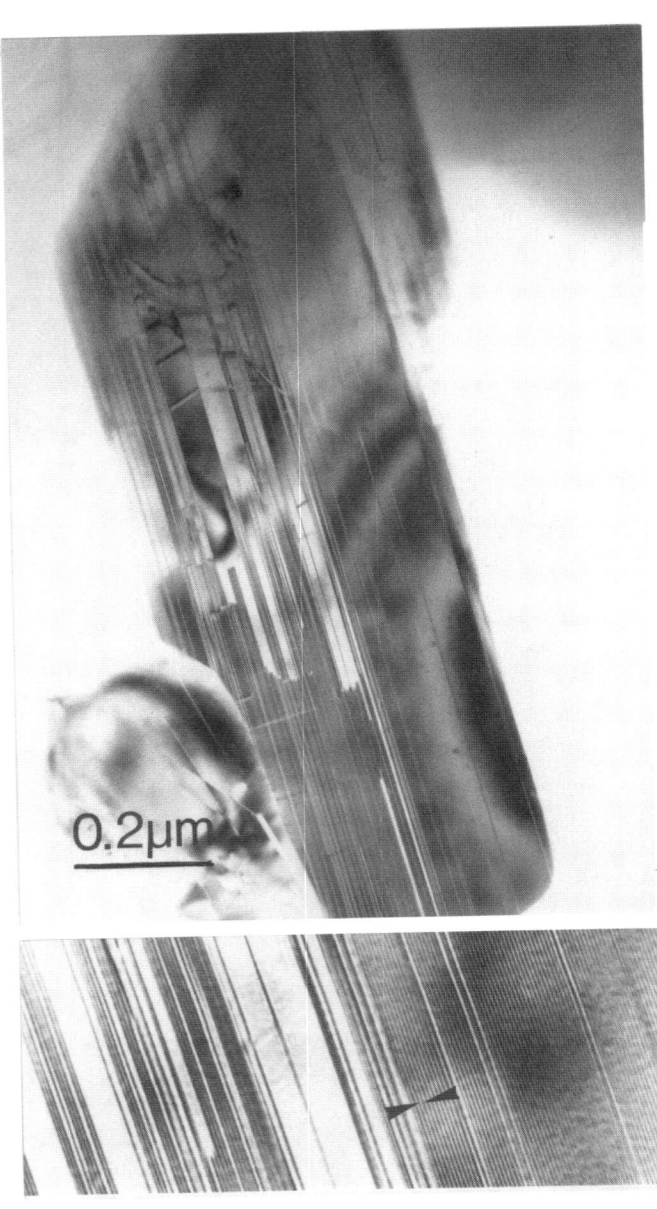

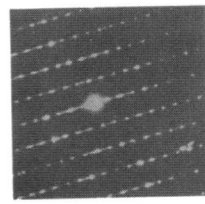

Figure 7: P phase precipitate in RR2071 aged at 950°C and orientated on the $[010]_P$ pole. The inset shows the fringes resulting from the inclusion within the aperture of the very closely spaced diffraction spots from the $(100)_P$ planes spaced at 1.69nm. Note the twinned areas which do not show fringes.

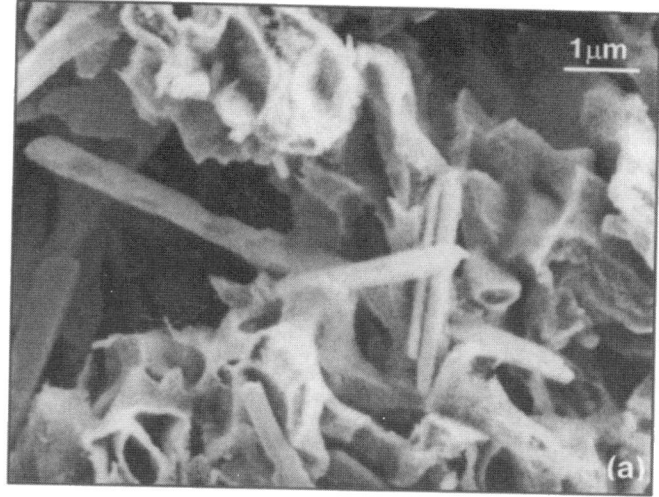

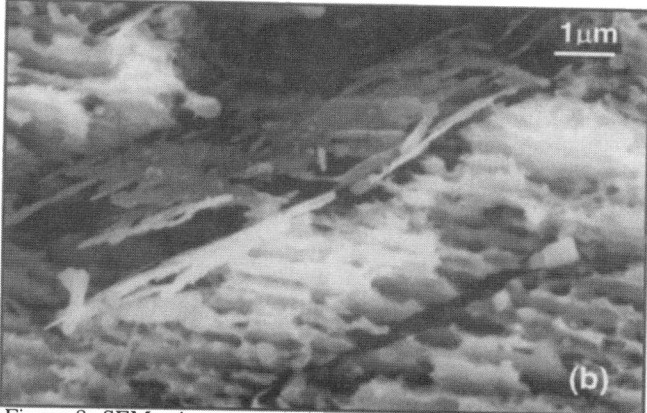

Figure 8: SEM micrographs of the TCPs extracted from RR2071 crept specimens: (a) Branching rods of P phase forming three dimensional network in a specimen crept at 1000°C for 671h; (b) Planar mats of μ and planar σ in specimen crept at 850°C for 1972h.

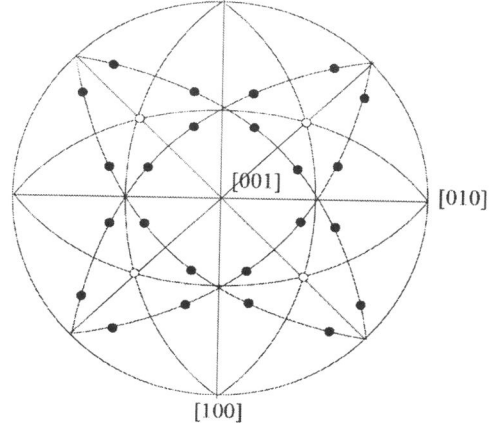

● - long axis of P-phase; O -{111} poles in γ/γ'.
Figure 9: Stereogram plotting the possible fast growth direction for the rods of the P-phase relative to the matrix.

stereogram for the γ/γ' matrix identifying all the matrix directions in which the long axes of P phase precipitates can lie. These are grouped some 15° from the <011>$_\gamma$ directions. As they grow, the precipitates can branch and intercept to form a three-dimensional network of rods. With prolonged ageing the rods thicken and segment. The result is the apparently random distribution of TCPs seen at higher temperatures and after prolonged heating at 1000°C.

The predominance of the μ phase at low temperature (<950°C) and the P phase at high temperature (>950°C) was confirmed by X-ray analysis of the extracted TCP residues from specimens crept for 1972h at 850°C and 671h at 1000°C and reported in detail in ref [18]. The lattice parameters for the μ and P phases in RR2071 were determined from the X-ray spectra obtained from such residues and these are summarised in Table 3.

Hex axes	μ Reference - Fe$_7$W$_6$	μ in RR 2071
a	0.4755nm	0.4735nm
b	0.4755nm	0.4735nm
c	2.583nm	2.554nm

	P Reference – Cr$_9$Mo$_{21}$Ni$_{20}$	P in RR2071
a	1.698nm	1.690nm
b	0.475nm	0.471nm
c	0.907nm	0.904nm

Table 3. Lattice parameters of μ and P phase in RR2071

TCP Phase Composition

The compositions of the TCP phases have been determined by EDS analysis in the TEM, and these are given in Table 4. The values for the σ, μ and R phases were determined from specimens aged at 800 and 900°C and those for P phase at 1000°C. Annealing times were 1000-2000h. The composition of the μ phase present at several temperatures showed little variation with temperature or annealing time. Where possible, several precipitates have been examined and the average value given; in the case of μ over 20 precipitates were analysed. For R phase the composition of only one precipitate was measured and this value is quoted. The values for the very thin precipitates of σ contain variable amounts of γ' from the surrounding material and are therefore higher in aluminum and nickel. These values show a consistent pattern and a single example is quoted which shows the lowest value of the aluminum content and is hence presumed to contain the least γ'.

Figure 10 shows the phase compositions in at % plotted on the same axes and the 'signatures' of each of the phases can be seen. The compositions of the P and R phases are very similar and characterised by equal chromium and molybdenum concentrations. The μ phase compositions are distinctive, being high in molybdenum and low in chromium and rhenium. In contrast the σ phase shows high chromium and low molybdenum and rhenium concentrations.

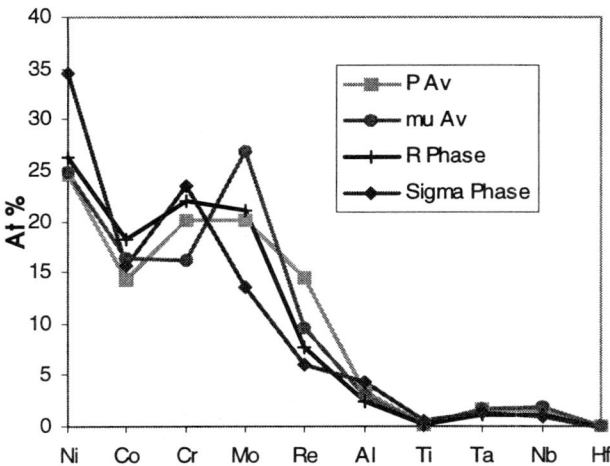

Figure 10 Graph of the compositions (at %) of the TCP phases formed in RR2071 determined by EDX analysis in the TEM.

At %	EXPERIMENTAL COMPOSITIONS			
	P phase 1000°C	μ phase 900°C	σ phase 800°C	R phase 800°C
Ni	24.6	24.7	34.6	26.4
Co	14.3	16.4	15.6	18.2
Cr	20.1	16.1	23.5	22.0
Mo	20.0	26.8	13.5	21.0
Re	14.5	9.5	5.9	7.8
Al	3.5	2.9	4.2	2.5
Ti	0.1	0.2	0.6	0.3
Ta	1.7	1.7	1.3	1.1
Nb	1.5	1.8	1.0	1.1
Hf	0.0	0.0	0.0	0.0

Table 4: Experimental compositions of TCP phases in RR2071, at %

At %	CALCULATED COMPOSITIONS			
	P phase 1000°C	μ phase 900°C	σ phase 800°C	R-phase 800°C
Ni	20.4	19.2	26.6	21.3
Co	13.6	13.3	13.8	12.7
Cr	25.6	17.1	37.0	24.1
Mo	25.8	34.2	19.6	27.1
Re	14.7	16.2	14.9	14.9

Table 5: Calculated compositions using the 'Super 5' database, at %

Comparison with Thermocalc predictions.

Using the results quoted here together with those from other alloys, the thermodynamic database used to predict the equilibrium structure of rhenium-containing alloys, was modified to give the version known as 'Super 5' [16]. The compositions of the TCP phases for the alloy RR2071 predicted using this database are given in Table 5. The 'signatures' of each of the phases are correctly predicted by the database but the experimental values for nickel are consistently higher, and the values for chromium and molybdenum lower, than predicted. At 1000°C and 2000h the average diffusion distance for the slowest element, rhenium, in γ is approximately 14μm, of the order of the precipitate spacing. This drops to 5μm at 900°C, [18, 20] and may explain the lower experimental rhenium values found for the lower temperature phases, μ, R and σ. The diffusion of the remaining TCP-forming elements is sufficiently rapid for equilibrium to be reached during the annealing time.

The relative stabilities of the various phases can be demonstrated by plotting the calculated phase occurrence as a function of the temperature for the alloy RR2071 where only one of the TCP phases is allowed to occur at any one time. Figure 11 shows that, for this alloy, the TCP phases are predicted to have very similar Gibbs free energies, and if this is indeed the case, it makes the prediction of the correct TCP phase extremely sensitive to the database parameters. Nevertheless, the predictions of the database were compared with the three similar rhenium-containing alloys for which detailed information on the TCP phase occurrence is available.

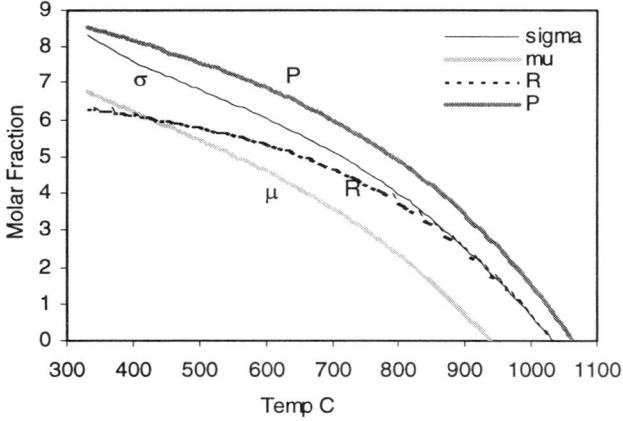

Figure 11: Calculated molar fraction of TCP phases predicted to form in the RR2071 alloy as a function of temperature with only one TCP phase being allowed to form at any one time showing the relative stability of the phases P, R, μ and σ.

The 'Super 5' database was used to predict the equilibrium TCP phases as a function of temperature for the alloys RR2071, CMSX-4 and Alloy 800 [15] and these are shown in Figure 12. RR2071 shows abundant TCP phase formation exhibiting all four of the major phases over the temperature range. The major phase at 800-900°C was μ with a small proportion of R and σ (the σ phase appears to be meta-stable). At 950°C and above, P phase is the majority phase with small amounts of μ at 950°C but not above that temperature. Although the database correctly predicts the formation of P phase in RR2071 at the higher temperatures, it underestimates the stability of the μ phase at the lower temperature. The solvus for P, predicted to be at 1060°C, is also lower than the experimental value of ~1150°C.

In the Alloy 800, studied by Darolia et. al.[15] the majority phases, P and σ, are reported to occur together up to at least 2100°F or 1150°C. The μ phase is also present in small amounts. The database predicts the occurrence of the phases P and σ, both having very similar stabilities, but, again, the predicted solvus for P at 1050°C clearly underestimates the solvus of the TCP phases. In contrast, CMSX-4 [7] shows the formation of the phases μ and R in small quantities. At the higher temperature of 1150°C, equal amounts of each phase are observed but at 1050°C μ is the most abundant phase. P phase has not been reported in CMSX-4 indicating that the replacement of tungsten by molybdenum stabilises P at the expense of R. However as pointed out by Saunders, [14] the free energies of the phases σ, R and P are particularly close in the alloy CMSX-4, and small changes in the rhenium, chromium or tungsten levels can cause any of the phases to become the dominant predicted phase. Neither P-phase nor μ is predicted by the database. Indeed the μ phase is present in all three alloys and is the majority phase in RR2071 and CMSX-4 at lower temperatures. The stability of the μ phase is under-estimated by the database. Attempts to remedy this within the existing model of the μ phase resulted in an unrealistically high rhenium content of the phase and indicate that the existing sub-lattice model of the μ phase requires further modification [16].

Prediction of the stability of rhenium-containing alloys with respect to TCP phases is not simply a function of equilibrium thermodynamics. The work reported here has highlighted the role that kinetics plays in the formation of TCPs. In the alloy RR2071, the precipitation of the σ phase at 900°C, and its subsequent disappearance, demonstrates that σ is not the equilibrium phase, but, although it is replaced after long annealing times by μ phase, the σ determines the morphology of the latter phase. The existence of a planar precipitate was suggested by Darolia et al to explain the inter-woven mats of P and σ precipitates parallel to {111} planes in Alloy 800. The difficulty of nucleating TCPs in CMSX-4 is remarked upon by Proctor [7], who observes that in CMSX-4 the μ phase precipitates are extremely large single needles up to 35μm in length. This is very much larger than any precipitates observed in RR2071. Thus the ease of nucleation of a particular phase can play an important role in the stability of an alloy in the short term and affect the morphology of the phases in the long term.

Despite the reservations outlined above, the level of detail and accuracy of the predictions from the rhenium-containing database exceeds that from other methods[21] and is in a form which can be integrated with diffusion data to provide a powerful modelling tool to incorporate kinetic effects [20, 22].

Summary and Conclusions

1. RR2071 exhibits a variety of TCP phases over the temperature range. The major phase at 800-900°C was μ with a small proportion of R and σ (the σ phase appears to be meta-stable at 900°C). At 950°C and above, P phase is the majority phase with small amounts of μ which vanish with increasing temperature.

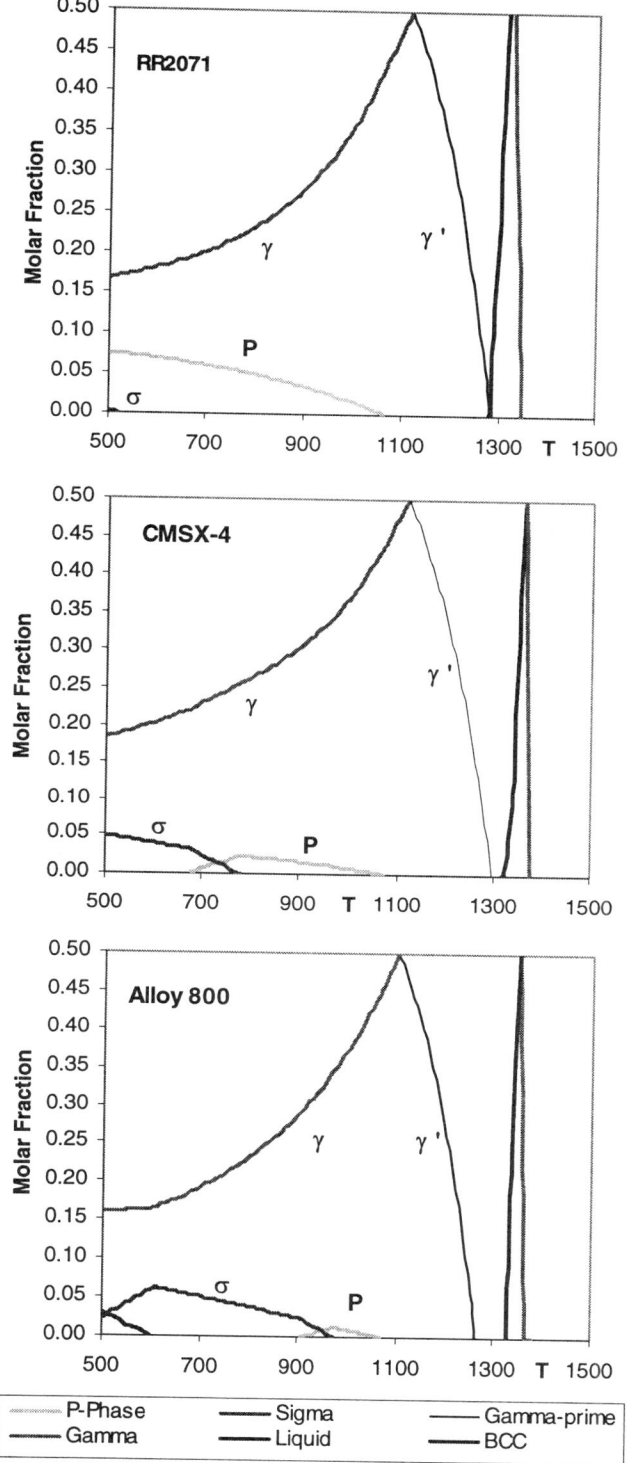

Figure 12: Equilibrium calculations using the Super 5 database to predict the phase molar fractions in RR2071, CMSX-4 and Alloy 800.

2. The TTT diagram for TCP formation has been determined; precipitation is most rapid at ~1025°C with significant occurrence of the P phase after 20 hours. No precipitation was observed at temperatures of 1150°C and above.

3. At lower temperatures meta-stable σ phase is readily nucleated as very thin polycrystalline sheets. It appears to be the pre-cursor for the equilibrium phases μ, and R.

4. At higher temperatures no evidence of σ was found and the equilibrium phase, P, grows throughout the matrix to form a 3-dimensional network of branching rods. The morphology which is exhibited by this phase has been rationalised.

5. The μ, σ and P phases are readily distinguished by composition: μ is rich in Mo (or W) σ is rich in chromium and low in molybdenum, and P and R both have roughly equal concentrations of molybdenum and chromium. Increasing the alloy molybdenum/tungsten ratio stabilises P at the expense of R.

6. The modified database for the rhenium-containing alloys has proved reasonably successful at predicting the occurrence of the P and σ phases but underestimates the stability of these phases. However the calculations using the current 'Super 5' database failed to predict the occurrence of the μ phase in any of the alloys examined, and the model for this phase requires further development.

Acknowledgements

The authors thank the Cambridge Commonwealth Trust, the Engineering Physical Sciences Research Council (EPSRC), Rolls-Royce plc and the Defence Evaluation & Research Agency (DERA) for sponsoring this work. We would also like to thank Nigel Saunders of Thermotech for many valuable discussions concerning the thermodynamic modelling.

References

1. A.F. Giamei and D.L. Anton, "Rhenium Additions to a Ni-base Superalloy: Effects on Microstructure", Metall Trans 16A, (1985), 1997-2005.

2. K. Harris, G.L. Erickson, S.L. Sikkenga, W. Brentnall, J.M. Aurrecoechea and K.G. Kubarych, "Development of the rhenium-containing superalloys CMSX-4 & CM186LC for single crystal blade and directionally solidified vane applications in advanced turbine engines", Superalloys, ed. S.D. Antolovitch et al., (Warrendale, PA, USA, 1992), 297-306.

3. A.K. Sinha, "Topologically Close-Packed Structures of Transition Metal Alloys", Progress in Materials Science, (1972), 79-185.

4. F.C. Frank and J.S. Kasper, Acta. Crystall., 12, (1959), 483.

5. S.T. Wlodek, Trans. of ASM, 57, (1964), 110-119.

6. E.W. Ross, Journal of Metals, 19, (1967), 12-14.

7. C. Proctor, "Formation and effects of intermetallics in the Re-containing superalloy CMSX-4" (PhD Thesis, University of Cambridge, 1993)

8. M. Pessah, P. Caron and T. Khan, "Effect of μ phase on the mechanical properties of a nickel-based single crystal superalloy" Superalloys (1992), ed. S.D. Antolovitch et al., (Warrendale, PA, USA,1992) 567-576.

9. T. Yokokawa, M. Osawa, H. Murakami, T. Kobayashi, Y. Koizumi, T. Yamagata and H. Harada, "Design of high Re containing single crystal superalloys for industrial gas turbines", Materials for Advanced Power Engineering, ed. J. Lecomte-Beckers, F Schubert and P.J. Ennis, Forschungszentrum, Julich, Germany, (1998), 1121-1128.

10. B. Sundman, B. Jansson and J. Andersson, "The Thermocalc Databank System", Calphad, 9 (1985), 153-190.

11. N. Saunders, Phil. Trans. Roy. Soc. Lond A, 351, (1995), 543-561.

12. N. Saunders, "Phase diagram calculations for Ni-based superalloys", Superalloys, ed. R.D. Kissinger, D.J. Deye, D.L. Anton, A.D. Cetel, M.V. Nathal, T.M. Pollock and D.M. Woodford, (Warrendale PA, USA,1996), 101-109.

13. N.J. Saunders and A.P. Miodownik, "Calphad, A Comprehensive Guide", Pergamon Press, (1998).

14. N. Saunders, "Modelling of Nickel-base Superalloys", Advanced Materials and Processes, 156, 9, (1999), 29-31.

15. R. Daralia, D.F. Lahrman and R.D. Field, "Formation of topologically close packed phases in Ni-based single crystals", Superalloys, ed. S. Reichman, D.N. Duhl, G. Maurer, S. Antolovich and C. Lund, (Warrendale PA, USA, 1988) 255-264.

16. N. Saunders, C.J. Small, "Computer modeling of Ni-based superalloy phase equilibria" Priv. Com. (1999) Thermotech plc, Surrey Technology Centre, 40 Occam Road, The Surrey Research Park, Guildford, Surrey GU2 5YG, UK.

17 R.C. Reed, M.P. Jackson and Y.S. Na, "Characterisation and Modelling of the Precipitation of the Sigma Phase in Udimet 720 and Udimet 720Li", Metallurgical and Materials Transactions, 30A, (1999), 521-533.

18 M.S.A. Karunaratne, "Diffusional phenomena in Ni-based superalloys", (PhD Thesis, University of Cambridge, 2000).

19. D.S. Zhou, H.Q. Ye and K.H. Kuo, "An HREM study of the intergrowth structures of σ-related phases and the μ phase", Phil. Mag. A, 57 (6), (1988), 907-922.

20. M.S.A. Karunaratne, P. Carter and R.C. Reed, "Interdiffusion in the FCC-A1 phase of the Ni-(Re,Ta,W) binary systems between 900°C and 1300°C', Mat. Sci. and Eng., A281, (1999), 229-233.

21. H. Harada, K. Ohno, T. Yamagata, T. Yokokawa and M. Yamazaki, "Phase calculation and its use in alloy design program for Ni-based superalloys", Superalloys, ed. S. Reichman, D.N. Duhl, G. Maurer, S. Antolovich and C. Lund, (Warrendale PA, USA, 1988), 733-742.

22. N..Matan, H. Winand, P. Carter, M. Karunaratne, P.Bogdanoff and R.C. Reed, "A coupled thermodynamic/kinetic model for diffusional processes in superalloys", Acta Mater., 46 (13), (1998), 4587-4600.

A LOW-COST SECOND GENERATION SINGLE CRYSTAL SUPERALLOY DD6

J. R. Li, Z. G. Zhong, D. Z. Tang, S. Z. Liu, P. Wei, P. Y. Wei, Z. T. Wu, D. Huang and M. Han

Beijing Institute of Aeronautical Materials

P. O. Box 81-1

Beijing 100095, China

Abstract

A low-cost second generation single crystal (SC) superalloy, designated DD6, has been developed for aeroengine turbine blade applications. DD6 contains 2 wt. % rhenium, which is about 2/3 of that of the second generation single crystal superalloys such as PWA1484, CMSX-4 and René N5. The alloy system employs the relatively high additive refractory element (tungsten, molybdenum, tantalum, rhenium and niobium) content of about 19.5 wt. %. A cost reduction of about 25% is expected for this alloy.

DD6 alloy has an approximate 40 ℃ improvement of creep strength relative to the first generation single crystal superalloys such as DD3, a Chinese first generation single crystal alloy whose creep rupture properties are comparable with PWA1480 alloy. The tensile properties and creep rupture properties of the alloy are comparable to those of the second generation single crystal alloys such as SC180, René N5, CMSX-4 and PWA1484. Most notably, the alloy provides superior oxidation resistance and good hot corrosion resistance. The advantage shown in the creep rupture properties over DD3 also generally hold for fatigue properties.

DD6 also has good microstructure stability, heat treatment characteristics and environmental properties.

Casting trials have been conducted on DD6 involving a great number of bars and some complex shaped hollow turbine blades. These trials demonstrated that DD6 possesses excellent single crystal castability.

Introduction

First generation and second generation single crystal alloys have been widely used for advanced commercial and military aeroengines since the 1980's[1-16], and third generation single crystal alloys have recently been developed[17-21]. The second generation single crystal alloys provide an approximate 30 ℃ improvement of creep strength relative to the first generation, while the third generation exhibit about 60 ℃ improvement of creep strength in comparison to the first. Table 1 presents the compositions of first, second and third generation single crystal alloys, including DD6 and DD3. A main distinction of the chemical compositions of the first, second and third generation single crystal superalloys is rhenium-free, 3 wt. % Re and 6 wt. % Re, respectively[1-21]. The effects of Re on the mechanical properties of single crystal superalloys are very significant, especially in improving the creep rupture life. However, Re is a rare element and the price of Re is very expensive. The aeroengine manufacturers all consider Re as a strategic element. Therefore, the development of low-Re second generation single crystal alloy is important. The goals of the present research were as follows: 1) develop a low cost second generation single crystal alloy, using low Re; 2) maintain the mechanical properties of the alloy at levels equivalent to those of the second generation single crystal alloys, such as SC180, René N5, CMSX-4 and PWA1484; 3) design a alloy combining high strengths with good environmental properties, microstructural stability, heat treatment characteristics and castability.

A low-cost second generation single crystal alloy, designated DD6, has been developed by Beijing Institute of Aeronautical Materials for aeroengine blade applications. Based on the study of single crystal superalloys for many years, with the help of the computer aided design of alloy compositions, the contents of tungsten, molybdenum, tantalum, rhenium and aluminum were judiciously balanced with the predominating rhenium requirement. A large number of trials were carried out, and the composition for DD6 alloy has been determined. The alloy contains 2 wt. % Re, and the alloy system employs the relatively high additive refractory element (W, Mo, Ta, Re and Nb) content of about 19.5 wt. %; this in comparison to DD3 which is about 9.5 wt. %. The Re content of DD6 alloy is about 2/3 of that of other second generation single crystal alloys such as PWA1484, CMSX-4 and René N5. Thus a cost reduction of about 25% for this alloy is expected in China.

The microstructural stability at temperatures above 1000°C was a

Superalloys 2000
Edited by T.M. Pollock, R.D. Kissinger, R.R. Bowman,
K.A. Green, M. McLean, S. Olson, and J.J. Schirra
TMS (The Minerals, Metals & Materials Society), 2000

key concern during the development of DD6. TCP phase formation is typically observed in many single crystal superalloys, especially those containing Re, although a small amount of TCP is not considered detrimental to creep rupture and other properties[19]. DD6 alloy exposed at 1093 ℃ for 1000 hours shows no TCP phases formation and the microstructure of the alloy possesses good stability.

Table I   Nominal compositions of three generations of single crystal superalloys (wt. %)

Element	Cr	Co	Mo	W	Ta	Re	V	Nb	Al	Ti	Hf	B	Ni	Density	Ref
Alloy	First Generation Single Crystals													Kg/cm³	
PWA1480	10	5	–	4	12	–	–	–	5.0	1.5	–	–	Bal.	8.70	1
Rene' N4	9	8	2	6	4	–	–	0.5	3.7	4.2	–	–	Bal.	8.56	2
SRR99	8	5	–	10	3	–	–	–	5.5	2.2	–	–	Bal.	8.56	3
RR2000	10	15	3	–	–	–	1	–	5.5	4.0	–	–	Bal.	7.87	3
AM1	8	6	2	6	9	–	–	–	5.2	1.2	–	–	Bal.	8.59	4
AM3	8	6	2	5	4	–	–	–	6.0	2.0	–	–	Bal.	8.25	5
CMSX-2	8	5	0.6	8	6	–	–	–	5.6	1.0	–	–	Bal.	8.56	6
CMSX-6	10	5	3	–	2	–	–	–	4.8	4.7	0.1	–	Bal.	7.98	7
CMSX-11B	12.5	7	0.5	5	5	–	–	0.1	3.6	4.2	0.04	–	Bal.	8.44	8, 9
CMSX-11C	14.9	3	0.4	4.5	5	–	–	0.1	3.4	4.2	0.04	–	Bal.	8.36	9
DD3	9.5	5	4	5.5	–	–	–	–	5.8	2	–	–	Bal.	8.2	10
	Second Generation Single Crystals														
PWA1484	5	10	2	6	9	3	–	–	5.6	–	0.1	–	Bal.	8.95	11, 12
Rene' N5	7	7.5	1.5	5	7	3	–	–	6.2	–	0.15	0.004	Bal.	8.63	13
CMSX-4	6.5	9	0.6	6	6.5	3	–	–	5.6	1	0.1	–	Bal.	8.70	14
SC180	5	10	2	5	8.5	3	–	–	5.2	1	0.1	–	Bal.	8.84	15
DD6	4.3	9	2	8	7.5	2	–	0.5	5.6	–	0.1	–	Bal.	8.83	16
	Third Generation Single Crystals														
Rene' N6	4.2	12.5	1.4	6	7.2	5.4	–	–	5.75	–	0.15	–	Bal.	8.98	17-19
CMSX-10	2	3	0.4	5	8	6	–	0.1	5.7	0.2	0.03	–	Bal.	9.05	20, 21

DD6 alloy has an approximate 40℃ improved creep strength relative to the first generation single crystal superalloys such as DD3 alloy. The tensile properties and creep rupture properties of DD6 are comparable to those of the second generation single crystal superalloys such as SC180, René N5, CMSX-4 and PWA1484. The yield strength of DD6 alloy at 760°C is approximately 981MPa. The creep rupture lives of DD6 at the conditions of 982°C/248.2MPa, 1070°C/160MPa and 1093°C/124MPa are about 307hours, 178hours and 484hours, respectively. Most notably, the alloy provides superior oxidation resistance and hot corrosion resistance[16]. The superior environmental properties of DD6 are attributable to its higher aluminum content and tantalum content, and the fact that it has optimum compositions. High cycle fatigue tests have been performed. The advantage shown in the creep rupture properties over DD3 generally hold for the fatigue properties.

Casting trials on DD6 alloy have been conducted involving a great number of slabs, bars and some complex shaped hollow turbine blades. These trials demonstrated that DD6 alloy is not prone to the formation of single crystal process defects such as freckles and slivers, and possesses excellent single crystal castability. The MCrAlX coating has been applied to DD6 alloy successfully.

Alloy Design

A common characteristic of the chemical compositions of the second generation single crystal superalloys widely used in the world is 3 wt. % Re[11-15]. Re has very important effects on improving the creep rupture life and it is the most expensive element in single crystal superalloys, so decreasing Re content will reduce alloy cost. The effects of Re on creep rupture life of a single crystal superalloy were studied in the period of DD6 development. It was found in the study that Re can greatly improve the creep rupture life, however the total level of refractory elements such as W, Mo, Nb, Ta and Re, also obviously affects the creep rupture life in single crystal superalloys[22]. Therefore, it is suggested that not only the Re content must be emphasized but also the total refractory element level in the alloy should be selected correctly. The Re content in DD6 alloy system was thereby designed to be about 2 wt. %.

Generally, the second generation single crystal alloys contain 8-9 wt.% of W+Re. On the basis of containing 2 wt. % Re, relatively high W+Re levels are employed to attain the high creep strength exhibited by DD6, particularly at elevated temperatures. Moreover, W and Cr tend to be involved in TCP phase formation. Utilizing high W content requires relatively low chromium alloying.

Since tantalum is a more efficient hardener and has beneficial effect on single crystal castability in reducing alloy freckle formation, plus it positively influences environmental properties, a Ta level of about 7.5 wt. % is employed in DD6 alloy system. Based on the mechanical properties, TCP and microstructural stability, environmental resistance and castability of the candidate alloys, the nominal composition of DD6 alloy was finally selected.

The results of phase calculations of DD6 alloy are listed in Table II. The values of the $\overline{Mdt}$ and $\overline{Nv}$ of DD6 are smaller than the critical values for alloy microstructure stability, respectively[23, 24]. The P value is a parameter which predicts the overall merit of the composition. The compositions with high P values will have high strength in combination with stability, heat treatability and resistance to oxidation and corrosion. For an optimum alloy, the most useful composition currently known by the inventors, the P value will be slightly in excess of 3940[12]. The various parameters in Table II show that DD6 alloy should has high strength, good microstructure stability, heat treatment characteristic and environmental properties.

Table II   Results of the phase calculation of DD6 alloy [12, 23, 24]

Design parameters	$\overline{Mdt}$ of alloy	$\overline{Nv}$ of γ matrix	P Value	Misfit of γ′ and γ, %
Calculated value	0.969	1.869	3965	- 0.09
Critical value	0.985	2.3	3940	–

Alloy Manufacture

Commercially pure raw materials for DD6 alloy were used for experimental study and commercial applications. The master alloy heat of DD6 was melted in a vacuum induction furnace according to the process set down by many experiments. The test bars, slabs and blades of DD6 alloy were cast in a directional solidified vacuum furnace.

Microstructure and Heat Treatment

The heat treatment regime for DD6 alloy was carefully studied to provide acceptable microstructure and optimal mechanical properties. The test results show that DD6 provides attractive solution heat treatment capability. The alloy provides a 20℃ heat treatment window (numerical difference in ℃ between the γ′ solvus and incipient melting point) and is thereby able to be fully solutioned. The heat treatment regime of DD6 alloy consists of a three-step process employing solution, primary aging and secondary aging heat treatments. The solution heat treatment, a 1315℃ for 4 hours cycle, dissolves the coarse as-cast γ′ and eutectic γ-γ′ present in the alloy for reprecipitation into more useful fine γ′. Following the solution heat treatment cycle, the primary aging treatment, a 1120℃ for 4 hours cycle, is used to bond the coating to the alloy, as well as to produce an optimum γ′ size and distribution, and the secondary aging treatment is performed at 870℃ for 32 hours to precipitate ultra fine γ′ between the larger particles.

The microstructures of DD6 alloy in the as-cast, as-solutioned and fully heat treated conditions are shown in Figure 1. The fully heat treated microstructure of DD6 alloy consists of approximately 65 vol. % of cuboidal γ′ precipitates in γ matrix, and the average width of γ′ is about 0.4μm.

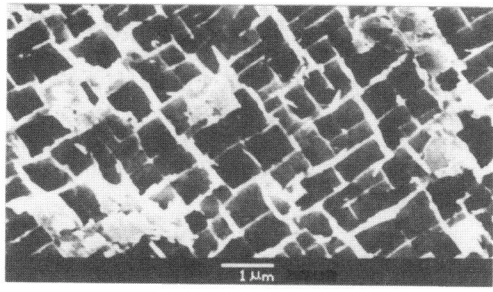

(a)   As-cast.

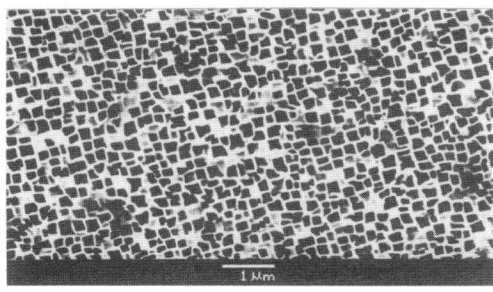

(b)   1315℃/ 4 Hrs./AC.

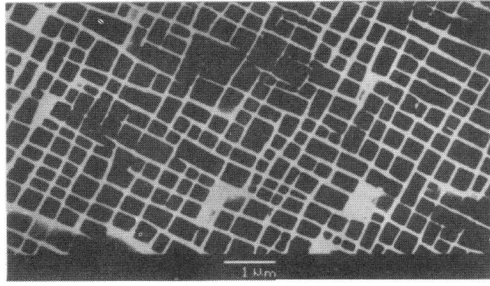

(c) 1315℃/4 Hrs./AC. + 1120℃/4 Hrs./AC. + 870℃/32 Hrs./AC.

(d) 1315℃/4 Hrs./AC. + 1120℃/4 Hrs./AC. + 870℃/32 Hrs./AC.

Figure 1: Microstructures of DD6 alloy in the (a) as-cast, (b) as-solutioned, (c) and (d) fully heat treated.

Microstructural Stability

The microstructural stability at temperatures above 1000℃ was a key concern in the development of DD6 alloy. TCP phase formation is typically observed in many single crystal superalloys, especially those containing Re, although a small amount of TCP is not considered detrimental to creep rupture and other properties[19]. DD6 alloy exposed at 1093℃ for 1000 hours shows no TCP phases formation and the microstructure of the alloy possesses good stability. The microstructural stability of DD6 is maintained by the reduction of chromium content and the increase of cobalt content. It can be found from Figure 2 that coarsening and rafting of γ′ precipitates occurred in the exposed specimens of DD6 alloy.

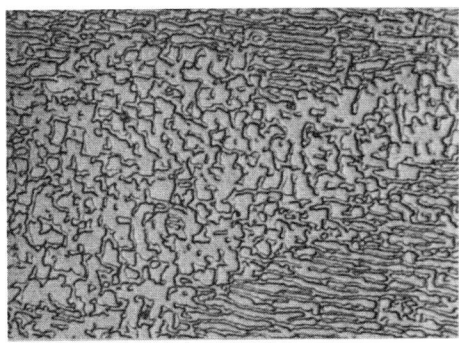

Figure 2: Microstructure of DD6 exposed at 1093℃ for 1000 hours.

Mechanical Properties

Tensile Properties

The tensile properties of DD6 have been extensively characterized in a series of tests between room temperature and 1070℃ and are shown in Figure 3. It can be seen that the ultimate tensile strength and yield strength of DD6 alloy peaks at about 760℃ and 850℃ respectively, while relatively good ductility is maintained throughout the tested regime from room temperature to 1070℃. The ultimate tensile strength and yield strength of DD6 at 760°C are approximately 1164MPa and 981MPa, respectively. Table III shows a comparison of the yield strength of DD6 alloy with other single crystal superalloys. The yield strength of DD6 at room temperature is equivalent to that of the first generation single crystal DD3, while DD6 offers an improvement compared with that of DD3 at middle and high temperatures. The yield strengths of DD6 alloy at 20℃, 650℃ and 980℃ are similar to those of other second generation single crystal alloys.

Table III Yield strength of DD6 alloy and other single crystal superalloys[11, 14, 25, 26]

Alloy	Yield strength, MPa				
	20℃	650℃	850℃	871℃	980℃
PWA1484		965			
CMSX-4	971	947		899	704
DD3	927		862		523
DD6	929	965	1035		721

Creep Rupture Properties

The creep rupture properties of DD6 alloy were studied from 650 ℃ to 1100℃. The creep rupture lives of DD6 alloy at the conditions of 982°C/248.2MPa, 1070°C/160MPa, 1093 ℃ /124MPa and 1100°C/140MPa are about 307 hours, 178 hours, 484 hours and 148 hours, respectively. The creep rupture properties are summarized in Figure 4 showing a Larson Miller Parameter comparison between DD6 and DD3. Figure 5 illustrates the 0.2% creep strength of DD6 and DD3 at 850℃. The second generation single crystal alloys generally exhibit enhanced creep strength relative to the first generation single crystal alloys. DD6 alloy possesses greater creep strength than DD3 alloy from

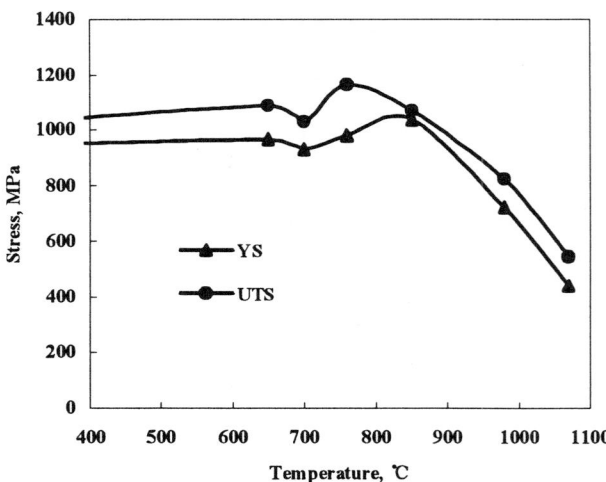

Figure 3: Yield strength and ultimate tensile strength of DD6.

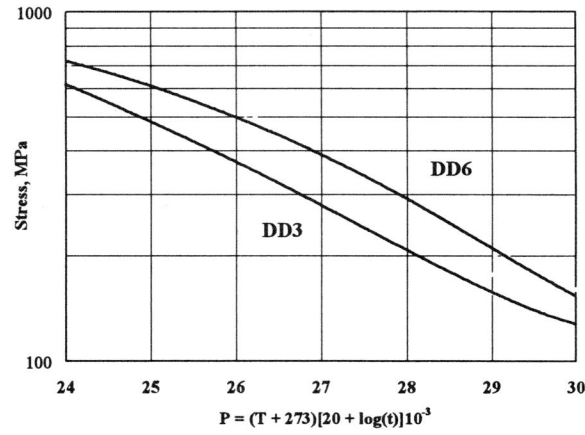

Figure 4: Larson-Miller creep strength of DD6 and DD3[26].

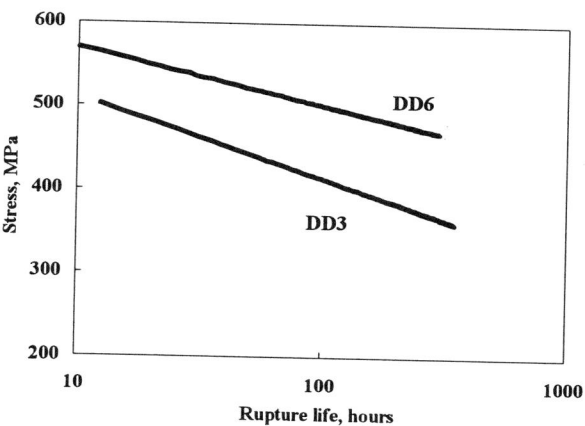

Figure 5: 850℃ 0.2% creep strength of DD6 and DD3[26].

760℃ to 1100℃. The relative improvement depends upon the alloy systems compared, the advantage of DD6 alloy in temperature capability is about 40℃ relative to DD3 alloy. The 40℃ advantage over DD3 alloy is maintained even at the higher temperatures and longer times due to the good microstructural stability of DD6 alloy. The high creep rupture properties of DD6 alloy are mainly attributable to the refractory element contents.

The creep rupture lives of DD6 and other second generation single crystal superalloys at 982 ℃ /248.2MPa and 1093 ℃ /124MPa are shown in Figure 6 and Figure 7, respectively. The creep strengths of DD6 alloy are at least equivalent to those of other second generation single crystal superalloys widely used in the world, such as CMSX-4, Rene'N5 and SC180, even better in certain tests.

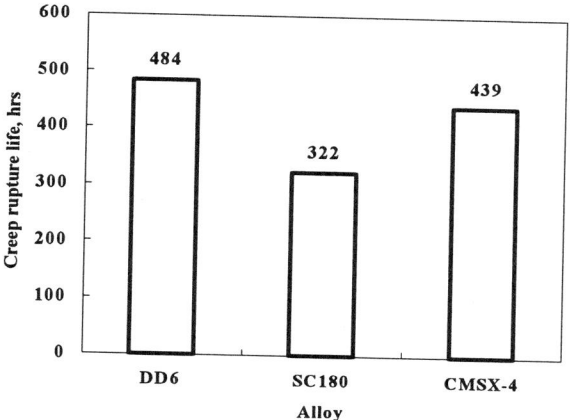

Figure 7: Creep rupture lives of DD6 and other second generation single crystal superalloys at 1093℃/124MPa[14, 15].

Fatigue

High cycle fatigue tests have been performed for DD6 alloy. The advantage shown in the creep rupture properties over DD3 generally hold for the fatigue properties. The smooth high cycle fatigue life of DD6 at 700℃ in the uncoated condition is shown in Figure 8. For 10^7 cycle smooth high cycle fatigue (HCF), DD6 alloy provides a 20MPa advantage relative to DD3 alloy at 700℃ in the uncoated condition. DD6 alloy exhibits excellent HCF capability overall especially at higher temperatures (between 800-950℃). The low cycle fatigue properties for DD6 alloy are being studied.

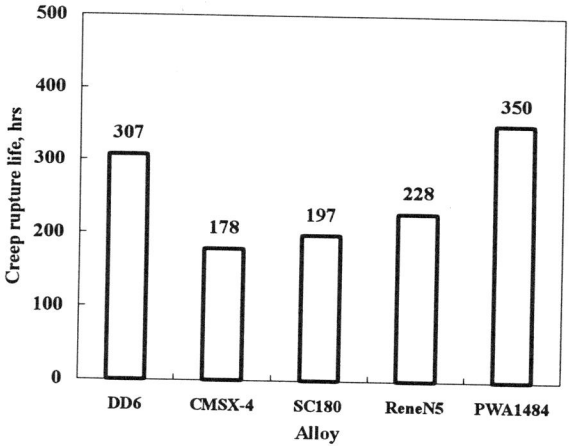

Figure 6: Creep rupture lives of DD6 and other second generation single crystal superalloys at 982℃/248.2MPa[11, 13-15].

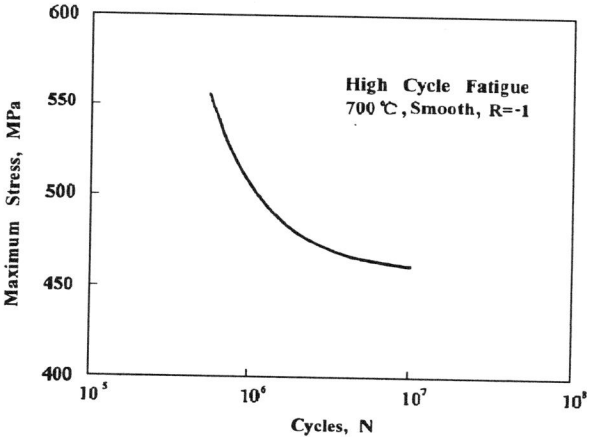

Figure 8: High cycle fatigue of DD6 at 700℃ in uncoated condition.

Environmental Properties

Environmental tests performed on superalloys often give rise to varied results, depending on methods employed. Tests of the environmental resistance of DD6 alloy were undertaken by both burner rig and crucible (static) evaluation methods. The tests of the oxidation resistance of DD6 alloy were performed at 1000°C and 1100°C. Weight change rates of the oxidation resistance tests of DD6 and DD3 alloy at 1100°C for 100 hours, in the uncoated conditions by the crucible method, are shown in Figure 9[16, 26]. The oxidation resistance of DD3 alloy is low, while DD6 alloy exhibits very good oxidation resistance for the duration of the test, ie., 100 hours, which is about 4-6 times improvement relative to DD3. The excellent oxidation resistance of DD6 is attributable to its Ta content compared to DD3 that contains no Ta, and the fact that it contains no Ti, despite the Cr level of DD6 being 50% lower than DD3.

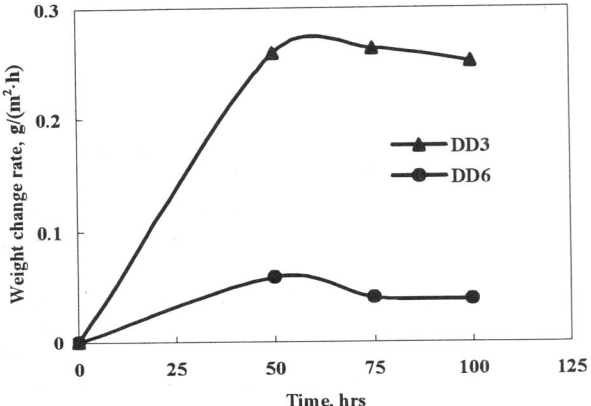

Figure 9: Oxidation resistance properties of DD6 and DD3 at 1100°C on uncoated specimens by crucible method[16, 26].

Table IV lists the hot corrosion resistance properties of DD6 and DD3 alloy at 900°C for tested 100 hours in uncoated conditions by the burner rig method [16, 26]. The actual test results are presented in terms of the weight change rates of the specimen. In this table, the salt content in the hot corrosion resistance test for DD6 is 0.002% while that of DD3 is 0.0005%, but the weight change rate of DD6 is 10 times improvement compared to DD3. The specimen appearances tested for the hot corrosion resistance of DD6 and DD3 at 900°C for 100 hours, in uncoated conditions at 0.002% salt content by burner rig method, are shown in Figure 10.

Table IV   Hot corrosion resistance of DD6 and DD3
at 900°C for 100 hours [16, 26]

Alloy	DD6	DD3
Weight change rate, g/(m²·h)	0.053	0.595

* The salt content in the hot corrosion resistance test of DD6 is 0.002%, while that of DD3 is 0.0005%.

Therefore, DD6 alloy provides surprisingly good oxidation characteristics, as confirmed through both burner rig and crucible tests. At the same time, similar multiple source testing confirms that DD6 exhibits excellent hot corrosion capabilities.

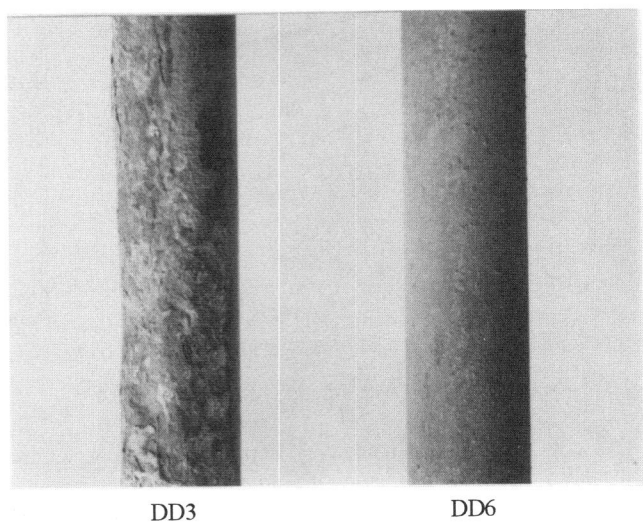

DD3   DD6

Figure 10: Appearances of the specimens of the hot corrosion resistance for DD6 and DD3 tested by burner rig method at 900°C for 100 hours, in uncoated condition at 0.002% salt content.

The MCrAlX coating was applied to DD6 alloy successfully. For a 10^7 cycle at 700°C, the smooth high cycle fatigue life of DD6 alloy with MCrAlX coating is approximately equal to that of the alloy in uncoated condition. The applications of environmental coating may further improve the oxidation and hot corrosion resistance of DD6 alloy.

Castability

Casting trials have been conducted on DD6 alloy involving a great number of test bars, test slabs and some complex shaped hollow turbine blades. The results of the detailed inspections for the crystal quality and casting yields obtained in these trials demonstrate that DD6 possesses excellent castability.

The castability of DD6 alloy is enhanced by alloy design. It is considered that addition of Ta in alloy has beneficial effect on single crystal casting process in reducing alloy freckle formation[21]. Thus significant Ta element is employed to DD6 alloy and the Ta:W ratio was regulated for beneficial effects.

Summary

A low-cost second generation single crystal superalloy, designated DD6, has been developed. With the help of computer aided design alloy composition, the contents of tungsten, molybdenum, tantalum and rhenium were judiciously balanced with the predominating rhenium requirement. The alloy contains 2 wt.% of Re, only 2/3 of other second generation single crystal superalloys, such as PWA1484, CMSX-4 and Rene′N5. Therefore, a significant cost reduction for DD6 alloy is expected.

DD6 alloy has an approximate 40 ℃ improvement of creep strength relative to the first generation single crystal alloy DD3. The tensile properties, creep rupture properties of DD6 alloy are comparable to those of other second generation single crystal superalloys used in commercial applications. DD6 alloy also provides excellent oxidation resistance and hot corrosion resistance. In addition, the microstructural stability of DD6 at elevated temperature after long term exposure is maintained. DD6 alloy also has satisfactory heat treatment characteristics, good single crystal castability and coatability.

References

1. M. Gell, D. N. Duhl and A. F. Giamei, "The Development of Single Crystal Superalloy Turbine Blades," Superalloys 1980, (Warrendale, PA: TMS, 1980), 205-214.
2. J. W. Holmes and K. S. O′Hara, ASTM STP 942, (Philadelphia, PA: ASTM, 1988), 672-691.
3. D. A. Ford and R. P. Arthey, " Development of Single Crystal Alloys for Specific Engine Application," Superalloys 1984, ed. M. Gell et al., (Warrendale, PA: TMS, 1984), 115-124.
4. E. Bachelet and G. Lamanthe, (Paper Presented at the National Symposium on SX Superalloys, Viallard-de-Lans, France, 26-28 February 1986).
5. T. Khan and M. Brun, (Paper Presented at the Symposium on SX Alloys, Munich, Germany, MTU/SMCT, June 1989).
6. K. Harris and G. L. Erickson, Cannon-Muskegon Corporation, U. S. patient 4,582,548–CMSX-2 Alloy.
7. K. Harris and G. L. Erickson, Cannon-Muskegon Corporation, U. S. patient 4,721,540–CMSX-6 Alloy.
8. G. L. Erickson, Cannon-Muskegon Corporation, U. S. patient 5,489,346–CMSX-11B Alloy.
9. G. L. Erickson, "The Development of the CMSX-11B and CMSX-11C Alloys for Industrial Gas Turbine Application," Superalloys 1996, ed. R. D. Kissing et al., (Warrendale, PA: TMS, 1996), 45-52
10. Wu Zhongtang, Wen Zhongyuan and Chen Dehou, "Composition Design and Experimental Study of Single Crystal Alloy DD3," Acta Metallurgica Sinica, 23(4) (1987), B171-B178.
11. A. D. Cetel and D. N. Duhl, "Second Generation Nickel-Base Single Crystal Superalloy," Superalloys 1988, ed. S. Reichman et al., (Warrendale, PA: TMS, 1988), 235-244.
12. D. N. Duhl and A. D. Cetel, United Technologies Corporation, U. S. patient 4,719,080– PWA1484 Alloy.
13. C. S. Wukusick and L. Buchakjian, U. K. Patent Appl. GB2235697, "Improved Property Balanced Nickel-base Superalloys for Producing Single Crystal Articles"–Rene′ N5 Alloy.
14. K. Harris and G. L. Erickson, Cannon-Muskegon Corporation, U. S. patient 4,643,782–CMSX-4 Alloy.
15. X. Nguyen-Dinh, Allied-Signal, Inc., U. S. patent 4,935,072–SC 180 Alloy.
16. Li Jiarong, Zhong Zhengang, Tang Dingzhong et al., "Low Cost Second Generation Single Crystal Superalloy DD6," Acta Metallurgica Sinica, 35(Suppl.2) (1999), S266-S269.
17. W. S. Walston, E. W. Ross, T. M. Pollock and K. S. O′Hara, General Electric Company, U. S. patent 5,455,120–Rene′ N6 Alloy
18. W. S. Walston, E. W. Ross, K. S. O′Hara and T. M. Pollock, General Electric Company, U. S. patent 5,270,123–Rene′ N6 Alloy.
19. W. S. Walston, K. S. O′Hara, E. W. Ross, T. M. Pollock and W. H. Murphy, "Rene′N6: Third Generation Single Crystal Superalloy," Superalloys 1996, ed. R. D. Kissinger et al., (Warrendale, PA: TMS, 1996), 27-34.
20. G.L.Erickson, Cannon-Muskegon Corporation, U.S. patent 5,366,695–CMSX-10 Alloy.
21. G.L.Erickson, "The Development and Application of CMSX-10," Superalloys 1996, ed. R. D. Kissinger et al., (Warrendale, PA: TMS, 1996), 35-44.
22. Jiarong Li, Dingzhong Tang, Riling Lao et al., "Effects of Rhenium on Creep Rupture Life of a Single Crystal Superalloy," Journal of Materials Science and Technology, 15(1) (1999), 53-57.
23. K. Matsugi, Y. Murata and M. Morinaga, "An Electronic Approach to Alloy Design and Its Application to Nickel-Based Single Crystal Superalloys," Materials Science and Engineering, 172A(1993), 101-110.
24. R. G. Barrows and J. B. Newkirk, "A Modified System for Predicting σ Formation," Metallurgical Transactions, 3(11) (1972), 2889-2893.
25. A.Sengupta, S.K.Putatunda, L.Bartosiewicz et al., "Tensile Behavior of a New Single Crystal Nickel-Based Superalloy (CMSX-4) at Room and Elevated Temperatures," Journal of Materials Engineering and Performance, 3(5) (1994), 664-672.
26. Wu Zhongtang, "DD3 alloy," Chines Aeronautical Materials Handbook, vol. 2, ed. Yan Minggao, (Beijing, China Standard Press, 1989), 868-881.

THE DEVELOPMENT OF IMPROVED PERFORMANCE PM UDIMET® 720 TURBINE DISKS

S. K. Jain, B. A. Ewing, C. A. Yin

Rolls-Royce Corporation, Indianapolis, Indiana

Abstract

Udimet® 720 is an important alloy because it exhibits an outstanding balance of strength, temperature, and defect tolerance characteristics (Ref. 1). In its cast and wrought alloy form, it is employed for turbine disk components utilized in a large number of civil and military propulsion systems. In its powder metal (PM) alloy form, which has been shown to be very cost competitive with its cast and wrought counterpart, Udimet 720 exhibits superior alloy homogeneity that provides an opportunity to develop the uniform and controlled microstructures desired for advanced designs (Ref. 2). Because it was recognized that Udimet 720 compositions developed for cast and wrought applications were not necessarily the best for PM processing, it was decided to study PM Udimet 720 chemistry, processing, and mechanical property relationships. The goal of this effort was to determine if an improved balance of performance characteristics could be developed for advanced turbine disk applications. For this work, four Udimet 720 chemistry modifications involving boron, zirconium, and hafnium additions were made to the baseline composition. Also represented in the program, for reference, was a contemporary baseline PM Udimet 720 composition.

To produce materials for evaluation, powder representative of each of the five compositions was argon atomized and screened to a -150 mesh powder fraction and consolidated by hot isostatic pressing (HIP) technique into 30 lb billet preforms. The HIP processed billets were then isothermally forged at two different forging temperatures into pancakes. After forging, selected pancakes were subsolvus and supersolvus solution heat treated to achieve target grain sizes, respectively, of ASTM 11 and ASTM 9. Cut-up evaluations were then performed on fully heat treated disks representative of each grain size and chemistry combination. This included tensile, creep rupture, fatigue crack growth (FCG), and low cycle fatigue (LCF) testing at temperatures and conditions of interest for gas turbine disk applications.

Test results showed supersolvus processing to an ASTM 9 target grain size resulted in attractive reductions in 1200°F FCG rate relative to both the subsolvus processed ASTM 11 PM alloys and cast and wrought Udimet 720 disk product subsolvus processed. Significant improvements in both 1200°F LCF life and 1350°F creep rupture capability were also achieved for material processed to the ASTM 9 grain size. It was observed that the higher boron and boron/zirconium levels in the coarse grained PM alloys enhanced these properties. By comparison, tensile testing did not show trends that could be attributed to the chemistry modifications, but 0.2% tensile yield strengths for the ASTM 9 material were reduced on the order of 10 ksi relative to the ASTM 11 material.

From this work it was found that selected chemistry modifications, when combined with appropriate grain size control, can yield an improved balance of FCG resistance, LCF capability, creep strength, and tensile properties relative to the baseline PM alloy as well as traditional cast and wrought Udimet 720 that is subsolvus processed. In particular, this work has shown promise for PM Udimet 720 alloy forms exhibiting boron and boron/zirconium levels that exceed baseline PM and cast and wrought Udimet 720 target levels. Based on these results, it should be possible to greatly extend the performance capabilities of PM Udimet 720 turbine disks.

*® Udimet is a registered trademark of Special Metals Corporation.

Superalloys 2000
Edited by T.M. Pollock, R.D. Kissinger, R.R. Bowman,
K.A. Green, M. McLean, S. Olson, and J.J. Schirra
TMS (The Minerals, Metals & Materials Society), 2000

Introduction

Early cast and wrought Udimet 720 compositions used for disk forgings were susceptible to chemical segregation, causing wide variability in grain size and heat treatment response. As a result, extensive homogenization and billet working practices had to be developed to improve alloy homogeneity. Problems were also encountered with the formation of boride and carbide stringers that can act as nucleation sites, which lead to early fatigue cracking and premature component failure. These stringer problems were related to the alloying constituents required to promote grain boundary strengthening.

Due to these difficulties, several changes were made to the melt practices and chemistry used for Udimet 720 disk components. As an example, melt practices were changed from a vacuum induction melt plus vacuum arc remelt double melt practice to a vacuum induction melt plus electroslag remelt plus vacuum arc remelt triple melt practice to improve cleanliness and structure. Elements that can lead to stringer formation were also adjusted; specifically, carbon and boron were reduced. Due to the lower carbon levels, chromium content was reduced as well to develop a chromium/carbon balance to minimize, if not eliminate, the formation of a deleterious chromium rich sigma phase. A comparison of first and current generation cast and wrought compositions reflecting these changes is presented in Table I.

Overall, the cast and wrought processing methods developed for Udimet 720 have been very successful in managing the segregation related issues associated with the manufacture of relatively small cast and wrought disks. These disks typically require forging multiple weights of up to 150 lb, and utilize 4 to 6-in. diameter billet. As a result, 4 and 6-in. diameter cast and wrought Udimet 720 billet is routinely being used to produce turbine disk forgings used in a number of Rolls-Royce Corporation engine applications. These include AE 3007 turbofan engines powering the Embraer RJ 135/145 and Citation X aircraft, the AE 2100 turboprop engine powering the C-130J and Saab 2000 aircraft, the AE 1107 turboshaft engine used on the V-22 Osprey tiltrotor, and the Light Helicopter Turbine Engine Company (LHTEC) T800 turboshaft engine powering the Comanche RAH-66 helicopter.

Unfortunately, as disk size increases beyond that accommodated by 6 to 8-in. diameter billet, segregation problems and related microstructural control issues become significantly more difficult to manage. For example, turbine disks for turbofan engines powering modern wide-bodied aircraft require large 1000-lb 12-in. diameter billet. As a result of increased billet size, the microstructural control and quality levels desired are very difficult to achieve using conventional cast and wrought methods

and processing complexity is dramatically increased. For these applications the use of PM processing affords the opportunity to produce cost-effective, high quality homogeneous billet product with uniform microstructural characteristics (Ref. 3).

Because of the opportunities PM billet processing provides for the improved microstructural control demanded by new disk designs, as well as the recognition that the compositions developed for cast and wrought applications were not necessarily the best for PM processing, it was decided to study PM Udimet 720 chemistry, processing, and mechanical property relationships. The goal of this effort was to determine if an improved balance of performance characteristics could be developed for advanced turbine disk applications.

To achieve an improved balance of defect tolerance, LCF resistance, creep strength, and tensile strength relative to cast and wrought Udimet 720, the work focused on two key aspects of disk manufacturing: a) chemistry modifications and b) isothermal forging/heat treat practices. The approach involved evaluating subscale disk forgings produced by a HIP consolidation and isothermal forging processing route. Crucible Materials Corporation produced and consolidated the powders and the Ladish Company produced and heat treated the forgings to target grain sizes of ASTM 9 and ASTM 11.

Udimet 720 Chemistry Modification

For this work, four Udimet 720 chemistry modifications involving boron, zirconium, and hafnium additions were made to the baseline composition (Table II). Also represented in the program, for reference, was a standard PM Udimet 720 composition (Alloy 1). Of the four compositions, Alloy 2 was designed to evaluate hafnium effects relative to the baseline Alloy 1, whereas Alloys 3 and 5 were formulated to evaluate the effect of higher boron. Alloy 4 was designed to evaluate the effect of increased boron and zirconium. This alloying approach was taken as a result of previous work performed by both Rolls-Royce Corporation and Rolls-Royce plc.

Powder Production and HIP Compaction

The powder for the program was produced at the Crucible Research Division of the Crucible Materials Corporation by conventional vacuum induction melting and argon atomization. Two heats were made for each composition. The heats were analyzed individually for major alloying elements and then blended. Powder characterization included scanning electron microscope (SEM) examination of the surface characteristics and optical examination of the as-atomized powder microstructures.

Table I. Evolution of cast and wrought Udimet 720 alloy chemistries.

Alloy	Alloy compositions (weight percent)										
	C	Cr	Co	Mo	Ti	Al	W	Zr	B	Hf	Ni
First generation	0.035	18.0	14.75	3.0	5.0	2.5	1.25	0.035	0.035	-	Bal.
Current generation	0.015	16.0	14.75	3.0	5.0	2.5	1.25	0.035	0.017	-	Bal.

Table II. Target alloy compositions for PM Udimet 720 alloy/process modification program.

Alloy	Target alloy compositions (weight percent)										
	C	Cr	Co	Mo	Ti	Al	W	Zr	B	Hf	Ni
Alloy 1 (baseline Udimet 720)	0.025	16.0	14.75	3.0	5.0	2.5	1.25	0.035	0.02	-	Bal.
Alloy 2	0.025	16.0	14.75	3.0	5.0	2.5	1.25	0.035	0.02	0.75	Bal.
Alloy 3	0.025	16.0	14.75	3.0	5.0	2.5	1.25	0.035	0.03	-	Bal.
Alloy 4	0.025	16.0	14.75	3.0	5.0	2.5	1.25	0.070	0.03	-	Bal.
Alloy 5	0.025	16.0	14.75	3.0	5.0	2.5	1.25	0.035	0.04	-	Bal.

Water elutriation was also conducted on -150 mesh powder from each of the compositions. The results of these evaluations indicated chemistry requirements were satisfied. SEM and water elutriation results were acceptable and considered as typical of Crucible's powder product produced by their laboratory equipment.

Nominal inside dimensions for the steel HIP containers were 4 5/8 in. diameter by 8 3/8 in. long. The cans filled with -150 powder were first outgassed at room temperature and then heated to 350°F under vacuum the ensure removal of water vapor. After outgassing, the cans were sealed off by pressure welding the stems and tungsten inert gas (TIG) welding the ends. The parameters used for HIP consolidation of the compacts were 2065°F and 15 ksi for 4 hr. Following HIP, the can material was removed, compacts were centerless ground, and were ultrasonically inspected. No anomalies were found in either the ultrasonic inspection results or the microstructural reviews conducted on the material.

The thermal induced porosity (TIP) response of the baseline Udimet 720 composition was determined by exposing the 50 g samples at temperature for 1 hr and measuring densities. The exposure at 2160°F resulted in a density decrease of 0.03 to 0.10% and at 2260°F resulted in a density decrease of 0.48 to 0.55%. These results are typical of -150 mesh powder.

Forging and Heat Treatment Development

Prior to isothermal forging of the HIP consolidated compacts, the forging characteristics of as-HIP material were evaluated by performing compression tests on baseline Udimet 720 material. The testing involved isothermally upsetting 3/8-in. diameter by 5/8-in. button specimens at six temperatures (Table III). The strain rate used for all tests was 0.003/in./in./sec. All specimens were forged to 70% upset. Selection of these parameters was based on previous Ladish experience with PM superalloys. Following upset, the specimens were cut into four quadrants. One was saved to preserve the forged microstructure and two were, respectively, solution heat treated to temperatures 115°F (2000°F) and 50°F (2065°F) below the 2115°F gamma prime solvus temperature for this material. One of the quadrants was also solution heat treated to 25°F (2140°F) above the gamma prime solvus temperature. All four quadrants were prepared for microstructural and grain size evaluations. Grain sizes varied from as fine as ASTM 13 to as coarse as ASTM 6 (Table III).

In general, there was little significant influence of forging/upset temperature on as-forged grain size. Relative to heat treatment, there was little grain size difference between the as-forged structure and the 2000°F heat treated material. Comparing the 2000 to the 2065°F heat treatment showed coarsening into the grain size regime of interest. Comparing the 2065°F heat treated

material to the 2140°F material showed additional coarsening, with material from the lower forging temperatures showing the best overall coarsening response.

Following a review of the microstructures and grain sizes obtained, two forging temperatures were selected for further study. Specifically selected were 2065°F, which is near the 2115°F solvus temperature, and 1990°F, which is near the current forging temperature. The 2065°F forgings were solutioned at 2065°F to obtain a target subsolvus grain size of ASTM 11 and the 1990°F forgings at 2140°F to obtain a target grain size of ASTM 9. Interest in these grain sizes was based upon previous work that showed an attractive balance of mechanical properties in forgings processed to these grain size conditions.

For each alloy, one 9-in. diameter by 1.25-in. thick pancake forging was made at 2065°F and another at 1990°F. Forging response of the alloys was generally good. Forgings made at 2065°F received a subsolvus solution heat treatment of 2065°F/2 hr and forgings made at 1990°F received a supersolvus solution heat treatment of 2140°F/4 hr. Cooling from solution heat treatments was by fan air quenching to develop an approximate cooling rate of 275°F/min. Further heat treatment included stabilization at 1400°F/8 hr/air cooled (AC) and aging at 1200°F/24 hr/AC for both subsolvus and supersolvus materials. Subsequently, it was determined that the subsolvus processing resulted in a relatively uniform grain size of ASTM 11 and the supersolvus processing resulted in a relatively uniform grain size of ASTM 9.

Mechanical Property Evaluations

Since the purpose of this investigation was to examine the effects of boron, zirconium, and hafnium additions and thermal-mechanical processes on the mechanical property behavior of PM Udimet 720, a wide range of mechanical property tests were performed. As shown in Table IV, these included tensile, creep rupture, LCF, and FCG rate tests at temperatures and conditions of interest for gas turbine disk applications. By selecting these conditions, it was also possible to compare the results to cast and wrought Udimet 720 properties from a Rolls-Royce Corporation database for full-scale disks representative of several production heats that have been subsolvus processed to a predominant ASTM10 grain size condition and oil quenched to maximize tensile capabilities.

Average tensile results of duplicate tests for subsolvus and supersolvus PM alloys are listed in Tables V and VI, respectively, along with cast and wrought Udimet 720 properties. Tensile testing did not show trends that could be attributed to the chemistry modifications; however, 0.2% tensile yield strengths of the supersolvus ASTM 9 materials were on the order of 10 ksi lower than subsolvus ASTM 11 material. The lower tensile yield strengths offered by supersolvus coarse grained material is an expected superalloy material behavior. By comparison to subsolvus processed cast and wrought Udimet 720, the subsolvus processed PM alloys generally exhibited slightly lower yield

Table III. As-forged and heat treated PM Udimet 720 grain sizes.

Forging* temp—°F	Predominant ASTM grain sizes as a function of heat treatment			
	As-forged	2000°F/ 2 hr	2065°F/ 4 hr	2140°F/ 2 hr
2085	12	11	10.5	6
2065	12	12	11	6
2040	13	12.5	11	7.5
2015	13	13	11.5	8.5
1990	12	12.5	11	8.5
1965	12	12	11.5	8.5

*Gamma prime solvus temperature is 2115°F

Table IV. Test Matrix for PM Udimet 720 alloy modifications.

Property	Test conditions
Tensile	Room temperature (RT), 800°F, and 1200°F
0.2% creep-rupture	1250°F/115 ksi and 1350°F/70 ksi
FCG	800°F and 1200°F/R = 0.05/with and without a 5- min dwell
LCF	800°F,R = 0.0,strain range = 0.9% 1200°F,R = 0.0, strain range = 0.8%

strengths. This is attributable to the faster cooling rates associated with the manufacture of the disks used to generate the cast and wrought database.

Creep capability of the candidate compositions was evaluated by testing duplicate specimens and comparing the time to 0.2% creep strain at 1250°F/115 ksi and 1350°F/70 ksi test conditions (Figures 1 and 2). As expected, the coarser grain supersolvus processed alloys showed superiority over the finer grain subsolvus processed alloys. It ranged from an approximate 4X advantage in life at the 1250°F test condition to as much as 30X advantage in life at the 1350°F test condition. Relative to cast and wrought performance, the subsolvus processed PM alloys exhibited creep characteristics that were slightly better at 1250°F and comparable at 1350°F. From a PM alloy compositional viewpoint, the subsolvus processed material appeared to be relatively insensitive to compositional variations irrespective of test conditions. However, for the supersolvus processed material there appeared to be a significant influence of boron and boron/zirconium additions on creep life for the specimens tested at 1350°F. The improved creep capability at 1350°F for supersolvus processed Alloys 3, 4, and 5 is likely associated with the beneficial effects of these additions on grain boundary strength. With respect to hafnium, its addition appeared to be most beneficial to stress rupture life of subsolvus material tested at both 1250 and 1350°F (Figure 3). Tests on supersolvus treated material were discontinued after 0.2% creep strain, therefore, stress rupture lives could not be plotted.

The elevated temperature LCF capability of the candidate compositions was evaluated by comparing the cycles to crack initiation on four to six specimens tested at a strain range of 0.9% with 20 cpm and R = 0 at 800°F (Figure 4) and for a strain range of 0.8% with 20 cpm and R = 0 at 1200°F (Figure 5). The results of the LCF testing showed a significant effect of both grain size and composition at both 800 and 1200°F test conditions.

In particular for the 800°F condition, it was shown that LCF lives for the subsolvus processed material were superior to those for supersolvus processed material. In addition, at the 800°F test condition, a trend developed for the subsolvus processed material in which LCF lives appeared to increase with increasing boron and boron/zirconium content (Alloys 1, 3, and 4), with a peak in performance being reached for Alloy 4 (0.03B/0.07Zr). However, at 800°F this effect was not evident in the supersolvus processed material. Relative to subsolvus processed cast and wrought material, the 800°F capabilities of the PM alloys were very attractive, with subsolvus processed Alloy 4 exhibiting an approximate 6X advantage in cyclic life.

Relative to the 1200°F testing, it was shown that the supersolvus processed alloys generally showed higher cyclic lives than the subsolvus processed alloys. This was the opposite of what was observed in the 800°F testing. One exception to this was Alloy 2, the hafnium bearing alloy. After reviewing individual specimen results and identities, it is still unclear as to why the supersolvus form of this alloy did not perform better. The 1200°F testing also showed trends similar to those observed in the 800°F testing in that cyclic life appeared to increase as a function of increasing boron and boron/zirconium content to a peak in performance for Alloy 4 (0.03B/0.07Zr). However, the alloying trend was far more pronounced in the supersolvus processed material than in

Table V. Subsolvus heat treated tensile data.

Forging serial no.		C/W Udimet 720	PM Alloy 1 Udimet 720	PM Alloy 2	PM Alloy 3	PM Alloy 4	PM Alloy 5
Solution heat treat ASTM grain size		Subsolvus 10	Subsolvus 11	Subsolvus 11	Subsolvus 11	Subsolvus 11	Subsolvus 11
Temp—°F	Properties						
RT	U.T.S.	233.9	239.3	243.3	240.3	240.0	240.3
RT	0.2% Y.S.	175.9	170.7	170.8	170.5	170.7	170.7
RT	% El.	17.1	19.0	19.0	20.7	18.0	19.3
RT	% RA	18.4	20.7	20.0	22.3	18.3	21.0
800	U.T.S.	224.4	226.0	228.5	226.5	227.5	226.0
800	0.2% Y.S.	167.6	163.5	164.0	164.0	164.0	163.5
800	% El.	18.2	18.0	20.0	20.5	18.0	19.5
800	% RA	18.7	20.0	23.0	23.5	21.0	21.5
1200	U.T.S.	214.0	198.7	202.7	199.7	199.3	200.3
1200	0.2% Y.S.	165.6	155.3	157.4	156.5	156.5	157.0
1200	% El.	18.1	21.3	27.3	31.0	19.0	17.7
1200	% RA	19.4	22.0	26.0	32.0	20.0	19.0

Table VI. Supersolvus heat treated tensile data.

Forging serial no.		PM Alloy 1 Udimet 720	PM Alloy 2	PM Alloy 3	PM Alloy 4	PM Alloy 5
Solution heat treat ASTM grain size		Supersolvus 9	Supersolvus 9	Supersolvus 9	Supersolvus 9	Supersolvus 9
Temp—°F	Properties					
RT	U.T.S.	230.0	234.7	229.3	229.7	230.0
RT	0.2% Y.S.	160.7	159.0	158.4	158.7	157.7
RT	% El.	15.3	17.3	17.3	17.0	17.3
RT	% RA	17.3	18.7	18.7	17.3	19.3
800	U.T.S.	219.0	221.5	221.0	220.0	217.5
800	0.2% Y.S.	155.5	154.0	155.5	155.0	155.5
800	% El.	13.5	16.5	15.5	16.0	16.5
800	% RA	19.5	18.5	16.5	17.0	16.0
1200	U.T.S.	207.0	207.3	205.7	207.3	207.7
1200	0.2% Y.S.	146.0	145.3	143.7	146.3	145.7
1200	% El.	25.3	31.0	28.0	23.7	23.3
1200	% RA	25.7	29.7	27.7	24.3	25.0

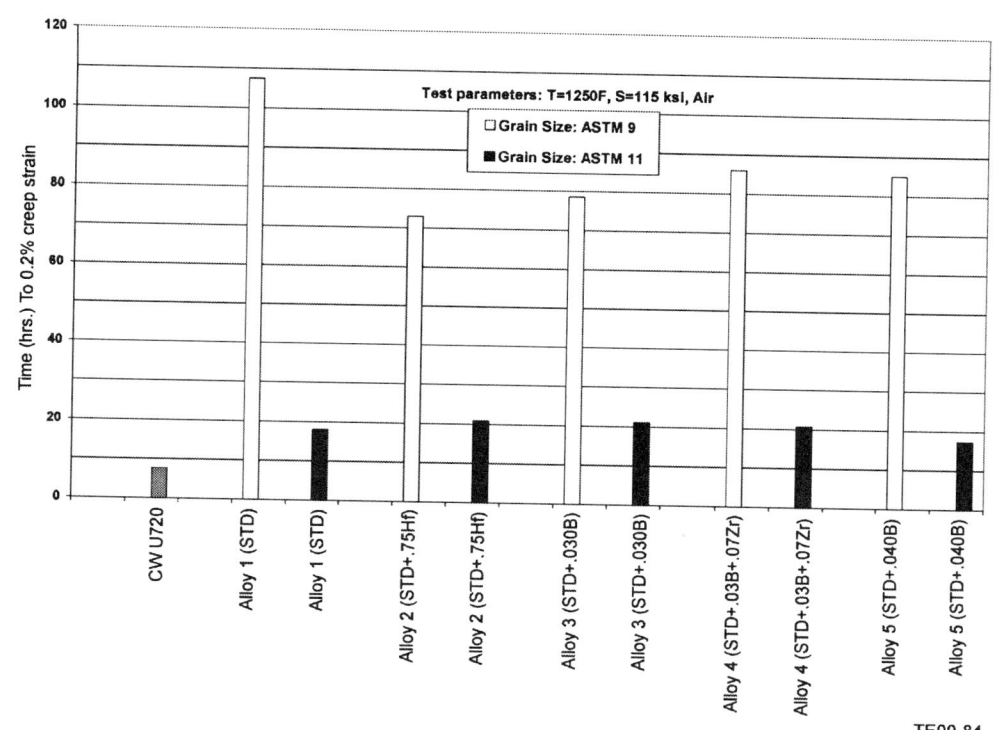

Figure 1. Effect of PM Udimet 720 alloy modifications and solution temperatures (grain sizes ASTM 9 and 11) on 1250°F/115 ksi creep behavior.

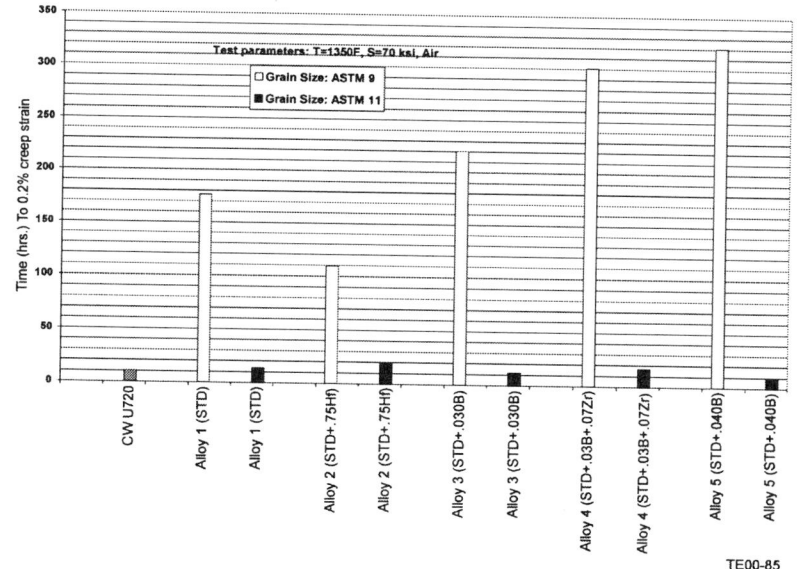

Figure 2. Effect of PM Udimet 720 alloy modifications and solution temperatures (grain sizes ASTM 9 and 11) on 1350°F/70 ksi creep behavior.

the subsolvus material. By comparison to subsolvus processed cast and wrought Udimet 720, all of the PM compositions offered superior lives, with the greatest advantage being noted for supersolvus processed materials at 1200°F. For example, supersolvus processed Alloy 4 exhibited 70,000 cycles average life versus 11,000 cycles average life for subsolvus processed cast and wrought Udimet 720.

To develop an improved understanding of the LCF test results, the fracture surfaces of failed LCF bars were examined; they generally showed the 800°F bars to be transgranular in nature and the 1200°F bars to be intergranular in failure mode. As might be expected, there were occasional failure origins associated with the presence of ceramic particles, although little scatter was noted on the multiple tests conducted.

To evaluate the crack growth rate capability of the candidate compositions, 1200°F fatigue crack growth tests were conducted on compact tension specimens with 5-min dwell and without dwell at peak load (Figures 6 and 7) and without dwell at 800°F

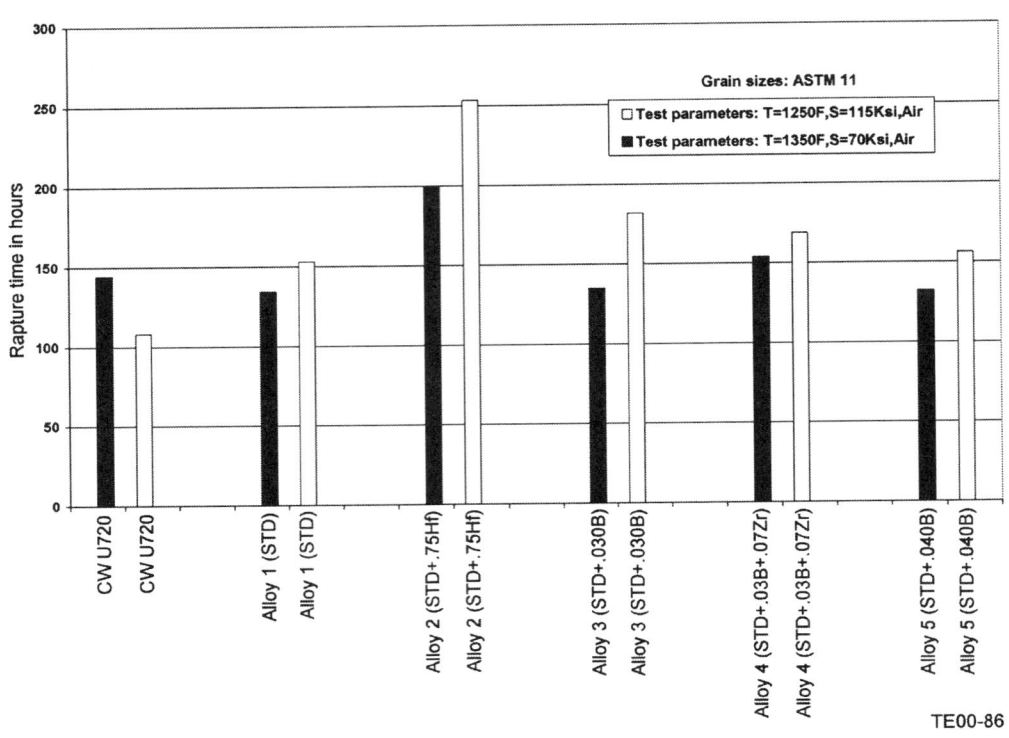

Figure 3. Effect of PM Udimet 720 alloy modifications on 1250°F/115 ksi and 1350°F/70 ksi stress rupture behavior of subsolvus heat treated (grains size ASTM 11) material.

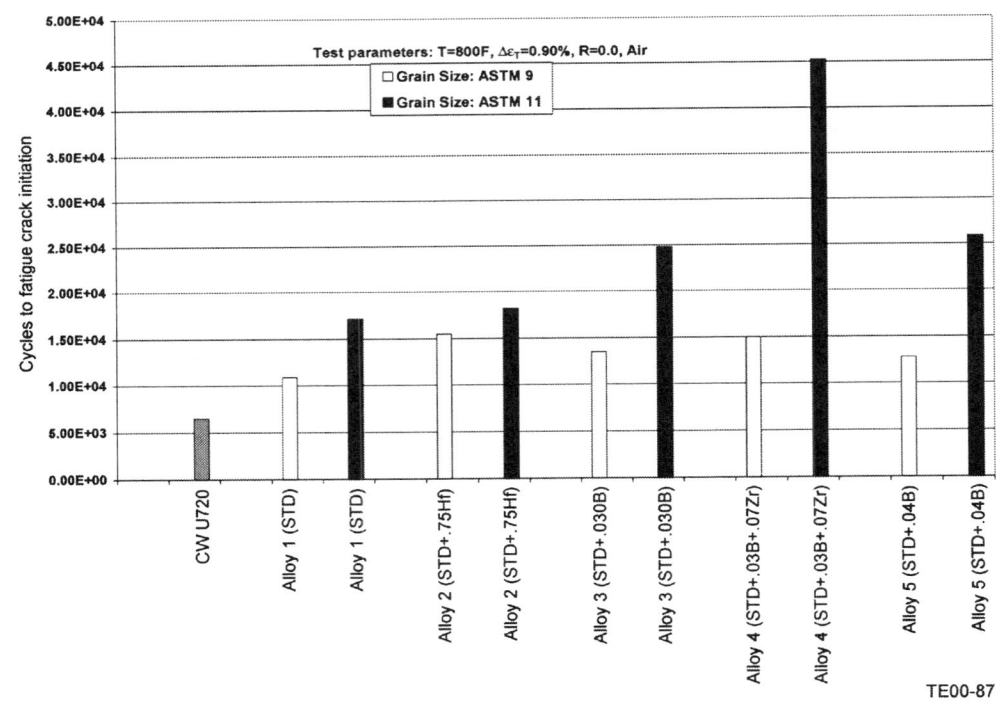

Figure 4. Effect of PM Udimet 720 alloy modifications and solution temperatures (grain sizes ASTM 9 and 11) on 800°F strain controlled LCF behavior ($\Delta\varepsilon_T$=0.9%, R=0.0, F=20CPM).

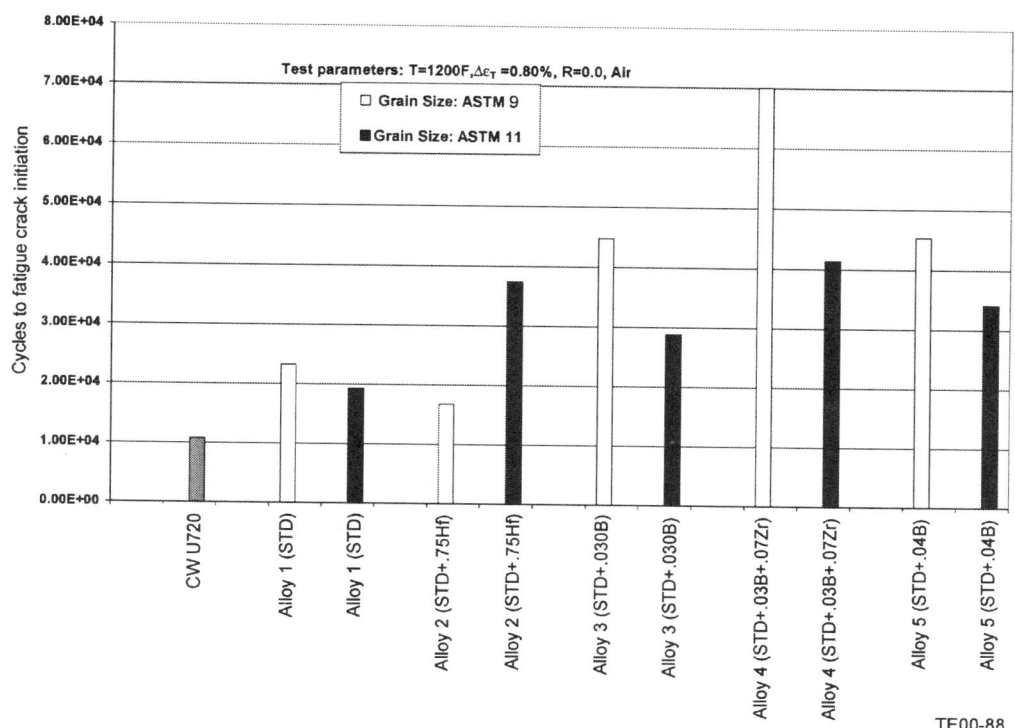

Figure 5. Effect of PM Udimet 720 alloy modifications and solution temperatures (grain sizes ASTM 9 and 11) on 1200°F strain controlled LCF behavior ($\Delta\varepsilon_T$=0.8%, R=0.0, F=20CPM).

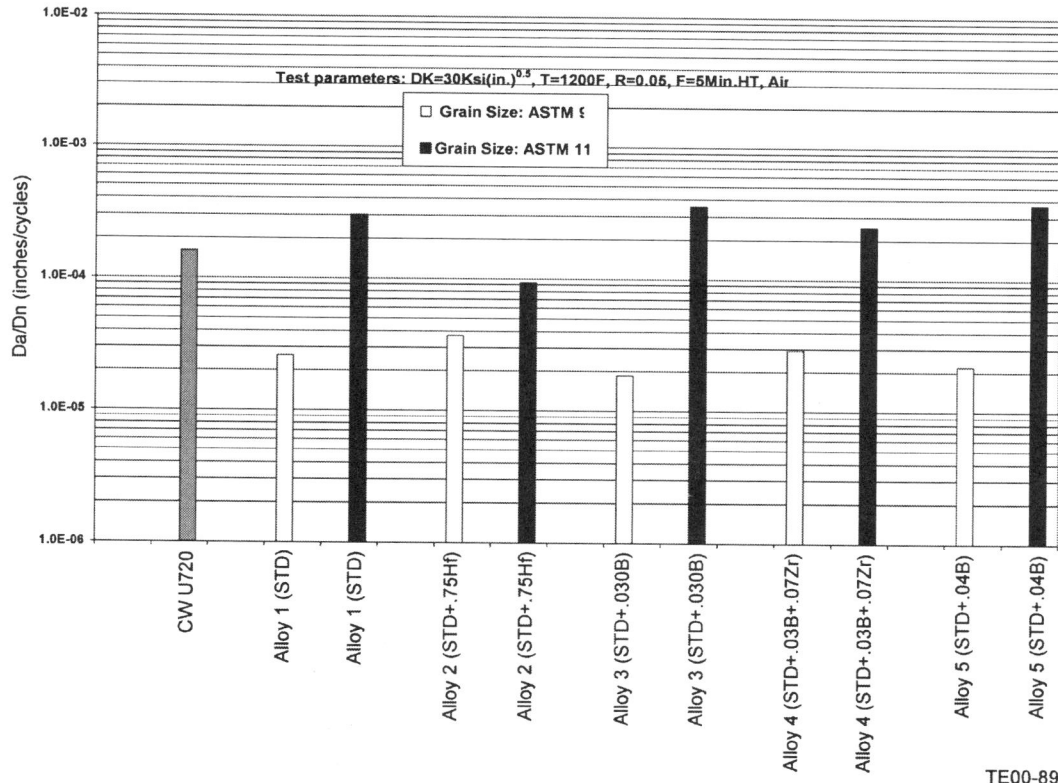

Figure 6. Effect of PM Udimet 720 alloy modifications and solution temperatures (grain sizes ASTM 9 and 11) on 1200°F/5-min hold FCG rate (Da/Dn) behavior (ΔK=30 ksi-in$^{1/2}$ R=0.05).

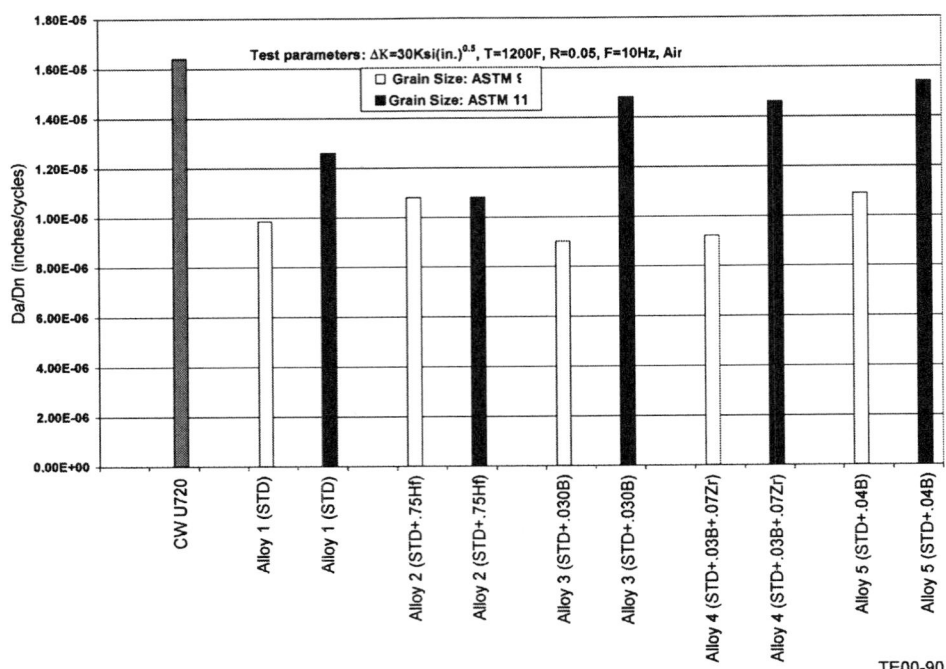

Figure 7. Effect of PM Udimet 720 alloy modifications and solution temperatures (grain sizes ASTM 9 and 11) on 1200°F FCG rate (Da/Dn) behavior (ΔK=30 ksi-in$^{1/2}$ R=0.05, Freq=10Hz).

(Figure 8). Duplicate tests were conducted on both subsolvus and supersolvus processed materials and average crack growth rates were plotted at ΔK = 30 ksi-in.$^{1/2}$

The dwell test is important because the rim of a turbine disk is at high temperature for several minutes when the engine is at takeoff power. Thus, the crack growth rate due to high temperature dwell must be minimized to ensure disk damage tolerance. As shown in Figure 6, the supersolvus processed candidate alloys had greatly decelerated crack growth rates when tested under dwell conditions at 1200°F. It was also observed that, with the exception of subsolvus processed hafnium modified Alloy 2, the crack growth rate characteristics of the modified alloys were similar. In the case of subsolvus processed Alloy 2, the hafnium addition appeared to offer an advantage over the other subsolvus processed nonhafnium bearing alloys.

With respect to 1200°F and 800°F nondwell tests, the data (Figures 7 and 8) indicated an overall trend in which the crack growth resistance characteristics of supersolvus processed alloys were improved over the subsolvus processed alloys. It was also noted that the crack growth resistance seemed to be relatively insensitive to compositional modifications.

Microstructural Observations

Microstructural review verified the target grain sizes of ASTM 11 for the subsolvus processed material and an ASTM 9 for the supersolvus processed material were achieved. It was also found that the grain sizes were very uniform from location to location within a forging. Additionally, microstructural appearances for a given target grain size were generally very similar from alloy to alloy. Photomicrographs representative of the PM Udimet 720 modifications are shown in Figure 9.

Examination of the microstructures representative of the ASTM 11 and ASTM 9 processed material found the

grain boundaries were decorated with gramma prime, M_3B_2, and $M_{23}C_6$ particles (Figure 10). Also, random ZrO_2 and HfO_2 particles were found in matrix areas of both subsolvus and supersolvus processed material away from grain boundaries. Relative to the morphology of the gamma prime in the alloys, there was evidence of residual overaged gamma prime phase in the subsolvus processed material. Also, the aging gamma prime in the supersolvus processed material tended to be slightly coarser than that seen in the subsolvus material (Figure 11).

Overall, it was not possible to clearly pinpoint grain boundary decoration characteristics and matrix microstructural features as a function of the chemistry modifications made in this program. However, a significant influence of both chemistry and grain size on mechanical properties was observed. Future work is needed to thoroughly understand the 800 and 1200°F LCF, 1200°F FCG, and 1350°F creep property relationships involved with the boron, zirconium, and hafnium chemistry modifications evaluated in this program.

Summary of Results

- Zirconium and boron modifications resulted in significant improvements in 800 and 1200°F LCF life:

  ♦ 0.03B/0.07Zr LCF life was approximately 2.5X better than baseline PM Udimet 720 at 800°F for subsolvus processed material and approximately 3X better for supersolvus processed material at 1200°F.

  ♦ 0.03B/0.07Zr LCF life was approximately 6X greater than cast and wrought Udimet 720 at both 800°F for subsolvus processed and 1200°F for supersolvus processed material.

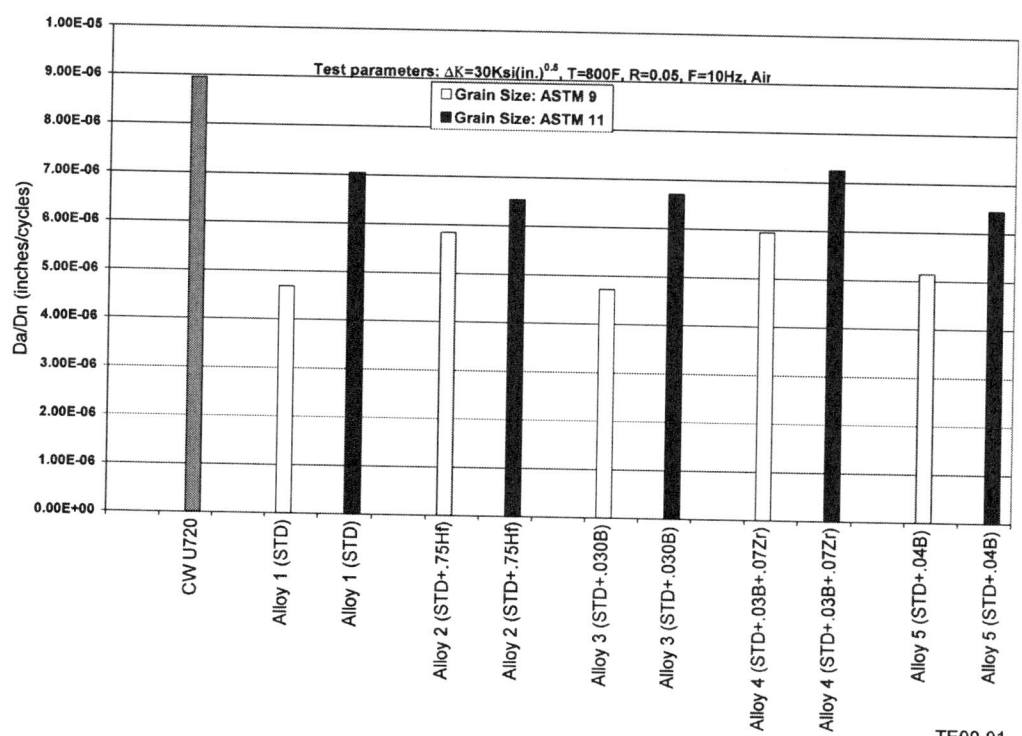

Figure 8. Effect of PM Udimet 720 alloy modifications and solution temperatures (grain sizes ASTM 9 and 11) on 800°F FCG rate (Da/Dn) behavior (ΔK=30 ksi-in$^{1/2}$ R=0.05, Freq=10Hz).

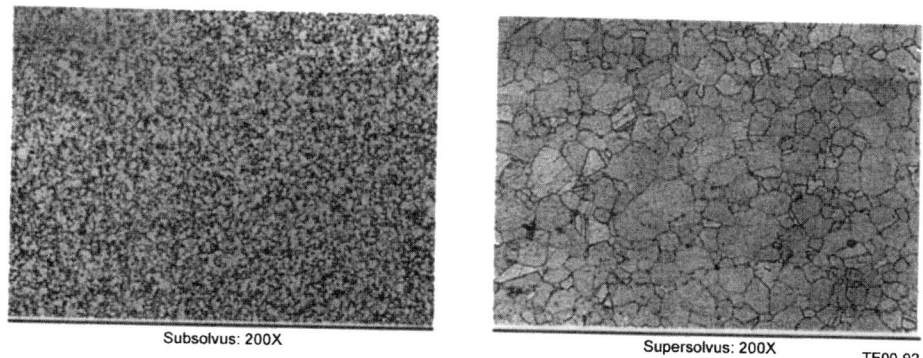

Figure 9. Typical low magnification microstructures observed for subsolvus and supersolvus processed PM Udimet 720 alloys.

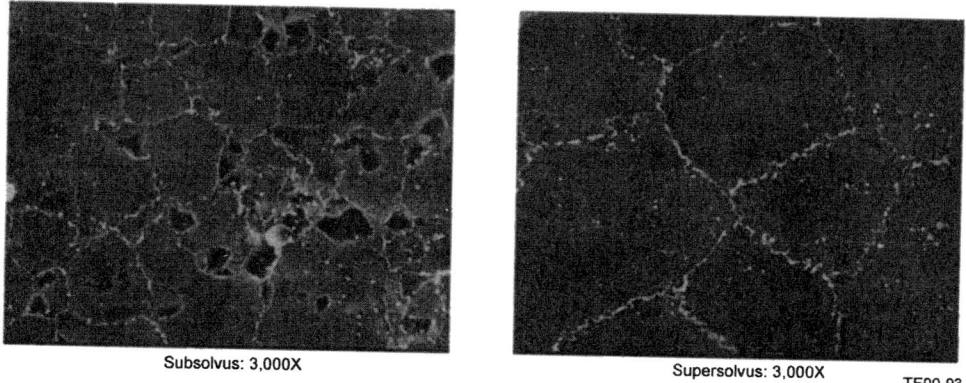

Figure 10. Typical grain boundary microstructures observed for subsolvus and supersolvus processed PM Udimet 720 alloys.

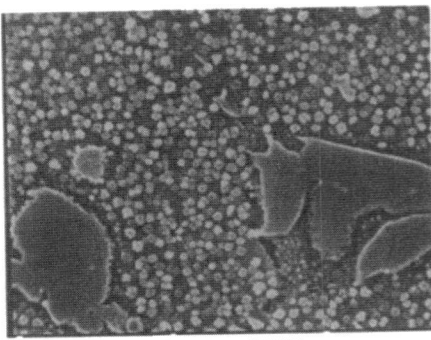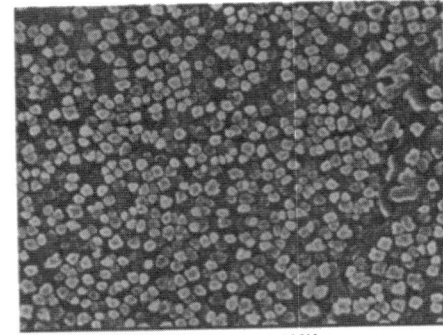

Subsolvus: 10,000X   Supersolvus: 10,000X   TE00-94

Figure 11. Typical gamma prime microstructures observed for subsolvus and supersolvus processed PM Udimet 720 alloys.

- Subsolvus processed (ASTM 11) material generally yielded the best 800°F LCF results; supersolvus material generally yielded the best 1200°F LCF results.

- Grain size had the greatest impact on FCG resistance at 800 and 1200°F:

  ♦ Supersolvus (ASTM 9) material was significantly better than subsolvus processed material.

  ♦ The FCG rates of the modified alloys were similar to one another; however, they had better (lower) FCG rates than cast and wrought Udimet 720 at 800 and 1200°F.

- Of the modified compositions in the subsolvus processed condition, the hafnium modified alloy appeared to offer the best overall FCG resistance at 1200°F.

- No significant effect of Udimet 720 alloy modifications was noted in tensile results:

- The 0.2% tensile yield strength of supersolvus processed material was reduced on the order of 10 ksi compared to subsolvus processed material.

- Little influence of alloy modifications was observed for the 1250°F creep results, but strong boron and boron/zirconium trends were noted for the 1350°F tests:

  • The best 1350°F results were noted for 0.03B/ 0.07Zr modifications.

Conclusions and Recommendations

The results of this program have shown that the compositions developed to manage segregation issues with cast and wrought Udimet 720 are nonoptimum from a mechanical property viewpoint for PM Udimet 720 alloy forms. This is because PM processing offers the ability to increase alloying elements such as boron and zirconium to levels beyond those that can ordinarily be tolerated with traditional cast and wrought processing. Additional work is recommended to further define the optimum alloy composition for PM Udimet 720 and to scale the effort to larger commercial heat sizes.

Additionally, the concept of developing a dual microstructure PM Udimet 720 disk featuring a coarse grained rim and a fine grained bore should be considered. This is because the results of this work have shown the coarse grained microstructures can develop very attractive creep rupture, crack growth rate, and LCF capabilities at elevated temperature conditions such as might be required by rim locations in advanced disk designs. Similarly, the fine grained structures investigated in this program have shown

tensile and LCF capabilities that are attractive for lower temperature bore regions. Therefore, by combining these two structures into a single disk design, it may be possible to develop an overall performance capability greater than what could be developed by a disk processed to a single microstructural condition.

Acknowledgments

The authors wish to acknowledge the cooperative spirit and technical support provided by the Crucible Materials Research Division, who provided the as-HIP PM billet material used in the program, and the Ladish Company, who conducted the isothermal forging and heat treatment of the PM billet product. Particular attention is directed to John Moll and Fred Yolton of Crucible Materials Research Division and David Furrer and Joe Lemsky of the Ladish Company. Without their advice and expertise in supplying the wrought product that was the subject of this program, the outcome of the effort may not have been nearly as positive. In addition, John Radavich of Micro-Met Laboratories is acknowledged for his electron microscopy work, phase extraction, and X-ray diffraction efforts. His work has helped to characterize the structures developed, and his comments and advice, as always, were found to be helpful and informative.

References

1) F. E. Sczerzenie and G. E. Maurer, "Development of Udimet 720 for High Strength Disk Applications," Superalloys 1984 (Warrendale, PA; The Metallurgical Society, 1984), 573-582.

2) K. R. Bain et al., "Development of Damage Tolerant Microstructures in Udimet 720," Superalloys 1988 (Warrendale, PA; The Metallurgical Society, 1988), 13-22.

3) K. A. Green, J. A. Lemsky, R. M. Gasior, "Development of Isothermally Forged P/M Udimet 720 for Turbine Disk Applications," Superalloys 1996 (Warrendale, PA; The Metallurgical Society, 1996), 697-703.

MICROSTRUCTURAL STABILITY AND CRACK GROWTH BEHAVIOUR OF A POLYCRYSTALLINE NICKEL-BASE SUPERALLOY.

D. W. Hunt, D. K. Skelton, and D. M. Knowles.

Rolls Royce University Technology Centre
Dept. Materials Science & Metallurgy
University of Cambridge.

Abstract

In this paper the microstructural instability of a recently developed powder-processed alloy, RR1000, is investigated and the corresponding influence on the creep-fatigue crack growth behaviour scrutinised. Equilibrium phase calculations of the alloy composition predict the presence of the TCP (Topologically Close Packed) phase sigma at the intended operating temperature of the alloy. Subsequent exposures at elevated temperatures reveal intergranular sigma phase precipitation. The kinetics of sigma phase formation has been determined experimentally by quantitative x-ray diffraction of exposed specimens. The effect of this grain boundary modification during prolonged exposure on the mechanical properties of the alloy was then investigated by testing specimens with two thermal histories: as-heat-treated (with a standard solution and ageing heat treatment developed for RR1000) and exposed to contain 1 wt% sigma phase on the grain boundaries.

To enable the discrimination of creep and environmental damage mechanisms during crack growth, fatigue and sustained load crack growth tests were performed in both vacuum and air. The influence of exposure on tensile and creep rupture behaviour was also investigated. In order to provide a suitable reference, a single fatigue crack propagation test was performed on the as-heat-treated material at room temperature. All other tests were performed at 725 °C. From the crack growth rates recorded and fractographic studies it is concluded creep and environmental mechanisms are responsible for increases in crack growth rates observed when the temperature is increased from 20 °C to 725 °C. Sigma phase precipitation during alloy exposure increases the severity of creep and environmental processes leading to increases in crack growth rates.

Superalloys 2000
Edited by T.M. Pollock, R.D. Kissinger, R.R. Bowman,
K.A. Green, M. McLean, S. Olson, and J.J. Schirra
TMS (The Minerals, Metals & Materials Society), 2000

Introduction

The demand for increased service lives and operating temperatures of gas turbine engines has led to the design and use of turbine disc alloys of increasing chemical complexity. Many latest generation alloys are increasingly unstable at the intended operating temperatures. In order to maximise the engineering performance of such alloys a comprehensive understanding of the influences of microstructural instability on the mechanical properties is required.

In this study a recently developed disc alloy, RR1000[1], has been assessed. Thermo-calc software[2] together with a thermodynamic databank[3] was used to predict the equilibrium phase composition of the alloy, which includes the TCP (Topologically Close Packed) phase sigma. The presence and effect of sigma in other nickel-base superalloys has been noted by various researchers[4-8]. Substantial decreases in ductility and creep rupture lives have been reported. The severity of property degradation previously observed appears to be sensitive to the morphology of sigma phase[4]. Needles, platelets or continuous intergranular films are proposed to be more deleterious than a discrete, more equiaxed intragranular morphology. Some mechanisms accounting for observed property losses have been suggested[8]. A hard sigma phase, especially a plate-like or needle morphology on grain boundaries offer ideal sites for crack initiation. TCP phases such as sigma are also rich in refractory elements like chromium, cobalt, molybdenum and tungsten. Precipitation of these phases will result in solute depletion in the surrounding γ matrix leading to localised compositional weakening. The susceptibility to environmental attack may likewise be increased if the formation of a protective Cr_2O_3 scale is inhibited through chromium depletion.

The effect of sigma phase precipitation on fatigue crack growth, particularly at elevated temperatures, is less clear. In order to assess crack growth at elevated temperatures, it is essential to consider damage mechanisms resulting from fatigue, creep and environmental processes. The relative contribution of these mechanisms will also be influenced by the microstructural state of the alloy.

Materials and Methods

The composition of RR1000 as well as other well-known disc alloys is presented in table I.

Table I. Composition (wt%, balance nickel) of RR1000 and other disc alloys [1,8].

Element	RR1000	U720Li	Waspaloy
Cr	14.35 – 15.15	16	19.5
Co	14.0 – 19.0	15	13.5
Mo	4.25 – 5.25	3.0	4.3
W	-	1.25	-
Al	2.85 – 3.15	2.5	1.3
Ti	3.45 – 4.15	5.0	3.0
Ta	1.35 – 2.15	-	-
Hf	0.0 – 1.0	-	-
Zr	0.05 – 0.07	0.035	-
C	0.012 – 0.033	0.015	0.08
B	0.01 – 0.025	0.015	0.006

Quantitative analysis of sigma phase precipitation kinetics was achieved using bulk electrolytic phase extraction of thermally exposed specimens. The powder-like extracts were examined by X-ray diffraction. Diffraction data was quantified as mass fraction phase using Quasar software[9] coupled to a crystallographic database. This technique is based on the Rietveld Method[10]. Data gathered by this procedure was used to construct TTT curves for sigma phase precipitation. The effect of the measured microstructural instability on the creep-fatigue behaviour of RR1000 was assessed by testing material in the as-heat-treated condition in comparison to exposed test pieces. In order to precipitate sigma phase within convenient time periods, specimens were exposed above the perceived operating temperature of RR1000. The exposed material contained 1 wt% sigma phase following a heat treatment equivalent to the nose position of the TTT curve determined as described. In order to differentiate fatigue, creep and environmental damage mechanisms occurring during crack growth at elevated temperatures, fatigue and sustained load crack propagation tests were carried out in vacuum and air at 725 °C. A 1–1–1–1 trapezoidal waveform with R = 0.5 was used for fatigue tests. SENB (single edge notch bend) specimens were used for both fatigue and sustained load testing. The crack length was monitored using the d.c. potential drop method[11]. The influence of exposure on the creep rupture and tensile behaviour of the alloy was also investigated at 725 °C.

Microstructural Stability

Examination of the microstructure after solutioning and ageing heat treatments reveals a fine grain size (typically $8 - 12$ μm) with a blocky primary γ' situated on the grain boundaries and a fine intergranular secondary γ' with an average size of 225 nm[12]. A general view of the microstructure, showing primary γ' and grain structure, is shown in figure 1. No sigma phase can be detected in the as-heat-treated alloy.

Subsequent prolonged exposure at elevated temperatures leads to the intergranular precipitation of sigma phase, developing extensive grain boundary networks after lengthy treatments. This microstructural degradation of RR1000 is shown in figures 2 and 3. The 'nose' of the TTT curve, illustrated in figure 4, was found to be significantly higher than the perceived maximum operating temperature of RR1000.

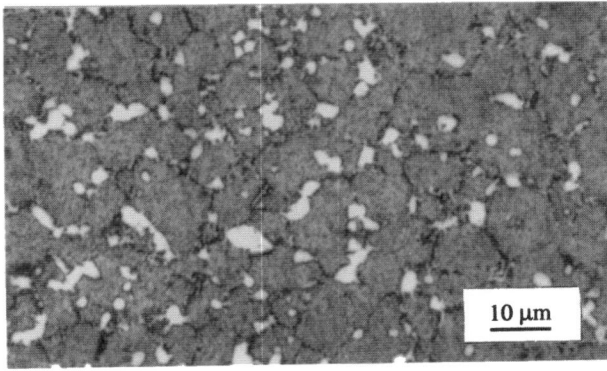

Figure 1. Optical micrograph of RR1000 showing an equiaxed grain structure (size 8 – 12 μm) with primary γ' situated on the grain boundaries. The secondary intragranular γ' may only discerned at higher magnification.

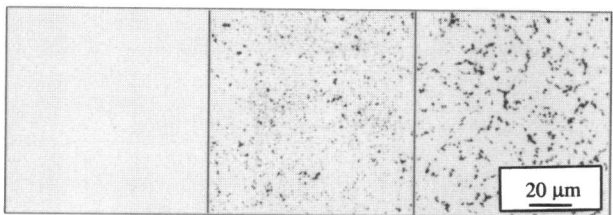

Figure 2. Optical micrographs similarly etched to reveal the extent of sigma phase precipitation during prolonged exposures at elevated temperatures. Image on left shows the as-heat-treated condition. No sigma phase is detected either through microscopical examination or x-ray diffraction analysis.

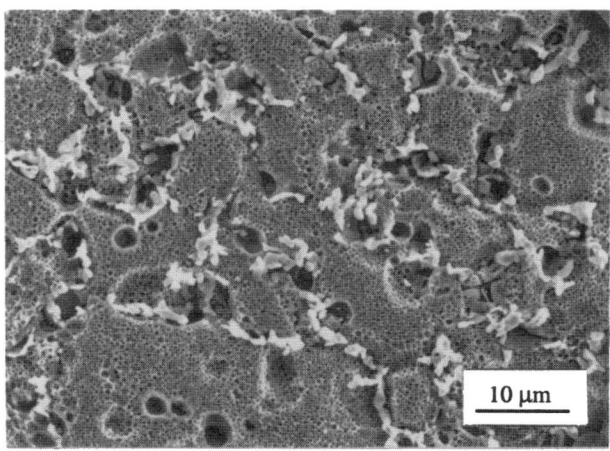

Figure 3. Deep-etched micrograph showing white sigma phase precipitates along the grain boundaries.

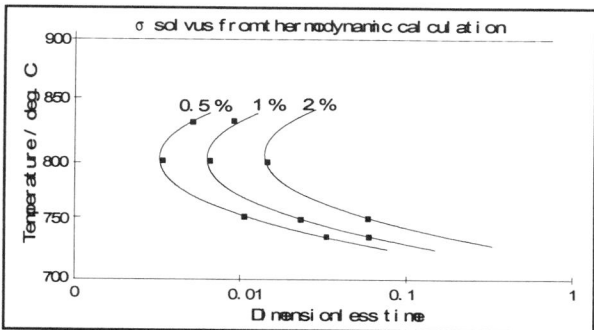

Figure 4. Experimentally determined TTT curves for sigma phase precipitation in RR1000.

Apart from intergranular precipitation of sigma phase it is possible that prolonged elevated ageing treatments may result in coarsening of fine secondary γ' precipitates. Examination of γ' using carbon extract replicas was carried out of a specimen exposed to approximately the nose position of the 2 wt% curve in figure 4. The average γ' particle size over this period was found to be

228 nm (average size in the as-heat-treated alloy was measured as 225 nm).

Mechanical Properties

Tensile Strength

The tensile behaviour as a function of temperature and alloy exposure is plotted on normalised axes in figure 5.

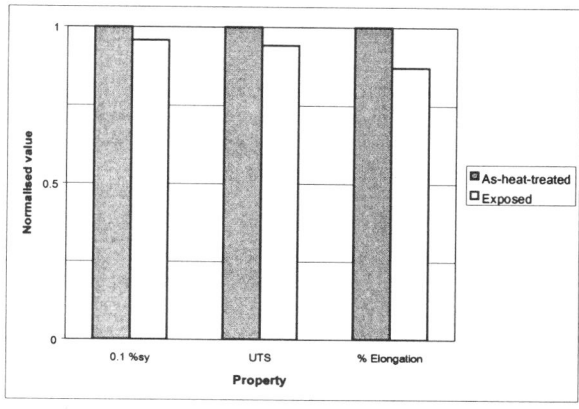

Figure 5. Tensile properties (0.1% σ_y, UTS and elongation) of RR1000 at 725 °C as a function alloy exposure.

Yield strength, UTS and ductility are reduced for the exposed condition. The effect on ductility is notable with the elongation being decreased by 13% whilst the yield strength is reduced by just 4%. The ultimate tensile strength of the exposed test-piece is decreased by 6%.

Creep Rupture

The effect of alloy exposure on the creep properties of RR1000 is shown in figures 6 and 7.

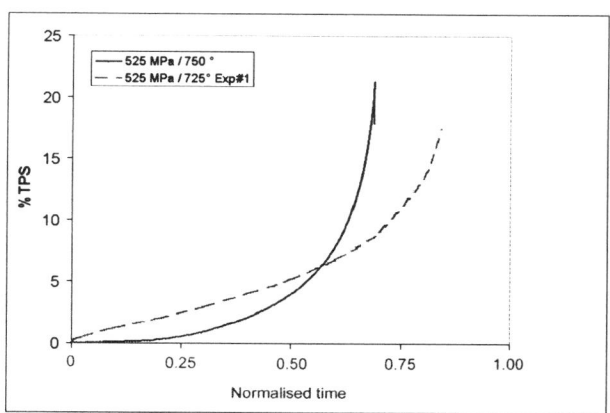

Figure 6. Creep curves demonstrating modification in strain evolution of the exposed alloy. It should be noted that in this case, creep data for the as-heat-treated alloy is for 750 °C not 725 °C.

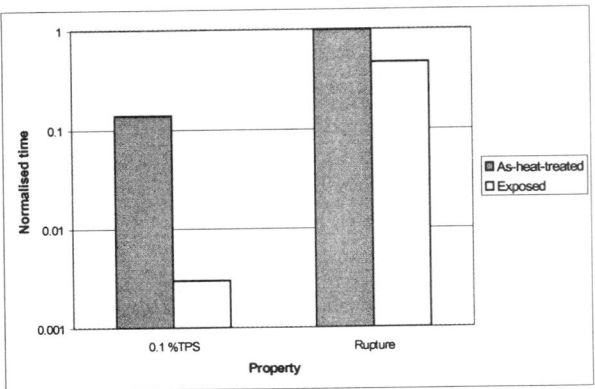

Figure 7. The influence of exposure on the creep properties of RR1000. In this instance data at 725 °C is used for the as-heat-treated alloy. Notice the significant effect on the creep strain rate at low TPS (total plastic strain).

Whilst the rupture life by is reduced by 53%, it is the effect on early strain rates which is most conspicuous with the time to 0.1 %TPS decreased by 97%. The higher secondary creep rates are clearly visible in figure 6.

Fatigue Crack Growth

The crack growth rates recorded during fatigue testing RR1000 are shown in figure 8.

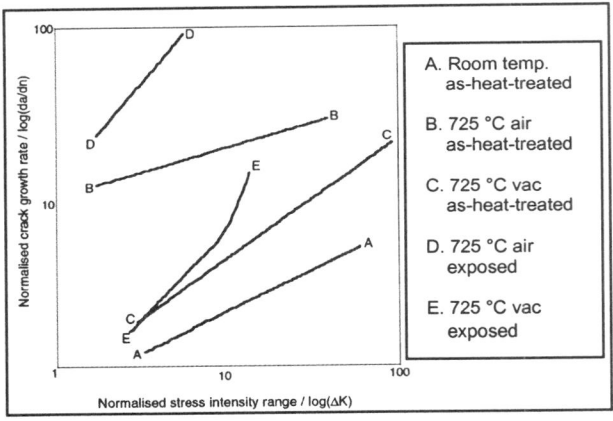

Figure 8. Fatigue crack propagation in RR1000 as a function of temperature, environment and microstructure. A balanced trapezoidal waveform (1-1-1-1, 0.25 Hz, R=0.5) was utilised.

The effect of temperature and environment is striking during fatigue crack propagation in RR1000. An order of magnitude increase in crack growth rate is observed when the temperature is increased to 725 °C (curves A and B). The influence of environment, particularly at low ΔK, is also dramatic (curves B and C). Exposure of RR1000 results in an extensive increase in crack growth rate, especially in air (curves B and D). The crack growth rates recorded in the exposed alloy in vacuum at low ΔK are similar to the as-heat-treated specimen (curves C and E). However, the Paris slope in vacuum is significantly higher and the crack growth rates diverge substantially.

Sustained Load Crack Growth

Crack growth resulting from a sustained load is shown in figure 9.

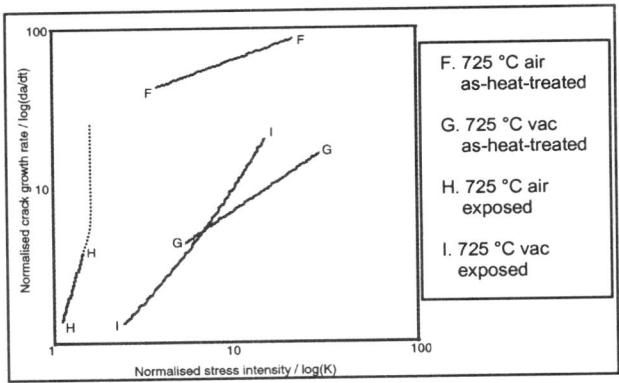

Figure 9. The influence of environment and exposure on the sustained load crack growth rate in RR1000. It should be noted that the extreme sensitivity of the exposed specimen in air resulted in rapid failure at relatively low stress intensities and the very high crack growth rates attained could not be readily recorded.

The influence of environment is also apparent during sustained load crack growth. The presence of sigma phase in the exposed specimens has a dramatic effect, particularly in an oxidising environment. The effect of sigma phase precipitation in air is extreme with very little crack growth and rapid failure over the stress intensity range tested. In vacuum, the exposed specimen has similar crack growth rates at lower values of K although initiating crack growth at very low stress intensities proved to be difficult to achieve with test-pieces in the as-heat-treated condition.

Discussion

Microstructural Degradation

Prolonged exposure of RR1000 over the temperature range examined results in a transformation of the grain boundary microstructure. No significant change in the average γ' size could be detected up to an exposure resulting in 2 wt% sigma phase (i.e. significantly greater than that of specimens used for mechanical property degradation assessment). The discrete intergranular precipitation of sigma phase may therefore said to be associated with the property changes which were recorded in RR1000.

Crack Growth Behaviour

The test results gathered during this study validate the proposition that crack growth in RR1000 at 725 °C can arise due to fatigue, creep and environmental mechanisms. The grain boundary modification brought about by sigma phase precipitation during prolonged exposure alters the relative contribution of these. The environmental and creep contributions may be distinguished by consideration / comparison of the crack growth tests performed in vacuum and air at 725 °C. It is assumed that the level of vacuum achieved during the tests (typically $pO_2 = 10^{-7}$ mbar), whilst from a thermodynamic consideration not low enough to preclude Cr_2O_3 or NiO formation, is sufficient to suppress the environmental effects observed during air tests.

798

During vacuum tests of the as-heat-treated material the extent of creep damage is sensitive to the stress intensity range, ΔK, in fatigue tests or stress intensity, K, in sustained load tests. This is supported by the difference in fracture morphology demonstrated at room temperature and 725 °C vacuum as shown in figures 10 and 11.

Figure 10. Fracture surface of room temperature fatigue specimen, demonstrating transgranular failure at all ΔK.

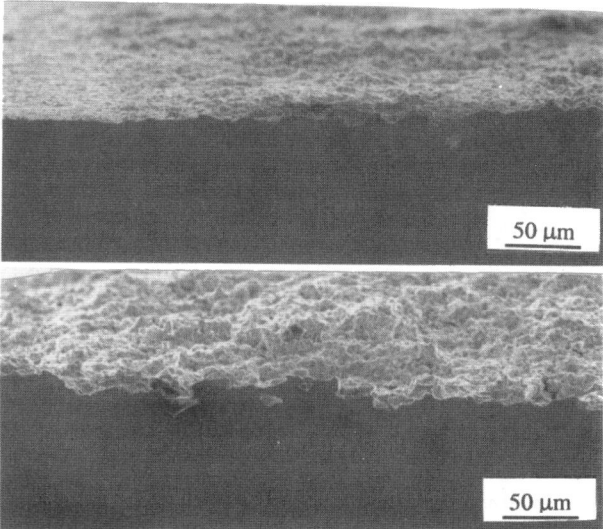

Figure 11. Fatigue fracture surfaces developed during crack growth in vacuum. Top shows low ΔK with pre-crack, high ΔK is shown in the bottom image.

At room temperature in an ambient environment crack growth may be said to be due to purely fatigue mechanisms, and the fracture morphology is transgranular at all values of ΔK (figure 10). At 725 °C in vacuum, a progressive transition to intergranular fracture occurs as ΔK (and K_{max}) increases whilst the crack growth rates recorded demonstrate an increased sensitivity to ΔK. During sustained load crack growth, where the fatigue contribution is removed, fracture proceeds in an intergranular manner. Furthermore, creep strain investigations performed elsewhere[12] have shown that creep damage in RR1000 develops via intergranular cavitation. The increase in Paris slope during crack growth in vacuum is notable. At low ΔK the absolute crack growth rates are similar to that observed at room temperature where damage resulting in crack growth originates from purely fatigue

processes. At 725 °C, creep mechanisms become significant. Static, or time dependent events contributing to crack growth have been studied elsewhere [13-15]. The creep contribution observed in crack growth in RR1000 at 725 °C may be similarly rationalised as increasing the Paris exponent, m.

The effect of intergranular sigma phase precipitation during alloy exposure on the creep contribution is quite dramatic. Evidence from tensile tests of the exposed material suggest that whilst yield strength is not greatly influenced, there is an appreciable degree of embrittlement which is evident from the reduction in tensile elongation. The strain evolution during creep testing is also significant. Given the modification in creep behaviour, an acceleration of crack growth rates in vacuum is anticipated. Figure 12 shows the fracture morphology of the exposed alloy when fatigue tested in vacuum. Although the absolute values of crack growth rate at low ΔK in the exposed material (figure 8) is similar to that recorded in the as-heat-treated alloy, the extent of creep damage appears to be increased. Fracture appears to be intergranular at all values of ΔK and the crack growth rate displays an increased sensitivity to ΔK. This is also evident in the crack growth rates recorded under sustained load in vacuum as shown in figure 9.

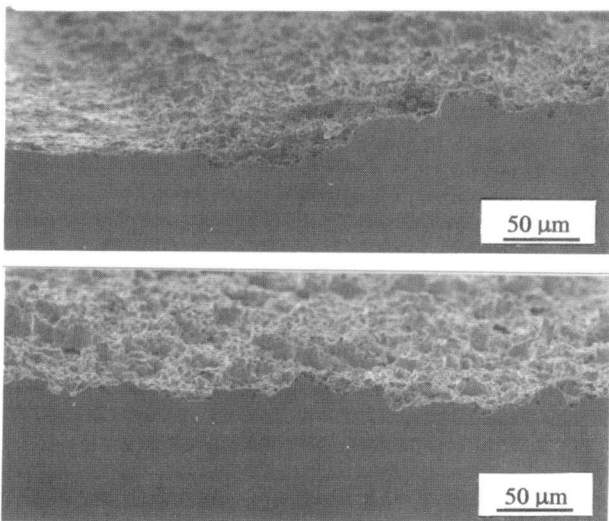

Figure 12. Fracture morphology developed in exposed specimens when fatigue tested in vacuum. The degree of creep damage is increased with intergranular fracture apparent at low ΔK as well as high ΔK (top and bottom).

Close examination of the crack tip region reveals the nature of damage which leads to the increased sensitivity to creep crack growth. Figure 13 shows a crack tip in the exposed alloy developed under sustained load in vacuum. Failure appears to occur discretely ahead of the main crack tip. These small cracks evolve resulting in fracture as shown in previous figures. In the exposed alloy localised failure appears to occur at the interfaces of sigma phase. An example of this behaviour is shown in figure 14. The influence of grain boundary particles on cavity nucleation during creep has been studied extensively[16], as has the effect of sigma precipitation on creep ductility in other alloys[4,5,8]. The mechanisms of local compositional weakening and damage initiation previously proposed[8] are in agreement with the crack growth processes observed in vacuum in RR1000.

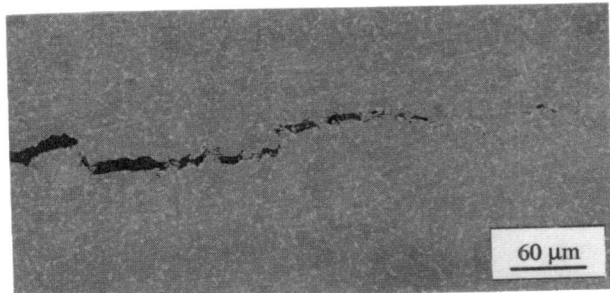

Figure 13. Crack tip from sustained load crack growth under vacuum of the exposed material.

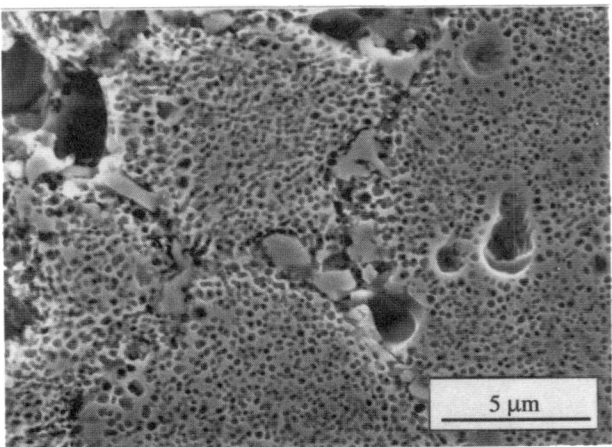

Figure 14. Local failure ahead of the crack tip. Shown is an etched surface to reveal the location of sigma phase. Microstructure appears porous due to the preferential dissolution of γ' during etching. Sigma phase is visible as unetched discrete precipitates approximately 2 μm in size, centrally positioned in the field of view. Short cracks are apparent at the sigma phase interface.

The damage mechanisms leading to crack growth are markedly different during air tests at 725 °C. The environmental effects observed result in substantial increases in crack growth rates. The increases in crack growth rate are concurrent with a transition to intergranular fracture. The fatigue fracture morphology of the as-heat-treated material is shown in figure 15.

Crack growth in air at 725 °C may also be distinguished from the vacuum behaviour by the presence of secondary cracking observed in the test-pieces. This occurred along grain boundaries adjacent the fracture surface. An example of this phenomenon is shown in figure 16 and suggests severe grain boundary embrittlement.

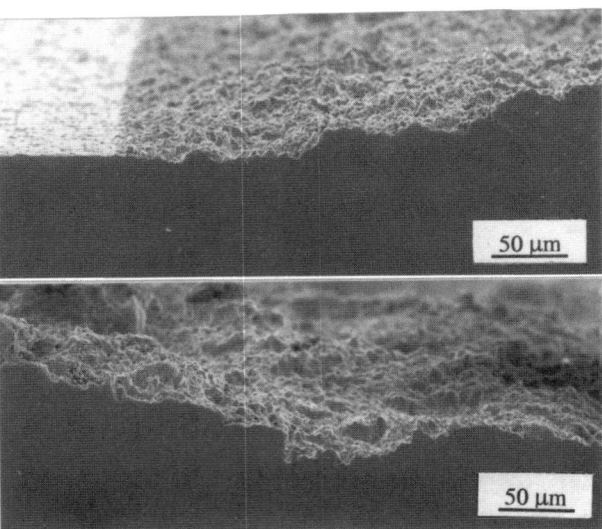

Figure 15. Fatigue fracture surfaces (low and high ΔK top and bottom respectively) of as-heat-treated RR1000 fatigue tested in air at 725 °C.

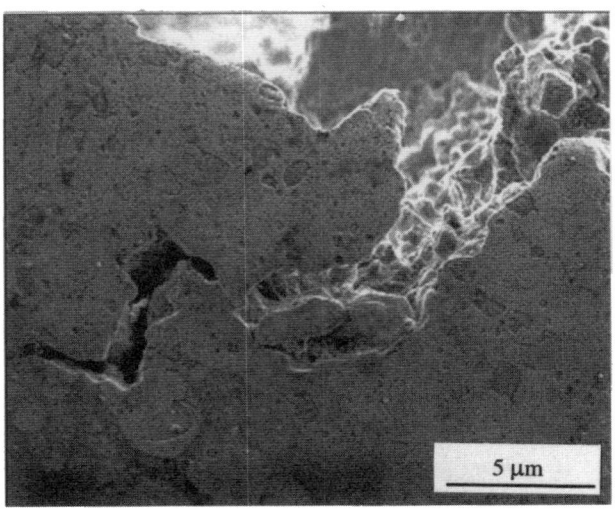

Figure 16. Secondary intergranular cracking observed in specimens tested in air at 725 °C. This phenomenon was not found in vacuum test-pieces.

The dramatic and detrimental effect of an oxidising environment on crack growth has been investigated in other alloy systems[17-20]. In particular, the behaviour of alloy 718 has been widely documented. Three mechanisms of oxygen induced embrittlement have been proposed. Short-range oxygen diffusion along grain boundaries may result in oxide intrusions that easily rupture under the constrained geometry of the crack tip and accelerate crack growth. Localised oxygen induced damage may also occur ahead of the crack tip if oxygen is transported to greater distances along the boundaries leading to internal grain boundary oxidation. A third environmental damage mechanism[21] has been proposed involving cavitation ahead of the crack tip. This process may arise due to the oxidation of carbides and the subsequent release of carbon monoxide and/or carbon dioxide. Other investigators[22] have reasoned that this mechanism may be inhibited in alloys containing significant quantities of chromium due to the preferential formation of Cr_2O_3.

The dramatic environmental assisted acceleration is also observed during sustained load crack growth in air at 725 °C. The mechanistic influence of sigma phase from an environmental consideration is less clear than that of creep. Precipitation of sigma phase dramatically increases the fatigue crack growth rates recorded in air. Again, this increase is associated with further intergranular embrittlement. The fatigue fracture surfaces of the exposed test-pieces are shown in figure 17. Although increasingly subtle, more extensive intergranular embrittlement and fracture is discernible. The extent of secondary cracking as shown in figure 16 is also significantly increased supporting this observation.

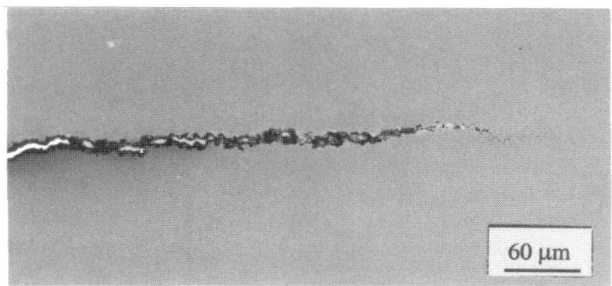

Figure 18. Crack tip in the exposed material due sustained load in air at 725 °C

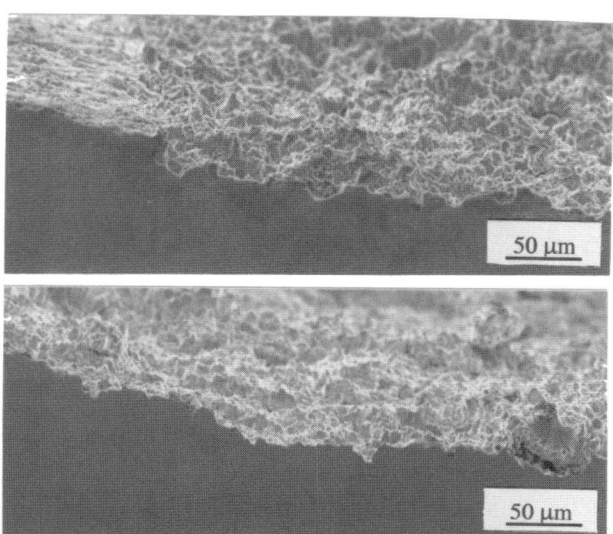

Figure 17. Fatigue fracture surfaces of exposed alloy in air at 725 °C. The extent of intergranular embrittlement is increases over that observed in the as-heat-treated material.

The crack growth rates recorded from the sustained load test of the exposed alloy (figure 9, curve H) are the cause of some confusion. The absolute values of crack growth rate recorded are quite low. However, the rapid fracture which occurred in this specimen prevented accurate measurement of the crack growth rates achieved. It seems that this test was conducted very close to a critical stress intensity for the exposed alloy at this temperature and requires further investigation.

Unlike the localised creep damage ahead of the crack tip observed in vacuum, the environmental mechanism leading to accelerated crack growth in air appears to be one of intergranular oxidation at the crack tip. The crack tip from an interrupted sustained load test of the exposed alloy is shown in figures 18 and 19. This clearly indicates the extensive oxidation which occurs at the crack tip and dominates the crack growth rates recorded. This contrasts sharply with the observations in vacuum voiding ahead of the crack tip was observed.

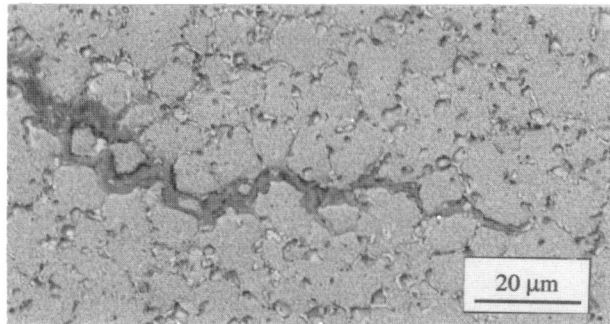

Figure 19. Extensive intergranular oxidation ahead of the crack tip.

The mechanism of oxidation observed appears to be in broad agreement with models of intergranular oxidation previously proposed for crack tip oxidation. No evidence of discrete oxidation ahead of the crack tip was found during this investigation. The precise mechanism by which sigma phase contributes to this intergranular oxidation is not certain as of yet, however it seems likely that solute (in particular chromium) depletion effects are responsible for the increased susceptibility to environmental attack.

Conclusions

Prolonged thermal exposure of RR1000 leads to intergranular precipitation of sigma phase which has been experimentally characterised in the form of TTT curves. This grain boundary modification leads to quite dramatic changes in the mechanical properties of the alloy.

At 725 °C crack growth due to time dependent mechanisms of creep and oxidation result in a modification in fracture behaviour, namely a transition from ductile transgranular fracture associated with pure fatigue to intergranular fracture. These transient effects have been investigated by examining the crack growth behaviour during sustained load tests. The creep mechanism displays an increased sensitivity to crack tip stress over oxidation processes. However, in terms of the contribution to absolute crack growth rates over the stress intensity range tested environmental mechanisms dominate at 725 °C. Sigma precipitation increases the severity of both creep and environmental mechanisms leading to increased crack growth rates. Creep damage in exposed specimens is associated with localised failure at the sigma interface. In air at 725 °C intergranular oxidation appears to be responsible for the crack growth rate increases recorded.

Acknowledgements

The authors gratefully acknowledge the Engineering and Physical Science Research Council, Rolls Royce plc. and DERA for their financial support and provisional of material.

References

1. S. J. Hessell et al, "Nickel Alloy for Turbine Engine Components", United States Patent, Number 5897718 (27 April 1999).

2. Bo Sundmann, Thermo-calc Users' Guide (Stockholm, Sweden: Division of Computational Thermodynamics, Department of Materials Science & Engineering, Royal Institute of Technology, Stockholm, Jan. 1997.)

3. Rolls Royce Aerospace Group, Thermodynamic Database for Nickel-base Superalloys, Rolls Royce plc, Derby U.K., 1999.

4. J. R. Mihalisin, C. G. Bieber, and R. T. Grant, "Sigma-Its Occurrence, Effect, and Control in Nickel-Base Superalloys", TMS AIME, 242, Dec 1968, 2399-2414.

5. G. Chen et al, "Grain Boundary Embrittlement by μ and σ Phases in Iron-Base Superalloys", Proceedings of the 4th International Conference on Superalloys, 1980, Ed. J. K. Tien et al, TMS, Seven Springs PA, (1980), 323-333.

6. P. W. Keefe, S. O. Mancuso, and G. E. Maurer, "Effects of Heat Treatment and Chemistry on the Long-Term Phase Stability of a High Strength Nickel-Based Superalloy", Proceedings of the 7th International Symposium on Superalloys, Ed. S. D. Antolovich et al, TMS, Seven Springs PA, (1992), 487-496.

7. G. F. Vander Voort and H. M. James, "Wrought Heat-Resistant Alloys", Metals Handbook, vol. 9, (American Society for Metals, 1985), 308-313.

8. C. T. Sims, N. S. Stoloff, and W. C. Hagel, eds., Superalloys II. High Temperature Materials for Aerospace & Industrial Power (New York, NY: John Wiley & Sons inc., 1987), 221-226.

9. Philips Electronics N.V., Quasar Version 1.0. Software for Quantitative Standardless Analysis using the Rietveld Method. User's Guide, December 1994.

10. D. L. Bish and S. A. Howard, "Quantitative Phase Analysis Using the Rietveld Method," J. Appl. Cryst., 21, (1988), 86-91.

11. M. A. Hicks and A. C. Pickard, "A Comparison of theoretical and experimental methods of calibrating the electrical potential drop technique for crack length determination," Int. J. Fracture, 20, (1982), 91-101.

12. A. J. Manning, Development of a Polycrystalline Ni Base Superalloy for Gas Turbine Disc Application, (PhD Thesis, University of Cambridge, U.K., 1999).

13. D. K. Skelton and D. M. Knowles, "The Creep Fatigue of Turbine Disc Alloys," Proceedings of the 7th International Fatigue Conference (June 1999, Beijing, P. R. China), 2111-2116.

14. J. F. Knott, "Models of Fatigue Crack Growth", Proceedings of a Conference on Fatigue Crack Growth, 1984, Ed. R. A. Smith, Cambridge, U.K. (September 1984), 31-52.

15. H. Riedel, Fracture at High Temperatures (Berlin, Germany: Springer-Verlag, 1987), 274-277.

16. H. Riedel, Fracture at High Temperatures (Berlin, Germany: Springer-Verlag, 1987), 59-60, 112-114.

17. H. Ghonem and D. Zheng, "Depth of intergranular oxygen diffusion during environmental-dependent fatigue crack growth in alloy 718," Mat. Sci. Eng., A150, (1992), 151-160.

18. H. Ghonem, T. Nicholas, and A. Pineau, "Elevated Temperature Fatigue Crack Growth in Alloy 718 – Part II: Effects of Environmental and Material Variables," Fatigue Fract. Engng. Mater. Struct., 16, (6), (1993), 577-590.

19. E. Andrieu et al, "Intergranular Crack Tip Oxidation Mechanism in a Nickel-Based Superalloy", Mat. Sci. Eng., A154, (1992), 21-28.

20. K. D. Challenger, R. P. Skelton, and J. S. Kamen, "The Effect of Oxidation on Fatigue Crack Growth in 2.25Cr-1Mo Steel at 525 °C: A Metallographic Examination," Mat. Sci. Eng., 91, (1987), 1-6.

21. R. H. Bricknell, and D. A. Woodford, Acta Metall., "The Mechanism of Cavity Formation During high Temperature Oxidation of Nickel", 44, 1982, 257-264.

22. H. M. Lu et al, "Environmentally-Enhanced Cavity Growth in Nickel and Nickel-Based Alloys," Acta Mater., 44, (8), (1996), 3259-3266.

THE APPLICATION OF CALPHAD CALCULATIONS TO NI-BASED SUPERALLOYS

N. Saunders✣, M. Fahrmann★ and C. J. Small⌃

✣Thermotech Ltd., Surrey Technology Centre, The Surrey Research Park, Guildford, Surrey GU2 5YG, U. K.

★Special Metals Corporation, 3200 Riverside Drive, Huntington, WV 25705-1771, U. S. A.

⌃Rolls-Royce plc., PO Box 31, Derby DE24 8BJ, U. K.

Abstract

In recent years thermodynamic modelling via the CALPHAD method has been extensively applied to industrial alloys of many types. Although pertaining to equilibrium conditions, use has shown that valuable information can be gained for a variety of practical applications. A paper presented at the last Seven Springs meeting gave some theoretical background to the CALPHAD method and described the development of the methodology to Ni-based superalloys. The main purpose of that paper was to provide validation of results against an extensive experimental literature which, at the time, concentrated on γ/γ' equilibria and liquid/solid equilibria. The present paper will present an extension of the validation process to take into account η formation and provide a review and examples of the practical application of the CALPHAD method to industrial alloys. It will expand on some of the topics briefly raised in the previous paper and demonstrate that the CALPHAD route is readily extendable to conditions that depart from equilibrium. It will also be shown that it can be used to provide fundamental input for calculations of physical and mechanical properties.

Introduction

In a previous paper the construction of a thermodynamic database for calculation of multi-component phase equilibria in Ni-based superalloys was decscribed[1]. In conjunction with an appropriate software package, such as Thermo-Calc[2] (used in the present studies), phase equilibria for multi-component Ni-based superalloys can be readily calculated. Numerous comparisons between calculation and experiment were given for γ/γ' phase equilibria and liquid/solid equilibria, and general phase equilibria calculations for some commercial alloys were shown. The paper concentrated on γ/γ' alloys touched on aspects to do with the

extension of the CALPHAD method to non-equilibrium transformation and also the prediction of fundamental properties such as anti-phase domain boundary (APB) and stacking fault (SFE) energies. Since this previous paper, the database has undergone substantial development, with the inclusion of new models, and its capabilities extended to new types of alloys, in particular to general NiFe-based superalloys and Re-containing single crystal alloys. As part of this process, validation work has been undertaken by companies such as Special Metals Corporation. and Rolls-Royce plc.

It has now also been used extensively world-wide and examples of its use are becoming well documented[3,4,5,6,7,8]. The purpose of the present paper is to first describe the extension of capability to NiFe-based superalloys, with emphasis on the formation of η and γ''. A review of examples of its application will then be made, a number of which emphasise that the CALPHAD methodology can be extended to areas outside of the field of equilibrium studies. Finally, the recent extension of the database to include Re will be discussed

NiFe-based Superalloys

NiFe-based superalloys, such as alloy 706 and alloy 718, can exhibit substantially more complex phase behaviour than γ/γ' type superalloys. They can be characterised by the formation of δ-Ni_3Nb, γ'', Laves, η as well as the usual types of TCP phases such as μ and σ. Transformation behaviour is more complex with the metastable γ'' being one of the primary hardening phases and a series of metastable states can arise during heat treatment.

While the CALPHAD route straightforwardly provides an equilibrium calculation, it can also be used to calculate various metastable states that arise if certain phases do not form for

Superalloys 2000
Edited by T.M. Pollock, R.D. Kissinger, R.R. Bowman,
K.A. Green, M. McLean, S. Olson, and J.J. Schirra
TMS (The Minerals, Metals & Materials Society), 2000

kinetic reasons. This approach has been classically used in steels where cementite is formed in preference to graphite during solid state transformations. In this case, graphite is "suspended" from the calculation by simply not considering it in the thermodynamic calculation; equilibrium is then calculated with cementite instead. Such metastable calculations have been made for numerous other material types, often with excellent success[3]. In a similar fashion such metastable calculations can be made for NiFe-based superalloys, particularly (i) with respect to the formation of γ" and (ii) alloys where γ' and η compete. It is also possible to take into account kinetic effects using a combined thermodynamic and kinetic approach and work has been done on solidification modelling for NiFe-based superalloys.

γ" Formation.

γ" is a metastable phase and in equilibrium conditions would not form. *Figure 1* shows a calculated phase % vs. Temperature plot for alloy 718. In the solid state the stable phases are γ, δ-Ni₃Nb, γ', sigma, MC (which at low temperatures transforms to $M_{23}C_6$) and at close to 600°C, α-Cr becomes stable (denoted BCC in the figure). However, after solution annealing and subsequent heat treatment at temperatures well below the δ solvus, the γ" transformation is favoured over formation of stable δ because of its comparatively rapid transformation rate. In this circumstance, a metastable equilibrium is formed with γ" instead of the equilibrium δ, and *Figure 2* shows a phase % vs. Temperature plot for this case.

γ" will transform to δ if left at temperature for sufficiently long times, as is well known from experiment, and *Figure 3* shows the experimentally established TTT diagram for alloy 718[9]. Although the equilibrium calculation does not predict transformation rates, certain critical features are well matched. For example the stable δ solvus and the solvus temperatures for γ and γ" in the absence of δ are well matched, and very close to each other as would be expected in this alloy which is γ'/γ" hardened.

The TTT diagram also shows the formation of σ and α-Cr at low temperatures after long anneals. The case of α-Cr formation is interesting as its solvus temperature is actually very close to that of σ. If its transformation rate is accelerated, for example by the application of stress or strain due to prior cold work, or if, in general, its transformation rate is faster than σ, there is the clear potential for it to form in accord with experimental studies[10]. In other γ" hardened alloys such as 625 similar calculations can be made for the equilibrium and metastable state and good agreement with experiment is found.

η Formation.

In NiFe-based superalloys there is a tendency to design alloys with lower Al concentrations than in γ/γ' hardened alloys. This promotes the formation of δ if Nb is present but, if Ti is also present, the Al/Ti ratio can become sufficiently low such that η forms in preference to γ' as the equilibrium phase. This phenomenon is well known in certain γ/γ' alloys, such as some Nimonic types, but in an analogous fashion to the γ"/δ case the faster kinetics of γ' growth permits it to form in preference to η. The γ' will eventually transform to η if left at temperature for sufficiently long periods, but if used at low temperatures the

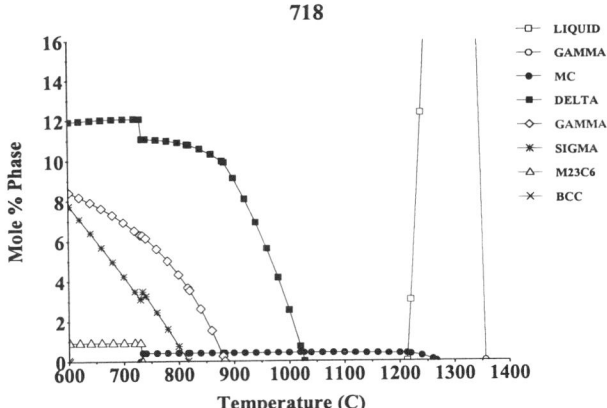

Figure 1 Calculated phase % vs. Temp plot for alloy 718

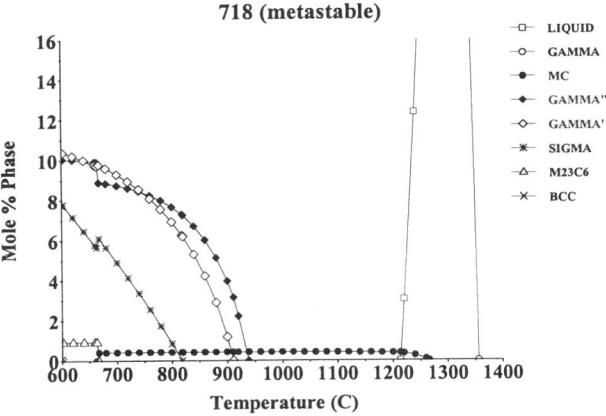

Figure 2: Calculated phase % vs. Temp. plot for alloy 718 with delta phase suppressed

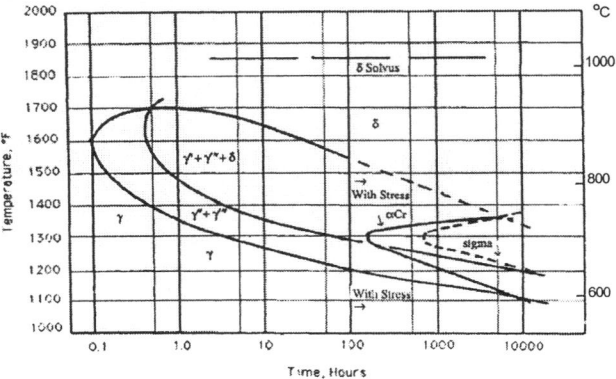

Figure 3: TTT diagram for alloy 718 from Basile Radavich[9]

γ'→η transformation is suppressed and "normal" γ' strengthening can be achieved.

For such alloys it is valuable to have some knowledge of both the η and γ' solvus temperatures. As part of a validation process at Special Metals a wide-ranging comparison was made with known behaviour of alloys and Table 1 shows results of the comparison.

Whenever the presence of η is noted, it was confirmed by phase extraction techniques combined with x-ray diffraction analyses. The absence of η was usually established in laboratory-type annealing studies of up to several thousand hours of exposure.

Table 1: Comparison between observed & calculated η formation‡.

Ni-base alloys	Obs.	Calc.	η Solvus Obs.(°C)	η Solvus Calc. (°C)
NIMONIC alloy 80A	No	No	-	-
INCONEL Alloy X-750	Yes	Yes	900	910
NiCo-base alloys				
Exp. alloy HV8964	Yes (850°C)	Yes	-	-
NIMONIC alloy 90	No	No	-	-
NIMONIC alloy 105	No	No	-	-
NIMONIC alloy 263	Yes (900°C)	Yes	-	-
NIMONIC alloy 115	No	No	-	-
NIMONIC alloy PK33	No	No	-	-
NIMONIC alloy EPK57	Yes (950°C)	Yes	-	-
NiFe-base alloys				
INCOLOY alloy 908	Yes (850°C)	Yes	-	-
NI-SPAN-C alloy 902	Yes	Yes	870	900
INCONEL alloy 706	Yes	Yes	950	980
NIMONIC alloy 901	Yes	Yes	980	1000
NIMONIC alloy PE11	Yes	Yes	>900	800
NIMONIC alloy PE16	No	No	-	-

It is clear that the database is providing high quality information on γ'/η phase competition and in a later section this capability was utilised to help design a new alloy for superheater tubes in advanced power plants.

Solidification Modelling.

A great deal of interest surrounds the solidification behaviour of NiFe-based superalloys. Unlike many γ/γ' types, a Laves phase can form during the final part of solidification. This phase is often metastable and can be removed by high temperature solution treatment, but in some cases it remains stable in the solid state. The CALPHAD method can be used to model aspects of solidification[11,12] that are otherwise quite difficult or time consuming to measure, e.g., segregation patterns and heat evolution. It is also possible to directly relate measured thermal effects to particular transformations.

Boutwell et al.[8] used an early version of the Ni-database to model the solidification behaviour of a 706 alloy. *Figure 4* shows a Fraction Solid vs. Temperature plot calculated using the so-called "Scheil-Gulliver" model. This model assumes little or no back diffusion occurs in the solid phase during solidification and has been used with excellent success for a wide range of alloy types[8].

Five critical points are calculated. The onset of solidification with the formation of γ is marked A, B marks the formation of

‡ INCONEL, INCOLOY, NIMONIC and Ni-SPAN-C are trademarks of the Special Metals Corporation family of companies.

MC carbide, C the Laves phase start, D the η phase start and finally solidification is complete at point E where the fraction solid equals one. Experimental DTA confirmed that the calculated transformation temperatures were in good agreement with experiment (Table 2).

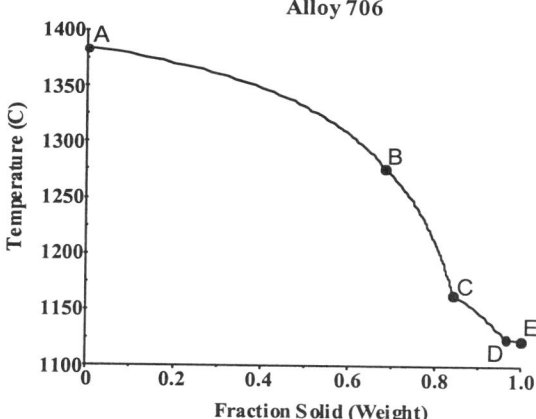

Figure 4: Fraction solid vs. Temperature plot for "Scheil-Gulliver" solidification of a 706 alloy

Table 2: Comparison between experimentally determined DTA results and a Scheil-Gulliver simulation for a 706 alloy[8]

	Liquidus	MC start	Laves start	η start	Solidific-ation end
Centre (DTA)	1381	1240	1164		
Edge (DTA)	1388	1261			
Calculated	1385	1277	1168	1126	1125

The segregation patterns were calculated and *Figure 5* shows the composition of γ as a function of solid transformed. As well as providing results pertaining to the physical metallurgy of the casting, it is also possible to obtain thermo-physical data that can be used for casting simulations. *Figure 6:* shows such a plot for the alloy investigated by Boutwell et al.[8].

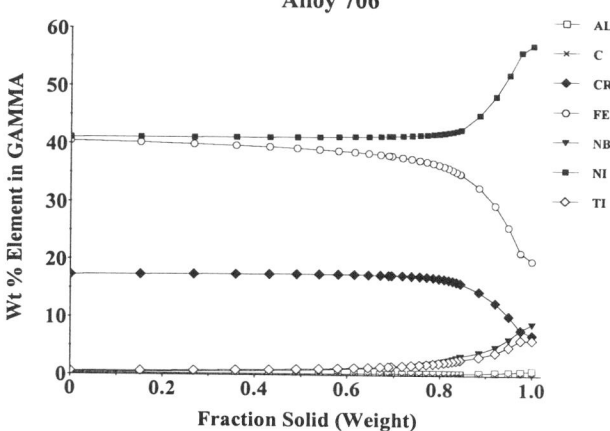

Figure 5: calculated segregation patterns in the γ phase after solidification of a 706 alloy

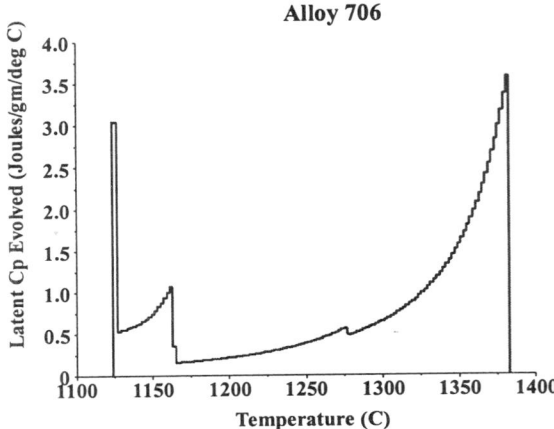

Alloy 706

Figure 6: Plot of latent Cp of solidification vs. Temperature for a 706 alloy during "Scheil-Gulliver" solidification.

Alloy Design and Development.

Design of an Affordable New Alloy for Superheater Tubing in Advanced Power Plants.

The material requirements for new, advanced power plants means that older, more established alloys, such as Fe-based austenitic and ferritic steels will not match long term service requirements. To this end a design programme was instigated to produce a new Ni-based alloy that could be used for critical components and the following design targets were set[6].

1. A metal loss of <2mm in 200,000 hrs. of operation in a hot flue gas/coal ash environment.
2. A minimum 100 MPa rupture stress for 100,000 hrs at 750°C.
3. Metallurgical stability at 750°C for the same period of time.
4. Manufacturability in various sizes and lengths.
5. Ease of field installation, meeting weldability and bending requirements.

A baseline composition for the new allow was derived from the well established NIMONIC alloy 263 and further specifications added to it.

1. The Cr content was raised from 20 to 24wt% to increase coal ash corrosion resistance.
2. The Mo level was drastically dropped from 5.8wt% to <1wt%. This was partly done to counteract the increased Cr stabilising σ and μ. But corrosion testing in a simulated boiler environment also suggested that Mo could significantly reduce corrosion resistance.
3. Nb was added to improve weldability.
4. The alloy be γ'-hardened with levels of up to 15-20 vol.% γ'.

Using a "traditional" approach to alloy development would have normally mean that a series of alloys would have to be melted, tested and examined so that the effect of all of these changes could be understood. However, the use of the CALPHAD route meant that the change in phase "make-up" of the alloy could be easily calculated.

The development programme proceeded in a series of steps. The first step was to look at the effect of adding Al and Ti to the base alloy. The main aim of this exercise was to see how the

competition between δ, η and γ' was affected by various levels of Al and Ti and, also, their ratio. So that the phase fields involving δ, η and γ' could be evaluated, an isothermal section of the base alloy at 750°C with Al and Ti as the axes was calculated. In this case 1 wt%Nb was added to improve weldability of the final alloy.

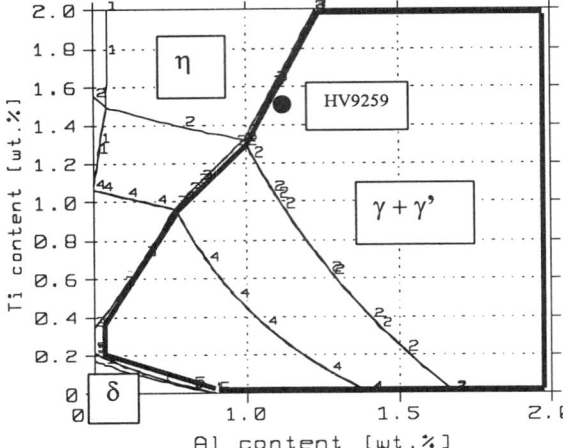

Figure 7: Calculated isothermal section at 750°C of a baseline alloy 1wt%Nb with Al and Ti as axes. μ and δ labels show areas where these phases are predicted to form. The desired γ+γ' area is outlined within the bold line. For the sake of clarity carbide formation is not shown (HV9259 depicts the composition of an experimental heat)

It was clear that the alloy could be γ'-hardened across a wide range of Al and Ti levels and the next step was to define the Al and Ti levels such that 15-20 vol.% γ' was produced. This was done by traversing the section shown in Figure 8 and constraining the amount of η to be zero and varying the amount of Al.

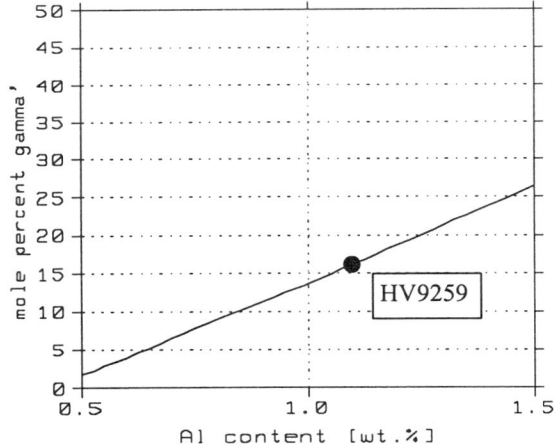

Figure 8: : Amount of γ' as a function of Al content added to the baseline alloy with Nb=1wt% calculated at 750°C. The Ti-content follows the γ'/γ'+η phase boundary in Figure 7.

This provided the maximum permissible Ti:Al ratio to strengthen γ' without compromising phase stability. The flexibility of the Thermo-Calc software allowed this to be done and at the same time a plot of mole % γ' could be plotted against the Al level in the alloy. When this level was between 15 and 20% the range of Al and Ti content was defined.

After these steps were complete, the effect of Cr and Mo was more closely examined. *Figure 9* shows an isothermal section of the alloy at a constant 1.1wt%Al, 1.5wt%Ti and 1.0wt%Nb level with Cr and Mo levels varied. The section clearly shows that, as Cr levels rise above 23wt%, σ stability would be greatly enhanced and the level of Mo was kept well below the critical level of 3wt% for an alloy with 24wt%Cr.

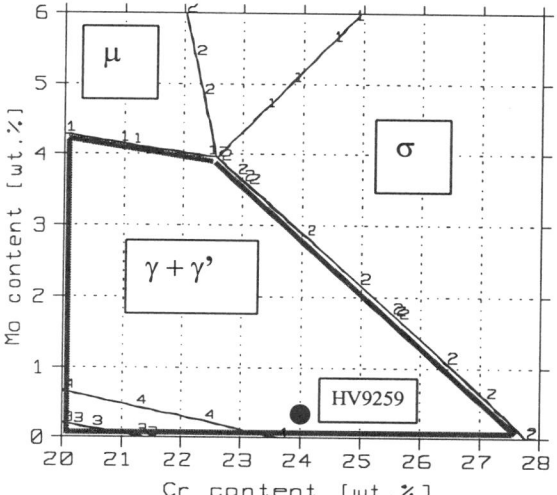

Figure 9: Calculated isothermal section at 750°C of a baseline alloy + 1.1%Al + 1.5%Ti + 1%Nb (wt%) with Cr and Mo as axes. μ and σ labels shows areas were these phases appear and the desired γ+γ' area is outlined within the bold line.

To confirm that alloys lying outside the calculated ranges of stability did indeed form deleterious phases, critical experiments were made. These supported the calculated phase diagrams. The ease of using the calculation route clearly enhanced the alloy design programme, which was both aided and expedited by having calculated information on phase amounts, at hand. This alleviated the necessity for a wide-ranging testing programme, significantly cutting costs and shortening development cycle times.

Design of a New Gas Turbine Disc Alloy.

In a recent paper, two of the present authors described the background to an alloy development programme for a new gas turbine disc alloy[4]. The paper emphasised the way a CALPHAD approach to predicting phase equilibria could accelerate the whole development programme by providing key input parameters for the design concepts. The paper mainly discussed issues associated with σ and μ stability. In itself the input of the phase equilibrium input was decisive in the decision making process by which the chemistry of the final alloy was decided. As in the previous example, the CALPHAD route effectively provided a rapid experimentation route that allowed critical features of the alloy to be quantitatively designed on the computer and used in the alloy development process. The test

matrix of experimental design alloys was then much reduced and, with features such as the γ'_s and amount of γ' under control, common thermo-mechanical treatment schedules were maintained for the experimental alloys.

As part of the development programme, the CALPHAD route was extended to predict anti-phase domain boundary (APB) energies. This information was used to maximise target mechanical properties so that design goals could be reached. The approach to predicting APB energies using a CALPHAD approach has been described in detail previously[13] and can be summarised as follows.

In a perfectly ordered superlattice, such as the $L1_2$, the position of the various unlike and like atoms are prescribed by the ordering of the superlattice. When dislocations pass through this ordered structure a breakdown of local chemical order ensues with the subsequent creation of an APB. This boundary has a characteristic energy dependent on the change in the number of like and unlike bonds across the boundary. The number of such bonds across the APB is known from crystallographic considerations and, if the energy of the various bonds can be calculated, the APB energy can also be calculated.

If it is considered that the APB energy is predominantly controlled by the first nearest neighbours, i.e. second and higher order nearest neighbours are relatively small in comparison, the APB energy is expressed only in terms of first nearest neighbour interactions, W_{AA}, W_{BB} and W_{AB}. This leads to the following expressions[14],

$$W_1 = W_{AB} - \frac{1}{2}(W_{AA} - W_{BB}) \qquad (1)$$

where W_1 is the first nearer neighbour interaction energy. The APB energy is then given by

$$APB[hkl] = W_1 a^{-2} \left[\frac{h}{\sqrt{(h^2 + k^2 + l^2)}} \right]. \qquad (2)$$

Eq.2 implies that APB[001] is equal to zero. However, experimental results suggest that the APB energy for the {001} planes in Ni_3Al is of the same order of magnitude as the {111}. It is therefore necessary to take into account energies up to at least the third nearest neighbour (W_3). The relevant equations for APB energies then take the form[15,16]

$$APB[111] = a^{-2}\left(\frac{1}{3}\right)^{-\frac{1}{2}}[W_1 - 3W_2 + 4W_3 - ...] \qquad (3)$$

and

$$APB[001] = a^{-2}[-W_2 + 4W_3 - ...] \qquad (4)$$

To utilise these equations it is necessary to link the values of W_1, W_2 and W_3 with the Ni-database, which can be done using the Bragg-Williams-Gorsky ordering model[13]. In this case energy of formation of the γ phase and the subsequent ordering energy to γ' are used as the critical input.

The model was tested on binary $L1_2$ compounds before extending it to multi-component alloys. The end result was that a series of calculations for multi-component Ni-based superalloys were made and validated against experimental measurements. *Figure 10* shows the comparison for this validation process and the agreement is very good.

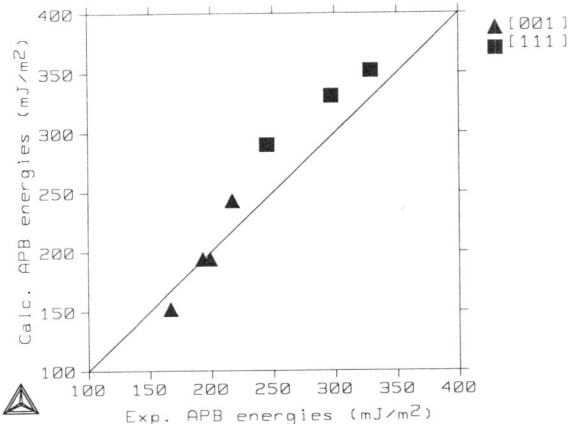

Figure 10: Comparison of calculated and experimentally measured[17] [111] and [001] APB energies for some superalloys

The advantage of using a calculation route is great. Firstly, the experimental measurement of APB energies is a taxing process, especially if the fault size is small. It requires great skill and expertise and, very importantly, the accuracy of measurement may be as much reliant on resolution of the electron microscope. Additionally, advances have been made in the correction methods applied to the raw measured data[18]. As a consequence the measured value of the APB energy has tended to increase with time (see for example Fig.3 in ref.13). The use of older measured values in theoretical equations may therefore incur inherent inaccuracy and it is useful to have a method of deriving values that is both internally self-consistent and can be verified by independent recent experiments. The procedure for calculating the requisite energies is very simple and was routinely performed during the development of the new disc alloy.

One of the design criteria of the alloy was to meet a specified low cycle fatigue life while meeting all of the other mechanical property requirements in an acceptably stable alloy. To this end Mo levels were increased to strengthen the γ matrix and Ta and Ti used to strengthen the γ'. However, the alloying additions had to be balanced to ensure that (i) the amounts and solvus temperatures of the TCP phases σ and μ were kept under control, (ii) the target 0.2% proof stress was achieved and (iii) the APB energy was kept sufficiently low for the deformation mechanism in the alloy to be cutting not climb. The latter provided planar slip rather than wavy slip and ensured that the low cycle fatigue life at high temperatures matched the design target.

The use of APB energy calculations to help control mechanical property requirements provided a clear enhancement to the value of a CALPHAD augmented alloy development approach. The design programme was progressed rapidly without the commitment of large amounts of time and resource in a 'traditional' experimental route. Further details of the alloy

development programme will be given in another presentation at the current meeting[19].

TCP formation in Re-containing Single Crystal alloys.

Recent alloy development of HP blades has almost exclusively relied on the addition of Re. Typically levels of between 2-6wt% are now added to single crystal alloys. Because Re is such a heavy element, this relates to a very small addition in atomic terms, but the effect on properties such as creep and strength is very pronounced. Re also has a very profound effect on the solvus temperature for TCP formation, raising it substantially[20,21]. The reason for this is not readily understood in terms of a PHACOMP approach[21], and it is therefore interesting to look at how CALPHAD methods can deal with Re.

In thermodynamic terms, Re is quite different to other elements that have a pronounced effect on TCP formation (e.g. Cr and Mo). It bears similarity to Cr in that a simple phase diagram between Ni and Re is formed with significant solubility of Re in Ni. However, in thermodynamic terms, Re forms very stable σ phases with elements such as Mo, Cr and W. Further, these σ phases contain high levels of Re (50-70at%Re). In contrast a "traditional" σ phase containing Ni, Co, Cr, Mo and W does not exhibit very negative heats of formation. In fact it is not always stable in the binary systems, only appearing in ternary and higher alloys. The corollary of both of these affects this is that Re strongly stabilises σ in Ni-alloys, substantially more so than elements such Cr, Mo and W and the partition coefficient, $x_{Re}^{\sigma} / x_{Re}^{\gamma}$ is very high.

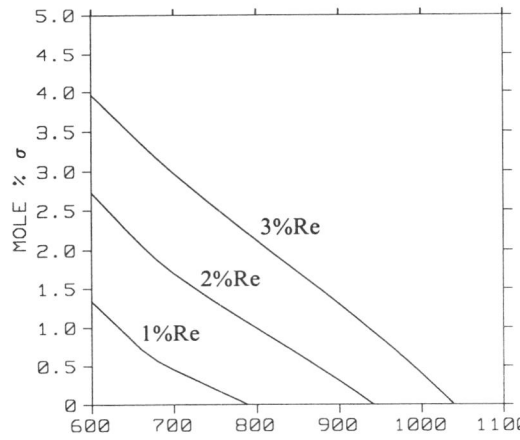

Figure 11: Predicted effect of Re on σ formation in CMSX-4[22]

Re has now been added to the Ni-database whereby the phase behaviour of commercial single crystal alloys can be calculated. The profound effect that Re has on σ stability can be demonstrated by using a CMSX-4 alloy. CMSX-4 typically has the composition in wt% Ni-6.3Cr-9Co-0.6Mo-6W-6.5Ta-3Re-1Ti-0.1Hf. *Figure 11* shows the effect on sigma formation when the Re concentration is changed from 3 to 1wt% (at 0%Re no sigma is predicted). It is clear that the effect of Re is extremely strong, especially when considering that such a dramatic change in sigma stability is produced by an addition of only 1at%Re. However, although clearly enhancing the formation of TCP phases the amount in the alloy is restricted by the total level of Re because σ is so rich in Re. This has the corollary that Re containing single crystal alloys, such as CMSX-4, may be relatively tolerant to TCP formation even though their temperatures of formation can be very high.

In the present meeting Rae et al.[7] report work from an experimental programme looking at a Re-containing single crystal superalloy, RR2071. In this alloy 4 different TCP phases can form depending on the heat treatment. The composition of RR2071 is (in wt%) Ni-9.5Co-6.6Cr-4.5Mo-2.8Re-5.6Al-1.3Ti-7.3Ta-0.3Nb-0.1Hf. The alloy was annealed at various temperatures for times of between 20 and 2600hrs. The results[7] are summarised below.

i)   At 1000°C the alloy was observed to form only P.
ii) At 950°C some μ is observed as well as P.
iii) At 900°C σ forms in the early stages before transforming to μ.
iv) At 800°C the alloy is predominantly μ with some σ.

R was occasionally observed at the lower temperatures. It was noted that at 900°C and below σ was the first phase to form, before transforming to the predominant μ phase. This feature was attributed to the more rapid transformation kinetics of σ, potentially arising from a favoured low energy $(1\,\bar{1}\,1)_\gamma$ interface.

Rae et al.[7] analysed their experimental results using a previous version of the Ni-database. Although providing valuable information on stability of single crystal alloys, the predictions from this version were incorrect in terms of which phases would form. For example, only P was predicted to be stable. However, it was clear that the stability of all the TCP phases was very close and small changes in Gibbs energy would then shift the balance of stability meaning that one phase would then predominate over the other.

The database used in the present paper has since been extensively re-developed and includes new models for σ and μ[23]. One of the effects of the new development work is that predictions for the equilibrium phases are now much closer to that observed. *Figure 12* shows a calculated phase % vs. Temperature plot for the RR2071 alloy using the new database. All features now appear

accurately predicted, with the exception of the occasional observation of R.

At 1000°C the plot correctly shows the formation of P phase. At 950°C some small amount of μ is predicted in addition to P. At 900°C μ has become the predominant phase, with only trace amounts of P. Just below 900°C, P disappears and σ forms. The σ phase then remains as a minor phase as the temperature is lowered. The absence of the σ phase at 900°C is in accord with the observation by Rae et al.[7] that it is probably metastable at this temperature.

It is interesting to now look at two other single crystal alloys where quantitative experimental assessment of the phases has been performed. Following the example of Rae et al.[7], CMSX4 and a Re-containing experimental "Alloy 800", studied by Proctor[24] and Darolia et al.[21] respectively, have been used here.

Proctor[24] studied samples annealed at 1050 and 1150°C and found that, at these temperatures, CMSX4 is a μ stable alloy, with small amounts of R present at 1150°C. At temperatures between 870 and 1150°C, Darolia et al.[21] found that Alloy 800 contains predominantly P and σ, with some μ also present. *Figure 13 & Figure 14* show calculated phase % vs. Temperature plots for both alloys.

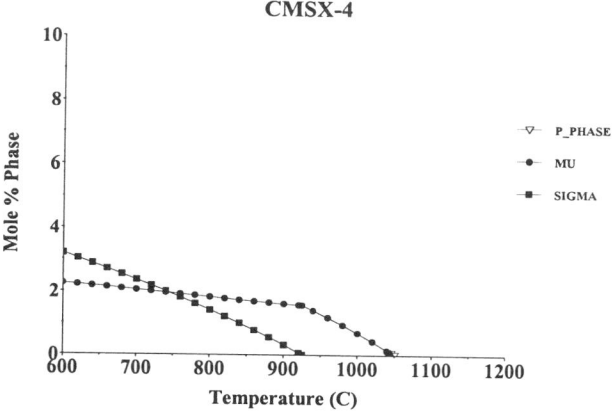

Figure 13 Calculated phase % vs. Temperature plot for a CMSX-4 single crystal alloy

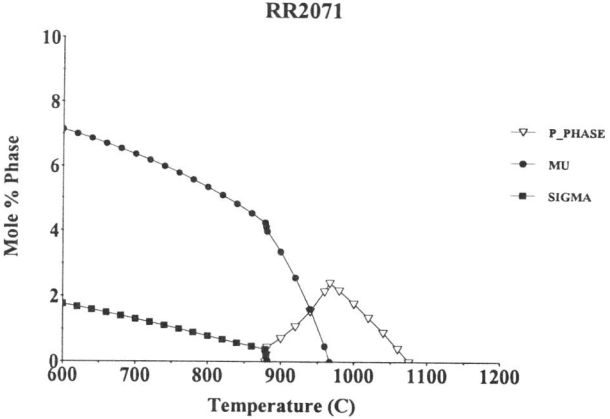

Figure 12: Calculated phase % vs. Temperature plot for the experimental single crystal alloy RR2071

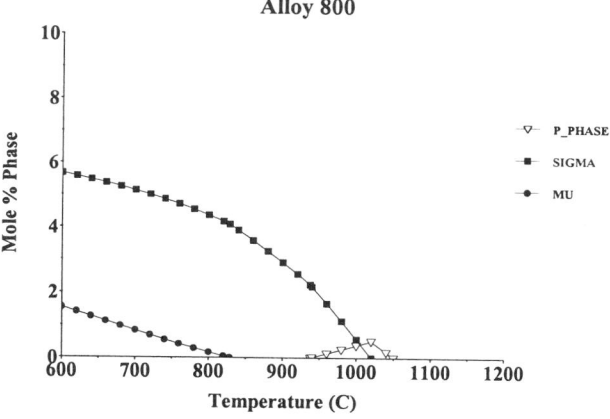

Figure 14: Calculated phase % vs. Temperature plot for the experimental single crystal alloy A800

CMSX-4 is correctly calculated to be μ dominant at higher temperatures with a small amount of P forming just below the TCP solvus temperature. The calculations also predict that at lower temperatures σ would form. Alloy 800 is predicted to be predominantly P and σ stable at the temperatures reported by Darolia et al.[21], with μ forming at lower temperatures.

It is clear that the new database provides very good predictions for the TCP phases formed in single crystal alloys. The alloys used for comparison purposes exhibit distinctly different behaviour and the calculations represent almost exactly the phases that are observed. The comparison between experiment and calculation does however suggest that the calculated TCP solvus temperatures may be too low. For example,

(i)   in CMSX-4 μ and R are observed at 1150°C while the highest TCP solvus temperature is calculated as 1050°C,
(ii) in RR2071 P is observed at 1100°C while the calculated solvus temperature is 1075°C and
(iii) the maximum rate of TCP formation for alloy 800 is 1090°C while the calculated P solvus is 1050°C.

It is, as yet, uncertain if this is due to a fault with the calculations or whether segregation remaining from casting is causing an enhanced susceptibility to TCP formation. Further work will be undertaken to resolve this question in the future.

Summary and Conclusions

The present paper has reported on recent developments to a database that can be utilised for CALPHAD-type thermodynamic calculations for Ni-based superalloys. These calculations allow features of both stable and metastable phase equilibria to be calculated for a wide range of Ni-based superalloy types. Examples of its application to NiFe-based superalloys, disc alloys and single crystal alloys are given. It can also be used in the calculation of non-equilibrium transformations as demonstrated for a case of solidification modelling. Emphasis has been placed both on the validation of the database against experiment and the practical use of the calculations. As part of the present paper, the extension to predicting fundamental physical properties and mechanical behaviour is discussed and an example of its practical use in such circumstances presented. It is clear that the CALPHAD methodology has a role to play in many aspects of Ni-based superalloys use and development.

References

1. N. Saunders, Superalloys 1996 eds. R. D. Kissinger, D. J. Deye, D. L. Anton, A. D. Cetel, M. V. Nathal, T. M. Pollock, and D. A. Woodford (Warrendale, PA: TMS, 1996), 101
2. B. Sundman, "Thermo-Calc, a General Tool for Phase Diagram Calculations and Manipulations", User Aspects of Phase Diagrams, ed. F. H. Hayes, (London: Inst.Metals, 1991) 130
3. N. Saunders and A. P. Miodownik, CALPHAD – Calculation of Phase Diagrams, Pergamon Materials Series vol.1, ed. R. W. Cahn, (Oxford: Elsevier Science, 1998)
4. C. J. Small and N. Saunders, "Development of a New Gas Turbine Disc Alloy", MRS Bulletin, 24, (1999), 22
5. N. Saunders, "Modelling of Nickel-Base Superalloys", Advanced Materials & Processes, 156, (1999), 29
6. M. G. Fahrmann and G. D. Smith, "Capitalizing on Computational Tools in Alloy and Process Development" to be published in Proc.Symp.Advanced Technologies for Superalloy Affordability (Warrendale, PA: TMS, 2000)
7. C. M. F. Rae, M. Karunatatne, C. J. Small, R. W. Broomfield, C. N. Jones and R. C. Reed, "Topologically Close Packed Phases in an Experimnetal Re-containing Single Crystal Superalloy", presented at 9th International Symposium on Superalloys, Seven Springs, PA, 2000
8. B. A. Boutwell, R. G. Thompson, N. Saunders, S. K. Mannan, and J. J. deBarbadillo, "Phase Formation Modeling of an Alloy 706 Casting Using Computational Thermodynamics", Superalloys 718, 625, 706 and Various Derivatives, ed. E. A. Loria, (Warrendale, PA: TMS, 1996), 99
9. A. O Basile and J. F. Radavich "A Current TTT Diagram for Wrought Alloy 718", Superalloy 718, 625 & Various Derivatives, ed. E. A Loria, (Warrendale, PA:TMS, 1991), 325
10. B. A. Lindsley, X. Pierron, G. E. Maurer, J. Radavich, "α-Cr Formation in Alloy 718 During Long Term Exposure: The Effects of Chemistry and Deformation", Long Term Stability of High Temperature Materials, eds.G. E Fuchs, K. A. Dannemann and T. C. Deragon, (Warrendale, PA: TMS, 1999.) 123
11. W. J. Boettinger, U. R. Kattner, S. R. Coriell, Y. A. Chang and B. A. Mueller, "Development of Multi-Component Solidification Micromodels Using a Thermodynamic Phase Diagram Database" Modeling of Casting, Welding and Advanced Solidification Processes, VII, eds. M. Cross and J. Campbell, J. (Warrendale, PA: TMS 1995) 649
12. N. Saunders, "Applicability of the Equilibrium and 'Scheil Model' to Solidification in MultiComponent Alloys", Solidification Processing 1997" eds. J. Beech and H. Jones (Univ.Sheffield: 1997) 362
13. A. P. Miodownik and N. Saunders "The calculation of APB Energies in $L1_2$ Compounds Using a Thermodynamic Database", Applications of Thermodynamics in the Synthesis and Processing of Materials, P. Nash and B. Sundman, (Warrendale, PA: TMS, 1995), 91
14. P. A. Flinn, "Theory of Deformation in Superlattices", Trans.Met.Soc.AIME, 218, (1960), 145
15. G. Inden, S. Bruns and H. Ackermann, "The Energy of Mechanically Produced Antiphase Boundaries (APB) in $L1_2$ and $L1_0$ Ordered FCC Alloys", Phil.Mag.A, 53, (1986) 87
16. A. G. Khachaturyan and J. W. Morris Jr., "The Interfacial Tension of Antiphase Domain Boundaries", Phil.Mag.A., 56, (1987), 517
17. W. M. Stobbs, "Measurement of APB Energy in a Series of Ni-based Alloys", Univ.Cambridge report to Rolls-Royce plc., Derby, U.K. 1993
18. N. Baluc, R. Schaublin and K. J. Hemper, "Methods for Determining Precise Values of Antiphase Boundary Energies in Ni_3Al", Phil.Mag.Lett., 64, (1991), 327
19. M.Hardy, C.J.Small, S.A.Franklin and D.J.Bryant, "The development of a new Ni-alloy for disc applications up to 725°C", presented at 9th International Symposium on Superalloys, Seven Springs, PA, 2000
20. G. L. Erickson, K. Harris and R. E. Schwer, "The Development of CMSX-5 a 3rd Generation High Strength Single Crystal Alloy" presented at the TMS Annual Meeting, New York, 1985
21. R. Darolia, D. F. Lahrman R. D. Field, "Formation of Topologically Closed Packed Phases in Nickel Base Single Crystal Superalloys", Superalloys 1988, eds. S. Reichman, D.N.

Duhl, G. Maurer, S. Antolovich and C. Lund (Warrendale, PA: TMS, 1988) 255

22. N. Saunders, "Calculated Phase Equilibria and its Use in Predicting Long Thermal Stability in Complex Alloys", Long Term Thermal Stability of High Temperature Materials, eds. G. E Fuchs, K. A. Dannemann and T. C. Deragon. (Warrendale, PA: TMS, 1999), 81

23. N. Saunders, Ni-DATA ver.4, Thermotech Ltd., The Surrey Research Park, Guildford, Surrey, UK

24. C. S. Proctor, "Formation and Effects of Intermetallics in the Rhenieum-containing Nickel-base Superalloy CMSX-4", (Ph. D. Thesis, University of Cambridge, 1993)

FORMATION OF A Pt$_2$Mo TYPE PHASE IN LONG-TERM AGED INCONEL® ALLOY 686

Michael G. Fahrmann and James R. Crum

Special Metals Corporation
3200 Riverside Drive
Huntington, WV 25705

Abstract

INCONEL[1] alloy 686 is a newly developed highly corrosion-resistant Ni-Cr-Mo superalloy, intended for usage at moderate temperatures. The alloy is ordinarily single-phase austenitic in the as-fabricated, i.e., mill-annealed condition. The metallurgical stability of this alloy upon extended aging at two different temperatures of 450 °C (842 °F) and 500 °C (932 °F) for times up to 5,000 hours was investigated. The material originated from a commercial heat. Samples were aged in both the mill-annealed and mill-annealed plus 15 % cold worked conditions, and subsequently examined by differential thermal analysis, optical and electron microscopy. Evidence for the formation of a low-temperature intermetallic Pt$_2$Mo type phase is provided. The implications of this metallurgical reaction for ductility and corrosion resistance are discussed.

Introduction

INCONEL alloy 686 is a newly developed Ni-Cr-Mo superalloy designed to be highly corrosion resistant to both oxidizing and reducing acids [1]. In the as-fabricated, i.e., mill-annealed condition the alloy is ordinarily single-phase austenitic. Due to heavy alloying with Cr and, in particular, Mo, it is not surprising that this alloy is prone to μ / P phase formation at temperatures exceeding approximately 760 °C (1500 °F) [2]. Such exposure temperatures are, however, well above the intended rather moderate usage temperatures.

Despite the relatively low usage temperatures of 550 °C (1000 °F) or less, the excessive lifetimes required from the material in some applications warranted a study of its metallurgical stability. At the same time, the retention of the corrosion resistance and selected mechanical properties such as tensile ductility upon extended aging were investigated and correlated with any microstructural changes observed.

Experiment

For this study material was available from a commercial heat of INCONEL alloy 686 whose composition is shown in Table I. Note the very low C and Si contents typical of this family of highly corrosion-resistant alloys.

Table I : Chemical composition (in wt.%) of the INCONEL alloy 686 material studied.

Ni	Cr	Mo	W	Fe	Mn	Si	Co	Al	C
bal	20.4	16.3	3.9	0.4	0.2	0.01	0.04	0.2	0.002

The sample blanks originated from mill-annealed, approximately 25 mm (1") thick hot rolled plate (ASTM grain size # 1 1/2). A portion of this plate was cold rolled (15% reduction in thickness) prior to aging. Both, blanks from the mill-annealed and mill-annealed plus cold worked material were exposed to temperatures of 450 °C (842 °F) and 500 °C (932 °F), respectively, for times up to 5,000 hours in air. Samples were pulled after 200 h, 2,000 h, and 5,000 h.

Differential thermal analyses (DTA) were performed on 200 mg specimens prepared from the aged samples. The dynamic atmosphere was He, and alumina powder was used as a reference sample. The heating rate was set to 40 °C /min, and the cooling rate upon reaching 1200 °C to 20 °C /min. From a few selected aged conditions specimens were prepared for transmission electron microscopy following standard procedures, i.e., jet polishing in a solution of 10% perchloric acid in methanol at − 45 °C. A Philips EM 400 operated at 100 kV was employed for the microscopy.

Tensile and corrosion testing were performed on the aged material in conformance with ASTM standards E-8 and G28 practice A (sulfuric acid + ferric sulfate).

---

[1] INCONEL is a registered trademark of the Special Metals Corporation family of companies

Superalloys 2000
Edited by T.M. Pollock, R.D.,Kissinger, R.R. Bowman,
K.A. Green, M. McLean, S. Olson, and J.J. Schirra
TMS (The Minerals, Metals & Materials Society), 2000

Results

Thermal Analysis

Typical DTA traces of the as-fabricated and long-term aged material are shown in Figs. 1 a and 1b, respectively. Note the strong endothermic reaction that took place at around 700 °C (1300 °F) in the aged sample upon heating. This reaction is virtually non-existent in the mill-annealed condition.

No complementary exothermic reaction was detected upon cooling, nor was the endothermic reaction present upon second heating. This is indicative of the dissolution of a low-enthalpy phase upon first heating, which does not re-form in any appreciable amounts upon cooling at the cooling rates employed in this study.

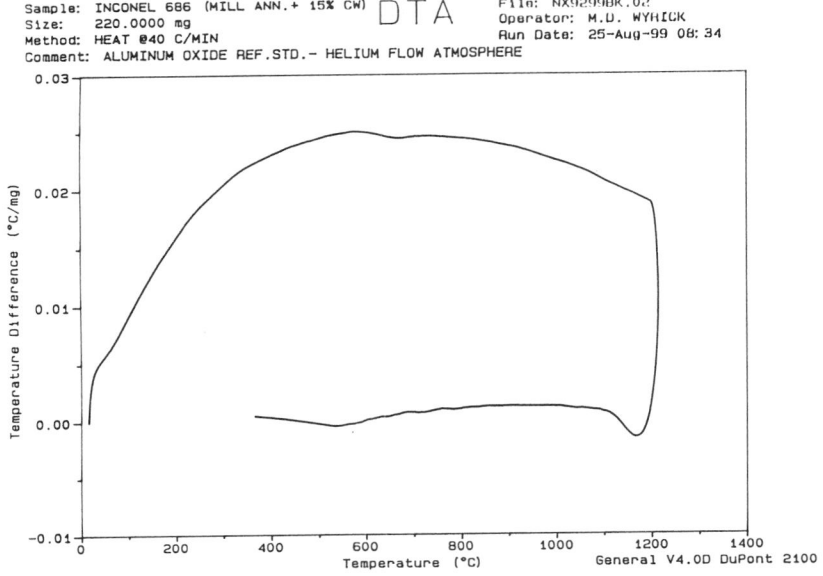

Fig. 1 a : Typical DTA trace of mill-annealed samples of INCONEL alloy 686.

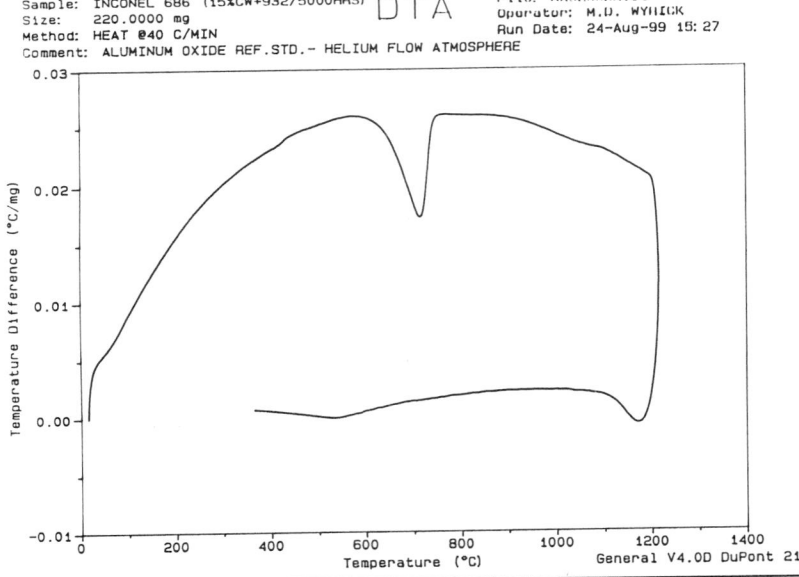

Fig. 1 b : DTA trace of a sample of INCONEL alloy 686, mill-annealed plus 15% cold worked, and aged for 5,000 hours at 500 °C (932 °F).

A strong dependence of the magnitude of the endothermic reaction on the aging conditions was observed. This relationship is semi-quantitatively reflected in Fig. 2, in which the latent heat of dissolution of that phase was expressed by the area under the respective DTA master curve. The reproducibility of the quoted numbers is about ± 1 unit. Although not necessarily accurate in absolute terms, this parameter does reflect trends since the same conditions (heating rate, gas flow etc.) were employed in the thermal analysis.

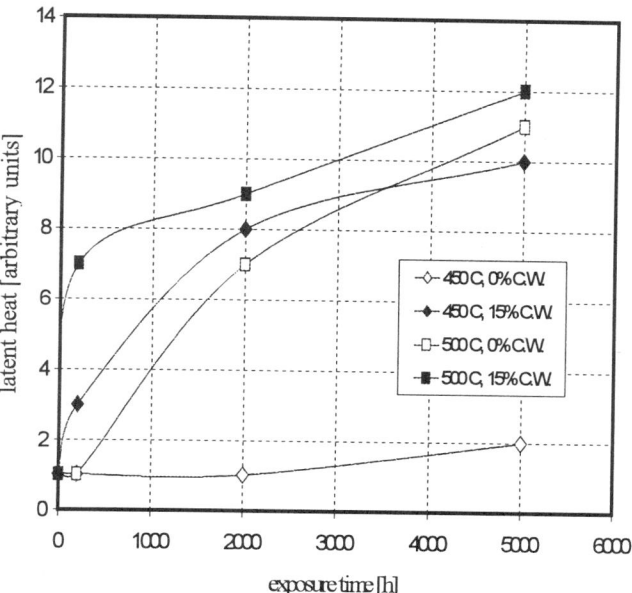

Figure 2 : The latent heat of dissolution (in arbitrary units) of the phase(s) formed upon long-term aging in samples of INCONEL alloy 686 is plotted as a function of aging time. Curve parameters are the aging temperature and the amount of cold work prior to aging.

The magnitude of the effect increases continuously with aging time and temperature, indicative of a diffusion-controlled phase transformation. For longer aging times the effect seems to saturate. Evidently, prior cold work greatly enhances the kinetics of formation of the phase(s) as most compelling for the 450 °C exposures.

Aged Microstructure

TEM selected area diffraction patterns (Fig. 3) of the long-term aged samples revealed consistently well-resolved superlattice reflections at every 1/3 <220>, 1/3 <420>, and 1/3 <311> type reciprocal lattice vectors for various beam directions. This was true regardless of the amount of prior cold work and the actual aging temperature.

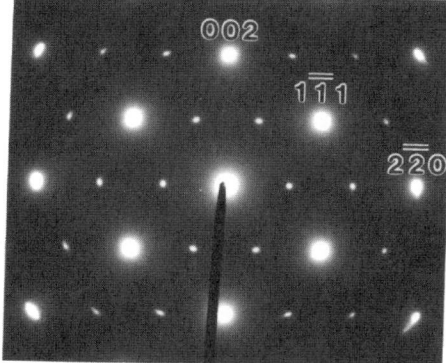

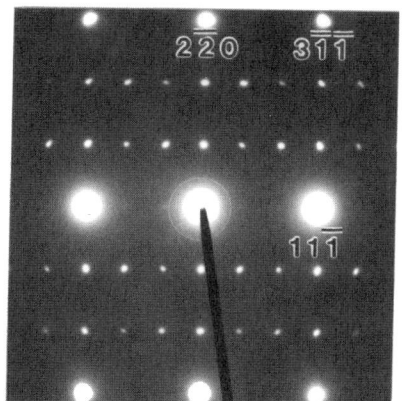

Figure 3 : TEM selected area diffraction patterns (750 mm camera length) obtained from long-term aged samples of INCONEL alloy 686 for various zone axes : (a) [100], (b) [110], (c) [111], and (d) [112]. The fundamental reflections were indexed in terms of fcc notations.

815

These superlattice reflections and their location with respect to the fundamental reflections (the location of the fundamental diffraction spots agreed well with the location of the diffraction spots of the matrix in the mill-annealed condition) indicate the presence of a second phase whose structure is derived from the matrix phase by some ordering reaction. The ordered phase was subsequently examined using dark-field techniques employing the various superlattice reflections for imaging.

Fig. 4 shows the typical intragranular precipitation of the ordered phase. Note the uniformity of the dispersion and the extremely fine length scale involved : the precipitate size is of the order of 10 nm. Such features cannot be resolved with a light microscope nor a scanning electron microscope. Also note the large apparent volume fraction of the ordered phase.

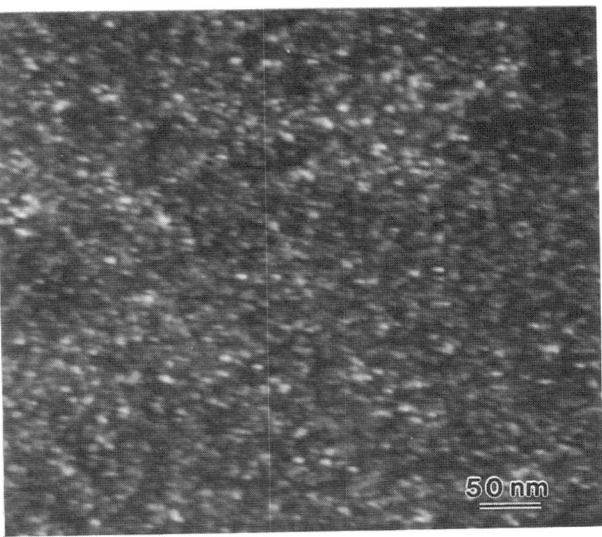

Fig. 5 a : Dark-field image ($g = 1/3 <311>$) of the microstructure obtained after aging mill-annealed plus 15% cold worked samples of INCONEL alloy 686 for 5,000 hours at 450 °C (magnification 170,000x). The beam direction is close to a <114> zone axis.

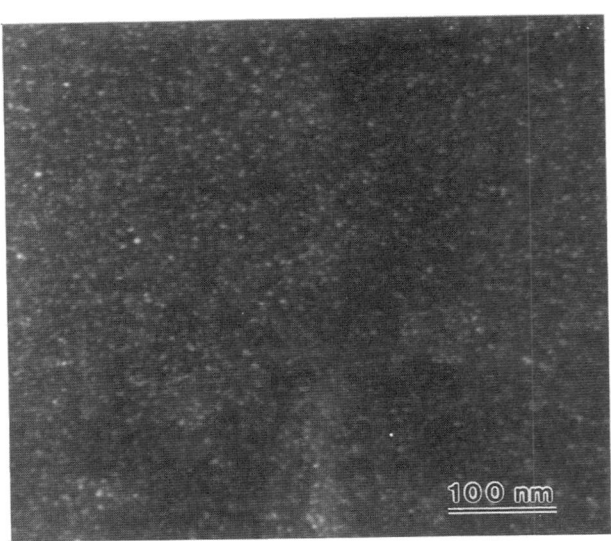

Figure 4 : Typical dark-field image ($g = 1/3 <311>$) of the microstructure obtained after long-term aging of samples of INCONEL alloy 686 at 450 °C (magnification 135,000x). The beam direction is close to a <114> zone axis.

At even higher magnifications the morphology of the ordered precipitates (Fig. 5) becomes more apparent : their shape is predominantly disk-like for both aging temperatures studied.

The size of the ordered features appears to be markedly bigger in the sample aged at 500 °C in comparison with the sample aged at 450 °C for the same aging time of 5,000 hours (Fig. 5). A comparison of the volume fractions is problematic due to the generally low-contrast images and their sensitivity to foil bending. Moreover, different type g vectors were employed for imaging in the micrographs of Fig. 5.

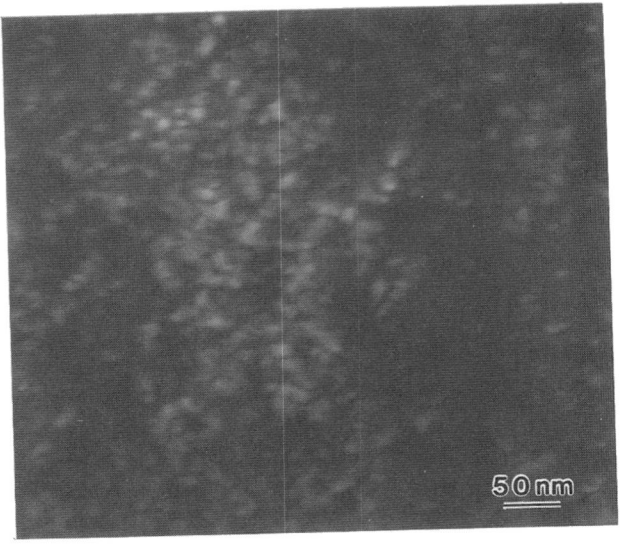

Fig. 5 b : Dark-field image ($g = 1/3 <220>$) of the microstructure obtained after aging mill-annealed plus 15% cold worked samples of INCONEL alloy 686 for 5,000 hours at 500 °C (magnification 170,000x). The beam direction is close to a <114> zone axis.

Surprisingly, a relatively low dislocation density was observed in all TEM specimens prepared from the 15% cold worked material even though the data in Fig. 2 suggests a marked effect of prior cold work on the kinetics of formation of the ordered phase. In fact, fairly large areas where the ordered phase had apparently homogeneously nucleated were found to be devoid of dislocations (Fig. 6).

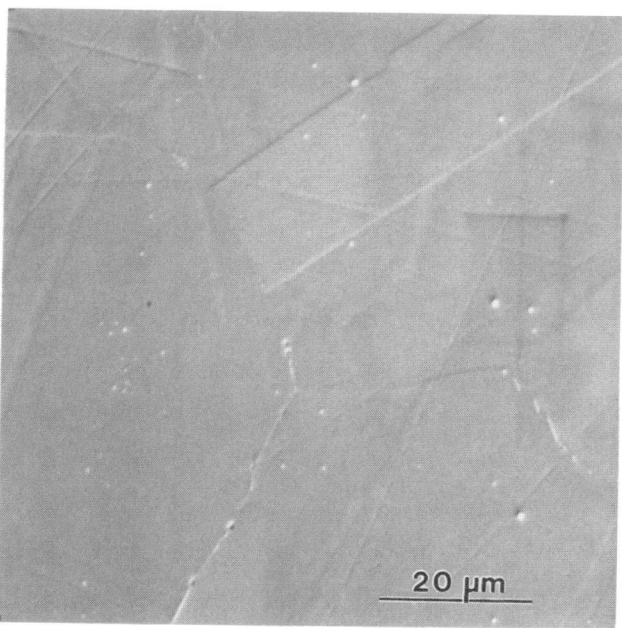

Fig. 7 a : SEM image of the typical microstructure of INCONEL alloy 686 in the mill-annealed plus 15% cold worked condition. The sample was etched electrolytically in 10 % chromic acid.

Fig. 6 : Bright-field image (two-beam condition, $\mathbf{g} = <220>$) of the microstructure obtained after aging mill-annealed plus 15% cold worked samples of INCONEL alloy 686 for 5,000 hours at 450 °C (magnification 135,000x).

The grain boundary microstructure is usually of great concern when retention of ductility and corrosion resistance (in particular, susceptibility to intergranular attack) upon aging are required.

Examination on both the light microscopy and scanning electron microscopy (SEM) levels did not reveal any marked precipitation in the grain boundaries of the aged samples in comparison with the mill-annealed samples (Fig. 7).

Examination of a TEM specimen which happened to contain a grain or twin boundary (the exact nature of this boundary was not elucidated) confirmed these findings : the boundary appears to be devoid of any gross precipitation (Fig. 8). The ordered phase nucleated homogeneously even in very close proximity to the boundary, and no denuded zones were observed (Fig. 9).

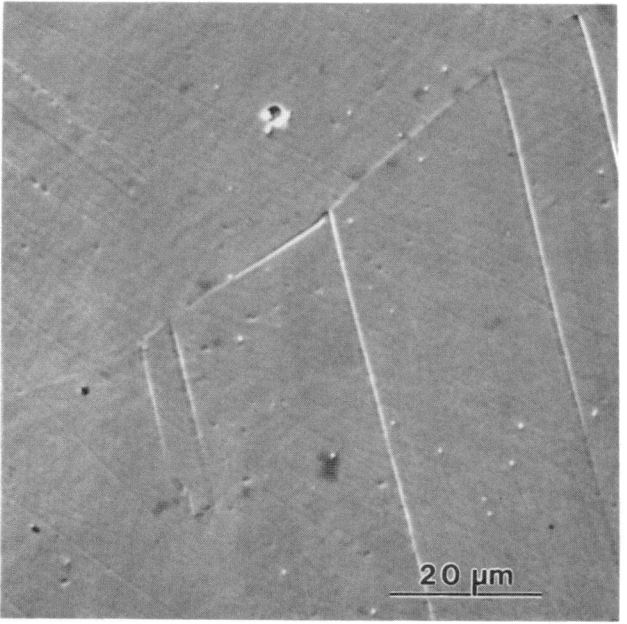

Fig. 7 b : SEM image of the microstructure obtained after aging mill-annealed plus 15% cold worked samples of INCONEL alloy 686 for 5,000 hours at 500 °C. The sample was etched electrolytically in 10 % chromic acid.

Fig. 8 : Bright-field image of a grain or twin boundary after aging mill-annealed plus 15% cold worked samples of INCONEL alloy 686 for 5,000 hours at 450 °C (magnification 22,000x).

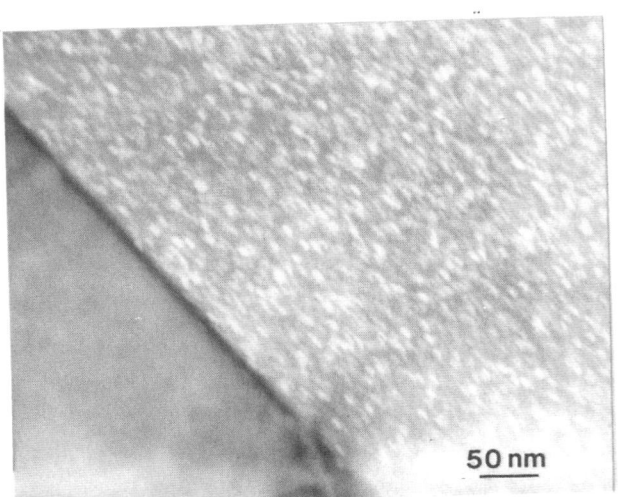

Fig. 9 : Dark-field image (**g** = 1/3<220>) of the microstructure in the vicinity of the boundary shown in Fig. 8 (magnification 170,000x). The beam direction is close to a <112> zone axis.

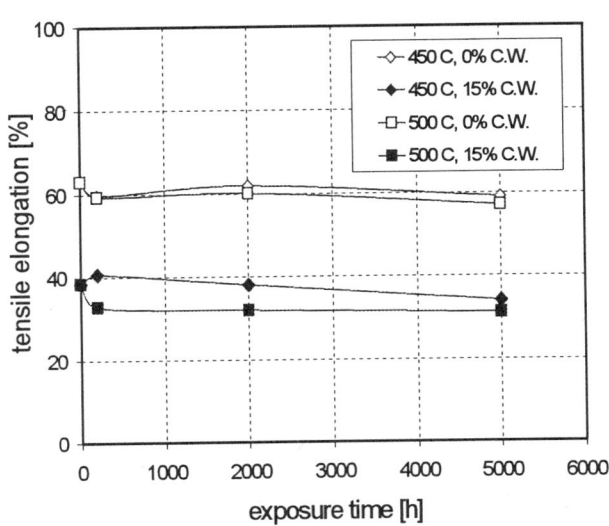

Fig. 10 : The tensile ductility of aged samples of INCONEL alloy 686 is plotted as a function of aging time. Curve parameters are the aging temperature and the amount of prior cold work.

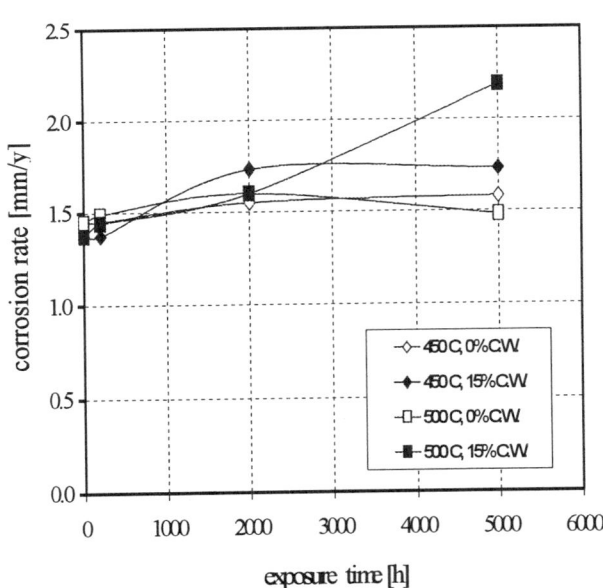

Fig. 11 : The corrosion rate per ASTM G28A of aged samples of INCONEL alloy 686 is plotted as a function of aging time. Curve parameters are the aging temperature and the amount of prior cold work.

Retention of tensile ductility and corrosion resistance upon aging is shown in Figs. 10 and 11, respectively. Except for the expected drop in ductility due to cold work, aging by itself did not lead to any significant ductility losses for the times and temperatures studied. The same is true for the corrosion resistance : within the scatter of the individual tests no significant deterioration of corrosion resistance was observed in this particular environment.

Discussion

The combination of thermal analysis and electron diffraction provides compelling evidence for the formation of an ordered, low-enthalpy phase in mill-annealed INCONEL alloy 686 upon aging at moderately elevated temperatures.

The observed diffraction patterns are identical with those reported for Ni_2Cr and (metastable) Ni_2Mo in the respective binary systems [3,4], and a presumed $Ni_2(Cr,Mo)$ phase in HASTELLOY[2] alloys S, C-4, and C-276 [5,6]. The patterns are typical of the Pt_2Mo structure type (orthorhombic crystal lattice) which can be derived from the A1 matrix structure by a peculiar stacking of the {220} planes, i.e., the alternation of two planes of predominantly Ni atoms followed by one plane of predominantly Cr / Mo atoms [4]. The length scale of the extremely fine and uniform intragranular distribution of the low-temperature phase found in this study (Fig. 5) is very similar to that observed in other Ni-Cr-Mo alloys aged at comparable temperatures [5,6].

The combined atomic fraction of Cr and Mo in INCONEL alloy 686, equaling roughly 30 at.%, leads one to believe that an ordering mechanism was active (the composition range for Ni_2Cr is quoted to be 25 to 36 at.% [3]). Such ordering only requires short-range lattice diffusion (characteristic diffusion length of the size of the unit cell) and can, thus, proceed at a reasonable pace even at temperatures as low as 450 °C. Moreover, it would promote a uniform dispersion since no nucleation event in the classical sense is needed.

Following this line of reasoning one would expect the long-term aged material to consist of 100 % ordered phase. This seems to be in contradiction to the apparent volume (area) fraction displayed in the dark-field images in Fig. 5. It has to be recalled, however, that, depending on the particular **g** vector selected, only a fraction of the six possible orientational variants [4] will be in contrast. For instance, when employing **g** = 1/3 [220], only one variant will be in contrast. Hence, the actual volume fraction of the ordered phase will be quite high after 5,000 hours of aging, and the reaction kinetics will have slowed down already considerably at this point as suggested by Fig. 2.

A poorly understood aspect of the ordering reaction concerns the accelerating effect of prior cold work on the kinetics of formation of this phase. Although well documented in Fig. 2, no direct microscopic evidence could be obtained. It is speculated that the generally higher vacancy concentration resulting from cold work might have enhanced the kinetics.

Notwithstanding the detailed nature of the mechanism, the extremely small size and, more importantly, uniform distribution of the ordered phase up to grain boundaries (Fig. 9) are thought to be primarily responsible for the retention of corrosion resistance. This reasoning is based on the apparent absence of zones deprived of the alloying elements, in particular Cr and Mo, that made the aged material virtually immune to intergranular attack. In contrast, "sensitized" microstructures, i.e., microstructures exhibiting heavy precipitation of μ /P phases in the grain boundaries [2], are generally susceptible to massive intergranular corrosive attack. At the same time, the "clean" grain boundaries (being devoid of gross precipitation of continuous films like the ones reported for HASTELLOY alloy C-276 [6] after long-term aging at 537 °C) are believed to have been instrumental in retaining tensile ductility.

It is interesting to note that the magnitude of the endothermic DTA peak associated with the dissolution of the ordered phase (Fig. 2) upon heating is much greater than the magnitude of the endothermic DTA peak associated with the dissolution of μ / P phase in a sample exhibiting massive precipitation thereof. Hence, in addition to facilitating kinetics, the ordered phase also seems to be favored thermodynamically at the temperatures studied.

Summary

The microstructural stability of INCONEL alloy 686, a newly developed highly corrosion-resistant Ni-Cr-Mo superalloy, was studied at two temperatures of 450 °C (842 °F) and 500 °C (932 °F) for aging times up to 5,000 hours.

Employing differential thermal analysis and electron microscopy, it was found that the austenitic matrix undergoes an ordering reaction at these rather moderate temperatures, resulting in the formation of a phase isomorph to Pt_2Mo. Cold working prior to aging greatly enhanced its kinetics of formation.

The extremely fine-scale and, more importantly, apparently uniform dispersion of this phase are believed to be the main cause for the excellent retention of resistance against intergranular corrosion and the retention of tensile ductility upon aging.

Acknowledgments

Helpful discussions with T. Summers (Lawrence Livermoore National Laboratories) are acknowledged.

---

[2] HASTELLOY is a registered trademark of HAYNES INTERNATIONAL

References

[1] J. R. Crum, J. M. Poole, and E. L. Hibner,
     U. S. patent 5,019,184.

[2] J. R. Crum and L. E. Shoemaker,
     NACE Annual Conference and Corrosion Show,
     Corrosion'93, paper no. 423.

[3] T. B. Massalski, Binary Alloy Phase Diagrams,
     Vol. 1, (ASM,1986), p. 843.

[4] S. K. Das and G. Thomas, phys. stat. sol.(a) 21 (1974)
     177.

[5] H. M. Tawancy, Met. Trans. 11 A (1980) 1764.

[6] H. M. Tawancy, J. Mat. Sci. 16 (1981) 2883.

DEVELOPMENT OF NEW NITRIDED NICKEL-BASE ALLOYS FOR HIGH TEMPERATURE APPLICATIONS

Claudio D. Penna

Institute of Metallurgy
Swiss Federal Institute of Technology - Zentrum
8092 Zurich - Switzerland

Abstract

The development strategy for what we believe to be a highly promising new family of superalloys is presented. Most superalloys reach their upper limit for useful service above 900°C: age hardened alloys encounter a dramatic drop of their very high strength because precipitates are no longer thermally and/or thermodynamically stable. Solid solution alloys still provide useful strengths slightly above 1000°C. In the temperature range 800 to 1100°C other strengthening mechanisms and phases must be used: examples are ODS alloys. The development presented consists in the production of a fairly fine grain structure (~50µm) in which nitride particles (with a size of 1-5 µm) are dispersed in order to avoid grain growth at high temperatures, and in the use of solid solution additions to promote additional high temperature strength and creep resistance. The useful temperature of this new alloy is situated in a range above 800°, up to 1100°C; below 800°C age hardened alloys are much stronger and therefore more suitable.

The alloys were produced by nitriding a Ni-Cr melt under high nitrogen partial pressure. For typical content of 0.5-0.8N during alloy solidification 10-20vol.% nitrides were formed, and homogeneously dispersed in the matrix after thermomechanical treatment. A grain size close to 50 µm was obtained. The nitridation and thermomechanical treatments have been carried out for different alloy compositions with a common Ni-Cr-N-base. The microstructure was shown to have a fairly good stability at 1100°C.

Small grains are believed to be deleterious for creep resistance, because they can lead to accelerated creep by grain boundary sliding: for this reason many high temperature alloys do not contain grain boundaries or only very low angle ones and some others (ODS or DS) have large elongated grains. However, good creep resistance is believed to be achievable by stabilizing grain boundaries by the nitrides.

In the alloys at temperatures below 1120°C the π nitride was observed; its precipitation was accompanied by a volume variation, which was measured by dilatometry. W, Mo and Ta were considerated as possible solid solution strengtheners and their effect tested on the alloys. W was identified as best candidate. The π nitride was found to be greatly stabilized by additions of Mo and W.

High temperature tensile tests revealed that the presence of nitrogen and tungsten induces a remarkable strengthening as well as the appearance of a pronounced yield phenomenon in the stress-strain behavior. Strength levels above 150 MPa at 1000°C were reached already with nitride-free Ni-Cr-N-W alloys.

Superalloys 2000
Edited by T.M. Pollock, R.D. Kissinger, R.R. Bowman,
K.A. Green, M. McLean, S. Olson, and J.J. Schirra
TMS (The Minerals, Metals & Materials Society), 2000

Introduction

A fundamental way of reaching higher efficiencies in a gas turbine is to raise the service temperature. As a consequence, there is a constant demand, from strength and corrosion points of view, for materials suitable for ever higher service temperatures. Motivated by this challenge we are developing a concept to obtain elevated temperature strength in nickel-base alloys.

When considering the use of alloys at high homologous temperatures ($T/Tm >0.5$) thermally activated processes start to play an important role; new glide systems may become active and diffusion causes microstructure evolutions, additionally vacancy diffusion assisted creep, dislocation climb as well as grain boundary migration and grain boundary sliding can occur.

Superalloys found today in high temperature applications, such as gas turbines, are based mainly on γ' precipitation strengthening, solid solution hardening or particles dispersion strengthening (or a combination of these). If specific mechanical properties at service temperatures above 900°C are required, precipitation strengthening might no longer be suitable because precipitates become unstable or rapidly coarse due to coalescence. Most superalloys reach therefore their upper limit for useful service above 900°C: age hardened alloys such as Nimonic 80A or Waspaloy encounter a dramatic drop of their very high strength. Solid solution alloys such as Hastelloy W or Haynes 230 still provide useful strengths slightly above 1000°C. In the temperature range 800 to 1100°C other strengthening mechanisms and phases must be used: examples are ODS alloys. Oxides dispersion (ODS) alloys provide reliable strength up 1100°C; however, their anisotropy, limited workability, relatively expensive cost and small rupture strain restrict their application. Superalloys development for service temperatures above 900°C, up to 1100°C, in the last decade has been characterized by strong solid solution strengthening additions together with the appearance of relatively fine-grained microstructures. A nickel matrix containing between 20 and 30 percent chrome is a common feature of these alloys (Haynes alloy HR120 and HR230, Inco Alloy 601GC and others [1,2, 3]) taking advantage of the high temperature stability of the cfc-γ phase and the high chromium content to increase the resistance to oxidation.

The development strategy for what we believe to be a highly promising new group of Ni-base superalloys is based on strengthening a Ni-Cr matrix with nitrogen and some substitutional solid solution strengtheners on one side; and by creating a fine grain microstructure stabilized against grain growth by the precipitation of nitrides, on the other side.

Solid solution strengthening

At high temperature solid solution strengthening is mainly due to elastic (misfit) and modulus interactions between matrix and foreign atoms, while other contributions can be ignored [4]; interstitials usually have a stronger strengthening effect than substituted atoms (relative per unit concentration), due to the tendency to produce non-spherical distortion in the lattice.

In Table I a few elements, typically found in superalloys, are listed together with their atom radii and elastic properties. In a Ni-30Cr matrix best candidates based on differences in size and moduli are primarily Re, W and also Nb, Mo and Ta, as also found in ref. [2-5] Each increment of 10 wt.% of tungsten and molybdenum the creep rate decreases by an order of magnitude, where tungsten strengthens to a somewhat larger extent than does molybdenum [3], which

is probably due to the difference in their elastic properties. Re as been also studied extensively as solid solution strengthener in nickel base alloys; it is also found to dramatically slow down diffusion at high temperatures [6]. Ta and Nb are often alloyed in order to precipitate intermetallic precipitates, as in Inconel 718.

Table I Relevant properties* of some elements for solid solution strengthening.

	Atomic radius [Å]	Young's modulus [GPa]	Shear modulus [GPa]
Ni	1.62	207	76
N	0.75	-	-
Co	1.67	211	83
Fe	1.72	200	81
Al	1.82	68	25
Cr	1.85	279	115
Re	1.97	469	176
Mo	2.01	330	120
W	2.02	400	175
Ta	2.09	185	69
Nb	2.09	103	37

* Data from: Metals Handbook, Vol.2, 10th ed., ASM Int. 1990.

The strengthening by interstitial nitrogen in steel is remarkable at room temperature [7], and a similar effect should be expected in nickel based alloys.

If these elements are to be used as solid solution strengtheners then it also important to consider the relative stability of their nitrides. For instance Ta and Nb form very stable nitrides. While Mo and W form stable carbides, they are not very strong nitride formers [8]. In the case of Ta, for instance, in a Ni30Cr alloy containing nitrogen the matrix will be depleted of this element because of the precipitation of nitrides.

Grain boundary strengthening

For high temperature applications of polycristalline alloys, grain boundary sliding (GBS) has been identified as a major deformation mechanism [9, 10, 11]. In the creep deformation mechanism maps for pure nickel [12] large portions of the map are dominated by GBS. Because of their sliding at high temperature, grain boundaries have been avoided, partially or completely, by production of single crystals or directionally solidified alloys. In polycristal alloys a dispersion of sufficiently stable particles have been shown, theoretically and experimentally, to significantly reduce grain boundary sliding. In Ni-base superalloys, grain boundary carbides have been shown to decrease the creep rate by suppressing GBS [9, 10 13].

Of course avoiding grain boundaries, or using a coarse grains structure, will exclude grain boundary strengthening. It is not clear whether grain boundary strengthening can be effective at temperatures above 900°C, although it has been shown that the Petch slope k_y varies weakly in the 77 to 300 °K temperature range for a series of binary Cu-Ni, Cu-Zn and Cu-Al alloys [14]. Grain boundary strengthening is known to be very effective in ferritic steels, but less so in austenitic iron based alloys [15].

Grain growth

In a similar way as with grain boundary sliding, particles are able to impede grain coarsening under certain conditions [16]. According to Ashby [17], at high homologous temperatures, for particles smaller than $r_0 = 3/2 \, f\gamma/P$ (where γ: boundary energy, P: driving pressure, f: volume fraction of particles), the boundary mobility decreases with increasing particle size. At homologous temperatures of 0.8 the mobility is in most cases controlled by matrix or interface diffusion. However, for particles larger than r_0 the boundary can escape and its intrinsic mobility is the controlling factor; in this range the model predicts an increasing mobility with increasing particle radius, calculated to be roughly of two orders of magnitude larger than for radii smaller than r_0. The lowest mobility should be obtained with a radius slightly smaller than r_0.

The Ni-Cr-N system

The solubility of nitrogen in iron-base melts is substantially higher than in a similar nickel-base alloys. The nitrogen solubility in a nickel-base melt can be raised by using higher partial pressures and alloying with Cr. With nitrogen partial pressure up to 50 bar, one can put more than 1% N in solution in the melt. The nitrogen solubility in the solid state for the same system is not as high, and primary nitrides precipitate. In this system four phases can be found depending on the temperature and composition: γ (fcc matrix), an hexagonal chrome nitride Cr_2N (ε), a cubic chrome nitride CrN and a mixed nitride: the π phase [2, 18, 19]. It has been concluded that the π phase has the structure of ß-manganese, and a stoichiometry close to $Cr_{13}Ni_7N_4$ [18]. At 1000°C the π phase is formed through the peritectoid reaction:

$$\gamma + Cr_2N \rightarrow \pi \qquad (1)$$

The phase was found to be unstable above 1100°C.

We have calculated the pseudo-binary phase diagram of the system Ni-Cr-N, for Cr content varying from 20 to 30 percent (using the software program Thermocalc®). In Figure 1 the pseudo binary phase diagram for Ni30Cr alloyed with nitrogen is presented. It shows the temperature and composition range for the π phase and of the ε nitride. The calculation correlates fairly well with information found in the literature. Lowering the chromium content results in the appearance of the cubic nitride, „coming from the right" in the diagram, and shifting to lower nitrogen concentration as the Cr content decreases further. Also the stability of the π phase is decreased with lower chromium concentrations.

With the same program the influence of Mo, Ta, Co and W on the

pseudo-binary phase diagram of Ni30Cr-N can be tested. The calculations show that Co, Mo and W have extended solubilities in γ. Cobalt was considered as a candidate for partial substitution of nickel to control alloy nitrogen solubility without influencing other properties. The thermodynamical calculation for tantalum predicts no solubility in γ and the precipitation of an intermetallic N-free Ta-Ni-Cr phase and the π nitride. Because of the stability of TaN [7] this prediction appears suspicious. In the case of tungsten, cobalt and molybdenum, the calculation shows that the π-phase can contain some W, Co and Mo respectively, by substitution of part of the Cr. The solubility limit of W in Ni20wt.%Cr is 17.5wt.% [2].

Alloy development strategy

The system Ni-30Cr shows very interesting features for high temperature applications. The presence of chromium allows the precipitation of nitrides; which, according to the calculated phase diagram for nitrogen concentration lower than ~0.7wt.%, are not stable in the melt. It is therefore possible to precipitate a nitride in the γ phase during the solidification of the alloy. If precipitation occurs in the melt, then some of the nitrides would be segregated either to the top or to the bottom of the melt. In the case of CrN and Cr_2N, for instance, having densities of 6.1 and 6.5 g/cm^3 respectively, the precipitates would segregate in the slag (nickel base superalloys have densities of 8-10 g/cm^3). A high chromium content also provides good high temperature oxidation resistance, which can be further improved with small additions of Y [20].

Interesting mechanical properties have been recognized to the Ni-20Cr alloys strengthened with W and Mo [2,3] for high temperature applications. For instance Haynes®230® provides a 0.2% yield strength of 145 MPa at 982°; this alloys contains 14 wt.% W and 2 wt.% Mo. Good strength and creep resistance have also been obtained with the alloy Ni30Cr(0.6-1)N [19]; the 0.2 yield strength at 1000°C was 85 MPa, together with a 10,000 hours creep rupture strength higher than 12 MPa.

The development strategy for what we believe to be a highly promising new group of Ni-base superalloys is based on strengthening a Ni matrix with 20 to 30% Cr with nitrogen and some substitutional solid solution strengtheners such as W, Mo and Ta on one side; and by creating a fine grain microstructure stabilized against grain growth by the precipitation of nitrides, on the other side. From a manufacturing point of view, this alloy can be produced by nitriding the melt, avoiding processing by powder metallurgy.

The development of this material demands fundamental investigations on several aspects: the phases that can occur as well as their properties, the role of interstitial nitrogen, the importance of its solid solution strengthening and grain boundary strengthening at high temperatures, the stability of the nitrides, the stability of the microstructure as well as the corrosion behavior. This paper presents the first part of the development dealing with the investigation on the Ni-Cr-N system and the influence of W, Mo and Ta as well as on the role of interstitial nitrogen on mechanical properties.

Experimental procedure

Alloy manufacturing

Alloys used in this work were produced in a vacuum induction furnace using pure Ni (99.95%), Co (99.9%), Cr (99.91%) and binary alloys of 55.2%Ni-44.7%W, 56.2%Ni-43.6%Ta and 55.4%Ni-44.1%Mo (compositions in wt.%).

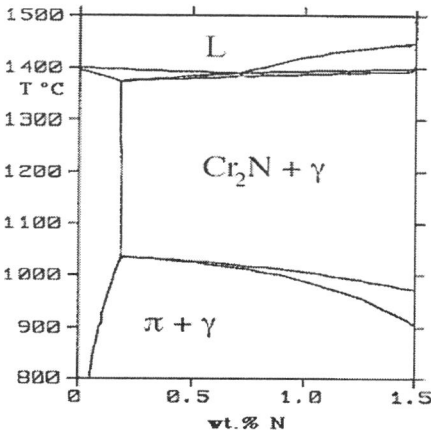

Figure 1: Ni30Cr-N pseudo binary equilibrium phase diagram calculated with Thermocalc®.

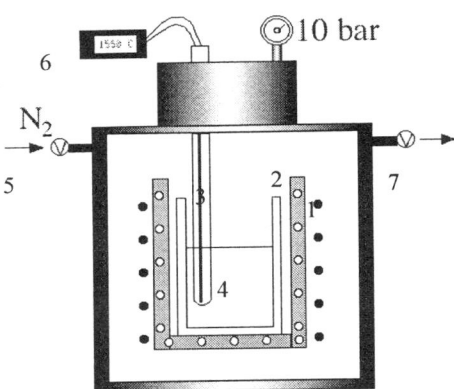

Figure 2: Schematic representation of the nitriding furnace. 1) Induction heating. 2) External crucible with cooling system. 3) Thermocouple. 4) Melt. 5) Nitrogen entry-valve. 6) Temperature indicator. 7) Exit valve. In order to evacuate the air prior to nitridation, the chamber is rinsed twice with at least 20 bar of nitrogen.

The N-free alloys were then remelted in our in-house developed high nitrogen pressure induction furnace (see Figure 2). The applied nitrogen pressure can be set between 1 and 200 bar, depending on alloy composition and the amount of nitrogen needed. The nitriding time and temperature were always 60 minutes and 1550°C, respectively. Typical pressures for Ni30Cr-based alloys in order to obtain 0.5-0.8 wt.% nitrogen are 15-25 bar. This values vary quite strongly with the Cr-content, as well as with the nature of other alloying elements. Once a 60 minutes nitriding time is reached, the melt is rapidly cooled down (~ 67°C/min.) by applying an approximately 8 times higher N_2 pressure, avoiding at the same time the danger of melt boiling. In Table II the compositions of the alloys discussed in this paper are given. The nitrided cylinders with a 50 mm diameter and a length of 90 mm were then homogenized for 24 hours at 1250°C. In order to obtain a fine and homogenous dispersion of the nitrides, the specimen were forged in the temperature range 1150-1200°C to 15x90xYmm blocks, which were then reheated to 1200°C before quenching. Since alloys must not be brought to forging temperature too quickly, otherwise microcraks develop and the material loses its workability, the blocks were first preheated in 30 minutes to 700°C and then transferred to the oven at 1200°C.

Table II Chemical composition of tested alloys (wt.%)

Cr	Mo	W	Ta	Co	Al	V	N	P(N_2)
30	-	-	-	-	-	-	0.89	20
30	-	5	-	-	-	0.2	0.17	10
30	-	5	-	-	-	-	0.75	20
30	4	-	-	-	-	-	0.7	10
30	8	-	-	-	-	-	0.38	6
30	8	-	-	-	-	-	0.67	10
25	-	-	6	-	-	-	0.63	25
30	-	-	-	6	-	-	0.58	20
25	-	5	6	-	0.5	-	0.50	15
25	-	8	-	-	0.5	-	0.12	10

Ni-Cr-N system

A peritectoid reaction has been observed by dilatometry in the alloys Ni30Cr0.9N, Ni30Cr4Mo0.7N, Ni30Cr8Mo0.7N and Ni30Cr5W0.75N. Dilatometry was carried out in an argon atmosphere with a constant heating and cooling rate of 0.2 °C/s, in the 900 to 1300°C temperature range, where the reaction was expected.

High temperature properties

Two alloys were selected for testing the high temperature effect of the nitrides at 1100°C: Ni30Cr6Co0.6N and Ni30Cr8Mo0.4N. At this temperature in the former the equilibrium phases are Cr_2N and γ, while in the latter are π and γ. The alloys were aged at 1100°C in air for 835 hours.

High temperature tensile tests in air were carried out with a Schenk hydraulic machine at 1000°C at a constant rate of 2 mm/min. A first series of alloys was produced in order to study the effect of nitrogen in solid solution in the Ni30Cr5W alloy. Specimen were tested after 45-60 stabilization at 1000°C ± 0.5°. The temperature was monitored directly from a thermocouple placed on the specimen surface. Standard M-10 tensile specimens with a diameter of 6 mm and a gage length of 30 mm were used.

Other characterization

Chemical analysis were conducted with a EDX analyzer, for metallic elements, on a CamScan scanning electron microscope, working at 20KV. Nitrogen content is measured with a LECO TC-436 Nitrogen/Oxygen Determinator, which uses the inert gas fusion principle.

Results and discussion

Alloys microstructure

The microstructures of a Ni30Cr6Co0.6N after casting and nitridation, after homogenizing and after forging at 1200°C are shown in Figure 3 to Figure 5 respectively. The Co-addition did not affect the nitrogen content obtained after nitriding. The cooling after nitriding is fast enough to avoid the formation of the π nitride. Cr_2N forms as eutectic lamella during solidification. Homogenizing at 1250°C causes the thin nitride lamella to spheroidise, in order to lower interfacial energy. After homogenizing, the nitride particles have a

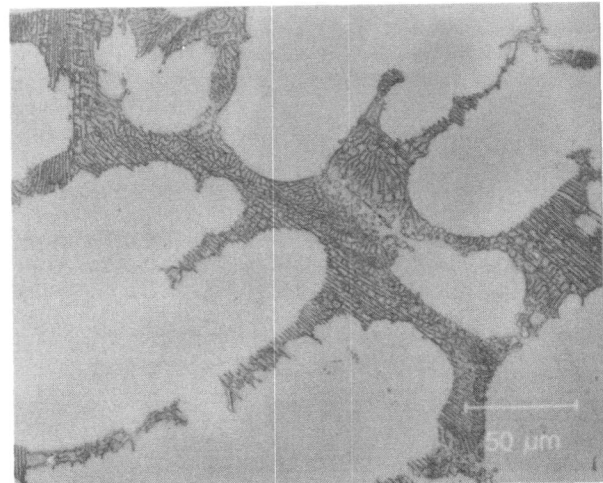

Figure 3: Microstructure of a Ni30Cr6Co0.6N alloy, after nitriding showing the Cr_2N lamella.

size of 1 to 5 µm.

The hot workability of the alloys, including the alloys containing Mo, W or Ta, is good. The presence of the nitride particles does not affect the workability of the materials for alloys containing less than 20 vol.%. Specimen containing more than 20 vol.% showed poor workability (cracking). The microstructure has a grain size

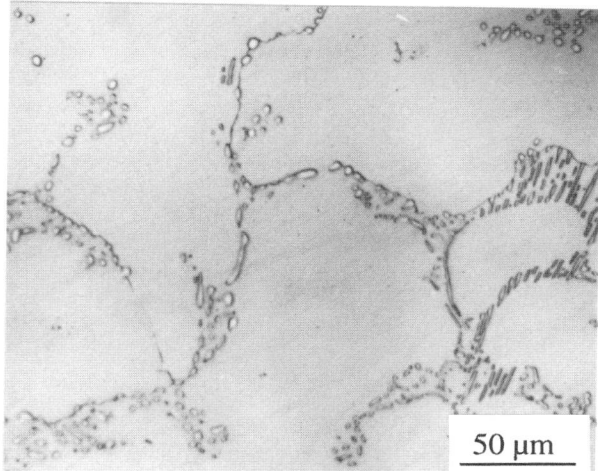

Figure 4: Microstructure of Ni30Cr6Co0.6N after 24 hours at 1250°C.

close to 50µm, and a fairly homogeneous distribution of nitride particles. The nitride volume fraction of the specimen shown in (Figure 5) is 0.12, as determined by quantitative image analysis.

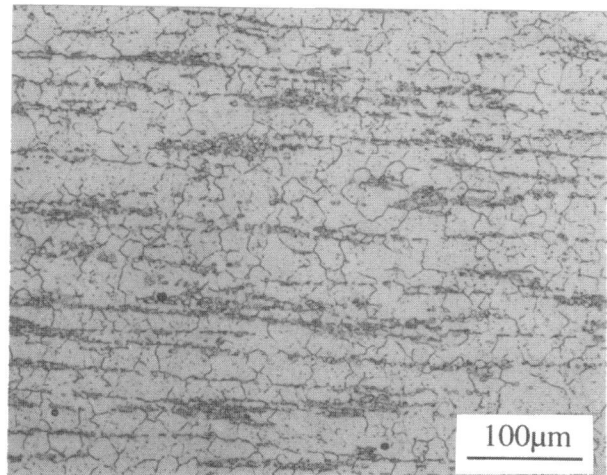

Figure 5: Microstructure of Ni30Cr6Co0.6N after forging.

The same manufacturing process was tested on similar alloys alloyed with Ta, Mo and W (Figure 6). The nitriding was successful in all cases, even though variations in the nitrogen content occurred depending on alloy composition. As the alloy in the nitriding furnace was cooled down rapidly, the microstructure was composed in all cases of γ and chromium nitride lamella. The lamella contain a non negligible amount of Ta (30wt.%), in the alloy where this element is present. After aging at 1100°C for 835 hours, this alloy developed a microstructure composed of γ, a Ni-Ta-Cr phase and the

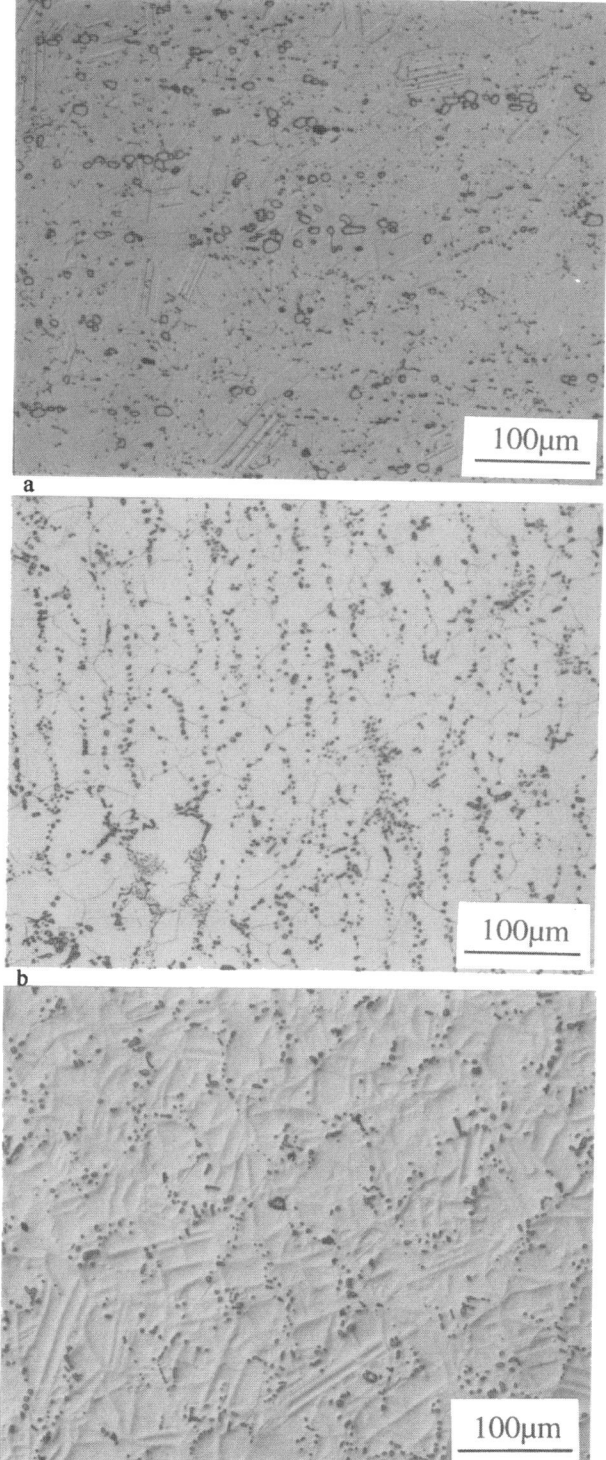

Figure 6: Examples of nitrided, homogenized and forged alloys: (a) Ni30Cr8Mo0.6N, (b) Ni25Cr6Ta0.6N and (c) Ni25Cr4W6Ta0.5N.

π nitride, as predicted by the thermodynamical calculations. The Mo-alloyed specimen after homogenization was composed of γ and π, while the W- and Co-alloyed specimen were composed of γ and Cr_2N. In all cases the hot workability was fairly good, and the nitrides were dispersed quite uniformly throughout the volume of the material. Many annealing twins could be observed in the alloys containing tungsten, tantalum and molybdenum additions.

The peritectoid reaction

The phase transformation between Cr_2N, γ and π phase (see equation 1) revealed a measurable volume decrease in the material of the order of 0.1%. The change was detected by dilatometry, during heating and cooling at a constant rate. It was therefore possible to study this reaction and the effect of alloying W and Mo on the material. Figure 7 shows an example of a dilatometry test run on the alloy Ni30Cr0.9N. The measured relative change in length of the specimen was converted into relative rate of length change, which would give a peak in the temperature range of the reaction. Of course kinetics effect must be taken into account and the estimated temperatures are slightly shifted to lower temperatures, if measured during heating, and to higher temperatures, if measured during cooling.

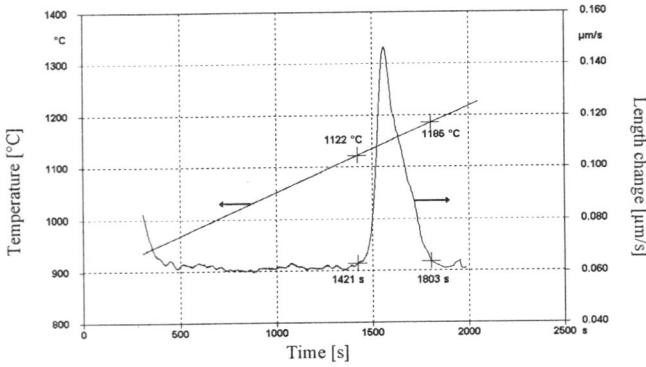

Figure 7: Dilatometry result of test ran on Ni30Cr0.9N.

The results obtained with different specimens (Table III) show the influence of W and Mo on the reaction, i.e. relative stability of the nitrides. As it can be seen both W and Mo cause a shift to higher temperatures; two facts can be responsible. On one side, the presence of atoms with a much larger atomic radius, and a lower mobility, can lower the nitrogen diffusivity, and therefore the „incubation" time of the reaction. On the other side, W and Mo can substitute for part of the Cr in the π phase, as revealed by EDX, and therefore have a stabilizing effect, shifting the reaction to higher temperatures. The π nitride appears to be greatly stabilized by Mo,

Table III Effect of Mo- and W-additions on $\gamma + Cr_2N \rightarrow \pi$ reaction temperatures.

Ni30Cr-	0.9N	4Mo 0.7N	8Mo 0.7N	5W 0.7N
ΔT (up)	1120°-1185°	1210°-1280°	>1300°	1160°-1180°
ΔT(down)	1180°-1195°	1260°-1280°	>1300°	1180°-1240°
π compos. [wt.%]	Ni 42 Cr 58	Ni 41 Cr 51 Mo 8	Ni 43 Cr 45 Mo 11	Ni 41 Cr 53 W 4.5
γ comp. [wt.%]	Ni 77 Cr 23	Ni 69 Cr 28 Mo 3.5	Ni 65 Cr 29 Mo 6.5	Ni 68 Cr 26 W 5

as also confirmed in reference 18 to a point that in the specimen containing 8wt.% Mo, Cr_2N can only be observed in a fast cooled

melt, while after homogenization at 1250°C π nitrides only are found. The tungsten addition has a similar effect, even though less dramatic. The effects of alloying Mo and W are for both heating and cooling measured reaction temperature ranges very similar. It can be concluded therefore, that Mo and W effectively stabilizes the π nitride shifting the reaction to higher temperatures

High temperature properties

In order to monitor the high temperature stability of the microstructure two alloys were aged at 1100°C, which is the temperature considered as the upper limit service temperature for the present development. The compositions Ni30Cr6Co0.6N and

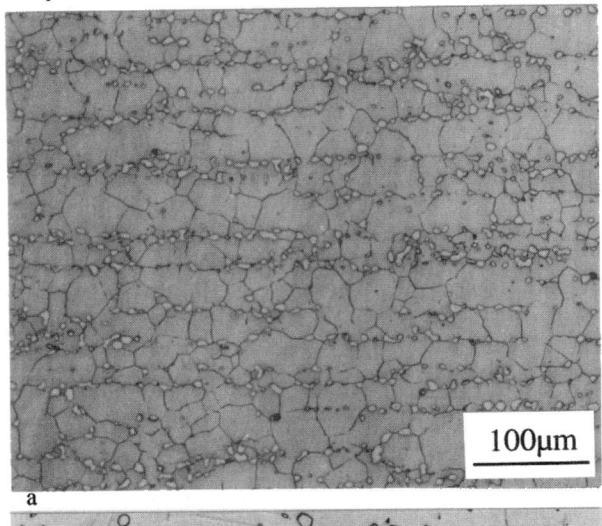

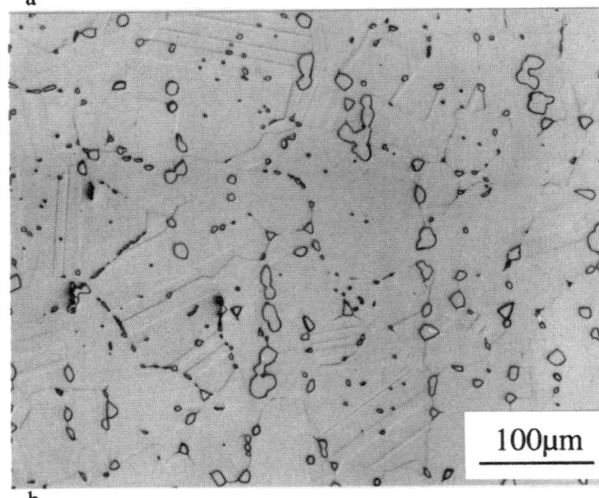

Figure 8: Alloys Ni30Cr6Co0.6N (a) and Ni30Cr8Mo0.7N (b) after 835 hours at 1100°C.

Ni30Cr8Mo0.4N were chosen in order to study the microstructure stability for both Cr_2N and π nitrides. Figure 8 shows the microstructure of the two alloys after 835 hours aging at 1100°C (see for comparison the microstructures after forging in Figure 5 and Figure 6a). As it can be observed, a certain change in the distribution and size of the particles has occurred, while the grain size has increased only very slightly, remaining close to 50-100 µm. This increase in nitride volume fraction probably takes place during annealing at the expense of submicron nitride particle, residues after forging, and of the slightly nitrogen supersaturated γ. This is also confirmed by the change in hardness measured after annealing in γ in the Ni30Cr6Co0.6N alloy dropping from 235 ±7 HV$^{0.05}$ to 208 ± 3 HV$^{0.05}$.

Mechanical properties

Tensile tests at 1000°C were run on specimens of composition Ni-30Cr-5W, Ni-30Cr-5W-0.17N and Ni-25Cr-8W-0.12N. In both nitrided alloys all the nitrogen is in solid solution; in both cases after annealing at 900°C only a minor amount (<0.5vol.%) of nitride precipitation was observed on the grain boundaries.

The results in Figure 9 show remarkable differences in both strength and stress-strain behavior between the nitrogen-free and the nitrided specimens. The strength is basically doubled, and in the nitrided case a pronounced yield point appears (see Figure 9 and Figure 10). Moreover, if the tensile test is unloaded for a few seconds, a new yield point appears, showing a strain aging phenomenon (Figure 10). These tested specimens had a quite large grain size between 300 and 400 μm. These results clearly indicate a major role played by nitrogen.

The strain aging effect has two possible interpretations. On one side, solid solution strengtheners, on temporarily unloading, are able to diffuse over short distances to dislocation cores anchoring them, resulting in the appearance of a new yield point. The phenomenon is well known: for instance in steels [4] even at 127°C interstitial nitrogen and carbon cause strain aging on reloading after

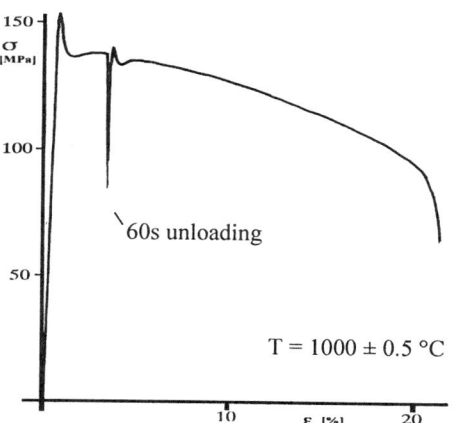

Figure 10: Tensile stress-strain curve for alloy Ni30Cr8W0.12N

placement, which would cause a second yield point. TEM investigations are necessary for confirmation of this point.

On the same alloy, but with a higher nitrogen content (0.75wt.%), x-ray diffraction phase identification analysis was carried out. The diffraction spectrum, shown in Figure 11, clearly reveals the presence of nickel, α-chromium, traces of Cr$_2$N as well as four peaks which can be attributed to the π phase ((013), (123), (134) and (125) planes) with the ß-manganese structure and a lattice parameter of 0.624 nm. At small diffraction angles a wide peak is detected, which can be related to the presence of clusters in the matrix (as indicated by the arrow in the figure).

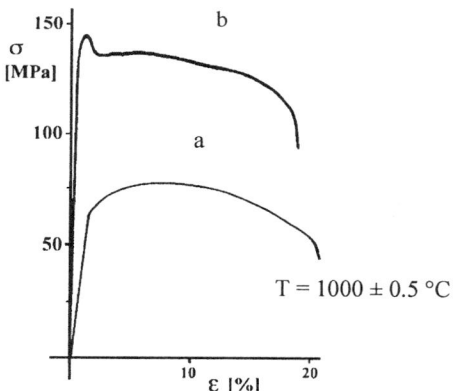

Figure 9: Tensile stress-strain curves for Ni30Cr5W N-free (a) and alloyed with 0.17wt.%N (b).

several days. At 1000°C even W atoms may have high enough mobility to participate in the pronounced yield strength, while strain aging, for unloading times of the order of one minute is not likely to be caused by W. Therefore the most likely interpretation is that strain aging is caused by nitrogen, which can diffuse rapidly to dislocation cores.

Considering the strong affinity between chromium and nitrogen, clustering or the formation of short range ordering is possible. Then a second possible interpretation is that dislocations are (also) anchored by clusters and under loading once a critical stress is reached brake away or unlocking [21] occurs and a yield point appears. At temperatures high enough, thermal activation allows dislocations to climb and, on unloading, regain an anchored

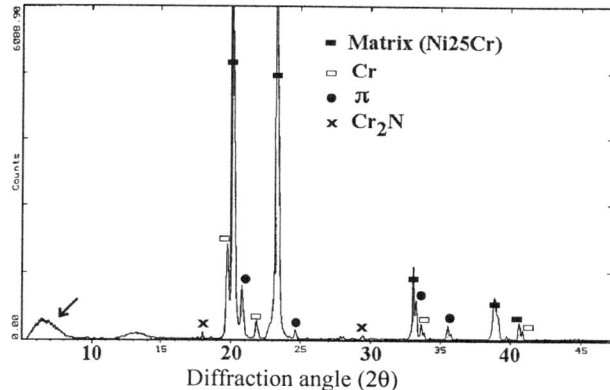

Figure 11: X-ray diffraction spectrum for alloy Ni30Cr5W0.6N (wavelength 0.7092 Å, Mo-k$_\alpha$2, CT: 6.5s).

The presence of nitrogen remarkably increases the strength of the material at 1000°C, while an increase of tungsten, from 5 to 8wt.%, has a less strong effect; the alloy contains though 0.12wt.% N (0.05 less than Ni30Cr5W). In this case, the first yield point is sharper, confirming that tungsten also contributed to it.

Concluding remarks

The stability of the microstructure and the good high temperature strength achieved with the combination of Ni-Cr-N with solid solution strengtheners is expected to be the base of a new high temperature alloys family. Strengths above 150 MPa at 1000°C should be reached, combining the strengthening effects of nitrogen, substituted elements, nitrides and grain boundaries. Creep tests need to be carried out, in order to confirm the inhibition of grain boundary sliding by the presence of nitride particles.

Summary of results

Ni-Cr-base alloys were nitrided in the liquid phase under high nitrogen partial pressures. For supersaturated alloys, lamellar nitrides are formed during the solidification; these can be broken up and homogeneously disperse in the matrix by means of thermomechanical treatments.

In the Ni-Cr-N system Cr_2N as well as the π nitride can be formed. The π phase become unstable above 1180°C but it is strongly stabilized by additions of Mo and, less dramatically however, by addition of W. Under 1180°C the π phase appears through a peritectoid reaction accompanied by a volume change of the order of 0.1%.

With a dispersion of fairly coarse nitrides a grain size close to 50 μm was maintained after 835 hours at 1100°C.

The strengthening by nitrogen and tungsten allowed to reach tensile strengths above 150 MPa at 1000°C. Evidence for cluster formation induced by the presence of nitrogen at 1000°C was found, and the appearance of a pronounced yield point and strain aging can, at least partially, be related to it.

References

1. U. Brill and D. C. Agarwal, „Alloy 2100GT: a New Ta-fortified Ni-Cr-Al-Alloy for Land Gas Turbines", Corrosion 99 (NACE, Houston TX, 1999), paper No.57.

2. T. Matsuo, M. Kikuchi and M. Takeyama, „Strengthening Mechanisms of Ni-Cr-W Based Superalloys for Very High Temperature Gas Cooled Reactors", Proc. First Int. Conf. of Heat-Resistant Materials, (ASM International, Ohio, USA, 1991), 601-614.

3. T. Matsuo et al.,"Strengthening of Nickel-Base Superalloys for Nuclear Heat Exchanger Applications", J. of Material Science 22 (1987), 1901-1907.

4. George D. Dieter, Mechanical Metallurgy (London, McGraw-Hill Book Company, UK, 1988), 203.

5. A. R.Braun and J. F. Radavich, „Microstructural and Mechanical Properties Comparison of P/M 718 and P/M TA 718", Superalloy 718 - Metallurgy and Applications (Warrendale PA, E.A. Loria, Metals and Materials Society, 1989), 623-629.

6. H. S. Ko, K.W. Paik et al., „Influence of Rhenium on the Microstructures and Mechanical Properties of a Mechanically Alloyed Oxide Dispersion-Strengthened Nickel-Base Superalloy", J. of Mat. Science 33 (13) (1998), 3361-3370.

7. V. G. Gavriljuk and H. Berns, „High Nitrogen Steels, Structure, Properties, Manufacturing, Applications" (Springer-Verlag Berlin, 1999), 10.

8. T. Rosenqvist, Principles of Extractive Metallurgy (London, McGraw Hill Book Company, 1974), 250.

9. X. J. Wu and A. K. Koul, „Grain Boundary Sliding in the Presence of Grain Boundary Precipitates during Transient Creep", Metall. and Mat. Trans. A 26A (1995), 905-914.

10. F. T. Furillo, J. M. Davidson and J. K. Tien, „ The Effect of Grain Boundary Carbides on the Creep and Back Stress of a Nickel-Base Superalloy", Materials Science and Eng. 39 (1979), 267-273.

11. X. J. Wu and A. K. Koul, „Modeling Creep in Complex Engineering Alloys", Creep and Stress Relaxation in Miniature Structures and Components, ed. H. D. Merchant (Warrendale PA, The Minerals, Metals & Materials Society, 1997), 3-17.

12. H. Lüthy, R. A. White and O. D. Sherby, „Grain Boundary Sliding and Deformation Mechanism Maps", Materials Science and Eng. 39 (1979), 211-216.

13. R. Castillo, A. K. Koul and J-P. A. Immarigeon, „The Effect of Service Exposure on the Creep Properties of Cast IN-738LC Subjected to Low Stress High Temperature Creep Conditions", Superalloys 1988 (Warrendale PA, S. Reichman, D.N. Duhl, G. Maurer, S. Antolovich and C. Lund, The Metallurgical Society, 1988), 805-814.

14. H. Suzuki and K. Nakanishi, „A Theory of the Grain Size Dependence of the Yield Stress in Face-Centered Cubic Alloys", Trans. JIM 16 (1975), 17-27.

15. K.J. Irvine, T. Gladman and F.B. Pichering, „The Strength of Austenitic Stainless Steels", JISI 199 (1969), 1017-1028.

16. M. F. Ashby, J. Harper and J. Lewis, „ The Interaction of Crystal Boundaries with Second-Phase Particles", Trans. Metallurgical Society of AIME, 245 (1969), 413-420.

17. M. F. Ashby, „The Influence of Particles on Boundary Mobility", Recrystallization and Grain Growth of Multi-Phase and Particle Containing Materials, Proc. Conf. 1st Int. Symp. on Metallurgy and Materials Science (ed. N. Hansen, A. R. Jones and T. Leffers, Riso National Lab., Roskilde, Denmark, 1980), 325-336.

18. N. Ono, M. Kajihara, and M. Kikuchi, „Formation and Stability of a Nitride with the Structure of Beta Manganese in Ni-Cr-N Ternary System", Metallurgical Transactions A, 23A (1992), 1389-1393.

19. U. Brill, D. C. Agarwal, „ Properties and Corrosion Resistance Behavior of a New Nitride-Strengthened Nickel-Chromium Superalloy" , Corrosion 98 (NACE, Houston TX, 1998), paper No.437.

20. Y. Zhang, D. Zhu and D. A. Shores, „Effect of Yttrium on the Oxidation behavior of Cast Ni-30Cr Alloy", Acta metall. mater., 43 (1995), 4015-4025.

21. R. W. Balluffi and A. V. Granato, „Dislocations, Vacancies and Interstitials", Dislocations in Solids, ed. F. R. N. Nabarro (Oxford, North-Holland Publishing Company, 1983), 32-35.

MC-NG: A 4th GENERATION SINGLE-CRYSTAL SUPERALLOY FOR FUTURE AERONAUTICAL TURBINE BLADES AND VANES

Didier Argence*, Cyril Vernault**, Yves Desvallées* and Dominique Fournier**

*SNECMA, Materials and Processing Department, Villaroche center 77556 Moissy-Cramayel Cedex, France
**TURBOMECA, Materials and Coatings Department, 64511 Bordes Cedex, France

Abstract

The growing demand for an increasing turbine inlet temperature has led both French engine manufacturers Snecma and Turbomeca to take into account the attractive potential of alloys containing rhenium and ruthenium. The MC-NG is a 4th generation single-crystal superalloy developed and patented by Onera.

This paper discusses both microstructural aspects and mechanical properties obtained on the first MC-NG alloy elaborated in industrial conditions. Results demonstrate that there is a promising alternative to the third generation superalloys to increase the temperature capability of the single-crystal turbine blades without the drawback of a too high density or microstructural instability features.

Introduction

It is clearly established and fully documented that aeronautical turbine blades are submitted to creep and fatigue at high temperature. In such conditions fatigue and creep tests performed in laboratories have shown that grain boundaries constitute preferential sites for crack initiation and propagation. Hence, the continuous need for an increasing turbine inlet temperature has led engine manufacturers to use materials which have a low intergranular sensibility.

Considering this context and including casting progress in directional solidification, gas turbine manufacturers who used, in the past, cast polycrystalline superalloys have moved towards columnar structures and finally single crystals. By developing nickel base superalloy chemistries, alloy designers have largely contributed to the improvement of high temperature mechanical properties of this kind of materials. As a result, the latest single-crystal superalloys constitute the 4th generation. This new alloys family is characterised by high rhenium and ruthenium contents. These alloying elements were indeed shown to have a strong beneficial effect on the long-term microstructural destabilisation especially during high temperature creep.

Currently, calculations for future engine programs based on behaviour and damage laws obtained for single crystals in service (AM1, AM3 and MC2 used for the M88 Rafale fighter's engine or Arriel 2 and Arrius 2 for Eurocopter and Sikorsky helicopter applications) show the advantages of the new single-crystal generation.

For both French engine manufacturers Snecma and Turbomeca, these new objectives can be reached with the manufacturing of MC-NG (Mono Cristal – Nouvelle Génération) blades which are intended for commercial and military engines. The MC-NG is a single-crystal superalloy developed and patented by Onera [1]. Before being used for operating engines, it is necessary to fully characterise the single crystal superalloy in order to obtain the basic properties and to identify its behaviour laws and damage models.

This paper deals with preliminary results regarding metallurgical characteristics, mechanical and oxidation-corrosion properties obtained on the first MC-NG alloy elaborated in industrial conditions.

Metallurgical study

As-cast material and solutioning treatment analysis

The composition and density of MC-NG are indicated in Table I together with chemical compositions and densities of AM1, AM3, MC2 and CMSX-4 alloys. It is to be noted that first (AM1 and AM3) and second (MC2) single-crystal generations are composed of rather similar elements whereas MC-NG exhibits an innovative composition with rhenium and ruthenium contents. Since these elements are relatively heavy, MC-NG density is higher than current single-crystal alloys, but as the same level of density than CMSX-4.

Furthermore, some differences in microstructure of the as-cast material can be expected in MC-NG as compared to AM1, AM3 and MC2. A metallurgical study on small bars of 10 and 14 mm diameter allowed us to quantify the eutectic content, by using an image analysis method. This method consists in examining 50 non-consecutive fields at a magnification of 250. The average fraction of eutectic phase counted on MC-NG is 2.8%. As compared to AM1 alloy (3%) the eutectic percentage is similar.

An other target set for the as-cast MC-NG alloy study was to determine the aspect and the size of the gamma prime phase. Two populations of γ' can be observed. The first one is located inside inter-dendritic areas where precipitates exhibit a square or circular features and a size in the range 0.6 – 0.7 µm (Figure 1). Furthermore, in dendrite's bulk, γ' looks like squares or stars of around 0.5 µm (Figure 2). However, a similar aspect has already been observed on current single crystals.

The definition of the solution treatment depends also on the burning temperature of the alloy. Hence, isothermal burning tests were performed on MC-NG at intervals of 10°C between 1310°C and 1360°C for 1 hour. Few burn points appeared at 1340°C. After a new test at 1335°C, the burning temperature of the MC-NG alloy has been ultimately defined at 1340°C. Quantitative studies on eutectic content and burning tests justify that the MC-NG superalloy solution treatment consist of a 10-hour dwell time at 1340°C (close to the CMSX-10 superalloy solution treatment) after a very slow rate heating (3°C per hour) between 1310°C and 1340°C. Indeed, this heat treatment is necessary to remove all the eutectic phases and to homogenise the material. It is to be noticed that the solution treatment is longer than on current single crystals since the dwell time is only 3 hours at 1300°C for AM1, AM3 and MC2 superalloys.

Table I Chemical compositions and densities of considered single-crystal superalloys

Alloy	Cr	Co	Mo	W	Ta	Re	Ru	Al	Ti	Hf	Ni	Density
AM1	8	6	2	6	9	-	-	5.2	1.2	-	bal	8.6
AM3	8	5.5	2.2	5	3.5	-	-	6	2	-	bal	8.25
MC2	8	5	2	8	6	-	-	5	1.5	-	bal	8.63
CMSX-4	6.5	9	0.6	6	6.5	3	-	5.6	1	0.1	bal	8.7
MC-NG[†]	4	<0.2	1	5	5	4	4	6	0.5	0.1	bal	8.75

[†]: MC-NG data correspond to the MC-NG 544 grade elaborated and patented by Onera.

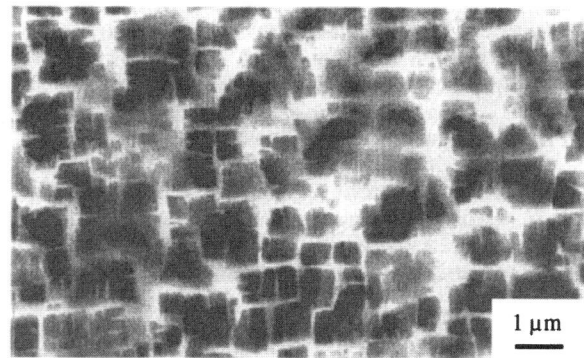

Figure 1: γ' aspect inside inter-dendritic zone.

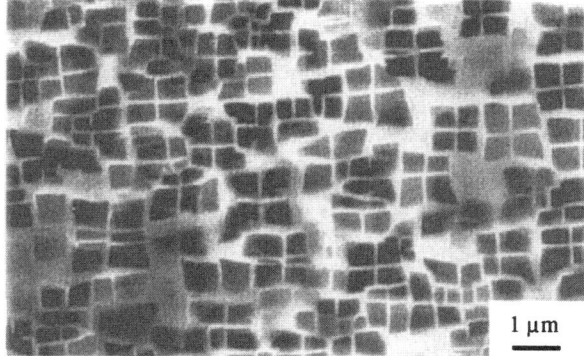

Figure 2: γ' aspect in dendrite's bulk.

Ageing treatment definition

Different heat treatments were applied in order to define the optimum conditions to provide expected the mechanical properties. Since it is necessary to use coatings in service, the considered ageing treatment, consists of two successive thermal treatments; the first one is aimed at ensuring coating diffusion whereas the second one stabilises the microstructure of the alloy.

Moreover, a brazing heat treatment simulation has also been considered. Table II synthesizes all explored conditions.

For each condition, the tensile properties at 750°C and γ' size were determined in order to choose the adequate ageing treatment.

The results are also indicated in Table II.

Tensile tests reveal the important role of the stabilising treatment (R2). For instance, reducing the dwell time from 16 hours to 4 hours significantly decreases the mechanical resistance.

A secondary role can be attributed to the primary γ' size. For instance, the increase of their size induces an improvement of tensile strength. Nevertheless, this effect disappears when the stabilising treatment duration is sufficient (16 hours). Even if it is not evidenced, the drastic effect of the stabilising dwell time is supposed to be partly due to a secondary γ' phase precipitation.

It ought to also be noted that the brazing heat treatment simulation does not modify the tensile properties of MC-NG alloy at 750°C. Hence, the thermal treatment sequence 1100°C/4hrs + 870°C/16 hrs has been chosen as the basic one.

In addition to defined conditions for obtaining a fully heat-treated alloy, the influence of heat treatment on microstructure stability was analysed. It appears that the γ' phase in MC-NG is less sensitive to a dwell time at very high temperature (1225°C) than in AM3 and MC2 alloys. While γ' precipitates are solution-treated for current single crystals (AM3, MC2), their aspect and distribution in MC-NG are nearly unchanged at this temperature. This behaviour must be correlated to the presence of alloying elements Re and Ru which reinforce the γ' phase. As a consequence, an improvement of the mechanical properties at high temperature is expected.

Table II MC-NG heat-treatment conditions explored and associated tensile properties at 750°C

Heat treatment	Brazing treatment (B)	Diffusion treatment (R1)	Stabilisation treatment (R2)	Ultimate Tensile Stress (MPa)	Yield Stress (MPa)	Elongation A%	γ' size (μm)
N° 1	-	1100°C/4 hrs	870°C/4 hrs	1038	809	11.6	540
N° 2	-	1150°C/4 hrs	870°C/4 hrs	1054	826	13.2	610
N° 3	-	1100°C/7 hrs	870°C/4 hrs	1044	818	12.4	640
N° 4	-	1100°C/4 hrs	870°C/16 hrs	1225	915	7.7	550
N° 5	-	1150°C/4 hrs	870°C/16 hrs	1105	827	10.2	700
N° 6	1220°C/23 min	1100°C/4 hrs	870°C/16 hrs	1191	904	7.3	720

Mechanical properties

Mechanical properties of MC-NG superalloy have been characterised though tensile, creep and low-cycle fatigue tests. Specimens were machined from fully heat-treated bars of 14 mm diameter. The disorientation between the <001> axis of the crystal and the test bar axis was, in all cases, less than 10 degrees.

Tensile properties

Tensile tests were performed on smooth samples from room temperature up to 1150°C. Figure 3 represents the ultimate tensile stress and the yield stress as a function of temperature.

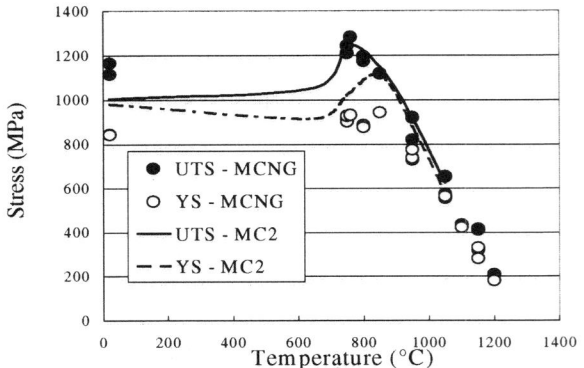

Figure 3: Tensile properties as a function of temperature.

Roughly tensile data are above current single-crystal data up to 1150°C. Only the yield strength is slightly below between room temperature and 800°C. This behaviour is not yet understood.

Creep properties

Creep tests were performed in the temperature range of 900°C to 1200°C. Since a Larson-Miller diagram does not allow the illustration of all test conditions when controlling strain or when damage mode is changed, MC-NG creep strength is represented in two figures. The first one shows a classic Larson-Miller curve where all test results obtained up to 1150°C are collected (Figure 4).

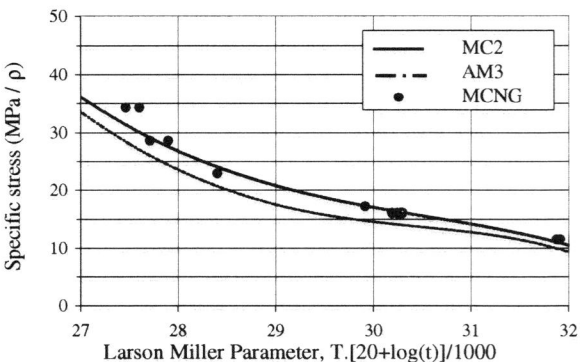

Figure 4: Creep-rupture results - Larson – Miller diagram.

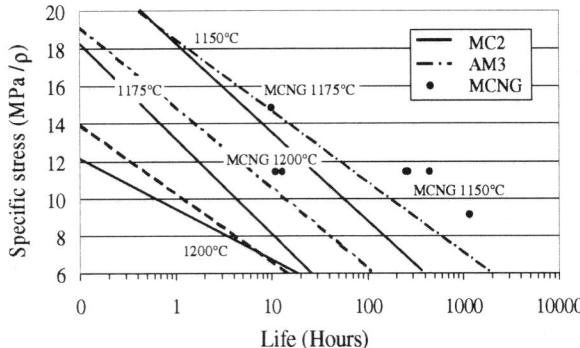

Figure 5: Creep results – time to rupture as a function of temperature.

Indeed this temperature was identified to be the transition between two behaviour modes. Figure 5 illustrates, for different temperatures, times to rupture as a function of the density normalised specific stress.

Considering both figures it can be concluded that the MC-NG creep strength is remarkable. In a large range of temperatures, the average time to rupture of this alloy is higher than on current single crystals. This difference is even more important for medium (~ 900°C) and very high temperatures (≥ 1150°C). For instance at 1200°C, improvements by a factor 25 and 60 on creep rupture times are observed in comparison with AM3 and MC2 alloys respectively. However, MC2 remains the reference material for creep properties in the temperature range of 1000°C-1050°C.

The significant higher creep resistance of MC-NG for temperatures above 1150°C can be explained by the increase of the residual fraction of γ' phase in this temperature range as compared to the case of MC2. This behaviour has been previously discussed in this paper, in terms of the delaying effect on microstructural destabilisation associated to the Re and Ru presence.

Low-cycle fatigue properties

Cyclic tests were carried out under different levels of stress amplitude at 950°C and 1100°C. Low-cycle fatigue tests were performed with and without dwell time (trapezoid cycle 10-90-10 or sinusoidal cycle at 0.5 Hz).

Figure 6 illustrates the MC-NG number of cycles to failure as a function of the normalised stress amplitude for all fatigue tests. Open symbols represent tests with dwell time at both temperatures. Comparable published data on AM1 alloys [2] are included.

Considering pure low cycle fatigue at 950°C, it appears that MC-NG properties are very promising. For instance, an improvement by a factor 6 on the fatigue life is observed in comparison with AM1.

Introducing a dwell time during cycling induces a decrease of the fatigue strength. This negative effect is equivalent for both test temperatures and roughly represents a factor 40 on the number of cycles to failure.

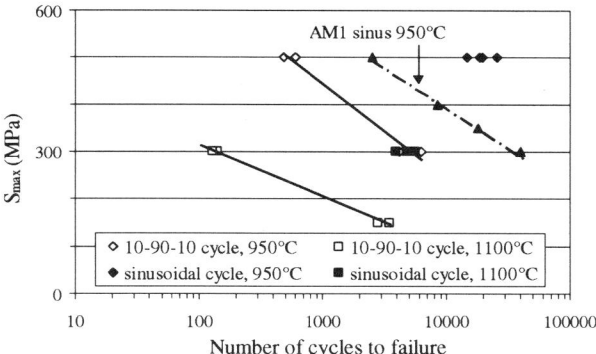

Figure 6: MC-NG low-cycle fatigue results – Number of cycles to failure as a function of normalised stress amplitude.

SEM analyses of fracture surfaces have shown, for tests performed without dwell time, that fatigue damage mechanisms are different for both test temperatures.

Indeed at 950°C, only one microcrack initiates on a subsurface shrinkage and leads to failure. On the contrary, at 1100°C, even if the thick oxide layer on the fracture surface does not allow the identification of initiation sites, the specimens' surface exhibits numerous uniformly distributed cracks.

As regard fatigue tests with dwell time, numerous cracks are also observed on samples' surface. Microcracks are initiated from the thick superficial oxide layer.

For both temperatures 950°C and 1100°C, microstructural investigations carried out from fracture area show the formation of a typical raft structure. This structure appears to be more evident at higher temperature.

Furthermore, observations of the specimens' longitudinal sections reveal the large population of micropores aligned along interdendritic spaces. Such behaviour is associated with creep damage during dwell time.

Environmental effect and microstructural stability

The behaviour of MC-NG in relation to environment has been investigated with air oxidation and sulfidation corrosion experiments under cyclic temperature. In order to prevent the damage development, the specimens were coated with a protective layer. This layer also plays the role of bondcoat for the thermal barrier coating. However the protection can induce the formation of a Secondary Reaction Zone (SRZ) within the underlying material [3,4]. This zone contains plate-like shaped Topologically Closed Packed (TCP) precipitates which could affect the mechanical properties [5,6].

Oxidation resistance

Several cyclic oxidation tests with a dwell time at 1100°C were performed on uncoated AM1 and MC-NG specimens. The classic curve of mass evolution (denoted Δm) is shown in Figure 7:

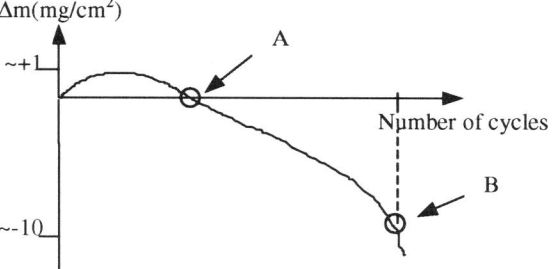

Figure 7: Standard curve of mass evolution during oxidation test.

At the beginning of the test, the specimen weight increases which is progressively balanced by scaling of the oxidation layer. The A (Δm=0) and B (Δm=-10) values are used for data analysis. They allow the comparison of the specimen oxidation resistance. Below -10mg/cm^2 we consider that the oxide is entirely spalled. The tests performed on uncoated specimens show that the MC-NG is clearly more resistant to oxidation than AM1 (Table III). For the same number of cycles the mass reduction of AM1 is close to -10mg/cm^2 while it is still positive for MC-NG.

Table III Oxidation resistance of uncoated
AM1 and MC-NG specimens

Alloy	Δm=0 mg/cm^2 (number of cycles)	Δm=-10 mg/cm^2 (number of cycles)
AM1	150	500
MC-NG	400	>1300

This difference of oxidation behaviour can be explained by the beneficial influence of alloying elements like rhenium and ruthenium which are present in the MC-NG alloy. However, other studies have shown that the difference is strongly linked with the alloy cleanness and particularly to the segregating elements which inhibit oxide scale adhesion [7,8].

Corrosion resistance

Snecma carried out several cyclic corrosion tests, at 850-900°C, on uncoated AM1 and MC-NG specimens.

During the 45 minute dwell time, a salt mist is introduced in the chamber with kerosene to obtain specific conditions (1mg/cm^2 of Na_2SO_4 for 100h exposure). For the data analysis, the number of cycles to initiation (pits) and to complete corrosion of the surface are counted. The results (Table IV) show that the corrosion resistance of MC-NG is better than the AM1 one.

Table IV Corrosion resistance of uncoated
AM1 and MC-NG specimens

Alloy	Initiation (number of cycles)	End (number of cycles)
AM1	30-50	100-150
MC-NG	140-240	200-300

The intrinsic corrosion resistance of MC-NG was also investigated by Turbomeca with other cyclic corrosion tests at 900°C. Results are similar to those obtained on AM3 but are lower than on CMSX-11 and MC2. Further analysis has still to be carried out to explain these differences.

Microstructural stability

Subsurface microstructure The MC-NG alloy is sensitive to the precipitation of TCP phases in relation to its high content of refractory element like rhenium. The present study concerns the metallurgical characterization of these phases and their influence on the low-cycle fatigue properties. The SRZ is generated during coating processes with the formation of Ni-Al protective layer. The chemical diffusion at high temperature generates a needle-shape type P TCP phase in a γ' matrix. The thickness of this zone ranges from 5 to 40μm (Figure 8).

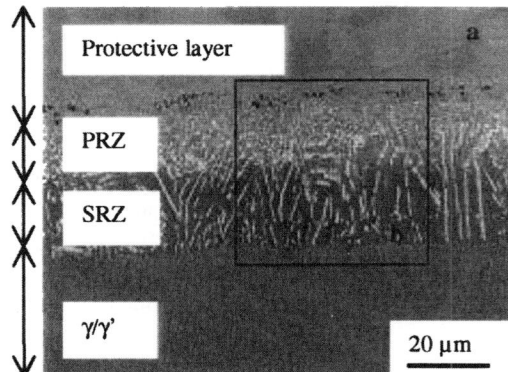

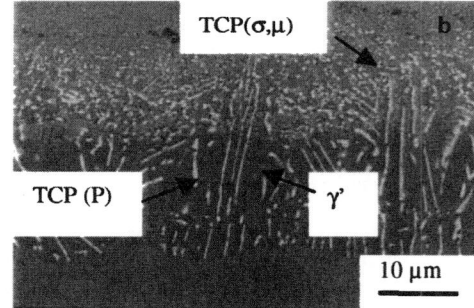

Figure 8: SRZ description (a) with detail (b).

The surface residual stresses have a catalytic effect on the SRZ development. This observation has already been detailed in reference [3]. The removal by an electrolytic polishing of a surface layer (a few microns thick) prevents the SRZ formation; an ageing treatment after coating generates the precipitation of TCP phases again but with a much slower kinetics.

To evaluate its impact on low cycle fatigue properties, several tests have been performed at 950°C and 1100°C on coated and uncoated MC-NG specimens (Table V).

Table V Comparison of MC-NG life
of coated and uncoated MC-NG specimens

Coating	Temperature (°C)	Stress (MPa)	Cycles to failure
No	950	500	18622
No	950	500	14791
Yes	950	500	6150
Yes	950	500	9687
No	1100	300	4784
No	1100	300	3880
Yes	1100	300	2928

We observe that the fatigue life decreases by a factor 2 for the coated specimens. Two reasons can explain this decrease: the effect of the protective layer and/or the effect of SRZ alone. We have been able to prove that this difference is mainly related to the protective layer independently of SRZ formation. Indeed, if we compare the fatigue life of uncoated and coated MC-NG to that of uncoated and coated AM1 (where there is no SRZ formation) we obtain an equivalent reduction ratio (1.7). Thus it can be assumed that the SRZ alone does not have a detrimental effect on the fatigue properties at 950°C and 1100°C. Moreover the fatigue life remains acceptable with regard to the specifications of the MC-NG damage law (see § Identification and validation of the damage model).

Furthermore a protective coating has also been applied by Turbomeca with a diffusionless MCrAlY type layer. No TCP phase precipitation has been observed with this type of coating, which is consistent with the involved diffusion mechanism for SRZ generation.

Volume microstructure Several creep tests were also performed at 950°C - 1050°C and 1150°C to evaluate the MC-NG sensitivity to the volume TCP phase formation. The results show that a local coalescence of γ' precipitates is particularly pronounced with increasing temperature (rafting formation) but does not affect the creep life in comparison with other single crystals (see § Creep properties). Moreover no TCP phase precipitation has been observed for long time exposures of up to 1665 hours.

Component life prediction

In addition to these experimental studies, a specific original approach is implemented to use all these results for the design and life calculations of turbine blades in running conditions. This approach consists in determining constitutive equations which are introduced in finite element analysis codes for part life prediction [9-11]. The behaviour model allows us to describe:
- the existing relation between stresses and strains in elasto-viscoplastic conditions,
- the evolution of these mechanical properties related to a finite volume of blade submitted to an external complex loading.
Moreover a lifing model describes the damage evolution laws within the material versus time to exposure. The time integration of these equations allows us to predict the time to

failure for blades and vanes which are submitted to a complex loading with the interaction between fatigue, creep and environmental effects.

MC-NG constitutive equations

Hypothesis Constitutive equations link stress, strain and time. The model is based on the continuous medium mechanical theory. The material is assumed to be homogeneous on the volume element scale. In comparison to the volume element, the heterogeneity is too small, so that the microscopic behaviour is considered equivalent to the mean macroscopic behaviour.

Expression of constitutive equations To describe this behaviour, elastic and plastic strains are distinguished. The elastic strain is given by the Hooke relation whereas the plastic strain is given by the Onera model [9,10] using seven parameters. This law is labeled 'double viscosity' since it takes into account the material behaviour for fast and slow deformation rates (denoted respectively ε_{pr} and ε_{pl}). This is quite representative of monotonic tensile and creep loadings. So the proposed model is quite accurate to describe the dwell time effects which are observed during engine running. The expression of the law is given by the following equations:

$$\dot{\varepsilon}_{pr} = \left\langle \frac{|\sigma - X| - R}{Kr} \right\rangle^{nr} \qquad \dot{X} = C\left(a\,\dot{\varepsilon}_{pr} - X\,\dot{p} \right)$$

$$\dot{\varepsilon}_{p_l} = \left(\frac{\sigma}{K_l} \right)^{nl} \qquad R = 0 \qquad (1)$$

Ki and ni are respectively the consistency and the hardening parameters in relation to the strain rate and the temperature.
R is the isotropic hardening (homogeneous increasing of the elasticity domain in relation to the elastic strain).
X is the kinematic hardening (translation of the elasticity domain).
C and a relate to the derivative of X in relation to the time.

Identification and validation of the constitutive equations Tensile and fatigue tests at several temperatures were performed to identify the constitutive equation parameters. This method allows the determination of the parameters for each temperature. The parameter smoothing on all the temperature range is then obtained with a specific optimization software developed by the Ecole des Mines de Paris and the Ecole Normale Supérieure de Cachan (SiDoLo ™).
After optimization, the validation of the model is carried out with creep-fatigue tests at 950°C and 1050°C. Figures 9a and 9b show a rather good fit between MC-NG experimental data and simulation.

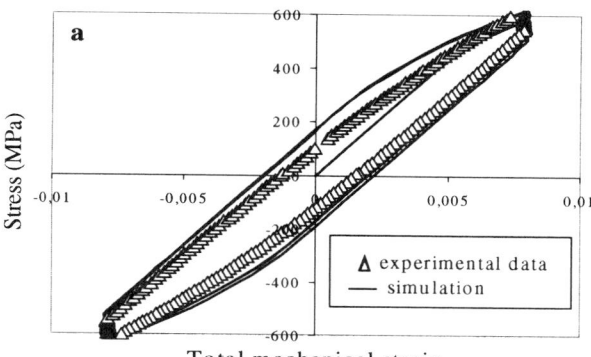

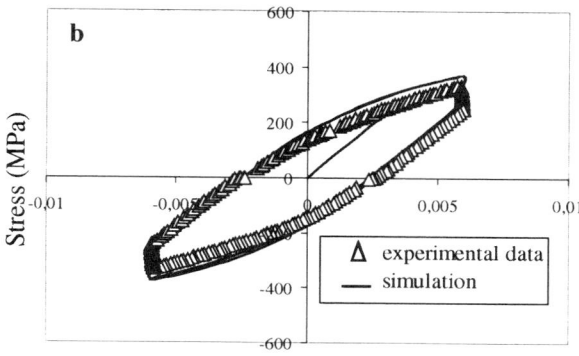

Figure 9: Comparison of MC-NG experimental data and simulation on creep-fatigue tests at 950°C (a) and 1050°C (b).

However, thermomechanical tests (cyclic strain and temperature) were performed to determine the limits of the model. This type of loading is more discriminating since it is closer to the actual engine running conditions [12].

MC-NG damage model

Hypothesis The absence of damage corresponds to a material free of cracking and cavity on the microscopic scale. The continuous damage theory describes the evolution between initiation and macroscopic cracking (a few millimeters). In the model we take into account:
- the fatigue damage (D_f) which is a function of the number of cycles,
- the creep damage (D_c) which is a function of the time.
Both damages are scalar values and can be added.
The environment effects are not directly considered, but are integrated within the D_c variable.

Expression of the damage model The expression of the damage model requires the knowledge of the parameter D. This parameter describes the material damage level. We consider it by introducing the effective stress concept:

$$\sigma_{eff} = \sigma/(1 - D) \qquad (2)$$

The fatigue damage in anisothermal conditions is given by the Chaboche model [9] with the following equations:

$$\delta D_f = \left[1 - (1-D)^{\beta+1}\right]^{\alpha}\left(\frac{\sigma_{max} - \overline{\sigma}}{M(1-D)}\right)^{\beta}\delta N \qquad (3)$$

With:
$$\begin{cases} M = M_0\left(1 - b_1\overline{\sigma}\right) \\ \alpha = 1 - a\left\langle\dfrac{\sigma_{max} - \sigma_l}{1 - \sigma_{max}}\right\rangle \\ \sigma_l = \sigma_{l0}\left(1 - b_2\sigma_{l0}\right) + \overline{\sigma} \end{cases} \qquad (4)$$

The coefficients β, a, σ_{l0}, M_0, b_1 and b_2 are unique for a given material.
The Rabotnov model [11] describes the creep damage. Its expression is as follows:

$$\delta D_c = \left(\frac{\sigma}{A}\right)^r (1 - D_c)^{-k}\delta t \qquad (5)$$

The coefficients A, r and k depend on the alloy and the temperature.
At high temperatures fatigue and creep damage can interact. To simulate this phenomenon we simply add the elementary damage values D_f and D_c. We therefore take into consideration the physical effects such as:
- the increasing of the fatigue crack propagation rate due to the creep cavities,
- the increasing of the creep cavities volume due to the concentration of stress at the crack tip.
This method leads to a non linear damage evolution which is well assessed by the experimental data.

Identification and validation of the damage model The identification of the damage model is based on several tensile, fatigue and creep tests. Up to now the number of tests performed on the MC-NG are not sufficient to identify all the parameters. However, some referring fatigue and creep tests (Figures 5 and 6) show that the MC-NG mean life is twice greater than that of the AM1. Therefore, in a first approach the damage model parameters were determined by using the AM1 curves with a factor 2 on the number of cycles or the time to failure.

Application of the constitutive equations and the damage model for the part life prediction

The life of a high pressure turbine blade of the M88 military engine (Figure 10) has been calculated with MC-NG behaviour and damage laws.

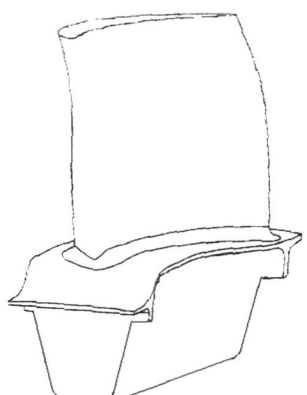

Figure 10: Design of a high pressure turbine blade of the M88 military engine.

The part life prediction method involves the following stages:
- 1D calculation using the elasto-viscoplastic behaviour established in the radial direction (direction <100> of the single crystal),
- stress calculation in different sections perpendicular to the blade axis,
- prediction of the number of cycles to failure using the damage model with the creep and fatigue interactions.
Then some overstress coefficients are locally applied to take into account:
- the disparity due to the anisotropy. This difference is determined by the comparison of 1D and 3D calculations which are performed in a localised area of the part,
- the stress concentrations due to the geometrical singularities, like holes.
The external loadings used in the numerical prediction are representative of an endurance cycle similar to the flying conditions. This complex cycle unites the temperature and stress variations. The results are presented for the critical section of the blade (Figure 11). The plotted factors represent the ratios between the MC-NG and AM1 number of cycles to failure. In this section the maximum temperature is 1050°C.

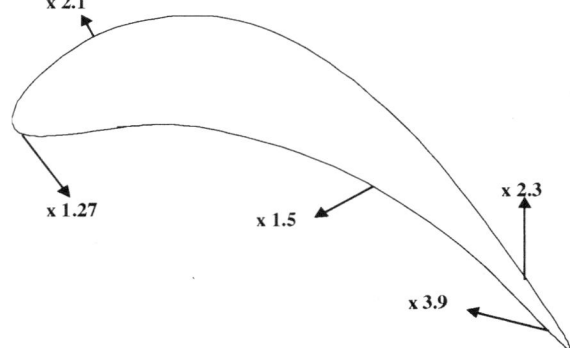

Figure 11: Predicted gain of life between AM1 and MC-NG for the same solicitations.

These results lead to the following remarks:
- in the creep predominant areas (with 950°C<T<1050°C) the MC-NG gain ranges between 1.3 and 2.1. In these conditions the MC-NG contribution is limited,
- in the fatigue predominant areas the gain ranges between 2.3 and 3.9. This is due to the more favourable behaviour of MC-NG as compared to AM1. In these conditions the stress relaxation is more important than for AM1.
The data analysis enables the modification of the blade shape to decrease the critical area solicitations. After design optimization for MC-NG, the life gain would be better than a factor 2 throughout the part compared to the AM1 blade.

Conclusion

This paper synthesizes the main basic works performed on the new generation single crystal MC-NG by Snecma and Turbomeca. It presents the different stages for the implementation of a new material which are the development, the characterization and the industrial design of parts. With its outstanding mechanical properties at high temperature and good environment resistance, MC-NG appears to be quite a valuable candidate for turbine blades in future turboshaft engines. This result is the consequence of a beneficial collaboration with the Onera research laboratory. Today Snecma and Turbomeca have produced some MC-NG blades for high pressure turbines. Both engine manufacturers will shortly test those parts on their experimental engines in very severe loading and temperature conditions. In parallel the evaluation of implementation processes (machining, brazing) and repairing is in progress.

Acknowledgements

The authors are grateful to the Service des Programmes Aéronautiques for their financial support of the MC-NG programme and acknowledge Laurence Potez, Alain Lyoret and Jean-Louis Ragot for their scientific participation in this research work.

References

1. P. Caron, "High γ' Solvus New Generation Nickel-Based Superalloys for Single Crystal Turbine Blade Applications" (Paper will be presented at the 9th International Symposium on Superalloys, Champion, Pennsylvania, 17-21 September 2000).

2. Philippe Perruchaut, Patrick Villechaise, and José Mendez, "Some Aspects of Environmental Effects on the Fatigue Damage of the AM1 Single Cristal Superalloy at High Temperature", Corrosion-Déformation Interactions (Nice, France: Thierry Magnin, 1996), 342.

3. W.S. Walston, J.C. Schaeffer and W.H. Murphy, "A New Type of Microstructural Instability in Superalloys – SRZ", Superalloys 1996, ed. R. D. Kissinger et al. (Warrendale, PA: The Minerals, Metals and Materials Society, 1996), 9-18.

4. W.S. Walston and al., "René N6: Third Generation Single Crystal Superalloy", Superalloys 1996, ed. R. D. Kissinger et al. (Warrendale, PA: The Minerals, Metals and Materials Society, 1996), 27-34.

5. David N. Duhl, Alloy Phase Stability and Design (Pittsburgh, PA: Materials Research Society, 1991), 389-399.

6. "Development and Turbine Engine Performance of Advanced Rhenium Containing Superalloys for Single Crystal and Directionnally Solidified Airfoils", Rhenium and Rhenium Alloys, ed. B. D. Bryskin (Warrendale, PA: The Minerals, Metals and Materials Society, 1997),731-754.

7. J. L. Smialek et al., "Effects of Hydrogen Annealing, Sulfur Segregation and Diffusion on the Cyclic Oxidation Resistance of Superalloys: a Review," Thin Solid Films, 253 (1994), 285-292.

8. Y. Zhang et al., "Synthesis and Cyclic Oxidation Behavior of a (Ni,Pt)Al Coating on a Desulfurized Ni-Based Superalloy", Metallurgical and Materials Transaction A, 30A (1999), 2679-2687.

9. Jean-Louis Chaboche, and Jean Lemaitre, Mécanique des Matériaux Solides (Paris, France: Dunod, 1985).

10. P. Poubanne, R. Krafft, and J. P. Mascarell, "Modélisation du Comportement Sous Sollicitation de Fluage Cyclique (Double Viscosité)" (Report YLEC 137/90, Snecma, 1990).

11. Y.N. Rabotnov, Creep Problem In Structural Members (North Holland Publishing Company, 1969).

12. J.Y. Guedou and Y. Honnorat, "Thermomechanical Fatigue of Turbo-Engine Blade Superalloys," Thermomechanical Fatigue Behavior of Materials, ed. Huseyin Sehitoglu (Philadelphia, PA: ASTM STP 1186, 1993),157-175.

Color plates of figures to the paper, Predicting Grain Size Evolution of UDIMET® alloy 718 During the "Cogging" Process Through the Use of Numerical Analysis (pages 39-47).

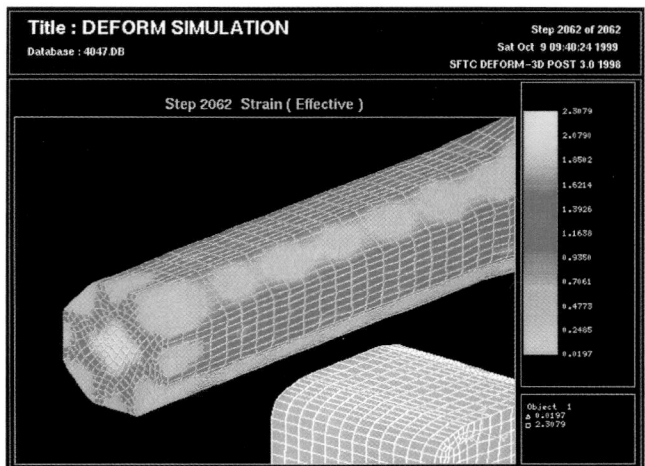

Figure 5: Practice 1: Final state of strain.

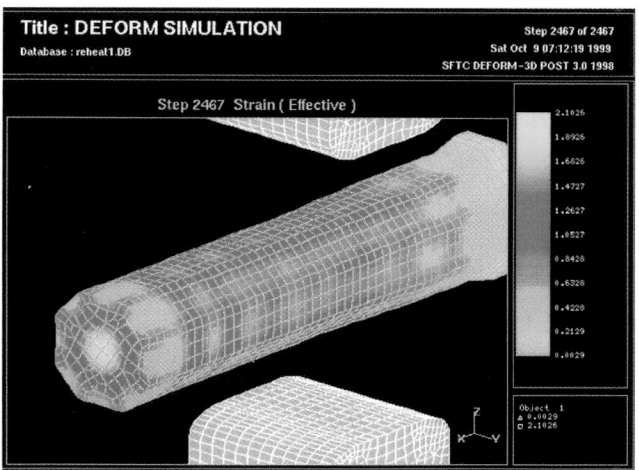

Figure 6: Practice 2: Final state of strain.

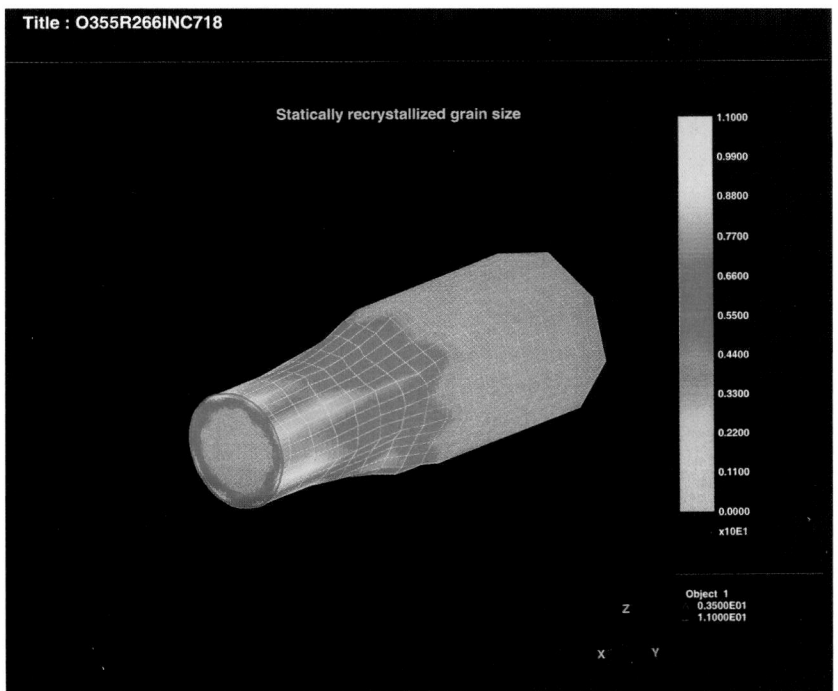

Figure 14: Statically recrystallized grain size. Predicted grain size ranges from ASTM 5.5 at center to 6.0 near edge.

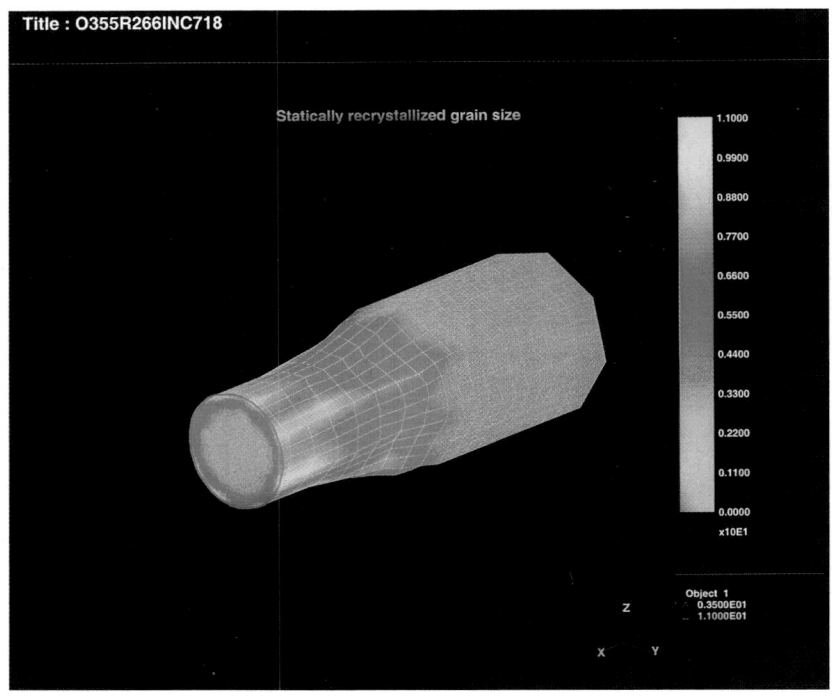

Figure 15: Dynamically recrystallized grain size. Predicted grain size ranges from ASTM 3.5 at center to 7.0 near edge where grain size measurements are made.

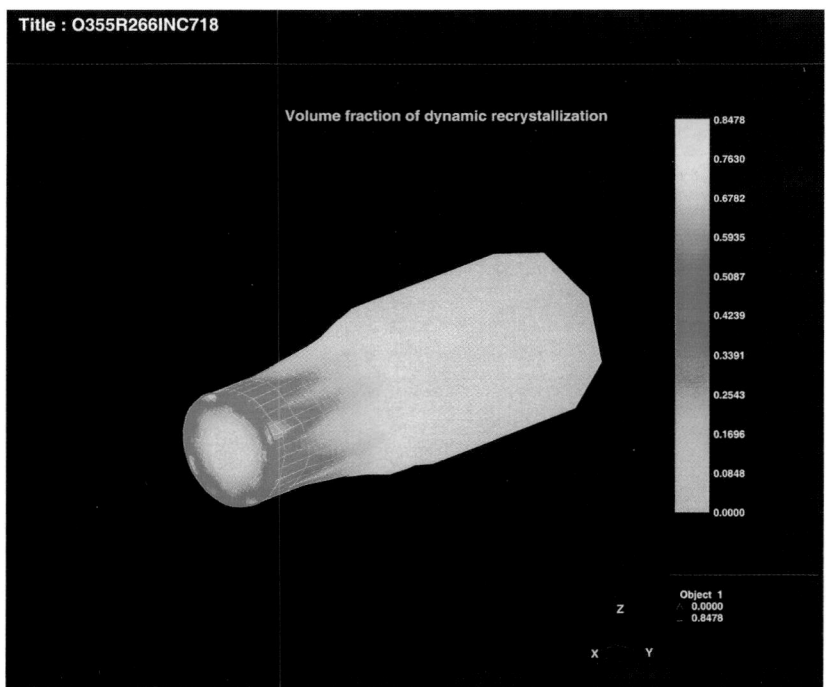

Figure 16: Dynamically recrystallized grain size fraction Rx.

failure for blades and vanes which are submitted to a complex loading with the interaction between fatigue, creep and environmental effects.

MC-NG constitutive equations

Hypothesis Constitutive equations link stress, strain and time. The model is based on the continuous medium mechanical theory. The material is assumed to be homogeneous on the volume element scale. In comparison to the volume element, the heterogeneity is too small, so that the microscopic behaviour is considered equivalent to the mean macroscopic behaviour.

Expression of constitutive equations To describe this behaviour, elastic and plastic strains are distinguished. The elastic strain is given by the Hooke relation whereas the plastic strain is given by the Onera model [9,10] using seven parameters. This law is labeled 'double viscosity' since it takes into account the material behaviour for fast and slow deformation rates (denoted respectively ε_{pr} and ε_{pl}). This is quite representative of monotonic tensile and creep loadings. So the proposed model is quite accurate to describe the dwell time effects which are observed during engine running. The expression of the law is given by the following equations:

$$\dot{\varepsilon}_{pr} = \left\langle \frac{|\sigma - X| - R}{Kr} \right\rangle^{nr} \qquad \dot{X} = C\left(a\dot{\varepsilon}_{pr} - X\dot{p} \right)$$

$$\dot{\varepsilon}_{pl} = \left(\frac{\sigma}{K_l} \right)^{nl} \qquad R = 0 \qquad (1)$$

Ki and ni are respectively the consistency and the hardening parameters in relation to the strain rate and the temperature.
R is the isotropic hardening (homogeneous increasing of the elasticity domain in relation to the elastic strain).
X is the kinematic hardening (translation of the elasticity domain).
C and a relate to the derivative of X in relation to the time.

Identification and validation of the constitutive equations
Tensile and fatigue tests at several temperatures were performed to identify the constitutive equation parameters. This method allows the determination of the parameters for each temperature. The parameter smoothing on all the temperature range is then obtained with a specific optimization software developed by the Ecole des Mines de Paris and the Ecole Normale Supérieure de Cachan (SiDoLo ™).
After optimization, the validation of the model is carried out with creep-fatigue tests at 950°C and 1050°C. Figures 9a and 9b show a rather good fit between MC-NG experimental data and simulation.

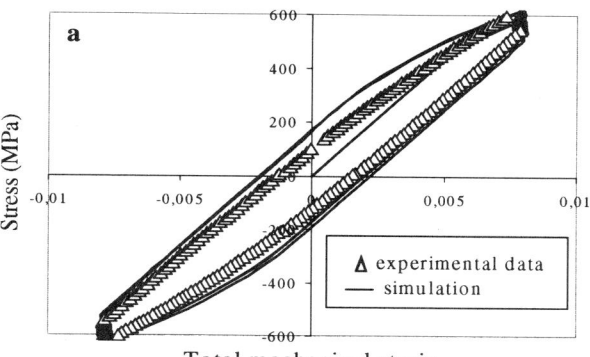

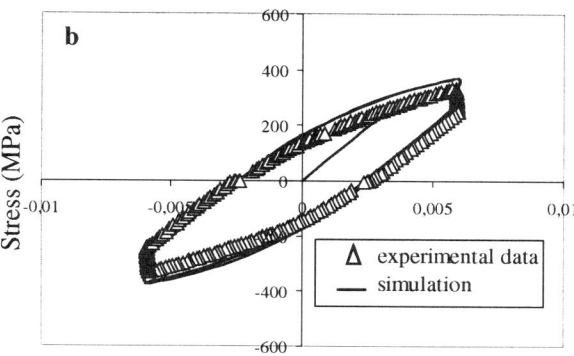

Figure 9: Comparison of MC-NG experimental data and simulation on creep-fatigue tests at 950°C (a) and 1050°C (b).

However, thermomechanical tests (cyclic strain and temperature) were performed to determine the limits of the model. This type of loading is more discriminating since it is closer to the actual engine running conditions [12].

MC-NG damage model

Hypothesis The absence of damage corresponds to a material free of cracking and cavity on the microscopic scale. The continuous damage theory describes the evolution between initiation and macroscopic cracking (a few millimeters). In the model we take into account:
- the fatigue damage (D_f) which is a function of the number of cycles,
- the creep damage (D_c) which is a function of the time.
Both damages are scalar values and can be added.
The environment effects are not directly considered, but are integrated within the D_c variable.

Expression of the damage model The expression of the damage model requires the knowledge of the parameter D. This parameter describes the material damage level. We consider it by introducing the effective stress concept:

$$\sigma_{eff} = \sigma / (1 - D) \qquad (2)$$

The fatigue damage in anisothermal conditions is given by the Chaboche model [9] with the following equations:

$$\delta D_f = \left[1 - (1-D)^{\beta+1}\right]^{\alpha} \left(\frac{\sigma_{max} - \overline{\sigma}}{M(1-D)}\right)^{\beta} \delta N \qquad (3)$$

With:
$$\begin{cases} M = M_0\left(1 - b_1\overline{\sigma}\right) \\ \alpha = 1 - a\left\langle\dfrac{\sigma_{max} - \sigma_l}{1 - \sigma_{max}}\right\rangle \\ \sigma_l = \sigma_{l0}\left(1 - b_2\sigma_{l0}\right) + \overline{\sigma} \end{cases} \qquad (4)$$

The coefficients β, a, σ_{l0}, M_0, b_1 and b_2 are unique for a given material.

The Rabotnov model [11] describes the creep damage. Its expression is as follows:

$$\delta D_c = \left(\frac{\sigma}{A}\right)^r (1 - D_c)^{-k} \delta t \qquad (5)$$

The coefficients A, r and k depend on the alloy and the temperature.

At high temperatures fatigue and creep damage can interact. To simulate this phenomenon we simply add the elementary damage values D_f and D_c. We therefore take into consideration the physical effects such as:
- the increasing of the fatigue crack propagation rate due to the creep cavities,
- the increasing of the creep cavities volume due to the concentration of stress at the crack tip.

This method leads to a non linear damage evolution which is well assessed by the experimental data.

Identification and validation of the damage model The identification of the damage model is based on several tensile, fatigue and creep tests. Up to now the number of tests performed on the MC-NG are not sufficient to identify all the parameters. However, some referring fatigue and creep tests (Figures 5 and 6) show that the MC-NG mean life is twice greater than that of the AM1. Therefore, in a first approach the damage model parameters were determined by using the AM1 curves with a factor 2 on the number of cycles or the time to failure.

Application of the constitutive equations and the damage model for the part life prediction

The life of a high pressure turbine blade of the M88 military engine (Figure 10) has been calculated with MC-NG behaviour and damage laws.

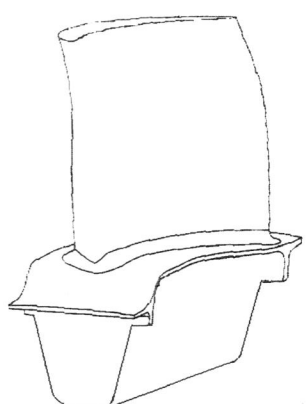

Figure 10: Design of a high pressure turbine blade of the M88 military engine.

The part life prediction method involves the following stages:
- 1D calculation using the elasto-viscoplastic behaviour established in the radial direction (direction <100> of the single crystal),
- stress calculation in different sections perpendicular to the blade axis,
- prediction of the number of cycles to failure using the damage model with the creep and fatigue interactions.

Then some overstress coefficients are locally applied to take into account:
- the disparity due to the anisotropy. This difference is determined by the comparison of 1D and 3D calculations which are performed in a localised area of the part,
- the stress concentrations due to the geometrical singularities, like holes.

The external loadings used in the numerical prediction are representative of an endurance cycle similar to the flying conditions. This complex cycle unites the temperature and stress variations. The results are presented for the critical section of the blade (Figure 11). The plotted factors represent the ratios between the MC-NG and AM1 number of cycles to failure. In this section the maximum temperature is 1050°C.

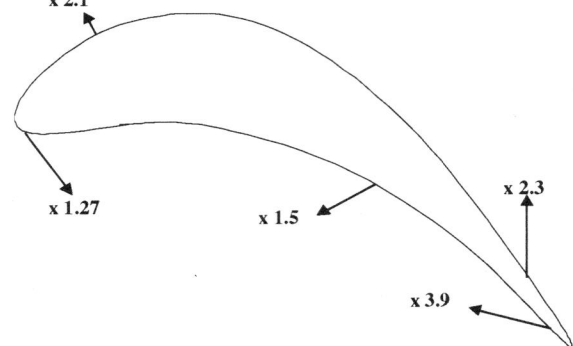

Figure 11: Predicted gain of life between AM1 and MC-NG for the same solicitations.

Color plates of figures to the paper, Sub-Solvus Recrystallization Mechanisms in UDIMET® Alloy 720LI (pages 59-68).

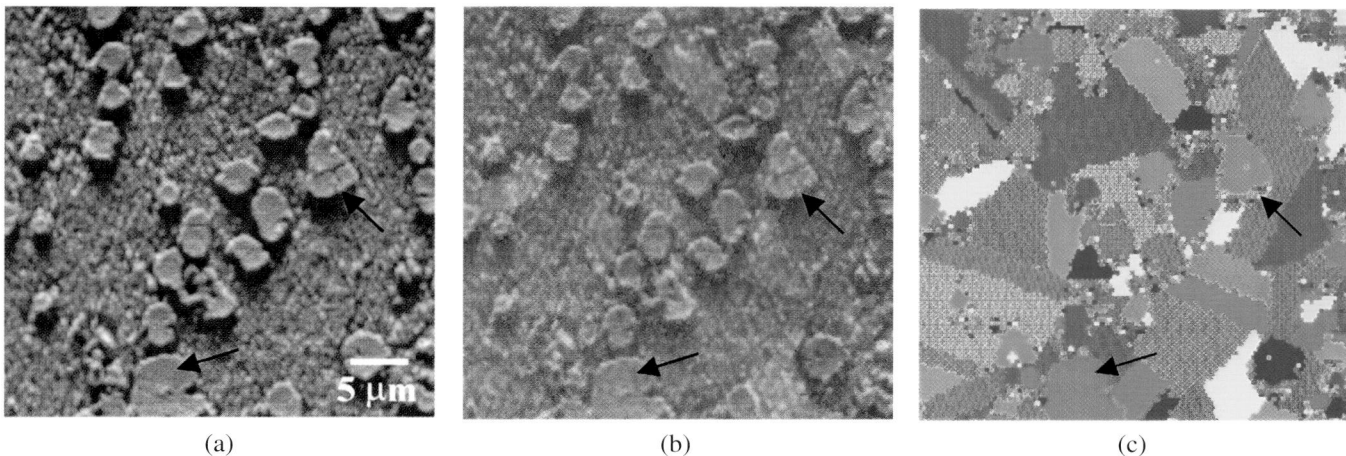

(a)	(b)	(c)

Figure 1. SEM image and EBSD orientation map of material A. (a) and (c) are SEM and orientation maps of the same region respectively. (b) is a superposition of image (a) and (c). The arrows indicate some γ' precipitates misoriented with the surrounding grains.

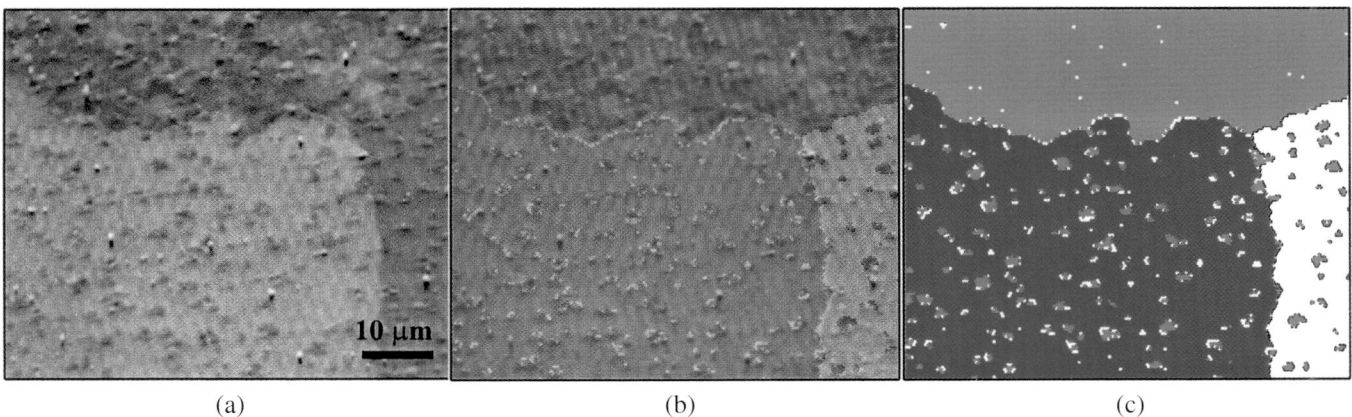

(a)	(b)	(c)

Figure 7. SEM image and EBSD orientation map of dynamically recrystallized material C. (a) and (c) are SEM and orientation maps of the same region respectively. (b) is a superposition of image (a) and (c). The red region is the original, unrecrystallized grain.

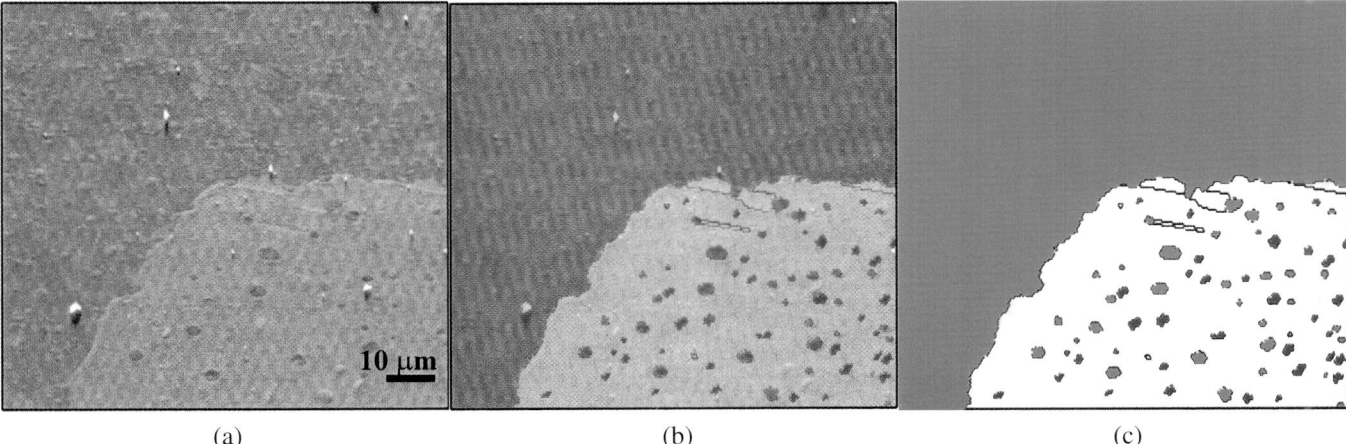

(a)	(b)	(c)

Figure 8. SEM image and EBSD orientation map of statically recrystallized material C. (a) and (c) are SEM and orientation maps of the same region respectively. (b) is a superposition of image (a) and (c). The red region is the original, unrecrystallized grain.

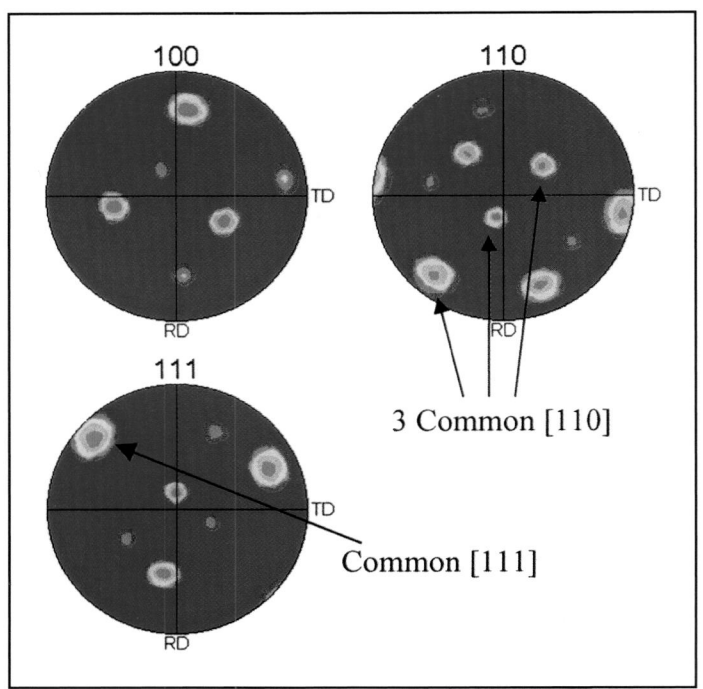

Figure 9. Pole figures for the recrystallized grain and the blue γ'
in Figure 8c.

Subject Index

849

Alloy Index

Author Index